MW01273664

DISCARD

Plants of British Columbia

Plants of British Columbia

Scientific and Common Names of Vascular Plants, Bryophytes, and Lichens

Hong Qian and Karel Klinka

UBC PRESS / VANCOUVER

Printed in Canada by Friesens on acid-free paper ∞

ISBN 0-7748-0652-4

Canadian Cataloguing in Publication Data

Qian, Hong, 1957–
Plants of British Columbia

ISBN 0-7748-0652-4

1. Botany – British Columbia – Nomenclature. I. Klinka, K., 1937– II. Title.
QK203.B7Q25 1997 581.9711'01'4 C97-910833-0

UBC Press gratefully acknowledges the ongoing support to its publishing program from the Canada Council for the Arts, the British Columbia Arts Council, and the Department of Canadian Heritage of the Government of Canada.

UBC Press
University of British Columbia
6344 Memorial Road
Vancouver, BC V6T 1Z2
(604) 822-5959
Fax: 1-800-668-0821
E-mail: orders@ubcpress.ubc.ca
http://www.ubcpress.ubc.ca

To my father, Zhong Qian; my mother, Youju Zhang;

and my wife, Luqiong Ling;

and to my teachers, Zhan Wang and Shuchun Li

— Hong Qian

Contents

Acknowledgements

Many people have contributed to this book, to whom we offer our thanks. We particularly would like to acknowledge W.B. Schofield (University of British Columbia) for reviewing our provincial bryophyte list and providing us with information on recently recorded bryophytes in the province, and C. Chourmouzis (University of British Columbia) for linguistic editing of part of the manuscript.

Our special appreciation goes to Luqiong Ling (University of British Columbia), who has played a very important role in this project. She, as a volunteer, has unselfishly devoted countless hours to data entry, checking spellings, and manuscript editing for the last five years. This book would not have been possible without her assistance, continuing encouragement, and support.

Introduction

British Columbia, the westernmost province of Canada, has an area of 948,600 square kilometres and is located between 48°19' and 60° N latitude and 114°03' and 139°03' W longitude within an elevation range of 0 to 4,663 metres (Farley 1979). The province is predominantly forested and mountainous, but some areas feature plateaus, plains, and basins. Although the area of the province represents only 9.5% of Canada's surface, its flora accounts for over 60% of all species in Canada. The rich floral presence can be attributed to the great diversity of climates, landforms, and soils. Considering the glacial history, geology, physiography, and the adjacency to the Pacific Ocean in the west and the continental mass in the east, a great variety of ecosystems have developed throughout the province. These range from the coastal rain forest in the west, to the semiarid grasslands in the southern interior, to the boreal forest in the north, and to the subalpine forest and alpine tundra in both the coastal and interior mountain ranges.

In British Columbia, the development and application of biogeoclimatic ecosystem classification to natural resources management (Krajina 1969; Meidinger and Pojar 1991; MacKinnon *et al.* 1992), the comprehensive natural resources inventory (Resources Inventory Committee 1995), and the concerns regarding biodiversity and ecosystem sustainability have coincided with an increased interest in plant and vegetation sciences. There are several guides and technical manuals that support these endeavours. The guides describe and illustrate common vascular plants, mosses, liverworts, and lichens for particular regions of the province, and the manuals are authoritative taxonomical reference works that include detailed descriptions, nomenclature, and identification keys. Examples of some of the recently published guides include Klinka *et al.* (1989), MacKinnon *et al.* (1992), Pojar and MacKinnon (1994), and Parish *et al.* (1996). Examples of some of the technical manuals pertinent to the province include Hultén (1968), Hitchcock *et al.* (1955-1969), Hitchcock and Cronquist (1973), Taylor and MacBryde (1977), Scoggan (1978, 1979), and Douglas *et al.* (1989, 1990, 1991, 1994) for vascular plants; Ireland *et al.* (1987), Schofield (1969, 1992), Vitt *et al.* (1988), Stotler and Crandall-Stotler (1977), Anderson (1990), and Anderson *et al.* (1990) for bryophytes; and Esslinger and Egan (1995), Noble *et al.* (1987), and Goward *et al.* (1994) for lichens.

Overall, there are sufficient botanical references pertinent to the province and these include guides, books, manuals, keys, and floras. While this ever-growing information base expands our knowledge of the flora, it often creates taxonomic difficulties, inconsistencies, and confusion. This is due to different personal preferences in using some plant names and the growth of taxonomic knowledge, which has resulted in changes of nomenclature. For example:

- Separate references can use different names for the same species (e.g., *Oxycoccus oxycoccos* [in Douglas *et al.* 1990, Pojar and MacKinnon 1994, and Parish *et al.* 1996] versus *Vaccinium oxycoccos* [in Klinka *et al.* 1989 and MacKinnon *et al.* 1992] versus *Oxycoccus microcarpus* [in Johnson *et al.* 1995]; *Eurynchium oreganum* [in McMinn 1957] versus *Kindbergia oregana* [in Ireland 1987, Vitt *et al.* 1988, Klinka *et al.* 1989, Schofield 1992, and Pojar and MacKinnon 1994] versus *Stokesiella oregana* [in Klinka 1976]).
- The same reference may use a certain name as the accepted name for one species and the same name as a synonym for another (e.g., in Douglas *et al.* [1994: 34-35], *Carex hindsii* is listed as an accepted name but also listed as a synonym for *Carex lenticularis* var. *limnophila*).
- Some names can be misapplied (e.g., *Adiantum pedatum* [in Vitt *et al.* 1988, Klinka *et al.* 1989, Pojar and MacKinnon 1994, and Parish *et al.* 1996], which occurs only in eastern North America, has been misapplied to *Adiantum aleuticum* [Douglas *et al.* 1991 and Flora of North America Editorial Committee 1993], which occurs in British Columbia).

Further, all checklists published to date for the overall flora of the province are for separate plant groups – vascular plants, bryophytes, and lichens.

Every living and non-living thing ought to be given a single name because we think, learn, and communicate by using the names of things that surround us. Using more than one name for one thing would make our thinking, learning, and communicating cumbersome and our knowledge about the things more difficult to convey to others. Although, for some, the word "plants" is satisfactory when referring to many individual plants, the need to give each plant identity, i.e., a unique (certain and definite) name, has been recognized for several centuries. Consequently, each plant has been given a scientific or Latin name, according to the *International Code of Botanical Nomenclature* (Greuter *et al.* 1988), that is used by professionals around the world. Each plant also has a common or vernacular name that is used by both laypeople and professionals as a common language when referring to particular plants. Due to the fact that not all the plant forms in existence have yet been described and that not all the described plants have been known in terms of their systematic relationships with others, botanists have been continually making changes to plant nomenclature and taxonomy. Further, no rules exist for recognizing one accepted common name. One of the major goals of plant scientists worldwide is to standardize plant nomenclature for a region, a province, a country, a continent, and eventually the globe, so that people in different places are able to communicate and share information using the same standardized "language" and the same standardized format. The standardization of plant nomenclature is particularly important for biodiversity inventory, assessment, and conservation.

Plant scientists, specifically taxonomists dedicated to the systematic study of flora, recognize the need for compiling, from time to time, a reference – either a checklist or catalogue – of the flora within a given area. This reference would summarize additions and alterations to the nomenclature resulting from continuing taxonomic investigations (e.g., Kartesz 1994; Brako *et al.* 1995; Czerepanov 1995; Vitt *et al.* 1988; Stotler and Crandall-Stotler 1977; Anderson *et al.* 1990; Santesson 1993). The purpose of these checklists is to establish a taxonomic and nomenclatural framework, in accordance with current botanical nomenclature, that would accommodate the needs of those involved in various studies of flora and vegetation. Similar to the function of a dictionary, a checklist would aid authors in detecting improper, misspelled, or misapplied names that would unnecessarily diminish the merit of their work. A typical checklist for a plant species would give an accepted (valid) scientific name, a synonym(s), a family name, and a common name(s) (if available), but may include some other floristic data such as native range, plant duration, plant habit, and plant habitat (*op. cit.*). However, every checklist must be updated from time to time to take into consideration changes in nomenclature due to new knowledge on plant systematics and distributions. For example, the previous checklist for lichens of British Columbia included 1,013 species (Nobel *et al.* 1987), whereas now over 1,285 species are known to occur in the province (see Table 1).

This volume provides a complete, accurate, and up-to-date checklist of all known vascular plants, bryophytes, and lichens (including lichenicolous fungi), both native and naturalized to British Columbia. The novelty of this checklist is its combination of all the three major plant groups (vascular plants, bryophytes, and lichens) in a single volume. It was designed to be a concise and coherent reference work for a wide range of users, including the general public, students, and professionals. To provide an effective way of sharing botanical information and communication within the province, Canada, and North America, the nomenclature used in this book follows the most recently updated North American Standards (Stotler and Crandall-Stotler 1977; Anderson 1990; Anderson *et al.* 1990; Kartesz 1994; Esslinger and Egan 1995). However, due to the large size of the province and to the remoteness of some regions, it is conceivable that for some groups of plants, such as lichens, the checklist may be incomplete. But as Kartesz (1994:ix) stated: "Any attempt to develop a … floristic summary … can only represent a provisional best effort toward an accurate account of our present floristic knowledge, and each attempt must be viewed as another cycle to improve our understanding of the massive taxonomic and nomenclatural data that have accrued."

Sources of Data and Nomenclature

We compiled the checklist after an extensive review of literature that included over 300 publications, including regional guides and manuals, provincial species checklists, theses, monographs, and journal articles. We have not made any nomenclatural changes; in general, the nomenclature follows Kartesz (1994) for vascular plants, Anderson (1990) and Anderson *et al.* (1990) for mosses, Stotler and Crandall-Stotler (1977) for hepatics, and Esslinger and Egan (1995) for lichens, except for a few names which were published after the works cited above were published. For example, we recognized *Sinosenecio newcombe* (Janovec and Barkley 1996) as an accepted name instead of *Senecio newcombe* (Kartesz 1994).

Although common names are not used in scientific and international communications, there is the need to relate the scientific names to their common names for all those interested in plants. However, ignoring language difference, the common names vary with region, time, and personal preferences. When we compiled the list of common names from the literature, we found that some species have several common names (e.g., six common names [cattail, common cattail, broad-leaved cattail, common bulrush, reedmace, and nail-rod] have been used for *Typha latifolia*), and that the same common name has been used for more than one species (e.g., black-eyed rosette has been used for both *Physcia stellaris* and *Physcia phace*). For those species that have more than one common name, we have selected one as the accepted common name based on the frequency of use in the literature; for those species that have their common name duplicated, we assigned that name to one selected species to avoid confusion. For vascular plants, most of the common names we accepted were adopted from the first common name for each species given in Douglas *et al.* (1989, 1990, 1991, 1994).

Format

This book consists of three parts and an Appendix. Part 1 contains systematic lists of scientific names, Part 2 provides an alphabetical list of scientific names, and Part 3 presents an alphabetical list of common names. The Appendix contains a list of excluded names.

Part 1 includes three lists of scientific names, one for each of the three major plant groups: vascular plants, bryophytes, and lichens. Within each major plant group, families are organized alphabetically, and within each family, the scientific names of genera and species are listed alphabetically. The systematic arrangement of genera into families follows (i) Flora of North America Editorial Committee (1993) for pteridophytes and gymnosperms; (ii) Cronquist (1988) for angiosperms (which has been followed in the Flora of North America North of Mexico); (iii) Anderson (1990) and Anderson *et al.* (1990) for mosses, Stotler and Crandall-Stotler (1977) for liverworts and hornworts; and (iv) Eriksson and Hawksworth (1993) and Greuter *et al.* (1993) for lichens (including lichenicolous fungi). Twenty genera of lichens (e.g., *Abrothallus*) are listed in "uncertain-family" because their taxonomic status at the family rank has not yet been determined.

The family name is followed by a one-letter abbreviation in parentheses differentiating the major plant groups (V = vascular plants, B = bryophytes, and L = lichens). Vascular plants and bryophytes are further divided into minor plant groups: vascular plants into pteridophytes (F), gymnosperms (G), dicots (D), and monocots (M); and bryophytes into mosses (M) and hepatics (H). A one-letter abbreviation for the minor plant groups is given after the abbreviation for the major plant groups and a colon (:) as a separator. For example, "V" and "D" in **Aceraceae** (V:D) indicate that the major plant group is vascular plants and that the minor plant group is dicots.

The names in **boldface** or in normal print are accepted names, while those printed in *italic* are synonyms. Each accepted name below the family rank is followed by an authority, an acronym (in square brackets), and, if available, a common name (in parentheses), e.g., **Acer glabrum** var. **douglasii** (Hook.) Dippel [ACERGLA1] (Douglas maple).

Acronyms are unique codes. Acronyms for genera consist of the first seven letters of the generic name plus a dollar sign ($) (e.g., the acronym for the genus **Adiantum** is ADIANTU$), except for a few acronyms for which their seventh letter was derived from the eighth, ninth, or tenth letter of the generic name to avoid duplication. Acronyms for binomial names were framed by taking the first four letters of the generic name and the first three letters of species epithet, and then combining the two parts. Similar to the acronyms for the genera, if two or more acronyms for binomial names were the same, we used the fourth, fifth, or sixth letter of species epithets to avoid duplications. An acronym for an infraspecific taxon (i.e., subspecies, variety, or form) was framed by placing a number (from 0 to 9) after the acronym of its binomial name. The number "0" was reserved for the typical infraspecific taxon, e.g., the acronyms for **Artemisia tilesii** ssp. **elatior** and **Artemisia tilesii** ssp. **tilesii** are ARTETIL1 and ARTETIL0, respectively. Thus, there is no acronym or common name duplicated for any of the three major plant groups. Taxa with an asterisk (*) printed in front of the accepted name are exotic but naturalized to British Columbia, e.g., ***Acer platanoides** L. [ACERPLA] (Norway maple).

If the accepted scientific name had a synonym(s) associated with it, its synonym(s) was listed in the following indented line(s) of the text with the abbreviation "SY:". For example, **Acer glabrum** var. **douglasii** has two synonyms; they were listed as

Acer glabrum var. **douglasii** (Hook.) Dippel
[ACERGLA1] (Douglas maple)
SY: *Acer douglasiii* Hook.
SY: *Acer glabrum* ssp. *douglasii* (Hook.) Wesmael

All synonym(s) listed under an accepted name are also listed independently in the proper location (i.e., in the alphabetical order according to their family) without the indent and abbreviation "SY" but followed by an equal sign (=) and the accepted name plus acronym in square brackets, e.g., *Acer douglasii* Hook. = Acer glabrum var. douglasii [ACERGLA1].

The format of Part 2 is similar to Part 1 except (i) scientific names (both accepted names and synonyms) are listed alphabetically regardless of plant groups and families; (ii) common names are not included; (iii) family names follow acronyms in square brackets and, in turn, are followed by one-letter abbreviations for the major and, if available, the minor plant groups, with a colon as a separator between them; and (iv) synonyms are listed independently, i.e., for each accepted scientific name that has a synonym(s) the synonym name(s) is not listed immediately following the accepted name. For example:

Acer glabrum Torr. [ACERGLA] Aceraceae (V:D)
Acer glabrum ssp. *Douglasii* (Hook.) Wesmael = Acer glabrum var. douglasii [ACERGLA1] Aceraceae (V:D)
Acer glabrum var. **douglasii** (Hook.) Dippel [ACERGLA1] Aceraceae (V:D)
Acer macrophyllum Pursh [ACERMAC] Aceraceae (V:D)

In Part 3 common names are listed alphabetically. Each common name is followed by an equal sign (=) and the accepted scientific name with abbreviations for the major and, if available, the minor plant group in parentheses. Those common names printed in **boldface** are the names recommended to be used for the taxon whose accepted scientific name appears on the right side of the equal sign, e.g., **Douglas maple** = Acer glabrum var. douglasii (V:D). Each common name is also indexed on its keyword. In the example given above, the common name **Douglas maple** is also indexed under "maple, Douglas" followed by a right-pointed arrow (→) and the common name with every word in its regular order (i.e., maple, Douglas → Douglas maple). Thus both common name and its keyword can be used to locate the scientific name(s) associated with that common name.

In the Appendix a total of 1,090 excluded names are listed. These are names present in the botanical literature that have been misapplied to plants in the province and for which correct names have not been determined. The users who fail to locate a specific Latin name in Part 1 or Part 2 should check the Appendix.

Symbols and Abbreviations

- * an exotic taxon
- = a name on the left side of this symbol is a synonym of the name on the right side of the symbol
- SY: an abbreviation of the word "synonym"; a name following this abbreviation is a synonym of the name above it in boldface
- → see under the name on the right side of the symbol.

Use of the Book

This book is designed to determine, for any included plant species, the accepted scientific name, synonym(s), common name, and family name. Users not familiar with plant taxonomy and nomenclature can start their search by using either Part 2 or Part 3, depending on whether they are searching with a common or scientific name. In both situations, the search is facilitated by the alphabetical order of names. Part 2 is used to confirm from a given scientific name of a species its valid scientific name, authority, acronym, family, and plant group; Part 3 is used to locate from a given common name of a species its valid scientific name and plant group. Part 1 provides the authority, acronym, recommended common name and family name, and lists all synonyms of the valid scientific name according to the major plant groups and families.

Summary of the Taxa in the Book

This book includes 4,349 species and 604 infraspecific taxa belonging to a total of 1,128 genera and 304 families of native plants (Table 1) and 582 introduced (exotic but naturalized) taxa. We treated a total of 16,919 scientific names, 3,976 common names, with 2,960 recommended for use, and 1,090 excluded names.

Table 1

Number of families, genera, species, and infraspecific taxa (subspecies and varieties) of vascular plants, bryophytes, and lichens native to British Columbia.

Plant group	*Families*	*Genera*	*Species*	*Infraspecific taxa*
Vascular plants	**125**	**565**	**2,105**	**487**
Pteridophytes	15	31	102	12
Gymnosperms	3	10	28	3
Dicots	87	405	1,418	350
Monocots	20	119	557	122
Bryophytes	**96**	**284**	**955**	**66**
Hepatics	40	76	237	18
Mosses	56	208	718	48
Lichens	**83**	**279**	**1,289**	**51**

Literature Cited

Anderson, L.E. 1990. A checklist of *Sphagnum* in North America north of Mexico. Bryologist 93: 500-501.

Anderson, L.E., Crum, H.A., and Buck, W.R. 1990. List of the mosses of North America north of Mexico. Bryologist 93: 448-499.

Brako, L., Rossman, A.Y., and Farr, D.F. 1995. Scientific and common names of 7,000 vascular plants in the United States. The American Phytopathological Society Press, St. Paul, Minnesota.

Cronquist, A. 1988. The evolution and classification of flowering plants. 2nd ed. New York Botanical Garden, Bronx, New York.

Czerepanov, S.K. 1995. Vascular plants of Russia and adjacent states (the former USSR). Cambridge University Press, Cambridge.

Douglas, G.W., Straley, G.B., and Meidinger, D. 1989, 1990, 1991, 1994. The vascular plants of British Columbia. Vols. 1-4. B.C. Min. For., Victoria, British Columbia.

Eriksson, O.V., and Hawksworth, D.L. 1993. Outline of the ascomycetes - 1993. Systema Ascomycetum 12: 51-257.

Esslinger, T.L., and Egan, R.S. 1995. A sixth checklist of lichen-forming, lichenicolous, and allied fungi of the continental United States and Canada. Bryologist 98: 467-549.

Farley, A.L. 1979. Atlas of British Columbia: people, environment, and resource use. University of British Columbia Press, Vancouver, British Columbia.

Flora of North America Editorial Committee. 1993. Flora of North America north of Mexico. Vol. 2. Oxford University Press, New York.

Goward, T., McCune, B., and Meidinger, D. 1994. The lichens of British Columbia, illustrated keys. Part 1. Foliose and squamulose species. Research Branch, B.C. Min. For., Victoria, British Columbia.

Greuter, W., Burdet, H.M., Chaloner, W.G., Demoulin, V., Grolle, R., Hawksworth, D.L., Nicolson, D.H., Silva, P.C., Stafleu, F.A., Voss, E.G., and McNeill, J. (eds.) 1988. International code of botanical nomenclature: adopted by the fourteenth International Botanical Congress, Berlin, July-August 1987. Koeltz Scientific Books, Königstein, Germany.

Greuter, W., Brummitt, R.K., Farr, E., Kilian, N., Kirk, P.M., and Silva, P.C. 1993. Names in current use for extant plant genera. Koeltz Scientific Books, Königstein, Germany.

Hitchcock, C.L., and Cronquist, A. 1973. Flora of the Pacific Northwest: an illustrated manual. University of Washington Press, Seattle, Washington.

Hitchcock, C.L., Cronquist, A., Ownbey, M., and Thompson, J.W. 1955-1969. Vascular plants of the Pacific Northwest. Vols. 1-5. University of Washington Press, Seattle, Washington.

Holland, S.S. 1976. Landforms of British Columbia: a physiographic outline. 2nd ed. B.C. Dept. Mines Pet. Resources, Bull. No. 48. Victoria, British Columbia.

Hultén, E. 1968. Flora of Alaska and neighboring territories: a manual of the vascular plants. Stanford University Press, Stanford, California.

Ireland, R.R., Brassard, G.R., Schofield, W.B., and Vitt, D.H. 1987. Checklist of the mosses of Canada II. Lindbergia 13: 1-62.

Janovec, J.P., and Barkley, T.M. 1996. *Sinosenecio newcombei* (Asteraceae: Senecioneae): a new combination for a North American plant in an Asiatic genus. Novon 6: 265-267.

Johnson, D., Kershaw, L., MacKinnon, A., and Pojar, J. 1995. Plants of the western boreal forest and aspen parkland. Lone Pine Publishing, Edmonton, Alberta.

Kartesz, J.T. 1994. A synonymized checklist of the vascular flora of the United States, Canada, and Greenland. 2nd ed. Vols. 1 & 2. Timber Press, Portland, Oregon.

Klinka, K. 1976. Ecosystem units, their classification, interpretation, and mapping in UBC Research Forest. Ph.D. thesis, Fac. For., University of British Columbia, Vancouver, British Columbia.

Klinka, K., Krajina, V.J., Ceska, A., and Scagel, A.M. 1989. Indicator plants of coastal British Columbia. University of British Columbia Press, Vancouver, British Columbia.

Krajina, V.J. 1969. Ecology of forest trees in British Columbia. Ecol. West. N. Amer. 2: 1-146.

MacKinnon, A., Pojar, J., and Coupé, R. (eds.) 1992. Plants of northern British Columbia. Lone Pine Publishing, Edmonton, Alberta.

McMinn, R.G. 1957. Water relations in the Douglas-fir region on Vancouver Island. Ph.D. thesis, Dept. Biol. & Bot., University of British Columbia, Vancouver, British Columbia.

Meidinger, D.V., and Pojar, J. (eds.) 1991. Ecosystems of British Columbia. Special Report Series No. 6, Victoria, British Columbia.

Noble, W.J., Ahti, T., Otto, G.F., and Brodo, I.M. 1987. A second checklist and bibliography of the lichens and allied fungi of British Columbia. Syllogeus 61: 1-95.

Parish, R., Coupé, R., and Lloyd, D. (eds.) 1996. Plants of southern interior British Columbia. Lone Pine Publishing, Vancouver, British Columbia.

Pojar, J., and MacKinnon, A. (eds.) 1994. Plants of coastal British Columbia including Washington, Oregon and Alaska. Lone Pine Publishing, Edmonton, Alberta.

Resources Inventory Committee. 1995. Standards for terrestrial ecosystem mapping in B.C. Ecosystems Working Group, Resources Inventory Committee, Victoria, British Columbia.

Santesson, R. 1993. The lichens and lichenicolous fungi of Sweden and Norway. Botanical Museum, Uppsala.

Schofield, W.B. 1969. A checklist of Hepaticae and Anthocerotae of British Columbia. Syesis 1: 157-162.

Schofield, W.B. 1992. Some common mosses of British Columbia. Royal British Columbia Museum, Victoria, British Columbia.

Scoggan, H.J. 1978, 1979. The flora of Canada. Vols. 1-4. National Museums of Canada, Ottawa.

Stotler, R., and Crandall-Stotler, B. 1977. A checklist of the liverworts and hornworts of North America. Bryologist 80: 405-428.

Taylor, R.L., and MacBryde, B. 1977. Vascular plants of British Columbia. Tech. Bull. No. 4. University of British Columbia Press, Vancouver, British Columbia.

Vitt, D.H., Marsh, J.E., and Bovey, R.B. 1988. Mosses, lichens and ferns of northwest North America. Lone Pine Publishing, Edmonton, Alberta.

Part 1

Systematic Lists of Scientific Names

Vascular Plants

ACERACEAE (V:D)

Acer L. [ACER$]
Acer circinatum Pursh [ACERCIR] (vine maple)
Acer douglasii Hook. = Acer glabrum var. douglasii [ACERGLA1]
Acer glabrum Torr. [ACERGLA] (Rocky Mountain maple)
Acer glabrum ssp. *douglasii* (Hook.) Wesmael = Acer glabrum var. douglasii [ACERGLA1]
Acer glabrum var. **douglasii** (Hook.) Dippel [ACERGLA1] (Douglas maple)
SY: *Acer douglasii* Hook.
SY: *Acer glabrum* ssp. *douglasii* (Hook.) Wesmael
Acer macrophyllum Pursh [ACERMAC] (bigleaf maple)
*****Acer negundo** L. [ACERNEG] (box elder)
*****Acer platanoides** L. [ACERPLA] (Norway maple)
SY: *Acer platanoides* var. *schwedleri* Nichols.
Acer platanoides var. *schwedleri* Nichols. = Acer platanoides [ACERPLA]
*****Acer pseudoplatanus** L. [ACERPSE] (sycamore maple)

ACORACEAE (V:M)

Acorus L. [ACORUS$]
Acorus americanus (Raf.) Raf. [ACORAME] (American sweet-flag)
SY: *Acorus calamus* var. *americanus* (Raf.) H.D. Wulff.
SY: *Acorus calamus* auct.
Acorus calamus auct. = Acorus americanus [ACORAME]
Acorus calamus var. *americanus* (Raf.) H.D. Wulff. = Acorus americanus [ACORAME]

ADOXACEAE (V:D)

Adoxa L. [ADOXA$]
Adoxa moschatellina L. [ADOXMOS] (moschatel)

ALISMATACEAE (V:M)

Alisma L. [ALISMA$]
Alisma brevipes Greene = Alisma triviale [ALISTRI]
Alisma geyeri Torr. = Alisma gramineum [ALISGRA]
Alisma gramineum Lej. [ALISGRA] (narrow-leaved water-plantain)
SY: *Alisma geyeri* Torr.
SY: *Alisma gramineum* var. *angustissimum* (DC.) Hendricks
SY: *Alisma gramineum* var. *geyeri* (Torr.) Lam.
SY: *Alisma gramineum* var. *graminifolium* (Wahlenb.) Hendricks
SY: *Alisma lanceolatum* Gray
Alisma gramineum var. *angustissimum* (DC.) Hendricks = Alisma gramineum [ALISGRA]
Alisma gramineum var. *geyeri* (Torr.) Lam. = Alisma gramineum [ALISGRA]
Alisma gramineum var. *graminifolium* (Wahlenb.) Hendricks = Alisma gramineum [ALISGRA]
Alisma lanceolatum Gray = Alisma gramineum [ALISGRA]
Alisma plantago-aquatica ssp. *brevipes* (Greene) Samuelsson = Alisma triviale [ALISTRI]
Alisma plantago-aquatica var. *americanum* J.A. Schultes = Alisma triviale [ALISTRI]
Alisma plantago-aquatica var. *brevipes* (Greene) Victorin = Alisma triviale [ALISTRI]
Alisma plantago-aquatica var. *triviale* (Britt., Sterns & Pogg.) Far. = Alisma triviale [ALISTRI]
Alisma triviale Pursh [ALISTRI]
SY: *Alisma brevipes* Greene
SY: *Alisma plantago-aquatica* var. *americanum* J.A. Schultes
SY: *Alisma plantago-aquatica* ssp. *brevipes* (Greene) Samuelsson
SY: *Alisma plantago-aquatica* var. *brevipes* (Greene) Victorin
SY: *Alisma plantago-aquatica* var. *triviale* (Britt., Sterns & Pogg.) Far.
Sagittaria L. [SAGITTA$]
Sagittaria arifolia Nutt. ex J.G. Sm. = Sagittaria cuneata [SAGICUN]
Sagittaria cuneata Sheldon [SAGICUN] (arum-leaved arrowhead)
SY: *Sagittaria arifolia* Nutt. ex J.G. Sm.
Sagittaria engelmanniana ssp. *longirostra* (Micheli) Bogin = Sagittaria latifolia [SAGILAT]
Sagittaria esculenta T.J. Howell = Sagittaria latifolia [SAGILAT]
Sagittaria latifolia Willd. [SAGILAT] (wapato)
SY: *Sagittaria engelmanniana* ssp. *longirostra* (Micheli) Bogin
SY: *Sagittaria esculenta* T.J. Howell
SY: *Sagittaria latifolia* var. *obtusa* (Engelm.) Wieg.
SY: *Sagittaria longirostra* (Micheli) J.G. Sm.
SY: *Sagittaria obtusa* Muhl. ex Willd.
SY: *Sagittaria ornithorhyncha* Small
SY: *Sagittaria planipes* Fern.
SY: *Sagittaria variabilis* var. *obtusa* Engelm.
SY: *Sagittaria viscosa* C. Mohr p.p.
Sagittaria latifolia var. *obtusa* (Engelm.) Wieg. = Sagittaria latifolia [SAGILAT]
Sagittaria longirostra (Micheli) J.G. Sm. = Sagittaria latifolia [SAGILAT]
Sagittaria obtusa Muhl. ex Willd. = Sagittaria latifolia [SAGILAT]
Sagittaria ornithorhyncha Small = Sagittaria latifolia [SAGILAT]
Sagittaria planipes Fern. = Sagittaria latifolia [SAGILAT]
Sagittaria variabilis var. *obtusa* Engelm. = Sagittaria latifolia [SAGILAT]
Sagittaria viscosa C. Mohr p.p. = Sagittaria latifolia [SAGILAT]

AMARANTHACEAE (V:D)

Amaranthus L. [AMARANT$]
*****Amaranthus albus** L. [AMARALB] (tumbleweed)
SY: *Amaranthus albus* var. *pubescens* (Uline & Bray) Fern.
SY: *Amaranthus graecizans* var. *pubescens* Uline & Bray
SY: *Amaranthus pubescens* (Uline & Bray) Rydb.
Amaranthus albus var. *pubescens* (Uline & Bray) Fern. = Amaranthus albus [AMARALB]
*****Amaranthus blitoides** S. Wats. [AMARBLI] (prostrate pigweed)
Amaranthus bouchonii Thellung = Amaranthus powellii [AMARPOW]
Amaranthus bracteosus Uline & Bray = Amaranthus powellii [AMARPOW]
Amaranthus graecizans var. *pubescens* Uline & Bray = Amaranthus albus [AMARALB]
Amaranthus powellii S. Wats. [AMARPOW] (Powell's amaranth)
SY: *Amaranthus bouchonii* Thellung
SY: *Amaranthus bracteosus* Uline & Bray
SY: *Amaranthus retroflexus* var. *powellii* (S. Wats.) Boivin
Amaranthus pubescens (Uline & Bray) Rydb. = Amaranthus albus [AMARALB]
*****Amaranthus retroflexus** L. [AMARRET] (rough pigweed)
SY: *Amaranthus retroflexus* var. *salicifolius* I.M. Johnston
Amaranthus retroflexus var. *powellii* (S. Wats.) Boivin = Amaranthus powellii [AMARPOW]
Amaranthus retroflexus var. *salicifolius* I.M. Johnston = Amaranthus retroflexus [AMARRET]

ANACARDIACEAE (V:D)

Rhus L. [RHUS$]

Rhus borealis Greene = Rhus glabra [RHUSGLA]
Rhus calophylla Greene = Rhus glabra [RHUSGLA]
Rhus diversiloba Torr. & Gray = Toxicodendron diversilobum [TOXIDIV]
Rhus glabra L. [RHUSGLA] (smooth sumac)
SY: *Rhus borealis* Greene
SY: *Rhus calophylla* Greene
SY: *Rhus glabra* var. *laciniata* Carr.
SY: *Rhus glabra* var. *occidentalis* Torr.
Rhus glabra var. *laciniata* Carr. = Rhus glabra [RHUSGLA]
Rhus glabra var. *occidentalis* Torr. = Rhus glabra [RHUSGLA]
Rhus radicans var. *rydbergii* (Small ex Rydb.) Rehd. = Toxicodendron rydbergii [TOXIRYD]
Rhus radicans var. *vulgaris* (Michx.) DC. = Toxicodendron rydbergii [TOXIRYD]
Rhus toxicodendron var. *vulgaris* Michx. = Toxicodendron rydbergii [TOXIRYD]
Toxicodendron P. Mill. [TOXICOD$]
Toxicodendron desertorum Lunell = Toxicodendron rydbergii [TOXIRYD]
Toxicodendron diversilobum (Torr. & Gray) Greene [TOXIDIV] (poison-oak)
SY: *Rhus diversiloba* Torr. & Gray
SY: *Toxicodendron radicans* ssp. *diversilobum* (Torr. & Gray) Thorne
Toxicodendron radicans ssp. *diversilobum* (Torr. & Gray) Thorne = Toxicodendron diversilobum [TOXIDIV]
Toxicodendron radicans var. *rydbergii* (Small ex Rydb.) Erskine = Toxicodendron rydbergii [TOXIRYD]
Toxicodendron rydbergii (Small ex Rydb.) Greene [TOXIRYD] (poison-ivy)
SY: *Rhus radicans* var. *rydbergii* (Small ex Rydb.) Rehd.
SY: *Rhus radicans* var. *vulgaris* (Michx.) DC.
SY: *Rhus toxicodendron* var. *vulgaris* Michx.
SY: *Toxicodendron desertorum* Lunell
SY: *Toxicodendron radicans* var. *rydbergii* (Small ex Rydb.) Erskine

APIACEAE (V:D)

Aegopodium L. [AEGOPOD$]
***Aegopodium podagraria** L. [AEGOPOA] (goutweed)
SY: *Aegopodium podagraria* var. *variegatum* Bailey
Aegopodium podagraria var. *variegatum* Bailey = Aegopodium podagraria [AEGOPOA]
Anethum L. [ANETHUM$]
***Anethum graveolens** L. [ANETGRA] (common dill)
Angelica L. [ANGELIC$]
Angelica arguta Nutt. [ANGEARG] (sharptooth angelica)
SY: *Angelica lyallii* S. Wats.
Angelica dawsonii S. Wats. [ANGEDAW] (Dawson's angelica)
Angelica genuflexa Nutt. [ANGEGEN] (kneeling angelica)
Angelica lucida L. [ANGELUC] (seacoast angelica)
SY: *Coelopleurum actaeifolium* (Michx.) Coult. & Rose
SY: *Coelopleurum gmelinii* (DC.) Ledeb.
SY: *Coelopleurum lucidum* (L.) Fern.
SY: *Coelopleurum lucidum* ssp. *gmelinii* (DC.) A. & D. Löve
Angelica lyallii S. Wats. = Angelica arguta [ANGEARG]
Anthriscus Pers. [ANTHRIC$]
***Anthriscus caucalis** Bieb. [ANTHCAU] (bur chervil)
SY: *Anthriscus neglecta* var. *scandix* (Scop.) Hyl.
SY: *Anthriscus scandicina* (Weber ex Wiggers) Mansf.
Anthriscus neglecta var. *scandix* (Scop.) Hyl. = Anthriscus caucalis [ANTHCAU]
Anthriscus scandicina (Weber ex Wiggers) Mansf. = Anthriscus caucalis [ANTHCAU]
Berula Bess. ex W.D.J. Koch [BERULA$]
Berula erecta (Huds.) Coville [BERUERE] (cut-leaved water-parsnip)
SY: *Berula erecta* var. *incisa* (Torr.) Cronq.
SY: *Berula incisa* (Torr.) G.N. Jones
SY: *Berula pusilla* Fern.
SY: *Siella erecta* (Huds.) M. Pimen.
SY: *Sium incisum* Torr.
Berula erecta var. *incisa* (Torr.) Cronq. = Berula erecta [BERUERE]
Berula incisa (Torr.) G.N. Jones = Berula erecta [BERUERE]
Berula pusilla Fern. = Berula erecta [BERUERE]
Bupleurum L. [BUPLEUR$]
Bupleurum americanum Coult. & Rose [BUPLAME] (American thorough-wax)
SY: *Bupleurum triradiatum* ssp. *arcticum* (Regel) Hultén
Bupleurum triradiatum ssp. *arcticum* (Regel) Hultén = Bupleurum americanum [BUPLAME]
Carum L. [CARUM$]
***Carum carvi** L. [CARUCAR] (caraway)
Caucalis microcarpa Hook. & Arn. = Yabea microcarpa [YABEMIC]
Cicuta L. [CICUTA$]
Cicuta bulbifera L. [CICUBUL] (bulbous water-hemlock)
Cicuta douglasii (DC.) Coult. & Rose [CICUDOU] (Douglas' water-hemlock)
SY: *Cicuta maculata* var. *californica* (Gray) Boivin
Cicuta mackenzieana Raup = Cicuta virosa [CICUVIR]
Cicuta maculata L. [CICUMAU] (spotted water-hemlock)
Cicuta maculata var. **angustifolia** Hook. [CICUMAU1] (spotted cowbane)
SY: *Cicuta occidentalis* Greene
Cicuta maculata var. *californica* (Gray) Boivin = Cicuta douglasii [CICUDOU]
Cicuta occidentalis Greene = Cicuta maculata var. angustifolia [CICUMAU1]
Cicuta virosa L. [CICUVIR] (European water-hemlock)
SY: *Cicuta mackenzieana* Raup
Cnidium Cusson ex Juss. [CNIDIUM$]
Cnidium cnidiifolium (Turcz.) Schischkin [CNIDCNI] (hemlock-parsley)
SY: *Conioselinum cnidiifolium* (Turcz.) Porsild
Coelopleurum actaeifolium (Michx.) Coult. & Rose = Angelica lucida [ANGELUC]
Coelopleurum gmelinii (DC.) Ledeb. = Angelica lucida [ANGELUC]
Coelopleurum lucidum (L.) Fern. = Angelica lucida [ANGELUC]
Coelopleurum lucidum ssp. *gmelinii* (DC.) A. & D. Löve = Angelica lucida [ANGELUC]
Cogswellia foeniculacea (Nutt.) Coult. & Rose = Lomatium foeniculaceum [LOMAFOE]
Cogswellia macrocarpa (Hook. & Arn.) M.E. Jones = Lomatium macrocarpum [LOMAMAC]
Cogswellia nudicaulis (Pursh) M.E. Jones = Lomatium nudicaule [LOMANUD]
Cogswellia villosa (Raf.) J.A. Schultes = Lomatium foeniculaceum [LOMAFOE]
Conioselinum Hoffmann [CONIOSE$]
Conioselinum chinense var. *pacificum* (S. Wats.) Boivin = Conioselinum gmelinii [CONIGME]
Conioselinum cnidiifolium (Turcz.) Porsild = Cnidium cnidiifolium [CNIDCNI]
Conioselinum gmelinii (Cham. & Schlecht.) Steud. [CONIGME]
SY: *Conioselinum chinense* var. *pacificum* (S. Wats.) Boivin
SY: *Conioselinum pacificum* (S. Wats.) Coult. & Rose
Conioselinum pacificum (S. Wats.) Coult. & Rose = Conioselinum gmelinii [CONIGME]
Conium L. [CONIUM$]
***Conium maculatum** L. [CONIMAC] (poison-hemlock)
Cynomarathrum brandegei Coult. & Rose = Lomatium brandegei [LOMABRA]
Daucus L. [DAUCUS$]
***Daucus carota** L. [DAUCCAR] (wild carrot)
Daucus pusillus Michx. [DAUCPUS] (American wild carrot)
Eryngium L. [ERYNGIU$]
***Eryngium planum** L. [ERYNPLA] (plains eryngo)
Ferula macrocarpa Hook. & Arn. = Lomatium macrocarpum [LOMAMAC]
Foeniculum P. Mill. [FOENICU$]
Foeniculum foeniculum (L.) Karst. = Foeniculum vulgare [FOENVUL]
***Foeniculum vulgare** P. Mill. [FOENVUL] (sweet fennel)

SY: *Foeniculum foeniculum* (L.) Karst.
Glehnia F. Schmidt ex Miq. [GLEHNIA$]
Glehnia leiocarpa Mathias = Glehnia littoralis ssp. leiocarpa [GLEHLIT1]
Glehnia littoralis F. Schmidt ex Miq. [GLEHLIT] (American silvertop)
Glehnia littoralis ssp. **leiocarpa** (Mathias) Hultén [GLEHLIT1] (beach-carrot)
SY: *Glehnia leiocarpa* Mathias
SY: *Glehnia littoralis* var. *leiocarpa* (Mathias) Boivin
Glehnia littoralis var. *leiocarpa* (Mathias) Boivin = Glehnia littoralis ssp. leiocarpa [GLEHLIT1]
Glycosma occidentalis Nutt. ex Torr. & Gray = Osmorhiza occidentalis [OSMOOCC]
Haloscias hultenii (Fern.) Holub = Ligusticum scoticum ssp. hultenii [LIGUSCO1]
Heracleum L. [HERACLE$]
Heracleum lanatum Michx. = Heracleum maximum [HERAMAX]
*__Heracleum mantegazzianum__ Sommier & Levier [HERAMAN] (giant cow-parsnip)
Heracleum maximum Bartr. [HERAMAX]
SY: *Heracleum lanatum* Michx.
SY: *Heracleum sphondylium* ssp. *lanatum* (Michx.) A. & D. Löve
SY: *Heracleum sphondylium* var. *lanatum* (Michx.) Dorn
SY: *Heracleum sphondylium* ssp. *montanum* (Schleich. ex Gaudin) Briq.
Heracleum sphondylium ssp. *lanatum* (Michx.) A. & D. Löve = Heracleum maximum [HERAMAX]
Heracleum sphondylium ssp. *montanum* (Schleich. ex Gaudin) Briq. = Heracleum maximum [HERAMAX]
Heracleum sphondylium var. *lanatum* (Michx.) Dorn = Heracleum maximum [HERAMAX]
Hydrocotyle L. [HYDROCO$]
Hydrocotyle ranunculoides L. f. [HYDRRAN] (floating marsh pennywort)
Hydrocotyle verticillata Thunb. [HYDRVER] (whorled marsh pennywort)
Leptotaenia dissecta Nutt. = Lomatium dissectum var. dissectum [LOMADIS0]
Leptotaenia multifida Nutt. = Lomatium dissectum var. multifidum [LOMADIS2]
Ligusticum L. [LIGUSTI$]
Ligusticum caeruleimontanum St. John = Ligusticum canbyi [LIGUCAN]
Ligusticum calderi Mathias & Constance [LIGUCAL] (Calder's lovage)
Ligusticum canbyi Coult. & Rose [LIGUCAN] (Canby's lovage)
SY: *Ligusticum caeruleimontanum* St. John
SY: *Ligusticum leibergii* Coult. & Rose
Ligusticum hultenii Fern. = Ligusticum scoticum ssp. hultenii [LIGUSCO1]
Ligusticum leibergii Coult. & Rose = Ligusticum canbyi [LIGUCAN]
Ligusticum scoticum L. [LIGUSCO] (Scottish wild lovage)
Ligusticum scoticum ssp. **hultenii** (Fern.) Calder & Taylor [LIGUSCO1] (beach lovage)
SY: *Haloscias hultenii* (Fern.) Holub
SY: *Ligusticum hultenii* Fern.
SY: *Ligusticum scoticum* var. *hultenii* (Fern.) Boivin
Ligusticum scoticum var. *hultenii* (Fern.) Boivin = Ligusticum scoticum ssp. hultenii [LIGUSCO1]
Ligusticum verticillatum (Hook.) Coult. & Rose ex Rose [LIGUVER] (verticillate-umbel lovage)
Lilaeopsis Greene [LILAEOP$]
Lilaeopsis occidentalis Coult. & Rose [LILAOCC] (western lilaeopsis)
Lomatium Raf. [LOMATIU$]
Lomatium ambiguum (Nutt.) Coult. & Rose [LOMAAMB] (swale desert-parsley)
SY: *Peucedanum ambiguum* (Nutt.) Nutt. ex Torr. & Gray
Lomatium angustatum (Coult. & Rose) St. John = Lomatium martindalei [LOMAMAR]
Lomatium angustatum var. *flavum* G.N. Jones = Lomatium martindalei [LOMAMAR]
Lomatium brandegei (Coult. & Rose) J.F. Macbr. [LOMABRA] (Brandegee's lomatium)
SY: *Cynomarathrum brandegei* Coult. & Rose
Lomatium dissectum (Nutt.) Mathias & Constance [LOMADIS] (fern-leaved desert-parsley)
Lomatium dissectum var. **dissectum** [LOMADIS0]
SY: *Leptotaenia dissecta* Nutt.
Lomatium dissectum var. *eatonii* (Coult. & Rose) Cronq. = Lomatium dissectum var. multifidum [LOMADIS2]
Lomatium dissectum var. **multifidum** (Nutt.) Mathias & Constance [LOMADIS2]
SY: *Leptotaenia multifida* Nutt.
SY: *Lomatium dissectum* var. *eatonii* (Coult. & Rose) Cronq.
Lomatium foeniculaceum (Nutt.) Coult. & Rose [LOMAFOE] (carrot-leaf desert-parsley)
SY: *Cogswellia foeniculacea* (Nutt.) Coult. & Rose
SY: *Cogswellia villosa* (Raf.) J.A. Schultes
Lomatium geyeri (S. Wats.) Coult. & Rose [LOMAGEY] (Geyer's desert-parsley)
Lomatium grayi (Coult. & Rose) Coult. & Rose [LOMAGRA] (Gray's desert-parsley)
Lomatium macrocarpum (Nutt. ex Torr. & Gray) Coult. & Rose [LOMAMAC] (large-fruited desert-parsley)
SY: *Cogswellia macrocarpa* (Hook. & Arn.) M.E. Jones
SY: *Ferula macrocarpa* Hook. & Arn.
SY: *Lomatium macrocarpum* var. *artemisiarum* Piper
SY: *Lomatium macrocarpum* var. *ellipticum* (Torr. & Gray) Jepson
SY: *Peucedanum macrocarpum* Nutt. ex Torr. & Gray
Lomatium macrocarpum var. *artemisiarum* Piper = Lomatium macrocarpum [LOMAMAC]
Lomatium macrocarpum var. *ellipticum* (Torr. & Gray) Jepson = Lomatium macrocarpum [LOMAMAC]
Lomatium martindalei (Coult. & Rose) Coult. & Rose [LOMAMAR] (Martindale's lomatium)
SY: *Lomatium angustatum* (Coult. & Rose) St. John
SY: *Lomatium angustatum* var. *flavum* G.N. Jones
SY: *Lomatium martindalei* var. *angustatum* (Coult. & Rose) Coult. & Rose
SY: *Lomatium martindalei* var. *flavum* (G.N. Jones) Cronq.
Lomatium martindalei var. *angustatum* (Coult. & Rose) Coult. & Rose = Lomatium martindalei [LOMAMAR]
Lomatium martindalei var. *flavum* (G.N. Jones) Cronq. = Lomatium martindalei [LOMAMAR]
Lomatium nudicaule (Pursh) Coult. & Rose [LOMANUD] (bare-stem desert-parsley)
SY: *Cogswellia nudicaulis* (Pursh) M.E. Jones
Lomatium platycarpum (Torr.) Coult. & Rose = Lomatium simplex [LOMASIM]
Lomatium sandbergii (Coult. & Rose) Coult. & Rose [LOMASAN] (Sandberg's desert-parsley)
SY: *Peucedanum sandbergii* Coult. & Rose
Lomatium simplex (Nutt.) J.F. Macbr. [LOMASIM]
SY: *Lomatium platycarpum* (Torr.) Coult. & Rose
SY: *Lomatium triternatum* ssp. *platycarpum* (Torr.) Cronq.
Lomatium triternatum (Pursh) Coult. & Rose [LOMATRI] (nine-leaved desert-parsley)
Lomatium triternatum ssp. *platycarpum* (Torr.) Cronq. = Lomatium simplex [LOMASIM]
Lomatium utriculatum (Nutt. ex Torr. & Gray) Coult. & Rose [LOMAUTR] (spring gold)
SY: *Lomatium vaseyi* (Coult. & Rose) Coult. & Rose
Lomatium vaseyi (Coult. & Rose) Coult. & Rose = Lomatium utriculatum [LOMAUTR]
Oenanthe L. [OENANTH$]
Oenanthe sarmentosa K. Presl ex DC. [OENASAR] (Pacific water-parsley)
Osmorhiza Raf. [OSMORHI$]
Osmorhiza berteroi DC. [OSMOBER]
SY: *Osmorhiza chilensis* Hook. & Arn.
SY: *Osmorhiza divaricata* (Britt.) Suksdorf
SY: *Osmorhiza nuda* Torr.
SY: *Washingtonia divaricata* Britt.

Osmorhiza chilensis Hook. & Arn. = Osmorhiza berteroi [OSMOBER]
Osmorhiza chilensis var. *cupressimontana* (Boivin) Boivin = Osmorhiza depauperata [OSMODEP]
Osmorhiza chilensis var. *purpurea* (Coult. & Rose) Boivin = Osmorhiza purpurea [OSMOPUR]
Osmorhiza depauperata Phil. [OSMODEP] (blunt-fruited sweet-cicely)
SY: *Osmorhiza chilensis* var. *cupressimontana* (Boivin) Boivin
SY: *Osmorhiza obtusa* (Coult. & Rose) Fern.
SY: *Washingtonia obtusa* Coult. & Rose
Osmorhiza divaricata (Britt.) Suksdorf = Osmorhiza berteroi [OSMOBER]
Osmorhiza nuda Torr. = Osmorhiza berteroi [OSMOBER]
Osmorhiza obtusa (Coult. & Rose) Fern. = Osmorhiza depauperata [OSMODEP]
Osmorhiza occidentalis (Nutt. ex Torr. & Gray) Torr. [OSMOOCC] (western sweet-cicely)
SY: *Glycosma occidentalis* Nutt. ex Torr. & Gray
Osmorhiza purpurea (Coult. & Rose) Suksdorf [OSMOPUR] (purple sweet-cicely)
SY: *Osmorhiza chilensis* var. *purpurea* (Coult. & Rose) Boivin
SY: *Washingtonia purpurea* Coult. & Rose
Pastinaca L. [PASTINA$]
***Pastinaca sativa** L. [PASTSAT] (common parsnip)
Perideridia Reichenb. [PERIDER$]
Perideridia gairdneri (Hook. & Arn.) Mathias [PERIGAI] (Gairdner's yampah)
Peucedanum ambiguum (Nutt.) Nutt. ex Torr. & Gray = Lomatium ambiguum [LOMAAMB]
Peucedanum macrocarpum Nutt. ex Torr. & Gray = Lomatium macrocarpum [LOMAMAC]
Peucedanum sandbergii Coult. & Rose = Lomatium sandbergii [LOMASAN]
Sanicula L. [SANICUL$]
Sanicula arctopoides Hook. & Arn. [SANIARC] (snake-root)
SY: *Sanicula crassicaulis* var. *howellii* (Coult. & Rose) Mathias
SY: *Sanicula* × *howellii* (Coult. & Rose) Shan & Constance
Sanicula bipinnatifida Dougl. ex Hook. [SANIBIP] (purple sanicle)
SY: *Sanicula bipinnatifida* var. *flava* Jepson
Sanicula bipinnatifida var. *flava* Jepson = Sanicula bipinnatifida [SANIBIP]
Sanicula crassicaulis Poepp. ex DC. [SANICRA] (Pacific sanicle)
Sanicula crassicaulis var. *howellii* (Coult. & Rose) Mathias = Sanicula arctopoides [SANIARC]
Sanicula graveolens Poepp. ex DC. [SANIGRA] (Sierra sanicle)
SY: *Sanicula graveolens* var. *septentrionalis* (Greene) St. John
SY: *Sanicula nevadensis* S. Wats.
SY: *Sanicula nevadensis* var. *septentrionalis* (Greene) Mathias
Sanicula graveolens var. *septentrionalis* (Greene) St. John = Sanicula graveolens [SANIGRA]
Sanicula × *howellii* (Coult. & Rose) Shan & Constance = Sanicula arctopoides [SANIARC]
Sanicula marilandica L. [SANIMAR] (black sanicle)
SY: *Sanicula marilandica* var. *petiolulata* Fern.
Sanicula marilandica var. *petiolulata* Fern. = Sanicula marilandica [SANIMAR]
Sanicula nevadensis S. Wats. = Sanicula graveolens [SANIGRA]
Sanicula nevadensis var. *septentrionalis* (Greene) Mathias = Sanicula graveolens [SANIGRA]
Scandix L. [SCANDIX$]
***Scandix pecten-veneris** L. [SCANPEC] (Venus'-comb)
Siella erecta (Huds.) M. Pimen. = Berula erecta [BERUERE]
Sium L. [SIUM$]
Sium cicutifolium Schrank = Sium suave [SIUMSUA]
Sium floridanum Small = Sium suave [SIUMSUA]
Sium incisum Torr. = Berula erecta [BERUERE]
Sium suave Walt. [SIUMSUA] (hemlock water-parsnip)
SY: *Sium cicutifolium* Schrank
SY: *Sium floridanum* Small
SY: *Sium suave* var. *floridanum* (Small) C.F. Reed
Sium suave var. *floridanum* (Small) C.F. Reed = Sium suave [SIUMSUA]
Washingtonia divaricata Britt. = Osmorhiza berteroi [OSMOBER]
Washingtonia obtusa Coult. & Rose = Osmorhiza depauperata [OSMODEP]
Washingtonia purpurea Coult. & Rose = Osmorhiza purpurea [OSMOPUR]
Yabea K.-Pol. [YABEA$]
Yabea microcarpa (Hook. & Arn.) K.-Pol. [YABEMIC]
SY: *Caucalis microcarpa* Hook. & Arn.
Zizia W.D.J. Koch [ZIZIA$]
Zizia aptera (Gray) Fern. [ZIZIAPT] (heart-leaved Alexanders)
SY: *Zizia aptera* var. *occidentalis* Fern.
SY: *Zizia cordata* W.D.J. Koch ex DC.
Zizia aptera var. *occidentalis* Fern. = Zizia aptera [ZIZIAPT]
Zizia cordata W.D.J. Koch ex DC. = Zizia aptera [ZIZIAPT]

APOCYNACEAE (V:D)

Apocynum L. [APOCYNU$]
Apocynum androsaemifolium L. [APOCAND] (spreading dogbane)
Apocynum androsaemifolium ssp. **androsaemifolium** [APOCAND0]
Apocynum androsaemifolium ssp. **pumilum** (Gray) Boivin [APOCAND1]
SY: *Apocynum androsaemifolium* var. *pumilum* Gray
Apocynum androsaemifolium var. *pumilum* Gray = Apocynum androsaemifolium ssp. pumilum [APOCAND1]
Apocynum cannabinum L. [APOCCAN] (Indian-hemp)
SY: *Apocynum cannabinum* var. *angustifolium* (Woot.) N. Holmgren
SY: *Apocynum cannabinum* var. *glaberrimum* A. DC.
SY: *Apocynum cannabinum* var. *greeneanum* (Bég. & Bel.) Woods.
SY: *Apocynum cannabinum* var. *hypericifolium* Gray
SY: *Apocynum cannabinum* var. *nemorale* (G.S. Mill.) Fern.
SY: *Apocynum cannabinum* var. *pubescens* (Mitchell ex R. Br.) Woods.
SY: *Apocynum cannabinum* var. *suksdorfii* (Greene) Bég. & Bel.
SY: *Apocynum hypericifolium* Ait.
SY: *Apocynum pubescens* Mitchell ex R. Br.
SY: *Apocynum sibiricum* Jacq.
SY: *Apocynum sibiricum* var. *cordigerum* (Greene) Fern.
SY: *Apocynum sibiricum* var. *farwellii* (Greene) Fern.
SY: *Apocynum sibiricum* var. *salignum* (Greene) Fern.
SY: *Apocynum suksdorfii* Greene
SY: *Apocynum suksdorfii* var. *angustifolium* (Woot.) Woods.
Apocynum cannabinum var. *angustifolium* (Woot.) N. Holmgren = Apocynum cannabinum [APOCCAN]
Apocynum cannabinum var. *glaberrimum* A. DC. = Apocynum cannabinum [APOCCAN]
Apocynum cannabinum var. *greeneanum* (Bég. & Bel.) Woods. = Apocynum cannabinum [APOCCAN]
Apocynum cannabinum var. *hypericifolium* Gray = Apocynum cannabinum [APOCCAN]
Apocynum cannabinum var. *nemorale* (G.S. Mill.) Fern. = Apocynum cannabinum [APOCCAN]
Apocynum cannabinum var. *pubescens* (Mitchell ex R. Br.) Woods. = Apocynum cannabinum [APOCCAN]
Apocynum cannabinum var. *suksdorfii* (Greene) Bég. & Bel. = Apocynum cannabinum [APOCCAN]
Apocynum × **floribundum** Greene (pro sp.) [APOCFLO]
SY: *Apocynum jonesii* Woods.
SY: *Apocynum medium* Greene
SY: *Apocynum medium* var. *floribundum* (Greene) Woods.
SY: *Apocynum medium* var. *leuconeuron* (Greene) Woods.
SY: *Apocynum medium* var. *lividum* (Greene) Woods.
SY: *Apocynum medium* var. *sarniense* (Greene) Woods.
SY: *Apocynum medium* var. *vestitum* (Greene) Woods.
SY: *Apocynum milleri* Britt.
Apocynum hypericifolium Ait. = Apocynum cannabinum [APOCCAN]
Apocynum jonesii Woods. = Apocynum × floribundum [APOCFLO]
Apocynum medium Greene = Apocynum × floribundum [APOCFLO]

Apocynum medium var. *floribundum* (Greene) Woods. = Apocynum × floribundum [APOCFLO]
Apocynum medium var. *leuconeuron* (Greene) Woods. = Apocynum × floribundum [APOCFLO]
Apocynum medium var. *lividum* (Greene) Woods. = Apocynum × floribundum [APOCFLO]
Apocynum medium var. *sarniense* (Greene) Woods. = Apocynum × floribundum [APOCFLO]
Apocynum medium var. *vestitum* (Greene) Woods. = Apocynum × floribundum [APOCFLO]
Apocynum milleri Britt. = Apocynum × floribundum [APOCFLO]
Apocynum pubescens Mitchell ex R. Br. = Apocynum cannabinum [APOCCAN]
Apocynum sibiricum Jacq. = Apocynum cannabinum [APOCCAN]
Apocynum sibiricum var. *cordigerum* (Greene) Fern. = Apocynum cannabinum [APOCCAN]
Apocynum sibiricum var. *farwellii* (Greene) Fern. = Apocynum cannabinum [APOCCAN]
Apocynum sibiricum var. *salignum* (Greene) Fern. = Apocynum cannabinum [APOCCAN]
Apocynum suksdorfii Greene = Apocynum cannabinum [APOCCAN]
Apocynum suksdorfii var. *angustifolium* (Woot.) Woods. = Apocynum cannabinum [APOCCAN]
Vinca L. [VINCA$]
•**Vinca major** L. [VINCMAJ] (large periwinkle)
SY: *Vinca major* var. *variegata* Loud.
Vinca major var. *variegata* Loud. = Vinca major [VINCMAJ]
•**Vinca minor** L. [VINCMIN] (common periwinkle)

AQUIFOLIACEAE (V:D)

Ilex L. [ILEX$]
•**Ilex aquifolium** L. [ILEXAQU] (English holly)

ARACEAE (V:M)

Calla L. [CALLA$]
Calla palustris L. [CALLPAL] (water arum)
Lysichiton Schott [LYSICHI$]
Lysichiton americanum Hultén & St. John [LYSIAME] (skunk cabbage)

ARALIACEAE (V:D)

Aralia L. [ARALIA$]
Aralia nudicaulis L. [ARALNUD] (wild sarsaparilla)
Echinopanax horridus (Sm.) Dcne. & Planch. ex H.A.T. Harms = Oplopanax horridus [OPLOHOR]
Hedera L. [HEDERA$]
•**Hedera helix** L. [HEDEHEL] (English ivy)
Oplopanax (Torr. & Gray) Miq. [OPLOPAN$]
Oplopanax horridus Miq. [OPLOHOR] (devil's club)
SY: *Echinopanax horridus* (Sm.) Dcne. & Planch. ex H.A.T. Harms

ARISTOLOCHIACEAE (V:D)

Asarum L. [ASARUM$]
Asarum caudatum Lindl. [ASARCAU] (wild ginger)

ASCLEPIADACEAE (V:D)

Asclepias L. [ASCLEPI$]
Asclepias ovalifolia Dcne. [ASCLOVA] (oval-leaved milkweed)
Asclepias speciosa Torr. [ASCLSPE] (showy milkweed)

ASPLENIACEAE (V:F)

Asplenium L. [ASPLENI$]
Asplenium adulterinum Milde [ASPLADU] (corrupt spleenwort)
Asplenium melanocaulon Willd. = Asplenium trichomanes [ASPLTRI]
Asplenium trichomanes L. [ASPLTRI] (maidenhair spleenwort)
SY: *Asplenium melanocaulon* Willd.
Asplenium trichomanes-ramosum L. [ASPLTRC]
SY: *Asplenium viride* Huds.
Asplenium viride Huds. = Asplenium trichomanes-ramosum [ASPLTRC]

ASTERACEAE (V:D)

Achillea L. [ACHILLE$]
Achillea alpicola (Rydb.) Rydb. = Achillea millefolium var. alpicola [ACHIMIL1]
Achillea angustissima Rydb. = Achillea millefolium var. occidentalis [ACHIMIL8]
Achillea borealis Bong. = Achillea millefolium var. borealis [ACHIMIL2]
Achillea borealis ssp. *typica* Keck = Achillea millefolium var. borealis [ACHIMIL2]
Achillea eradiata Piper = Achillea millefolium var. occidentalis [ACHIMIL8]
Achillea fusca Rydb. = Achillea millefolium var. alpicola [ACHIMIL1]
Achillea gracilis Raf. = Achillea millefolium var. occidentalis [ACHIMIL8]
Achillea lanulosa Nutt. = Achillea millefolium var. occidentalis [ACHIMIL8]
Achillea lanulosa ssp. *alpicola* (Rydb.) Keck = Achillea millefolium var. alpicola [ACHIMIL1]
Achillea lanulosa ssp. *typica* Keck = Achillea millefolium var. occidentalis [ACHIMIL8]
Achillea lanulosa var. *alpicola* Rydb. = Achillea millefolium var. alpicola [ACHIMIL1]
Achillea lanulosa var. *arachnoidea* Lunell = Achillea millefolium var. occidentalis [ACHIMIL8]
Achillea lanulosa var. *eradiata* (Piper) M.E. Peck = Achillea millefolium var. occidentalis [ACHIMIL8]
Achillea laxiflora Pollard & Cockerell = Achillea millefolium var. occidentalis [ACHIMIL8]
•**Achillea millefolium** L. [ACHIMIL] (common yarrow)
Achillea millefolium ssp. *atrotegula* Boivin = Achillea millefolium var. borealis [ACHIMIL2]
Achillea millefolium ssp. *borealis* (Bong.) Breitung = Achillea millefolium var. borealis [ACHIMIL2]
Achillea millefolium ssp. *lanulosa* (Nutt.) Piper = Achillea millefolium var. occidentalis [ACHIMIL8]
Achillea millefolium ssp. *occidentalis* (DC.) Hyl. = Achillea millefolium var. occidentalis [ACHIMIL8]
Achillea millefolium ssp. *pallidotegula* Boivin = Achillea millefolium var. occidentalis [ACHIMIL8]
•**Achillea millefolium** var. **alpicola** (Rydb.) Garrett [ACHIMIL1]
SY: *Achillea alpicola* (Rydb.) Rydb.
SY: *Achillea fusca* Rydb.
SY: *Achillea lanulosa* ssp. *alpicola* (Rydb.) Keck
SY: *Achillea lanulosa* var. *alpicola* Rydb.
SY: *Achillea millefolium* var. *fusca* (Rydb.) G.N. Jones
Achillea millefolium var. *aspleniifolia* (Vent.) Farw. = Achillea millefolium var. occidentalis [ACHIMIL8]
•**Achillea millefolium** var. **borealis** (Bong.) Farw. [ACHIMIL2]
SY: *Achillea borealis* Bong.
SY: *Achillea borealis* ssp. *typica* Keck
SY: *Achillea millefolium* ssp. *atrotegula* Boivin
SY: *Achillea millefolium* ssp. *borealis* (Bong.) Breitung
SY: *Achillea millefolium* var. *fulva* Boivin
SY: *Achillea millefolium* var. *parviligula* Boivin
SY: *Achillea millefolium* var. *parvula* Boivin
Achillea millefolium var. *fulva* Boivin = Achillea millefolium var. borealis [ACHIMIL2]
Achillea millefolium var. *fusca* (Rydb.) G.N. Jones = Achillea millefolium var. alpicola [ACHIMIL1]
Achillea millefolium var. *gracilis* Raf. ex DC. = Achillea millefolium var. occidentalis [ACHIMIL8]

Achillea millefolium var. *lanulosa* (Nutt.) Piper = Achillea millefolium var. occidentalis [ACHIMIL8]
•**Achillea millefolium** var. **occidentalis** DC. [ACHIMIL8]
SY: *Achillea angustissima* Rydb.
SY: *Achillea eradiata* Piper
SY: *Achillea gracilis* Raf.
SY: *Achillea lanulosa* Nutt.
SY: *Achillea lanulosa* var. *arachnoidea* Lunell
SY: *Achillea lanulosa* var. *eradiata* (Piper) M.E. Peck
SY: *Achillea lanulosa* ssp. *typica* Keck
SY: *Achillea laxiflora* Pollard & Cockerell
SY: *Achillea millefolium* var. *aspleniifolia* (Vent.) Farw.
SY: *Achillea millefolium* var. *gracilis* Raf. ex DC.
SY: *Achillea millefolium* ssp. *lanulosa* (Nutt.) Piper
SY: *Achillea millefolium* var. *lanulosa* (Nutt.) Piper
SY: *Achillea millefolium* ssp. *occidentalis* (DC.) Hyl.
SY: *Achillea millefolium* ssp. *pallidotegula* Boivin
SY: *Achillea millefolium* var. *rosea* (Desf.) Torr. & Gray
SY: *Achillea millefolium* var. *russeolata* Boivin
SY: *Achillea occidentalis* (DC.) Raf. ex Rydb.
SY: *Achillea rosea* Desf.
SY: *Achillea tomentosa* Pursh
•**Achillea millefolium** var. **pacifica** (Rydb.) G.N. Jones [ACHIMIL6]
SY: *Achillea pacifica* Rydb.
Achillea millefolium var. *parviligula* Boivin = Achillea millefolium var. borealis [ACHIMIL2]
Achillea millefolium var. *parvula* Boivin = Achillea millefolium var. borealis [ACHIMIL2]
Achillea millefolium var. *rosea* (Desf.) Torr. & Gray = Achillea millefolium var. occidentalis [ACHIMIL8]
Achillea millefolium var. *russeolata* Boivin = Achillea millefolium var. occidentalis [ACHIMIL8]
Achillea multiflora Hook. = Achillea sibirica [ACHISIB]
Achillea occidentalis (DC.) Raf. ex Rydb. = Achillea millefolium var. occidentalis [ACHIMIL8]
Achillea pacifica Rydb. = Achillea millefolium var. pacifica [ACHIMIL6]
Achillea rosea Desf. = Achillea millefolium var. occidentalis [ACHIMIL8]
Achillea sibirica Ledeb. [ACHISIB] (Siberian yarrow)
SY: *Achillea multiflora* Hook.
Achillea tomentosa Pursh = Achillea millefolium var. occidentalis [ACHIMIL8]
Acosta diffusa (Lam.) Soják = Centaurea diffusa [CENTDIF]
Acroptilon Cass. [ACROPTI$]
•**Acroptilon repens** (L.) DC. [ACROREP]
SY: *Centaurea picris* Pallas ex Willd.
SY: *Centaurea repens* L.
Adenocaulon Hook. [ADENOCA$]
Adenocaulon bicolor Hook. [ADENBIC] (pathfinder)
Agoseris Raf. [AGOSERI$]
Agoseris aurantiaca (Hook.) Greene [AGOSAUR] (orange agoseris)
SY: *Agoseris graminifolia* Greene
SY: *Agoseris rostrata* Rydb.
SY: *Troximon aurantiacum* Hook.
Agoseris glauca (Pursh) Raf. [AGOSGLA] (pale agoseris)
Agoseris glauca var. *asper* (Rydb.) Cronq. = Agoseris glauca var. dasycephala [AGOSGLA1]
Agoseris glauca var. **dasycephala** (Torr. & Gray) Jepson [AGOSGLA1] (short-beaked agoseris)
SY: *Agoseris glauca* var. *asper* (Rydb.) Cronq.
SY: *Agoseris glauca* var. *pumila* (Nutt.) Garrett
SY: *Agoseris scorzonerifolia* (Schrad.) Greene
SY: *Troximon glaucum* var. *dasycephalum* Torr. & Gray
Agoseris glauca var. *pumila* (Nutt.) Garrett = Agoseris glauca var. dasycephala [AGOSGLA1]
Agoseris graminifolia Greene = Agoseris aurantiaca [AGOSAUR]
Agoseris grandiflora (Nutt.) Greene [AGOSGRA] (large-flowered agoseris)
SY: *Agoseris laciniata* (Nutt.) Greene
SY: *Agoseris plebeja* (Greene) Greene
SY: *Troximon grandiflorum* Nutt.
Agoseris heterophylla (Nutt.) Greene [AGOSHET] (annual goat-chicory)
SY: *Agoseris heterophylla* ssp. *californica* (Nutt.) Piper
SY: *Agoseris heterophylla* var. *cryptopleura* Jepson
SY: *Agoseris heterophylla* var. *kymapleura* Greene
SY: *Agoseris heterophylla* ssp. *normalis* Piper
SY: *Macrorhynchus heterophyllus* Nutt.
Agoseris heterophylla ssp. *californica* (Nutt.) Piper = Agoseris heterophylla [AGOSHET]
Agoseris heterophylla ssp. *normalis* Piper = Agoseris heterophylla [AGOSHET]
Agoseris heterophylla var. *cryptopleura* Jepson = Agoseris heterophylla [AGOSHET]
Agoseris heterophylla var. *kymapleura* Greene = Agoseris heterophylla [AGOSHET]
Agoseris laciniata (Nutt.) Greene = Agoseris grandiflora [AGOSGRA]
Agoseris lackschewitzii D. Henderson & R. Moseley [AGOSLAC] (pink agoseris)
Agoseris plebeja (Greene) Greene = Agoseris grandiflora [AGOSGRA]
Agoseris rostrata Rydb. = Agoseris aurantiaca [AGOSAUR]
Agoseris scorzonerifolia (Schrad.) Greene = Agoseris glauca var. dasycephala [AGOSGLA1]
Ambrosia L. [AMBROSI$]
•**Ambrosia artemisiifolia** L. [AMBRART] (annual ragweed)
Ambrosia artemisiifolia var. **elatior** (L.) Descourtils [AMBRART1]
SY: *Ambrosia elatior* L.
Ambrosia chamissonis (Less.) Greene [AMBRCHA] (silver burweed)
SY: *Ambrosia chamissonis* var. *bipinnatisecta* (Less.) J.T. Howell
SY: *Franseria chamissonis* Less.
SY: *Franseria chamissonis* ssp. *bipinnatisecta* (Less.) Wiggins & Stockwell
SY: *Franseria chamissonis* var. *bipinnatisecta* Less.
Ambrosia chamissonis var. *bipinnatisecta* (Less.) J.T. Howell = Ambrosia chamissonis [AMBRCHA]
Ambrosia coronopifolia Torr. & Gray [AMBRCOR]
SY: *Ambrosia psilostachya* var. *coronopifolia* (Torr. & Gray) Farw.
Ambrosia elatior L. = Ambrosia artemisiifolia var. elatior [AMBRART1]
Ambrosia psilostachya var. *coronopifolia* (Torr. & Gray) Farw. = Ambrosia coronopifolia [AMBRCOR]
Anaphalis DC. [ANAPHAL$]
Anaphalis margaritacea (L.) Benth. & Hook. f. [ANAPMAR] (pearly everlasting)
SY: *Anaphalis margaritacea* var. *angustior* (Miq.) Nakai
SY: *Anaphalis margaritacea* var. *intercedens* Hara
SY: *Anaphalis margaritacea* var. *occidentalis* Greene
SY: *Anaphalis margaritacea* var. *revoluta* Suksdorf
SY: *Anaphalis margaritacea* var. *subalpina* Gray
SY: *Anaphalis occidentalis* (Greene) Heller
SY: *Gnaphalium margaritaceum* L.
Anaphalis margaritacea var. *angustior* (Miq.) Nakai = Anaphalis margaritacea [ANAPMAR]
Anaphalis margaritacea var. *intercedens* Hara = Anaphalis margaritacea [ANAPMAR]
Anaphalis margaritacea var. *occidentalis* Greene = Anaphalis margaritacea [ANAPMAR]
Anaphalis margaritacea var. *revoluta* Suksdorf = Anaphalis margaritacea [ANAPMAR]
Anaphalis margaritacea var. *subalpina* Gray = Anaphalis margaritacea [ANAPMAR]
Anaphalis occidentalis (Greene) Heller = Anaphalis margaritacea [ANAPMAR]
Antennaria Gaertn. [ANTENNA$]
Antennaria aizoides Greene = Antennaria umbrinella [ANTEUMB]
Antennaria alborosea Porsild ex M.P. Porsild = Antennaria rosea [ANTEROS]
Antennaria alpina var. *glabrata* J. Vahl = Antennaria monocephala [ANTEMON]
Antennaria alpina var. *intermedia* Rosenv. = Antennaria umbrinella [ANTEUMB]

Antennaria alpina var. *media* (Greene) Jepson = Antennaria media [ANTEMED]
Antennaria anaphaloides Rydb. [ANTEANA]
SY: *Antennaria pulcherrima* ssp. *anaphaloides* (Rydb.) W.A. Weber
SY: *Antennaria pulcherrima* var. *anaphaloides* (Rydb.) G.W. Douglas
Antennaria angustata Greene = Antennaria monocephala [ANTEMON]
Antennaria angustiarum Lunell = Antennaria neglecta [ANTENEG]
Antennaria aprica Greene = Antennaria parvifolia [ANTEPAR]
Antennaria aprica var. *minuscula* (Boivin) Boivin = Antennaria parvifolia [ANTEPAR]
Antennaria argentea ssp. *aberrans* E. Nels. = Antennaria luzuloides [ANTELUZ]
Antennaria athabascensis Greene = Antennaria neglecta [ANTENEG]
Antennaria aureola Lunell = Antennaria parvifolia [ANTEPAR]
Antennaria austromontana E. Nels. = Antennaria media [ANTEMED]
Antennaria burwellensis Malte = Antennaria monocephala [ANTEMON]
Antennaria callilepis Greene = Antennaria howellii ssp. howellii [ANTEHOW0]
Antennaria campestris Rydb. = Antennaria neglecta [ANTENEG]
Antennaria campestris var. *athabascensis* (Greene) Boivin = Antennaria neglecta [ANTENEG]
Antennaria candida Greene = Antennaria media [ANTEMED]
Antennaria carpathica var. *lanata* Hook. = Antennaria lanata [ANTELAN]
Antennaria carpathica var. *pulcherrima* Hook. = Antennaria pulcherrima [ANTEPUL]
Antennaria chelonica Lunell = Antennaria neglecta [ANTENEG]
Antennaria congesta Malte = Antennaria monocephala [ANTEMON]
Antennaria densa Greene = Antennaria media [ANTEMED]
Antennaria dimorpha (Nutt.) Torr. & Gray [ANTEDIM] (low pussytoes)
SY: *Antennaria dimorpha* var. *integra* Henderson
SY: *Antennaria dimorpha* var. *latisquama* (Piper) M.E. Peck
SY: *Antennaria dimorpha* var. *macrocephala* D.C. Eat.
SY: *Antennaria latisquama* Piper
SY: *Antennaria macrocephala* (D.C. Eat.) Rydb.
SY: *Gnaphalium dimorphum* Nutt.
Antennaria dimorpha var. *integra* Henderson = Antennaria dimorpha [ANTEDIM]
Antennaria dimorpha var. *latisquama* (Piper) M.E. Peck = Antennaria dimorpha [ANTEDIM]
Antennaria dimorpha var. *macrocephala* D.C. Eat. = Antennaria dimorpha [ANTEDIM]
Antennaria erosa Greene = Antennaria neglecta [ANTENEG]
Antennaria exilis Greene = Antennaria monocephala [ANTEMON]
Antennaria eximia Greene = Antennaria howellii ssp. howellii [ANTEHOW0]
Antennaria fernaldiana Polunin = Antennaria monocephala [ANTEMON]
Antennaria flavescens Rydb. = Antennaria umbrinella [ANTEUMB]
Antennaria fusca E. Nels. = Antennaria parvifolia [ANTEPAR]
Antennaria glabrata (J. Vahl) Greene = Antennaria monocephala [ANTEMON]
Antennaria gormanii St. John = Antennaria media [ANTEMED]
Antennaria holmii Greene = Antennaria parvifolia [ANTEPAR]
Antennaria howellii Greene [ANTEHOW]
Antennaria howellii ssp. **howellii** [ANTEHOW0]
SY: *Antennaria callilepis* Greene
SY: *Antennaria eximia* Greene
SY: *Antennaria neglecta* ssp. *howellii* (Greene) Hultén
SY: *Antennaria neglecta* var. *howellii* (Greene) Cronq.
SY: *Antennaria neodioica* ssp. *howellii* (Greene) Bayer
Antennaria howellii ssp. **neodioica** (Greene) Bayer [ANTEHOW1]
SY: *Antennaria neglecta* var. *attenuata* (Fern.) Cronq.
SY: *Antennaria neglecta* var. *neodioica* (Greene) Cronq.
SY: *Antennaria neodioica* Greene
SY: *Antennaria neodioica* var. *attenuata* Fern.
SY: *Antennaria neodioica* var. *chlorophylla* Fern.
SY: *Antennaria neodioica* var. *grandis* Fern.
SY: *Antennaria neodioica* var. *interjecta* Fern.
SY: *Antennaria rhodantha* Fern.
SY: *Antennaria rupicola* Fern.
SY: *Antennaria russellii* Boivin
Antennaria howellii var. *athabascensis* (Greene) Boivin = Antennaria neglecta [ANTENEG]
Antennaria howellii var. *campestris* (Rydb.) Boivin = Antennaria neglecta [ANTENEG]
Antennaria hudsonica Malte = Antennaria monocephala [ANTEMON]
Antennaria intermedia (Rosenv.) M.P. Porsild = Antennaria umbrinella [ANTEUMB]
Antennaria lanata (Hook.) Greene [ANTELAN] (woolly pussytoes)
SY: *Antennaria carpathica* var. *lanata* Hook.
Antennaria latisquama Piper = Antennaria dimorpha [ANTEDIM]
Antennaria longifolia Greene = Antennaria neglecta [ANTENEG]
Antennaria lunellii Greene = Antennaria neglecta [ANTENEG]
Antennaria luzuloides Torr. & Gray [ANTELUZ] (woodrush pussytoes)
SY: *Antennaria argentea* ssp. *aberrans* E. Nels.
SY: *Antennaria luzuloides* var. *oblanceolata* (Rydb.) M.E. Peck
SY: *Antennaria microcephala* Gray
SY: *Antennaria oblanceolata* Rydb.
SY: *Antennaria pyramidata* Greene
Antennaria luzuloides var. *oblanceolata* (Rydb.) M.E. Peck = Antennaria luzuloides [ANTELUZ]
Antennaria macrocephala (D.C. Eat.) Rydb. = Antennaria dimorpha [ANTEDIM]
Antennaria media Greene [ANTEMED]
SY: *Antennaria alpina* var. *media* (Greene) Jepson
SY: *Antennaria austromontana* E. Nels.
SY: *Antennaria candida* Greene
SY: *Antennaria densa* Greene
SY: *Antennaria gormanii* St. John
SY: *Antennaria modesta* Greene
SY: *Antennaria mucronata* E. Nels.
Antennaria microcephala Gray = Antennaria luzuloides [ANTELUZ]
Antennaria microphylla Rydb. [ANTEMIC] (small-leaved pussytoes)
SY: *Antennaria nitida* Greene
SY: *Antennaria rosea* var. *nitida* (Greene) Breitung
SY: *Antennaria solstitialis* Greene
SY: *Antennaria parvifolia* auct.
Antennaria minuscula Boivin = Antennaria parvifolia [ANTEPAR]
Antennaria modesta Greene = Antennaria media [ANTEMED]
Antennaria monocephala DC. [ANTEMON] (one-headed pussytoes)
SY: *Antennaria alpina* var. *glabrata* J. Vahl
SY: *Antennaria angustata* Greene
SY: *Antennaria burwellensis* Malte
SY: *Antennaria congesta* Malte
SY: *Antennaria exilis* Greene
SY: *Antennaria fernaldiana* Polunin
SY: *Antennaria glabrata* (J. Vahl) Greene
SY: *Antennaria hudsonica* Malte
SY: *Antennaria monocephala* ssp. *angustata* (Greene) Hultén
SY: *Antennaria monocephala* var. *exilis* (Greene) Hultén
SY: *Antennaria monocephala* var. *latisquamea* Hultén
SY: *Antennaria monocephala* ssp. *philonipha* (Porsild) Hultén
SY: *Antennaria nitens* Greene
SY: *Antennaria philonipha* Porsild
SY: *Antennaria shumaginensis* Porsild
SY: *Antennaria tansleyi* Polunin
SY: *Antennaria tweedsmurii* Polunin
Antennaria monocephala ssp. *angustata* (Greene) Hultén = Antennaria monocephala [ANTEMON]
Antennaria monocephala ssp. *philonipha* (Porsild) Hultén = Antennaria monocephala [ANTEMON]
Antennaria monocephala var. *exilis* (Greene) Hultén = Antennaria monocephala [ANTEMON]
Antennaria monocephala var. *latisquamea* Hultén = Antennaria monocephala [ANTEMON]
Antennaria mucronata E. Nels. = Antennaria media [ANTEMED]
Antennaria nebraskensis Greene = Antennaria neglecta [ANTENEG]

Antennaria neglecta Greene [ANTENEG] (broad-leaved pussytoes)
SY: *Antennaria angustiarum* Lunell
SY: *Antennaria athabascensis* Greene
SY: *Antennaria campestris* Rydb.
SY: *Antennaria campestris* var. *athabascensis* (Greene) Boivin
SY: *Antennaria chelonica* Lunell
SY: *Antennaria erosa* Greene
SY: *Antennaria howellii* var. *athabascensis* (Greene) Boivin
SY: *Antennaria howellii* var. *campestris* (Rydb.) Boivin
SY: *Antennaria longifolia* Greene
SY: *Antennaria lunellii* Greene
SY: *Antennaria nebraskensis* Greene
SY: *Antennaria neglecta* var. *athabascensis* (Greene) Taylor & MacBryde
SY: *Antennaria neglecta* var. *campestris* (Rydb.) Steyermark
SY: *Antennaria parvula* Greene
SY: *Antennaria wilsonii* Greene
Antennaria neglecta ssp. *howellii* (Greene) Hultén = Antennaria howellii ssp. howellii [ANTEHOW0]
Antennaria neglecta var. *athabascensis* (Greene) Taylor & MacBryde = Antennaria neglecta [ANTENEG]
Antennaria neglecta var. *attenuata* (Fern.) Cronq. = Antennaria howellii ssp. neodioica [ANTEHOW1]
Antennaria neglecta var. *campestris* (Rydb.) Steyermark = Antennaria neglecta [ANTENEG]
Antennaria neglecta var. *howellii* (Greene) Cronq. = Antennaria howellii ssp. howellii [ANTEHOW0]
Antennaria neglecta var. *neodioica* (Greene) Cronq. = Antennaria howellii ssp. neodioica [ANTEHOW1]
Antennaria neodioica Greene = Antennaria howellii ssp. neodioica [ANTEHOW1]
Antennaria neodioica ssp. *howellii* (Greene) Bayer = Antennaria howellii ssp. howellii [ANTEHOW0]
Antennaria neodioica var. *attenuata* Fern. = Antennaria howellii ssp. neodioica [ANTEHOW1]
Antennaria neodioica var. *chlorophylla* Fern. = Antennaria howellii ssp. neodioica [ANTEHOW1]
Antennaria neodioica var. *grandis* Fern. = Antennaria howellii ssp. neodioica [ANTEHOW1]
Antennaria neodioica var. *interjecta* Fern. = Antennaria howellii ssp. neodioica [ANTEHOW1]
Antennaria nitens Greene = Antennaria monocephala [ANTEMON]
Antennaria nitida Greene = Antennaria microphylla [ANTEMIC]
Antennaria oblanceolata Rydb. = Antennaria luzuloides [ANTELUZ]
Antennaria oblancifolia E. Nels. = Antennaria racemosa [ANTERAC]
Antennaria obtusata Greene = Antennaria parvifolia [ANTEPAR]
Antennaria parvifolia Nutt. [ANTEPAR] (Nuttall's pussytoes)
SY: *Antennaria aprica* Greene
SY: *Antennaria aprica* var. *minuscula* (Boivin) Boivin
SY: *Antennaria aureola* Lunell
SY: *Antennaria fusca* E. Nels.
SY: *Antennaria holmii* Greene
SY: *Antennaria minuscula* Boivin
SY: *Antennaria obtusata* Greene
SY: *Antennaria pumila* Greene
SY: *Antennaria recurva* Greene
SY: *Antennaria rhodanthus* Suksdorf
Antennaria parvifolia auct. = Antennaria microphylla [ANTEMIC]
Antennaria parvula Greene = Antennaria neglecta [ANTENEG]
Antennaria pedicellata Greene = Antennaria racemosa [ANTERAC]
Antennaria philonipha Porsild = Antennaria monocephala [ANTEMON]
Antennaria piperi Rydb. = Antennaria racemosa [ANTERAC]
Antennaria pulcherrima (Hook.) Greene [ANTEPUL] (showy pussytoes)
SY: *Antennaria carpathica* var. *pulcherrima* Hook.
SY: *Antennaria pulcherrima* var. *angustisquama* Porsild
SY: *Antennaria pulcherrima* var. *sordida* Boivin
Antennaria pulcherrima ssp. *anaphaloides* (Rydb.) W.A. Weber = Antennaria anaphaloides [ANTEANA]
Antennaria pulcherrima var. *anaphaloides* (Rydb.) G.W. Douglas = Antennaria anaphaloides [ANTEANA]
Antennaria pulcherrima var. *angustisquama* Porsild = Antennaria pulcherrima [ANTEPUL]
Antennaria pulcherrima var. *sordida* Boivin = Antennaria pulcherrima [ANTEPUL]
Antennaria pumila Greene = Antennaria parvifolia [ANTEPAR]
Antennaria pyramidata Greene = Antennaria luzuloides [ANTELUZ]
Antennaria racemosa Hook. [ANTERAC] (racemose pussytoes)
SY: *Antennaria oblancifolia* E. Nels.
SY: *Antennaria pedicellata* Greene
SY: *Antennaria piperi* Rydb.
Antennaria recurva Greene = Antennaria parvifolia [ANTEPAR]
Antennaria reflexa E. Nels. = Antennaria umbrinella [ANTEUMB]
Antennaria rhodantha Fern. = Antennaria howellii ssp. neodioica [ANTEHOW1]
Antennaria rhodanthus Suksdorf = Antennaria parvifolia [ANTEPAR]
Antennaria rosea Greene [ANTEROS]
SY: *Antennaria alborosea* Porsild ex M.P. Porsild
Antennaria rosea var. *nitida* (Greene) Breitung = Antennaria microphylla [ANTEMIC]
Antennaria rupicola Fern. = Antennaria howellii ssp. neodioica [ANTEHOW1]
Antennaria russellii Boivin = Antennaria howellii ssp. neodioica [ANTEHOW1]
Antennaria shumaginensis Porsild = Antennaria monocephala [ANTEMON]
Antennaria solstitialis Greene = Antennaria microphylla [ANTEMIC]
Antennaria tansleyi Polunin = Antennaria monocephala [ANTEMON]
Antennaria tweedsmurii Polunin = Antennaria monocephala [ANTEMON]
Antennaria umbrinella Rydb. [ANTEUMB] (umber pussytoes)
SY: *Antennaria aizoides* Greene
SY: *Antennaria alpina* var. *intermedia* Rosenv.
SY: *Antennaria flavescens* Rydb.
SY: *Antennaria intermedia* (Rosenv.) M.P. Porsild
SY: *Antennaria reflexa* E. Nels.
Antennaria wilsonii Greene = Antennaria neglecta [ANTENEG]
Anthemis L. [ANTHEMI$]
*__Anthemis arvensis__ L. [ANTHARV] (corn chamomile)
*__Anthemis cotula__ L. [ANTHCOT] (stinking chamomile)
SY: *Maruta cotula* (L.) DC.
*__Anthemis tinctoria__ L. [ANTHTIN] (yellow chamomile)
SY: *Cota tinctoria* (L.) J. Gay
Apargidium boreale (Bong.) Torr. & Gray = Microseris borealis [MICRBOR]
Arctium L. [ARCTIUM$]
*__Arctium lappa__ L. [ARCTLAP] (great burdock)
*__Arctium minus__ Bernh. [ARCTMIN] (common burdock)
SY: *Lappa minor* Hill
Arnica L. [ARNICA$]
Arnica alpina ssp. *angustifolia* (Vahl) Maguire = Arnica angustifolia ssp. angustifolia [ARNIANG0]
Arnica alpina ssp. *attenuata* (Greene) Maguire = Arnica angustifolia ssp. angustifolia [ARNIANG0]
Arnica alpina ssp. *lonchophylla* (Greene) Taylor & MacBryde = Arnica lonchophylla [ARNILOG]
Arnica alpina ssp. *sornborgeri* (Fern.) Maguire = Arnica angustifolia ssp. angustifolia [ARNIANG0]
Arnica alpina ssp. *tomentosa* (Macoun) Maguire = Arnica angustifolia ssp. tomentosa [ARNIANG4]
Arnica alpina var. *angustifolia* (Vahl) Fern. = Arnica angustifolia ssp. angustifolia [ARNIANG0]
Arnica alpina var. *attenuata* (Greene) Ediger & Barkl. = Arnica angustifolia ssp. angustifolia [ARNIANG0]
Arnica alpina var. *linearis* Hultén = Arnica angustifolia ssp. angustifolia [ARNIANG0]
Arnica alpina var. *lonchophylla* (Greene) Welsh = Arnica lonchophylla [ARNILOG]
Arnica alpina var. *plantaginea* (Pursh) Ediger & Barkl. = Arnica angustifolia ssp. angustifolia [ARNIANG0]
Arnica alpina var. *tomentosa* (Macoun) Cronq. = Arnica angustifolia ssp. tomentosa [ARNIANG4]

Arnica alpina var. *ungavensis* Boivin = Arnica angustifolia ssp. angustifolia [ARNIANG0]
Arnica alpina var. *vestita* Hultén = Arnica angustifolia ssp. angustifolia [ARNIANG0]
Arnica amplexicaulis Nutt. [ARNIAMP] (streambank arnica)
SY: *Arnica amplexicaulis* ssp. *genuina* Maguire
SY: *Arnica amplexicaulis* var. *piperi* St. John & Warren
SY: *Arnica amplexicaulis* ssp. *prima* (Maguire) Maguire
SY: *Arnica amplexicaulis* var. *prima* (Maguire) Boivin
SY: *Arnica amplexifolia* Rydb.
SY: *Arnica mollis* var. *aspera* (Greene) Boivin
Arnica amplexicaulis ssp. *genuina* Maguire = Arnica amplexicaulis [ARNIAMP]
Arnica amplexicaulis ssp. *prima* (Maguire) Maguire = Arnica amplexicaulis [ARNIAMP]
Arnica amplexicaulis var. *piperi* St. John & Warren = Arnica amplexicaulis [ARNIAMP]
Arnica amplexicaulis var. *prima* (Maguire) Boivin = Arnica amplexicaulis [ARNIAMP]
Arnica amplexifolia Rydb. = Arnica amplexicaulis [ARNIAMP]
Arnica angustifolia Vahl [ARNIANG] (alpine arnica)
Arnica angustifolia ssp. **angustifolia** [ARNIANG0]
SY: *Arnica alpina* ssp. *angustifolia* (Vahl) Maguire
SY: *Arnica alpina* var. *angustifolia* (Vahl) Fern.
SY: *Arnica alpina* ssp. *attenuata* (Greene) Maguire
SY: *Arnica alpina* var. *attenuata* (Greene) Ediger & Barkl.
SY: *Arnica alpina* var. *linearis* Hultén
SY: *Arnica alpina* var. *plantaginea* (Pursh) Ediger & Barkl.
SY: *Arnica alpina* ssp. *sornborgeri* (Fern.) Maguire
SY: *Arnica alpina* var. *ungavensis* Boivin
SY: *Arnica alpina* var. *vestita* Hultén
SY: *Arnica angustifolia* ssp. *attenuata* (Greene) G.W. Douglas
SY: *Arnica plantaginea* Pursh
SY: *Arnica sornborgeri* Fern.
SY: *Arnica terrae-novae* Fern.
Arnica angustifolia ssp. *attenuata* (Greene) G.W. Douglas = Arnica angustifolia ssp. angustifolia [ARNIANG0]
Arnica angustifolia ssp. *lonchophylla* (Greene) G.W. & G.R. Dougl. = Arnica lonchophylla [ARNILOG]
Arnica angustifolia ssp. **tomentosa** (Macoun) G.W. & G.R. Dougl. [ARNIANG4]
SY: *Arnica alpina* ssp. *tomentosa* (Macoun) Maguire
SY: *Arnica alpina* var. *tomentosa* (Macoun) Cronq.
SY: *Arnica tomentosa* Macoun
Arnica cascadensis St. John = Arnica rydbergii [ARNIRYD]
Arnica chamissonis Less. [ARNICHA] (meadow arnica)
Arnica chamissonis ssp. **chamissonis** [ARNICHA0]
Arnica chamissonis ssp. **foliosa** (Nutt.) Maguire [ARNICHA2]
Arnica chamissonis ssp. **foliosa** var. **incana** (Gray) Hultén [ARNICHA1]
SY: *Arnica chamissonis* ssp. *incana* (Gray) Maguire
SY: *Arnica foliosa* var. *incana* Gray
Arnica chamissonis ssp. *incana* (Gray) Maguire = Arnica chamissonis ssp. foliosa var. incana [ARNICHA1]
Arnica chionopappa Fern. = Arnica lonchophylla [ARNILOG]
Arnica cordifolia Hook. [ARNICOR] (heart-leaved arnica)
SY: *Arnica cordifolia* ssp. *genuina* Maguire
SY: *Arnica cordifolia* var. *humilis* (Rydb.) Maguire
SY: *Arnica cordifolia* var. *pumila* (Rydb.) Maguire
SY: *Arnica cordifolia* var. *whitneyi* (Fern.) Maguire
SY: *Arnica hardinae* St. John
SY: *Arnica humilis* Rydb.
SY: *Arnica paniculata* A. Nels.
SY: *Arnica whitneyi* Fern.
Arnica cordifolia ssp. *genuina* Maguire = Arnica cordifolia [ARNICOR]
Arnica cordifolia var. *humilis* (Rydb.) Maguire = Arnica cordifolia [ARNICOR]
Arnica cordifolia var. *pumila* (Rydb.) Maguire = Arnica cordifolia [ARNICOR]
Arnica cordifolia var. *whitneyi* (Fern.) Maguire = Arnica cordifolia [ARNICOR]
Arnica × diversifolia Greene (pro sp.) [ARNIDIV]
SY: *Arnica diversifolia* Greene
Arnica diversifolia Greene = Arnica × diversifolia [ARNIDIV]
Arnica foliosa var. *incana* Gray = Arnica chamissonis ssp. foliosa var. incana [ARNICHA1]
Arnica frigida C.A. Mey. ex Iljin [ARNIFRI] (northern arnica)
SY: *Arnica louiseana* ssp. *frigida* (C.A. Mey. ex Iljin) Maguire
Arnica fulgens Pursh [ARNIFUL] (orange arnica)
Arnica fulgens var. *sororia* (Greene) G.W. & G.R. Dougl. = Arnica sororia [ARNISOR]
Arnica gaspensis Fern. = Arnica lonchophylla [ARNILOG]
Arnica gracilis Rydb. [ARNIGRA] (high mountain arnica)
SY: *Arnica latifolia* var. *gracilis* (Rydb.) Cronq.
Arnica hardinae St. John = Arnica cordifolia [ARNICOR]
Arnica humilis Rydb. = Arnica cordifolia [ARNICOR]
Arnica latifolia Bong. [ARNILAT] (mountain arnica)
SY: *Arnica latifolia* var. *angustifolia* Herder
Arnica latifolia var. *angustifolia* Herder = Arnica latifolia [ARNILAT]
Arnica latifolia var. *gracilis* (Rydb.) Cronq. = Arnica gracilis [ARNIGRA]
Arnica lessingii (Torr. & Gray) Greene [ARNILES] (purple arnica)
SY: *Arnica lessingii* ssp. *norbergii* Hultén & Maguire
Arnica lessingii ssp. *norbergii* Hultén & Maguire = Arnica lessingii [ARNILES]
Arnica lonchophylla Greene [ARNILOG] (spear-leaved arnica)
SY: *Arnica alpina* ssp. *lonchophylla* (Greene) Taylor & MacBryde
SY: *Arnica alpina* var. *lonchophylla* (Greene) Welsh
SY: *Arnica angustifolia* ssp. *lonchophylla* (Greene) G.W. & G.R. Dougl.
SY: *Arnica chionopappa* Fern.
SY: *Arnica gaspensis* Fern.
SY: *Arnica lonchophylla* ssp. *chionopappa* (Fern.) Maguire
Arnica lonchophylla ssp. *chionopappa* (Fern.) Maguire = Arnica lonchophylla [ARNILOG]
Arnica louiseana ssp. *frigida* (C.A. Mey. ex Iljin) Maguire = Arnica frigida [ARNIFRI]
Arnica mollis Hook. [ARNIMOL] (hairy arnica)
SY: *Arnica mollis* var. *silvatica* (Greene) Maguire
Arnica mollis var. *aspera* (Greene) Boivin = Arnica amplexicaulis [ARNIAMP]
Arnica mollis var. *silvatica* (Greene) Maguire = Arnica mollis [ARNIMOL]
Arnica paniculata A. Nels. = Arnica cordifolia [ARNICOR]
Arnica parryi Gray [ARNIPAR] (Parry's arnica)
Arnica plantaginea Pursh = Arnica angustifolia ssp. angustifolia [ARNIANG0]
Arnica rydbergii Greene [ARNIRYD] (Rydberg's arnica)
SY: *Arnica cascadensis* St. John
Arnica sornborgeri Fern. = Arnica angustifolia ssp. angustifolia [ARNIANG0]
Arnica sororia Greene [ARNISOR]
SY: *Arnica fulgens* var. *sororia* (Greene) G.W. & G.R. Dougl.
Arnica terrae-novae Fern. = Arnica angustifolia ssp. angustifolia [ARNIANG0]
Arnica tomentosa Macoun = Arnica angustifolia ssp. tomentosa [ARNIANG4]
Arnica whitneyi Fern. = Arnica cordifolia [ARNICOR]
Artemisia L. [ARTEMIS$]
***Artemisia absinthium** L. [ARTEABS] (wormwood)
Artemisia alaskana Rydb. [ARTEALA] (Alaska sagebrush)
SY: *Artemisia tyrellii* Rydb.
Artemisia angustifolia (Gray) Rydb. = Artemisia tridentata ssp. tridentata [ARTETRI0]
Artemisia arctica Less. [ARTEARC] (mountain sagewort)
SY: *Artemisia norvegica* var. *piceetorum* Welsh & Goodrich
SY: *Artemisia norvegica* ssp. *saxatilis* (Bess.) Hall & Clements
SY: *Artemisia norvegica* var. *saxatilis* (Bess.) Jepson
***Artemisia biennis** Willd. [ARTEBIE] (biennial sagewort)
Artemisia borealis Pallas = Artemisia campestris ssp. borealis [ARTECAM1]
Artemisia campestris L. [ARTECAM] (northern wormwood)
Artemisia campestris ssp. **borealis** (Pallas) Hall & Clements [ARTECAM1]

SY: *Artemisia borealis* Pallas
Artemisia campestris ssp. **pacifica** (Nutt.) Hall & Clements [ARTECAM2]
SY: *Artemisia campestris* var. *douglasiana* (Bess.) Boivin
SY: *Artemisia campestris* var. *pacifica* (Nutt.) M.E. Peck
SY: *Artemisia campestris* var. *scouleriana* (Bess.) Cronq.
SY: *Artemisia campestris* var. *strutzia* Welsh
SY: *Artemisia camporum* Rydb.
SY: *Artemisia pacifica* Nutt.
SY: *Oligosporus campestris* ssp. *pacificus* (Nutt.) W.A. Weber
Artemisia campestris var. *douglasiana* (Bess.) Boivin = Artemisia campestris ssp. pacifica [ARTECAM2]
Artemisia campestris var. *pacifica* (Nutt.) M.E. Peck = Artemisia campestris ssp. pacifica [ARTECAM2]
Artemisia campestris var. *scouleriana* (Bess.) Cronq. = Artemisia campestris ssp. pacifica [ARTECAM2]
Artemisia campestris var. *strutzia* Welsh = Artemisia campestris ssp. pacifica [ARTECAM2]
Artemisia camporum Rydb. = Artemisia campestris ssp. pacifica [ARTECAM2]
Artemisia cana Pursh [ARTECAN] (silver sagebrush)
Artemisia candicans Rydb. = Artemisia ludoviciana ssp. candicans [ARTELUD1]
Artemisia diversifolia Rydb. = Artemisia ludoviciana ssp. ludoviciana [ARTELUD0]
Artemisia dracunculus L. [ARTEDRA] (tarragon)
Artemisia elatior (Torr. & Gray) Rydb. = Artemisia tilesii ssp. elatior [ARTETIL1]
Artemisia frigida Willd. [ARTEFRI] (prairie sagewort)
Artemisia furcata Bieb. [ARTEFUR] (three-fork wormwood)
SY: *Artemisia trifurcata* Steph. ex Spreng.
Artemisia furcata var. **heterophylla** (Bess.) Hultén [ARTEFUR1] (three-forked mugwort)
SY: *Artemisia heterophylla* Bess.
Artemisia gnaphalodes Nutt. = Artemisia ludoviciana ssp. ludoviciana [ARTELUD0]
Artemisia herriotii Rydb. = Artemisia ludoviciana ssp. ludoviciana [ARTELUD0]
Artemisia heterophylla Bess. = Artemisia furcata var. heterophylla [ARTEFUR1]
Artemisia incompta Nutt. = Artemisia ludoviciana ssp. incompta [ARTELUD2]
Artemisia lindleyana Bess. [ARTELIN] (Columbia River mugwort)
SY: *Artemisia prescottiana* Bess.
Artemisia longifolia Nutt. [ARTELON] (long-leaved mugwort)
Artemisia ludoviciana Nutt. [ARTELUD] (western mugwort)
Artemisia ludoviciana ssp. **candicans** (Rydb.) Keck [ARTELUD1]
SY: *Artemisia candicans* Rydb.
SY: *Artemisia ludoviciana* var. *latiloba* Nutt.
Artemisia ludoviciana ssp. **incompta** (Nutt.) Keck [ARTELUD2]
SY: *Artemisia incompta* Nutt.
SY: *Artemisia ludoviciana* var. *incompta* (Nutt.) Cronq.
Artemisia ludoviciana ssp. **ludoviciana** [ARTELUD0]
SY: *Artemisia diversifolia* Rydb.
SY: *Artemisia gnaphalodes* Nutt.
SY: *Artemisia herriotii* Rydb.
SY: *Artemisia ludoviciana* var. *americana* (Bess.) Fern.
SY: *Artemisia ludoviciana* var. *brittonii* (Rydb.) Fern.
SY: *Artemisia ludoviciana* var. *gnaphalodes* (Nutt.) Torr. & Gray
SY: *Artemisia ludoviciana* var. *latifolia* (Bess.) Torr. & Gray
SY: *Artemisia ludoviciana* var. *pabularis* (A. Nels.) Fern.
SY: *Artemisia ludoviciana* ssp. *typica* Keck
SY: *Artemisia pabularis* (A. Nels.) Rydb.
SY: *Artemisia purshiana* Bess.
SY: *Artemisia vulgaris* ssp. *ludoviciana* (Nutt.) Hall & Clements
SY: *Artemisia vulgaris* var. *ludoviciana* (Nutt.) Kuntze
Artemisia ludoviciana ssp. *typica* Keck = Artemisia ludoviciana ssp. ludoviciana [ARTELUD0]
Artemisia ludoviciana var. *americana* (Bess.) Fern. = Artemisia ludoviciana ssp. ludoviciana [ARTELUD0]
Artemisia ludoviciana var. *brittonii* (Rydb.) Fern. = Artemisia ludoviciana ssp. ludoviciana [ARTELUD0]
Artemisia ludoviciana var. *gnaphalodes* (Nutt.) Torr. & Gray = Artemisia ludoviciana ssp. ludoviciana [ARTELUD0]
Artemisia ludoviciana var. *incompta* (Nutt.) Cronq. = Artemisia ludoviciana ssp. incompta [ARTELUD2]
Artemisia ludoviciana var. *latifolia* (Bess.) Torr. & Gray = Artemisia ludoviciana ssp. ludoviciana [ARTELUD0]
Artemisia ludoviciana var. *latiloba* Nutt. = Artemisia ludoviciana ssp. candicans [ARTELUD1]
Artemisia ludoviciana var. *pabularis* (A. Nels.) Fern. = Artemisia ludoviciana ssp. ludoviciana [ARTELUD0]
Artemisia michauxiana Bess. [ARTEMIC] (Michaux's mugwort)
SY: *Artemisia vulgaris* ssp. *michauxiana* (Bess.) St. John
Artemisia norvegica ssp. *saxatilis* (Bess.) Hall & Clements = Artemisia arctica [ARTEARC]
Artemisia norvegica var. *piceetorum* Welsh & Goodrich = Artemisia arctica [ARTEARC]
Artemisia norvegica var. *saxatilis* (Bess.) Jepson = Artemisia arctica [ARTEARC]
Artemisia pabularis (A. Nels.) Rydb. = Artemisia ludoviciana ssp. ludoviciana [ARTELUD0]
Artemisia pacifica Nutt. = Artemisia campestris ssp. pacifica [ARTECAM2]
Artemisia prescottiana Bess. = Artemisia lindleyana [ARTELIN]
Artemisia purshiana Bess. = Artemisia ludoviciana ssp. ludoviciana [ARTELUD0]
Artemisia suksdorfii Piper [ARTESUK] (Suksdorf's mugwort)
Artemisia tilesii Ledeb. [ARTETIL] (Aleutian mugwort)
Artemisia tilesii ssp. **elatior** (Torr. & Gray) Hultén [ARTETIL1]
SY: *Artemisia elatior* (Torr. & Gray) Rydb.
SY: *Artemisia tilesii* var. *elatior* Torr. & Gray
Artemisia tilesii ssp. **tilesii** [ARTETIL0]
Artemisia tilesii ssp. **unalaschcensis** (Bess.) Hultén [ARTETIL2]
SY: *Artemisia tilesii* var. *unalaschcensis* Bess.
Artemisia tilesii var. *elatior* Torr. & Gray = Artemisia tilesii ssp. elatior [ARTETIL1]
Artemisia tilesii var. *unalaschcensis* Bess. = Artemisia tilesii ssp. unalaschcensis [ARTETIL2]
Artemisia tridentata Nutt. [ARTETRI] (big sagebrush)
Artemisia tridentata ssp. *parishii* (Gray) Hall & Clements = Artemisia tridentata ssp. tridentata [ARTETRI0]
Artemisia tridentata ssp. **tridentata** [ARTETRI0]
SY: *Artemisia angustifolia* (Gray) Rydb.
SY: *Artemisia tridentata* var. *angustifolia* Gray
SY: *Artemisia tridentata* ssp. *parishii* (Gray) Hall & Clements
SY: *Artemisia tridentata* var. *parishii* (Gray) Jepson
SY: *Seriphidium tridentatum* (Nutt.) W.A. Weber
SY: *Seriphidium tridentatum* ssp. *parishii* (Gray) W.A. Weber
Artemisia tridentata ssp. **vaseyana** (Rydb.) Beetle [ARTETRI2]
SY: *Artemisia tridentata* var. *pauciflora* Winward & Goodrich
SY: *Artemisia tridentata* var. *vaseyana* (Rydb.) Boivin
SY: *Artemisia vaseyana* Rydb.
SY: *Seriphidium tridentatum* ssp. *vaseyanum* (Rydb.) W.A. Weber
SY: *Seriphidium vaseyanum* (Rydb.) W.A. Weber
Artemisia tridentata var. *angustifolia* Gray = Artemisia tridentata ssp. tridentata [ARTETRI0]
Artemisia tridentata var. *parishii* (Gray) Jepson = Artemisia tridentata ssp. tridentata [ARTETRI0]
Artemisia tridentata var. *pauciflora* Winward & Goodrich = Artemisia tridentata ssp. vaseyana [ARTETRI2]
Artemisia tridentata var. *vaseyana* (Rydb.) Boivin = Artemisia tridentata ssp. vaseyana [ARTETRI2]
Artemisia trifurcata Steph. ex Spreng. = Artemisia furcata [ARTEFUR]
Artemisia tripartita Rydb. [ARTETRP] (threetip sagebrush)
Artemisia tyrellii Rydb. = Artemisia alaskana [ARTEALA]
Artemisia vaseyana Rydb. = Artemisia tridentata ssp. vaseyana [ARTETRI2]
***Artemisia vulgaris** L. [ARTEVUL] (common mugwort)
Artemisia vulgaris ssp. *ludoviciana* (Nutt.) Hall & Clements = Artemisia ludoviciana ssp. ludoviciana [ARTELUD0]
Artemisia vulgaris ssp. *michauxiana* (Bess.) St. John = Artemisia michauxiana [ARTEMIC]

Artemisia vulgaris var. *ludoviciana* (Nutt.) Kuntze = Artemisia ludoviciana ssp. ludoviciana [ARTELUD0]
Askellia elegans (Hook.) W.A. Weber = Crepis elegans [CREPELE]
Askellia nana ssp. *ramosa* (Babcock) W.A. Weber = Crepis nana ssp. ramosa [CREPNAN2]
Aster L. [ASTER$]
Aster adscendens Lindl. = Aster ascendens [ASTEASC]
Aster alpinus L. [ASTEALP] (alpine aster)
Aster alpinus ssp. *vierhapperi* Onno = Aster alpinus var. vierhapperi [ASTEALP1]
Aster alpinus var. **vierhapperi** (Onno) Cronq. [ASTEALP1]
SY: *Aster alpinus* ssp. *vierhapperi* Onno
Aster ascendens Lindl. [ASTEASC] (long-leaved aster)
SY: *Aster adscendens* Lindl.
SY: *Aster chilensis* ssp. *adscendens* (Lindl.) Cronq.
SY: *Aster macounii* Rydb.
SY: *Aster subgriseus* Rydb.
SY: *Virgulaster ascendens* (Lindl.) Semple
Aster borealis (Torr. & Gray) Prov. [ASTEBOR] (rush aster)
SY: *Aster franklinianus* Rydb.
SY: *Aster junciformis* Rydb.
SY: *Aster laxifolius* var. *borealis* Torr. & Gray
SY: *Symphyotrichum boreale* (Torr. & Gray) A. & D. Löve
Aster brachyactis Blake = Brachyactis ciliata ssp. angusta [BRACCIL1]
Aster × bracteolatus Nutt. (pro sp.) [ASTEBRA]
SY: *Aster bracteolatus* Nutt.
Aster bracteolatus Nutt. = Aster × bracteolatus [ASTEBRA]
Aster campestris Nutt. [ASTECAM] (meadow aster)
SY: *Virgulus campestris* (Nutt.) Reveal & Keener
Aster canescens Pursh = Machaeranthera canescens [MACHCAN]
Aster chilensis ssp. *adscendens* (Lindl.) Cronq. = Aster ascendens [ASTEASC]
Aster ciliolatus Lindl. [ASTECIL] (fringed aster)
SY: *Aster lindleyanus* Torr. & Gray
Aster conspicuus Lindl. [ASTECON] (showy aster)
Aster curtus Cronq. [ASTECUR] (white-top aster)
SY: *Sericocarpus rigidus* Lindl.
Aster douglasii Lindl. = Aster subspicatus [ASTESUB]
Aster eatonii (Gray) T.J. Howell [ASTEEAT] (Eaton's aster)
SY: *Aster foliaceus* var. *eatonii* Gray
SY: *Aster mearnsii* Rydb.
Aster elegans var. *engelmannii* D.C. Eat. = Aster engelmannii [ASTEENG]
Aster elegantulus Porsild = Aster falcatus [ASTEFAL]
Aster eliasii A. Nels. = Aster radulinus [ASTERAD]
Aster engelmannii (D.C. Eat.) Gray [ASTEENG] (Engelmann's aster)
SY: *Aster elegans* var. *engelmannii* D.C. Eat.
SY: *Eucephalus engelmannii* (D.C. Eat.) Greene
Aster engelmannii var. *paucicapitatus* B.L. Robins. = Aster paucicapitatus [ASTEPAU]
Aster ericoides L. [ASTEERI] (white heath aster)
Aster ericoides ssp. *pansus* (Blake) A.G. Jones = Aster ericoides var. pansus [ASTEERI1]
Aster ericoides var. *commutatus* (Torr. & Gray) Boivin p.p. = Aster falcatus [ASTEFAL]
Aster ericoides var. **pansus** (Blake) Boivin [ASTEERI1]
SY: *Aster ericoides* ssp. *pansus* (Blake) A.G. Jones
SY: *Aster multiflorus* var. *pansus* Blake
SY: *Aster pansus* (Blake) Cronq.
Aster exscapus Richards. = Townsendia exscapa [TOWNEXS]
Aster falcatus Lindl. [ASTEFAL] (rough white prairie aster)
SY: *Aster elegantulus* Porsild
SY: *Aster ericoides* var. *commutatus* (Torr. & Gray) Boivin p.p.
SY: *Aster ramulosus* Lindl.
SY: *Lasallea falcata* (Lindl.) Semple & Brouillet
SY: *Virgulus falcatus* (Lindl.) Reveal & Keener
Aster foliaceus Lindl. ex DC. [ASTEFOL] (leafy aster)
Aster foliaceus var. *eatonii* Gray = Aster eatonii [ASTEEAT]
Aster franklinianus Rydb. = Aster borealis [ASTEBOR]
Aster geyeri (Gray) T.J. Howell = Aster laevis var. geyeri [ASTELAE1]
Aster hesperius Gray = Aster lanceolatus ssp. hesperius [ASTELAN1]
Aster hesperius var. *laetevirens* (Greene) Cronq. = Aster lanceolatus ssp. hesperius [ASTELAN1]
Aster hesperius var. *wootonii* Greene = Aster lanceolatus ssp. hesperius [ASTELAN1]
Aster junciformis Rydb. = Aster borealis [ASTEBOR]
Aster laetevirens Greene = Aster lanceolatus ssp. hesperius [ASTELAN1]
Aster laevis L. [ASTELAE] (smooth blue aster)
Aster laevis ssp. *geyeri* (Gray) Piper = Aster laevis var. geyeri [ASTELAE1]
Aster laevis var. **geyeri** Gray [ASTELAE1]
SY: *Aster geyeri* (Gray) T.J. Howell
SY: *Aster laevis* ssp. *geyeri* (Gray) Piper
SY: *Aster pickettianus* Suksdorf
Aster lanceolatus Willd. [ASTELAN]
Aster lanceolatus ssp. **hesperius** (Gray) Semple & Chmielewski [ASTELAN1]
SY: *Aster hesperius* Gray
SY: *Aster hesperius* var. *laetevirens* (Greene) Cronq.
SY: *Aster hesperius* var. *wootonii* Greene
SY: *Aster laetevirens* Greene
SY: *Aster osterhoutii* Rydb.
SY: *Aster wootonii* (Greene) Greene
SY: *Symphyotrichum hesperium* (Gray) A. & D. Löve
Aster latissimifolius var. *serotinus* Kuntze = Solidago gigantea [SOLIGIG]
Aster laxifolius var. *borealis* Torr. & Gray = Aster borealis [ASTEBOR]
Aster lindleyanus Torr. & Gray = Aster ciliolatus [ASTECIL]
Aster macounii Rydb. = Aster ascendens [ASTEASC]
Aster major (Hook.) Porter = Aster modestus [ASTEMOD]
Aster mearnsii Rydb. = Aster eatonii [ASTEEAT]
Aster meritus A. Nels. = Aster sibiricus var. meritus [ASTESIB1]
Aster modestus Lindl. [ASTEMOD] (great northern aster)
SY: *Aster major* (Hook.) Porter
SY: *Aster unalaschkensis* var. *major* Hook.
SY: *Weberaster modestus* (Lindl.) A. & D. Löve
Aster multiflorus var. *pansus* Blake = Aster ericoides var. pansus [ASTEERI1]
Aster occidentalis (Nutt.) Torr. & Gray [ASTEOCC] (western mountain aster)
Aster occidentalis var. **intermedius** Gray [ASTEOCC2]
Aster occidentalis var. **occidentalis** [ASTEOCC0]
SY: *Aster spathulatus* Lindl.
SY: *Tripolium occidentale* Nutt.
Aster osterhoutii Rydb. = Aster lanceolatus ssp. hesperius [ASTELAN1]
Aster pansus (Blake) Cronq. = Aster ericoides var. pansus [ASTEERI1]
Aster paucicapitatus (B.L. Robins.) B.L. Robins. [ASTEPAU] (Olympic Mountain aster)
SY: *Aster engelmannii* var. *paucicapitatus* B.L. Robins.
Aster pickettianus Suksdorf = Aster laevis var. geyeri [ASTELAE1]
Aster radulinus Gray [ASTERAD] (rough-leaved aster)
SY: *Aster eliasii* A. Nels.
SY: *Weberaster radulinus* (Gray) A. & D. Löve
Aster ramulosus Lindl. = Aster falcatus [ASTEFAL]
Aster sibiricus L. [ASTESIB] (arctic aster)
Aster sibiricus ssp. *meritus* (A. Nels.) G.W. Doug. = Aster sibiricus var. meritus [ASTESIB1]
Aster sibiricus var. **meritus** (A. Nels.) Raup [ASTESIB1]
SY: *Aster meritus* A. Nels.
SY: *Aster sibiricus* ssp. *meritus* (A. Nels.) G.W. Doug.
Aster spathulatus Lindl. = Aster occidentalis var. occidentalis [ASTEOCC0]
Aster stenomeres Gray = Ionactis stenomeres [IONASTE]
Aster subgriseus Rydb. = Aster ascendens [ASTEASC]
Aster subspicatus Nees [ASTESUB] (Douglas' aster)
SY: *Aster douglasii* Lindl.
Aster unalaschkensis var. *major* Hook. = Aster modestus [ASTEMOD]
Aster wootonii (Greene) Greene = Aster lanceolatus ssp. hesperius [ASTELAN1]

Baeria maritima Gray = Lasthenia maritima [LASTMAR]
Baeria minor ssp. *maritima* (Gray) Ferris = Lasthenia maritima [LASTMAR]
Balsamorhiza Nutt. [BALSAMO$]
Balsamorhiza deltoidea Nutt. [BALSDEL] (deltoid balsamroot)
Balsamorhiza sagittata (Pursh) Nutt. [BALSSAG] (arrow-leaved balsamroot)
Bellis L. [BELLIS$]
•**Bellis perennis** L. [BELLPER] (English daisy)
Bidens L. [BIDENS$]
Bidens amplissima Greene [BIDEAMP] (Vancouver Island beggarticks)
Bidens beckii Torr. ex Spreng. = Megalodonta beckii [MEGABEC]
Bidens cernua L. [BIDECER] (nodding beggarticks)
SY: *Bidens cernua* var. *dentata* (Nutt.) Boivin
SY: *Bidens cernua* var. *elliptica* Wieg.
SY: *Bidens cernua* var. *integra* Wieg.
SY: *Bidens cernua* var. *minima* (Huds.) Pursh
SY: *Bidens cernua* var. *oligodonta* Fern. & St. John
SY: *Bidens cernua* var. *radiata* DC.
SY: *Bidens glaucescens* Greene
Bidens cernua var. *dentata* (Nutt.) Boivin = Bidens cernua [BIDECER]
Bidens cernua var. *elliptica* Wieg. = Bidens cernua [BIDECER]
Bidens cernua var. *integra* Wieg. = Bidens cernua [BIDECER]
Bidens cernua var. *minima* (Huds.) Pursh = Bidens cernua [BIDECER]
Bidens cernua var. *oligodonta* Fern. & St. John = Bidens cernua [BIDECER]
Bidens cernua var. *radiata* DC. = Bidens cernua [BIDECER]
Bidens frondosa L. [BIDEFRO] (common beggarticks)
SY: *Bidens frondosa* var. *anomala* Porter ex Fern.
SY: *Bidens frondosa* var. *caudata* Sherff
SY: *Bidens frondosa* var. *pallida* Wieg.
SY: *Bidens frondosa* var. *puberula* Wieg.
SY: *Bidens frondosa* var. *stenodonta* Fern. & St. John
Bidens frondosa var. *anomala* Porter ex Fern. = Bidens frondosa [BIDEFRO]
Bidens frondosa var. *caudata* Sherff = Bidens frondosa [BIDEFRO]
Bidens frondosa var. *pallida* Wieg. = Bidens frondosa [BIDEFRO]
Bidens frondosa var. *puberula* Wieg. = Bidens frondosa [BIDEFRO]
Bidens frondosa var. *stenodonta* Fern. & St. John = Bidens frondosa [BIDEFRO]
Bidens glaucescens Greene = Bidens cernua [BIDECER]
Bidens puberula Wieg. = Bidens vulgata [BIDEVUL]
Bidens vulgata Greene [BIDEVUL] (tall beggarticks)
SY: *Bidens puberula* Wieg.
SY: *Bidens vulgata* var. *puberula* (Wieg.) Greene
SY: *Bidens vulgata* var. *schizantha* Lunell
Bidens vulgata var. *puberula* (Wieg.) Greene = Bidens vulgata [BIDEVUL]
Bidens vulgata var. *schizantha* Lunell = Bidens vulgata [BIDEVUL]
Brachyactis Ledeb. [BRACHYA$]
Brachyactis angusta (Lindl.) Britt. = Brachyactis ciliata ssp. angusta [BRACCIL1]
Brachyactis ciliata (Ledeb.) Ledeb. [BRACCIL]
Brachyactis ciliata ssp. **angusta** (Lindl.) A.G. Jones [BRACCIL1]
SY: *Aster brachyactis* Blake
SY: *Brachyactis angusta* (Lindl.) Britt.
SY: *Tripolium angustum* Lindl.
Brickellia Ell. [BRICKEL$]
Brickellia grandiflora (Hook.) Nutt. [BRICGRA] (large-flowered brickellia)
SY: *Brickellia grandiflora* var. *petiolaris* Gray
SY: *Coleosanthus grandiflorus* (Hook.) Kuntze
Brickellia grandiflora var. *petiolaris* Gray = Brickellia grandiflora [BRICGRA]
Brickellia oblongifolia Nutt. [BRICOBL] (narrow-leaved brickellia)
SY: *Brickellia oblongifolia* var. *typica* B.L. Robins.
Brickellia oblongifolia var. *typica* B.L. Robins. = Brickellia oblongifolia [BRICOBL]
Cacalia nardosmia Gray = Cacaliopsis nardosmia [CACANAR]
Cacalia nardosmia var. *glabrata* (Piper) Boivin = Cacaliopsis nardosmia [CACANAR]
Cacaliopsis Gray [CACALIO$]
Cacaliopsis nardosmia (Gray) Gray [CACANAR] (silvercrown)
SY: *Cacalia nardosmia* Gray
SY: *Cacalia nardosmia* var. *glabrata* (Piper) Boivin
SY: *Cacaliopsis nardosmia* ssp. *glabrata* Piper
SY: *Cacaliopsis nardosmia* var. *glabrata* Piper
SY: *Luina nardosmia* (Gray) Cronq.
SY: *Luina nardosmia* var. *glabrata* (Piper) Cronq.
Cacaliopsis nardosmia ssp. *glabrata* Piper = Cacaliopsis nardosmia [CACANAR]
Cacaliopsis nardosmia var. *glabrata* Piper = Cacaliopsis nardosmia [CACANAR]
Carduus L. [CARDUUS$]
•**Carduus acanthoides** L. [CARDACA] (plumeless thistle)
Carduus arvensis (L.) Robson = Cirsium arvense [CIRSARV]
•**Carduus crispus** L. [CARDCRI] (curled thistle)
Carduus lanceolatus L. = Cirsium vulgare [CIRSVUL]
Carduus macounii Greene = Cirsium edule [CIRSEDU]
Carduus marianus L. = Silybum marianum [SILYMAR]
•**Carduus nutans** L. [CARDNUA]
•**Carduus nutans** ssp. **leiophyllus** (Petrovic) Stojanov & Stef. [CARDNUA1] (nodding thistle)
SY: *Carduus nutans* var. *leiophyllus* (Petrovic) Arènes
SY: *Carduus nutans* var. *vestitus* (Hallier) Boivin
Carduus nutans var. *leiophyllus* (Petrovic) Arènes = Carduus nutans ssp. leiophyllus [CARDNUA1]
Carduus nutans var. *vestitus* (Hallier) Boivin = Carduus nutans ssp. leiophyllus [CARDNUA1]
Carduus vulgaris Savi = Cirsium vulgare [CIRSVUL]
Carthamus L. [CARTHAM$]
Carthamus baeticus (Boiss. & Reut.) Lara = Carthamus lanatus ssp. baeticus [CARTLAN1]
•**Carthamus lanatus** L. [CARTLAN] (woolly distaff-thistle)
•**Carthamus lanatus** ssp. **baeticus** (Boiss. & Reut.) Nyman [CARTLAN1] (distaff thistle)
SY: *Carthamus baeticus* (Boiss. & Reut.) Lara
Centaurea L. [CENTAUR$]
•**Centaurea biebersteinii** DC. [CENTBIE]
SY: *Centaurea maculosa* auct.
•**Centaurea cyanus** L. [CENTCYA] (cornflower)
SY: *Leucacantha cyanus* (L.) Nieuwl. & Lunell
Centaurea debeauxii Gren. & Godr. [CENTDEB]
•**Centaurea debeauxii** ssp. **thuillieri** Dostál [CENTDEB1]
SY: *Centaurea pratensis* Thuill.
•**Centaurea diffusa** Lam. [CENTDIF] (diffuse knapweed)
SY: *Acosta diffusa* (Lam.) Soják
Centaurea dubia ssp. *vochinensis* (Bernh. ex Reichenb.) Hayek = Centaurea nigrescens [CENTNIR]
Centaurea maculosa auct. = Centaurea biebersteinii [CENTBIE]
•**Centaurea melitensis** L. [CENTMEL] (maltese star-thistle)
•**Centaurea montana** L. [CENTMON] (mountain bluet)
•**Centaurea nigrescens** Willd. [CENTNIR] (short-fringed knotweed)
SY: *Centaurea dubia* ssp. *vochinensis* (Bernh. ex Reichenb.) Hayek
SY: *Centaurea vochinensis* Bernh. ex Reichenb.
•**Centaurea paniculata** L. [CENTPAN] (jersey knapweed)
Centaurea picris Pallas ex Willd. = Acroptilon repens [ACROREP]
Centaurea pratensis Thuill. = Centaurea debeauxii ssp. thuillieri [CENTDEB1]
Centaurea repens L. = Acroptilon repens [ACROREP]
Centaurea vochinensis Bernh. ex Reichenb. = Centaurea nigrescens [CENTNIR]
Chaenactis DC. [CHAENAC$]
Chaenactis alpina (Gray) M.E. Jones = Chaenactis douglasii var. alpina [CHAEDOU4]
Chaenactis angustifolia Greene = Chaenactis douglasii var. douglasii [CHAEDOU0]
Chaenactis brachiata Greene = Chaenactis douglasii var. douglasii [CHAEDOU0]
Chaenactis brachiata var. *stansburiana* Stockwell = Chaenactis douglasii var. douglasii [CHAEDOU0]

Chaenactis cineria Stockwell = Chaenactis douglasii var. douglasii [CHAEDOU0]
Chaenactis douglasii (Hook.) Hook. & Arn. [CHAEDOU] (hoary false yarrow)
Chaenactis douglasii var. *achilleifolia* (Hook. & Arn.) Gray = Chaenactis douglasii var. douglasii [CHAEDOU0]
Chaenactis douglasii var. **alpina** Gray [CHAEDOU4]
SY: *Chaenactis alpina* (Gray) M.E. Jones
SY: *Chaenactis minuscula* Greene
Chaenactis douglasii var. **douglasii** [CHAEDOU0]
SY: *Chaenactis angustifolia* Greene
SY: *Chaenactis brachiata* Greene
SY: *Chaenactis brachiata* var. *stansburiana* Stockwell
SY: *Chaenactis cineria* Stockwell
SY: *Chaenactis douglasii* var. *achilleifolia* (Hook. & Arn.) Gray
SY: *Chaenactis douglasii* var. *glandulosa* Cronq.
SY: *Chaenactis douglasii* var. *montana* M.E. Jones
SY: *Chaenactis douglasii* var. *nana* Stockwell
SY: *Chaenactis douglasii* var. *rubricaulis* Rydb.
SY: *Chaenactis douglasii* var. *typicus* Cronq.
SY: *Chaenactis humilis* Rydb.
SY: *Chaenactis panamintensis* Stockwell
SY: *Chaenactis ramosa* Stockwell
SY: *Chaenactis rubricaulis* Rydb.
SY: *Chaenactis suksdorfii* Stockwell
Chaenactis douglasii var. *glandulosa* Cronq. = Chaenactis douglasii var. douglasii [CHAEDOU0]
Chaenactis douglasii var. *montana* M.E. Jones = Chaenactis douglasii var. douglasii [CHAEDOU0]
Chaenactis douglasii var. *nana* Stockwell = Chaenactis douglasii var. douglasii [CHAEDOU0]
Chaenactis douglasii var. *rubricaulis* Rydb. = Chaenactis douglasii var. douglasii [CHAEDOU0]
Chaenactis douglasii var. *typicus* Cronq. = Chaenactis douglasii var. douglasii [CHAEDOU0]
Chaenactis humilis Rydb. = Chaenactis douglasii var. douglasii [CHAEDOU0]
Chaenactis minuscula Greene = Chaenactis douglasii var. alpina [CHAEDOU4]
Chaenactis panamintensis Stockwell = Chaenactis douglasii var. douglasii [CHAEDOU0]
Chaenactis ramosa Stockwell = Chaenactis douglasii var. douglasii [CHAEDOU0]
Chaenactis rubricaulis Rydb. = Chaenactis douglasii var. douglasii [CHAEDOU0]
Chaenactis suksdorfii Stockwell = Chaenactis douglasii var. douglasii [CHAEDOU0]
Chamomilla chamomilla (L.) Rydb. = Matricaria recutita [MATRREC]
Chamomilla inodora (L.) Gilib. = Matricaria perforata [MATRPER]
Chamomilla recutita (L.) Rauschert = Matricaria recutita [MATRREC]
Chamomilla suaveolens (Pursh) Rydb. = Matricaria discoidea [MATRDIS]
Chlorocrepis albiflora (Hook.) W.A. Weber = Hieracium albiflorum [HIERALI]
Chondrilla L. [CHONDRI$]
***Chondrilla juncea** L. [CHONJUN] (gum-succory)
Chrysanthemum arcticum auct. = Dendranthema arcticum ssp. polare [DENDARC1]
Chrysanthemum arcticum var. *polare* (Hultén) Boivin = Dendranthema arcticum ssp. polare [DENDARC1]
Chrysanthemum bipinnatum ssp. *huronense* (Nutt.) Hultén = Tanacetum bipinnatum ssp. huronense [TANABIP1]
Chrysanthemum integrifolium Richards. = Leucanthemum integrifolium [LEUCINT]
Chrysanthemum leucanthemum L. = Leucanthemum vulgare [LEUCVUL]
Chrysanthemum leucanthemum var. *boecheri* Boivin = Leucanthemum vulgare [LEUCVUL]
Chrysanthemum leucanthemum var. *pinnatifidum* Lecoq & Lamotte = Leucanthemum vulgare [LEUCVUL]
Chrysanthemum parthenium (L.) Bernh. = Tanacetum parthenium [TANAPAR]
Chrysanthemum uliginosum Pers. = Tanacetum vulgare [TANAVUL]
Chrysanthemum vulgare (L.) Bernh. = Tanacetum vulgare [TANAVUL]
Chrysopsis angustifolia Rydb. = Heterotheca villosa var. villosa [HETEVIL0]
Chrysopsis arida A. Nels. = Heterotheca villosa var. hispida [HETEVIL1]
Chrysopsis bakeri Greene = Heterotheca villosa var. villosa [HETEVIL0]
Chrysopsis ballardii Rydb. = Heterotheca villosa var. villosa [HETEVIL0]
Chrysopsis butleri Rydb. = Heterotheca villosa var. hispida [HETEVIL1]
Chrysopsis canescens var. *nana* Gray = Heterotheca villosa var. hispida [HETEVIL1]
Chrysopsis caudata Rydb. = Heterotheca villosa var. villosa [HETEVIL0]
Chrysopsis columbiana Greene = Heterotheca villosa var. hispida [HETEVIL1]
Chrysopsis foliosa (Nutt.) Shinners = Heterotheca villosa var. villosa [HETEVIL0]
Chrysopsis hirsutissima Greene = Heterotheca villosa var. hispida [HETEVIL1]
Chrysopsis hispida (Hook.) DC. = Heterotheca villosa var. hispida [HETEVIL1]
Chrysopsis horrida Rydb. = Heterotheca villosa var. hispida [HETEVIL1]
Chrysopsis imbricata A. Nels. = Heterotheca villosa var. villosa [HETEVIL0]
Chrysopsis mollis Nutt. = Heterotheca villosa var. villosa [HETEVIL0]
Chrysopsis nitidula Woot. & Standl. = Heterotheca villosa var. villosa [HETEVIL0]
Chrysopsis pedunculata Greene = Heterotheca villosa var. villosa [HETEVIL0]
Chrysopsis villosa (Pursh) Nutt. ex DC. = Heterotheca villosa var. villosa [HETEVIL0]
Chrysopsis villosa var. *angustifolia* (Rydb.) Cronq. = Heterotheca villosa var. villosa [HETEVIL0]
Chrysopsis villosa var. *foliosa* (Nutt.) D.C. Eat. = Heterotheca villosa var. villosa [HETEVIL0]
Chrysopsis villosa var. *hispida* (Hook.) Gray = Heterotheca villosa var. hispida [HETEVIL1]
Chrysopsis viscida ssp. *cinerascens* Blake = Heterotheca villosa var. hispida [HETEVIL1]
Chrysopsis wisconsinensis Shinners = Heterotheca villosa var. hispida [HETEVIL1]
Chrysothamnus Nutt. [CHRYSOT$]
Chrysothamnus nauseosus (Pallas ex Pursh) Britt. [CHRYNAU] (common rabbit-brush)
Chrysothamnus nauseosus ssp. **albicaulis** (Nutt.) Hall & Clements [CHRYNAU1]
SY: *Chrysothamnus nauseosus* var. *albicaulis* (Nutt.) Rydb.
SY: *Chrysothamnus nauseosus* ssp. *speciosus* (Nutt.) Hall & Clements
SY: *Chrysothamnus nauseosus* var. *speciosus* (Nutt.) Hall
Chrysothamnus nauseosus ssp. *speciosus* (Nutt.) Hall & Clements = Chrysothamnus nauseosus ssp. albicaulis [CHRYNAU1]
Chrysothamnus nauseosus var. *albicaulis* (Nutt.) Rydb. = Chrysothamnus nauseosus ssp. albicaulis [CHRYNAU1]
Chrysothamnus nauseosus var. *speciosus* (Nutt.) Hall = Chrysothamnus nauseosus ssp. albicaulis [CHRYNAU1]
Chrysothamnus viscidiflorus (Hook.) Nutt. [CHRYVIS] (green rabbit-brush)
Chrysothamnus viscidiflorus ssp. *elegans* (Greene) Hall & Clements = Chrysothamnus viscidiflorus ssp. lanceolatus [CHRYVIS1]
Chrysothamnus viscidiflorus ssp. **lanceolatus** (Nutt.) Hall & Clements [CHRYVIS1]
SY: *Chrysothamnus viscidiflorus* ssp. *elegans* (Greene) Hall & Clements
SY: *Chrysothamnus viscidiflorus* var. *elegans* (Greene) Blake

SY: *Chrysothamnus viscidiflorus* var. *lanceolatus* (Nutt.) Greene
Chrysothamnus viscidiflorus var. *elegans* (Greene) Blake = Chrysothamnus viscidiflorus ssp. lanceolatus [CHRYVIS1]
Chrysothamnus viscidiflorus var. *lanceolatus* (Nutt.) Greene = Chrysothamnus viscidiflorus ssp. lanceolatus [CHRYVIS1]
Cichorium L. [CICHORI$]
*__Cichorium intybus__ L. [CICHINT] (chicory)
Cirsium P. Mill. [CIRSIUM$]
*__Cirsium arvense__ (L.) Scop. [CIRSARV] (Canada thistle)
SY: *Carduus arvensis* (L.) Robson
SY: *Cirsium arvense* var. *argenteum* (Vest) Fiori
SY: *Cirsium arvense* var. *horridum* Wimmer & Grab.
SY: *Cirsium arvense* var. *integrifolium* Wimmer & Grab.
SY: *Cirsium arvense* var. *mite* Wimmer & Grab.
SY: *Cirsium arvense* var. *vestitum* Wimmer & Grab.
SY: *Cirsium incanum* (Gmel.) Fisch.
SY: *Cirsium setosum* (Willd.) Bess. ex Bieb.
SY: *Serratula arvensis* L.
Cirsium arvense var. *argenteum* (Vest) Fiori = Cirsium arvense [CIRSARV]
Cirsium arvense var. *horridum* Wimmer & Grab. = Cirsium arvense [CIRSARV]
Cirsium arvense var. *integrifolium* Wimmer & Grab. = Cirsium arvense [CIRSARV]
Cirsium arvense var. *mite* Wimmer & Grab. = Cirsium arvense [CIRSARV]
Cirsium arvense var. *vestitum* Wimmer & Grab. = Cirsium arvense [CIRSARV]
Cirsium brevistylum Cronq. [CIRSBRE] (short-styled thistle)
Cirsium coccinatum Osterhout = Cirsium drummondii [CIRSDRU]
Cirsium drummondii Torr. & Gray [CIRSDRU] (Drummond's thistle)
SY: *Cirsium coccinatum* Osterhout
Cirsium edule Nutt. [CIRSEDU] (edible thistle)
SY: *Carduus macounii* Greene
SY: *Cirsium macounii* (Greene) Rydb.
Cirsium foliosum (Hook.) DC. [CIRSFOL] (leafy thistle)
SY: *Cirsium foliosum* var. *minganense* (Victorin) Boivin
Cirsium foliosum var. *minganense* (Victorin) Boivin = Cirsium foliosum [CIRSFOL]
Cirsium hookerianum Nutt. [CIRSHOO] (Hooker's thistle)
Cirsium incanum (Gmel.) Fisch. = Cirsium arvense [CIRSARV]
Cirsium lanceolatum (L.) Scop. = Cirsium vulgare [CIRSVUL]
Cirsium lanceolatum var. *hypoleucum* DC. = Cirsium vulgare [CIRSVUL]
Cirsium macounii (Greene) Rydb. = Cirsium edule [CIRSEDU]
*__Cirsium palustre__ (L.) Scop. [CIRSPAL] (marsh thistle)
Cirsium scariosum Nutt. [CIRSSCA] (elk thistle)
Cirsium setosum (Willd.) Bess. ex Bieb. = Cirsium arvense [CIRSARV]
Cirsium undulatum (Nutt.) Spreng. [CIRSUND] (wavy-leaved thistle)
*__Cirsium vulgare__ (Savi) Ten. [CIRSVUL] (bull thistle)
SY: *Carduus lanceolatus* L.
SY: *Carduus vulgaris* Savi
SY: *Cirsium lanceolatum* (L.) Scop.
SY: *Cirsium lanceolatum* var. *hypoleucum* DC.
Coleosanthus grandiflorus (Hook.) Kuntze = Brickellia grandiflora [BRICGRA]
Conyza Less. [CONYZA$]
*__Conyza canadensis__ (L.) Cronq. [CONYCAN]
SY: *Erigeron canadensis* L.
*__Conyza canadensis__ var. **glabrata** (Gray) Cronq. [CONYCAN2] (horseweed)
SY: *Erigeron canadensis* var. *glabratus* Gray
Coreopsis L. [COREOPS$]
Coreopsis atkinsoniana Dougl. ex Lindl. = Coreopsis tinctoria var. atkinsoniana [CORETIN1]
Coreopsis crassifolia Ait. = Coreopsis lanceolata [CORELAN]
Coreopsis heterogyna Fern. = Coreopsis lanceolata [CORELAN]
*__Coreopsis lanceolata__ L. [CORELAN] (garden coreopsis)
SY: *Coreopsis crassifolia* Ait.
SY: *Coreopsis heterogyna* Fern.
SY: *Coreopsis lanceolata* var. *villosa* Michx.
Coreopsis lanceolata var. *villosa* Michx. = Coreopsis lanceolata [CORELAN]
Coreopsis tinctoria Nutt. [CORETIN]
Coreopsis tinctoria var. **atkinsoniana** (Dougl. ex Lindl.) H.M. Parker [CORETIN1]
SY: *Coreopsis atkinsoniana* Dougl. ex Lindl.
Cota tinctoria (L.) J. Gay = Anthemis tinctoria [ANTHTIN]
Cotula L. [COTULA$]
*__Cotula coronopifolia__ L. [COTUCOR] (brass buttons)
Crepis L. [CREPIS$]
Crepis alpicola (Rydb.) A. Nels. = Crepis runcinata [CREPRUN]
Crepis atrabarba Heller [CREPATR] (slender hawksbeard)
SY: *Crepis exilis* Osterhout
Crepis atrabarba ssp. **atrabarba** [CREPATR0]
SY: *Crepis atribarba* ssp. *cytotaxonomicorum* (Boivin) W.A. Weber
SY: *Crepis atribarba* var. *cytotaxonomicorum* Boivin
SY: *Crepis atribarba* ssp. *typicus* Babcock & Stebbins
SY: *Psilochenia atribarba* (Heller) W.A. Weber
Crepis atrabarba ssp. **originalis** (Babcock & Stebbins) Babcock & Stebbins [CREPATR2]
SY: *Crepis barbigera* Leib. ex Coville
Crepis atribarba ssp. *cytotaxonomicorum* (Boivin) W.A. Weber = Crepis atrabarba ssp. atrabarba [CREPATR0]
Crepis atribarba ssp. *typicus* Babcock & Stebbins = Crepis atrabarba ssp. atrabarba [CREPATR0]
Crepis atribarba var. *cytotaxonomicorum* Boivin = Crepis atrabarba ssp. atrabarba [CREPATR0]
Crepis barbigera Leib. ex Coville = Crepis atrabarba ssp. originalis [CREPATR2]
*__Crepis capillaris__ (L.) Wallr. [CREPCAP] (smooth hawksbeard)
Crepis elegans Hook. [CREPELE] (elegant hawksbeard)
SY: *Askellia elegans* (Hook.) W.A. Weber
Crepis exilis Osterhout = Crepis atrabarba [CREPATR]
Crepis glaucella Rydb. = Crepis runcinata [CREPRUN]
Crepis intermedia Gray [CREPINT] (gray hawksbeard)
SY: *Psilochenia intermedia* (Gray) W.A. Weber
Crepis modocensis Greene [CREPMOD] (low hawksbeard)
Crepis modocensis ssp. **modocensis** [CREPMOD0]
SY: *Crepis modocensis* ssp. *typica* Babcock & Stebbins
SY: *Psilochenia modocensis* (Greene) W.A. Weber
Crepis modocensis ssp. **rostrata** (Coville) Babcock & Stebbins [CREPMOD2]
SY: *Crepis modocensis* var. *rostrata* (Coville) Boivin
SY: *Psilochenia modocensis* ssp. *rostrata* (Coville) W.A. Weber
Crepis modocensis ssp. *typica* Babcock & Stebbins = Crepis modocensis ssp. modocensis [CREPMOD0]
Crepis modocensis var. *rostrata* (Coville) Boivin = Crepis modocensis ssp. rostrata [CREPMOD2]
Crepis nana Richards. [CREPNAN] (dwarf hawksbeard)
Crepis nana ssp. **ramosa** Babcock [CREPNAN2]
SY: *Askellia nana* ssp. *ramosa* (Babcock) W.A. Weber
Crepis neomexicana Woot. & Standl. = Crepis runcinata [CREPRUN]
*__Crepis nicaeensis__ Balbis ex Pers. [CREPNIC] (French hawksbeard)
Crepis occidentalis Nutt. [CREPOCC] (western hawksbeard)
Crepis occidentalis ssp. **costata** (Gray) Babcock & Stebbins [CREPOCC1]
SY: *Crepis occidentalis* var. *costata* Gray
SY: *Psilochenia occidentalis* ssp. *costata* (Gray) W.A. Weber
Crepis occidentalis ssp. **occidentalis** [CREPOCC0]
SY: *Crepis occidentalis* ssp. *typica* Babcock & Stebbins
SY: *Psilochenia occidentalis* (Nutt.) Nutt.
Crepis occidentalis ssp. **pumila** (Rydb.) Babcock & Stebbins [CREPOCC3]
SY: *Crepis occidentalis* var. *pumila* (Rydb.) Babcock & Stebbins
Crepis occidentalis ssp. *typica* Babcock & Stebbins = Crepis occidentalis ssp. occidentalis [CREPOCC0]
Crepis occidentalis var. *costata* Gray = Crepis occidentalis ssp. costata [CREPOCC1]
Crepis occidentalis var. *pumila* (Rydb.) Babcock & Stebbins = Crepis occidentalis ssp. pumila [CREPOCC3]
Crepis perplexans Rydb. = Crepis runcinata [CREPRUN]

Crepis runcinata (James) Torr. & Gray [CREPRUN] (dandelion hawksbeard)
SY: *Crepis alpicola* (Rydb.) A. Nels.
SY: *Crepis glaucella* Rydb.
SY: *Crepis neomexicana* Woot. & Standl.
SY: *Crepis perplexans* Rydb.
SY: *Crepis runcinata* ssp. *typica* Babcock & Stebbins
SY: *Psilochenia runcinata* (James) A. & D. Löve
Crepis runcinata ssp. *typica* Babcock & Stebbins = Crepis runcinata [CREPRUN]
*****Crepis tectorum** L. [CREPTEC] (annual hawksbeard)
*****Crepis vesicaria** L. [CREPVES] (beaked hawksbeard)
*****Crepis vesicaria** ssp. **haenseleri** (Boiss. ex DC.) P.D. Sell [CREPVES1]
SY: *Crepis vesicaria* ssp. *taraxacifolia* (Thuill.) Thellung ex Schinz & R. Keller
SY: *Crepis vesicaria* var. *taraxacifolia* (Thuill.) Boivin
Crepis vesicaria ssp. *taraxacifolia* (Thuill.) Thellung ex Schinz & R. Keller = Crepis vesicaria ssp. haenseleri [CREPVES1]
Crepis vesicaria var. *taraxacifolia* (Thuill.) Boivin = Crepis vesicaria ssp. haenseleri [CREPVES1]
Crocidium Hook. [CROCIDI$]
Crocidium multicaule Hook. [CROCMUL] (gold star)
Cyclachaena xanthifolia (Nutt.) Fresen. = Iva xanthifolia [IVA XAN]
Dendranthema (DC.) Des Moulins [DENDRAN$]
Dendranthema arcticum (L.) Tzvelev [DENDARC]
Dendranthema arcticum ssp. **polare** (Hultén) Heywood [DENDARC1]
SY: *Chrysanthemum arcticum* var. *polare* (Hultén) Boivin
SY: *Dendranthema hultenii* (A. & D. Löve) Tzvelev
SY: *Chrysanthemum arcticum* auct.
SY: *Leucanthemum arcticum* auct.
Dendranthema hultenii (A. & D. Löve) Tzvelev = Dendranthema arcticum ssp. polare [DENDARC1]
Doronicum L. [DORONIC$]
*****Doronicum pardalianches** L. [DOROPAR] (great leopard's-bane)
Ericameria Nutt. [ERICAME$]
Ericameria bloomeri (Gray) J.F. Macbr. [ERICBLO]
SY: *Haplopappus bloomeri* Gray
SY: *Haplopappus bloomeri* var. *angustatus* Gray
SY: *Haplopappus bloomeri* ssp. *sonnei* (Gray) Hall
SY: *Haplopappus bloomeri* var. *sonnei* (Gray) Greene
Erigeron L. [ERIGERO$]
Erigeron acris L. = Trimorpha acris var. asteroides [TRIMACR1]
Erigeron acris ssp. *debilis* (Gray) Piper = Trimorpha acris var. debilis [TRIMACR2]
Erigeron acris ssp. *politus* (Fries) Schinz & R. Keller = Trimorpha acris var. asteroides [TRIMACR1]
Erigeron acris var. *asteroides* (Andrz. ex Bess.) DC. = Trimorpha acris var. asteroides [TRIMACR1]
Erigeron acris var. *debilis* Gray = Trimorpha acris var. debilis [TRIMACR2]
Erigeron acris var. *elatus* (Hook.) Cronq. = Trimorpha elata [TRIMELA]
Erigeron alpinus var. *elatus* Hook. = Trimorpha elata [TRIMELA]
Erigeron angulosus ssp. *debilis* (Gray) Piper = Trimorpha acris var. debilis [TRIMACR2]
*****Erigeron annuus** (L.) Pers. [ERIGANN] (annual fleabane)
SY: *Erigeron annuus* var. *discoideus* (Victorin & Rouss.) Cronq.
Erigeron annuus ssp. *strigosus* (Muhl. ex Willd.) Wagenitz = Erigeron strigosus [ERIGSTR]
Erigeron annuus var. *discoideus* (Victorin & Rouss.) Cronq. = Erigeron annuus [ERIGANN]
Erigeron anodonta Lunell = Erigeron glabellus var. pubescens [ERIGGLA1]
Erigeron asperum var. *pubescens* (Hook.) Breitung = Erigeron glabellus var. pubescens [ERIGGLA1]
Erigeron aureus Greene [ERIGAUR] (golden fleabane)
Erigeron caespitosus Nutt. [ERIGCAE] (tufted fleabane)
SY: *Erigeron caespitosus* var. *laccoliticus* M.E. Jones
Erigeron caespitosus var. *laccoliticus* M.E. Jones = Erigeron caespitosus [ERIGCAE]
Erigeron canadensis L. = Conyza canadensis [CONYCAN]
Erigeron canadensis var. *glabratus* Gray = Conyza canadensis var. glabrata [CONYCAN2]
Erigeron compositus Pursh [ERIGCOM] (dwarf mountain fleabane)
SY: *Erigeron compositus* var. *discoideus* Gray
SY: *Erigeron compositus* var. *glabratus* Macoun
SY: *Erigeron compositus* var. *multifidus* (Rydb.) J.F. Macbr. & Payson
SY: *Erigeron compositus* var. *typicus* Hook.
Erigeron compositus var. *discoideus* Gray = Erigeron compositus [ERIGCOM]
Erigeron compositus var. *glabratus* Macoun = Erigeron compositus [ERIGCOM]
Erigeron compositus var. *multifidus* (Rydb.) J.F. Macbr. & Payson = Erigeron compositus [ERIGCOM]
Erigeron compositus var. *typicus* Hook. = Erigeron compositus [ERIGCOM]
Erigeron conspicuus Rydb. = Erigeron subtrinervis var. conspicuus [ERIGSUB1]
Erigeron corymbosus Nutt. [ERIGCOR] (long-leaved fleabane)
Erigeron debilis (Gray) Rydb. = Trimorpha acris var. debilis [TRIMACR2]
Erigeron divergens Torr. & Gray [ERIGDIV] (spreading fleabane)
SY: *Erigeron divergens* var. *typicus* Cronq.
Erigeron divergens var. *typicus* Cronq. = Erigeron divergens [ERIGDIV]
Erigeron droebachianus O.F. Muell. ex Retz. = Trimorpha acris var. asteroides [TRIMACR1]
Erigeron elatus (Hook.) Greene = Trimorpha elata [TRIMELA]
Erigeron elatus var. *oligocephalus* (Fern. & Wieg.) Fern. = Trimorpha elata [TRIMELA]
Erigeron elongatus Ledeb. = Trimorpha acris var. asteroides [TRIMACR1]
Erigeron eriocephalus J. Vahl = Erigeron uniflorus ssp. eriocephalus [ERIGUNI1]
Erigeron filifolius (Hook.) Nutt. [ERIGFIL] (thread-leaved fleabane)
SY: *Erigeron filifolius* var. *typicus* Cronq.
Erigeron filifolius var. *typicus* Cronq. = Erigeron filifolius [ERIGFIL]
Erigeron flagellaris Gray [ERIGFLA] (trailing fleabane)
SY: *Erigeron flagellaris* var. *typicus* Cronq.
SY: *Erigeron nudiflorus* Buckl.
Erigeron flagellaris var. *typicus* Cronq. = Erigeron flagellaris [ERIGFLA]
Erigeron glabellus Nutt. [ERIGGLA] (streamside fleabane)
Erigeron glabellus ssp. *pubescens* (Hook.) Cronq. = Erigeron glabellus var. pubescens [ERIGGLA1]
Erigeron glabellus var. **pubescens** Hook. [ERIGGLA1]
SY: *Erigeron anodonta* Lunell
SY: *Erigeron asperum* var. *pubescens* (Hook.) Breitung
SY: *Erigeron glabellus* ssp. *pubescens* (Hook.) Cronq.
SY: *Erigeron oligodontus* Lunell
Erigeron grandiflorus Hook. [ERIGGRA] (large-flowered fleabane)
Erigeron humilis Graham [ERIGHUM] (arctic daisy)
SY: *Erigeron unalaschkensis* (DC.) Vierh.
SY: *Erigeron uniflorus* var. *unalaschkensis* (DC.) Ostenf.
Erigeron jucundus Greene = Trimorpha acris var. debilis [TRIMACR2]
Erigeron lanatus Hook. [ERIGLAN] (woolly daisy)
Erigeron leibergii Piper [ERIGLEI] (Leiberg's fleabane)
Erigeron linearis (Hook.) Piper [ERIGLIN] (line-leaved daisy)
SY: *Erigeron peucephyllus* Gray
Erigeron lonchophyllus Hook. = Trimorpha lonchophylla [TRIMLON]
Erigeron lonchophyllus var. *laurentianus* Victorin = Trimorpha lonchophylla [TRIMLON]
Erigeron minor (Hook.) Rydb. = Trimorpha lonchophylla [TRIMLON]
Erigeron nivalis Nutt. = Trimorpha acris var. debilis [TRIMACR2]
Erigeron nudiflorus Buckl. = Erigeron flagellaris [ERIGFLA]
Erigeron oligodontus Lunell = Erigeron glabellus var. pubescens [ERIGGLA1]
Erigeron pallens Cronq. [ERIGPAL]
SY: *Erigeron purpuratus* ssp. *pallens* (Cronq.) G.W. Douglas

Erigeron peregrinus (Banks ex Pursh) Greene [ERIGPER] (subalpine daisy)
Erigeron peregrinus ssp. **callianthemus** (Greene) Cronq. [ERIGPER2]
SY: *Erigeron peregrinus* ssp. *callianthemus* var. *angustifolius* (Gray) Cronq.
SY: *Erigeron peregrinus* ssp. *callianthemus* var. *scaposus* (Torr. & Gray) Cronq.
Erigeron peregrinus ssp. *callianthemus* var. *angustifolius* (Gray) Cronq. = Erigeron peregrinus ssp. callianthemus [ERIGPER2]
Erigeron peregrinus ssp. *callianthemus* var. *scaposus* (Torr. & Gray) Cronq. = Erigeron peregrinus ssp. callianthemus [ERIGPER2]
Erigeron peregrinus ssp. **peregrinus** [ERIGPER0]
SY: *Erigeron peregrinus* var. *thompsonii* (Blake ex J.W. Thompson) Cronq.
Erigeron peregrinus ssp. **peregrinus** var. **dawsonii** Greene [ERIGPER1]
Erigeron peregrinus var. *thompsonii* (Blake ex J.W. Thompson) Cronq. = Erigeron peregrinus ssp. peregrinus [ERIGPER0]
Erigeron peucephyllus Gray = Erigeron linearis [ERIGLIN]
Erigeron philadelphicus L. [ERIGPHI] (Philadelphia fleabane)
Erigeron poliospermus Gray [ERIGPOL] (cushion fleabane)
SY: *Erigeron poliospermus* var. *typicus* Cronq.
Erigeron poliospermus var. *typicus* Cronq. = Erigeron poliospermus [ERIGPOL]
Erigeron politus Fries = Trimorpha acris var. asteroides [TRIMACR1]
Erigeron pumilus Nutt. [ERIGPUM] (shaggy fleabane)
Erigeron pumilus ssp. **intermedius** Cronq. [ERIGPUM1]
Erigeron purpuratus Greene [ERIGPUR] (purple daisy)
Erigeron purpuratus ssp. *pallens* (Cronq.) G.W. Douglas = Erigeron pallens [ERIGPAL]
Erigeron ramosus (Walt.) B.S.P. = Erigeron strigosus [ERIGSTR]
Erigeron salishii G.W. Douglas & Packer [ERIGSAL] (Salish daisy)
Erigeron speciosus (Lindl.) DC. [ERIGSPE] (showy fleabane)
SY: *Erigeron speciosus* var. *typicus* Cronq.
Erigeron speciosus var. *conspicuus* (Rydb.) Breitung = Erigeron subtrinervis var. conspicuus [ERIGSUB1]
Erigeron speciosus var. *typicus* Cronq. = Erigeron speciosus [ERIGSPE]
Erigeron strigosus Muhl. ex Willd. [ERIGSTR] (prairie fleabane)
SY: *Erigeron annuus* ssp. *strigosus* (Muhl. ex Willd.) Wagenitz
SY: *Erigeron ramosus* (Walt.) B.S.P.
SY: *Erigeron strigosus* var. *discoideus* Robbins ex Gray
SY: *Erigeron strigosus* var. *eligulatus* Cronq.
SY: *Erigeron strigosus* var. *typicus* Cronq.
SY: *Erigeron traversii* Shinners
Erigeron strigosus var. *discoideus* Robbins ex Gray = Erigeron strigosus [ERIGSTR]
Erigeron strigosus var. *eligulatus* Cronq. = Erigeron strigosus [ERIGSTR]
Erigeron strigosus var. *typicus* Cronq. = Erigeron strigosus [ERIGSTR]
Erigeron subtrinervis Rydb. ex Porter & Britt. [ERIGSUB] (three-nerve fleabane)
Erigeron subtrinervis ssp. *conspicuus* (Rydb.) Cronq. = Erigeron subtrinervis var. conspicuus [ERIGSUB1]
Erigeron subtrinervis var. **conspicuus** (Rydb.) Cronq. [ERIGSUB1]
SY: *Erigeron conspicuus* Rydb.
SY: *Erigeron speciosus* var. *conspicuus* (Rydb.) Breitung
SY: *Erigeron subtrinervis* ssp. *conspicuus* (Rydb.) Cronq.
Erigeron traversii Shinners = Erigeron strigosus [ERIGSTR]
Erigeron trifidus Hook. [ERIGTRI] (three-lobed daisy)
Erigeron unalaschkensis (DC.) Vierh. = Erigeron humilis [ERIGHUM]
Erigeron uniflorus L. [ERIGUNI] (one-flowered fleabane)
Erigeron uniflorus ssp. **eriocephalus** (J. Vahl) Cronq. [ERIGUNI1]
SY: *Erigeron eriocephalus* J. Vahl
SY: *Erigeron uniflorus* var. *eriocephalus* (J. Vahl) Boivin
Erigeron uniflorus var. *eriocephalus* (J. Vahl) Boivin = Erigeron uniflorus ssp. eriocephalus [ERIGUNI1]
Erigeron uniflorus var. *unalaschkensis* (DC.) Ostenf. = Erigeron humilis [ERIGHUM]
Eriophyllum Lag. [ERIOPHY$]
Eriophyllum lanatum (Pursh) Forbes [ERIOLAN] (woolly eriophyllum)
SY: *Eriophyllum lanatum* var. *typicum* Constance
Eriophyllum lanatum var. *typicum* Constance = Eriophyllum lanatum [ERIOLAN]
Eucephalus engelmannii (D.C. Eat.) Greene = Aster engelmannii [ASTEENG]
Eupatoriadelphus maculatus var. *bruneri* (Gray) King & H.E. Robins. = Eupatorium maculatum var. bruneri [EUPAMAC1]
Eupatorium L. [EUPATOR$]
Eupatorium bruneri Gray = Eupatorium maculatum var. bruneri [EUPAMAC1]
Eupatorium maculatum L. [EUPAMAC] (spotted joe-pye-weed)
Eupatorium maculatum ssp. *bruneri* (Gray) G.W. Douglas = Eupatorium maculatum var. bruneri [EUPAMAC1]
Eupatorium maculatum var. **bruneri** (Gray) Breitung [EUPAMAC1]
SY: *Eupatoriadelphus maculatus* var. *bruneri* (Gray) King & H.E. Robins.
SY: *Eupatorium bruneri* Gray
SY: *Eupatorium maculatum* ssp. *bruneri* (Gray) G.W. Douglas
Euthamia Nutt. ex Cass. [EUTHAMI$]
Euthamia californica Gandog. = Euthamia occidentalis [EUTHOCC]
Euthamia graminifolia (L.) Nutt. [EUTHGRA] (flat-top goldentop)
SY: *Euthamia graminifolia* var. *major* (Michx.) Moldenke
SY: *Solidago graminifolia* (L.) Salisb.
SY: *Solidago graminifolia* var. *major* (Michx.) Fern.
Euthamia graminifolia var. *major* (Michx.) Moldenke = Euthamia graminifolia [EUTHGRA]
Euthamia linearifolia Gandog. = Euthamia occidentalis [EUTHOCC]
Euthamia occidentalis Nutt. [EUTHOCC] (western goldenrod)
SY: *Euthamia californica* Gandog.
SY: *Euthamia linearifolia* Gandog.
SY: *Solidago occidentalis* (Nutt.) Torr. & Gray
Filaginella palustris (Nutt.) Holub = Gnaphalium palustre [GNAPPAL]
Filaginella uliginosa (L.) Opiz = Gnaphalium uliginosum [GNAPULI]
Filago L. [FILAGO$]
***Filago arvensis** L. [FILAARV] (field filago)
SY: *Gnaphalium arvense* L.
SY: *Logfia arvensis* (L.) Holub
SY: *Oglifa arvensis* (L.) Cass.
Filago germanica L. = Filago vulgaris [FILAVUL]
***Filago vulgaris** Lam. [FILAVUL] (common filago)
SY: *Filago germanica* L.
SY: *Gifola germanica* Dumort.
SY: *Gnaphalium germanicum* Scopoli
Franseria chamissonis Less. = Ambrosia chamissonis [AMBRCHA]
Franseria chamissonis ssp. *bipinnatisecta* (Less.) Wiggins & Stockwell = Ambrosia chamissonis [AMBRCHA]
Franseria chamissonis var. *bipinnatisecta* Less. = Ambrosia chamissonis [AMBRCHA]
Gaillardia Foug. [GAILLAR$]
Gaillardia aristata Pursh [GAILARI] (brown-eyed Susan)
Galinsoga Ruiz & Pavón [GALINSO$]
Galinsoga aristulata Bickn. = Galinsoga quadriradiata [GALIQUA]
Galinsoga bicolorata St. John & White = Galinsoga quadriradiata [GALIQUA]
Galinsoga caracasana (DC.) Schultz-Bip. = Galinsoga quadriradiata [GALIQUA]
Galinsoga ciliata (Raf.) Blake = Galinsoga quadriradiata [GALIQUA]
***Galinsoga parviflora** Cav. [GALIPAR] (small-flowered galinsoga)
SY: *Galinsoga parviflora* var. *semicalva* Gray
SY: *Galinsoga semicalva* (Gray) St. John & White
SY: *Galinsoga semicalva* var. *percalva* Blake
Galinsoga parviflora var. *semicalva* Gray = Galinsoga parviflora [GALIPAR]
***Galinsoga quadriradiata** Ruiz & Pavón [GALIQUA]
SY: *Galinsoga aristulata* Bickn.
SY: *Galinsoga bicolorata* St. John & White
SY: *Galinsoga caracasana* (DC.) Schultz-Bip.
SY: *Galinsoga ciliata* (Raf.) Blake

Galinsoga semicalva (Gray) St. John & White = Galinsoga parviflora [GALIPAR]
Galinsoga semicalva var. *percalva* Blake = Galinsoga parviflora [GALIPAR]
Gamochaeta Weddell [GAMOCHA$]
*•**Gamochaeta purpurea** (L.) Cabrera [GAMOPUR]
SY: *Gnaphalium purpureum* L.
Gifola germanica Dumort. = Filago vulgaris [FILAVUL]
Gnaphalium L. [GNAPHAL$]
Gnaphalium arvense L. = Filago arvensis [FILAARV]
Gnaphalium beneolens A. Davids. = Gnaphalium microcephalum [GNAPMIC]
Gnaphalium chilense Spreng. = Gnaphalium stramineum [GNAPSTR]
Gnaphalium chilense var. *confertifolium* Greene = Gnaphalium stramineum [GNAPSTR]
Gnaphalium decurrens Ives = Gnaphalium viscosum [GNAPVIS]
Gnaphalium dimorphum Nutt. = Antennaria dimorpha [ANTEDIM]
Gnaphalium germanicum Scopoli = Filago vulgaris [FILAVUL]
Gnaphalium macounii Greene = Gnaphalium viscosum [GNAPVIS]
Gnaphalium margaritaceum L. = Anaphalis margaritacea [ANAPMAR]
Gnaphalium microcephalum Nutt. [GNAPMIC] (slender cudweed)
SY: *Gnaphalium beneolens* A. Davids.
SY: *Gnaphalium microcephalum* ssp. *thermale* (E. Nels.) G.W. Douglas
SY: *Gnaphalium microcephalum* var. *thermale* (E. Nels.) Cronq.
SY: *Gnaphalium thermale* E. Nels.
Gnaphalium microcephalum ssp. *thermale* (E. Nels.) G.W. Douglas = Gnaphalium microcephalum [GNAPMIC]
Gnaphalium microcephalum var. *thermale* (E. Nels.) Cronq. = Gnaphalium microcephalum [GNAPMIC]
Gnaphalium palustre Nutt. [GNAPPAL] (lowland cudweed)
SY: *Filaginella palustris* (Nutt.) Holub
Gnaphalium purpureum L. = Gamochaeta purpurea [GAMOPUR]
Gnaphalium stramineum Kunth [GNAPSTR]
SY: *Gnaphalium chilense* Spreng.
SY: *Gnaphalium chilense* var. *confertifolium* Greene
SY: *Pseudognaphalium stramineum* (Kunth) W.A. Weber
Gnaphalium sylvaticum L. = Omalotheca sylvatica [OMALSYL]
Gnaphalium thermale E. Nels. = Gnaphalium microcephalum [GNAPMIC]
*•**Gnaphalium uliginosum** L. [GNAPULI] (marsh cudweed)
SY: *Filaginella uliginosa* (L.) Opiz
Gnaphalium viscosum Kunth [GNAPVIS] (sticky cudweed)
SY: *Gnaphalium decurrens* Ives
SY: *Gnaphalium macounii* Greene
SY: *Pseudognaphalium viscosum* (Kunth) W.A. Weber
Grindelia Willd. [GRINDEL$]
Grindelia arenicola Steyermark = Grindelia integrifolia var. macrophylla [GRININT1]
Grindelia integrifolia DC. [GRININT] (entire-leaved gumweed)
Grindelia integrifolia var. **macrophylla** (Greene) Cronq. [GRININT1]
SY: *Grindelia arenicola* Steyermark
SY: *Grindelia macrophylla* Greene
SY: *Grindelia stricta* DC.
SY: *Grindelia stricta* ssp. *blakei* (Steyermark) Keck
SY: *Grindelia stricta* ssp. *venulosa* (Jepson) Keck
Grindelia macrophylla Greene = Grindelia integrifolia var. macrophylla [GRININT1]
Grindelia perennis A. Nels. = Grindelia squarrosa var. quasiperennis [GRINSQU1]
Grindelia serrulata Rydb. = Grindelia squarrosa var. serrulata [GRINSQU2]
Grindelia squarrosa (Pursh) Dunal [GRINSQU] (curly-cup gumweed)
Grindelia squarrosa var. **quasiperennis** Lunell [GRINSQU1]
SY: *Grindelia perennis* A. Nels.
*•**Grindelia squarrosa** var. **serrulata** (Rydb.) Steyermark [GRINSQU2]
SY: *Grindelia serrulata* Rydb.
*•**Grindelia squarrosa** var. **squarrosa** [GRINSQU0]
Grindelia stricta DC. = Grindelia integrifolia var. macrophylla [GRININT1]
Grindelia stricta ssp. *blakei* (Steyermark) Keck = Grindelia integrifolia var. macrophylla [GRININT1]
Grindelia stricta ssp. *venulosa* (Jepson) Keck = Grindelia integrifolia var. macrophylla [GRININT1]
Haplopappus bloomeri Gray = Ericameria bloomeri [ERICBLO]
Haplopappus bloomeri ssp. *sonnei* (Gray) Hall = Ericameria bloomeri [ERICBLO]
Haplopappus bloomeri var. *angustatus* Gray = Ericameria bloomeri [ERICBLO]
Haplopappus bloomeri var. *sonnei* (Gray) Greene = Ericameria bloomeri [ERICBLO]
Haplopappus carthamoides (Hook.) Gray = Pyrrocoma carthamoides [PYRRCAR]
Haplopappus carthamoides ssp. *rigidus* (Rydb.) Hall = Pyrrocoma carthamoides [PYRRCAR]
Haplopappus carthamoides var. *erythropappus* (Rydb.) St. John = Pyrrocoma carthamoides [PYRRCAR]
Haplopappus carthamoides var. *rigidus* (Rydb.) M.E. Peck = Pyrrocoma carthamoides [PYRRCAR]
Haplopappus carthamoides var. *typicus* (Hall) Cronq. = Pyrrocoma carthamoides [PYRRCAR]
Haplopappus lyallii Gray = Tonestus lyallii [TONELYA]
Helenium L. [HELENIU$]
Helenium autumnale L. [HELEAUT] (mountain sneezeweed)
Helenium autumnale var. **grandiflorum** (Nutt.) Torr. & Gray [HELEAUT1]
SY: *Helenium macranthum* Rydb.
Helenium autumnale var. **montanum** (Nutt.) Fern. [HELEAUT2]
SY: *Helenium montanum* Nutt.
Helenium macranthum Rydb. = Helenium autumnale var. grandiflorum [HELEAUT1]
Helenium montanum Nutt. = Helenium autumnale var. montanum [HELEAUT2]
Helianthella Torr. & Gray [HELIANT$]
Helianthella uniflora (Nutt.) Torr. & Gray [HELIUNI] (Rocky Mountain helianthella)
Helianthella uniflora var. **douglasii** (Torr. & Gray) W.A. Weber [HELIUNI1] (Douglas' helianthella)
Helianthus L. [HELIANH$]
*•**Helianthus annuus** L. [HELIANN] (common sunflower)
SY: *Helianthus annuus* ssp. *jaegeri* (Heiser) Heiser
SY: *Helianthus annuus* ssp. *lenticularis* (Dougl. ex Lindl.) Cockerell
SY: *Helianthus annuus* var. *lenticularis* (Dougl. ex Lindl.) Steyermark
SY: *Helianthus annuus* var. *macrocarpus* (DC.) Cockerell
SY: *Helianthus annuus* ssp. *texanus* Heiser
SY: *Helianthus annuus* var. *texanus* (Heiser) Shinners
SY: *Helianthus aridus* Rydb.
SY: *Helianthus lenticularis* Dougl. ex Lindl.
Helianthus annuus ssp. *jaegeri* (Heiser) Heiser = Helianthus annuus [HELIANN]
Helianthus annuus ssp. *lenticularis* (Dougl. ex Lindl.) Cockerell = Helianthus annuus [HELIANN]
Helianthus annuus ssp. *texanus* Heiser = Helianthus annuus [HELIANN]
Helianthus annuus var. *lenticularis* (Dougl. ex Lindl.) Steyermark = Helianthus annuus [HELIANN]
Helianthus annuus var. *macrocarpus* (DC.) Cockerell = Helianthus annuus [HELIANN]
Helianthus annuus var. *texanus* (Heiser) Shinners = Helianthus annuus [HELIANN]
Helianthus aridus Rydb. = Helianthus annuus [HELIANN]
Helianthus fascicularis Greene = Helianthus nuttallii [HELINUT]
Helianthus laetiflorus var. *subrhomboideus* (Rydb.) Fern. = Helianthus pauciflorus ssp. subrhomboideus [HELIPAU1]
Helianthus lenticularis Dougl. ex Lindl. = Helianthus annuus [HELIANN]
Helianthus nuttallii Torr. & Gray [HELINUT] (Nuttall's sunflower)
SY: *Helianthus fascicularis* Greene
SY: *Helianthus nuttallii* ssp. *canadensis* R.W. Long

SY: *Helianthus nuttallii* ssp. *coloradensis* (Cockerell) R.W. Long
Helianthus nuttallii ssp. *canadensis* R.W. Long = Helianthus nuttallii [HELINUT]
Helianthus nuttallii ssp. *coloradensis* (Cockerell) R.W. Long = Helianthus nuttallii [HELINUT]
Helianthus pauciflorus Nutt. [HELIPAU]
Helianthus pauciflorus ssp. **subrhomboideus** (Rydb.) O. Spring & E. Schilling [HELIPAU1]
SY: *Helianthus laetiflorus* var. *subrhomboideus* (Rydb.) Fern.
SY: *Helianthus pauciflorus* var. *subrhomboideus* (Rydb.) Cronq.
SY: *Helianthus rigidus* ssp. *laetiflorus* (Rydb.) Heiser
SY: *Helianthus rigidus* ssp. *subrhomboideus* (Rydb.) Heiser
SY: *Helianthus rigidus* var. *subrhomboideus* (Rydb.) Cronq.
SY: *Helianthus subrhomboideus* Rydb.
Helianthus pauciflorus var. *subrhomboideus* (Rydb.) Cronq. = Helianthus pauciflorus ssp. subrhomboideus [HELIPAU1]
Helianthus rigidus ssp. *laetiflorus* (Rydb.) Heiser = Helianthus pauciflorus ssp. subrhomboideus [HELIPAU1]
Helianthus rigidus ssp. *subrhomboideus* (Rydb.) Heiser = Helianthus pauciflorus ssp. subrhomboideus [HELIPAU1]
Helianthus rigidus var. *subrhomboideus* (Rydb.) Cronq. = Helianthus pauciflorus ssp. subrhomboideus [HELIPAU1]
Helianthus subrhomboideus Rydb. = Helianthus pauciflorus ssp. subrhomboideus [HELIPAU1]
Heterotheca Cass. [HETEROT$]
Heterotheca horrida (Rydb.) Harms = Heterotheca villosa var. hispida [HETEVIL1]
Heterotheca horrida ssp. *cinerascens* (Blake) Semple = Heterotheca villosa var. hispida [HETEVIL1]
Heterotheca villosa (Pursh) Shinners [HETEVIL] (golden-aster)
Heterotheca villosa var. *angustifolia* (Rydb.) Harms = Heterotheca villosa var. villosa [HETEVIL0]
Heterotheca villosa var. *foliosa* (Nutt.) Harms = Heterotheca villosa var. villosa [HETEVIL0]
Heterotheca villosa var. **hispida** (Hook.) Harms [HETEVIL1]
SY: *Chrysopsis arida* A. Nels.
SY: *Chrysopsis butleri* Rydb.
SY: *Chrysopsis canescens* var. *nana* Gray
SY: *Chrysopsis columbiana* Greene
SY: *Chrysopsis hirsutissima* Greene
SY: *Chrysopsis hispida* (Hook.) DC.
SY: *Chrysopsis horrida* Rydb.
SY: *Chrysopsis villosa* var. *hispida* (Hook.) Gray
SY: *Chrysopsis viscida* ssp. *cinerascens* Blake
SY: *Chrysopsis wisconsinensis* Shinners
SY: *Heterotheca horrida* (Rydb.) Harms
SY: *Heterotheca horrida* ssp. *cinerascens* (Blake) Semple
SY: *Heterotheca wisconsinensis* (Shinners) Shinners
Heterotheca villosa var. *pedunculata* (Greene) Harms ex Semple = Heterotheca villosa var. villosa [HETEVIL0]
Heterotheca villosa var. **villosa** [HETEVIL0]
SY: *Chrysopsis angustifolia* Rydb.
SY: *Chrysopsis bakeri* Greene
SY: *Chrysopsis ballardii* Rydb.
SY: *Chrysopsis caudata* Rydb.
SY: *Chrysopsis foliosa* (Nutt.) Shinners
SY: *Chrysopsis imbricata* A. Nels.
SY: *Chrysopsis mollis* Nutt.
SY: *Chrysopsis nitidula* Woot. & Standl.
SY: *Chrysopsis pedunculata* Greene
SY: *Chrysopsis villosa* (Pursh) Nutt. ex DC.
SY: *Chrysopsis villosa* var. *angustifolia* (Rydb.) Cronq.
SY: *Chrysopsis villosa* var. *foliosa* (Nutt.) D.C. Eat.
SY: *Heterotheca villosa* var. *angustifolia* (Rydb.) Harms
SY: *Heterotheca villosa* var. *foliosa* (Nutt.) Harms
SY: *Heterotheca villosa* var. *pedunculata* (Greene) Harms ex Semple
Heterotheca wisconsinensis (Shinners) Shinners = Heterotheca villosa var. hispida [HETEVIL1]
Hieracium L. [HIERACI$]
Hieracium albertinum Farr = Hieracium cynoglossoides [HIERCYN]
Hieracium albiflorum Hook. [HIERALI] (white-flowered hawkweed)
SY: *Chlorocrepis albiflora* (Hook.) W.A. Weber
SY: *Hieracium helleri* Gandog.
•**Hieracium aurantiacum** L. [HIERAUR] (orange hawkweed)
Hieracium chapacanum Zahn = Hieracium scouleri [HIERSCO]
Hieracium cusickii Gandog. = Hieracium cynoglossoides [HIERCYN]
Hieracium cynoglossoides Arv.-Touv. [HIERCYN] (hound's-tongue hawkweed)
SY: *Hieracium albertinum* Farr
SY: *Hieracium cusickii* Gandog.
SY: *Hieracium scouleri* var. *albertinum* (Farr) G.W. Douglas & G.A. Allen
SY: *Hieracium scouleri* var. *griseum* (Rydb.) A. Nels.
Hieracium florentinum All. = Hieracium piloselloides [HIERPIO]
Hieracium gracile Hook. [HIERGRA] (slender hawkweed)
SY: *Hieracium triste* var. *gracile* (Hook.) Gray
Hieracium helleri Gandog. = Hieracium albiflorum [HIERALI]
Hieracium lachenalii K.C. Gmel. [HIERLAC]
SY: *Hieracium vulgatum* Fries
•**Hieracium murorum** L. [HIERMUR] (wall hawkweed)
Hieracium parryi Zahn = Hieracium scouleri [HIERSCO]
•**Hieracium pilosella** L. [HIERPIL] (mouse-ear hawkweed)
•**Hieracium piloselloides** Vill. [HIERPIO] (tall hawkweed)
SY: *Hieracium florentinum* All.
Hieracium scabriusculum Schwein. = Hieracium umbellatum [HIERUMB]
Hieracium scabriusculum var. *perhirsutum* Lepage = Hieracium umbellatum [HIERUMB]
Hieracium scabriusculum var. *saximontanum* Lepage = Hieracium umbellatum [HIERUMB]
Hieracium scabriusculum var. *scabrum* (Schwein.) Lepage = Hieracium umbellatum [HIERUMB]
Hieracium scouleri Hook. [HIERSCO] (Scouler's hawkweed)
SY: *Hieracium chapacanum* Zahn
SY: *Hieracium parryi* Zahn
Hieracium scouleri var. *albertinum* (Farr) G.W. Douglas & G.A. Allen = Hieracium cynoglossoides [HIERCYN]
Hieracium scouleri var. *griseum* (Rydb.) A. Nels. = Hieracium cynoglossoides [HIERCYN]
Hieracium triste Willd. ex Spreng. [HIERTRI] (woolly hawkweed)
Hieracium triste var. *gracile* (Hook.) Gray = Hieracium gracile [HIERGRA]
•**Hieracium umbellatum** L. [HIERUMB] (narrow-leaved hawkweed)
SY: *Hieracium scabriusculum* Schwein.
SY: *Hieracium scabriusculum* var. *perhirsutum* Lepage
SY: *Hieracium scabriusculum* var. *saximontanum* Lepage
SY: *Hieracium scabriusculum* var. *scabrum* (Schwein.) Lepage
Hieracium vulgatum Fries = Hieracium lachenalii [HIERLAC]
Hyoseris virginica L. = Krigia virginica [KRIGVIR]
Hypochoeris L. = Hypochaeris [HYPOCHA$]
Hypochaeris L. [HYPOCHA$]
•**Hypochaeris glabra** L. [HYPOGLA] (smooth cat's-ear)
•**Hypochaeris radicata** L. [HYPORAD] (hairy cat's-ear)
Inula L. [INULA$]
•**Inula helenium** L. [INULHEL] (elecampane)
Ionactis Greene [IONACTI$]
Ionactis stenomeres (Gray) Greene [IONASTE]
SY: *Aster stenomeres* Gray
Iva L. [IVA$]
Iva axillaris Pursh [IVA AXI] (deer-root)
SY: *Iva axillaris* ssp. *robustior* (Hook.) Bassett
SY: *Iva axillaris* var. *robustior* Hook.
Iva axillaris ssp. *robustior* (Hook.) Bassett = Iva axillaris [IVA AXI]
Iva axillaris var. *robustior* Hook. = Iva axillaris [IVA AXI]
•**Iva xanthifolia** Nutt. [IVA XAN] (marsh-elder)
SY: *Cyclachaena xanthifolia* (Nutt.) Fresen.
Jaumea Pers. [JAUMEA$]
Jaumea carnosa (Less.) Gray [JAUMCAR] (fleshy jaumea)
Krigia Schreb. [KRIGIA$]
•**Krigia virginica** (L.) Willd. [KRIGVIR] (virginia dwarf dandelion)
SY: *Hyoseris virginica* L.
Lactuca L. [LACTUCA$]
Lactuca biennis (Moench) Fern. [LACTBIE] (tall blue lettuce)
SY: *Lactuca spicata* var. *integrifolia* (Torr. & Gray) Britt.
SY: *Mulgedium spicatum* var. *integrifolium* (Torr. & Gray) Small

*•**Lactuca canadensis** L. [LACTCAN] (Florida blue lettuce)
SY: *Lactuca canadensis* var. *integrifolia* (Bigelow) Torr. & Gray
SY: *Lactuca canadensis* var. *typica* Wieg.
SY: *Lactuca sagittifolia* Ell.
Lactuca canadensis var. *integrifolia* (Bigelow) Torr. & Gray = Lactuca canadensis [LACTCAN]
Lactuca canadensis var. *typica* Wieg. = Lactuca canadensis [LACTCAN]
Lactuca muralis (L.) Fresen. = Mycelis muralis [MYCEMUR]
Lactuca oblongifolia Nutt. = Lactuca tatarica var. pulchella [LACTTAT1]
Lactuca pulchella (Pursh) DC. = Lactuca tatarica var. pulchella [LACTTAT1]
Lactuca sagittifolia Ell. = Lactuca canadensis [LACTCAN]
Lactuca scariola L. = Lactuca serriola [LACTSER]
•**Lactuca serriola** L. [LACTSER] (prickly lettuce)
SY: *Lactuca scariola* L.
Lactuca spicata var. *integrifolia* (Torr. & Gray) Britt. = Lactuca biennis [LACTBIE]
Lactuca tatarica (L.) C.A. Mey. [LACTTAT] (common blue lettuce)
Lactuca tatarica ssp. *pulchella* (Pursh) Stebbins = Lactuca tatarica var. pulchella [LACTTAT1]
Lactuca tatarica var. *heterophylla* (Nutt.) Boivin = Lactuca tatarica var. pulchella [LACTTAT1]
Lactuca tatarica var. **pulchella** (Pursh) Breitung [LACTTAT1]
SY: *Lactuca oblongifolia* Nutt.
SY: *Lactuca pulchella* (Pursh) DC.
SY: *Lactuca tatarica* var. *heterophylla* (Nutt.) Boivin
SY: *Lactuca tatarica* ssp. *pulchella* (Pursh) Stebbins
SY: *Sonchus tataricus* L. p.p.
Lappa minor Hill = Arctium minus [ARCTMIN]
Lapsana L. [LAPSANA$]
•**Lapsana communis** L. [LAPSCOM] (nipplewort)
Lasallea falcata (Lindl.) Semple & Brouillet = Aster falcatus [ASTEFAL]
Lasthenia Cass. [LASTHEN$]
Lasthenia maritima (Gray) Ornduff [LASTMAR] (hairy goldfields)
SY: *Baeria maritima* Gray
SY: *Baeria minor* ssp. *maritima* (Gray) Ferris
SY: *Lasthenia minor* ssp. *maritima* (Gray) Ornduff
SY: *Lasthenia minor* var. *maritima* (Gray) Cronq.
Lasthenia minor ssp. *maritima* (Gray) Ornduff = Lasthenia maritima [LASTMAR]
Lasthenia minor var. *maritima* (Gray) Cronq. = Lasthenia maritima [LASTMAR]
Leontodon L. [LEONTOD$]
•**Leontodon autumnalis** L. [LEONAUT] (autumn hawkbit)
Leontodon erythrospermum (Andrz. ex Bess.) Britt. = Taraxacum laevigatum [TARALAE]
Leontodon leysseri (Wallr.) G. Beck = Leontodon taraxacoides [LEONTAR]
Leontodon nudicaulis ssp. *taraxacoides* (Vill.) Schinz & Thellung = Leontodon taraxacoides [LEONTAR]
Leontodon saxatilis ssp. *taraxacoides* (Vill.) Holub & Moravec = Leontodon taraxacoides [LEONTAR]
•**Leontodon taraxacoides** (Vill.) Mérat [LEONTAR] (hairy hawkbit)
SY: *Leontodon leysseri* (Wallr.) G. Beck
SY: *Leontodon nudicaulis* ssp. *taraxacoides* (Vill.) Schinz & Thellung
SY: *Leontodon saxatilis* ssp. *taraxacoides* (Vill.) Holub & Moravec
Lepachys columnifera (Nutt.) Rydb. = Ratibida columnifera [RATICOL]
Lepidanthus suaveolens (Pursh) Nutt. = Matricaria discoidea [MATRDIS]
Lepidotheca suaveolens (Pursh) Nutt. = Matricaria discoidea [MATRDIS]
Leucacantha cyanus (L.) Nieuwl. & Lunell = Centaurea cyanus [CENTCYA]
Leucanthemum P. Mill. [LEUCANT$]
Leucanthemum arcticum auct. = Dendranthema arcticum ssp. polare [DENDARC1]
Leucanthemum integrifolium (Richards.) DC. [LEUCINT] (entire-leaved daisy)
SY: *Chrysanthemum integrifolium* Richards.
Leucanthemum leucanthemum (L.) Rydb. = Leucanthemum vulgare [LEUCVUL]
•**Leucanthemum vulgare** Lam. [LEUCVUL] (oxeye daisy)
SY: *Chrysanthemum leucanthemum* L.
SY: *Chrysanthemum leucanthemum* var. *boecheri* Boivin
SY: *Chrysanthemum leucanthemum* var. *pinnatifidum* Lecoq & Lamotte
SY: *Leucanthemum leucanthemum* (L.) Rydb.
SY: *Leucanthemum vulgare* var. *pinnatifidum* (Lecoq & Lamotte) Moldenke
Leucanthemum vulgare var. *pinnatifidum* (Lecoq & Lamotte) Moldenke = Leucanthemum vulgare [LEUCVUL]
Logfia arvensis (L.) Holub = Filago arvensis [FILAARV]
Luina Benth. [LUINA$]
Luina hypoleuca Benth. [LUINHYP] (silverback luina)
Luina nardosmia (Gray) Cronq. = Cacaliopsis nardosmia [CACANAR]
Luina nardosmia var. *glabrata* (Piper) Cronq. = Cacaliopsis nardosmia [CACANAR]
Lygodesmia D. Don [LYGODES$]
Lygodesmia juncea (Pursh) D. Don ex Hook. [LYGOJUN] (rush-like skeleton-plant)
Machaeranthera Nees [MACHAER$]
Machaeranthera canescens (Pursh) Gray [MACHCAN] (hoary aster)
SY: *Aster canescens* Pursh
Macrorhynchus heterophyllus Nutt. = Agoseris heterophylla [AGOSHET]
Madia Molina [MADIA$]
Madia capitata Nutt. = Madia sativa [MADISAT]
Madia dissitiflora (Nutt.) Torr. & Gray = Madia gracilis [MADIGRA]
Madia exigua (Sm.) Gray [MADIEXI] (little tarweed)
Madia glomerata Hook. [MADIGLO] (clustered tarweed)
Madia gracilis (Sm.) Keck & J. Clausen ex Applegate [MADIGRA] (slender tarweed)
SY: *Madia dissitiflora* (Nutt.) Torr. & Gray
SY: *Madia gracilis* ssp. *collina* Keck
SY: *Madia gracilis* ssp. *pilosa* Keck
Madia gracilis ssp. *collina* Keck = Madia gracilis [MADIGRA]
Madia gracilis ssp. *pilosa* Keck = Madia gracilis [MADIGRA]
Madia madioides (Nutt.) Greene [MADIMAD] (woodland tarweed)
Madia minima (Gray) Keck [MADIMIN] (small-headed tarweed)
Madia sativa Molina [MADISAT] (Chilean tarweed)
SY: *Madia capitata* Nutt.
SY: *Madia sativa* ssp. *capitata* (Nutt.) Piper
SY: *Madia sativa* var. *congesta* Torr. & Gray
Madia sativa ssp. *capitata* (Nutt.) Piper = Madia sativa [MADISAT]
Madia sativa var. *congesta* Torr. & Gray = Madia sativa [MADISAT]
Mariana mariana (L.) Hill = Silybum marianum [SILYMAR]
Maruta cotula (L.) DC. = Anthemis cotula [ANTHCOT]
Matricaria L. [MATRICA$]
Matricaria chamomilla L. 1755 & 1763, non 1753 = Matricaria recutita [MATRREC]
Matricaria chamomilla var. *coronata* (J. Gay) Coss. & Germ. = Matricaria recutita [MATRREC]
Matricaria discoidea DC. [MATRDIS] (pineapple-weed)
SY: *Chamomilla suaveolens* (Pursh) Rydb.
SY: *Lepidanthus suaveolens* (Pursh) Nutt.
SY: *Lepidotheca suaveolens* (Pursh) Nutt.
SY: *Matricaria suaveolens* (Pursh) Buch.
SY: *Santolina suaveolens* Pursh
SY: *Tanacetum suaveolens* (Pursh) Hook.
SY: *Matricaria matricarioides* auct.
Matricaria inodora L. = Matricaria perforata [MATRPER]
Matricaria inodora var. *agrestis* (Knaf.) Willmot = Matricaria perforata [MATRPER]
Matricaria maritima ssp. *inodora* (L.) Clapham = Matricaria perforata [MATRPER]
Matricaria maritima var. *agrestis* (Knaf) Wilmott = Matricaria perforata [MATRPER]
Matricaria matricarioides auct. = Matricaria discoidea [MATRDIS]

Matricaria parthenium L. = Tanacetum parthenium [TANAPAR]
*__Matricaria perforata__ Mérat [MATRPER] (scentless mayweed)
SY: *Chamomilla inodora* (L.) Gilib.
SY: *Matricaria inodora* L.
SY: *Matricaria inodora* var. *agrestis* (Knaf.) Willmot
SY: *Matricaria maritima* var. *agrestis* (Knaf) Wilmott
SY: *Matricaria maritima* ssp. *inodora* (L.) Clapham
SY: *Tripleurospermum inodorum* (L.) Schultz-Bip.
*__Matricaria recutita__ L. [MATRREC] (wild chamomile)
SY: *Chamomilla chamomilla* (L.) Rydb.
SY: *Chamomilla recutita* (L.) Rauschert
SY: *Matricaria chamomilla* L. 1755 & 1763, non 1753
SY: *Matricaria chamomilla* var. *coronata* (J. Gay) Coss. & Germ.
SY: *Matricaria suaveolens* L.
Matricaria suaveolens (Pursh) Buch. = Matricaria discoidea [MATRDIS]
Matricaria suaveolens L. = Matricaria recutita [MATRREC]
Megalodonta Greene [MEGALOD$]
Megalodonta beckii (Torr. ex Spreng.) Greene [MEGABEC]
SY: *Bidens beckii* Torr. ex Spreng.
Microseris D. Don [MICROSE$]
Microseris bigelovii (Gray) Schultz-Bip. [MICRBIG] (coastal microseris)
Microseris borealis (Bong.) Schultz-Bip. [MICRBOR] (apargidium)
SY: *Apargidium boreale* (Bong.) Torr. & Gray
Microseris nutans (Hook.) Schultz-Bip. [MICRNUT] (nodding microseris)
SY: *Ptilocalais nutans* (Hook.) Greene
SY: *Scorzonella nutans* Hook.
SY: *Scorzonella nutans* var. *major* (Gray) M.E. Peck
Microseris troximoides Gray = Nothocalais troximoides [NOTHTRO]
Mulgedium spicatum var. *integrifolium* (Torr. & Gray) Small = Lactuca biennis [LACTBIE]
Mycelis Cass. [MYCELIS$]
*__Mycelis muralis__ (L.) Dumort. [MYCEMUR] (wall-lettuce)
SY: *Lactuca muralis* (L.) Fresen.
Nardosmia arctica (Porsild) A. & D. Löve = Petasites frigidus var. palmatus [PETAFRI3]
Nardosmia japonica Sieb. & Zucc. = Petasites japonicus [PETAJAP]
Nothocalais (Gray) Greene [NOTHOCA$]
Nothocalais troximoides (Gray) Greene [NOTHTRO] (false-agoseris)
SY: *Microseris troximoides* Gray
SY: *Scorzonella troximoides* (Gray) Jepson
Oglifa arvensis (L.) Cass. = Filago arvensis [FILAARV]
Oligosporus campestris ssp. *pacificus* (Nutt.) W.A. Weber = Artemisia campestris ssp. pacifica [ARTECAM2]
Omalotheca Cass. [OMALOTH$]
*__Omalotheca sylvatica__ (L.) Schultz-Bip. & F.W. Schultz [OMALSYL]
SY: *Gnaphalium sylvaticum* L.
Onopordum L. [ONOPORD$]
*__Onopordum acanthium__ L. [ONOPACA] (Scotch thistle)
Packera cana (Hook.) W.A. Weber & A. Löve = Senecio canus [SENECAN]
Packera cymbalarioides (Buek) W.A. Weber & A. Löve = Senecio cymbalarioides [SENECYM]
Packera hyperborealis (Greenm.) A. & D. Löve = Senecio hyperborealis [SENEHYP]
Packera indecora (Greene) A. & D. Löve = Senecio indecorus [SENEIND]
Packera ogotorukensis (Packer) A. & D. Löve = Senecio ogotorukensis [SENEOGO]
Packera pauciflora (Pursh) A. & D. Löve = Senecio pauciflorus [SENEPAU]
Packera paupercula (Michx.) A. & D. Löve = Senecio pauperculus [SENEPAP]
Packera pseudaurea (Rydb.) W.A. Weber & A. Löve = Senecio pseudaureus [SENEPSE]
Petasites P. Mill. [PETASIT$]
Petasites alaskanus Rydb. = Petasites frigidus var. frigidus [PETAFRI0]
Petasites arcticus Porsild = Petasites frigidus var. palmatus [PETAFRI3]
Petasites corymbosus (R. Br.) Rydb. = Petasites frigidus var. frigidus [PETAFRI0]
Petasites dentatus Blank. = Petasites sagittatus [PETASAG]
Petasites frigidus (L.) Fries [PETAFRI] (sweet coltsfoot)
Petasites frigidus var. *corymbosus* (R. Br.) Cronq. = Petasites frigidus var. frigidus [PETAFRI0]
Petasites frigidus var. **frigidus** [PETAFRI0]
SY: *Petasites alaskanus* Rydb.
SY: *Petasites corymbosus* (R. Br.) Rydb.
SY: *Petasites frigidus* var. *corymbosus* (R. Br.) Cronq.
SY: *Petasites frigidus* var. *hyperboreoides* Hultén
SY: *Petasites gracilis* Britt.
Petasites frigidus var. *hyperboreoides* Hultén = Petasites frigidus var. frigidus [PETAFRI0]
Petasites frigidus var. **nivalis** (Greene) Cronq. [PETAFRI2]
SY: *Petasites hyperboreus* Rydb.
SY: *Petasites nivalis* Greene
SY: *Petasites palmatus* var. *frigidus* Macoun
SY: *Petasites vitifolius* Greene
Petasites frigidus var. **palmatus** (Ait.) Cronq. [PETAFRI3]
SY: *Nardosmia arctica* (Porsild) A. & D. Löve
SY: *Petasites arcticus* Porsild
SY: *Petasites hookerianus* (Nutt.) Rydb.
SY: *Petasites palmatus* (Ait.) Gray
SY: *Petasites speciosus* (Nutt.) Piper
Petasites gracilis Britt. = Petasites frigidus var. frigidus [PETAFRI0]
Petasites hookerianus (Nutt.) Rydb. = Petasites frigidus var. palmatus [PETAFRI3]
Petasites hyperboreus Rydb. = Petasites frigidus var. nivalis [PETAFRI2]
*__Petasites japonicus__ (Sieb. & Zucc.) Maxim. [PETAJAP] (Japanese coltsfoot)
SY: *Nardosmia japonica* Sieb. & Zucc.
Petasites nivalis Greene = Petasites frigidus var. nivalis [PETAFRI2]
Petasites palmatus (Ait.) Gray = Petasites frigidus var. palmatus [PETAFRI3]
Petasites palmatus var. *frigidus* Macoun = Petasites frigidus var. nivalis [PETAFRI2]
Petasites sagittatus (Banks ex Pursh) Gray [PETASAG] (arrow-leaved coltsfoot)
SY: *Petasites dentatus* Blank.
Petasites speciosus (Nutt.) Piper = Petasites frigidus var. palmatus [PETAFRI3]
Petasites vitifolius Greene = Petasites frigidus var. nivalis [PETAFRI2]
Prenanthes L. [PRENANT$]
Prenanthes alata (Hook.) D. Dietr. [PRENALA] (western rattlesnake-root)
SY: *Prenanthes lessingii* Hultén
Prenanthes lessingii Hultén = Prenanthes alata [PRENALA]
Prenanthes racemosa Michx. [PRENRAC] (purple rattlesnake-root)
Prenanthes racemosa ssp. **multiflora** Cronq. [PRENRAC1]
SY: *Prenanthes racemosa* var. *multiflora* (Cronq.) Dorn
Prenanthes racemosa var. *multiflora* (Cronq.) Dorn = Prenanthes racemosa ssp. multiflora [PRENRAC1]
Pseudognaphalium stramineum (Kunth) W.A. Weber = Gnaphalium stramineum [GNAPSTR]
Pseudognaphalium viscosum (Kunth) W.A. Weber = Gnaphalium viscosum [GNAPVIS]
Psilocarphus Nutt. [PSILOCA$]
Psilocarphus elatior (Gray) Gray [PSILELA] (tall woolly-heads)
Psilocarphus tenellus Nutt. [PSILTEN] (slender woolly-heads)
Psilochenia atribarba (Heller) W.A. Weber = Crepis atrabarba ssp. atrabarba [CREPATR0]
Psilochenia intermedia (Gray) W.A. Weber = Crepis intermedia [CREPINT]
Psilochenia modocensis (Greene) W.A. Weber = Crepis modocensis ssp. modocensis [CREPMOD0]
Psilochenia modocensis ssp. *rostrata* (Coville) W.A. Weber = Crepis modocensis ssp. rostrata [CREPMOD2]
Psilochenia occidentalis (Nutt.) Nutt. = Crepis occidentalis ssp. occidentalis [CREPOCC0]

Psilochenia occidentalis ssp. *costata* (Gray) W.A. Weber = Crepis occidentalis ssp. costata [CREPOCC1]
Psilochenia runcinata (James) A. & D. Löve = Crepis runcinata [CREPRUN]
Ptilocalais nutans (Hook.) Greene = Microseris nutans [MICRNUT]
Pyrrocoma Hook. [PYRROCO$]
Pyrrocoma carthamoides Hook. [PYRRCAR]
SY: *Haplopappus carthamoides* (Hook.) Gray
SY: *Haplopappus carthamoides* var. *erythropappus* (Rydb.) St. John
SY: *Haplopappus carthamoides* ssp. *rigidus* (Rydb.) Hall
SY: *Haplopappus carthamoides* var. *rigidus* (Rydb.) M.E. Peck
SY: *Haplopappus carthamoides* var. *typicus* (Hall) Cronq.
Ratibida Raf. [RATIBID$]
Ratibida columnaris (Sims) D. Don = Ratibida columnifera [RATICOL]
Ratibida columnaris var. *pulcherrima* (DC.) D. Don = Ratibida columnifera [RATICOL]
Ratibida columnifera (Nutt.) Woot. & Standl. [RATICOL] (prairie coneflower)
SY: *Lepachys columnifera* (Nutt.) Rydb.
SY: *Ratibida columnaris* (Sims) D. Don
SY: *Ratibida columnaris* var. *pulcherrima* (DC.) D. Don
SY: *Rudbeckia columnaris* Sims
SY: *Rudbeckia columnaris* Pursh
SY: *Rudbeckia columnifera* Nutt.
Rudbeckia L. [RUDBECK$]
Rudbeckia columnaris Sims = Ratibida columnifera [RATICOL]
Rudbeckia columnaris Pursh = Ratibida columnifera [RATICOL]
Rudbeckia columnifera Nutt. = Ratibida columnifera [RATICOL]
***Rudbeckia hirta** L. [RUDBHIR] (black-eyed Susan)
Santolina suaveolens Pursh = Matricaria discoidea [MATRDIS]
Saussurea DC. [SAUSSUR$]
Saussurea americana D.C. Eat. [SAUSAME] (American sawwort)
Saussurea angustifolia (Willd.) DC. [SAUSANG] (northern sawwort)
Saussurea densa (Hook.) Rydb. [SAUSDEN]
SY: *Saussurea nuda* ssp. *densa* (Hook.) G.W. Douglas
SY: *Saussurea nuda* var. *densa* (Hook.) Hultén
SY: *Saussurea weberi* Hultén
Saussurea nuda ssp. *densa* (Hook.) G.W. Douglas = Saussurea densa [SAUSDEN]
Saussurea nuda var. *densa* (Hook.) Hultén = Saussurea densa [SAUSDEN]
Saussurea weberi Hultén = Saussurea densa [SAUSDEN]
Scorzonella nutans Hook. = Microseris nutans [MICRNUT]
Scorzonella nutans var. *major* (Gray) M.E. Peck = Microseris nutans [MICRNUT]
Scorzonella troximoides (Gray) Jepson = Nothocalais troximoides [NOTHTRO]
Senecio L. [SENECIO$]
Senecio alaskanus Hultén = Senecio yukonensis [SENEYUK]
Senecio atropurpureus (Ledeb.) Fedtsch. [SENEATR] (purple-haired groundsel)
SY: *Senecio atropurpureus* var. *dentatus* (Gray) Hultén
SY: *Senecio atropurpureus* ssp. *frigidus* (Richards.) Hultén
SY: *Senecio atropurpureus* ssp. *tomentosus* (Kjellm.) Hultén
SY: *Senecio atropurpureus* var. *ulmeri* (Steffen) Porsild
SY: *Senecio frigidus* (Richards.) Less.
SY: *Senecio kjellmanii* Porsild
SY: *Tephroseris atropurpurea* ssp. *frigida* (Richards.) A. & D. Löve
SY: *Tephroseris atropurpurea* ssp. *tomentosa* (Kjellm.) A. & D. Löve
SY: *Tephroseris integrifolia* ssp. *atropurpurea* (Ledeb.) B. Nordenstam
Senecio atropurpureus ssp. *frigidus* (Richards.) Hultén = Senecio atropurpureus [SENEATR]
Senecio atropurpureus ssp. *tomentosus* (Kjellm.) Hultén = Senecio atropurpureus [SENEATR]
Senecio atropurpureus var. *dentatus* (Gray) Hultén = Senecio atropurpureus [SENEATR]
Senecio atropurpureus var. *ulmeri* (Steffen) Porsild = Senecio atropurpureus [SENEATR]
Senecio balsamitae Muhl. ex Willd. = Senecio pauperculus [SENEPAP]
Senecio canus Hook. [SENECAN] (woolly groundsel)
SY: *Packera cana* (Hook.) W.A. Weber & A. Löve
SY: *Senecio convallium* Greenm.
SY: *Senecio hallii* Britt.
SY: *Senecio hallii* var. *discoidea* W.A. Weber
SY: *Senecio harbourii* Rydb.
SY: *Senecio howellii* Greene
SY: *Senecio purshianus* Nutt.
Senecio columbianus Greene = Senecio integerrimus var. exaltatus [SENEINT1]
Senecio congestus (R. Br.) DC. [SENECON] (marsh fleabane)
SY: *Senecio congestus* var. *palustris* (L.) Fern.
SY: *Senecio congestus* ssp. *tonsus* (Fern.) A. & D. Löve
SY: *Senecio congestus* var. *tonsus* Fern.
SY: *Senecio palustris* (L.) Hook.
Senecio congestus ssp. *tonsus* (Fern.) A. & D. Löve = Senecio congestus [SENECON]
Senecio congestus var. *palustris* (L.) Fern. = Senecio congestus [SENECON]
Senecio congestus var. *tonsus* Fern. = Senecio congestus [SENECON]
Senecio conterminus Greenm. [SENECOT] (high alpine butterweed)
Senecio convallium Greenm. = Senecio canus [SENECAN]
Senecio crawfordii (Britt.) G.W. & G.R. Dougl. = Senecio pauperculus [SENEPAP]
Senecio crepidineus Greene = Senecio elmeri [SENEELM]
Senecio cymbalarioides Buek [SENECYM] (alpine meadow butterweed)
SY: *Packera cymbalarioides* (Buek) W.A. Weber & A. Löve
SY: *Senecio cymbalarioides* ssp. *moresbiensis* Calder & Taylor
SY: *Senecio cymbalarioides* var. *moresbiensis* Calder & Taylor
SY: *Senecio moresbiensis* (Calder & Taylor) Dougl. & Ruyle-Dougl.
SY: *Senecio streptanthifolius* var. *moresbiensis* (Calder & Taylor) Boivin
SY: *Senecio subnudus* DC.
Senecio cymbalarioides Nutt. = Senecio streptanthifolius [SENESTR]
Senecio cymbalarioides ssp. *moresbiensis* Calder & Taylor = Senecio cymbalarioides [SENECYM]
Senecio cymbalarioides var. *moresbiensis* Calder & Taylor = Senecio cymbalarioides [SENECYM]
Senecio discoideus (Hook.) Britt. = Senecio pauciflorus [SENEPAU]
Senecio elmeri Piper [SENEELM] (Elmer's butterweed)
SY: *Senecio crepidineus* Greene
Senecio eremophilus Richards. [SENEERE] (desert ragwort)
SY: *Senecio glaucifolius* Rydb.
Senecio exaltatus Nutt. = Senecio integerrimus var. exaltatus [SENEINT1]
Senecio foetidus J.T. Howell = Senecio hydrophiloides [SENEHYR]
Senecio foetidus var. *hydrophiloides* (Rydb.) T.M. Barkl. ex Cronq. = Senecio hydrophiloides [SENEHYR]
Senecio fremontii Torr. & Gray [SENEFRE] (dwarf mountain butterweed)
Senecio frigidus (Richards.) Less. = Senecio atropurpureus [SENEATR]
Senecio fuscatus Hayek [SENEFUS]
SY: *Senecio lindstroemii* (Ostenf.) Porsild
SY: *Senecio tundricola* Tolm.
Senecio gaspensis Greenm. = Senecio pauperculus [SENEPAP]
Senecio gaspensis var. *firmifolius* (Greenm.) Fern. = Senecio pauperculus [SENEPAP]
Senecio gibbsonsii Greene = Senecio triangularis [SENETRI]
Senecio glaucifolius Rydb. = Senecio eremophilus [SENEERE]
Senecio hallii Britt. = Senecio canus [SENECAN]
Senecio hallii var. *discoidea* W.A. Weber = Senecio canus [SENECAN]
Senecio harbourii Rydb. = Senecio canus [SENECAN]
Senecio hookeri Torr. & Gray = Senecio integerrimus var. exaltatus [SENEINT1]
Senecio howellii Greene = Senecio canus [SENECAN]

Senecio hydrophiloides Rydb. [SENEHYR]
SY: *Senecio foetidus* J.T. Howell
SY: *Senecio foetidus* var. *hydrophiloides* (Rydb.) T.M. Barkl. ex Cronq.
SY: *Senecio pereziifolius* Rydb.
Senecio hydrophilus Nutt. [SENEHYD] (alkali-marsh butterweed)
Senecio hyperborealis Greenm. [SENEHYP]
SY: *Packera hyperborealis* (Greenm.) A. & D. Löve
Senecio indecorus Greene [SENEIND] (rayless ragwort)
SY: *Packera indecora* (Greene) A. & D. Löve
SY: *Senecio pauciflorus* var. *fallax* (Greenm.) Greenm.
Senecio integerrimus Nutt. [SENEINT] (western groundsel)
Senecio integerrimus var. **exaltatus** (Nutt.) Cronq. [SENEINT1]
SY: *Senecio columbianus* Greene
SY: *Senecio exaltatus* Nutt.
SY: *Senecio hookeri* Torr. & Gray
SY: *Senecio integerrimus* var. *vaseyi* (Greenm.) Cronq.
SY: *Senecio vaseyi* Greenm.
Senecio integerrimus var. *lugens* (Richards.) Boivin = Senecio lugens [SENELUG]
Senecio integerrimus var. **ochroleucus** (Gray) Cronq. [SENEINT3]
Senecio integerrimus var. *vaseyi* (Greenm.) Cronq. = Senecio integerrimus var. exaltatus [SENEINT1]
*Senecio jacobaea** L. [SENEJAC] (tansy ragwort)
Senecio kjellmanii Porsild = Senecio atropurpureus [SENEATR]
Senecio ligulifolius Greene = Senecio macounii [SENEMAC]
Senecio lindstroemii (Ostenf.) Porsild = Senecio fuscatus [SENEFUS]
Senecio lugens Richards. [SENELUG] (black-tipped groundsel)
SY: *Senecio integerrimus* var. *lugens* (Richards.) Boivin
Senecio macounii Greene [SENEMAC] (Macoun's groundsel)
SY: *Senecio ligulifolius* Greene
Senecio megacephalus Nutt. [SENEMEG] (large-headed groundsel)
Senecio moresbiensis (Calder & Taylor) Dougl. & Ruyle-Dougl. = Senecio cymbalarioides [SENECYM]
Senecio newcombei Greene = Sinosenecio newcombei [SINONEW]
Senecio ogotorukensis Packer [SENEOGO] (Ogotoruk Creek butterweed)
SY: *Packera ogotorukensis* (Packer) A. & D. Löve
Senecio palustris (L.) Hook. = Senecio congestus [SENECON]
Senecio pauciflorus Pursh [SENEPAU] (rayless alpine butterweed)
SY: *Packera pauciflora* (Pursh) A. & D. Löve
SY: *Senecio discoideus* (Hook.) Britt.
Senecio pauciflorus var. *fallax* (Greenm.) Greenm. = Senecio indecorus [SENEIND]
Senecio pauciflorus var. *jucundulus* Jepson = Senecio pseudaureus [SENEPSE]
Senecio pauperculus Michx. [SENEPAP] (Canadian butterweed)
SY: *Packera paupercula* (Michx.) A. & D. Löve
SY: *Senecio balsamitae* Muhl. ex Willd.
SY: *Senecio crawfordii* (Britt.) G.W. & G.R. Dougl.
SY: *Senecio gaspensis* Greenm.
SY: *Senecio gaspensis* var. *firmifolius* (Greenm.) Fern.
SY: *Senecio pauperculus* var. *balsamitae* (Muhl. ex Willd.) Fern.
SY: *Senecio pauperculus* var. *crawfordii* (Britt.) T.M. Barkl.
SY: *Senecio pauperculus* var. *firmifolius* (Greenm.) Greenm.
SY: *Senecio pauperculus* var. *neoscoticus* Fern.
SY: *Senecio pauperculus* var. *praelongus* (Greenm.) House
SY: *Senecio pauperculus* var. *thompsoniensis* (Greenm.) Boivin
SY: *Senecio tweedyi* Rydb.
Senecio pauperculus var. *balsamitae* (Muhl. ex Willd.) Fern. = Senecio pauperculus [SENEPAP]
Senecio pauperculus var. *crawfordii* (Britt.) T.M. Barkl. = Senecio pauperculus [SENEPAP]
Senecio pauperculus var. *firmifolius* (Greenm.) Greenm. = Senecio pauperculus [SENEPAP]
Senecio pauperculus var. *neoscoticus* Fern. = Senecio pauperculus [SENEPAP]
Senecio pauperculus var. *praelongus* (Greenm.) House = Senecio pauperculus [SENEPAP]
Senecio pauperculus var. *thompsoniensis* (Greenm.) Boivin = Senecio pauperculus [SENEPAP]
Senecio pereziifolius Rydb. = Senecio hydrophiloides [SENEHYR]
Senecio plattensis Nutt. [SENEPLA] (plains butterweed)
SY: *Senecio pseudotomentosus* Mackenzie & Bush
Senecio pseudaureus Rydb. [SENEPSE] (streambank butterweed)
SY: *Packera pseudaurea* (Rydb.) W.A. Weber & A. Löve
SY: *Senecio pauciflorus* var. *jucundulus* Jepson
Senecio pseudoarnica Less. [SENEPSU] (beach groundsel)
Senecio pseudotomentosus Mackenzie & Bush = Senecio plattensis [SENEPLA]
Senecio purshianus Nutt. = Senecio canus [SENECAN]
Senecio serra Hook. [SENESER] (tall butterweed)
Senecio sheldonensis Porsild [SENESHE] (Mount Sheldon butterweed)
Senecio streptanthifolius Greene [SENESTR] (Rocky Mountain butterweed)
SY: *Senecio cymbalarioides* Nutt.
Senecio streptanthifolius var. *moresbiensis* (Calder & Taylor) Boivin = Senecio cymbalarioides [SENECYM]
Senecio subnudus DC. = Senecio cymbalarioides [SENECYM]
*Senecio sylvaticus** L. [SENESYL] (wood groundsel)
Senecio triangularis Hook. [SENETRI] (arrow-leaved groundsel)
SY: *Senecio gibbsonsii* Greene
SY: *Senecio triangularis* var. *angustifolius* G.N. Jones
Senecio triangularis var. *angustifolius* G.N. Jones = Senecio triangularis [SENETRI]
Senecio tundricola Tolm. = Senecio fuscatus [SENEFUS]
Senecio tweedyi Rydb. = Senecio pauperculus [SENEPAP]
Senecio vaseyi Greenm. = Senecio integerrimus var. exaltatus [SENEINT1]
*Senecio viscosus** L. [SENEVIS] (sticky ragwort)
*Senecio vulgaris** L. [SENEVUL] (common groundsel)
Senecio yukonensis Porsild [SENEYUK] (Yukon groundsel)
SY: *Senecio alaskanus* Hultén
Sericocarpus rigidus Lindl. = Aster curtus [ASTECUR]
Seriphidium tridentatum (Nutt.) W.A. Weber = Artemisia tridentata ssp. tridentata [ARTETRI0]
Seriphidium tridentatum ssp. *parishii* (Gray) W.A. Weber = Artemisia tridentata ssp. tridentata [ARTETRI0]
Seriphidium tridentatum ssp. *vaseyanum* (Rydb.) W.A. Weber = Artemisia tridentata ssp. vaseyana [ARTETRI2]
Seriphidium vaseyanum (Rydb.) W.A. Weber = Artemisia tridentata ssp. vaseyana [ARTETRI2]
Serratula arvensis L. = Cirsium arvense [CIRSARV]
Silybum Adans. [SILYBUM$]
*Silybum marianum** (L.) Gaertn. [SILYMAR] (milk thistle)
SY: *Carduus marianus* L.
SY: *Mariana mariana* (L.) Hill
Sinosenecio B. Nord. [SINOSEN$]
Sinosenecio newcombei (Greene) J.P. Janovec & T.M. Barkley [SINONEW] (Newcombe's butterweed)
SY: *Senecio newcombei* Greene
Solidago L. [SOLIDAG$]
Solidago altissima var. *gilvocanescens* (Rydb.) Semple = Solidago canadensis var. gilvocanescens [SOLICAN1]
Solidago canadensis L. [SOLICAN] (Canada goldenrod)
Solidago canadensis ssp. *elongata* (Nutt.) Keck = Solidago canadensis var. salebrosa [SOLICAN2]
Solidago canadensis ssp. *salebrosa* (Piper) Keck = Solidago canadensis var. salebrosa [SOLICAN2]
Solidago canadensis var. *elongata* (Nutt.) M.E. Peck = Solidago canadensis var. salebrosa [SOLICAN2]
Solidago canadensis var. **gilvocanescens** Rydb. [SOLICAN1]
SY: *Solidago altissima* var. *gilvocanescens* (Rydb.) Semple
SY: *Solidago gilvocanescens* (Rydb.) Smyth
SY: *Solidago pruinosa* Greene
Solidago canadensis var. **salebrosa** (Piper) M.E. Jones [SOLICAN2]
SY: *Solidago canadensis* ssp. *elongata* (Nutt.) Keck
SY: *Solidago canadensis* var. *elongata* (Nutt.) M.E. Peck
SY: *Solidago canadensis* ssp. *salebrosa* (Piper) Keck
SY: *Solidago elongata* Nutt.
SY: *Solidago lepida* var. *elongata* (Nutt.) Fern.
SY: *Solidago lepida* var. *fallax* Fern.
Solidago canadensis var. **subserrata** (DC.) Cronq. [SOLICAN3]
SY: *Solidago lepida* DC.
SY: *Solidago lepida* var. *molina* Fern.

Solidago decemflora DC. = Solidago nemoralis var. longipetiolata [SOLINEM1]
Solidago elongata Nutt. = Solidago canadensis var. salebrosa [SOLICAN2]
Solidago gigantea Ait. [SOLIGIG] (late goldenrod)
SY: *Aster latissimifolius* var. *serotinus* Kuntze
SY: *Solidago gigantea* var. *leiophylla* Fern.
SY: *Solidago gigantea* var. *pitcheri* (Nutt.) Shinners
SY: *Solidago gigantea* ssp. *serotina* (Kuntze) McNeill
SY: *Solidago gigantea* var. *serotina* (Kuntze) Cronq.
SY: *Solidago gigantea* var. *shinnersii* Beaudry
SY: *Solidago × leiophallax* Friesner
SY: *Solidago pitcheri* Nutt.
SY: *Solidago serotina* Ait.
SY: *Solidago serotinoides* A. & D. Löve
Solidago gigantea ssp. *serotina* (Kuntze) McNeill = Solidago gigantea [SOLIGIG]
Solidago gigantea var. *leiophylla* Fern. = Solidago gigantea [SOLIGIG]
Solidago gigantea var. *pitcheri* (Nutt.) Shinners = Solidago gigantea [SOLIGIG]
Solidago gigantea var. *serotina* (Kuntze) Cronq. = Solidago gigantea [SOLIGIG]
Solidago gigantea var. *shinnersii* Beaudry = Solidago gigantea [SOLIGIG]
Solidago gilvocanescens (Rydb.) Smyth = Solidago canadensis var. gilvocanescens [SOLICAN1]
Solidago graminifolia (L.) Salisb. = Euthamia graminifolia [EUTHGRA]
Solidago graminifolia var. *major* (Michx.) Fern. = Euthamia graminifolia [EUTHGRA]
Solidago × leiophallax Friesner = Solidago gigantea [SOLIGIG]
Solidago lepida DC. = Solidago canadensis var. subserrata [SOLICAN3]
Solidago lepida var. *elongata* (Nutt.) Fern. = Solidago canadensis var. salebrosa [SOLICAN2]
Solidago lepida var. *fallax* Fern. = Solidago canadensis var. salebrosa [SOLICAN2]
Solidago lepida var. *molina* Fern. = Solidago canadensis var. subserrata [SOLICAN3]
Solidago longipetiolata Mackenzie & Bush = Solidago nemoralis var. longipetiolata [SOLINEM1]
Solidago missouriensis Nutt. [SOLIMIS] (Missouri goldenrod)
SY: *Solidago missouriensis* var. *montana* Gray
Solidago missouriensis var. *montana* Gray = Solidago missouriensis [SOLIMIS]
Solidago multiradiata Ait. [SOLIMUL] (northern goldenrod)
Solidago nemoralis Ait. [SOLINEM] (gray goldenrod)
Solidago nemoralis ssp. *decemflora* (DC.) Brammall = Solidago nemoralis var. longipetiolata [SOLINEM1]
Solidago nemoralis ssp. *longipetiolata* (Mackenzie & Bush) G.W. Douglas = Solidago nemoralis var. longipetiolata [SOLINEM1]
Solidago nemoralis var. *decemflora* (DC.) Fern. = Solidago nemoralis var. longipetiolata [SOLINEM1]
Solidago nemoralis var. **longipetiolata** (Mackenzie & Bush) Palmer & Steyermark [SOLINEM1]
SY: *Solidago decemflora* DC.
SY: *Solidago longipetiolata* Mackenzie & Bush
SY: *Solidago nemoralis* ssp. *decemflora* (DC.) Brammall
SY: *Solidago nemoralis* var. *decemflora* (DC.) Fern.
SY: *Solidago nemoralis* ssp. *longipetiolata* (Mackenzie & Bush) G.W. Douglas
SY: *Solidago pulcherrima* A. Nels.
Solidago neomexicana (Gray) Woot. & Standl. = Solidago spathulata var. neomexicana [SOLISPA5]
Solidago occidentalis (Nutt.) Torr. & Gray = Euthamia occidentalis [EUTHOCC]
Solidago pitcheri Nutt. = Solidago gigantea [SOLIGIG]
Solidago pruinosa Greene = Solidago canadensis var. gilvocanescens [SOLICAN1]
Solidago pulcherrima A. Nels. = Solidago nemoralis var. longipetiolata [SOLINEM1]
Solidago serotina Ait. = Solidago gigantea [SOLIGIG]
Solidago serotinoides A. & D. Löve = Solidago gigantea [SOLIGIG]
Solidago simplex Kunth [SOLISIM]
Solidago spathulata DC. [SOLISPA] (spikelike goldenrod)
Solidago spathulata var. **neomexicana** (Gray) Cronq. [SOLISPA5]
SY: *Solidago neomexicana* (Gray) Woot. & Standl.
Sonchus L. [SONCHUS$]
***Sonchus arvensis** L. [SONCARV] (perennial sow-thistle)
***Sonchus arvensis** ssp. **arvensis** [SONCARV0]
SY: *Sonchus arvensis* var. *shumovichii* Boivin
***Sonchus arvensis** ssp. **uliginosus** (Bieb.) Nyman [SONCARV1]
SY: *Sonchus arvensis* var. *glabrescens* Guenth., Grab. & Wimmer
SY: *Sonchus uliginosus* Bieb.
Sonchus arvensis var. *glabrescens* Guenth., Grab. & Wimmer = Sonchus arvensis ssp. uliginosus [SONCARV1]
Sonchus arvensis var. *shumovichii* Boivin = Sonchus arvensis ssp. arvensis [SONCARV0]
***Sonchus asper** (L.) Hill [SONCASP] (prickly sow-thistle)
***Sonchus oleraceus** L. [SONCOLE] (common sow-thistle)
Sonchus tataricus L. p.p. = Lactuca tatarica var. pulchella [LACTTAT1]
Sonchus uliginosus Bieb. = Sonchus arvensis ssp. uliginosus [SONCARV1]
Stephanomeria Nutt. [STEPHAN$]
Stephanomeria tenuifolia (Raf.) Hall [STEPTEN] (narrow-leaved stephanomeria)
Symphyotrichum boreale (Torr. & Gray) A. & D. Löve = Aster borealis [ASTEBOR]
Symphyotrichum hesperium (Gray) A. & D. Löve = Aster lanceolatus ssp. hesperius [ASTELAN1]
Tanacetum L. [TANACET$]
Tanacetum bipinnatum (L.) Schultz-Bip. [TANABIP] (dune tansy)
Tanacetum bipinnatum ssp. **huronense** (Nutt.) Breitung [TANABIP1]
SY: *Chrysanthemum bipinnatum* ssp. *huronense* (Nutt.) Hultén
SY: *Tanacetum douglasii* DC.
SY: *Tanacetum huronense* Nutt.
SY: *Tanacetum huronense* var. *bifarium* Fern.
SY: *Tanacetum huronense* var. *floccosum* Raup
SY: *Tanacetum huronense* var. *johannense* Fern.
SY: *Tanacetum huronense* var. *terrae-novae* Fern.
Tanacetum douglasii DC. = Tanacetum bipinnatum ssp. huronense [TANABIP1]
Tanacetum huronense Nutt. = Tanacetum bipinnatum ssp. huronense [TANABIP1]
Tanacetum huronense var. *bifarium* Fern. = Tanacetum bipinnatum ssp. huronense [TANABIP1]
Tanacetum huronense var. *floccosum* Raup = Tanacetum bipinnatum ssp. huronense [TANABIP1]
Tanacetum huronense var. *johannense* Fern. = Tanacetum bipinnatum ssp. huronense [TANABIP1]
Tanacetum huronense var. *terrae-novae* Fern. = Tanacetum bipinnatum ssp. huronense [TANABIP1]
***Tanacetum parthenium** (L.) Schultz-Bip. [TANAPAR] (feverfew)
SY: *Chrysanthemum parthenium* (L.) Bernh.
SY: *Matricaria parthenium* L.
Tanacetum suaveolens (Pursh) Hook. = Matricaria discoidea [MATRDIS]
***Tanacetum vulgare** L. [TANAVUL] (common tansy)
SY: *Chrysanthemum uliginosum* Pers.
SY: *Chrysanthemum vulgare* (L.) Bernh.
SY: *Tanacetum vulgare* var. *crispum* DC.
Tanacetum vulgare var. *crispum* DC. = Tanacetum vulgare [TANAVUL]
Taraxacum G.H. Weber ex Wiggers [TARAXAC$]
Taraxacum ambigens Fern. = Taraxacum officinale ssp. ceratophorum [TARAOFF1]
Taraxacum ambigens var. *flutius* Fern. = Taraxacum officinale ssp. ceratophorum [TARAOFF1]
Taraxacum amphiphron Böcher = Taraxacum officinale ssp. ceratophorum [TARAOFF1]
Taraxacum arctogenum Dahlst. = Taraxacum officinale ssp. ceratophorum [TARAOFF1]

Taraxacum brachyceras Dahlst. = Taraxacum officinale ssp. ceratophorum [TARAOFF1]
Taraxacum carneocoloratum A. Nels. = Taraxacum officinale ssp. ceratophorum [TARAOFF1]
Taraxacum carthamopsis Porsild = Taraxacum officinale ssp. ceratophorum [TARAOFF1]
Taraxacum ceratophorum (Ledeb.) DC. = Taraxacum officinale ssp. ceratophorum [TARAOFF1]
Taraxacum disseminatum Hagl. = Taraxacum laevigatum [TARALAE]
Taraxacum dumetorum Greene = Taraxacum officinale ssp. ceratophorum [TARAOFF1]
Taraxacum eriophorum Rydb. [TARAERI]
Taraxacum erythrospermum Andrz. ex Bess. = Taraxacum laevigatum [TARALAE]
Taraxacum eurylepium Dahlst. = Taraxacum officinale ssp. ceratophorum [TARAOFF1]
Taraxacum hyperboreum Dahlst. = Taraxacum officinale ssp. ceratophorum [TARAOFF1]
Taraxacum integratum Hagl. = Taraxacum officinale ssp. ceratophorum [TARAOFF1]
Taraxacum kamtschaticum Dahlst. = Taraxacum lyratum [TARALYR]
Taraxacum lacerum Greene = Taraxacum officinale ssp. ceratophorum [TARAOFF1]
Taraxacum lacistophyllum (Dahlst.) Raunk. = Taraxacum laevigatum [TARALAE]
*Taraxacum laevigatum** (Willd.) DC. [TARALAE] (red-seeded dandelion)
SY: *Leontodon erythrospermum* (Andrz. ex Bess.) Britt.
SY: *Taraxacum disseminatum* Hagl.
SY: *Taraxacum erythrospermum* Andrz. ex Bess.
SY: *Taraxacum lacistophyllum* (Dahlst.) Raunk.
SY: *Taraxacum scanicum* Dahlst.
Taraxacum lapponicum Kihlm. ex Hand.-Maz. = Taraxacum officinale ssp. ceratophorum [TARAOFF1]
Taraxacum latispinulosum M.P. Christens. = Taraxacum officinale ssp. ceratophorum [TARAOFF1]
Taraxacum laurentianum Fern. = Taraxacum officinale ssp. ceratophorum [TARAOFF1]
Taraxacum longii Fern. = Taraxacum officinale ssp. ceratophorum [TARAOFF1]
Taraxacum lyratum (Ledeb.) DC. [TARALYR] (horned dandelion)
SY: *Taraxacum kamtschaticum* Dahlst.
SY: *Taraxacum scopulorum* (Gray) Rydb.
SY: *Taraxacum sibiricum* Dahlst.
Taraxacum malteanum Dahlst. = Taraxacum officinale ssp. ceratophorum [TARAOFF1]
Taraxacum maurolepium Hagl. = Taraxacum officinale ssp. ceratophorum [TARAOFF1]
Taraxacum mitratum Hagl. = Taraxacum officinale ssp. ceratophorum [TARAOFF1]
Taraxacum multesimum Hagl. = Taraxacum officinale ssp. ceratophorum [TARAOFF1]
Taraxacum naevosum Dahlst. = Taraxacum officinale ssp. ceratophorum [TARAOFF1]
***Taraxacum officinale** G.H. Weber ex Wiggers [TARAOFF] (common dandelion)
Taraxacum officinale ssp. **ceratophorum** (Ledeb.) Schinz ex Thellung [TARAOFF1]
SY: *Taraxacum ambigens* Fern.
SY: *Taraxacum ambigens* var. *flutius* Fern.
SY: *Taraxacum amphiphron* Böcher
SY: *Taraxacum arctogenum* Dahlst.
SY: *Taraxacum brachyceras* Dahlst.
SY: *Taraxacum carneocoloratum* A. Nels.
SY: *Taraxacum carthamopsis* Porsild
SY: *Taraxacum ceratophorum* (Ledeb.) DC.
SY: *Taraxacum dumetorum* Greene
SY: *Taraxacum eurylepium* Dahlst.
SY: *Taraxacum hyperboreum* Dahlst.
SY: *Taraxacum integratum* Hagl.
SY: *Taraxacum lacerum* Greene
SY: *Taraxacum lapponicum* Kihlm. ex Hand.-Maz.
SY: *Taraxacum latispinulosum* M.P. Christens.
SY: *Taraxacum laurentianum* Fern.
SY: *Taraxacum longii* Fern.
SY: *Taraxacum malteanum* Dahlst.
SY: *Taraxacum maurolepium* Hagl.
SY: *Taraxacum mitratum* Hagl.
SY: *Taraxacum multesimum* Hagl.
SY: *Taraxacum naevosum* Dahlst.
SY: *Taraxacum ovinum* Rydb.
SY: *Taraxacum paucisquamosum* M.E. Peck
SY: *Taraxacum pellianum* Porsild
SY: *Taraxacum pseudonorvegicum* Dahlst.
SY: *Taraxacum purpuridens* Dahlst.
SY: *Taraxacum torngatense* Fern.
SY: *Taraxacum trigonolobum* Dahlst.
SY: *Taraxacum umbrinum* Dahlst.
Taraxacum ovinum Rydb. = Taraxacum officinale ssp. ceratophorum [TARAOFF1]
Taraxacum paucisquamosum M.E. Peck = Taraxacum officinale ssp. ceratophorum [TARAOFF1]
Taraxacum pellianum Porsild = Taraxacum officinale ssp. ceratophorum [TARAOFF1]
Taraxacum pseudonorvegicum Dahlst. = Taraxacum officinale ssp. ceratophorum [TARAOFF1]
Taraxacum purpuridens Dahlst. = Taraxacum officinale ssp. ceratophorum [TARAOFF1]
Taraxacum scanicum Dahlst. = Taraxacum laevigatum [TARALAE]
Taraxacum scopulorum (Gray) Rydb. = Taraxacum lyratum [TARALYR]
Taraxacum sibiricum Dahlst. = Taraxacum lyratum [TARALYR]
Taraxacum torngatense Fern. = Taraxacum officinale ssp. ceratophorum [TARAOFF1]
Taraxacum trigonolobum Dahlst. = Taraxacum officinale ssp. ceratophorum [TARAOFF1]
Taraxacum umbrinum Dahlst. = Taraxacum officinale ssp. ceratophorum [TARAOFF1]
Tephroseris atropurpurea ssp. *frigida* (Richards.) A. & D. Löve = Senecio atropurpureus [SENEATR]
Tephroseris atropurpurea ssp. *tomentosa* (Kjellm.) A. & D. Löve = Senecio atropurpureus [SENEATR]
Tephroseris integrifolia ssp. *atropurpurea* (Ledeb.) B. Nordenstam = Senecio atropurpureus [SENEATR]
Tetradymia DC. [TETRADY$]
Tetradymia canescens DC. [TETRCAN] (grey horsebrush)
SY: *Tetradymia canescens* var. *inermis* (Rydb.) Payson
Tetradymia canescens var. *inermis* (Rydb.) Payson = Tetradymia canescens [TETRCAN]
Tonestus A. Nels. [TONESTU$]
Tonestus lyallii (Gray) A. Nels. [TONELYA]
SY: *Haplopappus lyallii* Gray
Townsendia Hook. [TOWNSEN$]
Townsendia exscapa (Richards.) Porter [TOWNEXS] (easter daisy)
SY: *Aster exscapus* Richards.
SY: *Townsendia intermedia* Rydb.
SY: *Townsendia sericea* Hook.
Townsendia hookeri Beaman [TOWNHOO] (Hooker's townsendia)
Townsendia intermedia Rydb. = Townsendia exscapa [TOWNEXS]
Townsendia parryi D.C. Eat. [TOWNPAR] (Parry's townsendia)
Townsendia sericea Hook. = Townsendia exscapa [TOWNEXS]
Tragopogon L. [TRAGOPO$]
***Tragopogon dubius** Scop. [TRAGDUB] (yellow salsify)
SY: *Tragopogon dubius* ssp. *major* (Jacq.) Voll.
SY: *Tragopogon major* Jacq.
Tragopogon dubius ssp. *major* (Jacq.) Voll. = Tragopogon dubius [TRAGDUB]
Tragopogon major Jacq. = Tragopogon dubius [TRAGDUB]
***Tragopogon porrifolius** L. [TRAGPOR] (common salsify)
***Tragopogon pratensis** L. [TRAGPRA] (meadow salsify)
Trimorpha Cass. [TRIMORP$]
Trimorpha acris (L.) Nesom [TRIMACR]
Trimorpha acris var. **asteroides** (Andrz. ex Bess.) Nesom [TRIMACR1]

SY: *Erigeron acris* L.
SY: *Erigeron acris* var. *asteroides* (Andrz. ex Bess.) DC.
SY: *Erigeron acris* ssp. *politus* (Fries) Schinz & R. Keller
SY: *Erigeron droebachianus* O.F. Muell. ex Retz.
SY: *Erigeron elongatus* Ledeb.
SY: *Erigeron politus* Fries
Trimorpha acris var. **debilis** (Gray) Nesom [TRIMACR2]
SY: *Erigeron acris* ssp. *debilis* (Gray) Piper
SY: *Erigeron acris* var. *debilis* Gray
SY: *Erigeron angulosus* ssp. *debilis* (Gray) Piper
SY: *Erigeron debilis* (Gray) Rydb.
SY: *Erigeron jucundus* Greene
SY: *Erigeron nivalis* Nutt.
Trimorpha elata (Hook.) Nesom [TRIMELA]
SY: *Erigeron acris* var. *elatus* (Hook.) Cronq.
SY: *Erigeron alpinus* var. *elatus* Hook.
SY: *Erigeron elatus* (Hook.) Greene
SY: *Erigeron elatus* var. *oligocephalus* (Fern. & Wieg.) Fern.
Trimorpha lonchophylla (Hook.) Nesom [TRIMLON]
SY: *Erigeron lonchophyllus* Hook.
SY: *Erigeron lonchophyllus* var. *laurentianus* Victorin
SY: *Erigeron minor* (Hook.) Rydb.
Tripleurospermum inodorum (L.) Schultz-Bip. = Matricaria perforata [MATRPER]
Tripolium angustum Lindl. = Brachyactis ciliata ssp. angusta [BRACCIL1]
Tripolium occidentale Nutt. = Aster occidentalis var. occidentalis [ASTEOCC0]
Troximon aurantiacum Hook. = Agoseris aurantiaca [AGOSAUR]
Troximon glaucum var. *dasycephalum* Torr. & Gray = Agoseris glauca var. dasycephala [AGOSGLA1]
Troximon grandiflorum Nutt. = Agoseris grandiflora [AGOSGRA]
Tussilago L. [TUSSILA$]
*****Tussilago farfara** L. [TUSSFAR] (coltsfoot)
Virgulaster ascendens (Lindl.) Semple = Aster ascendens [ASTEASC]
Virgulus campestris (Nutt.) Reveal & Keener = Aster campestris [ASTECAM]
Virgulus falcatus (Lindl.) Reveal & Keener = Aster falcatus [ASTEFAL]
Weberaster modestus (Lindl.) A. & D. Löve = Aster modestus [ASTEMOD]
Weberaster radulinus (Gray) A. & D. Löve = Aster radulinus [ASTERAD]
Xanthium L. [XANTHIU$]
Xanthium acerosum Greene = Xanthium strumarium var. canadense [XANTSTR1]
Xanthium californicum Greene = Xanthium strumarium var. canadense [XANTSTR1]
Xanthium californicum var. *rotundifolium* Widder = Xanthium strumarium var. canadense [XANTSTR1]
Xanthium campestre Greene = Xanthium strumarium var. canadense [XANTSTR1]
Xanthium canadense P. Mill. = Xanthium strumarium var. canadense [XANTSTR1]
Xanthium cavanillesii Schouw = Xanthium strumarium var. canadense [XANTSTR1]
Xanthium cenchroides Millsp. & Sherff = Xanthium strumarium var. canadense [XANTSTR1]
Xanthium commune Britt. = Xanthium strumarium var. canadense [XANTSTR1]
Xanthium echinatum Murr. = Xanthium strumarium var. canadense [XANTSTR1]
Xanthium glanduliferum Greene = Xanthium strumarium var. canadense [XANTSTR1]
Xanthium italicum Moretti = Xanthium strumarium var. canadense [XANTSTR1]
Xanthium macounii Britt. = Xanthium strumarium var. canadense [XANTSTR1]
Xanthium oligacanthum Piper = Xanthium strumarium var. canadense [XANTSTR1]
Xanthium oviforme Wallr. = Xanthium strumarium var. canadense [XANTSTR1]
Xanthium pensylvanicum Wallr. = Xanthium strumarium var. canadense [XANTSTR1]
Xanthium saccharatum Wallr. = Xanthium strumarium var. canadense [XANTSTR1]
Xanthium speciosum Kearney = Xanthium strumarium var. canadense [XANTSTR1]
*****Xanthium strumarium** L. [XANTSTR] (rough cocklebur)
Xanthium strumarium ssp. *italicum* (Moretti) D. Löve = Xanthium strumarium var. canadense [XANTSTR1]
*****Xanthium strumarium** var. **canadense** (P. Mill.) Torr. & Gray [XANTSTR1] (common cocklebur)
SY: *Xanthium acerosum* Greene
SY: *Xanthium californicum* Greene
SY: *Xanthium californicum* var. *rotundifolium* Widder
SY: *Xanthium campestre* Greene
SY: *Xanthium canadense* P. Mill.
SY: *Xanthium cavanillesii* Schouw
SY: *Xanthium cenchroides* Millsp. & Sherff
SY: *Xanthium commune* Britt.
SY: *Xanthium echinatum* Murr.
SY: *Xanthium glanduliferum* Greene
SY: *Xanthium italicum* Moretti
SY: *Xanthium macounii* Britt.
SY: *Xanthium oligacanthum* Piper
SY: *Xanthium oviforme* Wallr.
SY: *Xanthium pensylvanicum* Wallr.
SY: *Xanthium saccharatum* Wallr.
SY: *Xanthium speciosum* Kearney
SY: *Xanthium strumarium* ssp. *italicum* (Moretti) D. Löve
SY: *Xanthium strumarium* var. *oviforme* (Wallr.) M.E. Peck
SY: *Xanthium strumarium* var. *pensylvanicum* (Wallr.) M.E. Peck
SY: *Xanthium varians* Greene
Xanthium strumarium var. *oviforme* (Wallr.) M.E. Peck = Xanthium strumarium var. canadense [XANTSTR1]
Xanthium strumarium var. *pensylvanicum* (Wallr.) M.E. Peck = Xanthium strumarium var. canadense [XANTSTR1]
Xanthium varians Greene = Xanthium strumarium var. canadense [XANTSTR1]

AZOLLACEAE (V:F)

Azolla Lam. [AZOLLA$]
Azolla filiculoides Lam. [AZOLFIL] (large mosquito fern)
Azolla mexicana Schlecht. & Cham. ex K. Presl [AZOLMEX] (Mexican mosquito fern)

BALSAMINACEAE (V:D)

Impatiens L. [IMPATIE$]
Impatiens aurella Rydb. [IMPAAUR] (orange touch-me-not)
Impatiens biflora Walt. = Impatiens capensis [IMPACAP]
Impatiens biflora var. *ecalcarata* (Blank) M.E. Jones = Impatiens ecalcarata [IMPAECA]
Impatiens capensis Meerb. [IMPACAP] (spotted touch-me-not)
SY: *Impatiens biflora* Walt.
SY: *Impatiens fulva* Nutt.
SY: *Impatiens noli-tangere* ssp. *biflora* (Walt.) Hultén
SY: *Impatiens nortonii* Rydb.
Impatiens ecalcarata Blank. [IMPAECA] (spurless touch-me-not)
SY: *Impatiens biflora* var. *ecalcarata* (Blank) M.E. Jones
Impatiens fulva Nutt. = Impatiens capensis [IMPACAP]
*****Impatiens glandulifera** Royle [IMPAGLA] (policeman's helmet)
SY: *Impatiens roylei* Walp.
*****Impatiens noli-tangere** L. [IMPANOL] (common touch-me-not)
SY: *Impatiens occidentalis* Rydb.
Impatiens noli-tangere ssp. *biflora* (Walt.) Hultén = Impatiens capensis [IMPACAP]
Impatiens nortonii Rydb. = Impatiens capensis [IMPACAP]
Impatiens occidentalis Rydb. = Impatiens noli-tangere [IMPANOL]
*****Impatiens parviflora** DC. [IMPAPAR] (small touch-me-not)
Impatiens roylei Walp. = Impatiens glandulifera [IMPAGLA]

BERBERIDACEAE (V:D)

Achlys DC. [ACHLYS$]
Achlys triphylla (Sm.) DC. [ACHLTRI] (vanilla-leaf)
Berberis L. [BERBERI$]
Berberis aquifolium Pursh = Mahonia aquifolium [MAHOAQU]
Berberis aquifolium var. *repens* (Lindl.) Scoggan = Mahonia repens [MAHOREP]
Berberis nervosa Pursh = Mahonia nervosa [MAHONER]
Berberis repens Lindl. = Mahonia repens [MAHOREP]
Berberis sonnei (Abrams) McMinn = Mahonia repens [MAHOREP]
*__Berberis vulgaris__ L. [BERBVUL] (common barberry)
Mahonia Nutt. [MAHONIA$]
Mahonia aquifolium (Pursh) Nutt. [MAHOAQU] (tall Oregon-grape)
SY: *Berberis aquifolium* Pursh
SY: *Odostemon aquifolium* (Pursh) Rydb.
Mahonia nervosa (Pursh) Nutt. [MAHONER] (dull Oregon-grape)
SY: *Berberis nervosa* Pursh
Mahonia repens (Lindl.) G. Don [MAHOREP] (creeping Oregon-grape)
SY: *Berberis aquifolium* var. *repens* (Lindl.) Scoggan
SY: *Berberis repens* Lindl.
SY: *Berberis sonnei* (Abrams) McMinn
SY: *Mahonia sonnei* Abrams
SY: *Odostemon repens* (Lindl.) Cockerell
Mahonia sonnei Abrams = Mahonia repens [MAHOREP]
Odostemon aquifolium (Pursh) Rydb. = Mahonia aquifolium [MAHOAQU]
Odostemon repens (Lindl.) Cockerell = Mahonia repens [MAHOREP]

BETULACEAE (V:D)

Alnus P. Mill. [ALNUS$]
Alnus alnobetula (Ehrh.) K. Koch p.p. = Alnus viridis ssp. sinuata [ALNUVIR2]
Alnus crispa (Ait.) Pursh = Alnus viridis ssp. crispa [ALNUVIR5]
Alnus crispa ssp. *laciniata* Hultén = Alnus viridis ssp. sinuata [ALNUVIR2]
Alnus crispa ssp. *sinuata* (Regel) Hultén = Alnus viridis ssp. sinuata [ALNUVIR2]
Alnus crispa var. *elongata* Raup = Alnus viridis ssp. crispa [ALNUVIR5]
Alnus crispa var. *mollis* (Fern.) Fern. = Alnus viridis ssp. crispa [ALNUVIR5]
Alnus × hultenii Murai = Alnus viridis ssp. crispa [ALNUVIR5]
Alnus incana (L.) Moench [ALNUINC] (gray alder)
Alnus incana ssp. *rugosa* var. *occidentalis* (Dippel) C.L. Hitchc. = Alnus incana ssp. tenuifolia [ALNUINC2]
Alnus incana ssp. **tenuifolia** (Nutt.) Breitung [ALNUINC2] (mountain alder)
SY: *Alnus incana* var. *occidentalis* (Dippel) C.L. Hitchc.
SY: *Alnus incana* ssp. *rugosa* var. *occidentalis* (Dippel) C.L. Hitchc.
SY: *Alnus incana* var. *virescens* S. Wats.
SY: *Alnus × purpusii* Callier
SY: *Alnus tenuifolia* Nutt.
SY: *Alnus tenuifolia* var. *occidentalis* (Dippel) Collier
Alnus incana var. *occidentalis* (Dippel) C.L. Hitchc. = Alnus incana ssp. tenuifolia [ALNUINC2]
Alnus incana var. *virescens* S. Wats. = Alnus incana ssp. tenuifolia [ALNUINC2]
Alnus oregona Nutt. = Alnus rubra [ALNURUB]
Alnus oregona var. *pinnatisecta* Starker = Alnus rubra [ALNURUB]
Alnus × purpusii Callier = Alnus incana ssp. tenuifolia [ALNUINC2]
Alnus rubra Bong. [ALNURUB] (red alder)
SY: *Alnus oregona* Nutt.
SY: *Alnus oregona* var. *pinnatisecta* Starker
Alnus sinuata (Regel) Rydb. = Alnus viridis ssp. sinuata [ALNUVIR2]
Alnus tenuifolia Nutt. = Alnus incana ssp. tenuifolia [ALNUINC2]
Alnus tenuifolia var. *occidentalis* (Dippel) Collier = Alnus incana ssp. tenuifolia [ALNUINC2]
Alnus viridis (Vill.) Lam. & DC. [ALNUVIR] (green alder)
Alnus viridis ssp. **crispa** (Ait.) Turrill [ALNUVIR5]
SY: *Alnus crispa* (Ait.) Pursh
SY: *Alnus crispa* var. *elongata* Raup
SY: *Alnus crispa* var. *mollis* (Fern.) Fern.
SY: *Alnus × hultenii* Murai
SY: *Alnus viridis* var. *crispa* (Ait.) House
SY: *Alnus viridis* ssp. *fruticosa* (Rupr.) Nyman
SY: *Duschekia viridis* (Chaix) Opiz
Alnus viridis ssp. *fruticosa* (Rupr.) Nyman = Alnus viridis ssp. crispa [ALNUVIR5]
Alnus viridis ssp. **sinuata** (Regel) A. & D. Löve [ALNUVIR2] (Sitka alder)
SY: *Alnus alnobetula* (Ehrh.) K. Koch p.p.
SY: *Alnus crispa* ssp. *laciniata* Hultén
SY: *Alnus crispa* ssp. *sinuata* (Regel) Hultén
SY: *Alnus sinuata* (Regel) Rydb.
SY: *Alnus viridis* var. *sinuata* Regel
SY: *Duschekia sinuata* (Regel) Pouzar
Alnus viridis var. *crispa* (Ait.) House = Alnus viridis ssp. crispa [ALNUVIR5]
Alnus viridis var. *sinuata* Regel = Alnus viridis ssp. sinuata [ALNUVIR2]
Betula L. [BETULA$]
Betula alaskana Sarg. = Betula neoalaskana [BETUNEO]
Betula alba L. = Betula pubescens [BETUPUB]
Betula alba var. *commutata* Regel = Betula papyrifera var. commutata [BETUPAP4]
Betula beeniana A. Nels. = Betula occidentalis [BETUOCC]
Betula exilis Sukatschev = Betula nana [BETUNAN]
Betula fontinalis Sarg. = Betula occidentalis [BETUOCC]
Betula glandulifera (Regel) Butler = Betula pumila var. glandulifera [BETUPUM1]
Betula glandulosa Michx. = Betula nana [BETUNAN]
Betula glandulosa var. *glandulifera* (Regel) Gleason = Betula pumila var. glandulifera [BETUPUM1]
Betula glandulosa var. *hallii* (T.J. Howell) C.L. Hitchc. = Betula nana [BETUNAN]
Betula glandulosa var. *sibirica* (Ledeb.) Schneid. = Betula nana [BETUNAN]
Betula michauxii Sarg. = Betula nana [BETUNAN]
Betula nana L. [BETUNAN] (swamp birch)
SY: *Betula exilis* Sukatschev
SY: *Betula glandulosa* Michx.
SY: *Betula glandulosa* var. *hallii* (T.J. Howell) C.L. Hitchc.
SY: *Betula glandulosa* var. *sibirica* (Ledeb.) Schneid.
SY: *Betula michauxii* Sarg.
SY: *Betula nana* ssp. *exilis* (Sukatschev) Hultén
SY: *Betula nana* var. *sibirica* Ledeb.
SY: *Betula terrae-novae* Fern.
Betula nana ssp. *exilis* (Sukatschev) Hultén = Betula nana [BETUNAN]
Betula nana var. *glandulifera* (Regel) Boivin = Betula pumila var. glandulifera [BETUPUM1]
Betula nana var. *sibirica* Ledeb. = Betula nana [BETUNAN]
Betula neoalaskana Sarg. [BETUNEO] (Alaska paper birch)
SY: *Betula alaskana* Sarg.
SY: *Betula papyrifera* ssp. *humilis* (Regel) Hultén
SY: *Betula papyrifera* var. *humilis* (Regel) Fern. & Raup
SY: *Betula papyrifera* var. *neoalaskana* (Sarg.) Raup
SY: *Betula resinifera* Britt.
Betula occidentalis Hook. [BETUOCC] (water birch)
SY: *Betula beeniana* A. Nels.
SY: *Betula fontinalis* Sarg.
SY: *Betula occidentalis* var. *inopina* (Jepson) C.L. Hitchc.
SY: *Betula papyrifera* ssp. *occidentalis* (Hook.) Hultén
SY: *Betula papyrifera* var. *occidentalis* (Hook.) Sarg.
Betula occidentalis var. *inopina* (Jepson) C.L. Hitchc. = Betula occidentalis [BETUOCC]
Betula papyrifera Marsh. [BETUPAP] (paper birch)
Betula papyrifera ssp. *humilis* (Regel) Hultén = Betula neoalaskana [BETUNEO]

Betula papyrifera ssp. *occidentalis* (Hook.) Hultén = Betula occidentalis [BETUOCC]
Betula papyrifera var. **commutata** (Regel) Fern. [BETUPAP4]
SY: *Betula alba* var. *commutata* Regel
Betula papyrifera var. *elobata* (Fern.) Sarg. = Betula papyrifera var. papyrifera [BETUPAP0]
Betula papyrifera var. *humilis* (Regel) Fern. & Raup = Betula neoalaskana [BETUNEO]
Betula papyrifera var. *macrostachya* Fern. = Betula papyrifera var. papyrifera [BETUPAP0]
Betula papyrifera var. *neoalaskana* (Sarg.) Raup = Betula neoalaskana [BETUNEO]
Betula papyrifera var. *occidentalis* (Hook.) Sarg. = Betula occidentalis [BETUOCC]
Betula papyrifera var. **papyrifera** [BETUPAP0]
SY: *Betula papyrifera* var. *elobata* (Fern.) Sarg.
SY: *Betula papyrifera* var. *macrostachya* Fern.
SY: *Betula papyrifera* var. *pensilis* Fern.
Betula papyrifera var. *pensilis* Fern. = Betula papyrifera var. papyrifera [BETUPAP0]
***Betula pendula** Roth [BETUPEN] (European birch)
SY: *Betula verrucosa* Ehrh.
Betula piperi Britt. = Betula × utahensis [BETUXUT]
Betula × piperi Britt. = Betula × utahensis [BETUXUT]
***Betula pubescens** Ehrh. [BETUPUB] (silver birch)
SY: *Betula alba* L.
Betula pumila L. [BETUPUM] (bog birch)
Betula pumila var. **glandulifera** Regel [BETUPUM1] (low birch)
SY: *Betula glandulifera* (Regel) Butler
SY: *Betula glandulosa* var. *glandulifera* (Regel) Gleason
SY: *Betula nana* var. *glandulifera* (Regel) Boivin
SY: *Betula pumila* var. *glandulifera* f. *hallii* (Howell) Brayshaw
Betula pumila var. *glandulifera* f. *hallii* (Howell) Brayshaw = Betula pumila var. glandulifera [BETUPUM1]
Betula resinifera Britt. = Betula neoalaskana [BETUNEO]
Betula terrae-novae Fern. = Betula nana [BETUNAN]
Betula × utahensis Britt. (pro sp.) [BETUXUT]
SY: *Betula piperi* Britt.
SY: *Betula × piperi* Britt.
Betula verrucosa Ehrh. = Betula pendula [BETUPEN]
Betula × winteri Dugle [BETUXWI] (Alaska × paper birch hybrid)
Corylus L. [CORYLUS$]
***Corylus avellana** L. [CORYAVE] (hazelnut)
Corylus californica (A. DC.) Rose = Corylus cornuta var. californica [CORYCOR1]
Corylus cornuta Marsh. [CORYCOR] (beaked hazelnut)
Corylus cornuta var. **californica** (A. DC.) Sharp [CORYCOR1] (California hazelnut)
SY: *Corylus californica* (A. DC.) Rose
Corylus cornuta var. **cornuta** [CORYCOR0]
SY: *Corylus rostrata* Ait.
Corylus cornuta var. **glandulosa** Boivin [CORYCOR3]
Corylus rostrata Ait. = Corylus cornuta var. cornuta [CORYCOR0]
Duschekia sinuata (Regel) Pouzar = Alnus viridis ssp. sinuata [ALNUVIR2]
Duschekia viridis (Chaix) Opiz = Alnus viridis ssp. crispa [ALNUVIR5]

BLECHNACEAE (V:F)

Blechnum L. [BLECHNU$]
Blechnum doodioides Hook. = Blechnum spicant [BLECSPI]
Blechnum spicant (L.) Roth [BLECSPI] (deer fern)
SY: *Blechnum doodioides* Hook.
SY: *Blechnum spicant* var. *elongatum* (Hook.) Boivin
SY: *Lomaria spicant* (L.) Desv.
SY: *Osmunda spicant* L.
SY: *Struthiopteris spicant* (L.) Weiss
Blechnum spicant var. *elongatum* (Hook.) Boivin = Blechnum spicant [BLECSPI]
Lomaria spicant (L.) Desv. = Blechnum spicant [BLECSPI]
Osmunda spicant L. = Blechnum spicant [BLECSPI]
Struthiopteris spicant (L.) Weiss = Blechnum spicant [BLECSPI]
Woodwardia Sm. [WOODWAR$]
Woodwardia chamissoi Brack. = Woodwardia fimbriata [WOODFIM]
Woodwardia fimbriata Sm. [WOODFIM] (giant chain fern)
SY: *Woodwardia chamissoi* Brack.

BORAGINACEAE (V:D)

Amsinckia Lehm. [AMSINCK$]
Amsinckia arizonica Suksdorf = Amsinckia intermedia [AMSIINT]
Amsinckia barbata Greene = Amsinckia lycopsoides [AMSILYC]
Amsinckia demissa Suksdorf = Amsinckia intermedia [AMSIINT]
Amsinckia echinata Gray = Amsinckia intermedia [AMSIINT]
Amsinckia hispida (Ruiz & Pavón) I.M. Johnston = Amsinckia lycopsoides [AMSILYC]
Amsinckia idahoensis M.E. Jones = Amsinckia lycopsoides [AMSILYC]
Amsinckia intactilis J.F. Macbr. = Amsinckia intermedia [AMSIINT]
Amsinckia intermedia Fisch. & C.A. Mey. [AMSIINT] (common fiddleneck)
SY: *Amsinckia arizonica* Suksdorf
SY: *Amsinckia demissa* Suksdorf
SY: *Amsinckia echinata* Gray
SY: *Amsinckia intactilis* J.F. Macbr.
SY: *Amsinckia intermedia* var. *echinata* (Gray) Wiggins
SY: *Amsinckia microphylla* Suksdorf
SY: *Amsinckia nana* Suksdorf
SY: *Amsinckia rigida* Suksdorf
Amsinckia intermedia var. *echinata* (Gray) Wiggins = Amsinckia intermedia [AMSIINT]
Amsinckia lycopsoides Lehm. [AMSILYC] (bugloss fiddleneck)
SY: *Amsinckia barbata* Greene
SY: *Amsinckia hispida* (Ruiz & Pavón) I.M. Johnston
SY: *Amsinckia idahoensis* M.E. Jones
SY: *Amsinckia parviflora* Heller
SY: *Benthamia lycopsoides* (Lehm.) Lindl. ex Druce
Amsinckia menziesii (Lehm.) A. Nels. & J.F. Macbr. [AMSIMEN] (small-flowered fiddleneck)
SY: *Amsinckia micrantha* Suksdorf
SY: *Echium menziesii* Lehm.
Amsinckia micrantha Suksdorf = Amsinckia menziesii [AMSIMEN]
Amsinckia microphylla Suksdorf = Amsinckia intermedia [AMSIINT]
Amsinckia nana Suksdorf = Amsinckia intermedia [AMSIINT]
Amsinckia parviflora Heller = Amsinckia lycopsoides [AMSILYC]
Amsinckia retrorsa Suksdorf [AMSIRET] (rigid fiddleneck)
SY: *Amsinckia rugosa* Rydb.
Amsinckia rigida Suksdorf = Amsinckia intermedia [AMSIINT]
Amsinckia rugosa Rydb. = Amsinckia retrorsa [AMSIRET]
Amsinckia scouleri I.M. Johnston = Amsinckia spectabilis [AMSISPE]
Amsinckia spectabilis Fisch. & C.A. Mey. [AMSISPE] (seaside fiddleneck)
SY: *Amsinckia scouleri* I.M. Johnston
SY: *Amsinckia spectabilis* var. *bracteosa* (Gray) Boivin
SY: *Amsinckia spectabilis* var. *microcarpa* (Greene) Jepson & Hoover
SY: *Amsinckia spectabilis* var. *nicolai* (Jepson) I.M. Johnston ex Munz
Amsinckia spectabilis var. *bracteosa* (Gray) Boivin = Amsinckia spectabilis [AMSISPE]
Amsinckia spectabilis var. *microcarpa* (Greene) Jepson & Hoover = Amsinckia spectabilis [AMSISPE]
Amsinckia spectabilis var. *nicolai* (Jepson) I.M. Johnston ex Munz = Amsinckia spectabilis [AMSISPE]
Anchusa L. [ANCHUSA$]
***Anchusa officinalis** L. [ANCHOFF] (alkanet bugloss)
SY: *Anchusa procera* Bess. ex Link
Anchusa procera Bess. ex Link = Anchusa officinalis [ANCHOFF]
Asperugo L. [ASPERUG$]
***Asperugo procumbens** L. [ASPEPRO] (madwort)
Batschia linearifolia (Goldie) Small = Lithospermum incisum [LITHINC]

Benthamia lycopsoides (Lehm.) Lindl. ex Druce = Amsinckia lycopsoides [AMSILYC]
Borago L. [BORAGO$]
*****Borago officinalis** L. [BORAOFF] (common borage)
Buglossoides Moench [BUGLOSS$]
*****Buglossoides arvensis** (L.) I.M. Johnston [BUGLARV]
SY: *Lithospermum arvense* L.
Cryptantha Lehm. ex G. Don [CRYPTAN$]
Cryptantha affinis (Gray) Greene [CRYPAFF] (common cryptantha)
Cryptantha ambigua (Gray) Greene [CRYPAMB] (obscure cryptantha)
Cryptantha barbigera var. *fergusoniae* J.F. Macbr. = Cryptantha intermedia [CRYPINT]
Cryptantha bradburiana Payson = Cryptantha celosioides [CRYPCEL]
Cryptantha calycosa (Gray) Rydb. = Cryptantha torreyana [CRYPTOR]
Cryptantha celosioides (Eastw.) Payson [CRYPCEL] (cockscomb cryptantha)
SY: *Cryptantha bradburiana* Payson
SY: *Cryptantha macounii* (Eastw.) Payson
SY: *Cryptantha nubigena* var. *celosioides* (Eastw.) Boivin
SY: *Cryptantha nubigena* var. *macounii* (Eastw.) Boivin
SY: *Cryptantha sheldonii* (Brand) Payson
SY: *Oreocarya celosioides* Eastw.
SY: *Oreocarya glomerata* (Pursh) Greene
SY: *Oreocarya macounii* Eastw.
SY: *Oreocarya sheldonii* Brand
Cryptantha eastwoodiae St. John = Cryptantha torreyana [CRYPTOR]
Cryptantha fendleri (Gray) Greene [CRYPFEN] (Fendler's cryptantha)
SY: *Cryptantha pattersonii* (Gray) Greene
Cryptantha flexulosa (A. Nels.) A. Nels. = Cryptantha torreyana [CRYPTOR]
Cryptantha fragilis M.E. Peck = Cryptantha intermedia [CRYPINT]
Cryptantha grandiflora Rydb. = Cryptantha intermedia [CRYPINT]
Cryptantha hendersonii (A. Nels.) Piper ex J.C. Nels. = Cryptantha intermedia [CRYPINT]
Cryptantha intermedia (Gray) Greene [CRYPINT]
SY: *Cryptantha barbigera* var. *fergusoniae* J.F. Macbr.
SY: *Cryptantha fragilis* M.E. Peck
SY: *Cryptantha grandiflora* Rydb.
SY: *Cryptantha hendersonii* (A. Nels.) Piper ex J.C. Nels.
SY: *Cryptantha intermedia* var. *grandiflora* (Rydb.) Cronq.
Cryptantha intermedia var. *grandiflora* (Rydb.) Cronq. = Cryptantha intermedia [CRYPINT]
Cryptantha macounii (Eastw.) Payson = Cryptantha celosioides [CRYPCEL]
Cryptantha nubigena (Greene) Payson [CRYPNUB] (Sierra cryptantha)
SY: *Oreocarya nubigena* Greene
Cryptantha nubigena var. *celosioides* (Eastw.) Boivin = Cryptantha celosioides [CRYPCEL]
Cryptantha nubigena var. *macounii* (Eastw.) Boivin = Cryptantha celosioides [CRYPCEL]
Cryptantha pattersonii (Gray) Greene = Cryptantha fendleri [CRYPFEN]
Cryptantha sheldonii (Brand) Payson = Cryptantha celosioides [CRYPCEL]
Cryptantha torreyana (Gray) Greene [CRYPTOR] (Torrey's cryptantha)
SY: *Cryptantha calycosa* (Gray) Rydb.
SY: *Cryptantha eastwoodiae* St. John
SY: *Cryptantha flexulosa* (A. Nels.) A. Nels.
SY: *Cryptantha torreyana* var. *calycosa* (Gray) Greene
SY: *Cryptantha torreyana* var. *pumila* (Heller) I.M. Johnston
Cryptantha torreyana var. *calycosa* (Gray) Greene = Cryptantha torreyana [CRYPTOR]
Cryptantha torreyana var. *pumila* (Heller) I.M. Johnston = Cryptantha torreyana [CRYPTOR]
Cryptantha watsonii (Gray) Greene [CRYPWAT] (Watson's cryptantha)
Cynoglossum L. [CYNOGLO$]
Cynoglossum boreale Fern. = Cynoglossum virginianum var. boreale [CYNOVIR1]
*****Cynoglossum officinale** L. [CYNOOFF] (common hound's-tongue)
Cynoglossum penicillatum Hook. & Arn. = Pectocarya penicillata [PECTPEN]
Cynoglossum virginianum L. [CYNOVIR]
Cynoglossum virginianum var. **boreale** (Fern.) Cooperrider [CYNOVIR1]
SY: *Cynoglossum boreale* Fern.
Echium L. [ECHIUM$]
Echium menziesii Lehm. = Amsinckia menziesii [AMSIMEN]
*****Echium vulgare** L. [ECHIVUL] (viper's bugloss)
Hackelia Opiz [HACKELI$]
Hackelia americana (Gray) Fern. = Hackelia deflexa var. americana [HACKDEF2]
Hackelia ciliata (Dougl. ex Lehm.) I.M. Johnston [HACKCIL] (Okanogan stickseed)
Hackelia deflexa (Wahlenb.) Opiz [HACKDEF] (nodding stickseed)
Hackelia deflexa ssp. *americana* (Gray) A. & D. Löve = Hackelia deflexa var. americana [HACKDEF2]
Hackelia deflexa var. **americana** (Gray) Fern. & I.M. Johnston [HACKDEF2]
SY: *Hackelia americana* (Gray) Fern.
SY: *Hackelia deflexa* ssp. *americana* (Gray) A. & D. Löve
SY: *Lappula americana* (Gray) Rydb.
SY: *Lappula deflexa* (Wahlenb.) Garcke p.p.
SY: *Lappula deflexa* ssp. *americana* (Gray) Hultén
SY: *Lappula deflexa* var. *americana* (Gray) Greene
Hackelia diffusa (Lehm.) I.M. Johnston [HACKDIF] (spreading stickseed)
Hackelia floribunda (Lehm.) I.M. Johnston [HACKFLO] (many-flowered stickseed)
SY: *Hackelia leptophylla* (Rydb.) I.M. Johnston
SY: *Lappula floribunda* (Lehm.) Greene
Hackelia jessicae (McGregor) Brand = Hackelia micrantha [HACKMIC]
Hackelia leptophylla (Rydb.) I.M. Johnston = Hackelia floribunda [HACKFLO]
Hackelia micrantha (Eastw.) J.L. Gentry [HACKMIC] (blue stickseed)
SY: *Hackelia jessicae* (McGregor) Brand
Lappula Moench [LAPPULA$]
Lappula americana (Gray) Rydb. = Hackelia deflexa var. americana [HACKDEF2]
Lappula deflexa (Wahlenb.) Garcke p.p. = Hackelia deflexa var. americana [HACKDEF2]
Lappula deflexa ssp. *americana* (Gray) Hultén = Hackelia deflexa var. americana [HACKDEF2]
Lappula deflexa var. *americana* (Gray) Greene = Hackelia deflexa var. americana [HACKDEF2]
Lappula echinata Gilib. = Lappula squarrosa [LAPPSQU]
Lappula echinata var. *occidentalis* (S. Wats.) Boivin = Lappula occidentalis var. occidentalis [LAPPOCC0]
Lappula erecta A. Nels. = Lappula squarrosa [LAPPSQU]
Lappula floribunda (Lehm.) Greene = Hackelia floribunda [HACKFLO]
Lappula fremontii (Torr.) Greene = Lappula squarrosa [LAPPSQU]
Lappula lappula (L.) Karst. = Lappula squarrosa [LAPPSQU]
Lappula myosotis Moench = Lappula squarrosa [LAPPSQU]
Lappula occidentalis (S. Wats.) Greene [LAPPOCC]
Lappula occidentalis var. **cupulata** (Gray) Higgins [LAPPOCC1]
SY: *Lappula redowskii* var. *cupulata* (Gray) M.E. Jones
SY: *Lappula redowskii* var. *texana* (Scheele) Brand
SY: *Lappula texana* (Scheele) Britt.
SY: *Lappula texana* var. *coronata* (Greene) A. Nels. & J.F. Macbr.
SY: *Lappula texana* var. *heterosperma* (Greene) A. Nels. & J.F. Macbr.
SY: *Lappula texana* var. *homosperma* (A. Nels.) A. Nels. & J.F. Macbr.
Lappula occidentalis var. **occidentalis** [LAPPOCC0]
SY: *Lappula echinata* var. *occidentalis* (S. Wats.) Boivin
SY: *Lappula redowskii* var. *desertorum* (Greene) I.M. Johnston

SY: *Lappula redowskii* var. *occidentalis* (S. Wats.) Rydb.
SY: *Lappula redowskii* auct.
Lappula redowskii auct. = Lappula occidentalis var. occidentalis [LAPPOCC0]
Lappula redowskii var. *cupulata* (Gray) M.E. Jones = Lappula occidentalis var. cupulata [LAPPOCC1]
Lappula redowskii var. *desertorum* (Greene) I.M. Johnston = Lappula occidentalis var. occidentalis [LAPPOCC0]
Lappula redowskii var. *occidentalis* (S. Wats.) Rydb. = Lappula occidentalis var. occidentalis [LAPPOCC0]
Lappula redowskii var. *texana* (Scheele) Brand = Lappula occidentalis var. cupulata [LAPPOCC1]
*__Lappula squarrosa__ (Retz.) Dumort. [LAPPSQU] (bristly stickseed)
SY: *Lappula echinata* Gilib.
SY: *Lappula erecta* A. Nels.
SY: *Lappula fremontii* (Torr.) Greene
SY: *Lappula lappula* (L.) Karst.
SY: *Lappula myosotis* Moench
SY: *Lappula squarrosa* var. *erecta* (A. Nels.) Dorn
Lappula squarrosa var. *erecta* (A. Nels.) Dorn = Lappula squarrosa [LAPPSQU]
Lappula texana (Scheele) Britt. = Lappula occidentalis var. cupulata [LAPPOCC1]
Lappula texana var. *coronata* (Greene) A. Nels. & J.F. Macbr. = Lappula occidentalis var. cupulata [LAPPOCC1]
Lappula texana var. *heterosperma* (Greene) A. Nels. & J.F. Macbr. = Lappula occidentalis var. cupulata [LAPPOCC1]
Lappula texana var. *homosperma* (A. Nels.) A. Nels. & J.F. Macbr. = Lappula occidentalis var. cupulata [LAPPOCC1]
Lithospermum L. [LITHOSP$]
Lithospermum angustifolium Michx. = Lithospermum incisum [LITHINC]
Lithospermum arvense L. = Buglossoides arvensis [BUGLARV]
Lithospermum incisum Lehm. [LITHINC] (yellow gromwell)
SY: *Batschia linearifolia* (Goldie) Small
SY: *Lithospermum angustifolium* Michx.
SY: *Lithospermum linearifolium* Goldie
SY: *Lithospermum mandanense* Spreng.
Lithospermum linearifolium Goldie = Lithospermum incisum [LITHINC]
Lithospermum mandanense Spreng. = Lithospermum incisum [LITHINC]
Lithospermum pilosum Nutt. = Lithospermum ruderale [LITHRUD]
Lithospermum ruderale Dougl. ex Lehm. [LITHRUD] (lemonweed gromwell)
SY: *Lithospermum pilosum* Nutt.
Mertensia Roth [MERTENS$]
Mertensia longiflora Greene [MERTLON] (long-flowered bluebells)
Mertensia maritima (L.) S.F. Gray [MERTMAR] (sea bluebells)
Mertensia palmeri A. Nels. & J.F. Macbr. = Mertensia paniculata var. paniculata [MERTPAN0]
Mertensia paniculata (Ait.) G. Don [MERTPAN] (tall bluebells)
Mertensia paniculata var. **borealis** (J.F. Macbr.) L.O. Williams [MERTPAN1]
Mertensia paniculata var. **paniculata** [MERTPAN0]
SY: *Mertensia palmeri* A. Nels. & J.F. Macbr.
Myosotis L. [MYOSOTI$]
Myosotis alpestris auct. = Myosotis asiatica [MYOSASI]
Myosotis alpestris ssp. *asiatica* Vesterg. = Myosotis asiatica [MYOSASI]
*__Myosotis arvensis__ (L.) Hill [MYOSARV] (field forget-me-not)
SY: *Myosotis scorpioides* var. *arvensis* L.
Myosotis asiatica (Vesterg.) Schischkin & Sergievskaja [MYOSASI] (mountain forget-me-not)
SY: *Myosotis alpestris* ssp. *asiatica* Vesterg.
SY: *Myosotis alpestris* auct.
SY: *Myosotis sylvatica* var. *alpestris* auct.
*__Myosotis discolor__ Pers. [MYOSDIS] (common forget-me-not)
SY: *Myosotis versicolor* (Pers.) Sm.
Myosotis laxa Lehm. [MYOSLAX] (small-flowered forget-me-not)
Myosotis micrantha auct. = Myosotis stricta [MYOSSTR]
Myosotis palustris (L.) Hill = Myosotis scorpioides [MYOSSCO]
*__Myosotis scorpioides__ L. [MYOSSCO] (forget-me-not)
SY: *Myosotis palustris* (L.) Hill
Myosotis scorpioides var. *arvensis* L. = Myosotis arvensis [MYOSARV]
*__Myosotis stricta__ Link ex Roemer & J.A. Schultes [MYOSSTR] (blue forget-me-not)
SY: *Myosotis micrantha* auct.
*__Myosotis sylvatica__ Ehrh. ex Hoffmann [MYOSSYL] (wood forget-me-not)
Myosotis sylvatica var. *alpestris* auct. = Myosotis asiatica [MYOSASI]
*__Myosotis verna__ Nutt. [MYOSVER] (spring forget-me-not)
Myosotis versicolor (Pers.) Sm. = Myosotis discolor [MYOSDIS]
Oreocarya celosioides Eastw. = Cryptantha celosioides [CRYPCEL]
Oreocarya glomerata (Pursh) Greene = Cryptantha celosioides [CRYPCEL]
Oreocarya macounii Eastw. = Cryptantha celosioides [CRYPCEL]
Oreocarya nubigena Greene = Cryptantha nubigena [CRYPNUB]
Oreocarya sheldonii Brand = Cryptantha celosioides [CRYPCEL]
Pectocarya DC. ex Meisn. [PECTOCA$]
Pectocarya linearis var. *penicillata* (Hook. & Arn.) M.E. Jones = Pectocarya penicillata [PECTPEN]
Pectocarya penicillata (Hook. & Arn.) A. DC. [PECTPEN] (winged combseed)
SY: *Cynoglossum penicillatum* Hook. & Arn.
SY: *Pectocarya linearis* var. *penicillata* (Hook. & Arn.) M.E. Jones
Plagiobothrys Fisch. & C.A. Mey. [PLAGIOB$]
Plagiobothrys asper Greene = Plagiobothrys tenellus [PLAGTEN]
Plagiobothrys figuratus (Piper) I.M. Johnston ex M.E. Peck [PLAGFIG] (fragrant popcornflower)
Plagiobothrys scouleri (Hook. & Arn.) I.M. Johnston [PLAGSCO] (Scouler's popcornflower)
Plagiobothrys tenellus (Nutt. ex Hook.) Gray [PLAGTEN] (slender popcornflower)
SY: *Plagiobothrys asper* Greene
Symphytum L. [SYMPHYT$]
Symphytum asperrimum Donn ex Sims = Symphytum asperum [SYMPASP]
*__Symphytum asperum__ Lepechin [SYMPASP] (rough comfrey)
SY: *Symphytum asperrimum* Donn ex Sims
*__Symphytum officinale__ L. [SYMPOFF] (common comfrey)
SY: *Symphytum officinale* ssp. *uliginosum* (Kern.) Nyman
SY: *Symphytum uliginosum* Kern.
Symphytum officinale ssp. *uliginosum* (Kern.) Nyman = Symphytum officinale [SYMPOFF]
Symphytum uliginosum Kern. = Symphytum officinale [SYMPOFF]

BRASSICACEAE (V:D)

Alliaria Heister ex Fabr. [ALLIARI$]
Alliaria alliaria (L.) Britt. = Alliaria petiolata [ALLIPET]
Alliaria officinalis Andrz. ex Bieb. = Alliaria petiolata [ALLIPET]
*__Alliaria petiolata__ (Bieb.) Cavara & Grande [ALLIPET] (garlic mustard)
SY: *Alliaria alliaria* (L.) Britt.
SY: *Alliaria officinalis* Andrz. ex Bieb.
SY: *Erysimum alliaria* L.
SY: *Sisymbrium alliaria* (L.) Scop.
Alyssum L. [ALYSSUM$]
*__Alyssum alyssoides__ (L.) L. [ALYSALY] (pale alyssum)
SY: *Alyssum calycinum* L.
SY: *Clypeola alyssoides* L.
Alyssum calycinum L. = Alyssum alyssoides [ALYSALY]
*__Alyssum desertorum__ Stapf [ALYSDES] (desert alyssum)
Alyssum incanum L. = Berteroa incana [BERTINC]
Alyssum maritimum (L.) Lam. = Lobularia maritima [LOBUMAR]
*__Alyssum murale__ Waldst. & Kit. [ALYSMUR] (wall alyssum)
Aphragmus Andrz. ex DC. [APHRAGM$]
Aphragmus eschscholtzianus Andrz. ex DC. [APHRESC] (Eschscholtz's little nightmare)
Arabidopsis Heynh. [ARABIDO$]

Arabidopsis glauca (Nutt.) Rydb. = Arabidopsis salsuginea [ARABSAL]
Arabidopsis novae-angliae (Rydb.) Britt. = Braya humilis [BRAYHUM]
Arabidopsis richardsonii Rydb. = Braya humilis [BRAYHUM]
Arabidopsis salsuginea (Pallas) N. Busch [ARABSAL]
SY: *Arabidopsis glauca* (Nutt.) Rydb.
SY: *Sisymbrium salsugineum* Pallas
SY: *Thellungiella salsuginea* (Pallas) O.E. Schulz
***Arabidopsis thaliana** (L.) Heynh. [ARABTHA] (mouse-ear cress)
SY: *Arabis thaliana* L.
SY: *Sisymbrium thalianum* (L.) J. Gay & Monn.
Arabis L. [ARABIS$]
Arabis acutina Greene = Arabis × divaricarpa [ARABDIV]
Arabis arcuata var. *secunda* (T.J. Howell) Robertson = Arabis holboellii var. retrofracta [ARABHOL5]
Arabis bourgovii Rydb. = Arabis holboellii var. collinsii [ARABHOL1]
Arabis brachycarpa (Torr. & Gray) Britt. = Arabis drummondii [ARABDRU]
Arabis bridgeri M.E. Jones = Arabis nuttallii [ARABNUT]
Arabis caduca A. Nels. = Arabis holboellii var. retrofracta [ARABHOL5]
Arabis collinsii Fern. = Arabis holboellii var. collinsii [ARABHOL1]
Arabis confinis S. Wats. = Arabis drummondii [ARABDRU]
Arabis confinis var. *interposita* (Greene) Welsh & Reveal = Arabis × divaricarpa [ARABDIV]
Arabis connexa Greene = Arabis drummondii [ARABDRU]
Arabis consanguinea Greene = Arabis holboellii var. retrofracta [ARABHOL5]
Arabis dacotica Greene = Arabis holboellii var. collinsii [ARABHOL1]
Arabis × divaricarpa A. Nels. (pro sp.) [ARABDIV]
SY: *Arabis acutina* Greene
SY: *Arabis confinis* var. *interposita* (Greene) Welsh & Reveal
SY: *Arabis divaricarpa* A. Nels.
SY: *Arabis divaricarpa* var. *dechamplainii* Boivin
SY: *Arabis divaricarpa* var. *interposita* (Greene) Rollins
SY: *Arabis divaricarpa* var. *stenocarpa* M. Hopkins
SY: *Arabis divaricarpa* var. *typica* Rollins
SY: *Arabis interposita* Greene
SY: *Boechera divaricarpa* (A. Nels.) A. & D. Löve
Arabis divaricarpa A. Nels. = Arabis × divaricarpa [ARABDIV]
Arabis divaricarpa var. *dacotica* (Greene) Boivin = Arabis holboellii var. collinsii [ARABHOL1]
Arabis divaricarpa var. *dechamplainii* Boivin = Arabis × divaricarpa [ARABDIV]
Arabis divaricarpa var. *interposita* (Greene) Rollins = Arabis × divaricarpa [ARABDIV]
Arabis divaricarpa var. *stenocarpa* M. Hopkins = Arabis × divaricarpa [ARABDIV]
Arabis divaricarpa var. *typica* Rollins = Arabis × divaricarpa [ARABDIV]
Arabis drummondii Gray [ARABDRU] (Drummond's rockcress)
SY: *Arabis brachycarpa* (Torr. & Gray) Britt.
SY: *Arabis confinis* S. Wats.
SY: *Arabis connexa* Greene
SY: *Arabis drummondii* var. *connexa* (Greene) Fern.
SY: *Arabis drummondii* var. *oxyphylla* (Greene) M. Hopkins
SY: *Arabis oxyphylla* Greene
SY: *Boechera drummondii* (Gray) A. & D. Löve
SY: *Erysimum drummondii* (Gray) Kuntze
SY: *Turritis drummondii* (Gray) Lunell
Arabis drummondii var. *connexa* (Greene) Fern. = Arabis drummondii [ARABDRU]
Arabis drummondii var. *oxyphylla* (Greene) M. Hopkins = Arabis drummondii [ARABDRU]
Arabis exilis A. Nels. = Arabis holboellii var. retrofracta [ARABHOL5]
Arabis glabra (L.) Bernh. [ARABGLA] (tower mustard)
SY: *Turritis glabra* L.
Arabis hirsuta (L.) Scop. [ARABHIR] (hairy rockcress)
Arabis hirsuta ssp. *eschscholtziana* (Andrz.) Hultén = Arabis hirsuta var. eschscholtziana [ARABHIR5]
Arabis hirsuta ssp. *pycnocarpa* (M. Hopkins) Hultén = Arabis hirsuta var. pycnocarpa [ARABHIR7]
Arabis hirsuta var. **eschscholtziana** (Andrz.) Rollins [ARABHIR5]
SY: *Arabis hirsuta* ssp. *eschscholtziana* (Andrz.) Hultén
Arabis hirsuta var. **glabrata** Torr. & Gray [ARABHIR6]
Arabis hirsuta var. **pycnocarpa** (M. Hopkins) Rollins [ARABHIR7]
SY: *Arabis hirsuta* ssp. *pycnocarpa* (M. Hopkins) Hultén
SY: *Arabis pycnocarpa* M. Hopkins
Arabis holboellii Hornem. [ARABHOL] (Holboell's rockcress)
Arabis holboellii var. **collinsii** (Fern.) Rollins [ARABHOL1]
SY: *Arabis bourgovii* Rydb.
SY: *Arabis collinsii* Fern.
SY: *Arabis dacotica* Greene
SY: *Arabis divaricarpa* var. *dacotica* (Greene) Boivin
SY: *Arabis retrofracta* var. *collinsii* (Fern.) Boivin
SY: *Boechera collinsii* (Fern.) A. & D. Löve
Arabis holboellii var. **holboellii** [ARABHOL0]
SY: *Boechera holboellii* (Hornem.) A. & D. Löve
Arabis holboellii var. **pendulocarpa** (A. Nels.) Rollins [ARABHOL3]
SY: *Arabis pendulocarpa* A. Nels.
Arabis holboellii var. **pinetorum** (Tidestrom) Rollins [ARABHOL4]
SY: *Arabis pinetorum* Tidestrom
Arabis holboellii var. **retrofracta** (Graham) Rydb. [ARABHOL5]
SY: *Arabis arcuata* var. *secunda* (T.J. Howell) Robertson
SY: *Arabis caduca* A. Nels.
SY: *Arabis consanguinea* Greene
SY: *Arabis exilis* A. Nels.
SY: *Arabis holboellii* var. *secunda* (T.J. Howell) Jepson
SY: *Arabis holboellii* var. *tenuis* Böcher
SY: *Arabis kochii* Blank.
SY: *Arabis lignipes* A. Nels.
SY: *Arabis mcdougalii* Rydb.
SY: *Arabis polyantha* Greene
SY: *Arabis retrofracta* Graham
SY: *Arabis retrofracta* var. *multicaulis* Boivin
SY: *Arabis rhodanthus* Greene
SY: *Arabis secunda* T.J. Howell
SY: *Arabis tenuis* Greene
SY: *Boechera retrofracta* (Graham) A. & D. Löve
SY: *Boechera tenuis* (Böcher) A. & D. Löve
Arabis holboellii var. *secunda* (T.J. Howell) Jepson = Arabis holboellii var. retrofracta [ARABHOL5]
Arabis holboellii var. *tenuis* Böcher = Arabis holboellii var. retrofracta [ARABHOL5]
Arabis hookeri Lange = Halimolobos mollis [HALIMOI]
Arabis interposita Greene = Arabis × divaricarpa [ARABDIV]
Arabis kochii Blank. = Arabis holboellii var. retrofracta [ARABHOL5]
Arabis lemmonii S. Wats. [ARABLEM] (Lemmon's rockcress)
Arabis lignipes A. Nels. = Arabis holboellii var. retrofracta [ARABHOL5]
Arabis lyallii S. Wats. [ARABLYA] (Lyall's rockcress)
Arabis lyrata L. [ARABLYR] (lyre-leaved rockcress)
Arabis lyrata ssp. *kamchatica* (Fisch. ex DC.) Hultén = Arabis lyrata var. kamchatica [ARABLYR2]
Arabis lyrata var. **kamchatica** Fisch. ex DC. [ARABLYR2] (Kamchatka lyre-leaved rockcress)
SY: *Arabis lyrata* ssp. *kamchatica* (Fisch. ex DC.) Hultén
SY: *Cardaminopsis kamchatica* (Fisch. ex DC.) O.E. Schulz
Arabis macella Piper = Arabis nuttallii [ARABNUT]
Arabis mcdougalii Rydb. = Arabis holboellii var. retrofracta [ARABHOL5]
Arabis microphylla Nutt. [ARABMIC] (small-leaved rockcress)
SY: *Arabis tenuicula* Greene
Arabis nuttallii B.L. Robins. [ARABNUT] (Nuttall's rockcress)
SY: *Arabis bridgeri* M.E. Jones
SY: *Arabis macella* Piper
Arabis oxyphylla Greene = Arabis drummondii [ARABDRU]
Arabis pendulocarpa A. Nels. = Arabis holboellii var. pendulocarpa [ARABHOL3]

Arabis pinetorum Tidestrom = Arabis holboellii var. pinetorum [ARABHOL4]
Arabis polyantha Greene = Arabis holboellii var. retrofracta [ARABHOL5]
Arabis pycnocarpa M. Hopkins = Arabis hirsuta var. pycnocarpa [ARABHIR7]
Arabis retrofracta Graham = Arabis holboellii var. retrofracta [ARABHOL5]
Arabis retrofracta var. *collinsii* (Fern.) Boivin = Arabis holboellii var. collinsii [ARABHOL1]
Arabis retrofracta var. *multicaulis* Boivin = Arabis holboellii var. retrofracta [ARABHOL5]
Arabis rhodanthus Greene = Arabis holboellii var. retrofracta [ARABHOL5]
Arabis secunda T.J. Howell = Arabis holboellii var. retrofracta [ARABHOL5]
Arabis sparsiflora Nutt. [ARABSPA] (sickle-pod rockcress)
Arabis tenuicula Greene = Arabis microphylla [ARABMIC]
Arabis tenuis Greene = Arabis holboellii var. retrofracta [ARABHOL5]
Arabis thaliana L. = Arabidopsis thaliana [ARABTHA]
Arabis whitedii Piper = Halimolobos whitedii [HALIWHI]
Armoracia P.G. Gaertn., B. Mey. & Scherb. [ARMORAC$]
Armoracia armoracia (L.) Britt. = Armoracia rusticana [ARMORUS]
Armoracia lapathifolia Gilib. = Armoracia rusticana [ARMORUS]
Armoracia rusticana P.G. Gaertn., B. Mey. & Scherb. [ARMORUS] (common horseradish)
SY: *Armoracia armoracia* (L.) Britt.
SY: *Armoracia lapathifolia* Gilib.
SY: *Cochlearia armoracia* L.
SY: *Radicula armoracia* (L.) B.L. Robins.
SY: *Rorippa armoracia* (L.) A.S. Hitchc.
Athysanus Greene [ATHYSAN$]
Athysanus pusillus (Hook.) Greene [ATHYPUS] (common sandweed)
SY: *Athysanus pusillus* var. *glabrior* S. Wats.
SY: *Thysanocarpus pusillus* Hook.
Athysanus pusillus var. *glabrior* S. Wats. = Athysanus pusillus [ATHYPUS]
Barbarea Ait. f. [BARBARE$]
Barbarea americana Rydb. = Barbarea orthoceras [BARBORT]
Barbarea arcuata (Opiz ex J.& K. Presl) Reichenb. = Barbarea vulgaris [BARBVUL]
Barbarea orthoceras Ledeb. [BARBORT] (American winter cress)
SY: *Barbarea americana* Rydb.
SY: *Barbarea orthoceras* var. *dolichocarpa* Fern.
Barbarea orthoceras var. *dolichocarpa* Fern. = Barbarea orthoceras [BARBORT]
***Barbarea verna** (P. Mill.) Aschers. [BARBVER] (early winter cress)
SY: *Campe verna* (P. Mill.) Heller
SY: *Erysimum vernum* P. Mill.
***Barbarea vulgaris** Ait. f. [BARBVUL] (bitter winter cress)
SY: *Barbarea arcuata* (Opiz ex J.& K. Presl) Reichenb.
SY: *Barbarea vulgaris* var. *arcuata* (Opiz ex J.& K. Presl) Fries
SY: *Barbarea vulgaris* var. *brachycarpa* Rouy & Foucaud
SY: *Barbarea vulgaris* var. *longisiliquosa* Carion
SY: *Barbarea vulgaris* var. *sylvestris* Fries
SY: *Campe barbarea* (L.) W. Wight ex Piper
Barbarea vulgaris var. *arcuata* (Opiz ex J.& K. Presl) Fries = Barbarea vulgaris [BARBVUL]
Barbarea vulgaris var. *brachycarpa* Rouy & Foucaud = Barbarea vulgaris [BARBVUL]
Barbarea vulgaris var. *longisiliquosa* Carion = Barbarea vulgaris [BARBVUL]
Barbarea vulgaris var. *sylvestris* Fries = Barbarea vulgaris [BARBVUL]
Berteroa DC. [BERTERO$]
***Berteroa incana** (L.) DC. [BERTINC] (hoary alyssum)
SY: *Alyssum incanum* L.
Boechera collinsii (Fern.) A. & D. Löve = Arabis holboellii var. collinsii [ARABHOL1]
Boechera divaricarpa (A. Nels.) A. & D. Löve = Arabis × divaricarpa [ARABDIV]
Boechera drummondii (Gray) A. & D. Löve = Arabis drummondii [ARABDRU]
Boechera holboellii (Hornem.) A. & D. Löve = Arabis holboellii var. holboellii [ARABHOL0]
Boechera retrofracta (Graham) A. & D. Löve = Arabis holboellii var. retrofracta [ARABHOL5]
Boechera tenuis (Böcher) A. & D. Löve = Arabis holboellii var. retrofracta [ARABHOL5]
Brassica L. [BRASSIC$]
Brassica alba Rabenh. = Sinapis alba [SINAALB]
Brassica arvensis Rabenh. = Sinapis arvensis [SINAARV]
Brassica campestris L. = Brassica rapa [BRASRAP]
Brassica campestris var. *rapa* (L.) Hartman = Brassica rapa [BRASRAP]
Brassica erucastrum L. = Erucastrum gallicum [ERUCGAL]
Brassica hirta Moench = Sinapis alba [SINAALB]
Brassica integrifolia (Vahl) Schulz = Brassica juncea [BRASJUN]
Brassica integrifolia Rupr. = Brassica juncea [BRASJUN]
Brassica japonica Thunb. = Brassica juncea [BRASJUN]
***Brassica juncea** (L.) Czern. [BRASJUN] (Indian mustard)
SY: *Brassica integrifolia* (Vahl) Schulz
SY: *Brassica integrifolia* Rupr.
SY: *Brassica japonica* Thunb.
SY: *Brassica juncea* var. *crispifolia* Bailey
SY: *Brassica juncea* var. *japonica* (Thunb.) Bailey
SY: *Brassica willdenowii* Boiss.
SY: *Sinapis juncea* L.
Brassica juncea var. *crispifolia* Bailey = Brassica juncea [BRASJUN]
Brassica juncea var. *japonica* (Thunb.) Bailey = Brassica juncea [BRASJUN]
Brassica kaber (DC.) L.C. Wheeler = Sinapis arvensis [SINAARV]
Brassica kaber var. *pinnatifida* (Stokes) L.C. Wheeler = Sinapis arvensis [SINAARV]
Brassica kaber var. *schkuhriana* (Reichenb.) L.C. Wheeler = Sinapis arvensis [SINAARV]
Brassica napobrassica (L.) P. Mill. = Brassica napus [BRASNAP]
***Brassica napus** L. [BRASNAP] (turnip)
SY: *Brassica napobrassica* (L.) P. Mill.
SY: *Brassica napus* var. *napobrassica* (L.) Reichenb.
Brassica napus var. *napobrassica* (L.) Reichenb. = Brassica napus [BRASNAP]
***Brassica nigra** (L.) W.D.J. Koch [BRASNIG] (black mustard)
SY: *Sinapis nigra* L.
Brassica orientalis L. = Conringia orientalis [CONRORI]
***Brassica rapa** L. [BRASRAP] (bird rape mustard)
SY: *Brassica campestris* L.
SY: *Brassica campestris* var. *rapa* (L.) Hartman
SY: *Brassica rapa* ssp. *campestris* (L.) Clapham
SY: *Brassica rapa* var. *campestris* (L.) W.D.J. Koch
SY: *Brassica rapa* ssp. *olifera* DC.
SY: *Brassica rapa* ssp. *sylvestris* Janchen
SY: *Caulanthus sulfureus* Payson
Brassica rapa ssp. *campestris* (L.) Clapham = Brassica rapa [BRASRAP]
Brassica rapa ssp. *olifera* DC. = Brassica rapa [BRASRAP]
Brassica rapa ssp. *sylvestris* Janchen = Brassica rapa [BRASRAP]
Brassica rapa var. *campestris* (L.) W.D.J. Koch = Brassica rapa [BRASRAP]
Brassica willdenowii Boiss. = Brassica juncea [BRASJUN]
Braya Sternb. & Hoppe [BRAYA$]
Braya alpina var. *americana* (Hook.) S. Wats. = Braya glabella [BRAYGLA]
Braya alpina var. *glabella* (Richards.) S. Wats. = Braya glabella [BRAYGLA]
Braya americana (Hook.) Fern. = Braya glabella [BRAYGLA]
Braya arctica Hook. = Braya glabella [BRAYGLA]
Braya bartlettiana Jordal = Braya glabella [BRAYGLA]
Braya glabella Richards. [BRAYGLA]
SY: *Braya alpina* var. *americana* (Hook.) S. Wats.
SY: *Braya alpina* var. *glabella* (Richards.) S. Wats.
SY: *Braya americana* (Hook.) Fern.
SY: *Braya arctica* Hook.
SY: *Braya bartlettiana* Jordal

SY: *Braya henryae* Raup
SY: *Braya humilis* var. *americana* (Hook.) Boivin
SY: *Braya humilis* var. *glabella* (Richards.) Boivin
Braya henryae Raup = Braya glabella [BRAYGLA]
Braya humilis (C.A. Mey.) B.L. Robins. [BRAYHUM] (dwarf braya)
SY: *Arabidopsis novae-angliae* (Rydb.) Britt.
SY: *Arabidopsis richardsonii* Rydb.
SY: *Braya humilis* ssp. *abbei* Böcher
SY: *Braya humilis* var. *abbei* (Böcher) Boivin
SY: *Braya humilis* ssp. *arctica* (Böcher) Rollins
SY: *Braya humilis* var. *arctica* (Böcher) Boivin
SY: *Braya humilis* var. *interior* (Böcher) Boivin
SY: *Braya humilis* var. *laurentiana* (Böcher) Boivin
SY: *Braya humilis* var. *leiocarpa* (Trautv.) Fern.
SY: *Braya humilis* var. *novae-angliae* (Rydb.) Fern.
SY: *Braya humilis* ssp. *richardsonii* (Rydb.) Hultén
SY: *Braya humilis* ssp. *ventosa* Rollins
SY: *Braya humilis* var. *ventosa* (Rollins) Boivin
SY: *Braya intermedia* Sorensen
SY: *Braya novae-angliae* (Rydb.) Sorensen
SY: *Braya novae-angliae* var. *interior* Böcher
SY: *Braya novae-angliae* var. *laurentiana* Böcher
SY: *Braya richardsonii* (Rydb.) Fern.
SY: *Pilosella novae-angliae* Rydb.
SY: *Pilosella richardsonii* Rydb.
SY: *Sisymbrium humile* C.A. Mey.
SY: *Torularia arctica* (Böcher) A. & D. Löve
SY: *Torularia humilis* (C.A. Mey.) O.E. Schulz
SY: *Torularia humilis* ssp. *arctica* Böcher
Braya humilis ssp. *abbei* Böcher = Braya humilis [BRAYHUM]
Braya humilis ssp. *arctica* (Böcher) Rollins = Braya humilis [BRAYHUM]
Braya humilis ssp. *richardsonii* (Rydb.) Hultén = Braya humilis [BRAYHUM]
Braya humilis ssp. *ventosa* Rollins = Braya humilis [BRAYHUM]
Braya humilis var. *abbei* (Böcher) Boivin = Braya humilis [BRAYHUM]
Braya humilis var. *americana* (Hook.) Boivin = Braya glabella [BRAYGLA]
Braya humilis var. *arctica* (Böcher) Boivin = Braya humilis [BRAYHUM]
Braya humilis var. *glabella* (Richards.) Boivin = Braya glabella [BRAYGLA]
Braya humilis var. *interior* (Böcher) Boivin = Braya humilis [BRAYHUM]
Braya humilis var. *laurentiana* (Böcher) Boivin = Braya humilis [BRAYHUM]
Braya humilis var. *leiocarpa* (Trautv.) Fern. = Braya humilis [BRAYHUM]
Braya humilis var. *novae-angliae* (Rydb.) Fern. = Braya humilis [BRAYHUM]
Braya humilis var. *ventosa* (Rollins) Boivin = Braya humilis [BRAYHUM]
Braya intermedia Sorensen = Braya humilis [BRAYHUM]
Braya novae-angliae (Rydb.) Sorensen = Braya humilis [BRAYHUM]
Braya novae-angliae var. *interior* Böcher = Braya humilis [BRAYHUM]
Braya novae-angliae var. *laurentiana* Böcher = Braya humilis [BRAYHUM]
Braya purpurascens (R. Br.) Bunge [BRAYPUR] (purple braya)
Braya richardsonii (Rydb.) Fern. = Braya humilis [BRAYHUM]
Bursa bursa-pastoris (L.) Britt. = Capsella bursa-pastoris [CAPSBUR]
Bursa bursa-pastoris var. *bifida* Crépin = Capsella bursa-pastoris [CAPSBUR]
Bursa gracilis Gren. = Capsella bursa-pastoris [CAPSBUR]
Bursa rubella Reut. = Capsella bursa-pastoris [CAPSBUR]
Cakile P. Mill. [CAKILE$]
Cakile cakile (L.) Karst. = Cakile maritima [CAKIMAR]
Cakile edentula (Bigelow) Hook. [CAKIEDE] (American searocket)
SY: *Cakile edentula* var. *californica* (Heller) Fern.
Cakile edentula var. *californica* (Heller) Fern. = Cakile edentula [CAKIEDE]
***Cakile maritima** Scop. [CAKIMAR] (European searocket)
SY: *Cakile cakile* (L.) Karst.
Camelina Crantz [CAMELIN$]
***Camelina microcarpa** DC. [CAMEMIC] (littlepod flax)
SY: *Camelina sativa* ssp. *microcarpa* (DC.) E. Schmid
***Camelina sativa** (L.) Crantz [CAMESAT] (falseflax)
Camelina sativa ssp. *microcarpa* (DC.) E. Schmid = Camelina microcarpa [CAMEMIC]
Campe barbarea (L.) W. Wight ex Piper = Barbarea vulgaris [BARBVUL]
Campe verna (P. Mill.) Heller = Barbarea verna [BARBVER]
Capsella Medik. [CAPSELL$]
***Capsella bursa-pastoris** (L.) Medik. [CAPSBUR] (shepherd's purse)
SY: *Bursa bursa-pastoris* (L.) Britt.
SY: *Bursa bursa-pastoris* var. *bifida* Crépin
SY: *Bursa gracilis* Gren.
SY: *Bursa rubella* Reut.
SY: *Thlaspi bursa-pastoris* L.
Carara didyma (L.) Britt. = Coronopus didymus [CORODID]
Cardamine L. [CARDAMI$]
Cardamine angulata Hook. [CARDANG] (angled bitter-cress)
Cardamine barbareifolia DC. = Rorippa barbareifolia [RORIBAR]
Cardamine bellidifolia L. [CARDBEL] (alpine bitter-cress)
SY: *Cardamine bellidifolia* var. *pinnatifida* Hultén
SY: *Cardamine bellidifolia* var. *sinuata* (J. Vahl) Lange
Cardamine bellidifolia var. *pinnatifida* Hultén = Cardamine bellidifolia [CARDBEL]
Cardamine bellidifolia var. *sinuata* (J. Vahl) Lange = Cardamine bellidifolia [CARDBEL]
Cardamine breweri S. Wats. [CARDBRE] (Brewer's bitter-cress)
Cardamine breweri var. **breweri** [CARDBRE0]
SY: *Cardamine hederifolia* Greene
SY: *Cardamine oregona* Piper
Cardamine breweri var. **orbicularis** (Greene) Detling [CARDBRE2]
SY: *Cardamine orbicularis* Greene
Cardamine cordifolia var. *lyallii* (S. Wats.) A. Nels. & J.F. Macbr. = Cardamine lyallii [CARDLYA]
Cardamine hederifolia Greene = Cardamine breweri var. breweri [CARDBRE0]
***Cardamine hirsuta** L. [CARDHIR] (hairy bitter-cress)
Cardamine hirsuta var. *kamtschatica* (Regel) O.E. Schulz = Cardamine oligosperma var. kamtschatica [CARDOLI1]
Cardamine kamtschatica (Regel) Piper = Cardamine oligosperma var. kamtschatica [CARDOLI1]
Cardamine lyallii S. Wats. [CARDLYA]
SY: *Cardamine cordifolia* var. *lyallii* (S. Wats.) A. Nels. & J.F. Macbr.
Cardamine nuttallii Greene [CARDNUT] (Nuttall's bitter-cress)
SY: *Cardamine pulcherrima* var. *tenella* (Pursh) C.L. Hitchc.
SY: *Dentaria tenella* Pursh
SY: *Dentaria tenella* var. *palmata* Detling
Cardamine nymanii Gandog. = Cardamine pratensis var. angustifolia [CARDPRA2]
Cardamine occidentalis (S. Wats. ex B.L. Robins.) T.J. Howell [CARDOCC] (western bitter-cress)
SY: *Cardamine pratensis* var. *occidentalis* S. Wats. ex B.L. Robins.
Cardamine oligosperma Nutt. [CARDOLI] (little western bitter-cress)
Cardamine oligosperma var. **kamtschatica** (Regel) Detling [CARDOLI1]
SY: *Cardamine hirsuta* var. *kamtschatica* (Regel) O.E. Schulz
SY: *Cardamine kamtschatica* (Regel) Piper
SY: *Cardamine umbellata* Greene
Cardamine orbicularis Greene = Cardamine breweri var. orbicularis [CARDBRE2]
Cardamine oregona Piper = Cardamine breweri var. breweri [CARDBRE0]
Cardamine parviflora L. [CARDPAR] (small-flowered bitter-cress)
Cardamine pensylvanica Muhl. ex Willd. [CARDPEN] (Pennsylvanian bitter-cress)

SY: *Cardamine pensylvanica* var. *brittoniana* Farw.
Cardamine pensylvanica var. *brittoniana* Farw. = Cardamine pensylvanica [CARDPEN]
Cardamine pratensis L. [CARDPRA] (cuckoo bitter-cress)
Cardamine pratensis ssp. *angustifolia* (Hook.) O.E. Schulz = Cardamine pratensis var. angustifolia [CARDPRA2]
Cardamine pratensis var. **angustifolia** Hook. [CARDPRA2]
SY: *Cardamine nymanii* Gandog.
SY: *Cardamine pratensis* ssp. *angustifolia* (Hook.) O.E. Schulz
Cardamine pratensis var. *occidentalis* S. Wats. ex B.L. Robins. = Cardamine occidentalis [CARDOCC]
Cardamine pulcherrima var. *tenella* (Pursh) C.L. Hitchc. = Cardamine nuttallii [CARDNUT]
Cardamine umbellata Greene = Cardamine oligosperma var. kamtschatica [CARDOLI1]
Cardaminopsis kamchatica (Fisch. ex DC.) O.E. Schulz = Arabis lyrata var. kamchatica [ARABLYR2]
Cardaria Desv. [CARDARI$]
Cardaria chalapensis (L.) Hand.-Maz. = Cardaria draba ssp. chalapensis [CARDDRA1]
*__Cardaria draba__ (L.) Desv. [CARDDRA] (heart-podded hoary-cress)
*__Cardaria draba__ ssp. **chalapensis** (L.) O.E. Schulz [CARDDRA1]
SY: *Cardaria chalapensis* (L.) Hand.-Maz.
SY: *Cardaria draba* var. *repens* (Schrenk) O.E. Schulz
SY: *Lepidium repens* (Schrenk) Boiss.
Cardaria draba var. *repens* (Schrenk) O.E. Schulz = Cardaria draba ssp. chalapensis [CARDDRA1]
*__Cardaria pubescens__ (C.A. Mey.) Jarmolenko [CARDPUB] (globe-pod hoary-cress)
SY: *Cardaria pubescens* var. *elongata* Rollins
SY: *Hymenophysa pubescens* C.A. Mey.
Cardaria pubescens var. *elongata* Rollins = Cardaria pubescens [CARDPUB]
Caulanthus sulfureus Payson = Brassica rapa [BRASRAP]
Cheiranthus cheiri L. = Erysimum cheiri [ERYSCHI]
Cheirinia cheiranthoides (L.) Link = Erysimum cheiranthoides [ERYSCHE]
Chorispora DC. [CHORISP$]
*__Chorispora tenella__ (Pallas) DC. [CHORTEN] (blue mustard)
Clypeola alyssoides L. = Alyssum alyssoides [ALYSALY]
Clypeola maritima L. = Lobularia maritima [LOBUMAR]
Cochlearia L. [COCHLEA$]
Cochlearia armoracia L. = Armoracia rusticana [ARMORUS]
Cochlearia groenlandica L. [COCHGRO]
SY: *Cochlearia officinalis* ssp. *arctica* (Schlecht.) Hultén
SY: *Cochlearia officinalis* var. *arctica* (Schlecht.) Gelert ex Anders. & Hessel
SY: *Cochlearia officinalis* ssp. *groenlandica* (L.) Porsild
SY: *Cochlearia officinalis* ssp. *oblongifolia* (DC.) Hultén
SY: *Cochleariopsis groenlandica* (L.) A. & D. Löve
SY: *Cochleariopsis groenlandica* ssp. *arctica* (Schlecht.) A. & D. Löve
SY: *Cochleariopsis groenlandica* ssp. *oblongifolia* (DC.) A. & D. Löve
Cochlearia officinalis ssp. *arctica* (Schlecht.) Hultén = Cochlearia groenlandica [COCHGRO]
Cochlearia officinalis ssp. *groenlandica* (L.) Porsild = Cochlearia groenlandica [COCHGRO]
Cochlearia officinalis ssp. *oblongifolia* (DC.) Hultén = Cochlearia groenlandica [COCHGRO]
Cochlearia officinalis var. *arctica* (Schlecht.) Gelert ex Anders. & Hessel = Cochlearia groenlandica [COCHGRO]
Cochleariopsis groenlandica (L.) A. & D. Löve = Cochlearia groenlandica [COCHGRO]
Cochleariopsis groenlandica ssp. *arctica* (Schlecht.) A. & D. Löve = Cochlearia groenlandica [COCHGRO]
Cochleariopsis groenlandica ssp. *oblongifolia* (DC.) A. & D. Löve = Cochlearia groenlandica [COCHGRO]
Conringia Heister ex Fabr. [CONRING$]
*__Conringia orientalis__ (L.) Andrz. [CONRORI] (hare's-ear mustard)
SY: *Brassica orientalis* L.
Coronopus Zinn [CORONOP$]
*__Coronopus didymus__ (L.) Sm. [CORODID] (lesser swine-cress)
SY: *Carara didyma* (L.) Britt.
SY: *Lepidium didymum* L.
Dentaria tenella Pursh = Cardamine nuttallii [CARDNUT]
Dentaria tenella var. *palmata* Detling = Cardamine nuttallii [CARDNUT]
Descurainia Webb & Berth. [DESCURA$]
Descurainia incana (Bernh. ex Fisch. & C.A. Mey.) Dorn [DESCINC]
SY: *Descurainia incana* var. *major* (Hook.) Dorn
SY: *Descurainia richardsonii* O.E. Schulz
SY: *Sisymbrium incanum* Bernh. ex Fisch. & C.A. Mey.
SY: *Sophia richardsonii* (O.E. Schulz) Rydb.
Descurainia incana ssp. **viscosa** (Rydb.) Kartesz & Gandhi [DESCINC1]
SY: *Descurainia incana* var. *viscosa* (Rydb.) Dorn
SY: *Descurainia richardsonii* ssp. *viscosa* (Rydb.) Detling
SY: *Descurainia richardsonii* var. *viscosa* (Rydb.) M.E. Peck
SY: *Sisymbrium viscosum* (Rydb.) Blank.
SY: *Sophia viscosa* Rydb.
Descurainia incana var. *major* (Hook.) Dorn = Descurainia incana [DESCINC]
Descurainia incana var. *viscosa* (Rydb.) Dorn = Descurainia incana ssp. viscosa [DESCINC1]
Descurainia intermedia (Rydb.) Daniels = Descurainia pinnata ssp. intermedia [DESCPIN2]
Descurainia pinnata (Walt.) Britt. [DESCPIN] (western tansymustard)
Descurainia pinnata ssp. **filipes** (Gray) Detling [DESCPIN1]
SY: *Descurainia pinnata* var. *filipes* (Gray) M.E. Peck
SY: *Sisymbrium incisum* var. *filipes* Gray
SY: *Sisymbrium longipedicellatum* Fourn.
SY: *Sophia filipes* (Gray) Heller
Descurainia pinnata ssp. **intermedia** (Rydb.) Detling [DESCPIN2]
SY: *Descurainia intermedia* (Rydb.) Daniels
SY: *Descurainia pinnata* var. *intermedia* (Rydb.) C.L. Hitchc.
SY: *Sophia intermedia* Rydb.
Descurainia pinnata var. *filipes* (Gray) M.E. Peck = Descurainia pinnata ssp. filipes [DESCPIN1]
Descurainia pinnata var. *intermedia* (Rydb.) C.L. Hitchc. = Descurainia pinnata ssp. intermedia [DESCPIN2]
Descurainia richardsonii O.E. Schulz = Descurainia incana [DESCINC]
Descurainia richardsonii ssp. *viscosa* (Rydb.) Detling = Descurainia incana ssp. viscosa [DESCINC1]
Descurainia richardsonii var. *viscosa* (Rydb.) M.E. Peck = Descurainia incana ssp. viscosa [DESCINC1]
*__Descurainia sophia__ (L.) Webb ex Prantl [DESCSOP] (flixweed)
SY: *Sisymbrium sophia* L.
SY: *Sophia sophia* (L.) Britt.
Descurainia sophioides (Fisch. ex Hook.) O.E. Schulz [DESCSOH] (northern tansymustard)
SY: *Sisymbrium sophioides* Fisch. ex Hook.
Draba L. [DRABA$]
Draba albertina Greene [DRABALB] (slender whitlow-grass)
SY: *Draba nitida* Greene
SY: *Draba stenoloba* var. *nana* (O.E. Schulz) C.L. Hitchc.
Draba allenii Fern. = Draba lactea [DRABLAC]
Draba alpina L. [DRABALP] (alpine whitlow-grass)
SY: *Draba alpina* var. *nana* Hook.
SY: *Draba alpina* var. *pilosa* (M.F. Adams ex DC.) Regel
SY: *Draba eschscholtzii* Pohle ex N. Busch
SY: *Draba micropetala* Hook.
SY: *Draba pilosa* M.F. Adams ex DC.
Draba alpina var. *nana* Hook. = Draba alpina [DRABALP]
Draba alpina var. *pilosa* (M.F. Adams ex DC.) Regel = Draba alpina [DRABALP]
Draba arabisans var. *canadensis* (Burnet) Fern. & Knowlt. = Draba glabella [DRABGLA]
Draba aurea Vahl ex Hornem. [DRABAUR] (golden whitlow-grass)
SY: *Draba aurea* var. *leiocarpa* (Payson & St. John) C.L. Hitchc.
SY: *Draba aurea* var. *neomexicana* (Greene) Tidestrom
SY: *Draba minganensis* (Victorin) Fern.
SY: *Draba neomexicana* Greene

Draba aurea var. *leiocarpa* (Payson & St. John) C.L. Hitchc. = Draba aurea [DRABAUR]
Draba aurea var. *neomexicana* (Greene) Tidestrom = Draba aurea [DRABAUR]
Draba bellii Holm = Draba corymbosa [DRABCOR]
Draba borealis DC. [DRABBOR] (northern whitlow-grass)
SY: *Draba mccallae* Rydb.
Draba caeruleomontana Payson & St. John = Draba densifolia [DRABDEN]
Draba cana Rydb. [DRABCAN] (lance-leaved whitlow-grass)
SY: *Draba stylaris* J. Gay ex W.D.J. Koch
Draba caroliniana Walt. = Draba reptans [DRABREP]
Draba cascadensis Payson & St. John = Draba praealta [DRABPRA]
Draba cinerea M.F. Adams [DRABCIN] (gray-leaved whitlow-grass)
Draba corymbosa R. Br. ex DC. [DRABCOR] (Baffin's Bay whitlow-grass)
SY: *Draba bellii* Holm
Draba crassifolia Graham [DRABCRA] (thick-leaved whitlow-grass)
Draba densifolia Nutt. [DRABDEN] (Nuttall's whitlow-grass)
SY: *Draba caeruleomontana* Payson & St. John
SY: *Draba nelsonii* J.F. Macbr. & Payson
SY: *Draba sphaerula* J.F. Macbr. & Payson
Draba eschscholtzii Pohle ex N. Busch = Draba alpina [DRABALP]
Draba exalata Ekman = Draba ruaxes [DRABRUA]
Draba fladnizensis Wulfen [DRABFLA] (Austrian whitlow-grass)
SY: *Draba tschuktschorum* Trautv.
Draba fladnizensis var. *heterotricha* (Lindbl.) J. Ball = Draba lactea [DRABLAC]
Draba glabella Pursh [DRABGLA] (smooth whitlow-grass)
SY: *Draba arabisans* var. *canadensis* (Burnet) Fern. & Knowlt.
Draba hatchiae Mulligan = Draba hyperborea [DRABHYP]
Draba hyperborea (L.) Desv. [DRABHYP] (North Pacific whitlow-grass)
SY: *Draba hatchiae* Mulligan
Draba incerta Payson [DRABINC] (Yellowstone whitlow-grass)
SY: *Draba laevicapsula* Payson
SY: *Draba peasei* Fern.
Draba juniperina Dorn = Draba oligosperma [DRABOLI]
Draba kananaskis Mulligan = Draba longipes [DRABLOG]
Draba lactea M.F. Adams [DRABLAC] (milky whitlow-grass)
SY: *Draba allenii* Fern.
SY: *Draba fladnizensis* var. *heterotricha* (Lindbl.) J. Ball
Draba laevicapsula Payson = Draba incerta [DRABINC]
Draba lonchocarpa Rydb. [DRABLON] (lance-fruited whitlow-grass)
Draba lonchocarpa var. *exigua* O.E. Schulz = Draba lonchocarpa var. lonchocarpa [DRABLON0]
Draba lonchocarpa var. **lonchocarpa** [DRABLON0]
SY: *Draba lonchocarpa* var. *exigua* O.E. Schulz
SY: *Draba nivalis* var. *elongata* S. Wats.
SY: *Draba nivalis* var. *exigua* (O.E. Schulz) C.L. Hitchc.
SY: *Draba nivalis* ssp. *lonchocarpa* (Rydb.) Hultén
Draba lonchocarpa var. **thompsonii** (C.L. Hitchc.) Rollins [DRABLON2]
SY: *Draba nivalis* var. *thompsonii* C.L. Hitchc.
Draba lonchocarpa var. **vestita** O.E. Schulz [DRABLON3]
Draba longipes Raup [DRABLOG] (long-stalked whitlow-grass)
SY: *Draba kananaskis* Mulligan
Draba macounii O.E. Schulz [DRABMAC] (Macoun's whitlow-grass)
Draba mccallae Rydb. = Draba borealis [DRABBOR]
Draba micrantha Nutt. = Draba reptans [DRABREP]
Draba micropetala Hook. = Draba alpina [DRABALP]
Draba minganensis (Victorin) Fern. = Draba aurea [DRABAUR]
Draba nelsonii J.F. Macbr. & Payson = Draba densifolia [DRABDEN]
Draba nemorosa L. [DRABNEM] (woods whitlow-grass)
Draba neomexicana Greene = Draba aurea [DRABAUR]
Draba nitida Greene = Draba albertina [DRABALB]
Draba nivalis Lilj. [DRABNIV] (snow whitlow-grass)
Draba nivalis ssp. *lonchocarpa* (Rydb.) Hultén = Draba lonchocarpa var. lonchocarpa [DRABLON0]
Draba nivalis var. *brevicula* Rollins = Draba porsildii [DRABPOR]
Draba nivalis var. *elongata* S. Wats. = Draba lonchocarpa var. lonchocarpa [DRABLON0]
Draba nivalis var. *exigua* (O.E. Schulz) C.L. Hitchc. = Draba lonchocarpa var. lonchocarpa [DRABLON0]
Draba nivalis var. *thompsonii* C.L. Hitchc. = Draba lonchocarpa var. thompsonii [DRABLON2]
Draba oligosperma Hook. [DRABOLI] (few-seeded whitlow-grass)
SY: *Draba juniperina* Dorn
SY: *Draba oligosperma* var. *juniperina* (Dorn) Welsh
SY: *Draba oligosperma* var. *subsessilis* (S. Wats.) O.E. Schulz
SY: *Draba subsessilis* S. Wats.
Draba oligosperma var. *juniperina* (Dorn) Welsh = Draba oligosperma [DRABOLI]
Draba oligosperma var. *subsessilis* (S. Wats.) O.E. Schulz = Draba oligosperma [DRABOLI]
Draba paysonii J.F. Macbr. [DRABPAY] (Payson's whitlow-grass)
Draba paysonii var. **treleasii** (O.E. Schulz) C.L. Hitchc. [DRABPAY1]
Draba peasei Fern. = Draba incerta [DRABINC]
Draba pilosa M.F. Adams ex DC. = Draba alpina [DRABALP]
Draba porsildii Mulligan [DRABPOR] (Porsild's whitlow-grass)
SY: *Draba nivalis* var. *brevicula* Rollins
Draba praealta Greene [DRABPRA] (tall whitlow-grass)
SY: *Draba cascadensis* Payson & St. John
Draba praecox Stev. = Draba verna [DRABVER]
Draba reptans (Lam.) Fern. [DRABREP] (Carolina whitlow-grass)
SY: *Draba caroliniana* Walt.
SY: *Draba micrantha* Nutt.
SY: *Draba reptans* var. *micrantha* (Nutt.) Fern.
SY: *Draba reptans* ssp. *stellifera* (O.E. Schulz) Abrams
SY: *Draba reptans* var. *stellifera* (O.E. Schulz) C.L. Hitchc.
SY: *Draba reptans* var. *typica* C.L. Hitchc.
Draba reptans ssp. *stellifera* (O.E. Schulz) Abrams = Draba reptans [DRABREP]
Draba reptans var. *micrantha* (Nutt.) Fern. = Draba reptans [DRABREP]
Draba reptans var. *stellifera* (O.E. Schulz) C.L. Hitchc. = Draba reptans [DRABREP]
Draba reptans var. *typica* C.L. Hitchc. = Draba reptans [DRABREP]
Draba ruaxes Payson & St. John [DRABRUA] (Coast Mountain whitlow-grass)
SY: *Draba exalata* Ekman
SY: *Draba ventosa* var. *ruaxes* (Payson & St. John) C.L. Hitchc.
Draba sphaerula J.F. Macbr. & Payson = Draba densifolia [DRABDEN]
Draba stenoloba Ledeb. [DRABSTE] (Alaska whitlow-grass)
Draba stenoloba var. *nana* (O.E. Schulz) C.L. Hitchc. = Draba albertina [DRABALB]
Draba stylaris J. Gay ex W.D.J. Koch = Draba cana [DRABCAN]
Draba subsessilis S. Wats. = Draba oligosperma [DRABOLI]
Draba tschuktschorum Trautv. = Draba fladnizensis [DRABFLA]
Draba ventosa var. *ruaxes* (Payson & St. John) C.L. Hitchc. = Draba ruaxes [DRABRUA]
***Draba verna** L. [DRABVER] (common whitlow-grass)
SY: *Draba praecox* Stev.
SY: *Draba verna* var. *aestivalis* Lej.
SY: *Draba verna* var. *boerhaavii* van Hall
SY: *Draba verna* var. *major* Stur
SY: *Erophila spathulata* A.F. Lang
SY: *Erophila verna* (L.) Bess.
SY: *Erophila verna* ssp. *praecox* (Stev.) S.M. Walters
SY: *Erophila verna* ssp. *spathulata* (A.F. Lang) S.M. Walters
Draba verna var. *aestivalis* Lej. = Draba verna [DRABVER]
Draba verna var. *boerhaavii* van Hall = Draba verna [DRABVER]
Draba verna var. *major* Stur = Draba verna [DRABVER]
Erophila spathulata A.F. Lang = Draba verna [DRABVER]
Erophila verna (L.) Bess. = Draba verna [DRABVER]
Erophila verna ssp. *praecox* (Stev.) S.M. Walters = Draba verna [DRABVER]
Erophila verna ssp. *spathulata* (A.F. Lang) S.M. Walters = Draba verna [DRABVER]
Erucastrum K. Presl [ERUCAST$]

*·**Erucastrum gallicum** (Willd.) O.E. Schulz [ERUCGAL] (dog mustard)
SY: *Brassica erucastrum* L.
Erysimum L. [ERYSIMU$]
Erysimum alliaria L. = Alliaria petiolata [ALLIPET]
Erysimum arenicola S. Wats. [ERYSARE] (cascade wallflower)
Erysimum arenicola var. **torulosum** (Piper) C.L. Hitchc. [ERYSARE1] (sand-dwelling wallflower)
SY: *Erysimum torulosum* Piper
Erysimum capitatum (Dougl. ex Hook.) Greene [ERYSCAP]
·**Erysimum cheiranthoides** L. [ERYSCHE] (wormseed mustard)
SY: *Cheirinia cheiranthoides* (L.) Link
SY: *Erysimum cheiranthoides* ssp. *altum* Ahti
Erysimum cheiranthoides ssp. *altum* Ahti = Erysimum cheiranthoides [ERYSCHE]
·**Erysimum cheiri** (L.) Crantz [ERYSCHI] (common wallflower)
SY: *Cheiranthus cheiri* L.
Erysimum drummondii (Gray) Kuntze = Arabis drummondii [ARABDRU]
Erysimum inconspicuum (S. Wats.) MacM. [ERYSINC] (small wallflower)
Erysimum officinale L. = Sisymbrium officinale [SISYOFF]
Erysimum pallasii (Pursh) Fern. [ERYSPAL] (Pallas' wallflower)
Erysimum torulosum Piper = Erysimum arenicola var. torulosum [ERYSARE1]
Erysimum vernum P. Mill. = Barbarea verna [BARBVER]
Eutrema R. Br. [EUTREMA$]
Eutrema edwardsii R. Br. [EUTREDW] (Edwards' wallflower)
Halimolobos Tausch [HALIMOL$]
Halimolobos mollis (Hook.) Rollins [HALIMOI] (soft halimolobos)
SY: *Arabis hookeri* Lange
SY: *Turritis mollis* Hook.
Halimolobos whitedii (Piper) Rollins [HALIWHI] (Whited's halimolobos)
SY: *Arabis whitedii* Piper
Hesperis L. [HESPERI$]
·**Hesperis matronalis** L. [HESPMAT] (dame's-violet)
Hutchinsia Ait. f. [HUTCHIN$]
Hutchinsia procumbens (L.) Desv. [HUTCPRO] (hutchinsia)
SY: *Hymenolobus procumbens* (L.) Nutt. ex Schinz & Thellung
Hymenolobus procumbens (L.) Nutt. ex Schinz & Thellung = Hutchinsia procumbens [HUTCPRO]
Hymenophysa pubescens C.A. Mey. = Cardaria pubescens [CARDPUB]
Iberis nudicaulis L. = Teesdalia nudicaulis [TEESNUD]
Idahoa A. Nels. & J.F. Macbr. [IDAHOA$]
Idahoa scapigera (Hook.) A. Nels. & J.F. Macbr. [IDAHSCA] (scalepod)
Isatis L. [ISATIS$]
·**Isatis tinctoria** L. [ISATTIN] (dyer's woad)
Koniga maritima (L.) R. Br. = Lobularia maritima [LOBUMAR]
Lepidium L. [LEPIDIU$]
Lepidium bourgeauanum Thellung [LEPIBOU] (branched pepper-grass)
SY: *Lepidium densiflorum* var. *bourgeauanum* (Thellung) C.L. Hitchc.
·**Lepidium campestre** (L.) Ait. f. [LEPICAM] (field pepper-grass)
SY: *Neolepia campestris* (L.) W.A. Weber
SY: *Thlaspi campestre* L.
Lepidium densiflorum Schrad. [LEPIDEN] (prairie pepper-grass)
Lepidium densiflorum var. *bourgeauanum* (Thellung) C.L. Hitchc. = Lepidium bourgeauanum [LEPIBOU]
Lepidium densiflorum var. **densiflorum** [LEPIDEN0]
SY: *Lepidium densiflorum* var. *typicum* Thellung
SY: *Lepidium neglectum* Thellung
SY: *Lepidium texanum* Buckl.
Lepidium densiflorum var. **elongatum** (Rydb.) Thellung [LEPIDEN2]
SY: *Lepidium elongatum* Rydb.
Lepidium densiflorum var. **macrocarpum** Mulligan [LEPIDEN3]
Lepidium densiflorum var. **pubicarpum** (A. Nels.) Thellung [LEPIDEN4]
SY: *Lepidium pubecarpum* A. Nels.
Lepidium densiflorum var. *typicum* Thellung = Lepidium densiflorum var. densiflorum [LEPIDEN0]
Lepidium didymum L. = Coronopus didymus [CORODID]
Lepidium divergens Osterhout = Lepidium ramosissimum [LEPIRAM]
Lepidium elongatum Rydb. = Lepidium densiflorum var. elongatum [LEPIDEN2]
·**Lepidium heterophyllum** (DC.) Benth. [LEPIHET] (Smith's pepper-grass)
SY: *Lepidium smithii* Hook.
Lepidium neglectum Thellung = Lepidium densiflorum var. densiflorum [LEPIDEN0]
·**Lepidium perfoliatum** L. [LEPIPER] (clasping-leaved pepper-grass)
Lepidium pubecarpum A. Nels. = Lepidium densiflorum var. pubicarpum [LEPIDEN4]
Lepidium ramosissimum A. Nels. [LEPIRAM]
SY: *Lepidium divergens* Osterhout
Lepidium repens (Schrenk) Boiss. = Cardaria draba ssp. chalapensis [CARDDRA1]
·**Lepidium sativum** L. [LEPISAT] (garden cress)
Lepidium smithii Hook. = Lepidium heterophyllum [LEPIHET]
Lepidium texanum Buckl. = Lepidium densiflorum var. densiflorum [LEPIDEN0]
Lepidium virginicum L. [LEPIVIR] (tall pepper-grass)
Lesquerella S. Wats. [LESQUER$]
Lesquerella arctica (Wormsk. ex Hornem.) S. Wats. [LESQARC] (arctic bladderpod)
SY: *Lesquerella arctica* ssp. *purshii* (S. Wats.) Porsild
SY: *Lesquerella arctica* var. *purshii* S. Wats.
SY: *Lesquerella arctica* var. *scammaniae* Rollins
SY: *Lesquerella purshii* (S. Wats.) Fern.
Lesquerella arctica ssp. *purshii* (S. Wats.) Porsild = Lesquerella arctica [LESQARC]
Lesquerella arctica var. *purshii* S. Wats. = Lesquerella arctica [LESQARC]
Lesquerella arctica var. *scammaniae* Rollins = Lesquerella arctica [LESQARC]
Lesquerella douglasii S. Wats. [LESQDOU] (Columbia bladderpod)
Lesquerella purshii (S. Wats.) Fern. = Lesquerella arctica [LESQARC]
Lobularia Desv. [LOBULAR$]
·**Lobularia maritima** (L.) Desv. [LOBUMAR] (sweet alyssum)
SY: *Alyssum maritimum* (L.) Lam.
SY: *Clypeola maritima* L.
SY: *Koniga maritima* (L.) R. Br.
Lunaria L. [LUNARIA$]
·**Lunaria annua** L. [LUNAANN] (honesty)
Myagrum paniculatum L. = Neslia paniculata [NESLPAN]
Nasturtium microphyllum Boenn. ex Reichenb. = Rorippa microphylla [RORIMIR]
Nasturtium officinale Ait. f. = Rorippa nasturtium-aquaticum [RORINAS]
Nasturtium officinale var. *microphyllum* (Boenn. ex Reichenb.) Thellung = Rorippa microphylla [RORIMIR]
Nasturtium officinale var. *siifolium* (Reichenb.) W.D.J. Koch = Rorippa nasturtium-aquaticum [RORINAS]
Neolepia campestris (L.) W.A. Weber = Lepidium campestre [LEPICAM]
Neslia Desv. [NESLIA$]
·**Neslia paniculata** (L.) Desv. [NESLPAN] (ball mustard)
SY: *Myagrum paniculatum* L.
Neuroloma nudicaule (L.) DC. = Parrya nudicaulis [PARRNUD]
Norta altissima (L.) Britt. = Sisymbrium altissimum [SISYALT]
Parrya R. Br. [PARRYA$]
Parrya nudicaulis (L.) Boiss. [PARRNUD] (northern parrya)
SY: *Neuroloma nudicaule* (L.) DC.
SY: *Parrya nudicaulis* var. *grandiflora* Hultén
SY: *Parrya nudicaulis* ssp. *interior* Hultén
SY: *Parrya nudicaulis* var. *interior* (Hultén) Boivin
SY: *Parrya nudicaulis* ssp. *septentrionalis* Hultén
SY: *Parrya platycarpa* Rydb.
SY: *Parrya rydbergii* Botsch.

Parrya nudicaulis ssp. *interior* Hultén = Parrya nudicaulis [PARRNUD]
Parrya nudicaulis ssp. *septentrionalis* Hultén = Parrya nudicaulis [PARRNUD]
Parrya nudicaulis var. *grandiflora* Hultén = Parrya nudicaulis [PARRNUD]
Parrya nudicaulis var. *interior* (Hultén) Boivin = Parrya nudicaulis [PARRNUD]
Parrya platycarpa Rydb. = Parrya nudicaulis [PARRNUD]
Parrya rydbergii Botsch. = Parrya nudicaulis [PARRNUD]
Physaria (Nutt. ex Torr. & Gray) Gray [PHYSARI$]
Physaria didymocarpa (Hook.) Gray [PHYSDID] (common twinpod)
SY: *Physaria didymocarpa* var. *normalis* Kuntze
Physaria didymocarpa var. *normalis* Kuntze = Physaria didymocarpa [PHYSDID]
Pilosella novae-angliae Rydb. = Braya humilis [BRAYHUM]
Pilosella richardsonii Rydb. = Braya humilis [BRAYHUM]
Radicula armoracia (L.) B.L. Robins. = Armoracia rusticana [ARMORUS]
Radicula sylvestris (L.) Druce = Rorippa sylvestris [RORISYL]
Raphanus L. [RAPHANU$]
***Raphanus raphanistrum** L. [RAPHRAP] (wild radish)
Raphanus raphanistrum var. *sativus* (L.) G. Beck = Raphanus sativus [RAPHSAT]
***Raphanus sativus** L. [RAPHSAT] (garden radish)
SY: *Raphanus raphanistrum* var. *sativus* (L.) G. Beck
Rorippa Scop. [RORIPPA$]
Rorippa armoracia (L.) A.S. Hitchc. = Armoracia rusticana [ARMORUS]
Rorippa barbareifolia (DC.) Kitagawa [RORIBAR] (hoary yellow cress)
SY: *Cardamine barbareifolia* DC.
SY: *Rorippa hispida* var. *barbareifolia* (DC.) Hultén
SY: *Rorippa islandica* var. *barbareifolia* (DC.) Welsh
Rorippa curvipes Greene [RORICUR] (blunt-leaved yellow cress)
Rorippa curvipes var. **integra** (Rydb.) R. Stuckey [RORICUR2]
SY: *Rorippa obtusa* var. *integra* (Rydb.) Victorin
Rorippa curvisiliqua (Hook.) Bess. ex Britt. [RORICUV] (western yellow cress)
Rorippa hispida var. *barbareifolia* (DC.) Hultén = Rorippa barbareifolia [RORIBAR]
Rorippa islandica var. *barbareifolia* (DC.) Welsh = Rorippa barbareifolia [RORIBAR]
***Rorippa microphylla** (Boenn. ex Reichenb.) Hyl. ex A. & D. Löve [RORIMIR]
SY: *Nasturtium microphyllum* Boenn. ex Reichenb.
SY: *Nasturtium officinale* var. *microphyllum* (Boenn. ex Reichenb.) Thellung
SY: *Rorippa nasturtium-aquaticum* var. *longisiliqua* (Irmisch) Boivin
***Rorippa nasturtium-aquaticum** (L.) Hayek [RORINAS]
SY: *Nasturtium officinale* Ait. f.
SY: *Nasturtium officinale* var. *siifolium* (Reichenb.) W.D.J. Koch
SY: *Sisymbrium nasturtium-aquaticum* L.
Rorippa nasturtium-aquaticum var. *longisiliqua* (Irmisch) Boivin = Rorippa microphylla [RORIMIR]
Rorippa obtusa var. *integra* (Rydb.) Victorin = Rorippa curvipes var. integra [RORICUR2]
Rorippa palustris (L.) Bess. [RORIPAL] (marsh yellow cress)
***Rorippa sylvestris** (L.) Bess. [RORISYL] (creeping yellow cress)
SY: *Radicula sylvestris* (L.) Druce
Schoenocrambe Greene [SCHOENO$]
Schoenocrambe linifolia (Nutt.) Greene [SCHOLIN] (plains mustard)
SY: *Sisymbrium linifolium* (Nutt.) Nutt. ex Torr. & Gray
Sinapis L. [SINAPIS$]
***Sinapis alba** L. [SINAALB]
SY: *Brassica alba* Rabenh.
SY: *Brassica hirta* Moench
***Sinapis arvensis** L. [SINAARV]
SY: *Brassica arvensis* Rabenh.
SY: *Brassica kaber* (DC.) L.C. Wheeler
SY: *Brassica kaber* var. *pinnatifida* (Stokes) L.C. Wheeler
SY: *Brassica kaber* var. *schkuhriana* (Reichenb.) L.C. Wheeler
Sinapis juncea L. = Brassica juncea [BRASJUN]
Sinapis nigra L. = Brassica nigra [BRASNIG]
Sisymbrium L. [SISYMBR$]
Sisymbrium alliaria (L.) Scop. = Alliaria petiolata [ALLIPET]
***Sisymbrium altissimum** L. [SISYALT] (tall tumble-mustard)
SY: *Norta altissima* (L.) Britt.
Sisymbrium humile C.A. Mey. = Braya humilis [BRAYHUM]
Sisymbrium incanum Bernh. ex Fisch. & C.A. Mey. = Descurainia incana [DESCINC]
Sisymbrium incisum var. *filipes* Gray = Descurainia pinnata ssp. filipes [DESCPIN1]
Sisymbrium linifolium (Nutt.) Nutt. ex Torr. & Gray = Schoenocrambe linifolia [SCHOLIN]
***Sisymbrium loeselii** L. [SISYLOE] (Loesel's tumble-mustard)
Sisymbrium longipedicellatum Fourn. = Descurainia pinnata ssp. filipes [DESCPIN1]
Sisymbrium nasturtium-aquaticum L. = Rorippa nasturtium-aquaticum [RORINAS]
***Sisymbrium officinale** (L.) Scop. [SISYOFF] (hedge mustard)
SY: *Erysimum officinale* L.
SY: *Sisymbrium officinale* var. *leiocarpum* DC.
Sisymbrium officinale var. *leiocarpum* DC. = Sisymbrium officinale [SISYOFF]
Sisymbrium salsugineum Pallas = Arabidopsis salsuginea [ARABSAL]
Sisymbrium sophia L. = Descurainia sophia [DESCSOP]
Sisymbrium sophioides Fisch. ex Hook. = Descurainia sophioides [DESCSOH]
Sisymbrium thalianum (L.) J. Gay & Monn. = Arabidopsis thaliana [ARABTHA]
Sisymbrium viscosum (Rydb.) Blank. = Descurainia incana ssp. viscosa [DESCINC1]
Smelowskia C.A. Mey. [SMELOWS$]
Smelowskia calycina (Steph. ex Willd.) C.A. Mey. [SMELCAL] (alpine smelowskia)
Smelowskia ovalis M.E. Jones [SMELOVA] (short-fruited smelowskia)
Sophia filipes (Gray) Heller = Descurainia pinnata ssp. filipes [DESCPIN1]
Sophia intermedia Rydb. = Descurainia pinnata ssp. intermedia [DESCPIN2]
Sophia richardsonii (O.E. Schulz) Rydb. = Descurainia incana [DESCINC]
Sophia sophia (L.) Britt. = Descurainia sophia [DESCSOP]
Sophia viscosa Rydb. = Descurainia incana ssp. viscosa [DESCINC1]
Subularia L. [SUBULAR$]
Subularia aquatica L. [SUBUAQU] (awlwort)
Subularia aquatica ssp. *americana* Mulligan & Calder = Subularia aquatica var. americana [SUBUAQU2]
Subularia aquatica var. **americana** (Mulligan & Calder) Boivin [SUBUAQU2]
SY: *Subularia aquatica* ssp. *americana* Mulligan & Calder
Teesdalia Ait. f. [TEESDAL$]
***Teesdalia nudicaulis** (L.) Ait. f. [TEESNUD] (shepherd's cress)
SY: *Iberis nudicaulis* L.
Thellungiella salsuginea (Pallas) O.E. Schulz = Arabidopsis salsuginea [ARABSAL]
Thelypodium Endl. [THELYPO$]
Thelypodium laciniatum (Hook.) Endl. ex Walp. [THELLAC] (thick-leaved thelypody)
SY: *Thelypodium laciniatum* var. *streptanthoides* (Leib. ex Piper) Payson
SY: *Thelypodium streptanthoides* Leib. ex Piper
Thelypodium laciniatum var. *milleflorum* (A. Nels.) Payson = Thelypodium milleflorum [THELMIL]
Thelypodium laciniatum var. *streptanthoides* (Leib. ex Piper) Payson = Thelypodium laciniatum [THELLAC]
Thelypodium milleflorum A. Nels. [THELMIL]
SY: *Thelypodium laciniatum* var. *milleflorum* (A. Nels.) Payson
Thelypodium streptanthoides Leib. ex Piper = Thelypodium laciniatum [THELLAC]

Thlaspi L. [THLASPI$]
*Thlaspi arvense** L. [THLAARV] (field pennycress)
Thlaspi bursa-pastoris L. = Capsella bursa-pastoris [CAPSBUR]
Thlaspi campestre L. = Lepidium campestre [LEPICAM]
Thysanocarpus Hook. [THYSANO$]
Thysanocarpus amplectens Greene = Thysanocarpus curvipes [THYSCUR]
Thysanocarpus curvipes Hook. [THYSCUR] (sand lacepod)
SY: *Thysanocarpus amplectens* Greene
SY: *Thysanocarpus curvipes* var. *elegans* (Fisch. & C.A. Mey.) B.L. Robins.
SY: *Thysanocarpus curvipes* var. *eradiatus* Jepson
SY: *Thysanocarpus curvipes* var. *longistylus* Jepson
SY: *Thysanocarpus elegans* Fisch. & C.A. Mey.
Thysanocarpus curvipes var. *elegans* (Fisch. & C.A. Mey.) B.L. Robins. = Thysanocarpus curvipes [THYSCUR]
Thysanocarpus curvipes var. *eradiatus* Jepson = Thysanocarpus curvipes [THYSCUR]
Thysanocarpus curvipes var. *longistylus* Jepson = Thysanocarpus curvipes [THYSCUR]
Thysanocarpus elegans Fisch. & C.A. Mey. = Thysanocarpus curvipes [THYSCUR]
Thysanocarpus pusillus Hook. = Athysanus pusillus [ATHYPUS]
Torularia arctica (Böcher) A. & D. Löve = Braya humilis [BRAYHUM]
Torularia humilis (C.A. Mey.) O.E. Schulz = Braya humilis [BRAYHUM]
Torularia humilis ssp. *arctica* Böcher = Braya humilis [BRAYHUM]
Turritis drummondii (Gray) Lunell = Arabis drummondii [ARABDRU]
Turritis glabra L. = Arabis glabra [ARABGLA]
Turritis mollis Hook. = Halimolobos mollis [HALIMOI]

BUDDLEJACEAE (V:D)

Buddleja L. [BUDDLEJ$]
***Buddleja davidii** Franch. [BUDDDAV] (butterfly-bush)

BUTOMACEAE (V:M)

Butomus L. [BUTOMUS$]
***Butomus umbellatus** L. [BUTOUMB] (flowering-rush)

CABOMBACEAE (V:D)

Brasenia Schreb. [BRASENI$]
Brasenia peltata Pursh = Brasenia schreberi [BRASSCH]
Brasenia schreberi J.F. Gmel. [BRASSCH] (water-shield)
SY: *Brasenia peltata* Pursh

CACTACEAE (V:D)

Opuntia P. Mill. [OPUNTIA$]
Opuntia fragilis (Nutt.) Haw. [OPUNFRA] (brittle prickly-pear cactus)
Opuntia polyacantha Haw. [OPUNPOL] (plains prickly-pear cactus)

CALLITRICHACEAE (V:D)

Callitriche L. [CALLITR$]
Callitriche anceps Fern. = Callitriche heterophylla ssp. heterophylla [CALLHET0]
Callitriche autumnalis L. = Callitriche hermaphroditica [CALLHER]
Callitriche bolanderi Hegelm. = Callitriche heterophylla ssp. bolanderi [CALLHET1]
Callitriche hermaphroditica L. [CALLHER] (northern water-starwort)
SY: *Callitriche autumnalis* L.
Callitriche heterophylla Pursh [CALLHET] (greater water-starwort)
Callitriche heterophylla ssp. **bolanderi** (Hegelm.) Calder & Taylor [CALLHET1] (diverse-leaved water-starwort)
SY: *Callitriche bolanderi* Hegelm.
SY: *Callitriche heterophylla* var. *bolanderi* (Hegelm.) Fassett
Callitriche heterophylla ssp. **heterophylla** [CALLHET0]
SY: *Callitriche anceps* Fern.
Callitriche heterophylla var. *bolanderi* (Hegelm.) Fassett = Callitriche heterophylla ssp. bolanderi [CALLHET1]
Callitriche marginata Torr. [CALLMAR] (winged water-starwort)
SY: *Callitriche sepulta* S. Wats.
Callitriche palustris L. [CALLPAU] (spring water-starwort)
SY: *Callitriche palustris* var. *verna* (L.) Fenley ex Jepson
SY: *Callitriche verna* L.
Callitriche palustris var. *verna* (L.) Fenley ex Jepson = Callitriche palustris [CALLPAU]
Callitriche sepulta S. Wats. = Callitriche marginata [CALLMAR]
***Callitriche stagnalis** Scop. [CALLSTA] (pond water-starwort)
Callitriche verna L. = Callitriche palustris [CALLPAU]

CAMPANULACEAE (V:D)

Campanula L. [CAMPANU$]
Campanula alaskana (Gray) W. Wight ex J.P. Anders. = Campanula rotundifolia [CAMPROT]
Campanula aurita Greene [CAMPAUR] (Alaskan harebell)
Campanula dubia A. DC. = Campanula rotundifolia [CAMPROT]
Campanula heterodoxa Bong. = Campanula rotundifolia [CAMPROT]
Campanula intercedens Witasek = Campanula rotundifolia [CAMPROT]
Campanula lasiocarpa Cham. [CAMPLAS] (mountain harebell)
SY: *Campanula lasiocarpa* ssp. *latisepala* (Hultén) Hultén
SY: *Campanula latisepala* Hultén
SY: *Campanula latisepala* var. *dubia* Hultén
Campanula lasiocarpa ssp. *latisepala* (Hultén) Hultén = Campanula lasiocarpa [CAMPLAS]
Campanula latisepala Hultén = Campanula lasiocarpa [CAMPLAS]
Campanula latisepala var. *dubia* Hultén = Campanula lasiocarpa [CAMPLAS]
***Campanula medium** L. [CAMPMED] (canterbury-bells)
***Campanula persicifolia** L. [CAMPPER] (peach-leaved bellflower)
SY: *Campanula persicifolia* var. *alba* Horton
Campanula persicifolia var. *alba* Horton = Campanula persicifolia [CAMPPER]
Campanula petiolata A. DC. = Campanula rotundifolia [CAMPROT]
***Campanula rapunculoides** L. [CAMPRAP] (creeping bellflower)
SY: *Campanula rapunculoides* var. *ucranica* (Bess.) K. Koch
Campanula rapunculoides var. *ucranica* (Bess.) K. Koch = Campanula rapunculoides [CAMPRAP]
Campanula rotundifolia L. [CAMPROT] (common harebell)
SY: *Campanula alaskana* (Gray) W. Wight ex J.P. Anders.
SY: *Campanula dubia* A. DC.
SY: *Campanula heterodoxa* Bong.
SY: *Campanula intercedens* Witasek
SY: *Campanula petiolata* A. DC.
SY: *Campanula rotundifolia* var. *alaskana* Gray
SY: *Campanula rotundifolia* var. *intercedens* (Witasek) Farw.
SY: *Campanula rotundifolia* var. *lancifolia* Mert. & Koch
SY: *Campanula rotundifolia* var. *petiolata* (A. DC.) J.K. Henry
SY: *Campanula rotundifolia* var. *velutina* A. DC.
SY: *Campanula sacajaweana* M.E. Peck
Campanula rotundifolia var. *alaskana* Gray = Campanula rotundifolia [CAMPROT]
Campanula rotundifolia var. *intercedens* (Witasek) Farw. = Campanula rotundifolia [CAMPROT]
Campanula rotundifolia var. *lancifolia* Mert. & Koch = Campanula rotundifolia [CAMPROT]
Campanula rotundifolia var. *petiolata* (A. DC.) J.K. Henry = Campanula rotundifolia [CAMPROT]
Campanula rotundifolia var. *velutina* A. DC. = Campanula rotundifolia [CAMPROT]
Campanula sacajaweana M.E. Peck = Campanula rotundifolia [CAMPROT]

Campanula scouleri Hook. ex A. DC. [CAMPSCO] (Scouler's harebell)
Campanula uniflora L. [CAMPUNI] (arctic harebell)
Downingia Torr. [DOWINGI$]
Downingia elegans (Dougl. ex Lindl.) Torr. [DOWIELE] (common downingia)
Githopsis Nutt. [GITHOPS$]
Githopsis calycina Benth. = Githopsis specularioides [GITHSPE]
Githopsis calycina var. *hirsuta* Benth. = Githopsis specularioides [GITHSPE]
Githopsis specularioides Nutt. [GITHSPE] (common bluecup)
SY: *Githopsis calycina* Benth.
SY: *Githopsis calycina* var. *hirsuta* Benth.
SY: *Githopsis specularioides* var. *hirsuta* Nutt.
Githopsis specularioides var. *hirsuta* Nutt. = Githopsis specularioides [GITHSPE]
Heterocodon Nutt. [HETEROC$]
Heterocodon rariflorum Nutt. [HETERAR] (heterocodon)
SY: *Specularia rariflora* (Nutt.) McVaugh
Legousia perfoliata (L.) Britt. = Triodanis perfoliata [TRIOPER]
Lobelia L. [LOBELIA$]
Lobelia dortmanna L. [LOBEDOR] (water lobelia)
Lobelia inflata L. [LOBEINF] (Indian-tobacco)
Lobelia kalmii L. [LOBEKAL] (Kalm's lobelia)
SY: *Lobelia kalmii* var. *strictiflora* Rydb.
SY: *Lobelia strictiflora* (Rydb.) Lunell
Lobelia kalmii var. *strictiflora* Rydb. = Lobelia kalmii [LOBEKAL]
Lobelia strictiflora (Rydb.) Lunell = Lobelia kalmii [LOBEKAL]
Specularia perfoliata (L.) A. DC. = Triodanis perfoliata [TRIOPER]
Specularia rariflora (Nutt.) McVaugh = Heterocodon rariflorum [HETERAR]
Triodanis Raf. ex Greene [TRIODAN$]
Triodanis perfoliata (L.) Nieuwl. [TRIOPER]
SY: *Legousia perfoliata* (L.) Britt.
SY: *Specularia perfoliata* (L.) A. DC.

CANNABACEAE (V:D)

Humulus L. [HUMULUS$]
***Humulus lupulus** L. [HUMULUP] (common hop)

CAPPARACEAE (V:D)

Cleome L. [CLEOME$]
Cleome serrulata Pursh [CLEOSER] (stinking-clover bee-plant)
SY: *Cleome serrulata* var. *angusta* (M.E. Jones) Tidestrom
Cleome serrulata var. *angusta* (M.E. Jones) Tidestrom = Cleome serrulata [CLEOSER]

CAPRIFOLIACEAE (V:D)

Distegia involucrata (Banks ex Spreng.) Cockerell = Lonicera involucrata [LONIINV]
Linnaea L. [LINNAEA$]
Linnaea americana Forbes = Linnaea borealis ssp. longiflora [LINNBOR2]
Linnaea borealis L. [LINNBOR] (twinflower)
Linnaea borealis ssp. *americana* (Forbes) Hultén ex Clausen = Linnaea borealis ssp. longiflora [LINNBOR2]
Linnaea borealis ssp. **borealis** [LINNBOR0]
Linnaea borealis ssp. **longiflora** (Torr.) Hultén [LINNBOR2]
SY: *Linnaea americana* Forbes
SY: *Linnaea borealis* ssp. *americana* (Forbes) Hultén ex Clausen
SY: *Linnaea borealis* var. *americana* (Forbes) Rehd.
SY: *Linnaea borealis* var. *longiflora* Torr.
Linnaea borealis var. *americana* (Forbes) Rehd. = Linnaea borealis ssp. longiflora [LINNBOR2]
Linnaea borealis var. *longiflora* Torr. = Linnaea borealis ssp. longiflora [LINNBOR2]
Lonicera L. [LONICER$]
***Lonicera caerulea** L. [LONICAE] (bluefly honeysuckle)
Lonicera ciliosa (Pursh) Poir. ex DC. [LONICIL] (western trumpet honeysuckle)
Lonicera dioica L. [LONIDIO] (limber honeysuckle)
Lonicera dioica var. **glaucescens** (Rydb.) Butters [LONIDIO1] (glaucous-leaved honeysuckle)
SY: *Lonicera glaucescens* (Rydb.) Rydb.
***Lonicera etrusca** Santi [LONIETR] (Etruscan honeysuckle)
Lonicera glaucescens (Rydb.) Rydb. = Lonicera dioica var. glaucescens [LONIDIO1]
Lonicera hispidula (Lindl.) Dougl. ex Torr. & Gray [LONIHIS] (hairy honeysuckle)
Lonicera involucrata Banks ex Spreng. [LONIINV] (bracked honeysuckle)
SY: *Distegia involucrata* (Banks ex Spreng.) Cockerell
SY: *Lonicera involucrata* var. *flavescens* (Dippel) Rehd.
SY: *Xylosteum involucratum* (Banks ex Spreng.) Richards.
Lonicera involucrata var. *flavescens* (Dippel) Rehd. = Lonicera involucrata [LONIINV]
Lonicera utahensis S. Wats. [LONIUTA] (Utah honeysuckle)
Sambucus L. [SAMBUCU$]
Sambucus callicarpa Greene = Sambucus racemosa ssp. pubens var. arborescens [SAMBRAC2]
Sambucus cerulea Raf. [SAMBCER] (blue elder)
Sambucus melanocarpa Gray = Sambucus racemosa ssp. pubens var. melanocarpa [SAMBRAC4]
Sambucus pubens Michx. = Sambucus racemosa ssp. pubens var. pubens [SAMBRAC5]
Sambucus pubens var. *arborescens* Torr. & Gray = Sambucus racemosa ssp. pubens var. arborescens [SAMBRAC2]
Sambucus racemosa L. [SAMBRAC] (red elderberry)
Sambucus racemosa ssp. **pubens** (Michx.) House [SAMBRAC1]
Sambucus racemosa ssp. **pubens** var. **arborescens** (Torr. & Gray) Gray [SAMBRAC2] (coastal red elder)
SY: *Sambucus callicarpa* Greene
SY: *Sambucus pubens* var. *arborescens* Torr. & Gray
Sambucus racemosa ssp. *pubens* var. *leucocarpa* (Torr. & Gray) Cronq. = Sambucus racemosa ssp. pubens var. pubens [SAMBRAC5]
Sambucus racemosa ssp. **pubens** var. **melanocarpa** (Gray) McMinn [SAMBRAC4] (black elder)
SY: *Sambucus melanocarpa* Gray
Sambucus racemosa ssp. **pubens** var. **pubens** (Michx.) Koehne [SAMBRAC5]
SY: *Sambucus pubens* Michx.
SY: *Sambucus racemosa* var. *leucocarpa* (Torr. & Gray) Cronq.
SY: *Sambucus racemosa* ssp. *pubens* var. *leucocarpa* (Torr. & Gray) Cronq.
Sambucus racemosa var. *leucocarpa* (Torr. & Gray) Cronq. = Sambucus racemosa ssp. pubens var. pubens [SAMBRAC5]
Symphoricarpos Duham. [SYMPHOR$]
Symphoricarpos albus (L.) Blake [SYMPALB] (common snowberry)
Symphoricarpos albus ssp. *laevigatus* (Fern.) Hultén = Symphoricarpos albus var. laevigatus [SYMPALB2]
Symphoricarpos albus var. **albus** [SYMPALB0]
SY: *Symphoricarpos albus* var. *pauciflorus* (W.J. Robins. ex Gray) Blake
SY: *Symphoricarpos pauciflorus* W.J. Robins. ex Gray
SY: *Symphoricarpos racemosus* Michx.
Symphoricarpos albus var. **laevigatus** (Fern.) Blake [SYMPALB2]
SY: *Symphoricarpos albus* ssp. *laevigatus* (Fern.) Hultén
SY: *Symphoricarpos rivularis* Suksdorf
Symphoricarpos albus var. *pauciflorus* (W.J. Robins. ex Gray) Blake = Symphoricarpos albus var. albus [SYMPALB0]
Symphoricarpos hesperius G.N. Jones [SYMPHES]
SY: *Symphoricarpos mollis* ssp. *hesperius* (G.N. Jones) Abrams ex Ferris
SY: *Symphoricarpos mollis* var. *hesperius* (G.N. Jones) Cronq.
Symphoricarpos mollis ssp. *hesperius* (G.N. Jones) Abrams ex Ferris = Symphoricarpos hesperius [SYMPHES]
Symphoricarpos mollis var. *hesperius* (G.N. Jones) Cronq. = Symphoricarpos hesperius [SYMPHES]
Symphoricarpos occidentalis Hook. [SYMPOCC] (western snowberry)

Symphoricarpos oreophilus Gray [SYMPORE] (mountain snowberry)
Symphoricarpos oreophilus var. **utahensis** (Rydb.) A. Nels. [SYMPORE1] (Utah mountain snowberry)
SY: *Symphoricarpos tetonensis* A. Nels.
SY: *Symphoricarpos utahensis* Rydb.
SY: *Symphoricarpos vaccinioides* Rydb.
Symphoricarpos pauciflorus W.J. Robins. ex Gray = Symphoricarpos albus var. albus [SYMPALB0]
Symphoricarpos racemosus Michx. = Symphoricarpos albus var. albus [SYMPALB0]
Symphoricarpos rivularis Suksdorf = Symphoricarpos albus var. laevigatus [SYMPALB2]
Symphoricarpos tetonensis A. Nels. = Symphoricarpos oreophilus var. utahensis [SYMPORE1]
Symphoricarpos utahensis Rydb. = Symphoricarpos oreophilus var. utahensis [SYMPORE1]
Symphoricarpos vaccinioides Rydb. = Symphoricarpos oreophilus var. utahensis [SYMPORE1]
Viburnum L. [VIBURNU$]
Viburnum edule (Michx.) Raf. [VIBUEDU] (low bush-cranberry)
SY: *Viburnum pauciflorum* La Pylaie ex Torr. & Gray
Viburnum opulus L. [VIBUOPU] (American bush-cranberry)
Viburnum opulus ssp. *trilobum* (Marsh.) Clausen = Viburnum opulus var. americanum [VIBUOPU1]
Viburnum opulus var. **americanum** Ait. [VIBUOPU1]
SY: *Viburnum opulus* ssp. *trilobum* (Marsh.) Clausen
SY: *Viburnum trilobum* Marsh.
Viburnum pauciflorum La Pylaie ex Torr. & Gray = Viburnum edule [VIBUEDU]
Viburnum trilobum Marsh. = Viburnum opulus var. americanum [VIBUOPU1]
Xylosteum involucratum (Banks ex Spreng.) Richards. = Lonicera involucrata [LONIINV]

CARYOPHYLLACEAE (V:D)

Agrostemma coronaria L. = Lychnis coronaria [LYCHCOR]
Alsinanthe elegans (Cham. & Schlecht.) A. & D. Löve = Minuartia elegans [MINUELE]
Alsinanthe stricta ssp. *dawsonensis* (Britt.) A. & D. Löve = Minuartia dawsonensis [MINUDAW]
Alsine americana (Porter ex B.L. Robins.) Rydb. = Stellaria americana [STELAME]
Alsine aquatica (L.) Britt. = Myosoton aquaticum [MYOSAQU]
Alsine baicalensis Coville = Stellaria umbellata [STELUMB]
Alsine bongardiana (Fern.) A. Davids. & Moxley = Stellaria borealis ssp. sitchana [STELBOR1]
Alsine brachypetala (Bong.) T.J. Howell = Stellaria borealis ssp. sitchana [STELBOR1]
Alsine calycantha (Ledeb.) Rydb. = Stellaria calycantha [STELCAL]
Alsine crispa (Cham. & Schlecht.) Holz. = Stellaria crispa [STELCRI]
Alsine elegans (Cham. & Schlecht.) Fenzl = Minuartia elegans [MINUELE]
Alsine humifusa (Rottb.) Britt. = Stellaria humifusa [STELHUM]
Alsine obtusa (Engelm.) Rose = Stellaria obtusa [STELOBT]
Alsine simcoei T.J. Howell = Stellaria calycantha [STELCAL]
Alsine uliginosa (Murr.) Britt. = Stellaria alsine [STELALS]
Alsine viridula Piper = Stellaria obtusa [STELOBT]
Alsine washingtoniana (B.L. Robins.) Heller = Stellaria obtusa [STELOBT]
Alsinopsis obtusiloba Rydb. = Minuartia obtusiloba [MINUOBT]
Alsinopsis occidentalis Heller = Minuartia nuttallii [MINUNUT]
Anotites viscosa Greene = Silene menziesii var. viscosa [SILEMEN2]
Arenaria L. [ARENARI$]
Arenaria capillaris Poir. [ARENCAP] (slender mountain sandwort)
Arenaria capillaris ssp. **americana** Maguire [ARENCAP1] (thread-leaved sandwort)
SY: *Arenaria capillaris* var. *americana* (Maguire) R.J. Davis
SY: *Arenaria formosa* (Fisch.) Regel
SY: *Eremogone americana* (Maguire) S. Ikonnikov
Arenaria capillaris var. *americana* (Maguire) R.J. Davis = Arenaria capillaris ssp. americana [ARENCAP1]
Arenaria dawsonensis Britt. = Minuartia dawsonensis [MINUDAW]
Arenaria elegans Cham. & Schlecht. = Minuartia elegans [MINUELE]
Arenaria elegans var. *columbiana* Raup = Minuartia elegans [MINUELE]
Arenaria formosa (Fisch.) Regel = Arenaria capillaris ssp. americana [ARENCAP1]
Arenaria lateriflora L. = Moehringia lateriflora [MOEHLAT]
Arenaria lateriflora var. *angustifolia* (Regel) St. John = Moehringia lateriflora [MOEHLAT]
Arenaria lateriflora var. *tayloriae* St. John = Moehringia lateriflora [MOEHLAT]
Arenaria litorea Fern. = Minuartia dawsonensis [MINUDAW]
Arenaria longipedunculata Hultén [ARENLON] (low sandwort)
Arenaria macrophylla Hook. = Moehringia macrophylla [MOEHMAC]
Arenaria nuttallii Pax = Minuartia nuttallii [MINUNUT]
Arenaria obtusiloba (Rydb.) Fern. = Minuartia obtusiloba [MINUOBT]
Arenaria peploides var. *major* Hook. = Honkenya peploides ssp. major [HONKPEP1]
Arenaria peploides var. *maxima* Fern. = Honkenya peploides ssp. major [HONKPEP1]
Arenaria propinqua Richards. = Minuartia rubella [MINURUB]
Arenaria pungens Nutt. = Minuartia nuttallii [MINUNUT]
Arenaria pusilla S. Wats. = Minuartia pusilla [MINUPUS]
Arenaria rossii ssp. *columbiana* (Raup) Maguire = Minuartia elegans [MINUELE]
Arenaria rossii ssp. *elegans* (Cham. & Schlecht.) Maguire = Minuartia elegans [MINUELE]
Arenaria rossii var. *apetala* Maguire = Minuartia elegans [MINUELE]
Arenaria rossii var. *columbiana* (Raup) Maguire = Minuartia elegans [MINUELE]
Arenaria rossii var. *corollina* Fenzl = Minuartia elegans [MINUELE]
Arenaria rossii var. *elegans* (Cham. & Schlecht.) Welsh = Minuartia elegans [MINUELE]
Arenaria rubella (Wahlenb.) Sm. = Minuartia rubella [MINURUB]
Arenaria sajanensis Willd. ex Schlecht. = Minuartia biflora [MINUBIF]
*__Arenaria serpyllifolia__ L. [ARENSER] (thyme-leaved sandwort)
Arenaria stephaniana var. *americana* (Porter ex B.L. Robins.) Shinners = Stellaria americana [STELAME]
Arenaria stricta ssp. *dawsonensis* (Britt.) Maguire = Minuartia dawsonensis [MINUDAW]
Arenaria stricta var. *dawsonensis* (Britt.) Scoggan = Minuartia dawsonensis [MINUDAW]
Arenaria stricta var. *litorea* (Fern.) Britt. = Minuartia dawsonensis [MINUDAW]
Arenaria tenella Nutt. = Minuartia tenella [MINUTEN]
Arenaria verna var. *propinqua* (Richards.) Fern. = Minuartia rubella [MINURUB]
Arenaria verna var. *rubella* (Wahlenb.) S. Wats. = Minuartia rubella [MINURUB]
Cerastium L. [CERASTI$]
Cerastium acutatum Suksdorf = Cerastium glomeratum [CERAGLO]
Cerastium adsurgens Greene = Cerastium fontanum ssp. vulgare [CERAFON1]
Cerastium alpinum var. *capillare* (Fern. & Wieg.) Boivin = Cerastium beeringianum ssp. earlei [CERABEE2]
Cerastium aquaticum L. = Myosoton aquaticum [MYOSAQU]
Cerastium arvense L. [CERAARV] (field chickweed)
Cerastium beeringianum Cham. & Schlecht. [CERABEE] (Bering chickweed)
Cerastium beeringianum ssp. **beeringianum** [CERABEE0]
Cerastium beeringianum ssp. **earlei** (Rydb.) Hultén [CERABEE2]
SY: *Cerastium alpinum* var. *capillare* (Fern. & Wieg.) Boivin
SY: *Cerastium beeringianum* var. *capillare* Fern. & Wieg.
SY: *Cerastium earlei* Rydb.
Cerastium beeringianum var. *capillare* Fern. & Wieg. = Cerastium beeringianum ssp. earlei [CERABEE2]
Cerastium earlei Rydb. = Cerastium beeringianum ssp. earlei [CERABEE2]

Cerastium fischerianum Ser. [CERAFIS] (Fischer's chickweed)
*•**Cerastium fontanum** Baumg. [CERAFON] (mouse-ear chickweed)
Cerastium fontanum ssp. *triviale* (Link) Jalas = Cerastium fontanum ssp. vulgare [CERAFON1]
*•**Cerastium fontanum** ssp. **vulgare** (Hartman) Greuter & Burdet [CERAFON1]
SY: *Cerastium adsurgens* Greene
SY: *Cerastium fontanum* ssp. *triviale* (Link) Jalas
SY: *Cerastium holosteoides* var. *vulgare* (Hartman) Hyl.
SY: *Cerastium triviale* Link
SY: *Cerastium vulgatum* L. 1762, non 1755
SY: *Cerastium vulgatum* var. *hirsutum* Fries
*•**Cerastium glomeratum** Thuill. [CERAGLO] (sticky chickweed)
SY: *Cerastium acutatum* Suksdorf
SY: *Cerastium glomeratum* var. *apetalum* (Dumort.) Fenzl
SY: *Cerastium viscosum* auct.
Cerastium glomeratum var. *apetalum* (Dumort.) Fenzl = Cerastium glomeratum [CERAGLO]
Cerastium holosteoides var. *vulgare* (Hartman) Hyl. = Cerastium fontanum ssp. vulgare [CERAFON1]
Cerastium nutans Raf. [CERANUT] (nodding chickweed)
*•**Cerastium semidecandrum** L. [CERASEM] (little chickweed)
*•**Cerastium tomentosum** L. [CERATOM] (snow-in-summer)
Cerastium triviale Link = Cerastium fontanum ssp. vulgare [CERAFON1]
Cerastium viscosum auct. = Cerastium glomeratum [CERAGLO]
Cerastium vulgatum L. 1762, non 1755 = Cerastium fontanum ssp. vulgare [CERAFON1]
Cerastium vulgatum var. *hirsutum* Fries = Cerastium fontanum ssp. vulgare [CERAFON1]
Coronaria coriacea (Moench) Schischkin & Gorschk. = Lychnis coronaria [LYCHCOR]
Corrigiola L. [CORRIGI$]
*•**Corrigiola litoralis** L. [CORRLIT] (strapwort)
Dianthus L. [DIANTHU$]
*•**Dianthus armeria** L. [DIANARM] (Deptford pink)
*•**Dianthus barbatus** L. [DIANBAR] (sweet William)
*•**Dianthus deltoides** L. [DIANDEL] (maiden pink)
Dianthus prolifer L. = Petrorhagia prolifera [PETRPRO]
Eremogone americana (Maguire) S. Ikonnikov = Arenaria capillaris ssp. americana [ARENCAP1]
Gastrolychnis affinis (J. Vahl ex Fries) Tolm. & Kozh. = Silene involucrata [SILEINV]
Gastrolychnis drummondii (Hook.) A. & D. Löve = Silene drummondii [SILEDRU]
Gastrolychnis involucrata (Cham. & Schlecht.) A. & D. Löve = Silene involucrata [SILEINV]
Gastrolychnis ostenfeldii (Porsild) Petrovsky = Silene taimyrensis [SILETAI]
Gastrolychnis taimyrensis (Tolm.) S.K. Czer. = Silene taimyrensis [SILETAI]
Gastrolychnis triflora ssp. *dawsonii* (B.L. Robins.) A. & D. Löve = Silene taimyrensis [SILETAI]
Gypsophila L. [GYPSOPH$]
*•**Gypsophila paniculata** L. [GYPSPAN] (baby's breath)
Holosteum L. [HOLOSTE$]
*•**Holosteum umbellatum** L. [HOLOUMB] (umbellate chickweed)
Honkenya Ehrh. [HONKENY$]
Honkenya oblongifolia Torr. & Gray = Honkenya peploides ssp. major [HONKPEP1]
Honkenya peploides (L.) Ehrh. [HONKPEP] (seabeach sandwort)
Honkenya peploides ssp. **major** (Hook.) Hultén [HONKPEP1]
SY: *Arenaria peploides* var. *major* Hook.
SY: *Arenaria peploides* var. *maxima* Fern.
SY: *Honkenya oblongifolia* Torr. & Gray
SY: *Honkenya peploides* var. *major* (Hook.) Abrams
Honkenya peploides var. *major* (Hook.) Abrams = Honkenya peploides ssp. major [HONKPEP1]
Lidia biflora (L.) A. & D. Löve = Minuartia biflora [MINUBIF]
Lidia obtusiloba (Rydb.) A. & D. Löve = Minuartia obtusiloba [MINUOBT]
Lychnis L. [LYCHNIS$]
Lychnis affinis J. Vahl ex Fries = Silene involucrata [SILEINV]
Lychnis alba P. Mill. = Silene latifolia ssp. alba [SILELAT1]
Lychnis apetala ssp. *attenuata* (Farr) Maguire = Silene uralensis ssp. attenuata [SILEURA1]
Lychnis apetala var. *attenuata* (Farr) C.L. Hitchc. = Silene uralensis ssp. attenuata [SILEURA1]
Lychnis attenuata Farr = Silene uralensis ssp. attenuata [SILEURA1]
*•**Lychnis coronaria** (L.) Desr. [LYCHCOR] (rose campion)
SY: *Agrostemma coronaria* L.
SY: *Coronaria coriacea* (Moench) Schischkin & Gorschk.
SY: *Silene coronaria* (L.) Clairville
Lychnis dawsonii (B.L. Robins.) J.P. Andrs. = Silene taimyrensis [SILETAI]
Lychnis dioica L. = Silene dioica [SILEDIO]
Lychnis drummondii (Hook.) S. Wats. = Silene drummondii [SILEDRU]
Lychnis furcata (Raf.) Fern. = Silene involucrata [SILEINV]
Lychnis gillettii Boivin = Silene involucrata [SILEINV]
Lychnis loveae Boivin = Silene latifolia ssp. alba [SILELAT1]
Lychnis × *loveae* Boivin = Silene latifolia ssp. alba [SILELAT1]
Lychnis pudica Boivin = Silene drummondii [SILEDRU]
Lychnis saponaria Jessen = Saponaria officinalis [SAPOOFF]
Lychnis taimyrensis (Tolm.) Polunin = Silene taimyrensis [SILETAI]
Lychnis triflora ssp. *dawsonii* (B.L. Robins.) Maguire = Silene taimyrensis [SILETAI]
Lychnis triflora var. *dawsonii* B.L. Robins. = Silene taimyrensis [SILETAI]
Lychnis vespertina Sibthorp = Silene latifolia ssp. alba [SILELAT1]
Malachium aquaticum (L.) Fries = Myosoton aquaticum [MYOSAQU]
Melandrium affine (J. Vahl ex Fries) J. Vahl = Silene involucrata [SILEINV]
Melandrium album (P. Mill.) Garcke = Silene latifolia ssp. alba [SILELAT1]
Melandrium apetalum ssp. *attenuatum* (Farr) Hara = Silene uralensis ssp. attenuata [SILEURA1]
Melandrium dioicum (L.) Coss. & Germ. = Silene dioica [SILEDIO]
Melandrium dioicum ssp. *rubrum* (Wieg.) D. Löve = Silene dioica [SILEDIO]
Melandrium drummondii (Hook.) Hultén = Silene drummondii [SILEDRU]
Melandrium furcatum (Raf.) Hadac = Silene involucrata [SILEINV]
Melandrium noctiflorum (L.) Fries = Silene noctiflora [SILENOC]
Melandrium ostenfeldii Porsild = Silene taimyrensis [SILETAI]
Melandrium taimyrense Tolm. = Silene taimyrensis [SILETAI]
Minuartia L. [MINUART$]
Minuartia austromontana S.J. Wolf & Packer [MINUAUS] (Rocky Mountain sandwort)
Minuartia biflora (L.) Schinz & Thellung [MINUBIF] (mountain sandwort)
SY: *Arenaria sajanensis* Willd. ex Schlecht.
SY: *Lidia biflora* (L.) A. & D. Löve
SY: *Stellaria biflora* L.
Minuartia dawsonensis (Britt.) House [MINUDAW] (rock sandwort)
SY: *Alsinanthe stricta* ssp. *dawsonensis* (Britt.) A. & D. Löve
SY: *Arenaria dawsonensis* Britt.
SY: *Arenaria litorea* Fern.
SY: *Arenaria stricta* ssp. *dawsonensis* (Britt.) Maguire
SY: *Arenaria stricta* var. *dawsonensis* (Britt.) Scoggan
SY: *Arenaria stricta* var. *litorea* (Fern.) Britt.
Minuartia elegans (Cham. & Schlecht.) Schischkin [MINUELE] (northern sandwort)
SY: *Alsinanthe elegans* (Cham. & Schlecht.) A. & D. Löve
SY: *Alsine elegans* (Cham. & Schlecht.) Fenzl
SY: *Arenaria elegans* Cham. & Schlecht.
SY: *Arenaria elegans* var. *columbiana* Raup
SY: *Arenaria rossii* var. *apetala* Maguire
SY: *Arenaria rossii* ssp. *columbiana* (Raup) Maguire
SY: *Arenaria rossii* var. *columbiana* (Raup) Maguire
SY: *Arenaria rossii* var. *corollina* Fenzl
SY: *Arenaria rossii* ssp. *elegans* (Cham. & Schlecht.) Maguire
SY: *Arenaria rossii* var. *elegans* (Cham. & Schlecht.) Welsh
SY: *Minuartia orthotrichoides* Schischkin
SY: *Minuartia rossii* var. *elegans* (Cham. & Schlecht.) Hultén

SY: *Minuartia rossii* var. *orthotrichoides* (Schischkin) Hultén
Minuartia nuttallii (Pax) Briq. [MINUNUT] (Nuttall's sandwort)
SY: *Alsinopsis occidentalis* Heller
SY: *Arenaria nuttallii* Pax
SY: *Arenaria pungens* Nutt.
SY: *Minuartia pungens* (Nutt.) Mattf.
SY: *Minuopsis nuttallii* (Pax) W.A. Weber
Minuartia obtusiloba (Rydb.) House [MINUOBT] (alpine sandwort)
SY: *Alsinopsis obtusiloba* Rydb.
SY: *Arenaria obtusiloba* (Rydb.) Fern.
SY: *Lidia obtusiloba* (Rydb.) A. & D. Löve
Minuartia orthotrichoides Schischkin = Minuartia elegans [MINUELE]
Minuartia pungens (Nutt.) Mattf. = Minuartia nuttallii [MINUNUT]
Minuartia pusilla (S. Wats.) Mattf. [MINUPUS] (dwarf sandwort)
SY: *Arenaria pusilla* S. Wats.
Minuartia rossii var. *elegans* (Cham. & Schlecht.) Hultén = Minuartia elegans [MINUELE]
Minuartia rossii var. *orthotrichoides* (Schischkin) Hultén = Minuartia elegans [MINUELE]
Minuartia rubella (Wahlenb.) Hiern [MINURUB] (boreal sandwort)
SY: *Arenaria propinqua* Richards.
SY: *Arenaria rubella* (Wahlenb.) Sm.
SY: *Arenaria verna* var. *propinqua* (Richards.) Fern.
SY: *Arenaria verna* var. *rubella* (Wahlenb.) S. Wats.
SY: *Tryphane rubella* (Wahlenb.) Reichenb.
Minuartia tenella (Nutt.) Mattf. [MINUTEN] (slender sandwort)
SY: *Arenaria tenella* Nutt.
Minuopsis nuttallii (Pax) W.A. Weber = Minuartia nuttallii [MINUNUT]
Moehringia L. [MOEHRIN$]
Moehringia lateriflora (L.) Fenzl [MOEHLAT] (blunt-leaved sandwort)
SY: *Arenaria lateriflora* L.
SY: *Arenaria lateriflora* var. *angustifolia* (Regel) St. John
SY: *Arenaria lateriflora* var. *tayloriae* St. John
Moehringia macrophylla (Hook.) Fenzl [MOEHMAC] (big-leaved sandwort)
SY: *Arenaria macrophylla* Hook.
Moenchia Ehrh. [MOENCHI$]
*__Moenchia erecta__ (L.) P.G. Gaertn., B. Mey. & Scherb. [MOENERE] (upright chickweed)
Mollugo tetraphylla L. = Polycarpon tetraphyllum [POLYTET]
Myosoton Moench [MYOSOTO$]
*__Myosoton aquaticum__ (L.) Moench [MYOSAQU] (water chickweed)
SY: *Alsine aquatica* (L.) Britt.
SY: *Cerastium aquaticum* L.
SY: *Malachium aquaticum* (L.) Fries
SY: *Stellaria aquatica* (L.) Scop.
Oberna commutata (Guss.) S. Ikonnikov = Silene vulgaris [SILEVUL]
Petrorhagia (Ser.) Link [PETRORH$]
*__Petrorhagia prolifera__ (L.) P.W. Ball & Heywood [PETRPRO] (petrorhagia)
SY: *Dianthus prolifer* L.
SY: *Tunica prolifera* (L.) Scop.
*__Petrorhagia saxifraga__ (L.) Link [PETRSAX] (tunic flower)
SY: *Tunica saxifraga* (L.) Scop.
Polycarpon L. [POLYCAR$]
*__Polycarpon tetraphyllum__ (L.) L. [POLYTET] (four-leaved all-seed)
SY: *Mollugo tetraphylla* L.
Sagina L. [SAGINA$]
Sagina crassicaulis S. Wats. = Sagina maxima ssp. crassicaulis [SAGIMAX1]
Sagina crassicaulis var. *littoralis* (Hultén) Hultén = Sagina maxima ssp. maxima [SAGIMAX0]
Sagina decumbens (Ell.) Torr. & Gray [SAGIDEC] (trailing pearlwort)
Sagina decumbens ssp. **occidentalis** (S. Wats.) Crow [SAGIDEC1] (western pearlwort)
SY: *Sagina occidentalis* S. Wats.
Sagina intermedia Fenzl = Sagina nivalis [SAGINIV]
*__Sagina japonica__ (Sw.) Ohwi [SAGIJAP] (Japanese pearlwort)
Sagina linnaei K. Presl = Sagina saginoides [SAGISAG]
Sagina littoralis Hultén = Sagina maxima ssp. maxima [SAGIMAX0]
Sagina maxima Gray [SAGIMAX] (coastal pearlwort)
Sagina maxima ssp. **crassicaulis** (S. Wats.) Crow [SAGIMAX1]
SY: *Sagina crassicaulis* S. Wats.
Sagina maxima ssp. **maxima** [SAGIMAX0]
SY: *Sagina crassicaulis* var. *littoralis* (Hultén) Hultén
SY: *Sagina littoralis* Hultén
SY: *Sagina maxima* var. *littorea* (Mackenzie) Hara
Sagina maxima var. *littorea* (Mackenzie) Hara = Sagina maxima ssp. maxima [SAGIMAX0]
Sagina micrantha (Bunge) Fern. = Sagina saginoides [SAGISAG]
Sagina nivalis (Lindbl.) Fries [SAGINIV] (snow pearlwort)
SY: *Sagina intermedia* Fenzl
SY: *Spergella intermedia* (Fenzl) A. & D. Löve
Sagina occidentalis S. Wats. = Sagina decumbens ssp. occidentalis [SAGIDEC1]
Sagina procumbens L. [SAGIPRO] (bird's-eye pearlwort)
SY: *Sagina procumbens* var. *compacta* Lange
Sagina procumbens var. *compacta* Lange = Sagina procumbens [SAGIPRO]
Sagina saginoides (L.) Karst. [SAGISAG] (arctic pearlwort)
SY: *Sagina linnaei* K. Presl
SY: *Sagina micrantha* (Bunge) Fern.
SY: *Sagina saginoides* var. *hesperia* Fern.
SY: *Spergella saginoides* (L.) Reichenb.
Sagina saginoides var. *hesperia* Fern. = Sagina saginoides [SAGISAG]
Saponaria L. [SAPONAR$]
*__Saponaria officinalis__ L. [SAPOOFF] (bouncing-bet)
SY: *Lychnis saponaria* Jessen
Saponaria vaccaria L. = Vaccaria hispanica [VACCHIS]
Scleranthus L. [SCLERAN$]
*__Scleranthus annuus__ L. [SCLEANN] (annual knawel)
Silene L. [SILENE$]
Silene acaulis (L.) Jacq. [SILEACA] (moss campion)
Silene acaulis ssp. *subacaulescens* (F.N. Williams) Hultén = Silene acaulis var. subacaulescens [SILEACA5]
Silene acaulis var. **acaulis** [SILEACA0]
Silene acaulis var. **subacaulescens** (F.N. Williams) Fern. & St. John [SILEACA5]
SY: *Silene acaulis* ssp. *subacaulescens* (F.N. Williams) Hultén
Silene alba (P. Mill.) Krause = Silene latifolia ssp. alba [SILELAT1]
Silene anglica L. = Silene gallica [SILEGAL]
Silene antirrhina L. [SILEANT] (sleepy catchfly)
SY: *Silene antirrhina* var. *confinis* Fern.
SY: *Silene antirrhina* var. *depauperata* Rydb.
SY: *Silene antirrhina* var. *divaricata* B.L. Robins.
SY: *Silene antirrhina* var. *laevigata* Engelm. & Gray
SY: *Silene antirrhina* var. *subglaber* Engelm. & Gray
SY: *Silene antirrhina* var. *vaccarifolia* Rydb.
Silene antirrhina var. *confinis* Fern. = Silene antirrhina [SILEANT]
Silene antirrhina var. *depauperata* Rydb. = Silene antirrhina [SILEANT]
Silene antirrhina var. *divaricata* B.L. Robins. = Silene antirrhina [SILEANT]
Silene antirrhina var. *laevigata* Engelm. & Gray = Silene antirrhina [SILEANT]
Silene antirrhina var. *subglaber* Engelm. & Gray = Silene antirrhina [SILEANT]
Silene antirrhina var. *vaccarifolia* Rydb. = Silene antirrhina [SILEANT]
*__Silene armeria__ L. [SILEARM] (sweet William catchfly)
Silene attenuata (Farr) Bocquet = Silene uralensis ssp. attenuata [SILEURA1]
Silene coronaria (L.) Clairville = Lychnis coronaria [LYCHCOR]
*__Silene csereii__ Baumg. [SILECSE] (biennial campion)
Silene cucubalus Wibel = Silene vulgaris [SILEVUL]
Silene cucubalus var. *latifolia* (P. Mill.) G. Beck = Silene vulgaris [SILEVUL]
*__Silene dichotoma__ Ehrh. [SILEDIC] (forked catchfly)
*__Silene dioica__ (L.) Clairville [SILEDIO] (red campion)
SY: *Lychnis dioica* L.
SY: *Melandrium dioicum* (L.) Coss. & Germ.

SY: *Melandrium dioicum* ssp. *rubrum* (Wieg.) D. Löve
Silene douglasii Hook. [SILEDOU] (Douglas' campion)
SY: *Silene douglasii* var. *villosa* C.L. Hitchc. & Maguire
SY: *Silene lyallii* S. Wats.
Silene douglasii var. *macounii* (S. Wats.) B.L. Robins. = Silene parryi [SILEPAR]
Silene douglasii var. *villosa* C.L. Hitchc. & Maguire = Silene douglasii [SILEDOU]
Silene drummondii Hook. [SILEDRU] (Drummond's campion)
SY: *Gastrolychnis drummondii* (Hook.) A. & D. Löve
SY: *Lychnis drummondii* (Hook.) S. Wats.
SY: *Lychnis pudica* Boivin
SY: *Melandrium drummondii* (Hook.) Hultén
SY: *Wahlbergella drummondii* (Hook.) Rydb.
Silene furcata Raf. = Silene involucrata [SILEINV]
***Silene gallica** L. [SILEGAL] (small-flowered catchfly)
SY: *Silene anglica* L.
Silene grandis Eastw. = Silene scouleri [SILESCO]
Silene inflata Sm. = Silene vulgaris [SILEVUL]
Silene involucrata (Cham. & Schlecht.) Bocquet [SILEINV] (arctic campion)
SY: *Gastrolychnis affinis* (J. Vahl ex Fries) Tolm. & Kozh.
SY: *Gastrolychnis involucrata* (Cham. & Schlecht.) A. & D. Löve
SY: *Lychnis affinis* J. Vahl ex Fries
SY: *Lychnis furcata* (Raf.) Fern.
SY: *Lychnis gillettii* Boivin
SY: *Melandrium affine* (J. Vahl ex Fries) J. Vahl
SY: *Melandrium furcatum* (Raf.) Hadac
SY: *Silene furcata* Raf.
Silene latifolia Poir. [SILELAT]
Silene latifolia (P. Mill.) Britten & Rendle = Silene vulgaris [SILEVUL]
***Silene latifolia** ssp. **alba** (P. Mill.) Greuter & Burdet [SILELAT1]
SY: *Lychnis alba* P. Mill.
SY: *Lychnis loveae* Boivin
SY: *Lychnis* × *loveae* Boivin
SY: *Lychnis vespertina* Sibthorp
SY: *Melandrium album* (P. Mill.) Garcke
SY: *Silene alba* (P. Mill.) Krause
SY: *Silene pratensis* (Rafn) Godr. & Gren.
Silene lyallii S. Wats. = Silene douglasii [SILEDOU]
Silene macounii S. Wats. = Silene parryi [SILEPAR]
Silene menziesii Hook. [SILEMEN] (Menzies' campion)
Silene menziesii var. **viscosa** (Greene) C.L. Hitchc. & Maguire [SILEMEN2]
SY: *Anotites viscosa* Greene
***Silene noctiflora** L. [SILENOC] (night-flowering catchfly)
SY: *Melandrium noctiflorum* (L.) Fries
Silene pacifica Eastw. = Silene scouleri ssp. scouleri var. pacifica [SILESCO0]
Silene parryi (S. Wats.) C.L. Hitchc. & Maguire [SILEPAR] (Parry's campion)
SY: *Silene douglasii* var. *macounii* (S. Wats.) B.L. Robins.
SY: *Silene macounii* S. Wats.
SY: *Silene scouleri* var. *macounii* (S. Wats.) Boivin
SY: *Silene tetonensis* A. Nels.
SY: *Wahlbergella parryi* (S. Wats.) Rydb.
Silene pratensis (Rafn) Godr. & Gren. = Silene latifolia ssp. alba [SILELAT1]
Silene repens Patrin ex Pers. [SILEREP] (pink campion)
Silene repens var. *costata* (Williams) Boivin = Silene scouleri [SILESCO]
Silene scouleri Hook. [SILESCO] (Scouler's campion)
SY: *Silene grandis* Eastw.
SY: *Silene repens* var. *costata* (Williams) Boivin
SY: *Silene scouleri* ssp. *grandis* (Eastw.) C.L. Hitchc. & Maguire
Silene scouleri ssp. *grandis* (Eastw.) C.L. Hitchc. & Maguire = Silene scouleri [SILESCO]
Silene scouleri ssp. **scouleri** var. **pacifica** (Eastw.) C.L. Hitchc. [SILESCO0]
SY: *Silene pacifica* Eastw.
SY: *Silene scouleri* var. *pacifica* (Eastw.) C.L. Hitchc.
Silene scouleri var. *macounii* (S. Wats.) Boivin = Silene parryi [SILEPAR]
Silene scouleri var. *pacifica* (Eastw.) C.L. Hitchc. = Silene scouleri ssp. scouleri var. pacifica [SILESCO0]
Silene taimyrensis (Tolm.) Bocquet [SILETAI] (Taimyr campion)
SY: *Gastrolychnis ostenfeldii* (Porsild) Petrovsky
SY: *Gastrolychnis taimyrensis* (Tolm.) S.K. Czer.
SY: *Gastrolychnis triflora* ssp. *dawsonii* (B.L. Robins.) A. & D. Löve
SY: *Lychnis dawsonii* (B.L. Robins.) J.P. Andrs.
SY: *Lychnis taimyrensis* (Tolm.) Polunin
SY: *Lychnis triflora* ssp. *dawsonii* (B.L. Robins.) Maguire
SY: *Lychnis triflora* var. *dawsonii* B.L. Robins.
SY: *Melandrium ostenfeldii* Porsild
SY: *Melandrium taimyrense* Tolm.
Silene tetonensis A. Nels. = Silene parryi [SILEPAR]
Silene uralensis (Rupr.) Bocquet [SILEURA] (apetalous catchfly)
Silene uralensis ssp. **attenuata** (Farr) McNeill [SILEURA1] (apetalous campion)
SY: *Lychnis apetala* ssp. *attenuata* (Farr) Maguire
SY: *Lychnis apetala* var. *attenuata* (Farr) C.L. Hitchc.
SY: *Lychnis attenuata* Farr
SY: *Melandrium apetalum* ssp. *attenuatum* (Farr) Hara
SY: *Silene attenuata* (Farr) Bocquet
SY: *Silene wahlbergella* ssp. *attenuata* (Farr) Hultén
SY: *Wahlbergella attenuata* (Farr) Rydb.
***Silene vulgaris** (Moench) Garcke [SILEVUL] (bladder campion)
SY: *Oberna commutata* (Guss.) S. Ikonnikov
SY: *Silene cucubalus* Wibel
SY: *Silene cucubalus* var. *latifolia* (P. Mill.) G. Beck
SY: *Silene inflata* Sm.
SY: *Silene latifolia* (P. Mill.) Britten & Rendle
Silene wahlbergella ssp. *attenuata* (Farr) Hultén = Silene uralensis ssp. attenuata [SILEURA1]
Spergella intermedia (Fenzl) A. & D. Löve = Sagina nivalis [SAGINIV]
Spergella saginoides (L.) Reichenb. = Sagina saginoides [SAGISAG]
Spergula L. [SPERGUL$]
***Spergula arvensis** L. [SPERARV] (corn-spurry)
Spergularia (Pers.) J.& K. Presl [SPERGUA$]
Spergularia alata Wieg. = Spergularia salina [SPERSAL]
Spergularia canadensis (Pers.) G. Don [SPERCAN] (Canadian sand-spurry)
Spergularia canadensis var. **canadensis** [SPERCAN0]
SY: *Tissa canadensis* (Pers.) Britt.
Spergularia canadensis var. **occidentalis** R.P. Rossb. [SPERCAN2]
Spergularia leiosperma (Kindb.) F. Schmidt = Spergularia salina [SPERSAL]
Spergularia macrotheca (Hornem.) Heynh. [SPERMAC] (beach sand-spurry)
Spergularia marina (L.) Griseb. = Spergularia salina [SPERSAL]
Spergularia marina var. *leiosperma* (Kindb.) Guerke = Spergularia salina [SPERSAL]
Spergularia marina var. *simonii* O.& I. Deg. = Spergularia salina [SPERSAL]
***Spergularia rubra** (L.) J.& K. Presl [SPERRUB] (red sand-spurry)
SY: *Spergularia rubra* var. *perennans* (Kindb.) B.L. Robins.
SY: *Tissa rubra* (L.) Britt.
Spergularia rubra var. *perennans* (Kindb.) B.L. Robins. = Spergularia rubra [SPERRUB]
Spergularia salina J.& K. Presl [SPERSAL]
SY: *Spergularia alata* Wieg.
SY: *Spergularia leiosperma* (Kindb.) F. Schmidt
SY: *Spergularia marina* (L.) Griseb.
SY: *Spergularia marina* var. *leiosperma* (Kindb.) Guerke
SY: *Spergularia marina* var. *simonii* O.& I. Deg.
SY: *Spergularia sparsiflora* (Greene) A. Nels.
SY: *Tissa marina* (L.) Britt.
Spergularia sparsiflora (Greene) A. Nels. = Spergularia salina [SPERSAL]
Stellaria L. [STELLAR$]
Stellaria alaskana Hultén [STELALA]
Stellaria alsine Grimm [STELALS] (bog starwort)

SY: *Alsine uliginosa* (Murr.) Britt.
SY: *Stellaria uliginosa* Murr.
Stellaria americana (Porter ex B.L. Robins.) Standl. [STELAME] (American starwort)
SY: *Alsine americana* (Porter ex B.L. Robins.) Rydb.
SY: *Arenaria stephaniana* var. *americana* (Porter ex B.L. Robins.) Shinners
SY: *Stellaria dichotoma* var. *americana* Porter ex B.L. Robins.
Stellaria aquatica (L.) Scop. = Myosoton aquaticum [MYOSAQU]
Stellaria biflora L. = Minuartia biflora [MINUBIF]
Stellaria borealis Bigelow [STELBOR]
SY: *Stellaria calycantha* ssp. *interior* Hultén
SY: *Stellaria calycantha* var. *isophylla* (Fern.) Fern.
Stellaria borealis ssp. *bongardiana* (Fern.) Piper & Beattie = Stellaria borealis ssp. sitchana [STELBOR1]
Stellaria borealis ssp. **sitchana** (Steud.) Piper [STELBOR1]
SY: *Alsine bongardiana* (Fern.) A. Davids. & Moxley
SY: *Alsine brachypetala* (Bong.) T.J. Howell
SY: *Stellaria borealis* ssp. *bongardiana* (Fern.) Piper & Beattie
SY: *Stellaria borealis* var. *bongardiana* Fern.
SY: *Stellaria borealis* var. *sitchana* (Steud.) Fern.
SY: *Stellaria brachypetala* var. *bongardiana* Fern.
SY: *Stellaria calycantha* var. *bongardiana* (Fern.) Fern.
SY: *Stellaria calycantha* var. *sitchana* (Steud.) Fern.
SY: *Stellaria sitchana* Steud.
SY: *Stellaria sitchana* var. *bongardiana* (Fern.) Hultén
Stellaria borealis var. *bongardiana* Fern. = Stellaria borealis ssp. sitchana [STELBOR1]
Stellaria borealis var. *crispa* (Cham. & Schlecht.) Fenzl ex Torr. & Gray = Stellaria crispa [STELCRI]
Stellaria borealis var. *simcoei* (T.J. Howell) Fern. = Stellaria calycantha [STELCAL]
Stellaria borealis var. *sitchana* (Steud.) Fern. = Stellaria borealis ssp. sitchana [STELBOR1]
Stellaria brachypetala var. *bongardiana* Fern. = Stellaria borealis ssp. sitchana [STELBOR1]
Stellaria calycantha (Ledeb.) Bong. [STELCAL] (northern starwort)
SY: *Alsine calycantha* (Ledeb.) Rydb.
SY: *Alsine simcoei* T.J. Howell
SY: *Stellaria borealis* var. *simcoei* (T.J. Howell) Fern.
SY: *Stellaria calycantha* var. *simcoei* (T.J. Howell) Fern.
SY: *Stellaria simcoei* (T.J. Howell) C.L. Hitchc.
Stellaria calycantha ssp. *interior* Hultén = Stellaria borealis [STELBOR]
Stellaria calycantha var. *bongardiana* (Fern.) Fern. = Stellaria borealis ssp. sitchana [STELBOR1]
Stellaria calycantha var. *isophylla* (Fern.) Fern. = Stellaria borealis [STELBOR]
Stellaria calycantha var. *simcoei* (T.J. Howell) Fern. = Stellaria calycantha [STELCAL]
Stellaria calycantha var. *sitchana* (Steud.) Fern. = Stellaria borealis ssp. sitchana [STELBOR1]
Stellaria crassifolia Ehrh. [STELCRA] (thick-leaved starwort)
Stellaria crispa Cham. & Schlecht. [STELCRI] (crisp starwort)
SY: *Alsine crispa* (Cham. & Schlecht.) Holz.
SY: *Stellaria borealis* var. *crispa* (Cham. & Schlecht.) Fenzl ex Torr. & Gray
Stellaria dichotoma var. *americana* Porter ex B.L. Robins. = Stellaria americana [STELAME]
Stellaria edwardsii R. Br. = Stellaria longipes [STELLOG]
Stellaria gonomischa Boivin = Stellaria umbellata [STELUMB]
***Stellaria graminea** L. [STELGRA] (grass-leaved starwort)
Stellaria humifusa Rottb. [STELHUM] (salt marsh starwort)
SY: *Alsine humifusa* (Rottb.) Britt.
SY: *Stellaria humifusa* var. *oblongifolia* Fenzl
SY: *Stellaria humifusa* var. *suberecta* Boivin
Stellaria humifusa var. *oblongifolia* Fenzl = Stellaria humifusa [STELHUM]
Stellaria humifusa var. *suberecta* Boivin = Stellaria humifusa [STELHUM]
Stellaria laeta Richards. = Stellaria longipes [STELLOG]
Stellaria longipes Goldie [STELLOG] (long-stalked starwort)
SY: *Stellaria edwardsii* R. Br.
SY: *Stellaria laeta* Richards.
SY: *Stellaria longipes* var. *altocaulis* (Hultén) C.L. Hitchc.
SY: *Stellaria longipes* var. *edwardsii* (R. Br.) Gray
SY: *Stellaria longipes* var. *laeta* (Richards.) S. Wats.
SY: *Stellaria longipes* var. *subvestita* (Greene) Polunin
SY: *Stellaria monantha* Hultén
SY: *Stellaria monantha* var. *altocaulis* Hultén
SY: *Stellaria stricta* Richards.
SY: *Stellaria subvestita* Greene
Stellaria longipes var. *altocaulis* (Hultén) C.L. Hitchc. = Stellaria longipes [STELLOG]
Stellaria longipes var. *edwardsii* (R. Br.) Gray = Stellaria longipes [STELLOG]
Stellaria longipes var. *laeta* (Richards.) S. Wats. = Stellaria longipes [STELLOG]
Stellaria longipes var. *subvestita* (Greene) Polunin = Stellaria longipes [STELLOG]
***Stellaria media** (L.) Vill. [STELMED] (common starwort)
Stellaria monantha Hultén = Stellaria longipes [STELLOG]
Stellaria monantha var. *altocaulis* Hultén = Stellaria longipes [STELLOG]
Stellaria nitens Nutt. [STELNIT] (shiny starwort)
SY: *Stellaria praecox* A. Nels.
Stellaria obtusa Engelm. [STELOBT] (blunt-sepaled starwort)
SY: *Alsine obtusa* (Engelm.) Rose
SY: *Alsine viridula* Piper
SY: *Alsine washingtoniana* (B.L. Robins.) Heller
SY: *Stellaria viridula* (Piper) St. John
SY: *Stellaria washingtoniana* B.L. Robins.
Stellaria praecox A. Nels. = Stellaria nitens [STELNIT]
Stellaria simcoei (T.J. Howell) C.L. Hitchc. = Stellaria calycantha [STELCAL]
Stellaria sitchana Steud. = Stellaria borealis ssp. sitchana [STELBOR1]
Stellaria sitchana var. *bongardiana* (Fern.) Hultén = Stellaria borealis ssp. sitchana [STELBOR1]
Stellaria stricta Richards. = Stellaria longipes [STELLOG]
Stellaria subvestita Greene = Stellaria longipes [STELLOG]
Stellaria uliginosa Murr. = Stellaria alsine [STELALS]
Stellaria umbellata Turcz. ex Kar. & Kir. [STELUMB] (umbellate starwort)
SY: *Alsine baicalensis* Coville
SY: *Stellaria gonomischa* Boivin
SY: *Stellaria weberi* Boivin
Stellaria viridula (Piper) St. John = Stellaria obtusa [STELOBT]
Stellaria washingtoniana B.L. Robins. = Stellaria obtusa [STELOBT]
Stellaria weberi Boivin = Stellaria umbellata [STELUMB]
Tissa canadensis (Pers.) Britt. = Spergularia canadensis var. canadensis [SPERCAN0]
Tissa marina (L.) Britt. = Spergularia salina [SPERSAL]
Tissa rubra (L.) Britt. = Spergularia rubra [SPERRUB]
Tryphane rubella (Wahlenb.) Reichenb. = Minuartia rubella [MINURUB]
Tunica prolifera (L.) Scop. = Petrorhagia prolifera [PETRPRO]
Tunica saxifraga (L.) Scop. = Petrorhagia saxifraga [PETRSAX]
Vaccaria von Wolf [VACCARI$]
***Vaccaria hispanica** (P. Mill.) Rauschert [VACCHIS]
SY: *Saponaria vaccaria* L.
SY: *Vaccaria pyramidata* Medik.
SY: *Vaccaria segetalis* Garcke ex Aschers.
SY: *Vaccaria vaccaria* (L.) Britt.
SY: *Vaccaria vulgaris* Host
Vaccaria pyramidata Medik. = Vaccaria hispanica [VACCHIS]
Vaccaria segetalis Garcke ex Aschers. = Vaccaria hispanica [VACCHIS]
Vaccaria vaccaria (L.) Britt. = Vaccaria hispanica [VACCHIS]
Vaccaria vulgaris Host = Vaccaria hispanica [VACCHIS]
Wahlbergella attenuata (Farr) Rydb. = Silene uralensis ssp. attenuata [SILEURA1]
Wahlbergella drummondii (Hook.) Rydb. = Silene drummondii [SILEDRU]
Wahlbergella parryi (S. Wats.) Rydb. = Silene parryi [SILEPAR]

CELASTRACEAE (V:D)

Euonymus L. [EUONYMU$]
Euonymus occidentalis Nutt. ex Torr. [EUONOCC] (western wahoo)
SY: *Evonymus occidentalis* Nutt. ex Torr.
Evonymus occidentalis Nutt. ex Torr. = Euonymus occidentalis [EUONOCC]
Pachistima myrsinites (Pursh) Raf. = Paxistima myrsinites [PAXIMYR]
Paxistima Raf. [PAXISTI$]
Paxistima myrsinites (Pursh) Raf. [PAXIMYR] (falsebox)
SY: *Pachistima myrsinites* (Pursh) Raf.

CERATOPHYLLACEAE (V:D)

Ceratophyllum L. [CERATOP$]
Ceratophyllum apiculatum Cham. = Ceratophyllum demersum [CERADEM]
Ceratophyllum demersum L. [CERADEM] (common hornwort)
SY: *Ceratophyllum apiculatum* Cham.
SY: *Ceratophyllum demersum* var. *apiculatum* (Cham.) Garcke
SY: *Ceratophyllum demersum* var. *apiculatum* (Cham.) Aschers.
Ceratophyllum demersum var. *apiculatum* (Cham.) Aschers. = Ceratophyllum demersum [CERADEM]
Ceratophyllum demersum var. *apiculatum* (Cham.) Garcke = Ceratophyllum demersum [CERADEM]
Ceratophyllum demersum var. *echinatum* (Gray) Gray = Ceratophyllum echinatum [CERAECH]
Ceratophyllum echinatum Gray [CERAECH] (spring hornwort)
SY: *Ceratophyllum demersum* var. *echinatum* (Gray) Gray
SY: *Ceratophyllum submersum* var. *echinatum* (Gray) Wilmot-Dear
Ceratophyllum submersum var. *echinatum* (Gray) Wilmot-Dear = Ceratophyllum echinatum [CERAECH]

CHENOPODIACEAE (V:D)

Atriplex L. [ATRIPLE$]
Atriplex alaskensis S. Wats. [ATRIALA] (Alaska orache)
SY: *Atriplex patula* ssp. *alaskensis* (S. Wats.) Hall & Clements
SY: *Atriplex patula* var. *alaskensis* (S. Wats.) Welsh
Atriplex argentea Nutt. [ATRIARG] (silverscale)
Atriplex buxifolia Rydb. = Atriplex nuttallii [ATRINUT]
Atriplex gmelinii C.A. Mey. ex Bong. [ATRIGME] (Gmelin's orache)
SY: *Atriplex gmelinii* var. *zosterifolia* (Hook.) Moq.
SY: *Atriplex patula* ssp. *obtusa* (Cham.) Hall & Clements
SY: *Atriplex patula* var. *obtusa* (Cham.) M.E. Peck
SY: *Atriplex patula* ssp. *zosterifolia* (Hook.) Hall & Clements
SY: *Atriplex patula* var. *zosterifolia* (Hook.) C.L. Hitchc.
Atriplex gmelinii var. *zosterifolia* (Hook.) Moq. = Atriplex gmelinii [ATRIGME]
Atriplex heterosperma Bunge = Atriplex micrantha [ATRIMIC]
***Atriplex hortensis** L. [ATRIHOR] (garden orache)
SY: *Atriplex hortensis* var. *atrosanguinea* hort.
Atriplex hortensis var. *atrosanguinea* hort. = Atriplex hortensis [ATRIHOR]
Atriplex longipes Drej. = Atriplex nudicaulis [ATRINUD]
Atriplex longipes ssp. *praecox* (Hülphers) Tresson = Atriplex nudicaulis [ATRINUD]
***Atriplex micrantha** C.A. Mey. [ATRIMIC]
SY: *Atriplex heterosperma* Bunge
***Atriplex nudicaulis** Boguslaw [ATRINUD]
SY: *Atriplex longipes* Drej.
SY: *Atriplex longipes* ssp. *praecox* (Hülphers) Tresson
SY: *Atriplex praecox* Hülphers
Atriplex nuttallii S. Wats. [ATRINUT] (Nuttall's orache)
SY: *Atriplex buxifolia* Rydb.
SY: *Atriplex nuttallii* ssp. *buxifolia* (Rydb.) Hall & Clements
Atriplex nuttallii ssp. *buxifolia* (Rydb.) Hall & Clements = Atriplex nuttallii [ATRINUT]
***Atriplex oblongifolia** Waldst. & Kit. [ATRIOBL] (oblong-leaved orache)
Atriplex patula L. [ATRIPAT] (common orache)
Atriplex patula ssp. *alaskensis* (S. Wats.) Hall & Clements = Atriplex alaskensis [ATRIALA]
Atriplex patula ssp. *hastata* Hall & Clements p.p. = Atriplex subspicata [ATRISUB]
Atriplex patula ssp. *obtusa* (Cham.) Hall & Clements = Atriplex gmelinii [ATRIGME]
Atriplex patula ssp. *subspicata* (Nutt.) Fosberg = Atriplex subspicata [ATRISUB]
Atriplex patula ssp. *zosterifolia* (Hook.) Hall & Clements = Atriplex gmelinii [ATRIGME]
Atriplex patula var. *alaskensis* (S. Wats.) Welsh = Atriplex alaskensis [ATRIALA]
Atriplex patula var. *obtusa* (Cham.) M.E. Peck = Atriplex gmelinii [ATRIGME]
Atriplex patula var. *subspicata* (Nutt.) S. Wats. = Atriplex subspicata [ATRISUB]
Atriplex patula var. *zosterifolia* (Hook.) C.L. Hitchc. = Atriplex gmelinii [ATRIGME]
Atriplex praecox Hülphers = Atriplex nudicaulis [ATRINUD]
***Atriplex rosea** L. [ATRIROS] (red orache)
Atriplex subspicata (Nutt.) Rydb. [ATRISUB] (saline orache)
SY: *Atriplex patula* ssp. *hastata* Hall & Clements p.p.
SY: *Atriplex patula* ssp. *subspicata* (Nutt.) Fosberg
SY: *Atriplex patula* var. *subspicata* (Nutt.) S. Wats.
Atriplex truncata (Torr. ex S. Wats.) Gray [ATRITRU] (wedgescalf orache)
Axyris L. [AXYRIS$]
***Axyris amaranthoides** L. [AXYRAMA] (Russian pigweed)
Bassia All. [BASSIA$]
***Bassia hyssopifolia** (Pallas) Volk. [BASSHYS] (five-hooked bassia)
SY: *Echinopsilon hyssopifolius* (Pallas) Moq.
SY: *Kochia hyssopifolia* (Pallas) Schrad.
Bassia sieversiana (Pallas) W.A. Weber = Kochia scoparia [KOCHSCO]
Blitum capitatum L. = Chenopodium capitatum [CHENCAP]
Blitum chenopodioides L. = Chenopodium botryodes [CHENBOR]
Botrydium botrys (L.) Small = Chenopodium botrys [CHENBOT]
Chenopodium L. [CHENOPO$]
***Chenopodium album** L. [CHENALB] (lamb's-quarters)
SY: *Chenopodium album* var. *lanceolatum* (Muhl. ex Willd.) Coss. & Germ.
SY: *Chenopodium album* var. *polymorphum* Aellen
SY: *Chenopodium amaranticolor* Coste & Reyn.
SY: *Chenopodium giganteum* D. Don
SY: *Chenopodium lanceolatum* Muhl. ex Willd.
SY: *Chenopodium suecicum* J. Murr.
Chenopodium album ssp. *striatum* (Krasan) J. Murr = Chenopodium strictum ssp. striatum [CHENSTR2]
Chenopodium album var. *lanceolatum* (Muhl. ex Willd.) Coss. & Germ. = Chenopodium album [CHENALB]
Chenopodium album var. *polymorphum* Aellen = Chenopodium album [CHENALB]
Chenopodium amaranticolor Coste & Reyn. = Chenopodium album [CHENALB]
Chenopodium aridum A. Nels. = Chenopodium atrovirens [CHENATR]
Chenopodium atrovirens Rydb. [CHENATR] (dark lamb's-quarters)
SY: *Chenopodium aridum* A. Nels.
SY: *Chenopodium fremontii* var. *atrovirens* (Rydb.) Fosberg
SY: *Chenopodium incognitum* H.A. Wahl p.p.
SY: *Chenopodium wolfii* Rydb.
Chenopodium botryodes Sm. [CHENBOR]
SY: *Blitum chenopodioides* L.
SY: *Chenopodium chenopodioides* (L.) Aellen
SY: *Chenopodium chenopodioides* var. *degenianum* (Aellen) Aellen
SY: *Chenopodium chenopodioides* var. *lengyelianum* (Aellen) Aellen
SY: *Chenopodium rubrum* var. *glomeratum* Wallr.
SY: *Chenopodium rubrum* var. *humile* auct.

*__Chenopodium botrys__ L. [CHENBOT] (Jerusalem-oak)
SY: *Botrydium botrys* (L.) Small
SY: *Teloxys botrys* (L.) W.A. Weber
Chenopodium capitatum (L.) Aschers. [CHENCAP] (strawberry-blite)
SY: *Blitum capitatum* L.
Chenopodium chenopodioides (L.) Aellen = Chenopodium botryodes [CHENBOR]
Chenopodium chenopodioides var. *degenianum* (Aellen) Aellen = Chenopodium botryodes [CHENBOR]
Chenopodium chenopodioides var. *lengyelianum* (Aellen) Aellen = Chenopodium botryodes [CHENBOR]
Chenopodium desiccatum A. Nels. [CHENDES]
SY: *Chenopodium leptophyllum* var. *oblongifolium* S. Wats.
SY: *Chenopodium oblongifolium* (S. Wats.) Rydb.
SY: *Chenopodium pratericola* ssp. *desiccatum* (A. Nels.) Aellen
SY: *Chenopodium pratericola* var. *oblongifolium* (S. Wats.) H.A. Wahl
Chenopodium fremontii var. *atrovirens* (Rydb.) Fosberg = Chenopodium atrovirens [CHENATR]
Chenopodium giganteum D. Don = Chenopodium album [CHENALB]
Chenopodium gigantospermum Aellen = Chenopodium simplex [CHENSIM]
Chenopodium glaucum ssp. *salinum* (Standl.) Aellen = Chenopodium salinum [CHENSAL]
Chenopodium glaucum var. *pulchrum* Aellen = Chenopodium salinum [CHENSAL]
Chenopodium glaucum var. *salinum* (Standl.) Boivin = Chenopodium salinum [CHENSAL]
Chenopodium hybridum auct. = Chenopodium simplex [CHENSIM]
Chenopodium hybridum ssp. *gigantospermum* (Aellen) Hultén = Chenopodium simplex [CHENSIM]
Chenopodium hybridum var. *gigantospermum* (Aellen) Rouleau = Chenopodium simplex [CHENSIM]
Chenopodium hybridum var. *simplex* Torr. = Chenopodium simplex [CHENSIM]
Chenopodium incognitum H.A. Wahl p.p. = Chenopodium atrovirens [CHENATR]
Chenopodium lanceolatum Muhl. ex Willd. = Chenopodium album [CHENALB]
Chenopodium leptophyllum var. *oblongifolium* S. Wats. = Chenopodium desiccatum [CHENDES]
Chenopodium oblongifolium (S. Wats.) Rydb. = Chenopodium desiccatum [CHENDES]
Chenopodium pratericola ssp. *desiccatum* (A. Nels.) Aellen = Chenopodium desiccatum [CHENDES]
Chenopodium pratericola var. *oblongifolium* (S. Wats.) H.A. Wahl = Chenopodium desiccatum [CHENDES]
Chenopodium rubrum L. [CHENRUB] (red goosefoot)
Chenopodium rubrum var. *glomeratum* Wallr. = Chenopodium botryodes [CHENBOR]
Chenopodium rubrum var. *humile* auct. = Chenopodium botryodes [CHENBOR]
Chenopodium salinum Standl. [CHENSAL]
SY: *Chenopodium glaucum* var. *pulchrum* Aellen
SY: *Chenopodium glaucum* ssp. *salinum* (Standl.) Aellen
SY: *Chenopodium glaucum* var. *salinum* (Standl.) Boivin
*__Chenopodium simplex__ (Torr.) Raf. [CHENSIM]
SY: *Chenopodium gigantospermum* Aellen
SY: *Chenopodium hybridum* ssp. *gigantospermum* (Aellen) Hultén
SY: *Chenopodium hybridum* var. *gigantospermum* (Aellen) Rouleau
SY: *Chenopodium hybridum* var. *simplex* Torr.
SY: *Chenopodium hybridum* auct.
Chenopodium strictum Roth [CHENSTR]
Chenopodium strictum ssp. **glaucophyllum** (Aellen) Aellen & Just. [CHENSTR1]
SY: *Chenopodium strictum* var. *glaucophyllum* (Aellen) H.A. Wahl
*__Chenopodium strictum__ ssp. **striatum** (Krasan) Aellen & Iljin [CHENSTR2]
SY: *Chenopodium album* ssp. *striatum* (Krasan) J. Murr
Chenopodium strictum var. *glaucophyllum* (Aellen) H.A. Wahl = Chenopodium strictum ssp. glaucophyllum [CHENSTR1]
Chenopodium suecicum J. Murr. = Chenopodium album [CHENALB]
*__Chenopodium urbicum__ L. [CHENURB] (upright goosefoot)
SY: *Chenopodium urbicum* var. *intermedium* (Mert. & Koch) W.D.J. Koch
Chenopodium urbicum var. *intermedium* (Mert. & Koch) W.D.J. Koch = Chenopodium urbicum [CHENURB]
Chenopodium wolfii Rydb. = Chenopodium atrovirens [CHENATR]
Corispermum L. [CORISPE$]
Corispermum americanum (Nutt.) Nutt. = Corispermum hyssopifolium [CORIHYS]
Corispermum hyssopifolium L. [CORIHYS] (bugseed)
SY: *Corispermum americanum* (Nutt.) Nutt.
SY: *Corispermum hyssopifolium* var. *rubricaule* Hook.
SY: *Corispermum marginale* Rydb.
SY: *Corispermum simplicissimum* Lunell
Corispermum hyssopifolium var. *rubricaule* Hook. = Corispermum hyssopifolium [CORIHYS]
Corispermum marginale Rydb. = Corispermum hyssopifolium [CORIHYS]
Corispermum simplicissimum Lunell = Corispermum hyssopifolium [CORIHYS]
Echinopsilon hyssopifolius (Pallas) Moq. = Bassia hyssopifolia [BASSHYS]
Kochia Roth [KOCHIA$]
Kochia alata Bates = Kochia scoparia [KOCHSCO]
Kochia hyssopifolia (Pallas) Schrad. = Bassia hyssopifolia [BASSHYS]
*__Kochia scoparia__ (L.) Schrad. [KOCHSCO] (summer-cypress)
SY: *Bassia sieversiana* (Pallas) W.A. Weber
SY: *Kochia alata* Bates
SY: *Kochia scoparia* var. *culta* Farw.
SY: *Kochia scoparia* var. *pubescens* Fenzl
SY: *Kochia scoparia* var. *subvillosa* Moq.
SY: *Kochia scoparia* var. *trichophila* (Stapf) Bailey
SY: *Kochia sieversiana* (Pallas) C.A. Mey.
SY: *Kochia trichophila* Stapf
Kochia scoparia var. *culta* Farw. = Kochia scoparia [KOCHSCO]
Kochia scoparia var. *pubescens* Fenzl = Kochia scoparia [KOCHSCO]
Kochia scoparia var. *subvillosa* Moq. = Kochia scoparia [KOCHSCO]
Kochia scoparia var. *trichophila* (Stapf) Bailey = Kochia scoparia [KOCHSCO]
Kochia sieversiana (Pallas) C.A. Mey. = Kochia scoparia [KOCHSCO]
Kochia trichophila Stapf = Kochia scoparia [KOCHSCO]
Monolepis Schrad. [MONOLEP$]
*__Monolepis nuttalliana__ (J.A. Schultes) Greene [MONONUT] (poverty weed)
Polycnemum L. [POLYCNE$]
*__Polycnemum arvense__ L. [POLYARV] (crunch weed)
Salicornia L. [SALICOR$]
Salicornia depressa Standl. = Salicornia virginica [SALIVIR]
Salicornia europaea auct. = Salicornia maritima [SALIMAR]
Salicornia europaea ssp. *rubra* (A. Nels.) Breitung = Salicornia rubra [SALIRUB]
Salicornia europaea var. *prona* (Lunell) Boivin = Salicornia rubra [SALIRUB]
Salicornia maritima Wolff & Jefferies [SALIMAR]
SY: *Salicornia europaea* auct.
Salicornia pacifica Standl. = Sarcocornia pacifica [SARCPAC]
Salicornia rubra A. Nels. [SALIRUB]
SY: *Salicornia europaea* var. *prona* (Lunell) Boivin
SY: *Salicornia europaea* ssp. *rubra* (A. Nels.) Breitung
Salicornia virginica L. [SALIVIR] (American glasswort)
SY: *Salicornia depressa* Standl.
Salsola L. [SALSOLA$]
*__Salsola kali__ L. [SALSKAL] (Russian thistle)
Sarcocornia A.J. Scott [SARCOCO$]
Sarcocornia pacifica (Standl.) A.J. Scott [SARCPAC]
SY: *Salicornia pacifica* Standl.

Schobera occidentalis S. Wats. = Suaeda calceoliformis [SUAECAL]
Suaeda Forsk. ex Scop. [SUAEDA$]
Suaeda americana (Pers.) Fern. = Suaeda calceoliformis [SUAECAL]
Suaeda calceoliformis (Hook.) Moq. [SUAECAL] (seablite)
SY: *Schobera occidentalis* S. Wats.
SY: *Suaeda americana* (Pers.) Fern.
SY: *Suaeda depressa* var. *erecta* S. Wats.
SY: *Suaeda maritima* var. *americana* (Pers.) Boivin
SY: *Suaeda minutiflora* S. Wats.
SY: *Suaeda occidentalis* (S. Wats.) S. Wats.
SY: *Suaeda depressa* auct.
Suaeda depressa auct. = Suaeda calceoliformis [SUAECAL]
Suaeda depressa var. *erecta* S. Wats. = Suaeda calceoliformis [SUAECAL]
Suaeda maritima var. *americana* (Pers.) Boivin = Suaeda calceoliformis [SUAECAL]
Suaeda minutiflora S. Wats. = Suaeda calceoliformis [SUAECAL]
Suaeda occidentalis (S. Wats.) S. Wats. = Suaeda calceoliformis [SUAECAL]
Teloxys botrys (L.) W.A. Weber = Chenopodium botrys [CHENBOT]

CLUSIACEAE (V:D)

Hypericum L. [HYPERIC$]
Hypericum anagalloides Cham. & Schlecht. [HYPEANA] (bog St. John's-wort)
Hypericum canadense var. *majus* Gray = Hypericum majus [HYPEMAJ]
Hypericum formosum auct. = Hypericum scouleri [HYPESCO]
Hypericum formosum ssp. *scouleri* (Hook.) C.L. Hitchc. = Hypericum scouleri ssp. scouleri [HYPESCO0]
Hypericum formosum var. *nortoniae* (M.E. Jones) C.L. Hitchc. = Hypericum scouleri ssp. nortoniae [HYPESCO4]
Hypericum formosum var. *scouleri* (Hook.) Coult. = Hypericum scouleri ssp. scouleri [HYPESCO0]
Hypericum majus (Gray) Britt. [HYPEMAJ] (large Canadian St. John's-wort)
SY: *Hypericum canadense* var. *majus* Gray
Hypericum nortoniae M.E. Jones = Hypericum scouleri ssp. nortoniae [HYPESCO4]
*__Hypericum perforatum__ L. [HYPEPER] (common St. John's-wort)
Hypericum scouleri Hook. [HYPESCO] (Scouler's St. John's-wort)
SY: *Hypericum formosum* auct.
Hypericum scouleri ssp. **nortoniae** (M.E. Jones) C.L. Hitchc. [HYPESCO4]
SY: *Hypericum formosum* var. *nortoniae* (M.E. Jones) C.L. Hitchc.
SY: *Hypericum nortoniae* M.E. Jones
Hypericum scouleri ssp. **scouleri** [HYPESCO0]
SY: *Hypericum formosum* ssp. *scouleri* (Hook.) C.L. Hitchc.
SY: *Hypericum formosum* var. *scouleri* (Hook.) Coult.

CONVOLVULACEAE (V:D)

Calystegia R. Br. [CALYSTE$]
*__Calystegia sepium__ (L.) R. Br. [CALYSEP] (hedge bindweed)
SY: *Convolvulus nashii* House
SY: *Convolvulus sepium* L.
SY: *Convolvulus sepium* var. *communis* R. Tryon
Calystegia soldanella (L.) R. Br. ex Roemer & J.A. Schultes [CALYSOL] (beach bindweed)
SY: *Convolvulus soldanella* L.
Convolvulus L. [CONVOLV$]
Convolvulus ambigens House = Convolvulus arvensis [CONVARV]
*__Convolvulus arvensis__ L. [CONVARV] (field bindweed)
SY: *Convolvulus ambigens* House
SY: *Strophocaulos arvensis* (L.) Small
Convolvulus nashii House = Calystegia sepium [CALYSEP]
Convolvulus sepium L. = Calystegia sepium [CALYSEP]
Convolvulus sepium var. *communis* R. Tryon = Calystegia sepium [CALYSEP]
Convolvulus soldanella L. = Calystegia soldanella [CALYSOL]
Strophocaulos arvensis (L.) Small = Convolvulus arvensis [CONVARV]

CORNACEAE (V:D)

Chamaepericlymenum canadense (L.) Aschers. & Graebn. = Cornus canadensis [CORNCAN]
Chamaepericlymenum suecicum (L.) Aschers. & Graebn. = Cornus suecica [CORNSUE]
Chamaepericlymenum unalaschkense (Ledeb.) Rydb. = Cornus unalaschkensis [CORNUNA]
Cornella canadensis (L.) Rydb. = Cornus canadensis [CORNCAN]
Cornella suecica (L.) Rydb. = Cornus suecica [CORNSUE]
Cornus L. [CORNUS$]
Cornus alba L. p.p. = Cornus sericea [CORNSER]
Cornus alba ssp. *stolonifera* (Michx.) Wangerin = Cornus sericea [CORNSER]
Cornus alba var. *baileyi* (Coult. & Evans) Boivin = Cornus sericea [CORNSER]
Cornus alba var. *californica* (C.A. Mey.) Boivin = Cornus sericea [CORNSER]
Cornus alba var. *interior* (Rydb.) Boivin = Cornus sericea [CORNSER]
Cornus baileyi Coult. & Evans = Cornus sericea [CORNSER]
Cornus × *californica* C.A. Mey. = Cornus sericea [CORNSER]
Cornus canadensis L. [CORNCAN] (bunchberry)
SY: *Chamaepericlymenum canadense* (L.) Aschers. & Graebn.
SY: *Cornella canadensis* (L.) Rydb.
SY: *Cornus canadensis* var. *dutillyi* (Lepage) Boivin
Cornus canadensis var. *dutillyi* (Lepage) Boivin = Cornus canadensis [CORNCAN]
Cornus instolonea A. Nels. = Cornus sericea [CORNSER]
Cornus interior (Rydb.) N. Petersen = Cornus sericea [CORNSER]
Cornus nuttallii Audubon ex Torr. & Gray [CORNNUT] (western dogwood)
Cornus sericea L. [CORNSER] (red-osier dogwood)
SY: *Cornus alba* L. p.p.
SY: *Cornus alba* var. *baileyi* (Coult. & Evans) Boivin
SY: *Cornus alba* var. *californica* (C.A. Mey.) Boivin
SY: *Cornus alba* var. *interior* (Rydb.) Boivin
SY: *Cornus alba* ssp. *stolonifera* (Michx.) Wangerin
SY: *Cornus baileyi* Coult. & Evans
SY: *Cornus* × *californica* C.A. Mey.
SY: *Cornus instolonea* A. Nels.
SY: *Cornus interior* (Rydb.) N. Petersen
SY: *Cornus sericea* var. *interior* (Rydb.) St. John
SY: *Cornus sericea* ssp. *stolonifera* (Michx.) Fosberg
SY: *Cornus stolonifera* Michx.
SY: *Cornus stolonifera* var. *baileyi* (Coult. & Evans) Drescher
SY: *Cornus stolonifera* var. *interior* (Rydb.) St. John
SY: *Swida instolonea* (A. Nels.) Rydb.
SY: *Swida stolonifera* (Michx.) Rydb.
Cornus sericea ssp. *stolonifera* (Michx.) Fosberg = Cornus sericea [CORNSER]
Cornus sericea var. *interior* (Rydb.) St. John = Cornus sericea [CORNSER]
Cornus stolonifera Michx. = Cornus sericea [CORNSER]
Cornus stolonifera var. *baileyi* (Coult. & Evans) Drescher = Cornus sericea [CORNSER]
Cornus stolonifera var. *interior* (Rydb.) St. John = Cornus sericea [CORNSER]
Cornus suecica L. [CORNSUE] (dwarf dog bunchberry)
SY: *Chamaepericlymenum suecicum* (L.) Aschers. & Graebn.
SY: *Cornella suecica* (L.) Rydb.
SY: *Swida suecica* (L.) Holub
Cornus unalaschkensis Ledeb. [CORNUNA] (Cordilleran bunchberry)
SY: *Chamaepericlymenum unalaschkense* (Ledeb.) Rydb.
Swida instolonea (A. Nels.) Rydb. = Cornus sericea [CORNSER]
Swida stolonifera (Michx.) Rydb. = Cornus sericea [CORNSER]
Swida suecica (L.) Holub = Cornus suecica [CORNSUE]

CRASSULACEAE (V:D)

Amerosedum divergens (S. Wats.) A. & D. Löve = Sedum divergens [SEDUDIV]
Amerosedum lanceolatum (Torr.) A. & D. Löve = Sedum lanceolatum ssp. lanceolatum [SEDULAN0]
Amerosedum nesioticum (G.N. Jones) A. & D. Löve = Sedum lanceolatum ssp. nesioticum [SEDULAN1]
Bulliarda aquatica (L.) DC. = Crassula aquatica [CRASAQU]
Crassula L. [CRASSUL$]
Crassula aquatica (L.) Schoenl. [CRASAQU] (pigmyweed)
SY: *Bulliarda aquatica* (L.) DC.
SY: *Hydrophila aquatica* (L.) House
SY: *Tillaea angustifolia* Nutt.
SY: *Tillaea aquatica* L.
SY: *Tillaea ascendens* Eat.
SY: *Tillaeastrum aquaticum* (L.) Britt.
Crassula connata (Ruiz & Pavón) Berger [CRASCON]
SY: *Crassula erecta* (Hook. & Arn.) Berger
SY: *Tillaea erecta* Hook. & Arn.
SY: *Tillaea minima* Gay
SY: *Tillaea minima* Miers ex Hook. & Arn.
SY: *Tillaea rubescens* Kunth
Crassula erecta (Hook. & Arn.) Berger = Crassula connata [CRASCON]
Gordonia spathulifolia (Hook.) A. & D. Löve = Sedum spathulifolium ssp. spathulifolium [SEDUSPA0]
Hydrophila aquatica (L.) House = Crassula aquatica [CRASAQU]
Oreosedum album (L.) Grulich = Sedum album [SEDUALB]
Rhodiola integrifolia Raf. = Sedum integrifolium [SEDUINT]
Sedum L. [SEDUM$]
***Sedum acre** L. [SEDUACR] (goldmoss stonecrop)
Sedum alaskanum (Rose) Rose ex Hutchinson = Sedum integrifolium [SEDUINT]
***Sedum album** L. [SEDUALB] (white stonecrop)
SY: *Oreosedum album* (L.) Grulich
Sedum divergens S. Wats. [SEDUDIV] (spreading stonecrop)
SY: *Amerosedum divergens* (S. Wats.) A. & D. Löve
Sedum douglasii Hook. = Sedum stenopetalum [SEDUSTE]
Sedum frigidum Rydb. = Sedum integrifolium [SEDUINT]
Sedum integrifolium (Raf.) A. Nels. [SEDUINT] (entire-leaf stonecrop)
SY: *Rhodiola integrifolia* Raf.
SY: *Sedum alaskanum* (Rose) Rose ex Hutchinson
SY: *Sedum frigidum* Rydb.
SY: *Sedum rosea* var. *alaskanum* (Rose) Berger
SY: *Sedum rosea* var. *frigidum* (Rydb.) Hultén
SY: *Sedum rosea* ssp. *integrifolium* (Raf.) Hultén
SY: *Sedum rosea* var. *integrifolium* (Raf.) Berger
SY: *Tolmachevia integrifolia* (Raf.) A. & D. Löve
Sedum lanceolatum Torr. [SEDULAN] (lance-leaved stonecrop)
Sedum lanceolatum ssp. **lanceolatum** [SEDULAN0]
SY: *Amerosedum lanceolatum* (Torr.) A. & D. Löve
Sedum lanceolatum ssp. **nesioticum** (G.N. Jones) Clausen [SEDULAN1]
SY: *Amerosedum nesioticum* (G.N. Jones) A. & D. Löve
SY: *Sedum lanceolatum* var. *nesioticum* (G.N. Jones) C.L. Hitchc.
Sedum lanceolatum var. *nesioticum* (G.N. Jones) C.L. Hitchc. = Sedum lanceolatum ssp. nesioticum [SEDULAN1]
Sedum oreganum Nutt. [SEDUORE] (Oregon stonecrop)
Sedum pruinosum Britt. = Sedum spathulifolium ssp. pruinosum [SEDUSPA1]
Sedum rosea ssp. *integrifolium* (Raf.) Hultén = Sedum integrifolium [SEDUINT]
Sedum rosea var. *alaskanum* (Rose) Berger = Sedum integrifolium [SEDUINT]
Sedum rosea var. *frigidum* (Rydb.) Hultén = Sedum integrifolium [SEDUINT]
Sedum rosea var. *integrifolium* (Raf.) Berger = Sedum integrifolium [SEDUINT]
Sedum spathulifolium Hook. [SEDUSPA] (broad-leaved stonecrop)
Sedum spathulifolium ssp. *anomalum* (Britt.) Clausen & Uhl = Sedum spathulifolium ssp. spathulifolium [SEDUSPA0]
Sedum spathulifolium ssp. **pruinosum** (Britt.) Clausen & Uhl [SEDUSPA1]
SY: *Sedum pruinosum* Britt.
SY: *Sedum spathulifolium* var. *pruinosum* (Britt.) Boivin
Sedum spathulifolium ssp. **spathulifolium** [SEDUSPA0]
SY: *Gordonia spathulifolia* (Hook.) A. & D. Löve
SY: *Sedum spathulifolium* ssp. *anomalum* (Britt.) Clausen & Uhl
Sedum spathulifolium var. *pruinosum* (Britt.) Boivin = Sedum spathulifolium ssp. pruinosum [SEDUSPA1]
Sedum stenopetalum Pursh [SEDUSTE] (worm-leaved stonecrop)
SY: *Sedum douglasii* Hook.
Tillaea angustifolia Nutt. = Crassula aquatica [CRASAQU]
Tillaea aquatica L. = Crassula aquatica [CRASAQU]
Tillaea ascendens Eat. = Crassula aquatica [CRASAQU]
Tillaea erecta Hook. & Arn. = Crassula connata [CRASCON]
Tillaea minima Gay = Crassula connata [CRASCON]
Tillaea minima Miers ex Hook. & Arn. = Crassula connata [CRASCON]
Tillaea rubescens Kunth = Crassula connata [CRASCON]
Tillaeastrum aquaticum (L.) Britt. = Crassula aquatica [CRASAQU]
Tolmachevia integrifolia (Raf.) A. & D. Löve = Sedum integrifolium [SEDUINT]

CUCURBITACEAE (V:D)

Echinocystis Torr. & Gray [ECHINOY$]
***Echinocystis lobata** (Michx.) Torr. & Gray [ECHILOB] (wild cucumber)
SY: *Micrampelis lobata* (Michx.) Greene
SY: *Sicyos lobata* Michx.
Echinocystis oregana (Torr. ex S. Wats.) Cogn. = Marah oreganus [MARAORE]
Marah Kellogg [MARAH$]
Marah oreganus (Torr. ex S. Wats.) T.J. Howell [MARAORE] (manroot)
SY: *Echinocystis oregana* (Torr. ex S. Wats.) Cogn.
Micrampelis lobata (Michx.) Greene = Echinocystis lobata [ECHILOB]
Sicyos lobata Michx. = Echinocystis lobata [ECHILOB]

CUPRESSACEAE (V:G)

Chamaecyparis Spach [CHAMAEC$]
Chamaecyparis nootkatensis (D. Don) Spach [CHAMNOO] (yellow cedar)
SY: *Cupressus nootkatensis* D. Don
Cupressus nootkatensis D. Don = Chamaecyparis nootkatensis [CHAMNOO]
Juniperus L. [JUNIPER$]
Juniperus alpina (Sm.) S.F. Gray = Juniperus communis var. montana [JUNICOM1]
Juniperus communis L. [JUNICOM] (common juniper)
Juniperus communis ssp. *alpina* (Sm.) Celak. = Juniperus communis var. montana [JUNICOM1]
Juniperus communis ssp. *nana* (Willd.) Syme = Juniperus communis var. montana [JUNICOM1]
Juniperus communis ssp. *saxitilis* (Pallas) E. Murr. = Juniperus communis var. montana [JUNICOM1]
Juniperus communis var. *alpina* Sm. = Juniperus communis var. montana [JUNICOM1]
Juniperus communis var. *jackii* Rehd. = Juniperus communis var. montana [JUNICOM1]
Juniperus communis var. **montana** Ait. [JUNICOM1]
SY: *Juniperus alpina* (Sm.) S.F. Gray
SY: *Juniperus communis* ssp. *alpina* (Sm.) Celak.
SY: *Juniperus communis* var. *alpina* Sm.
SY: *Juniperus communis* var. *jackii* Rehd.
SY: *Juniperus communis* ssp. *nana* (Willd.) Syme
SY: *Juniperus communis* var. *saxatilis* Pallas
SY: *Juniperus communis* ssp. *saxitilis* (Pallas) E. Murr.
SY: *Juniperus nana* Willd.
SY: *Juniperus sibirica* Burgsd.

Juniperus communis var. *saxatilis* Pallas = Juniperus communis var. montana [JUNICOM1]
Juniperus horizontalis Moench [JUNIHOR] (creeping juniper)
SY: *Juniperus horizontalis* var. *douglasii* hort.
SY: *Juniperus horizontalis* var. *variegata* Beissn.
SY: *Juniperus hudsonica* Forbes
SY: *Juniperus prostrata* Pers.
SY: *Juniperus repens* Nutt.
SY: *Juniperus virginiana* var. *prostrata* (Pers.) Torr.
SY: *Sabina horizontalis* (Moench) Rydb.
SY: *Sabina prostrata* (Pers.) Antoine
Juniperus horizontalis var. *douglasii* hort. = Juniperus horizontalis [JUNIHOR]
Juniperus horizontalis var. *variegata* Beissn. = Juniperus horizontalis [JUNIHOR]
Juniperus hudsonica Forbes = Juniperus horizontalis [JUNIHOR]
Juniperus nana Willd. = Juniperus communis var. montana [JUNICOM1]
Juniperus prostrata Pers. = Juniperus horizontalis [JUNIHOR]
Juniperus repens Nutt. = Juniperus horizontalis [JUNIHOR]
Juniperus scopulorum Sarg. [JUNISCO] (Rocky Mountain juniper)
SY: *Juniperus scopulorum* var. *columnaris* Fassett
SY: *Juniperus virginiana* var. *montana* Vasey
SY: *Juniperus virginiana* ssp. *scopulorum* (Sarg.) E. Murr.
SY: *Juniperus virginiana* var. *scopulorum* (Sarg.) Lemmon
SY: *Sabina scopulorum* (Sarg.) Rydb.
Juniperus scopulorum var. *columnaris* Fassett = Juniperus scopulorum [JUNISCO]
Juniperus sibirica Burgsd. = Juniperus communis var. montana [JUNICOM1]
Juniperus virginiana ssp. *scopulorum* (Sarg.) E. Murr. = Juniperus scopulorum [JUNISCO]
Juniperus virginiana var. *montana* Vasey = Juniperus scopulorum [JUNISCO]
Juniperus virginiana var. *prostrata* (Pers.) Torr. = Juniperus horizontalis [JUNIHOR]
Juniperus virginiana var. *scopulorum* (Sarg.) Lemmon = Juniperus scopulorum [JUNISCO]
Sabina horizontalis (Moench) Rydb. = Juniperus horizontalis [JUNIHOR]
Sabina prostrata (Pers.) Antoine = Juniperus horizontalis [JUNIHOR]
Sabina scopulorum (Sarg.) Rydb. = Juniperus scopulorum [JUNISCO]
Thuja L. [THUJA$]
Thuja plicata Donn ex D. Don [THUJPLI] (western redcedar)

CUSCUTACEAE (V:D)

Cuscuta L. [CUSCUTA$]
***Cuscuta approximata** Bab. [CUSCAPP] (clustered dodder)
Cuscuta cephalanthi Engelm. [CUSCCEP] (button-bush dodder)
SY: *Grammica cephalanthi* (Engelm.) Hadac & Chrtek
***Cuscuta epithymum** L. [CUSCEPI] (common dodder)
Cuscuta pentagona Engelm. [CUSCPEN] (field dodder)
SY: *Grammica pentagona* (Engelm.) W.A. Weber
Cuscuta salina Engelm. [CUSCSAL] (salt marsh dodder)
SY: *Grammica salina* (Engelm.) Taylor & MacBryde
Grammica cephalanthi (Engelm.) Hadac & Chrtek = Cuscuta cephalanthi [CUSCCEP]
Grammica pentagona (Engelm.) W.A. Weber = Cuscuta pentagona [CUSCPEN]
Grammica salina (Engelm.) Taylor & MacBryde = Cuscuta salina [CUSCSAL]

CYPERACEAE (V:M)

Amphiscirpus nevadensis (S. Wats.) Oteng Yeboah = Scirpus nevadensis [SCIRNEV]
Baeothryon alpinum (L.) Egor. = Eriophorum alpinum [ERIOALP]
Baeothryon cespitosum (L.) A. Dietr. = Scirpus cespitosus [SCIRCES]
Bolboschoenus maritimus (L.) Palla = Scirpus maritimus [SCIRMAR]
Bolboschoenus maritimus ssp. *paludosus* (A. Nels.) A. & D. Löve = Scirpus maritimus [SCIRMAR]
Bolboschoenus paludosus (A. Nels.) Soó = Scirpus maritimus [SCIRMAR]
Carex L. [CAREX$]
Carex ablata Bailey = Carex luzulina var. ablata [CARELUZ1]
Carex accedens Holm = Carex aperta [CAREAPE]
Carex acutina var. *tenuior* Bailey = Carex aperta [CAREAPE]
Carex acutinella Mackenzie = Carex aquatilis var. aquatilis [CAREAQU0]
Carex aenea Fern. [CAREAEN] (bronze sedge)
Carex albonigra Mackenzie [CAREALB] (two-toned sedge)
Carex ambusta Boott = Carex saxatilis var. major [CARESAX2]
Carex amphigena (Fern.) Mackenzie = Carex glareosa var. amphigens [CAREGLR2]
Carex amplifolia Boott [CAREAMP] (bigleaf sedge)
Carex angarae Steud. = Carex norvegica ssp. inferalpina [CARENOR1]
Carex angustior Mackenzie = Carex echinata ssp. echinata [CAREECH0]
Carex angustior var. *gracilenta* Clausen & H.A. Wahl = Carex echinata ssp. echinata [CAREECH0]
Carex anthericoides Presl = Carex macrocephala [CAREMAR]
Carex anthoxanthea J.& K. Presl [CAREANT] (yellow-flowered sedge)
Carex aperta Boott [CAREAPE] (Columbia sedge)
SY: *Carex accedens* Holm
SY: *Carex acutina* var. *tenuior* Bailey
SY: *Carex stylosa* var. *virens* Bailey
SY: *Carex turgidula* Bailey
Carex apoda Clokey = Carex atrosquama [CAREATO]
Carex aquatilis Wahlenb. [CAREAQU] (water sedge)
Carex aquatilis ssp. *altior* (Rydb.) Hultén = Carex aquatilis var. aquatilis [CAREAQU0]
Carex aquatilis ssp. *stans* (Drej.) Hultén = Carex aquatilis var. stans [CAREAQU3]
Carex aquatilis var. *altior* (Rydb.) Fern. = Carex aquatilis var. aquatilis [CAREAQU0]
Carex aquatilis var. **aquatilis** [CAREAQU0]
SY: *Carex acutinella* Mackenzie
SY: *Carex aquatilis* ssp. *altior* (Rydb.) Hultén
SY: *Carex aquatilis* var. *altior* (Rydb.) Fern.
SY: *Carex aquatilis* var. *substricta* Kükenth.
SY: *Carex substricta* (Kükenth.) Mackenzie
SY: *Carex suksdorfii* Kükenth.
Carex aquatilis var. **dives** (Holm) Kükenth. [CAREAQU6]
SY: *Carex aquatilis* var. *sitchensis* (Prescott ex Bong.) Kelso
SY: *Carex howellii* Bailey
SY: *Carex panda* C.B. Clarke
SY: *Carex sitchensis* Prescott ex Bong.
Carex aquatilis var. *sitchensis* (Prescott ex Bong.) Kelso = Carex aquatilis var. dives [CAREAQU6]
Carex aquatilis var. **stans** (Drej.) Boott [CAREAQU3]
SY: *Carex aquatilis* ssp. *stans* (Drej.) Hultén
SY: *Carex stans* Drej.
Carex aquatilis var. *substricta* Kükenth. = Carex aquatilis var. aquatilis [CAREAQU0]
Carex arcta Boott [CAREARC] (northern clustered sedge)
Carex arctaeformis Mackenzie [CAREART]
SY: *Carex canescens* ssp. *arctaeformis* (Mackenzie) Calder & Taylor
Carex arenicola F. Schmidt = Carex pansa [CAREPAN]
Carex arenicola ssp. *pansa* (Bailey) T. Koyama & Calder = Carex pansa [CAREPAN]
Carex athabascensis F.J. Herm. = Carex scirpoidea [CARESCI]
Carex atherodes Spreng. [CAREATH] (awned sedge)
Carex athrostachya Olney [CAREATR] (slender-beaked sedge)
Carex atrata L. = Carex atratiformis [CAREATA]
Carex atrata ssp. *atrosquama* (Mackenzie) Hultén = Carex atrosquama [CAREATO]
Carex atrata var. *atrosquama* (Mackenzie) Cronq. = Carex atrosquama [CAREATO]
Carex atrata var. *erecta* W. Boott = Carex heteroneura var. epapillosa [CAREHET1]
Carex atratiformis Britt. [CAREATA] (black sedge)

SY: *Carex atrata* L.
Carex atratiformis ssp. *raymondii* (Calder) Porsild = Carex raymondii [CARERAM]
Carex atrosquama Mackenzie [CAREATO] (black-scaled sedge)
SY: *Carex apoda* Clokey
SY: *Carex atrata* ssp. *atrosquama* (Mackenzie) Hultén
SY: *Carex atrata* var. *atrosquama* (Mackenzie) Cronq.
Carex aurea Nutt. [CAREAUR] (golden sedge)
Carex backii var. *saximontana* (Mackenzie) Boivin = Carex saximontana [CARESAI]
Carex bebbii Olney ex Fern. [CAREBEB] (Bebb's sedge)
Carex beringensis C.B. Clarke = Carex podocarpa [CAREPOD]
Carex bicolor Bellardi ex All. [CAREBIC] (two-coloured sedge)
Carex bigelowii Torr. ex Schwein. [CAREBIG] (Bigelow's sedge)
SY: *Carex bigelowii* ssp. *hyperborea* (Drej.) Böcher
SY: *Carex consimilis* Holm
Carex bigelowii ssp. *hyperborea* (Drej.) Böcher = Carex bigelowii [CAREBIG]
Carex bipartita All. [CAREBIP] (two-parted sedge)
SY: *Carex bipartita* var. *austromontana* F.J. Herm.
SY: *Carex lachenalii* Schkuhr
SY: *Carex lagopina* Wahlenb.
SY: *Carex tripartita* auct.
Carex bipartita var. *amphigena* (Fern.) Polunin = Carex glareosa var. amphigens [CAREGLR2]
Carex bipartita var. *austromontana* F.J. Herm. = Carex bipartita [CAREBIP]
Carex bolanderi Olney [CAREBOL] (Bolander's sedge)
SY: *Carex deweyana* var. *bolanderi* (Olney) W. Boott
Carex brevicaulis Mackenzie [CAREBRE] (short-stemmed sedge)
SY: *Carex deflexa* var. *brevicaulis* (Mackenzie) Boivin
Carex brevior (Dewey) Mackenzie ex Lunell [CAREBRV] (short-beaked sedge)
SY: *Carex festucacea* var. *brevior* (Dewey) Fern.
Carex brevipes W. Boott = Carex rossii [CAREROS]
Carex breweri var. *paddoensis* (Suksdorf) Cronq. = Carex engelmannii [CAREENG]
Carex brunnescens (Pers.) Poir. [CAREBRU] (brownish sedge)
SY: *Carex sphaerostachya* (Tuckerman) Dewey
Carex brunnescens ssp. **alaskana** Kalela [CAREBRU1]
Carex brunnescens ssp. **pacifica** Kalela [CAREBRU2]
Carex brunnescens ssp. **sphaerostachya** (Tuckerman) Kalela [CAREBRU3]
SY: *Carex brunnescens* var. *sphaerostachya* (Tuckerman) Kükenth.
Carex brunnescens var. *sphaerostachya* (Tuckerman) Kükenth. = Carex brunnescens ssp. sphaerostachya [CAREBRU3]
Carex buxbaumii Wahlenb. [CAREBUX] (Buxbaum's sedge)
Carex camporum Mackenzie = Carex praegracilis [CAREPRE]
Carex campylocarpa Holm = Carex scopulorum var. bracteosa [CARESCP1]
Carex campylocarpa ssp. *affinis* Maguire & A. Holmgren = Carex scopulorum var. bracteosa [CARESCP1]
Carex canescens L. [CARECAN] (grey sedge)
SY: *Carex cinerea* Poll.
SY: *Carex curta* Goodenough
Carex canescens ssp. *arctaeformis* (Mackenzie) Calder & Taylor = Carex arctaeformis [CAREART]
Carex canescens var. *subloliacea* (Laestad.) Hartman = Carex lapponica [CARELAP]
Carex capillaris L. [CARECAP] (hair-like sedge)
SY: *Carex capillaris* ssp. *chlorostachys* (Stev.) A. & D. Löve & Raymond
SY: *Carex capillaris* var. *elongata* Olney ex Fern.
SY: *Carex capillaris* var. *major* Blytt
SY: *Carex capillaris* ssp. *robustior* (Drej. ex Lange) Böcher
SY: *Carex chlorostachys* Stev.
SY: *Carex fuscidula* Krecz. ex Egor.
Carex capillaris ssp. *chlorostachys* (Stev.) A. & D. Löve & Raymond = Carex capillaris [CARECAP]
Carex capillaris ssp. *krausei* (Boeckl.) Böcher = Carex krausei [CAREKRA]
Carex capillaris ssp. *robustior* (Drej. ex Lange) Böcher = Carex capillaris [CARECAP]
Carex capillaris var. *elongata* Olney ex Fern. = Carex capillaris [CARECAP]
Carex capillaris var. *krausei* (Boeckl.) Crantz = Carex krausei [CAREKRA]
Carex capillaris var. *major* Blytt = Carex capillaris [CARECAP]
Carex capitata L. [CARECAI] (capitate sedge)
Carex cephalantha (Bailey) Bickn. = Carex echinata ssp. echinata [CAREECH0]
Carex chlorostachys Stev. = Carex capillaris [CARECAP]
Carex chordorrhiza Ehrh. ex L. f. [CARECHO] (cordroot sedge)
Carex cinerea Poll. = Carex canescens [CARECAN]
Carex circinata C.A. Mey. [CARECIR] (coiled sedge)
Carex comosa Boott [CARECOM] (bearded sedge)
Carex concinna R. Br. [CARECON] (beautiful sedge)
Carex concinnoides Mackenzie [CARECOC] (northwestern sedge)
Carex consimilis Holm = Carex bigelowii [CAREBIG]
Carex crawei Dewey [CARECRA] (Crawe's sedge)
Carex crawfordii Fern. [CARECRW] (Crawford's sedge)
SY: *Carex crawfordii* var. *vigens* Fern.
Carex crawfordii var. *vigens* Fern. = Carex crawfordii [CARECRW]
Carex cryptocarpa C.A. Mey. = Carex lyngbyei [CARELYN]
Carex cryptochlaena Holm = Carex lyngbyei [CARELYN]
Carex curta Goodenough = Carex canescens [CARECAN]
Carex cusickii Mackenzie ex Piper & Beattie [CARECUS] (Cusick's sedge)
SY: *Carex diandra* var. *ampla* (Bailey) Kukenth.
SY: *Carex obovoidea* Cronq.
Carex deflexa Hornem. [CAREDEF] (bent sedge)
Carex deflexa var. *brevicaulis* (Mackenzie) Boivin = Carex brevicaulis [CAREBRE]
Carex deflexa var. *rossii* (Boott) Bailey = Carex rossii [CAREROS]
Carex deweyana Schwein. [CAREDEW] (Dewey's sedge)
Carex deweyana ssp. *leptopoda* (Mackenzie) Calder & Taylor = Carex leptopoda [CARELEO]
Carex deweyana var. *bolanderi* (Olney) W. Boott = Carex bolanderi [CAREBOL]
Carex deweyana var. *leptopoda* (Mackenzie) Boivin = Carex leptopoda [CARELEO]
Carex diandra Schrank [CAREDIA] (lesser panicled sedge)
Carex diandra var. *ampla* (Bailey) Kukenth. = Carex cusickii [CARECUS]
Carex dioica ssp. *gynocrates* (Wormsk. ex Drej.) Hultén = Carex gynocrates [CAREGYN]
Carex dioica var. *gynocrates* (Wormsk. ex Drej.) Ostenf. = Carex gynocrates [CAREGYN]
Carex disperma Dewey [CAREDIS] (two-seeded sedge)
SY: *Carex tenella* Schukuhr
Carex diversistylis Roach = Carex rossii [CAREROS]
Carex douglasii Boott [CAREDOU] (Douglas' sedge)
Carex drummondiana Dewey = Carex rupestris var. drummondiana [CARERUP1]
Carex duriuscula C.A. Mey. [CAREDUR]
SY: *Carex eleocharis* Bailey
SY: *Carex stenophylla* ssp. *eleocharis* (Bailey) Hultén
SY: *Carex stenophylla* var. *enervis* Kükenth.
Carex eastwoodiana Stacey = Carex phaeocephala [CAREPHA]
Carex eburnea Boott [CAREEBU] (bristle-leaf sedge)
Carex echinata Murr. [CAREECH] (star sedge)
Carex echinata ssp. **echinata** [CAREECH0]
SY: *Carex angustior* Mackenzie
SY: *Carex angustior* var. *gracilenta* Clausen & H.A. Wahl
SY: *Carex cephalantha* (Bailey) Bickn.
SY: *Carex echinata* var. *angustata* (Carey) Bailey
SY: *Carex hawaiiensis* St. John
SY: *Carex josselynii* (Fern.) Mackenzie ex Pease
SY: *Carex laricina* Mackenzie ex Bright
SY: *Carex leersii* Willd.
SY: *Carex muricata* var. *angustata* (Carey) Carey ex Gleason
SY: *Carex muricata* var. *cephalantha* (Bailey) Wieg. & Eames
SY: *Carex muricata* var. *laricina* (Mackenzie ex Bright) Gleason
SY: *Carex ormantha* (Fern.) Mackenzie

SY: *Carex phyllomanica* var. *angustata* (Carey) Boivin
SY: *Carex phyllomanica* var. *ormantha* (Fern.) Boivin
SY: *Carex svensonis* Skottsberg
Carex echinata ssp. **phyllomanica** (W. Boott) Reznicek [CAREECH2] (coastal stellate sedge)
SY: *Carex phyllomanica* W. Boott
Carex echinata var. *angustata* (Carey) Bailey = Carex echinata ssp. echinata [CAREECH0]
Carex eleocharis Bailey = Carex duriuscula [CAREDUR]
Carex elyniformis Porsild = Carex filifolia [CAREFIL]
Carex enanderi Holm = Carex lenticularis var. dolia [CARELEN1]
Carex engelmannii Bailey [CAREENG] (Engelmann's sedge)
SY: *Carex breweri* var. *paddoensis* (Suksdorf) Cronq.
Carex epapillosa Mackenzie = Carex heteroneura var. epapillosa [CAREHET1]
Carex erxlebeniana L. Kelso = Carex inops ssp. heliophila [CAREINO1]
Carex eurystachya F.J. Herm. = Carex lenticularis var. dolia [CARELEN1]
Carex exsiccata Bailey [CAREEXS] (inflated sedge)
SY: *Carex vesicaria* var. *major* Boott
Carex festiva var. *decumbens* Holm = Carex haydeniana [CAREHAY]
Carex festiva var. *gracilis* Olney = Carex pachystachya [CAREPAC]
Carex festiva var. *pachystachya* (Cham. ex Steud.) Bailey = Carex pachystachya [CAREPAC]
Carex festivella Mackenzie = Carex microptera [CAREMIO]
Carex festucacea var. *brevior* (Dewey) Fern. = Carex brevior [CAREBRV]
Carex feta Bailey [CAREFET] (green-sheathed sedge)
SY: *Carex straminea* var. *mixta* Bailey
Carex filifolia Nutt. [CAREFIL] (thread-leaved sedge)
SY: *Carex elyniformis* Porsild
Carex flava L. [CAREFLA] (yellow sedge)
SY: *Carex flava* var. *fertilis* Peck
SY: *Carex flava* var. *gaspensis* Fern.
SY: *Carex flava* var. *graminis* Bailey
SY: *Carex flava* var. *laxior* (Kükenth.) Gleason
SY: *Carex flava* var. *rectirostra* Gaudin
SY: *Carex nevadensis* ssp. *flavella* (Krecz.) Janchen
Carex flava var. *fertilis* Peck = Carex flava [CAREFLA]
Carex flava var. *gaspensis* Fern. = Carex flava [CAREFLA]
Carex flava var. *graminis* Bailey = Carex flava [CAREFLA]
Carex flava var. *laxior* (Kükenth.) Gleason = Carex flava [CAREFLA]
Carex flava var. *rectirostra* Gaudin = Carex flava [CAREFLA]
Carex foenea Willd. [CAREFOE] (hay sedge)
SY: *Carex siccata* Dewey
Carex franklinii Boott [CAREFRA] (Franklin's sedge)
SY: *Carex petricosa* var. *franklinii* (Boott) Boivin
Carex fuliginosa ssp. *misandra* (R. Br.) Nyman = Carex misandra [CAREMIS]
Carex fuscidula Krecz. ex Egor. = Carex capillaris [CARECAP]
Carex garberi Fern. [CAREGAR] (Garber's sedge)
SY: *Carex garberi* ssp. *bifaria* (Fern.) Hultén
SY: *Carex garberi* var. *bifaria* Fern.
Carex garberi ssp. *bifaria* (Fern.) Hultén = Carex garberi [CAREGAR]
Carex garberi var. *bifaria* Fern. = Carex garberi [CAREGAR]
Carex geyeri Boott [CAREGEY] (elk sedge)
Carex glacialis Mackenzie [CAREGLA] (glacier sedge)
Carex glareosa Schkuhr ex Wahlenb. [CAREGLR] (lesser saltmarsh sedge)
Carex glareosa var. **amphigens** Fern. [CAREGLR2]
SY: *Carex amphigena* (Fern.) Mackenzie
SY: *Carex bipartita* var. *amphigena* (Fern.) Polunin
Carex gmelinii Hook. & Arn. [CAREGME] (Gmelin's sedge)
Carex gymnoclada Holm = Carex scopulorum var. bracteosa [CARESCP1]
Carex gynocrates Wormsk. ex Drej. [CAREGYN] (yellow bog sedge)
SY: *Carex dioica* ssp. *gynocrates* (Wormsk. ex Drej.) Hultén
SY: *Carex dioica* var. *gynocrates* (Wormsk. ex Drej.) Ostenf.
Carex hawaiiensis St. John = Carex echinata ssp. echinata [CAREECH0]
Carex haydeniana Olney [CAREHAY] (Hayden's sedge)
SY: *Carex festiva* var. *decumbens* Holm
SY: *Carex macloviana* ssp. *haydeniana* (Olney) Taylor & MacBryde
SY: *Carex nubicola* Mackenzie
Carex heleonastes L. f. [CAREHEL] (Hudson Bay sedge)
Carex heliophila Mackenzie = Carex inops ssp. heliophila [CAREINO1]
Carex hendersonii Bailey [CAREHEN] (Henderson's sedge)
Carex hepburnii Boott = Carex nardina var. hepburnii [CARENAR1]
Carex heteroneura W. Boott [CAREHET]
Carex heteroneura var. **epapillosa** (Mackenzie) F.J. Herm. [CAREHET1]
SY: *Carex atrata* var. *erecta* W. Boott
SY: *Carex epapillosa* Mackenzie
Carex hindsii C.B. Clarke = Carex lenticularis var. limnophila [CARELEN4]
Carex hoodii Boott [CAREHOO] (Hood's sedge)
Carex howellii Bailey = Carex aquatilis var. dives [CAREAQU6]
Carex hystericina Muhl. ex Willd. [CAREHYS] (porcupine sedge)
Carex illota Bailey [CAREILL] (sheep sedge)
Carex incurva Lightf. = Carex maritima [CAREMAT]
Carex incurviformis Mackenzie [CAREINC]
SY: *Carex maritima* var. *incurviformis* (Mackenzie) Boivin
Carex inflata var. *utriculata* (Boott) Druce = Carex utriculata [CAREUTR]
Carex inops Bailey [CAREINO] (long-stoloned sedge)
Carex inops ssp. **heliophila** (Mackenzie) Crins [CAREINO1]
SY: *Carex erxlebeniana* L. Kelso
SY: *Carex heliophila* Mackenzie
SY: *Carex pensylvanica* var. *digyna* Boeckl.
SY: *Carex pensylvanica* ssp. *heliophila* (Mackenzie) W.A. Weber
Carex inops ssp. **inops** [CAREINO0]
Carex interior Bailey [CAREINT] (inland sedge)
SY: *Carex interior* ssp. *charlestonensis* Clokey
SY: *Carex interior* var. *keweenawensis* F.J. Herm.
Carex interior ssp. *charlestonensis* Clokey = Carex interior [CAREINT]
Carex interior var. *keweenawensis* F.J. Herm. = Carex interior [CAREINT]
Carex interrupta Boeckl. [CAREINE] (green-fruited sedge)
SY: *Carex interrupta* var. *distenta* Kükenth.
Carex interrupta var. *distenta* Kükenth. = Carex interrupta [CAREINE]
Carex invisa Bailey = Carex spectabilis [CARESPE]
Carex irrigua Wahlenb. = Carex magellanica ssp. irrigua [CAREMAG1]
Carex jacobi-peteri Hultén = Carex pyrenaica ssp. micropoda [CAREPYR1]
Carex jimescalderi Boivin = Carex leptalea ssp. pacifica [CARELET2]
Carex josselynii (Fern.) Mackenzie ex Pease = Carex echinata ssp. echinata [CAREECH0]
Carex kelloggii W. Boott = Carex lenticularis var. lipocarpa [CARELEN5]
Carex krausei Boeckl. [CAREKRA] (Krause's sedge)
SY: *Carex capillaris* ssp. *krausei* (Boeckl.) Böcher
SY: *Carex capillaris* var. *krausei* (Boeckl.) Crantz
Carex lachenalii Schkuhr = Carex bipartita [CAREBIP]
Carex laeviculmis Meinsh. [CARELAE] (smooth-stemmed sedge)
Carex lagopina Wahlenb. = Carex bipartita [CAREBIP]
Carex lanuginosa Michx. [CARELAN] (woolly sedge)
SY: *Carex lasiocarpa* var. *latifolia* (Boeckl.) Gilly
Carex lanuginosa var. *americana* (Fern.) Boivin = Carex lasiocarpa var. americana [CARELAS1]
Carex lapponica O.F. Lang [CARELAP] (Lapland sedge)
SY: *Carex canescens* var. *subloliacea* (Laestad.) Hartman
Carex laricina Mackenzie ex Bright = Carex echinata ssp. echinata [CAREECH0]
Carex lasiocarpa Ehrh. [CARELAS] (hairy-fruited sedge)

Carex lasiocarpa ssp. *americana* (Fern.) Love & Bernard = Carex lasiocarpa var. americana [CARELAS1]
Carex lasiocarpa var. **americana** Fern. [CARELAS1]
SY: *Carex lanuginosa* var. *americana* (Fern.) Boivin
SY: *Carex lasiocarpa* ssp. *americana* (Fern.) Love & Bernard
Carex lasiocarpa var. *latifolia* (Boeckl.) Gilly = Carex lanuginosa [CARELAN]
Carex leersii Willd. = Carex echinata ssp. echinata [CAREECH0]
Carex lenticularis Michx. [CARELEN] (lens-fruited sedge)
Carex lenticularis var. *albimontana* Dewey = Carex lenticularis var. lenticularis [CARELEN0]
Carex lenticularis var. *blakei* Dewey = Carex lenticularis var. lenticularis [CARELEN0]
Carex lenticularis var. **dolia** (M.E. Jones) L.A. Standley [CARELEN1]
SY: *Carex enanderi* Holm
SY: *Carex eurystachya* F.J. Herm.
SY: *Carex plectocarpa* F.J. Herm.
Carex lenticularis var. *eucycla* Fern. = Carex lenticularis var. lenticularis [CARELEN0]
Carex lenticularis var. **impressa** (Bailey) L.A. Standley [CARELEN2]
SY: *Carex limnaea* Holm
SY: *Carex paucicostata* Mackenzie
Carex lenticularis var. **lenticularis** [CARELEN0]
SY: *Carex lenticularis* var. *albimontana* Dewey
SY: *Carex lenticularis* var. *blakei* Dewey
SY: *Carex lenticularis* var. *eucycla* Fern.
SY: *Carex lenticularis* var. *merens* Howe
Carex lenticularis var. **limnophila** (Holm) Cronq. [CARELEN4]
SY: *Carex hindsii* C.B. Clarke
Carex lenticularis var. **lipocarpa** (Holm) L.A. Standley [CARELEN5]
SY: *Carex kelloggii* W. Boott
SY: *Carex lenticularis* var. *pallida* (W. Boott) Dorn
Carex lenticularis var. *merens* Howe = Carex lenticularis var. lenticularis [CARELEN0]
Carex lenticularis var. *pallida* (W. Boott) Dorn = Carex lenticularis var. lipocarpa [CARELEN5]
Carex leporina L. [CARELEP]
SY: *Carex tracyi* Mackenzie
Carex leptalea Wahlenb. [CARELET] (bristle-stalked sedge)
Carex leptalea ssp. **leptalea** [CARELET0]
Carex leptalea ssp. **pacifica** Calder & Taylor [CARELET2]
SY: *Carex jimescalderi* Boivin
Carex leptopoda Mackenzie [CARELEO]
SY: *Carex deweyana* ssp. *leptopoda* (Mackenzie) Calder & Taylor
SY: *Carex deweyana* var. *leptopoda* (Mackenzie) Boivin
Carex liddonii Boott = Carex petasata [CAREPET]
Carex limnaea Holm = Carex lenticularis var. impressa [CARELEN2]
Carex limnophila F.J. Herm. = Carex microptera [CAREMIO]
Carex limosa L. [CARELIM] (shore sedge)
Carex livida (Wahlenb.) Willd. [CARELIV]
Carex livida var. *grayana* (Dewey) Fern. = Carex livida var. radicaulis [CARELIV1]
Carex livida var. **radicaulis** Paine [CARELIV1]
SY: *Carex livida* var. *grayana* (Dewey) Fern.
Carex loliacea L. [CARELOL] (ryegrass sedge)
Carex luzulina Olney [CARELUZ]
Carex luzulina var. **ablata** (Bailey) F.J. Herm. [CARELUZ1]
SY: *Carex ablata* Bailey
Carex lyallii Boott = Carex raynoldsii [CARERAY]
Carex lyngbyei Hornem. [CARELYN] (Lingbye's sedge)
SY: *Carex cryptocarpa* C.A. Mey.
SY: *Carex cryptochlaena* Holm
SY: *Carex lyngbyei* var. *robusta* (Bailey) Cronq.
Carex lyngbyei var. *robusta* (Bailey) Cronq. = Carex lyngbyei [CARELYN]
Carex macloviana d'Urv. [CAREMAL] (Falkland Island sedge)
Carex macloviana ssp. *festivella* (Mackenzie) A. & D. Löve = Carex microptera [CAREMIO]
Carex macloviana ssp. *haydeniana* (Olney) Taylor & MacBryde = Carex haydeniana [CAREHAY]
Carex macloviana ssp. *pachystachya* (Cham. ex Steud.) Hultén = Carex pachystachya [CAREPAC]
Carex macloviana var. *microptera* (Mackenzie) Boivin = Carex microptera [CAREMIO]
Carex macloviana var. *pachystachya* (Cham. ex Steud.) Kükenth. = Carex pachystachya [CAREPAC]
Carex macrocephala Willd. ex Spreng. [CAREMAR] (large-headed sedge)
SY: *Carex anthericoides* Presl
SY: *Carex macrocephala* ssp. *anthericoides* (Presl) Hultén
Carex macrocephala ssp. *anthericoides* (Presl) Hultén = Carex macrocephala [CAREMAR]
Carex macrochaeta C.A. Mey. [CAREMAO] (large-awned sedge)
Carex macrogyna Turcz. ex Steud. = Carex petricosa [CAREPER]
Carex magellanica Lam. [CAREMAG] (poor sedge)
SY: *Carex paupercula* Michx.
Carex magellanica ssp. **irrigua** (Wahlenb.) Hultén [CAREMAG1]
SY: *Carex irrigua* Wahlenb.
SY: *Carex magellanica* var. *irrigua* (Wahlenb.) B.S.P.
SY: *Carex paupercula* var. *irrigua* (Wahlenb.) Fern.
Carex magellanica var. *irrigua* (Wahlenb.) B.S.P. = Carex magellanica ssp. irrigua [CAREMAG1]
Carex magnifica Dewey ex Piper = Carex obnupta [CAREOBN]
Carex marina Dewey [CAREMAI]
Carex maritima Gunn. [CAREMAT] (curved-spiked sedge)
SY: *Carex incurva* Lightf.
Carex maritima var. *incurviformis* (Mackenzie) Boivin = Carex incurviformis [CAREINC]
Carex media R. Br. = Carex norvegica ssp. inferalpina [CARENOR1]
Carex membranacea Hook. [CAREMEM] (fragile sedge)
SY: *Carex membranopacta* Bailey
Carex membranopacta Bailey = Carex membranacea [CAREMEM]
Carex mertensii Prescott ex Bong. [CAREMER] (Mertens' sedge)
Carex microchaeta Holm [CAREMIC] (small-awned sedge)
Carex microglochin Wahlenb. [CAREMIR] (few-seeded bog sedge)
Carex micropoda C.A. Mey. = Carex pyrenaica ssp. micropoda [CAREPYR1]
Carex microptera Mackenzie [CAREMIO] (small-winged sedge)
SY: *Carex festivella* Mackenzie
SY: *Carex limnophila* F.J. Herm.
SY: *Carex macloviana* ssp. *festivella* (Mackenzie) A. & D. Löve
SY: *Carex macloviana* var. *microptera* (Mackenzie) Boivin
SY: *Carex microptera* var. *crassinervia* F.J. Herm.
SY: *Carex microptera* var. *limnophila* (F.J. Herm.) Dorn
Carex microptera var. *crassinervia* F.J. Herm. = Carex microptera [CAREMIO]
Carex microptera var. *limnophila* (F.J. Herm.) Dorn = Carex microptera [CAREMIO]
Carex misandra R. Br. [CAREMIS] (short-leaved sedge)
SY: *Carex fuliginosa* ssp. *misandra* (R. Br.) Nyman
Carex miserabilis Mackenzie = Carex scopulorum var. prionophylla [CARESCP2]
Carex montanensis Bailey = Carex podocarpa [CAREPOD]
Carex multimoda Bailey = Carex pachystachya [CAREPAC]
Carex muricata var. *angustata* (Carey) Carey ex Gleason = Carex echinata ssp. echinata [CAREECH0]
Carex muricata var. *cephalantha* (Bailey) Wieg. & Eames = Carex echinata ssp. echinata [CAREECH0]
Carex muricata var. *laricina* (Mackenzie ex Bright) Gleason = Carex echinata ssp. echinata [CAREECH0]
Carex nardina Fries [CARENAR] (spikenard sedge)
Carex nardina ssp. *hepburnii* (Boott) A. & D. Löve & Kapoor = Carex nardina var. hepburnii [CARENAR1]
Carex nardina var. **hepburnii** (Boott) Kükenth. [CARENAR1]
SY: *Carex hepburnii* Boott
SY: *Carex nardina* ssp. *hepburnii* (Boott) A. & D. Löve & Kapoor
Carex nevadensis ssp. *flavella* (Krecz.) Janchen = Carex flava [CAREFLA]
Carex nigella Boott = Carex spectabilis [CARESPE]
Carex nigricans C.A. Mey. [CARENIG] (black alpine sedge)
Carex nigromarginata var. *elliptica* (Boott) Gleason = Carex peckii [CAREPEC]

Carex nigromarginata var. *minor* (Boott) Gleason = Carex peckii [CAREPEC]
Carex norvegica Retz. [CARENOR]
Carex norvegica ssp. **inferalpina** (Wahlenb.) Hultén [CARENOR1]
SY: *Carex angarae* Steud.
SY: *Carex media* R. Br.
SY: *Carex norvegica* var. *inferalpina* (Wahlenb.) Boivin
SY: *Carex vahlii* var. *inferalpina* Wahlenb.
Carex norvegica var. *inferalpina* (Wahlenb.) Boivin = Carex norvegica ssp. inferalpina [CARENOR1]
Carex nubicola Mackenzie = Carex haydeniana [CAREHAY]
Carex obnupta Bailey [CAREOBN] (slough sedge)
SY: *Carex magnifica* Dewey ex Piper
Carex obovoidea Cronq. = Carex cusickii [CARECUS]
Carex obtusata Lilj. [CAREOBT] (blunt sedge)
Carex oederi var. *viridula* (Michx.) Hultén = Carex viridula [CAREVIR]
Carex ormantha (Fern.) Mackenzie = Carex echinata ssp. echinata [CAREECH0]
Carex pachystachya Cham. ex Steud. [CAREPAC] (thick-headed sedge)
SY: *Carex festiva* var. *gracilis* Olney
SY: *Carex festiva* var. *pachystachya* (Cham. ex Steud.) Bailey
SY: *Carex macloviana* ssp. *pachystachya* (Cham. ex Steud.) Hultén
SY: *Carex macloviana* var. *pachystachya* (Cham. ex Steud.) Kükenth.
SY: *Carex multimoda* Bailey
SY: *Carex pachystachya* var. *gracilis* (Olney) Mackenzie
SY: *Carex pachystachya* var. *monds-coulteri* L. Kelso
SY: *Carex pyrophila* Gandog.
Carex pachystachya var. *gracilis* (Olney) Mackenzie = Carex pachystachya [CAREPAC]
Carex pachystachya var. *monds-coulteri* L. Kelso = Carex pachystachya [CAREPAC]
*__Carex pallescens__ L. [CAREPAL] (pale sedge)
SY: *Carex pallescens* var. *neogaea* Fern.
Carex pallescens var. *neogaea* Fern. = Carex pallescens [CAREPAL]
Carex panda C.B. Clarke = Carex aquatilis var. dives [CAREAQU6]
Carex pansa Bailey [CAREPAN] (sand-dune sedge)
SY: *Carex arenicola* F. Schmidt
SY: *Carex arenicola* ssp. *pansa* (Bailey) T. Koyama & Calder
Carex parryana Dewey [CAREPAR] (Parry's sedge)
Carex paucicostata Mackenzie = Carex lenticularis var. impressa [CARELEN2]
Carex pauciflora Lightf. [CAREPAU] (few-flowered sedge)
Carex paupercula Michx. = Carex magellanica [CAREMAG]
Carex paupercula var. *irrigua* (Wahlenb.) Fern. = Carex magellanica ssp. irrigua [CAREMAG1]
Carex paysonis Clokey [CAREPAY] (Payson's sedge)
Carex peckii Howe [CAREPEC] (Peck's sedge)
SY: *Carex nigromarginata* var. *elliptica* (Boott) Gleason
SY: *Carex nigromarginata* var. *minor* (Boott) Gleason
Carex pedunculata Muhl. ex Willd. [CAREPED] (peduncled sedge)
Carex pensylvanica ssp. *heliophila* (Mackenzie) W.A. Weber = Carex inops ssp. heliophila [CAREINO1]
Carex pensylvanica var. *digyna* Boeckl. = Carex inops ssp. heliophila [CAREINO1]
Carex petasata Dewey [CAREPET] (pasture sedge)
SY: *Carex liddonii* Boott
Carex petricosa Dewey [CAREPER] (rock-dwelling sedge)
SY: *Carex macrogyna* Turcz. ex Steud.
SY: *Carex petricosa* var. *distichiflora* (Boivin) Boivin
Carex petricosa var. *distichiflora* (Boivin) Boivin = Carex petricosa [CAREPER]
Carex petricosa var. *franklinii* (Boott) Boivin = Carex franklinii [CAREFRA]
Carex phaeocephala Piper [CAREPHA] (dunhead sedge)
SY: *Carex eastwoodiana* Stacey
Carex phyllomanica W. Boott = Carex echinata ssp. phyllomanica [CAREECH2]
Carex phyllomanica var. *angustata* (Carey) Boivin = Carex echinata ssp. echinata [CAREECH0]
Carex phyllomanica var. *ormantha* (Fern.) Boivin = Carex echinata ssp. echinata [CAREECH0]
Carex physocarpa J.& K. Presl = Carex saxatilis var. major [CARESAX2]
Carex plectocarpa F.J. Herm. = Carex lenticularis var. dolia [CARELEN1]
Carex pluriflora Hultén [CAREPLU] (many-flowered sedge)
SY: *Carex rariflora* ssp. *pluriflora* (Hultén) Egor.
SY: *Carex rariflora* var. *pluriflora* (Hultén) Boivin
SY: *Carex stygia* auct.
Carex podocarpa R. Br. [CAREPOD] (graceful mountain sedge)
SY: *Carex beringensis* C.B. Clarke
SY: *Carex montanensis* Bailey
SY: *Carex venustula* Holm
Carex praegracilis W. Boott [CAREPRE] (field sedge)
SY: *Carex camporum* Mackenzie
Carex prairea Dewey ex Wood [CAREPRR] (prairie sedge)
Carex praticola Rydb. [CAREPRT] (meadow sedge)
SY: *Carex praticola* var. *subcoriacea* F.J. Herm.
Carex praticola var. *subcoriacea* F.J. Herm. = Carex praticola [CAREPRT]
Carex preslii Steud. [CAREPRS] (Presl's sedge)
Carex prionophylla Holm = Carex scopulorum var. prionophylla [CARESCP2]
Carex pseudoscirpoidea Rydb. [CAREPSE]
SY: *Carex scirpoidea* var. *pseudoscirpoidea* (Rydb.) Cronq.
Carex pyrenaica Wahlenb. [CAREPYR] (Pyrenean sedge)
Carex pyrenaica ssp. **micropoda** (C.A. Mey.) Hultén [CAREPYR1]
SY: *Carex jacobi-peteri* Hultén
SY: *Carex micropoda* C.A. Mey.
Carex pyrophila Gandog. = Carex pachystachya [CAREPAC]
Carex rariflora ssp. *pluriflora* (Hultén) Egor. = Carex pluriflora [CAREPLU]
Carex rariflora var. *pluriflora* (Hultén) Boivin = Carex pluriflora [CAREPLU]
Carex raymondii Calder [CARERAM]
SY: *Carex atratiformis* ssp. *raymondii* (Calder) Porsild
Carex raynoldsii Dewey [CARERAY] (Raynolds' sedge)
SY: *Carex lyallii* Boott
Carex retrorsa Schwein. [CARERET] (long-bracted sedge)
Carex rhynchosphysa Fisch., C.A. Mey. & Avé-Lall. = Carex utriculata [CAREUTR]
Carex richardsonii R. Br. [CARERIC] (Richardson's sedge)
Carex rossii Boott [CAREROS] (Ross' sedge)
SY: *Carex brevipes* W. Boott
SY: *Carex deflexa* var. *rossii* (Boott) Bailey
SY: *Carex diversistylis* Roach
Carex rostrata Stokes [CAREROR] (beaked sedge)
SY: *Carex rostrata* var. *ambigens* Fern.
Carex rostrata var. *ambigens* Fern. = Carex rostrata [CAREROR]
Carex rostrata var. *utriculata* (Boott) Bailey = Carex utriculata [CAREUTR]
Carex rugosperma var. *tonsa* (Fern.) E.G. Voss = Carex tonsa [CARETON]
Carex rupestris All. [CARERUP] (curly sedge)
Carex rupestris ssp. *drummondiana* (Dewey) Holub = Carex rupestris var. drummondiana [CARERUP1]
Carex rupestris var. **drummondiana** (Dewey) Bailey [CARERUP1]
SY: *Carex drummondiana* Dewey
SY: *Carex rupestris* ssp. *drummondiana* (Dewey) Holub
Carex rupestris var. **rupestris** [CARERUP0]
Carex saltuensis Bailey = Carex vaginata [CAREVAG]
Carex sartwellii Dewey [CARESAR] (Sartwell's sedge)
Carex saxatilis L. [CARESAX] (russet sedge)
Carex saxatilis ssp. *laxa* (Trautv.) Kalela = Carex saxatilis var. saxatilis [CARESAX0]
Carex saxatilis var. **major** Olney [CARESAX2]
SY: *Carex ambusta* Boott
SY: *Carex physocarpa* J.& K. Presl
Carex saxatilis var. **saxatilis** [CARESAX0]
SY: *Carex saxatilis* ssp. *laxa* (Trautv.) Kalela
Carex saximontana Mackenzie [CARESAI] (Rocky Mountain sedge)
SY: *Carex backii* var. *saximontana* (Mackenzie) Boivin

Carex scirpiformis Mackenzie = Carex scirpoidea [CARESCI]
Carex scirpoidea Michx. [CARESCI] (single-spiked sedge)
SY: *Carex athabascensis* F.J. Herm.
SY: *Carex scirpiformis* Mackenzie
SY: *Carex scirpoidea* var. *convoluta* Kükenth.
SY: *Carex scirpoidea* var. *scirpiformis* (Mackenzie) O'Neill & Duman
SY: *Carex scirpoidea* var. *stenochlaena* Holm
SY: *Carex stenochlaena* (Holm) Mackenzie
Carex scirpoidea var. *convoluta* Kükenth. = Carex scirpoidea [CARESCI]
Carex scirpoidea var. *pseudoscirpoidea* (Rydb.) Cronq. = Carex pseudoscirpoidea [CAREPSE]
Carex scirpoidea var. *scirpiformis* (Mackenzie) O'Neill & Duman = Carex scirpoidea [CARESCI]
Carex scirpoidea var. *stenochlaena* Holm = Carex scirpoidea [CARESCI]
Carex scoparia Schkuhr ex Willd. [CARESCO] (pointed broom sedge)
Carex scopulorum Holm [CARESCP] (Holm's Rocky Mountain sedge)
Carex scopulorum var. **bracteosa** (Bailey) F.J. Herm. [CARESCP1]
SY: *Carex campylocarpa* Holm
SY: *Carex campylocarpa* ssp. *affinis* Maguire & A. Holmgren
SY: *Carex gymnoclada* Holm
Carex scopulorum var. **prionophylla** (Holm) L.A. Standley [CARESCP2]
SY: *Carex miserabilis* Mackenzie
SY: *Carex prionophylla* Holm
Carex siccata Dewey = Carex foenea [CAREFOE]
Carex simulata Mackenzie [CARESIM] (analogue sedge)
Carex sitchensis Prescott ex Bong. = Carex aquatilis var. dives [CAREAQU6]
Carex spaniocarpa Steud. = Carex supina var. spaniocarpa [CARESUP1]
Carex spectabilis Dewey [CARESPE] (showy sedge)
SY: *Carex invisa* Bailey
SY: *Carex nigella* Boott
SY: *Carex tolmiei* Boott
Carex sphaerostachya (Tuckerman) Dewey = Carex brunnescens [CAREBRU]
Carex sprengelii Dewey ex Spreng. [CARESPR] (Sprengel's sedge)
Carex stans Drej. = Carex aquatilis var. stans [CAREAQU3]
Carex stenochlaena (Holm) Mackenzie = Carex scirpoidea [CARESCI]
Carex stenophylla ssp. *eleocharis* (Bailey) Hultén = Carex duriuscula [CAREDUR]
Carex stenophylla var. *enervis* Kükenth. = Carex duriuscula [CAREDUR]
Carex stipata Muhl. ex Willd. [CARESTI] (awl-fruited sedge)
Carex straminea var. *mixta* Bailey = Carex feta [CAREFET]
Carex stygia auct. = Carex pluriflora [CAREPLU]
Carex stylosa C.A. Mey. [CARESTY] (long-styled sedge)
SY: *Carex stylosa* var. *nigritella* (Drej.) Fern.
Carex stylosa var. *nigritella* (Drej.) Fern. = Carex stylosa [CARESTY]
Carex stylosa var. *virens* Bailey = Carex aperta [CAREAPE]
Carex substricta (Kükenth.) Mackenzie = Carex aquatilis var. aquatilis [CAREAQU0]
Carex suksdorfii Kükenth. = Carex aquatilis var. aquatilis [CAREAQU0]
Carex supina Willd. ex Wahlenb. [CARESUP] (weak arctic sedge)
Carex supina ssp. *spaniocarpa* (Steud.) Hultén = Carex supina var. spaniocarpa [CARESUP1]
Carex supina var. **spaniocarpa** (Steud.) Boivin [CARESUP1]
SY: *Carex spaniocarpa* Steud.
SY: *Carex supina* ssp. *spaniocarpa* (Steud.) Hultén
Carex svensonis Skottsberg = Carex echinata ssp. echinata [CAREECH0]
Carex sychnocephala Carey [CARESYC] (many-headed sedge)
Carex tenella Schukuhr = Carex disperma [CAREDIS]
Carex tenera Dewey [CARETEE] (tender sedge)
SY: *Carex tenera* var. *echinodes* (Fern.) Wieg.
Carex tenera var. *echinodes* (Fern.) Wieg. = Carex tenera [CARETEE]
Carex tenuiflora Wahlenb. [CARETEN] (sparse-leaved sedge)
Carex tolmiei Boott = Carex spectabilis [CARESPE]
Carex tonsa (Fern.) Bickn. [CARETON]
SY: *Carex rugosperma* var. *tonsa* (Fern.) E.G. Voss
SY: *Carex umbellata* var. *tonsa* Fern.
Carex tracyi Mackenzie = Carex leporina [CARELEP]
Carex tripartita auct. = Carex bipartita [CAREBIP]
Carex trisperma Dewey [CARETRS] (three-seeded sedge)
Carex turgidula Bailey = Carex aperta [CAREAPE]
Carex umbellata var. *tonsa* Fern. = Carex tonsa [CARETON]
Carex unilateralis Mackenzie [CAREUNI] (one-sided sedge)
Carex utriculata Boott [CAREUTR] (bottle sedge)
SY: *Carex inflata* var. *utriculata* (Boott) Druce
SY: *Carex rhynchosphysa* Fisch., C.A. Mey. & Avé-Lall.
SY: *Carex rostrata* var. *utriculata* (Boott) Bailey
Carex vaginata Tausch [CAREVAG] (sheathed sedge)
SY: *Carex saltuensis* Bailey
Carex vahlii var. *inferalpina* Wahlenb. = Carex norvegica ssp. inferalpina [CARENOR1]
Carex venustula Holm = Carex podocarpa [CAREPOD]
Carex vesicaria L. [CAREVES]
Carex vesicaria var. *major* Boott = Carex exsiccata [CAREEXS]
Carex viridula Michx. [CAREVIR] (green sedge)
SY: *Carex oederi* var. *viridula* (Michx.) Hultén
Carex vulpinoidea Michx. [CAREVUL] (fox sedge)
Carex xerantica Bailey [CAREXER] (dryland sedge)
Cyperus L. [CYPERUS$]
Cyperus aristatus Rottb. = Cyperus squarrosus [CYPESQU]
Cyperus aristatus var. *runyonii* O'Neill = Cyperus squarrosus [CYPESQU]
Cyperus inflexus Muhl. = Cyperus squarrosus [CYPESQU]
Cyperus squarrosus L. [CYPESQU]
SY: *Cyperus aristatus* Rottb.
SY: *Cyperus aristatus* var. *runyonii* O'Neill
SY: *Cyperus inflexus* Muhl.
Dulichium Pers. [DULICHI$]
Dulichium arundinaceum (L.) Britt. [DULIARU] (three-way sedge)
Eleocharis R. Br. [ELEOCHA$]
Eleocharis acicularis (L.) Roemer & J.A. Schultes [ELEOACI] (needle spike-rush)
Eleocharis acuminata (Muhl.) Nees = Eleocharis compressa [ELEOCOM]
Eleocharis atropurpurea (Retz.) J.& K. Presl [ELEOATR] (purple spike-rush)
SY: *Eleocharis multiflora* Chapman
SY: *Scirpus atropurpureus* Retz.
Eleocharis bernardina Munz & Johnston = Eleocharis quinqueflora [ELEOQUI]
Eleocharis calva var. *australis* (Nees) St. John = Eleocharis palustris [ELEOPAL]
Eleocharis capitata var. *borealis* Svens. = Eleocharis elliptica [ELEOELL]
Eleocharis coloradoensis (Britt.) Gilly = Eleocharis parvula [ELEOPAR]
Eleocharis compressa Sullivant [ELEOCOM]
SY: *Eleocharis acuminata* (Muhl.) Nees
SY: *Eleocharis compressa* var. *atrata* Svens.
SY: *Eleocharis elliptica* var. *compressa* (Sullivant) Drapalik & Mohlenbrock
SY: *Eleocharis tenuis* var. *atrata* (Svens.) Boivin
Eleocharis compressa var. *atrata* Svens. = Eleocharis compressa [ELEOCOM]
Eleocharis diandra C. Wright = Eleocharis obtusa [ELEOOBT]
Eleocharis elliptica Kunth [ELEOELL]
SY: *Eleocharis capitata* var. *borealis* Svens.
SY: *Eleocharis tenuis* var. *borealis* (Svens.) Gleason
Eleocharis elliptica var. *compressa* (Sullivant) Drapalik & Mohlenbrock = Eleocharis compressa [ELEOCOM]
Eleocharis kamtschatica (C.A. Mey.) Komarov [ELEOKAM] (Kamchatka spike-rush)
SY: *Scirpus kamtschaticus* C.A. Mey.

Eleocharis leptos (Steud.) Svens. = Eleocharis parvula [ELEOPAR]
Eleocharis leptos var. *coloradoensis* (Britt.) Svens. = Eleocharis parvula [ELEOPAR]
Eleocharis leptos var. *johnstonii* Svens. = Eleocharis parvula [ELEOPAR]
Eleocharis macounii Fern. = Eleocharis obtusa [ELEOOBT]
Eleocharis macrostachya Britt. = Eleocharis palustris [ELEOPAL]
Eleocharis membranacea Gilly = Eleocharis parvula [ELEOPAR]
Eleocharis multiflora Chapman = Eleocharis atropurpurea [ELEOATR]
Eleocharis nitida var. *borealis* (Svens.) Gleason = Eleocharis tenuis [ELEOTEN]
Eleocharis obtusa (Willd.) J.A. Schultes [ELEOOBT]
SY: *Eleocharis diandra* C. Wright
SY: *Eleocharis macounii* Fern.
SY: *Eleocharis obtusa* var. *ellipsoidalis* Fern.
SY: *Eleocharis obtusa* var. *gigantea* (C.B. Clarke) Fern.
SY: *Eleocharis obtusa* var. *jejuna* Fern.
SY: *Eleocharis obtusa* var. *peasei* Svens.
SY: *Eleocharis ovata* var. *obtusa* (Willd.) Kükenth. ex Skottsberg
SY: *Scirpus obtusus* Willd.
Eleocharis obtusa var. *ellipsoidalis* Fern. = Eleocharis obtusa [ELEOOBT]
Eleocharis obtusa var. *gigantea* (C.B. Clarke) Fern. = Eleocharis obtusa [ELEOOBT]
Eleocharis obtusa var. *jejuna* Fern. = Eleocharis obtusa [ELEOOBT]
Eleocharis obtusa var. *peasei* Svens. = Eleocharis obtusa [ELEOOBT]
Eleocharis ovata var. *obtusa* (Willd.) Kükenth. ex Skottsberg = Eleocharis obtusa [ELEOOBT]
Eleocharis palustris (L.) Roemer & J.A. Schultes [ELEOPAL] (creeping spike-rush)
SY: *Eleocharis calva* var. *australis* (Nees) St. John
SY: *Eleocharis macrostachya* Britt.
SY: *Eleocharis palustris* var. *australis* Nees
SY: *Eleocharis palustris* var. *major* Sonder
SY: *Eleocharis perlonga* Fern. & Brack.
SY: *Eleocharis smallii* var. *major* (Sonder) Seymour
SY: *Eleocharis xyridiformis* Fern. & Brack.
Eleocharis palustris var. *australis* Nees = Eleocharis palustris [ELEOPAL]
Eleocharis palustris var. *major* Sonder = Eleocharis palustris [ELEOPAL]
Eleocharis parvula (Roemer & J.A. Schultes) Link ex Bluff, Nees & Schauer [ELEOPAR] (small spike-rush)
SY: *Eleocharis coloradoensis* (Britt.) Gilly
SY: *Eleocharis leptos* (Steud.) Svens.
SY: *Eleocharis leptos* var. *coloradoensis* (Britt.) Svens.
SY: *Eleocharis leptos* var. *johnstonii* Svens.
SY: *Eleocharis membranacea* Gilly
SY: *Eleocharis parvula* var. *anachaeta* (Torr.) Svens.
SY: *Eleocharis parvula* var. *coloradoensis* (Britt.) Beetle
Eleocharis parvula var. *anachaeta* (Torr.) Svens. = Eleocharis parvula [ELEOPAR]
Eleocharis parvula var. *coloradoensis* (Britt.) Beetle = Eleocharis parvula [ELEOPAR]
Eleocharis pauciflora (Lightf.) Link = Eleocharis quinqueflora [ELEOQUI]
Eleocharis pauciflora var. *bernardina* (Munz & Johnston) Svens. = Eleocharis quinqueflora [ELEOQUI]
Eleocharis pauciflora var. *fernaldii* Svens. = Eleocharis quinqueflora [ELEOQUI]
Eleocharis pauciflora var. *suksdorfiana* (Beauv.) Svens. = Eleocharis quinqueflora [ELEOQUI]
Eleocharis perlonga Fern. & Brack. = Eleocharis palustris [ELEOPAL]
Eleocharis quinqueflora (F.X. Hartmann) Schwarz [ELEOQUI] (few-flowered spike-rush)
SY: *Eleocharis bernardina* Munz & Johnston
SY: *Eleocharis pauciflora* (Lightf.) Link
SY: *Eleocharis pauciflora* var. *bernardina* (Munz & Johnston) Svens.
SY: *Eleocharis pauciflora* var. *fernaldii* Svens.
SY: *Eleocharis pauciflora* var. *suksdorfiana* (Beauv.) Svens.
SY: *Eleocharis quinqueflora* ssp. *fernaldii* (Svens.) Hultén
SY: *Eleocharis quinqueflora* ssp. *suksdorfiana* (Beauv.) Hultén
SY: *Eleocharis quinqueflora* var. *suksdorfiana* (Beauv.) J.T. Howell
SY: *Eleocharis suksdorfiana* Beauv.
SY: *Scirpus nanus* Spreng.
SY: *Scirpus pauciflorus* Lightf.
SY: *Scirpus quinqueflorus* F.X. Hartmann
Eleocharis quinqueflora ssp. *fernaldii* (Svens.) Hultén = Eleocharis quinqueflora [ELEOQUI]
Eleocharis quinqueflora ssp. *suksdorfiana* (Beauv.) Hultén = Eleocharis quinqueflora [ELEOQUI]
Eleocharis quinqueflora var. *suksdorfiana* (Beauv.) J.T. Howell = Eleocharis quinqueflora [ELEOQUI]
Eleocharis rostellata (Torr.) Torr. [ELEOROS] (beaked spike-rush)
SY: *Eleocharis rostellata* var. *congdonii* Jepson
SY: *Eleocharis rostellata* var. *occidentalis* S. Wats.
SY: *Scirpus rostellatus* Torr.
Eleocharis rostellata var. *congdonii* Jepson = Eleocharis rostellata [ELEOROS]
Eleocharis rostellata var. *occidentalis* S. Wats. = Eleocharis rostellata [ELEOROS]
Eleocharis smallii var. *major* (Sonder) Seymour = Eleocharis palustris [ELEOPAL]
Eleocharis suksdorfiana Beauv. = Eleocharis quinqueflora [ELEOQUI]
Eleocharis tenuis (Willd.) J.A. Schultes [ELEOTEN] (slender spike-rush)
SY: *Eleocharis nitida* var. *borealis* (Svens.) Gleason
Eleocharis tenuis var. *atrata* (Svens.) Boivin = Eleocharis compressa [ELEOCOM]
Eleocharis tenuis var. *borealis* (Svens.) Gleason = Eleocharis elliptica [ELEOELL]
Eleocharis uniglumis (Link) J.A. Schultes [ELEOUNI]
SY: *Scirpus uniglumis* Link
Eleocharis xyridiformis Fern. & Brack. = Eleocharis palustris [ELEOPAL]
Elyna bellardii (All.) Degl. = Kobresia myosuroides [KOBRMYO]
Elyna sibirica Turcz. ex Ledeb. = Kobresia sibirica [KOBRSIB]
Eriophorum L. [ERIOPHO$]
Eriophorum alpinum L. [ERIOALP]
SY: *Baeothryon alpinum* (L.) Egor.
SY: *Leucocoma alpina* (L.) Rydb.
SY: *Scirpus hudsonianus* (Michx.) Fern.
SY: *Trichophorum alpinum* (L.) Pers.
Eriophorum altaicum Meinsh. [ERIOALT]
Eriophorum altaicum var. **neogaeum** Raymond [ERIOALT1]
Eriophorum angustifolium Honckeny [ERIOANG] (narrow-leaved cotton-grass)
Eriophorum angustifolium ssp. **scabriusculum** Hultén [ERIOANG1]
Eriophorum angustifolium ssp. **subarcticum** (Vassiljev) Hultén ex Kartesz & Gandhi [ERIOANG2]
Eriophorum angustifolium ssp. **triste** (T. Fries) Hultén [ERIOANG3]
SY: *Eriophorum triste* (T. Fries) Hadac & A. Löve
Eriophorum brachyantherum Trautv. & C.A. Mey. [ERIOBRA] (close-sheathed cotton-grass)
Eriophorum callitrix Cham. ex C.A. Mey. [ERIOCAL] (arctic cotton-grass)
Eriophorum chamissonis C.A. Mey. [ERIOCHA] (Chamisso's cotton-grass)
SY: *Eriophorum russeolum* ssp. *rufescens* (E. Anders.) Hyl.
Eriophorum gracile W.D.J. Koch [ERIOGRA] (slender cotton-grass)
Eriophorum polystachion var. *viridicarinatum* Engelm. = Eriophorum viridicarinatum [ERIOVIR]
Eriophorum russeolum Fries ex Hartman [ERIORUS]
Eriophorum russeolum ssp. *rufescens* (E. Anders.) Hyl. = Eriophorum chamissonis [ERIOCHA]
Eriophorum scheuchzeri Hoppe [ERIOSCH] (Scheuchzer's cotton-grass)
SY: *Eriophorum scheuchzeri* var. *tenuifolium* Ohwi

Eriophorum scheuchzeri var. *tenuifolium* Ohwi = Eriophorum scheuchzeri [ERIOSCH]
Eriophorum spissum Fern. = Eriophorum vaginatum var. spissum [ERIOVAG2]
Eriophorum spissum var. *erubescens* (Fern.) Fern. = Eriophorum vaginatum var. spissum [ERIOVAG2]
Eriophorum triste (T. Fries) Hadac & A. Löve = Eriophorum angustifolium ssp. triste [ERIOANG3]
Eriophorum vaginatum L. [ERIOVAG] (sheathed cotton-grass)
Eriophorum vaginatum ssp. *spissum* (Fern.) Hultén = Eriophorum vaginatum var. spissum [ERIOVAG2]
Eriophorum vaginatum var. **spissum** (Fern.) Boivin [ERIOVAG2]
SY: *Eriophorum spissum* Fern.
SY: *Eriophorum spissum* var. *erubescens* (Fern.) Fern.
SY: *Eriophorum vaginatum* ssp. *spissum* (Fern.) Hultén
*****Eriophorum virginicum** L. [ERIOVIG] (tawny cotton-grass)
Eriophorum viridicarinatum (Engelm.) Fern. [ERIOVIR] (green-keeled cotton-grass)
SY: *Eriophorum polystachion* var. *viridicarinatum* Engelm.
Hemicarpha micrantha (Vahl) Pax = Lipocarpha micrantha [LIPOMIC]
Hemicarpha micrantha var. *minor* (Schrad.) Friedland = Lipocarpha micrantha [LIPOMIC]
Isolepis cernua (Vahl) Roemer & J.A. Schultes p.p. = Scirpus cernuus var. californicus [SCIRCER1]
Isolepis setaceus (L.) R. Br. = Scirpus setaceus [SCIRSET]
Kobresia Willd. [KOBRESI$]
Kobresia bellardii (All.) K. Koch = Kobresia myosuroides [KOBRMYO]
Kobresia bellardii var. *macrocarpa* (Clokey ex Mackenzie) Harrington = Kobresia sibirica [KOBRSIB]
Kobresia bipartita (All.) Dalla Torre = Kobresia simpliciuscula [KOBRSIM]
Kobresia hyperborea Porsild = Kobresia sibirica [KOBRSIB]
Kobresia hyperborea var. *alaskana* Duman = Kobresia sibirica [KOBRSIB]
Kobresia macrocarpa Clokey ex Mackenzie = Kobresia sibirica [KOBRSIB]
Kobresia myosuroides (Vill.) Fiori [KOBRMYO] (Bellard's kobresia)
SY: *Elyna bellardii* (All.) Degl.
SY: *Kobresia bellardii* (All.) K. Koch
Kobresia schoenoides (C.A. Mey.) Steud. p.p. = Kobresia sibirica [KOBRSIB]
Kobresia schoenoides var. *lepagei* (Duman) Boivin = Kobresia sibirica [KOBRSIB]
Kobresia sibirica (Turcz. ex Ledeb.) Boeckl. [KOBRSIB] (Siberian kobresia)
SY: *Elyna sibirica* Turcz. ex Ledeb.
SY: *Kobresia bellardii* var. *macrocarpa* (Clokey ex Mackenzie) Harrington
SY: *Kobresia hyperborea* Porsild
SY: *Kobresia hyperborea* var. *alaskana* Duman
SY: *Kobresia macrocarpa* Clokey ex Mackenzie
SY: *Kobresia schoenoides* (C.A. Mey.) Steud. p.p.
SY: *Kobresia schoenoides* var. *lepagei* (Duman) Boivin
Kobresia simpliciuscula (Wahlenb.) Mackenzie [KOBRSIM] (simple kobresia)
SY: *Kobresia bipartita* (All.) Dalla Torre
SY: *Kobresia simpliciuscula* var. *americana* Duman
Kobresia simpliciuscula var. *americana* Duman = Kobresia simpliciuscula [KOBRSIM]
Leucocoma alpina (L.) Rydb. = Eriophorum alpinum [ERIOALP]
Lipocarpha R. Br. [LIPOCAR$]
Lipocarpha micrantha (Vahl) G. Tucker [LIPOMIC]
SY: *Hemicarpha micrantha* (Vahl) Pax
SY: *Hemicarpha micrantha* var. *minor* (Schrad.) Friedland
SY: *Scirpus micranthus* Vahl
Rhynchospora Vahl [RHYNCHO$]
Rhynchospora alba (L.) Vahl [RHYNALB] (white beak-rush)
SY: *Rhynchospora luquillensis* Britt.
Rhynchospora capillacea Torr. [RHYNCAP] (brown beak-rush)
SY: *Rhynchospora capillacea* var. *leviseta* E.J. Hill ex Gray
SY: *Rhynchospora smallii* Britt.
Rhynchospora capillacea var. *leviseta* E.J. Hill ex Gray = Rhynchospora capillacea [RHYNCAP]
Rhynchospora luquillensis Britt. = Rhynchospora alba [RHYNALB]
Rhynchospora smallii Britt. = Rhynchospora capillacea [RHYNCAP]
Schoenoplectus americanus (Pers.) Volk. ex Schinz & R. Keller = Scirpus americanus [SCIRAME]
Schoenoplectus lacustris ssp. *acutus* (Muhl. ex Bigelow) A. & D. Löve = Scirpus acutus [SCIRACU]
Schoenoplectus lacustris ssp. *creber* (Fern.) A. & D. Löve = Scirpus tabernaemontani [SCIRTAB]
Schoenoplectus lacustris ssp. *validus* (Vahl) T. Koyama = Scirpus tabernaemontani [SCIRTAB]
Schoenoplectus maritimus (L.) Lye = Scirpus maritimus [SCIRMAR]
Schoenoplectus pungens (Vahl) Palla = Scirpus pungens [SCIRPUN]
Schoenoplectus validus (Vahl) A. & D. Löve = Scirpus tabernaemontani [SCIRTAB]
Scirpus L. [SCIRPUS$]
Scirpus acutus Muhl. ex Bigelow [SCIRACU] (hard-stemmed bulrush)
SY: *Schoenoplectus lacustris* ssp. *acutus* (Muhl. ex Bigelow) A. & D. Löve
SY: *Scirpus lacustris* L. p.p.
SY: *Scirpus occidentalis* (S. Wats.) Chase
Scirpus americanus Pers. [SCIRAME] (American bulrush)
SY: *Schoenoplectus americanus* (Pers.) Volk. ex Schinz & R. Keller
SY: *Scirpus americanus* var. *monophyllus* (J.& K. Presl) T. Koyama
SY: *Scirpus chilensis* Nees & Meyen ex Kunth
SY: *Scirpus conglomeratus* Kunth
SY: *Scirpus monophyllus* J.& K. Presl
SY: *Scirpus olneyi* Gray
SY: *Scirpus pungens* ssp. *monophyllus* (J.& K. Presl) Taylor & MacBryde
Scirpus americanus var. *longispicatus* Britt. = Scirpus pungens [SCIRPUN]
Scirpus americanus var. *monophyllus* (J.& K. Presl) T. Koyama = Scirpus americanus [SCIRAME]
Scirpus americanus var. *polyphyllus* (Boeckl.) Beetle = Scirpus pungens [SCIRPUN]
Scirpus atrocinctus Fern. [SCIRATR]
SY: *Scirpus cyperinus* var. *brachypodus* (Fern.) Gilly
Scirpus atropurpureus Retz. = Eleocharis atropurpurea [ELEOATR]
Scirpus atrovirens var. *pallidus* Britt. = Scirpus pallidus [SCIRPAI]
Scirpus bergsonii Schuyler = Scirpus saximontanus [SCIRSAX]
Scirpus cernuus Vahl [SCIRCER] (low clubrush)
Scirpus cernuus ssp. *californicus* (Torr.) Thorne = Scirpus cernuus var. californicus [SCIRCER1]
Scirpus cernuus var. **californicus** (Torr.) Beetle [SCIRCER1]
SY: *Isolepis cernua* (Vahl) Roemer & J.A. Schultes p.p.
SY: *Scirpus cernuus* ssp. *californicus* (Torr.) Thorne
Scirpus cespitosus L. [SCIRCES]
SY: *Baeothryon cespitosum* (L.) A. Dietr.
SY: *Scirpus cespitosus* var. *austriacus* (Pallas) Aschers. & Graebn.
SY: *Scirpus cespitosus* var. *callosus* Bigelow
SY: *Scirpus cespitosus* var. *delicatulus* Fern.
SY: *Trichophorum cespitosum* (L.) Hartman
Scirpus cespitosus var. *austriacus* (Pallas) Aschers. & Graebn. = Scirpus cespitosus [SCIRCES]
Scirpus cespitosus var. *callosus* Bigelow = Scirpus cespitosus [SCIRCES]
Scirpus cespitosus var. *delicatulus* Fern. = Scirpus cespitosus [SCIRCES]
Scirpus chilensis Nees & Meyen ex Kunth = Scirpus americanus [SCIRAME]
Scirpus conglomeratus Kunth = Scirpus americanus [SCIRAME]
Scirpus cyperinus var. *brachypodus* (Fern.) Gilly = Scirpus atrocinctus [SCIRATR]
Scirpus fernaldii Bickn. = Scirpus maritimus [SCIRMAR]
Scirpus fluviatilis (Torr.) Gray [SCIRFLU] (river bulrush)
Scirpus hudsonianus (Michx.) Fern. = Eriophorum alpinum [ERIOALP]

Scirpus kamtschaticus C.A. Mey. = Eleocharis kamtschatica [ELEOKAM]
Scirpus lacustris L. p.p. = Scirpus acutus [SCIRACU]
Scirpus lacustris ssp. *creber* (Fern.) T. Koyama = Scirpus tabernaemontani [SCIRTAB]
Scirpus lacustris ssp. *glaucus* (Reichenb.) Hartman = Scirpus tabernaemontani [SCIRTAB]
Scirpus lacustris ssp. *tabernaemontani* (K.C. Gmel.) Syme = Scirpus tabernaemontani [SCIRTAB]
Scirpus lacustris ssp. *validus* (Vahl) T. Koyama = Scirpus tabernaemontani [SCIRTAB]
Scirpus longispicatus (Britt.) Smyth = Scirpus pungens [SCIRPUN]
Scirpus maritimus L. [SCIRMAR] (saltmarsh bulrush)
SY: *Bolboschoenus maritimus* (L.) Palla
SY: *Bolboschoenus maritimus* ssp. *paludosus* (A. Nels.) A. & D. Löve
SY: *Bolboschoenus paludosus* (A. Nels.) Soó
SY: *Schoenoplectus maritimus* (L.) Lye
SY: *Scirpus fernaldii* Bickn.
SY: *Scirpus maritimus* var. *fernaldii* (Bickn.) Beetle
SY: *Scirpus maritimus* var. *paludosus* (A. Nels.) Kükenth.
SY: *Scirpus pacificus* Britt.
SY: *Scirpus paludosus* A. Nels.
SY: *Scirpus paludosus* var. *atlanticus* Fern.
Scirpus maritimus var. *fernaldii* (Bickn.) Beetle = Scirpus maritimus [SCIRMAR]
Scirpus maritimus var. *paludosus* (A. Nels.) Kükenth. = Scirpus maritimus [SCIRMAR]
Scirpus micranthus Vahl = Lipocarpha micrantha [LIPOMIC]
Scirpus microcarpus J.& K. Presl [SCIRMIC] (small-flowered bulrush)
SY: *Scirpus microcarpus* var. *longispicatus* M.E. Peck
SY: *Scirpus microcarpus* var. *rubrotinctus* (Fern.) M.E. Jones
SY: *Scirpus rubrotinctus* Fern.
SY: *Scirpus sylvaticus* ssp. *digynus* (Boeckl.) Koyama
Scirpus microcarpus var. *longispicatus* M.E. Peck = Scirpus microcarpus [SCIRMIC]
Scirpus microcarpus var. *rubrotinctus* (Fern.) M.E. Jones = Scirpus microcarpus [SCIRMIC]
Scirpus monophyllus J.& K. Presl = Scirpus americanus [SCIRAME]
Scirpus nanus Spreng. = Eleocharis quinqueflora [ELEOQUI]
Scirpus nevadensis S. Wats. [SCIRNEV] (Nevada bulrush)
SY: *Amphiscirpus nevadensis* (S. Wats.) Oteng Yeboah
Scirpus obtusus Willd. = Eleocharis obtusa [ELEOOBT]
Scirpus occidentalis (S. Wats.) Chase = Scirpus acutus [SCIRACU]
Scirpus olneyi Gray = Scirpus americanus [SCIRAME]
Scirpus pacificus Britt. = Scirpus maritimus [SCIRMAR]
Scirpus pallidus (Britt.) Fern. [SCIRPAI] (pale bulrush)
SY: *Scirpus atrovirens* var. *pallidus* Britt.
Scirpus paludosus A. Nels. = Scirpus maritimus [SCIRMAR]
Scirpus paludosus var. *atlanticus* Fern. = Scirpus maritimus [SCIRMAR]
Scirpus pauciflorus Lightf. = Eleocharis quinqueflora [ELEOQUI]
Scirpus pumilus auct. = Scirpus rollandii [SCIRROL]
Scirpus pumilus ssp. *rollandii* (Fern.) Raymond = Scirpus rollandii [SCIRROL]
Scirpus pumilus var. *rollandii* (Fern.) Beetle = Scirpus rollandii [SCIRROL]
Scirpus pungens Vahl [SCIRPUN]
SY: *Schoenoplectus pungens* (Vahl) Palla
SY: *Scirpus americanus* var. *longispicatus* Britt.
SY: *Scirpus americanus* var. *polyphyllus* (Boeckl.) Beetle
SY: *Scirpus longispicatus* (Britt.) Smyth
SY: *Scirpus pungens* var. *longisetus* Benth. & F. Muell.
SY: *Scirpus pungens* var. *longispicatus* (Britt.) Taylor & MacBryde
SY: *Scirpus pungens* var. *polyphyllus* Boeckl.
Scirpus pungens ssp. *monophyllus* (J.& K. Presl) Taylor & MacBryde = Scirpus americanus [SCIRAME]
Scirpus pungens var. *longisetus* Benth. & F. Muell. = Scirpus pungens [SCIRPUN]
Scirpus pungens var. *longispicatus* (Britt.) Taylor & MacBryde = Scirpus pungens [SCIRPUN]
Scirpus pungens var. *polyphyllus* Boeckl. = Scirpus pungens [SCIRPUN]
Scirpus quinqueflorus F.X. Hartmann = Eleocharis quinqueflora [ELEOQUI]
Scirpus rollandii Fern. [SCIRROL]
SY: *Scirpus pumilus* ssp. *rollandii* (Fern.) Raymond
SY: *Scirpus pumilus* var. *rollandii* (Fern.) Beetle
SY: *Trichophorum pumilum* ssp. *rollandii* (Fern.) Taylor & MacBryde
SY: *Trichophorum pumilum* var. *rollandii* (Fern.) Hultén
SY: *Trichophorum rollandii* (Fern.) Hultén
SY: *Scirpus pumilus* auct.
SY: *Trichophorum pumilum* auct.
Scirpus rostellatus Torr. = Eleocharis rostellata [ELEOROS]
Scirpus rubrotinctus Fern. = Scirpus microcarpus [SCIRMIC]
Scirpus saximontanus Fern. [SCIRSAX]
SY: *Scirpus bergsonii* Schuyler
SY: *Scirpus supinus* var. *saximontanus* (Fern.) T. Koyama
*__Scirpus setaceus__ L. [SCIRSET] (bristle clubrush)
SY: *Isolepis setaceus* (L.) R. Br.
Scirpus steinmetzii Fern. = Scirpus tabernaemontani [SCIRTAB]
Scirpus subterminalis Torr. [SCIRSUB] (water clubrush)
Scirpus supinus var. *saximontanus* (Fern.) T. Koyama = Scirpus saximontanus [SCIRSAX]
Scirpus sylvaticus ssp. *digynus* (Boeckl.) Koyama = Scirpus microcarpus [SCIRMIC]
Scirpus tabernaemontani K.C. Gmel. [SCIRTAB]
SY: *Schoenoplectus lacustris* ssp. *creber* (Fern.) A. & D. Löve
SY: *Schoenoplectus lacustris* ssp. *validus* (Vahl) T. Koyama
SY: *Schoenoplectus validus* (Vahl) A. & D. Löve
SY: *Scirpus lacustris* ssp. *creber* (Fern.) T. Koyama
SY: *Scirpus lacustris* ssp. *glaucus* (Reichenb.) Hartman
SY: *Scirpus lacustris* ssp. *tabernaemontani* (K.C. Gmel.) Syme
SY: *Scirpus lacustris* ssp. *validus* (Vahl) T. Koyama
SY: *Scirpus steinmetzii* Fern.
SY: *Scirpus validus* Vahl
SY: *Scirpus validus* var. *creber* Fern.
Scirpus uniglumis Link = Eleocharis uniglumis [ELEOUNI]
Scirpus validus Vahl = Scirpus tabernaemontani [SCIRTAB]
Scirpus validus var. *creber* Fern. = Scirpus tabernaemontani [SCIRTAB]
Trichophorum alpinum (L.) Pers. = Eriophorum alpinum [ERIOALP]
Trichophorum cespitosum (L.) Hartman = Scirpus cespitosus [SCIRCES]
Trichophorum pumilum auct. = Scirpus rollandii [SCIRROL]
Trichophorum pumilum ssp. *rollandii* (Fern.) Taylor & MacBryde = Scirpus rollandii [SCIRROL]
Trichophorum pumilum var. *rollandii* (Fern.) Hultén = Scirpus rollandii [SCIRROL]
Trichophorum rollandii (Fern.) Hultén = Scirpus rollandii [SCIRROL]

DENNSTAEDTIACEAE (V:F)

Pteridium Gleditsch ex Scop. [PTERIDI$]
Pteridium aquilinum (L.) Kuhn [PTERAQU] (bracken fern)
Pteridium aquilinum ssp. *lanuginosum* (Bong.) Hultén = Pteridium aquilinum var. pubescens [PTERAQU7]
Pteridium aquilinum ssp. *latiusculum* (Desv.) C.N. Page = Pteridium aquilinum var. latiusculum [PTERAQU6]
Pteridium aquilinum var. *lanuginosum* (Bong.) Fern. = Pteridium aquilinum var. pubescens [PTERAQU7]
Pteridium aquilinum var. **latiusculum** (Desv.) Underwood ex Heller [PTERAQU6]
SY: *Pteridium aquilinum* ssp. *latiusculum* (Desv.) C.N. Page
SY: *Pteridium latiusculum* (Desv.) Hieron.
Pteridium aquilinum var. **pubescens** Underwood [PTERAQU7]
SY: *Pteridium aquilinum* ssp. *lanuginosum* (Bong.) Hultén
SY: *Pteridium aquilinum* var. *lanuginosum* (Bong.) Fern.
Pteridium latiusculum (Desv.) Hieron. = Pteridium aquilinum var. latiusculum [PTERAQU6]

DIAPENSIACEAE (V:D)

Diapensia L. [DIAPENS$]
Diapensia lapponica L. [DIAPLAP] (diapensia)
Diapensia lapponica ssp. *obovata* (F. Schmidt) Hultén = Diapensia lapponica var. obovata [DIAPLAP1]
Diapensia lapponica var. **obovata** F. Schmidt [DIAPLAP1]
SY: *Diapensia lapponica* ssp. *obovata* (F. Schmidt) Hultén
SY: *Diapensia lapponica* var. *rosea* Hultén
SY: *Diapensia obovata* (F. Schmidt) Nakai
Diapensia lapponica var. *rosea* Hultén = Diapensia lapponica var. obovata [DIAPLAP1]
Diapensia obovata (F. Schmidt) Nakai = Diapensia lapponica var. obovata [DIAPLAP1]

DIPSACACEAE (V:D)

Dipsacus L. [DIPSACU$]
***Dipsacus fullonum** L. [DIPSFUL] (Fuller's teasel)
Knautia L. [KNAUTIA$]
***Knautia arvensis** (L.) Coult. [KNAUARV] (field scabious)
SY: *Scabiosa arvensis* L.
Scabiosa L. [SCABIOS$]
Scabiosa arvensis L. = Knautia arvensis [KNAUARV]
***Scabiosa ochroleuca** L. [SCABOCH] (yellow scabious)

DROSERACEAE (V:D)

Drosera L. [DROSERA$]
Drosera anglica Huds. [DROSANG] (oblong-leaved sundew)
SY: *Drosera longifolia* L.
Drosera longifolia L. = Drosera anglica [DROSANG]
Drosera rotundifolia L. [DROSROT] (round-leaved sundew)

DRYOPTERIDACEAE (V:F)

Athyrium Roth [ATHYRIU$]
Athyrium alpestre (Hoppe) Milde = Athyrium americanum [ATHYAME]
Athyrium alpestre ssp. *americanum* (Butters) Lellinger = Athyrium americanum [ATHYAME]
Athyrium alpestre var. *americanum* Butters = Athyrium americanum [ATHYAME]
Athyrium alpestre var. *cyclosorum* (Rupr.) T. Moore = Athyrium filix-femina ssp. cyclosorum [ATHYFIL1]
Athyrium alpestre var. *gaspense* Fern. = Athyrium americanum [ATHYAME]
Athyrium americanum (Butters) Maxon [ATHYAME]
SY: *Athyrium alpestre* (Hoppe) Milde
SY: *Athyrium alpestre* ssp. *americanum* (Butters) Lellinger
SY: *Athyrium alpestre* var. *americanum* Butters
SY: *Athyrium alpestre* var. *gaspense* Fern.
SY: *Athyrium distentifolium* ssp. *americanum* (Butters) Hultén
SY: *Athyrium distentifolium* var. *americanum* (Butters) Boivin
Athyrium angustum var. *boreale* Jennings = Athyrium filix-femina ssp. cyclosorum [ATHYFIL1]
Athyrium angustum var. *elatius* (Link) Butters = Athyrium filix-femina ssp. cyclosorum [ATHYFIL1]
Athyrium cyclosorum Rupr. = Athyrium filix-femina ssp. cyclosorum [ATHYFIL1]
Athyrium distentifolium ssp. *americanum* (Butters) Hultén = Athyrium americanum [ATHYAME]
Athyrium distentifolium var. *americanum* (Butters) Boivin = Athyrium americanum [ATHYAME]
Athyrium filix-femina (L.) Roth [ATHYFIL] (lady fern)
Athyrium filix-femina ssp. **cyclosorum** (Rupr.) C. Christens. [ATHYFIL1]
SY: *Athyrium alpestre* var. *cyclosorum* (Rupr.) T. Moore
SY: *Athyrium angustum* var. *boreale* Jennings
SY: *Athyrium angustum* var. *elatius* (Link) Butters
SY: *Athyrium cyclosorum* Rupr.
SY: *Athyrium filix-femina* var. *californicum* Butters
SY: *Athyrium filix-femina* var. *cyclosorum* (Rupr.) Ledeb.
SY: *Athyrium filix-femina* var. *sitchense* (Rupr.) Ledeb.
Athyrium filix-femina var. *californicum* Butters = Athyrium filix-femina ssp. cyclosorum [ATHYFIL1]
Athyrium filix-femina var. *cyclosorum* (Rupr.) Ledeb. = Athyrium filix-femina ssp. cyclosorum [ATHYFIL1]
Athyrium filix-femina var. *sitchense* (Rupr.) Ledeb. = Athyrium filix-femina ssp. cyclosorum [ATHYFIL1]
Cystopteris Bernh. [CYSTOPT$]
Cystopteris dickieana Sim = Cystopteris fragilis [CYSTFRA]
Cystopteris fragilis (L.) Bernh. [CYSTFRA] (fragile bladder fern)
SY: *Cystopteris dickieana* Sim
SY: *Cystopteris fragilis* var. *angustata* Lawson
SY: *Cystopteris fragilis* ssp. *dickieana* (Sim) Hyl.
SY: *Cystopteris fragilis* var. *woodsioides* Christ
SY: *Filix fragilis* (L.) Underwood
Cystopteris fragilis ssp. *dickieana* (Sim) Hyl. = Cystopteris fragilis [CYSTFRA]
Cystopteris fragilis var. *angustata* Lawson = Cystopteris fragilis [CYSTFRA]
Cystopteris fragilis var. *woodsioides* Christ = Cystopteris fragilis [CYSTFRA]
Cystopteris montana (Lam.) Bernh. ex Desv. [CYSTMON] (mountain bladder fern)
SY: *Filix montana* (Lam.) Underwood
SY: *Rhizomatopteris montana* (Lam.) Khokhr.
Dryopteris Adans. [DRYOPTE$]
Dryopteris arguta (Kaulfuss) Watt [DRYOARG] (coastal wood fern)
Dryopteris assimilis S. Walker = Dryopteris expansa [DRYOEXP]
Dryopteris austriaca var. *spinulosa* (O.F. Muell.) Fisch. = Dryopteris carthusiana [DRYOCAR]
Dryopteris carthusiana (Vill.) H.P. Fuchs [DRYOCAR] (toothed wood fern)
SY: *Dryopteris austriaca* var. *spinulosa* (O.F. Muell.) Fisch.
SY: *Dryopteris spinulosa* (O.F. Muell.) Watt
Dryopteris cristata (L.) Gray [DRYOCRI] (crested wood fern)
Dryopteris dilatata auct. = Dryopteris expansa [DRYOEXP]
Dryopteris disjuncta (Rupr.) Morton = Gymnocarpium disjunctum [GYMNDIS]
Dryopteris dryopteris (L.) Britt. = Gymnocarpium dryopteris [GYMNDRY]
Dryopteris expansa (K. Presl) Fraser-Jenkins & Jermy [DRYOEXP] (spiny wood fern)
SY: *Dryopteris assimilis* S. Walker
SY: *Dryopteris spinulosa* ssp. *assimilis* (Walker) Schidlay
SY: *Dryopteris dilatata* auct.
Dryopteris filix-mas (L.) Schott [DRYOFIL] (male fern)
Dryopteris fragrans (L.) Schott [DRYOFRA] (fragrant wood fern)
Dryopteris linnaeana C. Christens. = Gymnocarpium dryopteris [GYMNDRY]
Dryopteris marginalis (L.) Gray [DRYOMAR] (marginal wood fern)
Dryopteris spinulosa (O.F. Muell.) Watt = Dryopteris carthusiana [DRYOCAR]
Dryopteris spinulosa ssp. *assimilis* (Walker) Schidlay = Dryopteris expansa [DRYOEXP]
Filix fragilis (L.) Underwood = Cystopteris fragilis [CYSTFRA]
Filix montana (Lam.) Underwood = Cystopteris montana [CYSTMON]
Gymnocarpium Newman [GYMNOCA$]
Gymnocarpium continentale (V. Petrov) Pojark. = Gymnocarpium jessoense ssp. parvulum [GYMNJES1]
Gymnocarpium disjunctum (Rupr.) Ching [GYMNDIS] (western oak fern)
SY: *Dryopteris disjuncta* (Rupr.) Morton
SY: *Gymnocarpium dryopteris* ssp. *disjunctum* (Rupr.) Sarvela
SY: *Gymnocarpium dryopteris* var. *disjunctum* (Rupr.) Ching
Gymnocarpium dryopteris (L.) Newman [GYMNDRY] (oak fern)
SY: *Dryopteris dryopteris* (L.) Britt.
SY: *Dryopteris linnaeana* C. Christens.
SY: *Phegopteris dryopteris* (L.) Fée
SY: *Thelypteris dryopteris* (L.) Slosson
Gymnocarpium dryopteris ssp. *disjunctum* (Rupr.) Sarvela = Gymnocarpium disjunctum [GYMNDIS]

Gymnocarpium dryopteris var. *disjunctum* (Rupr.) Ching = Gymnocarpium disjunctum [GYMNDIS]
Gymnocarpium jessoense (Koidzumi) Koidzumi [GYMNJES] (Asian oak fern)
Gymnocarpium jessoense ssp. **parvulum** Sarvela [GYMNJES1] (Nahanni oak fern)
SY: *Gymnocarpium continentale* (V. Petrov) Pojark.
Matteuccia Todaro [MATTEUC$]
Matteuccia pensylvanica (Willd.) Raymond = Matteuccia struthiopteris [MATTSTR]
Matteuccia struthiopteris (L.) Todaro [MATTSTR] (ostrich fern)
SY: *Matteuccia pensylvanica* (Willd.) Raymond
SY: *Matteuccia struthiopteris* var. *pensylvanica* (Willd.) Morton
SY: *Matteuccia struthiopteris* var. *pubescens* (Terry) Clute
SY: *Onoclea struthiopteris* (L.) Hoffmann p.p.
SY: *Onoclea struthiopteris* var. *pensylvanica* (Willd.) Boivin
SY: *Pteretis nodulosa* (Michx.) Nieuwl.
SY: *Pteretis pensylvanica* (Willd.) Fern.
Matteuccia struthiopteris var. *pensylvanica* (Willd.) Morton = Matteuccia struthiopteris [MATTSTR]
Matteuccia struthiopteris var. *pubescens* (Terry) Clute = Matteuccia struthiopteris [MATTSTR]
Onoclea struthiopteris (L.) Hoffmann p.p. = Matteuccia struthiopteris [MATTSTR]
Onoclea struthiopteris var. *pensylvanica* (Willd.) Boivin = Matteuccia struthiopteris [MATTSTR]
Phegopteris dryopteris (L.) Fée = Gymnocarpium dryopteris [GYMNDRY]
Polystichum Roth [POLYSTI$]
Polystichum alaskense Maxon = Polystichum setigerum [POLYSET]
Polystichum andersonii Hopkins [POLYAND] (Anderson's holly fern)
SY: *Polystichum braunii* ssp. *andersonii* (Hopkins) Calder & Taylor
Polystichum braunii (Spenner) Fée [POLYBRA] (Braun's holly fern)
SY: *Polystichum braunii* ssp. *purshii* (Fern.) Calder & Taylor
SY: *Polystichum braunii* var. *purshii* Fern.
Polystichum braunii ssp. *alaskense* (Maxon) Calder & Taylor = Polystichum setigerum [POLYSET]
Polystichum braunii ssp. *andersonii* (Hopkins) Calder & Taylor = Polystichum andersonii [POLYAND]
Polystichum braunii ssp. *purshii* (Fern.) Calder & Taylor = Polystichum braunii [POLYBRA]
Polystichum braunii var. *alaskense* (Maxon) Hultén = Polystichum setigerum [POLYSET]
Polystichum braunii var. *purshii* Fern. = Polystichum braunii [POLYBRA]
Polystichum imbricans (D.C. Eat.) D.H. Wagner [POLYIMB] (narrow-leaved sword fern)
Polystichum kruckebergii W.H. Wagner [POLYKRU] (Kruckeberg's holly fern)
Polystichum lemmonii Underwood [POLYLEM] (Lemmon's holly fern)
SY: *Polystichum mohrioides* var. *lemmonii* (Underwood) Fern.
Polystichum lonchitis (L.) Roth [POLYLOC] (mountain holly fern)
Polystichum mohrioides var. *lemmonii* (Underwood) Fern. = Polystichum lemmonii [POLYLEM]
Polystichum mohrioides var. *scopulinum* (D.C. Eat.) Fern. = Polystichum scopulinum [POLYSCP]
Polystichum munitum (Kaulfuss) K. Presl [POLYMUN] (sword fern)
SY: *Polystichum munitum* var. *incisoserratum* (D.C. Eat.) Underwood
Polystichum munitum var. *incisoserratum* (D.C. Eat.) Underwood = Polystichum munitum [POLYMUN]
Polystichum scopulinum (D.C. Eat.) Maxon [POLYSCP] (crag holly fern)
SY: *Polystichum mohrioides* var. *scopulinum* (D.C. Eat.) Fern.
Polystichum setigerum (K. Presl) K. Presl [POLYSET] (Alaska holly fern)
SY: *Polystichum alaskense* Maxon
SY: *Polystichum braunii* ssp. *alaskense* (Maxon) Calder & Taylor
SY: *Polystichum braunii* var. *alaskense* (Maxon) Hultén
Pteretis nodulosa (Michx.) Nieuwl. = Matteuccia struthiopteris [MATTSTR]
Pteretis pensylvanica (Willd.) Fern. = Matteuccia struthiopteris [MATTSTR]
Rhizomatopteris montana (Lam.) Khokhr. = Cystopteris montana [CYSTMON]
Thelypteris dryopteris (L.) Slosson = Gymnocarpium dryopteris [GYMNDRY]
Woodsia R. Br. [WOODSIA$]
Woodsia alpina (Bolton) S.F. Gray [WOODALP] (alpine cliff fern)
SY: *Woodsia alpina* var. *bellii* Lawson
SY: *Woodsia glabella* var. *bellii* (Lawson) Lawson
Woodsia alpina var. *bellii* Lawson = Woodsia alpina [WOODALP]
Woodsia glabella R. Br. ex Richards. [WOODGLA] (smooth cliff fern)
Woodsia glabella var. *bellii* (Lawson) Lawson = Woodsia alpina [WOODALP]
Woodsia ilvensis (L.) R. Br. [WOODILV] (rusty cliff fern)
Woodsia oregana D.C. Eat. [WOODORE] (western cliff fern)
Woodsia scopulina D.C. Eat. [WOODSCO] (mountain cliff fern)

ELAEAGNACEAE (V:D)

Elaeagnus L. [ELAEAGN$]
*__Elaeagnus angustifolia__ L. [ELAEANG] (Russian olive)
Elaeagnus argentea Pursh = Elaeagnus commutata [ELAECOM]
Elaeagnus canadensis (L.) A. Nels. = Shepherdia canadensis [SHEPCAN]
Elaeagnus commutata Bernh. ex Rydb. [ELAECOM] (silverberry)
SY: *Elaeagnus argentea* Pursh
Elaeagnus utilis A. Nels. = Shepherdia argentea [SHEPARG]
Lepargyrea argentea (Pursh) Greene = Shepherdia argentea [SHEPARG]
Lepargyrea canadensis (L.) Greene = Shepherdia canadensis [SHEPCAN]
Shepherdia Nutt. [SHEPHER$]
Shepherdia argentea (Pursh) Nutt. [SHEPARG] (thorny buffalo-berry)
SY: *Elaeagnus utilis* A. Nels.
SY: *Lepargyrea argentea* (Pursh) Greene
Shepherdia canadensis (L.) Nutt. [SHEPCAN] (Canada buffalo-berry)
SY: *Elaeagnus canadensis* (L.) A. Nels.
SY: *Lepargyrea canadensis* (L.) Greene

ELATINACEAE (V:D)

Elatine L. [ELATINE$]
Elatine rubella Rydb. [ELATRUB] (three-flowered waterwort)

EMPETRACEAE (V:D)

Empetrum L. [EMPETRU$]
Empetrum nigrum L. [EMPENIG] (crowberry)

EQUISETACEAE (V:F)

Equisetum L. [EQUISET$]
Equisetum affine Engelm. = Equisetum hyemale var. affine [EQUIHYE2]
Equisetum arvense L. [EQUIARV] (common horsetail)
SY: *Equisetum arvense* var. *alpestre* Wahlenb.
SY: *Equisetum arvense* var. *boreale* (Bong.) Rupr.
SY: *Equisetum arvense* var. *riparium* Farw.
SY: *Equisetum calderi* Boivin
Equisetum arvense var. *alpestre* Wahlenb. = Equisetum arvense [EQUIARV]
Equisetum arvense var. *boreale* (Bong.) Rupr. = Equisetum arvense [EQUIARV]
Equisetum arvense var. *riparium* Farw. = Equisetum arvense [EQUIARV]

Equisetum braunii Milde = Equisetum telmateia var. braunii [EQUITEL1]
Equisetum calderi Boivin = Equisetum arvense [EQUIARV]
Equisetum fluviatile L. [EQUIFLU] (swamp horsetail)
SY: *Equisetum fluviatile* var. *limosum* (L.) Gilbert
SY: *Equisetum limosum* L.
Equisetum fluviatile var. *limosum* (L.) Gilbert = Equisetum fluviatile [EQUIFLU]
Equisetum funstonii A.A. Eat. = Equisetum laevigatum [EQUILAE]
Equisetum hyemale L. [EQUIHYE] (common scouring-rush)
Equisetum hyemale ssp. *affine* (Engelm.) Calder & Taylor = Equisetum hyemale var. affine [EQUIHYE2]
Equisetum hyemale var. **affine** (Engelm.) A.A. Eat. [EQUIHYE2]
SY: *Equisetum affine* Engelm.
SY: *Equisetum hyemale* ssp. *affine* (Engelm.) Calder & Taylor
SY: *Equisetum hyemale* var. *californicum* Milde
SY: *Equisetum hyemale* var. *elatum* (Engelm.) Morton
SY: *Equisetum hyemale* var. *pseudohyemale* (Farw.) Morton
SY: *Equisetum hyemale* var. *robustum* (A. Braun) A.A. Eat.
SY: *Equisetum praealtum* Raf.
SY: *Equisetum robustum* A. Braun
SY: *Hippochaete hyemalis* (L.) Brubin
SY: *Hippochaete hyemalis* ssp. *affinis* (Engelm.) W.A. Weber
Equisetum hyemale var. *californicum* Milde = Equisetum hyemale var. affine [EQUIHYE2]
Equisetum hyemale var. *elatum* (Engelm.) Morton = Equisetum hyemale var. affine [EQUIHYE2]
Equisetum hyemale var. *pseudohyemale* (Farw.) Morton = Equisetum hyemale var. affine [EQUIHYE2]
Equisetum hyemale var. *robustum* (A. Braun) A.A. Eat. = Equisetum hyemale var. affine [EQUIHYE2]
Equisetum kansanum Schaffn. = Equisetum laevigatum [EQUILAE]
Equisetum laevigatum A. Braun [EQUILAE] (smooth scouring-rush)
SY: *Equisetum funstonii* A.A. Eat.
SY: *Equisetum kansanum* Schaffn.
SY: *Equisetum laevigatum* ssp. *funstonii* (A.A. Eat.) Hartman
SY: *Hippochaete laevigata* (A. Braun) Farw.
Equisetum laevigatum ssp. *funstonii* (A.A. Eat.) Hartman = Equisetum laevigatum [EQUILAE]
Equisetum limosum L. = Equisetum fluviatile [EQUIFLU]
Equisetum × mackaii (Newm.) Brichan [EQUIMAC]
SY: *Equisetum trachyodon* (A. Braun) W.D.J. Koch
SY: *Equisetum × trachyodon* A. Br.
SY: *Equisetum variegatum* var. *jesupii* A.A. Eat.
Equisetum maximum auct. = Equisetum telmateia var. braunii [EQUITEL1]
Equisetum palustre L. [EQUIPAL] (marsh horsetail)
SY: *Equisetum palustre* var. *americanum* Victorin
SY: *Equisetum palustre* var. *simplicissimum* A. Braun
Equisetum palustre var. *americanum* Victorin = Equisetum palustre [EQUIPAL]
Equisetum palustre var. *simplicissimum* A. Braun = Equisetum palustre [EQUIPAL]
Equisetum praealtum Raf. = Equisetum hyemale var. affine [EQUIHYE2]
Equisetum pratense Ehrh. [EQUIPRA] (meadow horsetail)
Equisetum robustum A. Braun = Equisetum hyemale var. affine [EQUIHYE2]
Equisetum scirpoides Michx. [EQUISCI] (dwarf scouring-rush)
Equisetum sylvaticum L. [EQUISYL] (wood horsetail)
SY: *Equisetum sylvaticum* var. *multiramosum* (Fern.) Wherry
SY: *Equisetum sylvaticum* var. *pauciramosum* Milde
Equisetum sylvaticum var. *multiramosum* (Fern.) Wherry = Equisetum sylvaticum [EQUISYL]
Equisetum sylvaticum var. *pauciramosum* Milde = Equisetum sylvaticum [EQUISYL]
Equisetum telmateia Ehrh. [EQUITEL] (giant horsetail)
Equisetum telmateia ssp. *braunii* (Milde) Hauke = Equisetum telmateia var. braunii [EQUITEL1]
Equisetum telmateia var. **braunii** (Milde) Milde [EQUITEL1]
SY: *Equisetum braunii* Milde
SY: *Equisetum telmateia* ssp. *braunii* (Milde) Hauke
SY: *Equisetum maximum* auct.
Equisetum trachyodon (A. Braun) W.D.J. Koch = Equisetum × mackaii [EQUIMAC]
Equisetum × trachyodon A. Br. = Equisetum × mackaii [EQUIMAC]
Equisetum variegatum Schleich. ex F. Weber & D.M.H. Mohr [EQUIVAR] (northern scouring-rush)
Equisetum variegatum ssp. *alaskanum* (A.A. Eat.) Hultén = Equisetum variegatum var. alaskanum [EQUIVAR1]
Equisetum variegatum var. **alaskanum** A.A. Eat. [EQUIVAR1]
SY: *Equisetum variegatum* ssp. *alaskanum* (A.A. Eat.) Hultén
Equisetum variegatum var. *anceps* Milde = Equisetum variegatum var. variegatum [EQUIVAR0]
Equisetum variegatum var. *jesupii* A.A. Eat. = Equisetum × mackaii [EQUIMAC]
Equisetum variegatum var. **variegatum** [EQUIVAR0]
SY: *Equisetum variegatum* var. *anceps* Milde
SY: *Hippochaete variegata* (Schleich. ex F. Weber & D.M.H. Mohr) Bruhin
Hippochaete hyemalis (L.) Brubin = Equisetum hyemale var. affine [EQUIHYE2]
Hippochaete hyemalis ssp. *affinis* (Engelm.) W.A. Weber = Equisetum hyemale var. affine [EQUIHYE2]
Hippochaete laevigata (A. Braun) Farw. = Equisetum laevigatum [EQUILAE]
Hippochaete variegata (Schleich. ex F. Weber & D.M.H. Mohr) Bruhin = Equisetum variegatum var. variegatum **[EQUIVAR0]**

ERICACEAE (V:D)

Andromeda L. [ANDROME$]
Andromeda polifolia L. [ANDRPOL] (bog-rosemary)
Arbutus L. [ARBUTUS$]
Arbutus alpina L. = Arctostaphylos alpina [ARCTALP]
Arbutus menziesii Pursh [ARBUMEN] (Pacific madrone)
Arctostaphylos Adans. [ARCTOST$]
Arctostaphylos adenotricha (Fern. & J.F. Macbr.) A. & D. Löve & Kapoor = Arctostaphylos uva-ursi [ARCTUVA]
Arctostaphylos alpina (L.) Spreng. [ARCTALP] (alpine bearberry)
SY: *Arbutus alpina* L.
SY: *Arctous alpina* (L.) Niedenzu
SY: *Mairania alpina* (L.) Desv.
Arctostaphylos alpina ssp. *rubra* (Rehd. & Wilson) Hultén = Arctostaphylos rubra [ARCTRUB]
Arctostaphylos alpina var. *rubra* (Rehd. & Wilson) Bean = Arctostaphylos rubra [ARCTRUB]
Arctostaphylos columbiana Piper [ARCTCOL] (hairy manzanita)
Arctostaphylos rubra (Rehd. & Wilson) Fern. [ARCTRUB] (red bearberry)
SY: *Arctostaphylos alpina* ssp. *rubra* (Rehd. & Wilson) Hultén
SY: *Arctostaphylos alpina* var. *rubra* (Rehd. & Wilson) Bean
SY: *Arctous erythrocarpa* Small
SY: *Arctous rubra* (Rehd. & Wilson) Nakai
Arctostaphylos uva-ursi (L.) Spreng. [ARCTUVA] (kinnikinnick)
SY: *Arctostaphylos adenotricha* (Fern. & J.F. Macbr.) A. & D. Löve & Kapoor
SY: *Arctostaphylos uva-ursi* ssp. *adenotricha* (Fern. & J.F. Macbr.) Calder & Taylor
SY: *Arctostaphylos uva-ursi* var. *adenotricha* Fern. & J.F. Macbr.
SY: *Arctostaphylos uva-ursi* ssp. *coactilis* (Fern. & J.F. Macbr.) A. & D. Löve & Kapoor
SY: *Arctostaphylos uva-ursi* var. *coactilis* Fern. & J.F. Macbr.
SY: *Arctostaphylos uva-ursi* var. *leobreweri* J.B. Roof
SY: *Arctostaphylos uva-ursi* ssp. *longipilosa* Packer & Denford
SY: *Arctostaphylos uva-ursi* var. *marinensis* J.B. Roof
SY: *Arctostaphylos uva-ursi* ssp. *monoensis* J.B. Roof
SY: *Arctostaphylos uva-ursi* var. *pacifica* Hultén
SY: *Arctostaphylos uva-ursi* ssp. *stipitata* Packer & Denford
SY: *Arctostaphylos uva-ursi* var. *stipitata* (Packer & Denford) Dorn
SY: *Arctostaphylos uva-ursi* var. *suborbiculata* W. Knight
SY: *Uva-Ursi uva-ursi* (L.) Britt.
Arctostaphylos uva-ursi ssp. *adenotricha* (Fern. & J.F. Macbr.) Calder & Taylor = Arctostaphylos uva-ursi [ARCTUVA]

Arctostaphylos uva-ursi ssp. *coactilis* (Fern. & J.F. Macbr.) A. & D. Löve & Kapoor = Arctostaphylos uva-ursi [ARCTUVA]
Arctostaphylos uva-ursi ssp. *longipilosa* Packer & Denford = Arctostaphylos uva-ursi [ARCTUVA]
Arctostaphylos uva-ursi ssp. *monoensis* J.B. Roof = Arctostaphylos uva-ursi [ARCTUVA]
Arctostaphylos uva-ursi ssp. *stipitata* Packer & Denford = Arctostaphylos uva-ursi [ARCTUVA]
Arctostaphylos uva-ursi var. *adenotricha* Fern. & J.F. Macbr. = Arctostaphylos uva-ursi [ARCTUVA]
Arctostaphylos uva-ursi var. *coactilis* Fern. & J.F. Macbr. = Arctostaphylos uva-ursi [ARCTUVA]
Arctostaphylos uva-ursi var. *leobreweri* J.B. Roof = Arctostaphylos uva-ursi [ARCTUVA]
Arctostaphylos uva-ursi var. *marinensis* J.B. Roof = Arctostaphylos uva-ursi [ARCTUVA]
Arctostaphylos uva-ursi var. *pacifica* Hultén = Arctostaphylos uva-ursi [ARCTUVA]
Arctostaphylos uva-ursi var. *stipitata* (Packer & Denford) Dorn = Arctostaphylos uva-ursi [ARCTUVA]
Arctostaphylos uva-ursi var. *suborbiculata* W. Knight = Arctostaphylos uva-ursi [ARCTUVA]
Arctous alpina (L.) Niedenzu = Arctostaphylos alpina [ARCTALP]
Arctous erythrocarpa Small = Arctostaphylos rubra [ARCTRUB]
Arctous rubra (Rehd. & Wilson) Nakai = Arctostaphylos rubra [ARCTRUB]
Azalea procumbens L. = Loiseleuria procumbens [LOISPRO]
Azaleastrum albiflorum (Hook.) Rydb. = Rhododendron albiflorum [RHODALB]
Calluna Salisb. [CALLUNA$]
***Calluna vulgaris** (L.) Hull [CALLVUL] (heather)
Cassiope D. Don [CASSIOP$]
Cassiope lycopodioides (Pallas) D. Don [CASSLYC] (club-moss mountain-heather)
Cassiope lycopodioides ssp. *cristipilosa* Calder & Taylor = Cassiope lycopodioides var. cristipilosa [CASSLYC1]
Cassiope lycopodioides var. **cristipilosa** (Calder & Taylor) Boivin [CASSLYC1]
SY: *Cassiope lycopodioides* ssp. *cristipilosa* Calder & Taylor
Cassiope mertensiana (Bong.) D. Don [CASSMER] (white mountain-heather)
Cassiope stelleriana (Pallas) DC. = Harrimanella stelleriana [HARRSTE]
Cassiope tetragona (L.) D. Don [CASSTET] (four-angled mountain-heather)
Cassiope tetragona ssp. *saximontana* (Small) Porsild = Cassiope tetragona var. saximontana [CASSTET1]
Cassiope tetragona var. **saximontana** (Small) C.L. Hitchc. [CASSTET1]
SY: *Cassiope tetragona* ssp. *saximontana* (Small) Porsild
Cassiope tetragona var. **tetragona** [CASSTET0]
Chamaecistus procumbens (L.) Kuntze = Loiseleuria procumbens [LOISPRO]
Chamaedaphne Moench [CHAMAED$]
Chamaedaphne calyculata (L.) Moench [CHAMCAL] (leatherleaf)
Chiogenes hispidula (L.) Torr. & Gray = Gaultheria hispidula [GAULHIS]
Cladothamnus pyroliflorus Bong. = Elliottia pyroliflorus [ELLIPYR]
Cyanococcus canadensis (Kalm ex A. Rich.) Rydb. = Vaccinium myrtilloides [VACCMYR]
Elliottia Muhl. ex Ell. [ELLIOTT$]
Elliottia pyroliflorus (Bong.) S.W. Brim & P.F. Stevens [ELLIPYR]
SY: *Cladothamnus pyroliflorus* Bong.
Gaultheria L. [GAULTHE$]
Gaultheria hispidula (L.) Muhl. ex Bigelow [GAULHIS] (creeping snowberry)
SY: *Chiogenes hispidula* (L.) Torr. & Gray
Gaultheria humifusa (Graham) Rydb. [GAULHUM] (alpine wintergreen)
Gaultheria ovatifolia Gray [GAULOVA] (western teaberry)
Gaultheria shallon Pursh [GAULSHA] (salal)
Harrimanella Coville [HARRIMA$]
Harrimanella stelleriana (Pallas) Coville [HARRSTE]
SY: *Cassiope stelleriana* (Pallas) DC.
Kalmia L. [KALMIA$]
Kalmia microphylla (Hook.) Heller [KALMMIC] (western bog-laurel)
SY: *Kalmia microphylla* ssp. *occidentalis* (Small) Taylor & MacBryde
SY: *Kalmia microphylla* var. *occidentalis* (Small) Ebinger
SY: *Kalmia occidentalis* Small
SY: *Kalmia polifolia* ssp. *microphylla* (Hook.) Calder & Taylor
SY: *Kalmia polifolia* var. *microphylla* (Hook.) Hall
SY: *Kalmia polifolia* ssp. *occidentalis* (Small) Abrams
Kalmia microphylla ssp. *occidentalis* (Small) Taylor & MacBryde = Kalmia microphylla [KALMMIC]
Kalmia microphylla var. *occidentalis* (Small) Ebinger = Kalmia microphylla [KALMMIC]
Kalmia occidentalis Small = Kalmia microphylla [KALMMIC]
Kalmia polifolia ssp. *microphylla* (Hook.) Calder & Taylor = Kalmia microphylla [KALMMIC]
Kalmia polifolia ssp. *occidentalis* (Small) Abrams = Kalmia microphylla [KALMMIC]
Kalmia polifolia var. *microphylla* (Hook.) Hall = Kalmia microphylla [KALMMIC]
Ledum L. [LEDUM$]
Ledum decumbens (Ait.) Lodd. ex Steud. = Ledum palustre ssp. decumbens [LEDUPAL1]
Ledum glandulosum Nutt. [LEDUGLA] (trapper's tea)
SY: *Ledum glandulosum* var. *californicum* (Kellogg) C.L. Hitchc.
Ledum glandulosum var. *californicum* (Kellogg) C.L. Hitchc. = Ledum glandulosum [LEDUGLA]
Ledum groenlandicum Oeder [LEDUGRO] (Labrador tea)
SY: *Ledum palustre* ssp. *groenlandicum* (Oeder) Hultén
SY: *Ledum palustre* var. *latifolium* (Jacq.) Michx.
SY: *Rhododendron groenlandicum* (Oeder) Kron & Judd
Ledum palustre L. [LEDUPAL] (marsh Labrador-tea)
Ledum palustre ssp. **decumbens** (Ait.) Hultén [LEDUPAL1] (northern Labrador tea)
SY: *Ledum decumbens* (Ait.) Lodd. ex Steud.
SY: *Ledum palustre* var. *decumbens* Ait.
SY: *Rhododendron tomentosum* ssp. *subarcticum* (Harmaja) G. Wallace
Ledum palustre ssp. *groenlandicum* (Oeder) Hultén = Ledum groenlandicum [LEDUGRO]
Ledum palustre var. *decumbens* Ait. = Ledum palustre ssp. decumbens [LEDUPAL1]
Ledum palustre var. *latifolium* (Jacq.) Michx. = Ledum groenlandicum [LEDUGRO]
Loiseleuria Desv. [LOISELE$]
Loiseleuria procumbens (L.) Desv. [LOISPRO] (alpine-azalea)
SY: *Azalea procumbens* L.
SY: *Chamaecistus procumbens* (L.) Kuntze
Mairania alpina (L.) Desv. = Arctostaphylos alpina [ARCTALP]
Menziesia Sm. [MENZIES$]
Menziesia ferruginea Sm. [MENZFER] (false azalea)
SY: *Menziesia ferruginea* ssp. *glabella* (Gray) Calder & Taylor
SY: *Menziesia ferruginea* var. *glabella* (Gray) M.E. Peck
SY: *Menziesia glabella* Gray
Menziesia ferruginea ssp. *glabella* (Gray) Calder & Taylor = Menziesia ferruginea [MENZFER]
Menziesia ferruginea var. *glabella* (Gray) M.E. Peck = Menziesia ferruginea [MENZFER]
Menziesia glabella Gray = Menziesia ferruginea [MENZFER]
Oxycoccus hagerupii A. & D. Löve = Vaccinium oxycoccos [VACCOXY]
Oxycoccus intermedius (Gray) Rydb. = Vaccinium oxycoccos [VACCOXY]
Oxycoccus microcarpos Turcz. ex Rupr. = Vaccinium oxycoccos [VACCOXY]
Oxycoccus ovalifolius (Michx.) Porsild = Vaccinium oxycoccos [VACCOXY]
Oxycoccus oxycoccos (L.) Adolphi = Vaccinium oxycoccos [VACCOXY]
Oxycoccus oxycoccos (L.) MacM. = Vaccinium oxycoccos [VACCOXY]

Oxycoccus palustris Pers. = Vaccinium oxycoccos [VACCOXY]
Oxycoccus palustris ssp. *microphyllus* (Lange) A. & D. Löve = Vaccinium oxycoccos [VACCOXY]
Oxycoccus palustris var. *intermedius* (Gray) T.J. Howell = Vaccinium oxycoccos [VACCOXY]
Oxycoccus palustris var. *ovalifolius* (Michx.) Seymour = Vaccinium oxycoccos [VACCOXY]
Oxycoccus quadripetalus Gilib. = Vaccinium oxycoccos [VACCOXY]
Oxycoccus quadripetalus var. *microphyllus* (Lange) Porsild = Vaccinium oxycoccos [VACCOXY]
Phyllodoce Salisb. [PHYLLOD$]
Phyllodoce aleutica ssp. *glanduliflora* (Hook.) Hultén = Phyllodoce glanduliflora [PHYLGLA]
Phyllodoce empetriformis (Sm.) D. Don [PHYLEMP] (pink mountain-heather)
Phyllodoce glanduliflora (Hook.) Coville [PHYLGLA] (yellow mountain-heather)
SY: *Phyllodoce aleutica* ssp. *glanduliflora* (Hook.) Hultén
Rhododendron L. [RHODODE$]
Rhododendron albiflorum Hook. [RHODALB] (white-flowered rhododendron)
SY: *Azaleastrum albiflorum* (Hook.) Rydb.
Rhododendron californicum Hook. = Rhododendron macrophyllum [RHODMAC]
Rhododendron groenlandicum (Oeder) Kron & Judd = Ledum groenlandicum [LEDUGRO]
Rhododendron lapponicum (L.) Wahlenb. [RHODLAP] (Lapland rosebay)
Rhododendron macrophyllum D. Don ex G. Don [RHODMAC] (Pacific rhododendron)
SY: *Rhododendron californicum* Hook.
Rhododendron tomentosum ssp. *subarcticum* (Harmaja) G. Wallace = Ledum palustre ssp. decumbens [LEDUPAL1]
Uva-Ursi uva-ursi (L.) Britt. = Arctostaphylos uva-ursi [ARCTUVA]
Vaccinium L. [VACCINI$]
Vaccinium alaskaense T.J. Howell = Vaccinium ovalifolium [VACCOVA]
Vaccinium angustifolium var. *myrtilloides* (Michx.) House = Vaccinium myrtilloides [VACCMYR]
Vaccinium caespitosum Michx. [VACCCAE] (dwarf blueberry)
Vaccinium canadense Kalm ex A. Rich. = Vaccinium myrtilloides [VACCMYR]
Vaccinium coccineum Piper = Vaccinium membranaceum [VACCMEM]
Vaccinium deliciosum Piper [VACCDEL] (blue-leaved huckleberry)
Vaccinium globulare Rydb. = Vaccinium membranaceum [VACCMEM]
Vaccinium membranaceum Dougl. ex Torr. [VACCMEM] (black huckleberry)
SY: *Vaccinium coccineum* Piper
SY: *Vaccinium globulare* Rydb.
SY: *Vaccinium membranaceum* var. *rigidum* (Hook.) Fern.
Vaccinium membranaceum var. *rigidum* (Hook.) Fern. = Vaccinium membranaceum [VACCMEM]
Vaccinium microcarpum (Turcz. ex Rupr.) Schmalh. = Vaccinium oxycoccos [VACCOXY]
Vaccinium myrtilloides Michx. [VACCMYR] (velvet-leaved blueberry)
SY: *Cyanococcus canadensis* (Kalm ex A. Rich.) Rydb.
SY: *Vaccinium angustifolium* var. *myrtilloides* (Michx.) House
SY: *Vaccinium canadense* Kalm ex A. Rich.
Vaccinium myrtillus L. [VACCMYT] (low bilberry)
Vaccinium occidentale Gray = Vaccinium uliginosum [VACCULI]
Vaccinium ovalifolium Sm. [VACCOVA] (oval-leaved blueberry)
SY: *Vaccinium alaskaense* T.J. Howell
Vaccinium ovatum Pursh [VACCOVT] (evergreen huckleberry)
Vaccinium oxycoccos L. [VACCOXY] (bog cranberry)
SY: *Oxycoccus hagerupii* A. & D. Löve
SY: *Oxycoccus intermedius* (Gray) Rydb.
SY: *Oxycoccus microcarpos* Turcz. ex Rupr.
SY: *Oxycoccus ovalifolius* (Michx.) Porsild
SY: *Oxycoccus oxycoccos* (L.) MacM.
SY: *Oxycoccus oxycoccos* (L.) Adolphi
SY: *Oxycoccus palustris* Pers.
SY: *Oxycoccus palustris* var. *intermedius* (Gray) T.J. Howell
SY: *Oxycoccus palustris* ssp. *microphyllus* (Lange) A. & D. Löve
SY: *Oxycoccus palustris* var. *ovalifolius* (Michx.) Seymour
SY: *Oxycoccus quadripetalus* Gilib.
SY: *Oxycoccus quadripetalus* var. *microphyllus* (Lange) Porsild
SY: *Vaccinium microcarpum* (Turcz. ex Rupr.) Schmalh.
SY: *Vaccinium oxycoccos* var. *intermedium* Gray
SY: *Vaccinium oxycoccos* var. *microphyllum* (Lange) Rouss. & Raymond
SY: *Vaccinium oxycoccos* var. *ovalifolium* Michx.
Vaccinium oxycoccos var. *intermedium* Gray = Vaccinium oxycoccos [VACCOXY]
Vaccinium oxycoccos var. *microphyllum* (Lange) Rouss. & Raymond = Vaccinium oxycoccos [VACCOXY]
Vaccinium oxycoccos var. *ovalifolium* Michx. = Vaccinium oxycoccos [VACCOXY]
Vaccinium parvifolium Sm. [VACCPAR] (red huckleberry)
Vaccinium salicinum Cham. = Vaccinium uliginosum [VACCULI]
Vaccinium scoparium Leib. ex Coville [VACCSCO] (grouseberry)
Vaccinium uliginosum L. [VACCULI] (bog blueberry)
SY: *Vaccinium occidentale* Gray
SY: *Vaccinium salicinum* Cham.
SY: *Vaccinium uliginosum* ssp. *alpinum* (Bigelow) Hultén
SY: *Vaccinium uliginosum* var. *alpinum* Bigelow
SY: *Vaccinium uliginosum* ssp. *gaultherioides* (Bigelow) S.B. Young
SY: *Vaccinium uliginosum* var. *langeanum* Malte
SY: *Vaccinium uliginosum* ssp. *microphyllum* Lange
SY: *Vaccinium uliginosum* ssp. *occidentale* (Gray) Hultén
SY: *Vaccinium uliginosum* var. *occidentale* (Gray) Hara
SY: *Vaccinium uliginosum* ssp. *pedris* (Harshberger) S.B. Young
SY: *Vaccinium uliginosum* ssp. *pubescens* (Wormsk. ex Hornem.) S.B. Young
SY: *Vaccinium uliginosum* var. *salicinum* (Cham.) Hultén
Vaccinium uliginosum ssp. *alpinum* (Bigelow) Hultén = Vaccinium uliginosum [VACCULI]
Vaccinium uliginosum ssp. *gaultherioides* (Bigelow) S.B. Young = Vaccinium uliginosum [VACCULI]
Vaccinium uliginosum ssp. *microphyllum* Lange = Vaccinium uliginosum [VACCULI]
Vaccinium uliginosum ssp. *occidentale* (Gray) Hultén = Vaccinium uliginosum [VACCULI]
Vaccinium uliginosum ssp. *pedris* (Harshberger) S.B. Young = Vaccinium uliginosum [VACCULI]
Vaccinium uliginosum ssp. *pubescens* (Wormsk. ex Hornem.) S.B. Young = Vaccinium uliginosum [VACCULI]
Vaccinium uliginosum var. *alpinum* Bigelow = Vaccinium uliginosum [VACCULI]
Vaccinium uliginosum var. *langeanum* Malte = Vaccinium uliginosum [VACCULI]
Vaccinium uliginosum var. *occidentale* (Gray) Hara = Vaccinium uliginosum [VACCULI]
Vaccinium uliginosum var. *salicinum* (Cham.) Hultén = Vaccinium uliginosum [VACCULI]
Vaccinium vitis-idaea L. [VACCVIT] (lingonberry)
Vaccinium vitis-idaea ssp. **minus** (Lodd.) Hultén [VACCVIT1]
SY: *Vaccinium vitis-idaea* var. *minus* Lodd.
SY: *Vaccinium vitis-idaea* var. *punctatum* Moench
Vaccinium vitis-idaea var. *minus* Lodd. = Vaccinium vitis-idaea ssp. minus [VACCVIT1]
Vaccinium vitis-idaea var. *punctatum* Moench = Vaccinium vitis-idaea ssp. minus [VACCVIT1]

EUPHORBIACEAE (V:D)

Chamaesyce S.F. Gray [CHAMAES$]
Chamaesyce albicaulis (Rydb.) Rydb. = Chamaesyce serpyllifolia [CHAMSER]
Chamaesyce glyptosperma (Engelm.) Small [CHAMGLY]
SY: *Euphorbia glyptosperma* Engelm.
*****Chamaesyce maculata** (L.) Small [CHAMMAC]

SY: *Chamaesyce mathewsii* Small
SY: *Chamaesyce supina* (Raf.) Moldenke
SY: *Chamaesyce tracyi* Small
SY: *Euphorbia maculata* L.
SY: *Euphorbia supina* Raf.
Chamaesyce mathewsii Small = Chamaesyce maculata [CHAMMAC]
Chamaesyce neomexicana (Greene) Standl. = Chamaesyce serpyllifolia [CHAMSER]
Chamaesyce serpyllifolia (Pers.) Small [CHAMSER]
SY: *Chamaesyce albicaulis* (Rydb.) Rydb.
SY: *Chamaesyce neomexicana* (Greene) Standl.
SY: *Euphorbia neomexicana* Greene
SY: *Euphorbia serpyllifolia* Pers.
Chamaesyce supina (Raf.) Moldenke = Chamaesyce maculata [CHAMMAC]
Chamaesyce tracyi Small = Chamaesyce maculata [CHAMMAC]
Euphorbia L. [EUPHORB$]
*Euphorbia cyparissias** L. [EUPHCYP] (cypress spurge)
SY: *Galarhoeus cyparissias* (L.) Small ex Rydb.
SY: *Tithymalus cyparissias* (L.) Hill
*Euphorbia esula** L. [EUPHESU] (leafy spurge)
*Euphorbia exigua** L. [EUPHEXI] (dwarf spurge)
Euphorbia glyptosperma Engelm. = Chamaesyce glyptosperma [CHAMGLY]
*Euphorbia helioscopia** L. [EUPHHEL] (summer spurge)
SY: *Galarhoeus helioscopius* (L.) Haw.
SY: *Tithymalus helioscopius* (L.) Hill
*Euphorbia lathyris** L. [EUPHLAT] (caper spurge)
SY: *Galarhoeus lathyris* (L.) Haw.
SY: *Tithymalus lathyris* (L.) Hill
Euphorbia maculata L. = Chamaesyce maculata [CHAMMAC]
Euphorbia neomexicana Greene = Chamaesyce serpyllifolia [CHAMSER]
*Euphorbia peplus** L. [EUPHPEP] (petty spurge)
SY: *Galarhoeus peplus* (L.) Haw.
SY: *Tithymalus peplus* (L.) Hill
Euphorbia serpyllifolia Pers. = Chamaesyce serpyllifolia [CHAMSER]
Euphorbia supina Raf. = Chamaesyce maculata [CHAMMAC]
Galarhoeus cyparissias (L.) Small ex Rydb. = Euphorbia cyparissias [EUPHCYP]
Galarhoeus helioscopius (L.) Haw. = Euphorbia helioscopia [EUPHHEL]
Galarhoeus lathyris (L.) Haw. = Euphorbia lathyris [EUPHLAT]
Galarhoeus peplus (L.) Haw. = Euphorbia peplus [EUPHPEP]
Tithymalus cyparissias (L.) Hill = Euphorbia cyparissias [EUPHCYP]
Tithymalus helioscopius (L.) Hill = Euphorbia helioscopia [EUPHHEL]
Tithymalus lathyris (L.) Hill = Euphorbia lathyris [EUPHLAT]
Tithymalus peplus (L.) Hill = Euphorbia peplus [EUPHPEP]

FABACEAE (V:D)

Acmispon americanum (Nutt.) Rydb. = Lotus unifoliolatus [LOTUUNI]
Aragallus splendens (Dougl. ex Hook.) Greene = Oxytropis splendens [OXYTSPL]
Aragallus viscidulus Rydb. = Oxytropis viscida [OXYTVIS]
Astragalus L. [ASTRAGA$]
Astragalus aboriginum Richards. = Astragalus australis [ASTRAUS]
Astragalus aboriginum var. *glabriusculus* (Hook.) Rydb. = Astragalus australis [ASTRAUS]
Astragalus aboriginum var. *lepagei* (Hultén) Boivin = Astragalus australis [ASTRAUS]
Astragalus aboriginum var. *richardsonii* (Sheldon) Boivin = Astragalus australis [ASTRAUS]
Astragalus adsurgens Pallas [ASTRADS] (standing milk-vetch)
Astragalus adsurgens ssp. *robustior* (Hook.) Welsh = Astragalus adsurgens var. robustior [ASTRADS1]
Astragalus adsurgens var. **robustior** Hook. [ASTRADS1] (ascending purple milk-vetch)
SY: *Astragalus adsurgens* ssp. *robustior* (Hook.) Welsh
SY: *Astragalus striatus* Nutt.
SY: *Astragalus sulphurescens* Rydb.
Astragalus agrestis Dougl. ex G. Don [ASTRAGR] (purple milk-vetch)
SY: *Astragalus danicus* var. *dasyglottis* (Fisch. ex DC.) Boivin
SY: *Astragalus dasyglottis* Fisch. ex DC.
SY: *Astragalus goniatus* Nutt.
SY: *Astragalus hypoglottis* Hook.
Astragalus alpinus L. [ASTRALP] (alpine milk-vetch)
SY: *Astragalus alpinus* ssp. *alaskanus* Hultén
SY: *Astragalus alpinus* ssp. *arcticus* Hultén
Astragalus alpinus ssp. *alaskanus* Hultén = Astragalus alpinus [ASTRALP]
Astragalus alpinus ssp. *arcticus* Hultén = Astragalus alpinus [ASTRALP]
Astragalus americanus (Hook.) M.E. Jones [ASTRAME] (American milk-vetch)
SY: *Astragalus frigidus* (L.) Gray p.p.
SY: *Astragalus frigidus* var. *americanus* (Hook.) S. Wats.
SY: *Astragalus frigidus* var. *gaspensis* (Rouss.) Fern.
SY: *Phaca americana* (Hook.) Rydb. ex Small
Astragalus australis (L.) Lam. [ASTRAUS]
SY: *Astragalus aboriginum* Richards.
SY: *Astragalus aboriginum* var. *glabriusculus* (Hook.) Rydb.
SY: *Astragalus aboriginum* var. *lepagei* (Hultén) Boivin
SY: *Astragalus aboriginum* var. *richardsonii* (Sheldon) Boivin
SY: *Astragalus australis* var. *glabriusculus* (Hook.) Isely
SY: *Astragalus australis* var. *major* (Gray) Isely
SY: *Astragalus forwoodii* S. Wats.
SY: *Astragalus forwoodii* var. *wallowensis* (Rydb.) M.E. Peck
SY: *Astragalus glabriusculus* var. *major* Gray
SY: *Astragalus linearis* (Rydb.) Porsild
SY: *Astragalus richardsonii* Sheldon
SY: *Astragalus scrupulicola* Fern. & Weatherby
SY: *Atelophragma aboriginorum* (Richards.) Rydb.
Astragalus australis var. *glabriusculus* (Hook.) Isely = Astragalus australis [ASTRAUS]
Astragalus australis var. *major* (Gray) Isely = Astragalus australis [ASTRAUS]
Astragalus beckwithii Torr. & Gray [ASTRBEC] (Beckwith's milk-vetch)
Astragalus beckwithii var. **weiserensis** M.E. Jones [ASTRBEC1] (Weiser milk-vetch)
SY: *Astragalus weiserensis* (M.E. Jones) Abrams
Astragalus blakei Egglest. = Astragalus robbinsii var. minor [ASTRROB1]
Astragalus bourgovii Gray [ASTRBOU] (Bourgeau's milk-vetch)
Astragalus brevidens (Gandog.) Rydb. = Astragalus canadensis var. brevidens [ASTRCAN1]
Astragalus canadensis L. [ASTRCAN] (Canadian milk-vetch)
Astragalus canadensis var. **brevidens** (Gandog.) Barneby [ASTRCAN1]
SY: *Astragalus brevidens* (Gandog.) Rydb.
Astragalus canadensis var. **canadensis** [ASTRCAN0]
SY: *Astragalus canadensis* var. *carolinianus* (L.) M.E. Jones
SY: *Astragalus canadensis* var. *longilobus* Fassett
SY: *Astragalus carolinianus* L.
SY: *Astragalus halei* Rydb.
Astragalus canadensis var. *carolinianus* (L.) M.E. Jones = Astragalus canadensis var. canadensis [ASTRCAN0]
Astragalus canadensis var. *longilobus* Fassett = Astragalus canadensis var. canadensis [ASTRCAN0]
Astragalus canadensis var. **mortonii** (Nutt.) S. Wats. [ASTRCAN3]
SY: *Astragalus mortonii* Nutt.
Astragalus carolinianus L. = Astragalus canadensis var. canadensis [ASTRCAN0]
*Astragalus cicer** L. [ASTRCIC] (chick-pea milk-vetch)
Astragalus collieri (Rydb.) Porsild = Astragalus robbinsii var. minor [ASTRROB1]
Astragalus collinus (Hook.) Dougl. ex G. Don [ASTRCOL] (hillside milk-vetch)
Astragalus convallarius Greene [ASTRCON] (lesser rushy milk-vetch)
Astragalus crassicarpus Nutt. [ASTRCRA] (ground plum)

Astragalus danicus var. *dasyglottis* (Fisch. ex DC.) Boivin = Astragalus agrestis [ASTRAGR]
Astragalus dasyglottis Fisch. ex DC. = Astragalus agrestis [ASTRAGR]
Astragalus decumbens var. *serotinus* (Gray ex Cooper) M.E. Jones = Astragalus miser var. serotinus [ASTRMIS2]
Astragalus drummondii Dougl. ex Hook. [ASTRDRU] (Drummond's milk-vetch)
SY: *Tium drummondii* (Dougl. ex Hook.) Rydb.
Astragalus eucosmus B.L. Robins. [ASTREUC] (elegant milk-vetch)
SY: *Astragalus eucosmus* var. *facinorum* Fern.
SY: *Astragalus eucosmus* ssp. *sealei* (Lepage) Hultén
SY: *Astragalus parviflorus* (Pursh) MacM.
SY: *Astragalus sealei* Lepage
SY: *Atelophragma elegans* (Hook.) Rydb.
Astragalus eucosmus ssp. *sealei* (Lepage) Hultén = Astragalus eucosmus [ASTREUC]
Astragalus eucosmus var. *facinorum* Fern. = Astragalus eucosmus [ASTREUC]
Astragalus filipes Torr. ex Gray [ASTRFIL] (threadstalk milk-vetch)
SY: *Astragalus filipes* var. *residuus* Jepson
SY: *Astragalus macgregorii* (Rydb.) Tidestrom
SY: *Astragalus stenophyllus* Torr. & Gray
SY: *Astragalus stenophyllus* var. *filipes* (Torr. ex Gray) Tidestrom
Astragalus filipes var. *residuus* Jepson = Astragalus filipes [ASTRFIL]
Astragalus forwoodii S. Wats. = Astragalus australis [ASTRAUS]
Astragalus forwoodii var. *wallowensis* (Rydb.) M.E. Peck = Astragalus australis [ASTRAUS]
Astragalus frigidus (L.) Gray p.p. = Astragalus americanus [ASTRAME]
Astragalus frigidus var. *americanus* (Hook.) S. Wats. = Astragalus americanus [ASTRAME]
Astragalus frigidus var. *gaspensis* (Rouss.) Fern. = Astragalus americanus [ASTRAME]
Astragalus glabriusculus var. *major* Gray = Astragalus australis [ASTRAUS]
Astragalus glareosus Dougl. ex Hook. = Astragalus purshii var. glareosus [ASTRPUR1]
Astragalus goniatus Nutt. = Astragalus agrestis [ASTRAGR]
Astragalus halei Rydb. = Astragalus canadensis var. canadensis [ASTRCAN0]
Astragalus hypoglottis Hook. = Astragalus agrestis [ASTRAGR]
Astragalus incurvus (Rydb.) Abrams = Astragalus purshii var. purshii [ASTRPUR0]
Astragalus lentiginosus Dougl. ex Hook. [ASTRLEN] (freckled milk-vetch)
Astragalus lentiginosus var. *macrolobus* (Rydb.) Barneby = Astragalus lentiginosus var. salinus [ASTRLEN2]
Astragalus lentiginosus var. **salinus** (T.J. Howell) Barneby [ASTRLEN2]
SY: *Astragalus lentiginosus* var. *macrolobus* (Rydb.) Barneby
SY: *Astragalus salinus* T.J. Howell
Astragalus linearis (Rydb.) Porsild = Astragalus australis [ASTRAUS]
Astragalus lotiflorus Hook. [ASTRLOT] (lotus milk-vetch)
SY: *Astragalus lotiflorus* var. *nebraskensis* Bates
SY: *Astragalus lotiflorus* var. *reverchonii* (Gray) M.E. Jones
SY: *Batidophaca lotiflorus* (Hook.) Rydb.
Astragalus lotiflorus var. *nebraskensis* Bates = Astragalus lotiflorus [ASTRLOT]
Astragalus lotiflorus var. *reverchonii* (Gray) M.E. Jones = Astragalus lotiflorus [ASTRLOT]
Astragalus macgregorii (Rydb.) Tidestrom = Astragalus filipes [ASTRFIL]
Astragalus macounii Rydb. = Astragalus robbinsii var. minor [ASTRROB1]
Astragalus microcystis Gray [ASTRMIC] (least bladdery milk-vetch)
Astragalus miser Dougl. [ASTRMIS] (timber milk-vetch)
Astragalus miser var. **miser** [ASTRMIS0]
SY: *Astragalus strigosus* Coult. & Fisher
Astragalus miser var. **serotinus** (Gray ex Cooper) Barneby [ASTRMIS2]
SY: *Astragalus decumbens* var. *serotinus* (Gray ex Cooper) M.E. Jones
SY: *Astragalus serotinus* Gray ex Cooper
Astragalus mortonii Nutt. = Astragalus canadensis var. mortonii [ASTRCAN3]
Astragalus nutzotinensis Rouss. [ASTRNUT] (Nutzotin milk-vetch)
Astragalus parviflorus (Pursh) MacM. = Astragalus eucosmus [ASTREUC]
Astragalus purshii Dougl. ex Hook. [ASTRPUR] (woollypod milk-vetch)
Astragalus purshii var. **glareosus** (Dougl. ex Hook.) Barneby [ASTRPUR1]
SY: *Astragalus glareosus* Dougl. ex Hook.
SY: *Astragalus ventosus* Suksdorf ex Rydb.
Astragalus purshii var. *interior* M.E. Jones = Astragalus purshii var. purshii [ASTRPUR0]
Astragalus purshii var. **purshii** [ASTRPUR0]
SY: *Astragalus incurvus* (Rydb.) Abrams
SY: *Astragalus purshii* var. *interior* M.E. Jones
Astragalus richardsonii Sheldon = Astragalus australis [ASTRAUS]
Astragalus robbinsii (Oakes) Gray [ASTRROB] (Robbins' milk-vetch)
Astragalus robbinsii var. *blakei* (Egglest.) Barneby = Astragalus robbinsii var. minor [ASTRROB1]
Astragalus robbinsii var. **minor** (Hook.) Barneby [ASTRROB1]
SY: *Astragalus blakei* Egglest.
SY: *Astragalus collieri* (Rydb.) Porsild
SY: *Astragalus macounii* Rydb.
SY: *Astragalus robbinsii* var. *blakei* (Egglest.) Barneby
Astragalus salinus T.J. Howell = Astragalus lentiginosus var. salinus [ASTRLEN2]
Astragalus sclerocarpus Gray [ASTRSCL] (The Dalles milk-vetch)
Astragalus scrupulicola Fern. & Weatherby = Astragalus australis [ASTRAUS]
Astragalus sealei Lepage = Astragalus eucosmus [ASTREUC]
Astragalus serotinus Gray ex Cooper = Astragalus miser var. serotinus [ASTRMIS2]
Astragalus spaldingii Gray [ASTRSPA] (Spalding's milk-vetch)
Astragalus splendens (Dougl. ex Hook.) Tidestrom = Oxytropis splendens [OXYTSPL]
Astragalus stenophyllus Torr. & Gray = Astragalus filipes [ASTRFIL]
Astragalus stenophyllus var. *filipes* (Torr. ex Gray) Tidestrom = Astragalus filipes [ASTRFIL]
Astragalus striatus Nutt. = Astragalus adsurgens var. robustior [ASTRADS1]
Astragalus strigosus Coult. & Fisher = Astragalus miser var. miser [ASTRMIS0]
Astragalus sulphurescens Rydb. = Astragalus adsurgens var. robustior [ASTRADS1]
Astragalus tenellus Pursh [ASTRTEN] (pulse milk-vetch)
SY: *Astragalus tenellus* var. *strigulosus* (Rydb.) F.J. Herm.
SY: *Homalobus tenellus* (Pursh) Britt.
Astragalus tenellus var. *strigulosus* (Rydb.) F.J. Herm. = Astragalus tenellus [ASTRTEN]
Astragalus umbellatus Bunge [ASTRUMB] (tundra milk-vetch)
Astragalus ventosus Suksdorf ex Rydb. = Astragalus purshii var. glareosus [ASTRPUR1]
Astragalus weiserensis (M.E. Jones) Abrams = Astragalus beckwithii var. weiserensis [ASTRBEC1]
Atelophragma aboriginorum (Richards.) Rydb. = Astragalus australis [ASTRAUS]
Atelophragma elegans (Hook.) Rydb. = Astragalus eucosmus [ASTREUC]
Batidophaca lotiflorus (Hook.) Rydb. = Astragalus lotiflorus [ASTRLOT]
Coronilla L. [CORONIL$]
***Coronilla varia** L. [COROVAR] (common crown-vetch)
SY: *Securigera varia* (L.) Lassen
Cytisus L. [CYTISUS$]
***Cytisus scoparius** (L.) Link [CYTISCO] (Scotch broom)
Glycyrrhiza L. [GLYCYRR$]
Glycyrrhiza glutinosa Nutt. = Glycyrrhiza lepidota [GLYCLEI]
Glycyrrhiza lepidota Pursh [GLYCLEI] (American licorice)

SY: *Glycyrrhiza glutinosa* Nutt.
SY: *Glycyrrhiza lepidota* var. *glutinosa* (Nutt.) S. Wats.
Glycyrrhiza lepidota var. *glutinosa* (Nutt.) S. Wats. = Glycyrrhiza lepidota [GLYCLEI]
Hedysarum L. [HEDYSAR$]
Hedysarum alpinum L. [HEDYALP] (alpine hedysarum)
Hedysarum alpinum ssp. *americanum* (Michx.) Fedtsch. = Hedysarum alpinum var. americanum [HEDYALP1]
Hedysarum alpinum var. **americanum** Michx. [HEDYALP1]
SY: *Hedysarum alpinum* ssp. *americanum* (Michx.) Fedtsch.
SY: *Hedysarum americanum* (Michx.) Britt.
Hedysarum americanum (Michx.) Britt. = Hedysarum alpinum var. americanum [HEDYALP1]
Hedysarum boreale Nutt. [HEDYBOR] (northern hedysarum)
Hedysarum boreale ssp. **boreale** [HEDYBOR0]
Hedysarum boreale ssp. **mackenzii** (Richards.) Welsh [HEDYBOR2]
SY: *Hedysarum boreale* var. *mackenzii* (Richards.) C.L. Hitchc.
SY: *Hedysarum mackenzii* Richards.
Hedysarum boreale var. *mackenzii* (Richards.) C.L. Hitchc. = Hedysarum boreale ssp. mackenzii [HEDYBOR2]
Hedysarum mackenzii Richards. = Hedysarum boreale ssp. mackenzii [HEDYBOR2]
Hedysarum occidentale Greene [HEDYOCC] (western hedysarum)
SY: *Hedysarum occidentale* var. *canone* Welsh
SY: *Hedysarum uintahense* A. Nels.
Hedysarum occidentale var. *canone* Welsh = Hedysarum occidentale [HEDYOCC]
Hedysarum sulphurescens Rydb. [HEDYSUL] (yellow hedysarum)
Hedysarum uintahense A. Nels. = Hedysarum occidentale [HEDYOCC]
Homalobus tenellus (Pursh) Britt. = Astragalus tenellus [ASTRTEN]
Hosackia americana (Nutt.) Piper = Lotus unifoliolatus [LOTUUNI]
Hosackia decumbens Benth. = Lotus nevadensis var. douglasii [LOTUNEV1]
Hosackia denticulata E. Drew = Lotus denticulatus [LOTUDEN]
Hosackia gracilis Benth. = Lotus formosissimus [LOTUFOR]
Hosackia parviflora Benth. = Lotus micranthus [LOTUMIC]
Hosackia pinnata (Hook.) Abrams = Lotus pinnatus [LOTUPIN]
Laburnum Medik. [LABURNU$]
***Laburnum anagyroides** Medik. [LABUANA] (laburnum)
Lathyrus L. [LATHYRU$]
Lathyrus bijugatus White [LATHBIJ] (pinewood peavine)
SY: *Lathyrus bijugatus* var. *sandbergii* White
Lathyrus bijugatus var. *sandbergii* White = Lathyrus bijugatus [LATHBIJ]
Lathyrus japonicus Willd. [LATHJAP] (beach pea)
Lathyrus japonicus ssp. *maritimus* (L.) P.W. Ball = Lathyrus japonicus var. maritimus [LATHJAP1]
Lathyrus japonicus var. *glaber* (Ser.) Fern. = Lathyrus japonicus var. maritimus [LATHJAP1]
Lathyrus japonicus var. **maritimus** (L.) Kartesz & Gandhi [LATHJAP1]
SY: *Lathyrus japonicus* var. *glaber* (Ser.) Fern.
SY: *Lathyrus japonicus* ssp. *maritimus* (L.) P.W. Ball
SY: *Lathyrus maritimus* Bigelow
SY: *Lathyrus maritimus* var. *glaber* (Ser.) Eames
SY: *Pisum maritimum* L.
SY: *Pisum maritimum* var. *glabrum* Ser.
***Lathyrus latifolius** L. [LATHLAT] (broad-leaved peavine)
Lathyrus littoralis (Nutt.) Endl. [LATHLIT] (grey beach peavine)
Lathyrus maritimus Bigelow = Lathyrus japonicus var. maritimus [LATHJAP1]
Lathyrus maritimus var. *glaber* (Ser.) Eames = Lathyrus japonicus var. maritimus [LATHJAP1]
Lathyrus myrtifolius Muhl. ex Willd. = Lathyrus palustris [LATHPAL]
Lathyrus nevadensis S. Wats. [LATHNEV] (purple peavine)
Lathyrus nevadensis ssp. **lanceolatus** var. **pilosellus** (M.E. Peck) C.L. Hitchc. [LATHNEV2]
SY: *Lathyrus nevadensis* var. *pilosellus* (Peck) C.L. Hitchc.
Lathyrus nevadensis var. *pilosellus* (Peck) C.L. Hitchc. = Lathyrus nevadensis ssp. lanceolatus var. pilosellus [LATHNEV2]
Lathyrus ochroleucus Hook. [LATHOCH] (creamy peavine)
Lathyrus palustris L. [LATHPAL] (marsh peavine)
SY: *Lathyrus myrtifolius* Muhl. ex Willd.
SY: *Lathyrus palustris* var. *linearifolius* Ser.
SY: *Lathyrus palustris* var. *macranthus* (White) Fern.
SY: *Lathyrus palustris* var. *meridionalis* Butters & St. John
SY: *Lathyrus palustris* var. *myrtifolius* (Muhl. ex Willd.) Gray
SY: *Lathyrus palustris* ssp. *pilosus* (Cham.) Hultén
SY: *Lathyrus palustris* var. *pilosus* (Cham.) Ledeb.
SY: *Lathyrus palustris* var. *retusus* Fern. & St. John
SY: *Orobus myrtifolius* (Muhl. ex Willd.) A. Hall
Lathyrus palustris ssp. *pilosus* (Cham.) Hultén = Lathyrus palustris [LATHPAL]
Lathyrus palustris var. *linearifolius* Ser. = Lathyrus palustris [LATHPAL]
Lathyrus palustris var. *macranthus* (White) Fern. = Lathyrus palustris [LATHPAL]
Lathyrus palustris var. *meridionalis* Butters & St. John = Lathyrus palustris [LATHPAL]
Lathyrus palustris var. *myrtifolius* (Muhl. ex Willd.) Gray = Lathyrus palustris [LATHPAL]
Lathyrus palustris var. *pilosus* (Cham.) Ledeb. = Lathyrus palustris [LATHPAL]
Lathyrus palustris var. *retusus* Fern. & St. John = Lathyrus palustris [LATHPAL]
***Lathyrus pratensis** L. [LATHPRA] (meadow peavine)
***Lathyrus sphaericus** Retz. [LATHSPH] (grass peavine)
***Lathyrus sylvestris** L. [LATHSYL] (narrow-leaved peavine)
Lotus L. [LOTUS$]
Lotus americanus (Nutt.) Bisch. = Lotus unifoliolatus [LOTUUNI]
***Lotus corniculatus** L. [LOTUCOR] (birds-foot trefoil)
SY: *Lotus corniculatus* var. *arvensis* (Schkuhr) Ser. ex DC.
Lotus corniculatus var. *arvensis* (Schkuhr) Ser. ex DC. = Lotus corniculatus [LOTUCOR]
Lotus corniculatus var. *tenuifolius* L. = Lotus tenuis [LOTUTEN]
Lotus denticulatus (E. Drew) Greene [LOTUDEN] (meadow birds-foot trefoil)
SY: *Hosackia denticulata* E. Drew
Lotus douglasii Greene = Lotus nevadensis var. douglasii [LOTUNEV1]
Lotus formosissimus Greene [LOTUFOR] (seaside birds-foot trefoil)
SY: *Hosackia gracilis* Benth.
Lotus micranthus Benth. [LOTUMIC] (small-flowered birds-foot trefoil)
SY: *Hosackia parviflora* Benth.
Lotus nevadensis (S. Wats.) Greene [LOTUNEV] (Nevada birds-foot trefoil)
Lotus nevadensis var. **douglasii** (Greene) Ottley [LOTUNEV1] (Nevada birds-food trefoil)
SY: *Hosackia decumbens* Benth.
SY: *Lotus douglasii* Greene
***Lotus pedunculatus** Cav. [LOTUPED] (pedunculate birds-foot trefoil)
SY: *Lotus uliginosus* Schkuhr
Lotus pinnatus Hook. [LOTUPIN] (bog birds-foot trefoil)
SY: *Hosackia pinnata* (Hook.) Abrams
Lotus purshianus F.E. & E.G. Clem. = Lotus unifoliolatus [LOTUUNI]
Lotus purshianus var. *glaber* (Nutt.) Munz = Lotus unifoliolatus [LOTUUNI]
Lotus sericeus Pursh = Lotus unifoliolatus [LOTUUNI]
***Lotus tenuis** Waldst. & Kit. ex Willd. [LOTUTEN] (narrow-leaved birds-foot trefoil)
SY: *Lotus corniculatus* var. *tenuifolius* L.
Lotus uliginosus Schkuhr = Lotus pedunculatus [LOTUPED]
Lotus unifoliolatus (Hook.) Benth. [LOTUUNI]
SY: *Acmispon americanum* (Nutt.) Rydb.
SY: *Hosackia americana* (Nutt.) Piper
SY: *Lotus americanus* (Nutt.) Bisch.
SY: *Lotus purshianus* F.E. & E.G. Clem.
SY: *Lotus purshianus* var. *glaber* (Nutt.) Munz
SY: *Lotus sericeus* Pursh

Lupinaster wormskioldii (Lehm.) K. Presl = Trifolium wormskjoldii [TRIFWOR]
Lupinus L. [LUPINUS$]
Lupinus albertensis C.P. Sm. = Lupinus nootkatensis var. nootkatensis [LUPINOO0]
Lupinus amniculi-putori C.P. Sm. = Lupinus arbustus ssp. neolaxiflorus [LUPIARU1]
Lupinus apricus Greene = Lupinus vallicola ssp. apricus [LUPIVAL1]
Lupinus arboreus Sims [LUPIARB] (tree lupine)
Lupinus arboreus var. *fruticosus* (Sims) S. Wats. = Lupinus nootkatensis var. fruticosus [LUPINOO1]
Lupinus arbustus Dougl. ex Lindl. [LUPIARU]
Lupinus arbustus ssp. **neolaxiflorus** D. Dunn [LUPIARU1] (spurred lupine)
SY: *Lupinus amniculi-putori* C.P. Sm.
SY: *Lupinus augustii* C.P. Sm.
SY: *Lupinus caudatus* var. *submanens* C.P. Sm.
SY: *Lupinus laxiflorus* var. *lyleanus* C.P. Sm.
SY: *Lupinus lyleanus* C.P. Sm.
SY: *Lupinus mackeyi* C.P. Sm.
SY: *Lupinus standingii* C.P. Sm.
SY: *Lupinus stipaphilus* C.P. Sm.
SY: *Lupinus stockii* C.P. Sm.
SY: *Lupinus wenachensis* Eastw.
SY: *Lupinus yakimensis* C.P. Sm.
Lupinus arbustus ssp. **pseudoparviflorus** (Rydb.) D. Dunn [LUPIARU2] (Montana lupine)
SY: *Lupinus laxiflorus* var. *elmerianus* C.P. Sm.
SY: *Lupinus laxiflorus* var. *pseudoparviflorus* (Rydb.) C.P. Sm. & St. John
SY: *Lupinus laxispicatus* Rydb.
SY: *Lupinus laxispicatus* var. *whithamii* C.P. Sm.
SY: *Lupinus mucronulatus* var. *umatillensis* C.P. Sm.
SY: *Lupinus scheuberae* Rydb.
Lupinus arcticus S. Wats. [LUPIARC] (arctic lupine)
Lupinus arcticus ssp. **arcticus** [LUPIARC0]
SY: *Lupinus borealis* Heller
SY: *Lupinus donnellyensis* C.P. Sm.
SY: *Lupinus gakonensis* C.P. Sm.
SY: *Lupinus multicaulis* C.P. Sm.
SY: *Lupinus multifolius* C.P. Sm.
SY: *Lupinus nootkatensis* var. *kiellmannii* Ostenf.
SY: *Lupinus polyphyllus* ssp. *arcticus* (S. Wats.) L. Phillips
SY: *Lupinus relictus* Hultén
SY: *Lupinus yukonensis* Greene
Lupinus arcticus ssp. **canadensis** (C.P. Sm.) D. Dunn [LUPIARC2]
SY: *Lupinus latifolius* var. *canadensis* C.P. Sm.
Lupinus arcticus ssp. **subalpinus** (Piper & B.L. Robins.) D. Dunn [LUPIARC3]
SY: *Lupinus arcticus* var. *subalpinus* (Piper & B.L. Robins.) C.P. Sm.
SY: *Lupinus glacialis* C.P. Sm.
SY: *Lupinus latifolius* var. *subalpinus* (Piper & B.L. Robins.) C.P. Sm.
SY: *Lupinus subalpinus* Piper & B.L. Robins.
SY: *Lupinus volcanicus* Greene
SY: *Lupinus volcanicus* var. *rupestricola* C.P. Sm.
Lupinus arcticus var. *subalpinus* (Piper & B.L. Robins.) C.P. Sm. = Lupinus arcticus ssp. subalpinus [LUPIARC3]
Lupinus argenteus Pursh [LUPIARG] (silvery lupine)
Lupinus argenteus ssp. **argenteus** [LUPIARG0]
Lupinus augustii C.P. Sm. = Lupinus arbustus ssp. neolaxiflorus [LUPIARU1]
Lupinus bicolor Lindl. [LUPIBIC] (two-coloured lupine)
SY: *Lupinus micranthus* var. *bicolor* (Lindl.) S. Wats.
Lupinus bingenensis Suksdorf [LUPIBIN]
Lupinus bingenensis var. **subsaccatus** Suksdorf [LUPIBIN1]
SY: *Lupinus sulphureus* ssp. *subsaccatus* (Suksdorf) L. Phillips
SY: *Lupinus sulphureus* var. *subsaccatus* (Suksdorf) C.L. Hitchc.
Lupinus borealis Heller = Lupinus arcticus ssp. arcticus [LUPIARC0]
Lupinus burkei S. Wats. [LUPIBUR] (bigleaf lupine)
SY: *Lupinus polyphyllus* var. *burkei* (S. Wats.) C.L. Hitchc.
Lupinus caudatus var. *submanens* C.P. Sm. = Lupinus arbustus ssp. neolaxiflorus [LUPIARU1]
Lupinus densiflorus Benth. [LUPIDEN] (white-whorl lupine)
SY: *Lupinus densiflorus* var. *latilabris* C.P. Sm.
SY: *Lupinus densiflorus* var. *scopulorum* C.P. Sm.
SY: *Lupinus densiflorus* var. *stenopetalus* C.P. Sm.
SY: *Lupinus densiflorus* var. *tracyi* C.P. Sm.
SY: *Lupinus microcarpus* ssp. *scopulorum* (C.P. Sm.) C.P. Sm.
SY: *Lupinus microcarpus* var. *scopulorum* C.P. Sm.
Lupinus densiflorus var. *latilabris* C.P. Sm. = Lupinus densiflorus [LUPIDEN]
Lupinus densiflorus var. *scopulorum* C.P. Sm. = Lupinus densiflorus [LUPIDEN]
Lupinus densiflorus var. *stenopetalus* C.P. Sm. = Lupinus densiflorus [LUPIDEN]
Lupinus densiflorus var. *tracyi* C.P. Sm. = Lupinus densiflorus [LUPIDEN]
Lupinus donnellyensis C.P. Sm. = Lupinus arcticus ssp. arcticus [LUPIARC0]
Lupinus gakonensis C.P. Sm. = Lupinus arcticus ssp. arcticus [LUPIARC0]
Lupinus glacialis C.P. Sm. = Lupinus arcticus ssp. subalpinus [LUPIARC3]
Lupinus kiskensis C.P. Sm. = Lupinus nootkatensis var. nootkatensis [LUPINOO0]
Lupinus kuschei Eastw. [LUPIKUS] (Yukon lupine)
SY: *Lupinus sericeus* var. *kuschei* (Eastw.) Boivin
Lupinus latifolius var. *canadensis* C.P. Sm. = Lupinus arcticus ssp. canadensis [LUPIARC2]
Lupinus latifolius var. *subalpinus* (Piper & B.L. Robins.) C.P. Sm. = Lupinus arcticus ssp. subalpinus [LUPIARC3]
Lupinus laxiflorus var. *elmerianus* C.P. Sm. = Lupinus arbustus ssp. pseudoparviflorus [LUPIARU2]
Lupinus laxiflorus var. *lyleanus* C.P. Sm. = Lupinus arbustus ssp. neolaxiflorus [LUPIARU1]
Lupinus laxiflorus var. *pseudoparviflorus* (Rydb.) C.P. Sm. & St. John = Lupinus arbustus ssp. pseudoparviflorus [LUPIARU2]
Lupinus laxispicatus Rydb. = Lupinus arbustus ssp. pseudoparviflorus [LUPIARU2]
Lupinus laxispicatus var. *whithamii* C.P. Sm. = Lupinus arbustus ssp. pseudoparviflorus [LUPIARU2]
Lupinus lepidus Dougl. ex Lindl. [LUPILEP] (prairie lupine)
Lupinus lepidus ssp. *lyallii* (Gray) Detling = Lupinus lyallii [LUPILYA]
Lupinus leucophyllus Dougl. ex Lindl. [LUPILEU] (velvet lupine)
Lupinus lignipes Heller = Lupinus rivularis [LUPIRIV]
Lupinus littoralis Dougl. [LUPILIT] (seashore lupine)
Lupinus lyallii Gray [LUPILYA] (dwarf mountain lupine)
SY: *Lupinus lepidus* ssp. *lyallii* (Gray) Detling
Lupinus lyleanus C.P. Sm. = Lupinus arbustus ssp. neolaxiflorus [LUPIARU1]
Lupinus mackeyi C.P. Sm. = Lupinus arbustus ssp. neolaxiflorus [LUPIARU1]
Lupinus micranthus Dougl. = Lupinus polycarpus [LUPIPOL]
Lupinus micranthus var. *bicolor* (Lindl.) S. Wats. = Lupinus bicolor [LUPIBIC]
Lupinus microcarpus ssp. *scopulorum* (C.P. Sm.) C.P. Sm. = Lupinus densiflorus [LUPIDEN]
Lupinus microcarpus var. *scopulorum* C.P. Sm. = Lupinus densiflorus [LUPIDEN]
Lupinus minimus Dougl. ex Hook. [LUPIMIN]
SY: *Lupinus ovinus* Greene
SY: *Lupinus piperi* B.L. Robins.
Lupinus mucronulatus var. *umatillensis* C.P. Sm. = Lupinus arbustus ssp. pseudoparviflorus [LUPIARU2]
Lupinus multicaulis C.P. Sm. = Lupinus arcticus ssp. arcticus [LUPIARC0]
Lupinus multifolius C.P. Sm. = Lupinus arcticus ssp. arcticus [LUPIARC0]
Lupinus nanus var. *apricus* (Greene) C.P. Sm. = Lupinus vallicola ssp. apricus [LUPIVAL1]
Lupinus nootkatensis Donn ex Sims [LUPINOO] (Nootka lupine)

Lupinus nootkatensis var. *ethel-looffii* C.P. Sm. = Lupinus nootkatensis var. nootkatensis [LUPINOO0]
Lupinus nootkatensis var. **fruticosus** Sims [LUPINOO1]
SY: *Lupinus arboreus* var. *fruticosus* (Sims) S. Wats.
SY: *Lupinus nootkatensis* var. *glaber* Hook.
SY: *Lupinus nootkatensis* var. *unalaskensis* S. Wats.
Lupinus nootkatensis var. *glaber* Hook. = Lupinus nootkatensis var. fruticosus [LUPINOO1]
Lupinus nootkatensis var. *henry-looffii* C.P. Sm. = Lupinus nootkatensis var. nootkatensis [LUPINOO0]
Lupinus nootkatensis var. *kiellmannii* Ostenf. = Lupinus arcticus ssp. arcticus [LUPIARC0]
Lupinus nootkatensis var. **nootkatensis** [LUPINOO0]
SY: *Lupinus albertensis* C.P. Sm.
SY: *Lupinus kiskensis* C.P. Sm.
SY: *Lupinus nootkatensis* var. *ethel-looffii* C.P. Sm.
SY: *Lupinus nootkatensis* var. *henry-looffii* C.P. Sm.
SY: *Lupinus nootkatensis* var. *perlanatus* C.P. Sm.
SY: *Lupinus perennis* ssp. *nootkatensis* (Donn ex Sims) L. Phillips
SY: *Lupinus trifurcatus* C.P. Sm.
Lupinus nootkatensis var. *perlanatus* C.P. Sm. = Lupinus nootkatensis var. nootkatensis [LUPINOO0]
Lupinus nootkatensis var. *unalaskensis* S. Wats. = Lupinus nootkatensis var. fruticosus [LUPINOO1]
Lupinus oreganus Heller [LUPIORE]
Lupinus oreganus var. **kincaidii** C.P. Sm. [LUPIORE1]
SY: *Lupinus sulphureus* ssp. *kincaidii* (C.P. Sm.) L. Phillips
SY: *Lupinus sulphureus* var. *kincaidii* (C.P. Sm.) C.L. Hitchc.
Lupinus ovinus Greene = Lupinus minimus [LUPIMIN]
Lupinus perennis ssp. *nootkatensis* (Donn ex Sims) L. Phillips = Lupinus nootkatensis var. nootkatensis [LUPINOO0]
Lupinus piperi B.L. Robins. = Lupinus minimus [LUPIMIN]
Lupinus polycarpus Greene [LUPIPOL] (small-flowered lupine)
SY: *Lupinus micranthus* Dougl.
Lupinus polyphyllus Lindl. [LUPIPOY] (large-leaved lupine)
Lupinus polyphyllus ssp. *arcticus* (S. Wats.) L. Phillips = Lupinus arcticus ssp. arcticus [LUPIARC0]
Lupinus polyphyllus var. *burkei* (S. Wats.) C.L. Hitchc. = Lupinus burkei [LUPIBUR]
Lupinus relictus Hultén = Lupinus arcticus ssp. arcticus [LUPIARC0]
Lupinus rivularis Dougl. ex Lindl. [LUPIRIV] (stream-bank lupine)
SY: *Lupinus lignipes* Heller
Lupinus scheuberae Rydb. = Lupinus arbustus ssp. pseudoparviflorus [LUPIARU2]
Lupinus sericeus Pursh [LUPISER] (silky lupine)
Lupinus sericeus var. *kuschei* (Eastw.) Boivin = Lupinus kuschei [LUPIKUS]
Lupinus standingii C.P. Sm. = Lupinus arbustus ssp. neolaxiflorus [LUPIARU1]
Lupinus stipaphilus C.P. Sm. = Lupinus arbustus ssp. neolaxiflorus [LUPIARU1]
Lupinus stockii C.P. Sm. = Lupinus arbustus ssp. neolaxiflorus [LUPIARU1]
Lupinus subalpinus Piper & B.L. Robins. = Lupinus arcticus ssp. subalpinus [LUPIARC3]
Lupinus sulphureus Dougl. ex Hook. [LUPISUL] (sulphur lupine)
SY: *Lupinus sulphureus* var. *applegateanus* C.P. Sm.
SY: *Lupinus sulphureus* var. *echleranus* C.P. Sm.
Lupinus sulphureus ssp. *kincaidii* (C.P. Sm.) L. Phillips = Lupinus oreganus var. kincaidii [LUPIORE1]
Lupinus sulphureus ssp. *subsaccatus* (Suksdorf) L. Phillips = Lupinus bingenensis var. subsaccatus [LUPIBIN1]
Lupinus sulphureus var. *applegateanus* C.P. Sm. = Lupinus sulphureus [LUPISUL]
Lupinus sulphureus var. *echleranus* C.P. Sm. = Lupinus sulphureus [LUPISUL]
Lupinus sulphureus var. *kincaidii* (C.P. Sm.) C.L. Hitchc. = Lupinus oreganus var. kincaidii [LUPIORE1]
Lupinus sulphureus var. *subsaccatus* (Suksdorf) C.L. Hitchc. = Lupinus bingenensis var. subsaccatus [LUPIBIN1]
Lupinus trifurcatus C.P. Sm. = Lupinus nootkatensis var. nootkatensis [LUPINOO0]
Lupinus vallicola Heller [LUPIVAL]
Lupinus vallicola ssp. **apricus** (Greene) D. Dunn [LUPIVAL1]
SY: *Lupinus apricus* Greene
SY: *Lupinus nanus* var. *apricus* (Greene) C.P. Sm.
SY: *Lupinus vallicola* var. *apricus* (Greene) C.P. Sm.
Lupinus vallicola var. *apricus* (Greene) C.P. Sm. = Lupinus vallicola ssp. apricus [LUPIVAL1]
Lupinus volcanicus Greene = Lupinus arcticus ssp. subalpinus [LUPIARC3]
Lupinus volcanicus var. *rupestricola* C.P. Sm. = Lupinus arcticus ssp. subalpinus [LUPIARC3]
Lupinus wenachensis Eastw. = Lupinus arbustus ssp. neolaxiflorus [LUPIARU1]
Lupinus wyethii S. Wats. [LUPIWYE] (Wyeth's lupine)
Lupinus yakimensis C.P. Sm. = Lupinus arbustus ssp. neolaxiflorus [LUPIARU1]
Lupinus yukonensis Greene = Lupinus arcticus ssp. arcticus [LUPIARC0]
Medicago L. [MEDICAG$]
***Medicago arabica** (L.) Huds. [MEDIARA] (spotted medic)
SY: *Medicago arabica* ssp. *inermis* Ricker
Medicago arabica ssp. *inermis* Ricker = Medicago arabica [MEDIARA]
Medicago falcata L. = Medicago sativa ssp. falcata [MEDISAT1]
Medicago hispida Gaertn. = Medicago polymorpha [MEDIPOL]
Medicago hispida var. *apiculata* (Willd.) Urban = Medicago polymorpha [MEDIPOL]
Medicago hispida var. *confinis* (W.D.J. Koch) Burnat = Medicago polymorpha [MEDIPOL]
***Medicago lupulina** L. [MEDILUP] (black medic)
SY: *Medicago lupulina* var. *cupaniana* (Guss.) Boiss.
SY: *Medicago lupulina* var. *glandulosa* Neilr.
Medicago lupulina var. *cupaniana* (Guss.) Boiss. = Medicago lupulina [MEDILUP]
Medicago lupulina var. *glandulosa* Neilr. = Medicago lupulina [MEDILUP]
***Medicago polymorpha** L. [MEDIPOL] (bur-clover)
SY: *Medicago hispida* Gaertn.
SY: *Medicago hispida* var. *apiculata* (Willd.) Urban
SY: *Medicago hispida* var. *confinis* (W.D.J. Koch) Burnat
SY: *Medicago polymorpha* var. *brevispina* (Benth.) Heyn
SY: *Medicago polymorpha* var. *ciliaris* (Ser.) Shinners
SY: *Medicago polymorpha* var. *nigra* L.
SY: *Medicago polymorpha* var. *polygyra* (Urban) Shinners
SY: *Medicago polymorpha* var. *tricycla* (Gren. & Godr.) Shinners
SY: *Medicago polymorpha* var. *vulgaris* (Benth.) Shinners
Medicago polymorpha var. *brevispina* (Benth.) Heyn = Medicago polymorpha [MEDIPOL]
Medicago polymorpha var. *ciliaris* (Ser.) Shinners = Medicago polymorpha [MEDIPOL]
Medicago polymorpha var. *nigra* L. = Medicago polymorpha [MEDIPOL]
Medicago polymorpha var. *polygyra* (Urban) Shinners = Medicago polymorpha [MEDIPOL]
Medicago polymorpha var. *tricycla* (Gren. & Godr.) Shinners = Medicago polymorpha [MEDIPOL]
Medicago polymorpha var. *vulgaris* (Benth.) Shinners = Medicago polymorpha [MEDIPOL]
***Medicago sativa** L. [MEDISAT] (alfalfa)
***Medicago sativa** ssp. **falcata** (L.) Arcang. [MEDISAT1]
SY: *Medicago falcata* L.
Melilotus P. Mill. [MELILOT$]
Melilotus alba Desr. = Melilotus officinalis [MELIOFF]
Melilotus albus Medik. = Melilotus officinalis [MELIOFF]
Melilotus albus var. *annuus* Coe = Melilotus officinalis [MELIOFF]
***Melilotus officinalis** (L.) Lam. [MELIOFF] (yellow sweet-clover)
SY: *Melilotus alba* Desr.
SY: *Melilotus albus* Medik.
SY: *Melilotus albus* var. *annuus* Coe
Onobrychis P. Mill. [ONOBRYC$]
***Onobrychis viciifolia** Scop. [ONOBVIC] (Sainfoin)
Orobus myrtifolius (Muhl. ex Willd.) A. Hall = Lathyrus palustris [LATHPAL]
Oxytropis DC. [OXYTROP$]

Oxytropis alaskana A. Nels. = Oxytropis campestris var. varians [OXYTCAM7]
Oxytropis alpicola (Rydb.) M.E. Jones = Oxytropis campestris var. cusickii [OXYTCAM2]
Oxytropis arctica R. Br. [OXYTARC] (arctic locoweed)
Oxytropis campestris (L.) DC. [OXYTCAM] (field locoweed)
Oxytropis campestris ssp. *gracilis* (A. Nels.) Hultén = Oxytropis monticola [OXYTMON]
Oxytropis campestris ssp. *jordalii* (Porsild) Hultén = Oxytropis jordalii ssp. jordalii [OXYTJOR0]
Oxytropis campestris ssp. *melanocephala* Hook. = Oxytropis maydelliana [OXYTMAY]
Oxytropis campestris var. *cervinus* (Greene) Boivin = Oxytropis monticola [OXYTMON]
Oxytropis campestris var. **columbiana** (St. John) Barneby [OXYTCAM8]
SY: *Oxytropis columbiana* St. John
Oxytropis campestris var. **cusickii** (Greenm.) Barneby [OXYTCAM2]
SY: *Oxytropis alpicola* (Rydb.) M.E. Jones
SY: *Oxytropis campestris* var. *rydbergii* (A. Nels.) R.J. Davis
SY: *Oxytropis cusickii* Greenm.
SY: *Oxytropis rydbergii* A. Nels.
Oxytropis campestris var. *davisii* Welsh = Oxytropis jordalii ssp. davisii [OXYTJOR1]
Oxytropis campestris var. *glabrata* Hook. = Oxytropis maydelliana [OXYTMAY]
Oxytropis campestris var. *gracilis* (A. Nels.) Barneby = Oxytropis monticola [OXYTMON]
Oxytropis campestris var. *jordalii* (Porsild) Welsh = Oxytropis jordalii ssp. jordalii [OXYTJOR0]
Oxytropis campestris var. *rydbergii* (A. Nels.) R.J. Davis = Oxytropis campestris var. cusickii [OXYTCAM2]
Oxytropis campestris var. **varians** (Rydb.) Barneby [OXYTCAM7]
SY: *Oxytropis alaskana* A. Nels.
SY: *Oxytropis hyperborea* Porsild
SY: *Oxytropis varians* (Rydb.) K. Schum.
Oxytropis columbiana St. John = Oxytropis campestris var. columbiana [OXYTCAM8]
Oxytropis cusickii Greenm. = Oxytropis campestris var. cusickii [OXYTCAM2]
Oxytropis deflexa (Pallas) DC. [OXYTDEF] (pendant-pod locoweed)
Oxytropis glabrata (Hook.) A. Nels. = Oxytropis maydelliana [OXYTMAY]
Oxytropis gracilis (A. Nels.) K. Schum. = Oxytropis monticola [OXYTMON]
Oxytropis huddelsonii Porsild [OXYTHUD] (Huddelson's locoweed)
Oxytropis hyperborea Porsild = Oxytropis campestris var. varians [OXYTCAM7]
Oxytropis jordalii Porsild [OXYTJOR] (Jordal's locoweed)
Oxytropis jordalii ssp. **davisii** (Welsh) Elisens & Packer [OXYTJOR1]
SY: *Oxytropis campestris* var. *davisii* Welsh
Oxytropis jordalii ssp. **jordalii** [OXYTJOR0]
SY: *Oxytropis campestris* ssp. *jordalii* (Porsild) Hultén
SY: *Oxytropis campestris* var. *jordalii* (Porsild) Welsh
Oxytropis leucantha var. *depressus* (Rydb.) Boivin = Oxytropis viscida [OXYTVIS]
Oxytropis luteola (Greene) Piper & Beattie = Oxytropis monticola [OXYTMON]
Oxytropis macounii (Greene) Rydb. = Oxytropis sericea var. spicata [OXYTSER1]
Oxytropis maydelliana Trautv. [OXYTMAY] (Maydell's locoweed)
SY: *Oxytropis campestris* var. *glabrata* Hook.
SY: *Oxytropis campestris* ssp. *melanocephala* Hook.
SY: *Oxytropis glabrata* (Hook.) A. Nels.
SY: *Oxytropis maydelliana* ssp. *melanocephala* (Hook.) Porsild
Oxytropis maydelliana ssp. *melanocephala* (Hook.) Porsild = Oxytropis maydelliana [OXYTMAY]
Oxytropis monticola Gray [OXYTMON] (yellow-flower locoweed)
SY: *Oxytropis campestris* var. *cervinus* (Greene) Boivin
SY: *Oxytropis campestris* ssp. *gracilis* (A. Nels.) Hultén
SY: *Oxytropis campestris* var. *gracilis* (A. Nels.) Barneby
SY: *Oxytropis gracilis* (A. Nels.) K. Schum.
SY: *Oxytropis luteola* (Greene) Piper & Beattie
SY: *Oxytropis villosa* (Rydb.) K. Schum.
Oxytropis nigrescens (Pallas) Fisch. ex DC. [OXYTNIG] (blackish locoweed)
Oxytropis podocarpa Gray [OXYTPOD] (stalked-pod locoweed)
SY: *Oxytropis podocarpa* var. *inflata* (Hook.) Boivin
Oxytropis podocarpa var. *inflata* (Hook.) Boivin = Oxytropis podocarpa [OXYTPOD]
Oxytropis richardsonii (Hook.) K. Schum. = Oxytropis splendens [OXYTSPL]
Oxytropis rydbergii A. Nels. = Oxytropis campestris var. cusickii [OXYTCAM2]
Oxytropis scammaniana Hultén [OXYTSCA] (Scamman's locoweed)
Oxytropis sericea Nutt. [OXYTSER] (Rocky Mountain locoweed)
Oxytropis sericea var. **spicata** (Hook.) Barneby [OXYTSER1] (silky locoweed)
SY: *Oxytropis macounii* (Greene) Rydb.
SY: *Oxytropis spicata* (Hook.) Standl.
Oxytropis spicata (Hook.) Standl. = Oxytropis sericea var. spicata [OXYTSER1]
Oxytropis splendens Dougl. ex Hook. [OXYTSPL] (showy locoweed)
SY: *Aragallus splendens* (Dougl. ex Hook.) Greene
SY: *Astragalus splendens* (Dougl. ex Hook.) Tidestrom
SY: *Oxytropis richardsonii* (Hook.) K. Schum.
SY: *Oxytropis splendens* var. *richardsonii* Hook.
SY: *Oxytropis splendens* var. *vestita* Hook.
Oxytropis splendens var. *richardsonii* Hook. = Oxytropis splendens [OXYTSPL]
Oxytropis splendens var. *vestita* Hook. = Oxytropis splendens [OXYTSPL]
Oxytropis varians (Rydb.) K. Schum. = Oxytropis campestris var. varians [OXYTCAM7]
Oxytropis villosa (Rydb.) K. Schum. = Oxytropis monticola [OXYTMON]
Oxytropis viscida Nutt. [OXYTVIS] (sticky locoweed)
SY: *Aragallus viscidulus* Rydb.
SY: *Oxytropis leucantha* var. *depressus* (Rydb.) Boivin
Phaca americana (Hook.) Rydb. ex Small = Astragalus americanus [ASTRAME]
Pisum maritimum L. = Lathyrus japonicus var. maritimus [LATHJAP1]
Pisum maritimum var. *glabrum* Ser. = Lathyrus japonicus var. maritimus [LATHJAP1]
Psoralea physodes Dougl. ex Hook. = Rupertia physodes [RUPEPHY]
Robinia L. [ROBINIA$]
***Robinia pseudoacacia** L. [ROBIPSE] (black locust)
SY: *Robinia pseudoacacia* var. *rectissima* (L.) Raber
Robinia pseudoacacia var. *rectissima* (L.) Raber = Robinia pseudoacacia [ROBIPSE]
Rupertia J. Grimes [RUPERTI$]
Rupertia physodes (Dougl. ex Hook.) J. Grimes [RUPEPHY]
SY: *Psoralea physodes* Dougl. ex Hook.
Securigera varia (L.) Lassen = Coronilla varia [COROVAR]
Thermopsis R. Br. ex Ait. f. [THERMOP$]
Thermopsis montana Nutt. = Thermopsis rhombifolia var. montana [THERRHO1]
Thermopsis pinetorum Greene = Thermopsis rhombifolia var. montana [THERRHO1]
Thermopsis rhombifolia (Nutt. ex Pursh) Nutt. ex Richards. [THERRHO] (prairie golden bean)
Thermopsis rhombifolia var. **montana** (Nutt.) Isely [THERRHO1]
SY: *Thermopsis montana* Nutt.
SY: *Thermopsis pinetorum* Greene
Tium drummondii (Dougl. ex Hook.) Rydb. = Astragalus drummondii [ASTRDRU]
Trifolium L. [TRIFOLI$]
Trifolium agrarium L. = Trifolium aureum [TRIFAUR]
Trifolium albopurpureum var. *dichotomum* (Hook. & Arn.) Isely = Trifolium dichotomum [TRIFDIC]

Trifolium appendiculatum Loja. = Trifolium variegatum [TRIFVAR]
*·**Trifolium arvense*** L. [TRIFARV] (hare's-foot clover)
·**Trifolium aureum** Pollich [TRIFAUR] (yellow clover)
SY: *Trifolium agrarium* L.
Trifolium bicephalum Elmer = Trifolium macraei [TRIFMAC]
Trifolium bifidum Gray [TRIFBIF] (pinole clover)
SY: *Trifolium bifidum* var. *decipiens* Greene
Trifolium bifidum var. *decipiens* Greene = Trifolium bifidum [TRIFBIF]
·**Trifolium campestre** Schreb. [TRIFCAM] (low hop-clover)
SY: *Trifolium procumbens* L. 1755, non 1753
Trifolium catalinae S. Wats. = Trifolium macraei [TRIFMAC]
Trifolium cyathiferum Lindl. [TRIFCYA] (cup clover)
Trifolium depauperatum Desv. [TRIFDEP] (pverty clover)
Trifolium dianthum Greene = Trifolium variegatum [TRIFVAR]
Trifolium dichotomum Hook. & Arn. [TRIFDIC]
SY: *Trifolium albopurpureum* var. *dichotomum* (Hook. & Arn.) Isely
SY: *Trifolium dichotomum* var. *turbinatum* Jepson
SY: *Trifolium macraei* var. *dichotomum* (Hook. & Arn.) Brewer ex S. Wats.
Trifolium dichotomum var. *turbinatum* Jepson = Trifolium dichotomum [TRIFDIC]
·**Trifolium dubium** Sibthorp [TRIFDUB] (small hop-clover)
Trifolium elegans Savi = Trifolium hybridum [TRIFHYB]
Trifolium fendleri Greene = Trifolium wormskjoldii [TRIFWOR]
Trifolium fimbriatum Lindl. = Trifolium wormskjoldii [TRIFWOR]
·**Trifolium fragiferum** L. [TRIFFRA] (strawberry clover)
SY: *Trifolium fragiferum* ssp. *bonannii* (K. Presl) Soják
Trifolium fragiferum ssp. *bonannii* (K. Presl) Soják = Trifolium fragiferum [TRIFFRA]
Trifolium geminiflorum Greene = Trifolium variegatum [TRIFVAR]
Trifolium heterodon Torr. & Gray = Trifolium wormskjoldii [TRIFWOR]
·**Trifolium hybridum** L. [TRIFHYB] (alsike clover)
SY: *Trifolium elegans* Savi
SY: *Trifolium hybridum* ssp. *elegans* (Savi) Aschers. & Graebn.
SY: *Trifolium hybridum* var. *elegans* (Savi) Boiss.
SY: *Trifolium hybridum* var. *pratense* Rabenh.
Trifolium hybridum ssp. *elegans* (Savi) Aschers. & Graebn. = Trifolium hybridum [TRIFHYB]
Trifolium hybridum var. *elegans* (Savi) Boiss. = Trifolium hybridum [TRIFHYB]
Trifolium hybridum var. *pratense* Rabenh. = Trifolium hybridum [TRIFHYB]
·**Trifolium incarnatum** L. [TRIFINC] (crimson clover)
SY: *Trifolium incarnatum* var. *elatius* Gibelli & Belli
Trifolium incarnatum var. *elatius* Gibelli & Belli = Trifolium incarnatum [TRIFINC]
Trifolium involucratum var. *fendleri* (Greene) McDermott = Trifolium wormskjoldii [TRIFWOR]
Trifolium involucratum var. *fimbriatum* (Lindl.) McDermott = Trifolium wormskjoldii [TRIFWOR]
Trifolium involucratum var. *heterodon* (Torr. & Gray) S. Wats. = Trifolium wormskjoldii [TRIFWOR]
Trifolium involucratum var. *kennedianum* McDermott = Trifolium wormskjoldii [TRIFWOR]
Trifolium kennedianum (McDermott) A. Nels. & J.F. Macbr. = Trifolium wormskjoldii [TRIFWOR]
Trifolium macraei Hook. & Arn. [TRIFMAC] (Macrae's clover)
SY: *Trifolium bicephalum* Elmer
SY: *Trifolium catalinae* S. Wats.
SY: *Trifolium mercedense* Kennedy
SY: *Trifolium traskiae* Kennedy
Trifolium macraei var. *dichotomum* (Hook. & Arn.) Brewer ex S. Wats. = Trifolium dichotomum [TRIFDIC]
Trifolium melananthum Hook. & Arn. = Trifolium variegatum [TRIFVAR]
Trifolium mercedense Kennedy = Trifolium macraei [TRIFMAC]
Trifolium microcephalum Pursh [TRIFMIC] (small-headed clover)
Trifolium microdon Hook. & Arn. [TRIFMIR] (thimble clover)
SY: *Trifolium microdon* var. *pilosum* Eastw.
Trifolium microdon var. *pilosum* Eastw. = Trifolium microdon [TRIFMIR]
Trifolium oliganthum Steud. [TRIFOLG] (few-flowered clover)
SY: *Trifolium pauciflorum* Nutt.
SY: *Trifolium variegatum* var. *pauciflorum* (Nutt.) McDermott
Trifolium pauciflorum Nutt. = Trifolium oliganthum [TRIFOLG]
·**Trifolium pratense** L. [TRIFPRA] (red clover)
SY: *Trifolium pratense* var. *sativum* (P. Mill.) Schreb.
Trifolium pratense var. *sativum* (P. Mill.) Schreb. = Trifolium pratense [TRIFPRA]
Trifolium procumbens L. 1755, non 1753 = Trifolium campestre [TRIFCAM]
·**Trifolium repens** L. [TRIFREP] (white clover)
Trifolium spinulosum Dougl. ex Hook. = Trifolium wormskjoldii [TRIFWOR]
·**Trifolium subterraneum** L. [TRIFSUB] (subterranean clover)
Trifolium traskiae Kennedy = Trifolium macraei [TRIFMAC]
Trifolium tridentatum Lindl. = Trifolium willdenowii [TRIFWIL]
Trifolium tridentatum var. *aciculare* (Nutt.) McDermott = Trifolium willdenowii [TRIFWIL]
Trifolium trilobatum Jepson = Trifolium variegatum [TRIFVAR]
Trifolium variegatum Nutt. [TRIFVAR] (white-tipped clover)
SY: *Trifolium appendiculatum* Loja.
SY: *Trifolium dianthum* Greene
SY: *Trifolium geminiflorum* Greene
SY: *Trifolium melananthum* Hook. & Arn.
SY: *Trifolium trilobatum* Jepson
SY: *Trifolium variegatum* var. *major* Loja.
SY: *Trifolium variegatum* var. *melananthum* (Hook. & Arn.) Greene
SY: *Trifolium variegatum* var. *rostratum* (Greene) C.L. Hitchc.
SY: *Trifolium variegatum* var. *trilobatum* (Jepson) McDermott
Trifolium variegatum var. *major* Loja. = Trifolium variegatum [TRIFVAR]
Trifolium variegatum var. *melananthum* (Hook. & Arn.) Greene = Trifolium variegatum [TRIFVAR]
Trifolium variegatum var. *pauciflorum* (Nutt.) McDermott = Trifolium oliganthum [TRIFOLG]
Trifolium variegatum var. *rostratum* (Greene) C.L. Hitchc. = Trifolium variegatum [TRIFVAR]
Trifolium variegatum var. *trilobatum* (Jepson) McDermott = Trifolium variegatum [TRIFVAR]
Trifolium willdenowii Spreng. [TRIFWIL]
SY: *Trifolium tridentatum* Lindl.
SY: *Trifolium tridentatum* var. *aciculare* (Nutt.) McDermott
Trifolium willdenowii var. *fimbriatum* (Lindl.) Ewan = Trifolium wormskjoldii [TRIFWOR]
Trifolium willdenowii var. *kennedianum* (McDermott) Ewan = Trifolium wormskjoldii [TRIFWOR]
Trifolium wormskjoldii Lehm. [TRIFWOR] (springbank clover)
SY: *Lupinaster wormskioldii* (Lehm.) K. Presl
SY: *Trifolium fendleri* Greene
SY: *Trifolium fimbriatum* Lindl.
SY: *Trifolium heterodon* Torr. & Gray
SY: *Trifolium involucratum* var. *fendleri* (Greene) McDermott
SY: *Trifolium involucratum* var. *fimbriatum* (Lindl.) McDermott
SY: *Trifolium involucratum* var. *heterodon* (Torr. & Gray) S. Wats.
SY: *Trifolium involucratum* var. *kennedianum* McDermott
SY: *Trifolium kennedianum* (McDermott) A. Nels. & J.F. Macbr.
SY: *Trifolium spinulosum* Dougl. ex Hook.
SY: *Trifolium willdenowii* var. *fimbriatum* (Lindl.) Ewan
SY: *Trifolium willdenowii* var. *kennedianum* (McDermott) Ewan
SY: *Trifolium wormskjoldii* var. *fimbriatum* (Lindl.) Jepson
SY: *Trifolium wormskjoldii* var. *kennedianum* (McDermott) Jepson
Trifolium wormskjoldii var. *fimbriatum* (Lindl.) Jepson = Trifolium wormskjoldii [TRIFWOR]
Trifolium wormskjoldii var. *kennedianum* (McDermott) Jepson = Trifolium wormskjoldii [TRIFWOR]
Ulex L. [ULEX$]
·**Ulex europaeus** L. [ULEXEUR] (gorse)
Vicia L. [VICIA$]

Vicia americana Muhl. ex Willd. [VICIAME] (American vetch)
Vicia angustifolia L. = Vicia sativa ssp. nigra [VICISAT1]
Vicia angustifolia var. *segetalis* (Thuill.) W.D.J. Koch = Vicia sativa ssp. nigra [VICISAT1]
Vicia angustifolia var. *uncinata* (Desv.) Rouy = Vicia sativa ssp. nigra [VICISAT1]
*•**Vicia cracca** L. [VICICRA] (tufted vetch)
Vicia gigantea Hook. = Vicia nigricans ssp. gigantea [VICINIG1]
•**Vicia hirsuta** (L.) S.F. Gray [VICIHIR] (tiny vetch)
Vicia nigricans Hook. & Arn. [VICINIG]
Vicia nigricans ssp. **gigantea** (Hook.) Lassetter & Gunn. [VICINIG1]
SY: *Vicia gigantea* Hook.
•**Vicia sativa** L. [VICISAT] (common vetch)
•**Vicia sativa** ssp. **nigra** (L.) Ehrh. [VICISAT1]
SY: *Vicia angustifolia* L.
SY: *Vicia angustifolia* var. *segetalis* (Thuill.) W.D.J. Koch
SY: *Vicia angustifolia* var. *uncinata* (Desv.) Rouy
SY: *Vicia sativa* var. *angustifolia* (L.) Ser.
SY: *Vicia sativa* var. *nigra* L.
SY: *Vicia sativa* var. *segetalis* (Thuill.) Ser.
•**Vicia sativa** ssp. **sativa** [VICISAT0]
SY: *Vicia sativa* var. *linearis* Lange
Vicia sativa var. *angustifolia* (L.) Ser. = Vicia sativa ssp. nigra [VICISAT1]
Vicia sativa var. *linearis* Lange = Vicia sativa ssp. sativa [VICISAT0]
Vicia sativa var. *nigra* L. = Vicia sativa ssp. nigra [VICISAT1]
Vicia sativa var. *segetalis* (Thuill.) Ser. = Vicia sativa ssp. nigra [VICISAT1]
•**Vicia tetrasperma** (L.) Schreb. [VICITET] (slender vetch)
SY: *Vicia tetrasperma* var. *tenuissima* Druce
Vicia tetrasperma var. *tenuissima* Druce = Vicia tetrasperma [VICITET]
•**Vicia villosa** Roth [VICIVIL] (hairy vetch)

FAGACEAE (V:D)

Quercus L. [QUERCUS$]
Quercus garryana Dougl. ex Hook. [QUERGAR] (Garry oak)
•**Quercus robur** L. [QUERROB] (English oak)

FUMARIACEAE (V:D)

Capnoides aureum (Willd.) Kuntze = Corydalis aurea [CORYAUR]
Capnoides sempervirens (L.) Borkh. = Corydalis sempervirens [CORYSEM]
Corydalis Vent. [CORYDAL$]
Corydalis aurea Willd. [CORYAUR] (golden corydalis)
SY: *Capnoides aureum* (Willd.) Kuntze
SY: *Corydalis washingtoniana* Fedde
Corydalis pauciflora (Steph.) Pers. [CORYPAU] (few-flowered corydalis)
SY: *Corydalis pauciflora* var. *albiflora* Porsild
SY: *Corydalis pauciflora* var. *chamissonis* Fedde
Corydalis pauciflora var. *albiflora* Porsild = Corydalis pauciflora [CORYPAU]
Corydalis pauciflora var. *chamissonis* Fedde = Corydalis pauciflora [CORYPAU]
Corydalis scouleri Hook. [CORYSCO] (Scouler's corydalis)
Corydalis sempervirens (L.) Pers. [CORYSEM] (pink corydalis)
SY: *Capnoides sempervirens* (L.) Borkh.
Corydalis washingtoniana Fedde = Corydalis aurea [CORYAUR]
Dicentra Bernh. [DICENTR$]
Dicentra formosa (Andr.) Walp. [DICEFOR] (Pacific bleeding heart)
Dicentra uniflora Kellogg [DICEUNI] (steer's head)
Fumaria L. [FUMARIA$]
•**Fumaria bastardii** Boreau [FUMABAS] (bastard fumatory)
•**Fumaria officinalis** L. [FUMAOFF] (common fumatory)

GENTIANACEAE (V:D)

Amarella acuta (Michx.) Raf. = Gentianella amarella ssp. acuta [GENTAMA1]
Amarella plebeja (Ledeb. ex Spreng.) Greene = Gentianella amarella ssp. acuta [GENTAMA1]
Amarella plebeja var. *holmii* (Wettst.) Rydb. = Gentianella amarella ssp. acuta [GENTAMA1]
Amarella strictiflora (Rydb.) Greene = Gentianella amarella ssp. acuta [GENTAMA1]
Anthopogon tonsum (Lunell) Rydb. = Gentianopsis macounii [GENTMAC]
Centaurium Hill [CENTAUI$]
Centaurium curvistamineum (Wittr.) Abrams = Centaurium muhlenbergii [CENTMUH]
•**Centaurium erythraea** Rafn [CENTERY] (common centaury)
SY: *Centaurium umbellatum* auct.
Centaurium exaltatum (Griseb.) W. Wight ex Piper [CENTEXA] (western centaury)
SY: *Centaurium namophilum* var. *nevadense* Broome
SY: *Centaurium nuttallii* (S. Wats.) Heller
SY: *Cicendia exaltata* Griseb.
Centaurium muhlenbergii (Griseb.) W. Wight ex Piper [CENTMUH] (Muhlenberg's centaury)
SY: *Centaurium curvistamineum* (Wittr.) Abrams
Centaurium namophilum var. *nevadense* Broome = Centaurium exaltatum [CENTEXA]
Centaurium nuttallii (S. Wats.) Heller = Centaurium exaltatum [CENTEXA]
Centaurium umbellatum auct. = Centaurium erythraea [CENTERY]
Cicendia exaltata Griseb. = Centaurium exaltatum [CENTEXA]
Ciminalis prostrata (Haenke) A. & D. Löve = Gentiana prostrata [GENTPRO]
Comastoma tenella (Rottb.) Toyokuni = Gentianella tenella [GENTTEN]
Dasystephana affinis (Griseb.) Rydb. = Gentiana affinis [GENTAFF]
Dasystephana glauca (Pallas) Rydb. = Gentiana glauca [GENTGLA]
Dasystephana interrupta (Greene) Rydb. = Gentiana affinis [GENTAFF]
Gentiana L. [GENTIAN$]
Gentiana acuta Michx. = Gentianella amarella ssp. acuta [GENTAMA1]
Gentiana affinis Griseb. [GENTAFF] (prairie gentian)
SY: *Dasystephana affinis* (Griseb.) Rydb.
SY: *Dasystephana interrupta* (Greene) Rydb.
SY: *Gentiana affinis* var. *bigelovii* (Gray) Kusnez.
SY: *Gentiana affinis* var. *forwoodii* (Gray) Kusnez.
SY: *Gentiana affinis* var. *major* A. Nels. & Kennedy
SY: *Gentiana affinis* var. *ovata* Gray
SY: *Gentiana affinis* var. *parvidentata* Kusnez.
SY: *Gentiana bigelovii* Gray
SY: *Gentiana forwoodii* Gray
SY: *Gentiana interrupta* Greene
SY: *Gentiana oregana* Engelm. ex Gray
SY: *Gentiana rusbyi* Greene
SY: *Pneumonanthe affinis* (Griseb.) W.A. Weber
Gentiana affinis var. *bigelovii* (Gray) Kusnez. = Gentiana affinis [GENTAFF]
Gentiana affinis var. *forwoodii* (Gray) Kusnez. = Gentiana affinis [GENTAFF]
Gentiana affinis var. *major* A. Nels. & Kennedy = Gentiana affinis [GENTAFF]
Gentiana affinis var. *ovata* Gray = Gentiana affinis [GENTAFF]
Gentiana affinis var. *parvidentata* Kusnez. = Gentiana affinis [GENTAFF]
Gentiana amarella L. auct. p.p. = Gentianella amarella ssp. acuta [GENTAMA1]
Gentiana amarella ssp. *acuta* (Michx.) Hultén = Gentianella amarella ssp. acuta [GENTAMA1]
Gentiana amarella var. *acuta* (Michx.) Herder = Gentianella amarella ssp. acuta [GENTAMA1]
Gentiana amarella var. *plebeja* (Ledeb. ex Spreng.) Hultén = Gentianella amarella ssp. acuta [GENTAMA1]
Gentiana amarella var. *stricta* (Griseb.) S. Wats. = Gentianella amarella ssp. acuta [GENTAMA1]

Gentiana arctophila Griseb. = Gentianella propinqua [GENTPRP]
Gentiana bigelovii Gray = Gentiana affinis [GENTAFF]
Gentiana calycosa Griseb. [GENTCAL] (mountain bog gentian)
SY: *Gentiana calycosa* var. *obtusiloba* (Rydb.) C.L. Hitchc.
SY: *Gentiana calycosa* var. *xantha* A. Nels.
SY: *Pneumonanthe calycosa* (Griseb.) Greene
Gentiana calycosa var. *obtusiloba* (Rydb.) C.L. Hitchc. = Gentiana calycosa [GENTCAL]
Gentiana calycosa var. *xantha* A. Nels. = Gentiana calycosa [GENTCAL]
Gentiana crinita var. *tonsa* (Lunell) Boivin = Gentianopsis macounii [GENTMAC]
Gentiana douglasiana Bong. [GENTDOU] (swamp gentian)
Gentiana forwoodii Gray = Gentiana affinis [GENTAFF]
Gentiana gaspensis Victorin = Gentianopsis macounii [GENTMAC]
Gentiana glauca Pallas [GENTGLA] (glaucous gentian)
SY: *Dasystephana glauca* (Pallas) Rydb.
SY: *Gentianodes glauca* (Pallas) A. & D. Löve
Gentiana interrupta Greene = Gentiana affinis [GENTAFF]
Gentiana macounii Holm = Gentianopsis macounii [GENTMAC]
Gentiana menziesii Griseb. = Gentiana sceptrum [GENTSCE]
Gentiana oregana Engelm. ex Gray = Gentiana affinis [GENTAFF]
Gentiana platypetala Griseb. [GENTPLA] (broad-petalled gentian)
Gentiana plebeja Ledeb. ex Spreng. = Gentianella amarella ssp. acuta [GENTAMA1]
Gentiana plebeja var. *holmii* Wettst. = Gentianella amarella ssp. acuta [GENTAMA1]
Gentiana propinqua Richards. = Gentianella propinqua [GENTPRP]
Gentiana propinqua ssp. *arctophila* (Griseb.) Hultén = Gentianella propinqua [GENTPRP]
Gentiana prostrata Haenke [GENTPRO] (moss gentian)
SY: *Ciminalis prostrata* (Haenke) A. & D. Löve
SY: *Gentiana prostrata* var. *americana* Engelm.
Gentiana prostrata var. *americana* Engelm. = Gentiana prostrata [GENTPRO]
Gentiana rusbyi Greene = Gentiana affinis [GENTAFF]
Gentiana sceptrum Griseb. [GENTSCE] (king gentian)
SY: *Gentiana menziesii* Griseb.
SY: *Gentiana sceptrum* var. *cascadensis* M.E. Peck
SY: *Gentiana sceptrum* var. *humilis* Engelm. ex Gray
Gentiana sceptrum var. *cascadensis* M.E. Peck = Gentiana sceptrum [GENTSCE]
Gentiana sceptrum var. *humilis* Engelm. ex Gray = Gentiana sceptrum [GENTSCE]
Gentiana strictiflora (Rydb.) A. Nels. = Gentianella amarella ssp. acuta [GENTAMA1]
Gentiana tenella Rottb. = Gentianella tenella [GENTTEN]
Gentiana tonsa (Lunell) Victorin = Gentianopsis macounii [GENTMAC]
Gentianella Moench [GENTIAE$]
Gentianella acuta (Michx.) Hiitonen = Gentianella amarella ssp. acuta [GENTAMA1]
Gentianella amarella (L.) Boerner [GENTAMA] (northern gentian)
Gentianella amarella ssp. **acuta** (Michx.) J. Gillett [GENTAMA1]
SY: *Amarella acuta* (Michx.) Raf.
SY: *Amarella plebeja* (Ledeb. ex Spreng.) Greene
SY: *Amarella plebeja* var. *holmii* (Wettst.) Rydb.
SY: *Amarella strictiflora* (Rydb.) Greene
SY: *Gentiana acuta* Michx.
SY: *Gentiana amarella* L. auct. p.p.
SY: *Gentiana amarella* ssp. *acuta* (Michx.) Hultén
SY: *Gentiana amarella* var. *acuta* (Michx.) Herder
SY: *Gentiana amarella* var. *plebeja* (Ledeb. ex Spreng.) Hultén
SY: *Gentiana amarella* var. *stricta* (Griseb.) S. Wats.
SY: *Gentiana plebeja* Ledeb. ex Spreng.
SY: *Gentiana plebeja* var. *holmii* Wettst.
SY: *Gentiana strictiflora* (Rydb.) A. Nels.
SY: *Gentianella acuta* (Michx.) Hiitonen
SY: *Gentianella amarella* var. *acuta* (Michx.) Herder
SY: *Gentianella strictiflora* (Rydb.) W.A. Weber
Gentianella amarella var. *acuta* (Michx.) Herder = Gentianella amarella ssp. acuta [GENTAMA1]
Gentianella crinita ssp. *macounii* (Holm) J. Gillett = Gentianopsis macounii [GENTMAC]
Gentianella propinqua (Richards.) J. Gillett [GENTPRP] (four-parted gentian)
SY: *Gentiana arctophila* Griseb.
SY: *Gentiana propinqua* Richards.
SY: *Gentiana propinqua* ssp. *arctophila* (Griseb.) Hultén
Gentianella strictiflora (Rydb.) W.A. Weber = Gentianella amarella ssp. acuta [GENTAMA1]
Gentianella tenella (Rottb.) Boerner [GENTTEN] (dane's dwarf-gentian)
SY: *Comastoma tenella* (Rottb.) Toyokuni
SY: *Gentiana tenella* Rottb.
SY: *Lomatogonium tenellum* (Rottb.) A. Löve & D. Löve
Gentianodes glauca (Pallas) A. & D. Löve = Gentiana glauca [GENTGLA]
Gentianopsis Ma [GENTIAO$]
Gentianopsis macounii (Holm) Iltis [GENTMAC]
SY: *Anthopogon tonsum* (Lunell) Rydb.
SY: *Gentiana crinita* var. *tonsa* (Lunell) Boivin
SY: *Gentiana gaspensis* Victorin
SY: *Gentiana macounii* Holm
SY: *Gentiana tonsa* (Lunell) Victorin
SY: *Gentianella crinita* ssp. *macounii* (Holm) J. Gillett
SY: *Gentianopsis procera* ssp. *macounii* (Holm) Iltis
Gentianopsis procera ssp. *macounii* (Holm) Iltis = Gentianopsis macounii [GENTMAC]
Halenia Borkh. [HALENIA$]
Halenia deflexa (Sm.) Griseb. [HALEDEF] (spurred gentian)
Lomatogonium A. Braun [LOMATOG$]
Lomatogonium rotatum (L.) Fries ex Fern. [LOMAROT] (marsh felwort)
SY: *Lomatogonium rotatum* ssp. *tenuifolium* (Griseb.) Porsild
SY: *Pleurogyne rotata* (L.) Griseb.
Lomatogonium rotatum ssp. *tenuifolium* (Griseb.) Porsild = Lomatogonium rotatum [LOMAROT]
Lomatogonium tenellum (Rottb.) A. Löve & D. Löve = Gentianella tenella [GENTTEN]
Pleurogyne rotata (L.) Griseb. = Lomatogonium rotatum [LOMAROT]
Pneumonanthe affinis (Griseb.) W.A. Weber = Gentiana affinis [GENTAFF]
Pneumonanthe calycosa (Griseb.) Greene = Gentiana calycosa [GENTCAL]
Swertia L. [SWERTIA$]
Swertia perennis L. [SWERPER] (alpine bog swertia)
SY: *Swertia perennis* var. *obtusa* (Ledeb.) Ledeb. ex Griseb.
Swertia perennis var. *obtusa* (Ledeb.) Ledeb. ex Griseb. = Swertia perennis [SWERPER]

GERANIACEAE (V:D)

Erodium L'Hér. ex Ait. [ERODIUM$]
***Erodium cicutarium** (L.) L'Hér. ex Ait. [ERODCIC] (common stork's-bill)
Geranium L. [GERANIU$]
Geranium bicknellii Britt. [GERABIC] (Bicknell's geranium)
Geranium carolinianum L. [GERACAR] (Carolina geranium)
Geranium carolinianum var. **sphaerospermum** (Fern.) Breitung [GERACAR2]
SY: *Geranium sphaerospermum* Fern.
***Geranium dissectum** L. [GERADIS] (cut-leaved geranium)
SY: *Geranium laxum* Hanks
Geranium erianthum DC. [GERAERI] (northern geranium)
SY: *Geranium pratense* var. *erianthum* (DC.) Boivin
Geranium laxum Hanks = Geranium dissectum [GERADIS]
***Geranium molle** L. [GERAMOL] (dovefoot geranium)
Geranium pratense var. *erianthum* (DC.) Boivin = Geranium erianthum [GERAERI]
***Geranium pusillum** L. [GERAPUS] (small-flowered geranium)
Geranium richardsonii Fisch. & Trautv. [GERARIC] (Richardson's geranium)

Geranium robertianum L. [GERAROB] (Robert geranium)
SY: *Robertiella robertiana* (L.) Hanks
Geranium sphaerospermum Fern. = Geranium carolinianum var. sphaerospermum [GERACAR2]
Geranium viscosissimum Fisch. & C.A. Mey. ex C.A. Mey. [GERAVIS] (sticky purple geranium)
Robertiella robertiana (L.) Hanks = Geranium robertianum [GERAROB]

GROSSULARIACEAE (V:D)

Grossularia cognata (Greene) Coville & Britt. = Ribes oxyacanthoides ssp. cognatum [RIBEOXY1]
Grossularia irrigua (Dougl.) Coville & Britt. = Ribes oxyacanthoides ssp. irriguum [RIBEOXY2]
Grossularia lobbii (Gray) Coville & Britt. = Ribes lobbii [RIBELOB]
Grossularia nonscripta Berger = Ribes oxyacanthoides ssp. irriguum [RIBEOXY2]
Grossularia oxyacanthoides (L.) P. Mill. = Ribes oxyacanthoides ssp. oxyacanthoides [RIBEOXY0]
Limnobotrya lacustris (Pers.) Rydb. = Ribes lacustre [RIBELAC]
Limnobotrya montigena (McClatchie) Rydb. = Ribes montigenum [RIBEMON]
Ribes L. [RIBES$]
Ribes acerifolium T.J. Howell [RIBEACE]
SY: *Ribes howellii* Greene
Ribes bracteosum Dougl. ex Hook. [RIBEBRA] (stink currant)
Ribes cereum Dougl. [RIBECER] (squaw currant)
SY: *Ribes reniforme* Nutt.
SY: *Ribes viscidulum* Berger
Ribes cognatum Greene = Ribes oxyacanthoides ssp. cognatum [RIBEOXY1]
Ribes divaricatum Dougl. [RIBEDIV] (wild gooseberry)
Ribes divaricatum var. *irriguum* (Dougl.) Gray = Ribes oxyacanthoides ssp. irriguum [RIBEOXY2]
Ribes glandulosum Grauer [RIBEGLA] (skunk currant)
SY: *Ribes prostratum* L'Hér.
SY: *Ribes resinosum* Pursh
Ribes howellii Greene = Ribes acerifolium [RIBEACE]
Ribes hudsonianum Richards. [RIBEHUD] (northern blackcurrant)
Ribes hudsonianum var. **hudsonianum** [RIBEHUD0]
Ribes hudsonianum var. **petiolare** (Dougl.) Jancz. [RIBEHUD2]
SY: *Ribes petiolare* Dougl.
Ribes inerme Rydb. [RIBEINE] (white-stemmed gooseberry)
Ribes irriguum Dougl. = Ribes oxyacanthoides ssp. irriguum [RIBEOXY2]
Ribes lacustre (Pers.) Poir. [RIBELAC] (black gooseberry)
SY: *Limnobotrya lacustris* (Pers.) Rydb.
SY: *Ribes lacustre* var. *parvulum* Gray
SY: *Ribes oxycanthoides* var. *lacustre* Pers.
Ribes lacustre var. *molle* Gray = Ribes montigenum [RIBEMON]
Ribes lacustre var. *parvulum* Gray = Ribes lacustre [RIBELAC]
Ribes laxiflorum Pursh [RIBELAX] (trailing black currant)
Ribes lentum Coville & Rose = Ribes montigenum [RIBEMON]
Ribes leucoderme Heller = Ribes oxyacanthoides ssp. irriguum [RIBEOXY2]
Ribes lobbii Gray [RIBELOB] (gummy gooseberry)
SY: *Grossularia lobbii* (Gray) Coville & Britt.
Ribes montigenum McClatchie [RIBEMON] (mountain gooseberry)
SY: *Limnobotrya montigena* (McClatchie) Rydb.
SY: *Ribes lacustre* var. *molle* Gray
SY: *Ribes lentum* Coville & Rose
SY: *Ribes nubigenum* McClatchie
Ribes nonscripta (Berger) Standl. = Ribes oxyacanthoides ssp. irriguum [RIBEOXY2]
Ribes nubigenum McClatchie = Ribes montigenum [RIBEMON]
Ribes oxyacanthoides L. [RIBEOXY] (northern gooseberry)
Ribes oxyacanthoides ssp. **cognatum** (Greene) Sinnott [RIBEOXY1]
SY: *Grossularia cognata* (Greene) Coville & Britt.
SY: *Ribes cognatum* Greene
Ribes oxyacanthoides ssp. **irriguum** (Dougl.) Sinnott [RIBEOXY2]
SY: *Grossularia irrigua* (Dougl.) Coville & Britt.
SY: *Grossularia nonscripta* Berger
SY: *Ribes divaricatum* var. *irriguum* (Dougl.) Gray
SY: *Ribes irriguum* Dougl.
SY: *Ribes leucoderme* Heller
SY: *Ribes nonscripta* (Berger) Standl.
SY: *Ribes oxyacanthoides* var. *irriguum* (Dougl.) Jancz.
SY: *Ribes oxyacanthoides* var. *leucoderme* (Heller) Jancz.
Ribes oxyacanthoides ssp. **oxyacanthoides** [RIBEOXY0]
SY: *Grossularia oxyacanthoides* (L.) P. Mill.
Ribes oxyacanthoides var. *irriguum* (Dougl.) Jancz. = Ribes oxyacanthoides ssp. irriguum [RIBEOXY2]
Ribes oxyacanthoides var. *leucoderme* (Heller) Jancz. = Ribes oxyacanthoides ssp. irriguum [RIBEOXY2]
Ribes oxycanthoides var. *lacustre* Pers. = Ribes lacustre [RIBELAC]
Ribes petiolare Dougl. = Ribes hudsonianum var. petiolare [RIBEHUD2]
Ribes prostratum L'Hér. = Ribes glandulosum [RIBEGLA]
Ribes reniforme Nutt. = Ribes cereum [RIBECER]
Ribes resinosum Pursh = Ribes glandulosum [RIBEGLA]
Ribes rubrum var. *alaskanum* (Berger) Boivin = Ribes triste [RIBETRI]
Ribes rubrum var. *propinquum* (Turcz.) Trautv. & C.A. Mey. = Ribes triste [RIBETRI]
Ribes sanguineum Pursh [RIBESAN] (red-flowering currant)
Ribes triste Pallas [RIBETRI] (swamp red currant)
SY: *Ribes rubrum* var. *alaskanum* (Berger) Boivin
SY: *Ribes rubrum* var. *propinquum* (Turcz.) Trautv. & C.A. Mey.
SY: *Ribes triste* var. *albinervium* (Michx.) Fern.
Ribes triste var. *albinervium* (Michx.) Fern. = Ribes triste [RIBETRI]
Ribes viscidulum Berger = Ribes cereum [RIBECER]
Ribes viscosissimum Pursh [RIBEVIS] (sticky currant)
SY: *Ribes viscosissimum* var. *hallii* Jancz.
Ribes viscosissimum var. *hallii* Jancz. = Ribes viscosissimum [RIBEVIS]

HALORAGACEAE (V:D)

Enydria aquatica Vell. = Myriophyllum aquaticum [MYRIAQU]
Myriophyllum L. [MYRIOPH$]
***Myriophyllum aquaticum** (Vell.) Verdc. [MYRIAQU] (Brazilian water-milfoil)
SY: *Enydria aquatica* Vell.
SY: *Myriophyllum brasiliense* Camb.
SY: *Myriophyllum proserpinacoides* Gillies ex Hook. & Arn.
Myriophyllum brasiliense Camb. = Myriophyllum aquaticum [MYRIAQU]
Myriophyllum elatinoides Gaud. = Myriophyllum quitense [MYRIQUI]
Myriophyllum exalbescens Fern. = Myriophyllum sibiricum [MYRISIB]
Myriophyllum exalbescens var. *magdalenense* (Fern.) A. Löve = Myriophyllum sibiricum [MYRISIB]
Myriophyllum farwellii Morong [MYRIFAR] (Farwell's water-milfoil)
***Myriophyllum heterophyllum** Michx. [MYRIHET] (vari-leaved water-milfoil)
Myriophyllum hippuroides Nutt. ex Torr. & Gray [MYRIHIP] (western water-milfoil)
Myriophyllum magdalenense Fern. = Myriophyllum sibiricum [MYRISIB]
Myriophyllum proserpinacoides Gillies ex Hook. & Arn. = Myriophyllum aquaticum [MYRIAQU]
Myriophyllum quitense Kunth [MYRIQUI] (waterwort water-milfoil)
SY: *Myriophyllum elatinoides* Gaud.
Myriophyllum sibiricum Komarov [MYRISIB] (Siberian water-milfoil)
SY: *Myriophyllum exalbescens* Fern.
SY: *Myriophyllum exalbescens* var. *magdalenense* (Fern.) A. Löve
SY: *Myriophyllum magdalenense* Fern.
SY: *Myriophyllum spicatum* var. *capillaceum* Lange
SY: *Myriophyllum spicatum* ssp. *exalbescens* (Fern.) Hultén
SY: *Myriophyllum spicatum* var. *exalbescens* (Fern.) Jepson

SY: *Myriophyllum spicatum* ssp. *squamosum* Laestad. ex Hartman
SY: *Myriophyllum spicatum* var. *squamosum* (Laestad. ex Hartman) Hartman
*__Myriophyllum spicatum__ L. [MYRISPI] (Eurasian water-milfoil)
Myriophyllum spicatum ssp. *exalbescens* (Fern.) Hultén = Myriophyllum sibiricum [MYRISIB]
Myriophyllum spicatum ssp. *squamosum* Laestad. ex Hartman = Myriophyllum sibiricum [MYRISIB]
Myriophyllum spicatum var. *capillaceum* Lange = Myriophyllum sibiricum [MYRISIB]
Myriophyllum spicatum var. *exalbescens* (Fern.) Jepson = Myriophyllum sibiricum [MYRISIB]
Myriophyllum spicatum var. *squamosum* (Laestad. ex Hartman) Hartman = Myriophyllum sibiricum [MYRISIB]
Myriophyllum ussuriense (Regel) Maxim. [MYRIUSS] (Ussurian water-milfoil)
*__Myriophyllum verticillatum__ L. [MYRIVER] (verticillate water-milfoil)
SY: *Myriophyllum verticillatum* var. *cheneyi* Fassett
SY: *Myriophyllum verticillatum* var. *intermedium* W.D.J. Koch
SY: *Myriophyllum verticillatum* var. *pectinatum* Wallr.
SY: *Myriophyllum verticillatum* var. *pinnatifidum* Wallr.
Myriophyllum verticillatum var. *cheneyi* Fassett = Myriophyllum verticillatum [MYRIVER]
Myriophyllum verticillatum var. *intermedium* W.D.J. Koch = Myriophyllum verticillatum [MYRIVER]
Myriophyllum verticillatum var. *pectinatum* Wallr. = Myriophyllum verticillatum [MYRIVER]
Myriophyllum verticillatum var. *pinnatifidum* Wallr. = Myriophyllum verticillatum [MYRIVER]

HIPPURIDACEAE (V:D)

Hippuris L. [HIPPURI$]
Hippuris montana Ledeb. [HIPPMON] (mountain mare's-tail)
Hippuris tetraphylla L. f. [HIPPTET] (four-leaved mare's-tail)
Hippuris vulgaris L. [HIPPVUL] (common mare's-tail)

HYDRANGEACEAE (V:D)

Philadelphus L. [PHILADE$]
Philadelphus lewisii Pursh [PHILLEW] (mock-orange)

HYDROCHARITACEAE (V:M)

Anacharis canadensis (Michx.) Planch. = Elodea canadensis [ELODCAN]
Anacharis canadensis var. *planchonii* (Caspary) Victorin = Elodea canadensis [ELODCAN]
Anacharis densa (Planch.) Victorin = Egeria densa [EGERDEN]
Anacharis nuttallii Planch. = Elodea nuttallii [ELODNUT]
Anacharis occidentalis (Pursh) Victorin = Elodea nuttallii [ELODNUT]
Egeria Planch. [EGERIA$]
*__Egeria densa__ Planch. [EGERDEN] (Brazilian waterweed)
SY: *Anacharis densa* (Planch.) Victorin
SY: *Elodea densa* (Planch.) Caspary
SY: *Philotria densa* (Planch.) Small & St. John
Elodea Michx. [ELODEA$]
Elodea brandegae St. John = Elodea canadensis [ELODCAN]
Elodea canadensis Michx. [ELODCAN] (Canadian waterweed)
SY: *Anacharis canadensis* (Michx.) Planch.
SY: *Anacharis canadensis* var. *planchonii* (Caspary) Victorin
SY: *Elodea brandegae* St. John
SY: *Elodea ioensis* Wylie
SY: *Elodea linearis* (Rydb.) St. John
SY: *Elodea planchonii* Caspary
SY: *Philotria canadensis* (Michx.) Britt.
SY: *Philotria linearis* Rydb.
Elodea columbiana St. John = Elodea nuttallii [ELODNUT]
Elodea densa (Planch.) Caspary = Egeria densa [EGERDEN]
Elodea ioensis Wylie = Elodea canadensis [ELODCAN]
Elodea linearis (Rydb.) St. John = Elodea canadensis [ELODCAN]
Elodea minor (Engelm. ex Caspary) Farw. = Elodea nuttallii [ELODNUT]
Elodea nuttallii (Planch.) St. John [ELODNUT] (Nuttall's waterweed)
SY: *Anacharis nuttallii* Planch.
SY: *Anacharis occidentalis* (Pursh) Victorin
SY: *Elodea columbiana* St. John
SY: *Elodea minor* (Engelm. ex Caspary) Farw.
SY: *Elodea occidentalis* (Pursh) St. John
SY: *Philotria angustifolia* (Muhl.) Britt. ex Rydb.
SY: *Philotria minor* (Engelm. ex Caspary) Small
SY: *Philotria nuttallii* (Planch.) Rydb.
SY: *Philotria occidentalis* (Pursh) House
SY: *Udora verticillata* var. *minor* Engelm. ex Caspary
Elodea occidentalis (Pursh) St. John = Elodea nuttallii [ELODNUT]
Elodea planchonii Caspary = Elodea canadensis [ELODCAN]
Philotria angustifolia (Muhl.) Britt. ex Rydb. = Elodea nuttallii [ELODNUT]
Philotria canadensis (Michx.) Britt. = Elodea canadensis [ELODCAN]
Philotria densa (Planch.) Small & St. John = Egeria densa [EGERDEN]
Philotria linearis Rydb. = Elodea canadensis [ELODCAN]
Philotria minor (Engelm. ex Caspary) Small = Elodea nuttallii [ELODNUT]
Philotria nuttallii (Planch.) Rydb. = Elodea nuttallii [ELODNUT]
Philotria occidentalis (Pursh) House = Elodea nuttallii [ELODNUT]
Udora verticillata var. *minor* Engelm. ex Caspary = Elodea nuttallii [ELODNUT]
Vallisneria L. [VALLISN$]
*__Vallisneria americana__ Michx. [VALLAME] (American tapegrass)
SY: *Vallisneria asiatica* Michx.
SY: *Vallisneria neotropicalis* Victorin
SY: *Vallisneria spiralis* var. *asiatica* (Michx.) Torr.
Vallisneria asiatica Michx. = Vallisneria americana [VALLAME]
Vallisneria neotropicalis Victorin = Vallisneria americana [VALLAME]
Vallisneria spiralis var. *asiatica* (Michx.) Torr. = Vallisneria americana [VALLAME]

HYDROPHYLLACEAE (V:D)

Ellisia L. [ELLISIA$]
Ellisia nyctelea (L.) L. [ELLINYC] (ellisia)
SY: *Ellisia nyctelea* var. *coloradensis* Brand
SY: *Nyctelea nyctelea* (L.) Britt.
Ellisia nyctelea var. *coloradensis* Brand = Ellisia nyctelea [ELLINYC]
Hesperochiron S. Wats. [HESPERO$]
Hesperochiron pumilus (Dougl. ex Griseb.) Porter [HESPPUM] (dwarf hesperochiron)
SY: *Hesperochiron villosulus* (Greene) Suksdorf
Hesperochiron villosulus (Greene) Suksdorf = Hesperochiron pumilus [HESPPUM]
Hydrophyllum L. [HYDROPH$]
Hydrophyllum capitatum Dougl. ex Benth. [HYDRCAP] (cat's-breeches)
Hydrophyllum fendleri (Gray) Heller [HYDRFEN] (Fendler's waterleaf)
Hydrophyllum fendleri var. **albifrons** (Heller) J.F. Macbr. [HYDRFEN1]
Hydrophyllum tenuipes Heller [HYDRTEN] (Pacific waterleaf)
Nemophila Nutt. [NEMOPHI$]
Nemophila breviflora Gray [NEMOBRE] (Great Basin nemophila)
Nemophila parviflora Dougl. ex Benth. [NEMOPAR] (small-flowered nemophila)
Nemophila pedunculata Dougl. ex Benth. [NEMOPED] (meadow nemophila)
Nyctelea nyctelea (L.) Britt. = Ellisia nyctelea [ELLINYC]
Phacelia Juss. [PHACELI$]
Phacelia franklinii (R. Br.) Gray [PHACFRA] (Franklin's phacelia)

Phacelia hastata Dougl. ex Lehm. [PHACHAS] (silverleaf phacelia)
SY: *Phacelia hastata* var. *leucophylla* (Torr.) Cronq.
Phacelia hastata var. *leptosepala* (Rydb.) Cronq. = Phacelia leptosepala [PHACLEP]
Phacelia hastata var. *leucophylla* (Torr.) Cronq. = Phacelia hastata [PHACHAS]
Phacelia leptosepala Rydb. [PHACLEP] (narrow-sepaled phacelia)
SY: *Phacelia hastata* var. *leptosepala* (Rydb.) Cronq.
Phacelia linearis (Pursh) Holz. [PHACLIN] (thread-leaved phacelia)
Phacelia lyallii (Gray) Rydb. [PHACLYA] (Lyall's phacelia)
Phacelia mollis J.F. Macbr. [PHACMOL] (MacBryde's phacelia)
Phacelia ramosissima Dougl. ex Lehm. [PHACRAM] (branched phacelia)
Phacelia sericea (Graham) Gray [PHACSER] (silky phacelia)
Romanzoffia Cham. [ROMANZO$]
Romanzoffia sitchensis Bong. [ROMASIT] (Sitka romanzoffia)
Romanzoffia tracyi Jepson [ROMATRA] (Tracy's romanzoffia)

HYMENOPHYLLACEAE (V:F)

Hymenophyllum Sm. [HYMENOP$]
Hymenophyllum wrightii Bosch [HYMEWRI]
SY: *Mecodium wrightii* (Bosch) Copeland
Mecodium wrightii (Bosch) Copeland = Hymenophyllum wrightii [HYMEWRI]

IRIDACEAE (V:M)

Crocosmia Planch. [CROCOSM$]
*__Crocosmia × crocosmiiflora__ (V. Lemoine ex E. Morr.) N.E. Br. [CROCXCR] (montbretia)
SY: *Tritonia × crocosmiiflora* (V. Lemoine ex E. Morr.) Nichols.
Hydastylus borealis Bickn. = Sisyrinchium californicum [SISYCAL]
Hydastylus brachypus Bickn. = Sisyrinchium californicum [SISYCAL]
Hydastylus californicus (Ker-Gawl. ex Sims) Salisb. = Sisyrinchium californicum [SISYCAL]
Iris L. [IRIS$]
Iris longipetala Herbert = Iris missouriensis [IRISMIS]
Iris missouriensis Nutt. [IRISMIS] (western blue iris)
SY: *Iris longipetala* Herbert
SY: *Iris missouriensis* var. *arizonica* (Dykes) R.C. Foster
SY: *Iris missouriensis* var. *pelogonus* (Goodding) R.C. Foster
SY: *Iris pariensis* Welsh
SY: *Iris tolmieana* Herbert
Iris missouriensis var. *arizonica* (Dykes) R.C. Foster = Iris missouriensis [IRISMIS]
Iris missouriensis var. *pelogonus* (Goodding) R.C. Foster = Iris missouriensis [IRISMIS]
Iris pariensis Welsh = Iris missouriensis [IRISMIS]
*__Iris pseudacorus__ L. [IRISPSE] (yellow iris)
*__Iris sibirica__ L. [IRISSIB] (Siberian iris)
Iris tolmieana Herbert = Iris missouriensis [IRISMIS]
Olsynium Raf. [OLSYNIU$]
Olsynium douglasii (A. Dietr.) Bickn. [OLSYDOU]
Olsynium douglasii var. **douglasii** [OLSYDOU0] (satinflower)
SY: *Sisyrinchium douglasii* A. Dietr.
SY: *Sisyrinchium inalatum* A. Nels.
Olsynium douglasii var. **inflatum** (Suksdorf) Cholewa & D. Henderson [OLSYDOU2] (purple blue-eyed-grass)
SY: *Sisyrinchium douglasii* var. *inflatum* (Suksdorf) P. Holmgren
SY: *Sisyrinchium inflatum* (Suksdorf) St. John
Sisyrinchium L. [SISYRIN$]
Sisyrinchium angustifolium P. Mill. [SISYANG]
SY: *Sisyrinchium graminoides* Bickn.
Sisyrinchium birameum Piper = Sisyrinchium idahoense var. idahoense [SISYIDA0]
Sisyrinchium boreale (Bickn.) Henry = Sisyrinchium californicum [SISYCAL]
Sisyrinchium californicum (Ker-Gawl. ex Sims) Ait. [SISYCAL] (golden-eyed-grass)
SY: *Hydastylus borealis* Bickn.
SY: *Hydastylus brachypus* Bickn.
SY: *Hydastylus californicus* (Ker-Gawl. ex Sims) Salisb.
SY: *Sisyrinchium boreale* (Bickn.) Henry
Sisyrinchium douglasii A. Dietr. = Olsynium douglasii var. douglasii [OLSYDOU0]
Sisyrinchium douglasii var. *inflatum* (Suksdorf) P. Holmgren = Olsynium douglasii var. inflatum [OLSYDOU2]
Sisyrinchium graminoides Bickn. = Sisyrinchium angustifolium [SISYANG]
Sisyrinchium idahoense Bickn. [SISYIDA] (Idaho blue-eyed-grass)
Sisyrinchium idahoense var. **idahoense** [SISYIDA0]
SY: *Sisyrinchium birameum* Piper
Sisyrinchium idahoense var. **macounii** (Bickn.) D. Henderson [SISYIDA2] (blue-eyed-grass)
SY: *Sisyrinchium macounii* Bickn.
Sisyrinchium inalatum A. Nels. = Olsynium douglasii var. douglasii [OLSYDOU0]
Sisyrinchium inflatum (Suksdorf) St. John = Olsynium douglasii var. inflatum [OLSYDOU2]
Sisyrinchium littorale Greene [SISYLIT] (shore blue-eyed-grass)
Sisyrinchium macounii Bickn. = Sisyrinchium idahoense var. macounii [SISYIDA2]
Sisyrinchium montanum Greene [SISYMON] (mountain blue-eyed-grass)
Sisyrinchium septentrionale Bickn. [SISYSEP] (northern blue-eyed-grass)
Tritonia × crocosmiiflora (V. Lemoine ex E. Morr.) Nichols. = Crocosmia × crocosmiiflora [CROCXCR]

ISOETACEAE (V:F)

Isoetes L. [ISOETES$]
Isoetes beringensis Komarov = Isoetes maritima [ISOEMAR]
Isoetes bolanderi Engelm. [ISOEBOL] (Bolander's quillwort)
Isoetes braunii Durieu = Isoetes echinospora [ISOEECH]
Isoetes echinospora Durieu [ISOEECH] (bristle-like quillwort)
SY: *Isoetes braunii* Durieu
SY: *Isoetes echinospora* ssp. *asiatica* (Makino) A. Löve
SY: *Isoetes echinospora* var. *asiatica* Makino
SY: *Isoetes echinospora* var. *braunii* (Durieu) Engelm.
SY: *Isoetes echinospora* var. *hesperia* (C.F. Reed) A. Löve
SY: *Isoetes echinospora* ssp. *muricata* (Durieu) A. & D. Löve
SY: *Isoetes echinospora* var. *muricata* (Durieu) Engelm.
SY: *Isoetes echinospora* var. *robusta* Engelm.
SY: *Isoetes echinospora* var. *savilei* Boivin
SY: *Isoetes muricata* Durieu
SY: *Isoetes muricata* var. *braunii* (Durieu) C.F. Reed
SY: *Isoetes muricata* var. *hesperia* C.F. Reed
SY: *Isoetes setacea* Lam. p.p.
SY: *Isoetes setacea* ssp. *muricata* (Durieu) Holub
Isoetes echinospora ssp. *asiatica* (Makino) A. Löve = Isoetes echinospora [ISOEECH]
Isoetes echinospora ssp. *maritima* (Underwood) A. Löve = Isoetes maritima [ISOEMAR]
Isoetes echinospora ssp. *muricata* (Durieu) A. & D. Löve = Isoetes echinospora [ISOEECH]
Isoetes echinospora var. *asiatica* Makino = Isoetes echinospora [ISOEECH]
Isoetes echinospora var. *braunii* (Durieu) Engelm. = Isoetes echinospora [ISOEECH]
Isoetes echinospora var. *hesperia* (C.F. Reed) A. Löve = Isoetes echinospora [ISOEECH]
Isoetes echinospora var. *maritima* (Underwood) A.A. Eat. = Isoetes maritima [ISOEMAR]
Isoetes echinospora var. *muricata* (Durieu) Engelm. = Isoetes echinospora [ISOEECH]
Isoetes echinospora var. *robusta* Engelm. = Isoetes echinospora [ISOEECH]
Isoetes echinospora var. *savilei* Boivin = Isoetes echinospora [ISOEECH]
Isoetes flettii (A.A. Eat.) N.E. Pfeiffer [ISOEFLE]
Isoetes howellii Engelm. [ISOEHOW] (Howell's quillwort)
SY: *Isoetes howellii* var. *minima* (A.A. Eat.) Pfeiffer

Isoetes howellii var. *minima* (A.A. Eat.) Pfeiffer = Isoetes howellii [ISOEHOW]
Isoetes lacustris ssp. *paupercula* (Engelm.) J. Feilberg = Isoetes occidentalis [ISOEOCC]
Isoetes lacustris var. *paupercula* Engelm. = Isoetes occidentalis [ISOEOCC]
Isoetes macounii A.A. Eaton = Isoetes maritima [ISOEMAR]
Isoetes maritima Underwood [ISOEMAR] (coastal quillwort)
SY: *Isoetes beringensis* Komarov
SY: *Isoetes echinospora* ssp. *maritima* (Underwood) A. Löve
SY: *Isoetes echinospora* var. *maritima* (Underwood) A.A. Eat.
SY: *Isoetes macounii* A.A. Eaton
SY: *Isoetes muricata* ssp. *maritima* (Underwood) Hultén
Isoetes muricata Durieu = Isoetes echinospora [ISOEECH]
Isoetes muricata ssp. *maritima* (Underwood) Hultén = Isoetes maritima [ISOEMAR]
Isoetes muricata var. *braunii* (Durieu) C.F. Reed = Isoetes echinospora [ISOEECH]
Isoetes muricata var. *hesperia* C.F. Reed = Isoetes echinospora [ISOEECH]
Isoetes nuttallii A. Braun ex Engelm. [ISOENUT] (Nuttall's quillwort)
Isoetes occidentalis Henderson [ISOEOCC] (western quillwort)
SY: *Isoetes lacustris* ssp. *paupercula* (Engelm.) J. Feilberg
SY: *Isoetes lacustris* var. *paupercula* Engelm.
SY: *Isoetes paupercula* (Engelm.) A.A. Eat.
SY: *Isoetes piperi* A.A. Eat.
Isoetes paupercula (Engelm.) A.A. Eat. = Isoetes occidentalis [ISOEOCC]
Isoetes piperi A.A. Eat. = Isoetes occidentalis [ISOEOCC]
Isoetes × pseudotruncata D.M. Britton & D.F. Brunt. [ISOEPSE]
Isoetes setacea Lam. p.p. = Isoetes echinospora [ISOEECH]
Isoetes setacea ssp. *muricata* (Durieu) Holub = Isoetes echinospora [ISOEECH]
Isoetes × truncata (A.A. Eat.) Clute (pro sp.) [ISOETRU]
SY: *Isoetes truncata* (A.A. Eat.) Clute
Isoetes truncata (A.A. Eat.) Clute = Isoetes × truncata [ISOETRU]

JUNCACEAE (V:M)

Juncoides hyperboreum (R. Br.) Sheldon p.p. = Luzula confusa [LUZUCON]
Juncoides piperi Coville = Luzula piperi [LUZUPIP]
Juncoides spicatum (L.) Kuntze = Luzula spicata [LUZUSPI]
Juncus L. [JUNCUS$]
Juncus acuminatus Michx. [JUNCACU] (tapered rush)
Juncus albescens (Lange) Fern. [JUNCALB] (whitish rush)
SY: *Juncus triglumis* ssp. *albescens* (Lange) Hultén
SY: *Juncus triglumis* var. *albescens* Lange
Juncus alpinoarticulatus Chaix [JUNCALP] (alpine rush)
SY: *Juncus geniculatus* Schrank
Juncus arcticus Willd. [JUNCARC] (arctic rush)
Juncus arcticus ssp. **alaskanus** Hultén [JUNCARC1]
SY: *Juncus arcticus* var. *alaskanus* (Hultén) Welsh
SY: *Juncus balticus* var. *alaskanus* (Hultén) Porsild
Juncus arcticus ssp. *ater* (Rydb.) Hultén = Juncus balticus var. montanus [JUNCBAL1]
Juncus arcticus ssp. *balticus* (Willd.) Hyl. = Juncus balticus [JUNCBAL]
Juncus arcticus ssp. *littoralis* (Engelm.) Hultén = Juncus balticus var. littoralis [JUNCBAL2]
Juncus arcticus ssp. *sitchensis* Engelm. = Juncus haenkei [JUNCHAE]
Juncus arcticus var. *alaskanus* (Hultén) Welsh = Juncus arcticus ssp. alaskanus [JUNCARC1]
Juncus arcticus var. *montanus* (Engelm.) Welsh = Juncus balticus var. montanus [JUNCBAL1]
Juncus arcticus var. *sitchensis* Engelm. = Juncus haenkei [JUNCHAE]
Juncus articulatus L. [JUNCART] (jointed rush)
SY: *Juncus articulatus* var. *obtusatus* Engelm.
SY: *Luzula hyperborea* R. Br. p.p.
Juncus articulatus var. *obtusatus* Engelm. = Juncus articulatus [JUNCART]
Juncus ater Rydb. = Juncus balticus var. montanus [JUNCBAL1]
Juncus balticus Willd. [JUNCBAL] (Baltic rush)
SY: *Juncus arcticus* ssp. *balticus* (Willd.) Hyl.
SY: *Juncus vallicola* Rydb.
Juncus balticus var. *alaskanus* (Hultén) Porsild = Juncus arcticus ssp. alaskanus [JUNCARC1]
Juncus balticus var. *haenkei* (E. Mey.) Buch. = Juncus haenkei [JUNCHAE]
Juncus balticus var. **littoralis** Engelm. [JUNCBAL2]
SY: *Juncus arcticus* ssp. *littoralis* (Engelm.) Hultén
Juncus balticus var. **montanus** Engelm. [JUNCBAL1]
SY: *Juncus arcticus* ssp. *ater* (Rydb.) Hultén
SY: *Juncus arcticus* var. *montanus* (Engelm.) Welsh
SY: *Juncus ater* Rydb.
Juncus biglumis L. [JUNCBIG] (two-flowered rush)
Juncus bolanderi Engelm. [JUNCBOL] (Bolander's rush)
Juncus brachystylus (Engelm.) Piper = Juncus kelloggii [JUNCKEL]
Juncus brevicaudatus (Engelm.) Fern. [JUNCBRE] (short-tailed rush)
Juncus breweri Engelm. [JUNCBRW] (Brewer's rush)
Juncus bufonius L. [JUNCBUF] (toad rush)
***Juncus bulbosus** L. [JUNCBUL] (bulbous rush)
SY: *Juncus supinus* Moench
Juncus castaneus Sm. [JUNCCAS] (chestnut rush)
Juncus castaneus ssp. **castaneus** [JUNCCAS0]
Juncus castaneus ssp. **leucochlamys** (Zing. ex Krecz.) Hultén [JUNCCAS2]
SY: *Juncus leucochlamys* Zing. ex Krecz.
Juncus confusus Coville [JUNCCON] (Colorado rush)
SY: *Juncus exilis* Osterhout
Juncus conglomeratus L. = Juncus effusus var. conglomeratus [JUNCEFF6]
Juncus covillei Piper [JUNCCOV] (Coville's rush)
Juncus drummondii E. Mey. [JUNCDRU] (Drummond's rush)
Juncus drummondii var. **drummondii** [JUNCDRU0]
Juncus drummondii var. *logifructus* St. John = Juncus drummondii var. subtriflorus [JUNCDRU2]
Juncus drummondii var. **subtriflorus** (E. Mey.) C.L. Hitchc. [JUNCDRU2]
SY: *Juncus drummondii* var. *logifructus* St. John
Juncus effusus L. [JUNCEFF] (common rush)
Juncus effusus var. **brunneus** Engelm. [JUNCEFF1]
Juncus effusus var. *caeruleomontanus* St. John = Juncus effusus var. conglomeratus [JUNCEFF6]
Juncus effusus var. *compactus* auct. = Juncus effusus var. conglomeratus [JUNCEFF6]
***Juncus effusus** var. **conglomeratus** (L.) Engelm. [JUNCEFF6]
SY: *Juncus conglomeratus* L.
SY: *Juncus effusus* var. *caeruleomontanus* St. John
SY: *Juncus effusus* var. *compactus* auct.
Juncus effusus var. **gracilis** Hook. [JUNCEFF2]
Juncus effusus var. **pacificus** Fern. & Wieg. [JUNCEFF3]
Juncus ensifolius Wikstr. [JUNCENS] (dagger-leaved rush)
SY: *Juncus ensifolius* var. *major* Hook.
SY: *Juncus xiphioides* var. *triandrus* Engelm.
Juncus ensifolius var. *brunnescens* (Rydb.) Cronq. = Juncus saximontanus [JUNCSAX]
Juncus ensifolius var. *major* Hook. = Juncus ensifolius [JUNCENS]
Juncus ensifolius var. *montanus* (Engelm.) C.L. Hitchc. = Juncus saximontanus [JUNCSAX]
Juncus exilis Osterhout = Juncus confusus [JUNCCON]
Juncus falcatus E. Mey. [JUNCFAL] (sickle-leaved rush)
Juncus falcatus ssp. *sitchensis* (Buch.) Hultén = Juncus prominens [JUNCPRO]
Juncus falcatus var. *sitchensis* Buch. = Juncus prominens [JUNCPRO]
Juncus filiformis L. [JUNCFIL] (thread rush)
Juncus geniculatus Schrank = Juncus alpinoarticulatus [JUNCALP]
Juncus gerardii Loisel. [JUNCGER] (mud rush)
Juncus greenei var. *vaseyi* (Engelm.) Boivin = Juncus vaseyi [JUNCVAS]

Juncus haenkei E. Mey. [JUNCHAE]
SY: *Juncus arcticus* ssp. *sitchensis* Engelm.
SY: *Juncus arcticus* var. *sitchensis* Engelm.
SY: *Juncus balticus* var. *haenkei* (E. Mey.) Buch.
Juncus jonesii Rydb. = Juncus regelii [JUNCREG]
Juncus kelloggii Engelm. [JUNCKEL] (Kellogg's rush)
SY: *Juncus brachystylus* (Engelm.) Piper
Juncus lesuerii Boland. [JUNCLES]
Juncus lesuerii var. *tracyi* Jepson = Juncus tracyi [JUNCTRA]
Juncus leucochlamys Zing. ex Krecz. = Juncus castaneus ssp. leucochlamys [JUNCCAS2]
Juncus longistylis Torr. [JUNCLON] (long-styled rush)
Juncus macer S.F. Gray = Juncus tenuis [JUNCTEN]
Juncus melanocarpus Michx. = Luzula parviflora ssp. melanocarpa [LUZUPAR4]
Juncus mertensianus Bong. [JUNCMER] (Mertens' rush)
SY: *Juncus slwookoorum* S.B. Young
Juncus nevadensis S. Wats. [JUNCNEV] (Sierra rush)
Juncus nodosus L. [JUNCNOD] (tuberous rush)
Juncus occidentalis Wieg. [JUNCOCC]
SY: *Juncus tenuis* var. *congestus* Engelm.
Juncus oreganus S. Wats. = Juncus supiniformis [JUNCSUP]
Juncus oxymeris Engelm. [JUNCOXY] (pointed rush)
Juncus parryi Engelm. [JUNCPAR] (Parry's rush)
Juncus prominens (Buch.) Miyabe & Kudo [JUNCPRO]
SY: *Juncus falcatus* ssp. *sitchensis* (Buch.) Hultén
SY: *Juncus falcatus* var. *sitchensis* Buch.
Juncus regelii Buch. [JUNCREG] (Regel's rush)
SY: *Juncus jonesii* Rydb.
Juncus saximontanus A. Nels. [JUNCSAX]
SY: *Juncus ensifolius* var. *brunnescens* (Rydb.) Cronq.
SY: *Juncus ensifolius* var. *montanus* (Engelm.) C.L. Hitchc.
SY: *Juncus saximontanus* var. *robustior* M.E. Peck
SY: *Juncus xiphioides* var. *macranthus* Engelm.
SY: *Juncus xiphioides* var. *montanus* Engelm.
Juncus saximontanus var. *robustior* M.E. Peck = Juncus saximontanus [JUNCSAX]
Juncus slwookoorum S.B. Young = Juncus mertensianus [JUNCMER]
Juncus stygius L. [JUNCSTY] (bog rush)
Juncus stygius ssp. **americanus** (Buch.) Hultén [JUNCSTY1]
SY: *Juncus stygius* var. *americanus* Buch.
Juncus stygius var. *americanus* Buch. = Juncus stygius ssp. americanus [JUNCSTY1]
Juncus supiniformis Engelm. [JUNCSUP] (spreading rush)
SY: *Juncus oreganus* S. Wats.
Juncus supinus Moench = Juncus bulbosus [JUNCBUL]
Juncus tenuis Willd. [JUNCTEN] (slender rush)
SY: *Juncus macer* S.F. Gray
SY: *Juncus tenuis* var. *anthelatus* Wieg.
SY: *Juncus tenuis* var. *multicornis* E. Mey.
SY: *Juncus tenuis* var. *williamsii* Fern.
Juncus tenuis var. *anthelatus* Wieg. = Juncus tenuis [JUNCTEN]
Juncus tenuis var. *congestus* Engelm. = Juncus occidentalis [JUNCOCC]
Juncus tenuis var. *multicornis* E. Mey. = Juncus tenuis [JUNCTEN]
Juncus tenuis var. *williamsii* Fern. = Juncus tenuis [JUNCTEN]
Juncus torreyi Coville [JUNCTOR] (Torrey's rush)
Juncus tracyi Rydb. [JUNCTRA]
SY: *Juncus lesuerii* var. *tracyi* Jepson
Juncus triglumis L. [JUNCTRI] (three-flowered rush)
Juncus triglumis ssp. *albescens* (Lange) Hultén = Juncus albescens [JUNCALB]
Juncus triglumis var. *albescens* Lange = Juncus albescens [JUNCALB]
Juncus vallicola Rydb. = Juncus balticus [JUNCBAL]
Juncus vaseyi Engelm. [JUNCVAS] (Vasey's rush)
SY: *Juncus greenei* var. *vaseyi* (Engelm.) Boivin
Juncus xiphioides var. *macranthus* Engelm. = Juncus saximontanus [JUNCSAX]
Juncus xiphioides var. *montanus* Engelm. = Juncus saximontanus [JUNCSAX]
Juncus xiphioides var. *triandrus* Engelm. = Juncus ensifolius [JUNCENS]
Luzula DC. [LUZULA$]
Luzula arctica Blytt [LUZUARC] (arctic woodrush)
SY: *Luzula nivalis* (Laestad.) Beurling
Luzula arcuata (Wahlenb.) Sw. [LUZUARU] (curved alpine woodrush)
Luzula arcuata ssp. **unalaschcensis** (Buch.) Hultén [LUZUARU1]
SY: *Luzula arcuata* var. *unalaschcensis* Buch.
SY: *Luzula unalaschcensis* (Buch.) Satake
Luzula arcuata var. *unalaschcensis* Buch. = Luzula arcuata ssp. unalaschcensis [LUZUARU1]
Luzula campestris var. *congesta* (Thuill.) E. Mey. = Luzula congesta [LUZUCOG]
Luzula campestris var. *frigida* Buch. = Luzula multiflora ssp. frigida [LUZUMUL7]
Luzula campestris var. *multiflora* (Ehrh.) Celak. = Luzula multiflora [LUZUMUL]
Luzula comosa E. Mey. = Luzula congesta [LUZUCOG]
Luzula comosa var. *congesta* (Thuill.) S. Wats. = Luzula congesta [LUZUCOG]
Luzula confusa Lindeberg [LUZUCON] (confused woodrush)
SY: *Juncoides hyperboreum* (R. Br.) Sheldon p.p.
Luzula congesta (Thuill.) Lej. [LUZUCOG]
SY: *Luzula campestris* var. *congesta* (Thuill.) E. Mey.
SY: *Luzula comosa* E. Mey.
SY: *Luzula comosa* var. *congesta* (Thuill.) S. Wats.
SY: *Luzula intermedia* (Thuill.) A. Nels.
SY: *Luzula multiflora* var. *acadiensis* Fern.
SY: *Luzula multiflora* ssp. *comosa* (E. Mey.) Hultén
SY: *Luzula multiflora* var. *comosa* (E. Mey.) Fern. & Wieg.
SY: *Luzula multiflora* ssp. *congesta* (Thuill.) Hyl.
SY: *Luzula multiflora* var. *congesta* (Thuill.) Koch
SY: *Luzula orestera* C.W. Sharsmith
SY: *Luzula spicata* ssp. *saximontana* A. & D. Löve
Luzula fastigiata E. Mey. = Luzula parviflora ssp. fastigiata [LUZUPAR1]
Luzula frigida (Buch.) Samuelsson = Luzula multiflora ssp. frigida [LUZUMUL7]
Luzula glabrata (Hoppe ex Rostk.) Desv. [LUZUGLA]
Luzula glabrata var. **hitchcockii** (Hämet-Ahti) Dorn [LUZUGLA1]
SY: *Luzula hitchcockii* Hämet-Ahti
Luzula groenlandica Böcher [LUZUGRO] (Greenland woodrush)
Luzula hitchcockii Hämet-Ahti = Luzula glabrata var. hitchcockii [LUZUGLA1]
Luzula hyperborea R. Br. p.p. = Juncus articulatus [JUNCART]
Luzula intermedia (Thuill.) A. Nels. = Luzula congesta [LUZUCOG]
Luzula melanocarpa (Michx.) Tolm. = Luzula parviflora ssp. melanocarpa [LUZUPAR4]
Luzula multiflora (Ehrh.) Lej. [LUZUMUL] (many-flowered woodrush)
SY: *Luzula campestris* var. *multiflora* (Ehrh.) Celak.
Luzula multiflora ssp. *comosa* (E. Mey.) Hultén = Luzula congesta [LUZUCOG]
Luzula multiflora ssp. *congesta* (Thuill.) Hyl. = Luzula congesta [LUZUCOG]
Luzula multiflora ssp. **frigida** (Buch.) Krecz. [LUZUMUL7]
SY: *Luzula campestris* var. *frigida* Buch.
SY: *Luzula frigida* (Buch.) Samuelsson
SY: *Luzula multiflora* var. *frigida* (Buch.) Samuelsson
SY: *Luzula multiflora* var. *fusconigra* Celak.
SY: *Luzula sudetica* var. *frigida* (Buch.) Fern.
Luzula multiflora ssp. *kjellmaniana* (Miyabe & Kudo) Tolm. = Luzula multiflora ssp. multiflora var. kjellmanioides **[LUZUMUL4]**
Luzula multiflora ssp. **multiflora** [LUZUMUL0]
Luzula multiflora ssp. **multiflora** var. **kjellmanioides** Taylor & MacBryde [LUZUMUL4]
SY: *Luzula multiflora* ssp. *kjellmaniana* (Miyabe & Kudo) Tolm.
Luzula multiflora var. *acadiensis* Fern. = Luzula congesta [LUZUCOG]
Luzula multiflora var. *comosa* (E. Mey.) Fern. & Wieg. = Luzula congesta [LUZUCOG]

Luzula multiflora var. *congesta* (Thuill.) Koch = Luzula congesta [LUZUCOG]
Luzula multiflora var. *frigida* (Buch.) Samuelsson = Luzula multiflora ssp. frigida [LUZUMUL7]
Luzula multiflora var. *fusconigra* Celak. = Luzula multiflora ssp. frigida [LUZUMUL7]
Luzula nivalis (Laestad.) Beurling = Luzula arctica [LUZUARC]
Luzula orestera C.W. Sharsmith = Luzula congesta [LUZUCOG]
Luzula parviflora (Ehrh.) Desv. [LUZUPAR] (small-flowered woodrush)
Luzula parviflora ssp. **fastigiata** (E. Mey.) Hämet-Ahti [LUZUPAR1]
SY: *Luzula fastigiata* E. Mey.
Luzula parviflora ssp. **melanocarpa** (Michx.) Hämet-Ahti [LUZUPAR4]
SY: *Juncus melanocarpus* Michx.
SY: *Luzula melanocarpa* (Michx.) Tolm.
SY: *Luzula parviflora* ssp. *melanocarpa* (Michx.) Tolm.
SY: *Luzula parviflora* var. *melanocarpa* (Michx.) Buch.
Luzula parviflora ssp. *melanocarpa* (Michx.) Tolm. = Luzula parviflora ssp. melanocarpa [LUZUPAR4]
Luzula parviflora var. *melanocarpa* (Michx.) Buch. = Luzula parviflora ssp. melanocarpa [LUZUPAR4]
Luzula pilosa var. *rufescens* (Fisch. ex E. Mey.) Boivin = Luzula rufescens [LUZURUF]
Luzula piperi (Coville) M.E. Jones [LUZUPIP] (Piper's woodrush)
SY: *Juncoides piperi* Coville
SY: *Luzula wahlenbergii* ssp. *piperi* (Coville) Hultén
Luzula rufescens Fisch. ex E. Mey. [LUZURUF] (rusty woodrush)
SY: *Luzula pilosa* var. *rufescens* (Fisch. ex E. Mey.) Boivin
Luzula spicata (L.) DC. [LUZUSPI] (spiked woodrush)
SY: *Juncoides spicatum* (L.) Kuntze
Luzula spicata ssp. *saximontana* A. & D. Löve = Luzula congesta [LUZUCOG]
Luzula sudetica var. *frigida* (Buch.) Fern. = Luzula multiflora ssp. frigida [LUZUMUL7]
Luzula unalaschcensis (Buch.) Satake = Luzula arcuata ssp. unalaschcensis [LUZUARU1]
Luzula wahlenbergii ssp. *piperi* (Coville) Hultén = Luzula piperi [LUZUPIP]

JUNCAGINACEAE (V:M)

Lilaea Bonpl. [LILAEA$]
Lilaea scilloides (Poir.) Hauman [LILASCI] (flowering quillwort)
SY: *Lilaea subulata* Bonpl.
Lilaea subulata Bonpl. = Lilaea scilloides [LILASCI]
Triglochin L. [TRIGLOC$]
Triglochin concinnum Burtt-Davy [TRIGCON] (graceful arrow-grass)
Triglochin concinnum var. *debile* (M.E. Jones) J.T. Howell = Triglochin maritimum [TRIGMAR]
Triglochin debile (M.E. Jones) A. & D. Löve = Triglochin maritimum [TRIGMAR]
Triglochin elatum Nutt. = Triglochin maritimum [TRIGMAR]
Triglochin maritimum L. [TRIGMAR] (seaside arrow-grass)
SY: *Triglochin concinnum* var. *debile* (M.E. Jones) J.T. Howell
SY: *Triglochin debile* (M.E. Jones) A. & D. Löve
SY: *Triglochin elatum* Nutt.
SY: *Triglochin maritimum* var. *elatum* (Nutt.) Gray
Triglochin maritimum var. *elatum* (Nutt.) Gray = Triglochin maritimum [TRIGMAR]
Triglochin palustre L. [TRIGPAL] (marsh arrow-grass)

LAMIACEAE (V:D)

Acinos P. Mill. [ACINOS$]
*__Acinos arvensis__ (Lam.) Dandy [ACINARV]
SY: *Acinos thymoides* (L.) Moench
SY: *Calamintha acinos* (L.) Clairville ex Gaud.
SY: *Clinopodium acinos* (L.) Kuntze
SY: *Satureja acinos* (L.) Scheele
Acinos thymoides (L.) Moench = Acinos arvensis [ACINARV]
Agastache Clayton ex Gronov. [AGASTAC$]
Agastache anethiodora (Nutt.) Britt. = Agastache foeniculum [AGASFOE]
Agastache foeniculum (Pursh) Kuntze [AGASFOE] (giant-hyssop)
SY: *Agastache anethiodora* (Nutt.) Britt.
Agastache urticifolia (Benth.) Kuntze [AGASURT] (nettle-leaved giant-hyssop)
Ajuga L. [AJUGA$]
*__Ajuga reptans__ L. [AJUGREP] (bugle-weed)
Calamintha acinos (L.) Clairville ex Gaud. = Acinos arvensis [ACINARV]
Calamintha clinopodium Benth = Clinopodium vulgare [CLINVUL]
Clinopodium L. [CLINOPO$]
Clinopodium acinos (L.) Kuntze = Acinos arvensis [ACINARV]
*__Clinopodium vulgare__ L. [CLINVUL]
SY: *Calamintha clinopodium* Benth
SY: *Clinopodium vulgare* var. *neogaea* (Fern.) C.F. Reed
SY: *Satureja vulgaris* (L.) Fritsch
SY: *Satureja vulgaris* var. *diminuta* (Simon) Fern. & Wieg.
SY: *Satureja vulgaris* var. *neogaea* Fern.
Clinopodium vulgare var. *neogaea* (Fern.) C.F. Reed = Clinopodium vulgare [CLINVUL]
Dracocephalum L. [DRACOCE$]
Dracocephalum nuttallii Britt. = Physostegia parviflora [PHYSPAR]
Dracocephalum parviflorum Nutt. [DRACPAR] (American dragonhead)
SY: *Moldavica parviflora* (Nutt.) Britt.
*__Dracocephalum thymiflorum__ L. [DRACTHY] (Eurasian dragonhead)
SY: *Moldavica thymiflora* (L.) Rydb.
Galeopsis L. [GALEOPS$]
*__Galeopsis bifida__ Boenn. [GALEBIF]
SY: *Galeopsis tetrahit* var. *bifida* (Boenn.) Lej. & Court.
*__Galeopsis tetrahit__ L. [GALETET] (hemp-nettle)
Galeopsis tetrahit var. *bifida* (Boenn.) Lej. & Court. = Galeopsis bifida [GALEBIF]
Glechoma L. [GLECHOM$]
*__Glechoma hederacea__ L. [GLECHED] (ground-ivy)
Lamium L. [LAMIUM$]
*__Lamium amplexicaule__ L. [LAMIAMP] (common dead-nettle)
SY: *Lamium amplexicaule* var. *album* A.L. & M.C. Pickens
Lamium amplexicaule var. *album* A.L. & M.C. Pickens = Lamium amplexicaule [LAMIAMP]
Lamium hybridum auct. = Lamium purpureum var. incisum [LAMIPUR1]
*__Lamium maculatum__ L. [LAMIMAC] (spotted dead-nettle)
*__Lamium purpureum__ L. [LAMIPUR] (purple dead-nettle)
*__Lamium purpureum__ var. **incisum** (Willd.) Pers. [LAMIPUR1]
SY: *Lamium hybridum* auct.
Leonurus L. [LEONURU$]
*__Leonurus cardiaca__ L. [LEONCAR] (common motherwort)
Lycopus L. [LYCOPUS$]
Lycopus americanus Muhl. ex W. Bart. [LYCOAME] (cut-leaved water horehound)
SY: *Lycopus americanus* var. *longii* Benner
SY: *Lycopus americanus* var. *scabrifolius* Fern.
SY: *Lycopus sinuatus* Ell.
Lycopus americanus var. *longii* Benner = Lycopus americanus [LYCOAME]
Lycopus americanus var. *scabrifolius* Fern. = Lycopus americanus [LYCOAME]
Lycopus asper Greene [LYCOASP] (rough water horehound)
SY: *Lycopus lucidus* ssp. *americanus* (Gray) Hultén
SY: *Lycopus lucidus* var. *americanus* Gray
Lycopus lucidus ssp. *americanus* (Gray) Hultén = Lycopus asper [LYCOASP]
Lycopus lucidus var. *americanus* Gray = Lycopus asper [LYCOASP]
Lycopus sinuatus Ell. = Lycopus americanus [LYCOAME]
Lycopus uniflorus Michx. [LYCOUNI] (northern water horehound)
Marrubium L. [MARRUBI$]
*__Marrubium vulgare__ L. [MARRVUL] (common horehound)
Melissa L. [MELISSA$]

*__Melissa officinalis__ L. [MELIOFI] (lemon balm)
Mentha L. [MENTHA$]
*__Mentha aquatica__ L. [MENTAQU]
SY: *Mentha citrata* Ehrh.
SY: *Mentha piperita* ssp. *citrata* (Ehrh.) Briq.
SY: *Mentha* × *piperita* var. *citrata* (Ehrh.) Boivin (pro nm.)
Mentha aquatica var. *crispa* (L.) Benth. = Mentha × piperita [MENTXPI]
Mentha arvensis L. [MENTARV] (wild mint)
SY: *Mentha gentilis* L.
SY: *Mentha* × *gentilis* L.
Mentha arvensis ssp. *borealis* (Michx.) Taylor & MacBryde = Mentha canadensis [MENTCAN]
Mentha arvensis ssp. *haplocalyx* Briq. = Mentha canadensis [MENTCAN]
Mentha arvensis var. *canadensis* (L.) Kuntze = Mentha canadensis [MENTCAN]
Mentha arvensis var. *glabrata* (Benth.) Fern. = Mentha canadensis [MENTCAN]
Mentha arvensis var. *lanata* Piper = Mentha canadensis [MENTCAN]
Mentha arvensis var. *villosa* (Benth.) S.R. Stewart = Mentha canadensis [MENTCAN]
Mentha canadensis L. [MENTCAN]
SY: *Mentha arvensis* ssp. *borealis* (Michx.) Taylor & MacBryde
SY: *Mentha arvensis* var. *canadensis* (L.) Kuntze
SY: *Mentha arvensis* var. *glabrata* (Benth.) Fern.
SY: *Mentha arvensis* ssp. *haplocalyx* Briq.
SY: *Mentha arvensis* var. *lanata* Piper
SY: *Mentha arvensis* var. *villosa* (Benth.) S.R. Stewart
SY: *Mentha glabrior* (Hook.) Rydb.
SY: *Mentha penardii* (Briq.) Rydb.
Mentha citrata Ehrh. = Mentha aquatica [MENTAQU]
Mentha crispa L. = Mentha × piperita [MENTXPI]
Mentha dumetorum Schultes = Mentha × piperita [MENTXPI]
Mentha gentilis L. = Mentha arvensis [MENTARV]
Mentha × *gentilis* L. = Mentha arvensis [MENTARV]
Mentha glabrior (Hook.) Rydb. = Mentha canadensis [MENTCAN]
Mentha penardii (Briq.) Rydb. = Mentha canadensis [MENTCAN]
*__Mentha × piperita__ L. (pro sp.) [MENTXPI] (peppermint)
SY: *Mentha aquatica* var. *crispa* (L.) Benth.
SY: *Mentha crispa* L.
SY: *Mentha dumetorum* Schultes
Mentha piperita ssp. *citrata* (Ehrh.) Briq. = Mentha aquatica [MENTAQU]
Mentha × *piperita* var. *citrata* (Ehrh.) Boivin (pro nm.) = Mentha aquatica [MENTAQU]
*__Mentha pulegium__ L. [MENTPUL] (pennyroyal)
Mentha rotundifolia auct. = Mentha suaveolens [MENTSUA]
*__Mentha spicata__ L. [MENTSPI] (spearmint)
SY: *Mentha viridis* L.
*__Mentha suaveolens__ Ehrh. [MENTSUA] (applemint)
SY: *Mentha rotundifolia* auct.
Mentha viridis L. = Mentha spicata [MENTSPI]
Micromeria chamissonis (Benth.) Greene = Satureja douglasii [SATUDOU]
Moldavica parviflora (Nutt.) Britt. = Dracocephalum parviflorum [DRACPAR]
Moldavica thymiflora (L.) Rydb. = Dracocephalum thymiflorum [DRACTHY]
Monarda L. [MONARDA$]
Monarda fistulosa L. [MONAFIS] (wild bergamot)
Monarda fistulosa var. **menthifolia** (Graham) Fern. [MONAFIS2]
SY: *Monarda menthifolia* Graham
Monarda fistulosa var. **mollis** (L.) Benth. [MONAFIS1]
SY: *Monarda mollis* L.
SY: *Monarda scabra* Beck
Monarda menthifolia Graham = Monarda fistulosa var. menthifolia [MONAFIS2]
Monarda mollis L. = Monarda fistulosa var. mollis [MONAFIS1]
Monarda scabra Beck = Monarda fistulosa var. mollis [MONAFIS1]
Monardella Benth. [MONARDE$]
Monardella odoratissima Benth. [MONAODO] (monardella)
Nepeta L. [NEPETA$]
*__Nepeta cataria__ L. [NEPECAT] (catnip)
Origanum L. [ORIGANU$]
*__Origanum vulgare__ L. [ORIGVUL] (wild marjoram)
Physostegia Benth. [PHYSOST$]
Physostegia nuttallii (Britt.) Fassett = Physostegia parviflora [PHYSPAR]
Physostegia parviflora Nutt. ex Gray [PHYSPAR] (purple dragonhead)
SY: *Dracocephalum nuttallii* Britt.
SY: *Physostegia nuttallii* (Britt.) Fassett
SY: *Physostegia virginiana* var. *parviflora* (Nutt. ex Gray) Boivin
Physostegia virginiana var. *parviflora* (Nutt. ex Gray) Boivin = Physostegia parviflora [PHYSPAR]
Prunella L. [PRUNELL$]
Prunella vulgaris L. [PRUNVUL] (self-heal)
Prunella vulgaris ssp. **lanceolata** (W. Bart.) Hultén [PRUNVUL1]
SY: *Prunella vulgaris* var. *elongata* Benth.
SY: *Prunella vulgaris* var. *lanceolata* (W. Bart.) Fern.
*__Prunella vulgaris__ ssp. **vulgaris** [PRUNVUL0]
SY: *Prunella vulgaris* var. *atropurpurea* Fern.
SY: *Prunella vulgaris* var. *calvescens* Fern.
SY: *Prunella vulgaris* var. *hispida* Benth.
SY: *Prunella vulgaris* var. *minor* Sm.
SY: *Prunella vulgaris* var. *nana* Clute
SY: *Prunella vulgaris* var. *parviflora* (Poir.) DC.
SY: *Prunella vulgaris* var. *rouleauiana* Victorin
Prunella vulgaris var. *atropurpurea* Fern. = Prunella vulgaris ssp. vulgaris [PRUNVUL0]
Prunella vulgaris var. *calvescens* Fern. = Prunella vulgaris ssp. vulgaris [PRUNVUL0]
Prunella vulgaris var. *elongata* Benth. = Prunella vulgaris ssp. lanceolata [PRUNVUL1]
Prunella vulgaris var. *hispida* Benth. = Prunella vulgaris ssp. vulgaris [PRUNVUL0]
Prunella vulgaris var. *lanceolata* (W. Bart.) Fern. = Prunella vulgaris ssp. lanceolata [PRUNVUL1]
Prunella vulgaris var. *minor* Sm. = Prunella vulgaris ssp. vulgaris [PRUNVUL0]
Prunella vulgaris var. *nana* Clute = Prunella vulgaris ssp. vulgaris [PRUNVUL0]
Prunella vulgaris var. *parviflora* (Poir.) DC. = Prunella vulgaris ssp. vulgaris [PRUNVUL0]
Prunella vulgaris var. *rouleauiana* Victorin = Prunella vulgaris ssp. vulgaris [PRUNVUL0]
Salvia L. [SALVIA$]
*__Salvia nemorosa__ L. [SALVNEM] (wood sage)
Satureja L. [SATUREJ$]
Satureja acinos (L.) Scheele = Acinos arvensis [ACINARV]
Satureja chamissonis (Benth.) Briq. = Satureja douglasii [SATUDOU]
Satureja douglasii (Benth.) Briq. [SATUDOU] (yerba buena)
SY: *Micromeria chamissonis* (Benth.) Greene
SY: *Satureja chamissonis* (Benth.) Briq.
Satureja vulgaris (L.) Fritsch = Clinopodium vulgare [CLINVUL]
Satureja vulgaris var. *diminuta* (Simon) Fern. & Wieg. = Clinopodium vulgare [CLINVUL]
Satureja vulgaris var. *neogaea* Fern. = Clinopodium vulgare [CLINVUL]
Scutellaria L. [SCUTELL$]
Scutellaria angustifolia Pursh [SCUTANG] (narrow-leaved skullcap)
Scutellaria epilobiifolia A. Hamilton = Scutellaria galericulata [SCUTGAL]
Scutellaria galericulata L. [SCUTGAL] (marsh skullcap)
SY: *Scutellaria epilobiifolia* A. Hamilton
SY: *Scutellaria galericulata* var. *epilobiifolia* (A. Hamilton) Jordal
SY: *Scutellaria galericulata* ssp. *pubescens* (Benth.) A. & D. Löve
SY: *Scutellaria galericulata* var. *pubescens* Benth.
Scutellaria galericulata ssp. *pubescens* (Benth.) A. & D. Löve = Scutellaria galericulata [SCUTGAL]
Scutellaria galericulata var. *epilobiifolia* (A. Hamilton) Jordal = Scutellaria galericulata [SCUTGAL]
Scutellaria galericulata var. *pubescens* Benth. = Scutellaria galericulata [SCUTGAL]

Scutellaria lateriflora L. [SCUTLAT] (blue skullcap)
Stachys L. [STACHYS$]
*****Stachys arvensis** (L.) L. [STACARV] (field hedge-nettle)
Stachys asperrima Rydb. = Stachys palustris ssp. pilosa [STACPAL1]
Stachys borealis Rydb. = Stachys palustris ssp. pilosa [STACPAL1]
*****Stachys byzantina** K. Koch ex Scheele [STACBYZ] (lamb's-ear)
SY: *Stachys lanata* Jacq.
SY: *Stachys olympica* Poir.
Stachys ciliata Epling [STACCIL]
SY: *Stachys cooleyae* Heller
Stachys cooleyae Heller = Stachys ciliata [STACCIL]
Stachys lanata Jacq. = Stachys byzantina [STACBYZ]
Stachys olympica Poir. = Stachys byzantina [STACBYZ]
Stachys palustris L. [STACPAL] (marsh hedge-nettle)
Stachys palustris ssp. **pilosa** (Nutt.) Epling [STACPAL1] (swamp hedge-nettle)
SY: *Stachys asperrima* Rydb.
SY: *Stachys borealis* Rydb.
SY: *Stachys palustris* var. *elliptica* Clos
SY: *Stachys palustris* var. *petiolata* Clos
SY: *Stachys palustris* var. *pilosa* (Nutt.) Fern.
SY: *Stachys palustris* var. *segetum* (Mutel) Grogn.
SY: *Stachys pilosa* Nutt.
SY: *Stachys rigida* Nutt. ex Benth.
SY: *Stachys rigida* ssp. *lanata* Epling
SY: *Stachys rigida* ssp. *quercetorum* (Heller) Epling
SY: *Stachys rigida* ssp. *rivularis* (Heller) Epling
SY: *Stachys scopulorum* Greene
SY: *Stachys teucriifolia* Rydb.
SY: *Stachys teucriiformis* Rydb.
Stachys palustris var. *elliptica* Clos = Stachys palustris ssp. pilosa [STACPAL1]
Stachys palustris var. *petiolata* Clos = Stachys palustris ssp. pilosa [STACPAL1]
Stachys palustris var. *pilosa* (Nutt.) Fern. = Stachys palustris ssp. pilosa [STACPAL1]
Stachys palustris var. *segetum* (Mutel) Grogn. = Stachys palustris ssp. pilosa [STACPAL1]
Stachys pilosa Nutt. = Stachys palustris ssp. pilosa [STACPAL1]
Stachys rigida Nutt. ex Benth. = Stachys palustris ssp. pilosa [STACPAL1]
Stachys rigida ssp. *lanata* Epling = Stachys palustris ssp. pilosa [STACPAL1]
Stachys rigida ssp. *quercetorum* (Heller) Epling = Stachys palustris ssp. pilosa [STACPAL1]
Stachys rigida ssp. *rivularis* (Heller) Epling = Stachys palustris ssp. pilosa [STACPAL1]
Stachys scopulorum Greene = Stachys palustris ssp. pilosa [STACPAL1]
Stachys teucriifolia Rydb. = Stachys palustris ssp. pilosa [STACPAL1]
Stachys teucriiformis Rydb. = Stachys palustris ssp. pilosa [STACPAL1]
Teucrium L. [TEUCRIU$]
Teucrium boreale Bickn. = Teucrium canadense var. occidentale [TEUCCAN2]
Teucrium canadense L. [TEUCCAN] (American germander)
Teucrium canadense ssp. *occidentale* (Gray) W.A. Weber = Teucrium canadense var. occidentale [TEUCCAN2]
Teucrium canadense ssp. *viscidum* (Piper) Taylor & MacBryde = Teucrium canadense var. occidentale [TEUCCAN2]
Teucrium canadense var. *boreale* (Bickn.) Shinners = Teucrium canadense var. occidentale [TEUCCAN2]
Teucrium canadense var. **occidentale** (Gray) McClintock & Epling [TEUCCAN2]
SY: *Teucrium boreale* Bickn.
SY: *Teucrium canadense* var. *boreale* (Bickn.) Shinners
SY: *Teucrium canadense* ssp. *occidentale* (Gray) W.A. Weber
SY: *Teucrium canadense* ssp. *viscidum* (Piper) Taylor & MacBryde
SY: *Teucrium occidentale* Gray
SY: *Teucrium occidentale* var. *boreale* (Bickn.) Fern.
Teucrium occidentale Gray = Teucrium canadense var. occidentale [TEUCCAN2]
Teucrium occidentale var. *boreale* (Bickn.) Fern. = Teucrium canadense var. occidentale [TEUCCAN2]
Thymus L. [THYMUS$]
Thymus arcticus (Dur.) Ronniger = Thymus praecox ssp. arcticus [THYMPRA1]
Thymus praecox Opiz [THYMPRA]
*****Thymus praecox** ssp. **arcticus** (Dur.) Jalas [THYMPRA1]
SY: *Thymus arcticus* (Dur.) Ronniger
SY: *Thymus serpyllum* auct.
Thymus serpyllum auct. = Thymus praecox ssp. arcticus [THYMPRA1]

LEMNACEAE (V:M)

Bruniera columbiana (Karst.) Nieuwl. = Wolffia columbiana [WOLFCOL]
Lemna L. [LEMNA$]
Lemna cyclostasa Ell. ex Schleid. = Lemna minor [LEMNMIN]
Lemna minima Chev. ex Schleid. = Lemna minor [LEMNMIN]
Lemna minor L. [LEMNMIN] (common duckweed)
SY: *Lemna cyclostasa* Ell. ex Schleid.
SY: *Lemna minima* Chev. ex Schleid.
Lemna trisulca L. [LEMNTRI] (ivy-leaved duckweed)
Lemna turionifera Landolt [LEMNTUR]
Spirodela Schleid. [SPIRODE$]
Spirodela polyrhiza (L.) Schleid. [SPIRPOL] (great duckmeat)
SY: *Spirodela polyrrhiza* var. *masonii* Daubs
Spirodela polyrrhiza var. *masonii* Daubs = Spirodela polyrhiza [SPIRPOL]
Wolffia Horkel ex Schleid. [WOLFFIA$]
Wolffia borealis (Engelm. ex Hegelm.) Landolt ex Landolt & Wildi [WOLFBOR] (northern water-meal)
SY: *Wolffia punctata* auct.
Wolffia columbiana Karst. [WOLFCOL] (water-meal)
SY: *Bruniera columbiana* (Karst.) Nieuwl.
Wolffia punctata auct. = Wolffia borealis [WOLFBOR]

LENTIBULARIACEAE (V:D)

Pinguicula L. [PINGUIC$]
Pinguicula arctica Eastw. = Pinguicula macroceras [PINGMAC]
Pinguicula macroceras Link [PINGMAC]
SY: *Pinguicula arctica* Eastw.
SY: *Pinguicula vulgaris* ssp. *macroceras* (Link) Calder & Taylor
SY: *Pinguicula vulgaris* var. *macroceras* (Link) Herder
Pinguicula villosa L. [PINGVIL] (hairy butterwort)
Pinguicula vulgaris L. [PINGVUL] (common butterwort)
SY: *Pinguicula vulgaris* var. *americana* Gray
Pinguicula vulgaris ssp. *macroceras* (Link) Calder & Taylor = Pinguicula macroceras [PINGMAC]
Pinguicula vulgaris var. *americana* Gray = Pinguicula vulgaris [PINGVUL]
Pinguicula vulgaris var. *macroceras* (Link) Herder = Pinguicula macroceras [PINGMAC]
Utricularia L. [UTRICUL$]
Utricularia biflora Lam. = Utricularia gibba [UTRIGIB]
Utricularia gibba L. [UTRIGIB] (humped bladderwort)
SY: *Utricularia biflora* Lam.
SY: *Utricularia obtusa* Sw.
SY: *Utricularia pumila* Walt.
Utricularia intermedia Hayne [UTRIINT] (flat-leaved bladderwort)
Utricularia macrorhiza Le Conte [UTRIMAC]
SY: *Utricularia vulgaris* L.
SY: *Utricularia vulgaris* var. *americana* Gray
SY: *Utricularia vulgaris* ssp. *macrorhiza* (Le Conte) Clausen
Utricularia minor L. [UTRIMIN] (lesser bladderwort)
Utricularia obtusa Sw. = Utricularia gibba [UTRIGIB]
Utricularia pumila Walt. = Utricularia gibba [UTRIGIB]
Utricularia vulgaris L. = Utricularia macrorhiza [UTRIMAC]

Utricularia vulgaris ssp. *macrorhiza* (Le Conte) Clausen = Utricularia macrorhiza [UTRIMAC]
Utricularia vulgaris var. *americana* Gray = Utricularia macrorhiza [UTRIMAC]

LILIACEAE (V:M)

Allium L. [ALLIUM$]
Allium acuminatum Hook. [ALLIACU] (Hooker's onion)
SY: *Allium acuminatum* var. *cuspidatum* Fern.
Allium acuminatum var. *cuspidatum* Fern. = Allium acuminatum [ALLIACU]
Allium amplectens Torr. [ALLIAMP] (slimleaf onion)
SY: *Allium attenuifolium* Kellogg
SY: *Allium monospermum* Jepson
SY: *Allium occidentale* Gray
Allium arenicola Osterhout = Allium geyeri var. tenerum [ALLIGEY2]
Allium attenuifolium Kellogg = Allium amplectens [ALLIAMP]
Allium cascadense M.E. Peck = Allium crenulatum [ALLICRE]
Allium cernuum Roth [ALLICER] (nodding onion)
Allium crenulatum Wieg. [ALLICRE] (Olympic onion)
SY: *Allium cascadense* M.E. Peck
SY: *Allium vancouverense* Macoun
SY: *Allium watsonii* T.J. Howell
Allium dictyotum Greene = Allium geyeri var. geyeri [ALLIGEY0]
Allium fibrosum Rydb. = Allium geyeri var. tenerum [ALLIGEY2]
Allium geyeri S. Wats. [ALLIGEY] (Geyer's onion)
Allium geyeri ssp. *tenerum* (M.E. Jones) Traub & Ownbey = Allium geyeri var. tenerum [ALLIGEY2]
Allium geyeri var. **geyeri** [ALLIGEY0]
SY: *Allium dictyotum* Greene
Allium geyeri var. *graniferum* Henderson = Allium geyeri var. tenerum [ALLIGEY2]
Allium geyeri var. **tenerum** M.E. Jones [ALLIGEY2]
SY: *Allium arenicola* Osterhout
SY: *Allium fibrosum* Rydb.
SY: *Allium geyeri* var. *graniferum* Henderson
SY: *Allium geyeri* ssp. *tenerum* (M.E. Jones) Traub & Ownbey
SY: *Allium rubrum* Osterhout
SY: *Allium rydbergii* J.F. Macbr.
SY: *Allium sabulicola* Osterhout
Allium monospermum Jepson = Allium amplectens [ALLIAMP]
Allium occidentale Gray = Allium amplectens [ALLIAMP]
Allium rubrum Osterhout = Allium geyeri var. tenerum [ALLIGEY2]
Allium rydbergii J.F. Macbr. = Allium geyeri var. tenerum [ALLIGEY2]
Allium sabulicola Osterhout = Allium geyeri var. tenerum [ALLIGEY2]
Allium schoenoprasum L. [ALLISCH] (wild chives)
Allium schoenoprasum ssp. *sibiricum* (L.) Celak. = Allium schoenoprasum var. sibiricum [ALLISCH1]
Allium schoenoprasum var. *laurentianum* Fern. = Allium schoenoprasum var. sibiricum [ALLISCH1]
Allium schoenoprasum var. **sibiricum** (L.) Hartman [ALLISCH1]
SY: *Allium schoenoprasum* var. *laurentianum* Fern.
SY: *Allium schoenoprasum* ssp. *sibiricum* (L.) Celak.
SY: *Allium sibiricum* L.
Allium sibiricum L. = Allium schoenoprasum var. sibiricum [ALLISCH1]
Allium validum S. Wats. [ALLIVAL] (swamp onion)
Allium vancouverense Macoun = Allium crenulatum [ALLICRE]
*****Allium vineale** L. [ALLIVIN] (field garlic)
Allium watsonii T.J. Howell = Allium crenulatum [ALLICRE]
Asparagus L. [ASPARAG$]
*****Asparagus officinalis** L. [ASPAOFF] (garden asparagus)
Brodiaea Sm. [BRODIAE$]
Brodiaea coronaria (Salisb.) Engl. [BRODCOR] (harvest brodiaea)
Brodiaea douglasii S. Wats. = Triteleia grandiflora [TRITGRA]
Brodiaea douglasii var. *howellii* (S. Wats.) M.E. Peck = Triteleia howellii [TRITHOW]
Brodiaea grandiflora (Lindl.) J.F. Macbr. = Triteleia grandiflora [TRITGRA]
Brodiaea howellii S. Wats. = Triteleia howellii [TRITHOW]
Brodiaea hyacinthina (Lindl.) Baker = Triteleia hyacinthina [TRITHYA]
Bulbocodium serotinum L. = Lloydia serotina ssp. serotina [LLOYSER0]
Calochortus Pursh [CALOCHO$]
Calochortus apiculatus Baker [CALOAPI] (three-spot mariposa lily)
Calochortus lyallii Baker [CALOLYA] (Lyall's mariposa lily)
Calochortus macrocarpus Dougl. [CALOMAC] (sagebrush mariposa lily)
Camassia Lindl. [CAMASSI$]
Camassia leichtlinii (Baker) S. Wats. [CAMALEI] (great camas)
Camassia quamash (Pursh) Greene [CAMAQUA] (common camas)
Camassia quamash ssp. **maxima** Gould [CAMAQUA1]
SY: *Camassia quamash* var. *maxima* (Gould) Boivin
Camassia quamash ssp. **quamash** [CAMAQUA0]
SY: *Camassia quamash* ssp. *teapeae* (St. John) St. John
SY: *Phalangium quamash* Pursh
Camassia quamash ssp. *teapeae* (St. John) St. John = Camassia quamash ssp. quamash [CAMAQUA0]
Camassia quamash var. *maxima* (Gould) Boivin = Camassia quamash ssp. maxima [CAMAQUA1]
Clintonia Raf. [CLINTON$]
Clintonia uniflora (Menzies ex J.A. & J.H. Schultes) Kunth [CLINUNI] (queen's cup)
SY: *Smilacina borealis* var. *uniflora* Menzies ex J.A. & J.H. Schultes
Convallaria stellata L. = Maianthemum stellatum [MAIASTE]
Convallaria trifolia L. = Maianthemum trifolium [MAIATRI]
Disporum Salisb. ex D. Don [DISPORU$]
Disporum hookeri (Torr.) Nichols. [DISPHOO] (Hooker's fairybells)
Disporum hookeri var. **oreganum** (S. Wats.) Q. Jones [DISPHOO1] (Oregon fairybells)
SY: *Disporum oreganum* (S. Wats.) W. Mill.
Disporum oreganum (S. Wats.) W. Mill. = Disporum hookeri var. oreganum [DISPHOO1]
Disporum smithii (Hook.) Piper [DISPSMI] (Smith's fairybells)
SY: *Uvularia smithii* Hook.
Disporum trachycarpum (S. Wats.) Benth. & Hook. f. [DISPTRA] (rough-fruited fairybells)
SY: *Disporum trachycarpum* var. *subglabrum* E.H. Kelso
SY: *Prosartes trachycarpa* S. Wats.
Disporum trachycarpum var. *subglabrum* E.H. Kelso = Disporum trachycarpum [DISPTRA]
Endymion hispanicus (P. Mill.) Chouard = Hyacinthoides hispanica [HYACHIS]
Erythronium L. [ERYTHRO$]
Erythronium grandiflorum Pursh [ERYTGRA] (yellow glacier lily)
Erythronium montanum S. Wats. [ERYTMON] (white glacier lily)
Erythronium oregonum Applegate [ERYTORE] (white fawn lily)
Erythronium revolutum Sm. [ERYTREV] (pink fawn lily)
Fritillaria L. [FRITILL$]
Fritillaria affinis (Schultes) Sealy = Fritillaria lanceolata [FRITLAN]
Fritillaria camschatcensis (L.) Ker-Gawl. [FRITCAM] (northern rice-root)
SY: *Lilium camschatcense* L.
Fritillaria camschatcensis var. *floribunda* (Benth.) Boivin = Fritillaria lanceolata [FRITLAN]
Fritillaria eximia Eastw. = Fritillaria lanceolata [FRITLAN]
Fritillaria lanceolata Pursh [FRITLAN] (chocolate lily)
SY: *Fritillaria affinis* (Schultes) Sealy
SY: *Fritillaria camschatcensis* var. *floribunda* (Benth.) Boivin
SY: *Fritillaria eximia* Eastw.
SY: *Fritillaria lanceolata* var. *tristulis* A.L. Grant
SY: *Fritillaria mutica* Lindl.
SY: *Fritillaria mutica* var. *gracilis* (S. Wats.) Jepson
Fritillaria lanceolata var. *tristulis* A.L. Grant = Fritillaria lanceolata [FRITLAN]
Fritillaria mutica Lindl. = Fritillaria lanceolata [FRITLAN]
Fritillaria mutica var. *gracilis* (S. Wats.) Jepson = Fritillaria lanceolata [FRITLAN]

Fritillaria pudica (Pursh) Spreng. [FRITPUD] (yellow bell)
SY: *Lilium pudicum* Pursh
SY: *Ochrocodon pudicus* (Pursh) Rydb.
Helonias tenax Pursh = Xerophyllum tenax [XEROTEN]
Hyacinthoides Medik. [HYACINT$]
***Hyacinthoides hispanica** (P. Mill.) Rothm. [HYACHIS] (Spanish bluebells)
SY: *Endymion hispanicus* (P. Mill.) Chouard
SY: *Scilla hispanica* P. Mill.
Kruhsea streptopoides (Ledeb.) Kearney p.p. = Streptopus streptopoides var. brevipes [STRESTR1]
Lilium L. [LILIUM$]
Lilium andinum Nutt. = Lilium philadelphicum var. andinum [LILIPHI1]
Lilium camschatcense L. = Fritillaria camschatcensis [FRITCAM]
Lilium canadense var. *parviflorum* Hook. = Lilium columbianum [LILICOL]
Lilium columbianum hort. ex Baker [LILICOL] (tiger lily)
SY: *Lilium canadense* var. *parviflorum* Hook.
Lilium montanum A. Nels. = Lilium philadelphicum var. andinum [LILIPHI1]
Lilium philadelphicum L. [LILIPHI] (wood lily)
Lilium philadelphicum var. **andinum** (Nutt.) Ker-Gawl. [LILIPHI1]
SY: *Lilium andinum* Nutt.
SY: *Lilium montanum* A. Nels.
SY: *Lilium philadelphicum* var. *montanum* (A. Nels.) Wherry
SY: *Lilium umbellatum* Pursh
Lilium philadelphicum var. *montanum* (A. Nels.) Wherry = Lilium philadelphicum var. andinum [LILIPHI1]
Lilium pudicum Pursh = Fritillaria pudica [FRITPUD]
Lilium umbellatum Pursh = Lilium philadelphicum var. andinum [LILIPHI1]
Lloydia Salisb. ex Reichenb. [LLOYDIA$]
Lloydia serotina (L.) Reichenb. [LLOYSER] (alp lily)
Lloydia serotina ssp. **flava** Calder & Taylor [LLOYSER1]
SY: *Lloydia serotina* var. *flava* (Calder & Taylor) Boivin
Lloydia serotina ssp. **serotina** [LLOYSER0]
SY: *Bulbocodium serotinum* L.
Lloydia serotina var. *flava* (Calder & Taylor) Boivin = Lloydia serotina ssp. flava [LLOYSER1]
Maianthemum G.H. Weber ex Wiggers [MAIANTH$]
Maianthemum amplexicaule (Nutt.) W.A. Weber = Maianthemum racemosum ssp. amplexicaule [MAIARAC1]
Maianthemum bifolium ssp. *kamtschaticum* (J.F. Gmel. ex Cham.) E. Murr. = Maianthemum dilatatum [MAIADIL]
Maianthemum bifolium var. *kamtschaticum* (J.F. Gmel. ex Cham.) Trautv. & C.A. Mey. = Maianthemum dilatatum [MAIADIL]
Maianthemum canadense Desf. [MAIACAN] (Canadian mayflower)
SY: *Maianthemum canadense* var. *interius* Fern.
SY: *Maianthemum canadense* var. *pubescens* Gates & Ehlers
SY: *Unifolium canadense* (Desf.) Greene
Maianthemum canadense var. *interius* Fern. = Maianthemum canadense [MAIACAN]
Maianthemum canadense var. *pubescens* Gates & Ehlers = Maianthemum canadense [MAIACAN]
Maianthemum dilatatum (Wood) A. Nels. & J.F. Macbr. [MAIADIL] (false lily-of-the-valley)
SY: *Maianthemum bifolium* ssp. *kamtschaticum* (J.F. Gmel. ex Cham.) E. Murr.
SY: *Maianthemum bifolium* var. *kamtschaticum* (J.F. Gmel. ex Cham.) Trautv. & C.A. Mey.
SY: *Maianthemum kamtschaticum* (J.F. Gmel. ex Cham.) Nakai
Maianthemum kamtschaticum (J.F. Gmel. ex Cham.) Nakai = Maianthemum dilatatum [MAIADIL]
Maianthemum racemosum (L.) Link [MAIARAC]
Maianthemum racemosum ssp. **amplexicaule** (Nutt.) LaFrankie [MAIARAC1]
SY: *Maianthemum amplexicaule* (Nutt.) W.A. Weber
SY: *Maianthemum racemosum* var. *amplexicaule* (Nutt.) Dorn
SY: *Smilacina amplexicaulis* Nutt.
SY: *Smilacina amplexicaulis* var. *glabra* J.F. Macbr.
SY: *Smilacina amplexicaulis* var. *jenkinsii* Boivin
SY: *Smilacina amplexicaulis* var. *ovata* Boivin
SY: *Smilacina racemosa* var. *amplexicaulis* (Nutt.) S. Wats.
SY: *Smilacina racemosa* var. *brachystyla* G. Henderson
SY: *Smilacina racemosa* var. *glabra* (J.F. Macbr.) St. John
SY: *Smilacina racemosa* var. *jenkinsii* (Boivin) Boivin
SY: *Vagnera amplexicaulis* (Nutt.) Greene
SY: *Vagnera amplexicaulis* var. *glabra* (J.F. Macbr.) Abrams
Maianthemum racemosum var. *amplexicaule* (Nutt.) Dorn = Maianthemum racemosum ssp. amplexicaule [MAIARAC1]
Maianthemum stellatum (L.) Link [MAIASTE]
SY: *Convallaria stellata* L.
SY: *Smilacina liliacea* (Greene) Wynd
SY: *Smilacina sessilifolia* Nutt. ex Baker
SY: *Smilacina stellata* (L.) Desf.
SY: *Smilacina stellata* var. *crassa* Victorin
SY: *Smilacina stellàta* var. *mollis* Farw.
SY: *Smilacina stellata* var. *sessilifolia* (Nutt. ex Baker) G. Henderson
SY: *Smilacina stellata* var. *sylvatica* Victorin & Rouss.
SY: *Vagnera liliacea* (Greene) Rydb.
SY: *Vagnera sessilifolia* (Nutt. ex Baker) Greene
SY: *Vagnera stellata* (L.) Morong
Maianthemum trifolium (L.) Sloboda [MAIATRI]
SY: *Convallaria trifolia* L.
SY: *Smilacina trifolia* (L.) Desf.
SY: *Tovaria trifolia* (L.) Neck. ex Baker
SY: *Vagnera trifolia* (L.) Morong
Narcissus L. [NARCISS$]
***Narcissus poeticus** L. [NARCPOE] (poet's narcissus)
***Narcissus pseudonarcissus** L. [NARCPSE] (daffodil)
Narthecium pusillum Michx. = Tofieldia pusilla [TOFIPUS]
Ochrocodon pudicus (Pursh) Rydb. = Fritillaria pudica [FRITPUD]
Ornithogalum L. [ORNITHO$]
***Ornithogalum umbellatum** L. [ORNIUMB] (star-of-Bethlehem)
Phalangium quamash Pursh = Camassia quamash ssp. quamash [CAMAQUA0]
Prosartes trachycarpa S. Wats. = Disporum trachycarpum [DISPTRA]
Scilla hispanica P. Mill. = Hyacinthoides hispanica [HYACHIS]
Smilacina amplexicaulis Nutt. = Maianthemum racemosum ssp. amplexicaule [MAIARAC1]
Smilacina amplexicaulis var. *glabra* J.F. Macbr. = Maianthemum racemosum ssp. amplexicaule [MAIARAC1]
Smilacina amplexicaulis var. *jenkinsii* Boivin = Maianthemum racemosum ssp. amplexicaule [MAIARAC1]
Smilacina amplexicaulis var. *ovata* Boivin = Maianthemum racemosum ssp. amplexicaule [MAIARAC1]
Smilacina borealis var. *uniflora* Menzies ex J.A. & J.H. Schultes = Clintonia uniflora [CLINUNI]
Smilacina liliacea (Greene) Wynd = Maianthemum stellatum [MAIASTE]
Smilacina racemosa var. *amplexicaulis* (Nutt.) S. Wats. = Maianthemum racemosum ssp. amplexicaule [MAIARAC1]
Smilacina racemosa var. *brachystyla* G. Henderson = Maianthemum racemosum ssp. amplexicaule [MAIARAC1]
Smilacina racemosa var. *glabra* (J.F. Macbr.) St. John = Maianthemum racemosum ssp. amplexicaule [MAIARAC1]
Smilacina racemosa var. *jenkinsii* (Boivin) Boivin = Maianthemum racemosum ssp. amplexicaule [MAIARAC1]
Smilacina sessilifolia Nutt. ex Baker = Maianthemum stellatum [MAIASTE]
Smilacina stellata (L.) Desf. = Maianthemum stellatum [MAIASTE]
Smilacina stellata var. *crassa* Victorin = Maianthemum stellatum [MAIASTE]
Smilacina stellata var. *mollis* Farw. = Maianthemum stellatum [MAIASTE]
Smilacina stellata var. *sessilifolia* (Nutt. ex Baker) G. Henderson = Maianthemum stellatum [MAIASTE]
Smilacina stellata var. *sylvatica* Victorin & Rouss. = Maianthemum stellatum [MAIASTE]
Smilacina trifolia (L.) Desf. = Maianthemum trifolium [MAIATRI]
Stenanthella occidentalis (Gray) Rydb. = Stenanthium occidentale [STENOCC]
Stenanthium (Gray) Kunth [STENANT$]

Stenanthium occidentale Gray [STENOCC] (western mountainbells)
SY: *Stenanthella occidentalis* (Gray) Rydb.
Streptopus Michx. [STREPTO$]
Streptopus amplexifolius (L.) DC. [STREAMP] (clasping twistedstalk)
Streptopus amplexifolius var. *americanus* J.A. & J.H. Schultes = Streptopus amplexifolius var. amplexifolius [STREAMP0]
Streptopus amplexifolius var. **amplexifolius** [STREAMP0]
SY: *Streptopus amplexifolius* var. *americanus* J.A. & J.H. Schultes
SY: *Streptopus amplexifolius* var. *denticulatus* Fassett
SY: *Streptopus amplexifolius* var. *grandiflorus* Fassett
SY: *Tortipes amplexifolius* (L.) Small
SY: *Uvularia amplexifolia* L.
Streptopus amplexifolius var. **chalazatus** Fassett [STREAMP2]
SY: *Streptopus fassettii* A. & D. Löve
Streptopus amplexifolius var. *denticulatus* Fassett = Streptopus amplexifolius var. amplexifolius [STREAMP0]
Streptopus amplexifolius var. *grandiflorus* Fassett = Streptopus amplexifolius var. amplexifolius [STREAMP0]
Streptopus curvipes Vail = Streptopus roseus var. curvipes [STREROS1]
Streptopus fassettii A. & D. Löve = Streptopus amplexifolius var. chalazatus [STREAMP2]
Streptopus roseus Michx. [STREROS] (rosy twistedstalk)
Streptopus roseus ssp. *curvipes* (Vail) Hultén = Streptopus roseus var. curvipes [STREROS1]
Streptopus roseus var. **curvipes** (Vail) Fassett [STREROS1]
SY: *Streptopus curvipes* Vail
SY: *Streptopus roseus* ssp. *curvipes* (Vail) Hultén
Streptopus streptopoides (Ledeb.) Frye & Rigg [STRESTR] (small twistedstalk)
Streptopus streptopoides ssp. *brevipes* (Baker) Calder & Taylor = Streptopus streptopoides var. brevipes [STRESTR1]
Streptopus streptopoides var. **brevipes** (Baker) Fassett [STRESTR1]
SY: *Kruhsea streptopoides* (Ledeb.) Kearney p.p.
SY: *Streptopus streptopoides* ssp. *brevipes* (Baker) Calder & Taylor
Tofieldia Huds. [TOFIELD$]
Tofieldia borealis Wahlenb. = Tofieldia pusilla [TOFIPUS]
Tofieldia coccinea Richards. [TOFICOC] (northern false asphodel)
Tofieldia glutinosa (Michx.) Pers. [TOFIGLU] (sticky false asphodel)
Tofieldia pusilla (Michx.) Pers. [TOFIPUS] (common false asphodel)
SY: *Narthecium pusillum* Michx.
SY: *Tofieldia borealis* Wahlenb.
Tortipes amplexifolius (L.) Small = Streptopus amplexifolius var. amplexifolius [STREAMP0]
Tovaria trifolia (L.) Neck. ex Baker = Maianthemum trifolium [MAIATRI]
Toxicoscordion gramineum (Rydb.) Rydb. = Zigadenus venenosus var. gramineus [ZIGAVEN1]
Toxicoscordion venenosum (S. Wats.) Rydb. = Zigadenus venenosus var. venenosus [ZIGAVEN0]
Trillium L. [TRILLIU$]
Trillium ovatum Pursh [TRILOVA] (western trillium)
Triteleia Dougl. ex Lindl. [TRITELE$]
Triteleia bicolor (Suksdorf) Abrams = Triteleia howellii [TRITHOW]
Triteleia grandiflora Lindl. [TRITGRA] (large-flowered triteleia)
SY: *Brodiaea douglasii* S. Wats.
SY: *Brodiaea grandiflora* (Lindl.) J.F. Macbr.
Triteleia grandiflora var. *howellii* (S. Wats.) Hoover = Triteleia howellii [TRITHOW]
Triteleia howellii (S. Wats.) Greene [TRITHOW] (Howell's triteleia)
SY: *Brodiaea douglasii* var. *howellii* (S. Wats.) M.E. Peck
SY: *Brodiaea howellii* S. Wats.
SY: *Triteleia bicolor* (Suksdorf) Abrams
SY: *Triteleia grandiflora* var. *howellii* (S. Wats.) Hoover
Triteleia hyacinthina (Lindl.) Greene [TRITHYA] (white triteleia)
SY: *Brodiaea hyacinthina* (Lindl.) Baker
Unifolium canadense (Desf.) Greene = Maianthemum canadense [MAIACAN]
Uvularia amplexifolia L. = Streptopus amplexifolius var. amplexifolius [STREAMP0]
Uvularia smithii Hook. = Disporum smithii [DISPSMI]
Vagnera amplexicaulis (Nutt.) Greene = Maianthemum racemosum ssp. amplexicaule [MAIARAC1]
Vagnera amplexicaulis var. *glabra* (J.F. Macbr.) Abrams = Maianthemum racemosum ssp. amplexicaule [MAIARAC1]
Vagnera liliacea (Greene) Rydb. = Maianthemum stellatum [MAIASTE]
Vagnera sessilifolia (Nutt. ex Baker) Greene = Maianthemum stellatum [MAIASTE]
Vagnera stellata (L.) Morong = Maianthemum stellatum [MAIASTE]
Vagnera trifolia (L.) Morong = Maianthemum trifolium [MAIATRI]
Veratrum L. [VERATRU$]
Veratrum eschscholtzianum (J.A. & J.H. Schultes) Rydb. ex Heller = Veratrum viride [VERAVIR]
Veratrum eschscholtzii Gray = Veratrum viride [VERAVIR]
Veratrum eschscholtzii var. *incriminatum* Boivin = Veratrum viride [VERAVIR]
Veratrum viride Ait. [VERAVIR] (Indian hellebore)
SY: *Veratrum eschscholtzianum* (J.A. & J.H. Schultes) Rydb. ex Heller
SY: *Veratrum eschscholtzii* Gray
SY: *Veratrum eschscholtzii* var. *incriminatum* Boivin
SY: *Veratrum viride* ssp. *eschscholtzii* (Gray) A. & D. Löve
SY: *Veratrum viride* var. *eschscholtzii* (Gray) Breitung
Veratrum viride ssp. *eschscholtzii* (Gray) A. & D. Löve = Veratrum viride [VERAVIR]
Veratrum viride var. *eschscholtzii* (Gray) Breitung = Veratrum viride [VERAVIR]
Xerophyllum Michx. [XEROPHY$]
Xerophyllum tenax (Pursh) Nutt. [XEROTEN] (bear-grass)
SY: *Helonias tenax* Pursh
Zygadenus Endl. = Zigadenus [ZIGADEN$]
Zigadenus Michx. [ZIGADEN$]
Zigadenus elegans Pursh [ZIGAELE] (mountain death-camas)
Zigadenus gramineus Rydb. = Zigadenus venenosus var. gramineus [ZIGAVEN1]
Zigadenus intermedius Rydb. = Zigadenus venenosus var. gramineus [ZIGAVEN1]
Zigadenus venenosus S. Wats. [ZIGAVEN] (meadow death-camas)
Zigadenus venenosus var. **gramineus** (Rydb.) Walsh ex M.E. Peck [ZIGAVEN1]
SY: *Toxicoscordion gramineum* (Rydb.) Rydb.
SY: *Zigadenus gramineus* Rydb.
SY: *Zigadenus intermedius* Rydb.
Zigadenus venenosus var. **venenosus** [ZIGAVEN0]
SY: *Toxicoscordion venenosum* (S. Wats.) Rydb.

LIMNANTHACEAE (V:D)

Limnanthes R. Br. [LIMNANT$]
Limnanthes macounii Trel. [LIMNMAC] (Macoun's meadow-foam)

LINACEAE (V:D)

Adenolinum lewisii (Pursh) A. & D. Löve = Linum lewisii [LINULEW]
Linum L. [LINUM$]
Linum angustifolium Huds. = Linum bienne [LINUBIE]
***Linum bienne** P. Mill. [LINUBIE] (pale flax)
SY: *Linum angustifolium* Huds.
Linum humile P. Mill. = Linum usitatissimum [LINUUSI]
Linum lewisii Pursh [LINULEW] (wild blue flax)
SY: *Adenolinum lewisii* (Pursh) A. & D. Löve
SY: *Linum perenne* ssp. *lewisii* (Pursh) Hultén
SY: *Linum perenne* var. *lewisii* (Pursh) Eat. & J. Wright
Linum perenne L. [LINUPER] (blue flax)
Linum perenne ssp. *lewisii* (Pursh) Hultén = Linum lewisii [LINULEW]
Linum perenne var. *lewisii* (Pursh) Eat. & J. Wright = Linum lewisii [LINULEW]
***Linum usitatissimum** L. [LINUUSI] (common flax)
SY: *Linum humile* P. Mill.
SY: *Linum usitatissimum* var. *humile* (P. Mill.) Pers.

Linum usitatissimum var. *humile* (P. Mill.) Pers. = Linum usitatissimum [LINUUSI]

LOASACEAE (V:D)

Acrolasia albicaulis (Dougl. ex Hook.) Rydb. = Mentzelia albicaulis [MENTALB]
Mentzelia L. [MENTZEL$]
Mentzelia albicaulis (Dougl. ex Hook.) Dougl. ex Torr. & Gray [MENTALB] (small-flowered evening star)
SY: *Acrolasia albicaulis* (Dougl. ex Hook.) Rydb.
SY: *Mentzelia albicaulis* var. *ctenophora* (Rydb.) St. John
SY: *Mentzelia albicaulis* var. *gracilis* J. Darl.
SY: *Mentzelia albicaulis* var. *tenerrima* (Rydb.) St. John
SY: *Mentzelia gracilis* H.J. Thompson & Lewis
Mentzelia albicaulis var. *ctenophora* (Rydb.) St. John = Mentzelia albicaulis [MENTALB]
Mentzelia albicaulis var. *gracilis* J. Darl. = Mentzelia albicaulis [MENTALB]
Mentzelia albicaulis var. *tenerrima* (Rydb.) St. John = Mentzelia albicaulis [MENTALB]
Mentzelia brandegei S. Wats. = Mentzelia laevicaulis var. parviflora [MENTLAE2]
Mentzelia dispersa S. Wats. [MENTDIS] (bushy mentzelia)
Mentzelia douglasii St. John = Mentzelia laevicaulis var. parviflora [MENTLAE2]
Mentzelia gracilis H.J. Thompson & Lewis = Mentzelia albicaulis [MENTALB]
Mentzelia laevicaulis (Dougl. ex Hook.) Torr. & Gray [MENTLAE] (blazing-star)
Mentzelia laevicaulis var. *acuminata* (Rydb.) A. Nels. & J.F. Macbr. = Mentzelia laevicaulis var. laevicaulis [MENTLAE0]
Mentzelia laevicaulis var. **laevicaulis** [MENTLAE0]
SY: *Mentzelia laevicaulis* var. *acuminata* (Rydb.) A. Nels. & J.F. Macbr.
SY: *Nuttallia laevicaulis* (Dougl. ex Hook.) Greene
Mentzelia laevicaulis var. **parviflora** (Dougl. ex Hook.) C.L. Hitchc. [MENTLAE2]
SY: *Mentzelia brandegei* S. Wats.
SY: *Mentzelia douglasii* St. John
Nuttallia laevicaulis (Dougl. ex Hook.) Greene = Mentzelia laevicaulis var. laevicaulis [MENTLAE0]

LYCOPODIACEAE (V:F)

Diphasiastrum alpinum (L.) Holub = Lycopodium alpinum [LYCOALP]
Diphasiastrum complanatum (L.) Holub = Lycopodium complanatum [LYCOCOM]
Diphasiastrum sitchense (Rupr.) Holub = Lycopodium sitchense [LYCOSIT]
Diphasium alpinum (L.) Rothm. = Lycopodium alpinum [LYCOALP]
Diphasium anceps (Wallr.) A. & D. Löve = Lycopodium complanatum [LYCOCOM]
Diphasium complanatum (L.) Rothm. = Lycopodium complanatum [LYCOCOM]
Diphasium complanatum ssp. *montellii* Kukkonen = Lycopodium complanatum [LYCOCOM]
Diphasium sitchense (Rupr.) A. & D. Löve = Lycopodium sitchense [LYCOSIT]
Diphasium wallrothii H.P. Fuchs = Lycopodium complanatum [LYCOCOM]
Huperzia Bernh. [HUPERZI$]
Huperzia chinensis (Christ) Czern. [HUPECHI]
SY: *Huperzia miyoshiana* (Makino) Ching
SY: *Huperzia selago* ssp. *chinensis* (Christ) A. & D. Löve
SY: *Huperzia selago* var. *miyoshiana* (Makino) Taylor & MacBryde
SY: *Lycopodium chinense* Christ
SY: *Lycopodium miyoshianum* Makino
SY: *Lycopodium selago* ssp. *chinense* (Christ) Hultén
SY: *Lycopodium selago* ssp. *miyoshianum* (Makino) Calder & Taylor
SY: *Lycopodium selago* var. *miyoshianum* (Makino) Makino
SY: *Urostachys chinensis* (Christ) Herter ex Nessel
SY: *Urostachys miyoshiana* (Makino) Herter ex Nessel
Huperzia haleakalae (Brack.) Holub [HUPEHAL] (Haleakala fir clubmoss)
SY: *Lycopodium haleakalae* Brack.
SY: *Urostachys haleakalae* (Brack.) Herter
Huperzia miyoshiana (Makino) Ching = Huperzia chinensis [HUPECHI]
Huperzia occidentalis (Clute) Kartesz & Gandhi [HUPEOCC] (western fir clubmoss)
SY: *Lycopodium lucidulum* var. *occidentale* (Clute) L.R. Wilson
SY: *Lycopodium lucidulum* var. *tryonii* Mohlenbrock
Huperzia selago ssp. *chinensis* (Christ) A. & D. Löve = Huperzia chinensis [HUPECHI]
Huperzia selago var. *miyoshiana* (Makino) Taylor & MacBryde = Huperzia chinensis [HUPECHI]
Lepidotis inundata (L.) C. Borner = Lycopodiella inundata [LYCOINU]
Lycopodiella Holub [LYCOPOD$]
Lycopodiella inundata (L.) Holub [LYCOINU] (bog clubmoss)
SY: *Lepidotis inundata* (L.) C. Borner
SY: *Lycopodium inundatum* L.
Lycopodium L. [LYCOPOI$]
Lycopodium alpinum L. [LYCOALP] (alpine clubmoss)
SY: *Diphasiastrum alpinum* (L.) Holub
SY: *Diphasium alpinum* (L.) Rothm.
Lycopodium anceps Wallr. = Lycopodium complanatum [LYCOCOM]
Lycopodium annotinum L. [LYCOANN] (stiff clubmoss)
SY: *Lycopodium pungens* La Pylaie ex Iljin
Lycopodium chinense Christ = Huperzia chinensis [HUPECHI]
Lycopodium clavatum L. [LYCOCLA] (running clubmoss)
Lycopodium clavatum ssp. *megastachyon* (Fern. & Bissell) Selin = Lycopodium clavatum var. monostachyon [LYCOCLA3]
Lycopodium clavatum ssp. *monostachyon* (Grev. & Hook.) Seland. = Lycopodium clavatum var. monostachyon [LYCOCLA3]
Lycopodium clavatum var. **clavatum** [LYCOCLA0]
SY: *Lycopodium clavatum* var. *laurentianum* Victorin
SY: *Lycopodium clavatum* var. *subremotum* Victorin
SY: *Lycopodium clavatum* var. *tristachyum* Hook.
Lycopodium clavatum var. **integerrimum** Spring [LYCOCLA2]
Lycopodium clavatum var. *laurentianum* Victorin = Lycopodium clavatum var. clavatum [LYCOCLA0]
Lycopodium clavatum var. *megastachyon* Fern. & Bissell = Lycopodium clavatum var. monostachyon [LYCOCLA3]
Lycopodium clavatum var. **monostachyon** Grev. & Hook. [LYCOCLA3]
SY: *Lycopodium clavatum* ssp. *megastachyon* (Fern. & Bissell) Selin
SY: *Lycopodium clavatum* var. *megastachyon* Fern. & Bissell
SY: *Lycopodium clavatum* ssp. *monostachyon* (Grev. & Hook.) Seland.
SY: *Lycopodium lagopus* (Laestad.) Zinserl. ex Kuzen
Lycopodium clavatum var. *subremotum* Victorin = Lycopodium clavatum var. clavatum [LYCOCLA0]
Lycopodium clavatum var. *tristachyum* Hook. = Lycopodium clavatum var. clavatum [LYCOCLA0]
Lycopodium complanatum L. [LYCOCOM] (ground-cedar)
SY: *Diphasiastrum complanatum* (L.) Holub
SY: *Diphasium anceps* (Wallr.) A. & D. Löve
SY: *Diphasium complanatum* (L.) Rothm.
SY: *Diphasium complanatum* ssp. *montellii* Kukkonen
SY: *Diphasium wallrothii* H.P. Fuchs
SY: *Lycopodium anceps* Wallr.
SY: *Lycopodium complanatum* ssp. *anceps* (Wallr.) Aschers.
SY: *Lycopodium complanatum* var. *canadense* Victorin
Lycopodium complanatum ssp. *anceps* (Wallr.) Aschers. = Lycopodium complanatum [LYCOCOM]
Lycopodium complanatum var. *canadense* Victorin = Lycopodium complanatum [LYCOCOM]

Lycopodium dendroideum Michx. [LYCODEN] (ground-pine)
SY: *Lycopodium obscurum* var. *dendroideum* (Michx.) D.C. Eat.
SY: *Lycopodium obscurum* var. *hybridum* Farw.
Lycopodium haleakalae Brack. = Huperzia haleakalae [HUPEHAL]
Lycopodium inundatum L. = Lycopodiella inundata [LYCOINU]
Lycopodium lagopus (Laestad.) Zinserl. ex Kuzen = Lycopodium clavatum var. monostachyon [LYCOCLA3]
Lycopodium lucidulum var. *occidentale* (Clute) L.R. Wilson = Huperzia occidentalis [HUPEOCC]
Lycopodium lucidulum var. *tryonii* Mohlenbrock = Huperzia occidentalis [HUPEOCC]
Lycopodium miyoshianum Makino = Huperzia chinensis [HUPECHI]
Lycopodium obscurum var. *dendroideum* (Michx.) D.C. Eat. = Lycopodium dendroideum [LYCODEN]
Lycopodium obscurum var. *hybridum* Farw. = Lycopodium dendroideum [LYCODEN]
Lycopodium pungens La Pylaie ex Iljin = Lycopodium annotinum [LYCOANN]
Lycopodium sabinifolium ssp. *sitchense* (Rupr.) Calder & Taylor = Lycopodium sitchense [LYCOSIT]
Lycopodium sabinifolium var. *sitchense* (Rupr.) Fern. = Lycopodium sitchense [LYCOSIT]
Lycopodium selago ssp. *chinense* (Christ) Hultén = Huperzia chinensis [HUPECHI]
Lycopodium selago ssp. *miyoshianum* (Makino) Calder & Taylor = Huperzia chinensis [HUPECHI]
Lycopodium selago var. *miyoshianum* (Makino) Makino = Huperzia chinensis [HUPECHI]
Lycopodium sitchense Rupr. [LYCOSIT] (Alaska clubmoss)
SY: *Diphasiastrum sitchense* (Rupr.) Holub
SY: *Diphasium sitchense* (Rupr.) A. & D. Löve
SY: *Lycopodium sabinifolium* ssp. *sitchense* (Rupr.) Calder & Taylor
SY: *Lycopodium sabinifolium* var. *sitchense* (Rupr.) Fern.
Urostachys chinensis (Christ) Herter ex Nessel = Huperzia chinensis [HUPECHI]
Urostachys haleakalae (Brack.) Herter = Huperzia haleakalae [HUPEHAL]
Urostachys miyoshiana (Makino) Herter ex Nessel = Huperzia chinensis [HUPECHI]

LYTHRACEAE (V:D)

Lythrum L. [LYTHRUM$]
Lythrum alatum Pursh [LYTHALA] (winged loosestrife)
***Lythrum portula** (L.) D.A. Webber [LYTHPOR]
SY: *Peplis portula* L.
***Lythrum salicaria** L. [LYTHSAL] (purple loosestrife)
SY: *Lythrum salicaria* var. *gracilior* Turcz.
SY: *Lythrum salicaria* var. *tomentosum* (P. Mill.) DC.
SY: *Lythrum salicaria* var. *vulgare* DC.
Lythrum salicaria var. *gracilior* Turcz. = Lythrum salicaria [LYTHSAL]
Lythrum salicaria var. *tomentosum* (P. Mill.) DC. = Lythrum salicaria [LYTHSAL]
Lythrum salicaria var. *vulgare* DC. = Lythrum salicaria [LYTHSAL]
Peplis portula L. = Lythrum portula [LYTHPOR]
Rotala L. [ROTALA$]
Rotala catholica (Cham. & Schlecht.) van Leeuwen = Rotala ramosior [ROTARAM]
Rotala dentifera (Gray) Koehne = Rotala ramosior [ROTARAM]
Rotala ramosior (L.) Koehne [ROTARAM] (toothcup meadow-foam)
SY: *Rotala catholica* (Cham. & Schlecht.) van Leeuwen
SY: *Rotala dentifera* (Gray) Koehne
SY: *Rotala ramosior* var. *interior* Fern. & Grisc.
SY: *Rotala ramosior* var. *typica* Fern. & Grisc.
Rotala ramosior var. *interior* Fern. & Grisc. = Rotala ramosior [ROTARAM]
Rotala ramosior var. *typica* Fern. & Grisc. = Rotala ramosior [ROTARAM]

MALVACEAE (V:D)

Abutilon P. Mill. [ABUTILO$]
Abutilon abutilon (L.) Rusby = Abutilon theophrasti [ABUTTHE]
Abutilon avicennae Gaertn. = Abutilon theophrasti [ABUTTHE]
***Abutilon theophrasti** Medik. [ABUTTHE] (velvet-leaf)
SY: *Abutilon abutilon* (L.) Rusby
SY: *Abutilon avicennae* Gaertn.
Iliamna Greene [ILIAMNA$]
Iliamna rivularis (Dougl. ex Hook.) Greene [ILIARIV] (streambank globe-mallow)
SY: *Sphaeralcea rivularis* (Dougl. ex Hook.) Torr.
Malva L. [MALVA$]
Malva mauritiana L. = Malva sylvestris [MALVSYL]
***Malva moschata** L. [MALVMOS] (musk mallow)
***Malva neglecta** Wallr. [MALVNEG] (dwarf mallow)
***Malva parviflora** L. [MALVPAR] (small-flowered mallow)
Malva pusilla auct. = Malva rotundifolia [MALVROT]
***Malva rotundifolia** L. [MALVROT]
SY: *Malva pusilla* auct.
***Malva sylvestris** L. [MALVSYL] (common mallow)
SY: *Malva mauritiana* L.
SY: *Malva sylvestris* ssp. *mauritiana* (L.) Thellung
SY: *Malva sylvestris* var. *mauritiana* (L.) Boiss.
Malva sylvestris ssp. *mauritiana* (L.) Thellung = Malva sylvestris [MALVSYL]
Malva sylvestris var. *mauritiana* (L.) Boiss. = Malva sylvestris [MALVSYL]
Sidalcea Gray [SIDALCE$]
Sidalcea hendersonii S. Wats. [SIDAHEN] (Henderson's checker-mallow)
Sidalcea oregana (Nutt. ex Torr. & Gray) Gray [SIDAORE] (Oregon checker-mallow)
Sidalcea oregana var. **procera** C.L. Hitchc. [SIDAORE1]
Sphaeralcea St.-Hil. [SPHAERA$]
Sphaeralcea coccinea (Nutt.) Rydb. [SPHACOC] (scarlet globe-mallow)
Sphaeralcea munroana (Dougl. ex Lindl.) Spach ex Gray [SPHAMUN] (Munroe's globe-mallow)
Sphaeralcea rivularis (Dougl. ex Hook.) Torr. = Iliamna rivularis [ILIARIV]

MARSILEACEAE (V:F)

Marsilea L. [MARSILE$]
Marsilea vestita Hook. & Grev. [MARSVES] (hairy water-clover)

MENYANTHACEAE (V:D)

Fauria Franch. [FAURIA$]
Fauria crista-galli (Menzies ex Hook.) Makino [FAURCRI] (deer-cabbage)
SY: *Nephrophyllidium crista-galli* (Menzies ex Hook.) Gilg
Menyanthes L. [MENYANT$]
Menyanthes trifoliata L. [MENYTRI] (buckbean)
SY: *Menyanthes trifoliata* var. *minor* Raf.
Menyanthes trifoliata var. *minor* Raf. = Menyanthes trifoliata [MENYTRI]
Nephrophyllidium crista-galli (Menzies ex Hook.) Gilg = Fauria crista-galli [FAURCRI]

MOLLUGINACEAE (V:D)

Mollugo L. [MOLLUGO$]
Mollugo berteriana Ser. = Mollugo verticillata [MOLLVER]
***Mollugo verticillata** L. [MOLLVER] (common carpetweed)
SY: *Mollugo berteriana* Ser.

MONOTROPACEAE (V:D)

Allotropa Torr. & Gray [ALLOTRO$]

Allotropa virgata Torr. & Gray ex Gray [ALLOVIR] (candystick)
Hemitomes Gray [HEMITOM$]
Hemitomes congestum Gray [HEMICON] (gnome-plant)
Hypopitys americana (DC.) Small = Monotropa hypopithys [MONOHYP]
Hypopitys fimbriata (Gray) T.J. Howell = Monotropa hypopithys [MONOHYP]
Hypopitys insignata Bickn. = Monotropa hypopithys [MONOHYP]
Hypopitys lanuginosa (Michx.) Nutt. = Monotropa hypopithys [MONOHYP]
Hypopitys latisquama Rydb. = Monotropa hypopithys [MONOHYP]
Hypopitys monotropa Crantz = Monotropa hypopithys [MONOHYP]
Monotropa L. [MONOTRO$]
Monotropa brittonii Small = Monotropa uniflora [MONOUNI]
Monotropa hypopithys L. [MONOHYP]
SY: *Hypopitys americana* (DC.) Small
SY: *Hypopitys fimbriata* (Gray) T.J. Howell
SY: *Hypopitys insignata* Bickn.
SY: *Hypopitys lanuginosa* (Michx.) Nutt.
SY: *Hypopitys latisquama* Rydb.
SY: *Hypopitys monotropa* Crantz
SY: *Monotropa hypopithys* var. *americana* (DC.) Domin
SY: *Monotropa hypopithys* ssp. *lanuginosa* (Michx.) Hara
SY: *Monotropa hypopithys* var. *latisquama* (Rydb.) Kearney & Peebles
SY: *Monotropa hypopithys* var. *rubra* (Torr.) Farw.
SY: *Monotropa lanuginosa* Michx.
SY: *Monotropa latisquama* (Rydb.) Hultén
Monotropa hypopithys ssp. *lanuginosa* (Michx.) Hara = Monotropa hypopithys [MONOHYP]
Monotropa hypopithys var. *americana* (DC.) Domin = Monotropa hypopithys [MONOHYP]
Monotropa hypopithys var. *latisquama* (Rydb.) Kearney & Peebles = Monotropa hypopithys [MONOHYP]
Monotropa hypopithys var. *rubra* (Torr.) Farw. = Monotropa hypopithys [MONOHYP]
Monotropa lanuginosa Michx. = Monotropa hypopithys [MONOHYP]
Monotropa latisquama (Rydb.) Hultén = Monotropa hypopithys [MONOHYP]
Monotropa uniflora L. [MONOUNI] (Indian-pipe)
SY: *Monotropa brittonii* Small
Pleuricospora Gray [PLEURIC$]
Pleuricospora fimbriolata Gray [PLEUFIM]
SY: *Pleuricospora longipetala* T.J. Howell
Pleuricospora longipetala T.J. Howell = Pleuricospora fimbriolata [PLEUFIM]
Pterospora Nutt. [PTEROSP$]
Pterospora andromedea Nutt. [PTERAND] (pinedrops)

MORACEAE (V:D)

Morus L. [MORUS$]
***Morus alba** L. [MORUALB] (white mulberry)
SY: *Morus alba* var. *tatarica* (L.) Ser.
SY: *Morus tatarica* L.
Morus alba var. *tatarica* (L.) Ser. = Morus alba [MORUALB]
Morus tatarica L. = Morus alba [MORUALB]

MYRICACEAE (V:D)

Myrica L. [MYRICA$]
Myrica californica Cham. [MYRICAL] (California wax-myrtle)
Myrica gale L. [MYRIGAL] (sweet gale)

NAJADACEAE (V:M)

Caulinia flexilis Willd. = Najas flexilis [NAJAFLE]
Najas L. [NAJAS$]
Najas caespitosus (Maguire) Reveal = Najas flexilis [NAJAFLE]
Najas flexilis (Willd.) Rostk. & Schmidt [NAJAFLE] (wavy water nymph)
SY: *Caulinia flexilis* Willd.
SY: *Najas caespitosus* (Maguire) Reveal
SY: *Najas flexilis* ssp. *caespitosus* Maguire
SY: *Najas flexilis* var. *congesta* Farw.
SY: *Najas flexilis* var. *robusta* Morong
Najas flexilis ssp. *caespitosus* Maguire = Najas flexilis [NAJAFLE]
Najas flexilis var. *congesta* Farw. = Najas flexilis [NAJAFLE]
Najas flexilis var. *robusta* Morong = Najas flexilis [NAJAFLE]

NYCTAGINACEAE (V:D)

Abronia Juss. [ABRONIA$]
Abronia latifolia Eschsch. [ABROLAT] (yellow sand-verbena)
Allionia hirsuta Pursh = Mirabilis hirsuta [MIRAHIR]
Allionia nyctaginea Michx. = Mirabilis nyctaginea [MIRANYC]
Mirabilis L. [MIRABIL$]
Mirabilis hirsuta (Pursh) MacM. [MIRAHIR] (hairy umbrellawort)
SY: *Allionia hirsuta* Pursh
SY: *Oxybaphus hirsutus* (Pursh) Sweet
***Mirabilis nyctaginea** (Michx.) MacM. [MIRANYC] (umbrellawort)
SY: *Allionia nyctaginea* Michx.
SY: *Oxybaphus nyctagineus* (Michx.) Sweet
Oxybaphus hirsutus (Pursh) Sweet = Mirabilis hirsuta [MIRAHIR]
Oxybaphus nyctagineus (Michx.) Sweet = Mirabilis nyctaginea [MIRANYC]

NYMPHAEACEAE (V:D)

Castalia flava (Leitner) Greene = Nymphaea mexicana [NYMPMEX]
Castalia lekophylla Small = Nymphaea odorata [NYMPODO]
Castalia minor (Sims) Nyar = Nymphaea odorata [NYMPODO]
Castalia odorata (Ait.) Wood = Nymphaea odorata [NYMPODO]
Castalia reniformis DC. = Nymphaea odorata [NYMPODO]
Castalia tetragona (Georgi) Lawson = Nymphaea tetragona [NYMPTET]
Castalia tuberosa (Paine) Greene = Nymphaea odorata [NYMPODO]
Nuphar Sm. [NUPHAR$]
Nuphar advena var. *fraterna* (Mill. & Standl.) Standl. = Nuphar lutea ssp. variegata [NUPHLUT2]
Nuphar lutea (L.) Sm. [NUPHLUT] (yellow water-lily)
Nuphar lutea ssp. **polysepala** (Engelm.) E.O. Beal [NUPHLUT1]
SY: *Nuphar polysepala* Engelm.
SY: *Nymphaea polysepala* (Engelm.) Greene
SY: *Nymphozanthus polysepalus* (Engelm.) Fern.
Nuphar lutea ssp. **variegata** (Dur.) E.O. Beal [NUPHLUT2]
SY: *Nuphar advena* var. *fraterna* (Mill. & Standl.) Standl.
SY: *Nuphar variegata* Dur.
SY: *Nymphaea fraterna* Mill. & Standl.
Nuphar polysepala Engelm. = Nuphar lutea ssp. polysepala [NUPHLUT1]
Nuphar variegata Dur. = Nuphar lutea ssp. variegata [NUPHLUT2]
Nymphaea L. [NYMPHAE$]
***Nymphaea alba** L. [NYMPALB] (European white water-lily)
Nymphaea fraterna Mill. & Standl. = Nuphar lutea ssp. variegata [NUPHLUT2]
***Nymphaea mexicana** Zucc. [NYMPMEX] (banana water-lily)
SY: *Castalia flava* (Leitner) Greene
Nymphaea minor (Sims) DC. = Nymphaea odorata [NYMPODO]
***Nymphaea odorata** Ait. [NYMPODO] (fragrant water-lily)
SY: *Castalia lekophylla* Small
SY: *Castalia minor* (Sims) Nyar
SY: *Castalia odorata* (Ait.) Wood
SY: *Castalia reniformis* DC.
SY: *Castalia tuberosa* (Paine) Greene
SY: *Nymphaea minor* (Sims) DC.
SY: *Nymphaea odorata* var. *gigantea* Tricker
SY: *Nymphaea odorata* var. *godfreyi* Ward
SY: *Nymphaea odorata* var. *maxima* (Conrad) Boivin
SY: *Nymphaea odorata* var. *minor* Sims
SY: *Nymphaea odorata* var. *rosea* Pursh

SY: *Nymphaea odorata* var. *stenopetala* Fern.
SY: *Nymphaea odorata* var. *villosa* Caspary
SY: *Nymphaea tuberosa* Paine
Nymphaea odorata var. *gigantea* Tricker = Nymphaea odorata [NYMPODO]
Nymphaea odorata var. *godfreyi* Ward = Nymphaea odorata [NYMPODO]
Nymphaea odorata var. *maxima* (Conrad) Boivin = Nymphaea odorata [NYMPODO]
Nymphaea odorata var. *minor* Sims = Nymphaea odorata [NYMPODO]
Nymphaea odorata var. *rosea* Pursh = Nymphaea odorata [NYMPODO]
Nymphaea odorata var. *stenopetala* Fern. = Nymphaea odorata [NYMPODO]
Nymphaea odorata var. *villosa* Caspary = Nymphaea odorata [NYMPODO]
Nymphaea polysepala (Engelm.) Greene = Nuphar lutea ssp. polysepala [NUPHLUT1]
Nymphaea tetragona Georgi [NYMPTET] (pygmy water-lily)
SY: *Castalia tetragona* (Georgi) Lawson
SY: *Nymphaea tetragona* ssp. *leibergii* (Morong) Porsild
SY: *Nymphaea tetragona* var. *leibergii* (Morong) Boivin
Nymphaea tetragona ssp. *leibergii* (Morong) Porsild = Nymphaea tetragona [NYMPTET]
Nymphaea tetragona var. *leibergii* (Morong) Boivin = Nymphaea tetragona [NYMPTET]
Nymphaea tuberosa Paine = Nymphaea odorata [NYMPODO]
Nymphozanthus polysepalus (Engelm.) Fern. = Nuphar lutea ssp. polysepala [NUPHLUT1]

OLEACEAE (V:D)

Ligustrum L. [LIGUSTR$]
****Ligustrum vulgare** L. [LIGUVUL] (common privet)

ONAGRACEAE (V:D)

Boisduvalia Spach [BOISDUV$]
Boisduvalia densiflora (Lindl.) S. Wats. [BOISDEN] (dense spike-primrose)
SY: *Boisduvalia densiflora* var. *pallescens* Suksdorf
SY: *Boisduvalia densiflora* var. *salicina* (Rydb.) Munz
SY: *Boisduvalia salicina* Rydb.
SY: *Oenothera densiflora* Lindl.
Boisduvalia densiflora var. *pallescens* Suksdorf = Boisduvalia densiflora [BOISDEN]
Boisduvalia densiflora var. *salicina* (Rydb.) Munz = Boisduvalia densiflora [BOISDEN]
Boisduvalia glabella (Nutt.) Walp. [BOISGLA] (smooth spike-primrose)
SY: *Boisduvalia glabella* var. *campestris* (Jepson) Jepson
SY: *Oenothera glabella* Nutt.
Boisduvalia glabella var. *campestris* (Jepson) Jepson = Boisduvalia glabella [BOISGLA]
Boisduvalia salicina Rydb. = Boisduvalia densiflora [BOISDEN]
Boisduvalia stricta (Gray) Greene [BOISSTR] (brook spike-primrose)
SY: *Gayophytum strictum* Gray
Camissonia Link [CAMISSO$]
Camissonia andina (Nutt.) Raven [CAMIAND] (Andean evening-primrose)
SY: *Oenothera andina* Nutt.
Camissonia breviflora (Torr. & Gray) Raven [CAMIBRE] (short-flowered evening-primrose)
SY: *Oenothera breviflora* Torr. & Gray
SY: *Taraxia breviflora* (Torr. & Gray) Nutt. ex Small
Camissonia contorta (Dougl. ex Lehm.) Kearney [CAMICON] (contorted-podded evening-primrose)
SY: *Oenothera contorta* Dougl. ex Lehm.
SY: *Oenothera cruciata* (S. Wats.) Munz
Chamaenerion angustifolium (L.) Scop. = Epilobium angustifolium ssp. angustifolium [EPILANG0]
Chamaenerion latifolium (L.) Sweet = Epilobium latifolium [EPILLAT]
Chamerion angustifolium (L.) Holub = Epilobium angustifolium ssp. angustifolium [EPILANG0]
Chamerion danielsii D. Löve = Epilobium angustifolium ssp. circumvagum [EPILANG2]
Chamerion latifolium (L.) Holub = Epilobium latifolium [EPILLAT]
Chamerion platyphyllum (Daniels) A. & D. Löve = Epilobium angustifolium ssp. circumvagum [EPILANG2]
Chamerion spicatum (Lam.) S.F. Gray = Epilobium angustifolium ssp. angustifolium [EPILANG0]
Chamerion subdentatum (Rydb.) A. & D. Löve = Epilobium latifolium [EPILLAT]
Circaea L. [CIRCAEA$]
Circaea alpina L. [CIRCALP] (alpine enchanter's-nightshade)
Circaea alpina ssp. **alpina** [CIRCALP0]
Circaea alpina ssp. **pacifica** (Aschers. & Magnus) Raven [CIRCALP2] (Pacific enchanter's-nightshade)
SY: *Circaea alpina* var. *pacifica* (Aschers. & Magnus) M.E. Jones
SY: *Circaea pacifica* Aschers. & Magnus
Circaea alpina var. *pacifica* (Aschers. & Magnus) M.E. Jones = Circaea alpina ssp. pacifica [CIRCALP2]
Circaea pacifica Aschers. & Magnus = Circaea alpina ssp. pacifica [CIRCALP2]
Clarkia Pursh [CLARKIA$]
Clarkia amoena (Lehm.) A. Nels. & J.F. Macbr. [CLARAMO] (farewell-to-spring)
Clarkia amoena ssp. **caurina** (Abrams ex Piper) H.F. & M.E. Lewis [CLARAMO1]
SY: *Clarkia amoena* var. *caurina* (Abrams ex Piper) C.L. Hitchc.
SY: *Clarkia amoena* var. *pacifica* (M.E. Peck) C.L. Hitchc.
SY: *Godetia pacifica* M.E. Peck
Clarkia amoena ssp. **lindleyi** (Dougl.) H.F. & M.E. Lewis [CLARAMO2]
SY: *Clarkia amoena* var. *lindleyi* (Dougl.) C.L. Hitchc.
Clarkia amoena var. *caurina* (Abrams ex Piper) C.L. Hitchc. = Clarkia amoena ssp. caurina [CLARAMO1]
Clarkia amoena var. *lindleyi* (Dougl.) C.L. Hitchc. = Clarkia amoena ssp. lindleyi [CLARAMO2]
Clarkia amoena var. *pacifica* (M.E. Peck) C.L. Hitchc. = Clarkia amoena ssp. caurina [CLARAMO1]
Clarkia pulchella Pursh [CLARPUL] (pink fairies)
Clarkia rhomboidea Dougl. ex Hook. [CLARRHO] (common clarkia)
Epilobium L. [EPILOBI$]
Epilobium adenocaulon Hausskn. = Epilobium ciliatum ssp. ciliatum [EPILCIL0]
Epilobium adenocaulon var. *cinerascens* (Piper) M.E. Peck = Epilobium ciliatum ssp. glandulosum [EPILCIL2]
Epilobium adenocaulon var. *ecomosum* (Fassett) Munz = Epilobium ciliatum ssp. ciliatum [EPILCIL0]
Epilobium adenocaulon var. *holosericeum* (Trel.) Munz = Epilobium ciliatum ssp. ciliatum [EPILCIL0]
Epilobium adenocaulon var. *occidentale* Trel. = Epilobium ciliatum ssp. glandulosum [EPILCIL2]
Epilobium adenocaulon var. *parishii* (Trel.) Munz = Epilobium ciliatum ssp. ciliatum [EPILCIL0]
Epilobium adenocaulon var. *perplexans* Trel. = Epilobium ciliatum ssp. ciliatum [EPILCIL0]
Epilobium alpinum L. p.p. = Epilobium anagallidifolium [EPILANA]
Epilobium alpinum var. *albiflorum* (Suksdorf) C.L. Hitchc. = Epilobium clavatum [EPILCLA]
Epilobium alpinum var. *clavatum* (Trel.) C.L. Hitchc. = Epilobium clavatum [EPILCLA]
Epilobium alpinum var. *gracillimum* (Trel.) C.L. Hitchc. = Epilobium oregonense [EPILORE]
Epilobium alpinum var. *lactiflorum* (Hausskn.) C.L. Hitchc. = Epilobium lactiflorum [EPILLAC]
Epilobium alpinum var. *nutans* Hornem. = Epilobium hornemannii ssp. hornemannii [EPILHOR0]

Epilobium alpinum var. *sertulatum* (Hausskn.) Welsh = Epilobium hornemannii ssp. behringianum [EPILHOR1]
Epilobium americanum Hausskn. = Epilobium ciliatum ssp. ciliatum [EPILCIL0]
Epilobium anagallidifolium Lam. [EPILANA] (alpine willowherb)
SY: *Epilobium alpinum* L. p.p.
SY: *Epilobium anagallidifolium* var. *pseudoscaposum* (Hausskn.) Hultén
Epilobium anagallidifolium var. *pseudoscaposum* (Hausskn.) Hultén = Epilobium anagallidifolium [EPILANA]
Epilobium angustifolium L. [EPILANG] (fireweed)
Epilobium angustifolium ssp. **angustifolium** [EPILANG0]
SY: *Chamaenerion angustifolium* (L.) Scop.
SY: *Chamerion angustifolium* (L.) Holub
SY: *Chamerion spicatum* (Lam.) S.F. Gray
SY: *Epilobium angustifolium* var. *intermedium* (Lange) Fern.
Epilobium angustifolium ssp. **circumvagum** Mosquin [EPILANG2]
SY: *Chamerion danielsii* D. Löve
SY: *Chamerion platyphyllum* (Daniels) A. & D. Löve
SY: *Epilobium angustifolium* var. *abbreviatum* (Lunell) Munz
SY: *Epilobium angustifolium* var. *canescens* Wood
SY: *Epilobium angustifolium* ssp. *macrophyllum* (Hausskn.) Hultén
SY: *Epilobium angustifolium* var. *macrophyllum* (Hausskn.) Fern.
SY: *Epilobium angustifolium* var. *platyphyllum* (Daniels) Fern.
Epilobium angustifolium ssp. *macrophyllum* (Hausskn.) Hultén = Epilobium angustifolium ssp. circumvagum [EPILANG2]
Epilobium angustifolium var. *abbreviatum* (Lunell) Munz = Epilobium angustifolium ssp. circumvagum [EPILANG2]
Epilobium angustifolium var. *canescens* Wood = Epilobium angustifolium ssp. circumvagum [EPILANG2]
Epilobium angustifolium var. *intermedium* (Lange) Fern. = Epilobium angustifolium ssp. angustifolium [EPILANG0]
Epilobium angustifolium var. *macrophyllum* (Hausskn.) Fern. = Epilobium angustifolium ssp. circumvagum [EPILANG2]
Epilobium angustifolium var. *platyphyllum* (Daniels) Fern. = Epilobium angustifolium ssp. circumvagum [EPILANG2]
Epilobium behringianum Hausskn. = Epilobium hornemannii ssp. behringianum [EPILHOR1]
Epilobium boreale Hausskn. = Epilobium ciliatum ssp. glandulosum [EPILCIL2]
Epilobium brachycarpum K. Presl [EPILBRA] (tall annual willowherb)
SY: *Epilobium paniculatum* Nutt. ex Torr. & Gray
SY: *Epilobium paniculatum* var. *hammondii* (T.J. Howell) M.E. Peck
SY: *Epilobium paniculatum* var. *jucundum* (Gray) Trel.
SY: *Epilobium paniculatum* var. *laevicaule* (Rydb.) Munz
SY: *Epilobium paniculatum* var. *subulatum* (Hausskn.) Fern.
SY: *Epilobium paniculatum* var. *tracyi* (Rydb.) Munz
Epilobium brevistylum Barbey = Epilobium ciliatum ssp. ciliatum [EPILCIL0]
Epilobium brevistylum var. *subfalcatum* (Trel.) Munz = Epilobium halleanum [EPILHAL]
Epilobium brevistylum var. *tenue* (Trel.) Jepson = Epilobium halleanum [EPILHAL]
Epilobium brevistylum var. *ursinum* (Parish ex Trel.) Jepson = Epilobium ciliatum ssp. ciliatum [EPILCIL0]
Epilobium californicum Hausskn. = Epilobium ciliatum ssp. ciliatum [EPILCIL0]
Epilobium californicum var. *holosericeum* (Trel.) Munz = Epilobium ciliatum ssp. ciliatum [EPILCIL0]
Epilobium ciliatum Raf. [EPILCIL] (purple-leaved willowherb)
Epilobium ciliatum ssp. **ciliatum** [EPILCIL0]
SY: *Epilobium adenocaulon* Hausskn.
SY: *Epilobium adenocaulon* var. *ecomosum* (Fassett) Munz
SY: *Epilobium adenocaulon* var. *holosericeum* (Trel.) Munz
SY: *Epilobium adenocaulon* var. *parishii* (Trel.) Munz
SY: *Epilobium adenocaulon* var. *perplexans* Trel.
SY: *Epilobium americanum* Hausskn.
SY: *Epilobium brevistylum* Barbey
SY: *Epilobium brevistylum* var. *ursinum* (Parish ex Trel.) Jepson
SY: *Epilobium californicum* Hausskn.
SY: *Epilobium californicum* var. *holosericeum* (Trel.) Munz
SY: *Epilobium ciliatum* var. *ecomosum* (Fassett) Boivin
SY: *Epilobium delicatum* Trel.
SY: *Epilobium ecomosum* (Fassett) Fern.
SY: *Epilobium glandulosum* var. *adenocaulon* (Hausskn.) Fern.
SY: *Epilobium glandulosum* var. *macounii* (Trel.) C.L. Hitchc.
SY: *Epilobium leptocarpum* var. *macounii* Trel.
SY: *Epilobium ursinum* Parish ex Trel.
SY: *Epilobium watsonii* var. *parishii* (Trel.) C.L. Hitchc.
Epilobium ciliatum ssp. **glandulosum** (Lehm.) Hoch & Raven [EPILCIL2]
SY: *Epilobium adenocaulon* var. *cinerascens* (Piper) M.E. Peck
SY: *Epilobium adenocaulon* var. *occidentale* Trel.
SY: *Epilobium boreale* Hausskn.
SY: *Epilobium ciliatum* var. *glandulosum* (Lehm.) Dorn
SY: *Epilobium glandulosum* Lehm.
SY: *Epilobium glandulosum* var. *cardiophyllum* Fern.
SY: *Epilobium glandulosum* var. *occidentale* (Trel.) Fern.
SY: *Epilobium watsonii* var. *occidentale* (Trel.) C.L. Hitchc.
Epilobium ciliatum ssp. **watsonii** (Barbey) Hoch & Raven [EPILCIL3]
SY: *Epilobium franciscanum* Barbey
SY: *Epilobium watsonii* Barbey
SY: *Epilobium watsonii* var. *franciscanum* (Barbey) Jepson
Epilobium ciliatum var. *ecomosum* (Fassett) Boivin = Epilobium ciliatum ssp. ciliatum [EPILCIL0]
Epilobium ciliatum var. *glandulosum* (Lehm.) Dorn = Epilobium ciliatum ssp. glandulosum [EPILCIL2]
Epilobium clavatum Trel. [EPILCLA] (club-fruited willowherb)
SY: *Epilobium alpinum* var. *albiflorum* (Suksdorf) C.L. Hitchc.
SY: *Epilobium alpinum* var. *clavatum* (Trel.) C.L. Hitchc.
SY: *Epilobium clavatum* var. *glareosum* (G.N. Jones) Munz
SY: *Epilobium glareosum* G.N. Jones
Epilobium clavatum var. *glareosum* (G.N. Jones) Munz = Epilobium clavatum [EPILCLA]
Epilobium davuricum Fisch. ex Hornem. [EPILDAV] (swamp willowherb)
SY: *Epilobium palustre* var. *davuricum* (Fisch. ex Hornem.) Welsh
Epilobium delicatum Trel. = Epilobium ciliatum ssp. ciliatum [EPILCIL0]
Epilobium drummondii Hausskn. = Epilobium saximontanum [EPILSAX]
Epilobium ecomosum (Fassett) Fern. = Epilobium ciliatum ssp. ciliatum [EPILCIL0]
Epilobium foliosum (Torr. & Gray) Suksdorf [EPILFOL] (foliose willowherb)
SY: *Epilobium minutum* var. *foliosum* Torr. & Gray
Epilobium franciscanum Barbey = Epilobium ciliatum ssp. watsonii [EPILCIL3]
Epilobium glaberrimum Barbey [EPILGLA] (smooth willowherb)
Epilobium glaberrimum ssp. **fastigiatum** (Nutt.) Hoch & Raven [EPILGLA1]
SY: *Epilobium glaberrimum* var. *fastigiatum* (Nutt.) Trel. ex Jepson
SY: *Epilobium platyphyllum* Rydb.
Epilobium glaberrimum var. *fastigiatum* (Nutt.) Trel. ex Jepson = Epilobium glaberrimum ssp. fastigiatum [EPILGLA1]
Epilobium glandulosum Lehm. = Epilobium ciliatum ssp. glandulosum [EPILCIL2]
Epilobium glandulosum var. *adenocaulon* (Hausskn.) Fern. = Epilobium ciliatum ssp. ciliatum [EPILCIL0]
Epilobium glandulosum var. *brionense* Fern. = Epilobium saximontanum [EPILSAX]
Epilobium glandulosum var. *cardiophyllum* Fern. = Epilobium ciliatum ssp. glandulosum [EPILCIL2]
Epilobium glandulosum var. *macounii* (Trel.) C.L. Hitchc. = Epilobium ciliatum ssp. ciliatum [EPILCIL0]
Epilobium glandulosum var. *occidentale* (Trel.) Fern. = Epilobium ciliatum ssp. glandulosum [EPILCIL2]
Epilobium glandulosum var. *tenue* (Trel.) C.L. Hitchc. = Epilobium halleanum [EPILHAL]
Epilobium glareosum G.N. Jones = Epilobium clavatum [EPILCLA]
Epilobium halleanum Hausskn. [EPILHAL] (Hall's willowherb)

SY: *Epilobium brevistylum* var. *subfalcatum* (Trel.) Munz
SY: *Epilobium brevistylum* var. *tenue* (Trel.) Jepson
SY: *Epilobium glandulosum* var. *tenue* (Trel.) C.L. Hitchc.
SY: *Epilobium pringleanum* Hausskn.
SY: *Epilobium pringleanum* var. *tenue* (Trel.) Munz
***Epilobium hirsutum** L. [EPILHIR] (hairy willowherb)
Epilobium hornemannii Reichenb. [EPILHOR] (Hornemann's willowherb)
Epilobium hornemannii ssp. **behringianum** (Hausskn.) Hoch & Raven [EPILHOR1]
SY: *Epilobium alpinum* var. *sertulatum* (Hausskn.) Welsh
SY: *Epilobium behringianum* Hausskn.
SY: *Epilobium sertulatum* Hausskn.
Epilobium hornemannii ssp. **hornemannii** [EPILHOR0]
SY: *Epilobium alpinum* var. *nutans* Hornem.
Epilobium hornemannii var. *lactiflorum* (Hausskn.) D. Löve = Epilobium lactiflorum [EPILLAC]
Epilobium lactiflorum Hausskn. [EPILLAC] (white-flowered willowherb)
SY: *Epilobium alpinum* var. *lactiflorum* (Hausskn.) C.L. Hitchc.
SY: *Epilobium hornemannii* var. *lactiflorum* (Hausskn.) D. Löve
Epilobium latifolium L. [EPILLAT] (broad-leaved willowherb)
SY: *Chamaenerion latifolium* (L.) Sweet
SY: *Chamerion latifolium* (L.) Holub
SY: *Chamerion subdentatum* (Rydb.) A. & D. Löve
Epilobium leptocarpum Hausskn. [EPILLEP]
Epilobium leptocarpum var. *macounii* Trel. = Epilobium ciliatum ssp. ciliatum [EPILCIL0]
Epilobium leptophyllum Raf. [EPILLET] (narrow-leaved willowherb)
SY: *Epilobium nesophilum* (Fern.) Fern.
SY: *Epilobium nesophilum* var. *sabulonense* Fern.
SY: *Epilobium palustre* var. *gracile* (Farw.) Dorn
SY: *Epilobium palustre* var. *sabulonense* (Fern.) Boivin
SY: *Epilobium rosmarinifolium* Pursh
Epilobium lineare Muhl. = Epilobium palustre [EPILPAL]
Epilobium luteum Pursh [EPILLUT] (yellow willowherb)
SY: *Epilobium treleaseanum* Levl. p.p.
Epilobium minutum Lindl. ex Lehm. [EPILMIN] (small-flowered willowherb)
Epilobium minutum var. *foliosum* Torr. & Gray = Epilobium foliosum [EPILFOL]
Epilobium mirabile Trel. ex Piper [EPILMIR] (hairy-stemmed willowherb)
Epilobium nesophilum (Fern.) Fern. = Epilobium leptophyllum [EPILLET]
Epilobium nesophilum var. *sabulonense* Fern. = Epilobium leptophyllum [EPILLET]
Epilobium oliganthum Michx. = Epilobium palustre [EPILPAL]
Epilobium oregonense Hausskn. [EPILORE] (Oregon willowherb)
SY: *Epilobium alpinum* var. *gracillimum* (Trel.) C.L. Hitchc.
Epilobium palustre L. [EPILPAL]
SY: *Epilobium lineare* Muhl.
SY: *Epilobium oliganthum* Michx.
SY: *Epilobium palustre* var. *grammadophyllum* Hausskn.
SY: *Epilobium palustre* var. *labradoricum* Hausskn.
SY: *Epilobium palustre* var. *lapponicum* Wahlenb.
SY: *Epilobium palustre* var. *longirameum* Fern. & Wieg.
SY: *Epilobium palustre* var. *oliganthum* (Michx.) Fern.
SY: *Epilobium pylaieanum* Fern.
SY: *Epilobium wyomingense* A. Nels.
Epilobium palustre var. *davuricum* (Fisch. ex Hornem.) Welsh = Epilobium davuricum [EPILDAV]
Epilobium palustre var. *gracile* (Farw.) Dorn = Epilobium leptophyllum [EPILLET]
Epilobium palustre var. *grammadophyllum* Hausskn. = Epilobium palustre [EPILPAL]
Epilobium palustre var. *labradoricum* Hausskn. = Epilobium palustre [EPILPAL]
Epilobium palustre var. *lapponicum* Wahlenb. = Epilobium palustre [EPILPAL]
Epilobium palustre var. *longirameum* Fern. & Wieg. = Epilobium palustre [EPILPAL]
Epilobium palustre var. *oliganthum* (Michx.) Fern. = Epilobium palustre [EPILPAL]
Epilobium palustre var. *sabulonense* (Fern.) Boivin = Epilobium leptophyllum [EPILLET]
Epilobium paniculatum Nutt. ex Torr. & Gray = Epilobium brachycarpum [EPILBRA]
Epilobium paniculatum var. *hammondii* (T.J. Howell) M.E. Peck = Epilobium brachycarpum [EPILBRA]
Epilobium paniculatum var. *juncundum* (Gray) Trel. = Epilobium brachycarpum [EPILBRA]
Epilobium paniculatum var. *laevicaule* (Rydb.) Munz = Epilobium brachycarpum [EPILBRA]
Epilobium paniculatum var. *subulatum* (Hausskn.) Fern. = Epilobium brachycarpum [EPILBRA]
Epilobium paniculatum var. *tracyi* (Rydb.) Munz = Epilobium brachycarpum [EPILBRA]
Epilobium platyphyllum Rydb. = Epilobium glaberrimum ssp. fastigiatum [EPILGLA1]
Epilobium pringleanum Hausskn. = Epilobium halleanum [EPILHAL]
Epilobium pringleanum var. *tenue* (Trel.) Munz = Epilobium halleanum [EPILHAL]
Epilobium pylaieanum Fern. = Epilobium palustre [EPILPAL]
Epilobium rosmarinifolium Pursh = Epilobium leptophyllum [EPILLET]
Epilobium saximontanum Hausskn. [EPILSAX] (Rocky Mountain willowherb)
SY: *Epilobium drummondii* Hausskn.
SY: *Epilobium glandulosum* var. *brionense* Fern.
SY: *Epilobium scalare* Fern.
SY: *Epilobium steckerianum* Fern.
Epilobium scalare Fern. = Epilobium saximontanum [EPILSAX]
Epilobium sertulatum Hausskn. = Epilobium hornemannii ssp. behringianum [EPILHOR1]
Epilobium steckerianum Fern. = Epilobium saximontanum [EPILSAX]
Epilobium treleaseanum Levl. p.p. = Epilobium luteum [EPILLUT]
Epilobium ursinum Parish ex Trel. = Epilobium ciliatum ssp. ciliatum [EPILCIL0]
Epilobium watsonii Barbey = Epilobium ciliatum ssp. watsonii [EPILCIL3]
Epilobium watsonii var. *franciscanum* (Barbey) Jepson = Epilobium ciliatum ssp. watsonii [EPILCIL3]
Epilobium watsonii var. *occidentale* (Trel.) C.L. Hitchc. = Epilobium ciliatum ssp. glandulosum [EPILCIL2]
Epilobium watsonii var. *parishii* (Trel.) C.L. Hitchc. = Epilobium ciliatum ssp. ciliatum [EPILCIL0]
Epilobium wyomingense A. Nels. = Epilobium palustre [EPILPAL]
Gaura L. [GAURA$]
Gaura coccinea Nutt. ex Pursh [GAURCOC] (scarlet gaura)
SY: *Gaura coccinea* var. *arizonica* Munz
SY: *Gaura coccinea* var. *epilobioides* (Kunth) Munz
SY: *Gaura coccinea* var. *glabra* (Lehm.) Munz
SY: *Gaura coccinea* var. *parvifolia* (Torr.) Rickett
SY: *Gaura coccinea* var. *typica* Munz
SY: *Gaura glabra* Lehm.
SY: *Gaura odorata* Sessé ex Lag.
Gaura coccinea var. *arizonica* Munz = Gaura coccinea [GAURCOC]
Gaura coccinea var. *epilobioides* (Kunth) Munz = Gaura coccinea [GAURCOC]
Gaura coccinea var. *glabra* (Lehm.) Munz = Gaura coccinea [GAURCOC]
Gaura coccinea var. *parvifolia* (Torr.) Rickett = Gaura coccinea [GAURCOC]
Gaura coccinea var. *typica* Munz = Gaura coccinea [GAURCOC]
Gaura glabra Lehm. = Gaura coccinea [GAURCOC]
Gaura odorata Sessé ex Lag. = Gaura coccinea [GAURCOC]
Gayophytum A. Juss. [GAYOPHY$]
Gayophytum diffusum Torr. & Gray [GAYODIF] (spreading groundsmoke)
Gayophytum diffusum ssp. **parviflorum** Lewis & Szweykowski [GAYODIF1]
SY: *Gayophytum diffusum* var. *strictipes* (Hook.) Dorn
SY: *Gayophytum helleri* var. *erosulatum* Jepson

SY: *Gayophytum intermedium* Rydb.
SY: *Gayophytum lasiospermum* Greene
SY: *Gayophytum lasiospermum* var. *hoffmannii* Munz
SY: *Gayophytum nuttallii* var. *abramsii* Munz
SY: *Gayophytum nuttallii* var. *intermedium* (Rydb.) Munz
Gayophytum diffusum var. *strictipes* (Hook.) Dorn = Gayophytum diffusum ssp. parviflorum [GAYODIF1]
Gayophytum helleri var. *erosulatum* Jepson = Gayophytum diffusum ssp. parviflorum [GAYODIF1]
Gayophytum humile Juss. [GAYOHUM] (dwarf groundsmoke)
SY: *Gayophytum nuttallii* Torr. & Gray
Gayophytum intermedium Rydb. = Gayophytum diffusum ssp. parviflorum [GAYODIF1]
Gayophytum lasiospermum Greene = Gayophytum diffusum ssp. parviflorum [GAYODIF1]
Gayophytum lasiospermum var. *hoffmannii* Munz = Gayophytum diffusum ssp. parviflorum [GAYODIF1]
Gayophytum nuttallii Torr. & Gray = Gayophytum humile [GAYOHUM]
Gayophytum nuttallii var. *abramsii* Munz = Gayophytum diffusum ssp. parviflorum [GAYODIF1]
Gayophytum nuttallii var. *intermedium* (Rydb.) Munz = Gayophytum diffusum ssp. parviflorum [GAYODIF1]
Gayophytum ramosissimum Torr. & Gray [GAYORAM] (hairstem groundsmoke)
Gayophytum strictum Gray = Boisduvalia stricta [BOISSTR]
Godetia pacifica M.E. Peck = Clarkia amoena ssp. caurina [CLARAMO1]
Isnardia palustris L. = Ludwigia palustris [LUDWPAL]
Kneiffia perennis (L.) Pennell = Oenothera perennis [OENOPER]
Kneiffia pumila (L.) Spach = Oenothera perennis [OENOPER]
Ludwigia L. [LUDWIGI$]
Ludwigia palustris (L.) Ell. [LUDWPAL] (water-purslane)
SY: *Isnardia palustris* L.
SY: *Ludwigia palustris* var. *americana* (DC.) Fern. & Grisc.
SY: *Ludwigia palustris* var. *nana* Fern. & Grisc.
SY: *Ludwigia palustris* var. *pacifica* Fern. & Grisc.
Ludwigia palustris var. *americana* (DC.) Fern. & Grisc. = Ludwigia palustris [LUDWPAL]
Ludwigia palustris var. *nana* Fern. & Grisc. = Ludwigia palustris [LUDWPAL]
Ludwigia palustris var. *pacifica* Fern. & Grisc. = Ludwigia palustris [LUDWPAL]
Oenothera L. [OENOTHE$]
Oenothera andina Nutt. = Camissonia andina [CAMIAND]
***Oenothera biennis** L. [OENOBIE] (common evening-primrose)
SY: *Oenothera biennis* ssp. *caeciarum* Munz
SY: *Oenothera biennis* ssp. *centralis* Munz
SY: *Oenothera biennis* var. *pycnocarpa* (Atkinson & Bartlett) Wieg.
SY: *Oenothera muricata* L.
SY: *Oenothera pycnocarpa* Atkinson & Bartlett
Oenothera biennis ssp. *caeciarum* Munz = Oenothera biennis [OENOBIE]
Oenothera biennis ssp. *centralis* Munz = Oenothera biennis [OENOBIE]
Oenothera biennis var. *pycnocarpa* (Atkinson & Bartlett) Wieg. = Oenothera biennis [OENOBIE]
Oenothera biennis var. *strigosa* (Rydb.) Piper = Oenothera villosa ssp. strigosa [OENOVIL1]
Oenothera breviflora Torr. & Gray = Camissonia breviflora [CAMIBRE]
Oenothera cheradophila Bartlett = Oenothera villosa ssp. strigosa [OENOVIL1]
Oenothera contorta Dougl. ex Lehm. = Camissonia contorta [CAMICON]
Oenothera cruciata (S. Wats.) Munz = Camissonia contorta [CAMICON]
Oenothera densiflora Lindl. = Boisduvalia densiflora [BOISDEN]
Oenothera depressa ssp. *strigosa* (Rydb.) Taylor & MacBryde = Oenothera villosa ssp. strigosa [OENOVIL1]
Oenothera erythrosepala Borbás = Oenothera glazioviana [OENOGLA]
Oenothera glabella Nutt. = Boisduvalia glabella [BOISGLA]
***Oenothera glazioviana** Micheli [OENOGLA] (red-sepaled evening-primrose)
SY: *Oenothera erythrosepala* Borbás
Oenothera muricata L. = Oenothera biennis [OENOBIE]
Oenothera pallida Lindl. [OENOPAL] (white-pole evening-primrose)
SY: *Oenothera pallida* var. *idahoensis* Munz
SY: *Oenothera pallida* var. *typica* Munz
Oenothera pallida var. *idahoensis* Munz = Oenothera pallida [OENOPAL]
Oenothera pallida var. *typica* Munz = Oenothera pallida [OENOPAL]
***Oenothera perennis** L. [OENOPER] (perennial sundrops)
SY: *Kneiffia perennis* (L.) Pennell
SY: *Kneiffia pumila* (L.) Spach
SY: *Oenothera perennis* var. *rectipilis* (Blake) Blake
SY: *Oenothera perennis* var. *typica* Munz
Oenothera perennis var. *rectipilis* (Blake) Blake = Oenothera perennis [OENOPER]
Oenothera perennis var. *typica* Munz = Oenothera perennis [OENOPER]
Oenothera procera Woot. & Standl. = Oenothera villosa ssp. strigosa [OENOVIL1]
Oenothera pycnocarpa Atkinson & Bartlett = Oenothera biennis [OENOBIE]
Oenothera rydbergii House = Oenothera villosa ssp. strigosa [OENOVIL1]
Oenothera strigosa (Rydb.) Mackenzie & Bush = Oenothera villosa ssp. strigosa [OENOVIL1]
Oenothera strigosa ssp. *cheradophila* (Bartlett) Munz = Oenothera villosa ssp. strigosa [OENOVIL1]
Oenothera villosa Thunb. [OENOVIL] (yellow evening-primrose)
Oenothera villosa ssp. *cheradophila* (Bartlett) W. Dietr. & Raven = Oenothera villosa ssp. strigosa [OENOVIL1]
Oenothera villosa ssp. **strigosa** (Rydb.) W. Dietr. & Raven [OENOVIL1]
SY: *Oenothera biennis* var. *strigosa* (Rydb.) Piper
SY: *Oenothera cheradophila* Bartlett
SY: *Oenothera depressa* ssp. *strigosa* (Rydb.) Taylor & MacBryde
SY: *Oenothera procera* Woot. & Standl.
SY: *Oenothera rydbergii* House
SY: *Oenothera strigosa* (Rydb.) Mackenzie & Bush
SY: *Oenothera strigosa* ssp. *cheradophila* (Bartlett) Munz
SY: *Oenothera villosa* ssp. *cheradophila* (Bartlett) W. Dietr. & Raven
SY: *Oenothera villosa* var. *strigosa* (Rydb.) Dorn
Oenothera villosa var. *strigosa* (Rydb.) Dorn = Oenothera villosa ssp. strigosa [OENOVIL1]
Taraxia breviflora (Torr. & Gray) Nutt. ex Small = Camissonia breviflora [CAMIBRE]

OPHIOGLOSSACEAE (V:F)

Botrychium Sw. [BOTRICH$]
Botrychium boreale auct. = Botrychium pinnatum [BOTRPIN]
Botrychium boreale var. *obtusilobum* auct. = Botrychium pinnatum [BOTRPIN]
Botrychium californicum Underwood = Botrychium multifidum [BOTRMUL]
Botrychium coulteri Underwood = Botrychium multifidum [BOTRMUL]
Botrychium lanceolatum (Gmel.) Angstr. [BOTRLAN] (lance-leaved moonwort)
Botrychium lunaria (L.) Sw. [BOTRLUN] (common moonwort)
SY: *Botrychium lunaria* var. *onondagense* (Underwood) House
SY: *Botrychium onondagense* Underwood
Botrychium lunaria ssp. *minganense* (Victorin) Calder & Taylor = Botrychium minganense [BOTRMIN]
Botrychium lunaria var. *minganense* (Victorin) Dole = Botrychium minganense [BOTRMIN]
Botrychium lunaria var. *onondagense* (Underwood) House = Botrychium lunaria [BOTRLUN]

Botrychium matricariae (Schrank) Spreng. = Botrychium multifidum [BOTRMUL]
Botrychium minganense Victorin [BOTRMIN] (Mingan moonwort)
SY: *Botrychium lunaria* ssp. *minganense* (Victorin) Calder & Taylor
SY: *Botrychium lunaria* var. *minganense* (Victorin) Dole
Botrychium multifidum (Gmel.) Trev. [BOTRMUL] (leathery grape fern)
SY: *Botrychium californicum* Underwood
SY: *Botrychium coulteri* Underwood
SY: *Botrychium matricariae* (Schrank) Spreng.
SY: *Botrychium multifidum* ssp. *californicum* (Underwood) Clausen
SY: *Botrychium multifidum* var. *californicum* (Underwood) Broun
SY: *Botrychium multifidum* ssp. *coulteri* (Underwood) Clausen
SY: *Botrychium multifidum* var. *coulteri* (Underwood) Broun
SY: *Botrychium multifidum* var. *intermedium* (D.C. Eat.) Farw.
SY: *Botrychium multifidum* var. *robustum* (Rupr.) C. Christens.
SY: *Botrychium multifidum* ssp. *silaifolium* (K. Presl) Clausen
SY: *Botrychium multifidum* var. *silaifolium* (K. Presl) Broun
SY: *Botrychium occidentale* Underwood
SY: *Botrychium silaifolium* K. Presl
SY: *Botrychium silaifolium* var. *coulteri* (Underwood) Jepson
SY: *Botrychium ternatum* var. *intermedium* Eaton
SY: *Sceptridium multifidum* (Gmel.) Tagawa
Botrychium multifidum ssp. *californicum* (Underwood) Clausen = Botrychium multifidum [BOTRMUL]
Botrychium multifidum ssp. *coulteri* (Underwood) Clausen = Botrychium multifidum [BOTRMUL]
Botrychium multifidum ssp. *silaifolium* (K. Presl) Clausen = Botrychium multifidum [BOTRMUL]
Botrychium multifidum var. *californicum* (Underwood) Broun = Botrychium multifidum [BOTRMUL]
Botrychium multifidum var. *coulteri* (Underwood) Broun = Botrychium multifidum [BOTRMUL]
Botrychium multifidum var. *intermedium* (D.C. Eat.) Farw. = Botrychium multifidum [BOTRMUL]
Botrychium multifidum var. *robustum* (Rupr.) C. Christens. = Botrychium multifidum [BOTRMUL]
Botrychium multifidum var. *silaifolium* (K. Presl) Broun = Botrychium multifidum [BOTRMUL]
Botrychium occidentale Underwood = Botrychium multifidum [BOTRMUL]
Botrychium onondagense Underwood = Botrychium lunaria [BOTRLUN]
Botrychium paradoxum W.H. Wagner [BOTRPAR] (two-spiked moonwort)
Botrychium pinnatum St. John [BOTRPIN] (northwestern moonwort)
SY: *Botrychium boreale* auct.
SY: *Botrychium boreale* var. *obtusilobum* auct.
Botrychium silaifolium K. Presl = Botrychium multifidum [BOTRMUL]
Botrychium silaifolium var. *coulteri* (Underwood) Jepson = Botrychium multifidum [BOTRMUL]
Botrychium simplex E. Hitchc. [BOTRSIM] (least moonwort)
Botrychium ternatum var. *intermedium* Eaton = Botrychium multifidum [BOTRMUL]
Botrychium virginianum (L.) Sw. [BOTRVIR] (Virginia grape fern)
SY: *Botrychium virginianum* ssp. *europaeum* (Angstr.) Jáv.
SY: *Botrychium virginianum* var. *europaeum* Angstr.
SY: *Botrypus virginianus* (L.) Michx.
Botrychium virginianum ssp. *europaeum* (Angstr.) Jáv. = Botrychium virginianum [BOTRVIR]
Botrychium virginianum var. *europaeum* Angstr. = Botrychium virginianum [BOTRVIR]
Botrypus virginianus (L.) Michx. = Botrychium virginianum [BOTRVIR]
Ophioglossum L. [OPHIOGL$]
Ophioglossum alaskanum Britton = Ophioglossum pusillum [OPHIPUS]
Ophioglossum pusillum Raf. [OPHIPUS] (northern adder's-tongue)
SY: *Ophioglossum alaskanum* Britton
SY: *Ophioglossum vulgatum* var. *alaskanum* (E.G. Britt.) C. Christens.
SY: *Ophioglossum vulgatum* var. *pseudopodum* (Blake) Farw.
SY: *Ophioglossum vulgatum* auct.
Ophioglossum vulgatum auct. = Ophioglossum pusillum [OPHIPUS]
Ophioglossum vulgatum var. *alaskanum* (E.G. Britt.) C. Christens. = Ophioglossum pusillum [OPHIPUS]
Ophioglossum vulgatum var. *pseudopodum* (Blake) Farw. = Ophioglossum pusillum [OPHIPUS]
Sceptridium multifidum (Gmel.) Tagawa = Botrychium multifidum [BOTRMUL]

ORCHIDACEAE (V:M)

Amerorchis Hultén [AMERORC$]
Amerorchis rotundifolia (Banks ex Pursh) Hultén [AMERROT] (round-leaved orchis)
SY: *Orchis rotundifolia* Banks ex Pursh
SY: *Orchis rotundifolia* var. *lineata* Mousley
Calypso Salisb. [CALYPSO$]
Calypso bulbosa (L.) Oakes [CALYBUL] (Venus'-slipper)
Calypso bulbosa ssp. *occidentalis* (Holz.) Calder & Taylor = Calypso bulbosa var. occidentalis [CALYBUL3]
Calypso bulbosa var. **occidentalis** (Holz.) Boivin [CALYBUL3]
SY: *Calypso bulbosa* ssp. *occidentalis* (Holz.) Calder & Taylor
Cephalanthera L.C. Rich. [CEPHALA$]
Cephalanthera austiniae (Gray) Heller [CEPHAUS] (phantom orchid)
SY: *Eburophyton austiniae* (Gray) Heller
Coeloglossum Hartman [COELOGL$]
Coeloglossum bracteatum (Muhl. ex Willd.) Parl. = Coeloglossum viride var. virescens [COELVIR1]
Coeloglossum viride (L.) Hartman [COELVIR] (long-bracted frog orchid)
Coeloglossum viride ssp. *bracteatum* (Muhl. ex Willd.) Hultén = Coeloglossum viride var. virescens [COELVIR1]
Coeloglossum viride var. **virescens** (Muhl. ex Willd.) Luer [COELVIR1]
SY: *Coeloglossum bracteatum* (Muhl. ex Willd.) Parl.
SY: *Coeloglossum viride* ssp. *bracteatum* (Muhl. ex Willd.) Hultén
SY: *Habenaria bracteata* (Muhl. ex Willd.) R. Br. ex Ait. f.
SY: *Habenaria viridis* var. *bracteata* (Muhl. ex Willd.) Reichenb. ex Gray
Corallorrhiza Gagnebin [CORALLO$]
Corallorrhiza corallorrhiza (L.) Karst. = Corallorrhiza trifida [CORATRI]
Corallorrhiza maculata (Raf.) Raf. [CORAMAC] (spotted coralroot)
SY: *Corallorrhiza maculata* var. *flavida* (M.E. Peck) Cockerell
SY: *Corallorrhiza maculata* var. *immaculata* M.E. Peck
SY: *Corallorrhiza maculata* var. *intermedia* Farw.
SY: *Corallorrhiza maculata* var. *occidentalis* (Lindl.) Cockerell
SY: *Corallorrhiza maculata* var. *punicea* Bartlett
SY: *Corallorrhiza multiflora* Nutt.
Corallorrhiza maculata ssp. *mertensiana* (Bong.) Calder & Taylor = Corallorrhiza mertensiana [CORAMER]
Corallorrhiza maculata var. *flavida* (M.E. Peck) Cockerell = Corallorrhiza maculata [CORAMAC]
Corallorrhiza maculata var. *immaculata* M.E. Peck = Corallorrhiza maculata [CORAMAC]
Corallorrhiza maculata var. *intermedia* Farw. = Corallorrhiza maculata [CORAMAC]
Corallorrhiza maculata var. *occidentalis* (Lindl.) Cockerell = Corallorrhiza maculata [CORAMAC]
Corallorrhiza maculata var. *punicea* Bartlett = Corallorrhiza maculata [CORAMAC]
Corallorrhiza mertensiana Bong. [CORAMER]
SY: *Corallorrhiza maculata* ssp. *mertensiana* (Bong.) Calder & Taylor
Corallorrhiza multiflora Nutt. = Corallorrhiza maculata [CORAMAC]
Corallorrhiza striata Lindl. [CORASTR] (striped coralroot)
Corallorrhiza trifida Chatelain [CORATRI] (yellow coralroot)
SY: *Corallorrhiza corallorrhiza* (L.) Karst.

SY: *Corallorrhiza trifida* var. *verna* (Nutt.) Fern.
Corallorrhiza trifida var. *verna* (Nutt.) Fern. = Corallorrhiza trifida [CORATRI]
Cypripedium L. [CYPRIPE$]
Cypripedium calceolus ssp. *parviflorum* (Salisb.) Hultén = Cypripedium parviflorum [CYPRPAR]
Cypripedium calceolus var. *parviflorum* (Salisb.) Fern. = Cypripedium parviflorum [CYPRPAR]
Cypripedium montanum Dougl. ex Lindl. [CYPRMON] (mountain lady's-slipper)
Cypripedium parviflorum Salisb. [CYPRPAR]
SY: *Cypripedium calceolus* ssp. *parviflorum* (Salisb.) Hultén
SY: *Cypripedium calceolus* var. *parviflorum* (Salisb.) Fern.
Cypripedium passerinum Richards. [CYPRPAS] (sparrow's-egg lady's-slipper)
SY: *Cypripedium passerinum* var. *minganense* Victorin
Cypripedium passerinum var. *minganense* Victorin = Cypripedium passerinum [CYPRPAS]
Eburophyton austiniae (Gray) Heller = Cephalanthera austiniae [CEPHAUS]
Epipactis Zinn [EPIPACT$]
Epipactis gigantea Dougl. ex Hook. [EPIPGIG] (giant helleborine)
*****Epipactis helleborine** (L.) Crantz [EPIPHEL] (helleborine)
SY: *Epipactis latifolia* (L.) All.
SY: *Serapias helleborine* L.
Epipactis latifolia (L.) All. = Epipactis helleborine [EPIPHEL]
Goodyera R. Br. ex Ait. f. [GOODYER$]
Goodyera decipiens (Hook.) F.T. Hubbard = Goodyera oblongifolia [GOODOBL]
Goodyera oblongifolia Raf. [GOODOBL] (rattlesnake-plantain)
SY: *Goodyera decipiens* (Hook.) F.T. Hubbard
SY: *Goodyera oblongifolia* var. *reticulata* Boivin
SY: *Peramium decipiens* (Hook.) Piper
Goodyera oblongifolia var. *reticulata* Boivin = Goodyera oblongifolia [GOODOBL]
Goodyera ophioides (Fern.) Rydb. = Goodyera repens [GOODREP]
Goodyera repens (L.) R. Br. ex Ait. f. [GOODREP] (lesser rattlesnake-plantain)
SY: *Goodyera ophioides* (Fern.) Rydb.
SY: *Goodyera repens* ssp. *ophioides* (Fern.) A. Löve & Simon
SY: *Goodyera repens* var. *ophioides* Fern.
SY: *Peramium ophioides* (Fern.) Rydb.
Goodyera repens ssp. *ophioides* (Fern.) A. Löve & Simon = Goodyera repens [GOODREP]
Goodyera repens var. *ophioides* Fern. = Goodyera repens [GOODREP]
Habenaria bracteata (Muhl. ex Willd.) R. Br. ex Ait. f. = Coeloglossum viride var. virescens [COELVIR1]
Habenaria chorisiana Cham. = Platanthera chorisiana [PLATCHO]
Habenaria dilatata (Pursh) Hook. = Platanthera dilatata var. dilatata [PLATDIL0]
Habenaria dilatata var. *albiflora* (Cham.) Correll = Platanthera dilatata var. albiflora [PLATDIL1]
Habenaria dilatata var. *leucostachys* (Lindl.) Ames = Platanthera leucostachys [PLATLEU]
Habenaria elegans (Lindl.) Boland. = Piperia elegans [PIPEELE]
Habenaria elegans var. *maritima* (Greene) Ames = Piperia elegans [PIPEELE]
Habenaria greenei Jepson = Piperia elegans [PIPEELE]
Habenaria hyperborea (L.) R. Br. ex Ait. f. = Platanthera hyperborea [PLATHYP]
Habenaria leucostachys (Lindl.) S. Wats. = Platanthera leucostachys [PLATLEU]
Habenaria maritima Greene = Piperia elegans [PIPEELE]
Habenaria obtusata (Banks ex Pursh) Richards. = Platanthera obtusata [PLATOBT]
Habenaria obtusata var. *collectanea* Fern. = Platanthera obtusata [PLATOBT]
Habenaria orbiculata (Pursh) Torr. = Platanthera orbiculata [PLATORB]
Habenaria saccata Greene = Platanthera stricta [PLATSTR]
Habenaria stricta (Lindl.) Rydb. = Platanthera stricta [PLATSTR]
Habenaria unalascensis (Spreng.) S. Wats. = Piperia unalascensis [PIPEUNA]
Habenaria unalascensis var. *maritima* (Greene) Correll = Piperia elegans [PIPEELE]
Habenaria viridis var. *bracteata* (Muhl. ex Willd.) Reichenb. ex Gray = Coeloglossum viride var. virescens [COELVIR1]
Hammarbya paludosa (L.) Kuntze = Malaxis paludosa [MALAPAL]
Ibidium strictum (Rydb.) House = Spiranthes romanzoffiana [SPIRROM]
Limnorchis dilatata (Pursh) Rydb. = Platanthera dilatata var. dilatata [PLATDIL0]
Limnorchis dilatata ssp. *albiflora* (Cham.) A. Löve & Simon = Platanthera dilatata var. albiflora [PLATDIL1]
Limnorchis graminifolia Rydb. = Platanthera leucostachys [PLATLEU]
Limnorchis saccata (Greene) A. Löve & Simon = Platanthera stricta [PLATSTR]
Limnorchis stricta (Lindl.) Rydb. = Platanthera stricta [PLATSTR]
Liparis L.C. Rich. [LIPARIS$]
Liparis loeselii (L.) L.C. Rich. [LIPALOE] (Loesel's liparis)
Listera R. Br. ex Ait. f. [LISTERA$]
Listera banksiana auct. = Listera caurina [LISTCAU]
Listera borealis Morong [LISTBOR] (northern twayblade)
Listera caurina Piper [LISTCAU] (northwestern twayblade)
SY: *Ophrys caurina* (Piper) Rydb.
SY: *Listera banksiana* auct.
Listera convallarioides (Sw.) Nutt. ex Ell. [LISTCON] (broad-leaved twayblade)
SY: *Ophrys convallarioides* (Sw.) W. Wight
Listera cordata (L.) R. Br. ex Ait. f. [LISTCOR] (heart-leaved twayblade)
Lysiella obtusata (Banks ex Pursh) Rydb. = Platanthera obtusata [PLATOBT]
Malaxis Soland. ex Sw. [MALAXIS$]
Malaxis brachypoda (Gray) Fern. [MALABRA]
SY: *Malaxis monophyllos* ssp. *brachypoda* (Gray) A. & D. Löve
SY: *Malaxis monophyllos* var. *brachypoda* (Gray) F. Morris & Eames
SY: *Malaxis monophyllos* auct.
Malaxis diphyllos Cham. [MALADIP]
SY: *Malaxis monophyllos* var. *diphyllos* (Cham.) Luer
Malaxis monophyllos auct. = Malaxis brachypoda [MALABRA]
Malaxis monophyllos ssp. *brachypoda* (Gray) A. & D. Löve = Malaxis brachypoda [MALABRA]
Malaxis monophyllos var. *brachypoda* (Gray) F. Morris & Eames = Malaxis brachypoda [MALABRA]
Malaxis monophyllos var. *diphyllos* (Cham.) Luer = Malaxis diphyllos [MALADIP]
Malaxis paludosa (L.) Sw. [MALAPAL]
SY: *Hammarbya paludosa* (L.) Kuntze
Ophrys caurina (Piper) Rydb. = Listera caurina [LISTCAU]
Ophrys convallarioides (Sw.) W. Wight = Listera convallarioides [LISTCON]
Orchis rotundifolia Banks ex Pursh = Amerorchis rotundifolia [AMERROT]
Orchis rotundifolia var. *lineata* Mousley = Amerorchis rotundifolia [AMERROT]
Peramium decipiens (Hook.) Piper = Goodyera oblongifolia [GOODOBL]
Peramium ophioides (Fern.) Rydb. = Goodyera repens [GOODREP]
Piperia Rydb. [PIPERIA$]
Piperia elegans (Lindl.) Rydb. [PIPEELE] (elegant rein orchid)
SY: *Habenaria elegans* (Lindl.) Boland.
SY: *Habenaria elegans* var. *maritima* (Greene) Ames
SY: *Habenaria greenei* Jepson
SY: *Habenaria maritima* Greene
SY: *Habenaria unalascensis* var. *maritima* (Greene) Correll
SY: *Piperia maritima* (Greene) Rydb.
SY: *Piperia multiflora* Rydb.
SY: *Platanthera unalascensis* ssp. *maritima* (Greene) de Filipps
SY: *Platanthera unalascensis* var. *maritima* (Greene) Correll
Piperia maritima (Greene) Rydb. = Piperia elegans [PIPEELE]
Piperia multiflora Rydb. = Piperia elegans [PIPEELE]

Piperia unalascensis (Spreng.) Rydb. [PIPEUNA] (Alaska rein orchid)
SY: *Habenaria unalascensis* (Spreng.) S. Wats.
SY: *Platanthera cooperi* (S. Wats.) Rydb.
SY: *Platanthera unalascensis* (Spreng.) Kurtz
SY: *Spiranthes unalascensis* Spreng.
Platanthera L.C. Rich. [PLATANT$]
Platanthera chorisiana (Cham.) Reichenb. [PLATCHO] (small bog orchid)
SY: *Habenaria chorisiana* Cham.
Platanthera cooperi (S. Wats.) Rydb. = Piperia unalascensis [PIPEUNA]
Platanthera dilatata (Pursh) Lindl. ex Beck [PLATDIL] (white bog orchid)
Platanthera dilatata var. **albiflora** (Cham.) Ledeb. [PLATDIL1]
SY: *Habenaria dilatata* var. *albiflora* (Cham.) Correll
SY: *Limnorchis dilatata* ssp. *albiflora* (Cham.) A. Löve & Simon
Platanthera dilatata var. *angustifolia* Hook. = Platanthera dilatata var. dilatata [PLATDIL0]
Platanthera dilatata var. **dilatata** [PLATDIL0]
SY: *Habenaria dilatata* (Pursh) Hook.
SY: *Limnorchis dilatata* (Pursh) Rydb.
SY: *Platanthera dilatata* var. *angustifolia* Hook.
Platanthera dilatata var. *gracilis* Ledeb. = Platanthera stricta [PLATSTR]
Platanthera dilatata var. *leucostachys* (Lindl.) Luer = Platanthera leucostachys [PLATLEU]
Platanthera hyperborea (L.) Lindl. [PLATHYP] (green-flowered bog orchid)
SY: *Habenaria hyperborea* (L.) R. Br. ex Ait. f.
Platanthera hyperborea var. *purpurascens* (Rydb.) Luer = Platanthera stricta [PLATSTR]
Platanthera leucostachys Lindl. [PLATLEU]
SY: *Habenaria dilatata* var. *leucostachys* (Lindl.) Ames
SY: *Habenaria leucostachys* (Lindl.) S. Wats.
SY: *Limnorchis graminifolia* Rydb.
SY: *Platanthera dilatata* var. *leucostachys* (Lindl.) Luer
Platanthera obtusata (Banks ex Pursh) Lindl. [PLATOBT] (one-leaved rein orchid)
SY: *Habenaria obtusata* (Banks ex Pursh) Richards.
SY: *Habenaria obtusata* var. *collectanea* Fern.
SY: *Lysiella obtusata* (Banks ex Pursh) Rydb.
SY: *Platanthera obtusata* ssp. *oligantha* (Turcz.) Hultén
Platanthera obtusata ssp. *oligantha* (Turcz.) Hultén = Platanthera obtusata [PLATOBT]
Platanthera orbiculata (Pursh) Lindl. [PLATORB] (round-leaved rein orchid)
SY: *Habenaria orbiculata* (Pursh) Torr.
Platanthera saccata (Greene) Hultén = Platanthera stricta [PLATSTR]
Platanthera stricta Lindl. [PLATSTR] (slender bog orchid)
SY: *Habenaria saccata* Greene
SY: *Habenaria stricta* (Lindl.) Rydb.
SY: *Limnorchis saccata* (Greene) A. Löve & Simon
SY: *Limnorchis stricta* (Lindl.) Rydb.
SY: *Platanthera dilatata* var. *gracilis* Ledeb.
SY: *Platanthera hyperborea* var. *purpurascens* (Rydb.) Luer
SY: *Platanthera saccata* (Greene) Hultén
Platanthera unalascensis (Spreng.) Kurtz = Piperia unalascensis [PIPEUNA]
Platanthera unalascensis ssp. *maritima* (Greene) de Filipps = Piperia elegans [PIPEELE]
Platanthera unalascensis var. *maritima* (Greene) Correll = Piperia elegans [PIPEELE]
Serapias helleborine L. = Epipactis helleborine [EPIPHEL]
Spiranthes L.C. Rich. [SPIRANT$]
Spiranthes romanzoffiana Cham. [SPIRROM] (hooded ladies' tresses)
SY: *Ibidium strictum* (Rydb.) House
SY: *Spiranthes stricta* Rydb.
Spiranthes stricta Rydb. = Spiranthes romanzoffiana [SPIRROM]
Spiranthes unalascensis Spreng. = Piperia unalascensis [PIPEUNA]

OROBANCHACEAE (V:D)

Anoplanthus fasciculatus (Nutt.) Walp. = Orobanche fasciculata [OROBFAS]
Aphyllon pinorum (Geyer ex Hook.) Gray = Orobanche pinorum [OROBPIN]
Boschniakia C.A. Mey. ex Bong. [BOSCHNI$]
Boschniakia hookeri Walp. [BOSCHOO] (Vancouver groundcone)
Boschniakia rossica (Cham. & Schlecht.) Fedtsch. [BOSCROS] (northern groundcone)
Boschniakia strobilacea Gray [BOSCSTR]
Myzorrhiza californica (Cham. & Schlecht.) Rydb. = Orobanche californica [OROBCAL]
Myzorrhiza pinorum (Geyer ex Hook.) Rydb. = Orobanche pinorum [OROBPIN]
Orobanche L. [OROBANC$]
Orobanche californica Cham. & Schlecht. [OROBCAL] (California broomrape)
SY: *Myzorrhiza californica* (Cham. & Schlecht.) Rydb.
SY: *Orobanche grayana* var. *nelsonii* Munz
SY: *Orobanche grayana* var. *violacea* (Eastw.) Munz
Orobanche corymbosa (Rydb.) Ferris [OROBCOR] (flat-topped broomrape)
Orobanche corymbosa ssp. **mutabilis** Heckard [OROBCOR1]
Orobanche fasciculata Nutt. [OROBFAS] (clustered broomrape)
SY: *Anoplanthus fasciculatus* (Nutt.) Walp.
SY: *Orobanche fasciculata* var. *franciscana* Achey
SY: *Orobanche fasciculata* var. *lutea* (Parry) Achey
SY: *Orobanche fasciculata* var. *subulata* Goodman
SY: *Orobanche fasciculata* var. *typica* Achey
SY: *Thalesia fasciculata* (Nutt.) Britt.
SY: *Thalesia lutea* (Parry) Rydb.
Orobanche fasciculata var. *franciscana* Achey = Orobanche fasciculata [OROBFAS]
Orobanche fasciculata var. *lutea* (Parry) Achey = Orobanche fasciculata [OROBFAS]
Orobanche fasciculata var. *subulata* Goodman = Orobanche fasciculata [OROBFAS]
Orobanche fasciculata var. *typica* Achey = Orobanche fasciculata [OROBFAS]
Orobanche grayana var. *nelsonii* Munz = Orobanche californica [OROBCAL]
Orobanche grayana var. *violacea* (Eastw.) Munz = Orobanche californica [OROBCAL]
Orobanche pinorum Geyer ex Hook. [OROBPIN] (pine broomrape)
SY: *Aphyllon pinorum* (Geyer ex Hook.) Gray
SY: *Myzorrhiza pinorum* (Geyer ex Hook.) Rydb.
SY: *Phelipaea pinorum* (Geyer ex Hook.) Gray
Orobanche porphyrantha G. Beck = Orobanche uniflora [OROBUNI]
Orobanche purpurea Jacq. = Orobanche uniflora [OROBUNI]
Orobanche terrae-novae Fern. = Orobanche uniflora [OROBUNI]
Orobanche uniflora L. [OROBUNI] (naked broomrape)
SY: *Orobanche porphyrantha* G. Beck
SY: *Orobanche purpurea* Jacq.
SY: *Orobanche terrae-novae* Fern.
SY: *Orobanche uniflora* var. *minuta* (Suksdorf) G. Beck
SY: *Orobanche uniflora* ssp. *occidentalis* (Greene) Abrams ex Ferris
SY: *Orobanche uniflora* var. *occidentalis* (Greene) Taylor & MacBryde
SY: *Orobanche uniflora* var. *purpurea* (Heller) Achey
SY: *Orobanche uniflora* var. *sedii* (Suksdorf) Achey
SY: *Orobanche uniflora* var. *terrae-novae* (Fern.) Munz
SY: *Orobanche uniflora* var. *typica* Achey
SY: *Thalesia uniflora* (L.) Britt.
Orobanche uniflora ssp. *occidentalis* (Greene) Abrams ex Ferris = Orobanche uniflora [OROBUNI]
Orobanche uniflora var. *minuta* (Suksdorf) G. Beck = Orobanche uniflora [OROBUNI]
Orobanche uniflora var. *occidentalis* (Greene) Taylor & MacBryde = Orobanche uniflora [OROBUNI]
Orobanche uniflora var. *purpurea* (Heller) Achey = Orobanche uniflora [OROBUNI]

Orobanche uniflora var. *sedii* (Suksdorf) Achey = Orobanche uniflora [OROBUNI]
Orobanche uniflora var. *terrae-novae* (Fern.) Munz = Orobanche uniflora [OROBUNI]
Orobanche uniflora var. *typica* Achey = Orobanche uniflora [OROBUNI]
Phelipaea pinorum (Geyer ex Hook.) Gray = Orobanche pinorum [OROBPIN]
Thalesia fasciculata (Nutt.) Britt. = Orobanche fasciculata [OROBFAS]
Thalesia lutea (Parry) Rydb. = Orobanche fasciculata [OROBFAS]
Thalesia uniflora (L.) Britt. = Orobanche uniflora [OROBUNI]

OXALIDACEAE (V:D)

Acetosella corniculata (L.) Kuntze = Oxalis corniculata [OXALCOR]
Ceratoxalis coloradensis (Rydb.) Lunell = Oxalis stricta [OXALSTR]
Ceratoxalis cymosa (Small) Lunell = Oxalis stricta [OXALSTR]
Oxalis L. [OXALIS$]
Oxalis acetosella ssp. *oregana* (Nutt.) D. Löve = Oxalis oregana [OXALORE]
Oxalis bushii (Small) Small = Oxalis stricta [OXALSTR]
Oxalis coloradensis Rydb. = Oxalis stricta [OXALSTR]
***Oxalis corniculata** L. [OXALCOR] (yellow oxalis)
SY: *Acetosella corniculata* (L.) Kuntze
SY: *Oxalis corniculata* var. *atropurpurea* Planch.
SY: *Oxalis corniculata* var. *langloisii* (Small) Wieg.
SY: *Oxalis corniculata* var. *lupulina* (R. Knuth) Zucc.
SY: *Oxalis corniculata* var. *macrophylla* Arsene ex R. Knuth
SY: *Oxalis corniculata* var. *minor* Laing
SY: *Oxalis corniculata* var. *reptans* Laing
SY: *Oxalis corniculata* var. *villosa* (Bieb.) Hohen.
SY: *Oxalis corniculata* var. *viscidula* Wieg.
SY: *Oxalis langloisii* (Small) Fedde
SY: *Oxalis pusilla* Salisb.
SY: *Oxalis repens* Thunb.
SY: *Oxalis villosa* Bieb.
SY: *Xanthoxalis corniculata* (L.) Small
SY: *Xanthoxalis corniculata* var. *atropurpurea* (Planch.) Moldenke
SY: *Xanthoxalis langloisii* Small
SY: *Xanthoxalis repens* (Thunb.) Moldenke
Oxalis corniculata var. *atropurpurea* Planch. = Oxalis corniculata [OXALCOR]
Oxalis corniculata var. *langloisii* (Small) Wieg. = Oxalis corniculata [OXALCOR]
Oxalis corniculata var. *lupulina* (R. Knuth) Zucc. = Oxalis corniculata [OXALCOR]
Oxalis corniculata var. *macrophylla* Arsene ex R. Knuth = Oxalis corniculata [OXALCOR]
Oxalis corniculata var. *minor* Laing = Oxalis corniculata [OXALCOR]
Oxalis corniculata var. *reptans* Laing = Oxalis corniculata [OXALCOR]
Oxalis corniculata var. *villosa* (Bieb.) Hohen. = Oxalis corniculata [OXALCOR]
Oxalis corniculata var. *viscidula* Wieg. = Oxalis corniculata [OXALCOR]
Oxalis cymosa Small = Oxalis stricta [OXALSTR]
Oxalis europaea Jord. = Oxalis stricta [OXALSTR]
Oxalis europaea var. *bushii* (Small) Wieg. = Oxalis stricta [OXALSTR]
Oxalis europaea var. *rufa* (Small) Young = Oxalis stricta [OXALSTR]
Oxalis fontana Bunge = Oxalis stricta [OXALSTR]
Oxalis fontana var. *bushii* (Small) Hara = Oxalis stricta [OXALSTR]
Oxalis interior (Small) Fedde = Oxalis stricta [OXALSTR]
Oxalis langloisii (Small) Fedde = Oxalis corniculata [OXALCOR]
Oxalis oregana Nutt. [OXALORE] (redwood sorrel)
SY: *Oxalis acetosella* ssp. *oregana* (Nutt.) D. Löve
SY: *Oxalis oregana* var. *smallii* (R. Knuth) M.E. Peck
Oxalis oregana var. *smallii* (R. Knuth) M.E. Peck = Oxalis oregana [OXALORE]
Oxalis pusilla Salisb. = Oxalis corniculata [OXALCOR]
Oxalis repens Thunb. = Oxalis corniculata [OXALCOR]
Oxalis rufa Small = Oxalis stricta [OXALSTR]
***Oxalis stricta** L. [OXALSTR] (upright yellow oxalis)
SY: *Ceratoxalis coloradensis* (Rydb.) Lunell
SY: *Ceratoxalis cymosa* (Small) Lunell
SY: *Oxalis bushii* (Small) Small
SY: *Oxalis coloradensis* Rydb.
SY: *Oxalis cymosa* Small
SY: *Oxalis europaea* Jord.
SY: *Oxalis europaea* var. *bushii* (Small) Wieg.
SY: *Oxalis europaea* var. *rufa* (Small) Young
SY: *Oxalis fontana* Bunge
SY: *Oxalis fontana* var. *bushii* (Small) Hara
SY: *Oxalis interior* (Small) Fedde
SY: *Oxalis rufa* Small
SY: *Oxalis stricta* var. *decumbens* Bitter
SY: *Oxalis stricta* var. *piletocarpa* Wieg.
SY: *Oxalis stricta* var. *rufa* (Small) Farw.
SY: *Oxalis stricta* var. *villicaulis* (Wieg.) Farw.
SY: *Xanthoxalis bushii* Small
SY: *Xanthoxalis coloradensis* (Rydb.) Rydb.
SY: *Xanthoxalis cymosa* (Small) Small
SY: *Xanthoxalis dillenii* var. *piletocarpa* (Wieg.) Holub
SY: *Xanthoxalis interior* Small
SY: *Xanthoxalis rufa* (Small) Small
SY: *Xanthoxalis stricta* (L.) Small
SY: *Xanthoxalis stricta* var. *piletocarpa* (Wieg.) Moldenke
Oxalis stricta var. *decumbens* Bitter = Oxalis stricta [OXALSTR]
Oxalis stricta var. *piletocarpa* Wieg. = Oxalis stricta [OXALSTR]
Oxalis stricta var. *rufa* (Small) Farw. = Oxalis stricta [OXALSTR]
Oxalis stricta var. *villicaulis* (Wieg.) Farw. = Oxalis stricta [OXALSTR]
Oxalis villosa Bieb. = Oxalis corniculata [OXALCOR]
Xanthoxalis bushii Small = Oxalis stricta [OXALSTR]
Xanthoxalis coloradensis (Rydb.) Rydb. = Oxalis stricta [OXALSTR]
Xanthoxalis corniculata (L.) Small = Oxalis corniculata [OXALCOR]
Xanthoxalis corniculata var. *atropurpurea* (Planch.) Moldenke = Oxalis corniculata [OXALCOR]
Xanthoxalis cymosa (Small) Small = Oxalis stricta [OXALSTR]
Xanthoxalis dillenii var. *piletocarpa* (Wieg.) Holub = Oxalis stricta [OXALSTR]
Xanthoxalis interior Small = Oxalis stricta [OXALSTR]
Xanthoxalis langloisii Small = Oxalis corniculata [OXALCOR]
Xanthoxalis repens (Thunb.) Moldenke = Oxalis corniculata [OXALCOR]
Xanthoxalis rufa (Small) Small = Oxalis stricta [OXALSTR]
Xanthoxalis stricta (L.) Small = Oxalis stricta [OXALSTR]
Xanthoxalis stricta var. *piletocarpa* (Wieg.) Moldenke = Oxalis stricta [OXALSTR]

PAPAVERACEAE (V:D)

Chelidonium L. [CHELIDO$]
***Chelidonium majus** L. [CHELMAJ] (celandine)
Eschscholzia Cham. [ESCHSCH$]
***Eschscholzia californica** Cham. [ESCHCAL] (California poppy)
Meconella Nutt. [MECONEL$]
Meconella oregana Nutt. [MECOORE] (white meconella)
Papaver L. [PAPAVER$]
Papaver alaskanum Hultén = Papaver lapponicum ssp. occidentale [PAPALAP1]
Papaver alaskanum var. *macranthum* Hultén = Papaver macounii [PAPAMAC]
Papaver alboroseum Hultén [PAPAALB] (pale poppy)
Papaver alpinum L. [PAPAALP] (dwarf poppy)
SY: *Papaver nudicaule* var. *pseudocorylatifolium* Fedde
Papaver cornwallisense D. Löve = Papaver lapponicum ssp. occidentale [PAPALAP1]
Papaver denalii Gjaerevoll = Papaver lapponicum ssp. occidentale [PAPALAP1]
Papaver freedmanianum D. Löve = Papaver lapponicum ssp. occidentale [PAPALAP1]

Papaver hultenii Knaben = Papaver macounii [PAPAMAC]
Papaver hultenii var. *salmonicolor* Hultén = Papaver macounii [PAPAMAC]
Papaver keelei Porsild = Papaver macounii [PAPAMAC]
Papaver kluanense D. Löve = Papaver lapponicum ssp. occidentale [PAPALAP1]
Papaver lapponicum (Tolm.) Nordh. [PAPALAP] (Lapland poppy)
Papaver lapponicum ssp. **occidentale** (Lundstr.) Knaben [PAPALAP1] (arctic poppy)
SY: *Papaver alaskanum* Hultén
SY: *Papaver cornwallisense* D. Löve
SY: *Papaver denalii* Gjaerevoll
SY: *Papaver freedmanianum* D. Löve
SY: *Papaver kluanense* D. Löve
SY: *Papaver lapponicum* ssp. *porsildii* Knaben
SY: *Papaver nigroflavum* D. Löve
SY: *Papaver nudicaule* var. *coloradense* Fedde
SY: *Papaver nudicaule* var. *columbianum* Fedde
SY: *Papaver radicatum* ssp. *lapponicum* Tolm.
SY: *Papaver radicatum* ssp. *occidentale* Lundstr.
SY: *Papaver radicatum* ssp. *porsildii* (Knaben) D. Löve
Papaver lapponicum ssp. *porsildii* Knaben = Papaver lapponicum ssp. occidentale [PAPALAP1]
Papaver macounii Greene [PAPAMAC] (Macoun's poppy)
SY: *Papaver alaskanum* var. *macranthum* Hultén
SY: *Papaver hultenii* Knaben
SY: *Papaver hultenii* var. *salmonicolor* Hultén
SY: *Papaver keelei* Porsild
SY: *Papaver macounii* var. *discolor* Hultén
SY: *Papaver scammianum* D. Löve
Papaver macounii var. *discolor* Hultén = Papaver macounii [PAPAMAC]
Papaver nigroflavum D. Löve = Papaver lapponicum ssp. occidentale [PAPALAP1]
***Papaver nudicaule** L. [PAPANUD] (Iceland poppy)
Papaver nudicaule var. *coloradense* Fedde = Papaver lapponicum ssp. occidentale [PAPALAP1]
Papaver nudicaule var. *columbianum* Fedde = Papaver lapponicum ssp. occidentale [PAPALAP1]
Papaver nudicaule var. *pseudocorylatifolium* Fedde = Papaver alpinum [PAPAALP]
Papaver pygmaeum Rydb. [PAPAPYG]
SY: *Papaver radicatum* var. *pygmaeum* (Rydb.) Welsh
Papaver radicatum ssp. *lapponicum* Tolm. = Papaver lapponicum ssp. occidentale [PAPALAP1]
Papaver radicatum ssp. *occidentale* Lundstr. = Papaver lapponicum ssp. occidentale [PAPALAP1]
Papaver radicatum ssp. *porsildii* (Knaben) D. Löve = Papaver lapponicum ssp. occidentale [PAPALAP1]
Papaver radicatum var. *pygmaeum* (Rydb.) Welsh = Papaver pygmaeum [PAPAPYG]
***Papaver rhoeas** L. [PAPARHO] (corn poppy)
Papaver scammianum D. Löve = Papaver macounii [PAPAMAC]
***Papaver somniferum** L. [PAPASOM] (opium poppy)

PINACEAE (V:G)

Abies P. Mill. [ABIES$]
Abies amabilis (Dougl. ex Loud.) Dougl. ex Forbes [ABIEAMA] (amabilis fir)
Abies balsemea (L.) P. Mill. [ABIEBAL] (balsam fir)
SY: *Pinus balsamea* L.
Abies excelsior Franco = Abies grandis [ABIEGRA]
Abies grandis (Dougl. ex D. Don) Lindl. [ABIEGRA] (grand fir)
SY: *Abies excelsior* Franco
Abies lasiocarpa (Hook.) Nutt. [ABIELAS] (subalpine fir)
Abies nobilis (Dougl. ex D. Don) Lindl. = Abies procera [ABIEPRO]
***Abies procera** Rehd. [ABIEPRO] (noble fir)
SY: *Abies nobilis* (Dougl. ex D. Don) Lindl.
Larix P. Mill. [LARIX$]
Larix alaskensis W. Wight = Larix laricina [LARILAR]
Larix laricina (Du Roi) K. Koch [LARILAR] (tamarack)
SY: *Larix alaskensis* W. Wight
SY: *Larix laricina* var. *alaskensis* (W. Wight) Raup
Larix laricina var. *alaskensis* (W. Wight) Raup = Larix laricina [LARILAR]
Larix lyallii Parl. [LARILYA] (subalpine larch)
Larix occidentalis Nutt. [LARIOCC] (western larch)
Picea A. Dietr. [PICEA$]
Picea canadensis (P. Mill.) B.S.P. = Picea glauca [PICEGLA]
Picea engelmannii Parry ex Engelm. [PICEENG] (Engelmann spruce)
SY: *Picea glauca* ssp. *engelmannii* (Parry ex Engelm.) T.M.C. Taylor
SY: *Picea glauca* var. *engelmannii* (Parry ex Engelm.) Boivin
Picea engelmannii × glauca [PICEENE] (hybrid white spruce)
Picea glauca (Moench) Voss [PICEGLA] (white spruce)
SY: *Picea canadensis* (P. Mill.) B.S.P.
SY: *Picea glauca* var. *albertiana* (S. Br.) Sarg.
SY: *Picea glauca* var. *densata* Bailey
SY: *Picea glauca* var. *porsildii* Raup
Picea glauca ssp. *engelmannii* (Parry ex Engelm.) T.M.C. Taylor = Picea engelmannii [PICEENG]
Picea glauca var. *albertiana* (S. Br.) Sarg. = Picea glauca [PICEGLA]
Picea glauca var. *densata* Bailey = Picea glauca [PICEGLA]
Picea glauca var. *engelmannii* (Parry ex Engelm.) Boivin = Picea engelmannii [PICEENG]
Picea glauca var. *porsildii* Raup = Picea glauca [PICEGLA]
Picea × lutzii Little [PICEXLU] (Roche spruce
Picea mariana (P. Mill.) B.S.P. [PICEMAR] (black spruce)
Picea sitchensis (Bong.) Carr. [PICESIT] (Sitka spruce)
Pinus L. [PINUS$]
Pinus albicaulis Engelm. [PINUALB] (whitebark pine)
Pinus balsamea L. = Abies balsemea [ABIEBAL]
Pinus banksiana Lamb. [PINUBAN] (jack pine)
SY: *Pinus divaricata* (Ait.) Dum.-Cours.
Pinus contorta Dougl. ex Loud. [PINUCON] (lodgepole pine)
Pinus contorta ssp. *latifolia* (Engelm. ex S. Wats.) Critchfield = Pinus contorta var. latifolia [PINUCON2]
Pinus contorta var. **contorta** [PINUCON0] (shore pine)
Pinus contorta var. **latifolia** Engelm. ex S. Wats. [PINUCON2]
SY: *Pinus contorta* ssp. *latifolia* (Engelm. ex S. Wats.) Critchfield
SY: *Pinus divaricata* var. *hendersonii* (Lemmon) Boivin
SY: *Pinus divaricata* var. *latifolia* (Engelm. ex S. Wats.) Boivin
Pinus divaricata (Ait.) Dum.-Cours. = Pinus banksiana [PINUBAN]
Pinus divaricata var. *hendersonii* (Lemmon) Boivin = Pinus contorta var. latifolia [PINUCON2]
Pinus divaricata var. *latifolia* (Engelm. ex S. Wats.) Boivin = Pinus contorta var. latifolia [PINUCON2]
Pinus divaricata var. × *musci* Boivin = Pinus × murraybanksiana [PINUXMU]
Pinus flexilis James [PINUFLE] (limber pine)
Pinus monticola Dougl. ex D. Don [PINUMON] (western white pine)
SY: *Pinus strobus* var. *monticola* (Dougl. ex D. Don) Nutt.
Pinus × murraybanksiana Righter & Stockwell [PINUXMU] (lodgepole × jack pine hybrid)
SY: *Pinus divaricata* var. × *musci* Boivin
Pinus ponderosa P.& C. Lawson [PINUPON] (ponderosa pine)
Pinus strobus var. *monticola* (Dougl. ex D. Don) Nutt. = Pinus monticola [PINUMON]
Pseudotsuga Carr. [PSEUDOT$]
Pseudotsuga menziesii (Mirbel) Franco [PSEUMEN] (Douglas-fir)
Pseudotsuga menziesii var. **glauca** (Beissn.) Franco [PSEUMEN1] (Rocky Mountain Douglas-fir)
SY: *Pseudotsuga taxifolia* var. *glauca* (Beissn.) Sudworth
Pseudotsuga menziesii var. **menziesii** [PSEUMEN0] (coast Douglas-fir)
SY: *Pseudotsuga taxifolia* (Lamb.) Britt.
Pseudotsuga taxifolia (Lamb.) Britt. = Pseudotsuga menziesii var. menziesii [PSEUMEN0]
Pseudotsuga taxifolia var. *glauca* (Beissn.) Sudworth = Pseudotsuga menziesii var. glauca [PSEUMEN1]
Tsuga Carr. [TSUGA$]
Tsuga heterophylla (Raf.) Sarg. [TSUGHET] (western hemlock)
Tsuga heterophylla × mertensiana [TSUGHEE] (mountain × western hemlock hybrid)
Tsuga mertensiana (Bong.) Carr. [TSUGMER] (mountain hemlock)

PLANTAGINACEAE (V:D)

Plantago L. [PLANTAG$]
Plantago arenaria Waldst. & Kit. = Plantago psyllium [PLANPSY]
Plantago bigelovii Gray [PLANBIG]
Plantago canescens M.F. Adams [PLANCAN] (arctic plantain)
SY: *Plantago canescens* var. *cylindrica* (Macoun) Boivin
SY: *Plantago septata* Morris ex Rydb.
Plantago canescens var. *cylindrica* (Macoun) Boivin = Plantago canescens [PLANCAN]
Plantago coronopus L. [PLANCOR] (buck's-horn plantain)
SY: *Plantago coronopus* ssp. *commutata* (Guss.) Pilger
Plantago coronopus ssp. *commutata* (Guss.) Pilger = Plantago coronopus [PLANCOR]
Plantago elongata Pursh [PLANELO] (slender plantain)
Plantago elongata ssp. **pentasperma** Bassett [PLANELO2]
Plantago eriopoda Torr. [PLANERI] (alkali plantain)
SY: *Plantago shastensis* Greene
Plantago gnaphalioides Nutt. = Plantago patagonica [PLANPAT]
Plantago indica L. = Plantago psyllium [PLANPSY]
Plantago juncoides Lam. = Plantago maritima var. juncoides [PLANMAR1]
Plantago juncoides var. *decipiens* (Barneoud) Fern. = Plantago maritima var. juncoides [PLANMAR1]
Plantago juncoides var. *glauca* (Hornem.) Fern. = Plantago maritima var. juncoides [PLANMAR1]
Plantago juncoides var. *laurentiana* Fern. = Plantago maritima var. juncoides [PLANMAR1]
*Plantago lanceolata** L. [PLANLAN] (ribwort plantain)
SY: *Plantago lanceolata* var. *sphaerostachya* Mert. & Koch
Plantago lanceolata var. *sphaerostachya* Mert. & Koch = Plantago lanceolata [PLANLAN]
Plantago macrocarpa Cham. & Schlecht. [PLANMAC] (Alaska plantain)
*Plantago major** L. [PLANMAJ] (common plantain)
Plantago maritima L. [PLANMAR] (goosetongue)
Plantago maritima ssp. *borealis* (Lange) Blytt & O. Dahl = Plantago maritima var. juncoides [PLANMAR1]
Plantago maritima ssp. *juncoides* (Lam.) Hultén = Plantago maritima var. juncoides [PLANMAR1]
Plantago maritima var. **juncoides** (Lam.) Gray [PLANMAR1]
SY: *Plantago juncoides* Lam.
SY: *Plantago juncoides* var. *decipiens* (Barneoud) Fern.
SY: *Plantago juncoides* var. *glauca* (Hornem.) Fern.
SY: *Plantago juncoides* var. *laurentiana* Fern.
SY: *Plantago maritima* ssp. *borealis* (Lange) Blytt & O. Dahl
SY: *Plantago maritima* ssp. *juncoides* (Lam.) Hultén
SY: *Plantago oliganthos* Roemer & J.A. Schultes
SY: *Plantago oliganthos* var. *fallax* Fern.
*Plantago media** L. [PLANMED] (hoary plantain)
SY: *Plantago media* var. *monnieri* (Giraud.) Roug.
Plantago media var. *monnieri* (Giraud.) Roug. = Plantago media [PLANMED]
Plantago oliganthos Roemer & J.A. Schultes = Plantago maritima var. juncoides [PLANMAR1]
Plantago oliganthos var. *fallax* Fern. = Plantago maritima var. juncoides [PLANMAR1]
Plantago patagonica Jacq. [PLANPAT] (woolly plantain)
SY: *Plantago gnaphalioides* Nutt.
SY: *Plantago patagonica* var. *breviscapa* (Shinners) Shinners
SY: *Plantago patagonica* var. *gnaphalioides* (Nutt.) Gray
SY: *Plantago patagonica* var. *oblonga* (Morris) Shinners
SY: *Plantago patagonica* var. *spinulosa* (Dcne.) Gray
SY: *Plantago picta* Morris
SY: *Plantago purshii* Roemer & J.A. Schultes
SY: *Plantago purshii* var. *breviscapa* Shinners
SY: *Plantago purshii* var. *oblonga* (Morris) Shinners
SY: *Plantago purshii* var. *picta* Pilger
SY: *Plantago purshii* var. *spinulosa* (Dcne.) Shinners
SY: *Plantago spinulosa* Dcne.
SY: *Plantago wyomingensis* Gandog.
Plantago patagonica var. *breviscapa* (Shinners) Shinners = Plantago patagonica [PLANPAT]
Plantago patagonica var. *gnaphalioides* (Nutt.) Gray = Plantago patagonica [PLANPAT]
Plantago patagonica var. *oblonga* (Morris) Shinners = Plantago patagonica [PLANPAT]
Plantago patagonica var. *spinulosa* (Dcne.) Gray = Plantago patagonica [PLANPAT]
Plantago picta Morris = Plantago patagonica [PLANPAT]
*Plantago psyllium** L. [PLANPSY] (whorled plantain)
SY: *Plantago arenaria* Waldst. & Kit.
SY: *Plantago indica* L.
SY: *Plantago scabra* Moench
Plantago purshii Roemer & J.A. Schultes = Plantago patagonica [PLANPAT]
Plantago purshii var. *breviscapa* Shinners = Plantago patagonica [PLANPAT]
Plantago purshii var. *oblonga* (Morris) Shinners = Plantago patagonica [PLANPAT]
Plantago purshii var. *picta* Pilger = Plantago patagonica [PLANPAT]
Plantago purshii var. *spinulosa* (Dcne.) Shinners = Plantago patagonica [PLANPAT]
Plantago scabra Moench = Plantago psyllium [PLANPSY]
Plantago septata Morris ex Rydb. = Plantago canescens [PLANCAN]
Plantago shastensis Greene = Plantago eriopoda [PLANERI]
Plantago spinulosa Dcne. = Plantago patagonica [PLANPAT]
Plantago wyomingensis Gandog. = Plantago patagonica [PLANPAT]

PLUMBAGINACEAE (V:D)

Armeria (DC.) Willd. [ARMERIA$]
Armeria arctica (Cham.) Wallr. = Armeria maritima ssp. purpurea [ARMEMAR3]
Armeria arctica ssp. *californica* (Boiss.) Abrams = Armeria maritima ssp. californica [ARMEMAR2]
Armeria californica Boiss. = Armeria maritima ssp. californica [ARMEMAR2]
Armeria maritima (P. Mill.) Willd. [ARMEMAR] (thrift)
Armeria maritima ssp. *arctica* (Cham.) Hultén = Armeria maritima ssp. purpurea [ARMEMAR3]
Armeria maritima ssp. **californica** (Boiss.) Porsild [ARMEMAR2]
SY: *Armeria arctica* ssp. *californica* (Boiss.) Abrams
SY: *Armeria californica* Boiss.
SY: *Armeria maritima* var. *californica* (Boiss.) G.H.M. Lawrence
Armeria maritima ssp. **purpurea** (W.D.J. Koch) A. & D. Löve [ARMEMAR3]
SY: *Armeria arctica* (Cham.) Wallr.
SY: *Armeria maritima* ssp. *arctica* (Cham.) Hultén
SY: *Armeria maritima* var. *purpurea* (W.D.J. Koch) G.H.M. Lawrence
Armeria maritima var. *californica* (Boiss.) G.H.M. Lawrence = Armeria maritima ssp. californica [ARMEMAR2]
Armeria maritima var. *purpurea* (W.D.J. Koch) G.H.M. Lawrence = Armeria maritima ssp. purpurea [ARMEMAR3]

POACEAE (V:M)

× *Agrohordeum macounii* (Vasey) Lepage = × Elyhordeum macounii [ELYHMAC]
× *Agrohordeum macounii* var. *valencianum* Bowden = × Elyhordeum macounii [ELYHMAC]
Agropyron Gaertn. [AGROPYR$]
Agropyron alaskanum Scribn. & Merr. = Elymus alaskanus [ELYMALA]
Agropyron albicans Scribn. & J.G. Sm. = Elymus lanceolatus ssp. albicans [ELYMLAN1]
Agropyron albicans var. *griffithii* (Scribn. & J.G. Sm. ex Piper) Beetle = Elymus lanceolatus ssp. albicans [ELYMLAN1]
Agropyron brevifolium Scribn. = Elymus trachycaulus ssp. trachycaulus [ELYMTRA0]
Agropyron × *brevifolium* Scribn. = Elymus trachycaulus ssp. trachycaulus [ELYMTRA0]

Agropyron caninum ssp. *majus* (Vasey) C.L. Hitchc. = Elymus trachycaulus ssp. trachycaulus [ELYMTRA0]
Agropyron caninum var. *andinum* (Scribn. & J.G. Sm.) C.L. Hitchc. = Elymus trachycaulus ssp. trachycaulus [ELYMTRA0]
Agropyron caninum var. *hornemannii* (Koch) Pease & Moore = Elymus trachycaulus ssp. trachycaulus [ELYMTRA0]
Agropyron caninum var. *mitchellii* Welsh = Elymus trachycaulus ssp. trachycaulus [ELYMTRA0]
Agropyron caninum var. *unilaterale* (Cassidy) C.L. Hitchc. = Elymus trachycaulus ssp. subsecundus [ELYMTRA1]
Agropyron caninum var. *latiglume* (Scribn. & J.G. Sm.) Pease & Moore = Elymus alaskanus ssp. latiglumis [ELYMALA1]
*****Agropyron cristatum** (L.) Gaertn. [AGROCRI] (crested wheatgrass)
Agropyron cristatum ssp. *desertorum* (Fisch. ex Link) A. Löve = Agropyron desertorum [AGRODES]
Agropyron cristatum ssp. *fragile* (Roth) A. Löve p.p. = Agropyron fragile ssp. sibiricum [AGROFRA1]
Agropyron cristatum var. *desertorum* (Fisch. ex Link) Dorn = Agropyron desertorum [AGRODES]
Agropyron cristatum var. *fragile* (Roth) Dorn p.p. = Agropyron fragile ssp. sibiricum [AGROFRA1]
Agropyron dasystachyum (Hook.) Scribn. & J.G. Sm. = Elymus lanceolatus ssp. lanceolatus [ELYMLAN0]
Agropyron dasystachyum ssp. *albicans* (Scribn. & J.G. Sm.) Dewey = Elymus lanceolatus ssp. albicans [ELYMLAN1]
Agropyron dasystachyum ssp. *yukonense* (Scribn. & Merr.) Dewey = Elymus × yukonensis [ELYMYUK]
Agropyron dasystachyum var. *riparum* (Scribn. & J.G. Sm.) Bowden = Elymus lanceolatus ssp. lanceolatus [ELYMLAN0]
*****Agropyron desertorum** (Fisch. ex Link) J.A. Schultes [AGRODES] (desert wheatgrass)
SY: *Agropyron cristatum* ssp. *desertorum* (Fisch. ex Link) A. Löve
SY: *Agropyron cristatum* var. *desertorum* (Fisch. ex Link) Dorn
Agropyron elmeri Scribn. = Elymus lanceolatus ssp. lanceolatus [ELYMLAN0]
*****Agropyron fragile** (Roth) P. Candargy [AGROFRA] (Siberian wheatgrass)
Agropyron fragile ssp. **sibiricum** (Willd.) Melderis [AGROFRA1]
SY: *Agropyron cristatum* ssp. *fragile* (Roth) A. Löve p.p.
SY: *Agropyron cristatum* var. *fragile* (Roth) Dorn p.p.
SY: *Agropyron fragile* var. *sibiricum* (Willd.) Tzvelev
SY: *Agropyron sibiricum* (Willd.) Beauv.
Agropyron fragile var. *sibiricum* (Willd.) Tzvelev = Agropyron fragile ssp. sibiricum [AGROFRA1]
Agropyron griffithii Scribn. & J.G. Sm. ex Piper = Elymus lanceolatus ssp. albicans [ELYMLAN1]
Agropyron inerme (Scribn. & J.G. Sm.) Rydb. = Pseudoroegneria spicata ssp. inermis [PSEUSPI1]
Agropyron intermedium (Host) Beauv. = Elytrigia intermedia [ELYTINT]
Agropyron intermedium var. *trichophorum* (Link) Halac. = Elytrigia intermedia [ELYTINT]
Agropyron lanceolatum Scribn. & J.G. Sm. = Elymus lanceolatus ssp. lanceolatus [ELYMLAN0]
Agropyron latiglume (Scribn. & J.G. Sm.) Rydb. = Elymus alaskanus ssp. latiglumis [ELYMALA1]
Agropyron molle (Scribn. & J.G. Sm.) Rydb. = Pascopyrum smithii [PASCSMI]
Agropyron novae-angliae Scribn. = Elymus trachycaulus ssp. trachycaulus [ELYMTRA0]
Agropyron pauciflorum (Schwein.) A.S. Hitchc. ex Silveus = Elymus trachycaulus ssp. trachycaulus [ELYMTRA0]
Agropyron pauciflorum ssp. *majus* (Vasey) Melderis = Elymus trachycaulus ssp. trachycaulus [ELYMTRA0]
Agropyron pauciflorum ssp. *novae-angliae* (Scribn.) Melderis = Elymus trachycaulus ssp. trachycaulus [ELYMTRA0]
Agropyron pauciflorum ssp. *teslinense* (Porsild & Senn) Melderis = Elymus trachycaulus ssp. trachycaulus [ELYMTRA0]
Agropyron pauciflorum var. *glaucum* (Pease & Moore) Taylor = Elymus trachycaulus ssp. subsecundus [ELYMTRA1]
Agropyron pauciflorum var. *novae-angliae* (Scribn.) Taylor & MacBryde = Elymus trachycaulus ssp. trachycaulus [ELYMTRA0]
Agropyron pauciflorum var. *unilaterale* (Vasey) Taylor & MacBryde = Elymus trachycaulus ssp. subsecundus [ELYMTRA1]
Agropyron prostratum (L. f.) Beauv. = Eremopyrum triticeum [EREMTRI]
Agropyron repens (L.) Beauv. = Elytrigia repens [ELYTREP]
Agropyron repens var. *subulatum* (Schreb.) Roemer & J.A. Schultes = Elytrigia repens [ELYTREP]
Agropyron riparum Scribn. & J.G. Sm. = Elymus lanceolatus ssp. lanceolatus [ELYMLAN0]
Agropyron saundersii (Vasey) A.S. Hitchc. = Elymus × saundersii [ELYMXSA]
Agropyron scribneri Vasey = Elymus scribneri [ELYMSCR]
Agropyron sibiricum (Willd.) Beauv. = Agropyron fragile ssp. sibiricum [AGROFRA1]
Agropyron smithii Rydb. = Pascopyrum smithii [PASCSMI]
Agropyron smithii var. *molle* (Scribn. & J.G. Sm.) M.E. Jones = Pascopyrum smithii [PASCSMI]
Agropyron smithii var. *palmeri* (Scribn. & J.G. Sm.) Heller = Pascopyrum smithii [PASCSMI]
Agropyron spicatum Pursh = Pseudoroegneria spicata [PSEUSPI]
Agropyron spicatum var. *inerme* (Scribn. & J.G. Sm.) Heller = Pseudoroegneria spicata ssp. inermis [PSEUSPI1]
Agropyron spicatum var. *pubescens* Elmer = Pseudoroegneria spicata [PSEUSPI]
Agropyron subsecundum (Link) A.S. Hitchc. = Elymus trachycaulus ssp. subsecundus [ELYMTRA1]
Agropyron subsecundum var. *andinum* (Scribn. & J.G. Sm.) A.S. Hitchc. = Elymus trachycaulus ssp. trachycaulus [ELYMTRA0]
Agropyron tenerum Vasey = Elymus trachycaulus ssp. trachycaulus [ELYMTRA0]
Agropyron teslinense Porsild & Senn = Elymus trachycaulus ssp. trachycaulus [ELYMTRA0]
Agropyron trachycaulum (Link) Malte ex H.F. Lewis = Elymus trachycaulus ssp. trachycaulus [ELYMTRA0]
Agropyron trachycaulum var. *ciliatum* (Scribn. & J.G. Sm.) Gleason = Elymus trachycaulus ssp. subsecundus [ELYMTRA1]
Agropyron trachycaulum var. *glaucum* (Pease & Moore) Malte = Elymus trachycaulus ssp. subsecundus [ELYMTRA1]
Agropyron trachycaulum var. *latiglume* (Scribn. & J.G. Sm.) Beetle = Elymus alaskanus ssp. latiglumis [ELYMALA1]
Agropyron trachycaulum var. *majus* (Vasey) Fern. = Elymus trachycaulus ssp. trachycaulus [ELYMTRA0]
Agropyron trachycaulum var. *novae-angliae* (Scribn.) Fern. = Elymus trachycaulus ssp. trachycaulus [ELYMTRA0]
Agropyron trachycaulum var. *unilaterale* (Cassidy) Malte = Elymus trachycaulus ssp. subsecundus [ELYMTRA1]
Agropyron trichophorum (Link) Richter = Elytrigia intermedia [ELYTINT]
Agropyron triticeum Gaertn. = Eremopyrum triticeum [EREMTRI]
Agropyron varnense (Velen.) Hayek = Elytrigia pontica [ELYTPON]
Agropyron vaseyi Scribn. & J.G. Sm. = Pseudoroegneria spicata [PSEUSPI]
Agropyron violaceum (Hornem.) Lange = Elymus alaskanus [ELYMALA]
Agropyron violaceum ssp. *andinum* (Scribn. & J.G. Sm.) Melderis = Elymus trachycaulus ssp. trachycaulus [ELYMTRA0]
Agropyron violaceum var. *andinum* Scribn. & J.G. Sm. = Elymus trachycaulus ssp. trachycaulus [ELYMTRA0]
Agropyron yukonense Scribn. & Merr. = Elymus × yukonensis [ELYMYUK]
× *Agrositanion saundersii* (Vasey) Bowden = Elymus × saundersii [ELYMXSA]
Agrostis L. [AGROSTI$]
Agrostis aequivalvis (Trin.) Trin. [AGROAEQ] (Alaska bentgrass)
SY: *Podagrostis aequivalvis* (Trin.) Scribn. & Merr.
Agrostis airoides Torr. = Sporobolus airoides [SPORAIR]
Agrostis alba auct. = Agrostis gigantea [AGROGIG]
Agrostis alba var. *palustris* (Huds.) Pers. = Agrostis stolonifera [AGROSTO]
Agrostis alba var. *stolonifera* (L.) Sm. = Agrostis stolonifera [AGROSTO]
Agrostis borealis Hartman = Agrostis mertensii [AGROMER]

Agrostis borealis var. *americana* (Scribn.) Fern. = Agrostis mertensii [AGROMER]
Agrostis borealis var. *paludosa* (Scribn.) Fern. = Agrostis mertensii [AGROMER]
*__Agrostis capillaris__ L. [AGROCAP] (colonial bentgrass)
SY: *Agrostis sylvatica* Huds.
SY: *Agrostis tenuis* Sibthorp
SY: *Agrostis tenuis* var. *aristata* (Parnell) Druce
SY: *Agrostis tenuis* var. *hispida* (Willd.) Philipson
SY: *Agrostis tenuis* var. *pumila* (L.) Druce
Agrostis cryptandra Torr. = Sporobolus cryptandrus [SPORCRY]
Agrostis diegoensis Vasey [AGRODIE] (thin bentgrass)
SY: *Agrostis pallens* var. *vaseyi* St. John
Agrostis exarata Trin. [AGROEXA] (spike bentgrass)
Agrostis exarata ssp. *minor* (Hook.) C.L. Hitchc. = Agrostis exarata var. minor [AGROEXA5]
Agrostis exarata var. **minor** Hook. [AGROEXA5]
SY: *Agrostis exarata* ssp. *minor* (Hook.) C.L. Hitchc.
SY: *Agrostis exarata* var. *purpurascens* Hultén
Agrostis exarata var. **monolepis** (Torr.) A.S. Hitchc. [AGROEXA6]
Agrostis exarata var. *purpurascens* Hultén = Agrostis exarata var. minor [AGROEXA5]
*__Agrostis gigantea__ Roth [AGROGIG] (redtop)
SY: *Agrostis gigantea* var. *dispar* (Michx.) Philipson
SY: *Agrostis nigra* With.
SY: *Agrostis stolonifera* ssp. *gigantea* (Roth) Schuebl. & Martens
SY: *Agrostis stolonifera* var. *major* (Gaudin) Farw.
SY: *Agrostis alba* auct.
Agrostis gigantea var. *dispar* (Michx.) Philipson = Agrostis gigantea [AGROGIG]
Agrostis inflata Scribn. [AGROINF]
Agrostis interrupta L. = Apera interrupta [APERINT]
Agrostis maritima Lam. = Agrostis stolonifera [AGROSTO]
Agrostis mertensii Trin. [AGROMER] (northern bentgrass)
SY: *Agrostis borealis* Hartman
SY: *Agrostis borealis* var. *americana* (Scribn.) Fern.
SY: *Agrostis borealis* var. *paludosa* (Scribn.) Fern.
SY: *Agrostis mertensii* ssp. *borealis* (Hartman) Tzvelev
Agrostis mertensii ssp. *borealis* (Hartman) Tzvelev = Agrostis mertensii [AGROMER]
Agrostis mexicana L. = Muhlenbergia mexicana [MUHLMEX]
Agrostis microphylla Steud. [AGROMIC] (small-leaved bentgrass)
Agrostis nigra With. = Agrostis gigantea [AGROGIG]
Agrostis oregonensis Vasey [AGROORE] (Oregon bentgrass)
Agrostis pallens Trin. [AGROPAL] (dune bentgrass)
Agrostis pallens var. *vaseyi* St. John = Agrostis diegoensis [AGRODIE]
Agrostis palustris Huds. = Agrostis stolonifera [AGROSTO]
Agrostis racemosa Michx. = Muhlenbergia racemosa [MUHLRAC]
Agrostis scabra Willd. [AGROSCA] (hair bentgrass)
*__Agrostis stolonifera__ L. [AGROSTO] (creeping bentgrass)
SY: *Agrostis alba* var. *palustris* (Huds.) Pers.
SY: *Agrostis alba* var. *stolonifera* (L.) Sm.
SY: *Agrostis maritima* Lam.
SY: *Agrostis palustris* Huds.
SY: *Agrostis stolonifera* var. *compacta* Hartman
SY: *Agrostis stolonifera* var. *palustris* (Huds.) Farw.
Agrostis stolonifera ssp. *gigantea* (Roth) Schuebl. & Martens = Agrostis gigantea [AGROGIG]
Agrostis stolonifera var. *compacta* Hartman = Agrostis stolonifera [AGROSTO]
Agrostis stolonifera var. *major* (Gaudin) Farw. = Agrostis gigantea [AGROGIG]
Agrostis stolonifera var. *palustris* (Huds.) Farw. = Agrostis stolonifera [AGROSTO]
Agrostis sylvatica Huds. = Agrostis capillaris [AGROCAP]
Agrostis tenuis Sibthorp = Agrostis capillaris [AGROCAP]
Agrostis tenuis var. *aristata* (Parnell) Druce = Agrostis capillaris [AGROCAP]
Agrostis tenuis var. *hispida* (Willd.) Philipson = Agrostis capillaris [AGROCAP]
Agrostis tenuis var. *pumila* (L.) Druce = Agrostis capillaris [AGROCAP]
Agrostis variabilis Rydb. [AGROVAR] (mountain bentgrass)
Aira L. [AIRA$]
Aira atropurpurea Wahlenb. = Vahlodea atropurpurea [VAHLATR]
*__Aira caryophyllea__ L. [AIRACAR] (silver hairgrass)
SY: *Aspris caryophyllea* (L.) Nash
Aira cespitosa L. = Deschampsia cespitosa ssp. cespitosa [DESCCES0]
Aira danthonioides Trin. = Deschampsia danthonioides [DESCDAN]
Aira elongata Hook. = Deschampsia elongata [DESCELO]
Aira holciformis (J. Presl) Steud. = Deschampsia cespitosa ssp. holciformis [DESCCES3]
Aira obtusata Michx. = Sphenopholis obtusata [SPHEOBT]
*__Aira praecox__ L. [AIRAPRA] (early hairgrass)
SY: *Aspris praecox* (L.) Nash
Aira spicata L. = Trisetum spicatum [TRISSPI]
Alopecurus L. [ALOPECU$]
Alopecurus aequalis Sobol. [ALOPAEQ] (little meadow-foxtail)
Alopecurus alpinus Sm. = Alopecurus borealis [ALOPBOR]
Alopecurus alpinus ssp. *glaucus* (Less.) Hultén = Alopecurus borealis [ALOPBOR]
Alopecurus alpinus ssp. *stejnegeri* (Vasey) Hultén = Alopecurus borealis [ALOPBOR]
Alopecurus alpinus var. *glaucus* (Less.) Krylov = Alopecurus borealis [ALOPBOR]
Alopecurus alpinus var. *stejnegeri* (Vasey) Hultén = Alopecurus borealis [ALOPBOR]
Alopecurus borealis Trin. [ALOPBOR]
SY: *Alopecurus alpinus* Sm.
SY: *Alopecurus alpinus* ssp. *glaucus* (Less.) Hultén
SY: *Alopecurus alpinus* var. *glaucus* (Less.) Krylov
SY: *Alopecurus alpinus* ssp. *stejnegeri* (Vasey) Hultén
SY: *Alopecurus alpinus* var. *stejnegeri* (Vasey) Hultén
SY: *Alopecurus glaucus* Less.
SY: *Alopecurus occidentalis* Scribn. & Tweedy
Alopecurus carolinianus Walt. [ALOPCAR] (carolina meadow-foxtail)
SY: *Alopecurus macounii* Vasey
SY: *Alopecurus ramosus* Poir.
*__Alopecurus geniculatus__ L. [ALOPGEN] (water meadow-foxtail)
SY: *Alopecurus pallescens* Piper
Alopecurus glaucus Less. = Alopecurus borealis [ALOPBOR]
Alopecurus macounii Vasey = Alopecurus carolinianus [ALOPCAR]
Alopecurus monspeliensis L. = Polypogon monspeliensis [POLYMOS]
Alopecurus occidentalis Scribn. & Tweedy = Alopecurus borealis [ALOPBOR]
Alopecurus pallescens Piper = Alopecurus geniculatus [ALOPGEN]
*__Alopecurus pratensis__ L. [ALOPPRA] (meadow-foxtail)
Alopecurus ramosus Poir. = Alopecurus carolinianus [ALOPCAR]
Ammophila Host [AMMOPHI$]
*__Ammophila arenaria__ (L.) Link [AMMOARE] (European beachgrass)
Andropogon scoparius Michx. = Schizachyrium scoparium [SCHISCO]
Aneurolepidium piperi (Bowden) Baum = Leymus cinereus [LEYMCIN]
Anisantha diandra (Roth) Tutin ex Tzvelev = Bromus diandrus [BROMDIA]
Anisantha sterilis (L.) Nevski = Bromus sterilis [BROMSTE]
Anthoxanthum L. [ANTHOXA$]
*__Anthoxanthum aristatum__ Boiss. [ANTHARI]
SY: *Anthoxanthum odoratum* var. *puelii* (Lecoq & Lamotte) Coss. & Durieu
SY: *Anthoxanthum puelii* Lecoq & Lamotte
Anthoxanthum nitens (Weber) Y. Schouten & Veldkamp = Hierochloe odorata [HIERODO]
*__Anthoxanthum odoratum__ L. [ANTHODO] (sweet vernalgrass)
Anthoxanthum odoratum var. *puelii* (Lecoq & Lamotte) Coss. & Durieu = Anthoxanthum aristatum [ANTHARI]
Anthoxanthum puelii Lecoq & Lamotte = Anthoxanthum aristatum [ANTHARI]
Apera Adans. [APERA$]
*__Apera interrupta__ (L.) Beauv. [APERINT] (dense silky bentgrass)

SY: *Agrostis interrupta* L.
Arctagrostis Griseb. [ARCTAGR$]
Arctagrostis angustifolia Nash = Arctagrostis latifolia ssp. arundinacea [ARCTLAT1]
Arctagrostis angustifolia var. *crassispica* Bowden = Arctagrostis latifolia ssp. arundinacea [ARCTLAT1]
Arctagrostis arundinacea (Trin.) Beal = Arctagrostis latifolia ssp. arundinacea [ARCTLAT1]
Arctagrostis latifolia (R. Br.) Griseb. [ARCTLAT] (polargrass)
Arctagrostis latifolia ssp. **arundinacea** (Trin.) Tzvelev [ARCTLAT1]
SY: *Arctagrostis angustifolia* Nash
SY: *Arctagrostis angustifolia* var. *crassispica* Bowden
SY: *Arctagrostis arundinacea* (Trin.) Beal
SY: *Arctagrostis latifolia* var. *angustifolia* (Nash) Hultén
SY: *Arctagrostis latifolia* var. *arundinacea* (Trin.) Griseb.
SY: *Arctagrostis poaeoides* Nash
Arctagrostis latifolia var. *angustifolia* (Nash) Hultén = Arctagrostis latifolia ssp. arundinacea [ARCTLAT1]
Arctagrostis latifolia var. *arundinacea* (Trin.) Griseb. = Arctagrostis latifolia ssp. arundinacea [ARCTLAT1]
Arctagrostis poaeoides Nash = Arctagrostis latifolia ssp. arundinacea [ARCTLAT1]
Arctophila Rupr. ex Anderss. [ARCTOPH$]
Arctophila fulva (Trin.) Rupr. ex Anderss. [ARCTFUL] (pendantgrass)
SY: *Colpodium fulvum* (Trin.) Griseb.
Arctopoa eminens (J. Presl) Probat. = Poa eminens [POA EMI]
Aristida L. [ARISTID$]
Aristida longiseta Steud. = Aristida purpurea var. longiseta [ARISPUR1]
Aristida longiseta var. *rariflora* A.S. Hitchc. = Aristida purpurea var. longiseta [ARISPUR1]
Aristida longiseta var. *robusta* Merr. = Aristida purpurea var. longiseta [ARISPUR1]
Aristida oligantha Michx. [ARISOLI] (prairie three-awn)
Aristida purpurea Nutt. [ARISPUR]
Aristida purpurea var. **longiseta** (Steud.) Vasey [ARISPUR1]
SY: *Aristida longiseta* Steud.
SY: *Aristida longiseta* var. *rariflora* A.S. Hitchc.
SY: *Aristida longiseta* var. *robusta* Merr.
SY: *Aristida purpurea* var. *robusta* (Merr.) Piper
Aristida purpurea var. *robusta* (Merr.) Piper = Aristida purpurea var. longiseta [ARISPUR1]
Arrhenatherum Beauv. [ARRHENA$]
***Arrhenatherum elatius** (L.) J.& K. Presl [ARRHELA] (tall oatgrass)
Arundo festucacea Willd. = Scolochloa festucacea [SCOLFES]
Aspris caryophyllea (L.) Nash = Aira caryophyllea [AIRACAR]
Aspris praecox (L.) Nash = Aira praecox [AIRAPRA]
Avena L. [AVENA$]
Avena byzantina K. Koch = Avena sativa [AVENSAT]
Avena dubia Leers = Ventenata dubia [VENTDUB]
***Avena fatua** L. [AVENFAT] (wild oat)
SY: *Avena fatua* var. *glabrata* Peterm.
SY: *Avena fatua* var. *vilis* (Wallr.) Hausskn.
Avena fatua var. *glabrata* Peterm. = Avena fatua [AVENFAT]
Avena fatua var. *sativa* (L.) Hausskn. = Avena sativa [AVENSAT]
Avena fatua var. *vilis* (Wallr.) Hausskn. = Avena fatua [AVENFAT]
Avena hookeri Scribn. = Helictotrichon hookeri [HELIHOO]
***Avena sativa** L. [AVENSAT] (common oat)
SY: *Avena byzantina* K. Koch
SY: *Avena fatua* var. *sativa* (L.) Hausskn.
SY: *Avena sativa* var. *orientalis* (Schreb.) Alef.
Avena sativa var. *orientalis* (Schreb.) Alef. = Avena sativa [AVENSAT]
Avena smithii Porter ex Gray = Melica smithii [MELISMI]
Avena torreyi Nash = Schizachne purpurascens [SCHIPUR]
Avenochloa hookeri (Scribn.) Holub = Helictotrichon hookeri [HELIHOO]
Avenula hookeri (Scribn.) Holub = Helictotrichon hookeri [HELIHOO]
Beckmannia Host [BECKMAN$]
Beckmannia eruciformis auct. = Beckmannia syzigachne [BECKSYZ]
Beckmannia eruciformis ssp. *baicalensis* (Kusnez.) Hultén = Beckmannia syzigachne [BECKSYZ]
Beckmannia eruciformis var. *uniflora* Scribn. ex Gray = Beckmannia syzigachne [BECKSYZ]
Beckmannia syzigachne (Steud.) Fern. [BECKSYZ] (American sloughgrass)
SY: *Beckmannia eruciformis* ssp. *baicalensis* (Kusnez.) Hultén
SY: *Beckmannia eruciformis* var. *uniflora* Scribn. ex Gray
SY: *Beckmannia syzigachne* ssp. *baicalensis* (Kusnez.) Koyama & Kawano
SY: *Beckmannia syzigachne* var. *uniflora* (Scribn. ex Gray) Boivin
SY: *Beckmannia eruciformis* auct.
Beckmannia syzigachne ssp. *baicalensis* (Kusnez.) Koyama & Kawano = Beckmannia syzigachne [BECKSYZ]
Beckmannia syzigachne var. *uniflora* (Scribn. ex Gray) Boivin = Beckmannia syzigachne [BECKSYZ]
Bouteloua Lag. [BOUTELO$]
Bouteloua gracilis (Willd. ex Kunth) Lag. ex Griffiths [BOUTGRA] (blue grama)
SY: *Bouteloua gracilis* var. *stricta* (Vasey) A.S. Hitchc.
SY: *Bouteloua oligostachya* (Nutt.) Torr. ex Gray
SY: *Chondrosum gracile* Willd. ex Kunth
SY: *Chondrosum oligostachyum* (Nutt.) Torr.
Bouteloua gracilis var. *stricta* (Vasey) A.S. Hitchc. = Bouteloua gracilis [BOUTGRA]
Bouteloua oligostachya (Nutt.) Torr. ex Gray = Bouteloua gracilis [BOUTGRA]
Briza L. [BRIZA$]
***Briza maxima** L. [BRIZMAX] (big quaking grass)
***Briza minor** L. [BRIZMIN] (little quaking grass)
Bromelica bulbosa (Geyer ex Porter & Coult.) W.A. Weber = Melica bulbosa [MELIBUL]
Bromelica smithii (Porter ex Gray) Farw. = Melica smithii [MELISMI]
Bromelica spectabilis (Scribn.) W.A. Weber = Melica spectabilis [MELISPE]
Bromopsis anomala (Rupr. ex Fourn.) Holub = Bromus anomalus [BROMANO]
Bromopsis canadensis (Michx.) Holub = Bromus canadensis [BROMCAN]
Bromopsis ciliata (L.) Holub = Bromus ciliatus [BROMCIL]
Bromopsis dicksonii (Mitchell & Wilton) A. & D. Löve = Bromus inermis ssp. pumpellianus var. pumpellianus **[BROMINE3]**
Bromopsis inermis ssp. *pumpelliana* (Scribn.) W.A. Weber = Bromus inermis ssp. pumpellianus var. pumpellianus [BROMINE3]
Bromopsis pacifica (Shear) Holub = Bromus pacificus [BROMPAC]
Bromopsis porteri Coult. = Bromus anomalus [BROMANO]
Bromopsis pumpelliana (Scribn.) Holub = Bromus inermis ssp. pumpellianus var. pumpellianus [BROMINE3]
Bromopsis richardsonii (Link) Holub = Bromus canadensis [BROMCAN]
Bromopsis vulgaris (Hook.) Holub = Bromus vulgaris [BROMVUL]
Bromus L. [BROMUS$]
Bromus aleutensis Trin. ex Griseb. = Bromus sitchensis var. aleutensis [BROMSIT1]
Bromus anomalus Rupr. ex Fourn. [BROMANO] (nodding brome)
SY: *Bromopsis anomala* (Rupr. ex Fourn.) Holub
SY: *Bromopsis porteri* Coult.
SY: *Bromus ciliatus* var. *porteri* (Coult.) Rydb.
SY: *Bromus porteri* (Coult.) Nash
***Bromus briziformis** Fisch. & C.A. Mey. [BROMBRI] (rattle brome)
Bromus canadensis Michx. [BROMCAN]
SY: *Bromopsis canadensis* (Michx.) Holub
SY: *Bromopsis richardsonii* (Link) Holub
SY: *Bromus ciliatus* var. *richardsonii* (Link) Boivin
SY: *Bromus dudleyi* Fern.
SY: *Bromus richardsonii* Link
SY: *Bromus richardsonii* var. *pallidus* (Hook.) Shear
Bromus carinatus Hook. & Arn. [BROMCAR] (California brome)
SY: *Bromus carinatus* var. *californicus* (Nutt. ex Buckl.) Shear
SY: *Bromus carinatus* var. *hookerianus* (Thurb.) Shear
SY: *Bromus hookerianus* Thurb.
SY: *Ceratochloa carinata* (Hook. & Arn.) Tutin

Bromus carinatus var. *californicus* (Nutt. ex Buckl.) Shear = Bromus carinatus [BROMCAR]
Bromus carinatus var. *hookerianus* (Thurb.) Shear = Bromus carinatus [BROMCAR]
Bromus carinatus var. *marginatus* (Nees ex Steud.) A.S. Hitchc. = Bromus marginatus [BROMMAR]
Bromus ciliatus L. [BROMCIL] (fringed brome)
SY: *Bromopsis ciliata* (L.) Holub
SY: *Bromus ciliatus* var. *genuinus* Fern.
Bromus ciliatus var. *coloradensis* Vasey ex Beal = Bromus inermis ssp. pumpellianus var. pumpellianus [BROMINE3]
Bromus ciliatus var. *genuinus* Fern. = Bromus ciliatus [BROMCIL]
Bromus ciliatus var. *porteri* (Coult.) Rydb. = Bromus anomalus [BROMANO]
Bromus ciliatus var. *richardsonii* (Link) Boivin = Bromus canadensis [BROMCAN]
*__Bromus commutatus__ Schrad. [BROMCOM] (meadow brome)
SY: *Bromus commutatus* var. *apricorum* Simonkai
Bromus commutatus var. *apricorum* Simonkai = Bromus commutatus [BROMCOM]
Bromus dertonensis All. = Vulpia bromoides [VULPBRO]
*__Bromus diandrus__ Roth [BROMDIA]
SY: *Anisantha diandra* (Roth) Tutin ex Tzvelev
SY: *Bromus gussonei* Parl.
SY: *Bromus rigidus* var. *gussonei* (Parl.) Coss. & Durieu
SY: *Bromus villosus* Forsk.
SY: *Bromus maximus* auct.
SY: *Bromus rigidus* auct.
Bromus dudleyi Fern. = Bromus canadensis [BROMCAN]
Bromus eximius (Shear) Piper = Bromus vulgaris [BROMVUL]
Bromus gussonei Parl. = Bromus diandrus [BROMDIA]
Bromus hookerianus Thurb. = Bromus carinatus [BROMCAR]
*__Bromus hordeaceus__ L. [BROMHOR] (soft brome)
*__Bromus hordeaceus__ ssp. **hordeaceus** [BROMHOR0]
SY: *Bromus mollis* auct.
*__Bromus hordeaceus__ ssp. **thominei** (Hardham ex Nyman) Victorin & Weiller [BROMHOR2]
SY: *Bromus thominei* Hardham ex Nyman
Bromus inermis Leyss. [BROMINE] (smooth brome)
*__Bromus inermis__ ssp. **inermis** [BROMINE0]
Bromus inermis ssp. **pumpellianus** (Scribn.) Wagnon [BROMINE2] (pumpelly brome)
Bromus inermis ssp. **pumpellianus** var. **pumpellianus** (Scribn.) C.L. Hitchc. [BROMINE3]
SY: *Bromopsis dicksonii* (Mitchell & Wilton) A. & D. Löve
SY: *Bromopsis inermis* ssp. *pumpelliana* (Scribn.) W.A. Weber
SY: *Bromopsis pumpelliana* (Scribn.) Holub
SY: *Bromus ciliatus* var. *coloradensis* Vasey ex Beal
SY: *Bromus pumpellianus* Scribn.
SY: *Bromus pumpellianus* ssp. *dicksonii* Mitchell & Wilton
SY: *Bromus pumpellianus* var. *villosissimus* Hultén
Bromus inermis ssp. **pumpellianus** var. **purpurascens** (Hook.) Wagnon [BROMINE1]
SY: *Bromus inermis* var. *tweedyi* (Scribn. ex Beal) C.L. Hitchc.
SY: *Bromus pumpellianus* var. *tweedyi* Scribn. ex Beal
Bromus inermis var. *tweedyi* (Scribn. ex Beal) C.L. Hitchc. = Bromus inermis ssp. pumpellianus var. purpurascens **[BROMINE1]**
*__Bromus japonicus__ Thunb. ex Murr. [BROMJAP] (Japanese brome)
SY: *Bromus japonicus* var. *porrectus* Hack.
SY: *Bromus patulus* Mert. & Koch
Bromus japonicus var. *porrectus* Hack. = Bromus japonicus [BROMJAP]
Bromus marginatus Nees ex Steud. [BROMMAR]
SY: *Bromus carinatus* var. *marginatus* (Nees ex Steud.) A.S. Hitchc.
SY: *Bromus sitchensis* var. *marginatus* (Nees ex Steud.) Boivin
Bromus maximus auct. = Bromus diandrus [BROMDIA]
Bromus mollis auct. = Bromus hordeaceus ssp. hordeaceus [BROMHOR0]
Bromus pacificus Shear [BROMPAC] (Pacific brome)
SY: *Bromopsis pacifica* (Shear) Holub
Bromus patulus Mert. & Koch = Bromus japonicus [BROMJAP]
Bromus porteri (Coult.) Nash = Bromus anomalus [BROMANO]
Bromus pumpellianus Scribn. = Bromus inermis ssp. pumpellianus var. pumpellianus [BROMINE3]
Bromus pumpellianus ssp. *dicksonii* Mitchell & Wilton = Bromus inermis ssp. pumpellianus var. pumpellianus [BROMINE3]
Bromus pumpellianus var. *tweedyi* Scribn. ex Beal = Bromus inermis ssp. pumpellianus var. purpurascens [BROMINE1]
Bromus pumpellianus var. *villosissimus* Hultén = Bromus inermis ssp. pumpellianus var. pumpellianus [BROMINE3]
*__Bromus racemosus__ L. [BROMRAC] (European smooth brome)
Bromus richardsonii Link = Bromus canadensis [BROMCAN]
Bromus richardsonii var. *pallidus* (Hook.) Shear = Bromus canadensis [BROMCAN]
Bromus rigidus auct. = Bromus diandrus [BROMDIA]
Bromus rigidus var. *gussonei* (Parl.) Coss. & Durieu = Bromus diandrus [BROMDIA]
*__Bromus secalinus__ L. [BROMSEC] (rye brome)
SY: *Bromus secalinus* var. *hirsutus* Kindb.
SY: *Bromus secalinus* var. *hirtus* (F.W. Schultz) Hegi
Bromus secalinus var. *hirsutus* Kindb. = Bromus secalinus [BROMSEC]
Bromus secalinus var. *hirtus* (F.W. Schultz) Hegi = Bromus secalinus [BROMSEC]
Bromus sitchensis Trin. [BROMSIT] (Alaska brome)
Bromus sitchensis var. **aleutensis** (Trin. ex Griseb.) Hultén [BROMSIT1]
SY: *Bromus aleutensis* Trin. ex Griseb.
Bromus sitchensis var. *marginatus* (Nees ex Steud.) Boivin = Bromus marginatus [BROMMAR]
Bromus sitchensis var. **sitchensis** [BROMSIT0]
*__Bromus squarrosus__ L. [BROMSQU] (corn brome)
*__Bromus sterilis__ L. [BROMSTE] (barren brome)
SY: *Anisantha sterilis* (L.) Nevski
*__Bromus tectorum__ L. [BROMTEC] (cheatgrass)
Bromus thominei Hardham ex Nyman = Bromus hordeaceus ssp. thominei [BROMHOR2]
Bromus villosus Forsk. = Bromus diandrus [BROMDIA]
Bromus vulgaris (Hook.) Shear [BROMVUL] (Columbia brome)
SY: *Bromopsis vulgaris* (Hook.) Holub
SY: *Bromus eximius* (Shear) Piper
SY: *Bromus vulgaris* var. *eximius* Shear
SY: *Bromus vulgaris* var. *robustus* Shear
Bromus vulgaris var. *eximius* Shear = Bromus vulgaris [BROMVUL]
Bromus vulgaris var. *robustus* Shear = Bromus vulgaris [BROMVUL]
Calamagrostis Adans. [CALAMAG$]
Calamagrostis anomala Suksdorf = Calamagrostis canadensis var. canadensis [CALACAN0]
Calamagrostis atropurpurea Nash = Calamagrostis canadensis var. canadensis [CALACAN0]
Calamagrostis californica Kearney = Calamagrostis stricta ssp. inexpansa [CALASTR2]
Calamagrostis canadensis (Michx.) Beauv. [CALACAN] (bluejoint)
Calamagrostis canadensis ssp. *langsdorfii* (Link) Hultén = Calamagrostis canadensis var. langsdorfii [CALACAN7]
Calamagrostis canadensis var. *acuminata* Vasey ex Shear & Rydb. = Calamagrostis canadensis var. canadensis [CALACAN0]
Calamagrostis canadensis var. **canadensis** [CALACAN0]
SY: *Calamagrostis anomala* Suksdorf
SY: *Calamagrostis atropurpurea* Nash
SY: *Calamagrostis canadensis* var. *acuminata* Vasey ex Shear & Rydb.
SY: *Calamagrostis canadensis* var. *pallida* (Vasey & Scribn.) Stebbins
SY: *Calamagrostis canadensis* var. *robusta* Vasey
SY: *Calamagrostis canadensis* var. *typica* Stebbins
SY: *Calamagrostis expansa* var. *robusta* (Vasey) Stebbins
SY: *Calamagrostis inexpansa* var. *cuprea* Kearney
SY: *Calamagrostis inexpansa* var. *robusta* (Vasey) Stebbins
SY: *Calamagrostis pallida* Vasey & Scribn. ex Vasey, non Nutt. ex Gray
Calamagrostis canadensis var. **langsdorfii** (Link) Inman [CALACAN7]
SY: *Calamagrostis canadensis* ssp. *langsdorfii* (Link) Hultén
SY: *Calamagrostis canadensis* var. *scabra* (J. Presl) A.S. Hitchc.

SY: *Calamagrostis langsdorfii* (Link) Trin.
SY: *Calamagrostis nubila* Louis-Marie
SY: *Calamagrostis scabra* J. Presl
Calamagrostis canadensis var. *pallida* (Vasey & Scribn.) Stebbins = Calamagrostis canadensis var. canadensis [CALACAN0]
Calamagrostis canadensis var. *robusta* Vasey = Calamagrostis canadensis var. canadensis [CALACAN0]
Calamagrostis canadensis var. *scabra* (J. Presl) A.S. Hitchc. = Calamagrostis canadensis var. langsdorfii [CALACAN7]
Calamagrostis canadensis var. *typica* Stebbins = Calamagrostis canadensis var. canadensis [CALACAN0]
Calamagrostis chordorrhiza Porsild = Calamagrostis stricta ssp. inexpansa [CALASTR2]
Calamagrostis crassiglumis Thurb. [CALACRA]
Calamagrostis expansa Rickett & Gilly = Calamagrostis stricta ssp. inexpansa [CALASTR2]
Calamagrostis expansa var. *robusta* (Vasey) Stebbins = Calamagrostis canadensis var. canadensis [CALACAN0]
Calamagrostis fasciculata Kearney = Calamagrostis rubescens [CALARUB]
Calamagrostis fernaldii Louis-Marie = Calamagrostis stricta ssp. inexpansa [CALASTR2]
Calamagrostis hyperborea var. *americana* (Vasey) Kearney = Calamagrostis stricta ssp. inexpansa [CALASTR2]
Calamagrostis hyperborea var. *elongata* Kearney = Calamagrostis stricta ssp. inexpansa [CALASTR2]
Calamagrostis hyperborea var. *stenodes* Kearney = Calamagrostis stricta ssp. inexpansa [CALASTR2]
Calamagrostis inexpansa Gray = Calamagrostis stricta ssp. inexpansa [CALASTR2]
Calamagrostis inexpansa var. *barbulata* Kearney = Calamagrostis stricta ssp. inexpansa [CALASTR2]
Calamagrostis inexpansa var. *brevior* (Vasey) Stebbins = Calamagrostis stricta ssp. inexpansa [CALASTR2]
Calamagrostis inexpansa var. *cuprea* Kearney = Calamagrostis canadensis var. canadensis [CALACAN0]
Calamagrostis inexpansa var. *novae-angliae* Stebbins = Calamagrostis stricta ssp. inexpansa [CALASTR2]
Calamagrostis inexpansa var. *robusta* (Vasey) Stebbins = Calamagrostis canadensis var. canadensis [CALACAN0]
Calamagrostis labradorica Kearney = Calamagrostis stricta ssp. inexpansa [CALASTR2]
Calamagrostis lacustris (Kearney) Nash = Calamagrostis stricta ssp. inexpansa [CALASTR2]
Calamagrostis langsdorfii (Link) Trin. = Calamagrostis canadensis var. langsdorfii [CALACAN7]
Calamagrostis lapponica (Wahlenb.) Hartman [CALALAP] (Lapland reedgrass)
SY: *Calamagrostis lapponica* var. *alpina* Hartman
SY: *Calamagrostis lapponica* var. *groenlandica* Lange
SY: *Calamagrostis lapponica* var. *nearctica* Porsild
Calamagrostis lapponica var. *alpina* Hartman = Calamagrostis lapponica [CALALAP]
Calamagrostis lapponica var. *brevipilis* Stebbins = Calamagrostis stricta ssp. inexpansa [CALASTR2]
Calamagrostis lapponica var. *groenlandica* Lange = Calamagrostis lapponica [CALALAP]
Calamagrostis lapponica var. *nearctica* Porsild = Calamagrostis lapponica [CALALAP]
Calamagrostis luxurians Rydb. = Calamagrostis rubescens [CALARUB]
Calamagrostis montanensis Scribn. ex Vasey [CALAMON] (plains reedgrass)
Calamagrostis neglecta (Ehrh.) P.G. Gaertn., B. Mey. & Scherb. = Calamagrostis stricta [CALASTR]
Calamagrostis nubila Louis-Marie = Calamagrostis canadensis var. langsdorfii [CALACAN7]
Calamagrostis nutkaensis (J. Presl) J. Presl ex Steud. [CALANUT] (Pacific reedgrass)
SY: *Deyeuxia nutkaensis* J. Presl
Calamagrostis pallida Vasey & Scribn. ex Vasey, non Nutt. ex Gray = Calamagrostis canadensis var. canadensis **[CALACAN0]**
Calamagrostis pickeringii var. *lacustris* (Kearney) A.S. Hitchc. = Calamagrostis stricta ssp. inexpansa [CALASTR2]
Calamagrostis purpurascens R. Br. [CALAPUR] (purple reedgrass)
SY: *Calamagrostis purpurascens* ssp. *tasuensis* Calder & Taylor
Calamagrostis purpurascens ssp. *tasuensis* Calder & Taylor = Calamagrostis purpurascens [CALAPUR]
Calamagrostis rubescens Buckl. [CALARUB] (pinegrass)
SY: *Calamagrostis fasciculata* Kearney
SY: *Calamagrostis luxurians* Rydb.
Calamagrostis scabra J. Presl = Calamagrostis canadensis var. langsdorfii [CALACAN7]
Calamagrostis scopulorum var. *bakeri* Stebbins = Calamagrostis stricta ssp. inexpansa [CALASTR2]
Calamagrostis scribneri Beal [CALASCR] (Scribner's reedgrass)
Calamagrostis sesquiflora (Trin.) Tzvelev [CALASES] (one-and-a-half-flowered reedgrass)
Calamagrostis stricta (Timm) Koel. [CALASTR] (slimstem reedgrass)
SY: *Calamagrostis neglecta* (Ehrh.) P.G. Gaertn., B. Mey. & Scherb.
Calamagrostis stricta ssp. **inexpansa** (Gray) C.W. Greene [CALASTR2]
SY: *Calamagrostis californica* Kearney
SY: *Calamagrostis chordorrhiza* Porsild
SY: *Calamagrostis expansa* Rickett & Gilly
SY: *Calamagrostis fernaldii* Louis-Marie
SY: *Calamagrostis hyperborea* var. *americana* (Vasey) Kearney
SY: *Calamagrostis hyperborea* var. *elongata* Kearney
SY: *Calamagrostis hyperborea* var. *stenodes* Kearney
SY: *Calamagrostis inexpansa* Gray
SY: *Calamagrostis inexpansa* var. *barbulata* Kearney
SY: *Calamagrostis inexpansa* var. *brevior* (Vasey) Stebbins
SY: *Calamagrostis inexpansa* var. *novae-angliae* Stebbins
SY: *Calamagrostis labradorica* Kearney
SY: *Calamagrostis lacustris* (Kearney) Nash
SY: *Calamagrostis lapponica* var. *brevipilis* Stebbins
SY: *Calamagrostis pickeringii* var. *lacustris* (Kearney) A.S. Hitchc.
SY: *Calamagrostis scopulorum* var. *bakeri* Stebbins
SY: *Calamagrostis stricta* var. *brevior* Vasey
SY: *Calamagrostis stricta* var. *lacustris* (Kearney) C.W. Greene
Calamagrostis stricta ssp. **stricta** [CALASTR0]
Calamagrostis stricta var. *brevior* Vasey = Calamagrostis stricta ssp. inexpansa [CALASTR2]
Calamagrostis stricta var. *lacustris* (Kearney) C.W. Greene = Calamagrostis stricta ssp. inexpansa [CALASTR2]
Calamovilfa (Gray) Hack. ex Scribn. & Southworth [CALAMOV$]
Calamovilfa longifolia (Hook.) Scribn. [CALALON] (prairie sandgrass)
Catabrosa Beauv. [CATABRO$]
Catabrosa aquatica (L.) Beauv. [CATAAQU] (water hairgrass)
Cenchrus L. [CENCHRU$]
Cenchrus longispinus (Hack.) Fern. [CENCLON] (burgrass)
Ceratochloa carinata (Hook. & Arn.) Tutin = Bromus carinatus [BROMCAR]
Chaetochloa glauca (L.) Scribn. = Setaria glauca [SETAGLA]
Chaetochloa italica (L.) Scribn. = Setaria italica [SETAITA]
Chaetochloa lutescens (Weigel) Stuntz = Setaria glauca [SETAGLA]
Chondrosum gracile Willd. ex Kunth = Bouteloua gracilis [BOUTGRA]
Chondrosum oligostachyum (Nutt.) Torr. = Bouteloua gracilis [BOUTGRA]
Cinna L. [CINNA$]
Cinna latifolia (Trev. ex Goepp.) Griseb. [CINNLAT] (wood reedgrass)
Coleanthus Seidel [COLEANT$]
*****Coleanthus subtilis** (Tratt.) Seidel [COLESUB] (moss grass)
Colpodium fulvum (Trin.) Griseb. = Arctophila fulva [ARCTFUL]
Critesion brachyantherum (Nevski) Barkworth & Dewey = Hordeum brachyantherum [HORDBRA]
Critesion geniculatum (All.) A. Lövc = Hordeum marinum ssp. gussonianum [HORDMAR1]

Critesion hystrix (Roth) A. Löve = Hordeum marinum ssp. gussonianum [HORDMAR1]
Critesion jubatum (L.) Nevski = Hordeum jubatum ssp. jubatum [HORDJUB0]
Critesion jubatum ssp. *breviaristatum* (Bowden) A. & D. Löve = Hordeum jubatum ssp. breviaristatum [HORDJUB1]
Critesion marinum ssp. *gussonianum* (Parl.) Barkworth & Dewey = Hordeum marinum ssp. gussonianum [HORDMAR1]
Critesion murinum (L.) A. Löve = Hordeum murinum ssp. murinum [HORDMUR0]
Critesion murinum ssp. *leporinum* (Link) A. Löve = Hordeum murinum ssp. leporinum [HORDMUR1]
Cynodon L.C. Rich. [CYNODON$]
•**Cynodon dactylon** (L.) Pers. [CYNODAC] (Bermuda grass)
Cynosurus L. [CYNOSUR$]
•**Cynosurus cristatus** L. [CYNOCRI] (crested dogtail)
•**Cynosurus echinatus** L. [CYNOECH] (hedgehog dogtail)
Dactylis L. [DACTYLI$]
•**Dactylis glomerata** L. [DACTGLO] (orchardgrass)
Danthonia DC. [DANTHON$]
Danthonia americana Scribn. = Danthonia californica [DANTCAL]
Danthonia californica Boland. [DANTCAL] (California oatgrass)
SY: *Danthonia americana* Scribn.
SY: *Danthonia californica* var. *americana* (Scribn.) A.S. Hitchc.
SY: *Danthonia californica* var. *palousensis* St. John
SY: *Danthonia californica* var. *piperi* St. John
SY: *Danthonia macounii* A.S. Hitchc.
Danthonia californica var. *americana* (Scribn.) A.S. Hitchc. = Danthonia californica [DANTCAL]
Danthonia californica var. *palousensis* St. John = Danthonia californica [DANTCAL]
Danthonia californica var. *piperi* St. John = Danthonia californica [DANTCAL]
Danthonia canadensis Baum & Findlay = Danthonia intermedia [DANTINT]
Danthonia cusickii (T.A. Williams) A.S. Hitchc. = Danthonia intermedia [DANTINT]
•**Danthonia decumbens** (L.) DC. [DANTDEC] (heather-grass)
SY: *Festuca decumbens* L.
SY: *Sieglingia decumbens* (L.) Bernh.
Danthonia intermedia Vasey [DANTINT] (timber oatgrass)
SY: *Danthonia canadensis* Baum & Findlay
SY: *Danthonia cusickii* (T.A. Williams) A.S. Hitchc.
SY: *Danthonia intermedia* var. *cusickii* T.A. Williams
Danthonia intermedia var. *cusickii* T.A. Williams = Danthonia intermedia [DANTINT]
Danthonia macounii A.S. Hitchc. = Danthonia californica [DANTCAL]
Danthonia pinetorum Piper = Danthonia spicata [DANTSPI]
Danthonia spicata (L.) Beauv. ex Roemer & J.A. Schultes [DANTSPI] (poverty oatgrass)
SY: *Danthonia pinetorum* Piper
SY: *Danthonia spicata* var. *longipila* Scribn. & Merr.
SY: *Danthonia spicata* var. *pinetorum* Piper
SY: *Danthonia thermalis* Scribn.
Danthonia spicata var. *longipila* Scribn. & Merr. = Danthonia spicata [DANTSPI]
Danthonia spicata var. *pinetorum* Piper = Danthonia spicata [DANTSPI]
Danthonia thermalis Scribn. = Danthonia spicata [DANTSPI]
Danthonia unispicata (Thurb.) Munro ex Macoun [DANTUNI] (one-spike oatgrass)
Deschampsia Beauv. [DESCHAM$]
Deschampsia atropurpurea (Wahlenb.) Scheele = Vahlodea atropurpurea [VAHLATR]
Deschampsia atropurpurea var. *latifolia* (Hook.) Scribn. ex Macoun = Vahlodea atropurpurea [VAHLATR]
Deschampsia atropurpurea var. *paramushirensis* Kudo = Vahlodea atropurpurea [VAHLATR]
Deschampsia atropurpurea var. *payettii* Lepage = Vahlodea atropurpurea [VAHLATR]
Deschampsia beringensis Hultén = Deschampsia cespitosa ssp. beringensis [DESCCES1]
Deschampsia calycina J. Presl = Deschampsia danthonioides [DESCDAN]
Deschampsia cespitosa (L.) Beauv. [DESCCES] (tufted hairgrass)
Deschampsia cespitosa ssp. **beringensis** (Hultén) W.E. Lawrence [DESCCES1]
SY: *Deschampsia beringensis* Hultén
SY: *Deschampsia cespitosa* var. *arctica* Vasey
Deschampsia cespitosa ssp. **cespitosa** [DESCCES0]
SY: *Aira cespitosa* L.
SY: *Deschampsia cespitosa* var. *abbei* Boivin
SY: *Deschampsia cespitosa* var. *alpicola* (Rydb.) A. & D. Löve & Kapoor
SY: *Deschampsia cespitosa* ssp. *genuina* (Reichenb.) Volk.
SY: *Deschampsia cespitosa* var. *intercotidalis* Boivin
SY: *Deschampsia cespitosa* var. *littoralis* (Gaudin) Richter
SY: *Deschampsia cespitosa* var. *longiflora* Beal
SY: *Deschampsia cespitosa* var. *maritima* Vasey
Deschampsia cespitosa ssp. *genuina* (Reichenb.) Volk. = Deschampsia cespitosa ssp. cespitosa [DESCCES0]
Deschampsia cespitosa ssp. **holciformis** (J. Presl) W.E. Lawrence [DESCCES3]
SY: *Aira holciformis* (J. Presl) Steud.
SY: *Deschampsia holciformis* J. Presl
Deschampsia cespitosa var. *abbei* Boivin = Deschampsia cespitosa ssp. cespitosa [DESCCES0]
Deschampsia cespitosa var. *alpicola* (Rydb.) A. & D. Löve & Kapoor = Deschampsia cespitosa ssp. cespitosa [DESCCES0]
Deschampsia cespitosa var. *arctica* Vasey = Deschampsia cespitosa ssp. beringensis [DESCCES1]
Deschampsia cespitosa var. *intercotidalis* Boivin = Deschampsia cespitosa ssp. cespitosa [DESCCES0]
Deschampsia cespitosa var. *littoralis* (Gaudin) Richter = Deschampsia cespitosa ssp. cespitosa [DESCCES0]
Deschampsia cespitosa var. *longiflora* Beal = Deschampsia cespitosa ssp. cespitosa [DESCCES0]
Deschampsia cespitosa var. *maritima* Vasey = Deschampsia cespitosa ssp. cespitosa [DESCCES0]
Deschampsia ciliata (Vasey ex Beal) Rydb. = Deschampsia elongata [DESCELO]
Deschampsia danthonioides (Trin.) Munro [DESCDAN] (annual hairgrass)
SY: *Aira danthonioides* Trin.
SY: *Deschampsia calycina* J. Presl
SY: *Deschampsia danthonioides* var. *gracilis* (Vasey) Munz
Deschampsia danthonioides var. *gracilis* (Vasey) Munz = Deschampsia danthonioides [DESCDAN]
Deschampsia elongata (Hook.) Munro [DESCELO] (slender hairgrass)
SY: *Aira elongata* Hook.
SY: *Deschampsia ciliata* (Vasey ex Beal) Rydb.
•**Deschampsia flexuosa** (L.) Trin. [DESCFLE] (wavy hairgrass)
Deschampsia holciformis J. Presl = Deschampsia cespitosa ssp. holciformis [DESCCES3]
Deschampsia pacifica Tatew. & Ohwi = Vahlodea atropurpurea [VAHLATR]
Deyeuxia nutkaensis J. Presl = Calamagrostis nutkaensis [CALANUT]
Dichanthelium (A.S. Hitchc. & Chase) Gould [DICHANT$]
Dichanthelium acuminatum (Sw.) Gould & C.A. Clark [DICHACU]
Dichanthelium acuminatum var. **fasciculatum** (Torr.) Freckmann [DICHACU1]
SY: *Dichanthelium acuminatum* var. *implicatum* (Scribn.) Gould & C.A. Clark
SY: *Dichanthelium lanuginosum* (Ell.) Gould
SY: *Dichanthelium lanuginosum* var. *fasciculatum* (Torr.) Spellenberg
SY: *Panicum acuminatum* var. *fasciculatum* (Torr.) Lelong
SY: *Panicum acuminatum* var. *implicatum* (Scribn.) C.F. Reed
SY: *Panicum brodiei* St. John
SY: *Panicum curtifolium* Nash
SY: *Panicum glutinoscabrum* Fern.
SY: *Panicum huachucae* Ashe
SY: *Panicum huachucae* var. *fasciculatum* (Torr.) F.T. Hubbard
SY: *Panicum implicatum* Scribn.

SY: *Panicum languidum* A.S. Hitchc.
SY: *Panicum lanuginosum* Ell.
SY: *Panicum lanuginosum* var. *fasciculatum* (Torr.) Fern.
SY: *Panicum lanuginosum* var. *huachucae* (Ashe) A.S. Hitchc.
SY: *Panicum lanuginosum* var. *implicatum* (Scribn.) Fern.
SY: *Panicum lanuginosum* var. *tennesseense* (Ashe) Gleason
SY: *Panicum lassenianum* Schmoll
SY: *Panicum lindheimeri* var. *fasciculatum* (Torr.) Fern.
SY: *Panicum occidentale* Scribn.
SY: *Panicum pacificum* A.S. Hitchc. & Chase
SY: *Panicum subvillosum* Ashe
SY: *Panicum tennesseense* Ashe
Dichanthelium acuminatum var. *implicatum* (Scribn.) Gould & C.A. Clark = Dichanthelium acuminatum var. fasciculatum [DICHACU1]
Dichanthelium acuminatum var. **thermale** (Boland.) Freckmann [DICHACU2]
SY: *Dichanthelium lanuginosum* var. *thermale* (Boland.) Spellenberg
SY: *Panicum thermale* Boland.
Dichanthelium lanuginosum (Ell.) Gould = Dichanthelium acuminatum var. fasciculatum [DICHACU1]
Dichanthelium lanuginosum var. *fasciculatum* (Torr.) Spellenberg = Dichanthelium acuminatum var. fasciculatum [DICHACU1]
Dichanthelium lanuginosum var. *thermale* (Boland.) Spellenberg = Dichanthelium acuminatum var. thermale [DICHACU2]
Dichanthelium oligosanthes (J.A. Schultes) Gould [DICHOLI]
Dichanthelium oligosanthes var. *helleri* (Nash) Mohlenbrock = Dichanthelium oligosanthes var. scribnerianum [DICHOLI1]
Dichanthelium oligosanthes var. **scribnerianum** (Nash) Gould [DICHOLI1]
SY: *Dichanthelium oligosanthes* var. *helleri* (Nash) Mohlenbrock
SY: *Panicum helleri* Nash
SY: *Panicum macrocarpon* Le Conte ex Torr.
SY: *Panicum oligosanthes* var. *helleri* (Nash) Fern.
SY: *Panicum oligosanthes* var. *scribnerianum* (Nash) Fern.
SY: *Panicum scoparium* S. Wats. ex Nash
SY: *Panicum scribnerianum* Nash
Digitaria Haller [DIGITAR$]
*****Digitaria ischaemum** (Schreb.) Muhl. [DIGIISC] (smooth crabgrass)
*****Digitaria sanguinalis** (L.) Scop. [DIGISAN] (hairy crabgrass)
SY: *Panicum sanguinale* L.
SY: *Syntherisma sanguinalis* (L.) Dulac
Distichlis Raf. [DISTICH$]
Distichlis dentata Rydb. = Distichlis spicata [DISTSPI]
Distichlis spicata (L.) Greene [DISTSPI] (seashore saltgrass)
SY: *Distichlis dentata* Rydb.
SY: *Distichlis spicata* var. *borealis* (J. Presl) Beetle
SY: *Distichlis spicata* var. *divaricata* Beetle
SY: *Distichlis spicata* var. *nana* Beetle
SY: *Distichlis spicata* var. *stolonifera* Beetle
SY: *Distichlis spicata* ssp. *stricta* (Torr.) Thorne
SY: *Distichlis spicata* var. *stricta* (Torr.) Scribn.
SY: *Distichlis stricta* (Torr.) Rydb.
SY: *Distichlis stricta* var. *dentata* (Rydb.) C.L. Hitchc.
SY: *Uniola spicata* L.
Distichlis spicata ssp. *stricta* (Torr.) Thorne = Distichlis spicata [DISTSPI]
Distichlis spicata var. *borealis* (J. Presl) Beetle = Distichlis spicata [DISTSPI]
Distichlis spicata var. *divaricata* Beetle = Distichlis spicata [DISTSPI]
Distichlis spicata var. *nana* Beetle = Distichlis spicata [DISTSPI]
Distichlis spicata var. *stolonifera* Beetle = Distichlis spicata [DISTSPI]
Distichlis spicata var. *stricta* (Torr.) Scribn. = Distichlis spicata [DISTSPI]
Distichlis stricta (Torr.) Rydb. = Distichlis spicata [DISTSPI]
Distichlis stricta var. *dentata* (Rydb.) C.L. Hitchc. = Distichlis spicata [DISTSPI]
Echinochloa Beauv. [ECHINOC$]
*****Echinochloa crusgalli** (L.) Beauv. [ECHICRU] (large barnyard-grass)

× **Elyhordeum** Mansf. ex Zizin & Petrowa [ELYHORD$]
× **Elyhordeum macounii** (Vasey) Barkworth & Dewey [ELYHMAC]
SY: × *Agrohordeum macounii* (Vasey) Lepage
SY: × *Agrohordeum macounii* var. *valencianum* Bowden
SY: × *Elymordeum macounii* (Vasey) Barkworth
SY: *Elymus macounii* Vasey
SY: × *Elytesion macounii* (Vasey) Barkworth & Dewey
× *Elymordeum macounii* (Vasey) Barkworth = × Elyhordeum macounii [ELYHMAC]
Elymus L. [ELYMUS$]
Elymus alaskanus (Scribn. & Merr.) A. Löve [ELYMALA] (Alaska wildrye)
SY: *Agropyron alaskanum* Scribn. & Merr.
SY: *Agropyron violaceum* (Hornem.) Lange
SY: *Elymus trachycaulus* ssp. *violaceus* (Hornem.) A. & D. Löve
Elymus alaskanus ssp. **latiglumis** (Scribn. & J.G. Sm.) A. Löve [ELYMALA1]
SY: *Agropyron caninum* var. *latiglume* (Scribn. & J.G. Sm.) Pease & Moore
SY: *Agropyron latiglume* (Scribn. & J.G. Sm.) Rydb.
SY: *Agropyron trachycaulum* var. *latiglume* (Scribn. & J.G. Sm.) Beetle
SY: *Elymus trachycaulus* ssp. *latiglumis* (Scribn. & J.G. Sm.) Barkworth & Dewey
SY: *Elymus trachycaulus* var. *latiglumis* (Scribn. & J.G. Sm.) Beetle
Elymus albicans (Scribn. & J.G. Sm.) A. Löve = Elymus lanceolatus ssp. albicans [ELYMLAN1]
Elymus albicans var. *griffithii* (Scribn. & J.G. Sm. ex Piper) Dorn = Elymus lanceolatus ssp. albicans [ELYMLAN1]
Elymus brachystachys Scribn. & Ball = Elymus canadensis [ELYMCAN]
Elymus brownii Scribn. & J.G. Sm. = Leymus innovatus [LEYMINN]
Elymus canadensis L. [ELYMCAN] (Canada wildrye)
SY: *Elymus brachystachys* Scribn. & Ball
Elymus cinereus Scribn. & Merr. = Leymus cinereus [LEYMCIN]
Elymus cinereus var. *pubens* (Piper) C.L. Hitchc. = Leymus cinereus [LEYMCIN]
Elymus condensatus var. *pubens* Piper = Leymus cinereus [LEYMCIN]
Elymus condensatus var. *triticoides* (Buckl.) Thurb. = Leymus triticoides [LEYMTRI]
Elymus elongatus ssp. *ponticus* (Podp.) Melderis = Elytrigia pontica [ELYTPON]
Elymus elongatus var. *ponticus* (Podp.) Dorn = Elytrigia pontica [ELYTPON]
Elymus elymoides (Raf.) Swezey [ELYMELY] (squirreltail grass)
SY: *Elymus elymoides* var. *brevifolius* (J.G. Sm.) Dorn
SY: *Elymus elymoides* ssp. *californicus* Barkworth
SY: *Elymus longifolius* (J.G. Sm.) Gould
SY: *Elymus sitanion* J.A. Schultes
SY: *Sitanion elymoides* Raf.
SY: *Sitanion hystrix* (Nutt.) J.G. Sm.
SY: *Sitanion hystrix* var. *brevifolium* (J.G. Sm.) C.L. Hitchc.
SY: *Sitanion hystrix* var. *californicum* (J.G. Sm.) F.D. Wilson
SY: *Sitanion longifolium* J.G. Sm.
Elymus elymoides ssp. *californicus* Barkworth = Elymus elymoides [ELYMELY]
Elymus elymoides var. *brevifolius* (J.G. Sm.) Dorn = Elymus elymoides [ELYMELY]
Elymus glaucus Buckl. [ELYMGLA] (blue wildrye)
Elymus glaucus ssp. **glaucus** [ELYMGLA0]
SY: *Elymus glaucus* var. *breviaristatus* Burtt-Davy
Elymus glaucus ssp. **virescens** (Piper) Gould [ELYMGLA5]
SY: *Elymus glaucus* var. *virescens* (Piper) Bowden
SY: *Elymus virescens* Piper
Elymus glaucus var. *breviaristatus* Burtt-Davy = Elymus glaucus ssp. glaucus [ELYMGLA0]
Elymus glaucus var. *virescens* (Piper) Bowden = Elymus glaucus ssp. virescens [ELYMGLA5]
Elymus griffithii (Scribn. & J.G. Sm. ex Piper) A. Löve = Elymus lanceolatus ssp. albicans [ELYMLAN1]
Elymus hirsutus J. Presl [ELYMHIR] (hairy wildrye)

Elymus hispidus (Opiz) Melderis = Elytrigia intermedia [ELYTINT]
Elymus hispidus ssp. *barbulatus* (Schur) Melderis = Elytrigia intermedia [ELYTINT]
Elymus hispidus var. *ruthenicus* (Griseb.) Dorn = Elytrigia intermedia [ELYTINT]
Elymus innovatus Beal = Leymus innovatus [LEYMINN]
Elymus innovatus var. *glabratus* Bowden = Leymus innovatus [LEYMINN]
Elymus innovatus var. *velutinus* Bowden = Leymus innovatus [LEYMINN]
Elymus lanceolatus (Scribn. & J.G. Sm.) Gould [ELYMLAN] (thickspike wildrye)
Elymus lanceolatus ssp. **albicans** (Scribn. & J.G. Sm.) Barkworth & Dewey [ELYMLAN1]
SY: *Agropyron albicans* Scribn. & J.G. Sm.
SY: *Agropyron albicans* var. *griffithii* (Scribn. & J.G. Sm. ex Piper) Beetle
SY: *Agropyron dasystachyum* ssp. *albicans* (Scribn. & J.G. Sm.) Dewey
SY: *Agropyron griffithii* Scribn. & J.G. Sm. ex Piper
SY: *Elymus albicans* (Scribn. & J.G. Sm.) A. Löve
SY: *Elymus albicans* var. *griffithii* (Scribn. & J.G. Sm. ex Piper) Dorn
SY: *Elymus griffithii* (Scribn. & J.G. Sm. ex Piper) A. Löve
SY: *Elytrigia dasystachya* ssp. *albicans* (Scribn. & J.G. Sm.) Dewey
SY: *Roegneria albicans* (Scribn. & J.G. Sm.) Beetle
SY: *Roegneria albicans* var. *griffithii* (Scribn. & J.G. Sm. ex Piper) Beetle
Elymus lanceolatus ssp. **lanceolatus** [ELYMLAN0]
SY: *Agropyron dasystachyum* (Hook.) Scribn. & J.G. Sm.
SY: *Agropyron dasystachyum* var. *riparum* (Scribn. & J.G. Sm.) Bowden
SY: *Agropyron elmeri* Scribn.
SY: *Agropyron lanceolatum* Scribn. & J.G. Sm.
SY: *Agropyron riparum* Scribn. & J.G. Sm.
SY: *Elymus lanceolatus* var. *riparius* (Scribn. & J.G. Sm.) Dorn
SY: *Elytrigia dasystachya* (Hook.) A. & D. Löve
SY: *Elytrigia ripara* (Scribn. & J.G. Sm.) Beetle
Elymus lanceolatus ssp. *yukonensis* (Scribn. & Merr.) A. Löve = Elymus × yukonensis [ELYMYUK]
Elymus lanceolatus var. *riparius* (Scribn. & J.G. Sm.) Dorn = Elymus lanceolatus ssp. lanceolatus [ELYMLAN0]
Elymus longifolius (J.G. Sm.) Gould = Elymus elymoides [ELYMELY]
Elymus macounii Vasey = × Elyhordeum macounii [ELYHMAC]
Elymus mollis Trin. = Leymus mollis [LEYMMOL]
Elymus orcuttianus Vasey = Leymus triticoides [LEYMTRI]
Elymus pauciflorum var. *subsecundus* Gould = Elymus trachycaulus ssp. subsecundus [ELYMTRA1]
Elymus pauciflorus (Schwein.) Gould = Elymus trachycaulus ssp. trachycaulus [ELYMTRA0]
Elymus piperi Bowden = Leymus cinereus [LEYMCIN]
Elymus repens (L.) Gould = Elytrigia repens [ELYTREP]
Elymus × saundersii Vasey (pro sp.) [ELYMXSA] (Saunder's wildrye)
SY: *Agropyron saundersii* (Vasey) A.S. Hitchc.
SY: × *Agrositanion saundersii* (Vasey) Bowden
SY: *Elymus saundersii* var. *californicus* Hoover
Elymus saundersii var. *californicus* Hoover = Elymus × saundersii [ELYMXSA]
Elymus scribneri (Vasey) M.E. Jones [ELYMSCR]
SY: *Agropyron scribneri* Vasey
Elymus sitanion J.A. Schultes = Elymus elymoides [ELYMELY]
Elymus smithii (Rydb.) Gould = Pascopyrum smithii [PASCSMI]
Elymus spicatus (Pursh) Gould = Pseudoroegneria spicata [PSEUSPI]
Elymus trachycaulus (Link) Gould ex Shinners [ELYMTRA] (slender wheatgrass)
Elymus trachycaulus ssp. *andinus* (Scribn. & J.G. Sm.) A. & D. Löve = Elymus trachycaulus ssp. trachycaulus [ELYMTRA0]
Elymus trachycaulus ssp. *latiglumis* (Scribn. & J.G. Sm.) Barkworth & Dewey = Elymus alaskanus ssp. latiglumis [ELYMALA1]
Elymus trachycaulus ssp. *novae-angliae* (Scribn.) Tzvelev = Elymus trachycaulus ssp. trachycaulus [ELYMTRA0]
Elymus trachycaulus ssp. **subsecundus** (Link) A. & D. Löve [ELYMTRA1]
SY: *Agropyron caninum* var. *unilaterale* (Cassidy) C.L. Hitchc.
SY: *Agropyron pauciflorum* var. *glaucum* (Pease & Moore) Taylor
SY: *Agropyron pauciflorum* var. *unilaterale* (Vasey) Taylor & MacBryde
SY: *Agropyron subsecundum* (Link) A.S. Hitchc.
SY: *Agropyron trachycaulum* var. *ciliatum* (Scribn. & J.G. Sm.) Gleason
SY: *Agropyron trachycaulum* var. *glaucum* (Pease & Moore) Malte
SY: *Agropyron trachycaulum* var. *unilaterale* (Cassidy) Malte
SY: *Elymus pauciflorum* var. *subsecundus* Gould
SY: *Elymus trachycaulus* var. *unilateralis* (Cassidy) Beetle
Elymus trachycaulus ssp. *teslinensis* (Porsild & Senn) A. Löve = Elymus trachycaulus ssp. trachycaulus [ELYMTRA0]
Elymus trachycaulus ssp. **trachycaulus** [ELYMTRA0]
SY: *Agropyron brevifolium* Scribn.
SY: *Agropyron* × *brevifolium* Scribn.
SY: *Agropyron caninum* var. *andinum* (Scribn. & J.G. Sm.) C.L. Hitchc.
SY: *Agropyron caninum* var. *hornemannii* (Koch) Pease & Moore
SY: *Agropyron caninum* ssp. *majus* (Vasey) C.L. Hitchc.
SY: *Agropyron caninum* var. *mitchellii* Welsh
SY: *Agropyron novae-angliae* Scribn.
SY: *Agropyron pauciflorum* (Schwein.) A.S. Hitchc. ex Silveus
SY: *Agropyron pauciflorum* ssp. *majus* (Vasey) Melderis
SY: *Agropyron pauciflorum* ssp. *novae-angliae* (Scribn.) Melderis
SY: *Agropyron pauciflorum* var. *novae-angliae* (Scribn.) Taylor & MacBryde
SY: *Agropyron pauciflorum* ssp. *teslinense* (Porsild & Senn) Melderis
SY: *Agropyron subsecundum* var. *andinum* (Scribn. & J.G. Sm.) A.S. Hitchc.
SY: *Agropyron tenerum* Vasey
SY: *Agropyron teslinense* Porsild & Senn
SY: *Agropyron trachycaulum* (Link) Malte ex H.F. Lewis
SY: *Agropyron trachycaulum* var. *majus* (Vasey) Fern.
SY: *Agropyron trachycaulum* var. *novae-angliae* (Scribn.) Fern.
SY: *Agropyron violaceum* ssp. *andinum* (Scribn. & J.G. Sm.) Melderis
SY: *Agropyron violaceum* var. *andinum* Scribn. & J.G. Sm.
SY: *Elymus pauciflorus* (Schwein.) Gould
SY: *Elymus trachycaulus* ssp. *andinus* (Scribn. & J.G. Sm.) A. & D. Löve
SY: *Elymus trachycaulus* var. *andinus* (Scribn. & J.G. Sm.) Dorn
SY: *Elymus trachycaulus* var. *majus* (Vasey) Beetle
SY: *Elymus trachycaulus* ssp. *novae-angliae* (Scribn.) Tzvelev
SY: *Elymus trachycaulus* ssp. *teslinensis* (Porsild & Senn) A. Löve
SY: *Roegneria pauciflora* (Schwein.) Hyl.
SY: *Roegneria trachycaula* (Link) Nevski
SY: *Triticum trachycaulum* Link
Elymus trachycaulus ssp. *violaceus* (Hornem.) A. & D. Löve = Elymus alaskanus [ELYMALA]
Elymus trachycaulus var. *andinus* (Scribn. & J.G. Sm.) Dorn = Elymus trachycaulus ssp. trachycaulus [ELYMTRA0]
Elymus trachycaulus var. *latiglumis* (Scribn. & J.G. Sm.) Beetle = Elymus alaskanus ssp. latiglumis [ELYMALA1]
Elymus trachycaulus var. *majus* (Vasey) Beetle = Elymus trachycaulus ssp. trachycaulus [ELYMTRA0]
Elymus trachycaulus var. *unilateralis* (Cassidy) Beetle = Elymus trachycaulus ssp. subsecundus [ELYMTRA1]
Elymus triticoides Buckl. = Leymus triticoides [LEYMTRI]
Elymus triticoides var. *pubescens* A.S. Hitchc. = Leymus triticoides [LEYMTRI]
Elymus varnensis (Velen.) Runemark = Elytrigia pontica [ELYTPON]
Elymus virescens Piper = Elymus glaucus ssp. virescens [ELYMGLA5]
Elymus × yukonensis (Scribn. & Merr.) A. Löve (pro sp.) [ELYMYUK]

SY: *Agropyron dasystachyum* ssp. *yukonense* (Scribn. & Merr.) Dewey
SY: *Agropyron yukonense* Scribn. & Merr.
SY: *Elymus lanceolatus* ssp. *yukonensis* (Scribn. & Merr.) A. Löve
SY: *Elymus yukonensis* (Scribn. & Merr.) A. Löve
SY: *Elytrigia dasystachya* ssp. *yukonensis* (Scribn. & Merr.) Dewey
Elymus yukonensis (Scribn. & Merr.) A. Löve = Elymus × yukonensis [ELYMYUK]
× *Elytesion macounii* (Vasey) Barkworth & Dewey = × Elyhordeum macounii [ELYHMAC]
Elytrigia Desv. [ELYTRIG$]
Elytrigia dasystachya (Hook.) A. & D. Löve = Elymus lanceolatus ssp. lanceolatus [ELYMLAN0]
Elytrigia dasystachya ssp. *albicans* (Scribn. & J.G. Sm.) Dewey = Elymus lanceolatus ssp. albicans [ELYMLAN1]
Elytrigia dasystachya ssp. *yukonensis* (Scribn. & Merr.) Dewey = Elymus × yukonensis [ELYMYUK]
*__Elytrigia intermedia__ (Host) Nevski [ELYTINT]
SY: *Agropyron intermedium* (Host) Beauv.
SY: *Agropyron intermedium* var. *trichophorum* (Link) Halac.
SY: *Agropyron trichophorum* (Link) Richter
SY: *Elymus hispidus* (Opiz) Melderis
SY: *Elymus hispidus* ssp. *barbulatus* (Schur) Melderis
SY: *Elymus hispidus* var. *ruthenicus* (Griseb.) Dorn
SY: *Elytrigia intermedia* ssp. *barbulata* (Schur) A. Löve
SY: *Thinopyrum intermedium* (Host) Barkworth & Dewey
SY: *Thinopyrum intermedium* ssp. *barbulatum* (Schur) Barkworth & Dewey
Elytrigia intermedia ssp. *barbulata* (Schur) A. Löve = Elytrigia intermedia [ELYTINT]
*__Elytrigia pontica__ (Podp.) Holub [ELYTPON]
SY: *Agropyron varnense* (Velen.) Hayek
SY: *Elymus elongatus* ssp. *ponticus* (Podp.) Melderis
SY: *Elymus elongatus* var. *ponticus* (Podp.) Dorn
SY: *Elymus varnensis* (Velen.) Runemark
SY: *Thinopyrum ponticum* (Podp.) Barkworth & Dewey
*__Elytrigia repens__ (L.) Desv. ex B.D. Jackson [ELYTREP]
SY: *Agropyron repens* (L.) Beauv.
SY: *Agropyron repens* var. *subulatum* (Schreb.) Roemer & J.A. Schultes
SY: *Elymus repens* (L.) Gould
SY: *Triticum repens* L.
Elytrigia ripara (Scribn. & J.G. Sm.) Beetle = Elymus lanceolatus ssp. lanceolatus [ELYMLAN0]
Elytrigia smithii (Rydb.) Nevski = Pascopyrum smithii [PASCSMI]
Elytrigia smithii var. *mollis* (Scribn. & J.G. Sm.) Beetle = Pascopyrum smithii [PASCSMI]
Elytrigia spicata (Pursh) Dewey = Pseudoroegneria spicata [PSEUSPI]
Eragrostis von Wolf [ERAGROS$]
*__Eragrostis cilianensis__ (All.) Lut. ex Janchen [ERAGCIL] (stinkgrass)
SY: *Eragrostis major* Host
SY: *Eragrostis megastachya* (Koel.) Link
SY: *Poa cilianensis* All.
Eragrostis eragrostis (L.) Beauv. = Eragrostis minor [ERAGMIN]
Eragrostis major Host = Eragrostis cilianensis [ERAGCIL]
Eragrostis megastachya (Koel.) Link = Eragrostis cilianensis [ERAGCIL]
*__Eragrostis minor__ Host [ERAGMIN] (little lovegrass)
SY: *Eragrostis eragrostis* (L.) Beauv.
SY: *Eragrostis poaeoides* Beauv. ex Roemer & J.A. Schultes
Eragrostis pectinacea (Michx.) Nees ex Steud. [ERAGPEC] (tufted lovegrass)
*__Eragrostis pilosa__ (L.) Beauv. [ERAGPIL] (Indian lovegrass)
Eragrostis poaeoides Beauv. ex Roemer & J.A. Schultes = Eragrostis minor [ERAGMIN]
Eremopyrum (Ledeb.) Jaubert & Spach [EREMOPY$]
*__Eremopyrum triticeum__ (Gaertn.) Nevski [EREMTRI] (annual wheatgrass)
SY: *Agropyron prostratum* (L. f.) Beauv.
SY: *Agropyron triticeum* Gaertn.
Eriocoma cuspidata Nutt. = Oryzopsis hymenoides [ORYZHYM]
Festuca L. [FESTUCA$]
Festuca altaica Trin. [FESTALT] (Altai fescue)
Festuca altaica ssp. *scabrella* (Torr. ex Hook.) Hultén = Festuca campestris [FESTCAM]
Festuca altaica var. *major* (Vasey) Gleason = Festuca campestris [FESTCAM]
Festuca altaica var. *scabrella* (Torr. ex Hook.) Breitung = Festuca campestris [FESTCAM]
*__Festuca arundinacea__ Schreb. [FESTARU] (tall fescue)
SY: *Festuca elatior* ssp. *arundinacea* (Schreb.) Hack.
SY: *Festuca elatior* var. *arundinacea* (Schreb.) C.F.H. Wimmer
Festuca aucta Krecz. & Bobr. = Festuca rubra ssp. aucta [FESTRUB5]
Festuca baffinensis Polunin [FESTBAF] (Baffin fescue)
Festuca brachyphylla J.A. Schultes ex J.A. & J.H. Schultes [FESTBRA] (alpine fescue)
SY: *Festuca ovina* var. *brachyphylla* (J.A. Schultes ex J.A. & J.H. Schultes) Piper ex A.S. Hitchc.
SY: *Festuca ovina* var. *brevifolia* S. Wats.
Festuca brachyphylla var. *rydbergii* (St.-Yves) Cronq. = Festuca saximontana [FESTSAX]
Festuca brevifolia var. *utahensis* St.-Yves = Festuca minutiflora [FESTMIN]
Festuca brevipila Tracey = Festuca trachyphylla [FESTTRA]
Festuca bromoides L. = Vulpia bromoides [VULPBRO]
Festuca campestris Rydb. [FESTCAM] (rough fescue)
SY: *Festuca altaica* var. *major* (Vasey) Gleason
SY: *Festuca altaica* ssp. *scabrella* (Torr. ex Hook.) Hultén
SY: *Festuca altaica* var. *scabrella* (Torr. ex Hook.) Breitung
SY: *Festuca scabrella* Torr. ex Hook.
SY: *Festuca scabrella* var. *major* Vasey
Festuca capillata Lam. = Festuca filiformis [FESTFIL]
Festuca decumbens L. = Danthonia decumbens [DANTDEC]
Festuca densiuscula (Hack. ex Piper) Alexeev = Festuca rubra ssp. pruinosa [FESTRUB1]
Festuca detonensis (All.) Aschers. & Graebn. = Vulpia bromoides [VULPBRO]
Festuca duriuscula auct. = Festuca trachyphylla [FESTTRA]
Festuca elatior L. = Festuca pratensis [FESTPRA]
Festuca elatior ssp. *arundinacea* (Schreb.) Hack. = Festuca arundinacea [FESTARU]
Festuca elatior var. *arundinacea* (Schreb.) C.F.H. Wimmer = Festuca arundinacea [FESTARU]
*__Festuca filiformis__ Pourret [FESTFIL] (hair fescue)
SY: *Festuca capillata* Lam.
SY: *Festuca ovina* var. *capillata* (Lam.) Alef.
SY: *Festuca ovina* var. *tenuifolia* (Sibthorp) Sm.
SY: *Festuca tenuifolia* Sibthorp
Festuca gracilenta Buckl. = Vulpia octoflora var. glauca [VULPOCT1]
Festuca howellii Hack. ex Beal = Festuca viridula [FESTVIR]
Festuca idahoensis Elmer [FESTIDA] (Idaho fescue)
Festuca idahoensis var. **roemeri** Pavlick [FESTIDA2]
SY: *Festuca roemeri* (Pavlick) Alexeev
Festuca megalura Nutt. = Vulpia myuros [VULPMYU]
Festuca megalura var. *hirsuta* (Hack.) Aschers. & Graebn. = Vulpia myuros [VULPMYU]
Festuca microstachys var. *pauciflora* Scribn. ex Beal = Vulpia microstachys var. pauciflora [VULPMIC1]
Festuca microstachys var. *simulans* (Hoover) Hoover = Vulpia microstachys var. pauciflora [VULPMIC1]
Festuca minutiflora Rydb. [FESTMIN] (little fescue)
SY: *Festuca brevifolia* var. *utahensis* St.-Yves
SY: *Festuca ovina* var. *minutiflora* (Rydb.) J.T. Howell
Festuca myuros L. = Vulpia myuros [VULPMYU]
Festuca occidentalis Hook. [FESTOCC] (western fescue)
SY: *Festuca ovina* var. *polyphylla* Vasey ex Beal
Festuca octoflora Walt. = Vulpia octoflora var. octoflora [VULPOCT0]
Festuca octoflora ssp. *hirtella* Piper = Vulpia octoflora var. hirtella [VULPOCT2]

Festuca octoflora var. *aristulata* Torr. ex L.H. Dewey = Vulpia octoflora var. octoflora [VULPOCT0]
Festuca octoflora var. *glauca* (Nutt.) Fern. = Vulpia octoflora var. glauca [VULPOCT1]
Festuca octoflora var. *hirtella* (Piper) Piper ex A.S. Hitchc. = Vulpia octoflora var. hirtella [VULPOCT2]
Festuca octoflora var. *tenella* (Willd.) Fern. = Vulpia octoflora var. glauca [VULPOCT1]
Festuca ovina var. *brachyphylla* (J.A. Schultes ex J.A. & J.H. Schultes) Piper ex A.S. Hitchc. = Festuca brachyphylla **[FESTBRA]**
Festuca ovina var. *brevifolia* S. Wats. = Festuca brachyphylla [FESTBRA]
Festuca ovina var. *capillata* (Lam.) Alef. = Festuca filiformis [FESTFIL]
Festuca ovina var. *duriuscula* auct. = Festuca trachyphylla [FESTTRA]
Festuca ovina var. *minutiflora* (Rydb.) J.T. Howell = Festuca minutiflora [FESTMIN]
Festuca ovina var. *polyphylla* Vasey ex Beal = Festuca occidentalis [FESTOCC]
Festuca ovina var. *purpusiana* St.-Yves = Festuca saximontana var. purpusiana [FESTSAX1]
Festuca ovina var. *saximontana* (Rydb.) Gleason = Festuca saximontana [FESTSAX]
Festuca ovina var. *tenuifolia* (Sibthorp) Sm. = Festuca filiformis [FESTFIL]
Festuca pacifica Piper = Vulpia microstachys var. pauciflora [VULPMIC1]
Festuca pacifica var. *simulans* Hoover = Vulpia microstachys var. pauciflora [VULPMIC1]
*****Festuca pratensis*** Huds. [FESTPRA] (meadow fescue)
SY: *Festuca elatior* L.
SY: *Festuca shortii* Kunth ex Wood
Festuca prolifera (Piper) Fern. = Festuca rubra ssp. arctica [FESTRUB3]
Festuca prolifera var. *lasiolepis* Fern. = Festuca rubra ssp. arctica [FESTRUB3]
Festuca reflexa Buckl. = Vulpia microstachys var. pauciflora [VULPMIC1]
Festuca richardsonii Hook. = Festuca rubra ssp. arctica [FESTRUB3]
Festuca richardsonii ssp. *cryophila* (Krecz. & Bobr.) A. & D. Löve = Festuca rubra ssp. arctica [FESTRUB3]
Festuca roemeri (Pavlick) Alexeev = Festuca idahoensis var. roemeri [FESTIDA2]
Festuca rubra L. [FESTRUB] (red fescue)
Festuca rubra ssp. **arctica** (Hack.) Govor. [FESTRUB3]
SY: *Festuca prolifera* (Piper) Fern.
SY: *Festuca prolifera* var. *lasiolepis* Fern.
SY: *Festuca richardsonii* Hook.
SY: *Festuca richardsonii* ssp. *cryophila* (Krecz. & Bobr.) A. & D. Löve
SY: *Festuca rubra* ssp. *cryophila* (Krecz. & Bobr.) Hultén
SY: *Festuca rubra* var. *mutica* Hartman
SY: *Festuca rubra* var. *prolifera* (Piper) Piper
SY: *Festuca rubra* ssp. *richardsonii* (Hook.) Hultén
SY: *Festuca rubra* var. *richardsonii* (Hook.) Hultén
Festuca rubra ssp. **arnicola** Alexeev [FESTRUB4]
Festuca rubra ssp. **aucta** (Krecz. & Bobr.) Hultén [FESTRUB5]
SY: *Festuca aucta* Krecz. & Bobr.
SY: *Festuca rubra* ssp. *aucta* f. *pseudovivipara* Pavl.
Festuca rubra ssp. *aucta* f. *pseudovivipara* Pavl. = Festuca rubra ssp. aucta [FESTRUB5]
Festuca rubra ssp. *cryophila* (Krecz. & Bobr.) Hultén = Festuca rubra ssp. arctica [FESTRUB3]
Festuca rubra ssp. *densiuscula* Hack. ex Piper = Festuca rubra ssp. pruinosa [FESTRUB1]
Festuca rubra ssp. **pruinosa** (Hack.) Piper [FESTRUB1]
SY: *Festuca densiuscula* (Hack. ex Piper) Alexeev
SY: *Festuca rubra* ssp. *densiuscula* Hack. ex Piper
Festuca rubra ssp. *richardsonii* (Hook.) Hultén = Festuca rubra ssp. arctica [FESTRUB3]
Festuca rubra ssp. **secunda** (J. Presl) Pavlick [FESTRUB7]
Festuca rubra ssp. **secunda** var. **mediana** Pavlick [FESTRUB8]
SY: *Festuca rubra* var. *littoralis* Vasey ex Beal
Festuca rubra ssp. **vallicola** (Rydb.) Pavlick [FESTRUB9]
SY: *Festuca vallicola* Rydb.
Festuca rubra var. *littoralis* Vasey ex Beal = Festuca rubra ssp. secunda var. mediana [FESTRUB8]
Festuca rubra var. *mutica* Hartman = Festuca rubra ssp. arctica [FESTRUB3]
Festuca rubra var. *prolifera* (Piper) Piper = Festuca rubra ssp. arctica [FESTRUB3]
Festuca rubra var. *richardsonii* (Hook.) Hultén = Festuca rubra ssp. arctica [FESTRUB3]
Festuca saximontana Rydb. [FESTSAX] (Rocky Mountain fescue)
SY: *Festuca brachyphylla* var. *rydbergii* (St.-Yves) Cronq.
SY: *Festuca ovina* var. *saximontana* (Rydb.) Gleason
Festuca saximontana var. **purpusiana** (St.-Yves) Frederiksen & Pavlick [FESTSAX1]
SY: *Festuca ovina* var. *purpusiana* St.-Yves
Festuca saximontana var. **robertsiana** Pavlick [FESTSAX2]
Festuca scabrella Torr. ex Hook. = Festuca campestris [FESTCAM]
Festuca scabrella var. *major* Vasey = Festuca campestris [FESTCAM]
Festuca shortii Kunth ex Wood = Festuca pratensis [FESTPRA]
Festuca subulata Trin. [FESTSUB] (bearded fescue)
Festuca subuliflora Scribn. [FESTSUU] (crinkle-awned fescue)
Festuca tenella Willd. = Vulpia octoflora var. glauca [VULPOCT1]
Festuca tenella var. *glauca* Nutt. = Vulpia octoflora var. glauca [VULPOCT1]
Festuca tenuifolia Sibthorp = Festuca filiformis [FESTFIL]
*****Festuca trachyphylla*** (Hack.) Krajina [FESTTRA] (hard fescue)
SY: *Festuca brevipila* Tracey
SY: *Festuca duriuscula* auct.
SY: *Festuca ovina* var. *duriuscula* auct.
Festuca vallicola Rydb. = Festuca rubra ssp. vallicola [FESTRUB9]
Festuca viridula Vasey [FESTVIR] (green fescue)
SY: *Festuca howellii* Hack. ex Beal
Festuca vivipara (L.) Sm. [FESTVIV]
Festuca vivipara ssp. **glabra** Frederiksen [FESTVIV2]
Festuca viviparoidea Krajina ex Pavlick [FESTVII] (viviparous fescue)
Festuca viviparoidea var. **krajinae** Pavlick [FESTVII1] (Krajina's fescue)
Fluminia festucacea (Willd.) A.S. Hitchc. = Scolochloa festucacea [SCOLFES]
Glyceria R. Br. [GLYCERI$]
Glyceria borealis (Nash) Batchelder [GLYCBOR] (northern mannagrass)
SY: *Panicularia borealis* Nash
Glyceria canadensis (Michx.) Trin. [GLYCCAN] (rattlesnake-grass)
SY: *Panicularia canadensis* (Michx.) Kuntze
Glyceria elata (Nash ex Rydb.) M.E. Jones [GLYCELA] (tall mannagrass)
Glyceria fernaldii (A.S. Hitchc.) St. John = Torreyochloa pallida var. fernaldii [TORRPAL1]
Glyceria grandis S. Wats. [GLYCGRA] (reed mannagrass)
Glyceria leptostachya Buckl. [GLYCLEP] (slender-spiked mannagrass)
*****Glyceria maxima*** (Hartman) Holmb. [GLYCMAX] (giant mannagrass)
SY: *Glyceria spectabilis* Mert. & Koch
SY: *Molinia maxima* Hartman
Glyceria occidentalis (Piper) J.C. Nels. [GLYCOCC] (western mannagrass)
Glyceria pallida (Torr.) Trin. = Torreyochloa pallida [TORRPAL]
Glyceria pallida var. *fernaldii* A.S. Hitchc. = Torreyochloa pallida var. fernaldii [TORRPAL1]
Glyceria pauciflora J. Presl = Torreyochloa pallida var. pauciflora [TORRPAL2]
Glyceria spectabilis Mert. & Koch = Glyceria maxima [GLYCMAX]
Glyceria striata (Lam.) A.S. Hitchc. [GLYCSTR] (fowl mannagrass)
Glyceria striata ssp. *stricta* (Scribn.) Hultén = Glyceria striata var. stricta [GLYCSTR1]
Glyceria striata var. **stricta** (Scribn.) Fern. [GLYCSTR1]

SY: *Glyceria striata* ssp. *stricta* (Scribn.) Hultén
Helictotrichon Bess. ex J.A. & J.H. Schultes [HELICTO$]
Helictotrichon hookeri (Scribn.) Henr. [HELIHOO] (spike-oat)
SY: *Avena hookeri* Scribn.
SY: *Avenochloa hookeri* (Scribn.) Holub
SY: *Avenula hookeri* (Scribn.) Holub
Hierochloe R. Br. [HIEROCH$]
Hierochloe alpina (Sw. ex Willd.) Roemer & J.A. Schultes [HIERALP] (alpine sweetgrass)
Hierochloe odorata (L.) Beauv. [HIERODO] (common sweetgrass)
SY: *Anthoxanthum nitens* (Weber) Y. Schouten & Veldkamp
Holcus L. [HOLCUS$]
*Holcus lanatus** L. [HOLCLAN] (common velvet-grass)
SY: *Nothoholcus lanatus* (L.) Nash
*Holcus mollis** L. [HOLCMOL] (creeping softgrass)
Homalocenchrus oryzoides (L.) Pollich = Leersia oryzoides [LEERORY]
Hordeum L. [HORDEUM$]
Hordeum aegiceras Nees ex Royle = Hordeum vulgare [HORDVUL]
Hordeum boreale Scribn. & J.G. Sm. = Hordeum brachyantherum [HORDBRA]
Hordeum brachyantherum Nevski [HORDBRA] (meadow barley)
SY: *Critesion brachyantherum* (Nevski) Barkworth & Dewey
SY: *Hordeum boreale* Scribn. & J.G. Sm.
SY: *Hordeum jubatum* var. *boreale* (Scribn. & J.G. Sm.) Boivin
SY: *Hordeum nodosum* L.
SY: *Hordeum nodosum* var. *boreale* (Scribn. & J.G. Sm.) A.S. Hitchc.
Hordeum caespitosum Scribn. ex Pammel = Hordeum jubatum ssp. jubatum [HORDJUB0]
Hordeum × *caespitosum* Scribn. ex Pammel = Hordeum jubatum ssp. jubatum [HORDJUB0]
Hordeum distichon L. = Hordeum vulgare [HORDVUL]
Hordeum geniculatum All. = Hordeum marinum ssp. gussonianum [HORDMAR1]
Hordeum gussonianum Parl. = Hordeum marinum ssp. gussonianum [HORDMAR1]
Hordeum hexastichum L. = Hordeum vulgare [HORDVUL]
Hordeum hystrix Roth = Hordeum marinum ssp. gussonianum [HORDMAR1]
Hordeum jubatum L. [HORDJUB] (foxtail barley)
Hordeum jubatum ssp. **breviaristatum** Bowden [HORDJUB1]
SY: *Critesion jubatum* ssp. *breviaristatum* (Bowden) A. & D. Löve
SY: *Hordeum jubatum* var. *breviaristatum* (Nevski) Bowden
Hordeum jubatum ssp. × **intermedium** Bowden [HORDJUB2]
Hordeum jubatum ssp. **jubatum** [HORDJUB0]
SY: *Critesion jubatum* (L.) Nevski
SY: *Hordeum caespitosum* Scribn. ex Pammel
SY: *Hordeum* × *caespitosum* Scribn. ex Pammel
SY: *Hordeum jubatum* var. *caespitosum* (Scribn. ex Pammel) A.S. Hitchc.
Hordeum jubatum var. *boreale* (Scribn. & J.G. Sm.) Boivin = Hordeum brachyantherum [HORDBRA]
Hordeum jubatum var. *breviaristatum* (Nevski) Bowden = Hordeum jubatum ssp. breviaristatum [HORDJUB1]
Hordeum jubatum var. *caespitosum* (Scribn. ex Pammel) A.S. Hitchc. = Hordeum jubatum ssp. jubatum [HORDJUB0]
Hordeum leporinum Link = Hordeum murinum ssp. leporinum [HORDMUR1]
*Hordeum marinum** Huds. [HORDMAR] (seaside barley)
*Hordeum marinum** ssp. **gussonianum** (Parl.) Thellung [HORDMAR1] (Mediterranean barley)
SY: *Critesion geniculatum* (All.) A. Löve
SY: *Critesion hystrix* (Roth) A. Löve
SY: *Critesion marinum* ssp. *gussonianum* (Parl.) Barkworth & Dewey
SY: *Hordeum geniculatum* All.
SY: *Hordeum gussonianum* Parl.
SY: *Hordeum hystrix* Roth
*Hordeum murinum** L. [HORDMUR] (wall barley)
*Hordeum murinum** ssp. **leporinum** (Link) Arcang. [HORDMUR1]
SY: *Critesion murinum* ssp. *leporinum* (Link) A. Löve
SY: *Hordeum leporinum* Link
*Hordeum murinum** ssp. **murinum** [HORDMUR0]
SY: *Critesion murinum* (L.) A. Löve
Hordeum nodosum L. = Hordeum brachyantherum [HORDBRA]
Hordeum nodosum var. *boreale* (Scribn. & J.G. Sm.) A.S. Hitchc. = Hordeum brachyantherum [HORDBRA]
*Hordeum vulgare** L. [HORDVUL] (common barley)
SY: *Hordeum aegiceras* Nees ex Royle
SY: *Hordeum distichon* L.
SY: *Hordeum hexastichum* L.
SY: *Hordeum vulgare* var. *trifurcatum* (Schlecht.) Alef.
Hordeum vulgare var. *trifurcatum* (Schlecht.) Alef. = Hordeum vulgare [HORDVUL]
Koeleria Pers. [KOELERI$]
Koeleria cristata auct. = Koeleria macrantha [KOELMAC]
Koeleria cristata var. *longifolia* Vasey ex Burtt-Davy = Koeleria macrantha [KOELMAC]
Koeleria cristata var. *pinetorum* Abrams = Koeleria macrantha [KOELMAC]
Koeleria gracilis Pers. = Koeleria macrantha [KOELMAC]
Koeleria macrantha (Ledeb.) J.A. Schultes [KOELMAC] (Junegrass)
SY: *Koeleria cristata* var. *longifolia* Vasey ex Burtt-Davy
SY: *Koeleria cristata* var. *pinetorum* Abrams
SY: *Koeleria gracilis* Pers.
SY: *Koeleria nitida* Nutt.
SY: *Koeleria yukonensis* Hultén
SY: *Koeleria cristata* auct.
SY: *Koeleria pyramidata* auct.
Koeleria nitida Nutt. = Koeleria macrantha [KOELMAC]
Koeleria pyramidata auct. = Koeleria macrantha [KOELMAC]
Koeleria yukonensis Hultén = Koeleria macrantha [KOELMAC]
Leersia Sw. [LEERSIA$]
Leersia oryzoides (L.) Sw. [LEERORY] (rice cutgrass)
SY: *Homalocenchrus oryzoides* (L.) Pollich
SY: *Phalaris oryzoides* L.
Leptochloa Beauv. [LEPTOCH$]
*Leptochloa fascicularis** (Lam.) Gray [LEPTFAS] (sprangletop)
Leymus Hochst. [LEYMUS$]
Leymus cinereus (Scribn. & Merr.) A. Löve [LEYMCIN] (giant wildrye)
SY: *Aneurolepidium piperi* (Bowden) Baum
SY: *Elymus cinereus* Scribn. & Merr.
SY: *Elymus cinereus* var. *pubens* (Piper) C.L. Hitchc.
SY: *Elymus condensatus* var. *pubens* Piper
SY: *Elymus piperi* Bowden
Leymus innovatus (Beal) Pilger [LEYMINN] (fuzzy-spiked wildrye)
SY: *Elymus brownii* Scribn. & J.G. Sm.
SY: *Elymus innovatus* Beal
SY: *Elymus innovatus* var. *glabratus* Bowden
SY: *Elymus innovatus* var. *velutinus* Bowden
SY: *Leymus velutinus* (Bowden) A. & D. Löve
Leymus mollis (Trin.) Hara [LEYMMOL] (dune wildrye)
SY: *Elymus mollis* Trin.
Leymus triticoides (Buckl.) Pilger [LEYMTRI] (creeping wildrye)
SY: *Elymus condensatus* var. *triticoides* (Buckl.) Thurb.
SY: *Elymus orcuttianus* Vasey
SY: *Elymus triticoides* Buckl.
SY: *Elymus triticoides* var. *pubescens* A.S. Hitchc.
Leymus velutinus (Bowden) A. & D. Löve = Leymus innovatus [LEYMINN]
Lolium L. [LOLIUM$]
Lolium arvense With. = Lolium temulentum [LOLITEM]
Lolium multiflorum Lam. = Lolium perenne ssp. multiflorum [LOLIPER1]
Lolium multiflorum var. *diminutum* Mutel = Lolium perenne ssp. multiflorum [LOLIPER1]
Lolium multiflorum var. *muticum* DC. = Lolium perenne ssp. multiflorum [LOLIPER1]
Lolium multiflorum var. *ramosum* Guss. ex Arcang. = Lolium perenne ssp. multiflorum [LOLIPER1]
*Lolium perenne** L. [LOLIPER] (perennial ryegrass)
*Lolium perenne** ssp. **multiflorum** (Lam.) Husnot [LOLIPER1]
SY: *Lolium multiflorum* Lam.
SY: *Lolium multiflorum* var. *diminutum* Mutel

SY: *Lolium multiflorum* var. *muticum* DC.
SY: *Lolium multiflorum* var. *ramosum* Guss. ex Arcang.
SY: *Lolium perenne* var. *multiflorum* (Lam.) Parnell
Lolium perenne var. *multiflorum* (Lam.) Parnell = Lolium perenne ssp. multiflorum [LOLIPER1]
***Lolium temulentum** L. [LOLITEM] (bearded ryegrass)
SY: *Lolium arvense* With.
SY: *Lolium temulentum* var. *arvense* (With.) Lilja
SY: *Lolium temulentum* var. *leptochaeton* A. Braun
SY: *Lolium temulentum* var. *macrochaeton* A. Braun
Lolium temulentum var. *arvense* (With.) Lilja = Lolium temulentum [LOLITEM]
Lolium temulentum var. *leptochaeton* A. Braun = Lolium temulentum [LOLITEM]
Lolium temulentum var. *macrochaeton* A. Braun = Lolium temulentum [LOLITEM]
Lophochlaena Nees [LOPHOCH$]
Lophochlaena refracta Gray [LOPHREF]
SY: *Pleuropogon refractus* (Gray) Benth. ex Vasey
Melica L. [MELICA$]
Melica bella Piper = Melica bulbosa [MELIBUL]
Melica bella ssp. *intonsa* Piper = Melica bulbosa [MELIBUL]
Melica bulbosa Geyer ex Porter & Coult. [MELIBUL] (oniongrass)
SY: *Bromelica bulbosa* (Geyer ex Porter & Coult.) W.A. Weber
SY: *Melica bella* Piper
SY: *Melica bella* ssp. *intonsa* Piper
SY: *Melica bulbosa* var. *intonsa* (Piper) M.E. Peck
Melica bulbosa var. *intonsa* (Piper) M.E. Peck = Melica bulbosa [MELIBUL]
Melica harfordii Boland. [MELIHAR] (Harford's melic)
SY: *Melica harfordii* var. *minor* Vasey
Melica harfordii var. *minor* Vasey = Melica harfordii [MELIHAR]
Melica purpurascens (Torr.) A.S. Hitchc. = Schizachne purpurascens [SCHIPUR]
Melica smithii (Porter ex Gray) Vasey [MELISMI] (Smith's melic)
SY: *Avena smithii* Porter ex Gray
SY: *Bromelica smithii* (Porter ex Gray) Farw.
Melica spectabilis Scribn. [MELISPE] (purple oniongrass)
SY: *Bromelica spectabilis* (Scribn.) W.A. Weber
Melica subulata (Griseb.) Scribn. [MELISUB] (Alaska oniongrass)
Molinia maxima Hartman = Glyceria maxima [GLYCMAX]
Muhlenbergia Schreb. [MUHLENB$]
Muhlenbergia ambigua Torr. = Muhlenbergia mexicana [MUHLMEX]
Muhlenbergia andina (Nutt.) A.S. Hitchc. [MUHLAND] (foxtail muhly)
SY: *Muhlenbergia comata* (Thurb.) Thurb. ex Benth.
Muhlenbergia asperifolia (Nees & Meyen ex Trin.) Parodi [MUHLASP] (alkali muhly)
SY: *Sporobolus asperifolius* (Nees & Meyen ex Trin.) Nees
Muhlenbergia comata (Thurb.) Thurb. ex Benth. = Muhlenbergia andina [MUHLAND]
Muhlenbergia filiformis (Thurb. ex S. Wats.) Rydb. [MUHLFIL] (slender muhly)
SY: *Muhlenbergia filiformis* var. *fortis* E.H. Kelso
SY: *Muhlenbergia idahoensis* St. John
SY: *Muhlenbergia simplex* (Scribn.) Rydb.
Muhlenbergia filiformis var. *fortis* E.H. Kelso = Muhlenbergia filiformis [MUHLFIL]
Muhlenbergia foliosa (Roemer & J.A. Schultes) Trin. = Muhlenbergia mexicana [MUHLMEX]
Muhlenbergia foliosa ssp. *ambigua* (Torr.) Scribn. = Muhlenbergia mexicana [MUHLMEX]
Muhlenbergia foliosa ssp. *setiglumis* (S. Wats.) Scribn. = Muhlenbergia mexicana [MUHLMEX]
Muhlenbergia glomerata (Willd.) Trin. [MUHLGLO] (marsh muhly)
SY: *Muhlenbergia glomerata* var. *cinnoides* (Link) F.J. Herm.
SY: *Muhlenbergia racemosa* var. *cinnoides* (Link) Boivin
Muhlenbergia glomerata var. *cinnoides* (Link) F.J. Herm. = Muhlenbergia glomerata [MUHLGLO]
Muhlenbergia idahoensis St. John = Muhlenbergia filiformis [MUHLFIL]
Muhlenbergia mexicana (L.) Trin. [MUHLMEX] (wirestem muhly)
SY: *Agrostis mexicana* L.
SY: *Muhlenbergia ambigua* Torr.
SY: *Muhlenbergia foliosa* (Roemer & J.A. Schultes) Trin.
SY: *Muhlenbergia foliosa* ssp. *ambigua* (Torr.) Scribn.
SY: *Muhlenbergia foliosa* ssp. *setiglumis* (S. Wats.) Scribn.
SY: *Muhlenbergia mexicana* var. *filiformis* (Willd.) Scribn.
Muhlenbergia mexicana var. *filiformis* (Willd.) Scribn. = Muhlenbergia mexicana [MUHLMEX]
Muhlenbergia racemosa (Michx.) B.S.P. [MUHLRAC] (satin grass)
SY: *Agrostis racemosa* Michx.
Muhlenbergia racemosa var. *cinnoides* (Link) Boivin = Muhlenbergia glomerata [MUHLGLO]
Muhlenbergia richardsonis (Trin.) Rydb. [MUHLRIC] (mat muhly)
SY: *Muhlenbergia squarrosa* (Trin.) Rydb.
Muhlenbergia simplex (Scribn.) Rydb. = Muhlenbergia filiformis [MUHLFIL]
Muhlenbergia squarrosa (Trin.) Rydb. = Muhlenbergia richardsonis [MUHLRIC]
Nassella (Trin.) Desv. [NASSELL$]
Nassella viridula (Trin.) Barkworth [NASSVIR] (green needlegrass)
SY: *Stipa viridula* Trin.
Nothoholcus lanatus (L.) Nash = Holcus lanatus [HOLCLAN]
Oryzopsis Michx. [ORYZOPS$]
Oryzopsis asperifolia Michx. [ORYZASP] (rough-leaved ricegrass)
Oryzopsis canadensis (Poir.) Torr. [ORYZCAN] (Canada ryegrass)
SY: *Stipa canadensis* Poir.
Oryzopsis exigua Thurb. [ORYZEXI] (little ricegrass)
Oryzopsis hymenoides (Roemer & J.A. Schultes) Ricker ex Piper [ORYZHYM]
SY: *Eriocoma cuspidata* Nutt.
SY: *Stipa hymenoides* Roemer & J.A. Schultes
Oryzopsis micrantha (Trin. & Rupr.) Thurb. [ORYZMIC] (small-flowered ricegrass)
Oryzopsis pungens (Torr. ex Spreng.) A.S. Hitchc. [ORYZPUN] (short-awned ricegrass)
Panicularia borealis Nash = Glyceria borealis [GLYCBOR]
Panicularia canadensis (Michx.) Kuntze = Glyceria canadensis [GLYCCAN]
Panicum L. [PANICUM$]
Panicum acuminatum var. *fasciculatum* (Torr.) Lelong = Dichanthelium acuminatum var. fasciculatum [DICHACU1]
Panicum acuminatum var. *implicatum* (Scribn.) C.F. Reed = Dichanthelium acuminatum var. fasciculatum [DICHACU1]
Panicum barbipulvinatum Nash = Panicum capillare [PANICAP]
Panicum brodiei St. John = Dichanthelium acuminatum var. fasciculatum [DICHACU1]
Panicum capillare L. [PANICAP] (common witchgrass)
SY: *Panicum barbipulvinatum* Nash
SY: *Panicum capillare* var. *agreste* Gattinger
SY: *Panicum capillare* ssp. *barbipulvinatum* (Nash) Tzvelev
SY: *Panicum capillare* var. *barbipulvinatum* (Nash) R.L. McGregor
SY: *Panicum capillare* var. *brevifolium* Vasey ex Rydb. & Shear
SY: *Panicum capillare* var. *occidentale* Rydb.
Panicum capillare ssp. *barbipulvinatum* (Nash) Tzvelev = Panicum capillare [PANICAP]
Panicum capillare var. *agreste* Gattinger = Panicum capillare [PANICAP]
Panicum capillare var. *barbipulvinatum* (Nash) R.L. McGregor = Panicum capillare [PANICAP]
Panicum capillare var. *brevifolium* Vasey ex Rydb. & Shear = Panicum capillare [PANICAP]
Panicum capillare var. *occidentale* Rydb. = Panicum capillare [PANICAP]
Panicum curtifolium Nash = Dichanthelium acuminatum var. fasciculatum [DICHACU1]
***Panicum dichotomiflorum** Michx. [PANIDIC] (smooth witchgrass)
Panicum glaucum L. = Setaria glauca [SETAGLA]
Panicum glutinoscabrum Fern. = Dichanthelium acuminatum var. fasciculatum [DICHACU1]
Panicum helleri Nash = Dichanthelium oligosanthes var. scribnerianum [DICHOLI1]

Panicum huachucae Ashe = Dichanthelium acuminatum var. fasciculatum [DICHACU1]
Panicum huachucae var. *fasciculatum* (Torr.) F.T. Hubbard = Dichanthelium acuminatum var. fasciculatum [DICHACU1]
Panicum implicatum Scribn. = Dichanthelium acuminatum var. fasciculatum [DICHACU1]
Panicum italicum L. = Setaria italica [SETAITA]
Panicum languidum A.S. Hitchc. = Dichanthelium acuminatum var. fasciculatum [DICHACU1]
Panicum lanuginosum Ell. = Dichanthelium acuminatum var. fasciculatum [DICHACU1]
Panicum lanuginosum var. *fasciculatum* (Torr.) Fern. = Dichanthelium acuminatum var. fasciculatum [DICHACU1]
Panicum lanuginosum var. *huachucae* (Ashe) A.S. Hitchc. = Dichanthelium acuminatum var. fasciculatum [DICHACU1]
Panicum lanuginosum var. *implicatum* (Scribn.) Fern. = Dichanthelium acuminatum var. fasciculatum [DICHACU1]
Panicum lanuginosum var. *tennesseense* (Ashe) Gleason = Dichanthelium acuminatum var. fasciculatum [DICHACU1]
Panicum lassenianum Schmoll = Dichanthelium acuminatum var. fasciculatum [DICHACU1]
Panicum lindheimeri var. *fasciculatum* (Torr.) Fern. = Dichanthelium acuminatum var. fasciculatum [DICHACU1]
Panicum macrocarpon Le Conte ex Torr. = Dichanthelium oligosanthes var. scribnerianum [DICHOLI1]
*__Panicum miliaceum__ L. [PANIMIL] (broom-corn millet)
Panicum occidentale Scribn. = Dichanthelium acuminatum var. fasciculatum [DICHACU1]
Panicum oligosanthes var. *helleri* (Nash) Fern. = Dichanthelium oligosanthes var. scribnerianum [DICHOLI1]
Panicum oligosanthes var. *scribnerianum* (Nash) Fern. = Dichanthelium oligosanthes var. scribnerianum [DICHOLI1]
Panicum pacificum A.S. Hitchc. & Chase = Dichanthelium acuminatum var. fasciculatum [DICHACU1]
Panicum sanguinale L. = Digitaria sanguinalis [DIGISAN]
Panicum scoparium S. Wats. ex Nash = Dichanthelium oligosanthes var. scribnerianum [DICHOLI1]
Panicum scribnerianum Nash = Dichanthelium oligosanthes var. scribnerianum [DICHOLI1]
Panicum subvillosum Ashe = Dichanthelium acuminatum var. fasciculatum [DICHACU1]
Panicum tennesseense Ashe = Dichanthelium acuminatum var. fasciculatum [DICHACU1]
Panicum thermale Boland. = Dichanthelium acuminatum var. thermale [DICHACU2]
Panicum verticillatum var. *ambiguum* Guss. = Setaria verticillata var. ambigua [SETAVER1]
Pascopyrum A. Löve [PASCOPY$]
Pascopyrum smithii (Rydb.) A. Löve [PASCSMI]
SY: *Agropyron molle* (Scribn. & J.G. Sm.) Rydb.
SY: *Agropyron smithii* Rydb.
SY: *Agropyron smithii* var. *molle* (Scribn. & J.G. Sm.) M.E. Jones
SY: *Agropyron smithii* var. *palmeri* (Scribn. & J.G. Sm.) Heller
SY: *Elymus smithii* (Rydb.) Gould
SY: *Elytrigia smithii* (Rydb.) Nevski
SY: *Elytrigia smithii* var. *mollis* (Scribn. & J.G. Sm.) Beetle
Pennisetum glaucum (L.) R. Br. = Setaria glauca [SETAGLA]
Phalaris L. [PHALARI$]
Phalaris arundinacea L. [PHALARU] (reed canarygrass)
SY: *Phalaris arundinacea* var. *picta* L.
SY: *Phalaroides arundinacea* (L.) Raeusch.
SY: *Phalaroides arundinacea* var. *picta* (L.) Tzvelev
Phalaris arundinacea var. *picta* L. = Phalaris arundinacea [PHALARU]
*__Phalaris canariensis__ L. [PHALCAN] (canarygrass)
Phalaris oryzoides L. = Leersia oryzoides [LEERORY]
Phalaroides arundinacea (L.) Raeusch. = Phalaris arundinacea [PHALARU]
Phalaroides arundinacea var. *picta* (L.) Tzvelev = Phalaris arundinacea [PHALARU]
Phippsia arctica (Hook.) A. & D. Löve = Puccinellia arctica [PUCCARC]
Phippsia borealis (Swallen) A. & D. Löve = Puccinellia arctica [PUCCARC]
Phippsia interior (Sorensen) A. & D. Löve = Puccinellia interior [PUCCINT]
Phippsia nutkaensis (J. Presl) A. & D. Löve = Puccinellia nutkaensis [PUCCNUT]
Phleum L. [PHLEUM$]
Phleum alpinum L. [PHLEALP] (alpine timothy)
SY: *Phleum alpinum* var. *commutatum* (Gaudin) Griseb.
SY: *Phleum commutatum* Gaudin
SY: *Phleum commutatum* var. *americanum* (Fourn.) Hultén
Phleum alpinum var. *commutatum* (Gaudin) Griseb. = Phleum alpinum [PHLEALP]
Phleum commutatum Gaudin = Phleum alpinum [PHLEALP]
Phleum commutatum var. *americanum* (Fourn.) Hultén = Phleum alpinum [PHLEALP]
*__Phleum pratense__ L. [PHLEPRA] (common timothy)
Phragmites Adans. [PHRAGMI$]
Phragmites australis (Cav.) Trin. ex Steud. [PHRAAUS] (common reedgrass)
SY: *Phragmites australis* var. *berlandieri* (Fourn.) C.F. Reed
SY: *Phragmites communis* Trin.
SY: *Phragmites communis* ssp. *berlandieri* (Fourn.) A. & D. Löve
SY: *Phragmites communis* var. *berlandieri* (Fourn.) Fern.
SY: *Phragmites phragmites* (L.) Karst.
Phragmites australis var. *berlandieri* (Fourn.) C.F. Reed = Phragmites australis [PHRAAUS]
Phragmites communis Trin. = Phragmites australis [PHRAAUS]
Phragmites communis ssp. *berlandieri* (Fourn.) A. & D. Löve = Phragmites australis [PHRAAUS]
Phragmites communis var. *berlandieri* (Fourn.) Fern. = Phragmites australis [PHRAAUS]
Phragmites phragmites (L.) Karst. = Phragmites australis [PHRAAUS]
Pleuropogon refractus (Gray) Benth. ex Vasey = Lophochlaena refracta [LOPHREF]
Poa L. [POA$]
Poa abbreviata R. Br. [POA ABB] (northern bluegrass)
Poa abbreviata ssp. *jordalii* (Porsild) Hultén = Poa abbreviata ssp. pattersonii [POA ABB3]
Poa abbreviata ssp. **pattersonii** (Vasey) A. & D. Löve & Kapoor [POA ABB3] (abbreviated bluegrass)
SY: *Poa abbreviata* ssp. *jordalii* (Porsild) Hultén
SY: *Poa jordalii* Porsild
SY: *Poa pattersonii* Vasey
Poa agassizensis Boivin & D. Löve = Poa pratensis [POA PRA]
Poa alpigena (Fries ex Blytt) Lindm. f. = Poa pratensis [POA PRA]
Poa alpigena ssp. *colpodea* (Fries ex Blytt) Tzvelev = Poa pratensis [POA PRA]
Poa alpigena var. *colpodea* (Fries ex Blytt) Schol. = Poa pratensis [POA PRA]
Poa alpina L. [POA ALP] (alpine bluegrass)
Poa alpina var. *purpurascens* Vassey = Poa fendleriana [POA FEN]
Poa ampla Merr. = Poa secunda [POA SEC]
Poa angustifolia L. = Poa pratensis [POA PRA]
*__Poa annua__ L. [POA ANN] (annual bluegrass)
SY: *Poa annua* var. *aquatica* Aschers.
SY: *Poa annua* var. *reptans* Hausskn.
Poa annua var. *aquatica* Aschers. = Poa annua [POA ANN]
Poa annua var. *reptans* Hausskn. = Poa annua [POA ANN]
Poa arctica R. Br. [POA ARC] (arctic bluegrass)
Poa arctica ssp. **arctica** [POA ARC0]
SY: *Poa arctica* var. *glabriflora* Rosh.
SY: *Poa arctica* var. *vivipara* Hook.
SY: *Poa cenisia* var. *arctica* (R. Br.) Richter
SY: *Poa longipila* Nash
Poa arctica ssp. **lanata** (Scribn. & Merr.) Soreng [POA ARC5]
SY: *Poa arctica* var. *lanata* (Scribn. & Merr.) Bovin
SY: *Poa arctica* var. *vivipara* Hultén
SY: *Poa lanata* Scribn. & Merr.
Poa arctica ssp. **longiculmis** Hultén [POA ARC3]
Poa arctica ssp. **williamsii** (Nash) Hultén [POA ARC4]
SY: *Poa williamsii* Nash

Poa arctica var. *glabriflora* Rosh. = Poa arctica ssp. arctica [POA ARC0]
Poa arctica var. *lanata* (Scribn. & Merr.) Bovin = Poa arctica ssp. lanata [POA ARC5]
Poa arctica var. *vivipara* Hultén = Poa arctica ssp. lanata [POA ARC5]
Poa arctica var. *vivipara* Hook. = Poa arctica ssp. arctica [POA ARC0]
Poa bolanderi Vasey [POA BOL]
Poa bolanderi ssp. **howellii** (Vasey & Scribn.) Keck [POA BOL1]
SY: *Poa bolanderi* var. *howellii* (Vasey & Scribn.) M.E. Jones
SY: *Poa howellii* Vasey & Scribn.
Poa bolanderi var. *howellii* (Vasey & Scribn.) M.E. Jones = Poa bolanderi ssp. howellii [POA BOL1]
Poa brachyanthera Hultén = Poa pseudoabbreviata [POA PSE]
Poa brachyglossa Piper = Poa secunda [POA SEC]
Poa buckleyana Nash = Poa secunda [POA SEC]
*__Poa bulbosa__ L. [POA BUL] (bulbous bluegrass)
Poa canbyi (Scribn.) T.J. Howell = Poa secunda [POA SEC]
Poa cenisia var. *arctica* (R. Br.) Richter = Poa arctica ssp. arctica [POA ARC0]
Poa cilianensis All. = Eragrostis cilianensis [ERAGCIL]
*__Poa compressa__ L. [POA COM] (Canada bluegrass)
Poa confinis Vasey [POA CON] (beach bluegrass)
Poa confusa Rydb. = Poa secunda [POA SEC]
Poa crocata Michx. = Poa palustris [POA PAL]
Poa cusickii Vasey = Poa fendleriana [POA FEN]
Poa cusickii ssp. *epilis* (Scribn.) W.A. Weber = Poa fendleriana [POA FEN]
Poa cusickii ssp. *pallida* Soreng = Poa fendleriana [POA FEN]
Poa cusickii ssp. *pubens* Keck = Poa fendleriana [POA FEN]
Poa cusickii ssp. *purpurascens* (Vasey) Soreng = Poa fendleriana [POA FEN]
Poa cusickii var. *epilis* (Scribn.) C.L. Hitchc. = Poa fendleriana [POA FEN]
Poa cusickii var. *purpurascens* (Vasey) C.L. Hitchc. = Poa fendleriana [POA FEN]
Poa douglasii Nees [POA DOU]
Poa douglasii ssp. **macrantha** (Vasey) Keck [POA DOU1]
SY: *Poa douglasii* var. *macrantha* (Vasey) Boivin
SY: *Poa macrantha* Vasey
Poa douglasii var. *macrantha* (Vasey) Boivin = Poa douglasii ssp. macrantha [POA DOU1]
Poa eminens J. Presl [POA EMI] (eminent bluegrass)
SY: *Arctopoa eminens* (J. Presl) Probat.
Poa englishii St. John & Hardin = Poa secunda [POA SEC]
Poa epilis Scribn. = Poa fendleriana [POA FEN]
Poa fendleriana (Steud.) Vasey [POA FEN] (Fendler's bluegrass)
SY: *Poa alpina* var. *purpurascens* Vassey
SY: *Poa cusickii* Vasey
SY: *Poa cusickii* ssp. *epilis* (Scribn.) W.A. Weber
SY: *Poa cusickii* var. *epilis* (Scribn.) C.L. Hitchc.
SY: *Poa cusickii* ssp. *pallida* Soreng
SY: *Poa cusickii* ssp. *pubens* Keck
SY: *Poa cusickii* ssp. *purpurascens* (Vasey) Soreng
SY: *Poa cusickii* var. *purpurascens* (Vasey) C.L. Hitchc.
SY: *Poa epilis* Scribn.
SY: *Poa fendleriana* ssp. *longiligula* (Scribn. & Williams) Soreng
SY: *Poa fendleriana* var. *longiligula* (Scribn. & Williams) Gould
SY: *Poa fendleriana* var. *wyomingensis* T.A. Williams
SY: *Poa longiligula* Scribn. & Williams
SY: *Poa nematophylla* Rydb.
SY: *Poa purpurascens* Vasey
SY: *Poa purpurascens* var. *epilis* (Scribn.) M.E. Jones
SY: *Poa subaristata* Scribn.
Poa fendleriana ssp. *longiligula* (Scribn. & Williams) Soreng = Poa fendleriana [POA FEN]
Poa fendleriana var. *longiligula* (Scribn. & Williams) Gould = Poa fendleriana [POA FEN]
Poa fendleriana var. *wyomingensis* T.A. Williams = Poa fendleriana [POA FEN]
Poa filifolia Vasey = Poa secunda [POA SEC]
Poa glacialis A. Hitchc. = Poa leptocoma ssp. paucispicula [POA LEP3]
Poa glauca Vahl [POA GLA] (glaucous bluegrass)
Poa glauca ssp. *conferta* (Blytt) Lindm. = Poa glauca ssp. glauca [POA GLA0]
Poa glauca ssp. **glauca** [POA GLA0]
SY: *Poa glauca* ssp. *conferta* (Blytt) Lindm.
SY: *Poa glauca* var. *conferta* (Blytt) Nannf.
SY: *Poa glauca* var. *laxiuscula* (Blytt) Lindm.
Poa glauca ssp. **rupicola** (Nash ex Rydb.) W.A. Weber [POA GLA2]
SY: *Poa glauca* var. *rupicola* (Nash ex Rydb.) Boivin
SY: *Poa rupicola* Nash ex Rydb.
Poa glauca var. *conferta* (Blytt) Nannf. = Poa glauca ssp. glauca [POA GLA0]
Poa glauca var. *laxiuscula* (Blytt) Lindm. = Poa glauca ssp. glauca [POA GLA0]
Poa glauca var. *rupicola* (Nash ex Rydb.) Boivin = Poa glauca ssp. rupicola [POA GLA2]
Poa gracillima Vasey = Poa secunda [POA SEC]
Poa gracillima var. *multnomae* (Piper) C.L. Hitchc. = Poa secunda [POA SEC]
Poa howellii Vasey & Scribn. = Poa bolanderi ssp. howellii [POA BOL1]
Poa incurva Scribn. & Williams = Poa secunda [POA SEC]
*__Poa infirma__ Kunth [POA INF] (diploid annual bluegrass)
Poa interior Rydb., p.p. [POA INT] (inland bluegrass)
SY: *Poa nemoralis* ssp. *interior* (Rydb.) W.A. Weber
SY: *Poa nemoralis* var. *interior* (Rydb.) Butters & Abbe
Poa irrigata Lindm. = Poa subcaerulea [POA SUB]
Poa jordalii Porsild = Poa abbreviata ssp. pattersonii [POA ABB3]
Poa juncifolia Scribn. = Poa secunda [POA SEC]
Poa juncifolia ssp. *porteri* Keck = Poa secunda [POA SEC]
Poa juncifolia var. *ampla* (Merr.) Dorn = Poa secunda [POA SEC]
Poa laevigata Scribn. = Poa secunda [POA SEC]
Poa lanata Scribn. & Merr. = Poa arctica ssp. lanata [POA ARC5]
Poa laxa Haenke [POA LAA] (Mt. Washington bluegrass)
Poa laxa ssp. **banffiana** Soreng [POA LAA1] (Banff bluegrass)
Poa laxiflora Buckl. [POA LAX] (lax-flowered bluegrass)
Poa leptocoma Trin. [POA LEP] (bog bluegrass)
Poa leptocoma ssp. **paucispicula** (Scribn. & Merr.) Tzvelev [POA LEP3]
SY: *Poa glacialis* A. Hitchc.
SY: *Poa leptocoma* var. *paucispicula* (Scribn. & Merr.) C.L. Hitchc.
SY: *Poa merrilliana* Scribn. & Merr.
SY: *Poa paucispicula* Scribn. & Merr.
Poa leptocoma var. *paucispicula* (Scribn. & Merr.) C.L. Hitchc. = Poa leptocoma ssp. paucispicula [POA LEP3]
Poa lettermanii Vasey [POA LET] (Letterman's bluegrass)
SY: *Poa montevansii* L. Kelso
SY: *Puccinellia lettermanii* (Vasey) Ponert
Poa longiligula Scribn. & Williams = Poa fendleriana [POA FEN]
Poa longipila Nash = Poa arctica ssp. arctica [POA ARC0]
Poa macrantha Vasey = Poa douglasii ssp. macrantha [POA DOU1]
Poa marcida A.S. Hitchc. [POA MAR] (weeping bluegrass)
SY: *Poa saltuensis* var. *marcida* (A.S. Hitchc.) Boivin
Poa merrilliana Scribn. & Merr. = Poa leptocoma ssp. paucispicula [POA LEP3]
Poa montevansii L. Kelso = Poa lettermanii [POA LET]
Poa nematophylla Rydb. = Poa fendleriana [POA FEN]
*__Poa nemoralis__ L. [POA NEM] (wood bluegrass)
Poa nemoralis ssp. *interior* (Rydb.) W.A. Weber = Poa interior [POA INT]
Poa nemoralis var. *interior* (Rydb.) Butters & Abbe = Poa interior [POA INT]
Poa nervosa (Hook.) Vasey [POA NER] (coastal bluegrass)
Poa nervosa var. **wheeleri** (Vasey) C.L. Hitchc. [POA NER1]
SY: *Poa wheeleri* Vasey
Poa nevadensis Vasey ex Scribn. = Poa secunda [POA SEC]
Poa nevadensis var. *juncifolia* (Scribn.) Beetle = Poa secunda [POA SEC]
Poa orcuttiana Vasey = Poa secunda [POA SEC]
Poa palustris L. [POA PAL] (fowl bluegrass)

SY: *Poa crocata* Michx.
SY: *Poa triflora* Gilib.
Poa pattersonii Vasey = Poa abbreviata ssp. pattersonii [POA ABB3]
Poa paucispicula Scribn. & Merr. = Poa leptocoma ssp. paucispicula [POA LEP3]
*Poa pratensis** L. [POA PRA] (Kentucky bluegrass)
SY: *Poa agassizensis* Boivin & D. Löve
SY: *Poa alpigena* (Fries ex Blytt) Lindm. f.
SY: *Poa alpigena* ssp. *colpodea* (Fries ex Blytt) Tzvelev
SY: *Poa alpigena* var. *colpodea* (Fries ex Blytt) Schol.
SY: *Poa angustifolia* L.
SY: *Poa pratensis* ssp. *agassizensis* (Boivin & D. Löve) Taylor & MacBryde
SY: *Poa pratensis* ssp. *alpigena* (Fries ex Blytt) Hiitonen
SY: *Poa pratensis* ssp. *angustifolia* (L.) Lej.
SY: *Poa pratensis* var. *angustifolia* (L.) Gaudin
SY: *Poa pratensis* var. *apligena* Fries ex Blytt
SY: *Poa pratensis* ssp. *colpodea* (Fries ex Blytt) Tzvelev
SY: *Poa pratensis* var. *colpodea* (Fries ex Blytt) Soreng
SY: *Poa pratensis* var. *domestica* Laestad.
SY: *Poa pratensis* var. *gelida* (Roemer & J.A. Schultes) Böcher
SY: *Poa pratensis* var. *iantha* Wahlenb.
SY: *Poa pratensis* var. *vivipara* (Malmgr.) Boivin
Poa pratensis ssp. *agassizensis* (Boivin & D. Löve) Taylor & MacBryde = Poa pratensis [POA PRA]
Poa pratensis ssp. *alpigena* (Fries ex Blytt) Hiitonen = Poa pratensis [POA PRA]
Poa pratensis ssp. *angustifolia* (L.) Lej. = Poa pratensis [POA PRA]
Poa pratensis ssp. *colpodea* (Fries ex Blytt) Tzvelev = Poa pratensis [POA PRA]
Poa pratensis ssp. *irrigata* (Lindm.) Lindb. f. = Poa subcaerulea [POA SUB]
Poa pratensis ssp. *rigens* (Hartman) Tzvelev = Poa subcaerulea [POA SUB]
Poa pratensis var. *angustifolia* (L.) Gaudin = Poa pratensis [POA PRA]
Poa pratensis var. *apligena* Fries ex Blytt = Poa pratensis [POA PRA]
Poa pratensis var. *colpodea* (Fries ex Blytt) Soreng = Poa pratensis [POA PRA]
Poa pratensis var. *domestica* Laestad. = Poa pratensis [POA PRA]
Poa pratensis var. *gelida* (Roemer & J.A. Schultes) Böcher = Poa pratensis [POA PRA]
Poa pratensis var. *iantha* Wahlenb. = Poa pratensis [POA PRA]
Poa pratensis var. *rigens* (Hartman) Wahlenb. = Poa subcaerulea [POA SUB]
Poa pratensis var. *vivipara* (Malmgr.) Boivin = Poa pratensis [POA PRA]
Poa pseudoabbreviata Rosh. [POA PSE] (polar bluegrass)
SY: *Poa brachyanthera* Hultén
Poa purpurascens Vasey = Poa fendleriana [POA FEN]
Poa purpurascens var. *epilis* (Scribn.) M.E. Jones = Poa fendleriana [POA FEN]
Poa rigens Hartman = Poa subcaerulea [POA SUB]
Poa rupicola Nash ex Rydb. = Poa glauca ssp. rupicola [POA GLA2]
Poa saltuensis var. *marcida* (A.S. Hitchc.) Boivin = Poa marcida [POA MAR]
Poa sandbergii Vasey = Poa secunda [POA SEC]
Poa scabrella (Thurb.) Benth. ex Vasey = Poa secunda [POA SEC]
Poa secunda J. Presl [POA SEC] (Sandberg's bluegrass)
SY: *Poa ampla* Merr.
SY: *Poa brachyglossa* Piper
SY: *Poa buckleyana* Nash
SY: *Poa canbyi* (Scribn.) T.J. Howell
SY: *Poa confusa* Rydb.
SY: *Poa englishii* St. John & Hardin
SY: *Poa filifolia* Vasey
SY: *Poa gracillima* Vasey
SY: *Poa gracillima* var. *multnomae* (Piper) C.L. Hitchc.
SY: *Poa incurva* Scribn. & Williams
SY: *Poa juncifolia* Scribn.
SY: *Poa juncifolia* var. *ampla* (Merr.) Dorn
SY: *Poa juncifolia* ssp. *porteri* Keck
SY: *Poa laevigata* Scribn.
SY: *Poa nevadensis* Vasey ex Scribn.
SY: *Poa nevadensis* var. *juncifolia* (Scribn.) Beetle
SY: *Poa orcuttiana* Vasey
SY: *Poa sandbergii* Vasey
SY: *Poa scabrella* (Thurb.) Benth. ex Vasey
SY: *Poa secunda* var. *elongata* (Vasey) Dorn
SY: *Poa secunda* var. *incurva* (Scribn. & Williams) Beetle
SY: *Poa secunda* ssp. *juncifolia* (Scribn.) Soreng
SY: *Poa secunda* var. *stenophylla* (Vasey) Beetle
SY: *Poa tenerrima* Scribn.
Poa secunda ssp. *juncifolia* (Scribn.) Soreng = Poa secunda [POA SEC]
Poa secunda var. *elongata* (Vasey) Dorn = Poa secunda [POA SEC]
Poa secunda var. *incurva* (Scribn. & Williams) Beetle = Poa secunda [POA SEC]
Poa secunda var. *stenophylla* (Vasey) Beetle = Poa secunda [POA SEC]
Poa stenantha Trin. [POA STE] (narrow-flowered bluegrass)
Poa subaristata Scribn. = Poa fendleriana [POA FEN]
***Poa subcaerulea** Sm. [POA SUB]
SY: *Poa irrigata* Lindm.
SY: *Poa pratensis* ssp. *irrigata* (Lindm.) Lindb. f.
SY: *Poa pratensis* ssp. *rigens* (Hartman) Tzvelev
SY: *Poa pratensis* var. *rigens* (Hartman) Wahlenb.
SY: *Poa rigens* Hartman
Poa tenerrima Scribn. = Poa secunda [POA SEC]
Poa triflora Gilib. = Poa palustris [POA PAL]
***Poa trivialis** L. [POA TRI] (rough bluegrass)
Poa wheeleri Vasey = Poa nervosa var. wheeleri [POA NER1]
Poa williamsii Nash = Poa arctica ssp. williamsii [POA ARC4]
Podagrostis aequivalvis (Trin.) Scribn. & Merr. = Agrostis aequivalvis [AGROAEQ]
Polypogon Desf. [POLYPOG$]
***Polypogon monspeliensis** (L.) Desf. [POLYMOS] (rabbitfoot polypogon)
SY: *Alopecurus monspeliensis* L.
Pseudoroegneria (Nevski) A. Löve [PSEUDOR$]
Pseudoroegneria spicata (Pursh) A. Löve [PSEUSPI]
SY: *Agropyron spicatum* Pursh
SY: *Agropyron spicatum* var. *pubescens* Elmer
SY: *Agropyron vaseyi* Scribn. & J.G. Sm.
SY: *Elymus spicatus* (Pursh) Gould
SY: *Elytrigia spicata* (Pursh) Dewey
SY: *Roegneria spicata* (Pursh) Beetle
Pseudoroegneria spicata ssp. **inermis** (Scribn. & J.G. Sm.) A. Löve [PSEUSPI1]
SY: *Agropyron inerme* (Scribn. & J.G. Sm.) Rydb.
SY: *Agropyron spicatum* var. *inerme* (Scribn. & J.G. Sm.) Heller
Puccinellia Parl. [PUCCINE$]
Puccinellia airoides (Nutt.) S. Wats. & Coult. = Puccinellia nuttalliana [PUCCNUA]
Puccinellia arctica (Hook.) Fern. & Weatherby [PUCCARC]
SY: *Phippsia arctica* (Hook.) A. & D. Löve
SY: *Phippsia borealis* (Swallen) A. & D. Löve
SY: *Puccinellia borealis* Swallen
Puccinellia borealis Swallen = Puccinellia arctica [PUCCARC]
Puccinellia cusickii Weatherby = Puccinellia nuttalliana [PUCCNUA]
***Puccinellia distans** (Jacq.) Parl. [PUCCDIS] (weeping alkaligrass)
Puccinellia fernaldii (A.S. Hitchc.) E.G. Voss = Torreyochloa pallida var. fernaldii [TORRPAL1]
Puccinellia grandis Swallen [PUCCGRA]
Puccinellia interior Sorensen [PUCCINT] (inland alkaligrass)
SY: *Phippsia interior* (Sorensen) A. & D. Löve
Puccinellia kurilensis (Takeda) Honda [PUCCKUR]
SY: *Puccinellia pumila* (Vasey) A.S. Hitchc.
Puccinellia lettermanii (Vasey) Ponert = Poa lettermanii [POA LET]
Puccinellia nutkaensis (J. Presl) Fern. & Weatherby [PUCCNUT] (Pacific alkaligrass)
SY: *Phippsia nutkaensis* (J. Presl) A. & D. Löve
Puccinellia nuttalliana (J.A. Schultes) A.S. Hitchc. [PUCCNUA] (Nuttall's alkaligrass)
SY: *Puccinellia airoides* (Nutt.) S. Wats. & Coult.
SY: *Puccinellia cusickii* Weatherby

Puccinellia pauciflora (J. Presl) Munz = Torreyochloa pallida var. pauciflora [TORRPAL2]
Puccinellia pauciflora var. *holmii* (Beal) C.L. Hitchc. = Torreyochloa pallida var. pauciflora [TORRPAL2]
Puccinellia pauciflora var. *microtheca* (Buckl.) C.L. Hitchc. = Torreyochloa pallida var. pauciflora [TORRPAL2]
Puccinellia pumila (Vasey) A.S. Hitchc. = Puccinellia kurilensis [PUCCKUR]
Roegneria albicans (Scribn. & J.G. Sm.) Beetle = Elymus lanceolatus ssp. albicans [ELYMLAN1]
Roegneria albicans var. *griffithii* (Scribn. & J.G. Sm. ex Piper) Beetle = Elymus lanceolatus ssp. albicans [ELYMLAN1]
Roegneria pauciflora (Schwein.) Hyl. = Elymus trachycaulus ssp. trachycaulus [ELYMTRA0]
Roegneria spicata (Pursh) Beetle = Pseudoroegneria spicata [PSEUSPI]
Roegneria trachycaula (Link) Nevski = Elymus trachycaulus ssp. trachycaulus [ELYMTRA0]
Schizachne Hack. [SCHIZAC$]
Schizachne purpurascens (Torr.) Swallen [SCHIPUR] (false melic)
SY: *Avena torreyi* Nash
SY: *Melica purpurascens* (Torr.) A.S. Hitchc.
SY: *Schizachne purpurascens* var. *pubescens* Dore
SY: *Schizachne stricta* (Michx.) Hultén
SY: *Trisetum purpurascens* Torr.
Schizachne purpurascens var. *pubescens* Dore = Schizachne purpurascens [SCHIPUR]
Schizachne stricta (Michx.) Hultén = Schizachne purpurascens [SCHIPUR]
Schizachyrium Nees [SCHIZAH$]
Schizachyrium scoparium (Michx.) Nash [SCHISCO] (little bluestem)
SY: *Andropogon scoparius* Michx.
Scolochloa Link [SCOLOCH$]
Scolochloa festucacea (Willd.) Link [SCOLFES] (rivergrass)
SY: *Arundo festucacea* Willd.
SY: *Fluminia festucacea* (Willd.) A.S. Hitchc.
Secale L. [SECALE$]
*__Secale cereale__ L. [SECACER] (rye)
SY: *Triticum cereale* (L.) Salisb.
Setaria Beauv. [SETARIA$]
Setaria decipiens Schimp. ex Nyman = Setaria verticillata var. ambigua [SETAVER1]
*__Setaria glauca__ (L.) Beauv. [SETAGLA] (yellow bristlegrass)
SY: *Chaetochloa glauca* (L.) Scribn.
SY: *Chaetochloa lutescens* (Weigel) Stuntz
SY: *Panicum glaucum* L.
SY: *Pennisetum glaucum* (L.) R. Br.
SY: *Setaria lutescens* (Weigel) F.T. Hubbard
*__Setaria italica__ (L.) Beauv. [SETAITA] (foxtail bristlegrass)
SY: *Chaetochloa italica* (L.) Scribn.
SY: *Panicum italicum* L.
SY: *Setaria italica subvar. metzeri* (Koern.) F.T. Hubbard
SY: *Setaria italica* var. *metzeri* (Koern.) Jáv.
SY: *Setaria italica* var. *stramineofructa* (F.T. Hubbard) Bailey
Setaria italica subvar. metzeri (Koern.) F.T. Hubbard = Setaria italica [SETAITA]
Setaria italica var. *metzeri* (Koern.) Jáv. = Setaria italica [SETAITA]
Setaria italica var. *stramineofructa* (F.T. Hubbard) Bailey = Setaria italica [SETAITA]
Setaria lutescens (Weigel) F.T. Hubbard = Setaria glauca [SETAGLA]
*__Setaria verticillata__ (L.) Beauv. [SETAVER] (bur bristlegrass)
Setaria verticillata f. *ambigua* (Guss.) Boivin = Setaria verticillata var. ambigua [SETAVER1]
Setaria verticillata var. **ambigua** (Guss.) Parl. [SETAVER1]
SY: *Panicum verticillatum* var. *ambiguum* Guss.
SY: *Setaria decipiens* Schimp. ex Nyman
SY: *Setaria verticillata* f. *ambigua* (Guss.) Boivin
SY: *Setaria viridis* var. *ambigua* (Guss.) Coss. & Durieu
*__Setaria viridis__ (L.) Beauv. [SETAVIR] (green bristlegrass)
Setaria viridis var. *ambigua* (Guss.) Coss. & Durieu = Setaria verticillata var. ambigua [SETAVER1]
Sieglingia decumbens (L.) Bernh. = Danthonia decumbens [DANTDEC]
Sitanion elymoides Raf. = Elymus elymoides [ELYMELY]
Sitanion hystrix (Nutt.) J.G. Sm. = Elymus elymoides [ELYMELY]
Sitanion hystrix var. *brevifolium* (J.G. Sm.) C.L. Hitchc. = Elymus elymoides [ELYMELY]
Sitanion hystrix var. *californicum* (J.G. Sm.) F.D. Wilson = Elymus elymoides [ELYMELY]
Sitanion longifolium J.G. Sm. = Elymus elymoides [ELYMELY]
Spartina Schreb. [SPARTIN$]
Spartina gracilis Trin. [SPARGRA] (alkali cordgrass)
*__Spartina patens__ (Ait.) Muhl. [SPARPAT] (salt meadowgrass)
SY: *Spartina patens* var. *juncea* (Michx.) A.S. Hitchc.
SY: *Spartina patens* var. *monogyna* (M.A. Curtis) Fern.
Spartina patens var. *juncea* (Michx.) A.S. Hitchc. = Spartina patens [SPARPAT]
Spartina patens var. *monogyna* (M.A. Curtis) Fern. = Spartina patens [SPARPAT]
Sphenopholis Scribn. [SPHENOP$]
Sphenopholis intermedia (Rydb.) Rydb. [SPHEINT]
SY: *Sphenopholis intermedia* var. *pilosa* Dore
SY: *Sphenopholis obtusata* var. *major* (Torr.) K.S. Erdman
Sphenopholis intermedia var. *pilosa* Dore = Sphenopholis intermedia [SPHEINT]
Sphenopholis obtusata (Michx.) Scribn. [SPHEOBT] (prairie wedgegrass)
SY: *Aira obtusata* Michx.
SY: *Sphenopholis obtusata* var. *lobata* (Trin.) Scribn.
SY: *Sphenopholis obtusata* var. *pubescens* (Scribn. & Merr.) Scribn.
Sphenopholis obtusata var. *lobata* (Trin.) Scribn. = Sphenopholis obtusata [SPHEOBT]
Sphenopholis obtusata var. *major* (Torr.) K.S. Erdman = Sphenopholis intermedia [SPHEINT]
Sphenopholis obtusata var. *pubescens* (Scribn. & Merr.) Scribn. = Sphenopholis obtusata [SPHEOBT]
Sporobolus R. Br. [SPOROBO$]
Sporobolus airoides (Torr.) Torr. [SPORAIR] (hairgrass dropseed)
SY: *Agrostis airoides* Torr.
Sporobolus asper (Beauv.) Kunth = Sporobolus compositus [SPORCOM]
Sporobolus asper var. *hookeri* (Trin.) Vasey = Sporobolus compositus [SPORCOM]
Sporobolus asperifolius (Nees & Meyen ex Trin.) Nees = Muhlenbergia asperifolia [MUHLASP]
Sporobolus compositus (Poir.) Merr. [SPORCOM]
SY: *Sporobolus asper* (Beauv.) Kunth
SY: *Sporobolus asper* var. *hookeri* (Trin.) Vasey
Sporobolus cryptandrus (Torr.) Gray [SPORCRY] (sand dropseed)
SY: *Agrostis cryptandra* Torr.
SY: *Sporobolus cryptandrus* ssp. *fuscicola* (Hook.) E.K. Jones & Fassett
SY: *Sporobolus cryptandrus* var. *fuscicola* (Hook.) Pohl
SY: *Sporobolus cryptandrus* var. *occidentalis* E.K. Jones & Fassett
SY: *Sporobolus cryptandrus* ssp. *typicus* var. *occidentalis* E.K. Jones & Fassett
Sporobolus cryptandrus ssp. *fuscicola* (Hook.) E.K. Jones & Fassett = Sporobolus cryptandrus [SPORCRY]
Sporobolus cryptandrus ssp. *typicus* var. *occidentalis* E.K. Jones & Fassett = Sporobolus cryptandrus [SPORCRY]
Sporobolus cryptandrus var. *fuscicola* (Hook.) Pohl = Sporobolus cryptandrus [SPORCRY]
Sporobolus cryptandrus var. *occidentalis* E.K. Jones & Fassett = Sporobolus cryptandrus [SPORCRY]
*__Sporobolus vaginiflorus__ (Torr. ex Gray) Wood [SPORVAG] (poverty grass)
SY: *Sporobolus vaginiflorus* var. *inaequalis* Fern.
Sporobolus vaginiflorus var. *inaequalis* Fern. = Sporobolus vaginiflorus [SPORVAG]
Stipa L. [STIPA$]
Stipa canadensis Poir. = Oryzopsis canadensis [ORYZCAN]
Stipa columbiana auct. = Stipa nelsonii ssp. dorei [STIPNEL1]
Stipa comata Trin. & Rupr. [STIPCOM] (needle-and-thread grass)

SY: *Stipa tweedyi* Scribn.
Stipa comata var. **intermedia** Scribn. & Tweedy [STIPCOM1]
Stipa curtiseta (A.S. Hitchc.) Barkworth [STIPCUR] (short-awned porcupinegrass)
SY: *Stipa spartea* var. *curtiseta* A.S. Hitchc.
Stipa elmeri Piper & Brodie ex Scribn. = Stipa occidentalis var. pubescens [STIPOCC3]
Stipa hymenoides Roemer & J.A. Schultes = Oryzopsis hymenoides [ORYZHYM]
Stipa lemmonii (Vasey) Scribn. [STIPLEM] (Lemmon's needlegrass)
SY: *Stipa lemmonii* var. *jonesii* Scribn.
SY: *Stipa lemmonii* var. *pubescens* Crampton
Stipa lemmonii var. *jonesii* Scribn. = Stipa lemmonii [STIPLEM]
Stipa lemmonii var. *pubescens* Crampton = Stipa lemmonii [STIPLEM]
Stipa minor (Vasey) Scribn. = Stipa nelsonii ssp. dorei [STIPNEL1]
Stipa nelsonii Scribn. [STIPNEL] (Nelson's needlegrass)
Stipa nelsonii ssp. **dorei** Barkworth & Maze [STIPNEL1] (Columbian needlegrass)
SY: *Stipa minor* (Vasey) Scribn.
SY: *Stipa nelsonii* var. *dorei* (Barkworth & Maze) Dorn
SY: *Stipa occidentalis* var. *minor* (Vasey) C.L. Hitchc.
SY: *Stipa columbiana* auct.
Stipa nelsonii var. *dorei* (Barkworth & Maze) Dorn = Stipa nelsonii ssp. dorei [STIPNEL1]
Stipa occidentalis Thurb. ex S. Wats. [STIPOCC] (western needle grass)
Stipa occidentalis var. *minor* (Vasey) C.L. Hitchc. = Stipa nelsonii ssp. dorei [STIPNEL1]
Stipa occidentalis var. **pubescens** (Vasey) Maze, Taylor & MacBryde [STIPOCC3] (stiff needlegrass)
SY: *Stipa elmeri* Piper & Brodie ex Scribn.
Stipa richardsonii Link [STIPRIC] (spreading needlegrass)
Stipa spartea Trin. [STIPSPA] (porcupinegrass)
Stipa spartea var. *curtiseta* A.S. Hitchc. = Stipa curtiseta [STIPCUR]
Stipa tweedyi Scribn. = Stipa comata [STIPCOM]
Stipa viridula Trin. = Nassella viridula [NASSVIR]
Syntherisma sanguinalis (L.) Dulac = Digitaria sanguinalis [DIGISAN]
Thinopyrum intermedium (Host) Barkworth & Dewey = Elytrigia intermedia [ELYTINT]
Thinopyrum intermedium ssp. *barbulatum* (Schur) Barkworth & Dewey = Elytrigia intermedia [ELYTINT]
Thinopyrum ponticum (Podp.) Barkworth & Dewey = Elytrigia pontica [ELYTPON]
Torreyochloa Church [TORREYO$]
Torreyochloa fernaldii (A.S. Hitchc.) Church = Torreyochloa pallida var. fernaldii [TORRPAL1]
Torreyochloa pallida (Torr.) Church [TORRPAL] (Fernald's false-manna)
SY: *Glyceria pallida* (Torr.) Trin.
Torreyochloa pallida var. **fernaldii** (A.S. Hitchc.) Dore ex Koyama & Kawano [TORRPAL1]
SY: *Glyceria fernaldii* (A.S. Hitchc.) St. John
SY: *Glyceria pallida* var. *fernaldii* A.S. Hitchc.
SY: *Puccinellia fernaldii* (A.S. Hitchc.) E.G. Voss
SY: *Torreyochloa fernaldii* (A.S. Hitchc.) Church
Torreyochloa pallida var. **pauciflora** (J. Presl) J.I. Davis [TORRPAL2]
SY: *Glyceria pauciflora* J. Presl
SY: *Puccinellia pauciflora* (J. Presl) Munz
SY: *Puccinellia pauciflora* var. *holmii* (Beal) C.L. Hitchc.
SY: *Puccinellia pauciflora* var. *microtheca* (Buckl.) C.L. Hitchc.
SY: *Torreyochloa pauciflora* (J. Presl) Church
SY: *Torreyochloa pauciflora* var. *holmii* (Beal) Taylor & MacBryde
SY: *Torreyochloa pauciflora* var. *microtheca* (Buckl.) Taylor & MacBryde
Torreyochloa pauciflora (J. Presl) Church = Torreyochloa pallida var. pauciflora [TORRPAL2]
Torreyochloa pauciflora var. *holmii* (Beal) Taylor & MacBryde = Torreyochloa pallida var. pauciflora [TORRPAL2]
Torreyochloa pauciflora var. *microtheca* (Buckl.) Taylor & MacBryde = Torreyochloa pallida var. pauciflora [TORRPAL2]
Trisetum Pers. [TRISETU$]
Trisetum canescens Buckl. = Trisetum cernuum var. canescens [TRISCER3]
Trisetum cernuum Trin. [TRISCER] (nodding trisetum)
Trisetum cernuum ssp. *canescens* (Buckl.) Calder & Taylor = Trisetum cernuum var. canescens [TRISCER3]
Trisetum cernuum var. **canescens** (Buckl.) Beal [TRISCER3] (tall trisetum)
SY: *Trisetum canescens* Buckl.
SY: *Trisetum cernuum* ssp. *canescens* (Buckl.) Calder & Taylor
Trisetum cernuum var. **cernuum** [TRISCER0]
Trisetum purpurascens Torr. = Schizachne purpurascens [SCHIPUR]
Trisetum spicatum (L.) Richter [TRISSPI] (spike trisetum)
SY: *Aira spicata* L.
SY: *Trisetum spicatum* ssp. *alaskanum* (Nash) Hultén
SY: *Trisetum spicatum* var. *alaskanum* (Nash) Malte ex Louis-Marie
SY: *Trisetum spicatum* ssp. *congdonii* (Scribn. & Merr.) Hultén
SY: *Trisetum spicatum* var. *congdonii* (Scribn. & Merr.) A.S. Hitchc.
SY: *Trisetum spicatum* var. *maidenii* (Gandog.) Fern.
SY: *Trisetum spicatum* ssp. *majus* (Rydb.) Hultén
SY: *Trisetum spicatum* var. *majus* (Rydb.) Farw.
SY: *Trisetum spicatum* ssp. *molle* (Kunth) Hultén
SY: *Trisetum spicatum* var. *molle* (Kunth) Beal
SY: *Trisetum spicatum* ssp. *pilosiglume* (Fern.) Hultén
SY: *Trisetum spicatum* var. *pilosiglume* Fern.
SY: *Trisetum spicatum* var. *spicatiforme* Hultén
SY: *Trisetum spicatum* var. *villosissimum* (Lange) Louis-Marie
SY: *Trisetum subspicatum* (L.) Beauv.
SY: *Trisetum triflorum* (Bigelow) A. & D. Löve
SY: *Trisetum triflorum* ssp. *molle* (Kunth) A. & D. Löve
SY: *Trisetum villosissimum* (Lange) Louis-Marie
Trisetum spicatum ssp. *alaskanum* (Nash) Hultén = Trisetum spicatum [TRISSPI]
Trisetum spicatum ssp. *congdonii* (Scribn. & Merr.) Hultén = Trisetum spicatum [TRISSPI]
Trisetum spicatum ssp. *majus* (Rydb.) Hultén = Trisetum spicatum [TRISSPI]
Trisetum spicatum ssp. *molle* (Kunth) Hultén = Trisetum spicatum [TRISSPI]
Trisetum spicatum ssp. *pilosiglume* (Fern.) Hultén = Trisetum spicatum [TRISSPI]
Trisetum spicatum var. *alaskanum* (Nash) Malte ex Louis-Marie = Trisetum spicatum [TRISSPI]
Trisetum spicatum var. *congdonii* (Scribn. & Merr.) A.S. Hitchc. = Trisetum spicatum [TRISSPI]
Trisetum spicatum var. *maidenii* (Gandog.) Fern. = Trisetum spicatum [TRISSPI]
Trisetum spicatum var. *majus* (Rydb.) Farw. = Trisetum spicatum [TRISSPI]
Trisetum spicatum var. *molle* (Kunth) Beal = Trisetum spicatum [TRISSPI]
Trisetum spicatum var. *pilosiglume* Fern. = Trisetum spicatum [TRISSPI]
Trisetum spicatum var. *spicatiforme* Hultén = Trisetum spicatum [TRISSPI]
Trisetum spicatum var. *villosissimum* (Lange) Louis-Marie = Trisetum spicatum [TRISSPI]
Trisetum subspicatum (L.) Beauv. = Trisetum spicatum [TRISSPI]
Trisetum triflorum (Bigelow) A. & D. Löve = Trisetum spicatum [TRISSPI]
Trisetum triflorum ssp. *molle* (Kunth) A. & D. Löve = Trisetum spicatum [TRISSPI]
Trisetum villosissimum (Lange) Louis-Marie = Trisetum spicatum [TRISSPI]
Trisetum wolfii Vasey [TRISWOL] (Wolf's trisetum)
Triticum L. [TRITICU$]
*__Triticum aestivum__ L. [TRITAES] (wheat)
SY: *Triticum hybernum* L.
SY: *Triticum macha* Dekap. & Menab.

SY: *Triticum sativum* Lam.
SY: *Triticum sphaerococcum* Percival
SY: *Triticum vulgare* Vill.
Triticum cereale (L.) Salisb. = Secale cereale [SECACER]
Triticum hybernum L. = Triticum aestivum [TRITAES]
Triticum macha Dekap. & Menab. = Triticum aestivum [TRITAES]
Triticum repens L. = Elytrigia repens [ELYTREP]
Triticum sativum Lam. = Triticum aestivum [TRITAES]
Triticum sphaerococcum Percival = Triticum aestivum [TRITAES]
Triticum trachycaulum Link = Elymus trachycaulus ssp. trachycaulus [ELYMTRA0]
Triticum vulgare Vill. = Triticum aestivum [TRITAES]
Uniola spicata L. = Distichlis spicata [DISTSPI]
Vahlodea Fries [VAHLODE$]
Vahlodea atropurpurea (Wahlenb.) Fries ex Hartman [VAHLATR] (mountain hairgrass)
SY: *Aira atropurpurea* Wahlenb.
SY: *Deschampsia atropurpurea* (Wahlenb.) Scheele
SY: *Deschampsia atropurpurea* var. *latifolia* (Hook.) Scribn. ex Macoun
SY: *Deschampsia atropurpurea* var. *paramushirensis* Kudo
SY: *Deschampsia atropurpurea* var. *payettii* Lepage
SY: *Deschampsia pacifica* Tatew. & Ohwi
SY: *Vahlodea atropurpurea* ssp. *latifolia* (Hook.) Porsild
SY: *Vahlodea atropurpurea* ssp. *paramushirensis* (Kudo) Hultén
SY: *Vahlodea flexuosa* (Honda) Ohwi
SY: *Vahlodea latifolia* (Hook.) Hultén
Vahlodea atropurpurea ssp. *latifolia* (Hook.) Porsild = Vahlodea atropurpurea [VAHLATR]
Vahlodea atropurpurea ssp. *paramushirensis* (Kudo) Hultén = Vahlodea atropurpurea [VAHLATR]
Vahlodea flexuosa (Honda) Ohwi = Vahlodea atropurpurea [VAHLATR]
Vahlodea latifolia (Hook.) Hultén = Vahlodea atropurpurea [VAHLATR]
Ventenata Koel. [VENTENA$]
Ventenata avenacea Koel. = Ventenata dubia [VENTDUB]
*Ventenata dubia** (Leers) Coss. & Durieu [VENTDUB] (ventenata)
SY: *Avena dubia* Leers
SY: *Ventenata avenacea* Koel.
Vulpia K.C. Gmel. [VULPIA$]
*Vulpia bromoides** (L.) S.F. Gray [VULPBRO] (barren fescue)
SY: *Bromus dertonensis* All.
SY: *Festuca bromoides* L.
SY: *Festuca detonensis* (All.) Aschers. & Graebn.
SY: *Vulpia dertonensis* (All.) Gola
Vulpia dertonensis (All.) Gola = Vulpia bromoides [VULPBRO]
Vulpia megalura (Nutt.) Rydb. = Vulpia myuros [VULPMYU]
Vulpia microstachys (Nutt.) Munro [VULPMIC] (small fescue)
Vulpia microstachys var. **pauciflora** (Scribn. ex Beal) Lonard & Gould [VULPMIC1]
SY: *Festuca microstachys* var. *pauciflora* Scribn. ex Beal
SY: *Festuca microstachys* var. *simulans* (Hoover) Hoover
SY: *Festuca pacifica* Piper
SY: *Festuca pacifica* var. *simulans* Hoover
SY: *Festuca reflexa* Buckl.
SY: *Vulpia pacifica* (Piper) Rydb.
SY: *Vulpia reflexa* (Buckl.) Rydb.
Vulpia myuros (L.) K.C. Gmel. [VULPMYU] (rattail fescue)
SY: *Festuca megalura* Nutt.
SY: *Festuca megalura* var. *hirsuta* (Hack.) Aschers. & Graebn.
SY: *Festuca myuros* L.
SY: *Vulpia megalura* (Nutt.) Rydb.
SY: *Vulpia myuros* var. *hirsuta* Hack.
Vulpia myuros var. *hirsuta* Hack. = Vulpia myuros [VULPMYU]
Vulpia octoflora (Walt.) Rydb. [VULPOCT] (six-weeks grass)
Vulpia octoflora var. **glauca** (Nutt.) Fern. [VULPOCT1]
SY: *Festuca gracilenta* Buckl.
SY: *Festuca octoflora* var. *glauca* (Nutt.) Fern.
SY: *Festuca octoflora* var. *tenella* (Willd.) Fern.
SY: *Festuca tenella* Willd.
SY: *Festuca tenella* var. *glauca* Nutt.
SY: *Vulpia octoflora* var. *tenella* (Willd.) Fern.
Vulpia octoflora var. **hirtella** (Piper) Henr. [VULPOCT2]
SY: *Festuca octoflora* ssp. *hirtella* Piper
SY: *Festuca octoflora* var. *hirtella* (Piper) Piper ex A.S. Hitchc.
Vulpia octoflora var. **octoflora** [VULPOCT0]
SY: *Festuca octoflora* Walt.
SY: *Festuca octoflora* var. *aristulata* Torr. ex L.H. Dewey
Vulpia octoflora var. *tenella* (Willd.) Fern. = Vulpia octoflora var. glauca [VULPOCT1]
Vulpia pacifica (Piper) Rydb. = Vulpia microstachys var. pauciflora [VULPMIC1]
Vulpia reflexa (Buckl.) Rydb. = Vulpia microstachys var. pauciflora [VULPMIC1]

POLEMONIACEAE (V:D)

Collomia Nutt. [COLLOMI$]
Collomia grandiflora Dougl. ex Lindl. [COLLGRN] (large-flowered collomia)
Collomia heterophylla Dougl. ex Hook. [COLLHET] (vari-leaved collomia)
Collomia linearis Nutt. [COLLLIN] (narrow-leaved collomia)
Gilia Ruiz & Pavón [GILIA$]
Gilia aggregata (Pursh) Spreng. = Ipomopsis aggregata [IPOMAGG]
Gilia capitata Sims [GILICAP] (globe gilia)
Gilia gracilis Hook. = Phlox gracilis [PHLOGRA]
Gilia hallii Parish = Leptodactylon pungens [LEPTPUN]
Gilia inconspicua var. *sinuata* (Dougl. ex Benth.) Gray = Gilia sinuata [GILISIN]
Gilia minutiflora Benth. = Ipomopsis minutiflora [IPOMMIN]
Gilia pungens (Torr.) Benth. = Leptodactylon pungens [LEPTPUN]
Gilia pungens var. *hookeri* (Dougl. ex Hook.) Gray = Leptodactylon pungens [LEPTPUN]
Gilia septentrionalis (Mason) St. John = Linanthus septentrionalis [LINASEP]
Gilia sinuata Dougl. ex Benth. [GILISIN] (shy gilia)
SY: *Gilia inconspicua* var. *sinuata* (Dougl. ex Benth.) Gray
Gilia squarrosa (Eschsch.) Hook. & Arn. = Navarretia squarrosa [NAVASQU]
Ipomopsis Michx. [IPOMOPS$]
Ipomopsis aggregata (Pursh) V. Grant [IPOMAGG] (scarlet gilia)
SY: *Gilia aggregata* (Pursh) Spreng.
Ipomopsis minutiflora (Benth.) V. Grant [IPOMMIN] (small-flowered ipomopsis)
SY: *Gilia minutiflora* Benth.
Leptodactylon Hook. & Arn. [LEPTODA$]
Leptodactylon hazeliae M.E. Peck = Leptodactylon pungens [LEPTPUN]
Leptodactylon lilacinum Greene ex Baker = Leptodactylon pungens [LEPTPUN]
Leptodactylon pungens (Torr.) Torr. ex Nutt. [LEPTPUN] (prickly phlox)
SY: *Gilia hallii* Parish
SY: *Gilia pungens* (Torr.) Benth.
SY: *Gilia pungens* var. *hookeri* (Dougl. ex Hook.) Gray
SY: *Leptodactylon hazeliae* M.E. Peck
SY: *Leptodactylon lilacinum* Greene ex Baker
SY: *Leptodactylon pungens* ssp. *brevifolium* (Rydb.) Wherry
SY: *Leptodactylon pungens* ssp. *eupungens* (Brand) Wherry
SY: *Leptodactylon pungens* ssp. *hallii* (Parish) Mason
SY: *Leptodactylon pungens* var. *hallii* (Parish) Jepson
SY: *Leptodactylon pungens* ssp. *hazeliae* (M.E. Peck) Meinke
SY: *Leptodactylon pungens* ssp. *hookeri* (Dougl. ex Hook.) Wherry
SY: *Leptodactylon pungens* var. *hookeri* (Dougl. ex Hook.) Jepson
SY: *Leptodactylon pungens* ssp. *pulchriflorum* (Brand) Mason
SY: *Leptodactylon pungens* ssp. *squarrosum* (Gray) Tidestrom
Leptodactylon pungens ssp. *brevifolium* (Rydb.) Wherry = Leptodactylon pungens [LEPTPUN]
Leptodactylon pungens ssp. *eupungens* (Brand) Wherry = Leptodactylon pungens [LEPTPUN]
Leptodactylon pungens ssp. *hallii* (Parish) Mason = Leptodactylon pungens [LEPTPUN]

Leptodactylon pungens ssp. *hazeliae* (M.E. Peck) Meinke = Leptodactylon pungens [LEPTPUN]
Leptodactylon pungens ssp. *hookeri* (Dougl. ex Hook.) Wherry = Leptodactylon pungens [LEPTPUN]
Leptodactylon pungens ssp. *pulchriflorum* (Brand) Mason = Leptodactylon pungens [LEPTPUN]
Leptodactylon pungens ssp. *squarrosum* (Gray) Tidestrom = Leptodactylon pungens [LEPTPUN]
Leptodactylon pungens var. *hallii* (Parish) Jepson = Leptodactylon pungens [LEPTPUN]
Leptodactylon pungens var. *hookeri* (Dougl. ex Hook.) Jepson = Leptodactylon pungens [LEPTPUN]
Linanthus Benth. [LINANTH$]
Linanthus bicolor (Nutt.) Greene [LINABIC] (bicolored flaxflower)
Linanthus harknessii (Curran) Greene [LINAHAR] (Harkness' flaxflower)
Linanthus harknessii var. *septentrionalis* (Mason) Jepson & V. Bailey = Linanthus septentrionalis [LINASEP]
Linanthus septentrionalis Mason [LINASEP] (northern flaxflower)
SY: *Gilia septentrionalis* (Mason) St. John
SY: *Linanthus harknessii* var. *septentrionalis* (Mason) Jepson & V. Bailey
Microsteris gracilis (Hook.) Greene = Phlox gracilis [PHLOGRA]
Navarretia Ruiz & Pavón [NAVARRE$]
Navarretia intertexta (Benth.) Hook. [NAVAINT] (needle-leaved navarretia)
SY: *Navarretia minima* var. *intertexta* (Benth.) Boivin
Navarretia minima var. *intertexta* (Benth.) Boivin = Navarretia intertexta [NAVAINT]
Navarretia squarrosa (Eschsch.) Hook. & Arn. [NAVASQU] (skunkweed)
SY: *Gilia squarrosa* (Eschsch.) Hook. & Arn.
Phlox L. [PHLOX$]
Phlox alyssifolia Greene [PHLOALY] (alyssum-leaved phlox)
Phlox caespitosa Nutt. [PHLOCAE] (tufyed phlox)
SY: *Phlox caespitosa* ssp. *eucaespitosa* Brand
SY: *Phlox douglasii* Hook.
SY: *Phlox douglasii* ssp. *eudouglasii* Brand
Phlox caespitosa ssp. *eucaespitosa* Brand = Phlox caespitosa [PHLOCAE]
Phlox diffusa Benth. [PHLODIF] (spreading phlox)
Phlox diffusa ssp. **longistylis** Wherry [PHLODIF1]
SY: *Phlox diffusa* var. *longistylis* (Wherry) M.E. Peck
Phlox diffusa var. *longistylis* (Wherry) M.E. Peck = Phlox diffusa ssp. longistylis [PHLODIF1]
Phlox douglasii Hook. = Phlox caespitosa [PHLOCAE]
Phlox douglasii ssp. *eudouglasii* Brand = Phlox caespitosa [PHLOCAE]
Phlox gracilis (Hook.) Greene [PHLOGRA]
SY: *Gilia gracilis* Hook.
SY: *Microsteris gracilis* (Hook.) Greene
Phlox hoodii Richards. [PHLOHOO] (Hood's phlox)
Phlox longifolia Nutt. [PHLOLON] (long-leaved phlox)
Phlox speciosa Pursh [PHLOSPE] (showy phlox)
Polemoniella micrantha (Benth.) Heller = Polemonium micranthum [POLEMIC]
Polemonium L. [POLEMON$]
Polemonium acutiflorum Willd. ex Roemer & J.A. Schultes [POLEACU]
SY: *Polemonium caeruleum* ssp. *villosum* (J.H. Rudolph ex Georgi) Brand
Polemonium boreale M.F. Adams [POLEBOR] (northern Jacob's-ladder)
***Polemonium caeruleum** L. [POLECAE] (tall Jacob's-ladder)
Polemonium caeruleum ssp. *amygdalinum* (Wherry) Munz = Polemonium occidentale [POLEOCC]
Polemonium caeruleum ssp. *occidentale* (Greene) J.F. Davids. = Polemonium occidentale [POLEOCC]
Polemonium caeruleum ssp. *villosum* (J.H. Rudolph ex Georgi) Brand = Polemonium acutiflorum [POLEACU]
Polemonium caeruleum var. *pterospermum* Benth. = Polemonium occidentale [POLEOCC]
Polemonium elegans Greene [POLEELE] (elegant Jacob's-ladder)
Polemonium helleri Brand = Polemonium occidentale [POLEOCC]
Polemonium intermedium (Brand) Rydb. = Polemonium occidentale [POLEOCC]
Polemonium micranthum Benth. [POLEMIC] (littlebells polemonium)
SY: *Polemoniella micrantha* (Benth.) Heller
Polemonium occidentale Greene [POLEOCC]
SY: *Polemonium caeruleum* ssp. *amygdalinum* (Wherry) Munz
SY: *Polemonium caeruleum* ssp. *occidentale* (Greene) J.F. Davids.
SY: *Polemonium caeruleum* var. *pterospermum* Benth.
SY: *Polemonium helleri* Brand
SY: *Polemonium intermedium* (Brand) Rydb.
SY: *Polemonium occidentale* ssp. *amygdalium* Wherry
SY: *Polemonium occidentale* ssp. *typicum* Wherry
Polemonium occidentale ssp. *amygdalium* Wherry = Polemonium occidentale [POLEOCC]
Polemonium occidentale ssp. *typicum* Wherry = Polemonium occidentale [POLEOCC]
Polemonium pulcherrimum Hook. [POLEPUL] (showy Jacob's-ladder)
Polemonium viscosum Nutt. [POLEVIS] (skunk Jacob's-ladder)
SY: *Polemonium viscosum* ssp. *genuinum* Wherry
SY: *Polemonium viscosum* ssp. *lemmonii* (Brand) Wherry
Polemonium viscosum ssp. *genuinum* Wherry = Polemonium viscosum [POLEVIS]
Polemonium viscosum ssp. *lemmonii* (Brand) Wherry = Polemonium viscosum [POLEVIS]

POLYGALACEAE (V:D)

Polygala L. [POLYGAL$]
Polygala senega L. [POLYSEN] (seneca-root)

POLYGONACEAE (V:D)

Acetosa alpestris (Jacq.) A. Löve = Rumex acetosa ssp. alpestris [RUMEACE3]
Acetosa oblongifolia (L.) A. & D. Löve = Rumex obtusifolius [RUMEOBT]
Acetosa pratensis ssp. *alpestris* (Jacq.) A. Löve = Rumex acetosa ssp. alpestris [RUMEACE3]
Acetosella acetosella (L.) Small = Rumex acetosella [RUMEACT]
Acetosella tenuifolia (Wallr.) A. Löve = Rumex acetosella [RUMEACT]
Acetosella vulgaris (Koch) Fourr. = Rumex acetosella [RUMEACT]
Aconogonum polystachyum (Wallich ex Meisn.) Haraldson = Polygonum polystachyum [POLYPOY]
Bilderdykia convolvulus (L.) Dumort. = Polygonum convolvulus [POLYCON]
Bistorta bistortoides (Pursh) Small = Polygonum bistortoides [POLYBIS]
Bistorta vivipara (L.) S.F. Gray = Polygonum viviparum [POLYVIV]
Eriogonum Michx. [ERIOGON$]
Eriogonum androsaceum Benth. [ERIOAND] (androsace buckwheat)
SY: *Eriogonum flavum* var. *androsaceum* (Benth.) M.E. Jones
Eriogonum angustifolium Nutt. = Eriogonum heracleoides var. angustifolium [ERIOHER1]
Eriogonum depauperatum Small = Eriogonum pauciflorum [ERIOPAU]
Eriogonum flavum Nutt. [ERIOFLA] (alpine golden wild buckwheat)
Eriogonum flavum ssp. *piperi* (Greene) S. Stokes = Eriogonum flavum var. piperi [ERIOFLA1]
Eriogonum flavum var. *androsaceum* (Benth.) M.E. Jones = Eriogonum androsaceum [ERIOAND]
Eriogonum flavum var. *linguifolium* Gandog. = Eriogonum flavum var. piperi [ERIOFLA1]
Eriogonum flavum var. **piperi** (Greene) M.E. Jones [ERIOFLA1]
SY: *Eriogonum flavum* var. *linguifolium* Gandog.
SY: *Eriogonum flavum* ssp. *piperi* (Greene) S. Stokes
SY: *Eriogonum piperi* Greene

Eriogonum heracleoides Nutt. [ERIOHER] (parsnip-flowered buckwheat)
Eriogonum heracleoides var. **angustifolium** (Nutt.) Torr. & Gray [ERIOHER1]
SY: *Eriogonum angustifolium* Nutt.
Eriogonum heracleoides var. *subalpinum* (Greene) S. Stokes = Eriogonum umbellatum var. majus [ERIOUMB1]
Eriogonum multiceps Nees = Eriogonum pauciflorum [ERIOPAU]
Eriogonum nivale Canby = Eriogonum ovalifolium var. nivale [ERIOOVA3]
Eriogonum niveum Dougl. ex Benth. [ERIONIV] (snow buckwheat)
SY: *Eriogonum niveum* ssp. *decumbens* (Benth.) S. Stokes
SY: *Eriogonum niveum* var. *decumbens* (Benth.) Torr. & Gray
SY: *Eriogonum niveum* var. *dichotomum* (Dougl. ex Benth.) M.E. Jones
SY: *Eriogonum strictum* var. *lachnostegium* Benth.
Eriogonum niveum ssp. *decumbens* (Benth.) S. Stokes = Eriogonum niveum [ERIONIV]
Eriogonum niveum var. *decumbens* (Benth.) Torr. & Gray = Eriogonum niveum [ERIONIV]
Eriogonum niveum var. *dichotomum* (Dougl. ex Benth.) M.E. Jones = Eriogonum niveum [ERIONIV]
Eriogonum ovalifolium Nutt. [ERIOOVA] (cushion buckwheat)
Eriogonum ovalifolium var. **nivale** (Canby) M.E. Jones [ERIOOVA3]
SY: *Eriogonum nivale* Canby
SY: *Eriogonum rhodanthum* A. Nels. & Kennedy
Eriogonum pauciflorum Pursh [ERIOPAU] (few-flowered buckwheat)
SY: *Eriogonum depauperatum* Small
SY: *Eriogonum multiceps* Nees
Eriogonum piperi Greene = Eriogonum flavum var. piperi [ERIOFLA1]
Eriogonum pyrolifolium Hook. [ERIOPYR] (alpine buckwheat)
Eriogonum pyrolifolium var. *bellingerianum* M.E. Peck = Eriogonum pyrolifolium var. coryphaeum [ERIOPYR1]
Eriogonum pyrolifolium var. **coryphaeum** Torr. & Gray [ERIOPYR1]
SY: *Eriogonum pyrolifolium* var. *bellingerianum* M.E. Peck
Eriogonum rhodanthum A. Nels. & Kennedy = Eriogonum ovalifolium var. nivale [ERIOOVA3]
Eriogonum strictum Benth. [ERIOSTR] (strict buckwheat)
Eriogonum strictum ssp. **proliferum** (Torr. & Gray) S. Stokes [ERIOSTR1]
Eriogonum strictum var. *lachnostegium* Benth. = Eriogonum niveum [ERIONIV]
Eriogonum subalpinum Greene = Eriogonum umbellatum var. majus [ERIOUMB1]
Eriogonum umbellatum Torr. [ERIOUMB] (sulphur buckwheat)
Eriogonum umbellatum ssp. *majus* (Hook.) Piper = Eriogonum umbellatum var. majus [ERIOUMB1]
Eriogonum umbellatum var. **majus** Hook. [ERIOUMB1]
SY: *Eriogonum heracleoides* var. *subalpinum* (Greene) S. Stokes
SY: *Eriogonum subalpinum* Greene
SY: *Eriogonum umbellatum* ssp. *majus* (Hook.) Piper
SY: *Eriogonum umbellatum* var. *subalpinum* (Greene) M.E. Jones
Eriogonum umbellatum var. *subalpinum* (Greene) M.E. Jones = Eriogonum umbellatum var. majus [ERIOUMB1]
Eriogonum umbellatum var. **umbellatum** [ERIOUMB0]
Fagopyrum P. Mill. [FAGOPYR$]
*__Fagopyrum esculentum__ Moench [FAGOESC] (buckwheat)
SY: *Fagopyrum fagopyrum* (L.) Karst.
SY: *Fagopyrum sagittatum* Gilib.
SY: *Fagopyrum vulgare* Hill
SY: *Polygonum fagopyrum* L.
Fagopyrum fagopyrum (L.) Karst. = Fagopyrum esculentum [FAGOESC]
Fagopyrum sagittatum Gilib. = Fagopyrum esculentum [FAGOESC]
Fagopyrum vulgare Hill = Fagopyrum esculentum [FAGOESC]
Fallopia convolvulus (L.) A. Löve = Polygonum convolvulus [POLYCON]
Koenigia L. [KOENIGI$]
Koenigia islandica L. [KOENISL] (island koenigia)
SY: *Macounastrum islandicum* (L.) Small
Macounastrum islandicum (L.) Small = Koenigia islandica [KOENISL]
Oxyria Hill [OXYRIA$]
Oxyria digyna (L.) Hill [OXYRDIG] (mountain sorrel)
SY: *Rumex digyna* L.
Persicaria amphibia (L.) S.F. Gray p.p. = Polygonum amphibium var. emersum [POLYAMP1]
Persicaria amphibia var. *emersa* (Michx.) Hickman = Polygonum amphibium var. emersum [POLYAMP1]
Persicaria amphibia var. *stipulacea* (Coleman) Hara = Polygonum amphibium var. stipulaceum [POLYAMP2]
Persicaria coccinea (Muhl. ex Willd.) Greene = Polygonum amphibium var. emersum [POLYAMP1]
Persicaria hydropiper (L.) Opiz = Polygonum hydropiper [POLYHYD]
Persicaria hydropiperoides (Michx.) Small = Polygonum hydropiperoides [POLYHYR]
Persicaria hydropiperoides var. *breviciliata* (Fern.) C.F. Reed = Polygonum hydropiperoides [POLYHYR]
Persicaria hydropiperoides var. *euronotora* (Fern.) C.F. Reed = Polygonum hydropiperoides [POLYHYR]
Persicaria hydropiperoides var. *opelousana* (Riddell ex Small) J.S. Wilson = Polygonum hydropiperoides [POLYHYR]
Persicaria lapathifolia (L.) S.F. Gray = Polygonum lapathifolium [POLYLAP]
Persicaria maculata (Raf.) S.F. Gray = Polygonum persicaria [POLYPES]
Persicaria mesochora Greene = Polygonum amphibium var. stipulaceum [POLYAMP2]
Persicaria muehlenbergii (S. Wats.) Small = Polygonum amphibium var. emersum [POLYAMP1]
Persicaria nebraskensis Greene = Polygonum amphibium var. stipulaceum [POLYAMP2]
Persicaria nepalense (Meisn.) Miyabe = Polygonum nepalense [POLYNEP]
Persicaria opelousana (Riddell ex Small) Small = Polygonum hydropiperoides [POLYHYR]
Persicaria paludicola Small = Polygonum hydropiperoides [POLYHYR]
Persicaria persicaria (L.) Small = Polygonum persicaria [POLYPES]
Persicaria persicarioides (Kunth) Small = Polygonum hydropiperoides [POLYHYR]
Persicaria punctata (Ell.) Small = Polygonum punctatum [POLYPUN]
Persicaria ruderalis (Salisb.) C.F. Reed = Polygonum persicaria [POLYPES]
Persicaria ruderalis var. *vulgaris* (Webb & Moq.) C.F. Reed = Polygonum persicaria [POLYPES]
Persicaria vulgaris Webb & Moq. = Polygonum persicaria [POLYPES]
Pleuropterus cuspidatus (Sieb. & Zucc.) Moldenke = Polygonum cuspidatum [POLYCUS]
Pleuropterus zuccarinii (Small) Small = Polygonum cuspidatum [POLYCUS]
Polygonum L. [POLYGON$]
Polygonum achoreum Blake [POLYACH] (Blake's knotweed)
SY: *Polygonum erectum* ssp. *achoreum* Blaka
Polygonum aequale Lindm. = Polygonum arenastrum [POLYARE]
Polygonum alatum Hamilton ex D. Don = Polygonum nepalense [POLYNEP]
Polygonum amphibium L. [POLYAMP] (water smartweed)
Polygonum amphibium ssp. *laevimarginatum* Hultén = Polygonum amphibium var. stipulaceum [POLYAMP2]
Polygonum amphibium var. **emersum** Michx. [POLYAMP1]
SY: *Persicaria amphibia* (L.) S.F. Gray p.p.
SY: *Persicaria amphibia* var. *emersa* (Michx.) Hickman
SY: *Persicaria coccinea* (Muhl. ex Willd.) Greene
SY: *Persicaria muehlenbergii* (S. Wats.) Small
SY: *Polygonum coccineum* Muhl. ex Willd.
SY: *Polygonum coccineum* var. *pratincola* (Greene) Stanford
SY: *Polygonum coccineum* var. *terrestre* Willd.
SY: *Polygonum muehlenbergii* S. Wats.

SY: *Polygonum muehlenbergii* var. *terrestre* (Willd.) Trel.
Polygonum amphibium var. *hartwrightii* (Gray) Bissell = Polygonum amphibium var. stipulaceum [POLYAMP2]
Polygonum amphibium var. **stipulaceum** Coleman [POLYAMP2]
SY: *Persicaria amphibia* var. *stipulacea* (Coleman) Hara
SY: *Persicaria mesochora* Greene
SY: *Persicaria nebraskensis* Greene
SY: *Polygonum amphibium* var. *hartwrightii* (Gray) Bissell
SY: *Polygonum amphibium* ssp. *laevimarginatum* Hultén
SY: *Polygonum coccineum* var. *rigidulum* (Sheldon) Stanford
SY: *Polygonum fluitans* Eat.
SY: *Polygonum hartwrightii* Gray
SY: *Polygonum inundatum* Raf.
SY: *Polygonum natans* Eat.
*__Polygonum arenastrum__ Jord. ex Boreau [POLYARE] (oval-leaved knotweed)
SY: *Polygonum aequale* Lindm.
SY: *Polygonum aviculare* var. *arenastrum* (Jord. ex Boreau) Rouy
SY: *Polygonum montereyense* Brenckle
Polygonum austiniae Greene = Polygonum douglasii ssp. austiniae [POLYDOU3]
*__Polygonum aviculare__ L. [POLYAVI] (common knotweed)
SY: *Polygonum aviculare* var. *vegetum* Ledeb.
SY: *Polygonum heterophyllum* Lindl.
SY: *Polygonum monspeliense* Pers.
Polygonum aviculare var. *arenastrum* (Jord. ex Boreau) Rouy = Polygonum arenastrum [POLYARE]
Polygonum aviculare var. *littorale* (Link) Mert. = Polygonum buxiforme [POLYBUX]
Polygonum aviculare var. *vegetum* Ledeb. = Polygonum aviculare [POLYAVI]
Polygonum bistortoides Pursh [POLYBIS] (American bistort)
SY: *Bistorta bistortoides* (Pursh) Small
SY: *Polygonum bistortoides* var. *linearifolium* (S. Wats.) Small
SY: *Polygonum bistortoides* var. *oblongifolium* (Meisn.) St. John
SY: *Polygonum cephalophorum* Greene
SY: *Polygonum glastifolium* Greene
SY: *Polygonum vulcanicum* Greene
Polygonum bistortoides var. *linearifolium* (S. Wats.) Small = Polygonum bistortoides [POLYBIS]
Polygonum bistortoides var. *oblongifolium* (Meisn.) St. John = Polygonum bistortoides [POLYBIS]
Polygonum buxifolium Nutt. ex Bong. = Polygonum fowleri [POLYFOW]
*__Polygonum buxiforme__ Small [POLYBUX] (eastern knotweed)
SY: *Polygonum aviculare* var. *littorale* (Link) Mert.
SY: *Polygonum littorale* Link
Polygonum buxiforme var. *montanum* (Small) R.J. Davis = Polygonum douglasii ssp. douglasii [POLYDOU0]
Polygonum cephalophorum Greene = Polygonum bistortoides [POLYBIS]
Polygonum coarctatum var. *majus* Meisn. = Polygonum douglasii ssp. majus [POLYDOU5]
Polygonum coccineum Muhl. ex Willd. = Polygonum amphibium var. emersum [POLYAMP1]
Polygonum coccineum var. *pratincola* (Greene) Stanford = Polygonum amphibium var. emersum [POLYAMP1]
Polygonum coccineum var. *rigidulum* (Sheldon) Stanford = Polygonum amphibium var. stipulaceum [POLYAMP2]
Polygonum coccineum var. *terrestre* Willd. = Polygonum amphibium var. emersum [POLYAMP1]
*__Polygonum convolvulus__ L. [POLYCON] (black bindweed)
SY: *Bilderdykia convolvulus* (L.) Dumort.
SY: *Fallopia convolvulus* (L.) A. Löve
*__Polygonum cuspidatum__ Sieb. & Zucc. [POLYCUS] (Japanese knotweed)
SY: *Pleuropterus cuspidatus* (Sieb. & Zucc.) Moldenke
SY: *Pleuropterus zuccarinii* (Small) Small
SY: *Polygonum cuspidatum* var. *compactum* (Hook f.) Bailey
SY: *Polygonum zuccarinii* Small
SY: *Reynoutria japonica* Houtt.
Polygonum cuspidatum var. *compactum* (Hook f.) Bailey = Polygonum cuspidatum [POLYCUS]
Polygonum douglasii Greene [POLYDOU] (Douglas' knotweed)
Polygonum douglasii ssp. **austiniae** (Greene) E. Murr. [POLYDOU3] (Austin's knotweed)
SY: *Polygonum austiniae* Greene
SY: *Polygonum douglasii* var. *austiniae* (Greene) M.E. Jones
Polygonum douglasii ssp. **douglasii** [POLYDOU0]
SY: *Polygonum buxiforme* var. *montanum* (Small) R.J. Davis
SY: *Polygonum douglasii* var. *latifolium* (Engelm.) Greene
SY: *Polygonum emaciatum* A. Nels.
SY: *Polygonum montanum* Small
Polygonum douglasii ssp. **engelmannii** (Greene) Kartesz & Gandhi [POLYDOU1]
SY: *Polygonum douglasii* var. *microspermum* (Engelm.) Dorn
SY: *Polygonum engelmannii* Greene
SY: *Polygonum microspermum* (Engelm.) Small
SY: *Polygonum tenue* var. *microspermum* Engelm.
Polygonum douglasii ssp. **majus** (Meisn.) Hickman [POLYDOU5] (wiry knotweed)
SY: *Polygonum coarctatum* var. *majus* Meisn.
SY: *Polygonum majus* (Meisn.) Piper
Polygonum douglasii ssp. **nuttallii** (Small) Hickman [POLYDOU6] (Nuttall's knotweed)
SY: *Polygonum nuttallii* Small
Polygonum douglasii ssp. **spergulariiforme** (Meisn. ex Small) Hickman [POLYDOU7] (spurry knotweed)
SY: *Polygonum spergulariiforme* Meisn. ex Small
Polygonum douglasii var. *austiniae* (Greene) M.E. Jones = Polygonum douglasii ssp. austiniae [POLYDOU3]
Polygonum douglasii var. *latifolium* (Engelm.) Greene = Polygonum douglasii ssp. douglasii [POLYDOU0]
Polygonum douglasii var. *microspermum* (Engelm.) Dorn = Polygonum douglasii ssp. engelmannii [POLYDOU1]
Polygonum dubium Stein = Polygonum persicaria [POLYPES]
Polygonum emaciatum A. Nels. = Polygonum douglasii ssp. douglasii [POLYDOU0]
Polygonum engelmannii Greene = Polygonum douglasii ssp. engelmannii [POLYDOU1]
Polygonum erectum ssp. *achoreum* Blaka = Polygonum achoreum [POLYACH]
Polygonum fagopyrum L. = Fagopyrum esculentum [FAGOESC]
Polygonum fluitans Eat. = Polygonum amphibium var. stipulaceum [POLYAMP2]
Polygonum fowleri B.L. Robins. [POLYFOW] (Fowler's knotweed)
SY: *Polygonum buxifolium* Nutt. ex Bong.
Polygonum fugax Small = Polygonum viviparum [POLYVIV]
Polygonum glastifolium Greene = Polygonum bistortoides [POLYBIS]
Polygonum hartwrightii Gray = Polygonum amphibium var. stipulaceum [POLYAMP2]
Polygonum heterophyllum Lindl. = Polygonum aviculare [POLYAVI]
*__Polygonum hydropiper__ L. [POLYHYD] (marshpepper smartweed)
SY: *Persicaria hydropiper* (L.) Opiz
SY: *Polygonum hydropiper* var. *projectum* Stanford
Polygonum hydropiper var. *projectum* Stanford = Polygonum hydropiper [POLYHYD]
Polygonum hydropiperoides Michx. [POLYHYR] (water-pepper)
SY: *Persicaria hydropiperoides* (Michx.) Small
SY: *Persicaria hydropiperoides* var. *breviciliata* (Fern.) C.F. Reed
SY: *Persicaria hydropiperoides* var. *euronotora* (Fern.) C.F. Reed
SY: *Persicaria hydropiperoides* var. *opelousana* (Riddell ex Small) J.S. Wilson
SY: *Persicaria opelousana* (Riddell ex Small) Small
SY: *Persicaria paludicola* Small
SY: *Persicaria persicarioides* (Kunth) Small
SY: *Polygonum hydropiperoides* var. *adenocalyx* (Stanford) Gleason
SY: *Polygonum hydropiperoides* var. *asperifolium* Stanford
SY: *Polygonum hydropiperoides* var. *breviciliatum* Fern.
SY: *Polygonum hydropiperoides* var. *bushianum* Stanford
SY: *Polygonum hydropiperoides* var. *digitatum* Fern.
SY: *Polygonum hydropiperoides* var. *euronotorum* Fern.
SY: *Polygonum hydropiperoides* var. *opelousanum* (Riddell ex Small) Riddell ex W. Stone
SY: *Polygonum hydropiperoides* var. *psilostachyum* St. John

SY: *Polygonum hydropiperoides* var. *strigosum* (Small) Stanford
SY: *Polygonum opelousanum* Riddell ex Small
SY: *Polygonum opelousanum* var. *adenocalyx* Stanford
SY: *Polygonum persicarioides* Kunth
Polygonum hydropiperoides var. *adenocalyx* (Stanford) Gleason = Polygonum hydropiperoides [POLYHYR]
Polygonum hydropiperoides var. *asperifolium* Stanford = Polygonum hydropiperoides [POLYHYR]
Polygonum hydropiperoides var. *breviciliatum* Fern. = Polygonum hydropiperoides [POLYHYR]
Polygonum hydropiperoides var. *bushianum* Stanford = Polygonum hydropiperoides [POLYHYR]
Polygonum hydropiperoides var. *digitatum* Fern. = Polygonum hydropiperoides [POLYHYR]
Polygonum hydropiperoides var. *euronotorum* Fern. = Polygonum hydropiperoides [POLYHYR]
Polygonum hydropiperoides var. *opelousanum* (Riddell ex Small) Riddell ex W. Stone = Polygonum hydropiperoides [POLYHYR]
Polygonum hydropiperoides var. *psilostachyum* St. John = Polygonum hydropiperoides [POLYHYR]
Polygonum hydropiperoides var. *strigosum* (Small) Stanford = Polygonum hydropiperoides [POLYHYR]
Polygonum inundatum Raf. = Polygonum amphibium var. stipulaceum [POLYAMP2]
Polygonum kelloggii Greene = Polygonum polygaloides ssp. kelloggii [POLYPOL1]
Polygonum lapathifolium L. [POLYLAP] (willow weed)
SY: *Persicaria lapathifolia* (L.) S.F. Gray
SY: *Polygonum scabrum* Moench
Polygonum littorale Link = Polygonum buxiforme [POLYBUX]
Polygonum macounii Small ex Macoun = Polygonum viviparum [POLYVIV]
Polygonum majus (Meisn.) Piper = Polygonum douglasii ssp. majus [POLYDOU5]
Polygonum microspermum (Engelm.) Small = Polygonum douglasii ssp. engelmannii [POLYDOU1]
Polygonum minimum S. Wats. [POLYMIN] (leafy dwarf knotweed)
Polygonum minus auct. = Polygonum persicaria [POLYPES]
Polygonum minus var. *subcontinuum* (Meisn.) Fern. = Polygonum persicaria [POLYPES]
Polygonum monspeliense Pers. = Polygonum aviculare [POLYAVI]
Polygonum montanum Small = Polygonum douglasii ssp. douglasii [POLYDOU0]
Polygonum montereyense Brenckle = Polygonum arenastrum [POLYARE]
Polygonum muehlenbergii S. Wats. = Polygonum amphibium var. emersum [POLYAMP1]
Polygonum muehlenbergii var. *terrestre* (Willd.) Trel. = Polygonum amphibium var. emersum [POLYAMP1]
Polygonum natans Eat. = Polygonum amphibium var. stipulaceum [POLYAMP2]
Polygonum nepalense* Meisn. [POLYNEP] (Nepalese knotweed)
SY: *Persicaria nepalense* (Meisn.) Miyabe
SY: *Polygonum alatum* Hamilton ex D. Don
Polygonum nuttallii Small = Polygonum douglasii ssp. nuttallii [POLYDOU6]
Polygonum opelousanum Riddell ex Small = Polygonum hydropiperoides [POLYHYR]
Polygonum opelousanum var. *adenocalyx* Stanford = Polygonum hydropiperoides [POLYHYR]
Polygonum paronychia Cham. & Schlecht. [POLYPAR] (black knotweed)
*__Polygonum persicaria__ L. [POLYPES] (lady's-thumb)
SY: *Persicaria maculata* (Raf.) S.F. Gray
SY: *Persicaria persicaria* (L.) Small
SY: *Persicaria ruderalis* (Salisb.) C.F. Reed
SY: *Persicaria ruderalis* var. *vulgaris* (Webb & Moq.) C.F. Reed
SY: *Persicaria vulgaris* Webb & Moq.
SY: *Polygonum dubium* Stein
SY: *Polygonum minus* var. *subcontinuum* (Meisn.) Fern.
SY: *Polygonum persicaria* var. *angustifolium* Beckh.
SY: *Polygonum persicaria* var. *ruderale* (Salisb.) Meisn.
SY: *Polygonum puritanorum* Fern.
SY: *Polygonum minus* auct.
Polygonum persicaria var. *angustifolium* Beckh. = Polygonum persicaria [POLYPES]
Polygonum persicaria var. *ruderale* (Salisb.) Meisn. = Polygonum persicaria [POLYPES]
Polygonum persicarioides Kunth = Polygonum hydropiperoides [POLYHYR]
Polygonum polygaloides Meisn. [POLYPOL] (white-margin knotweed)
Polygonum polygaloides ssp. **kelloggii** (Greene) Hickman [POLYPOL1] (Kellogg's knotweed)
SY: *Polygonum kelloggii* Greene
SY: *Polygonum unifolium* Greene
*__Polygonum polystachyum__ Wallich ex Meisn. [POLYPOY] (Himalayan knotweed)
SY: *Aconogonum polystachyum* (Wallich ex Meisn.) Haraldson
Polygonum prolificum (Small) B.L. Robins. = Polygonum ramosissimum var. prolificum [POLYRAM1]
Polygonum punctatum Ell. [POLYPUN] (dotted smartweed)
SY: *Persicaria punctata* (Ell.) Small
Polygonum puritanorum Fern. = Polygonum persicaria [POLYPES]
Polygonum ramosissimum Michx. [POLYRAM] (bushy knotweed)
Polygonum ramosissimum var. **prolificum** Small [POLYRAM1]
SY: *Polygonum prolificum* (Small) B.L. Robins.
*__Polygonum sachalinense__ F. Schmidt ex Maxim. [POLYSAC] (giant knotweed)
SY: *Reynoutria sachalinensis* (F. Schmidt ex Maxim.) Nakai
Polygonum scabrum Moench = Polygonum lapathifolium [POLYLAP]
Polygonum spergulariiforme Meisn. ex Small = Polygonum douglasii ssp. spergulariiforme [POLYDOU7]
Polygonum tenue var. *microspermum* Engelm. = Polygonum douglasii ssp. engelmannii [POLYDOU1]
Polygonum unifolium Greene = Polygonum polygaloides ssp. kelloggii [POLYPOL1]
Polygonum viviparum L. [POLYVIV] (alpine bistort)
SY: *Bistorta vivipara* (L.) S.F. Gray
SY: *Polygonum fugax* Small
SY: *Polygonum macounii* Small ex Macoun
SY: *Polygonum viviparum* var. *alpinum* Wahlenb.
SY: *Polygonum viviparum* var. *macounii* (Small ex Macoun) Hultén
Polygonum viviparum var. *alpinum* Wahlenb. = Polygonum viviparum [POLYVIV]
Polygonum viviparum var. *macounii* (Small ex Macoun) Hultén = Polygonum viviparum [POLYVIV]
Polygonum vulcanicum Greene = Polygonum bistortoides [POLYBIS]
Polygonum zuccarinii Small = Polygonum cuspidatum [POLYCUS]
Reynoutria japonica Houtt. = Polygonum cuspidatum [POLYCUS]
Reynoutria sachalinensis (F. Schmidt ex Maxim.) Nakai = Polygonum sachalinense [POLYSAC]
Rumex L. [RUMEX$]
Rumex acetosa L. [RUMEACE] (green sorrel)
*__Rumex acetosa__ ssp. **acetosa** [RUMEACE0]
Rumex acetosa ssp. **alpestris** (Jacq.) A. Löve [RUMEACE3]
SY: *Acetosa alpestris* (Jacq.) A. Löve
SY: *Acetosa pratensis* ssp. *alpestris* (Jacq.) A. Löve
SY: *Rumex alpestris* Jacq.
Rumex acetosa ssp. **arifolius** (All.) Blytt & O. Dahl [RUMEACE2]
*__Rumex acetosella__ L. [RUMEACT] (sheep sorrel)
SY: *Acetosella acetosella* (L.) Small
SY: *Acetosella tenuifolia* (Wallr.) A. Löve
SY: *Acetosella vulgaris* (Koch) Fourr.
SY: *Rumex acetosella* ssp. *angiocarpus* (Murb.) Murb.
SY: *Rumex acetosella* var. *pyrenaeus* (Pourret) Timbal-Lagrave
SY: *Rumex acetosella* var. *tenuifolius* Wallr.
SY: *Rumex angiocarpus* Murb.
SY: *Rumex tenuifolius* (Wallr.) A. Löve
Rumex acetosella ssp. *angiocarpus* (Murb.) Murb. = Rumex acetosella [RUMEACT]
Rumex acetosella var. *pyrenaeus* (Pourret) Timbal-Lagrave = Rumex acetosella [RUMEACT]

Rumex acetosella var. *tenuifolius* Wallr. = Rumex acetosella [RUMEACT]
Rumex alpestris Jacq. = Rumex acetosa ssp. alpestris [RUMEACE3]
Rumex angiocarpus Murb. = Rumex acetosella [RUMEACT]
Rumex aquaticus L. [RUMEAQU]
Rumex aquaticus ssp. *fenestratus* (Greene) Hultén = Rumex aquaticus var. fenestratus [RUMEAQU1]
Rumex aquaticus ssp. *occidentalis* (S. Wats.) Hultén = Rumex aquaticus var. fenestratus [RUMEAQU1]
Rumex aquaticus var. **fenestratus** (Greene) Dorn [RUMEAQU1]
SY: *Rumex aquaticus* ssp. *fenestratus* (Greene) Hultén
SY: *Rumex aquaticus* ssp. *occidentalis* (S. Wats.) Hultén
SY: *Rumex fenestratus* Greene
SY: *Rumex occidentalis* S. Wats.
SY: *Rumex occidentalis* var. *fenestratus* (Greene) Lepage
SY: *Rumex occidentalis* var. *labradoricus* (Rech. f.) Lepage
SY: *Rumex occidentalis* var. *procerus* (Greene) J.T. Howell
Rumex arcticus Trautv. [RUMEARC] (arctic dock)
•**Rumex conglomeratus** Murr. [RUMECON] (clustered dock)
•**Rumex crispus** L. [RUMECRI] (curled dock)
Rumex digyna L. = Oxyria digyna [OXYRDIG]
Rumex fenestratus Greene = Rumex aquaticus var. fenestratus [RUMEAQU1]
Rumex fueginus Phil. = Rumex maritimus [RUMEMAR]
Rumex maritimus L. [RUMEMAR] (golden dock)
SY: *Rumex fueginus* Phil.
SY: *Rumex maritimus* var. *athrix* St. John
SY: *Rumex maritimus* ssp. *fueginus* (Phil.) Hultén
SY: *Rumex maritimus* var. *fueginus* (Phil.) Dusen
SY: *Rumex maritimus* var. *persicarioides* (L.) R.S. Mitchell
SY: *Rumex persicarioides* L.
Rumex maritimus ssp. *fueginus* (Phil.) Hultén = Rumex maritimus [RUMEMAR]
Rumex maritimus var. *athrix* St. John = Rumex maritimus [RUMEMAR]
Rumex maritimus var. *fueginus* (Phil.) Dusen = Rumex maritimus [RUMEMAR]
Rumex maritimus var. *persicarioides* (L.) R.S. Mitchell = Rumex maritimus [RUMEMAR]
Rumex mexicanus Meisn. = Rumex salicifolius var. mexicanus [RUMESAL1]
Rumex mexicanus var. *angustifolius* (Meisn.) Boivin = Rumex salicifolius var. mexicanus [RUMESAL1]
Rumex mexicanus var. *subarcticus* (Lepage) Boivin = Rumex salicifolius var. mexicanus [RUMESAL1]
Rumex mexicanus var. *transitorius* (Rech. f.) Boivin = Rumex salicifolius var. transitorius [RUMESAL2]
•**Rumex obtusifolius** L. [RUMEOBT] (bitter dock)
SY: *Acetosa oblongifolia* (L.) A. & D. Löve
SY: *Rumex obtusifolius* ssp. *agrestis* (Fries) Danser
SY: *Rumex obtusifolius* ssp. *sylvestris* (Wallr.) Rech. f.
SY: *Rumex obtusifolius* var. *sylvestris* (Wallr.) Koch
Rumex obtusifolius ssp. *agrestis* (Fries) Danser = Rumex obtusifolius [RUMEOBT]
Rumex obtusifolius ssp. *sylvestris* (Wallr.) Rech. f. = Rumex obtusifolius [RUMEOBT]
Rumex obtusifolius var. *sylvestris* (Wallr.) Koch = Rumex obtusifolius [RUMEOBT]
Rumex occidentalis S. Wats. = Rumex aquaticus var. fenestratus [RUMEAQU1]
Rumex occidentalis var. *fenestratus* (Greene) Lepage = Rumex aquaticus var. fenestratus [RUMEAQU1]
Rumex occidentalis var. *labradoricus* (Rech. f.) Lepage = Rumex aquaticus var. fenestratus [RUMEAQU1]
Rumex occidentalis var. *procerus* (Greene) J.T. Howell = Rumex aquaticus var. fenestratus [RUMEAQU1]
•**Rumex patientia** L. [RUMEPAT] (patience dock)
Rumex paucifolius Nutt. [RUMEPAU] (alpine sorrel)
Rumex persicarioides L. = Rumex maritimus [RUMEMAR]
Rumex quadrangulivalvis (Danser) Rech. f. = Rumex salicifolius var. mexicanus [RUMESAL1]
Rumex salicifolius Weinm. [RUMESAL] (willow dock)
Rumex salicifolius ssp. *triangulivalvis* Danser = Rumex salicifolius var. mexicanus [RUMESAL1]
Rumex salicifolius ssp. *triangulivalvis* var. *mexicanus* (Meisn.) C.L. Hitchc. = Rumex salicifolius var. mexicanus [RUMESAL1]
Rumex salicifolius var. **mexicanus** (Meisn.) C.L. Hitchc. [RUMESAL1]
SY: *Rumex mexicanus* Meisn.
SY: *Rumex mexicanus* var. *angustifolius* (Meisn.) Boivin
SY: *Rumex mexicanus* var. *subarcticus* (Lepage) Boivin
SY: *Rumex quadrangulivalvis* (Danser) Rech. f.
SY: *Rumex salicifolius* ssp. *triangulivalvis* Danser
SY: *Rumex salicifolius* var. *triangulivalvis* (Danser) C.L. Hitchc.
SY: *Rumex salicifolius* ssp. *triangulivalvis* var. *mexicanus* (Meisn.) C.L. Hitchc.
SY: *Rumex triangulivalvis* (Danser) Rech. f.
SY: *Rumex triangulivalvis* var. *oreolapathum* Rech. f.
Rumex salicifolius var. **salicifolius** [RUMESAL0]
Rumex salicifolius var. **transitorius** (Rech. f.) Hickman [RUMESAL2]
SY: *Rumex mexicanus* var. *transitorius* (Rech. f.) Boivin
SY: *Rumex transitorius* Rech. f.
Rumex salicifolius var. *triangulivalvis* (Danser) C.L. Hitchc. = Rumex salicifolius var. mexicanus [RUMESAL1]
Rumex tenuifolius (Wallr.) A. Löve = Rumex acetosella [RUMEACT]
Rumex transitorius Rech. f. = Rumex salicifolius var. transitorius [RUMESAL2]
Rumex triangulivalvis (Danser) Rech. f. = Rumex salicifolius var. mexicanus [RUMESAL1]
Rumex triangulivalvis var. *oreolapathum* Rech. f. = Rumex salicifolius var. mexicanus [RUMESAL1]

POLYPODIACEAE (V:F)

Polypodium L. [POLYPOD$]
Polypodium amorphum Suksdorf [POLYAMO] (irregular polypody)
SY: *Polypodium montense* F.A. Lang
Polypodium glycyrrhiza D.C. Eat. [POLYGLY] (licorice fern)
SY: *Polypodium occidentale* (Hook.) Maxon
SY: *Polypodium vulgare* var. *commune* Milde
SY: *Polypodium vulgare* ssp. *occidentale* (Hook.) Hultén
SY: *Polypodium vulgare* var. *occidentale* Hook.
Polypodium hesperium Maxon [POLYHES] (western polypody)
SY: *Polypodium vulgare* ssp. *columbianum* (Gilbert) Hultén
SY: *Polypodium vulgare* var. *columbianum* Gilbert
SY: *Polypodium vulgare* var. *hesperium* (Maxon) A. Nels. & J.F. Macbr.
Polypodium montense F.A. Lang = Polypodium amorphum [POLYAMO]
Polypodium occidentale (Hook.) Maxon = Polypodium glycyrrhiza [POLYGLY]
Polypodium scouleri Hook. & Grev. [POLYSCO] (leathery polypody)
Polypodium sibiricum Sipl. [POLYSIB]
Polypodium vulgare ssp. *columbianum* (Gilbert) Hultén = Polypodium hesperium [POLYHES]
Polypodium vulgare ssp. *occidentale* (Hook.) Hultén = Polypodium glycyrrhiza [POLYGLY]
Polypodium vulgare var. *columbianum* Gilbert = Polypodium hesperium [POLYHES]
Polypodium vulgare var. *commune* Milde = Polypodium glycyrrhiza [POLYGLY]
Polypodium vulgare var. *hesperium* (Maxon) A. Nels. & J.F. Macbr. = Polypodium hesperium [POLYHES]
Polypodium vulgare var. *occidentale* Hook. = Polypodium glycyrrhiza [POLYGLY]

PORTULACACEAE (V:D)

Calandrinia Kunth [CALANDR$]
Calandrinia ciliata (Ruiz & Pavón) DC. [CALACIL] (desert rock purslane)
SY: *Calandrinia ciliata* var. *menziesii* (Hook.) J.F. Macbr.

Calandrinia ciliata var. *menziesii* (Hook.) J.F. Macbr. = Calandrinia ciliata [CALACIL]
Calandrinia tweedyi Gray = Cistanthe tweedyi [CISTTWE]
Calyptridium umbellatum var. *caudiciferum* (Gray) Jepson = Cistanthe umbellata var. caudicifera [CISTUMB1]
Cistanthe Spach [CISTANT$]
Cistanthe tweedyi (Gray) Hershkovitz [CISTTWE]
SY: *Calandrinia tweedyi* Gray
SY: *Lewisia tweedyi* (Gray) B.L. Robins.
Cistanthe umbellata (Torr.) Hershkovitz [CISTUMB]
SY: *Spraguea umbellata* Torr.
Cistanthe umbellata var. **caudicifera** (Gray) Kartesz & Gandhi [CISTUMB1]
SY: *Calyptridium umbellatum* var. *caudiciferum* (Gray) Jepson
SY: *Spraguea umbellata* var. *caudicifera* Gray
Claytonia L. [CLAYTON$]
Claytonia asarifolia auct. = Claytonia cordifolia [CLAYCOR]
Claytonia bostockii Porsild = Montia bostockii [MONTBOS]
Claytonia caroliniana var. *lanceolata* (Pursh) S. Wats. = Claytonia lanceolata [CLAYLAN]
Claytonia caroliniana var. *tuberosa* (Pallas ex J.A. Schultes) Boivin = Claytonia tuberosa [CLAYTUB]
Claytonia chamissoi Ledeb. ex Spreng. = Montia chamissoi [MONTCHA]
Claytonia cordifolia S. Wats. [CLAYCOR] (heart-leaved spring-beauty)
SY: *Claytonia sibirica* var. *cordifolia* (S. Wats.) R.J. Davis
SY: *Montia cordifolia* (S. Wats.) Pax & K. Hoffmann
SY: *Claytonia asarifolia* auct.
Claytonia dichotoma Nutt. = Montia dichotoma [MONTDIC]
Claytonia diffusa Nutt. = Montia diffusa [MONTDIF]
Claytonia flagellaris Bong. = Montia parvifolia ssp. flagellaris [MONTPAR1]
Claytonia heterophylla (Torr. & Gray) Swanson [CLAYHET]
SY: *Claytonia sibirica* var. *heterophylla* (Torr. & Gray) Gray
SY: *Montia heterophylla* (Torr. & Gray) Jepson
SY: *Montia sibirica* var. *heterophylla* (Torr. & Gray) B.L. Robins.
Claytonia howellii (S. Wats.) Piper = Montia howellii [MONTHOW]
Claytonia lanceolata Pursh [CLAYLAN] (western spring-beauty)
SY: *Claytonia caroliniana* var. *lanceolata* (Pursh) S. Wats.
Claytonia lanceolata var. **pacifica** McNeill [CLAYLAN2]
Claytonia linearis Dougl. ex Hook. = Montia linearis [MONTLIN]
Claytonia megarhiza (Gray) Parry ex S. Wats. [CLAYMEG] (alpine spring-beauty)
Claytonia parviflora Dougl. ex Hook. [CLAYPAR]
SY: *Claytonia perfoliata* var. *parviflora* (Dougl. ex Hook.) Torr.
SY: *Montia parviflora* (Dougl. ex Hook.) T.J. Howell
SY: *Montia perfoliata* var. *parviflora* (Dougl. ex Hook.) Jepson
Claytonia parviflora var. *depressa* Gray = Claytonia rubra [CLAYRUB]
Claytonia parviflora var. *glauca* Nutt. ex Torr. & Gray = Claytonia rubra [CLAYRUB]
Claytonia parvifolia Moc. ex DC. = Montia parvifolia ssp. parvifolia [MONTPAR0]
Claytonia parvifolia ssp. *flagellaris* (Bong.) Hultén = Montia parvifolia ssp. flagellaris [MONTPAR1]
Claytonia parvifolia var. *flagellaris* (Bong.) R.J. Davis = Montia parvifolia ssp. flagellaris [MONTPAR1]
Claytonia perfoliata Donn ex Willd. [CLAYPER] (miner's-lettuce)
SY: *Montia perfoliata* (Donn ex Willd.) T.J. Howell
Claytonia perfoliata var. *depressa* (Gray) Poelln. = Claytonia rubra [CLAYRUB]
Claytonia perfoliata var. *parviflora* (Dougl. ex Hook.) Torr. = Claytonia parviflora [CLAYPAR]
Claytonia rubra (T.J. Howell) Tidestrom [CLAYRUB]
SY: *Claytonia parviflora* var. *depressa* Gray
SY: *Claytonia parviflora* var. *glauca* Nutt. ex Torr. & Gray
SY: *Claytonia perfoliata* var. *depressa* (Gray) Poelln.
SY: *Montia perfoliata* var. *depressa* (Gray) Jepson
SY: *Montia perfoliata* ssp. *glauca* (Nutt. ex Torr. & Gray) Ferris
Claytonia sarmentosa C.A. Mey. [CLAYSAR] (Alaska spring-beauty)
SY: *Montia sarmentosa* (C.A. Mey.) B.L. Robins.
Claytonia sibirica L. [CLAYSIB] (Siberian miner's-lettuce)
SY: *Montia sibirica* (L.) T.J. Howell
Claytonia sibirica var. *cordifolia* (S. Wats.) R.J. Davis = Claytonia cordifolia [CLAYCOR]
Claytonia sibirica var. *heterophylla* (Torr. & Gray) Gray = Claytonia heterophylla [CLAYHET]
Claytonia spathulata Dougl. ex Hook. [CLAYSPA] (pale spring-beauty)
SY: *Montia spathulata* (Dougl. ex Hook.) T.J. Howell
Claytonia triphylla S. Wats. = Lewisia triphylla [LEWITRI]
Claytonia tuberosa Pallas ex J.A. Schultes [CLAYTUB] (tuberous spring-beauty)
SY: *Claytonia caroliniana* var. *tuberosa* (Pallas ex J.A. Schultes) Boivin
Crunocallis chamissoi (Ledeb. ex Spreng.) Rydb. = Montia chamissoi [MONTCHA]
Erocallis triphylla (S. Wats.) Rydb. = Lewisia triphylla [LEWITRI]
Lewisia Pursh [LEWISIA$]
Lewisia columbiana (T.J. Howell ex Gray) B.L. Robins. [LEWICOL] (Columbia lewisia)
Lewisia glandulosa (Rydb.) Dempster = Lewisia pygmaea [LEWIPYG]
Lewisia minima (A. Nels.) A. Nels. = Lewisia pygmaea [LEWIPYG]
Lewisia pygmaea (Gray) B.L. Robins. [LEWIPYG] (alpine lewisia)
SY: *Lewisia glandulosa* (Rydb.) Dempster
SY: *Lewisia minima* (A. Nels.) A. Nels.
SY: *Lewisia pygmaea* var. *aridorum* Bartlett
SY: *Lewisia pygmaea* ssp. *glandulosa* (Rydb.) Ferris
SY: *Oreobroma pygmaeum* (Gray) T.J. Howell
SY: *Talinum pygmaeum* Gray
Lewisia pygmaea ssp. *glandulosa* (Rydb.) Ferris = Lewisia pygmaea [LEWIPYG]
Lewisia pygmaea var. *aridorum* Bartlett = Lewisia pygmaea [LEWIPYG]
Lewisia rediviva Pursh [LEWIRED] (bitterroot)
Lewisia triphylla (S. Wats.) B.L. Robins. [LEWITRI] (three-leaved lewisia)
SY: *Claytonia triphylla* S. Wats.
SY: *Erocallis triphylla* (S. Wats.) Rydb.
Lewisia tweedyi (Gray) B.L. Robins. = Cistanthe tweedyi [CISTTWE]
Montia L. [MONTIA$]
Montia bostockii (Porsild) Welsh [MONTBOS] (Bostock's montia)
SY: *Claytonia bostockii* Porsild
Montia chamissoi (Ledeb. ex Spreng.) Greene [MONTCHA] (Chamisso's montia)
SY: *Claytonia chamissoi* Ledeb. ex Spreng.
SY: *Crunocallis chamissoi* (Ledeb. ex Spreng.) Rydb.
Montia cordifolia (S. Wats.) Pax & K. Hoffmann = Claytonia cordifolia [CLAYCOR]
Montia dichotoma (Nutt.) T.J. Howell [MONTDIC] (dwarf montia)
SY: *Claytonia dichotoma* Nutt.
SY: *Montiastrum dichotomum* (Nutt.) Rydb.
Montia diffusa (Nutt.) Greene [MONTDIF] (spreading montia)
SY: *Claytonia diffusa* Nutt.
Montia flagellaris (Bong.) B.L. Robins. = Montia parvifolia ssp. flagellaris [MONTPAR1]
Montia fontana L. [MONTFON] (blinks chickweed)
SY: *Montia fontana* var. *tenerrima* (Gray) Fern. & Wieg.
SY: *Montia lamprosperma* Cham.
SY: *Montia fontana* ssp. *amporitana* auct.
Montia fontana ssp. *amporitana* auct. = Montia fontana [MONTFON]
Montia fontana ssp. **variabilis** S.M. Walters [MONTFON3]
SY: *Montia funstonii* Rydb.
Montia fontana var. *tenerrima* (Gray) Fern. & Wieg. = Montia fontana [MONTFON]
Montia funstonii Rydb. = Montia fontana ssp. variabilis [MONTFON3]
Montia heterophylla (Torr. & Gray) Jepson = Claytonia heterophylla [CLAYHET]
Montia howellii S. Wats. [MONTHOW] (Howell's montia)
SY: *Claytonia howellii* (S. Wats.) Piper
Montia lamprosperma Cham. = Montia fontana [MONTFON]

Montia linearis (Dougl. ex Hook.) Greene [MONTLIN] (narrow-leaved montia)
SY: *Claytonia linearis* Dougl. ex Hook.
SY: *Montiastrum lineare* (Dougl. ex Hook.) Rydb.
Montia parviflora (Dougl. ex Hook.) T.J. Howell = Claytonia parviflora [CLAYPAR]
Montia parvifolia (Moc. ex DC.) Greene [MONTPAR] (small-leaved montia)
Montia parvifolia ssp. **flagellaris** (Bong.) Ferris [MONTPAR1]
SY: *Claytonia flagellaris* Bong.
SY: *Claytonia parvifolia* ssp. *flagellaris* (Bong.) Hultén
SY: *Claytonia parvifolia* var. *flagellaris* (Bong.) R.J. Davis
SY: *Montia flagellaris* (Bong.) B.L. Robins.
SY: *Montia parvifolia* var. *flagellaris* (Bong.) C.L. Hitchc.
Montia parvifolia ssp. **parvifolia** [MONTPAR0]
SY: *Claytonia parvifolia* Moc. ex DC.
Montia parvifolia var. *flagellaris* (Bong.) C.L. Hitchc. = Montia parvifolia ssp. flagellaris [MONTPAR1]
Montia perfoliata (Donn ex Willd.) T.J. Howell = Claytonia perfoliata [CLAYPER]
Montia perfoliata ssp. *glauca* (Nutt. ex Torr. & Gray) Ferris = Claytonia rubra [CLAYRUB]
Montia perfoliata var. *depressa* (Gray) Jepson = Claytonia rubra [CLAYRUB]
Montia perfoliata var. *parviflora* (Dougl. ex Hook.) Jepson = Claytonia parviflora [CLAYPAR]
Montia sarmentosa (C.A. Mey.) B.L. Robins. = Claytonia sarmentosa [CLAYSAR]
Montia sibirica (L.) T.J. Howell = Claytonia sibirica [CLAYSIB]
Montia sibirica var. *heterophylla* (Torr. & Gray) B.L. Robins. = Claytonia heterophylla [CLAYHET]
Montia spathulata (Dougl. ex Hook.) T.J. Howell = Claytonia spathulata [CLAYSPA]
Montiastrum dichotomum (Nutt.) Rydb. = Montia dichotoma [MONTDIC]
Montiastrum lineare (Dougl. ex Hook.) Rydb. = Montia linearis [MONTLIN]
Oreobroma pygmaeum (Gray) T.J. Howell = Lewisia pygmaea [LEWIPYG]
Portulaca L. [PORTULA$]
*Portulaca oleracea** L. [PORTOLE] (common purslane)
Spraguea umbellata Torr. = Cistanthe umbellata [CISTUMB]
Spraguea umbellata var. *caudicifera* Gray = Cistanthe umbellata var. caudicifera [CISTUMB1]
Talinum Adans. [TALINUM$]
Talinum okanoganense English = Talinum sediforme [TALISED]
Talinum pygmaeum Gray = Lewisia pygmaea [LEWIPYG]
Talinum sediforme Poelln. [TALISED] (Okanogan fameflower)
SY: *Talinum okanoganense* English

POTAMOGETONACEAE (V:M)

Potamogeton L. [POTAMOG$]
Potamogeton alpinus Balbis [POTAALP] (northern pondweed)
SY: *Potamogeton alpinus* var. *subellipticus* (Fern.) Ogden
SY: *Potamogeton alpinus* ssp. *tenuifolius* (Raf.) Hultén
SY: *Potamogeton alpinus* var. *tenuifolius* (Raf.) Ogden
SY: *Potamogeton tenuifolius* Raf.
Potamogeton alpinus ssp. *tenuifolius* (Raf.) Hultén = Potamogeton alpinus [POTAALP]
Potamogeton alpinus var. *subellipticus* (Fern.) Ogden = Potamogeton alpinus [POTAALP]
Potamogeton alpinus var. *tenuifolius* (Raf.) Ogden = Potamogeton alpinus [POTAALP]
Potamogeton americanus Cham. & Schlecht. = Potamogeton nodosus [POTANOD]
Potamogeton amplexicaulis Kar. = Potamogeton perfoliatus [POTAPER]
Potamogeton amplifolius Tuckerman [POTAAMP] (large-leaved pondweed)
Potamogeton angustifolius Bercht. & K. Presl = Potamogeton illinoensis [POTAILL]
Potamogeton berchtoldii Fieber = Potamogeton pusillus var. tenuissimus [POTAPUS2]
Potamogeton berchtoldii var. *acuminatus* Fieber = Potamogeton pusillus var. tenuissimus [POTAPUS2]
Potamogeton berchtoldii var. *colpophilus* (Fern.) Fern. = Potamogeton pusillus var. tenuissimus [POTAPUS2]
Potamogeton berchtoldii var. *lacunatus* (Hagstr.) Fern. = Potamogeton pusillus var. tenuissimus [POTAPUS2]
Potamogeton berchtoldii var. *mucronatus* Fieber = Potamogeton pusillus var. tenuissimus [POTAPUS2]
Potamogeton berchtoldii var. *polyphyllus* (Morong) Fern. = Potamogeton pusillus var. tenuissimus [POTAPUS2]
Potamogeton berchtoldii var. *tenuissimus* (Mert. & Koch) Fern. = Potamogeton pusillus var. tenuissimus [POTAPUS2]
Potamogeton bupleuroides Fern. = Potamogeton perfoliatus [POTAPER]
Potamogeton compressus auct. = Potamogeton zosteriformis [POTAZOS]
*Potamogeton crispus** L. [POTACRI] (curled pondweed)
Potamogeton epihydrus Raf. [POTAEPI] (ribbon-leaved pondweed)
SY: *Potamogeton epihydrus* ssp. *nuttallii* (Cham. & Schlecht.) Calder & Taylor
SY: *Potamogeton epihydrus* var. *nuttallii* (Cham. & Schlecht.) Fern.
SY: *Potamogeton epihydrus* var. *ramosus* (Peck) House
SY: *Potamogeton nuttallii* Cham. & Schlecht.
Potamogeton epihydrus ssp. *nuttallii* (Cham. & Schlecht.) Calder & Taylor = Potamogeton epihydrus [POTAEPI]
Potamogeton epihydrus var. *nuttallii* (Cham. & Schlecht.) Fern. = Potamogeton epihydrus [POTAEPI]
Potamogeton epihydrus var. *ramosus* (Peck) House = Potamogeton epihydrus [POTAEPI]
Potamogeton filiformis Pers. [POTAFIL] (slender-leaved pondweed)
Potamogeton fluitans Roth = Potamogeton nodosus [POTANOD]
Potamogeton foliosus Raf. [POTAFOL] (closed-leaved pondweed)
SY: *Potamogeton foliosus* var. *macellus* Fern.
Potamogeton foliosus var. *macellus* Fern. = Potamogeton foliosus [POTAFOL]
Potamogeton friesii Rupr. [POTAFRI] (flat-stalked pondweed)
Potamogeton gramineus L. [POTAGRA] (grass-leaved pondweed)
SY: *Potamogeton gramineus* var. *graminifolius* Fries
SY: *Potamogeton gramineus* var. *maximus* Morong
SY: *Potamogeton gramineus* var. *myriophyllus* J.W. Robbins
SY: *Potamogeton gramineus* var. *typicus* Ogden
Potamogeton gramineus var. *graminifolius* Fries = Potamogeton gramineus [POTAGRA]
Potamogeton gramineus var. *maximus* Morong = Potamogeton gramineus [POTAGRA]
Potamogeton gramineus var. *myriophyllus* J.W. Robbins = Potamogeton gramineus [POTAGRA]
Potamogeton gramineus var. *typicus* Ogden = Potamogeton gramineus [POTAGRA]
Potamogeton heterophyllus Schreb. = Potamogeton illinoensis [POTAILL]
Potamogeton illinoensis Morong [POTAILL] (Illinois pondweed)
SY: *Potamogeton angustifolius* Bercht. & K. Presl
SY: *Potamogeton heterophyllus* Schreb.
Potamogeton interruptus Kit. = Potamogeton vaginatus [POTAVAG]
Potamogeton longiligulatus Fern. = Potamogeton strictifolius [POTASTR]
Potamogeton natans L. [POTANAT] (floating-leaved pondweed)
Potamogeton nodosus Poir. [POTANOD] (long-leaved pondweed)
SY: *Potamogeton americanus* Cham. & Schlecht.
SY: *Potamogeton fluitans* Roth
Potamogeton nuttallii Cham. & Schlecht. = Potamogeton epihydrus [POTAEPI]
Potamogeton oakesianus J.W. Robbins [POTAOAK] (Oakes' pondweed)
Potamogeton obtusifolius Mert. & Koch [POTAOBT] (blunt-leaved pondweed)
Potamogeton panormitanus Biv. = Potamogeton pusillus [POTAPUS]
Potamogeton pectinatus L. [POTAPEC] (fennel-leaved pondweed)
Potamogeton perfoliatus L. [POTAPER] (perfoliate pondweed)

SY: *Potamogeton amplexicaulis* Kar.
SY: *Potamogeton bupleuroides* Fern.
SY: *Potamogeton perfoliatus* ssp. *bupleuroides* (Fern.) Hultén
SY: *Potamogeton perfoliatus* var. *bupleuroides* (Fern.) Farw.
Potamogeton perfoliatus ssp. *bupleuroides* (Fern.) Hultén = Potamogeton perfoliatus [POTAPER]
Potamogeton perfoliatus ssp. *richardsonii* (Benn.) Hultén = Potamogeton richardsonii [POTARIC]
Potamogeton perfoliatus var. *bupleuroides* (Fern.) Farw. = Potamogeton perfoliatus [POTAPER]
Potamogeton perfoliatus var. *richardsonii* Benn. = Potamogeton richardsonii [POTARIC]
Potamogeton praelongus Wulfen [POTAPRA] (long-stalked pondweed)
SY: *Potamogeton praelongus* var. *angustifolius* Graebn.
Potamogeton praelongus var. *angustifolius* Graebn. = Potamogeton praelongus [POTAPRA]
Potamogeton pusillus L. [POTAPUS] (small-leaved pondweed)
SY: *Potamogeton panormitanus* Biv.
SY: *Potamogeton pusillus* var. *minor* (Biv.) Fern. & Schub.
Potamogeton pusillus var. *minor* (Biv.) Fern. & Schub. = Potamogeton pusillus [POTAPUS]
Potamogeton pusillus var. *mucronatus* (Fieber) Graebn. = Potamogeton pusillus var. tenuissimus [POTAPUS2]
Potamogeton pusillus var. *rutiloides* (Fern.) Boivin = Potamogeton strictifolius [POTASTR]
Potamogeton pusillus var. **tenuissimus** Mert. & Koch [POTAPUS2] (small pondweed)
SY: *Potamogeton berchtoldii* Fieber
SY: *Potamogeton berchtoldii* var. *acuminatus* Fieber
SY: *Potamogeton berchtoldii* var. *colpophilus* (Fern.) Fern.
SY: *Potamogeton berchtoldii* var. *lacunatus* (Hagstr.) Fern.
SY: *Potamogeton berchtoldii* var. *mucronatus* Fieber
SY: *Potamogeton berchtoldii* var. *polyphyllus* (Morong) Fern.
SY: *Potamogeton berchtoldii* var. *tenuissimus* (Mert. & Koch) Fern.
SY: *Potamogeton pusillus* var. *mucronatus* (Fieber) Graebn.
Potamogeton richardsonii (Benn.) Rydb. [POTARIC] (Richardson's pondweed)
SY: *Potamogeton perfoliatus* ssp. *richardsonii* (Benn.) Hultén
SY: *Potamogeton perfoliatus* var. *richardsonii* Benn.
Potamogeton robbinsii Oakes [POTAROB] (Robbins' pondweed)
Potamogeton strictifolius Benn. [POTASTR] (stiff-leaved pondweed)
SY: *Potamogeton longiligulatus* Fern.
SY: *Potamogeton pusillus* var. *rutiloides* (Fern.) Boivin
SY: *Potamogeton strictifolius* var. *rutiloides* Fern.
SY: *Potamogeton strictifolius* var. *typicus* Fern.
Potamogeton strictifolius var. *rutiloides* Fern. = Potamogeton strictifolius [POTASTR]
Potamogeton strictifolius var. *typicus* Fern. = Potamogeton strictifolius [POTASTR]
Potamogeton tenuifolius Raf. = Potamogeton alpinus [POTAALP]
Potamogeton vaginatus Turcz. [POTAVAG] (large-sheathed pondweed)
SY: *Potamogeton interruptus* Kit.
Potamogeton zosterifolius ssp. *zosteriformis* (Fern.) Hultén = Potamogeton zosteriformis [POTAZOS]
Potamogeton zosteriformis Fern. [POTAZOS] (eel-grass pondweed)
SY: *Potamogeton zosterifolius* ssp. *zosteriformis* (Fern.) Hultén
SY: *Potamogeton compressus* auct.

PRIMULACEAE (V:D)

Anagallis L. [ANAGALL$]
***Anagallis arvensis** L. [ANAGARV] (scarlet pimpernel)
Anagallis minima (L.) Krause [ANAGMIN]
SY: *Centunculus minimus* L.
Androsace L. [ANDROSA$]
Androsace alaskana Coville & Standl. ex Hultén = Douglasia alaskana [DOUGALA]
Androsace alaskana var. *reediae* Welsh & Goodrich = Douglasia alaskana [DOUGALA]
Androsace arizonica (Gray) Derganc = Androsace occidentalis [ANDROCC]
Androsace chamaejasme Wulfen [ANDRCHA] (sweet-flowered fairy-candelabra)
Androsace chamaejasme ssp. **lehmanniana** (Spreng.) Hultén [ANDRCHA1]
SY: *Androsace chamaejasme* var. *arctica* R. Knuth
SY: *Androsace lehmanniana* Spreng.
Androsace chamaejasme var. *arctica* R. Knuth = Androsace chamaejasme ssp. lehmanniana [ANDRCHA1]
Androsace diffusa Small = Androsace septentrionalis ssp. subulifera [ANDRSEP2]
Androsace filiformis Retz. [ANDRFIL] (slender-flowered fairy-candelabra)
Androsace lehmanniana Spreng. = Androsace chamaejasme ssp. lehmanniana [ANDRCHA1]
Androsace occidentalis Pursh [ANDROCC] (western fairy-candelabra)
SY: *Androsace arizonica* (Gray) Derganc
SY: *Androsace occidentalis* var. *arizonica* (Gray) St. John
SY: *Androsace occidentalis* var. *simplex* (Rydb.) St. John
Androsace occidentalis var. *arizonica* (Gray) St. John = Androsace occidentalis [ANDROCC]
Androsace occidentalis var. *simplex* (Rydb.) St. John = Androsace occidentalis [ANDROCC]
Androsace puberulenta Rydb. = Androsace septentrionalis ssp. puberulenta [ANDRSEP1]
Androsace septentrionalis L. [ANDRSEP] (northern fairy-candelabra)
Androsace septentrionalis ssp. **puberulenta** (Rydb.) G.T. Robbins [ANDRSEP1]
SY: *Androsace puberulenta* Rydb.
SY: *Androsace septentrionalis* var. *puberulenta* (Rydb.) R. Knuth
Androsace septentrionalis ssp. **septentrionalis** [ANDRSEP0]
Androsace septentrionalis ssp. **subulifera** (Gray) G.T. Robbins [ANDRSEP2]
SY: *Androsace diffusa* Small
SY: *Androsace septentrionalis* var. *diffusa* (Small) R. Knuth
SY: *Androsace septentrionalis* var. *subulifera* Gray
Androsace septentrionalis ssp. **subumbellata** (A. Nels.) G.T. Robbins [ANDRSEP3]
SY: *Androsace septentrionalis* var. *subumbellata* A. Nels.
Androsace septentrionalis var. *diffusa* (Small) R. Knuth = Androsace septentrionalis ssp. subulifera [ANDRSEP2]
Androsace septentrionalis var. *puberulenta* (Rydb.) R. Knuth = Androsace septentrionalis ssp. puberulenta [ANDRSEP1]
Androsace septentrionalis var. *subulifera* Gray = Androsace septentrionalis ssp. subulifera [ANDRSEP2]
Androsace septentrionalis var. *subumbellata* A. Nels. = Androsace septentrionalis ssp. subumbellata [ANDRSEP3]
Centunculus minimus L. = Anagallis minima [ANAGMIN]
Dodecatheon L. [DODECAT$]
Dodecatheon conjugens Greene [DODECON] (slimpod shootingstar)
Dodecatheon conjugens ssp. **conjugens** [DODECON0]
SY: *Dodecatheon conjugens* var. *beamishii* Boivin
SY: *Dodecatheon conjugens* ssp. *leptophyllum* (Suksdorf) Piper
SY: *Dodecatheon cylindrocarpum* Rydb.
Dodecatheon conjugens ssp. *leptophyllum* (Suksdorf) Piper = Dodecatheon conjugens ssp. conjugens [DODECON0]
Dodecatheon conjugens ssp. **viscidum** (Piper) H.J. Thompson [DODECON1]
SY: *Dodecatheon conjugens* var. *viscidum* (Piper) Mason ex St. John
SY: *Dodecatheon viscidum* Piper
Dodecatheon conjugens var. *beamishii* Boivin = Dodecatheon conjugens ssp. conjugens [DODECON0]
Dodecatheon conjugens var. *viscidum* (Piper) Mason ex St. John = Dodecatheon conjugens ssp. viscidum [DODECON1]
Dodecatheon cusickii Greene = Dodecatheon pulchellum ssp. cusickii [DODEPUL1]
Dodecatheon cusickii var. *album* Suksdorf = Dodecatheon pulchellum ssp. cusickii [DODEPUL1]

Dodecatheon cylindrocarpum Rydb. = Dodecatheon conjugens ssp. conjugens [DODECON0]
Dodecatheon dentatum Hook. [DODEDEN] (white shootingstar)
Dodecatheon frigidum Cham. & Schlecht. [DODEFRI] (northern shootingstar)
Dodecatheon hendersonii Gray [DODEHEN] (broad-leaved shootingstar)
Dodecatheon jeffreyi Van Houtte [DODEJEF] (Jeffrey's shootingstar)
Dodecatheon pauciflorum var. *alaskanum* (Hultén) C.L. Hitchc. = Dodecatheon pulchellum ssp. macrocarpum [DODEPUL2]
Dodecatheon pauciflorum var. *cusickii* (Greene) Mason ex St. John = Dodecatheon pulchellum ssp. cusickii [DODEPUL1]
Dodecatheon pauciflorum var. *salinum* (A. Nels.) R. Knuth = Dodecatheon pulchellum ssp. pulchellum [DODEPUL0]
Dodecatheon pauciflorum var. *watsonii* (Tidestrom) C.L. Hitchc. = Dodecatheon pulchellum ssp. pulchellum [DODEPUL0]
Dodecatheon puberulum (Nutt.) Piper = Dodecatheon pulchellum ssp. cusickii [DODEPUL1]
Dodecatheon pulchellum (Raf.) Merr. [DODEPUL] (few-flowered shootingstar)
Dodecatheon pulchellum ssp. *alaskanum* (Hultén) Hultén = Dodecatheon pulchellum ssp. macrocarpum [DODEPUL2]
Dodecatheon pulchellum ssp. **cusickii** (Greene) Calder & Taylor [DODEPUL1]
SY: *Dodecatheon cusickii* Greene
SY: *Dodecatheon cusickii* var. *album* Suksdorf
SY: *Dodecatheon pauciflorum* var. *cusickii* (Greene) Mason ex St. John
SY: *Dodecatheon puberulum* (Nutt.) Piper
SY: *Dodecatheon pulchellum* var. *album* (Suksdorf) Boivin
SY: *Dodecatheon pulchellum* var. *cusickii* (Greene) Reveal
Dodecatheon pulchellum ssp. **macrocarpum** (Gray) Taylor & MacBryde [DODEPUL2]
SY: *Dodecatheon pauciflorum* var. *alaskanum* (Hultén) C.L. Hitchc.
SY: *Dodecatheon pulchellum* ssp. *alaskanum* (Hultén) Hultén
SY: *Dodecatheon pulchellum* var. *alaskanum* (Hultén) Reveal
SY: *Dodecatheon pulchellum* var. *alaskanum* (Hultén) Boivin
SY: *Dodecatheon pulchellum* ssp. *superbum* (Pennell & Stair) Hultén
SY: *Dodecatheon radicatum* ssp. *macrocarpum* (Gray) Beamish
Dodecatheon pulchellum ssp. **pulchellum** [DODEPUL0]
SY: *Dodecatheon pauciflorum* var. *salinum* (A. Nels.) R. Knuth
SY: *Dodecatheon pauciflorum* var. *watsonii* (Tidestrom) C.L. Hitchc.
SY: *Dodecatheon pulchellum* var. *watsonii* (Tidestrom) Boivin
SY: *Dodecatheon pulchellum* var. *watsonii* (Tidestrom) C.L. Hitchc.
SY: *Dodecatheon pulchellum* var. *watsonii* (Tidestrom) Reveal
SY: *Dodecatheon pulchellum* var. *zionense* (Eastw.) Welsh
SY: *Dodecatheon radicatum* Greene
SY: *Dodecatheon radicatum* ssp. *watsonii* (Tidestrom) H.J. Thompson
SY: *Dodecatheon salinum* A. Nels.
SY: *Dodecatheon zionense* Eastw.
Dodecatheon pulchellum ssp. *superbum* (Pennell & Stair) Hultén = Dodecatheon pulchellum ssp. macrocarpum [DODEPUL2]
Dodecatheon pulchellum var. *alaskanum* (Hultén) Reveal = Dodecatheon pulchellum ssp. macrocarpum [DODEPUL2]
Dodecatheon pulchellum var. *alaskanum* (Hultén) Boivin = Dodecatheon pulchellum ssp. macrocarpum [DODEPUL2]
Dodecatheon pulchellum var. *album* (Suksdorf) Boivin = Dodecatheon pulchellum ssp. cusickii [DODEPUL1]
Dodecatheon pulchellum var. *cusickii* (Greene) Reveal = Dodecatheon pulchellum ssp. cusickii [DODEPUL1]
Dodecatheon pulchellum var. *watsonii* (Tidestrom) C.L. Hitchc. = Dodecatheon pulchellum ssp. pulchellum [DODEPUL0]
Dodecatheon pulchellum var. *watsonii* (Tidestrom) Boivin = Dodecatheon pulchellum ssp. pulchellum [DODEPUL0]
Dodecatheon pulchellum var. *watsonii* (Tidestrom) Reveal = Dodecatheon pulchellum ssp. pulchellum [DODEPUL0]
Dodecatheon pulchellum var. *zionense* (Eastw.) Welsh = Dodecatheon pulchellum ssp. pulchellum [DODEPUL0]
Dodecatheon radicatum Greene = Dodecatheon pulchellum ssp. pulchellum [DODEPUL0]
Dodecatheon radicatum ssp. *macrocarpum* (Gray) Beamish = Dodecatheon pulchellum ssp. macrocarpum [DODEPUL2]
Dodecatheon radicatum ssp. *watsonii* (Tidestrom) H.J. Thompson = Dodecatheon pulchellum ssp. pulchellum [DODEPUL0]
Dodecatheon salinum A. Nels. = Dodecatheon pulchellum ssp. pulchellum [DODEPUL0]
Dodecatheon viscidum Piper = Dodecatheon conjugens ssp. viscidum [DODECON1]
Dodecatheon zionense Eastw. = Dodecatheon pulchellum ssp. pulchellum [DODEPUL0]
Douglasia Lindl. [DOUGLAS$]
Douglasia alaskana (Coville & Standl. ex Hultén) S. Kelso [DOUGALA]
SY: *Androsace alaskana* Coville & Standl. ex Hultén
SY: *Androsace alaskana* var. *reediae* Welsh & Goodrich
Douglasia arctica var. *gormanii* (Constance) Boivin = Douglasia gormanii [DOUGGOR]
Douglasia biflora A. Nels. = Douglasia montana [DOUGMON]
Douglasia gormanii Constance [DOUGGOR] (Gorman's douglasia)
SY: *Douglasia arctica* var. *gormanii* (Constance) Boivin
SY: *Douglasia ochotensis* ssp. *gormanii* (Constance) A. & D. Löve
Douglasia laevigata Gray [DOUGLAE] (smooth douglasia)
Douglasia laevigata ssp. *ciliolata* (Constance) Calder & Taylor = Douglasia laevigata var. ciliolata [DOUGLAE2]
Douglasia laevigata var. **ciliolata** Constance [DOUGLAE2]
SY: *Douglasia laevigata* ssp. *ciliolata* (Constance) Calder & Taylor
Douglasia montana Gray [DOUGMON] (Rocky Mountain douglasia)
SY: *Douglasia biflora* A. Nels.
SY: *Douglasia montana* var. *biflora* (A. Nels.) R. Knuth
SY: *Gregoria montana* (Gray) House
Douglasia montana var. *biflora* (A. Nels.) R. Knuth = Douglasia montana [DOUGMON]
Douglasia nivalis Lindl. [DOUGNIV] (snow douglasia)
Douglasia ochotensis ssp. *gormanii* (Constance) A. & D. Löve = Douglasia gormanii [DOUGGOR]
Glaux L. [GLAUX$]
Glaux maritima L. [GLAUMAR] (sea-milkwort)
Glaux maritima ssp. **maritima** [GLAUMAR0]
Glaux maritima ssp. **obtusifolia** (Fern.) Boivin [GLAUMAR2]
SY: *Glaux maritima* var. *angustifolia* Boivin
SY: *Glaux maritima* var. *macrophylla* Boivin
SY: *Glaux maritima* var. *obtusifolia* Fern.
Glaux maritima var. *angustifolia* Boivin = Glaux maritima ssp. obtusifolia [GLAUMAR2]
Glaux maritima var. *macrophylla* Boivin = Glaux maritima ssp. obtusifolia [GLAUMAR2]
Glaux maritima var. *obtusifolia* Fern. = Glaux maritima ssp. obtusifolia [GLAUMAR2]
Gregoria montana (Gray) House = Douglasia montana [DOUGMON]
Lysimachia L. [LYSIMAC$]
Lysimachia ciliata L. [LYSICIL] (fringed loosestrife)
SY: *Steironema ciliatum* (L.) Baudo
SY: *Steironema pumilum* Greene
•**Lysimachia nummularia** L. [LYSINUM] (creeping loosestrife)
•**Lysimachia punctata** L. [LYSIPUN] (spotted loosestrife)
SY: *Lysimachia punctata* var. *verticillata* (Bieb.) Klatt
Lysimachia punctata var. *verticillata* (Bieb.) Klatt = Lysimachia punctata [LYSIPUN]
•**Lysimachia terrestris** (L.) B.S.P. [LYSITER] (bog loosestrife)
SY: *Lysimachia terrestris* var. *ovata* (Rand & Redf.) Fern.
Lysimachia terrestris var. *ovata* (Rand & Redf.) Fern. = Lysimachia terrestris [LYSITER]
Lysimachia thyrsiflora L. [LYSITHY] (tufted loosestrife)
SY: *Naumburgia thyrsiflora* (L.) Duby
•**Lysimachia vulgaris** L. [LYSIVUL] (yellow loosestrife)
Naumburgia thyrsiflora (L.) Duby = Lysimachia thyrsiflora [LYSITHY]

Primula L. [PRIMULA$]
Primula clusiana auct. = Primula nutans [PRIMNUT]
Primula cuneifolia Ledeb. [PRIMCUN] (pixie-eyes)
Primula cuneifolia ssp. *saxifragifolia* (Lehm.) W.W. Sm. & G. Forrest = Primula cuneifolia var. saxifragifolia [PRIMCUN1]
Primula cuneifolia var. **saxifragifolia** (Lehm.) Pax & R. Knuth [PRIMCUN1]
SY: *Primula cuneifolia* ssp. *saxifragifolia* (Lehm.) W.W. Sm. & G. Forrest
Primula egaliksensis Wormsk. ex Hornem. [PRIMEGA] (Greenland primrose)
SY: *Primula groenlandica* (Warming) W.W. Sm. & G. Forrest
Primula farinosa var. *incana* Fern. = Primula incana [PRIMINC]
Primula groenlandica (Warming) W.W. Sm. & G. Forrest = Primula egaliksensis [PRIMEGA]
Primula incana M.E. Jones [PRIMINC] (mealy primrose)
SY: *Primula farinosa* var. *incana* Fern.
Primula intercedens Fern. = Primula mistassinica [PRIMMIS]
Primula maccalliana Wieg. = Primula mistassinica [PRIMMIS]
Primula mistassinica Michx. [PRIMMIS] (Mistassini primrose)
SY: *Primula intercedens* Fern.
SY: *Primula maccalliana* Wieg.
SY: *Primula mistassinica* var. *intercedens* (Fern.) Boivin
SY: *Primula mistassinica* var. *noveboracensis* Fern.
Primula mistassinica var. *intercedens* (Fern.) Boivin = Primula mistassinica [PRIMMIS]
Primula mistassinica var. *noveboracensis* Fern. = Primula mistassinica [PRIMMIS]
Primula nutans Georgi [PRIMNUT]
SY: *Primula sibirica* Jacq.
SY: *Primula clusiana* auct.
Primula sibirica Jacq. = Primula nutans [PRIMNUT]
Primula stricta Hornem. [PRIMSTR] (upright primrose)
Samolus L. [SAMOLUS$]
Samolus floribundus Kunth = Samolus valerandi ssp. parviflorus [SAMOVAL1]
Samolus parviflorus Raf. = Samolus valerandi ssp. parviflorus [SAMOVAL1]
Samolus valerandi L. [SAMOVAL] (brookweed)
Samolus valerandi ssp. **parviflorus** (Raf.) Hultén [SAMOVAL1]
SY: *Samolus floribundus* Kunth
SY: *Samolus parviflorus* Raf.
Steironema ciliatum (L.) Baudo = Lysimachia ciliata [LYSICIL]
Steironema pumilum Greene = Lysimachia ciliata [LYSICIL]
Trientalis L. [TRIENTA$]
Trientalis arctica Fisch. ex Hook. = Trientalis europaea ssp. arctica [TRIEEUR1]
Trientalis borealis Raf. [TRIEBOR] (northern starflower)
Trientalis borealis ssp. **latifolia** (Hook.) Hultén [TRIEBOR1]
SY: *Trientalis europaea* var. *latifolia* (Hook.) Torr.
SY: *Trientalis latifolia* Hook.
Trientalis europaea L. [TRIEEUR] (European starflower)
Trientalis europaea ssp. **arctica** (Fisch. ex Hook.) Hultén [TRIEEUR1]
SY: *Trientalis arctica* Fisch. ex Hook.
SY: *Trientalis europaea* var. *arctica* (Fisch. ex Hook.) Ledeb.
Trientalis europaea var. *arctica* (Fisch. ex Hook.) Ledeb. = Trientalis europaea ssp. arctica [TRIEEUR1]
Trientalis europaea var. *latifolia* (Hook.) Torr. = Trientalis borealis ssp. latifolia [TRIEBOR1]
Trientalis latifolia Hook. = Trientalis borealis ssp. latifolia [TRIEBOR1]

PTERIDACEAE (V:F)

Adiantum L. [ADIANTU$]
Adiantum aleuticum (Rupr.) Paris [ADIAALE] (maidenhair fern)
SY: *Adiantum pedatum* ssp. *aleuticum* (Rupr.) Calder & Taylor
SY: *Adiantum pedatum* var. *aleuticum* Rupr.
SY: *Adiantum pedatum* ssp. *calderi* Cody
SY: *Adiantum pedatum* ssp. *subpumilum* (W.H. Wagner) Lellinger
SY: *Adiantum pedatum* var. *subpumilum* W.H. Wagner
SY: *Adiantum pedatum* auct.
Adiantum capillus-veneris L. [ADIACAP] (Venus-hair fern)
SY: *Adiantum capillus-veneris* var. *modestum* (Underwood) Fern.
SY: *Adiantum capillus-veneris* var. *protrusum* Fern.
SY: *Adiantum modestum* Underwood
Adiantum capillus-veneris var. *modestum* (Underwood) Fern. = Adiantum capillus-veneris [ADIACAP]
Adiantum capillus-veneris var. *protrusum* Fern. = Adiantum capillus-veneris [ADIACAP]
Adiantum modestum Underwood = Adiantum capillus-veneris [ADIACAP]
Adiantum pedatum auct. = Adiantum aleuticum [ADIAALE]
Adiantum pedatum ssp. *aleuticum* (Rupr.) Calder & Taylor = Adiantum aleuticum [ADIAALE]
Adiantum pedatum ssp. *calderi* Cody = Adiantum aleuticum [ADIAALE]
Adiantum pedatum ssp. *subpumilum* (W.H. Wagner) Lellinger = Adiantum aleuticum [ADIAALE]
Adiantum pedatum var. *aleuticum* Rupr. = Adiantum aleuticum [ADIAALE]
Adiantum pedatum var. *subpumilum* W.H. Wagner = Adiantum aleuticum [ADIAALE]
Aspidotis (Nutt. ex Hook.) Copeland [ASPIDOT$]
Aspidotis densa (Brack.) Lellinger [ASPIDEN] (Indian's-dream fern)
SY: *Cheilanthes densa* (Brack.) St. John
SY: *Cheilanthes siliquosa* Maxon
SY: *Cryptogramma densa* (Brack.) Diels
SY: *Onychium densum* Brack.
SY: *Pellaea densa* (Brack.) Hook.
Cheilanthes Sw. [CHEILAN$]
Cheilanthes densa (Brack.) St. John = Aspidotis densa [ASPIDEN]
Cheilanthes feei T. Moore [CHEIFEE] (slender lip fern)
Cheilanthes gracillima D.C. Eat. [CHEIGRA] (lace fern)
SY: *Cheilanthes gracillima* var. *aberrans* M.E. Jones
Cheilanthes gracillima var. *aberrans* M.E. Jones = Cheilanthes gracillima [CHEIGRA]
Cheilanthes siliquosa Maxon = Aspidotis densa [ASPIDEN]
Cryptogramma R. Br. [CRYPTOG$]
Cryptogramma acrostichoides R. Br. [CRYPACR] (parsley fern)
SY: *Cryptogramma crispa* ssp. *acrostichoides* (R. Br.) Hultén
SY: *Cryptogramma crispa* var. *macrostichoides* (R. Br.) C.B. Clarke
Cryptogramma acrostichoides var. *sitchensis* (Rupr.) C. Christ. = Cryptogramma sitchensis [CRYPSIT]
Cryptogramma cascadensis E.R. Alverson [CRYPCAS] (Cascade parsley fern)
Cryptogramma crispa ssp. *acrostichoides* (R. Br.) Hultén = Cryptogramma acrostichoides [CRYPACR]
Cryptogramma crispa var. *macrostichoides* (R. Br.) C.B. Clarke = Cryptogramma acrostichoides [CRYPACR]
Cryptogramma crispa var. *sitchensis* (Rupr.) C. Christens. = Cryptogramma sitchensis [CRYPSIT]
Cryptogramma densa (Brack.) Diels = Aspidotis densa [ASPIDEN]
Cryptogramma sitchensis (Rupr.) T. Moore [CRYPSIT] (Sitka parsley fern)
SY: *Cryptogramma acrostichoides* var. *sitchensis* (Rupr.) C. Christ.
SY: *Cryptogramma crispa* var. *sitchensis* (Rupr.) C. Christens.
Cryptogramma stelleri (Gmel.) Prantl [CRYPSTE] (slender rock-brake)
Onychium densum Brack. = Aspidotis densa [ASPIDEN]
Pellaea Link [PELLAEA$]
Pellaea atropurpurea var. *simplex* (Butters) Morton = Pellaea glabella ssp. simplex [PELLGLA2]
Pellaea densa (Brack.) Hook. = Aspidotis densa [ASPIDEN]
Pellaea glabella Mett. ex Kuhn [PELLGLA] (smooth cliff-brake)
Pellaea glabella ssp. **occidentalis** (E. Nels.) Windham [PELLGLA1]
SY: *Pellaea glabella* var. *nana* (L.C. Rich.) Cody
SY: *Pellaea glabella* var. *occidentalis* (E. Nels.) Butters
SY: *Pellaea occidentalis* (E. Nels.) Rydb.
SY: *Pellaea pumila* Rydb.
Pellaea glabella ssp. **simplex** (Butters) A. & D. Löve [PELLGLA2]
SY: *Pellaea atropurpurea* var. *simplex* (Butters) Morton

SY: *Pellaea glabella* var. *simplex* Butters
SY: *Pellaea occidentalis* ssp. *simplex* (Butters) Gastony
SY: *Pellaea suksdorfiana* Butters
Pellaea glabella var. *nana* (L.C. Rich.) Cody = Pellaea glabella ssp. occidentalis [PELLGLA1]
Pellaea glabella var. *occidentalis* (E. Nels.) Butters = Pellaea glabella ssp. occidentalis [PELLGLA1]
Pellaea glabella var. *simplex* Butters = Pellaea glabella ssp. simplex [PELLGLA2]
Pellaea occidentalis (E. Nels.) Rydb. = Pellaea glabella ssp. occidentalis [PELLGLA1]
Pellaea occidentalis ssp. *simplex* (Butters) Gastony = Pellaea glabella ssp. simplex [PELLGLA2]
Pellaea pumila Rydb. = Pellaea glabella ssp. occidentalis [PELLGLA1]
Pellaea suksdorfiana Butters = Pellaea glabella ssp. simplex [PELLGLA2]
Pentagramma Yatsk. [PENTAGR$]
Pentagramma triangularis (Kaulfuss) Yatskievych, Windham & Wollenweber [PENTTRI] (goldenback fern)
SY: *Pityrogramma triangularis* (Kaulfuss) Maxon
Pityrogramma triangularis (Kaulfuss) Maxon = Pentagramma triangularis [PENTTRI]

PYROLACEAE (V:D)

Braxilia minor (L.) House = Pyrola minor [PYROMIN]
Chimaphila Pursh [CHIMAPH$]
Chimaphila menziesii (R. Br. ex D. Don) Spreng. [CHIMMEN] (Menzies' pipsissewa)
Chimaphila occidentalis Rydb. = Chimaphila umbellata ssp. occidentalis [CHIMUMB1]
Chimaphila umbellata (L.) W. Bart. [CHIMUMB] (prince's pine)
Chimaphila umbellata ssp. **occidentalis** (Rydb.) Hultén [CHIMUMB1]
SY: *Chimaphila occidentalis* Rydb.
SY: *Chimaphila umbellata* var. *occidentalis* (Rydb.) Blake
Chimaphila umbellata var. *occidentalis* (Rydb.) Blake = Chimaphila umbellata ssp. occidentalis [CHIMUMB1]
Erxlebenia minor (L.) Rydb. = Pyrola minor [PYROMIN]
Moneses Salisb. ex S.F. Gray [MONESES$]
Moneses uniflora (L.) Gray [MONEUNI] (single delight)
SY: *Pyrola uniflora* L.
Orthilia Raf. [ORTHILI$]
Orthilia secunda (L.) House [ORTHSEC] (one-sided wintergreen)
SY: *Orthilia secunda* ssp. *obtusata* (Turcz.) Böcher
SY: *Orthilia secunda* var. *obtusata* (Turcz.) House
SY: *Pyrola secunda* L.
SY: *Pyrola secunda* ssp. *obtusata* (Turcz.) Hultén
SY: *Pyrola secunda* var. *obtusata* Turcz.
SY: *Ramischia elatior* Rydb.
SY: *Ramischia secunda* (L.) Garcke
Orthilia secunda ssp. *obtusata* (Turcz.) Böcher = Orthilia secunda [ORTHSEC]
Orthilia secunda var. *obtusata* (Turcz.) House = Orthilia secunda [ORTHSEC]
Pyrola L. [PYROLA$]
Pyrola aphylla Sm. = Pyrola picta [PYROPIC]
Pyrola aphylla var. *leptosepala* Nutt. = Pyrola picta [PYROPIC]
Pyrola aphylla var. *paucifolia* T.J. Howell = Pyrola picta [PYROPIC]
Pyrola asarifolia Michx. [PYROASA] (pink wintergreen)
SY: *Pyrola asarifolia* var. *incarnata* (DC.) Fern.
SY: *Pyrola asarifolia* var. *ovata* Farw.
SY: *Pyrola asarifolia* var. *purpurea* (Bunge) Fern.
SY: *Pyrola californica* Krísa
SY: *Pyrola elata* Nutt.
SY: *Pyrola rotundifolia* ssp. *asarifolia* (Michx.) A. & D. Löve
SY: *Pyrola uliginosa* Torr. & Gray ex Torr.
SY: *Pyrola uliginosa* var. *gracilis* Jennings
Pyrola asarifolia ssp. **bracteata** (Hook.) Haber [PYROASA1]
SY: *Pyrola asarifolia* var. *bracteata* (Hook.) Jepson
SY: *Pyrola bracteata* Hook.
SY: *Pyrola bracteata* var. *hillii* J.K. Henry
Pyrola asarifolia var. *bracteata* (Hook.) Jepson = Pyrola asarifolia ssp. bracteata [PYROASA1]
Pyrola asarifolia var. *incarnata* (DC.) Fern. = Pyrola asarifolia [PYROASA]
Pyrola asarifolia var. *ovata* Farw. = Pyrola asarifolia [PYROASA]
Pyrola asarifolia var. *purpurea* (Bunge) Fern. = Pyrola asarifolia [PYROASA]
Pyrola blanda Andres = Pyrola picta [PYROPIC]
Pyrola borealis Rydb. = Pyrola grandiflora [PYROGRA]
Pyrola bracteata Hook. = Pyrola asarifolia ssp. bracteata [PYROASA1]
Pyrola bracteata var. *hillii* J.K. Henry = Pyrola asarifolia ssp. bracteata [PYROASA1]
Pyrola californica Krísa = Pyrola asarifolia [PYROASA]
Pyrola canadensis Andres = Pyrola grandiflora [PYROGRA]
Pyrola chlorantha Sw. [PYROCHL] (green wintergreen)
SY: *Pyrola chlorantha* var. *convoluta* (W. Bart.) Fern.
SY: *Pyrola chlorantha* var. *paucifolia* Fern.
SY: *Pyrola chlorantha* var. *revoluta* Jennings
SY: *Pyrola convoluta* W. Bart.
SY: *Pyrola oxypetala* Austin ex Gray
SY: *Pyrola virens* Schreb.
SY: *Pyrola virens* var. *convoluta* (W. Bart.) Fern.
SY: *Pyrola virens* var. *saximontana* Fern.
Pyrola chlorantha var. *convoluta* (W. Bart.) Fern. = Pyrola chlorantha [PYROCHL]
Pyrola chlorantha var. *paucifolia* Fern. = Pyrola chlorantha [PYROCHL]
Pyrola chlorantha var. *revoluta* Jennings = Pyrola chlorantha [PYROCHL]
Pyrola compacta Jennings = Pyrola elliptica [PYROELL]
Pyrola conardiana Andres = Pyrola picta [PYROPIC]
Pyrola convoluta W. Bart. = Pyrola chlorantha [PYROCHL]
Pyrola dentata Sm. = Pyrola picta [PYROPIC]
Pyrola dentata var. *apophylla* Copeland = Pyrola picta [PYROPIC]
Pyrola dentata var. *integra* Gray = Pyrola picta [PYROPIC]
Pyrola elata Nutt. = Pyrola asarifolia [PYROASA]
Pyrola elliptica Nutt. [PYROELL] (white wintergreen)
SY: *Pyrola compacta* Jennings
Pyrola gormanii Rydb. = Pyrola grandiflora [PYROGRA]
Pyrola grandiflora Radius [PYROGRA] (arctic wintergreen)
SY: *Pyrola borealis* Rydb.
SY: *Pyrola canadensis* Andres
SY: *Pyrola gormanii* Rydb.
SY: *Pyrola grandiflora* var. *canadensis* (Andres) Porsild
SY: *Pyrola occidentalis* R. Br. ex D. Don
Pyrola grandiflora var. *canadensis* (Andres) Porsild = Pyrola grandiflora [PYROGRA]
Pyrola minor L. [PYROMIN] (lesser wintergreen)
SY: *Braxilia minor* (L.) House
SY: *Erxlebenia minor* (L.) Rydb.
SY: *Pyrola minor* var. *parviflora* Boivin
Pyrola minor var. *parviflora* Boivin = Pyrola minor [PYROMIN]
Pyrola occidentalis R. Br. ex D. Don = Pyrola grandiflora [PYROGRA]
Pyrola oxypetala Austin ex Gray = Pyrola chlorantha [PYROCHL]
Pyrola pallida Greene = Pyrola picta [PYROPIC]
Pyrola paradoxa Andres = Pyrola picta [PYROPIC]
Pyrola picta Sm. [PYROPIC] (white-veined wintergreen)
SY: *Pyrola aphylla* Sm.
SY: *Pyrola aphylla* var. *leptosepala* Nutt.
SY: *Pyrola aphylla* var. *paucifolia* T.J. Howell
SY: *Pyrola blanda* Andres
SY: *Pyrola conardiana* Andres
SY: *Pyrola dentata* Sm.
SY: *Pyrola dentata* var. *apophylla* Copeland
SY: *Pyrola dentata* var. *integra* Gray
SY: *Pyrola pallida* Greene
SY: *Pyrola paradoxa* Andres
SY: *Pyrola picta* ssp. *dentata* (Sm.) Piper
SY: *Pyrola picta* var. *dentata* (Sm.) Dorn
SY: *Pyrola picta* ssp. *integra* (Gray) Piper
SY: *Pyrola picta* ssp. *pallida* Andres

SY: *Pyrola septentrionalis* Andres
SY: *Pyrola sparsifolia* Suksdorf
Pyrola picta ssp. *dentata* (Sm.) Piper = Pyrola picta [PYROPIC]
Pyrola picta ssp. *integra* (Gray) Piper = Pyrola picta [PYROPIC]
Pyrola picta ssp. *pallida* Andres = Pyrola picta [PYROPIC]
Pyrola picta var. *dentata* (Sm.) Dorn = Pyrola picta [PYROPIC]
Pyrola rotundifolia ssp. *asarifolia* (Michx.) A. & D. Löve = Pyrola asarifolia [PYROASA]
Pyrola secunda L. = Orthilia secunda [ORTHSEC]
Pyrola secunda ssp. *obtusata* (Turcz.) Hultén = Orthilia secunda [ORTHSEC]
Pyrola secunda var. *obtusata* Turcz. = Orthilia secunda [ORTHSEC]
Pyrola septentrionalis Andres = Pyrola picta [PYROPIC]
Pyrola sparsifolia Suksdorf = Pyrola picta [PYROPIC]
Pyrola uliginosa Torr. & Gray ex Torr. = Pyrola asarifolia [PYROASA]
Pyrola uliginosa var. *gracilis* Jennings = Pyrola asarifolia [PYROASA]
Pyrola uniflora L. = Moneses uniflora [MONEUNI]
Pyrola virens Schreb. = Pyrola chlorantha [PYROCHL]
Pyrola virens var. *convoluta* (W. Bart.) Fern. = Pyrola chlorantha [PYROCHL]
Pyrola virens var. *saximontana* Fern. = Pyrola chlorantha [PYROCHL]
Ramischia elatior Rydb. = Orthilia secunda [ORTHSEC]
Ramischia secunda (L.) Garcke = Orthilia secunda [ORTHSEC]

RANUNCULACEAE (V:D)

Aconitum L. [ACONITU$]
Aconitum columbianum Nutt. [ACONCOL] (Columbian monkshood)
SY: *Aconitum columbianum* var. *bakeri* (Greene) Harrington
SY: *Aconitum columbianum* ssp. *pallidum* Piper
Aconitum columbianum ssp. *pallidum* Piper = Aconitum columbianum [ACONCOL]
Aconitum columbianum var. *bakeri* (Greene) Harrington = Aconitum columbianum [ACONCOL]
Aconitum delphiniifolium DC. [ACONDEL] (mountain monkshood)
SY: *Aconitum delphiniifolium* var. *albiflorum* Porsild
Aconitum delphiniifolium var. *albiflorum* Porsild = Aconitum delphiniifolium [ACONDEL]
Actaea L. [ACTAEA$]
Actaea arguta Nutt. = Actaea rubra ssp. arguta [ACTARUB1]
Actaea arguta var. *viridiflora* (Greene) Tidestrom = Actaea rubra ssp. arguta [ACTARUB1]
Actaea rubra (Ait.) Willd. [ACTARUB] (baneberry)
Actaea rubra ssp. **arguta** (Nutt.) Hultén [ACTARUB1]
SY: *Actaea arguta* Nutt.
SY: *Actaea arguta* var. *viridiflora* (Greene) Tidestrom
SY: *Actaea rubra* var. *arguta* (Nutt.) Lawson
SY: *Actaea spicata* var. *arguta* (Nutt.) Torr.
Actaea rubra var. *arguta* (Nutt.) Lawson = Actaea rubra ssp. arguta [ACTARUB1]
Actaea spicata var. *arguta* (Nutt.) Torr. = Actaea rubra ssp. arguta [ACTARUB1]
Anemone L. [ANEMONE$]
Anemone canadensis L. [ANEMCAN] (Canada anemone)
Anemone cylindrica Gray [ANEMCYL] (long-fruited anemone)
Anemone drummondii S. Wats. [ANEMDRU] (alpine anemone)
Anemone drummondii var. *lithophila* (Rydb.) C.L. Hitchc. = Anemone lithophila [ANEMLIT]
Anemone lithophila Rydb. [ANEMLIT]
SY: *Anemone drummondii* var. *lithophila* (Rydb.) C.L. Hitchc.
Anemone ludoviciana Nutt. = Pulsatilla patens ssp. multifida [PULSPAT1]
Anemone lyallii Britt. [ANEMLYA] (Lyall's anemone)
SY: *Anemone nemorosa* var. *lyallii* (Britt.) Ulbr.
SY: *Anemone quinquefolia* var. *lyallii* (Britt.) B.L. Robins.
Anemone multifida Poir. [ANEMMUL] (cut-leaved anemone)
SY: *Anemone multifida* var. *saxicola* f. *hirsuta* (C.L. Hitchc.) T.C. Brayshaw
Anemone multifida (Pritz.) Zamels = Pulsatilla patens ssp. multifida [PULSPAT1]
Anemone multifida var. **hirsuta** C.L. Hitchc. [ANEMMUL1]
Anemone multifida var. **richardsiana** Fern. [ANEMMUL3]
Anemone multifida var. *saxicola* f. *hirsuta* (C.L. Hitchc.) T.C. Brayshaw = Anemone multifida [ANEMMUL]
Anemone narcissiflora L. [ANEMNAR] (narcissus anemone)
Anemone narcissiflora ssp. **alaskana** Hultén [ANEMNAR1]
SY: *Anemone narcissiflora* var. *alaskana* (Hultén) Boivin
Anemone narcissiflora ssp. **interior** Hultén [ANEMNAR2]
SY: *Anemone narcissiflora* var. *interior* (Hultén) Boivin
Anemone narcissiflora var. *alaskana* (Hultén) Boivin = Anemone narcissiflora ssp. alaskana [ANEMNAR1]
Anemone narcissiflora var. *interior* (Hultén) Boivin = Anemone narcissiflora ssp. interior [ANEMNAR2]
Anemone nemorosa var. *lyallii* (Britt.) Ulbr. = Anemone lyallii [ANEMLYA]
Anemone nuttalliana DC. = Pulsatilla patens ssp. multifida [PULSPAT1]
Anemone occidentalis S. Wats. = Pulsatilla occidentalis [PULSOCC]
Anemone parviflora Michx. [ANEMPAR] (small-flowered anemone)
Anemone patens ssp. *multifida* (Pritzel) Hultén = Pulsatilla patens ssp. multifida [PULSPAT1]
Anemone patens var. *multifida* Pritz. = Pulsatilla patens ssp. multifida [PULSPAT1]
Anemone patens var. *nuttalliana* (DC.) Gray = Pulsatilla patens ssp. multifida [PULSPAT1]
Anemone patens var. *wolfgangiana* (Bess.) K. Koch = Pulsatilla patens ssp. multifida [PULSPAT1]
Anemone piperi Britt. ex Rydb. [ANEMPIP] (Piper's anemone)
Anemone quinquefolia var. *lyallii* (Britt.) B.L. Robins. = Anemone lyallii [ANEMLYA]
Anemone richardsonii Hook. [ANEMRIC] (yellow anemone)
SY: *Jurtsevia richardsonii* (Hook.) A. & D. Löve
Anemone riparia Fern. = Anemone virginiana var. riparia [ANEMVIR2]
Anemone virginiana L. [ANEMVIR]
Anemone virginiana var. **riparia** (Fern.) Boivin [ANEMVIR2]
SY: *Anemone riparia* Fern.
Aquilegia L. [AQUILEG$]
Aquilegia brevistyla Hook. [AQUIBRE] (blue columbine)
Aquilegia flavescens S. Wats. [AQUIFLA] (yellow columbine)
Aquilegia formosa Fisch. ex DC. [AQUIFOR] (red columbine)
Batrachium aquatile (L.) Dumort. = Ranunculus aquatilis [RANUAQU]
Batrachium circinatum ssp. *subrigidum* (W. Drew) A. & D. Löve = Ranunculus longirostris [RANULON]
Batrachium flaccidum (Pers.) Rupr. = Ranunculus trichophyllus [RANUTRI]
Batrachium longirostre (Godr.) F.W. Schultz = Ranunculus longirostris [RANULON]
Batrachium porteri (Britt.) Britt. = Ranunculus trichophyllus [RANUTRI]
Batrachium trichophyllum (Chaix) F.W. Schultz = Ranunculus trichophyllus [RANUTRI]
Caltha L. [CALTHA$]
Caltha asarifolia DC. = Caltha palustris [CALTPAL]
Caltha biflora DC. = Caltha leptosepala ssp. howellii [CALTLEP3]
Caltha biflora ssp. *howellii* (Huth) Abrams = Caltha leptosepala ssp. howellii [CALTLEP3]
Caltha biflora var. *rotundifolia* (Huth) C.L. Hitchc. = Caltha leptosepala [CALTLEP]
Caltha howellii Huth = Caltha leptosepala ssp. howellii [CALTLEP3]
Caltha leptosepala DC. [CALTLEP] (white marsh-marigold)
SY: *Caltha biflora* var. *rotundifolia* (Huth) C.L. Hitchc.
Caltha leptosepala ssp. *biflora* (DC.) P.G. Sm. = Caltha leptosepala ssp. howellii [CALTLEP3]
Caltha leptosepala ssp. **howellii** (Huth) P.G. Sm. [CALTLEP3]
SY: *Caltha biflora* DC.
SY: *Caltha biflora* ssp. *howellii* (Huth) Abrams
SY: *Caltha howellii* Huth
SY: *Caltha leptosepala* ssp. *biflora* (DC.) P.G. Sm.
SY: *Caltha leptosepala* var. *biflora* (DC.) Lawson

Caltha leptosepala var. *biflora* (DC.) Lawson = Caltha leptosepala ssp. howellii [CALTLEP3]
Caltha natans Pallas ex Georgi [CALTNAT] (floating marsh-marigold)
SY: *Thacla natans* (Pallas ex Georgi) Deyl & Soják
Caltha natans var. *asarifolia* (DC.) Huth = Caltha palustris [CALTPAL]
Caltha palustris L. [CALTPAL] (yellow marsh-marigold)
SY: *Caltha asarifolia* DC.
SY: *Caltha natans* var. *asarifolia* (DC.) Huth
SY: *Caltha palustris* ssp. *asarifolia* (DC.) Hultén
SY: *Caltha palustris* var. *asarifolia* (DC.) Rothrock
SY: *Caltha palustris* var. *flabellifolia* (Pursh) Torr. & Gray
Caltha palustris ssp. *asarifolia* (DC.) Hultén = Caltha palustris [CALTPAL]
Caltha palustris var. *asarifolia* (DC.) Rothrock = Caltha palustris [CALTPAL]
Caltha palustris var. *flabellifolia* (Pursh) Torr. & Gray = Caltha palustris [CALTPAL]
Ceratocephala Moench
Ceratocephala orthoceras DC. = Ceratocephala testiculatus [CERATES]
*__Ceratocephala testiculatus__ (Crantz) Roth [CERATES]
SY: *Ceratocephala orthoceras* DC.
SY: *Ranunculus testiculatus* Crantz
Cimicifuga Wernischeck [CIMICIF$]
Cimicifuga elata Nutt. [CIMIELA] (tall bugbane)
Clematis L. [CLEMATI$]
Clematis columbiana (Nutt.) Torr. & Gray [CLEMCOL]
SY: *Clematis verticillaris* var. *columbiana* (Nutt.) Gray
Clematis ligusticifolia Nutt. [CLEMLIG] (white clematis)
Clematis occidentalis (Hornem.) DC. [CLEMOCC] (Columbia clematis)
Clematis occidentalis ssp. *grosseserrata* (Rydb.) Taylor & MacBryde = Clematis occidentalis var. grosseserrata [CLEMOCC2]
Clematis occidentalis var. **grosseserrata** (Rydb.) J. Pringle [CLEMOCC2]
SY: *Clematis occidentalis* ssp. *grosseserrata* (Rydb.) Taylor & MacBryde
Clematis orientalis var. *tangutica* Maxim. = Clematis tangutica [CLEMTAN]
*__Clematis tangutica__ (Maxim.) Korsh. [CLEMTAN] (golden clematis)
SY: *Clematis orientalis* var. *tangutica* Maxim.
Clematis verticillaris var. *columbiana* (Nutt.) Gray = Clematis columbiana [CLEMCOL]
*__Clematis vitalba__ L. [CLEMVIT] (traveler's joy)
Consolida S.F. Gray [CONSOLI$]
*__Consolida ajacis__ (L.) Schur [CONSAJA]
SY: *Consolida ambigua* (L.) P.W. Ball & Heywood
SY: *Delphinium ajacis* L.
SY: *Delphinium ambiguum* L.
Consolida ambigua (L.) P.W. Ball & Heywood = Consolida ajacis [CONSAJA]
Coptidium lapponicum (L.) Gandog. = Ranunculus lapponicus [RANULAP]
Coptis Salisb. [COPTIS$]
Coptis aspleniifolia Salisb. [COPTASP] (fern-leaved goldthread)
Coptis trifolia (L.) Salisb. [COPTTRI] (three-leaved goldthread)
Delphinium L. [DELPHIN$]
Delphinium ajacis L. = Consolida ajacis [CONSAJA]
Delphinium ambiguum L. = Consolida ajacis [CONSAJA]
Delphinium bicolor Nutt. [DELPBIC] (Montana larkspur)
Delphinium brownii Rydb. = Delphinium glaucum [DELPGLA]
Delphinium burkei Greene [DELPBUR] (Burke's larkspur)
SY: *Delphinium burkei* ssp. *distichiflorum* (Hook.) Ewan
SY: *Delphinium burkei* var. *distichiflorum* (Hook.) St. John
SY: *Delphinium × diversicolor* Rydb.
SY: *Delphinium strictum* A. Nels.
SY: *Delphinium strictum* var. *distichiflorum* (Hook.) St. John
Delphinium burkei ssp. *distichiflorum* (Hook.) Ewan = Delphinium burkei [DELPBUR]
Delphinium burkei var. *distichiflorum* (Hook.) St. John = Delphinium burkei [DELPBUR]
Delphinium depauperatum Nutt. [DELPDEP] (slim larkspur)
Delphinium × diversicolor Rydb. = Delphinium burkei [DELPBUR]
Delphinium glaucum S. Wats. [DELPGLA] (tall larkspur)
SY: *Delphinium brownii* Rydb.
SY: *Delphinium scopulorum* var. *glaucum* (S. Wats.) Gray
Delphinium menziesii DC. [DELPMEN] (Menzies' larkspur)
Delphinium menziesii ssp. **menziesii** [DELPMEN0]
Delphinium nelsonii Greene = Delphinium nuttallianum [DELPNUT]
Delphinium nelsonii ssp. *utahense* (S. Wats.) Ewan = Delphinium nuttallianum [DELPNUT]
Delphinium nuttallianum Pritz. ex Walp. [DELPNUT] (upland larkspur)
SY: *Delphinium nelsonii* Greene
SY: *Delphinium nelsonii* ssp. *utahense* (S. Wats.) Ewan
SY: *Delphinium nuttallianum* var. *levicaule* C.L. Hitchc.
Delphinium nuttallianum var. *levicaule* C.L. Hitchc. = Delphinium nuttallianum [DELPNUT]
Delphinium scopulorum var. *glaucum* (S. Wats.) Gray = Delphinium glaucum [DELPGLA]
Delphinium strictum A. Nels. = Delphinium burkei [DELPBUR]
Delphinium strictum var. *distichiflorum* (Hook.) St. John = Delphinium burkei [DELPBUR]
Enemion Raf. [ENEMION$]
Enemion savilei (Calder & Taylor) Keener [ENEMSAV]
SY: *Isopyrum savilei* Calder & Taylor
Isopyrum savilei Calder & Taylor = Enemion savilei [ENEMSAV]
Jurtsevia richardsonii (Hook.) A. & D. Löve = Anemone richardsonii [ANEMRIC]
Kumlienia Greene [KUMLIEN$]
Kumlienia cooleyae (Vasey & Rose) Greene [KUMLCOO]
SY: *Ranunculus cooleyae* Vasey & Rose
Myosurus L. [MYOSURU$]
Myosurus aristatus Benth. [MYOSARI] (bristly mousetail)
SY: *Myosurus minimus* var. *aristatus* (Benth.) Boivin
Myosurus minimus L. [MYOSMIN] (tiny mousetail)
Myosurus minimus var. *aristatus* (Benth.) Boivin = Myosurus aristatus [MYOSARI]
Pulsatilla P. Mill. [PULSATI$]
Pulsatilla hirsutissima (Pursh) Britt. = Pulsatilla patens ssp. multifida [PULSPAT1]
Pulsatilla ludoviciana Heller = Pulsatilla patens ssp. multifida [PULSPAT1]
Pulsatilla nuttalliana (DC.) Spreng. = Pulsatilla patens ssp. multifida [PULSPAT1]
Pulsatilla occidentalis (S. Wats.) Freyn [PULSOCC] (western pasqueflower)
SY: *Anemone occidentalis* S. Wats.
Pulsatilla patens (L.) P. Mill. [PULSPAT]
Pulsatilla patens ssp. *hirsutissima* (Pursh) Zamels = Pulsatilla patens ssp. multifida [PULSPAT1]
Pulsatilla patens ssp. **multifida** (Pritz.) Zamels [PULSPAT1] (prairie crocus)
SY: *Anemone ludoviciana* Nutt.
SY: *Anemone multifida* (Pritz.) Zamels
SY: *Anemone nuttalliana* DC.
SY: *Anemone patens* ssp. *multifida* (Pritzel) Hultén
SY: *Anemone patens* var. *multifida* Pritz.
SY: *Anemone patens* var. *nuttalliana* (DC.) Gray
SY: *Anemone patens* var. *wolfgangiana* (Bess.) K. Koch
SY: *Pulsatilla hirsutissima* (Pursh) Britt.
SY: *Pulsatilla ludoviciana* Heller
SY: *Pulsatilla nuttalliana* (DC.) Spreng.
SY: *Pulsatilla patens* ssp. *hirsutissima* (Pursh) Zamels
Ranunculus L. [RANUNCU$]
Ranunculus abortivus L. [RANUABO] (kidney-leaved buttercup)
SY: *Ranunculus abortivus* ssp. *acrolasius* (Fern.) Kapoor & A. & D. Löve
SY: *Ranunculus abortivus* var. *acrolasius* Fern.
SY: *Ranunculus abortivus* var. *eucyclus* Fern.
SY: *Ranunculus abortivus* var. *indivisus* Fern.
SY: *Ranunculus abortivus* var. *typicus* Fern.
Ranunculus abortivus ssp. *acrolasius* (Fern.) Kapoor & A. & D. Löve = Ranunculus abortivus [RANUABO]

Ranunculus abortivus var. *acrolasius* Fern. = Ranunculus abortivus [RANUABO]
Ranunculus abortivus var. *eucyclus* Fern. = Ranunculus abortivus [RANUABO]
Ranunculus abortivus var. *indivisus* Fern. = Ranunculus abortivus [RANUABO]
Ranunculus abortivus var. *typicus* Fern. = Ranunculus abortivus [RANUABO]
*__Ranunculus acris__ L. [RANUACR] (meadow buttercup)
Ranunculus affinis R. Br. = Ranunculus pedatifidus var. affinis [RANUPED1]
Ranunculus alismifolius Geyer ex Benth. [RANUALI] (water-plantain buttercup)
SY: *Ranunculus alismifolius* var. *hartwegii* (Greene) Jepson
SY: *Ranunculus alismifolius* var. *lemmonii* (Gray) L. Benson
SY: *Ranunculus alismifolius* var. *typicus* L. Benson
SY: *Ranunculus hartwegii* Greene
SY: *Ranunculus lemmonii* Gray
Ranunculus alismifolius var. *hartwegii* (Greene) Jepson = Ranunculus alismifolius [RANUALI]
Ranunculus alismifolius var. *lemmonii* (Gray) L. Benson = Ranunculus alismifolius [RANUALI]
Ranunculus alismifolius var. *typicus* L. Benson = Ranunculus alismifolius [RANUALI]
Ranunculus alpeophilus A. Nels. = Ranunculus inamoenus var. alpeophilus [RANUINA1]
Ranunculus amphibius James = Ranunculus longirostris [RANULON]
Ranunculus aquatilis L. [RANUAQU] (white water-buttercup)
SY: *Batrachium aquatile* (L.) Dumort.
SY: *Ranunculus aquatilis* var. *hispidulus* E. Drew
SY: *Ranunculus aquatilis* var. *typicus* L. Benson
SY: *Ranunculus trichophyllus* var. *hispidulus* (E. Drew) W. Drew
Ranunculus aquatilis var. *calvescens* (W. Drew) L. Benson = Ranunculus trichophyllus [RANUTRI]
Ranunculus aquatilis var. *capillaceus* (Thuill.) DC. = Ranunculus trichophyllus [RANUTRI]
Ranunculus aquatilis var. *harrisii* L. Benson = Ranunculus trichophyllus [RANUTRI]
Ranunculus aquatilis var. *hispidulus* E. Drew = Ranunculus aquatilis [RANUAQU]
Ranunculus aquatilis var. *lalondei* L. Benson = Ranunculus trichophyllus [RANUTRI]
Ranunculus aquatilis var. *lobbii* (Hiern) S. Wats. = Ranunculus lobbii [RANULOB]
Ranunculus aquatilis var. *longirostris* (Godr.) Lawson = Ranunculus longirostris [RANULON]
Ranunculus aquatilis var. *porteri* (Britt.) L. Benson = Ranunculus trichophyllus [RANUTRI]
Ranunculus aquatilis var. *subrigidus* (W. Drew) Breitung = Ranunculus longirostris [RANULON]
Ranunculus aquatilis var. *trichophyllus* (Chaix) Gray = Ranunculus trichophyllus [RANUTRI]
Ranunculus aquatilis var. *typicus* L. Benson = Ranunculus aquatilis [RANUAQU]
Ranunculus bongardii Greene = Ranunculus uncinatus var. parviflorus [RANUUNC1]
Ranunculus californicus Benth. [RANUCAL] (California buttercup)
SY: *Ranunculus californicus* var. *austromontanus* L. Benson
SY: *Ranunculus californicus* var. *cuneatus* Greene
SY: *Ranunculus californicus* var. *gratus* Jepson
SY: *Ranunculus californicus* var. *rugulosus* (Greene) L. Benson
SY: *Ranunculus californicus* var. *typicus* L. Benson
SY: *Ranunculus rugulosus* Greene
Ranunculus californicus var. *austromontanus* L. Benson = Ranunculus californicus [RANUCAL]
Ranunculus californicus var. *cuneatus* Greene = Ranunculus californicus [RANUCAL]
Ranunculus californicus var. *gratus* Jepson = Ranunculus californicus [RANUCAL]
Ranunculus californicus var. *rugulosus* (Greene) L. Benson = Ranunculus californicus [RANUCAL]
Ranunculus californicus var. *typicus* L. Benson = Ranunculus californicus [RANUCAL]
Ranunculus cardiophyllus Hook. [RANUCAR] (heart-leaved buttercup)
SY: *Ranunculus cardiophyllus* var. *subsagittatus* (Gray) L. Benson
SY: *Ranunculus cardiophyllus* var. *typicus* L. Benson
SY: *Ranunculus pedatifidus* var. *cardiophyllus* (Hook.) Britt.
Ranunculus cardiophyllus var. *subsagittatus* (Gray) L. Benson = Ranunculus cardiophyllus [RANUCAR]
Ranunculus cardiophyllus var. *typicus* L. Benson = Ranunculus cardiophyllus [RANUCAR]
Ranunculus circinatus auct. = Ranunculus longirostris [RANULON]
Ranunculus circinatus var. *subrigidus* (W. Drew) L. Benson = Ranunculus longirostris [RANULON]
Ranunculus cooleyae Vasey & Rose = Kumlienia cooleyae [KUMLCOO]
Ranunculus cymbalaria Pursh [RANUCYM] (shore buttercup)
Ranunculus cymbalaria ssp. *saximontanus* (Fern.) Thorne = Ranunculus cymbalaria var. saximontanus [RANUCYM2]
Ranunculus cymbalaria var. **saximontanus** Fern. [RANUCYM2]
SY: *Ranunculus cymbalaria* ssp. *saximontanus* (Fern.) Thorne
Ranunculus delphiniifolius Torr. ex Eat. = Ranunculus flabellaris [RANUFLA]
Ranunculus ellipticus Greene = Ranunculus glaberrimus var. ellipticus [RANUGLA1]
Ranunculus eschscholtzii Schlecht. [RANUESC] (subalpine buttercup)
SY: *Ranunculus eschscholtzii* var. *typicus* L. Benson
SY: *Ranunculus nivalis* var. *eschscholtzii* (Schlecht.) S. Wats.
Ranunculus eschscholtzii var. *suksdorfii* (Gray) L. Benson = Ranunculus suksdorfii [RANUSUK]
Ranunculus eschscholtzii var. *typicus* L. Benson = Ranunculus eschscholtzii [RANUESC]
*__Ranunculus ficaria__ L. [RANUFIC] (lesser celandine)
Ranunculus filiformis Michx. = Ranunculus flammula var. filiformis [RANUFLM1]
Ranunculus flabellaris Raf. [RANUFLA] (yellow water-buttercup)
SY: *Ranunculus delphiniifolius* Torr. ex Eat.
Ranunculus flaccidus Pers. = Ranunculus trichophyllus [RANUTRI]
Ranunculus flammula L. [RANUFLM] (lesser spearwort)
SY: *Ranunculus flammula* var. *ovalis* (Bigelow) L. Benson
Ranunculus flammula var. **filiformis** (Michx.) Hook. [RANUFLM1]
SY: *Ranunculus filiformis* Michx.
SY: *Ranunculus flammula* var. *reptans* (L.) E. Mey.
SY: *Ranunculus reptans* L.
SY: *Ranunculus reptans* var. *filiformis* (Michx.) DC.
SY: *Ranunculus reptans* var. *intermedius* (Hook.) Torr. & Gray
Ranunculus flammula var. *ovalis* (Bigelow) L. Benson = Ranunculus flammula [RANUFLM]
Ranunculus flammula var. *reptans* (L.) E. Mey. = Ranunculus flammula var. filiformis [RANUFLM1]
Ranunculus gelidus Kar. & Kir. = Ranunculus karelinii [RANUKAR]
Ranunculus gelidus ssp. *grayi* (Britt.) Hultén = Ranunculus karelinii [RANUKAR]
Ranunculus gelidus var. *shumaginensis* Hultén = Ranunculus karelinii [RANUKAR]
Ranunculus glaberrimus Hook. [RANUGLA] (sagebrush buttercup)
Ranunculus glaberrimus var. *buddii* Boivin = Ranunculus glaberrimus var. ellipticus [RANUGLA1]
Ranunculus glaberrimus var. **ellipticus** (Greene) Greene [RANUGLA1]
SY: *Ranunculus ellipticus* Greene
SY: *Ranunculus glaberrimus* var. *buddii* Boivin
Ranunculus glaberrimus var. **glaberrimus** [RANUGLA0]
SY: *Ranunculus glaberrimus* var. *typicus* L. Benson
Ranunculus glaberrimus var. *typicus* L. Benson = Ranunculus glaberrimus var. glaberrimus [RANUGLA0]
Ranunculus gmelinii DC. [RANUGME] (small yellow water-buttercup)
Ranunculus gmelinii ssp. *purshii* (Richards.) Hultén = Ranunculus gmelinii var. purshii [RANUGME1]
Ranunculus gmelinii var. *hookeri* (D. Don) L. Benson = Ranunculus gmelinii var. purshii [RANUGME1]
Ranunculus gmelinii var. **limosus** (Nutt.) Hara [RANUGME4]
SY: *Ranunculus limosus* Nutt.

Ranunculus gmelinii var. *prolificus* (Fern.) Hara = Ranunculus gmelinii var. purshii [RANUGME1]
Ranunculus gmelinii var. **purshii** (Richards.) Hara [RANUGME1]
SY: *Ranunculus gmelinii* var. *hookeri* (D. Don) L. Benson
SY: *Ranunculus gmelinii* var. *prolificus* (Fern.) Hara
SY: *Ranunculus gmelinii* ssp. *purshii* (Richards.) Hultén
SY: *Ranunculus gmelinii* var. *terrestris* (Ledeb.) L. Benson
SY: *Ranunculus purshii* Richards.
Ranunculus gmelinii var. *terrestris* (Ledeb.) L. Benson = Ranunculus gmelinii var. purshii [RANUGME1]
Ranunculus grayi Britt. = Ranunculus karelinii [RANUKAR]
Ranunculus hartwegii Greene = Ranunculus alismifolius [RANUALI]
Ranunculus hexasepalus (L. Benson) L. Benson [RANUHEX]
SY: *Ranunculus occidentalis* var. *hexasepalus* L. Benson
Ranunculus hyperboreus Rottb. [RANUHYP] (far-northern buttercup)
SY: *Ranunculus natans* var. *intertextus* (Greene) L. Benson
Ranunculus inamoenus Greene [RANUINA] (unlovely buttercup)
Ranunculus inamoenus var. **alpeophilus** (A. Nels.) L. Benson [RANUINA1]
SY: *Ranunculus alpeophilus* A. Nels.
Ranunculus karelinii Czern. [RANUKAR]
SY: *Ranunculus gelidus* Kar. & Kir.
SY: *Ranunculus gelidus* ssp. *grayi* (Britt.) Hultén
SY: *Ranunculus gelidus* var. *shumaginensis* Hultén
SY: *Ranunculus grayi* Britt.
Ranunculus lapponicus L. [RANULAP] (Lapland buttercup)
SY: *Coptidium lapponicum* (L.) Gandog.
Ranunculus lemmonii Gray = Ranunculus alismifolius [RANUALI]
Ranunculus limosus Nutt. = Ranunculus gmelinii var. limosus [RANUGME4]
Ranunculus lobbii (Hiern) Gray [RANULOB] (Lobb's water-buttercup)
SY: *Ranunculus aquatilis* var. *lobbii* (Hiern) S. Wats.
Ranunculus longirostris Godr. [RANULON]
SY: *Batrachium circinatum* ssp. *subrigidum* (W. Drew) A. & D. Löve
SY: *Batrachium longirostre* (Godr.) F.W. Schultz
SY: *Ranunculus amphibius* James
SY: *Ranunculus aquatilis* var. *longirostris* (Godr.) Lawson
SY: *Ranunculus aquatilis* var. *subrigidus* (W. Drew) Breitung
SY: *Ranunculus circinatus* var. *subrigidus* (W. Drew) L. Benson
SY: *Ranunculus subrigidus* W. Drew
SY: *Ranunculus usneoides* Greene
SY: *Ranunculus circinatus* auct.
Ranunculus macounii Britt. [RANUMAC] (Macoun's buttercup)
SY: *Ranunculus macounii* var. *oreganus* (Gray) Davis
Ranunculus macounii var. *oreganus* (Gray) Davis = Ranunculus macounii [RANUMAC]
Ranunculus natans var. *intertextus* (Greene) L. Benson = Ranunculus hyperboreus [RANUHYP]
Ranunculus nivalis L. [RANUNIV] (snow buttercup)
Ranunculus nivalis var. *eschscholtzii* (Schlecht.) S. Wats. = Ranunculus eschscholtzii [RANUESC]
Ranunculus nivalis var. *sulphureus* (Soland. ex Phipps) Wahlenb. = Ranunculus sulphureus [RANUSUL]
Ranunculus occidentalis Nutt. [RANUOCC] (western buttercup)
Ranunculus occidentalis var. **brevistylis** Greene [RANUOCC2]
Ranunculus occidentalis var. *hexasepalus* L. Benson = Ranunculus hexasepalus [RANUHEX]
Ranunculus occidentalis var. *parviflorus* Torr. = Ranunculus uncinatus var. parviflorus [RANUUNC1]
Ranunculus orthorhynchus Hook. [RANUORT] (straight-beaked buttercup)
SY: *Ranunculus orthorhynchus* ssp. *alaschensis* (L. Benson) Hultén
SY: *Ranunculus orthorhynchus* var. *alaschensis* L. Benson
SY: *Ranunculus orthorhynchus* var. *hallii* Jepson
SY: *Ranunculus orthorhynchus* ssp. *platyphyllus* (Gray) Taylor & MacBryde
SY: *Ranunculus orthorhynchus* var. *platyphyllus* Gray
SY: *Ranunculus orthorhynchus* var. *typicus* L. Benson
Ranunculus orthorhynchus ssp. *alaschensis* (L. Benson) Hultén = Ranunculus orthorhynchus [RANUORT]
Ranunculus orthorhynchus ssp. *platyphyllus* (Gray) Taylor & MacBryde = Ranunculus orthorhynchus [RANUORT]
Ranunculus orthorhynchus var. *alaschensis* L. Benson = Ranunculus orthorhynchus [RANUORT]
Ranunculus orthorhynchus var. *hallii* Jepson = Ranunculus orthorhynchus [RANUORT]
Ranunculus orthorhynchus var. *platyphyllus* Gray = Ranunculus orthorhynchus [RANUORT]
Ranunculus orthorhynchus var. *typicus* L. Benson = Ranunculus orthorhynchus [RANUORT]
Ranunculus ovalis Raf. = Ranunculus rhomboideus [RANURHO]
Ranunculus parvulus L. = Ranunculus sardous [RANUSAR]
Ranunculus pedatifidus Sm. [RANUPED] (birdfoot buttercup)
Ranunculus pedatifidus ssp. *affinis* (R. Br.) Hultén = Ranunculus pedatifidus var. affinis [RANUPED1]
Ranunculus pedatifidus var. **affinis** (R. Br.) L. Benson [RANUPED1]
SY: *Ranunculus affinis* R. Br.
SY: *Ranunculus pedatifidus* ssp. *affinis* (R. Br.) Hultén
SY: *Ranunculus pedatifidus* var. *leiocarpus* (Trautv.) Fern.
Ranunculus pedatifidus var. *cardiophyllus* (Hook.) Britt. = Ranunculus cardiophyllus [RANUCAR]
Ranunculus pedatifidus var. *leiocarpus* (Trautv.) Fern. = Ranunculus pedatifidus var. affinis [RANUPED1]
Ranunculus pensylvanicus L. f. [RANUPEN] (Pennsylvania buttercup)
Ranunculus purshii Richards. = Ranunculus gmelinii var. purshii [RANUGME1]
Ranunculus pygmaeus Wahlenb. [RANUPYG] (pygmy buttercup)
*__Ranunculus repens__ L. [RANUREP] (creeping buttercup)
SY: *Ranunculus repens* var. *villosus* Lamotte
Ranunculus repens var. **degeneratus** Schur [RANUREP2]
SY: *Ranunculus repens* var. *pleniflorus* Fern.
Ranunculus repens var. *erectus* DC. = Ranunculus repens var. glabratus [RANUREP1]
Ranunculus repens var. **glabratus** DC. [RANUREP1]
SY: *Ranunculus repens* var. *erectus* DC.
Ranunculus repens var. *pleniflorus* Fern. = Ranunculus repens var. degeneratus [RANUREP2]
Ranunculus repens var. *villosus* Lamotte = Ranunculus repens [RANUREP]
Ranunculus reptans L. = Ranunculus flammula var. filiformis [RANUFLM1]
Ranunculus reptans var. *filiformis* (Michx.) DC. = Ranunculus flammula var. filiformis [RANUFLM1]
Ranunculus reptans var. *intermedius* (Hook.) Torr. & Gray = Ranunculus flammula var. filiformis [RANUFLM1]
Ranunculus rhomboideus Goldie [RANURHO] (prairie buttercup)
SY: *Ranunculus ovalis* Raf.
Ranunculus rugulosus Greene = Ranunculus californicus [RANUCAL]
*__Ranunculus sardous__ Crantz [RANUSAR] (hairy buttercup)
SY: *Ranunculus parvulus* L.
Ranunculus sceleratus L. [RANUSCE] (celery-leaved buttercup)
Ranunculus sceleratus ssp. *multifidus* (Nutt.) Hultén = Ranunculus sceleratus var. multifidus [RANUSCE1]
Ranunculus sceleratus var. **multifidus** Nutt. [RANUSCE1]
SY: *Ranunculus sceleratus* ssp. *multifidus* (Nutt.) Hultén
Ranunculus subrigidus W. Drew = Ranunculus longirostris [RANULON]
Ranunculus suksdorfii Gray [RANUSUK]
SY: *Ranunculus eschscholtzii* var. *suksdorfii* (Gray) L. Benson
Ranunculus sulphureus Soland. ex Phipps [RANUSUL] (sulphur buttercup)
SY: *Ranunculus nivalis* var. *sulphureus* (Soland. ex Phipps) Wahlenb.
SY: *Ranunculus sulphureus* var. *intercedens* Hultén
Ranunculus sulphureus var. *intercedens* Hultén = Ranunculus sulphureus [RANUSUL]
Ranunculus testiculatus Crantz = Ceratocephala testiculatus [CERATES]

Ranunculus trichophyllus Chaix [RANUTRI]
SY: *Batrachium flaccidum* (Pers.) Rupr.
SY: *Batrachium porteri* (Britt.) Britt.
SY: *Batrachium trichophyllum* (Chaix) F.W. Schultz
SY: *Ranunculus aquatilis* var. *calvescens* (W. Drew) L. Benson
SY: *Ranunculus aquatilis* var. *capillaceus* (Thuill.) DC.
SY: *Ranunculus aquatilis* var. *harrisii* L. Benson
SY: *Ranunculus aquatilis* var. *lalondei* L. Benson
SY: *Ranunculus aquatilis* var. *porteri* (Britt.) L. Benson
SY: *Ranunculus aquatilis* var. *trichophyllus* (Chaix) Gray
SY: *Ranunculus flaccidus* Pers.
SY: *Ranunculus trichophyllus* var. *calvescens* W. Drew
SY: *Ranunculus trichophyllus* var. *typicus* W. Drew
Ranunculus trichophyllus var. *calvescens* W. Drew = Ranunculus trichophyllus [RANUTRI]
Ranunculus trichophyllus var. *hispidulus* (E. Drew) W. Drew = Ranunculus aquatilis [RANUAQU]
Ranunculus trichophyllus var. *typicus* W. Drew = Ranunculus trichophyllus [RANUTRI]
Ranunculus uncinatus D. Don ex G. Don [RANUUNC] (little buttercup)
Ranunculus uncinatus var. **parviflorus** (Torr.) L. Benson [RANUUNC1]
SY: *Ranunculus bongardii* Greene
SY: *Ranunculus occidentalis* var. *parviflorus* Torr.
Ranunculus usneoides Greene = Ranunculus longirostris [RANULON]
Ranunculus verecundus B.L. Robins. ex Piper [RANUVER] (modest buttercup)
Thacla natans (Pallas ex Georgi) Deyl & Soják = Caltha natans [CALTNAT]
Thalictrum L. [THALICT$]
Thalictrum alpinum L. [THALALP] (alpine meadowrue)
SY: *Thalictrum alpinum* var. *hebetum* Boivin
Thalictrum alpinum var. *hebetum* Boivin = Thalictrum alpinum [THALALP]
Thalictrum breitungii Boivin = Thalictrum occidentale var. occidentale [THALOCC0]
Thalictrum columbianum Rydb. = Thalictrum venulosum [THALVEN]
Thalictrum confine Fern. = Thalictrum venulosum [THALVEN]
Thalictrum confine var. *columbianum* (Rydb.) Boivin = Thalictrum venulosum [THALVEN]
Thalictrum confine var. *greeneanum* Boivin = Thalictrum venulosum [THALVEN]
Thalictrum dasycarpum Fisch. & Avé-Lall. [THALDAS] (purple meadowrue)
SY: *Thalictrum dasycarpum* var. *hypoglaucum* (Rydb.) Boivin
SY: *Thalictrum hypoglaucum* Rydb.
Thalictrum dasycarpum var. *hypoglaucum* (Rydb.) Boivin = Thalictrum dasycarpum [THALDAS]
Thalictrum hypoglaucum Rydb. = Thalictrum dasycarpum [THALDAS]
Thalictrum occidentale Gray [THALOCC] (western meadowrue)
Thalictrum occidentale var. *breitungii* (Bovin) T.C. Brayshaw = Thalictrum occidentale var. occidentale [THALOCC0]
Thalictrum occidentale var. *columbianum* (Rydb.) M.E. Peck = Thalictrum venulosum [THALVEN]
Thalictrum occidentale var. **macounii** Boivin [THALOCC4]
Thalictrum occidentale var. *megacarpum* (Torr.) St. John = Thalictrum occidentale var. occidentale [THALOCC0]
Thalictrum occidentale var. **occidentale** [THALOCC0]
SY: *Thalictrum breitungii* Boivin
SY: *Thalictrum occidentale* var. *breitungii* (Bovin) T.C. Brayshaw
SY: *Thalictrum occidentale* var. *megacarpum* (Torr.) St. John
SY: *Thalictrum occidentale* var. *palousense* St. John
Thalictrum occidentale var. *palousense* St. John = Thalictrum occidentale var. occidentale [THALOCC0]
Thalictrum sparsiflorum Turcz. ex Fisch. & C.A. Mey. [THALSPA] (few-flowered meadowrue)
Thalictrum sparsiflorum var. **richardsonii** (Gray) Boivin [THALSPA1] (Richardson's meadowrue)
Thalictrum turneri Boivin = Thalictrum venulosum [THALVEN]
Thalictrum venulosum Trel. [THALVEN] (veiny meadowrue)
SY: *Thalictrum columbianum* Rydb.
SY: *Thalictrum confine* Fern.
SY: *Thalictrum confine* var. *columbianum* (Rydb.) Boivin
SY: *Thalictrum confine* var. *greeneanum* Boivin
SY: *Thalictrum occidentale* var. *columbianum* (Rydb.) M.E. Peck
SY: *Thalictrum turneri* Boivin
SY: *Thalictrum venulosum* var. *confine* (Fern.) Boivin
SY: *Thalictrum venulosum* var. *fissum* (Greene) Boivin
SY: *Thalictrum venulosum* var. *lunellii* (Greene) Boivin
SY: *Thalictrum venulosum* var. *turneri* (Boivin) Boivin
Thalictrum venulosum var. *confine* (Fern.) Boivin = Thalictrum venulosum [THALVEN]
Thalictrum venulosum var. *fissum* (Greene) Boivin = Thalictrum venulosum [THALVEN]
Thalictrum venulosum var. *lunellii* (Greene) Boivin = Thalictrum venulosum [THALVEN]
Thalictrum venulosum var. *turneri* (Boivin) Boivin = Thalictrum venulosum [THALVEN]
Trautvetteria Fisch. & C.A. Mey. [TRAUTVE$]
Trautvetteria caroliniensis (Walt.) Vail [TRAUCAR] (false bugbane)
Trollius L. [TROLLIU$]
Trollius albiflorus (Gray) Rydb. = Trollius laxus ssp. albiflorus [TROLLAX1]
Trollius laxus Salisb. [TROLLAX] (globeflower)
Trollius laxus ssp. **albiflorus** (Gray) A. & D. Löve & Kapoor [TROLLAX1]
SY: *Trollius albiflorus* (Gray) Rydb.
SY: *Trollius laxus* var. *albiflorus* Gray
Trollius laxus var. *albiflorus* Gray = Trollius laxus ssp. albiflorus [TROLLAX1]

RESEDACEAE (V:D)

Reseda L. [RESEDA$]
***Reseda alba** L. [RESEALB] (white mignonette)
***Reseda lutea** L. [RESELUT] (yellow mignonette)

RHAMNACEAE (V:D)

Ceanothus L. [CEANOTH$]
Ceanothus oreganus Nutt. = Ceanothus sanguineus [CEANSAN]
Ceanothus sanguineus Pursh [CEANSAN] (redstem ceanothus)
SY: *Ceanothus oreganus* Nutt.
Ceanothus velutinus Dougl. ex Hook. [CEANVEL] (snowbrush)
Ceanothus velutinus ssp. **hookeri** (M.C. Johnston) C. Schmidt [CEANVEL1]
SY: *Ceanothus velutinus* var. *hookeri* M.C. Johnston
SY: *Ceanothus velutinus* var. *laevigatus* Torr. & Gray
Ceanothus velutinus ssp. **velutinus** [CEANVEL0]
Ceanothus velutinus var. *hookeri* M.C. Johnston = Ceanothus velutinus ssp. hookeri [CEANVEL1]
Ceanothus velutinus var. *laevigatus* Torr. & Gray = Ceanothus velutinus ssp. hookeri [CEANVEL1]
Frangula P. Mill. [FRANGUL$]
Frangula purshiana (DC.) Cooper [FRANPUR]
SY: *Rhamnus purshiana* DC.
Rhamnus L. [RHAMNUS$]
Rhamnus alnifolia L'Hér. [RHAMALN] (alder-leaved buckthorn)
Rhamnus purshiana DC. = Frangula purshiana [FRANPUR]

ROSACEAE (V:D)

Acomastylis calthifolia (Menzies ex Sm.) Bolle = Geum calthifolium [GEUMCAL]
Acomastylis macrantha (Kearney) Bolle = Geum × macranthum [GEUMMAR]
Agrimonia L. [AGRIMON$]
Agrimonia gryposepala Wallr. [AGRIGRY] (common agrimony)
Agrimonia striata Michx. [AGRISTR] (grooved agrimony)

Alchemilla L. [ALCHEMI$]
Alchemilla arvensis (L.) Scop. = Aphanes arvensis [APHAARV]
Alchemilla cuneifolia Nutt. = Aphanes arvensis [APHAARV]
Alchemilla microcarpa Boiss. & Reut. = Aphanes microcarpa [APHAMIC]
Alchemilla occidentalis Nutt. = Aphanes arvensis [APHAARV]
*****Alchemilla subcrenata** Buser [ALCHSUB] (lady's-mantle)
Amelanchier Medik. [AMELANC$]
Amelanchier alnifolia (Nutt.) Nutt. ex M. Roemer [AMELALN] (Saskatoon)
Amelanchier alnifolia ssp. *florida* (Lindl.) Hultén = Amelanchier alnifolia var. semiintegrifolia [AMELALN4]
Amelanchier alnifolia var. **alnifolia** [AMELALN0]
Amelanchier alnifolia var. **cusickii** (Fern.) C.L. Hitchc. [AMELALN2]
SY: *Amelanchier cusickii* Fern.
SY: *Amelanchier florida* var. *cusickii* (Fern.) M.E. Peck
Amelanchier alnifolia var. **humptulipensis** (G.N. Jones) C.L. Hitchc. [AMELALN3]
SY: *Amelanchier florida* var. *humptulipensis* G.N. Jones
Amelanchier alnifolia var. **semiintegrifolia** (Hook.) C.L. Hitchc. [AMELALN4]
SY: *Amelanchier alnifolia* ssp. *florida* (Lindl.) Hultén
SY: *Amelanchier florida* Lindl.
Amelanchier cusickii Fern. = Amelanchier alnifolia var. cusickii [AMELALN2]
Amelanchier florida Lindl. = Amelanchier alnifolia var. semiintegrifolia [AMELALN4]
Amelanchier florida var. *cusickii* (Fern.) M.E. Peck = Amelanchier alnifolia var. cusickii [AMELALN2]
Amelanchier florida var. *humptulipensis* G.N. Jones = Amelanchier alnifolia var. humptulipensis [AMELALN3]
Aphanes L. [APHANES$]
*****Aphanes arvensis** L. [APHAARV] (field parsley-piert)
SY: *Alchemilla arvensis* (L.) Scop.
SY: *Alchemilla cuneifolia* Nutt.
SY: *Alchemilla occidentalis* Nutt.
SY: *Aphanes occidentalis* (Nutt.) Rydb.
Aphanes australis Rydb. = Aphanes microcarpa [APHAMIC]
*****Aphanes microcarpa** (Boiss. & Reut.) Rothm. [APHAMIC] (small-fruited parsley-piert)
SY: *Alchemilla microcarpa* Boiss. & Reut.
SY: *Aphanes australis* Rydb.
Aphanes occidentalis (Nutt.) Rydb. = Aphanes arvensis [APHAARV]
Argentina Hill [ARGENTI$]
Argentina anserina (L.) Rydb. [ARGEANS]
SY: *Argentina anserina* var. *concolor* Rydb.
SY: *Argentina argentea* (L.) Rydb.
SY: *Potentilla anserina* L.
SY: *Potentilla anserina* var. *concolor* Ser.
SY: *Potentilla anserina* var. *sericea* (L.) Hayne
SY: *Potentilla anserina* var. *yukonensis* (Hultén) Boivin
SY: *Potentilla egedii* ssp. *yukonensis* (Hultén) Hultén
SY: *Potentilla yukonensis* Hultén
Argentina anserina var. *concolor* Rydb. = Argentina anserina [ARGEANS]
Argentina argentea (L.) Rydb. = Argentina anserina [ARGEANS]
Argentina egedii (Wormsk.) Rydb. [ARGEEGE]
SY: *Potentilla anserina* ssp. *egedii* (Wormsk.) Hiitonen
SY: *Potentilla anserina* var. *grandis* Torr. & Gray
SY: *Potentilla anserina* var. *lanata* Boivin
SY: *Potentilla anserina* ssp. *pacifica* (T.J. Howell) Rousi
SY: *Potentilla anserina* var. *rolandii* (Boivin) Boivin
SY: *Potentilla egedii* Wormsk.
SY: *Potentilla egedii* ssp. *grandis* (Torr. & Gray) Hultén
SY: *Potentilla egedii* var. *grandis* (Torr. & Gray) J.T. Howell
SY: *Potentilla pacifica* T.J. Howell
SY: *Potentilla rolandii* Boivin
Aruncus L. [ARUNCUS$]
Aruncus aruncus (L.) Karst. = Aruncus dioicus var. vulgaris [ARUNDIO1]
Aruncus dioicus (Walt.) Fern. [ARUNDIO] (goat's-beard)
Aruncus dioicus var. **vulgaris** (Maxim.) Hara [ARUNDIO1]
SY: *Aruncus aruncus* (L.) Karst.
SY: *Aruncus sylvester* Kostel. ex Maxim.
SY: *Aruncus vulgaris* (Maxim.) Raf. ex Hara
SY: *Spiraea aruncus* L.
Aruncus sylvester Kostel. ex Maxim. = Aruncus dioicus var. vulgaris [ARUNDIO1]
Aruncus vulgaris (Maxim.) Raf. ex Hara = Aruncus dioicus var. vulgaris [ARUNDIO1]
Cerasus avium (L.) Moench = Prunus avium [PRUNAVI]
Cerasus laurocerasus (L.) Loisel. = Prunus laurocerasus [PRUNLAU]
Chamaerhodos Bunge [CHAMAER$]
Chamaerhodos erecta (L.) Bunge [CHAMERE] (little-rose)
Chamaerhodos erecta ssp. **nuttallii** (Pickering ex Rydb.) Hultén [CHAMERE1] (American chamaerhodos)
SY: *Chamaerhodos erecta* var. *parviflora* (Nutt.) C.L. Hitchc.
SY: *Chamaerhodos nuttallii* Pickering ex Rydb.
SY: *Chamaerhodos nuttallii* var. *keweenawensis* Fern.
Chamaerhodos erecta var. *parviflora* (Nutt.) C.L. Hitchc. = Chamaerhodos erecta ssp. nuttallii [CHAMERE1]
Chamaerhodos nuttallii Pickering ex Rydb. = Chamaerhodos erecta ssp. nuttallii [CHAMERE1]
Chamaerhodos nuttallii var. *keweenawensis* Fern. = Chamaerhodos erecta ssp. nuttallii [CHAMERE1]
Comarum L. [COMARUM$]
Comarum palustre L. [COMAPAL]
SY: *Potentilla palustris* (L.) Scop.
SY: *Potentilla palustris* var. *parvifolia* (Raf.) Fern. & Long
SY: *Potentilla palustris* var. *villosa* (Pers.) Lehm.
Cotoneaster Medik. [COTONEA$]
*****Cotoneaster bullatus** Boiss. [COTOBUL] (bullate-leaved cotoneaster)
*****Cotoneaster horizontalis** Dcne. [COTOHOR] (rock cotoneaster)
*****Cotoneaster simonsii** Baker [COTOSIM] (Simons' cotoneaster)
Crataegus L. [CRATAEG$]
Crataegus columbiana T.J. Howell [CRATCOL] (red hawthorn)
Crataegus douglasii Lindl. [CRATDOU] (black hawthorn)
Crataegus douglasii var. *suksdorfii* Sarg. = Crataegus suksdorfii [CRATSUK]
*****Crataegus monogyna** Jacq. [CRATMON] (common hawthorn)
Crataegus suksdorfii (Sarg.) Kruschke [CRATSUK] (Suksdorf's hawthorn)
SY: *Crataegus douglasii* var. *suksdorfii* Sarg.
Cylactis arctica ssp. *acaulis* (Michx.) W.A. Weber = Rubus arcticus ssp. acaulis [RUBUARC1]
Dryas L. [DRYAS$]
Dryas alaskensis Porsild = Dryas octopetala ssp. alaskensis [DRYAOCT1]
Dryas drummondii Richards. ex Hook. [DRYADRU] (yellow mountain-avens)
Dryas drummondii var. **drummondii** [DRYADRU0]
Dryas drummondii var. **eglandulosa** Porsild [DRYADRU2]
Dryas drummondii var. **tomentosa** (Farr) L.O. Williams [DRYADRU3]
Dryas hookeriana Juz. = Dryas octopetala ssp. hookeriana [DRYAOCT2]
Dryas integrifolia Vahl [DRYAINT] (entire-leaved mountain-avens)
Dryas integrifolia ssp. **integrifolia** [DRYAINT0]
SY: *Dryas integrifolia* var. *canescens* Simm.
SY: *Dryas octopetala* var. *integrifolia* (Vahl) Hook. f.
Dryas integrifolia ssp. **sylvatica** (Hultén) Hultén [DRYAINT2]
SY: *Dryas sylvatica* (Hultén) Porsild
Dryas integrifolia var. *canescens* Simm. = Dryas integrifolia ssp. integrifolia [DRYAINT0]
Dryas octopetala L. [DRYAOCT] (white mountain-avens)
Dryas octopetala ssp. **alaskensis** (Porsild) Hultén [DRYAOCT1]
SY: *Dryas alaskensis* Porsild
Dryas octopetala ssp. **hookeriana** (Juz.) Hultén [DRYAOCT2]
SY: *Dryas hookeriana* Juz.
SY: *Dryas octopetala* var. *hookeriana* (Juz.) Breitung
Dryas octopetala ssp. **octopetala** [DRYAOCT0]
Dryas octopetala var. *hookeriana* (Juz.) Breitung = Dryas octopetala ssp. hookeriana [DRYAOCT2]

Dryas octopetala var. *integrifolia* (Vahl) Hook. f. = Dryas integrifolia ssp. integrifolia [DRYAINT0]
Dryas sylvatica (Hultén) Porsild = Dryas integrifolia ssp. sylvatica [DRYAINT2]
Drymocallis glandulosa (Lindl.) Rydb. = Potentilla glandulosa ssp. glandulosa [POTEGLA0]
Duchesnea Sm. [DUCHESN$]
***Duchesnea indica** (Andr.) Focke [DUCHIND] (Indian strawberry)
SY: *Fragaria indica* Andr.
Erythrocoma ciliata Pursh = Geum triflorum var. ciliatum [GEUMTRI1]
Erythrocoma triflora (Pursh) Greene = Geum triflorum var. triflorum [GEUMTRI0]
Fragaria L. [FRAGARI$]
Fragaria americana (Porter) Britt. = Fragaria vesca ssp. americana [FRAGVES1]
Fragaria bracteata Heller = Fragaria vesca ssp. bracteata [FRAGVES2]
Fragaria chiloensis (L.) P. Mill. [FRAGCHI] (coastal strawberry)
Fragaria chiloensis ssp. **lucida** (Vilm.) Staudt [FRAGCHI1]
Fragaria chiloensis ssp. **pacifica** Staudt [FRAGCHI2]
Fragaria glauca (S. Wats.) Rydb. = Fragaria virginiana ssp. glauca [FRAGVIR1]
Fragaria helleri Holz. = Fragaria vesca ssp. bracteata [FRAGVES2]
Fragaria indica Andr. = Duchesnea indica [DUCHIND]
Fragaria pauciflora Rydb. = Fragaria virginiana ssp. glauca [FRAGVIR1]
Fragaria platypetala Rydb. = Fragaria virginiana ssp. platypetala [FRAGVIR2]
Fragaria platypetala var. *sibbaldifolia* (Rydb.) Jepson = Fragaria virginiana ssp. platypetala [FRAGVIR2]
Fragaria sibbaldifolia Rydb. = Fragaria virginiana ssp. platypetala [FRAGVIR2]
Fragaria suksdorfii Rydb. = Fragaria virginiana ssp. platypetala [FRAGVIR2]
Fragaria truncata Rydb. = Fragaria virginiana ssp. platypetala [FRAGVIR2]
Fragaria vesca L. [FRAGVES] (wood strawberry)
Fragaria vesca ssp. **americana** (Porter) Staudt [FRAGVES1]
SY: *Fragaria americana* (Porter) Britt.
SY: *Fragaria vesca* var. *americana* Porter
Fragaria vesca ssp. **bracteata** (Heller) Staudt [FRAGVES2]
SY: *Fragaria bracteata* Heller
SY: *Fragaria helleri* Holz.
SY: *Fragaria vesca* var. *bracteata* (Heller) R.J. Davis
Fragaria vesca var. *americana* Porter = Fragaria vesca ssp. americana [FRAGVES1]
Fragaria vesca var. *bracteata* (Heller) R.J. Davis = Fragaria vesca ssp. bracteata [FRAGVES2]
Fragaria virginiana Duchesne [FRAGVIR] (wild strawberry)
Fragaria virginiana ssp. **glauca** (S. Wats.) Staudt [FRAGVIR1] (blue-leaved strawberry)
SY: *Fragaria glauca* (S. Wats.) Rydb.
SY: *Fragaria pauciflora* Rydb.
SY: *Fragaria virginiana* var. *glauca* S. Wats.
SY: *Fragaria virginiana* var. *terrae-novae* (Rydb.) Fern. & Wieg.
Fragaria virginiana ssp. **platypetala** (Rydb.) Staudt [FRAGVIR2]
SY: *Fragaria platypetala* Rydb.
SY: *Fragaria platypetala* var. *sibbaldifolia* (Rydb.) Jepson
SY: *Fragaria sibbaldifolia* Rydb.
SY: *Fragaria suksdorfii* Rydb.
SY: *Fragaria truncata* Rydb.
SY: *Fragaria virginiana* var. *platypetala* (Rydb.) Hall
Fragaria virginiana var. *glauca* S. Wats. = Fragaria virginiana ssp. glauca [FRAGVIR1]
Fragaria virginiana var. *platypetala* (Rydb.) Hall = Fragaria virginiana ssp. platypetala [FRAGVIR2]
Fragaria virginiana var. *terrae-novae* (Rydb.) Fern. & Wieg. = Fragaria virginiana ssp. glauca [FRAGVIR1]
Geum L. [GEUM$]
Geum aleppicum Jacq. [GEUMALE] (yellow avens)
SY: *Geum aleppicum* ssp. *strictum* (Ait.) Clausen
SY: *Geum aleppicum* var. *strictum* (Ait.) Fern.
SY: *Geum strictum* Ait.
SY: *Geum strictum* var. *decurrens* (Rydb.) Kearney & Peebles
Geum aleppicum ssp. *strictum* (Ait.) Clausen = Geum aleppicum [GEUMALE]
Geum aleppicum var. *strictum* (Ait.) Fern. = Geum aleppicum [GEUMALE]
Geum calthifolium Menzies ex Sm. [GEUMCAL] (caltha-leaved avens)
SY: *Acomastylis calthifolia* (Menzies ex Sm.) Bolle
Geum ciliatum Pursh = Geum triflorum var. ciliatum [GEUMTRI1]
Geum ciliatum var. *griseum* (Greene) Kearney & Peebles = Geum triflorum var. triflorum [GEUMTRI0]
Geum × macranthum (Kearney) Boivin [GEUMMAR]
SY: *Acomastylis macrantha* (Kearney) Bolle
SY: *Geum schofieldii* Calder & Taylor
SY: *Sieversia × macrantha* Kearney
Geum macrophyllum Willd. [GEUMMAC] (large-leaved avens)
Geum macrophyllum ssp. *perincisum* (Rydb.) Hultén = Geum macrophyllum var. perincisum [GEUMMAC1]
Geum macrophyllum var. **macrophyllum** [GEUMMAC0]
Geum macrophyllum var. **perincisum** (Rydb.) Raup [GEUMMAC1]
SY: *Geum macrophyllum* ssp. *perincisum* (Rydb.) Hultén
SY: *Geum macrophyllum* var. *rydbergii* Farw.
SY: *Geum oregonense* (Scheutz) Rydb.
SY: *Geum perincisum* Rydb.
SY: *Geum perincisum* var. *intermedium* Boivin
Geum macrophyllum var. *rydbergii* Farw. = Geum macrophyllum var. perincisum [GEUMMAC1]
Geum oregonense (Scheutz) Rydb. = Geum macrophyllum var. perincisum [GEUMMAC1]
Geum perincisum Rydb. = Geum macrophyllum var. perincisum [GEUMMAC1]
Geum perincisum var. *intermedium* Boivin = Geum macrophyllum var. perincisum [GEUMMAC1]
Geum rivale L. [GEUMRIV] (water avens)
Geum rossii (R. Br.) Ser. [GEUMROS] (Ross' avens)
Geum schofieldii Calder & Taylor = Geum × macranthum [GEUMMAR]
Geum strictum Ait. = Geum aleppicum [GEUMALE]
Geum strictum var. *decurrens* (Rydb.) Kearney & Peebles = Geum aleppicum [GEUMALE]
Geum triflorum Pursh [GEUMTRI] (three-flowered avens)
Geum triflorum var. **ciliatum** (Pursh) Fassett [GEUMTRI1]
SY: *Erythrocoma ciliata* Pursh
SY: *Geum ciliatum* Pursh
SY: *Sieversia ciliata* (Pursh) G. Don
Geum triflorum var. **triflorum** [GEUMTRI0]
SY: *Erythrocoma triflora* (Pursh) Greene
SY: *Geum ciliatum* var. *griseum* (Greene) Kearney & Peebles
SY: *Sieversia triflora* (Pursh) R. Br.
Holodiscus (K. Koch) Maxim. [HOLODIS$]
Holodiscus boursieri (Carr.) Rehd. = Holodiscus discolor [HOLODIC]
Holodiscus discolor (Pursh) Maxim. [HOLODIC] (ocean-spray)
SY: *Holodiscus boursieri* (Carr.) Rehd.
SY: *Holodiscus discolor* var. *delnortensis* Ley
SY: *Holodiscus discolor* ssp. *franciscanus* (Rydb.) Taylor & MacBryde
SY: *Holodiscus discolor* var. *franciscanus* (Rydb.) Jepson
SY: *Holodiscus discolor* var. *glabrescens* (Greenm.) Jepson
SY: *Holodiscus dumosus* var. *australis* (Heller) Ley
SY: *Holodiscus dumosus* var. *glabrescens* (Greenm.) C.L. Hichc.
SY: *Holodiscus dumosus* ssp. *saxicola* (Heller) Abrams
SY: *Holodiscus glabrescens* (Greenm.) Heller ex Jepson
SY: *Holodiscus microphyllus* Rydb.
SY: *Holodiscus microphyllus* var. *glabrescens* (Greenm.) Ley
SY: *Holodiscus microphyllus* var. *sericeus* Ley
SY: *Holodiscus microphyllus* var. *typicus* Ley
SY: *Sericotheca discolor* (Pursh) Rydb.
SY: *Spiraea discolor* Pursh
Holodiscus discolor ssp. *franciscanus* (Rydb.) Taylor & MacBryde = Holodiscus discolor [HOLODIC]

Holodiscus discolor var. *delnortensis* Ley = Holodiscus discolor [HOLODIC]
Holodiscus discolor var. *franciscanus* (Rydb.) Jepson = Holodiscus discolor [HOLODIC]
Holodiscus discolor var. *glabrescens* (Greenm.) Jepson = Holodiscus discolor [HOLODIC]
Holodiscus dumosus ssp. *saxicola* (Heller) Abrams = Holodiscus discolor [HOLODIC]
Holodiscus dumosus var. *australis* (Heller) Ley = Holodiscus discolor [HOLODIC]
Holodiscus dumosus var. *glabrescens* (Greenm.) C.L. Hichc. = Holodiscus discolor [HOLODIC]
Holodiscus glabrescens (Greenm.) Heller ex Jepson = Holodiscus discolor [HOLODIC]
Holodiscus microphyllus Rydb. = Holodiscus discolor [HOLODIC]
Holodiscus microphyllus var. *glabrescens* (Greenm.) Ley = Holodiscus discolor [HOLODIC]
Holodiscus microphyllus var. *sericeus* Ley = Holodiscus discolor [HOLODIC]
Holodiscus microphyllus var. *typicus* Ley = Holodiscus discolor [HOLODIC]
Luetkea Bong. [LUETKEA$]
Luetkea pectinata (Pursh) Kuntze [LUETPEC] (partridgefoot)
SY: *Saxifraga pectinata* Pursh
Malus P. Mill. [MALUS$]
Malus communis Poir. = Malus pumila [MALUPUM]
Malus diversifolia (Bong.) M. Roemer = Malus fusca [MALUFUS]
Malus domestica (Borkh.) Borkh. = Malus pumila [MALUPUM]
Malus fusca (Raf.) Schneid. [MALUFUS] (Pacific crab apple)
SY: *Malus diversifolia* (Bong.) M. Roemer
SY: *Malus fusca* var. *levipes* (Nutt.) Schneid.
SY: *Pyrus diversifolia* Bong.
SY: *Pyrus fusca* Raf.
SY: *Pyrus rivularis* Dougl. ex Hook.
Malus fusca var. *levipes* (Nutt.) Schneid. = Malus fusca [MALUFUS]
Malus malus (L.) Britt. = Malus sylvestris [MALUSYL]
*__Malus pumila__ P. Mill. [MALUPUM] (cultivated apple)
SY: *Malus communis* Poir.
SY: *Malus domestica* (Borkh.) Borkh.
SY: *Pyrus pumila* (P. Mill.) K. Koch
Malus sylvestris P. Mill. [MALUSYL]
SY: *Malus malus* (L.) Britt.
SY: *Pyrus malus* L.
Nuttallia cerasiformis Torr. & Gray ex Hook. & Arn. = Oemleria cerasiformis [OEMLCER]
Oemleria Reichenb. [OEMLERI$]
Oemleria cerasiformis (Torr. & Gray ex Hook. & Arn.) Landon [OEMLCER] (Indian-plum)
SY: *Nuttallia cerasiformis* Torr. & Gray ex Hook. & Arn.
SY: *Osmaronia cerasiformis* (Torr. & Gray ex Hook. & Arn.) Greene
Osmaronia cerasiformis (Torr. & Gray ex Hook. & Arn.) Greene = Oemleria cerasiformis [OEMLCER]
Padus melanocarpa (A. Nels.) Shafer = Prunus virginiana var. melanocarpa [PRUNVIR2]
Padus virginiana ssp. *melanocarpa* (A. Nels.) W.A. Weber = Prunus virginiana var. melanocarpa [PRUNVIR2]
Pentaphylloides Duham. [PENTAPH$]
Pentaphylloides floribunda (Pursh) A. Löve [PENTFLO]
SY: *Potentilla floribunda* Pursh
SY: *Potentilla fruticosa* ssp. *floribunda* (Pursh) Elkington
SY: *Potentilla fruticosa* var. *tenuifolia* Lehm.
SY: *Potentilla fruticosa* auct.
Physocarpus Maxim. [PHYSOCA$]
Physocarpus capitatus (Pursh) Kuntze [PHYSCAP] (Pacific ninebark)
SY: *Physocarpus opulifolius* var. *tomentellus* (Ser.) Boivin
Physocarpus malvaceus (Greene) Kuntze [PHYSMAL] (mallow ninebark)
Physocarpus opulifolius var. *tomentellus* (Ser.) Boivin = Physocarpus capitatus [PHYSCAP]
Potentilla L. [POTENTI$]
Potentilla altaica Bunge = Potentilla nivea var. pentaphylla [POTENIV1]
Potentilla angustata Rydb. = Potentilla gracilis var. nuttallii [POTEGRA1]
Potentilla anomalofolia M.E. Peck = Potentilla drummondii [POTEDRU]
Potentilla anserina L. = Argentina anserina [ARGEANS]
Potentilla anserina ssp. *egedii* (Wormsk.) Hiitonen = Argentina egedii [ARGEEGE]
Potentilla anserina ssp. *pacifica* (T.J. Howell) Rousi = Argentina egedii [ARGEEGE]
Potentilla anserina var. *concolor* Ser. = Argentina anserina [ARGEANS]
Potentilla anserina var. *grandis* Torr. & Gray = Argentina egedii [ARGEEGE]
Potentilla anserina var. *lanata* Boivin = Argentina egedii [ARGEEGE]
Potentilla anserina var. *rolandii* (Boivin) Boivin = Argentina egedii [ARGEEGE]
Potentilla anserina var. *sericea* (L.) Hayne = Argentina anserina [ARGEANS]
Potentilla anserina var. *yukonensis* (Hultén) Boivin = Argentina anserina [ARGEANS]
*__Potentilla argentea__ L. [POTEARG] (silvery cinquefoil)
Potentilla arguta Pursh [POTEARU] (white cinquefoil)
Potentilla arguta ssp. **convallaria** (Rydb.) Keck [POTEARU2]
SY: *Potentilla arguta* var. *convallaria* (Rydb.) T. Wolf
SY: *Potentilla convallaria* Rydb.
Potentilla arguta var. *convallaria* (Rydb.) T. Wolf = Potentilla arguta ssp. convallaria [POTEARU2]
Potentilla biennis Greene [POTEBIE] (biennial cinquefoil)
Potentilla biflora Willd. ex Schlecht. [POTEBIF] (two-flowered cinquefoil)
Potentilla bipinnatifida Dougl. ex Hook. [POTEBIP] (bipinnate cinquefoil)
SY: *Potentilla pensylvanica* var. *bipinnatifida* (Dougl. ex Hook.) Torr. & Gray
Potentilla bipinnatifida var. **glabrata** (Lehm. ex Hook.) Kohli & Packer [POTEBIP2]
SY: *Potentilla glabella* Rydb.
SY: *Potentilla pensylvanica* var. *glabrata* (Lehm. ex Hook.) S. Wats.
Potentilla blasckeana Turcz. ex Lehm. = Potentilla gracilis var. nuttallii [POTEGRA1]
Potentilla blasckeana var. *permollis* (Rydb.) Wolf = Potentilla gracilis var. nuttallii [POTEGRA1]
Potentilla camporum Rydb. = Potentilla pulcherrima [POTEPUC]
Potentilla convallaria Rydb. = Potentilla arguta ssp. convallaria [POTEARU2]
Potentilla diversifolia Lehm. [POTEDIV] (diverse-leaved cinquefoil)
Potentilla diversifolia ssp. *glaucophylla* Lehm. = Potentilla diversifolia var. diversifolia [POTEDIV0]
Potentilla diversifolia var. **diversifolia** [POTEDIV0]
SY: *Potentilla diversifolia* ssp. *glaucophylla* Lehm.
SY: *Potentilla diversifolia* var. *glaucophylla* (Lehm.) S. Wats.
Potentilla diversifolia var. *glaucophylla* (Lehm.) S. Wats. = Potentilla diversifolia var. diversifolia [POTEDIV0]
Potentilla diversifolia var. **perdissecta** (Rydb.) C.L. Hitchc. [POTEDIV2]
Potentilla drummondii Lehm. [POTEDRU] (Drummond's cinquefoil)
SY: *Potentilla anomalofolia* M.E. Peck
Potentilla egedii Wormsk. = Argentina egedii [ARGEEGE]
Potentilla egedii ssp. *grandis* (Torr. & Gray) Hultén = Argentina egedii [ARGEEGE]
Potentilla egedii ssp. *yukonensis* (Hultén) Hultén = Argentina anserina [ARGEANS]
Potentilla egedii var. *grandis* (Torr. & Gray) J.T. Howell = Argentina egedii [ARGEEGE]
Potentilla elegans Cham. & Schlecht. [POTEELE] (elegant cinquefoil)
Potentilla emarginata Pursh = Potentilla nana [POTENAN]

Potentilla etomentosa Rydb. = Potentilla gracilis var. nuttallii [POTEGRA1]
Potentilla etomentosa var. *hallii* (Rydb.) Abrams = Potentilla gracilis var. nuttallii [POTEGRA1]
Potentilla fastigiata Nutt. = Potentilla gracilis var. nuttallii [POTEGRA1]
Potentilla flabellifolia Hook. ex Torr. & Gray [POTEFLA] (fan-leaved cinquefoil)
Potentilla flabellifolia var. *emarginata* (Pursh) Boivin = Potentilla nana [POTENAN]
Potentilla flabelliformis Lehm. [POTEFLB]
SY: *Potentilla gracilis* var. *flabelliformis* (Lehm.) Nutt. ex Torr. & Gray
Potentilla floribunda Pursh = Pentaphylloides floribunda [PENTFLO]
Potentilla fruticosa auct. = Pentaphylloides floribunda [PENTFLO]
Potentilla fruticosa ssp. *floribunda* (Pursh) Elkington = Pentaphylloides floribunda [PENTFLO]
Potentilla fruticosa var. *tenuifolia* Lehm. = Pentaphylloides floribunda [PENTFLO]
Potentilla glabella Rydb. = Potentilla bipinnatifida var. glabrata [POTEBIP2]
Potentilla glandulosa Lindl. [POTEGLA] (sticky cinquefoil)
Potentilla glandulosa ssp. **glandulosa** [POTEGLA0]
SY: *Drymocallis glandulosa* (Lindl.) Rydb.
SY: *Potentilla glandulosa* var. *campanulata* C.L. Hitchc.
SY: *Potentilla glandulosa* var. *incisa* Lindl.
SY: *Potentilla glandulosa* ssp. *typica* Keck
SY: *Potentilla rhomboidea* Rydb.
Potentilla glandulosa ssp. **pseudorupestris** (Rydb.) Keck [POTEGLA2]
SY: *Potentilla glandulosa* var. *pseudorupestris* (Rydb.) Breitung
Potentilla glandulosa ssp. *typica* Keck = Potentilla glandulosa ssp. glandulosa [POTEGLA0]
Potentilla glandulosa var. *campanulata* C.L. Hitchc. = Potentilla glandulosa ssp. glandulosa [POTEGLA0]
Potentilla glandulosa var. *incisa* Lindl. = Potentilla glandulosa ssp. glandulosa [POTEGLA0]
Potentilla glandulosa var. *pseudorupestris* (Rydb.) Breitung = Potentilla glandulosa ssp. pseudorupestris [POTEGLA2]
Potentilla glomerata A. Nels. = Potentilla gracilis var. nuttallii [POTEGRA1]
Potentilla gracilis Dougl. ex Hook. [POTEGRA] (graceful cinquefoil)
Potentilla gracilis ssp. *nuttallii* (Lehm.) Keck = Potentilla gracilis var. nuttallii [POTEGRA1]
Potentilla gracilis var. *blasckeana* (Turcz. ex Lehm.) Jepson = Potentilla gracilis var. nuttallii [POTEGRA1]
Potentilla gracilis var. *flabelliformis* (Lehm.) Nutt. ex Torr. & Gray = Potentilla flabelliformis [POTEFLB]
Potentilla gracilis var. *glabrata* (Lehm.) C.L. Hitchc. = Potentilla gracilis var. nuttallii [POTEGRA1]
Potentilla gracilis var. **gracilis** [POTEGRA0]
SY: *Potentilla longipedunculata* Rydb.
SY: *Potentilla macropetala* Rydb.
Potentilla gracilis var. **nuttallii** (Lehm.) Sheldon [POTEGRA1]
SY: *Potentilla angustata* Rydb.
SY: *Potentilla blasckeana* Turcz. ex Lehm.
SY: *Potentilla blasckeana* var. *permollis* (Rydb.) Wolf
SY: *Potentilla etomentosa* Rydb.
SY: *Potentilla etomentosa* var. *hallii* (Rydb.) Abrams
SY: *Potentilla fastigiata* Nutt.
SY: *Potentilla glomerata* A. Nels.
SY: *Potentilla gracilis* var. *blasckeana* (Turcz. ex Lehm.) Jepson
SY: *Potentilla gracilis* var. *glabrata* (Lehm.) C.L. Hitchc.
SY: *Potentilla gracilis* ssp. *nuttallii* (Lehm.) Keck
SY: *Potentilla gracilis* var. *permollis* (Rydb.) C.L. Hitchc.
SY: *Potentilla gracilis* var. *rigida* S. Wats.
SY: *Potentilla indiges* M.E. Peck
SY: *Potentilla jucunda* A. Nels.
SY: *Potentilla nuttallii* Lehm.
SY: *Potentilla permollis* Rydb.
SY: *Potentilla rectiformis* Rydb.
SY: *Potentilla viridescens* Rydb.
Potentilla gracilis var. *permollis* (Rydb.) C.L. Hitchc. = Potentilla gracilis var. nuttallii [POTEGRA1]
Potentilla gracilis var. *pulcherrima* (Lehm.) Fern. = Potentilla pulcherrima [POTEPUC]
Potentilla gracilis var. *rigida* S. Wats. = Potentilla gracilis var. nuttallii [POTEGRA1]
Potentilla hippiana Lehm. [POTEHIP] (woolly cinquefoil)
Potentilla hookeriana Lehm. [POTEHOO] (Hooker's cinquefoil)
Potentilla hyparctica Malte = Potentilla nana [POTENAN]
Potentilla hyparctica ssp. *nana* (Willd. ex Schlecht.) Hultén = Potentilla nana [POTENAN]
Potentilla hyparctica var. *elatior* (Abrom.) Fern. = Potentilla nana [POTENAN]
Potentilla indiges M.E. Peck = Potentilla gracilis var. nuttallii [POTEGRA1]
Potentilla jucunda A. Nels. = Potentilla gracilis var. nuttallii [POTEGRA1]
Potentilla ledebouriana Porsild = Potentilla uniflora [POTEUNI]
Potentilla longipedunculata Rydb. = Potentilla gracilis var. gracilis [POTEGRA0]
Potentilla macropetala Rydb. = Potentilla gracilis var. gracilis [POTEGRA0]
Potentilla nana Willd. ex Schlecht. [POTENAN]
SY: *Potentilla emarginata* Pursh
SY: *Potentilla flabellifolia* var. *emarginata* (Pursh) Boivin
SY: *Potentilla hyparctica* Malte
SY: *Potentilla hyparctica* var. *elatior* (Abrom.) Fern.
SY: *Potentilla hyparctica* ssp. *nana* (Willd. ex Schlecht.) Hultén
Potentilla nicolletii (S. Wats.) Sheldon = Potentilla paradoxa [POTEPAR]
Potentilla nivea L. [POTENIV] (snow cinquefoil)
Potentilla nivea ssp. *chionodes* Hiitonen = Potentilla nivea var. pentaphylla [POTENIV1]
Potentilla nivea ssp. *subquinata* (Lange) Hultén = Potentilla nivea var. pentaphylla [POTENIV1]
Potentilla nivea var. *macrophylla* Ser. = Potentilla nivea var. pentaphylla [POTENIV1]
Potentilla nivea var. **pentaphylla** Lehm. [POTENIV1]
SY: *Potentilla altaica* Bunge
SY: *Potentilla nivea* ssp. *chionodes* Hiitonen
SY: *Potentilla nivea* var. *macrophylla* Ser.
SY: *Potentilla nivea* ssp. *subquinata* (Lange) Hultén
SY: *Potentilla nivea* var. *subquinata* Lange
SY: *Potentilla quinquefolia* (Rydb.) Rydb.
Potentilla nivea var. *subquinata* Lange = Potentilla nivea var. pentaphylla [POTENIV1]
Potentilla nivea var. *villosa* (Pallas ex Pursh) Regel & Tiling = Potentilla villosa [POTEVIL]
Potentilla norvegica L. [POTENOR] (Norwegian cinquefoil)
Potentilla nuttallii Lehm. = Potentilla gracilis var. nuttallii [POTEGRA1]
Potentilla ovina Macoun [POTEOVI] (sheep cinquefoil)
Potentilla pacifica T.J. Howell = Argentina egedii [ARGEEGE]
Potentilla palustris (L.) Scop. = Comarum palustre [COMAPAL]
Potentilla palustris var. *parvifolia* (Raf.) Fern. & Long = Comarum palustre [COMAPAL]
Potentilla palustris var. *villosa* (Pers.) Lehm. = Comarum palustre [COMAPAL]
Potentilla paradoxa Nutt. [POTEPAR] (bushy cinquefoil)
SY: *Potentilla nicolletii* (S. Wats.) Sheldon
SY: *Potentilla supina* ssp. *paradoxa* (Nutt.) Soják
Potentilla pensylvanica L. [POTEPEN] (Pennsylvania cinquefoil)
Potentilla pensylvanica var. *bipinnatifida* (Dougl. ex Hook.) Torr. & Gray = Potentilla bipinnatifida [POTEBIP]
Potentilla pensylvanica var. *glabrata* (Lehm. ex Hook.) S. Wats. = Potentilla bipinnatifida var. glabrata [POTEBIP2]
Potentilla permollis Rydb. = Potentilla gracilis var. nuttallii [POTEGRA1]
Potentilla pulcherrima Lehm. [POTEPUC]
SY: *Potentilla camporum* Rydb.
SY: *Potentilla gracilis* var. *pulcherrima* (Lehm.) Fern.
Potentilla quinquefolia (Rydb.) Rydb. = Potentilla nivea var. pentaphylla [POTENIV1]

*•**Potentilla recta** L. [POTEREC] (sulphur cinquefoil)
SY: *Potentilla recta* var. *obscura* (Nestler) W.D.J. Koch
SY: *Potentilla recta* var. *pilosa* (Willd.) Ledeb.
SY: *Potentilla recta* var. *sulphurea* (Lam. & DC.) Peyr.
Potentilla recta var. *obscura* (Nestler) W.D.J. Koch = Potentilla recta [POTEREC]
Potentilla recta var. *pilosa* (Willd.) Ledeb. = Potentilla recta [POTEREC]
Potentilla recta var. *sulphurea* (Lam. & DC.) Peyr. = Potentilla recta [POTEREC]
Potentilla rectiformis Rydb. = Potentilla gracilis var. nuttallii [POTEGRA1]
Potentilla rhomboidea Rydb. = Potentilla glandulosa ssp. glandulosa [POTEGLA0]
Potentilla rivalis Nutt. [POTERIV] (brook cinquefoil)
Potentilla rolandii Boivin = Argentina egedii [ARGEEGE]
Potentilla sibbaldii Haller f. = Sibbaldia procumbens [SIBBPRO]
Potentilla supina ssp. *paradoxa* (Nutt.) Soják = Potentilla paradoxa [POTEPAR]
Potentilla uniflora Ledeb. [POTEUNI] (one-flowered cinquefoil)
SY: *Potentilla ledebouriana* Porsild
Potentilla villosa Pallas ex Pursh [POTEVIL] (villous cinquefoil)
SY: *Potentilla nivea* var. *villosa* (Pallas ex Pursh) Regel & Tiling
SY: *Potentilla villosa* var. *parviflora* C.L. Hitchc.
Potentilla villosa var. *parviflora* C.L. Hitchc. = Potentilla villosa [POTEVIL]
Potentilla viridescens Rydb. = Potentilla gracilis var. nuttallii [POTEGRA1]
Potentilla yukonensis Hultén = Argentina anserina [ARGEANS]
Prunus L. [PRUNUS$]
•**Prunus avium** (L.) L. [PRUNAVI] (sweet cherry)
SY: *Cerasus avium* (L.) Moench
Prunus demissa (Nutt.) Walp. = Prunus virginiana var. demissa [PRUNVIR1]
Prunus emarginata (Dougl. ex Hook.) Walp. [PRUNEMA] (bitter cherry)
•**Prunus laurocerasus** L. [PRUNLAU] (cherry-laurel)
SY: *Cerasus laurocerasus* (L.) Loisel.
•**Prunus mahaleb** L. [PRUNMAH] (mahaleb cherry)
Prunus melanocarpa (A. Nels.) Rydb. = Prunus virginiana var. melanocarpa [PRUNVIR2]
Prunus pensylvanica L. f. [PRUNPEN] (pin cherry)
•**Prunus spinosa** L. [PRUNSPI] (sloe)
Prunus virginiana L. [PRUNVIR] (choke cherry)
Prunus virginiana ssp. *demissa* (Nutt.) Taylor & MacBryde = Prunus virginiana var. demissa [PRUNVIR1]
Prunus virginiana ssp. *melanocarpa* (A. Nels.) Taylor & MacBryde = Prunus virginiana var. melanocarpa [PRUNVIR2]
Prunus virginiana var. **demissa** (Nutt.) Torr. [PRUNVIR1]
SY: *Prunus demissa* (Nutt.) Walp.
SY: *Prunus virginiana* ssp. *demissa* (Nutt.) Taylor & MacBryde
Prunus virginiana var. **melanocarpa** (A. Nels.) Sarg. [PRUNVIR2]
SY: *Padus melanocarpa* (A. Nels.) Shafer
SY: *Padus virginiana* ssp. *melanocarpa* (A. Nels.) W.A. Weber
SY: *Prunus melanocarpa* (A. Nels.) Rydb.
SY: *Prunus virginiana* ssp. *melanocarpa* (A. Nels.) Taylor & MacBryde
Purshia DC. ex Poir. [PURSHIA$]
Purshia tridentata (Pursh) DC. [PURSTRI] (antelope-brush)
Pyrus L. [PYRUS$]
Pyrus aucuparia (L.) Gaertn. = Sorbus aucuparia [SORBAUC]
•**Pyrus communis** L. [PYRUCOM] (common pear)
Pyrus diversifolia Bong. = Malus fusca [MALUFUS]
Pyrus fusca Raf. = Malus fusca [MALUFUS]
Pyrus malus L. = Malus sylvestris [MALUSYL]
Pyrus pumila (P. Mill.) K. Koch = Malus pumila [MALUPUM]
Pyrus rivularis Dougl. ex Hook. = Malus fusca [MALUFUS]
Pyrus sitchensis (M. Roemer) Piper = Sorbus sitchensis var. sitchensis [SORBSIT0]
Rosa L. [ROSA$]
Rosa acicularis Lindl. [ROSAACI] (prickly rose)
Rosa acicularis ssp. **sayi** (Schwein.) W.H. Lewis [ROSAACI1]
SY: *Rosa acicularis* var. *bourgeauiana* (Crépin) Crépin
SY: *Rosa acicularis* var. *sayana* Erlanson
SY: *Rosa bourgeauiana* Crépin
SY: *Rosa collaris* Rydb.
SY: *Rosa engelmannii* S. Wats.
SY: *Rosa sayi* Schwein.
Rosa acicularis var. *bourgeauiana* (Crépin) Crépin = Rosa acicularis ssp. sayi [ROSAACI1]
Rosa acicularis var. *sayana* Erlanson = Rosa acicularis ssp. sayi [ROSAACI1]
Rosa anatonensis St. John = Rosa nutkana var. hispida [ROSANUT1]
Rosa arizonica Rydb. = Rosa woodsii var. ultramontana [ROSAWOO1]
Rosa arizonica var. *granulifera* (Rydb.) Kearney & Peebles = Rosa woodsii var. ultramontana [ROSAWOO1]
Rosa arkansana Porter [ROSAARK] (Arkansas rose)
Rosa bourgeauiana Crépin = Rosa acicularis ssp. sayi [ROSAACI1]
Rosa caeruleimontana St. John = Rosa nutkana var. hispida [ROSANUT1]
•**Rosa canina** L. [ROSACAN] (dog rose)
SY: *Rosa canina* var. *dumetorum* Baker
Rosa canina var. *dumetorum* Baker = Rosa canina [ROSACAN]
Rosa collaris Rydb. = Rosa acicularis ssp. sayi [ROSAACI1]
Rosa covillei Greene = Rosa woodsii var. ultramontana [ROSAWOO1]
Rosa durandii Crépin = Rosa nutkana var. nutkana [ROSANUT0]
•**Rosa eglanteria** L. [ROSAEGL] (sweetbrier)
SY: *Rosa rubiginosa* L.
Rosa engelmannii S. Wats. = Rosa acicularis ssp. sayi [ROSAACI1]
Rosa gymnocarpa Nutt. [ROSAGYM] (baldhip rose)
Rosa jonesii St. John = Rosa nutkana var. hispida [ROSANUT1]
Rosa lapwaiensis St. John = Rosa woodsii var. ultramontana [ROSAWOO1]
Rosa macdougalii Holz. = Rosa nutkana var. hispida [ROSANUT1]
Rosa macounii Greene = Rosa woodsii var. ultramontana [ROSAWOO1]
Rosa megalantha G.N. Jones = Rosa nutkana var. hispida [ROSANUT1]
Rosa nutkana K. Presl [ROSANUT] (Nootka rose)
Rosa nutkana var. **hispida** Fern. [ROSANUT1]
SY: *Rosa anatonensis* St. John
SY: *Rosa caeruleimontana* St. John
SY: *Rosa jonesii* St. John
SY: *Rosa macdougalii* Holz.
SY: *Rosa megalantha* G.N. Jones
SY: *Rosa spaldingii* Crépin
SY: *Rosa spaldingii* var. *alta* (Suksdorf) G.N. Jones
SY: *Rosa spaldingii* var. *hispida* (Fern.) G.N. Jones
SY: *Rosa spaldingii* var. *parkeri* (S. Wats.) St. John
Rosa nutkana var. **nutkana** [ROSANUT0]
SY: *Rosa durandii* Crépin
Rosa pecosensis Cockerell = Rosa woodsii var. ultramontana [ROSAWOO1]
Rosa pisocarpa Gray [ROSAPIS] (clustered wild rose)
SY: *Rosa pisocarpa* var. *rivalis* (Eastw.) Jepson
SY: *Rosa rivalis* Eastw.
Rosa pisocarpa var. *rivalis* (Eastw.) Jepson = Rosa pisocarpa [ROSAPIS]
Rosa rivalis Eastw. = Rosa pisocarpa [ROSAPIS]
Rosa rubiginosa L. = Rosa eglanteria [ROSAEGL]
Rosa sayi Schwein. = Rosa acicularis ssp. sayi [ROSAACI1]
Rosa spaldingii Crépin = Rosa nutkana var. hispida [ROSANUT1]
Rosa spaldingii var. *alta* (Suksdorf) G.N. Jones = Rosa nutkana var. hispida [ROSANUT1]
Rosa spaldingii var. *hispida* (Fern.) G.N. Jones = Rosa nutkana var. hispida [ROSANUT1]
Rosa spaldingii var. *parkeri* (S. Wats.) St. John = Rosa nutkana var. hispida [ROSANUT1]
Rosa ultramontana (S. Wats.) Heller = Rosa woodsii var. ultramontana [ROSAWOO1]
Rosa woodsii Lindl. [ROSAWOO] (Woods' rose)
Rosa woodsii ssp. *ultramontana* (S. Wats.) Taylor & MacBryde = Rosa woodsii var. ultramontana [ROSAWOO1]

Rosa woodsii var. *arizonica* (Rydb.) W.C. Martin & C.R. Hutchins = Rosa woodsii var. ultramontana [ROSAWOO1]
Rosa woodsii var. *granulifera* (Rydb.) W.C. Martin & C.R. Hutchins = Rosa woodsii var. ultramontana [ROSAWOO1]
Rosa woodsii var. *macounii* (Greene) W.C. Martin & C.R. Hutchins = Rosa woodsii var. ultramontana [ROSAWOO1]
Rosa woodsii var. **ultramontana** (S. Wats.) Jepson [ROSAWOO1]
SY: *Rosa arizonica* Rydb.
SY: *Rosa arizonica* var. *granulifera* (Rydb.) Kearney & Peebles
SY: *Rosa covillei* Greene
SY: *Rosa lapwaiensis* St. John
SY: *Rosa macounii* Greene
SY: *Rosa pecosensis* Cockerell
SY: *Rosa ultramontana* (S. Wats.) Heller
SY: *Rosa woodsii* var. *arizonica* (Rydb.) W.C. Martin & C.R. Hutchins
SY: *Rosa woodsii* var. *granulifera* (Rydb.) W.C. Martin & C.R. Hutchins
SY: *Rosa woodsii* var. *macounii* (Greene) W.C. Martin & C.R. Hutchins
SY: *Rosa woodsii* ssp. *ultramontana* (S. Wats.) Taylor & MacBryde
Rubus L. [RUBUS$]
Rubus acaulis Michx. = Rubus arcticus ssp. acaulis [RUBUARC1]
*__Rubus allegheniensis__ Porter [RUBUALL] (Allegheny blackberry)
Rubus arcticus L. [RUBUARC] (nagoonberry)
Rubus arcticus ssp. **acaulis** (Michx.) Focke [RUBUARC1] (dwarf raspberry)
SY: *Cylactis arctica* ssp. *acaulis* (Michx.) W.A. Weber
SY: *Rubus acaulis* Michx.
SY: *Rubus arcticus* var. *acaulis* (Michx.) Boivin
Rubus arcticus ssp. **stellatus** (Sm.) Boivin [RUBUARC2]
SY: *Rubus arcticus* var. *stellatus* (Sm.) Boivin
SY: *Rubus stellatus* Sm.
Rubus arcticus var. *acaulis* (Michx.) Boivin = Rubus arcticus ssp. acaulis [RUBUARC1]
Rubus arcticus var. *stellatus* (Sm.) Boivin = Rubus arcticus ssp. stellatus [RUBUARC2]
Rubus carolinianus Rydb. = Rubus idaeus ssp. strigosus [RUBUIDA1]
Rubus chamaemorus L. [RUBUCHA] (cloudberry)
*__Rubus discolor__ Weihe & Nees [RUBUDIS] (Himalayan blackberry)
SY: *Rubus procerus* P.J. Muell.
Rubus idaeus L. [RUBUIDA] (red raspberry)
Rubus idaeus ssp. *melanolasius* (Dieck) Focke = Rubus idaeus ssp. strigosus [RUBUIDA1]
Rubus idaeus ssp. *sachalinensis* (Levl.) Focke = Rubus idaeus ssp. strigosus [RUBUIDA1]
Rubus idaeus ssp. **strigosus** (Michx.) Focke [RUBUIDA1]
SY: *Rubus carolinianus* Rydb.
SY: *Rubus idaeus* var. *aculeatissimus* Regel & Tiling
SY: *Rubus idaeus* var. *canadensis* Richards.
SY: *Rubus idaeus* var. *gracilipes* M.E. Jones
SY: *Rubus idaeus* ssp. *melanolasius* (Dieck) Focke
SY: *Rubus idaeus* var. *melanolasius* (Dieck) R.J. Davis
SY: *Rubus idaeus* var. *melanotrachys* (Focke) Fern.
SY: *Rubus idaeus* ssp. *sachalinensis* (Levl.) Focke
SY: *Rubus idaeus* var. *strigosus* (Michx.) Maxim.
SY: *Rubus melanolasius* Dieck
SY: *Rubus neglectus* Peck
SY: *Rubus strigosus* Michx.
SY: *Rubus strigosus* var. *acalyphacea* (Greene) Bailey
SY: *Rubus strigosus* var. *arizonicus* (Greene) Kearney & Peebles
SY: *Rubus strigosus* var. *canadensis* (Richards.) House
Rubus idaeus var. *aculeatissimus* Regel & Tiling = Rubus idaeus ssp. strigosus [RUBUIDA1]
Rubus idaeus var. *canadensis* Richards. = Rubus idaeus ssp. strigosus [RUBUIDA1]
Rubus idaeus var. *gracilipes* M.E. Jones = Rubus idaeus ssp. strigosus [RUBUIDA1]
Rubus idaeus var. *melanolasius* (Dieck) R.J. Davis = Rubus idaeus ssp. strigosus [RUBUIDA1]
Rubus idaeus var. *melanotrachys* (Focke) Fern. = Rubus idaeus ssp. strigosus [RUBUIDA1]
Rubus idaeus var. *strigosus* (Michx.) Maxim. = Rubus idaeus ssp. strigosus [RUBUIDA1]
*__Rubus laciniatus__ Willd. [RUBULAC] (evergreen blackberry)
Rubus lasiococcus Gray [RUBULAS] (dwarf bramble)
Rubus leucodermis Dougl. ex Torr. & Gray [RUBULEU] (black raspberry)
Rubus macropetalus Dougl. ex Hook. = Rubus ursinus ssp. macropetalus [RUBUURS1]
Rubus melanolasius Dieck = Rubus idaeus ssp. strigosus [RUBUIDA1]
Rubus neglectus Peck = Rubus idaeus ssp. strigosus [RUBUIDA1]
Rubus nivalis Dougl. ex Hook. [RUBUNIV] (snow bramble)
Rubus parviflorus Nutt. [RUBUPAR] (thimbleberry)
Rubus pedatus Sm. [RUBUPED] (five-leaved bramble)
Rubus procerus P.J. Muell. = Rubus discolor [RUBUDIS]
Rubus pubescens Raf. [RUBUPUB] (trailing raspberry)
Rubus spectabilis Pursh [RUBUSPE] (salmonberry)
Rubus stellatus Sm. = Rubus arcticus ssp. stellatus [RUBUARC2]
Rubus strigosus Michx. = Rubus idaeus ssp. strigosus [RUBUIDA1]
Rubus strigosus var. *acalyphacea* (Greene) Bailey = Rubus idaeus ssp. strigosus [RUBUIDA1]
Rubus strigosus var. *arizonicus* (Greene) Kearney & Peebles = Rubus idaeus ssp. strigosus [RUBUIDA1]
Rubus strigosus var. *canadensis* (Richards.) House = Rubus idaeus ssp. strigosus [RUBUIDA1]
Rubus ursinus Cham. & Schlecht. [RUBUURS] (trailing blackberry)
Rubus ursinus ssp. **macropetalus** (Dougl. ex Hook.) Taylor & MacBryde [RUBUURS1]
SY: *Rubus macropetalus* Dougl. ex Hook.
SY: *Rubus ursinus* var. *macropetalus* (Dougl. ex Hook.) S.W. Br.
Rubus ursinus var. *macropetalus* (Dougl. ex Hook.) S.W. Br. = Rubus ursinus ssp. macropetalus [RUBUURS1]
Sanguisorba L. [SANGUIS$]
Sanguisorba canadensis L. [SANGCAN] (Sitka burnet)
SY: *Sanguisorba canadensis* ssp. *latifolia* (Hook.) Calder & Taylor
SY: *Sanguisorba canadensis* var. *latifolia* Hook.
SY: *Sanguisorba sitchensis* C.A. Mey.
SY: *Sanguisorba stipulata* Raf.
Sanguisorba canadensis ssp. *latifolia* (Hook.) Calder & Taylor = Sanguisorba canadensis [SANGCAN]
Sanguisorba canadensis var. *latifolia* Hook. = Sanguisorba canadensis [SANGCAN]
Sanguisorba menziesii Rydb. [SANGMEN] (Menzies' burnet)
Sanguisorba microcephala K. Presl = Sanguisorba officinalis [SANGOFF]
*__Sanguisorba minor__ Scop. [SANGMIN] (salad burnet)
Sanguisorba occidentalis Nutt. [SANGOCC] (western burnet)
Sanguisorba officinalis L. [SANGOFF] (great burnet)
SY: *Sanguisorba microcephala* K. Presl
SY: *Sanguisorba officinalis* ssp. *microcephala* (K. Presl) Calder & Taylor
SY: *Sanguisorba officinalis* var. *polygama* (W. Nyl.) Mela & Caj.
Sanguisorba officinalis ssp. *microcephala* (K. Presl) Calder & Taylor = Sanguisorba officinalis [SANGOFF]
Sanguisorba officinalis var. *polygama* (W. Nyl.) Mela & Caj. = Sanguisorba officinalis [SANGOFF]
Sanguisorba sitchensis C.A. Mey. = Sanguisorba canadensis [SANGCAN]
Sanguisorba stipulata Raf. = Sanguisorba canadensis [SANGCAN]
Saxifraga pectinata Pursh = Luetkea pectinata [LUETPEC]
Sericotheca discolor (Pursh) Rydb. = Holodiscus discolor [HOLODIC]
Sibbaldia L. [SIBBALD$]
Sibbaldia procumbens L. [SIBBPRO] (sibbaldia)
SY: *Potentilla sibbaldii* Haller f.
Sieversia ciliata (Pursh) G. Don = Geum triflorum var. ciliatum [GEUMTRI1]
Sieversia × *macrantha* Kearney = Geum × macranthum [GEUMMAR]

Sieversia triflora (Pursh) R. Br. = Geum triflorum var. triflorum [GEUMTRI0]
Sorbus L. [SORBUS$]
*****Sorbus aucuparia** L. [SORBAUC] (European mountain-ash)
SY: *Pyrus aucuparia* (L.) Gaertn.
Sorbus cascadensis G.N. Jones = Sorbus scopulina var. cascadensis [SORBSCO1]
Sorbus occidentalis (S. Wats.) Greene = Sorbus sitchensis var. grayi [SORBSIT3]
Sorbus scopulina Greene [SORBSCO] (western mountain-ash)
Sorbus scopulina var. **cascadensis** (G.N. Jones) C.L. Hitchc. [SORBSCO1]
SY: *Sorbus cascadensis* G.N. Jones
Sorbus scopulina var. **scopulina** [SORBSCO0]
Sorbus sitchensis M. Roemer [SORBSIT] (Sitka mountain-ash)
Sorbus sitchensis ssp. *grayi* (Wenzig) Calder & Taylor = Sorbus sitchensis var. grayi [SORBSIT3]
Sorbus sitchensis var. **grayi** (Wenzig) C.L. Hitchc. [SORBSIT3]
SY: *Sorbus occidentalis* (S. Wats.) Greene
SY: *Sorbus sitchensis* ssp. *grayi* (Wenzig) Calder & Taylor
Sorbus sitchensis var. **sitchensis** [SORBSIT0]
SY: *Pyrus sitchensis* (M. Roemer) Piper
Spiraea L. [SPIRAEA$]
Spiraea aruncus L. = Aruncus dioicus var. vulgaris [ARUNDIO1]
Spiraea beauverdiana auct. = Spiraea stevenii [SPIRSTE]
Spiraea beauverdiana var. *stevenii* Schneid. = Spiraea stevenii [SPIRSTE]
Spiraea betulifolia Pallas [SPIRBET] (birch-leaved spirea)
Spiraea betulifolia ssp. *lucida* (Dougl. ex Greene) Taylor & MacBryde = Spiraea betulifolia var. lucida [SPIRBET2]
Spiraea betulifolia var. **lucida** (Dougl. ex Greene) C.L. Hitchc. [SPIRBET2]
SY: *Spiraea betulifolia* ssp. *lucida* (Dougl. ex Greene) Taylor & MacBryde
SY: *Spiraea lucida* Dougl. ex Greene
Spiraea densiflora Nutt. ex Greenm. = Spiraea splendens [SPIRSPL]
Spiraea densiflora ssp. *splendens* (Baumann ex K. Koch) Abrams = Spiraea splendens [SPIRSPL]
Spiraea densiflora var. *splendens* (Baumann ex K. Koch) C.L. Hitchc. = Spiraea splendens [SPIRSPL]
Spiraea discolor Pursh = Holodiscus discolor [HOLODIC]
Spiraea douglasii Hook. [SPIRDOU] (hardhack)
Spiraea douglasii ssp. *menziesii* (Hook.) Calder & Taylor = Spiraea douglasii var. menziesii [SPIRDOU1]
Spiraea douglasii var. **douglasii** [SPIRDOU0]
SY: *Spiraea douglasii* var. *roseata* (Rydb.) C.L. Hitchc.
Spiraea douglasii var. **menziesii** (Hook.) K. Presl [SPIRDOU1]
SY: *Spiraea douglasii* ssp. *menziesii* (Hook.) Calder & Taylor
SY: *Spiraea menziesii* Hook.
SY: *Spiraea subvillosa* Rydb.
Spiraea douglasii var. *roseata* (Rydb.) C.L. Hitchc. = Spiraea douglasii var. douglasii [SPIRDOU0]
Spiraea lucida Dougl. ex Greene = Spiraea betulifolia var. lucida [SPIRBET2]
Spiraea menziesii Hook. = Spiraea douglasii var. menziesii [SPIRDOU1]
Spiraea × pyramidata Greene (pro sp.) [SPIRPYR]
SY: *Spiraea pyramidata* Greene
SY: *Spiraea tomentulosa* Rydb.
Spiraea pyramidata Greene = Spiraea × pyramidata [SPIRPYR]
Spiraea splendens Baumann ex K. Koch [SPIRSPL]
SY: *Spiraea densiflora* Nutt. ex Greenm.
SY: *Spiraea densiflora* ssp. *splendens* (Baumann ex K. Koch) Abrams
SY: *Spiraea densiflora* var. *splendens* (Baumann ex K. Koch) C.L. Hitchc.
Spiraea stevenii (Schneid.) Rydb. [SPIRSTE] (Steven's spirea)
SY: *Spiraea beauverdiana* var. *stevenii* Schneid.
SY: *Spiraea beauverdiana* auct.
Spiraea subvillosa Rydb. = Spiraea douglasii var. menziesii [SPIRDOU1]
Spiraea tomentulosa Rydb. = Spiraea × pyramidata [SPIRPYR]

RUBIACEAE (V:D)

Asperula odorata L. = Galium odoratum [GALIODO]
Galium L. [GALIUM$]
Galium agreste var. *echinospermum* Wallr. = Galium aparine [GALIAPA]
Galium aparine L. [GALIAPA] (cleavers)
SY: *Galium agreste* var. *echinospermum* Wallr.
SY: *Galium aparine* var. *echinospermum* (Wallr.) Farw.
SY: *Galium aparine* var. *intermedium* (Merr.) Briq.
SY: *Galium aparine* var. *minor* Hook.
SY: *Galium aparine* var. *vaillantii* (DC.) Koch
SY: *Galium spurium* var. *echinospermum* (Wallr.) Hayek
SY: *Galium spurium* var. *vaillantii* (DC.) Gren. & Godr.
SY: *Galium spurium* var. *vaillantii* (DC.) G. Beck
SY: *Galium vaillantii* DC.
Galium aparine var. *echinospermum* (Wallr.) Farw. = Galium aparine [GALIAPA]
Galium aparine var. *intermedium* (Merr.) Briq. = Galium aparine [GALIAPA]
Galium aparine var. *minor* Hook. = Galium aparine [GALIAPA]
Galium aparine var. *vaillantii* (DC.) Koch = Galium aparine [GALIAPA]
Galium asperrimum var. *asperulum* Gray = Galium mexicanum ssp. asperulum [GALIMEX1]
Galium asperulum (Gray) Rydb. = Galium mexicanum ssp. asperulum [GALIMEX1]
Galium bifolium S. Wats. [GALIBIF] (thin-leaved bedstraw)
Galium boreale L. [GALIBOR] (northern bedstraw)
SY: *Galium boreale* var. *hyssopifolium* (Hoffmann) DC.
SY: *Galium boreale* var. *intermedium* DC.
SY: *Galium boreale* var. *linearifolium* Rydb.
SY: *Galium boreale* var. *scabrum* DC.
SY: *Galium boreale* ssp. *septentrionale* (Roemer & J.A. Schultes) Iltis
SY: *Galium boreale* ssp. *septentrionale* (Roemer & J.A. Schultes) Hara
SY: *Galium boreale* var. *typicum* G. Beck
SY: *Galium hyssopifolium* Hoffmann
SY: *Galium septentrionale* Roemer & J.A. Schultes
SY: *Galium strictum* Torr.
Galium boreale ssp. *septentrionale* (Roemer & J.A. Schultes) Iltis = Galium boreale [GALIBOR]
Galium boreale ssp. *septentrionale* (Roemer & J.A. Schultes) Hara = Galium boreale [GALIBOR]
Galium boreale var. *hyssopifolium* (Hoffmann) DC. = Galium boreale [GALIBOR]
Galium boreale var. *intermedium* DC. = Galium boreale [GALIBOR]
Galium boreale var. *linearifolium* Rydb. = Galium boreale [GALIBOR]
Galium boreale var. *scabrum* DC. = Galium boreale [GALIBOR]
Galium boreale var. *typicum* G. Beck = Galium boreale [GALIBOR]
Galium brachiatum Pursh = Galium triflorum [GALITRF]
Galium brandegei Gray p.p. = Galium trifidum ssp. trifidum [GALITRI0]
Galium claytonii var. *subbiflorum* (Wieg.) Wieg. = Galium trifidum ssp. subbiflorum [GALITRI4]
Galium columbianum Rydb. = Galium trifidum ssp. columbianum [GALITRI3]
Galium cymosum Wieg. = Galium trifidum ssp. columbianum [GALITRI3]
Galium erectum Huds. = Galium mollugo [GALIMOL]
Galium hyssopifolium Hoffmann = Galium boreale [GALIBOR]
Galium kamtschaticum Steller ex J.A. & J.H. Schultes [GALIKAM] (boreal bedstraw)
Galium mexicanum Kunth [GALIMEX] (Mexican bedstraw)
Galium mexicanum ssp. **asperulum** (Gray) Dempster [GALIMEX1] (rough bedstraw)
SY: *Galium asperrimum* var. *asperulum* Gray
SY: *Galium asperulum* (Gray) Rydb.
SY: *Galium mexicanum* var. *asperulum* (Gray) Dempster
Galium mexicanum var. *asperulum* (Gray) Dempster = Galium mexicanum ssp. asperulum [GALIMEX1]

*Galium mollugo L. [GALIMOL] (white bedstraw)
SY: *Galium erectum* Huds.
SY: *Galium mollugo* ssp. *erectum* (Huds.) Briq.
SY: *Galium mollugo* var. *erectum* (Huds.) Domin
Galium mollugo ssp. *erectum* (Huds.) Briq. = Galium mollugo [GALIMOL]
Galium mollugo var. *erectum* (Huds.) Domin = Galium mollugo [GALIMOL]
***Galium odoratum** (L.) Scop. [GALIODO] (sweet woodruff)
SY: *Asperula odorata* L.
Galium pennsylvanicum W. Bart. = Galium triflorum [GALITRF]
Galium septentrionale Roemer & J.A. Schultes = Galium boreale [GALIBOR]
Galium spurium var. *echinospermum* (Wallr.) Hayek = Galium aparine [GALIAPA]
Galium spurium var. *vaillantii* (DC.) G. Beck = Galium aparine [GALIAPA]
Galium spurium var. *vaillantii* (DC.) Gren. & Godr. = Galium aparine [GALIAPA]
Galium strictum Torr. = Galium boreale [GALIBOR]
Galium subbiflorum (Wieg.) Rydb. = Galium trifidum ssp. subbiflorum [GALITRI4]
Galium tinctorium var. *subbiflorum* (Wieg.) Fern. = Galium trifidum ssp. subbiflorum [GALITRI4]
Galium trifidum L. [GALITRI] (small bedstraw)
Galium trifidum ssp. **columbianum** (Rydb.) Hultén [GALITRI3]
SY: *Galium columbianum* Rydb.
SY: *Galium cymosum* Wieg.
SY: *Galium trifidum* ssp. *pacificum* (Wieg.) Piper
SY: *Galium trifidum* var. *pacificum* Wieg.
Galium trifidum ssp. *pacificum* (Wieg.) Piper = Galium trifidum ssp. columbianum [GALITRI3]
Galium trifidum ssp. **subbiflorum** (Wieg.) Piper [GALITRI4]
SY: *Galium claytonii* var. *subbiflorum* (Wieg.) Wieg.
SY: *Galium subbiflorum* (Wieg.) Rydb.
SY: *Galium tinctorium* var. *subbiflorum* (Wieg.) Fern.
SY: *Galium trifidum* var. *pusillum* Gray
SY: *Galium trifidum* var. *subbiflorum* Wieg.
Galium trifidum ssp. **trifidum** [GALITRI0]
SY: *Galium brandegei* Gray p.p.
Galium trifidum var. *pacificum* Wieg. = Galium trifidum ssp. columbianum [GALITRI3]
Galium trifidum var. *pusillum* Gray = Galium trifidum ssp. subbiflorum [GALITRI4]
Galium trifidum var. *subbiflorum* Wieg. = Galium trifidum ssp. subbiflorum [GALITRI4]
Galium triflorum Michx. [GALITRF] (sweet-scented bedstraw)
SY: *Galium brachiatum* Pursh
SY: *Galium pennsylvanicum* W. Bart.
SY: *Galium triflorum* var. *asprelliforme* Fern.
SY: *Galium triflorum* var. *viridiflorum* DC.
Galium triflorum var. *asprelliforme* Fern. = Galium triflorum [GALITRF]
Galium triflorum var. *viridiflorum* DC. = Galium triflorum [GALITRF]
Galium vaillantii DC. = Galium aparine [GALIAPA]
***Galium verum** L. [GALIVER] (yellow bedstraw)
Sherardia L. [SHERARD$]
***Sherardia arvensis** L. [SHERARV] (field madder)

RUPPIACEAE (V:M)

Ruppia L. [RUPPIA$]
Ruppia cirrhosa ssp. *occidentalis* (S. Wats.) A. & D. Löve = Ruppia maritima [RUPPMAR]
Ruppia maritima L. [RUPPMAR] (ditch-grass)
SY: *Ruppia cirrhosa* ssp. *occidentalis* (S. Wats.) A. & D. Löve
SY: *Ruppia maritima* var. *brevirostris* Agardh
SY: *Ruppia maritima* var. *exigua* Fern. & Wieg.
SY: *Ruppia maritima* var. *intermedia* (Thed.) Aschers. & Graebn.
SY: *Ruppia maritima* var. *longipes* Hagstr.
SY: *Ruppia maritima* var. *obliqua* (Schur) Aschers. & Graebn.
SY: *Ruppia maritima* var. *occidentalis* (S. Wats.) Graebn.
SY: *Ruppia maritima* var. *pacifica* St. John & Fosberg
SY: *Ruppia maritima* var. *rostrata* Agardh
SY: *Ruppia maritima* var. *spiralis* Morris
SY: *Ruppia maritima* var. *subcapitata* Fern. & Wieg.
SY: *Ruppia occidentalis* S. Wats.
SY: *Ruppia pectinata* Rydb.
Ruppia maritima var. *brevirostris* Agardh = Ruppia maritima [RUPPMAR]
Ruppia maritima var. *exigua* Fern. & Wieg. = Ruppia maritima [RUPPMAR]
Ruppia maritima var. *intermedia* (Thed.) Aschers. & Graebn. = Ruppia maritima [RUPPMAR]
Ruppia maritima var. *longipes* Hagstr. = Ruppia maritima [RUPPMAR]
Ruppia maritima var. *obliqua* (Schur) Aschers. & Graebn. = Ruppia maritima [RUPPMAR]
Ruppia maritima var. *occidentalis* (S. Wats.) Graebn. = Ruppia maritima [RUPPMAR]
Ruppia maritima var. *pacifica* St. John & Fosberg = Ruppia maritima [RUPPMAR]
Ruppia maritima var. *rostrata* Agardh = Ruppia maritima [RUPPMAR]
Ruppia maritima var. *spiralis* Morris = Ruppia maritima [RUPPMAR]
Ruppia maritima var. *subcapitata* Fern. & Wieg. = Ruppia maritima [RUPPMAR]
Ruppia occidentalis S. Wats. = Ruppia maritima [RUPPMAR]
Ruppia pectinata Rydb. = Ruppia maritima [RUPPMAR]

SALICACEAE (V:D)

Monilistus monilifera (Ait.) Raf. ex B.D. Jackson = Populus deltoides ssp. monilifera [POPUDEL2]
Populus L. [POPULUS$]
Populus angulata Ait. = Populus deltoides ssp. deltoides [POPUDEL0]
Populus angulata var. *missouriensis* A. Henry = Populus deltoides ssp. deltoides [POPUDEL0]
Populus aurea Tidestrom = Populus tremuloides [POPUTRE]
Populus balsamifera L. [POPUBAL] (balsam poplar)
Populus balsamifera ssp. **balsamifera** [POPUBAL0]
SY: *Populus balsamifera* var. *candicans* (Ait.) Gray
SY: *Populus balsamifera* var. *fernaldiana* Rouleau
SY: *Populus balsamifera* var. *lanceolata* Marsh.
SY: *Populus balsamifera* var. *michauxii* (Dode) A. Henry
SY: *Populus balsamifera* var. *subcordata* Hyl.
SY: *Populus candicans* Ait.
SY: *Populus michauxii* Dode
SY: *Populus tacamahaca* P. Mill.
SY: *Populus tacamahaca* var. *candicans* (Ait.) Stout
SY: *Populus tacamahaca* var. *lanceolata* (Marsh.) Farw.
SY: *Populus tacamahaca* var. *michauxii* (Dode) Farw.
Populus balsamifera ssp. **trichocarpa** (Torr. & Gray ex Hook.) Brayshaw [POPUBAL2] (black cottonwood)
SY: *Populus balsamifera* var. *californica* S. Wats.
SY: *Populus hastata* Dode p.p.
SY: *Populus trichocarpa* Torr. & Gray ex Hook.
SY: *Populus trichocarpa* var. *cupulata* S. Wats.
SY: *Populus trichocarpa* ssp. *hastata* (Dode) Dode p.p.
SY: *Populus trichocarpa* var. *hastata* (Dode) A. Henry p.p.
SY: *Populus trichocarpa* var. *ingrata* (Jepson) Jepson
Populus balsamifera var. *californica* S. Wats. = Populus balsamifera ssp. trichocarpa [POPUBAL2]
Populus balsamifera var. *candicans* (Ait.) Gray = Populus balsamifera ssp. balsamifera [POPUBAL0]
Populus balsamifera var. *fernaldiana* Rouleau = Populus balsamifera ssp. balsamifera [POPUBAL0]
Populus balsamifera var. *lanceolata* Marsh. = Populus balsamifera ssp. balsamifera [POPUBAL0]
Populus balsamifera var. *michauxii* (Dode) A. Henry = Populus balsamifera ssp. balsamifera [POPUBAL0]
Populus balsamifera var. *missouriensis* (A. Henry) Rehd. = Populus deltoides ssp. deltoides [POPUDEL0]

Populus balsamifera var. *pilosa* Sarg. = Populus deltoides ssp. deltoides [POPUDEL0]
Populus balsamifera var. *subcordata* Hyl. = Populus balsamifera ssp. balsamifera [POPUBAL0]
Populus balsamifera var. *virginiana* (Foug.) Sarg. = Populus deltoides ssp. deltoides [POPUDEL0]
Populus besseyana Dode = Populus deltoides ssp. monilifera [POPUDEL2]
Populus canadensis var. *virginiana* (Foug.) Fiori = Populus deltoides ssp. deltoides [POPUDEL0]
Populus candicans Ait. = Populus balsamifera ssp. balsamifera [POPUBAL0]
Populus cercidiphylla Britt. = Populus tremuloides [POPUTRE]
*__Populus deltoides__ Bartr. ex Marsh. [POPUDEL]
*__Populus deltoides__ ssp. **deltoides** [POPUDEL0] (southern cottonwood)
SY: *Populus angulata* Ait.
SY: *Populus angulata* var. *missouriensis* A. Henry
SY: *Populus balsamifera* var. *missouriensis* (A. Henry) Rehd.
SY: *Populus balsamifera* var. *pilosa* Sarg.
SY: *Populus balsamifera* var. *virginiana* (Foug.) Sarg.
SY: *Populus canadensis* var. *virginiana* (Foug.) Fiori
SY: *Populus deltoides* var. *angulata* (Ait.) Sarg.
SY: *Populus deltoides* var. *missouriensis* (A. Henry) A. Henry
SY: *Populus deltoides* var. *pilosa* (Sarg.) Sudworth
SY: *Populus deltoides* var. *virginiana* (Foug.) Sudworth
SY: *Populus nigra* var. *virginiana* (Foug.) Castigl.
SY: *Populus palmeri* Sarg.
SY: *Populus virginiana* Foug.
SY: *Populus virginiana* var. *pilosa* (Sarg.) F.C. Gates
*__Populus deltoides__ ssp. **monilifera** (Ait.) Eckenwalder [POPUDEL2] (plains cottonwood)
SY: *Monilistus monilifera* (Ait.) Raf. ex B.D. Jackson
SY: *Populus besseyana* Dode
SY: *Populus deltoides* var. *occidentalis* Rydb.
SY: *Populus monilifera* Ait.
SY: *Populus occidentalis* (Rydb.) Britt. ex Rydb.
SY: *Populus sargentii* Dode
SY: *Populus sargentii* var. *texana* (Sarg.) Correll
SY: *Populus texana* Sarg.
Populus deltoides var. *angulata* (Ait.) Sarg. = Populus deltoides ssp. deltoides [POPUDEL0]
Populus deltoides var. *missouriensis* (A. Henry) A. Henry = Populus deltoides ssp. deltoides [POPUDEL0]
Populus deltoides var. *occidentalis* Rydb. = Populus deltoides ssp. monilifera [POPUDEL2]
Populus deltoides var. *pilosa* (Sarg.) Sudworth = Populus deltoides ssp. deltoides [POPUDEL0]
Populus deltoides var. *virginiana* (Foug.) Sudworth = Populus deltoides ssp. deltoides [POPUDEL0]
Populus hastata Dode p.p. = Populus balsamifera ssp. trichocarpa [POPUBAL2]
Populus michauxii Dode = Populus balsamifera ssp. balsamifera [POPUBAL0]
Populus monilifera Ait. = Populus deltoides ssp. monilifera [POPUDEL2]
Populus nigra var. *virginiana* (Foug.) Castigl. = Populus deltoides ssp. deltoides [POPUDEL0]
Populus occidentalis (Rydb.) Britt. ex Rydb. = Populus deltoides ssp. monilifera [POPUDEL2]
Populus palmeri Sarg. = Populus deltoides ssp. deltoides [POPUDEL0]
Populus × polygonifolia Bernard = Populus tremuloides [POPUTRE]
Populus sargentii Dode = Populus deltoides ssp. monilifera [POPUDEL2]
Populus sargentii var. *texana* (Sarg.) Correll = Populus deltoides ssp. monilifera [POPUDEL2]
Populus tacamahaca P. Mill. = Populus balsamifera ssp. balsamifera [POPUBAL0]
Populus tacamahaca var. *candicans* (Ait.) Stout = Populus balsamifera ssp. balsamifera [POPUBAL0]
Populus tacamahaca var. *lanceolata* (Marsh.) Farw. = Populus balsamifera ssp. balsamifera [POPUBAL0]
Populus tacamahaca var. *michauxii* (Dode) Farw. = Populus balsamifera ssp. balsamifera [POPUBAL0]
Populus texana Sarg. = Populus deltoides ssp. monilifera [POPUDEL2]
Populus tremula ssp. *tremuloides* (Michx.) A. & D. Löve = Populus tremuloides [POPUTRE]
Populus tremuloides Michx. [POPUTRE] (trembling aspen)
SY: *Populus aurea* Tidestrom
SY: *Populus cercidiphylla* Britt.
SY: *Populus × polygonifolia* Bernard
SY: *Populus tremula* ssp. *tremuloides* (Michx.) A. & D. Löve
SY: *Populus tremuloides* var. *aurea* (Tidestrom) Daniels
SY: *Populus tremuloides* var. *cercidiphylla* (Britt.) Sudworth
SY: *Populus tremuloides* var. *intermedia* Victorin
SY: *Populus tremuloides* var. *magnifica* Victorin
SY: *Populus tremuloides* var. *rhomboidea* Victorin
SY: *Populus tremuloides* var. *vancouveriana* (Trel.) Sarg.
SY: *Populus vancouveriana* Trel.
Populus tremuloides var. *aurea* (Tidestrom) Daniels = Populus tremuloides [POPUTRE]
Populus tremuloides var. *cercidiphylla* (Britt.) Sudworth = Populus tremuloides [POPUTRE]
Populus tremuloides var. *intermedia* Victorin = Populus tremuloides [POPUTRE]
Populus tremuloides var. *magnifica* Victorin = Populus tremuloides [POPUTRE]
Populus tremuloides var. *rhomboidea* Victorin = Populus tremuloides [POPUTRE]
Populus tremuloides var. *vancouveriana* (Trel.) Sarg. = Populus tremuloides [POPUTRE]
Populus trichocarpa Torr. & Gray ex Hook. = Populus balsamifera ssp. trichocarpa [POPUBAL2]
Populus trichocarpa ssp. *hastata* (Dode) Dode p.p. = Populus balsamifera ssp. trichocarpa [POPUBAL2]
Populus trichocarpa var. *cupulata* S. Wats. = Populus balsamifera ssp. trichocarpa [POPUBAL2]
Populus trichocarpa var. *hastata* (Dode) A. Henry p.p. = Populus balsamifera ssp. trichocarpa [POPUBAL2]
Populus trichocarpa var. *ingrata* (Jepson) Jepson = Populus balsamifera ssp. trichocarpa [POPUBAL2]
Populus vancouveriana Trel. = Populus tremuloides [POPUTRE]
Populus virginiana Foug. = Populus deltoides ssp. deltoides [POPUDEL0]
Populus virginiana var. *pilosa* (Sarg.) F.C. Gates = Populus deltoides ssp. deltoides [POPUDEL0]
Salix L. [SALIX$]
Salix alaxensis (Anderss.) Coville [SALIALA] (Alaska willow)
Salix alaxensis ssp. *longistylis* (Rydb.) Hultén = Salix alaxensis var. longistylis [SALIALA2]
Salix alaxensis var. **alaxensis** [SALIALA0]
SY: *Salix alaxensis* var. *obovalifolia* Ball
SY: *Salix speciosa* Hook. & Arn.
SY: *Salix speciosa* var. *alaxensis* Anderss.
Salix alaxensis var. **longistylis** (Rydb.) Schneid. [SALIALA2]
SY: *Salix alaxensis* ssp. *longistylis* (Rydb.) Hultén
SY: *Salix longistylis* Rydb.
Salix alaxensis var. *obovalifolia* Ball = Salix alaxensis var. alaxensis [SALIALA0]
*__Salix alba__ L. [SALIALB] (white willow)
Salix alba ssp. *vitellina* (L.) Arcang. = Salix alba var. vitellina [SALIALB1]
*__Salix alba__ var. **vitellina** (L.) Stokes [SALIALB1]
SY: *Salix alba* ssp. *vitellina* (L.) Arcang.
SY: *Salix vitellina* L.
Salix albertana Rowlee = Salix barrattiana [SALIBAA]
Salix aliena Flod. = Salix setchelliana [SALISET]
Salix amplifolia Coville = Salix hookeriana [SALIHOO]
Salix amygdaloides Anderss. [SALIAMY] (peach-leaf willow)
SY: *Salix amygdaloides* var. *wrightii* (Anderss.) Schneid.
SY: *Salix nigra* var. *amygdaloides* (Anderss.) Anderss.
SY: *Salix nigra* var. *wrightii* (Anderss.) Anderss.
SY: *Salix wrightii* Anderss.

Salix amygdaloides var. *wrightii* (Anderss.) Schneid. = Salix amygdaloides [SALIAMY]
Salix ancorifera Fern. = Salix discolor [SALIDIS]
Salix anglorum var. *antiplasta* Schneid. = Salix arctica [SALIARC]
Salix anglorum var. *araioclada* Schneid. = Salix arctica [SALIARC]
Salix anglorum var. *kophophylla* Schneid. = Salix arctica [SALIARC]
Salix arbusculoides Anderss. [SALIARB] (northern bush willow)
SY: *Salix humillima* Anderss.
SY: *Salix humillima* var. *puberula* (Anderss.) Anderss.
SY: *Salix saskatchevana* von Seem.
Salix arbusculoides var. *glabra* (Anderss.) Anderss. ex Schneid. = Salix planifolia ssp. pulchra [SALIPLA2]
Salix arctica Pallas [SALIARC] (arctic willow)
SY: *Salix anglorum* var. *antiplasta* Schneid.
SY: *Salix anglorum* var. *araioclada* Schneid.
SY: *Salix anglorum* var. *kophophylla* Schneid.
SY: *Salix arctica* R. Br. ex Richards.
SY: *Salix arctica* var. *antiplasta* (Schneid.) Fern.
SY: *Salix arctica* var. *araioclada* (Schneid.) Raup
SY: *Salix arctica* var. *brownei* Anderss.
SY: *Salix arctica* var. *caespitosa* (Kennedy) L. Kelso
SY: *Salix arctica* ssp. *crassijulis* (Trautv.) Skvort.
SY: *Salix arctica* var. *graminifolia* (E.H. Kelso) L. Kelso
SY: *Salix arctica* var. *kophophylla* (Schneid.) Polunin
SY: *Salix arctica* var. *pallasii* (Anderss.) Kurtz
SY: *Salix arctica* ssp. *petraea* (Anderss.) A. & D. Löve & Kapoor
SY: *Salix arctica* var. *petraea* Anderss.
SY: *Salix arctica* var. *petrophila* (Rydb.) L. Kelso
SY: *Salix arctica* ssp. *tortulosa* (Trautv.) Hultén
SY: *Salix arctica* var. *tortulosa* (Trautv.) Raup
SY: *Salix brownei* (Anderss.) Bebb
SY: *Salix brownei* var. *petraea* (Anderss.) Bebb
SY: *Salix caespitosa* Kennedy
SY: *Salix crassijulis* Trautv.
SY: *Salix hudsonensis* Schneid.
SY: *Salix pallasii* Anderss.
SY: *Salix pallasii* var. *crassijulis* (Trautv.) Anderss.
SY: *Salix petrophila* Rydb.
SY: *Salix petrophila* var. *caespitosa* (Kennedy) Schneid.
SY: *Salix tortulosa* Trautv.
Salix arctica R. Br. ex Richards. = Salix arctica [SALIARC]
Salix arctica ssp. *crassijulis* (Trautv.) Skvort. = Salix arctica [SALIARC]
Salix arctica ssp. *petraea* (Anderss.) A. & D. Löve & Kapoor = Salix arctica [SALIARC]
Salix arctica ssp. *tortulosa* (Trautv.) Hultén = Salix arctica [SALIARC]
Salix arctica var. *antiplasta* (Schneid.) Fern. = Salix arctica [SALIARC]
Salix arctica var. *araioclada* (Schneid.) Raup = Salix arctica [SALIARC]
Salix arctica var. *brownei* Anderss. = Salix arctica [SALIARC]
Salix arctica var. *caespitosa* (Kennedy) L. Kelso = Salix arctica [SALIARC]
Salix arctica var. *graminifolia* (E.H. Kelso) L. Kelso = Salix arctica [SALIARC]
Salix arctica var. *kophophylla* (Schneid.) Polunin = Salix arctica [SALIARC]
Salix arctica var. *pallasii* (Anderss.) Kurtz = Salix arctica [SALIARC]
Salix arctica var. *petraea* Anderss. = Salix arctica [SALIARC]
Salix arctica var. *petrophila* (Rydb.) L. Kelso = Salix arctica [SALIARC]
Salix arctica var. *subcordata* (Anderss.) Schneid. = Salix glauca var. villosa [SALIGLA2]
Salix arctica var. *tortulosa* (Trautv.) Raup = Salix arctica [SALIARC]
Salix argophylla Nutt. = Salix exigua [SALIEXI]
Salix arguta Anderss. = Salix lucida ssp. lasiandra [SALILUC2]
Salix arguta var. *alpigena* Anderss. = Salix serissima [SALISER]
Salix arguta var. *erythrocoma* (Anderss.) Anderss. = Salix lucida ssp. lasiandra [SALILUC2]
Salix arguta var. *lasiandra* (Benth.) Anderss. = Salix lucida ssp. lasiandra [SALILUC2]
Salix arguta var. *pallescens* Anderss. = Salix serissima [SALISER]
Salix athabascensis Raup [SALIATH] (Athabasca willow)
SY: *Salix fallax* Raup
SY: *Salix pedicellaris* var. *athabascensis* (Raup) Boivin
Salix austinae Bebb = Salix lemmonii [SALILEM]
Salix balsamifera (Hook.) Barratt ex Anderss. = Salix pyrifolia [SALIPYR]
Salix balsamifera var. *alpestris* Bebb = Salix pyrifolia [SALIPYR]
Salix balsamifera var. *lanceolata* Bebb = Salix pyrifolia [SALIPYR]
Salix balsamifera var. *vegeta* Bebb = Salix pyrifolia [SALIPYR]
Salix barclayi Anderss. [SALIBAR] (Barclay's willow)
SY: *Salix barclayi* var. *angustifolia* (Anderss.) Anderss. ex Schneid.
SY: *Salix barclayi* var. *conjuncta* (Bebb) Ball ex Schneid.
SY: *Salix conjuncta* Bebb
SY: *Salix hoyeriana* Dieck
SY: *Salix pyrolifolia* var. *hoyeriana* (Dieck) Dippel
SY: *Salix regelii* Anderss.
Salix barclayi var. *angustifolia* (Anderss.) Anderss. ex Schneid. = Salix barclayi [SALIBAR]
Salix barclayi var. *commutata* (Bebb) L. Kelso = Salix commutata [SALICOM]
Salix barclayi var. *conjuncta* (Bebb) Ball ex Schneid. = Salix barclayi [SALIBAR]
Salix barclayi var. *hebecarpa* Anderss. = Salix planifolia ssp. pulchra [SALIPLA2]
Salix barclayi var. *pseudomonticola* (Ball) L. Kelso = Salix pseudomonticola [SALIPSE]
Salix barrattiana Hook. [SALIBAA] (Barratt's willow)
SY: *Salix albertana* Rowlee
SY: *Salix barrattiana* var. *angustifolia* Anderss.
SY: *Salix barrattiana* var. *latifolia* Anderss.
SY: *Salix barrattiana* var. *marcescens* Raup
Salix barrattiana var. *angustifolia* Anderss. = Salix barrattiana [SALIBAA]
Salix barrattiana var. *latifolia* Anderss. = Salix barrattiana [SALIBAA]
Salix barrattiana var. *marcescens* Raup = Salix barrattiana [SALIBAA]
Salix barrattiana var. *tweedyi* Bebb ex Rose = Salix tweedyi [SALITWE]
Salix bebbiana Sarg. [SALIBEB] (Bebb's willow)
SY: *Salix bebbiana* var. *capreifolia* (Fern.) Fern.
SY: *Salix bebbiana* var. *depilis* Raup
SY: *Salix bebbiana* var. *luxurians* (Fern.) Fern.
SY: *Salix bebbiana* var. *perrostrata* (Rydb.) Schneid.
SY: *Salix bebbiana* var. *projecta* (Fern.) Schneid.
SY: *Salix depressa* ssp. *rostrata* (Richards.) Hiitonen
SY: *Salix livida* var. *occidentalis* (Anderss.) Gray
SY: *Salix livida* var. *rostrata* (Richards.) Dippel
SY: *Salix perrostrata* Rydb.
SY: *Salix rostrata* Richards.
SY: *Salix rostrata* var. *capreifolia* Fern.
SY: *Salix rostrata* var. *luxurians* Fern.
SY: *Salix rostrata* var. *perrostrata* (Rydb.) Fern.
SY: *Salix rostrata* var. *projecta* Fern.
SY: *Salix starkeana* ssp. *bebbiana* (Sarg.) Youngberg
SY: *Salix vagans* var. *occidentalis* Anderss.
SY: *Salix vagans* var. *rostrata* (Richards.) Anderss.
Salix bebbiana var. *capreifolia* (Fern.) Fern. = Salix bebbiana [SALIBEB]
Salix bebbiana var. *depilis* Raup = Salix bebbiana [SALIBEB]
Salix bebbiana var. *luxurians* (Fern.) Fern. = Salix bebbiana [SALIBEB]
Salix bebbiana var. *perrostrata* (Rydb.) Schneid. = Salix bebbiana [SALIBEB]
Salix bebbiana var. *projecta* (Fern.) Schneid. = Salix bebbiana [SALIBEB]
Salix bella Piper = Salix drummondiana [SALIDRU]
Salix bolanderiana Rowlee = Salix melanopsis [SALIMEL]
Salix boothii Dorn [SALIBOO] (Booth's willow)
SY: *Salix pseudocordata* var. *aequalis* (Anderss.) Ball ex Schneid.
SY: *Salix pseudomyrsinites* var. *aequalis* (Anderss.) Anderss. ex Ball

SY: *Salix pseudocordata* auct.
SY: *Salix pseudomyrsinites* auct.
Salix brachycarpa Nutt. [SALIBRA] (short-fruited willow)
Salix brachycarpa ssp. **brachycarpa** [SALIBRA0]
Salix brachycarpa ssp. **niphoclada** (Rydb.) Argus [SALIBRA2] (snow willow)
SY: *Salix brachycarpa* var. *mexiae* Ball
SY: *Salix glauca* var. *niphoclada* (Rydb.) Wiggins
SY: *Salix muriei* Hultén
SY: *Salix niphoclada* Rydb.
SY: *Salix niphoclada* var. *mexiae* (Ball) Hultén
SY: *Salix niphoclada* var. *muriei* (Hultén) Raup
Salix brachycarpa var. *mexiae* Ball = Salix brachycarpa ssp. niphoclada [SALIBRA2]
Salix brachystachys Benth. = Salix scouleriana [SALISCO]
Salix brachystachys var. *scouleriana* (Barratt ex Hook.) Anderss. = Salix scouleriana [SALISCO]
Salix brownei (Anderss.) Bebb = Salix arctica [SALIARC]
Salix brownei var. *petraea* (Anderss.) Bebb = Salix arctica [SALIARC]
Salix brownii var. *tenera* (Anderss.) M.E. Jones = Salix cascadensis [SALICAS]
Salix caespitosa Kennedy = Salix arctica [SALIARC]
Salix candida Fluegge ex Willd. [SALICAN] (hoary willow)
SY: *Salix candida* var. *denudata* Anderss.
SY: *Salix candida* var. *tomentosa* Anderss.
SY: *Salix candidula* Nieuwl.
Salix candida var. *denudata* Anderss. = Salix candida [SALICAN]
Salix candida var. *tomentosa* Anderss. = Salix candida [SALICAN]
Salix candidula Nieuwl. = Salix candida [SALICAN]
Salix capreoides Anderss. = Salix scouleriana [SALISCO]
Salix cascadensis Cockerell [SALICAS] (Cascade willow)
SY: *Salix brownii* var. *tenera* (Anderss.) M.E. Jones
SY: *Salix cascadensis* var. *thompsonii* Brayshaw
SY: *Salix tenera* Anderss.
Salix cascadensis var. *thompsonii* Brayshaw = Salix cascadensis [SALICAS]
Salix caudata (Nutt.) Heller = Salix lucida ssp. caudata [SALILUC1]
Salix caudata var. *bryantiana* Ball & Braceline = Salix lucida ssp. caudata [SALILUC1]
Salix caudata var. *parvifolia* Ball = Salix lucida ssp. caudata [SALILUC1]
Salix chlorophylla Anderss. = Salix planifolia ssp. planifolia [SALIPLA0]
Salix chlorophylla var. *monica* (Bebb) Flod. = Salix planifolia ssp. planifolia [SALIPLA0]
Salix chlorophylla var. *nelsonii* (Ball) Flod. = Salix planifolia ssp. planifolia [SALIPLA0]
Salix chlorophylla var. *pychnocarpa* (Anderss.) Anderss. = Salix planifolia ssp. planifolia [SALIPLA0]
Salix commutata Bebb [SALICOM] (variable willow)
SY: *Salix barclayi* var. *commutata* (Bebb) L. Kelso
SY: *Salix commutata* var. *denudata* Bebb
SY: *Salix commutata* var. *mixta* Piper
SY: *Salix commutata* var. *puberula* Bebb
SY: *Salix commutata* var. *sericea* Bebb
Salix commutata var. *denudata* Bebb = Salix commutata [SALICOM]
Salix commutata var. *mixta* Piper = Salix commutata [SALICOM]
Salix commutata var. *puberula* Bebb = Salix commutata [SALICOM]
Salix commutata var. *sericea* Bebb = Salix commutata [SALICOM]
Salix conformis Forbes = Salix discolor [SALIDIS]
Salix conjuncta Bebb = Salix barclayi [SALIBAR]
Salix cordata var. *balsamifera* Hook. = Salix pyrifolia [SALIPYR]
Salix cordata var. *mackenzieana* Hook. = Salix prolixa [SALIPRO]
Salix coulteri Anderss. = Salix sitchensis [SALISIT]
Salix covillei Eastw. = Salix drummondiana [SALIDRU]
Salix crassa Barratt = Salix discolor [SALIDIS]
Salix crassijulis Trautv. = Salix arctica [SALIARC]
Salix cuneata Nutt. = Salix sitchensis [SALISIT]
Salix depressa ssp. *rostrata* (Richards.) Hiitonen = Salix bebbiana [SALIBEB]
Salix desertorum Richards. = Salix glauca var. villosa [SALIGLA2]
Salix desertorum var. *elata* Anderss. = Salix glauca var. villosa [SALIGLA2]
Salix discolor Muhl. [SALIDIS] (pussy willow)
SY: *Salix ancorifera* Fern.
SY: *Salix conformis* Forbes
SY: *Salix crassa* Barratt
SY: *Salix discolor* var. *overi* Ball
SY: *Salix discolor* var. *prinoides* (Pursh) Anderss.
SY: *Salix discolor* var. *rigidior* (Anderss.) Schneid.
SY: *Salix fuscata* Pursh
SY: *Salix prinoides* Pursh
SY: *Salix sensitiva* Barratt
SY: *Salix squamata* Rydb.
Salix discolor var. *overi* Ball = Salix discolor [SALIDIS]
Salix discolor var. *prinoides* (Pursh) Anderss. = Salix discolor [SALIDIS]
Salix discolor var. *rigidior* (Anderss.) Schneid. = Salix discolor [SALIDIS]
Salix drummondiana Barratt ex Hook. [SALIDRU] (Drummond's willow)
SY: *Salix bella* Piper
SY: *Salix covillei* Eastw.
SY: *Salix drummondiana* var. *bella* (Piper) Ball
SY: *Salix drummondiana* ssp. *subcoerulea* (Piper) E. Murr.
SY: *Salix drummondiana* var. *subcoerulea* (Piper) Ball
SY: *Salix pachnophora* Rydb.
SY: *Salix subcoerulea* Piper
Salix drummondiana ssp. *subcoerulea* (Piper) E. Murr. = Salix drummondiana [SALIDRU]
Salix drummondiana var. *bella* (Piper) Ball = Salix drummondiana [SALIDRU]
Salix drummondiana var. *subcoerulea* (Piper) Ball = Salix drummondiana [SALIDRU]
Salix eriocephala ssy. prolixa (Anderss.) Argus = Salix prolixa [SALIPRO]
Salix exigua Nutt. [SALIEXI] (sandbar willow)
SY: *Salix argophylla* Nutt.
SY: *Salix exigua* var. *angustissima* (Anderss.) Reveal & Broome
SY: *Salix exigua* var. *exterior* (Fern.) C.F. Reed
SY: *Salix exigua* ssp. *interior* (Rowlee) Cronq.
SY: *Salix exigua* var. *luteosericea* (Rydb.) Schneid.
SY: *Salix exigua* var. *nevadensis* (S. Wats.) Schneid.
SY: *Salix exigua* var. *pedicellata* (Anderss.) Cronq.
SY: *Salix exigua* var. *stenophylla* (Rydb.) Schneid.
SY: *Salix exigua* var. *virens* Rowlee
SY: *Salix fluviatilis* var. *argophylla* (Nutt.) Sarg.
SY: *Salix fluviatilis* var. *sericans* (Nees) Boivin
SY: *Salix hindsiana* var. *tenuifolia* (Anderss.) Anderss.
SY: *Salix interior* Rowlee
SY: *Salix interior* var. *angustissima* (Anderss.) Dayton
SY: *Salix interior* var. *exterior* Fern.
SY: *Salix interior* var. *luteosericea* (Rydb.) Schneid.
SY: *Salix interior* var. *pedicellata* (Anderss.) Ball
SY: *Salix interior* var. *wheeleri* Rowlee
SY: *Salix linearifolia* Rydb.
SY: *Salix longifolia* Muhl. non Lam.
SY: *Salix longifolia* var. *argophylla* (Nutt.) Anderss.
SY: *Salix longifolia* var. *exigua* (Nutt.) Bebb
SY: *Salix longifolia* var. *interior* (Rowlee) M.E. Jones
SY: *Salix longifolia* var. *opaca* Anderss.
SY: *Salix longifolia* var. *pedicellata* Anderss.
SY: *Salix longifolia* var. *sericans* Nees ex Wied-Neuw.
SY: *Salix longifolia* var. *wheeleri* (Rowlee) Schneid.
SY: *Salix luteosericea* Rydb.
SY: *Salix malacophylla* Nutt. ex Ball
SY: *Salix nevadensis* S. Wats.
SY: *Salix rubra* Richards.
SY: *Salix stenophylla* Rydb.
SY: *Salix thurberi* Rowlee
SY: *Salix wheeleri* (Rowlee) Rydb.
Salix exigua ssp. *interior* (Rowlee) Cronq. = Salix exigua [SALIEXI]
Salix exigua ssp. *melanopsis* (Nutt.) Cronq. = Salix melanopsis [SALIMEL]

Salix exigua var. *angustissima* (Anderss.) Reveal & Broome = Salix exigua [SALIEXI]
Salix exigua var. *exterior* (Fern.) C.F. Reed = Salix exigua [SALIEXI]
Salix exigua var. *gracilipes* (Ball) Cronq. = Salix melanopsis [SALIMEL]
Salix exigua var. *luteosericea* (Rydb.) Schneid. = Salix exigua [SALIEXI]
Salix exigua var. *nevadensis* (S. Wats.) Schneid. = Salix exigua [SALIEXI]
Salix exigua var. *parishiana* (Rowlee) Jepson = Salix sessilifolia [SALISES]
Salix exigua var. *pedicellata* (Anderss.) Cronq. = Salix exigua [SALIEXI]
Salix exigua var. *stenophylla* (Rydb.) Schneid. = Salix exigua [SALIEXI]
Salix exigua var. *tenerrima* (Henderson) Schneid. = Salix melanopsis [SALIMEL]
Salix exigua var. *virens* Rowlee = Salix exigua [SALIEXI]
Salix fallax Raup = Salix athabascensis [SALIATH]
Salix farriae Ball [SALIFAR] (Farr's willow)
SY: *Salix farriae* var. *microserrulata* Ball
SY: *Salix hastata* var. *farriae* (Ball) Hultén
Salix farriae var. *microserrulata* Ball = Salix farriae [SALIFAR]
Salix fendleriana Anderss. = Salix lucida ssp. caudata [SALILUC1]
Salix fernaldii Blank. = Salix vestita [SALIVES]
Salix flavescens Nutt. = Salix scouleriana [SALISCO]
Salix flavescens var. *capreoides* (Anderss.) Bebb = Salix scouleriana [SALISCO]
Salix flavescens var. *scouleriana* (Barratt ex Hook.) Bebb = Salix scouleriana [SALISCO]
Salix fluviatilis var. *argophylla* (Nutt.) Sarg. = Salix exigua [SALIEXI]
Salix fluviatilis var. *sericans* (Nees) Boivin = Salix exigua [SALIEXI]
Salix fluviatilis var. *sessilifolia* (Nutt.) Scoggan = Salix sessilifolia [SALISES]
Salix fluviatilis var. *tenerrima* (Henderson) T.J. Howell = Salix melanopsis [SALIMEL]
Salix fulcrata var. *subglauca* Anderss. = Salix planifolia ssp. pulchra [SALIPLA2]
Salix fuscata Pursh = Salix discolor [SALIDIS]
Salix fuscescens var. *hebecarpa* Fern. = Salix pedicellaris [SALIPED]
Salix geyeriana Anderss. [SALIGEY] (Geyer's willow)
SY: *Salix geyeriana* ssp. *argentea* (Bebb) E. Murr.
SY: *Salix geyeriana* var. *argentea* (Bebb) Schneid.
SY: *Salix geyeriana* var. *meleina* J.K. Henry
SY: *Salix macrocarpa* Nutt.
SY: *Salix macrocarpa* var. *argentea* Bebb
SY: *Salix meleina* (J.K. Henry) G.N. Jones
Salix geyeriana ssp. *argentea* (Bebb) E. Murr. = Salix geyeriana [SALIGEY]
Salix geyeriana var. *argentea* (Bebb) Schneid. = Salix geyeriana [SALIGEY]
Salix geyeriana var. *meleina* J.K. Henry = Salix geyeriana [SALIGEY]
Salix glauca L. [SALIGLA] (grey-leaved willow)
Salix glauca ssp. *acutifolia* (Hook.) Hultén = Salix glauca var. acutifolia [SALIGLA1]
Salix glauca ssp. *desertorum* (Richards.) Hultén = Salix glauca var. villosa [SALIGLA2]
Salix glauca ssp. *glabrescens* (Anderss.) Hultén = Salix glauca var. villosa [SALIGLA2]
Salix glauca var. **acutifolia** (Hook.) Schneid. [SALIGLA1]
SY: *Salix glauca* ssp. *acutifolia* (Hook.) Hultén
SY: *Salix glauca* var. *alicea* Ball
SY: *Salix glauca* var. *perstipulata* Raup
SY: *Salix glauca* var. *poliophylla* (Schneid.) Raup
SY: *Salix glauca* var. *seemanii* (Rydb.) Ostenf.
SY: *Salix seemanii* Rydb.
SY: *Salix villosa* var. *acutifolia* Hook.
Salix glauca var. *alicea* Ball = Salix glauca var. acutifolia [SALIGLA1]
Salix glauca var. *glabrescens* (Anderss.) Schneid. = Salix glauca var. villosa [SALIGLA2]
Salix glauca var. *kenosha* (L. Kelso) L. Kelso = Salix glauca var. villosa [SALIGLA2]
Salix glauca var. *niphoclada* (Rydb.) Wiggins = Salix brachycarpa ssp. niphoclada [SALIBRA2]
Salix glauca var. *perstipulata* Raup = Salix glauca var. acutifolia [SALIGLA1]
Salix glauca var. *poliophylla* (Schneid.) Raup = Salix glauca var. acutifolia [SALIGLA1]
Salix glauca var. *pseudolapponum* (von Seem.) L. Kelso = Salix glauca var. villosa [SALIGLA2]
Salix glauca var. *seemanii* (Rydb.) Ostenf. = Salix glauca var. acutifolia [SALIGLA1]
Salix glauca var. *sericea* Hultén = Salix glauca var. villosa [SALIGLA2]
Salix glauca var. *subincurva* (E.H. Kelso) L. Kelso = Salix glauca var. villosa [SALIGLA2]
Salix glauca var. **villosa** (D. Don ex Hook.) Anderss. [SALIGLA2]
SY: *Salix arctica* var. *subcordata* (Anderss.) Schneid.
SY: *Salix desertorum* Richards.
SY: *Salix desertorum* var. *elata* Anderss.
SY: *Salix glauca* ssp. *desertorum* (Richards.) Hultén
SY: *Salix glauca* ssp. *glabrescens* (Anderss.) Hultén
SY: *Salix glauca* var. *glabrescens* (Anderss.) Schneid.
SY: *Salix glauca* var. *kenosha* (L. Kelso) L. Kelso
SY: *Salix glauca* var. *pseudolapponum* (von Seem.) L. Kelso
SY: *Salix glauca* var. *sericea* Hultén
SY: *Salix glauca* var. *subincurva* (E.H. Kelso) L. Kelso
SY: *Salix glaucops* Anderss.
SY: *Salix* × *glaucops* Anderss.
SY: *Salix glaucops* var. *glabrescens* Anderss.
SY: *Salix glaucops* var. *villosa* (D. Don ex Hook.) Anderss.
SY: *Salix nudescens* Rydb.
SY: *Salix pseudolapponum* von Seem.
SY: *Salix pseudolapponum* var. *kenosha* L. Kelso
SY: *Salix pseudolapponum* var. *subincurva* E.H. Kelso
SY: *Salix subcordata* Anderss.
SY: *Salix villosa* D. Don ex Hook.
SY: *Salix wolfii* var. *pseudolapponum* (von Seem.) M.E. Jones
SY: *Salix wyomingensis* Rydb.
Salix glaucops Anderss. = Salix glauca var. villosa [SALIGLA2]
Salix × *glaucops* Anderss. = Salix glauca var. villosa [SALIGLA2]
Salix glaucops var. *glabrescens* Anderss. = Salix glauca var. villosa [SALIGLA2]
Salix glaucops var. *villosa* (D. Don ex Hook.) Anderss. = Salix glauca var. villosa [SALIGLA2]
Salix gracilis Anderss. = Salix petiolaris [SALIPET]
Salix gracilis var. *rosmarinoides* Anderss. = Salix petiolaris [SALIPET]
Salix gracilis var. *textoris* Fern. = Salix petiolaris [SALIPET]
Salix hastata var. *farriae* (Ball) Hultén = Salix farriae [SALIFAR]
Salix hebecarpa (Fern.) Fern. = Salix pedicellaris [SALIPED]
Salix hindsiana var. *parishiana* (Rowlee) Ball = Salix sessilifolia [SALISES]
Salix hindsiana var. *tenuifolia* (Anderss.) Anderss. = Salix exigua [SALIEXI]
Salix hookeriana Barratt ex Hook. [SALIHOO] (Hooker's willow)
SY: *Salix amplifolia* Coville
SY: *Salix hookeriana* var. *laurifolia* J.K. Henry
SY: *Salix hookeriana* var. *tomentosa* J.K. Henry ex Schneid.
SY: *Salix piperi* Bebb
Salix hookeriana var. *laurifolia* J.K. Henry = Salix hookeriana [SALIHOO]
Salix hookeriana var. *tomentosa* J.K. Henry ex Schneid. = Salix hookeriana [SALIHOO]
Salix hoyeriana Dieck = Salix barclayi [SALIBAR]
Salix hudsonensis Schneid. = Salix arctica [SALIARC]
Salix humillima Anderss. = Salix arbusculoides [SALIARB]
Salix humillima var. *puberula* (Anderss.) Anderss. = Salix arbusculoides [SALIARB]
Salix interior Rowlee = Salix exigua [SALIEXI]
Salix interior var. *angustissima* (Anderss.) Dayton = Salix exigua [SALIEXI]
Salix interior var. *exterior* Fern. = Salix exigua [SALIEXI]

Salix interior var. *luteosericea* (Rydb.) Schneid. = Salix exigua [SALIEXI]
Salix interior var. *pedicellata* (Anderss.) Ball = Salix exigua [SALIEXI]
Salix interior var. *wheeleri* Rowlee = Salix exigua [SALIEXI]
Salix lanata L. [SALILAN] (woolly willow)
Salix lanata ssp. **richardsonii** (Hook.) Skvort. [SALILAN1] (Richardson's willow)
SY: *Salix richardsonii* Hook.
SY: *Salix richardsonii* var. *mckeandii* Polunin
Salix lancifolia Anderss. = Salix lucida ssp. lasiandra [SALILUC2]
Salix lasiandra Benth. = Salix lucida ssp. lasiandra [SALILUC2]
Salix lasiandra ssp. *caudata* (Nutt.) E. Murr. = Salix lucida ssp. caudata [SALILUC1]
Salix lasiandra var. *abramsii* Ball = Salix lucida ssp. lasiandra [SALILUC2]
Salix lasiandra var. *caudata* (Nutt.) Sudworth = Salix lucida ssp. caudata [SALILUC1]
Salix lasiandra var. *fendleriana* (Anderss.) Bebb = Salix lucida ssp. caudata [SALILUC1]
Salix lasiandra var. *lancifolia* (Anderss.) Bebb = Salix lucida ssp. lasiandra [SALILUC2]
Salix lasiandra var. *lyallii* Sarg. = Salix lucida ssp. lasiandra [SALILUC2]
Salix lasiandra var. *macrophylla* (Anderss.) Little = Salix lucida ssp. lasiandra [SALILUC2]
Salix lasiandra var. *recomponens* Raup = Salix lucida ssp. lasiandra [SALILUC2]
Salix leiolepis Fern. = Salix vestita [SALIVES]
Salix lemmonii Bebb [SALILEM] (Lemmon's willow)
SY: *Salix austinae* Bebb
SY: *Salix lemmonii* var. *austinae* (Bebb) Schneid.
SY: *Salix lemmonii* var. *macrostachya* Bebb
SY: *Salix lemmonii* var. *melanopsis* Bebb
SY: *Salix lemmonii* var. *sphaerostachya* Bebb
Salix lemmonii var. *austinae* (Bebb) Schneid. = Salix lemmonii [SALILEM]
Salix lemmonii var. *macrostachya* Bebb = Salix lemmonii [SALILEM]
Salix lemmonii var. *melanopsis* Bebb = Salix lemmonii [SALILEM]
Salix lemmonii var. *sphaerostachya* Bebb = Salix lemmonii [SALILEM]
Salix linearifolia Rydb. = Salix exigua [SALIEXI]
Salix lingulata Anderss. = Salix myrtillifolia var. myrtillifolia [SALIMYR0]
Salix livida var. *occidentalis* (Anderss.) Gray = Salix bebbiana [SALIBEB]
Salix livida var. *rostrata* (Richards.) Dippel = Salix bebbiana [SALIBEB]
Salix longifolia Muhl. non Lam. = Salix exigua [SALIEXI]
Salix longifolia var. *argophylla* (Nutt.) Anderss. = Salix exigua [SALIEXI]
Salix longifolia var. *exigua* (Nutt.) Bebb = Salix exigua [SALIEXI]
Salix longifolia var. *interior* (Rowlee) M.E. Jones = Salix exigua [SALIEXI]
Salix longifolia var. *opaca* Anderss. = Salix exigua [SALIEXI]
Salix longifolia var. *pedicellata* Anderss. = Salix exigua [SALIEXI]
Salix longifolia var. *sericans* Nees ex Wied-Neuw. = Salix exigua [SALIEXI]
Salix longifolia var. *sessilifolia* (Nutt.) M.E. Jones = Salix sessilifolia [SALISES]
Salix longifolia var. *tenerrima* Henderson = Salix melanopsis [SALIMEL]
Salix longifolia var. *wheeleri* (Rowlee) Schneid. = Salix exigua [SALIEXI]
Salix longistylis Rydb. = Salix alaxensis var. longistylis [SALIALA2]
Salix lucida Muhl. [SALILUC] (shining willow)
Salix lucida ssp. **caudata** (Nutt.) E. Murr. [SALILUC1] (tail-leaved willow)
SY: *Salix caudata* (Nutt.) Heller
SY: *Salix caudata* var. *bryantiana* Ball & Braceline
SY: *Salix caudata* var. *parvifolia* Ball
SY: *Salix fendleriana* Anderss.
SY: *Salix lasiandra* ssp. *caudata* (Nutt.) E. Murr.
SY: *Salix lasiandra* var. *caudata* (Nutt.) Sudworth
SY: *Salix lasiandra* var. *fendleriana* (Anderss.) Bebb
SY: *Salix pentandra* var. *caudata* Nutt.
Salix lucida ssp. **lasiandra** (Benth.) E. Murr. [SALILUC2] (Pacific willow)
SY: *Salix arguta* Anderss.
SY: *Salix arguta* var. *erythrocoma* (Anderss.) Anderss.
SY: *Salix arguta* var. *lasiandra* (Benth.) Anderss.
SY: *Salix lancifolia* Anderss.
SY: *Salix lasiandra* Benth.
SY: *Salix lasiandra* var. *abramsii* Ball
SY: *Salix lasiandra* var. *lancifolia* (Anderss.) Bebb
SY: *Salix lasiandra* var. *lyallii* Sarg.
SY: *Salix lasiandra* var. *macrophylla* (Anderss.) Little
SY: *Salix lasiandra* var. *recomponens* Raup
SY: *Salix lucida* var. *macrophylla* Anderss.
SY: *Salix lyallii* (Sarg.) Heller
SY: *Salix speciosa* Nutt.
Salix lucida var. *macrophylla* Anderss. = Salix lucida ssp. lasiandra [SALILUC2]
Salix lucida var. *serissima* Bailey = Salix serissima [SALISER]
Salix luteosericea Rydb. = Salix exigua [SALIEXI]
Salix lyallii (Sarg.) Heller = Salix lucida ssp. lasiandra [SALILUC2]
Salix maccalliana Rowlee [SALIMAC] (MacCalla's willow)
Salix mackenzieana (Hook.) Barratt ex Anderss. = Salix prolixa [SALIPRO]
Salix mackenzieana var. *macrogemma* Ball = Salix prolixa [SALIPRO]
Salix macrocarpa Nutt. = Salix geyeriana [SALIGEY]
Salix macrocarpa var. *argentea* Bebb = Salix geyeriana [SALIGEY]
Salix macrostachya Nutt. = Salix sessilifolia [SALISES]
Salix macrostachya var. *cusickii* Rowlee = Salix sessilifolia [SALISES]
Salix malacophylla Nutt. ex Ball = Salix exigua [SALIEXI]
Salix melanopsis Nutt. [SALIMEL] (dusky willow)
SY: *Salix bolanderiana* Rowlee
SY: *Salix exigua* var. *gracilipes* (Ball) Cronq.
SY: *Salix exigua* ssp. *melanopsis* (Nutt.) Cronq.
SY: *Salix exigua* var. *tenerrima* (Henderson) Schneid.
SY: *Salix fluviatilis* var. *tenerrima* (Henderson) T.J. Howell
SY: *Salix longifolia* var. *tenerrima* Henderson
SY: *Salix melanopsis* var. *bolanderiana* (Rowlee) Schneid.
SY: *Salix melanopsis* var. *gracilipes* Ball
SY: *Salix melanopsis* var. *kronkheittii* L. Kelso
SY: *Salix melanopsis* var. *tenerrima* (Henderson) Ball
SY: *Salix parksiana* Ball
SY: *Salix sessilifolia* var. *vancouverensis* Brayshaw
SY: *Salix tenerrima* (Henderson) Heller
Salix melanopsis var. *bolanderiana* (Rowlee) Schneid. = Salix melanopsis [SALIMEL]
Salix melanopsis var. *gracilipes* Ball = Salix melanopsis [SALIMEL]
Salix melanopsis var. *kronkheittii* L. Kelso = Salix melanopsis [SALIMEL]
Salix melanopsis var. *tenerrima* (Henderson) Ball = Salix melanopsis [SALIMEL]
Salix meleina (J.K. Henry) G.N. Jones = Salix geyeriana [SALIGEY]
Salix monica Bebb = Salix planifolia ssp. planifolia [SALIPLA0]
Salix muriei Hultén = Salix brachycarpa ssp. niphoclada [SALIBRA2]
Salix myrsinites var. *curtifolia* Anderss. = Salix myrtillifolia var. myrtillifolia [SALIMYR0]
Salix myrtillifolia Anderss. [SALIMYR] (bilberry willow)
Salix myrtillifolia var. **cordata** (Anderss.) Dorn [SALIMYR1]
SY: *Salix myrtillifolia* var. *pseudomyrsinites* (Anderss.) Ball ex Hultén
SY: *Salix novae-angliae* var. *pseudomyrsinites* (Anderss.) Anderss.
SY: *Salix pseudocordata* var. *cordata* (Anderss.) Ball
SY: *Salix pseudomyrsinites* Anderss.
SY: *Salix novae-angliae* auct.
Salix myrtillifolia var. *curtifolia* (Anderss.) Bebb ex Rose = Salix myrtillifolia var. myrtillifolia [SALIMYR0]

Salix myrtillifolia var. *lingulata* (Anderss.) Ball = Salix myrtillifolia var. myrtillifolia [SALIMYR0]
Salix myrtillifolia var. **myrtillifolia** [SALIMYR0]
SY: *Salix lingulata* Anderss.
SY: *Salix myrsinites* var. *curtifolia* Anderss.
SY: *Salix myrtillifolia* var. *curtifolia* (Anderss.) Bebb ex Rose
SY: *Salix myrtillifolia* var. *lingulata* (Anderss.) Ball
SY: *Salix novae-angliae* var. *myrtillifolia* (Anderss.) Anderss.
SY: *Salix novae-angliae* var. *pseudocordata* Anderss.
SY: *Salix pseudocordata* (Anderss.) Rydb.
Salix myrtillifolia var. *pseudomyrsinites* (Anderss.) Ball ex Hultén = Salix myrtillifolia var. cordata [SALIMYR1]
Salix myrtilloides var. *hypoglauca* (Fern.) Ball = Salix pedicellaris [SALIPED]
Salix myrtilloides var. *pedicellaris* (Pursh) Anderss. = Salix pedicellaris [SALIPED]
Salix nelsonii Ball = Salix planifolia ssp. planifolia [SALIPLA0]
Salix neoforbesii Toepffer = Salix petiolaris [SALIPET]
Salix nevadensis S. Wats. = Salix exigua [SALIEXI]
Salix nigra var. *amygdaloides* (Anderss.) Anderss. = Salix amygdaloides [SALIAMY]
Salix nigra var. *wrightii* (Anderss.) Anderss. = Salix amygdaloides [SALIAMY]
Salix niphoclada Rydb. = Salix brachycarpa ssp. niphoclada [SALIBRA2]
Salix niphoclada var. *mexiae* (Ball) Hultén = Salix brachycarpa ssp. niphoclada [SALIBRA2]
Salix niphoclada var. *muriei* (Hultén) Raup = Salix brachycarpa ssp. niphoclada [SALIBRA2]
Salix nivalis Hook. = Salix reticulata ssp. nivalis [SALIRET2]
Salix nivalis var. *saximontana* (Rydb.) Schneid. = Salix reticulata ssp. nivalis [SALIRET2]
Salix novae-angliae auct. = Salix myrtillifolia var. cordata [SALIMYR1]
Salix novae-angliae var. *myrtillifolia* (Anderss.) Anderss. = Salix myrtillifolia var. myrtillifolia [SALIMYR0]
Salix novae-angliae var. *pseudocordata* Anderss. = Salix myrtillifolia var. myrtillifolia [SALIMYR0]
Salix novae-angliae var. *pseudomyrsinites* (Anderss.) Anderss. = Salix myrtillifolia var. cordata [SALIMYR1]
Salix nudescens Rydb. = Salix glauca var. villosa [SALIGLA2]
Salix nuttallii Sarg. = Salix scouleriana [SALISCO]
Salix nuttallii var. *capreoides* (Anderss.) Sarg. = Salix scouleriana [SALISCO]
Salix orbicularis Anderss. = Salix reticulata ssp. reticulata [SALIRET0]
Salix pachnophora Rydb. = Salix drummondiana [SALIDRU]
Salix pallasii Anderss. = Salix arctica [SALIARC]
Salix pallasii var. *crassijulis* (Trautv.) Anderss. = Salix arctica [SALIARC]
Salix parishiana Rowlee = Salix sessilifolia [SALISES]
Salix parksiana Ball = Salix melanopsis [SALIMEL]
Salix pedicellaris Pursh [SALIPED] (bog willow)
SY: *Salix fuscescens* var. *hebecarpa* Fern.
SY: *Salix hebecarpa* (Fern.) Fern.
SY: *Salix myrtilloides* var. *hypoglauca* (Fern.) Ball
SY: *Salix myrtilloides* var. *pedicellaris* (Pursh) Anderss.
SY: *Salix pedicellaris* var. *hypoglauca* Fern.
SY: *Salix pedicellaris* var. *tenuescens* Fern.
Salix pedicellaris var. *athabascensis* (Raup) Boivin = Salix athabascensis [SALIATH]
Salix pedicellaris var. *hypoglauca* Fern. = Salix pedicellaris [SALIPED]
Salix pedicellaris var. *tenuescens* Fern. = Salix pedicellaris [SALIPED]
Salix pennata Ball = Salix planifolia ssp. planifolia [SALIPLA0]
Salix pentandra var. *caudata* Nutt. = Salix lucida ssp. caudata [SALILUC1]
Salix perrostrata Rydb. = Salix bebbiana [SALIBEB]
Salix petiolaris Sm. [SALIPET] (meadow willow)
SY: *Salix gracilis* Anderss.
SY: *Salix gracilis* var. *rosmarinoides* Anderss.
SY: *Salix gracilis* var. *textoris* Fern.
SY: *Salix neoforbesii* Toepffer
SY: *Salix petiolaris* var. *angustifolia* Anderss.
SY: *Salix petiolaris* var. *gracilis* (Anderss.) Anderss.
SY: *Salix petiolaris* var. *rosmarinoides* (Anderss.) Schneid.
SY: *Salix petiolaris* var. *subsericea* Anderss.
SY: *Salix sericea* var. *subsericea* (Anderss.) Rydb.
SY: *Salix × subsericea* (Anderss.) Schneid.
Salix petiolaris var. *angustifolia* Anderss. = Salix petiolaris [SALIPET]
Salix petiolaris var. *gracilis* (Anderss.) Anderss. = Salix petiolaris [SALIPET]
Salix petiolaris var. *rosmarinoides* (Anderss.) Schneid. = Salix petiolaris [SALIPET]
Salix petiolaris var. *subsericea* Anderss. = Salix petiolaris [SALIPET]
Salix petrophila Rydb. = Salix arctica [SALIARC]
Salix petrophila var. *caespitosa* (Kennedy) Schneid. = Salix arctica [SALIARC]
Salix phylicifolia ssp. *planifolia* (Pursh) Hiitonen = Salix planifolia ssp. planifolia [SALIPLA0]
Salix phylicifolia ssp. *pulchra* (Cham.) Hultén = Salix planifolia ssp. pulchra [SALIPLA2]
Salix phylicifolia var. *monica* (Bebb) Jepson = Salix planifolia ssp. planifolia [SALIPLA0]
Salix phylicifolia var. *pennata* (Ball) Cronq. = Salix planifolia ssp. planifolia [SALIPLA0]
Salix phylicifolia var. *subglauca* (Anderss.) Boivin = Salix planifolia ssp. pulchra [SALIPLA2]
Salix phylicoides Anderss. = Salix planifolia ssp. pulchra [SALIPLA2]
Salix piperi Bebb = Salix hookeriana [SALIHOO]
Salix planifolia Pursh [SALIPLA] (tea-leaved willow)
Salix planifolia ssp. **planifolia** [SALIPLA0]
SY: *Salix chlorophylla* Anderss.
SY: *Salix chlorophylla* var. *monica* (Bebb) Flod.
SY: *Salix chlorophylla* var. *nelsonii* (Ball) Flod.
SY: *Salix chlorophylla* var. *pychnocarpa* (Anderss.) Anderss.
SY: *Salix monica* Bebb
SY: *Salix nelsonii* Ball
SY: *Salix pennata* Ball
SY: *Salix phylicifolia* var. *monica* (Bebb) Jepson
SY: *Salix phylicifolia* var. *pennata* (Ball) Cronq.
SY: *Salix phylicifolia* ssp. *planifolia* (Pursh) Hiitonen
SY: *Salix planifolia* var. *monica* (Bebb) Schneid.
SY: *Salix planifolia* var. *nelsonii* (Ball) Ball ex E.C. Sm.
SY: *Salix planifolia* var. *pennata* (Ball) Ball ex Dutilly, Lepage & Daman
SY: *Salix pychnocarpa* Anderss.
Salix planifolia ssp. **pulchra** (Cham.) Argus [SALIPLA2] (diamond-leaved willow)
SY: *Salix arbusculoides* var. *glabra* (Anderss.) Anderss. ex Schneid.
SY: *Salix barclayi* var. *hebecarpa* Anderss.
SY: *Salix fulcrata* var. *subglauca* Anderss.
SY: *Salix phylicifolia* ssp. *pulchra* (Cham.) Hultén
SY: *Salix phylicifolia* var. *subglauca* (Anderss.) Boivin
SY: *Salix phylicoides* Anderss.
SY: *Salix planifolia* var. *yukonensis* (Schneid.) Argus
SY: *Salix pulchra* Cham.
SY: *Salix pulchra* var. *looffiae* Ball
SY: *Salix pulchra* var. *palmeri* Ball
SY: *Salix pulchra* var. *yukonensis* Schneid.
Salix planifolia var. *monica* (Bebb) Schneid. = Salix planifolia ssp. planifolia [SALIPLA0]
Salix planifolia var. *nelsonii* (Ball) Ball ex E.C. Sm. = Salix planifolia ssp. planifolia [SALIPLA0]
Salix planifolia var. *pennata* (Ball) Ball ex Dutilly, Lepage & Daman = Salix planifolia ssp. planifolia [SALIPLA0]
Salix planifolia var. *yukonensis* (Schneid.) Argus = Salix planifolia ssp. pulchra [SALIPLA2]
Salix polaris Wahlenb. [SALIPOL] (polar willow)
SY: *Salix polaris* var. *glabrata* Hultén
SY: *Salix polaris* ssp. *pseudopolaris* (Flod.) Hultén
SY: *Salix polaris* var. *selwynensis* Raup
SY: *Salix pseudopolaris* Flod.

Salix polaris ssp. *pseudopolaris* (Flod.) Hultén = Salix polaris [SALIPOL]
Salix polaris var. *glabrata* Hultén = Salix polaris [SALIPOL]
Salix polaris var. *selwynensis* Raup = Salix polaris [SALIPOL]
Salix prinoides Pursh = Salix discolor [SALIDIS]
Salix prolixa Anderss. [SALIPRO] (Mackenzie's willow)
SY: *Salix cordata* var. *mackenzieana* Hook.
SY: *Salix eriocephala ssy. prolixa* (Anderss.) Argus
SY: *Salix mackenzieana* (Hook.) Barratt ex Anderss.
SY: *Salix mackenzieana* var. *macrogemma* Ball
SY: *Salix rigida* ssp. *mackenzieana* (Hook.) E. Murr.
SY: *Salix rigida* var. *mackenzieana* (Hook.) Cronq.
SY: *Salix rigida* var. *macrogemma* (Ball) Cronq.
Salix pseudocordata (Anderss.) Rydb. = Salix myrtillifolia var. myrtillifolia [SALIMYR0]
Salix pseudocordata auct. = Salix boothii [SALIBOO]
Salix pseudocordata var. *aequalis* (Anderss.) Ball ex Schneid. = Salix boothii [SALIBOO]
Salix pseudocordata var. *cordata* (Anderss.) Ball = Salix myrtillifolia var. cordata [SALIMYR1]
Salix pseudolapponum von Seem. = Salix glauca var. villosa [SALIGLA2]
Salix pseudolapponum var. *kenosha* L. Kelso = Salix glauca var. villosa [SALIGLA2]
Salix pseudolapponum var. *subincurva* E.H. Kelso = Salix glauca var. villosa [SALIGLA2]
Salix pseudomonticola Ball [SALIPSE] (mountain willow)
SY: *Salix barclayi* var. *pseudomonticola* (Ball) L. Kelso
Salix pseudomyrsinites Anderss. = Salix myrtillifolia var. cordata [SALIMYR1]
Salix pseudomyrsinites auct. = Salix boothii [SALIBOO]
Salix pseudomyrsinites var. *aequalis* (Anderss.) Anderss. ex Ball = Salix boothii [SALIBOO]
Salix pseudopolaris Flod. = Salix polaris [SALIPOL]
Salix pulchra Cham. = Salix planifolia ssp. pulchra [SALIPLA2]
Salix pulchra var. *looffiae* Ball = Salix planifolia ssp. pulchra [SALIPLA2]
Salix pulchra var. *palmeri* Ball = Salix planifolia ssp. pulchra [SALIPLA2]
Salix pulchra var. *yukonensis* Schneid. = Salix planifolia ssp. pulchra [SALIPLA2]
Salix pychnocarpa Anderss. = Salix planifolia ssp. planifolia [SALIPLA0]
Salix pyrifolia Anderss. [SALIPYR] (balsam willow)
SY: *Salix balsamifera* (Hook.) Barratt ex Anderss.
SY: *Salix balsamifera* var. *alpestris* Bebb
SY: *Salix balsamifera* var. *lanceolata* Bebb
SY: *Salix balsamifera* var. *vegeta* Bebb
SY: *Salix cordata* var. *balsamifera* Hook.
SY: *Salix pyrifolia* var. *lanceolata* (Bebb) Fern.
Salix pyrifolia var. *lanceolata* (Bebb) Fern. = Salix pyrifolia [SALIPYR]
Salix pyrolifolia var. *hoyeriana* (Dieck) Dippel = Salix barclayi [SALIBAR]
Salix raupii Argus [SALIRAU] (Raup's willow)
Salix regelii Anderss. = Salix barclayi [SALIBAR]
Salix reticulata L. [SALIRET] (netted willow)
Salix reticulata ssp. **glabellicarpa** Argus [SALIRET1] (glabrous dwarf willow)
Salix reticulata ssp. **nivalis** (Hook.) A. & D. Löve & Kapoor [SALIRET2] (dwarf snow willow)
SY: *Salix nivalis* Hook.
SY: *Salix nivalis* var. *saximontana* (Rydb.) Schneid.
SY: *Salix reticulata* var. *nana* (Hook.) Anderss.
SY: *Salix reticulata* var. *nivalis* (Hook.) Anderss.
SY: *Salix reticulata* var. *saximontana* (Rydb.) L. Kelso
SY: *Salix saximontana* Rydb.
SY: *Salix venusta* Anderss.
SY: *Salix vestita* var. *nana* Hook.
Salix reticulata ssp. *orbicularis* (Anderss.) Flod. = Salix reticulata ssp. reticulata [SALIRET0]
Salix reticulata ssp. **reticulata** [SALIRET0]
SY: *Salix orbicularis* Anderss.
SY: *Salix reticulata* var. *gigantifolia* Ball
SY: *Salix reticulata* var. *glabra* Trautv.
SY: *Salix reticulata* ssp. *orbicularis* (Anderss.) Flod.
SY: *Salix reticulata* var. *orbicularis* (Anderss.) Komarov
SY: *Salix reticulata* var. *semicalva* Fern.
Salix reticulata var. *gigantifolia* Ball = Salix reticulata ssp. reticulata [SALIRET0]
Salix reticulata var. *glabra* Trautv. = Salix reticulata ssp. reticulata [SALIRET0]
Salix reticulata var. *nana* (Hook.) Anderss. = Salix reticulata ssp. nivalis [SALIRET2]
Salix reticulata var. *nivalis* (Hook.) Anderss. = Salix reticulata ssp. nivalis [SALIRET2]
Salix reticulata var. *orbicularis* (Anderss.) Komarov = Salix reticulata ssp. reticulata [SALIRET0]
Salix reticulata var. *saximontana* (Rydb.) L. Kelso = Salix reticulata ssp. nivalis [SALIRET2]
Salix reticulata var. *semicalva* Fern. = Salix reticulata ssp. reticulata [SALIRET0]
Salix reticulata var. *vestita* (Pursh) Anderss. = Salix vestita [SALIVES]
Salix richardsonii Hook. = Salix lanata ssp. richardsonii [SALILAN1]
Salix richardsonii var. *mckeandii* Polunin = Salix lanata ssp. richardsonii [SALILAN1]
Salix rigida ssp. *mackenzieana* (Hook.) E. Murr. = Salix prolixa [SALIPRO]
Salix rigida var. *mackenzieana* (Hook.) Cronq. = Salix prolixa [SALIPRO]
Salix rigida var. *macrogemma* (Ball) Cronq. = Salix prolixa [SALIPRO]
Salix rostrata Richards. = Salix bebbiana [SALIBEB]
Salix rostrata var. *capreifolia* Fern. = Salix bebbiana [SALIBEB]
Salix rostrata var. *luxurians* Fern. = Salix bebbiana [SALIBEB]
Salix rostrata var. *perrostrata* (Rydb.) Fern. = Salix bebbiana [SALIBEB]
Salix rostrata var. *projecta* Fern. = Salix bebbiana [SALIBEB]
Salix rotundifolia Nutt. = Salix tweedyi [SALITWE]
Salix rubra Richards. = Salix exigua [SALIEXI]
Salix saskatchevana von Seem. = Salix arbusculoides [SALIARB]
Salix saximontana Rydb. = Salix reticulata ssp. nivalis [SALIRET2]
Salix scouleriana Barratt ex Hook. [SALISCO] (Scouler's willow)
SY: *Salix brachystachys* Benth.
SY: *Salix brachystachys* var. *scouleriana* (Barratt ex Hook.) Anderss.
SY: *Salix capreoides* Anderss.
SY: *Salix flavescens* Nutt.
SY: *Salix flavescens* var. *capreoides* (Anderss.) Bebb
SY: *Salix flavescens* var. *scouleriana* (Barratt ex Hook.) Bebb
SY: *Salix nuttallii* Sarg.
SY: *Salix nuttallii* var. *capreoides* (Anderss.) Sarg.
SY: *Salix scouleriana* var. *brachystachys* (Benth.) M.E. Jones
SY: *Salix scouleriana* var. *coetanea* Ball
SY: *Salix scouleriana* var. *crassijulis* (Anderss.) Schneid.
SY: *Salix scouleriana* var. *flavescens* (Nutt.) J.K. Henry
SY: *Salix scouleriana* f. *poikila* Schneid.
SY: *Salix scouleriana* var. *poikila* Schneid.
SY: *Salix scouleriana* var. *thompsonii* Ball
SY: *Salix stagnalis* Nutt.
Salix scouleriana f. *poikila* Schneid. = Salix scouleriana [SALISCO]
Salix scouleriana var. *brachystachys* (Benth.) M.E. Jones = Salix scouleriana [SALISCO]
Salix scouleriana var. *coetanea* Ball = Salix scouleriana [SALISCO]
Salix scouleriana var. *crassijulis* (Anderss.) Schneid. = Salix scouleriana [SALISCO]
Salix scouleriana var. *flavescens* (Nutt.) J.K. Henry = Salix scouleriana [SALISCO]
Salix scouleriana var. *poikila* Schneid. = Salix scouleriana [SALISCO]
Salix scouleriana var. *thompsonii* Ball = Salix scouleriana [SALISCO]
Salix seemanii Rydb. = Salix glauca var. acutifolia [SALIGLA1]
Salix sensitiva Barratt = Salix discolor [SALIDIS]

Salix sericea var. *subsericea* (Anderss.) Rydb. = Salix petiolaris [SALIPET]
Salix serissima (Bailey) Fern. [SALISER] (autumn willow)
SY: *Salix arguta* var. *alpigena* Anderss.
SY: *Salix arguta* var. *pallescens* Anderss.
SY: *Salix lucida* var. *serissima* Bailey
Salix sessilifolia Nutt. [SALISES] (soft-leaved sandbar willow)
SY: *Salix exigua* var. *parishiana* (Rowlee) Jepson
SY: *Salix fluviatilis* var. *sessilifolia* (Nutt.) Scoggan
SY: *Salix hindsiana* var. *parishiana* (Rowlee) Ball
SY: *Salix longifolia* var. *sessilifolia* (Nutt.) M.E. Jones
SY: *Salix macrostachya* Nutt.
SY: *Salix macrostachya* var. *cusickii* Rowlee
SY: *Salix parishiana* Rowlee
SY: *Salix sessilifolia* var. *villosa* Anderss.
Salix sessilifolia var. *vancouverensis* Brayshaw = Salix melanopsis [SALIMEL]
Salix sessilifolia var. *villosa* Anderss. = Salix sessilifolia [SALISES]
Salix setchelliana Ball [SALISET] (Setchell's willow)
SY: *Salix aliena* Flod.
Salix sitchensis Sanson ex Bong. [SALISIT] (Sitka willow)
SY: *Salix coulteri* Anderss.
SY: *Salix cuneata* Nutt.
SY: *Salix sitchensis* var. *congesta* (Anderss.) Anderss.
SY: *Salix sitchensis* var. *denudata* (Anderss.) Anderss.
SY: *Salix sitchensis* var. *parviflora* (Jepson) Jepson
SY: *Salix sitchensis* var. *ralphiana* (Jepson) Jepson
Salix sitchensis var. *congesta* (Anderss.) Anderss. = Salix sitchensis [SALISIT]
Salix sitchensis var. *denudata* (Anderss.) Anderss. = Salix sitchensis [SALISIT]
Salix sitchensis var. *parviflora* (Jepson) Jepson = Salix sitchensis [SALISIT]
Salix sitchensis var. *ralphiana* (Jepson) Jepson = Salix sitchensis [SALISIT]
Salix speciosa Hook. & Arn. = Salix alaxensis var. alaxensis [SALIALA0]
Salix speciosa Nutt. = Salix lucida ssp. lasiandra [SALILUC2]
Salix speciosa var. *alaxensis* Anderss. = Salix alaxensis var. alaxensis [SALIALA0]
Salix squamata Rydb. = Salix discolor [SALIDIS]
Salix stagnalis Nutt. = Salix scouleriana [SALISCO]
Salix starkeana ssp. *bebbiana* (Sarg.) Youngberg = Salix bebbiana [SALIBEB]
Salix stenophylla Rydb. = Salix exigua [SALIEXI]
Salix stolonifera Coville [SALISTO] (creeping willow)
Salix subcoerulea Piper = Salix drummondiana [SALIDRU]
Salix subcordata Anderss. = Salix glauca var. villosa [SALIGLA2]
Salix × *subsericea* (Anderss.) Schneid. = Salix petiolaris [SALIPET]
Salix tenera Anderss. = Salix cascadensis [SALICAS]
Salix tenerrima (Henderson) Heller = Salix melanopsis [SALIMEL]
Salix thurberi Rowlee = Salix exigua [SALIEXI]
Salix tortulosa Trautv. = Salix arctica [SALIARC]
Salix tweedyi (Bebb ex Rose) Ball [SALITWE] (Tweedy's willow)
SY: *Salix barrattiana* var. *tweedyi* Bebb ex Rose
SY: *Salix rotundifolia* Nutt.
Salix vagans var. *occidentalis* Anderss. = Salix bebbiana [SALIBEB]
Salix vagans var. *rostrata* (Richards.) Anderss. = Salix bebbiana [SALIBEB]
Salix venusta Anderss. = Salix reticulata ssp. nivalis [SALIRET2]
Salix vestita Pursh [SALIVES] (rock willow)
SY: *Salix fernaldii* Blank.
SY: *Salix leiolepis* Fern.
SY: *Salix reticulata* var. *vestita* (Pursh) Anderss.
SY: *Salix vestita* var. *erecta* Anderss.
SY: *Salix vestita* var. *humilior* Anderss.
SY: *Salix vestita* ssp. *leiolepis* (Fern.) Argus
SY: *Salix vestita* var. *psilophylla* Fern. & St. John
Salix vestita ssp. *leiolepis* (Fern.) Argus = Salix vestita [SALIVES]
Salix vestita var. *erecta* Anderss. = Salix vestita [SALIVES]
Salix vestita var. *humilior* Anderss. = Salix vestita [SALIVES]
Salix vestita var. *nana* Hook. = Salix reticulata ssp. nivalis [SALIRET2]
Salix vestita var. *psilophylla* Fern. & St. John = Salix vestita [SALIVES]
Salix villosa D. Don ex Hook. = Salix glauca var. villosa [SALIGLA2]
Salix villosa var. *acutifolia* Hook. = Salix glauca var. acutifolia [SALIGLA1]
Salix vitellina L. = Salix alba var. vitellina [SALIALB1]
Salix wheeleri (Rowlee) Rydb. = Salix exigua [SALIEXI]
Salix wolfii var. *pseudolapponum* (von Seem.) M.E. Jones = Salix glauca var. villosa [SALIGLA2]
Salix wrightii Anderss. = Salix amygdaloides [SALIAMY]
Salix wyomingensis Rydb. = Salix glauca var. villosa [SALIGLA2]

SANTALACEAE (V:D)

Comandra Nutt. [COMANDR$]
Comandra californica Eastw. ex Rydb. = Comandra umbellata ssp. californica [COMAUMB1]
Comandra lividum Richards. = Geocaulon lividum [GEOCLIV]
Comandra pallida A. DC. = Comandra umbellata ssp. pallida [COMAUMB2]
Comandra umbellata (L.) Nutt. [COMAUMB] (pale comandra)
Comandra umbellata ssp. **californica** (Eastw. ex Rydb.) Piehl [COMAUMB1]
SY: *Comandra californica* Eastw. ex Rydb.
SY: *Comandra umbellata* var. *californica* (Eastw. ex Rydb.) C.L. Hitchc.
Comandra umbellata ssp. **pallida** (A. DC.) Piehl [COMAUMB2]
SY: *Comandra pallida* A. DC.
SY: *Comandra umbellata* var. *angustifolia* (A. DC.) Torr.
SY: *Comandra umbellata* var. *pallida* (A. DC.) M.E. Jones
Comandra umbellata var. *angustifolia* (A. DC.) Torr. = Comandra umbellata ssp. pallida [COMAUMB2]
Comandra umbellata var. *californica* (Eastw. ex Rydb.) C.L. Hitchc. = Comandra umbellata ssp. californica [COMAUMB1]
Comandra umbellata var. *pallida* (A. DC.) M.E. Jones = Comandra umbellata ssp. pallida [COMAUMB2]
Geocaulon Fern. [GEOCAUL$]
Geocaulon lividum (Richards.) Fern. [GEOCLIV] (bastard toad-flax)
SY: *Comandra lividum* Richards.

SARRACENIACEAE (V:D)

Sarracenia L. [SARRACE$]
Sarracenia heterophylla Eat. = Sarracenia purpurea [SARRPUR]
Sarracenia purpurea L. [SARRPUR] (common pitcher-plant)
SY: *Sarracenia heterophylla* Eat.
SY: *Sarracenia purpurea* ssp. *gibbosa* (Raf.) Wherry
SY: *Sarracenia purpurea* ssp. *heterophylla* (Eat.) Torr.
SY: *Sarracenia purpurea* var. *ripicola* Boivin
SY: *Sarracenia purpurea* var. *stolonifera* Macfarlane & D.W. Steckbeck
SY: *Sarracenia purpurea* var. *terrae-novae* La Pylaie
Sarracenia purpurea ssp. *gibbosa* (Raf.) Wherry = Sarracenia purpurea [SARRPUR]
Sarracenia purpurea ssp. *heterophylla* (Eat.) Torr. = Sarracenia purpurea [SARRPUR]
Sarracenia purpurea var. *ripicola* Boivin = Sarracenia purpurea [SARRPUR]
Sarracenia purpurea var. *stolonifera* Macfarlane & D.W. Steckbeck = Sarracenia purpurea [SARRPUR]
Sarracenia purpurea var. *terrae-novae* La Pylaie = Sarracenia purpurea [SARRPUR]

SAXIFRAGACEAE (V:D)

Boykinia Nutt. [BOYKINI$]
Boykinia elata (Nutt.) Greene = Boykinia occidentalis [BOYKOCC]
Boykinia occidentalis Torr. & Gray [BOYKOCC] (coast boykinia)
SY: *Boykinia elata* (Nutt.) Greene
SY: *Boykinia vancouverensis* (Rydb.) Fedde
Boykinia vancouverensis (Rydb.) Fedde = Boykinia occidentalis [BOYKOCC]

Chrysosplenium L. [CHRYSOS$]
Chrysosplenium alternifolium ssp. *iowense* (Rydb.) Hultén = Chrysosplenium iowense [CHRYIOW]
Chrysosplenium alternifolium ssp. *tetrandrum* (Lund) Hultén = Chrysosplenium tetrandrum [CHRYTET]
Chrysosplenium alternifolium var. *iowense* (Rydb.) Boivin = Chrysosplenium iowense [CHRYIOW]
Chrysosplenium alternifolium var. *tetrandrum* Lund = Chrysosplenium tetrandrum [CHRYTET]
Chrysosplenium iowense Rydb. [CHRYIOW] (Iowa golden-saxifrage)
SY: *Chrysosplenium alternifolium* ssp. *iowense* (Rydb.) Hultén
SY: *Chrysosplenium alternifolium* var. *iowense* (Rydb.) Boivin
Chrysosplenium tetrandrum (Lund) Th. Fries [CHRYTET] (northern golden-saxifrage)
SY: *Chrysosplenium alternifolium* ssp. *tetrandrum* (Lund) Hultén
SY: *Chrysosplenium alternifolium* var. *tetrandrum* Lund
Chrysosplenium wrightii Franch. & Savigny [CHRYWRI] (Wright's golden-saxifrage)
Ciliaria austromontana (Wieg.) W.A. Weber = Saxifraga bronchialis ssp. austromontana [SAXIBRO1]
Ciliaria funstonii (Small) W.A. Weber = Saxifraga bronchialis ssp. funstonii [SAXIBRO2]
Elmera Rydb. [ELMERA$]
Elmera racemosa (S. Wats.) Rydb. [ELMERAC]
SY: *Heuchera racemosa* S. Wats.
Hemieva ranunculifolia (Hook.) Raf. = Suksdorfia ranunculifolia [SUKSRAN]
Heuchera L. [HEUCHER$]
Heuchera chlorantha Piper [HEUCCHL] (meadow alumroot)
Heuchera cylindrica Dougl. ex Hook. [HEUCCYL] (round-leaved alumroot)
Heuchera cylindrica var. **cylindrica** [HEUCCYL0]
SY: *Heuchera cylindrica* var. *suksdorfii* (Rydb.) Dorn
SY: *Heuchera saxicola* E. Nels.
SY: *Heuchera suksdorfii* Rydb.
Heuchera cylindrica var. **glabella** (Torr. & Gray) Wheelock [HEUCCYL2]
SY: *Heuchera glabella* Torr. & Gray
Heuchera cylindrica var. **orbicularis** (Rosendahl, Butters & Lakela) Calder & Savile [HEUCCYL3]
SY: *Heuchera ovalifolia* var. *orbicularis* Rosendahl, Butters & Lakela
Heuchera cylindrica var. **septentrionalis** Rosendahl, Butters & Lakela [HEUCCYL4]
Heuchera cylindrica var. *suksdorfii* (Rydb.) Dorn = Heuchera cylindrica var. cylindrica [HEUCCYL0]
Heuchera diversifolia Rydb. = Heuchera micrantha var. diversifolia [HEUCMIC1]
Heuchera glabella Torr. & Gray = Heuchera cylindrica var. glabella [HEUCCYL2]
Heuchera glabra Willd. ex Roemer & J.A. Schultes [HEUCGLA] (smooth alumroot)
Heuchera micrantha Dougl. ex Lindl. [HEUCMIC] (small-flowered alumroot)
Heuchera micrantha var. **diversifolia** (Rydb.) Rosendahl, Butters & Lakela [HEUCMIC1]
SY: *Heuchera diversifolia* Rydb.
Heuchera ovalifolia var. *orbicularis* Rosendahl, Butters & Lakela = Heuchera cylindrica var. orbicularis [HEUCCYL3]
Heuchera racemosa S. Wats. = Elmera racemosa [ELMERAC]
Heuchera richardsonii R. Br. [HEUCRIC] (Richardson's alumroot)
SY: *Heuchera richardsonii* var. *affinis* Rosendahl, Butters & Lakela
SY: *Heuchera richardsonii* var. *grayana* Rosendahl, Butters & Lakela
SY: *Heuchera richardsonii* var. *hispidior* Rosendahl, Butters & Lakela
Heuchera richardsonii var. *affinis* Rosendahl, Butters & Lakela = Heuchera richardsonii [HEUCRIC]
Heuchera richardsonii var. *grayana* Rosendahl, Butters & Lakela = Heuchera richardsonii [HEUCRIC]
Heuchera richardsonii var. *hispidior* Rosendahl, Butters & Lakela = Heuchera richardsonii [HEUCRIC]
Heuchera saxicola E. Nels. = Heuchera cylindrica var. cylindrica [HEUCCYL0]
Heuchera suksdorfii Rydb. = Heuchera cylindrica var. cylindrica [HEUCCYL0]
Hirculus serpyllifolius (Pursh) W.A. Weber = Saxifraga serpyllifolia [SAXISER]
Leptarrhena R. Br. [LEPTARR$]
Leptarrhena pyrolifolia (D. Don) R. Br. ex Ser. [LEPTPYR] (leatherleaf saxifrage)
SY: *Saxifraga pyrolifolia* D. Don
Leptasea aizoides (L.) Haw. = Saxifraga aizoides [SAXIAIZ]
Leptasea tricuspidata (Rottb.) Haw. = Saxifraga tricuspidata [SAXITRI]
Lithophragma (Nutt.) Torr. & Gray [LITHOPH$]
Lithophragma australe Rydb. = Lithophragma tenellum [LITHTEN]
Lithophragma brevilobum Rydb. = Lithophragma tenellum [LITHTEN]
Lithophragma bulbiferum Rydb. = Lithophragma glabrum [LITHGLA]
Lithophragma glabrum Nutt. [LITHGLA] (smooth woodland star)
SY: *Lithophragma bulbiferum* Rydb.
SY: *Lithophragma glabrum* var. *bulbiferum* (Rydb.) Jepson
SY: *Lithophragma glabrum* var. *ramulosum* (Suksdorf) Boivin
Lithophragma glabrum var. *bulbiferum* (Rydb.) Jepson = Lithophragma glabrum [LITHGLA]
Lithophragma glabrum var. *ramulosum* (Suksdorf) Boivin = Lithophragma glabrum [LITHGLA]
Lithophragma parviflorum (Hook.) Nutt. ex Torr. & Gray [LITHPAR] (small-flowered woodland star)
Lithophragma rupicola Greene = Lithophragma tenellum [LITHTEN]
Lithophragma tenellum Nutt. [LITHTEN] (slender woodland star)
SY: *Lithophragma australe* Rydb.
SY: *Lithophragma brevilobum* Rydb.
SY: *Lithophragma rupicola* Greene
SY: *Lithophragma tenellum* var. *thompsonii* (Hoover) C.L. Hitchc.
SY: *Lithophragma thompsonii* Hoover
Lithophragma tenellum var. *thompsonii* (Hoover) C.L. Hitchc. = Lithophragma tenellum [LITHTEN]
Lithophragma thompsonii Hoover = Lithophragma tenellum [LITHTEN]
Micranthes bidens Small = Saxifraga integrifolia [SAXIINT]
Micranthes lata Small = Saxifraga occidentalis [SAXIOCC]
Micranthes lyallii (Engl.) Small = Saxifraga lyallii ssp. lyallii [SAXILYA0]
Micranthes montana Small = Saxifraga nidifica [SAXINID]
Micranthes nivalis (L.) Small = Saxifraga nivalis [SAXINIV]
Micranthes occidentalis (S. Wats.) Small = Saxifraga occidentalis [SAXIOCC]
Micranthes odontoloma (Piper) W.A. Weber = Saxifraga odontoloma [SAXIODO]
Micranthes saximontana Small = Saxifraga occidentalis [SAXIOCC]
Mitella L. [MITELLA$]
Mitella breweri Gray [MITEBRE] (Brewer's mitrewort)
SY: *Pectiantia breweri* (Gray) Rydb.
Mitella caulescens Nutt. [MITECAU] (leafy mitrewort)
SY: *Mitellastra caulescens* (Nutt.) T.J. Howell
Mitella nuda L. [MITENUD] (common mitrewort)
Mitella ovalis Greene [MITEOVA] (oval-leaved mitrewort)
SY: *Pectiantia ovalis* (Greene) Rydb.
Mitella pentandra Hook. [MITEPEN] (five-stamened mitrewort)
SY: *Pectiantia pentandra* (Hook.) Rydb.
Mitella trifida Graham [MITETRI] (three-toothed mitrewort)
Mitellastra caulescens (Nutt.) T.J. Howell = Mitella caulescens [MITECAU]
Muscaria adscendens (L.) Small p.p. = Saxifraga adscendens ssp. oregonensis [SAXIADS1]
Muscaria micropetala Small = Saxifraga cespitosa ssp. monticola [SAXICES1]
Muscaria monticola Small = Saxifraga cespitosa ssp. monticola [SAXICES1]
Parnassia L. [PARNASS$]

Parnassia fimbriata Koenig [PARNFIM] (fringed grass-of-Parnassus)
Parnassia kotzebuei Cham. ex Spreng. [PARNKOT] (Kotzebue's grass-of-Parnassus)
SY: *Parnassia kotzebuei* var. *pumila* C.L. Hitchc. & Ownbey
Parnassia kotzebuei var. *pumila* C.L. Hitchc. & Ownbey = Parnassia kotzebuei [PARNKOT]
Parnassia montanensis Fern. & Rydb. ex Rydb. = Parnassia palustris var. montanensis [PARNPAL2]
Parnassia palustris L. [PARNPAL] (northern grass-of-Parnassus)
Parnassia palustris var. **montanensis** (Fern. & Rydb. ex Rydb.) C.L. Hitchc. [PARNPAL2]
SY: *Parnassia montanensis* Fern. & Rydb. ex Rydb.
Parnassia palustris var. *parviflora* (DC.) Boivin = Parnassia parviflora [PARNPAR]
Parnassia parviflora DC. [PARNPAR] (small-flowered grass-of-Parnassus)
SY: *Parnassia palustris* var. *parviflora* (DC.) Boivin
Pectiantia breweri (Gray) Rydb. = Mitella breweri [MITEBRE]
Pectiantia ovalis (Greene) Rydb. = Mitella ovalis [MITEOVA]
Pectiantia pentandra (Hook.) Rydb. = Mitella pentandra [MITEPEN]
Saxifraga L. [SAXIFRA$]
Saxifraga adscendens L. [SAXIADS] (wedge-leaved saxifrage)
Saxifraga adscendens ssp. **oregonensis** (Raf.) Bacig. [SAXIADS1] (Oregon saxifrage)
SY: *Muscaria adscendens* (L.) Small p.p.
SY: *Saxifraga adscendens* var. *oregonensis* (Raf.) Breitung
Saxifraga adscendens var. *oregonensis* (Raf.) Breitung = Saxifraga adscendens ssp. oregonensis [SAXIADS1]
Saxifraga aequidentata (Small) Rosendahl = Saxifraga rufidula [SAXIRUF]
Saxifraga aestivalis auct. = Saxifraga odontoloma [SAXIODO]
Saxifraga aizoides L. [SAXIAIZ] (evergreen saxifrage)
SY: *Leptasea aizoides* (L.) Haw.
Saxifraga arguta auct. = Saxifraga odontoloma [SAXIODO]
Saxifraga austromontana Wieg. = Saxifraga bronchialis ssp. austromontana [SAXIBRO1]
Saxifraga bracteosa Suksdorf = Saxifraga integrifolia [SAXIINT]
Saxifraga bracteosa var. *angustifolia* Suksdorf = Saxifraga integrifolia [SAXIINT]
Saxifraga bracteosa var. *micropetala* Suksdorf = Saxifraga nidifica [SAXINID]
Saxifraga bronchialis L. [SAXIBRO] (spotted saxifrage)
Saxifraga bronchialis ssp. **austromontana** (Wieg.) Piper [SAXIBRO1]
SY: *Ciliaria austromontana* (Wieg.) W.A. Weber
SY: *Saxifraga austromontana* Wieg.
SY: *Saxifraga bronchialis* var. *austromontana* (Wieg.) Piper ex G.N. Jones
Saxifraga bronchialis ssp. **funstonii** (Small) Hultén [SAXIBRO2]
SY: *Ciliaria funstonii* (Small) W.A. Weber
SY: *Saxifraga bronchialis* var. *purpureomaculata* Hultén
SY: *Saxifraga firma* Litv. ex Losinsk.
SY: *Saxifraga funstonii* (Small) Fedde
Saxifraga bronchialis var. *austromontana* (Wieg.) Piper ex G.N. Jones = Saxifraga bronchialis ssp. austromontana [SAXIBRO1]
Saxifraga bronchialis var. *purpureomaculata* Hultén = Saxifraga bronchialis ssp. funstonii [SAXIBRO2]
Saxifraga calycina Sternb. [SAXICAL]
SY: *Saxifraga davurica* ssp. *grandipetala* (Engl. & Irmsch.) Hultén
SY: *Saxifraga davurica* var. *grandipetala* (Engl. & Irmsch.) Boivin
Saxifraga cernua L. [SAXICER] (nodding saxifrage)
SY: *Saxifraga cernua* var. *exilioides* Polunin
Saxifraga cernua var. *exilioides* Polunin = Saxifraga cernua [SAXICER]
Saxifraga cespitosa L. [SAXICES] (tufted saxifrage)
Saxifraga cespitosa ssp. **monticola** (Small) Porsild [SAXICES1]
SY: *Muscaria micropetala* Small
SY: *Muscaria monticola* Small
SY: *Saxifraga cespitosa* var. *minima* Blank.
SY: *Saxifraga micropetala* (Small) Fedde
SY: *Saxifraga monticola* (Small) A. & D. Löve
Saxifraga cespitosa ssp. **sileneflora** (Sternb. ex Cham.) Hultén [SAXICES2]
SY: *Saxifraga sileneflora* Sternb. ex Cham.
Saxifraga cespitosa ssp. **subgemmifera** Engl. & Irmsch. [SAXICES3]
SY: *Saxifraga cespitosa* var. *subgemmifera* (Engl. & Irmsch.) C.L. Hitchc.
Saxifraga cespitosa var. **emarginata** (Small) Rosendahl [SAXICES4]
Saxifraga cespitosa var. *minima* Blank. = Saxifraga cespitosa ssp. monticola [SAXICES1]
Saxifraga cespitosa var. *subgemmifera* (Engl. & Irmsch.) C.L. Hitchc. = Saxifraga cespitosa ssp. subgemmifera [SAXICES3]
Saxifraga columbiana Piper = Saxifraga nidifica [SAXINID]
Saxifraga crenatifolia (Small) Fedde = Saxifraga nidifica [SAXINID]
Saxifraga davurica ssp. *grandipetala* (Engl. & Irmsch.) Hultén = Saxifraga calycina [SAXICAL]
Saxifraga davurica var. *grandipetala* (Engl. & Irmsch.) Boivin = Saxifraga calycina [SAXICAL]
Saxifraga debilis Engelm. ex Gray = Saxifraga rivularis [SAXIRIV]
Saxifraga ferruginea Graham [SAXIFER] (Alaska saxifrage)
SY: *Saxifraga newcombei* (Small) Engl. & Irmsch.
Saxifraga firma Litv. ex Losinsk. = Saxifraga bronchialis ssp. funstonii [SAXIBRO2]
Saxifraga flagellaris Willd. ex Sternb. [SAXIFLA] (whiplash saxifrage)
Saxifraga flagellaris ssp. **setigera** (Pursh) Tolm. [SAXIFLA1] (stoloniferous saxifrage)
SY: *Saxifraga setigera* Pursh
Saxifraga fragosa var. *leucandra* Suksdorf = Saxifraga integrifolia [SAXIINT]
Saxifraga funstonii (Small) Fedde = Saxifraga bronchialis ssp. funstonii [SAXIBRO2]
Saxifraga hieraciifolia Waldst. & Kit. [SAXIHIE] (hawkweed-leaved saxifrage)
Saxifraga hirculus L. [SAXIHIR] (yellow marsh saxifrage)
Saxifraga hyperborea R. Br. = Saxifraga rivularis [SAXIRIV]
Saxifraga hyperborea ssp. *debilis* (Engelm. ex Gray) A. & D. Löve & Kapoor = Saxifraga rivularis [SAXIRIV]
Saxifraga integrifolia Hook. [SAXIINT] (grassland saxifrage)
SY: *Micranthes bidens* Small
SY: *Saxifraga bracteosa* Suksdorf
SY: *Saxifraga bracteosa* var. *angustifolia* Suksdorf
SY: *Saxifraga fragosa* var. *leucandra* Suksdorf
SY: *Saxifraga laevicarpa* A.M. Johnson
Saxifraga integrifolia var. *columbiana* (Piper) C.L. Hitchc. = Saxifraga nidifica [SAXINID]
Saxifraga integrifolia var. *leptopetala* (Suksdorf) Engl. & Irmsch. = Saxifraga nidifica [SAXINID]
Saxifraga integrifolia var. *micropetala* (Suksdorf) Engl. & Irmsch. = Saxifraga nidifica [SAXINID]
Saxifraga klickitatensis A.M. Johnson = Saxifraga rufidula [SAXIRUF]
Saxifraga laevicarpa A.M. Johnson = Saxifraga integrifolia [SAXIINT]
Saxifraga lyallii Engl. [SAXILYA] (red-stemmed saxifrage)
Saxifraga lyallii ssp. **hultenii** (Calder & Savile) Calder & Taylor [SAXILYA1]
SY: *Saxifraga lyallii* var. *hultenii* Calder & Savile
Saxifraga lyallii ssp. **lyallii** [SAXILYA0]
SY: *Micranthes lyallii* (Engl.) Small
SY: *Saxifraga lyallii* var. *laxa* Engl.
Saxifraga lyallii var. *hultenii* Calder & Savile = Saxifraga lyallii ssp. hultenii [SAXILYA1]
Saxifraga lyallii var. *laxa* Engl. = Saxifraga lyallii ssp. lyallii [SAXILYA0]
Saxifraga mertensiana Bong. [SAXIMER] (wood saxifrage)
SY: *Saxifraga mertensiana* var. *eastwoodiae* (Small) Engl. & Irmsch.
Saxifraga mertensiana var. *eastwoodiae* (Small) Engl. & Irmsch. = Saxifraga mertensiana [SAXIMER]

Saxifraga micropetala (Small) Fedde = Saxifraga cespitosa ssp. monticola [SAXICES1]
Saxifraga montana (Small) Fedde = Saxifraga nidifica [SAXINID]
Saxifraga monticola (Small) A. & D. Löve = Saxifraga cespitosa ssp. monticola [SAXICES1]
Saxifraga nelsoniana D. Don [SAXINEL] (cordate-leaved saxifrage)
SY: *Saxifraga punctata* L.
Saxifraga nelsoniana ssp. **carlottae** (Calder & Savile) Hultén [SAXINEL1]
SY: *Saxifraga punctata* ssp. *carlottae* Calder & Savile
SY: *Saxifraga punctata* var. *carlottae* (Calder & Savile) Boivin
Saxifraga nelsoniana ssp. **cascadensis** (Calder & Savile) Hultén [SAXINEL2]
SY: *Saxifraga punctata* ssp. *cascadensis* Calder & Savile
SY: *Saxifraga punctata* var. *cascadensis* (Calder & Savile) Boivin
Saxifraga nelsoniana ssp. **pacifica** (Hultén) Hultén [SAXINEL3]
SY: *Saxifraga punctata* ssp. *pacifica* Hultén
Saxifraga nelsoniana ssp. **porsildiana** (Calder & Savile) Hultén [SAXINEL4]
SY: *Saxifraga porsildiana* (Calder & Savile) Jurtzev & Petrovsky
SY: *Saxifraga punctata* ssp. *porsildiana* Calder & Savile
SY: *Saxifraga punctata* var. *porsildiana* (Calder & Savile) Boivin
Saxifraga newcombei (Small) Engl. & Irmsch. = Saxifraga ferruginea [SAXIFER]
Saxifraga nidifica Greene [SAXINID] (peak saxifrage)
SY: *Micranthes montana* Small
SY: *Saxifraga bracteosa* var. *micropetala* Suksdorf
SY: *Saxifraga columbiana* Piper
SY: *Saxifraga crenatifolia* (Small) Fedde
SY: *Saxifraga integrifolia* var. *columbiana* (Piper) C.L. Hitchc.
SY: *Saxifraga integrifolia* var. *leptopetala* (Suksdorf) Engl. & Irmsch.
SY: *Saxifraga integrifolia* var. *micropetala* (Suksdorf) Engl. & Irmsch.
SY: *Saxifraga montana* (Small) Fedde
SY: *Saxifraga plantaginea* Small
Saxifraga nivalis L. [SAXINIV] (alpine saxifrage)
SY: *Micranthes nivalis* (L.) Small
Saxifraga occidentalis S. Wats. [SAXIOCC] (western saxifrage)
SY: *Micranthes lata* Small
SY: *Micranthes occidentalis* (S. Wats.) Small
SY: *Micranthes saximontana* Small
SY: *Saxifraga occidentalis* var. *allenii* (Small) C.L. Hitchc.
SY: *Saxifraga occidentalis* var. *wallowensis* M.E. Peck
SY: *Saxifraga reflexa* ssp. *occidentalis* (S. Wats.) Hultén
Saxifraga occidentalis ssp. *rufidula* (Small) Bacig. = Saxifraga rufidula [SAXIRUF]
Saxifraga occidentalis var. *aequidentata* (Small) M.E. Peck = Saxifraga rufidula [SAXIRUF]
Saxifraga occidentalis var. *allenii* (Small) C.L. Hitchc. = Saxifraga occidentalis [SAXIOCC]
Saxifraga occidentalis var. *rufidula* (Small) C.L. Hitchc. = Saxifraga rufidula [SAXIRUF]
Saxifraga occidentalis var. *wallowensis* M.E. Peck = Saxifraga occidentalis [SAXIOCC]
Saxifraga odontoloma Piper [SAXIODO] (stream saxifrage)
SY: *Micranthes odontoloma* (Piper) W.A. Weber
SY: *Saxifraga aestivalis* auct.
SY: *Saxifraga arguta* auct.
Saxifraga oppositifolia L. [SAXIOPP] (purple mountain saxifrage)
Saxifraga plantaginea Small = Saxifraga nidifica [SAXINID]
Saxifraga porsildiana (Calder & Savile) Jurtzev & Petrovsky = Saxifraga nelsoniana ssp. porsildiana **[SAXINEL4]**
Saxifraga punctata L. = Saxifraga nelsoniana [SAXINEL]
Saxifraga punctata ssp. *carlottae* Calder & Savile = Saxifraga nelsoniana ssp. carlottae [SAXINEL1]
Saxifraga punctata ssp. *cascadensis* Calder & Savile = Saxifraga nelsoniana ssp. cascadensis [SAXINEL2]
Saxifraga punctata ssp. *pacifica* Hultén = Saxifraga nelsoniana ssp. pacifica [SAXINEL3]
Saxifraga punctata ssp. *porsildiana* Calder & Savile = Saxifraga nelsoniana ssp. porsildiana [SAXINEL4]
Saxifraga punctata var. *carlottae* (Calder & Savile) Boivin = Saxifraga nelsoniana ssp. carlottae [SAXINEL1]
Saxifraga punctata var. *cascadensis* (Calder & Savile) Boivin = Saxifraga nelsoniana ssp. cascadensis [SAXINEL2]
Saxifraga punctata var. *porsildiana* (Calder & Savile) Boivin = Saxifraga nelsoniana ssp. porsildiana [SAXINEL4]
Saxifraga pyrolifolia D. Don = Leptarrhena pyrolifolia [LEPTPYR]
Saxifraga reflexa Hook. [SAXIREF] (Yukon saxifrage)
Saxifraga reflexa ssp. *occidentalis* (S. Wats.) Hultén = Saxifraga occidentalis [SAXIOCC]
Saxifraga rivularis L. [SAXIRIV] (brook saxifrage)
SY: *Saxifraga debilis* Engelm. ex Gray
SY: *Saxifraga hyperborea* R. Br.
SY: *Saxifraga hyperborea* ssp. *debilis* (Engelm. ex Gray) A. & D. Löve & Kapoor
SY: *Saxifraga rivularis* var. *debilis* (Engelm. ex Gray) Dorn
SY: *Saxifraga rivularis* var. *flexuosa* (Sternb.) Engl. & Irmsch.
SY: *Saxifraga rivularis* var. *hyperborea* (R. Br.) Dorn
Saxifraga rivularis var. *debilis* (Engelm. ex Gray) Dorn = Saxifraga rivularis [SAXIRIV]
Saxifraga rivularis var. *flexuosa* (Sternb.) Engl. & Irmsch. = Saxifraga rivularis [SAXIRIV]
Saxifraga rivularis var. *hyperborea* (R. Br.) Dorn = Saxifraga rivularis [SAXIRIV]
Saxifraga rufidula (Small) Macoun [SAXIRUF] (rusty-haired saxifrage)
SY: *Saxifraga aequidentata* (Small) Rosendahl
SY: *Saxifraga klickitatensis* A.M. Johnson
SY: *Saxifraga occidentalis* var. *aequidentata* (Small) M.E. Peck
SY: *Saxifraga occidentalis* ssp. *rufidula* (Small) Bacig.
SY: *Saxifraga occidentalis* var. *rufidula* (Small) C.L. Hitchc.
Saxifraga serpyllifolia Pursh [SAXISER] (thyme-leaved saxifrage)
SY: *Hirculus serpyllifolius* (Pursh) W.A. Weber
SY: *Saxifraga serpyllifolia* var. *purpurea* Hultén
Saxifraga serpyllifolia var. *purpurea* Hultén = Saxifraga serpyllifolia [SAXISER]
Saxifraga setigera Pursh = Saxifraga flagellaris ssp. setigera [SAXIFLA1]
Saxifraga sileneflora Sternb. ex Cham. = Saxifraga cespitosa ssp. sileneflora [SAXICES2]
Saxifraga taylori Calder & Savile [SAXITAY] (Taylor's saxifrage)
Saxifraga tischii Skelly [SAXITIS]
Saxifraga tolmiei Torr. & Gray [SAXITOL] (Tolmie's saxifrage)
SY: *Saxifraga tolmiei* var. *ledifolia* (Greene) Engl. & Irmsch.
Saxifraga tolmiei var. *ledifolia* (Greene) Engl. & Irmsch. = Saxifraga tolmiei [SAXITOL]
Saxifraga tricuspidata Rottb. [SAXITRI] (three-toothed saxifrage)
SY: *Leptasea tricuspidata* (Rottb.) Haw.
*__Saxifraga tridactylites__ L. [SAXITRD] (rue-leaved saxifrage)
Suksdorfia Gray [SUKSDOR$]
Suksdorfia ranunculifolia (Hook.) Engl. [SUKSRAN] (buttercup-leaved suksdorfia)
SY: *Hemieva ranunculifolia* (Hook.) Raf.
Suksdorfia violacea Gray [SUKSVIO] (violet suksdorfia)
Tellima R. Br. [TELLIMA$]
Tellima grandiflora (Pursh) Dougl. ex Lindl. [TELLGRA] (fringecup)
SY: *Tellima odorata* T.J. Howell
Tellima odorata T.J. Howell = Tellima grandiflora [TELLGRA]
Tiarella L. [TIARELL$]
Tiarella californica (Kellogg) Rydb. = Tiarella trifoliata var. laciniata [TIARTRI1]
Tiarella laciniata Hook. = Tiarella trifoliata var. laciniata [TIARTRI1]
Tiarella menziesii Pursh = Tolmiea menziesii [TOLMMEN]
Tiarella trifoliata L. [TIARTRI] (foamflower)
Tiarella trifoliata ssp. *unifoliata* (Hook.) Kern = Tiarella trifoliata var. unifoliata [TIARTRI2]
Tiarella trifoliata var. **laciniata** (Hook.) Wheelock [TIARTRI1]
SY: *Tiarella californica* (Kellogg) Rydb.
SY: *Tiarella laciniata* Hook.
SY: *Tiarella trifoliata* var. × *laciniata* (Hook.) Wheelock

Tiarella trifoliata var. × *laciniata* (Hook.) Wheelock = Tiarella trifoliata var. laciniata [TIARTRI1]
Tiarella trifoliata var. **trifoliata** [TIARTRI0] (three-leaved foamflower)
Tiarella trifoliata var. **unifoliata** (Hook.) Kurtz [TIARTRI2] (one-leaved foamflower)
SY: *Tiarella trifoliata* ssp. *unifoliata* (Hook.) Kern
SY: *Tiarella unifoliata* Hook.
Tiarella unifoliata Hook. = Tiarella trifoliata var. unifoliata [TIARTRI2]
Tolmiea Torr. & Gray [TOLMIEA$]
Tolmiea menziesii (Pursh) Torr. & Gray [TOLMMEN] (piggy-back plant)
SY: *Tiarella menziesii* Pursh

SCHEUCHZERIACEAE (V:M)

Scheuchzeria L. [SCHEUCH$]
Scheuchzeria americana (Fern.) G.N. Jones = Scheuchzeria palustris ssp. americana [SCHEPAL1]
Scheuchzeria palustris L. [SCHEPAL] (scheuchzeria)
Scheuchzeria palustris ssp. **americana** (Fern.) Hultén [SCHEPAL1]
SY: *Scheuchzeria americana* (Fern.) G.N. Jones
SY: *Scheuchzeria palustris* var. *americana* Fern.
Scheuchzeria palustris var. *americana* Fern. = Scheuchzeria palustris ssp. americana [SCHEPAL1]

SCROPHULARIACEAE (V:D)

Antirrhinum elatine L. = Kickxia elatine [KICKELA]
Antirrhinum tenellum Pursh = Collinsia parviflora var. grandiflora [COLLPAR1]
Besseya Rydb. [BESSEYA$]
Besseya cinerea (Raf.) Pennell = Besseya wyomingensis [BESSWYO]
Besseya wyomingensis (A. Nels.) Rydb. [BESSWYO] (Wyoming kitten-tails)
SY: *Besseya cinerea* (Raf.) Pennell
Castilleja Mutis ex L. f. [CASTILL$]
Castilleja ambigua Hook. & Arn. [CASTAMB]
SY: *Orthocarpus castillejoides* Benth.
Castilleja angustifolia var. *hispida* (Benth.) Fern. = Castilleja hispida [CASTHIS]
Castilleja angustifolia var. *whitedii* Piper = Castilleja elmeri [CASTELM]
Castilleja attenuata (Gray) Chuang & Heckard [CASTATT]
SY: *Orthocarpus attenuatus* Gray
Castilleja cervina Greenm. [CASTCER] (deer paintbrush)
Castilleja chrymactis Pennell = Castilleja miniata [CASTMIN]
Castilleja confusa Greene = Castilleja miniata [CASTMIN]
Castilleja cusickii Greenm. [CASTCUS] (Cusick's paintbrush)
SY: *Castilleja lutea* Heller
Castilleja elmeri Fern. [CASTELM] (Elmer's paintbrush)
SY: *Castilleja angustifolia* var. *whitedii* Piper
Castilleja exilis A. Nels. [CASTEXI] (annual paintbrush)
Castilleja fulva Pennell [CASTFUL] (boreal paintbrush)
Castilleja gracillima Rydb. = Castilleja miniata [CASTMIN]
Castilleja hispida Benth. [CASTHIS] (harsh paintbrush)
SY: *Castilleja angustifolia* var. *hispida* (Benth.) Fern.
SY: *Castilleja hispida* ssp. *abbreviata* (Fern.) Pennell
Castilleja hispida ssp. *abbreviata* (Fern.) Pennell = Castilleja hispida [CASTHIS]
Castilleja hyetophila Pennell [CASTHYE] (common red paintbrush)
Castilleja hyperborea Pennell [CASTHYP] (northern paintbrush)
SY: *Castilleja villosissima* Pennell
Castilleja inconstans Standl. = Castilleja miniata [CASTMIN]
Castilleja lauta A. Nels. = Castilleja rhexifolia [CASTRHE]
Castilleja leonardii Rydb. = Castilleja rhexifolia [CASTRHE]
Castilleja levisecta Greenm. [CASTLEV] (golden paintbrush)
Castilleja lutea Heller = Castilleja cusickii [CASTCUS]
Castilleja luteovirens Rydb. = Castilleja sulphurea [CASTSUL]
Castilleja lutescens (Greenm.) Rydb. [CASTLUT] (yellow paintbrush)
Castilleja miniata Dougl. ex Hook. [CASTMIN] (scarlet paintbrush)
SY: *Castilleja chrymactis* Pennell
SY: *Castilleja confusa* Greene
SY: *Castilleja gracillima* Rydb.
SY: *Castilleja inconstans* Standl.
SY: *Castilleja oblongifolia* Gray
Castilleja mogollonica Pennell = Castilleja sulphurea [CASTSUL]
Castilleja oblongifolia Gray = Castilleja miniata [CASTMIN]
Castilleja occidentalis Torr. [CASTOCC] (western paintbrush)
Castilleja oregonensis Gandog. = Castilleja rhexifolia [CASTRHE]
Castilleja oreopola ssp. *albida* Pennell = Castilleja parviflora var. albida [CASTPAR1]
Castilleja pallescens (Gray) Greenm. [CASTPAL] (palish paintbrush)
Castilleja parviflora Bong. [CASTPAR] (small-flowered paintbrush)
Castilleja parviflora var. **albida** (Pennell) Ownbey [CASTPAR1]
SY: *Castilleja oreopola* ssp. *albida* Pennell
Castilleja raupii Pennell [CASTRAU] (Raup's paintbrush)
Castilleja rhexifolia Rydb. [CASTRHE] (alpine paintbrush)
SY: *Castilleja lauta* A. Nels.
SY: *Castilleja leonardii* Rydb.
SY: *Castilleja oregonensis* Gandog.
Castilleja rhexifolia var. *sulphurea* (Rydb.) Atwood = Castilleja sulphurea [CASTSUL]
Castilleja rupicola Piper ex Fern. [CASTRUP] (cliff paintbrush)
Castilleja sulphurea Rydb. [CASTSUL] (sulphur paintbrush)
SY: *Castilleja luteovirens* Rydb.
SY: *Castilleja mogollonica* Pennell
SY: *Castilleja rhexifolia* var. *sulphurea* (Rydb.) Atwood
Castilleja tenuis (Heller) Chuang & Heckard [CASTTEN]
SY: *Orthocarpus hispidus* Benth.
Castilleja thompsonii Pennell [CASTTHO] (Thompson's paintbrush)
SY: *Castilleja villicaulis* Pennell & Ownbey
Castilleja unalaschcensis (Cham. & Schlecht.) Malte [CASTUNA] (unalaska paintbrush)
Castilleja villicaulis Pennell & Ownbey = Castilleja thompsonii [CASTTHO]
Castilleja villosissima Pennell = Castilleja hyperborea [CASTHYP]
Chaenorrhinum (DC. ex Duby) Reichenb. [CHAENOR$]
*__Chaenorrhinum minus__ (L.) Lange [CHAEMIN] (common snapdragon)
Collinsia Nutt. [COLLINS$]
Collinsia grandiflora Lindl. = Collinsia parviflora var. grandiflora [COLLPAR1]
Collinsia grandiflora var. *pusilla* Gray = Collinsia parviflora var. grandiflora [COLLPAR1]
Collinsia parviflora Lindl. [COLLPAR] (small-flowered blue-eyed Mary)
Collinsia parviflora var. **grandiflora** (Lindl.) Ganders & Krause [COLLPAR1]
SY: *Antirrhinum tenellum* Pursh
SY: *Collinsia grandiflora* Lindl.
SY: *Collinsia grandiflora* var. *pusilla* Gray
SY: *Collinsia tenella* (Pursh) Piper
Collinsia tenella (Pursh) Piper = Collinsia parviflora var. grandiflora [COLLPAR1]
Cymbalaria Hill [CYMBALA$]
*__Cymbalaria muralis__ P.G. Gaertn., B. Mey. & Scherb. [CYMBMUR] (ivy-leaved toadflax)
SY: *Linaria cymbalaria* (L.) P. Mill.
Digitalis L. [DIGITAL$]
*__Digitalis purpurea__ L. [DIGIPUR] (common foxglove)
Eunanus breweri Greene = Mimulus breweri [MIMUBRW]
Euphrasia L. [EUPHRAS$]
Euphrasia americana Wettst. = Euphrasia nemorosa [EUPHNEM]
Euphrasia arctica ssp. *borealis* (Townsend) Yeo = Euphrasia nemorosa [EUPHNEM]
Euphrasia arctica var. *disjuncta* (Fern. & Wieg.) Cronq. = Euphrasia disjuncta [EUPHDIS]
Euphrasia borealis (Townsend) Wettst. = Euphrasia nemorosa [EUPHNEM]
Euphrasia disjuncta Fern. & Wieg. [EUPHDIS]
SY: *Euphrasia arctica* var. *disjuncta* (Fern. & Wieg.) Cronq.
*__Euphrasia nemorosa__ (Pers.) Wallr. [EUPHNEM] (eastern eyebright)

SY: *Euphrasia americana* Wettst.
SY: *Euphrasia arctica* ssp. *borealis* (Townsend) Yeo
SY: *Euphrasia borealis* (Townsend) Wettst.
Gratiola L. [GRATIOL$]
Gratiola ebracteata Benth. ex A. DC. [GRATEBR] (bractless hedge-hyssop)
Gratiola neglecta Torr. [GRATNEG] (common American hedge-hyssop)
SY: *Gratiola neglecta* var. *glaberrima* Fern.
Gratiola neglecta var. *glaberrima* Fern. = Gratiola neglecta [GRATNEG]
Ilysanthes inequalis (Walt.) Pennell = Lindernia dubia var. anagallidea [LINDDUB1]
Kickxia Dumort. [KICKXIA$]
*__Kickxia elatine__ (L.) Dumort. [KICKELA] (sharp-leaved fluellen)
SY: *Antirrhinum elatine* L.
SY: *Linaria elatine* (L.) P. Mill.
Limosella L. [LIMOSEL$]
Limosella aquatica L. [LIMOAQU] (water mudwort)
Linaria P. Mill. [LINARIA$]
Linaria canadensis var. *texana* (Scheele) Pennell = Nuttallanthus texanus [NUTTTEX]
Linaria cymbalaria (L.) P. Mill. = Cymbalaria muralis [CYMBMUR]
*__Linaria dalmatica__ (L.) P. Mill. [LINADAL] (dalmatian toadflax)
SY: *Linaria genistifolia* ssp. *dalmatica* (L.) Maire & Petitm.
Linaria elatine (L.) P. Mill. = Kickxia elatine [KICKELA]
*__Linaria genistifolia__ (L.) P. Mill. [LINAGEN] (broom-leaf toadflax)
Linaria genistifolia ssp. *dalmatica* (L.) Maire & Petitm. = Linaria dalmatica [LINADAL]
Linaria linaria (L.) Karst. = Linaria vulgaris [LINAVUL]
*__Linaria purpurea__ (L.) P. Mill. [LINAPUR] (purple toadflax)
Linaria texana Scheele = Nuttallanthus texanus [NUTTTEX]
*__Linaria vulgaris__ P. Mill. [LINAVUL] (butter-and-eggs)
SY: *Linaria linaria* (L.) Karst.
Lindernia All. [LINDERN$]
Lindernia anagallidea (Michx.) Pennell = Lindernia dubia var. anagallidea [LINDDUB1]
Lindernia dubia (L.) Pennell [LINDDUB]
Lindernia dubia var. **anagallidea** (Michx.) Cooperrider [LINDDUB1]
SY: *Ilysanthes inequalis* (Walt.) Pennell
SY: *Lindernia anagallidea* (Michx.) Pennell
Melampyrum L. [MELAMPY$]
*__Melampyrum lineare__ Desr. [MELALIN] (cow-wheat)
SY: *Melampyrum lineare* var. *americanum* (Michx.) Beauverd
Melampyrum lineare var. *americanum* (Michx.) Beauverd = Melampyrum lineare [MELALIN]
Mimulus L. [MIMULUS$]
Mimulus alpinus (Gray) Piper = Mimulus tilingii var. caespitosus [MIMUTIL1]
Mimulus alsinoides Dougl. ex Benth. [MIMUALS] (chickweed monkey-flower)
Mimulus arvensis Greene = Mimulus guttatus [MIMUGUT]
Mimulus bakeri Gandog. = Mimulus guttatus [MIMUGUT]
Mimulus brachystylis Edwin = Mimulus guttatus [MIMUGUT]
Mimulus breviflorus Piper [MIMUBRE] (short-flowered monkey-flower)
SY: *Mimulus inflatulus* Suksdorf
Mimulus breweri (Greene) Coville [MIMUBRW] (Brewer's monkey-flower)
SY: *Eunanus breweri* Greene
SY: *Mimulus rubellus* var. *breweri* (Greene) Jepson
Mimulus caespitosus (Greene) Greene = Mimulus tilingii var. caespitosus [MIMUTIL1]
Mimulus caespitosus var. *implexus* (Greene) M.E. Peck = Mimulus tilingii var. caespitosus [MIMUTIL1]
Mimulus clementinus Greene = Mimulus guttatus [MIMUGUT]
Mimulus corallinus (Greene) A.L. Grant = Mimulus tilingii var. tilingii [MIMUTIL0]
Mimulus cordatus Greene = Mimulus guttatus [MIMUGUT]
Mimulus cuspidata Greene = Mimulus guttatus [MIMUGUT]
Mimulus decorus (A.L. Grant) Suksdorf = Mimulus guttatus [MIMUGUT]
Mimulus deltoides Gandog. = Mimulus floribundus [MIMUFLO]
Mimulus dentatus Nutt. ex Benth. [MIMUDEN] (toothed-leaved monkey-flower)
Mimulus equinnus Greene = Mimulus guttatus [MIMUGUT]
Mimulus floribundus Lindl. [MIMUFLO] (purple-stemmed monkey-flower)
SY: *Mimulus deltoides* Gandog.
SY: *Mimulus floribundus* var. *geniculatus* (Greene) A.L. Grant
SY: *Mimulus floribundus* var. *membranaceus* (A. Nels.) A.L. Grant
SY: *Mimulus floribundus* ssp. *moorei* Iltis
SY: *Mimulus floribundus* var. *subulatus* A.L. Grant
SY: *Mimulus geniculatus* Greene
SY: *Mimulus membranaceus* A. Nels.
SY: *Mimulus multiflorus* Pennell
SY: *Mimulus peduncularis* Dougl. ex Benth.
SY: *Mimulus pubescens* Benth.
SY: *Mimulus serotinus* Suksdorf
SY: *Mimulus subulatus* (A.L. Grant) Pennell
SY: *Mimulus trisulcatus* Pennell
Mimulus floribundus ssp. *moorei* Iltis = Mimulus floribundus [MIMUFLO]
Mimulus floribundus var. *geniculatus* (Greene) A.L. Grant = Mimulus floribundus [MIMUFLO]
Mimulus floribundus var. *membranaceus* (A. Nels.) A.L. Grant = Mimulus floribundus [MIMUFLO]
Mimulus floribundus var. *subulatus* A.L. Grant = Mimulus floribundus [MIMUFLO]
Mimulus geniculatus Greene = Mimulus floribundus [MIMUFLO]
Mimulus glabratus var. *ascendens* Gray = Mimulus guttatus [MIMUGUT]
Mimulus glareosus Greene = Mimulus guttatus [MIMUGUT]
Mimulus grandiflorus J.T. Howell = Mimulus guttatus [MIMUGUT]
Mimulus grandis (Greene) Heller = Mimulus guttatus [MIMUGUT]
Mimulus guttatus DC. [MIMUGUT] (yellow monkey-flower)
SY: *Mimulus arvensis* Greene
SY: *Mimulus bakeri* Gandog.
SY: *Mimulus brachystylis* Edwin
SY: *Mimulus clementinus* Greene
SY: *Mimulus cordatus* Greene
SY: *Mimulus cuspidata* Greene
SY: *Mimulus decorus* (A.L. Grant) Suksdorf
SY: *Mimulus equinnus* Greene
SY: *Mimulus glabratus* var. *ascendens* Gray
SY: *Mimulus glareosus* Greene
SY: *Mimulus grandiflorus* J.T. Howell
SY: *Mimulus grandis* (Greene) Heller
SY: *Mimulus guttatus* ssp. *arenicola* Pennell
SY: *Mimulus guttatus* ssp. *arvensis* (Greene) Munz
SY: *Mimulus guttatus* var. *arvensis* (Greene) A.L. Grant
SY: *Mimulus guttatus* var. *decorus* A.L. Grant
SY: *Mimulus guttatus* var. *depauperatus* (Gray) A.L. Grant
SY: *Mimulus guttatus* var. *gracilis* (Gray) Campbell
SY: *Mimulus guttatus* var. *grandis* Greene
SY: *Mimulus guttatus* ssp. *haidensis* Calder & Taylor
SY: *Mimulus guttatus* var. *hallii* (Greene) A.L. Grant
SY: *Mimulus guttatus* var. *insignis* Greene
SY: *Mimulus guttatus* var. *laxus* (Pennell ex M.E. Peck) M.E. Peck
SY: *Mimulus guttatus* ssp. *littoralis* Pennell
SY: *Mimulus guttatus* var. *lyratus* (Benth.) Pennell ex M.E. Peck
SY: *Mimulus guttatus* ssp. *micranthus* (Heller) Munz
SY: *Mimulus guttatus* var. *microphyllus* (Benth.) Pennell ex M.E. Peck
SY: *Mimulus guttatus* var. *nasutus* (Greene) Jepson
SY: *Mimulus guttatus* var. *puberulus* (Greene ex Rydb.) A.L. Grant
SY: *Mimulus guttatus* ssp. *scouleri* (Hook.) Pennell
SY: *Mimulus hallii* Greene
SY: *Mimulus hirsutus* J.T. Howell
SY: *Mimulus langsdorfii* Donn ex Greene
SY: *Mimulus langsdorfii* var. *argutus* Greene
SY: *Mimulus langsdorfii* var. *arvensis* (Greene) Jepson
SY: *Mimulus langsdorfii* var. *californicus* Jepson

SY: *Mimulus langsdorfii* var. *grandis* (Greene) Greene
SY: *Mimulus langsdorfii* var. *guttatus* (Fisch. ex DC.) Jepson
SY: *Mimulus langsdorfii* var. *insignis* (Greene) A.L. Grant
SY: *Mimulus langsdorfii* var. *microphyllus* (Benth.) A. Nels. & J.F. Macbr.
SY: *Mimulus langsdorfii* var. *minimus* Henry
SY: *Mimulus langsdorfii* var. *nasutus* (Greene) Jepson
SY: *Mimulus langsdorfii* var. *platyphyllus* Greene
SY: *Mimulus laxus* Pennell ex M.E. Peck
SY: *Mimulus longulus* Greene
SY: *Mimulus luteus* var. *depauperatus* Gray
SY: *Mimulus luteus* var. *gracilis* Gray
SY: *Mimulus lyratus* Benth.
SY: *Mimulus maguirei* Pennell
SY: *Mimulus marmoratus* Greene
SY: *Mimulus micranthus* Heller
SY: *Mimulus microphyllus* Benth.
SY: *Mimulus nasutus* Greene
SY: *Mimulus nasutus* var. *micranthus* (Heller) A.L. Grant
SY: *Mimulus paniculatus* Greene
SY: *Mimulus pardalis* Pennell
SY: *Mimulus parishii* Gandog.
SY: *Mimulus petiolaris* Greene
SY: *Mimulus prionophyllus* Greene
SY: *Mimulus procerus* Greene
SY: *Mimulus puberulus* Greene ex Rydb.
SY: *Mimulus puncticalyx* Gandog.
SY: *Mimulus rivularis* Nutt.
SY: *Mimulus scouleri* Hook.
SY: *Mimulus subreniformis* Greene
SY: *Mimulus tenellus* Nutt. ex Gray
SY: *Mimulus thermalis* A. Nels.
SY: *Mimulus unimaculatus* Pennell

Mimulus guttatus ssp. *arenicola* Pennell = Mimulus guttatus [MIMUGUT]
Mimulus guttatus ssp. *arvensis* (Greene) Munz = Mimulus guttatus [MIMUGUT]
Mimulus guttatus ssp. *haidensis* Calder & Taylor = Mimulus guttatus [MIMUGUT]
Mimulus guttatus ssp. *littoralis* Pennell = Mimulus guttatus [MIMUGUT]
Mimulus guttatus ssp. *micranthus* (Heller) Munz = Mimulus guttatus [MIMUGUT]
Mimulus guttatus ssp. *scouleri* (Hook.) Pennell = Mimulus guttatus [MIMUGUT]
Mimulus guttatus var. *arvensis* (Greene) A.L. Grant = Mimulus guttatus [MIMUGUT]
Mimulus guttatus var. *decorus* A.L. Grant = Mimulus guttatus [MIMUGUT]
Mimulus guttatus var. *depauperatus* (Gray) A.L. Grant = Mimulus guttatus [MIMUGUT]
Mimulus guttatus var. *gracilis* (Gray) Campbell = Mimulus guttatus [MIMUGUT]
Mimulus guttatus var. *grandis* Greene = Mimulus guttatus [MIMUGUT]
Mimulus guttatus var. *hallii* (Greene) A.L. Grant = Mimulus guttatus [MIMUGUT]
Mimulus guttatus var. *insignis* Greene = Mimulus guttatus [MIMUGUT]
Mimulus guttatus var. *laxus* (Pennell ex M.E. Peck) M.E. Peck = Mimulus guttatus [MIMUGUT]
Mimulus guttatus var. *lyratus* (Benth.) Pennell ex M.E. Peck = Mimulus guttatus [MIMUGUT]
Mimulus guttatus var. *microphyllus* (Benth.) Pennell ex M.E. Peck = Mimulus guttatus [MIMUGUT]
Mimulus guttatus var. *nasutus* (Greene) Jepson = Mimulus guttatus [MIMUGUT]
Mimulus guttatus var. *puberulus* (Greene ex Rydb.) A.L. Grant = Mimulus guttatus [MIMUGUT]
Mimulus hallii Greene = Mimulus guttatus [MIMUGUT]
Mimulus hirsutus J.T. Howell = Mimulus guttatus [MIMUGUT]
Mimulus implicatus Greene = Mimulus tilingii var. tilingii [MIMUTIL0]
Mimulus inflatulus Suksdorf = Mimulus breviflorus [MIMUBRE]
Mimulus langsdorfii Donn ex Greene = Mimulus guttatus [MIMUGUT]
Mimulus langsdorfii var. *argutus* Greene = Mimulus guttatus [MIMUGUT]
Mimulus langsdorfii var. *arvensis* (Greene) Jepson = Mimulus guttatus [MIMUGUT]
Mimulus langsdorfii var. *californicus* Jepson = Mimulus guttatus [MIMUGUT]
Mimulus langsdorfii var. *grandis* (Greene) Greene = Mimulus guttatus [MIMUGUT]
Mimulus langsdorfii var. *guttatus* (Fisch. ex DC.) Jepson = Mimulus guttatus [MIMUGUT]
Mimulus langsdorfii var. *insignis* (Greene) A.L. Grant = Mimulus guttatus [MIMUGUT]
Mimulus langsdorfii var. *microphyllus* (Benth.) A. Nels. & J.F. Macbr. = Mimulus guttatus [MIMUGUT]
Mimulus langsdorfii var. *minimus* Henry = Mimulus guttatus [MIMUGUT]
Mimulus langsdorfii var. *minor* (A. Nels.) Cockerell = Mimulus tilingii var. tilingii [MIMUTIL0]
Mimulus langsdorfii var. *nasutus* (Greene) Jepson = Mimulus guttatus [MIMUGUT]
Mimulus langsdorfii var. *platyphyllus* Greene = Mimulus guttatus [MIMUGUT]
Mimulus langsdorfii var. *tilingii* (Regel) Greene = Mimulus tilingii var. tilingii [MIMUTIL0]
Mimulus laxus Pennell ex M.E. Peck = Mimulus guttatus [MIMUGUT]
Mimulus lewisii Pursh [MIMULEW] (pink monkey-flower)
Mimulus longulus Greene = Mimulus guttatus [MIMUGUT]
Mimulus lucens Greene = Mimulus tilingii var. tilingii [MIMUTIL0]
Mimulus luteus var. *alpinus* Gray = Mimulus tilingii var. caespitosus [MIMUTIL1]
Mimulus luteus var. *depauperatus* Gray = Mimulus guttatus [MIMUGUT]
Mimulus luteus var. *gracilis* Gray = Mimulus guttatus [MIMUGUT]
Mimulus lyratus Benth. = Mimulus guttatus [MIMUGUT]
Mimulus maguirei Pennell = Mimulus guttatus [MIMUGUT]
Mimulus marmoratus Greene = Mimulus guttatus [MIMUGUT]
Mimulus membranaceus A. Nels. = Mimulus floribundus [MIMUFLO]
Mimulus micranthus Heller = Mimulus guttatus [MIMUGUT]
Mimulus microphyllus Benth. = Mimulus guttatus [MIMUGUT]
Mimulus minor A. Nels. = Mimulus tilingii var. tilingii [MIMUTIL0]
Mimulus minusculus Greene = Mimulus tilingii var. tilingii [MIMUTIL0]
Mimulus moschatus Dougl. ex Lindl. [MIMUMOS] (musk-flower)
Mimulus multiflorus Pennell = Mimulus floribundus [MIMUFLO]
Mimulus nasutus Greene = Mimulus guttatus [MIMUGUT]
Mimulus nasutus var. *micranthus* (Heller) A.L. Grant = Mimulus guttatus [MIMUGUT]
Mimulus paniculatus Greene = Mimulus guttatus [MIMUGUT]
Mimulus pardalis Pennell = Mimulus guttatus [MIMUGUT]
Mimulus parishii Gandog. = Mimulus guttatus [MIMUGUT]
Mimulus peduncularis Dougl. ex Benth. = Mimulus floribundus [MIMUFLO]
Mimulus petiolaris Greene = Mimulus guttatus [MIMUGUT]
Mimulus prionophyllus Greene = Mimulus guttatus [MIMUGUT]
Mimulus procerus Greene = Mimulus guttatus [MIMUGUT]
Mimulus puberulus Greene ex Rydb. = Mimulus guttatus [MIMUGUT]
Mimulus pubescens Benth. = Mimulus floribundus [MIMUFLO]
Mimulus puncticalyx Gandog. = Mimulus guttatus [MIMUGUT]
Mimulus rivularis Nutt. = Mimulus guttatus [MIMUGUT]
Mimulus rubellus var. *breweri* (Greene) Jepson = Mimulus breweri [MIMUBRW]
Mimulus scouleri Hook. = Mimulus guttatus [MIMUGUT]
Mimulus scouleri var. *caespitosus* Greene = Mimulus tilingii var. caespitosus [MIMUTIL1]
Mimulus serotinus Suksdorf = Mimulus floribundus [MIMUFLO]
Mimulus subreniformis Greene = Mimulus guttatus [MIMUGUT]

Mimulus subulatus (A.L. Grant) Pennell = Mimulus floribundus [MIMUFLO]
Mimulus tenellus Nutt. ex Gray = Mimulus guttatus [MIMUGUT]
Mimulus thermalis A. Nels. = Mimulus guttatus [MIMUGUT]
Mimulus tilingii Regel [MIMUTIL] (mountain monkey-flower)
Mimulus tilingii var. **caespitosus** (Greene) A.L. Grant [MIMUTIL1]
SY: *Mimulus alpinus* (Gray) Piper
SY: *Mimulus caespitosus* (Greene) Greene
SY: *Mimulus caespitosus* var. *implexus* (Greene) M.E. Peck
SY: *Mimulus luteus* var. *alpinus* Gray
SY: *Mimulus scouleri* var. *caespitosus* Greene
Mimulus tilingii var. *corallinus* (Greene) A.L. Grant = Mimulus tilingii var. tilingii [MIMUTIL0]
Mimulus tilingii var. **tilingii** [MIMUTIL0]
SY: *Mimulus corallinus* (Greene) A.L. Grant
SY: *Mimulus implicatus* Greene
SY: *Mimulus langsdorfii* var. *minor* (A. Nels.) Cockerell
SY: *Mimulus langsdorfii* var. *tilingii* (Regel) Greene
SY: *Mimulus lucens* Greene
SY: *Mimulus minor* A. Nels.
SY: *Mimulus minusculus* Greene
SY: *Mimulus tilingii* var. *corallinus* (Greene) A.L. Grant
SY: *Mimulus veronicifolius* Greene
Mimulus trisulcatus Pennell = Mimulus floribundus [MIMUFLO]
Mimulus unimaculatus Pennell = Mimulus guttatus [MIMUGUT]
Mimulus veronicifolius Greene = Mimulus tilingii var. tilingii [MIMUTIL0]
Nothochelone (Gray) Straw [NOTHOCH$]
Nothochelone nemorosa (Dougl. ex Lindl.) Straw [NOTHNEM] (woodland penstemon)
SY: *Penstemon nemorosus* (Dougl. ex Lindl.) Trautv.
Nuttallanthus D.A. Sutton [NUTTALL$]
Nuttallanthus texanus (Scheele) D.A. Sutton [NUTTTEX]
SY: *Linaria canadensis* var. *texana* (Scheele) Pennell
SY: *Linaria texana* Scheele
Orthocarpus Nutt. [ORTHOCA$]
Orthocarpus attenuatus Gray = Castilleja attenuata [CASTATT]
Orthocarpus bracteosus Benth. [ORTHBRA] (rosy owl-clover)
SY: *Orthocarpus bracteosus* var. *albus* Keck
Orthocarpus bracteosus var. *albus* Keck = Orthocarpus bracteosus [ORTHBRA]
Orthocarpus castillejoides Benth. = Castilleja ambigua [CASTAMB]
Orthocarpus faucibarbatus ssp. *albidus* Keck = Triphysaria versicolor [TRIPVER]
Orthocarpus faucibarbatus var. *albidus* (Keck) J.T. Howell = Triphysaria versicolor [TRIPVER]
Orthocarpus hispidus Benth. = Castilleja tenuis [CASTTEN]
Orthocarpus imbricatus Torr. ex S. Wats. [ORTHIMB] (mountain owl-clover)
Orthocarpus luteus Nutt. [ORTHLUT] (yellow owl-clover)
Orthocarpus pusillus Benth. = Triphysaria pusilla [TRIPPUS]
Orthocarpus pusillus var. *densiuscuus* (Gandog.) Keck = Triphysaria pusilla [TRIPPUS]
Orthocarpus tenuifolius (Pursh) Benth. [ORTHTEN] (thin-leaved owl-clover)
Parentucellia Viviani [PARENTU$]
***Parentucellia viscosa** (L.) Caruel [PAREVIS] (yellow parentucellia)
Pediculariopsis verticillata (L.) A. & D. Löve = Pedicularis verticillata [PEDIVER]
Pedicularis L. [PEDICUL$]
Pedicularis arctica R. Br. = Pedicularis langsdorfii ssp. arctica [PEDILAG1]
Pedicularis bracteosa Benth. [PEDIBRA] (bracted lousewort)
Pedicularis bracteosa var. **bracteosa** [PEDIBRA0]
SY: *Pedicularis montanensis* Rydb.
SY: *Pedicularis paddoensis* Pennell
SY: *Pedicularis thompsonii* Pennell
Pedicularis bracteosa var. **latifolia** (Pennell) Cronq. [PEDIBRA2]
SY: *Pedicularis latifolia* Pennell
Pedicularis capitata M.F. Adams [PEDICAP] (capitate lousewort)
SY: *Pedicularis nelsonii* R. Br.
Pedicularis contorta Benth. [PEDICON] (coil-beaked lousewort)
Pedicularis groenlandica Retz. [PEDIGRO] (elephant's-head lousewort)
Pedicularis hians Eastw. = Pedicularis langsdorfii ssp. arctica [PEDILAG1]
Pedicularis kanei Dur. = Pedicularis lanata [PEDILAN]
Pedicularis labradorica Wirsing [PEDILAB] (Labrador lousewort)
Pedicularis lanata Cham. & Schlecht. [PEDILAN] (woolly lousewort)
SY: *Pedicularis kanei* Dur.
SY: *Pedicularis willdenowii* Vved.
Pedicularis langsdorfii Fisch. ex Stev. [PEDILAG] (Langsdorf's lousewort)
Pedicularis langsdorfii ssp. **arctica** (R. Br.) Pennell [PEDILAG1] (arctic Langsdorf's lousewort)
SY: *Pedicularis arctica* R. Br.
SY: *Pedicularis hians* Eastw.
SY: *Pedicularis langsdorfii* var. *arctica* (R. Br.) Polunin
Pedicularis langsdorfii var. *arctica* (R. Br.) Polunin = Pedicularis langsdorfii ssp. arctica [PEDILAG1]
Pedicularis latifolia Pennell = Pedicularis bracteosa var. latifolia [PEDIBRA2]
Pedicularis macrodonta Richards. [PEDIMAC] (boreal lousewort)
SY: *Pedicularis parviflora* var. *macrodonta* (Richards.) Welsh
Pedicularis montanensis Rydb. = Pedicularis bracteosa var. bracteosa [PEDIBRA0]
Pedicularis nelsonii R. Br. = Pedicularis capitata [PEDICAP]
Pedicularis oederi Vahl ex Hornem. [PEDIOED] (Oeder's lousewort)
SY: *Pedicularis versicolor* Wahlenb.
Pedicularis ornithorhyncha Benth. [PEDIORN] (bird's-beak lousewort)
SY: *Pedicularis pedicellata* Bunge
SY: *Pedicularis subnuda* Benth.
Pedicularis paddoensis Pennell = Pedicularis bracteosa var. bracteosa [PEDIBRA0]
Pedicularis parviflora Sm. ex Rees [PEDIPAR] (small-flowered lousewort)
SY: *Pedicularis pennellii* ssp. *insularis* Calder & Taylor
SY: *Pedicularis pennellii* var. *insularis* (Calder & Taylor) Boivin
Pedicularis parviflora var. *macrodonta* (Richards.) Welsh = Pedicularis macrodonta [PEDIMAC]
Pedicularis pedicellata Bunge = Pedicularis ornithorhyncha [PEDIORN]
Pedicularis pennellii ssp. *insularis* Calder & Taylor = Pedicularis parviflora [PEDIPAR]
Pedicularis pennellii var. *insularis* (Calder & Taylor) Boivin = Pedicularis parviflora [PEDIPAR]
Pedicularis racemosa Dougl. ex Benth. [PEDIRAC] (sickletop lousewort)
Pedicularis subnuda Benth. = Pedicularis ornithorhyncha [PEDIORN]
Pedicularis sudetica Willd. [PEDISUD] (Sudeten lousewort)
Pedicularis sudetica ssp. **interior** (Hultén) Hultén [PEDISUD1]
SY: *Pedicularis sudetica* var. *gymnocephala* Trautv.
Pedicularis sudetica var. *gymnocephala* Trautv. = Pedicularis sudetica ssp. interior [PEDISUD1]
Pedicularis thompsonii Pennell = Pedicularis bracteosa var. bracteosa [PEDIBRA0]
Pedicularis versicolor Wahlenb. = Pedicularis oederi [PEDIOED]
Pedicularis verticillata L. [PEDIVER] (whorled lousewort)
SY: *Pediculariopsis verticillata* (L.) A. & D. Löve
Pedicularis willdenowii Vved. = Pedicularis lanata [PEDILAN]
Penstemon Schmidel [PENSTEM$]
Penstemon albertinus Greene [PENSALB] (Alberta penstemon)
Penstemon confertus Dougl. ex Lindl. [PENSCON] (yellow penstemon)
Penstemon confertus ssp. *procerus* (Dougl. ex Graham) D.V. Clark = Penstemon procerus var. procerus [PENSPRO0]
Penstemon confertus var. *procerus* (Dougl. ex Graham) Coville = Penstemon procerus var. procerus [PENSPRO0]
Penstemon cusickii Gray [PENSCUS] (Cusick's penstemon)
Penstemon davidsonii Greene [PENSDAV] (Davidson's penstemon)
Penstemon davidsonii ssp. *menziesii* Keck = Penstemon davidsonii var. menziesii [PENSDAV2]
Penstemon davidsonii var. **davidsonii** [PENSDAV0]

SY: *Penstemon menziesii* ssp. *davidsonii* (Greene) Piper
SY: *Penstemon menziesii* ssp. *thompsonii* Pennell
Penstemon davidsonii var. *ellipticus* (Coult. & Fisher) Boivin = Penstemon ellipticus [PENSELL]
Penstemon davidsonii var. **menziesii** (Keck) Cronq. [PENSDAV2]
SY: *Penstemon davidsonii* ssp. *menziesii* Keck
SY: *Penstemon menziesii* Hook. p.p.
Penstemon ellipticus Coult. & Fisher [PENSELL] (oval penstemon)
SY: *Penstemon davidsonii* var. *ellipticus* (Coult. & Fisher) Boivin
Penstemon eriantherus Pursh [PENSERI] (fuzzy-tongued penstemon)
Penstemon fruticosus (Pursh) Greene [PENSFRU] (shrubby penstemon)
Penstemon fruticosus ssp. *scouleri* (Lindl.) Pennell & Keck = Penstemon fruticosus var. scouleri [PENSFRU2]
Penstemon fruticosus var. **scouleri** (Lindl.) Cronq. [PENSFRU2]
SY: *Penstemon fruticosus* ssp. *scouleri* (Lindl.) Pennell & Keck
Penstemon gormanii Greene [PENSGOR] (Gorman's penstemon)
Penstemon gracilis Nutt. [PENSGRA] (slender penstemon)
Penstemon lyallii (Gray) Gray [PENSLYA] (Lyall's penstemon)
Penstemon menziesii Hook. p.p. = Penstemon davidsonii var. menziesii [PENSDAV2]
Penstemon menziesii ssp. *davidsonii* (Greene) Piper = Penstemon davidsonii var. davidsonii [PENSDAV0]
Penstemon menziesii ssp. *thompsonii* Pennell = Penstemon davidsonii var. davidsonii [PENSDAV0]
Penstemon nemorosus (Dougl. ex Lindl.) Trautv. = Nothochelone nemorosa [NOTHNEM]
Penstemon nitidus Dougl. ex Benth. [PENSNIT] (shining penstemon)
Penstemon ovatus Dougl. ex Hook. [PENSOVA] (broad-leaved penstemon)
Penstemon procerus Dougl. ex Graham [PENSPRO] (small-flowered penstemon)
Penstemon procerus var. **procerus** [PENSPRO0]
SY: *Penstemon confertus* ssp. *procerus* (Dougl. ex Graham) D.V. Clark
SY: *Penstemon confertus* var. *procerus* (Dougl. ex Graham) Coville
Penstemon procerus var. **tolmiei** (Hook.) Cronq. [PENSPRO2]
SY: *Penstemon tolmiei* Hook.
Penstemon pruinosus Dougl. ex Lindl. [PENSPRU] (Chelan penstemon)
Penstemon richardsonii Dougl. ex Lindl. [PENSRIC] (Richardson's penstemon)
Penstemon serrulatus Menzies ex Sm. [PENSSER] (coast penstemon)
Penstemon tolmiei Hook. = Penstemon procerus var. tolmiei [PENSPRO2]
Rhinanthus L. [RHINANT$]
Rhinanthus borealis ssp. *kyrolliae* (Chabert) Pennell = Rhinanthus minor [RHINMIN]
Rhinanthus crista-galli L. = Rhinanthus minor [RHINMIN]
Rhinanthus minor L. [RHINMIN] (yellow rattle)
SY: *Rhinanthus borealis* ssp. *kyrolliae* (Chabert) Pennell
SY: *Rhinanthus crista-galli* L.
SY: *Rhinanthus stenophyllus* (Schur) Schinz & Thellung
Rhinanthus stenophyllus (Schur) Schinz & Thellung = Rhinanthus minor [RHINMIN]
Scrophularia L. [SCROPHU$]
Scrophularia californica var. *oregana* (Pennell) Boivin = Scrophularia oregana [SCROORE]
Scrophularia californica var. *oregana* (Pennell) Cronq. = Scrophularia oregana [SCROORE]
Scrophularia lanceolata Pursh [SCROLAN] (lance-leaved figwort)
SY: *Scrophularia pectinata* Raf.
Scrophularia oregana Pennell [SCROORE]
SY: *Scrophularia californica* var. *oregana* (Pennell) Cronq.
SY: *Scrophularia californica* var. *oregana* (Pennell) Boivin
Scrophularia pectinata Raf. = Scrophularia lanceolata [SCROLAN]
Tonella Nutt. ex Gray [TONELLA$]
Tonella tenella (Benth.) Heller [TONETEN] (small-flowered tonella)
Triphysaria Fisch. & C.A. Mey. [TRIPHYS$]
Triphysaria pusilla (Benth.) Chuang & Heckard [TRIPPUS]
SY: *Orthocarpus pusillus* Benth.
SY: *Orthocarpus pusillus* var. *densiuscuus* (Gandog.) Keck
Triphysaria versicolor Fisch. & C.A. Mey. [TRIPVER]
SY: *Orthocarpus faucibarbatus* ssp. *albidus* Keck
SY: *Orthocarpus faucibarbatus* var. *albidus* (Keck) J.T. Howell
Verbascum L. [VERBASC$]
*__Verbascum blattaria__ L. [VERBBLA] (moth mullein)
*__Verbascum phlomoides__ L. [VERBPHL] (woolly mullein)
*__Verbascum thapsus__ L. [VERBTHA] (great mullein)
Veronica L. [VERONIC$]
Veronica alpina var. *alterniflora* Fern. = Veronica wormskjoldii [VEROWOR]
Veronica alpina var. *cascadensis* Fern. = Veronica wormskjoldii [VEROWOR]
Veronica alpina var. *geminiflora* Fern. = Veronica wormskjoldii [VEROWOR]
Veronica alpina var. *nutans* (Bong.) Boivin = Veronica wormskjoldii [VEROWOR]
Veronica alpina var. *terrae-novae* Fern. = Veronica wormskjoldii [VEROWOR]
Veronica alpina var. *unalaschcensis* Cham. & Schlecht. = Veronica wormskjoldii [VEROWOR]
Veronica alpina var. *wormskjoldii* Hook. = Veronica wormskjoldii [VEROWOR]
Veronica americana Schwein. ex Benth. [VEROAME] (American brooklime)
SY: *Veronica beccabunga* ssp. *americana* (Raf.) Sellers
Veronica anagallis L. = Veronica anagallis-aquatica [VEROANA]
*__Veronica anagallis-aquatica__ L. [VEROANA] (blue water speedwell)
SY: *Veronica anagallis* L.
SY: *Veronica catenata* Pennell
SY: *Veronica catenata* var. *glandulosa* (Farw.) Pennell
SY: *Veronica comosa* var. *glaberrima* (Pennell) Boivin
SY: *Veronica comosa* var. *glandulosa* (Farw.) Boivin
SY: *Veronica connata* ssp. *glaberrima* Pennell
SY: *Veronica connata* var. *glaberrima* (Pennell) Fern.
SY: *Veronica connata* var. *typica* Pennell
SY: *Veronica glandifera* Pennell
SY: *Veronica* × *lackschewitzii* Keller
*__Veronica arvensis__ L. [VEROARV] (wall speedwell)
Veronica beccabunga L. [VEROBEC] (European speedwell)
Veronica beccabunga ssp. *americana* (Raf.) Sellers = Veronica americana [VEROAME]
Veronica catenata Pennell = Veronica anagallis-aquatica [VEROANA]
Veronica catenata var. *glandulosa* (Farw.) Pennell = Veronica anagallis-aquatica [VEROANA]
*__Veronica chamaedrys__ L. [VEROCHA] (germander speedwell)
Veronica comosa var. *glaberrima* (Pennell) Boivin = Veronica anagallis-aquatica [VEROANA]
Veronica comosa var. *glandulosa* (Farw.) Boivin = Veronica anagallis-aquatica [VEROANA]
Veronica connata ssp. *glaberrima* Pennell = Veronica anagallis-aquatica [VEROANA]
Veronica connata var. *glaberrima* (Pennell) Fern. = Veronica anagallis-aquatica [VEROANA]
Veronica connata var. *typica* Pennell = Veronica anagallis-aquatica [VEROANA]
Veronica cusickii Gray [VEROCUS] (Cusick's speedwell)
*__Veronica filiformis__ Sm. [VEROFIL] (slender speedwell)
Veronica glandifera Pennell = Veronica anagallis-aquatica [VEROANA]
Veronica hederaefolia L. = Veronica hederifolia [VEROHED]
*__Veronica hederifolia__ L. [VEROHED]
SY: *Veronica hederaefolia* L.
Veronica humifusa Dickson = Veronica serpyllifolia ssp. humifusa [VEROSER1]
Veronica × *lackschewitzii* Keller = Veronica anagallis-aquatica [VEROANA]
Veronica nutans Bong. = Veronica wormskjoldii [VEROWOR]
*__Veronica officinalis__ L. [VEROOFF] (common speedwell)
Veronica peregrina L. [VEROPER] (purslane speedwell)

Veronica peregrina ssp. **peregrina** [VEROPER0]
SY: *Veronica peregrina* var. *typica* Pennell
Veronica peregrina ssp. **xalapensis** (Kunth) Pennell [VEROPER1]
SY: *Veronica peregrina* var. *xalapensis* (Kunth) Pennell
SY: *Veronica sherwoodii* M.E. Peck
SY: *Veronica xalapensis* Kunth
Veronica peregrina var. *typica* Pennell = Veronica peregrina ssp. peregrina [VEROPER0]
Veronica peregrina var. *xalapensis* (Kunth) Pennell = Veronica peregrina ssp. xalapensis [VEROPER1]
***Veronica persica** Poir. [VEROPES] (bird's-eye speedwell)
Veronica scutellata L. [VEROSCU] (marsh speedwell)
SY: *Veronica scutellata* var. *villosa* Schumacher
Veronica scutellata var. *villosa* Schumacher = Veronica scutellata [VEROSCU]
***Veronica serpyllifolia** L. [VEROSER] (thyme-leaved speedwell)
Veronica serpyllifolia ssp. **humifusa** (Dickson) Syme [VEROSER1]
SY: *Veronica humifusa* Dickson
SY: *Veronica serpyllifolia* var. *borealis* Laestad.
SY: *Veronica serpyllifolia* var. *humifusa* (Dickson) Vahl
SY: *Veronica tenella* All.
SY: *Veronicastrum serpyllifolium* ssp. *humifusum* (Dickson) W.A. Weber
Veronica serpyllifolia ssp. **serpyllifolia** [VEROSER0]
SY: *Veronicastrum serpyllifolium* (L.) Fourr.
Veronica serpyllifolia var. *borealis* Laestad. = Veronica serpyllifolia ssp. humifusa [VEROSER1]
Veronica serpyllifolia var. *humifusa* (Dickson) Vahl = Veronica serpyllifolia ssp. humifusa [VEROSER1]
Veronica sherwoodii M.E. Peck = Veronica peregrina ssp. xalapensis [VEROPER1]
Veronica stelleri Pallas ex Link = Veronica wormskjoldii [VEROWOR]
Veronica stelleri var. *glabrescens* Hultén = Veronica wormskjoldii [VEROWOR]
Veronica tenella All. = Veronica serpyllifolia ssp. humifusa [VEROSER1]
***Veronica verna** L. [VEROVER] (spring speedwell)
Veronica wormskjoldii Roemer & J.A. Schultes [VEROWOR] (alpine speedwell)
SY: *Veronica alpina* var. *alterniflora* Fern.
SY: *Veronica alpina* var. *cascadensis* Fern.
SY: *Veronica alpina* var. *geminiflora* Fern.
SY: *Veronica alpina* var. *nutans* (Bong.) Boivin
SY: *Veronica alpina* var. *terrae-novae* Fern.
SY: *Veronica alpina* var. *unalaschcensis* Cham. & Schlecht.
SY: *Veronica alpina* var. *wormskjoldii* Hook.
SY: *Veronica nutans* Bong.
SY: *Veronica stelleri* Pallas ex Link
SY: *Veronica stelleri* var. *glabrescens* Hultén
SY: *Veronica wormskjoldii* ssp. *alterniflora* (Fern.) Pennell
Veronica wormskjoldii ssp. *alterniflora* (Fern.) Pennell = Veronica wormskjoldii [VEROWOR]
Veronica xalapensis Kunth = Veronica peregrina ssp. xalapensis [VEROPER1]
Veronicastrum serpyllifolium (L.) Fourr. = Veronica serpyllifolia ssp. serpyllifolia [VEROSER0]
Veronicastrum serpyllifolium ssp. *humifusum* (Dickson) W.A. Weber = Veronica serpyllifolia ssp. humifusa [VEROSER1]

SELAGINELLACEAE (V:F)

Selaginella Beauv. [SELAGIN$]
Selaginella densa Rydb. [SELADEN] (compact selaginella)
Selaginella densa var. **scopulorum** (Maxon) R. Tryon [SELADEN2]
SY: *Selaginella engelmannii* var. *scopulorum* (Maxon) C.F. Reed
SY: *Selaginella scopulorum* Maxon
Selaginella engelmannii var. *scopulorum* (Maxon) C.F. Reed = Selaginella densa var. scopulorum [SELADEN2]
Selaginella oregana D.C. Eat. [SELAORE] (Oregon selaginella)
Selaginella scopulorum Maxon = Selaginella densa var. scopulorum [SELADEN2]
Selaginella selaginoides (L.) Link [SELASEL] (mountain-moss)
Selaginella sibirica (Milde) Hieron. [SELASIB] (northern selaginella)
Selaginella wallacei Hieron. [SELAWAL] (Wallace's selaginella)

SOLANACEAE (V:D)

Androcera rostrata (Dunal) Rydb. = Solanum rostratum [SOLAROS]
Datura L. [DATURA$]
***Datura stramonium** L. [DATUSTR] (jimsonweed)
SY: *Datura stramonium* var. *tatula* (L.) Torr.
SY: *Datura tatula* L.
Datura stramonium var. *tatula* (L.) Torr. = Datura stramonium [DATUSTR]
Datura tatula L. = Datura stramonium [DATUSTR]
Lycium L. [LYCIUM$]
***Lycium barbarum** L. [LYCIBAR]
SY: *Lycium halimifolium* P. Mill.
Lycium halimifolium P. Mill. = Lycium barbarum [LYCIBAR]
Solanum L. [SOLANUM$]
***Solanum americanum** P. Mill. [SOLAAME] (black nightshade)
SY: *Solanum americanum* var. *nodiflorum* (Jacq.) Edmonds
SY: *Solanum americanum* var. *patulum* (L.) Edmonds
SY: *Solanum fistulosum* Dunal ex Poir.
SY: *Solanum hermannii* Dunal
SY: *Solanum linnaeanum* Hepper & Jaeger
SY: *Solanum nigrum* L., p.p.
SY: *Solanum nigrum* var. *americanum* (P. Mill.) O.E. Schulz
SY: *Solanum nigrum* var. *virginicum* L.
SY: *Solanum nodiflorum* Jacq.
Solanum americanum var. *nodiflorum* (Jacq.) Edmonds = Solanum americanum [SOLAAME]
Solanum americanum var. *patulum* (L.) Edmonds = Solanum americanum [SOLAAME]
Solanum cornutum auct. = Solanum rostratum [SOLAROS]
***Solanum dulcamara** L. [SOLADUL] (European bittersweet)
Solanum fistulosum Dunal ex Poir. = Solanum americanum [SOLAAME]
Solanum hermannii Dunal = Solanum americanum [SOLAAME]
Solanum linnaeanum Hepper & Jaeger = Solanum americanum [SOLAAME]
Solanum nigrum L., p.p. = Solanum americanum [SOLAAME]
Solanum nigrum var. *americanum* (P. Mill.) O.E. Schulz = Solanum americanum [SOLAAME]
Solanum nigrum var. *virginicum* L. = Solanum americanum [SOLAAME]
Solanum nodiflorum Jacq. = Solanum americanum [SOLAAME]
***Solanum rostratum** Dunal [SOLAROS] (buffalo-bur)
SY: *Androcera rostrata* (Dunal) Rydb.
SY: *Solanum cornutum* auct.
***Solanum sarrachoides** Sendtner [SOLASAR] (hairy nightshade)
Solanum triflorum Nutt. [SOLATRI] (cut-leaved nightshade)

SPARGANIACEAE (V:M)

Sparganium L. [SPARGAN$]
Sparganium acaule (Beeby) Rydb. = Sparganium angustifolium [SPARANG]
Sparganium angustifolium Michx. [SPARANG] (narrow-leaved bur-reed)
SY: *Sparganium acaule* (Beeby) Rydb.
SY: *Sparganium angustifolium* ssp. *emersum* (Rehmann) T.C. Brayshaw
SY: *Sparganium angustifolium* var. *multipedunculatum* (Morong) T.C. Brayshaw
SY: *Sparganium chlorocarpum* var. *acaule* (Beeby) Fern.
SY: *Sparganium emersum* Rehmann
SY: *Sparganium emersum* var. *angustifolium* (Michx.) Taylor & MacBryde
SY: *Sparganium emersum* var. *multipedunculatum* (Morong) Reveal
SY: *Sparganium multipedunculatum* (Morong) Rydb.
SY: *Sparganium simplex* var. *multipedunculatum* Morong

Sparganium angustifolium ssp. *emersum* (Rehmann) T.C. Brayshaw = Sparganium angustifolium [SPARANG]
Sparganium angustifolium var. *multipedunculatum* (Morong) T.C. Brayshaw = Sparganium angustifolium [SPARANG]
Sparganium californicum Greene = Sparganium eurycarpum [SPAREUR]
Sparganium chlorocarpum var. *acaule* (Beeby) Fern. = Sparganium angustifolium [SPARANG]
Sparganium emersum Rehmann = Sparganium angustifolium [SPARANG]
Sparganium emersum var. *angustifolium* (Michx.) Taylor & MacBryde = Sparganium angustifolium [SPARANG]
Sparganium emersum var. *multipedunculatum* (Morong) Reveal = Sparganium angustifolium [SPARANG]
Sparganium eurycarpum Engelm. ex Gray [SPAREUR] (broad-fruited bur-reed)
SY: *Sparganium californicum* Greene
SY: *Sparganium eurycarpum* var. *greenei* (Morong) Graebn.
SY: *Sparganium greenei* Morong
Sparganium eurycarpum var. *greenei* (Morong) Graebn. = Sparganium eurycarpum [SPAREUR]
Sparganium fluctuans (Morong) B.L. Robins. [SPARFLU] (water bur-reed)
Sparganium glomeratum Laestad. ex Beurling [SPARGLO] (glomerate bur-reed)
Sparganium greenei Morong = Sparganium eurycarpum [SPAREUR]
Sparganium hyperboreum Laestad. ex Beurling [SPARHYP] (northern bur-reed)
Sparganium minimum (Hartman) Wallr. = Sparganium nutans [SPARNUT]
Sparganium multipedunculatum (Morong) Rydb. = Sparganium angustifolium [SPARANG]
Sparganium nutans L. [SPARNUT] (small bur-reed)
SY: *Sparganium minimum* (Hartman) Wallr.
Sparganium simplex var. *multipedunculatum* Morong = Sparganium angustifolium [SPARANG]

TAXACEAE (V:G)

Taxus L. [TAXUS$]
Taxus brevifolia Nutt. [TAXUBRE] (western yew)

THELYPTERIDACEAE (V:F)

Dryopteris nevadensis (Baker) Underwood = Thelypteris nevadensis [THELNEV]
Dryopteris oregana C. Christens. = Thelypteris nevadensis [THELNEV]
Dryopteris phegopteris (L.) C. Christens. = Phegopteris connectilis [PHEGCON]
Lastrea oregana (C. Christens.) Copeland = Thelypteris nevadensis [THELNEV]
Oreopteris quelpaertensis (Christ) Holub = Thelypteris quelpaertensis [THELQUE]
Phegopteris Fée [PHEGOPT$]
Phegopteris connectilis (Michx.) Watt [PHEGCON] (narrow beech fern)
SY: *Dryopteris phegopteris* (L.) C. Christens.
SY: *Phegopteris polypodioides* Fée
SY: *Thelypteris phegopteris* (L.) Slosson
Phegopteris polypodioides Fée = Phegopteris connectilis [PHEGCON]
Thelypteris Schmidel [THELYPT$]
Thelypteris limbosperma auct. = Thelypteris quelpaertensis [THELQUE]
Thelypteris nevadensis (Baker) Clute ex Morton [THELNEV] (Nevada marsh fern)
SY: *Dryopteris nevadensis* (Baker) Underwood
SY: *Dryopteris oregana* C. Christens.
SY: *Lastrea oregana* (C. Christens.) Copeland
Thelypteris phegopteris (L.) Slosson = Phegopteris connectilis [PHEGCON]
Thelypteris quelpaertensis (Christ) Ching [THELQUE] (mountain fern)
SY: *Oreopteris quelpaertensis* (Christ) Holub
SY: *Thelypteris limbosperma* auct.

THYMELAEACEAE (V:D)

Daphne L. [DAPHNE$]
*__Daphne laureola__ L. [DAPHLAU] (spurge-laurel)

TYPHACEAE (V:M)

Typha L. [TYPHA$]
Typha angustifolia L. [TYPHANG] (lesser cattail)
SY: *Typha angustifolia* var. *calumetensis* Peattie
SY: *Typha angustifolia* var. *elongata* (Dudley) Wieg.
Typha angustifolia var. *calumetensis* Peattie = Typha angustifolia [TYPHANG]
Typha angustifolia var. *elongata* (Dudley) Wieg. = Typha angustifolia [TYPHANG]
Typha latifolia L. [TYPHLAT] (broad-leaved cattail)

ULMACEAE (V:D)

Ulmus L. [ULMUS$]
*__Ulmus pumila__ L. [ULMUPUM] (Siberian elm)

URTICACEAE (V:D)

Parietaria L. [PARIETA$]
Parietaria pensylvanica Muhl. ex Willd. [PARIPEN] (Pennsylvania pellitory)
Urtica L. [URTICA$]
Urtica californica Greene = Urtica dioica ssp. gracilis [URTIDIO1]
Urtica cardiophylla Rydb. = Urtica dioica ssp. gracilis [URTIDIO1]
Urtica dioica L. [URTIDIO] (stinging nettle)
Urtica dioica ssp. **gracilis** (Ait.) Seland. [URTIDIO1]
SY: *Urtica californica* Greene
SY: *Urtica cardiophylla* Rydb.
SY: *Urtica dioica* var. *angustifolia* Schlecht.
SY: *Urtica dioica* var. *californica* (Greene) C.L. Hitchc.
SY: *Urtica dioica* var. *gracilis* (Ait.) C.L. Hitchc.
SY: *Urtica dioica* ssp. *gracilis* var. *lyallii* (S. Wats.) C.L. Hitchc.
SY: *Urtica dioica* var. *lyallii* (S. Wats.) C.L. Hitchc.
SY: *Urtica dioica* var. *procera* (Muhl. ex Willd.) Weddell
SY: *Urtica gracilis* Ait.
SY: *Urtica lyallii* S. Wats.
SY: *Urtica lyallii* var. *californica* (Greene) Jepson
SY: *Urtica major* H.P. Fuchs
SY: *Urtica procera* Muhl. ex Willd.
SY: *Urtica strigosissima* Rydb.
SY: *Urtica viridis* Rydb.
Urtica dioica ssp. *gracilis* var. *lyallii* (S. Wats.) C.L. Hitchc. = Urtica dioica ssp. gracilis [URTIDIO1]
Urtica dioica var. *angustifolia* Schlecht. = Urtica dioica ssp. gracilis [URTIDIO1]
Urtica dioica var. *californica* (Greene) C.L. Hitchc. = Urtica dioica ssp. gracilis [URTIDIO1]
Urtica dioica var. *gracilis* (Ait.) C.L. Hitchc. = Urtica dioica ssp. gracilis [URTIDIO1]
Urtica dioica var. *lyallii* (S. Wats.) C.L. Hitchc. = Urtica dioica ssp. gracilis [URTIDIO1]
Urtica dioica var. *procera* (Muhl. ex Willd.) Weddell = Urtica dioica ssp. gracilis [URTIDIO1]
Urtica gracilis Ait. = Urtica dioica ssp. gracilis [URTIDIO1]
Urtica lyallii S. Wats. = Urtica dioica ssp. gracilis [URTIDIO1]
Urtica lyallii var. *californica* (Greene) Jepson = Urtica dioica ssp. gracilis [URTIDIO1]
Urtica major H.P. Fuchs = Urtica dioica ssp. gracilis [URTIDIO1]
Urtica procera Muhl. ex Willd. = Urtica dioica ssp. gracilis [URTIDIO1]

Urtica strigosissima Rydb. = Urtica dioica ssp. gracilis [URTIDIO1]
***Urtica urens** L. [URTIURE] (dog nettle)
Urtica viridis Rydb. = Urtica dioica ssp. gracilis [URTIDIO1]

VALERIANACEAE (V:D)

Plectritis (Lindl.) DC. [PLECTRI$]
Plectritis anomala (Gray) Suksdorf = Plectritis congesta ssp. brachystemon [PLECCON1]
Plectritis anomala var. *gibbosa* (Suksdorf) Dyal = Plectritis congesta ssp. brachystemon [PLECCON1]
Plectritis aphanoptera (Gray) Suksdorf = Plectritis congesta ssp. brachystemon [PLECCON1]
Plectritis congesta (Lindl.) DC. [PLECCON] (sea blush)
Plectritis congesta ssp. **brachystemon** (Fisch. & C.A. Mey.) Morey [PLECCON1]
SY: *Plectritis anomala* (Gray) Suksdorf
SY: *Plectritis anomala* var. *gibbosa* (Suksdorf) Dyal
SY: *Plectritis aphanoptera* (Gray) Suksdorf
SY: *Plectritis congesta* var. *major* (Fisch. & C.A. Mey.) Dyal
SY: *Plectritis magna* (Greene) Suksdorf
SY: *Plectritis samolifolia* (DC.) Hoeck
SY: *Plectritis samolifolia* var. *involuta* (Suksdorf) Dyal
Plectritis congesta var. *major* (Fisch. & C.A. Mey.) Dyal = Plectritis congesta ssp. brachystemon [PLECCON1]
Plectritis macrocera Torr. & Gray [PLECMAC] (long-spurred plectritis)
Plectritis magna (Greene) Suksdorf = Plectritis congesta ssp. brachystemon [PLECCON1]
Plectritis samolifolia (DC.) Hoeck = Plectritis congesta ssp. brachystemon [PLECCON1]
Plectritis samolifolia var. *involuta* (Suksdorf) Dyal = Plectritis congesta ssp. brachystemon [PLECCON1]
Valeriana L. [VALERIA$]
Valeriana capitata Pallas ex Link [VALECAP] (capitate valerian)
Valeriana dioica L. [VALEDIO] (marsh valerian)
Valeriana dioica ssp. *sylvatica* (S. Wats.) F.G. Mey. = Valeriana dioica var. sylvatica [VALEDIO1]
Valeriana dioica var. **sylvatica** S. Wats. [VALEDIO1]
SY: *Valeriana dioica* ssp. *sylvatica* (S. Wats.) F.G. Mey.
SY: *Valeriana septentrionalis* Rydb.
SY: *Valeriana sylvatica* Soland. ex Richards.
Valeriana edulis Nutt. ex Torr. & Gray [VALEEDU] (edible valerian)
***Valeriana officinalis** L. [VALEOFF] (garden heliotrope)
Valeriana scouleri Rydb. [VALESCO] (Scouler's valerian)
SY: *Valeriana sitchensis* ssp. *scouleri* (Rydb.) F.G. Mey.
SY: *Valeriana sitchensis* var. *scouleri* (Rydb.) M.E. Jones
Valeriana septentrionalis Rydb. = Valeriana dioica var. sylvatica [VALEDIO1]
Valeriana sitchensis Bong. [VALESIT] (Sitka valerian)
Valeriana sitchensis ssp. *scouleri* (Rydb.) F.G. Mey. = Valeriana scouleri [VALESCO]
Valeriana sitchensis var. *scouleri* (Rydb.) M.E. Jones = Valeriana scouleri [VALESCO]
Valeriana sylvatica Soland. ex Richards. = Valeriana dioica var. sylvatica [VALEDIO1]
Valerianella P. Mill. [VALERIN$]
***Valerianella locusta** (L.) Lat. [VALELOC] (cornsalad)
SY: *Valerianella olitoria* (L.) Pollich
Valerianella olitoria (L.) Pollich = Valerianella locusta [VALELOC]

VERBENACEAE (V:D)

Verbena L. [VERBENA$]
Verbena bracteata Lag. & Rodr. [VERBBRA] (bracted vervain)
SY: *Verbena bracteosa* Michx.
SY: *Verbena imbricata* Woot. & Standl.
Verbena bracteosa Michx. = Verbena bracteata [VERBBRA]
Verbena hastata L. [VERBHAS] (blue vervain)
Verbena imbricata Woot. & Standl. = Verbena bracteata [VERBBRA]

VIOLACEAE (V:D)

Viola L. [VIOLA$]
Viola achyrophora Greene = Viola epipsila ssp. repens [VIOLEPI1]
Viola adunca Sm. [VIOLADU] (early blue violet)
***Viola arvensis** Murr. [VIOLARV] (European field pansy)
SY: *Viola tricolor* var. *arvensis* (Murr.) Boiss.
Viola biflora L. [VIOLBIF] (twinflower violet)
Viola biflora ssp. **carlottae** Calder & Taylor [VIOLBIF1] (Queen Charlotte twinflower violet)
SY: *Viola biflora* var. *carlottae* (Calder & Taylor) Boivin
Viola biflora var. *carlottae* (Calder & Taylor) Boivin = Viola biflora ssp. carlottae [VIOLBIF1]
Viola canadensis L. [VIOLCAN] (Canada violet)
Viola canadensis ssp. *rydbergii* (Greene) House = Viola canadensis var. corymbosa [VIOLCAN1]
Viola canadensis var. **corymbosa** Nutt. ex Torr. & Gray [VIOLCAN1]
SY: *Viola canadensis* ssp. *rydbergii* (Greene) House
Viola cognata Greene = Viola nephrophylla var. cognata [VIOLNEP1]
Viola epipsila Ledeb. [VIOLEPI] (dwarf marsh violet)
Viola epipsila ssp. **repens** Becker [VIOLEPI1]
SY: *Viola achyrophora* Greene
SY: *Viola epipsiloides* A. & D. Löve
SY: *Viola repens* Turcz. ex Trautv. & C.A. Mey.
Viola epipsiloides A. & D. Löve = Viola epipsila ssp. repens [VIOLEPI1]
Viola glabella Nutt. [VIOLGLA] (stream violet)
Viola howellii Gray [VIOLHOW] (Howell's violet)
***Viola lanceolata** L. [VIOLLAN] (lance-leaved violet)
Viola langsdorfii Fisch. ex Gingins [VIOLLAG] (Alaska violet)
SY: *Viola simulata* M.S. Baker
SY: *Viola superba* M.S. Baker
Viola maccabeiana M.S. Baker [VIOLMAC]
Viola macloskeyi Lloyd [VIOLMAL] (small white violet)
Viola macloskeyi ssp. **pallens** (Banks ex DC.) M.S. Baker [VIOLMAL2]
SY: *Viola macloskeyi* var. *pallens* (Banks ex DC.) C.L. Hitchc.
SY: *Viola pallens* (Banks ex DC.) Brainerd
SY: *Viola pallens* var. *subreptans* Rouss.
Viola macloskeyi var. *pallens* (Banks ex DC.) C.L. Hitchc. = Viola macloskeyi ssp. pallens [VIOLMAL2]
Viola nephrophylla Greene [VIOLNEP] (bog violet)
Viola nephrophylla var. **cognata** (Greene) C.L. Hitchc. [VIOLNEP1]
SY: *Viola cognata* Greene
Viola nephrophylla var. **nephrophylla** [VIOLNEP0]
Viola nuttallii ssp. *praemorsa* (Dougl. ex Lindl.) Piper = Viola praemorsa [VIOLPRA]
Viola nuttallii var. *major* Hook. = Viola vallicola var. major [VIOLVAL1]
Viola nuttallii var. *praemorsa* (Dougl. ex Lindl.) S. Wats. = Viola praemorsa [VIOLPRA]
Viola nuttallii var. *vallicola* (A. Nels.) St. John = Viola vallicola [VIOLVAL]
***Viola odorata** L. [VIOLODO] (sweet violet)
Viola orbiculata Geyer ex Holz. [VIOLORB] (rounded-leaved violet)
SY: *Viola sempervirens* var. *orbiculata* (Geyer ex Holz.) J.K. Henry
SY: *Viola sempervirens* var. *orbiculoides* M.S. Baker
Viola pallens (Banks ex DC.) Brainerd = Viola macloskeyi ssp. pallens [VIOLMAL2]
Viola pallens var. *subreptans* Rouss. = Viola macloskeyi ssp. pallens [VIOLMAL2]
Viola palustris L. [VIOLPAL] (marsh violet)
Viola praemorsa Dougl. ex Lindl. [VIOLPRA] (yellow montane violet)
SY: *Viola nuttallii* ssp. *praemorsa* (Dougl. ex Lindl.) Piper
SY: *Viola nuttallii* var. *praemorsa* (Dougl. ex Lindl.) S. Wats.
Viola praemorsa ssp. *major* (Hook.) M.S. Baker ex M.E. Peck = Viola vallicola var. major [VIOLVAL1]
Viola praemorsa var. *major* (Hook.) M.E. Peck = Viola vallicola var. major [VIOLVAL1]
Viola purpurea Kellogg [VIOLPUR] (goose-foot yellow violet)

Viola purpurea ssp. *atriplicifolia* (Greene) M.S. Baker & J.C. Clausen ex M.E. Peck = Viola purpurea ssp. venosa **[VIOLPUR1]**
Viola purpurea ssp. **venosa** (S. Wats.) M.S. Baker & J.C. Clausen [VIOLPUR1]
SY: *Viola purpurea* ssp. *atriplicifolia* (Greene) M.S. Baker & J.C. Clausen ex M.E. Peck
SY: *Viola purpurea* var. *atriplicifolia* (Greene) M.E. Peck
SY: *Viola purpurea* var. *venosa* (S. Wats.) Brainerd
Viola purpurea var. *atriplicifolia* (Greene) M.E. Peck = Viola purpurea ssp. venosa [VIOLPUR1]
Viola purpurea var. *venosa* (S. Wats.) Brainerd = Viola purpurea ssp. venosa [VIOLPUR1]
Viola renifolia Gray [VIOLREN] (kidney-leaved violet)
SY: *Viola renifolia* var. *brainerdii* (Greene) Fern.
Viola renifolia var. *brainerdii* (Greene) Fern. = Viola renifolia [VIOLREN]
Viola repens Turcz. ex Trautv. & C.A. Mey. = Viola epipsila ssp. repens [VIOLEPI1]
Viola sarmentosa Dougl. ex Hook. = Viola sempervirens [VIOLSEM]
Viola selkirkii Pursh ex Goldie [VIOLSEL] (Selkirk's violet)
Viola sempervirens Greene [VIOLSEM] (trailing yellow violet)
SY: *Viola sarmentosa* Dougl. ex Hook.
Viola sempervirens var. *orbiculata* (Geyer ex Holz.) J.K. Henry = Viola orbiculata [VIOLORB]
Viola sempervirens var. *orbiculoides* M.S. Baker = Viola orbiculata [VIOLORB]
Viola septentrionalis Greene [VIOLSEP] (northern violet)
Viola simulata M.S. Baker = Viola langsdorfii [VIOLLAG]
Viola superba M.S. Baker = Viola langsdorfii [VIOLLAG]
*__Viola tricolor__ L. [VIOLTRI] (European wild pansy)
Viola tricolor var. *arvensis* (Murr.) Boiss. = Viola arvensis [VIOLARV]
Viola vallicola A. Nels. [VIOLVAL] (valley violet)
SY: *Viola nuttallii* var. *vallicola* (A. Nels.) St. John
Viola vallicola var. **major** (Hook.) Fabijan [VIOLVAL1] (yellow sagebrush violet)
SY: *Viola nuttallii* var. *major* Hook.
SY: *Viola praemorsa* ssp. *major* (Hook.) M.S. Baker ex M.E. Peck
SY: *Viola praemorsa* var. *major* (Hook.) M.E. Peck

VISCACEAE (V:D)

Arceuthobium Bieb. [ARCEUTH$]
Arceuthobium americanum Nutt. ex Engelm. [ARCEAME] (American dwarf mistletoe)
Arceuthobium campylopodum Engelm. [ARCECAM] (western dwarf mistletoe)
Arceuthobium douglasii Engelm. [ARCEDOU] (Douglas' dwarf mistletoe)

ZANNICHELLIACEAE (V:M)

Zannichellia L. [ZANNICH$]
Zannichellia major (Hartman) Boenn. ex Reichenb. = Zannichellia palustris [ZANNPAL]
Zannichellia palustris L. [ZANNPAL] (horned pondweed)
SY: *Zannichellia major* (Hartman) Boenn. ex Reichenb.
SY: *Zannichellia palustris* var. *major* (Hartman) W.D.J. Koch
SY: *Zannichellia palustris* var. *stenophylla* Aschers. & Graebn.
Zannichellia palustris var. *major* (Hartman) W.D.J. Koch = Zannichellia palustris [ZANNPAL]
Zannichellia palustris var. *stenophylla* Aschers. & Graebn. = Zannichellia palustris [ZANNPAL]

ZOSTERACEAE (V:M)

Phyllospadix Hook. [PHYLLOS$]
Phyllospadix scouleri Hook. [PHYLSCO] (Scouler's surf-grass)
Phyllospadix serrulatus Rupr. ex Aschers. [PHYLSER] (toothed surf-grass)
Phyllospadix torreyi S. Wats. [PHYLTOR] (Torrey's surf-grass)
Zostera L. [ZOSTERA$]
Zostera americana den Hartog = Zostera japonica [ZOSTJAP]
*__Zostera japonica__ Aschers. & Graebn. [ZOSTJAP] (Japanese eel-grass)
SY: *Zostera americana* den Hartog
SY: *Zostera nana* Roth
SY: *Zostera noltii* auct.
Zostera marina L. [ZOSTMAR] (common eel-grass)
SY: *Zostera oregana* S. Wats.
SY: *Zostera stenophylla* Raf.
Zostera marina var. **latifolia** Morong [ZOSTMAR1]
Zostera marina var. **stenophylla** Aschers. & Graebn. [ZOSTMAR2]
Zostera nana Roth = Zostera japonica [ZOSTJAP]
Zostera noltii auct. = Zostera japonica [ZOSTJAP]
Zostera oregana S. Wats. = Zostera marina [ZOSTMAR]
Zostera stenophylla Raf. = Zostera marina [ZOSTMAR]

ZYGOPHYLLACEAE (V:D)

Tribulus L. [TRIBULU$]
*__Tribulus terrestris__ L. [TRIBTER] (puncture vine))

Bryophytes

ADELANTHACEAE (B:H)

Odontoschisma (Dum.) Dum. [ODONTOS$]
Odontoschisma denudatum (Nees ex Mart.) Dum. [ODONDEN]
SY: *Odontoschisma gibbsiae* Evans
SY: *Odontoschisma sphagni* auct.
Odontoschisma elongatum (Lindb.) Evans [ODONELO]
Odontoschisma gibbsiae Evans = Odontoschisma denudatum [ODONDEN]
Odontoschisma macounii (Aust.) Underw. [ODONMAC]
Odontoschisma sphagni auct. = Odontoschisma denudatum [ODONDEN]

AMBLYSTEGIACEAE (B:M)

Acrocladium cordifolium (Hedw.) P. Rich. & Wallace = Calliergon cordifolium [CALLCOR]
Acrocladium cuspidatum (Hedw.) Lindb. = Calliergonella cuspidata [CALLCUS]
Amblystegium Schimp. in B.S.G. [AMBLYST$]
Amblystegium americanum Grout = Conardia compacta [CONACOM]
Amblystegium brevipes Card. & Thér. ex Holz. = Leptodictyum riparium [LEPTRIP]
Amblystegium compactum (C. Müll.) Aust. = Conardia compacta [CONACOM]
Amblystegium fallax (Brid.) Milde = Cratoneuron filicinum [CRATFIL]
Amblystegium filicinum (Hedw.) De Not. = Cratoneuron filicinum [CRATFIL]
Amblystegium filicinum f. *marianopolitana* Dupret ex Moxley = Hygroamblystegium tenax [HYGRTEN]
Amblystegium fluviatile (Hedw.) Schimp. in B.S.G. = Hygroamblystegium fluviatile [HYGRFLU]
Amblystegium fluviatile var. *noterophilum* (Sull. & Lesq. in Sull.) Flow. = Hygroamblystegium noterophilum [HYGRNOT]
Amblystegium irriguum (Hook. & Wils.) B.S.G. = Hygroamblystegium tenax [HYGRTEN]
Amblystegium irriguum f. *marianopolitana* Dupret = Hygroamblystegium tenax [HYGRTEN]
Amblystegium juratzkanum Schimp. = Amblystegium serpens var. juratzkanum [AMBLSER1]
Amblystegium juratzkanum var. *giganteum* (Grout) Grout = Amblystegium serpens var. juratzkanum [AMBLSER1]
Amblystegium noterophilum (Sull. & Lesq. in Sull.) Holz. = Hygroamblystegium noterophilum [HYGRNOT]
Amblystegium polygamum Schimp. in B.S.G. = Campylium polygamum [CAMPPOL]
Amblystegium porphyrrhizon Schimp. = Amblystegium serpens var. juratzkanum [AMBLSER1]
Amblystegium radicale P. Beauv. = Campylium radicale [CAMPRAD]
Amblystegium riparium (Hedw.) Schimp. in B.S.G. = Leptodictyum riparium [LEPTRIP]
Amblystegium riparium f. *fluitans* (Lesq. & James) Flow. = Leptodictyum riparium [LEPTRIP]
Amblystegium riparium var. *abbreviatum* Schimp. in B.S.G. = Leptodictyum riparium [LEPTRIP]
Amblystegium riparium var. *flaccidum* (Lesq. & James) Ren. & Card. = Leptodictyum riparium [LEPTRIP]
Amblystegium riparium var. *fluitans* (Lesq. & James) Ren. & Card. = Leptodictyum riparium [LEPTRIP]
Amblystegium saxatile Schimp. = Campylium radicale [CAMPRAD]
Amblystegium serpens (Hedw.) Schimp. in B.S.G. [AMBLSER]
SY: *Amblystegium serpens* var. *beringianum* Card. & Thér.
SY: *Amblystegium serpens* var. *tenue* (Brid.) Schimp. in B.S.G.
SY: *Hypnum serpens* Hedw.
Amblystegium serpens var. *beringianum* Card. & Thér. = Amblystegium serpens [AMBLSER]
Amblystegium serpens var. *giganteum* Grout = Amblystegium serpens var. juratzkanum [AMBLSER1]
Amblystegium serpens var. **juratzkanum** (Schimp.) Rau & Herv. [AMBLSER1]
SY: *Amblystegium juratzkanum* Schimp.
SY: *Amblystegium juratzkanum* var. *giganteum* (Grout) Grout
SY: *Amblystegium porphyrrhizon* Schimp.
SY: *Amblystegium serpens* var. *giganteum* Grout
Amblystegium serpens var. **serpens** [AMBLSER0]
Amblystegium serpens var. *tenue* (Brid.) Schimp. in B.S.G. = Amblystegium serpens [AMBLSER]
Amblystegium sipho (P. Beauv.) Card. = Leptodictyum riparium [LEPTRIP]
Amblystegium stellatum (Hedw.) Lindb. = Campylium stellatum [CAMPSTE]
Amblystegium subcompactum C. Müll. & Kindb. in Mac. & Kindb. = Conardia compacta [CONACOM]
Amblystegium tenax (Hedw.) C. Jens. = Hygroamblystegium tenax [HYGRTEN]
Amblystegium vacillans Sull. = Leptodictyum riparium [LEPTRIP]
Amblystegium varium (Hedw.) Lindb. [AMBLVAR]
SY: *Amblystegium varium* var. *alaskanum* Card. & Thér.
SY: *Amblystegium varium* var. *lancifolium* Grout
SY: *Amblystegium varium* var. *ovatum* (Grout) Grout
SY: *Amblystegium varium* var. *parvulum* (Aust.) Grout
SY: *Hygroamblystegium varium* (Hedw.) Mönk.
SY: *Hypnum vacium* (Hedw.) P. Beauv.
Amblystegium varium var. *alaskanum* Card. & Thér. = Amblystegium varium [AMBLVAR]
Amblystegium varium var. *lancifolium* Grout = Amblystegium varium [AMBLVAR]
Amblystegium varium var. *ovatum* (Grout) Grout = Amblystegium varium [AMBLVAR]
Amblystegium varium var. *parvulum* (Aust.) Grout = Amblystegium varium [AMBLVAR]
Brachythecium edentatum Williams = Warnstorfia pseudostraminea [WARNPSE]
Brachythecium pennellii Bartr. = Leptodictyum riparium [LEPTRIP]
Calliergidium pseudostramineum (C. Müll.) Grout = Warnstorfia pseudostraminea [WARNPSE]
Calliergidium pseudostramineum var. *plesiostramineum* (Ren.) Grout = Warnstorfia pseudostraminea [WARNPSE]
Calliergon (Sull.) Kindb. [CALLIEG$]
Calliergon cordifolium (Hedw.) Kindb. [CALLCOR]
SY: *Acrocladium cordifolium* (Hedw.) P. Rich. & Wallace
SY: *Calliergon cordifolium* var. *angustifolium* G. Roth
SY: *Calliergon cordifolium* var. *fontinaloides* (J. Lange) G. Roth
SY: *Calliergon cordifolium* var. *intermedium* Mönk. in Bauer
SY: *Calliergon cordifolium* var. *lanutocaule* (Bryhn) Broth.
SY: *Calliergon cordifolium* var. *latifolium* Karcz.
SY: *Calliergon cordifolium* f. *laxum* (Röll) Podp.
SY: *Hypnum cordifolium* Hedw.
Calliergon cordifolium f. *laxum* (Röll) Podp. = Calliergon cordifolium [CALLCOR]
Calliergon cordifolium var. *angustifolium* G. Roth = Calliergon cordifolium [CALLCOR]
Calliergon cordifolium var. *fontinaloides* (J. Lange) G. Roth = Calliergon cordifolium [CALLCOR]
Calliergon cordifolium var. *intermedium* Mönk. in Bauer = Calliergon cordifolium [CALLCOR]

Calliergon cordifolium var. *lanutocaule* (Bryhn) Broth. = Calliergon cordifolium [CALLCOR]
Calliergon cordifolium var. *latifolium* Karcz. = Calliergon cordifolium [CALLCOR]
Calliergon cuspidata var. *pungens* (Schimp.) Warnst. = Calliergonella cuspidata [CALLCUS]
Calliergon cuspidatum (Hedw.) Kindb. = Calliergonella cuspidata [CALLCUS]
Calliergon cuspidatum f. *acuteramosum* Bauer ex Karcz. = Calliergonella cuspidata [CALLCUS]
Calliergon cuspidatum f. *turgescens* (Wheld.) Karcz. = Calliergonella cuspidata [CALLCUS]
Calliergon cuspidatum var. *brevifolium* Sanio ex Warnst. = Calliergonella cuspidata [CALLCUS]
Calliergon cuspidatum var. *umbrosum* (Loeske) Warnst. = Calliergonella cuspidata [CALLCUS]
Calliergon giganteum (Schimp.) Kindb. [CALLGIG] (giant water moss)
SY: *Calliergon giganteum* f. *crassicostatum* (Mik.) Karcz.
SY: *Calliergon giganteum* var. *cyclophyllotum* (Holz.) Grout
SY: *Calliergon giganteum* f. *decurrens* (Mik.) Karcz.
SY: *Calliergon giganteum* var. *dendroides* (Limpr.) G. Roth
SY: *Calliergon giganteum* var. *fluitans* (Klinggr.) G. Roth
SY: *Calliergon giganteum* var. *hystricosum* G. Roth & Bock in G. Roth
SY: *Calliergon giganteum* var. *immersum* Ruthe ex Karcz.
SY: *Calliergon giganteum* var. *labradorense* (Ren. & Card.) Grout
SY: *Calliergon giganteum* var. *pennatum* Karcz.
SY: *Calliergon giganteum* var. *tenue* Karcz.
SY: *Calliergon subsarmentosum* Kindb.
SY: *Hypnum giganteum* Schimp.
SY: *Hypnum giganteum* var. *fluitans* (Rabenh.) Heug.
Calliergon giganteum f. *crassicostatum* (Mik.) Karcz. = Calliergon giganteum [CALLGIG]
Calliergon giganteum f. *decurrens* (Mik.) Karcz. = Calliergon giganteum [CALLGIG]
Calliergon giganteum var. *cyclophyllotum* (Holz.) Grout = Calliergon giganteum [CALLGIG]
Calliergon giganteum var. *dendroides* (Limpr.) G. Roth = Calliergon giganteum [CALLGIG]
Calliergon giganteum var. *fluitans* (Klinggr.) G. Roth = Calliergon giganteum [CALLGIG]
Calliergon giganteum var. *hystricosum* G. Roth & Bock in G. Roth = Calliergon giganteum [CALLGIG]
Calliergon giganteum var. *immersum* Ruthe ex Karcz. = Calliergon giganteum [CALLGIG]
Calliergon giganteum var. *labradorense* (Ren. & Card.) Grout = Calliergon giganteum [CALLGIG]
Calliergon giganteum var. *pennatum* Karcz. = Calliergon giganteum [CALLGIG]
Calliergon giganteum var. *tenue* Karcz. = Calliergon giganteum [CALLGIG]
Calliergon pseudosarmentosum (Card. & Thér.) Broth. = Drepanocladus pseudosarmentosus [DREPPSE]
Calliergon richardsonii (Mitt.) Kindb. in Warnst. [CALLRIC] (Richardson's water moss)
SY: *Calliergon subgiganteum* Kindb.
SY: *Hypnum richardsonii* (Mitt.) Lesq. & James
SY: *Hypnum subgiganteum* Kindb. ex Grout
Calliergon sarmentosum (Wahlenb.) Kindb. = Sarmenthypnum sarmentosum [SARMSAR]
Calliergon sarmentosum f. *heterophyllum* Arnell & C. Jens. = Sarmenthypnum sarmentosum [SARMSAR]
Calliergon sarmentosum f. *homophyllum* Arnell & C. Jens. = Sarmenthypnum sarmentosum [SARMSAR]
Calliergon sarmentosum var. *beringianum* (Card. & Thér.) Grout = Sarmenthypnum sarmentosum [SARMSAR]
Calliergon sarmentosum var. *crispum* Karcz. = Sarmenthypnum sarmentosum [SARMSAR]
Calliergon sarmentosum var. *fallaciosum* (Milde) G. Roth = Sarmenthypnum sarmentosum [SARMSAR]
Calliergon sarmentosum var. *flagellare* Karcz. = Sarmenthypnum sarmentosum [SARMSAR]
Calliergon sarmentosum var. *fontinaloides* (Berggr.) G. Roth = Sarmenthypnum sarmentosum [SARMSAR]
Calliergon sarmentosum var. *subpinnatum* Warnst. = Sarmenthypnum sarmentosum [SARMSAR]
Calliergon stramineum (Brid.) Kindb. [CALLSTR] (straw-like feather moss)
SY: *Calliergon stramineum* var. *flagellaceum* (G. Roth & Bock in G. Roth) Karcz.
SY: *Calliergon stramineum* var. *laxifolium* (Kindb.) Karcz.
SY: *Calliergon stramineum* var. *nivale* (Lor.) G. Roth
SY: *Calliergon stramineum* var. *patens* (Lindb.) G. Roth
SY: *Calliergon stramineum* f. *subtrifarium* (Saelan) Warnst.
SY: *Hypnum stramineum* Dicks. ex Brid.
SY: *Hypnum stramineum* var. *patens* (Lindb.) Par.
Calliergon stramineum f. *subtrifarium* (Saelan) Warnst. = Calliergon stramineum [CALLSTR]
Calliergon stramineum var. *flagellaceum* (G. Roth & Bock in G. Roth) Karcz. = Calliergon stramineum [CALLSTR]
Calliergon stramineum var. *laxifolium* (Kindb.) Karcz. = Calliergon stramineum [CALLSTR]
Calliergon stramineum var. *nivale* (Lor.) G. Roth = Calliergon stramineum [CALLSTR]
Calliergon stramineum var. *patens* (Lindb.) G. Roth = Calliergon stramineum [CALLSTR]
Calliergon subgiganteum Kindb. = Calliergon richardsonii [CALLRIC]
Calliergon subsarmentosum Kindb. = Calliergon giganteum [CALLGIG]
Calliergon subsarmentosum Kindb. = Sarmenthypnum sarmentosum [SARMSAR]
Calliergon subturgescens Kindb. = Pseudocalliergon turgescens [PSEUTUR]
Calliergon trifarium (Web. & Mohr) Kindb. [CALLTRI]
SY: *Hypnum trifarium* Web. & Mohr
SY: *Scorpidium trifarium* (Web. & Mohr) Paul
Calliergon turgescens (T. Jens.) Kindb. = Pseudocalliergon turgescens [PSEUTUR]
Calliergon turgescens var. *patens* Karcz. = Pseudocalliergon turgescens [PSEUTUR]
Calliergon turgescens var. *tenue* (Berggr.) Karcz. = Pseudocalliergon turgescens [PSEUTUR]
Calliergon wickesiae Grout = Loeskypnum wickesiae [LOESWIC]
Calliergonella Loeske [CALLIEO$]
Calliergonella conardii Lawt. = Calliergonella cuspidata [CALLCUS]
Calliergonella cuspidata (Hedw.) Loeske [CALLCUS] (spear moss)
SY: *Acrocladium cuspidatum* (Hedw.) Lindb.
SY: *Calliergon cuspidata* var. *pungens* (Schimp.) Warnst.
SY: *Calliergon cuspidatum* (Hedw.) Kindb.
SY: *Calliergon cuspidatum* f. *acuteramosum* Bauer ex Karcz.
SY: *Calliergon cuspidatum* var. *brevifolium* Sanio ex Warnst.
SY: *Calliergon cuspidatum* f. *turgescens* (Wheld.) Karcz.
SY: *Calliergon cuspidatum* var. *umbrosum* (Loeske) Warnst.
SY: *Calliergonella conardii* Lawt.
SY: *Calliergonella cuspidata* f. *abbreviata* (Röll) Podp.
SY: *Calliergonella cuspidata* f. *acuteramosa* Bauer ex Karez.
SY: *Calliergonella cuspidata* f. *brevifolium* (Sanio in Warnst.) Podp.
SY: *Calliergonella cuspidata* var. *caespitosum* (Whiteh.) Karez.
SY: *Calliergonella cuspidata* var. *pungens* (Schimp.) Latz.
SY: *Calliergonella cuspidata* var. *umbrosus* (Loeske) Warnst.
SY: *Hypnum cuspidatum* Hedw.
Calliergonella cuspidata f. *abbreviata* (Röll) Podp. = Calliergonella cuspidata [CALLCUS]
Calliergonella cuspidata f. *acuteramosa* Bauer ex Karez. = Calliergonella cuspidata [CALLCUS]
Calliergonella cuspidata f. *brevifolium* (Sanio in Warnst.) Podp. = Calliergonella cuspidata [CALLCUS]
Calliergonella cuspidata var. *caespitosum* (Whiteh.) Karez. = Calliergonella cuspidata [CALLCUS]
Calliergonella cuspidata var. *pungens* (Schimp.) Latz. = Calliergonella cuspidata [CALLCUS]
Calliergonella cuspidata var. *umbrosus* (Loeske) Warnst. = Calliergonella cuspidata [CALLCUS]

Camptothecium paulianum Grout in Holz. & Frye = Sanionia uncinata var. uncinata [SANIUNC0]
Campylium (Sull.) Mitt. [CAMPYLI$]
Campylium calcareum Crundw. & Nyh. [CAMPCAL]
Campylium chrysophyllum (Brid.) J. Lange [CAMPCHR]
SY: *Campylium chrysophyllum* var. *brevifolium* (Ren. & Card.) Grout
SY: *Campylium chrysophyllum* var. *zemliae* (C. Jens.) Grout
SY: *Chryso-hypnum chrysophyllum* (Brid.) Loeske
SY: *Hypnum chrysophyllum* Brid.
SY: *Hypnum chrysophyllum* var. *brevifolium* Ren. & Card.
SY: *Hypnum sinuolatum* (Kindb.) Kindb. in Röll
SY: *Hypnum subsecundum* Kindb.
SY: *Hypnum unicostatum* C. Müll. & Kindb. in Mac. & Kindb.
Campylium chrysophyllum var. *brevifolium* (Ren. & Card.) Grout = Campylium chrysophyllum [CAMPCHR]
Campylium chrysophyllum var. *zemliae* (C. Jens.) Grout = Campylium chrysophyllum [CAMPCHR]
Campylium halleri (Hedw.) Lindb. [CAMPHAL]
SY: *Campylophyllum halleri* (Hedw.) Fleisch.
Campylium hispidulum (Brid.) Mitt. [CAMPHIS]
SY: *Campylium hispidulum* var. *cordatum* Grout
SY: *Campylium hispidulum* var. *sommerfeltii* (Myr.) Lindb.
SY: *Campylium sommerfeltii* (Myr.) J. Lange
SY: *Chryso-hypnum hispidulum* (Brid.) Roth
SY: *Hypnum byssirameum* C. Müll. & Kindb. in Mac. & Kindb.
SY: *Hypnum hispidulum* Brid.
SY: *Hypnum sommerfeltii* Myr.
SY: *Myurella squarrosa* Grout
Campylium hispidulum var. *cordatum* Grout = Campylium hispidulum [CAMPHIS]
Campylium hispidulum var. *sommerfeltii* (Myr.) Lindb. = Campylium hispidulum [CAMPHIS]
Campylium polygamum (Schimp. in B.S.G.) C. Jens. [CAMPPOL]
SY: *Amblystegium polygamum* Schimp. in B.S.G.
SY: *Campylium polygamum* var. *fluitans* Grout
SY: *Campylium polygamum* var. *minus* (Schimp.) G. Roth
SY: *Chryso-hypnum polygamum* (B.S.G.) Loeske
SY: *Hypnum polygamum* (B.S.G.) Wils.
Campylium polygamum var. *fluitans* Grout = Campylium polygamum [CAMPPOL]
Campylium polygamum var. *longinerve* (Ren. & Card.) Grout = Leptodictyum riparium [LEPTRIP]
Campylium polygamum var. *minus* (Schimp.) G. Roth = Campylium polygamum [CAMPPOL]
Campylium protensum (Brid.) Kindb. = Campylium stellatum var. protensum [CAMPSTE2]
Campylium radicale (P. Beauv.) Grout [CAMPRAD]
SY: *Amblystegium radicale* P. Beauv.
SY: *Amblystegium saxatile* Schimp.
SY: *Hypnum bergenense* Aust.
SY: *Hypnum decursivulum* C. Müll. & Kindb. in Mac. & Kindb.
SY: *Hypnum radicale* (P. Beauv.) Brid.
SY: *Leptodictyum radicale* (P. Beauv.) Kanda
Campylium sommerfeltii (Myr.) J. Lange = Campylium hispidulum [CAMPHIS]
Campylium stellatum (Hedw.) C. Jens. [CAMPSTE] (golden star moss)
SY: *Amblystegium stellatum* (Hedw.) Lindb.
SY: *Chryso-hypnum stellatum* (Hedw.) Loeske
SY: *Hypnum stellatum* Hedw.
Campylium stellatum var. **protensum** (Brid.) Bryhn [CAMPSTE2]
SY: *Campylium protensum* (Brid.) Kindb.
SY: *Hypnum stellatum* var. *protensum* (Brid.) Röhl.
Campylium stellatum var. **stellatum** [CAMPSTE0]
Campylophyllum halleri (Hedw.) Fleisch. = Campylium halleri [CAMPHAL]
Chryso-hypnum chrysophyllum (Brid.) Loeske = Campylium chrysophyllum [CAMPCHR]
Chryso-hypnum hispidulum (Brid.) Roth = Campylium hispidulum [CAMPHIS]
Chryso-hypnum polygamum (B.S.G.) Loeske = Campylium polygamum [CAMPPOL]
Chryso-hypnum stellatum (Hedw.) Loeske = Campylium stellatum [CAMPSTE]
Conardia Robins. [CONARDI$]
Conardia compacta (C. Müll.) Robins. [CONACOM]
SY: *Amblystegium americanum* Grout
SY: *Amblystegium compactum* (C. Müll.) Aust.
SY: *Amblystegium subcompactum* C. Müll. & Kindb. in Mac. & Kindb.
SY: *Hypnum compactum* (Hook.) C. Müll.
SY: *Rhynchostegiella compacta* (C. Müll.) Loeske
SY: *Rhynchostegiella compacta* var. *americana* (Grout) Crum, Steere & Anderson
Cratoneuron (Sull.) Spruce [CRATONE$]
Cratoneuron commutatum (Brid.) G. Roth = Palustriella commutata [PALUCOM]
Cratoneuron commutatum var. *falcatum* (Brid.) Mönk. = Palustriella commutata [PALUCOM]
Cratoneuron commutatum var. *sulcatum* (Lindb.) Mönk. = Palustriella commutata [PALUCOM]
Cratoneuron curvicaule (Jur.) G. Roth = Cratoneuron filicinum [CRATFIL]
Cratoneuron falcatum (Brid.) G. Roth = Palustriella commutata [PALUCOM]
Cratoneuron filicinum (Hedw.) Spruce [CRATFIL] (spring claw moss)
SY: *Amblystegium fallax* (Brid.) Milde
SY: *Amblystegium filicinum* (Hedw.) De Not.
SY: *Cratoneuron curvicaule* (Jur.) G. Roth
SY: *Cratoneuron filicinum* var. *aciculinum* (C. Müll. & Kindb. in Mac. & Kindb.) Grout
SY: *Cratoneuron filicinum* var. *curvicaule* (Jur.) Mönk.
SY: *Cratoneuron filicinum* var. *fallax* (Brid.) Roth
SY: *Cratoneuron filicinum* var. *filiforme* (Par.) Wijk & Marg.
SY: *Hygroamblystegium filicinum* (Hedw.) Loeske
SY: *Hypnum filicinum* Hedw.
SY: *Hypnum filicinum* var. *aciculinum* C. Müll. & Kindb. in Mac. & Kindb.
SY: *Hypnum filicinum* var. *filiforme* Berggr.
Cratoneuron filicinum var. *aciculinum* (C. Müll. & Kindb. in Mac. & Kindb.) Grout = Cratoneuron filicinum [CRATFIL]
Cratoneuron filicinum var. *curvicaule* (Jur.) Mönk. = Cratoneuron filicinum [CRATFIL]
Cratoneuron filicinum var. *fallax* (Brid.) Roth = Cratoneuron filicinum [CRATFIL]
Cratoneuron filicinum var. *filiforme* (Par.) Wijk & Marg. = Cratoneuron filicinum [CRATFIL]
Dichelyma longinerve Kindb. in Mac. = Warnstorfia exannulata var. exannulata [WARNEXA0]
Drepanocladus (C. Müll.) G. Roth (Amblystegiacea [DREPANO$]
Drepanocladus aduncus (Hedw.) Warnst. [DREPADU] (common hook moss)
SY: *Drepanocladus aduncus* f. *aquaticus* (Sanio) Mönk.
SY: *Drepanocladus aduncus* var. *aquaticus* (Sanio) Ren.
SY: *Drepanocladus aduncus* var. *intermedius* (Schimp.) Roth
SY: *Drepanocladus aduncus* var. *kneiffii* f. *intermedius* (B.S.G.) Mönk.
SY: *Drepanocladus aduncus* f. *laxus* Husn.
SY: *Drepanocladus aduncus* var. *paternus* (Sanio) Jel.
SY: *Hypnum aduncum* Hedw.
SY: *Hypnum aduncum* var. *aquaticum* (Sanio) Sanio
SY: *Hypnum aduncum* var. *platyphyllum* (Kindb.) Par.
SY: *Hypnum conflatum* C. Müll. & Kindb. in Mac. & Kindb.
SY: *Hypnum kneiffii* var. *platyphyllum* Kindb. in Mac. & Kindb.
Drepanocladus aduncus f. *aquaticus* (Sanio) Mönk. = Drepanocladus aduncus [DREPADU]
Drepanocladus aduncus f. *laxus* Husn. = Drepanocladus aduncus [DREPADU]
Drepanocladus aduncus var. **aduncus** [DREPADU0]
SY: *Drepanocladus aduncus* var. *pseudofluitans* (Sanio) Glow.
Drepanocladus aduncus var. *aquaticus* (Sanio) Ren. = Drepanocladus aduncus [DREPADU]
Drepanocladus aduncus var. *capillifolius* (Warnst.) Riehm. = Drepanocladus capillifolius [DREPCAP]

Drepanocladus aduncus var. *intermedius* (Schimp.) Roth = Drepanocladus aduncus [DREPADU]
Drepanocladus aduncus var. **kneiffii** (Schimp. in B.S.G.) Mönk. [DREPADU3]
SY: *Hypnum kneiffii* (B.S.G.) Schimp. in Wils.
Drepanocladus aduncus var. *kneiffii* f. *intermedius* (B.S.G.) Mönk. = Drepanocladus aduncus [DREPADU]
Drepanocladus aduncus var. *paternus* (Sanio) Jel. = Drepanocladus aduncus [DREPADU]
Drepanocladus aduncus var. **polycarpus** (Bland. ex Voit) G. Roth [DREPADU4]
SY: *Drepanocladus aduncus* var. *polycarpus* f. *gracilescens* (B.S.G.) Mönk.
SY: *Drepanocladus polycarpus* (Voit) Warnst.
SY: *Harpidium polycarpum* (Voit) Williams
Drepanocladus aduncus var. *polycarpus* f. *gracilescens* (B.S.G.) Mönk. = Drepanocladus aduncus var. polycarpus [DREPADU4]
Drepanocladus aduncus var. *pseudofluitans* (Sanio) Glow. = Drepanocladus aduncus var. aduncus [DREPADU0]
Drepanocladus aduncus var. *typicus* f. *capillifolius* Mönk. = Drepanocladus capillifolius [DREPCAP]
Drepanocladus badius (Hartm.) G. Roth = Loeskypnum badium [LOESBAD]
Drepanocladus berggrenii (C. Jens.) G. Roth = Warnstorfia fluitans var. fluitans [WARNFLU0]
Drepanocladus capillifolius (Warnst.) Warnst. [DREPCAP]
SY: *Drepanocladus aduncus* var. *capillifolius* (Warnst.) Riehm.
SY: *Drepanocladus aduncus* var. *typicus* f. *capillifolius* Mönk.
SY: *Drepanocladus capillifolius* f. *fallax* Ren.
SY: *Hypnum capillifolium* Warnst.
Drepanocladus capillifolius f. *fallax* Ren. = Drepanocladus capillifolius [DREPCAP]
Drepanocladus crassicostatus Janssens [DREPCRA]
SY: *Warnstorfia crassicostata* (Jansssens) Crum & Anderson
Drepanocladus exannulatus (Schimp. in B.S.G.) Warnst. = Warnstorfia exannulata var. exannulata [WARNEXA0]
Drepanocladus exannulatus f. *brachydictyus* (Ren.) Smirn. = Warnstorfia exannulata var. exannulata [WARNEXA0]
Drepanocladus exannulatus f. *falcifolius* Ren. = Warnstorfia exannulata var. exannulata [WARNEXA0]
Drepanocladus exannulatus f. *orthophyllus* (Milde) Mönk. = Warnstorfia exannulata var. exannulata [WARNEXA0]
Drepanocladus exannulatus f. *submersus* Mönk. = Warnstorfia exannulata var. exannulata [WARNEXA0]
Drepanocladus exannulatus f. *tundrae* (Arnell) Mönk. = Warnstorfia tundrae [WARNTUN]
Drepanocladus exannulatus var. *alpinus* (Grav.) Wijk & Marg. = Warnstorfia exannulata var. exannulata [WARNEXA0]
Drepanocladus exannulatus var. *brachydictyon* (Ren.) G. Roth = Warnstorfia exannulata var. exannulata [WARNEXA0]
Drepanocladus exannulatus var. *falcifolius* (Ren.) Grout = Warnstorfia exannulata var. exannulata [WARNEXA0]
Drepanocladus exannulatus var. *purpurascens* (Schimp.) Herz. = Warnstorfia exannulata var. purpurascens [WARNEXA1]
Drepanocladus exannulatus var. *rotae* (De Not.) Loeske = Warnstorfia exannulata var. exannulata [WARNEXA0]
Drepanocladus exannulatus var. *serratus* (Milde) Loeske = Warnstorfia exannulata var. exannulata [WARNEXA0]
Drepanocladus exannulatus var. *tundrae* (Arnell) Warnst. = Warnstorfia tundrae [WARNTUN]
Drepanocladus fluitans (Hedw.) Warnst. = Warnstorfia fluitans [WARNFLU]
Drepanocladus fluitans f. *gracilis* Boul. = Warnstorfia fluitans [WARNFLU]
Drepanocladus fluitans f. *jeanbernatii* (Ren.) Mönk. = Warnstorfia fluitans [WARNFLU]
Drepanocladus fluitans f. *setiformis* (Ren.) Mönk. = Warnstorfia fluitans [WARNFLU]
Drepanocladus fluitans var. *berggrenii* (C. Jens.) C. Jens. in Weim. = Warnstorfia fluitans [WARNFLU]
Drepanocladus fluitans var. *falcatus* (Sanio ex C. Jens.) G. Roth = Warnstorfia fluitans var. falcata [WARNFLU1]
Drepanocladus fluitans var. *falcifolius* f. *viridis* Boul. = Warnstorfia exannulata var. exannulata [WARNEXA0]
Drepanocladus fluitans var. *gracilis* Warnst. = Warnstorfia fluitans [WARNFLU]
Drepanocladus fluitans var. *jeanbernatii* (Ren.) Warnst. = Warnstorfia fluitans [WARNFLU]
Drepanocladus fluitans var. *submersus* (Schimp.) Loeske = Warnstorfia fluitans [WARNFLU]
Drepanocladus fluitans var. *uncatus* Crum, Steere & Anderson = Warnstorfia fluitans var. falcata [WARNFLU1]
Drepanocladus intermedius (Lindb.) Warnst. = Limprichtia cossonii [LIMPCOS]
Drepanocladus lapponicus (Norrl.) Smirn. = Hamatocaulis lapponicus [HAMALAP]
Drepanocladus latinervis Warnst. = Drepanocladus sendtneri [DREPSEN]
Drepanocladus polycarpus (Voit) Warnst. = Drepanocladus aduncus var. polycarpus [DREPADU4]
Drepanocladus procerus (Ren. & Arnell in Husn.) = Warnstorfia procera [WARNPRO]
Drepanocladus pseudosarmentosus (Card. & Thér.) Perss. [DREPPSE]
SY: *Calliergon pseudosarmentosum* (Card. & Thér.) Broth.
Drepanocladus pseudostramineus (C. Müll.) G. Roth = Warnstorfia pseudostraminea [WARNPSE]
Drepanocladus purpurascens (Schimp.) Loeske = Warnstorfia exannulata var. purpurascens [WARNEXA1]
Drepanocladus revolvens (Sw.) Warnst. = Limprichtia revolvens [LIMPREV]
Drepanocladus revolvens var. *intermedia* (Lindb.) Grout = Limprichtia cossonii [LIMPCOS]
Drepanocladus revolvens var. *miquelonensis* (Ren.) Grout = Scorpidium scorpioides [SCORSCO]
Drepanocladus schulzei G. Roth = Warnstorfia fluitans var. falcata [WARNFLU1]
Drepanocladus scorpioides (Hedw.) Warnst. = Scorpidium scorpioides [SCORSCO]
Drepanocladus sendtneri (Schimp.) Warnst. [DREPSEN]
SY: *Drepanocladus latinervis* Warnst.
SY: *Drepanocladus sendtneri* var. *wilsonii* (Lindb.) Warnst.
SY: *Hypnum falcatum* var. *microphyllum* Kindb. in Mac. & Kindb.
SY: *Hypnum sendtneri* Schimp. in H. Müll.
SY: *Hypnum wilsonii* (Lindb.) Schimp. ex Ren.
Drepanocladus sendtneri var. *wilsonii* (Lindb.) Warnst. = Drepanocladus sendtneri [DREPSEN]
Drepanocladus submersus (Schimp.) Warnst. = Warnstorfia fluitans [WARNFLU]
Drepanocladus trichophyllus (Warnst.) Podp. = Warnstorfia trichophylla [WARNTRI]
Drepanocladus tundrae (Arnell) Loeske = Warnstorfia tundrae [WARNTUN]
Drepanocladus uncinatus (Hedw.) Warnst. = Sanionia uncinata [SANIUNC]
Drepanocladus uncinatus var. *alpinus* (Ren.) Warnst. = Sanionia uncinata [SANIUNC]
Drepanocladus uncinatus var. *gracilescens* (B.S.G.) Warnst. = Sanionia uncinata [SANIUNC]
Drepanocladus uncinatus var. *gracillimus* (Berggr.) Storm. = Sanionia uncinata [SANIUNC]
Drepanocladus uncinatus var. *plumosus* (Schimp.) Warnst. = Sanionia uncinata var. uncinata [SANIUNC0]
Drepanocladus uncinatus var. *plumulosus* (Schimp. in B.S.G.) Warnst. = Sanionia uncinata var. uncinata [SANIUNC0]
Drepanocladus uncinatus var. *subjulaceus* (Schimp. in B.S.G.) Warnst. = Sanionia uncinata var. uncinata [SANIUNC0]
Drepanocladus uncinatus var. *symmetricus* (Ren. & Card.) Grout = Sanionia uncinata var. symmetrica [SANIUNC1]
Drepanocladus vernicosus (Mitt.) Warnst. = Hamatocaulis vernicosus [HAMAVER]
Hamatocaulis Hedenäs [HAMATOC$]
Hamatocaulis lapponicus (Norrl.) Hedenäs [HAMALAP]
SY: *Drepanocladus lapponicus* (Norrl.) Smirn.
Hamatocaulis vernicosus (Mitt.) Hedenäs [HAMAVER]

SY: *Drepanocladus vernicosus* (Mitt.) Warnst.
SY: *Hypnum vernicosus* Lindb. in Hartm.
SY: *Limprichtia vernicosa* (Mitt.) Loeske
SY: *Scorpidium vernicosum* (Mitt.) Tuom.
Harpidium polycarpum (Voit) Williams = Drepanocladus aduncus var. polycarpus [DREPADU4]
Hygroamblystegium Loeske [HYGROAM$]
Hygroamblystegium filicinum (Hedw.) Loeske = Cratoneuron filicinum [CRATFIL]
Hygroamblystegium fluviatile (Hedw.) Loeske [HYGRFLU]
SY: *Amblystegium fluviatile* (Hedw.) Schimp. in B.S.G.
SY: *Hygroamblystegium fluviatile* var. *ovatum* Grout
SY: *Hypnum fluviatile* Hedw.
Hygroamblystegium fluviatile var. *ovatum* Grout = Hygroamblystegium fluviatile [HYGRFLU]
Hygroamblystegium irriguum f. *marianopolitana* Dupret = Hygroamblystegium tenax [HYGRTEN]
Hygroamblystegium noterophilum (Sull. & Lesq. in Sull.) Warnst. [HYGRNOT]
SY: *Amblystegium fluviatile* var. *noterophilum* (Sull. & Lesq. in Sull.) Flow.
SY: *Amblystegium noterophilum* (Sull. & Lesq. in Sull.) Holz.
Hygroamblystegium tenax (Hedw.) Jenn. [HYGRTEN]
SY: *Amblystegium filicinum* f. *marianopolitana* Dupret ex Moxley
SY: *Amblystegium irriguum* (Hook. & Wils.) B.S.G.
SY: *Amblystegium irriguum* f. *marianopolitana* Dupret
SY: *Amblystegium tenax* (Hedw.) C. Jens.
SY: *Hygroamblystegium irriguum* f. *marianopolitana* Dupret
Hygroamblystegium varium (Hedw.) Mönk. = Amblystegium varium [AMBLVAR]
Hygrohypnum Lindb. [HYGROHY$]
Hygrohypnum alpestre (Hedw.) Loeske [HYGRALP]
SY: *Hypnum alpestre* Sw. ex Hedw.
Hygrohypnum alpinum (Lindb.) Loeske [HYGRALI]
Hygrohypnum arcticum (Somm.) Loeske = Hygrohypnum smithii [HYGRSMI]
Hygrohypnum bestii (Ren. & Bryhn in Ren.) Broth. [HYGRBES]
SY: *Hygrohypnum molle* var. *bestii* (Ren. & Bryhn) Hab.
Hygrohypnum cochlearifolium (Vent. ex De Not.) Broth. [HYGRCOC]
Hygrohypnum dilitatum (Wils.) Loeske = Hygrohypnum molle [HYGRMOL]
Hygrohypnum duriusculum (De Not.) Jamieson [HYGRDUR] (rigid brook moss)
Hygrohypnum luridum (Hedw.) Jenn. [HYGRLUR]
SY: *Hygrohypnum luridum* var. *ehlei* (Arnell) Wijk & Marg.
SY: *Hygrohypnum luridum* var. *julaceum* (Schimp. in B.S.G.) Podp.
SY: *Hygrohypnum luridum* ssp. *pseudomontanum* (Kindb.) Wijk & Marg.
SY: *Hygrohypnum luridum* var. *subsphaericarpon* (Brid.) C. Jens. in Podp.
SY: *Hygrohypnum palustre* Loeske
SY: *Hygrohypnum palustre* var. *ehlei* (Arnell) Grout
SY: *Hygrohypnum palustre* var. *julaceum* (Schimp. in B.S.G.) Loeske
SY: *Hygrohypnum palustre* var. *subsphaericarpon* (Brid.) Loeske
SY: *Hygrohypnum pseudomontanum* (Kindb.) Grout
SY: *Hygrohypnum subeugyrium* var. *occidentale* (Card. & Thér.) Grout
SY: *Hypnum columbicopalustre* C. Müll. & Kindb. in Mac. & Kindb.
SY: *Hypnum palustre* Brid.
SY: *Hypnum pseudo-arcticum* Kindb.
Hygrohypnum luridum ssp. *pseudomontanum* (Kindb.) Wijk & Marg. = Hygrohypnum luridum [HYGRLUR]
Hygrohypnum luridum var. *ehlei* (Arnell) Wijk & Marg. = Hygrohypnum luridum [HYGRLUR]
Hygrohypnum luridum var. *julaceum* (Schimp. in B.S.G.) Podp. = Hygrohypnum luridum [HYGRLUR]
Hygrohypnum luridum var. *subsphaericarpon* (Brid.) C. Jens. in Podp. = Hygrohypnum luridum [HYGRLUR]
Hygrohypnum micans (Mitt.) Broth. [HYGRMIC]
SY: *Hygrohypnum novae-caesareae* (Aust.) Grout
SY: *Rhaphidostegium novae-caesareae* (Aust.) Kindb.
SY: *Sematophyllum micans* (Mitt.) Braithw.
SY: *Sematophyllum novae-caesareae* (Aust.) Britt.
Hygrohypnum molle (Hedw.) Loeske [HYGRMOL]
SY: *Hygrohypnum dilitatum* (Wils.) Loeske
SY: *Hygrohypnum smithii* var. *goulardii* (Schimp.) Wijk & Marg.
SY: *Hypnum circulifolium* C. Müll. & Kindb. in Mac. & Kindb.
SY: *Hypnum dilatatum* Wils. ex Schimp.
Hygrohypnum molle var. *bestii* (Ren. & Bryhn) Hab. = Hygrohypnum bestii [HYGRBES]
Hygrohypnum norvegicum (Schimp. in B.S.G.) Amann [HYGRNOR]
Hygrohypnum novae-caesareae (Aust.) Grout = Hygrohypnum micans [HYGRMIC]
Hygrohypnum ochraceum (Turn. ex Wils.) Loeske [HYGROCH]
SY: *Hygrohypnum ochraceum* var. *filiforme* (Limpr.) Amann
SY: *Hygrohypnum ochraceum* var. *flaccidum* (Milde) Amann
SY: *Hygrohypnum ochraceum* var. *uncinatum* (Milde) Loeske
SY: *Hypnum exannulatum* ssp. *pseudolycopodioides* Kindb.
SY: *Hypnum ochraceum* Turn. ex Wils.
SY: *Hypnum pseudolycopodioides* Kindb. ex Nichols
Hygrohypnum ochraceum var. *filiforme* (Limpr.) Amann = Hygrohypnum ochraceum [HYGROCH]
Hygrohypnum ochraceum var. *flaccidum* (Milde) Amann = Hygrohypnum ochraceum [HYGROCH]
Hygrohypnum ochraceum var. *uncinatum* (Milde) Loeske = Hygrohypnum ochraceum [HYGROCH]
Hygrohypnum palustre Loeske = Hygrohypnum luridum [HYGRLUR]
Hygrohypnum palustre var. *ehlei* (Arnell) Grout = Hygrohypnum luridum [HYGRLUR]
Hygrohypnum palustre var. *julaceum* (Schimp. in B.S.G.) Loeske = Hygrohypnum luridum [HYGRLUR]
Hygrohypnum palustre var. *subsphaericarpon* (Brid.) Loeske = Hygrohypnum luridum [HYGRLUR]
Hygrohypnum polare (Lindb.) Loeske [HYGRPOL]
SY: *Hygrohypnum polare* var. *falcatum* (Bryhn) Broth.
SY: *Hypnum polare* Lindb.
Hygrohypnum polare var. *falcatum* (Bryhn) Broth. = Hygrohypnum polare [HYGRPOL]
Hygrohypnum pseudomontanum (Kindb.) Grout = Hygrohypnum luridum [HYGRLUR]
Hygrohypnum smithii (Sw. in Lilj.) Broth. [HYGRSMI]
SY: *Hygrohypnum arcticum* (Somm.) Loeske
SY: *Hypnum arcticum* Somm. in Wahlenb.
SY: *Hypnum goulardii* Schimp.
SY: *Hypnum torrentis* C. Müll. & Kindb. in Mac. & Kindb.
Hygrohypnum smithii var. *goulardii* (Schimp.) Wijk & Marg. = Hygrohypnum molle [HYGRMOL]
Hygrohypnum styriacum (Limpr.) Broth. [HYGRSTY]
Hygrohypnum subeugyrium var. *occidentale* (Card. & Thér.) Grout = Hygrohypnum luridum [HYGRLUR]
Hypnum aduncum Hedw. = Drepanocladus aduncus [DREPADU]
Hypnum aduncum var. *aquaticum* (Sanio) Sanio = Drepanocladus aduncus [DREPADU]
Hypnum aduncum var. *platyphyllum* (Kindb.) Par. = Drepanocladus aduncus [DREPADU]
Hypnum alpestre Sw. ex Hedw. = Hygrohypnum alpestre [HYGRALP]
Hypnum amblyphyllum Williams = Warnstorfia tundrae [WARNTUN]
Hypnum arcticum Somm. in Wahlenb. = Hygrohypnum smithii [HYGRSMI]
Hypnum badium C.J. Hartm. = Loeskypnum badium [LOESBAD]
Hypnum bergenense Aust. = Campylium radicale [CAMPRAD]
Hypnum byssirameum C. Müll. & Kindb. in Mac. & Kindb. = Campylium hispidulum [CAMPHIS]
Hypnum capillifolium Warnst. = Drepanocladus capillifolius [DREPCAP]
Hypnum chrysophyllum Brid. = Campylium chrysophyllum [CAMPCHR]
Hypnum chrysophyllum var. *brevifolium* Ren. & Card. = Campylium chrysophyllum [CAMPCHR]

Hypnum circulifolium C. Müll. & Kindb. in Mac. & Kindb. = Hygrohypnum molle [HYGRMOL]
Hypnum columbicopalustre C. Müll. & Kindb. in Mac. & Kindb. = Hygrohypnum luridum [HYGRLUR]
Hypnum commutatum Hedw. = Palustriella commutata [PALUCOM]
Hypnum commutatum ssp. *sulcatum* (Lindb.) Boul. = Palustriella commutata [PALUCOM]
Hypnum compactum (Hook.) C. Müll. = Conardia compacta [CONACOM]
Hypnum conflatum C. Müll. & Kindb. in Mac. & Kindb. = Drepanocladus aduncus [DREPADU]
Hypnum cordifolium Hedw. = Calliergon cordifolium [CALLCOR]
Hypnum cossonii Schimp. = Limprichtia revolvens [LIMPREV]
Hypnum cuspidatum Hedw. = Calliergonella cuspidata [CALLCUS]
Hypnum decursivulum C. Müll. & Kindb. in Mac. & Kindb. = Campylium radicale [CAMPRAD]
Hypnum dilatatum Wils. ex Schimp. = Hygrohypnum molle [HYGRMOL]
Hypnum exannulatum B.S.G. = Warnstorfia exannulata var. exannulata [WARNEXA0]
Hypnum exannulatum ssp. *pseudolycopodioides* Kindb. = Hygrohypnum ochraceum [HYGROCH]
Hypnum exannulatum ssp. *pseudostramineum* (C. Müll.) Kindb. = Warnstorfia pseudostraminea [WARNPSE]
Hypnum exannulatum var. *purpurascens* (Schimp.) Milde = Warnstorfia exannulata var. exannulata [WARNEXA0]
Hypnum falcatum Brid. = Palustriella commutata [PALUCOM]
Hypnum falcatum var. *microphyllum* Kindb. in Mac. & Kindb. = Drepanocladus sendtneri [DREPSEN]
Hypnum filicinum Hedw. = Cratoneuron filicinum [CRATFIL]
Hypnum filicinum var. *aciculinum* C. Müll. & Kindb. in Mac. & Kindb. = Cratoneuron filicinum [CRATFIL]
Hypnum filicinum var. *filiforme* Berggr. = Cratoneuron filicinum [CRATFIL]
Hypnum fluitans Hedw. = Warnstorfia fluitans [WARNFLU]
Hypnum fluitans var. *falcifolius* f. *viridis* Boul. = Warnstorfia exannulata var. exannulata [WARNEXA0]
Hypnum fluitans var. *gracilis* Boul. in Ren. = Warnstorfia fluitans [WARNFLU]
Hypnum fluitans var. *jeanbernatii* Ren. = Warnstorfia fluitans [WARNFLU]
Hypnum fluviatile Hedw. = Hygroamblystegium fluviatile [HYGRFLU]
Hypnum giganteum Schimp. = Calliergon giganteum [CALLGIG]
Hypnum giganteum var. *fluitans* (Rabenh.) Heug. = Calliergon giganteum [CALLGIG]
Hypnum goulardii Schimp. = Hygrohypnum smithii [HYGRSMI]
Hypnum hispidulum Brid. = Campylium hispidulum [CAMPHIS]
Hypnum kneiffii (B.S.G.) Schimp. in Wils. = Drepanocladus aduncus var. kneiffii [DREPADU3]
Hypnum kneiffii var. *platyphyllum* Kindb. in Mac. & Kindb. = Drepanocladus aduncus [DREPADU]
Hypnum longinerve (Kindb.) Kindb. = Warnstorfia exannulata var. exannulata [WARNEXA0]
Hypnum moseri Kindb. = Sanionia uncinata [SANIUNC]
Hypnum ochraceum Turn. ex Wils. = Hygrohypnum ochraceum [HYGROCH]
Hypnum orthothecioides (Lindb.) Par. = Sanionia uncinata [SANIUNC]
Hypnum palustre Brid. = Hygrohypnum luridum [HYGRLUR]
Hypnum polare Lindb. = Hygrohypnum polare [HYGRPOL]
Hypnum polygamum (B.S.G.) Wils. = Campylium polygamum [CAMPPOL]
Hypnum pseudo-arcticum Kindb. = Hygrohypnum luridum [HYGRLUR]
Hypnum pseudolycopodioides Kindb. ex Nichols = Hygrohypnum ochraceum [HYGROCH]
Hypnum radicale (P. Beauv.) Brid. = Campylium radicale [CAMPRAD]
Hypnum revolvens Sw. = Limprichtia revolvens [LIMPREV]
Hypnum richardsonii (Mitt.) Lesq. & James = Calliergon richardsonii [CALLRIC]
Hypnum riparium Hedw. = Leptodictyum riparium [LEPTRIP]
Hypnum sarmentosum Wahlenb. = Sarmenthypnum sarmentosum [SARMSAR]
Hypnum sarmentosum var. *acuminatum* Bryhn = Sarmenthypnum sarmentosum [SARMSAR]
Hypnum sarmentosum var. *fontinaloides* Berggr. = Sarmenthypnum sarmentosum [SARMSAR]
Hypnum scorpioides Hedw. = Scorpidium scorpioides [SCORSCO]
Hypnum scorpioides var. *gracilescens* Sanio in Klinggr. = Scorpidium scorpioides [SCORSCO]
Hypnum sendtneri Schimp. in H. Müll. = Drepanocladus sendtneri [DREPSEN]
Hypnum serpens Hedw. = Amblystegium serpens [AMBLSER]
Hypnum sinuolatum (Kindb.) Kindb. in Röll = Campylium chrysophyllum [CAMPCHR]
Hypnum sommerfeltii Myr. = Campylium hispidulum [CAMPHIS]
Hypnum stellatum Hedw. = Campylium stellatum [CAMPSTE]
Hypnum stellatum var. *protensum* (Brid.) Röhl. = Campylium stellatum var. protensum [CAMPSTE2]
Hypnum stramineum Dicks. ex Brid. = Calliergon stramineum [CALLSTR]
Hypnum stramineum var. *patens* (Lindb.) Par. = Calliergon stramineum [CALLSTR]
Hypnum subgiganteum Kindb. ex Grout = Calliergon richardsonii [CALLRIC]
Hypnum subsecundum Kindb. = Campylium chrysophyllum [CAMPCHR]
Hypnum torrentis C. Müll. & Kindb. in Mac. & Kindb. = Hygrohypnum smithii [HYGRSMI]
Hypnum trifarium Web. & Mohr = Calliergon trifarium [CALLTRI]
Hypnum turgescens T. Jens. = Pseudocalliergon turgescens [PSEUTUR]
Hypnum uncinatum Hedw. = Sanionia uncinata [SANIUNC]
Hypnum uncinatum var. *alpinum* (Ren.) Hamm. = Sanionia uncinata [SANIUNC]
Hypnum uncinatum var. *gracilescens* B.S.G. = Sanionia uncinata [SANIUNC]
Hypnum uncinatum var. *gracillimum* Berggr. = Sanionia uncinata [SANIUNC]
Hypnum uncinatum var. *micropterum* Kindb. in Mac. & Kindb. = Sanionia uncinata [SANIUNC]
Hypnum uncinatum var. *plumosum* B.S.G. = Sanionia uncinata [SANIUNC]
Hypnum uncinatum var. *symmetricum* (Ren. & Card.) Ren. & Card. = Sanionia uncinata var. symmetrica [SANIUNC1]
Hypnum unicostatum C. Müll. & Kindb. in Mac. & Kindb. = Campylium chrysophyllum [CAMPCHR]
Hypnum vacillans Lesq. & James = Leptodictyum riparium [LEPTRIP]
Hypnum vacium (Hedw.) P. Beauv. = Amblystegium varium [AMBLVAR]
Hypnum vernicosus Lindb. in Hartm. = Hamatocaulis vernicosus [HAMAVER]
Hypnum wilsonii (Lindb.) Schimp. ex Ren. = Drepanocladus sendtneri [DREPSEN]
Leptodictyum (Schimp.) Warnst. [LEPTODI$]
Leptodictyum brevipes (Card. & Thér. ex Holz.) Broth. = Leptodictyum riparium [LEPTRIP]
Leptodictyum laxirete (Card. & Thér.) Broth. = Leptodictyum riparium [LEPTRIP]
Leptodictyum pennellii (Bartr.) Robins. = Leptodictyum riparium [LEPTRIP]
Leptodictyum radicale (P. Beauv.) Kanda = Campylium radicale [CAMPRAD]
Leptodictyum riparium (Hedw.) Warnst. [LEPTRIP]
SY: *Amblystegium brevipes* Card. & Thér. ex Holz.
SY: *Amblystegium riparium* (Hedw.) Schimp. in B.S.G.
SY: *Amblystegium riparium* var. *abbreviatum* Schimp. in B.S.G.
SY: *Amblystegium riparium* var. *flaccidum* (Lesq. & James) Ren. & Card.
SY: *Amblystegium riparium* f. *fluitans* (Lesq. & James) Flow.
SY: *Amblystegium riparium* var. *fluitans* (Lesq. & James) Ren. & Card.
SY: *Amblystegium sipho* (P. Beauv.) Card.

SY: *Amblystegium vacillans* Sull.
SY: *Brachythecium pennellii* Bartr.
SY: *Campylium polygamum* var. *longinerve* (Ren. & Card.) Grout
SY: *Hypnum riparium* Hedw.
SY: *Hypnum vacillans* Lesq. & James
SY: *Leptodictyum brevipes* (Card. & Thér. ex Holz.) Broth.
SY: *Leptodictyum laxirete* (Card. & Thér.) Broth.
SY: *Leptodictyum pennellii* (Bartr.) Robins.
SY: *Leptodictyum riparium* f. *abbreviatum* (Bruch & Schimp.) Grout
SY: *Leptodictyum riparium* var. *abbreviatum* (Schimp. in B.S.G.) Grout
SY: *Leptodictyum riparium* var. *brachyphyllum* (Card. & Thér.) Grout
SY: *Leptodictyum riparium* var. *elongatum* (Schimp. in B.S.G.) Warnst.
SY: *Leptodictyum riparium* var. *flaccidum* (Lesq. & James) Grout
SY: *Leptodictyum riparium* f. *fluitans* (Lesq. & James) Grout
SY: *Leptodictyum riparium* f. *longifolium* (Schultz) Grout
SY: *Leptodictyum riparium* var. *longifolium* (Schultz) Warnst.
SY: *Leptodictyum riparium* var. *nigrescens* Wynne in E. Whiteh.
SY: *Leptodictyum riparium* f. *obtusum* (Grout) Grout
SY: *Leptodictyum riparium* var. *obtusum* (Grout) Grout
SY: *Leptodictyum sipho* (P. Beauv.) Broth.
SY: *Leptodictyum vacillans* (Sull.) Broth.
SY: *Rhynchostegiella georgiana* Dix. & Grout in Grout

Leptodictyum riparium f. *abbreviatum* (Bruch & Schimp.) Grout = Leptodictyum riparium [LEPTRIP]
Leptodictyum riparium f. *fluitans* (Lesq. & James) Grout = Leptodictyum riparium [LEPTRIP]
Leptodictyum riparium f. *longifolium* (Schultz) Grout = Leptodictyum riparium [LEPTRIP]
Leptodictyum riparium f. *obtusum* (Grout) Grout = Leptodictyum riparium [LEPTRIP]
Leptodictyum riparium var. *abbreviatum* (Schimp. in B.S.G.) Grout = Leptodictyum riparium [LEPTRIP]
Leptodictyum riparium var. *brachyphyllum* (Card. & Thér.) Grout = Leptodictyum riparium [LEPTRIP]
Leptodictyum riparium var. *elongatum* (Schimp. in B.S.G.) Warnst. = Leptodictyum riparium [LEPTRIP]
Leptodictyum riparium var. *flaccidum* (Lesq. & James) Grout = Leptodictyum riparium [LEPTRIP]
Leptodictyum riparium var. *longifolium* (Schultz) Warnst. = Leptodictyum riparium [LEPTRIP]
Leptodictyum riparium var. *nigrescens* Wynne in E. Whiteh. = Leptodictyum riparium [LEPTRIP]
Leptodictyum riparium var. *obtusum* (Grout) Grout = Leptodictyum riparium [LEPTRIP]
Leptodictyum sipho (P. Beauv.) Broth. = Leptodictyum riparium [LEPTRIP]
Leptodictyum vacillans (Sull.) Broth. = Leptodictyum riparium [LEPTRIP]

Limprichtia Loeske [LIMPRIC$]

Limprichtia cossonii (Schimp.) Anderson, Crum & Buck [LIMPCOS]
SY: *Drepanocladus intermedius* (Lindb.) Warnst.
SY: *Drepanocladus revolvens* var. *intermedia* (Lindb.) Grout
SY: *Limprichtia intermedia* (Lindb. in Hartm.) Loeske
SY: *Scorpidium cossonii* (Schimp.) Hedenäs

Limprichtia intermedia (Lindb. in Hartm.) Loeske = Limprichtia cossonii [LIMPCOS]

Limprichtia revolvens (Sw.) Loeske [LIMPREV] (red hook moss)
SY: *Drepanocladus revolvens* (Sw.) Warnst.
SY: *Hypnum cossonii* Schimp.
SY: *Hypnum revolvens* Sw.
SY: *Scorpidium revolvens* (Sw.) Hedenäs

Limprichtia vernicosa (Mitt.) Loeske = Hamatocaulis vernicosus [HAMAVER]

Loeskypnum Paul [LOESKYP$]

Loeskypnum badium (Hartm.) Paul [LOESBAD]
SY: *Drepanocladus badius* (Hartm.) G. Roth
SY: *Hypnum badium* C.J. Hartm.

Loeskypnum wickesiae (Grout) Tuom. [LOESWIC]
SY: *Calliergon wickesiae* Grout

Myurella squarrosa Grout = Campylium hispidulum [CAMPHIS]

Palustriella Ochyra [PALUSTR$]

Palustriella commutata (Brid.) Ochyra [PALUCOM]
SY: *Cratoneuron commutatum* (Brid.) G. Roth
SY: *Cratoneuron commutatum* var. *falcatum* (Brid.) Mönk.
SY: *Cratoneuron commutatum* var. *sulcatum* (Lindb.) Mönk.
SY: *Cratoneuron falcatum* (Brid.) G. Roth
SY: *Hypnum commutatum* Hedw.
SY: *Hypnum commutatum* ssp. *sulcatum* (Lindb.) Boul.
SY: *Hypnum falcatum* Brid.

Pseudocalliergon (Limpr.) Loeske [PSEUDOG$]

Pseudocalliergon turgescens (T. Jens.) Loeske [PSEUTUR]
SY: *Calliergon subturgescens* Kindb.
SY: *Calliergon turgescens* (T. Jens.) Kindb.
SY: *Calliergon turgescens* var. *patens* Karcz.
SY: *Calliergon turgescens* var. *tenue* (Berggr.) Karcz.
SY: *Hypnum turgescens* T. Jens.
SY: *Scorpidium turgescens* (T. Jens.) Loeske

Rhaphidostegium novae-caesareae (Aust.) Kindb. = Hygrohypnum micans [HYGRMIC]
Rhynchostegiella compacta (C. Müll.) Loeske = Conardia compacta [CONACOM]
Rhynchostegiella compacta var. *americana* (Grout) Crum, Steere & Anderson = Conardia compacta [CONACOM]
Rhynchostegiella georgiana Dix. & Grout in Grout = Leptodictyum riparium [LEPTRIP]

Sanionia Loeske [SANIONI$]

Sanionia uncinata (Hedw.) Loeske [SANIUNC] (sickle moss)
SY: *Drepanocladus uncinatus* (Hedw.) Warnst.
SY: *Drepanocladus uncinatus* var. *alpinus* (Ren.) Warnst.
SY: *Drepanocladus uncinatus* var. *gracilescens* (B.S.G.) Warnst.
SY: *Drepanocladus uncinatus* var. *gracillimus* (Berggr.) Storm.
SY: *Hypnum moseri* Kindb.
SY: *Hypnum orthothecioides* (Lindb.) Par.
SY: *Hypnum uncinatum* Hedw.
SY: *Hypnum uncinatum* var. *alpinum* (Ren.) Hamm.
SY: *Hypnum uncinatum* var. *gracilescens* B.S.G.
SY: *Hypnum uncinatum* var. *gracillimum* Berggr.
SY: *Hypnum uncinatum* var. *micropterum* Kindb. in Mac. & Kindb.
SY: *Hypnum uncinatum* var. *plumosum* B.S.G.

Sanionia uncinata var. **symmetrica** (Ren. & Card.) Crum & Anderson [SANIUNC1]
SY: *Drepanocladus uncinatus* var. *symmetricus* (Ren. & Card.) Grout
SY: *Hypnum uncinatum* var. *symmetricum* (Ren. & Card.) Ren. & Card.

Sanionia uncinata var. **uncinata** [SANIUNC0]
SY: *Camptothecium paulianum* Grout in Holz. & Frye
SY: *Drepanocladus uncinatus* var. *plumosus* (Schimp.) Warnst.
SY: *Drepanocladus uncinatus* var. *plumulosus* (Schimp. in B.S.G.) Warnst.
SY: *Drepanocladus uncinatus* var. *subjulaceus* (Schimp. in B.S.G.) Warnst.
SY: *Tomentypnum paulianum* (Grout) Broth. ex Robins.

Sarmenthypnum Tuom. & T. Kop. [SARMENT$]

Sarmenthypnum sarmentosum (Wahlenb.) Tuom. & T. Kop. [SARMSAR]
SY: *Calliergon sarmentosum* (Wahlenb.) Kindb.
SY: *Calliergon sarmentosum* var. *beringianum* (Card. & Thér.) Grout
SY: *Calliergon sarmentosum* var. *crispum* Karcz.
SY: *Calliergon sarmentosum* var. *fallaciosum* (Milde) G. Roth
SY: *Calliergon sarmentosum* var. *flagellare* Karcz.
SY: *Calliergon sarmentosum* var. *fontinaloides* (Berggr.) G. Roth
SY: *Calliergon sarmentosum* f. *heterophyllum* Arnell & C. Jens.
SY: *Calliergon sarmentosum* f. *homophyllum* Arnell & C. Jens.
SY: *Calliergon sarmentosum* var. *subpinnatum* Warnst.
SY: *Calliergon subsarmentosum* Kindb.
SY: *Hypnum sarmentosum* Wahlenb.
SY: *Hypnum sarmentosum* var. *acuminatum* Bryhn
SY: *Hypnum sarmentosum* var. *fontinaloides* Berggr.

Scorpidium (Schimp.) Limpr. [SCORPID$]
Scorpidium cossonii (Schimp.) Hedenäs = Limprichtia cossonii [LIMPCOS]
Scorpidium revolvens (Sw.) Hedenäs = Limprichtia revolvens [LIMPREV]
Scorpidium scorpioides (Hedw.) Limpr. [SCORSCO] (sausage moss)
SY: *Drepanocladus revolvens* var. *miquelonensis* (Ren.) Grout
SY: *Drepanocladus scorpioides* (Hedw.) Warnst.
SY: *Hypnum scorpioides* Hedw.
SY: *Hypnum scorpioides* var. *gracilescens* Sanio in Klinggr.
SY: *Scorpidium scorpioides* f. *gracilescens* (Sanio) C. Jens.
Scorpidium scorpioides f. *gracilescens* (Sanio) C. Jens. = Scorpidium scorpioides [SCORSCO]
Scorpidium trifarium (Web. & Mohr) Paul = Calliergon trifarium [CALLTRI]
Scorpidium turgescens (T. Jens.) Loeske = Pseudocalliergon turgescens [PSEUTUR]
Scorpidium vernicosum (Mitt.) Tuom. = Hamatocaulis vernicosus [HAMAVER]
Sematophyllum micans (Mitt.) Braithw. = Hygrohypnum micans [HYGRMIC]
Sematophyllum novae-caesareae (Aust.) Britt. = Hygrohypnum micans [HYGRMIC]
Tomentypnum paulianum (Grout) Broth. ex Robins. = Sanionia uncinata var. uncinata [SANIUNC0]
Warnstorfia Loeske [WARNSTO$]
Warnstorfia crassicostata (Jansssens) Crum & Anderson = Drepanocladus crassicostatus [DREPCRA]
Warnstorfia exannulata (Schimp. in B.S.G.) Loeske [WARNEXA]
Warnstorfia exannulata var. **exannulata** [WARNEXA0] (hooked spring moss)
SY: *Dichelyma longinerve* Kindb. in Mac.
SY: *Drepanocladus exannulatus* (Schimp. in B.S.G.) Warnst.
SY: *Drepanocladus exannulatus* var. *alpinus* (Grav.) Wijk & Marg.
SY: *Drepanocladus exannulatus* var. *brachydictyon* (Ren.) G. Roth
SY: *Drepanocladus exannulatus* f. *brachydictyus* (Ren.) Smirn.
SY: *Drepanocladus exannulatus* f. *falcifolius* Ren.
SY: *Drepanocladus exannulatus* var. *falcifolius* (Ren.) Grout
SY: *Drepanocladus exannulatus* f. *orthophyllus* (Milde) Mönk.
SY: *Drepanocladus exannulatus* var. *rotae* (De Not.) Loeske
SY: *Drepanocladus exannulatus* var. *serratus* (Milde) Loeske
SY: *Drepanocladus exannulatus* f. *submersus* Mönk.
SY: *Drepanocladus fluitans* var. *falcifolius* f. *viridis* Boul.
SY: *Hypnum exannulatum* B.S.G.
SY: *Hypnum exannulatum* var. *purpurascens* (Schimp.) Milde
SY: *Hypnum fluitans* var. *falcifolius* f. *viridis* Boul.
SY: *Hypnum longinerve* (Kindb.) Kindb.
Warnstorfia exannulata var. **purpurascens** (Schimp.) Tuom. & T. Kop. [WARNEXA1]
SY: *Drepanocladus exannulatus* var. *purpurascens* (Schimp.) Herz.
SY: *Drepanocladus purpurascens* (Schimp.) Loeske
Warnstorfia fluitans (Hedw.) Loeske [WARNFLU] (water hook moss)
SY: *Drepanocladus fluitans* (Hedw.) Warnst.
SY: *Drepanocladus fluitans* var. *berggrenii* (C. Jens.) C. Jens. in Weim.
SY: *Drepanocladus fluitans* f. *gracilis* Boul.
SY: *Drepanocladus fluitans* var. *gracilis* Warnst.
SY: *Drepanocladus fluitans* f. *jeanbernatii* (Ren.) Mönk.
SY: *Drepanocladus fluitans* var. *jeanbernatii* (Ren.) Warnst.
SY: *Drepanocladus fluitans* f. *setiformis* (Ren.) Mönk.
SY: *Drepanocladus fluitans* var. *submersus* (Schimp.) Loeske
SY: *Drepanocladus submersus* (Schimp.) Warnst.
SY: *Hypnum fluitans* Hedw.
SY: *Hypnum fluitans* var. *gracilis* Boul. in Ren.
SY: *Hypnum fluitans* var. *jeanbernatii* Ren.
Warnstorfia fluitans var. **falcata** (Sanio ex C. Jens.) Crum & Anderson [WARNFLU1]
SY: *Drepanocladus fluitans* var. *falcatus* (Sanio ex C. Jens.) G. Roth
SY: *Drepanocladus fluitans* var. *uncatus* Crum, Steere & Anderson
SY: *Drepanocladus schulzei* G. Roth
Warnstorfia fluitans var. **fluitans** [WARNFLU0]
SY: *Drepanocladus berggrenii* (C. Jens.) G. Roth
Warnstorfia procera (Ren. & Arnell in Husn.) Tuom. & T. Kop. [WARNPRO]
SY: *Drepanocladus procerus* (Ren. & Arnell in Husn.)
Warnstorfia pseudostraminea (C. Müll.) Tuom. & T. Kop. [WARNPSE]
SY: *Brachythecium edentatum* Williams
SY: *Calliergidium pseudostramineum* (C. Müll.) Grout
SY: *Calliergidium pseudostramineum* var. *plesiostramineum* (Ren.) Grout
SY: *Drepanocladus pseudostramineus* (C. Müll.) G. Roth
SY: *Hypnum exannulatum* ssp. *pseudostramineum* (C. Müll.) Kindb.
Warnstorfia trichophylla (Warnst.) Tuom. & T. Kop. [WARNTRI]
SY: *Drepanocladus trichophyllus* (Warnst.) Podp.
Warnstorfia tundrae (Arnell) Loeske [WARNTUN]
SY: *Drepanocladus exannulatus* f. *tundrae* (Arnell) Mönk.
SY: *Drepanocladus exannulatus* var. *tundrae* (Arnell) Warnst.
SY: *Drepanocladus tundrae* (Arnell) Loeske
SY: *Hypnum amblyphyllum* Williams

ANDREAEACEAE (B:M)

Andreaea Hedw. [ANDREAE$]
Andreaea alpestris (Thed.) Schimp. [ANDRALP]
SY: *Andreaea rupestris* var. *alpestris* (Thed.) Sharp in Grout
Andreaea angustata Lindb. ex Limpr. = Andreaea heinemannii [ANDRHEI]
Andreaea baileyi (Holz.) Holz. = Andreaea nivalis [ANDRNIV]
Andreaea blyttii Schimp. [ANDRBLY]
Andreaea blyttii ssp. *angustata* (Lindb. ex Limpr.) Schultze-Motel = Andreaea heinemannii [ANDRHEI]
Andreaea blyttii var. *obtusifolia* (Berggr.) Sharp = Andreaea heinemannii [ANDRHEI]
Andreaea crassinervia var. *obtusifolia* Berggr. = Andreaea heinemannii [ANDRHEI]
Andreaea heinemannii Hampe & C. Müll. [ANDRHEI]
SY: *Andreaea angustata* Lindb. ex Limpr.
SY: *Andreaea blyttii* ssp. *angustata* (Lindb. ex Limpr.) Schultze-Motel
SY: *Andreaea blyttii* var. *obtusifolia* (Berggr.) Sharp
SY: *Andreaea crassinervia* var. *obtusifolia* Berggr.
Andreaea macounii Kindb. = Andreaea nivalis [ANDRNIV]
Andreaea megistospora B. Murr. [ANDRMEG]
Andreaea megistospora ssp. *epapillosa* B. Murr. = Andreaea megistospora var. epapillosa [ANDRMEG1]
Andreaea megistospora var. **epapillosa** (B. Murr.) Crum & Anderson [ANDRMEG1]
SY: *Andreaea megistospora* ssp. *epapillosa* B. Murr.
Andreaea megistospora var. **megistospora** [ANDRMEG0]
Andreaea mutabilis Hook. f. & Wils. [ANDRMUT]
Andreaea nivalis Hook. [ANDRNIV]
SY: *Andreaea baileyi* (Holz.) Holz.
SY: *Andreaea macounii* Kindb.
Andreaea papillosa Lindb. = Andreaea rupestris var. papillosa [ANDRRUP3]
Andreaea parvifolia C. Müll. = Andreaea rupestris var. rupestris [ANDRRUP0]
Andreaea petrophila Ehrh. ex Fürnr. = Andreaea rupestris var. rupestris [ANDRRUP0]
Andreaea rothii Web. & Mohr [ANDRROT]
Andreaea rupestris Hedw. [ANDRRUP] (black rock moss)
Andreaea rupestris ssp. *papillosa* (Lindb.) C. Jens. = Andreaea rupestris var. papillosa [ANDRRUP3]
Andreaea rupestris var. *acuminata* auct. = Andreaea rupestris var. papillosa [ANDRRUP3]
Andreaea rupestris var. *alpestris* (Thed.) Sharp in Grout = Andreaea alpestris [ANDRALP]

Andreaea rupestris var. **papillosa** (Lindb.) Podp. [ANDRRUP3]
SY: *Andreaea papillosa* Lindb.
SY: *Andreaea rupestris* ssp. *papillosa* (Lindb.) C. Jens.
SY: *Andreaea rupestris* var. *sparsifolia* (Zett.) Sharp in Grout
SY: *Andreaea sparsifolia* var. *sublaevis* Kindb.
SY: *Andreaea rupestris* var. *acuminata* auct.
Andreaea rupestris var. **rupestris** [ANDRRUP0]
SY: *Andreaea parvifolia* C. Müll.
SY: *Andreaea petrophila* Ehrh. ex Fürnr.
Andreaea rupestris var. *sparsifolia* (Zett.) Sharp in Grout = Andreaea rupestris var. papillosa [ANDRRUP3]
Andreaea schofieldiana B. Murr. [ANDRSCH]
Andreaea sinuosa B. Murr. [ANDRSIN]
Andreaea sparsifolia var. *sublaevis* Kindb. = Andreaea rupestris var. papillosa [ANDRRUP3]

ANDREAEOBRYACEAE (B:M)

Andreaeobryum Steere & B. Murr. [ANDREAO$]
Andreaeobryum macrosporum Steere & B. Murr. [ANDRMAC]

ANEURACEAE (B:H)

Aneura Dum. [ANEURA$]
Aneura pinguis (L.) Dum. [ANEUPIN]
SY: *Riccardia pinguis* (L.) Gray
Riccardia Gray [RICCARD$]
Riccardia chamedryfolia (With.) Grolle [RICCCHA]
SY: *Riccardia sinuata* (Dicks.) Trev.
Riccardia latifrons Lindb. [RICCLAT]
Riccardia multifida (L.) Gray [RICCMUL] (comb liverwort)
Riccardia palmata (Hedw.) Carruth. [RICCPAL]
Riccardia pinguis (L.) Gray = Aneura pinguis [ANEUPIN]
Riccardia sinuata (Dicks.) Trev. = Riccardia chamedryfolia [RICCCHA]

ANOMODONTACEAE (B:M)

Pterogonium Sm. [PTEROGO$]
Pterogonium gracile (Hedw.) Sm. [PTERGRA]

ANTHELIACEAE (B:H)

Anthelia (Dum. emend. Schiffn.) Dum. [ANTHELI$]
Anthelia julacea (L.) Dum. [ANTHJUL]
Anthelia juratzkana (Limpr.) Trev. [ANTHJUR]

ANTHOCEROTACEAE (B:H)

Anthoceros L. emend. Prosk. [ANTHOCE$]
Anthoceros crispulus (Mont.) Douin = Anthoceros punctatus [ANTHPUN]
Anthoceros fusiformis Aust. [ANTHFUS]
Anthoceros hallii Aust. = Phaeoceros hallii [PHAEHAL]
Anthoceros phymatodes M.A. Howe = Phaeoceros hallii [PHAEHAL]
Anthoceros punctatus L. [ANTHPUN]
SY: *Anthoceros crispulus* (Mont.) Douin
Phaeoceros Prosk. [PHAEOCE$]
Phaeoceros hallii (Aust.) Prosk. [PHAEHAL]
SY: *Anthoceros hallii* Aust.
SY: *Anthoceros phymatodes* M.A. Howe
Phaeoceros laevis (L.) Prosk. [PHAELAE]

ARNELLIACEAE (B:H)

Arnellia Lindb. [ARNELLI$]
Arnellia fennica (Gott.) Lindb. [ARNEFEN]

AULACOMNIACEAE (B:M)

Aulacomnium Schwaegr. [AULACOM$]
Aulacomnium acuminatum (Lindb. & Arnell) Kindb. [AULAACU]
Aulacomnium androgynum (Hedw.) Schwaegr. [AULAAND] (lover's moss)
Aulacomnium palustre (Hedw.) Schwaegr. [AULAPAL] (ribbed bog moss)
SY: *Aulacomnium palustre* var. *congestum* Boul.
SY: *Aulacomnium palustre* var. *dimorphum* Card. & Thér.
SY: *Aulacomnium palustre* var. *fasciculare* (Brid.) B.S.G.
SY: *Aulacomnium palustre* var. *imbricatum* Bruch & Schimp. in B.S.G.
SY: *Aulacomnium palustre* f. *laxifolium* (Kindb. in Mac. & Kindb.) Podp.
SY: *Aulacomnium palustre* var. *laxifolium* Kindb. in Mac. & Kindb.
SY: *Aulacomnium palustre* var. *polycephalum* (Brid.) Hüb.
Aulacomnium palustre f. *laxifolium* (Kindb. in Mac. & Kindb.) Podp. = Aulacomnium palustre [AULAPAL]
Aulacomnium palustre var. *congestum* Boul. = Aulacomnium palustre [AULAPAL]
Aulacomnium palustre var. *dimorphum* Card. & Thér. = Aulacomnium palustre [AULAPAL]
Aulacomnium palustre var. *fasciculare* (Brid.) B.S.G. = Aulacomnium palustre [AULAPAL]
Aulacomnium palustre var. *imbricatum* Bruch & Schimp. in B.S.G. = Aulacomnium palustre [AULAPAL]
Aulacomnium palustre var. *laxifolium* Kindb. in Mac. & Kindb. = Aulacomnium palustre [AULAPAL]
Aulacomnium palustre var. *polycephalum* (Brid.) Hüb. = Aulacomnium palustre [AULAPAL]
Aulacomnium turgidum (Wahlenb.) Schwaegr. [AULATUR]

AYTONIACEAE (B:H)

Asterella P. Beauv. [ASTEREL$]
Asterella gracilis (Web.) Underw. [ASTEGRA]
Asterella lindenbergiana (Corda) Lindb. [ASTELIN]
Grimaldia barbifrons Bisch. = Mannia fragrans [MANNFRA]
Mannia Opiz [MANNIA$]
Mannia fragrans (Balbis) Frye & Clark [MANNFRA]
SY: *Grimaldia barbifrons* Bisch.
Reboulia Raddi [REBOULI$]
Reboulia hemisphaerica (L.) Raddi [REBOHEM]

BARTRAMIACEAE (B:M)

Anacolia Schimp. [ANACOLI$]
Anacolia menziesii (Turn.) Par. [ANACMEN]
SY: *Anacolia menziesii* f. *grandifolia* Flow. in Grout
SY: *Bartramia menziesii* Turn.
Anacolia menziesii f. *grandifolia* Flow. in Grout = Anacolia menziesii [ANACMEN]
Bartramia Hedw. [BARTRAM$]
Bartramia breviseta Lindb. = Bartramia ithyphylla [BARTITH]
Bartramia circinnulata C. Müll. & Kindb. in Mac. & Kindb. = Bartramia pomiformis [BARTPOM]
Bartramia crispa Brid. = Bartramia pomiformis [BARTPOM]
Bartramia fontana (Hedw.) Turn. = Philonotis fontana [PHILFON]
Bartramia glauco-viridis C. Müll. & Kindb. in Mac. & Kindb. = Bartramia pomiformis [BARTPOM]
Bartramia halleriana Hedw. [BARTHAL]
Bartramia ithyphylla Brid. [BARTITH]
SY: *Bartramia breviseta* Lindb.
SY: *Bartramia ithyphylla* var. *breviseta* (Lindb.) Kindb.
SY: *Bartramia ithyphylla* var. *fragilifolia* Card. & Thér.
SY: *Bartramia ithyphylla* var. *strigosa* (Wahlenb.) Hartm.
Bartramia ithyphylla var. *breviseta* (Lindb.) Kindb. = Bartramia ithyphylla [BARTITH]
Bartramia ithyphylla var. *fragilifolia* Card. & Thér. = Bartramia ithyphylla [BARTITH]

Bartramia ithyphylla var. *strigosa* (Wahlenb.) Hartm. = Bartramia ithyphylla [BARTITH]
Bartramia marchica (Hedw.) Sw. in Schrad. = Philonotis marchica [PHILMAR]
Bartramia menziesii Turn. = Anacolia menziesii [ANACMEN]
Bartramia oederi Brid. = Plagiopus oederiana [PLAGOED]
Bartramia oederiana Turn. = Plagiopus oederiana [PLAGOED]
Bartramia pomiformis Hedw. [BARTPOM] (apple moss)
SY: *Bartramia circinnulata* C. Müll. & Kindb. in Mac. & Kindb.
SY: *Bartramia crispa* Brid.
SY: *Bartramia glauco-viridis* C. Müll. & Kindb. in Mac. & Kindb.
SY: *Bartramia pomiformis* var. *crispa* (Brid.) Bruch & Schimp. in B.S.G.
SY: *Bartramia pomiformis* var. *elongata* Turn.
Bartramia pomiformis var. *crispa* (Brid.) Bruch & Schimp. in B.S.G. = Bartramia pomiformis [BARTPOM]
Bartramia pomiformis var. *elongata* Turn. = Bartramia pomiformis [BARTPOM]
Bartramia stricta Brid. [BARTSTR]
Conostomum Sw. in Web. & Mohr [CONOSTO$]
Conostomum boreale Sw. = Conostomum tetragonum [CONOTET]
Conostomum tetragonum (Hedw.) Lindb. [CONOTET]
SY: *Conostomum boreale* Sw.
Philonotis Brid. [PHILONO$]
Philonotis acutiflora Kindb. in Röll = Philonotis fontana var. pumila [PHILFON4]
Philonotis americana Dism. = Philonotis fontana var. americana [PHILFON1]
Philonotis americana var. *gracilescens* (Dism.) Flow. in Grout = Philonotis fontana var. americana [PHILFON1]
Philonotis americana var. *torquata* (Ren. & Geh. in Geh.) Flow. in Grout = Philonotis fontana var. americana [PHILFON1]
Philonotis arnellii Husn. = Philonotis capillaris [PHILCAP]
Philonotis calcarea f. *occidentalis* Flow. in Grout = Philonotis fontana var. fontana [PHILFON0]
Philonotis capillaris Lindb. in Hartm. [PHILCAP]
SY: *Philonotis arnellii* Husn.
SY: *Philonotis macounii* Lesq. & James
SY: *Philonotis vancouverensis* Kindb.
Philonotis fontana (Hedw.) Brid. [PHILFON] (swamp moss)
SY: *Bartramia fontana* (Hedw.) Turn.
SY: *Philonotis fontana* var. *microblasta* C. Müll. & Kindb. in Mac. & Kindb.
SY: *Philonotis fontana* var. *microthamnia* Kindb.
Philonotis fontana var. *adpressa* (Ferg.) Limpr. = Philonotis fontana var. fontana [PHILFON0]
Philonotis fontana var. **americana** (Dism.) Flow. [PHILFON1]
SY: *Philonotis americana* Dism.
SY: *Philonotis americana* var. *gracilescens* (Dism.) Flow. in Grout
SY: *Philonotis americana* var. *torquata* (Ren. & Geh. in Geh.) Flow. in Grout
SY: *Philonotis fontana* var. *seriata* Breidl.
SY: *Philonotis fontana* var. *serrata* Kindb. in Mac. & Kindb.
Philonotis fontana var. *borealis* Hag. = Philonotis fontana var. fontana [PHILFON0]
Philonotis fontana var. *brachyphylla* Kindb. in Mac. = Philonotis fontana var. fontana [PHILFON0]
Philonotis fontana var. *columbiae* Kindb. in Mac. & Kindb. = Philonotis fontana var. fontana [PHILFON0]
Philonotis fontana var. *falcata* (Hook.) Brid. = Philonotis fontana var. fontana [PHILFON0]
Philonotis fontana var. **fontana** [PHILFON0]
SY: *Philonotis calcarea* f. *occidentalis* Flow. in Grout
SY: *Philonotis fontana* var. *adpressa* (Ferg.) Limpr.
SY: *Philonotis fontana* var. *borealis* Hag.
SY: *Philonotis fontana* var. *brachyphylla* Kindb. in Mac.
SY: *Philonotis fontana* var. *columbiae* Kindb. in Mac. & Kindb.
SY: *Philonotis fontana* var. *falcata* (Hook.) Brid.
SY: *Philonotis fontana* var. *heterophylla* Dism.
SY: *Philonotis fontana* var. *laxa* Vent.
Philonotis fontana var. *heterophylla* Dism. = Philonotis fontana var. fontana [PHILFON0]
Philonotis fontana var. *laxa* Vent. = Philonotis fontana var. fontana [PHILFON0]
Philonotis fontana var. *microblasta* C. Müll. & Kindb. in Mac. & Kindb. = Philonotis fontana [PHILFON]
Philonotis fontana var. *microthamnia* Kindb. = Philonotis fontana [PHILFON]
Philonotis fontana var. **pumila** (Turn.) Brid. [PHILFON4]
SY: *Philonotis acutiflora* Kindb. in Röll
SY: *Philonotis tomentella* Mol. in Lor.
Philonotis fontana var. *seriata* (Mitt.) Kindb. = Philonotis yezoana [PHILYEZ]
Philonotis fontana var. *seriata* Breidl. = Philonotis fontana var. americana [PHILFON1]
Philonotis fontana var. *seriata* f. *dimorphophylla* Flow. in Grout = Philonotis yezoana [PHILYEZ]
Philonotis fontana var. *seriata* f. *occidentalis* Flow. in Grout = Philonotis yezoana [PHILYEZ]
Philonotis fontana var. *serrata* Kindb. in Mac. & Kindb. = Philonotis fontana var. americana [PHILFON1]
Philonotis macounii Lesq. & James = Philonotis capillaris [PHILCAP]
Philonotis marchica (Hedw.) Brid. [PHILMAR]
SY: *Bartramia marchica* (Hedw.) Sw. in Schrad.
SY: *Philonotis marchica* var. *laxa* (Limpr.) Loeske & Warnst.
Philonotis marchica var. *laxa* (Limpr.) Loeske & Warnst. = Philonotis marchica [PHILMAR]
Philonotis tomentella Mol. in Lor. = Philonotis fontana var. pumila [PHILFON4]
Philonotis vancouverensis Kindb. = Philonotis capillaris [PHILCAP]
Philonotis yezoana Besch. & Card. in Card. [PHILYEZ]
SY: *Philonotis fontana* var. *seriata* (Mitt.) Kindb.
SY: *Philonotis fontana* var. *seriata* f. *dimorphophylla* Flow. in Grout
SY: *Philonotis fontana* var. *seriata* f. *occidentalis* Flow. in Grout
Plagiopus Brid. [PLAGIOP$]
Plagiopus oederi (Brid.) Limpr. = Plagiopus oederiana [PLAGOED]
Plagiopus oederi f. *alpinus* (Schwaegr.) Hab. = Plagiopus oederiana [PLAGOED]
Plagiopus oederi var. *alpina* (Schwaegr.) Torre & Sarnth. = Plagiopus oederiana [PLAGOED]
Plagiopus oederiana (Sw.) Crum & Anderson [PLAGOED]
SY: *Bartramia oederi* Brid.
SY: *Bartramia oederiana* Turn.
SY: *Plagiopus oederi* (Brid.) Limpr.
SY: *Plagiopus oederi* var. *alpina* (Schwaegr.) Torre & Sarnth.
SY: *Plagiopus oederi* f. *alpinus* (Schwaegr.) Hab.

BLASIACEAE (B:H)

Blasia L. [BLASIA$]
Blasia pusilla L. [BLASPUS]

BRACHYTHECIACEAE (B:M)

Bestia breweriana (Lesq.) Grout = Isothecium cristatum [ISOTCRI]
Bestia breweriana var. *howei* (Kindb.) Grout = Isothecium cristatum [ISOTCRI]
Bestia breweriana var. *lutescens* (Lesq. & James) Grout = Isothecium cristatum [ISOTCRI]
Bestia cristata (Hampe) L. Koch = Isothecium cristatum [ISOTCRI]
Brachythecium Schimp. in B.S.G. [BRACHYT$]
Brachythecium acutum (Mitt.) Sull. [BRACACT]
SY: *Brachythecium salebrosum* ssp. *mamilligerum* Kindb. in Mac. & Kindb.
SY: *Rhynchostegium revelstokense* (Kindb.) Kindb.
Brachythecium albicans (Hedw.) Schimp. in B.S.G. [BRACALB] (lawn moss)
SY: *Brachythecium albicans* var. *occidentale* Ren. & Card.
SY: *Brachythecium beringianum* Card. & Thér.
SY: *Brachythecium pseudoalbicans* Kindb.
SY: *Brachythecium salebrosum* var. *waghornei* Ren. & Card.
SY: *Chamberlainia albicans* (Hedw.) Robins.
SY: *Hypnum albicans* Neck. ex Hedw.

Brachythecium albicans var. *occidentale* Ren. & Card. = Brachythecium albicans [BRACALB]
Brachythecium asperrimum (Mitt.) Sull. = Brachythecium frigidum [BRACFRI]
Brachythecium beringianum Card. & Thér. = Brachythecium albicans [BRACALB]
Brachythecium bestii Grout = Brachythecium reflexum var. pacificum [BRACREF1]
Brachythecium calcareum Kindb. [BRACCAL]
SY: *Brachythecium flexicaule* Ren. & Card. in Grout
SY: *Brachythecium labradoricum* (Kindb.) Par.
SY: *Brachythecium salebrosum* var. *densum* Schimp. in B.S.G.
SY: *Chamberlainia calcarea* (Kindb.) Robins.
SY: *Eurhynchium labradoricum* Kindb.
Brachythecium campestre (C. Müll.) Schimp. in B.S.G. [BRACCAM]
SY: *Chamberlainia campestris* (C. Müll.) Robins.
SY: *Hypnum campestre* (C. Müll.) Mitt.
Brachythecium cavernosum Kindb. = Brachythecium rutabulum [BRACRUT]
Brachythecium cirrosum (Schwaegr. in Schultes) Schimp. = Cirriphyllum cirrosum [CIRRCIR]
Brachythecium cirrosum var. *coloradense* (Aust.) Wijk & Marg. = Cirriphyllum cirrosum [CIRRCIR]
Brachythecium collinum (Schleich. ex C. Müll.) Schimp. in B.S.G. [BRACCOL]
SY: *Chamberlainia collina* (Schleich. ex C. Müll.) Robins.
Brachythecium collinum var. *idahense* (Ren. & Card.) Grout = Brachythecium velutinum var. velutinum [BRACVEL0]
Brachythecium columbico-rutabulum Kindb. in Mac. & Kindb. = Brachythecium frigidum [BRACFRI]
Brachythecium curtum (Lindb.) Limpr. = Brachythecium oedipodium [BRACOED]
Brachythecium erythrorrhizon Schimp. in B.S.G. [BRACERY]
SY: *Brachythecium harpidioides* C. Müll. & Kindb. in Mac. & Kindb.
SY: *Chamberlainia erythrorrhiza* (Schimp. in B.S.G.) Robins.
Brachythecium erythrorrhizon var. *suberythrorrhizon* (Ren. & Card.) Grout = Brachythecium velutinum var. venustum [BRACVEL1]
Brachythecium fendleri var. *idahense* (Ren. & Card.) Wijk & Marg. = Brachythecium velutinum var. velutinum [BRACVEL0]
Brachythecium flagellare (Hedw.) Jenn. = Brachythecium plumosum [BRACPLU]
Brachythecium flagellare var. *homomallum* (Schimp. in B.S.G.) Jenn. = Brachythecium plumosum [BRACPLU]
Brachythecium flagellare var. *pringlei* (Williams) Grout = Brachythecium plumosum [BRACPLU]
Brachythecium flagellare var. *roellii* (Ren. & Card.) Grout = Brachythecium plumosum [BRACPLU]
Brachythecium flexicaule Ren. & Card. in Grout = Brachythecium calcareum [BRACCAL]
Brachythecium frigidum (C. Müll.) Besch. [BRACFRI] (golden short-capsuled moss)
SY: *Brachythecium asperrimum* (Mitt.) Sull.
SY: *Brachythecium columbico-rutabulum* Kindb. in Mac. & Kindb.
SY: *Brachythecium gemmascens* C. Müll. & Kindb. in Mac. & Kindb.
SY: *Brachythecium lamprochryseum* C. Müll. & Kindb. in Mac. & Kindb.
SY: *Brachythecium lamprochryseum* var. *giganteum* Grout
SY: *Brachythecium lamprochryseum* var. *solfatarense* Grout
SY: *Brachythecium pacificum* Jenn.
SY: *Brachythecium pseudostarkei* Ren. & Card.
SY: *Brachythecium spurio-rutabulum* C. Müll. & Kindb. in Mac. & Kindb.
SY: *Brachythecium subasperrimum* Card. & Thér.
SY: *Brachythecium subintricatum* Kindb.
SY: *Brachythecium washingtonianum* D.C. Eat. in Grout
Brachythecium gemmascens C. Müll. & Kindb. in Mac. & Kindb. = Brachythecium frigidum [BRACFRI]
Brachythecium groenlandicum (C. Jens.) Schljak. [BRACGRO]
Brachythecium harpidioides C. Müll. & Kindb. in Mac. & Kindb. = Brachythecium erythrorrhizon [BRACERY]
Brachythecium holzingeri (Grout) Grout [BRACHOL]
SY: *Brachythecium tromsoense* auct.
Brachythecium hylotapetum B. Hig. & N. Hig. [BRACHYL] (woodsy ragged moss)
Brachythecium idahense Ren. & Card. = Brachythecium velutinum var. velutinum [BRACVEL0]
Brachythecium labradoricum (Kindb.) Par. = Brachythecium calcareum [BRACCAL]
Brachythecium laetum (Brid.) B.S.G. = Brachythecium oxycladon [BRACOXY]
Brachythecium laevisetum Kindb. = Brachythecium salebrosum [BRACSAL]
Brachythecium lamprochryseum C. Müll. & Kindb. in Mac. & Kindb. = Brachythecium frigidum [BRACFRI]
Brachythecium lamprochryseum var. *giganteum* Grout = Brachythecium frigidum [BRACFRI]
Brachythecium lamprochryseum var. *solfatarense* Grout = Brachythecium frigidum [BRACFRI]
Brachythecium latifolium Kindb. = Brachythecium nelsonii [BRACNEL]
Brachythecium leibergii Grout [BRACLEI]
SY: *Chamberlainia leibergii* (Grout) Robins.
Brachythecium leucoglaucum C. Müll. & Kindb. in Mac. & Kindb. = Brachythecium rutabulum [BRACRUT]
Brachythecium mirabundum C. Müll. & Kindb. in Mac. & Kindb. = Brachythecium rutabulum [BRACRUT]
Brachythecium nanopes C. Müll. & Kindb. in Mac. & Kindb. = Brachythecium populeum [BRACPOP]
Brachythecium nelsonii Grout [BRACNEL]
SY: *Brachythecium latifolium* Kindb.
Brachythecium oedipodium (Mitt.) Jaeg. [BRACOED] (short-leaved ragged moss)
SY: *Brachythecium curtum* (Lindb.) Limpr.
SY: *Brachythecium starkei* var. *curtum* (Lindb.) Warnst.
SY: *Brachythecium starkei* var. *explanatum* (Brid.) Mönk.
SY: *Brachythecium starkei* ssp. *oedipodium* (Mitt.) Ren. & Card.
SY: *Eurhynchium subintegrifolium* Kindb.
SY: *Rhynchostegium subintegrifolium* (Kindb.) Kindb.
Brachythecium oxycladon (Brid.) Jaeg. [BRACOXY]
SY: *Brachythecium laetum* (Brid.) B.S.G.
SY: *Brachythecium oxycladon* var. *dentatum* (Lesq. & James) Grout
SY: *Brachythecium spurio-acuminatum* C. Müll. & Kindb. in Mac. & Kindb.
SY: *Chamberlainia oxyclada* (Brid.) Robins.
Brachythecium oxycladon var. *dentatum* (Lesq. & James) Grout = Brachythecium oxycladon [BRACOXY]
Brachythecium pacificum Jenn. = Brachythecium frigidum [BRACFRI]
Brachythecium petrophilum Williams = Brachythecium velutinum var. venustum [BRACVEL1]
Brachythecium piliferum (Hedw.) Kindb. = Cirriphyllum piliferum [CIRRPIL]
Brachythecium platycladum C. Müll. & Kindb. in Mac. & Kindb. = Brachythecium rutabulum [BRACRUT]
Brachythecium plumosum (Hedw.) Schimp. in B.S.G. [BRACPLU]
SY: *Brachythecium flagellare* (Hedw.) Jenn.
SY: *Brachythecium flagellare* var. *homomallum* (Schimp. in B.S.G.) Jenn.
SY: *Brachythecium flagellare* var. *pringlei* (Williams) Grout
SY: *Brachythecium flagellare* var. *roellii* (Ren. & Card.) Grout
SY: *Brachythecium plumosum* var. *homomallum* Schimp. in B.S.G.
SY: *Brachythecium plumosum* var. *pringlei* (Williams) Grout
SY: *Brachythecium plumosum* var. *roellii* (Ren. & Card.) Grout
SY: *Brachythecium roellii* Ren. & Card.
SY: *Brachythecium rutabuliforme* Kindb. in Mac. & Kindb.
SY: *Eurhynchium semiasperum* C. Müll. & Kindb. in Mac. & Kindb.
SY: *Hypnum plumosum* Hedw.

Brachythecium plumosum var. *homomallum* Schimp. in B.S.G. = Brachythecium plumosum [BRACPLU]
Brachythecium plumosum var. *pringlei* (Williams) Grout = Brachythecium plumosum [BRACPLU]
Brachythecium plumosum var. *roellii* (Ren. & Card.) Grout = Brachythecium plumosum [BRACPLU]
Brachythecium populeum (Hedw.) Schimp. in B.S.G. [BRACPOP]
SY: *Brachythecium nanopes* C. Müll. & Kindb. in Mac. & Kindb.
SY: *Brachythecium populeum* var. *majus* Schimp. in B.S.G.
SY: *Brachythecium populeum* var. *ovatum* Grout
SY: *Brachythecium populeum* var. *rufescens* Schimp. in B.S.G.
SY: *Hypnum populeum* Hedw.
Brachythecium populeum var. *majus* Schimp. in B.S.G. = Brachythecium populeum [BRACPOP]
Brachythecium populeum var. *ovatum* Grout = Brachythecium populeum [BRACPOP]
Brachythecium populeum var. *rufescens* Schimp. in B.S.G. = Brachythecium populeum [BRACPOP]
Brachythecium pseudoalbicans Kindb. = Brachythecium albicans [BRACALB]
Brachythecium pseudostarkei Ren. & Card. = Brachythecium frigidum [BRACFRI]
Brachythecium reflexum (Starke in Web. & Mohr) Schimp. in B.S.G. [BRACREF]
SY: *Hypnum reflexum* Stark in Web. & Mohr
Brachythecium reflexum var. **pacificum** Ren. & Card. in Röll [BRACREF1]
SY: *Brachythecium bestii* Grout
SY: *Brachythecium starkei* var. *pacificum* (Ren. & Card. in Röll) Lawt.
Brachythecium reflexum var. **reflexum** [BRACREF0]
Brachythecium rivulare Schimp. in B.S.G. [BRACRIV] (waterside feather moss)
SY: *Brachythecium rivulare* var. *cataractarum* Saut.
SY: *Brachythecium rivulare* var. *lamoillense* Grout
SY: *Brachythecium rivulare* var. *laxum* Grout
SY: *Brachythecium rivulare* var. *noveboracense* (Grout) Robins.
SY: *Brachythecium rivulare* var. *obtusulum* Kindb. in Mac. & Kindb.
SY: *Hypnum rivulare* (B.S.G.) Bruch in Wils.
Brachythecium rivulare var. *cataractarum* Saut. = Brachythecium rivulare [BRACRIV]
Brachythecium rivulare var. *lamoillense* Grout = Brachythecium rivulare [BRACRIV]
Brachythecium rivulare var. *laxum* Grout = Brachythecium rivulare [BRACRIV]
Brachythecium rivulare var. *noveboracense* (Grout) Robins. = Brachythecium rivulare [BRACRIV]
Brachythecium rivulare var. *obtusulum* Kindb. in Mac. & Kindb. = Brachythecium rivulare [BRACRIV]
Brachythecium roellii Ren. & Card. = Brachythecium plumosum [BRACPLU]
Brachythecium rutabuliforme Kindb. in Mac. & Kindb. = Brachythecium plumosum [BRACPLU]
Brachythecium rutabulum (Hedw.) Schimp. in B.S.G. [BRACRUT]
SY: *Brachythecium cavernosum* Kindb.
SY: *Brachythecium leucoglaucum* C. Müll. & Kindb. in Mac. & Kindb.
SY: *Brachythecium mirabundum* C. Müll. & Kindb. in Mac. & Kindb.
SY: *Brachythecium platycladum* C. Müll. & Kindb. in Mac. & Kindb.
SY: *Brachythecium rutabulum* var. *flavescens* Schimp. in B.S.G.
SY: *Brachythecium rutabulum* var. *turgescens* Limpr.
SY: *Hypnum rutabulum* Hedw.
Brachythecium rutabulum var. *flavescens* Schimp. in B.S.G. = Brachythecium rutabulum [BRACRUT]
Brachythecium rutabulum var. *turgescens* Limpr. = Brachythecium rutabulum [BRACRUT]
Brachythecium salebrosum (Web. & Mohr) Schimp. in B.S.G. [BRACSAL] (golden ragged moss)
SY: *Brachythecium laevisetum* Kindb.
SY: *Chamberlainia salebrosa* (Web. & Mohr) Robins.
SY: *Hypnum salebrosum* Hoffm. ex Web. & Mohr
Brachythecium salebrosum ssp. *mamilligerum* Kindb. in Mac. & Kindb. = Brachythecium acutum [BRACACT]
Brachythecium salebrosum var. *densum* Schimp. in B.S.G. = Brachythecium calcareum [BRACCAL]
Brachythecium salebrosum var. *waghornei* Ren. & Card. = Brachythecium albicans [BRACALB]
Brachythecium serrulatum (Hedw.) Robins. = Steerecleus serrulatus [STEESER]
Brachythecium spurio-acuminatum C. Müll. & Kindb. in Mac. & Kindb. = Brachythecium oxycladon [BRACOXY]
Brachythecium spurio-rutabulum C. Müll. & Kindb. in Mac. & Kindb. = Brachythecium frigidum [BRACFRI]
Brachythecium starkei (Brid.) Schimp. in B.S.G. [BRACSTA]
SY: *Eurhynchium pseudoserrulatum* Kindb.
SY: *Hypnum starkei* Brid.
SY: *Rhynchostegium pseudo-serrulatum* (Kindb.) Kindb.
Brachythecium starkei ssp. *oedipodium* (Mitt.) Ren. & Card. = Brachythecium oedipodium [BRACOED]
Brachythecium starkei var. *curtum* (Lindb.) Warnst. = Brachythecium oedipodium [BRACOED]
Brachythecium starkei var. *explanatum* (Brid.) Mönk. = Brachythecium oedipodium [BRACOED]
Brachythecium starkei var. *pacificum* (Ren. & Card. in Röll) Lawt. = Brachythecium reflexum var. pacificum [BRACREF1]
Brachythecium subasperrimum Card. & Thér. = Brachythecium frigidum [BRACFRI]
Brachythecium suberythrorrhizon Ren. & Card. = Brachythecium velutinum var. venustum [BRACVEL1]
Brachythecium subintricatum Kindb. = Brachythecium frigidum [BRACFRI]
Brachythecium trachypodium (Brid.) Schimp. in B.S.G. [BRACTRA]
Brachythecium tromsoense auct. = Brachythecium holzingeri [BRACHOL]
Brachythecium turgidum (Hartm.) Kindb. [BRACTUR] (thick ragged moss)
SY: *Chamberlainia turgida* (Hartm.) Robins.
Brachythecium velutinum (Hedw.) Schimp. in B.S.G. [BRACVEL] (velvet feather moss)
SY: *Chamberlainia velutina* (Hedw.) Robins.
SY: *Hypnum velutinum* Hedw.
Brachythecium velutinum var. **velutinum** [BRACVEL0]
SY: *Brachythecium collinum* var. *idahense* (Ren. & Card.) Grout
SY: *Brachythecium fendleri* var. *idahense* (Ren. & Card.) Wijk & Marg.
SY: *Brachythecium idahense* Ren. & Card.
Brachythecium velutinum var. **venustum** (De Not.) Arc. [BRACVEL1]
SY: *Brachythecium erythrorrhizon* var. *suberythrorrhizon* (Ren. & Card.) Grout
SY: *Brachythecium petrophilum* Williams
SY: *Brachythecium suberythrorrhizon* Ren. & Card.
SY: *Chamberlainia collina* var. *suberythrorrhiza* (Ren. & Card.) Robins.
Brachythecium washingtonianum D.C. Eat. in Grout = Brachythecium frigidum [BRACFRI]
Bryhnia Kaur. [BRYHNIA$]
Bryhnia brittoniae (Grout) Robins. = Eurhynchium praelongum [EURHPRA]
Bryhnia hultenii Bartr. in Grout [BRYHHUL]
Bryhnia oregana (Sull.) Robins. = Eurhynchium oreganum [EURHORE]
Bryhnia stokesii (Turn.) Robins. = Eurhynchium praelongum [EURHPRA]
Camptothecium aeneum (Mitt.) Jaeg. = Homalothecium aeneum [HOMAAEN]
Camptothecium aeneum var. *dolosum* (Ren. & Card.) Grout = Homalothecium aeneum [HOMAAEN]
Camptothecium aeneum var. *robustum* Grout = Homalothecium aeneum [HOMAAEN]
Camptothecium alsioides Kindb. = Homalothecium arenarium [HOMAARE]

Camptothecium amesiae Ren. & Card. = Homalothecium pinnatifidum [HOMAPIN]
Camptothecium arenarium (Lesq.) Jaeg. = Homalothecium arenarium [HOMAARE]
Camptothecium hamatidens var. *tenue* Kindb. = Homalothecium nuttallii [HOMANUT]
Camptothecium hematidens (Kindb.) Kindb. = Homalothecium nuttallii [HOMANUT]
Camptothecium lutescens var. *occidentale* Ren. & Card. = Homalothecium fulgescens [HOMAFUL]
Camptothecium megaptilum Sull. = Trachybryum megaptilum [TRACMEG]
Camptothecium megaptilum var. *fosteri* Grout = Trachybryum megaptilum [TRACMEG]
Camptothecium nitens (Hedw.) Schimp. = Tomentypnum nitens [TOMENIT]
Camptothecium nitens var. *falcatum* Peck in Burnh. = Tomentypnum falcifolium [TOMEFAL]
Camptothecium nitens var. *falcifolium* Ren. ex Nichols = Tomentypnum falcifolium [TOMEFAL]
Camptothecium nitens var. *leucobaisi* Kindb. = Tomentypnum nitens [TOMENIT]
Camptothecium nitens var. *microtheca* Kindb. = Tomentypnum nitens [TOMENIT]
Camptothecium nuttallii (Wils.) Schimp. in B.S.G. = Homalothecium nuttallii [HOMANUT]
Camptothecium pinnatifidum (Sull. & Lesq.) Sull. = Homalothecium pinnatifidum [HOMAPIN]
Chamberlainia albicans (Hedw.) Robins. = Brachythecium albicans [BRACALB]
Chamberlainia calcarea (Kindb.) Robins. = Brachythecium calcareum [BRACCAL]
Chamberlainia campestris (C. Müll.) Robins. = Brachythecium campestre [BRACCAM]
Chamberlainia collina (Schleich. ex C. Müll.) Robins. = Brachythecium collinum [BRACCOL]
Chamberlainia collina var. *suberythrorrhiza* (Ren. & Card.) Robins. = Brachythecium velutinum var. venustum [BRACVEL1]
Chamberlainia erythrorrhiza (Schimp. in B.S.G.) Robins. = Brachythecium erythrorrhizon [BRACERY]
Chamberlainia leibergii (Grout) Robins. = Brachythecium leibergii [BRACLEI]
Chamberlainia oxyclada (Brid.) Robins. = Brachythecium oxycladon [BRACOXY]
Chamberlainia salebrosa (Web. & Mohr) Robins. = Brachythecium salebrosum [BRACSAL]
Chamberlainia turgida (Hartm.) Robins. = Brachythecium turgidum [BRACTUR]
Chamberlainia velutina (Hedw.) Robins. = Brachythecium velutinum [BRACVEL]
Cirriphyllum Grout [CIRRIPH$]
Cirriphyllum cirrosum (Schwaegr. in Schultes) Grout [CIRRCIR]
SY: *Brachythecium cirrosum* (Schwaegr. in Schultes) Schimp.
SY: *Brachythecium cirrosum* var. *coloradense* (Aust.) Wijk & Marg.
SY: *Cirriphyllum cirrosum* var. *coloradense* (Aust.) Grout
Cirriphyllum cirrosum var. *coloradense* (Aust.) Grout = Cirriphyllum cirrosum [CIRRCIR]
Cirriphyllum piliferum (Hedw.) Grout [CIRRPIL]
SY: *Brachythecium piliferum* (Hedw.) Kindb.
SY: *Eurhynchium piliferum* (Hedw.) B.S.G.
Eurhynchium Schimp. in B.S.G. [EURHYNC$]
Eurhynchium acutifolium Kindb. = Eurhynchium praelongum [EURHPRA]
Eurhynchium brittoniae Grout = Eurhynchium praelongum [EURHPRA]
Eurhynchium diversifolium Schimp. in B.S.G. = Eurhynchium pulchellum var. pulchellum [EURHPUL0]
Eurhynchium fallax (Ren. & Card.) Grout = Eurhynchium pulchellum var. pulchellum [EURHPUL0]
Eurhynchium fallax var. *taylorae* (Williams) Grout = Eurhynchium pulchellum var. barnesii [EURHPUL1]
Eurhynchium fasciculosum (Hedw.) Dix. = Eurhynchium pulchellum var. pulchellum [EURHPUL0]
Eurhynchium labradoricum Kindb. = Brachythecium calcareum [BRACCAL]
Eurhynchium macounii Kindb. = Scleropodium touretii var. colpophyllum [SCLETOU1]
Eurhynchium myosuroides (Brid.) Schimp. = Isothecium myosuroides [ISOTMYO]
Eurhynchium oreganum (Sull.) Jaeg. [EURHORE] (Oregon beaked moss)
SY: *Bryhnia oregana* (Sull.) Robins.
SY: *Hypnum oreganum* Sull.
SY: *Kindbergia oregana* (Sull.) Ochyra
SY: *Stokesiella oregana* (Sull.) Robins.
Eurhynchium piliferum (Hedw.) B.S.G. = Cirriphyllum piliferum [CIRRPIL]
Eurhynchium praecox (Hedw.) De Not. in Picc. = Eurhynchium pulchellum var. pulchellum [EURHPUL0]
Eurhynchium praelongum (Hedw.) Schimp. in B.S.G. [EURHPRA] (slender beaked moss)
SY: *Bryhnia brittoniae* (Grout) Robins.
SY: *Bryhnia stokesii* (Turn.) Robins.
SY: *Eurhynchium acutifolium* Kindb.
SY: *Eurhynchium brittoniae* Grout
SY: *Eurhynchium praelongum* var. *californicum* Grout
SY: *Eurhynchium praelongum* var. *stokesii* (Turn.) Dix.
SY: *Eurhynchium stokesii* (Turn.) Schimp. in B.S.G.
SY: *Eurhynchium stokesii* var. *californicum* (Grout) Grout
SY: *Eurhynchium stokesii* ssp. *pseudospeciosum* Kindb.
SY: *Hypnum praelongum* Hedw.
SY: *Hypnum stokessii* Turn.
SY: *Kindbergia brittoniae* (Grout) Ochyra
SY: *Kindbergia praelonga* (Hedw.) Ochyra
SY: *Kindbergia praelonga* var. *californica* (Grout) Ochyra
SY: *Kindbergia praelonga* var. *stokesii* (Turn.) Ochyra
SY: *Oxyrrhynchium praelongum* (Hedw.) Warnst.
SY: *Oxyrrhynchium praelongum* var. *californicum* (Grout) Wijk & Marg.
SY: *Oxyrrhynchium praelongum* var. *stokesii* (Turn.) Podp.
SY: *Plagiothecium bifariellum* Kindb.
SY: *Stokesiella brittoniae* (Grout) Robins.
SY: *Stokesiella praelonga* (Hedw.) Robins.
SY: *Stokesiella praelonga* var. *californicum* (Grout) Crum & Anderson
SY: *Stokesiella praelonga* var. *stokesii* (Turn.) Crum
Eurhynchium praelongum var. *californicum* Grout = Eurhynchium praelongum [EURHPRA]
Eurhynchium praelongum var. *stokesii* (Turn.) Dix. = Eurhynchium praelongum [EURHPRA]
Eurhynchium pseudoserrulatum Kindb. = Brachythecium starkei [BRACSTA]
Eurhynchium pulchellum (Hedw.) Jenn. [EURHPUL] (common beaked moss)
SY: *Eurhynchium pulchellum* var. *diversifolium* (B.S.G.) C. Jens.
SY: *Hypnum strigosum* Hoffm. ex Web. & Mohr
SY: *Rhynchostegium pulchellum* (Hedw.) Robins.
Eurhynchium pulchellum var. **barnesii** (Ren. & Card.) Grout [EURHPUL1]
SY: *Eurhynchium fallax* var. *taylorae* (Williams) Grout
SY: *Eurhynchium substrigosum* var. *taylorae* (Ren. & Card.) Grout
Eurhynchium pulchellum var. *diversifolium* (B.S.G.) C. Jens. = Eurhynchium pulchellum [EURHPUL]
Eurhynchium pulchellum var. *praecox* (Hedw.) Dix. = Eurhynchium pulchellum var. pulchellum [EURHPUL0]
Eurhynchium pulchellum var. **pulchellum** [EURHPUL0]
SY: *Eurhynchium diversifolium* Schimp. in B.S.G.
SY: *Eurhynchium fallax* (Ren. & Card.) Grout
SY: *Eurhynchium fasciculosum* (Hedw.) Dix.
SY: *Eurhynchium praecox* (Hedw.) De Not. in Picc.
SY: *Eurhynchium pulchellum* var. *praecox* (Hedw.) Dix.
SY: *Eurhynchium pulchellum* var. *robustum* (Röll) Amann
SY: *Eurhynchium strigosum* (Web. & Mohr) Schimp. in B.S.G.

SY: *Eurhynchium strigosum* var. *praecox* (Hedw.) Husn.
SY: *Eurhynchium strigosum* var. *robustum* Röll
SY: *Eurhynchium strigosum* var. *scabrisetum* Grout
SY: *Eurhynchium substrigosum* Kindb. in Mac. & Kindb.
SY: *Rhynchostegium strigosum* (Web. & Mohr) De Not.
Eurhynchium pulchellum var. *robustum* (Röll) Amann = Eurhynchium pulchellum var. pulchellum [EURHPUL0]
Eurhynchium riparioides (Hedw.) P. Rich. = Platyhypnidium riparioides [PLATRIP]
Eurhynchium rusciforme (Brid.) Milde = Platyhypnidium riparioides [PLATRIP]
Eurhynchium rusciforme var. *complanatum* H. Schultz ex Limpr. = Platyhypnidium riparioides [PLATRIP]
Eurhynchium semiasperum C. Müll. & Kindb. in Mac. & Kindb. = Brachythecium plumosum [BRACPLU]
Eurhynchium serrulatum (Hedw.) Kindb. = Steerecleus serrulatus [STEESER]
Eurhynchium serrulatum ssp. *ericense* Kindb. = Steerecleus serrulatus [STEESER]
Eurhynchium stokesii (Turn.) Schimp. in B.S.G. = Eurhynchium praelongum [EURHPRA]
Eurhynchium stokesii ssp. *pseudospeciosum* Kindb. = Eurhynchium praelongum [EURHPRA]
Eurhynchium stokesii var. *californicum* (Grout) Grout = Eurhynchium praelongum [EURHPRA]
Eurhynchium strigosum (Web. & Mohr) Schimp. in B.S.G. = Eurhynchium pulchellum var. pulchellum [EURHPUL0]
Eurhynchium strigosum var. *praecox* (Hedw.) Husn. = Eurhynchium pulchellum var. pulchellum [EURHPUL0]
Eurhynchium strigosum var. *robustum* Röll = Eurhynchium pulchellum var. pulchellum [EURHPUL0]
Eurhynchium strigosum var. *scabrisetum* Grout = Eurhynchium pulchellum var. pulchellum [EURHPUL0]
Eurhynchium subintegrifolium Kindb. = Brachythecium oedipodium [BRACOED]
Eurhynchium substrigosum Kindb. in Mac. & Kindb. = Eurhynchium pulchellum var. pulchellum [EURHPUL0]
Eurhynchium substrigosum var. *taylorae* (Ren. & Card.) Grout = Eurhynchium pulchellum var. barnesii [EURHPUL1]
Homalothecium Schimp. in B.S.G. [HOMALOH$]
Homalothecium aeneum (Mitt.) Lawt. [HOMAAEN] (golden curls moss)
SY: *Camptothecium aeneum* (Mitt.) Jaeg.
SY: *Camptothecium aeneum* var. *dolosum* (Ren. & Card.) Grout
SY: *Camptothecium aeneum* var. *robustum* Grout
Homalothecium arenarium (Lesq.) Lawt. [HOMAARE]
SY: *Camptothecium alsioides* Kindb.
SY: *Camptothecium arenarium* (Lesq.) Jaeg.
Homalothecium corticola Kindb. in Mac. & Kindb. = Homalothecium nevadense [HOMANEV]
Homalothecium fulgescens (Mitt. ex C. Müll.) Lawt. [HOMAFUL] (yellow moss)
SY: *Camptothecium lutescens* var. *occidentale* Ren. & Card.
SY: *Homalothecium pseudosericeum* (C. Müll.) Jaeg. & Sauerb.
Homalothecium megaptilum (Sull.) Robins. = Trachybryum megaptilum [TRACMEG]
Homalothecium nevadense (Lesq.) Ren. & Card. [HOMANEV]
SY: *Homalothecium corticola* Kindb. in Mac. & Kindb.
SY: *Homalothecium nevadense* var. *subulatum* Ren. & Card. in Röll
SY: *Homalothecium sericeoides* C. Müll. & Kindb. in Mac. & Kindb.
Homalothecium nevadense var. *subulatum* Ren. & Card. in Röll = Homalothecium nevadense [HOMANEV]
Homalothecium nitens (Hedw.) Robins. = Tomentypnum nitens [TOMENIT]
Homalothecium nitens var. *falcifolium* (Ren. ex Nichols) Wijk & Marg. = Tomentypnum falcifolium [TOMEFAL]
Homalothecium nuttallii (Wils.) Jaeg. [HOMANUT]
SY: *Camptothecium hamatidens* var. *tenue* Kindb.
SY: *Camptothecium hematidens* (Kindb.) Kindb.
SY: *Camptothecium nuttallii* (Wils.) Schimp. in B.S.G.
SY: *Homalothecium nuttallii* var. *hamatidens* (Kindb.) Grout
SY: *Homalothecium nuttallii* var. *stoloniferum* (Lesq.) L. Koch
SY: *Homalothecium nuttallii* var. *tenue* (Kindb.) Grout
Homalothecium nuttallii var. *hamatidens* (Kindb.) Grout = Homalothecium nuttallii [HOMANUT]
Homalothecium nuttallii var. *stoloniferum* (Lesq.) L. Koch = Homalothecium nuttallii [HOMANUT]
Homalothecium nuttallii var. *tenue* (Kindb.) Grout = Homalothecium nuttallii [HOMANUT]
Homalothecium pinnatifidum (Sull. & Lesq.) Lawt. [HOMAPIN]
SY: *Camptothecium amesiae* Ren. & Card.
SY: *Camptothecium pinnatifidum* (Sull. & Lesq.) Sull.
Homalothecium pseudosericeum (C. Müll.) Jaeg. & Sauerb. = Homalothecium fulgescens [HOMAFUL]
Homalothecium sericeoides C. Müll. & Kindb. in Mac. & Kindb. = Homalothecium nevadense [HOMANEV]
Hypnum albicans Neck. ex Hedw. = Brachythecium albicans [BRACALB]
Hypnum campestre (C. Müll.) Mitt. = Brachythecium campestre [BRACCAM]
Hypnum myosuroides Brid. = Isothecium myosuroides [ISOTMYO]
Hypnum nitens Hedw. = Tomentypnum nitens [TOMENIT]
Hypnum oreganum Sull. = Eurhynchium oreganum [EURHORE]
Hypnum plumosum Hedw. = Brachythecium plumosum [BRACPLU]
Hypnum populeum Hedw. = Brachythecium populeum [BRACPOP]
Hypnum praelongum Hedw. = Eurhynchium praelongum [EURHPRA]
Hypnum reflexum Stark in Web. & Mohr = Brachythecium reflexum [BRACREF]
Hypnum rivulare (B.S.G.) Bruch in Wils. = Brachythecium rivulare [BRACRIV]
Hypnum rusciforme Weiss ex Brid. = Platyhypnidium riparioides [PLATRIP]
Hypnum rutabulum Hedw. = Brachythecium rutabulum [BRACRUT]
Hypnum salebrosum Hoffm. ex Web. & Mohr = Brachythecium salebrosum [BRACSAL]
Hypnum serrulatum Hedw. = Steerecleus serrulatus [STEESER]
Hypnum starkei Brid. = Brachythecium starkei [BRACSTA]
Hypnum stokessii Turn. = Eurhynchium praelongum [EURHPRA]
Hypnum strigosum Hoffm. ex Web. & Mohr = Eurhynchium pulchellum [EURHPUL]
Hypnum velutinum Hedw. = Brachythecium velutinum [BRACVEL]
Isothecium Brid. [ISOTHEC$]
Isothecium acuticuspis (Mitt.) Mac. & Kindb. = Isothecium myosuroides [ISOTMYO]
Isothecium aggredatum (Mitt.) Jaeg. & Sauerb. = Isothecium cristatum [ISOTCRI]
Isothecium aplocladum (Mitt.) Kindb. = Scleropodium cespitans var. sublaeve [SCLECES2]
Isothecium brachycladon Kindb. = Isothecium myosuroides [ISOTMYO]
Isothecium brewerianum (Lesq.) Kindb. in Mac. & Kindb. = Isothecium cristatum [ISOTCRI]
Isothecium cardotii Kindb. = Isothecium myosuroides [ISOTMYO]
Isothecium cristatum (Hampe) Robins. [ISOTCRI]
SY: *Bestia breweriana* (Lesq.) Grout
SY: *Bestia breweriana* var. *howei* (Kindb.) Grout
SY: *Bestia breweriana* var. *lutescens* (Lesq. & James) Grout
SY: *Bestia cristata* (Hampe) L. Koch
SY: *Isothecium aggredatum* (Mitt.) Jaeg. & Sauerb.
SY: *Isothecium brewerianum* (Lesq.) Kindb. in Mac. & Kindb.
SY: *Isothecium cristatum* var. *howei* (Kindb.) Crum, Steere & Anderson
SY: *Isothecium cristatum* var. *lutescens* (Lesq. & James) Crum, Steere & Anderson
SY: *Isothecium hylocomioides* (Kindb.) Kindb.
SY: *Pterogonium peregrinum* Schimp.
Isothecium cristatum var. *howei* (Kindb.) Crum, Steere & Anderson = Isothecium cristatum [ISOTCRI]
Isothecium cristatum var. *lutescens* (Lesq. & James) Crum, Steere & Anderson = Isothecium cristatum [ISOTCRI]
Isothecium eumyosuroides Dix. = Isothecium myosuroides [ISOTMYO]

Isothecium eumyosuroides var. *cavernarum* (Mol.) Crum, Steere & Anderson = Isothecium myosuroides [ISOTMYO]
Isothecium hylocomioides (Kindb.) Kindb. = Isothecium cristatum [ISOTCRI]
Isothecium lentum (Mitt.) Kindb. = Scleropodium cespitans [SCLECES]
Isothecium myosuroides Brid. [ISOTMYO] (cat-tail moss)
SY: *Eurhynchium myosuroides* (Brid.) Schimp.
SY: *Hypnum myosuroides* Brid.
SY: *Isothecium acuticuspis* (Mitt.) Mac. & Kindb.
SY: *Isothecium brachycladon* Kindb.
SY: *Isothecium cardotii* Kindb.
SY: *Isothecium eumyosuroides* Dix.
SY: *Isothecium eumyosuroides* var. *cavernarum* (Mol.) Crum, Steere & Anderson
SY: *Isothecium myosuroides* var. *cavernarum* Mol.
SY: *Isothecium spiculiferum* (Mitt.) Ren. & Card.
SY: *Isothecium spiculiferum* var. *cardotii* (Kindb. in Mac. & Kindb.) Crum, Steere & Anderson
SY: *Isothecium spiculiferum* var. *myurellum* (Kindb.) Wijk & Marg.
SY: *Isothecium stoloniferum* Brid.
SY: *Isothecium stoloniferum* var. *cardotii* (Kindb. in Mac. & Kindb.) Wijk & Marg.
SY: *Isothecium stoloniferum* var. *myurellum* (Kindb.) Wijk & Marg.
SY: *Isothecium stoloniferum* var. *spiculiferum* (Mitt.) R.D. Williams & Schof.
SY: *Isothecium stoloniferum* var. *thamnoides* (Kindb.) Wijk & Marg.
SY: *Isothecium thamnoides* Kindb.
SY: *Pseudisothecium myosuroides* (Brid.) Grout
SY: *Pseudisothecium myosuroides* var. *filescens* (Ren.) Grout
SY: *Pseudisothecium stoloniferum* (Brid.) Grout
SY: *Pseudisothecium stoloniferum* var. *cardotii* (Kindb. in Mac. & Kindb.) Grout
SY: *Pseudisothecium stoloniferum* var. *myurellum* (Kindb.) Grout
Isothecium myosuroides var. *cavernarum* Mol. = Isothecium myosuroides [ISOTMYO]
Isothecium spiculiferum (Mitt.) Ren. & Card. = Isothecium myosuroides [ISOTMYO]
Isothecium spiculiferum var. *cardotii* (Kindb. in Mac. & Kindb.) Crum, Steere & Anderson = Isothecium myosuroides [ISOTMYO]
Isothecium spiculiferum var. *myurellum* (Kindb.) Wijk & Marg. = Isothecium myosuroides [ISOTMYO]
Isothecium stoloniferum Brid. = Isothecium myosuroides [ISOTMYO]
Isothecium stoloniferum var. *cardotii* (Kindb. in Mac. & Kindb.) Wijk & Marg. = Isothecium myosuroides [ISOTMYO]
Isothecium stoloniferum var. *myurellum* (Kindb.) Wijk & Marg. = Isothecium myosuroides [ISOTMYO]
Isothecium stoloniferum var. *spiculiferum* (Mitt.) R.D. Williams & Schof. = Isothecium myosuroides [ISOTMYO]
Isothecium stoloniferum var. *thamnoides* (Kindb.) Wijk & Marg. = Isothecium myosuroides [ISOTMYO]
Isothecium thamnoides Kindb. = Isothecium myosuroides [ISOTMYO]
Kindbergia brittoniae (Grout) Ochyra = Eurhynchium praelongum [EURHPRA]
Kindbergia oregana (Sull.) Ochyra = Eurhynchium oreganum [EURHORE]
Kindbergia praelonga (Hedw.) Ochyra = Eurhynchium praelongum [EURHPRA]
Kindbergia praelonga var. *californica* (Grout) Ochyra = Eurhynchium praelongum [EURHPRA]
Kindbergia praelonga var. *stokesii* (Turn.) Ochyra = Eurhynchium praelongum [EURHPRA]
Myrinia dieckii Ren. & Card. in Röll = Scleropodium cespitans var. sublaeve [SCLECES2]
Oxyrrhynchium praelongum (Hedw.) Warnst. = Eurhynchium praelongum [EURHPRA]
Oxyrrhynchium praelongum var. *californicum* (Grout) Wijk & Marg. = Eurhynchium praelongum [EURHPRA]
Oxyrrhynchium praelongum var. *stokesii* (Turn.) Podp. = Eurhynchium praelongum [EURHPRA]
Oxyrrhynchium riparioides (Hedw.) Jenn. = Platyhypnidium riparioides [PLATRIP]
Oxyrrhynchium rusciforme Warnst. = Platyhypnidium riparioides [PLATRIP]
Oxyrrhynchium rusciforme var. *complanatum* (Limpr.) Warnst. = Platyhypnidium riparioides [PLATRIP]
Plagiothecium bifariellum Kindb. = Eurhynchium praelongum [EURHPRA]
Platyhypnidium Fleisch. [PLATYHY$]
Platyhypnidium riparioides (Hedw.) Dix. [PLATRIP]
SY: *Eurhynchium riparioides* (Hedw.) P. Rich.
SY: *Eurhynchium rusciforme* (Brid.) Milde
SY: *Eurhynchium rusciforme* var. *complanatum* H. Schultz ex Limpr.
SY: *Hypnum rusciforme* Weiss ex Brid.
SY: *Oxyrrhynchium riparioides* (Hedw.) Jenn.
SY: *Oxyrrhynchium rusciforme* Warnst.
SY: *Oxyrrhynchium rusciforme* var. *complanatum* (Limpr.) Warnst.
SY: *Rhynchostegium riparioides* (Hedw.) Card. in Tourr.
SY: *Rhynchostegium rusciforme* (Brid.) Schimp. in B.S.G.
Pseudisothecium myosuroides (Brid.) Grout = Isothecium myosuroides [ISOTMYO]
Pseudisothecium myosuroides var. *filescens* (Ren.) Grout = Isothecium myosuroides [ISOTMYO]
Pseudisothecium stoloniferum (Brid.) Grout = Isothecium myosuroides [ISOTMYO]
Pseudisothecium stoloniferum var. *cardotii* (Kindb. in Mac. & Kindb.) Grout = Isothecium myosuroides [ISOTMYO]
Pseudisothecium stoloniferum var. *myurellum* (Kindb.) Grout = Isothecium myosuroides [ISOTMYO]
Pseudoscleropodium (Limpr.) Fleisch. in Broth. [PSEUDOS$]
***Pseudoscleropodium purum** (Hedw.) Fleisch. in Broth. [PSEUPUR]
Pterogonium peregrinum Schimp. = Isothecium cristatum [ISOTCRI]
Rhynchostegium pseudo-serrulatum (Kindb.) Kindb. = Brachythecium starkei [BRACSTA]
Rhynchostegium pulchellum (Hedw.) Robins. = Eurhynchium pulchellum [EURHPUL]
Rhynchostegium revelstokense (Kindb.) Kindb. = Brachythecium acutum [BRACACT]
Rhynchostegium riparioides (Hedw.) Card. in Tourr. = Platyhypnidium riparioides [PLATRIP]
Rhynchostegium rusciforme (Brid.) Schimp. in B.S.G. = Platyhypnidium riparioides [PLATRIP]
Rhynchostegium serrulatum (Hedw.) Jaeg. = Steerecleus serrulatus [STEESER]
Rhynchostegium serrulatum ssp. *eriense* (Kindb.) Kindb. = Steerecleus serrulatus [STEESER]
Rhynchostegium serrulatum ssp. *hispidifolium* (Kindb.) Kindb. = Steerecleus serrulatus [STEESER]
Rhynchostegium strigosum (Web. & Mohr) De Not. = Eurhynchium pulchellum var. pulchellum [EURHPUL0]
Rhynchostegium subintegrifolium (Kindb.) Kindb. = Brachythecium oedipodium [BRACOED]
Scleropodium Schimp. in B.S.G. [SCLEROP$]
Scleropodium apocladum (Mitt.) Grout = Scleropodium cespitans var. sublaeve [SCLECES2]
Scleropodium apocladum var. *obtusum* Grout = Scleropodium cespitans var. sublaeve [SCLECES2]
Scleropodium caespitosum Schimp. in B.S.G. = Scleropodium cespitans var. cespitans [SCLECES0]
Scleropodium caespitosum var. *sublaeve* Ren. & Card. in Röll = Scleropodium cespitans var. sublaeve [SCLECES2]
Scleropodium cespitans (C. Müll.) L. Koch [SCLECES]
SY: *Isothecium lentum* (Mitt.) Kindb.
Scleropodium cespitans var. **cespitans** [SCLECES0]
SY: *Scleropodium caespitosum* Schimp. in B.S.G.
Scleropodium cespitans var. *laeve* Ren. & Card. in Röll = Scleropodium cespitans var. sublaeve [SCLECES2]

Scleropodium cespitans var. **sublaeve** (Ren. & Card. in Röll) Wijk & Marg. [SCLECES2]
SY: *Isothecium aplocladum* (Mitt.) Kindb.
SY: *Myrinia dieckii* Ren. & Card. in Röll
SY: *Scleropodium apocladum* (Mitt.) Grout
SY: *Scleropodium apocladum* var. *obtusum* Grout
SY: *Scleropodium caespitosum* var. *sublaeve* Ren. & Card. in Röll
SY: *Scleropodium cespitans* var. *laeve* Ren. & Card. in Röll
Scleropodium colpophyllum (Sull.) Grout = Scleropodium touretii var. colpophyllum [SCLETOU1]
Scleropodium colpophyllum var. *attenuatum* Grout = Scleropodium touretii var. colpophyllum [SCLETOU1]
Scleropodium illecebrum Schimp. in B.S.G. = Scleropodium touretii [SCLETOU]
Scleropodium obtusifolium (Jaeg.) Kindb. in Mac. & Kindb. [SCLEOBT] (blunt-leaved moss)
Scleropodium touretii (Brid.) L. Koch [SCLETOU]
SY: *Scleropodium illecebrum* Schimp. in B.S.G.
Scleropodium touretii var. **colpophyllum** (Sull.) Lawt. ex Crum [SCLETOU1]
SY: *Eurhynchium macounii* Kindb.
SY: *Scleropodium colpophyllum* (Sull.) Grout
SY: *Scleropodium colpophyllum* var. *attenuatum* Grout
Scleropodium touretii var. **touretii** [SCLETOU0]
Steerecleus Robins. [STEEREC$]
Steerecleus serrulatus (Hedw.) Robins. [STEESER]
SY: *Brachythecium serrulatum* (Hedw.) Robins.
SY: *Eurhynchium serrulatum* (Hedw.) Kindb.
SY: *Eurhynchium serrulatum* ssp. *ericense* Kindb.
SY: *Hypnum serrulatum* Hedw.
SY: *Rhynchostegium serrulatum* (Hedw.) Jaeg.
SY: *Rhynchostegium serrulatum* ssp. *eriense* (Kindb.) Kindb.
SY: *Rhynchostegium serrulatum* ssp. *hispidifolium* (Kindb.) Kindb.
Stokesiella brittoniae (Grout) Robins. = Eurhynchium praelongum [EURHPRA]
Stokesiella oregana (Sull.) Robins. = Eurhynchium oreganum [EURHORE]
Stokesiella praelonga (Hedw.) Robins. = Eurhynchium praelongum [EURHPRA]
Stokesiella praelonga var. *californicum* (Grout) Crum & Anderson = Eurhynchium praelongum [EURHPRA]
Stokesiella praelonga var. *stokesii* (Turn.) Crum = Eurhynchium praelongum [EURHPRA]
Tomenthypnum Loeske = Tomentypnum [TOMENTY$]
Tomentypnum Loeske [TOMENTY$]
Tomentypnum falcifolium (Ren. ex Nichols) Tuom. in Ahti & Fagers. [TOMEFAL]
SY: *Camptothecium nitens* var. *falcatum* Peck in Burnh.
SY: *Camptothecium nitens* var. *falcifolium* Ren. ex Nichols
SY: *Homalothecium nitens* var. *falcifolium* (Ren. ex Nichols) Wijk & Marg.
SY: *Tomentypnum nitens* var. *falcifolium* Ren. ex Nichols
Tomentypnum nitens (Hedw.) Loeske [TOMENIT] (golden fuzzy fen moss)
SY: *Camptothecium nitens* (Hedw.) Schimp.
SY: *Camptothecium nitens* var. *leucobaisi* Kindb.
SY: *Camptothecium nitens* var. *microtheca* Kindb.
SY: *Homalothecium nitens* (Hedw.) Robins.
SY: *Hypnum nitens* Hedw.
SY: *Tomentypnum nitens* var. *insigne* (Milde) C. Jens.
SY: *Tomentypnum nitens* var. *involutum* (Limpr.) C. Jens.
Tomentypnum nitens var. *falcifolium* Ren. ex Nichols = Tomentypnum falcifolium [TOMEFAL]
Tomentypnum nitens var. *insigne* (Milde) C. Jens. = Tomentypnum nitens [TOMENIT]
Tomentypnum nitens var. *involutum* (Limpr.) C. Jens. = Tomentypnum nitens [TOMENIT]
Trachybryum (Broth.) Schof. [TRACHYB$]
Trachybryum megaptilum (Sull.) Schof. [TRACMEG]
SY: *Camptothecium megaptilum* Sull.
SY: *Camptothecium megaptilum* var. *fosteri* Grout
SY: *Homalothecium megaptilum* (Sull.) Robins.

BRUCHIACEAE (B:M)

Trematodon Michx. [TREMATO$]
Trematodon ambiguus (Hedw.) Hornsch. [TREMAMB]
Trematodon boasii Schof. [TREMBOA]
Trematodon montanus Belland & Brass. [TREMMON]

BRYACEAE (B:M)

Anomobryum Schimp. [ANOMOBR$]
Anomobryum concinnatum (Spruce) Lindb. = Anomobryum filiforme [ANOMFIL]
Anomobryum filiforme (Dicks.) Solms in Rabenh. [ANOMFIL]
SY: *Anomobryum concinnatum* (Spruce) Lindb.
SY: *Anomobryum filiforme* var. *concinnatum* (Spruce) Weis.
SY: *Anomobryum julaceum* (Brid.) Schimp.
SY: *Bryum bullatum* C. Müll.
SY: *Bryum filiforme* Dicks.
SY: *Pohlia filiformis* (Dicks.) Andrews in Grout
SY: *Pohlia filiformis* var. *concinnata* (Spruce) Grout
Anomobryum filiforme var. *concinnatum* (Spruce) Weis. = Anomobryum filiforme [ANOMFIL]
Anomobryum julaceum (Brid.) Schimp. = Anomobryum filiforme [ANOMFIL]
Bryum Hedw. [BRYUM$]
Bryum aciculinum Kindb. = Bryum pseudotriquetrum [BRYUPSE]
Bryum aciculinum (Hoppe & Hornsch.) B.S.G. = Pohlia elongata var. greenii [POHLELO1]
Bryum acutiusculum C. Müll. = Bryum amblyodon [BRYUAMB]
Bryum affine Lindb. & Arnell = Bryum lisae var. cuspidatum [BRYULIS1]
Bryum affine var. *cirrhatum* (Hüb.) Braithw. = Bryum lonchocaulon [BRYULON]
Bryum alaskanum Kindb. = Bryum amblyodon [BRYUAMB]
Bryum albicans (Wahlenb.) Röhl. = Pohlia wahlenbergii [POHLWAH]
Bryum algovicum Sendtn. ex C. Müll. [BRYUALG]
SY: *Bryum angustirete* ssp. *fridtzii* (Hag.) Wijk & Marg.
SY: *Bryum angustirete* ssp. *parvulum* (Kindb.) Wijk & Marg.
SY: *Bryum fridtzii* Hag.
SY: *Bryum pendulum* ssp. *nanum* Kindb.
Bryum alpinum Huds. ex With. [BRYUALP]
SY: *Bryum laurentianum* Card. & Thér.
Bryum amblyodon C. Müll. [BRYUAMB]
SY: *Bryum acutiusculum* C. Müll.
SY: *Bryum alaskanum* Kindb.
SY: *Bryum biddlecomiae* Aust.
SY: *Bryum edwardsianum* C. Müll. & Kindb. in Mac. & Kindb.
SY: *Bryum erubescens* Kindb. in Mac. & Kindb.
SY: *Bryum inclinatum* (Brid.) Bland.
SY: *Bryum stenotrichum* C. Müll.
SY: *Bryum stenotrichum* var. *biddlecomiae* (Aust.) Lawt.
Bryum angustirete ssp. *fridtzii* (Hag.) Wijk & Marg. = Bryum algovicum [BRYUALG]
Bryum angustirete ssp. *parvulum* (Kindb.) Wijk & Marg. = Bryum algovicum [BRYUALG]
Bryum anoectangiaceum C. Müll. & Kindb. in Mac. & Kindb. = Bryum pallens [BRYUPAL]
Bryum archangelicum auct. = Bryum salinum [BRYUSAL]
Bryum arcticum (R. Br.) Bruch & Schimp. in B.S.G. [BRYUARC]
SY: *Bryum arcticum* var. *arcuatum* (Limpr.) Podp.
SY: *Bryum arcticum* var. *micans* Amann
SY: *Bryum arcticum* var. *tomentosum* Joerg.
SY: *Bryum aurimontanum* Kindb.
SY: *Bryum micans* Limpr.
SY: *Bryum submicans* Kindb.
SY: *Bryum tomentosum* (Joerg.) Hag.
Bryum arcticum var. *arcuatum* (Limpr.) Podp. = Bryum arcticum [BRYUARC]
Bryum arcticum var. *micans* Amann = Bryum arcticum [BRYUARC]

Bryum arcticum var. *tomentosum* Joerg. = Bryum arcticum [BRYUARC]
Bryum argenteum Hedw. [BRYUARG] (silver moss)
SY: *Bryum argenteum* var. *insigne* Podp.
SY: *Bryum argenteum* var. *lanatum* (P. Beauv.) Hampe
SY: *Bryum argenteum* var. *majus* Schwaegr.
SY: *Bryum lanatum* (P. Beauv.) Brid.
Bryum argenteum var. *insigne* Podp. = Bryum argenteum [BRYUARG]
Bryum argenteum var. *lanatum* (P. Beauv.) Hampe = Bryum argenteum [BRYUARG]
Bryum argenteum var. *majus* Schwaegr. = Bryum argenteum [BRYUARG]
Bryum atropurpureum Bruch & Schimp. in B.S.G. = Bryum dichotomum [BRYUDIC]
Bryum atropurpureum Wahlenb. in Fürnr. = Pohlia atropurpurea [POHLATR]
Bryum atwateriae C. Müll. = Bryum miniatum [BRYUMIN]
Bryum aurimontanum Kindb. = Bryum arcticum [BRYUARC]
Bryum bicolor Dicks. = Bryum dichotomum [BRYUDIC]
Bryum biddlecomiae Aust. = Bryum amblyodon [BRYUAMB]
Bryum bimum (Brid.) Turn. = Bryum pseudotriquetrum [BRYUPSE]
Bryum bimum var. *angustifolium* Kindb. in Mac. & Kindb. = Bryum lisae var. cuspidatum [BRYULIS1]
Bryum blindii Bruch & Schimp. in B.S.G. [BRYUBLI]
Bryum bornholmense Winkelm. & Ruthe [BRYUBOR]
Bryum bullatum C. Müll. = Anomobryum filiforme [ANOMFIL]
Bryum caespiticium Hedw. [BRYUCAE]
SY: *Bryum microcephalum* C. Müll. & Kindb. in Mac. & Kindb.
SY: *Bryum oligochloron* C. Müll. & Kindb. in Mac. & Kindb.
SY: *Bryum submuticum* Philib.
SY: *Bryum synoico-caespiticium* C. Müll. & Kindb. in Mac. & Kindb.
SY: *Bryum vancouveriense* Kindb.
Bryum calobryoides Spence [BRYUCAO]
Bryum calophyllum R. Br. [BRYUCAL]
Bryum canariense Brid. [BRYUCAN]
SY: *Bryum hendersonii* Ren. & Card.
Bryum capillare Hedw. [BRYUCAP]
SY: *Bryum capillare* ssp. *heteroneuron* C. Müll. in Mac. & Kindb.
SY: *Bryum heteroneuron* (C. Müll. & Kindb. in Mac. & Kindb.) Ren. & Card.
SY: *Bryum heteroneuron* var. *brevicuspidatum* (Kindb.) Ren. & Card.
SY: *Bryum oreganum* Sull.
SY: *Bryum speirophyllum* Kindb.
SY: *Bryum streptophyllum* Kindb.
SY: *Bryum tomentosum* Kindb.
SY: *Bryum trichophorum* Kindb.
Bryum capillare ssp. *erythroloma* Kindb. = Bryum erythroloma [BRYUERY]
Bryum capillare ssp. *heteroneuron* C. Müll. in Mac. & Kindb. = Bryum capillare [BRYUCAP]
Bryum capillare var. *flaccidum* (Brid.) Bruch & Schimp. in B.S.G. = Bryum flaccidum [BRYUFLA]
Bryum capitellatum C. Müll. & Kindb. in Mac. & Kindb. = Bryum gemmiparum [BRYUGEM]
Bryum cernuum (Hedw.) Bruch & Schimp. in B.S.G. = Bryum uliginosum [BRYUULI]
Bryum cirrhatum Hornsch. = Bryum lonchocaulon [BRYULON]
Bryum columbico-caespiticum Kindb. = Bryum lonchocaulon [BRYULON]
Bryum crassirameum Ren. & Card. = Bryum pseudotriquetrum [BRYUPSE]
Bryum creberrimum Tayl. = Bryum lisae var. cuspidatum [BRYULIS1]
Bryum crispulum Hampe in Hag. = Bryum pseudotriquetrum [BRYUPSE]
Bryum crudum (Hedw.) Turn. = Pohlia cruda [POHLCRU]
Bryum cryophilum Mårt. = Bryum cyclophyllum [BRYUCYC]
Bryum cuspidatum (Bruch & Schimp. in B.S.G.) Schimp. = Bryum lisae var. cuspidatum [BRYULIS1]
Bryum cyclophylloides Kindb. = Bryum turbinatum [BRYUTUR]
Bryum cyclophyllum (Schwaegr.) Bruch & Schimp. in B.S.G. [BRYUCYC]
SY: *Bryum cryophilum* Mårt.
SY: *Bryum obtusifolium* Lindb.
SY: *Bryum tortifolium* Funck in Brid.
Bryum denticulatum Kindb. = Bryum turbinatum [BRYUTUR]
Bryum dichotomum Hedw. [BRYUDIC]
SY: *Bryum atropurpureum* Bruch & Schimp. in B.S.G.
SY: *Bryum bicolor* Dicks.
SY: *Bryum microglobum* C. Müll. & Kindb. in Mac. & Kindb.
SY: *Bryum occidentale* Sull.
Bryum duvalii Voit in Sturm = Bryum weigelii [BRYUWEI]
Bryum duvalii var. *gaspeanum* Kindb. in Mac. & Kindb. = Bryum turbinatum [BRYUTUR]
Bryum duvalii var. *latodecurrens* C. Müll. & Kindb. in Mac. & Kindb. = Bryum weigelii [BRYUWEI]
Bryum edwardsianum C. Müll. & Kindb. in Mac. & Kindb. = Bryum amblyodon [BRYUAMB]
Bryum elongatum (Hedw.) With. = Pohlia elongata [POHLELO]
Bryum erubescens Kindb. in Mac. & Kindb. = Bryum amblyodon [BRYUAMB]
Bryum erythroloma (Kindb.) Syed [BRYUERY]
SY: *Bryum capillare* ssp. *erythroloma* Kindb.
Bryum erythrophylloides Kindb. in Mac. & Kindb. = Bryum turbinatum [BRYUTUR]
Bryum extenuatum Ren. & Card. = Bryum turbinatum [BRYUTUR]
Bryum fallax Milde in Rabenh. = Bryum pallens [BRYUPAL]
Bryum filiforme Dicks. = Anomobryum filiforme [ANOMFIL]
Bryum flaccidum Brid. [BRYUFLA]
SY: *Bryum capillare* var. *flaccidum* (Brid.) Bruch & Schimp. in B.S.G.
SY: *Bryum laevifilum* Syed
Bryum flagellosum Kindb. = Bryum lisae var. cuspidatum [BRYULIS1]
Bryum flexuosum Aust. = Bryum gemmiparum [BRYUGEM]
Bryum fridtzii Hag. = Bryum algovicum [BRYUALG]
Bryum gemmiparum De Not. [BRYUGEM]
SY: *Bryum capitellatum* C. Müll. & Kindb. in Mac. & Kindb.
SY: *Bryum flexuosum* Aust.
Bryum grandirete Kindb. = Bryum turbinatum [BRYUTUR]
Bryum haematocarpum C. Müll. & Kindb. in Mac. & Kindb. = Bryum pseudotriquetrum [BRYUPSE]
Bryum haematophyllum Kindb. in Mac. & Kindb. = Bryum turbinatum [BRYUTUR]
Bryum hamicuspis Kindb. = Bryum lisae var. cuspidatum [BRYULIS1]
Bryum hendersonii Ren. & Card. = Bryum canariense [BRYUCAN]
Bryum heteroneuron (C. Müll. & Kindb. in Mac. & Kindb.) Ren. & Card. = Bryum capillare [BRYUCAP]
Bryum heteroneuron var. *brevicuspidatum* (Kindb.) Ren. & Card. = Bryum capillare [BRYUCAP]
Bryum hydrophilum Kindb. = Bryum pallens [BRYUPAL]
Bryum inclinatum (Brid.) Bland. = Bryum amblyodon [BRYUAMB]
Bryum intermedium auct. = Bryum lisae var. cuspidatum [BRYULIS1]
Bryum laevifilum Syed = Bryum flaccidum [BRYUFLA]
Bryum lanatum (P. Beauv.) Brid. = Bryum argenteum [BRYUARG]
Bryum languidum Hag. = Bryum turbinatum [BRYUTUR]
Bryum lapponicum auct. = Bryum salinum [BRYUSAL]
Bryum laurentianum Card. & Thér. = Bryum alpinum [BRYUALP]
Bryum lescurianum Sull. = Pohlia lescuriana [POHLLES]
Bryum leucolomatum C. Müll. & Kindb. in Mac. & Kindb. = Bryum lonchocaulon [BRYULON]
Bryum lisae De Not. [BRYULIS]
Bryum lisae var. **cuspidatum** (Bruch & Schimp. in B.S.G.) Marg. [BRYULIS1]
SY: *Bryum affine* Lindb. & Arnell
SY: *Bryum bimum* var. *angustifolium* Kindb. in Mac. & Kindb.
SY: *Bryum creberrimum* Tayl.
SY: *Bryum cuspidatum* (Bruch & Schimp. in B.S.G.) Schimp.
SY: *Bryum flagellosum* Kindb.
SY: *Bryum hamicuspis* Kindb.
SY: *Bryum pseudostirtonii* Philib. ex Card. & Thér.

SY: *Bryum intermedium* auct.
Bryum lonchocaulon C. Müll. [BRYULON]
SY: *Bryum affine* var. *cirrhatum* (Hüb.) Braithw.
SY: *Bryum cirrhatum* Hornsch.
SY: *Bryum columbico-caespiticum* Kindb.
SY: *Bryum leucolomatum* C. Müll. & Kindb. in Mac. & Kindb.
SY: *Bryum producticolle* Kindb.
SY: *Bryum revelstokense* Kindb.
Bryum meesioides Kindb. in Mac. [BRYUMEE]
SY: *Bryum pallens* var. *meesioides* (Kindb. in Mac.) Broth.
Bryum micans Limpr. = Bryum arcticum [BRYUARC]
Bryum microcephalum C. Müll. & Kindb. in Mac. & Kindb. = Bryum caespiticium [BRYUCAE]
Bryum microerythrocarpum C. Müll. & Kindb. in Mac. & Kindb. = Bryum subapiculatum [BRYUSUB]
Bryum microglobum C. Müll. & Kindb. in Mac. & Kindb. = Bryum dichotomum [BRYUDIC]
Bryum miniatum Lesq. [BRYUMIN] (red bryum)
SY: *Bryum atwateriae* C. Müll.
Bryum muehlenbeckii Bruch & Schimp. in B.S.G. [BRYUMUE]
SY: *Bryum percurrentinerve* Kindb. in Mac.
SY: *Bryum pygmaeo-alpinum* C. Müll. & Kindb. in Mac. & Kindb.
SY: *Bryum rauei* Aust.
SY: *Bryum rubricundulum* C. Müll. & Kindb. in Mac. & Kindb.
Bryum neodamense Itzig in C. Müll. = Bryum pseudotriquetrum [BRYUPSE]
Bryum neodamense var. *ovatum* Lindb. & Arnell = Bryum pseudotriquetrum [BRYUPSE]
Bryum nutans (Hedw.) Turn. = Pohlia nutans [POHLNUT]
Bryum obtusifolium Lindb. = Bryum cyclophyllum [BRYUCYC]
Bryum occidentale Sull. = Bryum dichotomum [BRYUDIC]
Bryum oligochloron C. Müll. & Kindb. in Mac. & Kindb. = Bryum caespiticium [BRYUCAE]
Bryum oreganum Sull. = Bryum capillare [BRYUCAP]
Bryum ovatum Jur. = Bryum pseudotriquetrum [BRYUPSE]
Bryum pallens (Brid.) Sw. in Röhl. [BRYUPAL]
SY: *Bryum anoectangiaceum* C. Müll. & Kindb. in Mac. & Kindb.
SY: *Bryum fallax* Milde in Rabenh.
SY: *Bryum hydrophilum* Kindb.
SY: *Bryum pallens* var. *alpinum* (B.S.G.) Podp.
SY: *Bryum pallens* var. *fallax* Jur.
SY: *Bryum subpurpurascens* Kindb. in Mac. & Kindb.
Bryum pallens var. *alpinum* (B.S.G.) Podp. = Bryum pallens [BRYUPAL]
Bryum pallens var. *fallax* Jur. = Bryum pallens [BRYUPAL]
Bryum pallens var. *meesioides* (Kindb. in Mac.) Broth. = Bryum meesioides [BRYUMEE]
Bryum pallescens Schleich. ex Schwaegr. [BRYUPAE]
SY: *Bryum pallescens* var. *laxifolium* Kindb. in Max.
SY: *Bryum pallescens* var. *longifolium* Kindb.
SY: *Bryum pallescens* var. *subrotundum* (Brid.) B.S.G.
SY: *Bryum subrotundum* Brid.
Bryum pallescens var. *grande* Kindb. = Bryum pseudotriquetrum [BRYUPSE]
Bryum pallescens var. *laxifolium* Kindb. in Max. = Bryum pallescens [BRYUPAE]
Bryum pallescens var. *longifolium* Kindb. = Bryum pallescens [BRYUPAE]
Bryum pallescens var. *subrotundum* (Brid.) B.S.G. = Bryum pallescens [BRYUPAE]
Bryum pendulum ssp. *nanum* Kindb. = Bryum algovicum [BRYUALG]
Bryum percurrentinerve Kindb. in Mac. = Bryum muehlenbeckii [BRYUMUE]
Bryum producticolle Kindb. = Bryum lonchocaulon [BRYULON]
Bryum pseudostirtonii Philib. ex Card. & Thér. = Bryum lisae var. cuspidatum [BRYULIS1]
Bryum pseudotriquetrum (Hedw.) Gaertn., Meyer & Scherb. [BRYUPSE] (tall clustered thread moss)
SY: *Bryum aciculinum* Kindb.
SY: *Bryum bimum* (Brid.) Turn.
SY: *Bryum crassirameum* Ren. & Card.
SY: *Bryum crispulum* Hampe in Hag.
SY: *Bryum haematocarpum* C. Müll. & Kindb. in Mac. & Kindb.
SY: *Bryum neodamense* Itzig in C. Müll.
SY: *Bryum neodamense* var. *ovatum* Lindb. & Arnell
SY: *Bryum ovatum* Jur.
SY: *Bryum pallescens* var. *grande* Kindb.
SY: *Bryum pseudotriquetrum* var. *bimum* (Brid.) Lilj.
SY: *Bryum pseudotriquetrum* var. *compactum* B.S.G.
SY: *Bryum pseudotriquetrum* var. *crassirameum* Ren. & Card.
SY: *Bryum pseudotriquetrum* ssp. *crispulum* (Roth) C. Jens.
SY: *Bryum pseudotriquetrum* var. *hyalodontium* C. Müll. & Kindb. in Mac. & Kindb.
SY: *Bryum subpercurrentinerve* Kindb.
SY: *Bryum ventricosum* Relh.
SY: *Bryum ventricosum* var. *compactum* (B.S.G.) Lindb.
Bryum pseudotriquetrum ssp. *crispulum* (Roth) C. Jens. = Bryum pseudotriquetrum [BRYUPSE]
Bryum pseudotriquetrum var. *bimum* (Brid.) Lilj. = Bryum pseudotriquetrum [BRYUPSE]
Bryum pseudotriquetrum var. *compactum* B.S.G. = Bryum pseudotriquetrum [BRYUPSE]
Bryum pseudotriquetrum var. *crassirameum* Ren. & Card. = Bryum pseudotriquetrum [BRYUPSE]
Bryum pseudotriquetrum var. *hyalodontium* C. Müll. & Kindb. in Mac. & Kindb. = Bryum pseudotriquetrum [BRYUPSE]
Bryum pygmaeo-alpinum C. Müll. & Kindb. in Mac. & Kindb. = Bryum muehlenbeckii [BRYUMUE]
Bryum pyriforme (Hedw.) Lam. & Cand. = Leptobryum pyriforme [LEPTPYI]
Bryum rauei Aust. = Bryum muehlenbeckii [BRYUMUE]
Bryum retusum Hag. = Bryum salinum [BRYUSAL]
Bryum revelstokense Kindb. = Bryum lonchocaulon [BRYULON]
Bryum roseum (Hedw.) Crom. = Rhodobryum roseum [RHODROS]
Bryum rubricundulum C. Müll. & Kindb. in Mac. & Kindb. = Bryum muehlenbeckii [BRYUMUE]
Bryum salinum Hag. ex Limpr. [BRYUSAL]
SY: *Bryum retusum* Hag.
SY: *Bryum subtumidum* Limpr. in Jaeg.
SY: *Bryum archangelicum* auct.
SY: *Bryum lapponicum* auct.
Bryum sandbergii Holz. = Roellia roellii [ROELROE]
Bryum schleicheri Schwaegr. [BRYUSCH]
Bryum simplex Kindb. in Mac. & Kindb. = Roellia roellii [ROELROE]
Bryum speirophyllum Kindb. = Bryum capillare [BRYUCAP]
Bryum stenotrichum C. Müll. = Bryum amblyodon [BRYUAMB]
Bryum stenotrichum var. *biddlecomiae* (Aust.) Lawt. = Bryum amblyodon [BRYUAMB]
Bryum streptophyllum Kindb. = Bryum capillare [BRYUCAP]
Bryum subapiculatum Hampe [BRYUSUB]
SY: *Bryum microerythrocarpum* C. Müll. & Kindb. in Mac. & Kindb.
Bryum submicans Kindb. = Bryum arcticum [BRYUARC]
Bryum submuticum Philib. = Bryum caespiticium [BRYUCAE]
Bryum suborbiculare Philib. = Bryum turbinatum [BRYUTUR]
Bryum subpercurrentinerve Kindb. = Bryum pseudotriquetrum [BRYUPSE]
Bryum subpurpurascens Kindb. in Mac. & Kindb. = Bryum pallens [BRYUPAL]
Bryum subrotundum Brid. = Bryum pallescens [BRYUPAE]
Bryum subtumidum Limpr. in Jaeg. = Bryum salinum [BRYUSAL]
Bryum synoico-caespiticium C. Müll. & Kindb. in Mac. & Kindb. = Bryum caespiticium [BRYUCAE]
Bryum tenuisetum Limpr. [BRYUTEN]
Bryum tomentosum (Joerg.) Hag. = Bryum arcticum [BRYUARC]
Bryum tomentosum Kindb. = Bryum capillare [BRYUCAP]
Bryum tortifolium Funck in Brid. = Bryum cyclophyllum [BRYUCYC]
Bryum trichophorum Kindb. = Bryum capillare [BRYUCAP]
Bryum turbinatum (Hedw.) Turn. [BRYUTUR]
SY: *Bryum cyclophylloides* Kindb.
SY: *Bryum denticulatum* Kindb.
SY: *Bryum duvalii* var. *gaspeanum* Kindb. in Mac. & Kindb.
SY: *Bryum erythrophylloides* Kindb. in Mac. & Kindb.

SY: *Bryum extenuatum* Ren. & Card.
SY: *Bryum grandirete* Kindb.
SY: *Bryum haematophyllum* Kindb. in Mac. & Kindb.
SY: *Bryum languidum* Hag.
SY: *Bryum suborbiculare* Philib.
SY: *Splachnobryum kieneri* Williams
Bryum uliginosum (Brid.) Bruch & Schimp. in B.S.G. [BRYUULI]
SY: *Bryum cernuum* (Hedw.) Bruch & Schimp. in B.S.G.
Bryum vancouveriense Kindb. = Bryum caespiticium [BRYUCAE]
Bryum ventricosum Relh. = Bryum pseudotriquetrum [BRYUPSE]
Bryum ventricosum var. *compactum* (B.S.G.) Lindb. = Bryum pseudotriquetrum [BRYUPSE]
Bryum violaceum Crundw. & Nyh. [BRYUVIO]
Bryum wahlenbergii (Web. & Mohr) Schwaegr. = Pohlia wahlenbergii [POHLWAH]
Bryum weigelii Spreng. in Biehler [BRYUWEI]
SY: *Bryum duvalii* Voit in Sturm
SY: *Bryum duvalii* var. *latodecurrens* C. Müll. & Kindb. in Mac. & Kindb.
Epipterygium Lindb. [EPIPTER$]
Epipterygium tozeri (Grev.) Lindb. [EPIPTOZ]
SY: *Pohlia tozeri* (Grev.) Delogne
SY: *Webera tozeri* (Grev.) Schimp.
Leptobryum (Bruch & Schimp. in B.S.G.) Wils. [LEPTOBR$]
Leptobryum pyriforme (Hedw.) Wils. [LEPTPYI] (pear moss)
SY: *Bryum pyriforme* (Hedw.) Lam. & Cand.
SY: *Leptobryum pyriforme* var. *flagelliferum* Holz.
Leptobryum pyriforme var. *flagelliferum* Holz. = Leptobryum pyriforme [LEPTPYI]
Mielichhoferia Nees & Hornsch. [MIELICH$]
Mielichhoferia macrocarpa (Hook. in Drumm.) Bruch & Schimp. ex Jaeg. [MIELMAC]
SY: *Mielichhoferia macrocarpa* var. *pungens* Bartr.
Mielichhoferia macrocarpa var. *pungens* Bartr. = Mielichhoferia macrocarpa [MIELMAC]
Mielichhoferia mielichhoferi (Funck ex Hook.) Wijk & Marg. = Mielichhoferia mielichhoferiana [MIELMIE]
Mielichhoferia mielichhoferiana (Funck in Hook.) Loeske [MIELMIE]
SY: *Mielichhoferia mielichhoferi* (Funck ex Hook.) Wijk & Marg.
Mniobryum albicans (Wahlenb.) Limpr. = Pohlia wahlenbergii [POHLWAH]
Mniobryum longibracteatum (Broth. in Röll) Broth. = Pohlia longibracteata [POHLLON]
Mniobryum wahlenbergii (Web. & Mohr) Jenn. = Pohlia wahlenbergii [POHLWAH]
Mniobryum wahlenbergii var. *glacialis* (Brid.) Wijk & Marg. = Pohlia wahlenbergii [POHLWAH]
Plagiobryum Lindb. [PLAGIOR$]
Plagiobryum demissum (Hook.) Lindb. [PLAGDEM]
SY: *Zieria demissa* (Hook.) Schimp.
Plagiobryum zieri (Hedw.) Lindb. [PLAGZIE]
SY: *Zieria julacea* Schimp.
Pohlia Hedw. [POHLIA$]
Pohlia acuminata Hoppe & Hornsch. = Pohlia elongata var. elongata [POHLELO0]
Pohlia albicans Lindb. = Pohlia wahlenbergii [POHLWAH]
Pohlia andalusica (Hohn.) Broth. [POHLAND]
SY: *Pohlia rothii* auct.
Pohlia annotina (Hedw.) Lindb. [POHLANN]
SY: *Pohlia annotina* var. *decipiens* Loeske
SY: *Pohlia annotina* var. *loeskei* Crum, Steere & Anderson
SY: *Pohlia camptotrachela* var. *decipiens* (Loeske) Nyh.
SY: *Pohlia grandiflora* ssp. *proligera* var. *decipiens* (Loeske) Kuc
SY: *Webera annotina* (Hedw.) Bruch in Schwaegr.
Pohlia annotina var. *decipiens* Loeske = Pohlia annotina [POHLANN]
Pohlia annotina var. *loeskei* Crum, Steere & Anderson = Pohlia annotina [POHLANN]
Pohlia atropurpurea (Wahlenb. in Fürnr.) H. Lindb. [POHLATR]
SY: *Bryum atropurpureum* Wahlenb. in Fürnr.
Pohlia bolanderi (Sull.) Broth. [POHLBOL]
Pohlia brevinervis Lindb. & Arnell [POHLBRE]
Pohlia bulbifera (Warnst.) Warnst. [POHLBUL]
SY: *Pohlia camptotrachela* var. *bulbifera* (Lam. & DC.) Wijk & Marg.
SY: *Webera pseudo-carnea* (Kindb.) Mac.
Pohlia camptotrachela (Ren. & Card.) Broth. [POHLCAM]
Pohlia camptotrachela var. *bulbifera* (Lam. & DC.) Wijk & Marg. = Pohlia bulbifera [POHLBUL]
Pohlia camptotrachela var. *decipiens* (Loeske) Nyh. = Pohlia annotina [POHLANN]
Pohlia cardotii (Ren. in Ren. & Card.) Broth. [POHLCAR]
Pohlia carnea (Schimp.) Lindb. = Pohlia melanodon [POHLMEL]
Pohlia columbica (Kindb. in Mac. & Kindb.) Andrews [POHLCOL]
SY: *Webera columbica* Kindb. in Mac. & Kindb.
Pohlia commutata (Schimp.) Lindb. = Pohlia drummondii [POHLDRU]
Pohlia commutata var. *filum* (Schimp.) Dus. = Pohlia filum [POHLFIU]
Pohlia cruda (Hedw.) Lindb. [POHLCRU]
SY: *Bryum crudum* (Hedw.) Turn.
SY: *Pohlia cruda* var. *oregonensis* (Par.) Wijk & Marg.
SY: *Webera cruda* (Hedw.) Fürnr.
SY: *Webera micro-apiculata* C. Müll. & Kindb. in Mac. & Kindb.
Pohlia cruda var. *oregonensis* (Par.) Wijk & Marg. = Pohlia cruda [POHLCRU]
Pohlia crudoides (Sull. & Lesq.) Broth. [POHLCRD]
SY: *Webera crudoides* (Sull. & Lesq.) Jaeg. & Sauerb.
Pohlia cucullata (Schwaegr.) Lindb. = Pohlia obtusifolia [POHLOBT]
Pohlia defecta (Sanio) Andrews in Grout = Pohlia erecta [POHLERE]
Pohlia delicatula (Hedw.) Grout = Pohlia melanodon [POHLMEL]
Pohlia drummondii (C. Müll.) Andrews [POHLDRU]
SY: *Pohlia commutata* (Schimp.) Lindb.
SY: *Webera commutata* Schimp.
SY: *Webera microdenticulata* C. Müll. & Kindb. in Mac. & Kindb.
SY: *Webera polymorphoides* Kindb. in Mac. & Kindb.
SY: *Webera pycnodecurrens* C. Müll. & Kindb. in Mac. & Kindb.
SY: *Webera subcucullata* C. Müll. & Kindb. in Mac. & Kindb.
SY: *Webera subpolymorpha* (Kindb.) Par.
Pohlia drummondii var. *gracilis* (Bruch & Schimp. in B.S.G.) Podp. = Pohlia filum [POHLFIU]
Pohlia elongata Hedw. [POHLELO]
SY: *Bryum elongatum* (Hedw.) With.
SY: *Webera acuminata* (Hoppe & Hornsch.) Schimp.
SY: *Webera elongata* (Hedw.) Schwaegr.
SY: *Webera microcaulon* C. Müll. & Kindb. in Mac. & Kindb.
Pohlia elongata var. **elongata** [POHLELO0]
SY: *Pohlia acuminata* Hoppe & Hornsch.
SY: *Pohlia minor* ssp. *acuminata* (Hoppe & Hornsch.) Wijk & Marg.
SY: *Pohlia minor* var. *brachycarpa* (Hoppe & Hornsch.) Wijk & Marg.
Pohlia elongata var. **greenii** (Brid.) Shaw [POHLELO1]
SY: *Bryum aciculinum* (Hoppe & Hornsch.) B.S.G.
SY: *Pohlia elongata* var. *minor* Hartm.
SY: *Pohlia minor* Schleich ex Schwaegr.
SY: *Webera polymorpha* (Hoppe & Hornsch.) Schimp.
Pohlia elongata var. *minor* Hartm. = Pohlia elongata var. greenii [POHLELO1]
Pohlia erecta Lindb. [POHLERE]
SY: *Pohlia defecta* (Sanio) Andrews in Grout
Pohlia filiformis (Dicks.) Andrews in Grout = Anomobryum filiforme [ANOMFIL]
Pohlia filiformis var. *concinnata* (Spruce) Grout = Anomobryum filiforme [ANOMFIL]
Pohlia filum (Schimp.) Mårt. [POHLFIU]
SY: *Pohlia commutata* var. *filum* (Schimp.) Dus.
SY: *Pohlia drummondii* var. *gracilis* (Bruch & Schimp. in B.S.G.) Podp.
SY: *Pohlia gracilis* (Bruch & Schimp. in B.S.G.) Lindb.
SY: *Pohlia rothii* (Corr. ex Limpr.) Broth.
SY: *Pohlia schleicheri* Crum
SY: *Webera commutata* var. *gracile* Schleich.
SY: *Webera gracilis* (B.S.G.) De Not.

Pohlia gracilis (Bruch & Schimp. in B.S.G.) Lindb. = Pohlia filum [POHLFIU]
Pohlia grandiflora ssp. *proligera* (Kindb. ex Breidl.) Kuc = Pohlia proligera [POHLPRO]
Pohlia grandiflora ssp. *proligera* var. *decipiens* (Loeske) Kuc = Pohlia annotina [POHLANN]
Pohlia lescuriana (Sull.) Grout [POHLLES]
SY: *Bryum lescurianum* Sull.
SY: *Pohlia pulchella* (Hedw.) Lindb.
SY: *Webera lescuriana* (Sull.) Jaeg. & Sauerb.
SY: *Webera pulchella* Schimp.
Pohlia longibracteata Broth. in Röll [POHLLON]
SY: *Mniobryum longibracteatum* (Broth. in Röll) Broth.
Pohlia longicolla (Hedw.) Lindb. [POHLLOG]
SY: *Webera longicolla* Hedw.
Pohlia ludwigii (Spreng. ex Schwaegr.) Broth. [POHLLUD]
SY: *Webera ludwigii* (Schwaegr.) Fürnr.
SY: *Webera ludwigii* var. *microphylla* Kindb. in Mac. & Kindb.
Pohlia melanodon (Brid.) Shaw [POHLMEL]
SY: *Pohlia carnea* (Schimp.) Lindb.
SY: *Pohlia delicatula* (Hedw.) Grout
SY: *Webera carnea* Schimp.
Pohlia minor Schleich ex Schwaegr. = Pohlia elongata var. greenii [POHLELO1]
Pohlia minor ssp. *acuminata* (Hoppe & Hornsch.) Wijk & Marg. = Pohlia elongata var. elongata [POHLELO0]
Pohlia minor var. *brachycarpa* (Hoppe & Hornsch.) Wijk & Marg. = Pohlia elongata var. elongata [POHLELO0]
Pohlia nutans (Hedw.) Lindb. [POHLNUT] (common nodding pohlia)
SY: *Bryum nutans* (Hedw.) Turn.
SY: *Pohlia nutans* ssp. *schimperi* (C. Müll.) Nyh.
SY: *Pohlia rutilans* (Schimp.) Lindb.
SY: *Pohlia schimperi* (C. Müll.) Andrews in Grout
SY: *Webera canaliculata* C. Müll. & Kindb. in Mac. & Kindb.
SY: *Webera nutans* Hedw.
SY: *Webera nutans* var. *macounii* (Kindb.) Mac.
SY: *Webera schimperi* (C. Müll.) Schimp.
Pohlia nutans ssp. *schimperi* (C. Müll.) Nyh. = Pohlia nutans [POHLNUT]
Pohlia obtusifolia (Brid.) L. Koch [POHLOBT]
SY: *Pohlia cucullata* (Schwaegr.) Lindb.
SY: *Webera cucullata* (Schwaegr.) Schimp.
Pohlia pacifica Shaw [POHLPAC]
Pohlia proligera (Kindb. ex Breidl.) Lindb. ex Arnell [POHLPRO]
SY: *Pohlia grandiflora* ssp. *proligera* (Kindb. ex Breidl.) Kuc
SY: *Webera proligera* Kindb.
Pohlia pulchella (Hedw.) Lindb. = Pohlia lescuriana [POHLLES]
Pohlia rothii (Corr. ex Limpr.) Broth. = Pohlia filum [POHLFIU]
Pohlia rothii auct. = Pohlia andalusica [POHLAND]
Pohlia rutilans (Schimp.) Lindb. = Pohlia nutans [POHLNUT]
Pohlia schimperi (C. Müll.) Andrews in Grout = Pohlia nutans [POHLNUT]
Pohlia schleicheri Crum = Pohlia filum [POHLFIU]
Pohlia sphagnicola (Bruch & Schimp.) Lindb. & Arnell [POHLSPH]
SY: *Webera sphagnicola* (B.S.G.) Schimp.
Pohlia tozeri (Grev.) Delogne = Epipterygium tozeri [EPIPTOZ]
Pohlia tundrae Shaw [POHLTUN]
Pohlia vexans (Limpr.) H. Lindb. [POHLVEX]
Pohlia wahlenbergii (Web. & Mohr) Andrews [POHLWAH]
SY: *Bryum albicans* (Wahlenb.) Röhl.
SY: *Bryum wahlenbergii* (Web. & Mohr) Schwaegr.
SY: *Mniobryum albicans* (Wahlenb.) Limpr.
SY: *Mniobryum wahlenbergii* (Web. & Mohr) Jenn.
SY: *Mniobryum wahlenbergii* var. *glacialis* (Brid.) Wijk & Marg.
SY: *Pohlia albicans* Lindb.
SY: *Webera albicans* Schimp.
Rhodobryum (Schimp.) Hampe [RHODOBR$]
Rhodobryum roseum (Hedw.) Limpr. [RHODROS]
SY: *Bryum roseum* (Hedw.) Crom.
Roellia Kindb. [ROELLIA$]
Roellia roellii (Broth. in Röll) Andrews ex Crum [ROELROE]
SY: *Bryum sandbergii* Holz.
SY: *Bryum simplex* Kindb. in Mac. & Kindb.
SY: *Roellia simplex* Kindb.
Roellia simplex Kindb. = Roellia roellii [ROELROE]
Splachnobryum kieneri Williams = Bryum turbinatum [BRYUTUR]
Webera acuminata (Hoppe & Hornsch.) Schimp. = Pohlia elongata [POHLELO]
Webera albicans Schimp. = Pohlia wahlenbergii [POHLWAH]
Webera annotina (Hedw.) Bruch in Schwaegr. = Pohlia annotina [POHLANN]
Webera canaliculata C. Müll. & Kindb. in Mac. & Kindb. = Pohlia nutans [POHLNUT]
Webera carnea Schimp. = Pohlia melanodon [POHLMEL]
Webera columbica Kindb. in Mac. & Kindb. = Pohlia columbica [POHLCOL]
Webera commutata Schimp. = Pohlia drummondii [POHLDRU]
Webera commutata var. *gracile* Schleich. = Pohlia filum [POHLFIU]
Webera cruda (Hedw.) Fürnr. = Pohlia cruda [POHLCRU]
Webera crudoides (Sull. & Lesq.) Jaeg. & Sauerb. = Pohlia crudoides [POHLCRD]
Webera cucullata (Schwaegr.) Schimp. = Pohlia obtusifolia [POHLOBT]
Webera elongata (Hedw.) Schwaegr. = Pohlia elongata [POHLELO]
Webera gracilis (B.S.G.) De Not. = Pohlia filum [POHLFIU]
Webera lescuriana (Sull.) Jaeg. & Sauerb. = Pohlia lescuriana [POHLLES]
Webera longicolla Hedw. = Pohlia longicolla [POHLLOG]
Webera ludwigii (Schwaegr.) Fürnr. = Pohlia ludwigii [POHLLUD]
Webera ludwigii var. *microphylla* Kindb. in Mac. & Kindb. = Pohlia ludwigii [POHLLUD]
Webera micro-apiculata C. Müll. & Kindb. in Mac. & Kindb. = Pohlia cruda [POHLCRU]
Webera microcaulon C. Müll. & Kindb. in Mac. & Kindb. = Pohlia elongata [POHLELO]
Webera microdenticulata C. Müll. & Kindb. in Mac. & Kindb. = Pohlia drummondii [POHLDRU]
Webera nutans Hedw. = Pohlia nutans [POHLNUT]
Webera nutans var. *macounii* (Kindb.) Mac. = Pohlia nutans [POHLNUT]
Webera polymorpha (Hoppe & Hornsch.) Schimp. = Pohlia elongata var. greenii [POHLELO1]
Webera polymorphoides Kindb. in Mac. & Kindb. = Pohlia drummondii [POHLDRU]
Webera proligera Kindb. = Pohlia proligera [POHLPRO]
Webera pseudo-carnea (Kindb.) Mac. = Pohlia bulbifera [POHLBUL]
Webera pulchella Schimp. = Pohlia lescuriana [POHLLES]
Webera pycnodecurrens C. Müll. & Kindb. in Mac. & Kindb. = Pohlia drummondii [POHLDRU]
Webera schimperi (C. Müll.) Schimp. = Pohlia nutans [POHLNUT]
Webera sphagnicola (B.S.G.) Schimp. = Pohlia sphagnicola [POHLSPH]
Webera subcucullata C. Müll. & Kindb. in Mac. & Kindb. = Pohlia drummondii [POHLDRU]
Webera subpolymorpha (Kindb.) Par. = Pohlia drummondii [POHLDRU]
Webera tozeri (Grev.) Schimp. = Epipterygium tozeri [EPIPTOZ]
Zieria demissa (Hook.) Schimp. = Plagiobryum demissum [PLAGDEM]
Zieria julacea Schimp. = Plagiobryum zieri [PLAGZIE]

BUXBAUMIACEAE (B:M)

Buxbaumia Hedw. [BUXBAUM$]
Buxbaumia aphylla Hedw. [BUXBAPH]
Buxbaumia indusiata Brid. = Buxbaumia viridis [BUXBVIR]
Buxbaumia piperi Best [BUXBPIP]
Buxbaumia viridis (DC.) Moug. & Nestl. [BUXBVIR]
SY: *Buxbaumia indusiata* Brid.
Diphyscium Mohr [DIPHYSC$]
Diphyscium foliosum (Hedw.) Mohr [DIPHFOL]
SY: *Diphyscium sessile* Lindb.
SY: *Webera sessilis* Lindb.
Diphyscium sessile Lindb. = Diphyscium foliosum [DIPHFOL]
Webera sessilis Lindb. = Diphyscium foliosum [DIPHFOL]

CALYPOGEIACEAE (B:H)

Calypogeia Raddi [CALYPOG$]
Calypogeia azurea Stotler & Crotz. [CALYAZU]
SY: *Calypogeia cyanophora* Schust.
SY: *Calypogeia trichomanis* auct.
Calypogeia cyanophora Schust. = Calypogeia azurea [CALYAZU]
Calypogeia fissa (L.) Raddi [CALYFIS]
Calypogeia integristipula Steph. [CALYINT]
SY: *Calypogeia meylanii* Buch
SY: *Calypogeia neesiana* var. *meylanii* (Buch) Schust.
Calypogeia meylanii Buch = Calypogeia integristipula [CALYINT]
Calypogeia muelleriana (Schiffn.) K. Müll. [CALYMUE]
Calypogeia muelleriana f. **muelleriana** [CALYMUE0]
Calypogeia muelleriana f. **schofieldii** Hong [CALYMUE1]
Calypogeia muelleriana f. **shieldii** Hong [CALYMUE2]
Calypogeia neesiana (Mass. & Carest.) K. Müll. [CALYNEE]
Calypogeia neesiana var. *meylanii* (Buch) Schust. = Calypogeia integristipula [CALYINT]
Calypogeia paludosa Warnst. = Calypogeia sphagnicola [CALYSPH]
Calypogeia sphagnicola (Arnell & J. Perss.) Warnst. & Loeske [CALYSPH]
SY: *Calypogeia paludosa* Warnst.
Calypogeia suecica (Arnell & J. Perss.) K. Müll. [CALYSUE]
Calypogeia trichomanis auct. = Calypogeia azurea [CALYAZU]
Calypogeja Raddi = Calypogeia [CALYPOG$]

CATOSCOPIACEAE (B:M)

Catoscopium Brid. [CATOSCO$]
Catoscopium nigritum (Hedw.) Brid. [CATONIG] (golf club moss)

CEPHALOZIACEAE (B:H)

Cephalozia (Dum. emend. Schiffn.) Dum. [CEPHALO$]
Cephalozia ambigua Mass. = Cephalozia bicuspidata ssp. ambigua [CEPHBIC1]
Cephalozia bicuspidata (L.) Dum. [CEPHBIC] (slender two-toothed wort)
Cephalozia bicuspidata ssp. **ambigua** (Mass.) Schust. [CEPHBIC1]
SY: *Cephalozia ambigua* Mass.
Cephalozia bicuspidata ssp. **bicuspidata** [CEPHBIC0]
Cephalozia catenulata (Hub.) Lindb. [CEPHCAT]
Cephalozia connivens (Dicks.) Lindb. [CEPHCON]
Cephalozia fluitans (Nees) Spruce = Cladopodiella fluitans [CLADFLU]
Cephalozia leucantha Spruce [CEPHLEU]
Cephalozia lunulifolia (Dum.) Dum. [CEPHLUN]
SY: *Cephalozia media* Lindb.
Cephalozia macounii (Aust.) Aust. [CEPHMAC]
Cephalozia media Lindb. = Cephalozia lunulifolia [CEPHLUN]
Cephalozia pleniceps (Aust.) Lindb. [CEPHPLE]
Cladopodiella Buch [CLADOPO$]
Cladopodiella fluitans (Nees) Joerg. [CLADFLU]
SY: *Cephalozia fluitans* (Nees) Spruce
Hygrobiella Spruce [HYGROBI$]
Hygrobiella laxifolia (Hook.) Spruce [HYGRLAX]
Pleuroclada albescens (Hook.) Spruce = Pleurocladula albescens [PLEUALB]
Pleuroclada albescens var. *islandica* (Nees) Spruce = Pleurocladula albescens [PLEUALB]
Pleurocladula Grolle [PLEUROL$]
Pleurocladula albescens (Hook.) Grolle [PLEUALB]
SY: *Pleuroclada albescens* (Hook.) Spruce
SY: *Pleuroclada albescens* var. *islandica* (Nees) Spruce
SY: *Pleurocladula islandica* (Nees) Grolle
Pleurocladula islandica (Nees) Grolle = Pleurocladula albescens [PLEUALB]
Schofieldia Godfr. [SCHOFIE$]
Schofieldia monticola Godfr. [SCHOMON]

CEPHALOZIELLACEAE (B:H)

Cephaloziella (Spruce) Steph. [CEPHALZ$]
Cephaloziella arctica Bryhn & Douin [CEPHARC]
SY: *Cephaloziella glacialis* Douin
Cephaloziella brinkmani Douin [CEPHBRI]
Cephaloziella byssacea (Roth) Warnst. = Cephaloziella divaricata [CEPHDIV]
Cephaloziella byssacea var. *scabra* (M.A. Howe) Schust. = Cephaloziella divaricata var. scabra [CEPHDIV2]
Cephaloziella divaricata (Sm.) Schiffn. [CEPHDIV]
SY: *Cephaloziella byssacea* (Roth) Warnst.
SY: *Cephaloziella starkei* (Funck ex Nees) Schiffn.
SY: *Jungermannia confervoides* Raddi
Cephaloziella divaricata var. **divaricata** [CEPHDIV0]
Cephaloziella divaricata var. **scabra** M.A. Howe [CEPHDIV2]
SY: *Cephaloziella byssacea* var. *scabra* (M.A. Howe) Schust.
SY: *Cephaloziella papilosa* (Douin) Schiffn.
Cephaloziella elachista (Jack) Schiffn. [CEPHELA]
Cephaloziella glacialis Douin = Cephaloziella arctica [CEPHARC]
Cephaloziella hampeana (Nees) Schiffn. [CEPHHAM]
Cephaloziella papilosa (Douin) Schiffn. = Cephaloziella divaricata var. scabra [CEPHDIV2]
Cephaloziella phyllacantha (Mass. & Carest.) K. Müll. [CEPHPHY]
Cephaloziella rubella (Nees) Warnst. [CEPHRUB]
Cephaloziella spinosa Douin = Cephaloziella subdentata [CEPHSUB]
Cephaloziella starkei (Funck ex Nees) Schiffn. = Cephaloziella divaricata [CEPHDIV]
Cephaloziella striatula (C. Jens.) Douin = Cephaloziella subdentata [CEPHSUB]
Cephaloziella subdentata Warnst. [CEPHSUB]
SY: *Cephaloziella spinosa* Douin
SY: *Cephaloziella striatula* (C. Jens.) Douin
Cephaloziella turneri (Hook.) K. Müll. [CEPHTUR]
SY: *Prionolobus turneri* (Hook.) Spruce
Jungermannia confervoides Raddi = Cephaloziella divaricata [CEPHDIV]
Prionolobus turneri (Hook.) Spruce = Cephaloziella turneri [CEPHTUR]

CLEVEACEAE (B:H)

Athalamia Falc. [ATHALAM$]
Athalamia hyalina (Sommerf.) Hatt. [ATHAHYA]
SY: *Clevea hyalina* (Somm.) Lindb.
Clevea hyalina (Somm.) Lindb. = Athalamia hyalina [ATHAHYA]
Peltolepis Lindb. [PELTOLE$]
Peltolepis grandis Lindb. = Peltolepis quadrata [PELTQUA]
Peltolepis quadrata (Saut.) K. Müll. [PELTQUA]
SY: *Peltolepis grandis* Lindb.
Sauteria Nees [SAUTERI$]
Sauteria alpina (Nees) Nees [SAUTALP]

CLIMACIACEAE (B:M)

Climacium Web. & Mohr [CLIMACI$]
Climacium dendroides (Hedw.) Web. & Mohr [CLIMDEN] (tree moss)
SY: *Climacium dendroides* var. *oregonensis* Ren. & Card.
Climacium dendroides var. *oregonensis* Ren. & Card. = Climacium dendroides [CLIMDEN]

CODONIACEAE (B:H)

Fossombronia Raddi [FOSSOMB$]
Fossombronia dumortieri (Hüb. & Genth.) Lindb. = Fossombronia foveolata [FOSSFOV]
Fossombronia foveolata Lindb. [FOSSFOV]
SY: *Fossombronia dumortieri* (Hüb. & Genth.) Lindb.
Fossombronia longiseta Aust. [FOSSLON]

CONOCEPHALACEAE (B:H)

Conocephalum Wigg. [CONOCEP$]
Conocephalum conicum (L.) Lindb. [CONOCON] (snake liverwort)

DALTONIACEAE (B:M)

Daltonia Hook. & Tayl. [DALTONI$]
Daltonia splachnoides (Sm. in Sm. & Sowerby) Hook. & Tayl. [DALTSPL]

DICRANACEAE (B:M)

Anisothecium grevilleanum (Brid.) Arnell & C. Jens. = Dicranella grevilleana [DICRGRE]
Anisothecium rufescens (With.) Lindb. = Dicranella rufescens [DICRRUF]
Anisothecium schreberianum (Hedw.) Dix. = Dicranella schreberiana var. schreberiana [DICRSCH0]
Anisothecium vaginale (Hedw.) Loeske = Dicranella crispa [DICRCRI]
Anisothecium varium (Hedw.) Mitt. = Dicranella varia [DICRVAR]
Aongstroemia Bruch & Schimp. in B.S.G. [AONGSTR$]
Aongstroemia longipes (Somm.) Bruch & Schimp. in B.S.G. [AONGLON]
SY: *Dicranum julaceum* Hook. & Wils. in Drumm.
Arctoa Bruch & Schimp. in B.S.G. [ARCTOA$]
Arctoa blyttii (Schimp.) Loeske = Kiaeria blyttii [KIAEBLY]
Arctoa falcata (Hedw.) Loeske = Kiaeria falcata [KIAEFAL]
Arctoa fulvella (Dicks.) Bruch & Schimp. in B.S.G. [ARCTFUV]
SY: *Dicranum fulvellum* (Dicks.) Sm.
Arctoa starkei (Web. & Mohr) Loeske = Kiaeria starkei [KIAESTA]
Bartleya ohioensis Robins. = Dicranella cerviculata [DICRCER]
Campylopus Brid. [CAMPYLO$]
Campylopus atrovirens De Not. [CAMPATR] (black fish hook moss)
Campylopus atrovirens var. **cucullatifolius** J.-P. Frahm [CAMPATR1]
Campylopus canadensis Kindb. = Paraleucobryum longifolium [PARALON]
Campylopus eberhardtii Par. = Campylopus japonicus [CAMPJAP]
Campylopus excurrens Dix. = Campylopus japonicus [CAMPJAP]
Campylopus flexuosus (Hedw.) Brid. [CAMPFLE]
SY: *Dicranum palustre* La Pyl. in Brid.
Campylopus fragilis (Brid.) Bruch & Schimp. in B.S.G. [CAMPFRA]
Campylopus introflexus (Hedw.) Brid. [CAMPINT]
Campylopus irrigatus Thér. = Campylopus japonicus [CAMPJAP]
Campylopus japonicus Broth. [CAMPJAP]
SY: *Campylopus eberhardtii* Par.
SY: *Campylopus excurrens* Dix.
SY: *Campylopus irrigatus* Thér.
SY: *Campylopus pseudomülleri* Card.
SY: *Campylopus saint-pierrei* Thér.
SY: *Campylopus uii* Broth.
Campylopus pseudomülleri Card. = Campylopus japonicus [CAMPJAP]
Campylopus saint-pierrei Thér. = Campylopus japonicus [CAMPJAP]
Campylopus schimperi Milde [CAMPSCI]
SY: *Campylopus subulatus* var. *schimperi* (Milde) Husn.
Campylopus schwarzii Schimp. [CAMPSCH]
Campylopus subulatus Schimp. in Rabenh. [CAMPSUB]
Campylopus subulatus var. *schimperi* (Milde) Husn. = Campylopus schimperi [CAMPSCI]
Campylopus uii Broth. = Campylopus japonicus [CAMPJAP]
Cnestrum alpestre (Wahlenb.) Nyh. ex Mog. = Cynodontium alpestre [CYNOALP]
Cnestrum glaucescens (Lindb. & Arnell) Holm. ex Mog. & Steere = Cynodontium glaucescens [CYNOGLA]
Cnestrum schisti (Web. & Mohr) Hag. = Cynodontium schisti [CYNOSCH]
Cynodontium Bruch & Schimp. ex Schimp. [CYNODOT$]
Cynodontium alpestre (Wahlenb.) Milde [CYNOALP]
SY: *Cnestrum alpestre* (Wahlenb.) Nyh. ex Mog.
SY: *Oncophorus alpestris* (Wahlenb.) Lindb.
SY: *Oncophorus strumulosus* (C. Müll. & Kindb.) Britt. in Williams
Cynodontium glaucescens (Lindb. & Arnell) Par. [CYNOGLA]
SY: *Cnestrum glaucescens* (Lindb. & Arnell) Holm. ex Mog. & Steere
SY: *Oncophorus glaucescens* Lindb. & Arnell
Cynodontium jenneri (Schimp. in Howie) Stirt. [CYNOJEN]
SY: *Cynodontium polycarpon* var. *laxirete* (Dix.) Dix.
SY: *Oncophorus jenneri* (Schimp. in Howie) Williams
Cynodontium polycarpon (Hedw.) Schimp. [CYNOPOL]
SY: *Dicranum polycarpum* (Hedw.) Web. & Mohr
SY: *Oncophorus polycarpus* (Hedw.) Brid.
Cynodontium polycarpon var. *laxirete* (Dix.) Dix. = Cynodontium jenneri [CYNOJEN]
Cynodontium schisti (Web. & Mohr) Lindb. [CYNOSCH]
SY: *Cnestrum schisti* (Web. & Mohr) Hag.
SY: *Oncophorus schisti* (Web. & Mohr) Lindb.
Cynodontium strumiferum (Hedw.) Lindb. [CYNOSTR]
SY: *Oncophorus polycarpus* var. *strumiferus* (Hedw.) Brid.
Cynodontium subalpestre Kindb. in Mac. & Kindb. = Cynodontium tenellum [CYNOTEN]
Cynodontium tenellum (Bruch & Schimp. in B.S.G.) Limpr. [CYNOTEN]
SY: *Cynodontium subalpestre* Kindb. in Mac. & Kindb.
SY: *Cynodontium torquescens* Limpr.
SY: *Oncophorus tenellus* (Bruch & Schimp. in B.S.G.) Williams
Cynodontium torquescens Limpr. = Cynodontium tenellum [CYNOTEN]
Cynodontium virens (Hedw.) Schimp. = Oncophorus virens [ONCOVIR]
Cynodontium virens var. *serratum* (B.S.G.) Schimp. = Oncophorus virens [ONCOVIR]
Cynodontium virens var. *wahlenbergii* (Brid.) Schimp. = Oncophorus wahlenbergii [ONCOWAH]
Cynodontium wahlenbergii (Brid.) Hartm. = Oncophorus wahlenbergii [ONCOWAH]
Cynodontium wahlenbergii var. *compactum* (B.S.G.) Mac. & Kindb. = Oncophorus wahlenbergii [ONCOWAH]
Cynodontium wahlenbergii var. *gracile* (Broth.) Mönk. = Oncophorus wahlenbergii [ONCOWAH]
Dichodontium Schimp. [DICHODO$]
Dichodontium flavescens (With.) Lindb. = Dichodontium pellucidum [DICHPEL]
Dichodontium olympicum Ren. & Card. [DICHOLY]
Dichodontium pellucidum (Hedw.) Schimp. [DICHPEL] (wet rock moss)
SY: *Dichodontium flavescens* (With.) Lindb.
SY: *Dichodontium pellucidum* ssp. *fagimontanum* (Brid.) Kindb.
SY: *Dichodontium pellucidum* var. *fagimontanum* (Brid.) Schimp.
SY: *Dichodontium pellucidum* var. *flavescens* (With.) Moore
Dichodontium pellucidum ssp. *fagimontanum* (Brid.) Kindb. = Dichodontium pellucidum [DICHPEL]
Dichodontium pellucidum var. *fagimontanum* (Brid.) Schimp. = Dichodontium pellucidum [DICHPEL]
Dichodontium pellucidum var. *flavescens* (With.) Moore = Dichodontium pellucidum [DICHPEL]
Dicranella (C. Müll.) Schimp. [DICRANE$]
Dicranella cerviculata (Hedw.) Schimp. [DICRCER]
SY: *Bartleya ohioensis* Robins.
SY: *Dicranella cerviculata* var. *americana* Grout
SY: *Dicranella ohioense* (Robins.) Crum
SY: *Dicranella polaris* Kindb.
SY: *Dicranum cerviculatum* Hedw.
Dicranella cerviculata var. *americana* Grout = Dicranella cerviculata [DICRCER]
Dicranella crispa (Hedw.) Schimp. [DICRCRI]
SY: *Anisothecium vaginale* (Hedw.) Loeske

Dicranella grevilleana (Brid.) Schimp. [DICRGRE]
SY: *Anisothecium grevilleanum* (Brid.) Arnell & C. Jens.
Dicranella heteromalla (Hedw.) Schimp. [DICRHET]
SY: *Dicranella heteromalla* var. *orthocarpa* (Hedw.) Jaeg.
SY: *Dicranella heteromalla* var. *sericea* (Schimp.) Pfeff.
SY: *Dicranum heteromallum* Hedw.
Dicranella heteromalla var. *orthocarpa* (Hedw.) Jaeg. = Dicranella heteromalla [DICRHET]
Dicranella heteromalla var. *sericea* (Schimp.) Pfeff. = Dicranella heteromalla [DICRHET]
Dicranella howei Ren. & Card. [DICRHOW]
Dicranella hutchinsonii Krajina = Dicranella rufescens [DICRRUF]
Dicranella ohioense (Robins.) Crum = Dicranella cerviculata [DICRCER]
Dicranella pacifica Schof. [DICRPAC]
Dicranella palustris (Dicks.) Crundw. ex Warb. [DICRPAL]
SY: *Dicranella squarrosa* (Stark) Schimp.
SY: *Dicranella squarrosa* f. *fluitans* Grout
Dicranella polaris Kindb. = Dicranella cerviculata [DICRCER]
Dicranella rubra Lindb. = Dicranella varia [DICRVAR]
Dicranella rufescens (With.) Schimp. [DICRRUF]
SY: *Anisothecium rufescens* (With.) Lindb.
SY: *Dicranella hutchinsonii* Krajina
SY: *Dicranum rufescens* (With.) Turn.
Dicranella schreberi Schimp. = Dicranella schreberiana [DICRSCH]
Dicranella schreberi var. *elata* Schimp. = Dicranella schreberiana var. robusta [DICRSCH1]
Dicranella schreberiana (Hedw.) Hilf. ex Crum & Anderson [DICRSCH]
SY: *Dicranella schreberi* Schimp.
SY: *Dicranum schreberi* Sw.
Dicranella schreberiana var. **robusta** (Schimp. ex Braithw.) Crum & Anderson [DICRSCH1]
SY: *Dicranella schreberi* var. *elata* Schimp.
Dicranella schreberiana var. **schreberiana** [DICRSCH0]
SY: *Anisothecium schreberianum* (Hedw.) Dix.
Dicranella squarrosa (Stark) Schimp. = Dicranella palustris [DICRPAL]
Dicranella squarrosa f. *fluitans* Grout = Dicranella palustris [DICRPAL]
Dicranella stickinensis Grout [DICRSTI]
Dicranella subulata (Hedw.) Schimp. [DICRSUB]
SY: *Dicranum subulatum* Hedw.
Dicranella varia (Hedw.) Schimp. [DICRVAR]
SY: *Anisothecium varium* (Hedw.) Mitt.
SY: *Dicranella rubra* Lindb.
SY: *Dicranum varium* Hedw.
Dicranodontium Bruch & Schimp. in B.S.G. [DICRANO$]
Dicranodontium asperulum (Mitt.) Broth. [DICRASP]
Dicranodontium denudatum (Brid.) Britt. in Williams [DICRDEN]
SY: *Dicranodontium longirostre* (Web. & Mohr) B.S.G.
Dicranodontium longirostre (Web. & Mohr) B.S.G. = Dicranodontium denudatum [DICRDEN]
Dicranodontium subporodictyon Broth. [DICRSUP]
Dicranodontium uncinatum (Harv. in Hook.) Jaeg. [DICRUNC]
Dicranoweisia Lindb. ex Milde [DICRANW$]
Dicranoweisia cirrata (Hedw.) Lindb. ex Milde [DICRCIR] (curly thatch moss)
SY: *Didymodon hinckleyi* Bartr.
Dicranoweisia contermina Ren. & Card. = Dicranoweisia crispula [DICRCRS]
Dicranoweisia crispula (Hedw.) Lindb. ex Milde [DICRCRS] (yellow-green cushion moss)
SY: *Dicranoweisia contermina* Ren. & Card.
SY: *Dicranoweisia crispula* var. *compacta* (Schleich. ex Schwaegr.) Lindb. in Kindb.
SY: *Dicranoweisia crispula* var. *contermina* (Ren. & Card.) Grout
SY: *Dicranoweisia crispula* var. *roellii* (Kindb. in Röll) Lawt.
SY: *Dicranoweisia roellii* Kindb. in Röll
Dicranoweisia crispula var. *compacta* (Schleich. ex Schwaegr.) Lindb. in Kindb. = Dicranoweisia crispula [DICRCRS]
Dicranoweisia crispula var. *contermina* (Ren. & Card.) Grout = Dicranoweisia crispula [DICRCRS]
Dicranoweisia crispula var. *roellii* (Kindb. in Röll) Lawt. = Dicranoweisia crispula [DICRCRS]
Dicranoweisia obliqua Kindb. = Kiaeria blyttii [KIAEBLY]
Dicranoweisia roellii Kindb. in Röll = Dicranoweisia crispula [DICRCRS]
Dicranoweisia subcompacta Card. & Thér. = Kiaeria starkei [KIAESTA]
Dicranum Hedw. [DICRANU$]
Dicranum acutifolium (Lindb. & Arnell) C. Jens. ex Weinm. [DICRACU]
SY: *Dicranum bergeri* ssp. *rupincola* Kindb.
SY: *Dicranum rupincola* (Kindb.) Perss.
Dicranum affine Funck = Dicranum undulatum [DICRUND]
Dicranum alatum (Barnes) Card. & Thér. = Dicranum scoparium [DICRSCO]
Dicranum albicans var. *denticulatum* Kindb. in Mac. & Kindb. = Paraleucobryum longifolium [PARALON]
Dicranum algidum Kindb. = Dicranum spadiceum [DICRSPA]
Dicranum angustum Lindb. [DICRANG]
SY: *Dicranum laevidens* Williams
Dicranum arcticum Schimp. = Kiaeria glacialis [KIAEGLA]
Dicranum bergeri Bland. in Sturm = Dicranum undulatum [DICRUND]
Dicranum bergeri ssp. *rupincola* Kindb. = Dicranum acutifolium [DICRACU]
Dicranum blyttii B.S.G. = Kiaeria blyttii [KIAEBLY]
Dicranum bonjeanii var. *alatum* Barnes in Röll = Dicranum scoparium [DICRSCO]
Dicranum brevifolium (Lindb.) Lindb. [DICRBRE]
SY: *Dicranum muehlenbeckii* var. *brevifolium* Lindb.
SY: *Dicranum muehlenbeckii* var. *cirratum* (Schimp.) Lindb. in Norrl.
Dicranum canadense Kindb. in Mac. = Dicranum scoparium [DICRSCO]
Dicranum cerviculatum Hedw. = Dicranella cerviculata [DICRCER]
Dicranum consobrinum Ren. & Card. = Dicranum scoparium [DICRSCO]
Dicranum crispulum C. Müll. & Kindb. in Mac. & Kindb. = Dicranum fuscescens [DICRFUS]
Dicranum elongatum Schleich. ex Schwaegr. [DICRELO]
SY: *Dicranum elongatum* ssp. *attenuatum* Kindb.
SY: *Dicranum elongatum* ssp. *subfragilifolium* Kindb.
SY: *Dicranum subflagellare* Card. & Thér.
Dicranum elongatum ssp. *attenuatum* Kindb. = Dicranum elongatum [DICRELO]
Dicranum elongatum ssp. *subfragilifolium* Kindb. = Dicranum elongatum [DICRELO]
Dicranum falcatum Hedw. = Kiaeria falcata [KIAEFAL]
Dicranum flagellare Hedw. [DICRFLA] (whip fork moss)
SY: *Dicranum flagellare* var. *minutissimum* Grout
SY: *Orthodicranum flagellare* (Hedw.) Loeske
Dicranum flagellare var. *minutissimum* Grout = Dicranum flagellare [DICRFLA]
Dicranum fragilifolium Lindb. [DICRFRA]
Dicranum fulvellum (Dicks.) Sm. = Arctoa fulvella [ARCTFUV]
Dicranum fuscescens Turn. [DICRFUS] (dusky fork moss)
SY: *Dicranum crispulum* C. Müll. & Kindb. in Mac. & Kindb.
SY: *Dicranum fuscescens* ssp. *subbrevifolium* Kindb.
SY: *Dicranum sulcatum* Kindb. in Mac.
Dicranum fuscescens ssp. *subbrevifolium* Kindb. = Dicranum fuscescens [DICRFUS]
Dicranum groenlandicum Brid. [DICRGRO]
Dicranum heteromallum Hedw. = Dicranella heteromalla [DICRHET]
Dicranum julaceum Hook. & Wils. in Drumm. = Aongstroemia longipes [AONGLON]
Dicranum laevidens Williams = Dicranum angustum [DICRANG]
Dicranum longifolium Hedw. = Paraleucobryum longifolium [PARALON]
Dicranum longifolium var. *subalpinum* Milde = Paraleucobryum longifolium [PARALON]
Dicranum majus Sm. [DICRMAJ]
Dicranum majus var. **majus** [DICRMAJ0]
Dicranum majus var. **orthophyllum** A. Br. ex Milde [DICRMAJ2]

Dicranum microcarpum Hook. in Drumm. = Oncophorus wahlenbergii [ONCOWAH]
Dicranum molle (Wils.) Lindb. = Kiaeria glacialis [KIAEGLA]
Dicranum montanum Hedw. [DICRMON]
SY: *Orthodicranum montanum* (Hedw.) Loeske
Dicranum muehlenbeckii Bruch & Schimp. in B.S.G. [DICRMUE]
Dicranum muehlenbeckii var. *brevifolium* Lindb. = Dicranum brevifolium [DICRBRE]
Dicranum muehlenbeckii var. *cirratum* (Schimp.) Lindb. in Norrl. = Dicranum brevifolium [DICRBRE]
Dicranum neglectum Jur. ex De Not. = Dicranum spadiceum [DICRSPA]
Dicranum pallidisetum (Bail. in Holz.) Irel. [DICRPAI] (pale-stalked broom moss)
Dicranum pallidum Bruch & Schimp. ex C. Müll. = Dicranum scoparium [DICRSCO]
Dicranum palustre La Pyl. in Brid. = Campylopus flexuosus [CAMPFLE]
Dicranum polycarpum (Hedw.) Web. & Mohr = Cynodontium polycarpon [CYNOPOL]
Dicranum polysetum Sw. [DICRPOL] (wavy-leaved moss)
SY: *Dicranum rugosum* (Funck) Hoffm. ex Brid.
SY: *Dicranum rugosum* var. *rugulosum* Kindb.
SY: *Dicranum schraderi* Wahlenb.
SY: *Dicranum stenodictyon* Kindb.
SY: *Dicranum undulatum* Ehrh. ex Web. & Mohr non Brid.
SY: *Dicranum undulatum* ssp. *ontariense* Kindb.
Dicranum pumilum Saut. = Kiaeria falcata [KIAEFAL]
Dicranum rufescens (With.) Turn. = Dicranella rufescens [DICRRUF]
Dicranum rugosum (Funck) Hoffm. ex Brid. = Dicranum polysetum [DICRPOL]
Dicranum rugosum var. *rugulosum* Kindb. = Dicranum polysetum [DICRPOL]
Dicranum rupincola (Kindb.) Perss. = Dicranum acutifolium [DICRACU]
Dicranum sauteri var. *pachytrichum* Kindb. in Mac. & Kindb. = Paraleucobryum longifolium [PARALON]
Dicranum schisti Lindb. = Kiaeria blyttii [KIAEBLY]
Dicranum schraderi Wahlenb. = Dicranum polysetum [DICRPOL]
Dicranum schreberi Sw. = Dicranella schreberiana [DICRSCH]
Dicranum scopariiforme Kindb. = Dicranum scoparium [DICRSCO]
Dicranum scoparium Hedw. [DICRSCO] (broom moss)
SY: *Dicranum alatum* (Barnes) Card. & Thér.
SY: *Dicranum bonjeanii* var. *alatum* Barnes in Röll
SY: *Dicranum canadense* Kindb. in Mac.
SY: *Dicranum consobrinum* Ren. & Card.
SY: *Dicranum pallidum* Bruch & Schimp. ex C. Müll.
SY: *Dicranum scopariiforme* Kindb.
SY: *Dicranum scoparium* ssp. *involutum* Kindb. in Röll
SY: *Dicranum scoparium* var. *nigrescens* Györf.
SY: *Dicranum scoparium* var. *orthophyllum* Brid.
SY: *Dicranum scoparium* var. *scopariiforme* (Kindb.) Kindb.
Dicranum scoparium ssp. *involutum* Kindb. in Röll = Dicranum scoparium [DICRSCO]
Dicranum scoparium var. *nigrescens* Györf. = Dicranum scoparium [DICRSCO]
Dicranum scoparium var. *orthophyllum* Brid. = Dicranum scoparium [DICRSCO]
Dicranum scoparium var. *scopariiforme* (Kindb.) Kindb. = Dicranum scoparium [DICRSCO]
Dicranum serratum Kindb. = Paraleucobryum longifolium [PARALON]
Dicranum spadiceum Zett. [DICRSPA]
SY: *Dicranum algidum* Kindb.
SY: *Dicranum neglectum* Jur. ex De Not.
Dicranum starkei Web. & Mohr = Kiaeria starkei [KIAESTA]
Dicranum stenodictyon Kindb. = Dicranum polysetum [DICRPOL]
Dicranum strictum Schleich. ex Mohr = Dicranum tauricum [DICRTAU]
Dicranum subflagellare Card. & Thér. = Dicranum elongatum [DICRELO]
Dicranum subulatum Hedw. = Dicranella subulata [DICRSUB]
Dicranum subulifolium Kindb. in Mac. = Paraleucobryum enerve [PARAENE]
Dicranum sulcatum Kindb. in Mac. = Dicranum fuscescens [DICRFUS]
Dicranum tauricum Sapenh. [DICRTAU] (broken-leaf moss)
SY: *Dicranum strictum* Schleich. ex Mohr
Dicranum undulatum Brid. [DICRUND] (wavy dicranum)
SY: *Dicranum affine* Funck
SY: *Dicranum bergeri* Bland. in Sturm
Dicranum undulatum Ehrh. ex Web. & Mohr non Brid. = Dicranum polysetum [DICRPOL]
Dicranum undulatum ssp. *ontariense* Kindb. = Dicranum polysetum [DICRPOL]
Dicranum varium Hedw. = Dicranella varia [DICRVAR]
Dicranum virens Hedw. = Oncophorus virens [ONCOVIR]
Dicranum virens var. *wahlenbergii* (Brid.) Hüb. = Oncophorus wahlenbergii [ONCOWAH]
Didymodon hinckleyi Bartr. = Dicranoweisia cirrata [DICRCIR]
Kiaeria Hag. [KIAERIA$]
Kiaeria blyttii (Schimp.) Broth. [KIAEBLY]
SY: *Arctoa blyttii* (Schimp.) Loeske
SY: *Dicranoweisia obliqua* Kindb.
SY: *Dicranum blyttii* B.S.G.
SY: *Dicranum schisti* Lindb.
SY: *Kiaeria blyttii* var. *hispidula* (Williams) Wijk & Marg.
Kiaeria blyttii var. *hispidula* (Williams) Wijk & Marg. = Kiaeria blyttii [KIAEBLY]
Kiaeria falcata (Hedw.) Hag. [KIAEFAL]
SY: *Arctoa falcata* (Hedw.) Loeske
SY: *Dicranum falcatum* Hedw.
SY: *Dicranum pumilum* Saut.
SY: *Kiaeria falcata* var. *pumila* (Saut.) Podp.
Kiaeria falcata var. *pumila* (Saut.) Podp. = Kiaeria falcata [KIAEFAL]
Kiaeria glacialis (Berggr.) Hag. [KIAEGLA]
SY: *Dicranum arcticum* Schimp.
SY: *Dicranum molle* (Wils.) Lindb.
Kiaeria starkei (Web. & Mohr) Hag. [KIAESTA]
SY: *Arctoa starkei* (Web. & Mohr) Loeske
SY: *Dicranoweisia subcompacta* Card. & Thér.
SY: *Dicranum starkei* Web. & Mohr
Oncophorus (Brid.) Brid. [ONCOPHO$]
Oncophorus alpestris (Wahlenb.) Lindb. = Cynodontium alpestre [CYNOALP]
Oncophorus glaucescens Lindb. & Arnell = Cynodontium glaucescens [CYNOGLA]
Oncophorus jenneri (Schimp. in Howie) Williams = Cynodontium jenneri [CYNOJEN]
Oncophorus polycarpus (Hedw.) Brid. = Cynodontium polycarpon [CYNOPOL]
Oncophorus polycarpus var. *strumiferus* (Hedw.) Brid. = Cynodontium strumiferum [CYNOSTR]
Oncophorus schisti (Web. & Mohr) Lindb. = Cynodontium schisti [CYNOSCH]
Oncophorus strumulosus (C. Müll. & Kindb.) Britt. in Williams = Cynodontium alpestre [CYNOALP]
Oncophorus tenellus (Bruch & Schimp. in B.S.G.) Williams = Cynodontium tenellum [CYNOTEN]
Oncophorus virens (Hedw.) Brid. [ONCOVIR]
SY: *Cynodontium virens* (Hedw.) Schimp.
SY: *Cynodontium virens* var. *serratum* (B.S.G.) Schimp.
SY: *Dicranum virens* Hedw.
SY: *Oncophorus virens* var. *nigrescens* Williams in Mac.
SY: *Oncophorus virens* var. *serratus* (Bruch & Schimp. in B.S.G.) Braithw.
Oncophorus virens var. *nigrescens* Williams in Mac. = Oncophorus virens [ONCOVIR]
Oncophorus virens var. *serratus* (Bruch & Schimp. in B.S.G.) Braithw. = Oncophorus virens [ONCOVIR]
Oncophorus wahlenbergii Brid. [ONCOWAH] (mountain curved-back moss)
SY: *Cynodontium virens* var. *wahlenbergii* (Brid.) Schimp.
SY: *Cynodontium wahlenbergii* (Brid.) Hartm.

SY: *Cynodontium wahlenbergii* var. *compactum* (B.S.G.) Mac. & Kindb.
SY: *Cynodontium wahlenbergii* var. *gracile* (Broth.) Mönk.
SY: *Dicranum microcarpum* Hook. in Drumm.
SY: *Dicranum virens* var. *wahlenbergii* (Brid.) Hüb.
SY: *Oncophorus wahlenbergii* var. *compactus* (Bruch & Schimp. in B.S.G.) Braithw.
SY: *Oncophorus wahlenbergii* var. *gracilis* (Broth.) Arnell & C. Jens.
Oncophorus wahlenbergii var. *compactus* (Bruch & Schimp. in B.S.G.) Braithw. = Oncophorus wahlenbergii [ONCOWAH]
Oncophorus wahlenbergii var. *gracilis* (Broth.) Arnell & C. Jens. = Oncophorus wahlenbergii [ONCOWAH]
Oreas Brid. [OREAS$]
Oreas martiana (Hoppe & Hornsch. in Hornsch.) Brid. [OREAMAR]
Orthodicranum flagellare (Hedw.) Loeske = Dicranum flagellare [DICRFLA]
Orthodicranum montanum (Hedw.) Loeske = Dicranum montanum [DICRMON]
Paraleucobryum (Lindb.) Loeske [PARALEU$]
Paraleucobryum enerve (Thed. in Hartm.) Loeske [PARAENE]
SY: *Dicranum subulifolium* Kindb. in Mac.
SY: *Paraleucobryum sauteri* (Bruch & Schimp. in B.S.G.) Loeske
Paraleucobryum longifolium (Hedw.) Loeske [PARALON]
SY: *Campylopus canadensis* Kindb.
SY: *Dicranum albicans* var. *denticulatum* Kindb. in Mac. & Kindb.
SY: *Dicranum longifolium* Hedw.
SY: *Dicranum longifolium* var. *subalpinum* Milde
SY: *Dicranum sauteri* var. *pachytrichum* Kindb. in Mac. & Kindb.
SY: *Dicranum serratum* Kindb.
SY: *Paraleucobryum longifolium* ssp. *serratum* (Kindb.) Podp.
SY: *Paraleucobryum longifolium* var. *subalpinum* (Milde) Cas.-Gil
Paraleucobryum longifolium ssp. *serratum* (Kindb.) Podp. = Paraleucobryum longifolium [PARALON]
Paraleucobryum longifolium var. *subalpinum* (Milde) Cas.-Gil = Paraleucobryum longifolium [PARALON]
Paraleucobryum sauteri (Bruch & Schimp. in B.S.G.) Loeske = Paraleucobryum enerve [PARAENE]
Pleuridium axillare (Sm.) Lindb. = Pseudephemerum nitidum [PSEUNIT]
Pseudephemerum (Lindb.) Loeske [PSEUDEP$]
***Pseudephemerum nitidum** (Hedw.) Loeske [PSEUNIT]
SY: *Pleuridium axillare* (Sm.) Lindb.
Rhabdoweisia Bruch & Schimp. in B.S.G. [RHABDOW$]
Rhabdoweisia crispata (With.) Lindb. [RHABCRI]
SY: *Rhabdoweisia denticulata* Bruch & Schimp. in B.S.G.
Rhabdoweisia denticulata Bruch & Schimp. in B.S.G. = Rhabdoweisia crispata [RHABCRI]

DISCELIACEAE (B:M)

Discelium Brid. [DISCELI$]
Discelium nudum (Dicks.) Brid. [DISCNUD]

DITRICHACEAE (B:M)

Ceratodon Brid. [CERATOD$]
Ceratodon columbiae Kindb. = Ceratodon purpureus [CERAPUR]
Ceratodon purpureus (Hedw.) Brid. [CERAPUR] (red roof moss)
SY: *Ceratodon columbiae* Kindb.
SY: *Ceratodon purpureus* f. *aristatus* Aust.
SY: *Mielichhoferia recurvifolium* Kindb.
Ceratodon purpureus f. *aristatus* Aust. = Ceratodon purpureus [CERAPUR]
Didymodon trifarius (Hedw.) Röhl. = Saelania glaucescens [SAELGLA]
Distichium Bruch & Schimp. in B.S.G. [DISTICI$]
Distichium capillaceum (Hedw.) Bruch & Schimp. in B.S.G. [DISTCAP] (erect-fruited iris moss)
SY: *Swartzia montana* Lindb.
Distichium inclinatum (Hedw.) Bruch & Schimp. in B.S.G. [DISTINC]
SY: *Swartzia inclinata* (Hedw.) P. Beauv.
Distichium macounii C. Müll. & Kindb. in Mac. & Kindb. = Ditrichum flexicaule [DITRFLE]
Ditrichum Hampe [DITRICH$]
Ditrichum ambiguum Best [DITRAMB]
SY: *Ditrichum tortuloides* Grout
Ditrichum crispatissimum (C. Müll.) Par. = Ditrichum flexicaule [DITRFLE]
Ditrichum cylindricum (Hedw.) Grout = Trichodon cylindricus [TRICCYL]
Ditrichum flexicaule (Schwaegr.) Hampe [DITRFLE] (slender-stemmed hair moss)
SY: *Distichium macounii* C. Müll. & Kindb. in Mac. & Kindb.
SY: *Ditrichum crispatissimum* (C. Müll.) Par.
SY: *Ditrichum flexicaule* f. *brevifolium* (Kindb.) Grout
SY: *Ditrichum flexicaule* var. *densum* (B.S.G.) Braithw.
SY: *Ditrichum flexicaule* f. *estellae* Hab.
SY: *Ditrichum flexicaule* f. *macounii* Mönk.
SY: *Ditrichum flexicaule* var. *sterile* (De Not.) Limpr.
SY: *Ditrichum giganteum* Williams
SY: *Leptotrichum brevifolium* Kindb. ex Par.
SY: *Leptotrichum brevifolium* var. *densum* Schimp.
Ditrichum flexicaule f. *brevifolium* (Kindb.) Grout = Ditrichum flexicaule [DITRFLE]
Ditrichum flexicaule f. *estellae* Hab. = Ditrichum flexicaule [DITRFLE]
Ditrichum flexicaule f. *macounii* Mönk. = Ditrichum flexicaule [DITRFLE]
Ditrichum flexicaule var. *densum* (B.S.G.) Braithw. = Ditrichum flexicaule [DITRFLE]
Ditrichum flexicaule var. *sterile* (De Not.) Limpr. = Ditrichum flexicaule [DITRFLE]
Ditrichum giganteum Williams = Ditrichum flexicaule [DITRFLE]
Ditrichum glaucescens (Hedw.) Hampe = Saelania glaucescens [SAELGLA]
Ditrichum heteromallum (Hedw.) Britt. [DITRHET]
SY: *Ditrichum homomallum* (Hedw.) Hampe
Ditrichum homomallum (Hedw.) Hampe = Ditrichum heteromallum [DITRHET]
Ditrichum montanum Leib. [DITRMON]
Ditrichum pusillum (Hedw.) Hampe [DITRPUS]
SY: *Ditrichum pusillum* var. *tortile* (Schrad.) Hag.
SY: *Ditrichum tortile* (Schrad.) Brockm.
SY: *Ditrichum tortile* ssp. *pusillum* Hedw.
SY: *Leptotrichum tortile* (Schrad.) Hampe in C. Müll.
SY: *Trichostomum tortile* Schrad.
Ditrichum pusillum var. *tortile* (Schrad.) Hag. = Ditrichum pusillum [DITRPUS]
Ditrichum schimperi (Lesq.) Kuntze [DITRSCH]
Ditrichum tenuifolium Lindb. = Trichodon cylindricus [TRICCYL]
Ditrichum tortile (Schrad.) Brockm. = Ditrichum pusillum [DITRPUS]
Ditrichum tortile ssp. *pusillum* Hedw. = Ditrichum pusillum [DITRPUS]
Ditrichum tortuloides Grout = Ditrichum ambiguum [DITRAMB]
Ditrichum zonatum (Brid.) Kindb. [DITRZON]
Ditrichum zonatum var. **scabrifolium** Dix. [DITRZON1]
Ditrichum zonatum var. **zonatum** [DITRZON0]
Leptotrichum brevifolium Kindb. ex Par. = Ditrichum flexicaule [DITRFLE]
Leptotrichum brevifolium var. *densum* Schimp. = Ditrichum flexicaule [DITRFLE]
Leptotrichum glaucescens (Hedw.) Hampe in Schimp. = Saelania glaucescens [SAELGLA]
Leptotrichum tortile (Schrad.) Hampe in C. Müll. = Ditrichum pusillum [DITRPUS]
Mielichhoferia recurvifolium Kindb. = Ceratodon purpureus [CERAPUR]
Pleuridium Rabenh. [PLEURID$]
Pleuridium alternifolium auct. = Pleuridium subulatum [PLEUSUB]
Pleuridium subulatum (Hedw.) Rabenh. [PLEUSUB]

SY: *Pleuridium alternifolium* auct.
Saelania Lindb. [SAELANI$]
Saelania glaucescens (Hedw.) Broth. in Bomanss. & Broth. [SAELGLA]
SY: *Didymodon trifarius* (Hedw.) Röhl.
SY: *Ditrichum glaucescens* (Hedw.) Hampe
SY: *Leptotrichum glaucescens* (Hedw.) Hampe in Schimp.
Swartzia inclinata (Hedw.) P. Beauv. = Distichium inclinatum [DISTINC]
Swartzia montana Lindb. = Distichium capillaceum [DISTCAP]
Trichodon Schimp. [TRICHOD$]
Trichodon cylindricus (Hedw.) Schimp. [TRICCYL]
SY: *Ditrichum cylindricum* (Hedw.) Grout
SY: *Ditrichum tenuifolium* Lindb.
Trichostomum tortile Schrad. = Ditrichum pusillum [DITRPUS]

ENCALYPTACEAE (B:M)

Bryobrittonia Williams [BRYOBRI$]
Bryobrittonia longipes (Williams) Horton [BRYOLON]
SY: *Bryobrittonia pellucida* Williams
Bryobrittonia pellucida Williams = Bryobrittonia longipes [BRYOLON]
Encalypta Hedw. [ENCALYP$]
Encalypta affinis Hedw. f. in Web. & Mohr [ENCAAFF]
Encalypta affinis ssp. *macounii* (Aust.) Horton = Encalypta affinis var. macounii [ENCAAFF1]
Encalypta affinis var. **affinis** [ENCAAFF0]
SY: *Encalypta apophysata* Nees & Hornsch. in Nees et al.
Encalypta affinis var. **macounii** (Aust.) Crum & Anderson [ENCAAFF1]
SY: *Encalypta affinis* ssp. *macounii* (Aust.) Horton
SY: *Encalypta leiocarpa* Kindb. in Mac.
SY: *Encalypta macounii* Aust.
Encalypta alpina Sm. [ENCAALP]
Encalypta apophysata Nees & Hornsch. = Encalypta affinis var. affinis [ENCAAFF0]
Encalypta brevicolla (Bruch & Schimp. in B.S.G.) Bruch ex Ångstr. [ENCABRE]
SY: *Encalypta labradorica* Kindb.
SY: *Encalypta subbrevicolla* Kindb.
Encalypta brevipes Schljak. [ENCABRV]
Encalypta ciliata Hedw. [ENCACIL]
SY: *Encalypta ciliata* var. *microstoma* Schimp.
SY: *Encalypta laciniata* Lindb.
Encalypta ciliata var. *microstoma* Schimp. = Encalypta ciliata [ENCACIL]
Encalypta intermedia Jur. in Jur. & Milde [ENCAINT]
Encalypta labradorica Kindb. = Encalypta brevicolla [ENCABRE]
Encalypta laciniata Lindb. = Encalypta ciliata [ENCACIL]
Encalypta leiocarpa Kindb. in Mac. = Encalypta affinis var. macounii [ENCAAFF1]
Encalypta longicolla Bruch [ENCALON]
Encalypta longipes Mitt. = Encalypta procera [ENCAPRO]
Encalypta macounii Aust. = Encalypta affinis var. macounii [ENCAAFF1]
Encalypta mutica Hag. [ENCAMUT]
Encalypta procera Bruch [ENCAPRO]
SY: *Encalypta longipes* Mitt.
SY: *Encalypta selwynii* Aust.
Encalypta rhaptocarpa Schwaegr. [ENCARHA] (grooved gnome-cap moss)
SY: *Encalypta rhaptocarpa* var. *leiomitra* Kindb.
SY: *Encalypta rhaptocarpa* var. *subspathulata* (C. Müll. & Kindb. in Mac. & Kindb.) Flow.
SY: *Encalypta subspathulata* C. Müll. & Kindb. in Mac. & Kindb.
SY: *Encalypta vulgaris* var. *rhaptocarpa* (Schwaegr.) Lawt.
Encalypta rhaptocarpa var. *leiomitra* Kindb. = Encalypta rhaptocarpa [ENCARHA]
Encalypta rhaptocarpa var. *subspathulata* (C. Müll. & Kindb. in Mac. & Kindb.) Flow. = Encalypta rhaptocarpa [ENCARHA]
Encalypta selwynii Aust. = Encalypta procera [ENCAPRO]
Encalypta spathulata C. Müll. [ENCASPA]
Encalypta subbrevicolla Kindb. = Encalypta brevicolla [ENCABRE]
Encalypta subspathulata C. Müll. & Kindb. in Mac. & Kindb. = Encalypta rhaptocarpa [ENCARHA]
Encalypta vulgaris Hedw. [ENCAVUL]
SY: *Encalypta vulgaris* var. *apiculata* Wahlenb.
SY: *Encalypta vulgaris* var. *mutica* Brid.
Encalypta vulgaris var. *apiculata* Wahlenb. = Encalypta vulgaris [ENCAVUL]
Encalypta vulgaris var. *mutica* Brid. = Encalypta vulgaris [ENCAVUL]
Encalypta vulgaris var. *rhaptocarpa* (Schwaegr.) Lawt. = Encalypta rhaptocarpa [ENCARHA]

ENTODONTACEAE (B:M)

Entodon C. Müll. [ENTODON$]
Entodon concinnus (De Not.) Par. [ENTOCON]
SY: *Entodon orthocarpus* (Brid.) Lindb.
Entodon orthocarpus (Brid.) Lindb. = Entodon concinnus [ENTOCON]

EPHEMERACEAE (B:M)

Micromitrium Aust. [MICROMI$]
***Micromitrium tenerum** (Bruch & Schimp. in B.S.G.) Crosby [MICRTEN]
SY: *Nanomitrium tenerum* (Bruch & Schimp. in B.S.G.) Lindb.
Nanomitrium tenerum (Bruch & Schimp. in B.S.G.) Lindb. = Micromitrium tenerum [MICRTEN]

FABRONIACEAE (B:M)

Fabronia Raddi [FABRONI$]
Fabronia pusilla Raddi [FABRPUS]

FISSIDENTACEAE (B:M)

Conomitrium julianum (Cand.) Mont. = Fissidens fontanus [FISSFON]
Fissidens Hedw. [FISSIDE$]
Fissidens adianthoides Hedw. [FISSADI]
SY: *Fissidens adianthoides* var. *immarginatus* Lindb. ex Lesq. & James
Fissidens adianthoides var. *immarginatus* Lindb. ex Lesq. & James = Fissidens adianthoides [FISSADI]
Fissidens andersonii Grout = Fissidens bryoides [FISSBRY]
Fissidens aphelotaxifolius Purs. [FISSAPH]
Fissidens bryoides Hedw. [FISSBRY]
SY: *Fissidens andersonii* Grout
SY: *Fissidens bryoides* var. *brevifolius* (Ren. & Card.) Wijk & Marg.
SY: *Fissidens bryoides* var. *incurvus* (Röhl.) Hüb.
SY: *Fissidens bryoides* var. *viridulus* (Sw.) Broth.
SY: *Fissidens exiguus* Sull.
SY: *Fissidens exiguus* var. *falcatulus* (Ren. & Card.) Grout
SY: *Fissidens falcatulus* Ren. & Card.
SY: *Fissidens incurvus* Starke ex Röhl.
SY: *Fissidens incurvus* var. *brevifolius* Ren. & Card.
SY: *Fissidens longifolius* Brid.
SY: *Fissidens minutulus* Sull.
SY: *Fissidens pusillus* (Wils.) Milde
SY: *Fissidens repandus* Wils.
SY: *Fissidens tamarindifolius* (Turn.) Brid.
SY: *Fissidens texanus* Lesq. & James
SY: *Fissidens tortilis* Hampe & C. Müll.
SY: *Fissidens viridulus* (Sw.) Wahlenb.
SY: *Fissidens viridulus* var. *brevifolius* (Ren. & Card.) Grout
SY: *Fissidens viridulus* var. *tamarindifolius* (Turn.) Grout
SY: *Fissidens viridulus* var. *texanus* (Lesq. & James) Grout

Fissidens bryoides var. *brevifolius* (Ren. & Card.) Wijk & Marg. = Fissidens bryoides [FISSBRY]
Fissidens bryoides var. *incurvus* (Röhl.) Hüb. = Fissidens bryoides [FISSBRY]
Fissidens bryoides var. *viridulus* (Sw.) Broth. = Fissidens bryoides [FISSBRY]
Fissidens debilis Schwaegr. = Fissidens fontanus [FISSFON]
Fissidens exiguus Sull. = Fissidens bryoides [FISSBRY]
Fissidens exiguus var. *falcatulus* (Ren. & Card.) Grout = Fissidens bryoides [FISSBRY]
Fissidens falcatulus Ren. & Card. = Fissidens bryoides [FISSBRY]
Fissidens fontanus (B. Pyl.) Steud. [FISSFON]
SY: *Conomitrium julianum* (Cand.) Mont.
SY: *Fissidens debilis* Schwaegr.
SY: *Fissidens julianus* (Savi ex DC.) Schimp.
SY: *Fissidens julianus* var. *americanus* Kindb.
Fissidens grandifrons Brid. [FISSGRA]
Fissidens incurvus Starke ex Röhl. = Fissidens bryoides [FISSBRY]
Fissidens incurvus var. *brevifolius* Ren. & Card. = Fissidens bryoides [FISSBRY]
Fissidens julianus (Savi ex DC.) Schimp. = Fissidens fontanus [FISSFON]
Fissidens julianus var. *americanus* Kindb. = Fissidens fontanus [FISSFON]
Fissidens limbatus Sull. [FISSLIM]
Fissidens longifolius Brid. = Fissidens bryoides [FISSBRY]
Fissidens minutulus Sull. = Fissidens bryoides [FISSBRY]
Fissidens osmundioides Hedw. [FISSOSM]
Fissidens pauperculus Howe [FISSPAU]
Fissidens pusillus (Wils.) Milde = Fissidens bryoides [FISSBRY]
Fissidens repandus Wils. = Fissidens bryoides [FISSBRY]
Fissidens rufulus auct. = Fissidens ventricosus [FISSVEN]
Fissidens tamarindifolius (Turn.) Brid. = Fissidens bryoides [FISSBRY]
Fissidens texanus Lesq. & James = Fissidens bryoides [FISSBRY]
Fissidens tortilis Hampe & C. Müll. = Fissidens bryoides [FISSBRY]
Fissidens ventricosus Lesq. [FISSVEN]
SY: *Fissidens rufulus* auct.
Fissidens viridulus (Sw.) Wahlenb. = Fissidens bryoides [FISSBRY]
Fissidens viridulus var. *brevifolius* (Ren. & Card.) Grout = Fissidens bryoides [FISSBRY]
Fissidens viridulus var. *tamarindifolius* (Turn.) Grout = Fissidens bryoides [FISSBRY]
Fissidens viridulus var. *texanus* (Lesq. & James) Grout = Fissidens bryoides [FISSBRY]

FONTINALACEAE (B:M)

Dichelyma Myr. [DICHELY$]
Dichelyma cylindricarpum Aust. = Dichelyma uncinatum [DICHUNC]
Dichelyma falcatum (Hedw.) Myr. [DICHFAL]
Dichelyma falcatum var. *uncinatum* (Mitt.) Lawt. = Dichelyma uncinatum [DICHUNC]
Dichelyma uncinatum Mitt. [DICHUNC]
SY: *Dichelyma cylindricarpum* Aust.
SY: *Dichelyma falcatum* var. *uncinatum* (Mitt.) Lawt.
Fontinalis Hedw. [FONTINA$]
Fontinalis allenii Card. in Nichols = Fontinalis antipyretica var. oregonensis [FONTANT3]
Fontinalis antipyretica Hedw. [FONTANT] (common water moss)
Fontinalis antipyretica var. **antipyretica** [FONTANT0]
SY: *Fontinalis antipyretica* var. *mollis* (C. Müll.) Welch in Grout
SY: *Fontinalis antipyretica* var. *patula* (Card.) Welch
SY: *Fontinalis patula* Card.
Fontinalis antipyretica var. **gigantea** (Sull.) Sull. [FONTANT1]
SY: *Fontinalis gigantea* Sull. in Sull. & Lesq.
Fontinalis antipyretica var. *mollis* (C. Müll.) Welch in Grout = Fontinalis antipyretica var. antipyretica [FONTANT0]
Fontinalis antipyretica var. **oregonensis** Ren. & Card. [FONTANT3]
SY: *Fontinalis allenii* Card. in Nichols
SY: *Fontinalis chrysophylla* Card.
SY: *Fontinalis howellii* Ren. & Card.
SY: *Fontinalis kindbergii* Ren. & Card.
Fontinalis antipyretica var. *patula* (Card.) Welch = Fontinalis antipyretica var. antipyretica [FONTANT0]
Fontinalis chrysophylla Card. = Fontinalis antipyretica var. oregonensis [FONTANT3]
Fontinalis gigantea Sull. in Sull. & Lesq. = Fontinalis antipyretica var. gigantea [FONTANT1]
Fontinalis howellii Ren. & Card. = Fontinalis antipyretica var. oregonensis [FONTANT3]
Fontinalis hypnoides Hartm. [FONTHYP]
Fontinalis kindbergii Ren. & Card. = Fontinalis antipyretica var. oregonensis [FONTANT3]
Fontinalis neomexicana Sull. & Lesq. [FONTNEO]
Fontinalis patula Card. = Fontinalis antipyretica var. antipyretica [FONTANT0]

FUNARIACEAE (B:M)

Aphanorrhegma patens (Hedw.) Lindb. = Physcomitrella patens [PHYSPAT]
Bryum pyriforme (Hedw.) With. = Physcomitrium pyriforme [PHYSPYR]
Entosthodon Schwaegr. [ENTOSTH$]
Entosthodon fascicularis (Hedw.) C. Müll. [ENTOFAS]
SY: *Entosthodon leibergii* Britt.
Entosthodon leibergii Britt. = Entosthodon fascicularis [ENTOFAS]
Entosthodon muhlenbergii (Turn.) Fife = Funaria muhlenbergii [FUNAMUH]
Entosthodon rubiginosus (Williams) Grout [ENTORUB]
Funaria Hedw. [FUNARIA$]
Funaria calcarea Wahlenb. = Funaria muhlenbergii [FUNAMUH]
Funaria hygrometrica Hedw. [FUNAHYG] (cord moss)
SY: *Funaria hygrometrica* var. *convoluta* (Hampe) Grout
SY: *Funaria hygrometrica* var. *patula* Bruch & Schimp. in B.S.G.
SY: *Funaria hygrometrica* var. *utahensis* Grout
Funaria hygrometrica var. *convoluta* (Hampe) Grout = Funaria hygrometrica [FUNAHYG]
Funaria hygrometrica var. *patula* Bruch & Schimp. in B.S.G. = Funaria hygrometrica [FUNAHYG]
Funaria hygrometrica var. *utahensis* Grout = Funaria hygrometrica [FUNAHYG]
Funaria mediterranea Lindb. = Funaria muhlenbergii [FUNAMUH]
Funaria muhlenbergii Turn. [FUNAMUH]
SY: *Entosthodon muhlenbergii* (Turn.) Fife
SY: *Funaria calcarea* Wahlenb.
SY: *Funaria mediterranea* Lindb.
SY: *Funaria muhlenbergii* var. *lineata* Grout
SY: *Funaria muhlenbergii* var. *patula* (Bruch & Schimp. in B.S.G.) Schimp.
Funaria muhlenbergii var. *lineata* Grout = Funaria muhlenbergii [FUNAMUH]
Funaria muhlenbergii var. *patula* (Bruch & Schimp. in B.S.G.) Schimp. = Funaria muhlenbergii [FUNAMUH]
Physcomitrella Bruch & Schimp. in B.S.G. [PHYSCOM$]
Physcomitrella patens (Hedw.) Bruch & Schimp. in B.S.G. [PHYSPAT]
SY: *Aphanorrhegma patens* (Hedw.) Lindb.
Physcomitrium (Brid.) Brid. [PHYSCOI$]
Physcomitrium australe Britt. = Physcomitrium pyriforme [PHYSPYR]
Physcomitrium californicum Britt. = Physcomitrium pyriforme [PHYSPYR]
Physcomitrium drummondii Britt. = Physcomitrium pyriforme [PHYSPYR]
Physcomitrium immersum Sull. [PHYSIMM]
Physcomitrium kellermanii Britt. = Physcomitrium pyriforme [PHYSPYR]
Physcomitrium kellermanii var. *drummondii* (Britt.) Grout = Physcomitrium pyriforme [PHYSPYR]
Physcomitrium megalocarpum Kindb. = Physcomitrium pyriforme [PHYSPYR]

Physcomitrium megalocarpum var. *californicum* (Britt.) Grout = Physcomitrium pyriforme [PHYSPYR]
Physcomitrium pyriforme (Hedw.) Hampe [PHYSPYR]
SY: *Bryum pyriforme* (Hedw.) With.
SY: *Physcomitrium australe* Britt.
SY: *Physcomitrium californicum* Britt.
SY: *Physcomitrium drummondii* Britt.
SY: *Physcomitrium kellermanii* Britt.
SY: *Physcomitrium kellermanii* var. *drummondii* (Britt.) Grout
SY: *Physcomitrium megalocarpum* Kindb.
SY: *Physcomitrium megalocarpum* var. *californicum* (Britt.) Grout
SY: *Physcomitrium pyriforme* var. *serratum* (Ren. & Card.) Crum & Anderson
SY: *Physcomitrium strangulatum* Kindb.
SY: *Physcomitrium turbinatum* C. Müll. ex Lesq. & James
SY: *Physcomitrium turbinatum* (Michx.) Britt.
SY: *Physcomitrium turbinatum* f. *australe* (Britt.) Grout
SY: *Physcomitrium turbinatum* var. *langloisii* (Ren. & Card.) Britt.
Physcomitrium pyriforme var. *serratum* (Ren. & Card.) Crum & Anderson = Physcomitrium pyriforme [PHYSPYR]
Physcomitrium strangulatum Kindb. = Physcomitrium pyriforme [PHYSPYR]
Physcomitrium turbinatum C. Müll. ex Lesq. & James = Physcomitrium pyriforme [PHYSPYR]
Physcomitrium turbinatum (Michx.) Britt. = Physcomitrium pyriforme [PHYSPYR]
Physcomitrium turbinatum f. *australe* (Britt.) Grout = Physcomitrium pyriforme [PHYSPYR]
Physcomitrium turbinatum var. *langloisii* (Ren. & Card.) Britt. = Physcomitrium pyriforme [PHYSPYR]

GEOCALYCACEAE (B:H)

Geocalyx Nees [GEOCALY$]
Geocalyx graveolens (Schrad.) Nees [GEOCGRA]
Harpanthus Nees [HARPANT$]
Harpanthus flotovianus (Nees) Nees [HARPFLO]

GRIMMIACEAE (B:M)

Coscinodon Spreng. [COSCINO$]
Coscinodon calyptratus (Hook. in Drumm.) C. Jens. ex Kindb. [COSCCAL] (steppe mouse-moss)
SY: *Grimmia calyptrata* Hook. in Drumm.
Coscinodon cribrosus (Hedw.) Spruce [COSCCRI]
SY: *Coscinodon pulvinatus* Spreng.
SY: *Grimmia cribrosa* Hedw.
Coscinodon pulvinatus Spreng. = Coscinodon cribrosus [COSCCRI]
Dryptodon Brid. [DRYPTOD$]
Dryptodon patens (Hedw.) Brid. [DRYPPAT]
SY: *Grimmia curvata* (Brid.) De Sloover
SY: *Grimmia patens* (Hedw.) Bruch & Schimp. in B.S.G.
SY: *Racomitrium patens* (Hedw.) Hüb.
Grimmia Hedw. [GRIMMIA$]
Grimmia affinis Hoppe & Hornsch. ex Hornsch. [GRIMAFF]
SY: *Grimmia arctophila* ssp. *labradorica* Kindb.
SY: *Grimmia ovalis* f. *affinis* (Hoppe & Hornsch. ex Hornsch.) G. Jones in Grout
Grimmia agassizii (Sull. & Lesq. in Sull.) Jaeg. = Schistidium agassizii [SCHIAGA]
Grimmia alpestris (Web. & Mohr) Schleich. = Grimmia tenerrima [GRIMTEE]
Grimmia alpestris var. *caespiticia* (Brid.) G. Jones in Grout = Grimmia tenerrima [GRIMTEE]
Grimmia alpestris var. *holzingeri* (Card. & Thér.) G. Jones in Grout = Grimmia tenerrima [GRIMTEE]
Grimmia alpestris var. *manniae* (C. Müll.) G. Jones in Grout = Grimmia tenerrima [GRIMTEE]
Grimmia alpestris var. *microstoma* Bruch & Schimp. in B.S.G. = Grimmia tenerrima [GRIMTEE]
Grimmia alpicola Sw. ex Hedw. = Schistidium agassizii [SCHIAGA]
Grimmia alpicola var. *dupretii* (Thér.) Crum = Schistidium apocarpum [SCHIAPO]
Grimmia alpicola var. *papillosa* (G. Jones) Hab. = Schistidium rivulare [SCHIRIV]
Grimmia alpicola var. *rivularis* f. *papillosa* G. Jones in Grout = Schistidium rivulare [SCHIRIV]
Grimmia angusta (Hag.) Broth. = Schistidium agassizii [SCHIAGA]
Grimmia anodon Bruch & Schimp. in B.S.G. [GRIMANO]
Grimmia anomala Hampe ex Schimp. [GRIMANM]
SY: *Grimmia hartmanii* var. *anomala* (Hampe ex Schimp.) Mönk.
Grimmia apocarpa Hedw. = Schistidium apocarpum [SCHIAPO]
Grimmia apocarpa f. *canadensis* Loeske = Schistidium apocarpum [SCHIAPO]
Grimmia apocarpa f. *epilosa* Warnst. = Schistidium apocarpum [SCHIAPO]
Grimmia apocarpa var. *alpicola* (Hedw.) Röhl. = Schistidium agassizii [SCHIAGA]
Grimmia apocarpa var. *ambigua* (Sull.) G. Jones in Grout = Schistidium apocarpum [SCHIAPO]
Grimmia apocarpa var. *atrofusca* (Schimp.) Husn. = Schistidium apocarpum [SCHIAPO]
Grimmia apocarpa var. *brunnescens* (Limpr.) Mönk. = Schistidium apocarpum [SCHIAPO]
Grimmia apocarpa var. *canadensis* Dupret = Schistidium apocarpum [SCHIAPO]
Grimmia apocarpa var. *conferta* (Funck) Spreng. = Schistidium apocarpum [SCHIAPO]
Grimmia apocarpa var. *conferta* f. *obtusifolia* (Schimp.) Mönk. = Schistidium apocarpum [SCHIAPO]
Grimmia apocarpa var. *dupretii* (Thér.) Sayre = Schistidium apocarpum [SCHIAPO]
Grimmia apocarpa var. *gracilis* Web. & Mohr = Schistidium apocarpum [SCHIAPO]
Grimmia apocarpa var. *nigrescens* Mol. = Schistidium apocarpum [SCHIAPO]
Grimmia apocarpa var. *obscuroviridis* Crum = Schistidium apocarpum [SCHIAPO]
Grimmia apocarpa var. *rivularis* (Brid.) Nees & Hornsch. = Schistidium rivulare [SCHIRIV]
Grimmia apocarpa var. *stricta* (Turn.) Hook. & Tayl. = Schistidium apocarpum [SCHIAPO]
Grimmia apocarpa var. *tenerrima* Nees & Hornsch. = Schistidium apocarpum [SCHIAPO]
Grimmia arctophila Kindb. = Grimmia laevigata [GRIMLAE]
Grimmia arctophila ssp. *labradorica* Kindb. = Grimmia affinis [GRIMAFF]
Grimmia arcuatifolia Kindb. = Racomitrium heterostichum [RACOHET]
Grimmia atricha C. Müll. & Kindb. in Mac. & Kindb. = Schistidium apocarpum [SCHIAPO]
Grimmia brittoniae Williams [GRIMBRI]
Grimmia calyptrata Hook. in Drumm. = Coscinodon calyptratus [COSCCAL]
Grimmia canadensis Kindb. = Grimmia trichophylla [GRIMTRI]
Grimmia chloroblasta Kindb. in Mac. & Kindb. = Schistidium heterophyllum [SCHIHET]
Grimmia coloradensis Aust. = Schistidium apocarpum [SCHIAPO]
Grimmia commutata Hüb. = Grimmia ovalis [GRIMOVA]
Grimmia conferta Funck = Schistidium apocarpum [SCHIAPO]
Grimmia conferta var. *compacta* Lesq. & James = Schistidium apocarpum [SCHIAPO]
Grimmia crassinervia (C. Müll.) Mac. & Kindb. = Schistidium maritimum [SCHIMAR]
Grimmia cribrosa Hedw. = Coscinodon cribrosus [COSCCRI]
Grimmia curvata (Brid.) De Sloover = Dryptodon patens [DRYPPAT]
Grimmia decipiens auct. = Grimmia trichophylla [GRIMTRI]
Grimmia densa Kindb. = Grimmia trichophylla [GRIMTRI]
Grimmia depilata Kindb. in Mac. & Kindb. = Grimmia trichophylla [GRIMTRI]
Grimmia donniana Sm. [GRIMDON]
SY: *Grimmia donniana* var. *triformis* (Carest. & De Not. in De Not.) Loeske
SY: *Grimmia manniae* C. Müll.

SY: *Grimmia microtricha* C. Müll. & Kindb. in Mac. & Kindb.
SY: *Grimmia nivalis* Kindb.
SY: *Grimmia tenella* (C. Müll.) Mac. & Kindb.
SY: *Grimmia velutina* Kindb.
Grimmia donniana var. *holzingeri* (Card. & Thér.) Wijk & Marg. = Grimmia tenerrima [GRIMTEE]
Grimmia donniana var. *manniae* (C. Müll.) Wijk & Marg. = Grimmia tenerrima [GRIMTEE]
Grimmia donniana var. *triformis* (Carest. & De Not. in De Not.) Loeske = Grimmia donniana [GRIMDON]
Grimmia dupretii Thér. = Schistidium apocarpum [SCHIAPO]
Grimmia elata Kindb. = Grimmia elatior [GRIMELA]
Grimmia elatior Bruch ex Bals. & De Not. [GRIMELA]
SY: *Grimmia elata* Kindb.
Grimmia elongata Kaulf. in Sturm [GRIMELO]
Grimmia glauca auct. = Grimmia laevigata [GRIMLAE]
Grimmia gracilis (Web. & Mohr ex Nees et al.) Schleich. ex Limpr. = Schistidium apocarpum [SCHIAPO]
Grimmia hartmanii var. *anomala* (Hampe ex Schimp.) Mönk. = Grimmia anomala [GRIMANM]
Grimmia heterophylla Kindb. in Mac. & Kindb. = Schistidium heterophyllum [SCHIHET]
Grimmia holzingeri Card. & Thér. = Grimmia tenerrima [GRIMTEE]
Grimmia laevigata (Brid.) Brid. [GRIMLAE]
SY: *Grimmia arctophila* Kindb.
SY: *Grimmia sarcocalyx* Kindb. in Mac. & Kindb.
SY: *Grimmia glauca* auct.
Grimmia leibergii Par. = Grimmia trichophylla [GRIMTRI]
Grimmia manniae C. Müll. = Grimmia donniana [GRIMDON]
Grimmia maritima Turn. = Schistidium maritimum [SCHIMAR]
Grimmia microtricha C. Müll. & Kindb. in Mac. & Kindb. = Grimmia donniana [GRIMDON]
Grimmia mollis Bruch & Schimp. in B.S.G. [GRIMMOL]
SY: *Hydrogrimmia mollis* (Bruch & Schimp. in B.S.G.) Loeske
Grimmia montana Bruch & Schimp. in B.S.G. [GRIMMON]
Grimmia nivalis Kindb. = Grimmia donniana [GRIMDON]
Grimmia ovalis (Hedw.) Lindb. [GRIMOVA]
SY: *Grimmia commutata* Hüb.
SY: *Grimmia ovata* Web. & Mohr
Grimmia ovalis f. *affinis* (Hoppe & Hornsch. ex Hornsch.) G. Jones in Grout = Grimmia affinis [GRIMAFF]
Grimmia ovata Web. & Mohr = Grimmia ovalis [GRIMOVA]
Grimmia patens (Hedw.) Bruch & Schimp. in B.S.G. = Dryptodon patens [DRYPPAT]
Grimmia plagiopodia Hedw. [GRIMPLA]
Grimmia plagiopodia ssp. *brandegei* (Aust.) Kindb. = Schistidium apocarpum [SCHIAPO]
Grimmia plagiopodia ssp. *pilifera* Lesq. & James = Schistidium apocarpum [SCHIAPO]
Grimmia procera Kindb. = Grimmia trichophylla [GRIMTRI]
Grimmia prolifera C. Müll. & Kindb. in Mac. & Kindb. = Grimmia torquata [GRIMTOR]
Grimmia pseudorivularis ssp. *lancifolia* Kindb. = Schistidium rivulare [SCHIRIV]
Grimmia pulvinata (Hedw.) Sm. [GRIMPUL]
Grimmia sarcocalyx Kindb. in Mac. & Kindb. = Grimmia laevigata [GRIMLAE]
Grimmia sessitana var. *subsulcata* (Limpr.) Breidl. = Grimmia tenerrima [GRIMTEE]
Grimmia stricta Turn. = Schistidium apocarpum [SCHIAPO]
Grimmia subpapillinervis Kindb. = Grimmia tenerrima [GRIMTEE]
Grimmia tenella (C. Müll.) Mac. & Kindb. = Grimmia donniana [GRIMDON]
Grimmia tenera Zett. = Schistidium tenerum [SCHITEN]
Grimmia tenerrima Ren. & Card. [GRIMTEE]
SY: *Grimmia alpestris* (Web. & Mohr) Schleich. ex Nees
SY: *Grimmia alpestris* var. *caespiticia* (Brid.) G. Jones in Grout
SY: *Grimmia alpestris* var. *holzingeri* (Card. & Thér.) G. Jones in Grout
SY: *Grimmia alpestris* var. *manniae* (C. Müll.) G. Jones in Grout
SY: *Grimmia alpestris* var. *microstoma* Bruch & Schimp. in B.S.G.
SY: *Grimmia donniana* var. *holzingeri* (Card. & Thér.) Wijk & Marg.
SY: *Grimmia donniana* var. *manniae* (C. Müll.) Wijk & Marg.
SY: *Grimmia holzingeri* Card. & Thér.
SY: *Grimmia sessitana* var. *subsulcata* (Limpr.) Breidl.
SY: *Grimmia subpapillinervis* Kindb.
SY: *Schistidium tenerrimum* (Ren. & Card.) G. Roth
Grimmia tenuicaulis Williams = Schistidium tenerum [SCHITEN]
Grimmia teretinervis Limpr. [GRIMTER]
SY: *Schistidium teretinerve* (Limpr.) Limpr.
Grimmia torquata Hornsch. in Grev. [GRIMTOR]
SY: *Grimmia prolifera* C. Müll. & Kindb. in Mac. & Kindb.
SY: *Grimmia tortifolia* Kindb.
SY: *Grimmia tortifolia* ssp. *pellucida* Kindb. in Röll
Grimmia tortifolia Kindb. = Grimmia torquata [GRIMTOR]
Grimmia tortifolia ssp. *pellucida* Kindb. in Röll = Grimmia torquata [GRIMTOR]
Grimmia trichophylla Grev. [GRIMTRI]
SY: *Grimmia canadensis* Kindb.
SY: *Grimmia densa* Kindb.
SY: *Grimmia depilata* Kindb. in Mac. & Kindb.
SY: *Grimmia leibergii* Par.
SY: *Grimmia procera* Kindb.
SY: *Grimmia trichophylla* var. *brachycarpa* De Not.
SY: *Grimmia trichophylla* var. *meridionalis* C. Müll.
SY: *Grimmia decipiens* auct.
Grimmia trichophylla var. *brachycarpa* De Not. = Grimmia trichophylla [GRIMTRI]
Grimmia trichophylla var. *meridionalis* C. Müll. = Grimmia trichophylla [GRIMTRI]
Grimmia unicolor Hook. in Grev. [GRIMUNI]
Grimmia velutina Kindb. = Grimmia donniana [GRIMDON]
Hydrogrimmia mollis (Bruch & Schimp. in B.S.G.) Loeske = Grimmia mollis [GRIMMOL]
Racomitrium Brid. [RACOMIT$]
Racomitrium aciculare (Hedw.) Brid. [RACOACI] (black-tufted rock moss)
SY: *Racomitrium neevii* (C. Müll.) Watts
Racomitrium affine (Schleich. ex Web. & Mohr) Lindb. [RACOAFF]
SY: *Racomitrium heterostichum* var. *affine* (Schleich. ex Web. & Mohr) Lesq.
SY: *Racomitrium heterostichum* var. *alopecurum* Hüb.
SY: *Racomitrium heterostichum* ssp. *micropoides* (Kindb.) Kindb.
Racomitrium alternuatum C. Müll. & Kindb. in Mac. & Kindb. = Racomitrium macounii [RACOMAC]
Racomitrium aquaticum (Brid. ex Schrad.) Brid. [RACOAQU]
SY: *Racomitrium protensum* (Braun) Hüb.
Racomitrium brevipes Kindb. in Mac. [RACOBRE]
SY: *Racomitrium sudeticum* f. *brevipes* (Kindb. in Mac.) Lawt.
Racomitrium canescens (Hedw.) Brid. [RACOCAN] (roadside rock moss)
Racomitrium canescens f. *epilosum* (H. Müll. ex Milde) G. Jones in Grout = Racomitrium ericoides [RACOERI]
Racomitrium canescens f. *ericoides* (Web. ex Brid.) Mönk. = Racomitrium ericoides [RACOERI]
Racomitrium canescens ssp. *latifolium* (C. Jens. in Lange & C. Jens.) Frisv. = Racomitrium canescens var. latifolium [RACOCAN1]
Racomitrium canescens var. **canescens** [RACOCAN0]
Racomitrium canescens var. *epilosum* H. Müll. ex Milde = Racomitrium ericoides [RACOERI]
Racomitrium canescens var. *ericoides* (Hedw.) Hampe = Racomitrium ericoides [RACOERI]
Racomitrium canescens var. **latifolium** C. Jens. [RACOCAN1]
SY: *Racomitrium canescens* ssp. *latifolium* (C. Jens. in Lange & C. Jens.) Frisv.
Racomitrium canescens var. *muticum* (Kindb. in Mac.) Mac. & Kindb. = Racomitrium muticum [RACOMUT]
Racomitrium canescens var. *strictum* Schlieph. in Limpr. = Racomitrium ericoides [RACOERI]
Racomitrium elongatum Ehrh. ex Frisv. [RACOELO]
Racomitrium ericoides (Web. ex Brid.) Brid. [RACOERI] (shaggy yellow sand moss)

SY: *Racomitrium canescens* f. *epilosum* (H. Müll. ex Milde) G. Jones in Grout
SY: *Racomitrium canescens* var. *epilosum* H. Müll. ex Milde
SY: *Racomitrium canescens* f. *ericoides* (Web. ex Brid.) Mönk.
SY: *Racomitrium canescens* var. *ericoides* (Hedw.) Hampe
SY: *Racomitrium canescens* var. *strictum* Schlieph. in Limpr.
Racomitrium fasciculare (Hedw.) Brid. [RACOFAS]
SY: *Racomitrium palmeri* (Kindb.) Kindb.
SY: *Racomitrium tenuinerve* Kindb.
Racomitrium heterostichum (Hedw.) Brid. [RACOHET] (yellow-green rock moss)
SY: *Grimmia arcuatifolia* Kindb.
SY: *Racomitrium heterostichum* var. *gracilescens* Bruch & Schimp. in B.S.G.
Racomitrium heterostichum ssp. *micropoides* (Kindb.) Kindb. = Racomitrium affine [RACOAFF]
Racomitrium heterostichum var. *affine* (Schleich. ex Web. & Mohr) Lesq. = Racomitrium affine [RACOAFF]
Racomitrium heterostichum var. *alopecurum* Hüb. = Racomitrium affine [RACOAFF]
Racomitrium heterostichum var. *gracilescens* Bruch & Schimp. in B.S.G. = Racomitrium heterostichum [RACOHET]
Racomitrium heterostichum var. *macounii* (Kindb. in Mac.) G. Jones in Grout = Racomitrium macounii [RACOMAC]
Racomitrium heterostichum var. *microcarpon* (Hedw.) Boul. = Racomitrium microcarpon [RACOMIC]
Racomitrium heterostichum var. *occidentale* Ren. & Card. = Racomitrium occidentale [RACOOCC]
Racomitrium heterostichum var. *ramulosum* (Lindb.) Corb. = Racomitrium microcarpon [RACOMIC]
Racomitrium heterostichum var. *sudeticum* (Funck) Bauer = Racomitrium sudeticum [RACOSUD]
Racomitrium hypnoides Lindb. = Racomitrium lanuginosum [RACOLAN]
Racomitrium lanuginosum (Hedw.) Brid. [RACOLAN] (hoary rock moss)
SY: *Racomitrium hypnoides* Lindb.
SY: *Racomitrium lanuginosum* var. *subimberbe* (Hartm.) Lindb.
Racomitrium lanuginosum var. *subimberbe* (Hartm.) Lindb. = Racomitrium lanuginosum [RACOLAN]
Racomitrium lawtonae Irel. [RACOLAW]
Racomitrium macounii Kindb. in Mac. [RACOMAC]
SY: *Racomitrium alternuatum* C. Müll. & Kindb. in Mac. & Kindb.
SY: *Racomitrium heterostichum* var. *macounii* (Kindb. in Mac.) G. Jones in Grout
SY: *Racomitrium macounii* ssp. *alpinum* (Lawt.) Frisv.
SY: *Racomitrium robustifolium* Kindb.
SY: *Racomitrium sudeticum* f. *alpinum* Lawt.
SY: *Racomitrium sudeticum* f. *americanum* Lawt.
SY: *Racomitrium sudeticum* f. *macounii* (Kindb. in Mac.) Lawt.
Racomitrium macounii ssp. *alpinum* (Lawt.) Frisv. = Racomitrium macounii [RACOMAC]
Racomitrium microcarpon (Hedw.) Brid. [RACOMIC]
SY: *Racomitrium heterostichum* var. *microcarpon* (Hedw.) Boul.
SY: *Racomitrium heterostichum* var. *ramulosum* (Lindb.) Corb.
SY: *Racomitrium ramulosum* Lindb.
Racomitrium muticum (Kindb. in Mac.) Frisv. [RACOMUT]
SY: *Racomitrium canescens* var. *muticum* (Kindb. in Mac.) Mac. & Kindb.
Racomitrium neevii (C. Müll.) Watts = Racomitrium aciculare [RACOACI]
Racomitrium obesum Frisv. [RACOOBE]
Racomitrium occidentale (Ren. & Card.) Ren. & Card. [RACOOCC]
SY: *Racomitrium heterostichum* var. *occidentale* Ren. & Card.
Racomitrium pacificum Irel. & Spence [RACOPAC]
Racomitrium palmeri (Kindb.) Kindb. = Racomitrium fasciculare [RACOFAS]
Racomitrium panschii (C. Müll.) Kindb. [RACOPAN]
Racomitrium patens (Hedw.) Hüb. = Dryptodon patens [DRYPPAT]
Racomitrium protensum (Braun) Hüb. = Racomitrium aquaticum [RACOAQU]
Racomitrium pygmaeum Frisv. [RACOPYG]
Racomitrium ramulosum Lindb. = Racomitrium microcarpon [RACOMIC]
Racomitrium robustifolium Kindb. = Racomitrium macounii [RACOMAC]
Racomitrium sudeticum (Funck) Bruch & Schimp. in B.S.G. [RACOSUD]
SY: *Racomitrium heterostichum* var. *sudeticum* (Funck) Bauer
SY: *Racomitrium sudeticum* var. *validius* Jur.
Racomitrium sudeticum f. *alpinum* Lawt. = Racomitrium macounii [RACOMAC]
Racomitrium sudeticum f. *americanum* Lawt. = Racomitrium macounii [RACOMAC]
Racomitrium sudeticum f. *brevipes* (Kindb. in Mac.) Lawt. = Racomitrium brevipes [RACOBRE]
Racomitrium sudeticum f. *macounii* (Kindb. in Mac.) Lawt. = Racomitrium macounii [RACOMAC]
Racomitrium sudeticum var. *validius* Jur. = Racomitrium sudeticum [RACOSUD]
Racomitrium tenuinerve Kindb. = Racomitrium fasciculare [RACOFAS]
Racomitrium varium (Mitt.) Jaeg. [RACOVAR]
Rhacomitrium Brid. =Racomitrium [RACOMIT$]
Schistidium Brid. [SCHISTI$]
Schistidium agassizii Sull. & Lesq. in Sull. [SCHIAGA]
SY: *Grimmia agassizii* (Sull. & Lesq. in Sull.) Jaeg.
SY: *Grimmia alpicola* Sw. ex Hedw.
SY: *Grimmia angusta* (Hag.) Broth.
SY: *Grimmia apocarpa* var. *alpicola* (Hedw.) Röhl.
SY: *Schistidium alpicola* (Hedw.) Limpr.
Schistidium alpicola (Hedw.) Limpr. = Schistidium agassizii [SCHIAGA]
Schistidium alpicola var. *dupretii* (Thér.) Crum = Schistidium apocarpum [SCHIAPO]
Schistidium ambiguum Sull. = Schistidium apocarpum [SCHIAPO]
Schistidium apocarpum (Hedw.) Bruch & Schimp. in B.S.G. [SCHIAPO] (common beard moss)
SY: *Grimmia alpicola* var. *dupretii* (Thér.) Crum
SY: *Grimmia apocarpa* Hedw.
SY: *Grimmia apocarpa* var. *ambigua* (Sull.) G. Jones in Grout
SY: *Grimmia apocarpa* var. *atrofusca* (Schimp.) Husn.
SY: *Grimmia apocarpa* var. *brunnescens* (Limpr.) Mönk.
SY: *Grimmia apocarpa* f. *canadensis* Loeske
SY: *Grimmia apocarpa* var. *canadensis* Dupret
SY: *Grimmia apocarpa* var. *conferta* (Funck) Spreng.
SY: *Grimmia apocarpa* var. *conferta* f. *obtusifolia* (Schimp.) Mönk.
SY: *Grimmia apocarpa* var. *dupretii* (Thér.) Sayre
SY: *Grimmia apocarpa* f. *epilosa* Warnst.
SY: *Grimmia apocarpa* var. *gracilis* Web. & Mohr ex Nees et al.
SY: *Grimmia apocarpa* var. *nigrescens* Mol.
SY: *Grimmia apocarpa* var. *obscuroviridis* Crum
SY: *Grimmia apocarpa* var. *stricta* (Turn.) Hook. & Tayl.
SY: *Grimmia apocarpa* var. *tenerrima* Nees & Hornsch.
SY: *Grimmia atricha* C. Müll. & Kindb. in Mac. & Kindb.
SY: *Grimmia coloradensis* Aust.
SY: *Grimmia conferta* Funck
SY: *Grimmia conferta* var. *compacta* Lesq. & James
SY: *Grimmia dupretii* Thér.
SY: *Grimmia gracilis* (Web. & Mohr ex Nees et al.) Schleich. ex Limpr.
SY: *Grimmia plagiopodia* ssp. *brandegei* (Aust.) Kindb.
SY: *Grimmia plagiopodia* ssp. *pilifera* Lesq. & James
SY: *Grimmia stricta* Turn.
SY: *Schistidium alpicola* var. *dupretii* (Thér.) Crum
SY: *Schistidium ambiguum* Sull.
SY: *Schistidium apocarpum* var. *ambiguum* (Sull.) G. Jones in Grout
SY: *Schistidium apocarpum* var. *atrofuscum* (Schimp.) C. Jens. ex Weim.
SY: *Schistidium apocarpum* var. *boreale* (Poelt) Duell
SY: *Schistidium apocarpum* var. *brunnescens* (Limpr.) Herz.
SY: *Schistidium boreale* Poelt
SY: *Schistidium apocarpum* var. *confertum* (Funck) Möll.

SY: *Schistidium apocarpum* var. *dupretii* (Thér.) Wijk & Marg.
SY: *Schistidium apocarpum* var. *gracile* (Web. & Mohr ex Nees et al.) Bruch & Schimp. in B.S.G.
SY: *Schistidium apocarpum* var. *nigrescens* (Mol.) Loeske
SY: *Schistidium apocarpum* var. *strictum* (Turn.) Moore
SY: *Schistidium atrofuscum* (Schimp.) Limpr.
SY: *Schistidium brunnescens* Limpr.
SY: *Schistidium confertum* (Funck) Bruch & Schimp. in B.S.G.
SY: *Schistidium dupretii* (Thér.) Web.
SY: *Schistidium gracile* (Web. & Mohr ex Nees et al.) Schleich. ex Limpr.
SY: *Schistidium papillosum* Culm.
SY: *Schistidium strictum* (Turn.) T. Kop. & Isov.
Schistidium apocarpum var. *ambiguum* (Sull.) G. Jones in Grout = Schistidium apocarpum [SCHIAPO]
Schistidium apocarpum var. *atrofuscum* (Schimp.) C. Jens. ex Weim. = Schistidium apocarpum [SCHIAPO]
Schistidium apocarpum var. *brunnescens* (Limpr.) Herz. = Schistidium apocarpum [SCHIAPO]
Schistidium apocarpum var. *boreale* (Poelt) Duell . = Schistidium apocarpum [SCHIAPO]
Schistidium apocarpum var. *confertum* (Funck) Möll. = Schistidium apocarpum [SCHIAPO]
Schistidium apocarpum var. *dupretii* (Thér.) Wijk & Marg. = Schistidium apocarpum [SCHIAPO]
Schistidium apocarpum var. *gracile* (Web. & Mohr ex Nees et al.) Bruch & Schimp. in B.S.G. = Schistidium apocarpum [SCHIAPO]
Schistidium apocarpum var. *nigrescens* (Mol.) Loeske = Schistidium apocarpum [SCHIAPO]
Schistidium apocarpum var. *strictum* (Turn.) Moore = Schistidium apocarpum [SCHIAPO]
Schistidium atrofuscum (Schimp.) Limpr. = Schistidium apocarpum [SCHIAPO]
Schistidium boreale Poelt = Schistidium apocarpum [SCHIAPO]
Schistidium brunnescens Limpr. = Schistidium apocarpum [SCHIAPO]
Schistidium confertum (Funck) Bruch & Schimp. in B.S.G. = Schistidium apocarpum [SCHIAPO]
Schistidium dupretii (Thér.) Web. = Schistidium apocarpum [SCHIAPO]
Schistidium gracile (Web. & Mohr ex Nees et al.) Schleich. ex Limpr. = Schistidium apocarpum [SCHIAPO]
Schistidium heterophyllum (Kindb. in Mac. & Kindb.) McIntosh [SCHIHET]
SY: *Grimmia chloroblasta* Kindb. in Mac. & Kindb.
SY: *Grimmia heterophylla* Kindb. in Mac. & Kindb.
Schistidium maritimum (Turn.) Bruch & Schimp. in B.S.G. [SCHIMAR]
SY: *Grimmia crassinervia* (C. Müll.) Mac. & Kindb.
SY: *Grimmia maritima* Turn.
Schistidium papillosum Culm. = Schistidium apocarpum [SCHIAPO]
Schistidium rivulare (Brid.) Podp. [SCHIRIV]
SY: *Grimmia alpicola* var. *papillosa* (G. Jones) Hab.
SY: *Grimmia alpicola* var. *rivularis* f. *papillosa* G. Jones in Grout
SY: *Grimmia apocarpa* var. *rivularis* (Brid.) Nees & Hornsch.
SY: *Grimmia pseudorivularis* ssp. *lancifolia* Kindb.
Schistidium strictum (Turn.) T. Kop. & Isov. = Schistidium apocarpum [SCHIAPO]
Schistidium tenerrimum (Ren. & Card.) G. Roth = Grimmia tenerrima [GRIMTEE]
Schistidium tenerum (Zett.) Nyh. [SCHITEN]
SY: *Grimmia tenera* Zett.
SY: *Grimmia tenuicaulis* Williams
Schistidium teretinerve (Limpr.) Limpr. = Grimmia teretinervis [GRIMTER]
Schistidium trichodon (Brid.) Poelt [SCHITRI]

GYMNOMITRIACEAE (B:H)

Eremonotus Lindb. & Kaal. ex Pears. [EREMONO$]
Eremonotus myriocarpus (Carring.) Pears. [EREMMYR]
Gymnomitrion Corda [GYMNOMI$]
Gymnomitrion apiculatum (Schiffn.) K. Müll. [GYMNAPI]
SY: *Marsupella apiculata* Schiffn.
Gymnomitrion concinnatum (Lightf.) Corda [GYMNCON]
Gymnomitrion corallioides Nees [GYMNCOR]
Gymnomitrion obtusum (Lindb.) Pears. [GYMNOBT]
Gymnomitrion pacificum Grolle [GYMNPAC]
SY: *Gymnomitrium crenulatum* auct.
Gymnomitrium crenulatum auct. = Gymnomitrion pacificum [GYMNPAC]
Gymnomitrium revolutum (Nees) Philib. = Marsupella revoluta [MARSREV]
Gymnomitrium varians (Lindb.) K. Müll. = Marsupella brevissima [MARSBRE]
Marsupella Dum. [MARSUPE$]
Marsupella alpina (Mass. & Carest.) H. Bern. [MARSALP]
Marsupella apiculata Schiffn. = Gymnomitrion apiculatum [GYMNAPI]
Marsupella aquatica (Lindb.) Schiffn. = Marsupella emarginata ssp. emarginata var. aquatica [MARSEMA2]
Marsupella boeckii (Aust.) Kaal. [MARSBOE]
SY: *Marsupella stableri* Spruce
Marsupella brevissima (Dum.) Grolle [MARSBRE]
SY: *Gymnomitrium varians* (Lindb.) K. Müll.
Marsupella commutata (Limpr.) H. Bern. [MARSCOM]
Marsupella condensata (Angstr.) Schiffn. [MARSCON]
Marsupella emarginata (Ehrh.) Dum. [MARSEMA] (rusty rock wort)
Marsupella emarginata ssp. **emarginata** [MARSEMA0]
Marsupella emarginata ssp. **emarginata** var. **aquatica** (Lindenb.) Dum. [MARSEMA2]
SY: *Marsupella aquatica* (Lindb.) Schiffn.
SY: *Marsupella emarginata* var. *pearsonii* (Schiffn.) Joerg.
Marsupella emarginata ssp. **emarginata** var. **emarginata** [MARSEMA3]
Marsupella emarginata ssp. **tubulosa** (Steph.) Kitag. [MARSEMA4]
Marsupella emarginata var. *pearsonii* (Schiffn.) Joerg. = Marsupella emarginata ssp. emarginata var. aquatica [MARSEMA2]
Marsupella revoluta (Nees) Dum. [MARSREV]
SY: *Gymnomitrium revolutum* (Nees) Philib.
Marsupella sparsifolia (Lindb.) Dum. [MARSSPA]
Marsupella sphacelata (Gieseke) Dum. [MARSSPH]
SY: *Marsupella sphacelata* var. *erythrorhiza* (Limpr.) Schiffn.
SY: *Marsupella sullivantii* (De Not.) Evans
Marsupella sphacelata f. **media** (Gott.) Schust. [MARSSPH1]
Marsupella sphacelata f. **sphacelata** [MARSSPH0]
Marsupella sphacelata var. *erythrorhiza* (Limpr.) Schiffn. = Marsupella sphacelata [MARSSPH]
Marsupella sprucei (Limpr.) H. Bern. = Marsupella ustulata var. sprucei [MARSUST1]
Marsupella stableri Spruce = Marsupella boeckii [MARSBOE]
Marsupella sullivantii (De Not.) Evans = Marsupella sphacelata [MARSSPH]
Marsupella ustulata (Hub.) Spruce [MARSUST]
Marsupella ustulata var. **sprucei** (Limpr.) Schust. [MARSUST1]
SY: *Marsupella sprucei* (Limpr.) H. Bern.

GYROTHYRACEAE (B:H)

Gyrothyra M.A. Howe [GYROTHY$]
Gyrothyra underwoodiana M.A. Howe [GYROUND]

HAPLOMITRIACEAE (B:H)

Haplomitrium Nees [HAPLOMI$]
Haplomitrium hookeri (Sm.) Nees [HAPLHOO]

HEDWIGIACEAE (B:M)

Braunia californica Lesq. = Pseudobraunia californica [PSEUCAL]
Hedwigia P. Beauv. [HEDWIGI$]
Hedwigia albicans Lindb. = Hedwigia ciliata [HEDWCIL]

Hedwigia albicans var. *subnuda* (Kindb.) C. Mohr = Hedwigia ciliata [HEDWCIL]
Hedwigia ciliata (Hedw.) P. Beauv. [HEDWCIL] (Hedwig's rock moss)
SY: *Hedwigia albicans* Lindb.
SY: *Hedwigia albicans* var. *subnuda* (Kindb.) C. Mohr
SY: *Hedwigia ciliata* var. *leucophaea* B.S.G.
SY: *Hedwigia ciliata* f. *secunda* (B.S.G.) G. Jones in Grout
SY: *Hedwigia ciliata* var. *secunda* B.S.G.
SY: *Hedwigia ciliata* ssp. *subnuda* Kindb. in Mac. & Kindb.
SY: *Hedwigia ciliata* var. *subnuda* Kindb. ex G. Jones in Grout
SY: *Hedwigia ciliata* f. *viridis* (B.S.G.) G. Jones in Grout
SY: *Hedwigia ciliata* var. *viridis* B.S.G.
Hedwigia ciliata f. *secunda* (B.S.G.) G. Jones in Grout = Hedwigia ciliata [HEDWCIL]
Hedwigia ciliata f. *viridis* (B.S.G.) G. Jones in Grout = Hedwigia ciliata [HEDWCIL]
Hedwigia ciliata ssp. *subnuda* Kindb. in Mac. & Kindb. = Hedwigia ciliata [HEDWCIL]
Hedwigia ciliata var. *leucophaea* B.S.G. = Hedwigia ciliata [HEDWCIL]
Hedwigia ciliata var. *secunda* B.S.G. = Hedwigia ciliata [HEDWCIL]
Hedwigia ciliata var. *subnuda* Kindb. ex G. Jones in Grout = Hedwigia ciliata [HEDWCIL]
Hedwigia ciliata var. *viridis* B.S.G. = Hedwigia ciliata [HEDWCIL]
Hedwigia stellata Hedenäs [HEDWSTE]
Pseudobraunia (Lesq. & James) Broth. [PSEUDOB$]
Pseudobraunia californica (Lesq.) Broth. [PSEUCAL]
SY: *Braunia californica* Lesq.

HELODIACEAE (B:M)

Elodium blandowii (Web. & Mohr) Broth. = Helodium blandowii [HELOBLA]
Helodium Warnst. [HELODIU$]
Helodium blandowii (Web. & Mohr) Warnst. [HELOBLA] (Blandow's feather moss)
SY: *Elodium blandowii* (Web. & Mohr) Broth.
SY: *Helodium lanatum* (Brid.) Broth.
SY: *Hypnum blandowii* Web. & Mohr
SY: *Thuidium blandowii* (Web. & Mohr) B.S.G.
SY: *Thuidium pseudoabietinum* Kindb.
Helodium lanatum (Brid.) Broth. = Helodium blandowii [HELOBLA]
Hypnum blandowii Web. & Mohr = Helodium blandowii [HELOBLA]
Thuidium blandowii (Web. & Mohr) B.S.G. = Helodium blandowii [HELOBLA]
Thuidium pseudoabietinum Kindb. = Helodium blandowii [HELOBLA]

HERBERTACEAE (B:H)

Herberta Gray = Herbertus [HERBERT$]
Herbertus Gray [HERBERT$]
Herbertus aduncus (Dicks.) Gray [HERBADU] (common scissor-leaf liverwort)
Herbertus hawaiiensis Mill. [HERBHAW]
Herbertus himalayanus auct. = Herbertus sakuraii [HERBSAK]
Herbertus sakuraii Steph. [HERBSAK]
SY: *Herbertus himalayanus* auct.
Herbertus sendtneri (Nees) Lindb. [HERBSEN]

HOOKERIACEAE (B:M)

Hookeria Sm. [HOOKERI$]
Hookeria acutifolia Hook. & Grev. [HOOKACU]
Hookeria lucens (Hedw.) Sm. [HOOKLUC] (clear moss)
SY: *Pterygophyllum lucens* (Hedw.) Brid.
Pterygophyllum lucens (Hedw.) Brid. = Hookeria lucens [HOOKLUC]

HYLOCOMIACEAE (B:M)

Calliergon schreberi (Brid.) Mitt. = Pleurozium schreberi [PLEUSCH]
Calliergonella schreberi (Brid.) Grout = Pleurozium schreberi [PLEUSCH]
Calliergonella schreberi var. *tananae* (Grout) Grout = Pleurozium schreberi [PLEUSCH]
Hylocomiastrum Fleisch. [HYLOCOI$]
Hylocomiastrum pyrenaicum (Spruce) Fleisch. in Broth. [HYLOPYE]
SY: *Hylocomium pyrenaicum* (Spruce) Lindb.
SY: *Hypnum oakesii* Sull. in Gray
SY: *Hypnum pyrenaicum* Spruce
Hylocomiastrum umbratum (Hedw.) Fleisch. in Broth. [HYLOUMR]
SY: *Hylocomium umbratum* (Hedw.) Schimp. in B.S.G.
SY: *Hypnum umbratum* Ehrh. ex Hedw.
Hylocomium Schimp. in B.S.G. [HYLOCOM$]
Hylocomium alaskanum (Lesq. & James) Aust. = Hylocomium splendens [HYLOSPL]
Hylocomium calvescens Lindb. = Rhytidiadelphus squarrosus [RHYTSQU]
Hylocomium giganteum Perss., nom. nud., non Bartr. = Hylocomium splendens [HYLOSPL]
Hylocomium loreum (Hedw.) B.S.G. = Rhytidiadelphus loreus [RHYTLOR]
Hylocomium proliferum (Brid.) Lindb. = Hylocomium splendens [HYLOSPL]
Hylocomium pyrenaicum (Spruce) Lindb. = Hylocomiastrum pyrenaicum [HYLOPYE]
Hylocomium robustum (Hook.) Kindb. = Rhytidiopsis robusta [RHYTROB]
Hylocomium splendens (Hedw.) Schimp. in B.S.G. [HYLOSPL] (step moss)
SY: *Hylocomium alaskanum* (Lesq. & James) Aust.
SY: *Hylocomium giganteum* Perss., nom. nud., non Bartr.
SY: *Hylocomium proliferum* (Brid.) Lindb.
SY: *Hylocomium splendens* var. *alaskanum* (Lesq. & James) Limpr.
SY: *Hylocomium splendens* var. *compactum* (Lesq. & James) Mac. & Kindb.
SY: *Hylocomium splendens* var. *gracilius* (Boul.) Husn.
SY: *Hylocomium splendens* var. *obtusifolium* (Geh.) Par.
SY: *Hypnum splendens* Hedw.
Hylocomium splendens var. *alaskanum* (Lesq. & James) Limpr. = Hylocomium splendens [HYLOSPL]
Hylocomium splendens var. *compactum* (Lesq. & James) Mac. & Kindb. = Hylocomium splendens [HYLOSPL]
Hylocomium splendens var. *gracilius* (Boul.) Husn. = Hylocomium splendens [HYLOSPL]
Hylocomium splendens var. *obtusifolium* (Geh.) Par. = Hylocomium splendens [HYLOSPL]
Hylocomium squarrosum ssp. *calvescens* Kindb. = Rhytidiadelphus squarrosus [RHYTSQU]
Hylocomium squarrosum var. *calvescens* (Kindb.) Hobk. = Rhytidiadelphus squarrosus [RHYTSQU]
Hylocomium triquetrum (Hedw.) B.S.G. = Rhytidiadelphus triquetrus [RHYTTRI]
Hylocomium umbratum (Hedw.) Schimp. in B.S.G. = Hylocomiastrum umbratum [HYLOUMR]
Hypnum oakesii Sull. in Gray = Hylocomiastrum pyrenaicum [HYLOPYE]
Hypnum pyrenaicum Spruce = Hylocomiastrum pyrenaicum [HYLOPYE]
Hypnum schreberi Brid. = Pleurozium schreberi [PLEUSCH]
Hypnum splendens Hedw. = Hylocomium splendens [HYLOSPL]
Hypnum squarrosum Hedw. = Rhytidiadelphus squarrosus [RHYTSQU]
Hypnum triquetrum Hedw. = Rhytidiadelphus triquetrus [RHYTTRI]
Hypnum umbratum Ehrh. ex Hedw. = Hylocomiastrum umbratum [HYLOUMR]
Pleurozium Mitt. [PLEUROI$]

Pleurozium schreberi (Brid.) Mitt. [PLEUSCH] (red-stemmed feathermoss)
SY: *Calliergon schreberi* (Brid.) Mitt.
SY: *Calliergonella schreberi* (Brid.) Grout
SY: *Calliergonella schreberi* var. *tananae* (Grout) Grout
SY: *Hypnum schreberi* Brid.
SY: *Pleurozium schreberi* var. *tananae* (Grout) Wijk & Marg.
Pleurozium schreberi var. *tananae* (Grout) Wijk & Marg. = Pleurozium schreberi [PLEUSCH]
Rhytidiadelphus (Lindb. ex Limpr.) Warnst. [RHYTIDI$]
Rhytidiadelphus calvescens (Kindb.) Broth. = Rhytidiadelphus squarrosus [RHYTSQU]
Rhytidiadelphus loreus (Hedw.) Warnst. [RHYTLOR] (lanky moss)
SY: *Hylocomium loreum* (Hedw.) B.S.G.
Rhytidiadelphus squarrosus (Hedw.) Warnst. [RHYTSQU] (bent-leaf moss)
SY: *Hylocomium calvescens* Lindb.
SY: *Hylocomium squarrosum* ssp. *calvescens* Kindb.
SY: *Hylocomium squarrosum* var. *calvescens* (Kindb.) Hobk.
SY: *Hypnum squarrosum* Hedw.
SY: *Rhytidiadelphus calvescens* (Kindb.) Broth.
SY: *Rhytidiadelphus squarrosus* var. *calvescens* (Kindb.) Warnst.
SY: *Rhytidiadelphus subpinnatus* (Lindb.) T. Kop.
Rhytidiadelphus squarrosus var. *calvescens* (Kindb.) Warnst. = Rhytidiadelphus squarrosus [RHYTSQU]
Rhytidiadelphus subpinnatus (Lindb.) T. Kop. = Rhytidiadelphus squarrosus [RHYTSQU]
Rhytidiadelphus triquetrus (Hedw.) Warnst. [RHYTTRI] (goose-necked moss)
SY: *Hylocomium triquetrum* (Hedw.) B.S.G.
SY: *Hypnum triquetrum* Hedw.
SY: *Rhytidiadelphus triquetrus* var. *beringianus* (Card. & Thér.) Grout
Rhytidiadelphus triquetrus var. *beringianus* (Card. & Thér.) Grout = Rhytidiadelphus triquetrus [RHYTTRI]
Rhytidiopsis Broth. [RHYTIDO$]
Rhytidiopsis robusta (Hook.) Broth. [RHYTROB] (pipecleaner moss)
SY: *Hylocomium robustum* (Hook.) Kindb.

HYPNACEAE (B:M)

Amblystegiella jungermannioides (Brid.) Giac. = Platydictya jungermannioides [PLATJUN]
Amblystegiella sprucei (Bruch) Loeske = Platydictya jungermannioides [PLATJUN]
Amblystegium jungermannoides (Brid.) A.J.E. Sm. = Platydictya jungermannioides [PLATJUN]
Amblystegium sprucei (Bruch) B.S.G. = Platydictya jungermannioides [PLATJUN]
Breidleria arcuata (Mol.) Loeske = Hypnum lindbergii [HYPNLIN]
Breidleria pratensis (Rabenh.) Loeske = Hypnum pratense [HYPNPRA]
Callicladium Crum [CALLICL$]
Callicladium haldanianum (Grev.) Crum [CALLHAL]
SY: *Heterophyllium haldanianum* (Grev.) Fleisch.
SY: *Hypnum flaccum* C. Müll. & Kindb. in Mac. & Kindb.
SY: *Hypnum haldanianum* Grev.
SY: *Plagiothecium brevipungens* Kindb. in Mac. & Kindb.
SY: *Robinsonia haldaniana* (Grev.) Crum
Campylium adscendens (Lindb.) Perss. in Perss. & Gjaer. = Herzogiella adscendens [HERZADS]
Campylium stellatum ssp. *treleasii* (Ren.) Grout = Herzogiella adscendens [HERZADS]
Campylium stellatum var. *adscendens* (Lindb.) Perss. = Herzogiella adscendens [HERZADS]
Campylium treleasii (Ren.) Broth. = Herzogiella adscendens [HERZADS]
Ctenidium (Schimp.) Mitt. [CTENIDI$]
Ctenidium procerrimum (De Not.) Lindb. = Hypnum procerrimum [HYPNPRO]
Ctenidium schofieldii Nish. [CTENSCH]
Cylindrothecium macounii (C. Müll. & Kindb. in Mac. & Kindb.) Ren. & Card. = Hypnum pratense [HYPNPRA]
Entodon expallens C. Müll. & Kindb. in Mac. & Kindb. = Hypnum pratense [HYPNPRA]
Entodon macounii C. Müll. & Kindb. in Mac. & Kindb. = Hypnum pratense [HYPNPRA]
Entodon repens var. *orthoclados* (Kindb.) Grout = Platygyrium repens [PLATREP]
Eurhynchium vaucheri (Lesq.) B.S.G. = Hypnum vaucheri [HYPNVAU]
Gollania Broth. [GOLLANI$]
Gollania densepinnata Dix. = Gollania turgens [GOLLTUR]
Gollania turgens (C. Müll.) Ando [GOLLTUR]
SY: *Gollania densepinnata* Dix.
Helodium pratense W. Koch ex Moxley = Hypnum pratense [HYPNPRA]
Helodium reptile Michx. ex Moxley = Hypnum pallescens [HYPNPAL]
Herzogiella Broth. [HERZOGI$]
Herzogiella adscendens (Lindb.) Iwats. & Schof. [HERZADS]
SY: *Campylium adscendens* (Lindb.) Perss. in Perss. & Gjaer.
SY: *Campylium stellatum* var. *adscendens* (Lindb.) Perss.
SY: *Campylium stellatum* ssp. *treleasii* (Ren.) Grout
SY: *Campylium treleasii* (Ren.) Broth.
SY: *Stereodon adscendens* Lindb.
Herzogiella seligeri (Brid.) Iwats. [HERZSEL]
SY: *Isopterygium seligeri* (Brid.) Dix. in C. Jens.
SY: *Isopterygium silesiacum* (Selig. ex P. Beauv.) Kindb.
SY: *Leskea seligeri* Brid.
SY: *Plagiothecium seligeri* (Brid.) Lindb.
SY: *Plagiothecium silesiacum* (Selig. ex P. Beauv.) Schimp. in B.S.G.
SY: *Sharpiella seligeri* (Brid.) Iwats.
Herzogiella striatella (Brid.) Iwats. [HERZSTR]
SY: *Hypnum muehlenbeckii* Schimp. ex Hartm.
SY: *Isopterygium striatellum* (Brid.) Loeske
SY: *Leskea striatella* Brid.
SY: *Plagiothecium muehlenbeckii* B.S.G.
SY: *Plagiothecium striatellum* (Brid.) Lindb.
SY: *Sharpiella striatella* (Brid.) Iwats.
Heterophyllium haldanianum (Grev.) Fleisch. = Callicladium haldanianum [CALLHAL]
Holmgrenia chrysea (Schwaegr. in Schultes) Lindb. = Orthothecium chryseum [ORTHCHR]
Holmgrenia intricata (Hartm.) Lindb. = Orthothecium intricatum [ORTHINT]
Holmgrenia stricta (Lor.) Grout = Orthothecium strictum [ORTHSTR]
Hypnum Hedw. [HYPNUM$]
Hypnum alaskae Kindb. = Hypnum callichroum [HYPNCAL]
Hypnum arcuatiforme Kindb. in Mac. & Kindb. = Hypnum lindbergii [HYPNLIN]
Hypnum arcuatum Lindb. = Hypnum lindbergii [HYPNLIN]
Hypnum arcuatum var. *americanum* Ren. & Card. = Hypnum lindbergii [HYPNLIN]
Hypnum arcuatum var. *demissum* Schimp. = Hypnum lindbergii [HYPNLIN]
Hypnum arcuatum var. *elatum* Schimp. = Hypnum lindbergii [HYPNLIN]
Hypnum bambergeri Schimp. [HYPNBAM]
Hypnum bridelianum Crum, Steere & Anderson = Hypnum recurvatum [HYPNREC]
Hypnum callichroum Funck ex Brid. [HYPNCAL]
SY: *Hypnum alaskae* Kindb.
SY: *Stereodon callichrous* (Brid.) Braithw.
Hypnum canadense Kindb. = Hypnum dieckii [HYPNDIE]
Hypnum circinale Hook. [HYPNCIR] (coiled-leaf moss)
SY: *Hypnum pseudorecurvans* Kindb.
SY: *Hypnum sequoieti* C. Müll.
Hypnum complexum (Mitt.) Jaeg. & Sauerb. = Hypnum cupressiforme [HYPNCUP]
Hypnum crista-castrensis Hedw. = Ptilium crista-castrensis [PTILCRI]

Hypnum cupressiforme Hedw. [HYPNCUP]
SY: *Hypnum complexum* (Mitt.) Jaeg. & Sauerb.
SY: *Hypnum pseudofastigiatum* C. Müll. & Kindb. in Mac. & Kindb.
SY: *Stereodon cupressiformis* (Hedw.) Brid. ex Mitt.
Hypnum cupressiforme var. *brevisetum* Schimp. = Hypnum cupressiforme var. cupressiforme [HYPNCUP0]
Hypnum cupressiforme var. **cupressiforme** [HYPNCUP0]
SY: *Hypnum cupressiforme* var. *brevisetum* Schimp.
Hypnum cupressiforme var. **filiforme** Brid. [HYPNCUP2]
Hypnum cupressiforme var. **lacunosum** Brid. [HYPNCUP1]
Hypnum cupressiforme var. **resupinatum** (Tayl.) Schimp. in Spruce [HYPNCUP3]
SY: *Hypnum pseudo-nemorosum* (Kindb.) Par.
Hypnum dieckii Ren. & Card. in Röll [HYPNDIE]
SY: *Hypnum canadense* Kindb.
Hypnum fastigiatum Brid. = Hypnum recurvatum [HYPNREC]
Hypnum flaccum C. Müll. & Kindb. in Mac. & Kindb. = Callicladium haldanianum [CALLHAL]
Hypnum geminum (Mitt.) Lesq. & James [HYPNGEM]
Hypnum haldanianum Grev. = Callicladium haldanianum [CALLHAL]
Hypnum hamulosum Schimp. in B.S.G. [HYPNHAM]
Hypnum implexum Ren. & Card. = Hypnum plicatulum [HYPNPLI]
Hypnum lindbergii Mitt. [HYPNLIN] (clay pigtail moss)
SY: *Breidleria arcuata* (Mol.) Loeske
SY: *Hypnum arcuatiforme* Kindb. in Mac. & Kindb.
SY: *Hypnum arcuatum* Lindb.
SY: *Hypnum arcuatum* var. *americanum* Ren. & Card.
SY: *Hypnum arcuatum* var. *demissum* Schimp.
SY: *Hypnum arcuatum* var. *elatum* Schimp.
SY: *Hypnum lindbergii* var. *americanum* (Ren. & Card.) E. Whiteh.
SY: *Hypnum lindbergii* var. *demissum* (Schimp.) Loeske
SY: *Hypnum lindbergii* var. *elatum* (Schimp.) Williams
SY: *Hypnum patientiae* Lindb. ex Milde
SY: *Hypnum patientiae* var. *americanum* (Ren. & Card.) Ren. & Card.
SY: *Hypnum patientiae* var. *demissum* (Schimp.) Jaeg.
SY: *Hypnum patientiae* var. *elatum* (Schimp.) Jaeg.
SY: *Hypnum pseudodrepanium* C. Müll. & Kindb. in Mac. & Kindb.
SY: *Hypnum renauldii* Kindb. in Mac. & Kindb.
SY: *Stereodon arcuatus* Lindb.
SY: *Stereodon lindbergii* (Mitt.) Braithw.
Hypnum lindbergii var. *americanum* (Ren. & Card.) E. Whiteh. = Hypnum lindbergii [HYPNLIN]
Hypnum lindbergii var. *demissum* (Schimp.) Loeske = Hypnum lindbergii [HYPNLIN]
Hypnum lindbergii var. *elatum* (Schimp.) Williams = Hypnum lindbergii [HYPNLIN]
Hypnum molluscoides Kindb. = Hypnum plicatulum [HYPNPLI]
Hypnum molluscum var. *molluscoides* (Par.) Grout = Hypnum plicatulum [HYPNPLI]
Hypnum muehlenbeckii Schimp. ex Hartm. = Herzogiella striatella [HERZSTR]
Hypnum nitidulum Wahlenb. = Isopterygiopsis pulchella [ISOPPUC]
Hypnum pallescens (Hedw.) P. Beauv. [HYPNPAL]
SY: *Helodium reptile* Michx. ex Moxley
SY: *Stereodon pallescens* (Hedw.) Mitt.
SY: *Stereodon reptilis* (Michx.) Mitt.
Hypnum passaicense (Aust.) Lesq. & James = Isopterygiopsis pulchella [ISOPPUC]
Hypnum patientiae Lindb. ex Milde = Hypnum lindbergii [HYPNLIN]
Hypnum patientiae var. *americanum* (Ren. & Card.) Ren. & Card. = Hypnum lindbergii [HYPNLIN]
Hypnum patientiae var. *demissum* (Schimp.) Jaeg. = Hypnum lindbergii [HYPNLIN]
Hypnum patientiae var. *elatum* (Schimp.) Jaeg. = Hypnum lindbergii [HYPNLIN]
Hypnum plicatile (Mitt.) Lesq. & James = Hypnum revolutum [HYPNREV]
Hypnum plicatulum (Lindb.) Jaeg. [HYPNPLI]
SY: *Hypnum implexum* Ren. & Card.
SY: *Hypnum molluscoides* Kindb.
SY: *Hypnum molluscum* var. *molluscoides* (Par.) Grout
SY: *Hypnum subplicatile* Limpr.
Hypnum pratense (Rabenh.) W. Koch ex Spruce [HYPNPRA]
SY: *Breidleria pratensis* (Rabenh.) Loeske
SY: *Cylindrothecium macounii* (C. Müll. & Kindb. in Mac. & Kindb.) Ren. & Card.
SY: *Entodon expallens* C. Müll. & Kindb. in Mac. & Kindb.
SY: *Entodon macounii* C. Müll. & Kindb. in Mac. & Kindb.
SY: *Helodium pratense* W. Koch ex Moxley
SY: *Hypnum pseudopratense* Kindb. in Mac. & Kindb.
SY: *Hypnum subflaccum* C. Müll. & Kindb. in Mac. & Kindb.
SY: *Stereodon pratensis* (Rabenh.) Warnst.
Hypnum procerrimum Mol. [HYPNPRO]
SY: *Ctenidium procerrimum* (De Not.) Lindb.
SY: *Pseudostereodon procerrimum* (Mol.) Fleisch. in Broth.
Hypnum pseudo-nemorosum (Kindb.) Par. = Hypnum cupressiforme var. resupinatum [HYPNCUP3]
Hypnum pseudodrepanium C. Müll. & Kindb. in Mac. & Kindb. = Hypnum lindbergii [HYPNLIN]
Hypnum pseudofastigiatum C. Müll. & Kindb. in Mac. & Kindb. = Hypnum cupressiforme [HYPNCUP]
Hypnum pseudopratense Kindb. in Mac. & Kindb. = Hypnum pratense [HYPNPRA]
Hypnum pseudorecurvans Kindb. = Hypnum circinale [HYPNCIR]
Hypnum pulchellum (Hedw.) Dicks. ex Brid. = Isopterygiopsis pulchella [ISOPPUC]
Hypnum ravaudii ssp. *fastigiatum* (Brid.) Wijk & Marg. = Hypnum recurvatum [HYPNREC]
Hypnum recurvatum (Lindb. & Arnell) Kindb. [HYPNREC]
SY: *Hypnum bridelianum* Crum, Steere & Anderson
SY: *Hypnum fastigiatum* Brid.
SY: *Hypnum ravaudii* ssp. *fastigiatum* (Brid.) Wijk & Marg.
Hypnum renauldii Kindb. in Mac. & Kindb. = Hypnum lindbergii [HYPNLIN]
Hypnum revolutum (Mitt.) Lindb. [HYPNREV] (rusty claw moss)
SY: *Hypnum plicatile* (Mitt.) Lesq. & James
SY: *Hypnum revolutum* var. *dolomiticum* (Milde) Mönk.
SY: *Hypnum revolutum* var. *subjulaceum* Bryhn
Hypnum revolutum var. *dolomiticum* (Milde) Mönk. = Hypnum revolutum [HYPNREV]
Hypnum revolutum var. *subjulaceum* Bryhn = Hypnum revolutum [HYPNREV]
Hypnum sequoieti C. Müll. = Hypnum circinale [HYPNCIR]
Hypnum subcomplexum Kindb. = Hypnum vaucheri [HYPNVAU]
Hypnum subflaccum C. Müll. & Kindb. in Mac. & Kindb. = Hypnum pratense [HYPNPRA]
Hypnum subimponens Lesq. [HYPNSUB] (curly hypnum)
SY: *Hypnum subimponens* var. *cristulum* Kindb. in Mac.
Hypnum subimponens var. *cristulum* Kindb. in Mac. = Hypnum subimponens [HYPNSUB]
Hypnum subplicatile Limpr. = Hypnum plicatulum [HYPNPLI]
Hypnum vaucheri Lesq. [HYPNVAU]
SY: *Eurhynchium vaucheri* (Lesq.) B.S.G.
SY: *Hypnum subcomplexum* Kindb.
Isopterygiopsis Iwats. [ISOPTER$]
Isopterygiopsis muelleriana (Schimp.) Iwats. [ISOPMUE]
SY: *Isopterygium muellerianum* (Schimp.) Jaeg.
SY: *Plagiothecium muellerianum* Schimp.
Isopterygiopsis pulchella (Hedw.) Iwats. [ISOPPUC]
SY: *Hypnum nitidulum* Wahlenb.
SY: *Hypnum passaicense* (Aust.) Lesq. & James
SY: *Hypnum pulchellum* (Hedw.) Dicks. ex Brid.
SY: *Isopterygium nitidulum* (Wahlenb.) Kindb.
SY: *Isopterygium nitidum* Lindb.
SY: *Isopterygium pulchellum* (Hedw.) Jaeg.
SY: *Isopterygium pulchellum* var. *nitidulum* (Wahlenb.) G. Roth
SY: *Plagiothecium geminum* (Mitt.) Jaeg. & Sauerb.
SY: *Plagiothecium passaicence* Aust.
SY: *Plagiothecium pseudolatebricola* Kindb. in Mac. & Kindb.
SY: *Plagiothecium pulchellum* (Hedw.) Schimp. in B.S.G.

SY: *Plagiothecium pulchellum* var. *nitidulum* (Wahlenb.) Ren. & Card.
Isopterygium borrerianum (C. Müll.) Lindb. = Pseudotaxiphyllum elegans [PSEUELE]
Isopterygium borrerianum var. *gracilens* (Aust. ex Grout) Crum, Steere & Anderson = Pseudotaxiphyllum elegans [PSEUELE]
Isopterygium borrerianum var. *terrestre* (Lindb.) Crum, Steere & Anderson = Pseudotaxiphyllum elegans [PSEUELE]
Isopterygium elegans (Brid.) Lindb. = Pseudotaxiphyllum elegans [PSEUELE]
Isopterygium elegans var. *gracilens* Aust. ex Grout = Pseudotaxiphyllum elegans [PSEUELE]
Isopterygium elegans var. *schimperi* (Jur. & Milde) Limpr. = Pseudotaxiphyllum elegans [PSEUELE]
Isopterygium elegans var. *terrestre* (Lindb.) Wijk & Marg. = Pseudotaxiphyllum elegans [PSEUELE]
Isopterygium muellerianum (Schimp.) Jaeg. = Isopterygiopsis muelleriana [ISOPMUE]
Isopterygium nitidulum (Wahlenb.) Kindb. = Isopterygiopsis pulchella [ISOPPUC]
Isopterygium nitidum Lindb. = Isopterygiopsis pulchella [ISOPPUC]
Isopterygium pulchellum (Hedw.) Jaeg. = Isopterygiopsis pulchella [ISOPPUC]
Isopterygium pulchellum var. *nitidulum* (Wahlenb.) G. Roth = Isopterygiopsis pulchella [ISOPPUC]
Isopterygium seligeri (Brid.) Dix. in C. Jens. = Herzogiella seligeri [HERZSEL]
Isopterygium silesiacum (Selig. ex P. Beauv.) Kindb. = Herzogiella seligeri [HERZSEL]
Isopterygium striatellum (Brid.) Loeske = Herzogiella striatella [HERZSTR]
Leskea seligeri Brid. = Herzogiella seligeri [HERZSEL]
Leskea striatella Brid. = Herzogiella striatella [HERZSTR]
Orthothecium Schimp. in B.S.G. [ORTHOTH$]
Orthothecium chryseum (Schwaegr. in Schultes) Schimp. in B.S.G. [ORTHCHR]
SY: *Holmgrenia chrysea* (Schwaegr. in Schultes) Lindb.
Orthothecium intricatum (Hartm.) Schimp. in B.S.G. [ORTHINT]
SY: *Holmgrenia intricata* (Hartm.) Lindb.
Orthothecium rubellum (Mitt.) Kindb. = Orthothecium strictum [ORTHSTR]
Orthothecium strictum Lor. [ORTHSTR]
SY: *Holmgrenia stricta* (Lor.) Grout
SY: *Orthothecium rubellum* (Mitt.) Kindb.
Plagiothecium brevipungens Kindb. in Mac. & Kindb. = Callicladium haldanianum [CALLHAL]
Plagiothecium elegans (Brid.) Sull. in Gray = Pseudotaxiphyllum elegans [PSEUELE]
Plagiothecium elegans var. *gracilens* Aust. ex Grout = Pseudotaxiphyllum elegans [PSEUELE]
Plagiothecium geminum (Mitt.) Jaeg. & Sauerb. = Isopterygiopsis pulchella [ISOPPUC]
Plagiothecium muehlenbeckii B.S.G. = Herzogiella striatella [HERZSTR]
Plagiothecium muellerianum Schimp. = Isopterygiopsis muelleriana [ISOPMUE]
Plagiothecium passaicence Aust. = Isopterygiopsis pulchella [ISOPPUC]
Plagiothecium pseudolatebricola Kindb. in Mac. & Kindb. = Isopterygiopsis pulchella [ISOPPUC]
Plagiothecium pulchellum (Hedw.) Schimp. in B.S.G. = Isopterygiopsis pulchella [ISOPPUC]
Plagiothecium pulchellum var. *nitidulum* (Wahlenb.) Ren. & Card. = Isopterygiopsis pulchella [ISOPPUC]
Plagiothecium seligeri (Brid.) Lindb. = Herzogiella seligeri [HERZSEL]
Plagiothecium silesiacum (Selig. ex P. Beauv.) Schimp. in B.S.G. = Herzogiella seligeri [HERZSEL]
Plagiothecium striatellum (Brid.) Lindb. = Herzogiella striatella [HERZSTR]
Platydictya Berk. [PLATYDI$]
Platydictya jungermannioides (Brid.) Crum [PLATJUN]
SY: *Amblystegiella jungermannioides* (Brid.) Giac.
SY: *Amblystegiella sprucei* (Bruch) Loeske
SY: *Amblystegium jungermannoides* (Brid.) A.J.E. Sm.
SY: *Amblystegium sprucei* (Bruch) B.S.G.
Platygyrium Schimp. in B.S.G. [PLATYGY$]
Platygyrium orthoclados Kindb. in Mac. & Kindb. = Platygyrium repens [PLATREP]
Platygyrium repens (Brid.) Schimp. in B.S.G. [PLATREP]
SY: *Entodon repens* var. *orthoclados* (Kindb.) Grout
SY: *Platygyrium orthoclados* Kindb. in Mac. & Kindb.
SY: *Platygyrium repens* var. *ascendens* (Schwaegr.) Grout
SY: *Platygyrium repens* var. *orthoclados* Kindb.
Platygyrium repens var. *ascendens* (Schwaegr.) Grout = Platygyrium repens [PLATREP]
Platygyrium repens var. *orthoclados* Kindb. = Platygyrium repens [PLATREP]
Pseudostereodon procerrimum (Mol.) Fleisch. in Broth. = Hypnum procerrimum [HYPNPRO]
Pseudotaxiphyllum Iwats. [PSEUDOH$]
Pseudotaxiphyllum elegans (Brid.) Iwats. [PSEUELE] (small flat moss)
SY: *Isopterygium borrerianum* (C. Müll.) Lindb.
SY: *Isopterygium borrerianum* var. *gracilens* (Aust. ex Grout) Crum, Steere & Anderson
SY: *Isopterygium borrerianum* var. *terrestre* (Lindb.) Crum, Steere & Anderson
SY: *Isopterygium elegans* (Brid.) Lindb.
SY: *Isopterygium elegans* var. *gracilens* Aust. ex Grout
SY: *Isopterygium elegans* var. *schimperi* (Jur. & Milde) Limpr.
SY: *Isopterygium elegans* var. *terrestre* (Lindb.) Wijk & Marg.
SY: *Plagiothecium elegans* (Brid.) Sull. in Gray
SY: *Plagiothecium elegans* var. *gracilens* Aust. ex Grout
Ptilium De Not. [PTILIUM$]
Ptilium crista-castrensis (Hedw.) De Not. [PTILCRI] (knight's plume)
SY: *Hypnum crista-castrensis* Hedw.
Pylaisia filari-acuminata C. Müll. & Kindb. in Mac. & Kindb. = Pylaisiella polyantha [PYLAPOL]
Pylaisia heteromalla Bruch & Schimp. = Pylaisiella polyantha [PYLAPOL]
Pylaisia intricata (Hedw.) Schimp. in B.S.G. = Pylaisiella intricata [PYLAINT]
Pylaisia jamesii Sull. & Lesq. = Pylaisiella polyantha [PYLAPOL]
Pylaisia ontariensis C. Müll. & Kindb. in Mac. & Kindb. = Pylaisiella polyantha [PYLAPOL]
Pylaisia polyantha (Hedw.) Schimp. in B.S.G. = Pylaisiella polyantha [PYLAPOL]
Pylaisia polyantha var. *brevifolia* (Lindb. & Arnell) Limpr. = Pylaisiella polyantha [PYLAPOL]
Pylaisia polyantha var. *jamesii* (Sull. & Lesq.) Rau & Herv. = Pylaisiella polyantha [PYLAPOL]
Pylaisia polyantha var. *pseudoplatygyria* (Kindb.) Grout = Pylaisiella polyantha [PYLAPOL]
Pylaisia subdenticulata Schimp. in B.S.G. = Pylaisiella polyantha [PYLAPOL]
Pylaisia velutina Schimp. in B.S.G. = Pylaisiella intricata [PYLAINT]
Pylaisiella Kindb. [PYLAISI$]
Pylaisiella intricata (Hedw.) Grout [PYLAINT]
SY: *Pylaisia intricata* (Hedw.) Schimp. in B.S.G.
SY: *Pylaisia velutina* Schimp. in B.S.G.
Pylaisiella polyantha (Hedw.) Grout [PYLAPOL] (aspen moss)
SY: *Pylaisia filari-acuminata* C. Müll. & Kindb. in Mac. & Kindb.
SY: *Pylaisia heteromalla* Bruch & Schimp.
SY: *Pylaisia jamesii* Sull. & Lesq.
SY: *Pylaisia ontariensis* C. Müll. & Kindb. in Mac. & Kindb.
SY: *Pylaisia polyantha* (Hedw.) Schimp. in B.S.G.
SY: *Pylaisia polyantha* var. *brevifolia* (Lindb. & Arnell) Limpr.
SY: *Pylaisia polyantha* var. *jamesii* (Sull. & Lesq.) Rau & Herv.
SY: *Pylaisia polyantha* var. *pseudoplatygyria* (Kindb.) Grout
SY: *Pylaisia subdenticulata* Schimp. in B.S.G.
SY: *Pylaisiella polyantha* var. *pseudoplatygyria* (Kindb.) Grout
Pylaisiella polyantha var. *pseudoplatygyria* (Kindb.) Grout = Pylaisiella polyantha [PYLAPOL]

Robinsonia haldaniana (Grev.) Crum = Callicladium haldanianum [CALLHAL]
Sharpiella seligeri (Brid.) Iwats. = Herzogiella seligeri [HERZSEL]
Sharpiella striatella (Brid.) Iwats. = Herzogiella striatella [HERZSTR]
Stereodon adscendens Lindb. = Herzogiella adscendens [HERZADS]
Stereodon arcuatus Lindb. = Hypnum lindbergii [HYPNLIN]
Stereodon callichrous (Brid.) Braithw. = Hypnum callichroum [HYPNCAL]
Stereodon cupressiformis (Hedw.) Brid. ex Mitt. = Hypnum cupressiforme [HYPNCUP]
Stereodon lindbergii (Mitt.) Braithw. = Hypnum lindbergii [HYPNLIN]
Stereodon pallescens (Hedw.) Mitt. = Hypnum pallescens [HYPNPAL]
Stereodon pratensis (Rabenh.) Warnst. = Hypnum pratense [HYPNPRA]
Stereodon reptilis (Michx.) Mitt. = Hypnum pallescens [HYPNPAL]
Tripterocladium (C. Müll.) Jaeg. [TRIPTER$]
Tripterocladium leucocladulum (C. Müll.) Jaeg. [TRIPLEU]
SY: *Tripterocladium rupestris* (Kindb.) Kindb.
Tripterocladium rupestris (Kindb.) Kindb. = Tripterocladium leucocladulum [TRIPLEU]

HYPOPTERYGIACEAE (B:M)

Hypopterygium Brid. [HYPOPTE$]
Hypopterygium canadense Kindb. = Hypopterygium fauriei [HYPOFAU]
Hypopterygium fauriei Besch. [HYPOFAU]
SY: *Hypopterygium canadense* Kindb.

JUBULACEAE (B:H)

Frullania Raddi [FRULLAN$]
Frullania bolanderi Aust. [FRULBOL]
Frullania californica (Aust.) Evans [FRULCAL]
Frullania franciscana M.A. Howe [FRULFRA]
Frullania hattoriana J.D. Godfr. & G.A. Godfr. [FRULHAT]
Frullania nisquallensis Sull. = Frullania tamarisci ssp. nisquallensis [FRULTAM2]
Frullania tamarisci (L.) Dum. [FRULTAM]
Frullania tamarisci ssp. **nisquallensis** (Sull.) Hatt. [FRULTAM2] (hanging millipede liverwort)
SY: *Frullania nisquallensis* Sull.
Frullania tamarisci ssp. **tamarisci** [FRULTAM0]

JUNGERMANNIACEAE (B:H)

Anastrepta (Lindb.) Schiffn. [ANASTRE$]
Anastrepta orcadensis (Hook) Schiffn. [ANASORC]
Anastrophyllum (Spruce) Steph. [ANASTRO$]
Anastrophyllum assimile (Mitt.) Steph. [ANASASS]
SY: *Anastrophyllum japonicum* var. *otianum* Hatt.
SY: *Anastrophyllum reichardtii* (Gott.) Steph.
Anastrophyllum donianum (Hook.) Steph. [ANASDON]
Anastrophyllum helleranum (Nees) Schust. [ANASHEL]
SY: *Anastrophyllum hellerianum* (Nees) Schust.
SY: *Crossocalyx hellerianus* (Nees) Meyl.
SY: *Eremonotus hellerianus* (Meyl.) Schust.
SY: *Isopaches hellerianus* (Nees) Buch
SY: *Sphenolobus helleranus* (Nees) Steph.
Anastrophyllum hellerianum (Nees) Schust. = Anastrophyllum helleranum [ANASHEL]
Anastrophyllum japonicum Steph. = Anastrophyllum michauxii [ANASMIC]
Anastrophyllum japonicum var. *otianum* Hatt. = Anastrophyllum assimile [ANASASS]
Anastrophyllum michauxii (Web.) Buch ex Evans [ANASMIC]
SY: *Anastrophyllum japonicum* Steph.
SY: *Anastrophyllum tamurae* Steph.
SY: *Sphenolobus michauxii* (Web.) Steph.
Anastrophyllum minutum (Schreb.) Schust. [ANASMIN]
SY: *Sphenolobus minutus* (Schreb.) Berggr.
Anastrophyllum minutum var. **grandis** (Gott. ex Lindb.) Schust. [ANASMIN1]
SY: *Cephaloziopsis saccatula* (Lindb.) Schiffn.
SY: *Sphenolobus minutus* var. *grandis* (Lindb.) Lindb. & S. Arnell
SY: *Sphenolobus saccatulus* (Lindb.) K. Müll.
Anastrophyllum minutum var. **minutum** [ANASMIN0]
SY: *Eremonotus minutus* (schreb.) Schust.
Anastrophyllum reichardtii (Gott.) Steph. = Anastrophyllum assimile [ANASASS]
Anastrophyllum saxicola (Schrad.) Schust. [ANASSAX]
SY: *Sphenolobus saxicolus* (Schrad.) Steph.
Anastrophyllum tamurae Steph. = Anastrophyllum michauxii [ANASMIC]
Aplozia sphaerocarpa (Hook.) Dum. = Jungermannia sphaerocarpa [JUNGSPH]
Barbilophozia Loeske [BARBILO$]
Barbilophozia attenuata (Mart.) Loeske [BARBATT]
SY: *Barbilophozia gracilis* (Schleich.) K. Müll.
SY: *Lophozia attenuata* (Mart.) Dum.
SY: *Lophozia gracilis* (Schleich.) Steph.
SY: *Orthocaulis attenuatus* (Mart.) Evans
SY: *Orthocaulis gracilis* (Schleich.) Buch
Barbilophozia barbata (Schmid. ex Schreb.) Loeske [BARBBAR]
SY: *Lophozia barbata* (Schmid.) Dum.
Barbilophozia binsteadii (Kaal.) Loeske [BARBBIN]
SY: *Lophozia binsteadii* (Kaal.) Evans
SY: *Orthocaulis binsteadii* (Kaal.) Buch
Barbilophozia floerkei (Web. & Mohr) Loeske [BARBFLO] (mountain leafy liverwort)
SY: *Lophozia floerkei* (Web. & Mohr) Schiffn.
SY: *Orthocaulis floerkei* (Web. & Mohr) Buch
Barbilophozia gracilis (Schleich.) K. Müll. = Barbilophozia attenuata [BARBATT]
Barbilophozia hatcheri (Evans) Loeske [BARBHAT] (Hatcher's fan wort)
SY: *Lophozia baueriana* Schiffn.
SY: *Lophozia hatcheri* (Evans) Steph.
Barbilophozia kunzeana (Hub.) Gams. [BARBKUN]
SY: *Lophozia kunzeana* (Hüb.) Buch
SY: *Lophozia rotundiloba* nom. herb.
SY: *Orthocaulis kunzeana* (Hüb.) Buch
Barbilophozia lycopodioides (Wallr.) Loeske [BARBLYC] (maple liverwort)
SY: *Lophozia lycopodioides* (Wallr.) Cogn.
Barbilophozia quadriloba (Lindb.) Loeske [BARBQUA]
SY: *Lophozia quadriloba* (Lindb.) Evans
SY: *Orthocaulis quadrilobus* (Lindb.) Buch
Cephaloziella pearsonii (Spruce) Douin = Sphenolobopsis pearsonii [SPHEPEA]
Cephaloziopsis pearsonii (Spruce) Schiffn. = Sphenolobopsis pearsonii [SPHEPEA]
Cephaloziopsis saccatula (Lindb.) Schiffn. = Anastrophyllum minutum var. grandis [ANASMIN1]
Chandonanthus Mitt. [CHANDON$]
Chandonanthus cavallii (Gola) S. Arnell = Tetralophozia setiformis [TETRSET]
Chandonanthus filiformis Steph. = Tetralophozia filiformis [TETRFIL]
Chandonanthus hirtellus (Web.) Mitt. [CHANHIR]
Chandonanthus pusillus Steph. = Tetralophozia filiformis [TETRFIL]
Chandonanthus quadrifidus Steph. = Tetralophozia setiformis [TETRSET]
Chandonanthus setiformis (Ehrh.) Lindb. = Tetralophozia setiformis [TETRSET]
Crossocalyx hellerianus (Nees) Meyl. = Anastrophyllum helleranum [ANASHEL]
Eremonotus hellerianus (Meyl.) Schust. = Anastrophyllum helleranum [ANASHEL]
Eremonotus minutus (schreb.) Schust. = Anastrophyllum minutum var. minutum [ANASMIN0]

Gymnocolea (Dum.) Dum. [GYMNOCO$]
Gymnocolea inflata (Huds.) Dum. [GYMNINF]
Isopaches bicrenatus (Schmid.) Buch = Lophozia bicrenata [LOPHBIC]
Isopaches decolorans Buch = Lophozia decolorans [LOPHDEC]
Isopaches hellerianus (Nees) Buch = Anastrophyllum helleranum [ANASHEL]
Jamesoniella (Spruce) Carring. [JAMESON$]
Jamesoniella autumnalis (DC.) Steph. [JAMEAUT] (Jameson's liverwort)
SY: *Jungermannia allenii* Clark
Jungermannia L. emend. Dum. [JUNGERM$]
Jungermannia allenii Clark = Jamesoniella autumnalis [JAMEAUT]
Jungermannia atrovirens Dum. [JUNGATR]
SY: *Jungermannia lanceolata* L.
SY: *Jungermannia tristis* Nees
SY: *Solenostoma atrovirens* (Dum.) K. Müll.
SY: *Solenostoma triste* (Nees) K. Müll
Jungermannia bolanderi Gott. ex Underw. = Jungermannia confertissima [JUNGCON]
Jungermannia borealis Damscholdt & Vana [JUNGBOR]
Jungermannia confertissima Nees [JUNGCON]
SY: *Jungermannia bolanderi* Gott. ex Underw.
SY: *Jungermannia danicola* Gott. ex Underw.
SY: *Solenostoma levieri* (Steph.) Steph.
SY: *Solenostoma pusillum* var. *vinaceum* Schust.
SY: *Jungermannia sphaerocarpa* auct.
Jungermannia danicola Gott. ex Underw. = Jungermannia confertissima [JUNGCON]
Jungermannia decolorans Limpr. = Lophozia decolorans [LOPHDEC]
Jungermannia exsertifolia Steph. [JUNGEXS]
Jungermannia exsertifolia ssp. **cordifolia** (Dum.) Vana [JUNGEXS1]
Jungermannia exsertifolia ssp. **exsertifolia** [JUNGEXS0]
Jungermannia hyalina Lyell [JUNGHYA]
SY: *Plectocolea hyalina* (Lyell) Mitt.
SY: *Solenostoma hyalinum* (Lyell) Mitt.
SY: *Solenostoma ontariensis* Schust.
Jungermannia lanceolata L. = Jungermannia atrovirens [JUNGATR]
Jungermannia lanceolata auct. = Jungermannia leiantha [JUNGLEI]
Jungermannia leiantha Grolle [JUNGLEI]
SY: *Jungermannia lanceolata* auct.
Jungermannia obovata Nees [JUNGOBO]
SY: *Plectocolea obovata* (Nees) Lindb.
SY: *Solenostoma obovatum* (Nees) Mass.
Jungermannia pumila With. [JUNGPUM]
SY: *Jungermannia pumila* var. *rivularis* (Schiffn.) Frye & Clark
SY: *Solenostoma pumilum* (With.) K. Müll
SY: *Solenostoma pumilum* ssp. *anomalum* Schust. & Damsh.
Jungermannia pumila var. *rivularis* (Schiffn.) Frye & Clark = Jungermannia pumila [JUNGPUM]
Jungermannia rubra Gott. ex Underw. [JUNGRUB]
SY: *Nardia rubra* (Gott.) Evans
SY: *Plectocolea rubra* (Gott. ex Underw.) Buch et al.
SY: *Solenostoma rubrum* (Gott. ex Underw.) Schust.
Jungermannia schusterana J.D. Godfr. & G.A. Godfr. [JUNGSCH]
Jungermannia sphaerocarpa Hook. [JUNGSPH]
SY: *Aplozia sphaerocarpa* (Hook.) Dum.
SY: *Jungermannia sphaerocarpa* var. *amplexicaulis* (Dum.) Frye & Clark
SY: *Jungermannia sphaerocarpa* var. *nana* (Ness) Frye & Clark
SY: *Solenostoma sphaerocarpa* (Hook.) Steph.
SY: *Solenostoma sphaerocarpa* var. *nana* (Nees) Schust.
Jungermannia sphaerocarpa auct. = Jungermannia confertissima [JUNGCON]
Jungermannia sphaerocarpa var. *amplexicaulis* (Dum.) Frye & Clark = Jungermannia sphaerocarpa [JUNGSPH]
Jungermannia sphaerocarpa var. *nana* (Ness) Frye & Clark = Jungermannia sphaerocarpa [JUNGSPH]
Jungermannia subelliptica (Lindb. ex Kaal.) Lev. [JUNGSUB]
SY: *Plectocolea subelliptica* (Lindb. ex Kaal.) Evans
SY: *Solenostoma subellipticum* (Lindb. ex Kaal.) Schust.
Jungermannia tristis Nees = Jungermannia atrovirens [JUNGATR]
Leiocolea badensis (Gott. ex Rabenh.) Joerg. = Lophozia badensis [LOPHBAD]
Leiocolea bantriensis (Hook.) Joerg. = Lophozia bantriensis [LOPHBAN]
Leiocolea gillmanii (Aust.) Evans = Lophozia gillmanii [LOPHGIL]
Leiocolea heterocolpa (Thed.) Buch = Lophozia heterocolpos [LOPHHEE]
Leiocolea heterocolpos (Thed.) Buch = Lophozia heterocolpos [LOPHHEE]
Leiocolea muelleri (Nees) Joerg. = Lophozia collaris [LOPHCOL]
Leiocolea obtusa (Lindb.) Buch = Lophozia obtusa [LOPHOBT]
Leiocolea rutheana (Lindb.) K. Müll. = Lophozia rutheana [LOPHRUT]
Lophozia (Dum.) Dum. [LOPHOZI$]
Lophozia alpestris (Schleich. ex Web.) Evans [LOPHALP]
Lophozia ascendens (Warnst.) Schust. [LOPHASC]
SY: *Lophozia gracillima* Buch
SY: *Sphenolobus ascendens* Warnst.
SY: *Lophozia longidens* auct.
Lophozia attenuata (Mart.) Dum. = Barbilophozia attenuata [BARBATT]
Lophozia badensis (Gott. ex Rabenh.) Schiffn. [LOPHBAD]
SY: *Leiocolea badensis* (Gott. ex Rabenh.) Joerg.
Lophozia bantriensis (Hook.) Steph. [LOPHBAN]
SY: *Leiocolea bantriensis* (Hook.) Joerg.
SY: *Lophozia hornschuchiana* Macoun
Lophozia barbata (Schmid.) Dum. = Barbilophozia barbata [BARBBAR]
Lophozia baueriana Schiffn. = Barbilophozia hatcheri [BARBHAT]
Lophozia bicrenata (Schmid. ex Hoffm.) Dum. [LOPHBIC]
SY: *Isopaches bicrenatus* (Schmid.) Buch
Lophozia binsteadii (Kaal.) Evans = Barbilophozia binsteadii [BARBBIN]
Lophozia collaris (Nees) Dum. [LOPHCOL]
SY: *Leiocolea muelleri* (Nees) Joerg.
SY: *Lophozia muelleri* (Nees) Dum.
Lophozia confertifolia Schiffn. = Lophozia wenzelii [LOPHWEN]
Lophozia decolorans (Limpr.) Steph. [LOPHDEC]
SY: *Isopaches decolorans* Buch
SY: *Jungermannia decolorans* Limpr.
Lophozia excisa (Dicks.) Dum. [LOPHEXC]
Lophozia fauriana Steph. = Lophozia guttulata [LOPHGUT]
Lophozia floerkei (Web. & Mohr) Schiffn. = Barbilophozia floerkei [BARBFLO]
Lophozia gillmanii (Aust.) Schust. [LOPHGIL]
SY: *Leiocolea gillmanii* (Aust.) Evans
SY: *Lophozia kaurini* (Limpr.) Steph.
Lophozia gracilis (Schleich.) Steph. = Barbilophozia attenuata [BARBATT]
Lophozia gracillima Buch = Lophozia ascendens [LOPHASC]
Lophozia grandiretis (Lindb.) Schiffn. [LOPHGRA]
SY: *Lophozia murmanica* Kaal.
Lophozia guttulata (Lindb. & Arnell) Evans [LOPHGUT]
SY: *Lophozia fauriana* Steph.
SY: *Lophozia porphyroleuca* (Nees) Schiffn.
Lophozia hatcheri (Evans) Steph. = Barbilophozia hatcheri [BARBHAT]
Lophozia heterocolpos (Thed.) M.A. Howe [LOPHHEE]
SY: *Leiocolea heterocolpa* (Thed.) Buch
SY: *Leiocolea heterocolpos* (Thed.) Buch
Lophozia hornschuchiana Macoun = Lophozia bantriensis [LOPHBAN]
Lophozia incisa (Schrad.) Dum. [LOPHINC]
Lophozia kaurini (Limpr.) Steph. = Lophozia gillmanii [LOPHGIL]
Lophozia kunzeana (Hüb.) Buch = Barbilophozia kunzeana [BARBKUN]
Lophozia longidens (Lindb.) Macoun [LOPHLON]
Lophozia longidens auct. = Lophozia ascendens [LOPHASC]
Lophozia longiflora (Nees) Schiffn. = Lophozia ventricosa var. longiflora [LOPHVEN3]
Lophozia lycopodioides (Wallr.) Cogn. = Barbilophozia lycopodioides [BARBLYC]

Lophozia muelleri (Nees) Dum. = Lophozia collaris [LOPHCOL]
Lophozia murmanica Kaal. = Lophozia grandiretis [LOPHGRA]
Lophozia obtusa (Lindb.) Evans [LOPHOBT]
SY: *Leiocolea obtusa* (Lindb.) Buch
SY: *Obtusifolium obtusum* (Lindb.) S. Arnell
Lophozia opacifolia Culm. [LOPHOPA]
Lophozia porphyroleuca (Nees) Schiffn. = Lophozia guttulata [LOPHGUT]
Lophozia quadriloba (Lindb.) Evans = Barbilophozia quadriloba [BARBQUA]
Lophozia quinquedentata (Huds.) Cogn. = Tritomaria quinquedentata [TRITQUI]
Lophozia rotundiloba nom. herb. = Barbilophozia kunzeana [BARBKUN]
Lophozia rutheana (Limpr.) M.A. Howe [LOPHRUT]
SY: *Leiocolea rutheana* (Lindb.) K. Müll.
SY: *Lophozia schultzii* (Nees) Schiffn.
Lophozia schultzii (Nees) Schiffn. = Lophozia rutheana [LOPHRUT]
Lophozia silvicola Buch = Lophozia ventricosa var. silvicola [LOPHVEN5]
Lophozia silvicoloides Kitag. = Lophozia ventricosa var. silvicola [LOPHVEN5]
Lophozia ventricosa (Dicks.) Dum. [LOPHVEN] (leafy liverwort)
Lophozia ventricosa var. **longiflora** (Nees) Macoun [LOPHVEN3]
SY: *Lophozia longiflora* (Nees) Schiffn.
Lophozia ventricosa var. **silvicola** (Buch) Jones [LOPHVEN5]
SY: *Lophozia silvicola* Buch
SY: *Lophozia silvicoloides* Kitag.
Lophozia ventricosa var. **ventricosa** [LOPHVEN0]
Lophozia wenzelii (Nees) Steph. [LOPHWEN]
SY: *Lophozia confertifolia* Schiffn.
Mylia Gray [MYLIA$]
Mylia anomala (Hook.) Gray [MYLIANO] (hard scale liverwort)
Mylia taylorii (Hook.) Gray [MYLITAY] (Taylor's liverwort)
Nardia Gray [NARDIA$]
Nardia breidleri (Limpr.) Lindb. [NARDBRE]
Nardia cavana Clark = Nardia scalaris [NARDSCA]
Nardia compressa (Hook.) Gray [NARDCOM]
Nardia geoscyphus (DeNot.) Lindb. [NARDGEO]
Nardia geoscyphus var. *insecta* (K. Müll.) Schiffn. = Nardia insecta [NARDINS]
Nardia insecta Lindb. [NARDINS]
SY: *Nardia geoscyphus* var. *insecta* (K. Müll.) Schiffn.
Nardia japonica Steph. [NARDJAP]
Nardia rubra (Gott.) Evans = Jungermannia rubra [JUNGRUB]
Nardia scalaris Gray [NARDSCA]
SY: *Nardia cavana* Clark
Obtusifolium obtusum (Lindb.) S. Arnell = Lophozia obtusa [LOPHOBT]
Orthocaulis attenuatus (Mart.) Evans = Barbilophozia attenuata [BARBATT]
Orthocaulis binsteadii (Kaal.) Buch = Barbilophozia binsteadii [BARBBIN]
Orthocaulis floerkei (Web. & Mohr) Buch = Barbilophozia floerkei [BARBFLO]
Orthocaulis gracilis (Schleich.) Buch = Barbilophozia attenuata [BARBATT]
Orthocaulis kunzeana (Hüb.) Buch = Barbilophozia kunzeana [BARBKUN]
Orthocaulis quadrilobus (Lindb.) Buch = Barbilophozia quadriloba [BARBQUA]
Plectocolea hyalina (Lyell) Mitt. = Jungermannia hyalina [JUNGHYA]
Plectocolea obovata (Nees) Lindb. = Jungermannia obovata [JUNGOBO]
Plectocolea rubra (Gott. ex Underw.) Buch et al. = Jungermannia rubra [JUNGRUB]
Plectocolea subelliptica (Lindb. ex Kaal.) Evans = Jungermannia subelliptica [JUNGSUB]
Saccobasis polita (Nees) Buch = Tritomaria polita [TRITPOL]
Solenostoma atrovirens (Dum.) K. Müll. = Jungermannia atrovirens [JUNGATR]
Solenostoma hyalinum (Lyell) Mitt. = Jungermannia hyalina [JUNGHYA]
Solenostoma levieri (Steph.) Steph. = Jungermannia confertissima [JUNGCON]
Solenostoma obovatum (Nees) Mass. = Jungermannia obovata [JUNGOBO]
Solenostoma ontariensis Schust. = Jungermannia hyalina [JUNGHYA]
Solenostoma pumilum (With.) K. Müll = Jungermannia pumila [JUNGPUM]
Solenostoma pumilum ssp. *anomalum* Schust. & Damsh. = Jungermannia pumila [JUNGPUM]
Solenostoma pusillum var. *vinaceum* Schust. = Jungermannia confertissima [JUNGCON]
Solenostoma rubrum (Gott. ex Underw.) Schust. = Jungermannia rubra [JUNGRUB]
Solenostoma sphaerocarpa (Hook.) Steph. = Jungermannia sphaerocarpa [JUNGSPH]
Solenostoma sphaerocarpa var. *nana* (Nees) Schust. = Jungermannia sphaerocarpa [JUNGSPH]
Solenostoma subellipticum (Lindb. ex Kaal.) Schust. = Jungermannia subelliptica [JUNGSUB]
Solenostoma triste (Nees) K. Müll = Jungermannia atrovirens [JUNGATR]
Sphenolobopsis Schust. & Kitag. [SPHENOL$]
Sphenolobopsis pearsonii (Spruce) Schust. [SPHEPEA]
SY: *Cephaloziella pearsonii* (Spruce) Douin
SY: *Cephaloziopsis pearsonii* (Spruce) Schiffn.
Sphenolobus ascendens Warnst. = Lophozia ascendens [LOPHASC]
Sphenolobus exsectiformis (Breidl.) K. Müll. = Tritomaria exsectiformis [TRITEXE]
Sphenolobus exsectus (Schmid. ex Schrad.) Steph. = Tritomaria exsecta [TRITEXS]
Sphenolobus helleranus (Nees) Steph. = Anastrophyllum helleranum [ANASHEL]
Sphenolobus michauxii (Web.) Steph. = Anastrophyllum michauxii [ANASMIC]
Sphenolobus minutus (Schreb.) Berggr. = Anastrophyllum minutum [ANASMIN]
Sphenolobus minutus var. *grandis* (Lindb.) Lindb. & S. Arnell = Anastrophyllum minutum var. grandis [ANASMIN1]
Sphenolobus politus Steph. = Tritomaria polita [TRITPOL]
Sphenolobus saccatulus (Lindb.) K. Müll. = Anastrophyllum minutum var. grandis [ANASMIN1]
Sphenolobus saxicolus (Schrad.) Steph. = Anastrophyllum saxicola [ANASSAX]
Temnoma setiforme (Ehrh.) M.A. Howe = Tetralophozia setiformis [TETRSET]
Tetralophozia (Schust.) Schljak. [TETRALO$]
Tetralophozia filiformis (Steph.) Lammes [TETRFIL]
SY: *Chandonanthus filiformis* Steph.
SY: *Chandonanthus pusillus* Steph.
Tetralophozia setiformis (Ehrh.) Schljak. [TETRSET]
SY: *Chandonanthus cavallii* (Gola) S. Arnell
SY: *Chandonanthus quadrifidus* Steph.
SY: *Chandonanthus setiformis* (Ehrh.) Lindb.
SY: *Temnoma setiforme* (Ehrh.) M.A. Howe
Tritomaria Schiffn. ex Loeske [TRITOMA$]
Tritomaria exsecta (Schmid. ex Schrad.) Loeske [TRITEXS]
SY: *Sphenolobus exsectus* (Schmid. ex Schrad.) Steph.
Tritomaria exsectiformis (Breidl.) Loeske [TRITEXE]
SY: *Sphenolobus exsectiformis* (Breidl.) K. Müll.
Tritomaria polita (Nees) Joerg. [TRITPOL]
SY: *Saccobasis polita* (Nees) Buch
SY: *Sphenolobus politus* Steph.
Tritomaria quinquedentata (Huds.) Buch [TRITQUI]
SY: *Lophozia quinquedentata* (Huds.) Cogn.
Tritomaria scitula (Tayl.) Joerg. [TRITSCI]

LEJEUNEACEAE (B:H)

Cololejeunea (Spruce) Schiffn. [COLOLEJ$]

Cololejeunea handelii (Herz.) Hatt. = Cololejeunea macounii [COLOMAC]
Cololejeunea macounii (Spruce ex Underw.) Evans [COLOMAC]
SY: *Cololejeunea handelii* (Herz.) Hatt.
SY: *Cololejeunea rupicola* (Steph.) Hatt.
Cololejeunea rupicola (Steph.) Hatt. = Cololejeunea macounii [COLOMAC]

LEPIDOZIACEAE (B:H)

Bazzania Gray [BAZZANI$]
Bazzania ambigua (Lindenb.) Trev. = Bazzania denudata [BAZZDEN]
Bazzania denudata (Torrey ex Gott. & al.) Trev. [BAZZDEN]
SY: *Bazzania ambigua* (Lindenb.) Trev.
Bazzania pearsonii Steph. [BAZZPEA]
Bazzania pearsonii f. **krajinai** Hong [BAZZPEA1]
Bazzania triangularis Lindb. = Bazzania tricrenata [BAZZTRI]
Bazzania tricrenata (Wahlenb.) Lindb. [BAZZTRI] (three-toothed whip liverwort)
SY: *Bazzania triangularis* Lindb.
Bazzania tricrenata f. **vittii** Hong [BAZZTRI2]
Bazzania tricrenata var. **fulfordiae** Hong [BAZZTRI1]
Bazzania tricrenata var. **tricrenata** [BAZZTRI0]
Bazzania trilobata (L.) Gray [BAZZTRL]
Dendrobazzania Schust. & Schof. [DENDROB$]
Dendrobazzania griffithiana (Steph.) Schust. & Schof. [DENDGRI]
Kurzia G. Martens [KURZIA$]
Kurzia makinoana (Steph.) Grolle = Kurzia sylvatica [KURZSYL]
Kurzia pauciflora (Dicks.) Grolle [KURZPAU]
Kurzia setacea (Web.) Grolle [KURZSET]
SY: *Leiocolea setacea* (Web.) Mitt.
SY: *Microlepidozia setacea* (Web.) Joerg.
Kurzia sylvatica (Evans) Grolle [KURZSYL]
SY: *Kurzia makinoana* (Steph.) Grolle
SY: *Leiocolea sylvatica* Evans
SY: *Microlepidozia makinoana* Hatt. & Mizut.
SY: *Microlepidozia sylvatica* (Evans) Joerg.
Kurzia trichoclados (K. Müll.) Grolle [KURZTRI]
Leiocolea setacea (Web.) Mitt. = Kurzia setacea [KURZSET]
Leiocolea sylvatica Evans = Kurzia sylvatica [KURZSYL]
Lepidozia (Dum.) Dum. [LEPIDOZ$]
Lepidozia filamentosa (Lehm. & Lindenb.) Lindenb. [LEPIFIL]
Lepidozia reptans (L.) Dum. [LEPIREP] (little hands liverwort)
Lepidozia sandvicensis Lindenb. [LEPISAN]
Microlepidozia makinoana Hatt. & Mizut. = Kurzia sylvatica [KURZSYL]
Microlepidozia setacea (Web.) Joerg. = Kurzia setacea [KURZSET]
Microlepidozia sylvatica (Evans) Joerg. = Kurzia sylvatica [KURZSYL]

LESKEACEAE (B:M)

Amblystegium montanae Bryhn in Holz. = Leskeella nervosa [LESKNER]
Anomodon heteroideus Kindb. in Mac. & Kindb. = Leskeella nervosa [LESKNER]
Anomodon subrigidulus Kindb. = Leskeella nervosa [LESKNER]
Bryohaplocladium Wat. & Iwats. [BRYOHAP$]
Bryohaplocladium microphyllum (Hedw.) Wat. & Iwats. [BRYOMIC]
SY: *Haplocladium microphyllum* (Hedw.) Broth.
SY: *Haplocladium microphyllum* var. *lignicola* (Kindb.) Reim.
SY: *Haplocladium microphyllum* var. *obtusum* (Grout) Crum
SY: *Hypnum gracile* Bruch & Schimp. in Hook.
SY: *Thuidium gracile* B.S.G.
SY: *Thuidium lignicola* Kindb.
SY: *Thuidium microphyllum* (Hedw.) Jaeg.
SY: *Thuidium microphyllum* var. *lignicola* (Kindb.) Best
SY: *Thuidium microphyllum* var. *obtusum* Grout
Claopodium (Lesq. & James) Ren. & Card. [CLAOPOD$]
Claopodium bolanderi Best [CLAOBOL]
SY: *Thuidium bolanderi* (Best) Par.
Claopodium crispifolium (Hook.) Ren. & Card. [CLAOCRI] (rough moss)
SY: *Thuidium crispifolium* (Hook.) Lindb.
Claopodium pellucinerve (Mitt.) Best [CLAOPEL]
SY: *Claopodium subpiliferum* (Lindb. & Arnell) Broth.
Claopodium subpiliferum (Lindb. & Arnell) Broth. = Claopodium pellucinerve [CLAOPEL]
Claopodium whippleanum (Sull. in Whipple & Ives) Ren. & Card. [CLAOWHI]
SY: *Claopodium whippleanum* var. *leuconeuron* (Sull. & Lesq.) Grout
SY: *Thuidium leuconeuron* (Sull. & Lesq.) Lesq.
SY: *Thuidium whippleanum* (Sull.) Jaeg. & Sauerb.
Claopodium whippleanum var. *leuconeuron* (Sull. & Lesq.) Grout = Claopodium whippleanum [CLAOWHI]
Haplocladium microphyllum (Hedw.) Broth. = Bryohaplocladium microphyllum [BRYOMIC]
Haplocladium microphyllum var. *lignicola* (Kindb.) Reim. = Bryohaplocladium microphyllum [BRYOMIC]
Haplocladium microphyllum var. *obtusum* (Grout) Crum = Bryohaplocladium microphyllum [BRYOMIC]
Heterocladium papillosum (Lindb.) Lindb. = Pseudoleskeella papillosa [PSEUPAP]
Hypnum atrovirens Dicks. = Pseudoleskea patens [PSEUPAT]
Hypnum gracile Bruch & Schimp. in Hook. = Bryohaplocladium microphyllum [BRYOMIC]
Lescuraea Schimp. in B.S.G. [LESCURA$]
Lescuraea atricha (Kindb. in Mac. & Kindb.) Lawt. = Pseudoleskea atricha [PSEUATR]
Lescuraea baileyi (Best & Grout in Grout) Lawt. = Pseudoleskea baileyi [PSEUBAI]
Lescuraea frigida Kindb. = Lescuraea saxicola [LESCSAX]
Lescuraea iliamniana Lawt. = Pseudoleskea julacea [PSEUJUL]
Lescuraea imperfecta C. Müll. & Kindb. in Mac. & Kindb. = Pseudoleskea stenophylla [PSEUSTE]
Lescuraea incurvata (Hedw.) Lawt. = Pseudoleskea incurvata [PSEUINC]
Lescuraea incurvata var. *gigantea* Lawt. = Pseudoleskea incurvata var. gigantea [PSEUINC1]
Lescuraea julacea Besch. & Card. in Card. = Pseudoleskea julacea [PSEUJUL]
Lescuraea mutabilis var. *saxicola* (Schimp. in B.S.G.) Hag. = Lescuraea saxicola [LESCSAX]
Lescuraea patens (Lindb.) Arnell & C. Jens. = Pseudoleskea patens [PSEUPAT]
Lescuraea radicosa (Mitt.) Mönk. = Pseudoleskea radicosa [PSEURAD]
Lescuraea radicosa var. *compacta* (Best) Lawt. = Pseudoleskea radicosa var. compacta [PSEURAD1]
Lescuraea radicosa var. *denudata* (Kindb. in Mac. & Kindb.) Lawt. = Pseudoleskea radicosa var. denudata [PSEURAD2]
Lescuraea radicosa var. *pallida* (Best) Lawt. = Pseudoleskea radicosa var. pallida [PSEURAD3]
Lescuraea saxicola (Schimp. in B.S.G.) Milde [LESCSAX]
SY: *Lescuraea frigida* Kindb.
SY: *Lescuraea mutabilis* var. *saxicola* (Schimp. in B.S.G.) Hag.
SY: *Pseudoleskea frigida* (Kindb.) Sharp in Grout
SY: *Pseudoleskea saxicola* (Schimp. in B.S.G.) Crum, Steere & Anderson
SY: *Pseudoleskea substriata* Best
Lescuraea stenophylla (Ren. & Card. in Röll) Kindb. = Pseudoleskea stenophylla [PSEUSTE]
Leskea Hedw. [LESKEA$]
Leskea arenicola Best = Leskea polycarpa [LESKPOL]
Leskea cyrtophylla Kindb. in Mac. & Kindb. = Pseudoleskeella tectorum [PSEUTEC]
Leskea montanae (Bryhn) Grout = Leskeella nervosa [LESKNER]
Leskea nervosa (Brid.) Myr. = Leskeella nervosa [LESKNER]
Leskea nervosa var. *bulbifera* (Brid.) Best in Mac. = Leskeella nervosa [LESKNER]
Leskea nervosa var. *flagellifera* Kindb. = Leskeella nervosa [LESKNER]

Leskea nervosa var. *nigrescens* (Kindb.) Best = Leskeella nervosa [LESKNER]
Leskea nigrescens Kindb. = Leskeella nervosa [LESKNER]
Leskea polycarpa Hedw. [LESKPOL]
SY: *Leskea arenicola* Best
SY: *Leskea polycarpa* var. *paludosa* (Hedw.) Schimp.
SY: *Leskea subobtusifolia* C. Müll. & Kindb. in Mac. & Kindb.
Leskea polycarpa var. *paludosa* (Hedw.) Schimp. = Leskea polycarpa [LESKPOL]
Leskea subobtusifolia C. Müll. & Kindb. in Mac. & Kindb. = Leskea polycarpa [LESKPOL]
Leskea tectorum (Funck ex Brid.) Lindb. = Pseudoleskeella tectorum [PSEUTEC]
Leskea tectorum var. *flagellifera* Best = Pseudoleskeella tectorum [PSEUTEC]
Leskea williamsii Best = Pseudoleskeella tectorum [PSEUTEC]
Leskea williamsii var. *filamentosa* Best = Pseudoleskeella tectorum [PSEUTEC]
Leskea wollei Aust. = Pseudoleskeella tectorum [PSEUTEC]
Leskeella (Limpr.) Loeske [LESKEEL$]
Leskeella nervosa (Brid.) Loeske [LESKNER]
SY: *Amblystegium montanae* Bryhn in Holz.
SY: *Anomodon heteroideus* Kindb. in Mac. & Kindb.
SY: *Anomodon subrigidulus* Kindb.
SY: *Leskea montanae* (Bryhn) Grout
SY: *Leskea nervosa* (Brid.) Myr.
SY: *Leskea nervosa* var. *bulbifera* (Brid.) Best in Mac.
SY: *Leskea nervosa* var. *flagellifera* Kindb.
SY: *Leskea nervosa* var. *nigrescens* (Kindb.) Best
SY: *Leskea nigrescens* Kindb.
SY: *Leskeella nervosa* f. *bulbifera* Möll.
SY: *Leskeella nervosa* var. *flagellifera* (Kindb.) Wijk & Marg.
SY: *Leskeella nervosa* var. *sibirica* (Arnell) Broth.
SY: *Leskeella nervosa* var. *subrigidula* (Kindb.) Podp.
SY: *Pseudoleskeella nervosa* (Brid.) Nyh.
Leskeella nervosa f. *bulbifera* Möll. = Leskeella nervosa [LESKNER]
Leskeella nervosa var. *flagellifera* (Kindb.) Wijk & Marg. = Leskeella nervosa [LESKNER]
Leskeella nervosa var. *sibirica* (Arnell) Broth. = Leskeella nervosa [LESKNER]
Leskeella nervosa var. *subrigidula* (Kindb.) Podp. = Leskeella nervosa [LESKNER]
Leskeella tectorum (Funck ex Brid.) Hag. = Pseudoleskeella tectorum [PSEUTEC]
Leskeella tectorum ssp. *cyrtophylla* (Kindb. in Mac. & Kindb.) Wijk & Marg. = Pseudoleskeella tectorum [PSEUTEC]
Pseudoleskea Schimp. in B.S.G. [PSEUDOL$]
Pseudoleskea atricha (Kindb. in Mac. & Kindb.) Kindb. [PSEUATR]
SY: *Lescuraea atricha* (Kindb. in Mac. & Kindb.) Lawt.
Pseudoleskea atrovirens (Brid.) Schimp. in B.S.G. = Pseudoleskea incurvata var. incurvata [PSEUINC0]
Pseudoleskea atrovirens auct. = Pseudoleskea patens [PSEUPAT]
Pseudoleskea baileyi Best & Grout in Grout [PSEUBAI] (rusty mountain heath moss)
SY: *Lescuraea baileyi* (Best & Grout in Grout) Lawt.
Pseudoleskea brachyclados (Schwaegr.) Kindb. = Pseudoleskea radicosa var. radicosa [PSEURAD0]
Pseudoleskea breidleri Kindb. = Pseudoleskea radicosa var. denudata [PSEURAD2]
Pseudoleskea denudata (Kindb. in Mac. & Kindb.) Kindb. in Best = Pseudoleskea radicosa var. denudata [PSEURAD2]
Pseudoleskea denudata var. *holzingeri* Best = Pseudoleskea radicosa var. radicosa [PSEURAD0]
Pseudoleskea falcicuspis C. Müll. & Kindb. in Mac. & Kindb. = Pseudoleskea incurvata var. incurvata [PSEUINC0]
Pseudoleskea frigida (Kindb.) Sharp in Grout = Lescuraea saxicola [LESCSAX]
Pseudoleskea howei (Best) L. Koch = Pseudoleskea radicosa var. radicosa [PSEURAD0]
Pseudoleskea iliamniana (Lawt.) Crum, Steere & Anderson = Pseudoleskea julacea [PSEUJUL]
Pseudoleskea incurvata (Hedw.) Loeske [PSEUINC]
SY: *Lescuraea incurvata* (Hedw.) Lawt.
Pseudoleskea incurvata var. **gigantea** (Lawt.) Crum, Steere & Anderson [PSEUINC1]
SY: *Lescuraea incurvata* var. *gigantea* Lawt.
Pseudoleskea incurvata var. **incurvata** [PSEUINC0]
SY: *Pseudoleskea atrovirens* (Brid.) Schimp. in B.S.G.
SY: *Pseudoleskea falcicuspis* C. Müll. & Kindb. in Mac. & Kindb.
SY: *Pseudoleskea oligoclada* Kindb.
Pseudoleskea julacea (Besch. & Card. in Card.) Crum, Steere & Anderson [PSEUJUL]
SY: *Lescuraea iliamniana* Lawt.
SY: *Lescuraea julacea* Besch. & Card. in Card.
SY: *Pseudoleskea iliamniana* (Lawt.) Crum, Steere & Anderson
Pseudoleskea oligoclada Kindb. = Pseudoleskea incurvata var. incurvata [PSEUINC0]
Pseudoleskea pallida Best = Pseudoleskea radicosa var. pallida [PSEURAD3]
Pseudoleskea patens (Lindb.) Kindb. [PSEUPAT]
SY: *Hypnum atrovirens* Dicks.
SY: *Lescuraea patens* (Lindb.) Arnell & C. Jens.
SY: *Pseudoleskea atrovirens* auct.
Pseudoleskea radicosa (Mitt.) Mac. & Kindb. [PSEURAD]
SY: *Lescuraea radicosa* (Mitt.) Mönk.
Pseudoleskea radicosa var. **compacta** Best [PSEURAD1]
SY: *Lescuraea radicosa* var. *compacta* (Best) Lawt.
Pseudoleskea radicosa var. **denudata** (Kindb. in Mac. & Kindb.) Wijk & Marg. [PSEURAD2]
SY: *Lescuraea radicosa* var. *denudata* (Kindb. in Mac. & Kindb.) Lawt.
SY: *Pseudoleskea breidleri* Kindb.
SY: *Pseudoleskea denudata* (Kindb. in Mac. & Kindb.) Kindb. in Best
SY: *Pseudoleskea sciuroides* var. *denudata* Kindb. in Mac. & Kindb.
Pseudoleskea radicosa var. *holzingeri* (Best) Hag. = Pseudoleskea radicosa var. radicosa [PSEURAD0]
Pseudoleskea radicosa var. **pallida** (Best) Crum, Steere & Anderson [PSEURAD3]
SY: *Lescuraea radicosa* var. *pallida* (Best) Lawt.
SY: *Pseudoleskea pallida* Best
Pseudoleskea radicosa var. **radicosa** [PSEURAD0]
SY: *Pseudoleskea brachyclados* (Schwaegr.) Kindb.
SY: *Pseudoleskea denudata* var. *holzingeri* Best
SY: *Pseudoleskea howei* (Best) L. Koch
SY: *Pseudoleskea radicosa* var. *holzingeri* (Best) Hag.
SY: *Pseudoleskea sciuroides* Kindb.
Pseudoleskea rigescens Lindb. = Pseudoleskea stenophylla [PSEUSTE]
Pseudoleskea saxicola (Schimp. in B.S.G.) Crum, Steere & Anderson = Lescuraea saxicola [LESCSAX]
Pseudoleskea sciuroides Kindb. = Pseudoleskea radicosa var. radicosa [PSEURAD0]
Pseudoleskea sciuroides var. *denudata* Kindb. in Mac. & Kindb. = Pseudoleskea radicosa var. denudata [PSEURAD2]
Pseudoleskea stenophylla Ren. & Card. in Röll [PSEUSTE]
SY: *Lescuraea imperfecta* C. Müll. & Kindb. in Mac. & Kindb.
SY: *Lescuraea stenophylla* (Ren. & Card. in Röll) Kindb.
SY: *Pseudoleskea rigescens* Lindb.
Pseudoleskea substriata Best = Lescuraea saxicola [LESCSAX]
Pseudoleskeella Kindb. [PSEUDOE$]
Pseudoleskeella nervosa (Brid.) Nyh. = Leskeella nervosa [LESKNER]
Pseudoleskeella tectorum (Funck ex Brid.) Kindb. in Broth. [PSEUTEC]
SY: *Leskea cyrtophylla* Kindb. in Mac. & Kindb.
SY: *Leskea tectorum* (Funck ex Brid.) Lindb.
SY: *Leskea tectorum* var. *flagellifera* Best
SY: *Leskea williamsii* Best
SY: *Leskea williamsii* var. *filamentosa* Best
SY: *Leskea wollei* Aust.
SY: *Leskeella tectorum* (Funck ex Brid.) Hag.
SY: *Leskeella tectorum* ssp. *cyrtophylla* (Kindb. in Mac. & Kindb.) Wijk & Marg.

SY: *Pseudoleskeella tectorum* var. *cyrtophylla* (Kindb. in Mac. & Kindb.) Crum, Steere & Anderson
SY: *Pseudoleskeella tectorum* var. *flagellifera* (Best) Amann
SY: *Pseudoleskeella williamsii* (Best) Crum, Steere & Anderson
SY: *Pseudoleskeella williamsii* var. *filamentosa* (Best) Crum, Steere & Anderson
Pseudoleskeella papillosa (Lindb.) Kindb [PSEUPAP]
SY: *Heterocladium papillosum* (Lindb.) Lindb.
Pseudoleskeella tectorum var. *cyrtophylla* (Kindb. in Mac. & Kindb.) Crum, Steere & Anderson = Pseudoleskeella tectorum [PSEUTEC]
Pseudoleskeella tectorum var. *flagellifera* (Best) Amann = Pseudoleskeella tectorum [PSEUTEC]
Pseudoleskeella williamsii (Best) Crum, Steere & Anderson = Pseudoleskeella tectorum [PSEUTEC]
Pseudoleskeella williamsii var. *filamentosa* (Best) Crum, Steere & Anderson = Pseudoleskeella tectorum [PSEUTEC]
Thuidium bolanderi (Best) Par. = Claopodium bolanderi [CLAOBOL]
Thuidium crispifolium (Hook.) Lindb. = Claopodium crispifolium [CLAOCRI]
Thuidium gracile B.S.G. = Bryohaplocladium microphyllum [BRYOMIC]
Thuidium leuconeuron (Sull. & Lesq.) Lesq. = Claopodium whippleanum [CLAOWHI]
Thuidium lignicola Kindb. = Bryohaplocladium microphyllum [BRYOMIC]
Thuidium microphyllum (Hedw.) Jaeg. = Bryohaplocladium microphyllum [BRYOMIC]
Thuidium microphyllum var. *lignicola* (Kindb.) Best = Bryohaplocladium microphyllum [BRYOMIC]
Thuidium microphyllum var. *obtusum* Grout = Bryohaplocladium microphyllum [BRYOMIC]
Thuidium whippleanum (Sull.) Jaeg. & Sauerb. = Claopodium whippleanum [CLAOWHI]

LEUCODONTACEAE (B:M)

Alsia Sull. [ALSIA$]
Alsia abietina (Hook.) Sull. = Dendroalsia abietina [DENDABI]
Alsia californica (Hook. & Arnott) Sull. [ALSICAL]
SY: *Alsia californica* var. *flagellifera* (Ren. & Card.) Sull.
Alsia californica var. *flagellifera* (Ren. & Card.) Sull. = Alsia californica [ALSICAL]
Antitrichia Brid. [ANTITRI$]
Antitrichia californica Sull. in Lesq. [ANTICAL]
SY: *Antitrichia tenella* Kindb.
Antitrichia curtipendula (Hedw.) Brid. [ANTICUR] (hanging moss)
Antitrichia tenella Kindb. = Antitrichia californica [ANTICAL]
Dendroalsia Britt. [DENDROA$]
Dendroalsia abietina (Hook.) Britt. [DENDABI] (plume moss)
SY: *Alsia abietina* (Hook.) Sull.

LOPHOCOLEACEAE (B:H)

Chiloscyphus Corda [CHILOSC$]
Chiloscyphus pallescens (Ehrh. ex Hoffm.) Dum. [CHILPAL]
Chiloscyphus polyanthos (L.) Corda [CHILPOL]
Chiloscyphus polyanthos var. **polyanthos** [CHILPOL0]
Chiloscyphus polyanthos var. **rivularis** (Schrad.) Nees [CHILPOL2]
SY: *Chiloscyphus rivularis* (Schrad.) Loeske
Chiloscyphus rivularis (Schrad.) Loeske = Chiloscyphus polyanthos var. rivularis [CHILPOL2]
Lophocolea (Dum.) Dum. [LOPHOCO$]
Lophocolea austinii Lindb. = Lophocolea heterophylla [LOPHHET]
Lophocolea bidentata (L.) Dum. [LOPHBID]
Lophocolea heterophylla (Schrad.) Dum. [LOPHHET]
SY: *Lophocolea austinii* Lindb.
SY: *Lophocolea macounii* Aust.
Lophocolea macounii Aust. = Lophocolea heterophylla [LOPHHET]
Lophocolea minor Nees [LOPHMIN]

LUNULARIACEAE (B:H)

Lunularia Adans. [LUNULAR$]
Lunularia cruciata (L.) Dum. [LUNUCRU]

MARCHANTIACEAE (B:H)

Bucegia Radian [BUCEGIA$]
Bucegia romanica Radian [BUCEROM]
Marchantia L. [MARCHAN$]
Marchantia alpestris (Nees) Burgeff [MARCALP]
Marchantia aquatica (Nees) Burgeff = Marchantia polymorpha [MARCPOL]
Marchantia polymorpha L. [MARCPOL] (lung liverwort)
SY: *Marchantia aquatica* (Nees) Burgeff
Preissia Corda [PREISSI$]
Preissia quadrata (Scop.) Nees [PREIQUA]
SY: *Pressia commutata* Nees
Pressia commutata Nees = Preissia quadrata [PREIQUA]

MASTIGOPHORACEAE (B:H)

Mastigophora Nees [MASTIGP$]
Mastigophora woodsii (Hook.) Nees [MASTWOO]

MEESIACEAE (B:M)

Amblyodon Bruch & Schimp. in B.S.G. [AMBLYOD$]
Amblyodon dealbatus (Hedw.) Bruch & Schimp. in B.S.G. [AMBLDEA]
SY: *Amblyodon dealbatus* var. *americanus* Ren. & Card.
Amblyodon dealbatus var. *americanus* Ren. & Card. = Amblyodon dealbatus [AMBLDEA]
Meesia Hedw. [MEESIA$]
Meesia albertini B.S.G. = Meesia longiseta [MEESLON]
Meesia hexasticha auct. = Meesia longiseta [MEESLON]
Meesia longiseta Hedw. [MEESLON]
SY: *Meesia albertini* B.S.G.
SY: *Meesia longiseta* var. *macounii* Grout
SY: *Meesia hexasticha* auct.
Meesia longiseta var. *macounii* Grout = Meesia longiseta [MEESLON]
Meesia trichodes Spruce = Meesia uliginosa [MEESULI]
Meesia trifaria Crum, Steere & Anderson = Meesia triquetra [MEESTRI]
Meesia triquetra (Richt.) Ångstr. [MEESTRI] (three-angled thread moss)
SY: *Meesia trifaria* Crum, Steere & Anderson
SY: *Meesia tristicha* Bruch
Meesia tristicha Bruch = Meesia triquetra [MEESTRI]
Meesia uliginosa Hedw. [MEESULI] (capillary thread moss)
SY: *Meesia trichodes* Spruce
SY: *Meesia uliginosa* var. *alpina* (Bruch) Hampe
SY: *Meesia uliginosa* var. *angustifolia* (Brid.) Hampe
SY: *Meesia uliginosa* var. *minor* (Brid.) Web. & Mohr
Meesia uliginosa var. *alpina* (Bruch) Hampe = Meesia uliginosa [MEESULI]
Meesia uliginosa var. *angustifolia* (Brid.) Hampe = Meesia uliginosa [MEESULI]
Meesia uliginosa var. *minor* (Brid.) Web. & Mohr = Meesia uliginosa [MEESULI]
Paludella Brid. [PALUDEL$]
Paludella squarrosa (Hedw.) Brid. [PALUSQU]

METZGERIACEAE (B:H)

Apometzgeria Kuwah. [APOMETZ$]
Apometzgeria pubescens (Schrank) Kuwah. [APOMPUB]
SY: *Metzgeria pubescens* (Schrank) Raddi
Metzgeria Raddi [METZGER$]
Metzgeria conjugata Lindb. [METZCON]

Metzgeria fruticulosa auct. = Metzgeria temperata [METZTEM]
Metzgeria hamata Lindb. = Metzgeria leptoneura [METZLEP]
Metzgeria leptoneura Spruce [METZLEP]
SY: *Metzgeria hamata* Lindb.
Metzgeria pubescens (Schrank) Raddi = Apometzgeria pubescens [APOMPUB]
Metzgeria temperata Kuwah. [METZTEM]
SY: *Metzgeria fruticulosa* auct.

MNIACEAE (B:M)

Cinclidium Sw. in Schrad. [CINCLID$]
Cinclidium arcticum Bruch & Schimp. in B.S.G. [CINCARC]
Cyrtomnium Holm. [CYRTOMN$]
Cyrtomnium hymenophylloides (Hüb.) Nyh. ex T. Kop. [CYRTHYM]
SY: *Mnium hymenophylloides* Hüb.
Cyrtomnium hymenophyllum (Bruch & Schimp. in B.S.G.) Holm. [CYRTHYE]
SY: *Mnium hymenophyllum* Bruch & Schimp. in B.S.G.
Leucolepis Lindb. [LEUCOLE$]
Leucolepis acanthoneuron (Schwaegr.) Lindb. [LEUCACA] (Menzies' tree moss)
SY: *Leucolepis menziesii* (Hook.) Steere in L. Koch
SY: *Mnium menziesii* Hook.
Leucolepis menziesii (Hook.) Steere in L. Koch = Leucolepis acanthoneuron [LEUCACA]
Mnium Hedw. [MNIUM$]
Mnium affine var. *ciliare* C. Müll. = Plagiomnium ciliare [PLAGCIL]
Mnium affine var. *integrifolium* (Lindb.) Milde = Plagiomnium ellipticum [PLAGELL]
Mnium affine var. *rugicum* (Laur.) Bruch & Schimp. in B.S.G. = Plagiomnium ellipticum [PLAGELL]
Mnium ambiguum H. Müll. [MNIUAMB]
SY: *Mnium inclinatum* Lindb.
SY: *Mnium lycopodioides* var. *inclinatum* (Lindb.) Wijk & Marg.
SY: *Mnium pseudolycopodioides* C. Müll. & Kindb. in Mac. & Kindb.
SY: *Mnium riparium* Mitt.
SY: *Mnium umbratile* Mitt.
Mnium arizonicum Amann [MNIUARI]
SY: *Mnium saximontanum* M. Bowers
Mnium blyttii Bruch & Schimp in B.S.G. [MNIUBLY]
SY: *Stellariomnium blyttii* (Bruch & Schimp. in B.S.G.) M. Bowers
Mnium ciliare (C. Müll.) Schimp. = Plagiomnium ciliare [PLAGCIL]
Mnium cinclidioides Hüb. = Pseudobryum cinclidioides [PSEUCIN]
Mnium cuspidatum Hedw. = Plagiomnium cuspidatum [PLAGCUS]
Mnium cuspidatum var. *tenellum* Kindb. in Mac. & Kindb. = Plagiomnium cuspidatum [PLAGCUS]
Mnium decurrens C. Müll. & Kindb. in Mac. & Kindb. = Mnium thomsonii [MNIUTHO]
Mnium drummondii Bruch & Schimp. = Plagiomnium drummondii [PLAGDRU]
Mnium ellipticum Brid. = Plagiomnium ellipticum [PLAGELL]
Mnium glabrescens Kindb. = Rhizomnium glabrescens [RHIZGLB]
Mnium glabrescens ssp. *chlorophyllosum* Kindb. = Rhizomnium punctatum [RHIZPUN]
Mnium gracile (T. Kop.) Crum & Anderson = Rhizomnium gracile [RHIZGRC]
Mnium hymenophylloides Hüb. = Cyrtomnium hymenophylloides [CYRTHYM]
Mnium hymenophyllum Bruch & Schimp. in B.S.G. = Cyrtomnium hymenophyllum [CYRTHYE]
Mnium inclinatum Lindb. = Mnium ambiguum [MNIUAMB]
Mnium insigne Mitt. = Plagiomnium insigne [PLAGINS]
Mnium insigne var. *intermedium* Kindb. in Mac. = Plagiomnium insigne [PLAGINS]
Mnium koponenii Crum = Rhizomnium gracile [RHIZGRC]
Mnium longirostre Brid. = Plagiomnium rostratum [PLAGROS]
Mnium lycopodioides var. *inclinatum* (Lindb.) Wijk & Marg. = Mnium ambiguum [MNIUAMB]
Mnium macrociliare C. Müll. & Kindb. in Mac. & Kindb. = Plagiomnium ciliare [PLAGCIL]
Mnium magnifolium Horik. = Rhizomnium magnifolium [RHIZMAG]
Mnium marginatum (With.) Brid. ex P. Beauv. [MNIUMAR]
SY: *Mnium serratum* Schrad. ex Brid.
Mnium medium Bruch & Schimp. in B.S.G. = Plagiomnium medium [PLAGMED]
Mnium menziesii Hook. = Leucolepis acanthoneuron [LEUCACA]
Mnium nudum Britt. & Williams = Rhizomnium nudum [RHIZNUD]
Mnium orthorrhynchum auct. = Mnium thomsonii [MNIUTHO]
Mnium pseudolycopodioides C. Müll. & Kindb. in Mac. & Kindb. = Mnium ambiguum [MNIUAMB]
Mnium pseudopunctatum Bruch & Schimp. = Rhizomnium pseudopunctatum [RHIZPSE]
Mnium punctatum Hedw. = Rhizomnium punctatum [RHIZPUN]
Mnium punctatum var. *elatum* Schimp. = Rhizomnium magnifolium [RHIZMAG]
Mnium riparium Mitt. = Mnium ambiguum [MNIUAMB]
Mnium robustum (Kindb.) Kindb. = Plagiomnium insigne [PLAGINS]
Mnium rostratum Schrad. = Plagiomnium rostratum [PLAGROS]
Mnium rugicum Laur. = Plagiomnium ellipticum [PLAGELL]
Mnium saximontanum M. Bowers = Mnium arizonicum [MNIUARI]
Mnium serratum Schrad. ex Brid. = Mnium marginatum [MNIUMAR]
Mnium serratum var. *macounii* Kindb. in Mac. & Kindb. = Mnium spinulosum [MNIUSPN]
Mnium spinulosum Bruch & Schimp. in B.S.G. [MNIUSPN] (Menzies' red-mouthed mnium)
SY: *Mnium serratum* var. *macounii* Kindb. in Mac. & Kindb.
Mnium subglobosum Bruch & Schimp. in B.S.G. = Rhizomnium pseudopunctatum [RHIZPSE]
Mnium sylvaticum Lindb. = Plagiomnium cuspidatum [PLAGCUS]
Mnium thomsonii Schimp. [MNIUTHO]
SY: *Mnium decurrens* C. Müll. & Kindb. in Mac. & Kindb.
SY: *Mnium orthorrhynchum* auct.
Mnium umbratile Mitt. = Mnium ambiguum [MNIUAMB]
Mnium venustum Mitt. = Plagiomnium venustum [PLAGVEN]
Plagiomnium T. Kop. [PLAGIOM$]
Plagiomnium ciliare (C. Müll.) T. Kop. [PLAGCIL]
SY: *Mnium affine* var. *ciliare* C. Müll.
SY: *Mnium ciliare* (C. Müll.) Schimp.
SY: *Mnium macrociliare* C. Müll. & Kindb. in Mac. & Kindb.
Plagiomnium cinclidioides (Hüb.) M. Bowers = Pseudobryum cinclidioides [PSEUCIN]
Plagiomnium cuspidatum (Hedw.) T. Kop. [PLAGCUS] (woodsy leafy moss)
SY: *Mnium cuspidatum* Hedw.
SY: *Mnium cuspidatum* var. *tenellum* Kindb. in Mac. & Kindb.
SY: *Mnium sylvaticum* Lindb.
Plagiomnium drummondii (Bruch & Schimp.) T. Kop. [PLAGDRU] (Drummond's leafy moss)
SY: *Mnium drummondii* Bruch & Schimp.
Plagiomnium ellipticum (Brid.) T. Kop. [PLAGELL] (marsh magnificent moss)
SY: *Mnium affine* var. *integrifolium* (Lindb.) Milde
SY: *Mnium affine* var. *rugicum* (Laur.) Bruch & Schimp. in B.S.G.
SY: *Mnium ellipticum* Brid.
SY: *Mnium rugicum* Laur.
SY: *Plagiomnium rugicum* (Laur.) T. Kop.
Plagiomnium insigne (Mitt.) T. Kop. [PLAGINS] (badge moss)
SY: *Mnium insigne* Mitt.
SY: *Mnium insigne* var. *intermedium* Kindb. in Mac.
SY: *Mnium robustum* (Kindb.) Kindb.
Plagiomnium medium (Bruch & Schimp. in B.S.G.) T. Kop. [PLAGMED] (trailing leafy moss)
SY: *Mnium medium* Bruch & Schimp. in B.S.G.
Plagiomnium rostratum (Schrad.) T. Kop. [PLAGROS]
SY: *Mnium longirostre* Brid.
SY: *Mnium rostratum* Schrad.
Plagiomnium rugicum (Laur.) T. Kop. = Plagiomnium ellipticum [PLAGELL]
Plagiomnium venustum (Mitt.) T. Kop. [PLAGVEN] (magnificent moss)

SY: *Mnium venustum* Mitt.
Pseudobryum (Kindb.) T. Kop. [PSEUDBY$]
Pseudobryum cinclidioides (Hüb.) T. Kop. [PSEUCIN]
SY: *Mnium cinclidioides* Hüb.
SY: *Plagiomnium cinclidioides* (Hüb.) M. Bowers
Rhizomnium (Broth.) T. Kop. [RHIZOMN$]
Rhizomnium glabrescens (Kindb.) T. Kop. [RHIZGLB] (fan moss)
SY: *Mnium glabrescens* Kindb.
Rhizomnium gracile T. Kop. [RHIZGRC]
SY: *Mnium gracile* (T. Kop.) Crum & Anderson
SY: *Mnium koponenii* Crum
Rhizomnium magnifolium (Horik.) T. Kop. [RHIZMAG] (hairy lantern moss)
SY: *Mnium magnifolium* Horik.
SY: *Mnium punctatum* var. *elatum* Schimp.
SY: *Rhizomnium perssonii* T. Kop
SY: *Rhizomnium punctatum* var. *elatum* (Schimp.) T. Kop.
Rhizomnium nudum (Britt. & Williams) T. Kop. [RHIZNUD]
SY: *Mnium nudum* Britt. & Williams
Rhizomnium perssonii T. Kop = Rhizomnium magnifolium [RHIZMAG]
Rhizomnium pseudopunctatum (Bruch & Schimp.) T. Kop. [RHIZPSE] (felt round moss)
SY: *Mnium pseudopunctatum* Bruch & Schimp.
SY: *Mnium subglobosum* Bruch & Schimp. in B.S.G.
Rhizomnium punctatum (Hedw.) T. Kop. [RHIZPUN]
SY: *Mnium glabrescens* ssp. *chlorophyllosum* Kindb.
SY: *Mnium punctatum* Hedw.
SY: *Rhizomnium punctatum* ssp. *chlorophyllosum* (Kindb.) T. Kop.
Rhizomnium punctatum ssp. *chlorophyllosum* (Kindb.) T. Kop. = Rhizomnium punctatum [RHIZPUN]
Rhizomnium punctatum var. *elatum* (Schimp.) T. Kop. = Rhizomnium magnifolium [RHIZMAG]
Stellariomnium blyttii (Bruch & Schimp. in B.S.G.) M. Bowers = Mnium blyttii [MNIUBLY]

MYRINIACEAE (B:M)

Leskea pulvinata Wahlenb. = Myrinia pulvinata [MYRIPUL]
Myrinia Schimp. [MYRINIA$]
Myrinia pulvinata (Wahlenb.) Schimp. [MYRIPUL]
SY: *Leskea pulvinata* Wahlenb.

NECKERACEAE (B:M)

Homalia (Brid.) Schimp. in B.S.G. [HOMALIA$]
Homalia gracilis James in Peck = Homalia trichomanoides [HOMATRI]
Homalia jamesii Schimp. = Homalia trichomanoides [HOMATRI]
Homalia jamesii var. *gracilis* (James in Peck) Wagn. = Homalia trichomanoides [HOMATRI]
Homalia macounii C. Müll. & Kindb. in Mac. & Kindb. = Homalia trichomanoides [HOMATRI]
Homalia trichomanoides (Hedw.) Schimp. in B.S.G. [HOMATRI]
SY: *Homalia gracilis* James in Peck
SY: *Homalia jamesii* Schimp.
SY: *Homalia jamesii* var. *gracilis* (James in Peck) Wagn.
SY: *Homalia macounii* C. Müll. & Kindb. in Mac. & Kindb.
SY: *Neckera gracilis* (James in Peck) Kindb.
SY: *Omalia trichomanoides* (Brid.) B.S.G.
Metaneckera Steere [METANEC$]
Metaneckera menziesii (Hook. in Drumm.) Steere [METAMEN] (Menzies' neckera)
SY: *Neckera menziesii* Hook. in Drumm.
SY: *Neckera menziesii* var. *amblyclada* Kindb. in Mac. & Kindb.
SY: *Neckera neomexicana* (Card.) Grout
SY: *Neckeradelphus menziesii* (Hook. in Drumm.) Steere
SY: *Porotrichum neomexicanum* (Card.) Wagn.
Neckera Hedw. [NECKERA$]
Neckera douglasii Hook. [NECKDOU] (Douglas' neckera)
SY: *Neckera douglasii* var. *macounii* Kindb.
Neckera douglasii var. *macounii* Kindb. = Neckera douglasii [NECKDOU]
Neckera gracilis (James in Peck) Kindb. = Homalia trichomanoides [HOMATRI]
Neckera menziesii Hook. in Drumm. = Metaneckera menziesii [METAMEN]
Neckera menziesii var. *amblyclada* Kindb. in Mac. & Kindb. = Metaneckera menziesii [METAMEN]
Neckera neomexicana (Card.) Grout = Metaneckera menziesii [METAMEN]
Neckera oligocarpa Bruch in Ångstr. = Neckera pennata [NECKPEN]
Neckera pennata Hedw. [NECKPEN]
SY: *Neckera oligocarpa* Bruch in Ångstr.
SY: *Neckera pennata* var. *oligocarpa* (Bruch in Ångstr.) C. Müll.
SY: *Neckera pennata* var. *tenera* C. Müll.
SY: *Neckera pterantha* (C. Müll. & Kindb.) Kindb.
Neckera pennata var. *oligocarpa* (Bruch in Ångstr.) C. Müll. = Neckera pennata [NECKPEN]
Neckera pennata var. *tenera* C. Müll. = Neckera pennata [NECKPEN]
Neckera pterantha (C. Müll. & Kindb.) Kindb. = Neckera pennata [NECKPEN]
Neckeradelphus menziesii (Hook. in Drumm.) Steere = Metaneckera menziesii [METAMEN]
Omalia trichomanoides (Brid.) B.S.G. = Homalia trichomanoides [HOMATRI]
Porotrichum neomexicanum (Card.) Wagn. = Metaneckera menziesii [METAMEN]

OEDIPODIACEAE (B:M)

Oedipodium Schwaegr. [OEDIPOD$]
Oedipodium griffithianum (Dicks.) Schwaegr. [OEDIGRI]

ORTHOTRICHACEAE (B:M)

Amphidium Schimp. [AMPHIDI$]
Amphidium californicum (Hampe ex C. Müll.) Broth. [AMPHCAL]
Amphidium lapponicum (Hedw.) Schimp. [AMPHLAP] (bottle moss)
SY: *Amphidium lapponicum* var. *crispatum* (Kindb.) Grout
SY: *Amphoridium lapponicum* (Hedw.) Schimp.
SY: *Zieria lapponicus* (Hedw.) B.S.G.
Amphidium lapponicum var. *crispatum* (Kindb.) Grout = Amphidium lapponicum [AMPHLAP]
Amphidium mougeotii (Bruch & Schimp. in B.S.G.) Schimp. [AMPHMOU]
SY: *Anoectangium mougeotii* (B.S.G.) Lindb.
SY: *Zieria mougeotii* B.S.G.
Amphoridium lapponicum (Hedw.) Schimp. = Amphidium lapponicum [AMPHLAP]
Amphoridium Schimp. = Amphidium [AMPHIDI$]
Anoectangium mougeotii (B.S.G.) Lindb. = Amphidium mougeotii [AMPHMOU]
Nyholmiella obtusifolia (Brid.) Holm. & Warncke = Orthotrichum obtusifolium [ORTHOBT]
Orthotrichum Hedw. [ORTHOTR$]
Orthotrichum affine Brid. [ORTHAFF]
Orthotrichum alpestre Hornsch. in B.S.G. [ORTHALP]
SY: *Orthotrichum alpestre* var. *majus* Lesq. & James
SY: *Orthotrichum alpestre* var. *occidentale* (James) Grout
SY: *Orthotrichum alpestre* var. *watsonii* (James) Grout
Orthotrichum alpestre var. *majus* Lesq. & James = Orthotrichum alpestre [ORTHALP]
Orthotrichum alpestre var. *occidentale* (James) Grout = Orthotrichum alpestre [ORTHALP]
Orthotrichum alpestre var. *watsonii* (James) Grout = Orthotrichum alpestre [ORTHALP]
Orthotrichum anomalum Hedw. [ORTHANO]
SY: *Orthotrichum anomalum* var. *americanum* Vent. in Mac. & Kindb.
SY: *Orthotrichum anomalum* var. *saxatile* Milde

SY: *Orthotrichum canadense* Bruch & Schimp.
SY: *Orthotrichum consimile* ssp. *anomaloides* Kindb.
Orthotrichum anomalum var. *americanum* Vent. in Mac. & Kindb. = Orthotrichum anomalum [ORTHANO]
Orthotrichum anomalum var. *saxatile* Milde = Orthotrichum anomalum [ORTHANO]
Orthotrichum arcticum Schimp. = Orthotrichum pylaisii [ORTHPYL]
Orthotrichum blyttii Schimp. = Orthotrichum pylaisii [ORTHPYL]
Orthotrichum canadense Bruch & Schimp. = Orthotrichum anomalum [ORTHANO]
Orthotrichum columbicum Mitt. = Orthotrichum consimile [ORTHCON]
Orthotrichum consimile Mitt. [ORTHCON]
SY: *Orthotrichum columbicum* Mitt.
SY: *Orthotrichum pulchellum* var. *columbicum* Grout
SY: *Orthotrichum pulchellum* ssp. *ulotaeforme* (Ren. & Card.) Kindb.
SY: *Orthotrichum winteri* Schimp.
Orthotrichum consimile ssp. *anomaloides* Kindb. = Orthotrichum anomalum [ORTHANO]
Orthotrichum cupulatum Brid. [ORTHCUP]
SY: *Orthotrichum nudum* var. *rudolphianum* Vent. in Husn.
Orthotrichum cupulatum var. *jamesianum* (Sull. in James in Watson) Lawt. = Orthotrichum pellucidum [ORTHPEU]
Orthotrichum cylindricarpum Lesq. in Jaeg. = Orthotrichum tenellum [ORTHTNE]
Orthotrichum diaphanum Brid. [ORTHDIA]
SY: *Orthotrichum garrettii* Grout & Flow. in Grout
Orthotrichum elegans Schwaegr. ex Grev. = Orthotrichum speciosum var. elegans [ORTHSPE1]
Orthotrichum euryphyllum Vent. in Röll = Orthotrichum rivulare [ORTHRIV]
Orthotrichum garrettii Grout & Flow. in Grout = Orthotrichum diaphanum [ORTHDIA]
Orthotrichum hallii Sull. & Lesq. in Sull. [ORTHHAL]
Orthotrichum jamesianum Sull. in James in Watson = Orthotrichum pellucidum [ORTHPEU]
Orthotrichum killiasii C. Müll. = Orthotrichum speciosum var. speciosum [ORTHSPE0]
Orthotrichum kingianum Lesq. = Orthotrichum laevigatum [ORTHLAE]
Orthotrichum laevigatum Zett. [ORTHLAE]
SY: *Orthotrichum kingianum* Lesq.
SY: *Orthotrichum laevigatum* var. *kingianum* (Lesq.) Grout
SY: *Orthotrichum laevigatum* f. *macounii* (Aust.) Lawt. & Vitt in Lawt.
SY: *Orthotrichum lonchothecium* C. Müll. & Kindb. in Mac. & Kindb.
SY: *Orthotrichum macounii* Aust.
SY: *Orthotrichum macounii* var. *lonchothecium* (C. Müll. & Kindb. in Mac. & Kindb.) Grout
SY: *Orthotrichum roellii* Vent. in Röll
SY: *Orthotrichum speciosum* var. *hainesiae* (Aust.) Par.
Orthotrichum laevigatum f. *macounii* (Aust.) Lawt. & Vitt in Lawt. = Orthotrichum laevigatum [ORTHLAE]
Orthotrichum laevigatum var. *kingianum* (Lesq.) Grout = Orthotrichum laevigatum [ORTHLAE]
Orthotrichum lonchothecium C. Müll. & Kindb. in Mac. & Kindb. = Orthotrichum laevigatum [ORTHLAE]
Orthotrichum lyellii Hook. & Tayl. [ORTHLYE] (lyell's bristle moss)
SY: *Orthotrichum lyellii* var. *papillosum* (Hampe) Sull.
SY: *Orthotrichum lyelloides* Kindb.
SY: *Orthotrichum papillosum* Hampe
SY: *Orthotrichum papillosum* ssp. *strictum* (Vent.) Kindb.
Orthotrichum lyellii var. *papillosum* (Hampe) Sull. = Orthotrichum lyellii [ORTHLYE]
Orthotrichum lyelloides Kindb. = Orthotrichum lyellii [ORTHLYE]
Orthotrichum macounii Aust. = Orthotrichum laevigatum [ORTHLAE]
Orthotrichum macounii var. *lonchothecium* (C. Müll. & Kindb. in Mac. & Kindb.) Grout = Orthotrichum laevigatum [ORTHLAE]
Orthotrichum microblepharum Schimp. = Orthotrichum pylaisii [ORTHPYL]
Orthotrichum nudum var. *rudolphianum* Vent. in Husn. = Orthotrichum cupulatum [ORTHCUP]
Orthotrichum obtusifolium Brid. [ORTHOBT] (blunt-leaved bristle moss)
SY: *Nyholmiella obtusifolia* (Brid.) Holm. & Warncke
Orthotrichum pallens Bruch ex Brid. [ORTHPAL]
Orthotrichum papillosum Hampe = Orthotrichum lyellii [ORTHLYE]
Orthotrichum papillosum ssp. *strictum* (Vent.) Kindb. = Orthotrichum lyellii [ORTHLYE]
Orthotrichum pellucidum Lindb. [ORTHPEU]
SY: *Orthotrichum cupulatum* var. *jamesianum* (Sull. in James in Watson) Lawt.
SY: *Orthotrichum jamesianum* Sull. in James in Watson
Orthotrichum psilothecium C. Müll. & Kindb. in Mac. & Kindb. = Orthotrichum speciosum var. elegans [ORTHSPE1]
Orthotrichum pulchellum Brunt. in Winch. & Gateh. [ORTHPUL]
SY: *Orthotrichum pulchellum* var. *groutii* Lawt.
Orthotrichum pulchellum ssp. *ulotaeforme* (Ren. & Card.) Kindb. = Orthotrichum consimile [ORTHCON]
Orthotrichum pulchellum var. *columbicum* Grout = Orthotrichum consimile [ORTHCON]
Orthotrichum pulchellum var. *groutii* Lawt. = Orthotrichum pulchellum [ORTHPUL]
Orthotrichum pylaisii Brid. [ORTHPYL]
SY: *Orthotrichum arcticum* Schimp.
SY: *Orthotrichum blyttii* Schimp.
SY: *Orthotrichum microblepharum* Schimp.
Orthotrichum rivulare Turn. [ORTHRIV]
SY: *Orthotrichum euryphyllum* Vent. in Röll
SY: *Orthotrichum sprucei* auct.
Orthotrichum roellii Vent. in Röll = Orthotrichum laevigatum [ORTHLAE]
Orthotrichum rupestre Schleich. ex Schwaegr. [ORTHRUP]
SY: *Orthotrichum rupestre* var. *globosum* (Lesq.) Grout
SY: *Orthotrichum rupestre* var. *macfaddenae* (Williams) Grout
SY: *Orthotrichum rupestre* ssp. *sturmii* (Hoppe & Hornsch.) Boul.
SY: *Orthotrichum texanum* Sull. & Lesq.
Orthotrichum rupestre ssp. *sturmii* (Hoppe & Hornsch.) Boul. = Orthotrichum rupestre [ORTHRUP]
Orthotrichum rupestre var. *globosum* (Lesq.) Grout = Orthotrichum rupestre [ORTHRUP]
Orthotrichum rupestre var. *macfaddenae* (Williams) Grout = Orthotrichum rupestre [ORTHRUP]
Orthotrichum speciosum Nees in Sturm [ORTHSPE] (hooded moss)
Orthotrichum speciosum var. **elegans** (Schwaegr. ex Hook. & Grev.) Warnst. [ORTHSPE1]
SY: *Orthotrichum elegans* Schwaegr. ex Grev.
SY: *Orthotrichum psilothecium* C. Müll. & Kindb. in Mac. & Kindb.
Orthotrichum speciosum var. *hainesiae* (Aust.) Par. = Orthotrichum laevigatum [ORTHLAE]
Orthotrichum speciosum var. *killiasii* (C. Müll.) Vent. = Orthotrichum speciosum var. speciosum [ORTHSPE0]
Orthotrichum speciosum var. **speciosum** [ORTHSPE0]
SY: *Orthotrichum killiasii* C. Müll.
SY: *Orthotrichum speciosum* var. *killiasii* (C. Müll.) Vent.
Orthotrichum sprucei auct. = Orthotrichum rivulare [ORTHRIV]
Orthotrichum striatum Hedw. [ORTHSTI]
Orthotrichum tenellum Bruch ex Brid. [ORTHTNE]
SY: *Orthotrichum cylindricarpum* Lesq. in Jaeg.
SY: *Orthotrichum tenellum* var. *coulteri* (Mitt.) Grout
Orthotrichum tenellum var. *coulteri* (Mitt.) Grout = Orthotrichum tenellum [ORTHTNE]
Orthotrichum texanum Sull. & Lesq. = Orthotrichum rupestre [ORTHRUP]
Orthotrichum winteri Schimp. = Orthotrichum consimile [ORTHCON]
Ulota Mohr [ULOTA$]
Ulota americanum Mitt. = Ulota curvifolia [ULOTCUR]
Ulota cirrata Grout = Ulota curvifolia [ULOTCUR]

Ulota crispa var. *alaskana* (Card. & Thér.) Grout = Ulota obtusiuscula [ULOTOBT]
Ulota curvifolia (Wahlenb.) Lilj. [ULOTCUR]
SY: *Ulota americanum* Mitt.
SY: *Ulota cirrata* Grout
SY: *Ulota scabrida* Kindb. in Mac. & Kindb.
SY: *Weissia curvifolia* (Wahlenb.) Lindb. in Braithw.
Ulota drummondii (Hook. & Grev. in Grev.) Brid. [ULOTDRU]
SY: *Ulota funstonii* Grout
Ulota funstonii Grout = Ulota drummondii [ULOTDRU]
Ulota maritima C. Müll. & Kindb. in Mac. & Kindb. = Ulota phyllantha [ULOTPHY]
Ulota megalospora Vent. in Röll [ULOTMEG]
SY: *Ulota subulata* Kindb. in Mac. & Kindb.
SY: *Ulota subulifolia* C. Müll. & Kindb. in Mac. & Kindb.
Ulota obtusiuscula C. Müll. & Kindb. in Mac. & Kindb. [ULOTOBT] (twisted ulota)
SY: *Ulota crispa* var. *alaskana* (Card. & Thér.) Grout
Ulota phyllantha Brid. [ULOTPHY]
SY: *Ulota maritima* C. Müll. & Kindb. in Mac. & Kindb.
SY: *Weissia maritima* (C. Müll. & Kindb.) Britt.
Ulota scabrida Kindb. in Mac. & Kindb. = Ulota curvifolia [ULOTCUR]
Ulota subulata Kindb. in Mac. & Kindb. = Ulota megalospora [ULOTMEG]
Ulota subulifolia C. Müll. & Kindb. in Mac. & Kindb. = Ulota megalospora [ULOTMEG]
Weissia curvifolia (Wahlenb.) Lindb. in Braithw. = Ulota curvifolia [ULOTCUR]
Weissia maritima (C. Müll. & Kindb.) Britt. = Ulota phyllantha [ULOTPHY]
Zieria lapponicus (Hedw.) B.S.G. = Amphidium lapponicum [AMPHLAP]
Zieria mougeotii B.S.G. = Amphidium mougeotii [AMPHMOU]
Zygodon Hook. & Tayl. [ZYGODON$]
Zygodon baumgartneri Malta = Zygodon viridissimus var. rupestris [ZYGOVIR2]
Zygodon gracilis Wils. in Berk. [ZYGOGRA]
Zygodon reinwardtii (Hornsch. in Reinw. & Hornsch.) A. Br. in Bruch & Schimp. [ZYGOREI]
SY: *Zygodon reinwardtii* var. *subintegrifolius* Malta
Zygodon reinwardtii var. *subintegrifolius* Malta = Zygodon reinwardtii [ZYGOREI]
Zygodon rufotomentosus Britt. ex Malta = Zygodon viridissimus var. rupestris [ZYGOVIR2]
Zygodon rupestris (Lindb. ex Hartm.) Lindb. ex Britt. = Zygodon viridissimus var. rupestris [ZYGOVIR2]
Zygodon viridissimus (Dicks.) Brid. [ZYGOVIR]
Zygodon viridissimus var. *dentatus* (Breidl.) Limpr. = Zygodon viridissimus var. viridissimus [ZYGOVIR0]
Zygodon viridissimus var. *rufotomentosus* (Britt. ex Malta) Grout = Zygodon viridissimus var. rupestris [ZYGOVIR2]
Zygodon viridissimus var. **rupestris** Lindb. ex Hartm. [ZYGOVIR2]
SY: *Zygodon baumgartneri* Malta
SY: *Zygodon rufotomentosus* Britt. ex Malta
SY: *Zygodon rupestris* (Lindb. ex Hartm.) Lindb. ex Britt.
SY: *Zygodon viridissimus* var. *rufotomentosus* (Britt. ex Malta) Grout
SY: *Zygodon viridissimus* var. *vulgaris* Malta
SY: *Zygodon vulgaris* (Malta) Nyh.
Zygodon viridissimus var. **viridissimus** [ZYGOVIR0]
SY: *Zygodon viridissimus* var. *dentatus* (Breidl.) Limpr.
Zygodon viridissimus var. *vulgaris* Malta = Zygodon viridissimus var. rupestris [ZYGOVIR2]
Zygodon vulgaris (Malta) Nyh. = Zygodon viridissimus var. rupestris [ZYGOVIR2]

PALLAVICINIACEAE (B:H)

Moerckia Gott. [MOERCKI$]
Moerckia blyttii (Moerck) Brockm. [MOERBLY]
Moerckia flotoviana (Nees) Schiffn. = Moerckia hibernica [MOERHIB]
Moerckia hibernica (Hook.) Gott. [MOERHIB]
SY: *Moerckia flotoviana* (Nees) Schiffn.

PELLIACEAE (B:H)

Calycularia Mitt. [CALYCUL$]
Calycularia crispula Mitt. [CALYCRI]
Pellia Raddi [PELLIA$]
Pellia borealis Lorbeer = Pellia epiphylla [PELLEPI]
Pellia columbiana Krajina & Brayshaw = Pellia neesiana [PELLNEE]
Pellia endiviifolia (Dicks.) Dum. [PELLEND]
SY: *Pellia fabbroniana* Raddi
Pellia epiphylla (L.) Corda [PELLEPI]
SY: *Pellia borealis* Lorbeer
Pellia fabbroniana Raddi = Pellia endiviifolia [PELLEND]
Pellia neesiana (Gott.) Limpr. [PELLNEE] (ring pellia)
SY: *Pellia columbiana* Krajina & Brayshaw

PLAGIOCHILACEAE (B:H)

Plagiochila (Dum.) Dum. [PLAGIOH$]
Plagiochila alaskana Evans = Plagiochila semidecurrens var. alaskana [PLAGSEM1]
Plagiochila arctica auct. = Plagiochila asplenioides [PLAGASL]
Plagiochila asplenioides (L.) Dum. [PLAGASL] (cedar-shake wort)
SY: *Plagiochila major* (Nees) S. Arnell
SY: *Plagiochila arctica* auct.
Plagiochila asplenioides ssp. *porelloides* (Torrey ex Nees) Schust. = Plagiochila porelloides [PLAGPOR]
Plagiochila firma ssp. *confusa* (Schust.) H. Inoue = Plagiochila schofieldiana [PLAGSCH]
Plagiochila fryei Evans = Plagiochila semidecurrens var. alaskana [PLAGSEM1]
Plagiochila major (Nees) S. Arnell = Plagiochila asplenioides [PLAGASL]
Plagiochila porelloides (Torrey ex Nees) Lindenb. [PLAGPOR] (cedar-shake liverwort)
SY: *Plagiochila asplenioides* ssp. *porelloides* (Torrey ex Nees) Schust.
Plagiochila rhizophora ssp. *confusa* Schust. = Plagiochila schofieldiana [PLAGSCH]
Plagiochila satoi Hatt. [PLAGSAT]
Plagiochila satoi var. **magnum** Schof. & Hong [PLAGSAT1]
Plagiochila satoi var. **satoi** [PLAGSAT0]
Plagiochila schofieldiana H. Inoue [PLAGSCH]
SY: *Plagiochila firma* ssp. *confusa* (Schust.) H. Inoue
SY: *Plagiochila rhizophora* ssp. *confusa* Schust.
Plagiochila semidecurrens Lehm. & Lindenb. [PLAGSEM]
Plagiochila semidecurrens ssp. *grossidens* (Herz.) Schust. p.p. = Plagiochila semidecurrens var. semidecurrens [PLAGSEM0]
Plagiochila semidecurrens ssp. *grossidens* (Herz.) Schust. p.p. = Plagiochila semidecurrens var. alaskana [PLAGSEM1]
Plagiochila semidecurrens var. **alaskana** (Evans) H. Inoue [PLAGSEM1]
SY: *Plagiochila alaskana* Evans
SY: *Plagiochila fryei* Evans
SY: *Plagiochila semidecurrens* ssp. *grossidens* (Herz.) Schust. p.p.
Plagiochila semidecurrens var. **semidecurrens** [PLAGSEM0]
SY: *Plagiochila semidecurrens* ssp. *grossidens* (Herz.) Schust. p.p.

PLAGIOTHECIACEAE (B:M)

Hypnum denticulatum Hedw. = Plagiothecium denticulatum [PLAGDEN]
Hypnum sullivantiae Schimp. ex Sull. = Plagiothecium cavifolium [PLAGCAV]
Isopterygium piliferum (Sw. ex Hartm.) Loeske = Plagiothecium piliferum [PLAGPIL]
Plagiotheciella pilifera (Sw. ex Hartm.) Fleisch. = Plagiothecium piliferum [PLAGPIL]
Plagiothecium Schimp. in B.S.G. [PLAGIOT$]

Plagiothecium aciculari-pungens C. Müll. & Kindb. in Mac. & Kindb. = Plagiothecium cavifolium [PLAGCAV]
Plagiothecium cavifolium (Brid.) Iwats. [PLAGCAV]
SY: *Hypnum sullivantiae* Schimp. ex Sull.
SY: *Plagiothecium aciculari-pungens* C. Müll. & Kindb. in Mac. & Kindb.
SY: *Plagiothecium cavifolium* var. *fallax* (Card. & Thér.) Iwats.
SY: *Plagiothecium fallax* Card. & Thér.
SY: *Plagiothecium neglectum* Mönk.
SY: *Plagiothecium roeseanum* Schimp. in B.S.G.
SY: *Plagiothecium sylvaticum* var. *orthocladium* (Schimp. in B.S.G.) Schimp.
SY: *Plagiothecium sylvaticum* var. *roeseanum* (Schimp. in B.S.G.) Lindb.
SY: *Plagiothecium platyphyllum* auct.
SY: *Plagiothecium succulentum* auct.
SY: *Plagiothecium sylvaticum* var. *succulentum* auct.
Plagiothecium cavifolium var. *fallax* (Card. & Thér.) Iwats. = Plagiothecium cavifolium [PLAGCAV]
Plagiothecium curvifolium auct. = Plagiothecium laetum [PLAGLAE]
Plagiothecium decursivifolium Kindb. in Mac. & Kindb. = Plagiothecium laetum [PLAGLAE]
Plagiothecium denticulatum (Hedw.) Schimp. in B.S.G. [PLAGDEN]
SY: *Hypnum denticulatum* Hedw.
SY: *Plagiothecium denticulatum* var. *bullulae* Grout
SY: *Plagiothecium denticulatum* var. *donii* Lindb.
SY: *Plagiothecium denticulatum* var. *majus* (Boul.) Delogne
SY: *Plagiothecium denticulatum* var. *obtusifolium* (Turn.) Moore
SY: *Plagiothecium denticulatum* var. *squarrosum* Kindb.
SY: *Plagiothecium denticulatum* var. *undulatum* Ruthe ex Geh.
SY: *Plagiothecium ruthei* auct.
Plagiothecium denticulatum var. *aptychus* (Spruce) Lees in Dix. = Plagiothecium laetum [PLAGLAE]
Plagiothecium denticulatum var. *bullulae* Grout = Plagiothecium denticulatum [PLAGDEN]
Plagiothecium denticulatum var. *donii* Lindb. = Plagiothecium denticulatum [PLAGDEN]
Plagiothecium denticulatum var. *laetum* (Schimp. in B.S.G.) Lindb. = Plagiothecium laetum [PLAGLAE]
Plagiothecium denticulatum var. *majus* (Boul.) Delogne = Plagiothecium denticulatum [PLAGDEN]
Plagiothecium denticulatum var. *obtusifolium* (Turn.) Moore = Plagiothecium denticulatum [PLAGDEN]
Plagiothecium denticulatum var. *squarrosum* Kindb. = Plagiothecium denticulatum [PLAGDEN]
Plagiothecium denticulatum var. *tenellum* Schimp. in B.S.G. = Plagiothecium laetum [PLAGLAE]
Plagiothecium denticulatum var. *undulatum* Ruthe ex Geh. = Plagiothecium denticulatum [PLAGDEN]
Plagiothecium fallax Card. & Thér. = Plagiothecium cavifolium [PLAGCAV]
Plagiothecium laetum Schimp. in B.S.G. [PLAGLAE]
SY: *Plagiothecium decursivifolium* Kindb. in Mac. & Kindb.
SY: *Plagiothecium denticulatum* var. *aptychus* (Spruce) Lees in Dix.
SY: *Plagiothecium denticulatum* var. *laetum* (Schimp. in B.S.G.) Lindb.
SY: *Plagiothecium denticulatum* var. *tenellum* Schimp. in B.S.G.
SY: *Plagiothecium curvifolium* auct.
Plagiothecium neglectum Mönk. = Plagiothecium cavifolium [PLAGCAV]
Plagiothecium piliferum (Sw. ex Hartm.) Schimp. in B.S.G. [PLAGPIL]
SY: *Isopterygium piliferum* (Sw. ex Hartm.) Loeske
SY: *Plagiotheciella pilifera* (Sw. ex Hartm.) Fleisch.
SY: *Plagiothecium trichophorum* (Spruce) Vent. & Bott.
Plagiothecium platyphyllum auct. = Plagiothecium cavifolium [PLAGCAV]
Plagiothecium roeseanum Schimp. in B.S.G. = Plagiothecium cavifolium [PLAGCAV]
Plagiothecium ruthei auct. = Plagiothecium denticulatum [PLAGDEN]
Plagiothecium succulentum auct. = Plagiothecium cavifolium [PLAGCAV]
Plagiothecium sylvaticum var. *orthocladium* (Schimp. in B.S.G.) Schimp. = Plagiothecium cavifolium [PLAGCAV]
Plagiothecium sylvaticum var. *roeseanum* (Schimp. in B.S.G.) Lindb. = Plagiothecium cavifolium [PLAGCAV]
Plagiothecium sylvaticum var. *succulentum* auct. = Plagiothecium cavifolium [PLAGCAV]
Plagiothecium trichophorum (Spruce) Vent. & Bott. = Plagiothecium piliferum [PLAGPIL]
Plagiothecium undulatum (Hedw.) Schimp. in B.S.G. [PLAGUND] (flat moss)
SY: *Plagiothecium undulatum* var. *myurum* Card. & Thér.
Plagiothecium undulatum var. *myurum* Card. & Thér. = Plagiothecium undulatum [PLAGUND]

PLEUROZIACEAE (B:H)

Pleurozia Dum. [PLEUROA$]
Pleurozia purpurea Lindb. [PLEUPUR] (purple-worm liverwort)

PLEUROZIOPSIDACEAE (B:M)

Girgensohnia ruthenica (Weinm.) Kindb. = Pleuroziopsis ruthenica [PLEURUT]
Pleuroziopsis Kindb. ex Britt. [PLEUROZ$]
Pleuroziopsis ruthenica (Weinm.) Kindb. ex Britt. [PLEURUT]
SY: *Girgensohnia ruthenica* (Weinm.) Kindb.

POLYTRICHACEAE (B:M)

Atrichum P. Beauv. [ATRICHU$]
Atrichum crispum var. *molle* (Holz.) Frye in Grout = Atrichum tenellum [ATRITEN]
Atrichum fertile Nawasch. = Atrichum haussknechtii [ATRIHAU]
Atrichum haussknechtii Jur. & Milde [ATRIHAU]
SY: *Atrichum fertile* Nawasch.
SY: *Atrichum undulatum* var. *gracilisetum* Besch.
SY: *Atrichum undulatum* var. *haussknechtii* (Jur. & Milde) Frye in Grout
SY: *Catharinea haussknechtii* (Jur. & Milde) Broth.
Atrichum leiophyllum Kindb. = Oligotrichum parallelum [OLIGPAR]
Atrichum papallelum Mitt. = Oligotrichum parallelum [OLIGPAR]
Atrichum rosulatum C. Müll. & Kindb. in Mac. & Kindb. = Atrichum selwynii [ATRISEL]
Atrichum selwynii Aust. [ATRISEL] (crane's-bill moss)
SY: *Atrichum rosulatum* C. Müll. & Kindb. in Mac. & Kindb.
SY: *Atrichum undulatum* var. *selwynii* (Aust.) Frye in Grout
SY: *Catharinea selwynii* (Aust.) Britt.
Atrichum tenellum (Röhl.) Bruch & Schimp. in B.S.G. [ATRITEN]
SY: *Atrichum crispum* var. *molle* (Holz.) Frye in Grout
SY: *Mnium orthorrhynchum* Brid.
Atrichum undulatum (Hedw.) P. Beauv. [ATRIUND]
SY: *Catharinea undulata* (Hedw.) Web. & Mohr
Atrichum undulatum var. *gracilisetum* Besch. = Atrichum haussknechtii [ATRIHAU]
Atrichum undulatum var. *haussknechtii* (Jur. & Milde) Frye in Grout = Atrichum haussknechtii [ATRIHAU]
Atrichum undulatum var. *selwynii* (Aust.) Frye in Grout = Atrichum selwynii [ATRISEL]
Bartramiopsis Kindb. [BARTRAO$]
Bartramiopsis lescurii (James) Kindb. [BARTLES]
SY: *Lyellia lescurii* (James) Salm.
Catharinea haussknechtii (Jur. & Milde) Broth. = Atrichum haussknechtii [ATRIHAU]
Catharinea selwynii (Aust.) Britt. = Atrichum selwynii [ATRISEL]
Catharinea undulata (Hedw.) Web. & Mohr = Atrichum undulatum [ATRIUND]
Lyellia lescurii (James) Salm. = Bartramiopsis lescurii [BARTLES]
Mnium orthorrhynchum Brid. = Atrichum tenellum [ATRITEN]
Oligotrichum Lam. & DC. [OLIGOTR$]

Oligotrichum aligerum Mitt. [OLIGALI] (small hair moss)
Oligotrichum hercynicum (Hedw.) Lam. & DC. [OLIGHER]
SY: *Oligotrichum hercynicum* var. *latifolium* C. Müll. & Kindb. in Mac. & Kindb.
Oligotrichum hercynicum var. *latifolium* C. Müll. & Kindb. in Mac. & Kindb. = Oligotrichum hercynicum [OLIGHER]
Oligotrichum laevigatum var. *cavifolium* (Wils.) Frye in Grout = Psilopilum cavifolium [PSILCAV]
Oligotrichum lyallii (Mitt.) Lindb. = Polytrichum lyallii [POLYLYA]
Oligotrichum parallelum (Mitt.) Kindb. [OLIGPAR] (large hair moss)
SY: *Atrichum leiophyllum* Kindb.
SY: *Atrichum papallelum* Mitt.
Pogonatum P. Beauv. [POGONAT$]
Pogonatum atrovirens Mitt. = Pogonatum contortum [POGOCON]
Pogonatum capillare (Michx.) Brid. = Pogonatum dentatum [POGODEN]
Pogonatum capillare var. *minus* (Wahlenb.) Par. = Pogonatum dentatum [POGODEN]
Pogonatum contortum (Brid.) Lesq. [POGOCON]
SY: *Pogonatum atrovirens* Mitt.
SY: *Pogonatum erythrodontium* Kindb. in Mac. & Kindb.
SY: *Pogonatum laterale* Schimp. in Jard.
Pogonatum dentatum (Brid.) Brid. [POGODEN]
SY: *Pogonatum capillare* (Michx.) Brid.
SY: *Pogonatum capillare* var. *minus* (Wahlenb.) Par.
SY: *Polytrichum capillare* Michx.
Pogonatum erythrodontium Kindb. in Mac. & Kindb. = Pogonatum contortum [POGOCON]
Pogonatum laterale Schimp. in Jard. = Pogonatum contortum [POGOCON]
Pogonatum sphaerothecium Besch. = Polytrichum sphaerothecium [POLYSPH]
Pogonatum urnigerum (Hedw.) P. Beauv. [POGOURN] (grey haircap moss)
SY: *Pogonatum urnigerum* var. *fasciculatum* (Michx.) Brid.
SY: *Pogonatum urnigerum* var. *humile* (Wahlenb.) Brid.
SY: *Pogonatum urnigerum* var. *subintegrifolium* (Arnell & C. Jens.) Möll.
Pogonatum urnigerum var. *fasciculatum* (Michx.) Brid. = Pogonatum urnigerum [POGOURN]
Pogonatum urnigerum var. *humile* (Wahlenb.) Brid. = Pogonatum urnigerum [POGOURN]
Pogonatum urnigerum var. *subintegrifolium* (Arnell & C. Jens.) Möll. = Pogonatum urnigerum [POGOURN]
Polytrichadelphus lyallii Mitt. = Polytrichum lyallii [POLYLYA]
Polytrichastrum G.L. Sm. [POLYTRC$]
Polytrichastrum alpinum (Hedw.) G.L. Sm. [POLYALP] (bristly haircap moss)
Polytrichastrum formosum (Hedw.) G.L. Sm. = Polytrichum formosum [POLYFOR]
Polytrichastrum longisetum (Brid.) G. L. Sm. = Polytrichum longisetum [POLYLOG]
Polytrichastrum lyallii (Mitt.) G.L. Sm. = Polytrichum lyallii [POLYLYA]
Polytrichastrum sexangulare (Brid.) G. L. Sm. = Polytrichum sexangulare [POLYSEX]
Polytrichum Hedw. [POLYTRI$]
Polytrichum affine Funck = Polytrichum strictum [POLYSTR]
Polytrichum alpestre Hoppe = Polytrichum strictum [POLYSTR]
Polytrichum anomalum (Milde) Milde = Polytrichum longisetum [POLYLOG]
Polytrichum apiculatum Kindb. = Polytrichum juniperinum [POLYJUN]
Polytrichum attenuatum Menz. ex Brid. = Polytrichum formosum [POLYFOR]
Polytrichum aurantiacum Hoppe ex Brid. = Polytrichum longisetum [POLYLOG]
Polytrichum capillare Michx. = Pogonatum dentatum [POGODEN]
Polytrichum commune Hedw. [POLYCOM] (common haircap moss)
SY: *Polytrichum commune* var. *canadense* Kindb. in Mac. & Kindb.
Polytrichum commune var. *canadense* Kindb. in Mac. & Kindb. = Polytrichum commune [POLYCOM]
Polytrichum commune var. **commune** [POLYCOM0]
SY: *Polytrichum commune* var. *maximoviczii* Lindb.
SY: *Polytrichum commune* var. *nigrescens* Warnst.
SY: *Polytrichum commune* var. *uliginosum* Wallr.
Polytrichum commune var. *maximoviczii* Lindb. = Polytrichum commune var. commune [POLYCOM0]
Polytrichum commune var. *nigrescens* Warnst. = Polytrichum commune var. commune [POLYCOM0]
Polytrichum commune var. **perigoniale** (Michx.) Hampe [POLYCOM4]
Polytrichum commune var. *uliginosum* Wallr. = Polytrichum commune var. commune [POLYCOM0]
Polytrichum conorhynchum Kindb. in Mac. & Kindb. = Polytrichum formosum [POLYFOR]
Polytrichum formosum Hedw. [POLYFOR]
SY: *Polytrichastrum formosum* (Hedw.) G.L. Sm.
SY: *Polytrichum attenuatum* Menz. ex Brid.
SY: *Polytrichum conorhynchum* Kindb. in Mac. & Kindb.
Polytrichum formosum var. *aurantiacum* (Hoppe ex Brid.) Hartm. = Polytrichum longisetum [POLYLOG]
Polytrichum gracile Bryhn = Polytrichum longisetum [POLYLOG]
Polytrichum gracile var. *anomalum* (Milde) Hag. = Polytrichum longisetum [POLYLOG]
Polytrichum juniperinum Hedw. [POLYJUN] (juniper haircap moss)
SY: *Polytrichum apiculatum* Kindb.
SY: *Polytrichum juniperinum* var. *waghornei* Kindb. in Mac.
Polytrichum juniperinum var. *affine* (Funck) Brid. = Polytrichum strictum [POLYSTR]
Polytrichum juniperinum var. *alpestre* (Hoppe) Röhl. = Polytrichum strictum [POLYSTR]
Polytrichum juniperinum var. *gracilius* Wahlenb. = Polytrichum strictum [POLYSTR]
Polytrichum juniperinum var. *waghornei* Kindb. in Mac. = Polytrichum juniperinum [POLYJUN]
Polytrichum longisetum Brid. [POLYLOG]
SY: *Polytrichastrum longisetum* (Brid.) G. L. Sm.
SY: *Polytrichum anomalum* (Milde) Milde
SY: *Polytrichum aurantiacum* Hoppe ex Brid.
SY: *Polytrichum formosum* var. *aurantiacum* (Hoppe ex Brid.) Hartm.
SY: *Polytrichum gracile* Bryhn
SY: *Polytrichum gracile* var. *anomalum* (Milde) Hag.
SY: *Polytrichum longisetum* var. *anomalum* (Milde) Hag.
SY: *Polytrichum longisetum* var. *aurantiacum* (Hoppe ex Brid.)
Polytrichum longisetum var. *anomalum* (Milde) Hag. = Polytrichum longisetum [POLYLOG]
Polytrichum longisetum var. *aurantiacum* (Hoppe ex Brid.) = Polytrichum longisetum [POLYLOG]
Polytrichum lyallii (Mitt.) Kindb. [POLYLYA]
SY: *Oligotrichum lyallii* (Mitt.) Lindb.
SY: *Polytrichadelphus lyallii* Mitt.
SY: *Polytrichastrum lyallii* (Mitt.) G.L. Sm.
Polytrichum norvegicum auct. = Polytrichum sexangulare [POLYSEX]
Polytrichum norvegicum var. *vulcanicum* (C. Jens.) Podp. = Polytrichum sphaerothecium [POLYSPH]
Polytrichum piliferum Hedw. [POLYPIL] (awned haircap moss)
Polytrichum sexangulare Brid. [POLYSEX]
SY: *Polytrichastrum sexangulare* (Brid.) G. L. Sm.
SY: *Polytrichum norvegicum* auct.
Polytrichum sexangulare var. *vulcanicum* C. Jens. = Polytrichum sphaerothecium [POLYSPH]
Polytrichum sphaerothecium (Besch.) C. Müll. [POLYSPH]
SY: *Pogonatum sphaerothecium* Besch.
SY: *Polytrichum norvegicum* var. *vulcanicum* (C. Jens.) Podp.
SY: *Polytrichum sexangulare* var. *vulcanicum* C. Jens.
Polytrichum strictum Brid. [POLYSTR] (slender haircap)
SY: *Polytrichum affine* Funck
SY: *Polytrichum alpestre* Hoppe
SY: *Polytrichum juniperinum* var. *affine* (Funck) Brid.

SY: *Polytrichum juniperinum* var. *alpestre* (Hoppe) Röhl.
SY: *Polytrichum juniperinum* var. *gracilius* Wahlenb.
Psilopilum Brid. [PSILOPI$]
Psilopilum cavifolium (Wils.) Hag. [PSILCAV]
SY: *Oligotrichum laevigatum* var. *cavifolium* (Wils.) Frye in Grout

PORELLACEAE (B:H)

Madotheca cordaeana (Hüb.) Dum. = Porella cordaeana [PORECOR]
Madotheca navicularis (Lehm. & Lindenb.) Dum. = Porella navicularis [PORENAV]
Madotheca roellii Steph. = Porella roellii [POREROE]
Porella L. [PORELLA$]
Porella cordaeana (Hub.) Moore [PORECOR] (dull scale feather wort)
SY: *Madotheca cordaeana* (Hüb.) Dum.
Porella navicularis (Lehm. & Lindenb.) Lindb. [PORENAV] (three-ruffle liverwort)
SY: *Madotheca navicularis* (Lehm. & Lindenb.) Dum.
Porella platyphylla (L.) Pfeiff. [POREPLA]
Porella platyphylla var. *platyphylloidea* (Schwein.) Frye & Clark = Porella platyphylloidea [POREPLT]
Porella platyphylloidea (Schwein.) Lindb. [POREPLT]
SY: *Porella platyphylla* var. *platyphylloidea* (Schwein.) Frye & Clark
Porella roellii Steph. [POREROE]
SY: *Madotheca roellii* Steph.
Porella roellii f. **crispata** Hong [POREROE1]
Porella roellii f. **schofieldiana** Hong [POREROE2]

POTTIACEAE (B:M)

Acaulon C. Müll. [ACAULON$]
Acaulon muticum (Hedw.) C. Müll. [ACAUMUT]
Acaulon muticum var. **rufescens** (Jaeg.) Crum [ACAUMUT1]
SY: *Acaulon rufescens* Jaeg.
Acaulon rufescens Jaeg. = Acaulon muticum var. rufescens [ACAUMUT1]
Aloina Kindb. [ALOINA$]
Aloina bifrons (De Not.) Delg. [ALOIBIF]
SY: *Aloina pilifera* (De Not.) Crum & Steere
SY: *Aloina rigida* var. *pilifera* (De Not.) Limpr.
Aloina brevirostris (Hook. & Grev.) Kindb. [ALOIBRE]
SY: *Barbula brevirostris* (Hook. & Grev.) Bruch in F. Müll.
SY: *Barbula macrorhyncha* Kindb. in Mac. & Kindb.
Aloina pilifera (De Not.) Crum & Steere = Aloina bifrons [ALOIBIF]
Aloina rigida (Hedw.) Limpr. [ALOIRIG]
SY: *Barbula rigida* Hedw.
Aloina rigida var. *pilifera* (De Not.) Limpr. = Aloina bifrons [ALOIBIF]
Anoectangium Schwaegr. [ANOECTA$]
Anoectangium aestivum (Hedw.) Mitt. [ANOEAES]
SY: *Anoectangium compactum* Schwaegr.
SY: *Anoectangium compactum* var. *alaskanum* Card. & Thér.
SY: *Anoectangium euchloron* (Schwaegr.) Mitt.
SY: *Anoectangium peckii* (Sull.) Sull. ex Aust.
Anoectangium arizonicum Bartr. = Gymnostomum aeruginosum [GYMNAER]
Anoectangium compactum Schwaegr. = Anoectangium aestivum [ANOEAES]
Anoectangium compactum var. *alaskanum* Card. & Thér. = Anoectangium aestivum [ANOEAES]
Anoectangium euchloron (Schwaegr.) Mitt. = Anoectangium aestivum [ANOEAES]
Anoectangium obtusifolium (Broth. & Par. in Card.) Grout = Molendoa sendtneriana [MOLESEN]
Anoectangium peckii (Sull.) Sull. ex Aust. = Anoectangium aestivum [ANOEAES]
Anoectangium sendtnerianum Bruch & Schimp. in B.S.G. = Molendoa sendtneriana [MOLESEN]
Anoectangium sendtnerianum var. *tenuinerve* (Limpr.) Mönk. = Anoectangium tenuinerve [ANOETEN]
Anoectangium tenuinerve (Limpr.) Par. [ANOETEN]
SY: *Anoectangium sendtnerianum* var. *tenuinerve* (Limpr.) Mönk.
Barbula Hedw. [BARBULA$]
Barbula aciphylla B.S.G. = Tortula norvegica [TORTNOR]
Barbula acuta (Brid.) Brid. = Didymodon rigidulus var. gracilis [DIDYRIG1]
Barbula acuta ssp. *icmadophila* (Schimp. ex C. Müll.) Amann = Didymodon rigidulus var. icmadophilus [DIDYRIG2]
Barbula acuta var. *bescherellei* (Sauerb. in Jaeg.) Crum = Didymodon rigidulus var. gracilis [DIDYRIG1]
Barbula acuta var. *icmadophila* (Schimp. ex C. Müll.) Crum = Didymodon rigidulus var. icmadophilus [DIDYRIG2]
Barbula amplexa Lesq. = Tortula amplexa [TORTAMP]
Barbula amplexifolia (Mitt.) Jaeg. [BARBAMP]
SY: *Barbula haringae* Crum
Barbula andreaeoides Kindb. = Didymodon subandreaeoides [DIDYSUB]
Barbula angustata (Schimp.) Mac. & Kindb. = Tortula subulata [TORTSUB]
Barbula asperifolia Mitt. = Didymodon asperifolius [DIDYASP]
Barbula bescherellei Sauerb. in Jaeg. = Didymodon rigidulus var. gracilis [DIDYRIG1]
Barbula brevirostris (Hook. & Grev.) Bruch in F. Müll. = Aloina brevirostris [ALOIBRE]
Barbula caespitosa Schwaegr. = Tortella humilis [TORTHUM]
Barbula circinnatula C. Müll. & Kindb. in Mac. & Kindb. = Didymodon vinealis var. flaccidus [DIDYVIN1]
Barbula columbiana (Herm. & Lawt.) Herm. & Lawt. = Bryoerythrophyllum columbianum [BRYOCOL]
Barbula convoluta Hedw. [BARBCON]
SY: *Streblotrichum convolutum* (Hedw.) P. Beauv.
Barbula convoluta var. **convoluta** [BARBCON0]
SY: *Barbula convoluta* var. *obtusata* C. Müll. & Kindb. in Mac. & Kindb.
Barbula convoluta var. **gallinula** Zand. [BARBCON1]
Barbula convoluta var. *obtusata* C. Müll. & Kindb. in Mac. & Kindb. = Barbula convoluta var. convoluta [BARBCON0]
Barbula cylindrica (Tayl.) Schimp. = Didymodon vinealis var. vinealis [DIDYVIN0]
Barbula eustegia Card. & Thér. [BARBEUS]
SY: *Barbula whitehouseae* Crum
Barbula fallax Hedw. = Didymodon fallax [DIDYFAL]
Barbula fallax var. *recurvifolia* (Wils.) Husn. = Didymodon fallax var. reflexus [DIDYFAL1]
Barbula ferruginascens Stirt. = Bryoerythrophyllum ferruginascens [BRYOFER]
Barbula fragilis (Hook. & Wils.) B.S.G. = Tortella fragilis [TORTFRA]
Barbula haringae Crum = Barbula amplexifolia [BARBAMP]
Barbula horridifolia C. Müll. & Kindb. in Mac. & Kindb. = Didymodon vinealis var. flaccidus [DIDYVIN1]
Barbula icmadophila Schimp. ex C. Müll. = Didymodon rigidulus var. icmadophilus [DIDYRIG2]
Barbula inclinata C. Müll. & Kindb. in Mac. & Kindb. = Tortella inclinata [TORTINC]
Barbula johansenii Williams = Didymodon johansenii [DIDYJOH]
Barbula laevipila (Brid.) Garov. = Tortula laevipila [TORTLAE]
Barbula laeviuscula Kindb. = Tortula ruralis [TORTRUA]
Barbula leptotricha C. Müll. & Kindb. in Mac. & Kindb. = Tortula ruralis [TORTRUA]
Barbula macrorhyncha Kindb. in Mac. & Kindb. = Aloina brevirostris [ALOIBRE]
Barbula megalocarpa Kindb. = Tortula princeps [TORTPRI]
Barbula melanocarpa C. Müll. & Kindb. in Mac. & Kindb. = Didymodon vinealis var. rubiginosus [DIDYVIN2]
Barbula michiganensis Steere = Didymodon michiganensis [DIDYMIC]
Barbula mucronifolia (Schwaegr.) Garov. = Tortula mucronifolia [TORTMUC]
Barbula nigrescens Mitt. = Didymodon nigrescens [DIDYNIG]
Barbula pseudorigidula Kindb. in Mac. & Kindb. = Didymodon vinealis var. flaccidus [DIDYVIN1]

Barbula recurvifolia Schimp. = Didymodon fallax var. reflexus [DIDYFAL1]
Barbula reflexa (Brid.) Brid. = Didymodon fallax var. reflexus [DIDYFAL1]
Barbula rigida Hedw. = Aloina rigida [ALOIRIG]
Barbula rigidula var. *icmadophila* Schimp. ex C. Müll. = Didymodon rigidulus var. icmadophilus [DIDYRIG2]
Barbula rigidula (Hedw.) Milde = Didymodon rigidulus [DIDYRIG]
Barbula rubella (Hüb.) Mitt. ex Lindb. = Bryoerythrophyllum recurvirostre [BRYOREC]
Barbula rubiginosa Mitt. = Didymodon vinealis var. rubiginosus [DIDYVIN2]
Barbula rufofusca Lawt. & Herm. = Didymodon nigrescens [DIDYNIG]
Barbula ruralis Hedw. = Tortula ruralis [TORTRUA]
Barbula subandreaeoides Kindb. = Didymodon subandreaeoides [DIDYSUB]
Barbula subgracilis C. Müll. & Kindb. in Mac. & Kindb. = Didymodon vinealis var. vinealis [DIDYVIN0]
Barbula subgracilis var. *viridior* Kindb. = Didymodon rigidulus [DIDYRIG]
Barbula subicmadophila C. Müll. & Kindb. in Mac. & Kindb. = Didymodon vinealis var. rubiginosus [DIDYVIN2]
Barbula subulata (Hedw.) P. Beauv. = Tortula subulata [TORTSUB]
Barbula tortuosa (Hedw.) Web. & Mohr = Tortella tortuosa [TORTTOR]
Barbula unguiculata Hedw. [BARBUNG]
SY: *Barbula unguiculata* f. *apiculata* (Hedw.) Mönk.
Barbula unguiculata f. *apiculata* (Hedw.) Mönk. = Barbula unguiculata [BARBUNG]
Barbula vinealis Brid. = Didymodon vinealis [DIDYVIN]
Barbula vinealis ssp. *cylindrica* (Tayl.) Podp. = Didymodon vinealis var. flaccidus [DIDYVIN1]
Barbula vinealis var. *flaccida* Bruch & Schimp. in B.S.G. = Didymodon vinealis var. flaccidus [DIDYVIN1]
Barbula whitehouseae Crum = Barbula eustegia [BARBEUS]
Bryoerythrophyllum Chen [BRYOERY$]
Bryoerythrophyllum alpigenum (Vent.) Chen = Bryoerythrophyllum recurvirostre [BRYOREC]
Bryoerythrophyllum columbianum (Herm. & Lawt.) Zand. [BRYOCOL]
SY: *Barbula columbiana* (Herm. & Lawt.) Herm. & Lawt.
SY: *Didymodon columbianus* Herm. & Lawt.
Bryoerythrophyllum ferruginascens (Stirt.) Giac. [BRYOFER]
SY: *Barbula ferruginascens* Stirt.
Bryoerythrophyllum recurvifolium (Tayl.) Zand. = Paraleptodontium recurvifolium [PARAREC]
Bryoerythrophyllum recurvirostre (Hedw.) Chen [BRYOREC]
SY: *Barbula rubella* (Hüb.) Mitt. ex Lindb.
SY: *Bryoerythrophyllum alpigenum* (Vent.) Chen
SY: *Bryoerythrophyllum recurvirostre* var. *dentatum* (Schimp.) Crum, Steere & Anderson
SY: *Didymodon alpigenum* Vent. in Jur.
SY: *Didymodon canadensis* Kindb. in Mac. & Kindb.
SY: *Didymodon recurvirostris* Hedw.
SY: *Didymodon recurvirostris* var. *dentatus* (Schimp.) Steere
SY: *Didymodon rubellus* (Bruch & Schimp. in B.S.G.) B.S.G.
SY: *Erythrobarbula recurvirostris* (Hedw.) Steere
Bryoerythrophyllum recurvirostre var. *dentatum* (Schimp.) Crum, Steere & Anderson = Bryoerythrophyllum recurvirostre [BRYOREC]
Crossidium Jur. [CROSSID$]
Crossidium aberrans Holz. & Bartr. [CROSABE]
SY: *Crossidium spatulaefolium* Holz. & Bartr.
Crossidium rosei Williams [CROSROS]
Crossidium seriatum Crum & Steere [CROSSER]
Crossidium spatulaefolium Holz. & Bartr. = Crossidium aberrans [CROSABE]
Crossidium squamiferum (Viv.) Jur. [CROSSQU]
Crumia Schof. [CRUMIA$]
Crumia latifolia (Kindb. in Mac.) Schof. [CRUMLAT]
SY: *Merceya latifolia* Kindb. in Mac.
SY: *Scopelophila latifolia* (Kindb. in Mac.) Ren. & Card.
Desmatodon Brid. [DESMATO$]
Desmatodon camptothecius Kindb. in Mac. & Kindb. = Desmatodon cernuus [DESMCER]
Desmatodon cernuus (Hüb.) Bruch & Schimp. in B.S.G. [DESMCER]
SY: *Desmatodon camptothecius* Kindb. in Mac. & Kindb.
SY: *Desmatodon cernuus* var. *xanthopus* Kindb.
Desmatodon cernuus var. *xanthopus* Kindb. = Desmatodon cernuus [DESMCER]
Desmatodon coloradensis Grout = Desmatodon obtusifolius [DESMOBT]
Desmatodon convolutus (Brid.) Grout [DESMCON]
SY: *Tortula atrovirens* (Sm.) Lindb.
Desmatodon guepinii Bruch & Schimp. in B.S.G. [DESMGUE]
Desmatodon heimii (Hedw.) Mitt. [DESMHEI]
Desmatodon hendersonii (Ren. & Card.) Williams in Millsp. & Nutt. = Didymodon tophaceus [DIDYTOP]
Desmatodon latifolius (Hedw.) Brid. [DESMLAT]
SY: *Desmatodon latifolius* var. *muticus* (Brid.) Brid.
Desmatodon latifolius var. *muticus* (Brid.) Brid. = Desmatodon latifolius [DESMLAT]
Desmatodon leucostoma (R. Br.) Berggr. [DESMLEU]
SY: *Desmatodon suberectus* (Hook.) Limpr.
Desmatodon obtusifolius (Schwaegr.) Schimp. [DESMOBT]
SY: *Desmatodon coloradensis* Grout
SY: *Desmatodon subtorquescens* C. Müll. & Kindb. in Mac. & Kindb.
SY: *Tortula obtusifolia* (Schwaegr.) Mathieu
Desmatodon randii (Kenn.) Laz. [DESMRAN]
SY: *Entosthodon neoscoticus* M.S. Brown
SY: *Pottia randii* Kenn.
Desmatodon suberectus (Hook.) Limpr. = Desmatodon leucostoma [DESMLEU]
Desmatodon subtorquescens C. Müll. & Kindb. in Mac. & Kindb. = Desmatodon obtusifolius [DESMOBT]
Didymodon Hedw. [DIDYMOD$]
Didymodon acutus (Brid.) Saito = Didymodon rigidulus var. gracilis [DIDYRIG1]
Didymodon acutus var. *ditrichoides* (Broth.) Zand. = Didymodon rigidulus var. gracilis [DIDYRIG1]
Didymodon alpigenum Vent. in Jur. = Bryoerythrophyllum recurvirostre [BRYOREC]
Didymodon asperifolius (Mitt.) Crum, Steere & Anderson [DIDYASP]
SY: *Barbula asperifolia* Mitt.
SY: *Didymodon rufus* Lor. in Rabenh.
Didymodon australasiae (Grev. & Hook.) Zand. = Trichostomopsis australasiae [TRICAUS]
Didymodon australasiae var. *umbrosus* (C. Müll.) Zand. = Trichostomopsis australasiae [TRICAUS]
Didymodon canadensis Kindb. in Mac. & Kindb. = Bryoerythrophyllum recurvirostre [BRYOREC]
Didymodon columbianus Herm. & Lawt. = Bryoerythrophyllum columbianum [BRYOCOL]
Didymodon cylindricus (Brid.) B.S.G. = Oxystegus tenuirostris [OXYSTEN]
Didymodon fallax (Hedw.) Zand. [DIDYFAL]
SY: *Barbula fallax* Hedw.
Didymodon fallax var. **reflexus** (Brid.) Zand. [DIDYFAL1]
SY: *Barbula fallax* var. *recurvifolia* (Wils.) Husn.
SY: *Barbula recurvifolia* Schimp.
SY: *Barbula reflexa* (Brid.) Brid.
SY: *Didymodon ferrugineus* (Schimp. ex Besch.) M.O. Hill
SY: *Didymodon rigidicaulis* (C. Müll.) Saito
SY: *Tortula reflexa* Brid.
Didymodon ferrugineus (Schimp. ex Besch.) M.O. Hill = Didymodon fallax var. reflexus [DIDYFAL1]
Didymodon fragilis Hook. & Wils. ex Drumm. = Tortella fragilis [TORTFRA]
Didymodon fuscoviridis Card. = Didymodon rigidulus var. rigidulus [DIDYRIG0]
Didymodon giganteus (Funck) Jur. = Geheebia gigantea [GEHEGIG]

Didymodon icmadophilus (Schimp. ex C. Müll.) Saito = Didymodon rigidulus var. icmadophilus [DIDYRIG2]
Didymodon insulanus (De Not.) M.O. Hill = Didymodon vinealis var. flaccidus [DIDYVIN1]
Didymodon johansenii (Williams) Crum [DIDYJOH]
SY: *Barbula johansenii* Williams
Didymodon luridus Hornsch. in Spreng. = Didymodon vinealis var. luridus [DIDYVIN3]
Didymodon michiganensis (Steere) Saito [DIDYMIC]
SY: *Barbula michiganensis* Steere
Didymodon nicholsonii Culm. = Didymodon vinealis var. nicholsonii [DIDYVIN4]
Didymodon nigrescens (Mitt.) Saito [DIDYNIG]
SY: *Barbula nigrescens* Mitt.
SY: *Barbula rufofusca* Lawt. & Herm.
Didymodon occidentalis Zand. = Didymodon vinealis var. rubiginosus [DIDYVIN2]
Didymodon recurvirostris Hedw. = Bryoerythrophyllum recurvirostre [BRYOREC]
Didymodon recurvirostris var. *dentatus* (Schimp.) Steere = Bryoerythrophyllum recurvirostre [BRYOREC]
Didymodon revolutus (Card.) Williams [DIDYREV]
SY: *Husnotiella revoluta* Card.
SY: *Husnotiella revoluta* var. *palmeri* (Card.) Bartr.
Didymodon rigidicaulis (C. Müll.) Saito = Didymodon fallax var. reflexus [DIDYFAL1]
Didymodon rigidulus Hedw. [DIDYRIG]
SY: *Barbula rigidula* (Hedw.) Milde
SY: *Barbula subgracilis* var. *viridior* Kindb.
Didymodon rigidulus var. **gracilis** (Schleich. ex Hook. & Grev.) Zand. [DIDYRIG1]
SY: *Barbula acuta* (Brid.) Brid.
SY: *Barbula acuta* var. *bescherellei* (Sauerb. in Jaeg.) Crum
SY: *Barbula bescherellei* Sauerb. in Jaeg.
SY: *Didymodon acutus* (Brid.) Saito
SY: *Didymodon acutus* var. *ditrichoides* (Broth.) Zand.
Didymodon rigidulus var. **icmadophilus** (Schimp. ex C. Müll.) Zand. [DIDYRIG2]
SY: *Barbula acuta* ssp. *icmadophila* (Schimp. ex C. Müll.) Amann
SY: *Barbula acuta* var. *icmadophila* (Schimp. ex C. Müll.) Crum
SY: *Barbula icmadophila* Schimp. ex C. Müll.
SY: *Barbula rigidula* var. *icmadophila* Schimp. ex C. Müll.
SY: *Didymodon icmadophilus* (Schimp. ex C. Müll.) Saito
Didymodon rigidulus var. **rigidulus** [DIDYRIG0]
SY: *Didymodon fuscoviridis* Card.
Didymodon rubellus (Bruch & Schimp. in B.S.G.) B.S.G. = Bryoerythrophyllum recurvirostre [BRYOREC]
Didymodon rufus Lor. in Rabenh. = Didymodon asperifolius [DIDYASP]
Didymodon subandreaeoides (Kindb.) Zand. [DIDYSUB]
SY: *Barbula andreaeoides* Kindb.
SY: *Barbula subandreaeoides* Kindb.
Didymodon tophaceus (Brid.) Lisa [DIDYTOP]
SY: *Desmatodon hendersonii* (Ren. & Card.) Williams in Millsp. & Nutt.
SY: *Didymodon tophaceus* var. *decurrens* Card. & Thér.
SY: *Husnotiella pringlei* (Card.) Grout
SY: *Trichostomum tophaceum* Brid.
Didymodon tophaceus var. *decurrens* Card. & Thér. = Didymodon tophaceus [DIDYTOP]
Didymodon trachyneuron Kindb. = Oxystegus tenuirostris [OXYSTEN]
Didymodon trifarius auct. = Didymodon vinealis var. luridus [DIDYVIN3]
Didymodon trifarius ssp. *nicholsonii* (Culm.) Wijk & Marg. = Didymodon vinealis var. nicholsonii [DIDYVIN4]
Didymodon umbrosus (C. Müll.) Zand. = Trichostomopsis australasiae [TRICAUS]
Didymodon vinealis (Brid.) Zand. [DIDYVIN]
SY: *Barbula vinealis* Brid.
Didymodon vinealis var. **flaccidus** (Bruch & Schimp. in Schimp.) Zand. [DIDYVIN1]
SY: *Barbula circinnatula* C. Müll. & Kindb. in Mac. & Kindb.
SY: *Barbula horridifolia* C. Müll. & Kindb. in Mac. & Kindb.
SY: *Barbula pseudorigidula* Kindb. in Mac. & Kindb.
SY: *Barbula vinealis* ssp. *cylindrica* (Tayl.) Podp.
SY: *Barbula vinealis* var. *flaccida* Bruch & Schimp. in B.S.G.
SY: *Didymodon insulanus* (De Not.) M.O. Hill
Didymodon vinealis var. **luridus** (Hornsch. in Spreng.) Zand. [DIDYVIN3]
SY: *Didymodon luridus* Hornsch. in Spreng.
SY: *Didymodon trifarius* auct.
Didymodon vinealis var. **nicholsonii** (Culm.) Zand. [DIDYVIN4]
SY: *Didymodon nicholsonii* Culm.
SY: *Didymodon trifarius* ssp. *nicholsonii* (Culm.) Wijk & Marg.
Didymodon vinealis var. **rubiginosus** (Mitt.) Zand. [DIDYVIN2]
SY: *Barbula melanocarpa* C. Müll. & Kindb. in Mac. & Kindb.
SY: *Barbula rubiginosa* Mitt.
SY: *Barbula subicmadophila* C. Müll. & Kindb. in Mac. & Kindb.
SY: *Didymodon occidentalis* Zand.
Didymodon vinealis var. **vinealis** [DIDYVIN0]
SY: *Barbula cylindrica* (Tayl.) Schimp.
SY: *Barbula subgracilis* C. Müll. & Kindb. in Mac. & Kindb.
Entosthodon neoscoticus M.S. Brown = Desmatodon randii [DESMRAN]
Erythrobarbula recurvirostris (Hedw.) Steere = Bryoerythrophyllum recurvirostre [BRYOREC]
Eucladium Bruch & Schimp. in B.S.G. [EUCLADI$]
Eucladium verticillatum (Brid.) Bruch & Schimp. in B.S.G. [EUCLVER]
Geheebia Schimp. [GEHEEBI$]
Geheebia gigantea (Funck) Boul. [GEHEGIG]
SY: *Didymodon giganteus* (Funck) Jur.
Gymnostomum Nees & Hornsch. [GYMNOSO$]
Gymnostomum aeruginosum Sm. [GYMNAER]
SY: *Anoectangium arizonicum* Bartr.
SY: *Gymnostomum rupestre* Schleich. ex Schwaegr.
SY: *Weissia rupestris* (Schwaegr.) C. Müll.
Gymnostomum curvirostre Hedw. ex Brid. = Hymenostylium recurvirostre [HYMEREC]
Gymnostomum curvirostre var. *commutatum* (Mitt.) Card. & Thér. = Hymenostylium recurvirostre [HYMEREC]
Gymnostomum insigne (Dix.) A.J.E. Sm. = Hymenostylium insigne [HYMEINS]
Gymnostomum latifolium (Zett.) Flow. in Crum = Hymenostylium recurvirostre [HYMEREC]
Gymnostomum recurvirostre Hedw. = Hymenostylium recurvirostre [HYMEREC]
Gymnostomum recurvirostre var. *commutatum* (Mitt.) Grout = Hymenostylium recurvirostre [HYMEREC]
Gymnostomum recurvirostre var. *latifolium* Zett. = Hymenostylium recurvirostre [HYMEREC]
Gymnostomum recurvirostre var. *scabrum* (Lindb.) Grout = Hymenostylium recurvirostre [HYMEREC]
Gymnostomum rupestre Schleich. ex Schwaegr. = Gymnostomum aeruginosum [GYMNAER]
Hilpertia scotteri (Zand. & Steere) Zand. = Tortula scotteri [TORTSCO]
Husnotiella pringlei (Card.) Grout = Didymodon tophaceus [DIDYTOP]
Husnotiella revoluta Card. = Didymodon revolutus [DIDYREV]
Husnotiella revoluta var. *palmeri* (Card.) Bartr. = Didymodon revolutus [DIDYREV]
Husnotiella torquescens (Card.) Bartr. = Trichostomopsis australasiae [TRICAUS]
Hymenostomum microstomum (Hedw.) R. Br. = Weissia hedwigii [WEISHED]
Hymenostylium Brid. [HYMENOS$]
Hymenostylium curvirostre Mitt. = Hymenostylium recurvirostre [HYMEREC]
Hymenostylium insigne (Dix.) Podp. [HYMEINS]
SY: *Gymnostomum insigne* (Dix.) A.J.E. Sm.
SY: *Hymenostylium recurvirostre* var. *insigne* (Dix.) Bartr.
Hymenostylium recurvirostre (Hedw.) Dix. [HYMEREC]
SY: *Gymnostomum curvirostre* Hedw. ex Brid.

SY: *Gymnostomum curvirostre* var. *commutatum* (Mitt.) Card. & Thér.
SY: *Gymnostomum latifolium* (Zett.) Flow. in Crum
SY: *Gymnostomum recurvirostre* Hedw.
SY: *Gymnostomum recurvirostre* var. *commutatum* (Mitt.) Grout
SY: *Gymnostomum recurvirostre* var. *latifolium* Zett.
SY: *Gymnostomum recurvirostre* var. *scabrum* (Lindb.) Grout
SY: *Hymenostylium curvirostre* Mitt.
SY: *Hymenostylium recurvirostre* var. *latifolium* (Zett.) Wijk & Marg.
SY: *Hymenostylium scabrum* (Lindb.) Loeske
SY: *Weissia curvirostris* C. Müll.
Hymenostylium recurvirostre var. *insigne* (Dix.) Bartr. = Hymenostylium insigne [HYMEINS]
Hymenostylium recurvirostre var. *latifolium* (Zett.) Wijk & Marg. = Hymenostylium recurvirostre [HYMEREC]
Hymenostylium scabrum (Lindb.) Loeske = Hymenostylium recurvirostre [HYMEREC]
Leptodontium recurvifolium (Tayl.) Lindb. = Paraleptodontium recurvifolium [PARAREC]
Merceya latifolia Kindb. in Mac. = Crumia latifolia [CRUMLAT]
Molendoa Lindb. [MOLENDO$]
Molendoa hornschuchiana (Hook.) Lindb. ex Limpr. [MOLEHOR]
Molendoa obtusifolia Broth. & Par. in Card. = Molendoa sendtneriana [MOLESEN]
Molendoa sendtneriana (Bruch & Schimp. in B.S.G.) Limpr. [MOLESEN]
SY: *Anoectangium obtusifolium* (Broth. & Par. in Card.) Grout
SY: *Anoectangium sendtnerianum* Bruch & Schimp. in B.S.G.
SY: *Molendoa obtusifolia* Broth. & Par. in Card.
Oxystegus (Limpr.) Hilp. [OXYSTEG$]
Oxystegus recurvifolius (Tayl.) Zand. = Paraleptodontium recurvifolium [PARAREC]
Oxystegus tenuirostris (Hook. & Tayl.) A. J. E. Sm. [OXYSTEN]
SY: *Didymodon cylindricus* (Brid.) B.S.G.
SY: *Didymodon trachyneuron* Kindb.
SY: *Trichostomum cylindricum* (Brid.) C. Müll.
SY: *Trichostomum tenuirostre* (Hook. & Tayl.) Lindb.
SY: *Trichostomum mollissimum* auct.
Paraleptodontium Long [PARALEP$]
Paraleptodontium recurvifolium (Tayl.) Long [PARAREC]
SY: *Bryoerythrophyllum recurvifolium* (Tayl.) Zand.
SY: *Leptodontium recurvifolium* (Tayl.) Lindb.
SY: *Oxystegus recurvifolius* (Tayl.) Zand.
Pharomitrium subsessile (Brid.) Schimp. = Pterygoneurum subsessile [PTERSUB]
Phascum Hedw. [PHASCUM$]
Phascum acaulon With. = Phascum cuspidatum [PHASCUS]
Phascum cuspidatum Hedw. [PHASCUS]
SY: *Phascum acaulon* With.
SY: *Phascum cuspidatum* var. *americanum* Ren. & Card.
SY: *Phascum cuspidatum* var. *henricii* (Ren. & Card.) Wijk & Marg.
SY: *Phascum cuspidatum* var. *piliferum* (Hedw.) Amann
SY: *Phascum cuspidatum* var. *schreberianum* (Dicks.) Brid.
SY: *Phascum piliferum* Schreb. ex Hedw.
Phascum cuspidatum var. *americanum* Ren. & Card. = Phascum cuspidatum [PHASCUS]
Phascum cuspidatum var. *henricii* (Ren. & Card.) Wijk & Marg. = Phascum cuspidatum [PHASCUS]
Phascum cuspidatum var. *piliferum* (Hedw.) Amann = Phascum cuspidatum [PHASCUS]
Phascum cuspidatum var. *schreberianum* (Dicks.) Brid. = Phascum cuspidatum [PHASCUS]
Phascum piliferum Schreb. ex Hedw. = Phascum cuspidatum [PHASCUS]
Phascum subexsertum Hook. in Drumm. = Pottia bryoides [POTTBRY]
Phascum vlassovii Laz. [PHASVLA]
Pottia (Reichenb.) Fürnr. [POTTIA$]
Pottia bryoides (Dicks.) Mitt. [POTTBRY]
SY: *Phascum subexsertum* Hook. in Drumm.
Pottia cavifolia Ehrh. ex Fürnr. = Pterygoneurum ovatum [PTEROVA]
Pottia latifolia (Schwaegr. in Schultes) C. Müll. = Stegonia latifolia [STEGLAT]
Pottia latifolia var. *pilifera* (Brid.) C. Müll. = Stegonia pilifera [STEGPIL]
Pottia littoralis Mitt. = Pottia truncata [POTTTRU]
Pottia nevadensis Card. & Thér. [POTTNEV]
Pottia randii Kenn. = Desmatodon randii [DESMRAN]
Pottia truncata (Hedw.) Fürnr. ex B.S.G. [POTTTRU]
SY: *Pottia littoralis* Mitt.
SY: *Pottia truncata* var. *littoralis* (Mitt.) Warnst.
SY: *Pottia truncatula* (With.) Buse
Pottia truncata var. *littoralis* (Mitt.) Warnst. = Pottia truncata [POTTTRU]
Pottia truncatula (With.) Buse = Pottia truncata [POTTTRU]
Pottia wilsonii (Hook.) Bruch & Schimp. in B.S.G. [POTTWIL]
Pseudocrossidium Williams [PSEUDOO$]
Pseudocrossidium hornschuchianum (Schultz) Zand. [PSEUHOR]
Pseudocrossidium revolutum (Brid. in Schrad.) Zand. [PSEUREV]
Pterygoneurum Jur. [PTERYGO$]
Pterygoneurum arcticum Steere = Pterygoneurum lamellatum [PTERLAM]
Pterygoneurum kozlovii Laz. [PTERKOZ]
SY: *Pterygoneurum smardaeanum* Vanek
Pterygoneurum lamellatum (Lindb.) Jur. [PTERLAM]
SY: *Pterygoneurum arcticum* Steere
Pterygoneurum ovatum (Hedw.) Dix. [PTEROVA]
SY: *Pottia cavifolia* Ehrh. ex Fürnr.
Pterygoneurum smardaeanum Vanek = Pterygoneurum kozlovii [PTERKOZ]
Pterygoneurum subsessile (Brid.) Jur. [PTERSUB] (woolly caterpillar moss)
SY: *Pharomitrium subsessile* (Brid.) Schimp.
SY: *Pterygoneurum subsessile* var. *henricii* (Rau) Wareh. in Grout
SY: *Pterygoneurum subsessile* var. *kieneri* Hab.
Pterygoneurum subsessile var. *henricii* (Rau) Wareh. in Grout = Pterygoneurum subsessile [PTERSUB]
Pterygoneurum subsessile var. *kieneri* Hab. = Pterygoneurum subsessile [PTERSUB]
Scopelophila latifolia (Kindb. in Mac.) Ren. & Card. = Crumia latifolia [CRUMLAT]
Stegonia Vent. [STEGONI$]
Stegonia latifolia (Schwaegr. in Schultes) Vent. ex Broth. [STEGLAT]
SY: *Pottia latifolia* (Schwaegr. in Schultes) C. Müll.
Stegonia latifolia var. *pilifera* (Brid.) Broth. = Stegonia pilifera [STEGPIL]
Stegonia pilifera (Brid.) Crum & Anderson [STEGPIL]
SY: *Pottia latifolia* var. *pilifera* (Brid.) C. Müll.
SY: *Stegonia latifolia* var. *pilifera* (Brid.) Broth.
Streblotrichum convolutum (Hedw.) P. Beauv. = Barbula convoluta [BARBCON]
Syntrichia ruralis (Hedw.) Web. & Mohr = Tortula ruralis [TORTRUA]
Timmiella (De Not.) Limpr. [TIMMIEL$]
Timmiella crassinervis (Hampe) L. Koch [TIMMCRA]
SY: *Timmiella flexiseta* var. *vancouveriensis* (Broth.) Grout
SY: *Trichostomum vancouveriense* (Broth.) Kindb.
SY: *Timmiella flexiseta* auct.
Timmiella flexiseta auct. = Timmiella crassinervis [TIMMCRA]
Timmiella flexiseta var. *vancouveriensis* (Broth.) Grout = Timmiella crassinervis [TIMMCRA]
Tortella (Lindb.) Limpr. [TORTELL$]
Tortella arctica (Arnell) Crundw. & Nyh. [TORTARC]
Tortella caespitosa (Schwaegr.) Limpr. = Tortella humilis [TORTHUM]
Tortella fragilis (Hook. & Wils. in Drumm.) Limpr. [TORTFRA] (fragile screw moss)
SY: *Barbula fragilis* (Hook. & Wils.) B.S.G.
SY: *Didymodon fragilis* Hook. & Wils. ex Drumm.
Tortella humilis (Hedw.) Jenn. [TORTHUM]
SY: *Barbula caespitosa* Schwaegr.

SY: *Tortella caespitosa* (Schwaegr.) Limpr.
Tortella inclinata (Hedw. f.) Limpr. [TORTINC]
SY: *Barbula inclinata* C. Müll. & Kindb. in Mac. & Kindb.
SY: *Tortella inclinatula* (C. Müll. & Kindb. in Mac. & Kindb.) Broth.
Tortella inclinatula (C. Müll. & Kindb. in Mac. & Kindb.) Broth. = Tortella inclinata [TORTINC]
Tortella tortuosa (Hedw.) Limpr. [TORTTOR]
SY: *Barbula tortuosa* (Hedw.) Web. & Mohr
Tortula Hedw. [TORTULA$]
Tortula amplexa (Lesq.) Steere [TORTAMP]
SY: *Barbula amplexa* Lesq.
Tortula atrovirens (Sm.) Lindb. = Desmatodon convolutus [DESMCON]
Tortula bistratosa Flow. = Tortula caninervis [TORTCAN]
Tortula bolanderi (Lesq.) Howe [TORTBOL]
Tortula brevipes (Lesq.) Broth. [TORTBRE]
Tortula calcicola Grebe. = Tortula ruralis [TORTRUA]
Tortula calcicolens W. Kramer = Tortula ruralis [TORTRUA]
Tortula caninervis (Mitt.) Broth. [TORTCAN]
SY: *Tortula bistratosa* Flow.
SY: *Tortula desertorum* Broth.
Tortula desertorum Broth. = Tortula caninervis [TORTCAN]
Tortula intermedia (Brid.) De Not. = Tortula ruralis [TORTRUA]
Tortula laevipila (Brid.) Schwaegr. [TORTLAE]
SY: *Barbula laevipila* (Brid.) Garov.
Tortula laevipila var. **laevipila** [TORTLAE0]
SY: *Tortula laevipila* var. *pagorum* Husn.
Tortula laevipila var. **meridionalis** (Schimp.) Wijk & Marg. [TORTLAE2]
Tortula laevipila var. *pagorum* Husn. = Tortula laevipila var. laevipila [TORTLAE0]
Tortula latifolia Bruch ex Hartm. [TORTLAT]
Tortula mucronifolia Schwaegr. [TORTMUC]
SY: *Barbula mucronifolia* (Schwaegr.) Garov.
Tortula muralis Hedw. [TORTMUR]
Tortula norvegica (Web.) Wahlenb. ex Lindb. [TORTNOR]
SY: *Barbula aciphylla* B.S.G.
Tortula obtusifolia (Schwaegr.) Mathieu = Desmatodon obtusifolius [DESMOBT]
Tortula princeps De Not. [TORTPRI]
SY: *Barbula megalocarpa* Kindb.
Tortula reflexa Brid. = Didymodon fallax var. reflexus [DIDYFAL1]
Tortula ruraliformis (Besch.) Ingh. = Tortula ruralis [TORTRUA]
Tortula ruralis (Hedw.) Gaertn., Meyer & Scherb. [TORTRUA] (hairy screw moss)
SY: *Barbula laeviuscula* Kindb.
SY: *Barbula leptotricha* C. Müll. & Kindb. in Mac. & Kindb.
SY: *Barbula ruralis* Hedw.
SY: *Syntrichia ruralis* (Hedw.) Web. & Mohr
SY: *Tortula calcicola* Grebe.
SY: *Tortula calcicolens* W. Kramer
SY: *Tortula intermedia* (Brid.) De Not.
SY: *Tortula ruraliformis* (Besch.) Ingh.
SY: *Tortula ruralis* var. *crinita* De Not.
Tortula ruralis var. *crinita* De Not. = Tortula ruralis [TORTRUA]
Tortula scotteri Zand. & Steere [TORTSCO]
SY: *Hilpertia scotteri* (Zand. & Steere) Zand.
Tortula subulata Hedw. [TORTSUB]
SY: *Barbula angustata* (Schimp.) Mac. & Kindb.
SY: *Barbula subulata* (Hedw.) P. Beauv.
SY: *Tortula subulata* var. *angustata* (Schimp.) Lindb.
Tortula subulata var. *angustata* (Schimp.) Lindb. = Tortula subulata [TORTSUB]
Tortula virescens (De Not.) De Not. [TORTVIR]
Trichostomopsis Card. [TRICHOS$]
Trichostomopsis australasiae (Grev. & Hook.) Robins. [TRICAUS]
SY: *Didymodon australasiae* (Grev. & Hook.) Zand.
SY: *Didymodon australasiae* var. *umbrosus* (C. Müll.) Zand.
SY: *Didymodon umbrosus* (C. Müll.) Zand.
SY: *Husnotiella torquescens* (Card.) Bartr.
SY: *Trichostomopsis australasiae* var. *umbrosus* (C. Müll.) Zand.
SY: *Trichostomopsis crispifolia* Card.
SY: *Trichostomopsis diaphanobasis* (Card.) Grout
SY: *Trichostomopsis fayae* Grout
SY: *Trichostomopsis umbrosus* (Card.) Robins.
Trichostomopsis australasiae var. *umbrosus* (C. Müll.) Zand. = Trichostomopsis australasiae [TRICAUS]
Trichostomopsis crispifolia Card. = Trichostomopsis australasiae [TRICAUS]
Trichostomopsis diaphanobasis (Card.) Grout = Trichostomopsis australasiae [TRICAUS]
Trichostomopsis fayae Grout = Trichostomopsis australasiae [TRICAUS]
Trichostomopsis umbrosus (Card.) Robins. = Trichostomopsis australasiae [TRICAUS]
Trichostomum Bruch [TRICHOT$]
Trichostomum arcticum Kaal. [TRICARC]
SY: *Trichostomum cuspidatissimum* Card. & Thér.
Trichostomum cuspidatissimum Card. & Thér. = Trichostomum arcticum [TRICARC]
Trichostomum cylindricum (Brid.) C. Müll. = Oxystegus tenuirostris [OXYSTEN]
Trichostomum mollissimum auct. = Oxystegus tenuirostris [OXYSTEN]
Trichostomum tenuirostre (Hook. & Tayl.) Lindb. = Oxystegus tenuirostris [OXYSTEN]
Trichostomum tophaceum Brid. = Didymodon tophaceus [DIDYTOP]
Trichostomum vancouveriense (Broth.) Kindb. = Timmiella crassinervis [TIMMCRA]
Weissia Hedw. [WEISSIA$]
Weissia andrewsii Bartr. = Weissia controversa [WEISCON]
Weissia brachycarpa auct. = Weissia hedwigii [WEISHED]
Weissia controversa Hedw. [WEISCON]
SY: *Weissia andrewsii* Bartr.
SY: *Weissia controversa* var. *longiseta* (Lesq. & James) Crum, Steere & Anderson
SY: *Weissia controversa* var. *wolfii* (Lesq. & James) Crum, Steere & Anderson
SY: *Weissia viridula* Hedw. ex Brid.
SY: *Weissia viridula* var. *australis* Aust.
SY: *Weissia viridula* var. *wolfii* Lesq. & James
SY: *Weissia wolfii* Lesq. & James
Weissia controversa var. *longiseta* (Lesq. & James) Crum, Steere & Anderson = Weissia controversa [WEISCON]
Weissia controversa var. *wolfii* (Lesq. & James) Crum, Steere & Anderson = Weissia controversa [WEISCON]
Weissia curvirostris C. Müll. = Hymenostylium recurvirostre [HYMEREC]
Weissia hedwigii Crum [WEISHED]
SY: *Hymenostomum microstomum* (Hedw.) R. Br.
SY: *Weissia microstoma* (Hedw.) C. Müll.
SY: *Weissia brachycarpa* auct.
Weissia microstoma (Hedw.) C. Müll. = Weissia hedwigii [WEISHED]
Weissia rupestris (Schwaegr.) C. Müll. = Gymnostomum aeruginosum [GYMNAER]
Weissia viridula Hedw. ex Brid. = Weissia controversa [WEISCON]
Weissia viridula var. *australis* Aust. = Weissia controversa [WEISCON]
Weissia viridula var. *wolfii* Lesq. & James = Weissia controversa [WEISCON]
Weissia wolfii Lesq. & James = Weissia controversa [WEISCON]

PSEUDOLEPICOLEACEAE (B:H)

Blepharostoma (Dum. emend. Lindb.) Dum. [BLEPHAR$]
Blepharostoma arachnoideum M.A. Howe [BLEPARA]
Blepharostoma arachnoideum var. *brevirete* (Bryhn & Kaal.) Frye & Clark = Blepharostoma trichophyllum ssp. brevirete [BLEPTRI1]
Blepharostoma trichophyllum (L.) Dum. [BLEPTRI]
Blepharostoma trichophyllum ssp. **brevirete** (Bryhn & Kaal.) Schust. [BLEPTRI1]
SY: *Blepharostoma arachnoideum* var. *brevirete* (Bryhn & Kaal.) Frye & Clark
Blepharostoma trichophyllum ssp. **trichophyllum** [BLEPTRI0]

PTERIGYNANDRACEAE (B:M)

Habrodon leucotrichus (Mitt.) Perss. = Iwatsukiella leucotricha [IWATLEU]
Heterocladium Schimp. in B.S.G. [HETEROL$]
Heterocladium aberrans Ren. & Card. = Heterocladium procurrens [HETEPRO]
Heterocladium dimorphum (Brid.) Schimp. in B.S.G. [HETEDIM]
SY: *Heterocladium squarrosulum* Lindb.
SY: *Hypnum dimorphum* Brid.
Heterocladium frullaniopsis C. Müll. & Kindb. in Mac. & Kindb. = Pterigynandrum filiforme [PTERFIL]
Heterocladium heteropteroides Best = Heterocladium macounii [HETEMAC]
Heterocladium heteropteroides var. *filescens* Best = Heterocladium macounii [HETEMAC]
Heterocladium heteropterum (Brid.) B.S.G. = Heterocladium macounii [HETEMAC]
Heterocladium macounii Best [HETEMAC]
SY: *Heterocladium heteropteroides* Best
SY: *Heterocladium heteropteroides* var. *filescens* Best
SY: *Heterocladium heteropterum* (Brid.) B.S.G.
Heterocladium procurrens (Mitt.) Jaeg. [HETEPRO] (tangle moss)
SY: *Heterocladium aberrans* Ren. & Card.
Heterocladium squarrosulum Lindb. = Heterocladium dimorphum [HETEDIM]
Hypnum dimorphum Brid. = Heterocladium dimorphum [HETEDIM]
Hypnum julaceum (Schwaegr.) Vill. ex Schwaegr. = Myurella julacea [MYURJUL]
Iwatsukiella Buck & Crum [IWATSUK$]
Iwatsukiella leucotricha (Mitt.) Buck & Crum [IWATLEU]
SY: *Habrodon leucotrichus* (Mitt.) Perss.
Myurella Schimp. in B.S.G. [MYURELL$]
Myurella apiculata (Somm.) Schimp. in B.S.G. = Myurella tenerrima [MYURTEN]
Myurella careyana Sull. in Sull. & Lesq. = Myurella sibirica [MYURSIB]
Myurella careyana var. *tenella* Hab. = Myurella sibirica [MYURSIB]
Myurella gracilis Lindb. = Myurella sibirica [MYURSIB]
Myurella julacea (Schwaegr.) Schimp. in B.S.G. [MYURJUL] (small mouse-tail moss)
SY: *Hypnum julaceum* (Schwaegr.) Vill. ex Schwaegr.
SY: *Myurella julacea* var. *scabrifolia* Lindb. ex Limpr.
Myurella julacea var. *scabrifolia* Lindb. ex Limpr. = Myurella julacea [MYURJUL]
Myurella sibirica (C. Müll.) Reim. [MYURSIB]
SY: *Myurella careyana* Sull. in Sull. & Lesq.
SY: *Myurella careyana* var. *tenella* Hab.
SY: *Myurella gracilis* Lindb.
SY: *Myurella sibirica* var. *tenella* (Hab.) Crum, Steere & Anderson
Myurella sibirica var. *tenella* (Hab.) Crum, Steere & Anderson = Myurella sibirica [MYURSIB]
Myurella tenerrima (Brid.) Lindb. [MYURTEN]
SY: *Myurella apiculata* (Somm.) Schimp. in B.S.G.
Pterigynandrum Hedw. [PTERIGY$]
Pterigynandrum filiforme Hedw. [PTERFIL]
SY: *Heterocladium frullaniopsis* C. Müll. & Kindb. in Mac. & Kindb.
SY: *Pterigynandrum filiforme* ssp. *decipiens* (Web. & Mohr) Kindb.
SY: *Pterigynandrum filiforme* var. *decipiens* (Web. & Mohr) Limpr.
SY: *Pterigynandrum filiforme* f. *majus* (De Not.) De Not.
SY: *Pterigynandrum filiforme* var. *majus* (De Not.) De Not.
SY: *Pterigynandrum filiforme* f. *minus* Lesq. & James
SY: *Pterigynandrum filiforme* var. *minus* Lesq. & James
SY: *Pterigynandrum filiforme* f. *papillosulum* (C. Müll. & Kindb. in Mac. & Kindb.) Grout
SY: *Pterigynandrum filiforme* var. *papillosulum* (C. Müll. & Kindb. in Mac. & Kindb.) Thér.
SY: *Pterigynandrum papillosulum* C. Müll. & Kindb. in Mac. & Kindb.
Pterigynandrum filiforme f. *majus* (De Not.) De Not. = Pterigynandrum filiforme [PTERFIL]
Pterigynandrum filiforme f. *minus* Lesq. & James = Pterigynandrum filiforme [PTERFIL]
Pterigynandrum filiforme f. *papillosulum* (C. Müll. & Kindb. in Mac. & Kindb.) Grout = Pterigynandrum filiforme [PTERFIL]
Pterigynandrum filiforme ssp. *decipiens* (Web. & Mohr) Kindb. = Pterigynandrum filiforme [PTERFIL]
Pterigynandrum filiforme var. *decipiens* (Web. & Mohr) Limpr. = Pterigynandrum filiforme [PTERFIL]
Pterigynandrum filiforme var. *majus* (De Not.) De Not. = Pterigynandrum filiforme [PTERFIL]
Pterigynandrum filiforme var. *minus* Lesq. & James = Pterigynandrum filiforme [PTERFIL]
Pterigynandrum filiforme var. *papillosulum* (C. Müll. & Kindb. in Mac. & Kindb.) Thér. = Pterigynandrum filiforme [PTERFIL]
Pterigynandrum papillosulum C. Müll. & Kindb. in Mac. & Kindb. = Pterigynandrum filiforme [PTERFIL]

PTILIDIACEAE (B:H)

Ptilidium Nees [PTILIDI$]
Ptilidium californicum (Aust.) Underw. [PTILCAL]
Ptilidium ciliare (L.) Hampe [PTILCIL] (northern naugehyde liverwort)
Ptilidium pulcherrimum (G. Web.) Hampe [PTILPUL] (naugehyde liverwort)

PTYCHOMITRIACEAE (B:M)

Brachysteleum polyphylloides C. Müll. = Ptychomitrium gardneri [PTYCGAR]
Ptychomitrium Fürnr. [PTYCHOM$]
Ptychomitrium gardneri Lesq. [PTYCGAR]
SY: *Brachysteleum polyphylloides* C. Müll.
SY: *Ptychomitrium kiusiuense* Sak.
SY: *Ptychomitrium longisetum* Reim. & Sak.
SY: *Ptychomitrium polyphylloides* (C. Müll.) Par.
SY: *Ptychomitrium robustum* Broth.
SY: *Ptychomitrium viride* Sak.
Ptychomitrium kiusiuense Sak. = Ptychomitrium gardneri [PTYCGAR]
Ptychomitrium longisetum Reim. & Sak. = Ptychomitrium gardneri [PTYCGAR]
Ptychomitrium polyphylloides (C. Müll.) Par. = Ptychomitrium gardneri [PTYCGAR]
Ptychomitrium robustum Broth. = Ptychomitrium gardneri [PTYCGAR]
Ptychomitrium viride Sak. = Ptychomitrium gardneri [PTYCGAR]

RADULACEAE (B:H)

Radula Dum. [RADULA$]
Radula auriculata Steph. [RADUAUR]
Radula bolanderi Gott. [RADUBOL]
Radula complanata (L.) Dum. [RADUCOM] (flat-leaved liverwort)
SY: *Radula hallii* Aust.
Radula hallii Aust. = Radula complanata [RADUCOM]
Radula obtusiloba Steph. [RADUOBT]
Radula obtusiloba ssp. **obtusiloba** [RADUOBT0]
Radula obtusiloba ssp. **polyclada** (Evans) Hatt. [RADUOBT1]
SY: *Radula polyclada* Evans
Radula polyclada Evans = Radula obtusiloba ssp. polyclada [RADUOBT1]
Radula prolifera S. Arnell [RADUPRO]

RHYTIDIACEAE (B:M)

Hypnum rugosum Hedw. = Rhytidium rugosum [RHYTRUG]

Rhytidium (Sull.) Kindb. [RHYTIDU$]
Rhytidium rugosum (Hedw.) Kindb. [RHYTRUG] (crumpled-leaf moss)
SY: *Hypnum rugosum* Hedw.

RICCIACEAE (B:H)

Riccia L. [RICCIA$]
Riccia beyrichiana Hampe ex Lehm. [RICCBEY]
Riccia cavernosa Hoffm. [RICCCAV]
SY: *Riccia crystallina* auct.
Riccia crystallina auct. = Riccia cavernosa [RICCCAV]
Riccia fluitans L. [RICCFLU]
SY: *Riccia rhenana* Lorbeer
Riccia rhenana Lorbeer = Riccia fluitans [RICCFLU]
Riccia sorocarpa Bisch. [RICCSOR]
Ricciocarpos Corda [RICCIOC$]
Ricciocarpos natans (L.) Corda [RICCNAT] (floating liverwort)

SCAPANIACEAE (B:H)

Diplophylleia albicans (L.) Trevis = Diplophyllum albicans [DIPLALB]
Diplophylleia plicata (Lindb.) Evans = Diplophyllum plicatum [DIPLPLI]
Diplophylleia taxifolia (Wahlenb.) Trevis. = Diplophyllum taxifolium [DIPLTAX]
Diplophyllum (Dum.) Dum. [DIPLOPH$]
Diplophyllum albicans (L.) Dum. [DIPLALB] (common fold-leaf liverwort)
SY: *Diplophylleia albicans* (L.) Trevis
Diplophyllum gymnostomophilum (Kaal.) Kaal. = Scapania gymnostomophila [SCAPGYM]
Diplophyllum hyalinum Brinkm. = Diplophyllum imbricatum [DIPLIMB]
Diplophyllum imbricatum (M.A. Howe) K. Müll. [DIPLIMB]
SY: *Diplophyllum hyalinum* Brinkm.
SY: *Macrodiplophyllum imbricatum* (M.A. Howe) Perss.
Diplophyllum incurvum Bryhn & Kaal. = Scapania gymnostomophila [SCAPGYM]
Diplophyllum obtusifolium (Hook.) Dum. [DIPLOBU]
Diplophyllum plicatum Lindb. [DIPLPLI]
SY: *Diplophylleia plicata* (Lindb.) Evans
SY: *Macrodiplophyllum plicatum* (Lindb.) Perss.
Diplophyllum taxifolium (Wahlenb.) Dum. [DIPLTAX] (yellow double-leaf wort)
SY: *Diplophylleia taxifolia* (Wahlenb.) Trevis.
Douinia (C. Jens.) Buch [DOUINIA$]
Douinia ovata (Dicks.) Buch [DOUIOVA]
Macrodiplophyllum imbricatum (M.A. Howe) Perss. = Diplophyllum imbricatum [DIPLIMB]
Macrodiplophyllum plicatum (Lindb.) Perss. = Diplophyllum plicatum [DIPLPLI]
Scapania (Dum.) Dum. [SCAPANI$]
Scapania americana K. Müll. [SCAPAME]
SY: *Scapania bolanderi* var. *americana* (K. Müll.) Frye & Clark
SY: *Scapania granulifera* Evans
Scapania apiculata Spruce [SCAPAPI]
Scapania bolanderi Aust. [SCAPBOL] (yellow-ladle liverwort)
Scapania bolanderi var. *americana* (K. Müll.) Frye & Clark = Scapania americana [SCAPAME]
Scapania convexula K. Müll. = Scapania spitzbergensis [SCAPSPI]
Scapania curta (Mart.) Dum. [SCAPCUR]
SY: *Scapania rosacea* (Corda) Dum.
Scapania cuspiduligera (Nees) K. Müll. [SCAPCUS]
Scapania glaucocephala (Tayl.) Aust. [SCAPGLA]
SY: *Scapaniella glaucocephala* (Tayl.) Evans
Scapania granulifera Evans = Scapania americana [SCAPAME]
Scapania gymnostomophila Kaal. [SCAPGYM]
SY: *Diplophyllum gymnostomophilum* (Kaal.) Kaal.
SY: *Diplophyllum incurvum* Bryhn & Kaal.
Scapania heterophylla M.A. Howe = Scapania undulata var. undulata [SCAPUND0]
Scapania hians Steph. [SCAPHIA]
Scapania hians ssp. **hians** [SCAPHIA0]
Scapania hians ssp. **salishensis** J. Godfrey & G. Godfrey [SCAPHIA1]
Scapania hollandiae Hong [SCAPHOL]
Scapania irrigua (Nees) Gott. & al. [SCAPIRR]
Scapania mucronata Buch [SCAPMUC]
Scapania oakesii Aust. = Scapania undulata var. oakesii [SCAPUND2]
Scapania obscura (Arnell & C. Jens.) Schiffn. [SCAPOBS]
Scapania paludicola Loeske & K. Müll. [SCAPPAL]
Scapania paludosa (K. Müll.) K. Müll. [SCAPPAU]
Scapania perlaxa Warnst. = Scapania subalpina [SCAPSUB]
Scapania rosacea (Corda) Dum. = Scapania curta [SCAPCUR]
Scapania scandica (Arnell & Buch) Macv. [SCAPSCA]
Scapania simmonsii Bryhn & Kaal. [SCAPSIM]
Scapania spitzbergensis (Lindb.) K. Müll. [SCAPSPI]
SY: *Scapania convexula* K. Müll.
Scapania subalpina (Nees) Dum. [SCAPSUB]
SY: *Scapania perlaxa* Warnst.
SY: *Scapania subalpina* var. *haynesiae* Frye & Clark
Scapania subalpina var. *haynesiae* Frye & Clark = Scapania subalpina [SCAPSUB]
Scapania uliginosa (Sw. ex Lindenb.) Dum. [SCAPULI]
Scapania umbrosa (Schrad.) Dum. [SCAPUMB]
Scapania undulata (L.) Dum. [SCAPUND] (water mitten-leaf wort)
Scapania undulata var. *heterophylla* Warnst. = Scapania undulata var. undulata [SCAPUND0]
Scapania undulata var. **oakesii** (Aust.) Buch [SCAPUND2]
SY: *Scapania oakesii* Aust.
Scapania undulata var. **undulata** [SCAPUND0]
SY: *Scapania heterophylla* M.A. Howe
SY: *Scapania undulata* var. *heterophylla* Warnst.
Scapaniella glaucocephala (Tayl.) Evans = Scapania glaucocephala [SCAPGLA]

SCHISTOSTEGACEAE (B:M)

Schistostega Mohr [SCHISTO$]
Schistostega osmundacea Mohr = Schistostega pennata [SCHIPEN]
Schistostega pennata (Hedw.) Web. & Mohr [SCHIPEN] (Goblin's gold)
SY: *Schistostega osmundacea* Mohr

SCOULERIACEAE (B:M)

Scouleria Hook. in Drumm. [SCOULER$]
Scouleria aquatica Hook. in Drumm. [SCOUAQU] (streamside moss)
Scouleria marginata Britt. [SCOUMAR]

SELIGERIACEAE (B:M)

Anodus donianus (Sm.) B.S.G. = Seligeria donniana [SELIDON]
Blindia Bruch & Schimp. in B.S.G. [BLINDIA$]
Blindia acuta (Hedw.) Bruch & Schimp. in B.S.G. [BLINACU] (silky tufted moss)
Brachydontium Fürnr. [BRACHYD$]
Brachydontium olympicum (Britt. in Frye) McIntosh & Spence [BRACOLY]
SY: *Grimmia olympica* Britt. in Frye
Brachyodus Fünr. ex Nees & Hornsch. = Brachydontium [BRACHYD$]
Grimmia olympica Britt. in Frye = Brachydontium olympicum [BRACOLY]
Seligeria Bruch & Schimp. in B.S.G. [SELIGER$]
Seligeria acutifolia Lindb. in Hartm. [SELIACU]
Seligeria campylopoda Kindb. in Mac. & Kindb. [SELICAM]
Seligeria careyana Vitt. & Schof. [SELICAR]

Seligeria donniana (Sm.) C. Müll. [SELIDON]
SY: *Anodus donianus* (Sm.) B.S.G.
SY: *Seligeria donnii* Lindb.
Seligeria donnii Lindb. = Seligeria donniana [SELIDON]
Seligeria recurvata (Hedw.) Bruch & Schimp. in B.S.G. [SELIREC]
Seligeria tristichoides Kindb. [SELITRI]

SEMATOPHYLLACEAE (B:M)

Acanthocladium carlottae Schof. = Wijkia carlottae [WIJKCAR]
Brotherella Loeske ex Fleisch. [BROTHEE$]
Brotherella canadensis Schof. [BROTCAN]
Brotherella roellii (Ren. & Card. in Röll) Fleisch. [BROTROE]
SY: *Pylaisiadelpha roellii* (Ren. & Card. in Röll) Buck
SY: *Rhaphidostegium roellii* Ren. & Card. in Röll
Pylaisiadelpha roellii (Ren. & Card. in Röll) Buck = Brotherella roellii [BROTROE]
Rhaphidostegium demissum (Wils.) De Not. = Sematophyllum demissum [SEMADEM]
Rhaphidostegium roellii Ren. & Card. in Röll = Brotherella roellii [BROTROE]
Sematophyllum Mitt. [SEMATOP$]
Sematophyllum carolinianum (C. Müll.) Britt. = Sematophyllum demissum [SEMADEM]
Sematophyllum demissum (Wils.) Mitt. [SEMADEM]
SY: *Rhaphidostegium demissum* (Wils.) De Not.
SY: *Sematophyllum carolinianum* (C. Müll.) Britt.
Wijkia Crum [WIJKIA$]
Wijkia carlottae (Schof.) Crum [WIJKCAR]
SY: *Acanthocladium carlottae* Schof.

SPHAEROCARPACEAE (B:H)

Sphaerocarpos Boehmer [SPHAERC$]
Sphaerocarpos texanus Aust. [SPHATEX]
SY: *Sphaerocarpus terrestris* Bisch.
Sphaerocarpus terrestris Bisch. = Sphaerocarpos texanus [SPHATEX]

SPHAGNACEAE (B:M)

Sphagnum L. [SPHAGNU$]
Sphagnum acutifolium Ehrh. ex Schrad. = Sphagnum capillifolium [SPHACAI]
Sphagnum acutifolium var. *intermedium* Aust. = Sphagnum warnstorfii [SPHAWAR]
Sphagnum acutifolium var. *pallescens* Warnst. = Sphagnum capillifolium [SPHACAI]
Sphagnum acutifolium var. *rubellum* (Wils.) Russ. = Sphagnum rubellum [SPHARUB]
Sphagnum acutifolium var. *rubrum* Brid. ex Warnst. = Sphagnum capillifolium [SPHACAI]
Sphagnum acutifolium var. *subtile* Russ. = Sphagnum subtile [SPHASUT]
Sphagnum acutifolium var. *tenellum* Schimp. = Sphagnum rubellum [SPHARUB]
Sphagnum acutifolium var. *versicolor* Warnst. = Sphagnum capillifolium [SPHACAI]
Sphagnum acutiforme var. *robustum* Warnst. = Sphagnum russowii [SPHARUS]
Sphagnum angustifolium (C. Jens. ex Russ.) C. Jens. in Tolf [SPHAANU] (yellow-green peat moss)
SY: *Sphagnum fallax* var. *angustifolium* (C. Jens. ex Russ.) Nyh.
SY: *Sphagnum flexuosum* var. *tenue* (Klinggr.) Pilous
SY: *Sphagnum parvifolium* (Sendtn.) Warnst.
SY: *Sphagnum recurvum* ssp. *angustifolium* C. Jens. ex Russ.
SY: *Sphagnum recurvum* var. *parvifolium* Sendtn. ex Warnst.
SY: *Sphagnum recurvum* var. *tenue* Klinggr.
Sphagnum annulatum var. *porosum* (Schlieph. & Warnst.) Maass & Isov. in Maass = Sphagnum jensenii [SPHAJEN]
Sphagnum aongstroemii Hartm. [SPHAAON]
Sphagnum apiculatum H. Lindb. in Bauer = Sphagnum fallax [SPHAFAL]
Sphagnum austinii Sull. in Aust. [SPHAAUS]
SY: *Sphagnum carlottae* Andrus, nomen nudum
SY: *Sphagnum imbricatum* ssp. *austinii* (Sull. in Aust.) Flatb.
Sphagnum balticum (Russ.) C. Jens. [SPHABAL]
Sphagnum capillaceum (Weiss.) Schrank = Sphagnum capillifolium [SPHACAI]
Sphagnum capillaceum var. *tenellum* (Schimp.) Andrews = Sphagnum rubellum [SPHARUB]
Sphagnum capillifolium (Ehrh.) Hedw. [SPHACAI] (small red peat moss)
SY: *Sphagnum acutifolium* Ehrh. ex Schrad.
SY: *Sphagnum acutifolium* var. *pallescens* Warnst.
SY: *Sphagnum acutifolium* var. *rubrum* Brid. ex Warnst.
SY: *Sphagnum acutifolium* var. *versicolor* Warnst.
SY: *Sphagnum capillaceum* (Weiss.) Schrank
SY: *Sphagnum intermedium* Hoffm.
SY: *Sphagnum nemoreum* Scop. auct. plur.
SY: *Sphagnum nemoreum* var. *alpinum* (Milde) Wijk & Marg.
SY: *Sphagnum nemoreum* var. *elegans* (Braithw.) Wijk & Marg.
SY: *Sphagnum nemoreum* f. *versicolor* (Warnst.) Podp.
Sphagnum capillifolium var. *tenellum* (Schimp.) Crum = Sphagnum rubellum [SPHARUB]
Sphagnum carlottae Andrus, nomen nudum = Sphagnum austinii [SPHAAUS]
Sphagnum centrale C. Jens. in Arnell & C. Jens. [SPHACEN]
SY: *Sphagnum papillosum* var. *intermedium* Warnst.
SY: *Sphagnum subbicolor* Hampe
Sphagnum compactum DC. in Lam. & DC. [SPHACOM]
SY: *Sphagnum compactum* var. *brachycladum* (Röll) Röll
SY: *Sphagnum compactum* var. *expositum* Maass
SY: *Sphagnum compactum* var. *imbricatum* Warnst.
SY: *Sphagnum compactum* var. *squarrosum* (Russ.) Warnst.
SY: *Sphagnum compactum* var. *subsquarrosum* Warnst.
SY: *Sphagnum rigidum* (Nees & Hornsch.) Schimp.
Sphagnum compactum var. *brachycladum* (Röll) Röll = Sphagnum compactum [SPHACOM]
Sphagnum compactum var. *expositum* Maass = Sphagnum compactum [SPHACOM]
Sphagnum compactum var. *imbricatum* Warnst. = Sphagnum compactum [SPHACOM]
Sphagnum compactum var. *squarrosum* (Russ.) Warnst. = Sphagnum compactum [SPHACOM]
Sphagnum compactum var. *subsquarrosum* Warnst. = Sphagnum compactum [SPHACOM]
Sphagnum contortum Schultz [SPHACON]
SY: *Sphagnum laricinum* (Wils.) Spruce
SY: *Sphagnum subsecundum* var. *contortum* (Schultz) Hüb.
Sphagnum cuspidatum Ehrh. ex Hoffm. [SPHACUS]
SY: *Sphagnum cuspidatum* var. *plumosum* Nees & Hornsch.
SY: *Sphagnum cuspidatum* var. *plumulosum* Schimp.
Sphagnum cuspidatum var. *dusenii* C. Jens. ex Warnst. = Sphagnum majus [SPHAMAJ]
Sphagnum cuspidatum var. *plumosum* Nees & Hornsch. = Sphagnum cuspidatum [SPHACUS]
Sphagnum cuspidatum var. *plumulosum* Schimp. = Sphagnum cuspidatum [SPHACUS]
Sphagnum cymbifolium (Ehrh.) Hedw. = Sphagnum palustre [SPHAPAL]
Sphagnum cymbifolium var. *laeve* Warnst. = Sphagnum palustre [SPHAPAL]
Sphagnum dusenii C. Jens. ex Russ. & Warnst. in Russ. = Sphagnum majus [SPHAMAJ]
Sphagnum fallax (Klinggr.) Klinggr. [SPHAFAL]
SY: *Sphagnum apiculatum* H. Lindb. in Bauer
SY: *Sphagnum flexuosum* var. *fallax* (Klinggr.) M.O. Hill ex A.J.E. Sm.
SY: *Sphagnum recurvum* var. *brevifolium* (Lindb. ex Braithw.) Warnst.
SY: *Sphagnum recurvum* var. *fallax* (Klinggr.) Paul in Koppe
SY: *Sphagnum recurvum* var. *mucronatum* (Russ.) Warnst.

Sphagnum fallax var. *angustifolium* (C. Jens. ex Russ.) Nyh. = Sphagnum angustifolium [SPHAANU]
Sphagnum fimbriatum Wils. in Wils. & Hook. f. in Hook. f. [SPHAFIM]
SY: *Sphagnum fimbriatum* var. *laxum* (Braithw.) Wijk & Marg.
SY: *Sphagnum fimbriatum* var. *robustum* Braithw. ex Warnst.
Sphagnum fimbriatum var. *laxum* (Braithw.) Wijk & Marg. = Sphagnum fimbriatum [SPHAFIM]
Sphagnum fimbriatum var. *robustum* Braithw. ex Warnst. = Sphagnum fimbriatum [SPHAFIM]
Sphagnum flexuosum var. *fallax* (Klinggr.) M.O. Hill ex A.J.E. Sm. = Sphagnum fallax [SPHAFAL]
Sphagnum flexuosum var. *tenue* (Klinggr.) Pilous = Sphagnum angustifolium [SPHAANU]
Sphagnum fuscum (Schimp.) Klinggr. [SPHAFUS] (rusty peat moss)
SY: *Sphagnum fuscum* f. *fuscescens* (Warnst.) Warnst.
SY: *Sphagnum fuscum* var. *fuscescens* Warnst.
SY: *Sphagnum fuscum* f. *pallescens* (Warnst.) Warnst.
SY: *Sphagnum fuscum* var. *pallescens* Warnst.
SY: *Sphagnum fuscum* f. *viride* (Warnst.) Warnst.
SY: *Sphagnum fuscum* var. *viride* Warnst.
SY: *Sphagnum tenuifolium* Warnst.
Sphagnum fuscum f. *fuscescens* (Warnst.) Warnst. = Sphagnum fuscum [SPHAFUS]
Sphagnum fuscum f. *pallescens* (Warnst.) Warnst. = Sphagnum fuscum [SPHAFUS]
Sphagnum fuscum f. *viride* (Warnst.) Warnst. = Sphagnum fuscum [SPHAFUS]
Sphagnum fuscum var. *fuscescens* Warnst. = Sphagnum fuscum [SPHAFUS]
Sphagnum fuscum var. *pallescens* Warnst. = Sphagnum fuscum [SPHAFUS]
Sphagnum fuscum var. *viride* Warnst. = Sphagnum fuscum [SPHAFUS]
Sphagnum girgensohnii Russ. [SPHAGIR] (white-toothed peat moss)
SY: *Sphagnum girgensohnii* var. *hygrophyllum* Russ.
SY: *Sphagnum girgensohnii* var. *stachyodes* Russ. in Warnst.
SY: *Sphagnum girgensohnii* var. *strictum* (Lindb.) Russ.
Sphagnum girgensohnii var. *hygrophyllum* Russ. = Sphagnum girgensohnii [SPHAGIR]
Sphagnum girgensohnii var. *stachyodes* Russ. in Warnst. = Sphagnum girgensohnii [SPHAGIR]
Sphagnum girgensohnii var. *strictum* (Lindb.) Russ. = Sphagnum girgensohnii [SPHAGIR]
Sphagnum henryense Warnst. [SPHAHEN]
Sphagnum imbricatum ssp. *austinii* (Sull. in Aust.) Flatb. = Sphagnum austinii [SPHAAUS]
Sphagnum intermedium Hoffm. = Sphagnum capillifolium [SPHACAI]
Sphagnum jensenii H. Lindb. [SPHAJEN]
SY: *Sphagnum annulatum* var. *porosum* (Schlieph. & Warnst.) Maass & Isov. in Maass
SY: *Sphagnum recurvum* var. *porosum* Schlieph. & Warnst. in Warnst.
Sphagnum junghuhnianum Dozy & Molk. [SPHAJUN]
Sphagnum junghuhnianum ssp. *pseudomolle* (Warnst.) Suz. = Sphagnum junghuhnianum var. pseudomolle [SPHAJUN1]
Sphagnum junghuhnianum var. **junghuhnianum** [SPHAJUN0]
Sphagnum junghuhnianum var. **pseudomolle** (Warnst.) Warnst. [SPHAJUN1]
SY: *Sphagnum junghuhnianum* ssp. *pseudomolle* (Warnst.) Suz.
Sphagnum laricinum (Wils.) Spruce = Sphagnum contortum [SPHACON]
Sphagnum lindbergii Schimp. in Lindb. [SPHALIN] (brown-stemmed bog moss)
SY: *Sphagnum lindbergii* var. *macrophyllum* Warnst.
SY: *Sphagnum lindbergii* var. *mesophyllum* Warnst.
SY: *Sphagnum lindbergii* var. *tenellum* (Lindb.) Röll
Sphagnum lindbergii var. *macrophyllum* Warnst. = Sphagnum lindbergii [SPHALIN]
Sphagnum lindbergii var. *mesophyllum* Warnst. = Sphagnum lindbergii [SPHALIN]
Sphagnum lindbergii var. *tenellum* (Lindb.) Röll = Sphagnum lindbergii [SPHALIN]
Sphagnum magellanicum Brid. [SPHAMAG] (midway peat moss)
SY: *Sphagnum magellanicum* var. *pallescens* (Warnst.) C. Jens.
SY: *Sphagnum magellanicum* var. *purpurascens* (Russ.) Warnst.
SY: *Sphagnum magellanicum* var. *roseum* (Röll) C. Jens.
SY: *Sphagnum medium* Limpr.
SY: *Sphagnum medium* var. *laeve* f. *purpurascens* Russ.
SY: *Sphagnum medium* var. *pallescens* (Warnst.) Warnst.
SY: *Sphagnum medium* var. *purpurascens* (Russ.) Warnst.
SY: *Sphagnum medium* var. *roseum* (Röll) Warnst.
Sphagnum magellanicum var. *pallescens* (Warnst.) C. Jens. = Sphagnum magellanicum [SPHAMAG]
Sphagnum magellanicum var. *purpurascens* (Russ.) Warnst. = Sphagnum magellanicum [SPHAMAG]
Sphagnum magellanicum var. *roseum* (Röll) C. Jens. = Sphagnum magellanicum [SPHAMAG]
Sphagnum majus (Russ.) C. Jens. [SPHAMAJ]
SY: *Sphagnum cuspidatum* var. *dusenii* C. Jens. ex Warnst.
SY: *Sphagnum dusenii* C. Jens. ex Russ. & Warnst. in Russ.
SY: *Sphagnum obtusum* var. *dusenii* C. Jens. ex Warnst.
Sphagnum medium Limpr. = Sphagnum magellanicum [SPHAMAG]
Sphagnum medium var. *laeve* f. *purpurascens* Russ. = Sphagnum magellanicum [SPHAMAG]
Sphagnum medium var. *pallescens* (Warnst.) Warnst. = Sphagnum magellanicum [SPHAMAG]
Sphagnum medium var. *purpurascens* (Russ.) Warnst. = Sphagnum magellanicum [SPHAMAG]
Sphagnum medium var. *roseum* (Röll) Warnst. = Sphagnum magellanicum [SPHAMAG]
Sphagnum mendocinum Sull. & Lesq. in Sull. [SPHAMEN]
Sphagnum molluscum Bruch = Sphagnum tenellum [SPHATEN]
Sphagnum nemoreum Scop. auct. plur. = Sphagnum capillifolium [SPHACAI]
Sphagnum nemoreum f. *versicolor* (Warnst.) Podp. = Sphagnum capillifolium [SPHACAI]
Sphagnum nemoreum var. *alpinum* (Milde) Wijk & Marg. = Sphagnum capillifolium [SPHACAI]
Sphagnum nemoreum var. *elegans* (Braithw.) Wijk & Marg. = Sphagnum capillifolium [SPHACAI]
Sphagnum nemoreum var. *subtile* Russ. = Sphagnum subtile [SPHASUT]
Sphagnum obtusum var. *dusenii* C. Jens. ex Warnst. = Sphagnum majus [SPHAMAJ]
Sphagnum pacificum Flatb. [SPHAPAC]
Sphagnum palustre L. [SPHAPAL]
SY: *Sphagnum cymbifolium* (Ehrh.) Hedw.
SY: *Sphagnum cymbifolium* var. *laeve* Warnst.
Sphagnum papillosum Lindb. [SPHAPAP] (fat bog moss)
SY: *Sphagnum papillosum* var. *laeve* Warnst.
SY: *Sphagnum papillosum* var. *sublaeve* Warnst.
Sphagnum papillosum var. *intermedium* Warnst. = Sphagnum centrale [SPHACEN]
Sphagnum papillosum var. *laeve* Warnst. = Sphagnum papillosum [SPHAPAP]
Sphagnum papillosum var. *sublaeve* Warnst. = Sphagnum papillosum [SPHAPAP]
Sphagnum parvifolium (Sendtn.) Warnst. = Sphagnum angustifolium [SPHAANU]
Sphagnum platyphyllum (Lindb. ex Braithw.) Sull. ex Warnst. [SPHAPLA]
SY: *Sphagnum subsecundum* var. *platyphyllum* (Lindb. ex Braithw.) Card.
Sphagnum plumulosum Röll = Sphagnum subnitens [SPHASUN]
Sphagnum pulchricoma C. Müll. = Sphagnum recurvum [SPHAREC]
Sphagnum pulchrum (Lindb. ex Braithw.) Warnst. [SPHAPUL]
SY: *Sphagnum pulchrum* var. *fusco-flavescens* (Warnst.) Warnst.
SY: *Sphagnum pulchrum* var. *nigricans* Warnst.
SY: *Sphagnum recurvum* var. *pulchrum* (Lindb.) Warnst.
Sphagnum pulchrum var. *fusco-flavescens* (Warnst.) Warnst. = Sphagnum pulchrum [SPHAPUL]
Sphagnum pulchrum var. *nigricans* Warnst. = Sphagnum pulchrum [SPHAPUL]

Sphagnum purpureum C. Jens. = Sphagnum rubellum [SPHARUB]
Sphagnum quinquefarium (Lindb. ex Braithw.) Warnst. [SPHAQUI]
Sphagnum recurvum P. Beauv. [SPHAREC]
SY: *Sphagnum pulchricoma* C. Müll.
Sphagnum recurvum ssp. *angustifolium* C. Jens. ex Russ. = Sphagnum angustifolium [SPHAANU]
Sphagnum recurvum var. *brevifolium* (Lindb. ex Braithw.) Warnst. = Sphagnum fallax [SPHAFAL]
Sphagnum recurvum var. *fallax* (Klinggr.) Paul in Koppe = Sphagnum fallax [SPHAFAL]
Sphagnum recurvum var. *mucronatum* (Russ.) Warnst. = Sphagnum fallax [SPHAFAL]
Sphagnum recurvum var. *parvifolium* Sendtn. ex Warnst. = Sphagnum angustifolium [SPHAANU]
Sphagnum recurvum var. *porosum* Schlieph. & Warnst. in Warnst. = Sphagnum jensenii [SPHAJEN]
Sphagnum recurvum var. *pulchrum* (Lindb.) Warnst. = Sphagnum pulchrum [SPHAPUL]
Sphagnum recurvum var. *tenue* Klinggr. = Sphagnum angustifolium [SPHAANU]
Sphagnum rigidum (Nees & Hornsch.) Schimp. = Sphagnum compactum [SPHACOM]
Sphagnum riparium Ångstr. [SPHARIP] (shore-growing peat moss)
Sphagnum robustum (Warnst.) Röll = Sphagnum russowii [SPHARUS]
Sphagnum rubellum Wils. [SPHARUB]
SY: *Sphagnum acutifolium* var. *rubellum* (Wils.) Russ.
SY: *Sphagnum acutifolium* var. *tenellum* Schimp.
SY: *Sphagnum capillaceum* var. *tenellum* (Schimp.) Andrews
SY: *Sphagnum capillifolium* var. *tenellum* (Schimp.) Crum
SY: *Sphagnum purpureum* C. Jens.
SY: *Sphagnum rubellum* var. *rubescens* Warnst.
SY: *Sphagnum rubellum* var. *rubrum* (Grav.) Horr.
SY: *Sphagnum rubellum* var. *violaceum* C. Jens.
SY: *Sphagnum tenellum* var. *rubellum* (Wils.) Klinggr.
SY: *Sphagnum tenellum* var. *rubrum* Grav. ex Warnst.
Sphagnum rubellum var. *rubescens* Warnst. = Sphagnum rubellum [SPHARUB]
Sphagnum rubellum var. *rubrum* (Grav.) Horr. = Sphagnum rubellum [SPHARUB]
Sphagnum rubellum var. *violaceum* C. Jens. = Sphagnum rubellum [SPHARUB]
Sphagnum rubiginosum Flatb. [SPHARUI]
Sphagnum russowii Warnst. [SPHARUS]
SY: *Sphagnum acutiforme* var. *robustum* Warnst.
SY: *Sphagnum robustum* (Warnst.) Röll
SY: *Sphagnum russowii* var. *girgensohnioides* Russ. ex Warnst.
SY: *Sphagnum russowii* var. *poecilum* Russ. ex Warnst.
SY: *Sphagnum russowii* var. *rhodochroum* Russ. in Warnst.
Sphagnum russowii var. *girgensohnioides* Russ. ex Warnst. = Sphagnum russowii [SPHARUS]
Sphagnum russowii var. *poecilum* Russ. ex Warnst. = Sphagnum russowii [SPHARUS]
Sphagnum russowii var. *rhodochroum* Russ. in Warnst. = Sphagnum russowii [SPHARUS]
Sphagnum schofieldii Crum [SPHASCH]
Sphagnum semisquarrosum Russ. = Sphagnum squarrosum [SPHASQU]
Sphagnum squarrosum Crome [SPHASQU] (spread-leaved peat moss)
SY: *Sphagnum semisquarrosum* Russ.
SY: *Sphagnum squarrosum* var. *cuspidatum* Warnst.
SY: *Sphagnum squarrosum* var. *imbricatum* Schimp.
SY: *Sphagnum squarrosum* var. *immersum* Beek ex Warnst.
SY: *Sphagnum squarrosum* var. *semisquarrosum* Russ. ex Warnst.
SY: *Sphagnum squarrosum* var. *spectabile* Russ. in Warnst.
SY: *Sphagnum squarrosum* var. *subsquarrosum* Russ. in Warnst.
Sphagnum squarrosum var. *cuspidatum* Warnst. = Sphagnum squarrosum [SPHASQU]
Sphagnum squarrosum var. *imbricatum* Schimp. = Sphagnum squarrosum [SPHASQU]
Sphagnum squarrosum var. *immersum* Beek ex Warnst. = Sphagnum squarrosum [SPHASQU]
Sphagnum squarrosum var. *semisquarrosum* Russ. ex Warnst. = Sphagnum squarrosum [SPHASQU]
Sphagnum squarrosum var. *spectabile* Russ. in Warnst. = Sphagnum squarrosum [SPHASQU]
Sphagnum squarrosum var. *subsquarrosum* Russ. in Warnst. = Sphagnum squarrosum [SPHASQU]
Sphagnum subbicolor Hampe = Sphagnum centrale [SPHACEN]
Sphagnum subnitens Russ. & Warnst. in Warnst. [SPHASUN]
SY: *Sphagnum plumulosum* Röll
Sphagnum subobesum Warnst. [SPHASUB]
SY: *Sphagnum subsecundum* var. *junsaiense* (Warnst.) Crum
Sphagnum subsecundum Nees in Sturm [SPHASUS]
Sphagnum subsecundum var. *contortum* (Schultz) Hüb. = Sphagnum contortum [SPHACON]
Sphagnum subsecundum var. *junsaiense* (Warnst.) Crum = Sphagnum subobesum [SPHASUB]
Sphagnum subsecundum var. *platyphyllum* (Lindb. ex Braithw.) Card. = Sphagnum platyphyllum [SPHAPLA]
Sphagnum subtile (Russ.) Warnst. [SPHASUT]
SY: *Sphagnum acutifolium* var. *subtile* Russ.
SY: *Sphagnum nemoreum* var. *subtile* Russ.
Sphagnum tenellum (Brid.) Bory [SPHATEN]
SY: *Sphagnum molluscum* Bruch
Sphagnum tenellum var. *rubellum* (Wils.) Klinggr. = Sphagnum rubellum [SPHARUB]
Sphagnum tenellum var. *rubrum* Grav. ex Warnst. = Sphagnum rubellum [SPHARUB]
Sphagnum tenuifolium Warnst. = Sphagnum fuscum [SPHAFUS]
Sphagnum teres (Schimp.) Ångstr. in Hartm. [SPHATER]
SY: *Sphagnum teres* var. *compactum* Warnst.
SY: *Sphagnum teres* var. *imbricatum* Warnst.
SY: *Sphagnum teres* var. *squarrosulum* (Schimp.) Warnst.
Sphagnum teres var. *compactum* Warnst. = Sphagnum teres [SPHATER]
Sphagnum teres var. *imbricatum* Warnst. = Sphagnum teres [SPHATER]
Sphagnum teres var. *squarrosulum* (Schimp.) Warnst. = Sphagnum teres [SPHATER]
Sphagnum warnstorfianum Du Rietz = Sphagnum warnstorfii [SPHAWAR]
Sphagnum warnstorfii Russ. [SPHAWAR] (Warnstorf's peat moss)
SY: *Sphagnum acutifolium* var. *intermedium* Aust.
SY: *Sphagnum warnstorfianum* Du Rietz
SY: *Sphagnum warnstorfii* var. *purpurascens* Russ. in Warnst.
SY: *Sphagnum warnstorfii* var. *viride* Russ. in Warnst.
Sphagnum warnstorfii var. *purpurascens* Russ. in Warnst. = Sphagnum warnstorfii [SPHAWAR]
Sphagnum warnstorfii var. *viride* Russ. in Warnst. = Sphagnum warnstorfii [SPHAWAR]
Sphagnum wilfii Crum [SPHAWIL]
Sphagnum wulfianum Girg. [SPHAWUL]
SY: *Sphagnum wulfianum* var. *squarrulosum* Russ.
SY: *Sphagnum wulfianum* var. *versicolor* Warnst.
SY: *Sphagnum wulfianum* var. *viride* Warnst.
Sphagnum wulfianum var. *squarrulosum* Russ. = Sphagnum wulfianum [SPHAWUL]
Sphagnum wulfianum var. *versicolor* Warnst. = Sphagnum wulfianum [SPHAWUL]
Sphagnum wulfianum var. *viride* Warnst. = Sphagnum wulfianum [SPHAWUL]

SPLACHNACEAE (B:M)

Dissodon splachnoides Grev. & Arnell = Tayloria serrata [TAYLSER]
Splachnum Hedw. [SPLACHU$]
Splachnum ampullaceum Hedw. [SPLAAMP]
Splachnum luteum Hedw. [SPLALUT] (fairy parasols)
Splachnum ovatum Dicks. ex Hedw. = Splachnum sphaericum [SPLASPH]
Splachnum rubrum Hedw. [SPLARUB]
Splachnum sphaericum Hedw. [SPLASPH]
SY: *Splachnum ovatum* Dicks. ex Hedw.
Splachnum vasculosum Hedw. [SPLAVAS]

Tayloria Hook. [TAYLORI$]
Tayloria froelichiana (Hedw.) Mitt. ex Broth. [TAYLFRO]
Tayloria lingulata (Dicks.) Lindb. [TAYLLIN]
Tayloria serrata (Hedw.) Bruch & Schimp. in B.S.G. [TAYLSER]
SY: *Dissodon splachnoides* Grev. & Arnell
SY: *Tayloria serrata* var. *flagellaris* (Brid.) Bruch & Schimp. in B.S.G.
SY: *Tayloria serrata* var. *tenuis* (With.) Bruch & Schimp. in B.S.G.
SY: *Tayloria tenuis* (With.) Schimp.
Tayloria serrata var. *flagellaris* (Brid.) Bruch & Schimp. in B.S.G. = Tayloria serrata [TAYLSER]
Tayloria serrata var. *tenuis* (With.) Bruch & Schimp. in B.S.G. = Tayloria serrata [TAYLSER]
Tayloria splachnoides (Schleich. ex Schwaegr.) Hook. [TAYLSPL]
Tayloria tenuis (With.) Schimp. = Tayloria serrata [TAYLSER]
Tetraplodon Bruch & Schimp. in B.S.G. [TETRAPL$]
Tetraplodon angustatus (Hedw.) Bruch & Schimp. in B.S.G. [TETRANG]
Tetraplodon blyttii Frisvoll [TETRBLY]
Tetraplodon bryoides Lindb. = Tetraplodon mnioides [TETRMNI]
Tetraplodon mnioides (Hedw.) Bruch & Schimp. in B.S.G. [TETRMNI] (common dung moss)
SY: *Tetraplodon bryoides* Lindb.
SY: *Tetraplodon mnioides* ssp. *breweri* var. *brevicollis* Kindb.
SY: *Tetraplodon mnioides* var. *brewerianus* (Hedw.) Bruch & Schimp. in B.S.G.
Tetraplodon mnioides ssp. *breweri* var. *brevicollis* Kindb. = Tetraplodon mnioides [TETRMNI]
Tetraplodon mnioides ssp. *pallidus* (Hag.) Kindb. = Tetraplodon pallidus [TETRPAL]
Tetraplodon mnioides var. *brewerianus* (Hedw.) Bruch & Schimp. in B.S.G. = Tetraplodon mnioides [TETRMNI]
Tetraplodon mnioides var. *cavifolius* Schimp. = Tetraplodon urceolatus [TETRURC]
Tetraplodon mnioides var. *urceolatus* (Hedw.) Steere in Polunin = Tetraplodon urceolatus [TETRURC]
Tetraplodon pallidus Hag. [TETRPAL]
SY: *Tetraplodon mnioides* ssp. *pallidus* (Hag.) Kindb.
Tetraplodon urceolatus (Hedw.) Bruch & Schimp. in Schimp. [TETRURC]
SY: *Tetraplodon mnioides* var. *cavifolius* Schimp.
SY: *Tetraplodon mnioides* var. *urceolatus* (Hedw.) Steere in Polunin

TAKAKIACEAE (B:M)

Takakia Hatt. & H. Inoue [TAKAKIA$]
Takakia lepidozioides Hatt. & H. Inoue [TAKALEP]

TARGIONIACEAE (B:H)

Targionia L. [TARGION$]
Targionia hypophylla L. [TARGHYP]

TETRAPHIDACEAE (B:M)

Georgia cuspidata Kindb. = Tetraphis pellucida [TETRPEL]
Georgia geniculata (Milde) Brockm. = Tetraphis geniculata [TETRGEN]
Georgia pellucida (Hedw.) Rabenh. = Tetraphis pellucida [TETRPEL]
Georgia trachypoda Kindb. = Tetraphis geniculata [TETRGEN]
Tetraphis Hedw. [TETRAPH$]
Tetraphis browniana var. *repanda* (Funck in Sturm) Hampe = Tetrodontium repandum [TETRREP]
Tetraphis brownianum (Dicks.) Grev. = Tetrodontium brownianum [TETRBRO]
Tetraphis geniculata Girg. ex Milde [TETRGEN]
SY: *Georgia geniculata* (Milde) Brockm.
SY: *Georgia trachypoda* Kindb.
Tetraphis pellucida Hedw. [TETRPEL] (four-toothed log moss)
SY: *Georgia cuspidata* Kindb.
SY: *Georgia pellucida* (Hedw.) Rabenh.
Tetrodontium Schwaegr. [TETRODO$]
Tetrodontium brownianum (Dicks.) Schwaegr. [TETRBRO]
SY: *Tetraphis brownianum* (Dicks.) Grev.
Tetrodontium repandum (Funck in Sturm) Schwaegr. [TETRREP]
SY: *Tetraphis browniana* var. *repanda* (Funck in Sturm) Hampe

THAMNOBRYACEAE (B:M)

Arbuscula leibergii (Britt.) Crum, Steere & Anderson = Thamnobryum neckeroides [THAMNEC]
Bestia holzingeri (Ren. & Card.) Broth. = Porotrichum vancouveriense [POROVAN]
Bestia occidentalis (Sull. & Lesq.) Grout = Porotrichum vancouveriense [POROVAN]
Bestia vancouveriensis (Kindb. in Mac.) Wijk & Marg. = Porotrichum vancouveriense [POROVAN]
Bryolawtonia vancouveriensis (Kindb. in Mac.) Enroth & Norris = Porotrichum vancouveriense [POROVAN]
Heterocladium vancouveriense (Kindb.) Kindb. in Mac. & Kindb. = Porotrichum vancouveriense [POROVAN]
Porothamnium bigelovii (Sull.) Fleisch. in Broth. = Porotrichum bigelovii [POROBIG]
Porotrichum (Brid.) Hampe [POROTRI$]
Porotrichum bigelovii (Sull.) Kindb. [POROBIG]
SY: *Porothamnium bigelovii* (Sull.) Fleisch. in Broth.
SY: *Thamnium bigelovii* (Sull.) Jaeg.
Porotrichum neckeroides (Hook.) Williams = Thamnobryum neckeroides [THAMNEC]
Porotrichum vancouveriense (Kindb. in Mac.) Crum [POROVAN]
SY: *Bestia holzingeri* (Ren. & Card.) Broth.
SY: *Bestia occidentalis* (Sull. & Lesq.) Grout
SY: *Bestia vancouveriensis* (Kindb. in Mac.) Wijk & Marg.
SY: *Bryolawtonia vancouveriensis* (Kindb. in Mac.) Enroth & Norris
SY: *Heterocladium vancouveriense* (Kindb.) Kindb. in Mac. & Kindb.
SY: *Pseudoleskeella occidentalis* (Kindb.) Kindb.
Pseudoleskeella occidentalis (Kindb.) Kindb. = Porotrichum vancouveriense [POROVAN]
Thamnium bigelovii (Sull.) Jaeg. = Porotrichum bigelovii [POROBIG]
Thamnobryum Nieuwl. [THAMNOB$]
Thamnobryum leibergii (Britt.) Ren. & Card. = Thamnobryum neckeroides [THAMNEC]
Thamnobryum neckeroides (Hook.) Lawt. [THAMNEC]
SY: *Arbuscula leibergii* (Britt.) Crum, Steere & Anderson
SY: *Porotrichum neckeroides* (Hook.) Williams
SY: *Thamnobryum leibergii* (Britt.) Ren. & Card.

THUIDIACEAE (B:M)

Abietinella C. Müll. [ABIETIN$]
Abietinella abietina (Hedw.) Fleisch. [ABIEABI] (wiry fern moss)
SY: *Hypnum abietinum* Hedw.
SY: *Thuidium abietinum* (Hedw.) Schimp. in B.S.G.
Hypnum abietinum Hedw. = Abietinella abietina [ABIEABI]
Hypnum recognitum Hedw. = Thuidium recognitum [THUIREC]
Thuidium Schimp. in B.S.G. [THUIDIU$]
Thuidium abietinum (Hedw.) Schimp. in B.S.G. = Abietinella abietina [ABIEABI]
Thuidium delicatulum var. *radicans* (Kindb.) Crum, Steere & Anderson = Thuidium philibertii [THUIPHI]
Thuidium philibertii Limpr. [THUIPHI]
SY: *Thuidium delicatulum* var. *radicans* (Kindb.) Crum, Steere & Anderson
Thuidium recognitum (Hedw.) Lindb. [THUIREC] (hook-leaf fern moss)
SY: *Hypnum recognitum* Hedw.

TIMMIACEAE (B:M)

Grevilleanum serratum Beck & Emmons = Timmia megapolitana var. megapolitana [TIMMMEG0]
Timmia Hedw. [TIMMIA$]
Timmia austriaca Hedw. [TIMMAUS] (false haircap moss)
SY: *Timmia austriaca* var. *arctica* (Lindb.) Arnell
Timmia austriaca var. *arctica* (Lindb.) Arnell = Timmia austriaca [TIMMAUS]
Timmia bavarica Hessl. = Timmia megapolitana var. bavarica [TIMMMEG1]
Timmia cucullata Michx. = Timmia megapolitana [TIMMMEG]
Timmia megapolitana Hedw. [TIMMMEG]
SY: *Timmia cucullata* Michx.
SY: *Timmia megapolitana* f. *cucullata* (Michx.) Sayre in Grout
Timmia megapolitana f. *cucullata* (Michx.) Sayre in Grout = Timmia megapolitana [TIMMMEG]
Timmia megapolitana ssp. *bavarica* (Hessl.) Brass. = Timmia megapolitana var. bavarica [TIMMMEG1]
Timmia megapolitana var. **bavarica** (Hessl.) Brid. [TIMMMEG1]
SY: *Timmia bavarica* Hessl.
SY: *Timmia megapolitana* ssp. *bavarica* (Hessl.) Brass.
Timmia megapolitana var. **megapolitana** [TIMMMEG0]
SY: *Grevilleanum serratum* Beck & Emmons
Timmia norvegica Zett. [TIMMNOR]
Timmia norvegica var. *crassiretis* Hess. = Timmia sibirica [TIMMSIB]
Timmia sibirica Lindb. & Arnell [TIMMSIB]
SY: *Timmia norvegica* var. *crassiretis* Hess.

TREUBIACEAE (B:H)

Apotreubia Hatt. & Mizut. [APOTREU$]
Apotreubia nana (Hatt. & H. Inoue) Hatt. & Mizut. [APOTNAN]
SY: *Treubia nana* Hatt. & H. Inoue
Treubia nana Hatt. & H. Inoue = Apotreubia nana [APOTNAN])

Lichens

ACAROSPORACEAE (L)

Acarospora A. Massal. [ACAROSP$]
Acarospora albida H. Magn. = Acarospora schleicheri [ACARSCH]
Acarospora amabilis H. Magn. = Acarospora schleicheri [ACARSCH]
Acarospora amphibola Wedd. = Acarospora smaragdula [ACARSMA]
Acarospora asahinae H. Magn. [ACARASA]
Acarospora badiofusca (Nyl.) Th. Fr. [ACARBAD]
SY: *Acarospora boulderensis* H. Magn.
Acarospora bella (Nyl.) Jatta = Acarospora schleicheri [ACARSCH]
Acarospora boulderensis H. Magn. = Acarospora badiofusca [ACARBAD]
Acarospora cartilaginea H. Magn. = Acarospora fuscata [ACARFUS]
Acarospora cervina A. Massal. [ACARCER]
Acarospora chlorophana (Wahlenb.) A. Massal. = Pleopsidium chlorophanum [PLEOCHL]
Acarospora chrysops (Tuck.) H. Magn. = Acarospora schleicheri [ACARSCH]
Acarospora contigua H. Magn. = Acarospora schleicheri [ACARSCH]
Acarospora dissipata H. Magn. = Acarospora schleicheri [ACARSCH]
Acarospora epilutescens Zahlbr. = Acarospora schleicheri [ACARSCH]
Acarospora erythrophora H. Magn. = Pleopsidium chlorophanum [PLEOCHL]
Acarospora evoluta H. Magn. = Acarospora schleicheri [ACARSCH]
Acarospora fuscata (Schrader) Arnold [ACARFUS]
SY: *Acarospora cartilaginea* H. Magn.
SY: *Acarospora squamulosa* (Schrader) Trevisan
Acarospora glaucocarpa (Ach.) Körber [ACARGLA]
Acarospora incertula H. Magn. = Pleopsidium chlorophanum [PLEOCHL]
Acarospora intercedens H. Magn. = Acarospora schleicheri [ACARSCH]
Acarospora macrospora (Hepp) Bagl. [ACARMAC]
SY: *Acarospora squamulosa* auct.
Acarospora novomexicana H. Magn. = Pleopsidium chlorophanum [PLEOCHL]
Acarospora ocellata H. Magn. = Acarospora schleicheri [ACARSCH]
Acarospora rhabarbarina Hue = Acarospora schleicheri [ACARSCH]
Acarospora rimulosa H. Magn. = Acarospora schleicheri [ACARSCH]
Acarospora rubicunda H. Magn. = Acarospora schleicheri [ACARSCH]
Acarospora rufescens (Ach.) Bausch = Acarospora smaragdula [ACARSMA]
Acarospora saxicola Fink = Glypholecia scabra [GLYPSCA]
Acarospora scabra (Pers.) Th. Fr. = Glypholecia scabra [GLYPSCA]
Acarospora schleicheri (Ach.) A. Massal. [ACARSCH]
SY: *Acarospora albida* H. Magn.
SY: *Acarospora amabilis* H. Magn.
SY: *Acarospora bella* (Nyl.) Jatta
SY: *Acarospora chrysops* (Tuck.) H. Magn.
SY: *Acarospora contigua* H. Magn.
SY: *Acarospora dissipata* H. Magn.
SY: *Acarospora epilutescens* Zahlbr.
SY: *Acarospora evoluta* H. Magn.
SY: *Acarospora intercedens* H. Magn.
SY: *Acarospora ocellata* H. Magn.
SY: *Acarospora rhabarbarina* Hue
SY: *Acarospora rimulosa* H. Magn.
SY: *Acarospora rubicunda* H. Magn.
SY: *Acarospora subalbida* H. Magn.
SY: *Acarospora subcontigua* H. Magn.
SY: *Acarospora xanthophana* (Nyl.) Jatta
Acarospora sinopica (Wahlenb.) Körber [ACARSIN]
Acarospora smaragdula (Wahlenb.) A. Massal. [ACARSMA]
SY: *Acarospora amphibola* Wedd.
SY: *Acarospora rufescens* (Ach.) Bausch
Acarospora squamulosa (Schrader) Trevisan = Acarospora fuscata [ACARFUS]
Acarospora squamulosa auct. = Acarospora macrospora [ACARMAC]
Acarospora subalbida H. Magn. = Acarospora schleicheri [ACARSCH]
Acarospora subcontigua H. Magn. = Acarospora schleicheri [ACARSCH]
Acarospora texana H. Magn. = Pleopsidium chlorophanum [PLEOCHL]
Acarospora veronensis A. Massal. [ACARVER]
Acarospora weldensis H. Magn. = Pleopsidium chlorophanum [PLEOCHL]
Acarospora xanthophana (Nyl.) Jatta = Acarospora schleicheri [ACARSCH]
Biatorella moriformis (Ach.) Th. Fr. = Strangospora moriformis [STRAMOR]
Biatorella pruinosa (Körber) Mudd = Sarcogyne regularis [SARCREG]
Biatorella simplex (Davies) Branth & Rostrup = Polysporina simplex [POLYSIM]
Biatorella testudinea (Ach.) A. Massal. = Sporastatia testudinea [SPORTES]
Glypholecia Nyl. [GLYPHOL$]
Glypholecia scabra (Pers.) Müll. Arg. [GLYPSCA]
SY: *Acarospora saxicola* Fink
SY: *Acarospora scabra* (Pers.) Th. Fr.
Pleopsidium Körber [PLEOPSI$]
Pleopsidium chlorophanum (Wahlenb.) Zopf [PLEOCHL]
SY: *Acarospora chlorophana* (Wahlenb.) A. Massal.
SY: *Acarospora erythrophora* H. Magn.
SY: *Acarospora incertula* H. Magn.
SY: *Acarospora novomexicana* H. Magn.
SY: *Acarospora texana* H. Magn.
SY: *Acarospora weldensis* H. Magn.
Polysporina Vezda [POLYSPO$]
Polysporina simplex (Davies) Vezda [POLYSIM]
SY: *Biatorella simplex* (Davies) Branth & Rostrup
SY: *Sarcogyne simplex* (Davies) Nyl.
Polysporina urceolata (Anzi) Brodo [POLYURC]
SY: *Sarcogyne urceolata* Anzi
Sarcogyne Flotow [SARCOGY$]
Sarcogyne pruinosa auct. = Sarcogyne regularis [SARCREG]
Sarcogyne regularis Körber [SARCREG]
SY: *Biatorella pruinosa* (Körber) Mudd
SY: *Sarcogyne pruinosa* auct.
Sarcogyne simplex (Davies) Nyl. = Polysporina simplex [POLYSIM]
Sarcogyne urceolata Anzi = Polysporina urceolata [POLYURC]
Sporastatia A. Massal. [SPORAST$]
Sporastatia cinerea (Schaerer) Körber = Sporastatia polyspora [SPORPOL]
Sporastatia polyspora (Nyl.) Grummann [SPORPOL]
SY: *Sporastatia cinerea* (Schaerer) Körber
Sporastatia testudinea (Ach.) A. Massal. [SPORTES]
SY: *Biatorella testudinea* (Ach.) A. Massal.
Strangospora Körber [STRANGO$]
Strangospora moriformis (Ach.) Stein [STRAMOR]
SY: *Biatorella moriformis* (Ach.) Th. Fr.

Strangospora pinicola (A. Massal.) Körber [STRAPIN]
Thelocarpon Nyl. ex Hue [THELOCA$]
Thelocarpon epibolum Nyl. [THELEPI]
Thelocarpon epibolum var. **epithallinum** (Leighton ex Nyl.) Salisb. [THELEPI1]

AGYRIACEAE (L)

Agyrium Fr. [AGYRIUM$]
Agyrium rufum (Pers.) Fr. [AGYRRUF]
Xylographa (Fr.) Fr. [XYLOGRA$]
Xylographa abietina (Pers.) Zahlbr. = Xylographa parallela [XYLOPAR]
Xylographa hians Tuck. [XYLOHIA]
SY: *Xylographa micrographa* G. Merr.
Xylographa micrographa G. Merr. = Xylographa hians [XYLOHIA]
Xylographa opegraphella Nyl. ex Rothr. [XYLOOPE]
Xylographa parallela (Ach.) Behlen & Desberger [XYLOPAR]
SY: *Xylographa abietina* (Pers.) Zahlbr.
Xylographa spilomatica (Anzi) Th. Fr. = Xylographa vitiligo [XYLOVIT]
Xylographa vitiligo (Ach.) J.R. Laundon [XYLOVIT]
SY: *Xylographa spilomatica* (Anzi) Th. Fr.

ALECTORIACEAE (L)

Alectoria Ach. [ALECTOR$]
Alectoria boryana Delise = Alectoria nigricans [ALECNIG]
Alectoria cincinnata (Fr.) Lynge = Alectoria sarmentosa ssp. vexillifera [ALECSAR2]
Alectoria imshaugii Brodo & D. Hawksw. [ALECIMS]
Alectoria luteola Mont. = Alectoria sarmentosa ssp. sarmentosa [ALECSAR0]
Alectoria nigricans (Ach.) Nyl. [ALECNIG]
SY: *Alectoria boryana* Delise
Alectoria ochroleuca (Hoffm.) A. Massal. [ALECOCH]
SY: *Alectoria ochroleuca* var. *rigida* Fr.
Alectoria ochroleuca var. *rigida* Fr. = Alectoria ochroleuca [ALECOCH]
Alectoria ochroleuca var. *sarmentosa* Nyl. = Alectoria sarmentosa ssp. sarmentosa [ALECSAR0]
Alectoria sarmentosa (Ach.) Ach. [ALECSAR] (common witch's hair)
Alectoria sarmentosa ssp. **sarmentosa** [ALECSAR0]
SY: *Alectoria luteola* Mont.
SY: *Alectoria ochroleuca* var. *sarmentosa* Nyl.
SY: *Alectoria sarmentosa* var. *gigantea* Räsänen
SY: *Alectoria stigmata* Bystrek
SY: *Alectoria subsarmentosa* Stirton
Alectoria sarmentosa ssp. **vexillifera** (Nyl.) D. Hawksw. [ALECSAR2]
SY: *Alectoria cincinnata* (Fr.) Lynge
SY: *Alectoria vexillifera* (Nyl.) Stizenb.
Alectoria sarmentosa var. *gigantea* Räsänen = Alectoria sarmentosa ssp. sarmentosa [ALECSAR0]
Alectoria stigmata Bystrek = Alectoria sarmentosa ssp. sarmentosa [ALECSAR0]
Alectoria subsarmentosa Stirton = Alectoria sarmentosa ssp. sarmentosa [ALECSAR0]
Alectoria vancouverensis (Gyelnik) Gyelnik ex Brodo & D. Hawksw. [ALECVAN]
Alectoria vexillifera (Nyl.) Stizenb. = Alectoria sarmentosa ssp. vexillifera [ALECSAR2]

ARCTOMIACEAE (L)

Arctomia Th. Fr. [ARCTOMI$]
Arctomia delicatula Th. Fr. [ARCTDEL]

ARTHONIACEAE (L)

Arthonia Ach. [ARTHONI$]
Arthonia arthonioides (Ach.) A.L. Sm. [ARTHART]
SY: *Arthonia aspersa* Leighton
Arthonia aspersa Leighton = Arthonia arthonioides [ARTHART]
Arthonia carneorufa Willey [ARTHCAN]
Arthonia dispersa (Schrader) Nyl. [ARTHDIS]
Arthonia epimela (Almq.) Lamb [ARTHEPM]
Arthonia epiphyscia Nyl. [ARTHEPI]
Arthonia fusca (A. Massal.) Hepp = Arthonia lapidicola [ARTHLAP]
Arthonia glebosa Tuck. [ARTHGLE]
Arthonia hypobela (Nyl.) Zahlbr. [ARTHHYP]
Arthonia ilicina Taylor [ARTHILI]
SY: *Arthothelium ilicinum* (Taylor) P. James
Arthonia intexta Almq. [ARTHINT]
Arthonia lapidicola (Taylor) Branth & Rostrup [ARTHLAP]
SY: *Arthonia fusca* (A. Massal.) Hepp
Arthonia macounii G. Merr. = Arthothelium macounii [ARTHMAC]
Arthonia patellulata Nyl. [ARTHPAT] (aspen comma)
Arthonia phaeobaea (Norman) Norman [ARTHPHA]
Arthonia polygramma Nyl. [ARTHPOL]
Arthonia radiata (Pers.) Ach. [ARTHRAD]
Arthonia vinosa Leighton [ARTHVIN]
Arthonia willeyi Tuck. [ARTHWIL]
Arthothelium A. Massal. [ARTHOTH$]
Arthothelium ilicinum (Taylor) P. James = Arthonia ilicina [ARTHILI]
Arthothelium macounii (G. Merr.) W. Noble [ARTHMAC]
SY: *Arthonia macounii* G. Merr.
Arthothelium spectabile A. Massal. [ARTHSPE]
Arthothelium violascens (Nyl.) Zahlbr. [ARTHVIO]

ARTHOPYRENIACEAE (L)

Arthopyrenia A. Massal. [ARTHOPY$]
Arthopyrenia antecellens (Nyl.) Arnold [ARTHANT]
Arthopyrenia cerasi (Schrader) A. Massal. [ARTHCER]
Arthopyrenia cinereopruinosa (Schaerer) A. Massal. [ARTHCIE]
SY: *Arthopyrenia pinicola* (Hepp) A. Massal.
Arthopyrenia epidermidis (DC.) A. Massal. = Arthopyrenia punctiformis [ARTHPUC]
Arthopyrenia fallax (Nyl.) Arnold = Arthopyrenia lapponina [ARTHLAO]
Arthopyrenia lapponina Anzi [ARTHLAO]
SY: *Arthopyrenia fallax* (Nyl.) Arnold
SY: *Polyblastiopsis fallax* (Nyl.) Fink
Arthopyrenia padi Rabenh. = Arthopyrenia punctiformis [ARTHPUC]
Arthopyrenia pinicola (Hepp) A. Massal. = Arthopyrenia cinereopruinosa [ARTHCIE]
Arthopyrenia plumbaria (Stizenb.) R.C. Harris [ARTHPLU]
SY: *Porina plumbaria* (Stizenb.) Hasse
SY: *Pyrenula herrei* Fink
SY: *Verrucaria plumbaria* Stizenb.
Arthopyrenia punctiformis (Pers.) A. Massal. [ARTHPUC]
SY: *Arthopyrenia epidermidis* (DC.) A. Massal.
SY: *Arthopyrenia padi* Rabenh.
Polyblastiopsis fallax (Nyl.) Fink = Arthopyrenia lapponina [ARTHLAO]
Porina plumbaria (Stizenb.) Hasse = Arthopyrenia plumbaria [ARTHPLU]
Pyrenula herrei Fink = Arthopyrenia plumbaria [ARTHPLU]
Tomasellia A. Massal. [TOMASEL$]
Tomasellia americana (Minks ex Willey) R.C. Harris [TOMAAME]
Verrucaria plumbaria Stizenb. = Arthopyrenia plumbaria [ARTHPLU]
Zwackhiomyces Grube & Hafellner [ZWACKHI$]
Zwackhiomyces coepulonus (Norman) Grube & R. Sant. [ZWACCOE]

ARTHRORHAPHIDACEAE (L)

Arthrorhaphis Th. Fr. [ARTHROR$]
Arthrorhaphis alpina (Schaerer) R. Sant. [ARTHALP]
SY: *Bacidia alpina* (Schaerer) Vainio

Arthrorhaphis citrinella (Ach.) Poelt [ARTHCIT]
SY: *Bacidia citrinella* (Ach.) Branth & Rostrup
SY: *Bacidia flavovirescens* (Dickson) Anzi
Bacidia alpina (Schaerer) Vainio = Arthrorhaphis alpina [ARTHALP]
Bacidia citrinella (Ach.) Branth & Rostrup = Arthrorhaphis citrinella [ARTHCIT]
Bacidia flavovirescens (Dickson) Anzi = Arthrorhaphis citrinella [ARTHCIT]

BACIDIACEAE (L)

Adelolecia Hertel & Hafellner [ADELOLE$]
Adelolecia pilati (Hepp) Hertel & Hafellner [ADELPIL]
SY: *Lecidea lyngeana* Zahlbr.
SY: *Lecidea pilati* (Hepp) Körber
SY: *Lecidea subauriculata* Lynge
Bacidia De Not. [BACIDIA$]
Bacidia abbrevians (Nyl.) Th. Fr. = Bacidia igniarii [BACIIGN]
Bacidia absistens (Nyl.) Arnold [BACIABS]
SY: *Lecidea absistens* Nyl.
Bacidia affinis (Stizenb.) Vainio = Bacidia subincompta [BACISUI]
Bacidia alaskensis (Nyl.) Zahlbr. [BACIALA]
SY: *Lecidea alaskensis* Nyl.
Bacidia arceutina (Ach.) Arnold [BACIARC]
Bacidia atrogrisea (Delise ex Hepp) Körber = Bacidia laurocerasi [BACILAU]
Bacidia auerswaldii (Hepp ex Stizenb.) Mig. [BACIAUE]
SY: *Bacidia effusella* Zahlbr., nom. ill.
SY: *Bilimbia effusa* Auersw. ex Rabenh.
SY: *Lecidea auerswaldii* Hepp ex Stizenb.
SY: *Lecidea effusa* (Auersw. ex Rabenh.) Stizenb.
Bacidia bacillifera (Nyl.) Arnold = Bacidia circumspecta [BACICIR]
Bacidia bagliettoana (Massal. & De Not.) Jatta [BACIBAG]
SY: *Bacidia muscorum* (Sw.) Mudd
Bacidia beckhausii Körber [BACIBEC]
SY: *Bacidia minuscula* Anzi
Bacidia carneoalbida (Müll. Arg.) Coppins = Biatora carneoalbida [BIATCAR]
Bacidia circumspecta (Nyl. ex Vainio) Malme [BACICIR]
SY: *Bacidia bacillifera* (Nyl.) Arnold
SY: *Lecidea bacillifera* f. *circumspecta* Nyl., nom. nud.
SY: *Lecidea bacillifera* var. *circumspecta* Nyl. ex Vainio
SY: *Lecidea circumspecta* (Nyl. ex Vainio) Hedl.
Bacidia effusella Zahlbr., nom. ill. = Bacidia auerswaldii [BACIAUE]
Bacidia endoleuca (Nyl.) Zahlbr. = Bacidia laurocerasi [BACILAU]
Bacidia friesiana (Hepp) Körber [BACIFRI]
Bacidia fuscorubella (Ach.) Bausch = Bacidia polychroa [BACIPOL]
Bacidia hegetschweileri (Hepp) Vainio = Bacidia subincompta [BACISUI]
Bacidia idahoensis H. Magn. = Bacidia laurocerasi ssp. idahoensis [BACILAU1]
Bacidia igniarii (Nyl.) Oksner [BACIIGN]
SY: *Bacidia abbrevians* (Nyl.) Th. Fr.
Bacidia invertens (Nyl.) Zahlbr. = Bacidia laurocerasi [BACILAU]
Bacidia laurocerasi (Delise ex Duby) Zahlbr. [BACILAU]
SY: *Bacidia atrogrisea* (Delise ex Hepp) Körber
SY: *Bacidia endoleuca* (Nyl.) Zahlbr.
SY: *Bacidia invertens* (Nyl.) Zahlbr.
SY: *Bacidia subacerina* Vainio
SY: *Biatora atrogrisea* Delise ex Hepp
SY: *Lecidea endoleucula* Nyl.
SY: *Lecidea invertens* Nyl.
SY: *Patellaria laurocerasi* Delise ex Duby
Bacidia laurocerasi ssp. **idahoensis** (H. Magn.) S. Ekman [BACILAU1]
SY: *Bacidia idahoensis* H. Magn.
Bacidia laurocerasi ssp. **laurocerasi** [BACILAU0]
Bacidia luteola (Schrad.) Mudd = Bacidia rubella [BACIRUB]
Bacidia luteola (Schrad.) Mudd = Bacidia rubella [BACIRUB]
Bacidia minuscula Anzi = Bacidia beckhausii [BACIBEC]
Bacidia muscorum (Sw.) Mudd = Bacidia bagliettoana [BACIBAG]
Bacidia naegelii (Hepp) Zahlbr. [BACINAE]
Bacidia nivalis Follmann [BACINIV]
Bacidia polychroa (Th. Fr.) Körber [BACIPOL]
SY: *Bacidia fuscorubella* (Ach.) Bausch
Bacidia rosella (Pers.) De Not. [BACIROS]
Bacidia rubella (Hoffm.) A. Massal. [BACIRUB]
SY: *Bacidia luteola* (Schrad.) Mudd
SY: *Bacidia luteola* (Schrad.) Mudd
SY: *Biatora luteola* (Schrad.) Fr.
SY: *Lecidea luteola* (Schrad.) Ach.
SY: *Lichen luteolus* Schrad.
SY: *Lichen lutereus* J.F. Gmel.
SY: *Lichen rubellus* Ehrh., nom. nud.
SY: *Verrucaria rubella* Hoffm.
Bacidia salmonea S. Ekman [BACISAL]
Bacidia scopulicola (Nyl.) A.L. Sm. [BACISCO]
Bacidia sphaeroides (Dickson) Zahlbr. = Biatora sphaeroides [BIATSPH]
Bacidia subacerina Vainio = Bacidia laurocerasi [BACILAU]
Bacidia subincompta (Nyl.) Arnold [BACISUI]
SY: *Bacidia affinis* (Stizenb.) Vainio
SY: *Bacidia hegetschweileri* (Hepp) Vainio
SY: *Biatora atrosanguinea* var. *hegetschweileri* Hepp
SY: *Lecidea hegetschweileri* Hepp
SY: *Lecidea separabilis* (Nyl.) Arnold
SY: *Lecidea subincompta* Nyl.
Bacidia vermifera (Nyl.) Th. Fr. [BACIVER]
Biatora Fr. [BIATORA$]
Biatora anthracophila (Nyl.) Hafellner [BIATANT]
SY: *Hypocenomyce anthracophila* (Nyl.) P. James & Gotth. Schneider
SY: *Lecidea anthracophila* Nyl.
SY: *Psora anthracophila* (Nyl.) Arnold
Biatora atrogrisea Delise ex Hepp = Bacidia laurocerasi [BACILAU]
Biatora atrosanguinea var. *hegetschweileri* Hepp = Bacidia subincompta [BACISUI]
Biatora botryosa Fr. [BIATBOT]
SY: *Lecidea botryosa* (Fr.) Th. Fr.
Biatora carneoalbida (Müll. Arg.) Coppins [BIATCAR]
SY: *Bacidia carneoalbida* (Müll. Arg.) Coppins
Biatora flavopunctata (Tønsberg) Hinteregger & Printzen [BIATFLA]
SY: *Lecanora flavopunctata* Tønsberg
Biatora helvola (Körber) Hellbom [BIATHEL]
SY: *Lecidea helvola* (Körber) Th. Fr.
Biatora luteola (Schrad.) Fr. = Bacidia rubella [BACIRUB]
Biatora pullata Norman [BIATPUL]
SY: *Lecidea pullata* (Norman) Th. Fr.
Biatora sphaeroides (Dickson) Körber [BIATSPH]
SY: *Bacidia sphaeroides* (Dickson) Zahlbr.
SY: *Catillaria sphaeroides* (A. Massal.) Schuler
Biatora vernalis (L.) Fr. [BIATVER] (dot lichen)
SY: *Lecidea vernalis* (L.) Ach.
Bilimbia effusa Auersw. ex Rabenh. = Bacidia auerswaldii [BACIAUE]
Catillaria atropurpurea (Schaerer) Th. Fr. = Catinaria atropurpurea [CATIATO]
Catillaria graniformis (K. Hagen) Vainio = Cliostomum corrugatum [CLIOCOR]
Catillaria griffithii (Sm.) Malme = Cliostomum griffithii [CLIOGRI]
Catillaria sphaeroides (A. Massal.) Schuler = Biatora sphaeroides [BIATSPH]
Catillaria tricolor auct. = Cliostomum griffithii [CLIOGRI]
Catinaria Vainio [CATINAR$]
Catinaria atropurpurea (Schaerer) Vezda & Poelt [CATIATO]
SY: *Catillaria atropurpurea* (Schaerer) Th. Fr.
Cliostomum Fr. [CLIOSTO$]
Cliostomum corrugatum (Ach.) Fr. [CLIOCOR]
SY: *Catillaria graniformis* (K. Hagen) Vainio
SY: *Cliostomum graniforme* (K. Hagen) Coppins
Cliostomum graniforme (K. Hagen) Coppins = Cliostomum corrugatum [CLIOCOR]
Cliostomum griffithii (Sm.) Coppins [CLIOGRI]
SY: *Catillaria griffithii* (Sm.) Malme

SY: *Catillaria tricolor* auct.
Cliostomum leprosum (Räsänen) Holien & Tønsberg [CLIOLEP]
SY: *Cliostomum luteolum* Gowan
Cliostomum luteolum Gowan = Cliostomum leprosum [CLIOLEP]
Hypocenomyce anthracophila (Nyl.) P. James & Gotth. Schneider = Biatora anthracophila [BIATANT]
Japewia Tønsberg [JAPEWIA$]
Japewia carrollii (Coppins & P. James) Tønsberg [JAPECAR]
Japewia subaurifera Muhr & Tønsberg [JAPESUB]
Japewia tornoënsis (Nyl.) Tønsberg [JUPETOR]
SY: *Lecidea tornoënsis* Nyl.
SY: *Mycoblastus tornoënsis* (Nyl.) R. Anderson
Lecania A. Massal. [LECANIA$]
Lecania cyrtella (Ach.) Th. Fr. [LECACYR]
SY: *Lecania cyrtellina* (Nyl.) Sandst.
Lecania cyrtellina (Nyl.) Sandst. = Lecania cyrtella [LECACYR]
Lecania dimera (Nyl.) Th. Fr. = Lecania dubitans [LECADUB]
Lecania dubitans (Nyl.) A.L. Sm. [LECADUB]
SY: *Lecania dimera* (Nyl.) Th. Fr.
Lecania dudleyi Herre [LECADUD]
Lecania erysibe (Ach.) Mudd [LECAERY]
Lecania fructigena Zahlbr. [LECAFRU]
Lecania hassei (Zahlbr.) W. Noble [LECAHAS]
SY: *Solenopsora hassei* (Zahlbr.) Zahlbr.
Lecania nylanderiana A. Massal. [LECANYL]
Lecanora atra (Hudson) Ach. = Tephromela atra [TEPHATR]
Lecanora flavopunctata Tønsberg = Biatora flavopunctata [BIATFLA]
Lecanora lentigera (Weber) Ach. = Squamarina lentigera [SQUALEN]
Lecidea absistens Nyl. = Bacidia absistens [BACIABS]
Lecidea alaskensis Nyl. = Bacidia alaskensis [BACIALA]
Lecidea anthracophila Nyl. = Biatora anthracophila [BIATANT]
Lecidea armeniaca (DC.) Fr. = Tephromela armeniaca [TEPHARM]
Lecidea auerswaldii Hepp ex Stizenb. = Bacidia auerswaldii [BACIAUE]
Lecidea bacillifera f. *circumspecta* Nyl., nom. nud. = Bacidia circumspecta [BACICIR]
Lecidea bacillifera var. *circumspecta* Nyl. ex Vainio = Bacidia circumspecta [BACICIR]
Lecidea botryosa (Fr.) Th. Fr. = Biatora botryosa [BIATBOT]
Lecidea circumspecta (Nyl. ex Vainio) Hedl. = Bacidia circumspecta [BACICIR]
Lecidea effusa (Auersw. ex Rabenh.) Stizenb. = Bacidia auerswaldii [BACIAUE]
Lecidea endoleucula Nyl. = Bacidia laurocerasi [BACILAU]
Lecidea hegetschweileri Hepp = Bacidia subincompta [BACISUI]
Lecidea helvola (Körber) Th. Fr. = Biatora helvola [BIATHEL]
Lecidea invertens Nyl. = Bacidia laurocerasi [BACILAU]
Lecidea luteola (Schrad.) Ach. = Bacidia rubella [BACIRUB]
Lecidea lyngeana Zahlbr. = Adelolecia pilati [ADELPIL]
Lecidea pilati (Hepp) Körber = Adelolecia pilati [ADELPIL]
Lecidea pullata (Norman) Th. Fr. = Biatora pullata [BIATPUL]
Lecidea separabilis (Nyl.) Arnold = Bacidia subincompta [BACISUI]
Lecidea subauriculata Lynge = Adelolecia pilati [ADELPIL]
Lecidea subincompta Nyl. = Bacidia subincompta [BACISUI]
Lecidea tornoënsis Nyl. = Japewia tornoënsis [JUPETOR]
Lecidea vernalis (L.) Ach. = Biatora vernalis [BIATVER]
Lichen luteolus Schrad. = Bacidia rubella [BACIRUB]
Lichen lutereus J.F. Gmel. = Bacidia rubella [BACIRUB]
Lichen rubellus Ehrh., nom. nud. = Bacidia rubella [BACIRUB]
Lopadium alpinum (Körber) R. Sant. = Schadonia alpina [SCHAALP]
Lopadium gemellum (Anzi) Stizenb. = Schadonia alpina [SCHAALP]
Mycoblastus tornoënsis (Nyl.) R. Anderson = Japewia tornoënsis [JUPETOR]
Patellaria laurocerasi Delise ex Duby = Bacidia laurocerasi [BACILAU]
Psora anthracophila (Nyl.) Arnold = Biatora anthracophila [BIATANT]
Schadonia Körber [SCHADON$]
Schadonia alpina Körber [SCHAALP]
SY: *Lopadium alpinum* (Körber) R. Sant.
SY: *Lopadium gemellum* (Anzi) Stizenb.
Solenopsora hassei (Zahlbr.) Zahlbr. = Lecania hassei [LECAHAS]
Squamarina Poelt [SQUAMAR$]
Squamarina cartilaginea (With.) P. James [SQUACAR]
SY: *Squamarina crassa* (Hudson) Poelt
Squamarina crassa (Hudson) Poelt = Squamarina cartilaginea [SQUACAR]
Squamarina lentigera (Weber) Poelt [SQUALEN]
SY: *Lecanora lentigera* (Weber) Ach.
Tephromela Choisy [TEPHROM$]
Tephromela armeniaca (DC.) Hertel & Rambold [TEPHARM]
SY: *Lecidea armeniaca* (DC.) Fr.
Tephromela atra (Hudson) Hafellner [TEPHATR]
SY: *Lecanora atra* (Hudson) Ach.
Verrucaria rubella Hoffm. = Bacidia rubella [BACIRUB]
Waynea Moberg [WAYNEA$]
Waynea californica Moberg = Waynea stoechadiana [WAYNSTO]
Waynea stoechadiana (Abbassi Maaf & Roux) Roux & Clerc [WAYNSTO] (knobbed scale)
SY: *Waynea californica* Moberg

BAEOMYCETACEAE (L)

Baeomyces Pers. [BAEOMYC$]
Baeomyces carneus Flörke [BAEOCAR]
Baeomyces fungoides auct. = Dibaeis baeomyces [DIBABAE]
Baeomyces placophyllus Ach. [BAEOPLA]
Baeomyces roseus Pers. = Dibaeis baeomyces [DIBABAE]
Baeomyces rufus (Hudson) Rebent. [BAEORUF]
Dibaeis Clem. [DIBAEIS$]
Dibaeis baeomyces (L. f.) Rambold & Hertel [DIBABAE]
SY: *Baeomyces roseus* Pers.
SY: *Dibaeis rosea* (Pers.) Clem.
SY: *Baeomyces fungoides* auct.
Dibaeis rosea (Pers.) Clem. = Dibaeis baeomyces [DIBABAE]
Icmadophila Trevisan [ICMADOP$]
Icmadophila ericetorum (L.) Zahlbr. [ICMAERI] (spraypaint lichen)

BIATORELLACEAE (L)

Biatorella De Not. [BIATORE$]
Biatorella fossarum (Dufour ex Fr.) Th. Fr. = Biatorella hemisphaerica [BIATHEM]
Biatorella hemisphaerica Anzi [BIATHEM]
SY: *Biatorella fossarum* (Dufour ex Fr.) Th. Fr.

BRIGANTIAEACEAE (L)

Brigantiaea Trevisan [BRIGANT$]
Brigantiaea purpurata (Zahlbr.) Haf. & Bellem. [BRIGPUR]

CALICIACEAE (L)

Acroscyphus Léveillé [ACROSCY$]
Acroscyphus sphaerophoroides Léveillé [ACROSPH]
Calicium Pers. [CALICIU$]
Calicium abietinum Pers. [CALIABI]
Calicium adaequatum Nyl. [CALIADA]
SY: *Calicium hemisphaericum* Howard
Calicium adspersum Pers. [CALIADS]
SY: *Calicium roscidum* (Ach.) Ach.
Calicium corynellum (Ach.) Ach. [CALICOR]
Calicium glaucellum Ach. [CALIGLA]
Calicium hemisphaericum Howard = Calicium adaequatum [CALIADA]
Calicium hyperellum (Ach.) Ach. = Calicium viride [CALIVIR]
Calicium lenticulare Ach. [CALILEN]
SY: *Calicium lentigerellum* Tuck.
SY: *Calicium subquercinum* Asah.
Calicium lentigerellum Tuck. = Calicium lenticulare [CALILEN]
Calicium lichenoides (L.) Schumacher = Calicium salicinum [CALISAL]

Calicium parvum Tibell [CALIPAR]
Calicium roscidum (Ach.) Ach. = Calicium adspersum [CALIADS]
Calicium roscidum var. *trabinellum* (Ach.) Schaerer = Calicium trabinellum [CALITRA]
Calicium salicinum Pers. [CALISAL]
SY: *Calicium lichenoides* (L.) Schumacher
SY: *Calicium sphaerocephalum* (L.) Ach.
SY: *Calicium trachelinum* Ach.
Calicium sphaerocephalum (L.) Ach. = Calicium salicinum [CALISAL]
Calicium subquercinum Asah. = Calicium lenticulare [CALILEN]
Calicium trabinellum (Ach.) Ach. [CALITRA]
SY: *Calicium roscidum* var. *trabinellum* (Ach.) Schaerer
Calicium trachelinum Ach. = Calicium salicinum [CALISAL]
Calicium viride Pers. [CALIVIR] (green stubble)
SY: *Calicium hyperellum* (Ach.) Ach.
Cypheliopsis bolanderi (Tuck.) Vainio = Thelomma mammosum [THELMAM]
Cyphelium Ach. [CYPHELU$]
Cyphelium caliciforme (Flotow) Zahlbr. = Thelomma occidentale [THELOCC]
Cyphelium inquinans (Sm.) Trevisan [CYPHINQ]
SY: *Cyphelium ventricosulum* (Müll. Arg.) Zahlbr.
Cyphelium karelicum (Vainio) Räsänen [CYPHKAR]
Cyphelium occidentale Herre = Thelomma occidentale [THELOCC]
Cyphelium ocellatum (Körber) Trevisan = Thelomma ocellatum [THELOCE]
Cyphelium pinicola Tibell [CYPHPIN]
Cyphelium tigillare (Ach.) Ach. [CYPHTIG]
Cyphelium trachylioides (Nyl. ex Branth & Rostrup) Erichsen [CYPHTRA]
Cyphelium ventricosulum (Müll. Arg.) Zahlbr. = Cyphelium inquinans [CYPHINQ]
Thelomma A. Massal. [THELOMM$]
Thelomma mammosum (Hepp) A. Massal. [THELMAM]
SY: *Cypheliopsis bolanderi* (Tuck.) Vainio
Thelomma occidentale (Herre) Tibell [THELOCC]
SY: *Cyphelium caliciforme* (Flotow) Zahlbr.
SY: *Cyphelium occidentale* Herre
Thelomma ocellatum (Körber) Tibell [THELOCE]
SY: *Cyphelium ocellatum* (Körber) Trevisan
Tholurna Norman [THOLURN$]
Tholurna dissimilis (Norman) Norman [THOLDIS]

CANDELARIACEAE (L)

Candelaria A. Massal. [CANDELA$]
Candelaria concolor (Dickson) Stein [CANDCON] (candleflame)
Candelariella Müll. Arg. [CANDELR$]
Candelariella athallina (Wedd.) Du Rietz [CANDATH]
Candelariella aurella (Hoffm.) Zahlbr. [CANDAUR]
SY: *Candelariella cerinella* (Flörke) Zahlbr.
SY: *Candelariella epixantha* auct.
Candelariella cerinella (Flörke) Zahlbr. = Candelariella aurella [CANDAUR]
Candelariella efflorescens R.C. Harris & W.R. Buck [CANDEFF]
Candelariella epixantha auct. = Candelariella aurella [CANDAUR]
Candelariella rosulans (Müll. Arg.) Zahlbr. [CANDROS]
Candelariella terrigena Räsänen [CANDTER]
Candelariella vitellina (Hoffm.) Müll. Arg. [CANDVIT]
Candelariella xanthostigma (Ach.) Lettau [CANDXAN]

CAPNODIACEAE (L)

Echinothecium Zopf [ECHINOT$]
Echinothecium reticulatum Zopf [ECHIRET]

CATILLARIACEAE (L)

Bacidia declinis (Tuck.) Zahlbr. = Catillaria nigroclavata [CATINIG]
Bacidia globulosa (Flörke) Hafellner & V. Wirth = Catillaria globulosa [CATIGLO]
Catillaria A. Massal. [CATILLA$]
Catillaria arctica Lynge = Toninia philippea [TONIPHI]
Catillaria athallina (Hepp) Hellbom = Toninia athallina [TONIATH]
Catillaria chalybeia (Borrer) A. Massal. [CATICHA]
Catillaria columbiana (G. Merr.) W. Noble [CATICOL]
Catillaria globulosa (Flörke) Th. Fr. [CATIGLO]
SY: *Bacidia globulosa* (Flörke) Hafellner & V. Wirth
Catillaria kansuensis H. Magn. = Toninia philippea [TONIPHI]
Catillaria nigroclavata (Nyl.) Schuler [CATINIG]
SY: *Bacidia declinis* (Tuck.) Zahlbr.
Catillaria subnitida Hellbom = Catillaria tristis [CATITRI]
Catillaria tristis (Müll. Arg.) Arnold [CATITRI]
SY: *Catillaria subnitida* Hellbom
SY: *Toninia caeruleonigricans* (Lightf.) Th. Fr.
Halecania M. Mayrh. [HALECAN$]
Halecania alpivaga (Th. Fr.) M. Mayrh. [HALEALP]
SY: *Lecania alpivaga* Th. Fr.
SY: *Lecania thallophila* H. Magn.
Kiliasia athallina (Hepp) Hafellner = Toninia athallina [TONIATH]
Kiliasia philippea (Mont.) Hafellner = Toninia philippea [TONIPHI]
Lecania alpivaga Th. Fr. = Halecania alpivaga [HALEALP]
Lecania thallophila H. Magn. = Halecania alpivaga [HALEALP]
Psora scholanderi (Lynge) R. Anderson = Toninia tristis [TONITRI]
Toninia A. Massal. [TONINIA$]
Toninia aromatica (Sm.) A. Massal. [TONIARO]
Toninia athallina (Hepp) Timdal [TONIATH]
SY: *Catillaria athallina* (Hepp) Hellbom
SY: *Kiliasia athallina* (Hepp) Hafellner
Toninia caeruleonigricans (Lightf.) Th. Fr. = Fuscopannaria praetermissa [FUSCPRE]
Toninia candida (Weber) Th. Fr. [TONICAN]
Toninia kolax Poelt = Toninia verrucarioides [TONIVER]
Toninia philippea (Mont.) Timdal [TONIPHI]
SY: *Catillaria arctica* Lynge
SY: *Catillaria kansuensis* H. Magn.
SY: *Kiliasia philippea* (Mont.) Hafellner
Toninia ruginosa (Tuck.) Herre [TONIRUG]
Toninia sedifolia (Scop.) Timdal [TONISED]
Toninia tabacina auct. = Toninia tristis [TONITRI]
Toninia tristis (Th. Fr.) Th. Fr. [TONITRI]
SY: *Psora scholanderi* (Lynge) R. Anderson
SY: *Toninia tabacina* auct.
Toninia tristis ssp. **asiae-centralis** (H. Magn.) Timdal [TONITRI1]
Toninia tristis ssp. **canadensis** Timdal [TONITRI2]
Toninia verrucarioides (Nyl.) Timdal [TONIVER]
SY: *Toninia kolax* Poelt

CHRYSOTRICHACEAE (L)

Chrysothrix Mont. [CHRYSOH$]
Chrysothrix candelaris (L.) J.R. Laundon [CHRYCAN]
SY: *Lepraria candelaris* (L.) Fr.
SY: *Lepraria citrina* (Schaerer) Rabenh.
SY: *Lepraria flava* (Schreber) Sm.
Chrysothrix chlorina (Ach.) J.R. Laundon [CHRYCHL] (lime dust)
SY: *Lepraria chlorina* (Ach.) Ach.
Lepraria candelaris (L.) Fr. = Chrysothrix candelaris [CHRYCAN]
Lepraria chlorina (Ach.) Ach. = Chrysothrix chlorina [CHRYCHL]
Lepraria citrina (Schaerer) Rabenh. = Chrysothrix candelaris [CHRYCAN]
Lepraria flava (Schreber) Sm. = Chrysothrix candelaris [CHRYCAN]

CLADONIACEAE (L)

Cladina Nyl. [CLADINA$]
Cladina aberrans (Abbayes) Hale & Culb. = Cladina stellaris var. aberrans [CLADSTE1]
Cladina alpestris (L.) Nyl. = Cladina stellaris [CLADSTE]
Cladina arbuscula (Wallr.) Hale & Culb. [CLADARB]
SY: *Cladonia arbuscula* (Wallr.) Flotow

SY: *Cladonia rangiferina* var. *sylvatica* L.
SY: *Cladonia silvatica* (L.) Räsänen
SY: *Cladonia sylvatica* L.
SY: *Cladonia sylvatica* auct.
Cladina arbuscula ssp. **arbuscula** [CLADARB0]
Cladina arbuscula ssp. *mitis* (Sandst.) Burgaz = Cladina mitis [CLADMIT]
Cladina arbuscula ssp. **beringiana** (Ahti) N.S. Golubk. [CLADARB1]
SY: *Cladina beringiana* (Ahti) Trass
SY: *Cladonia arbuscula* ssp. *beringiana* Ahti
Cladina beringiana (Ahti) Trass = Cladina arbuscula ssp. beringiana [CLADARB1]
Cladina ciliata (Stirton) Trass [CLADCIL]
SY: *Cladonia ciliata* Stirton
SY: *Cladonia tenuis* (Flörke) Harm.
Cladina ciliata f. *tenuis* (Flörke) Ahti = Cladina ciliata var. tenuis [CLADCIL3]
Cladina ciliata var. **ciliata** [CLADCIL0]
SY: *Cladina leucophaea* (Abbayes) Hale & Culb.
SY: *Cladonia leucophaea* Abbayes
SY: *Cladonia tenuis* var. *leucophaea* (Abbayes) Ahti
Cladina ciliata var. **tenuis** (Flörke) Ahti & M.J. Lai [CLADCIL3]
SY: *Cladina ciliata* f. *tenuis* (Flörke) Ahti
SY: *Cladina tenuis* (Flörke) Hale & Culb.
Cladina impexa de Lesd. = Cladina portentosa [CLADPOR]
Cladina leucophaea (Abbayes) Hale & Culb. = Cladina ciliata var. ciliata [CLADCIL0]
Cladina mitis (Sandst.) Hustich [CLADMIT] (green reindeer lichen)
SY: *Cladina arbuscula* ssp. *mitis* (Sandst.) Burgaz
SY: *Cladonia arbuscula* ssp. *mitis* (Sandst.) Ruoss
SY: *Cladonia mitis* Sandst.
SY: *Cladonia silvatica* var. *mitis* (Sandst.)Räsänen
Cladina pacifica (Ahti) Hale & Culb. = Cladina portentosa ssp. pacifica [CLADPOR1]
Cladina portentosa (Dufour) Follmann [CLADPOR] (coastal reindeer)
SY: *Cladina impexa* de Lesd.
SY: *Cladonia impexa* Harm.
Cladina portentosa ssp. **pacifica** f. **pacifica** [CLADPOR3]
Cladina portentosa ssp. **pacifica** (Ahti) Ahti [CLADPOR1]
SY: *Cladina pacifica* (Ahti) Hale & Culb.
SY: *Cladonia pacifica* Ahti
SY: *Cladonia portentosa* ssp. *pacifica* (Ahti) Ahti
Cladina portentosa ssp. **pacifica** f. **decolorans** (Ahti) Ahti [CLADPOR2]
SY: *Cladonia portentosa* ssp. *pacifica* f. *decolorans* Ahti
Cladina rangiferina (L.) Nyl. [CLADRAN] (reindeer lichen)
SY: *Cladonia rangiferina* (L.) F.H. Wigg.
SY: *Cladonia rangiferina* f. *nivea* Räsänen
Cladina rangiferina ssp. **grisea** (Ahti) Ahti & M.J. Lai [CLADRAN1]
SY: *Cladonia rangiferina* ssp. *grisea* Ahti
Cladina stellaris (Opiz) Brodo [CLADSTE] (northern reindeer lichen)
SY: *Cladina alpestris* (L.) Nyl.
SY: *Cladonia alpestris* (L.) Rabenh.
SY: *Cladonia rangiferina* var. *alpestris* Flörke
SY: *Cladonia stellaris* (Opiz) Pouzar & Vezda
SY: *Cladonia sylvatica* var. *alpestris* L.
Cladina stellaris var. **aberrans** (Abbayes) Ahti [CLADSTE1]
SY: *Cladina aberrans* (Abbayes) Hale & Culb.
SY: *Cladonia aberrans* (Abbayes) Stuckenb.
SY: *Cladonia alpestris* f. *aberrans* Abbayes
SY: *Cladonia alpestris* var. *aberrans* (Abbayes) Ahti
Cladina stellaris var. **stellaris** [CLADSTE0]
Cladina stygia (Fr.) Ahti [CLADSTY]
SY: *Cladonia stygia* (Fr.) Ruoss
Cladina tenuis (Flörke) Hale & Culb. = Cladina ciliata var. tenuis [CLADCIL3]
Cladonia P. Browne [CLADONI$]
Cladonia aberrans (Abbayes) Stuckenb. = Cladina stellaris var. aberrans [CLADSTE1]
Cladonia acuminata (Ach.) Norrlin [CLADACU]
Cladonia albonigra Brodo & Ahti [CLADALB]
Cladonia alpestris (L.) Rabenh. = Cladina stellaris [CLADSTE]
Cladonia alpestris f. *aberrans* Abbayes = Cladina stellaris var. aberrans [CLADSTE1]
Cladonia alpestris var. *aberrans* (Abbayes) Ahti = Cladina stellaris var. aberrans [CLADSTE1]
Cladonia alpicola (Flotow) Vainio = Cladonia macrophylla [CLADMAR]
Cladonia amaurocraea (Flörke) Schaerer [CLADAMA]
Cladonia arbuscula (Wallr.) Flotow = Cladina arbuscula [CLADARB]
Cladonia arbuscula ssp. *beringiana* Ahti = Cladina arbuscula ssp. beringiana [CLADARB1]
Cladonia arbuscula ssp. *mitis* (Sandst.) Ruoss = Cladina mitis [CLADMIT]
Cladonia artuata S. Hammer [CLADART]
Cladonia asahinae J.W. Thomson [CLADASA]
Cladonia bacillaris Nyl. = Cladonia macilenta var. bacillaris [CLADMAC1]
Cladonia bacilliformis (Nyl.) Glück [CLADBAI]
Cladonia balfourii Crombie = Cladonia macilenta [CLADMAC]
Cladonia bellidiflora (Ach.) Schaerer [CLADBEL]
SY: *Cladonia bellidiflora* var. *hookeri* (Tuck.) Nyl.
SY: *Cladonia hookeri* Tuck.
Cladonia bellidiflora var. *hookeri* (Tuck.) Nyl. = Cladonia bellidiflora [CLADBEL]
Cladonia blakei Robbins = Cladonia coccifera [CLADCOC]
Cladonia borealis S. Stenroos [CLADBOR] (red pixie-cup)
Cladonia botrytes (K. Hagen) Willd. [CLADBOY] (stump cladonia)
Cladonia caespiticia (Pers.) Flörke [CLADCAE]
Cladonia carassensis ssp. *japonica* (Vainio) Asah. = Cladonia crispata [CLADCRI]
Cladonia cariosa (Ach.) Sprengel [CLADCAI] (ribbed cladonia)
Cladonia carneola (Fr.) Fr. [CLADCAN]
Cladonia cenotea (Ach.) Schaerer [CLADCEN] (powdered funnel cladonia)
SY: *Cladonia cenotea* var. *crossota* (Ach.) Nyl.
Cladonia cenotea var. *crossota* (Ach.) Nyl. = Cladonia cenotea [CLADCEN]
Cladonia cervicornis (Ach.) Flotow [CLADCER]
Cladonia cervicornis ssp. **cervicornis** [CLADCER0]
SY: *Cladonia verticillata* var. *cervicornis* Flörke
Cladonia cervicornis ssp. **verticillata** (Hoffm.) Ahti [CLADCER1] (whorled cup lichen)
SY: *Cladonia gracilis* var. *verticillata* Fr.
SY: *Cladonia verticillata* (Hoffm.) Schaerer
Cladonia chlorophaea (Flörke ex Sommerf.) Sprengel [CLADCHL] (false pixie-cup)
Cladonia ciliata Stirton = Cladina ciliata [CLADCIL]
Cladonia coccifera (L.) Willd. [CLADCOC]
SY: *Cladonia blakei* Robbins
SY: *Cladonia cornucopioides* (L.) Fr.
SY: *Cladonia pseudodigitata* Gyelnik
Cladonia coccifera var. *pleurota* (Flörke) Schaerer = Cladonia pleurota [CLADPLE]
Cladonia coniocraea (Flörke) Sprengel [CLADCON] (tiny toothpick cladonia)
Cladonia conista A. Evans [CLADCOI]
SY: *Cladonia humilis* var. *bourgeanica* A.W. Archer
Cladonia conistea auct. = Cladonia humilis [CLADHUM]
Cladonia conoidea Ahti = Cladonia humilis [CLADHUM]
Cladonia cornucopioides (L.) Fr. = Cladonia coccifera [CLADCOC]
Cladonia cornuta (L.) Hoffm. [CLADCOR] (pioneer cladonia)
Cladonia cornuta ssp. **cornuta** [CLADCOR0]
Cladonia cornuta ssp. **groenlandica** (E. Dahl) Ahti [CLADCOR1]
SY: *Cladonia cornuta* var. *groenlandica* E. Dahl
SY: *Cladonia groenlandica* (E. Dahl) Trass
Cladonia cornuta var. *groenlandica* E. Dahl = Cladonia cornuta ssp. groenlandica [CLADCOR1]
Cladonia cornutoradiata (Navás) Sandst. = Cladonia subulata [CLADSUU]
Cladonia crispata (Ach.) Flotow [CLADCRI] (shrub funnel lichen)

SY: *Cladonia carassensis* ssp. *japonica* (Vainio) Asah.
SY: *Cladonia furcata* var. *crispata* Flörke
SY: *Cladonia japonica* Vainio
Cladonia crispata var. **cetrariiformis** (Delise) Vainio [CLADCRI2]
Cladonia crispata var. **crispata** [CLADCRI0]
Cladonia cryptochlorophaea Asah. [CLADCRY]
Cladonia cyanipes (Sommerf.) Nyl. [CLADCYA]
Cladonia cyathomorpha (A. Evans) A. Evans [CLADCYT]
Cladonia decorticata (Flörke) Sprengel [CLADDEC]
Cladonia deformis (L.) Hoffm. [CLADDEF]
Cladonia degenerans (Flörke) Sprengel = Cladonia phyllophora [CLADPHY]
Cladonia delessertii Vainio = Cladonia subfurcata [CLADSUF]
Cladonia delicata auct. = Cladonia parasitica [CLADPAR]
Cladonia digitata (L.) Hoffm. [CLADDIG]
Cladonia dimorpha S. Hammer [CLADDIM]
Cladonia ecmocyna Leighton [CLADECM] (orange-foot lichen)
Cladonia ecmocyna ssp. **occidentalis** Ahti [CLADECM3]
Cladonia ecmocyna ssp. **ecmocyna** [CLADECM0]
Cladonia ecmocyna ssp. **intermedia** (Robbins) Ahti [CLADECM1]
SY: *Cladonia ecmocyna* var. *intermedia* (Robbins) A. Evans
Cladonia ecmocyna var. *intermedia* (Robbins) A. Evans = Cladonia ecmocyna ssp. intermedia [CLADECM1]
Cladonia ecmocyna var. *macroceras* (Flörke) Ach. = Cladonia macroceras [CLADMAE]
Cladonia elongata (Jacq.) Hoffm. = Cladonia maxima [CLADMAX]
Cladonia elongata auct. = Cladonia macroceras [CLADMAE]
Cladonia fimbriata (L.) Fr. [CLADFIM]
SY: *Cladonia fimbriata* var. *costata* Flörke
SY: *Cladonia fimbriata* var. *prolifera* (Retz.) Mass.
SY: *Cladonia fimbriata* var. *simplex* f. *major* (K. Hagen) Vainio
SY: *Cladonia fimbriata* var. *simplex* f. *minor* (K. Hagen) Vainio
SY: *Cladonia fimbriata* var. *tubaeformis* Fr.
SY: *Cladonia major* (K. Hagen) Sandst.
Cladonia fimbriata var. *costata* Flörke = Cladonia fimbriata [CLADFIM]
Cladonia fimbriata var. *prolifera* (Retz.) Mass. = Cladonia fimbriata [CLADFIM]
Cladonia fimbriata var. *simplex* f. *major* (K. Hagen) Vainio = Cladonia fimbriata [CLADFIM]
Cladonia fimbriata var. *simplex* f. *minor* (K. Hagen) Vainio = Cladonia fimbriata [CLADFIM]
Cladonia fimbriata var. *subulata* Schaerer = Cladonia subulata [CLADSUU]
Cladonia fimbriata var. *tubaeformis* Fr. = Cladonia fimbriata [CLADFIM]
Cladonia flabelliformis Vainio = Cladonia polydactyla [CLADPOD]
Cladonia furcata (Hudson) Schrader [CLADFUR]
SY: *Cladonia furcata* var. *conspersa* Vainio
SY: *Cladonia furcata* var. *pinnata* (Flörke) Vainio
SY: *Cladonia furcata* var. *racemosa* Flörke
SY: *Cladonia furcata* var. *subulata* Flörke
SY: *Cladonia herrei* Fink ex J. Hedrick
SY: *Cladonia furcata* ssp. *subrangiformis* auct.
SY: *Cladonia subrangiformis* auct.
Cladonia furcata ssp. *subrangiformis* auct. = Cladonia furcata [CLADFUR]
Cladonia furcata var. *conspersa* Vainio = Cladonia furcata [CLADFUR]
Cladonia furcata var. *crispata* Flörke = Cladonia crispata [CLADCRI]
Cladonia furcata var. *pinnata* (Flörke) Vainio = Cladonia furcata [CLADFUR]
Cladonia furcata var. *racemosa* Flörke = Cladonia furcata [CLADFUR]
Cladonia furcata var. *subulata* Flörke = Cladonia furcata [CLADFUR]
Cladonia gonecha (Ach.) Asah. = Cladonia sulphurina [CLADSUL]
Cladonia gracilescens auct. = Cladonia stricta [CLADSTI]
Cladonia gracilis (L.) Willd. [CLADGRA]
SY: *Cladonia gracilis* f. *anthocephala* Flörke
Cladonia gracilis f. *anthocephala* Flörke = Cladonia gracilis [CLADGRA]
Cladonia gracilis ssp. *dilatata* (Hoffm.) Vainio = Cladonia gracilis ssp. turbinata [CLADGRA2]
Cladonia gracilis ssp. **elongata** (Jacq.) Vainio [CLADGRA1]
SY: *Cladonia gracilis* ssp. *nigripes* (Nyl.) Ahti
Cladonia gracilis ssp. *nigripes* (Nyl.) Ahti = Cladonia gracilis ssp. elongata [CLADGRA1]
Cladonia gracilis ssp. **turbinata** (Ach.) Ahti [CLADGRA2] (black-foot cladonia)
SY: *Cladonia gracilis* ssp. *dilatata* (Hoffm.) Vainio
Cladonia gracilis ssp. **vulnerata** Ahti [CLADGRA3]
Cladonia gracilis var. *elongata* Fr. = Cladonia maxima [CLADMAX]
Cladonia gracilis var. *verticillata* Fr. = Cladonia cervicornis ssp. verticillata [CLADCER1]
Cladonia granulans Vainio [CLADGRN]
Cladonia grayi G. Merr. ex Sandst. [CLADGRY]
Cladonia groenlandica (E. Dahl) Trass = Cladonia cornuta ssp. groenlandica [CLADCOR1]
Cladonia herrei Fink ex J. Hedrick = Cladonia furcata [CLADFUR]
Cladonia homosekikaica Nuno [CLADHOM]
Cladonia hookeri Tuck. = Cladonia bellidiflora [CLADBEL]
Cladonia humilis (With.) J.R. Laundon [CLADHUM]
SY: *Cladonia conoidea* Ahti
SY: *Cladonia conistea* auct.
Cladonia humilis var. *bourgeanica* A.W. Archer = Cladonia conista [CLADCOI]
Cladonia impexa Harm. = Cladina portentosa [CLADPOR]
Cladonia invisa Robbins = Cladonia ochrochlora [CLADOCH]
Cladonia japonica Vainio = Cladonia crispata [CLADCRI]
Cladonia kanewskii Oksner [CLADKAN]
Cladonia lepidota auct. = Cladonia stricta [CLADSTI]
Cladonia leucophaea Abbayes = Cladina ciliata var. ciliata [CLADCIL0]
Cladonia luteoalba Wheldon & A. Wilson [CLADLUT]
Cladonia macilenta Hoffm. [CLADMAC] (lipstick cladonia)
SY: *Cladonia balfourii* Crombie
Cladonia macilenta var. **bacillaris** (Genth) Schaerer [CLADMAC1]
SY: *Cladonia bacillaris* Nyl.
Cladonia macroceras (Delise) Hav. [CLADMAE]
SY: *Cladonia ecmocyna* var. *macroceras* (Flörke) Ach.
SY: *Cladonia elongata* auct.
Cladonia macrophylla (Schaerer) Stenh. [CLADMAR]
SY: *Cladonia alpicola* (Flotow) Vainio
Cladonia macrophyllodes Nyl. [CLADMAO]
Cladonia macroptera Räsänen = Cladonia scabriuscula [CLADSCA]
Cladonia major (K. Hagen) Sandst. = Cladonia fimbriata [CLADFIM]
Cladonia maxima (Asah.) Ahti [CLADMAX]
SY: *Cladonia elongata* (Jacq.) Hoffm.
SY: *Cladonia gracilis* var. *elongata* Fr.
Cladonia merochlorophaea Asah. [CLADMER]
Cladonia metacorallifera Asah. [CLADMET]
Cladonia mitis Sandst. = Cladina mitis [CLADMIT]
Cladonia multiformis G. Merr. [CLADMUL] (sieve cladonia)
Cladonia nemoxyna (Ach.) Arnold = Cladonia rei [CLADREI]
Cladonia nipponica Asah. [CLADNIP]
Cladonia norvegica Tønsberg & Holien [CLADNOR]
Cladonia novochlorophaea (Sipman) Brodo & Ahti [CLADNOV]
Cladonia ochrochlora Flörke [CLADOCH]
SY: *Cladonia invisa* Robbins
Cladonia pacifica Ahti = Cladina portentosa ssp. pacifica [CLADPOR1]
Cladonia parasitica (Hoffm.) Hoffm. [CLADPAR]
SY: *Cladonia delicata* auct.
Cladonia phyllophora Hoffm. [CLADPHY]
SY: *Cladonia degenerans* (Flörke) Sprengel
Cladonia pleurota (Flörke) Schaerer [CLADPLE]
SY: *Cladonia coccifera* var. *pleurota* (Flörke) Schaerer
Cladonia pocillum (Ach.) Grognot [CLADPOC]
SY: *Cladonia pyxidata* var. *pocillum* (Ach.) Flotow
Cladonia polydactyla (Flörke) Sprengel [CLADPOD]
SY: *Cladonia flabelliformis* Vainio
Cladonia poroscypha S. Hammer [CLADPOO]
Cladonia portentosa ssp. *pacifica* (Ahti) Ahti = Cladina portentosa ssp. pacifica [CLADPOR1]

Cladonia portentosa ssp. *pacifica* f. *decolorans* Ahti = Cladina portentosa ssp. pacifica f. decolorans [CLADPOR2]
Cladonia prolifica Ahti & S. Hammer [CLADPRO]
Cladonia pseudodigitata Gyelnik = Cladonia coccifera [CLADCOC]
Cladonia pseudorangiformis Asah. = Cladonia wainioi [CLADWAI]
Cladonia pseudostellata Asah. = Cladonia uncialis [CLADUNC]
Cladonia pyxidata (L.) Hoffm. [CLADPYX] (pixie-cup lichen)
SY: *Cladonia pyxidata* var. *neglecta* (Flörke) Mass.
Cladonia pyxidata var. *neglecta* (Flörke) Mass. = Cladonia pyxidata [CLADPYX]
Cladonia pyxidata var. *pocillum* (Ach.) Flotow = Cladonia pocillum [CLADPOC]
Cladonia rangiferina (L.) F.H. Wigg. = Cladina rangiferina [CLADRAN]
Cladonia rangiferina f. *nivea* Räsänen = Cladina rangiferina [CLADRAN]
Cladonia rangiferina ssp. *grisea* Ahti = Cladina rangiferina ssp. grisea [CLADRAN1]
Cladonia rangiferina var. *alpestris* Flörke = Cladina stellaris [CLADSTE]
Cladonia rangiferina var. *sylvatica* L. = Cladina arbuscula [CLADARB]
Cladonia rei Schaerer [CLADREI]
SY: *Cladonia nemoxyna* (Ach.) Arnold
Cladonia scabriuscula (Delise) Nyl. [CLADSCA] (shingled cladonia)
SY: *Cladonia macroptera* Räsänen
Cladonia silvatica (L.) Räsänen = Cladina arbuscula [CLADARB]
Cladonia silvatica var. *mitis* (Sandst.)Räsänen = Cladina mitis [CLADMIT]
Cladonia singularis S. Hammer [CLADSIN]
Cladonia squamosa Hoffm. [CLADSQU] (dragon cladonia)
Cladonia squamosa var. *muricella* (Del.) Vainio = Cladonia squamosa var. squamosa [CLADSQU0]
Cladonia squamosa var. **squamosa** [CLADSQU0]
SY: *Cladonia squamosa* var. *muricella* (Del.) Vainio
Cladonia squamosa var. **subsquamosa** (Nyl. ex Leighton) Vainio [CLADSQU1]
SY: *Cladonia subsquamosa* (Nyl. ex Leighton) Crombie
Cladonia stellaris (Opiz) Pouzar & Vezda = Cladina stellaris [CLADSTE]
Cladonia stricta (Nyl.) Nyl. [CLADSTI]
SY: *Cladonia gracilescens* auct.
SY: *Cladonia lepidota* auct.
Cladonia stygia (Fr.) Ruoss = Cladina stygia [CLADSTY]
Cladonia subfurcata (Nyl.) Arnold [CLADSUF]
SY: *Cladonia delessertii* Vainio
Cladonia subrangiformis auct. = Cladonia furcata [CLADFUR]
Cladonia subsquamosa (Nyl. ex Leighton) Crombie = Cladonia squamosa var. subsquamosa [CLADSQU1]
Cladonia subulata (L.) F.H. Wigg. [CLADSUU]
SY: *Cladonia cornutoradiata* (Navás) Sandst.
SY: *Cladonia fimbriata* var. *subulata* Schaerer
Cladonia sulphurina (Michaux) Fr. [CLADSUL] (sulphur cup)
SY: *Cladonia gonecha* (Ach.) Asah.
Cladonia sylvatica L. = Cladina arbuscula [CLADARB]
Cladonia sylvatica auct. = Cladina arbuscula [CLADARB]
Cladonia sylvatica var. *alpestris* L. = Cladina stellaris [CLADSTE]
Cladonia symphycarpa (Flörke) Fr. = Cladonia symphycarpia [CLADSYM]
Cladonia symphycarpia (Flörke) Fr. [CLADSYM]
SY: *Cladonia symphycarpa* (Flörke) Fr.
Cladonia tenuis (Flörke) Harm. = Cladina ciliata [CLADCIL]
Cladonia tenuis var. *leucophaea* (Abbayes) Ahti = Cladina ciliata var. ciliata [CLADCIL0]
Cladonia thomsonii Ahti [CLADTHO]
Cladonia transcendens (Vainio) Vainio [CLADTRA]
Cladonia umbricola Tønsberg & Ahti [CLADUMB]
Cladonia uncialis (L.) F.H. Wigg. [CLADUNC] (prickle cladonia)
SY: *Cladonia pseudostellata* Asah.
SY: *Cladonia uncialis* var. *adunca* Fr.
Cladonia uncialis var. *adunca* Fr. = Cladonia uncialis [CLADUNC]
Cladonia verruculosa (Vainio) Ahti [CLADVER]
Cladonia verticillata (Hoffm.) Schaerer = Cladonia cervicornis ssp. verticillata [CLADCER1]
Cladonia verticillata var. *cervicornis* Flörke = Cladonia cervicornis ssp. cervicornis [CLADCER0]
Cladonia wainii Savicz = Cladonia wainioi [CLADWAI]
Cladonia wainioi Savicz [CLADWAI]
SY: *Cladonia pseudorangiformis* Asah.
SY: *Cladonia wainii* Savicz

CLAVARIACEAE (L)

Multiclavula R. Petersen [MULTICL$]
Multiclavula corynoides (Peck) R. Petersen [MULTCOY]
Multiclavula sharpii R. Petersen [MULTSHA]

COCCOCARPIACEAE (L)

Spilonema Bornet [SPILONE$]
Spilonema revertens Nyl. [SPILREV]

COCCOTREMATACEAE (L)

Coccotrema Müll. Arg. [COCCOTR$]
Coccotrema maritimum Brodo [COCCMAR]
Coccotrema pocillarium (Cummings) Brodo [COCCPOC]
SY: *Ochrolechia pacifica* H. Magn.
SY: *Perforaria minuta* Degel.
Ochrolechia pacifica H. Magn. = Coccotrema pocillarium [COCCPOC]
Perforaria minuta Degel. = Coccotrema pocillarium [COCCPOC]

COLLEMATACEAE (L)

Collema F.H. Wigg. [COLLEMA$]
Collema arcticum Lynge = Collema ceraniscum [COLLCER]
Collema auriculatum Hoffm. = Collema auriforme [COLLAUI]
Collema auriforme (With.) Coppins & J.R. Laundon [COLLAUI] (jelly tarpaper)
SY: *Collema auriculatum* Hoffm.
SY: *Collema granosum* auct.
Collema bachmanianum (Fink) Degel. [COLLBAC] (tar tarpaper)
SY: *Collemodes bachmanianum* Fink
Collema callopismum A. Massal. [COLLCAO] (tripe tarpaper)
Collema callopismum var. **rhyparodes** (Nyl.) Degel. [COLLCAO1]
Collema ceraniscum Nyl. [COLLCER] (cushion tarpaper)
SY: *Collema arcticum* Lynge
Collema cheileum (Ach.) Ach. = Collema crispum [COLLCRI]
Collema coccophorum Tuck. [COLLCOC]
SY: *Collema dubium* de Lesd.
SY: *Collema novomexicanum* de Lesd.
Collema crispum (Hudson) F.H. Wigg. [COLLCRI] (papoose tarpaper)
SY: *Collema cheileum* (Ach.) Ach.
Collema cristatellum Tuck. = Collema tenax [COLLTEN]
Collema cristatum (L.) F.H. Wigg. [COLLCRS] (fingered tarpaper)
Collema cristatum var. **marginale** (Hudson) Degel. [COLLCRS1]
Collema dubium de Lesd. = Collema coccophorum [COLLCOC]
Collema fecundum Degel. [COLLFEC] (seaside tarpaper)
Collema flaccidum (Ach.) Ach. [COLLFLA] (butterfly tarpaper)
SY: *Synechoblastus rupestris* (Sw.) Trevisan
Collema furfuraceum (Arnold) Du Rietz [COLLFUR] (blistered tarpaper)
Collema furfuraceum var. **furfuraceum** [COLLFUR0]
Collema furfuraceum var. **luzonense** (Räsänen) Degel. [COLLFUR1]
SY: *Collema subfurfuraceum* Degel.
Collema furvum (Ach.) Ach. = Collema fuscovirens [COLLFUS]
Collema fuscovirens (With.) J.R. Laundon [COLLFUS] (bleb tarpaper)
SY: *Collema furvum* (Ach.) Ach.
SY: *Collema tuniforme* (Ach.) Ach.

Collema glebulentum (Nyl. ex Crombie) Degel. [COLLGLE] (amphibious tarpaper)
Collema granosum auct. = Collema auriforme [COLLAUI]
Collema multipartitum Sm. [COLLMUL] (protracted tarpaper)
Collema nigrescens (Hudson) DC. [COLLNIG] (broadleaf tarpaper)
Collema novomexicanum de Lesd. = Collema coccophorum [COLLCOC]
Collema polycarpon Hoffm. [COLLPOL] (shaly tarpaper)
SY: *Synechoblastus polycarpus* (Hoffm.) Dalla Torre & Sarnth.
SY: *Synechoblastus wyomingensis* Fink
Collema pulposum (Bernh.) Ach. = Collema tenax [COLLTEN]
Collema subflaccidum Degel. [COLLSUL] (moth tarpaper)
SY: *Collema subfurvum* auct.
Collema subfurfuraceum Degel. = Collema furfuraceum var. luzonense [COLLFUR1]
Collema subfurvum auct. = Collema subflaccidum [COLLSUL]
Collema subparvum Degel. [COLLSUB] (western tarpaper)
Collema tenax (Sw.) Ach. [COLLTEN]
SY: *Collema cristatellum* Tuck.
SY: *Collema pulposum* (Bernh.) Ach.
Collema tenax var. **corallinum** (A. Massal.) Degel. [COLLTEN1]
Collema tenax var. **crustaceum** (Kremp.) Degel. [COLLTEN2]
Collema tenax var. **tenax** [COLLTEN0]
Collema tuniforme (Ach.) Ach. = Collema fuscovirens [COLLFUS]
Collema undulatum Laurer ex Flotow [COLLUND] (protean tarpaper)
Collema undulatum var. **granulosum** Degel. [COLLUND1]
Collemodes bachmanianum Fink = Collema bachmanianum [COLLBAC]
Leciophysma Th. Fr. [LECIOPH$]
Leciophysma finmarkicum Th. Fr. [LECIFIN]
Leciophysma furfurascens (Nyl.) Gyelnik [LECIFUU]
Leptogium (Ach.) Gray [LEPTOGI$]
Leptogium brebissonii Mont. [LEPTBRE] (jellied vinyl)
Leptogium burgessii (L.) Mont. [LEPTBUR] (peppered vinyl)
SY: *Leptogium inflexum* Nyl.
Leptogium burnetiae C. W. Dodge [LEPTBUN]
Leptogium burnetiae var. *hirsutum* (Sierk) P.M. Jørg. = Leptogium hirsutum [LEPTHIR]
Leptogium caesium (Ach.) Vainio = Leptogium cyanescens [LEPTCYA]
Leptogium californicum Tuck. [LEPTCAL]
Leptogium corniculatum (Hoffm.) Minks [LEPTCOI] (antlered vinyl)
SY: *Leptogium palmatum* (Hudson) Mont.
Leptogium cyanescens (Rabenh.) Körber [LEPTCYA] (blue vinyl)
SY: *Leptogium caesium* (Ach.) Vainio
SY: *Leptogium tremelloides* auct.
Leptogium furfuraceum (Harm.) Sierk [LEPTFUR]
SY: *Leptogium papillosum* auct.
Leptogium gelatinosum (With.) J.R. Laundon [LEPTGEL] (petalled vinyl)
SY: *Leptogium sinuatum* (Hudson) A. Massal.
Leptogium hirsutum Sierk [LEPTHIR]
SY: *Leptogium burnetiae* var. *hirsutum* (Sierk) P.M. Jørg.
Leptogium inflexum Nyl. = Leptogium burgessii [LEPTBUR]
Leptogium lichenoides (L.) Zahlbr. [LEPTLIC] (tattered vinyl)
Leptogium lividofuscum (Florke ex Schlecht.) Flotow = Leptogium tenuissimum [LEPTTEN]
Leptogium minutissimum (Flörke) Fr. = Leptogium subtile [LEPTSUB]
Leptogium nanum Herre = Leptogium tenuissimum [LEPTTEN]
Leptogium palmatum (Hudson) Mont. = Leptogium corniculatum [LEPTCOI]
Leptogium papillosum auct. = Leptogium furfuraceum [LEPTFUR]
Leptogium perminutum Fink = Leptogium tenuissimum [LEPTTEN]
Leptogium platynum (Tuck.) Herre [LEPTPLA] (butterfly vinyl)
Leptogium polycarpum P.M. Jørg. & Goward [LEPTPOL] (peacock vinyl)
Leptogium rivale Tuck. [LEPTRIV] (streamside vinyl)
SY: *Polychidium rivale* (Tuck.) Fink
Leptogium saturninum (Dickson) Nyl. [LEPTSAT] (pappered vinyl)
Leptogium schraderi (Ach.) Nyl. [LEPTSCH] (wrinkled vinyl)
Leptogium sinuatum (Hudson) A. Massal. = Leptogium gelatinosum [LEPTGEL]
Leptogium subaridum Jørg. & Goward [LEPTSUA] (pincushion vinyl)
Leptogium subtile (Schrader) Torss. [LEPTSUB] (appressed vinyl)
SY: *Leptogium minutissimum* (Flörke) Fr.
Leptogium tenuissimum (Dickson) Körber [LEPTTEN] (lilliput vinyl)
SY: *Leptogium lividofuscum* (Florke ex Schlecht.) Flotow
SY: *Leptogium nanum* Herre
SY: *Leptogium perminutum* Fink
Leptogium teretiusculum (Wallr.) Arnold [LEPTTER] (shrubby vinyl)
Leptogium tremelloides auct. = Leptogium cyanescens [LEPTCYA]
Polychidium rivale (Tuck.) Fink = Leptogium rivale [LEPTRIV]
Synechoblastus polycarpus (Hoffm.) Dalla Torre & Sarnth. = Collema polycarpon [COLLPOL]
Synechoblastus rupestris (Sw.) Trevisan = Collema flaccidum [COLLFLA]
Synechoblastus wyomingensis Fink = Collema polycarpon [COLLPOL]

CONIOCYBACEAE (L)

Chaenotheca Th. Fr. [CHAENOT$]
Chaenotheca brachypoda (Ach.) Tibell [CHAEBRA]
SY: *Chaenotheca sulphurea* (Retz.) Middleborg & J.-E. Mattsson
SY: *Coniocybe sulphurea* (Retz.) Nyl.
Chaenotheca brunneola (Ach.) Müll. Arg. [CHAEBRU]
SY: *Chaenotheca hygrophila* Tibell
Chaenotheca carthusiae (Harm.) Lettau = Chaenotheca chlorella [CHAECHL]
Chaenotheca chlorella (Ach.) Müll. Arg. [CHAECHL]
SY: *Chaenotheca carthusiae* (Harm.) Lettau
Chaenotheca chrysocephala (Turner ex Ach.) Th. Fr. [CHAECHR]
Chaenotheca cinerea (Pers.) Tibell [CHAECIN]
SY: *Chaenotheca schaereri* (De Not.) Zahlbr.
SY: *Chaenotheca trichialis* var. *cinerea* (Pers.) Blomb. & Forss.
Chaenotheca ferruginea (Turner & Borrer) Mig. [CHAEFER]
SY: *Chaenotheca melanophaea* (Ach.) Zwackh
Chaenotheca furfuracea (L.) Tibell [CHAEFUR] (sulphur stubble)
SY: *Coniocybe furfuracea* (L.) Ach.
Chaenotheca gracillima (Vainio) Tibell [CHAEGRA]
SY: *Coniocybe gracillima* Vainio
Chaenotheca hispidula (Ach.) Zahlbr. [CHAEHIS]
Chaenotheca hygrophila Tibell = Chaenotheca brunneola [CHAEBRU]
Chaenotheca laevigata Nádv. [CHAELAE]
Chaenotheca melanophaea (Ach.) Zwackh = Chaenotheca ferruginea [CHAEFER]
Chaenotheca schaereri (De Not.) Zahlbr. = Chaenotheca cinerea [CHAECIN]
Chaenotheca stemonea (Ach.) Müll. Arg. [CHAESTE]
Chaenotheca subroscida (Eitner) Zahlbr. [CHAESUB]
Chaenotheca sulphurea (Retz.) Middleborg & J.-E. Mattsson = Chaenotheca brachypoda [CHAEBRA]
Chaenotheca trichialis (Ach.) Th. Fr. [CHAETRI]
Chaenotheca trichialis var. *cinerea* (Pers.) Blomb. & Forss. = Chaenotheca cinerea [CHAECIN]
Chaenotheca xyloxena Nádv. [CHAEXYL]
Coniocybe furfuracea (L.) Ach. = Chaenotheca furfuracea [CHAEFUR]
Coniocybe gracillima Vainio = Chaenotheca gracillima [CHAEGRA]
Coniocybe sulphurea (Retz.) Nyl. = Chaenotheca brachypoda [CHAEBRA]
Sclerophora Chevall. [SCLEROI$]
Sclerophora amabilis (Tibell) Tibell [SCLEAMA]
Sclerophora coniophaea (Norman) J. Mattsson & Middelb. [SCLECON]
Sclerophora farinacea (Chevall.) Chevall. [SCLEFAR]
Sclerophora peronella (Ach.) Tibell [SCLEPER]

DACAMPIACEAE (L)

Clypeococcum D. Hawksw. [CLYPEOC$]
Clypeococcum grossum (Körber) D. Hawksw. [CLYPGRO]
Clypeococcum hypocenomycis D. Hawksw. [CLYPHYP]

DACTYLOSPORACEAE (L)

Buelliella nuttallii (Calk. & Nyl.) Fink = Dactylospora lobariella [DACTLOB]
Dactylospora Körber [DACTYLO$]
Dactylospora lobariella (Nyl.) Hafellner [DACTLOB]
SY: *Buelliella nuttallii* (Calk. & Nyl.) Fink

ECTOLECHIACEAE (L)

Lopadium Körber [LOPADIU$]
Lopadium disciforme (Flotow) Kullhem [LOPADIS]
SY: *Lopadium pezizoideum* var. *disciforme* (Flotow) Körber
Lopadium pezizoideum (Ach.) Körber [LOPAPEZ]
Lopadium pezizoideum var. *disciforme* (Flotow) Körber = Lopadium disciforme [LOPADIS]

FUSCIDEACEAE (L)

Buellia atrata (Sm.) Anzi = Orphniospora moriopsis [ORPHMOR]
Buellia coracina (Nyl.) Körber = Orphniospora moriopsis [ORPHMOR]
Buellia moriopsis (A. Massal.) Th. Fr. = Orphniospora moriopsis [ORPHMOR]
Fuscidea V. Wirth & Vezda [FUSCIDE$]
Fuscidea lightfootii (Sm.) Coppins & P. James [FUSCLIG]
Fuscidea mollis (Wahlenb.) V. Wirth & Vezda [FUSCMOL]
SY: *Lecidea mollis* (Wahlenb.) Nyl.
Fuscidea viridis Tønsberg = Ropalospora viridis [ROPAVIR]
Lecidea mollis (Wahlenb.) Nyl. = Fuscidea mollis [FUSCMOL]
Orphniospora Körber [ORPHNIO$]
Orphniospora atrata (Sm.) Poelt = Orphniospora moriopsis [ORPHMOR]
Orphniospora moriopsis (A. Massal.) D. Hawksw. [ORPHMOR]
SY: *Buellia atrata* (Sm.) Anzi
SY: *Buellia coracina* (Nyl.) Körber
SY: *Buellia moriopsis* (A. Massal.) Th. Fr.
SY: *Orphniospora atrata* (Sm.) Poelt
Ropalospora A. Massal. [ROPALOS$]
Ropalospora viridis (Tønsberg) Tønsberg [ROPAVIR]
SY: *Fuscidea viridis* Tønsberg

GOMPHILLACEAE (L)

Gyalideopsis Vezda [GYALIDO$]
Gyalideopsis alnicola W. Noble & Vezda = Gyalideopsis piceicola [GYALPIC]
Gyalideopsis anastomosans P. James & Vezda [GYALANA]
Gyalideopsis piceicola (Nyl.) Vezda [GYALPIC]
SY: *Gyalideopsis alnicola* W. Noble & Vezda

GRAPHIDACEAE (L)

Graphina Müll. Arg. [GRAPHIN$]
Graphina anguina (Mont.) Müll. Arg. [GRAPANG]
Graphis Adans. [GRAPHIS$]
Graphis elegans (Borrer ex Sm.) Ach. [GRAPELE]
Graphis scripta (L.) Ach. [GRAPSCR] (pencil scirpt)

GYALECTACEAE (L)

Dimerella Trevisan [DIMEREL$]
Dimerella diluta (Pers.) Trevisan = Dimerella pineti [DIMEPIN]
Dimerella lutea (Dickson) Trevisan [DIMELUT]
Dimerella pineti (Ach.) Vezda [DIMEPIN]
SY: *Dimerella diluta* (Pers.) Trevisan
Gyalecta Ach. [GYALECT$]
Gyalecta cupularis (Hedw.) Schaerer = Gyalecta jenensis [GYALJEN]
Gyalecta friesii Flotow ex Körber [GYALFRI]
Gyalecta jenensis (Batsch) Zahlbr. [GYALJEN]
SY: *Gyalecta cupularis* (Hedw.) Schaerer

GYPSOPLACACEAE (L)

Gypsoplaca Timdal [GYPSOPL$]
Gypsoplaca macrophylla (Zahlbr.) Timdal [GYPSMAC]

HAEMATOMMATACEAE (L)

Haematomma elatinum (Ach.) A. Massal. = Loxospora elatina [LOXOELA]
Haematomma ochrophaeum (Tuck.) A. Massal. = Loxospora ochrophaea [LOXOOCH]
Loxospora A. Massal. [LOXOSPO$]
Loxospora elatina (Ach.) A. Massal. [LOXOELA]
SY: *Haematomma elatinum* (Ach.) A. Massal.
Loxospora ochrophaea (Tuck.) R.C. Harris [LOXOOCH]
SY: *Haematomma ochrophaeum* (Tuck.) A. Massal.

HEPPIACEAE (L)

Heppia Nägeli [HEPPIA$]
Heppia despreauxii (Mont.) Tuck. = Heppia lutosa [HEPPLUT]
Heppia lutosa (Ach.) Nyl. [HEPPLUT] (soil ruby)
SY: *Heppia despreauxii* (Mont.) Tuck.
SY: *Heppia virescens* (Despr.) Nyl.
SY: *Solorinaria despreauxii* (Mont.) Fink
Heppia virescens (Despr.) Nyl. = Heppia lutosa [HEPPLUT]
Solorinaria despreauxii (Mont.) Fink = Heppia lutosa [HEPPLUT]

HYMENELIACEAE (L)

Agrestia J.W. Thomson [AGRESTI$]
Agrestia cyphellata J.W. Thomson = Agrestia hispida [AGREHIS]
Agrestia hispida (Mereschk.) Hale & Culb. [AGREHIS]
SY: *Agrestia cyphellata* J.W. Thomson
Aspicilia A. Massal. [ASPICIL$]
Aspicilia alphoplaca (Wahlenb.) Poelt & Leuckert = Lobothallia alphoplaca [LOBOALP]
Aspicilia anseris (Lynge) J.W. Thomson [ASPIANS]
SY: *Lecanora anseris* Lynge
Aspicilia aquatica Körber [ASPIAQU]
SY: *Lecanora aquatica* (Körber) Hepp
Aspicilia caesiocinerea (Nyl. ex Malbr.) Arnold [ASPICAE]
SY: *Lecanora caesiocinerea* Nyl. ex Malbr.
Aspicilia caesiopruinosa (H. Magn.) J.W. Thomson [ASPICAS]
SY: *Lecanora caesiopruinosa* H. Magn.
Aspicilia calcarea (L.) Mudd [ASPICAL]
SY: *Lecanora calcarea* (L.) Sommerf.
Aspicilia candida (Anzi) Hue [ASPICAN]
SY: *Lecanora candida* (Anzi) Nyl.
Aspicilia cinerea (L.) Körber [ASPICIN]
SY: *Lecanora cinerea* (L.) Sommerf.
Aspicilia contorta (Hoffm.) Kremp. [ASPICON]
SY: *Lecanora contorta* (Hoffm.) J. Steiner
Aspicilia desertorum (Kremp.) Mereschk. [ASPIDES]
SY: *Lecanora desertorum* Kremp.
Aspicilia disserpens (Zahlbr.) Räsänen [ASPIDIS]
SY: *Lecanora disserpens* (Zahlbr.) H. Magn.
Aspicilia fimbriata (H. Magn.) Clauzade & Rondon [ASPIFIM]
SY: *Lecanora fimbriata* H. Magn.
Aspicilia gibbosa (Ach.) Körber [ASPIGIB]
SY: *Aspicilia gibbosula* (H. Magn.) Oksner
SY: *Lecanora gibbosa* (Ach.) Nyl.

SY: *Lecanora gibbosula* H. Magn.
Aspicilia gibbosula (H. Magn.) Oksner = Aspicilia gibbosa [ASPIGIB]
Aspicilia karelica (H. Magn.) Oksner [ASPIKAR]
SY: *Lecanora karelica* H. Magn.
Aspicilia lacustris (With.) Th. Fr. = Ionaspis lacustris [IONALAC]
Aspicilia laxula (H. Magn.) Brodo [ASPILAX]
SY: *Lecanora laxula* H. Magn.
Aspicilia leprosescens (Sandst.) Hav. [ASPILEP]
SY: *Lecanora leprosescens* Sandst.
Aspicilia melanaspis (Ach.) Poelt & Leuckert = Lobothallia melanaspis [LOBOMEL]
Aspicilia myrinii (Fr.) Stein [ASPIMYR]
SY: *Lecanora myrinii* (Fr.) Tuck.
Aspicilia praeradiosa (Nyl.) Poelt & Leuckert = Lobothallia praeradiosa [LOBOPRA]
Aspicilia reptans (Looman) Wetmore [ASPIREP]
SY: *Lecanora reptans* Looman
Aspicilia ryrkaipiae (H. Magn.) Oksner [ASPIRYR]
SY: *Lecanora ryrkaipiae* H. Magn.
Aspicilia subplicigera (H. Magn.) Oksner [ASPISUB]
Aspicilia supertegens Arnold [ASPISUP]
SY: *Lecanora supertegens* (Arnold) Zahlbr.
Aspicilia verrucigera Hue [ASPIVER]
SY: *Lecanora verrucigera* (Hue) Zahlbr.
Aspicilia verrucosa (Ach.) Körker = Megaspora verrucosa [MEGAVEU]
Hymenelia Kremp. [HYMENEL$]
Hymenelia epulotica (Ach.) Lutzoni [HYMEEPU]
SY: *Hymenelia prevostii* (Duby) Kremp.
SY: *Ionaspis epulotica* (Ach.) Blomb. & Forssell
SY: *Lecanora epulotica* (Ach.) Nyl.
Hymenelia heteromorpha (Kremp.) Lutzoni [HYMEHET]
SY: *Ionaspis heteromorpha* (Kremp.) Arnold
SY: *Ionaspis ochracella* (Nyl.) H. Magn.
SY: *Ionaspis reducta* H. Magn.
SY: *Ionaspis schismatopsis* (Nyl.) Hue
Hymenelia lacustris (With.) Choisy = Ionaspis lacustris [IONALAC]
Hymenelia melanocarpa (Kremp.) Arnold [HYMEMEL]
SY: *Ionaspis melanocarpa* (Kremp.) Arnold
Hymenelia prevostii (Duby) Kremp. = Hymenelia epulotica [HYMEEPU]
Ionaspis Th. Fr. [IONASPI$]
Ionaspis epulotica (Ach.) Blomb. & Forssell = Hymenelia epulotica [HYMEEPU]
Ionaspis heteromorpha (Kremp.) Arnold = Hymenelia heteromorpha [HYMEHET]
Ionaspis lacustris (With.) Lutzoni [IONALAC]
SY: *Aspicilia lacustris* (With.) Th. Fr.
SY: *Hymenelia lacustris* (With.) Choisy
SY: *Lecanora lacustris* (With.) Nyl.
Ionaspis melanocarpa (Kremp.) Arnold = Hymenelia melanocarpa [HYMEMEL]
Ionaspis ochracella (Nyl.) H. Magn. = Hymenelia heteromorpha [HYMEHET]
Ionaspis odora (Ach.) Th. Fr. ex Stein [IONAODO]
Ionaspis reducta H. Magn. = Hymenelia heteromorpha [HYMEHET]
Ionaspis schismatopsis (Nyl.) Hue = Hymenelia heteromorpha [HYMEHET]
Lecanora alphoplaca (Wahlenb.) Ach. = Lobothallia alphoplaca [LOBOALP]
Lecanora anseris Lynge = Aspicilia anseris [ASPIANS]
Lecanora aquatica (Körber) Hepp = Aspicilia aquatica [ASPIAQU]
Lecanora caesiocinerea Nyl. ex Malbr. = Aspicilia caesiocinerea [ASPICAE]
Lecanora caesiopruinosa H. Magn. = Aspicilia caesiopruinosa [ASPICAS]
Lecanora calcarea (L.) Sommerf. = Aspicilia calcarea [ASPICAL]
Lecanora candida (Anzi) Nyl. = Aspicilia candida [ASPICAN]
Lecanora cinerea (L.) Sommerf. = Aspicilia cinerea [ASPICIN]
Lecanora contorta (Hoffm.) J. Steiner = Aspicilia contorta [ASPICON]
Lecanora desertorum Kremp. = Aspicilia desertorum [ASPIDES]
Lecanora disserpens (Zahlbr.) H. Magn. = Aspicilia disserpens [ASPIDIS]
Lecanora epulotica (Ach.) Nyl. = Hymenelia epulotica [HYMEEPU]
Lecanora fimbriata H. Magn. = Aspicilia fimbriata [ASPIFIM]
Lecanora gibbosa (Ach.) Nyl. = Aspicilia gibbosa [ASPIGIB]
Lecanora gibbosula H. Magn. = Aspicilia gibbosa [ASPIGIB]
Lecanora karelica H. Magn. = Aspicilia karelica [ASPIKAR]
Lecanora lacustris (With.) Nyl. = Ionaspis lacustris [IONALAC]
Lecanora laxula H. Magn. = Aspicilia laxula [ASPILAX]
Lecanora leprosescens Sandst. = Aspicilia leprosescens [ASPILEP]
Lecanora melanaspis (Ach.) Ach. = Lobothallia melanaspis [LOBOMEL]
Lecanora mutabilis (Ach.) Nyl. = Megaspora verrucosa [MEGAVEU]
Lecanora myrinii (Fr.) Tuck. = Aspicilia myrinii [ASPIMYR]
Lecanora praeradiosa Nyl. = Lobothallia praeradiosa [LOBOPRA]
Lecanora reptans Looman = Aspicilia reptans [ASPIREP]
Lecanora ryrkaipiae H. Magn. = Aspicilia ryrkaipiae [ASPIRYR]
Lecanora supertegens (Arnold) Zahlbr. = Aspicilia supertegens [ASPISUP]
Lecanora thamnoplaca Tuck. = Lobothallia alphoplaca [LOBOALP]
Lecanora urceolata (Fr.) Wetmore = Megaspora verrucosa [MEGAVEU]
Lecanora verrucigera (Hue) Zahlbr. = Aspicilia verrucigera [ASPIVER]
Lecanora verrucosa (Ach.) Laurer = Megaspora verrucosa [MEGAVEU]
Lecidea atrata (Ach.) Wahlenb. = Tremolecia atrata [TREMATR]
Lecidea dicksonii auct. = Tremolecia atrata [TREMATR]
Lobothallia (Clauzade & Roux) Hafellner [LOBOTHA$]
Lobothallia alphoplaca (Wahlenb.) Hafellner [LOBOALP]
SY: *Aspicilia alphoplaca* (Wahlenb.) Poelt & Leuckert
SY: *Lecanora alphoplaca* (Wahlenb.) Ach.
SY: *Lecanora thamnoplaca* Tuck.
Lobothallia melanaspis (Ach.) Hafellner [LOBOMEL]
SY: *Aspicilia melanaspis* (Ach.) Poelt & Leuckert
SY: *Lecanora melanaspis* (Ach.) Ach.
Lobothallia praeradiosa (Nyl.) Hafellner [LOBOPRA]
SY: *Aspicilia praeradiosa* (Nyl.) Poelt & Leuckert
SY: *Lecanora praeradiosa* Nyl.
Megaspora (Clauz. & Roux) Hafellner & V. Wirth [MEGASPO$]
Megaspora verrucosa (Ach.) Hafellner & V. Wirth [MEGAVEU]
SY: *Aspicilia verrucosa* (Ach.) Körker
SY: *Lecanora mutabilis* (Ach.) Nyl.
SY: *Lecanora urceolata* (Fr.) Wetmore
SY: *Lecanora verrucosa* (Ach.) Laurer
SY: *Pachyospora mutabilis* (Ach.) A. Massal.
SY: *Pachyospora verrucosa* (Ach.) A. Massal.
SY: *Pertusaria freyi* Erichsen
Pachyospora mutabilis (Ach.) A. Massal. = Megaspora verrucosa [MEGAVEU]
Pachyospora verrucosa (Ach.) A. Massal. = Megaspora verrucosa [MEGAVEU]
Pertusaria freyi Erichsen = Megaspora verrucosa [MEGAVEU]
Tremolecia Choisy [TREMOLE$]
Tremolecia atrata (Ach.) Hertel [TREMATR]
SY: *Lecidea atrata* (Ach.) Wahlenb.
SY: *Lecidea dicksonii* auct.

LECANORACEAE (L)

Abrothallus oxysporus Tul. = Phacopsis oxyspora [PHACOXY]
Bacidia arnoldiana Körber = Bacidina arnoldiana [BACIARN]
Bacidia arthoniza (Nyl.) Zahlbr. = Lecidella stigmatea [LECISTI]
SY: *Bacidia effusa* (Auersw. ex Rabenh.) Lettau
Bacidia chlorococca (Stenh.) Lettau = Scoliciosporum chlorococcum [SCOLCHL]
Bacidia effusa (Auersw. ex Rabenh.) Lettau = Bacidia auerswaldii [BACIAUE]
Bacidia egenula (Nyl.) Arnold = Bacidina egenula [BACIEGE]
Bacidia inundata (Fr.) Körber = Bacidina inundata [BACIINU]
Bacidia umbrina (Ach.) Bausch = Scoliciosporum umbrinum [SCOLUMB]
Bacidina Vezda [BACIDIN$]

Bacidina arnoldiana (Körber) Wirth & Vezda [BACIARN]
SY: *Bacidia arnoldiana* Körber
SY: *Woessia fusarioides* D. Hawksw.
Bacidina egenula (Nyl.) Vezda [BACIEGE]
SY: *Bacidia egenula* (Nyl.) Arnold
Bacidina inundata (Fr.) Vezda [BACIINU]
SY: *Bacidia inundata* (Fr.) Körber
Bacidina ramea S. Ekman [BACIRAM]
Bryonora Poelt [BRYONOA$]
Bryonora castanea (Hepp) Poelt [BRYOCAS]
SY: *Lecanora castanea* (Hepp) Th. Fr.
Bryonora curvescens (Mudd) Poelt [BRYOCUR]
SY: *Lecania curvescens* (Mudd) A.L. Sm.
SY: *Lecanora curvescens* (Mudd) Nyl.
Bryonora pruinosa (Th. Fr.) Holtan-Hartwig [BRYOPRU]
Carbonea (Hertel) Hertel [CARBONE$]
Carbonea aggregantula (Müll. Arg.) Diederich & Triebel [CARBAGG]
Carbonea vorticosa (Flörke) Hertel [CARBVOR]
SY: *Lecidea vorticosa* (Flörke) Körber
Catillaria bahusiensis (Blomb.) Th. Fr. = Tylothallia biformigera [TYLOBIF]
Catillaria biformigera (Leighton) H. Magn. = Tylothallia biformigera [TYLOBIF]
Glaucomaria rupicola (L.) Choisy = Lecanora rupicola [LECARUP]
Lecania curvescens (Mudd) A.L. Sm. = Bryonora curvescens [BRYOCUR]
Lecanora Ach. [LECANOR$]
Lecanora albescens (Hoffm.) Branth & Rostrup [LECAALE]
Lecanora allophana Nyl. [LECAALL]
SY: *Lecanora subfusca* (L.) Ach.
Lecanora argopholis (Ach.) Ach. [LECAARO]
SY: *Lecanora frustulosa* (Dickson) Ach.
SY: *Lecanora occidentalis* (Lynge) Lynge
SY: *Lecanora oregana* Tuck.
Lecanora atrynea (Ach.) Röhl. = Lecanora cenisia [LECACEN]
Lecanora badia (Hoffm.) Ach. = Protoparmelia badia [PROTBAD]
Lecanora beringii Nyl. [LECABER]
Lecanora boligera (Norman ex Th. Fr.) Hedl. [LECABOL]
Lecanora cadubriae (A. Massal.) Hedl. [LECACAD]
SY: *Lecidea cadubriae* (A. Massal.) Nyl.
SY: *Lecidea ramulicola* H. Magn.
Lecanora cascadensis H. Magn. = Lecanora garovaglii [LECAGAR]
Lecanora castanea (Hepp) Th. Fr. = Bryonora castanea [BRYOCAS]
Lecanora cateilea (Ach.) A. Massal. [LECACTE]
Lecanora cenisia Ach. [LECACEN]
SY: *Lecanora atrynea* (Ach.) Röhl.
Lecanora chlarona (Ach.) Nyl. = Lecanora pulicaris [LECAPUL]
Lecanora chlarotera Nyl. [LECACHA]
Lecanora chloropolia (Erichsen) Almb. = Lecanora impudens [LECAIMP]
Lecanora christoi W.A. Weber = Lecanora phaedrophthalma [LECAPHA]
Lecanora chrysoleuca (Sm.) Ach. = Rhizoplaca chrysoleuca [RHIZCHR]
Lecanora cinereofusca H. Magn. [LECACIN]
Lecanora circumborealis Brodo & Vitik. [LECACIC] (rim lichen)
SY: *Lecanora coilocarpa* auct.
Lecanora coilocarpa (Ach.) Nyl. = Lecanora pulicaris [LECAPUL]
Lecanora coilocarpa auct. = Lecanora circumborealis [LECACIC]
Lecanora confusa Almb. [LECACOF]
Lecanora congesta Lynge [LECACON]
Lecanora conizaea (Ach.) Nyl. ex Crombie = Lecanora expallens [LECAEXP]
Lècanora conizaea auct. = Lecanora strobilina [LECASTO]
Lecanora conizaeoides Nyl. ex Crombie [LECACOA]
Lecanora contractula Nyl. [LECACOR]
Lecanora curvescens (Mudd) Nyl. = Bryonora curvescens [BRYOCUR]
Lecanora demissa (Flotow) Zahlbr. [LECADEM]
SY: *Lecanora incusa* (Fr.) Vainio
SY: *Lecanora subolivascens* Nyl.
Lecanora diffracta Ach. = Lecanora muralis [LECAMUR]
Lecanora dispersa (Pers.) Sommerf. [LECADIP]
Lecanora effusa (Hoffm.) Ach. = Lecanora saligna [LECASAI]
Lecanora epanora (Ach.) Ach. [LECAEPA]
Lecanora epibryon (Ach.) Ach. [LECAEPI]
Lecanora expallens Ach. [LECAEXP]
SY: *Lecanora conizaea* (Ach.) Nyl. ex Crombie
Lecanora eyerdamii Herre = Lecanora xylophila [LECAXYL]
Lecanora frustulosa (Dickson) Ach. = Lecanora argopholis [LECAARO]
Lecanora fuscescens (Sommerf.) Nyl. [LECAFUE]
SY: *Lecidea fuscescens* Sommerf.
Lecanora garovaglii (Körber) Zahlbr. [LECAGAR]
SY: *Lecanora cascadensis* H. Magn.
SY: *Lecanora nevadensis* H. Magn.
Lecanora grandis H. Magn. = Protoparmelia badia [PROTBAD]
Lecanora grantii H. Magn. = Lecanora xylophila [LECAXYL]
Lecanora hagenii (Ach.) Ach. [LECAHAG]
Lecanora helicopis (Wahlenb.) Ach. [LECAHEL]
Lecanora hypopta (Ach.) Vainio [LECAHYA]
SY: *Lecidea hypopta* Ach.
Lecanora impudens Degel. [LECAIMP]
SY: *Lecanora chloropolia* (Erichsen) Almb.
SY: *Lecanora variolascens* auct.
Lecanora incusa (Fr.) Vainio = Lecanora demissa [LECADEM]
Lecanora intricata (Ach.) Ach. [LECAINT]
SY: *Lecanora mutabilis* Sommerf.
Lecanora intumescens (Rebent.) Rabenh. [LECAINU]
Lecanora luteovernalis Brodo [LECALUT]
Lecanora marginata (Schaerer) Hertel & Rambold [LECAMAG]
SY: *Lecidea elata* Schaerer
SY: *Lecidea marginata* Schaerer
SY: *Lecidea purissima* Darbish.
Lecanora melanophthalma (DC.) Ramond = Rhizoplaca melanophthalma [RHIZMEL]
Lecanora mughicola Nyl. [LECAMUG]
Lecanora muralis (Schreber) Rabenh. [LECAMUR]
SY: *Lecanora diffracta* Ach.
SY: *Protoparmeliopsis muralis* (Schreber) Choisy
Lecanora mutabilis Sommerf. = Lecanora intricata [LECAINT]
Lecanora nevadensis H. Magn. = Lecanora garovaglii [LECAGAR]
Lecanora occidentalis (Lynge) Lynge = Lecanora argopholis [LECAARO]
Lecanora ochrococca (Nyl.) Clauzade & Roux = Protoparmelia ochrococca [PROTOCH]
Lecanora orae-frigidae R. Sant. [LECAORA]
SY: *Lecidea sorediata* Lynge
Lecanora oregana Tuck. = Lecanora argopholis [LECAARO]
Lecanora pacifica Tuck. [LECAPAI]
SY: *Lecanora tetraspora* H. Magn.
Lecanora palanderi Vainio = Lecanora zosterae [LECAZOS]
Lecanora peltata (Ramond) Steudel = Rhizoplaca peltata [RHIZPEL]
Lecanora persimilis (Th. Fr.) Nyl. [LECAPER]
Lecanora phaedrophthalma Poelt [LECAPHA]
SY: *Lecanora christoi* W.A. Weber
Lecanora phaeobola Tuck. = Protoparmelia ochrococca [PROTOCH]
Lecanora pinastri (Schaerer) H. Magn. = Lecanora pulicaris [LECAPUL]
Lecanora polytropa (Hoffm.) Rabenh. [LECAPOY]
Lecanora pringlei (Tuck.) Lamb [LECAPRI]
SY: *Lecidea pringlei* Tuck.
Lecanora pulicaris (Pers.) Ach. [LECAPUL]
SY: *Lecanora chlarona* (Ach.) Nyl.
SY: *Lecanora coilocarpa* (Ach.) Nyl.
SY: *Lecanora pinastri* (Schaerer) H. Magn.
Lecanora reagens Norman [LECAREA]
Lecanora riparia G. Merr. = Lecanora xylophila [LECAXYL]
Lecanora rubina (Vill.) Ach. = Rhizoplaca chrysoleuca [RHIZCHR]
Lecanora rugosella Zahlbr. [LECARUG]
Lecanora rupicola (L.) Zahlbr. [LECARUP]
SY: *Glaucomaria rupicola* (L.) Choisy
SY: *Lecanora sordida* (Pers.) Th. Fr.
Lecanora salicicola H. Magn. [LECASAL]
Lecanora saligna (Schrader) Zahlbr. [LECASAI]

SY: *Lecanora effusa* (Hoffm.) Ach.
SY: *Lecanoropsis saligna* (Schrader) Choisy
Lecanora sambuci (Pers.) Nyl. [LECASAM]
Lecanora scotopholis (Tuck.) Timdal [LECASCO]
SY: *Lecidea scotopholis* (Tuck.) Herre
SY: *Psora scotopholia* (Tuck.) Fink
SY: *Psorula scotopholis* (Tuck.) Gotth. Schneider
Lecanora semitensis Tuck. [LECASEM]
Lecanora sordida (Pers.) Th. Fr. = Lecanora rupicola [LECARUP]
Lecanora straminea Ach. [LECASTR]
Lecanora strobilina (Sprengel) Kieffer [LECASTO]
SY: *Lecanora conizaea* auct.
Lecanora subfusca (L.) Ach. = Lecanora allophana [LECAALL]
Lecanora subintricata (Nyl.) Th. Fr. [LECASUI]
Lecanora subolivascens Nyl. = Lecanora demissa [LECADEM]
Lecanora symmicta (Ach.) Ach. [LECASYM]
SY: *Lecanora symmictera* Nyl.
SY: *Lecidea symmicta* (Ach.) Ach.
Lecanora symmictera Nyl. = Lecanora symmicta [LECASYM]
Lecanora tetraspora H. Magn. = Lecanora pacifica [LECAPAI]
Lecanora torrida Vainio [LECATOR]
Lecanora varia (Hoffm.) Ach. [LECAVAR]
Lecanora variolascens auct. = Lecanora impudens [LECAIMP]
Lecanora weberi B.D. Ryan [LECAWEB]
Lecanora wisconsinensis H. Magn. [LECAWIS]
Lecanora xylophila Hue [LECAXYL]
SY: *Lecanora eyerdamii* Herre
SY: *Lecanora grantii* H. Magn.
SY: *Lecanora riparia* G. Merr.
Lecanora zosterae (Ach.) Nyl. [LECAZOS]
SY: *Lecanora palanderi* Vainio
Lecanoropsis saligna (Schrader) Choisy = Lecanora saligna [LECASAI]
Lecidea acrocyanea (Th. Fr.) H. Magn. = Lecidella patavina [LECIPAA]
Lecidea alaiensis Vainio = Lecidella patavina [LECIPAA]
Lecidea brandegei Tuck. = Rhizoplaca melanophthalma [RHIZMEL]
Lecidea cadubriae (A. Massal.) Nyl. = Lecanora cadubriae [LECACAD]
Lecidea carpathica (Körber) Szat. = Lecidella carpathica [LECICAA]
Lecidea catalinaria Stizenb. = Lecidella asema [LECIASE]
Lecidea cinnabarina Sommerf. = Pyrrhospora cinnabarina [PYRRCIN]
Lecidea elabens Fr. = Pyrrhospora elabens [PYRRELA]
Lecidea elaeochroma (Ach.) Ach. = Lecidella elaeochroma [LECIELO]
Lecidea elata Schaerer = Lecanora marginata [LECAMAG]
Lecidea endolithea Lynge = Lecidella patavina [LECIPAA]
Lecidea euphorea (Flörke) Nyl. = Lecidella euphorea [LECIEUP]
Lecidea evansii H. Magn. = Lecidella carpathica [LECICAA]
Lecidea fuscescens Sommerf. = Lecanora fuscescens [LECAFUE]
Lecidea glomerulosa (DC.) Steudel = Lecidella euphorea [LECIEUP]
Lecidea heppii R. Anderson & W.A. Weber = Lecidella wulfenii [LECIWUL]
Lecidea hypopta Ach. = Lecanora hypopta [LECAHYA]
Lecidea lacus-crateris H. Magn. = Lecidella stigmatea [LECISTI]
Lecidea latypea auct. = Lecidella carpathica [LECICAA]
Lecidea latypiza Nyl. = Lecidella carpathica [LECICAA]
Lecidea limitata auct. = Lecidella elaeochroma [LECIELO]
Lecidea marginata Schaerer = Lecanora marginata [LECAMAG]
Lecidea melancheima Tuck. = Pyrrhospora elabens [PYRRELA]
Lecidea micacea Körber = Lecidella stigmatea [LECISTI]
Lecidea ochrococca Nyl. = Protoparmelia ochrococca [PROTOCH]
Lecidea olivacea (Hoffm.) A. Massal. = Lecidella elaeochroma [LECIELO]
Lecidea oxyspora (Tul.) Nyl. = Phacopsis oxyspora [PHACOXY]
Lecidea prasinula (Wedd.) de Lesd. = Lecidella scabra [LECISCB]
Lecidea pringlei Tuck. = Lecanora pringlei [LECAPRI]
Lecidea purissima Darbish. = Lecanora marginata [LECAMAG]
Lecidea quernea (Dickson) Ach. = Pyrrhospora quernea [PYRRQUE]
Lecidea ramulicola H. Magn. = Lecanora cadubriae [LECACAD]
Lecidea russula Ach. = Pyrrhospora russula [PYRRRUS]
Lecidea scabra Taylor = Lecidella scabra [LECISCB]
Lecidea scotopholis (Tuck.) Herre = Lecanora scotopholis [LECASCO]
Lecidea sorediata Lynge = Lecanora orae-frigidae [LECAORA]
Lecidea stigmatea Ach. = Lecidella stigmatea [LECISTI]
Lecidea subcinnabarina Tønsberg = Pyrrhospora subcinnabarina [PYRRSUB]
Lecidea subcontinuior de Lesd. = Lecidella carpathica [LECICAA]
Lecidea symmicta (Ach.) Ach. = Lecanora symmicta [LECASYM]
Lecidea varians Ach. = Pyrrhospora varians [PYRRVAR]
Lecidea vorticosa (Flörke) Körber = Carbonea vorticosa [CARBVOR]
Lecidea vulgata Zahlbr. = Lecidella stigmatea [LECISTI]
Lecidea wulfenii (Hepp) Arnold = Lecidella wulfenii [LECIWUL]
Lecidella Körber [LECIDLL$]
Lecidella alaiensis (Vainio) Hertel = Lecidella patavina [LECIPAA]
Lecidella asema (Nyl.) Knoph & Hertel [LECIASE]
SY: *Lecidea catalinaria* Stizenb.
SY: *Lecidella elaeochromoides* (Nyl.) Knoph & Hertel
SY: *Lecidella subincongrua* (Nyl.) Hertel & Leuckert
SY: *Lecidella subincongrua* var. *elaeochromoides* (Nyl.) Hertel & Leuckert
Lecidella carpathica Körber [LECICAA]
SY: *Lecidea carpathica* (Körber) Szat.
SY: *Lecidea evansii* H. Magn.
SY: *Lecidea latypiza* Nyl.
SY: *Lecidea subcontinuior* de Lesd.
SY: *Lecidea latypea* auct.
Lecidella elaeochroma (Ach.) Hazsl. [LECIELO]
SY: *Lecidea elaeochroma* (Ach.) Ach.
SY: *Lecidea olivacea* (Hoffm.) A. Massal.
SY: *Lecidea limitata* auct.
Lecidella elaeochromoides (Nyl.) Knoph & Hertel = Lecidella asema [LECIASE]
Lecidella euphorea (Flörke) Hertel [LECIEUP]
SY: *Lecidea euphorea* (Flörke) Nyl.
SY: *Lecidea glomerulosa* (DC.) Steudel
SY: *Lecidella glomerulosa* (DC.) Choisy
Lecidella glomerulosa (DC.) Choisy = Lecidella euphorea [LECIEUP]
Lecidella inamoena (Müll. Arg.) Hertel = Lecidella patavina [LECIPAA]
Lecidella laureri (Hepp) Körber [LECILAU]
Lecidella patavina (A. Massal.) Knoph & Leuckert [LECIPAA]
SY: *Lecidea acrocyanea* (Th. Fr.) H. Magn.
SY: *Lecidea alaiensis* Vainio
SY: *Lecidea endolithea* Lynge
SY: *Lecidella alaiensis* (Vainio) Hertel
SY: *Lecidella inamoena* (Müll. Arg.) Hertel
SY: *Lecidella spitsbergensis* (Lynge) Hertel & Leuckert
Lecidella prasinula (Wedd.) Hertel = Lecidella scabra [LECISCB]
Lecidella scabra (Taylor) Hertel & Leuckert [LECISCB]
SY: *Lecidea prasinula* (Wedd.) de Lesd.
SY: *Lecidea scabra* Taylor
SY: *Lecidella prasinula* (Wedd.) Hertel
Lecidella spitsbergensis (Lynge) Hertel & Leuckert = Lecidella patavina [LECIPAA]
Lecidella stigmatea (Ach.) Hertel & Leuckert [LECISTI]
SY: *Bacidia arthoniza* (Nyl.) Zahlbr.
SY: *Lecidea lacus-crateris* H. Magn.
SY: *Lecidea micacea* Körber
SY: *Lecidea stigmatea* Ach.
SY: *Lecidea vulgata* Zahlbr.
Lecidella subincongrua (Nyl.) Hertel & Leuckert = Lecidella asema [LECIASE]
Lecidella subincongrua var. *elaeochromoides* (Nyl.) Hertel & Leuckert = Lecidella asema [LECIASE]
Lecidella wulfenii (Hepp) Körber [LECIWUL]
SY: *Lecidea heppii* R. Anderson & W.A. Weber
SY: *Lecidea wulfenii* (Hepp) Arnold
Megalaria Hafellner [MEGALAR$]
Megalaria brodoana S. Ekman & Tønsberg [MEGABRO]
Nesolechia oxyspora (Tul.) A. Massal. = Phacopsis oxyspora [PHACOXY]
Phacopsis Tul. [PHACOPS$]
Phacopsis huuskonenii Räsänen [PHACHUU]

Phacopsis oxyspora (Tul.) Triebel & Rambold [PHACOXY]
SY: *Abrothallus oxysporus* Tul.
SY: *Lecidea oxyspora* (Tul.) Nyl.
SY: *Nesolechia oxyspora* (Tul.) A. Massal.
Phacopsis vulpina Tul. [PHACVUL]
Protoblastenia cinnabarina (Sommerf.) Räsänen = Pyrrhospora cinnabarina [PYRRCIN]
Protoblastenia quernea (Dickson) Clauzade = Pyrrhospora quernea [PYRRQUE]
Protoblastenia russula (Ach.) Räsänen = Pyrrhospora russula [PYRRRUS]
Protoparmelia Choisy [PROTOPA$]
Protoparmelia badia (Hoffm.) Hafellner [PROTBAD]
SY: *Lecanora badia* (Hoffm.) Ach.
SY: *Lecanora grandis* H. Magn.
Protoparmelia ochrococca (Nyl.) P.M. Jørg., Rambold & Hertel [PROTOCH]
SY: *Lecanora ochrococca* (Nyl.) Clauzade & Roux
SY: *Lecanora phaeobola* Tuck.
SY: *Lecidea ochrococca* Nyl.
Protoparmeliopsis muralis (Schreber) Choisy = Lecanora muralis [LECAMUR]
Psora scotopholia (Tuck.) Fink = Lecanora scotopholis [LECASCO]
Psorula scotopholis (Tuck.) Gotth. Schneider = Lecanora scotopholis [LECASCO]
Pyrrhospora Körber [PYRRHOS$]
Pyrrhospora cinnabarina (Sommerf.) Choisy [PYRRCIN]
SY: *Lecidea cinnabarina* Sommerf.
SY: *Protoblastenia cinnabarina* (Sommerf.) Räsänen
Pyrrhospora elabens (Fr.) Hafellner [PYRRELA]
SY: *Lecidea elabens* Fr.
SY: *Lecidea melancheima* Tuck.
Pyrrhospora quernea (Dickson) Körber [PYRRQUE]
SY: *Lecidea quernea* (Dickson) Ach.
SY: *Protoblastenia quernea* (Dickson) Clauzade
Pyrrhospora russula (Ach.) Hafellner [PYRRRUS]
SY: *Lecidea russula* Ach.
SY: *Protoblastenia russula* (Ach.) Räsänen
Pyrrhospora subcinnabarina (Tønsberg) Hafellner [PYRRSUB]
SY: *Lecidea subcinnabarina* Tønsberg
Pyrrhospora varians (Ach.) R.C. Harris [PYRRVAR]
SY: *Lecidea varians* Ach.
Rhizoplaca Zopf [RHIZOPL$]
Rhizoplaca chrysoleuca (Sm.) Zopf [RHIZCHR] (pink-eyed rockbright)
SY: *Lecanora chrysoleuca* (Sm.) Ach.
SY: *Lecanora rubina* (Vill.) Ach.
Rhizoplaca melanophthalma (DC.) Leuckert & Poelt [RHIZMEL] (black-eyed rockbright)
SY: *Lecanora melanophthalma* (DC.) Ramond
SY: *Lecidea brandegei* Tuck.
Rhizoplaca peltata (Ramond) Leuckert & Poelt [RHIZPEL] (brown-eyed rockbright)
SY: *Lecanora peltata* (Ramond) Steudel
Rhizoplaca subdiscrepans (Nyl.) R. Sant. [RHIZSUB]
Scoliciosporum A. Massal. [SCOLICI$]
Scoliciosporum chlorococcum (Stenh.) Vezda [SCOLCHL]
SY: *Bacidia chlorococca* (Stenh.) Lettau
Scoliciosporum umbrinum (Ach.) Arnold [SCOLUMB]
SY: *Bacidia umbrina* (Ach.) Bausch
Tylothallia P. James & R. Kilias [TYLOTHA$]
Tylothallia biformigera (Leighton) P. James & R. Kilias [TYLOBIF]
SY: *Catillaria bahusiensis* (Blomb.) Th. Fr.
SY: *Catillaria biformigera* (Leighton) H. Magn.
Woessia fusarioides D. Hawksw. = Bacidina arnoldiana [BACIARN]

LECIDEACEAE (L)

Hypocenomyce Choisy [HYPOCEN$]
Hypocenomyce castaneocinerea (Räsänen) Timdal [HYPOCAS] (charcoal turtle)
Hypocenomyce friesii (Ach.) P. James & Gotth. Schneider [HYPOFRI] (old-growth turtle)
SY: *Lecidea friesii* Ach.
SY: *Psora friesii* (Ach.) Hellbom
Hypocenomyce leucococca R. Sant. [HYPOLEU] (alder turtle)
Hypocenomyce scalaris (Ach.) Choisy [HYPOSCA] (common shingle)
SY: *Lecidea ostreata* (Hoffm.) Schaerer
SY: *Lecidea scalaris* (Ach. ex Lilj.) Ach.
SY: *Psora ostreata* Hoffm.
SY: *Psora scalaris* (Ach. ex Lilj.) Hook.
Lecidea Ach. [LECIDEA$]
Lecidea albofuscescens Nyl. [LECIALO]
Lecidea alpestris Sommerf. [LECIALP]
Lecidea arctica Sommerf. = Lecidea caesioatra [LECICAS]
Lecidea atrobrunnea (Ramond ex Lam. & DC.) Schaerer [LECIATR]
Lecidea atrolutescens Nyl. = Lecidea mannii [LECIMAN]
Lecidea atroviridis (Arnold) Th. Fr. [LECIATO]
Lecidea auriculata Th. Fr. [LECIAUR]
Lecidea caesioatra Schaerer [LECICAS]
SY: *Lecidea arctica* Sommerf.
Lecidea columbiana H. Magn. = Lecidea tessellata [LECITEL]
Lecidea confluens (Weber) Ach. [LECICOF]
Lecidea cruciaria Tuck. [LECICRU]
Lecidea cyanea (Ach.) Rohl. = Lecidea tessellata [LECITEL]
Lecidea cyanescens Lynge = Lecidea lapicida [LECILAP]
Lecidea friesii Ach. = Hypocenomyce friesii [HYPOFRI]
Lecidea fuscoatra (L.) Ach. [LECIFUA]
SY: *Lecidea grisella* Flörke ex Schaerer
Lecidea geophana Nyl. = Steinia geophana [STEIGEO]
Lecidea grisella Flörke ex Schaerer = Lecidea fuscoatra [LECIFUA]
Lecidea lactea Flörke ex Schaerer = Lecidea lapicida [LECILAP]
Lecidea lapicida (Ach.) Ach. [LECILAP]
SY: *Lecidea cyanescens* Lynge
SY: *Lecidea lactea* Flörke ex Schaerer
SY: *Lecidea pantherina* (Ach.) Th. Fr.
Lecidea latypea Ach. = Lecidea plana [LECIPLN]
Lecidea leprarioides Tønsberg [LECILEP]
Lecidea leucothallina Arnold [LECILEC]
Lecidea lithophila (Ach.) Ach. [LECILIT]
SY: *Lecidea pruinosa* Ach.
Lecidea lurida (Ach.) DC. [LECILUR]
SY: *Lecidea petri* (Tuck.) Zahlbr.
SY: *Psora lurida* (Ach.) DC.
SY: *Psora petri* (Tuck.) Fink
Lecidea mannii Tuck. [LECIMAN]
SY: *Lecidea atrolutescens* Nyl.
Lecidea nylanderi (Anzi) Th. Fr. [LECINYL]
Lecidea ostreata (Hoffm.) Schaerer = Hypocenomyce scalaris [HYPOSCA]
Lecidea pantherina (Ach.) Th. Fr. = Lecidea lapicida [LECILAP]
Lecidea paupercula Th. Fr. = Lecidea praenubila [LECIPRE]
Lecidea petri (Tuck.) Zahlbr. = Lecidea lurida [LECILUR]
Lecidea phaeops Nyl. [LECIPHA]
Lecidea plana (J. Lahm) Nyl. [LECIPLN]
SY: *Lecidea latypea* Ach.
Lecidea praenubila Nyl. [LECIPRE]
SY: *Lecidea paupercula* Th. Fr.
Lecidea pruinosa Ach. = Lecidea lithophila [LECILIT]
Lecidea rufofusca (Anzi) Nyl. [LECIRUF]
Lecidea scalaris (Ach. ex Lilj.) Ach. = Hypocenomyce scalaris [HYPOSCA]
Lecidea silacea Ach. [LECISIL]
Lecidea tessellata Flörke [LECITEL]
SY: *Lecidea columbiana* H. Magn.
SY: *Lecidea cyanea* (Ach.) Rohl.
Lecidea umbonata (Hepp) Mudd [LECIUMB]
Lecidea vacciniicola Tønsberg [LECIVAC]
Psora friesii (Ach.) Hellbom = Hypocenomyce friesii [HYPOFRI]
Psora lurida (Ach.) DC. = Lecidea lurida [LECILUR]
Psora ostreata Hoffm. = Hypocenomyce scalaris [HYPOSCA]
Psora petri (Tuck.) Fink = Lecidea lurida [LECILUR]

Psora scalaris (Ach. ex Lilj.) Hook. = Hypocenomyce scalaris [HYPOSCA]
Steinia Körber [STEINIA$]
Steinia geophana (Nyl.) Stein [STEIGEO]
SY: *Lecidea geophana* Nyl.

LICHENOTHELIACEAE (L)

Lichenothelia D. Hawksw. [LICHENT$]
Lichenothelia metzleri (J. Lahm) D. Hawksw. [LICHMET]
SY: *Microthelia metzleri* J. Lahm
Microthelia metzleri J. Lahm = Lichenothelia metzleri [LICHMET]

LICHINACEAE (L)

Ephebe Fr. [EPHEBE$]
Ephebe hispidula (Ach.) Horwood [EPHEHIS]
SY: *Ephebeia hispidula* (Ach.) Nyl.
Ephebe lanata (L.) Vainio [EPHELAN]
Ephebeia hispidula (Ach.) Nyl. = Ephebe hispidula [EPHEHIS]
Euopsis Nyl. [EUOPSIS$]
Euopsis pulvinata (Schaerer) Nyl. [EUOPPUL]
SY: *Pyrenopsis pulvinata* (Schaerer) Th. Fr.
Gonohymenia nigritella (Lettau) Henssen = Lichinella nigritella [LICHNIG]
Lempholemma Körber [LEMPHOL$]
Lempholemma isidiodes (Nyl. ex Arnold) H. Magn. [LEMPISI]
Lempholemma myriococcum (Ach.) Th. Fr. = Lempholemma polyanthes [LEMPPOL]
Lempholemma polyanthes (Bernh.) Malme [LEMPPOL]
SY: *Lempholemma myriococcum* (Ach.) Th. Fr.
Lempholemma radiatum (Sommerf.) Henssen [LEMPRAD]
Lichinella Nyl. [LICHINE$]
Lichinella nigritella (Lettau) Moreno & Egea [LICHNIG] (coal tarpaper)
SY: *Gonohymenia nigritella* (Lettau) Henssen
SY: *Thyrea nigritella* Lettau
Lichinodium Nyl. [LICHINO$]
Lichinodium canadense Henssen [LICHCAN]
Phylliscum Nyl. [PHYLLIS$]
Phylliscum demangeonii (Moug. & Mont.) Nyl. [PHYLDEM] (lizard tripe)
SY: *Thyrea demangeonii* (Moug. & Mont.) Fink
Porocyphus Körber [POROCYP$]
Porocyphus kenmorensis (Holl ex Nyl.) Henssen [POROKEN]
Pyrenopsis pulvinata (Schaerer) Th. Fr. = Euopsis pulvinata [EUOPPUL]
Thermutis Fr. [THERMUT$]
Thermutis velutina (Ach.) Flotow [THERVEL]
Thyrea demangeonii (Moug. & Mont.) Fink = Phylliscum demangeonii [PHYLDEM]
Thyrea nigritella Lettau = Lichinella nigritella [LICHNIG]
Zahlbrucknerella Herre [ZAHLBRU$]
Zahlbrucknerella calcarea (Herre) Herre [ZAHLCAL]

LOBARIACEAE (L)

Dendriscocaulon Nyl. [DENDRIS$]
Dendriscocaulon intricatulum (Nyl.) Henssen [DENDINT]
SY: *Leptogidium intricatulum* Nyl.
SY: *Polychidium intricatulum* (Nyl.) Henssen
Leptogidium intricatulum Nyl. = Dendriscocaulon intricatulum [DENDINT]
Lobaria (Schreber) Hoffm. [LOBARIA$]
Lobaria hallii (Tuck.) Zahlbr. [LOBAHAL] (iron lung)
SY: *Sticta hallii* Tuck.
Lobaria linita (Ach.) Rabenh. [LOBALIN] (cabbage lung)
SY: *Sticta linita* Ach.
Lobaria linita var. **linita** [LOBALIN0]
Lobaria linita var. **tenuior** (Del.) Asah. [LOBALIN1]
Lobaria oregana (Tuck.) Müll. Arg. [LOBAORE] (lettuce lung)
SY: *Sticta oregana* Tuck.
Lobaria pulmonaria (L.) Hoffm. [LOBAPUL] (lungwort)
SY: *Sticta pulmonaria* (L.) Biroli
Lobaria retigera (Bory) Trevisan [LOBARET] (smoker's lung)
Lobaria scrobiculata (Scop.) DC. [LOBASCR] (textured lung)
SY: *Lobaria verrucosa* (Hudson) Hoffm.
SY: *Sticta verrucosa* (Hudson) Fink
Lobaria verrucosa (Hudson) Hoffm. = Lobaria scrobiculata [LOBASCR]
Polychidium intricatulum (Nyl.) Henssen = Dendriscocaulon intricatulum [DENDINT]
Pseudocyphellaria Vainio [PSEUDOY$]
Pseudocyphellaria anomala Brodo & Ahti [PSEUANO] (netted specklebelly)
Pseudocyphellaria anthraspis (Ach.) H. Magn. [PSEUANT] (dimpled specklebelly)
SY: *Sticta anthraspis* Ach.
Pseudocyphellaria aurata (Ach.) Vainio [PSEUAUR] (yellow specklebelly)
SY: *Sticta aurata* Ach.
Pseudocyphellaria crocata (L.) Vainio [PSEUCRO]
SY: *Pseudocyphellaria mougeotiana* (Delise) Vainio
SY: *Sticta crocata* (L.) Ach.
Pseudocyphellaria mougeotiana (Delise) Vainio = Pseudocyphellaria crocata [PSEUCRO]
Pseudocyphellaria rainierensis Imshaug [PSEURAI] (old-growth specklebelly)
Sticta (Schreber) Ach. [STICTA$]
Sticta anthraspis Ach. = Pseudocyphellaria anthraspis [PSEUANT]
Sticta arctica Degel. [STICARC] (arctic moon)
Sticta aurata Ach. = Pseudocyphellaria aurata [PSEUAUR]
Sticta crocata (L.) Ach. = Pseudocyphellaria crocata [PSEUCRO]
Sticta fuliginosa (Hoffm.) Ach. [STICFUL] (peppered moon)
Sticta hallii Tuck. = Lobaria hallii [LOBAHAL]
Sticta limbata (Sm.) Ach. [STICLIM] (powdered moon)
Sticta linita Ach. = Lobaria linita [LOBALIN]
Sticta oregana Tuck. = Lobaria oregana [LOBAORE]
Sticta oroborealis Goward & Tonsberg [STICORO]
Sticta pulmonaria (L.) Biroli = Lobaria pulmonaria [LOBAPUL]
Sticta verrucosa (Hudson) Fink = Lobaria scrobiculata [LOBASCR]
Sticta weigelii (Ach.) Vainio [STICWEI] (fringed moon)
Sticta wrightii Tuck. [STICWRI] (green moon)

MASTODIACEAE (L)

Kohlmeyera Schatz [KOHLMEY$]
Kohlmeyera complicatula (Nyl.) Schatz [KOHLCOM]
SY: *Leptogiopsis complicatula* Nyl.
SY: *Mastodia tessellata* Hook. f. & Harvey
SY: *Turgidosculum complicatulum* (Nyl.) Kohlm. & E. Kohlm.
Leptogiopsis complicatula Nyl. = Kohlmeyera complicatula [KOHLCOM]
Mastodia tessellata Hook. f. & Harvey = Kohlmeyera complicatula [KOHLCOM]
Turgidosculum complicatulum (Nyl.) Kohlm. & E. Kohlm. = Kohlmeyera complicatula [KOHLCOM]

MICAREACEAE (L)

Bacidia melaena (Nyl.) Zahlbr. = Micarea melaena [MICAMEL]
Bacidia trisepta (Hellbom) Zahlbr. = Micarea peliocarpa [MICAPEL]
Catillaria micrococca (Körber) Th. Fr. = Micarea prasina [MICAPRA]
Catillaria prasina (Fr.) Th. Fr. = Micarea prasina [MICAPRA]
Helocarpon Th. Fr. [HELOCAR$]
Helocarpon crassipes Th. Fr. [HELOCRA]
SY: *Lecidea crassipes* (Th. Fr.) Nyl.
SY: *Micarea crassipes* (Th. Fr.) Coppins
Lecidea aniptiza Stirton = Micarea denigrata [MICADEN]
Lecidea assimilata Nyl. = Micarea assimilata [MICAASS]
Lecidea crassipes (Th. Fr.) Nyl. = Helocarpon crassipes [HELOCRA]
Lecidea lucida (Ach.) Ach. = Psilolechia lucida [PSILLUC]

Lecidea misella (Nyl.) Nyl. = Micarea misella [MICAMIS]
Lecidea sylvicola Flotow = Micarea sylvicola [MICASYL]
Micarea Fr. [MICAREA$]
Micarea assimilata (Nyl.) Coppins [MICAASS]
SY: *Lecidea assimilata* Nyl.
Micarea cinerea (Schaerer) Hedl. [MICACIN]
Micarea crassipes (Th. Fr.) Coppins = Helocarpon crassipes [HELOCRA]
Micarea denigrata (Fr.) Hedl. [MICADEN]
SY: *Lecidea aniptiza* Stirton
SY: *Micarea hemipoliella* (Nyl.) Vezda
Micarea globularis (Ach. ex Nyl.) Hedl. = Micarea misella [MICAMIS]
Micarea hemipoliella (Nyl.) Vezda = Micarea denigrata [MICADEN]
Micarea lithinella (Nyl.) Hedl. [MICALIT]
Micarea lutulata (Nyl.) Coppins [MICALUT]
Micarea melaena (Nyl.) Hedl. [MICAMEL]
SY: *Bacidia melaena* (Nyl.) Zahlbr.
Micarea misella (Nyl.) Hedl. [MICAMIS]
SY: *Lecidea misella* (Nyl.) Nyl.
SY: *Micarea globularis* (Ach. ex Nyl.) Hedl.
Micarea peliocarpa (Anzi) Coppins & R. Sant. [MICAPEL]
SY: *Bacidia trisepta* (Hellbom) Zahlbr.
SY: *Micarea trisepta* (Hellbom) Wetmore
SY: *Micarea violacea* (Crouan ex Nyl.) Hedl.
Micarea prasina Fr. [MICAPRA]
SY: *Catillaria micrococca* (Körber) Th. Fr.
SY: *Catillaria prasina* (Fr.) Th. Fr.
Micarea sylvicola (Flotow) Vezda & V. Wirth [MICASYL]
SY: *Lecidea sylvicola* Flotow
Micarea trisepta (Hellbom) Wetmore = Micarea peliocarpa [MICAPEL]
Micarea violacea (Crouan ex Nyl.) Hedl. = Micarea peliocarpa [MICAPEL]
Psilolechia A. Massal. [PSILOLE$]
Psilolechia lucida (Ach.) Choisy [PSILLUC]
SY: *Lecidea lucida* (Ach.) Ach.
Scutula Tul. [SCUTULA$]
Scutula miliaris (Wallr.) Trevisan [SCUTMIL]
SY: *Scutula tuberculosa* (Th. Fr.) Rehm
Scutula stereocaulorum (Anzi) Körber [SCUTSTE]
Scutula tuberculosa (Th. Fr.) Rehm = Scutula miliaris [SCUTMIL]

MICROCALICIACEAE (L)

Calicium disseminatum Ach. = Microcalicium disseminatum [MICRDIS]
Coniocybopsis arenaria (Hampe ex Massal.) Vainio = Microcalicium arenarium [MICRARE]
Microcalicium Vainio [MICROCA$]
Microcalicium arenarium (Hampe ex Massal.) Tibell [MICRARE]
SY: *Coniocybopsis arenaria* (Hampe ex Massal.) Vainio
Microcalicium disseminatum (Ach.) Vainio [MICRDIS]
SY: *Calicium disseminatum* Ach.
SY: *Microcalicium subpedicellatum* (Schaerer) Tibell
SY: *Mycocalicium disseminatum* (Ach.) Fink
Microcalicium subpedicellatum (Schaerer) Tibell = Microcalicium disseminatum [MICRDIS]
Mycocalicium disseminatum (Ach.) Fink = Microcalicium disseminatum [MICRDIS]

MONOBLASTIACEAE (L)

Acrocordia A. Massal. [ACROCOR$]
Acrocordia gemmata (Ach.) A. Massal. [ACROGEM]
SY: *Arthopyrenia alba* (Schrader) Zahlbr.
SY: *Arthopyrenia gemmata* (Ach.) A. Massal.
SY: *Arthopyrenia sphaeroides* (Wallr.) Zahlbr.
Anisomeridium (Müll. Arg.) Choisy [ANISOME$]
Anisomeridium biforme (Borrer) R.C. Harris [ANISBIF]
SY: *Arthopyrenia biformis* (Borrer) A. Massal.
SY: *Arthopyrenia parvula* Zahlbr.
SY: *Ditremis biformis* (Borrer) R.C. Harris
SY: *Arthopyrenia conformis* auct.
Arthopyrenia alba (Schrader) Zahlbr. = Acrocordia gemmata [ACROGEM]
Arthopyrenia biformis (Borrer) A. Massal. = Anisomeridium biforme [ANISBIF]
Arthopyrenia conformis auct. = Anisomeridium biforme [ANISBIF]
Arthopyrenia gemmata (Ach.) A. Massal. = Acrocordia gemmata [ACROGEM]
Arthopyrenia parvula Zahlbr. = Anisomeridium biforme [ANISBIF]
Arthopyrenia sphaeroides (Wallr.) Zahlbr. = Acrocordia gemmata [ACROGEM]
Ditremis biformis (Borrer) R.C. Harris = Anisomeridium biforme [ANISBIF]

MYCOBLASTACEAE (L)

Haematomma caesium Coppins & P. James = Mycoblastus caesius [MYCOCAE]
Lecidea glabrescens Nyl. = Mycoblastus affinis [MYCOAFF]
Mycoblastus Norman [MYCOBLA$]
Mycoblastus affinis (Schaerer) Schauer [MYCOAFF]
SY: *Lecidea glabrescens* Nyl.
SY: *Mycoblastus melinus* (Kremp. ex Nyl.) Hellbom
Mycoblastus alpinus (Fr.) Kernst. [MYCOALP]
Mycoblastus caesius (Coppins & P. James) Tønsberg [MYCOCAE]
SY: *Haematomma caesium* Coppins & P. James
Mycoblastus fucatus (Stirton) Zahlbr. [MYCOFUC]
SY: *Mycoblastus glabrescens* (Nyl.) Zahlbr.
Mycoblastus glabrescens (Nyl.) Zahlbr. = Mycoblastus fucatus [MYCOFUC]
Mycoblastus melinus (Kremp. ex Nyl.) Hellbom = Mycoblastus affinis [MYCOAFF]
Mycoblastus sanguinarius (L.) Norman [MYCOSAN] (bloody heart)
SY: *Mycoblastus sanguinarius* var. *dodgeanus* Räsänen
Mycoblastus sanguinarius var. *dodgeanus* Räsänen = Mycoblastus sanguinarius [MYCOSAN]

MYCOCALICIACEAE (L)

Calicium asikkalense Vainio = Chaenothecopsis pusilla [CHAEPUS]
Calicium floerkei Zahlbr. = Chaenothecopsis pusilla [CHAEPUS]
Calicium populneum Brond. ex Duby = Phaeocalicium populneum [PHAEPOP]
Calicium pusillum auct. = Chaenothecopsis pusilla [CHAEPUS]
Calicium pusiolum Ach. = Chaenothecopsis pusiola [CHAEPUI]
Calicium subpusillum Vainio = Chaenothecopsis pusilla [CHAEPUS]
Calicium subtile Pers. = Mycocalicium subtile [MYCOSUB]
Chaenothecopsis Vainio [CHAENOH$]
Chaenothecopsis consociata (Nádv.) A.F.W. Schmidt [CHAECON]
Chaenothecopsis debilis (Turner & Borrer ex Sm.) Tibell [CHAEDEB]
Chaenothecopsis epithallina Tibell [CHAEEPI]
Chaenothecopsis lignicola (Nádv.) A.F.W. Schmidt = Chaenothecopsis pusiola [CHAEPUI]
Chaenothecopsis nana Tibell [CHAENAN]
Chaenothecopsis pusilla (Ach.) A.F.W. Schmidt [CHAEPUS]
SY: *Calicium asikkalense* Vainio
SY: *Calicium floerkei* Zahlbr.
SY: *Calicium subpusillum* Vainio
SY: *Chaenothecopsis subpusilla* (Vainio) Tibell
SY: *Calicium pusillum* auct.
Chaenothecopsis pusiola (Ach.) Vainio [CHAEPUI]
SY: *Calicium pusiolum* Ach.
SY: *Chaenothecopsis lignicola* (Nádv.) A.F.W. Schmidt
SY: *Mycocalicium pusiolum* (Ach.) Räsänen
Chaenothecopsis subpusilla (Vainio) Tibell = Chaenothecopsis pusilla [CHAEPUS]
Chaenothecopsis viridialba (Kremp.) A.F.W. Schmidt [CHAEVIR]
Chaenothecopsis viridireagens (Nádv.) A.F.W. Schmidt [CHAEVII]
Mycocalicium Vainio [MYCOCAC$]

Mycocalicium compressulum Nyl. ex Szat. = Phaeocalicium compressulum [PHAECOM]
Mycocalicium parietinum (Ach. ex Schaerer) D. Hawksw. = Mycocalicium subtile [MYCOSUB]
Mycocalicium pusiolum (Ach.) Räsänen = Chaenothecopsis pusiola [CHAEPUI]
Mycocalicium subtile (Pers.) Szat. [MYCOSUB]
SY: *Calicium subtile* Pers.
SY: *Mycocalicium parietinum* (Ach. ex Schaerer) D. Hawksw.
Phaeocalicium A.F.W. Schmidt [PHAEOCA$]
Phaeocalicium compressulum (Nyl. ex Szat.) A.F.W. Schmidt [PHAECOM]
SY: *Mycocalicium compressulum* Nyl. ex Szat.
Phaeocalicium populneum (Brond. ex Duby) A.F.W. Schmidt [PHAEPOP]
SY: *Calicium populneum* Brond. ex Duby
Stenocybe (Nyl.) Körber [STENOCY$]
Stenocybe byssacea (Fr.) Körber = Stenocybe pullatula [STENPUL]
Stenocybe clavata Tibell [STENCLA]
Stenocybe euspora (Nyl.) Anzi = Stenocybe major [STENMAJ]
Stenocybe major (Nyl.) Körber [STENMAJ]
SY: *Stenocybe euspora* (Nyl.) Anzi
Stenocybe pullatula (Ach.) Stein [STENPUL]
SY: *Stenocybe byssacea* (Fr.) Körber

MYCOSPHAERELLACEAE (L)

Sphaerellothecium Zopf [SPHAERE$]
Sphaerellothecium araneosum (Rehm ex Arnold) Zopf [SPHAARA]

NEPHROMATACEAE (L)

Nephroma Ach. [NEPHROM$]
Nephroma arcticum (L.) Torss. [NEPHARC] (green paw)
Nephroma bellum (Sprengel) Tuck. [NEPHBEL] (cat paw)
SY: *Nephroma subtomentellum* (Nyl.) Gyelnik
SY: *Nephroma laevigatum* auct.
Nephroma expallidum (Nyl.) Nyl. [NEPHEXP] (alpine paw)
Nephroma helveticum Ach. [NEPHHEL] (dog paw)
Nephroma helveticum ssp. **helveticum** [NEPHHEL0]
Nephroma helveticum ssp. **sipeanum** (Gyelnik) Goward & Ahti [NEPHHEL1]
SY: *Nephroma helveticum* var. *sipeanum* (Gyelnik) Wetmore
SY: *Nephromium canadense* Räsänen
Nephroma helveticum var. *sipeanum* (Gyelnik) Wetmore = Nephroma helveticum ssp. sipeanum [NEPHHEL1]
Nephroma isidiosum (Nyl.) Gyelnik [NEPHISI] (pepper paw)
Nephroma laevigatum Ach. [NEPHLAE] (seaside paw)
SY: *Nephroma lusitanicum* auct.
Nephroma laevigatum auct. = Nephroma bellum [NEPHBEL]
Nephroma lusitanicum auct. = Nephroma laevigatum [NEPHLAE]
Nephroma occultum Wetmore [NEPHOCC] (cryptic paw)
Nephroma parile (Ach.) Ach. [NEPHPAR] (powder paw)
Nephroma resupinatum (L.) Ach. [NEPHRES] (blistered paw)
Nephroma silvae-veteris Goward & Goffinet [NEPHSIL] (old-growth paw)
Nephroma subtomentellum (Nyl.) Gyelnik = Nephroma bellum [NEPHBEL]
Nephromium canadense Räsänen = Nephroma helveticum ssp. sipeanum [NEPHHEL1]

ODONTOTREMATACEAE (L)

Lethariicola Grummann [LETHARC$]
Lethariicola cucularis (Norman) Lumbsch & D. Hawksw. [LETHCUC]
SY: *Lethariicola sipei* Grummann
Lethariicola sipei Grummann = Lethariicola cucularis [LETHCUC]

OPEGRAPHACEAE (L)

Bacidia akompsa (Tuck.) Fink = Lecanactis akompsa [LECAAKO]
Bactrospora A. Massal. [BACTROS$]
Bactrospora patellarioides (Nyl.) Vainio [BACTPAT]
SY: *Lecanactis patellarioides* (Nyl.) Vainio
Enterographa Fée [ENTEROG$]
Enterographa zonata (Körber) Källsten [ENTEZON]
SY: *Opegrapha zonata* Körber
Lecanactis Körber [LECANAC$]
Lecanactis abietina (Ach.) Körber [LECAABI]
SY: *Lecanactis megaspora* (G. Merr.) Brodo
Lecanactis akompsa (Tuck.) Essl. & Egan [LECAAKO]
SY: *Bacidia akompsa* (Tuck.) Fink
Lecanactis megaspora (G. Merr.) Brodo = Lecanactis abietina [LECAABI]
Lecanactis patellarioides (Nyl.) Vainio = Bactrospora patellarioides [BACTPAT]
Opegrapha Ach. [OPEGRAP$]
Opegrapha betulina Sm. = Opegrapha herbarum [OPEGHER]
Opegrapha diaphora (Ach.) Ach. = Opegrapha varia [OPEGVAR]
Opegrapha gyrocarpa Flotow [OPEGGYR]
Opegrapha herbarum Mont. [OPEGHER]
SY: *Opegrapha betulina* Sm.
Opegrapha herpetica (Ach.) Ach. = Opegrapha rufescens [OPEGRUF]
Opegrapha lichenoides Pers. = Opegrapha varia [OPEGVAR]
Opegrapha ochrocheila Nyl. [OPEGOCH]
Opegrapha pulicaris auct. = Opegrapha varia [OPEGVAR]
Opegrapha rimalis Pers. ex Ach. = Opegrapha varia [OPEGVAR]
Opegrapha rufescens Pers. [OPEGRUF]
SY: *Opegrapha herpetica* (Ach.) Ach.
Opegrapha sphaerophoricola Isbrand & Alstrup [OPEGSPH]
Opegrapha varia Pers. [OPEGVAR]
SY: *Opegrapha diaphora* (Ach.) Ach.
SY: *Opegrapha lichenoides* Pers.
SY: *Opegrapha rimalis* Pers. ex Ach.
SY: *Opegrapha pulicaris* auct.
Opegrapha zonata Körber = Enterographa zonata [ENTEZON]

OPHIOPARMACEAE (L)

Bacidia herrei Zahlbr. = Ophioparma rubricosa [OPHIRUB]
Bacidia rubricosa (Müll. Arg.) Zahlbr. = Ophioparma rubricosa [OPHIRUB]
Haematomma californicum Sigal & D. Toren = Ophioparma rubricosa [OPHIRUB]
Haematomma lapponicum Räsänen = Ophioparma lapponica [OPHILAP]
Haematomma ventosum (L.) A. Massal. = Ophioparma lapponica [OPHILAP]
Ophioparma Norman [OPHIOPA$]
Ophioparma lapponica (Räsänen) Hafellner & R.W. Rogers [OPHILAP]
SY: *Haematomma lapponicum* Räsänen
SY: *Haematomma ventosum* (L.) A. Massal.
Ophioparma rubricosa (Müll. Arg.) S. Ekman [OPHIRUB]
SY: *Bacidia herrei* Zahlbr.
SY: *Bacidia rubricosa* (Müll. Arg.) Zahlbr.
SY: *Haematomma californicum* Sigal & D. Toren

PANNARIACEAE (L)

Amphiloma lanuginosum (Hoffm.) Nyl. = Leproloma membranaceum [LEPRMEB]
Erioderma Fée [ERIODER$]
Erioderma mollissimum (Samp.) Du Rietz [ERIOMOL]
Erioderma sorediatum D.J. Galloway & P.M. Jørg. [ERIOSOR]
SY: *Pannaria ahlneri* P.M. Jørg.
SY: *Pannaria laceratula* Hue
SY: *Pannaria leucostictoides* Ohlsson
SY: *Pannaria maritima* P.M. Jørg.
SY: *Pannaria mediterranea* Tav.
SY: *Pannaria praetermissa* Nyl.

SY: *Pannaria saubinetii* (Mont.) Nyl.
Leioderma Nyl. [LEIODER$]
Leioderma sorediatum D.J. Galloway & P.M. Jørg. [LEIOSOR] (treepelt)
Lepraria arctica (Lynge) Wetmore = Leproloma vouauxii [LEPRVOU]
Lepraria membranacea (Dickson) Vainio = Leproloma membranaceum [LEPRMEB]
Lepraria vouauxii (Hue) R.C. Harris = Leproloma vouauxii [LEPRVOU]
Leproloma Nyl. ex Crombie [LEPROLO$]
Leproloma diffusum J.R. Laundon [LEPRDIF]
Leproloma diffusum var. **chrysodetoides** J.R. Laundon [LEPRDIF1]
Leproloma diffusum var. **diffusum** [LEPRDIF0]
Leproloma membranaceum (Dickson) Vainio [LEPRMEB]
SY: *Amphiloma lanuginosum* (Hoffm.) Nyl.
SY: *Lepraria membranacea* (Dickson) Vainio
Leproloma vouauxii (Hue) J.R. Laundon [LEPRVOU]
SY: *Lepraria arctica* (Lynge) Wetmore
SY: *Lepraria vouauxii* (Hue) R.C. Harris
Pannaria Delise [PANNARI$]
Pannaria ahlneri P.M. Jørg. = Fuscopannaria ahlneri [FUSCAHL]
Pannaria cheiroloba (Müll. Arg.) Essl. & Egan [PANNCHE] (rock mouse)
SY: *Parmeliella cheiroloba* Müll. Arg.
Pannaria cyanolepra Tuck. [PANNCYA]
Pannaria hypnorum (Vahl) Körber = Psoroma hypnorum [PSORHYP]
Pannaria laceratula Hue = Fuscopannaria laceratula [FUSCLAC]
Pannaria leucostictoides Ohlsson = Fuscopannaria leucostictoides [FUSCLEU]
Pannaria maritima P.M. Jørg. = Fuscopannaria maritima [FUSCMAR]
Pannaria mediterranea Tav. = Fuscopannaria mediterranea [FUSCMED]
Pannaria pezizoides (Weber) Trevisan [PANNPEZ] (peacock mouse)
Pannaria praetermissa Nyl. = Fuscopannaria praetermissa [FUSCPRE]
Pannaria rubiginosa (Ach.) Bory [PANNRUB]
Pannaria saubinetii (Mont.) Nyl. = Fuscopannaria saubinetii [FUSCSAU]
Parmeliella Müll. Arg. [PARMELL$]
Parmeliella cheiroloba Müll. Arg. = Pannaria cheiroloba [PANNCHE]
Parmeliella triptophylla (Ach.) Müll. Arg. [PARMTRI] (fingered mouse)
Psoroma Michaux [PSOROMA$]
Psoroma hypnorum (Vahl) Gray [PSORHYP] (green mouse)
SY: *Pannaria hypnorum* (Vahl) Körber

PARMELIACEAE (L)

Ahtiana Goward [AHTIANA$]
Ahtiana pallidula (Tuck. ex Riddle) Goward & Thell [AHTIPAL] (pallid ruffle)
SY: *Cetraria pallidula* Tuck. ex Riddle
SY: *Tuckermannopsis pallidula* (Tuck. ex Riddle) Hale
Ahtiana sphaerosporella (Müll. Arg.) Goward [AHTISPH] (whitebark candlewax)
SY: *Parmelia sphaerosporella* Müll. Arg.
Alectoria abbreviata (Müll. Arg.) R. Howe = Nodobryoria abbreviata [NODOABB]
Alectoria achariana Gyelnik = Bryoria pseudofuscescens [BRYOPSE]
Alectoria ambigua Mot. = Bryoria trichodes ssp. americana [BRYOTRI1]
Alectoria americana Mot. = Bryoria trichodes ssp. americana [BRYOTRI1]
Alectoria bicolor (Ehrh.) Nyl. = Bryoria bicolor [BRYOBIC]
Alectoria californica (Tuck.) G. Merr. = Kaernefeltia californica [KAERCAL]
Alectoria canadensis Mot. = Bryoria trichodes ssp. trichodes [BRYOTRI0]
Alectoria capillaris (Ach.) Crombie = Bryoria capillaris [BRYOCAP]
Alectoria cervinula Mot. = Bryoria cervinula [BRYOCER]
Alectoria cetrariza Nyl. = Kaernefeltia californica [KAERCAL]
Alectoria chalybeiformis (L.) Gray = Bryoria chalybeiformis [BRYOCHA]
Alectoria corneliae Gyelnik = Bryoria fremontii [BRYOFRE]
Alectoria delicata Mot. = Bryoria trichodes [BRYOTRI]
Alectoria divergens (Ach.) Nyl. = Bryocaulon divergens [BRYODIV]
Alectoria divergens f. *abbreviata* Müll. Arg. = Nodobryoria abbreviata [NODOABB]
Alectoria fremontii Tuck. = Bryoria fremontii [BRYOFRE]
Alectoria fremontii f. *perfertilis* Räsänen = Bryoria fremontii [BRYOFRE]
Alectoria fremontii ssp. *olivacea* Räsänen = Bryoria fremontii [BRYOFRE]
Alectoria fuscescens Gyelnik = Bryoria fuscescens [BRYOFUS]
Alectoria glabra Mot. = Bryoria glabra [BRYOGLA]
Alectoria implexa (Hoffm.) Nyl. = Bryoria implexa [BRYOIMP]
Alectoria implexa f. *fuscidula* Arnold = Bryoria pikei [BRYOPIK]
Alectoria irvingii Llano = Bryoria nitidula [BRYONIT]
Alectoria jubata var. *bicolor* (Ehrh.) Nyl. = Bryoria bicolor [BRYOBIC]
Alectoria jubata var. *implexa* (Hoffm.) Ach. = Bryoria implexa [BRYOIMP]
Alectoria jubata var. *vrangiana* (Gyelnik) Räsänen = Bryoria implexa [BRYOIMP]
Alectoria lanea auct. = Bryoria nitidula [BRYONIT]
Alectoria lanestris (Ach.) Gyelnik = Bryoria lanestris [BRYOLAN]
Alectoria minuscula (Nyl. ex Arnold) Degel. = Pseudephebe minuscula [PSEUMIN]
Alectoria nana Mot. = Bryoria simplicior [BRYOSIM]
Alectoria nitidula (Th. Fr.) Vainio = Bryoria nitidula [BRYONIT]
Alectoria norstictica Mot. = Bryoria pseudofuscescens [BRYOPSE]
Alectoria oregana Tuck. = Nodobryoria oregana [NODOORE]
Alectoria positiva (Gyelnik) Mot. = Bryoria fuscescens [BRYOFUS]
Alectoria pseudofuscescens Gyelnik = Bryoria pseudofuscescens [BRYOPSE]
Alectoria pubescens (L.) R. Howe = Pseudephebe pubescens [PSEUPUB]
Alectoria setacea (Ach.) Mot. = Bryoria capillaris [BRYOCAP]
Alectoria simplicior (Vainio) Lynge = Bryoria simplicior [BRYOSIM]
Alectoria subcana (Nyl. ex Stizenb.) Gyelnik = Bryoria subcana [BRYOSUB]
Alectoria subdivergens E. Dahl = Nodobryoria subdivergens [NODOSUB]
Alectoria subtilis Mot. = Bryoria pseudofuscescens [BRYOPSE]
Alectoria tenerrima Mot. = Bryoria fremontii [BRYOFRE]
Alectoria tenuis E. Dahl = Bryoria tenuis [BRYOTEN]
Alectoria tortuosa G. Merr. = Bryoria tortuosa [BRYOTOR]
Alectoria virens auct. = Bryoria tortuosa [BRYOTOR]
Alectoria vrangiana Gyelnik = Bryoria implexa [BRYOIMP]
Allantoparmelia (Vainio) Essl. [ALLANTO$]
Allantoparmelia almquistii (Vainio) Essl. [ALLAALM] (rockgrub)
SY: *Parmelia almquistii* Vainio
Allantoparmelia alpicola (Th. Fr.) Essl. [ALLAALP]
SY: *Parmelia alpicola* Th. Fr.
Allocetraria cucullata (Bellardi) Randlane & Saag = Flavocetraria cucullata [FLAVCUC]
Allocetraria nivalis (L.) Randlane & Saag = Flavocetraria nivalis [FLAVNIV]
Arctoparmelia Hale [ARCTOPA$]
Arctoparmelia aleuritica (Nyl.) Hale = Arctoparmelia centrifuga [ARCTCEN]
Arctoparmelia centrifuga (L.) Hale [ARCTCEN] (rippled rockfrog)
SY: *Arctoparmelia aleuritica* (Nyl.) Hale
SY: *Parmelia aleuritica* Nyl.
SY: *Parmelia centrifuga* (L.) Ach.
SY: *Xanthoparmelia centrifuga* (L.) Hale
Arctoparmelia incurva (Pers.) Hale [ARCTINC] (powdered rockfrog)
SY: *Parmelia incurva* (Pers.) Fr.
SY: *Xanthoparmelia incurva* (Pers.) Hale
Arctoparmelia separata (Th. Fr.) Hale [ARCTSEP]

SY: *Parmelia birulae* var. *grumosa* Llano
SY: *Parmelia separata* Th. Fr.
SY: *Xanthoparmelia separata* (Th. Fr.) Hale
Arctoparmelia subcentrifuga (Oksner) Hale [ARCTSUB] (dissolving rockfrog)
SY: *Parmelia subcentrifuga* Oksner
SY: *Xanthoparmelia subcentrifuga* (Oksner) Hale
Asahinea Culb. & C. Culb. [ASAHINE$]
Asahinea chrysantha (Tuck.) Culb. & C. Culb. [ASAHCHR]
SY: *Cetraria chrysantha* Tuck.
Asahinea scholanderi (Llano) Culb. & C. Culb. [ASAHSCH] (arctic rag)
SY: *Cetraria scholanderi* Llano
Brodoa Goward [BRODOA$]
Brodoa intestiniformis auct. = Brodoa oroarctica [BRODORO]
Brodoa oroarctica (Krog) Goward [BRODORO]
SY: *Hypogymnia oroarctica* Krog
SY: *Lichen intestiniformis* Vill.
SY: *Brodoa intestiniformis* auct.
Bryocaulon Kärnefelt [BRYOCAU$]
Bryocaulon divergens (Ach.) Kärnefelt [BRYODIV]
SY: *Alectoria divergens* (Ach.) Nyl.
SY: *Bryopogon divergens* (Ach.) Schwend.
SY: *Coelocaulon divergens* (Ach.) R. Howe
SY: *Cornicularia divergens* Ach.
Bryocaulon pseudosatoanum (Asah.) Kärnefelt [BRYOPSD]
SY: *Cornicularia pseudosatoana* Asah.
Bryopogon divergens (Ach.) Schwend. = Bryocaulon divergens [BRYODIV]
Bryopogon fremontii (Tuck.) Rabenh. = Bryoria fremontii [BRYOFRE]
Bryopogon jubata var. *chalybeiformis* (Ach.) Rabenh. = Bryoria chalybeiformis [BRYOCHA]
Bryopogon jubata var. *implexa* Fr. = Bryoria implexa [BRYOIMP]
Bryopogon negativus Gyelnik = Bryoria lanestris [BRYOLAN]
Bryopogon pacificus Gyelnik = Bryoria fuscescens [BRYOFUS]
Bryoria Brodo & D. Hawksw. [BRYORIA$]
Bryoria abbreviata (Müll. Arg.) Brodo & D. Hawksw. = Nodobryoria abbreviata [NODOABB]
Bryoria bicolor (Ehrh.) Brodo & D. Hawksw. [BRYOBIC]
SY: *Alectoria bicolor* (Ehrh.) Nyl.
SY: *Alectoria jubata* var. *bicolor* (Ehrh.) Nyl.
Bryoria capillaris (Ach.) Brodo & D. Hawksw. [BRYOCAP]
SY: *Alectoria capillaris* (Ach.) Crombie
SY: *Alectoria setacea* (Ach.) Mot.
Bryoria carlottae Brodo & D. Hawksw. [BRYOCAR]
Bryoria cervinula Mot. ex Brodo & D. Hawksw. [BRYOCER]
SY: *Alectoria cervinula* Mot.
Bryoria chalybeiformis (L.) Brodo & D. Hawksw. [BRYOCHA]
SY: *Alectoria chalybeiformis* (L.) Gray
SY: *Bryopogon jubata* var. *chalybeiformis* (Ach.) Rabenh.
Bryoria fremontii (Tuck.) Brodo & D. Hawksw. [BRYOFRE] (edible horsehair)
SY: *Alectoria corneliae* Gyelnik
SY: *Alectoria fremontii* Tuck.
SY: *Alectoria fremontii* ssp. *olivacea* Räsänen
SY: *Alectoria fremontii* f. *perfertilis* Räsänen
SY: *Alectoria tenerrima* Mot.
SY: *Bryopogon fremontii* (Tuck.) Rabenh.
Bryoria friabilis Brodo & D. Hawksw. [BRYOFRI]
Bryoria fuscescens (Gyelnik) Brodo & D. Hawksw. [BRYOFUS] (speckled horsehair)
SY: *Alectoria fuscescens* Gyelnik
SY: *Alectoria positiva* (Gyelnik) Mot.
SY: *Bryopogon pacificus* Gyelnik
Bryoria glabra (Mot.) Brodo & D. Hawksw. [BRYOGLA]
SY: *Alectoria glabra* Mot.
Bryoria implexa (Hoffm.) Brodo & D. Hawksw. [BRYOIMP]
SY: *Alectoria implexa* (Hoffm.) Nyl.
SY: *Alectoria jubata* var. *implexa* (Hoffm.) Ach.
SY: *Alectoria jubata* var. *vrangiana* (Gyelnik) Räsänen
SY: *Alectoria vrangiana* Gyelnik
SY: *Bryopogon jubata* var. *implexa* Fr.
SY: *Bryoria vrangiana* (Gyelnik) Brodo & D. Hawksw.
Bryoria lanestris (Ach.) Brodo & D. Hawksw. [BRYOLAN] (brittle horsehair)
SY: *Alectoria lanestris* (Ach.) Gyelnik
SY: *Bryopogon negativus* Gyelnik
Bryoria nitidula (Th. Fr.) Brodo & D. Hawksw. [BRYONIT]
SY: *Alectoria irvingii* Llano
SY: *Alectoria nitidula* (Th. Fr.) Vainio
SY: *Alectoria lanea* auct.
Bryoria oregana (Tuck. ex Willey) Brodo & D. Hawksw. = Nodobryoria oregana [NODOORE]
Bryoria pikei Brodo & D. Hawksw. [BRYOPIK]
SY: *Alectoria implexa* f. *fuscidula* Arnold
Bryoria pseudofuscescens (Gyelnik) Brodo & D. Hawksw. [BRYOPSE]
SY: *Alectoria achariana* Gyelnik
SY: *Alectoria norstictica* Mot.
SY: *Alectoria pseudofuscescens* Gyelnik
SY: *Alectoria subtilis* Mot.
Bryoria simplicior (Vainio) Brodo & D. Hawksw. [BRYOSIM] (simple horsehair)
SY: *Alectoria nana* Mot.
SY: *Alectoria simplicior* (Vainio) Lynge
Bryoria subcana (Nyl. ex Stizenb.) Brodo & D. Hawksw. [BRYOSUB]
SY: *Alectoria subcana* (Nyl. ex Stizenb.) Gyelnik
Bryoria subdivergens (E. Dahl) Brodo & D. Hawksw. = Nodobryoria subdivergens [NODOSUB]
Bryoria tenuis (E. Dahl) Brodo & D. Hawksw. [BRYOTEN]
SY: *Alectoria tenuis* E. Dahl
Bryoria tortuosa (G. Merr.) Brodo & D. Hawksw. [BRYOTOR]
SY: *Alectoria tortuosa* G. Merr.
SY: *Alectoria virens* auct.
Bryoria trichodes (Michaux) Brodo & D. Hawksw. [BRYOTRI]
SY: *Alectoria delicata* Mot.
Bryoria trichodes ssp. **americana** (Mot.) Brodo & D. Hawksw. [BRYOTRI1]
SY: *Alectoria ambigua* Mot.
SY: *Alectoria americana* Mot.
Bryoria trichodes ssp. **trichodes** [BRYOTRI0]
SY: *Alectoria canadensis* Mot.
Bryoria vrangiana (Gyelnik) Brodo & D. Hawksw. = Bryoria implexa [BRYOIMP]
Cavernularia Degel. [CAVERNU$]
Cavernularia hultenii Degel. [CAVEHUL] (powdered saguaro)
Cavernularia lophyrea (Ach.) Degel. [CAVELOP] (eyed saguaro)
Cetraria Ach. [CETRARI$]
Cetraria aculeata (Schreber) Fr. [CETRACU] (spiny heath lichen)
SY: *Coelocaulon aculeatum* (Schreber) Link
SY: *Cornicularia aculeata* (Schreber) Ach.
Cetraria agnata (Nyl.) Kristinsson = Melanelia agnata [MELAAGN]
Cetraria arborialis (Zahlbr.) Howard = Tuckermannopsis subalpina [TUCKSUB]
Cetraria arctica (Hook.) Tuck. = Dactylina arctica [DACTARC]
Cetraria californica Tuck. = Kaernefeltia californica [KAERCAL]
Cetraria canadensis (Räsänen) Räsänen = Vulpicida canadensis [VULPCAN]
Cetraria chlorophylla (Willd.) Vainio = Tuckermannopsis chlorophylla [TUCKCHL]
Cetraria chrysantha Tuck. = Asahinea chrysantha [ASAHCHR]
Cetraria ciliaris var. *halei* (Culb. & C. Culb.) Ahti = Tuckermannopsis americana [TUCKAME]
Cetraria commixta (Nyl.) Th. Fr. = Melanelia commixta [MELACOM]
Cetraria cucullata (Bellardi) Ach. = Flavocetraria cucullata [FLAVCUC]
Cetraria delisei (Bory ex Schaerer) Nyl. = Cetrariella delisei [CETRDEI]
Cetraria ericetorum Opiz [CETRERI]
Cetraria ericetorum ssp. **reticulata** (Räsänen) Kärnefelt [CETRERI1]
Cetraria fahlunensis (L.) Schreber = Melanelia commixta [MELACOM]

Cetraria glauca (L.) Ach. = Platismatia glauca [PLATGLA]
Cetraria halei Culb. & C. Culb. = Tuckermannopsis americana [TUCKAME]
Cetraria hepatizon (Ach.) Vainio = Melanelia hepatizon [MELAHEP]
Cetraria herrei Imshaug = Platismatia herrei [PLATHER]
Cetraria iberica Crespo & Barreno = Kaernefeltia merrillii [KAERMER]
Cetraria idahoensis Essl. = Esslingeriana idahoensis [ESSLIDA]
Cetraria islandica (L.) Ach. [CETRISL] (icelandmoss)
Cetraria islandica ssp. **crispiformis** (Räsänen) Kärnefelt [CETRISL1]
Cetraria islandica ssp. **islandica** [CETRISL0]
Cetraria islandica ssp. **orientalis** (Asah.) Kärnefelt [CETRISL2]
Cetraria juniperina auct. = Vulpicida canadensis [VULPCAN]
Cetraria lacunosa Ach. = Platismatia lacunosa [PLATLAC]
Cetraria lacunosa var. *macounii* Du Rietz = Platismatia lacunosa [PLATLAC]
Cetraria laevigata Rass. [CETRLAE]
Cetraria madreporiformis (Ach.) Müll. Arg. = Dactylina madreporiformis [DACTMAD]
Cetraria merrillii Du Rietz = Kaernefeltia merrillii [KAERMER]
Cetraria muricata (Ach.) Eckfeldt [CETRMUR]
SY: *Coelocaulon muricatum* (Ach.) J.R. Laundon
SY: *Cornicularia muricata* (Ach.) Ach.
Cetraria nigricans Nyl. [CETRNIG] (blackened icelandmoss)
Cetraria nivalis (L.) Ach. = Flavocetraria nivalis [FLAVNIV]
Cetraria norvegica (Lynge) Du Rietz = Platismatia norvegica [PLATNOR]
Cetraria orbata (Nyl.) Fink = Tuckermannopsis orbata [TUCKORB]
Cetraria pallidula Tuck. ex Riddle = Ahtiana pallidula [AHTIPAL]
Cetraria pinastri (Scop.) Gray = Vulpicida pinastri [VULPPIN]
Cetraria platyphylla Tuck. = Tuckermannopsis platyphylla [TUCKPLA]
Cetraria polyschiza (Nyl.) Jatta = Melanelia hepatizon [MELAHEP]
Cetraria richardsonii Hook. = Masonhalea richardsonii [MASORIC]
Cetraria scholanderi Llano = Asahinea scholanderi [ASAHSCH]
Cetraria scutata (Wulfen) Poetsch = Tuckermannopsis sepincola [TUCKSEP]
Cetraria scutata auct. = Tuckermannopsis chlorophylla [TUCKCHL]
Cetraria sepincola (Ehrh.) Ach. = Tuckermannopsis sepincola [TUCKSEP]
Cetraria stenophylla (Tuck.) G. Merr. = Platismatia stenophylla [PLATSTE]
Cetraria subalpina Imshaug = Tuckermannopsis subalpina [TUCKSUB]
Cetraria tilesii Ach. = Vulpicida tilesii [VULPTIL]
Cetraria tuckermanii Herre = Platismatia herrei [PLATHER]
SY: *Cetraria delisei* (Bory ex Schaerer) Nyl.
Cetrelia Culb. & C. Culb. [CETRELI$]
Cetrelia cetrarioides (Duby) Culb. & C. Culb. [CETRCET] (speckled rag)
SY: *Parmelia cetrarioides* (Delise ex Duby) Nyl.
Chlorea vulpina Nyl. = Letharia vulpina [LETHVUL]
Coelocaulon aculeatum (Schreber) Link = Cetraria aculeata [CETRACU]
Coelocaulon cetrariza (Nyl.) Gyelnik = Kaernefeltia californica [KAERCAL]
Coelocaulon divergens (Ach.) R. Howe = Bryocaulon divergens [BRYODIV]
Coelocaulon muricatum (Ach.) J.R. Laundon = Cetraria muricata [CETRMUR]
Cornicularia (Schreber) Hoffm. [CORNICU$]
Cornicularia aculeata (Schreber) Ach. = Cetraria aculeata [CETRACU]
Cornicularia californica (Tuck.) Du Rietz = Kaernefeltia californica [KAERCAL]
Cornicularia divergens Ach. = Bryocaulon divergens [BRYODIV]
Cornicularia muricata (Ach.) Ach. = Cetraria muricata [CETRMUR]
Cornicularia normoerica (Gunn.) Du Rietz [CORNNOR]
Cornicularia pseudosatoana Asah. = Bryocaulon pseudosatoanum [BRYOPSD]
Dactylina Nyl. [DACTYLN$]
Dactylina arctica (Richardson) Nyl. [DACTARC] (few-finger lichen)
SY: *Cetraria arctica* (Hook.) Tuck.
Dactylina arctica ssp. *beringica* (C.D. Bird & J.W. Thomson) Kärnefelt & Thell = Dactylina beringica [DACTBER]
Dactylina beringica C.D. Bird & J.W. Thomson [DACTBER]
SY: *Dactylina arctica* ssp. *beringica* (C.D. Bird & J.W. Thomson) Kärnefelt & Thell
Dactylina madreporiformis (Ach.) Tuck. [DACTMAD]
SY: *Cetraria madreporiformis* (Ach.) Müll. Arg.
SY: *Dufourea madreporiformis* (Ach.) Ach.
Dactylina ramulosa (Hook.) Tuck. [DACTRAM]
Dufourea madreporiformis (Ach.) Ach. = Dactylina madreporiformis [DACTMAD]
Ephebe pubescens (L.) Fr. = Pseudephebe pubescens [PSEUPUB]
Esslingeriana Hale & M.J. Lai [ESSLING$]
Esslingeriana idahoensis (Essl.) Hale & M.J. Lai [ESSLIDA] (yellow rag)
SY: *Cetraria idahoensis* Essl.
Evernia Ach. [EVERNIA$]
Evernia divaricata (L.) Ach. [EVERDIV]
Evernia mesomorpha Nyl. [EVERMES] (spruce moss)
SY: *Evernia thamnodes* (Flotow) Arnold
Evernia perfragilis Llano [EVERPER]
Evernia prunastri (L.) Ach. [EVERPRU] (antlered perfume)
Evernia thamnodes (Flotow) Arnold = Evernia mesomorpha [EVERMES]
Evernia vulpina (L.) Ach. = Letharia vulpina [LETHVUL]
Flavocetraria Kärnefelt & Thell [FLAVOCE$]
Flavocetraria cucullata (Bellardi) Kärnefelt & Thell [FLAVCUC] (furled paperdoll)
SY: *Allocetraria cucullata* (Bellardi) Randlane & Saag
SY: *Cetraria cucullata* (Bellardi) Ach.
Flavocetraria nivalis (L.) Kärnefelt & Thell [FLAVNIV] (flattened snow light)
SY: *Allocetraria nivalis* (L.) Randlane & Saag
SY: *Cetraria nivalis* (L.) Ach.
Flavopunctelia (Krog) Hale [FLAVOPU$]
Flavopunctelia flaventior (Stirton) Hale [FLAVFLA] (green speckleback)
SY: *Parmelia andreana* Müll. Arg.
SY: *Parmelia flaventior* Stirton
SY: *Parmelia kernstockii* (Lynge) Zahlbr.
SY: *Punctelia flaventior* (Stirton) Krog
Foraminella ambigua (Wulfen) S.F. Meyer = Parmeliopsis ambigua [PARMAMB]
Foraminella hyperopta (Ach.) S.F. Meyer = Parmeliopsis hyperopta [PARMHYE]
Hypogymnia (Nyl.) Nyl. [HYPOGYM$]
Hypogymnia apinnata Goward & McCune [HYPOAPI] (beaded bone)
Hypogymnia austerodes (Nyl.) Räsänen [HYPOAUS] (powdered bone)
SY: *Parmelia austerodes* Nyl.
Hypogymnia bitteri (Lynge) Ahti [HYPOBIT]
SY: *Parmelia bitteri* Lynge
Hypogymnia duplicata (Ach.) Rass. [HYPODUP] (tickertape bone)
SY: *Hypogymnia elongata* (Hillm.) Rass.
Hypogymnia elongata (Hillm.) Rass. = Hypogymnia duplicata [HYPODUP]
Hypogymnia enteromorpha (Ach.) Nyl. [HYPOENT]
Hypogymnia heterophylla L. Pike [HYPOHET] (seaside bone)
Hypogymnia imshaugii Krog [HYPOIMS]
Hypogymnia imshaugii var. *inactiva* Krog = Hypogymnia inactiva [HYPOINA]
Hypogymnia inactiva (Krog) Ohlsson [HYPOINA] (forking bone)
SY: *Hypogymnia imshaugii* var. *inactiva* Krog
Hypogymnia metaphysodes (Asah.) Rass. [HYPOMET] (deflated bone)
Hypogymnia occidentalis L. Pike [HYPOOCC] (lattice bone)
Hypogymnia oceanica Goward [HYPOOCE]
SY: *Hypogymnia pseudophysodes* auct.
Hypogymnia oroarctica Krog = Brodoa oroarctica [BRODORO]
Hypogymnia physodes (L.) Nyl. [HYPOPHY]
SY: *Parmelia duplicata* var. *douglasicola* Gyelnik

SY: *Parmelia oregana* Gyelnik
SY: *Parmelia physodes* (L.) Ach.
Hypogymnia pseudophysodes auct. = Hypogymnia oceanica [HYPOOCE]
Hypogymnia rugosa (G. Merr.) L. Pike [HYPORUG] (puckered bone)
Hypogymnia subobscura (Vainio) Poelt [HYPOSUO] (heath bone)
SY: *Parmelia subobscura* Vainio
Hypogymnia tubulosa (Schaerer) Hav. [HYPOTUB] (dog bone)
SY: *Parmelia tubulosa* (Schaerer) Bitter
Hypogymnia vittata (Ach.) Parrique [HYPOVIT] (monk's-hood)
SY: *Parmelia vittata* (Ach.) Nyl.
Hypotrachyna (Vainio) Hale [HYPOTRA$]
Hypotrachyna revoluta (Flörke) Hale [HYPOREV] (grey loop)
SY: *Parmelia revoluta* Flörke
Hypotrachyna sinuosa (Sm.) Hale [HYPOSIN] (green loop)
SY: *Parmelia sinuosa* (Sm.) Ach.
Imshaugia S.F. Meyer [IMSHAUG$]
Imshaugia aleurites (Ach.) S.F. Meyer [IMSHALE] (salted starburst)
SY: *Parmeliopsis aleurites* (Ach.) Nyl.
Kaernefeltia Thell & Goward [KAERNEF$]
Kaernefeltia californica (Tuck.) Thell & Goward [KAERCAL] (seaside thornbush)
SY: *Alectoria californica* (Tuck.) G. Merr.
SY: *Alectoria cetrariza* Nyl.
SY: *Coelocaulon cetrariza* (Nyl.) Gyelnik
SY: *Cornicularia californica* (Tuck.) Du Rietz
SY: *Cetraria californica* Tuck.
Kaernefeltia merrillii (Du Rietz) Thell & Goward [KAERMER] (blackened thornbush)
SY: *Cetraria iberica* Crespo & Barreno
SY: *Cetraria merrillii* Du Rietz
Letharia (Th. Fr.) Zahlbr. [LETHARI$]
Letharia californica (Lév.) Hue = Letharia columbiana [LETHCOL]
Letharia columbiana (Nutt.) J.W. Thomson [LETHCOL]
SY: *Letharia californica* (Lév.) Hue
SY: *Letharia vulpina* f. *californica* (Lév.) W.A. Weber
Letharia vulpina (L.) Hue [LETHVUL] (wolf lichen)
SY: *Chlorea vulpina* Nyl.
SY: *Evernia vulpina* (L.) Ach.
Letharia vulpina f. *californica* (Lév.) W.A. Weber = Letharia columbiana [LETHCOL]
Lichen intestiniformis Vill. = Brodoa oroarctica [BRODORO]
Masonhalea Kärnefelt [MASONHA$]
Masonhalea richardsonii (Hook.) Kärnefelt [MASORIC] (arctic tumbleweed)
SY: *Cetraria richardsonii* Hook.
Melanelia Essl. [MELANEL$]
Melanelia agnata (Nyl.) Thell [MELAAGN] (leather brown)
SY: *Cetraria agnata* (Nyl.) Kristinsson
Melanelia albertana (Ahti) Essl. [MELAALE]
SY: *Parmelia albertana* Ahti
Melanelia commixta (Nyl.) Thell [MELACOM] (rock brown)
SY: *Cetraria commixta* (Nyl.) Th. Fr.
SY: *Cetraria fahlunensis* (L.) Schreber
Melanelia disjuncta (Erichsen) Essl. [MELADIS] (powdered brown)
SY: *Melanelia granulosa* (Lynge) Essl.
SY: *Parmelia denalii* Krog
SY: *Parmelia disjuncta* Erichsen
SY: *Parmelia granulosa* Lynge
Melanelia elegantula (Zahlbr.) Essl. [MELAELE]
SY: *Melanelia incolorata* (Parr.) Essl.
SY: *Parmelia elegantula* (Zahlbr.) Szat.
SY: *Parmelia elegantula* var. *americana* Räsänen
Melanelia exasperatula (Nyl.) Essl. [MELAEXS] (lustrous brown)
SY: *Parmelia exasperatula* Nyl.
Melanelia fuliginosa (Fr. ex Duby) Essl. [MELAFUL]
SY: *Melanelia glabratula* (Lamy) Essl.
SY: *Parmelia glabratula* (Lamy) Nyl.
Melanelia glabratula (Lamy) Essl. = Melanelia fuliginosa [MELAFUL]
Melanelia granulosa (Lynge) Essl. = Melanelia disjuncta [MELADIS]
Melanelia hepatizon (Ach.) Thell [MELAHEP] (rock brow)
SY: *Cetraria hepatizon* (Ach.) Vainio
SY: *Cetraria polyschiza* (Nyl.) Jatta
Melanelia incolorata (Parr.) Essl. = Melanelia elegantula [MELAELE]
Melanelia infumata (Nyl.) Essl. [MELAINF] (elegant brown)
SY: *Parmelia infumata* Nyl.
Melanelia loxodes (Nyl.) Essl. = Neofuscelia loxodes [NEOFLOX]
Melanelia multispora (A. Schneider) Essl. [MELAMUL] (eyed brown)
SY: *Parmelia multispora* A. Schneider
Melanelia olivaceoides (Krog) Essl. [MELAOLI]
SY: *Parmelia olivaceoides* Krog
Melanelia panniformis (Nyl.) Essl. [MELAPAN] (lattice brown)
SY: *Parmelia panniformis* (Nyl.) Vainio
Melanelia septentrionalis (Lynge) Essl. [MELASET] (northern brown)
SY: *Parmelia septentrionalis* (Lynge) Ahti
Melanelia sorediata (Ach.) Goward & Ahti [MELASOR]
SY: *Melanelia sorediosa* (Almb.) Essl.
SY: *Parmelia sorediata* (Ach.) Th. Fr.
SY: *Parmelia sorediosa* Almb.
Melanelia sorediosa (Almb.) Essl. = Melanelia sorediata [MELASOR]
Melanelia stygia (L.) Essl. [MELASTY]
SY: *Parmelia stygia* (L.) Ach.
Melanelia subargentifera (Nyl.) Essl. [MELASUA]
SY: *Parmelia conspurcata* (Schaerer) Vainio
SY: *Parmelia subargentifera* Nyl.
Melanelia subaurifera (Nyl.) Essl. [MELASUU] (abraded brown)
SY: *Parmelia subaurifera* Nyl.
Melanelia subelegantula (Essl.) Essl. [MELASUB] (subelegant brown)
SY: *Parmelia subelegantula* Essl.
Melanelia subolivacea (Nyl.) Essl. [MELASUO]
SY: *Parmelia subolivacea* Nyl.
Melanelia substygia (Räsänen) Essl. = Melanelia tominii [MELATOM]
Melanelia tominii (Oksner) Essl. [MELATOM]
SY: *Melanelia substygia* (Räsänen) Essl.
SY: *Parmelia saximontana* R. Anderson & W.A. Weber
SY: *Parmelia substygia* Räsänen
Melanelia trabeculata (Ahti) Essl. [MELATRA] (baby brown)
SY: *Parmelia trabeculata* Ahti
Menegazzia A. Massal. [MENEGAZ$]
Menegazzia pertusa (Schrank) Stein = Menegazzia terebrata [MENETER]
Menegazzia terebrata (Hoffm.) A. Massal. [MENETER] (magic treeflute)
SY: *Menegazzia pertusa* (Schrank) Stein
SY: *Parmelia pertusa* (Schrank) Schaerer
SY: *Parmelia sipeana* Gyelnik
Neofuscelia Essl. [NEOFUSC$]
Neofuscelia loxodes (Nyl.) Essl. [NEOFLOX] (blistered brown)
SY: *Melanelia loxodes* (Nyl.) Essl.
SY: *Parmelia isidiotyla* Nyl.
SY: *Parmelia loxodes* Nyl.
Neofuscelia subhosseana (Essl.) Essl. [NEOFSUB]
SY: *Parmelia subhosseana* Essl.
Neofuscelia verruculifera (Nyl.) Essl. [NEOFVER]
SY: *Parmelia verruculifera* Nyl.
Nephromopsis platyphylla (Tuck.) Herre = Tuckermannopsis platyphylla [TUCKPLA]
Nodobryoria Common & Brodo [NODOBRY$]
Nodobryoria abbreviata (Müll. Arg.) Common & Brodo [NODOABB]
SY: *Alectoria abbreviata* (Müll. Arg.) R. Howe
SY: *Alectoria divergens* f. *abbreviata* Müll. Arg.
SY: *Bryoria abbreviata* (Müll. Arg.) Brodo & D. Hawksw.
Nodobryoria oregana (Tuck.) Common & Brodo [NODOORE]
SY: *Alectoria oregana* Tuck.
SY: *Bryoria oregana* (Tuck. ex Willey) Brodo & D. Hawksw.
Nodobryoria subdivergens (E. Dahl) Common & Brodo [NODOSUB]

SY: *Alectoria subdivergens* E. Dahl
SY: *Bryoria subdivergens* (E. Dahl) Brodo & D. Hawksw.
Parmelia Ach. [PARMELI$]
Parmelia albertana Ahti = Melanelia albertana [MELAALE]
Parmelia aleuritica Nyl. = Arctoparmelia centrifuga [ARCTCEN]
Parmelia almquistii Vainio = Allantoparmelia almquistii [ALLAALM]
Parmelia alpicola Th. Fr. = Allantoparmelia alpicola [ALLAALP]
Parmelia andreana Müll. Arg. = Flavopunctelia flaventior [FLAVFLA]
Parmelia arnoldii Du Rietz = Parmotrema arnoldii [PARMARO]
Parmelia austerodes Nyl. = Hypogymnia austerodes [HYPOAUS]
Parmelia birulae var. *grumosa* Llano = Arctoparmelia separata [ARCTSEP]
Parmelia bitteri Lynge = Hypogymnia bitteri [HYPOBIT]
Parmelia centrifuga (L.) Ach. = Arctoparmelia centrifuga [ARCTCEN]
Parmelia cetrarioides (Delise ex Duby) Nyl. = Cetrelia cetrarioides [CETRCET]
Parmelia conspurcata (Schaerer) Vainio = Melanelia subargentifera [MELASUA]
Parmelia crinita Ach. = Parmotrema crinitum [PARMCRN]
Parmelia cumberlandia (Gyelnik) Hale = Xanthoparmelia cumberlandia [XANTCUM]
Parmelia denalii Krog = Melanelia disjuncta [MELADIS]
Parmelia disjuncta Erichsen = Melanelia disjuncta [MELADIS]
Parmelia dubia (Wulfen) Schaerer = Punctelia subrudecta [PUNCSUR]
Parmelia duplicata var. *douglasicola* Gyelnik = Hypogymnia physodes [HYPOPHY]
Parmelia elegantula (Zahlbr.) Szat. = Melanelia elegantula [MELAELE]
Parmelia elegantula var. *americana* Räsänen = Melanelia elegantula [MELAELE]
Parmelia exasperatula Nyl. = Melanelia exasperatula [MELAEXS]
Parmelia flaventior Stirton = Flavopunctelia flaventior [FLAVFLA]
Parmelia fraudans (Nyl.) Nyl. [PARMFRA] (green shield)
Parmelia glabratula (Lamy) Nyl. = Melanelia fuliginosa [MELAFUL]
Parmelia graminicola de Lesd. = Punctelia subrudecta [PUNCSUR]
Parmelia granulosa Lynge = Melanelia disjuncta [MELADIS]
Parmelia hygrophila Goward & Ahti [PARMHYG]
Parmelia incurva (Pers.) Fr. = Arctoparmelia incurva [ARCTINC]
Parmelia infumata Nyl. = Melanelia infumata [MELAINF]
Parmelia isidiotyla Nyl. = Neofuscelia loxodes [NEOFLOX]
Parmelia kerguelensis A. Wilson = Parmelia saxatilis [PARMSAX]
Parmelia kerguelensis auct. = Parmelia pseudosulcata [PARMPSE]
Parmelia kernstockii (Lynge) Zahlbr. = Flavopunctelia flaventior [FLAVFLA]
Parmelia lanata (L.) Wallr. = Pseudephebe pubescens [PSEUPUB]
Parmelia lineola E.C. Berry = Xanthoparmelia lineola [XANTLIN]
Parmelia loxodes Nyl. = Neofuscelia loxodes [NEOFLOX]
Parmelia mexicana Gyelnik = Xanthoparmelia mexicana [XANTMEX]
Parmelia mougeotii Schaerer = Xanthoparmelia mougeotii [XANTMOU]
Parmelia multispora A. Schneider = Melanelia multispora [MELAMUL]
Parmelia neoconspersa Gyelnik = Xanthoparmelia neoconspersa [XANTNEO]
Parmelia olivaceoides Krog = Melanelia olivaceoides [MELAOLI]
Parmelia omphalodes (L.) Ach. [PARMOMP] (unsalted shield)
Parmelia oregana Gyelnik = Hypogymnia physodes [HYPOPHY]
Parmelia panniformis (Nyl.) Vainio = Melanelia panniformis [MELAPAN]
Parmelia perlata (Hudson) Ach. = Parmotrema chinense [PARMCHN]
Parmelia pertusa (Schrank) Schaerer = Menegazzia terebrata [MENETER]
Parmelia physodes (L.) Ach. = Hypogymnia physodes [HYPOPHY]
Parmelia plittii Gyelnik = Xanthoparmelia plittii [XANTPLI]
Parmelia pseudosulcata Gyelnik [PARMPSE]
SY: *Parmelia kerguelensis* auct.
Parmelia pubescens (L.) Vainio = Pseudephebe pubescens [PSEUPUB]
Parmelia pubescens var. *reticulata* Cromb. = Pseudephebe pubescens [PSEUPUB]
Parmelia revoluta Flörke = Hypotrachyna revoluta [HYPOREV]
Parmelia saxatilis (L.) Ach. [PARMSAX] (salted shield)
SY: *Parmelia kerguelensis* A. Wilson
Parmelia saxatilis var. *divaricata* (Nyl.) Hue = Parmelia squarrosa [PARMSQU]
Parmelia saximontana R. Anderson & W.A. Weber = Melanelia tominii [MELATOM]
Parmelia separata Th. Fr. = Arctoparmelia separata [ARCTSEP]
Parmelia septentrionalis (Lynge) Ahti = Melanelia septentrionalis [MELASET]
Parmelia sinuosa (Sm.) Ach. = Hypotrachyna sinuosa [HYPOSIN]
Parmelia sipeana Gyelnik = Menegazzia terebrata [MENETER]
Parmelia somloënsis Gyelnik = Xanthoparmelia somloënsis [XANTSOM]
Parmelia sorediata (Ach.) Th. Fr. = Melanelia sorediata [MELASOR]
Parmelia sorediosa Almb. = Melanelia sorediata [MELASOR]
Parmelia sphaerosporella Müll. Arg. = Ahtiana sphaerosporella [AHTISPH]
Parmelia squarrosa Hale [PARMSQU]
SY: *Parmelia saxatilis* var. *divaricata* (Nyl.) Hue
Parmelia stenophylla (Ach.) Du Rietz = Xanthoparmelia somloënsis [XANTSOM]
Parmelia stictica (Duby) Nyl. = Punctelia stictica [PUNCSTI]
Parmelia stygia (L.) Ach. = Melanelia stygia [MELASTY]
Parmelia subargentifera Nyl. = Melanelia subargentifera [MELASUA]
Parmelia subaurifera Nyl. = Melanelia subaurifera [MELASUU]
Parmelia subcentrifuga Oksner = Arctoparmelia subcentrifuga [ARCTSUB]
Parmelia subelegantula Essl. = Melanelia subelegantula [MELASUB]
Parmelia subhosseana Essl. = Neofuscelia subhosseana [NEOFSUB]
Parmelia subobscura Vainio = Hypogymnia subobscura [HYPOSUO]
Parmelia subolivacea Nyl. = Melanelia subolivacea [MELASUO]
Parmelia subrudecta Nyl. = Punctelia subrudecta [PUNCSUR]
Parmelia substygia Räsänen = Melanelia tominii [MELATOM]
Parmelia sulcata Taylor [PARMSUL] (powdered shield)
Parmelia tasmanica Hook. f. & Taylor = Xanthoparmelia tasmanica [XANTTAS]
Parmelia trabeculata Ahti = Melanelia trabeculata [MELATRA]
Parmelia trichotera Hue = Parmotrema chinense [PARMCHN]
Parmelia tubulosa (Schaerer) Bitter = Hypogymnia tubulosa [HYPOTUB]
Parmelia ulophylla (Ach.) G. Merr. = Punctelia subrudecta [PUNCSUR]
Parmelia verruculifera Nyl. = Neofuscelia verruculifera [NEOFVER]
Parmelia vittata (Ach.) Nyl. = Hypogymnia vittata [HYPOVIT]
Parmelia wyomingica (Gyelnik) Hale = Xanthoparmelia wyomingica [XANTWYO]
Parmeliopsis Nyl. [PARMELO$]
Parmeliopsis aleurites (Ach.) Nyl. = Imshaugia aleurites [IMSHALE]
Parmeliopsis ambigua (Wulfen) Nyl. [PARMAMB] (green starburst)
SY: *Foraminella ambigua* (Wulfen) S.F. Meyer
Parmeliopsis diffusa (Weber) Riddle = Parmeliopsis hyperopta [PARMHYE]
Parmeliopsis hyperopta (Ach.) Arnold [PARMHYE] (grey starburst)
SY: *Foraminella hyperopta* (Ach.) S.F. Meyer
SY: *Parmeliopsis diffusa* (Weber) Riddle
Parmotrema A. Massal. [PARMOTR$]
Parmotrema arnoldii (Du Rietz) Hale [PARMARO] (powdered scatter-rug)
SY: *Parmelia arnoldii* Du Rietz
Parmotrema chinense (Osbeck) Hale & Ahti [PARMCHN] (Chinese scatter-rug)
SY: *Parmelia perlata* (Hudson) Ach.
SY: *Parmelia trichotera* Hue
SY: *Parmotrema perlatum* (Hudson) Choisy
Parmotrema crinitum (Ach.) Choisy [PARMCRN] (salted scatter-rug)
SY: *Parmelia crinita* Ach.

Parmotrema perlatum (Hudson) Choisy = Parmotrema chinense [PARMCHN]
Parmotrema sinense (Osbeck) Hale & Ahti [PARMSIN]
Platismatia Culb. & C. Culb. [PLATISM$]
Platismatia glauca (L.) Culb. & C. Culb. [PLATGLA] (ragbag)
SY: *Cetraria glauca* (L.) Ach.
Platismatia herrei (Imshaug) Culb. & C. Culb. [PLATHER] (tattered rag)
SY: *Cetraria herrei* Imshaug
SY: *Cetraria tuckermanii* Herre
Platismatia lacunosa (Ach.) Culb. & C. Culb. [PLATLAC] (crinkled rag)
SY: *Cetraria lacunosa* Ach.
SY: *Cetraria lacunosa* var. *macounii* Du Rietz
SY: *Platismatia lacunosa* var. *macounii* (Du Rietz) Culb. & C. Culb.
Platismatia lacunosa var. *macounii* (Du Rietz) Culb. & C. Culb. = Platismatia lacunosa [PLATLAC]
Platismatia norvegica (Lynge) Culb. & C. Culb. [PLATNOR] (laundered rag)
SY: *Cetraria norvegica* (Lynge) Du Rietz
Platismatia stenophylla (Tuck.) Culb. & C. Culb. [PLATSTE] (ribbon rag)
SY: *Cetraria stenophylla* (Tuck.) G. Merr.
Pseudephebe Choisy [PSEUDEH$]
Pseudephebe minuscula (Nyl. ex Arnold) Brodo & D. Hawksw. [PSEUMIN]
SY: *Alectoria minuscula* (Nyl. ex Arnold) Degel.
Pseudephebe pubescens (L.) Choisy [PSEUPUB] (velcro lichen)
SY: *Alectoria pubescens* (L.) R. Howe
SY: *Ephebe pubescens* (L.) Fr.
SY: *Parmelia lanata* (L.) Wallr.
SY: *Parmelia pubescens* (L.) Vainio
SY: *Parmelia pubescens* var. *reticulata* Cromb.
Punctelia Krog [PUNCTEL$]
Punctelia flaventior (Stirton) Krog = Flavopunctelia flaventior [FLAVFLA]
Punctelia stictica (Duby) Krog [PUNCSTI] (seaside speckleback)
SY: *Parmelia stictica* (Duby) Nyl.
Punctelia subrudecta (Nyl.) Krog [PUNCSUR] (forest speckleback)
SY: *Parmelia dubia* (Wulfen) Schaerer
SY: *Parmelia graminicola* de Lesd.
SY: *Parmelia subrudecta* Nyl.
SY: *Parmelia ulophylla* (Ach.) G. Merr.
Tuckermannopsis Gyelnik [TUCKERM$]
Tuckermannopsis americana (Sprengel) Hale [TUCKAME] (fringed ruffle)
SY: *Cetraria ciliaris* var. *halei* (Culb. & C. Culb.) Ahti
SY: *Cetraria halei* Culb. & C. Culb.
SY: *Tuckermannopsis halei* (Culb. & C. Culb.) M.J. Lai
Tuckermannopsis canadensis (Räsänen) Hale = Vulpicida canadensis [VULPCAN]
Tuckermannopsis chlorophylla (Willd.) Hale [TUCKCHL] (shadow ruffle)
SY: *Cetraria chlorophylla* (Willd.) Vainio
SY: *Cetraria scutata* auct.
Tuckermannopsis halei (Culb. & C. Culb.) M.J. Lai = Tuckermannopsis americana [TUCKAME]
Tuckermannopsis juniperina auct. = Vulpicida canadensis [VULPCAN]
Tuckermannopsis orbata (Nyl.) M.J. Lai [TUCKORB] (variable ruffle)
SY: *Cetraria orbata* (Nyl.) Fink
Tuckermannopsis pallidula (Tuck. ex Riddle) Hale = Ahtiana pallidula [AHTIPAL]
Tuckermannopsis pinastri (Scop.) Hale = Vulpicida pinastri [VULPPIN]
Tuckermannopsis platyphylla (Tuck.) Hale [TUCKPLA] (weathered ruffle)
SY: *Cetraria platyphylla* Tuck.
SY: *Nephromopsis platyphylla* (Tuck.) Herre
Tuckermannopsis sepincola (Ehrh.) Hale [TUCKSEP] (eyed ruffle)
SY: *Cetraria scutata* (Wulfen) Poetsch
SY: *Cetraria sepincola* (Ehrh.) Ach.
Tuckermannopsis subalpina (Imshaug) Kärnefelt [TUCKSUB]
SY: *Cetraria arborialis* (Zahlbr.) Howard
SY: *Cetraria subalpina* Imshaug
Usnea Dill. ex Adans. [USNEA$]
Usnea alpina Mot. [USNEALP]
Usnea angulata Ach. [USNEANG]
Usnea barbata (L.) F.H. Wigg. [USNEBAR]
SY: *Usnea prostrata* Vainio ex Räsänen
Usnea barbata var. *ceratina* Schaerer = Usnea ceratina [USNECER]
Usnea barbata var. *dasypoga* Fr. = Usnea filipendula [USNEFIL]
Usnea barbata var. *florida* Fr. = Usnea florida [USNEFLO]
Usnea barbata var. *hirta* Fr. = Usnea hirta [USNEHIR]
Usnea betulina Mot. = Usnea glabrescens [USNEGLB]
Usnea capitata Mot. [USNECAP]
Usnea catenulata Mot. [USNECAT]
Usnea caucasica Vainio = Usnea filipendula [USNEFIL]
Usnea cavernosa Tuck. [USNECAV] (pitted beard)
Usnea ceratina Ach. [USNECER]
SY: *Usnea barbata* var. *ceratina* Schaerer
Usnea chaetophora Stirton [USNECHA]
Usnea comosa auct. = Usnea subfloridana [USNESUF]
Usnea comosa var. *stuppea* Räsänen = Usnea stuppea [USNESTU]
Usnea compacta (Räsänen) Mot. = Usnea glabrescens [USNEGLB]
Usnea cornuta Körber [USNECOR]
SY: *Usnea inflata* Delise ex Duby
SY: *Usnea intexta* Stirton
SY: *Usnea subpectinata* Stirton
Usnea dasypoga auct. = Usnea filipendula [USNEFIL]
Usnea dasypoga ssp. *melanopoga* Mot. = Usnea filipendula [USNEFIL]
Usnea dasypoga var. *subscabrata* Vainio = Usnea filipendula [USNEFIL]
Usnea diplotypus Vainio [USNEDIP]
Usnea distincta Mot. = Usnea glabrescens [USNEGLB]
Usnea esthonica Räsänen = Usnea filipendula [USNEFIL]
Usnea filipendula Stirton [USNEFIL]
SY: *Usnea barbata* var. *dasypoga* Fr.
SY: *Usnea caucasica* Vainio
SY: *Usnea dasypoga* ssp. *melanopoga* Mot.
SY: *Usnea dasypoga* var. *subscabrata* Vainio
SY: *Usnea esthonica* Räsänen
SY: *Usnea flagellata* Mot.
SY: *Usnea sublaxa* Vainio
SY: *Usnea dasypoga* auct.
Usnea flagellata Mot. = Usnea filipendula [USNEFIL]
Usnea florida (L.) F.H. Wigg. [USNEFLO]
SY: *Usnea barbata* var. *florida* Fr.
Usnea fragilescens Hav. ex Lynge [USNEFRA]
Usnea fragilescens var. **mollis** (Vainio) Clerc [USNEFRA1]
SY: *Usnea mollis* Stirton
Usnea fulvoreagens (Räsänen) Räsänen [USNEFUL]
Usnea glabrata (Ach.) Vainio [USNEGLA]
SY: *Usnea kujalae* Räsänen
SY: *Usnea sorediifera* (Arnold) Lynge
Usnea glabrescens (Nyl. ex Vainio) Vainio [USNEGLB]
SY: *Usnea betulina* Mot.
SY: *Usnea compacta* (Räsänen) Mot.
SY: *Usnea distincta* Mot.
Usnea graciosa Mot. [USNEGRA]
Usnea hesperina Mot. [USNEHES]
Usnea hirta (L.) F.H. Wigg. [USNEHIR] (sugared beard)
SY: *Usnea barbata* var. *hirta* Fr.
Usnea inflata Delise ex Duby = Usnea cornuta [USNECOR]
Usnea intexta Stirton = Usnea cornuta [USNECOR]
Usnea kujalae Räsänen = Usnea glabrata [USNEGLA]
Usnea lapponica Vainio [USNELAP] (powered beard)
SY: *Usnea laricina* Vainio ex Räsänen
SY: *Usnea sorediifera* auct.
Usnea laricina Vainio ex Räsänen = Usnea lapponica [USNELAP]
Usnea longissima Ach. [USNELON] (Methuselah's beard)
Usnea longissima var. **corticata** R.H. Howe [USNELON1]
Usnea longissima var. **longissima** [USNELON0]

Usnea madeirensis Mot. [USNEMAD]
Usnea merrillii Mot. [USNEMER]
Usnea mollis Stirton = Usnea fragilescens var. mollis [USNEFRA1]
Usnea nidulans Mot. [USNENID]
Usnea occidentalis Mot. [USNEOCC]
Usnea prostrata Vainio ex Räsänen = Usnea barbata [USNEBAR]
Usnea rigida (Ach.) Mot. [USNERIG]
Usnea rubicunda Stirton [USNERUB]
Usnea rugulosa Vainio = Usnea scabrata [USNESCB]
Usnea scabiosa Mot. [USNESCA]
Usnea scabrata Nyl. [USNESCB] (scruffy beard)
SY: *Usnea rugulosa* Vainio
SY: *Usnea scabrata* ssp. *nylanderiana* Mot.
Usnea scabrata ssp. *nylanderiana* Mot. = Usnea scabrata [USNESCB]
Usnea silesiaca Mot. [USNESIL]
Usnea similis Mot. ex Räsänen = Usnea subfloridana [USNESUF]
Usnea sorediifera (Arnold) Lynge = Usnea glabrata [USNEGLA]
Usnea sorediifera auct. = Usnea lapponica [USNELAP]
Usnea stuppea (Räsänen) Mot. [USNESTU]
SY: *Usnea comosa* var. *stuppea* Räsänen
Usnea subfloridana Stirton [USNESUF]
SY: *Usnea similis* Mot. ex Räsänen
SY: *Usnea comosa* auct.
Usnea sublaxa Vainio = Usnea filipendula [USNEFIL]
Usnea subpectinata Stirton = Usnea cornuta [USNECOR]
Usnea substerilis Mot. [USNESUB]
Usnea sylvatica Mot. [USNESYL]
Usnea trichodea Ach. [USNETRI]
Usnea wasmuthii Räsänen [USNEWAS]
Usnea wirthii Clerc [USNEWIR] (blood-spattered beard)
Vulpicida J.-E. Mattsson & M.J. Lai [VULPICI$]
Vulpicida canadensis (Räsänen) J.-E. Mattsson & M.J. Lai [VULPCAN] (brown-eyed sunshine)
SY: *Cetraria canadensis* (Räsänen) Räsänen
SY: *Tuckermannopsis canadensis* (Räsänen) Hale
SY: *Cetraria juniperina* auct.
SY: *Tuckermannopsis juniperina* auct.
Vulpicida pinastri (Scop.) J.-E. Mattsson & M.J. Lai [VULPPIN] (powdered sunshine)
SY: *Cetraria pinastri* (Scop.) Gray
SY: *Tuckermannopsis pinastri* (Scop.) Hale
Vulpicida tilesii (Ach.) J.-E. Mattsson & M.J. Lai [VULPTIL] (limestone sunshine)
SY: *Cetraria tilesii* Ach.
Xanthoparmelia (Vainio) Hale [XANTHOP$]
Xanthoparmelia angustiphylla (Gyelnik) Hale [XANTANG]
SY: *Xanthoparmelia hypopsila* auct.
Xanthoparmelia camtschadalis (Ach.) Hale [XANTCAM] (vagabond rockfrog)
Xanthoparmelia centrifuga (L.) Hale = Arctoparmelia centrifuga [ARCTCEN]
Xanthoparmelia coloradoensis (Gyelnik) Hale [XANTCOL] (Colorado rockfrog)
SY: *Xanthoparmelia taractica* auct.
Xanthoparmelia cumberlandia (Gyelnik) Hale [XANTCUM] (questionable rockfrog)
SY: *Parmelia cumberlandia* (Gyelnik) Hale
Xanthoparmelia hypopsila auct. = Xanthoparmelia angustiphylla [XANTANG]
Xanthoparmelia incurva (Pers.) Hale = Arctoparmelia incurva [ARCTINC]
Xanthoparmelia lineola (E.C. Berry) Hale [XANTLIN]
SY: *Parmelia lineola* E.C. Berry
Xanthoparmelia mexicana (Gyelnik) Hale [XANTMEX]
SY: *Parmelia mexicana* Gyelnik
Xanthoparmelia mougeotii (Schaerer) Hale [XANTMOU]
SY: *Parmelia mougeotii* Schaerer
Xanthoparmelia neoconspersa (Gyelnik) Hale [XANTNEO]
SY: *Parmelia neoconspersa* Gyelnik
Xanthoparmelia planilobata (Gyelnik) Hale [XANTPLA] (mini rockfrog)
Xanthoparmelia plittii (Gyelnk) Hale [XANTPLI] (salted rockfrog)
SY: *Parmelia plittii* Gyelnik
Xanthoparmelia separata (Th. Fr.) Hale = Arctoparmelia separata [ARCTSEP]
Xanthoparmelia somloënsis (Gyelnik) Hale [XANTSOM]
SY: *Parmelia somloënsis* Gyelnik
SY: *Parmelia stenophylla* (Ach.) Du Rietz
Xanthoparmelia subcentrifuga (Oksner) Hale = Arctoparmelia subcentrifuga [ARCTSUB]
Xanthoparmelia taractica auct. = Xanthoparmelia coloradoensis [XANTCOL]
Xanthoparmelia tasmanica (Hook. f. & Taylor) Hale [XANTTAS]
SY: *Parmelia tasmanica* Hook. f. & Taylor
Xanthoparmelia wyomingica (Gyelnik) Hale [XANTWYO] (variable rockfrog)
SY: *Parmelia wyomingica* (Gyelnik) Hale

PELTIGERACEAE (L)

Hydrothyria J.L. Russell [HYDROTH$]
Hydrothyria venosa J.L. Russell [HYDRVEN] (waterfan)
Massalongia Körber [MASSALO$]
Massalongia carnosa (Dickson) Körber [MASSCAR] (bluff mouse)
Massalongia microphylliza (Nyl. ex Hasse) Henssen [MASSMIC] (soil mouse)
SY: *Placynthium dubium* Herre
SY: *Placynthium microphyllizum* (Nyl. ex Hasse) Hasse
Peltigera Willd. [PELTIGE$]
Peltigera aphthosa (L.) Willd. [PELTAPH]
Peltigera aphthosa f. *complicata* (Th. Fr.) Zahlbr. = Peltigera leucophlebia [PELTLEU]
Peltigera aphthosa var. *variolosa* A. Massal. = Peltigera leucophlebia [PELTLEU]
Peltigera britannica (Gyelnik) Holt.-Hartw. & Tønsberg [PELTBRI]
Peltigera canina (L.) Willd. [PELTCAN]
Peltigera canina var. *extenuata* Nyl. ex Vainio = Peltigera didactyla var. extenuata [PELTDID1]
Peltigera canina var. *rufescens* (Weiss) Mudd = Peltigera rufescens [PELTRUF]
Peltigera canina var. *rufescens* f. *innovans* (Körber) J.W. Thomson = Peltigera praetextata [PELTPRA]
Peltigera canina var. *spuria* (Ach.) Schaerer = Peltigera didactyla [PELTDID]
Peltigera chionophila Goward [PELTCHI]
Peltigera cinnamomea Goward [PELTCIN]
Peltigera collina (Ach.) Schrader [PELTCOL] (tree pelt)
SY: *Peltigera scutata* (Dickson) Duby
Peltigera degenii Gyelnik [PELTDEG]
Peltigera didactyla (With.) J.R. Laundon [PELTDID] (temporary pelt)
SY: *Peltigera canina* var. *spuria* (Ach.) Schaerer
SY: *Peltigera erumpens* (Taylor) Elenkin
SY: *Peltigera spuria* (Ach.) DC.
Peltigera didactyla var. **didactyla** [PELTDID0]
Peltigera didactyla var. **extenuata** (Nyl. ex Vainio) Goffinet & Hastings [PELTDID1]
SY: *Peltigera canina* var. *extenuata* Nyl. ex Vainio
Peltigera elisabethae Gyelnik [PELTELI]
SY: *Peltigera horizontalis* f. *zopfii* auct.
Peltigera erumpens (Taylor) Elenkin = Peltigera didactyla [PELTDID]
Peltigera evansiana Gyelnik [PELTEVA] (peppered pelt)
Peltigera horizontalis (Hudson) Baumg. [PELTHOR] (concentric pelt)
Peltigera horizontalis f. *zopfii* auct. = Peltigera elisabethae [PELTELI]
Peltigera hymenina (Ach.) Delise [PELTHYM]
SY: *Peltigera lactucifolia* (With.) J.R. Laundon
SY: *Peltigera polydactyla* var. *hymenina* (Ach.) Flotow
Peltigera kristinssonii Vitik. [PELTKRI]
SY: *Peltigera occidentalis* auct.
Peltigera lactucifolia (With.) J.R. Laundon = Peltigera hymenina [PELTHYM]
Peltigera lepidophora (Vainio) Bitter [PELTLEP] (butterfly pelt)

Peltigera leucophlebia (Nyl.) Gyelnik [PELTLEU] (freckle pelt)
SY: *Peltigera aphthosa* f. *complicata* (Th. Fr.) Zahlbr.
SY: *Peltigera aphthosa* var. *variolosa* A. Massal.
SY: *Peltigera variolosa* (A. Massal.) Gyelnik
Peltigera malacea (Ach.) Funck [PELTMAL] (apple pelt)
Peltigera membranacea (Ach.) Nyl. [PELTMEM] (dog pelt)
Peltigera neckeri Hepp ex Müll. Arg. [PELTNEC]
Peltigera neopolydactyla (Gyelnik) Gyelnik [PELTNEO] (frog pelt)
SY: *Peltigera occidentalis* (E. Dahl) Kristinsson
SY: *Peltigera polydactyla* var. *neopolydactyla* Gyelnik
Peltigera occidentalis (E. Dahl) Kristinsson = Peltigera neopolydactyla [PELTNEO]
Peltigera occidentalis auct. = Peltigera kristinssonii [PELTKRI]
Peltigera pacifica Vitik. [PELTPAC]
Peltigera polydactyla (Necker) Hoffm. = Peltigera polydactylon [PELTPOL]
Peltigera polydactylon (Necker) Hoffm. [PELTPOL]
SY: *Peltigera polydactyla* (Necker) Hoffm.
Peltigera polydactyla var. *hymenina* (Ach.) Flotow = Peltigera hymenina [PELTHYM]
Peltigera polydactyla var. *neopolydactyla* Gyelnik = Peltigera neopolydactyla [PELTNEO]
Peltigera ponojensis Gyelnik [PELTPON] (felt pelt)
Peltigera praetextata (Flörke ex Sommerf.) Zopf [PELTPRA] (born-again pelt)
SY: *Peltigera canina* var. *rufescens* f. *innovans* (Körber) J.W. Thomson
Peltigera pulverulenta (Taylor) Kremp. = Peltigera scabrosa [PELTSCA]
Peltigera retifoveata Vitik. [PELTRET] (sponge pelt)
Peltigera rufescens (Weiss) Humb. [PELTRUF]
SY: *Peltigera canina* var. *rufescens* (Weiss) Mudd
Peltigera scabrosa Th. Fr. [PELTSCA] (toad pelt)
SY: *Peltigera pulverulenta* (Taylor) Kremp.
Peltigera scutata (Dickson) Duby = Peltigera collina [PELTCOL]
Peltigera spuria (Ach.) DC. = Peltigera didactyla [PELTDID]
Peltigera variolosa (A. Massal.) Gyelnik = Peltigera leucophlebia [PELTLEU]
Peltigera venosa (L.) Hoffm. [PELTVEN] (fan pelt)
Placynthium dubium Herre = Massalongia microphylliza [MASSMIC]
Placynthium microphyllizum (Nyl. ex Hasse) Hasse = Massalongia microphylliza [MASSMIC]
Solorina Ach. [SOLORIN$]
Solorina bispora Nyl. [SOLOBIS]
Solorina crocea (L.) Ach. [SOLOCRO] (chocolate chip)
Solorina octospora (Arnold) Arnold [SOLOOCT] (tundra owl)
Solorina saccata (L.) Ach. [SOLOSAC] (woodland owl)
Solorina spongiosa (Ach.) Anzi [SOLOSPO] (fringed owl)

PELTULACEAE (L)

Heppia euploca (Ach.) Vainio = Peltula euploca [PELTEUP]
Heppia guepinii (Delise) Nyl. = Peltula euploca [PELTEUP]
Heppia polyphylla de Lesd. = Peltula euploca [PELTEUP]
Peltula Nyl. [PELTULA$]
Peltula euploca (Ach.) Poelt [PELTEUP] (rock-olive)
SY: *Heppia euploca* (Ach.) Vainio
SY: *Heppia guepinii* (Delise) Nyl.
SY: *Heppia polyphylla* de Lesd.

PERTUSARIACEAE (L)

Lecanora glaucomela Tuck. = Pertusaria glaucomela [PERTGLA]
Ochrolechia A. Massal. [OCHROLE$]
Ochrolechia androgyna (Hoffm.) Arnold [OCHRAND]
SY: *Ochrolechia androgyna* var. *pergranulosa* Räsänen
SY: *Ochrolechia mahluensis* Räsänen
Ochrolechia androgyna var. *pergranulosa* Räsänen = Ochrolechia androgyna [OCHRAND]
Ochrolechia arborea (Kreyer) Almb. [OCHRARB]
SY: *Variolaria lactea* var. *arborea* Kreyer
Ochrolechia californica Vers. = Ochrolechia oregonensis [OCHRORE]
Ochrolechia elisabethae-kolae Vers. = Ochrolechia frigida [OCHRFRI]
Ochrolechia farinacea Howard [OCHRFAR]
Ochrolechia frigida (Sw.) Lynge [OCHRFRI]
SY: *Ochrolechia elisabethae-kolae* Vers.
SY: *Ochrolechia gonatodes* (Ach.) Räsänen
SY: *Ochrolechia pterulina* (Nyl.) Howard
Ochrolechia geminipara (Th. Fr.) Vainio = Pertusaria geminipara [PERTGEM]
Ochrolechia gonatodes (Ach.) Räsänen = Ochrolechia frigida [OCHRFRI]
Ochrolechia gowardii Brodo [OCHRGOW]
Ochrolechia gyalectina (Nyl.) Zahlbr. [OCHRGYA]
Ochrolechia juvenalis Brodo [OCHRJUV]
Ochrolechia laevigata (Räsänen) Vers. ex Brodo [OCHRLAE]
SY: *Ochrolechia pallescens* var. *laevigata* Räsänen
Ochrolechia mahluensis Räsänen = Ochrolechia androgyna [OCHRAND]
Ochrolechia montana Brodo [OCHRMON]
Ochrolechia oregonensis H. Magn. [OCHRORE]
SY: *Ochrolechia californica* Vers.
Ochrolechia pallescens var. *laevigata* Räsänen = Ochrolechia laevigata [OCHRLAE]
Ochrolechia parella (L.) A. Massal. [OCHRPAR]
Ochrolechia pterulina (Nyl.) Howard = Ochrolechia frigida [OCHRFRI]
Ochrolechia rhodoleuca (Th. Fr.) Brodo [OCHRRHO]
SY: *Pertusaria rhodoleuca* Th. Fr.
Ochrolechia sorediosa Howard = Ochrolechia szatalaënsis [OCHRSZA]
Ochrolechia subathallina H. Magn. [OCHRSUB]
Ochrolechia subpallescens Vers. [OCHRSUP]
Ochrolechia subplicans (Nyl.) Brodo [OCHRSUL]
SY: *Pertusaria subplicans* Nyl.
Ochrolechia subplicans ssp. **hultenii** (Erichsen) Brodo [OCHRSUL1]
SY: *Pertusaria hultenii* Erichsen
Ochrolechia szatalaënsis Vers. [OCHRSZA]
SY: *Ochrolechia sorediosa* Howard
Ochrolechia tartarea (L.) A. Massal. [OCHRTAR]
Ochrolechia turneri (Sm.) Hasselrot [OCHRTUR]
Ochrolechia upsaliensis (L.) A. Massal. [OCHRUPS]
Ochrolechia xanthostoma (Sommerf.) K. Schmitz & Lumbsch [OCHRXAN]
SY: *Pertusaria xanthostoma* (Sommerf.) Fr.
Pertusaria DC. [PERTUSA$]
Pertusaria alaskensis Erichsen [PERTALA]
SY: *Pertusaria aleutensis* Erichsen
Pertusaria albescens (Hudson) Choisy & Werner [PERTALB]
SY: *Pertusaria discoidea* (Pers.) Malme
Pertusaria aleutensis Erichsen = Pertusaria alaskensis [PERTALA]
Pertusaria amara (Ach.) Nyl. [PERTAMA]
Pertusaria ambigens (Nyl.) Tuck. = Pertusaria subambigens [PERTSUA]
Pertusaria borealis Erichsen [PERTBOR]
Pertusaria carneopallida (Nyl.) Anzi [PERTCAR]
SY: *Pertusaria protuberans* (Sommerf. ex Th. Fr.) Th. Fr.
Pertusaria chiodectonoides Bagl. ex A. Massal. [PERTCHI]
SY: *Pertusaria nolens* Nyl.
Pertusaria concentrica Erichsen = Pertusaria multipunctoides [PERTMUT]
Pertusaria coriacea (Th. Fr.) Th. Fr. [PERTCOR]
SY: *Pertusaria coriacea* var. *obducens* (Nyl.) Vainio
Pertusaria coriacea var. *obducens* (Nyl.) Vainio = Pertusaria coriacea [PERTCOR]
Pertusaria dactylina (Ach.) Nyl. [PERTDAC]
Pertusaria discoidea (Pers.) Malme = Pertusaria albescens [PERTALB]
Pertusaria geminipara (Th. Fr.) C. Knight ex Brodo [PERTGEM]
SY: *Ochrolechia geminipara* (Th. Fr.) Vainio
Pertusaria glaucomela (Tuck.) Nyl. [PERTGLA]

SY: *Lecanora glaucomela* Tuck.
SY: *Pertusaria subpupillaris* Vezda
Pertusaria gyalectina (Nyl.) Zahlbr. [PERTGYA]
Pertusaria hakkodensis Yasuda ex Räsänen [PERTHAK]
Pertusaria hultenii Erichsen = Ochrolechia subplicans ssp. hultenii [OCHRSUL1]
Pertusaria leioplaca DC. [PERTLEI]
SY: *Pertusaria leucostoma* A. Massal.
SY: *Pertusaria tabuliformis* Erichsen
Pertusaria leucostoma A. Massal. = Pertusaria leioplaca [PERTLEI]
Pertusaria multipunctoides Dibben [PERTMUT]
SY: *Pertusaria concentrica* Erichsen
Pertusaria nolens Nyl. = Pertusaria chiodectonoides [PERTCHI]
Pertusaria octomela (Norman) Erichsen [PERTOCT]
Pertusaria oculata (Dickson) Th. Fr. [PERTOCU]
Pertusaria ophthalmiza (Nyl.) Nyl. [PERTOPH]
Pertusaria panyrga (Ach.) A. Massal. [PERTPAN]
Pertusaria protuberans (Sommerf. ex Th. Fr.) Th. Fr. = Pertusaria carneopallida [PERTCAR]
Pertusaria pupillaris (Nyl.) Th. Fr. [PERTPUP]
Pertusaria rhodoleuca Th. Fr. = Ochrolechia rhodoleuca [OCHRRHO]
Pertusaria sommerfeltii (Flörke ex Sommerf.) Fr. [PERTSOM]
Pertusaria stenhammarii Hellb. [PERTSTE]
Pertusaria subambigens Dibben [PERTSUA]
SY: *Pertusaria ambigens* (Nyl.) Tuck.
Pertusaria subdactylina Nyl. [PERTSUB]
Pertusaria suboculata Brodo & Dibben [PERTSUO]
Pertusaria subplicans Nyl. = Ochrolechia subplicans [OCHRSUL]
Pertusaria subpupillaris Vezda = Pertusaria glaucomela [PERTGLA]
Pertusaria tabuliformis Erichsen = Pertusaria leioplaca [PERTLEI]
Pertusaria xanthodes Müll. Arg. [PERTXAN]
Pertusaria xanthostoma (Sommerf.) Fr. = Ochrolechia xanthostoma [OCHRXAN]
Varicellaria Nyl. [VARICEL$]
Varicellaria kemensis Räsänen = Varicellaria rhodocarpa [VARIRHO]
Varicellaria rhodocarpa (Körber) Th. Fr. [VARIRHO]
SY: *Varicellaria kemensis* Räsänen
Variolaria lactea var. *arborea* Kreyer = Ochrolechia arborea [OCHRARB]

PHLYCTIDACEAE (L)

Phlyctis Wallr. [PHLYCTS$]
Phlyctis argena (Sprengel) Flotow [PHLYARG]
Phlyctis speirea G. Merr. [PHLYSPE]

PHYLLACHORACEAE (L)

Lichenochora Hafellner [LICHENH$]
Lichenochora thallina (Cooke) Hafellner [LICHTHA]

PHYSCIACEAE (L)

Amandinea Choisy ex Scheid. & H. Mayrh. [AMANDIN$]
Amandinea coniops (Wahlenb.) Choisy ex Scheid. & H. Mayrh. [AMANCON]
SY: *Buellia coniops* (Wahlenb.) Th. Fr.
Amandinea punctata (Hoffm.) Coppins & Scheid. [AMANPUN] (button lichen)
SY: *Buellia myriocarpa* (DC.) De Not.
SY: *Buellia pullata* Tuck.
SY: *Buellia punctata* (Hoffm.) A. Massal.
Anaptychia Körber [ANAPTYC$]
Anaptychia kaspica Gyelnik = Anaptychia setifera [ANAPSET]
Anaptychia leucomelaena auct. = Heterodermia leucomelos [HETELEU]
Anaptychia pseudospeciosa var. *tremulans* (Müll. Arg.) Kurok. = Heterodermia speciosa [HETESPE]
Anaptychia setifera Räsänen [ANAPSET] (eyed centipede)
SY: *Anaptychia kaspica* Gyelnik
Anaptychia speciosa (Wulfen) A. Massal. = Heterodermia speciosa [HETESPE]
Anaptychia ulotrichoides (Vainio) Vainio [ANAPULO]
Buellia De Not. [BUELLIA$]
Buellia aethalea (Ach.) Th. Fr. [BUELAET]
SY: *Buellia aethaleoides* (Nyl.) H. Olivier
SY: *Buellia malmei* Lynge
SY: *Buellia verruculosa* (Sm.) Mudd
Buellia aethaleoides (Nyl.) H. Olivier = Buellia aethalea [BUELAET]
Buellia alboatra (Hoffm.) Th. Fr. = Diplotomma alboatrum [DIPLALO]
Buellia badia (Fr.) A. Massal. [BUELBAD]
Buellia callispora (C. Knight) J. Steiner = Hafellia callispora [HAFECAL]
Buellia chlorophaea (Hepp ex Leighton) Lettau = Diplotomma chlorophaeum [DIPLCHL]
Buellia coniops (Wahlenb.) Th. Fr. = Amandinea coniops [AMANCON]
Buellia crystallifera (Vainio) Hav. [BUELCRY]
Buellia curtisii (Tuck.) Imshaug [BUELCUT]
Buellia disciformis (Fr.) Mudd [BUELDIS]
SY: *Buellia parasema* (Ach.) De Not.
Buellia elegans Poelt [BUELELE]
Buellia erubescens Arnold [BUELERU]
SY: *Buellia zahlbruckneri* J. Steiner
Buellia fosteri Imshaug, nom. inval. = Hafellia fosteri [HAFEFOS]
Buellia geophila (Flörke ex Sommerf.) Lynge [BUELGEO]
Buellia griseovirens (Turner & Borrer ex Sm.) Almb. [BUELGRI]
SY: *Buellia hassei* Imshaug
Buellia halonia (Ach.) Tuck. [BUELHAL]
Buellia hassei Imshaug = Buellia griseovirens [BUELGRI]
Buellia insignis (Naeg. ex Hepp) Th. Fr. [BUELINS]
Buellia jugorum (Arnold) Arnold [BUELJUG]
Buellia lepidastra (Tuck.) Tuck. [BUELLEP]
Buellia malmei Lynge = Buellia aethalea [BUELAET]
Buellia myriocarpa (DC.) De Not. = Amandinea punctata [AMANPUN]
Buellia oidalea (Nyl.) Tuck. [BUELOID]
SY: *Rhizocarpon oidaleum* (Nyl.) Fink
Buellia papillata (Sommerf.) Tuck. [BUELPAP]
SY: *Buelliopsis papillata* (Sommerf.) Fink
Buellia parasema (Ach.) De Not. = Buellia disciformis [BUELDIS]
Buellia penichra (Tuck.) Hasse = Diplotomma penichrum [DIPLPEN]
Buellia pullata Tuck. = Amandinea punctata [AMANPUN]
Buellia pulverulenta (Anzi) Jatta [BUELPUL]
Buellia punctata (Hoffm.) A. Massal. = Amandinea punctata [AMANPUN]
Buellia restituta (Stirton) Zahlbr. = Hafellia callispora [HAFECAL]
Buellia stellulata (Taylor) Mudd [BUELSTE]
Buellia stillingiana J. Steiner [BUELSTL]
Buellia subconnexa (Stirton) Zahlbr. = Hafellia callispora [HAFECAL]
Buellia triphragmioides Anzi [BUELTRI]
Buellia turgescens Tuck. [BUELTUR]
SY: *Buellia turgescentoides* Fink
Buellia turgescentoides Fink = Buellia turgescens [BUELTUR]
Buellia verruculosa (Sm.) Mudd = Buellia aethalea [BUELAET]
Buellia zahlbruckneri J. Steiner = Buellia erubescens [BUELERU]
Buelliopsis papillata (Sommerf.) Fink = Buellia papillata [BUELPAP]
Dimelaena Norman [DIMELAE$]
Dimelaena novomexicana (de Lesd.) Hale & Culb. = Dimelaena oreina [DIMEORE]
Dimelaena oreina (Ach.) Norman [DIMEORE]
SY: *Dimelaena novomexicana* (de Lesd.) Hale & Culb.
SY: *Dimelaena suboreina* (de Lesd.) Hale & Culb.
SY: *Rinodina hueana* Vainio
SY: *Rinodina novomexicana* de Lesd.
SY: *Rinodina oreina* (Ach.) A. Massal.
SY: *Rinodina suboreina* de Lesd.
Dimelaena suboreina (de Lesd.) Hale & Culb. = Dimelaena oreina [DIMEORE]
Dimelaena thysanota (Tuck.) Hale & Culb. [DIMETHY]

SY: *Rinodina thysanota* Tuck.
Diplotomma Flotow [DIPLOTO$]
Diplotomma alboatrum (Hoffm.) Flotow [DIPLALO]
SY: *Buellia alboatra* (Hoffm.) Th. Fr.
SY: *Rhizocarpon alboatrum* (Hoffm.) Anzi
Diplotomma chlorophaeum (Hepp ex Leighton) Szat. [DIPLCHL]
SY: *Buellia chlorophaea* (Hepp ex Leighton) Lettau
SY: *Rhizocarpon chlorophaeum* Hepp ex Leighton
Diplotomma penichrum (Tuck.) Szat. [DIPLPEN]
SY: *Buellia penichra* (Tuck.) Hasse
SY: *Rhizocarpon penichrum* (Tuck.) G. Merr.
Hafellia Kalb, H. Mayrh. & Scheid. [HAFELLI$]
Hafellia callispora (C. Knight) H. Mayrh. & Sheard [HAFECAL]
SY: *Buellia callispora* (C. Knight) J. Steiner
SY: *Buellia restituta* (Stirton) Zahlbr.
SY: *Buellia subconnexa* (Stirton) Zahlbr.
SY: *Lecidea callispora* Knight
SY: *Lecidea restituta* Stirton
SY: *Lecidea subconnexa* Stirton
Hafellia fosteri Imshaug & Sheard [HAFEFOS]
SY: *Buellia fosteri* Imshaug, nom. inval.
Heterodermia Trevisan [HETEROD$]
Heterodermia leucomelaena (L.) Poelt = Heterodermia leucomelos [HETELEU]
Heterodermia leucomelos (L.) Poelt [HETELEU] (elegant centipede)
SY: *Heterodermia leucomelaena* (L.) Poelt
SY: *Anaptychia leucomelaena* auct.
Heterodermia sitchensis Goward & W. Noble [HETESIT] (seaside centipede)
Heterodermia speciosa (Wulfen) Trevisan [HETESPE] (powdered centipede)
SY: *Anaptychia pseudospeciosa* var. *tremulans* (Müll. Arg.) Kurok.
SY: *Anaptychia speciosa* (Wulfen) A. Massal.
SY: *Heterodermia tremulans* (Müll. Arg.) Culb.
Heterodermia tremulans (Müll. Arg.) Culb. = Heterodermia speciosa [HETESPE]
Lecidea callispora Knight = Hafellia callispora [HAFECAL]
Lecidea restituta Stirton = Hafellia callispora [HAFECAL]
Lecidea subconnexa Stirton = Hafellia callispora [HAFECAL]
Phaeophyscia Moberg [PHAEOPH$]
Phaeophyscia adiastola (Essl.) Essl. [PHAEADI]
SY: *Physcia adiastola* Essl.
Phaeophyscia ciliata (Hoffm.) Moberg [PHAECIL]
SY: *Physcia ciliata* (Hoffm.) Du Rietz
SY: *Physcia obscura* auct.
Phaeophyscia constipata (Norrlin & Nyl.) Moberg [PHAECON] (pincushion shadow)
SY: *Physcia constipata* Norrlin & Nyl.
Phaeophyscia endococcina (Körber) Moberg [PHAEEND] (starburst shadow)
SY: *Physcia columbiana* de Lesd.
SY: *Physcia endococcina* (Körber) Th. Fr.
SY: *Physcia lithotodes* Nyl.
Phaeophyscia hirsuta (Mereschk.) Essl. [PHAEHIR] (powdered shadow)
SY: *Physcia hirsuta* Mereschk.
Phaeophyscia hirtella Essl. [PHAEHIT]
Phaeophyscia hispidula (Ach.) Essl. [PHAEHIS] (whiskered shadow)
SY: *Physcia hispidula* (Ach.) Frey
SY: *Physcia setosa* (Ach.) Nyl.
Phaeophyscia hispidula ssp. **hispidula** [PHAEHIS0]
Phaeophyscia hispidula ssp. **limbata** Poelt [PHAEHIS1]
Phaeophyscia kairamoi (Vainio) Moberg [PHAEKAI]
SY: *Physcia kairamoi* Vainio
Phaeophyscia nigricans (Flörke) Moberg [PHAENIG] (one-horse shadow)
SY: *Physcia nigricans* (Flörke) Stizenb.
Phaeophyscia orbicularis (Necker) Moberg [PHAEORB] (granulated shadow)
SY: *Physcia orbicularis* (Necker) Poetsch
Phaeophyscia sciastra (Ach.) Moberg [PHAESCI] (five o'clock shadow)
SY: *Physcia lithotea* (Ach.) Nyl.
SY: *Physcia sciastra* (Ach.) Du Rietz
Phaeorrhiza H. Mayrh. & Poelt [PHAEORR$]
Phaeorrhiza nimbosa (Fr.) H. Mayrh. & Poelt [PHAENIM]
SY: *Rinodina nimbosa* (Fr.) Th. Fr.
SY: *Rinodina phaeocarpa* (Sommerf.) Vainio
Phaeorrhiza sareptana (Tomin) H. Mayrh. & Poelt [PHAESAR]
SY: *Rinodina sareptana* (Tomin) H. Magn.
Physcia (Schreber) Michaux [PHYSCIA$]
Physcia adiastola Essl. = Phaeophyscia adiastola [PHAEADI]
Physcia adscendens (Fr.) H. Olivier [PHYSADS] (hooded rosette)
Physcia aipolia (Ehrh. ex Humb.) Fürnr. [PHYSAIP] (grey-eyed rosette)
SY: *Physcia cainii* Räsänen
Physcia aipolia var. **aipolia** [PHYSAIP0]
Physcia aipolia var. **alnophila** (Vainio) Lynge [PHYSAIP1]
Physcia biziana (A. Massal.) Zahlbr. [PHYSBIZ]
Physcia caesia (Hoffm.) Fürnr. [PHYSCAE]
Physcia cainii Räsänen = Physcia aipolia [PHYSAIP]
Physcia callosa Nyl. [PHYSCAO] (beaded rosette)
Physcia cascadensis H. Magn. [PHYSCAS]
Physcia ciliata (Hoffm.) Du Rietz = Phaeophyscia ciliata [PHAECIL]
Physcia columbiana de Lesd. = Phaeophyscia endococcina [PHAEEND]
Physcia constipata Norrlin & Nyl. = Phaeophyscia constipata [PHAECON]
Physcia detersa (Nyl.) Nyl. = Physconia detersa [PHYSDET]
Physcia dimidiata (Arnold) Nyl. [PHYSDIM] (frosted rosette)
Physcia dubia (Hoffm.) Lettau [PHYSDUB] (powdered rosette)
SY: *Physcia intermedia* Vainio
SY: *Physcia teretiuscula* (Ach.) Lynge
Physcia endococcina (Körber) Th. Fr. = Phaeophyscia endococcina [PHAEEND]
Physcia hirsuta Mereschk. = Phaeophyscia hirsuta [PHAEHIR]
Physcia hispida auct. = Physcia tenella [PHYSTEN]
Physcia hispidula (Ach.) Frey = Phaeophyscia hispidula [PHAEHIS]
Physcia intermedia Vainio = Physcia dubia [PHYSDUB]
Physcia kairamoi Vainio = Phaeophyscia kairamoi [PHAEKAI]
Physcia leptalea (Ach.) DC. [PHYSLEP] (fringed rosette)
SY: *Physcia semipinnata* (J.F. Gmelin) Moberg
Physcia lithotea (Ach.) Nyl. = Phaeophyscia sciastra [PHAESCI]
Physcia lithotodes Nyl. = Phaeophyscia endococcina [PHAEEND]
Physcia melops Dufour = Physcia phaea [PHYSPHA]
Physcia muscigena (Ach.) Nyl. = Physconia muscigena [PHYSMUS]
Physcia nigricans (Flörke) Stizenb. = Phaeophyscia nigricans [PHAENIG]
Physcia obscura auct. = Phaeophyscia ciliata [PHAECIL]
Physcia orbicularis (Necker) Poetsch = Phaeophyscia orbicularis [PHAEORB]
Physcia phaea (Tuck.) J.W. Thomson [PHYSPHA] (black-eyed rosette)
SY: *Physcia melops* Dufour
Physcia sciastra (Ach.) Du Rietz = Phaeophyscia sciastra [PHAESCI]
Physcia semipinnata (J.F. Gmelin) Moberg = Physcia leptalea [PHYSLEP]
Physcia setosa (Ach.) Nyl. = Phaeophyscia hispidula [PHAEHIS]
Physcia stellaris (L.) Nyl. [PHYSSTE]
Physcia tenella (Scop.) DC. [PHYSTEN]
SY: *Physcia hispida* auct.
Physcia teretiuscula (Ach.) Lynge = Physcia dubia [PHYSDUB]
Physcia wainioi Räsänen [PHYSWAI]
Physconia Poelt [PHYSCOA$]
Physconia americana Essl. [PHYSAMR]
Physconia detersa (Nyl.) Poelt [PHYSDET] (bottlebrush frost)
SY: *Physcia detersa* (Nyl.) Nyl.
Physconia enteroxantha (Nyl.) Poelt [PHYSENT] (bordered frost)
Physconia farrea auct. = Physconia perisidiosa [PHYSPER]
Physconia muscigena (Ach.) Poelt [PHYSMUS] (ground frost)
SY: *Physcia muscigena* (Ach.) Nyl.
Physconia perisidiosa (Erichsen) Moberg [PHYSPER]
SY: *Physconia farrea* auct.

Physconia thomsonii Essl. [PHYSTHO]
Rhizocarpon alboatrum (Hoffm.) Anzi = Diplotomma alboatrum [DIPLALO]
Rhizocarpon chlorophaeum Hepp ex Leighton = Diplotomma chlorophaeum [DIPLCHL]
Rhizocarpon oidaleum (Nyl.) Fink = Buellia oidalea [BUELOID]
Rhizocarpon penichrum (Tuck.) G. Merr. = Diplotomma penichrum [DIPLPEN]
Rinodina (Ach.) Gray [RINODIN$]
Rinodina archaea (Ach.) Arnold [RINOARC]
Rinodina bolanderi H. Magn. [RINOBOL]
Rinodina calcigena (Th. Fr.) Lynge [RINOCAL]
SY: *Rinodina occidentalis* Lynge
Rinodina colobina (Ach.) Th. Fr. [RINOCOL]
Rinodina confragosa (Ach.) Körber [RINOCON]
Rinodina conradii Körber [RINOCOR]
SY: *Rinodina intermedia* Bagl.
SY: *Rinodina sabulosa* Tuck.
Rinodina disjuncta Sheard & Tønsberg [RINODIS]
Rinodina efflorescens Malme [RINOEFF]
Rinodina excrescens Vainio [RINOEXC]
Rinodina exigua (Ach.) Gray [RINOEXI]
Rinodina gennarii Bagl. [RINOGEN]
SY: *Rinodina salina* Degel.
Rinodina hallii Tuck. [RINOHAI]
Rinodina hueana Vainio = Dimelaena oreina [DIMEORE]
Rinodina intermedia Bagl. = Rinodina conradii [RINOCOR]
Rinodina marysvillensis H. Magn. [RINOMAR]
Rinodina mniaraea (Ach.) Körber [RINOMNI]
Rinodina mucronatula H. Magn. [RINOMUC]
Rinodina nimbosa (Fr.) Th. Fr. = Phaeorrhiza nimbosa [PHAENIM]
Rinodina novomexicana de Lesd. = Dimelaena oreina [DIMEORE]
Rinodina occidentalis Lynge = Rinodina calcigena [RINOCAL]
Rinodina orbata (Ach.) Vainio = Rinodina turfacea [RINOTUR]
Rinodina oreina (Ach.) A. Massal. = Dimelaena oreina [DIMEORE]
Rinodina phaeocarpa (Sommerf.) Vainio = Phaeorrhiza nimbosa [PHAENIM]
Rinodina pyrina (Ach.) Arnold [RINOPYR]
Rinodina roscida (Sommerf.) Arnold [RINOROS]
Rinodina sabulosa Tuck. = Rinodina conradii [RINOCOR]
Rinodina salina Degel. = Rinodina gennarii [RINOGEN]
Rinodina sareptana (Tomin) H. Magn. = Phaeorrhiza sareptana [PHAESAR]
Rinodina septentrionalis Malme [RINOSEP]
Rinodina suboreina de Lesd. = Dimelaena oreina [DIMEORE]
Rinodina thysanota Tuck. = Dimelaena thysanota [DIMETHY]
Rinodina turfacea (Wahlenb.) Körber [RINOTUR]
SY: *Rinodina orbata* (Ach.) Vainio

PILOCARPACEAE (L)

Byssoloma Trevisan [BYSSOLO$]
Byssoloma leucoblepharum (Nyl.) Vainio [BYSSLEU]
Byssoloma rotuliforme (Müll. Arg.) R. Sant. = Byssoloma subdiscordans [BYSSSUB]
Byssoloma subdiscordans (Nyl.) P. James [BYSSSUB]
SY: *Byssoloma rotuliforme* (Müll. Arg.) R. Sant.
Catillaria bouteillei (Desmaz.) Zahlbr. = Fellhanera bouteillei [FELLBOU]
Fellhanera Vezda [FELLHAN$]
Fellhanera bouteillei (Desmaz.) Vezda [FELLBOU]
SY: *Catillaria bouteillei* (Desmaz.) Zahlbr.
Fellhanera subtilis (Vezda) Diederich & Sérusiaux [FELLSUB]

PLACYNTHIACEAE (L)

Koerberia A. Massal. [KOERBER$]
Koerberia sonomensis (Tuck.) Henssen [KOERSON] (brownette)
SY: *Pannaria sonomensis* Tuck.
Leptochidium Choisy [LEPTOCI$]
Leptochidium albociliatum (Desmaz.) Choisy [LEPTALB] (whiskered tarpaper)
SY: *Leptogium albociliatum* Desmaz.
SY: *Leptogium pilosellum* G. Merr.
SY: *Polychidium albociliatum* (Desmaz.) Zahlbr.
Leptogidium dendriscum (Nyl.) Nyl. = Polychidium dendriscum [POLYDEN]
Leptogium albociliatum Desmaz. = Leptochidium albociliatum [LEPTALB]
Leptogium muscicola (Sw.) Fr. = Polychidium muscicola [POLYMUS]
Leptogium pilosellum G. Merr. = Leptochidium albociliatum [LEPTALB]
Pannaria isidiata Degel. = Vestergrenopsis isidiata [VESTISI]
Pannaria sonomensis Tuck. = Koerberia sonomensis [KOERSON]
Placynthium (Ach.) Gray [PLACYNT$]
Placynthium asperellum (Ach.) Trevisan [PLACASE]
SY: *Placynthium aspratile* (Ach.) Henssen
Placynthium aspratile (Ach.) Henssen = Placynthium asperellum [PLACASE]
Placynthium flabellosum (Tuck.) Zahlbr. [PLACFLA]
Placynthium nigrum (Hudson) Gray [PLACNIG] (quilted brownette)
Placynthium nigrum var. **nigrum** [PLACNIG0]
Placynthium nigrum var. **tantaleum** (Hepp) Arnold [PLACNIG1]
Placynthium stenophyllum (Tuck.) Fink [PLACSTE]
Placynthium stenophyllum var. **isidiatum** Henssen [PLACSTE1]
Placynthium subradiatum (Nyl.) Arnold [PLACSUR]
Polychidium (Ach.) Gray [POLYCHI$]
Polychidium albociliatum (Desmaz.) Zahlbr. = Leptochidium albociliatum [LEPTALB]
Polychidium contortum Henssen [POLYCOT]
Polychidium dendriscum (Nyl.) Henssen [POLYDEN]
SY: *Leptogidium dendriscum* (Nyl.) Nyl.
Polychidium muscicola (Sw.) Gray [POLYMUS]
SY: *Leptogium muscicola* (Sw.) Fr.
Vestergrenopsis Gyelnik [VESTERG$]
Vestergrenopsis elaeina (Wahlenb.) Gyelnik [VESTELA] (eyed brownette)
Vestergrenopsis isidiata (Degel.) E. Dahl [VESTISI] (peppered brownette)
SY: *Pannaria isidiata* Degel.

PORPIDIACEAE (L)

Amygdalaria Norman [AMYGDAL$]
Amygdalaria consentiens (Nyl.) Hertel, Brodo & Mas. Inoue [AMYGCON]
SY: *Huilia consentiens* (Nyl.) Hertel
Amygdalaria continua Brodo & Hertel [AMYGCOT]
Amygdalaria haidensis Brodo & Hertel [AMYGHAI]
Amygdalaria panaeola (Ach.) Hertel & Brodo [AMYGPAN]
SY: *Huilia panaeola* (Ach.) Hertel
SY: *Lecidea panaeola* (Ach.) Ach.
Amygdalaria pelobotryon (Wahlenb.) Norman [AMYGPEL]
SY: *Aspicilia pelobotrya* (Wahlenb.) Th. Fr.
SY: *Lecanora pelobotrya* (Wahlenb.) Sommerf.
SY: *Lecidea pelobotrya* (Wahlenb.) Leighton
Amygdalaria subdissentiens (Nyl.) Mas. Inoue & Brodo [AMYGSUB]
Aspicilia alpina (Sommerf.) Arnold = Bellemerea alpina [BELLALP]
Aspicilia cinereorufescens (Ach.) A. Massal. = Bellemerea cinereorufescens [BELLCIN]
Aspicilia diamarta (Ach.) Boistel = Bellemerea diamarta [BELLDIA]
Aspicilia pelobotrya (Wahlenb.) Th. Fr. = Amygdalaria pelobotryon [AMYGPEL]
Aspicilia subsorediza (Lynge) R. Sant. = Bellemerea subsorediza [BELLSUB]
Bacidia accedens (Arnold) Lettau = Mycobilimbia sabuletorum [MYCOSAB]
Bacidia fusca (A. Massal.) Du Rietz = Mycobilimbia tetramera [MYCOTET]
Bacidia hypnophila (Turner ex Ach.) Zahlbr. = Mycobilimbia sabuletorum [MYCOSAB]

Bacidia obscurata (Sommerf.) Zahlbr. = Mycobilimbia tetramera [MYCOTET]
Bacidia sabuletorum (Schreber) Lettau = Mycobilimbia sabuletorum [MYCOSAB]
Bacidia tetramera (De Not.) Coppins = Mycobilimbia tetramera [MYCOTET]
Bellemerea Hafellner & Roux [BELLEME$]
Bellemerea alpina (Sommerf.) Clauzade & Roux [BELLALP]
SY: *Aspicilia alpina* (Sommerf.) Arnold
SY: *Lecanora alpina* Sommerf.
Bellemerea cinereorufescens (Ach.) Clauzade & Roux [BELLCIN]
SY: *Aspicilia cinereorufescens* (Ach.) A. Massal.
SY: *Lecanora cinereorufescens* (Ach.) Hepp
Bellemerea diamarta (Ach.) Hafellner & Roux [BELLDIA]
SY: *Aspicilia diamarta* (Ach.) Boistel
SY: *Lecanora diamarta* (Ach.) Vainio
Bellemerea subsorediza (Lynge) R. Sant. [BELLSUB]
SY: *Aspicilia subsorediza* (Lynge) R. Sant.
SY: *Lecidea subsorediza* Lynge
Clauzadea Hafellner & Bellem. [CLAUZAD$]
Clauzadea monticola (Ach. ex Schaerer) Hafellner & Bellem. [CLAUMON]
SY: *Lecidea monticola* Ach.
SY: *Protoblastenia monticola* (Ach.) J. Steiner
Farnoldia Hertel [FARNOLD$]
Farnoldia hypocrita (A. Massal.) Frøberg [FARNHYP]
SY: *Lecidea hypocrita* A. Massal.
SY: *Lecidea lithospersa* Zahlbr.
SY: *Lecidea ypocrita* A. Massal.
Farnoldia jurana (Schaerer) Hertel [FARNJUR]
SY: *Lecidea albosuffusa* Th. Fr.
SY: *Lecidea jurana* Schaerer
SY: *Melanolecia jurana* (Schaerer) Hertel
SY: *Tremolecia jurana* (Schaerer) Hertel
Huilia albocaerulescens (Wulfen) Hertel = Porpidia albocaerulescens [PORPALB]
Huilia consentiens (Nyl.) Hertel = Amygdalaria consentiens [AMYGCON]
Huilia crustulata (Ach.) Hertel = Porpidia crustulata [PORPCRU]
Huilia macrocarpa (DC.) Hertel = Porpidia macrocarpa [PORPMAC]
Huilia nigrocruenta (Anzi) Hertel = Porpidia macrocarpa [PORPMAC]
Huilia panaeola (Ach.) Hertel = Amygdalaria panaeola [AMYGPAN]
Huilia soredizodes (Lamy ex Nyl.) Hertel = Porpidia soredizodes [PORPSOR]
Huilia tuberculosa (Sm.) P. James = Porpidia tuberculosa [PORPTUB]
Hymenelia ochrolemma (Vainio) Gowan & Ahti = Porpidia ochrolemma [PORPOCH]
Koerberiella Stein [KOERBEI$]
Koerberiella wimmeriana (Körber) Stein [KOERWIM]
Lecanora alpina Sommerf. = Bellemerea alpina [BELLALP]
Lecanora cinereorufescens (Ach.) Hepp = Bellemerea cinereorufescens [BELLCIN]
Lecanora diamarta (Ach.) Vainio = Bellemerea diamarta [BELLDIA]
Lecanora pelobotrya (Wahlenb.) Sommerf. = Amygdalaria pelobotryon [AMYGPEL]
Lecidea albocaerulescens (Wulfen) Ach. = Porpidia albocaerulescens [PORPALB]
Lecidea albosuffusa Th. Fr. = Farnoldia jurana [FARNJUR]
Lecidea atrofusca (Hepp) Mudd = Mycobilimbia hypnorum [MYCOHYP]
Lecidea berengeriana (A. Massal.) Nyl. = Mycobilimbia berengeriana [MYCOBER]
Lecidea contigua Fr. = Porpidia macrocarpa [PORPMAC]
Lecidea contraponenda Arnold = Porpidia contraponenda [PORPCON]
Lecidea crustulata (Ach.) Sprengel = Porpidia crustulata [PORPCRU]
Lecidea fusca (Schaerer) Th. Fr. = Mycobilimbia hypnorum [MYCOHYP]
Lecidea hypnorum Lib. = Mycobilimbia hypnorum [MYCOHYP]
Lecidea hypocrita A. Massal. = Farnoldia hypocrita [FARNHYP]
Lecidea jurana Schaerer = Farnoldia jurana [FARNJUR]
Lecidea lithospersa Zahlbr. = Farnoldia hypocrita [FARNHYP]
Lecidea macrocarpa (DC.) Steudel = Porpidia macrocarpa [PORPMAC]
Lecidea monticola Ach. = Clauzadea monticola [CLAUMON]
Lecidea nigrocruenta Anzi = Porpidia macrocarpa [PORPMAC]
Lecidea panaeola (Ach.) Ach. = Amygdalaria panaeola [AMYGPAN]
Lecidea parasema var. *crustulata* Ach. = Porpidia crustulata [PORPCRU]
Lecidea pelobotrya (Wahlenb.) Leighton = Amygdalaria pelobotryon [AMYGPEL]
Lecidea phylliscina Nyl. = Porpidia macrocarpa [PORPMAC]
Lecidea platycarpa Ach. = Porpidia macrocarpa [PORPMAC]
Lecidea sanguineoatra auct. = Mycobilimbia hypnorum [MYCOHYP]
Lecidea soredifera Lowe = Porpidia crustulata [PORPCRU]
Lecidea sorediza Nyl. = Porpidia tuberculosa [PORPTUB]
Lecidea soredizodes (Lamy ex Nyl.) Vainio = Porpidia soredizodes [PORPSOR]
Lecidea steriza (Ach.) Vainio = Porpidia macrocarpa [PORPMAC]
Lecidea subsorediza Lynge = Bellemerea subsorediza [BELLSUB]
Lecidea templetonii Taylor = Mycobilimbia hypnorum [MYCOHYP]
Lecidea tumida A. Massal. = Porpidia tuberculosa [PORPTUB]
Lecidea ypocrita A. Massal. = Farnoldia hypocrita [FARNHYP]
Lichen albocaerulescens Wulfen = Porpidia albocaerulescens [PORPALB]
Melanolecia jurana (Schaerer) Hertel = Farnoldia jurana [FARNJUR]
Mycobilimbia Rehm [MYCOBIL$]
Mycobilimbia accedens (Arnold) V. Wirth ex Hafellner = Mycobilimbia sabuletorum [MYCOSAB]
Mycobilimbia berengeriana (A. Massal.) Hafellner & V. Wirth [MYCOBER]
SY: *Lecidea berengeriana* (A. Massal.) Nyl.
Mycobilimbia fusca (A. Massal.) Hafellner & V. Wirth = Mycobilimbia tetramera [MYCOTET]
Mycobilimbia hypnorum (Lib.) Kalb & Hafellner [MYCOHYP]
SY: *Lecidea atrofusca* (Hepp) Mudd
SY: *Lecidea fusca* (Schaerer) Th. Fr.
SY: *Lecidea hypnorum* Lib.
SY: *Lecidea templetonii* Taylor
SY: *Lecidea sanguineoatra* auct.
Mycobilimbia lobulata (Sommerf.) Hafellner [MYCOLOB]
SY: *Toninia lobulata* (Sommerf.) Lynge
Mycobilimbia obscurata (Sommerf.) Rehm = Mycobilimbia tetramera [MYCOTET]
Mycobilimbia sabuletorum (Schreber) Hafellner [MYCOSAB]
SY: *Bacidia accedens* (Arnold) Lettau
SY: *Bacidia hypnophila* (Turner ex Ach.) Zahlbr.
SY: *Bacidia sabuletorum* (Schreber) Lettau
SY: *Mycobilimbia accedens* (Arnold) V. Wirth ex Hafellner
Mycobilimbia tetramera (De Not.) W. Brunnbauer [MYCOTET]
SY: *Bacidia fusca* (A. Massal.) Du Rietz
SY: *Bacidia obscurata* (Sommerf.) Zahlbr.
SY: *Bacidia tetramera* (De Not.) Coppins
SY: *Mycobilimbia fusca* (A. Massal.) Hafellner & V. Wirth
SY: *Mycobilimbia obscurata* (Sommerf.) Rehm
Patellaria macrocarpa DC. = Porpidia macrocarpa [PORPMAC]
Porpidia Körber [PORPIDI$]
Porpidia albocaerulescens (Wulfen) Hertel & Knoph [PORPALB]
SY: *Huilia albocaerulescens* (Wulfen) Hertel
SY: *Lecidea albocaerulescens* (Wulfen) Ach.
SY: *Lichen albocaerulescens* Wulfen
Porpidia carlottiana Gowan [PORPCAR]
Porpidia contraponenda (Arnold) Knoph & Hertel [PORPCON]
SY: *Lecidea contraponenda* Arnold
Porpidia crustulata (Ach.) Hertel & Knoph [PORPCRU]
SY: *Huilia crustulata* (Ach.) Hertel
SY: *Lecidea crustulata* (Ach.) Sprengel
SY: *Lecidea parasema* var. *crustulata* Ach.
SY: *Lecidea soredifera* Lowe
Porpidia macrocarpa (DC.) Hertel & A.J. Schwab [PORPMAC]
SY: *Huilia macrocarpa* (DC.) Hertel
SY: *Huilia nigrocruenta* (Anzi) Hertel
SY: *Lecidea contigua* Fr.

SY: *Lecidea macrocarpa* (DC.) Steudel
SY: *Lecidea nigrocruenta* Anzi
SY: *Lecidea phylliscina* Nyl.
SY: *Lecidea platycarpa* Ach.
SY: *Lecidea steriza* (Ach.) Vainio
SY: *Patellaria macrocarpa* DC.
SY: *Porpidia nigrocruenta* (Anzi) Diederich & Sérus.
Porpidia nigrocruenta (Anzi) Diederich & Sérus. = Porpidia macrocarpa [PORPMAC]
Porpidia ochrolemma (Vainio) Brodo & R. Sant. [PORPOCH]
SY: *Hymenelia ochrolemma* (Vainio) Gowan & Ahti
SY: *Porpidia pseudomelinodes* A.J. Schwab
Porpidia pseudomelinodes A.J. Schwab = Porpidia ochrolemma [PORPOCH]
Porpidia soredizodes (Lamy ex Nyl.) J.R. Laundon [PORPSOR]
SY: *Huilia soredizodes* (Lamy ex Nyl.) Hertel
SY: *Lecidea soredizodes* (Lamy ex Nyl.) Vainio
Porpidia thomsonii Gowan [PORPTHO]
Porpidia tuberculosa (Sm.) Hertel & Knoph [PORPTUB]
SY: *Huilia tuberculosa* (Sm.) P. James
SY: *Lecidea sorediza* Nyl.
SY: *Lecidea tumida* A. Massal.
SY: *Spilonema tuberculosa* Sm.
Protoblastenia monticola (Ach.) J. Steiner = Clauzadea monticola [CLAUMON]
Spilonema tuberculosa Sm. = Porpidia tuberculosa [PORPTUB]
Toninia lobulata (Sommerf.) Lynge = Mycobilimbia lobulata [MYCOLOB]
Tremolecia jurana (Schaerer) Hertel = Farnoldia jurana [FARNJUR]

PROTOTHELENELLACEAE (L)

Microglaena sphinctrinoides (Nyl.) Lönnr. = Protothelenella sphinctrinoides [PROTSPH]
Protothelenella Räsänen [PROTOTH$]
Protothelenella leucothelia (Nyl.) H. Mayrh. & Poelt [PROTLEU]
Protothelenella sphinctrinoides (Nyl.) H. Mayrh. & Poelt [PROTSPH]
SY: *Microglaena sphinctrinoides* (Nyl.) Lönnr.

PSORACEAE (L)

Baeomyces rubiformis Ach. = Psora rubiformis [PSORRUB]
Biatora himalayana Church. Bab. = Psora himalayana [PSORHIM]
Lecidea brouardii (de Lesd.) Zahlbr. = Psorula rufonigra [PSORRUF]
Lecidea decipiens (Hedwig) Ach. = Psora decipiens [PSORDEC]
Lecidea demissa (Rutstr.) Ach. = Lecidoma demissum [LECIDEU]
Lecidea globifera Ach. = Psora globifera [PSORGLO]
Lecidea luridella Tuck. = Psora luridella [PSORLUR]
Lecidea novomexicana (de Lesd.) W.A. Weber ex R. Anderson = Psora nipponica [PSORNIP]
Lecidea rubiformis (Ach.) Wahlenb. = Psora rubiformis [PSORRUB]
Lecidea rufonigra (Tuck.) Nyl. = Psorula rufonigra [PSORRUF]
Lecidea russellii Tuck. = Psora russellii [PSORRUS]
Lecidoma Gotth. Schneider & Hertel [LECIDOM$]
Lecidoma demissum (Rutstr.) Gotth. Schneider & Hertel [LECIDEU]
SY: *Lecidea demissa* (Rutstr.) Ach.
SY: *Lepidoma demissum* (Rutstr.) Choisy
SY: *Psora demissa* (Rutstr.) Hepp
Lepidoma demissum (Rutstr.) Choisy = Lecidoma demissum [LECIDEU]
Protoblastenia (Zahlbr.) J. Steiner [PROTOBL$]
Protoblastenia incrustans (DC.) J. Steiner [PROTINC]
Protoblastenia rupestris (Scop.) J. Steiner [PROTRUP]
Protoblastenia terricola (Anzi) Lynge [PROTTER]
Psora Hoffm. [PSORA$]
Psora cerebriformis W.A. Weber [PSORCER] (fissured scale)
Psora decipiens (Hedwig) Hoffm. [PSORDEC] (sockeye scale)
SY: *Lecidea decipiens* (Hedwig) Ach.
Psora demissa (Rutstr.) Hepp = Lecidoma demissum [LECIDEU]
Psora elenkinii Rass. = Psora himalayana [PSORHIM]
Psora globifera (Ach.) A. Massal. [PSORGLO] (blackberry)
SY: *Lecidea globifera* Ach.
Psora himalayana (Church. Bab.) Timdal [PSORHIM] (mountain scale)
SY: *Biatora himalayana* Church. Bab.
SY: *Psora elenkinii* Rass.
Psora luridella (Tuck.) Fink [PSORLUR]
SY: *Lecidea luridella* Tuck.
Psora montana Timdal [PSORMON] (brown-eyed scale)
Psora nicolai de Lesd. = Psora russellii [PSORRUS]
Psora nipponica (Zahlbr.) Gotth. Schneider [PSORNIP] (butterfly scale)
SY: *Lecidea novomexicana* (de Lesd.) W.A. Weber ex R. Anderson
SY: *Psora novomexicana* de Lesd.
Psora novomexicana de Lesd. = Psora nipponica [PSORNIP]
Psora rubiformis (Ach.) Hook. [PSORRUB]
SY: *Baeomyces rubiformis* Ach.
SY: *Lecidea rubiformis* (Ach.) Wahlenb.
Psora rufonigra (Tuck.) A. Schneider = Psorula rufonigra [PSORRUF]
Psora russellii (Tuck.) A. Schneider [PSORRUS]
SY: *Lecidea russellii* Tuck.
SY: *Psora nicolai* de Lesd.
Psora tuckermanii R. Anderson ex Timdal [PSORTUC]
Psorula Gotth. Schneider [PSORULA$]
Psorula rufonigra (Tuck.) Gotth. Schneider [PSORRUF]
SY: *Lecidea brouardii* (de Lesd.) Zahlbr.
SY: *Lecidea rufonigra* (Tuck.) Nyl.
SY: *Psora rufonigra* (Tuck.) A. Schneider

PYRENULACEAE (L)

Arthopyrenia halodytes (Nyl.) Arnold = Pyrenocollema halodytes [PYREHAL]
Eopyrenula R.C. Harris [EOPYREN$]
Eopyrenula intermedia Coppins [EOPYINT]
SY: *Pyrenula leucoplaca* var. *pluriloculata* Fink
Parathelium microcarpum Riddle = Pyrenula microtheca [PYREMIC]
Pyrenocollema Reinke [PYRENOC$]
Pyrenocollema halodytes (Nyl.) R.C. Harris [PYREHAL]
SY: *Arthopyrenia halodytes* (Nyl.) Arnold
Pyrenula A. Massal. [PYRENUL$]
Pyrenula acutispora Kalb & Hafellner [PYREACU]
Pyrenula leucoplaca var. *pluriloculata* Fink = Eopyrenula intermedia [EOPYINT]
Pyrenula microtheca R.C. Harris [PYREMIC]
SY: *Parathelium microcarpum* Riddle
Pyrenula neglecta R.C. Harris = Pyrenula pseudobufonia [PYREPSE]
Pyrenula occidentalis (R.C. Harris) R.C. Harris [PYREOCC]
Pyrenula pseudobufonia (Rehm) R.C. Harris [PYREPSE]
SY: *Pyrenula neglecta* R.C. Harris

RAMALINACEAE (L)

Alectoria crinalis Ach. = Ramalina thrausta [RAMATHR]
Alectoria thrausta Ach. = Ramalina thrausta [RAMATHR]
Desmazieria cephalota (Tuck.) Follmann & Huneck = Niebla cephalota [NIEBCEP]
Fistulariella dilacerata (Hoffm.) Bowler & Rundel = Ramalina dilacerata [RAMADIL]
Fistulariella geniculata (Hook. f. & Taylor) Bowler & Rundel = Ramalina geniculata [RAMAGEN]
Fistulariella inflata (Hook. f. & Taylor) Bowler & Rundel = Ramalina inflata [RAMAINF]
Fistulariella minuscula (Nyl.) Bowler & Rundel = Ramalina dilacerata [RAMADIL]
Fistulariella roesleri (Hochst. ex Schaerer) Bowler & Rundel = Ramalina roesleri [RAMAROE]
Niebla Rundel & Bowler [NIEBLA$]
Niebla cephalota (Tuck.) Rundel & Bowler [NIEBCEP]
SY: *Desmazieria cephalota* (Tuck.) Follmann & Huneck
SY: *Ramalina cephalota* Tuck.

Ramalina Ach. [RAMALIN$]
Ramalina calicaris var. *farinacea* Schaerer = Ramalina farinacea [RAMAFAR]
Ramalina cephalota Tuck. = Niebla cephalota [NIEBCEP]
Ramalina crinalis (Ach.) Gyelnik = Ramalina thrausta [RAMATHR]
Ramalina dilacerata (Hoffm.) Hoffm. [RAMADIL] (punctured gristle)
SY: *Fistulariella dilacerata* (Hoffm.) Bowler & Rundel
SY: *Fistulariella minuscula* (Nyl.) Bowler & Rundel
SY: *Ramalina minuscula* (Nyl.) Nyl.
Ramalina dilacerata f. **pollinariella** Arnold [RAMADIL1]
Ramalina farinacea (L.) Ach. [RAMAFAR]
SY: *Ramalina calicaris* var. *farinacea* Schaerer
SY: *Ramalina farinacea* Schaerer
SY: *Ramalina hypoprotocetrarica* Culb.
SY: *Ramalina reagens* (de Lesd.) Culb.
Ramalina farinacea Schaerer = Ramalina farinacea [RAMAFAR]
Ramalina geniculata Hook. f. & Taylor [RAMAGEN]
SY: *Fistulariella geniculata* (Hook. f. & Taylor) Bowler & Rundel
Ramalina hypoprotocetrarica Culb. = Ramalina farinacea [RAMAFAR]
Ramalina inflata (Hook. f. & Taylor) Hook. f. & Taylor [RAMAINF]
SY: *Fistulariella inflata* (Hook. f. & Taylor) Bowler & Rundel
Ramalina menziesii Taylor [RAMAMEN] (fishnet)
SY: *Ramalina reticulata* (Nöhden) Kremp.
Ramalina minuscula (Nyl.) Nyl. = Ramalina dilacerata [RAMADIL]
Ramalina obtusata (Arnold) Bitter [RAMAOBT]
Ramalina pollinaria (Westr.) Ach. [RAMAPOL] (dusty gristle)
Ramalina pollinariella Nyl. = Ramalina roesleri [RAMAROE]
Ramalina reagens (de Lesd.) Culb. = Ramalina farinacea [RAMAFAR]
Ramalina reticulata (Nöhden) Kremp. = Ramalina menziesii [RAMAMEN]
Ramalina roesleri (Hochst. ex Schaerer) Hue [RAMAROE]
SY: *Fistulariella roesleri* (Hochst. ex Schaerer) Bowler & Rundel
SY: *Ramalina pollinariella* Nyl.
Ramalina sinensis Jatta [RAMASIN]
Ramalina subleptocarpha Rundel & Bowler [RAMASUB]
Ramalina thrausta (Ach.) Nyl. [RAMATHR]
SY: *Alectoria crinalis* Ach.
SY: *Alectoria thrausta* Ach.
SY: *Ramalina crinalis* (Ach.) Gyelnik

RHIZOCARPACEAE (L)

Buellia badioatra (Flörke ex Sprengel) Körber = Rhizocarpon badioatrum [RHIZBAD]
Buellia colludens (Nyl.) Arnold = Rhizocarpon hochstetteri [RHIZHOC]
Buellia pulchella (Schrader) Tuck. = Catolechia wahlenbergii [CATOWAH]
Buellia wahlenbergii (Ach.) Sheard = Catolechia wahlenbergii [CATOWAH]
Catolechia Flotow [CATOLEC$]
Catolechia wahlenbergii (Ach.) Körber [CATOWAH]
SY: *Buellia pulchella* (Schrader) Tuck.
SY: *Buellia wahlenbergii* (Ach.) Sheard
Poeltinula Hafellner [POELTIN$]
Poeltinula cerebrina (DC.) Hafellner [POELCER]
Rhizocarpon Ramond ex DC. [RHIZOCA$]
Rhizocarpon albineum (Tuck.) Fink = Rhizocarpon obscuratum [RHIZOBS]
Rhizocarpon alpicola (Anzi) Rabenh. [RHIZALP]
SY: *Rhizocarpon chionophilum* Th. Fr.
SY: *Rhizocarpon oreites* (Vainio) Zahlbr.
Rhizocarpon ambiguum (Schaerer) Zahlbr. = Rhizocarpon distinctum [RHIZDIT]
Rhizocarpon atroalbescens (Nyl.) Zahlbr. = Rhizocarpon eupetraeoides [RHIZEUE]
Rhizocarpon atroflavescens Lynge [RHIZATR]
Rhizocarpon badioatrum (Flörke ex Sprengel) Th. Fr. [RHIZBAD]
SY: *Buellia badioatra* (Flörke ex Sprengel) Körber
Rhizocarpon bolanderi (Tuck.) Herre [RHIZBOL]
Rhizocarpon chioneum (Norman) Th. Fr. [RHIZCHI]
Rhizocarpon chionophilum Th. Fr. = Rhizocarpon alpicola [RHIZALP]
Rhizocarpon cinereonigrum Vainio [RHIZCIE]
Rhizocarpon cinereovirens (Müll. Arg.) Vainio [RHIZCIN]
Rhizocarpon concentricum (Davies) Beltr. [RHIZCON]
SY: *Rhizocarpon perlutum* (Nyl.) Zahlbr.
Rhizocarpon concretum (Ach.) Elenkin = Rhizocarpon geminatum [RHIZGEM]
Rhizocarpon crystalligenum Lynge = Rhizocarpon superficiale ssp. boreale [RHIZSUP1]
Rhizocarpon disporum auct. = Rhizocarpon geminatum [RHIZGEM]
Rhizocarpon distinctum Th. Fr. [RHIZDIT]
SY: *Rhizocarpon ambiguum* (Schaerer) Zahlbr.
Rhizocarpon eupetraeoides (Nyl.) Blomb. & Forss. [RHIZEUE]
SY: *Rhizocarpon atroalbescens* (Nyl.) Zahlbr.
Rhizocarpon eupetraeum (Nyl.) Arnold [RHIZEUP]
SY: *Rhizocarpon intermedium* Degel.
Rhizocarpon geminatum Körber [RHIZGEM]
SY: *Rhizocarpon concretum* (Ach.) Elenkin
SY: *Rhizocarpon disporum* auct.
Rhizocarpon geographicum (L.) DC. [RHIZGEO] (green map)
Rhizocarpon grande (Flörke ex Flotow) Arnold [RHIZGRA]
Rhizocarpon hensseniae Brodo [RHIZHEN]
Rhizocarpon hochstetteri (Körber) Vainio [RHIZHOC]
SY: *Buellia colludens* (Nyl.) Arnold
Rhizocarpon inarense (Vainio) Vainio [RHIZINA]
Rhizocarpon intermediellum Räsänen [RHIZINT]
Rhizocarpon intermedium Degel. = Rhizocarpon eupetraeum [RHIZEUP]
Rhizocarpon interponens (Nyl.) Zahlbr. = Rhizocarpon obscuratum [RHIZOBS]
Rhizocarpon lavatum (Fr.) Hazsl. [RHIZLAV]
Rhizocarpon lecanorinum Anders [RHIZLEC]
Rhizocarpon macrosporum Räsänen [RHIZMAC]
SY: *Rhizocarpon sphaerosporum* Räsänen
Rhizocarpon obscuratum (Ach.) A. Massal. [RHIZOBS]
SY: *Rhizocarpon albineum* (Tuck.) Fink
SY: *Rhizocarpon interponens* (Nyl.) Zahlbr.
SY: *Rhizocarpon permodestum* Arnold
SY: *Rhizocarpon reductum* Th. Fr.
Rhizocarpon occidentale Lynge = Rhizocarpon superficiale [RHIZSUP]
Rhizocarpon oreites (Vainio) Zahlbr. = Rhizocarpon alpicola [RHIZALP]
Rhizocarpon perlutum (Nyl.) Zahlbr. = Rhizocarpon concentricum [RHIZCON]
Rhizocarpon permodestum Arnold = Rhizocarpon obscuratum [RHIZOBS]
Rhizocarpon polare Räsänen = Rhizocarpon superficiale [RHIZSUP]
Rhizocarpon pusillum Runem. [RHIZPUS]
Rhizocarpon reductum Th. Fr. = Rhizocarpon obscuratum [RHIZOBS]
Rhizocarpon riparium Räsänen [RHIZRIP]
Rhizocarpon sphaerosporum Räsänen = Rhizocarpon macrosporum [RHIZMAC]
Rhizocarpon subtile Runem. = Rhizocarpon viridiatrum [RHIZVIR]
Rhizocarpon superficiale (Schaerer) Vainio [RHIZSUP]
SY: *Rhizocarpon occidentale* Lynge
SY: *Rhizocarpon polare* Räsänen
Rhizocarpon superficiale ssp. **boreale** Runem. [RHIZSUP1]
SY: *Rhizocarpon crystalligenum* Lynge
Rhizocarpon umbilicatum (Ramond) Flagey [RHIZUMB]
Rhizocarpon umbilicatum f. **pseudospeireum** (Th. Fr.) Magnusson [RHIZUMB]
Rhizocarpon viridiatrum (Wulfen) Körber [RHIZVIR]
SY: *Rhizocarpon subtile* Runem.

RIMULARIACEAE (L)

Lecidea insularis Nyl. = Rimularia insularis [RIMUINS]

Lecidea intumescens (Flotow) Nyl. = Rimularia insularis [RIMUINS]
Rimularia Nyl. [RIMULAR$]
Rimularia insularis (Nyl.) Rambold & Hertel [RIMUINS]
SY: *Lecidea insularis* Nyl.
SY: *Lecidea intumescens* (Flotow) Nyl.

SCHAERERIACEAE (L)

Schaereria Körber [SCHAERE$]
Schaereria corticola Muhr & Tønsberg [SCHACOR]

SIPHULACEAE (L)

Siphula Fr. [SIPHULA$]
Siphula ceratites (Wahlenb.) Fr. [SIPHCER] (waterworm)
SY: *Siphula simplex* (Taylor) Nyl.
Siphula simplex (Taylor) Nyl. = Siphula ceratites [SIPHCER]
Thamnolia Ach. ex Schaerer [THAMNOL$]
Thamnolia vermicularis (Sw.) Ach. ex Schaerer [THAMVER] (rock worm lichen)
Thamnolia vermicularis var. **subuliformis** (Ehrh.) Schaerer [THAMVER1]
Thamnolia vermicularis var. **vermicularis** [THAMVER0]

SOLORINELLACEAE (L)

Gyalidea Lettau ex Vezda [GYALIDE$]
Gyalidea dodgei Vezda = Gyalidea hyalinescens [GYALHYA]
Gyalidea hyalinescens (Nyl.) Vezda [GYALHYA]
SY: *Gyalidea dodgei* Vezda
Solorinella Anzi [SOLORIE$]
Solorinella asteriscus Anzi [SOLOAST]

SPHAEROPHORACEAE (L)

Bunodophoron A. Massal. [BUNODOP$]
Bunodophoron melanocarpum (Sw.) Wedin [BUNOMEL]
SY: *Sphaerophorus compressum* Ach.
SY: *Sphaerophorus melanocarpus* (Sw.) DC.
Sphaerophoron coralloides Pers. = Sphaerophorus globosus [SPHAGLO]
Sphaerophoron globiferum var. *congestum* Müll. Arg. = Sphaerophorus globosus [SPHAGLO]
Sphaerophorus Pers. [SPHAERO$]
Sphaerophorus compressum Ach. = Bunodophoron melanocarpum [BUNOMEL]
Sphaerophorus fragilis (L.) Pers. [SPHAFRA]
Sphaerophorus globiferis DC. = Sphaerophorus globosus [SPHAGLO]
Sphaerophorus globiferous (L.) DC. = Sphaerophorus globosus [SPHAGLO]
Sphaerophorus globiferus (L.) DC. = Sphaerophorus globosus [SPHAGLO]
Sphaerophorus globosus (Hudson) Vainio [SPHAGLO] (common Christmas-tree)
SY: *Sphaerophoron coralloides* Pers.
SY: *Sphaerophoron globiferum* var. *congestum* Müll. Arg.
SY: *Sphaerophorus globiferis* DC.
SY: *Sphaerophorus globiferous* (L.) DC.
SY: *Sphaerophorus globiferus* (L.) DC.
SY: *Sphaerophorus globosus* var. *lacunosus* Tuck.
Sphaerophorus globosus var. **globosus** [SPHAGLO0]
Sphaerophorus globosus var. **gracilis** (Müll. Arg.) Zahlbr. [SPHAGLO1]
SY: *Sphaerophorus tuckermanii* Räsänen
Sphaerophorus globosus var. *lacunosus* Tuck. = Sphaerophorus globosus [SPHAGLO]
Sphaerophorus melanocarpus (Sw.) DC. = Bunodophoron melanocarpum [BUNOMEL]
Sphaerophorus tuckermanii Räsänen = Sphaerophorus globosus var. gracilis [SPHAGLO1]

STEREOCAULACEAE (L)

Lecidea pallida Th. Fr. = Pilophorus dovrensis [PILODOV]
Pilophorus Th. Fr. [PILOPHO$]
Pilophorus acicularis (Ach.) Th. Fr. [PILOACI] (devil's matchstick)
SY: *Pilophorus cereolus* var. *acicularis* (Ach.) Tuck.
Pilophorus cereolus (Ach.) Th. Fr. [PILOCER]
Pilophorus cereolus var. *acicularis* (Ach.) Tuck. = Pilophorus acicularis [PILOACI]
Pilophorus clavatus Th. Fr. [PILOCLA]
SY: *Pilophorus hallii* (Tuck.) Vainio
Pilophorus dovrensis (Nyl.) Timdal, Hertel & Rambold [PILODOV]
SY: *Lecidea pallida* Th. Fr.
SY: *Pilophorus pallidus* (Th. Fr.) Timdal
Pilophorus hallii (Tuck.) Vainio = Pilophorus clavatus [PILOCLA]
Pilophorus nigricaulis Satô [PILONIG]
Pilophorus pallidus (Th. Fr.) Timdal = Pilophorus dovrensis [PILODOV]
Pilophorus vegae Krog [PILOVEG]
Stereocaulon Hoffm. [STEREOC$]
Stereocaulon alpinum Laurer ex Funck [STERALP]
Stereocaulon arenarium (Savicz) Lamb [STERARE]
Stereocaulon botryosum Ach. [STERBOT]
Stereocaulon condensatum Hoffm. [STERCON]
Stereocaulon coniophyllum Lamb [STERCOI]
Stereocaulon coralloides Fr. = Stereocaulon dactylophyllum [STERDAC]
Stereocaulon dactylophyllum Flörke [STERDAC]
SY: *Stereocaulon coralloides* Fr.
Stereocaulon denudatum Flörke = Stereocaulon vesuvianum [STERVES]
Stereocaulon evolutoides (H. Magn.) Frey = Stereocaulon saxatile [STERSAX]
Stereocaulon glareosum (Savicz) H. Magn. [STERGLA]
Stereocaulon grande (H. Magn.) H. Magn. [STERGRA]
Stereocaulon intermedium (Savicz) H. Magn. [STERINT]
SY: *Stereocaulon intermedium* f. *compactum* Lamb
Stereocaulon intermedium f. *compactum* Lamb = Stereocaulon intermedium [STERINT]
Stereocaulon myriocarpum Th. Fr. [STERMYR]
Stereocaulon paschale (L.) Hoffm. [STERPAS] (common coral lichen)
Stereocaulon rivulorum H. Magn. [STERRIV]
Stereocaulon sasakii Zahlbr. [STERSAS]
Stereocaulon sasakii var. **simplex** (Riddle) Lamb [STERSAS1]
SY: *Stereocaulon tomentosum* var. *simplex* Riddle
Stereocaulon sasakii var. **tomentosoides** Lamb [STERSAS2]
Stereocaulon saxatile H. Magn. [STERSAX]
SY: *Stereocaulon evolutoides* (H. Magn.) Frey
Stereocaulon spathuliferum Vainio [STERSPA]
Stereocaulon sterile (Savicz) Lamb ex Krog [STERSTE]
Stereocaulon subcoralloides (Nyl.) Nyl. [STERSUC]
Stereocaulon tomentosum Fr. [STERTOM] (woolly coral)
Stereocaulon tomentosum var. *simplex* Riddle = Stereocaulon sasakii var. simplex [STERSAS1]
Stereocaulon vesuvianum Pers. [STERVES]
SY: *Stereocaulon denudatum* Flörke
SY: *Stereocaulon vesuvianum* var. *denudatum* (Flörke) Lamb ex Poelt
SY: *Stereocaulon vesuvianum* var. *nodulosum* (Wallr.) Lamb
SY: *Stereocaulon vesuvianum* var. *nodulosum* f. *verrucosum* Lamb
Stereocaulon vesuvianum var. *denudatum* (Flörke) Lamb ex Poelt = Stereocaulon vesuvianum [STERVES]
Stereocaulon vesuvianum var. *nodulosum* (Wallr.) Lamb = Stereocaulon vesuvianum [STERVES]
Stereocaulon vesuvianum var. *nodulosum* f. *verrucosum* Lamb = Stereocaulon vesuvianum [STERVES]

STICTIDACEAE (L)

Nanostictis M.S. Christ. [NANOSTI$]

Nanostictis peltigerae Sherwood [NANOPEL]

STRIGULACEAE (L)

Arthopyrenia faginea (Schaerer) Swinscow = Strigula stigmatella [STRISTI]
Strigula Fr. [STRIGUL$]
Strigula stigmatella (Ach.) R.C. Harris [STRISTI]
SY: *Arthopyrenia faginea* (Schaerer) Swinscow

TELOSCHISTACEAE (L)

Blastenia luteominium (Tuck.) Hasse. = Caloplaca luteominia [CALOLUT]
Blastenia sinapisperma (Lam. & DC.) A. Massal. = Caloplaca sinapisperma [CALOSIN]
Caloplaca Th. Fr. [CALOPLA$]
Caloplaca ammiospila (Wahlenb.) H. Olivier [CALOAMM]
SY: *Caloplaca cinnamomea* (Th. Fr.) H. Olivier
SY: *Caloplaca discoidalis* (Vainio) Lynge
Caloplaca approximata (Lynge) H. Magn. [CALOAPP]
Caloplaca arenaria (Pers.) Müll. Arg. [CALOARE]
SY: *Caloplaca lamprocheila* (DC.) Flagey
Caloplaca atroflava (Turner) Mong. [CALOATR]
SY: *Lecidea atroflava* Turn.
Caloplaca atrosanguinea (G. Merr.) Lamb [CALOATS]
SY: *Caloplaca herrei* Hasse
SY: *Caloplaca subnigricans* H. Magn.
Caloplaca aurantiaca (Lightf.) Th. Fr. = Caloplaca flavorubescens [CALOFLV]
Caloplaca bolacina (Tuck.) Herre [CALOBOL]
SY: *Placodium bolacinum* Tuck.
Caloplaca borealis (Vainio) Poelt [CALOBOR]
Caloplaca bracteata (Hoffm.) Jatta = Fulgensia bracteata [FULGBRA]
Caloplaca caesiorufella (Nyl.) Zahlbr. = Caloplaca phaeocarpella [CALOPHA]
Caloplaca castellana (Räsänen) Poelt [CALOCAS]
SY: *Caloplaca invadens* Lynge
Caloplaca cerina (Ehrh. ex Hedwig) Th. Fr. [CALOCER] (crusted orange lichen)
SY: *Caloplaca gilva* (Hoffm.) Zahlbr.
SY: *Caloplaca stillicidiorum* (Vahl) Lynge
Caloplaca cerinelloides (Erichsen) Poelt [CALOCEI]
Caloplaca chlorina (Flotow) H. Olivier [CALOCHL]
SY: *Lecanora chlorina* (Flotow) Lamy
SY: *Zeora cerina* var. *chlorina* Flotow
Caloplaca chrysophthalma Degel. [CALOCHR]
Caloplaca cinnamomea (Th. Fr.) H. Olivier = Caloplaca ammiospila [CALOAMM]
Caloplaca citrina (Hoffm.) Th. Fr. [CALOCIT]
SY: *Verrucaria citrina* Hoffm.
Caloplaca diphyodes (Nyl.) Jatta [CALODIP]
SY: *Lecania arctica* Lynge
Caloplaca discoidalis (Vainio) Lynge = Caloplaca ammiospila [CALOAMM]
Caloplaca elegans (Link) Th. Fr. = Xanthoria elegans [XANTELE]
Caloplaca epithallina Lynge [CALOEPI]
Caloplaca erythrella (Ach.) Kieffer = Caloplaca flavovirescens [CALOFLA]
Caloplaca feracissima H. Magn. [CALOFER]
Caloplaca ferruginea (Hudson) Th. Fr. [CALOFEU]
Caloplaca flavogranulosa Arup [CALOFLO]
Caloplaca flavorubescens (Hudson) J.R. Laundon [CALOFLV]
SY: *Caloplaca aurantiaca* (Lightf.) Th. Fr.
Caloplaca flavovirescens (Wulfen) Dalla Torre & Sarnth. [CALOFLA]
SY: *Caloplaca erythrella* (Ach.) Kieffer
Caloplaca fraudans (Th. Fr.) H. Olivier [CALOFRA]
Caloplaca gilva (Hoffm.) Zahlbr. = Caloplaca cerina [CALOCER]
Caloplaca gloriae Werner & Llimona [CALOGLO]
Caloplaca herrei Hasse = Caloplaca atrosanguinea [CALOATS]
Caloplaca holocarpa (Hoffm. ex Ach.) M. Wade [CALOHOL]
SY: *Caloplaca pyracea* (Ach.) Th. Fr.
Caloplaca inconspecta Arup [CALOINC]
Caloplaca invadens Lynge = Caloplaca castellana [CALOCAS]
Caloplaca jungermanniae (Vahl) Th. Fr. [CALOJUN]
Caloplaca lamprocheila (DC.) Flagey = Caloplaca arenaria [CALOARE]
Caloplaca leucoraea (Ach. ex Flörke) Branth = Caloplaca sinapisperma [CALOSIN]
Caloplaca litoricola Brodo [CALOLIT]
Caloplaca livida (Hepp) Jatta [CALOLIV]
Caloplaca luteominia (Tuck.) Zahlbr. [CALOLUT]
SY: *Blastenia luteominium* (Tuck.) Hasse.
SY: *Placodium luteominium* Tuck.
Caloplaca marina (Wedd.) Zahlbr. [CALOMAR]
SY: *Lecanora marina* Wedd.
Caloplaca marina ssp. **americana** Arup [CALOMAR1]
Caloplaca microthallina (Wedd.) Zahlbr. [CALOMIC]
Caloplaca nivalis (Körber) Th. Fr. [CALONIV]
Caloplaca obscurella (Körber) Th. Fr. [CALOOBS]
SY: *Caloplaca sarcopisioides* (Körber) Zahlbr.
Caloplaca oleicola (Steiner) v.d. Boom & Breuss [CALOOLE]
Caloplaca oregona H. Magn. [CALOORE]
Caloplaca phaeocarpella (Nyl.) Zahlbr. [CALOPHA]
SY: *Caloplaca caesiorufella* (Nyl.) Zahlbr.
Caloplaca pyracea (Ach.) Th. Fr. = Caloplaca holocarpa [CALOHOL]
Caloplaca rosei Hasse [CALOROS]
Caloplaca sarcopisioides (Körber) Zahlbr. = Caloplaca obscurella [CALOOBS]
Caloplaca scopularis (Nyl.) Lettau [CALOSCO]
Caloplaca sinapisperma (Lam. & DC.) Maheu & A. Gillet [CALOSIN]
SY: *Blastenia sinapisperma* (Lam. & DC.) A. Massal.
SY: *Caloplaca leucoraea* (Ach. ex Flörke) Branth
Caloplaca sorediata (Vainio) Du Rietz = Xanthoria sorediata [XANTSOR]
Caloplaca sorocarpa (Vainio) Zahlbr. [CALOSOR]
Caloplaca stillicidiorum (Vahl) Lynge = Caloplaca cerina [CALOCER]
Caloplaca subnigricans H. Magn. = Caloplaca atrosanguinea [CALOATS]
Caloplaca teicholyta (Ach.) J. Steiner [CALOTEI]
Caloplaca tetraspora (Nyl.) H. Olivier [CALOTET]
Caloplaca tominii Savicz [CALOTOM]
Caloplaca verruculifera (Vainio) Zahlbr. [CALOVER]
Caloplaca vicaria H. Magn. [CALOVIC]
Fulgensia A. Massal. & De Not. [FULGENS$]
Fulgensia bracteata (Hoffm.) Räsänen [FULGBRA]
SY: *Caloplaca bracteata* (Hoffm.) Jatta
Fulgensia desertorum (Tomin) Poelt [FULGDES]
Lecania arctica Lynge = Caloplaca diphyodes [CALODIP]
Lecanora chlorina (Flotow) Lamy = Caloplaca chlorina [CALOCHL]
Lecanora marina Wedd. = Caloplaca marina [CALOMAR]
Lecidea atroflava Turn. = Caloplaca atroflava [CALOATR]
Placodium bolacinum Tuck. = Caloplaca bolacina [CALOBOL]
Placodium luteominium Tuck. = Caloplaca luteominia [CALOLUT]
Teloschistes Norman [TELOSCH$]
Teloschistes candelaris (L.) Fink = Xanthoria candelaria [XANTCAN]
Teloschistes contortuplicatus (Ach.) Clauzade & Rondon ex Vezda [TELOCON]
Teloschistes controversus (Mass.) Müll Arg. = Xanthoria candelaria [XANTCAN]
Teloschistes controversus var. *pygmaeus* (Bory) Müll. Arg. = Xanthoria candelaria [XANTCAN]
Teloschistes parietinus (L.) Norman = Xanthoria parietina [XANTPAR]
Teloschistes polycarpus (Hoffm.) Tuck. = Xanthoria polycarpa [XANTPOL]
Teloschistes ramulosus Tuck. = Xanthoria ramulosa [XANTRAM]
Verrucaria citrina Hoffm. = Caloplaca citrina [CALOCIT]
Xanthoria (Fr.) Th. Fr. [XANTHOR$]

Xanthoria candelaria (L.) Th. Fr. [XANTCAN] (shrubby orange)
SY: *Teloschistes candelaris* (L.) Fink
SY: *Teloschistes controversus* (Mass.) Müll Arg.
SY: *Teloschistes controversus* var. *pygmaeus* (Bory) Müll. Arg.
SY: *Xanthoria oregana* Gyelnik
Xanthoria elegans (Link) Th. Fr. [XANTELE] (elegant orange)
SY: *Caloplaca elegans* (Link) Th. Fr.
Xanthoria fallax (Hepp) Arnold [XANTFAL] (powdered orange)
Xanthoria fulva (Hoffm.) Poelt & Petutschnig [XANTFUL]
Xanthoria oregana Gyelnik = Xanthoria candelaria [XANTCAN]
Xanthoria parietina (L.) Th. Fr. [XANTPAR]
SY: *Teloschistes parietinus* (L.) Norman
Xanthoria polycarpa (Hoffm.) Rieber [XANTPOL] (pincushion orange)
SY: *Teloschistes polycarpus* (Hoffm.) Tuck.
Xanthoria ramulosa (Tuck.) Herre [XANTRAM]
SY: *Teloschistes ramulosus* Tuck.
Xanthoria sorediata (Vainio) Poelt [XANTSOR] (sugared orange)
SY: *Caloplaca sorediata* (Vainio) Du Rietz
Zeora cerina var. *chlorina* Flotow = Caloplaca chlorina [CALOCHL]

THELEPHORACEAE (L)

Dictyonema C. Agardh [DICTYON$]
Dictyonema moorei (Nyl.) Henssen [DICTMOO]

THELOTREMATACEAE (L)

Diploschistes Norman [DIPLOSC$]
Diploschistes bisporus (Bagl.) J. Steiner [DIPLBIS]
Diploschistes muscorum (Scop.) R. Sant. [DIPLMUS] (cow pie)
Diploschistes scruposus (Schreber) Norman [DIPLSCR]
SY: *Urceolaria scruposa* (Schreber) Ach.
Thelotrema Ach. [THELOTR$]
Thelotrema lepadinum (Ach.) Ach. [THELLEP] (bark barnacle)
Urceolaria scruposa (Schreber) Ach. = Diploschistes scruposus [DIPLSCR]

THROMBIACEAE (L)

Thrombium Wallr. [THROMBI$]
Thrombium epigaeum (Pers.) Wallr. [THROEPI]

TRAPELIACEAE (L)

Anzina Scheid. [ANZINA$]
Anzina carneonivea (Anzi) Scheid. [ANZICAR]
Lecanora gelida (L.) Ach. = Placopsis gelida [PLACGEL]
Lecidea aeruginosa Borrer = Trapeliopsis flexuosa [TRAPFLE]
Lecidea coarctata (Sm.) Nyl. = Trapelia coarctata [TRAPCOA]
Lecidea flexuosa (Fr.) Nyl. = Trapeliopsis flexuosa [TRAPFLE]
Lecidea gelatinosa Flörke = Trapeliopsis gelatinosa [TRAPGEL]
Lecidea glebulosa (Fr.) Clem. = Trapeliopsis wallrothii [TRAPWAL]
Lecidea granulosa (Hoffm.) Ach. = Trapeliopsis granulosa [TRAPGRA]
Lecidea gregaria G. Merr. = Trapelia involuta [TRAPINV]
Lecidea humosa (Hoffm.) Nyl. = Placynthiella uliginosa [PLACULI]
Lecidea obtegens Th. Fr. = Trapelia obtegens [TRAPOBT]
Lecidea oligotropha J.R. Laundon = Placynthiella oligotropha [PLACOLI]
Lecidea ornata (Sommerf.) Hue = Trapelia involuta [TRAPINV]
Lecidea quadricolor (Dickson) Borrer = Trapeliopsis granulosa [TRAPGRA]
Lecidea uliginosa (Schrader) Ach. = Placynthiella uliginosa [PLACULI]
Lecidea viridescens (Schrader) Ach. = Trapeliopsis viridescens [TRAPVIR]
Lecidea wallrothii Flörke ex Sprengel = Trapeliopsis wallrothii [TRAPWAL]
Micarea gelatinosa (Flörke) Brodo = Trapeliopsis gelatinosa [TRAPGEL]
Micarea viridescens (Schrader) Brodo = Trapeliopsis viridescens [TRAPVIR]
Placopsis (Nyl.) Lindsay [PLACOPS$]
Placopsis cribellans (Nyl.) Räsänen [PLACCRI]
Placopsis effusa Lamb [PLACEFF]
Placopsis gelida (L.) Lindsay [PLACGEL] (bull's-eye)
SY: *Lecanora gelida* (L.) Ach.
Placynthiella Elenkin [PLACYNH$]
Placynthiella icmalea (Ach.) Coppins & P. James [PLACICM]
SY: *Saccomorpha icmalea* (Ach.) Clauzade & Roux
Placynthiella oligotropha (J.R. Laundon) Coppins & P. James [PLACOLI]
SY: *Lecidea oligotropha* J.R. Laundon
SY: *Saccomorpha oligotropha* (J.R. Laundon) Clauzade & Roux
Placynthiella uliginosa (Schrader) Coppins & P. James [PLACULI]
SY: *Lecidea humosa* (Hoffm.) Nyl.
SY: *Lecidea uliginosa* (Schrader) Ach.
SY: *Saccomorpha uliginosa* (Schrader) Hafellner
Saccomorpha icmalea (Ach.) Clauzade & Roux = Placynthiella icmalea [PLACICM]
Saccomorpha oligotropha (J.R. Laundon) Clauzade & Roux = Placynthiella oligotropha [PLACOLI]
Saccomorpha uliginosa (Schrader) Hafellner = Placynthiella uliginosa [PLACULI]
Trapelia Choisy [TRAPELI$]
Trapelia coarctata (Sm.) Choisy [TRAPCOA]
SY: *Lecidea coarctata* (Sm.) Nyl.
Trapelia corticola Coppins & P. James [TRAPCOR]
Trapelia involuta (Taylor) Hertel [TRAPINV]
SY: *Lecidea gregaria* G. Merr.
SY: *Lecidea ornata* (Sommerf.) Hue
Trapelia obtegens (Th. Fr.) Hertel [TRAPOBT]
SY: *Lecidea obtegens* Th. Fr.
Trapelia placodioides Coppins & P. James [TRAPPLA]
Trapeliopsis Hertel & Gotth. Schneider [TRAPELO$]
Trapeliopsis aeneofusca (Flotow) Coppins & P. James [TRAPAEN]
Trapeliopsis flexuosa (Fr.) Coppins & P. James [TRAPFLE]
SY: *Lecidea aeruginosa* Borrer
SY: *Lecidea flexuosa* (Fr.) Nyl.
Trapeliopsis gelatinosa (Flörke) Coppins & P. James [TRAPGEL]
SY: *Lecidea gelatinosa* Flörke
SY: *Micarea gelatinosa* (Flörke) Brodo
Trapeliopsis granulosa (Hoffm.) Lumbsch [TRAPGRA]
SY: *Lecidea granulosa* (Hoffm.) Ach.
SY: *Lecidea quadricolor* (Dickson) Borrer
Trapeliopsis pseudogranulosa Coppins & P. James [TRAPPSE]
Trapeliopsis viridescens (Schrader) Coppins & P. James [TRAPVIR]
SY: *Lecidea viridescens* (Schrader) Ach.
SY: *Micarea viridescens* (Schrader) Brodo
Trapeliopsis wallrothii (Flörke) Hertel & Gotth. Schneider [TRAPWAL]
SY: *Lecidea glebulosa* (Fr.) Clem.
SY: *Lecidea wallrothii* Flörke ex Sprengel

TRICHOLOMATACEAE (L)

Botrydina botryoides (L.) Redhead & Kuyper = Omphalina umbellifera [OMPHUMB]
Botrydina viridis (Ach.) Redhead & Kuyper = Omphalina hudsoniana [OMPHHUD]
Botrydina vulgaris Bréb. = Omphalina umbellifera [OMPHUMB]
Coriscium viride (Ach.) Vainio = Omphalina hudsoniana [OMPHHUD]
Omphalina Quélet [OMPHALI$]
Omphalina ericetorum (Pers.) M.T. Lange = Omphalina umbellifera [OMPHUMB]
Omphalina hudsoniana (H.S. Jenn.) H.E. Bigelow [OMPHHUD]
SY: *Botrydina viridis* (Ach.) Redhead & Kuyper
SY: *Coriscium viride* (Ach.) Vainio
SY: *Phytoconis viridis* (Ach.) Redhead & Kuyper
Omphalina umbellifera (L.) Quélet [OMPHUMB]
SY: *Botrydina botryoides* (L.) Redhead & Kuyper

SY: *Botrydina vulgaris* Bréb.
SY: *Omphalina ericetorum* (Pers.) M.T. Lange
SY: *Phytoconis ericetorum* (Pers.) Redhead & Kuyper
Phytoconis ericetorum (Pers.) Redhead & Kuyper = Omphalina umbellifera [OMPHUMB]
Phytoconis viridis (Ach.) Redhead & Kuyper = Omphalina hudsoniana [OMPHHUD]

TRICHOTHELIACEAE (L)

Porina Müll. Arg. [PORINA$]
Porina aenea (Wallr.) Zahlbr. = Pseudosagedia aenea [PSEUAEN]
Porina carpinea (Pers. ex Ach.) Zahlbr. = Pseudosagedia aenea [PSEUAEN]
Porina chlorotica (Ach.) Müll. Arg. = Pseudosagedia chlorotica [PSEUCHL]
Porina guentheri (Flotow) Zahlbr. = Pseudosagedia guentheri [PSEUGUE]
Porina lectissima (Fr.) Zahlbr. [PORILEC]
Porina leptalea (Durieu & Mont.) A.L. Sm. [PORILEP]
Porina linearis (Leighton) Zahlbr. = Pseudosagedia linearis [PSEULIN]
Pseudosagedia (Müll. Arg.) Choisy [PSEUDOD$]
Pseudosagedia aenea (Wallr.) Hafellner & Kalb [PSEUAEN]
SY: *Porina aenea* (Wallr.) Zahlbr.
SY: *Porina carpinea* (Pers. ex Ach.) Zahlbr.
Pseudosagedia chlorotica (Ach.) Hafellner & Kalb [PSEUCHL]
SY: *Porina chlorotica* (Ach.) Müll. Arg.
Pseudosagedia guentheri (Flotow) Hafellner & Kalb [PSEUGUE]
SY: *Porina guentheri* (Flotow) Zahlbr.
Pseudosagedia linearis (Leighton) Hafellner & Kalb [PSEULIN]
SY: *Porina linearis* (Leighton) Zahlbr.

UMBILICARIACEAE (L)

Actinogyra muehlenbergii (Ach.) Schol. = Umbilicaria muehlenbergii [UMBIMUE]
Actinogyra polyrrhiza (L.) Schol. = Umbilicaria polyrrhiza [UMBIPOY]
Agyrophora leiocarpa (DC.) Gyelnik = Umbilicaria leiocarpa [UMBILEI]
Agyrophora lyngei (Schol.) Llano = Umbilicaria lyngei [UMBILYN]
Agyrophora rigida (Du Rietz) Llano = Umbilicaria rigida [UMBIRIG]
Gyrophora angulata (Tuck.) Herre = Umbilicaria angulata [UMBIANG]
Gyrophora anthracina (Wulfen) Körber = Umbilicaria rigida [UMBIRIG]
Gyrophora arctica Ach. = Umbilicaria arctica [UMBIARC]
Gyrophora cylindrica (L.) Ach. = Umbilicaria cylindrica [UMBICYL]
Gyrophora decussata (Vill.) Zahlbr. = Umbilicaria decussata [UMBIDEC]
Gyrophora deusta (L.) Ach. = Umbilicaria deusta [UMBIDEU]
Gyrophora erosa (G. Weber) Ach. = Umbilicaria torrefacta [UMBITOR]
Gyrophora hyperborea Ach. = Umbilicaria hyperborea var. hyperborea [UMBIHYP0]
Gyrophora muehlenbergii Ach. = Umbilicaria muehlenbergii [UMBIMUE]
Gyrophora phaea (Tuck.) Nyl. = Umbilicaria phaea [UMBIPHA]
Gyrophora polyphylla (L.) Fink = Umbilicaria polyphylla [UMBIPOL]
Gyrophora polyrrhiza (L.) Körber = Umbilicaria polyrrhiza [UMBIPOY]
Gyrophora vellea (L.) Ach. = Umbilicaria vellea [UMBIVEL]
Lasallia Mérat [LASALLI$]
Lasallia papulosa (Ach.) Llano [LASAPAP]
SY: *Lasallia pustulata* ssp. *papulosa* (Ach.) W.A. Weber
SY: *Umbilicaria papulosa* (Ach.) Nyl.
SY: *Umbilicaria pustulata* var. *papulosa* (Ach.) Tuck.
Lasallia pensylvanica (Hoffm.) Llano [LASAPEN] (blistered rocktripe)
SY: *Umbilicaria pensylvanica* Hoffm.
Lasallia pustulata ssp. *papulosa* (Ach.) W.A. Weber = Lasallia papulosa [LASAPAP]
Omphalodiscus decussatus (Vill.) Schol. = Umbilicaria decussata [UMBIDEC]
Omphalodiscus krascheninnikovii (Savicz) Schol. = Umbilicaria krascheninnikovii [UMBIKRA]
Omphalodiscus virginis (Schaerer) Schol. = Umbilicaria virginis [UMBIVIR]
Umbilicaria Hoffm. [UMBILIC$]
Umbilicaria americana Poelt & T. Nash [UMBIAME]
Umbilicaria angulata Tuck. [UMBIANG] (asterisk rocktripe)
SY: *Gyrophora angulata* (Tuck.) Herre
Umbilicaria aprina Nyl. [UMBIAPR] (ashen rocktripe)
Umbilicaria aprina var. **halei** Llano [UMBIAPR1]
Umbilicaria arctica (Ach.) Nyl. [UMBIARC]
SY: *Gyrophora arctica* Ach.
Umbilicaria cinereorufescens (Schaerer) Frey [UMBICIN] (doubtful rocktripe)
Umbilicaria coriacea Imshaug = Umbilicaria rigida [UMBIRIG]
Umbilicaria cylindrica (L.) Delise ex Duby [UMBICYL] (fringed rocktripe)
SY: *Gyrophora cylindrica* (L.) Ach.
Umbilicaria decussata (Vill.) Zahlbr. [UMBIDEC]
SY: *Gyrophora decussata* (Vill.) Zahlbr.
SY: *Omphalodiscus decussatus* (Vill.) Schol.
Umbilicaria deusta (L.) Baumg. [UMBIDEU] (peppered rocktripe)
SY: *Gyrophora deusta* (L.) Ach.
Umbilicaria havaasii Llano [UMBIHAV] (ragged rocktripe)
Umbilicaria hirsuta (Sw. ex Westr.) Hoffm. [UMBIHIR]
Umbilicaria hyperborea (Ach.) Hoffm. [UMBIHYP]
Umbilicaria hyperborea var. **coccinea** Llano [UMBIHYP1] (blistered rocktripe)
Umbilicaria hyperborea var. **hyperborea** [UMBIHYP0]
SY: *Gyrophora hyperborea* Ach.
Umbilicaria hyperborea var. **radicicula** (J.E. Zetterst.) Hasselrot [UMBIHYP2]
Umbilicaria krascheninnikovii (Savicz) Zahlbr. [UMBIKRA]
SY: *Omphalodiscus krascheninnikovii* (Savicz) Schol.
Umbilicaria lambii Imshaug [UMBILAM] (windward rocktripe)
Umbilicaria leiocarpa DC. [UMBILEI]
SY: *Agyrophora leiocarpa* (DC.) Gyelnik
Umbilicaria lyngei Schol. [UMBILYN]
SY: *Agyrophora lyngei* (Schol.) Llano
Umbilicaria muehlenbergii (Ach.) Tuck. [UMBIMUE] (plated rocktripe)
SY: *Actinogyra muehlenbergii* (Ach.) Schol.
SY: *Gyrophora muehlenbergii* Ach.
Umbilicaria nylanderiana (Zahlbr.) H. Magn. [UMBINYL]
Umbilicaria papulosa (Ach.) Nyl. = Lasallia papulosa [LASAPAP]
Umbilicaria pensylvanica Hoffm. = Lasallia pensylvanica [LASAPEN]
Umbilicaria phaea Tuck. [UMBIPHA] (emery rocktripe)
SY: *Gyrophora phaea* (Tuck.) Nyl.
Umbilicaria polyphylla (L.) Baumg. [UMBIPOL] (petalled rocktripe)
SY: *Gyrophora polyphylla* (L.) Fink
Umbilicaria polyrrhiza (L.) Fr. [UMBIPOY] (ballpoint rocktripe)
SY: *Actinogyra polyrrhiza* (L.) Schol.
SY: *Gyrophora polyrrhiza* (L.) Körber
Umbilicaria proboscidea (L.) Schrader [UMBIPRO] (netted rocktripe)
Umbilicaria pustulata var. *papulosa* (Ach.) Tuck. = Lasallia papulosa [LASAPAP]
Umbilicaria rigida (Du Rietz) Frey [UMBIRIG] (roughened rocktripe)
SY: *Agyrophora rigida* (Du Rietz) Llano
SY: *Gyrophora anthracina* (Wulfen) Körber
SY: *Umbilicaria coriacea* Imshaug
Umbilicaria torrefacta (Lightf.) Schrader [UMBITOR] (punctured rocktripe)
SY: *Gyrophora erosa* (G. Weber) Ach.
Umbilicaria vellea (L.) Hoffm. [UMBIVEL] (frosted rocktripe)
SY: *Gyrophora vellea* (L.) Ach.
Umbilicaria virginis Schaerer [UMBIVIR] (blushing rocktripe)

SY: *Omphalodiscus virginis* (Schaerer) Schol.

UNCERTAIN-FAMILY (L)

Abrothallus De Not. [ABROTHA$]
Abrothallus bertianus De Not. [ABROBER]
Abrothallus cetrariae Kotte [ABROCET]
Abrothallus parmeliarum (Sommerf.) Arnold [ABROPAR]
SY: *Buelliella parmeliarum* (Sommerf.) Fink
Abrothallus welwitschii Tul. [ABROWEL]
Asterophoma D. Hawksw. [ASTEROP$]
Asterophoma mazaediicola D. Hawksw. [ASTEMAZ]
Bachmanniomyces D. Hawksw. [BACHMAN$]
Bachmanniomyces uncialicola (Zopf) D. Hawksw. [BACHUNC]
Biatoropsis Räsänen [BIATORO$]
Biatoropsis usnearum Räsänen [BIATUSN]
Buelliella parmeliarum (Sommerf.) Fink = Abrothallus parmeliarum [ABROPAR]
Cetraria hiascens (Fr.) Th. Fr. = Cetrariella delisei [CETRDEI]
Cetrariella Kärnefelt & Thell [CETRARE$]
Cetrariella delisei (Bory ex Schaerer) Kärnefelt & Thell [CETRDEI]
SY: *Cetraria hiascens* (Fr.) Th. Fr.
Coniothyrium lecanoracearum Vouaux = Lichenoconium usneae [LICHUSN]
Corticifraga D. Hawksw. & R. Sant. [CORTICI$]
Corticifraga fuckelii (Rehm) D. Hawksw. & R. Sant. [CORTFUC]
SY: *Phragmonaevia fuckelii* Rehm
Crocynia aliciae Hue = Lepraria lobificans [LEPRLOB]
Crocynia neglecta (Nyl.) Hue = Lepraria neglecta [LEPRNEG]
Cystocoleus Thwaites [CYSTOCO$]
Cystocoleus ebeneus (Dillwyn) Thwaites [CYSTEBE]
Fuscopannaria P.M. Jørg. [FUSCOPA$]
Fuscopannaria ahlneri (P.M. Jørg.) P.M. Jørg. [FUSCAHL] (roughened mouse)
Fuscopannaria laceratula (Hue) P.M. Jørg. [FUSCLAC] (cushion mouse)
Fuscopannaria leucostictoides (Ohlsson) P.M. Jørg. [FUSCLEU] (petalled mouse)
Fuscopannaria maritima (P.M. Jørg.) P.M. Jørg. [FUSCMAR] (seaside mouse)
Fuscopannaria mediterranea (Tav.) P.M. Jørg. [FUSCMED] (blue-edged mouse)
Fuscopannaria praetermissa (Nyl.) P.M. Jørg. [FUSCPRE] (moss mouse)
SY: *Pannaria lepidiota* (Sommerf.) Th. Fr.
Fuscopannaria saubinetii (Mont.) P.M. Jørg. [FUSCSAU] (pink-eyed mouse)
SY: *Parmeliella saubinetii* (Mont.) Zahlbr.
Hawksworthiana U. Braun [HAWKSWO$]
Hawksworthiana peltigericola (D. Hawksw.) U. Braun [HAWKPEL]
SY: *Ramularia peltigericola* D. Hawksw.
Illosporium Martius [ILLOSPO$]
Illosporium carneum Fr. [ILLOCAR]
Lepraria Ach. [LEPRARI$]
Lepraria aeruginosa auct. = Lepraria incana [LEPRINC]
Lepraria cacuminum (A. Massal.) Lohtander [LEPRCAC]
SY: *Leproloma angardianum* (Ovstedal) J.R. Laundon
SY: *Leproloma cacuminum* (A. Massal.) J.R. Laundon
Lepraria crassissima (Hue) Lettau [LEPRCRA]
Lepraria eburnea J.R. Laundon [LEPREBU]
Lepraria frigida J.R. Laundon [LEPRFRI]
Lepraria incana (L.) Ach. [LEPRINC]
SY: *Lepraria aeruginosa* auct.
Lepraria lobificans Nyl. [LEPRLOB]
SY: *Crocynia aliciae* Hue
Lepraria neglecta (Nyl.) Erichsen [LEPRNEG]
SY: *Crocynia neglecta* (Nyl.) Hue
Lepraria rigidula (de Lesd.) Tønsberg [LEPRRIG]
Leprocaulon Nyl. ex Lamy [LEPROCA$]
Leprocaulon albicans (Th. Fr.) Nyl. ex Hue [LEPRALB]
SY: *Stereocaulon albicans* Th. Fr.
Leprocaulon microscopicum (Vill.) Gams ex D. Hawksw. [LEPRMIC]
SY: *Stereocaulon microscopicum* (Vill.) Frey
Leprocaulon subalbicans (Lamb) Lamb & Ward [LEPRSUB]
SY: *Stereocaulon pseudoarbuscula* Asah.
Leproloma angardianum (Ovstedal) J.R. Laundon = Lepraria cacuminum [LEPRCAC]
Leproloma cacuminum (A. Massal.) J.R. Laundon = Lepraria cacuminum [LEPRCAC]
Lichenoconium Petrak & H. Sydow [LICHENC$]
Lichenoconium usneae (Anzi) D. Hawksw. [LICHUSN]
SY: *Coniothyrium lecanoracearum* Vouaux
Lichenomyces lichenum (Sommerf.) R. Sant. = Plectocarpon lichenum [PLECLIC]
Lichenosticta Zopf [LICHENS$]
Lichenosticta alcicorniaria (Lindsay) D. Hawksw. [LICHALC]
Microlychnus A. Funk [MICROLY$]
Microlychnus epicorticis A. Funk [MICREPI]
Pannaria lepidiota (Sommerf.) Th. Fr. = Fuscopannaria praetermissa [FUSCPRE]
Parmeliella saubinetii (Mont.) Zahlbr. = Fuscopannaria saubinetii [FUSCSAU]
Phragmonaevia fuckelii Rehm = Corticifraga fuckelii [CORTFUC]
Plectocarpon Fée [PLECTOC$]
Plectocarpon lichenum (Sommerf.) D. Hawksw. [PLECLIC]
SY: *Lichenomyces lichenum* (Sommerf.) R. Sant.
Plectocarpon nephromeum (Norman) Sant. [PLECNEP]
Racodium Fr. [RACODIU$]
Racodium rupestre Pers. [RACORUP]
Ramularia peltigericola D. Hawksw. = Hawksworthiana peltigericola [HAWKPEL]
Refractohilum D. Hawksw. [REFRACT$]
Refractohilum peltigerae (Keissler) D. Hawksw. [REFRPEL]
Roselliniella Vainio [ROSELLI$]
Roselliniella nephromatis (Crouan) Matzer & Hafellner [ROSENEP]
Stereocaulon albicans Th. Fr. = Leprocaulon albicans [LEPRALB]
Stereocaulon microscopicum (Vill.) Frey = Leprocaulon microscopicum [LEPRMIC]
Stereocaulon pseudoarbuscula Asah. = Leprocaulon subalbicans [LEPRSUB]
Szczawinskia A. Funk [SZCZAWI$]
Szczawinskia tsugae A. Funk [SZCZTSU]

VERRUCARIACEAE (L)

Agonimia Zahlbr. [AGONIMI$]
Agonimia tristicula (Nyl.) Zahlbr. [AGONTRI] (moss trifle lichen)
SY: *Polyblastia tristicula* (Nyl.) Arnold
Catapyrenium Flotow [CATAPYR$]
Catapyrenium cinereum (Pers.) Körber [CATACIN]
SY: *Dermatocarpon cinereum* (Pers.) Th. Fr.
SY: *Dermatocarpon hepaticum* (Ach.) Th. Fr.
SY: *Endocarpon cinereum* Pers.
SY: *Endocarpon tephroides* (Ach.) Ach.
SY: *Lichen tephroides* Ach.
SY: *Sagedia cinerea* (Pers.) Fr.
Catapyrenium compactum (A. Massal.) R. Sant. [CATACOM]
SY: *Dermatocarpon compactum* (A. Massal.) Lettau
SY: *Dermatocarpon rupicola* (de Lesd.) Zahlbr.
SY: *Endopyrenium americanum* de Lesd.
SY: *Endopyrenium rupicola* de Lesd.
Catapyrenium daedaleum (Kremp.) Stein [CATADAE] (ashen stipplescale)
SY: *Dermatocarpon daedaleum* (Kremp.) Th. Fr.
Catapyrenium heppioides (Zahlbr.) J.W. Thomson [CATAHEP]
SY: *Dermatocarpon heppioides* Zahlbr.
Catapyrenium lachneum (Ach.) R. Sant. [CATALAC]
SY: *Dermatocarpon lachneum* (Ach.) A.L. Sm.
SY: *Endocarpon hepaticum* Ach.
SY: *Endocarpon lachneum* (Ach.) Ach.
SY: *Endopyrenium hepaticum* (Ach.) Körber
SY: *Lichen lachneus* Ach.

SY: *Placidium hepaticum* var. *pruinosum* de Lesd.
SY: *Placidium hepaticum* var. *reticulatum* de Lesd.
Catapyrenium norvegicum Breuss [CATANOR]
Catapyrenium squamulosum (Ach.) Breuss [CATASQA] (brown stipplescale)
SY: *Dermatocarpon hepaticum* auct.
Dermatocarpon Eschw. [DERMATO$]
Dermatocarpon aquaticum (Weiss) Zahlbr. = Dermatocarpon luridum [DERMLUR]
Dermatocarpon cinereum (Pers.) Th. Fr. = Catapyrenium cinereum [CATACIN]
Dermatocarpon compactum (A. Massal.) Lettau = Catapyrenium compactum [CATACOM]
Dermatocarpon daedaleum (Kremp.) Th. Fr. = Catapyrenium daedaleum [CATADAE]
Dermatocarpon fluviatile (Weber) Th. Fr. = Dermatocarpon luridum [DERMLUR]
Dermatocarpon hepaticum (Ach.) Th. Fr. = Catapyrenium cinereum [CATACIN]
Dermatocarpon hepaticum auct. = Catapyrenium squamulosum [CATASQA]
Dermatocarpon heppioides Zahlbr. = Catapyrenium heppioides [CATAHEP]
Dermatocarpon intestiniforme (Körber) Hasse [DERMINT] (fissured stippleback)
SY: *Dermatocarpon polyphyllum* (Wulfen) Dalla Torre & Sarnth.
Dermatocarpon lachneum (Ach.) A.L. Sm. = Catapyrenium lachneum [CATALAC]
Dermatocarpon linkolae Räsänen [DERMLIN]
Dermatocarpon luridum (With.) J.R. Laundon [DERMLUR]
SY: *Dermatocarpon aquaticum* (Weiss) Zahlbr.
SY: *Dermatocarpon fluviatile* (Weber) Th. Fr.
SY: *Dermatocarpon weberi* (Ach.) W. Mann
Dermatocarpon miniatum (L.) W. Mann [DERMMIN] (limy stippleback)
SY: *Dermatocarpon muehlenbergii* (Ach.) Müll. Arg.
Dermatocarpon moulinsii (Mont.) Zahlbr. [DERMMOU] (shag stippleback)
Dermatocarpon muehlenbergii (Ach.) Müll. Arg. = Dermatocarpon miniatum [DERMMIN]
Dermatocarpon polyphyllum (Wulfen) Dalla Torre & Sarnth. = Dermatocarpon intestiniforme [DERMINT]
Dermatocarpon reticulatum H. Magn. [DERMRET] (northwest stippleback)
SY: *Dermatocarpon vagans* Imshaug
Dermatocarpon rivulorum (Arnold) Dalla Torre & Sarnth. [DERMRIV] (streamside stippleback)
Dermatocarpon rupicola (de Lesd.) Zahlbr. = Catapyrenium compactum [CATACOM]
Dermatocarpon vagans Imshaug = Dermatocarpon reticulatum [DERMRET]
Dermatocarpon weberi (Ach.) W. Mann = Dermatocarpon luridum [DERMLUR]
Endocarpon Hedwig [ENDOCAR$]
Endocarpon cinereum Pers. = Catapyrenium cinereum [CATACIN]
Endocarpon drummondii (Tuck.) Choisy = Staurothele drummondii [STAUDRU]
Endocarpon hepaticum Ach. = Catapyrenium lachneum [CATALAC]
Endocarpon lachneum (Ach.) Ach. = Catapyrenium lachneum [CATALAC]
Endocarpon pallidum Ach. [ENDOPAL]
SY: *Endocarpon pusillum* var. *pallidum* (Ach.) Körber
Endocarpon pulvinatum Th. Fr. [ENDOPUL] (rock stippleback)
SY: *Pyrenothamnia brandegei* (Tuck.) Zahlbr.
SY: *Pyrenothamnia spraguei* Tuck.
Endocarpon pusillum Hedwig [ENDOPUS] (soil stippleback)
Endocarpon pusillum var. *pallidum* (Ach.) Körber = Endocarpon pallidum [ENDOPAL]
Endocarpon tephroides (Ach.) Ach. = Catapyrenium cinereum [CATACIN]
Endocarpon wilmsoides Zahlbr. = Staurothele drummondii [STAUDRU]
Endopyrenium americanum de Lesd. = Catapyrenium compactum [CATACOM]
Endopyrenium hepaticum (Ach.) Körber = Catapyrenium lachneum [CATALAC]
Endopyrenium rupicola de Lesd. = Catapyrenium compactum [CATACOM]
Lichen lachneus Ach. = Catapyrenium lachneum [CATALAC]
Lichen tephroides Ach. = Catapyrenium cinereum [CATACIN]
Muellerella Hepp ex Müll. Arg. [MUELLER$]
Muellerella pygmaea (Körber) D. Hawksw. [MUELPYG]
SY: *Tichothecium pygmaeum* Körber
Normandina Nyl. [NORMAND$]
Normandina pulchella (Borrer) Nyl. [NORMPUL] (elf-ear)
Placidiopsis Beltr. [PLACIDI$]
Placidiopsis pseudocinerea Breuss [PLACPSE]
Placidium hepaticum var. *pruinosum* de Lesd. = Catapyrenium lachneum [CATALAC]
Placidium hepaticum var. *reticulatum* de Lesd. = Catapyrenium lachneum [CATALAC]
Polyblastia A. Massal. [POLYBLA$]
Polyblastia gelatinosa (Ach.) Th. Fr. [POLYGEL]
Polyblastia hyperborea Th. Fr. [POLYHYP]
SY: *Polyblastia integrascens* (Nyl.) Vainio
Polyblastia integrascens (Nyl.) Vainio = Polyblastia hyperborea [POLYHYP]
Polyblastia obsoleta Arnold [POLYOBS]
Polyblastia rupifraga A. Massal. = Staurothele rupifraga [STAURUP]
Polyblastia sommerfeltii Lynge = Polyblastia terrestris [POLYTER]
Polyblastia terrestris Th. Fr. [POLYTER]
SY: *Polyblastia sommerfeltii* Lynge
Polyblastia theleodes (Sommerf.) Th. Fr. [POLYTHE]
Polyblastia tristicula (Nyl.) Arnold = Agonimia tristicula [AGONTRI]
Polyblastia umbrina var. *areolata* (Ach.) Boist. = Staurothele areolata [STAUARE]
Pyrenothamnia brandegei (Tuck.) Zahlbr. = Endocarpon pulvinatum [ENDOPUL]
Pyrenothamnia spraguei Tuck. = Endocarpon pulvinatum [ENDOPUL]
Pyrenula areolata Ach. = Staurothele areolata [STAUARE]
Sagedia cinerea (Pers.) Fr. = Catapyrenium cinereum [CATACIN]
Sphaeromphale areolata (Ach.) Arnold = Staurothele areolata [STAUARE]
Sphaeromphale clopima var. *areolata* (Ach.) Trev. = Staurothele areolata [STAUARE]
Sphaeromphale fissa (Tayl.) Körber = Staurothele fissa [STAUFIS]
Sphaeromphale hazslinszkyi Körber = Staurothele fissa [STAUFIS]
Staurothele Norman [STAUROT$]
Staurothele ambrosiana (A. Massal.) Zschacke = Staurothele drummondii [STAUDRU]
Staurothele areolata (Ach.) Lettau [STAUARE]
SY: *Polyblastia umbrina* var. *areolata* (Ach.) Boist.
SY: *Pyrenula areolata* Ach.
SY: *Sphaeromphale areolata* (Ach.) Arnold
SY: *Sphaeromphale clopima* var. *areolata* (Ach.) Trev.
SY: *Staurothele clopima* var. *areolata* (Ach.) Vainio
SY: *Staurothele clopima* var. *mamillata* Vainio
Staurothele circinata Tuck. = Staurothele fissa [STAUFIS]
Staurothele clopima var. *areolata* (Ach.) Vainio = Staurothele areolata [STAUARE]
Staurothele clopima var. *mamillata* Vainio = Staurothele areolata [STAUARE]
Staurothele diffusilis (Nyl.) Zahlbr. = Staurothele drummondii [STAUDRU]
Staurothele drummondii (Tuck.) Tuck. [STAUDRU]
SY: *Endocarpon drummondii* (Tuck.) Choisy
SY: *Endocarpon wilmsoides* Zahlbr.
SY: *Staurothele ambrosiana* (A. Massal.) Zschacke
SY: *Staurothele diffusilis* (Nyl.) Zahlbr.
SY: *Staurothele fuscocuprea* (Nyl.) Zschacke
SY: *Staurothele fuscocuprea* f. *rimosior* (Nyl.) Zsch.
SY: *Staurothele fuscocuprea* f. *rimoso-areolata* Zsch.
SY: *Staurothele perradiata* Lynge
SY: *Staurothele perradiata* f. *dissoluta* Lynge

SY: *Staurothele perradiata* var. *hypothallina* Lamb
SY: *Staurothele septentrionalis* Lynge
SY: *Staurothele succedens* (Rehm) Arnold
SY: *Verrucaria cuprea* var. *fuscocuprea* Nyl.
SY: *Verrucaria diffusilis* Nyl.
SY: *Verrucaria drummondii* Tuck.
SY: *Verrucaria fuscocuprea* (Nyl.) Nyl.
SY: *Verrucaria fuscocuprea* f. *rimosior* Nyl.
Staurothele fissa (Taylor) Zwackh [STAUFIS]
SY: *Sphaeromphale fissa* (Tayl.) Körber
SY: *Sphaeromphale hazslinszkyi* Körber
SY: *Staurothele circinata* Tuck.
SY: *Staurothele glacialis* Herre
SY: *Staurothele hazslinskyi* (Körber) Blomb. & Forss.
SY: *Staurothele inconversa* (Nyl.) Blomb. & Forss.
SY: *Staurothele umbrina* (Wahlenb.) Tuck.
SY: *Stigmatomma fissum* (Tayl.) Syd.
SY: *Verrucaria fissa* Tayl.
SY: *Verrucaria inconversa* Nyl.
SY: *Staurothele lithina* auct.
Staurothele fuscocuprea (Nyl.) Zschacke = Staurothele drummondii [STAUDRU]
Staurothele fuscocuprea f. *rimosior* (Nyl.) Zsch. = Staurothele drummondii [STAUDRU]
Staurothele fuscocuprea f. *rimoso-areolata* Zsch. = Staurothele drummondii [STAUDRU]
Staurothele glacialis Herre = Staurothele fissa [STAUFIS]
Staurothele hazslinskyi (Körber) Blomb. & Forss. = Staurothele fissa [STAUFIS]
Staurothele inconversa (Nyl.) Blomb. & Forss. = Staurothele fissa [STAUFIS]
Staurothele lithina auct. = Staurothele fissa [STAUFIS]
Staurothele perradiata Lynge = Staurothele drummondii [STAUDRU]
Staurothele perradiata f. *dissoluta* Lynge = Staurothele drummondii [STAUDRU]
Staurothele perradiata var. *hypothallina* Lamb = Staurothele drummondii [STAUDRU]
Staurothele rupifraga (A. Massal.) Arnold [STAURUP]
SY: *Polyblastia rupifraga* A. Massal.
Staurothele septentrionalis Lynge = Staurothele drummondii [STAUDRU]
Staurothele succedens (Rehm) Arnold = Staurothele drummondii [STAUDRU]
Staurothele umbrina (Wahlenb.) Tuck. = Staurothele fissa [STAUFIS]
Stigmatomma fissum (Tayl.) Syd. = Staurothele fissa [STAUFIS]
Thelidium A. Massal. [THELIDI$]
Thelidium absconditum (Hepp) Rabenh. [THELABS]
Thelidium decipiens (Nyl.) Kremp. [THELDEC]
Thelidium incavatum Nyl. ex Mudd [THELINC]
Tichothecium pygmaeum Körber = Muellerella pygmaea [MUELPYG]
Verrucaria Schrader [VERRUCA$]
Verrucaria aethiobola Wahlenb. [VERRAET]
SY: *Verrucaria laevata* Ach.
Verrucaria cuprea var. *fuscocuprea* Nyl. = Staurothele drummondii [STAUDRU]
Verrucaria devergens Nyl. [VERRDEV]
Verrucaria diffusilis Nyl. = Staurothele drummondii [STAUDRU]
Verrucaria drummondii Tuck. = Staurothele drummondii [STAUDRU]
Verrucaria erichsenii Zsch. [VERRERI]
Verrucaria fissa Tayl. = Staurothele fissa [STAUFIS]
Verrucaria funckii (Sprengel) Zahlbr. [VERRFUN]
Verrucaria fuscocuprea (Nyl.) Nyl. = Staurothele drummondii [STAUDRU]
Verrucaria fuscocuprea f. *rimosior* Nyl. = Staurothele drummondii [STAUDRU]
Verrucaria fusconigrescens Nyl. [VERRFUS]
Verrucaria inconversa Nyl. = Staurothele fissa [STAUFIS]
Verrucaria laevata Ach. = Verrucaria aethiobola [VERRAET]
Verrucaria macrostoma Dufour ex DC. [VERRMAC]
Verrucaria margacea (Wahlenb.) Wahlenb. [VERRMAR]
Verrucaria maura Wahlenb. [VERRMAU] (sea tar)
Verrucaria mucosa Wahlenb. [VERRMUC]
Verrucaria muralis Ach. [VERRMUR]
Verrucaria nigrescens Pers. [VERRNIG]
Verrucaria prominula Nyl. [VERRPRO]
Verrucaria rupestris Schrader [VERRRUP]
Verrucaria sphaerospora Anzi [VERRSPH]
Verrucaria tectorum (A. Massal.) Körber [VERRTEC]
Verrucaria viridula (Schrader) Ach. [VERRVII]

Part 2

Alphabetical List of Scientific Names

A

Abies P. Mill. [ABIES$] Pinaceae (V:G)
Abies amabilis (Dougl. ex Loud.) Dougl. ex Forbes [ABIEAMA] Pinaceae (V:G)
Abies balsemea (L.) P. Mill. [ABIEBAL] Pinaceae (V:G)
Abies excelsior Franco = Abies grandis [ABIEGRA] Pinaceae (V:G)
Abies grandis (Dougl. ex D. Don) Lindl. [ABIEGRA] Pinaceae (V:G)
Abies lasiocarpa (Hook.) Nutt. [ABIELAS] Pinaceae (V:G)
Abies nobilis (Dougl. ex D. Don) Lindl. = Abies procera [ABIEPRO] Pinaceae (V:G)
***Abies procera** Rehd. [ABIEPRO] Pinaceae (V:G)
Abietinella C. Müll. [ABIETIN$] Thuidiaceae (B:M)
Abietinella abietina (Hedw.) Fleisch. [ABIEABI] Thuidiaceae (B:M)
Abronia Juss. [ABRONIA$] Nyctaginaceae (V:D)
Abronia latifolia Eschsch. [ABROLAT] Nyctaginaceae (V:D)
Abrothallus De Not. [ABROTHA$] Uncertain-family (L)
Abrothallus bertianus De Not. [ABROBER] Uncertain-family (L)
Abrothallus cetrariae Kotte [ABROCET] Uncertain-family (L)
Abrothallus oxysporus Tul. = Phacopsis oxyspora [PHACOXY] Lecanoraceae (L)
Abrothallus parmeliarum (Sommerf.) Arnold [ABROPAR] Uncertain-family (L)
Abrothallus welwitschii Tul. [ABROWEL] Uncertain-family (L)
Abutilon P. Mill. [ABUTILO$] Malvaceae (V:D)
Abutilon abutilon (L.) Rusby = Abutilon theophrasti [ABUTTHE] Malvaceae (V:D)
Abutilon avicennae Gaertn. = Abutilon theophrasti [ABUTTHE] Malvaceae (V:D)
***Abutilon theophrasti** Medik. [ABUTTHE] Malvaceae (V:D)
Acanthocladium carlottae Schof. = Wijkia carlottae [WIJKCAR] Sematophyllaceae (B:M)
Acarospora A. Massal. [ACAROSP$] Acarosporaceae (L)
Acarospora albida H. Magn. = Acarospora schleicheri [ACARSCH] Acarosporaceae (L)
Acarospora amabilis H. Magn. = Acarospora schleicheri [ACARSCH] Acarosporaceae (L)
Acarospora amphibola Wedd. = Acarospora smaragdula [ACARSMA] Acarosporaceae (L)
Acarospora asahinae H. Magn. [ACARASA] Acarosporaceae (L)
Acarospora badiofusca (Nyl.) Th. Fr. [ACARBAD] Acarosporaceae (L)
Acarospora bella (Nyl.) Jatta = Acarospora schleicheri [ACARSCH] Acarosporaceae (L)
Acarospora boulderensis H. Magn. = Acarospora badiofusca [ACARBAD] Acarosporaceae (L)
Acarospora cartilaginea H. Magn. = Acarospora fuscata [ACARFUS] Acarosporaceae (L)
Acarospora cervina A. Massal. [ACARCER] Acarosporaceae (L)
Acarospora chlorophana (Wahlenb.) A. Massal. = Pleopsidium chlorophanum [PLEOCHL] Acarosporaceae (L)
Acarospora chrysops (Tuck.) H. Magn. = Acarospora schleicheri [ACARSCH] Acarosporaceae (L)
Acarospora contigua H. Magn. = Acarospora schleicheri [ACARSCH] Acarosporaceae (L)
Acarospora dissipata H. Magn. = Acarospora schleicheri [ACARSCH] Acarosporaceae (L)
Acarospora epilutescens Zahlbr. = Acarospora schleicheri [ACARSCH] Acarosporaceae (L)
Acarospora erythrophora H. Magn. = Pleopsidium chlorophanum [PLEOCHL] Acarosporaceae (L)
Acarospora evoluta H. Magn. = Acarospora schleicheri [ACARSCH] Acarosporaceae (L)
Acarospora fuscata (Schrader) Arnold [ACARFUS] Acarosporaceae (L)
Acarospora glaucocarpa (Ach.) Körber [ACARGLA] Acarosporaceae (L)
Acarospora incertula H. Magn. = Pleopsidium chlorophanum [PLEOCHL] Acarosporaceae (L)
Acarospora intercedens H. Magn. = Acarospora schleicheri [ACARSCH] Acarosporaceae (L)
Acarospora macrospora (Hepp) Bagl. [ACARMAC] Acarosporaceae (L)
Acarospora novomexicana H. Magn. = Pleopsidium chlorophanum [PLEOCHL] Acarosporaceae (L)
Acarospora ocellata H. Magn. = Acarospora schleicheri [ACARSCH] Acarosporaceae (L)
Acarospora rhabarbarina Hue = Acarospora schleicheri [ACARSCH] Acarosporaceae (L)
Acarospora rimulosa H. Magn. = Acarospora schleicheri [ACARSCH] Acarosporaceae (L)
Acarospora rubicunda H. Magn. = Acarospora schleicheri [ACARSCH] Acarosporaceae (L)
Acarospora rufescens (Ach.) Bausch = Acarospora smaragdula [ACARSMA] Acarosporaceae (L)
Acarospora saxicola Fink = Glypholecia scabra [GLYPSCA] Acarosporaceae (L)
Acarospora scabra (Pers.) Th. Fr. = Glypholecia scabra [GLYPSCA] Acarosporaceae (L)
Acarospora schleicheri (Ach.) A. Massal. [ACARSCH] Acarosporaceae (L)
Acarospora sinopica (Wahlenb.) Körber [ACARSIN] Acarosporaceae (L)
Acarospora smaragdula (Wahlenb.) A. Massal. [ACARSMA] Acarosporaceae (L)
Acarospora squamulosa (Schrader) Trevisan = Acarospora fuscata [ACARFUS] Acarosporaceae (L)
Acarospora squamulosa auct. = Acarospora macrospora [ACARMAC] Acarosporaceae (L)
Acarospora subalbida H. Magn. = Acarospora schleicheri [ACARSCH] Acarosporaceae (L)
Acarospora subcontigua H. Magn. = Acarospora schleicheri [ACARSCH] Acarosporaceae (L)
Acarospora texana H. Magn. = Pleopsidium chlorophanum [PLEOCHL] Acarosporaceae (L)
Acarospora veronensis A. Massal. [ACARVER] Acarosporaceae (L)
Acarospora weldensis H. Magn. = Pleopsidium chlorophanum [PLEOCHL] Acarosporaceae (L)
Acarospora xanthophana (Nyl.) Jatta = Acarospora schleicheri [ACARSCH] Acarosporaceae (L)
Acaulon C. Müll. [ACAULON$] Pottiaceae (B:M)
Acaulon muticum (Hedw.) C. Müll. [ACAUMUT] Pottiaceae (B:M)
Acaulon muticum var. **rufescens** (Jaeg.) Crum [ACAUMUT1] Pottiaceae (B:M)
Acaulon rufescens Jaeg. = Acaulon muticum var. rufescens [ACAUMUT1] Pottiaceae (B:M)
Acer L. [ACER$] Aceraceae (V:D)
Acer circinatum Pursh [ACERCIR] Aceraceae (V:D)
Acer douglasii Hook. = Acer glabrum var. douglasii [ACERGLA1] Aceraceae (V:D)
Acer glabrum Torr. [ACERGLA] Aceraceae (V:D)
Acer glabrum ssp. *douglasii* (Hook.) Wesmael = Acer glabrum var. douglasii [ACERGLA1] Aceraceae (V:D)
Acer glabrum var. **douglasii** (Hook.) Dippel [ACERGLA1] Aceraceae (V:D)
Acer macrophyllum Pursh [ACERMAC] Aceraceae (V:D)
***Acer negundo** L. [ACERNEG] Aceraceae (V:D)
***Acer platanoides** L. [ACERPLA] Aceraceae (V:D)
Acer platanoides var. *schwedleri* Nichols. = Acer platanoides [ACERPLA] Aceraceae (V:D)
***Acer pseudoplatanus** L. [ACERPSE] Aceraceae (V:D)
Acetosa alpestris (Jacq.) A. Löve = Rumex acetosa ssp. alpestris [RUMEACE3] Polygonaceae (V:D)

Acetosa oblongifolia (L.) A. & D. Löve = Rumex obtusifolius [RUMEOBT] Polygonaceae (V:D)
Acetosa pratensis ssp. *alpestris* (Jacq.) A. Löve = Rumex acetosa ssp. alpestris [RUMEACE3] Polygonaceae (V:D)
Acetosella acetosella (L.) Small = Rumex acetosella [RUMEACT] Polygonaceae (V:D)
Acetosella corniculata (L.) Kuntze = Oxalis corniculata [OXALCOR] Oxalidaceae (V:D)
Acetosella tenuifolia (Wallr.) A. Löve = Rumex acetosella [RUMEACT] Polygonaceae (V:D)
Acetosella vulgaris (Koch) Fourr. = Rumex acetosella [RUMEACT] Polygonaceae (V:D)
Achillea L. [ACHILLE$] Asteraceae (V:D)
Achillea alpicola (Rydb.) Rydb. = Achillea millefolium var. alpicola [ACHIMIL1] Asteraceae (V:D)
Achillea angustissima Rydb. = Achillea millefolium var. occidentalis [ACHIMIL8] Asteraceae (V:D)
Achillea borealis Bong. = Achillea millefolium var. borealis [ACHIMIL2] Asteraceae (V:D)
Achillea borealis ssp. *typica* Keck = Achillea millefolium var. borealis [ACHIMIL2] Asteraceae (V:D)
Achillea eradiata Piper = Achillea millefolium var. occidentalis [ACHIMIL8] Asteraceae (V:D)
Achillea fusca Rydb. = Achillea millefolium var. alpicola [ACHIMIL1] Asteraceae (V:D)
Achillea gracilis Raf. = Achillea millefolium var. occidentalis [ACHIMIL8] Asteraceae (V:D)
Achillea lanulosa Nutt. = Achillea millefolium var. occidentalis [ACHIMIL8] Asteraceae (V:D)
Achillea lanulosa ssp. *alpicola* (Rydb.) Keck = Achillea millefolium var. alpicola [ACHIMIL1] Asteraceae (V:D)
Achillea lanulosa ssp. *typica* Keck = Achillea millefolium var. occidentalis [ACHIMIL8] Asteraceae (V:D)
Achillea lanulosa var. *alpicola* Rydb. = Achillea millefolium var. alpicola [ACHIMIL1] Asteraceae (V:D)
Achillea lanulosa var. *arachnoidea* Lunell = Achillea millefolium var. occidentalis [ACHIMIL8] Asteraceae (V:D)
Achillea lanulosa var. *eradiata* (Piper) M.E. Peck = Achillea millefolium var. occidentalis [ACHIMIL8] Asteraceae (V:D)
Achillea laxiflora Pollard & Cockerell = Achillea millefolium var. occidentalis [ACHIMIL8] Asteraceae (V:D)
***Achillea millefolium** L. [ACHIMIL] Asteraceae (V:D)
Achillea millefolium ssp. *atrotegula* Boivin = Achillea millefolium var. borealis [ACHIMIL2] Asteraceae (V:D)
Achillea millefolium ssp. *borealis* (Bong.) Breitung = Achillea millefolium var. borealis [ACHIMIL2] Asteraceae (V:D)
Achillea millefolium ssp. *lanulosa* (Nutt.) Piper = Achillea millefolium var. occidentalis [ACHIMIL8] Asteraceae (V:D)
Achillea millefolium ssp. *occidentalis* (DC.) Hyl. = Achillea millefolium var. occidentalis [ACHIMIL8] Asteraceae (V:D)
Achillea millefolium ssp. *pallidotegula* Boivin = Achillea millefolium var. occidentalis [ACHIMIL8] Asteraceae (V:D)
***Achillea millefolium** var. **alpicola** (Rydb.) Garrett [ACHIMIL1] Asteraceae (V:D)
Achillea millefolium var. *aspleniifolia* (Vent.) Farw. = Achillea millefolium var. occidentalis [ACHIMIL8] Asteraceae (V:D)
***Achillea millefolium** var. **borealis** (Bong.) Farw. [ACHIMIL2] Asteraceae (V:D)
Achillea millefolium var. *fulva* Boivin = Achillea millefolium var. borealis [ACHIMIL2] Asteraceae (V:D)
Achillea millefolium var. *fusca* (Rydb.) G.N. Jones = Achillea millefolium var. alpicola [ACHIMIL1] Asteraceae (V:D)
Achillea millefolium var. *gracilis* Raf. ex DC. = Achillea millefolium var. occidentalis [ACHIMIL8] Asteraceae (V:D)
Achillea millefolium var. *lanulosa* (Nutt.) Piper = Achillea millefolium var. occidentalis [ACHIMIL8] Asteraceae (V:D)
***Achillea millefolium** var. **occidentalis** DC. [ACHIMIL8] Asteraceae (V:D)
***Achillea millefolium** var. **pacifica** (Rydb.) G.N. Jones [ACHIMIL6] Asteraceae (V:D)
Achillea millefolium var. *parviligula* Boivin = Achillea millefolium var. borealis [ACHIMIL2] Asteraceae (V:D)
Achillea millefolium var. *parvula* Boivin = Achillea millefolium var. borealis [ACHIMIL2] Asteraceae (V:D)
Achillea millefolium var. *rosea* (Desf.) Torr. & Gray = Achillea millefolium var. occidentalis [ACHIMIL8] Asteraceae (V:D)
Achillea millefolium var. *russeolata* Boivin = Achillea millefolium var. occidentalis [ACHIMIL8] Asteraceae (V:D)
Achillea multiflora Hook. = Achillea sibirica [ACHISIB] Asteraceae (V:D)
Achillea occidentalis (DC.) Raf. ex Rydb. = Achillea millefolium var. occidentalis [ACHIMIL8] Asteraceae (V:D)
Achillea pacifica Rydb. = Achillea millefolium var. pacifica [ACHIMIL6] Asteraceae (V:D)
Achillea rosea Desf. = Achillea millefolium var. occidentalis [ACHIMIL8] Asteraceae (V:D)
Achillea sibirica Ledeb. [ACHISIB] Asteraceae (V:D)
Achillea tomentosa Pursh = Achillea millefolium var. occidentalis [ACHIMIL8] Asteraceae (V:D)
Achlys DC. [ACHLYS$] Berberidaceae (V:D)
Achlys triphylla (Sm.) DC. [ACHLTRI] Berberidaceae (V:D)
Acinos P. Mill. [ACINOS$] Lamiaceae (V:D)
***Acinos arvensis** (Lam.) Dandy [ACINARV] Lamiaceae (V:D)
Acinos thymoides (L.) Moench = Acinos arvensis [ACINARV] Lamiaceae (V:D)
Acmispon americanum (Nutt.) Rydb. = Lotus unifoliolatus [LOTUUNI] Fabaceae (V:D)
Acomastylis calthifolia (Menzies ex Sm.) Bolle = Geum calthifolium [GEUMCAL] Rosaceae (V:D)
Acomastylis macrantha (Kearney) Bolle = Geum × macranthum [GEUMMAR] Rosaceae (V:D)
Aconitum L. [ACONITU$] Ranunculaceae (V:D)
Aconitum columbianum Nutt. [ACONCOL] Ranunculaceae (V:D)
Aconitum columbianum ssp. *pallidum* Piper = Aconitum columbianum [ACONCOL] Ranunculaceae (V:D)
Aconitum columbianum var. *bakeri* (Greene) Harrington = Aconitum columbianum [ACONCOL] Ranunculaceae (V:D)
Aconitum delphiniifolium DC. [ACONDEL] Ranunculaceae (V:D)
Aconitum delphiniifolium var. *albiflorum* Porsild = Aconitum delphiniifolium [ACONDEL] Ranunculaceae (V:D)
Aconogonum polystachyum (Wallich ex Meisn.) Haraldson = Polygonum polystachyum [POLYPOY] Polygonaceae (V:D)
Acorus L. [ACORUS$] Acoraceae (V:M)
Acorus americanus (Raf.) Raf. [ACORAME] Acoraceae (V:M)
Acorus calamus auct. = Acorus americanus [ACORAME] Acoraceae (V:M)
Acorus calamus var. *americanus* (Raf.) H.D. Wulff. = Acorus americanus [ACORAME] Acoraceae (V:M)
Acosta diffusa (Lam.) Soják = Centaurea diffusa [CENTDIF] Asteraceae (V:D)
Acrocladium cordifolium (Hedw.) P. Rich. & Wallace = Calliergon cordifolium [CALLCOR] Amblystegiaceae (B:M)
Acrocladium cuspidatum (Hedw.) Lindb. = Calliergonella cuspidata [CALLCUS] Amblystegiaceae (B:M)
Acrocordia A. Massal. [ACROCOR$] Monoblastiaceae (L)
Acrocordia gemmata (Ach.) A. Massal. [ACROGEM] Monoblastiaceae (L)
Acrolasia albicaulis (Dougl. ex Hook.) Rydb. = Mentzelia albicaulis [MENTALB] Loasaceae (V:D)
Acroptilon Cass. [ACROPTI$] Asteraceae (V:D)
***Acroptilon repens** (L.) DC. [ACROREP] Asteraceae (V:D)
Acroscyphus Léveillé [ACROSCY$] Caliciaceae (L)
Acroscyphus sphaerophoroides Léveillé [ACROSPH] Caliciaceae (L)
Actaea L. [ACTAEA$] Ranunculaceae (V:D)
Actaea arguta Nutt. = Actaea rubra ssp. arguta [ACTARUB1] Ranunculaceae (V:D)
Actaea arguta var. *viridiflora* (Greene) Tidestrom = Actaea rubra ssp. arguta [ACTARUB1] Ranunculaceae (V:D)
Actaea rubra (Ait.) Willd. [ACTARUB] Ranunculaceae (V:D)
Actaea rubra ssp. **arguta** (Nutt.) Hultén [ACTARUB1] Ranunculaceae (V:D)
Actaea rubra var. *arguta* (Nutt.) Lawson = Actaea rubra ssp. arguta [ACTARUB1] Ranunculaceae (V:D)

Actaea spicata var. *arguta* (Nutt.) Torr. = Actaea rubra ssp. arguta [ACTARUB1] Ranunculaceae (V:D)
Actinogyra muehlenbergii (Ach.) Schol. = Umbilicaria muehlenbergii [UMBIMUE] Umbilicariaceae (L)
Actinogyra polyrrhiza (L.) Schol. = Umbilicaria polyrrhiza [UMBIPOY] Umbilicariaceae (L)
Adelolecia Hertel & Hafellner [ADELOLE$] Bacidiaceae (L)
Adelolecia pilati (Hepp) Hertel & Hafellner [ADELPIL] Bacidiaceae (L)
Adenocaulon Hook. [ADENOCA$] Asteraceae (V:D)
Adenocaulon bicolor Hook. [ADENBIC] Asteraceae (V:D)
Adenolinum lewisii (Pursh) A. & D. Löve = Linum lewisii [LINULEW] Linaceae (V:D)
Adiantum L. [ADIANTU$] Pteridaceae (V:F)
Adiantum aleuticum (Rupr.) Paris [ADIAALE] Pteridaceae (V:F)
Adiantum capillus-veneris L. [ADIACAP] Pteridaceae (V:F)
Adiantum capillus-veneris var. *modestum* (Underwood) Fern. = Adiantum capillus-veneris [ADIACAP] Pteridaceae (V:F)
Adiantum capillus-veneris var. *protrusum* Fern. = Adiantum capillus-veneris [ADIACAP] Pteridaceae (V:F)
Adiantum modestum Underwood = Adiantum capillus-veneris [ADIACAP] Pteridaceae (V:F)
Adiantum pedatum auct. = Adiantum aleuticum [ADIAALE] Pteridaceae (V:F)
Adiantum pedatum ssp. *aleuticum* (Rupr.) Calder & Taylor = Adiantum aleuticum [ADIAALE] Pteridaceae (V:F)
Adiantum pedatum ssp. *calderi* Cody = Adiantum aleuticum [ADIAALE] Pteridaceae (V:F)
Adiantum pedatum ssp. *subpumilum* (W.H. Wagner) Lellinger = Adiantum aleuticum [ADIAALE] Pteridaceae (V:F)
Adiantum pedatum var. *aleuticum* Rupr. = Adiantum aleuticum [ADIAALE] Pteridaceae (V:F)
Adiantum pedatum var. *subpumilum* W.H. Wagner = Adiantum aleuticum [ADIAALE] Pteridaceae (V:F)
Adoxa L. [ADOXA$] Adoxaceae (V:D)
Adoxa moschatellina L. [ADOXMOS] Adoxaceae (V:D)
Aegopodium L. [AEGOPOD$] Apiaceae (V:D)
***Aegopodium podagraria** L. [AEGOPOA] Apiaceae (V:D)
Aegopodium podagraria var. *variegatum* Bailey = Aegopodium podagraria [AEGOPOA] Apiaceae (V:D)
Agastache Clayton ex Gronov. [AGASTAC$] Lamiaceae (V:D)
Agastache anethiodora (Nutt.) Britt. = Agastache foeniculum [AGASFOE] Lamiaceae (V:D)
Agastache foeniculum (Pursh) Kuntze [AGASFOE] Lamiaceae (V:D)
Agastache urticifolia (Benth.) Kuntze [AGASURT] Lamiaceae (V:D)
Agonimia Zahlbr. [AGONIMI$] Verrucariaceae (L)
Agonimia tristicula (Nyl.) Zahlbr. [AGONTRI] Verrucariaceae (L)
Agoseris Raf. [AGOSERI$] Asteraceae (V:D)
Agoseris aurantiaca (Hook.) Greene [AGOSAUR] Asteraceae (V:D)
Agoseris glauca (Pursh) Raf. [AGOSGLA] Asteraceae (V:D)
Agoseris glauca var. *asper* (Rydb.) Cronq. = Agoseris glauca var. dasycephala [AGOSGLA1] Asteraceae (V:D)
Agoseris glauca var. **dasycephala** (Torr. & Gray) Jepson [AGOSGLA1] Asteraceae (V:D)
Agoseris glauca var. *pumila* (Nutt.) Garrett = Agoseris glauca var. dasycephala [AGOSGLA1] Asteraceae (V:D)
Agoseris graminifolia Greene = Agoseris aurantiaca [AGOSAUR] Asteraceae (V:D)
Agoseris grandiflora (Nutt.) Greene [AGOSGRA] Asteraceae (V:D)
Agoseris heterophylla (Nutt.) Greene [AGOSHET] Asteraceae (V:D)
Agoseris heterophylla ssp. *californica* (Nutt.) Piper = Agoseris heterophylla [AGOSHET] Asteraceae (V:D)
Agoseris heterophylla ssp. *normalis* Piper = Agoseris heterophylla [AGOSHET] Asteraceae (V:D)
Agoseris heterophylla var. *cryptopleura* Jepson = Agoseris heterophylla [AGOSHET] Asteraceae (V:D)
Agoseris heterophylla var. *kymapleura* Greene = Agoseris heterophylla [AGOSHET] Asteraceae (V:D)
Agoseris laciniata (Nutt.) Greene = Agoseris grandiflora [AGOSGRA] Asteraceae (V:D)
Agoseris lackschewitzii D. Henderson & R. Moseley [AGOSLAC] Asteraceae (V:D)
Agoseris plebeja (Greene) Greene = Agoseris grandiflora [AGOSGRA] Asteraceae (V:D)
Agoseris rostrata Rydb. = Agoseris aurantiaca [AGOSAUR] Asteraceae (V:D)
Agoseris scorzonerifolia (Schrad.) Greene = Agoseris glauca var. dasycephala [AGOSGLA1] Asteraceae (V:D)
Agrestia J.W. Thomson [AGRESTI$] Hymeneliaceae (L)
Agrestia cyphellata J.W. Thomson = Agrestia hispida [AGREHIS] Hymeneliaceae (L)
Agrestia hispida (Mereschk.) Hale & Culb. [AGREHIS] Hymeneliaceae (L)
Agrimonia L. [AGRIMON$] Rosaceae (V:D)
Agrimonia gryposepala Wallr. [AGRIGRY] Rosaceae (V:D)
Agrimonia striata Michx. [AGRISTR] Rosaceae (V:D)
× *Agrohordeum macounii* (Vasey) Lepage = × Elyhordeum macounii [ELYHMAC] Poaceae (V:M)
× *Agrohordeum macounii* var. *valencianum* Bowden = × Elyhordeum macounii [ELYHMAC] Poaceae (V:M)
Agropyron Gaertn. [AGROPYR$] Poaceae (V:M)
Agropyron alaskanum Scribn. & Merr. = Elymus alaskanus [ELYMALA] Poaceae (V:M)
Agropyron albicans Scribn. & J.G. Sm. = Elymus lanceolatus ssp. albicans [ELYMLAN1] Poaceae (V:M)
Agropyron albicans var. *griffithii* (Scribn. & J.G. Sm. ex Piper) Beetle = Elymus lanceolatus ssp. albicans [ELYMLAN1] Poaceae (V:M)
Agropyron × *brevifolium* Scribn. = Elymus trachycaulus ssp. trachycaulus [ELYMTRA0] Poaceae (V:M)
Agropyron brevifolium Scribn. = Elymus trachycaulus ssp. trachycaulus [ELYMTRA0] Poaceae (V:M)
Agropyron caninum ssp. *majus* (Vasey) C.L. Hitchc. = Elymus trachycaulus ssp. trachycaulus [ELYMTRA0] Poaceae (V:M)
Agropyron caninum var. *andinum* (Scribn. & J.G. Sm.) C.L. Hitchc. = Elymus trachycaulus ssp. trachycaulus [ELYMTRA0] Poaceae (V:M)
Agropyron caninum var. *hornemannii* (Koch) Pease & Moore = Elymus trachycaulus ssp. trachycaulus [ELYMTRA0] Poaceae (V:M)
Agropyron caninum var. *mitchellii* Welsh = Elymus trachycaulus ssp. trachycaulus [ELYMTRA0] Poaceae (V:M)
Agropyron caninum var. *unilaterale* (Cassidy) C.L. Hitchc. = Elymus trachycaulus ssp. subsecundus [ELYMTRA1] Poaceae (V:M)
Agropyron caninum var. *latiglume* (Scribn. & J.G. Sm.) Pease & Moore = Elymus alaskanus ssp. latiglumis [ELYMALA1] Poaceae (V:M)
***Agropyron cristatum** (L.) Gaertn. [AGROCRI] Poaceae (V:M)
Agropyron cristatum ssp. *desertorum* (Fisch. ex Link) A. Löve = Agropyron desertorum [AGRODES] Poaceae (V:M)
Agropyron cristatum ssp. *fragile* (Roth) A. Löve p.p. = Agropyron fragile ssp. sibiricum [AGROFRA1] Poaceae (V:M)
Agropyron cristatum var. *desertorum* (Fisch. ex Link) Dorn = Agropyron desertorum [AGRODES] Poaceae (V:M)
Agropyron cristatum var. *fragile* (Roth) Dorn p.p. = Agropyron fragile ssp. sibiricum [AGROFRA1] Poaceae (V:M)
Agropyron dasystachyum (Hook.) Scribn. & J.G. Sm. = Elymus lanceolatus ssp. lanceolatus [ELYMLAN0] Poaceae (V:M)
Agropyron dasystachyum ssp. *albicans* (Scribn. & J.G. Sm.) Dewey = Elymus lanceolatus ssp. albicans [ELYMLAN1] Poaceae (V:M)
Agropyron dasystachyum ssp. *yukonense* (Scribn. & Merr.) Dewey = Elymus × yukonensis [ELYMYUK] Poaceae (V:M)
Agropyron dasystachyum var. *riparum* (Scribn. & J.G. Sm.) Bowden = Elymus lanceolatus ssp. lanceolatus [ELYMLAN0] Poaceae (V:M)
***Agropyron desertorum** (Fisch. ex Link) J.A. Schultes [AGRODES] Poaceae (V:M)
Agropyron elmeri Scribn. = Elymus lanceolatus ssp. lanceolatus [ELYMLAN0] Poaceae (V:M)
***Agropyron fragile** (Roth) P. Candargy [AGROFRA] Poaceae (V:M)
Agropyron fragile ssp. **sibiricum** (Willd.) Melderis [AGROFRA1] Poaceae (V:M)
Agropyron fragile var. *sibiricum* (Willd.) Tzvelev = Agropyron fragile ssp. sibiricum [AGROFRA1] Poaceae (V:M)
Agropyron griffithii Scribn. & J.G. Sm. ex Piper = Elymus lanceolatus ssp. albicans [ELYMLAN1] Poaceae (V:M)

Agropyron inerme (Scribn. & J.G. Sm.) Rydb. = Pseudoroegneria spicata ssp. inermis [PSEUSPI1] Poaceae (V:M)
Agropyron intermedium (Host) Beauv. = Elytrigia intermedia [ELYTINT] Poaceae (V:M)
Agropyron intermedium var. *trichophorum* (Link) Halac. = Elytrigia intermedia [ELYTINT] Poaceae (V:M)
Agropyron lanceolatum Scribn. & J.G. Sm. = Elymus lanceolatus ssp. lanceolatus [ELYMLAN0] Poaceae (V:M)
Agropyron latiglume (Scribn. & J.G. Sm.) Rydb. = Elymus alaskanus ssp. latiglumis [ELYMALA1] Poaceae (V:M)
Agropyron molle (Scribn. & J.G. Sm.) Rydb. = Pascopyrum smithii [PASCSMI] Poaceae (V:M)
Agropyron novae-angliae Scribn. = Elymus trachycaulus ssp. trachycaulus [ELYMTRA0] Poaceae (V:M)
Agropyron pauciflorum (Schwein.) A.S. Hitchc. ex Silveus = Elymus trachycaulus ssp. trachycaulus [ELYMTRA0] Poaceae (V:M)
Agropyron pauciflorum ssp. *majus* (Vasey) Melderis = Elymus trachycaulus ssp. trachycaulus [ELYMTRA0] Poaceae (V:M)
Agropyron pauciflorum ssp. *novae-angliae* (Scribn.) Melderis = Elymus trachycaulus ssp. trachycaulus [ELYMTRA0] Poaceae (V:M)
Agropyron pauciflorum ssp. *teslinense* (Porsild & Senn) Melderis = Elymus trachycaulus ssp. trachycaulus [ELYMTRA0] Poaceae (V:M)
Agropyron pauciflorum var. *glaucum* (Pease & Moore) Taylor = Elymus trachycaulus ssp. subsecundus [ELYMTRA1] Poaceae (V:M)
Agropyron pauciflorum var. *novae-angliae* (Scribn.) Taylor & MacBryde = Elymus trachycaulus ssp. trachycaulus [ELYMTRA0] Poaceae (V:M)
Agropyron pauciflorum var. *unilaterale* (Vasey) Taylor & MacBryde = Elymus trachycaulus ssp. subsecundus [ELYMTRA1] Poaceae (V:M)
Agropyron prostratum (L. f.) Beauv. = Eremopyrum triticeum [EREMTRI] Poaceae (V:M)
Agropyron repens (L.) Beauv. = Elytrigia repens [ELYTREP] Poaceae (V:M)
Agropyron repens var. *subulatum* (Schreb.) Roemer & J.A. Schultes = Elytrigia repens [ELYTREP] Poaceae (V:M)
Agropyron riparum Scribn. & J.G. Sm. = Elymus lanceolatus ssp. lanceolatus [ELYMLAN0] Poaceae (V:M)
Agropyron saundersii (Vasey) A.S. Hitchc. = Elymus × saundersii [ELYMXSA] Poaceae (V:M)
Agropyron scribneri Vasey = Elymus scribneri [ELYMSCR] Poaceae (V:M)
Agropyron sibiricum (Willd.) Beauv. = Agropyron fragile ssp. sibiricum [AGROFRA1] Poaceae (V:M)
Agropyron smithii Rydb. = Pascopyrum smithii [PASCSMI] Poaceae (V:M)
Agropyron smithii var. *molle* (Scribn. & J.G. Sm.) M.E. Jones = Pascopyrum smithii [PASCSMI] Poaceae (V:M)
Agropyron smithii var. *palmeri* (Scribn. & J.G. Sm.) Heller = Pascopyrum smithii [PASCSMI] Poaceae (V:M)
Agropyron spicatum Pursh = Pseudoroegneria spicata [PSEUSPI] Poaceae (V:M)
Agropyron spicatum var. *inerme* (Scribn. & J.G. Sm.) Heller = Pseudoroegneria spicata ssp. inermis [PSEUSPI1] Poaceae (V:M)
Agropyron spicatum var. *pubescens* Elmer = Pseudoroegneria spicata [PSEUSPI] Poaceae (V:M)
Agropyron subsecundum (Link) A.S. Hitchc. = Elymus trachycaulus ssp. subsecundus [ELYMTRA1] Poaceae (V:M)
Agropyron subsecundum var. *andinum* (Scribn. & J.G. Sm.) A.S. Hitchc. = Elymus trachycaulus ssp. trachycaulus [ELYMTRA0] Poaceae (V:M)
Agropyron tenerum Vasey = Elymus trachycaulus ssp. trachycaulus [ELYMTRA0] Poaceae (V:M)
Agropyron teslinense Porsild & Senn = Elymus trachycaulus ssp. trachycaulus [ELYMTRA0] Poaceae (V:M)
Agropyron trachycaulum (Link) Malte ex H.F. Lewis = Elymus trachycaulus ssp. trachycaulus [ELYMTRA0] Poaceae (V:M)
Agropyron trachycaulum var. *ciliatum* (Scribn. & J.G. Sm.) Gleason = Elymus trachycaulus ssp. subsecundus [ELYMTRA1] Poaceae (V:M)
Agropyron trachycaulum var. *glaucum* (Pease & Moore) Malte = Elymus trachycaulus ssp. subsecundus [ELYMTRA1] Poaceae (V:M)
Agropyron trachycaulum var. *latiglume* (Scribn. & J.G. Sm.) Beetle = Elymus alaskanus ssp. latiglumis [ELYMALA1] Poaceae (V:M)
Agropyron trachycaulum var. *majus* (Vasey) Fern. = Elymus trachycaulus ssp. trachycaulus [ELYMTRA0] Poaceae (V:M)
Agropyron trachycaulum var. *novae-angliae* (Scribn.) Fern. = Elymus trachycaulus ssp. trachycaulus [ELYMTRA0] Poaceae (V:M)
Agropyron trachycaulum var. *unilaterale* (Cassidy) Malte = Elymus trachycaulus ssp. subsecundus [ELYMTRA1] Poaceae (V:M)
Agropyron trichophorum (Link) Richter = Elytrigia intermedia [ELYTINT] Poaceae (V:M)
Agropyron triticeum Gaertn. = Eremopyrum triticeum [EREMTRI] Poaceae (V:M)
Agropyron varnense (Velen.) Hayek = Elytrigia pontica [ELYTPON] Poaceae (V:M)
Agropyron vaseyi Scribn. & J.G. Sm. = Pseudoroegneria spicata [PSEUSPI] Poaceae (V:M)
Agropyron violaceum (Hornem.) Lange = Elymus alaskanus [ELYMALA] Poaceae (V:M)
Agropyron violaceum ssp. *andinum* (Scribn. & J.G. Sm.) Melderis = Elymus trachycaulus ssp. trachycaulus [ELYMTRA0] Poaceae (V:M)
Agropyron violaceum var. *andinum* Scribn. & J.G. Sm. = Elymus trachycaulus ssp. trachycaulus [ELYMTRA0] Poaceae (V:M)
Agropyron yukonense Scribn. & Merr. = Elymus × yukonensis [ELYMYUK] Poaceae (V:M)
× *Agrositanion saundersii* (Vasey) Bowden = Elymus × saundersii [ELYMXSA] Poaceae (V:M)
Agrostemma coronaria L. = Lychnis coronaria [LYCHCOR] Caryophyllaceae (V:D)
Agrostis L. [AGROSTI$] Poaceae (V:M)
Agrostis aequivalvis (Trin.) Trin. [AGROAEQ] Poaceae (V:M)
Agrostis airoides Torr. = Sporobolus airoides [SPORAIR] Poaceae (V:M)
Agrostis alba auct. = Agrostis gigantea [AGROGIG] Poaceae (V:M)
Agrostis alba var. *palustris* (Huds.) Pers. = Agrostis stolonifera [AGROSTO] Poaceae (V:M)
Agrostis alba var. *stolonifera* (L.) Sm. = Agrostis stolonifera [AGROSTO] Poaceae (V:M)
Agrostis borealis Hartman = Agrostis mertensii [AGROMER] Poaceae (V:M)
Agrostis borealis var. *americana* (Scribn.) Fern. = Agrostis mertensii [AGROMER] Poaceae (V:M)
Agrostis borealis var. *paludosa* (Scribn.) Fern. = Agrostis mertensii [AGROMER] Poaceae (V:M)
*__Agrostis capillaris__ L. [AGROCAP] Poaceae (V:M)
Agrostis cryptandra Torr. = Sporobolus cryptandrus [SPORCRY] Poaceae (V:M)
Agrostis diegoensis Vasey [AGRODIE] Poaceae (V:M)
Agrostis exarata Trin. [AGROEXA] Poaceae (V:M)
Agrostis exarata ssp. *minor* (Hook.) C.L. Hitchc. = Agrostis exarata var. minor [AGROEXA5] Poaceae (V:M)
Agrostis exarata var. **minor** Hook. [AGROEXA5] Poaceae (V:M)
Agrostis exarata var. **monolepis** (Torr.) A.S. Hitchc. [AGROEXA6] Poaceae (V:M)
Agrostis exarata var. *purpurascens* Hultén = Agrostis exarata var. minor [AGROEXA5] Poaceae (V:M)
*__Agrostis gigantea__ Roth [AGROGIG] Poaceae (V:M)
Agrostis gigantea var. *dispar* (Michx.) Philipson = Agrostis gigantea [AGROGIG] Poaceae (V:M)
Agrostis inflata Scribn. [AGROINF] Poaceae (V:M)
Agrostis interrupta L. = Apera interrupta [APERINT] Poaceae (V:M)
Agrostis maritima Lam. = Agrostis stolonifera [AGROSTO] Poaceae (V:M)
Agrostis mertensii Trin. [AGROMER] Poaceae (V:M)
Agrostis mertensii ssp. *borealis* (Hartman) Tzvelev = Agrostis mertensii [AGROMER] Poaceae (V:M)
Agrostis mexicana L. = Muhlenbergia mexicana [MUHLMEX] Poaceae (V:M)
Agrostis microphylla Steud. [AGROMIC] Poaceae (V:M)
Agrostis nigra With. = Agrostis gigantea [AGROGIG] Poaceae (V:M)

Agrostis oregonensis Vasey [AGROORE] Poaceae (V:M)
Agrostis pallens Trin. [AGROPAL] Poaceae (V:M)
Agrostis pallens var. *vaseyi* St. John = Agrostis diegoensis [AGRODIE] Poaceae (V:M)
Agrostis palustris Huds. = Agrostis stolonifera [AGROSTO] Poaceae (V:M)
Agrostis racemosa Michx. = Muhlenbergia racemosa [MUHLRAC] Poaceae (V:M)
Agrostis scabra Willd. [AGROSCA] Poaceae (V:M)
*•**Agrostis stolonifera** L. [AGROSTO] Poaceae (V:M)
Agrostis stolonifera ssp. *gigantea* (Roth) Schuebl. & Martens = Agrostis gigantea [AGROGIG] Poaceae (V:M)
Agrostis stolonifera var. *compacta* Hartman = Agrostis stolonifera [AGROSTO] Poaceae (V:M)
Agrostis stolonifera var. *major* (Gaudin) Farw. = Agrostis gigantea [AGROGIG] Poaceae (V:M)
Agrostis stolonifera var. *palustris* (Huds.) Farw. = Agrostis stolonifera [AGROSTO] Poaceae (V:M)
Agrostis sylvatica Huds. = Agrostis capillaris [AGROCAP] Poaceae (V:M)
Agrostis tenuis Sibthorp = Agrostis capillaris [AGROCAP] Poaceae (V:M)
Agrostis tenuis var. *aristata* (Parnell) Druce = Agrostis capillaris [AGROCAP] Poaceae (V:M)
Agrostis tenuis var. *hispida* (Willd.) Philipson = Agrostis capillaris [AGROCAP] Poaceae (V:M)
Agrostis tenuis var. *pumila* (L.) Druce = Agrostis capillaris [AGROCAP] Poaceae (V:M)
Agrostis variabilis Rydb. [AGROVAR] Poaceae (V:M)
Agyrium Fr. [AGYRIUM$] Agyriaceae (L)
Agyrium rufum (Pers.) Fr. [AGYRRUF] Agyriaceae (L)
Agyrophora leiocarpa (DC.) Gyelnik = Umbilicaria leiocarpa [UMBILEI] Umbilicariaceae (L)
Agyrophora lyngei (Schol.) Llano = Umbilicaria lyngei [UMBILYN] Umbilicariaceae (L)
Agyrophora rigida (Du Rietz) Llano = Umbilicaria rigida [UMBIRIG] Umbilicariaceae (L)
Ahtiana Goward [AHTIANA$] Parmeliaceae (L)
Ahtiana pallidula (Tuck. ex Riddle) Goward & Thell [AHTIPAL] Parmeliaceae (L)
Ahtiana sphaerosporella (Müll. Arg.) Goward [AHTISPH] Parmeliaceae (L)
Aira L. [AIRA$] Poaceae (V:M)
Aira atropurpurea Wahlenb. = Vahlodea atropurpurea [VAHLATR] Poaceae (V:M)
•**Aira caryophyllea** L. [AIRACAR] Poaceae (V:M)
Aira cespitosa L. = Deschampsia cespitosa ssp. cespitosa [DESCCES0] Poaceae (V:M)
Aira danthonioides Trin. = Deschampsia danthonioides [DESCDAN] Poaceae (V:M)
Aira elongata Hook. = Deschampsia elongata [DESCELO] Poaceae (V:M)
Aira holciformis (J. Presl) Steud. = Deschampsia cespitosa ssp. holciformis [DESCCES3] Poaceae (V:M)
Aira obtusata Michx. = Sphenopholis obtusata [SPHEOBT] Poaceae (V:M)
•**Aira praecox** L. [AIRAPRA] Poaceae (V:M)
Aira spicata L. = Trisetum spicatum [TRISSPI] Poaceae (V:M)
Ajuga L. [AJUGA$] Lamiaceae (V:D)
•**Ajuga reptans** L. [AJUGREP] Lamiaceae (V:D)
Alchemilla L. [ALCHEMI$] Rosaceae (V:D)
Alchemilla arvensis (L.) Scop. = Aphanes arvensis [APHAARV] Rosaceae (V:D)
Alchemilla cuneifolia Nutt. = Aphanes arvensis [APHAARV] Rosaceae (V:D)
Alchemilla microcarpa Boiss. & Reut. = Aphanes microcarpa [APHAMIC] Rosaceae (V:D)
Alchemilla occidentalis Nutt. = Aphanes arvensis [APHAARV] Rosaceae (V:D)
•**Alchemilla subcrenata** Buser [ALCHSUB] Rosaceae (V:D)
Alectoria Ach. [ALECTOR$] Alectoriaceae (L)
Alectoria abbreviata (Müll. Arg.) R. Howe = Nodobryoria abbreviata [NODOABB] Parmeliaceae (L)
Alectoria achariana Gyelnik = Bryoria pseudofuscescens [BRYOPSE] Parmeliaceae (L)
Alectoria ambigua Mot. = Bryoria trichodes ssp. americana [BRYOTRI1] Parmeliaceae (L)
Alectoria americana Mot. = Bryoria trichodes ssp. americana [BRYOTRI1] Parmeliaceae (L)
Alectoria bicolor (Ehrh.) Nyl. = Bryoria bicolor [BRYOBIC] Parmeliaceae (L)
Alectoria boryana Delise = Alectoria nigricans [ALECNIG] Alectoriaceae (L)
Alectoria californica (Tuck.) G. Merr. = Kaernefeltia californica [KAERCAL] Parmeliaceae (L)
Alectoria canadensis Mot. = Bryoria trichodes ssp. trichodes [BRYOTRI0] Parmeliaceae (L)
Alectoria capillaris (Ach.) Crombie = Bryoria capillaris [BRYOCAP] Parmeliaceae (L)
Alectoria cervinula Mot. = Bryoria cervinula [BRYOCER] Parmeliaceae (L)
Alectoria cetrariza Nyl. = Kaernefeltia californica [KAERCAL] Parmeliaceae (L)
Alectoria chalybeiformis (L.) Gray = Bryoria chalybeiformis [BRYOCHA] Parmeliaceae (L)
Alectoria cincinnata (Fr.) Lynge = Alectoria sarmentosa ssp. vexillifera [ALECSAR2] Alectoriaceae (L)
Alectoria corneliae Gyelnik = Bryoria fremontii [BRYOFRE] Parmeliaceae (L)
Alectoria crinalis Ach. = Ramalina thrausta [RAMATHR] Ramalinaceae (L)
Alectoria delicata Mot. = Bryoria trichodes [BRYOTRI] Parmeliaceae (L)
Alectoria divergens (Ach.) Nyl. = Bryocaulon divergens [BRYODIV] Parmeliaceae (L)
Alectoria divergens f. *abbreviata* Müll. Arg. = Nodobryoria abbreviata [NODOABB] Parmeliaceae (L)
Alectoria fremontii Tuck. = Bryoria fremontii [BRYOFRE] Parmeliaceae (L)
Alectoria fremontii f. *perfertilis* Räsänen = Bryoria fremontii [BRYOFRE] Parmeliaceae (L)
Alectoria fremontii ssp. *olivacea* Räsänen = Bryoria fremontii [BRYOFRE] Parmeliaceae (L)
Alectoria fuscescens Gyelnik = Bryoria fuscescens [BRYOFUS] Parmeliaceae (L)
Alectoria glabra Mot. = Bryoria glabra [BRYOGLA] Parmeliaceae (L)
Alectoria implexa (Hoffm.) Nyl. = Bryoria implexa [BRYOIMP] Parmeliaceae (L)
Alectoria implexa f. *fuscidula* Arnold = Bryoria pikei [BRYOPIK] Parmeliaceae (L)
Alectoria imshaugii Brodo & D. Hawksw. [ALECIMS] Alectoriaceae (L)
Alectoria irvingii Llano = Bryoria nitidula [BRYONIT] Parmeliaceae (L)
Alectoria jubata var. *bicolor* (Ehrh.) Nyl. = Bryoria bicolor [BRYOBIC] Parmeliaceae (L)
Alectoria jubata var. *implexa* (Hoffm.) Ach. = Bryoria implexa [BRYOIMP] Parmeliaceae (L)
Alectoria jubata var. *vrangiana* (Gyelnik) Räsänen = Bryoria implexa [BRYOIMP] Parmeliaceae (L)
Alectoria lanea auct. = Bryoria nitidula [BRYONIT] Parmeliaceae (L)
Alectoria lanestris (Ach.) Gyelnik = Bryoria lanestris [BRYOLAN] Parmeliaceae (L)
Alectoria luteola Mont. = Alectoria sarmentosa ssp. sarmentosa [ALECSAR0] Alectoriaceae (L)
Alectoria minuscula (Nyl. ex Arnold) Degel. = Pseudephebe minuscula [PSEUMIN] Parmeliaceae (L)
Alectoria nana Mot. = Bryoria simplicior [BRYOSIM] Parmeliaceae (L)
Alectoria nigricans (Ach.) Nyl. [ALECNIG] Alectoriaceae (L)
Alectoria nitidula (Th. Fr.) Vainio = Bryoria nitidula [BRYONIT] Parmeliaceae (L)
Alectoria norstictica Mot. = Bryoria pseudofuscescens [BRYOPSE] Parmeliaceae (L)

Alectoria ochroleuca (Hoffm.) A. Massal. [ALECOCH] Alectoriaceae (L)
Alectoria ochroleuca var. *rigida* Fr. = Alectoria ochroleuca [ALECOCH] Alectoriaceae (L)
Alectoria ochroleuca var. *sarmentosa* Nyl. = Alectoria sarmentosa ssp. sarmentosa [ALECSAR0] Alectoriaceae (L)
Alectoria oregana Tuck. = Nodobryoria oregana [NODOORE] Parmeliaceae (L)
Alectoria positiva (Gyelnik) Mot. = Bryoria fuscescens [BRYOFUS] Parmeliaceae (L)
Alectoria pseudofuscescens Gyelnik = Bryoria pseudofuscescens [BRYOPSE] Parmeliaceae (L)
Alectoria pubescens (L.) R. Howe = Pseudephebe pubescens [PSEUPUB] Parmeliaceae (L)
Alectoria sarmentosa (Ach.) Ach. [ALECSAR] Alectoriaceae (L)
Alectoria sarmentosa ssp. **sarmentosa** [ALECSAR0] Alectoriaceae (L)
Alectoria sarmentosa ssp. **vexillifera** (Nyl.) D. Hawksw. [ALECSAR2] Alectoriaceae (L)
Alectoria sarmentosa var. *gigantea* Räsänen = Alectoria sarmentosa ssp. sarmentosa [ALECSAR0] Alectoriaceae (L)
Alectoria setacea (Ach.) Mot. = Bryoria capillaris [BRYOCAP] Parmeliaceae (L)
Alectoria simplicior (Vainio) Lynge = Bryoria simplicior [BRYOSIM] Parmeliaceae (L)
Alectoria stigmata Bystrek = Alectoria sarmentosa ssp. sarmentosa [ALECSAR0] Alectoriaceae (L)
Alectoria subcana (Nyl. ex Stizenb.) Gyelnik = Bryoria subcana [BRYOSUB] Parmeliaceae (L)
Alectoria subdivergens E. Dahl = Nodobryoria subdivergens [NODOSUB] Parmeliaceae (L)
Alectoria subsarmentosa Stirton = Alectoria sarmentosa ssp. sarmentosa [ALECSAR0] Alectoriaceae (L)
Alectoria subtilis Mot. = Bryoria pseudofuscescens [BRYOPSE] Parmeliaceae (L)
Alectoria tenerrima Mot. = Bryoria fremontii [BRYOFRE] Parmeliaceae (L)
Alectoria tenuis E. Dahl = Bryoria tenuis [BRYOTEN] Parmeliaceae (L)
Alectoria thrausta Ach. = Ramalina thrausta [RAMATHR] Ramalinaceae (L)
Alectoria tortuosa G. Merr. = Bryoria tortuosa [BRYOTOR] Parmeliaceae (L)
Alectoria vancouverensis (Gyelnik) Gyelnik ex Brodo & D. Hawksw. [ALECVAN] Alectoriaceae (L)
Alectoria vexillifera (Nyl.) Stizenb. = Alectoria sarmentosa ssp. vexillifera [ALECSAR2] Alectoriaceae (L)
Alectoria virens auct. = Bryoria tortuosa [BRYOTOR] Parmeliaceae (L)
Alectoria vrangiana Gyelnik = Bryoria implexa [BRYOIMP] Parmeliaceae (L)
Alisma L. [ALISMA$] Alismataceae (V:M)
Alisma brevipes Greene = Alisma triviale [ALISTRI] Alismataceae (V:M)
Alisma geyeri Torr. = Alisma gramineum [ALISGRA] Alismataceae (V:M)
Alisma gramineum Lej. [ALISGRA] Alismataceae (V:M)
Alisma gramineum var. *angustissimum* (DC.) Hendricks = Alisma gramineum [ALISGRA] Alismataceae (V:M)
Alisma gramineum var. *geyeri* (Torr.) Lam. = Alisma gramineum [ALISGRA] Alismataceae (V:M)
Alisma gramineum var. *graminifolium* (Wahlenb.) Hendricks = Alisma gramineum [ALISGRA] Alismataceae (V:M)
Alisma lanceolatum Gray = Alisma gramineum [ALISGRA] Alismataceae (V:M)
Alisma plantago-aquatica ssp. *brevipes* (Greene) Samuelsson = Alisma triviale [ALISTRI] Alismataceae (V:M)
Alisma plantago-aquatica var. *americanum* J.A. Schultes = Alisma triviale [ALISTRI] Alismataceae (V:M)
Alisma plantago-aquatica var. *brevipes* (Greene) Victorin = Alisma triviale [ALISTRI] Alismataceae (V:M)
Alisma plantago-aquatica var. *triviale* (Britt., Sterns & Pogg.) Far. = Alisma triviale [ALISTRI] Alismataceae (V:M)
Alisma triviale Pursh [ALISTRI] Alismataceae (V:M)
Allantoparmelia (Vainio) Essl. [ALLANTO$] Parmeliaceae (L)
Allantoparmelia almquistii (Vainio) Essl. [ALLAALM] Parmeliaceae (L)
Allantoparmelia alpicola (Th. Fr.) Essl. [ALLAALP] Parmeliaceae (L)
Alliaria Heister ex Fabr. [ALLIARI$] Brassicaceae (V:D)
Alliaria alliaria (L.) Britt. = Alliaria petiolata [ALLIPET] Brassicaceae (V:D)
Alliaria officinalis Andrz. ex Bieb. = Alliaria petiolata [ALLIPET] Brassicaceae (V:D)
***Alliaria petiolata** (Bieb.) Cavara & Grande [ALLIPET] Brassicaceae (V:D)
Allionia hirsuta Pursh = Mirabilis hirsuta [MIRAHIR] Nyctaginaceae (V:D)
Allionia nyctaginea Michx. = Mirabilis nyctaginea [MIRANYC] Nyctaginaceae (V:D)
Allium L. [ALLIUM$] Liliaceae (V:M)
Allium acuminatum Hook. [ALLIACU] Liliaceae (V:M)
Allium acuminatum var. *cuspidatum* Fern. = Allium acuminatum [ALLIACU] Liliaceae (V:M)
Allium amplectens Torr. [ALLIAMP] Liliaceae (V:M)
Allium arenicola Osterhout = Allium geyeri var. tenerum [ALLIGEY2] Liliaceae (V:M)
Allium attenuifolium Kellogg = Allium amplectens [ALLIAMP] Liliaceae (V:M)
Allium cascadense M.E. Peck = Allium crenulatum [ALLICRE] Liliaceae (V:M)
Allium cernuum Roth [ALLICER] Liliaceae (V:M)
Allium crenulatum Wieg. [ALLICRE] Liliaceae (V:M)
Allium dictyotum Greene = Allium geyeri var. geyeri [ALLIGEY0] Liliaceae (V:M)
Allium fibrosum Rydb. = Allium geyeri var. tenerum [ALLIGEY2] Liliaceae (V:M)
Allium geyeri S. Wats. [ALLIGEY] Liliaceae (V:M)
Allium geyeri ssp. *tenerum* (M.E. Jones) Traub & Ownbey = Allium geyeri var. tenerum [ALLIGEY2] Liliaceae (V:M)
Allium geyeri var. **geyeri** [ALLIGEY0] Liliaceae (V:M)
Allium geyeri var. *graniferum* Henderson = Allium geyeri var. tenerum [ALLIGEY2] Liliaceae (V:M)
Allium geyeri var. **tenerum** M.E. Jones [ALLIGEY2] Liliaceae (V:M)
Allium monospermum Jepson = Allium amplectens [ALLIAMP] Liliaceae (V:M)
Allium occidentale Gray = Allium amplectens [ALLIAMP] Liliaceae (V:M)
Allium rubrum Osterhout = Allium geyeri var. tenerum [ALLIGEY2] Liliaceae (V:M)
Allium rydbergii J.F. Macbr. = Allium geyeri var. tenerum [ALLIGEY2] Liliaceae (V:M)
Allium sabulicola Osterhout = Allium geyeri var. tenerum [ALLIGEY2] Liliaceae (V:M)
Allium schoenoprasum L. [ALLISCH] Liliaceae (V:M)
Allium schoenoprasum ssp. *sibiricum* (L.) Celak. = Allium schoenoprasum var. sibiricum [ALLISCH1] Liliaceae (V:M)
Allium schoenoprasum var. *laurentianum* Fern. = Allium schoenoprasum var. sibiricum [ALLISCH1] Liliaceae (V:M)
Allium schoenoprasum var. **sibiricum** (L.) Hartman [ALLISCH1] Liliaceae (V:M)
Allium sibiricum L. = Allium schoenoprasum var. sibiricum [ALLISCH1] Liliaceae (V:M)
Allium validum S. Wats. [ALLIVAL] Liliaceae (V:M)
Allium vancouverense Macoun = Allium crenulatum [ALLICRE] Liliaceae (V:M)
***Allium vineale** L. [ALLIVIN] Liliaceae (V:M)
Allium watsonii T.J. Howell = Allium crenulatum [ALLICRE] Liliaceae (V:M)
Allocetraria cucullata (Bellardi) Randlane & Saag = Flavocetraria cucullata [FLAVCUC] Parmeliaceae (L)
Allocetraria nivalis (L.) Randlane & Saag = Flavocetraria nivalis [FLAVNIV] Parmeliaceae (L)
Allotropa Torr. & Gray [ALLOTRO$] Monotropaceae (V:D)

Allotropa virgata Torr. & Gray ex Gray [ALLOVIR] Monotropaceae (V:D)
Alnus P. Mill. [ALNUS$] Betulaceae (V:D)
Alnus alnobetula (Ehrh.) K. Koch p.p. = Alnus viridis ssp. sinuata [ALNUVIR2] Betulaceae (V:D)
Alnus crispa (Ait.) Pursh = Alnus viridis ssp. crispa [ALNUVIR5] Betulaceae (V:D)
Alnus crispa ssp. *laciniata* Hultén = Alnus viridis ssp. sinuata [ALNUVIR2] Betulaceae (V:D)
Alnus crispa ssp. *sinuata* (Regel) Hultén = Alnus viridis ssp. sinuata [ALNUVIR2] Betulaceae (V:D)
Alnus crispa var. *elongata* Raup = Alnus viridis ssp. crispa [ALNUVIR5] Betulaceae (V:D)
Alnus crispa var. *mollis* (Fern.) Fern. = Alnus viridis ssp. crispa [ALNUVIR5] Betulaceae (V:D)
Alnus x *hultenii* Murai = Alnus viridis ssp. crispa [ALNUVIR5] Betulaceae (V:D)
Alnus incana (L.) Moench [ALNUINC] Betulaceae (V:D)
Alnus incana ssp. *rugosa* var. *occidentalis* (Dippel) C.L. Hitchc. = Alnus incana ssp. tenuifolia [ALNUINC2] Betulaceae (V:D)
Alnus incana ssp. **tenuifolia** (Nutt.) Breitung [ALNUINC2] Betulaceae (V:D)
Alnus incana var. *occidentalis* (Dippel) C.L. Hitchc. = Alnus incana ssp. tenuifolia [ALNUINC2] Betulaceae (V:D)
Alnus incana var. *virescens* S. Wats. = Alnus incana ssp. tenuifolia [ALNUINC2] Betulaceae (V:D)
Alnus oregona Nutt. = Alnus rubra [ALNURUB] Betulaceae (V:D)
Alnus oregona var. *pinnatisecta* Starker = Alnus rubra [ALNURUB] Betulaceae (V:D)
Alnus x *purpusii* Callier = Alnus incana ssp. tenuifolia [ALNUINC2] Betulaceae (V:D)
Alnus rubra Bong. [ALNURUB] Betulaceae (V:D)
Alnus sinuata (Regel) Rydb. = Alnus viridis ssp. sinuata [ALNUVIR2] Betulaceae (V:D)
Alnus tenuifolia Nutt. = Alnus incana ssp. tenuifolia [ALNUINC2] Betulaceae (V:D)
Alnus tenuifolia var. *occidentalis* (Dippel) Collier = Alnus incana ssp. tenuifolia [ALNUINC2] Betulaceae (V:D)
Alnus viridis (Vill.) Lam. & DC. [ALNUVIR] Betulaceae (V:D)
Alnus viridis ssp. **crispa** (Ait.) Turrill [ALNUVIR5] Betulaceae (V:D)
Alnus viridis ssp. *fruticosa* (Rupr.) Nyman = Alnus viridis ssp. crispa [ALNUVIR5] Betulaceae (V:D)
Alnus viridis ssp. **sinuata** (Regel) A. & D. Löve [ALNUVIR2] Betulaceae (V:D)
Alnus viridis var. *crispa* (Ait.) House = Alnus viridis ssp. crispa [ALNUVIR5] Betulaceae (V:D)
Alnus viridis var. *sinuata* Regel = Alnus viridis ssp. sinuata [ALNUVIR2] Betulaceae (V:D)
Aloina Kindb. [ALOINA$] Pottiaceae (B:M)
Aloina bifrons (De Not.) Delg. [ALOIBIF] Pottiaceae (B:M)
Aloina brevirostris (Hook. & Grev.) Kindb. [ALOIBRE] Pottiaceae (B:M)
Aloina pilifera (De Not.) Crum & Steere = Aloina bifrons [ALOIBIF] Pottiaceae (B:M)
Aloina rigida (Hedw.) Limpr. [ALOIRIG] Pottiaceae (B:M)
Aloina rigida var. *pilifera* (De Not.) Limpr. = Aloina bifrons [ALOIBIF] Pottiaceae (B:M)
Alopecurus L. [ALOPECU$] Poaceae (V:M)
Alopecurus aequalis Sobol. [ALOPAEQ] Poaceae (V:M)
Alopecurus alpinus Sm. = Alopecurus borealis [ALOPBOR] Poaceae (V:M)
Alopecurus alpinus ssp. *glaucus* (Less.) Hultén = Alopecurus borealis [ALOPBOR] Poaceae (V:M)
Alopecurus alpinus ssp. *stejnegeri* (Vasey) Hultén = Alopecurus borealis [ALOPBOR] Poaceae (V:M)
Alopecurus alpinus var. *glaucus* (Less.) Krylov = Alopecurus borealis [ALOPBOR] Poaceae (V:M)
Alopecurus alpinus var. *stejnegeri* (Vasey) Hultén = Alopecurus borealis [ALOPBOR] Poaceae (V:M)
Alopecurus borealis Trin. [ALOPBOR] Poaceae (V:M)
Alopecurus carolinianus Walt. [ALOPCAR] Poaceae (V:M)
***Alopecurus geniculatus** L. [ALOPGEN] Poaceae (V:M)
Alopecurus glaucus Less. = Alopecurus borealis [ALOPBOR] Poaceae (V:M)
Alopecurus macounii Vasey = Alopecurus carolinianus [ALOPCAR] Poaceae (V:M)
Alopecurus monspeliensis L. = Polypogon monspeliensis [POLYMOS] Poaceae (V:M)
Alopecurus occidentalis Scribn. & Tweedy = Alopecurus borealis [ALOPBOR] Poaceae (V:M)
Alopecurus pallescens Piper = Alopecurus geniculatus [ALOPGEN] Poaceae (V:M)
***Alopecurus pratensis** L. [ALOPPRA] Poaceae (V:M)
Alopecurus ramosus Poir. = Alopecurus carolinianus [ALOPCAR] Poaceae (V:M)
Alsia Sull. [ALSIA$] Leucodontaceae (B:M)
Alsia abietina (Hook.) Sull. = Dendroalsia abietina [DENDABI] Leucodontaceae (B:M)
Alsia californica (Hook. & Arnott) Sull. [ALSICAL] Leucodontaceae (B:M)
Alsia californica var. *flagellifera* (Ren. & Card.) Sull. = Alsia californica [ALSICAL] Leucodontaceae (B:M)
Alsinanthe elegans (Cham. & Schlecht.) A. & D. Löve = Minuartia elegans [MINUELE] Caryophyllaceae (V:D)
Alsinanthe stricta ssp. *dawsonensis* (Britt.) A. & D. Löve = Minuartia dawsonensis [MINUDAW] Caryophyllaceae (V:D)
Alsine americana (Porter ex B.L. Robins.) Rydb. = Stellaria americana [STELAME] Caryophyllaceae (V:D)
Alsine aquatica (L.) Britt. = Myosoton aquaticum [MYOSAQU] Caryophyllaceae (V:D)
Alsine baicalensis Coville = Stellaria umbellata [STELUMB] Caryophyllaceae (V:D)
Alsine bongardiana (Fern.) A. Davids. & Moxley = Stellaria borealis ssp. sitchana [STELBOR1] Caryophyllaceae (V:D)
Alsine brachypetala (Bong.) T.J. Howell = Stellaria borealis ssp. sitchana [STELBOR1] Caryophyllaceae (V:D)
Alsine calycantha (Ledeb.) Rydb. = Stellaria calycantha [STELCAL] Caryophyllaceae (V:D)
Alsine crispa (Cham. & Schlecht.) Holz. = Stellaria crispa [STELCRI] Caryophyllaceae (V:D)
Alsine elegans (Cham. & Schlecht.) Fenzl = Minuartia elegans [MINUELE] Caryophyllaceae (V:D)
Alsine humifusa (Rottb.) Britt. = Stellaria humifusa [STELHUM] Caryophyllaceae (V:D)
Alsine obtusa (Engelm.) Rose = Stellaria obtusa [STELOBT] Caryophyllaceae (V:D)
Alsine simcoei T.J. Howell = Stellaria calycantha [STELCAL] Caryophyllaceae (V:D)
Alsine uliginosa (Murr.) Britt. = Stellaria alsine [STELALS] Caryophyllaceae (V:D)
Alsine viridula Piper = Stellaria obtusa [STELOBT] Caryophyllaceae (V:D)
Alsine washingtoniana (B.L. Robins.) Heller = Stellaria obtusa [STELOBT] Caryophyllaceae (V:D)
Alsinopsis obtusiloba Rydb. = Minuartia obtusiloba [MINUOBT] Caryophyllaceae (V:D)
Alsinopsis occidentalis Heller = Minuartia nuttallii [MINUNUT] Caryophyllaceae (V:D)
Alyssum L. [ALYSSUM$] Brassicaceae (V:D)
***Alyssum alyssoides** (L.) L. [ALYSALY] Brassicaceae (V:D)
Alyssum calycinum L. = Alyssum alyssoides [ALYSALY] Brassicaceae (V:D)
***Alyssum desertorum** Stapf [ALYSDES] Brassicaceae (V:D)
Alyssum incanum L. = Berteroa incana [BERTINC] Brassicaceae (V:D)
Alyssum maritimum (L.) Lam. = Lobularia maritima [LOBUMAR] Brassicaceae (V:D)
***Alyssum murale** Waldst. & Kit. [ALYSMUR] Brassicaceae (V:D)
Amandinea Choisy ex Scheid. & H. Mayrh. [AMANDIN$] Physciaceae (L)
Amandinea coniops (Wahlenb.) Choisy ex Scheid. & H. Mayrh. [AMANCON] Physciaceae (L)
Amandinea punctata (Hoffm.) Coppins & Scheid. [AMANPUN] Physciaceae (L)
Amaranthus L. [AMARANT$] Amaranthaceae (V:D)

*•**Amaranthus albus** L. [AMARALB] Amaranthaceae (V:D)
Amaranthus albus var. *pubescens* (Uline & Bray) Fern. = Amaranthus albus [AMARALB] Amaranthaceae (V:D)
•**Amaranthus blitoides** S. Wats. [AMARBLI] Amaranthaceae (V:D)
Amaranthus bouchonii Thellung = Amaranthus powellii [AMARPOW] Amaranthaceae (V:D)
Amaranthus bracteosus Uline & Bray = Amaranthus powellii [AMARPOW] Amaranthaceae (V:D)
Amaranthus graecizans var. *pubescens* Uline & Bray = Amaranthus albus [AMARALB] Amaranthaceae (V:D)
Amaranthus powellii S. Wats. [AMARPOW] Amaranthaceae (V:D)
Amaranthus pubescens (Uline & Bray) Rydb. = Amaranthus albus [AMARALB] Amaranthaceae (V:D)
•**Amaranthus retroflexus** L. [AMARRET] Amaranthaceae (V:D)
Amaranthus retroflexus var. *powellii* (S. Wats.) Boivin = Amaranthus powellii [AMARPOW] Amaranthaceae (V:D)
Amaranthus retroflexus var. *salicifolius* I.M. Johnston = Amaranthus retroflexus [AMARRET] Amaranthaceae (V:D)
Amarella acuta (Michx.) Raf. = Gentianella amarella ssp. acuta [GENTAMA1] Gentianaceae (V:D)
Amarella plebeja (Ledeb. ex Spreng.) Greene = Gentianella amarella ssp. acuta [GENTAMA1] Gentianaceae (V:D)
Amarella plebeja var. *holmii* (Wettst.) Rydb. = Gentianella amarella ssp. acuta [GENTAMA1] Gentianaceae (V:D)
Amarella strictiflora (Rydb.) Greene = Gentianella amarella ssp. acuta [GENTAMA1] Gentianaceae (V:D)
Amblyodon Bruch & Schimp. in B.S.G. [AMBLYOD$] Meesiaceae (B:M)
Amblyodon dealbatus (Hedw.) Bruch & Schimp. in B.S.G. [AMBLDEA] Meesiaceae (B:M)
Amblyodon dealbatus var. *americanus* Ren. & Card. = Amblyodon dealbatus [AMBLDEA] Meesiaceae (B:M)
Amblystegiella jungermannioides (Brid.) Giac. = Platydictya jungermannioides [PLATJUN] Hypnaceae (B:M)
Amblystegiella sprucei (Bruch) Loeske = Platydictya jungermannioides [PLATJUN] Hypnaceae (B:M)
Amblystegium Schimp. in B.S.G. [AMBLYST$] Amblystegiaceae (B:M)
Amblystegium americanum Grout = Conardia compacta [CONACOM] Amblystegiaceae (B:M)
Amblystegium brevipes Card. & Thér. ex Holz. = Leptodictyum riparium [LEPTRIP] Amblystegiaceae (B:M)
Amblystegium compactum (C. Müll.) Aust. = Conardia compacta [CONACOM] Amblystegiaceae (B:M)
Amblystegium fallax (Brid.) Milde = Cratoneuron filicinum [CRATFIL] Amblystegiaceae (B:M)
Amblystegium filicinum (Hedw.) De Not. = Cratoneuron filicinum [CRATFIL] Amblystegiaceae (B:M)
Amblystegium filicinum f. *marianopolitana* Dupret ex Moxley = Hygroamblystegium tenax [HYGRTEN] Amblystegiaceae (B:M)
Amblystegium fluviatile (Hedw.) Schimp. in B.S.G. = Hygroamblystegium fluviatile [HYGRFLU] Amblystegiaceae (B:M)
Amblystegium fluviatile var. *noterophilum* (Sull. & Lesq. in Sull.) Flow. = Hygroamblystegium noterophilum [HYGRNOT] Amblystegiaceae (B:M)
Amblystegium irriguum (Hook. & Wils.) B.S.G. = Hygroamblystegium tenax [HYGRTEN] Amblystegiaceae (B:M)
Amblystegium irriguum f. *marianopolitana* Dupret = Hygroamblystegium tenax [HYGRTEN] Amblystegiaceae (B:M)
Amblystegium jungermannoides (Brid.) A.J.E. Sm. = Platydictya jungermannioides [PLATJUN] Hypnaceae (B:M)
Amblystegium juratzkanum Schimp. = Amblystegium serpens var. juratzkanum [AMBLSER1] Amblystegiaceae (B:M)
Amblystegium juratzkanum var. *giganteum* (Grout) Grout = Amblystegium serpens var. juratzkanum [AMBLSER1] Amblystegiaceae (B:M)
Amblystegium montanae Bryhn in Holz. = Leskeella nervosa [LESKNER] Leskeaceae (B:M)
Amblystegium noterophilum (Sull. & Lesq. in Sull.) Holz. = Hygroamblystegium noterophilum [HYGRNOT] Amblystegiaceae (B:M)
Amblystegium polygamum Schimp. in B.S.G. = Campylium polygamum [CAMPPOL] Amblystegiaceae (B:M)
Amblystegium porphyrrhizon Schimp. = Amblystegium serpens var. juratzkanum [AMBLSER1] Amblystegiaceae (B:M)
Amblystegium radicale P. Beauv. = Campylium radicale [CAMPRAD] Amblystegiaceae (B:M)
Amblystegium riparium (Hedw.) Schimp. in B.S.G. = Leptodictyum riparium [LEPTRIP] Amblystegiaceae (B:M)
Amblystegium riparium f. *fluitans* (Lesq. & James) Flow. = Leptodictyum riparium [LEPTRIP] Amblystegiaceae (B:M)
Amblystegium riparium var. *abbreviatum* Schimp. in B.S.G. = Leptodictyum riparium [LEPTRIP] Amblystegiaceae (B:M)
Amblystegium riparium var. *flaccidum* (Lesq. & James) Ren. & Card. = Leptodictyum riparium [LEPTRIP] Amblystegiaceae (B:M)
Amblystegium riparium var. *fluitans* (Lesq. & James) Ren. & Card. = Leptodictyum riparium [LEPTRIP] Amblystegiaceae (B:M)
Amblystegium saxatile Schimp. = Campylium radicale [CAMPRAD] Amblystegiaceae (B:M)
Amblystegium serpens (Hedw.) Schimp. in B.S.G. [AMBLSER] Amblystegiaceae (B:M)
Amblystegium serpens var. *beringianum* Card. & Thér. = Amblystegium serpens [AMBLSER] Amblystegiaceae (B:M)
Amblystegium serpens var. *giganteum* Grout = Amblystegium serpens var. juratzkanum [AMBLSER1] Amblystegiaceae (B:M)
Amblystegium serpens var. **juratzkanum** (Schimp.) Rau & Herv. [AMBLSER1] Amblystegiaceae (B:M)
Amblystegium serpens var. **serpens** [AMBLSER0] Amblystegiaceae (B:M)
Amblystegium serpens var. *tenue* (Brid.) Schimp. in B.S.G. = Amblystegium serpens [AMBLSER] Amblystegiaceae (B:M)
Amblystegium sipho (P. Beauv.) Card. = Leptodictyum riparium [LEPTRIP] Amblystegiaceae (B:M)
Amblystegium sprucei (Bruch) B.S.G. = Platydictya jungermannioides [PLATJUN] Hypnaceae (B:M)
Amblystegium stellatum (Hedw.) Lindb. = Campylium stellatum [CAMPSTE] Amblystegiaceae (B:M)
Amblystegium subcompactum C. Müll. & Kindb. in Mac. & Kindb. = Conardia compacta [CONACOM] Amblystegiaceae (B:M)
Amblystegium tenax (Hedw.) C. Jens. = Hygroamblystegium tenax [HYGRTEN] Amblystegiaceae (B:M)
Amblystegium vacillans Sull. = Leptodictyum riparium [LEPTRIP] Amblystegiaceae (B:M)
Amblystegium varium (Hedw.) Lindb. [AMBLVAR] Amblystegiaceae (B:M)
Amblystegium varium var. *alaskanum* Card. & Thér. = Amblystegium varium [AMBLVAR] Amblystegiaceae (B:M)
Amblystegium varium var. *lancifolium* Grout = Amblystegium varium [AMBLVAR] Amblystegiaceae (B:M)
Amblystegium varium var. *ovatum* (Grout) Grout = Amblystegium varium [AMBLVAR] Amblystegiaceae (B:M)
Amblystegium varium var. *parvulum* (Aust.) Grout = Amblystegium varium [AMBLVAR] Amblystegiaceae (B:M)
Ambrosia L. [AMBROSI$] Asteraceae (V:D)
•**Ambrosia artemisiifolia** L. [AMBRART] Asteraceae (V:D)
Ambrosia artemisiifolia var. **elatior** (L.) Descourtils [AMBRART1] Asteraceae (V:D)
Ambrosia chamissonis (Less.) Greene [AMBRCHA] Asteraceae (V:D)
Ambrosia chamissonis var. *bipinnatisecta* (Less.) J.T. Howell = Ambrosia chamissonis [AMBRCHA] Asteraceae (V:D)
Ambrosia coronopifolia Torr. & Gray [AMBRCOR] Asteraceae (V:D)
Ambrosia elatior L. = Ambrosia artemisiifolia var. elatior [AMBRART1] Asteraceae (V:D)
Ambrosia psilostachya var. *coronopifolia* (Torr. & Gray) Farw. = Ambrosia coronopifolia [AMBRCOR] Asteraceae (V:D)
Amelanchier Medik. [AMELANC$] Rosaceae (V:D)
Amelanchier alnifolia (Nutt.) Nutt. ex M. Roemer [AMELALN] Rosaceae (V:D)
Amelanchier alnifolia ssp. *florida* (Lindl.) Hultén = Amelanchier alnifolia var. semiintegrifolia [AMELALN4] Rosaceae (V:D)
Amelanchier alnifolia var. **alnifolia** [AMELALN0] Rosaceae (V:D)

Amelanchier alnifolia var. **cusickii** (Fern.) C.L. Hitchc. [AMELALN2] Rosaceae (V:D)
Amelanchier alnifolia var. **humptulipensis** (G.N. Jones) C.L. Hitchc. [AMELALN3] Rosaceae (V:D)
Amelanchier alnifolia var. **semiintegrifolia** (Hook.) C.L. Hitchc. [AMELALN4] Rosaceae (V:D)
Amelanchier cusickii Fern. = Amelanchier alnifolia var. cusickii [AMELALN2] Rosaceae (V:D)
Amelanchier florida Lindl. = Amelanchier alnifolia var. semiintegrifolia [AMELALN4] Rosaceae (V:D)
Amelanchier florida var. *cusickii* (Fern.) M.E. Peck = Amelanchier alnifolia var. cusickii [AMELALN2] Rosaceae (V:D)
Amelanchier florida var. *humptulipensis* G.N. Jones = Amelanchier alnifolia var. humptulipensis [AMELALN3] Rosaceae (V:D)
Amerorchis Hultén [AMERORC$] Orchidaceae (V:M)
Amerorchis rotundifolia (Banks ex Pursh) Hultén [AMERROT] Orchidaceae (V:M)
Amerosedum divergens (S. Wats.) A. & D. Löve = Sedum divergens [SEDUDIV] Crassulaceae (V:D)
Amerosedum lanceolatum (Torr.) A. & D. Löve = Sedum lanceolatum ssp. lanceolatum [SEDULAN0] Crassulaceae (V:D)
Amerosedum nesioticum (G.N. Jones) A. & D. Löve = Sedum lanceolatum ssp. nesioticum [SEDULAN1] Crassulaceae (V:D)
Ammophila Host [AMMOPHI$] Poaceae (V:M)
***Ammophila arenaria** (L.) Link [AMMOARE] Poaceae (V:M)
Amphidium Schimp. [AMPHIDI$] Orthotrichaceae (B:M)
Amphidium californicum (Hampe ex C. Müll.) Broth. [AMPHCAL] Orthotrichaceae (B:M)
Amphidium lapponicum (Hedw.) Schimp. [AMPHLAP] Orthotrichaceae (B:M)
Amphidium lapponicum var. *crispatum* (Kindb.) Grout = Amphidium lapponicum [AMPHLAP] Orthotrichaceae (B:M)
Amphidium mougeotii (Bruch & Schimp. in B.S.G.) Schimp. [AMPHMOU] Orthotrichaceae (B:M)
Amphiloma lanuginosum (Hoffm.) Nyl. = Leproloma membranaceum [LEPRMEB] Pannariaceae (L)
Amphiscirpus nevadensis (S. Wats.) Oteng Yeboah = Scirpus nevadensis [SCIRNEV] Cyperaceae (V:M)
Amphoridium Schimp. = Amphidium Schimp. [AMPHIDI$] Orthotrichaceae (B:M)
Amphoridium lapponicum (Hedw.) Schimp. = Amphidium lapponicum [AMPHLAP] Orthotrichaceae (B:M)
Amsinckia Lehm. [AMSINCK$] Boraginaceae (V:D)
Amsinckia arizonica Suksdorf = Amsinckia intermedia [AMSIINT] Boraginaceae (V:D)
Amsinckia barbata Greene = Amsinckia lycopsoides [AMSILYC] Boraginaceae (V:D)
Amsinckia demissa Suksdorf = Amsinckia intermedia [AMSIINT] Boraginaceae (V:D)
Amsinckia echinata Gray = Amsinckia intermedia [AMSIINT] Boraginaceae (V:D)
Amsinckia hispida (Ruiz & Pavón) I.M. Johnston = Amsinckia lycopsoides [AMSILYC] Boraginaceae (V:D)
Amsinckia idahoensis M.E. Jones = Amsinckia lycopsoides [AMSILYC] Boraginaceae (V:D)
Amsinckia intactilis J.F. Macbr. = Amsinckia intermedia [AMSIINT] Boraginaceae (V:D)
Amsinckia intermedia Fisch. & C.A. Mey. [AMSIINT] Boraginaceae (V:D)
Amsinckia intermedia var. *echinata* (Gray) Wiggins = Amsinckia intermedia [AMSIINT] Boraginaceae (V:D)
Amsinckia lycopsoides Lehm. [AMSILYC] Boraginaceae (V:D)
Amsinckia menziesii (Lehm.) A. Nels. & J.F. Macbr. [AMSIMEN] Boraginaceae (V:D)
Amsinckia micrantha Suksdorf = Amsinckia menziesii [AMSIMEN] Boraginaceae (V:D)
Amsinckia microphylla Suksdorf = Amsinckia intermedia [AMSIINT] Boraginaceae (V:D)
Amsinckia nana Suksdorf = Amsinckia intermedia [AMSIINT] Boraginaceae (V:D)
Amsinckia parviflora Heller = Amsinckia lycopsoides [AMSILYC] Boraginaceae (V:D)
Amsinckia retrorsa Suksdorf [AMSIRET] Boraginaceae (V:D)
Amsinckia rigida Suksdorf = Amsinckia intermedia [AMSIINT] Boraginaceae (V:D)
Amsinckia rugosa Rydb. = Amsinckia retrorsa [AMSIRET] Boraginaceae (V:D)
Amsinckia scouleri I.M. Johnston = Amsinckia spectabilis [AMSISPE] Boraginaceae (V:D)
Amsinckia spectabilis Fisch. & C.A. Mey. [AMSISPE] Boraginaceae (V:D)
Amsinckia spectabilis var. *bracteosa* (Gray) Boivin = Amsinckia spectabilis [AMSISPE] Boraginaceae (V:D)
Amsinckia spectabilis var. *microcarpa* (Greene) Jepson & Hoover = Amsinckia spectabilis [AMSISPE] Boraginaceae (V:D)
Amsinckia spectabilis var. *nicolai* (Jepson) I.M. Johnston ex Munz = Amsinckia spectabilis [AMSISPE] Boraginaceae (V:D)
Amygdalaria Norman [AMYGDAL$] Porpidiaceae (L)
Amygdalaria consentiens (Nyl.) Hertel, Brodo & Mas. Inoue [AMYGCON] Porpidiaceae (L)
Amygdalaria continua Brodo & Hertel [AMYGCOT] Porpidiaceae (L)
Amygdalaria haidensis Brodo & Hertel [AMYGHAI] Porpidiaceae (L)
Amygdalaria panaeola (Ach.) Hertel & Brodo [AMYGPAN] Porpidiaceae (L)
Amygdalaria pelobotryon (Wahlenb.) Norman [AMYGPEL] Porpidiaceae (L)
Amygdalaria subdissentiens (Nyl.) Mas. Inoue & Brodo [AMYGSUB] Porpidiaceae (L)
Anacharis canadensis (Michx.) Planch. = Elodea canadensis [ELODCAN] Hydrocharitaceae (V:M)
Anacharis canadensis var. *planchonii* (Caspary) Victorin = Elodea canadensis [ELODCAN] Hydrocharitaceae (V:M)
Anacharis densa (Planch.) Victorin = Egeria densa [EGERDEN] Hydrocharitaceae (V:M)
Anacharis nuttallii Planch. = Elodea nuttallii [ELODNUT] Hydrocharitaceae (V:M)
Anacharis occidentalis (Pursh) Victorin = Elodea nuttallii [ELODNUT] Hydrocharitaceae (V:M)
Anacolia Schimp. [ANACOLI$] Bartramiaceae (B:M)
Anacolia menziesii (Turn.) Par. [ANACMEN] Bartramiaceae (B:M)
Anacolia menziesii f. *grandifolia* Flow. in Grout = Anacolia menziesii [ANACMEN] Bartramiaceae (B:M)
Anagallis L. [ANAGALL$] Primulaceae (V:D)
***Anagallis arvensis** L. [ANAGARV] Primulaceae (V:D)
Anagallis minima (L.) Krause [ANAGMIN] Primulaceae (V:D)
Anaphalis DC. [ANAPHAL$] Asteraceae (V:D)
Anaphalis margaritacea (L.) Benth. & Hook. f. [ANAPMAR] Asteraceae (V:D)
Anaphalis margaritacea var. *angustior* (Miq.) Nakai = Anaphalis margaritacea [ANAPMAR] Asteraceae (V:D)
Anaphalis margaritacea var. *intercedens* Hara = Anaphalis margaritacea [ANAPMAR] Asteraceae (V:D)
Anaphalis margaritacea var. *occidentalis* Greene = Anaphalis margaritacea [ANAPMAR] Asteraceae (V:D)
Anaphalis margaritacea var. *revoluta* Suksdorf = Anaphalis margaritacea [ANAPMAR] Asteraceae (V:D)
Anaphalis margaritacea var. *subalpina* Gray = Anaphalis margaritacea [ANAPMAR] Asteraceae (V:D)
Anaphalis occidentalis (Greene) Heller = Anaphalis margaritacea [ANAPMAR] Asteraceae (V:D)
Anaptychia Körber [ANAPTYC$] Physciaceae (L)
Anaptychia kaspica Gyelnik = Anaptychia setifera [ANAPSET] Physciaceae (L)
Anaptychia leucomelaena auct. = Heterodermia leucomelos [HETELEU] Physciaceae (L)
Anaptychia pseudospeciosa var. *tremulans* (Müll. Arg.) Kurok. = Heterodermia speciosa [HETESPE] Physciaceae (L)
Anaptychia setifera Räsänen [ANAPSET] Physciaceae (L)
Anaptychia speciosa (Wulfen) A. Massal. = Heterodermia speciosa [HETESPE] Physciaceae (L)
Anaptychia ulotrichoides (Vainio) Vainio [ANAPULO] Physciaceae (L)
Anastrepta (Lindb.) Schiffn. [ANASTRE$] Jungermanniaceae (B:H)

Anastrepta orcadensis (Hook) Schiffn. [ANASORC] Jungermanniaceae (B:H)
Anastrophyllum (Spruce) Steph. [ANASTRO$] Jungermanniaceae (B:H)
Anastrophyllum assimile (Mitt.) Steph. [ANASASS] Jungermanniaceae (B:H)
Anastrophyllum donianum (Hook.) Steph. [ANASDON] Jungermanniaceae (B:H)
Anastrophyllum helleranum (Nees) Schust. [ANASHEL] Jungermanniaceae (B:H)
Anastrophyllum hellerianum (Nees) Schust. = Anastrophyllum helleranum [ANASHEL] Jungermanniaceae (B:H)
Anastrophyllum japonicum Steph. = Anastrophyllum michauxii [ANASMIC] Jungermanniaceae (B:H)
Anastrophyllum japonicum var. *otianum* Hatt. = Anastrophyllum assimile [ANASASS] Jungermanniaceae (B:H)
Anastrophyllum michauxii (Web.) Buch ex Evans [ANASMIC] Jungermanniaceae (B:H)
Anastrophyllum minutum (Schreb.) Schust. [ANASMIN] Jungermanniaceae (B:H)
Anastrophyllum minutum var. **grandis** (Gott. ex Lindb.) Schust. [ANASMIN1] Jungermanniaceae (B:H)
Anastrophyllum minutum var. **minutum** [ANASMIN0] Jungermanniaceae (B:H)
Anastrophyllum reichardtii (Gott.) Steph. = Anastrophyllum assimile [ANASASS] Jungermanniaceae (B:H)
Anastrophyllum saxicola (Schrad.) Schust. [ANASSAX] Jungermanniaceae (B:H)
Anastrophyllum tamurae Steph. = Anastrophyllum michauxii [ANASMIC] Jungermanniaceae (B:H)
Anchusa L. [ANCHUSA$] Boraginaceae (V:D)
***Anchusa officinalis** L. [ANCHOFF] Boraginaceae (V:D)
Anchusa procera Bess. ex Link = Anchusa officinalis [ANCHOFF] Boraginaceae (V:D)
Andreaea Hedw. [ANDREAE$] Andreaeaceae (B:M)
Andreaea alpestris (Thed.) Schimp. [ANDRALP] Andreaeaceae (B:M)
Andreaea angustata Lindb. ex Limpr. = Andreaea heinemannii [ANDRHEI] Andreaeaceae (B:M)
Andreaea baileyi (Holz.) Holz. = Andreaea nivalis [ANDRNIV] Andreaeaceae (B:M)
Andreaea blyttii Schimp. [ANDRBLY] Andreaeaceae (B:M)
Andreaea blyttii ssp. *angustata* (Lindb. ex Limpr.) Schultze-Motel = Andreaea heinemannii [ANDRHEI] Andreaeaceae (B:M)
Andreaea blyttii var. *obtusifolia* (Berggr.) Sharp = Andreaea heinemannii [ANDRHEI] Andreaeaceae (B:M)
Andreaea crassinervia var. *obtusifolia* Berggr. = Andreaea heinemannii [ANDRHEI] Andreaeaceae (B:M)
Andreaea heinemannii Hampe & C. Müll. [ANDRHEI] Andreaeaceae (B:M)
Andreaea macounii Kindb. = Andreaea nivalis [ANDRNIV] Andreaeaceae (B:M)
Andreaea megistospora B. Murr. [ANDRMEG] Andreaeaceae (B:M)
Andreaea megistospora ssp. *epapillosa* B. Murr. = Andreaea megistospora var. epapillosa [ANDRMEG1] Andreaeaceae (B:M)
Andreaea megistospora var. **epapillosa** (B. Murr.) Crum & Anderson [ANDRMEG1] Andreaeaceae (B:M)
Andreaea megistospora var. **megistospora** [ANDRMEG0] Andreaeaceae (B:M)
Andreaea mutabilis Hook. f. & Wils. [ANDRMUT] Andreaeaceae (B:M)
Andreaea nivalis Hook. [ANDRNIV] Andreaeaceae (B:M)
Andreaea papillosa Lindb. = Andreaea rupestris var. papillosa [ANDRRUP3] Andreaeaceae (B:M)
Andreaea parvifolia C. Müll. = Andreaea rupestris var. rupestris [ANDRRUP0] Andreaeaceae (B:M)
Andreaea petrophila Ehrh. ex Fürnr. = Andreaea rupestris var. rupestris [ANDRRUP0] Andreaeaceae (B:M)
Andreaea rothii Web. & Mohr [ANDRROT] Andreaeaceae (B:M)
Andreaea rupestris Hedw. [ANDRRUP] Andreaeaceae (B:M)
Andreaea rupestris ssp. *papillosa* (Lindb.) C. Jens. = Andreaea rupestris var. papillosa [ANDRRUP3] Andreaeaceae (B:M)
Andreaea rupestris var. *acuminata* auct. = Andreaea rupestris var. papillosa [ANDRRUP3] Andreaeaceae (B:M)
Andreaea rupestris var. *alpestris* (Thed.) Sharp in Grout = Andreaea alpestris [ANDRALP] Andreaeaceae (B:M)
Andreaea rupestris var. **papillosa** (Lindb.) Podp. [ANDRRUP3] Andreaeaceae (B:M)
Andreaea rupestris var. **rupestris** [ANDRRUP0] Andreaeaceae (B:M)
Andreaea rupestris var. *sparsifolia* (Zett.) Sharp in Grout = Andreaea rupestris var. papillosa [ANDRRUP3] Andreaeaceae (B:M)
Andreaea schofieldiana B. Murr. [ANDRSCH] Andreaeaceae (B:M)
Andreaea sinuosa B. Murr. [ANDRSIN] Andreaeaceae (B:M)
Andreaea sparsifolia var. *sublaevis* Kindb. = Andreaea rupestris var. papillosa [ANDRRUP3] Andreaeaceae (B:M)
Andreaeobryum Steere & B. Murr. [ANDREAO$] Andreaeobryaceae (B:M)
Andreaeobryum macrosporum Steere & B. Murr. [ANDRMAC] Andreaeobryaceae (B:M)
Androcera rostrata (Dunal) Rydb. = Solanum rostratum [SOLAROS] Solanaceae (V:D)
Andromeda L. [ANDROME$] Ericaceae (V:D)
Andromeda polifolia L. [ANDRPOL] Ericaceae (V:D)
Andropogon scoparius Michx. = Schizachyrium scoparium [SCHISCO] Poaceae (V:M)
Androsace L. [ANDROSA$] Primulaceae (V:D)
Androsace alaskana Coville & Standl. ex Hultén = Douglasia alaskana [DOUGALA] Primulaceae (V:D)
Androsace alaskana var. *reediae* Welsh & Goodrich = Douglasia alaskana [DOUGALA] Primulaceae (V:D)
Androsace arizonica (Gray) Derganc = Androsace occidentalis [ANDROCC] Primulaceae (V:D)
Androsace chamaejasme Wulfen [ANDRCHA] Primulaceae (V:D)
Androsace chamaejasme ssp. **lehmanniana** (Spreng.) Hultén [ANDRCHA1] Primulaceae (V:D)
Androsace chamaejasme var. *arctica* R. Knuth = Androsace chamaejasme ssp. lehmanniana [ANDRCHA1] Primulaceae (V:D)
Androsace diffusa Small = Androsace septentrionalis ssp. subulifera [ANDRSEP2] Primulaceae (V:D)
Androsace filiformis Retz. [ANDRFIL] Primulaceae (V:D)
Androsace lehmanniana Spreng. = Androsace chamaejasme ssp. lehmanniana [ANDRCHA1] Primulaceae (V:D)
Androsace occidentalis Pursh [ANDROCC] Primulaceae (V:D)
Androsace occidentalis var. *arizonica* (Gray) St. John = Androsace occidentalis [ANDROCC] Primulaceae (V:D)
Androsace occidentalis var. *simplex* (Rydb.) St. John = Androsace occidentalis [ANDROCC] Primulaceae (V:D)
Androsace puberulenta Rydb. = Androsace septentrionalis ssp. puberulenta [ANDRSEP1] Primulaceae (V:D)
Androsace septentrionalis L. [ANDRSEP] Primulaceae (V:D)
Androsace septentrionalis ssp. **puberulenta** (Rydb.) G.T. Robbins [ANDRSEP1] Primulaceae (V:D)
Androsace septentrionalis ssp. **septentrionalis** [ANDRSEP0] Primulaceae (V:D)
Androsace septentrionalis ssp. **subulifera** (Gray) G.T. Robbins [ANDRSEP2] Primulaceae (V:D)
Androsace septentrionalis ssp. **subumbellata** (A. Nels.) G.T. Robbins [ANDRSEP3] Primulaceae (V:D)
Androsace septentrionalis var. *diffusa* (Small) R. Knuth = Androsace septentrionalis ssp. subulifera [ANDRSEP2] Primulaceae (V:D)
Androsace septentrionalis var. *puberulenta* (Rydb.) R. Knuth = Androsace septentrionalis ssp. puberulenta [ANDRSEP1] Primulaceae (V:D)
Androsace septentrionalis var. *subulifera* Gray = Androsace septentrionalis ssp. subulifera [ANDRSEP2] Primulaceae (V:D)
Androsace septentrionalis var. *subumbellata* A. Nels. = Androsace septentrionalis ssp. subumbellata [ANDRSEP3] Primulaceae (V:D)
Anemone L. [ANEMONE$] Ranunculaceae (V:D)
Anemone canadensis L. [ANEMCAN] Ranunculaceae (V:D)
Anemone cylindrica Gray [ANEMCYL] Ranunculaceae (V:D)
Anemone drummondii S. Wats. [ANEMDRU] Ranunculaceae (V:D)
Anemone drummondii var. *lithophila* (Rydb.) C.L. Hitchc. = Anemone lithophila [ANEMLIT] Ranunculaceae (V:D)
Anemone lithophila Rydb. [ANEMLIT] Ranunculaceae (V:D)

Anemone ludoviciana Nutt. = Pulsatilla patens ssp. multifida [PULSPAT1] Ranunculaceae (V:D)
Anemone lyallii Britt. [ANEMLYA] Ranunculaceae (V:D)
Anemone multifida Poir. [ANEMMUL] Ranunculaceae (V:D)
Anemone multifida (Pritz.) Zamels = Pulsatilla patens ssp. multifida [PULSPAT1] Ranunculaceae (V:D)
Anemone multifida var. **hirsuta** C.L. Hitchc. [ANEMMUL1] Ranunculaceae (V:D)
Anemone multifida var. **richardsiana** Fern. [ANEMMUL3] Ranunculaceae (V:D)
Anemone multifida var. *saxicola* f. *hirsuta* (C.L. Hitchc.) T.C. Brayshaw = Anemone multifida [ANEMMUL] Ranunculaceae (V:D)
Anemone narcissiflora L. [ANEMNAR] Ranunculaceae (V:D)
Anemone narcissiflora ssp. **alaskana** Hultén [ANEMNAR1] Ranunculaceae (V:D)
Anemone narcissiflora ssp. **interior** Hultén [ANEMNAR2] Ranunculaceae (V:D)
Anemone narcissiflora var. *alaskana* (Hultén) Boivin = Anemone narcissiflora ssp. alaskana [ANEMNAR1] Ranunculaceae (V:D)
Anemone narcissiflora var. *interior* (Hultén) Boivin = Anemone narcissiflora ssp. interior [ANEMNAR2] Ranunculaceae (V:D)
Anemone nemorosa var. *lyallii* (Britt.) Ulbr. = Anemone lyallii [ANEMLYA] Ranunculaceae (V:D)
Anemone nuttalliana DC. = Pulsatilla patens ssp. multifida [PULSPAT1] Ranunculaceae (V:D)
Anemone occidentalis S. Wats. = Pulsatilla occidentalis [PULSOCC] Ranunculaceae (V:D)
Anemone parviflora Michx. [ANEMPAR] Ranunculaceae (V:D)
Anemone patens ssp. *multifida* (Pritzel) Hultén = Pulsatilla patens ssp. multifida [PULSPAT1] Ranunculaceae (V:D)
Anemone patens var. *multifida* Pritz. = Pulsatilla patens ssp. multifida [PULSPAT1] Ranunculaceae (V:D)
Anemone patens var. *nuttalliana* (DC.) Gray = Pulsatilla patens ssp. multifida [PULSPAT1] Ranunculaceae (V:D)
Anemone patens var. *wolfgangiana* (Bess.) K. Koch = Pulsatilla patens ssp. multifida [PULSPAT1] Ranunculaceae (V:D)
Anemone piperi Britt. ex Rydb. [ANEMPIP] Ranunculaceae (V:D)
Anemone quinquefolia var. *lyallii* (Britt.) B.L. Robins. = Anemone lyallii [ANEMLYA] Ranunculaceae (V:D)
Anemone richardsonii Hook. [ANEMRIC] Ranunculaceae (V:D)
Anemone riparia Fern. = Anemone virginiana var. riparia [ANEMVIR2] Ranunculaceae (V:D)
Anemone virginiana L. [ANEMVIR] Ranunculaceae (V:D)
Anemone virginiana var. **riparia** (Fern.) Boivin [ANEMVIR2] Ranunculaceae (V:D)
Anethum L. [ANETHUM$] Apiaceae (V:D)
*__Anethum graveolens__ L. [ANETGRA] Apiaceae (V:D)
Aneura Dum. [ANEURA$] Aneuraceae (B:H)
Aneura pinguis (L.) Dum. [ANEUPIN] Aneuraceae (B:H)
Aneurolepidium piperi (Bowden) Baum = Leymus cinereus [LEYMCIN] Poaceae (V:M)
Angelica L. [ANGELIC$] Apiaceae (V:D)
Angelica arguta Nutt. [ANGEARG] Apiaceae (V:D)
Angelica dawsonii S. Wats. [ANGEDAW] Apiaceae (V:D)
Angelica genuflexa Nutt. [ANGEGEN] Apiaceae (V:D)
Angelica lucida L. [ANGELUC] Apiaceae (V:D)
Angelica lyallii S. Wats. = Angelica arguta [ANGEARG] Apiaceae (V:D)
Anisantha diandra (Roth) Tutin ex Tzvelev = Bromus diandrus [BROMDIA] Poaceae (V:M)
Anisantha sterilis (L.) Nevski = Bromus sterilis [BROMSTE] Poaceae (V:M)
Anisomeridium (Müll. Arg.) Choisy [ANISOME$] Monoblastiaceae (L)
Anisomeridium biforme (Borrer) R.C. Harris [ANISBIF] Monoblastiaceae (L)
Anisothecium grevilleanum (Brid.) Arnell & C. Jens. = Dicranella grevilleana [DICRGRE] Dicranaceae (B:M)
Anisothecium rufescens (With.) Lindb. = Dicranella rufescens [DICRRUF] Dicranaceae (B:M)
Anisothecium schreberianum (Hedw.) Dix. = Dicranella schreberiana var. schreberiana [DICRSCH0] Dicranaceae (B:M)
Anisothecium vaginale (Hedw.) Loeske = Dicranella crispa [DICRCRI] Dicranaceae (B:M)
Anisothecium varium (Hedw.) Mitt. = Dicranella varia [DICRVAR] Dicranaceae (B:M)
Anodus donianus (Sm.) B.S.G. = Seligeria donniana [SELIDON] Seligeriaceae (B:M)
Anoectangium Schwaegr. [ANOECTA$] Pottiaceae (B:M)
Anoectangium aestivum (Hedw.) Mitt. [ANOEAES] Pottiaceae (B:M)
Anoectangium arizonicum Bartr. = Gymnostomum aeruginosum [GYMNAER] Pottiaceae (B:M)
Anoectangium compactum Schwaegr. = Anoectangium aestivum [ANOEAES] Pottiaceae (B:M)
Anoectangium compactum var. *alaskanum* Card. & Thér. = Anoectangium aestivum [ANOEAES] Pottiaceae (B:M)
Anoectangium euchloron (Schwaegr.) Mitt. = Anoectangium aestivum [ANOEAES] Pottiaceae (B:M)
Anoectangium mougeotii (B.S.G.) Lindb. = Amphidium mougeotii [AMPHMOU] Orthotrichaceae (B:M)
Anoectangium obtusifolium (Broth. & Par. in Card.) Grout = Molendoa sendtneriana [MOLESEN] Pottiaceae (B:M)
Anoectangium peckii (Sull.) Sull. ex Aust. = Anoectangium aestivum [ANOEAES] Pottiaceae (B:M)
Anoectangium sendtnerianum Bruch & Schimp. in B.S.G. = Molendoa sendtneriana [MOLESEN] Pottiaceae (B:M)
Anoectangium sendtnerianum var. *tenuinerve* (Limpr.) Mönk. = Anoectangium tenuinerve [ANOETEN] Pottiaceae (B:M)
Anoectangium tenuinerve (Limpr.) Par. [ANOETEN] Pottiaceae (B:M)
Anomobryum Schimp. [ANOMOBR$] Bryaceae (B:M)
Anomobryum concinnatum (Spruce) Lindb. = Anomobryum filiforme [ANOMFIL] Bryaceae (B:M)
Anomobryum filiforme (Dicks.) Solms in Rabenh. [ANOMFIL] Bryaceae (B:M)
Anomobryum filiforme var. *concinnatum* (Spruce) Weis. = Anomobryum filiforme [ANOMFIL] Bryaceae (B:M)
Anomobryum julaceum (Brid.) Schimp. = Anomobryum filiforme [ANOMFIL] Bryaceae (B:M)
Anomodon heteroideus Kindb. in Mac. & Kindb. = Leskeella nervosa [LESKNER] Leskeaceae (B:M)
Anomodon subrigidulus Kindb. = Leskeella nervosa [LESKNER] Leskeaceae (B:M)
Anoplanthus fasciculatus (Nutt.) Walp. = Orobanche fasciculata [OROBFAS] Orobanchaceae (V:D)
Anotites viscosa Greene = Silene menziesii var. viscosa [SILEMEN2] Caryophyllaceae (V:D)
Antennaria Gaertn. [ANTENNA$] Asteraceae (V:D)
Antennaria aizoides Greene = Antennaria umbrinella [ANTEUMB] Asteraceae (V:D)
Antennaria alborosea Porsild ex M.P. Porsild = Antennaria rosea [ANTEROS] Asteraceae (V:D)
Antennaria alpina var. *glabrata* J. Vahl = Antennaria monocephala [ANTEMON] Asteraceae (V:D)
Antennaria alpina var. *intermedia* Rosenv. = Antennaria umbrinella [ANTEUMB] Asteraceae (V:D)
Antennaria alpina var. *media* (Greene) Jepson = Antennaria media [ANTEMED] Asteraceae (V:D)
Antennaria anaphaloides Rydb. [ANTEANA] Asteraceae (V:D)
Antennaria angustata Greene = Antennaria monocephala [ANTEMON] Asteraceae (V:D)
Antennaria angustiarum Lunell = Antennaria neglecta [ANTENEG] Asteraceae (V:D)
Antennaria aprica Greene = Antennaria parvifolia [ANTEPAR] Asteraceae (V:D)
Antennaria aprica var. *minuscula* (Boivin) Boivin = Antennaria parvifolia [ANTEPAR] Asteraceae (V:D)
Antennaria argentea ssp. *aberrans* E. Nels. = Antennaria luzuloides [ANTELUZ] Asteraceae (V:D)
Antennaria athabascensis Greene = Antennaria neglecta [ANTENEG] Asteraceae (V:D)
Antennaria aureola Lunell = Antennaria parvifolia [ANTEPAR] Asteraceae (V:D)

Antennaria austromontana E. Nels. = Antennaria media [ANTEMED] Asteraceae (V:D)
Antennaria burwellensis Malte = Antennaria monocephala [ANTEMON] Asteraceae (V:D)
Antennaria callilepis Greene = Antennaria howellii ssp. howellii [ANTEHOW0] Asteraceae (V:D)
Antennaria campestris Rydb. = Antennaria neglecta [ANTENEG] Asteraceae (V:D)
Antennaria campestris var. *athabascensis* (Greene) Boivin = Antennaria neglecta [ANTENEG] Asteraceae (V:D)
Antennaria candida Greene = Antennaria media [ANTEMED] Asteraceae (V:D)
Antennaria carpathica var. *lanata* Hook. = Antennaria lanata [ANTELAN] Asteraceae (V:D)
Antennaria carpathica var. *pulcherrima* Hook. = Antennaria pulcherrima [ANTEPUL] Asteraceae (V:D)
Antennaria chelonica Lunell = Antennaria neglecta [ANTENEG] Asteraceae (V:D)
Antennaria congesta Malte = Antennaria monocephala [ANTEMON] Asteraceae (V:D)
Antennaria densa Greene = Antennaria media [ANTEMED] Asteraceae (V:D)
Antennaria dimorpha (Nutt.) Torr. & Gray [ANTEDIM] Asteraceae (V:D)
Antennaria dimorpha var. *integra* Henderson = Antennaria dimorpha [ANTEDIM] Asteraceae (V:D)
Antennaria dimorpha var. *latisquama* (Piper) M.E. Peck = Antennaria dimorpha [ANTEDIM] Asteraceae (V:D)
Antennaria dimorpha var. *macrocephala* D.C. Eat. = Antennaria dimorpha [ANTEDIM] Asteraceae (V:D)
Antennaria erosa Greene = Antennaria neglecta [ANTENEG] Asteraceae (V:D)
Antennaria exilis Greene = Antennaria monocephala [ANTEMON] Asteraceae (V:D)
Antennaria eximia Greene = Antennaria howellii ssp. howellii [ANTEHOW0] Asteraceae (V:D)
Antennaria fernaldiana Polunin = Antennaria monocephala [ANTEMON] Asteraceae (V:D)
Antennaria flavescens Rydb. = Antennaria umbrinella [ANTEUMB] Asteraceae (V:D)
Antennaria fusca E. Nels. = Antennaria parvifolia [ANTEPAR] Asteraceae (V:D)
Antennaria glabrata (J. Vahl) Greene = Antennaria monocephala [ANTEMON] Asteraceae (V:D)
Antennaria gormanii St. John = Antennaria media [ANTEMED] Asteraceae (V:D)
Antennaria holmii Greene = Antennaria parvifolia [ANTEPAR] Asteraceae (V:D)
Antennaria howellii Greene [ANTEHOW] Asteraceae (V:D)
Antennaria howellii ssp. **howellii** [ANTEHOW0] Asteraceae (V:D)
Antennaria howellii ssp. **neodioica** (Greene) Bayer [ANTEHOW1] Asteraceae (V:D)
Antennaria howellii var. *athabascensis* (Greene) Boivin = Antennaria neglecta [ANTENEG] Asteraceae (V:D)
Antennaria howellii var. *campestris* (Rydb.) Boivin = Antennaria neglecta [ANTENEG] Asteraceae (V:D)
Antennaria hudsonica Malte = Antennaria monocephala [ANTEMON] Asteraceae (V:D)
Antennaria intermedia (Rosenv.) M.P. Porsild = Antennaria umbrinella [ANTEUMB] Asteraceae (V:D)
Antennaria lanata (Hook.) Greene [ANTELAN] Asteraceae (V:D)
Antennaria latisquama Piper = Antennaria dimorpha [ANTEDIM] Asteraceae (V:D)
Antennaria longifolia Greene = Antennaria neglecta [ANTENEG] Asteraceae (V:D)
Antennaria lunellii Greene = Antennaria neglecta [ANTENEG] Asteraceae (V:D)
Antennaria luzuloides Torr. & Gray [ANTELUZ] Asteraceae (V:D)
Antennaria luzuloides var. *oblanceolata* (Rydb.) M.E. Peck = Antennaria luzuloides [ANTELUZ] Asteraceae (V:D)
Antennaria macrocephala (D.C. Eat.) Rydb. = Antennaria dimorpha [ANTEDIM] Asteraceae (V:D)
Antennaria media Greene [ANTEMED] Asteraceae (V:D)
Antennaria microcephala Gray = Antennaria luzuloides [ANTELUZ] Asteraceae (V:D)
Antennaria microphylla Rydb. [ANTEMIC] Asteraceae (V:D)
Antennaria minuscula Boivin = Antennaria parvifolia [ANTEPAR] Asteraceae (V:D)
Antennaria modesta Greene = Antennaria media [ANTEMED] Asteraceae (V:D)
Antennaria monocephala DC. [ANTEMON] Asteraceae (V:D)
Antennaria monocephala ssp. *angustata* (Greene) Hultén = Antennaria monocephala [ANTEMON] Asteraceae (V:D)
Antennaria monocephala ssp. *philonipha* (Porsild) Hultén = Antennaria monocephala [ANTEMON] Asteraceae (V:D)
Antennaria monocephala var. *exilis* (Greene) Hultén = Antennaria monocephala [ANTEMON] Asteraceae (V:D)
Antennaria monocephala var. *latisquamea* Hultén = Antennaria monocephala [ANTEMON] Asteraceae (V:D)
Antennaria mucronata E. Nels. = Antennaria media [ANTEMED] Asteraceae (V:D)
Antennaria nebraskensis Greene = Antennaria neglecta [ANTENEG] Asteraceae (V:D)
Antennaria neglecta Greene [ANTENEG] Asteraceae (V:D)
Antennaria neglecta ssp. *howellii* (Greene) Hultén = Antennaria howellii ssp. howellii [ANTEHOW0] Asteraceae (V:D)
Antennaria neglecta var. *athabascensis* (Greene) Taylor & MacBryde = Antennaria neglecta [ANTENEG] Asteraceae (V:D)
Antennaria neglecta var. *attenuata* (Fern.) Cronq. = Antennaria howellii ssp. neodioica [ANTEHOW1] Asteraceae (V:D)
Antennaria neglecta var. *campestris* (Rydb.) Steyermark = Antennaria neglecta [ANTENEG] Asteraceae (V:D)
Antennaria neglecta var. *howellii* (Greene) Cronq. = Antennaria howellii ssp. howellii [ANTEHOW0] Asteraceae (V:D)
Antennaria neglecta var. *neodioica* (Greene) Cronq. = Antennaria howellii ssp. neodioica [ANTEHOW1] Asteraceae (V:D)
Antennaria neodioica Greene = Antennaria howellii ssp. neodioica [ANTEHOW1] Asteraceae (V:D)
Antennaria neodioica ssp. *howellii* (Greene) Bayer = Antennaria howellii ssp. howellii [ANTEHOW0] Asteraceae (V:D)
Antennaria neodioica var. *attenuata* Fern. = Antennaria howellii ssp. neodioica [ANTEHOW1] Asteraceae (V:D)
Antennaria neodioica var. *chlorophylla* Fern. = Antennaria howellii ssp. neodioica [ANTEHOW1] Asteraceae (V:D)
Antennaria neodioica var. *grandis* Fern. = Antennaria howellii ssp. neodioica [ANTEHOW1] Asteraceae (V:D)
Antennaria neodioica var. *interjecta* Fern. = Antennaria howellii ssp. neodioica [ANTEHOW1] Asteraceae (V:D)
Antennaria nitens Greene = Antennaria monocephala [ANTEMON] Asteraceae (V:D)
Antennaria nitida Greene = Antennaria microphylla [ANTEMIC] Asteraceae (V:D)
Antennaria oblanceolata Rydb. = Antennaria luzuloides [ANTELUZ] Asteraceae (V:D)
Antennaria oblancifolia E. Nels. = Antennaria racemosa [ANTERAC] Asteraceae (V:D)
Antennaria obtusata Greene = Antennaria parvifolia [ANTEPAR] Asteraceae (V:D)
Antennaria parvifolia Nutt. [ANTEPAR] Asteraceae (V:D)
Antennaria parvifolia auct. = Antennaria microphylla [ANTEMIC] Asteraceae (V:D)
Antennaria parvula Greene = Antennaria neglecta [ANTENEG] Asteraceae (V:D)
Antennaria pedicellata Greene = Antennaria racemosa [ANTERAC] Asteraceae (V:D)
Antennaria philonipha Porsild = Antennaria monocephala [ANTEMON] Asteraceae (V:D)
Antennaria piperi Rydb. = Antennaria racemosa [ANTERAC] Asteraceae (V:D)
Antennaria pulcherrima (Hook.) Greene [ANTEPUL] Asteraceae (V:D)
Antennaria pulcherrima ssp. *anaphaloides* (Rydb.) W.A. Weber = Antennaria anaphaloides [ANTEANA] Asteraceae (V:D)
Antennaria pulcherrima var. *anaphaloides* (Rydb.) G.W. Douglas = Antennaria anaphaloides [ANTEANA] Asteraceae (V:D)

Antennaria pulcherrima var. *angustisquama* Porsild = Antennaria pulcherrima [ANTEPUL] Asteraceae (V:D)
Antennaria pulcherrima var. *sordida* Boivin = Antennaria pulcherrima [ANTEPUL] Asteraceae (V:D)
Antennaria pumila Greene = Antennaria parvifolia [ANTEPAR] Asteraceae (V:D)
Antennaria pyramidata Greene = Antennaria luzuloides [ANTELUZ] Asteraceae (V:D)
Antennaria racemosa Hook. [ANTERAC] Asteraceae (V:D)
Antennaria recurva Greene = Antennaria parvifolia [ANTEPAR] Asteraceae (V:D)
Antennaria reflexa E. Nels. = Antennaria umbrinella [ANTEUMB] Asteraceae (V:D)
Antennaria rhodantha Fern. = Antennaria howellii ssp. neodioica [ANTEHOW1] Asteraceae (V:D)
Antennaria rhodanthus Suksdorf = Antennaria parvifolia [ANTEPAR] Asteraceae (V:D)
Antennaria rosea Greene [ANTEROS] Asteraceae (V:D)
Antennaria rosea var. *nitida* (Greene) Breitung = Antennaria microphylla [ANTEMIC] Asteraceae (V:D)
Antennaria rupicola Fern. = Antennaria howellii ssp. neodioica [ANTEHOW1] Asteraceae (V:D)
Antennaria russellii Boivin = Antennaria howellii ssp. neodioica [ANTEHOW1] Asteraceae (V:D)
Antennaria shumaginensis Porsild = Antennaria monocephala [ANTEMON] Asteraceae (V:D)
Antennaria solstitialis Greene = Antennaria microphylla [ANTEMIC] Asteraceae (V:D)
Antennaria tansleyi Polunin = Antennaria monocephala [ANTEMON] Asteraceae (V:D)
Antennaria tweedsmurii Polunin = Antennaria monocephala [ANTEMON] Asteraceae (V:D)
Antennaria umbrinella Rydb. [ANTEUMB] Asteraceae (V:D)
Antennaria wilsonii Greene = Antennaria neglecta [ANTENEG] Asteraceae (V:D)
Anthelia (Dum. emend. Schiffn.) Dum. [ANTHELI$] Antheliaceae (B:H)
Anthelia julacea (L.) Dum. [ANTHJUL] Antheliaceae (B:H)
Anthelia juratzkana (Limpr.) Trev. [ANTHJUR] Antheliaceae (B:H)
Anthemis L. [ANTHEMI$] Asteraceae (V:D)
***Anthemis arvensis** L. [ANTHARV] Asteraceae (V:D)
***Anthemis cotula** L. [ANTHCOT] Asteraceae (V:D)
***Anthemis tinctoria** L. [ANTHTIN] Asteraceae (V:D)
Anthoceros L. emend. Prosk. [ANTHOCE$] Anthocerotaceae (B:H)
Anthoceros crispulus (Mont.) Douin = Anthoceros punctatus [ANTHPUN] Anthocerotaceae (B:H)
Anthoceros fusiformis Aust. [ANTHFUS] Anthocerotaceae (B:H)
Anthoceros hallii Aust. = Phaeoceros hallii [PHAEHAL] Anthocerotaceae (B:H)
Anthoceros phymatodes M.A. Howe = Phaeoceros hallii [PHAEHAL] Anthocerotaceae (B:H)
Anthoceros punctatus L. [ANTHPUN] Anthocerotaceae (B:H)
Anthopogon tonsum (Lunell) Rydb. = Gentianopsis macounii [GENTMAC] Gentianaceae (V:D)
Anthoxanthum L. [ANTHOXA$] Poaceae (V:M)
***Anthoxanthum aristatum** Boiss. [ANTHARI] Poaceae (V:M)
Anthoxanthum nitens (Weber) Y. Schouten & Veldkamp = Hierochloe odorata [HIERODO] Poaceae (V:M)
***Anthoxanthum odoratum** L. [ANTHODO] Poaceae (V:M)
Anthoxanthum odoratum var. *puelii* (Lecoq & Lamotte) Coss. & Durieu = Anthoxanthum aristatum [ANTHARI] Poaceae (V:M)
Anthoxanthum puelii Lecoq & Lamotte = Anthoxanthum aristatum [ANTHARI] Poaceae (V:M)
Anthriscus Pers. [ANTHRIC$] Apiaceae (V:D)
***Anthriscus caucalis** Bieb. [ANTHCAU] Apiaceae (V:D)
Anthriscus neglecta var. *scandix* (Scop.) Hyl. = Anthriscus caucalis [ANTHCAU] Apiaceae (V:D)
Anthriscus scandicina (Weber ex Wiggers) Mansf. = Anthriscus caucalis [ANTHCAU] Apiaceae (V:D)
Antirrhinum elatine L. = Kickxia elatine [KICKELA] Scrophulariaceae (V:D)
Antirrhinum tenellum Pursh = Collinsia parviflora var. grandiflora [COLLPAR1] Scrophulariaceae (V:D)
Antitrichia Brid. [ANTITRI$] Leucodontaceae (B:M)
Antitrichia californica Sull. in Lesq. [ANTICAL] Leucodontaceae (B:M)
Antitrichia curtipendula (Hedw.) Brid. [ANTICUR] Leucodontaceae (B:M)
Antitrichia tenella Kindb. = Antitrichia californica [ANTICAL] Leucodontaceae (B:M)
Anzina Scheid. [ANZINA$] Trapeliaceae (L)
Anzina carneonivea (Anzi) Scheid. [ANZICAR] Trapeliaceae (L)
Aongstroemia Bruch & Schimp. in B.S.G. [AONGSTR$] Dicranaceae (B:M)
Aongstroemia longipes (Somm.) Bruch & Schimp. in B.S.G. [AONGLON] Dicranaceae (B:M)
Apargidium boreale (Bong.) Torr. & Gray = Microseris borealis [MICRBOR] Asteraceae (V:D)
Apera Adans. [APERA$] Poaceae (V:M)
***Apera interrupta** (L.) Beauv. [APERINT] Poaceae (V:M)
Aphanes L. [APHANES$] Rosaceae (V:D)
***Aphanes arvensis** L. [APHAARV] Rosaceae (V:D)
Aphanes australis Rydb. = Aphanes microcarpa [APHAMIC] Rosaceae (V:D)
***Aphanes microcarpa** (Boiss. & Reut.) Rothm. [APHAMIC] Rosaceae (V:D)
Aphanes occidentalis (Nutt.) Rydb. = Aphanes arvensis [APHAARV] Rosaceae (V:D)
Aphanorrhegma patens (Hedw.) Lindb. = Physcomitrella patens [PHYSPAT] Funariaceae (B:M)
Aphragmus Andrz. ex DC. [APHRAGM$] Brassicaceae (V:D)
Aphragmus eschscholtzianus Andrz. ex DC. [APHRESC] Brassicaceae (V:D)
Aphyllon pinorum (Geyer ex Hook.) Gray = Orobanche pinorum [OROBPIN] Orobanchaceae (V:D)
Aplozia sphaerocarpa (Hook.) Dum. = Jungermannia sphaerocarpa [JUNGSPH] Jungermanniaceae (B:H)
Apocynum L. [APOCYNU$] Apocynaceae (V:D)
Apocynum androsaemifolium L. [APOCAND] Apocynaceae (V:D)
Apocynum androsaemifolium ssp. **androsaemifolium** [APOCAND0] Apocynaceae (V:D)
Apocynum androsaemifolium ssp. **pumilum** (Gray) Boivin [APOCAND1] Apocynaceae (V:D)
Apocynum androsaemifolium var. *pumilum* Gray = Apocynum androsaemifolium ssp. pumilum [APOCAND1] Apocynaceae (V:D)
Apocynum cannabinum L. [APOCCAN] Apocynaceae (V:D)
Apocynum cannabinum var. *angustifolium* (Woot.) N. Holmgren = Apocynum cannabinum [APOCCAN] Apocynaceae (V:D)
Apocynum cannabinum var. *glaberrimum* A. DC. = Apocynum cannabinum [APOCCAN] Apocynaceae (V:D)
Apocynum cannabinum var. *greeneanum* (Bég. & Bel.) Woods. = Apocynum cannabinum [APOCCAN] Apocynaceae (V:D)
Apocynum cannabinum var. *hypericifolium* Gray = Apocynum cannabinum [APOCCAN] Apocynaceae (V:D)
Apocynum cannabinum var. *nemorale* (G.S. Mill.) Fern. = Apocynum cannabinum [APOCCAN] Apocynaceae (V:D)
Apocynum cannabinum var. *pubescens* (Mitchell ex R. Br.) Woods. = Apocynum cannabinum [APOCCAN] Apocynaceae (V:D)
Apocynum cannabinum var. *suksdorfii* (Greene) Bég. & Bel. = Apocynum cannabinum [APOCCAN] Apocynaceae (V:D)
Apocynum × floribundum Greene (pro sp.) [APOCFLO] Apocynaceae (V:D)
Apocynum hypericifolium Ait. = Apocynum cannabinum [APOCCAN] Apocynaceae (V:D)
Apocynum jonesii Woods. = Apocynum × floribundum [APOCFLO] Apocynaceae (V:D)
Apocynum medium Greene = Apocynum × floribundum [APOCFLO] Apocynaceae (V:D)
Apocynum medium var. *floribundum* (Greene) Woods. = Apocynum × floribundum [APOCFLO] Apocynaceae (V:D)
Apocynum medium var. *leuconeuron* (Greene) Woods. = Apocynum × floribundum [APOCFLO] Apocynaceae (V:D)
Apocynum medium var. *lividum* (Greene) Woods. = Apocynum × floribundum [APOCFLO] Apocynaceae (V:D)

Apocynum medium var. *sarniense* (Greene) Woods. = Apocynum × floribundum [APOCFLO] Apocynaceae (V:D)
Apocynum medium var. *vestitum* (Greene) Woods. = Apocynum × floribundum [APOCFLO] Apocynaceae (V:D)
Apocynum milleri Britt. = Apocynum × floribundum [APOCFLO] Apocynaceae (V:D)
Apocynum pubescens Mitchell ex R. Br. = Apocynum cannabinum [APOCCAN] Apocynaceae (V:D)
Apocynum sibiricum Jacq. = Apocynum cannabinum [APOCCAN] Apocynaceae (V:D)
Apocynum sibiricum var. *cordigerum* (Greene) Fern. = Apocynum cannabinum [APOCCAN] Apocynaceae (V:D)
Apocynum sibiricum var. *farwellii* (Greene) Fern. = Apocynum cannabinum [APOCCAN] Apocynaceae (V:D)
Apocynum sibiricum var. *salignum* (Greene) Fern. = Apocynum cannabinum [APOCCAN] Apocynaceae (V:D)
Apocynum suksdorfii Greene = Apocynum cannabinum [APOCCAN] Apocynaceae (V:D)
Apocynum suksdorfii var. *angustifolium* (Woot.) Woods. = Apocynum cannabinum [APOCCAN] Apocynaceae (V:D)
Apometzgeria Kuwah. [APOMETZ$] Metzgeriaceae (B:H)
Apometzgeria pubescens (Schrank) Kuwah. [APOMPUB] Metzgeriaceae (B:H)
Apotreubia Hatt. & Mizut. [APOTREU$] Treubiaceae (B:H)
Apotreubia nana (Hatt. & H. Inoue) Hatt. & Mizut. [APOTNAN] Treubiaceae (B:H)
Aquilegia L. [AQUILEG$] Ranunculaceae (V:D)
Aquilegia brevistyla Hook. [AQUIBRE] Ranunculaceae (V:D)
Aquilegia flavescens S. Wats. [AQUIFLA] Ranunculaceae (V:D)
Aquilegia formosa Fisch. ex DC. [AQUIFOR] Ranunculaceae (V:D)
Arabidopsis Heynh. [ARABIDO$] Brassicaceae (V:D)
Arabidopsis glauca (Nutt.) Rydb. = Arabidopsis salsuginea [ARABSAL] Brassicaceae (V:D)
Arabidopsis novae-angliae (Rydb.) Britt. = Braya humilis [BRAYHUM] Brassicaceae (V:D)
Arabidopsis richardsonii Rydb. = Braya humilis [BRAYHUM] Brassicaceae (V:D)
Arabidopsis salsuginea (Pallas) N. Busch [ARABSAL] Brassicaceae (V:D)
*****Arabidopsis thaliana** (L.) Heynh. [ARABTHA] Brassicaceae (V:D)
Arabis L. [ARABIS$] Brassicaceae (V:D)
Arabis acutina Greene = Arabis × divaricarpa [ARABDIV] Brassicaceae (V:D)
Arabis arcuata var. *secunda* (T.J. Howell) Robertson = Arabis holboellii var. retrofracta [ARABHOL5] Brassicaceae (V:D)
Arabis bourgovii Rydb. = Arabis holboellii var. collinsii [ARABHOL1] Brassicaceae (V:D)
Arabis brachycarpa (Torr. & Gray) Britt. = Arabis drummondii [ARABDRU] Brassicaceae (V:D)
Arabis bridgeri M.E. Jones = Arabis nuttallii [ARABNUT] Brassicaceae (V:D)
Arabis caduca A. Nels. = Arabis holboellii var. retrofracta [ARABHOL5] Brassicaceae (V:D)
Arabis collinsii Fern. = Arabis holboellii var. collinsii [ARABHOL1] Brassicaceae (V:D)
Arabis confinis S. Wats. = Arabis drummondii [ARABDRU] Brassicaceae (V:D)
Arabis confinis var. *interposita* (Greene) Welsh & Reveal = Arabis × divaricarpa [ARABDIV] Brassicaceae (V:D)
Arabis connexa Greene = Arabis drummondii [ARABDRU] Brassicaceae (V:D)
Arabis consanguinea Greene = Arabis holboellii var. retrofracta [ARABHOL5] Brassicaceae (V:D)
Arabis dacotica Greene = Arabis holboellii var. collinsii [ARABHOL1] Brassicaceae (V:D)
Arabis × divaricarpa A. Nels. (pro sp.) [ARABDIV] Brassicaceae (V:D)
Arabis divaricarpa A. Nels. = Arabis × divaricarpa [ARABDIV] Brassicaceae (V:D)
Arabis divaricarpa var. *dacotica* (Greene) Boivin = Arabis holboellii var. collinsii [ARABHOL1] Brassicaceae (V:D)
Arabis divaricarpa var. *dechamplainii* Boivin = Arabis × divaricarpa [ARABDIV] Brassicaceae (V:D)
Arabis divaricarpa var. *interposita* (Greene) Rollins = Arabis × divaricarpa [ARABDIV] Brassicaceae (V:D)
Arabis divaricarpa var. *stenocarpa* M. Hopkins = Arabis × divaricarpa [ARABDIV] Brassicaceae (V:D)
Arabis divaricarpa var. *typica* Rollins = Arabis × divaricarpa [ARABDIV] Brassicaceae (V:D)
Arabis drummondii Gray [ARABDRU] Brassicaceae (V:D)
Arabis drummondii var. *connexa* (Greene) Fern. = Arabis drummondii [ARABDRU] Brassicaceae (V:D)
Arabis drummondii var. *oxyphylla* (Greene) M. Hopkins = Arabis drummondii [ARABDRU] Brassicaceae (V:D)
Arabis exilis A. Nels. = Arabis holboellii var. retrofracta [ARABHOL5] Brassicaceae (V:D)
Arabis glabra (L.) Bernh. [ARABGLA] Brassicaceae (V:D)
Arabis hirsuta (L.) Scop. [ARABHIR] Brassicaceae (V:D)
Arabis hirsuta ssp. *eschscholtziana* (Andrz.) Hultén = Arabis hirsuta var. eschscholtziana [ARABHIR5] Brassicaceae (V:D)
Arabis hirsuta ssp. *pycnocarpa* (M. Hopkins) Hultén = Arabis hirsuta var. pycnocarpa [ARABHIR7] Brassicaceae (V:D)
Arabis hirsuta var. **eschscholtziana** (Andrz.) Rollins [ARABHIR5] Brassicaceae (V:D)
Arabis hirsuta var. **glabrata** Torr. & Gray [ARABHIR6] Brassicaceae (V:D)
Arabis hirsuta var. **pycnocarpa** (M. Hopkins) Rollins [ARABHIR7] Brassicaceae (V:D)
Arabis holboellii Hornem. [ARABHOL] Brassicaceae (V:D)
Arabis holboellii var. **collinsii** (Fern.) Rollins [ARABHOL1] Brassicaceae (V:D)
Arabis holboellii var. **holboellii** [ARABHOL0] Brassicaceae (V:D)
Arabis holboellii var. **pendulocarpa** (A. Nels.) Rollins [ARABHOL3] Brassicaceae (V:D)
Arabis holboellii var. **pinetorum** (Tidestrom) Rollins [ARABHOL4] Brassicaceae (V:D)
Arabis holboellii var. **retrofracta** (Graham) Rydb. [ARABHOL5] Brassicaceae (V:D)
Arabis holboellii var. *secunda* (T.J. Howell) Jepson = Arabis holboellii var. retrofracta [ARABHOL5] Brassicaceae (V:D)
Arabis holboellii var. *tenuis* Böcher = Arabis holboellii var. retrofracta [ARABHOL5] Brassicaceae (V:D)
Arabis hookeri Lange = Halimolobos mollis [HALIMOI] Brassicaceae (V:D)
Arabis interposita Greene = Arabis × divaricarpa [ARABDIV] Brassicaceae (V:D)
Arabis kochii Blank. = Arabis holboellii var. retrofracta [ARABHOL5] Brassicaceae (V:D)
Arabis lemmonii S. Wats. [ARABLEM] Brassicaceae (V:D)
Arabis lignipes A. Nels. = Arabis holboellii var. retrofracta [ARABHOL5] Brassicaceae (V:D)
Arabis lyallii S. Wats. [ARABLYA] Brassicaceae (V:D)
Arabis lyrata L. [ARABLYR] Brassicaceae (V:D)
Arabis lyrata ssp. *kamchatica* (Fisch. ex DC.) Hultén = Arabis lyrata var. kamchatica [ARABLYR2] Brassicaceae (V:D)
Arabis lyrata var. **kamchatica** Fisch. ex DC. [ARABLYR2] Brassicaceae (V:D)
Arabis macella Piper = Arabis nuttallii [ARABNUT] Brassicaceae (V:D)
Arabis mcdougalii Rydb. = Arabis holboellii var. retrofracta [ARABHOL5] Brassicaceae (V:D)
Arabis microphylla Nutt. [ARABMIC] Brassicaceae (V:D)
Arabis nuttallii B.L. Robins. [ARABNUT] Brassicaceae (V:D)
Arabis oxyphylla Greene = Arabis drummondii [ARABDRU] Brassicaceae (V:D)
Arabis pendulocarpa A. Nels. = Arabis holboellii var. pendulocarpa [ARABHOL3] Brassicaceae (V:D)
Arabis pinetorum Tidestrom = Arabis holboellii var. pinetorum [ARABHOL4] Brassicaceae (V:D)
Arabis polyantha Greene = Arabis holboellii var. retrofracta [ARABHOL5] Brassicaceae (V:D)
Arabis pycnocarpa M. Hopkins = Arabis hirsuta var. pycnocarpa [ARABHIR7] Brassicaceae (V:D)
Arabis retrofracta Graham = Arabis holboellii var. retrofracta [ARABHOL5] Brassicaceae (V:D)

Arabis retrofracta var. *collinsii* (Fern.) Boivin = Arabis holboellii var. collinsii [ARABHOL1] Brassicaceae (V:D)
Arabis retrofracta var. *multicaulis* Boivin = Arabis holboellii var. retrofracta [ARABHOL5] Brassicaceae (V:D)
Arabis rhodanthus Greene = Arabis holboellii var. retrofracta [ARABHOL5] Brassicaceae (V:D)
Arabis secunda T.J. Howell = Arabis holboellii var. retrofracta [ARABHOL5] Brassicaceae (V:D)
Arabis sparsiflora Nutt. [ARABSPA] Brassicaceae (V:D)
Arabis tenuicula Greene = Arabis microphylla [ARABMIC] Brassicaceae (V:D)
Arabis tenuis Greene = Arabis holboellii var. retrofracta [ARABHOL5] Brassicaceae (V:D)
Arabis thaliana L. = Arabidopsis thaliana [ARABTHA] Brassicaceae (V:D)
Arabis whitedii Piper = Halimolobos whitedii [HALIWHI] Brassicaceae (V:D)
Aragallus splendens (Dougl. ex Hook.) Greene = Oxytropis splendens [OXYTSPL] Fabaceae (V:D)
Aragallus viscidulus Rydb. = Oxytropis viscida [OXYTVIS] Fabaceae (V:D)
Aralia L. [ARALIA$] Araliaceae (V:D)
Aralia nudicaulis L. [ARALNUD] Araliaceae (V:D)
Arbuscula leibergii (Britt.) Crum, Steere & Anderson = Thamnobryum neckeroides [THAMNEC] Thamnobryaceae (B:M)
Arbutus L. [ARBUTUS$] Ericaceae (V:D)
Arbutus alpina L. = Arctostaphylos alpina [ARCTALP] Ericaceae (V:D)
Arbutus menziesii Pursh [ARBUMEN] Ericaceae (V:D)
Arceuthobium Bieb. [ARCEUTH$] Viscaceae (V:D)
Arceuthobium americanum Nutt. ex Engelm. [ARCEAME] Viscaceae (V:D)
Arceuthobium campylopodum Engelm. [ARCECAM] Viscaceae (V:D)
Arceuthobium douglasii Engelm. [ARCEDOU] Viscaceae (V:D)
Arctagrostis Griseb. [ARCTAGR$] Poaceae (V:M)
Arctagrostis angustifolia Nash = Arctagrostis latifolia ssp. arundinacea [ARCTLAT1] Poaceae (V:M)
Arctagrostis angustifolia var. *crassispica* Bowden = Arctagrostis latifolia ssp. arundinacea [ARCTLAT1] Poaceae (V:M)
Arctagrostis arundinacea (Trin.) Beal = Arctagrostis latifolia ssp. arundinacea [ARCTLAT1] Poaceae (V:M)
Arctagrostis latifolia (R. Br.) Griseb. [ARCTLAT] Poaceae (V:M)
Arctagrostis latifolia ssp. **arundinacea** (Trin.) Tzvelev [ARCTLAT1] Poaceae (V:M)
Arctagrostis latifolia var. *angustifolia* (Nash) Hultén = Arctagrostis latifolia ssp. arundinacea [ARCTLAT1] Poaceae (V:M)
Arctagrostis latifolia var. *arundinacea* (Trin.) Griseb. = Arctagrostis latifolia ssp. arundinacea [ARCTLAT1] Poaceae (V:M)
Arctagrostis poaeoides Nash = Arctagrostis latifolia ssp. arundinacea [ARCTLAT1] Poaceae (V:M)
Arctium L. [ARCTIUM$] Asteraceae (V:D)
*__Arctium lappa__ L. [ARCTLAP] Asteraceae (V:D)
*__Arctium minus__ Bernh. [ARCTMIN] Asteraceae (V:D)
Arctoa Bruch & Schimp. in B.S.G. [ARCTOA$] Dicranaceae (B:M)
Arctoa blyttii (Schimp.) Loeske = Kiaeria blyttii [KIAEBLY] Dicranaceae (B:M)
Arctoa falcata (Hedw.) Loeske = Kiaeria falcata [KIAEFAL] Dicranaceae (B:M)
Arctoa fulvella (Dicks.) Bruch & Schimp. in B.S.G. [ARCTFUV] Dicranaceae (B:M)
Arctoa starkei (Web. & Mohr) Loeske = Kiaeria starkei [KIAESTA] Dicranaceae (B:M)
Arctomia Th. Fr. [ARCTOMI$] Arctomiaceae (L)
Arctomia delicatula Th. Fr. [ARCTDEL] Arctomiaceae (L)
Arctoparmelia Hale [ARCTOPA$] Parmeliaceae (L)
Arctoparmelia aleuritica (Nyl.) Hale = Arctoparmelia centrifuga [ARCTCEN] Parmeliaceae (L)
Arctoparmelia centrifuga (L.) Hale [ARCTCEN] Parmeliaceae (L)
Arctoparmelia incurva (Pers.) Hale [ARCTINC] Parmeliaceae (L)
Arctoparmelia separata (Th. Fr.) Hale [ARCTSEP] Parmeliaceae (L)
Arctoparmelia subcentrifuga (Oksner) Hale [ARCTSUB] Parmeliaceae (L)
Arctophila Rupr. ex Anderss. [ARCTOPH$] Poaceae (V:M)
Arctophila fulva (Trin.) Rupr. ex Anderss. [ARCTFUL] Poaceae (V:M)
Arctopoa eminens (J. Presl) Probat. = Poa eminens [POA EMI] Poaceae (V:M)
Arctostaphylos Adans. [ARCTOST$] Ericaceae (V:D)
Arctostaphylos adenotricha (Fern. & J.F. Macbr.) A. & D. Löve & Kapoor = Arctostaphylos uva-ursi [ARCTUVA] Ericaceae (V:D)
Arctostaphylos alpina (L.) Spreng. [ARCTALP] Ericaceae (V:D)
Arctostaphylos alpina ssp. *rubra* (Rehd. & Wilson) Hultén = Arctostaphylos rubra [ARCTRUB] Ericaceae (V:D)
Arctostaphylos alpina var. *rubra* (Rehd. & Wilson) Bean = Arctostaphylos rubra [ARCTRUB] Ericaceae (V:D)
Arctostaphylos columbiana Piper [ARCTCOL] Ericaceae (V:D)
Arctostaphylos rubra (Rehd. & Wilson) Fern. [ARCTRUB] Ericaceae (V:D)
Arctostaphylos uva-ursi (L.) Spreng. [ARCTUVA] Ericaceae (V:D)
Arctostaphylos uva-ursi ssp. *adenotricha* (Fern. & J.F. Macbr.) Calder & Taylor = Arctostaphylos uva-ursi [ARCTUVA] Ericaceae (V:D)
Arctostaphylos uva-ursi ssp. *coactilis* (Fern. & J.F. Macbr.) A. & D. Löve & Kapoor = Arctostaphylos uva-ursi [ARCTUVA] Ericaceae (V:D)
Arctostaphylos uva-ursi ssp. *longipilosa* Packer & Denford = Arctostaphylos uva-ursi [ARCTUVA] Ericaceae (V:D)
Arctostaphylos uva-ursi ssp. *monoensis* J.B. Roof = Arctostaphylos uva-ursi [ARCTUVA] Ericaceae (V:D)
Arctostaphylos uva-ursi ssp. *stipitata* Packer & Denford = Arctostaphylos uva-ursi [ARCTUVA] Ericaceae (V:D)
Arctostaphylos uva-ursi var. *adenotricha* Fern. & J.F. Macbr. = Arctostaphylos uva-ursi [ARCTUVA] Ericaceae (V:D)
Arctostaphylos uva-ursi var. *coactilis* Fern. & J.F. Macbr. = Arctostaphylos uva-ursi [ARCTUVA] Ericaceae (V:D)
Arctostaphylos uva-ursi var. *leobreweri* J.B. Roof = Arctostaphylos uva-ursi [ARCTUVA] Ericaceae (V:D)
Arctostaphylos uva-ursi var. *marinensis* J.B. Roof = Arctostaphylos uva-ursi [ARCTUVA] Ericaceae (V:D)
Arctostaphylos uva-ursi var. *pacifica* Hultén = Arctostaphylos uva-ursi [ARCTUVA] Ericaceae (V:D)
Arctostaphylos uva-ursi var. *stipitata* (Packer & Denford) Dorn = Arctostaphylos uva-ursi [ARCTUVA] Ericaceae (V:D)
Arctostaphylos uva-ursi var. *suborbiculata* W. Knight = Arctostaphylos uva-ursi [ARCTUVA] Ericaceae (V:D)
Arctous alpina (L.) Niedenzu = Arctostaphylos alpina [ARCTALP] Ericaceae (V:D)
Arctous erythrocarpa Small = Arctostaphylos rubra [ARCTRUB] Ericaceae (V:D)
Arctous rubra (Rehd. & Wilson) Nakai = Arctostaphylos rubra [ARCTRUB] Ericaceae (V:D)
Arenaria L. [ARENARI$] Caryophyllaceae (V:D)
Arenaria capillaris Poir. [ARENCAP] Caryophyllaceae (V:D)
Arenaria capillaris ssp. **americana** Maguire [ARENCAP1] Caryophyllaceae (V:D)
Arenaria capillaris var. *americana* (Maguire) R.J. Davis = Arenaria capillaris ssp. americana [ARENCAP1] Caryophyllaceae (V:D)
Arenaria dawsonensis Britt. = Minuartia dawsonensis [MINUDAW] Caryophyllaceae (V:D)
Arenaria elegans Cham. & Schlecht. = Minuartia elegans [MINUELE] Caryophyllaceae (V:D)
Arenaria elegans var. *columbiana* Raup = Minuartia elegans [MINUELE] Caryophyllaceae (V:D)
Arenaria formosa (Fisch.) Regel = Arenaria capillaris ssp. americana [ARENCAP1] Caryophyllaceae (V:D)
Arenaria lateriflora L. = Moehringia lateriflora [MOEHLAT] Caryophyllaceae (V:D)
Arenaria lateriflora var. *angustifolia* (Regel) St. John = Moehringia lateriflora [MOEHLAT] Caryophyllaceae (V:D)
Arenaria lateriflora var. *tayloriae* St. John = Moehringia lateriflora [MOEHLAT] Caryophyllaceae (V:D)
Arenaria litorea Fern. = Minuartia dawsonensis [MINUDAW] Caryophyllaceae (V:D)
Arenaria longipedunculata Hultén [ARENLON] Caryophyllaceae (V:D)

Arenaria macrophylla Hook. = Moehringia macrophylla [MOEHMAC] Caryophyllaceae (V:D)
Arenaria nuttallii Pax = Minuartia nuttallii [MINUNUT] Caryophyllaceae (V:D)
Arenaria obtusiloba (Rydb.) Fern. = Minuartia obtusiloba [MINUOBT] Caryophyllaceae (V:D)
Arenaria peploides var. *major* Hook. = Honkenya peploides ssp. major [HONKPEP1] Caryophyllaceae (V:D)
Arenaria peploides var. *maxima* Fern. = Honkenya peploides ssp. major [HONKPEP1] Caryophyllaceae (V:D)
Arenaria propinqua Richards. = Minuartia rubella [MINURUB] Caryophyllaceae (V:D)
Arenaria pungens Nutt. = Minuartia nuttallii [MINUNUT] Caryophyllaceae (V:D)
Arenaria pusilla S. Wats. = Minuartia pusilla [MINUPUS] Caryophyllaceae (V:D)
Arenaria rossii ssp. *columbiana* (Raup) Maguire = Minuartia elegans [MINUELE] Caryophyllaceae (V:D)
Arenaria rossii ssp. *elegans* (Cham. & Schlecht.) Maguire = Minuartia elegans [MINUELE] Caryophyllaceae (V:D)
Arenaria rossii var. *apetala* Maguire = Minuartia elegans [MINUELE] Caryophyllaceae (V:D)
Arenaria rossii var. *columbiana* (Raup) Maguire = Minuartia elegans [MINUELE] Caryophyllaceae (V:D)
Arenaria rossii var. *corollina* Fenzl = Minuartia elegans [MINUELE] Caryophyllaceae (V:D)
Arenaria rossii var. *elegans* (Cham. & Schlecht.) Welsh = Minuartia elegans [MINUELE] Caryophyllaceae (V:D)
Arenaria rubella (Wahlenb.) Sm. = Minuartia rubella [MINURUB] Caryophyllaceae (V:D)
Arenaria sajanensis Willd. ex Schlecht. = Minuartia biflora [MINUBIF] Caryophyllaceae (V:D)
***Arenaria serpyllifolia** L. [ARENSER] Caryophyllaceae (V:D)
Arenaria stephaniana var. *americana* (Porter ex B.L. Robins.) Shinners = Stellaria americana [STELAME] Caryophyllaceae (V:D)
Arenaria stricta ssp. *dawsonensis* (Britt.) Maguire = Minuartia dawsonensis [MINUDAW] Caryophyllaceae (V:D)
Arenaria stricta var. *dawsonensis* (Britt.) Scoggan = Minuartia dawsonensis [MINUDAW] Caryophyllaceae (V:D)
Arenaria stricta var. *litorea* (Fern.) Britt. = Minuartia dawsonensis [MINUDAW] Caryophyllaceae (V:D)
Arenaria tenella Nutt. = Minuartia tenella [MINUTEN] Caryophyllaceae (V:D)
Arenaria verna var. *propinqua* (Richards.) Fern. = Minuartia rubella [MINURUB] Caryophyllaceae (V:D)
Arenaria verna var. *rubella* (Wahlenb.) S. Wats. = Minuartia rubella [MINURUB] Caryophyllaceae (V:D)
Argentina Hill [ARGENTI$] Rosaceae (V:D)
Argentina anserina (L.) Rydb. [ARGEANS] Rosaceae (V:D)
Argentina anserina var. *concolor* Rydb. = Argentina anserina [ARGEANS] Rosaceae (V:D)
Argentina argentea (L.) Rydb. = Argentina anserina [ARGEANS] Rosaceae (V:D)
Argentina egedii (Wormsk.) Rydb. [ARGEEGE] Rosaceae (V:D)
Aristida L. [ARISTID$] Poaceae (V:M)
Aristida longiseta Steud. = Aristida purpurea var. longiseta [ARISPUR1] Poaceae (V:M)
Aristida longiseta var. *rariflora* A.S. Hitchc. = Aristida purpurea var. longiseta [ARISPUR1] Poaceae (V:M)
Aristida longiseta var. *robusta* Merr. = Aristida purpurea var. longiseta [ARISPUR1] Poaceae (V:M)
Aristida oligantha Michx. [ARISOLI] Poaceae (V:M)
Aristida purpurea Nutt. [ARISPUR] Poaceae (V:M)
Aristida purpurea var. **longiseta** (Steud.) Vasey [ARISPUR1] Poaceae (V:M)
Aristida purpurea var. *robusta* (Merr.) Piper = Aristida purpurea var. longiseta [ARISPUR1] Poaceae (V:M)
Armeria (DC.) Willd. [ARMERIA$] Plumbaginaceae (V:D)
Armeria arctica (Cham.) Wallr. = Armeria maritima ssp. purpurea [ARMEMAR3] Plumbaginaceae (V:D)
Armeria arctica ssp. *californica* (Boiss.) Abrams = Armeria maritima ssp. californica [ARMEMAR2] Plumbaginaceae (V:D)
Armeria californica Boiss. = Armeria maritima ssp. californica [ARMEMAR2] Plumbaginaceae (V:D)
Armeria maritima (P. Mill.) Willd. [ARMEMAR] Plumbaginaceae (V:D)
Armeria maritima ssp. *arctica* (Cham.) Hultén = Armeria maritima ssp. purpurea [ARMEMAR3] Plumbaginaceae (V:D)
Armeria maritima ssp. **californica** (Boiss.) Porsild [ARMEMAR2] Plumbaginaceae (V:D)
Armeria maritima ssp. **purpurea** (W.D.J. Koch) A. & D. Löve [ARMEMAR3] Plumbaginaceae (V:D)
Armeria maritima var. *californica* (Boiss.) G.H.M. Lawrence = Armeria maritima ssp. californica [ARMEMAR2] Plumbaginaceae (V:D)
Armeria maritima var. *purpurea* (W.D.J. Koch) G.H.M. Lawrence = Armeria maritima ssp. purpurea [ARMEMAR3] Plumbaginaceae (V:D)
Armoracia P.G. Gaertn., B. Mey. & Scherb. [ARMORAC$] Brassicaceae (V:D)
Armoracia armoracia (L.) Britt. = Armoracia rusticana [ARMORUS] Brassicaceae (V:D)
Armoracia lapathifolia Gilib. = Armoracia rusticana [ARMORUS] Brassicaceae (V:D)
***Armoracia rusticana** P.G. Gaertn., B. Mey. & Scherb. [ARMORUS] Brassicaceae (V:D)
Arnellia Lindb. [ARNELLI$] Arnelliaceae (B:H)
Arnellia fennica (Gott.) Lindb. [ARNEFEN] Arnelliaceae (B:H)
Arnica L. [ARNICA$] Asteraceae (V:D)
Arnica alpina ssp. *angustifolia* (Vahl) Maguire = Arnica angustifolia ssp. angustifolia [ARNIANG0] Asteraceae (V:D)
Arnica alpina ssp. *attenuata* (Greene) Maguire = Arnica angustifolia ssp. angustifolia [ARNIANG0] Asteraceae (V:D)
Arnica alpina ssp. *lonchophylla* (Greene) Taylor & MacBryde = Arnica lonchophylla [ARNILOG] Asteraceae (V:D)
Arnica alpina ssp. *sornborgeri* (Fern.) Maguire = Arnica angustifolia ssp. angustifolia [ARNIANG0] Asteraceae (V:D)
Arnica alpina ssp. *tomentosa* (Macoun) Maguire = Arnica angustifolia ssp. tomentosa [ARNIANG4] Asteraceae (V:D)
Arnica alpina var. *angustifolia* (Vahl) Fern. = Arnica angustifolia ssp. angustifolia [ARNIANG0] Asteraceae (V:D)
Arnica alpina var. *attenuata* (Greene) Ediger & Barkl. = Arnica angustifolia ssp. angustifolia [ARNIANG0] Asteraceae (V:D)
Arnica alpina var. *linearis* Hultén = Arnica angustifolia ssp. angustifolia [ARNIANG0] Asteraceae (V:D)
Arnica alpina var. *lonchophylla* (Greene) Welsh = Arnica lonchophylla [ARNILOG] Asteraceae (V:D)
Arnica alpina var. *plantaginea* (Pursh) Ediger & Barkl. = Arnica angustifolia ssp. angustifolia [ARNIANG0] Asteraceae (V:D)
Arnica alpina var. *tomentosa* (Macoun) Cronq. = Arnica angustifolia ssp. tomentosa [ARNIANG4] Asteraceae (V:D)
Arnica alpina var. *ungavensis* Boivin = Arnica angustifolia ssp. angustifolia [ARNIANG0] Asteraceae (V:D)
Arnica alpina var. *vestita* Hultén = Arnica angustifolia ssp. angustifolia [ARNIANG0] Asteraceae (V:D)
Arnica amplexicaulis Nutt. [ARNIAMP] Asteraceae (V:D)
Arnica amplexicaulis ssp. *genuina* Maguire = Arnica amplexicaulis [ARNIAMP] Asteraceae (V:D)
Arnica amplexicaulis ssp. *prima* (Maguire) Maguire = Arnica amplexicaulis [ARNIAMP] Asteraceae (V:D)
Arnica amplexicaulis var. *piperi* St. John & Warren = Arnica amplexicaulis [ARNIAMP] Asteraceae (V:D)
Arnica amplexicaulis var. *prima* (Maguire) Boivin = Arnica amplexicaulis [ARNIAMP] Asteraceae (V:D)
Arnica amplexifolia Rydb. = Arnica amplexicaulis [ARNIAMP] Asteraceae (V:D)
Arnica angustifolia Vahl [ARNIANG] Asteraceae (V:D)
Arnica angustifolia ssp. **angustifolia** [ARNIANG0] Asteraceae (V:D)
Arnica angustifolia ssp. *attenuata* (Greene) G.W. Douglas = Arnica angustifolia ssp. angustifolia [ARNIANG0] Asteraceae (V:D)
Arnica angustifolia ssp. *lonchophylla* (Greene) G.W. & G.R. Dougl. = Arnica lonchophylla [ARNILOG] Asteraceae (V:D)
Arnica angustifolia ssp. **tomentosa** (Macoun) G.W. & G.R. Dougl. [ARNIANG4] Asteraceae (V:D)

Arnica cascadensis St. John = Arnica rydbergii [ARNIRYD] Asteraceae (V:D)
Arnica chamissonis Less. [ARNICHA] Asteraceae (V:D)
Arnica chamissonis ssp. **chamissonis** [ARNICHA0] Asteraceae (V:D)
Arnica chamissonis ssp. **foliosa** (Nutt.) Maguire [ARNICHA2] Asteraceae (V:D)
Arnica chamissonis ssp. **foliosa** var. **incana** (Gray) Hultén [ARNICHA1] Asteraceae (V:D)
Arnica chamissonis ssp. *incana* (Gray) Maguire = Arnica chamissonis ssp. foliosa var. incana [ARNICHA1] Asteraceae (V:D)
Arnica chionopappa Fern. = Arnica lonchophylla [ARNILOG] Asteraceae (V:D)
Arnica cordifolia Hook. [ARNICOR] Asteraceae (V:D)
Arnica cordifolia ssp. *genuina* Maguire = Arnica cordifolia [ARNICOR] Asteraceae (V:D)
Arnica cordifolia var. *humilis* (Rydb.) Maguire = Arnica cordifolia [ARNICOR] Asteraceae (V:D)
Arnica cordifolia var. *pumila* (Rydb.) Maguire = Arnica cordifolia [ARNICOR] Asteraceae (V:D)
Arnica cordifolia var. *whitneyi* (Fern.) Maguire = Arnica cordifolia [ARNICOR] Asteraceae (V:D)
Arnica × diversifolia Greene (pro sp.) [ARNIDIV] Asteraceae (V:D)
Arnica diversifolia Greene = Arnica × diversifolia [ARNIDIV] Asteraceae (V:D)
Arnica foliosa var. *incana* Gray = Arnica chamissonis ssp. foliosa var. incana [ARNICHA1] Asteraceae (V:D)
Arnica frigida C.A. Mey. ex Iljin [ARNIFRI] Asteraceae (V:D)
Arnica fulgens Pursh [ARNIFUL] Asteraceae (V:D)
Arnica fulgens var. *sororia* (Greene) G.W. & G.R. Dougl. = Arnica sororia [ARNISOR] Asteraceae (V:D)
Arnica gaspensis Fern. = Arnica lonchophylla [ARNILOG] Asteraceae (V:D)
Arnica gracilis Rydb. [ARNIGRA] Asteraceae (V:D)
Arnica hardinae St. John = Arnica cordifolia [ARNICOR] Asteraceae (V:D)
Arnica humilis Rydb. = Arnica cordifolia [ARNICOR] Asteraceae (V:D)
Arnica latifolia Bong. [ARNILAT] Asteraceae (V:D)
Arnica latifolia var. *angustifolia* Herder = Arnica latifolia [ARNILAT] Asteraceae (V:D)
Arnica latifolia var. *gracilis* (Rydb.) Cronq. = Arnica gracilis [ARNIGRA] Asteraceae (V:D)
Arnica lessingii (Torr. & Gray) Greene [ARNILES] Asteraceae (V:D)
Arnica lessingii ssp. *norbergii* Hultén & Maguire = Arnica lessingii [ARNILES] Asteraceae (V:D)
Arnica lonchophylla Greene [ARNILOG] Asteraceae (V:D)
Arnica lonchophylla ssp. *chionopappa* (Fern.) Maguire = Arnica lonchophylla [ARNILOG] Asteraceae (V:D)
Arnica louiseana ssp. *frigida* (C.A. Mey. ex Iljin) Maguire = Arnica frigida [ARNIFRI] Asteraceae (V:D)
Arnica mollis Hook. [ARNIMOL] Asteraceae (V:D)
Arnica mollis var. *aspera* (Greene) Boivin = Arnica amplexicaulis [ARNIAMP] Asteraceae (V:D)
Arnica mollis var. *silvatica* (Greene) Maguire = Arnica mollis [ARNIMOL] Asteraceae (V:D)
Arnica paniculata A. Nels. = Arnica cordifolia [ARNICOR] Asteraceae (V:D)
Arnica parryi Gray [ARNIPAR] Asteraceae (V:D)
Arnica plantaginea Pursh = Arnica angustifolia ssp. angustifolia [ARNIANG0] Asteraceae (V:D)
Arnica rydbergii Greene [ARNIRYD] Asteraceae (V:D)
Arnica sornborgeri Fern. = Arnica angustifolia ssp. angustifolia [ARNIANG0] Asteraceae (V:D)
Arnica sororia Greene [ARNISOR] Asteraceae (V:D)
Arnica terrae-novae Fern. = Arnica angustifolia ssp. angustifolia [ARNIANG0] Asteraceae (V:D)
Arnica tomentosa Macoun = Arnica angustifolia ssp. tomentosa [ARNIANG4] Asteraceae (V:D)
Arnica whitneyi Fern. = Arnica cordifolia [ARNICOR] Asteraceae (V:D)
Arrhenatherum Beauv. [ARRHENA$] Poaceae (V:M)
***Arrhenatherum elatius** (L.) J.& K. Presl [ARRHELA] Poaceae (V:M)
Artemisia L. [ARTEMIS$] Asteraceae (V:D)
***Artemisia absinthium** L. [ARTEABS] Asteraceae (V:D)
Artemisia alaskana Rydb. [ARTEALA] Asteraceae (V:D)
Artemisia angustifolia (Gray) Rydb. = Artemisia tridentata ssp. tridentata [ARTETRI0] Asteraceae (V:D)
Artemisia arctica Less. [ARTEARC] Asteraceae (V:D)
***Artemisia biennis** Willd. [ARTEBIE] Asteraceae (V:D)
Artemisia borealis Pallas = Artemisia campestris ssp. borealis [ARTECAM1] Asteraceae (V:D)
Artemisia campestris L. [ARTECAM] Asteraceae (V:D)
Artemisia campestris ssp. **borealis** (Pallas) Hall & Clements [ARTECAM1] Asteraceae (V:D)
Artemisia campestris ssp. **pacifica** (Nutt.) Hall & Clements [ARTECAM2] Asteraceae (V:D)
Artemisia campestris var. *douglasiana* (Bess.) Boivin = Artemisia campestris ssp. pacifica [ARTECAM2] Asteraceae (V:D)
Artemisia campestris var. *pacifica* (Nutt.) M.E. Peck = Artemisia campestris ssp. pacifica [ARTECAM2] Asteraceae (V:D)
Artemisia campestris var. *scouleriana* (Bess.) Cronq. = Artemisia campestris ssp. pacifica [ARTECAM2] Asteraceae (V:D)
Artemisia campestris var. *strutzia* Welsh = Artemisia campestris ssp. pacifica [ARTECAM2] Asteraceae (V:D)
Artemisia camporum Rydb. = Artemisia campestris ssp. pacifica [ARTECAM2] Asteraceae (V:D)
Artemisia cana Pursh [ARTECAN] Asteraceae (V:D)
Artemisia candicans Rydb. = Artemisia ludoviciana ssp. candicans [ARTELUD1] Asteraceae (V:D)
Artemisia diversifolia Rydb. = Artemisia ludoviciana ssp. ludoviciana [ARTELUD0] Asteraceae (V:D)
Artemisia dracunculus L. [ARTEDRA] Asteraceae (V:D)
Artemisia elatior (Torr. & Gray) Rydb. = Artemisia tilesii ssp. elatior [ARTETIL1] Asteraceae (V:D)
Artemisia frigida Willd. [ARTEFRI] Asteraceae (V:D)
Artemisia furcata Bieb. [ARTEFUR] Asteraceae (V:D)
Artemisia furcata var. **heterophylla** (Bess.) Hultén [ARTEFUR1] Asteraceae (V:D)
Artemisia gnaphalodes Nutt. = Artemisia ludoviciana ssp. ludoviciana [ARTELUD0] Asteraceae (V:D)
Artemisia herriotii Rydb. = Artemisia ludoviciana ssp. ludoviciana [ARTELUD0] Asteraceae (V:D)
Artemisia heterophylla Bess. = Artemisia furcata var. heterophylla [ARTEFUR1] Asteraceae (V:D)
Artemisia incompta Nutt. = Artemisia ludoviciana ssp. incompta [ARTELUD2] Asteraceae (V:D)
Artemisia lindleyana Bess. [ARTELIN] Asteraceae (V:D)
Artemisia longifolia Nutt. [ARTELON] Asteraceae (V:D)
Artemisia ludoviciana Nutt. [ARTELUD] Asteraceae (V:D)
Artemisia ludoviciana ssp. **candicans** (Rydb.) Keck [ARTELUD1] Asteraceae (V:D)
Artemisia ludoviciana ssp. **incompta** (Nutt.) Keck [ARTELUD2] Asteraceae (V:D)
Artemisia ludoviciana ssp. **ludoviciana** [ARTELUD0] Asteraceae (V:D)
Artemisia ludoviciana ssp. *typica* Keck = Artemisia ludoviciana ssp. ludoviciana [ARTELUD0] Asteraceae (V:D)
Artemisia ludoviciana var. *americana* (Bess.) Fern. = Artemisia ludoviciana ssp. ludoviciana [ARTELUD0] Asteraceae (V:D)
Artemisia ludoviciana var. *brittonii* (Rydb.) Fern. = Artemisia ludoviciana ssp. ludoviciana [ARTELUD0] Asteraceae (V:D)
Artemisia ludoviciana var. *gnaphalodes* (Nutt.) Torr. & Gray = Artemisia ludoviciana ssp. ludoviciana [ARTELUD0] Asteraceae (V:D)
Artemisia ludoviciana var. *incompta* (Nutt.) Cronq. = Artemisia ludoviciana ssp. incompta [ARTELUD2] Asteraceae (V:D)
Artemisia ludoviciana var. *latifolia* (Bess.) Torr. & Gray = Artemisia ludoviciana ssp. ludoviciana [ARTELUD0] Asteraceae (V:D)
Artemisia ludoviciana var. *latiloba* Nutt. = Artemisia ludoviciana ssp. candicans [ARTELUD1] Asteraceae (V:D)
Artemisia ludoviciana var. *pabularis* (A. Nels.) Fern. = Artemisia ludoviciana ssp. ludoviciana [ARTELUD0] Asteraceae (V:D)
Artemisia michauxiana Bess. [ARTEMIC] Asteraceae (V:D)

Artemisia norvegica ssp. *saxatilis* (Bess.) Hall & Clements = Artemisia arctica [ARTEARC] Asteraceae (V:D)
Artemisia norvegica var. *piceetorum* Welsh & Goodrich = Artemisia arctica [ARTEARC] Asteraceae (V:D)
Artemisia norvegica var. *saxatilis* (Bess.) Jepson = Artemisia arctica [ARTEARC] Asteraceae (V:D)
Artemisia pabularis (A. Nels.) Rydb. = Artemisia ludoviciana ssp. ludoviciana [ARTELUD0] Asteraceae (V:D)
Artemisia pacifica Nutt. = Artemisia campestris ssp. pacifica [ARTECAM2] Asteraceae (V:D)
Artemisia prescottiana Bess. = Artemisia lindleyana [ARTELIN] Asteraceae (V:D)
Artemisia purshiana Bess. = Artemisia ludoviciana ssp. ludoviciana [ARTELUD0] Asteraceae (V:D)
Artemisia suksdorfii Piper [ARTESUK] Asteraceae (V:D)
Artemisia tilesii Ledeb. [ARTETIL] Asteraceae (V:D)
Artemisia tilesii ssp. **elatior** (Torr. & Gray) Hultén [ARTETIL1] Asteraceae (V:D)
Artemisia tilesii ssp. **tilesii** [ARTETIL0] Asteraceae (V:D)
Artemisia tilesii ssp. **unalaschcensis** (Bess.) Hultén [ARTETIL2] Asteraceae (V:D)
Artemisia tilesii var. *elatior* Torr. & Gray = Artemisia tilesii ssp. elatior [ARTETIL1] Asteraceae (V:D)
Artemisia tilesii var. *unalaschcensis* Bess. = Artemisia tilesii ssp. unalaschcensis [ARTETIL2] Asteraceae (V:D)
Artemisia tridentata Nutt. [ARTETRI] Asteraceae (V:D)
Artemisia tridentata ssp. *parishii* (Gray) Hall & Clements = Artemisia tridentata ssp. tridentata [ARTETRI0] Asteraceae (V:D)
Artemisia tridentata ssp. **tridentata** [ARTETRI0] Asteraceae (V:D)
Artemisia tridentata ssp. **vaseyana** (Rydb.) Beetle [ARTETRI2] Asteraceae (V:D)
Artemisia tridentata var. *angustifolia* Gray = Artemisia tridentata ssp. tridentata [ARTETRI0] Asteraceae (V:D)
Artemisia tridentata var. *parishii* (Gray) Jepson = Artemisia tridentata ssp. tridentata [ARTETRI0] Asteraceae (V:D)
Artemisia tridentata var. *pauciflora* Winward & Goodrich = Artemisia tridentata ssp. vaseyana [ARTETRI2] Asteraceae (V:D)
Artemisia tridentata var. *vaseyana* (Rydb.) Boivin = Artemisia tridentata ssp. vaseyana [ARTETRI2] Asteraceae (V:D)
Artemisia trifurcata Steph. ex Spreng. = Artemisia furcata [ARTEFUR] Asteraceae (V:D)
Artemisia tripartita Rydb. [ARTETRP] Asteraceae (V:D)
Artemisia tyrellii Rydb. = Artemisia alaskana [ARTEALA] Asteraceae (V:D)
Artemisia vaseyana Rydb. = Artemisia tridentata ssp. vaseyana [ARTETRI2] Asteraceae (V:D)
***Artemisia vulgaris** L. [ARTEVUL] Asteraceae (V:D)
Artemisia vulgaris ssp. *ludoviciana* (Nutt.) Hall & Clements = Artemisia ludoviciana ssp. ludoviciana [ARTELUD0] Asteraceae (V:D)
Artemisia vulgaris ssp. *michauxiana* (Bess.) St. John = Artemisia michauxiana [ARTEMIC] Asteraceae (V:D)
Artemisia vulgaris var. *ludoviciana* (Nutt.) Kuntze = Artemisia ludoviciana ssp. ludoviciana [ARTELUD0] Asteraceae (V:D)
Arthonia Ach. [ARTHONI$] Arthoniaceae (L)
Arthonia arthonioides (Ach.) A.L. Sm. [ARTHART] Arthoniaceae (L)
Arthonia aspersa Leighton = Arthonia arthonioides [ARTHART] Arthoniaceae (L)
Arthonia carneorufa Willey [ARTHCAN] Arthoniaceae (L)
Arthonia dispersa (Schrader) Nyl. [ARTHDIS] Arthoniaceae (L)
Arthonia epimela (Almq.) Lamb [ARTHEPM] Arthoniaceae (L)
Arthonia epiphyscia Nyl. [ARTHEPI] Arthoniaceae (L)
Arthonia fusca (A. Massal.) Hepp = Arthonia lapidicola [ARTHLAP] Arthoniaceae (L)
Arthonia glebosa Tuck. [ARTHGLE] Arthoniaceae (L)
Arthonia hypobela (Nyl.) Zahlbr. [ARTHHYP] Arthoniaceae (L)
Arthonia ilicina Taylor [ARTHILI] Arthoniaceae (L)
Arthonia intexta Almq. [ARTHINT] Arthoniaceae (L)
Arthonia lapidicola (Taylor) Branth & Rostrup [ARTHLAP] Arthoniaceae (L)
Arthonia macounii G. Merr. = Arthothelium macounii [ARTHMAC] Arthoniaceae (L)
Arthonia patellulata Nyl. [ARTHPAT] Arthoniaceae (L)
Arthonia phaeobaea (Norman) Norman [ARTHPHA] Arthoniaceae (L)
Arthonia polygramma Nyl. [ARTHPOL] Arthoniaceae (L)
Arthonia radiata (Pers.) Ach. [ARTHRAD] Arthoniaceae (L)
Arthonia vinosa Leighton [ARTHVIN] Arthoniaceae (L)
Arthonia willeyi Tuck. [ARTHWIL] Arthoniaceae (L)
Arthopyrenia A. Massal. [ARTHOPY$] Arthopyreniaceae (L)
Arthopyrenia alba (Schrader) Zahlbr. = Acrocordia gemmata [ACROGEM] Monoblastiaceae (L)
Arthopyrenia antecellens (Nyl.) Arnold [ARTHANT] Arthopyreniaceae (L)
Arthopyrenia biformis (Borrer) A. Massal. = Anisomeridium biforme [ANISBIF] Monoblastiaceae (L)
Arthopyrenia cerasi (Schrader) A. Massal. [ARTHCER] Arthopyreniaceae (L)
Arthopyrenia cinereopruinosa (Schaerer) A. Massal. [ARTHCIE] Arthopyreniaceae (L)
Arthopyrenia conformis auct. = Anisomeridium biforme [ANISBIF] Monoblastiaceae (L)
Arthopyrenia epidermidis (DC.) A. Massal. = Arthopyrenia punctiformis [ARTHPUC] Arthopyreniaceae (L)
Arthopyrenia faginea (Schaerer) Swinscow = Strigula stigmatella [STRISTI] Strigulaceae (L)
Arthopyrenia fallax (Nyl.) Arnold = Arthopyrenia lapponina [ARTHLAO] Arthopyreniaceae (L)
Arthopyrenia gemmata (Ach.) A. Massal. = Acrocordia gemmata [ACROGEM] Monoblastiaceae (L)
Arthopyrenia halodytes (Nyl.) Arnold = Pyrenocollema halodytes [PYREHAL] Pyrenulaceae (L)
Arthopyrenia lapponina Anzi [ARTHLAO] Arthopyreniaceae (L)
Arthopyrenia padi Rabenh. = Arthopyrenia punctiformis [ARTHPUC] Arthopyreniaceae (L)
Arthopyrenia parvula Zahlbr. = Anisomeridium biforme [ANISBIF] Monoblastiaceae (L)
Arthopyrenia pinicola (Hepp) A. Massal. = Arthopyrenia cinereopruinosa [ARTHCIE] Arthopyreniaceae (L)
Arthopyrenia plumbaria (Stizenb.) R.C. Harris [ARTHPLU] Arthopyreniaceae (L)
Arthopyrenia punctiformis (Pers.) A. Massal. [ARTHPUC] Arthopyreniaceae (L)
Arthopyrenia sphaeroides (Wallr.) Zahlbr. = Acrocordia gemmata [ACROGEM] Monoblastiaceae (L)
Arthothelium A. Massal. [ARTHOTH$] Arthoniaceae (L)
Arthothelium ilicinum (Taylor) P. James = Arthonia ilicina [ARTHILI] Arthoniaceae (L)
Arthothelium macounii (G. Merr.) W. Noble [ARTHMAC] Arthoniaceae (L)
Arthothelium spectabile A. Massal. [ARTHSPE] Arthoniaceae (L)
Arthothelium violascens (Nyl.) Zahlbr. [ARTHVIO] Arthoniaceae (L)
Arthrorhaphis Th. Fr. [ARTHROR$] Arthrorhaphidaceae (L)
Arthrorhaphis alpina (Schaerer) R. Sant. [ARTHALP] Arthrorhaphidaceae (L)
Arthrorhaphis citrinella (Ach.) Poelt [ARTHCIT] Arthrorhaphidaceae (L)
Aruncus L. [ARUNCUS$] Rosaceae (V:D)
Aruncus aruncus (L.) Karst. = Aruncus dioicus var. vulgaris [ARUNDIO1] Rosaceae (V:D)
Aruncus dioicus (Walt.) Fern. [ARUNDIO] Rosaceae (V:D)
Aruncus dioicus var. **vulgaris** (Maxim.) Hara [ARUNDIO1] Rosaceae (V:D)
Aruncus sylvester Kostel. ex Maxim. = Aruncus dioicus var. vulgaris [ARUNDIO1] Rosaceae (V:D)
Aruncus vulgaris (Maxim.) Raf. ex Hara = Aruncus dioicus var. vulgaris [ARUNDIO1] Rosaceae (V:D)
Arundo festucacea Willd. = Scolochloa festucacea [SCOLFES] Poaceae (V:M)
Asahinea Culb. & C. Culb. [ASAHINE$] Parmeliaceae (L)
Asahinea chrysantha (Tuck.) Culb. & C. Culb. [ASAHCHR] Parmeliaceae (L)
Asahinea scholanderi (Llano) Culb. & C. Culb. [ASAHSCH] Parmeliaceae (L)

Asarum L. [ASARUM$] Aristolochiaceae (V:D)
Asarum caudatum Lindl. [ASARCAU] Aristolochiaceae (V:D)
Asclepias L. [ASCLEPI$] Asclepiadaceae (V:D)
Asclepias ovalifolia Dcne. [ASCLOVA] Asclepiadaceae (V:D)
Asclepias speciosa Torr. [ASCLSPE] Asclepiadaceae (V:D)
Askellia elegans (Hook.) W.A. Weber = Crepis elegans [CREPELE] Asteraceae (V:D)
Askellia nana ssp. *ramosa* (Babcock) W.A. Weber = Crepis nana ssp. ramosa [CREPNAN2] Asteraceae (V:D)
Asparagus L. [ASPARAG$] Liliaceae (V:M)
***Asparagus officinalis** L. [ASPAOFF] Liliaceae (V:M)
Asperugo L. [ASPERUG$] Boraginaceae (V:D)
***Asperugo procumbens** L. [ASPEPRO] Boraginaceae (V:D)
Asperula odorata L. = Galium odoratum [GALIODO] Rubiaceae (V:D)
Aspicilia A. Massal. [ASPICIL$] Hymeneliaceae (L)
Aspicilia alphoplaca (Wahlenb.) Poelt & Leuckert = Lobothallia alphoplaca [LOBOALP] Hymeneliaceae (L)
Aspicilia alpina (Sommerf.) Arnold = Bellemerea alpina [BELLALP] Porpidiaceae (L)
Aspicilia anseris (Lynge) J.W. Thomson [ASPIANS] Hymeneliaceae (L)
Aspicilia aquatica Körber [ASPIAQU] Hymeneliaceae (L)
Aspicilia caesiocinerea (Nyl. ex Malbr.) Arnold [ASPICAE] Hymeneliaceae (L)
Aspicilia caesiopruinosa (H. Magn.) J.W. Thomson [ASPICAS] Hymeneliaceae (L)
Aspicilia calcarea (L.) Mudd [ASPICAL] Hymeneliaceae (L)
Aspicilia candida (Anzi) Hue [ASPICAN] Hymeneliaceae (L)
Aspicilia cinerea (L.) Körber [ASPICIN] Hymeneliaceae (L)
Aspicilia cinereorufescens (Ach.) A. Massal. = Bellemerea cinereorufescens [BELLCIN] Porpidiaceae (L)
Aspicilia contorta (Hoffm.) Kremp. [ASPICON] Hymeneliaceae (L)
Aspicilia desertorum (Kremp.) Mereschk. [ASPIDES] Hymeneliaceae (L)
Aspicilia diamarta (Ach.) Boistel = Bellemerea diamarta [BELLDIA] Porpidiaceae (L)
Aspicilia disserpens (Zahlbr.) Räsänen [ASPIDIS] Hymeneliaceae (L)
Aspicilia fimbriata (H. Magn.) Clauzade & Rondon [ASPIFIM] Hymeneliaceae (L)
Aspicilia gibbosa (Ach.) Körber [ASPIGIB] Hymeneliaceae (L)
Aspicilia gibbosula (H. Magn.) Oksner = Aspicilia gibbosa [ASPIGIB] Hymeneliaceae (L)
Aspicilia karelica (H. Magn.) Oksner [ASPIKAR] Hymeneliaceae (L)
Aspicilia lacustris (With.) Th. Fr. = Ionaspis lacustris [IONALAC] Hymeneliaceae (L)
Aspicilia laxula (H. Magn.) Brodo [ASPILAX] Hymeneliaceae (L)
Aspicilia leprosescens (Sandst.) Hav. [ASPILEP] Hymeneliaceae (L)
Aspicilia melanaspis (Ach.) Poelt & Leuckert = Lobothallia melanaspis [LOBOMEL] Hymeneliaceae (L)
Aspicilia myrinii (Fr.) Stein [ASPIMYR] Hymeneliaceae (L)
Aspicilia pelobotrya (Wahlenb.) Th. Fr. = Amygdalaria pelobotryon [AMYGPEL] Porpidiaceae (L)
Aspicilia praeradiosa (Nyl.) Poelt & Leuckert = Lobothallia praeradiosa [LOBOPRA] Hymeneliaceae (L)
Aspicilia reptans (Looman) Wetmore [ASPIREP] Hymeneliaceae (L)
Aspicilia ryrkaipiae (H. Magn.) Oksner [ASPIRYR] Hymeneliaceae (L)
Aspicilia subplicigera (H. Magn.) Oksner [ASPISUB] Hymeneliaceae (L)
Aspicilia subsorediza (Lynge) R. Sant. = Bellemerea subsorediza [BELLSUB] Porpidiaceae (L)
Aspicilia supertegens Arnold [ASPISUP] Hymeneliaceae (L)
Aspicilia verrucigera Hue [ASPIVER] Hymeneliaceae (L)
Aspicilia verrucosa (Ach.) Körker = Megaspora verrucosa [MEGAVEU] Hymeneliaceae (L)
Aspidotis (Nutt. ex Hook.) Copeland [ASPIDOT$] Pteridaceae (V:F)
Aspidotis densa (Brack.) Lellinger [ASPIDEN] Pteridaceae (V:F)
Asplenium L. [ASPLENI$] Aspleniaceae (V:F)
Asplenium adulterinum Milde [ASPLADU] Aspleniaceae (V:F)
Asplenium melanocaulon Willd. = Asplenium trichomanes [ASPLTRI] Aspleniaceae (V:F)
Asplenium trichomanes L. [ASPLTRI] Aspleniaceae (V:F)
Asplenium trichomanes-ramosum L. [ASPLTRC] Aspleniaceae (V:F)
Asplenium viride Huds. = Asplenium trichomanes-ramosum [ASPLTRC] Aspleniaceae (V:F)
Aspris caryophyllea (L.) Nash = Aira caryophyllea [AIRACAR] Poaceae (V:M)
Aspris praecox (L.) Nash = Aira praecox [AIRAPRA] Poaceae (V:M)
Aster L. [ASTER$] Asteraceae (V:D)
Aster adscendens Lindl. = Aster ascendens [ASTEASC] Asteraceae (V:D)
Aster alpinus L. [ASTEALP] Asteraceae (V:D)
Aster alpinus ssp. *vierhapperi* Onno = Aster alpinus var. vierhapperi [ASTEALP1] Asteraceae (V:D)
Aster alpinus var. **vierhapperi** (Onno) Cronq. [ASTEALP1] Asteraceae (V:D)
Aster ascendens Lindl. [ASTEASC] Asteraceae (V:D)
Aster borealis (Torr. & Gray) Prov. [ASTEBOR] Asteraceae (V:D)
Aster brachyactis Blake = Brachyactis ciliata ssp. angusta [BRACCIL1] Asteraceae (V:D)
Aster × bracteolatus Nutt. (pro sp.) [ASTEBRA] Asteraceae (V:D)
Aster bracteolatus Nutt. = Aster × bracteolatus [ASTEBRA] Asteraceae (V:D)
Aster campestris Nutt. [ASTECAM] Asteraceae (V:D)
Aster canescens Pursh = Machaeranthera canescens [MACHCAN] Asteraceae (V:D)
Aster chilensis ssp. *adscendens* (Lindl.) Cronq. = Aster ascendens [ASTEASC] Asteraceae (V:D)
Aster ciliolatus Lindl. [ASTECIL] Asteraceae (V:D)
Aster conspicuus Lindl. [ASTECON] Asteraceae (V:D)
Aster curtus Cronq. [ASTECUR] Asteraceae (V:D)
Aster douglasii Lindl. = Aster subspicatus [ASTESUB] Asteraceae (V:D)
Aster eatonii (Gray) T.J. Howell [ASTEEAT] Asteraceae (V:D)
Aster elegans var. *engelmannii* D.C. Eat. = Aster engelmannii [ASTEENG] Asteraceae (V:D)
Aster elegantulus Porsild = Aster falcatus [ASTEFAL] Asteraceae (V:D)
Aster eliasii A. Nels. = Aster radulinus [ASTERAD] Asteraceae (V:D)
Aster engelmannii (D.C. Eat.) Gray [ASTEENG] Asteraceae (V:D)
Aster engelmannii var. *paucicapitatus* B.L. Robins. = Aster paucicapitatus [ASTEPAU] Asteraceae (V:D)
Aster ericoides L. [ASTEERI] Asteraceae (V:D)
Aster ericoides ssp. *pansus* (Blake) A.G. Jones = Aster ericoides var. pansus [ASTEERI1] Asteraceae (V:D)
Aster ericoides var. *commutatus* (Torr. & Gray) Boivin p.p. = Aster falcatus [ASTEFAL] Asteraceae (V:D)
Aster ericoides var. **pansus** (Blake) Boivin [ASTEERI1] Asteraceae (V:D)
Aster exscapus Richards. = Townsendia exscapa [TOWNEXS] Asteraceae (V:D)
Aster falcatus Lindl. [ASTEFAL] Asteraceae (V:D)
Aster foliaceus Lindl. ex DC. [ASTEFOL] Asteraceae (V:D)
Aster foliaceus var. *eatonii* Gray = Aster eatonii [ASTEEAT] Asteraceae (V:D)
Aster franklinianus Rydb. = Aster borealis [ASTEBOR] Asteraceae (V:D)
Aster geyeri (Gray) T.J. Howell = Aster laevis var. geyeri [ASTELAE1] Asteraceae (V:D)
Aster hesperius Gray = Aster lanceolatus ssp. hesperius [ASTELAN1] Asteraceae (V:D)
Aster hesperius var. *laetevirens* (Greene) Cronq. = Aster lanceolatus ssp. hesperius [ASTELAN1] Asteraceae (V:D)
Aster hesperius var. *wootonii* Greene = Aster lanceolatus ssp. hesperius [ASTELAN1] Asteraceae (V:D)
Aster junciformis Rydb. = Aster borealis [ASTEBOR] Asteraceae (V:D)
Aster laetevirens Greene = Aster lanceolatus ssp. hesperius [ASTELAN1] Asteraceae (V:D)
Aster laevis L. [ASTELAE] Asteraceae (V:D)
Aster laevis ssp. *geyeri* (Gray) Piper = Aster laevis var. geyeri [ASTELAE1] Asteraceae (V:D)
Aster laevis var. **geyeri** Gray [ASTELAE1] Asteraceae (V:D)
Aster lanceolatus Willd. [ASTELAN] Asteraceae (V:D)

Aster lanceolatus ssp. **hesperius** (Gray) Semple & Chmielewski [ASTELAN1] Asteraceae (V:D)
Aster latissimifolius var. *serotinus* Kuntze = Solidago gigantea [SOLIGIG] Asteraceae (V:D)
Aster laxifolius var. *borealis* Torr. & Gray = Aster borealis [ASTEBOR] Asteraceae (V:D)
Aster lindleyanus Torr. & Gray = Aster ciliolatus [ASTECIL] Asteraceae (V:D)
Aster macounii Rydb. = Aster ascendens [ASTEASC] Asteraceae (V:D)
Aster major (Hook.) Porter = Aster modestus [ASTEMOD] Asteraceae (V:D)
Aster mearnsii Rydb. = Aster eatonii [ASTEEAT] Asteraceae (V:D)
Aster meritus A. Nels. = Aster sibiricus var. meritus [ASTESIB1] Asteraceae (V:D)
Aster modestus Lindl. [ASTEMOD] Asteraceae (V:D)
Aster multiflorus var. *pansus* Blake = Aster ericoides var. pansus [ASTEERI1] Asteraceae (V:D)
Aster occidentalis (Nutt.) Torr. & Gray [ASTEOCC] Asteraceae (V:D)
Aster occidentalis var. **intermedius** Gray [ASTEOCC2] Asteraceae (V:D)
Aster occidentalis var. **occidentalis** [ASTEOCC0] Asteraceae (V:D)
Aster osterhoutii Rydb. = Aster lanceolatus ssp. hesperius [ASTELAN1] Asteraceae (V:D)
Aster pansus (Blake) Cronq. = Aster ericoides var. pansus [ASTEERI1] Asteraceae (V:D)
Aster paucicapitatus (B.L. Robins.) B.L. Robins. [ASTEPAU] Asteraceae (V:D)
Aster pickettianus Suksdorf = Aster laevis var. geyeri [ASTELAE1] Asteraceae (V:D)
Aster radulinus Gray [ASTERAD] Asteraceae (V:D)
Aster ramulosus Lindl. = Aster falcatus [ASTEFAL] Asteraceae (V:D)
Aster sibiricus L. [ASTESIB] Asteraceae (V:D)
Aster sibiricus ssp. *meritus* (A. Nels.) G.W. Doug. = Aster sibiricus var. meritus [ASTESIB1] Asteraceae (V:D)
Aster sibiricus var. **meritus** (A. Nels.) Raup [ASTESIB1] Asteraceae (V:D)
Aster spathulatus Lindl. = Aster occidentalis var. occidentalis [ASTEOCC0] Asteraceae (V:D)
Aster stenomeres Gray = Ionactis stenomeres [IONASTE] Asteraceae (V:D)
Aster subgriseus Rydb. = Aster ascendens [ASTEASC] Asteraceae (V:D)
Aster subspicatus Nees [ASTESUB] Asteraceae (V:D)
Aster unalaschkensis var. *major* Hook. = Aster modestus [ASTEMOD] Asteraceae (V:D)
Aster wootonii (Greene) Greene = Aster lanceolatus ssp. hesperius [ASTELAN1] Asteraceae (V:D)
Asterella P. Beauv. [ASTEREL$] Aytoniaceae (B:H)
Asterella gracilis (Web.) Underw. [ASTEGRA] Aytoniaceae (B:H)
Asterella lindenbergiana (Corda) Lindb. [ASTELIN] Aytoniaceae (B:H)
Asterophoma D. Hawksw. [ASTEROP$] Uncertain-family (L)
Asterophoma mazaediicola D. Hawksw. [ASTEMAZ] Uncertain-family (L)
Astragalus L. [ASTRAGA$] Fabaceae (V:D)
Astragalus aboriginum Richards. = Astragalus australis [ASTRAUS] Fabaceae (V:D)
Astragalus aboriginum var. *glabriusculus* (Hook.) Rydb. = Astragalus australis [ASTRAUS] Fabaceae (V:D)
Astragalus aboriginum var. *lepagei* (Hultén) Boivin = Astragalus australis [ASTRAUS] Fabaceae (V:D)
Astragalus aboriginum var. *richardsonii* (Sheldon) Boivin = Astragalus australis [ASTRAUS] Fabaceae (V:D)
Astragalus adsurgens Pallas [ASTRADS] Fabaceae (V:D)
Astragalus adsurgens ssp. *robustior* (Hook.) Welsh = Astragalus adsurgens var. robustior [ASTRADS1] Fabaceae (V:D)
Astragalus adsurgens var. **robustior** Hook. [ASTRADS1] Fabaceae (V:D)
Astragalus agrestis Dougl. ex G. Don [ASTRAGR] Fabaceae (V:D)
Astragalus alpinus L. [ASTRALP] Fabaceae (V:D)
Astragalus alpinus ssp. *alaskanus* Hultén = Astragalus alpinus [ASTRALP] Fabaceae (V:D)
Astragalus alpinus ssp. *arcticus* Hultén = Astragalus alpinus [ASTRALP] Fabaceae (V:D)
Astragalus americanus (Hook.) M.E. Jones [ASTRAME] Fabaceae (V:D)
Astragalus australis (L.) Lam. [ASTRAUS] Fabaceae (V:D)
Astragalus australis var. *glabriusculus* (Hook.) Isely = Astragalus australis [ASTRAUS] Fabaceae (V:D)
Astragalus australis var. *major* (Gray) Isely = Astragalus australis [ASTRAUS] Fabaceae (V:D)
Astragalus beckwithii Torr. & Gray [ASTRBEC] Fabaceae (V:D)
Astragalus beckwithii var. **weiserensis** M.E. Jones [ASTRBEC1] Fabaceae (V:D)
Astragalus blakei Egglest. = Astragalus robbinsii var. minor [ASTRROB1] Fabaceae (V:D)
Astragalus bourgovii Gray [ASTRBOU] Fabaceae (V:D)
Astragalus brevidens (Gandog.) Rydb. = Astragalus canadensis var. brevidens [ASTRCAN1] Fabaceae (V:D)
Astragalus canadensis L. [ASTRCAN] Fabaceae (V:D)
Astragalus canadensis var. **brevidens** (Gandog.) Barneby [ASTRCAN1] Fabaceae (V:D)
Astragalus canadensis var. **canadensis** [ASTRCAN0] Fabaceae (V:D)
Astragalus canadensis var. *carolinianus* (L.) M.E. Jones = Astragalus canadensis var. canadensis [ASTRCAN0] Fabaceae (V:D)
Astragalus canadensis var. *longilobus* Fassett = Astragalus canadensis var. canadensis [ASTRCAN0] Fabaceae (V:D)
Astragalus canadensis var. **mortonii** (Nutt.) S. Wats. [ASTRCAN3] Fabaceae (V:D)
Astragalus carolinianus L. = Astragalus canadensis var. canadensis [ASTRCAN0] Fabaceae (V:D)
*__Astragalus cicer__ L. [ASTRCIC] Fabaceae (V:D)
Astragalus collieri (Rydb.) Porsild = Astragalus robbinsii var. minor [ASTRROB1] Fabaceae (V:D)
Astragalus collinus (Hook.) Dougl. ex G. Don [ASTRCOL] Fabaceae (V:D)
Astragalus convallarius Greene [ASTRCON] Fabaceae (V:D)
Astragalus crassicarpus Nutt. [ASTRCRA] Fabaceae (V:D)
Astragalus danicus var. *dasyglottis* (Fisch. ex DC.) Boivin = Astragalus agrestis [ASTRAGR] Fabaceae (V:D)
Astragalus dasyglottis Fisch. ex DC. = Astragalus agrestis [ASTRAGR] Fabaceae (V:D)
Astragalus decumbens var. *serotinus* (Gray ex Cooper) M.E. Jones = Astragalus miser var. serotinus [ASTRMIS2] Fabaceae (V:D)
Astragalus drummondii Dougl. ex Hook. [ASTRDRU] Fabaceae (V:D)
Astragalus eucosmus B.L. Robins. [ASTREUC] Fabaceae (V:D)
Astragalus eucosmus ssp. *sealei* (Lepage) Hultén = Astragalus eucosmus [ASTREUC] Fabaceae (V:D)
Astragalus eucosmus var. *facinorum* Fern. = Astragalus eucosmus [ASTREUC] Fabaceae (V:D)
Astragalus filipes Torr. ex Gray [ASTRFIL] Fabaceae (V:D)
Astragalus filipes var. *residuus* Jepson = Astragalus filipes [ASTRFIL] Fabaceae (V:D)
Astragalus forwoodii S. Wats. = Astragalus australis [ASTRAUS] Fabaceae (V:D)
Astragalus forwoodii var. *wallowensis* (Rydb.) M.E. Peck = Astragalus australis [ASTRAUS] Fabaceae (V:D)
Astragalus frigidus (L.) Gray p.p. = Astragalus americanus [ASTRAME] Fabaceae (V:D)
Astragalus frigidus var. *americanus* (Hook.) S. Wats. = Astragalus americanus [ASTRAME] Fabaceae (V:D)
Astragalus frigidus var. *gaspensis* (Rouss.) Fern. = Astragalus americanus [ASTRAME] Fabaceae (V:D)
Astragalus glabriusculus var. *major* Gray = Astragalus australis [ASTRAUS] Fabaceae (V:D)
Astragalus glareosus Dougl. ex Hook. = Astragalus purshii var. glareosus [ASTRPUR1] Fabaceae (V:D)
Astragalus goniatus Nutt. = Astragalus agrestis [ASTRAGR] Fabaceae (V:D)
Astragalus halei Rydb. = Astragalus canadensis var. canadensis [ASTRCAN0] Fabaceae (V:D)

Astragalus hypoglottis Hook. = Astragalus agrestis [ASTRAGR] Fabaceae (V:D)
Astragalus incurvus (Rydb.) Abrams = Astragalus purshii var. purshii [ASTRPUR0] Fabaceae (V:D)
Astragalus lentiginosus Dougl. ex Hook. [ASTRLEN] Fabaceae (V:D)
Astragalus lentiginosus var. *macrolobus* (Rydb.) Barneby = Astragalus lentiginosus var. salinus [ASTRLEN2] Fabaceae (V:D)
Astragalus lentiginosus var. **salinus** (T.J. Howell) Barneby [ASTRLEN2] Fabaceae (V:D)
Astragalus linearis (Rydb.) Porsild = Astragalus australis [ASTRAUS] Fabaceae (V:D)
Astragalus lotiflorus Hook. [ASTRLOT] Fabaceae (V:D)
Astragalus lotiflorus var. *nebraskensis* Bates = Astragalus lotiflorus [ASTRLOT] Fabaceae (V:D)
Astragalus lotiflorus var. *reverchonii* (Gray) M.E. Jones = Astragalus lotiflorus [ASTRLOT] Fabaceae (V:D)
Astragalus macgregorii (Rydb.) Tidestrom = Astragalus filipes [ASTRFIL] Fabaceae (V:D)
Astragalus macounii Rydb. = Astragalus robbinsii var. minor [ASTRROB1] Fabaceae (V:D)
Astragalus microcystis Gray [ASTRMIC] Fabaceae (V:D)
Astragalus miser Dougl. [ASTRMIS] Fabaceae (V:D)
Astragalus miser var. **miser** [ASTRMIS0] Fabaceae (V:D)
Astragalus miser var. **serotinus** (Gray ex Cooper) Barneby [ASTRMIS2] Fabaceae (V:D)
Astragalus mortonii Nutt. = Astragalus canadensis var. mortonii [ASTRCAN3] Fabaceae (V:D)
Astragalus nutzotinensis Rouss. [ASTRNUT] Fabaceae (V:D)
Astragalus parviflorus (Pursh) MacM. = Astragalus eucosmus [ASTREUC] Fabaceae (V:D)
Astragalus purshii Dougl. ex Hook. [ASTRPUR] Fabaceae (V:D)
Astragalus purshii var. **glareosus** (Dougl. ex Hook.) Barneby [ASTRPUR1] Fabaceae (V:D)
Astragalus purshii var. *interior* M.E. Jones = Astragalus purshii var. purshii [ASTRPUR0] Fabaceae (V:D)
Astragalus purshii var. **purshii** [ASTRPUR0] Fabaceae (V:D)
Astragalus richardsonii Sheldon = Astragalus australis [ASTRAUS] Fabaceae (V:D)
Astragalus robbinsii (Oakes) Gray [ASTRROB] Fabaceae (V:D)
Astragalus robbinsii var. *blakei* (Egglest.) Barneby = Astragalus robbinsii var. minor [ASTRROB1] Fabaceae (V:D)
Astragalus robbinsii var. **minor** (Hook.) Barneby [ASTRROB1] Fabaceae (V:D)
Astragalus salinus T.J. Howell = Astragalus lentiginosus var. salinus [ASTRLEN2] Fabaceae (V:D)
Astragalus sclerocarpus Gray [ASTRSCL] Fabaceae (V:D)
Astragalus scrupulicola Fern. & Weatherby = Astragalus australis [ASTRAUS] Fabaceae (V:D)
Astragalus sealei Lepage = Astragalus eucosmus [ASTREUC] Fabaceae (V:D)
Astragalus serotinus Gray ex Cooper = Astragalus miser var. serotinus [ASTRMIS2] Fabaceae (V:D)
Astragalus spaldingii Gray [ASTRSPA] Fabaceae (V:D)
Astragalus splendens (Dougl. ex Hook.) Tidestrom = Oxytropis splendens [OXYTSPL] Fabaceae (V:D)
Astragalus stenophyllus Torr. & Gray = Astragalus filipes [ASTRFIL] Fabaceae (V:D)
Astragalus stenophyllus var. *filipes* (Torr. ex Gray) Tidestrom = Astragalus filipes [ASTRFIL] Fabaceae (V:D)
Astragalus striatus Nutt. = Astragalus adsurgens var. robustior [ASTRADS1] Fabaceae (V:D)
Astragalus strigosus Coult. & Fisher = Astragalus miser var. miser [ASTRMIS0] Fabaceae (V:D)
Astragalus sulphurescens Rydb. = Astragalus adsurgens var. robustior [ASTRADS1] Fabaceae (V:D)
Astragalus tenellus Pursh [ASTRTEN] Fabaceae (V:D)
Astragalus tenellus var. *strigulosus* (Rydb.) F.J. Herm. = Astragalus tenellus [ASTRTEN] Fabaceae (V:D)
Astragalus umbellatus Bunge [ASTRUMB] Fabaceae (V:D)
Astragalus ventosus Suksdorf ex Rydb. = Astragalus purshii var. glareosus [ASTRPUR1] Fabaceae (V:D)
Astragalus weiserensis (M.E. Jones) Abrams = Astragalus beckwithii var. weiserensis [ASTRBEC1] Fabaceae (V:D)
Atelophragma aboriginorum (Richards.) Rydb. = Astragalus australis [ASTRAUS] Fabaceae (V:D)
Atelophragma elegans (Hook.) Rydb. = Astragalus eucosmus [ASTREUC] Fabaceae (V:D)
Athalamia Falc. [ATHALAM$] Cleveaceae (B:H)
Athalamia hyalina (Sommerf.) Hatt. [ATHAHYA] Cleveaceae (B:H)
Athyrium Roth [ATHYRIU$] Dryopteridaceae (V:F)
Athyrium alpestre (Hoppe) Milde = Athyrium americanum [ATHYAME] Dryopteridaceae (V:F)
Athyrium alpestre ssp. *americanum* (Butters) Lellinger = Athyrium americanum [ATHYAME] Dryopteridaceae (V:F)
Athyrium alpestre var. *americanum* Butters = Athyrium americanum [ATHYAME] Dryopteridaceae (V:F)
Athyrium alpestre var. *cyclosorum* (Rupr.) T. Moore = Athyrium filix-femina ssp. cyclosorum [ATHYFIL1] Dryopteridaceae (V:F)
Athyrium alpestre var. *gaspense* Fern. = Athyrium americanum [ATHYAME] Dryopteridaceae (V:F)
Athyrium americanum (Butters) Maxon [ATHYAME] Dryopteridaceae (V:F)
Athyrium angustum var. *boreale* Jennings = Athyrium filix-femina ssp. cyclosorum [ATHYFIL1] Dryopteridaceae (V:F)
Athyrium angustum var. *elatius* (Link) Butters = Athyrium filix-femina ssp. cyclosorum [ATHYFIL1] Dryopteridaceae (V:F)
Athyrium cyclosorum Rupr. = Athyrium filix-femina ssp. cyclosorum [ATHYFIL1] Dryopteridaceae (V:F)
Athyrium distentifolium ssp. *americanum* (Butters) Hultén = Athyrium americanum [ATHYAME] Dryopteridaceae (V:F)
Athyrium distentifolium var. *americanum* (Butters) Boivin = Athyrium americanum [ATHYAME] Dryopteridaceae (V:F)
Athyrium filix-femina (L.) Roth [ATHYFIL] Dryopteridaceae (V:F)
Athyrium filix-femina ssp. **cyclosorum** (Rupr.) C. Christens. [ATHYFIL1] Dryopteridaceae (V:F)
Athyrium filix-femina var. *californicum* Butters = Athyrium filix-femina ssp. cyclosorum [ATHYFIL1] Dryopteridaceae (V:F)
Athyrium filix-femina var. *cyclosorum* (Rupr.) Ledeb. = Athyrium filix-femina ssp. cyclosorum [ATHYFIL1] Dryopteridaceae (V:F)
Athyrium filix-femina var. *sitchense* (Rupr.) Ledeb. = Athyrium filix-femina ssp. cyclosorum [ATHYFIL1] Dryopteridaceae (V:F)
Athysanus Greene [ATHYSAN$] Brassicaceae (V:D)
Athysanus pusillus (Hook.) Greene [ATHYPUS] Brassicaceae (V:D)
Athysanus pusillus var. *glabrior* S. Wats. = Athysanus pusillus [ATHYPUS] Brassicaceae (V:D)
Atrichum P. Beauv. [ATRICHU$] Polytrichaceae (B:M)
Atrichum crispum var. *molle* (Holz.) Frye in Grout = Atrichum tenellum [ATRITEN] Polytrichaceae (B:M)
Atrichum fertile Nawasch. = Atrichum haussknechtii [ATRIHAU] Polytrichaceae (B:M)
Atrichum haussknechtii Jur. & Milde [ATRIHAU] Polytrichaceae (B:M)
Atrichum leiophyllum Kindb. = Oligotrichum parallelum [OLIGPAR] Polytrichaceae (B:M)
Atrichum papallelum Mitt. = Oligotrichum parallelum [OLIGPAR] Polytrichaceae (B:M)
Atrichum rosulatum C. Müll. & Kindb. in Mac. & Kindb. = Atrichum selwynii [ATRISEL] Polytrichaceae (B:M)
Atrichum selwynii Aust. [ATRISEL] Polytrichaceae (B:M)
Atrichum tenellum (Röhl.) Bruch & Schimp. in B.S.G. [ATRITEN] Polytrichaceae (B:M)
Atrichum undulatum (Hedw.) P. Beauv. [ATRIUND] Polytrichaceae (B:M)
Atrichum undulatum var. *gracilisetum* Besch. = Atrichum haussknechtii [ATRIHAU] Polytrichaceae (B:M)
Atrichum undulatum var. *haussknechtii* (Jur. & Milde) Frye in Grout = Atrichum haussknechtii [ATRIHAU] Polytrichaceae (B:M)
Atrichum undulatum var. *selwynii* (Aust.) Frye in Grout = Atrichum selwynii [ATRISEL] Polytrichaceae (B:M)
Atriplex L. [ATRIPLE$] Chenopodiaceae (V:D)
Atriplex alaskensis S. Wats. [ATRIALA] Chenopodiaceae (V:D)
Atriplex argentea Nutt. [ATRIARG] Chenopodiaceae (V:D)
Atriplex buxifolia Rydb. = Atriplex nuttallii [ATRINUT] Chenopodiaceae (V:D)

Atriplex gmelinii C.A. Mey. ex Bong. [ATRIGME] Chenopodiaceae (V:D)
Atriplex gmelinii var. *zosterifolia* (Hook.) Moq. = Atriplex gmelinii [ATRIGME] Chenopodiaceae (V:D)
Atriplex heterosperma Bunge = Atriplex micrantha [ATRIMIC] Chenopodiaceae (V:D)
*__Atriplex hortensis__ L. [ATRIHOR] Chenopodiaceae (V:D)
Atriplex hortensis var. *atrosanguinea* hort. = Atriplex hortensis [ATRIHOR] Chenopodiaceae (V:D)
Atriplex longipes Drej. = Atriplex nudicaulis [ATRINUD] Chenopodiaceae (V:D)
Atriplex longipes ssp. *praecox* (Hülphers) Tresson = Atriplex nudicaulis [ATRINUD] Chenopodiaceae (V:D)
*__Atriplex micrantha__ C.A. Mey. [ATRIMIC] Chenopodiaceae (V:D)
*__Atriplex nudicaulis__ Boguslaw [ATRINUD] Chenopodiaceae (V:D)
Atriplex nuttallii S. Wats. [ATRINUT] Chenopodiaceae (V:D)
Atriplex nuttallii ssp. *buxifolia* (Rydb.) Hall & Clements = Atriplex nuttallii [ATRINUT] Chenopodiaceae (V:D)
*__Atriplex oblongifolia__ Waldst. & Kit. [ATRIOBL] Chenopodiaceae (V:D)
Atriplex patula L. [ATRIPAT] Chenopodiaceae (V:D)
Atriplex patula ssp. *alaskensis* (S. Wats.) Hall & Clements = Atriplex alaskensis [ATRIALA] Chenopodiaceae (V:D)
Atriplex patula ssp. *hastata* Hall & Clements p.p. = Atriplex subspicata [ATRISUB] Chenopodiaceae (V:D)
Atriplex patula ssp. *obtusa* (Cham.) Hall & Clements = Atriplex gmelinii [ATRIGME] Chenopodiaceae (V:D)
Atriplex patula ssp. *subspicata* (Nutt.) Fosberg = Atriplex subspicata [ATRISUB] Chenopodiaceae (V:D)
Atriplex patula ssp. *zosterifolia* (Hook.) Hall & Clements = Atriplex gmelinii [ATRIGME] Chenopodiaceae (V:D)
Atriplex patula var. *alaskensis* (S. Wats.) Welsh = Atriplex alaskensis [ATRIALA] Chenopodiaceae (V:D)
Atriplex patula var. *obtusa* (Cham.) M.E. Peck = Atriplex gmelinii [ATRIGME] Chenopodiaceae (V:D)
Atriplex patula var. *subspicata* (Nutt.) S. Wats. = Atriplex subspicata [ATRISUB] Chenopodiaceae (V:D)
Atriplex patula var. *zosterifolia* (Hook.) C.L. Hitchc. = Atriplex gmelinii [ATRIGME] Chenopodiaceae (V:D)
Atriplex praecox Hülphers = Atriplex nudicaulis [ATRINUD] Chenopodiaceae (V:D)
*__Atriplex rosea__ L. [ATRIROS] Chenopodiaceae (V:D)
Atriplex subspicata (Nutt.) Rydb. [ATRISUB] Chenopodiaceae (V:D)
Atriplex truncata (Torr. ex S. Wats.) Gray [ATRITRU] Chenopodiaceae (V:D)
Aulacomnium Schwaegr. [AULACOM$] Aulacomniaceae (B:M)
Aulacomnium acuminatum (Lindb. & Arnell) Kindb. [AULAACU] Aulacomniaceae (B:M)
Aulacomnium androgynum (Hedw.) Schwaegr. [AULAAND] Aulacomniaceae (B:M)
Aulacomnium palustre (Hedw.) Schwaegr. [AULAPAL] Aulacomniaceae (B:M)
Aulacomnium palustre f. *laxifolium* (Kindb. in Mac. & Kindb.) Podp. = Aulacomnium palustre [AULAPAL] Aulacomniaceae (B:M)
Aulacomnium palustre var. *congestum* Boul. = Aulacomnium palustre [AULAPAL] Aulacomniaceae (B:M)
Aulacomnium palustre var. *dimorphum* Card. & Thér. = Aulacomnium palustre [AULAPAL] Aulacomniaceae (B:M)
Aulacomnium palustre var. *fasciculare* (Brid.) B.S.G. = Aulacomnium palustre [AULAPAL] Aulacomniaceae (B:M)
Aulacomnium palustre var. *imbricatum* Bruch & Schimp. in B.S.G. = Aulacomnium palustre [AULAPAL] Aulacomniaceae (B:M)
Aulacomnium palustre var. *laxifolium* Kindb. in Mac. & Kindb. = Aulacomnium palustre [AULAPAL] Aulacomniaceae (B:M)
Aulacomnium palustre var. *polycephalum* (Brid.) Hüb. = Aulacomnium palustre [AULAPAL] Aulacomniaceae (B:M)
Aulacomnium turgidum (Wahlenb.) Schwaegr. [AULATUR] Aulacomniaceae (B:M)
Avena L. [AVENA$] Poaceae (V:M)
Avena byzantina K. Koch = Avena sativa [AVENSAT] Poaceae (V:M)
Avena dubia Leers = Ventenata dubia [VENTDUB] Poaceae (V:M)
*__Avena fatua__ L. [AVENFAT] Poaceae (V:M)
Avena fatua var. *glabrata* Peterm. = Avena fatua [AVENFAT] Poaceae (V:M)
Avena fatua var. *sativa* (L.) Hausskn. = Avena sativa [AVENSAT] Poaceae (V:M)
Avena fatua var. *vilis* (Wallr.) Hausskn. = Avena fatua [AVENFAT] Poaceae (V:M)
Avena hookeri Scribn. = Helictotrichon hookeri [HELIHOO] Poaceae (V:M)
*__Avena sativa__ L. [AVENSAT] Poaceae (V:M)
Avena sativa var. *orientalis* (Schreb.) Alef. = Avena sativa [AVENSAT] Poaceae (V:M)
Avena smithii Porter ex Gray = Melica smithii [MELISMI] Poaceae (V:M)
Avena torreyi Nash = Schizachne purpurascens [SCHIPUR] Poaceae (V:M)
Avenochloa hookeri (Scribn.) Holub = Helictotrichon hookeri [HELIHOO] Poaceae (V:M)
Avenula hookeri (Scribn.) Holub = Helictotrichon hookeri [HELIHOO] Poaceae (V:M)
Axyris L. [AXYRIS$] Chenopodiaceae (V:D)
*__Axyris amaranthoides__ L. [AXYRAMA] Chenopodiaceae (V:D)
Azalea procumbens L. = Loiseleuria procumbens [LOISPRO] Ericaceae (V:D)
Azaleastrum albiflorum (Hook.) Rydb. = Rhododendron albiflorum [RHODALB] Ericaceae (V:D)
Azolla Lam. [AZOLLA$] Azollaceae (V:F)
Azolla filiculoides Lam. [AZOLFIL] Azollaceae (V:F)
Azolla mexicana Schlecht. & Cham. ex K. Presl [AZOLMEX] Azollaceae (V:F)

B

Bachmanniomyces D. Hawksw. [BACHMAN$] Uncertain-family (L)
Bachmanniomyces uncialicola (Zopf) D. Hawksw. [BACHUNC] Uncertain-family (L)
Bacidia De Not. [BACIDIA$] Bacidiaceae (L)
Bacidia abbrevians (Nyl.) Th. Fr. = Bacidia igniarii [BACIIGN] Bacidiaceae (L)
Bacidia absistens (Nyl.) Arnold [BACIABS] Bacidiaceae (L)
Bacidia accedens (Arnold) Lettau = Mycobilimbia sabuletorum [MYCOSAB] Porpidiaceae (L)
Bacidia affinis (Stizenb.) Vainio = Bacidia subincompta [BACISUI] Bacidiaceae (L)
Bacidia akompsa (Tuck.) Fink = Lecanactis akompsa [LECAAKO] Opegraphaceae (L)
Bacidia alaskensis (Nyl.) Zahlbr. [BACIALA] Bacidiaceae (L)
Bacidia alpina (Schaerer) Vainio = Arthrorhaphis alpina [ARTHALP] Arthrorhaphidaceae (L)
Bacidia arceutina (Ach.) Arnold [BACIARC] Bacidiaceae (L)
Bacidia arnoldiana Körber = Bacidina arnoldiana [BACIARN] Lecanoraceae (L)
Bacidia arthoniza (Nyl.) Zahlbr. = Lecidella stigmatea [LECISTI] Lecanoraceae (L)
Bacidia atrogrisea (Delise ex Hepp) Körber = Bacidia laurocerasi [BACILAU] Bacidiaceae (L)
Bacidia auerswaldii (Hepp ex Stizenb.) Mig. [BACIAUE] Bacidiaceae (L)
Bacidia bacillifera (Nyl.) Arnold = Bacidia circumspecta [BACICIR] Bacidiaceae (L)

Bacidia bagliettoana (Massal. & De Not.) Jatta [BACIBAG] Bacidiaceae (L)
Bacidia beckhausii Körber [BACIBEC] Bacidiaceae (L)
Bacidia carneoalbida (Müll. Arg.) Coppins = Biatora carneoalbida [BIATCAR] Bacidiaceae (L)
Bacidia chlorococca (Stenh.) Lettau = Scoliciosporum chlorococcum [SCOLCHL] Lecanoraceae (L)
Bacidia circumspecta (Nyl. ex Vainio) Malme [BACICIR] Bacidiaceae (L)
Bacidia citrinella (Ach.) Branth & Rostrup = Arthrorhaphis citrinella [ARTHCIT] Arthrorhaphidaceae (L)
Bacidia declinis (Tuck.) Zahlbr. = Catillaria nigroclavata [CATINIG] Catillariaceae (L)
Bacidia effusa (Auersw. ex Rabenh.) Lettau = Bacidia auerswaldii [BACIAUE] Lecanoraceae (L)
Bacidia effusella Zahlbr., nom. ill. = Bacidia auerswaldii [BACIAUE] Bacidiaceae (L)
Bacidia egenula (Nyl.) Arnold = Bacidina egenula [BACIEGE] Lecanoraceae (L)
Bacidia endoleuca (Nyl.) Zahlbr. = Bacidia laurocerasi [BACILAU] Bacidiaceae (L)
Bacidia flavovirescens (Dickson) Anzi = Arthrorhaphis citrinella [ARTHCIT] Arthrorhaphidaceae (L)
Bacidia friesiana (Hepp) Körber [BACIFRI] Bacidiaceae (L)
Bacidia fusca (A. Massal.) Du Rietz = Mycobilimbia tetramera [MYCOTET] Porpidiaceae (L)
Bacidia fuscorubella (Ach.) Bausch = Bacidia polychroa [BACIPOL] Bacidiaceae (L)
Bacidia globulosa (Flörke) Hafellner & V. Wirth = Catillaria globulosa [CATIGLO] Catillariaceae (L)
Bacidia hegetschweileri (Hepp) Vainio = Bacidia subincompta [BACISUI] Bacidiaceae (L)
Bacidia herrei Zahlbr. = Ophioparma rubricosa [OPHIRUB] Ophioparmaceae (L)
Bacidia hypnophila (Turner ex Ach.) Zahlbr. = Mycobilimbia sabuletorum [MYCOSAB] Porpidiaceae (L)
Bacidia idahoensis H. Magn. = Bacidia laurocerasi ssp. idahoensis [BACILAU1] Bacidiaceae (L)
Bacidia igniarii (Nyl.) Oksner [BACIIGN] Bacidiaceae (L)
Bacidia inundata (Fr.) Körber = Bacidina inundata [BACIINU] Lecanoraceae (L)
Bacidia invertens (Nyl.) Zahlbr. = Bacidia laurocerasi [BACILAU] Bacidiaceae (L)
Bacidia laurocerasi (Delise ex Duby) Zahlbr. [BACILAU] Bacidiaceae (L)
Bacidia laurocerasi ssp. **idahoensis** (H. Magn.) S. Ekman [BACILAU1] Bacidiaceae (L)
Bacidia laurocerasi ssp. **laurocerasi** [BACILAU0] Bacidiaceae (L)
Bacidia luteola (Schrad.) Mudd = Bacidia rubella [BACIRUB] Bacidiaceae (L)
Bacidia luteola (Schrad.) Mudd = Bacidia rubella [BACIRUB] Bacidiaceae (L)
Bacidia melaena (Nyl.) Zahlbr. = Micarea melaena [MICAMEL] Micareaceae (L)
Bacidia minuscula Anzi = Bacidia beckhausii [BACIBEC] Bacidiaceae (L)
Bacidia muscorum (Sw.) Mudd = Bacidia bagliettoana [BACIBAG] Bacidiaceae (L)
Bacidia naegelii (Hepp) Zahlbr. [BACINAE] Bacidiaceae (L)
Bacidia nivalis Follmann [BACINIV] Bacidiaceae (L)
Bacidia obscurata (Sommerf.) Zahlbr. = Mycobilimbia tetramera [MYCOTET] Porpidiaceae (L)
Bacidia polychroa (Th. Fr.) Körber [BACIPOL] Bacidiaceae (L)
Bacidia rosella (Pers.) De Not. [BACIROS] Bacidiaceae (L)
Bacidia rubella (Hoffm.) A. Massal. [BACIRUB] Bacidiaceae (L)
Bacidia rubricosa (Müll. Arg.) Zahlbr. = Ophioparma rubricosa [OPHIRUB] Ophioparmaceae (L)
Bacidia sabuletorum (Schreber) Lettau = Mycobilimbia sabuletorum [MYCOSAB] Porpidiaceae (L)
Bacidia salmonea S. Ekman [BACISAL] Bacidiaceae (L)
Bacidia scopulicola (Nyl.) A.L. Sm. [BACISCO] Bacidiaceae (L)
Bacidia sphaeroides (Dickson) Zahlbr. = Biatora sphaeroides [BIATSPH] Bacidiaceae (L)
Bacidia subacerina Vainio = Bacidia laurocerasi [BACILAU] Bacidiaceae (L)
Bacidia subincompta (Nyl.) Arnold [BACISUI] Bacidiaceae (L)
Bacidia tetramera (De Not.) Coppins = Mycobilimbia tetramera [MYCOTET] Porpidiaceae (L)
Bacidia trisepta (Hellbom) Zahlbr. = Micarea peliocarpa [MICAPEL] Micareaceae (L)
Bacidia umbrina (Ach.) Bausch = Scoliciosporum umbrinum [SCOLUMB] Lecanoraceae (L)
Bacidia vermifera (Nyl.) Th. Fr. [BACIVER] Bacidiaceae (L)
Bacidina Vezda [BACIDIN$] Lecanoraceae (L)
Bacidina arnoldiana (Körber) Wirth & Vezda [BACIARN] Lecanoraceae (L)
Bacidina egenula (Nyl.) Vezda [BACIEGE] Lecanoraceae (L)
Bacidina inundata (Fr.) Vezda [BACIINU] Lecanoraceae (L)
Bacidina ramea S. Ekman [BACIRAM] Lecanoraceae (L)
Bactrospora A. Massal. [BACTROS$] Opegraphaceae (L)
Bactrospora patellarioides (Nyl.) Vainio [BACTPAT] Opegraphaceae (L)
Baeomyces Pers. [BAEOMYC$] Baeomycetaceae (L)
Baeomyces carneus Flörke [BAEOCAR] Baeomycetaceae (L)
Baeomyces fungoides auct. = Dibaeis baeomyces [DIBABAE] Baeomycetaceae (L)
Baeomyces placophyllus Ach. [BAEOPLA] Baeomycetaceae (L)
Baeomyces roseus Pers. = Dibaeis baeomyces [DIBABAE] Baeomycetaceae (L)
Baeomyces rubiformis Ach. = Psora rubiformis [PSORRUB] Psoraceae (L)
Baeomyces rufus (Hudson) Rebent. [BAEORUF] Baeomycetaceae (L)
Baeothryon alpinum (L.) Egor. = Eriophorum alpinum [ERIOALP] Cyperaceae (V:M)
Baeothryon cespitosum (L.) A. Dietr. = Scirpus cespitosus [SCIRCES] Cyperaceae (V:M)
Baeria maritima Gray = Lasthenia maritima [LASTMAR] Asteraceae (V:D)
Baeria minor ssp. *maritima* (Gray) Ferris = Lasthenia maritima [LASTMAR] Asteraceae (V:D)
Balsamorhiza Nutt. [BALSAMO$] Asteraceae (V:D)
Balsamorhiza deltoidea Nutt. [BALSDEL] Asteraceae (V:D)
Balsamorhiza sagittata (Pursh) Nutt. [BALSSAG] Asteraceae (V:D)
Barbarea Ait. f. [BARBARE$] Brassicaceae (V:D)
Barbarea americana Rydb. = Barbarea orthoceras [BARBORT] Brassicaceae (V:D)
Barbarea arcuata (Opiz ex J.& K. Presl) Reichenb. = Barbarea vulgaris [BARBVUL] Brassicaceae (V:D)
Barbarea orthoceras Ledeb. [BARBORT] Brassicaceae (V:D)
Barbarea orthoceras var. *dolichocarpa* Fern. = Barbarea orthoceras [BARBORT] Brassicaceae (V:D)
***Barbarea verna** (P. Mill.) Aschers. [BARBVER] Brassicaceae (V:D)
***Barbarea vulgaris** Ait. f. [BARBVUL] Brassicaceae (V:D)
Barbarea vulgaris var. *arcuata* (Opiz ex J.& K. Presl) Fries = Barbarea vulgaris [BARBVUL] Brassicaceae (V:D)
Barbarea vulgaris var. *brachycarpa* Rouy & Foucaud = Barbarea vulgaris [BARBVUL] Brassicaceae (V:D)
Barbarea vulgaris var. *longisiliquosa* Carion = Barbarea vulgaris [BARBVUL] Brassicaceae (V:D)
Barbarea vulgaris var. *sylvestris* Fries = Barbarea vulgaris [BARBVUL] Brassicaceae (V:D)
Barbilophozia Loeske [BARBILO$] Jungermanniaceae (B:H)
Barbilophozia attenuata (Mart.) Loeske [BARBATT] Jungermanniaceae (B:H)
Barbilophozia barbata (Schmid. ex Schreb.) Loeske [BARBBAR] Jungermanniaceae (B:H)
Barbilophozia binsteadii (Kaal.) Loeske [BARBBIN] Jungermanniaceae (B:H)
Barbilophozia floerkei (Web. & Mohr) Loeske [BARBFLO] Jungermanniaceae (B:H)
Barbilophozia gracilis (Schleich.) K. Müll. = Barbilophozia attenuata [BARBATT] Jungermanniaceae (B:H)
Barbilophozia hatcheri (Evans) Loeske [BARBHAT] Jungermanniaceae (B:H)

Barbilophozia kunzeana (Hub.) Gams. [BARBKUN] Jungermanniaceae (B:H)
Barbilophozia lycopodioides (Wallr.) Loeske [BARBLYC] Jungermanniaceae (B:H)
Barbilophozia quadriloba (Lindb.) Loeske [BARBQUA] Jungermanniaceae (B:H)
Barbula Hedw. [BARBULA$] Pottiaceae (B:M)
Barbula aciphylla B.S.G. = Tortula norvegica [TORTNOR] Pottiaceae (B:M)
Barbula acuta (Brid.) Brid. = Didymodon rigidulus var. gracilis [DIDYRIG1] Pottiaceae (B:M)
Barbula acuta ssp. *icmadophila* (Schimp. ex C. Müll.) Amann = Didymodon rigidulus var. icmadophilus [DIDYRIG2] Pottiaceae (B:M)
Barbula acuta var. *bescherellei* (Sauerb. in Jaeg.) Crum = Didymodon rigidulus var. gracilis [DIDYRIG1] Pottiaceae (B:M)
Barbula acuta var. *icmadophila* (Schimp. ex C. Müll.) Crum = Didymodon rigidulus var. icmadophilus [DIDYRIG2] Pottiaceae (B:M)
Barbula amplexa Lesq. = Tortula amplexa [TORTAMP] Pottiaceae (B:M)
Barbula amplexifolia (Mitt.) Jaeg. [BARBAMP] Pottiaceae (B:M)
Barbula andreaeoides Kindb. = Didymodon subandreaeoides [DIDYSUB] Pottiaceae (B:M)
Barbula angustata (Schimp.) Mac. & Kindb. = Tortula subulata [TORTSUB] Pottiaceae (B:M)
Barbula asperifolia Mitt. = Didymodon asperifolius [DIDYASP] Pottiaceae (B:M)
Barbula bescherellei Sauerb. in Jaeg. = Didymodon rigidulus var. gracilis [DIDYRIG1] Pottiaceae (B:M)
Barbula brevirostris (Hook. & Grev.) Bruch in F. Müll. = Aloina brevirostris [ALOIBRE] Pottiaceae (B:M)
Barbula caespitosa Schwaegr. = Tortella humilis [TORTHUM] Pottiaceae (B:M)
Barbula circinnatula C. Müll. & Kindb. in Mac. & Kindb. = Didymodon vinealis var. flaccidus [DIDYVIN1] Pottiaceae (B:M)
Barbula columbiana (Herm. & Lawt.) Herm. & Lawt. = Bryoerythrophyllum columbianum [BRYOCOL] Pottiaceae (B:M)
Barbula convoluta Hedw. [BARBCON] Pottiaceae (B:M)
Barbula convoluta var. **convoluta** [BARBCON0] Pottiaceae (B:M)
Barbula convoluta var. **gallinula** Zand. [BARBCON1] Pottiaceae (B:M)
Barbula convoluta var. *obtusata* C. Müll. & Kindb. in Mac. & Kindb. = Barbula convoluta var. convoluta [BARBCON0] Pottiaceae (B:M)
Barbula cylindrica (Tayl.) Schimp. = Didymodon vinealis var. vinealis [DIDYVIN0] Pottiaceae (B:M)
Barbula eustegia Card. & Thér. [BARBEUS] Pottiaceae (B:M)
Barbula fallax Hedw. = Didymodon fallax [DIDYFAL] Pottiaceae (B:M)
Barbula fallax var. *recurvifolia* (Wils.) Husn. = Didymodon fallax var. reflexus [DIDYFAL1] Pottiaceae (B:M)
Barbula ferruginascens Stirt. = Bryoerythrophyllum ferruginascens [BRYOFER] Pottiaceae (B:M)
Barbula fragilis (Hook. & Wils.) B.S.G. = Tortella fragilis [TORTFRA] Pottiaceae (B:M)
Barbula haringae Crum = Barbula amplexifolia [BARBAMP] Pottiaceae (B:M)
Barbula horridifolia C. Müll. & Kindb. in Mac. & Kindb. = Didymodon vinealis var. flaccidus [DIDYVIN1] Pottiaceae (B:M)
Barbula icmadophila Schimp. ex C. Müll. = Didymodon rigidulus var. icmadophilus [DIDYRIG2] Pottiaceae (B:M)
Barbula inclinata C. Müll. & Kindb. in Mac. & Kindb. = Tortella inclinata [TORTINC] Pottiaceae (B:M)
Barbula johansenii Williams = Didymodon johansenii [DIDYJOH] Pottiaceae (B:M)
Barbula laevipila (Brid.) Garov. = Tortula laevipila [TORTLAE] Pottiaceae (B:M)
Barbula laeviuscula Kindb. = Tortula ruralis [TORTRUA] Pottiaceae (B:M)
Barbula leptotricha C. Müll. & Kindb. in Mac. & Kindb. = Tortula ruralis [TORTRUA] Pottiaceae (B:M)
Barbula macrorhyncha Kindb. in Mac. & Kindb. = Aloina brevirostris [ALOIBRE] Pottiaceae (B:M)
Barbula megalocarpa Kindb. = Tortula princeps [TORTPRI] Pottiaceae (B:M)
Barbula melanocarpa C. Müll. & Kindb. in Mac. & Kindb. = Didymodon vinealis var. rubiginosus [DIDYVIN2] Pottiaceae (B:M)
Barbula michiganensis Steere = Didymodon michiganensis [DIDYMIC] Pottiaceae (B:M)
Barbula mucronifolia (Schwaegr.) Garov. = Tortula mucronifolia [TORTMUC] Pottiaceae (B:M)
Barbula nigrescens Mitt. = Didymodon nigrescens [DIDYNIG] Pottiaceae (B:M)
Barbula pseudorigidula Kindb. in Mac. & Kindb. = Didymodon vinealis var. flaccidus [DIDYVIN1] Pottiaceae (B:M)
Barbula recurvifolia Schimp. = Didymodon fallax var. reflexus [DIDYFAL1] Pottiaceae (B:M)
Barbula reflexa (Brid.) Brid. = Didymodon fallax var. reflexus [DIDYFAL1] Pottiaceae (B:M)
Barbula rigida Hedw. = Aloina rigida [ALOIRIG] Pottiaceae (B:M)
Barbula rigidula var. *icmadophila* Schimp. ex C. Müll. = Didymodon rigidulus var. icmadophilus [DIDYRIG2] Pottiaceae (B:M)
Barbula rigidula (Hedw.) Milde = Didymodon rigidulus [DIDYRIG] Pottiaceae (B:M)
Barbula rubella (Hüb.) Mitt. ex Lindb. = Bryoerythrophyllum recurvirostre [BRYOREC] Pottiaceae (B:M)
Barbula rubiginosa Mitt. = Didymodon vinealis var. rubiginosus [DIDYVIN2] Pottiaceae (B:M)
Barbula rufofusca Lawt. & Herm. = Didymodon nigrescens [DIDYNIG] Pottiaceae (B:M)
Barbula ruralis Hedw. = Tortula ruralis [TORTRUA] Pottiaceae (B:M)
Barbula subandreaeoides Kindb. = Didymodon subandreaeoides [DIDYSUB] Pottiaceae (B:M)
Barbula subgracilis C. Müll. & Kindb. in Mac. & Kindb. = Didymodon vinealis var. vinealis [DIDYVIN0] Pottiaceae (B:M)
Barbula subgracilis var. *viridior* Kindb. = Didymodon rigidulus [DIDYRIG] Pottiaceae (B:M)
Barbula subicmadophila C. Müll. & Kindb. in Mac. & Kindb. = Didymodon vinealis var. rubiginosus [DIDYVIN2] Pottiaceae (B:M)
Barbula subulata (Hedw.) P. Beauv. = Tortula subulata [TORTSUB] Pottiaceae (B:M)
Barbula tortuosa (Hedw.) Web. & Mohr = Tortella tortuosa [TORTTOR] Pottiaceae (B:M)
Barbula unguiculata Hedw. [BARBUNG] Pottiaceae (B:M)
Barbula unguiculata f. *apiculata* (Hedw.) Mönk. = Barbula unguiculata [BARBUNG] Pottiaceae (B:M)
Barbula vinealis Brid. = Didymodon vinealis [DIDYVIN] Pottiaceae (B:M)
Barbula vinealis ssp. *cylindrica* (Tayl.) Podp. = Didymodon vinealis var. flaccidus [DIDYVIN1] Pottiaceae (B:M)
Barbula vinealis var. *flaccida* Bruch & Schimp. in B.S.G. = Didymodon vinealis var. flaccidus [DIDYVIN1] Pottiaceae (B:M)
Barbula whitehouseae Crum = Barbula eustegia [BARBEUS] Pottiaceae (B:M)
Bartleya ohioensis Robins. = Dicranella cerviculata [DICRCER] Dicranaceae (B:M)
Bartramia Hedw. [BARTRAM$] Bartramiaceae (B:M)
Bartramia breviseta Lindb. = Bartramia ithyphylla [BARTITH] Bartramiaceae (B:M)
Bartramia circinnulata C. Müll. & Kindb. in Mac. & Kindb. = Bartramia pomiformis [BARTPOM] Bartramiaceae (B:M)
Bartramia crispa Brid. = Bartramia pomiformis [BARTPOM] Bartramiaceae (B:M)
Bartramia fontana (Hedw.) Turn. = Philonotis fontana [PHILFON] Bartramiaceae (B:M)
Bartramia glauco-viridis C. Müll. & Kindb. in Mac. & Kindb. = Bartramia pomiformis [BARTPOM] Bartramiaceae (B:M)
Bartramia halleriana Hedw. [BARTHAL] Bartramiaceae (B:M)
Bartramia ithyphylla Brid. [BARTITH] Bartramiaceae (B:M)
Bartramia ithyphylla var. *breviseta* (Lindb.) Kindb. = Bartramia ithyphylla [BARTITH] Bartramiaceae (B:M)

Bartramia ithyphylla var. *fragilifolia* Card. & Thér. = Bartramia ithyphylla [BARTITH] Bartramiaceae (B:M)
Bartramia ithyphylla var. *strigosa* (Wahlenb.) Hartm. = Bartramia ithyphylla [BARTITH] Bartramiaceae (B:M)
Bartramia marchica (Hedw.) Sw. in Schrad. = Philonotis marchica [PHILMAR] Bartramiaceae (B:M)
Bartramia menziesii Turn. = Anacolia menziesii [ANACMEN] Bartramiaceae (B:M)
Bartramia oederi Brid. = Plagiopus oederiana [PLAGOED] Bartramiaceae (B:M)
Bartramia oederiana Turn. = Plagiopus oederiana [PLAGOED] Bartramiaceae (B:M)
Bartramia pomiformis Hedw. [BARTPOM] Bartramiaceae (B:M)
Bartramia pomiformis var. *crispa* (Brid.) Bruch & Schimp. in B.S.G. = Bartramia pomiformis [BARTPOM] Bartramiaceae (B:M)
Bartramia pomiformis var. *elongata* Turn. = Bartramia pomiformis [BARTPOM] Bartramiaceae (B:M)
Bartramia stricta Brid. [BARTSTR] Bartramiaceae (B:M)
Bartramiopsis Kindb. [BARTRAO$] Polytrichaceae (B:M)
Bartramiopsis lescurii (James) Kindb. [BARTLES] Polytrichaceae (B:M)
Bassia All. [BASSIA$] Chenopodiaceae (V:D)
*__Bassia hyssopifolia__ (Pallas) Volk. [BASSHYS] Chenopodiaceae (V:D)
Bassia sieversiana (Pallas) W.A. Weber = Kochia scoparia [KOCHSCO] Chenopodiaceae (V:D)
Batidophaca lotiflorus (Hook.) Rydb. = Astragalus lotiflorus [ASTRLOT] Fabaceae (V:D)
Batrachium aquatile (L.) Dumort. = Ranunculus aquatilis [RANUAQU] Ranunculaceae (V:D)
Batrachium circinatum ssp. *subrigidum* (W. Drew) A. & D. Löve = Ranunculus longirostris [RANULON] Ranunculaceae (V:D)
Batrachium flaccidum (Pers.) Rupr. = Ranunculus trichophyllus [RANUTRI] Ranunculaceae (V:D)
Batrachium longirostre (Godr.) F.W. Schultz = Ranunculus longirostris [RANULON] Ranunculaceae (V:D)
Batrachium porteri (Britt.) Britt. = Ranunculus trichophyllus [RANUTRI] Ranunculaceae (V:D)
Batrachium trichophyllum (Chaix) F.W. Schultz = Ranunculus trichophyllus [RANUTRI] Ranunculaceae (V:D)
Batschia linearifolia (Goldie) Small = Lithospermum incisum [LITHINC] Boraginaceae (V:D)
Bazzania Gray [BAZZANI$] Lepidoziaceae (B:H)
Bazzania ambigua (Lindenb.) Trev. = Bazzania denudata [BAZZDEN] Lepidoziaceae (B:H)
Bazzania denudata (Torrey ex Gott. & al.) Trev. [BAZZDEN] Lepidoziaceae (B:H)
Bazzania pearsonii Steph. [BAZZPEA] Lepidoziaceae (B:H)
Bazzania pearsonii f. **krajinai** Hong [BAZZPEA1] Lepidoziaceae (B:H)
Bazzania triangularis Lindb. = Bazzania tricrenata [BAZZTRI] Lepidoziaceae (B:H)
Bazzania tricrenata (Wahlenb.) Lindb. [BAZZTRI] Lepidoziaceae (B:H)
Bazzania tricrenata f. **vittii** Hong [BAZZTRI2] Lepidoziaceae (B:H)
Bazzania tricrenata var. **fulfordiae** Hong [BAZZTRI1] Lepidoziaceae (B:H)
Bazzania tricrenata var. **tricrenata** [BAZZTRI0] Lepidoziaceae (B:H)
Bazzania trilobata (L.) Gray [BAZZTRL] Lepidoziaceae (B:H)
Beckmannia Host [BECKMAN$] Poaceae (V:M)
Beckmannia eruciformis auct. = Beckmannia syzigachne [BECKSYZ] Poaceae (V:M)
Beckmannia eruciformis ssp. *baicalensis* (Kusnez.) Hultén = Beckmannia syzigachne [BECKSYZ] Poaceae (V:M)
Beckmannia eruciformis var. *uniflora* Scribn. ex Gray = Beckmannia syzigachne [BECKSYZ] Poaceae (V:M)
Beckmannia syzigachne (Steud.) Fern. [BECKSYZ] Poaceae (V:M)
Beckmannia syzigachne ssp. *baicalensis* (Kusnez.) Koyama & Kawano = Beckmannia syzigachne [BECKSYZ] Poaceae (V:M)
Beckmannia syzigachne var. *uniflora* (Scribn. ex Gray) Boivin = Beckmannia syzigachne [BECKSYZ] Poaceae (V:M)
Bellemerea Hafellner & Roux [BELLEME$] Porpidiaceae (L)
Bellemerea alpina (Sommerf.) Clauzade & Roux [BELLALP] Porpidiaceae (L)
Bellemerea cinereorufescens (Ach.) Clauzade & Roux [BELLCIN] Porpidiaceae (L)
Bellemerea diamarta (Ach.) Hafellner & Roux [BELLDIA] Porpidiaceae (L)
Bellemerea subsorediza (Lynge) R. Sant. [BELLSUB] Porpidiaceae (L)
Bellis L. [BELLIS$] Asteraceae (V:D)
*__Bellis perennis__ L. [BELLPER] Asteraceae (V:D)
Benthamia lycopsoides (Lehm.) Lindl. ex Druce = Amsinckia lycopsoides [AMSILYC] Boraginaceae (V:D)
Berberis L. [BERBERI$] Berberidaceae (V:D)
Berberis aquifolium Pursh = Mahonia aquifolium [MAHOAQU] Berberidaceae (V:D)
Berberis aquifolium var. *repens* (Lindl.) Scoggan = Mahonia repens [MAHOREP] Berberidaceae (V:D)
Berberis nervosa Pursh = Mahonia nervosa [MAHONER] Berberidaceae (V:D)
Berberis repens Lindl. = Mahonia repens [MAHOREP] Berberidaceae (V:D)
Berberis sonnei (Abrams) McMinn = Mahonia repens [MAHOREP] Berberidaceae (V:D)
*__Berberis vulgaris__ L. [BERBVUL] Berberidaceae (V:D)
Berteroa DC. [BERTERO$] Brassicaceae (V:D)
*__Berteroa incana__ (L.) DC. [BERTINC] Brassicaceae (V:D)
Berula Bess. ex W.D.J. Koch [BERULA$] Apiaceae (V:D)
Berula erecta (Huds.) Coville [BERUERE] Apiaceae (V:D)
Berula erecta var. *incisa* (Torr.) Cronq. = Berula erecta [BERUERE] Apiaceae (V:D)
Berula incisa (Torr.) G.N. Jones = Berula erecta [BERUERE] Apiaceae (V:D)
Berula pusilla Fern. = Berula erecta [BERUERE] Apiaceae (V:D)
Besseya Rydb. [BESSEYA$] Scrophulariaceae (V:D)
Besseya cinerea (Raf.) Pennell = Besseya wyomingensis [BESSWYO] Scrophulariaceae (V:D)
Besseya wyomingensis (A. Nels.) Rydb. [BESSWYO] Scrophulariaceae (V:D)
Bestia breweriana (Lesq.) Grout = Isothecium cristatum [ISOTCRI] Brachytheciaceae (B:M)
Bestia breweriana var. *howei* (Kindb.) Grout = Isothecium cristatum [ISOTCRI] Brachytheciaceae (B:M)
Bestia breweriana var. *lutescens* (Lesq. & James) Grout = Isothecium cristatum [ISOTCRI] Brachytheciaceae (B:M)
Bestia cristata (Hampe) L. Koch = Isothecium cristatum [ISOTCRI] Brachytheciaceae (B:M)
Bestia holzingeri (Ren. & Card.) Broth. = Porotrichum vancouveriense [POROVAN] Thamnobryaceae (B:M)
Bestia occidentalis (Sull. & Lesq.) Grout = Porotrichum vancouveriense [POROVAN] Thamnobryaceae (B:M)
Bestia vancouveriensis (Kindb. in Mac.) Wijk & Marg. = Porotrichum vancouveriense [POROVAN] Thamnobryaceae (B:M)
Betula L. [BETULA$] Betulaceae (V:D)
Betula alaskana Sarg. = Betula neoalaskana [BETUNEO] Betulaceae (V:D)
Betula alba L. = Betula pubescens [BETUPUB] Betulaceae (V:D)
Betula alba var. *commutata* Regel = Betula papyrifera var. commutata [BETUPAP4] Betulaceae (V:D)
Betula beeniana A. Nels. = Betula occidentalis [BETUOCC] Betulaceae (V:D)
Betula exilis Sukatschev = Betula nana [BETUNAN] Betulaceae (V:D)
Betula fontinalis Sarg. = Betula occidentalis [BETUOCC] Betulaceae (V:D)
Betula glandulifera (Regel) Butler = Betula pumila var. glandulifera [BETUPUM1] Betulaceae (V:D)
Betula glandulosa Michx. = Betula nana [BETUNAN] Betulaceae (V:D)
Betula glandulosa var. *glandulifera* (Regel) Gleason = Betula pumila var. glandulifera [BETUPUM1] Betulaceae (V:D)
Betula glandulosa var. *hallii* (T.J. Howell) C.L. Hitchc. = Betula nana [BETUNAN] Betulaceae (V:D)

Betula glandulosa var. *sibirica* (Ledeb.) Schneid. = Betula nana [BETUNAN] Betulaceae (V:D)
Betula michauxii Sarg. = Betula nana [BETUNAN] Betulaceae (V:D)
Betula nana L. [BETUNAN] Betulaceae (V:D)
Betula nana ssp. *exilis* (Sukatschev) Hultén = Betula nana [BETUNAN] Betulaceae (V:D)
Betula nana var. *glandulifera* (Regel) Boivin = Betula pumila var. glandulifera [BETUPUM1] Betulaceae (V:D)
Betula nana var. *sibirica* Ledeb. = Betula nana [BETUNAN] Betulaceae (V:D)
Betula neoalaskana Sarg. [BETUNEO] Betulaceae (V:D)
Betula occidentalis Hook. [BETUOCC] Betulaceae (V:D)
Betula occidentalis var. *inopina* (Jepson) C.L. Hitchc. = Betula occidentalis [BETUOCC] Betulaceae (V:D)
Betula papyrifera Marsh. [BETUPAP] Betulaceae (V:D)
Betula papyrifera ssp. *humilis* (Regel) Hultén = Betula neoalaskana [BETUNEO] Betulaceae (V:D)
Betula papyrifera ssp. *occidentalis* (Hook.) Hultén = Betula occidentalis [BETUOCC] Betulaceae (V:D)
Betula papyrifera var. **commutata** (Regel) Fern. [BETUPAP4] Betulaceae (V:D)
Betula papyrifera var. *elobata* (Fern.) Sarg. = Betula papyrifera var. papyrifera [BETUPAP0] Betulaceae (V:D)
Betula papyrifera var. *humilis* (Regel) Fern. & Raup = Betula neoalaskana [BETUNEO] Betulaceae (V:D)
Betula papyrifera var. *macrostachya* Fern. = Betula papyrifera var. papyrifera [BETUPAP0] Betulaceae (V:D)
Betula papyrifera var. *neoalaskana* (Sarg.) Raup = Betula neoalaskana [BETUNEO] Betulaceae (V:D)
Betula papyrifera var. *occidentalis* (Hook.) Sarg. = Betula occidentalis [BETUOCC] Betulaceae (V:D)
Betula papyrifera var. **papyrifera** [BETUPAP0] Betulaceae (V:D)
Betula papyrifera var. *pensilis* Fern. = Betula papyrifera var. papyrifera [BETUPAP0] Betulaceae (V:D)
***Betula pendula** Roth [BETUPEN] Betulaceae (V:D)
Betula piperi Britt. = Betula × utahensis [BETUXUT] Betulaceae (V:D)
***Betula pubescens** Ehrh. [BETUPUB] Betulaceae (V:D)
Betula pumila L. [BETUPUM] Betulaceae (V:D)
Betula pumila var. **glandulifera** Regel [BETUPUM1] Betulaceae (V:D)
Betula pumila var. *glandulifera* f. *hallii* (Howell) Brayshaw = Betula pumila var. glandulifera [BETUPUM1] Betulaceae (V:D)
Betula resinifera Britt. = Betula neoalaskana [BETUNEO] Betulaceae (V:D)
Betula terrae-novae Fern. = Betula nana [BETUNAN] Betulaceae (V:D)
Betula × utahensis Britt. (pro sp.) [BETUXUT] Betulaceae (V:D)
Betula verrucosa Ehrh. = Betula pendula [BETUPEN] Betulaceae (V:D)
Betula × winteri Dugle [BETUXWI] Betulaceae (V:D)
Betula × piperi Britt. = Betula × utahensis [BETUXUT] Betulaceae (V:D)
Biatora Fr. [BIATORA$] Bacidiaceae (L)
Biatora anthracophila (Nyl.) Hafellner [BIATANT] Bacidiaceae (L)
Biatora atrogrisea Delise ex Hepp = Bacidia laurocerasi [BACILAU] Bacidiaceae (L)
Biatora atrosanguinea var. *hegetschweileri* Hepp = Bacidia subincompta [BACISUI] Bacidiaceae (L)
Biatora botryosa Fr. [BIATBOT] Bacidiaceae (L)
Biatora carneoalbida (Müll. Arg.) Coppins [BIATCAR] Bacidiaceae (L)
Biatora flavopunctata (Tønsberg) Hinteregger & Printzen [BIATFLA] Bacidiaceae (L)
Biatora helvola (Körber) Hellbom [BIATHEL] Bacidiaceae (L)
Biatora himalayana Church. Bab. = Psora himalayana [PSORHIM] Psoraceae (L)
Biatora luteola (Schrad.) Fr. = Bacidia rubella [BACIRUB] Bacidiaceae (L)
Biatora pullata Norman [BIATPUL] Bacidiaceae (L)
Biatora sphaeroides (Dickson) Körber [BIATSPH] Bacidiaceae (L)
Biatora vernalis (L.) Fr. [BIATVER] Bacidiaceae (L)
Biatorella De Not. [BIATORE$] Biatorellaceae (L)
Biatorella fossarum (Dufour ex Fr.) Th. Fr. = Biatorella hemisphaerica [BIATHEM] Biatorellaceae (L)
Biatorella hemisphaerica Anzi [BIATHEM] Biatorellaceae (L)
Biatorella moriformis (Ach.) Th. Fr. = Strangospora moriformis [STRAMOR] Acarosporaceae (L)
Biatorella pruinosa (Körber) Mudd = Sarcogyne regularis [SARCREG] Acarosporaceae (L)
Biatorella simplex (Davies) Branth & Rostrup = Polysporina simplex [POLYSIM] Acarosporaceae (L)
Biatorella testudinea (Ach.) A. Massal. = Sporastatia testudinea [SPORTES] Acarosporaceae (L)
Biatoropsis Räsänen [BIATORO$] Uncertain-family (L)
Biatoropsis usnearum Räsänen [BIATUSN] Uncertain-family (L)
Bidens L. [BIDENS$] Asteraceae (V:D)
Bidens amplissima Greene [BIDEAMP] Asteraceae (V:D)
Bidens beckii Torr. ex Spreng. = Megalodonta beckii [MEGABEC] Asteraceae (V:D)
Bidens cernua L. [BIDECER] Asteraceae (V:D)
Bidens cernua var. *dentata* (Nutt.) Boivin = Bidens cernua [BIDECER] Asteraceae (V:D)
Bidens cernua var. *elliptica* Wieg. = Bidens cernua [BIDECER] Asteraceae (V:D)
Bidens cernua var. *integra* Wieg. = Bidens cernua [BIDECER] Asteraceae (V:D)
Bidens cernua var. *minima* (Huds.) Pursh = Bidens cernua [BIDECER] Asteraceae (V:D)
Bidens cernua var. *oligodonta* Fern. & St. John = Bidens cernua [BIDECER] Asteraceae (V:D)
Bidens cernua var. *radiata* DC. = Bidens cernua [BIDECER] Asteraceae (V:D)
Bidens frondosa L. [BIDEFRO] Asteraceae (V:D)
Bidens frondosa var. *anomala* Porter ex Fern. = Bidens frondosa [BIDEFRO] Asteraceae (V:D)
Bidens frondosa var. *caudata* Sherff = Bidens frondosa [BIDEFRO] Asteraceae (V:D)
Bidens frondosa var. *pallida* Wieg. = Bidens frondosa [BIDEFRO] Asteraceae (V:D)
Bidens frondosa var. *puberula* Wieg. = Bidens frondosa [BIDEFRO] Asteraceae (V:D)
Bidens frondosa var. *stenodonta* Fern. & St. John = Bidens frondosa [BIDEFRO] Asteraceae (V:D)
Bidens glaucescens Greene = Bidens cernua [BIDECER] Asteraceae (V:D)
Bidens puberula Wieg. = Bidens vulgata [BIDEVUL] Asteraceae (V:D)
Bidens vulgata Greene [BIDEVUL] Asteraceae (V:D)
Bidens vulgata var. *puberula* (Wieg.) Greene = Bidens vulgata [BIDEVUL] Asteraceae (V:D)
Bidens vulgata var. *schizantha* Lunell = Bidens vulgata [BIDEVUL] Asteraceae (V:D)
Bilderdykia convolvulus (L.) Dumort. = Polygonum convolvulus [POLYCON] Polygonaceae (V:D)
Bilimbia effusa Auersw. ex Rabenh. = Bacidia auerswaldii [BACIAUE] Bacidiaceae (L)
Bistorta bistortoides (Pursh) Small = Polygonum bistortoides [POLYBIS] Polygonaceae (V:D)
Bistorta vivipara (L.) S.F. Gray = Polygonum viviparum [POLYVIV] Polygonaceae (V:D)
Blasia L. [BLASIA$] Blasiaceae (B:H)
Blasia pusilla L. [BLASPUS] Blasiaceae (B:H)
Blastenia luteominium (Tuck.) Hasse. = Caloplaca luteominia [CALOLUT] Teloschistaceae (L)
Blastenia sinapisperma (Lam. & DC.) A. Massal. = Caloplaca sinapisperma [CALOSIN] Teloschistaceae (L)
Blechnum L. [BLECHNU$] Blechnaceae (V:F)
Blechnum doodioides Hook. = Blechnum spicant [BLECSPI] Blechnaceae (V:F)
Blechnum spicant (L.) Roth [BLECSPI] Blechnaceae (V:F)
Blechnum spicant var. *elongatum* (Hook.) Boivin = Blechnum spicant [BLECSPI] Blechnaceae (V:F)
Blepharostoma (Dum. emend. Lindb.) Dum. [BLEPHAR$] Pseudolepicoleaceae (B:H)

Blepharostoma arachnoideum M.A. Howe [BLEPARA] Pseudolepicoleaceae (B:H)
Blepharostoma arachnoideum var. *brevirete* (Bryhn & Kaal.) Frye & Clark = Blepharostoma trichophyllum ssp. brevirete [BLEPTRI1] Pseudolepicoleaceae (B:H)
Blepharostoma trichophyllum (L.) Dum. [BLEPTRI] Pseudolepicoleaceae (B:H)
Blepharostoma trichophyllum ssp. **brevirete** (Bryhn & Kaal.) Schust. [BLEPTRI1] Pseudolepicoleaceae (B:H)
Blepharostoma trichophyllum ssp. **trichophyllum** [BLEPTRI0] Pseudolepicoleaceae (B:H)
Blindia Bruch & Schimp. in B.S.G. [BLINDIA$] Seligeriaceae (B:M)
Blindia acuta (Hedw.) Bruch & Schimp. in B.S.G. [BLINACU] Seligeriaceae (B:M)
Blitum capitatum L. = Chenopodium capitatum [CHENCAP] Chenopodiaceae (V:D)
Blitum chenopodioides L. = Chenopodium botryodes [CHENBOR] Chenopodiaceae (V:D)
Boechera collinsii (Fern.) A. & D. Löve = Arabis holboellii var. collinsii [ARABHOL1] Brassicaceae (V:D)
Boechera divaricarpa (A. Nels.) A. & D. Löve = Arabis × divaricarpa [ARABDIV] Brassicaceae (V:D)
Boechera drummondii (Gray) A. & D. Löve = Arabis drummondii [ARABDRU] Brassicaceae (V:D)
Boechera holboellii (Hornem.) A. & D. Löve = Arabis holboellii var. holboellii [ARABHOL0] Brassicaceae (V:D)
Boechera retrofracta (Graham) A. & D. Löve = Arabis holboellii var. retrofracta [ARABHOL5] Brassicaceae (V:D)
Boechera tenuis (Böcher) A. & D. Löve = Arabis holboellii var. retrofracta [ARABHOL5] Brassicaceae (V:D)
Boisduvalia Spach [BOISDUV$] Onagraceae (V:D)
Boisduvalia densiflora (Lindl.) S. Wats. [BOISDEN] Onagraceae (V:D)
Boisduvalia densiflora var. *pallescens* Suksdorf = Boisduvalia densiflora [BOISDEN] Onagraceae (V:D)
Boisduvalia densiflora var. *salicina* (Rydb.) Munz = Boisduvalia densiflora [BOISDEN] Onagraceae (V:D)
Boisduvalia glabella (Nutt.) Walp. [BOISGLA] Onagraceae (V:D)
Boisduvalia glabella var. *campestris* (Jepson) Jepson = Boisduvalia glabella [BOISGLA] Onagraceae (V:D)
Boisduvalia salicina Rydb. = Boisduvalia densiflora [BOISDEN] Onagraceae (V:D)
Boisduvalia stricta (Gray) Greene [BOISSTR] Onagraceae (V:D)
Bolboschoenus maritimus (L.) Palla = Scirpus maritimus [SCIRMAR] Cyperaceae (V:M)
Bolboschoenus maritimus ssp. *paludosus* (A. Nels.) A. & D. Löve = Scirpus maritimus [SCIRMAR] Cyperaceae (V:M)
Bolboschoenus paludosus (A. Nels.) Soó = Scirpus maritimus [SCIRMAR] Cyperaceae (V:M)
Borago L. [BORAGO$] Boraginaceae (V:D)
***Borago officinalis** L. [BORAOFF] Boraginaceae (V:D)
Boschniakia C.A. Mey. ex Bong. [BOSCHNI$] Orobanchaceae (V:D)
Boschniakia hookeri Walp. [BOSCHOO] Orobanchaceae (V:D)
Boschniakia rossica (Cham. & Schlecht.) Fedtsch. [BOSCROS] Orobanchaceae (V:D)
Boschniakia strobilacea Gray [BOSCSTR] Orobanchaceae (V:D)
Botrychium Sw. [BOTRICH$] Ophioglossaceae (V:F)
Botrychium boreale auct. = Botrychium pinnatum [BOTRPIN] Ophioglossaceae (V:F)
Botrychium boreale var. *obtusilobum* auct. = Botrychium pinnatum [BOTRPIN] Ophioglossaceae (V:F)
Botrychium californicum Underwood = Botrychium multifidum [BOTRMUL] Ophioglossaceae (V:F)
Botrychium coulteri Underwood = Botrychium multifidum [BOTRMUL] Ophioglossaceae (V:F)
Botrychium lanceolatum (Gmel.) Angstr. [BOTRLAN] Ophioglossaceae (V:F)
Botrychium lunaria (L.) Sw. [BOTRLUN] Ophioglossaceae (V:F)
Botrychium lunaria ssp. *minganense* (Victorin) Calder & Taylor = Botrychium minganense [BOTRMIN] Ophioglossaceae (V:F)
Botrychium lunaria var. *minganense* (Victorin) Dole = Botrychium minganense [BOTRMIN] Ophioglossaceae (V:F)
Botrychium lunaria var. *onondagense* (Underwood) House = Botrychium lunaria [BOTRLUN] Ophioglossaceae (V:F)
Botrychium matricariae (Schrank) Spreng. = Botrychium multifidum [BOTRMUL] Ophioglossaceae (V:F)
Botrychium minganense Victorin [BOTRMIN] Ophioglossaceae (V:F)
Botrychium multifidum (Gmel.) Trev. [BOTRMUL] Ophioglossaceae (V:F)
Botrychium multifidum ssp. *californicum* (Underwood) Clausen = Botrychium multifidum [BOTRMUL] Ophioglossaceae (V:F)
Botrychium multifidum ssp. *coulteri* (Underwood) Clausen = Botrychium multifidum [BOTRMUL] Ophioglossaceae (V:F)
Botrychium multifidum ssp. *silaifolium* (K. Presl) Clausen = Botrychium multifidum [BOTRMUL] Ophioglossaceae (V:F)
Botrychium multifidum var. *californicum* (Underwood) Broun = Botrychium multifidum [BOTRMUL] Ophioglossaceae (V:F)
Botrychium multifidum var. *coulteri* (Underwood) Broun = Botrychium multifidum [BOTRMUL] Ophioglossaceae (V:F)
Botrychium multifidum var. *intermedium* (D.C. Eat.) Farw. = Botrychium multifidum [BOTRMUL] Ophioglossaceae (V:F)
Botrychium multifidum var. *robustum* (Rupr.) C. Christens. = Botrychium multifidum [BOTRMUL] Ophioglossaceae (V:F)
Botrychium multifidum var. *silaifolium* (K. Presl) Broun = Botrychium multifidum [BOTRMUL] Ophioglossaceae (V:F)
Botrychium occidentale Underwood = Botrychium multifidum [BOTRMUL] Ophioglossaceae (V:F)
Botrychium onondagense Underwood = Botrychium lunaria [BOTRLUN] Ophioglossaceae (V:F)
Botrychium paradoxum W.H. Wagner [BOTRPAR] Ophioglossaceae (V:F)
Botrychium pinnatum St. John [BOTRPIN] Ophioglossaceae (V:F)
Botrychium silaifolium K. Presl = Botrychium multifidum [BOTRMUL] Ophioglossaceae (V:F)
Botrychium silaifolium var. *coulteri* (Underwood) Jepson = Botrychium multifidum [BOTRMUL] Ophioglossaceae (V:F)
Botrychium simplex E. Hitchc. [BOTRSIM] Ophioglossaceae (V:F)
Botrychium ternatum var. *intermedium* Eaton = Botrychium multifidum [BOTRMUL] Ophioglossaceae (V:F)
Botrychium virginianum (L.) Sw. [BOTRVIR] Ophioglossaceae (V:F)
Botrychium virginianum ssp. *europaeum* (Angstr.) Jáv. = Botrychium virginianum [BOTRVIR] Ophioglossaceae (V:F)
Botrychium virginianum var. *europaeum* Angstr. = Botrychium virginianum [BOTRVIR] Ophioglossaceae (V:F)
Botrydina botryoides (L.) Redhead & Kuyper = Omphalina umbellifera [OMPHUMB] Tricholomataceae (L)
Botrydina viridis (Ach.) Redhead & Kuyper = Omphalina hudsoniana [OMPHHUD] Tricholomataceae (L)
Botrydina vulgaris Bréb. = Omphalina umbellifera [OMPHUMB] Tricholomataceae (L)
Botrydium botrys (L.) Small = Chenopodium botrys [CHENBOT] Chenopodiaceae (V:D)
Botrypus virginianus (L.) Michx. = Botrychium virginianum [BOTRVIR] Ophioglossaceae (V:F)
Bouteloua Lag. [BOUTELO$] Poaceae (V:M)
Bouteloua gracilis (Willd. ex Kunth) Lag. ex Griffiths [BOUTGRA] Poaceae (V:M)
Bouteloua gracilis var. *stricta* (Vasey) A.S. Hitchc. = Bouteloua gracilis [BOUTGRA] Poaceae (V:M)
Bouteloua oligostachya (Nutt.) Torr. ex Gray = Bouteloua gracilis [BOUTGRA] Poaceae (V:M)
Boykinia Nutt. [BOYKINI$] Saxifragaceae (V:D)
Boykinia elata (Nutt.) Greene = Boykinia occidentalis [BOYKOCC] Saxifragaceae (V:D)
Boykinia occidentalis Torr. & Gray [BOYKOCC] Saxifragaceae (V:D)
Boykinia vancouverensis (Rydb.) Fedde = Boykinia occidentalis [BOYKOCC] Saxifragaceae (V:D)
Brachyactis Ledeb. [BRACHYA$] Asteraceae (V:D)
Brachyactis angusta (Lindl.) Britt. = Brachyactis ciliata ssp. angusta [BRACCIL1] Asteraceae (V:D)
Brachyactis ciliata (Ledeb.) Ledeb. [BRACCIL] Asteraceae (V:D)

Brachyactis ciliata ssp. **angusta** (Lindl.) A.G. Jones [BRACCIL1] Asteraceae (V:D)

Brachydontium Fürnr. [BRACHYD$] Seligeriaceae (B:M)

Brachydontium olympicum (Britt. in Frye) McIntosh & Spence [BRACOLY] Seligeriaceae (B:M)

Brachyodus Fürnr ex Nees & Hornsch. = Brachydontium Fürnr. [BRACHYD$] Seligeriaceae (B:M)

Brachysteleum polyphylloides C. Müll. = Ptychomitrium gardneri [PTYCGAR] Ptychomitriaceae (B:M)

Brachythecium Schimp. in B.S.G. [BRACHYT$] Brachytheciaceae (B:M)

Brachythecium acutum (Mitt.) Sull. [BRACACT] Brachytheciaceae (B:M)

Brachythecium albicans (Hedw.) Schimp. in B.S.G. [BRACALB] Brachytheciaceae (B:M)

Brachythecium albicans var. *occidentale* Ren. & Card. = Brachythecium albicans [BRACALB] Brachytheciaceae (B:M)

Brachythecium asperrimum (Mitt.) Sull. = Brachythecium frigidum [BRACFRI] Brachytheciaceae (B:M)

Brachythecium beringianum Card. & Thér. = Brachythecium albicans [BRACALB] Brachytheciaceae (B:M)

Brachythecium bestii Grout = Brachythecium reflexum var. pacificum [BRACREF1] Brachytheciaceae (B:M)

Brachythecium calcareum Kindb. [BRACCAL] Brachytheciaceae (B:M)

Brachythecium campestre (C. Müll.) Schimp. in B.S.G. [BRACCAM] Brachytheciaceae (B:M)

Brachythecium cavernosum Kindb. = Brachythecium rutabulum [BRACRUT] Brachytheciaceae (B:M)

Brachythecium cirrosum (Schwaegr. in Schultes) Schimp. = Cirriphyllum cirrosum [CIRRCIR] Brachytheciaceae (B:M)

Brachythecium cirrosum var. *coloradense* (Aust.) Wijk & Marg. = Cirriphyllum cirrosum [CIRRCIR] Brachytheciaceae (B:M)

Brachythecium collinum (Schleich. ex C. Müll.) Schimp. in B.S.G. [BRACCOL] Brachytheciaceae (B:M)

Brachythecium collinum var. *idahense* (Ren. & Card.) Grout = Brachythecium velutinum var. velutinum [BRACVEL0] Brachytheciaceae (B:M)

Brachythecium columbico-rutabulum Kindb. in Mac. & Kindb. = Brachythecium frigidum [BRACFRI] Brachytheciaceae (B:M)

Brachythecium curtum (Lindb.) Limpr. = Brachythecium oedipodium [BRACOED] Brachytheciaceae (B:M)

Brachythecium edentatum Williams = Warnstorfia pseudostraminea [WARNPSE] Amblystegiaceae (B:M)

Brachythecium erythrorrhizon Schimp. in B.S.G. [BRACERY] Brachytheciaceae (B:M)

Brachythecium erythrorrhizon var. *suberythrorrhizon* (Ren. & Card.) Grout = Brachythecium velutinum var. venustum [BRACVEL1] Brachytheciaceae (B:M)

Brachythecium fendleri var. *idahense* (Ren. & Card.) Wijk & Marg. = Brachythecium velutinum var. velutinum [BRACVEL0] Brachytheciaceae (B:M)

Brachythecium flagellare (Hedw.) Jenn. = Brachythecium plumosum [BRACPLU] Brachytheciaceae (B:M)

Brachythecium flagellare var. *homomallum* (Schimp. in B.S.G.) Jenn. = Brachythecium plumosum [BRACPLU] Brachytheciaceae (B:M)

Brachythecium flagellare var. *pringlei* (Williams) Grout = Brachythecium plumosum [BRACPLU] Brachytheciaceae (B:M)

Brachythecium flagellare var. *roellii* (Ren. & Card.) Grout = Brachythecium plumosum [BRACPLU] Brachytheciaceae (B:M)

Brachythecium flexicaule Ren. & Card. in Grout = Brachythecium calcareum [BRACCAL] Brachytheciaceae (B:M)

Brachythecium frigidum (C. Müll.) Besch. [BRACFRI] Brachytheciaceae (B:M)

Brachythecium gemmascens C. Müll. & Kindb. in Mac. & Kindb. = Brachythecium frigidum [BRACFRI] Brachytheciaceae (B:M)

Brachythecium groenlandicum (C. Jens.) Schljak. [BRACGRO] Brachytheciaceae (B:M)

Brachythecium harpidioides C. Müll. & Kindb. in Mac. & Kindb. = Brachythecium erythrorrhizon [BRACERY] Brachytheciaceae (B:M)

Brachythecium holzingeri (Grout) Grout [BRACHOL] Brachytheciaceae (B:M)

Brachythecium hylotapetum B. Hig. & N. Hig. [BRACHYL] Brachytheciaceae (B:M)

Brachythecium idahense Ren. & Card. = Brachythecium velutinum var. velutinum [BRACVEL0] Brachytheciaceae (B:M)

Brachythecium labradoricum (Kindb.) Par. = Brachythecium calcareum [BRACCAL] Brachytheciaceae (B:M)

Brachythecium laetum (Brid.) B.S.G. = Brachythecium oxycladon [BRACOXY] Brachytheciaceae (B:M)

Brachythecium laevisetum Kindb. = Brachythecium salebrosum [BRACSAL] Brachytheciaceae (B:M)

Brachythecium lamprochryseum C. Müll. & Kindb. in Mac. & Kindb. = Brachythecium frigidum [BRACFRI] Brachytheciaceae (B:M)

Brachythecium lamprochryseum var. *giganteum* Grout = Brachythecium frigidum [BRACFRI] Brachytheciaceae (B:M)

Brachythecium lamprochryseum var. *solfatarense* Grout = Brachythecium frigidum [BRACFRI] Brachytheciaceae (B:M)

Brachythecium latifolium Kindb. = Brachythecium nelsonii [BRACNEL] Brachytheciaceae (B:M)

Brachythecium leibergii Grout [BRACLEI] Brachytheciaceae (B:M)

Brachythecium leucoglaucum C. Müll. & Kindb. in Mac. & Kindb. = Brachythecium rutabulum [BRACRUT] Brachytheciaceae (B:M)

Brachythecium mirabundum C. Müll. & Kindb. in Mac. & Kindb. = Brachythecium rutabulum [BRACRUT] Brachytheciaceae (B:M)

Brachythecium nanopes C. Müll. & Kindb. in Mac. & Kindb. = Brachythecium populeum [BRACPOP] Brachytheciaceae (B:M)

Brachythecium nelsonii Grout [BRACNEL] Brachytheciaceae (B:M)

Brachythecium oedipodium (Mitt.) Jaeg. [BRACOED] Brachytheciaceae (B:M)

Brachythecium oxycladon (Brid.) Jaeg. [BRACOXY] Brachytheciaceae (B:M)

Brachythecium oxycladon var. *dentatum* (Lesq. & James) Grout = Brachythecium oxycladon [BRACOXY] Brachytheciaceae (B:M)

Brachythecium pacificum Jenn. = Brachythecium frigidum [BRACFRI] Brachytheciaceae (B:M)

Brachythecium pennellii Bartr. = Leptodictyum riparium [LEPTRIP] Amblystegiaceae (B:M)

Brachythecium petrophilum Williams = Brachythecium velutinum var. venustum [BRACVEL1] Brachytheciaceae (B:M)

Brachythecium piliferum (Hedw.) Kindb. = Cirriphyllum piliferum [CIRRPIL] Brachytheciaceae (B:M)

Brachythecium platycladum C. Müll. & Kindb. in Mac. & Kindb. = Brachythecium rutabulum [BRACRUT] Brachytheciaceae (B:M)

Brachythecium plumosum (Hedw.) Schimp. in B.S.G. [BRACPLU] Brachytheciaceae (B:M)

Brachythecium plumosum var. *homomallum* Schimp. in B.S.G. = Brachythecium plumosum [BRACPLU] Brachytheciaceae (B:M)

Brachythecium plumosum var. *pringlei* (Williams) Grout = Brachythecium plumosum [BRACPLU] Brachytheciaceae (B:M)

Brachythecium plumosum var. *roellii* (Ren. & Card.) Grout = Brachythecium plumosum [BRACPLU] Brachytheciaceae (B:M)

Brachythecium populeum (Hedw.) Schimp. in B.S.G. [BRACPOP] Brachytheciaceae (B:M)

Brachythecium populeum var. *majus* Schimp. in B.S.G. = Brachythecium populeum [BRACPOP] Brachytheciaceae (B:M)

Brachythecium populeum var. *ovatum* Grout = Brachythecium populeum [BRACPOP] Brachytheciaceae (B:M)

Brachythecium populeum var. *rufescens* Schimp. in B.S.G. = Brachythecium populeum [BRACPOP] Brachytheciaceae (B:M)

Brachythecium pseudoalbicans Kindb. = Brachythecium albicans [BRACALB] Brachytheciaceae (B:M)

Brachythecium pseudostarkei Ren. & Card. = Brachythecium frigidum [BRACFRI] Brachytheciaceae (B:M)

Brachythecium reflexum (Starke in Web. & Mohr) Schimp. in B.S.G. [BRACREF] Brachytheciaceae (B:M)

Brachythecium reflexum var. **pacificum** Ren. & Card. in Röll [BRACREF1] Brachytheciaceae (B:M)

Brachythecium reflexum var. **reflexum** [BRACREF0] Brachytheciaceae (B:M)

Brachythecium rivulare Schimp. in B.S.G. [BRACRIV] Brachytheciaceae (B:M)

Brachythecium rivulare var. *cataractarum* Saut. = Brachythecium rivulare [BRACRIV] Brachytheciaceae (B:M)

Brachythecium rivulare var. *lamoillense* Grout = Brachythecium rivulare [BRACRIV] Brachytheciaceae (B:M)
Brachythecium rivulare var. *laxum* Grout = Brachythecium rivulare [BRACRIV] Brachytheciaceae (B:M)
Brachythecium rivulare var. *noveboracense* (Grout) Robins. = Brachythecium rivulare [BRACRIV] Brachytheciaceae (B:M)
Brachythecium rivulare var. *obtusulum* Kindb. in Mac. & Kindb. = Brachythecium rivulare [BRACRIV] Brachytheciaceae (B:M)
Brachythecium roellii Ren. & Card. = Brachythecium plumosum [BRACPLU] Brachytheciaceae (B:M)
Brachythecium rutabuliforme Kindb. in Mac. & Kindb. = Brachythecium plumosum [BRACPLU] Brachytheciaceae (B:M)
Brachythecium rutabulum (Hedw.) Schimp. in B.S.G. [BRACRUT] Brachytheciaceae (B:M)
Brachythecium rutabulum var. *flavescens* Schimp. in B.S.G. = Brachythecium rutabulum [BRACRUT] Brachytheciaceae (B:M)
Brachythecium rutabulum var. *turgescens* Limpr. = Brachythecium rutabulum [BRACRUT] Brachytheciaceae (B:M)
Brachythecium salebrosum (Web. & Mohr) Schimp. in B.S.G. [BRACSAL] Brachytheciaceae (B:M)
Brachythecium salebrosum ssp. *mamilligerum* Kindb. in Mac. & Kindb. = Brachythecium acutum [BRACACT] Brachytheciaceae (B:M)
Brachythecium salebrosum var. *densum* Schimp. in B.S.G. = Brachythecium calcareum [BRACCAL] Brachytheciaceae (B:M)
Brachythecium salebrosum var. *waghornei* Ren. & Card. = Brachythecium albicans [BRACALB] Brachytheciaceae (B:M)
Brachythecium serrulatum (Hedw.) Robins. = Steerecleus serrulatus [STEESER] Brachytheciaceae (B:M)
Brachythecium spurio-acuminatum C. Müll. & Kindb. in Mac. & Kindb. = Brachythecium oxycladon [BRACOXY] Brachytheciaceae (B:M)
Brachythecium spurio-rutabulum C. Müll. & Kindb. in Mac. & Kindb. = Brachythecium frigidum [BRACFRI] Brachytheciaceae (B:M)
Brachythecium starkei (Brid.) Schimp. in B.S.G. [BRACSTA] Brachytheciaceae (B:M)
Brachythecium starkei ssp. *oedipodium* (Mitt.) Ren. & Card. = Brachythecium oedipodium [BRACOED] Brachytheciaceae (B:M)
Brachythecium starkei var. *curtum* (Lindb.) Warnst. = Brachythecium oedipodium [BRACOED] Brachytheciaceae (B:M)
Brachythecium starkei var. *explanatum* (Brid.) Mönk. = Brachythecium oedipodium [BRACOED] Brachytheciaceae (B:M)
Brachythecium starkei var. *pacificum* (Ren. & Card. in Röll) Lawt. = Brachythecium reflexum var. pacificum [BRACREF1] Brachytheciaceae (B:M)
Brachythecium subasperrimum Card. & Thér. = Brachythecium frigidum [BRACFRI] Brachytheciaceae (B:M)
Brachythecium suberythrorrhizon Ren. & Card. = Brachythecium velutinum var. venustum [BRACVEL1] Brachytheciaceae (B:M)
Brachythecium subintricatum Kindb. = Brachythecium frigidum [BRACFRI] Brachytheciaceae (B:M)
Brachythecium trachypodium (Brid.) Schimp. in B.S.G. [BRACTRA] Brachytheciaceae (B:M)
Brachythecium tromsoense auct. = Brachythecium holzingeri [BRACHOL] Brachytheciaceae (B:M)
Brachythecium turgidum (Hartm.) Kindb. [BRACTUR] Brachytheciaceae (B:M)
Brachythecium velutinum (Hedw.) Schimp. in B.S.G. [BRACVEL] Brachytheciaceae (B:M)
Brachythecium velutinum var. **velutinum** [BRACVEL0] Brachytheciaceae (B:M)
Brachythecium velutinum var. **venustum** (De Not.) Arc. [BRACVEL1] Brachytheciaceae (B:M)
Brachythecium washingtonianum D.C. Eat. in Grout = Brachythecium frigidum [BRACFRI] Brachytheciaceae (B:M)
Brasenia Schreb. [BRASENI$] Cabombaceae (V:D)
Brasenia peltata Pursh = Brasenia schreberi [BRASSCH] Cabombaceae (V:D)
Brasenia schreberi J.F. Gmel. [BRASSCH] Cabombaceae (V:D)
Brassica L. [BRASSIC$] Brassicaceae (V:D)
Brassica alba Rabenh. = Sinapis alba [SINAALB] Brassicaceae (V:D)
Brassica arvensis Rabenh. = Sinapis arvensis [SINAARV] Brassicaceae (V:D)
Brassica campestris L. = Brassica rapa [BRASRAP] Brassicaceae (V:D)
Brassica campestris var. *rapa* (L.) Hartman = Brassica rapa [BRASRAP] Brassicaceae (V:D)
Brassica erucastrum L. = Erucastrum gallicum [ERUCGAL] Brassicaceae (V:D)
Brassica hirta Moench = Sinapis alba [SINAALB] Brassicaceae (V:D)
Brassica integrifolia Rupr. = Brassica juncea [BRASJUN] Brassicaceae (V:D)
Brassica integrifolia (Vahl) Schulz = Brassica juncea [BRASJUN] Brassicaceae (V:D)
Brassica japonica Thunb. = Brassica juncea [BRASJUN] Brassicaceae (V:D)
***Brassica juncea** (L.) Czern. [BRASJUN] Brassicaceae (V:D)
Brassica juncea var. *crispifolia* Bailey = Brassica juncea [BRASJUN] Brassicaceae (V:D)
Brassica juncea var. *japonica* (Thunb.) Bailey = Brassica juncea [BRASJUN] Brassicaceae (V:D)
Brassica kaber (DC.) L.C. Wheeler = Sinapis arvensis [SINAARV] Brassicaceae (V:D)
Brassica kaber var. *pinnatifida* (Stokes) L.C. Wheeler = Sinapis arvensis [SINAARV] Brassicaceae (V:D)
Brassica kaber var. *schkuhriana* (Reichenb.) L.C. Wheeler = Sinapis arvensis [SINAARV] Brassicaceae (V:D)
Brassica napobrassica (L.) P. Mill. = Brassica napus [BRASNAP] Brassicaceae (V:D)
***Brassica napus** L. [BRASNAP] Brassicaceae (V:D)
Brassica napus var. *napobrassica* (L.) Reichenb. = Brassica napus [BRASNAP] Brassicaceae (V:D)
***Brassica nigra** (L.) W.D.J. Koch [BRASNIG] Brassicaceae (V:D)
Brassica orientalis L. = Conringia orientalis [CONRORI] Brassicaceae (V:D)
***Brassica rapa** L. [BRASRAP] Brassicaceae (V:D)
Brassica rapa ssp. *campestris* (L.) Clapham = Brassica rapa [BRASRAP] Brassicaceae (V:D)
Brassica rapa ssp. *olifera* DC. = Brassica rapa [BRASRAP] Brassicaceae (V:D)
Brassica rapa ssp. *sylvestris* Janchen = Brassica rapa [BRASRAP] Brassicaceae (V:D)
Brassica rapa var. *campestris* (L.) W.D.J. Koch = Brassica rapa [BRASRAP] Brassicaceae (V:D)
Brassica willdenowii Boiss. = Brassica juncea [BRASJUN] Brassicaceae (V:D)
Braunia californica Lesq. = Pseudobraunia californica [PSEUCAL] Hedwigiaceae (B:M)
Braxilia minor (L.) House = Pyrola minor [PYROMIN] Pyrolaceae (V:D)
Braya Sternb. & Hoppe [BRAYA$] Brassicaceae (V:D)
Braya alpina var. *americana* (Hook.) S. Wats. = Braya glabella [BRAYGLA] Brassicaceae (V:D)
Braya alpina var. *glabella* (Richards.) S. Wats. = Braya glabella [BRAYGLA] Brassicaceae (V:D)
Braya americana (Hook.) Fern. = Braya glabella [BRAYGLA] Brassicaceae (V:D)
Braya arctica Hook. = Braya glabella [BRAYGLA] Brassicaceae (V:D)
Braya bartlettiana Jordal = Braya glabella [BRAYGLA] Brassicaceae (V:D)
Braya glabella Richards. [BRAYGLA] Brassicaceae (V:D)
Braya henryae Raup = Braya glabella [BRAYGLA] Brassicaceae (V:D)
Braya humilis (C.A. Mey.) B.L. Robins. [BRAYHUM] Brassicaceae (V:D)
Braya humilis ssp. *abbei* Böcher = Braya humilis [BRAYHUM] Brassicaceae (V:D)
Braya humilis ssp. *arctica* (Böcher) Rollins = Braya humilis [BRAYHUM] Brassicaceae (V:D)
Braya humilis ssp. *richardsonii* (Rydb.) Hultén = Braya humilis [BRAYHUM] Brassicaceae (V:D)

Braya humilis ssp. *ventosa* Rollins = Braya humilis [BRAYHUM] Brassicaceae (V:D)
Braya humilis var. *abbei* (Böcher) Boivin = Braya humilis [BRAYHUM] Brassicaceae (V:D)
Braya humilis var. *americana* (Hook.) Boivin = Braya glabella [BRAYGLA] Brassicaceae (V:D)
Braya humilis var. *arctica* (Böcher) Boivin = Braya humilis [BRAYHUM] Brassicaceae (V:D)
Braya humilis var. *glabella* (Richards.) Boivin = Braya glabella [BRAYGLA] Brassicaceae (V:D)
Braya humilis var. *interior* (Böcher) Boivin = Braya humilis [BRAYHUM] Brassicaceae (V:D)
Braya humilis var. *laurentiana* (Böcher) Boivin = Braya humilis [BRAYHUM] Brassicaceae (V:D)
Braya humilis var. *leiocarpa* (Trautv.) Fern. = Braya humilis [BRAYHUM] Brassicaceae (V:D)
Braya humilis var. *novae-angliae* (Rydb.) Fern. = Braya humilis [BRAYHUM] Brassicaceae (V:D)
Braya humilis var. *ventosa* (Rollins) Boivin = Braya humilis [BRAYHUM] Brassicaceae (V:D)
Braya intermedia Sorensen = Braya humilis [BRAYHUM] Brassicaceae (V:D)
Braya novae-angliae (Rydb.) Sorensen = Braya humilis [BRAYHUM] Brassicaceae (V:D)
Braya novae-angliae var. *interior* Böcher = Braya humilis [BRAYHUM] Brassicaceae (V:D)
Braya novae-angliae var. *laurentiana* Böcher = Braya humilis [BRAYHUM] Brassicaceae (V:D)
Braya purpurascens (R. Br.) Bunge [BRAYPUR] Brassicaceae (V:D)
Braya richardsonii (Rydb.) Fern. = Braya humilis [BRAYHUM] Brassicaceae (V:D)
Breidleria arcuata (Mol.) Loeske = Hypnum lindbergii [HYPNLIN] Hypnaceae (B:M)
Breidleria pratensis (Rabenh.) Loeske = Hypnum pratense [HYPNPRA] Hypnaceae (B:M)
Brickellia Ell. [BRICKEL$] Asteraceae (V:D)
Brickellia grandiflora (Hook.) Nutt. [BRICGRA] Asteraceae (V:D)
Brickellia grandiflora var. *petiolaris* Gray = Brickellia grandiflora [BRICGRA] Asteraceae (V:D)
Brickellia oblongifolia Nutt. [BRICOBL] Asteraceae (V:D)
Brickellia oblongifolia var. *typica* B.L. Robins. = Brickellia oblongifolia [BRICOBL] Asteraceae (V:D)
Brigantiaea Trevisan [BRIGANT$] Brigantiaeaceae (L)
Brigantiaea purpurata (Zahlbr.) Haf. & Bellem. [BRIGPUR] Brigantiaeaceae (L)
Briza L. [BRIZA$] Poaceae (V:M)
*__Briza maxima__ L. [BRIZMAX] Poaceae (V:M)
*__Briza minor__ L. [BRIZMIN] Poaceae (V:M)
Brodiaea Sm. [BRODIAE$] Liliaceae (V:M)
Brodiaea coronaria (Salisb.) Engl. [BRODCOR] Liliaceae (V:M)
Brodiaea douglasii S. Wats. = Triteleia grandiflora [TRITGRA] Liliaceae (V:M)
Brodiaea douglasii var. *howellii* (S. Wats.) M.E. Peck = Triteleia howellii [TRITHOW] Liliaceae (V:M)
Brodiaea grandiflora (Lindl.) J.F. Macbr. = Triteleia grandiflora [TRITGRA] Liliaceae (V:M)
Brodiaea howellii S. Wats. = Triteleia howellii [TRITHOW] Liliaceae (V:M)
Brodiaea hyacinthina (Lindl.) Baker = Triteleia hyacinthina [TRITHYA] Liliaceae (V:M)
Brodoa Goward [BRODOA$] Parmeliaceae (L)
Brodoa intestiniformis auct. = Brodoa oroarctica [BRODORO] Parmeliaceae (L)
Brodoa oroarctica (Krog) Goward [BRODORO] Parmeliaceae (L)
Bromelica bulbosa (Geyer ex Porter & Coult.) W.A. Weber = Melica bulbosa [MELIBUL] Poaceae (V:M)
Bromelica smithii (Porter ex Gray) Farw. = Melica smithii [MELISMI] Poaceae (V:M)
Bromelica spectabilis (Scribn.) W.A. Weber = Melica spectabilis [MELISPE] Poaceae (V:M)
Bromopsis anomala (Rupr. ex Fourn.) Holub = Bromus anomalus [BROMANO] Poaceae (V:M)
Bromopsis canadensis (Michx.) Holub = Bromus canadensis [BROMCAN] Poaceae (V:M)
Bromopsis ciliata (L.) Holub = Bromus ciliatus [BROMCIL] Poaceae (V:M)
Bromopsis dicksonii (Mitchell & Wilton) A. & D. Löve = Bromus inermis ssp. pumpellianus var. pumpellianus [BROMINE3] Poaceae (V:M)
Bromopsis inermis ssp. *pumpelliana* (Scribn.) W.A. Weber = Bromus inermis ssp. pumpellianus var. pumpellianus [BROMINE3] Poaceae (V:M)
Bromopsis pacifica (Shear) Holub = Bromus pacificus [BROMPAC] Poaceae (V:M)
Bromopsis porteri Coult. = Bromus anomalus [BROMANO] Poaceae (V:M)
Bromopsis pumpelliana (Scribn.) Holub = Bromus inermis ssp. pumpellianus var. pumpellianus [BROMINE3] Poaceae (V:M)
Bromopsis richardsonii (Link) Holub = Bromus canadensis [BROMCAN] Poaceae (V:M)
Bromopsis vulgaris (Hook.) Holub = Bromus vulgaris [BROMVUL] Poaceae (V:M)
Bromus L. [BROMUS$] Poaceae (V:M)
Bromus aleutensis Trin. ex Griseb. = Bromus sitchensis var. aleutensis [BROMSIT1] Poaceae (V:M)
Bromus anomalus Rupr. ex Fourn. [BROMANO] Poaceae (V:M)
*__Bromus briziformis__ Fisch. & C.A. Mey. [BROMBRI] Poaceae (V:M)
Bromus canadensis Michx. [BROMCAN] Poaceae (V:M)
Bromus carinatus Hook. & Arn. [BROMCAR] Poaceae (V:M)
Bromus carinatus var. *californicus* (Nutt. ex Buckl.) Shear = Bromus carinatus [BROMCAR] Poaceae (V:M)
Bromus carinatus var. *hookerianus* (Thurb.) Shear = Bromus carinatus [BROMCAR] Poaceae (V:M)
Bromus carinatus var. *marginatus* (Nees ex Steud.) A.S. Hitchc. = Bromus marginatus [BROMMAR] Poaceae (V:M)
Bromus ciliatus L. [BROMCIL] Poaceae (V:M)
Bromus ciliatus var. *coloradensis* Vasey ex Beal = Bromus inermis ssp. pumpellianus var. pumpellianus [BROMINE3] Poaceae (V:M)
Bromus ciliatus var. *genuinus* Fern. = Bromus ciliatus [BROMCIL] Poaceae (V:M)
Bromus ciliatus var. *porteri* (Coult.) Rydb. = Bromus anomalus [BROMANO] Poaceae (V:M)
Bromus ciliatus var. *richardsonii* (Link) Boivin = Bromus canadensis [BROMCAN] Poaceae (V:M)
*__Bromus commutatus__ Schrad. [BROMCOM] Poaceae (V:M)
Bromus commutatus var. *apricorum* Simonkai = Bromus commutatus [BROMCOM] Poaceae (V:M)
Bromus dertonensis All. = Vulpia bromoides [VULPBRO] Poaceae (V:M)
*__Bromus diandrus__ Roth [BROMDIA] Poaceae (V:M)
Bromus dudleyi Fern. = Bromus canadensis [BROMCAN] Poaceae (V:M)
Bromus eximius (Shear) Piper = Bromus vulgaris [BROMVUL] Poaceae (V:M)
Bromus gussonei Parl. = Bromus diandrus [BROMDIA] Poaceae (V:M)
Bromus hookerianus Thurb. = Bromus carinatus [BROMCAR] Poaceae (V:M)
*__Bromus hordeaceus__ L. [BROMHOR] Poaceae (V:M)
*__Bromus hordeaceus__ ssp. **hordeaceus** [BROMHOR0] Poaceae (V:M)
*__Bromus hordeaceus__ ssp. **thominei** (Hardham ex Nyman) Victorin & Weiller [BROMHOR2] Poaceae (V:M)
Bromus inermis Leyss. [BROMINE] Poaceae (V:M)
*__Bromus inermis__ ssp. **inermis** [BROMINE0] Poaceae (V:M)
Bromus inermis ssp. **pumpellianus** (Scribn.) Wagnon [BROMINE2] Poaceae (V:M)
Bromus inermis ssp. **pumpellianus** var. **pumpellianus** (Scribn.) C.L. Hitchc. [BROMINE3] Poaceae (V:M)
Bromus inermis ssp. **pumpellianus** var. **purpurascens** (Hook.) Wagnon [BROMINE1] Poaceae (V:M)
Bromus inermis var. *tweedyi* (Scribn. ex Beal) C.L. Hitchc. = Bromus inermis ssp. pumpellianus var. purpurascens [BROMINE1] Poaceae (V:M)

*****Bromus japonicus*** Thunb. ex Murr. [BROMJAP] Poaceae (V:M)
Bromus japonicus var. *porrectus* Hack. = Bromus japonicus [BROMJAP] Poaceae (V:M)
Bromus marginatus Nees ex Steud. [BROMMAR] Poaceae (V:M)
Bromus maximus auct. = Bromus diandrus [BROMDIA] Poaceae (V:M)
Bromus mollis auct. = Bromus hordeaceus ssp. hordeaceus [BROMHOR0] Poaceae (V:M)
Bromus pacificus Shear [BROMPAC] Poaceae (V:M)
Bromus patulus Mert. & Koch = Bromus japonicus [BROMJAP] Poaceae (V:M)
Bromus porteri (Coult.) Nash = Bromus anomalus [BROMANO] Poaceae (V:M)
Bromus pumpellianus Scribn. = Bromus inermis ssp. pumpellianus var. pumpellianus [BROMINE3] Poaceae (V:M)
Bromus pumpellianus ssp. *dicksonii* Mitchell & Wilton = Bromus inermis ssp. pumpellianus var. pumpellianus [BROMINE3] Poaceae (V:M)
Bromus pumpellianus var. *tweedyi* Scribn. ex Beal = Bromus inermis ssp. pumpellianus var. purpurascens [BROMINE1] Poaceae (V:M)
Bromus pumpellianus var. *villosissimus* Hultén = Bromus inermis ssp. pumpellianus var. pumpellianus [BROMINE3] Poaceae (V:M)
*****Bromus racemosus*** L. [BROMRAC] Poaceae (V:M)
Bromus richardsonii Link = Bromus canadensis [BROMCAN] Poaceae (V:M)
Bromus richardsonii var. *pallidus* (Hook.) Shear = Bromus canadensis [BROMCAN] Poaceae (V:M)
Bromus rigidus auct. = Bromus diandrus [BROMDIA] Poaceae (V:M)
Bromus rigidus var. *gussonei* (Parl.) Coss. & Durieu = Bromus diandrus [BROMDIA] Poaceae (V:M)
*****Bromus secalinus*** L. [BROMSEC] Poaceae (V:M)
Bromus secalinus var. *hirsutus* Kindb. = Bromus secalinus [BROMSEC] Poaceae (V:M)
Bromus secalinus var. *hirtus* (F.W. Schultz) Hegi = Bromus secalinus [BROMSEC] Poaceae (V:M)
Bromus sitchensis Trin. [BROMSIT] Poaceae (V:M)
Bromus sitchensis var. **aleutensis** (Trin. ex Griseb.) Hultén [BROMSIT1] Poaceae (V:M)
Bromus sitchensis var. *marginatus* (Nees ex Steud.) Boivin = Bromus marginatus [BROMMAR] Poaceae (V:M)
Bromus sitchensis var. **sitchensis** [BROMSIT0] Poaceae (V:M)
*****Bromus squarrosus*** L. [BROMSQU] Poaceae (V:M)
*****Bromus sterilis*** L. [BROMSTE] Poaceae (V:M)
*****Bromus tectorum*** L. [BROMTEC] Poaceae (V:M)
Bromus thominei Hardham ex Nyman = Bromus hordeaceus ssp. thominei [BROMHOR2] Poaceae (V:M)
Bromus villosus Forsk. = Bromus diandrus [BROMDIA] Poaceae (V:M)
Bromus vulgaris (Hook.) Shear [BROMVUL] Poaceae (V:M)
Bromus vulgaris var. *eximius* Shear = Bromus vulgaris [BROMVUL] Poaceae (V:M)
Bromus vulgaris var. *robustus* Shear = Bromus vulgaris [BROMVUL] Poaceae (V:M)
Brotherella Loeske ex Fleisch. [BROTHEE$] Sematophyllaceae (B:M)
Brotherella canadensis Schof. [BROTCAN] Sematophyllaceae (B:M)
Brotherella roellii (Ren. & Card. in Röll) Fleisch. [BROTROE] Sematophyllaceae (B:M)
Bruniera columbiana (Karst.) Nieuwl. = Wolffia columbiana [WOLFCOL] Lemnaceae (V:M)
Bryhnia Kaur. [BRYHNIA$] Brachytheciaceae (B:M)
Bryhnia brittoniae (Grout) Robins. = Eurhynchium praelongum [EURHPRA] Brachytheciaceae (B:M)
Bryhnia hultenii Bartr. in Grout [BRYHHUL] Brachytheciaceae (B:M)
Bryhnia oregana (Sull.) Robins. = Eurhynchium oreganum [EURHORE] Brachytheciaceae (B:M)
Bryhnia stokesii (Turn.) Robins. = Eurhynchium praelongum [EURHPRA] Brachytheciaceae (B:M)
Bryobrittonia Williams [BRYOBRI$] Encalyptaceae (B:M)
Bryobrittonia longipes (Williams) Horton [BRYOLON] Encalyptaceae (B:M)
Bryobrittonia pellucida Williams = Bryobrittonia longipes [BRYOLON] Encalyptaceae (B:M)
Bryocaulon Kärnefelt [BRYOCAU$] Parmeliaceae (L)
Bryocaulon divergens (Ach.) Kärnefelt [BRYODIV] Parmeliaceae (L)
Bryocaulon pseudosatoanum (Asah.) Kärnefelt [BRYOPSD] Parmeliaceae (L)
Bryoerythrophyllum Chen [BRYOERY$] Pottiaceae (B:M)
Bryoerythrophyllum alpigenum (Vent.) Chen = Bryoerythrophyllum recurvirostre [BRYOREC] Pottiaceae (B:M)
Bryoerythrophyllum columbianum (Herm. & Lawt.) Zand. [BRYOCOL] Pottiaceae (B:M)
Bryoerythrophyllum ferruginascens (Stirt.) Giac. [BRYOFER] Pottiaceae (B:M)
Bryoerythrophyllum recurvifolium (Tayl.) Zand. = Paraleptodontium recurvifolium [PARAREC] Pottiaceae (B:M)
Bryoerythrophyllum recurvirostre (Hedw.) Chen [BRYOREC] Pottiaceae (B:M)
Bryoerythrophyllum recurvirostre var. *dentatum* (Schimp.) Crum, Steere & Anderson = Bryoerythrophyllum recurvirostre [BRYOREC] Pottiaceae (B:M)
Bryohaplocladium Wat. & Iwats. [BRYOHAP$] Leskeaceae (B:M)
Bryohaplocladium microphyllum (Hedw.) Wat. & Iwats. [BRYOMIC] Leskeaceae (B:M)
Bryolawtonia vancouveriensis (Kindb. in Mac.) Enroth & Norris = Porotrichum vancouveriense [POROVAN] Thamnobryaceae (B:M)
Bryonora Poelt [BRYONOA$] Lecanoraceae (L)
Bryonora castanea (Hepp) Poelt [BRYOCAS] Lecanoraceae (L)
Bryonora curvescens (Mudd) Poelt [BRYOCUR] Lecanoraceae (L)
Bryonora pruinosa (Th. Fr.) Holtan-Hartwig [BRYOPRU] Lecanoraceae (L)
Bryopogon divergens (Ach.) Schwend. = Bryocaulon divergens [BRYODIV] Parmeliaceae (L)
Bryopogon fremontii (Tuck.) Rabenh. = Bryoria fremontii [BRYOFRE] Parmeliaceae (L)
Bryopogon jubata var. *chalybeiformis* (Ach.) Rabenh. = Bryoria chalybeiformis [BRYOCHA] Parmeliaceae (L)
Bryopogon jubata var. *implexa* Fr. = Bryoria implexa [BRYOIMP] Parmeliaceae (L)
Bryopogon negativus Gyelnik = Bryoria lanestris [BRYOLAN] Parmeliaceae (L)
Bryopogon pacificus Gyelnik = Bryoria fuscescens [BRYOFUS] Parmeliaceae (L)
Bryoria Brodo & D. Hawksw. [BRYORIA$] Parmeliaceae (L)
Bryoria abbreviata (Müll. Arg.) Brodo & D. Hawksw. = Nodobryoria abbreviata [NODOABB] Parmeliaceae (L)
Bryoria bicolor (Ehrh.) Brodo & D. Hawksw. [BRYOBIC] Parmeliaceae (L)
Bryoria capillaris (Ach.) Brodo & D. Hawksw. [BRYOCAP] Parmeliaceae (L)
Bryoria carlottae Brodo & D. Hawksw. [BRYOCAR] Parmeliaceae (L)
Bryoria cervinula Mot. ex Brodo & D. Hawksw. [BRYOCER] Parmeliaceae (L)
Bryoria chalybeiformis (L.) Brodo & D. Hawksw. [BRYOCHA] Parmeliaceae (L)
Bryoria fremontii (Tuck.) Brodo & D. Hawksw. [BRYOFRE] Parmeliaceae (L)
Bryoria friabilis Brodo & D. Hawksw. [BRYOFRI] Parmeliaceae (L)
Bryoria fuscescens (Gyelnik) Brodo & D. Hawksw. [BRYOFUS] Parmeliaceae (L)
Bryoria glabra (Mot.) Brodo & D. Hawksw. [BRYOGLA] Parmeliaceae (L)
Bryoria implexa (Hoffm.) Brodo & D. Hawksw. [BRYOIMP] Parmeliaceae (L)
Bryoria lanestris (Ach.) Brodo & D. Hawksw. [BRYOLAN] Parmeliaceae (L)
Bryoria nitidula (Th. Fr.) Brodo & D. Hawksw. [BRYONIT] Parmeliaceae (L)
Bryoria oregana (Tuck. ex Willey) Brodo & D. Hawksw. = Nodobryoria oregana [NODOORE] Parmeliaceae (L)
Bryoria pikei Brodo & D. Hawksw. [BRYOPIK] Parmeliaceae (L)

Bryoria pseudofuscescens (Gyelnik) Brodo & D. Hawksw. [BRYOPSE] Parmeliaceae (L)
Bryoria simplicior (Vainio) Brodo & D. Hawksw. [BRYOSIM] Parmeliaceae (L)
Bryoria subcana (Nyl. ex Stizenb.) Brodo & D. Hawksw. [BRYOSUB] Parmeliaceae (L)
Bryoria subdivergens (E. Dahl) Brodo & D. Hawksw. = Nodobryoria subdivergens [NODOSUB] Parmeliaceae (L)
Bryoria tenuis (E. Dahl) Brodo & D. Hawksw. [BRYOTEN] Parmeliaceae (L)
Bryoria tortuosa (G. Merr.) Brodo & D. Hawksw. [BRYOTOR] Parmeliaceae (L)
Bryoria trichodes (Michaux) Brodo & D. Hawksw. [BRYOTRI] Parmeliaceae (L)
Bryoria trichodes ssp. **americana** (Mot.) Brodo & D. Hawksw. [BRYOTRI1] Parmeliaceae (L)
Bryoria trichodes ssp. **trichodes** [BRYOTRI0] Parmeliaceae (L)
Bryoria vrangiana (Gyelnik) Brodo & D. Hawksw. = Bryoria implexa [BRYOIMP] Parmeliaceae (L)
Bryum Hedw. [BRYUM$] Bryaceae (B:M)
Bryum aciculinum (Hoppe & Hornsch.) B.S.G. = Pohlia elongata var. greenii [POHLELO1] Bryaceae (B:M)
Bryum aciculinum Kindb. = Bryum pseudotriquetrum [BRYUPSE] Bryaceae (B:M)
Bryum acutiusculum C. Müll. = Bryum amblyodon [BRYUAMB] Bryaceae (B:M)
Bryum affine Lindb. & Arnell = Bryum lisae var. cuspidatum [BRYULIS1] Bryaceae (B:M)
Bryum affine var. *cirrhatum* (Hüb.) Braithw. = Bryum lonchocaulon [BRYULON] Bryaceae (B:M)
Bryum alaskanum Kindb. = Bryum amblyodon [BRYUAMB] Bryaceae (B:M)
Bryum albicans (Wahlenb.) Röhl. = Pohlia wahlenbergii [POHLWAH] Bryaceae (B:M)
Bryum algovicum Sendtn. ex C. Müll. [BRYUALG] Bryaceae (B:M)
Bryum alpinum Huds. ex With. [BRYUALP] Bryaceae (B:M)
Bryum amblyodon C. Müll. [BRYUAMB] Bryaceae (B:M)
Bryum angustirete ssp. *fridtzii* (Hag.) Wijk & Marg. = Bryum algovicum [BRYUALG] Bryaceae (B:M)
Bryum angustirete ssp. *parvulum* (Kindb.) Wijk & Marg. = Bryum algovicum [BRYUALG] Bryaceae (B:M)
Bryum anoectangiaceum C. Müll. & Kindb. in Mac. & Kindb. = Bryum pallens [BRYUPAL] Bryaceae (B:M)
Bryum archangelicum auct. = Bryum salinum [BRYUSAL] Bryaceae (B:M)
Bryum arcticum (R. Br.) Bruch & Schimp. in B.S.G. [BRYUARC] Bryaceae (B:M)
Bryum arcticum var. *arcuatum* (Limpr.) Podp. = Bryum arcticum [BRYUARC] Bryaceae (B:M)
Bryum arcticum var. *micans* Amann = Bryum arcticum [BRYUARC] Bryaceae (B:M)
Bryum arcticum var. *tomentosum* Joerg. = Bryum arcticum [BRYUARC] Bryaceae (B:M)
Bryum argenteum Hedw. [BRYUARG] Bryaceae (B:M)
Bryum argenteum var. *insigne* Podp. = Bryum argenteum [BRYUARG] Bryaceae (B:M)
Bryum argenteum var. *lanatum* (P. Beauv.) Hampe = Bryum argenteum [BRYUARG] Bryaceae (B:M)
Bryum argenteum var. *majus* Schwaegr. = Bryum argenteum [BRYUARG] Bryaceae (B:M)
Bryum atropurpureum Bruch & Schimp. in B.S.G. = Bryum dichotomum [BRYUDIC] Bryaceae (B:M)
Bryum atropurpureum Wahlenb. in Fürnr. = Pohlia atropurpurea [POHLATR] Bryaceae (B:M)
Bryum atwateriae C. Müll. = Bryum miniatum [BRYUMIN] Bryaceae (B:M)
Bryum aurimontanum Kindb. = Bryum arcticum [BRYUARC] Bryaceae (B:M)
Bryum bicolor Dicks. = Bryum dichotomum [BRYUDIC] Bryaceae (B:M)
Bryum biddlecomiae Aust. = Bryum amblyodon [BRYUAMB] Bryaceae (B:M)
Bryum bimum (Brid.) Turn. = Bryum pseudotriquetrum [BRYUPSE] Bryaceae (B:M)
Bryum bimum var. *angustifolium* Kindb. in Mac. & Kindb. = Bryum lisae var. cuspidatum [BRYULIS1] Bryaceae (B:M)
Bryum blindii Bruch & Schimp. in B.S.G. [BRYUBLI] Bryaceae (B:M)
Bryum bornholmense Winkelm. & Ruthe [BRYUBOR] Bryaceae (B:M)
Bryum bullatum C. Müll. = Anomobryum filiforme [ANOMFIL] Bryaceae (B:M)
Bryum caespiticium Hedw. [BRYUCAE] Bryaceae (B:M)
Bryum calobryoides Spence [BRYUCAO] Bryaceae (B:M)
Bryum calophyllum R. Br. [BRYUCAL] Bryaceae (B:M)
Bryum canariense Brid. [BRYUCAN] Bryaceae (B:M)
Bryum capillare Hedw. [BRYUCAP] Bryaceae (B:M)
Bryum capillare ssp. *erythroloma* Kindb. = Bryum erythroloma [BRYUERY] Bryaceae (B:M)
Bryum capillare ssp. *heteroneuron* C. Müll. in Mac. & Kindb. = Bryum capillare [BRYUCAP] Bryaceae (B:M)
Bryum capillare var. *flaccidum* (Brid.) Bruch & Schimp. in B.S.G. = Bryum flaccidum [BRYUFLA] Bryaceae (B:M)
Bryum capitellatum C. Müll. & Kindb. in Mac. & Kindb. = Bryum gemmiparum [BRYUGEM] Bryaceae (B:M)
Bryum cernuum (Hedw.) Bruch & Schimp. in B.S.G. = Bryum uliginosum [BRYUULI] Bryaceae (B:M)
Bryum cirrhatum Hornsch. = Bryum lonchocaulon [BRYULON] Bryaceae (B:M)
Bryum columbico-caespiticum Kindb. = Bryum lonchocaulon [BRYULON] Bryaceae (B:M)
Bryum crassirameum Ren. & Card. = Bryum pseudotriquetrum [BRYUPSE] Bryaceae (B:M)
Bryum creberrimum Tayl. = Bryum lisae var. cuspidatum [BRYULIS1] Bryaceae (B:M)
Bryum crispulum Hampe in Hag. = Bryum pseudotriquetrum [BRYUPSE] Bryaceae (B:M)
Bryum crudum (Hedw.) Turn. = Pohlia cruda [POHLCRU] Bryaceae (B:M)
Bryum cryophilum Mårt. = Bryum cyclophyllum [BRYUCYC] Bryaceae (B:M)
Bryum cuspidatum (Bruch & Schimp. in B.S.G.) Schimp. = Bryum lisae var. cuspidatum [BRYULIS1] Bryaceae (B:M)
Bryum cyclophylloides Kindb. = Bryum turbinatum [BRYUTUR] Bryaceae (B:M)
Bryum cyclophyllum (Schwaegr.) Bruch & Schimp. in B.S.G. [BRYUCYC] Bryaceae (B:M)
Bryum denticulatum Kindb. = Bryum turbinatum [BRYUTUR] Bryaceae (B:M)
Bryum dichotomum Hedw. [BRYUDIC] Bryaceae (B:M)
Bryum duvalii Voit in Sturm = Bryum weigelii [BRYUWEI] Bryaceae (B:M)
Bryum duvalii var. *gaspeanum* Kindb. in Mac. & Kindb. = Bryum turbinatum [BRYUTUR] Bryaceae (B:M)
Bryum duvalii var. *latodecurrens* C. Müll. & Kindb. in Mac. & Kindb. = Bryum weigelii [BRYUWEI] Bryaceae (B:M)
Bryum edwardsianum C. Müll. & Kindb. in Mac. & Kindb. = Bryum amblyodon [BRYUAMB] Bryaceae (B:M)
Bryum elongatum (Hedw.) With. = Pohlia elongata [POHLELO] Bryaceae (B:M)
Bryum erubescens Kindb. in Mac. & Kindb. = Bryum amblyodon [BRYUAMB] Bryaceae (B:M)
Bryum erythroloma (Kindb.) Syed [BRYUERY] Bryaceae (B:M)
Bryum erythrophylloides Kindb. in Mac. & Kindb. = Bryum turbinatum [BRYUTUR] Bryaceae (B:M)
Bryum extenuatum Ren. & Card. = Bryum turbinatum [BRYUTUR] Bryaceae (B:M)
Bryum fallax Milde in Rabenh. = Bryum pallens [BRYUPAL] Bryaceae (B:M)
Bryum filiforme Dicks. = Anomobryum filiforme [ANOMFIL] Bryaceae (B:M)
Bryum flaccidum Brid. [BRYUFLA] Bryaceae (B:M)
Bryum flagellosum Kindb. = Bryum lisae var. cuspidatum [BRYULIS1] Bryaceae (B:M)

Bryum flexuosum Aust. = Bryum gemmiparum [BRYUGEM] Bryaceae (B:M)
Bryum fridtzii Hag. = Bryum algovicum [BRYUALG] Bryaceae (B:M)
Bryum gemmiparum De Not. [BRYUGEM] Bryaceae (B:M)
Bryum grandirete Kindb. = Bryum turbinatum [BRYUTUR] Bryaceae (B:M)
Bryum haematocarpum C. Müll. & Kindb. in Mac. & Kindb. = Bryum pseudotriquetrum [BRYUPSE] Bryaceae (B:M)
Bryum haematophyllum Kindb. in Mac. & Kindb. = Bryum turbinatum [BRYUTUR] Bryaceae (B:M)
Bryum hamicuspis Kindb. = Bryum lisae var. cuspidatum [BRYULIS1] Bryaceae (B:M)
Bryum hendersonii Ren. & Card. = Bryum canariense [BRYUCAN] Bryaceae (B:M)
Bryum heteroneuron (C. Müll. & Kindb. in Mac. & Kindb.) Ren. & Card. = Bryum capillare [BRYUCAP] Bryaceae (B:M)
Bryum heteroneuron var. *brevicuspidatum* (Kindb.) Ren. & Card. = Bryum capillare [BRYUCAP] Bryaceae (B:M)
Bryum hydrophilum Kindb. = Bryum pallens [BRYUPAL] Bryaceae (B:M)
Bryum inclinatum (Brid.) Bland. = Bryum amblyodon [BRYUAMB] Bryaceae (B:M)
Bryum intermedium auct. = Bryum lisae var. cuspidatum [BRYULIS1] Bryaceae (B:M)
Bryum laevifilum Syed = Bryum flaccidum [BRYUFLA] Bryaceae (B:M)
Bryum lanatum (P. Beauv.) Brid. = Bryum argenteum [BRYUARG] Bryaceae (B:M)
Bryum languidum Hag. = Bryum turbinatum [BRYUTUR] Bryaceae (B:M)
Bryum lapponicum auct. = Bryum salinum [BRYUSAL] Bryaceae (B:M)
Bryum laurentianum Card. & Thér. = Bryum alpinum [BRYUALP] Bryaceae (B:M)
Bryum lescurianum Sull. = Pohlia lescuriana [POHLLES] Bryaceae (B:M)
Bryum leucolomatum C. Müll. & Kindb. in Mac. & Kindb. = Bryum lonchocaulon [BRYULON] Bryaceae (B:M)
Bryum lisae De Not. [BRYULIS] Bryaceae (B:M)
Bryum lisae var. **cuspidatum** (Bruch & Schimp. in B.S.G.) Marg. [BRYULIS1] Bryaceae (B:M)
Bryum lonchocaulon C. Müll. [BRYULON] Bryaceae (B:M)
Bryum meesioides Kindb. in Mac. [BRYUMEE] Bryaceae (B:M)
Bryum micans Limpr. = Bryum arcticum [BRYUARC] Bryaceae (B:M)
Bryum microcephalum C. Müll. & Kindb. in Mac. & Kindb. = Bryum caespiticium [BRYUCAE] Bryaceae (B:M)
Bryum microerythrocarpum C. Müll. & Kindb. in Mac. & Kindb. = Bryum subapiculatum [BRYUSUB] Bryaceae (B:M)
Bryum microglobum C. Müll. & Kindb. in Mac. & Kindb. = Bryum dichotomum [BRYUDIC] Bryaceae (B:M)
Bryum miniatum Lesq. [BRYUMIN] Bryaceae (B:M)
Bryum muehlenbeckii Bruch & Schimp. in B.S.G. [BRYUMUE] Bryaceae (B:M)
Bryum neodamense Itzig in C. Müll. = Bryum pseudotriquetrum [BRYUPSE] Bryaceae (B:M)
Bryum neodamense var. *ovatum* Lindb. & Arnell = Bryum pseudotriquetrum [BRYUPSE] Bryaceae (B:M)
Bryum nutans (Hedw.) Turn. = Pohlia nutans [POHLNUT] Bryaceae (B:M)
Bryum obtusifolium Lindb. = Bryum cyclophyllum [BRYUCYC] Bryaceae (B:M)
Bryum occidentale Sull. = Bryum dichotomum [BRYUDIC] Bryaceae (B:M)
Bryum oligochloron C. Müll. & Kindb. in Mac. & Kindb. = Bryum caespiticium [BRYUCAE] Bryaceae (B:M)
Bryum oreganum Sull. = Bryum capillare [BRYUCAP] Bryaceae (B:M)
Bryum ovatum Jur. = Bryum pseudotriquetrum [BRYUPSE] Bryaceae (B:M)
Bryum pallens (Brid.) Sw. in Röhl. [BRYUPAL] Bryaceae (B:M)
Bryum pallens var. *alpinum* (B.S.G.) Podp. = Bryum pallens [BRYUPAL] Bryaceae (B:M)
Bryum pallens var. *fallax* Jur. = Bryum pallens [BRYUPAL] Bryaceae (B:M)
Bryum pallens var. *meesioides* (Kindb. in Mac.) Broth. = Bryum meesioides [BRYUMEE] Bryaceae (B:M)
Bryum pallescens Schleich. ex Schwaegr. [BRYUPAE] Bryaceae (B:M)
Bryum pallescens var. *grande* Kindb. = Bryum pseudotriquetrum [BRYUPSE] Bryaceae (B:M)
Bryum pallescens var. *laxifolium* Kindb. in Max. = Bryum pallescens [BRYUPAE] Bryaceae (B:M)
Bryum pallescens var. *longifolium* Kindb. = Bryum pallescens [BRYUPAE] Bryaceae (B:M)
Bryum pallescens var. *subrotundum* (Brid.) B.S.G. = Bryum pallescens [BRYUPAE] Bryaceae (B:M)
Bryum pendulum ssp. *nanum* Kindb. = Bryum algovicum [BRYUALG] Bryaceae (B:M)
Bryum percurrentinerve Kindb. in Mac. = Bryum muehlenbeckii [BRYUMUE] Bryaceae (B:M)
Bryum producticolle Kindb. = Bryum lonchocaulon [BRYULON] Bryaceae (B:M)
Bryum pseudostirtonii Philib. ex Card. & Thér. = Bryum lisae var. cuspidatum [BRYULIS1] Bryaceae (B:M)
Bryum pseudotriquetrum (Hedw.) Gaertn., Meyer & Scherb. [BRYUPSE] Bryaceae (B:M)
Bryum pseudotriquetrum ssp. *crispulum* (Roth) C. Jens. = Bryum pseudotriquetrum [BRYUPSE] Bryaceae (B:M)
Bryum pseudotriquetrum var. *bimum* (Brid.) Lilj. = Bryum pseudotriquetrum [BRYUPSE] Bryaceae (B:M)
Bryum pseudotriquetrum var. *compactum* B.S.G. = Bryum pseudotriquetrum [BRYUPSE] Bryaceae (B:M)
Bryum pseudotriquetrum var. *crassirameum* Ren. & Card. = Bryum pseudotriquetrum [BRYUPSE] Bryaceae (B:M)
Bryum pseudotriquetrum var. *hyalodontium* C. Müll. & Kindb. in Mac. & Kindb. = Bryum pseudotriquetrum [BRYUPSE] Bryaceae (B:M)
Bryum pygmaeo-alpinum C. Müll. & Kindb. in Mac. & Kindb. = Bryum muehlenbeckii [BRYUMUE] Bryaceae (B:M)
Bryum pyriforme (Hedw.) Lam. & Cand. = Leptobryum pyriforme [LEPTPYI] Bryaceae (B:M)
Bryum pyriforme (Hedw.) With. = Physcomitrium pyriforme [PHYSPYR] Funariaceae (B:M)
Bryum rauei Aust. = Bryum muehlenbeckii [BRYUMUE] Bryaceae (B:M)
Bryum retusum Hag. = Bryum salinum [BRYUSAL] Bryaceae (B:M)
Bryum revelstokense Kindb. = Bryum lonchocaulon [BRYULON] Bryaceae (B:M)
Bryum roseum (Hedw.) Crom. = Rhodobryum roseum [RHODROS] Bryaceae (B:M)
Bryum rubricundulum C. Müll. & Kindb. in Mac. & Kindb. = Bryum muehlenbeckii [BRYUMUE] Bryaceae (B:M)
Bryum salinum Hag. ex Limpr. [BRYUSAL] Bryaceae (B:M)
Bryum sandbergii Holz. = Roellia roellii [ROELROE] Bryaceae (B:M)
Bryum schleicheri Schwaegr. [BRYUSCH] Bryaceae (B:M)
Bryum simplex Kindb. in Mac. & Kindb. = Roellia roellii [ROELROE] Bryaceae (B:M)
Bryum speirophyllum Kindb. = Bryum capillare [BRYUCAP] Bryaceae (B:M)
Bryum stenotrichum C. Müll. = Bryum amblyodon [BRYUAMB] Bryaceae (B:M)
Bryum stenotrichum var. *biddlecomiae* (Aust.) Lawt. = Bryum amblyodon [BRYUAMB] Bryaceae (B:M)
Bryum streptophyllum Kindb. = Bryum capillare [BRYUCAP] Bryaceae (B:M)
Bryum subapiculatum Hampe [BRYUSUB] Bryaceae (B:M)
Bryum submicans Kindb. = Bryum arcticum [BRYUARC] Bryaceae (B:M)
Bryum submuticum Philib. = Bryum caespiticium [BRYUCAE] Bryaceae (B:M)
Bryum suborbiculare Philib. = Bryum turbinatum [BRYUTUR] Bryaceae (B:M)

Bryum subpercurrentinerve Kindb. = Bryum pseudotriquetrum [BRYUPSE] Bryaceae (B:M)
Bryum subpurpurascens Kindb. in Mac. & Kindb. = Bryum pallens [BRYUPAL] Bryaceae (B:M)
Bryum subrotundum Brid. = Bryum pallescens [BRYUPAE] Bryaceae (B:M)
Bryum subtumidum Limpr. in Jaeg. = Bryum salinum [BRYUSAL] Bryaceae (B:M)
Bryum synoico-caespiticium C. Müll. & Kindb. in Mac. & Kindb. = Bryum caespiticium [BRYUCAE] Bryaceae (B:M)
Bryum tenuisetum Limpr. [BRYUTEN] Bryaceae (B:M)
Bryum tomentosum (Joerg.) Hag. = Bryum arcticum [BRYUARC] Bryaceae (B:M)
Bryum tomentosum Kindb. = Bryum capillare [BRYUCAP] Bryaceae (B:M)
Bryum tortifolium Funck in Brid. = Bryum cyclophyllum [BRYUCYC] Bryaceae (B:M)
Bryum trichophorum Kindb. = Bryum capillare [BRYUCAP] Bryaceae (B:M)
Bryum turbinatum (Hedw.) Turn. [BRYUTUR] Bryaceae (B:M)
Bryum uliginosum (Brid.) Bruch & Schimp. in B.S.G. [BRYUULI] Bryaceae (B:M)
Bryum vancouveriense Kindb. = Bryum caespiticium [BRYUCAE] Bryaceae (B:M)
Bryum ventricosum Relh. = Bryum pseudotriquetrum [BRYUPSE] Bryaceae (B:M)
Bryum ventricosum var. *compactum* (B.S.G.) Lindb. = Bryum pseudotriquetrum [BRYUPSE] Bryaceae (B:M)
Bryum violaceum Crundw. & Nyh. [BRYUVIO] Bryaceae (B:M)
Bryum wahlenbergii (Web. & Mohr) Schwaegr. = Pohlia wahlenbergii [POHLWAH] Bryaceae (B:M)
Bryum weigelii Spreng. in Biehler [BRYUWEI] Bryaceae (B:M)
Bucegia Radian [BUCEGIA$] Marchantiaceae (B:H)
Bucegia romanica Radian [BUCEROM] Marchantiaceae (B:H)
Buddleja L. [BUDDLEJ$] Buddlejaceae (V:D)
*__Buddleja davidii__ Franch. [BUDDDAV] Buddlejaceae (V:D)
Buellia De Not. [BUELLIA$] Physciaceae (L)
Buellia aethalea (Ach.) Th. Fr. [BUELAET] Physciaceae (L)
Buellia aethaleoides (Nyl.) H. Olivier = Buellia aethalea [BUELAET] Physciaceae (L)
Buellia alboatra (Hoffm.) Th. Fr. = Diplotomma alboatrum [DIPLALO] Physciaceae (L)
Buellia atrata (Sm.) Anzi = Orphniospora moriopsis [ORPHMOR] Fuscideaceae (L)
Buellia badia (Fr.) A. Massal. [BUELBAD] Physciaceae (L)
Buellia badioatra (Flörke ex Sprengel) Körber = Rhizocarpon badioatrum [RHIZBAD] Rhizocarpaceae (L)
Buellia callispora (C. Knight) J. Steiner = Hafellia callispora [HAFECAL] Physciaceae (L)
Buellia chlorophaea (Hepp ex Leighton) Lettau = Diplotomma chlorophaeum [DIPLCHL] Physciaceae (L)
Buellia colludens (Nyl.) Arnold = Rhizocarpon hochstetteri [RHIZHOC] Rhizocarpaceae (L)
Buellia coniops (Wahlenb.) Th. Fr. = Amandinea coniops [AMANCON] Physciaceae (L)
Buellia coracina (Nyl.) Körber = Orphniospora moriopsis [ORPHMOR] Fuscideaceae (L)
Buellia crystallifera (Vainio) Hav. [BUELCRY] Physciaceae (L)
Buellia curtisii (Tuck.) Imshaug [BUELCUT] Physciaceae (L)
Buellia disciformis (Fr.) Mudd [BUELDIS] Physciaceae (L)
Buellia elegans Poelt [BUELELE] Physciaceae (L)
Buellia erubescens Arnold [BUELERU] Physciaceae (L)
Buellia fosteri Imshaug, nom. inval. = Hafellia fosteri [HAFEFOS] Physciaceae (L)
Buellia geophila (Flörke ex Sommerf.) Lynge [BUELGEO] Physciaceae (L)
Buellia griseovirens (Turner & Borrer ex Sm.) Almb. [BUELGRI] Physciaceae (L)
Buellia halonia (Ach.) Tuck. [BUELHAL] Physciaceae (L)
Buellia hassei Imshaug = Buellia griseovirens [BUELGRI] Physciaceae (L)
Buellia insignis (Naeg. ex Hepp) Th. Fr. [BUELINS] Physciaceae (L)
Buellia jugorum (Arnold) Arnold [BUELJUG] Physciaceae (L)
Buellia lepidastra (Tuck.) Tuck. [BUELLEP] Physciaceae (L)
Buellia malmei Lynge = Buellia aethalea [BUELAET] Physciaceae (L)
Buellia moriopsis (A. Massal.) Th. Fr. = Orphniospora moriopsis [ORPHMOR] Fuscideaceae (L)
Buellia myriocarpa (DC.) De Not. = Amandinea punctata [AMANPUN] Physciaceae (L)
Buellia oidalea (Nyl.) Tuck. [BUELOID] Physciaceae (L)
Buellia papillata (Sommerf.) Tuck. [BUELPAP] Physciaceae (L)
Buellia parasema (Ach.) De Not. = Buellia disciformis [BUELDIS] Physciaceae (L)
Buellia penichra (Tuck.) Hasse = Diplotomma penichrum [DIPLPEN] Physciaceae (L)
Buellia pulchella (Schrader) Tuck. = Catolechia wahlenbergii [CATOWAH] Rhizocarpaceae (L)
Buellia pullata Tuck. = Amandinea punctata [AMANPUN] Physciaceae (L)
Buellia pulverulenta (Anzi) Jatta [BUELPUL] Physciaceae (L)
Buellia punctata (Hoffm.) A. Massal. = Amandinea punctata [AMANPUN] Physciaceae (L)
Buellia restituta (Stirton) Zahlbr. = Hafellia callispora [HAFECAL] Physciaceae (L)
Buellia stellulata (Taylor) Mudd [BUELSTE] Physciaceae (L)
Buellia stillingiana J. Steiner [BUELSTL] Physciaceae (L)
Buellia subconnexa (Stirton) Zahlbr. = Hafellia callispora [HAFECAL] Physciaceae (L)
Buellia triphragmioides Anzi [BUELTRI] Physciaceae (L)
Buellia turgescens Tuck. [BUELTUR] Physciaceae (L)
Buellia turgescentoides Fink = Buellia turgescens [BUELTUR] Physciaceae (L)
Buellia verruculosa (Sm.) Mudd = Buellia aethalea [BUELAET] Physciaceae (L)
Buellia wahlenbergii (Ach.) Sheard = Catolechia wahlenbergii [CATOWAH] Rhizocarpaceae (L)
Buellia zahlbruckneri J. Steiner = Buellia erubescens [BUELERU] Physciaceae (L)
Buelliella nuttallii (Calk. & Nyl.) Fink = Dactylospora lobariella [DACTLOB] Dactylosporaceae (L)
Buelliella parmeliarum (Sommerf.) Fink = Abrothallus parmeliarum [ABROPAR] Uncertain-family (L)
Buelliopsis papillata (Sommerf.) Fink = Buellia papillata [BUELPAP] Physciaceae (L)
Buglossoides Moench [BUGLOSS$] Boraginaceae (V:D)
*__Buglossoides arvensis__ (L.) I.M. Johnston [BUGLARV] Boraginaceae (V:D)
Bulbocodium serotinum L. = Lloydia serotina ssp. serotina [LLOYSER0] Liliaceae (V:M)
Bulliarda aquatica (L.) DC. = Crassula aquatica [CRASAQU] Crassulaceae (V:D)
Bunodophoron A. Massal. [BUNODOP$] Sphaerophoraceae (L)
Bunodophoron melanocarpum (Sw.) Wedin [BUNOMEL] Sphaerophoraceae (L)
Bupleurum L. [BUPLEUR$] Apiaceae (V:D)
Bupleurum americanum Coult. & Rose [BUPLAME] Apiaceae (V:D)
Bupleurum triradiatum ssp. *arcticum* (Regel) Hultén = Bupleurum americanum [BUPLAME] Apiaceae (V:D)
Bursa bursa-pastoris (L.) Britt. = Capsella bursa-pastoris [CAPSBUR] Brassicaceae (V:D)
Bursa bursa-pastoris var. *bifida* Crépin = Capsella bursa-pastoris [CAPSBUR] Brassicaceae (V:D)
Bursa gracilis Gren. = Capsella bursa-pastoris [CAPSBUR] Brassicaceae (V:D)
Bursa rubella Reut. = Capsella bursa-pastoris [CAPSBUR] Brassicaceae (V:D)
Butomus L. [BUTOMUS$] Butomaceae (V:M)
*__Butomus umbellatus__ L. [BUTOUMB] Butomaceae (V:M)
Buxbaumia Hedw. [BUXBAUM$] Buxbaumiaceae (B:M)
Buxbaumia aphylla Hedw. [BUXBAPH] Buxbaumiaceae (B:M)
Buxbaumia indusiata Brid. = Buxbaumia viridis [BUXBVIR] Buxbaumiaceae (B:M)
Buxbaumia piperi Best [BUXBPIP] Buxbaumiaceae (B:M)

Buxbaumia viridis (DC.) Moug. & Nestl. [BUXBVIR] Buxbaumiaceae (B:M)
Byssoloma Trevisan [BYSSOLO$] Pilocarpaceae (L)
Byssoloma leucoblepharum (Nyl.) Vainio [BYSSLEU] Pilocarpaceae (L)
Byssoloma rotuliforme (Müll. Arg.) R. Sant. = Byssoloma subdiscordans [BYSSSUB] Pilocarpaceae (L)
Byssoloma subdiscordans (Nyl.) P. James [BYSSSUB] Pilocarpaceae (L)

C

Cacalia nardosmia Gray = Cacaliopsis nardosmia [CACANAR] Asteraceae (V:D)
Cacalia nardosmia var. *glabrata* (Piper) Boivin = Cacaliopsis nardosmia [CACANAR] Asteraceae (V:D)
Cacaliopsis Gray [CACALIO$] Asteraceae (V:D)
Cacaliopsis nardosmia (Gray) Gray [CACANAR] Asteraceae (V:D)
Cacaliopsis nardosmia ssp. *glabrata* Piper = Cacaliopsis nardosmia [CACANAR] Asteraceae (V:D)
Cacaliopsis nardosmia var. *glabrata* Piper = Cacaliopsis nardosmia [CACANAR] Asteraceae (V:D)
Cakile P. Mill. [CAKILE$] Brassicaceae (V:D)
Cakile cakile (L.) Karst. = Cakile maritima [CAKIMAR] Brassicaceae (V:D)
Cakile edentula (Bigelow) Hook. [CAKIEDE] Brassicaceae (V:D)
Cakile edentula var. *californica* (Heller) Fern. = Cakile edentula [CAKIEDE] Brassicaceae (V:D)
***Cakile maritima** Scop. [CAKIMAR] Brassicaceae (V:D)
Calamagrostis Adans. [CALAMAG$] Poaceae (V:M)
Calamagrostis anomala Suksdorf = Calamagrostis canadensis var. canadensis [CALACAN0] Poaceae (V:M)
Calamagrostis atropurpurea Nash = Calamagrostis canadensis var. canadensis [CALACAN0] Poaceae (V:M)
Calamagrostis californica Kearney = Calamagrostis stricta ssp. inexpansa [CALASTR2] Poaceae (V:M)
Calamagrostis canadensis (Michx.) Beauv. [CALACAN] Poaceae (V:M)
Calamagrostis canadensis ssp. *langsdorfii* (Link) Hultén = Calamagrostis canadensis var. langsdorfii [CALACAN7] Poaceae (V:M)
Calamagrostis canadensis var. *acuminata* Vasey ex Shear & Rydb. = Calamagrostis canadensis var. canadensis [CALACAN0] Poaceae (V:M)
Calamagrostis canadensis var. **canadensis** [CALACAN0] Poaceae (V:M)
Calamagrostis canadensis var. **langsdorfii** (Link) Inman [CALACAN7] Poaceae (V:M)
Calamagrostis canadensis var. *pallida* (Vasey & Scribn.) Stebbins = Calamagrostis canadensis var. canadensis [CALACAN0] Poaceae (V:M)
Calamagrostis canadensis var. *robusta* Vasey = Calamagrostis canadensis var. canadensis [CALACAN0] Poaceae (V:M)
Calamagrostis canadensis var. *scabra* (J. Presl) A.S. Hitchc. = Calamagrostis canadensis var. langsdorfii [CALACAN7] Poaceae (V:M)
Calamagrostis canadensis var. *typica* Stebbins = Calamagrostis canadensis var. canadensis [CALACAN0] Poaceae (V:M)
Calamagrostis chordorrhiza Porsild = Calamagrostis stricta ssp. inexpansa [CALASTR2] Poaceae (V:M)
Calamagrostis crassiglumis Thurb. [CALACRA] Poaceae (V:M)
Calamagrostis expansa Rickett & Gilly = Calamagrostis stricta ssp. inexpansa [CALASTR2] Poaceae (V:M)
Calamagrostis expansa var. *robusta* (Vasey) Stebbins = Calamagrostis canadensis var. canadensis [CALACAN0] Poaceae (V:M)
Calamagrostis fasciculata Kearney = Calamagrostis rubescens [CALARUB] Poaceae (V:M)
Calamagrostis fernaldii Louis-Marie = Calamagrostis stricta ssp. inexpansa [CALASTR2] Poaceae (V:M)
Calamagrostis hyperborea var. *americana* (Vasey) Kearney = Calamagrostis stricta ssp. inexpansa [CALASTR2] Poaceae (V:M)
Calamagrostis hyperborea var. *elongata* Kearney = Calamagrostis stricta ssp. inexpansa [CALASTR2] Poaceae (V:M)
Calamagrostis hyperborea var. *stenodes* Kearney = Calamagrostis stricta ssp. inexpansa [CALASTR2] Poaceae (V:M)
Calamagrostis inexpansa Gray = Calamagrostis stricta ssp. inexpansa [CALASTR2] Poaceae (V:M)
Calamagrostis inexpansa var. *barbulata* Kearney = Calamagrostis stricta ssp. inexpansa [CALASTR2] Poaceae (V:M)
Calamagrostis inexpansa var. *brevior* (Vasey) Stebbins = Calamagrostis stricta ssp. inexpansa [CALASTR2] Poaceae (V:M)
Calamagrostis inexpansa var. *cuprea* Kearney = Calamagrostis canadensis var. canadensis [CALACAN0] Poaceae (V:M)
Calamagrostis inexpansa var. *novae-angliae* Stebbins = Calamagrostis stricta ssp. inexpansa [CALASTR2] Poaceae (V:M)
Calamagrostis inexpansa var. *robusta* (Vasey) Stebbins = Calamagrostis canadensis var. canadensis [CALACAN0] Poaceae (V:M)
Calamagrostis labradorica Kearney = Calamagrostis stricta ssp. inexpansa [CALASTR2] Poaceae (V:M)
Calamagrostis lacustris (Kearney) Nash = Calamagrostis stricta ssp. inexpansa [CALASTR2] Poaceae (V:M)
Calamagrostis langsdorfii (Link) Trin. = Calamagrostis canadensis var. langsdorfii [CALACAN7] Poaceae (V:M)
Calamagrostis lapponica (Wahlenb.) Hartman [CALALAP] Poaceae (V:M)
Calamagrostis lapponica var. *alpina* Hartman = Calamagrostis lapponica [CALALAP] Poaceae (V:M)
Calamagrostis lapponica var. *brevipilis* Stebbins = Calamagrostis stricta ssp. inexpansa [CALASTR2] Poaceae (V:M)
Calamagrostis lapponica var. *groenlandica* Lange = Calamagrostis lapponica [CALALAP] Poaceae (V:M)
Calamagrostis lapponica var. *nearctica* Porsild = Calamagrostis lapponica [CALALAP] Poaceae (V:M)
Calamagrostis luxurians Rydb. = Calamagrostis rubescens [CALARUB] Poaceae (V:M)
Calamagrostis montanensis Scribn. ex Vasey [CALAMON] Poaceae (V:M)
Calamagrostis neglecta (Ehrh.) P.G. Gaertn., B. Mey. & Scherb. = Calamagrostis stricta [CALASTR] Poaceae (V:M)
Calamagrostis nubila Louis-Marie = Calamagrostis canadensis var. langsdorfii [CALACAN7] Poaceae (V:M)
Calamagrostis nutkaensis (J. Presl) J. Presl ex Steud. [CALANUT] Poaceae (V:M)
Calamagrostis pallida Vasey & Scribn. ex Vasey, non Nutt. ex Gray = Calamagrostis canadensis var. canadensis [CALACAN0] Poaceae (V:M)
Calamagrostis pickeringii var. *lacustris* (Kearney) A.S. Hitchc. = Calamagrostis stricta ssp. inexpansa [CALASTR2] Poaceae (V:M)
Calamagrostis purpurascens R. Br. [CALAPUR] Poaceae (V:M)
Calamagrostis purpurascens ssp. *tasuensis* Calder & Taylor = Calamagrostis purpurascens [CALAPUR] Poaceae (V:M)
Calamagrostis rubescens Buckl. [CALARUB] Poaceae (V:M)
Calamagrostis scabra J. Presl = Calamagrostis canadensis var. langsdorfii [CALACAN7] Poaceae (V:M)
Calamagrostis scopulorum var. *bakeri* Stebbins = Calamagrostis stricta ssp. inexpansa [CALASTR2] Poaceae (V:M)
Calamagrostis scribneri Beal [CALASCR] Poaceae (V:M)
Calamagrostis sesquiflora (Trin.) Tzvelev [CALASES] Poaceae (V:M)
Calamagrostis stricta (Timm) Koel. [CALASTR] Poaceae (V:M)

Calamagrostis stricta ssp. **inexpansa** (Gray) C.W. Greene [CALASTR2] Poaceae (V:M)
Calamagrostis stricta ssp. **stricta** [CALASTR0] Poaceae (V:M)
Calamagrostis stricta var. *brevior* Vasey = Calamagrostis stricta ssp. inexpansa [CALASTR2] Poaceae (V:M)
Calamagrostis stricta var. *lacustris* (Kearney) C.W. Greene = Calamagrostis stricta ssp. inexpansa [CALASTR2] Poaceae (V:M)
Calamintha acinos (L.) Clairville ex Gaud. = Acinos arvensis [ACINARV] Lamiaceae (V:D)
Calamintha clinopodium Benth = Clinopodium vulgare [CLINVUL] Lamiaceae (V:D)
Calamovilfa (Gray) Hack. ex Scribn. & Southworth [CALAMOV$] Poaceae (V:M)
Calamovilfa longifolia (Hook.) Scribn. [CALALON] Poaceae (V:M)
Calandrinia Kunth [CALANDR$] Portulacaceae (V:D)
Calandrinia ciliata (Ruiz & Pavón) DC. [CALACIL] Portulacaceae (V:D)
Calandrinia ciliata var. *menziesii* (Hook.) J.F. Macbr. = Calandrinia ciliata [CALACIL] Portulacaceae (V:D)
Calandrinia tweedyi Gray = Cistanthe tweedyi [CISTTWE] Portulacaceae (V:D)
Calicium Pers. [CALICIU$] Caliciaceae (L)
Calicium abietinum Pers. [CALIABI] Caliciaceae (L)
Calicium adaequatum Nyl. [CALIADA] Caliciaceae (L)
Calicium adspersum Pers. [CALIADS] Caliciaceae (L)
Calicium asikkalense Vainio = Chaenothecopsis pusilla [CHAEPUS] Mycocaliciaceae (L)
Calicium corynellum (Ach.) Ach. [CALICOR] Caliciaceae (L)
Calicium disseminatum Ach. = Microcalicium disseminatum [MICRDIS] Microcaliciaceae (L)
Calicium floerkei Zahlbr. = Chaenothecopsis pusilla [CHAEPUS] Mycocaliciaceae (L)
Calicium glaucellum Ach. [CALIGLA] Caliciaceae (L)
Calicium hemisphaericum Howard = Calicium adaequatum [CALIADA] Caliciaceae (L)
Calicium hyperellum (Ach.) Ach. = Calicium viride [CALIVIR] Caliciaceae (L)
Calicium lenticulare Ach. [CALILEN] Caliciaceae (L)
Calicium lentigerellum Tuck. = Calicium lenticulare [CALILEN] Caliciaceae (L)
Calicium lichenoides (L.) Schumacher = Calicium salicinum [CALISAL] Caliciaceae (L)
Calicium parvum Tibell [CALIPAR] Caliciaceae (L)
Calicium populneum Brond. ex Duby = Phaeocalicium populneum [PHAEPOP] Mycocaliciaceae (L)
Calicium pusillum auct. = Chaenothecopsis pusilla [CHAEPUS] Mycocaliciaceae (L)
Calicium pusiolum Ach. = Chaenothecopsis pusiola [CHAEPUI] Mycocaliciaceae (L)
Calicium roscidum (Ach.) Ach. = Calicium adspersum [CALIADS] Caliciaceae (L)
Calicium roscidum var. *trabinellum* (Ach.) Schaerer = Calicium trabinellum [CALITRA] Caliciaceae (L)
Calicium salicinum Pers. [CALISAL] Caliciaceae (L)
Calicium sphaerocephalum (L.) Ach. = Calicium salicinum [CALISAL] Caliciaceae (L)
Calicium subpusillum Vainio = Chaenothecopsis pusilla [CHAEPUS] Mycocaliciaceae (L)
Calicium subquercinum Asah. = Calicium lenticulare [CALILEN] Caliciaceae (L)
Calicium subtile Pers. = Mycocalicium subtile [MYCOSUB] Mycocaliciaceae (L)
Calicium trabinellum (Ach.) Ach. [CALITRA] Caliciaceae (L)
Calicium trachelinum Ach. = Calicium salicinum [CALISAL] Caliciaceae (L)
Calicium viride Pers. [CALIVIR] Caliciaceae (L)
Calla L. [CALLA$] Araceae (V:M)
Calla palustris L. [CALLPAL] Araceae (V:M)
Callicladium Crum [CALLICL$] Hypnaceae (B:M)
Callicladium haldanianum (Grev.) Crum [CALLHAL] Hypnaceae (B:M)
Calliergidium pseudostramineum (C. Müll.) Grout = Warnstorfia pseudostraminea [WARNPSE] Amblystegiaceae (B:M)
Calliergidium pseudostramineum var. *plesiostramineum* (Ren.) Grout = Warnstorfia pseudostraminea [WARNPSE] Amblystegiaceae (B:M)
Calliergon (Sull.) Kindb. [CALLIEG$] Amblystegiaceae (B:M)
Calliergon cordifolium (Hedw.) Kindb. [CALLCOR] Amblystegiaceae (B:M)
Calliergon cordifolium f. *laxum* (Röll) Podp. = Calliergon cordifolium [CALLCOR] Amblystegiaceae (B:M)
Calliergon cordifolium var. *angustifolium* G. Roth = Calliergon cordifolium [CALLCOR] Amblystegiaceae (B:M)
Calliergon cordifolium var. *fontinaloides* (J. Lange) G. Roth = Calliergon cordifolium [CALLCOR] Amblystegiaceae (B:M)
Calliergon cordifolium var. *intermedium* Mönk. in Bauer = Calliergon cordifolium [CALLCOR] Amblystegiaceae (B:M)
Calliergon cordifolium var. *lanutocaule* (Bryhn) Broth. = Calliergon cordifolium [CALLCOR] Amblystegiaceae (B:M)
Calliergon cordifolium var. *latifolium* Karcz. = Calliergon cordifolium [CALLCOR] Amblystegiaceae (B:M)
Calliergon cuspidata var. *pungens* (Schimp.) Warnst. = Calliergonella cuspidata [CALLCUS] Amblystegiaceae (B:M)
Calliergon cuspidatum (Hedw.) Kindb. = Calliergonella cuspidata [CALLCUS] Amblystegiaceae (B:M)
Calliergon cuspidatum f. *acuteramosum* Bauer ex Karcz. = Calliergonella cuspidata [CALLCUS] Amblystegiaceae (B:M)
Calliergon cuspidatum f. *turgescens* (Wheld.) Karcz. = Calliergonella cuspidata [CALLCUS] Amblystegiaceae (B:M)
Calliergon cuspidatum var. *brevifolium* Sanio ex Warnst. = Calliergonella cuspidata [CALLCUS] Amblystegiaceae (B:M)
Calliergon cuspidatum var. *umbrosum* (Loeske) Warnst. = Calliergonella cuspidata [CALLCUS] Amblystegiaceae (B:M)
Calliergon giganteum (Schimp.) Kindb. [CALLGIG] Amblystegiaceae (B:M)
Calliergon giganteum f. *crassicostatum* (Mik.) Karcz. = Calliergon giganteum [CALLGIG] Amblystegiaceae (B:M)
Calliergon giganteum f. *decurrens* (Mik.) Karcz. = Calliergon giganteum [CALLGIG] Amblystegiaceae (B:M)
Calliergon giganteum var. *cyclophyllotum* (Holz.) Grout = Calliergon giganteum [CALLGIG] Amblystegiaceae (B:M)
Calliergon giganteum var. *dendroides* (Limpr.) G. Roth = Calliergon giganteum [CALLGIG] Amblystegiaceae (B:M)
Calliergon giganteum var. *fluitans* (Klinggr.) G. Roth = Calliergon giganteum [CALLGIG] Amblystegiaceae (B:M)
Calliergon giganteum var. *hystricosum* G. Roth & Bock in G. Roth = Calliergon giganteum [CALLGIG] Amblystegiaceae (B:M)
Calliergon giganteum var. *immersum* Ruthe ex Karcz. = Calliergon giganteum [CALLGIG] Amblystegiaceae (B:M)
Calliergon giganteum var. *labradorense* (Ren. & Card.) Grout = Calliergon giganteum [CALLGIG] Amblystegiaceae (B:M)
Calliergon giganteum var. *pennatum* Karcz. = Calliergon giganteum [CALLGIG] Amblystegiaceae (B:M)
Calliergon giganteum var. *tenue* Karcz. = Calliergon giganteum [CALLGIG] Amblystegiaceae (B:M)
Calliergon pseudosarmentosum (Card. & Thér.) Broth. = Drepanocladus pseudosarmentosus [DREPPSE] Amblystegiaceae (B:M)
Calliergon richardsonii (Mitt.) Kindb. in Warnst. [CALLRIC] Amblystegiaceae (B:M)
Calliergon sarmentosum (Wahlenb.) Kindb. = Sarmenthypnum sarmentosum [SARMSAR] Amblystegiaceae (B:M)
Calliergon sarmentosum f. *heterophyllum* Arnell & C. Jens. = Sarmenthypnum sarmentosum [SARMSAR] Amblystegiaceae (B:M)
Calliergon sarmentosum f. *homophyllum* Arnell & C. Jens. = Sarmenthypnum sarmentosum [SARMSAR] Amblystegiaceae (B:M)
Calliergon sarmentosum var. *beringianum* (Card. & Thér.) Grout = Sarmenthypnum sarmentosum [SARMSAR] Amblystegiaceae (B:M)
Calliergon sarmentosum var. *crispum* Karcz. = Sarmenthypnum sarmentosum [SARMSAR] Amblystegiaceae (B:M)
Calliergon sarmentosum var. *fallaciosum* (Milde) G. Roth = Sarmenthypnum sarmentosum [SARMSAR] Amblystegiaceae (B:M)

Calliergon sarmentosum var. *flagellare* Karcz. = Sarmenthypnum sarmentosum [SARMSAR] Amblystegiaceae (B:M)
Calliergon sarmentosum var. *fontinaloides* (Berggr.) G. Roth = Sarmenthypnum sarmentosum [SARMSAR] Amblystegiaceae (B:M)
Calliergon sarmentosum var. *subpinnatum* Warnst. = Sarmenthypnum sarmentosum [SARMSAR] Amblystegiaceae (B:M)
Calliergon schreberi (Brid.) Mitt. = Pleurozium schreberi [PLEUSCH] Hylocomiaceae (B:M)
Calliergon stramineum (Brid.) Kindb. [CALLSTR] Amblystegiaceae (B:M)
Calliergon stramineum f. *subtrifarium* (Saelan) Warnst. = Calliergon stramineum [CALLSTR] Amblystegiaceae (B:M)
Calliergon stramineum var. *flagellaceum* (G. Roth & Bock in G. Roth) Karcz. = Calliergon stramineum [CALLSTR] Amblystegiaceae (B:M)
Calliergon stramineum var. *laxifolium* (Kindb.) Karcz. = Calliergon stramineum [CALLSTR] Amblystegiaceae (B:M)
Calliergon stramineum var. *nivale* (Lor.) G. Roth = Calliergon stramineum [CALLSTR] Amblystegiaceae (B:M)
Calliergon stramineum var. *patens* (Lindb.) G. Roth = Calliergon stramineum [CALLSTR] Amblystegiaceae (B:M)
Calliergon subgiganteum Kindb. = Calliergon richardsonii [CALLRIC] Amblystegiaceae (B:M)
Calliergon subsarmentosum Kindb. = Sarmenthypnum sarmentosum [SARMSAR] Amblystegiaceae (B:M)
Calliergon subsarmentosum Kindb. = Calliergon giganteum [CALLGIG] Amblystegiaceae (B:M)
Calliergon subturgescens Kindb. = Pseudocalliergon turgescens [PSEUTUR] Amblystegiaceae (B:M)
Calliergon trifarium (Web. & Mohr) Kindb. [CALLTRI] Amblystegiaceae (B:M)
Calliergon turgescens (T. Jens.) Kindb. = Pseudocalliergon turgescens [PSEUTUR] Amblystegiaceae (B:M)
Calliergon turgescens var. *patens* Karcz. = Pseudocalliergon turgescens [PSEUTUR] Amblystegiaceae (B:M)
Calliergon turgescens var. *tenue* (Berggr.) Karcz. = Pseudocalliergon turgescens [PSEUTUR] Amblystegiaceae (B:M)
Calliergon wickesiae Grout = Loeskypnum wickesiae [LOESWIC] Amblystegiaceae (B:M)
Calliergonella Loeske [CALLIEO$] Amblystegiaceae (B:M)
Calliergonella conardii Lawt. = Calliergonella cuspidata [CALLCUS] Amblystegiaceae (B:M)
Calliergonella cuspidata (Hedw.) Loeske [CALLCUS] Amblystegiaceae (B:M)
Calliergonella cuspidata f. *abbreviata* (Röll) Podp. = Calliergonella cuspidata [CALLCUS] Amblystegiaceae (B:M)
Calliergonella cuspidata f. *acuteramosa* Bauer ex Karez. = Calliergonella cuspidata [CALLCUS] Amblystegiaceae (B:M)
Calliergonella cuspidata f. *brevifolium* (Sanio in Warnst.) Podp. = Calliergonella cuspidata [CALLCUS] Amblystegiaceae (B:M)
Calliergonella cuspidata var. *caespitosum* (Whiteh.) Karez. = Calliergonella cuspidata [CALLCUS] Amblystegiaceae (B:M)
Calliergonella cuspidata var. *pungens* (Schimp.) Latz. = Calliergonella cuspidata [CALLCUS] Amblystegiaceae (B:M)
Calliergonella cuspidata var. *umbrosus* (Loeske) Warnst. = Calliergonella cuspidata [CALLCUS] Amblystegiaceae (B:M)
Calliergonella schreberi (Brid.) Grout = Pleurozium schreberi [PLEUSCH] Hylocomiaceae (B:M)
Calliergonella schreberi var. *tananae* (Grout) Grout = Pleurozium schreberi [PLEUSCH] Hylocomiaceae (B:M)
Callitriche L. [CALLITR$] Callitrichaceae (V:D)
Callitriche anceps Fern. = Callitriche heterophylla ssp. heterophylla [CALLHET0] Callitrichaceae (V:D)
Callitriche autumnalis L. = Callitriche hermaphroditica [CALLHER] Callitrichaceae (V:D)
Callitriche bolanderi Hegelm. = Callitriche heterophylla ssp. bolanderi [CALLHET1] Callitrichaceae (V:D)
Callitriche hermaphroditica L. [CALLHER] Callitrichaceae (V:D)
Callitriche heterophylla Pursh [CALLHET] Callitrichaceae (V:D)
Callitriche heterophylla ssp. **bolanderi** (Hegelm.) Calder & Taylor [CALLHET1] Callitrichaceae (V:D)
Callitriche heterophylla ssp. **heterophylla** [CALLHET0] Callitrichaceae (V:D)
Callitriche heterophylla var. *bolanderi* (Hegelm.) Fassett = Callitriche heterophylla ssp. bolanderi [CALLHET1] Callitrichaceae (V:D)
Callitriche marginata Torr. [CALLMAR] Callitrichaceae (V:D)
Callitriche palustris L. [CALLPAU] Callitrichaceae (V:D)
Callitriche palustris var. *verna* (L.) Fenley ex Jepson = Callitriche palustris [CALLPAU] Callitrichaceae (V:D)
Callitriche sepulta S. Wats. = Callitriche marginata [CALLMAR] Callitrichaceae (V:D)
*__Callitriche stagnalis__ Scop. [CALLSTA] Callitrichaceae (V:D)
Callitriche verna L. = Callitriche palustris [CALLPAU] Callitrichaceae (V:D)
Calluna Salisb. [CALLUNA$] Ericaceae (V:D)
*__Calluna vulgaris__ (L.) Hull [CALLVUL] Ericaceae (V:D)
Calochortus Pursh [CALOCHO$] Liliaceae (V:M)
Calochortus apiculatus Baker [CALOAPI] Liliaceae (V:M)
Calochortus lyallii Baker [CALOLYA] Liliaceae (V:M)
Calochortus macrocarpus Dougl. [CALOMAC] Liliaceae (V:M)
Caloplaca Th. Fr. [CALOPLA$] Teloschistaceae (L)
Caloplaca ammiospila (Wahlenb.) H. Olivier [CALOAMM] Teloschistaceae (L)
Caloplaca approximata (Lynge) H. Magn. [CALOAPP] Teloschistaceae (L)
Caloplaca arenaria (Pers.) Müll. Arg. [CALOARE] Teloschistaceae (L)
Caloplaca atroflava (Turner) Mong. [CALOATR] Teloschistaceae (L)
Caloplaca atrosanguinea (G. Merr.) Lamb [CALOATS] Teloschistaceae (L)
Caloplaca aurantiaca (Lightf.) Th. Fr. = Caloplaca flavorubescens [CALOFLV] Teloschistaceae (L)
Caloplaca bolacina (Tuck.) Herre [CALOBOL] Teloschistaceae (L)
Caloplaca borealis (Vainio) Poelt [CALOBOR] Teloschistaceae (L)
Caloplaca bracteata (Hoffm.) Jatta = Fulgensia bracteata [FULGBRA] Teloschistaceae (L)
Caloplaca caesiorufella (Nyl.) Zahlbr. = Caloplaca phaeocarpella [CALOPHA] Teloschistaceae (L)
Caloplaca castellana (Räsänen) Poelt [CALOCAS] Teloschistaceae (L)
Caloplaca cerina (Ehrh. ex Hedwig) Th. Fr. [CALOCER] Teloschistaceae (L)
Caloplaca cerinelloides (Erichsen) Poelt [CALOCEI] Teloschistaceae (L)
Caloplaca chlorina (Flotow) H. Olivier [CALOCHL] Teloschistaceae (L)
Caloplaca chrysophthalma Degel. [CALOCHR] Teloschistaceae (L)
Caloplaca cinnamomea (Th. Fr.) H. Olivier = Caloplaca ammiospila [CALOAMM] Teloschistaceae (L)
Caloplaca citrina (Hoffm.) Th. Fr. [CALOCIT] Teloschistaceae (L)
Caloplaca diphyodes (Nyl.) Jatta [CALODIP] Teloschistaceae (L)
Caloplaca discoidalis (Vainio) Lynge = Caloplaca ammiospila [CALOAMM] Teloschistaceae (L)
Caloplaca elegans (Link) Th. Fr. = Xanthoria elegans [XANTELE] Teloschistaceae (L)
Caloplaca epithallina Lynge [CALOEPI] Teloschistaceae (L)
Caloplaca erythrella (Ach.) Kieffer = Caloplaca flavovirescens [CALOFLA] Teloschistaceae (L)
Caloplaca feracissima H. Magn. [CALOFER] Teloschistaceae (L)
Caloplaca ferruginea (Hudson) Th. Fr. [CALOFEU] Teloschistaceae (L)
Caloplaca flavogranulosa Arup [CALOFLO] Teloschistaceae (L)
Caloplaca flavorubescens (Hudson) J.R. Laundon [CALOFLV] Teloschistaceae (L)
Caloplaca flavovirescens (Wulfen) Dalla Torre & Sarnth. [CALOFLA] Teloschistaceae (L)
Caloplaca fraudans (Th. Fr.) H. Olivier [CALOFRA] Teloschistaceae (L)
Caloplaca gilva (Hoffm.) Zahlbr. = Caloplaca cerina [CALOCER] Teloschistaceae (L)
Caloplaca gloriae Werner & Llimona [CALOGLO] Teloschistaceae (L)

Caloplaca herrei Hasse = Caloplaca atrosanguinea [CALOATS] Teloschistaceae (L)
Caloplaca holocarpa (Hoffm. ex Ach.) M. Wade [CALOHOL] Teloschistaceae (L)
Caloplaca inconspecta Arup [CALOINC] Teloschistaceae (L)
Caloplaca invadens Lynge = Caloplaca castellana [CALOCAS] Teloschistaceae (L)
Caloplaca jungermanniae (Vahl) Th. Fr. [CALOJUN] Teloschistaceae (L)
Caloplaca lamprocheila (DC.) Flagey = Caloplaca arenaria [CALOARE] Teloschistaceae (L)
Caloplaca leucoraea (Ach. ex Flörke) Branth = Caloplaca sinapisperma [CALOSIN] Teloschistaceae (L)
Caloplaca litoricola Brodo [CALOLIT] Teloschistaceae (L)
Caloplaca livida (Hepp) Jatta [CALOLIV] Teloschistaceae (L)
Caloplaca luteominia (Tuck.) Zahlbr. [CALOLUT] Teloschistaceae (L)
Caloplaca marina (Wedd.) Zahlbr. [CALOMAR] Teloschistaceae (L)
Caloplaca marina ssp. **americana** Arup [CALOMAR1] Teloschistaceae (L)
Caloplaca microthallina (Wedd.) Zahlbr. [CALOMIC] Teloschistaceae (L)
Caloplaca nivalis (Körber) Th. Fr. [CALONIV] Teloschistaceae (L)
Caloplaca obscurella (Körber) Th. Fr. [CALOOBS] Teloschistaceae (L)
Caloplaca oleicola (Steiner) v.d. Boom & Breuss [CALOOLE] Teloschistaceae (L)
Caloplaca oregona H. Magn. [CALOORE] Teloschistaceae (L)
Caloplaca phaeocarpella (Nyl.) Zahlbr. [CALOPHA] Teloschistaceae (L)
Caloplaca pyracea (Ach.) Th. Fr. = Caloplaca holocarpa [CALOHOL] Teloschistaceae (L)
Caloplaca rosei Hasse [CALOROS] Teloschistaceae (L)
Caloplaca sarcopisioides (Körber) Zahlbr. = Caloplaca obscurella [CALOOBS] Teloschistaceae (L)
Caloplaca scopularis (Nyl.) Lettau [CALOSCO] Teloschistaceae (L)
Caloplaca sinapisperma (Lam. & DC.) Maheu & A. Gillet [CALOSIN] Teloschistaceae (L)
Caloplaca sorediata (Vainio) Du Rietz = Xanthoria sorediata [XANTSOR] Teloschistaceae (L)
Caloplaca sorocarpa (Vainio) Zahlbr. [CALOSOR] Teloschistaceae (L)
Caloplaca stillicidiorum (Vahl) Lynge = Caloplaca cerina [CALOCER] Teloschistaceae (L)
Caloplaca subnigricans H. Magn. = Caloplaca atrosanguinea [CALOATS] Teloschistaceae (L)
Caloplaca teicholyta (Ach.) J. Steiner [CALOTEI] Teloschistaceae (L)
Caloplaca tetraspora (Nyl.) H. Olivier [CALOTET] Teloschistaceae (L)
Caloplaca tominii Savicz [CALOTOM] Teloschistaceae (L)
Caloplaca verruculifera (Vainio) Zahlbr. [CALOVER] Teloschistaceae (L)
Caloplaca vicaria H. Magn. [CALOVIC] Teloschistaceae (L)
Caltha L. [CALTHA$] Ranunculaceae (V:D)
Caltha asarifolia DC. = Caltha palustris [CALTPAL] Ranunculaceae (V:D)
Caltha biflora DC. = Caltha leptosepala ssp. howellii [CALTLEP3] Ranunculaceae (V:D)
Caltha biflora ssp. *howellii* (Huth) Abrams = Caltha leptosepala ssp. howellii [CALTLEP3] Ranunculaceae (V:D)
Caltha biflora var. *rotundifolia* (Huth) C.L. Hitchc. = Caltha leptosepala [CALTLEP] Ranunculaceae (V:D)
Caltha howellii Huth = Caltha leptosepala ssp. howellii [CALTLEP3] Ranunculaceae (V:D)
Caltha leptosepala DC. [CALTLEP] Ranunculaceae (V:D)
Caltha leptosepala ssp. *biflora* (DC.) P.G. Sm. = Caltha leptosepala ssp. howellii [CALTLEP3] Ranunculaceae (V:D)
Caltha leptosepala ssp. **howellii** (Huth) P.G. Sm. [CALTLEP3] Ranunculaceae (V:D)
Caltha leptosepala var. *biflora* (DC.) Lawson = Caltha leptosepala ssp. howellii [CALTLEP3] Ranunculaceae (V:D)
Caltha natans Pallas ex Georgi [CALTNAT] Ranunculaceae (V:D)
Caltha natans var. *asarifolia* (DC.) Huth = Caltha palustris [CALTPAL] Ranunculaceae (V:D)
Caltha palustris L. [CALTPAL] Ranunculaceae (V:D)
Caltha palustris ssp. *asarifolia* (DC.) Hultén = Caltha palustris [CALTPAL] Ranunculaceae (V:D)
Caltha palustris var. *asarifolia* (DC.) Rothrock = Caltha palustris [CALTPAL] Ranunculaceae (V:D)
Caltha palustris var. *flabellifolia* (Pursh) Torr. & Gray = Caltha palustris [CALTPAL] Ranunculaceae (V:D)
Calycularia Mitt. [CALYCUL$] Pelliaceae (B:H)
Calycularia crispula Mitt. [CALYCRI] Pelliaceae (B:H)
Calypogeia Raddi [CALYPOG$] Calypogeiaceae (B:H)
Calypogeia azurea Stotler & Crotz. [CALYAZU] Calypogeiaceae (B:H)
Calypogeia cyanophora Schust. = Calypogeia azurea [CALYAZU] Calypogeiaceae (B:H)
Calypogeia fissa (L.) Raddi [CALYFIS] Calypogeiaceae (B:H)
Calypogeia integristipula Steph. [CALYINT] Calypogeiaceae (B:H)
Calypogeia meylanii Buch = Calypogeia integristipula [CALYINT] Calypogeiaceae (B:H)
Calypogeia muelleriana (Schiffn.) K. Müll. [CALYMUE] Calypogeiaceae (B:H)
Calypogeia muelleriana f. **muelleriana** [CALYMUE0] Calypogeiaceae (B:H)
Calypogeia muelleriana f. **schofieldii** Hong [CALYMUE1] Calypogeiaceae (B:H)
Calypogeia muelleriana f. **shieldii** Hong [CALYMUE2] Calypogeiaceae (B:H)
Calypogeia neesiana (Mass. & Carest.) K. Müll. [CALYNEE] Calypogeiaceae (B:H)
Calypogeia neesiana var. *meylanii* (Buch) Schust. = Calypogeia integristipula [CALYINT] Calypogeiaceae (B:H)
Calypogeia paludosa Warnst. = Calypogeia sphagnicola [CALYSPH] Calypogeiaceae (B:H)
Calypogeia sphagnicola (Arnell & J. Perss.) Warnst. & Loeske [CALYSPH] Calypogeiaceae (B:H)
Calypogeia suecica (Arnell & J. Perss.) K. Müll. [CALYSUE] Calypogeiaceae (B:H)
Calypogeia trichomanis auct. = Calypogeia azurea [CALYAZU] Calypogeiaceae (B:H)
Calypogeja Raddi = Calypogeia [CALYPOG$] Calypogeiaceae (B:H)
Calypso Salisb. [CALYPSO$] Orchidaceae (V:M)
Calypso bulbosa (L.) Oakes [CALYBUL] Orchidaceae (V:M)
Calypso bulbosa ssp. *occidentalis* (Holz.) Calder & Taylor = Calypso bulbosa var. occidentalis [CALYBUL3] Orchidaceae (V:M)
Calypso bulbosa var. **occidentalis** (Holz.) Boivin [CALYBUL3] Orchidaceae (V:M)
Calyptridium umbellatum var. *caudiciferum* (Gray) Jepson = Cistanthe umbellata var. caudicifera [CISTUMB1] Portulacaceae (V:D)
Calystegia R. Br. [CALYSTE$] Convolvulaceae (V:D)
*****Calystegia sepium** (L.) R. Br. [CALYSEP] Convolvulaceae (V:D)
Calystegia soldanella (L.) R. Br. ex Roemer & J.A. Schultes [CALYSOL] Convolvulaceae (V:D)
Camassia Lindl. [CAMASSI$] Liliaceae (V:M)
Camassia leichtlinii (Baker) S. Wats. [CAMALEI] Liliaceae (V:M)
Camassia quamash (Pursh) Greene [CAMAQUA] Liliaceae (V:M)
Camassia quamash ssp. **maxima** Gould [CAMAQUA1] Liliaceae (V:M)
Camassia quamash ssp. **quamash** [CAMAQUA0] Liliaceae (V:M)
Camassia quamash ssp. *teapeae* (St. John) St. John = Camassia quamash ssp. quamash [CAMAQUA0] Liliaceae (V:M)
Camassia quamash var. *maxima* (Gould) Boivin = Camassia quamash ssp. maxima [CAMAQUA1] Liliaceae (V:M)
Camelina Crantz [CAMELIN$] Brassicaceae (V:D)
*****Camelina microcarpa** DC. [CAMEMIC] Brassicaceae (V:D)
*****Camelina sativa** (L.) Crantz [CAMESAT] Brassicaceae (V:D)
Camelina sativa ssp. *microcarpa* (DC.) E. Schmid = Camelina microcarpa [CAMEMIC] Brassicaceae (V:D)
Camissonia Link [CAMISSO$] Onagraceae (V:D)
Camissonia andina (Nutt.) Raven [CAMIAND] Onagraceae (V:D)
Camissonia breviflora (Torr. & Gray) Raven [CAMIBRE] Onagraceae (V:D)

Camissonia contorta (Dougl. ex Lehm.) Kearney [CAMICON] Onagraceae (V:D)
Campanula L. [CAMPANU$] Campanulaceae (V:D)
Campanula alaskana (Gray) W. Wight ex J.P. Anders. = Campanula rotundifolia [CAMPROT] Campanulaceae (V:D)
Campanula aurita Greene [CAMPAUR] Campanulaceae (V:D)
Campanula dubia A. DC. = Campanula rotundifolia [CAMPROT] Campanulaceae (V:D)
Campanula heterodoxa Bong. = Campanula rotundifolia [CAMPROT] Campanulaceae (V:D)
Campanula intercedens Witasek = Campanula rotundifolia [CAMPROT] Campanulaceae (V:D)
Campanula lasiocarpa Cham. [CAMPLAS] Campanulaceae (V:D)
Campanula lasiocarpa ssp. *latisepala* (Hultén) Hultén = Campanula lasiocarpa [CAMPLAS] Campanulaceae (V:D)
Campanula latisepala Hultén = Campanula lasiocarpa [CAMPLAS] Campanulaceae (V:D)
Campanula latisepala var. *dubia* Hultén = Campanula lasiocarpa [CAMPLAS] Campanulaceae (V:D)
*__Campanula medium__ L. [CAMPMED] Campanulaceae (V:D)
*__Campanula persicifolia__ L. [CAMPPER] Campanulaceae (V:D)
Campanula persicifolia var. *alba* Horton = Campanula persicifolia [CAMPPER] Campanulaceae (V:D)
Campanula petiolata A. DC. = Campanula rotundifolia [CAMPROT] Campanulaceae (V:D)
*__Campanula rapunculoides__ L. [CAMPRAP] Campanulaceae (V:D)
Campanula rapunculoides var. *ucranica* (Bess.) K. Koch = Campanula rapunculoides [CAMPRAP] Campanulaceae (V:D)
Campanula rotundifolia L. [CAMPROT] Campanulaceae (V:D)
Campanula rotundifolia var. *alaskana* Gray = Campanula rotundifolia [CAMPROT] Campanulaceae (V:D)
Campanula rotundifolia var. *intercedens* (Witasek) Farw. = Campanula rotundifolia [CAMPROT] Campanulaceae (V:D)
Campanula rotundifolia var. *lancifolia* Mert. & Koch = Campanula rotundifolia [CAMPROT] Campanulaceae (V:D)
Campanula rotundifolia var. *petiolata* (A. DC.) J.K. Henry = Campanula rotundifolia [CAMPROT] Campanulaceae (V:D)
Campanula rotundifolia var. *velutina* A. DC. = Campanula rotundifolia [CAMPROT] Campanulaceae (V:D)
Campanula sacajaweana M.E. Peck = Campanula rotundifolia [CAMPROT] Campanulaceae (V:D)
Campanula scouleri Hook. ex A. DC. [CAMPSCO] Campanulaceae (V:D)
Campanula uniflora L. [CAMPUNI] Campanulaceae (V:D)
Campe barbarea (L.) W. Wight ex Piper = Barbarea vulgaris [BARBVUL] Brassicaceae (V:D)
Campe verna (P. Mill.) Heller = Barbarea verna [BARBVER] Brassicaceae (V:D)
Camptothecium aeneum (Mitt.) Jaeg. = Homalothecium aeneum [HOMAAEN] Brachytheciaceae (B:M)
Camptothecium aeneum var. *dolosum* (Ren. & Card.) Grout = Homalothecium aeneum [HOMAAEN] Brachytheciaceae (B:M)
Camptothecium aeneum var. *robustum* Grout = Homalothecium aeneum [HOMAAEN] Brachytheciaceae (B:M)
Camptothecium alsioides Kindb. = Homalothecium arenarium [HOMAARE] Brachytheciaceae (B:M)
Camptothecium amesiae Ren. & Card. = Homalothecium pinnatifidum [HOMAPIN] Brachytheciaceae (B:M)
Camptothecium arenarium (Lesq.) Jaeg. = Homalothecium arenarium [HOMAARE] Brachytheciaceae (B:M)
Camptothecium hamatidens var. *tenue* Kindb. = Homalothecium nuttallii [HOMANUT] Brachytheciaceae (B:M)
Camptothecium hematidens (Kindb.) Kindb. = Homalothecium nuttallii [HOMANUT] Brachytheciaceae (B:M)
Camptothecium lutescens var. *occidentale* Ren. & Card. = Homalothecium fulgescens [HOMAFUL] Brachytheciaceae (B:M)
Camptothecium megaptilum Sull. = Trachybryum megaptilum [TRACMEG] Brachytheciaceae (B:M)
Camptothecium megaptilum var. *fosteri* Grout = Trachybryum megaptilum [TRACMEG] Brachytheciaceae (B:M)
Camptothecium nitens (Hedw.) Schimp. = Tomentypnum nitens [TOMENIT] Brachytheciaceae (B:M)
Camptothecium nitens var. *falcatum* Peck in Burnh. = Tomentypnum falcifolium [TOMEFAL] Brachytheciaceae (B:M)
Camptothecium nitens var. *falcifolium* Ren. ex Nichols = Tomentypnum falcifolium [TOMEFAL] Brachytheciaceae (B:M)
Camptothecium nitens var. *leucobaisi* Kindb. = Tomentypnum nitens [TOMENIT] Brachytheciaceae (B:M)
Camptothecium nitens var. *microtheca* Kindb. = Tomentypnum nitens [TOMENIT] Brachytheciaceae (B:M)
Camptothecium nuttallii (Wils.) Schimp. in B.S.G. = Homalothecium nuttallii [HOMANUT] Brachytheciaceae (B:M)
Camptothecium paulianum Grout in Holz. & Frye = Sanionia uncinata var. uncinata [SANIUNC0] Amblystegiaceae (B:M)
Camptothecium pinnatifidum (Sull. & Lesq.) Sull. = Homalothecium pinnatifidum [HOMAPIN] Brachytheciaceae (B:M)
Campylium (Sull.) Mitt. [CAMPYLI$] Amblystegiaceae (B:M)
Campylium adscendens (Lindb.) Perss. in Perss. & Gjaer. = Herzogiella adscendens [HERZADS] Hypnaceae (B:M)
Campylium calcareum Crundw. & Nyh. [CAMPCAL] Amblystegiaceae (B:M)
Campylium chrysophyllum (Brid.) J. Lange [CAMPCHR] Amblystegiaceae (B:M)
Campylium chrysophyllum var. *brevifolium* (Ren. & Card.) Grout = Campylium chrysophyllum [CAMPCHR] Amblystegiaceae (B:M)
Campylium chrysophyllum var. *zemliae* (C. Jens.) Grout = Campylium chrysophyllum [CAMPCHR] Amblystegiaceae (B:M)
Campylium halleri (Hedw.) Lindb. [CAMPHAL] Amblystegiaceae (B:M)
Campylium hispidulum (Brid.) Mitt. [CAMPHIS] Amblystegiaceae (B:M)
Campylium hispidulum var. *cordatum* Grout = Campylium hispidulum [CAMPHIS] Amblystegiaceae (B:M)
Campylium hispidulum var. *sommerfeltii* (Myr.) Lindb. = Campylium hispidulum [CAMPHIS] Amblystegiaceae (B:M)
Campylium polygamum (Schimp. in B.S.G.) C. Jens. [CAMPPOL] Amblystegiaceae (B:M)
Campylium polygamum var. *fluitans* Grout = Campylium polygamum [CAMPPOL] Amblystegiaceae (B:M)
Campylium polygamum var. *longinerve* (Ren. & Card.) Grout = Leptodictyum riparium [LEPTRIP] Amblystegiaceae (B:M)
Campylium polygamum var. *minus* (Schimp.) G. Roth = Campylium polygamum [CAMPPOL] Amblystegiaceae (B:M)
Campylium protensum (Brid.) Kindb. = Campylium stellatum var. protensum [CAMPSTE2] Amblystegiaceae (B:M)
Campylium radicale (P. Beauv.) Grout [CAMPRAD] Amblystegiaceae (B:M)
Campylium sommerfeltii (Myr.) J. Lange = Campylium hispidulum [CAMPHIS] Amblystegiaceae (B:M)
Campylium stellatum (Hedw.) C. Jens. [CAMPSTE] Amblystegiaceae (B:M)
Campylium stellatum ssp. *treleasii* (Ren.) Grout = Herzogiella adscendens [HERZADS] Hypnaceae (B:M)
Campylium stellatum var. *adscendens* (Lindb.) Perss. = Herzogiella adscendens [HERZADS] Hypnaceae (B:M)
Campylium stellatum var. **protensum** (Brid.) Bryhn [CAMPSTE2] Amblystegiaceae (B:M)
Campylium stellatum var. **stellatum** [CAMPSTE0] Amblystegiaceae (B:M)
Campylium treleasii (Ren.) Broth. = Herzogiella adscendens [HERZADS] Hypnaceae (B:M)
Campylophyllum halleri (Hedw.) Fleisch. = Campylium halleri [CAMPHAL] Amblystegiaceae (B:M)
Campylopus Brid. [CAMPYLO$] Dicranaceae (B:M)
Campylopus atrovirens De Not. [CAMPATR] Dicranaceae (B:M)
Campylopus atrovirens var. **cucullatifolius** J.-P. Frahm [CAMPATR1] Dicranaceae (B:M)
Campylopus canadensis Kindb. = Paraleucobryum longifolium [PARALON] Dicranaceae (B:M)
Campylopus eberhardtii Par. = Campylopus japonicus [CAMPJAP] Dicranaceae (B:M)
Campylopus excurrens Dix. = Campylopus japonicus [CAMPJAP] Dicranaceae (B:M)
Campylopus flexuosus (Hedw.) Brid. [CAMPFLE] Dicranaceae (B:M)

Campylopus fragilis (Brid.) Bruch & Schimp. in B.S.G. [CAMPFRA] Dicranaceae (B:M)
Campylopus introflexus (Hedw.) Brid. [CAMPINT] Dicranaceae (B:M)
Campylopus irrigatus Thér. = Campylopus japonicus [CAMPJAP] Dicranaceae (B:M)
Campylopus japonicus Broth. [CAMPJAP] Dicranaceae (B:M)
Campylopus pseudomülleri Card. = Campylopus japonicus [CAMPJAP] Dicranaceae (B:M)
Campylopus saint-pierrei Thér. = Campylopus japonicus [CAMPJAP] Dicranaceae (B:M)
Campylopus schimperi Milde [CAMPSCI] Dicranaceae (B:M)
Campylopus schwarzii Schimp. [CAMPSCH] Dicranaceae (B:M)
Campylopus subulatus Schimp. in Rabenh. [CAMPSUB] Dicranaceae (B:M)
Campylopus subulatus var. *schimperi* (Milde) Husn. = Campylopus schimperi [CAMPSCI] Dicranaceae (B:M)
Campylopus uii Broth. = Campylopus japonicus [CAMPJAP] Dicranaceae (B:M)
Candelaria A. Massal. [CANDELA$] Candelariaceae (L)
Candelaria concolor (Dickson) Stein [CANDCON] Candelariaceae (L)
Candelariella Müll. Arg. [CANDELR$] Candelariaceae (L)
Candelariella athallina (Wedd.) Du Rietz [CANDATH] Candelariaceae (L)
Candelariella aurella (Hoffm.) Zahlbr. [CANDAUR] Candelariaceae (L)
Candelariella cerinella (Flörke) Zahlbr. = Candelariella aurella [CANDAUR] Candelariaceae (L)
Candelariella efflorescens R.C. Harris & W.R. Buck [CANDEFF] Candelariaceae (L)
Candelariella epixantha auct. = Candelariella aurella [CANDAUR] Candelariaceae (L)
Candelariella rosulans (Müll. Arg.) Zahlbr. [CANDROS] Candelariaceae (L)
Candelariella terrigena Räsänen [CANDTER] Candelariaceae (L)
Candelariella vitellina (Hoffm.) Müll. Arg. [CANDVIT] Candelariaceae (L)
Candelariella xanthostigma (Ach.) Lettau [CANDXAN] Candelariaceae (L)
Capnoides aureum (Willd.) Kuntze = Corydalis aurea [CORYAUR] Fumariaceae (V:D)
Capnoides sempervirens (L.) Borkh. = Corydalis sempervirens [CORYSEM] Fumariaceae (V:D)
Capsella Medik. [CAPSELL$] Brassicaceae (V:D)
***Capsella bursa-pastoris** (L.) Medik. [CAPSBUR] Brassicaceae (V:D)
Carara didyma (L.) Britt. = Coronopus didymus [CORODID] Brassicaceae (V:D)
Carbonea (Hertel) Hertel [CARBONE$] Lecanoraceae (L)
Carbonea aggregantula (Müll. Arg.) Diederich & Triebel [CARBAGG] Lecanoraceae (L)
Carbonea vorticosa (Flörke) Hertel [CARBVOR] Lecanoraceae (L)
Cardamine L. [CARDAMI$] Brassicaceae (V:D)
Cardamine angulata Hook. [CARDANG] Brassicaceae (V:D)
Cardamine barbareifolia DC. = Rorippa barbareifolia [RORIBAR] Brassicaceae (V:D)
Cardamine bellidifolia L. [CARDBEL] Brassicaceae (V:D)
Cardamine bellidifolia var. *pinnatifida* Hultén = Cardamine bellidifolia [CARDBEL] Brassicaceae (V:D)
Cardamine bellidifolia var. *sinuata* (J. Vahl) Lange = Cardamine bellidifolia [CARDBEL] Brassicaceae (V:D)
Cardamine breweri S. Wats. [CARDBRE] Brassicaceae (V:D)
Cardamine breweri var. **breweri** [CARDBRE0] Brassicaceae (V:D)
Cardamine breweri var. **orbicularis** (Greene) Detling [CARDBRE2] Brassicaceae (V:D)
Cardamine cordifolia var. *lyallii* (S. Wats.) A. Nels. & J.F. Macbr. = Cardamine lyallii [CARDLYA] Brassicaceae (V:D)
Cardamine hederifolia Greene = Cardamine breweri var. breweri [CARDBRE0] Brassicaceae (V:D)
***Cardamine hirsuta** L. [CARDHIR] Brassicaceae (V:D)
Cardamine hirsuta var. *kamtschatica* (Regel) O.E. Schulz = Cardamine oligosperma var. kamtschatica [CARDOLI1] Brassicaceae (V:D)
Cardamine kamtschatica (Regel) Piper = Cardamine oligosperma var. kamtschatica [CARDOLI1] Brassicaceae (V:D)
Cardamine lyallii S. Wats. [CARDLYA] Brassicaceae (V:D)
Cardamine nuttallii Greene [CARDNUT] Brassicaceae (V:D)
Cardamine nymanii Gandog. = Cardamine pratensis var. angustifolia [CARDPRA2] Brassicaceae (V:D)
Cardamine occidentalis (S. Wats. ex B.L. Robins.) T.J. Howell [CARDOCC] Brassicaceae (V:D)
Cardamine oligosperma Nutt. [CARDOLI] Brassicaceae (V:D)
Cardamine oligosperma var. **kamtschatica** (Regel) Detling [CARDOLI1] Brassicaceae (V:D)
Cardamine orbicularis Greene = Cardamine breweri var. orbicularis [CARDBRE2] Brassicaceae (V:D)
Cardamine oregona Piper = Cardamine breweri var. breweri [CARDBRE0] Brassicaceae (V:D)
Cardamine parviflora L. [CARDPAR] Brassicaceae (V:D)
Cardamine pensylvanica Muhl. ex Willd. [CARDPEN] Brassicaceae (V:D)
Cardamine pensylvanica var. *brittoniana* Farw. = Cardamine pensylvanica [CARDPEN] Brassicaceae (V:D)
Cardamine pratensis L. [CARDPRA] Brassicaceae (V:D)
Cardamine pratensis ssp. *angustifolia* (Hook.) O.E. Schulz = Cardamine pratensis var. angustifolia [CARDPRA2] Brassicaceae (V:D)
Cardamine pratensis var. **angustifolia** Hook. [CARDPRA2] Brassicaceae (V:D)
Cardamine pratensis var. *occidentalis* S. Wats. ex B.L. Robins. = Cardamine occidentalis [CARDOCC] Brassicaceae (V:D)
Cardamine pulcherrima var. *tenella* (Pursh) C.L. Hitchc. = Cardamine nuttallii [CARDNUT] Brassicaceae (V:D)
Cardamine umbellata Greene = Cardamine oligosperma var. kamtschatica [CARDOLI1] Brassicaceae (V:D)
Cardaminopsis kamchatica (Fisch. ex DC.) O.E. Schulz = Arabis lyrata var. kamchatica [ARABLYR2] Brassicaceae (V:D)
Cardaria Desv. [CARDARI$] Brassicaceae (V:D)
Cardaria chalapensis (L.) Hand.-Maz. = Cardaria draba ssp. chalapensis [CARDDRA1] Brassicaceae (V:D)
***Cardaria draba** (L.) Desv. [CARDDRA] Brassicaceae (V:D)
***Cardaria draba** ssp. **chalapensis** (L.) O.E. Schulz [CARDDRA1] Brassicaceae (V:D)
Cardaria draba var. *repens* (Schrenk) O.E. Schulz = Cardaria draba ssp. chalapensis [CARDDRA1] Brassicaceae (V:D)
***Cardaria pubescens** (C.A. Mey.) Jarmolenko [CARDPUB] Brassicaceae (V:D)
Cardaria pubescens var. *elongata* Rollins = Cardaria pubescens [CARDPUB] Brassicaceae (V:D)
Carduus L. [CARDUUS$] Asteraceae (V:D)
***Carduus acanthoides** L. [CARDACA] Asteraceae (V:D)
Carduus arvensis (L.) Robson = Cirsium arvense [CIRSARV] Asteraceae (V:D)
***Carduus crispus** L. [CARDCRI] Asteraceae (V:D)
Carduus lanceolatus L. = Cirsium vulgare [CIRSVUL] Asteraceae (V:D)
Carduus macounii Greene = Cirsium edule [CIRSEDU] Asteraceae (V:D)
Carduus marianus L. = Silybum marianum [SILYMAR] Asteraceae (V:D)
***Carduus nutans** L. [CARDNUA] Asteraceae (V:D)
***Carduus nutans** ssp. **leiophyllus** (Petrovic) Stojanov & Stef. [CARDNUA1] Asteraceae (V:D)
Carduus nutans var. *leiophyllus* (Petrovic) Arènes = Carduus nutans ssp. leiophyllus [CARDNUA1] Asteraceae (V:D)
Carduus nutans var. *vestitus* (Hallier) Boivin = Carduus nutans ssp. leiophyllus [CARDNUA1] Asteraceae (V:D)
Carduus vulgaris Savi = Cirsium vulgare [CIRSVUL] Asteraceae (V:D)
Carex L. [CAREX$] Cyperaceae (V:M)
Carex ablata Bailey = Carex luzulina var. ablata [CARELUZ1] Cyperaceae (V:M)
Carex accedens Holm = Carex aperta [CAREAPE] Cyperaceae (V:M)

Carex acutina var. *tenuior* Bailey = Carex aperta [CAREAPE] Cyperaceae (V:M)
Carex acutinella Mackenzie = Carex aquatilis var. aquatilis [CAREAQU0] Cyperaceae (V:M)
Carex aenea Fern. [CAREAEN] Cyperaceae (V:M)
Carex albonigra Mackenzie [CAREALB] Cyperaceae (V:M)
Carex ambusta Boott = Carex saxatilis var. major [CARESAX2] Cyperaceae (V:M)
Carex amphigena (Fern.) Mackenzie = Carex glareosa var. amphigens [CAREGLR2] Cyperaceae (V:M)
Carex amplifolia Boott [CAREAMP] Cyperaceae (V:M)
Carex angarae Steud. = Carex norvegica ssp. inferalpina [CARENOR1] Cyperaceae (V:M)
Carex angustior Mackenzie = Carex echinata ssp. echinata [CAREECH0] Cyperaceae (V:M)
Carex angustior var. *gracilenta* Clausen & H.A. Wahl = Carex echinata ssp. echinata [CAREECH0] Cyperaceae (V:M)
Carex anthericoides Presl = Carex macrocephala [CAREMAR] Cyperaceae (V:M)
Carex anthoxanthea J.& K. Presl [CAREANT] Cyperaceae (V:M)
Carex aperta Boott [CAREAPE] Cyperaceae (V:M)
Carex apoda Clokey = Carex atrosquama [CAREATO] Cyperaceae (V:M)
Carex aquatilis Wahlenb. [CAREAQU] Cyperaceae (V:M)
Carex aquatilis ssp. *altior* (Rydb.) Hultén = Carex aquatilis var. aquatilis [CAREAQU0] Cyperaceae (V:M)
Carex aquatilis ssp. *stans* (Drej.) Hultén = Carex aquatilis var. stans [CAREAQU3] Cyperaceae (V:M)
Carex aquatilis var. *altior* (Rydb.) Fern. = Carex aquatilis var. aquatilis [CAREAQU0] Cyperaceae (V:M)
Carex aquatilis var. **aquatilis** [CAREAQU0] Cyperaceae (V:M)
Carex aquatilis var. **dives** (Holm) Kükenth. [CAREAQU6] Cyperaceae (V:M)
Carex aquatilis var. *sitchensis* (Prescott ex Bong.) Kelso = Carex aquatilis var. dives [CAREAQU6] Cyperaceae (V:M)
Carex aquatilis var. **stans** (Drej.) Boott [CAREAQU3] Cyperaceae (V:M)
Carex aquatilis var. *substricta* Kükenth. = Carex aquatilis var. aquatilis [CAREAQU0] Cyperaceae (V:M)
Carex arcta Boott [CAREARC] Cyperaceae (V:M)
Carex arctaeformis Mackenzie [CAREART] Cyperaceae (V:M)
Carex arenicola F. Schmidt = Carex pansa [CAREPAN] Cyperaceae (V:M)
Carex arenicola ssp. *pansa* (Bailey) T. Koyama & Calder = Carex pansa [CAREPAN] Cyperaceae (V:M)
Carex athabascensis F.J. Herm. = Carex scirpoidea [CARESCI] Cyperaceae (V:M)
Carex atherodes Spreng. [CAREATH] Cyperaceae (V:M)
Carex athrostachya Olney [CAREATR] Cyperaceae (V:M)
Carex atrata L. = Carex atratiformis [CAREATA] Cyperaceae (V:M)
Carex atrata ssp. *atrosquama* (Mackenzie) Hultén = Carex atrosquama [CAREATO] Cyperaceae (V:M)
Carex atrata var. *atrosquama* (Mackenzie) Cronq. = Carex atrosquama [CAREATO] Cyperaceae (V:M)
Carex atrata var. *erecta* W. Boott = Carex heteroneura var. epapillosa [CAREHET1] Cyperaceae (V:M)
Carex atratiformis Britt. [CAREATA] Cyperaceae (V:M)
Carex atratiformis ssp. *raymondii* (Calder) Porsild = Carex raymondii [CARERAM] Cyperaceae (V:M)
Carex atrosquama Mackenzie [CAREATO] Cyperaceae (V:M)
Carex aurea Nutt. [CAREAUR] Cyperaceae (V:M)
Carex backii var. *saximontana* (Mackenzie) Boivin = Carex saximontana [CARESAI] Cyperaceae (V:M)
Carex bebbii Olney ex Fern. [CAREBEB] Cyperaceae (V:M)
Carex beringensis C.B. Clarke = Carex podocarpa [CAREPOD] Cyperaceae (V:M)
Carex bicolor Bellardi ex All. [CAREBIC] Cyperaceae (V:M)
Carex bigelowii Torr. ex Schwein. [CAREBIG] Cyperaceae (V:M)
Carex bigelowii ssp. *hyperborea* (Drej.) Böcher = Carex bigelowii [CAREBIG] Cyperaceae (V:M)
Carex bipartita All. [CAREBIP] Cyperaceae (V:M)
Carex bipartita var. *amphigena* (Fern.) Polunin = Carex glareosa var. amphigens [CAREGLR2] Cyperaceae (V:M)
Carex bipartita var. *austromontana* F.J. Herm. = Carex bipartita [CAREBIP] Cyperaceae (V:M)
Carex bolanderi Olney [CAREBOL] Cyperaceae (V:M)
Carex brevicaulis Mackenzie [CAREBRE] Cyperaceae (V:M)
Carex brevior (Dewey) Mackenzie ex Lunell [CAREBRV] Cyperaceae (V:M)
Carex brevipes W. Boott = Carex rossii [CAREROS] Cyperaceae (V:M)
Carex breweri var. *paddoensis* (Suksdorf) Cronq. = Carex engelmannii [CAREENG] Cyperaceae (V:M)
Carex brunnescens (Pers.) Poir. [CAREBRU] Cyperaceae (V:M)
Carex brunnescens ssp. **alaskana** Kalela [CAREBRU1] Cyperaceae (V:M)
Carex brunnescens ssp. **pacifica** Kalela [CAREBRU2] Cyperaceae (V:M)
Carex brunnescens ssp. **sphaerostachya** (Tuckerman) Kalela [CAREBRU3] Cyperaceae (V:M)
Carex brunnescens var. *sphaerostachya* (Tuckerman) Kükenth. = Carex brunnescens ssp. sphaerostachya [CAREBRU3] Cyperaceae (V:M)
Carex buxbaumii Wahlenb. [CAREBUX] Cyperaceae (V:M)
Carex camporum Mackenzie = Carex praegracilis [CAREPRE] Cyperaceae (V:M)
Carex campylocarpa Holm = Carex scopulorum var. bracteosa [CARESCP1] Cyperaceae (V:M)
Carex campylocarpa ssp. *affinis* Maguire & A. Holmgren = Carex scopulorum var. bracteosa [CARESCP1] Cyperaceae (V:M)
Carex canescens L. [CARECAN] Cyperaceae (V:M)
Carex canescens ssp. *arctaeformis* (Mackenzie) Calder & Taylor = Carex arctaeformis [CAREART] Cyperaceae (V:M)
Carex canescens var. *subloliacea* (Laestad.) Hartman = Carex lapponica [CARELAP] Cyperaceae (V:M)
Carex capillaris L. [CARECAP] Cyperaceae (V:M)
Carex capillaris ssp. *chlorostachys* (Stev.) A. & D. Löve & Raymond = Carex capillaris [CARECAP] Cyperaceae (V:M)
Carex capillaris ssp. *krausei* (Boeckl.) Böcher = Carex krausei [CAREKRA] Cyperaceae (V:M)
Carex capillaris ssp. *robustior* (Drej. ex Lange) Böcher = Carex capillaris [CARECAP] Cyperaceae (V:M)
Carex capillaris var. *elongata* Olney ex Fern. = Carex capillaris [CARECAP] Cyperaceae (V:M)
Carex capillaris var. *krausei* (Boeckl.) Crantz = Carex krausei [CAREKRA] Cyperaceae (V:M)
Carex capillaris var. *major* Blytt = Carex capillaris [CARECAP] Cyperaceae (V:M)
Carex capitata L. [CARECAI] Cyperaceae (V:M)
Carex cephalantha (Bailey) Bickn. = Carex echinata ssp. echinata [CAREECH0] Cyperaceae (V:M)
Carex chlorostachys Stev. = Carex capillaris [CARECAP] Cyperaceae (V:M)
Carex chordorrhiza Ehrh. ex L. f. [CARECHO] Cyperaceae (V:M)
Carex cinerea Poll. = Carex canescens [CARECAN] Cyperaceae (V:M)
Carex circinata C.A. Mey. [CARECIR] Cyperaceae (V:M)
Carex comosa Boott [CARECOM] Cyperaceae (V:M)
Carex concinna R. Br. [CARECON] Cyperaceae (V:M)
Carex concinnoides Mackenzie [CARECOC] Cyperaceae (V:M)
Carex consimilis Holm = Carex bigelowii [CAREBIG] Cyperaceae (V:M)
Carex crawei Dewey [CARECRA] Cyperaceae (V:M)
Carex crawfordii Fern. [CARECRW] Cyperaceae (V:M)
Carex crawfordii var. *vigens* Fern. = Carex crawfordii [CARECRW] Cyperaceae (V:M)
Carex cryptocarpa C.A. Mey. = Carex lyngbyei [CARELYN] Cyperaceae (V:M)
Carex cryptochlaena Holm = Carex lyngbyei [CARELYN] Cyperaceae (V:M)
Carex curta Goodenough = Carex canescens [CARECAN] Cyperaceae (V:M)
Carex cusickii Mackenzie ex Piper & Beattie [CARECUS] Cyperaceae (V:M)
Carex deflexa Hornem. [CAREDEF] Cyperaceae (V:M)

Carex deflexa var. *brevicaulis* (Mackenzie) Boivin = Carex brevicaulis [CAREBRE] Cyperaceae (V:M)
Carex deflexa var. *rossii* (Boott) Bailey = Carex rossii [CAREROS] Cyperaceae (V:M)
Carex deweyana Schwein. [CAREDEW] Cyperaceae (V:M)
Carex deweyana ssp. *leptopoda* (Mackenzie) Calder & Taylor = Carex leptopoda [CARELEO] Cyperaceae (V:M)
Carex deweyana var. *bolanderi* (Olney) W. Boott = Carex bolanderi [CAREBOL] Cyperaceae (V:M)
Carex deweyana var. *leptopoda* (Mackenzie) Boivin = Carex leptopoda [CARELEO] Cyperaceae (V:M)
Carex diandra Schrank [CAREDIA] Cyperaceae (V:M)
Carex diandra var. *ampla* (Bailey) Kukenth. = Carex cusickii [CARECUS] Cyperaceae (V:M)
Carex dioica ssp. *gynocrates* (Wormsk. ex Drej.) Hultén = Carex gynocrates [CAREGYN] Cyperaceae (V:M)
Carex dioica var. *gynocrates* (Wormsk. ex Drej.) Ostenf. = Carex gynocrates [CAREGYN] Cyperaceae (V:M)
Carex disperma Dewey [CAREDIS] Cyperaceae (V:M)
Carex diversistylis Roach = Carex rossii [CAREROS] Cyperaceae (V:M)
Carex douglasii Boott [CAREDOU] Cyperaceae (V:M)
Carex drummondiana Dewey = Carex rupestris var. drummondiana [CARERUP1] Cyperaceae (V:M)
Carex duriuscula C.A. Mey. [CAREDUR] Cyperaceae (V:M)
Carex eastwoodiana Stacey = Carex phaeocephala [CAREPHA] Cyperaceae (V:M)
Carex eburnea Boott [CAREEBU] Cyperaceae (V:M)
Carex echinata Murr. [CAREECH] Cyperaceae (V:M)
Carex echinata ssp. **echinata** [CAREECH0] Cyperaceae (V:M)
Carex echinata ssp. **phyllomanica** (W. Boott) Reznicek [CAREECH2] Cyperaceae (V:M)
Carex echinata var. *angustata* (Carey) Bailey = Carex echinata ssp. echinata [CAREECH0] Cyperaceae (V:M)
Carex eleocharis Bailey = Carex duriuscula [CAREDUR] Cyperaceae (V:M)
Carex elyniformis Porsild = Carex filifolia [CAREFIL] Cyperaceae (V:M)
Carex enanderi Holm = Carex lenticularis var. dolia [CARELEN1] Cyperaceae (V:M)
Carex engelmannii Bailey [CAREENG] Cyperaceae (V:M)
Carex epapillosa Mackenzie = Carex heteroneura var. epapillosa [CAREHET1] Cyperaceae (V:M)
Carex erxlebeniana L. Kelso = Carex inops ssp. heliophila [CAREINO1] Cyperaceae (V:M)
Carex eurystachya F.J. Herm. = Carex lenticularis var. dolia [CARELEN1] Cyperaceae (V:M)
Carex exsiccata Bailey [CAREEXS] Cyperaceae (V:M)
Carex festiva var. *decumbens* Holm = Carex haydeniana [CAREHAY] Cyperaceae (V:M)
Carex festiva var. *gracilis* Olney = Carex pachystachya [CAREPAC] Cyperaceae (V:M)
Carex festiva var. *pachystachya* (Cham. ex Steud.) Bailey = Carex pachystachya [CAREPAC] Cyperaceae (V:M)
Carex festivella Mackenzie = Carex microptera [CAREMIO] Cyperaceae (V:M)
Carex festucacea var. *brevior* (Dewey) Fern. = Carex brevior [CAREBRV] Cyperaceae (V:M)
Carex feta Bailey [CAREFET] Cyperaceae (V:M)
Carex filifolia Nutt. [CAREFIL] Cyperaceae (V:M)
Carex flava L. [CAREFLA] Cyperaceae (V:M)
Carex flava var. *fertilis* Peck = Carex flava [CAREFLA] Cyperaceae (V:M)
Carex flava var. *gaspensis* Fern. = Carex flava [CAREFLA] Cyperaceae (V:M)
Carex flava var. *graminis* Bailey = Carex flava [CAREFLA] Cyperaceae (V:M)
Carex flava var. *laxior* (Kükenth.) Gleason = Carex flava [CAREFLA] Cyperaceae (V:M)
Carex flava var. *rectirostra* Gaudin = Carex flava [CAREFLA] Cyperaceae (V:M)
Carex foenea Willd. [CAREFOE] Cyperaceae (V:M)
Carex franklinii Boott [CAREFRA] Cyperaceae (V:M)
Carex fuliginosa ssp. *misandra* (R. Br.) Nyman = Carex misandra [CAREMIS] Cyperaceae (V:M)
Carex fuscidula Krecz. ex Egor. = Carex capillaris [CARECAP] Cyperaceae (V:M)
Carex garberi Fern. [CAREGAR] Cyperaceae (V:M)
Carex garberi ssp. *bifaria* (Fern.) Hultén = Carex garberi [CAREGAR] Cyperaceae (V:M)
Carex garberi var. *bifaria* Fern. = Carex garberi [CAREGAR] Cyperaceae (V:M)
Carex geyeri Boott [CAREGEY] Cyperaceae (V:M)
Carex glacialis Mackenzie [CAREGLA] Cyperaceae (V:M)
Carex glareosa Schkuhr ex Wahlenb. [CAREGLR] Cyperaceae (V:M)
Carex glareosa var. **amphigens** Fern. [CAREGLR2] Cyperaceae (V:M)
Carex gmelinii Hook. & Arn. [CAREGME] Cyperaceae (V:M)
Carex gymnoclada Holm = Carex scopulorum var. bracteosa [CARESCP1] Cyperaceae (V:M)
Carex gynocrates Wormsk. ex Drej. [CAREGYN] Cyperaceae (V:M)
Carex hawaiiensis St. John = Carex echinata ssp. echinata [CAREECH0] Cyperaceae (V:M)
Carex haydeniana Olney [CAREHAY] Cyperaceae (V:M)
Carex heleonastes L. f. [CAREHEL] Cyperaceae (V:M)
Carex heliophila Mackenzie = Carex inops ssp. heliophila [CAREINO1] Cyperaceae (V:M)
Carex hendersonii Bailey [CAREHEN] Cyperaceae (V:M)
Carex hepburnii Boott = Carex nardina var. hepburnii [CARENAR1] Cyperaceae (V:M)
Carex heteroneura W. Boott [CAREHET] Cyperaceae (V:M)
Carex heteroneura var. **epapillosa** (Mackenzie) F.J. Herm. [CAREHET1] Cyperaceae (V:M)
Carex hindsii C.B. Clarke = Carex lenticularis var. limnophila [CARELEN4] Cyperaceae (V:M)
Carex hoodii Boott [CAREHOO] Cyperaceae (V:M)
Carex howellii Bailey = Carex aquatilis var. dives [CAREAQU6] Cyperaceae (V:M)
Carex hystericina Muhl. ex Willd. [CAREHYS] Cyperaceae (V:M)
Carex illota Bailey [CAREILL] Cyperaceae (V:M)
Carex incurva Lightf. = Carex maritima [CAREMAT] Cyperaceae (V:M)
Carex incurviformis Mackenzie [CAREINC] Cyperaceae (V:M)
Carex inflata var. *utriculata* (Boott) Druce = Carex utriculata [CAREUTR] Cyperaceae (V:M)
Carex inops Bailey [CAREINO] Cyperaceae (V:M)
Carex inops ssp. **heliophila** (Mackenzie) Crins [CAREINO1] Cyperaceae (V:M)
Carex inops ssp. **inops** [CAREINO0] Cyperaceae (V:M)
Carex interior Bailey [CAREINT] Cyperaceae (V:M)
Carex interior ssp. *charlestonensis* Clokey = Carex interior [CAREINT] Cyperaceae (V:M)
Carex interior var. *keweenawensis* F.J. Herm. = Carex interior [CAREINT] Cyperaceae (V:M)
Carex interrupta Boeckl. [CAREINE] Cyperaceae (V:M)
Carex interrupta var. *distenta* Kükenth. = Carex interrupta [CAREINE] Cyperaceae (V:M)
Carex invisa Bailey = Carex spectabilis [CARESPE] Cyperaceae (V:M)
Carex irrigua Wahlenb. = Carex magellanica ssp. irrigua [CAREMAG1] Cyperaceae (V:M)
Carex jacobi-peteri Hultén = Carex pyrenaica ssp. micropoda [CAREPYR1] Cyperaceae (V:M)
Carex jimescalderi Boivin = Carex leptalea ssp. pacifica [CARELET2] Cyperaceae (V:M)
Carex josselynii (Fern.) Mackenzie ex Pease = Carex echinata ssp. echinata [CAREECH0] Cyperaceae (V:M)
Carex kelloggii W. Boott = Carex lenticularis var. lipocarpa [CARELEN5] Cyperaceae (V:M)
Carex krausei Boeckl. [CAREKRA] Cyperaceae (V:M)
Carex lachenalii Schkuhr = Carex bipartita [CAREBIP] Cyperaceae (V:M)
Carex laeviculmis Meinsh. [CARELAE] Cyperaceae (V:M)
Carex lagopina Wahlenb. = Carex bipartita [CAREBIP] Cyperaceae (V:M)

Carex lanuginosa Michx. [CARELAN] Cyperaceae (V:M)
Carex lanuginosa var. *americana* (Fern.) Boivin = Carex lasiocarpa var. americana [CARELAS1] Cyperaceae (V:M)
Carex lapponica O.F. Lang [CARELAP] Cyperaceae (V:M)
Carex laricina Mackenzie ex Bright = Carex echinata ssp. echinata [CAREECH0] Cyperaceae (V:M)
Carex lasiocarpa Ehrh. [CARELAS] Cyperaceae (V:M)
Carex lasiocarpa ssp. *americana* (Fern.) Love & Bernard = Carex lasiocarpa var. americana [CARELAS1] Cyperaceae (V:M)
Carex lasiocarpa var. **americana** Fern. [CARELAS1] Cyperaceae (V:M)
Carex lasiocarpa var. *latifolia* (Boeckl.) Gilly = Carex lanuginosa [CARELAN] Cyperaceae (V:M)
Carex leersii Willd. = Carex echinata ssp. echinata [CAREECH0] Cyperaceae (V:M)
Carex lenticularis Michx. [CARELEN] Cyperaceae (V:M)
Carex lenticularis var. *albimontana* Dewey = Carex lenticularis var. lenticularis [CARELEN0] Cyperaceae (V:M)
Carex lenticularis var. *blakei* Dewey = Carex lenticularis var. lenticularis [CARELEN0] Cyperaceae (V:M)
Carex lenticularis var. **dolia** (M.E. Jones) L.A. Standley [CARELEN1] Cyperaceae (V:M)
Carex lenticularis var. *eucycla* Fern. = Carex lenticularis var. lenticularis [CARELEN0] Cyperaceae (V:M)
Carex lenticularis var. **impressa** (Bailey) L.A. Standley [CARELEN2] Cyperaceae (V:M)
Carex lenticularis var. **lenticularis** [CARELEN0] Cyperaceae (V:M)
Carex lenticularis var. **limnophila** (Holm) Cronq. [CARELEN4] Cyperaceae (V:M)
Carex lenticularis var. **lipocarpa** (Holm) L.A. Standley [CARELEN5] Cyperaceae (V:M)
Carex lenticularis var. *merens* Howe = Carex lenticularis var. lenticularis [CARELEN0] Cyperaceae (V:M)
Carex lenticularis var. *pallida* (W. Boott) Dorn = Carex lenticularis var. lipocarpa [CARELEN5] Cyperaceae (V:M)
Carex leporina L. [CARELEP] Cyperaceae (V:M)
Carex leptalea Wahlenb. [CARELET] Cyperaceae (V:M)
Carex leptalea ssp. **leptalea** [CARELET0] Cyperaceae (V:M)
Carex leptalea ssp. **pacifica** Calder & Taylor [CARELET2] Cyperaceae (V:M)
Carex leptopoda Mackenzie [CARELEO] Cyperaceae (V:M)
Carex liddonii Boott = Carex petasata [CAREPET] Cyperaceae (V:M)
Carex limnaea Holm = Carex lenticularis var. impressa [CARELEN2] Cyperaceae (V:M)
Carex limnophila F.J. Herm. = Carex microptera [CAREMIO] Cyperaceae (V:M)
Carex limosa L. [CARELIM] Cyperaceae (V:M)
Carex livida (Wahlenb.) Willd. [CARELIV] Cyperaceae (V:M)
Carex livida var. *grayana* (Dewey) Fern. = Carex livida var. radicaulis [CARELIV1] Cyperaceae (V:M)
Carex livida var. **radicaulis** Paine [CARELIV1] Cyperaceae (V:M)
Carex loliacea L. [CARELOL] Cyperaceae (V:M)
Carex luzulina Olney [CARELUZ] Cyperaceae (V:M)
Carex luzulina var. **ablata** (Bailey) F.J. Herm. [CARELUZ1] Cyperaceae (V:M)
Carex lyallii Boott = Carex raynoldsii [CARERAY] Cyperaceae (V:M)
Carex lyngbyei Hornem. [CARELYN] Cyperaceae (V:M)
Carex lyngbyei var. *robusta* (Bailey) Cronq. = Carex lyngbyei [CARELYN] Cyperaceae (V:M)
Carex macloviana d'Urv. [CAREMAL] Cyperaceae (V:M)
Carex macloviana ssp. *festivella* (Mackenzie) A. & D. Löve = Carex microptera [CAREMIO] Cyperaceae (V:M)
Carex macloviana ssp. *haydeniana* (Olney) Taylor & MacBryde = Carex haydeniana [CAREHAY] Cyperaceae (V:M)
Carex macloviana ssp. *pachystachya* (Cham. ex Steud.) Hultén = Carex pachystachya [CAREPAC] Cyperaceae (V:M)
Carex macloviana var. *microptera* (Mackenzie) Boivin = Carex microptera [CAREMIO] Cyperaceae (V:M)
Carex macloviana var. *pachystachya* (Cham. ex Steud.) Kükenth. = Carex pachystachya [CAREPAC] Cyperaceae (V:M)
Carex macrocephala Willd. ex Spreng. [CAREMAR] Cyperaceae (V:M)
Carex macrocephala ssp. *anthericoides* (Presl) Hultén = Carex macrocephala [CAREMAR] Cyperaceae (V:M)
Carex macrochaeta C.A. Mey. [CAREMAO] Cyperaceae (V:M)
Carex macrogyna Turcz. ex Steud. = Carex petricosa [CAREPER] Cyperaceae (V:M)
Carex magellanica Lam. [CAREMAG] Cyperaceae (V:M)
Carex magellanica ssp. **irrigua** (Wahlenb.) Hultén [CAREMAG1] Cyperaceae (V:M)
Carex magellanica var. *irrigua* (Wahlenb.) B.S.P. = Carex magellanica ssp. irrigua [CAREMAG1] Cyperaceae (V:M)
Carex magnifica Dewey ex Piper = Carex obnupta [CAREOBN] Cyperaceae (V:M)
Carex marina Dewey [CAREMAI] Cyperaceae (V:M)
Carex maritima Gunn. [CAREMAT] Cyperaceae (V:M)
Carex maritima var. *incurviformis* (Mackenzie) Boivin = Carex incurviformis [CAREINC] Cyperaceae (V:M)
Carex media R. Br. = Carex norvegica ssp. inferalpina [CARENOR1] Cyperaceae (V:M)
Carex membranacea Hook. [CAREMEM] Cyperaceae (V:M)
Carex membranopacta Bailey = Carex membranacea [CAREMEM] Cyperaceae (V:M)
Carex mertensii Prescott ex Bong. [CAREMER] Cyperaceae (V:M)
Carex microchaeta Holm [CAREMIC] Cyperaceae (V:M)
Carex microglochin Wahlenb. [CAREMIR] Cyperaceae (V:M)
Carex micropoda C.A. Mey. = Carex pyrenaica ssp. micropoda [CAREPYR1] Cyperaceae (V:M)
Carex microptera Mackenzie [CAREMIO] Cyperaceae (V:M)
Carex microptera var. *crassinervia* F.J. Herm. = Carex microptera [CAREMIO] Cyperaceae (V:M)
Carex microptera var. *limnophila* (F.J. Herm.) Dorn = Carex microptera [CAREMIO] Cyperaceae (V:M)
Carex misandra R. Br. [CAREMIS] Cyperaceae (V:M)
Carex miserabilis Mackenzie = Carex scopulorum var. prionophylla [CARESCP2] Cyperaceae (V:M)
Carex montanensis Bailey = Carex podocarpa [CAREPOD] Cyperaceae (V:M)
Carex multimoda Bailey = Carex pachystachya [CAREPAC] Cyperaceae (V:M)
Carex muricata var. *angustata* (Carey) Carey ex Gleason = Carex echinata ssp. echinata [CAREECH0] Cyperaceae (V:M)
Carex muricata var. *cephalantha* (Bailey) Wieg. & Eames = Carex echinata ssp. echinata [CAREECH0] Cyperaceae (V:M)
Carex muricata var. *laricina* (Mackenzie ex Bright) Gleason = Carex echinata ssp. echinata [CAREECH0] Cyperaceae (V:M)
Carex nardina Fries [CARENAR] Cyperaceae (V:M)
Carex nardina ssp. *hepburnii* (Boott) A. & D. Löve & Kapoor = Carex nardina var. hepburnii [CARENAR1] Cyperaceae (V:M)
Carex nardina var. **hepburnii** (Boott) Kükenth. [CARENAR1] Cyperaceae (V:M)
Carex nevadensis ssp. *flavella* (Krecz.) Janchen = Carex flava [CAREFLA] Cyperaceae (V:M)
Carex nigella Boott = Carex spectabilis [CARESPE] Cyperaceae (V:M)
Carex nigricans C.A. Mey. [CARENIG] Cyperaceae (V:M)
Carex nigromarginata var. *elliptica* (Boott) Gleason = Carex peckii [CAREPEC] Cyperaceae (V:M)
Carex nigromarginata var. *minor* (Boott) Gleason = Carex peckii [CAREPEC] Cyperaceae (V:M)
Carex norvegica Retz. [CARENOR] Cyperaceae (V:M)
Carex norvegica ssp. **inferalpina** (Wahlenb.) Hultén [CARENOR1] Cyperaceae (V:M)
Carex norvegica var. *inferalpina* (Wahlenb.) Boivin = Carex norvegica ssp. inferalpina [CARENOR1] Cyperaceae (V:M)
Carex nubicola Mackenzie = Carex haydeniana [CAREHAY] Cyperaceae (V:M)
Carex obnupta Bailey [CAREOBN] Cyperaceae (V:M)
Carex obovoidea Cronq. = Carex cusickii [CARECUS] Cyperaceae (V:M)
Carex obtusata Lilj. [CAREOBT] Cyperaceae (V:M)
Carex oederi var. *viridula* (Michx.) Hultén = Carex viridula [CAREVIR] Cyperaceae (V:M)
Carex ormantha (Fern.) Mackenzie = Carex echinata ssp. echinata [CAREECH0] Cyperaceae (V:M)

Carex pachystachya Cham. ex Steud. [CAREPAC] Cyperaceae (V:M)
Carex pachystachya var. *gracilis* (Olney) Mackenzie = Carex pachystachya [CAREPAC] Cyperaceae (V:M)
Carex pachystachya var. *monds-coulteri* L. Kelso = Carex pachystachya [CAREPAC] Cyperaceae (V:M)
***Carex pallescens** L. [CAREPAL] Cyperaceae (V:M)
Carex pallescens var. *neogaea* Fern. = Carex pallescens [CAREPAL] Cyperaceae (V:M)
Carex panda C.B. Clarke = Carex aquatilis var. dives [CAREAQU6] Cyperaceae (V:M)
Carex pansa Bailey [CAREPAN] Cyperaceae (V:M)
Carex parryana Dewey [CAREPAR] Cyperaceae (V:M)
Carex paucicostata Mackenzie = Carex lenticularis var. impressa [CARELEN2] Cyperaceae (V:M)
Carex pauciflora Lightf. [CAREPAU] Cyperaceae (V:M)
Carex paupercula Michx. = Carex magellanica [CAREMAG] Cyperaceae (V:M)
Carex paupercula var. *irrigua* (Wahlenb.) Fern. = Carex magellanica ssp. irrigua [CAREMAG1] Cyperaceae (V:M)
Carex paysonis Clokey [CAREPAY] Cyperaceae (V:M)
Carex peckii Howe [CAREPEC] Cyperaceae (V:M)
Carex pedunculata Muhl. ex Willd. [CAREPED] Cyperaceae (V:M)
Carex pensylvanica ssp. *heliophila* (Mackenzie) W.A. Weber = Carex inops ssp. heliophila [CAREINO1] Cyperaceae (V:M)
Carex pensylvanica var. *digyna* Boeckl. = Carex inops ssp. heliophila [CAREINO1] Cyperaceae (V:M)
Carex petasata Dewey [CAREPET] Cyperaceae (V:M)
Carex petricosa Dewey [CAREPER] Cyperaceae (V:M)
Carex petricosa var. *distichiflora* (Boivin) Boivin = Carex petricosa [CAREPER] Cyperaceae (V:M)
Carex petricosa var. *franklinii* (Boott) Boivin = Carex franklinii [CAREFRA] Cyperaceae (V:M)
Carex phaeocephala Piper [CAREPHA] Cyperaceae (V:M)
Carex phyllomanica W. Boott = Carex echinata ssp. phyllomanica [CAREECH2] Cyperaceae (V:M)
Carex phyllomanica var. *angustata* (Carey) Boivin = Carex echinata ssp. echinata [CAREECH0] Cyperaceae (V:M)
Carex phyllomanica var. *ormantha* (Fern.) Boivin = Carex echinata ssp. echinata [CAREECH0] Cyperaceae (V:M)
Carex physocarpa J.& K. Presl = Carex saxatilis var. major [CARESAX2] Cyperaceae (V:M)
Carex plectocarpa F.J. Herm. = Carex lenticularis var. dolia [CARELEN1] Cyperaceae (V:M)
Carex pluriflora Hultén [CAREPLU] Cyperaceae (V:M)
Carex podocarpa R. Br. [CAREPOD] Cyperaceae (V:M)
Carex praegracilis W. Boott [CAREPRE] Cyperaceae (V:M)
Carex prairea Dewey ex Wood [CAREPRR] Cyperaceae (V:M)
Carex praticola Rydb. [CAREPRT] Cyperaceae (V:M)
Carex praticola var. *subcoriacea* F.J. Herm. = Carex praticola [CAREPRT] Cyperaceae (V:M)
Carex preslii Steud. [CAREPRS] Cyperaceae (V:M)
Carex prionophylla Holm = Carex scopulorum var. prionophylla [CARESCP2] Cyperaceae (V:M)
Carex pseudoscirpoidea Rydb. [CAREPSE] Cyperaceae (V:M)
Carex pyrenaica Wahlenb. [CAREPYR] Cyperaceae (V:M)
Carex pyrenaica ssp. **micropoda** (C.A. Mey.) Hultén [CAREPYR1] Cyperaceae (V:M)
Carex pyrophila Gandog. = Carex pachystachya [CAREPAC] Cyperaceae (V:M)
Carex rariflora ssp. *pluriflora* (Hultén) Egor. = Carex pluriflora [CAREPLU] Cyperaceae (V:M)
Carex rariflora var. *pluriflora* (Hultén) Boivin = Carex pluriflora [CAREPLU] Cyperaceae (V:M)
Carex raymondii Calder [CARERAM] Cyperaceae (V:M)
Carex raynoldsii Dewey [CARERAY] Cyperaceae (V:M)
Carex retrorsa Schwein. [CARERET] Cyperaceae (V:M)
Carex rhynchosphysa Fisch., C.A. Mey. & Avé-Lall. = Carex utriculata [CAREUTR] Cyperaceae (V:M)
Carex richardsonii R. Br. [CARERIC] Cyperaceae (V:M)
Carex rossii Boott [CAREROS] Cyperaceae (V:M)
Carex rostrata Stokes [CAREROR] Cyperaceae (V:M)
Carex rostrata var. *ambigens* Fern. = Carex rostrata [CAREROR] Cyperaceae (V:M)
Carex rostrata var. *utriculata* (Boott) Bailey = Carex utriculata [CAREUTR] Cyperaceae (V:M)
Carex rugosperma var. *tonsa* (Fern.) E.G. Voss = Carex tonsa [CARETON] Cyperaceae (V:M)
Carex rupestris All. [CARERUP] Cyperaceae (V:M)
Carex rupestris ssp. *drummondiana* (Dewey) Holub = Carex rupestris var. drummondiana [CARERUP1] Cyperaceae (V:M)
Carex rupestris var. **drummondiana** (Dewey) Bailey [CARERUP1] Cyperaceae (V:M)
Carex rupestris var. **rupestris** [CARERUP0] Cyperaceae (V:M)
Carex saltuensis Bailey = Carex vaginata [CAREVAG] Cyperaceae (V:M)
Carex sartwellii Dewey [CARESAR] Cyperaceae (V:M)
Carex saxatilis L. [CARESAX] Cyperaceae (V:M)
Carex saxatilis ssp. *laxa* (Trautv.) Kalela = Carex saxatilis var. saxatilis [CARESAX0] Cyperaceae (V:M)
Carex saxatilis var. **major** Olney [CARESAX2] Cyperaceae (V:M)
Carex saxatilis var. **saxatilis** [CARESAX0] Cyperaceae (V:M)
Carex saximontana Mackenzie [CARESAI] Cyperaceae (V:M)
Carex scirpiformis Mackenzie = Carex scirpoidea [CARESCI] Cyperaceae (V:M)
Carex scirpoidea Michx. [CARESCI] Cyperaceae (V:M)
Carex scirpoidea var. *convoluta* Kükenth. = Carex scirpoidea [CARESCI] Cyperaceae (V:M)
Carex scirpoidea var. *pseudoscirpoidea* (Rydb.) Cronq. = Carex pseudoscirpoidea [CAREPSE] Cyperaceae (V:M)
Carex scirpoidea var. *scirpiformis* (Mackenzie) O'Neill & Duman = Carex scirpoidea [CARESCI] Cyperaceae (V:M)
Carex scirpoidea var. *stenochlaena* Holm = Carex scirpoidea [CARESCI] Cyperaceae (V:M)
Carex scoparia Schkuhr ex Willd. [CARESCO] Cyperaceae (V:M)
Carex scopulorum Holm [CARESCP] Cyperaceae (V:M)
Carex scopulorum var. **bracteosa** (Bailey) F.J. Herm. [CARESCP1] Cyperaceae (V:M)
Carex scopulorum var. **prionophylla** (Holm) L.A. Standley [CARESCP2] Cyperaceae (V:M)
Carex siccata Dewey = Carex foenea [CAREFOE] Cyperaceae (V:M)
Carex simulata Mackenzie [CARESIM] Cyperaceae (V:M)
Carex sitchensis Prescott ex Bong. = Carex aquatilis var. dives [CAREAQU6] Cyperaceae (V:M)
Carex spaniocarpa Steud. = Carex supina var. spaniocarpa [CARESUP1] Cyperaceae (V:M)
Carex spectabilis Dewey [CARESPE] Cyperaceae (V:M)
Carex sphaerostachya (Tuckerman) Dewey = Carex brunnescens [CAREBRU] Cyperaceae (V:M)
Carex sprengelii Dewey ex Spreng. [CARESPR] Cyperaceae (V:M)
Carex stans Drej. = Carex aquatilis var. stans [CAREAQU3] Cyperaceae (V:M)
Carex stenochlaena (Holm) Mackenzie = Carex scirpoidea [CARESCI] Cyperaceae (V:M)
Carex stenophylla ssp. *eleocharis* (Bailey) Hultén = Carex duriuscula [CAREDUR] Cyperaceae (V:M)
Carex stenophylla var. *enervis* Kükenth. = Carex duriuscula [CAREDUR] Cyperaceae (V:M)
Carex stipata Muhl. ex Willd. [CARESTI] Cyperaceae (V:M)
Carex straminea var. *mixta* Bailey = Carex feta [CAREFET] Cyperaceae (V:M)
Carex stygia auct. = Carex pluriflora [CAREPLU] Cyperaceae (V:M)
Carex stylosa C.A. Mey. [CARESTY] Cyperaceae (V:M)
Carex stylosa var. *nigritella* (Drej.) Fern. = Carex stylosa [CARESTY] Cyperaceae (V:M)
Carex stylosa var. *virens* Bailey = Carex aperta [CAREAPE] Cyperaceae (V:M)
Carex substricta (Kükenth.) Mackenzie = Carex aquatilis var. aquatilis [CAREAQU0] Cyperaceae (V:M)
Carex suksdorfii Kükenth. = Carex aquatilis var. aquatilis [CAREAQU0] Cyperaceae (V:M)
Carex supina Willd. ex Wahlenb. [CARESUP] Cyperaceae (V:M)
Carex supina ssp. *spaniocarpa* (Steud.) Hultén = Carex supina var. spaniocarpa [CARESUP1] Cyperaceae (V:M)

Carex supina var. **spaniocarpa** (Steud.) Boivin [CARESUP1] Cyperaceae (V:M)
Carex svensonis Skottsberg = Carex echinata ssp. echinata [CAREECH0] Cyperaceae (V:M)
Carex sychnocephala Carey [CARESYC] Cyperaceae (V:M)
Carex tenella Schukuhr = Carex disperma [CAREDIS] Cyperaceae (V:M)
Carex tenera Dewey [CARETEE] Cyperaceae (V:M)
Carex tenera var. *echinodes* (Fern.) Wieg. = Carex tenera [CARETEE] Cyperaceae (V:M)
Carex tenuiflora Wahlenb. [CARETEN] Cyperaceae (V:M)
Carex tolmiei Boott = Carex spectabilis [CARESPE] Cyperaceae (V:M)
Carex tonsa (Fern.) Bickn. [CARETON] Cyperaceae (V:M)
Carex tracyi Mackenzie = Carex leporina [CARELEP] Cyperaceae (V:M)
Carex tripartita auct. = Carex bipartita [CAREBIP] Cyperaceae (V:M)
Carex trisperma Dewey [CARETRS] Cyperaceae (V:M)
Carex turgidula Bailey = Carex aperta [CAREAPE] Cyperaceae (V:M)
Carex umbellata var. *tonsa* Fern. = Carex tonsa [CARETON] Cyperaceae (V:M)
Carex unilateralis Mackenzie [CAREUNI] Cyperaceae (V:M)
Carex utriculata Boott [CAREUTR] Cyperaceae (V:M)
Carex vaginata Tausch [CAREVAG] Cyperaceae (V:M)
Carex vahlii var. *inferalpina* Wahlenb. = Carex norvegica ssp. inferalpina [CARENOR1] Cyperaceae (V:M)
Carex venustula Holm = Carex podocarpa [CAREPOD] Cyperaceae (V:M)
Carex vesicaria L. [CAREVES] Cyperaceae (V:M)
Carex vesicaria var. *major* Boott = Carex exsiccata [CAREEXS] Cyperaceae (V:M)
Carex viridula Michx. [CAREVIR] Cyperaceae (V:M)
Carex vulpinoidea Michx. [CAREVUL] Cyperaceae (V:M)
Carex xerantica Bailey [CAREXER] Cyperaceae (V:M)
Carthamus L. [CARTHAM$] Asteraceae (V:D)
Carthamus baeticus (Boiss. & Reut.) Lara = Carthamus lanatus ssp. baeticus [CARTLAN1] Asteraceae (V:D)
***Carthamus lanatus** L. [CARTLAN] Asteraceae (V:D)
***Carthamus lanatus** ssp. **baeticus** (Boiss. & Reut.) Nyman [CARTLAN1] Asteraceae (V:D)
Carum L. [CARUM$] Apiaceae (V:D)
***Carum carvi** L. [CARUCAR] Apiaceae (V:D)
Cassiope D. Don [CASSIOP$] Ericaceae (V:D)
Cassiope lycopodioides (Pallas) D. Don [CASSLYC] Ericaceae (V:D)
Cassiope lycopodioides ssp. *cristipilosa* Calder & Taylor = Cassiope lycopodioides var. cristipilosa [CASSLYC1] Ericaceae (V:D)
Cassiope lycopodioides var. **cristipilosa** (Calder & Taylor) Boivin [CASSLYC1] Ericaceae (V:D)
Cassiope mertensiana (Bong.) D. Don [CASSMER] Ericaceae (V:D)
Cassiope stelleriana (Pallas) DC. = Harrimanella stelleriana [HARRSTE] Ericaceae (V:D)
Cassiope tetragona (L.) D. Don [CASSTET] Ericaceae (V:D)
Cassiope tetragona ssp. *saximontana* (Small) Porsild = Cassiope tetragona var. saximontana [CASSTET1] Ericaceae (V:D)
Cassiope tetragona var. **saximontana** (Small) C.L. Hitchc. [CASSTET1] Ericaceae (V:D)
Cassiope tetragona var. **tetragona** [CASSTET0] Ericaceae (V:D)
Castalia flava (Leitner) Greene = Nymphaea mexicana [NYMPMEX] Nymphaeaceae (V:D)
Castalia lekophylla Small = Nymphaea odorata [NYMPODO] Nymphaeaceae (V:D)
Castalia minor (Sims) Nyar = Nymphaea odorata [NYMPODO] Nymphaeaceae (V:D)
Castalia odorata (Ait.) Wood = Nymphaea odorata [NYMPODO] Nymphaeaceae (V:D)
Castalia reniformis DC. = Nymphaea odorata [NYMPODO] Nymphaeaceae (V:D)
Castalia tetragona (Georgi) Lawson = Nymphaea tetragona [NYMPTET] Nymphaeaceae (V:D)
Castalia tuberosa (Paine) Greene = Nymphaea odorata [NYMPODO] Nymphaeaceae (V:D)
Castilleja Mutis ex L. f. [CASTILL$] Scrophulariaceae (V:D)
Castilleja ambigua Hook. & Arn. [CASTAMB] Scrophulariaceae (V:D)
Castilleja angustifolia var. *hispida* (Benth.) Fern. = Castilleja hispida [CASTHIS] Scrophulariaceae (V:D)
Castilleja angustifolia var. *whitedii* Piper = Castilleja elmeri [CASTELM] Scrophulariaceae (V:D)
Castilleja attenuata (Gray) Chuang & Heckard [CASTATT] Scrophulariaceae (V:D)
Castilleja cervina Greenm. [CASTCER] Scrophulariaceae (V:D)
Castilleja chrymactis Pennell = Castilleja miniata [CASTMIN] Scrophulariaceae (V:D)
Castilleja confusa Greene = Castilleja miniata [CASTMIN] Scrophulariaceae (V:D)
Castilleja cusickii Greenm. [CASTCUS] Scrophulariaceae (V:D)
Castilleja elmeri Fern. [CASTELM] Scrophulariaceae (V:D)
Castilleja exilis A. Nels. [CASTEXI] Scrophulariaceae (V:D)
Castilleja fulva Pennell [CASTFUL] Scrophulariaceae (V:D)
Castilleja gracillima Rydb. = Castilleja miniata [CASTMIN] Scrophulariaceae (V:D)
Castilleja hispida Benth. [CASTHIS] Scrophulariaceae (V:D)
Castilleja hispida ssp. *abbreviata* (Fern.) Pennell = Castilleja hispida [CASTHIS] Scrophulariaceae (V:D)
Castilleja hyetophila Pennell [CASTHYE] Scrophulariaceae (V:D)
Castilleja hyperborea Pennell [CASTHYP] Scrophulariaceae (V:D)
Castilleja inconstans Standl. = Castilleja miniata [CASTMIN] Scrophulariaceae (V:D)
Castilleja lauta A. Nels. = Castilleja rhexifolia [CASTRHE] Scrophulariaceae (V:D)
Castilleja leonardii Rydb. = Castilleja rhexifolia [CASTRHE] Scrophulariaceae (V:D)
Castilleja levisecta Greenm. [CASTLEV] Scrophulariaceae (V:D)
Castilleja lutea Heller = Castilleja cusickii [CASTCUS] Scrophulariaceae (V:D)
Castilleja luteovirens Rydb. = Castilleja sulphurea [CASTSUL] Scrophulariaceae (V:D)
Castilleja lutescens (Greenm.) Rydb. [CASTLUT] Scrophulariaceae (V:D)
Castilleja miniata Dougl. ex Hook. [CASTMIN] Scrophulariaceae (V:D)
Castilleja mogollonica Pennell = Castilleja sulphurea [CASTSUL] Scrophulariaceae (V:D)
Castilleja oblongifolia Gray = Castilleja miniata [CASTMIN] Scrophulariaceae (V:D)
Castilleja occidentalis Torr. [CASTOCC] Scrophulariaceae (V:D)
Castilleja oregonensis Gandog. = Castilleja rhexifolia [CASTRHE] Scrophulariaceae (V:D)
Castilleja oreopola ssp. *albida* Pennell = Castilleja parviflora var. albida [CASTPAR1] Scrophulariaceae (V:D)
Castilleja pallescens (Gray) Greenm. [CASTPAL] Scrophulariaceae (V:D)
Castilleja parviflora Bong. [CASTPAR] Scrophulariaceae (V:D)
Castilleja parviflora var. **albida** (Pennell) Ownbey [CASTPAR1] Scrophulariaceae (V:D)
Castilleja raupii Pennell [CASTRAU] Scrophulariaceae (V:D)
Castilleja rhexifolia Rydb. [CASTRHE] Scrophulariaceae (V:D)
Castilleja rhexifolia var. *sulphurea* (Rydb.) Atwood = Castilleja sulphurea [CASTSUL] Scrophulariaceae (V:D)
Castilleja rupicola Piper ex Fern. [CASTRUP] Scrophulariaceae (V:D)
Castilleja sulphurea Rydb. [CASTSUL] Scrophulariaceae (V:D)
Castilleja tenuis (Heller) Chuang & Heckard [CASTTEN] Scrophulariaceae (V:D)
Castilleja thompsonii Pennell [CASTTHO] Scrophulariaceae (V:D)
Castilleja unalaschcensis (Cham. & Schlecht.) Malte [CASTUNA] Scrophulariaceae (V:D)
Castilleja villicaulis Pennell & Ownbey = Castilleja thompsonii [CASTTHO] Scrophulariaceae (V:D)
Castilleja villosissima Pennell = Castilleja hyperborea [CASTHYP] Scrophulariaceae (V:D)
Catabrosa Beauv. [CATABRO$] Poaceae (V:M)

Catabrosa aquatica (L.) Beauv. [CATAAQU] Poaceae (V:M)
Catapyrenium Flotow [CATAPYR$] Verrucariaceae (L)
Catapyrenium cinereum (Pers.) Körber [CATACIN] Verrucariaceae (L)
Catapyrenium compactum (A. Massal.) R. Sant. [CATACOM] Verrucariaceae (L)
Catapyrenium daedaleum (Kremp.) Stein [CATADAE] Verrucariaceae (L)
Catapyrenium heppioides (Zahlbr.) J.W. Thomson [CATAHEP] Verrucariaceae (L)
Catapyrenium lachneum (Ach.) R. Sant. [CATALAC] Verrucariaceae (L)
Catapyrenium norvegicum Breuss [CATANOR] Verrucariaceae (L)
Catapyrenium squamulosum (Ach.) Breuss [CATASQA] Verrucariaceae (L)
Catharinea haussknechtii (Jur. & Milde) Broth. = Atrichum haussknechtii [ATRIHAU] Polytrichaceae (B:M)
Catharinea selwynii (Aust.) Britt. = Atrichum selwynii [ATRISEL] Polytrichaceae (B:M)
Catharinea undulata (Hedw.) Web. & Mohr = Atrichum undulatum [ATRIUND] Polytrichaceae (B:M)
Catillaria A. Massal. [CATILLA$] Catillariaceae (L)
Catillaria arctica Lynge = Toninia philippea [TONIPHI] Catillariaceae (L)
Catillaria athallina (Hepp) Hellbom = Toninia athallina [TONIATH] Catillariaceae (L)
Catillaria atropurpurea (Schaerer) Th. Fr. = Catinaria atropurpurea [CATIATO] Bacidiaceae (L)
Catillaria bahusiensis (Blomb.) Th. Fr. = Tylothallia biformigera [TYLOBIF] Lecanoraceae (L)
Catillaria biformigera (Leighton) H. Magn. = Tylothallia biformigera [TYLOBIF] Lecanoraceae (L)
Catillaria bouteillei (Desmaz.) Zahlbr. = Fellhanera bouteillei [FELLBOU] Pilocarpaceae (L)
Catillaria chalybeia (Borrer) A. Massal. [CATICHA] Catillariaceae (L)
Catillaria columbiana (G. Merr.) W. Noble [CATICOL] Catillariaceae (L)
Catillaria globulosa (Flörke) Th. Fr. [CATIGLO] Catillariaceae (L)
Catillaria graniformis (K. Hagen) Vainio = Cliostomum corrugatum [CLIOCOR] Bacidiaceae (L)
Catillaria griffithii (Sm.) Malme = Cliostomum griffithii [CLIOGRI] Bacidiaceae (L)
Catillaria kansuensis H. Magn. = Toninia philippea [TONIPHI] Catillariaceae (L)
Catillaria micrococca (Körber) Th. Fr. = Micarea prasina [MICAPRA] Micareaceae (L)
Catillaria nigroclavata (Nyl.) Schuler [CATINIG] Catillariaceae (L)
Catillaria prasina (Fr.) Th. Fr. = Micarea prasina [MICAPRA] Micareaceae (L)
Catillaria sphaeroides (A. Massal.) Schuler = Biatora sphaeroides [BIATSPH] Bacidiaceae (L)
Catillaria subnitida Hellbom = Catillaria tristis [CATITRI] Catillariaceae (L)
Catillaria tricolor auct. = Cliostomum griffithii [CLIOGRI] Bacidiaceae (L)
Catillaria tristis (Müll. Arg.) Arnold [CATITRI] Catillariaceae (L)
Catinaria Vainio [CATINAR$] Bacidiaceae (L)
Catinaria atropurpurea (Schaerer) Vezda & Poelt [CATIATO] Bacidiaceae (L)
Catolechia Flotow [CATOLEC$] Rhizocarpaceae (L)
Catolechia wahlenbergii (Ach.) Körber [CATOWAH] Rhizocarpaceae (L)
Catoscopium Brid. [CATOSCO$] Catoscopiaceae (B:M)
Catoscopium nigritum (Hedw.) Brid. [CATONIG] Catoscopiaceae (B:M)
Caucalis microcarpa Hook. & Arn. = Yabea microcarpa [YABEMIC] Apiaceae (V:D)
Caulanthus sulfureus Payson = Brassica rapa [BRASRAP] Brassicaceae (V:D)
Caulinia flexilis Willd. = Najas flexilis [NAJAFLE] Najadaceae (V:M)
Cavernularia Degel. [CAVERNU$] Parmeliaceae (L)
Cavernularia hultenii Degel. [CAVEHUL] Parmeliaceae (L)
Cavernularia lophyrea (Ach.) Degel. [CAVELOP] Parmeliaceae (L)
Ceanothus L. [CEANOTH$] Rhamnaceae (V:D)
Ceanothus oreganus Nutt. = Ceanothus sanguineus [CEANSAN] Rhamnaceae (V:D)
Ceanothus sanguineus Pursh [CEANSAN] Rhamnaceae (V:D)
Ceanothus velutinus Dougl. ex Hook. [CEANVEL] Rhamnaceae (V:D)
Ceanothus velutinus ssp. **hookeri** (M.C. Johnston) C. Schmidt [CEANVEL1] Rhamnaceae (V:D)
Ceanothus velutinus ssp. **velutinus** [CEANVEL0] Rhamnaceae (V:D)
Ceanothus velutinus var. *hookeri* M.C. Johnston = Ceanothus velutinus ssp. hookeri [CEANVEL1] Rhamnaceae (V:D)
Ceanothus velutinus var. *laevigatus* Torr. & Gray = Ceanothus velutinus ssp. hookeri [CEANVEL1] Rhamnaceae (V:D)
Cenchrus L. [CENCHRU$] Poaceae (V:M)
Cenchrus longispinus (Hack.) Fern. [CENCLON] Poaceae (V:M)
Centaurea L. [CENTAUR$] Asteraceae (V:D)
*__Centaurea biebersteinii__ DC. [CENTBIE] Asteraceae (V:D)
*__Centaurea cyanus__ L. [CENTCYA] Asteraceae (V:D)
Centaurea debeauxii Gren. & Godr. [CENTDEB] Asteraceae (V:D)
*__Centaurea debeauxii__ ssp. **thuillieri** Dostál [CENTDEB1] Asteraceae (V:D)
*__Centaurea diffusa__ Lam. [CENTDIF] Asteraceae (V:D)
Centaurea dubia ssp. *vochinensis* (Bernh. ex Reichenb.) Hayek = Centaurea nigrescens [CENTNIR] Asteraceae (V:D)
Centaurea maculosa auct. = Centaurea biebersteinii [CENTBIE] Asteraceae (V:D)
*__Centaurea melitensis__ L. [CENTMEL] Asteraceae (V:D)
*__Centaurea montana__ L. [CENTMON] Asteraceae (V:D)
*__Centaurea nigrescens__ Willd. [CENTNIR] Asteraceae (V:D)
*__Centaurea paniculata__ L. [CENTPAN] Asteraceae (V:D)
Centaurea picris Pallas ex Willd. = Acroptilon repens [ACROREP] Asteraceae (V:D)
Centaurea pratensis Thuill. = Centaurea debeauxii ssp. thuillieri [CENTDEB1] Asteraceae (V:D)
Centaurea repens L. = Acroptilon repens [ACROREP] Asteraceae (V:D)
Centaurea vochinensis Bernh. ex Reichenb. = Centaurea nigrescens [CENTNIR] Asteraceae (V:D)
Centaurium Hill [CENTAUI$] Gentianaceae (V:D)
Centaurium curvistamineum (Wittr.) Abrams = Centaurium muhlenbergii [CENTMUH] Gentianaceae (V:D)
*__Centaurium erythraea__ Rafn [CENTERY] Gentianaceae (V:D)
Centaurium exaltatum (Griseb.) W. Wight ex Piper [CENTEXA] Gentianaceae (V:D)
Centaurium muhlenbergii (Griseb.) W. Wight ex Piper [CENTMUH] Gentianaceae (V:D)
Centaurium namophilum var. *nevadense* Broome = Centaurium exaltatum [CENTEXA] Gentianaceae (V:D)
Centaurium nuttallii (S. Wats.) Heller = Centaurium exaltatum [CENTEXA] Gentianaceae (V:D)
Centaurium umbellatum auct. = Centaurium erythraea [CENTERY] Gentianaceae (V:D)
Centunculus minimus L. = Anagallis minima [ANAGMIN] Primulaceae (V:D)
Cephalanthera L.C. Rich. [CEPHALA$] Orchidaceae (V:M)
Cephalanthera austiniae (Gray) Heller [CEPHAUS] Orchidaceae (V:M)
Cephalozia (Dum. emend. Schiffn.) Dum. [CEPHALO$] Cephaloziaceae (B:H)
Cephalozia ambigua Mass. = Cephalozia bicuspidata ssp. ambigua [CEPHBIC1] Cephaloziaceae (B:H)
Cephalozia bicuspidata (L.) Dum. [CEPHBIC] Cephaloziaceae (B:H)
Cephalozia bicuspidata ssp. **ambigua** (Mass.) Schust. [CEPHBIC1] Cephaloziaceae (B:H)
Cephalozia bicuspidata ssp. **bicuspidata** [CEPHBIC0] Cephaloziaceae (B:H)
Cephalozia catenulata (Hub.) Lindb. [CEPHCAT] Cephaloziaceae (B:H)

Cephalozia connivens (Dicks.) Lindb. [CEPHCON] Cephaloziaceae (B:H)
Cephalozia fluitans (Nees) Spruce = Cladopodiella fluitans [CLADFLU] Cephaloziaceae (B:H)
Cephalozia leucantha Spruce [CEPHLEU] Cephaloziaceae (B:H)
Cephalozia lunulifolia (Dum.) Dum. [CEPHLUN] Cephaloziaceae (B:H)
Cephalozia macounii (Aust.) Aust. [CEPHMAC] Cephaloziaceae (B:H)
Cephalozia media Lindb. = Cephalozia lunulifolia [CEPHLUN] Cephaloziaceae (B:H)
Cephalozia pleniceps (Aust.) Lindb. [CEPHPLE] Cephaloziaceae (B:H)
Cephaloziella (Spruce) Steph. [CEPHALZ$] Cephaloziellaceae (B:H)
Cephaloziella arctica Bryhn & Douin [CEPHARC] Cephaloziellaceae (B:H)
Cephaloziella brinkmani Douin [CEPHBRI] Cephaloziellaceae (B:H)
Cephaloziella byssacea (Roth) Warnst. = Cephaloziella divaricata [CEPHDIV] Cephaloziellaceae (B:H)
Cephaloziella byssacea var. *scabra* (M.A. Howe) Schust. = Cephaloziella divaricata var. scabra [CEPHDIV2] Cephaloziellaceae (B:H)
Cephaloziella divaricata (Sm.) Schiffn. [CEPHDIV] Cephaloziellaceae (B:H)
Cephaloziella divaricata var. **divaricata** [CEPHDIV0] Cephaloziellaceae (B:H)
Cephaloziella divaricata var. **scabra** M.A. Howe [CEPHDIV2] Cephaloziellaceae (B:H)
Cephaloziella elachista (Jack) Schiffn. [CEPHELA] Cephaloziellaceae (B:H)
Cephaloziella glacialis Douin = Cephaloziella arctica [CEPHARC] Cephaloziellaceae (B:H)
Cephaloziella hampeana (Nees) Schiffn. [CEPHHAM] Cephaloziellaceae (B:H)
Cephaloziella papilosa (Douin) Schiffn. = Cephaloziella divaricata var. scabra [CEPHDIV2] Cephaloziellaceae (B:H)
Cephaloziella pearsonii (Spruce) Douin = Sphenolobopsis pearsonii [SPHEPEA] Jungermanniaceae (B:H)
Cephaloziella phyllacantha (Mass. & Carest.) K. Müll. [CEPHPHY] Cephaloziellaceae (B:H)
Cephaloziella rubella (Nees) Warnst. [CEPHRUB] Cephaloziellaceae (B:H)
Cephaloziella spinosa Douin = Cephaloziella subdentata [CEPHSUB] Cephaloziellaceae (B:H)
Cephaloziella starkei (Funck ex Nees) Schiffn. = Cephaloziella divaricata [CEPHDIV] Cephaloziellaceae (B:H)
Cephaloziella striatula (C. Jens.) Douin = Cephaloziella subdentata [CEPHSUB] Cephaloziellaceae (B:H)
Cephaloziella subdentata Warnst. [CEPHSUB] Cephaloziellaceae (B:H)
Cephaloziella turneri (Hook.) K. Müll. [CEPHTUR] Cephaloziellaceae (B:H)
Cephaloziopsis pearsonii (Spruce) Schiffn. = Sphenolobopsis pearsonii [SPHEPEA] Jungermanniaceae (B:H)
Cephaloziopsis saccatula (Lindb.) Schiffn. = Anastrophyllum minutum var. grandis [ANASMIN1] Jungermanniaceae (B:H)
Cerastium L. [CERASTI$] Caryophyllaceae (V:D)
Cerastium acutatum Suksdorf = Cerastium glomeratum [CERAGLO] Caryophyllaceae (V:D)
Cerastium adsurgens Greene = Cerastium fontanum ssp. vulgare [CERAFON1] Caryophyllaceae (V:D)
Cerastium alpinum var. *capillare* (Fern. & Wieg.) Boivin = Cerastium beeringianum ssp. earlei [CERABEE2] Caryophyllaceae (V:D)
Cerastium aquaticum L. = Myosoton aquaticum [MYOSAQU] Caryophyllaceae (V:D)
Cerastium arvense L. [CERAARV] Caryophyllaceae (V:D)
Cerastium beeringianum Cham. & Schlecht. [CERABEE] Caryophyllaceae (V:D)
Cerastium beeringianum ssp. **beeringianum** [CERABEE0] Caryophyllaceae (V:D)
Cerastium beeringianum ssp. **earlei** (Rydb.) Hultén [CERABEE2] Caryophyllaceae (V:D)
Cerastium beeringianum var. *capillare* Fern. & Wieg. = Cerastium beeringianum ssp. earlei [CERABEE2] Caryophyllaceae (V:D)
Cerastium earlei Rydb. = Cerastium beeringianum ssp. earlei [CERABEE2] Caryophyllaceae (V:D)
Cerastium fischerianum Ser. [CERAFIS] Caryophyllaceae (V:D)
***Cerastium fontanum** Baumg. [CERAFON] Caryophyllaceae (V:D)
Cerastium fontanum ssp. *triviale* (Link) Jalas = Cerastium fontanum ssp. vulgare [CERAFON1] Caryophyllaceae (V:D)
***Cerastium fontanum** ssp. **vulgare** (Hartman) Greuter & Burdet [CERAFON1] Caryophyllaceae (V:D)
***Cerastium glomeratum** Thuill. [CERAGLO] Caryophyllaceae (V:D)
Cerastium glomeratum var. *apetalum* (Dumort.) Fenzl = Cerastium glomeratum [CERAGLO] Caryophyllaceae (V:D)
Cerastium holosteoides var. *vulgare* (Hartman) Hyl. = Cerastium fontanum ssp. vulgare [CERAFON1] Caryophyllaceae (V:D)
Cerastium nutans Raf. [CERANUT] Caryophyllaceae (V:D)
***Cerastium semidecandrum** L. [CERASEM] Caryophyllaceae (V:D)
***Cerastium tomentosum** L. [CERATOM] Caryophyllaceae (V:D)
Cerastium triviale Link = Cerastium fontanum ssp. vulgare [CERAFON1] Caryophyllaceae (V:D)
Cerastium viscosum auct. = Cerastium glomeratum [CERAGLO] Caryophyllaceae (V:D)
Cerastium vulgatum L. 1762, non 1755 = Cerastium fontanum ssp. vulgare [CERAFON1] Caryophyllaceae (V:D)
Cerastium vulgatum var. *hirsutum* Fries = Cerastium fontanum ssp. vulgare [CERAFON1] Caryophyllaceae (V:D)
Cerasus avium (L.) Moench = Prunus avium [PRUNAVI] Rosaceae (V:D)
Cerasus laurocerasus (L.) Loisel. = Prunus laurocerasus [PRUNLAU] Rosaceae (V:D)
Ceratocephala Moench Ranunculaceae (V:D)
Ceratocephala orthoceras DC. = Ceratocephala testiculatus [CERATES] Ranunculaceae (V:D)
***Ceratocephala testiculatus** (Crantz) Roth [CERATES] Ranunculaceae (V:D)
Ceratochloa carinata (Hook. & Arn.) Tutin = Bromus carinatus [BROMCAR] Poaceae (V:M)
Ceratodon Brid. [CERATOD$] Ditrichaceae (B:M)
Ceratodon columbiae Kindb. = Ceratodon purpureus [CERAPUR] Ditrichaceae (B:M)
Ceratodon purpureus (Hedw.) Brid. [CERAPUR] Ditrichaceae (B:M)
Ceratodon purpureus f. *aristatus* Aust. = Ceratodon purpureus [CERAPUR] Ditrichaceae (B:M)
Ceratophyllum L. [CERATOP$] Ceratophyllaceae (V:D)
Ceratophyllum apiculatum Cham. = Ceratophyllum demersum [CERADEM] Ceratophyllaceae (V:D)
Ceratophyllum demersum L. [CERADEM] Ceratophyllaceae (V:D)
Ceratophyllum demersum var. *apiculatum* (Cham.) Garcke = Ceratophyllum demersum [CERADEM] Ceratophyllaceae (V:D)
Ceratophyllum demersum var. *apiculatum* (Cham.) Aschers. = Ceratophyllum demersum [CERADEM] Ceratophyllaceae (V:D)
Ceratophyllum demersum var. *echinatum* (Gray) Gray = Ceratophyllum echinatum [CERAECH] Ceratophyllaceae (V:D)
Ceratophyllum echinatum Gray [CERAECH] Ceratophyllaceae (V:D)
Ceratophyllum submersum var. *echinatum* (Gray) Wilmot-Dear = Ceratophyllum echinatum [CERAECH] Ceratophyllaceae (V:D)
Ceratoxalis coloradensis (Rydb.) Lunell = Oxalis stricta [OXALSTR] Oxalidaceae (V:D)
Ceratoxalis cymosa (Small) Lunell = Oxalis stricta [OXALSTR] Oxalidaceae (V:D)
Cetraria Ach. [CETRARI$] Parmeliaceae (L)
Cetraria aculeata (Schreber) Fr. [CETRACU] Parmeliaceae (L)
Cetraria agnata (Nyl.) Kristinsson = Melanelia agnata [MELAAGN] Parmeliaceae (L)
Cetraria arborialis (Zahlbr.) Howard = Tuckermannopsis subalpina [TUCKSUB] Parmeliaceae (L)
Cetraria arctica (Hook.) Tuck. = Dactylina arctica [DACTARC] Parmeliaceae (L)
Cetraria californica Tuck. = Kaernefeltia californica [KAERCAL] Parmeliaceae (L)

Cetraria canadensis (Räsänen) Räsänen = Vulpicida canadensis [VULPCAN] Parmeliaceae (L)
Cetraria chlorophylla (Willd.) Vainio = Tuckermannopsis chlorophylla [TUCKCHL] Parmeliaceae (L)
Cetraria chrysantha Tuck. = Asahinea chrysantha [ASAHCHR] Parmeliaceae (L)
Cetraria ciliaris var. *halei* (Culb. & C. Culb.) Ahti = Tuckermannopsis americana [TUCKAME] Parmeliaceae (L)
Cetraria commixta (Nyl.) Th. Fr. = Melanelia commixta [MELACOM] Parmeliaceae (L)
Cetraria cucullata (Bellardi) Ach. = Flavocetraria cucullata [FLAVCUC] Parmeliaceae (L)
Cetraria delisei (Bory ex Schaerer) Nyl. = Cetrariella delisei [CETRDEI] Parmeliaceae (L)
Cetraria ericetorum Opiz [CETRERI] Parmeliaceae (L)
Cetraria ericetorum ssp. **reticulata** (Räsänen) Kärnefelt [CETRERI1] Parmeliaceae (L)
Cetraria fahlunensis (L.) Schreber = Melanelia commixta [MELACOM] Parmeliaceae (L)
Cetraria glauca (L.) Ach. = Platismatia glauca [PLATGLA] Parmeliaceae (L)
Cetraria halei Culb. & C. Culb. = Tuckermannopsis americana [TUCKAME] Parmeliaceae (L)
Cetraria hepatizon (Ach.) Vainio = Melanelia hepatizon [MELAHEP] Parmeliaceae (L)
Cetraria herrei Imshaug = Platismatia herrei [PLATHER] Parmeliaceae (L)
Cetraria hiascens (Fr.) Th. Fr. = Cetrariella delisei [CETRDEI] Uncertain-family (L)
Cetraria iberica Crespo & Barreno = Kaernefeltia merrillii [KAERMER] Parmeliaceae (L)
Cetraria idahoensis Essl. = Esslingeriana idahoensis [ESSLIDA] Parmeliaceae (L)
Cetraria islandica (L.) Ach. [CETRISL] Parmeliaceae (L)
Cetraria islandica ssp. **crispiformis** (Räsänen) Kärnefelt [CETRISL1] Parmeliaceae (L)
Cetraria islandica ssp. **islandica** [CETRISL0] Parmeliaceae (L)
Cetraria islandica ssp. **orientalis** (Asah.) Kärnefelt [CETRISL2] Parmeliaceae (L)
Cetraria juniperina auct. = Vulpicida canadensis [VULPCAN] Parmeliaceae (L)
Cetraria lacunosa Ach. = Platismatia lacunosa [PLATLAC] Parmeliaceae (L)
Cetraria lacunosa var. *macounii* Du Rietz = Platismatia lacunosa [PLATLAC] Parmeliaceae (L)
Cetraria laevigata Rass. [CETRLAE] Parmeliaceae (L)
Cetraria madreporiformis (Ach.) Müll. Arg. = Dactylina madreporiformis [DACTMAD] Parmeliaceae (L)
Cetraria merrillii Du Rietz = Kaernefeltia merrillii [KAERMER] Parmeliaceae (L)
Cetraria muricata (Ach.) Eckfeldt [CETRMUR] Parmeliaceae (L)
Cetraria nigricans Nyl. [CETRNIG] Parmeliaceae (L)
Cetraria nivalis (L.) Ach. = Flavocetraria nivalis [FLAVNIV] Parmeliaceae (L)
Cetraria norvegica (Lynge) Du Rietz = Platismatia norvegica [PLATNOR] Parmeliaceae (L)
Cetraria orbata (Nyl.) Fink = Tuckermannopsis orbata [TUCKORB] Parmeliaceae (L)
Cetraria pallidula Tuck. ex Riddle = Ahtiana pallidula [AHTIPAL] Parmeliaceae (L)
Cetraria pinastri (Scop.) Gray = Vulpicida pinastri [VULPPIN] Parmeliaceae (L)
Cetraria platyphylla Tuck. = Tuckermannopsis platyphylla [TUCKPLA] Parmeliaceae (L)
Cetraria polyschiza (Nyl.) Jatta = Melanelia hepatizon [MELAHEP] Parmeliaceae (L)
Cetraria richardsonii Hook. = Masonhalea richardsonii [MASORIC] Parmeliaceae (L)
Cetraria scholanderi Llano = Asahinea scholanderi [ASAHSCH] Parmeliaceae (L)
Cetraria scutata (Wulfen) Poetsch = Tuckermannopsis sepincola [TUCKSEP] Parmeliaceae (L)
Cetraria scutata auct. = Tuckermannopsis chlorophylla [TUCKCHL] Parmeliaceae (L)
Cetraria sepincola (Ehrh.) Ach. = Tuckermannopsis sepincola [TUCKSEP] Parmeliaceae (L)
Cetraria stenophylla (Tuck.) G. Merr. = Platismatia stenophylla [PLATSTE] Parmeliaceae (L)
Cetraria subalpina Imshaug = Tuckermannopsis subalpina [TUCKSUB] Parmeliaceae (L)
Cetraria tilesii Ach. = Vulpicida tilesii [VULPTIL] Parmeliaceae (L)
Cetraria tuckermanii Herre = Platismatia herrei [PLATHER] Parmeliaceae (L)
Cetrariella Kärnefelt & Thell [CETRARE$] Uncertain-family (L)
Cetrariella delisei (Bory ex Schaerer) Kärnefelt & Thell [CETRDEI] Uncertain-family (L)
Cetrelia Culb. & C. Culb. [CETRELI$] Parmeliaceae (L)
Cetrelia cetrarioides (Duby) Culb. & C. Culb. [CETRCET] Parmeliaceae (L)
Chaenactis DC. [CHAENAC$] Asteraceae (V:D)
Chaenactis alpina (Gray) M.E. Jones = Chaenactis douglasii var. alpina [CHAEDOU4] Asteraceae (V:D)
Chaenactis angustifolia Greene = Chaenactis douglasii var. douglasii [CHAEDOU0] Asteraceae (V:D)
Chaenactis brachiata Greene = Chaenactis douglasii var. douglasii [CHAEDOU0] Asteraceae (V:D)
Chaenactis brachiata var. *stansburiana* Stockwell = Chaenactis douglasii var. douglasii [CHAEDOU0] Asteraceae (V:D)
Chaenactis cineria Stockwell = Chaenactis douglasii var. douglasii [CHAEDOU0] Asteraceae (V:D)
Chaenactis douglasii (Hook.) Hook. & Arn. [CHAEDOU] Asteraceae (V:D)
Chaenactis douglasii var. *achilleifolia* (Hook. & Arn.) Gray = Chaenactis douglasii var. douglasii [CHAEDOU0] Asteraceae (V:D)
Chaenactis douglasii var. **alpina** Gray [CHAEDOU4] Asteraceae (V:D)
Chaenactis douglasii var. **douglasii** [CHAEDOU0] Asteraceae (V:D)
Chaenactis douglasii var. *glandulosa* Cronq. = Chaenactis douglasii var. douglasii [CHAEDOU0] Asteraceae (V:D)
Chaenactis douglasii var. *montana* M.E. Jones = Chaenactis douglasii var. douglasii [CHAEDOU0] Asteraceae (V:D)
Chaenactis douglasii var. *nana* Stockwell = Chaenactis douglasii var. douglasii [CHAEDOU0] Asteraceae (V:D)
Chaenactis douglasii var. *rubricaulis* Rydb. = Chaenactis douglasii var. douglasii [CHAEDOU0] Asteraceae (V:D)
Chaenactis douglasii var. *typicus* Cronq. = Chaenactis douglasii var. douglasii [CHAEDOU0] Asteraceae (V:D)
Chaenactis humilis Rydb. = Chaenactis douglasii var. douglasii [CHAEDOU0] Asteraceae (V:D)
Chaenactis minuscula Greene = Chaenactis douglasii var. alpina [CHAEDOU4] Asteraceae (V:D)
Chaenactis panamintensis Stockwell = Chaenactis douglasii var. douglasii [CHAEDOU0] Asteraceae (V:D)
Chaenactis ramosa Stockwell = Chaenactis douglasii var. douglasii [CHAEDOU0] Asteraceae (V:D)
Chaenactis rubricaulis Rydb. = Chaenactis douglasii var. douglasii [CHAEDOU0] Asteraceae (V:D)
Chaenactis suksdorfii Stockwell = Chaenactis douglasii var. douglasii [CHAEDOU0] Asteraceae (V:D)
Chaenorrhinum (DC. ex Duby) Reichenb. [CHAENOR$] Scrophulariaceae (V:D)
***Chaenorrhinum minus** (L.) Lange [CHAEMIN] Scrophulariaceae (V:D)
Chaenotheca Th. Fr. [CHAENOT$] Coniocybaceae (L)
Chaenotheca brachypoda (Ach.) Tibell [CHAEBRA] Coniocybaceae (L)
Chaenotheca brunneola (Ach.) Müll. Arg. [CHAEBRU] Coniocybaceae (L)
Chaenotheca carthusiae (Harm.) Lettau = Chaenotheca chlorella [CHAECHL] Coniocybaceae (L)
Chaenotheca chlorella (Ach.) Müll. Arg. [CHAECHL] Coniocybaceae (L)
Chaenotheca chrysocephala (Turner ex Ach.) Th. Fr. [CHAECHR] Coniocybaceae (L)

Chaenotheca cinerea (Pers.) Tibell [CHAECIN] Coniocybaceae (L)
Chaenotheca ferruginea (Turner & Borrer) Mig. [CHAEFER] Coniocybaceae (L)
Chaenotheca furfuracea (L.) Tibell [CHAEFUR] Coniocybaceae (L)
Chaenotheca gracillima (Vainio) Tibell [CHAEGRA] Coniocybaceae (L)
Chaenotheca hispidula (Ach.) Zahlbr. [CHAEHIS] Coniocybaceae (L)
Chaenotheca hygrophila Tibell = Chaenotheca brunneola [CHAEBRU] Coniocybaceae (L)
Chaenotheca laevigata Nádv. [CHAELAE] Coniocybaceae (L)
Chaenotheca melanophaea (Ach.) Zwackh = Chaenotheca ferruginea [CHAEFER] Coniocybaceae (L)
Chaenotheca schaereri (De Not.) Zahlbr. = Chaenotheca cinerea [CHAECIN] Coniocybaceae (L)
Chaenotheca stemonea (Ach.) Müll. Arg. [CHAESTE] Coniocybaceae (L)
Chaenotheca subroscida (Eitner) Zahlbr. [CHAESUB] Coniocybaceae (L)
Chaenotheca sulphurea (Retz.) Middleborg & J.-E. Mattsson = Chaenotheca brachypoda [CHAEBRA] Coniocybaceae (L)
Chaenotheca trichialis (Ach.) Th. Fr. [CHAETRI] Coniocybaceae (L)
Chaenotheca trichialis var. *cinerea* (Pers.) Blomb. & Forss. = Chaenotheca cinerea [CHAECIN] Coniocybaceae (L)
Chaenotheca xyloxena Nádv. [CHAEXYL] Coniocybaceae (L)
Chaenothecopsis Vainio [CHAENOH$] Mycocaliciaceae (L)
Chaenothecopsis consociata (Nádv.) A.F.W. Schmidt [CHAECON] Mycocaliciaceae (L)
Chaenothecopsis debilis (Turner & Borrer ex Sm.) Tibell [CHAEDEB] Mycocaliciaceae (L)
Chaenothecopsis epithallina Tibell [CHAEEPI] Mycocaliciaceae (L)
Chaenothecopsis lignicola (Nádv.) A.F.W. Schmidt = Chaenothecopsis pusiola [CHAEPUI] Mycocaliciaceae (L)
Chaenothecopsis nana Tibell [CHAENAN] Mycocaliciaceae (L)
Chaenothecopsis pusilla (Ach.) A.F.W. Schmidt [CHAEPUS] Mycocaliciaceae (L)
Chaenothecopsis pusiola (Ach.) Vainio [CHAEPUI] Mycocaliciaceae (L)
Chaenothecopsis subpusilla (Vainio) Tibell = Chaenothecopsis pusilla [CHAEPUS] Mycocaliciaceae (L)
Chaenothecopsis viridialba (Kremp.) A.F.W. Schmidt [CHAEVIR] Mycocaliciaceae (L)
Chaenothecopsis viridireagens (Nádv.) A.F.W. Schmidt [CHAEVII] Mycocaliciaceae (L)
Chaetochloa glauca (L.) Scribn. = Setaria glauca [SETAGLA] Poaceae (V:M)
Chaetochloa italica (L.) Scribn. = Setaria italica [SETAITA] Poaceae (V:M)
Chaetochloa lutescens (Weigel) Stuntz = Setaria glauca [SETAGLA] Poaceae (V:M)
Chamaecistus procumbens (L.) Kuntze = Loiseleuria procumbens [LOISPRO] Ericaceae (V:D)
Chamaecyparis Spach [CHAMAEC$] Cupressaceae (V:G)
Chamaecyparis nootkatensis (D. Don) Spach [CHAMNOO] Cupressaceae (V:G)
Chamaedaphne Moench [CHAMAED$] Ericaceae (V:D)
Chamaedaphne calyculata (L.) Moench [CHAMCAL] Ericaceae (V:D)
Chamaenerion angustifolium (L.) Scop. = Epilobium angustifolium ssp. angustifolium [EPILANG0] Onagraceae (V:D)
Chamaenerion latifolium (L.) Sweet = Epilobium latifolium [EPILLAT] Onagraceae (V:D)
Chamaepericlymenum canadense (L.) Aschers. & Graebn. = Cornus canadensis [CORNCAN] Cornaceae (V:D)
Chamaepericlymenum suecicum (L.) Aschers. & Graebn. = Cornus suecica [CORNSUE] Cornaceae (V:D)
Chamaepericlymenum unalaschkense (Ledeb.) Rydb. = Cornus unalaschkensis [CORNUNA] Cornaceae (V:D)
Chamaerhodos Bunge [CHAMAER$] Rosaceae (V:D)
Chamaerhodos erecta (L.) Bunge [CHAMERE] Rosaceae (V:D)
Chamaerhodos erecta ssp. **nuttallii** (Pickering ex Rydb.) Hultén [CHAMERE1] Rosaceae (V:D)
Chamaerhodos erecta var. *parviflora* (Nutt.) C.L. Hitchc. = Chamaerhodos erecta ssp. nuttallii [CHAMERE1] Rosaceae (V:D)
Chamaerhodos nuttallii Pickering ex Rydb. = Chamaerhodos erecta ssp. nuttallii [CHAMERE1] Rosaceae (V:D)
Chamaerhodos nuttallii var. *keweenawensis* Fern. = Chamaerhodos erecta ssp. nuttallii [CHAMERE1] Rosaceae (V:D)
Chamaesyce S.F. Gray [CHAMAES$] Euphorbiaceae (V:D)
Chamaesyce albicaulis (Rydb.) Rydb. = Chamaesyce serpyllifolia [CHAMSER] Euphorbiaceae (V:D)
Chamaesyce glyptosperma (Engelm.) Small [CHAMGLY] Euphorbiaceae (V:D)
*__Chamaesyce maculata__ (L.) Small [CHAMMAC] Euphorbiaceae (V:D)
Chamaesyce mathewsii Small = Chamaesyce maculata [CHAMMAC] Euphorbiaceae (V:D)
Chamaesyce neomexicana (Greene) Standl. = Chamaesyce serpyllifolia [CHAMSER] Euphorbiaceae (V:D)
Chamaesyce serpyllifolia (Pers.) Small [CHAMSER] Euphorbiaceae (V:D)
Chamaesyce supina (Raf.) Moldenke = Chamaesyce maculata [CHAMMAC] Euphorbiaceae (V:D)
Chamaesyce tracyi Small = Chamaesyce maculata [CHAMMAC] Euphorbiaceae (V:D)
Chamberlainia albicans (Hedw.) Robins. = Brachythecium albicans [BRACALB] Brachytheciaceae (B:M)
Chamberlainia calcarea (Kindb.) Robins. = Brachythecium calcareum [BRACCAL] Brachytheciaceae (B:M)
Chamberlainia campestris (C. Müll.) Robins. = Brachythecium campestre [BRACCAM] Brachytheciaceae (B:M)
Chamberlainia collina (Schleich. ex C. Müll.) Robins. = Brachythecium collinum [BRACCOL] Brachytheciaceae (B:M)
Chamberlainia collina var. *suberythrorrhiza* (Ren. & Card.) Robins. = Brachythecium velutinum var. venustum [BRACVEL1] Brachytheciaceae (B:M)
Chamberlainia erythrorrhiza (Schimp. in B.S.G.) Robins. = Brachythecium erythrorrhizon [BRACERY] Brachytheciaceae (B:M)
Chamberlainia leibergii (Grout) Robins. = Brachythecium leibergii [BRACLEI] Brachytheciaceae (B:M)
Chamberlainia oxyclada (Brid.) Robins. = Brachythecium oxycladon [BRACOXY] Brachytheciaceae (B:M)
Chamberlainia salebrosa (Web. & Mohr) Robins. = Brachythecium salebrosum [BRACSAL] Brachytheciaceae (B:M)
Chamberlainia turgida (Hartm.) Robins. = Brachythecium turgidum [BRACTUR] Brachytheciaceae (B:M)
Chamberlainia velutina (Hedw.) Robins. = Brachythecium velutinum [BRACVEL] Brachytheciaceae (B:M)
Chamerion angustifolium (L.) Holub = Epilobium angustifolium ssp. angustifolium [EPILANG0] Onagraceae (V:D)
Chamerion danielsii D. Löve = Epilobium angustifolium ssp. circumvagum [EPILANG2] Onagraceae (V:D)
Chamerion latifolium (L.) Holub = Epilobium latifolium [EPILLAT] Onagraceae (V:D)
Chamerion platyphyllum (Daniels) A. & D. Löve = Epilobium angustifolium ssp. circumvagum [EPILANG2] Onagraceae (V:D)
Chamerion spicatum (Lam.) S.F. Gray = Epilobium angustifolium ssp. angustifolium [EPILANG0] Onagraceae (V:D)
Chamerion subdentatum (Rydb.) A. & D. Löve = Epilobium latifolium [EPILLAT] Onagraceae (V:D)
Chamomilla chamomilla (L.) Rydb. = Matricaria recutita [MATRREC] Asteraceae (V:D)
Chamomilla inodora (L.) Gilib. = Matricaria perforata [MATRPER] Asteraceae (V:D)
Chamomilla recutita (L.) Rauschert = Matricaria recutita [MATRREC] Asteraceae (V:D)
Chamomilla suaveolens (Pursh) Rydb. = Matricaria discoidea [MATRDIS] Asteraceae (V:D)
Chandonanthus Mitt. [CHANDON$] Jungermanniaceae (B:H)
Chandonanthus cavallii (Gola) S. Arnell = Tetralophozia setiformis [TETRSET] Jungermanniaceae (B:H)
Chandonanthus filiformis Steph. = Tetralophozia filiformis [TETRFIL] Jungermanniaceae (B:H)

Chandonanthus hirtellus (Web.) Mitt. [CHANHIR] Jungermanniaceae (B:H)
Chandonanthus pusillus Steph. = Tetralophozia filiformis [TETRFIL] Jungermanniaceae (B:H)
Chandonanthus quadrifidus Steph. = Tetralophozia setiformis [TETRSET] Jungermanniaceae (B:H)
Chandonanthus setiformis (Ehrh.) Lindb. = Tetralophozia setiformis [TETRSET] Jungermanniaceae (B:H)
Cheilanthes Sw. [CHEILAN$] Pteridaceae (V:F)
Cheilanthes densa (Brack.) St. John = Aspidotis densa [ASPIDEN] Pteridaceae (V:F)
Cheilanthes feei T. Moore [CHEIFEE] Pteridaceae (V:F)
Cheilanthes gracillima D.C. Eat. [CHEIGRA] Pteridaceae (V:F)
Cheilanthes gracillima var. *aberrans* M.E. Jones = Cheilanthes gracillima [CHEIGRA] Pteridaceae (V:F)
Cheilanthes siliquosa Maxon = Aspidotis densa [ASPIDEN] Pteridaceae (V:F)
Cheiranthus cheiri L. = Erysimum cheiri [ERYSCHI] Brassicaceae (V:D)
Cheirinia cheiranthoides (L.) Link = Erysimum cheiranthoides [ERYSCHE] Brassicaceae (V:D)
Chelidonium L. [CHELIDO$] Papaveraceae (V:D)
*__Chelidonium majus__ L. [CHELMAJ] Papaveraceae (V:D)
Chenopodium L. [CHENOPO$] Chenopodiaceae (V:D)
*__Chenopodium album__ L. [CHENALB] Chenopodiaceae (V:D)
Chenopodium album ssp. *striatum* (Krasan) J. Murr = Chenopodium strictum ssp. striatum [CHENSTR2] Chenopodiaceae (V:D)
Chenopodium album var. *lanceolatum* (Muhl. ex Willd.) Coss. & Germ. = Chenopodium album [CHENALB] Chenopodiaceae (V:D)
Chenopodium album var. *polymorphum* Aellen = Chenopodium album [CHENALB] Chenopodiaceae (V:D)
Chenopodium amaranticolor Coste & Reyn. = Chenopodium album [CHENALB] Chenopodiaceae (V:D)
Chenopodium aridum A. Nels. = Chenopodium atrovirens [CHENATR] Chenopodiaceae (V:D)
Chenopodium atrovirens Rydb. [CHENATR] Chenopodiaceae (V:D)
Chenopodium botryodes Sm. [CHENBOR] Chenopodiaceae (V:D)
*__Chenopodium botrys__ L. [CHENBOT] Chenopodiaceae (V:D)
Chenopodium capitatum (L.) Aschers. [CHENCAP] Chenopodiaceae (V:D)
Chenopodium chenopodioides (L.) Aellen = Chenopodium botryodes [CHENBOR] Chenopodiaceae (V:D)
Chenopodium chenopodioides var. *degenianum* (Aellen) Aellen = Chenopodium botryodes [CHENBOR] Chenopodiaceae (V:D)
Chenopodium chenopodioides var. *lengyelianum* (Aellen) Aellen = Chenopodium botryodes [CHENBOR] Chenopodiaceae (V:D)
Chenopodium desiccatum A. Nels. [CHENDES] Chenopodiaceae (V:D)
Chenopodium fremontii var. *atrovirens* (Rydb.) Fosberg = Chenopodium atrovirens [CHENATR] Chenopodiaceae (V:D)
Chenopodium giganteum D. Don = Chenopodium album [CHENALB] Chenopodiaceae (V:D)
Chenopodium gigantospermum Aellen = Chenopodium simplex [CHENSIM] Chenopodiaceae (V:D)
Chenopodium glaucum ssp. *salinum* (Standl.) Aellen = Chenopodium salinum [CHENSAL] Chenopodiaceae (V:D)
Chenopodium glaucum var. *pulchrum* Aellen = Chenopodium salinum [CHENSAL] Chenopodiaceae (V:D)
Chenopodium glaucum var. *salinum* (Standl.) Boivin = Chenopodium salinum [CHENSAL] Chenopodiaceae (V:D)
Chenopodium hybridum auct. = Chenopodium simplex [CHENSIM] Chenopodiaceae (V:D)
Chenopodium hybridum ssp. *gigantospermum* (Aellen) Hultén = Chenopodium simplex [CHENSIM] Chenopodiaceae (V:D)
Chenopodium hybridum var. *gigantospermum* (Aellen) Rouleau = Chenopodium simplex [CHENSIM] Chenopodiaceae (V:D)
Chenopodium hybridum var. *simplex* Torr. = Chenopodium simplex [CHENSIM] Chenopodiaceae (V:D)
Chenopodium incognitum H.A. Wahl p.p. = Chenopodium atrovirens [CHENATR] Chenopodiaceae (V:D)
Chenopodium lanceolatum Muhl. ex Willd. = Chenopodium album [CHENALB] Chenopodiaceae (V:D)
Chenopodium leptophyllum var. *oblongifolium* S. Wats. = Chenopodium desiccatum [CHENDES] Chenopodiaceae (V:D)
Chenopodium oblongifolium (S. Wats.) Rydb. = Chenopodium desiccatum [CHENDES] Chenopodiaceae (V:D)
Chenopodium pratericola ssp. *desiccatum* (A. Nels.) Aellen = Chenopodium desiccatum [CHENDES] Chenopodiaceae (V:D)
Chenopodium pratericola var. *oblongifolium* (S. Wats.) H.A. Wahl = Chenopodium desiccatum [CHENDES] Chenopodiaceae (V:D)
Chenopodium rubrum L. [CHENRUB] Chenopodiaceae (V:D)
Chenopodium rubrum var. *glomeratum* Wallr. = Chenopodium botryodes [CHENBOR] Chenopodiaceae (V:D)
Chenopodium rubrum var. *humile* auct. = Chenopodium botryodes [CHENBOR] Chenopodiaceae (V:D)
Chenopodium salinum Standl. [CHENSAL] Chenopodiaceae (V:D)
*__Chenopodium simplex__ (Torr.) Raf. [CHENSIM] Chenopodiaceae (V:D)
Chenopodium strictum Roth [CHENSTR] Chenopodiaceae (V:D)
Chenopodium strictum ssp. **glaucophyllum** (Aellen) Aellen & Just. [CHENSTR1] Chenopodiaceae (V:D)
*__Chenopodium strictum__ ssp. **striatum** (Krasan) Aellen & Iljin [CHENSTR2] Chenopodiaceae (V:D)
Chenopodium strictum var. *glaucophyllum* (Aellen) H.A. Wahl = Chenopodium strictum ssp. glaucophyllum [CHENSTR1] Chenopodiaceae (V:D)
Chenopodium suecicum J. Murr. = Chenopodium album [CHENALB] Chenopodiaceae (V:D)
*__Chenopodium urbicum__ L. [CHENURB] Chenopodiaceae (V:D)
Chenopodium urbicum var. *intermedium* (Mert. & Koch) W.D.J. Koch = Chenopodium urbicum [CHENURB] Chenopodiaceae (V:D)
Chenopodium wolfii Rydb. = Chenopodium atrovirens [CHENATR] Chenopodiaceae (V:D)
Chiloscyphus Corda [CHILOSC$] Lophocoleaceae (B:H)
Chiloscyphus pallescens (Ehrh. ex Hoffm.) Dum. [CHILPAL] Lophocoleaceae (B:H)
Chiloscyphus polyanthos (L.) Corda [CHILPOL] Lophocoleaceae (B:H)
Chiloscyphus polyanthos var. **polyanthos** [CHILPOL0] Lophocoleaceae (B:H)
Chiloscyphus polyanthos var. **rivularis** (Schrad.) Nees [CHILPOL2] Lophocoleaceae (B:H)
Chiloscyphus rivularis (Schrad.) Loeske = Chiloscyphus polyanthos var. rivularis [CHILPOL2] Lophocoleaceae (B:H)
Chimaphila Pursh [CHIMAPH$] Pyrolaceae (V:D)
Chimaphila menziesii (R. Br. ex D. Don) Spreng. [CHIMMEN] Pyrolaceae (V:D)
Chimaphila occidentalis Rydb. = Chimaphila umbellata ssp. occidentalis [CHIMUMB1] Pyrolaceae (V:D)
Chimaphila umbellata (L.) W. Bart. [CHIMUMB] Pyrolaceae (V:D)
Chimaphila umbellata ssp. **occidentalis** (Rydb.) Hultén [CHIMUMB1] Pyrolaceae (V:D)
Chimaphila umbellata var. *occidentalis* (Rydb.) Blake = Chimaphila umbellata ssp. occidentalis [CHIMUMB1] Pyrolaceae (V:D)
Chiogenes hispidula (L.) Torr. & Gray = Gaultheria hispidula [GAULHIS] Ericaceae (V:D)
Chlorea vulpina Nyl. = Letharia vulpina [LETHVUL] Parmeliaceae (L)
Chlorocrepis albiflora (Hook.) W.A. Weber = Hieracium albiflorum [HIERALI] Asteraceae (V:D)
Chondrilla L. [CHONDRI$] Asteraceae (V:D)
*__Chondrilla juncea__ L. [CHONJUN] Asteraceae (V:D)
Chondrosum gracile Willd. ex Kunth = Bouteloua gracilis [BOUTGRA] Poaceae (V:M)
Chondrosum oligostachyum (Nutt.) Torr. = Bouteloua gracilis [BOUTGRA] Poaceae (V:M)
Chorispora DC. [CHORISP$] Brassicaceae (V:D)
*__Chorispora tenella__ (Pallas) DC. [CHORTEN] Brassicaceae (V:D)
Chrysanthemum arcticum auct. = Dendranthema arcticum ssp. polare [DENDARC1] Asteraceae (V:D)
Chrysanthemum arcticum var. *polare* (Hultén) Boivin = Dendranthema arcticum ssp. polare [DENDARC1] Asteraceae (V:D)

Chrysanthemum bipinnatum ssp. *huronense* (Nutt.) Hultén = Tanacetum bipinnatum ssp. huronense [TANABIP1] Asteraceae (V:D)
Chrysanthemum integrifolium Richards. = Leucanthemum integrifolium [LEUCINT] Asteraceae (V:D)
Chrysanthemum leucanthemum L. = Leucanthemum vulgare [LEUCVUL] Asteraceae (V:D)
Chrysanthemum leucanthemum var. *boecheri* Boivin = Leucanthemum vulgare [LEUCVUL] Asteraceae (V:D)
Chrysanthemum leucanthemum var. *pinnatifidum* Lecoq & Lamotte = Leucanthemum vulgare [LEUCVUL] Asteraceae (V:D)
Chrysanthemum parthenium (L.) Bernh. = Tanacetum parthenium [TANAPAR] Asteraceae (V:D)
Chrysanthemum uliginosum Pers. = Tanacetum vulgare [TANAVUL] Asteraceae (V:D)
Chrysanthemum vulgare (L.) Bernh. = Tanacetum vulgare [TANAVUL] Asteraceae (V:D)
Chryso-hypnum chrysophyllum (Brid.) Loeske = Campylium chrysophyllum [CAMPCHR] Amblystegiaceae (B:M)
Chryso-hypnum hispidulum (Brid.) Roth = Campylium hispidulum [CAMPHIS] Amblystegiaceae (B:M)
Chryso-hypnum polygamum (B.S.G.) Loeske = Campylium polygamum [CAMPPOL] Amblystegiaceae (B:M)
Chryso-hypnum stellatum (Hedw.) Loeske = Campylium stellatum [CAMPSTE] Amblystegiaceae (B:M)
Chrysopsis angustifolia Rydb. = Heterotheca villosa var. villosa [HETEVIL0] Asteraceae (V:D)
Chrysopsis arida A. Nels. = Heterotheca villosa var. hispida [HETEVIL1] Asteraceae (V:D)
Chrysopsis bakeri Greene = Heterotheca villosa var. villosa [HETEVIL0] Asteraceae (V:D)
Chrysopsis ballardii Rydb. = Heterotheca villosa var. villosa [HETEVIL0] Asteraceae (V:D)
Chrysopsis butleri Rydb. = Heterotheca villosa var. hispida [HETEVIL1] Asteraceae (V:D)
Chrysopsis canescens var. *nana* Gray = Heterotheca villosa var. hispida [HETEVIL1] Asteraceae (V:D)
Chrysopsis caudata Rydb. = Heterotheca villosa var. villosa [HETEVIL0] Asteraceae (V:D)
Chrysopsis columbiana Greene = Heterotheca villosa var. hispida [HETEVIL1] Asteraceae (V:D)
Chrysopsis foliosa (Nutt.) Shinners = Heterotheca villosa var. villosa [HETEVIL0] Asteraceae (V:D)
Chrysopsis hirsutissima Greene = Heterotheca villosa var. hispida [HETEVIL1] Asteraceae (V:D)
Chrysopsis hispida (Hook.) DC. = Heterotheca villosa var. hispida [HETEVIL1] Asteraceae (V:D)
Chrysopsis horrida Rydb. = Heterotheca villosa var. hispida [HETEVIL1] Asteraceae (V:D)
Chrysopsis imbricata A. Nels. = Heterotheca villosa var. villosa [HETEVIL0] Asteraceae (V:D)
Chrysopsis mollis Nutt. = Heterotheca villosa var. villosa [HETEVIL0] Asteraceae (V:D)
Chrysopsis nitidula Woot. & Standl. = Heterotheca villosa var. villosa [HETEVIL0] Asteraceae (V:D)
Chrysopsis pedunculata Greene = Heterotheca villosa var. villosa [HETEVIL0] Asteraceae (V:D)
Chrysopsis villosa (Pursh) Nutt. ex DC. = Heterotheca villosa var. villosa [HETEVIL0] Asteraceae (V:D)
Chrysopsis villosa var. *angustifolia* (Rydb.) Cronq. = Heterotheca villosa var. villosa [HETEVIL0] Asteraceae (V:D)
Chrysopsis villosa var. *foliosa* (Nutt.) D.C. Eat. = Heterotheca villosa var. villosa [HETEVIL0] Asteraceae (V:D)
Chrysopsis villosa var. *hispida* (Hook.) Gray = Heterotheca villosa var. hispida [HETEVIL1] Asteraceae (V:D)
Chrysopsis viscida ssp. *cinerascens* Blake = Heterotheca villosa var. hispida [HETEVIL1] Asteraceae (V:D)
Chrysopsis wisconsinensis Shinners = Heterotheca villosa var. hispida [HETEVIL1] Asteraceae (V:D)
Chrysosplenium L. [CHRYSOS$] Saxifragaceae (V:D)
Chrysosplenium alternifolium ssp. *iowense* (Rydb.) Hultén = Chrysosplenium iowense [CHRYIOW] Saxifragaceae (V:D)
Chrysosplenium alternifolium ssp. *tetrandrum* (Lund) Hultén = Chrysosplenium tetrandrum [CHRYTET] Saxifragaceae (V:D)
Chrysosplenium alternifolium var. *iowense* (Rydb.) Boivin = Chrysosplenium iowense [CHRYIOW] Saxifragaceae (V:D)
Chrysosplenium alternifolium var. *tetrandrum* Lund = Chrysosplenium tetrandrum [CHRYTET] Saxifragaceae (V:D)
Chrysosplenium iowense Rydb. [CHRYIOW] Saxifragaceae (V:D)
Chrysosplenium tetrandrum (Lund) Th. Fries [CHRYTET] Saxifragaceae (V:D)
Chrysosplenium wrightii Franch. & Savigny [CHRYWRI] Saxifragaceae (V:D)
Chrysothamnus Nutt. [CHRYSOT$] Asteraceae (V:D)
Chrysothamnus nauseosus (Pallas ex Pursh) Britt. [CHRYNAU] Asteraceae (V:D)
Chrysothamnus nauseosus ssp. **albicaulis** (Nutt.) Hall & Clements [CHRYNAU1] Asteraceae (V:D)
Chrysothamnus nauseosus ssp. *speciosus* (Nutt.) Hall & Clements = Chrysothamnus nauseosus ssp. albicaulis [CHRYNAU1] Asteraceae (V:D)
Chrysothamnus nauseosus var. *albicaulis* (Nutt.) Rydb. = Chrysothamnus nauseosus ssp. albicaulis [CHRYNAU1] Asteraceae (V:D)
Chrysothamnus nauseosus var. *speciosus* (Nutt.) Hall = Chrysothamnus nauseosus ssp. albicaulis [CHRYNAU1] Asteraceae (V:D)
Chrysothamnus viscidiflorus (Hook.) Nutt. [CHRYVIS] Asteraceae (V:D)
Chrysothamnus viscidiflorus ssp. *elegans* (Greene) Hall & Clements = Chrysothamnus viscidiflorus ssp. lanceolatus [CHRYVIS1] Asteraceae (V:D)
Chrysothamnus viscidiflorus ssp. **lanceolatus** (Nutt.) Hall & Clements [CHRYVIS1] Asteraceae (V:D)
Chrysothamnus viscidiflorus var. *elegans* (Greene) Blake = Chrysothamnus viscidiflorus ssp. lanceolatus [CHRYVIS1] Asteraceae (V:D)
Chrysothamnus viscidiflorus var. *lanceolatus* (Nutt.) Greene = Chrysothamnus viscidiflorus ssp. lanceolatus [CHRYVIS1] Asteraceae (V:D)
Chrysothrix Mont. [CHRYSOH$] Chrysotrichaceae (L)
Chrysothrix candelaris (L.) J.R. Laundon [CHRYCAN] Chrysotrichaceae (L)
Chrysothrix chlorina (Ach.) J.R. Laundon [CHRYCHL] Chrysotrichaceae (L)
Cicendia exaltata Griseb. = Centaurium exaltatum [CENTEXA] Gentianaceae (V:D)
Cichorium L. [CICHORI$] Asteraceae (V:D)
***Cichorium intybus** L. [CICHINT] Asteraceae (V:D)
Cicuta L. [CICUTA$] Apiaceae (V:D)
Cicuta bulbifera L. [CICUBUL] Apiaceae (V:D)
Cicuta douglasii (DC.) Coult. & Rose [CICUDOU] Apiaceae (V:D)
Cicuta mackenzieana Raup = Cicuta virosa [CICUVIR] Apiaceae (V:D)
Cicuta maculata L. [CICUMAU] Apiaceae (V:D)
Cicuta maculata var. **angustifolia** Hook. [CICUMAU1] Apiaceae (V:D)
Cicuta maculata var. *californica* (Gray) Boivin = Cicuta douglasii [CICUDOU] Apiaceae (V:D)
Cicuta occidentalis Greene = Cicuta maculata var. angustifolia [CICUMAU1] Apiaceae (V:D)
Cicuta virosa L. [CICUVIR] Apiaceae (V:D)
Ciliaria austromontana (Wieg.) W.A. Weber = Saxifraga bronchialis ssp. austromontana [SAXIBRO1] Saxifragaceae (V:D)
Ciliaria funstonii (Small) W.A. Weber = Saxifraga bronchialis ssp. funstonii [SAXIBRO2] Saxifragaceae (V:D)
Cimicifuga Wernischeck [CIMICIF$] Ranunculaceae (V:D)
Cimicifuga elata Nutt. [CIMIELA] Ranunculaceae (V:D)
Ciminalis prostrata (Haenke) A. & D. Löve = Gentiana prostrata [GENTPRO] Gentianaceae (V:D)
Cinclidium Sw. in Schrad. [CINCLID$] Mniaceae (B:M)
Cinclidium arcticum Bruch & Schimp. in B.S.G. [CINCARC] Mniaceae (B:M)
Cinna L. [CINNA$] Poaceae (V:M)
Cinna latifolia (Trev. ex Goepp.) Griseb. [CINNLAT] Poaceae (V:M)

Circaea L. [CIRCAEA$] Onagraceae (V:D)
Circaea alpina L. [CIRCALP] Onagraceae (V:D)
Circaea alpina ssp. **alpina** [CIRCALP0] Onagraceae (V:D)
Circaea alpina ssp. **pacifica** (Aschers. & Magnus) Raven [CIRCALP2] Onagraceae (V:D)
Circaea alpina var. *pacifica* (Aschers. & Magnus) M.E. Jones = Circaea alpina ssp. pacifica [CIRCALP2] Onagraceae (V:D)
Circaea pacifica Aschers. & Magnus = Circaea alpina ssp. pacifica [CIRCALP2] Onagraceae (V:D)
Cirriphyllum Grout [CIRRIPH$] Brachytheciaceae (B:M)
Cirriphyllum cirrosum (Schwaegr. in Schultes) Grout [CIRRCIR] Brachytheciaceae (B:M)
Cirriphyllum cirrosum var. *coloradense* (Aust.) Grout = Cirriphyllum cirrosum [CIRRCIR] Brachytheciaceae (B:M)
Cirriphyllum piliferum (Hedw.) Grout [CIRRPIL] Brachytheciaceae (B:M)
Cirsium P. Mill. [CIRSIUM$] Asteraceae (V:D)
•**Cirsium arvense** (L.) Scop. [CIRSARV] Asteraceae (V:D)
Cirsium arvense var. *argenteum* (Vest) Fiori = Cirsium arvense [CIRSARV] Asteraceae (V:D)
Cirsium arvense var. *horridum* Wimmer & Grab. = Cirsium arvense [CIRSARV] Asteraceae (V:D)
Cirsium arvense var. *integrifolium* Wimmer & Grab. = Cirsium arvense [CIRSARV] Asteraceae (V:D)
Cirsium arvense var. *mite* Wimmer & Grab. = Cirsium arvense [CIRSARV] Asteraceae (V:D)
Cirsium arvense var. *vestitum* Wimmer & Grab. = Cirsium arvense [CIRSARV] Asteraceae (V:D)
Cirsium brevistylum Cronq. [CIRSBRE] Asteraceae (V:D)
Cirsium coccinatum Osterhout = Cirsium drummondii [CIRSDRU] Asteraceae (V:D)
Cirsium drummondii Torr. & Gray [CIRSDRU] Asteraceae (V:D)
Cirsium edule Nutt. [CIRSEDU] Asteraceae (V:D)
Cirsium foliosum (Hook.) DC. [CIRSFOL] Asteraceae (V:D)
Cirsium foliosum var. *minganense* (Victorin) Boivin = Cirsium foliosum [CIRSFOL] Asteraceae (V:D)
Cirsium hookerianum Nutt. [CIRSHOO] Asteraceae (V:D)
Cirsium incanum (Gmel.) Fisch. = Cirsium arvense [CIRSARV] Asteraceae (V:D)
Cirsium lanceolatum (L.) Scop. = Cirsium vulgare [CIRSVUL] Asteraceae (V:D)
Cirsium lanceolatum var. *hypoleucum* DC. = Cirsium vulgare [CIRSVUL] Asteraceae (V:D)
Cirsium macounii (Greene) Rydb. = Cirsium edule [CIRSEDU] Asteraceae (V:D)
•**Cirsium palustre** (L.) Scop. [CIRSPAL] Asteraceae (V:D)
Cirsium scariosum Nutt. [CIRSSCA] Asteraceae (V:D)
Cirsium setosum (Willd.) Bess. ex Bieb. = Cirsium arvense [CIRSARV] Asteraceae (V:D)
Cirsium undulatum (Nutt.) Spreng. [CIRSUND] Asteraceae (V:D)
•**Cirsium vulgare** (Savi) Ten. [CIRSVUL] Asteraceae (V:D)
Cistanthe Spach [CISTANT$] Portulacaceae (V:D)
Cistanthe tweedyi (Gray) Hershkovitz [CISTTWE] Portulacaceae (V:D)
Cistanthe umbellata (Torr.) Hershkovitz [CISTUMB] Portulacaceae (V:D)
Cistanthe umbellata var. **caudicifera** (Gray) Kartesz & Gandhi [CISTUMB1] Portulacaceae (V:D)
Cladina Nyl. [CLADINA$] Cladoniaceae (L)
Cladina aberrans (Abbayes) Hale & Culb. = Cladina stellaris var. aberrans [CLADSTE1] Cladoniaceae (L)
Cladina alpestris (L.) Nyl. = Cladina stellaris [CLADSTE] Cladoniaceae (L)
Cladina arbuscula (Wallr.) Hale & Culb. [CLADARB] Cladoniaceae (L)
Cladina arbuscula ssp. **arbuscula** [CLADARB0] Cladoniaceae (L)
Cladina arbuscula ssp. *mitis* (Sandst.) Burgaz = Cladina mitis [CLADMIT] Cladoniaceae (L)
Cladina arbuscula ssp. **beringiana** (Ahti) N.S. Golubk. [CLADARB1] Cladoniaceae (L)
Cladina beringiana (Ahti) Trass = Cladina arbuscula ssp. beringiana [CLADARB1] Cladoniaceae (L)
Cladina ciliata (Stirton) Trass [CLADCIL] Cladoniaceae (L)
Cladina ciliata f. *tenuis* (Flörke) Ahti = Cladina ciliata var. tenuis [CLADCIL3] Cladoniaceae (L)
Cladina ciliata var. **ciliata** [CLADCIL0] Cladoniaceae (L)
Cladina ciliata var. **tenuis** (Flörke) Ahti & M.J. Lai [CLADCIL3] Cladoniaceae (L)
Cladina impexa de Lesd. = Cladina portentosa [CLADPOR] Cladoniaceae (L)
Cladina leucophaea (Abbayes) Hale & Culb. = Cladina ciliata var. ciliata [CLADCIL0] Cladoniaceae (L)
Cladina mitis (Sandst.) Hustich [CLADMIT] Cladoniaceae (L)
Cladina pacifica (Ahti) Hale & Culb. = Cladina portentosa ssp. pacifica [CLADPOR1] Cladoniaceae (L)
Cladina portentosa (Dufour) Follmann [CLADPOR] Cladoniaceae (L)
Cladina portentosa ssp. **pacifica** f. **pacifica** [CLADPOR3] Cladoniaceae (L)
Cladina portentosa ssp. **pacifica** (Ahti) Ahti [CLADPOR1] Cladoniaceae (L)
Cladina portentosa ssp. **pacifica** f. **decolorans** (Ahti) Ahti [CLADPOR2] Cladoniaceae (L)
Cladina rangiferina (L.) Nyl. [CLADRAN] Cladoniaceae (L)
Cladina rangiferina ssp. **grisea** (Ahti) Ahti & M.J. Lai [CLADRAN1] Cladoniaceae (L)
Cladina stellaris (Opiz) Brodo [CLADSTE] Cladoniaceae (L)
Cladina stellaris var. **aberrans** (Abbayes) Ahti [CLADSTE1] Cladoniaceae (L)
Cladina stellaris var. **stellaris** [CLADSTE0] Cladoniaceae (L)
Cladina stygia (Fr.) Ahti [CLADSTY] Cladoniaceae (L)
Cladina tenuis (Flörke) Hale & Culb. = Cladina ciliata var. tenuis [CLADCIL3] Cladoniaceae (L)
Cladonia P. Browne [CLADONI$] Cladoniaceae (L)
Cladonia aberrans (Abbayes) Stuckenb. = Cladina stellaris var. aberrans [CLADSTE1] Cladoniaceae (L)
Cladonia acuminata (Ach.) Norrlin [CLADACU] Cladoniaceae (L)
Cladonia albonigra Brodo & Ahti [CLADALB] Cladoniaceae (L)
Cladonia alpestris (L.) Rabenh. = Cladina stellaris [CLADSTE] Cladoniaceae (L)
Cladonia alpestris f. *aberrans* Abbayes = Cladina stellaris var. aberrans [CLADSTE1] Cladoniaceae (L)
Cladonia alpestris var. *aberrans* (Abbayes) Ahti = Cladina stellaris var. aberrans [CLADSTE1] Cladoniaceae (L)
Cladonia alpicola (Flotow) Vainio = Cladonia macrophylla [CLADMAR] Cladoniaceae (L)
Cladonia amaurocraea (Flörke) Schaerer [CLADAMA] Cladoniaceae (L)
Cladonia arbuscula (Wallr.) Flotow = Cladina arbuscula [CLADARB] Cladoniaceae (L)
Cladonia arbuscula ssp. *beringiana* Ahti = Cladina arbuscula ssp. beringiana [CLADARB1] Cladoniaceae (L)
Cladonia arbuscula ssp. *mitis* (Sandst.) Ruoss = Cladina mitis [CLADMIT] Cladoniaceae (L)
Cladonia artuata S. Hammer [CLADART] Cladoniaceae (L)
Cladonia asahinae J.W. Thomson [CLADASA] Cladoniaceae (L)
Cladonia bacillaris Nyl. = Cladonia macilenta var. bacillaris [CLADMAC1] Cladoniaceae (L)
Cladonia bacilliformis (Nyl.) Glück [CLADBAI] Cladoniaceae (L)
Cladonia balfourii Crombie = Cladonia macilenta [CLADMAC] Cladoniaceae (L)
Cladonia bellidiflora (Ach.) Schaerer [CLADBEL] Cladoniaceae (L)
Cladonia bellidiflora var. *hookeri* (Tuck.) Nyl. = Cladonia bellidiflora [CLADBEL] Cladoniaceae (L)
Cladonia blakei Robbins = Cladonia coccifera [CLADCOC] Cladoniaceae (L)
Cladonia borealis S. Stenroos [CLADBOR] Cladoniaceae (L)
Cladonia botrytes (K. Hagen) Willd. [CLADBOY] Cladoniaceae (L)
Cladonia caespiticia (Pers.) Flörke [CLADCAE] Cladoniaceae (L)
Cladonia carassensis ssp. *japonica* (Vainio) Asah. = Cladonia crispata [CLADCRI] Cladoniaceae (L)
Cladonia cariosa (Ach.) Sprengel [CLADCAI] Cladoniaceae (L)
Cladonia carneola (Fr.) Fr. [CLADCAN] Cladoniaceae (L)
Cladonia cenotea (Ach.) Schaerer [CLADCEN] Cladoniaceae (L)
Cladonia cenotea var. *crossota* (Ach.) Nyl. = Cladonia cenotea [CLADCEN] Cladoniaceae (L)

Cladonia cervicornis (Ach.) Flotow [CLADCER] Cladoniaceae (L)
Cladonia cervicornis ssp. **cervicornis** [CLADCER0] Cladoniaceae (L)
Cladonia cervicornis ssp. **verticillata** (Hoffm.) Ahti [CLADCER1] Cladoniaceae (L)
Cladonia chlorophaea (Flörke ex Sommerf.) Sprengel [CLADCHL] Cladoniaceae (L)
Cladonia ciliata Stirton = Cladina ciliata [CLADCIL] Cladoniaceae (L)
Cladonia coccifera (L.) Willd. [CLADCOC] Cladoniaceae (L)
Cladonia coccifera var. *pleurota* (Flörke) Schaerer = Cladonia pleurota [CLADPLE] Cladoniaceae (L)
Cladonia coniocraea (Flörke) Sprengel [CLADCON] Cladoniaceae (L)
Cladonia conista A. Evans [CLADCOI] Cladoniaceae (L)
Cladonia conistea auct. = Cladonia humilis [CLADHUM] Cladoniaceae (L)
Cladonia conoidea Ahti = Cladonia humilis [CLADHUM] Cladoniaceae (L)
Cladonia cornucopioides (L.) Fr. = Cladonia coccifera [CLADCOC] Cladoniaceae (L)
Cladonia cornuta (L.) Hoffm. [CLADCOR] Cladoniaceae (L)
Cladonia cornuta ssp. **cornuta** [CLADCOR0] Cladoniaceae (L)
Cladonia cornuta ssp. **groenlandica** (E. Dahl) Ahti [CLADCOR1] Cladoniaceae (L)
Cladonia cornuta var. *groenlandica* E. Dahl = Cladonia cornuta ssp. groenlandica [CLADCOR1] Cladoniaceae (L)
Cladonia cornutoradiata (Navás) Sandst. = Cladonia subulata [CLADSUU] Cladoniaceae (L)
Cladonia crispata (Ach.) Flotow [CLADCRI] Cladoniaceae (L)
Cladonia crispata var. **cetrariiformis** (Delise) Vainio [CLADCRI2] Cladoniaceae (L)
Cladonia crispata var. **crispata** [CLADCRI0] Cladoniaceae (L)
Cladonia cryptochlorophaea Asah. [CLADCRY] Cladoniaceae (L)
Cladonia cyanipes (Sommerf.) Nyl. [CLADCYA] Cladoniaceae (L)
Cladonia cyathomorpha (A. Evans) A. Evans [CLADCYT] Cladoniaceae (L)
Cladonia decorticata (Flörke) Sprengel [CLADDEC] Cladoniaceae (L)
Cladonia deformis (L.) Hoffm. [CLADDEF] Cladoniaceae (L)
Cladonia degenerans (Flörke) Sprengel = Cladonia phyllophora [CLADPHY] Cladoniaceae (L)
Cladonia delessertii Vainio = Cladonia subfurcata [CLADSUF] Cladoniaceae (L)
Cladonia delicata auct. = Cladonia parasitica [CLADPAR] Cladoniaceae (L)
Cladonia digitata (L.) Hoffm. [CLADDIG] Cladoniaceae (L)
Cladonia dimorpha S. Hammer [CLADDIM] Cladoniaceae (L)
Cladonia ecmocyna Leighton [CLADECM] Cladoniaceae (L)
Cladonia ecmocyna ssp. **occidentalis** Ahti [CLADECM3] Cladoniaceae (L)
Cladonia ecmocyna ssp. **ecmocyna** [CLADECM0] Cladoniaceae (L)
Cladonia ecmocyna ssp. **intermedia** (Robbins) Ahti [CLADECM1] Cladoniaceae (L)
Cladonia ecmocyna var. *intermedia* (Robbins) A. Evans = Cladonia ecmocyna ssp. intermedia [CLADECM1] Cladoniaceae (L)
Cladonia ecmocyna var. *macroceras* (Flörke) Ach. = Cladonia macroceras [CLADMAE] Cladoniaceae (L)
Cladonia elongata (Jacq.) Hoffm. = Cladonia maxima [CLADMAX] Cladoniaceae (L)
Cladonia elongata auct. = Cladonia macroceras [CLADMAE] Cladoniaceae (L)
Cladonia fimbriata (L.) Fr. [CLADFIM] Cladoniaceae (L)
Cladonia fimbriata var. *costata* Flörke = Cladonia fimbriata [CLADFIM] Cladoniaceae (L)
Cladonia fimbriata var. *prolifera* (Retz.) Mass. = Cladonia fimbriata [CLADFIM] Cladoniaceae (L)
Cladonia fimbriata var. *simplex* f. *major* (K. Hagen) Vainio = Cladonia fimbriata [CLADFIM] Cladoniaceae (L)
Cladonia fimbriata var. *simplex* f. *minor* (K. Hagen) Vainio = Cladonia fimbriata [CLADFIM] Cladoniaceae (L)
Cladonia fimbriata var. *subulata* Schaerer = Cladonia subulata [CLADSUU] Cladoniaceae (L)
Cladonia fimbriata var. *tubaeformis* Fr. = Cladonia fimbriata [CLADFIM] Cladoniaceae (L)
Cladonia flabelliformis Vainio = Cladonia polydactyla [CLADPOD] Cladoniaceae (L)
Cladonia furcata (Hudson) Schrader [CLADFUR] Cladoniaceae (L)
Cladonia furcata ssp. *subrangiformis* auct. = Cladonia furcata [CLADFUR] Cladoniaceae (L)
Cladonia furcata var. *conspersa* Vainio = Cladonia furcata [CLADFUR] Cladoniaceae (L)
Cladonia furcata var. *crispata* Flörke = Cladonia crispata [CLADCRI] Cladoniaceae (L)
Cladonia furcata var. *pinnata* (Flörke) Vainio = Cladonia furcata [CLADFUR] Cladoniaceae (L)
Cladonia furcata var. *racemosa* Flörke = Cladonia furcata [CLADFUR] Cladoniaceae (L)
Cladonia furcata var. *subulata* Flörke = Cladonia furcata [CLADFUR] Cladoniaceae (L)
Cladonia gonecha (Ach.) Asah. = Cladonia sulphurina [CLADSUL] Cladoniaceae (L)
Cladonia gracilescens auct. = Cladonia stricta [CLADSTI] Cladoniaceae (L)
Cladonia gracilis (L.) Willd. [CLADGRA] Cladoniaceae (L)
Cladonia gracilis f. *anthocephala* Flörke = Cladonia gracilis [CLADGRA] Cladoniaceae (L)
Cladonia gracilis ssp. *dilatata* (Hoffm.) Vainio = Cladonia gracilis ssp. turbinata [CLADGRA2] Cladoniaceae (L)
Cladonia gracilis ssp. **elongata** (Jacq.) Vainio [CLADGRA1] Cladoniaceae (L)
Cladonia gracilis ssp. *nigripes* (Nyl.) Ahti = Cladonia gracilis ssp. elongata [CLADGRA1] Cladoniaceae (L)
Cladonia gracilis ssp. **turbinata** (Ach.) Ahti [CLADGRA2] Cladoniaceae (L)
Cladonia gracilis ssp. **vulnerata** Ahti [CLADGRA3] Cladoniaceae (L)
Cladonia gracilis var. *elongata* Fr. = Cladonia maxima [CLADMAX] Cladoniaceae (L)
Cladonia gracilis var. *verticillata* Fr. = Cladonia cervicornis ssp. verticillata [CLADCER1] Cladoniaceae (L)
Cladonia granulans Vainio [CLADGRN] Cladoniaceae (L)
Cladonia grayi G. Merr. ex Sandst. [CLADGRY] Cladoniaceae (L)
Cladonia groenlandica (E. Dahl) Trass = Cladonia cornuta ssp. groenlandica [CLADCOR1] Cladoniaceae (L)
Cladonia herrei Fink ex J. Hedrick = Cladonia furcata [CLADFUR] Cladoniaceae (L)
Cladonia homosekikaica Nuno [CLADHOM] Cladoniaceae (L)
Cladonia hookeri Tuck. = Cladonia bellidiflora [CLADBEL] Cladoniaceae (L)
Cladonia humilis (With.) J.R. Laundon [CLADHUM] Cladoniaceae (L)
Cladonia humilis var. *bourgeanica* A.W. Archer = Cladonia conista [CLADCOI] Cladoniaceae (L)
Cladonia impexa Harm. = Cladina portentosa [CLADPOR] Cladoniaceae (L)
Cladonia invisa Robbins = Cladonia ochrochlora [CLADOCH] Cladoniaceae (L)
Cladonia japonica Vainio = Cladonia crispata [CLADCRI] Cladoniaceae (L)
Cladonia kanewskii Oksner [CLADKAN] Cladoniaceae (L)
Cladonia lepidota auct. = Cladonia stricta [CLADSTI] Cladoniaceae (L)
Cladonia leucophaea Abbayes = Cladina ciliata var. ciliata [CLADCIL0] Cladoniaceae (L)
Cladonia luteoalba Wheldon & A. Wilson [CLADLUT] Cladoniaceae (L)
Cladonia macilenta Hoffm. [CLADMAC] Cladoniaceae (L)
Cladonia macilenta var. **bacillaris** (Genth) Schaerer [CLADMAC1] Cladoniaceae (L)
Cladonia macroceras (Delise) Hav. [CLADMAE] Cladoniaceae (L)
Cladonia macrophylla (Schaerer) Stenh. [CLADMAR] Cladoniaceae (L)
Cladonia macrophyllodes Nyl. [CLADMAO] Cladoniaceae (L)
Cladonia macroptera Räsänen = Cladonia scabriuscula [CLADSCA] Cladoniaceae (L)

Cladonia major (K. Hagen) Sandst. = Cladonia fimbriata [CLADFIM] Cladoniaceae (L)
Cladonia maxima (Asah.) Ahti [CLADMAX] Cladoniaceae (L)
Cladonia merochlorophaea Asah. [CLADMER] Cladoniaceae (L)
Cladonia metacorallifera Asah. [CLADMET] Cladoniaceae (L)
Cladonia mitis Sandst. = Cladina mitis [CLADMIT] Cladoniaceae (L)
Cladonia multiformis G. Merr. [CLADMUL] Cladoniaceae (L)
Cladonia nemoxyna (Ach.) Arnold = Cladonia rei [CLADREI] Cladoniaceae (L)
Cladonia nipponica Asah. [CLADNIP] Cladoniaceae (L)
Cladonia norvegica Tønsberg & Holien [CLADNOR] Cladoniaceae (L)
Cladonia novochlorophaea (Sipman) Brodo & Ahti [CLADNOV] Cladoniaceae (L)
Cladonia ochrochlora Flörke [CLADOCH] Cladoniaceae (L)
Cladonia pacifica Ahti = Cladina portentosa ssp. pacifica [CLADPOR1] Cladoniaceae (L)
Cladonia parasitica (Hoffm.) Hoffm. [CLADPAR] Cladoniaceae (L)
Cladonia phyllophora Hoffm. [CLADPHY] Cladoniaceae (L)
Cladonia pleurota (Flörke) Schaerer [CLADPLE] Cladoniaceae (L)
Cladonia pocillum (Ach.) Grognot [CLADPOC] Cladoniaceae (L)
Cladonia polydactyla (Flörke) Sprengel [CLADPOD] Cladoniaceae (L)
Cladonia poroscypha S. Hammer [CLADPOO] Cladoniaceae (L)
Cladonia portentosa ssp. *pacifica* (Ahti) Ahti = Cladina portentosa ssp. pacifica [CLADPOR1] Cladoniaceae (L)
Cladonia portentosa ssp. *pacifica* f. *decolorans* Ahti = Cladina portentosa ssp. pacifica f. decolorans [CLADPOR2] Cladoniaceae (L)
Cladonia prolifica Ahti & S. Hammer [CLADPRO] Cladoniaceae (L)
Cladonia pseudodigitata Gyelnik = Cladonia coccifera [CLADCOC] Cladoniaceae (L)
Cladonia pseudorangiformis Asah. = Cladonia wainioi [CLADWAI] Cladoniaceae (L)
Cladonia pseudostellata Asah. = Cladonia uncialis [CLADUNC] Cladoniaceae (L)
Cladonia pyxidata (L.) Hoffm. [CLADPYX] Cladoniaceae (L)
Cladonia pyxidata var. *neglecta* (Flörke) Mass. = Cladonia pyxidata [CLADPYX] Cladoniaceae (L)
Cladonia pyxidata var. *pocillum* (Ach.) Flotow = Cladonia pocillum [CLADPOC] Cladoniaceae (L)
Cladonia rangiferina (L.) F.H. Wigg. = Cladina rangiferina [CLADRAN] Cladoniaceae (L)
Cladonia rangiferina f. *nivea* Räsänen = Cladina rangiferina [CLADRAN] Cladoniaceae (L)
Cladonia rangiferina ssp. *grisea* Ahti = Cladina rangiferina ssp. grisea [CLADRAN1] Cladoniaceae (L)
Cladonia rangiferina var. *alpestris* Flörke = Cladina stellaris [CLADSTE] Cladoniaceae (L)
Cladonia rangiferina var. *sylvatica* L. = Cladina arbuscula [CLADARB] Cladoniaceae (L)
Cladonia rei Schaerer [CLADREI] Cladoniaceae (L)
Cladonia scabriuscula (Delise) Nyl. [CLADSCA] Cladoniaceae (L)
Cladonia silvatica (L.) Räsänen = Cladina arbuscula [CLADARB] Cladoniaceae (L)
Cladonia silvatica var. *mitis* (Sandst.)Räsänen = Cladina mitis [CLADMIT] Cladoniaceae (L)
Cladonia singularis S. Hammer [CLADSIN] Cladoniaceae (L)
Cladonia squamosa Hoffm. [CLADSQU] Cladoniaceae (L)
Cladonia squamosa var. *muricella* (Del.) Vainio = Cladonia squamosa var. squamosa [CLADSQU0] Cladoniaceae (L)
Cladonia squamosa var. **squamosa** [CLADSQU0] Cladoniaceae (L)
Cladonia squamosa var. **subsquamosa** (Nyl. ex Leighton) Vainio [CLADSQU1] Cladoniaceae (L)
Cladonia stellaris (Opiz) Pouzar & Vezda = Cladina stellaris [CLADSTE] Cladoniaceae (L)
Cladonia stricta (Nyl.) Nyl. [CLADSTI] Cladoniaceae (L)
Cladonia stygia (Fr.) Ruoss = Cladina stygia [CLADSTY] Cladoniaceae (L)
Cladonia subfurcata (Nyl.) Arnold [CLADSUF] Cladoniaceae (L)
Cladonia subrangiformis auct. = Cladonia furcata [CLADFUR] Cladoniaceae (L)
Cladonia subsquamosa (Nyl. ex Leighton) Crombie = Cladonia squamosa var. subsquamosa [CLADSQU1] Cladoniaceae (L)
Cladonia subulata (L.) F.H. Wigg. [CLADSUU] Cladoniaceae (L)
Cladonia sulphurina (Michaux) Fr. [CLADSUL] Cladoniaceae (L)
Cladonia sylvatica L. = Cladina arbuscula [CLADARB] Cladoniaceae (L)
Cladonia sylvatica auct. = Cladina arbuscula [CLADARB] Cladoniaceae (L)
Cladonia sylvatica var. *alpestris* L. = Cladina stellaris [CLADSTE] Cladoniaceae (L)
Cladonia symphycarpa (Flörke) Fr. = Cladonia symphycarpon [CLADSYM] Cladoniaceae (L)
Cladonia symphycarpon (Flörke) Fr. [CLADSYM] Cladoniaceae (L)
Cladonia tenuis (Flörke) Harm. = Cladina ciliata [CLADCIL] Cladoniaceae (L)
Cladonia tenuis var. *leucophaea* (Abbayes) Ahti = Cladina ciliata var. ciliata [CLADCIL0] Cladoniaceae (L)
Cladonia thomsonii Ahti [CLADTHO] Cladoniaceae (L)
Cladonia transcendens (Vainio) Vainio [CLADTRA] Cladoniaceae (L)
Cladonia umbricola Tønsberg & Ahti [CLADUMB] Cladoniaceae (L)
Cladonia uncialis (L.) F.H. Wigg. [CLADUNC] Cladoniaceae (L)
Cladonia uncialis var. *adunca* Fr. = Cladonia uncialis [CLADUNC] Cladoniaceae (L)
Cladonia verruculosa (Vainio) Ahti [CLADVER] Cladoniaceae (L)
Cladonia verticillata (Hoffm.) Schaerer = Cladonia cervicornis ssp. verticillata [CLADCER1] Cladoniaceae (L)
Cladonia verticillata var. *cervicornis* Flörke = Cladonia cervicornis ssp. cervicornis [CLADCER0] Cladoniaceae (L)
Cladonia wainii Savicz = Cladonia wainioi [CLADWAI] Cladoniaceae (L)
Cladonia wainioi Savicz [CLADWAI] Cladoniaceae (L)
Cladopodiella Buch [CLADOPO$] Cephaloziaceae (B:H)
Cladopodiella fluitans (Nees) Joerg. [CLADFLU] Cephaloziaceae (B:H)
Cladothamnus pyroliflorus Bong. = Elliottia pyroliflorus [ELLIPYR] Ericaceae (V:D)
Claopodium (Lesq. & James) Ren. & Card. [CLAOPOD$] Leskeaceae (B:M)
Claopodium bolanderi Best [CLAOBOL] Leskeaceae (B:M)
Claopodium crispifolium (Hook.) Ren. & Card. [CLAOCRI] Leskeaceae (B:M)
Claopodium pellucinerve (Mitt.) Best [CLAOPEL] Leskeaceae (B:M)
Claopodium subpiliferum (Lindb. & Arnell) Broth. = Claopodium pellucinerve [CLAOPEL] Leskeaceae (B:M)
Claopodium whippleanum (Sull. in Whipple & Ives) Ren. & Card. [CLAOWHI] Leskeaceae (B:M)
Claopodium whippleanum var. *leuconeuron* (Sull. & Lesq.) Grout = Claopodium whippleanum [CLAOWHI] Leskeaceae (B:M)
Clarkia Pursh [CLARKIA$] Onagraceae (V:D)
Clarkia amoena (Lehm.) A. Nels. & J.F. Macbr. [CLARAMO] Onagraceae (V:D)
Clarkia amoena ssp. **caurina** (Abrams ex Piper) H.F. & M.E. Lewis [CLARAMO1] Onagraceae (V:D)
Clarkia amoena ssp. **lindleyi** (Dougl.) H.F. & M.E. Lewis [CLARAMO2] Onagraceae (V:D)
Clarkia amoena var. *caurina* (Abrams ex Piper) C.L. Hitchc. = Clarkia amoena ssp. caurina [CLARAMO1] Onagraceae (V:D)
Clarkia amoena var. *lindleyi* (Dougl.) C.L. Hitchc. = Clarkia amoena ssp. lindleyi [CLARAMO2] Onagraceae (V:D)
Clarkia amoena var. *pacifica* (M.E. Peck) C.L. Hitchc. = Clarkia amoena ssp. caurina [CLARAMO1] Onagraceae (V:D)
Clarkia pulchella Pursh [CLARPUL] Onagraceae (V:D)
Clarkia rhomboidea Dougl. ex Hook. [CLARRHO] Onagraceae (V:D)
Clauzadea Hafellner & Bellem. [CLAUZAD$] Porpidiaceae (L)
Clauzadea monticola (Ach. ex Schaerer) Hafellner & Bellem. [CLAUMON] Porpidiaceae (L)
Claytonia L. [CLAYTON$] Portulacaceae (V:D)
Claytonia asarifolia auct. = Claytonia cordifolia [CLAYCOR] Portulacaceae (V:D)

Claytonia bostockii Porsild = Montia bostockii [MONTBOS] Portulacaceae (V:D)
Claytonia caroliniana var. *lanceolata* (Pursh) S. Wats. = Claytonia lanceolata [CLAYLAN] Portulacaceae (V:D)
Claytonia caroliniana var. *tuberosa* (Pallas ex J.A. Schultes) Boivin = Claytonia tuberosa [CLAYTUB] Portulacaceae (V:D)
Claytonia chamissoi Ledeb. ex Spreng. = Montia chamissoi [MONTCHA] Portulacaceae (V:D)
Claytonia cordifolia S. Wats. [CLAYCOR] Portulacaceae (V:D)
Claytonia dichotoma Nutt. = Montia dichotoma [MONTDIC] Portulacaceae (V:D)
Claytonia diffusa Nutt. = Montia diffusa [MONTDIF] Portulacaceae (V:D)
Claytonia flagellaris Bong. = Montia parvifolia ssp. flagellaris [MONTPAR1] Portulacaceae (V:D)
Claytonia heterophylla (Torr. & Gray) Swanson [CLAYHET] Portulacaceae (V:D)
Claytonia howellii (S. Wats.) Piper = Montia howellii [MONTHOW] Portulacaceae (V:D)
Claytonia lanceolata Pursh [CLAYLAN] Portulacaceae (V:D)
Claytonia lanceolata var. **pacifica** McNeill [CLAYLAN2] Portulacaceae (V:D)
Claytonia linearis Dougl. ex Hook. = Montia linearis [MONTLIN] Portulacaceae (V:D)
Claytonia megarhiza (Gray) Parry ex S. Wats. [CLAYMEG] Portulacaceae (V:D)
Claytonia parviflora Dougl. ex Hook. [CLAYPAR] Portulacaceae (V:D)
Claytonia parviflora var. *depressa* Gray = Claytonia rubra [CLAYRUB] Portulacaceae (V:D)
Claytonia parviflora var. *glauca* Nutt. ex Torr. & Gray = Claytonia rubra [CLAYRUB] Portulacaceae (V:D)
Claytonia parvifolia Moc. ex DC. = Montia parvifolia ssp. parvifolia [MONTPAR0] Portulacaceae (V:D)
Claytonia parvifolia ssp. *flagellaris* (Bong.) Hultén = Montia parvifolia ssp. flagellaris [MONTPAR1] Portulacaceae (V:D)
Claytonia parvifolia var. *flagellaris* (Bong.) R.J. Davis = Montia parvifolia ssp. flagellaris [MONTPAR1] Portulacaceae (V:D)
Claytonia perfoliata Donn ex Willd. [CLAYPER] Portulacaceae (V:D)
Claytonia perfoliata var. *depressa* (Gray) Poelln. = Claytonia rubra [CLAYRUB] Portulacaceae (V:D)
Claytonia perfoliata var. *parviflora* (Dougl. ex Hook.) Torr. = Claytonia parviflora [CLAYPAR] Portulacaceae (V:D)
Claytonia rubra (T.J. Howell) Tidestrom [CLAYRUB] Portulacaceae (V:D)
Claytonia sarmentosa C.A. Mey. [CLAYSAR] Portulacaceae (V:D)
Claytonia sibirica L. [CLAYSIB] Portulacaceae (V:D)
Claytonia sibirica var. *cordifolia* (S. Wats.) R.J. Davis = Claytonia cordifolia [CLAYCOR] Portulacaceae (V:D)
Claytonia sibirica var. *heterophylla* (Torr. & Gray) Gray = Claytonia heterophylla [CLAYHET] Portulacaceae (V:D)
Claytonia spathulata Dougl. ex Hook. [CLAYSPA] Portulacaceae (V:D)
Claytonia triphylla S. Wats. = Lewisia triphylla [LEWITRI] Portulacaceae (V:D)
Claytonia tuberosa Pallas ex J.A. Schultes [CLAYTUB] Portulacaceae (V:D)
Clematis L. [CLEMATI$] Ranunculaceae (V:D)
Clematis columbiana (Nutt.) Torr. & Gray [CLEMCOL] Ranunculaceae (V:D)
Clematis ligusticifolia Nutt. [CLEMLIG] Ranunculaceae (V:D)
Clematis occidentalis (Hornem.) DC. [CLEMOCC] Ranunculaceae (V:D)
Clematis occidentalis ssp. *grosseserrata* (Rydb.) Taylor & MacBryde = Clematis occidentalis var. grosseserrata [CLEMOCC2] Ranunculaceae (V:D)
Clematis occidentalis var. **grosseserrata** (Rydb.) J. Pringle [CLEMOCC2] Ranunculaceae (V:D)
Clematis orientalis var. *tangutica* Maxim. = Clematis tangutica [CLEMTAN] Ranunculaceae (V:D)
*****Clematis tangutica** (Maxim.) Korsh. [CLEMTAN] Ranunculaceae (V:D)
Clematis verticillaris var. *columbiana* (Nutt.) Gray = Clematis columbiana [CLEMCOL] Ranunculaceae (V:D)
*****Clematis vitalba** L. [CLEMVIT] Ranunculaceae (V:D)
Cleome L. [CLEOME$] Capparaceae (V:D)
Cleome serrulata Pursh [CLEOSER] Capparaceae (V:D)
Cleome serrulata var. *angusta* (M.E. Jones) Tidestrom = Cleome serrulata [CLEOSER] Capparaceae (V:D)
Clevea hyalina (Somm.) Lindb. = Athalamia hyalina [ATHAHYA] Cleveaceae (B:H)
Climacium Web. & Mohr [CLIMACI$] Climaciaceae (B:M)
Climacium dendroides (Hedw.) Web. & Mohr [CLIMDEN] Climaciaceae (B:M)
Climacium dendroides var. *oregonensis* Ren. & Card. = Climacium dendroides [CLIMDEN] Climaciaceae (B:M)
Clinopodium L. [CLINOPO$] Lamiaceae (V:D)
Clinopodium acinos (L.) Kuntze = Acinos arvensis [ACINARV] Lamiaceae (V:D)
*****Clinopodium vulgare** L. [CLINVUL] Lamiaceae (V:D)
Clinopodium vulgare var. *neogaea* (Fern.) C.F. Reed = Clinopodium vulgare [CLINVUL] Lamiaceae (V:D)
Clintonia Raf. [CLINTON$] Liliaceae (V:M)
Clintonia uniflora (Menzies ex J.A. & J.H. Schultes) Kunth [CLINUNI] Liliaceae (V:M)
Cliostomum Fr. [CLIOSTO$] Bacidiaceae (L)
Cliostomum corrugatum (Ach.) Fr. [CLIOCOR] Bacidiaceae (L)
Cliostomum graniforme (K. Hagen) Coppins = Cliostomum corrugatum [CLIOCOR] Bacidiaceae (L)
Cliostomum griffithii (Sm.) Coppins [CLIOGRI] Bacidiaceae (L)
Cliostomum leprosum (Räsänen) Holien & Tønsberg [CLIOLEP] Bacidiaceae (L)
Cliostomum luteolum Gowan = Cliostomum leprosum [CLIOLEP] Bacidiaceae (L)
Clypeococcum D. Hawksw. [CLYPEOC$] Dacampiaceae (L)
Clypeococcum grossum (Körber) D. Hawksw. [CLYPGRO] Dacampiaceae (L)
Clypeococcum hypocenomycis D. Hawksw. [CLYPHYP] Dacampiaceae (L)
Clypeola alyssoides L. = Alyssum alyssoides [ALYSALY] Brassicaceae (V:D)
Clypeola maritima L. = Lobularia maritima [LOBUMAR] Brassicaceae (V:D)
Cnestrum alpestre (Wahlenb.) Nyh. ex Mog. = Cynodontium alpestre [CYNOALP] Dicranaceae (B:M)
Cnestrum glaucescens (Lindb. & Arnell) Holm. ex Mog. & Steere = Cynodontium glaucescens [CYNOGLA] Dicranaceae (B:M)
Cnestrum schisti (Web. & Mohr) Hag. = Cynodontium schisti [CYNOSCH] Dicranaceae (B:M)
Cnidium Cusson ex Juss. [CNIDIUM$] Apiaceae (V:D)
Cnidium cnidiifolium (Turcz.) Schischkin [CNIDCNI] Apiaceae (V:D)
Coccotrema Müll. Arg. [COCCOTR$] Coccotremataceae (L)
Coccotrema maritimum Brodo [COCCMAR] Coccotremataceae (L)
Coccotrema pocillarium (Cummings) Brodo [COCCPOC] Coccotremataceae (L)
Cochlearia L. [COCHLEA$] Brassicaceae (V:D)
Cochlearia armoracia L. = Armoracia rusticana [ARMORUS] Brassicaceae (V:D)
Cochlearia groenlandica L. [COCHGRO] Brassicaceae (V:D)
Cochlearia officinalis ssp. *arctica* (Schlecht.) Hultén = Cochlearia groenlandica [COCHGRO] Brassicaceae (V:D)
Cochlearia officinalis ssp. *groenlandica* (L.) Porsild = Cochlearia groenlandica [COCHGRO] Brassicaceae (V:D)
Cochlearia officinalis ssp. *oblongifolia* (DC.) Hultén = Cochlearia groenlandica [COCHGRO] Brassicaceae (V:D)
Cochlearia officinalis var. *arctica* (Schlecht.) Gelert ex Anders. & Hessel = Cochlearia groenlandica [COCHGRO] Brassicaceae (V:D)
Cochleariopsis groenlandica (L.) A. & D. Löve = Cochlearia groenlandica [COCHGRO] Brassicaceae (V:D)
Cochleariopsis groenlandica ssp. *arctica* (Schlecht.) A. & D. Löve = Cochlearia groenlandica [COCHGRO] Brassicaceae (V:D)
Cochleariopsis groenlandica ssp. *oblongifolia* (DC.) A. & D. Löve = Cochlearia groenlandica [COCHGRO] Brassicaceae (V:D)

Coelocaulon aculeatum (Schreber) Link = Cetraria aculeata [CETRACU] Parmeliaceae (L)
Coelocaulon cetrariza (Nyl.) Gyelnik = Kaernefeltia californica [KAERCAL] Parmeliaceae (L)
Coelocaulon divergens (Ach.) R. Howe = Bryocaulon divergens [BRYODIV] Parmeliaceae (L)
Coelocaulon muricatum (Ach.) J.R. Laundon = Cetraria muricata [CETRMUR] Parmeliaceae (L)
Coeloglossum Hartman [COELOGL$] Orchidaceae (V:M)
Coeloglossum bracteatum (Muhl. ex Willd.) Parl. = Coeloglossum viride var. virescens [COELVIR1] Orchidaceae (V:M)
Coeloglossum viride (L.) Hartman [COELVIR] Orchidaceae (V:M)
Coeloglossum viride ssp. *bracteatum* (Muhl. ex Willd.) Hultén = Coeloglossum viride var. virescens [COELVIR1] Orchidaceae (V:M)
Coeloglossum viride var. **virescens** (Muhl. ex Willd.) Luer [COELVIR1] Orchidaceae (V:M)
Coelopleurum actaeifolium (Michx.) Coult. & Rose = Angelica lucida [ANGELUC] Apiaceae (V:D)
Coelopleurum gmelinii (DC.) Ledeb. = Angelica lucida [ANGELUC] Apiaceae (V:D)
Coelopleurum lucidum (L.) Fern. = Angelica lucida [ANGELUC] Apiaceae (V:D)
Coelopleurum lucidum ssp. *gmelinii* (DC.) A. & D. Löve = Angelica lucida [ANGELUC] Apiaceae (V:D)
Cogswellia foeniculacea (Nutt.) Coult. & Rose = Lomatium foeniculaceum [LOMAFOE] Apiaceae (V:D)
Cogswellia macrocarpa (Hook. & Arn.) M.E. Jones = Lomatium macrocarpum [LOMAMAC] Apiaceae (V:D)
Cogswellia nudicaulis (Pursh) M.E. Jones = Lomatium nudicaule [LOMANUD] Apiaceae (V:D)
Cogswellia villosa (Raf.) J.A. Schultes = Lomatium foeniculaceum [LOMAFOE] Apiaceae (V:D)
Coleanthus Seidel [COLEANT$] Poaceae (V:M)
*__Coleanthus subtilis__ (Tratt.) Seidel [COLESUB] Poaceae (V:M)
Coleosanthus grandiflorus (Hook.) Kuntze = Brickellia grandiflora [BRICGRA] Asteraceae (V:D)
Collema F.H. Wigg. [COLLEMA$] Collemataceae (L)
Collema arcticum Lynge = Collema ceraniscum [COLLCER] Collemataceae (L)
Collema auriculatum Hoffm. = Collema auriforme [COLLAUI] Collemataceae (L)
Collema auriforme (With.) Coppins & J.R. Laundon [COLLAUI] Collemataceae (L)
Collema bachmanianum (Fink) Degel. [COLLBAC] Collemataceae (L)
Collema callopismum A. Massal. [COLLCAO] Collemataceae (L)
Collema callopismum var. **rhyparodes** (Nyl.) Degel. [COLLCAO1] Collemataceae (L)
Collema ceraniscum Nyl. [COLLCER] Collemataceae (L)
Collema cheileum (Ach.) Ach. = Collema crispum [COLLCRI] Collemataceae (L)
Collema coccophorum Tuck. [COLLCOC] Collemataceae (L)
Collema crispum (Hudson) F.H. Wigg. [COLLCRI] Collemataceae (L)
Collema cristatellum Tuck. = Collema tenax [COLLTEN] Collemataceae (L)
Collema cristatum (L.) F.H. Wigg. [COLLCRS] Collemataceae (L)
Collema cristatum var. **marginale** (Hudson) Degel. [COLLCRS1] Collemataceae (L)
Collema dubium de Lesd. = Collema coccophorum [COLLCOC] Collemataceae (L)
Collema fecundum Degel. [COLLFEC] Collemataceae (L)
Collema flaccidum (Ach.) Ach. [COLLFLA] Collemataceae (L)
Collema furfuraceum (Arnold) Du Rietz [COLLFUR] Collemataceae (L)
Collema furfuraceum var. **furfuraceum** [COLLFUR0] Collemataceae (L)
Collema furfuraceum var. **luzonense** (Räsänen) Degel. [COLLFUR1] Collemataceae (L)
Collema furvum (Ach.) Ach. = Collema fuscovirens [COLLFUS] Collemataceae (L)
Collema fuscovirens (With.) J.R. Laundon [COLLFUS] Collemataceae (L)
Collema glebulentum (Nyl. ex Crombie) Degel. [COLLGLE] Collemataceae (L)
Collema granosum auct. = Collema auriforme [COLLAUI] Collemataceae (L)
Collema multipartitum Sm. [COLLMUL] Collemataceae (L)
Collema nigrescens (Hudson) DC. [COLLNIG] Collemataceae (L)
Collema novomexicanum de Lesd. = Collema coccophorum [COLLCOC] Collemataceae (L)
Collema polycarpon Hoffm. [COLLPOL] Collemataceae (L)
Collema pulposum (Bernh.) Ach. = Collema tenax [COLLTEN] Collemataceae (L)
Collema subflaccidum Degel. [COLLSUL] Collemataceae (L)
Collema subfurfuraceum Degel. = Collema furfuraceum var. luzonense [COLLFUR1] Collemataceae (L)
Collema subfurvum auct. = Collema subflaccidum [COLLSUL] Collemataceae (L)
Collema subparvum Degel. [COLLSUB] Collemataceae (L)
Collema tenax (Sw.) Ach. [COLLTEN] Collemataceae (L)
Collema tenax var. **corallinum** (A. Massal.) Degel. [COLLTEN1] Collemataceae (L)
Collema tenax var. **crustaceum** (Kremp.) Degel. [COLLTEN2] Collemataceae (L)
Collema tenax var. **tenax** [COLLTEN0] Collemataceae (L)
Collema tuniforme (Ach.) Ach. = Collema fuscovirens [COLLFUS] Collemataceae (L)
Collema undulatum Laurer ex Flotow [COLLUND] Collemataceae (L)
Collema undulatum var. **granulosum** Degel. [COLLUND1] Collemataceae (L)
Collemodes bachmanianum Fink = Collema bachmanianum [COLLBAC] Collemataceae (L)
Collinsia Nutt. [COLLINS$] Scrophulariaceae (V:D)
Collinsia grandiflora Lindl. = Collinsia parviflora var. grandiflora [COLLPAR1] Scrophulariaceae (V:D)
Collinsia grandiflora var. *pusilla* Gray = Collinsia parviflora var. grandiflora [COLLPAR1] Scrophulariaceae (V:D)
Collinsia parviflora Lindl. [COLLPAR] Scrophulariaceae (V:D)
Collinsia parviflora var. **grandiflora** (Lindl.) Ganders & Krause [COLLPAR1] Scrophulariaceae (V:D)
Collinsia tenella (Pursh) Piper = Collinsia parviflora var. grandiflora [COLLPAR1] Scrophulariaceae (V:D)
Collomia Nutt. [COLLOMI$] Polemoniaceae (V:D)
Collomia grandiflora Dougl. ex Lindl. [COLLGRN] Polemoniaceae (V:D)
Collomia heterophylla Dougl. ex Hook. [COLLHET] Polemoniaceae (V:D)
Collomia linearis Nutt. [COLLLIN] Polemoniaceae (V:D)
Cololejeunea (Spruce) Schiffn. [COLOLEJ$] Lejeuneaceae (B:H)
Cololejeunea handelii (Herz.) Hatt. = Cololejeunea macounii [COLOMAC] Lejeuneaceae (B:H)
Cololejeunea macounii (Spruce ex Underw.) Evans [COLOMAC] Lejeuneaceae (B:H)
Cololejeunea rupicola (Steph.) Hatt. = Cololejeunea macounii [COLOMAC] Lejeuneaceae (B:H)
Colpodium fulvum (Trin.) Griseb. = Arctophila fulva [ARCTFUL] Poaceae (V:M)
Comandra Nutt. [COMANDR$] Santalaceae (V:D)
Comandra californica Eastw. ex Rydb. = Comandra umbellata ssp. californica [COMAUMB1] Santalaceae (V:D)
Comandra lividum Richards. = Geocaulon lividum [GEOCLIV] Santalaceae (V:D)
Comandra pallida A. DC. = Comandra umbellata ssp. pallida [COMAUMB2] Santalaceae (V:D)
Comandra umbellata (L.) Nutt. [COMAUMB] Santalaceae (V:D)
Comandra umbellata ssp. **californica** (Eastw. ex Rydb.) Piehl [COMAUMB1] Santalaceae (V:D)
Comandra umbellata ssp. **pallida** (A. DC.) Piehl [COMAUMB2] Santalaceae (V:D)
Comandra umbellata var. *angustifolia* (A. DC.) Torr. = Comandra umbellata ssp. pallida [COMAUMB2] Santalaceae (V:D)

Comandra umbellata var. *californica* (Eastw. ex Rydb.) C.L. Hitchc. = Comandra umbellata ssp. californica [COMAUMB1] Santalaceae (V:D)
Comandra umbellata var. *pallida* (A. DC.) M.E. Jones = Comandra umbellata ssp. pallida [COMAUMB2] Santalaceae (V:D)
Comarum L. [COMARUM$] Rosaceae (V:D)
Comarum palustre L. [COMAPAL] Rosaceae (V:D)
Comastoma tenella (Rottb.) Toyokuni = Gentianella tenella [GENTTEN] Gentianaceae (V:D)
Conardia Robins. [CONARDI$] Amblystegiaceae (B:M)
Conardia compacta (C. Müll.) Robins. [CONACOM] Amblystegiaceae (B:M)
Coniocybe furfuracea (L.) Ach. = Chaenotheca furfuracea [CHAEFUR] Coniocybaceae (L)
Coniocybe gracillima Vainio = Chaenotheca gracillima [CHAEGRA] Coniocybaceae (L)
Coniocybe sulphurea (Retz.) Nyl. = Chaenotheca brachypoda [CHAEBRA] Coniocybaceae (L)
Coniocybopsis arenaria (Hampe ex Massal.) Vainio = Microcalicium arenarium [MICRARE] Microcaliciaceae (L)
Conioselinum Hoffmann [CONIOSE$] Apiaceae (V:D)
Conioselinum chinense var. *pacificum* (S. Wats.) Boivin = Conioselinum gmelinii [CONIGME] Apiaceae (V:D)
Conioselinum cnidiifolium (Turcz.) Porsild = Cnidium cnidiifolium [CNIDCNI] Apiaceae (V:D)
Conioselinum gmelinii (Cham. & Schlecht.) Steud. [CONIGME] Apiaceae (V:D)
Conioselinum pacificum (S. Wats.) Coult. & Rose = Conioselinum gmelinii [CONIGME] Apiaceae (V:D)
Coniothyrium lecanoracearum Vouaux = Lichenoconium usneae [LICHUSN] Uncertain-family (L)
Conium L. [CONIUM$] Apiaceae (V:D)
*Conium maculatum** L. [CONIMAC] Apiaceae (V:D)
Conocephalum Wigg. [CONOCEP$] Conocephalaceae (B:H)
Conocephalum conicum (L.) Lindb. [CONOCON] Conocephalaceae (B:H)
Conomitrium julianum (Cand.) Mont. = Fissidens fontanus [FISSFON] Fissidentaceae (B:M)
Conostomum Sw. in Web. & Mohr [CONOSTO$] Bartramiaceae (B:M)
Conostomum boreale Sw. = Conostomum tetragonum [CONOTET] Bartramiaceae (B:M)
Conostomum tetragonum (Hedw.) Lindb. [CONOTET] Bartramiaceae (B:M)
Conringia Heister ex Fabr. [CONRING$] Brassicaceae (V:D)
*Conringia orientalis** (L.) Andrz. [CONRORI] Brassicaceae (V:D)
Consolida S.F. Gray [CONSOLI$] Ranunculaceae (V:D)
*Consolida ajacis** (L.) Schur [CONSAJA] Ranunculaceae (V:D)
Consolida ambigua (L.) P.W. Ball & Heywood = Consolida ajacis [CONSAJA] Ranunculaceae (V:D)
Convallaria stellata L. = Maianthemum stellatum [MAIASTE] Liliaceae (V:M)
Convallaria trifolia L. = Maianthemum trifolium [MAIATRI] Liliaceae (V:M)
Convolvulus L. [CONVOLV$] Convolvulaceae (V:D)
Convolvulus ambigens House = Convolvulus arvensis [CONVARV] Convolvulaceae (V:D)
*Convolvulus arvensis** L. [CONVARV] Convolvulaceae (V:D)
Convolvulus nashii House = Calystegia sepium [CALYSEP] Convolvulaceae (V:D)
Convolvulus sepium L. = Calystegia sepium [CALYSEP] Convolvulaceae (V:D)
Convolvulus sepium var. *communis* R. Tryon = Calystegia sepium [CALYSEP] Convolvulaceae (V:D)
Convolvulus soldanella L. = Calystegia soldanella [CALYSOL] Convolvulaceae (V:D)
Conyza Less. [CONYZA$] Asteraceae (V:D)
*Conyza canadensis** (L.) Cronq. [CONYCAN] Asteraceae (V:D)
*Conyza canadensis** var. **glabrata** (Gray) Cronq. [CONYCAN2] Asteraceae (V:D)
Coptidium lapponicum (L.) Gandog. = Ranunculus lapponicus [RANULAP] Ranunculaceae (V:D)
Coptis Salisb. [COPTIS$] Ranunculaceae (V:D)
Coptis aspleniifolia Salisb. [COPTASP] Ranunculaceae (V:D)
Coptis trifolia (L.) Salisb. [COPTTRI] Ranunculaceae (V:D)
Corallorrhiza Gagnebin [CORALLO$] Orchidaceae (V:M)
Corallorrhiza corallorrhiza (L.) Karst. = Corallorrhiza trifida [CORATRI] Orchidaceae (V:M)
Corallorrhiza maculata (Raf.) Raf. [CORAMAC] Orchidaceae (V:M)
Corallorrhiza maculata ssp. *mertensiana* (Bong.) Calder & Taylor = Corallorrhiza mertensiana [CORAMER] Orchidaceae (V:M)
Corallorrhiza maculata var. *flavida* (M.E. Peck) Cockerell = Corallorrhiza maculata [CORAMAC] Orchidaceae (V:M)
Corallorrhiza maculata var. *immaculata* M.E. Peck = Corallorrhiza maculata [CORAMAC] Orchidaceae (V:M)
Corallorrhiza maculata var. *intermedia* Farw. = Corallorrhiza maculata [CORAMAC] Orchidaceae (V:M)
Corallorrhiza maculata var. *occidentalis* (Lindl.) Cockerell = Corallorrhiza maculata [CORAMAC] Orchidaceae (V:M)
Corallorrhiza maculata var. *punicea* Bartlett = Corallorrhiza maculata [CORAMAC] Orchidaceae (V:M)
Corallorrhiza mertensiana Bong. [CORAMER] Orchidaceae (V:M)
Corallorrhiza multiflora Nutt. = Corallorrhiza maculata [CORAMAC] Orchidaceae (V:M)
Corallorrhiza striata Lindl. [CORASTR] Orchidaceae (V:M)
Corallorrhiza trifida Chatelain [CORATRI] Orchidaceae (V:M)
Corallorrhiza trifida var. *verna* (Nutt.) Fern. = Corallorrhiza trifida [CORATRI] Orchidaceae (V:M)
Coreopsis L. [COREOPS$] Asteraceae (V:D)
Coreopsis atkinsoniana Dougl. ex Lindl. = Coreopsis tinctoria var. atkinsoniana [CORETIN1] Asteraceae (V:D)
Coreopsis crassifolia Ait. = Coreopsis lanceolata [CORELAN] Asteraceae (V:D)
Coreopsis heterogyna Fern. = Coreopsis lanceolata [CORELAN] Asteraceae (V:D)
*Coreopsis lanceolata** L. [CORELAN] Asteraceae (V:D)
Coreopsis lanceolata var. *villosa* Michx. = Coreopsis lanceolata [CORELAN] Asteraceae (V:D)
Coreopsis tinctoria Nutt. [CORETIN] Asteraceae (V:D)
Coreopsis tinctoria var. **atkinsoniana** (Dougl. ex Lindl.) H.M. Parker [CORETIN1] Asteraceae (V:D)
Coriscium viride (Ach.) Vainio = Omphalina hudsoniana [OMPHHUD] Tricholomataceae (L)
Corispermum L. [CORISPE$] Chenopodiaceae (V:D)
Corispermum americanum (Nutt.) Nutt. = Corispermum hyssopifolium [CORIHYS] Chenopodiaceae (V:D)
Corispermum hyssopifolium L. [CORIHYS] Chenopodiaceae (V:D)
Corispermum hyssopifolium var. *rubricaule* Hook. = Corispermum hyssopifolium [CORIHYS] Chenopodiaceae (V:D)
Corispermum marginale Rydb. = Corispermum hyssopifolium [CORIHYS] Chenopodiaceae (V:D)
Corispermum simplicissimum Lunell = Corispermum hyssopifolium [CORIHYS] Chenopodiaceae (V:D)
Cornella canadensis (L.) Rydb. = Cornus canadensis [CORNCAN] Cornaceae (V:D)
Cornella suecica (L.) Rydb. = Cornus suecica [CORNSUE] Cornaceae (V:D)
Cornicularia (Schreber) Hoffm. [CORNICU$] Parmeliaceae (L)
Cornicularia aculeata (Schreber) Ach. = Cetraria aculeata [CETRACU] Parmeliaceae (L)
Cornicularia californica (Tuck.) Du Rietz = Kaernefeltia californica [KAERCAL] Parmeliaceae (L)
Cornicularia divergens Ach. = Bryocaulon divergens [BRYODIV] Parmeliaceae (L)
Cornicularia muricata (Ach.) Ach. = Cetraria muricata [CETRMUR] Parmeliaceae (L)
Cornicularia normoerica (Gunn.) Du Rietz [CORNNOR] Parmeliaceae (L)
Cornicularia pseudosatoana Asah. = Bryocaulon pseudosatoanum [BRYOPSD] Parmeliaceae (L)
Cornus L. [CORNUS$] Cornaceae (V:D)
Cornus alba L. p.p. = Cornus sericea [CORNSER] Cornaceae (V:D)
Cornus alba ssp. *stolonifera* (Michx.) Wangerin = Cornus sericea [CORNSER] Cornaceae (V:D)

Cornus alba var. *baileyi* (Coult. & Evans) Boivin = Cornus sericea [CORNSER] Cornaceae (V:D)
Cornus alba var. *californica* (C.A. Mey.) Boivin = Cornus sericea [CORNSER] Cornaceae (V:D)
Cornus alba var. *interior* (Rydb.) Boivin = Cornus sericea [CORNSER] Cornaceae (V:D)
Cornus baileyi Coult. & Evans = Cornus sericea [CORNSER] Cornaceae (V:D)
Cornus × *californica* C.A. Mey. = Cornus sericea [CORNSER] Cornaceae (V:D)
Cornus canadensis L. [CORNCAN] Cornaceae (V:D)
Cornus canadensis var. *dutillyi* (Lepage) Boivin = Cornus canadensis [CORNCAN] Cornaceae (V:D)
Cornus instolonea A. Nels. = Cornus sericea [CORNSER] Cornaceae (V:D)
Cornus interior (Rydb.) N. Petersen = Cornus sericea [CORNSER] Cornaceae (V:D)
Cornus nuttallii Audubon ex Torr. & Gray [CORNNUT] Cornaceae (V:D)
Cornus sericea L. [CORNSER] Cornaceae (V:D)
Cornus sericea ssp. *stolonifera* (Michx.) Fosberg = Cornus sericea [CORNSER] Cornaceae (V:D)
Cornus sericea var. *interior* (Rydb.) St. John = Cornus sericea [CORNSER] Cornaceae (V:D)
Cornus stolonifera Michx. = Cornus sericea [CORNSER] Cornaceae (V:D)
Cornus stolonifera var. *baileyi* (Coult. & Evans) Drescher = Cornus sericea [CORNSER] Cornaceae (V:D)
Cornus stolonifera var. *interior* (Rydb.) St. John = Cornus sericea [CORNSER] Cornaceae (V:D)
Cornus suecica L. [CORNSUE] Cornaceae (V:D)
Cornus unalaschkensis Ledeb. [CORNUNA] Cornaceae (V:D)
Coronaria coriacea (Moench) Schischkin & Gorschk. = Lychnis coronaria [LYCHCOR] Caryophyllaceae (V:D)
Coronilla L. [CORONIL$] Fabaceae (V:D)
*__Coronilla varia__ L. [COROVAR] Fabaceae (V:D)
Coronopus Zinn [CORONOP$] Brassicaceae (V:D)
*__Coronopus didymus__ (L.) Sm. [CORODID] Brassicaceae (V:D)
Corrigiola L. [CORRIGI$] Caryophyllaceae (V:D)
*__Corrigiola litoralis__ L. [CORRLIT] Caryophyllaceae (V:D)
Corticifraga D. Hawksw. & R. Sant. [CORTICI$] Uncertain-family (L)
Corticifraga fuckelii (Rehm) D. Hawksw. & R. Sant. [CORTFUC] Uncertain-family (L)
Corydalis Vent. [CORYDAL$] Fumariaceae (V:D)
Corydalis aurea Willd. [CORYAUR] Fumariaceae (V:D)
Corydalis pauciflora (Steph.) Pers. [CORYPAU] Fumariaceae (V:D)
Corydalis pauciflora var. *albiflora* Porsild = Corydalis pauciflora [CORYPAU] Fumariaceae (V:D)
Corydalis pauciflora var. *chamissonis* Fedde = Corydalis pauciflora [CORYPAU] Fumariaceae (V:D)
Corydalis scouleri Hook. [CORYSCO] Fumariaceae (V:D)
Corydalis sempervirens (L.) Pers. [CORYSEM] Fumariaceae (V:D)
Corydalis washingtoniana Fedde = Corydalis aurea [CORYAUR] Fumariaceae (V:D)
Corylus L. [CORYLUS$] Betulaceae (V:D)
*__Corylus avellana__ L. [CORYAVE] Betulaceae (V:D)
Corylus californica (A. DC.) Rose = Corylus cornuta var. californica [CORYCOR1] Betulaceae (V:D)
Corylus cornuta Marsh. [CORYCOR] Betulaceae (V:D)
Corylus cornuta var. **californica** (A. DC.) Sharp [CORYCOR1] Betulaceae (V:D)
Corylus cornuta var. **cornuta** [CORYCOR0] Betulaceae (V:D)
Corylus cornuta var. **glandulosa** Boivin [CORYCOR3] Betulaceae (V:D)
Corylus rostrata Ait. = Corylus cornuta var. cornuta [CORYCOR0] Betulaceae (V:D)
Coscinodon Spreng. [COSCINO$] Grimmiaceae (B:M)
Coscinodon calyptratus (Hook. in Drumm.) C. Jens. ex Kindb. [COSCCAL] Grimmiaceae (B:M)
Coscinodon cribrosus (Hedw.) Spruce [COSCCRI] Grimmiaceae (B:M)
Coscinodon pulvinatus Spreng. = Coscinodon cribrosus [COSCCRI] Grimmiaceae (B:M)
Cota tinctoria (L.) J. Gay = Anthemis tinctoria [ANTHTIN] Asteraceae (V:D)
Cotoneaster Medik. [COTONEA$] Rosaceae (V:D)
*__Cotoneaster bullatus__ Boiss. [COTOBUL] Rosaceae (V:D)
*__Cotoneaster horizontalis__ Dcne. [COTOHOR] Rosaceae (V:D)
*__Cotoneaster simonsii__ Baker [COTOSIM] Rosaceae (V:D)
Cotula L. [COTULA$] Asteraceae (V:D)
*__Cotula coronopifolia__ L. [COTUCOR] Asteraceae (V:D)
Crassula L. [CRASSUL$] Crassulaceae (V:D)
Crassula aquatica (L.) Schoenl. [CRASAQU] Crassulaceae (V:D)
Crassula connata (Ruiz & Pavón) Berger [CRASCON] Crassulaceae (V:D)
Crassula erecta (Hook. & Arn.) Berger = Crassula connata [CRASCON] Crassulaceae (V:D)
Crataegus L. [CRATAEG$] Rosaceae (V:D)
Crataegus columbiana T.J. Howell [CRATCOL] Rosaceae (V:D)
Crataegus douglasii Lindl. [CRATDOU] Rosaceae (V:D)
Crataegus douglasii var. *suksdorfii* Sarg. = Crataegus suksdorfii [CRATSUK] Rosaceae (V:D)
*__Crataegus monogyna__ Jacq. [CRATMON] Rosaceae (V:D)
Crataegus suksdorfii (Sarg.) Kruschke [CRATSUK] Rosaceae (V:D)
Cratoneuron (Sull.) Spruce [CRATONE$] Amblystegiaceae (B:M)
Cratoneuron commutatum (Brid.) G. Roth = Palustriella commutata [PALUCOM] Amblystegiaceae (B:M)
Cratoneuron commutatum var. *falcatum* (Brid.) Mönk. = Palustriella commutata [PALUCOM] Amblystegiaceae (B:M)
Cratoneuron commutatum var. *sulcatum* (Lindb.) Mönk. = Palustriella commutata [PALUCOM] Amblystegiaceae (B:M)
Cratoneuron curvicaule (Jur.) G. Roth = Cratoneuron filicinum [CRATFIL] Amblystegiaceae (B:M)
Cratoneuron falcatum (Brid.) G. Roth = Palustriella commutata [PALUCOM] Amblystegiaceae (B:M)
Cratoneuron filicinum (Hedw.) Spruce [CRATFIL] Amblystegiaceae (B:M)
Cratoneuron filicinum var. *aciculinum* (C. Müll. & Kindb. in Mac. & Kindb.) Grout = Cratoneuron filicinum [CRATFIL] Amblystegiaceae (B:M)
Cratoneuron filicinum var. *curvicaule* (Jur.) Mönk. = Cratoneuron filicinum [CRATFIL] Amblystegiaceae (B:M)
Cratoneuron filicinum var. *fallax* (Brid.) Roth = Cratoneuron filicinum [CRATFIL] Amblystegiaceae (B:M)
Cratoneuron filicinum var. *filiforme* (Par.) Wijk & Marg. = Cratoneuron filicinum [CRATFIL] Amblystegiaceae (B:M)
Crepis L. [CREPIS$] Asteraceae (V:D)
Crepis alpicola (Rydb.) A. Nels. = Crepis runcinata [CREPRUN] Asteraceae (V:D)
Crepis atrabarba Heller [CREPATR] Asteraceae (V:D)
Crepis atrabarba ssp. **atrabarba** [CREPATR0] Asteraceae (V:D)
Crepis atrabarba ssp. **originalis** (Babcock & Stebbins) Babcock & Stebbins [CREPATR2] Asteraceae (V:D)
Crepis atribarba ssp. *cytotaxonomicorum* (Boivin) W.A. Weber = Crepis atrabarba ssp. atrabarba [CREPATR0] Asteraceae (V:D)
Crepis atribarba ssp. *typicus* Babcock & Stebbins = Crepis atrabarba ssp. atrabarba [CREPATR0] Asteraceae (V:D)
Crepis atribarba var. *cytotaxonomicorum* Boivin = Crepis atrabarba ssp. atrabarba [CREPATR0] Asteraceae (V:D)
Crepis barbigera Leib. ex Coville = Crepis atrabarba ssp. originalis [CREPATR2] Asteraceae (V:D)
*__Crepis capillaris__ (L.) Wallr. [CREPCAP] Asteraceae (V:D)
Crepis elegans Hook. [CREPELE] Asteraceae (V:D)
Crepis exilis Osterhout = Crepis atrabarba [CREPATR] Asteraceae (V:D)
Crepis glaucella Rydb. = Crepis runcinata [CREPRUN] Asteraceae (V:D)
Crepis intermedia Gray [CREPINT] Asteraceae (V:D)
Crepis modocensis Greene [CREPMOD] Asteraceae (V:D)
Crepis modocensis ssp. **modocensis** [CREPMOD0] Asteraceae (V:D)
Crepis modocensis ssp. **rostrata** (Coville) Babcock & Stebbins [CREPMOD2] Asteraceae (V:D)
Crepis modocensis ssp. *typica* Babcock & Stebbins = Crepis modocensis ssp. modocensis [CREPMOD0] Asteraceae (V:D)

Crepis modocensis var. *rostrata* (Coville) Boivin = Crepis modocensis ssp. rostrata [CREPMOD2] Asteraceae (V:D)
Crepis nana Richards. [CREPNAN] Asteraceae (V:D)
Crepis nana ssp. **ramosa** Babcock [CREPNAN2] Asteraceae (V:D)
Crepis neomexicana Woot. & Standl. = Crepis runcinata [CREPRUN] Asteraceae (V:D)
*****Crepis nicaeensis** Balbis ex Pers. [CREPNIC] Asteraceae (V:D)
Crepis occidentalis Nutt. [CREPOCC] Asteraceae (V:D)
Crepis occidentalis ssp. **costata** (Gray) Babcock & Stebbins [CREPOCC1] Asteraceae (V:D)
Crepis occidentalis ssp. **occidentalis** [CREPOCC0] Asteraceae (V:D)
Crepis occidentalis ssp. **pumila** (Rydb.) Babcock & Stebbins [CREPOCC3] Asteraceae (V:D)
Crepis occidentalis ssp. *typica* Babcock & Stebbins = Crepis occidentalis ssp. occidentalis [CREPOCC0] Asteraceae (V:D)
Crepis occidentalis var. *costata* Gray = Crepis occidentalis ssp. costata [CREPOCC1] Asteraceae (V:D)
Crepis occidentalis var. *pumila* (Rydb.) Babcock & Stebbins = Crepis occidentalis ssp. pumila [CREPOCC3] Asteraceae (V:D)
Crepis perplexans Rydb. = Crepis runcinata [CREPRUN] Asteraceae (V:D)
Crepis runcinata (James) Torr. & Gray [CREPRUN] Asteraceae (V:D)
Crepis runcinata ssp. *typica* Babcock & Stebbins = Crepis runcinata [CREPRUN] Asteraceae (V:D)
*****Crepis tectorum** L. [CREPTEC] Asteraceae (V:D)
*****Crepis vesicaria** L. [CREPVES] Asteraceae (V:D)
*****Crepis vesicaria** ssp. **haenseleri** (Boiss. ex DC.) P.D. Sell [CREPVES1] Asteraceae (V:D)
Crepis vesicaria ssp. *taraxacifolia* (Thuill.) Thellung ex Schinz & R. Keller = Crepis vesicaria ssp. haenseleri [CREPVES1] Asteraceae (V:D)
Crepis vesicaria var. *taraxacifolia* (Thuill.) Boivin = Crepis vesicaria ssp. haenseleri [CREPVES1] Asteraceae (V:D)
Critesion brachyantherum (Nevski) Barkworth & Dewey = Hordeum brachyantherum [HORDBRA] Poaceae (V:M)
Critesion geniculatum (All.) A. Löve = Hordeum marinum ssp. gussonianum [HORDMAR1] Poaceae (V:M)
Critesion hystrix (Roth) A. Löve = Hordeum marinum ssp. gussonianum [HORDMAR1] Poaceae (V:M)
Critesion jubatum (L.) Nevski = Hordeum jubatum ssp. jubatum [HORDJUB0] Poaceae (V:M)
Critesion jubatum ssp. *breviaristatum* (Bowden) A. & D. Löve = Hordeum jubatum ssp. breviaristatum [HORDJUB1] Poaceae (V:M)
Critesion marinum ssp. *gussonianum* (Parl.) Barkworth & Dewey = Hordeum marinum ssp. gussonianum [HORDMAR1] Poaceae (V:M)
Critesion murinum (L.) A. Löve = Hordeum murinum ssp. murinum [HORDMUR0] Poaceae (V:M)
Critesion murinum ssp. *leporinum* (Link) A. Löve = Hordeum murinum ssp. leporinum [HORDMUR1] Poaceae (V:M)
Crocidium Hook. [CROCIDI$] Asteraceae (V:D)
Crocidium multicaule Hook. [CROCMUL] Asteraceae (V:D)
Crocosmia Planch. [CROCOSM$] Iridaceae (V:M)
*****Crocosmia × crocosmiiflora** (V. Lemoine ex E. Morr.) N.E. Br. [CROCXCR] Iridaceae (V:M)
Crocynia aliciae Hue = Lepraria lobificans [LEPRLOB] Uncertain-family (L)
Crocynia neglecta (Nyl.) Hue = Lepraria neglecta [LEPRNEG] Uncertain-family (L)
Crossidium Jur. [CROSSID$] Pottiaceae (B:M)
Crossidium aberrans Holz. & Bartr. [CROSABE] Pottiaceae (B:M)
Crossidium rosei Williams [CROSROS] Pottiaceae (B:M)
Crossidium seriatum Crum & Steere [CROSSER] Pottiaceae (B:M)
Crossidium spatulaefolium Holz. & Bartr. = Crossidium aberrans [CROSABE] Pottiaceae (B:M)
Crossidium squamiferum (Viv.) Jur. [CROSSQU] Pottiaceae (B:M)
Crossocalyx hellerianus (Nees) Meyl. = Anastrophyllum helleranum [ANASHEL] Jungermanniaceae (B:H)
Crumia Schof. [CRUMIA$] Pottiaceae (B:M)
Crumia latifolia (Kindb. in Mac.) Schof. [CRUMLAT] Pottiaceae (B:M)
Crunocallis chamissoi (Ledeb. ex Spreng.) Rydb. = Montia chamissoi [MONTCHA] Portulacaceae (V:D)
Cryptantha Lehm. ex G. Don [CRYPTAN$] Boraginaceae (V:D)
Cryptantha affinis (Gray) Greene [CRYPAFF] Boraginaceae (V:D)
Cryptantha ambigua (Gray) Greene [CRYPAMB] Boraginaceae (V:D)
Cryptantha barbigera var. *fergusoniae* J.F. Macbr. = Cryptantha intermedia [CRYPINT] Boraginaceae (V:D)
Cryptantha bradburiana Payson = Cryptantha celosioides [CRYPCEL] Boraginaceae (V:D)
Cryptantha calycosa (Gray) Rydb. = Cryptantha torreyana [CRYPTOR] Boraginaceae (V:D)
Cryptantha celosioides (Eastw.) Payson [CRYPCEL] Boraginaceae (V:D)
Cryptantha eastwoodiae St. John = Cryptantha torreyana [CRYPTOR] Boraginaceae (V:D)
Cryptantha fendleri (Gray) Greene [CRYPFEN] Boraginaceae (V:D)
Cryptantha flexulosa (A. Nels.) A. Nels. = Cryptantha torreyana [CRYPTOR] Boraginaceae (V:D)
Cryptantha fragilis M.E. Peck = Cryptantha intermedia [CRYPINT] Boraginaceae (V:D)
Cryptantha grandiflora Rydb. = Cryptantha intermedia [CRYPINT] Boraginaceae (V:D)
Cryptantha hendersonii (A. Nels.) Piper ex J.C. Nels. = Cryptantha intermedia [CRYPINT] Boraginaceae (V:D)
Cryptantha intermedia (Gray) Greene [CRYPINT] Boraginaceae (V:D)
Cryptantha intermedia var. *grandiflora* (Rydb.) Cronq. = Cryptantha intermedia [CRYPINT] Boraginaceae (V:D)
Cryptantha macounii (Eastw.) Payson = Cryptantha celosioides [CRYPCEL] Boraginaceae (V:D)
Cryptantha nubigena (Greene) Payson [CRYPNUB] Boraginaceae (V:D)
Cryptantha nubigena var. *celosioides* (Eastw.) Boivin = Cryptantha celosioides [CRYPCEL] Boraginaceae (V:D)
Cryptantha nubigena var. *macounii* (Eastw.) Boivin = Cryptantha celosioides [CRYPCEL] Boraginaceae (V:D)
Cryptantha pattersonii (Gray) Greene = Cryptantha fendleri [CRYPFEN] Boraginaceae (V:D)
Cryptantha sheldonii (Brand) Payson = Cryptantha celosioides [CRYPCEL] Boraginaceae (V:D)
Cryptantha torreyana (Gray) Greene [CRYPTOR] Boraginaceae (V:D)
Cryptantha torreyana var. *calycosa* (Gray) Greene = Cryptantha torreyana [CRYPTOR] Boraginaceae (V:D)
Cryptantha torreyana var. *pumila* (Heller) I.M. Johnston = Cryptantha torreyana [CRYPTOR] Boraginaceae (V:D)
Cryptantha watsonii (Gray) Greene [CRYPWAT] Boraginaceae (V:D)
Cryptogramma R. Br. [CRYPTOG$] Pteridaceae (V:F)
Cryptogramma acrostichoides R. Br. [CRYPACR] Pteridaceae (V:F)
Cryptogramma acrostichoides var. *sitchensis* (Rupr.) C. Christ. = Cryptogramma sitchensis [CRYPSIT] Pteridaceae (V:F)
Cryptogramma cascadensis E.R. Alverson [CRYPCAS] Pteridaceae (V:F)
Cryptogramma crispa ssp. *acrostichoides* (R. Br.) Hultén = Cryptogramma acrostichoides [CRYPACR] Pteridaceae (V:F)
Cryptogramma crispa var. *macrostichoides* (R. Br.) C.B. Clarke = Cryptogramma acrostichoides [CRYPACR] Pteridaceae (V:F)
Cryptogramma crispa var. *sitchensis* (Rupr.) C. Christens. = Cryptogramma sitchensis [CRYPSIT] Pteridaceae (V:F)
Cryptogramma densa (Brack.) Diels = Aspidotis densa [ASPIDEN] Pteridaceae (V:F)
Cryptogramma sitchensis (Rupr.) T. Moore [CRYPSIT] Pteridaceae (V:F)
Cryptogramma stelleri (Gmel.) Prantl [CRYPSTE] Pteridaceae (V:F)
Ctenidium (Schimp.) Mitt. [CTENIDI$] Hypnaceae (B:M)
Ctenidium procerrimum (De Not.) Lindb. = Hypnum procerrimum [HYPNPRO] Hypnaceae (B:M)
Ctenidium schofieldii Nish. [CTENSCH] Hypnaceae (B:M)

Cupressus nootkatensis D. Don = Chamaecyparis nootkatensis [CHAMNOO] Cupressaceae (V:G)
Cuscuta L. [CUSCUTA$] Cuscutaceae (V:D)
***Cuscuta approximata** Bab. [CUSCAPP] Cuscutaceae (V:D)
Cuscuta cephalanthi Engelm. [CUSCCEP] Cuscutaceae (V:D)
***Cuscuta epithymum** L. [CUSCEPI] Cuscutaceae (V:D)
Cuscuta pentagona Engelm. [CUSCPEN] Cuscutaceae (V:D)
Cuscuta salina Engelm. [CUSCSAL] Cuscutaceae (V:D)
Cyanococcus canadensis (Kalm ex A. Rich.) Rydb. = Vaccinium myrtilloides [VACCMYR] Ericaceae (V:D)
Cyclachaena xanthifolia (Nutt.) Fresen. = Iva xanthifolia [IVA XAN] Asteraceae (V:D)
Cylactis arctica ssp. *acaulis* (Michx.) W.A. Weber = Rubus arcticus ssp. acaulis [RUBUARC1] Rosaceae (V:D)
Cylindrothecium macounii (C. Müll. & Kindb. in Mac. & Kindb.) Ren. & Card. = Hypnum pratense [HYPNPRA] Hypnaceae (B:M)
Cymbalaria Hill [CYMBALA$] Scrophulariaceae (V:D)
***Cymbalaria muralis** P.G. Gaertn., B. Mey. & Scherb. [CYMBMUR] Scrophulariaceae (V:D)
Cynodon L.C. Rich. [CYNODON$] Poaceae (V:M)
***Cynodon dactylon** (L.) Pers. [CYNODAC] Poaceae (V:M)
Cynodontium Bruch & Schimp. ex Schimp. [CYNODOT$] Dicranaceae (B:M)
Cynodontium alpestre (Wahlenb.) Milde [CYNOALP] Dicranaceae (B:M)
Cynodontium glaucescens (Lindb. & Arnell) Par. [CYNOGLA] Dicranaceae (B:M)
Cynodontium jenneri (Schimp. in Howie) Stirt. [CYNOJEN] Dicranaceae (B:M)
Cynodontium polycarpon (Hedw.) Schimp. [CYNOPOL] Dicranaceae (B:M)
Cynodontium polycarpon var. *laxirete* (Dix.) Dix. = Cynodontium jenneri [CYNOJEN] Dicranaceae (B:M)
Cynodontium schisti (Web. & Mohr) Lindb. [CYNOSCH] Dicranaceae (B:M)
Cynodontium strumiferum (Hedw.) Lindb. [CYNOSTR] Dicranaceae (B:M)
Cynodontium subalpestre Kindb. in Mac. & Kindb. = Cynodontium tenellum [CYNOTEN] Dicranaceae (B:M)
Cynodontium tenellum (Bruch & Schimp. in B.S.G.) Limpr. [CYNOTEN] Dicranaceae (B:M)
Cynodontium torquescens Limpr. = Cynodontium tenellum [CYNOTEN] Dicranaceae (B:M)
Cynodontium virens (Hedw.) Schimp. = Oncophorus virens [ONCOVIR] Dicranaceae (B:M)
Cynodontium virens var. *serratum* (B.S.G.) Schimp. = Oncophorus virens [ONCOVIR] Dicranaceae (B:M)
Cynodontium virens var. *wahlenbergii* (Brid.) Schimp. = Oncophorus wahlenbergii [ONCOWAH] Dicranaceae (B:M)
Cynodontium wahlenbergii (Brid.) Hartm. = Oncophorus wahlenbergii [ONCOWAH] Dicranaceae (B:M)
Cynodontium wahlenbergii var. *compactum* (B.S.G.) Mac. & Kindb. = Oncophorus wahlenbergii [ONCOWAH] Dicranaceae (B:M)
Cynodontium wahlenbergii var. *gracile* (Broth.) Mönk. = Oncophorus wahlenbergii [ONCOWAH] Dicranaceae (B:M)
Cynoglossum L. [CYNOGLO$] Boraginaceae (V:D)
Cynoglossum boreale Fern. = Cynoglossum virginianum var. boreale [CYNOVIR1] Boraginaceae (V:D)
***Cynoglossum officinale** L. [CYNOOFF] Boraginaceae (V:D)
Cynoglossum penicillatum Hook. & Arn. = Pectocarya penicillata [PECTPEN] Boraginaceae (V:D)
Cynoglossum virginianum L. [CYNOVIR] Boraginaceae (V:D)
Cynoglossum virginianum var. **boreale** (Fern.) Cooperrider [CYNOVIR1] Boraginaceae (V:D)
Cynomarathrum brandegei Coult. & Rose = Lomatium brandegei [LOMABRA] Apiaceae (V:D)
Cynosurus L. [CYNOSUR$] Poaceae (V:M)
***Cynosurus cristatus** L. [CYNOCRI] Poaceae (V:M)
***Cynosurus echinatus** L. [CYNOECH] Poaceae (V:M)
Cyperus L. [CYPERUS$] Cyperaceae (V:M)
Cyperus aristatus Rottb. = Cyperus squarrosus [CYPESQU] Cyperaceae (V:M)
Cyperus aristatus var. *runyonii* O'Neill = Cyperus squarrosus [CYPESQU] Cyperaceae (V:M)
Cyperus inflexus Muhl. = Cyperus squarrosus [CYPESQU] Cyperaceae (V:M)
Cyperus squarrosus L. [CYPESQU] Cyperaceae (V:M)
Cypheliopsis bolanderi (Tuck.) Vainio = Thelomma mammosum [THELMAM] Caliciaceae (L)
Cyphelium Ach. [CYPHELU$] Caliciaceae (L)
Cyphelium caliciforme (Flotow) Zahlbr. = Thelomma occidentale [THELOCC] Caliciaceae (L)
Cyphelium inquinans (Sm.) Trevisan [CYPHINQ] Caliciaceae (L)
Cyphelium karelicum (Vainio) Räsänen [CYPHKAR] Caliciaceae (L)
Cyphelium occidentale Herre = Thelomma occidentale [THELOCC] Caliciaceae (L)
Cyphelium ocellatum (Körber) Trevisan = Thelomma ocellatum [THELOCE] Caliciaceae (L)
Cyphelium pinicola Tibell [CYPHPIN] Caliciaceae (L)
Cyphelium tigillare (Ach.) Ach. [CYPHTIG] Caliciaceae (L)
Cyphelium trachylioides (Nyl. ex Branth & Rostrup) Erichsen [CYPHTRA] Caliciaceae (L)
Cyphelium ventricosulum (Müll. Arg.) Zahlbr. = Cyphelium inquinans [CYPHINQ] Caliciaceae (L)
Cypripedium L. [CYPRIPE$] Orchidaceae (V:M)
Cypripedium calceolus ssp. *parviflorum* (Salisb.) Hultén = Cypripedium parviflorum [CYPRPAR] Orchidaceae (V:M)
Cypripedium calceolus var. *parviflorum* (Salisb.) Fern. = Cypripedium parviflorum [CYPRPAR] Orchidaceae (V:M)
Cypripedium montanum Dougl. ex Lindl. [CYPRMON] Orchidaceae (V:M)
Cypripedium parviflorum Salisb. [CYPRPAR] Orchidaceae (V:M)
Cypripedium passerinum Richards. [CYPRPAS] Orchidaceae (V:M)
Cypripedium passerinum var. *minganense* Victorin = Cypripedium passerinum [CYPRPAS] Orchidaceae (V:M)
Cyrtomnium Holm. [CYRTOMN$] Mniaceae (B:M)
Cyrtomnium hymenophylloides (Hüb.) Nyh. ex T. Kop. [CYRTHYM] Mniaceae (B:M)
Cyrtomnium hymenophyllum (Bruch & Schimp. in B.S.G.) Holm. [CYRTHYE] Mniaceae (B:M)
Cystocoleus Thwaites [CYSTOCO$] Uncertain-family (L)
Cystocoleus ebeneus (Dillwyn) Thwaites [CYSTEBE] Uncertain-family (L)
Cystopteris Bernh. [CYSTOPT$] Dryopteridaceae (V:F)
Cystopteris dickieana Sim = Cystopteris fragilis [CYSTFRA] Dryopteridaceae (V:F)
Cystopteris fragilis (L.) Bernh. [CYSTFRA] Dryopteridaceae (V:F)
Cystopteris fragilis ssp. *dickieana* (Sim) Hyl. = Cystopteris fragilis [CYSTFRA] Dryopteridaceae (V:F)
Cystopteris fragilis var. *angustata* Lawson = Cystopteris fragilis [CYSTFRA] Dryopteridaceae (V:F)
Cystopteris fragilis var. *woodsioides* Christ = Cystopteris fragilis [CYSTFRA] Dryopteridaceae (V:F)
Cystopteris montana (Lam.) Bernh. ex Desv. [CYSTMON] Dryopteridaceae (V:F)
Cytisus L. [CYTISUS$] Fabaceae (V:D)
***Cytisus scoparius** (L.) Link [CYTISCO] Fabaceae (V:D)

D

Dactylina Nyl. [DACTYLN$] Parmeliaceae (L)
Dactylina arctica (Richardson) Nyl. [DACTARC] Parmeliaceae (L)

Dactylina arctica ssp. *beringica* (C.D. Bird & J.W. Thomson) Kärnefelt & Thell = Dactylina beringica [DACTBER] Parmeliaceae (L)
Dactylina beringica C.D. Bird & J.W. Thomson [DACTBER] Parmeliaceae (L)
Dactylina madreporiformis (Ach.) Tuck. [DACTMAD] Parmeliaceae (L)
Dactylina ramulosa (Hook.) Tuck. [DACTRAM] Parmeliaceae (L)
Dactylis L. [DACTYLI$] Poaceae (V:M)
*Dactylis glomerata** L. [DACTGLO] Poaceae (V:M)
Dactylospora Körber [DACTYLO$] Dactylosporaceae (L)
Dactylospora lobariella (Nyl.) Hafellner [DACTLOB] Dactylosporaceae (L)
Daltonia Hook. & Tayl. [DALTONI$] Daltoniaceae (B:M)
Daltonia splachnoides (Sm. in Sm. & Sowerby) Hook. & Tayl. [DALTSPL] Daltoniaceae (B:M)
Danthonia DC. [DANTHON$] Poaceae (V:M)
Danthonia americana Scribn. = Danthonia californica [DANTCAL] Poaceae (V:M)
Danthonia californica Boland. [DANTCAL] Poaceae (V:M)
Danthonia californica var. *americana* (Scribn.) A.S. Hitchc. = Danthonia californica [DANTCAL] Poaceae (V:M)
Danthonia californica var. *palousensis* St. John = Danthonia californica [DANTCAL] Poaceae (V:M)
Danthonia californica var. *piperi* St. John = Danthonia californica [DANTCAL] Poaceae (V:M)
Danthonia canadensis Baum & Findlay = Danthonia intermedia [DANTINT] Poaceae (V:M)
Danthonia cusickii (T.A. Williams) A.S. Hitchc. = Danthonia intermedia [DANTINT] Poaceae (V:M)
*Danthonia decumbens** (L.) DC. [DANTDEC] Poaceae (V:M)
Danthonia intermedia Vasey [DANTINT] Poaceae (V:M)
Danthonia intermedia var. *cusickii* T.A. Williams = Danthonia intermedia [DANTINT] Poaceae (V:M)
Danthonia macounii A.S. Hitchc. = Danthonia californica [DANTCAL] Poaceae (V:M)
Danthonia pinetorum Piper = Danthonia spicata [DANTSPI] Poaceae (V:M)
Danthonia spicata (L.) Beauv. ex Roemer & J.A. Schultes [DANTSPI] Poaceae (V:M)
Danthonia spicata var. *longipila* Scribn. & Merr. = Danthonia spicata [DANTSPI] Poaceae (V:M)
Danthonia spicata var. *pinetorum* Piper = Danthonia spicata [DANTSPI] Poaceae (V:M)
Danthonia thermalis Scribn. = Danthonia spicata [DANTSPI] Poaceae (V:M)
Danthonia unispicata (Thurb.) Munro ex Macoun [DANTUNI] Poaceae (V:M)
Daphne L. [DAPHNE$] Thymelaeaceae (V:D)
*Daphne laureola** L. [DAPHLAU] Thymelaeaceae (V:D)
Dasystephana affinis (Griseb.) Rydb. = Gentiana affinis [GENTAFF] Gentianaceae (V:D)
Dasystephana glauca (Pallas) Rydb. = Gentiana glauca [GENTGLA] Gentianaceae (V:D)
Dasystephana interrupta (Greene) Rydb. = Gentiana affinis [GENTAFF] Gentianaceae (V:D)
Datura L. [DATURA$] Solanaceae (V:D)
*Datura stramonium** L. [DATUSTR] Solanaceae (V:D)
Datura stramonium var. *tatula* (L.) Torr. = Datura stramonium [DATUSTR] Solanaceae (V:D)
Datura tatula L. = Datura stramonium [DATUSTR] Solanaceae (V:D)
Daucus L. [DAUCUS$] Apiaceae (V:D)
*Daucus carota** L. [DAUCCAR] Apiaceae (V:D)
Daucus pusillus Michx. [DAUCPUS] Apiaceae (V:D)
Delphinium L. [DELPHIN$] Ranunculaceae (V:D)
Delphinium ajacis L. = Consolida ajacis [CONSAJA] Ranunculaceae (V:D)
Delphinium ambiguum L. = Consolida ajacis [CONSAJA] Ranunculaceae (V:D)
Delphinium bicolor Nutt. [DELPBIC] Ranunculaceae (V:D)
Delphinium brownii Rydb. = Delphinium glaucum [DELPGLA] Ranunculaceae (V:D)
Delphinium burkei Greene [DELPBUR] Ranunculaceae (V:D)
Delphinium burkei ssp. *distichiflorum* (Hook.) Ewan = Delphinium burkei [DELPBUR] Ranunculaceae (V:D)
Delphinium burkei var. *distichiflorum* (Hook.) St. John = Delphinium burkei [DELPBUR] Ranunculaceae (V:D)
Delphinium depauperatum Nutt. [DELPDEP] Ranunculaceae (V:D)
Delphinium × diversicolor Rydb. = Delphinium burkei [DELPBUR] Ranunculaceae (V:D)
Delphinium glaucum S. Wats. [DELPGLA] Ranunculaceae (V:D)
Delphinium menziesii DC. [DELPMEN] Ranunculaceae (V:D)
Delphinium menziesii ssp. **menziesii** [DELPMEN0] Ranunculaceae (V:D)
Delphinium nelsonii Greene = Delphinium nuttallianum [DELPNUT] Ranunculaceae (V:D)
Delphinium nelsonii ssp. *utahense* (S. Wats.) Ewan = Delphinium nuttallianum [DELPNUT] Ranunculaceae (V:D)
Delphinium nuttallianum Pritz. ex Walp. [DELPNUT] Ranunculaceae (V:D)
Delphinium nuttallianum var. *levicaule* C.L. Hitchc. = Delphinium nuttallianum [DELPNUT] Ranunculaceae (V:D)
Delphinium scopulorum var. *glaucum* (S. Wats.) Gray = Delphinium glaucum [DELPGLA] Ranunculaceae (V:D)
Delphinium strictum A. Nels. = Delphinium burkei [DELPBUR] Ranunculaceae (V:D)
Delphinium strictum var. *distichiflorum* (Hook.) St. John = Delphinium burkei [DELPBUR] Ranunculaceae (V:D)
Dendranthema (DC.) Des Moulins [DENDRAN$] Asteraceae (V:D)
Dendranthema arcticum (L.) Tzvelev [DENDARC] Asteraceae (V:D)
Dendranthema arcticum ssp. **polare** (Hultén) Heywood [DENDARC1] Asteraceae (V:D)
Dendranthema hultenii (A. & D. Löve) Tzvelev = Dendranthema arcticum ssp. polare [DENDARC1] Asteraceae (V:D)
Dendriscocaulon Nyl. [DENDRIS$] Lobariaceae (L)
Dendriscocaulon intricatulum (Nyl.) Henssen [DENDINT] Lobariaceae (L)
Dendroalsia Britt. [DENDROA$] Leucodontaceae (B:M)
Dendroalsia abietina (Hook.) Britt. [DENDABI] Leucodontaceae (B:M)
Dendrobazzania Schust. & Schof. [DENDROB$] Lepidoziaceae (B:H)
Dendrobazzania griffithiana (Steph.) Schust. & Schof. [DENDGRI] Lepidoziaceae (B:H)
Dentaria tenella Pursh = Cardamine nuttallii [CARDNUT] Brassicaceae (V:D)
Dentaria tenella var. *palmata* Detling = Cardamine nuttallii [CARDNUT] Brassicaceae (V:D)
Dermatocarpon Eschw. [DERMATO$] Verrucariaceae (L)
Dermatocarpon aquaticum (Weiss) Zahlbr. = Dermatocarpon luridum [DERMLUR] Verrucariaceae (L)
Dermatocarpon cinereum (Pers.) Th. Fr. = Catapyrenium cinereum [CATACIN] Verrucariaceae (L)
Dermatocarpon compactum (A. Massal.) Lettau = Catapyrenium compactum [CATACOM] Verrucariaceae (L)
Dermatocarpon daedaleum (Kremp.) Th. Fr. = Catapyrenium daedaleum [CATADAE] Verrucariaceae (L)
Dermatocarpon fluviatile (Weber) Th. Fr. = Dermatocarpon luridum [DERMLUR] Verrucariaceae (L)
Dermatocarpon hepaticum (Ach.) Th. Fr. = Catapyrenium cinereum [CATACIN] Verrucariaceae (L)
Dermatocarpon hepaticum auct. = Catapyrenium squamulosum [CATASQA] Verrucariaceae (L)
Dermatocarpon heppioides Zahlbr. = Catapyrenium heppioides [CATAHEP] Verrucariaceae (L)
Dermatocarpon intestiniforme (Körber) Hasse [DERMINT] Verrucariaceae (L)
Dermatocarpon lachneum (Ach.) A.L. Sm. = Catapyrenium lachneum [CATALAC] Verrucariaceae (L)
Dermatocarpon linkolae Räsänen [DERMLIN] Verrucariaceae (L)
Dermatocarpon luridum (With.) J.R. Laundon [DERMLUR] Verrucariaceae (L)
Dermatocarpon miniatum (L.) W. Mann [DERMMIN] Verrucariaceae (L)

Dermatocarpon moulinsii (Mont.) Zahlbr. [DERMMOU] Verrucariaceae (L)
Dermatocarpon muehlenbergii (Ach.) Müll. Arg. = Dermatocarpon miniatum [DERMMIN] Verrucariaceae (L)
Dermatocarpon polyphyllum (Wulfen) Dalla Torre & Sarnth. = Dermatocarpon intestiniforme [DERMINT] Verrucariaceae (L)
Dermatocarpon reticulatum H. Magn. [DERMRET] Verrucariaceae (L)
Dermatocarpon rivulorum (Arnold) Dalla Torre & Sarnth. [DERMRIV] Verrucariaceae (L)
Dermatocarpon rupicola (de Lesd.) Zahlbr. = Catapyrenium compactum [CATACOM] Verrucariaceae (L)
Dermatocarpon vagans Imshaug = Dermatocarpon reticulatum [DERMRET] Verrucariaceae (L)
Dermatocarpon weberi (Ach.) W. Mann = Dermatocarpon luridum [DERMLUR] Verrucariaceae (L)
Deschampsia Beauv. [DESCHAM$] Poaceae (V:M)
Deschampsia atropurpurea (Wahlenb.) Scheele = Vahlodea atropurpurea [VAHLATR] Poaceae (V:M)
Deschampsia atropurpurea var. *latifolia* (Hook.) Scribn. ex Macoun = Vahlodea atropurpurea [VAHLATR] Poaceae (V:M)
Deschampsia atropurpurea var. *paramushirensis* Kudo = Vahlodea atropurpurea [VAHLATR] Poaceae (V:M)
Deschampsia atropurpurea var. *payettii* Lepage = Vahlodea atropurpurea [VAHLATR] Poaceae (V:M)
Deschampsia beringensis Hultén = Deschampsia cespitosa ssp. beringensis [DESCCES1] Poaceae (V:M)
Deschampsia calycina J. Presl = Deschampsia danthonioides [DESCDAN] Poaceae (V:M)
Deschampsia cespitosa (L.) Beauv. [DESCCES] Poaceae (V:M)
Deschampsia cespitosa ssp. **beringensis** (Hultén) W.E. Lawrence [DESCCES1] Poaceae (V:M)
Deschampsia cespitosa ssp. **cespitosa** [DESCCES0] Poaceae (V:M)
Deschampsia cespitosa ssp. *genuina* (Reichenb.) Volk. = Deschampsia cespitosa ssp. cespitosa [DESCCES0] Poaceae (V:M)
Deschampsia cespitosa ssp. **holciformis** (J. Presl) W.E. Lawrence [DESCCES3] Poaceae (V:M)
Deschampsia cespitosa var. *abbei* Boivin = Deschampsia cespitosa ssp. cespitosa [DESCCES0] Poaceae (V:M)
Deschampsia cespitosa var. *alpicola* (Rydb.) A. & D. Löve & Kapoor = Deschampsia cespitosa ssp. cespitosa [DESCCES0] Poaceae (V:M)
Deschampsia cespitosa var. *arctica* Vasey = Deschampsia cespitosa ssp. beringensis [DESCCES1] Poaceae (V:M)
Deschampsia cespitosa var. *intercotidalis* Boivin = Deschampsia cespitosa ssp. cespitosa [DESCCES0] Poaceae (V:M)
Deschampsia cespitosa var. *littoralis* (Gaudin) Richter = Deschampsia cespitosa ssp. cespitosa [DESCCES0] Poaceae (V:M)
Deschampsia cespitosa var. *longiflora* Beal = Deschampsia cespitosa ssp. cespitosa [DESCCES0] Poaceae (V:M)
Deschampsia cespitosa var. *maritima* Vasey = Deschampsia cespitosa ssp. cespitosa [DESCCES0] Poaceae (V:M)
Deschampsia ciliata (Vasey ex Beal) Rydb. = Deschampsia elongata [DESCELO] Poaceae (V:M)
Deschampsia danthonioides (Trin.) Munro [DESCDAN] Poaceae (V:M)
Deschampsia danthonioides var. *gracilis* (Vasey) Munz = Deschampsia danthonioides [DESCDAN] Poaceae (V:M)
Deschampsia elongata (Hook.) Munro [DESCELO] Poaceae (V:M)
•**Deschampsia flexuosa** (L.) Trin. [DESCFLE] Poaceae (V:M)
Deschampsia holciformis J. Presl = Deschampsia cespitosa ssp. holciformis [DESCCES3] Poaceae (V:M)
Deschampsia pacifica Tatew. & Ohwi = Vahlodea atropurpurea [VAHLATR] Poaceae (V:M)
Descurainia Webb & Berth. [DESCURA$] Brassicaceae (V:D)
Descurainia incana (Bernh. ex Fisch. & C.A. Mey.) Dorn [DESCINC] Brassicaceae (V:D)
Descurainia incana ssp. **viscosa** (Rydb.) Kartesz & Gandhi [DESCINC1] Brassicaceae (V:D)
Descurainia incana var. *major* (Hook.) Dorn = Descurainia incana [DESCINC] Brassicaceae (V:D)
Descurainia incana var. *viscosa* (Rydb.) Dorn = Descurainia incana ssp. viscosa [DESCINC1] Brassicaceae (V:D)
Descurainia intermedia (Rydb.) Daniels = Descurainia pinnata ssp. intermedia [DESCPIN2] Brassicaceae (V:D)
Descurainia pinnata (Walt.) Britt. [DESCPIN] Brassicaceae (V:D)
Descurainia pinnata ssp. **filipes** (Gray) Detling [DESCPIN1] Brassicaceae (V:D)
Descurainia pinnata ssp. **intermedia** (Rydb.) Detling [DESCPIN2] Brassicaceae (V:D)
Descurainia pinnata var. *filipes* (Gray) M.E. Peck = Descurainia pinnata ssp. filipes [DESCPIN1] Brassicaceae (V:D)
Descurainia pinnata var. *intermedia* (Rydb.) C.L. Hitchc. = Descurainia pinnata ssp. intermedia [DESCPIN2] Brassicaceae (V:D)
Descurainia richardsonii O.E. Schulz = Descurainia incana [DESCINC] Brassicaceae (V:D)
Descurainia richardsonii ssp. *viscosa* (Rydb.) Detling = Descurainia incana ssp. viscosa [DESCINC1] Brassicaceae (V:D)
Descurainia richardsonii var. *viscosa* (Rydb.) M.E. Peck = Descurainia incana ssp. viscosa [DESCINC1] Brassicaceae (V:D)
•**Descurainia sophia** (L.) Webb ex Prantl [DESCSOP] Brassicaceae (V:D)
Descurainia sophioides (Fisch. ex Hook.) O.E. Schulz [DESCSOH] Brassicaceae (V:D)
Desmatodon Brid. [DESMATO$] Pottiaceae (B:M)
Desmatodon camptothecius Kindb. in Mac. & Kindb. = Desmatodon cernuus [DESMCER] Pottiaceae (B:M)
Desmatodon cernuus (Hüb.) Bruch & Schimp. in B.S.G. [DESMCER] Pottiaceae (B:M)
Desmatodon cernuus var. *xanthopus* Kindb. = Desmatodon cernuus [DESMCER] Pottiaceae (B:M)
Desmatodon coloradensis Grout = Desmatodon obtusifolius [DESMOBT] Pottiaceae (B:M)
Desmatodon convolutus (Brid.) Grout [DESMCON] Pottiaceae (B:M)
Desmatodon guepinii Bruch & Schimp. in B.S.G. [DESMGUE] Pottiaceae (B:M)
Desmatodon heimii (Hedw.) Mitt. [DESMHEI] Pottiaceae (B:M)
Desmatodon hendersonii (Ren. & Card.) Williams in Millsp. & Nutt. = Didymodon tophaceus [DIDYTOP] Pottiaceae (B:M)
Desmatodon latifolius (Hedw.) Brid. [DESMLAT] Pottiaceae (B:M)
Desmatodon latifolius var. *muticus* (Brid.) Brid. = Desmatodon latifolius [DESMLAT] Pottiaceae (B:M)
Desmatodon leucostoma (R. Br.) Berggr. [DESMLEU] Pottiaceae (B:M)
Desmatodon obtusifolius (Schwaegr.) Schimp. [DESMOBT] Pottiaceae (B:M)
Desmatodon randii (Kenn.) Laz. [DESMRAN] Pottiaceae (B:M)
Desmatodon suberectus (Hook.) Limpr. = Desmatodon leucostoma [DESMLEU] Pottiaceae (B:M)
Desmatodon subtorquescens C. Müll. & Kindb. in Mac. & Kindb. = Desmatodon obtusifolius [DESMOBT] Pottiaceae (B:M)
Desmazieria cephalota (Tuck.) Follmann & Huneck = Niebla cephalota [NIEBCEP] Ramalinaceae (L)
Deyeuxia nutkaensis J. Presl = Calamagrostis nutkaensis [CALANUT] Poaceae (V:M)
Dianthus L. [DIANTHU$] Caryophyllaceae (V:D)
•**Dianthus armeria** L. [DIANARM] Caryophyllaceae (V:D)
•**Dianthus barbatus** L. [DIANBAR] Caryophyllaceae (V:D)
•**Dianthus deltoides** L. [DIANDEL] Caryophyllaceae (V:D)
Dianthus prolifer L. = Petrorhagia prolifera [PETRPRO] Caryophyllaceae (V:D)
Diapensia L. [DIAPENS$] Diapensiaceae (V:D)
Diapensia lapponica L. [DIAPLAP] Diapensiaceae (V:D)
Diapensia lapponica ssp. *obovata* (F. Schmidt) Hultén = Diapensia lapponica var. obovata [DIAPLAP1] Diapensiaceae (V:D)
Diapensia lapponica var. **obovata** F. Schmidt [DIAPLAP1] Diapensiaceae (V:D)
Diapensia lapponica var. *rosea* Hultén = Diapensia lapponica var. obovata [DIAPLAP1] Diapensiaceae (V:D)
Diapensia obovata (F. Schmidt) Nakai = Diapensia lapponica var. obovata [DIAPLAP1] Diapensiaceae (V:D)
Dibaeis Clem. [DIBAEIS$] Baeomycetaceae (L)

Dibaeis baeomyces (L. f.) Rambold & Hertel [DIBABAE] Baeomycetaceae (L)
Dibaeis rosea (Pers.) Clem. = Dibaeis baeomyces [DIBABAE] Baeomycetaceae (L)
Dicentra Bernh. [DICENTR$] Fumariaceae (V:D)
Dicentra formosa (Andr.) Walp. [DICEFOR] Fumariaceae (V:D)
Dicentra uniflora Kellogg [DICEUNI] Fumariaceae (V:D)
Dichanthelium (A.S. Hitchc. & Chase) Gould [DICHANT$] Poaceae (V:M)
Dichanthelium acuminatum (Sw.) Gould & C.A. Clark [DICHACU] Poaceae (V:M)
Dichanthelium acuminatum var. **fasciculatum** (Torr.) Freckmann [DICHACU1] Poaceae (V:M)
Dichanthelium acuminatum var. *implicatum* (Scribn.) Gould & C.A. Clark = Dichanthelium acuminatum var. fasciculatum [DICHACU1] Poaceae (V:M)
Dichanthelium acuminatum var. **thermale** (Boland.) Freckmann [DICHACU2] Poaceae (V:M)
Dichanthelium lanuginosum (Ell.) Gould = Dichanthelium acuminatum var. fasciculatum [DICHACU1] Poaceae (V:M)
Dichanthelium lanuginosum var. *fasciculatum* (Torr.) Spellenberg = Dichanthelium acuminatum var. fasciculatum [DICHACU1] Poaceae (V:M)
Dichanthelium lanuginosum var. *thermale* (Boland.) Spellenberg = Dichanthelium acuminatum var. thermale [DICHACU2] Poaceae (V:M)
Dichanthelium oligosanthes (J.A. Schultes) Gould [DICHOLI] Poaceae (V:M)
Dichanthelium oligosanthes var. *helleri* (Nash) Mohlenbrock = Dichanthelium oligosanthes var. scribnerianum [DICHOLI1] Poaceae (V:M)
Dichanthelium oligosanthes var. **scribnerianum** (Nash) Gould [DICHOLI1] Poaceae (V:M)
Dichelyma Myr. [DICHELY$] Fontinalaceae (B:M)
Dichelyma cylindricarpum Aust. = Dichelyma uncinatum [DICHUNC] Fontinalaceae (B:M)
Dichelyma falcatum (Hedw.) Myr. [DICHFAL] Fontinalaceae (B:M)
Dichelyma falcatum var. *uncinatum* (Mitt.) Lawt. = Dichelyma uncinatum [DICHUNC] Fontinalaceae (B:M)
Dichelyma longinerve Kindb. in Mac. = Warnstorfia exannulata var. exannulata [WARNEXA0] Amblystegiaceae (B:M)
Dichelyma uncinatum Mitt. [DICHUNC] Fontinalaceae (B:M)
Dichodontium Schimp. [DICHODO$] Dicranaceae (B:M)
Dichodontium flavescens (With.) Lindb. = Dichodontium pellucidum [DICHPEL] Dicranaceae (B:M)
Dichodontium olympicum Ren. & Card. [DICHOLY] Dicranaceae (B:M)
Dichodontium pellucidum (Hedw.) Schimp. [DICHPEL] Dicranaceae (B:M)
Dichodontium pellucidum ssp. *fagimontanum* (Brid.) Kindb. = Dichodontium pellucidum [DICHPEL] Dicranaceae (B:M)
Dichodontium pellucidum var. *fagimontanum* (Brid.) Schimp. = Dichodontium pellucidum [DICHPEL] Dicranaceae (B:M)
Dichodontium pellucidum var. *flavescens* (With.) Moore = Dichodontium pellucidum [DICHPEL] Dicranaceae (B:M)
Dicranella (C. Müll.) Schimp. [DICRANE$] Dicranaceae (B:M)
Dicranella cerviculata (Hedw.) Schimp. [DICRCER] Dicranaceae (B:M)
Dicranella cerviculata var. *americana* Grout = Dicranella cerviculata [DICRCER] Dicranaceae (B:M)
Dicranella crispa (Hedw.) Schimp. [DICRCRI] Dicranaceae (B:M)
Dicranella grevilleana (Brid.) Schimp. [DICRGRE] Dicranaceae (B:M)
Dicranella heteromalla (Hedw.) Schimp. [DICRHET] Dicranaceae (B:M)
Dicranella heteromalla var. *orthocarpa* (Hedw.) Jaeg. = Dicranella heteromalla [DICRHET] Dicranaceae (B:M)
Dicranella heteromalla var. *sericea* (Schimp.) Pfeff. = Dicranella heteromalla [DICRHET] Dicranaceae (B:M)
Dicranella howei Ren. & Card. [DICRHOW] Dicranaceae (B:M)
Dicranella hutchinsonii Krajina = Dicranella rufescens [DICRRUF] Dicranaceae (B:M)
Dicranella ohioense (Robins.) Crum = Dicranella cerviculata [DICRCER] Dicranaceae (B:M)
Dicranella pacifica Schof. [DICRPAC] Dicranaceae (B:M)
Dicranella palustris (Dicks.) Crundw. ex Warb. [DICRPAL] Dicranaceae (B:M)
Dicranella polaris Kindb. = Dicranella cerviculata [DICRCER] Dicranaceae (B:M)
Dicranella rubra Lindb. = Dicranella varia [DICRVAR] Dicranaceae (B:M)
Dicranella rufescens (With.) Schimp. [DICRRUF] Dicranaceae (B:M)
Dicranella schreberi Schimp. = Dicranella schreberiana [DICRSCH] Dicranaceae (B:M)
Dicranella schreberi var. *elata* Schimp. = Dicranella schreberiana var. robusta [DICRSCH1] Dicranaceae (B:M)
Dicranella schreberiana (Hedw.) Hilf. ex Crum & Anderson [DICRSCH] Dicranaceae (B:M)
Dicranella schreberiana var. **robusta** (Schimp. ex Braithw.) Crum & Anderson [DICRSCH1] Dicranaceae (B:M)
Dicranella schreberiana var. **schreberiana** [DICRSCH0] Dicranaceae (B:M)
Dicranella squarrosa (Stark) Schimp. = Dicranella palustris [DICRPAL] Dicranaceae (B:M)
Dicranella squarrosa f. *fluitans* Grout = Dicranella palustris [DICRPAL] Dicranaceae (B:M)
Dicranella stickinensis Grout [DICRSTI] Dicranaceae (B:M)
Dicranella subulata (Hedw.) Schimp. [DICRSUB] Dicranaceae (B:M)
Dicranella varia (Hedw.) Schimp. [DICRVAR] Dicranaceae (B:M)
Dicranodontium Bruch & Schimp. in B.S.G. [DICRANO$] Dicranaceae (B:M)
Dicranodontium asperulum (Mitt.) Broth. [DICRASP] Dicranaceae (B:M)
Dicranodontium denudatum (Brid.) Britt. in Williams [DICRDEN] Dicranaceae (B:M)
Dicranodontium longirostre (Web. & Mohr) B.S.G. = Dicranodontium denudatum [DICRDEN] Dicranaceae (B:M)
Dicranodontium subporodictyon Broth. [DICRSUP] Dicranaceae (B:M)
Dicranodontium uncinatum (Harv. in Hook.) Jaeg. [DICRUNC] Dicranaceae (B:M)
Dicranoweisia Lindb. ex Milde [DICRANW$] Dicranaceae (B:M)
Dicranoweisia cirrata (Hedw.) Lindb. ex Milde [DICRCIR] Dicranaceae (B:M)
Dicranoweisia contermina Ren. & Card. = Dicranoweisia crispula [DICRCRS] Dicranaceae (B:M)
Dicranoweisia crispula (Hedw.) Lindb. ex Milde [DICRCRS] Dicranaceae (B:M)
Dicranoweisia crispula var. *compacta* (Schleich. ex Schwaegr.) Lindb. in Kindb. = Dicranoweisia crispula [DICRCRS] Dicranaceae (B:M)
Dicranoweisia crispula var. *contermina* (Ren. & Card.) Grout = Dicranoweisia crispula [DICRCRS] Dicranaceae (B:M)
Dicranoweisia crispula var. *roellii* (Kindb. in Röll) Lawt. = Dicranoweisia crispula [DICRCRS] Dicranaceae (B:M)
Dicranoweisia obliqua Kindb. = Kiaeria blyttii [KIAEBLY] Dicranaceae (B:M)
Dicranoweisia roellii Kindb. in Röll = Dicranoweisia crispula [DICRCRS] Dicranaceae (B:M)
Dicranoweisia subcompacta Card. & Thér. = Kiaeria starkei [KIAESTA] Dicranaceae (B:M)
Dicranum Hedw. [DICRANU$] Dicranaceae (B:M)
Dicranum acutifolium (Lindb. & Arnell) C. Jens. ex Weinm. [DICRACU] Dicranaceae (B:M)
Dicranum affine Funck = Dicranum undulatum [DICRUND] Dicranaceae (B:M)
Dicranum alatum (Barnes) Card. & Thér. = Dicranum scoparium [DICRSCO] Dicranaceae (B:M)
Dicranum albicans var. *denticulatum* Kindb. in Mac. & Kindb. = Paraleucobryum longifolium [PARALON] Dicranaceae (B:M)
Dicranum algidum Kindb. = Dicranum spadiceum [DICRSPA] Dicranaceae (B:M)
Dicranum angustum Lindb. [DICRANG] Dicranaceae (B:M)

Dicranum arcticum Schimp. = Kiaeria glacialis [KIAEGLA] Dicranaceae (B:M)
Dicranum bergeri Bland. in Sturm = Dicranum undulatum [DICRUND] Dicranaceae (B:M)
Dicranum bergeri ssp. *rupincola* Kindb. = Dicranum acutifolium [DICRACU] Dicranaceae (B:M)
Dicranum blyttii B.S.G. = Kiaeria blyttii [KIAEBLY] Dicranaceae (B:M)
Dicranum bonjeanii var. *alatum* Barnes in Röll = Dicranum scoparium [DICRSCO] Dicranaceae (B:M)
Dicranum brevifolium (Lindb.) Lindb. [DICRBRE] Dicranaceae (B:M)
Dicranum canadense Kindb. in Mac. = Dicranum scoparium [DICRSCO] Dicranaceae (B:M)
Dicranum cerviculatum Hedw. = Dicranella cerviculata [DICRCER] Dicranaceae (B:M)
Dicranum consobrinum Ren. & Card. = Dicranum scoparium [DICRSCO] Dicranaceae (B:M)
Dicranum crispulum C. Müll. & Kindb. in Mac. & Kindb. = Dicranum fuscescens [DICRFUS] Dicranaceae (B:M)
Dicranum elongatum Schleich. ex Schwaegr. [DICRELO] Dicranaceae (B:M)
Dicranum elongatum ssp. *attenuatum* Kindb. = Dicranum elongatum [DICRELO] Dicranaceae (B:M)
Dicranum elongatum ssp. *subfragilifolium* Kindb. = Dicranum elongatum [DICRELO] Dicranaceae (B:M)
Dicranum falcatum Hedw. = Kiaeria falcata [KIAEFAL] Dicranaceae (B:M)
Dicranum flagellare Hedw. [DICRFLA] Dicranaceae (B:M)
Dicranum flagellare var. *minutissimum* Grout = Dicranum flagellare [DICRFLA] Dicranaceae (B:M)
Dicranum fragilifolium Lindb. [DICRFRA] Dicranaceae (B:M)
Dicranum fulvellum (Dicks.) Sm. = Arctoa fulvella [ARCTFUV] Dicranaceae (B:M)
Dicranum fuscescens Turn. [DICRFUS] Dicranaceae (B:M)
Dicranum fuscescens ssp. *subbrevifolium* Kindb. = Dicranum fuscescens [DICRFUS] Dicranaceae (B:M)
Dicranum groenlandicum Brid. [DICRGRO] Dicranaceae (B:M)
Dicranum heteromallum Hedw. = Dicranella heteromalla [DICRHET] Dicranaceae (B:M)
Dicranum julaceum Hook. & Wils. in Drumm. = Aongstroemia longipes [AONGLON] Dicranaceae (B:M)
Dicranum laevidens Williams = Dicranum angustum [DICRANG] Dicranaceae (B:M)
Dicranum longifolium Hedw. = Paraleucobryum longifolium [PARALON] Dicranaceae (B:M)
Dicranum longifolium var. *subalpinum* Milde = Paraleucobryum longifolium [PARALON] Dicranaceae (B:M)
Dicranum majus Sm. [DICRMAJ] Dicranaceae (B:M)
Dicranum majus var. **majus** [DICRMAJ0] Dicranaceae (B:M)
Dicranum majus var. **orthophyllum** A. Br. ex Milde [DICRMAJ2] Dicranaceae (B:M)
Dicranum microcarpum Hook. in Drumm. = Oncophorus wahlenbergii [ONCOWAH] Dicranaceae (B:M)
Dicranum molle (Wils.) Lindb. = Kiaeria glacialis [KIAEGLA] Dicranaceae (B:M)
Dicranum montanum Hedw. [DICRMON] Dicranaceae (B:M)
Dicranum muehlenbeckii Bruch & Schimp. in B.S.G. [DICRMUE] Dicranaceae (B:M)
Dicranum muehlenbeckii var. *brevifolium* Lindb. = Dicranum brevifolium [DICRBRE] Dicranaceae (B:M)
Dicranum muehlenbeckii var. *cirratum* (Schimp.) Lindb. in Norrl. = Dicranum brevifolium [DICRBRE] Dicranaceae (B:M)
Dicranum neglectum Jur. ex De Not. = Dicranum spadiceum [DICRSPA] Dicranaceae (B:M)
Dicranum pallidisetum (Bail. in Holz.) Irel. [DICRPAI] Dicranaceae (B:M)
Dicranum pallidum Bruch & Schimp. ex C. Müll. = Dicranum scoparium [DICRSCO] Dicranaceae (B:M)
Dicranum palustre La Pyl. in Brid. = Campylopus flexuosus [CAMPFLE] Dicranaceae (B:M)
Dicranum polycarpum (Hedw.) Web. & Mohr = Cynodontium polycarpon [CYNOPOL] Dicranaceae (B:M)
Dicranum polysetum Sw. [DICRPOL] Dicranaceae (B:M)
Dicranum pumilum Saut. = Kiaeria falcata [KIAEFAL] Dicranaceae (B:M)
Dicranum rufescens (With.) Turn. = Dicranella rufescens [DICRRUF] Dicranaceae (B:M)
Dicranum rugosum (Funck) Hoffm. ex Brid. = Dicranum polysetum [DICRPOL] Dicranaceae (B:M)
Dicranum rugosum var. *rugulosum* Kindb. = Dicranum polysetum [DICRPOL] Dicranaceae (B:M)
Dicranum rupincola (Kindb.) Perss. = Dicranum acutifolium [DICRACU] Dicranaceae (B:M)
Dicranum sauteri var. *pachytrichum* Kindb. in Mac. & Kindb. = Paraleucobryum longifolium [PARALON] Dicranaceae (B:M)
Dicranum schisti Lindb. = Kiaeria blyttii [KIAEBLY] Dicranaceae (B:M)
Dicranum schraderi Wahlenb. = Dicranum polysetum [DICRPOL] Dicranaceae (B:M)
Dicranum schreberi Sw. = Dicranella schreberiana [DICRSCH] Dicranaceae (B:M)
Dicranum scopariiforme Kindb. = Dicranum scoparium [DICRSCO] Dicranaceae (B:M)
Dicranum scoparium Hedw. [DICRSCO] Dicranaceae (B:M)
Dicranum scoparium ssp. *involutum* Kindb. in Röll = Dicranum scoparium [DICRSCO] Dicranaceae (B:M)
Dicranum scoparium var. *nigrescens* Györf. = Dicranum scoparium [DICRSCO] Dicranaceae (B:M)
Dicranum scoparium var. *orthophyllum* Brid. = Dicranum scoparium [DICRSCO] Dicranaceae (B:M)
Dicranum scoparium var. *scopariiforme* (Kindb.) Kindb. = Dicranum scoparium [DICRSCO] Dicranaceae (B:M)
Dicranum serratum Kindb. = Paraleucobryum longifolium [PARALON] Dicranaceae (B:M)
Dicranum spadiceum Zett. [DICRSPA] Dicranaceae (B:M)
Dicranum starkei Web. & Mohr = Kiaeria starkei [KIAESTA] Dicranaceae (B:M)
Dicranum stenodictyon Kindb. = Dicranum polysetum [DICRPOL] Dicranaceae (B:M)
Dicranum strictum Schleich. ex Mohr = Dicranum tauricum [DICRTAU] Dicranaceae (B:M)
Dicranum subflagellare Card. & Thér. = Dicranum elongatum [DICRELO] Dicranaceae (B:M)
Dicranum subulatum Hedw. = Dicranella subulata [DICRSUB] Dicranaceae (B:M)
Dicranum subulifolium Kindb. in Mac. = Paraleucobryum enerve [PARAENE] Dicranaceae (B:M)
Dicranum sulcatum Kindb. in Mac. = Dicranum fuscescens [DICRFUS] Dicranaceae (B:M)
Dicranum tauricum Sapenh. [DICRTAU] Dicranaceae (B:M)
Dicranum undulatum Brid. [DICRUND] Dicranaceae (B:M)
Dicranum undulatum Ehrh. ex Web. & Mohr non Brid. = Dicranum polysetum [DICRPOL] Dicranaceae (B:M)
Dicranum undulatum ssp. *ontariense* Kindb. = Dicranum polysetum [DICRPOL] Dicranaceae (B:M)
Dicranum varium Hedw. = Dicranella varia [DICRVAR] Dicranaceae (B:M)
Dicranum virens Hedw. = Oncophorus virens [ONCOVIR] Dicranaceae (B:M)
Dicranum virens var. *wahlenbergii* (Brid.) Hüb. = Oncophorus wahlenbergii [ONCOWAH] Dicranaceae (B:M)
Dictyonema C. Agardh [DICTYON$] Thelephoraceae (L)
Dictyonema moorei (Nyl.) Henssen [DICTMOO] Thelephoraceae (L)
Didymodon Hedw. [DIDYMOD$] Pottiaceae (B:M)
Didymodon acutus (Brid.) Saito = Didymodon rigidulus var. gracilis [DIDYRIG1] Pottiaceae (B:M)
Didymodon acutus var. *ditrichoides* (Broth.) Zand. = Didymodon rigidulus var. gracilis [DIDYRIG1] Pottiaceae (B:M)
Didymodon alpigenum Vent. in Jur. = Bryoerythrophyllum recurvirostre [BRYOREC] Pottiaceae (B:M)
Didymodon asperifolius (Mitt.) Crum, Steere & Anderson [DIDYASP] Pottiaceae (B:M)
Didymodon australasiae (Grev. & Hook.) Zand. = Trichostomopsis australasiae [TRICAUS] Pottiaceae (B:M)

Didymodon australasiae var. *umbrosus* (C. Müll.) Zand. = Trichostomopsis australasiae [TRICAUS] Pottiaceae (B:M)
Didymodon canadensis Kindb. in Mac. & Kindb. = Bryoerythrophyllum recurvirostre [BRYOREC] Pottiaceae (B:M)
Didymodon columbianus Herm. & Lawt. = Bryoerythrophyllum columbianum [BRYOCOL] Pottiaceae (B:M)
Didymodon cylindricus (Brid.) B.S.G. = Oxystegus tenuirostris [OXYSTEN] Pottiaceae (B:M)
Didymodon fallax (Hedw.) Zand. [DIDYFAL] Pottiaceae (B:M)
Didymodon fallax var. **reflexus** (Brid.) Zand. [DIDYFAL1] Pottiaceae (B:M)
Didymodon ferrugineus (Schimp. ex Besch.) M.O. Hill = Didymodon fallax var. reflexus [DIDYFAL1] Pottiaceae (B:M)
Didymodon fragilis Hook. & Wils. ex Drumm. = Tortella fragilis [TORTFRA] Pottiaceae (B:M)
Didymodon fuscoviridis Card. = Didymodon rigidulus var. rigidulus [DIDYRIG0] Pottiaceae (B:M)
Didymodon giganteus (Funck) Jur. = Geheebia gigantea [GEHEGIG] Pottiaceae (B:M)
Didymodon hinckleyi Bartr. = Dicranoweisia cirrata [DICRCIR] Dicranaceae (B:M)
Didymodon icmadophilus (Schimp. ex C. Müll.) Saito = Didymodon rigidulus var. icmadophilus [DIDYRIG2] Pottiaceae (B:M)
Didymodon insulanus (De Not.) M.O. Hill = Didymodon vinealis var. flaccidus [DIDYVIN1] Pottiaceae (B:M)
Didymodon johansenii (Williams) Crum [DIDYJOH] Pottiaceae (B:M)
Didymodon luridus Hornsch. in Spreng. = Didymodon vinealis var. luridus [DIDYVIN3] Pottiaceae (B:M)
Didymodon michiganensis (Steere) Saito [DIDYMIC] Pottiaceae (B:M)
Didymodon nicholsonii Culm. = Didymodon vinealis var. nicholsonii [DIDYVIN4] Pottiaceae (B:M)
Didymodon nigrescens (Mitt.) Saito [DIDYNIG] Pottiaceae (B:M)
Didymodon occidentalis Zand. = Didymodon vinealis var. rubiginosus [DIDYVIN2] Pottiaceae (B:M)
Didymodon recurvirostris Hedw. = Bryoerythrophyllum recurvirostre [BRYOREC] Pottiaceae (B:M)
Didymodon recurvirostris var. *dentatus* (Schimp.) Steere = Bryoerythrophyllum recurvirostre [BRYOREC] Pottiaceae (B:M)
Didymodon revolutus (Card.) Williams [DIDYREV] Pottiaceae (B:M)
Didymodon rigidicaulis (C. Müll.) Saito = Didymodon fallax var. reflexus [DIDYFAL1] Pottiaceae (B:M)
Didymodon rigidulus Hedw. [DIDYRIG] Pottiaceae (B:M)
Didymodon rigidulus var. **gracilis** (Schleich. ex Hook. & Grev.) Zand. [DIDYRIG1] Pottiaceae (B:M)
Didymodon rigidulus var. **icmadophilus** (Schimp. ex C. Müll.) Zand. [DIDYRIG2] Pottiaceae (B:M)
Didymodon rigidulus var. **rigidulus** [DIDYRIG0] Pottiaceae (B:M)
Didymodon rubellus (Bruch & Schimp. in B.S.G.) B.S.G. = Bryoerythrophyllum recurvirostre [BRYOREC] Pottiaceae (B:M)
Didymodon rufus Lor. in Rabenh. = Didymodon asperifolius [DIDYASP] Pottiaceae (B:M)
Didymodon subandreaeoides (Kindb.) Zand. [DIDYSUB] Pottiaceae (B:M)
Didymodon tophaceus (Brid.) Lisa [DIDYTOP] Pottiaceae (B:M)
Didymodon tophaceus var. *decurrens* Card. & Thér. = Didymodon tophaceus [DIDYTOP] Pottiaceae (B:M)
Didymodon trachyneuron Kindb. = Oxystegus tenuirostris [OXYSTEN] Pottiaceae (B:M)
Didymodon trifarius (Hedw.) Röhl. = Saelania glaucescens [SAELGLA] Ditrichaceae (B:M)
Didymodon trifarius auct. = Didymodon vinealis var. luridus [DIDYVIN3] Pottiaceae (B:M)
Didymodon trifarius ssp. *nicholsonii* (Culm.) Wijk & Marg. = Didymodon vinealis var. nicholsonii [DIDYVIN4] Pottiaceae (B:M)
Didymodon umbrosus (C. Müll.) Zand. = Trichostomopsis australasiae [TRICAUS] Pottiaceae (B:M)
Didymodon vinealis (Brid.) Zand. [DIDYVIN] Pottiaceae (B:M)
Didymodon vinealis var. **flaccidus** (Bruch & Schimp. in Schimp.) Zand. [DIDYVIN1] Pottiaceae (B:M)
Didymodon vinealis var. **luridus** (Hornsch. in Spreng.) Zand. [DIDYVIN3] Pottiaceae (B:M)
Didymodon vinealis var. **nicholsonii** (Culm.) Zand. [DIDYVIN4] Pottiaceae (B:M)
Didymodon vinealis var. **rubiginosus** (Mitt.) Zand. [DIDYVIN2] Pottiaceae (B:M)
Didymodon vinealis var. **vinealis** [DIDYVIN0] Pottiaceae (B:M)
Digitalis L. [DIGITAL$] Scrophulariaceae (V:D)
***Digitalis purpurea** L. [DIGIPUR] Scrophulariaceae (V:D)
Digitaria Haller [DIGITAR$] Poaceae (V:M)
***Digitaria ischaemum** (Schreb.) Muhl. [DIGIISC] Poaceae (V:M)
***Digitaria sanguinalis** (L.) Scop. [DIGISAN] Poaceae (V:M)
Dimelaena Norman [DIMELAE$] Physciaceae (L)
Dimelaena novomexicana (de Lesd.) Hale & Culb. = Dimelaena oreina [DIMEORE] Physciaceae (L)
Dimelaena oreina (Ach.) Norman [DIMEORE] Physciaceae (L)
Dimelaena suboreina (de Lesd.) Hale & Culb. = Dimelaena oreina [DIMEORE] Physciaceae (L)
Dimelaena thysanota (Tuck.) Hale & Culb. [DIMETHY] Physciaceae (L)
Dimerella Trevisan [DIMEREL$] Gyalectaceae (L)
Dimerella diluta (Pers.) Trevisan = Dimerella pineti [DIMEPIN] Gyalectaceae (L)
Dimerella lutea (Dickson) Trevisan [DIMELUT] Gyalectaceae (L)
Dimerella pineti (Ach.) Vezda [DIMEPIN] Gyalectaceae (L)
Diphasiastrum alpinum (L.) Holub = Lycopodium alpinum [LYCOALP] Lycopodiaceae (V:F)
Diphasiastrum complanatum (L.) Holub = Lycopodium complanatum [LYCOCOM] Lycopodiaceae (V:F)
Diphasiastrum sitchense (Rupr.) Holub = Lycopodium sitchense [LYCOSIT] Lycopodiaceae (V:F)
Diphasium alpinum (L.) Rothm. = Lycopodium alpinum [LYCOALP] Lycopodiaceae (V:F)
Diphasium anceps (Wallr.) A. & D. Löve = Lycopodium complanatum [LYCOCOM] Lycopodiaceae (V:F)
Diphasium complanatum (L.) Rothm. = Lycopodium complanatum [LYCOCOM] Lycopodiaceae (V:F)
Diphasium complanatum ssp. *montellii* Kukkonen = Lycopodium complanatum [LYCOCOM] Lycopodiaceae (V:F)
Diphasium sitchense (Rupr.) A. & D. Löve = Lycopodium sitchense [LYCOSIT] Lycopodiaceae (V:F)
Diphasium wallrothii H.P. Fuchs = Lycopodium complanatum [LYCOCOM] Lycopodiaceae (V:F)
Diphyscium Mohr [DIPHYSC$] Buxbaumiaceae (B:M)
Diphyscium foliosum (Hedw.) Mohr [DIPHFOL] Buxbaumiaceae (B:M)
Diphyscium sessile Lindb. = Diphyscium foliosum [DIPHFOL] Buxbaumiaceae (B:M)
Diplophylleia albicans (L.) Trevis = Diplophyllum albicans [DIPLALB] Scapaniaceae (B:H)
Diplophylleia plicata (Lindb.) Evans = Diplophyllum plicatum [DIPLPLI] Scapaniaceae (B:H)
Diplophylleia taxifolia (Wahlenb.) Trevis. = Diplophyllum taxifolium [DIPLTAX] Scapaniaceae (B:H)
Diplophyllum (Dum.) Dum. [DIPLOPH$] Scapaniaceae (B:H)
Diplophyllum albicans (L.) Dum. [DIPLALB] Scapaniaceae (B:H)
Diplophyllum gymnostomophilum (Kaal.) Kaal. = Scapania gymnostomophila [SCAPGYM] Scapaniaceae (B:H)
Diplophyllum hyalinum Brinkm. = Diplophyllum imbricatum [DIPLIMB] Scapaniaceae (B:H)
Diplophyllum imbricatum (M.A. Howe) K. Müll. [DIPLIMB] Scapaniaceae (B:H)
Diplophyllum incurvum Bryhn & Kaal. = Scapania gymnostomophila [SCAPGYM] Scapaniaceae (B:H)
Diplophyllum obtusifolium (Hook.) Dum. [DIPLOBU] Scapaniaceae (B:H)
Diplophyllum plicatum Lindb. [DIPLPLI] Scapaniaceae (B:H)
Diplophyllum taxifolium (Wahlenb.) Dum. [DIPLTAX] Scapaniaceae (B:H)
Diploschistes Norman [DIPLOSC$] Thelotremataceae (L)
Diploschistes bisporus (Bagl.) J. Steiner [DIPLBIS] Thelotremataceae (L)

Diploschistes muscorum (Scop.) R. Sant. [DIPLMUS] Thelotremataceae (L)
Diploschistes scruposus (Schreber) Norman [DIPLSCR] Thelotremataceae (L)
Diplotomma Flotow [DIPLOTO$] Physciaceae (L)
Diplotomma alboatrum (Hoffm.) Flotow [DIPLALO] Physciaceae (L)
Diplotomma chlorophaeum (Hepp ex Leighton) Szat. [DIPLCHL] Physciaceae (L)
Diplotomma penichrum (Tuck.) Szat. [DIPLPEN] Physciaceae (L)
Dipsacus L. [DIPSACU$] Dipsacaceae (V:D)
***Dipsacus fullonum** L. [DIPSFUL] Dipsacaceae (V:D)
Discelium Brid. [DISCELI$] Disceliaceae (B:M)
Discelium nudum (Dicks.) Brid. [DISCNUD] Disceliaceae (B:M)
Disporum Salisb. ex D. Don [DISPORU$] Liliaceae (V:M)
Disporum hookeri (Torr.) Nichols. [DISPHOO] Liliaceae (V:M)
Disporum hookeri var. **oreganum** (S. Wats.) Q. Jones [DISPHOO1] Liliaceae (V:M)
Disporum oreganum (S. Wats.) W. Mill. = Disporum hookeri var. oreganum [DISPHOO1] Liliaceae (V:M)
Disporum smithii (Hook.) Piper [DISPSMI] Liliaceae (V:M)
Disporum trachycarpum (S. Wats.) Benth. & Hook. f. [DISPTRA] Liliaceae (V:M)
Disporum trachycarpum var. *subglabrum* E.H. Kelso = Disporum trachycarpum [DISPTRA] Liliaceae (V:M)
Dissodon splachnoides Grev. & Arnell = Tayloria serrata [TAYLSER] Splachnaceae (B:M)
Distegia involucrata (Banks ex Spreng.) Cockerell = Lonicera involucrata [LONIINV] Caprifoliaceae (V:D)
Distichium Bruch & Schimp. in B.S.G. [DISTICI$] Ditrichaceae (B:M)
Distichium capillaceum (Hedw.) Bruch & Schimp. in B.S.G. [DISTCAP] Ditrichaceae (B:M)
Distichium inclinatum (Hedw.) Bruch & Schimp. in B.S.G. [DISTINC] Ditrichaceae (B:M)
Distichium macounii C. Müll. & Kindb. in Mac. & Kindb. = Ditrichum flexicaule [DITRFLE] Ditrichaceae (B:M)
Distichlis Raf. [DISTICH$] Poaceae (V:M)
Distichlis dentata Rydb. = Distichlis spicata [DISTSPI] Poaceae (V:M)
Distichlis spicata (L.) Greene [DISTSPI] Poaceae (V:M)
Distichlis spicata ssp. *stricta* (Torr.) Thorne = Distichlis spicata [DISTSPI] Poaceae (V:M)
Distichlis spicata var. *borealis* (J. Presl) Beetle = Distichlis spicata [DISTSPI] Poaceae (V:M)
Distichlis spicata var. *divaricata* Beetle = Distichlis spicata [DISTSPI] Poaceae (V:M)
Distichlis spicata var. *nana* Beetle = Distichlis spicata [DISTSPI] Poaceae (V:M)
Distichlis spicata var. *stolonifera* Beetle = Distichlis spicata [DISTSPI] Poaceae (V:M)
Distichlis spicata var. *stricta* (Torr.) Scribn. = Distichlis spicata [DISTSPI] Poaceae (V:M)
Distichlis stricta (Torr.) Rydb. = Distichlis spicata [DISTSPI] Poaceae (V:M)
Distichlis stricta var. *dentata* (Rydb.) C.L. Hitchc. = Distichlis spicata [DISTSPI] Poaceae (V:M)
Ditremis biformis (Borrer) R.C. Harris = Anisomeridium biforme [ANISBIF] Monoblastiaceae (L)
Ditrichum Hampe [DITRICH$] Ditrichaceae (B:M)
Ditrichum ambiguum Best [DITRAMB] Ditrichaceae (B:M)
Ditrichum crispatissimum (C. Müll.) Par. = Ditrichum flexicaule [DITRFLE] Ditrichaceae (B:M)
Ditrichum cylindricum (Hedw.) Grout = Trichodon cylindricus [TRICCYL] Ditrichaceae (B:M)
Ditrichum flexicaule (Schwaegr.) Hampe [DITRFLE] Ditrichaceae (B:M)
Ditrichum flexicaule f. *brevifolium* (Kindb.) Grout = Ditrichum flexicaule [DITRFLE] Ditrichaceae (B:M)
Ditrichum flexicaule f. *estellae* Hab. = Ditrichum flexicaule [DITRFLE] Ditrichaceae (B:M)
Ditrichum flexicaule f. *macounii* Mönk. = Ditrichum flexicaule [DITRFLE] Ditrichaceae (B:M)
Ditrichum flexicaule var. *densum* (B.S.G.) Braithw. = Ditrichum flexicaule [DITRFLE] Ditrichaceae (B:M)
Ditrichum flexicaule var. *sterile* (De Not.) Limpr. = Ditrichum flexicaule [DITRFLE] Ditrichaceae (B:M)
Ditrichum giganteum Williams = Ditrichum flexicaule [DITRFLE] Ditrichaceae (B:M)
Ditrichum glaucescens (Hedw.) Hampe = Saelania glaucescens [SAELGLA] Ditrichaceae (B:M)
Ditrichum heteromallum (Hedw.) Britt. [DITRHET] Ditrichaceae (B:M)
Ditrichum homomallum (Hedw.) Hampe = Ditrichum heteromallum [DITRHET] Ditrichaceae (B:M)
Ditrichum montanum Leib. [DITRMON] Ditrichaceae (B:M)
Ditrichum pusillum (Hedw.) Hampe [DITRPUS] Ditrichaceae (B:M)
Ditrichum pusillum var. *tortile* (Schrad.) Hag. = Ditrichum pusillum [DITRPUS] Ditrichaceae (B:M)
Ditrichum schimperi (Lesq.) Kuntze [DITRSCH] Ditrichaceae (B:M)
Ditrichum tenuifolium Lindb. = Trichodon cylindricus [TRICCYL] Ditrichaceae (B:M)
Ditrichum tortile (Schrad.) Brockm. = Ditrichum pusillum [DITRPUS] Ditrichaceae (B:M)
Ditrichum tortile ssp. *pusillum* Hedw. = Ditrichum pusillum [DITRPUS] Ditrichaceae (B:M)
Ditrichum tortuloides Grout = Ditrichum ambiguum [DITRAMB] Ditrichaceae (B:M)
Ditrichum zonatum (Brid.) Kindb. [DITRZON] Ditrichaceae (B:M)
Ditrichum zonatum var. **scabrifolium** Dix. [DITRZON1] Ditrichaceae (B:M)
Ditrichum zonatum var. **zonatum** [DITRZON0] Ditrichaceae (B:M)
Dodecatheon L. [DODECAT$] Primulaceae (V:D)
Dodecatheon conjugens Greene [DODECON] Primulaceae (V:D)
Dodecatheon conjugens ssp. **conjugens** [DODECON0] Primulaceae (V:D)
Dodecatheon conjugens ssp. *leptophyllum* (Suksdorf) Piper = Dodecatheon conjugens ssp. conjugens [DODECON0] Primulaceae (V:D)
Dodecatheon conjugens ssp. **viscidum** (Piper) H.J. Thompson [DODECON1] Primulaceae (V:D)
Dodecatheon conjugens var. *beamishii* Boivin = Dodecatheon conjugens ssp. conjugens [DODECON0] Primulaceae (V:D)
Dodecatheon conjugens var. *viscidum* (Piper) Mason ex St. John = Dodecatheon conjugens ssp. viscidum [DODECON1] Primulaceae (V:D)
Dodecatheon cusickii Greene = Dodecatheon pulchellum ssp. cusickii [DODEPUL1] Primulaceae (V:D)
Dodecatheon cusickii var. *album* Suksdorf = Dodecatheon pulchellum ssp. cusickii [DODEPUL1] Primulaceae (V:D)
Dodecatheon cylindrocarpum Rydb. = Dodecatheon conjugens ssp. conjugens [DODECON0] Primulaceae (V:D)
Dodecatheon dentatum Hook. [DODEDEN] Primulaceae (V:D)
Dodecatheon frigidum Cham. & Schlecht. [DODEFRI] Primulaceae (V:D)
Dodecatheon hendersonii Gray [DODEHEN] Primulaceae (V:D)
Dodecatheon jeffreyi Van Houtte [DODEJEF] Primulaceae (V:D)
Dodecatheon pauciflorum var. *alaskanum* (Hultén) C.L. Hitchc. = Dodecatheon pulchellum ssp. macrocarpum [DODEPUL2] Primulaceae (V:D)
Dodecatheon pauciflorum var. *cusickii* (Greene) Mason ex St. John = Dodecatheon pulchellum ssp. cusickii [DODEPUL1] Primulaceae (V:D)
Dodecatheon pauciflorum var. *salinum* (A. Nels.) R. Knuth = Dodecatheon pulchellum ssp. pulchellum [DODEPUL0] Primulaceae (V:D)
Dodecatheon pauciflorum var. *watsonii* (Tidestrom) C.L. Hitchc. = Dodecatheon pulchellum ssp. pulchellum [DODEPUL0] Primulaceae (V:D)
Dodecatheon puberulum (Nutt.) Piper = Dodecatheon pulchellum ssp. cusickii [DODEPUL1] Primulaceae (V:D)
Dodecatheon pulchellum (Raf.) Merr. [DODEPUL] Primulaceae (V:D)
Dodecatheon pulchellum ssp. *alaskanum* (Hultén) Hultén = Dodecatheon pulchellum ssp. macrocarpum [DODEPUL2] Primulaceae (V:D)

Dodecatheon pulchellum ssp. **cusickii** (Greene) Calder & Taylor [DODEPUL1] Primulaceae (V:D)
Dodecatheon pulchellum ssp. **macrocarpum** (Gray) Taylor & MacBryde [DODEPUL2] Primulaceae (V:D)
Dodecatheon pulchellum ssp. **pulchellum** [DODEPUL0] Primulaceae (V:D)
Dodecatheon pulchellum ssp. *superbum* (Pennell & Stair) Hultén = Dodecatheon pulchellum ssp. macrocarpum [DODEPUL2] Primulaceae (V:D)
Dodecatheon pulchellum var. *alaskanum* (Hultén) Boivin = Dodecatheon pulchellum ssp. macrocarpum [DODEPUL2] Primulaceae (V:D)
Dodecatheon pulchellum var. *alaskanum* (Hultén) Reveal = Dodecatheon pulchellum ssp. macrocarpum [DODEPUL2] Primulaceae (V:D)
Dodecatheon pulchellum var. *album* (Suksdorf) Boivin = Dodecatheon pulchellum ssp. cusickii [DODEPUL1] Primulaceae (V:D)
Dodecatheon pulchellum var. *cusickii* (Greene) Reveal = Dodecatheon pulchellum ssp. cusickii [DODEPUL1] Primulaceae (V:D)
Dodecatheon pulchellum var. *watsonii* (Tidestrom) C.L. Hitchc. = Dodecatheon pulchellum ssp. pulchellum [DODEPUL0] Primulaceae (V:D)
Dodecatheon pulchellum var. *watsonii* (Tidestrom) Boivin = Dodecatheon pulchellum ssp. pulchellum [DODEPUL0] Primulaceae (V:D)
Dodecatheon pulchellum var. *watsonii* (Tidestrom) Reveal = Dodecatheon pulchellum ssp. pulchellum [DODEPUL0] Primulaceae (V:D)
Dodecatheon pulchellum var. *zionense* (Eastw.) Welsh = Dodecatheon pulchellum ssp. pulchellum [DODEPUL0] Primulaceae (V:D)
Dodecatheon radicatum Greene = Dodecatheon pulchellum ssp. pulchellum [DODEPUL0] Primulaceae (V:D)
Dodecatheon radicatum ssp. *macrocarpum* (Gray) Beamish = Dodecatheon pulchellum ssp. macrocarpum [DODEPUL2] Primulaceae (V:D)
Dodecatheon radicatum ssp. *watsonii* (Tidestrom) H.J. Thompson = Dodecatheon pulchellum ssp. pulchellum [DODEPUL0] Primulaceae (V:D)
Dodecatheon salinum A. Nels. = Dodecatheon pulchellum ssp. pulchellum [DODEPUL0] Primulaceae (V:D)
Dodecatheon viscidum Piper = Dodecatheon conjugens ssp. viscidum [DODECON1] Primulaceae (V:D)
Dodecatheon zionense Eastw. = Dodecatheon pulchellum ssp. pulchellum [DODEPUL0] Primulaceae (V:D)
Doronicum L. [DORONIC$] Asteraceae (V:D)
***Doronicum pardalianches** L. [DOROPAR] Asteraceae (V:D)
Douglasia Lindl. [DOUGLAS$] Primulaceae (V:D)
Douglasia alaskana (Coville & Standl. ex Hultén) S. Kelso [DOUGALA] Primulaceae (V:D)
Douglasia arctica var. *gormanii* (Constance) Boivin = Douglasia gormanii [DOUGGOR] Primulaceae (V:D)
Douglasia biflora A. Nels. = Douglasia montana [DOUGMON] Primulaceae (V:D)
Douglasia gormanii Constance [DOUGGOR] Primulaceae (V:D)
Douglasia laevigata Gray [DOUGLAE] Primulaceae (V:D)
Douglasia laevigata ssp. *ciliolata* (Constance) Calder & Taylor = Douglasia laevigata var. ciliolata [DOUGLAE2] Primulaceae (V:D)
Douglasia laevigata var. **ciliolata** Constance [DOUGLAE2] Primulaceae (V:D)
Douglasia montana Gray [DOUGMON] Primulaceae (V:D)
Douglasia montana var. *biflora* (A. Nels.) R. Knuth = Douglasia montana [DOUGMON] Primulaceae (V:D)
Douglasia nivalis Lindl. [DOUGNIV] Primulaceae (V:D)
Douglasia ochotensis ssp. *gormanii* (Constance) A. & D. Löve = Douglasia gormanii [DOUGGOR] Primulaceae (V:D)
Douinia (C. Jens.) Buch [DOUINIA$] Scapaniaceae (B:H)
Douinia ovata (Dicks.) Buch [DOUIOVA] Scapaniaceae (B:H)
Downingia Torr. [DOWINGI$] Campanulaceae (V:D)
Downingia elegans (Dougl. ex Lindl.) Torr. [DOWIELE] Campanulaceae (V:D)
Draba L. [DRABA$] Brassicaceae (V:D)
Draba albertina Greene [DRABALB] Brassicaceae (V:D)
Draba allenii Fern. = Draba lactea [DRABLAC] Brassicaceae (V:D)
Draba alpina L. [DRABALP] Brassicaceae (V:D)
Draba alpina var. *nana* Hook. = Draba alpina [DRABALP] Brassicaceae (V:D)
Draba alpina var. *pilosa* (M.F. Adams ex DC.) Regel = Draba alpina [DRABALP] Brassicaceae (V:D)
Draba arabisans var. *canadensis* (Burnet) Fern. & Knowlt. = Draba glabella [DRABGLA] Brassicaceae (V:D)
Draba aurea Vahl ex Hornem. [DRABAUR] Brassicaceae (V:D)
Draba aurea var. *leiocarpa* (Payson & St. John) C.L. Hitchc. = Draba aurea [DRABAUR] Brassicaceae (V:D)
Draba aurea var. *neomexicana* (Greene) Tidestrom = Draba aurea [DRABAUR] Brassicaceae (V:D)
Draba bellii Holm = Draba corymbosa [DRABCOR] Brassicaceae (V:D)
Draba borealis DC. [DRABBOR] Brassicaceae (V:D)
Draba caeruleomontana Payson & St. John = Draba densifolia [DRABDEN] Brassicaceae (V:D)
Draba cana Rydb. [DRABCAN] Brassicaceae (V:D)
Draba caroliniana Walt. = Draba reptans [DRABREP] Brassicaceae (V:D)
Draba cascadensis Payson & St. John = Draba praealta [DRABPRA] Brassicaceae (V:D)
Draba cinerea M.F. Adams [DRABCIN] Brassicaceae (V:D)
Draba corymbosa R. Br. ex DC. [DRABCOR] Brassicaceae (V:D)
Draba crassifolia Graham [DRABCRA] Brassicaceae (V:D)
Draba densifolia Nutt. [DRABDEN] Brassicaceae (V:D)
Draba eschscholtzii Pohle ex N. Busch = Draba alpina [DRABALP] Brassicaceae (V:D)
Draba exalata Ekman = Draba ruaxes [DRABRUA] Brassicaceae (V:D)
Draba fladnizensis Wulfen [DRABFLA] Brassicaceae (V:D)
Draba fladnizensis var. *heterotricha* (Lindbl.) J. Ball = Draba lactea [DRABLAC] Brassicaceae (V:D)
Draba glabella Pursh [DRABGLA] Brassicaceae (V:D)
Draba hatchiae Mulligan = Draba hyperborea [DRABHYP] Brassicaceae (V:D)
Draba hyperborea (L.) Desv. [DRABHYP] Brassicaceae (V:D)
Draba incerta Payson [DRABINC] Brassicaceae (V:D)
Draba juniperina Dorn = Draba oligosperma [DRABOLI] Brassicaceae (V:D)
Draba kananaskis Mulligan = Draba longipes [DRABLOG] Brassicaceae (V:D)
Draba lactea M.F. Adams [DRABLAC] Brassicaceae (V:D)
Draba laevicapsula Payson = Draba incerta [DRABINC] Brassicaceae (V:D)
Draba lonchocarpa Rydb. [DRABLON] Brassicaceae (V:D)
Draba lonchocarpa var. *exigua* O.E. Schulz = Draba lonchocarpa var. lonchocarpa [DRABLON0] Brassicaceae (V:D)
Draba lonchocarpa var. **lonchocarpa** [DRABLON0] Brassicaceae (V:D)
Draba lonchocarpa var. **thompsonii** (C.L. Hitchc.) Rollins [DRABLON2] Brassicaceae (V:D)
Draba lonchocarpa var. **vestita** O.E. Schulz [DRABLON3] Brassicaceae (V:D)
Draba longipes Raup [DRABLOG] Brassicaceae (V:D)
Draba macounii O.E. Schulz [DRABMAC] Brassicaceae (V:D)
Draba mccallae Rydb. = Draba borealis [DRABBOR] Brassicaceae (V:D)
Draba micrantha Nutt. = Draba reptans [DRABREP] Brassicaceae (V:D)
Draba micropetala Hook. = Draba alpina [DRABALP] Brassicaceae (V:D)
Draba minganensis (Victorin) Fern. = Draba aurea [DRABAUR] Brassicaceae (V:D)
Draba nelsonii J.F. Macbr. & Payson = Draba densifolia [DRABDEN] Brassicaceae (V:D)
Draba nemorosa L. [DRABNEM] Brassicaceae (V:D)
Draba neomexicana Greene = Draba aurea [DRABAUR] Brassicaceae (V:D)
Draba nitida Greene = Draba albertina [DRABALB] Brassicaceae (V:D)

Draba nivalis Lilj. [DRABNIV] Brassicaceae (V:D)
Draba nivalis ssp. *lonchocarpa* (Rydb.) Hultén = Draba lonchocarpa var. lonchocarpa [DRABLON0] Brassicaceae (V:D)
Draba nivalis var. *brevicula* Rollins = Draba porsildii [DRABPOR] Brassicaceae (V:D)
Draba nivalis var. *elongata* S. Wats. = Draba lonchocarpa var. lonchocarpa [DRABLON0] Brassicaceae (V:D)
Draba nivalis var. *exigua* (O.E. Schulz) C.L. Hitchc. = Draba lonchocarpa var. lonchocarpa [DRABLON0] Brassicaceae (V:D)
Draba nivalis var. *thompsonii* C.L. Hitchc. = Draba lonchocarpa var. thompsonii [DRABLON2] Brassicaceae (V:D)
Draba oligosperma Hook. [DRABOLI] Brassicaceae (V:D)
Draba oligosperma var. *juniperina* (Dorn) Welsh = Draba oligosperma [DRABOLI] Brassicaceae (V:D)
Draba oligosperma var. *subsessilis* (S. Wats.) O.E. Schulz = Draba oligosperma [DRABOLI] Brassicaceae (V:D)
Draba paysonii J.F. Macbr. [DRABPAY] Brassicaceae (V:D)
Draba paysonii var. **treleasii** (O.E. Schulz) C.L. Hitchc. [DRABPAY1] Brassicaceae (V:D)
Draba peasei Fern. = Draba incerta [DRABINC] Brassicaceae (V:D)
Draba pilosa M.F. Adams ex DC. = Draba alpina [DRABALP] Brassicaceae (V:D)
Draba porsildii Mulligan [DRABPOR] Brassicaceae (V:D)
Draba praealta Greene [DRABPRA] Brassicaceae (V:D)
Draba praecox Stev. = Draba verna [DRABVER] Brassicaceae (V:D)
Draba reptans (Lam.) Fern. [DRABREP] Brassicaceae (V:D)
Draba reptans ssp. *stellifera* (O.E. Schulz) Abrams = Draba reptans [DRABREP] Brassicaceae (V:D)
Draba reptans var. *micrantha* (Nutt.) Fern. = Draba reptans [DRABREP] Brassicaceae (V:D)
Draba reptans var. *stellifera* (O.E. Schulz) C.L. Hitchc. = Draba reptans [DRABREP] Brassicaceae (V:D)
Draba reptans var. *typica* C.L. Hitchc. = Draba reptans [DRABREP] Brassicaceae (V:D)
Draba ruaxes Payson & St. John [DRABRUA] Brassicaceae (V:D)
Draba sphaerula J.F. Macbr. & Payson = Draba densifolia [DRABDEN] Brassicaceae (V:D)
Draba stenoloba Ledeb. [DRABSTE] Brassicaceae (V:D)
Draba stenoloba var. *nana* (O.E. Schulz) C.L. Hitchc. = Draba albertina [DRABALB] Brassicaceae (V:D)
Draba stylaris J. Gay ex W.D.J. Koch = Draba cana [DRABCAN] Brassicaceae (V:D)
Draba subsessilis S. Wats. = Draba oligosperma [DRABOLI] Brassicaceae (V:D)
Draba tschuktschorum Trautv. = Draba fladnizensis [DRABFLA] Brassicaceae (V:D)
Draba ventosa var. *ruaxes* (Payson & St. John) C.L. Hitchc. = Draba ruaxes [DRABRUA] Brassicaceae (V:D)
***Draba verna** L. [DRABVER] Brassicaceae (V:D)
Draba verna var. *aestivalis* Lej. = Draba verna [DRABVER] Brassicaceae (V:D)
Draba verna var. *boerhaavii* van Hall = Draba verna [DRABVER] Brassicaceae (V:D)
Draba verna var. *major* Stur = Draba verna [DRABVER] Brassicaceae (V:D)
Dracocephalum L. [DRACOCE$] Lamiaceae (V:D)
Dracocephalum nuttallii Britt. = Physostegia parviflora [PHYSPAR] Lamiaceae (V:D)
Dracocephalum parviflorum Nutt. [DRACPAR] Lamiaceae (V:D)
***Dracocephalum thymiflorum** L. [DRACTHY] Lamiaceae (V:D)
Drepanocladus (C. Müll.) G. Roth (Amblystegiacea [DREPANO$] Amblystegiaceae (B:M)
Drepanocladus aduncus (Hedw.) Warnst. [DREPADU] Amblystegiaceae (B:M)
Drepanocladus aduncus f. *aquaticus* (Sanio) Mönk. = Drepanocladus aduncus [DREPADU] Amblystegiaceae (B:M)
Drepanocladus aduncus f. *laxus* Husn. = Drepanocladus aduncus [DREPADU] Amblystegiaceae (B:M)
Drepanocladus aduncus var. **aduncus** [DREPADU0] Amblystegiaceae (B:M)
Drepanocladus aduncus var. *aquaticus* (Sanio) Ren. = Drepanocladus aduncus [DREPADU] Amblystegiaceae (B:M)
Drepanocladus aduncus var. *capillifolius* (Warnst.) Riehm. = Drepanocladus capillifolius [DREPCAP] Amblystegiaceae (B:M)
Drepanocladus aduncus var. *intermedius* (Schimp.) Roth = Drepanocladus aduncus [DREPADU] Amblystegiaceae (B:M)
Drepanocladus aduncus var. **kneiffii** (Schimp. in B.S.G.) Mönk. [DREPADU3] Amblystegiaceae (B:M)
Drepanocladus aduncus var. *kneiffii* f. *intermedius* (B.S.G.) Mönk. = Drepanocladus aduncus [DREPADU] Amblystegiaceae (B:M)
Drepanocladus aduncus var. *paternus* (Sanio) Jel. = Drepanocladus aduncus [DREPADU] Amblystegiaceae (B:M)
Drepanocladus aduncus var. **polycarpus** (Bland. ex Voit) G. Roth [DREPADU4] Amblystegiaceae (B:M)
Drepanocladus aduncus var. *polycarpus* f. *gracilescens* (B.S.G.) Mönk. = Drepanocladus aduncus var. polycarpus [DREPADU4] Amblystegiaceae (B:M)
Drepanocladus aduncus var. *pseudofluitans* (Sanio) Glow. = Drepanocladus aduncus var. aduncus [DREPADU0] Amblystegiaceae (B:M)
Drepanocladus aduncus var. *typicus* f. *capillifolius* Mönk. = Drepanocladus capillifolius [DREPCAP] Amblystegiaceae (B:M)
Drepanocladus badius (Hartm.) G. Roth = Loeskypnum badium [LOESBAD] Amblystegiaceae (B:M)
Drepanocladus berggrenii (C. Jens.) G. Roth = Warnstorfia fluitans var. fluitans [WARNFLU0] Amblystegiaceae (B:M)
Drepanocladus capillifolius (Warnst.) Warnst. [DREPCAP] Amblystegiaceae (B:M)
Drepanocladus capillifolius f. *fallax* Ren. = Drepanocladus capillifolius [DREPCAP] Amblystegiaceae (B:M)
Drepanocladus crassicostatus Janssens [DREPCRA] Amblystegiaceae (B:M)
Drepanocladus exannulatus (Schimp. in B.S.G.) Warnst. = Warnstorfia exannulata var. exannulata [WARNEXA0] Amblystegiaceae (B:M)
Drepanocladus exannulatus f. *brachydictyus* (Ren.) Smirn. = Warnstorfia exannulata var. exannulata [WARNEXA0] Amblystegiaceae (B:M)
Drepanocladus exannulatus f. *falcifolius* Ren. = Warnstorfia exannulata var. exannulata [WARNEXA0] Amblystegiaceae (B:M)
Drepanocladus exannulatus f. *orthophyllus* (Milde) Mönk. = Warnstorfia exannulata var. exannulata [WARNEXA0] Amblystegiaceae (B:M)
Drepanocladus exannulatus f. *submersus* Mönk. = Warnstorfia exannulata var. exannulata [WARNEXA0] Amblystegiaceae (B:M)
Drepanocladus exannulatus f. *tundrae* (Arnell) Mönk. = Warnstorfia tundrae [WARNTUN] Amblystegiaceae (B:M)
Drepanocladus exannulatus var. *alpinus* (Grav.) Wijk & Marg. = Warnstorfia exannulata var. exannulata [WARNEXA0] Amblystegiaceae (B:M)
Drepanocladus exannulatus var. *brachydictyon* (Ren.) G. Roth = Warnstorfia exannulata var. exannulata [WARNEXA0] Amblystegiaceae (B:M)
Drepanocladus exannulatus var. *falcifolius* (Ren.) Grout = Warnstorfia exannulata var. exannulata [WARNEXA0] Amblystegiaceae (B:M)
Drepanocladus exannulatus var. *purpurascens* (Schimp.) Herz. = Warnstorfia exannulata var. purpurascens [WARNEXA1] Amblystegiaceae (B:M)
Drepanocladus exannulatus var. *rotae* (De Not.) Loeske = Warnstorfia exannulata var. exannulata [WARNEXA0] Amblystegiaceae (B:M)
Drepanocladus exannulatus var. *serratus* (Milde) Loeske = Warnstorfia exannulata var. exannulata [WARNEXA0] Amblystegiaceae (B:M)
Drepanocladus exannulatus var. *tundrae* (Arnell) Warnst. = Warnstorfia tundrae [WARNTUN] Amblystegiaceae (B:M)
Drepanocladus fluitans (Hedw.) Warnst. = Warnstorfia fluitans [WARNFLU] Amblystegiaceae (B:M)
Drepanocladus fluitans f. *gracilis* Boul. = Warnstorfia fluitans [WARNFLU] Amblystegiaceae (B:M)
Drepanocladus fluitans f. *jeanbernatii* (Ren.) Mönk. = Warnstorfia fluitans [WARNFLU] Amblystegiaceae (B:M)
Drepanocladus fluitans f. *setiformis* (Ren.) Mönk. = Warnstorfia fluitans [WARNFLU] Amblystegiaceae (B:M)

Drepanocladus fluitans var. *berggrenii* (C. Jens.) C. Jens. in Weim. = Warnstorfia fluitans [WARNFLU] Amblystegiaceae (B:M)
Drepanocladus fluitans var. *falcatus* (Sanio ex C. Jens.) G. Roth = Warnstorfia fluitans var. falcata [WARNFLU1] Amblystegiaceae (B:M)
Drepanocladus fluitans var. *falcifolius* f. *viridis* Boul. = Warnstorfia exannulata var. exannulata [WARNEXA0] Amblystegiaceae (B:M)
Drepanocladus fluitans var. *gracilis* Warnst. = Warnstorfia fluitans [WARNFLU] Amblystegiaceae (B:M)
Drepanocladus fluitans var. *jeanbernatii* (Ren.) Warnst. = Warnstorfia fluitans [WARNFLU] Amblystegiaceae (B:M)
Drepanocladus fluitans var. *submersus* (Schimp.) Loeske = Warnstorfia fluitans [WARNFLU] Amblystegiaceae (B:M)
Drepanocladus fluitans var. *uncatus* Crum, Steere & Anderson = Warnstorfia fluitans var. falcata [WARNFLU1] Amblystegiaceae (B:M)
Drepanocladus intermedius (Lindb.) Warnst. = Limprichtia cossonii [LIMPCOS] Amblystegiaceae (B:M)
Drepanocladus lapponicus (Norrl.) Smirn. = Hamatocaulis lapponicus [HAMALAP] Amblystegiaceae (B:M)
Drepanocladus latinervis Warnst. = Drepanocladus sendtneri [DREPSEN] Amblystegiaceae (B:M)
Drepanocladus polycarpus (Voit) Warnst. = Drepanocladus aduncus var. polycarpus [DREPADU4] Amblystegiaceae (B:M)
Drepanocladus procerus (Ren. & Arnell in Husn.) = Warnstorfia procera [WARNPRO] Amblystegiaceae (B:M)
Drepanocladus pseudosarmentosus (Card. & Thér.) Perss. [DREPPSE] Amblystegiaceae (B:M)
Drepanocladus pseudostramineus (C. Müll.) G. Roth = Warnstorfia pseudostraminea [WARNPSE] Amblystegiaceae (B:M)
Drepanocladus purpurascens (Schimp.) Loeske = Warnstorfia exannulata var. purpurascens [WARNEXA1] Amblystegiaceae (B:M)
Drepanocladus revolvens (Sw.) Warnst. = Limprichtia revolvens [LIMPREV] Amblystegiaceae (B:M)
Drepanocladus revolvens var. *intermedia* (Lindb.) Grout = Limprichtia cossonii [LIMPCOS] Amblystegiaceae (B:M)
Drepanocladus revolvens var. *miquelonensis* (Ren.) Grout = Scorpidium scorpioides [SCORSCO] Amblystegiaceae (B:M)
Drepanocladus schulzei G. Roth = Warnstorfia fluitans var. falcata [WARNFLU1] Amblystegiaceae (B:M)
Drepanocladus scorpioides (Hedw.) Warnst. = Scorpidium scorpioides [SCORSCO] Amblystegiaceae (B:M)
Drepanocladus sendtneri (Schimp.) Warnst. [DREPSEN] Amblystegiaceae (B:M)
Drepanocladus sendtneri var. *wilsonii* (Lindb.) Warnst. = Drepanocladus sendtneri [DREPSEN] Amblystegiaceae (B:M)
Drepanocladus submersus (Schimp.) Warnst. = Warnstorfia fluitans [WARNFLU] Amblystegiaceae (B:M)
Drepanocladus trichophyllus (Warnst.) Podp. = Warnstorfia trichophylla [WARNTRI] Amblystegiaceae (B:M)
Drepanocladus tundrae (Arnell) Loeske = Warnstorfia tundrae [WARNTUN] Amblystegiaceae (B:M)
Drepanocladus uncinatus (Hedw.) Warnst. = Sanionia uncinata [SANIUNC] Amblystegiaceae (B:M)
Drepanocladus uncinatus var. *alpinus* (Ren.) Warnst. = Sanionia uncinata [SANIUNC] Amblystegiaceae (B:M)
Drepanocladus uncinatus var. *gracilescens* (B.S.G.) Warnst. = Sanionia uncinata [SANIUNC] Amblystegiaceae (B:M)
Drepanocladus uncinatus var. *gracillimus* (Berggr.) Storm. = Sanionia uncinata [SANIUNC] Amblystegiaceae (B:M)
Drepanocladus uncinatus var. *plumosus* (Schimp.) Warnst. = Sanionia uncinata var. uncinata [SANIUNC0] Amblystegiaceae (B:M)
Drepanocladus uncinatus var. *plumulosus* (Schimp. in B.S.G.) Warnst. = Sanionia uncinata var. uncinata [SANIUNC0] Amblystegiaceae (B:M)
Drepanocladus uncinatus var. *subjulaceus* (Schimp. in B.S.G.) Warnst. = Sanionia uncinata var. uncinata [SANIUNC0] Amblystegiaceae (B:M)
Drepanocladus uncinatus var. *symmetricus* (Ren. & Card.) Grout = Sanionia uncinata var. symmetrica [SANIUNC1] Amblystegiaceae (B:M)
Drepanocladus vernicosus (Mitt.) Warnst. = Hamatocaulis vernicosus [HAMAVER] Amblystegiaceae (B:M)
Drosera L. [DROSERA$] Droseraceae (V:D)
Drosera anglica Huds. [DROSANG] Droseraceae (V:D)
Drosera longifolia L. = Drosera anglica [DROSANG] Droseraceae (V:D)
Drosera rotundifolia L. [DROSROT] Droseraceae (V:D)
Dryas L. [DRYAS$] Rosaceae (V:D)
Dryas alaskensis Porsild = Dryas octopetala ssp. alaskensis [DRYAOCT1] Rosaceae (V:D)
Dryas drummondii Richards. ex Hook. [DRYADRU] Rosaceae (V:D)
Dryas drummondii var. **drummondii** [DRYADRU0] Rosaceae (V:D)
Dryas drummondii var. **eglandulosa** Porsild [DRYADRU2] Rosaceae (V:D)
Dryas drummondii var. **tomentosa** (Farr) L.O. Williams [DRYADRU3] Rosaceae (V:D)
Dryas hookeriana Juz. = Dryas octopetala ssp. hookeriana [DRYAOCT2] Rosaceae (V:D)
Dryas integrifolia Vahl [DRYAINT] Rosaceae (V:D)
Dryas integrifolia ssp. **integrifolia** [DRYAINT0] Rosaceae (V:D)
Dryas integrifolia ssp. **sylvatica** (Hultén) Hultén [DRYAINT2] Rosaceae (V:D)
Dryas integrifolia var. *canescens* Simm. = Dryas integrifolia ssp. integrifolia [DRYAINT0] Rosaceae (V:D)
Dryas octopetala L. [DRYAOCT] Rosaceae (V:D)
Dryas octopetala ssp. **alaskensis** (Porsild) Hultén [DRYAOCT1] Rosaceae (V:D)
Dryas octopetala ssp. **hookeriana** (Juz.) Hultén [DRYAOCT2] Rosaceae (V:D)
Dryas octopetala ssp. **octopetala** [DRYAOCT0] Rosaceae (V:D)
Dryas octopetala var. *hookeriana* (Juz.) Breitung = Dryas octopetala ssp. hookeriana [DRYAOCT2] Rosaceae (V:D)
Dryas octopetala var. *integrifolia* (Vahl) Hook. f. = Dryas integrifolia ssp. integrifolia [DRYAINT0] Rosaceae (V:D)
Dryas sylvatica (Hultén) Porsild = Dryas integrifolia ssp. sylvatica [DRYAINT2] Rosaceae (V:D)
Drymocallis glandulosa (Lindl.) Rydb. = Potentilla glandulosa ssp. glandulosa [POTEGLA0] Rosaceae (V:D)
Dryopteris Adans. [DRYOPTE$] Dryopteridaceae (V:F)
Dryopteris arguta (Kaulfuss) Watt [DRYOARG] Dryopteridaceae (V:F)
Dryopteris assimilis S. Walker = Dryopteris expansa [DRYOEXP] Dryopteridaceae (V:F)
Dryopteris austriaca var. *spinulosa* (O.F. Muell.) Fisch. = Dryopteris carthusiana [DRYOCAR] Dryopteridaceae (V:F)
Dryopteris carthusiana (Vill.) H.P. Fuchs [DRYOCAR] Dryopteridaceae (V:F)
Dryopteris cristata (L.) Gray [DRYOCRI] Dryopteridaceae (V:F)
Dryopteris dilatata auct. = Dryopteris expansa [DRYOEXP] Dryopteridaceae (V:F)
Dryopteris disjuncta (Rupr.) Morton = Gymnocarpium disjunctum [GYMNDIS] Dryopteridaceae (V:F)
Dryopteris dryopteris (L.) Britt. = Gymnocarpium dryopteris [GYMNDRY] Dryopteridaceae (V:F)
Dryopteris expansa (K. Presl) Fraser-Jenkins & Jermy [DRYOEXP] Dryopteridaceae (V:F)
Dryopteris filix-mas (L.) Schott [DRYOFIL] Dryopteridaceae (V:F)
Dryopteris fragrans (L.) Schott [DRYOFRA] Dryopteridaceae (V:F)
Dryopteris linnaeana C. Christens. = Gymnocarpium dryopteris [GYMNDRY] Dryopteridaceae (V:F)
Dryopteris marginalis (L.) Gray [DRYOMAR] Dryopteridaceae (V:F)
Dryopteris nevadensis (Baker) Underwood = Thelypteris nevadensis [THELNEV] Thelypteridaceae (V:F)
Dryopteris oregana C. Christens. = Thelypteris nevadensis [THELNEV] Thelypteridaceae (V:F)
Dryopteris phegopteris (L.) C. Christens. = Phegopteris connectilis [PHEGCON] Thelypteridaceae (V:F)
Dryopteris spinulosa (O.F. Muell.) Watt = Dryopteris carthusiana [DRYOCAR] Dryopteridaceae (V:F)

Dryopteris spinulosa ssp. *assimilis* (Walker) Schidlay = Dryopteris expansa [DRYOEXP] Dryopteridaceae (V:F)
Dryptodon Brid. [DRYPTOD$] Grimmiaceae (B:M)
Dryptodon patens (Hedw.) Brid. [DRYPPAT] Grimmiaceae (B:M)
Duchesnea Sm. [DUCHESN$] Rosaceae (V:D)
***Duchesnea indica** (Andr.) Focke [DUCHIND] Rosaceae (V:D)
Dufourea madreporiformis (Ach.) Ach. = Dactylina madreporiformis [DACTMAD] Parmeliaceae (L)
Dulichium Pers. [DULICHI$] Cyperaceae (V:M)
Dulichium arundinaceum (L.) Britt. [DULIARU] Cyperaceae (V:M)
Duschekia sinuata (Regel) Pouzar = Alnus viridis ssp. sinuata [ALNUVIR2] Betulaceae (V:D)
Duschekia viridis (Chaix) Opiz = Alnus viridis ssp. crispa [ALNUVIR5] Betulaceae (V:D)

E

Eburophyton austiniae (Gray) Heller = Cephalanthera austiniae [CEPHAUS] Orchidaceae (V:M)
Echinochloa Beauv. [ECHINOC$] Poaceae (V:M)
***Echinochloa crusgalli** (L.) Beauv. [ECHICRU] Poaceae (V:M)
Echinocystis Torr. & Gray [ECHINOY$] Cucurbitaceae (V:D)
***Echinocystis lobata** (Michx.) Torr. & Gray [ECHILOB] Cucurbitaceae (V:D)
Echinocystis oregana (Torr. ex S. Wats.) Cogn. = Marah oreganus [MARAORE] Cucurbitaceae (V:D)
Echinopanax horridus (Sm.) Dcne. & Planch. ex H.A.T. Harms = Oplopanax horridus [OPLOHOR] Araliaceae (V:D)
Echinopsilon hyssopifolius (Pallas) Moq. = Bassia hyssopifolia [BASSHYS] Chenopodiaceae (V:D)
Echinothecium Zopf [ECHINOT$] Capnodiaceae (L)
Echinothecium reticulatum Zopf [ECHIRET] Capnodiaceae (L)
Echium L. [ECHIUM$] Boraginaceae (V:D)
Echium menziesii Lehm. = Amsinckia menziesii [AMSIMEN] Boraginaceae (V:D)
***Echium vulgare** L. [ECHIVUL] Boraginaceae (V:D)
Egeria Planch. [EGERIA$] Hydrocharitaceae (V:M)
***Egeria densa** Planch. [EGERDEN] Hydrocharitaceae (V:M)
Elaeagnus L. [ELAEAGN$] Elaeagnaceae (V:D)
***Elaeagnus angustifolia** L. [ELAEANG] Elaeagnaceae (V:D)
Elaeagnus argentea Pursh = Elaeagnus commutata [ELAECOM] Elaeagnaceae (V:D)
Elaeagnus canadensis (L.) A. Nels. = Shepherdia canadensis [SHEPCAN] Elaeagnaceae (V:D)
Elaeagnus commutata Bernh. ex Rydb. [ELAECOM] Elaeagnaceae (V:D)
Elaeagnus utilis A. Nels. = Shepherdia argentea [SHEPARG] Elaeagnaceae (V:D)
Elatine L. [ELATINE$] Elatinaceae (V:D)
Elatine rubella Rydb. [ELATRUB] Elatinaceae (V:D)
Eleocharis R. Br. [ELEOCHA$] Cyperaceae (V:M)
Eleocharis acicularis (L.) Roemer & J.A. Schultes [ELEOACI] Cyperaceae (V:M)
Eleocharis acuminata (Muhl.) Nees = Eleocharis compressa [ELEOCOM] Cyperaceae (V:M)
Eleocharis atropurpurea (Retz.) J.& K. Presl [ELEOATR] Cyperaceae (V:M)
Eleocharis bernardina Munz & Johnston = Eleocharis quinqueflora [ELEOQUI] Cyperaceae (V:M)
Eleocharis calva var. *australis* (Nees) St. John = Eleocharis palustris [ELEOPAL] Cyperaceae (V:M)
Eleocharis capitata var. *borealis* Svens. = Eleocharis elliptica [ELEOELL] Cyperaceae (V:M)
Eleocharis coloradoensis (Britt.) Gilly = Eleocharis parvula [ELEOPAR] Cyperaceae (V:M)
Eleocharis compressa Sullivant [ELEOCOM] Cyperaceae (V:M)
Eleocharis compressa var. *atrata* Svens. = Eleocharis compressa [ELEOCOM] Cyperaceae (V:M)
Eleocharis diandra C. Wright = Eleocharis obtusa [ELEOOBT] Cyperaceae (V:M)
Eleocharis elliptica Kunth [ELEOELL] Cyperaceae (V:M)
Eleocharis elliptica var. *compressa* (Sullivant) Drapalik & Mohlenbrock = Eleocharis compressa [ELEOCOM] Cyperaceae (V:M)
Eleocharis kamtschatica (C.A. Mey.) Komarov [ELEOKAM] Cyperaceae (V:M)
Eleocharis leptos (Steud.) Svens. = Eleocharis parvula [ELEOPAR] Cyperaceae (V:M)
Eleocharis leptos var. *coloradoensis* (Britt.) Svens. = Eleocharis parvula [ELEOPAR] Cyperaceae (V:M)
Eleocharis leptos var. *johnstonii* Svens. = Eleocharis parvula [ELEOPAR] Cyperaceae (V:M)
Eleocharis macounii Fern. = Eleocharis obtusa [ELEOOBT] Cyperaceae (V:M)
Eleocharis macrostachya Britt. = Eleocharis palustris [ELEOPAL] Cyperaceae (V:M)
Eleocharis membranacea Gilly = Eleocharis parvula [ELEOPAR] Cyperaceae (V:M)
Eleocharis multiflora Chapman = Eleocharis atropurpurea [ELEOATR] Cyperaceae (V:M)
Eleocharis nitida var. *borealis* (Svens.) Gleason = Eleocharis tenuis [ELEOTEN] Cyperaceae (V:M)
Eleocharis obtusa (Willd.) J.A. Schultes [ELEOOBT] Cyperaceae (V:M)
Eleocharis obtusa var. *ellipsoidalis* Fern. = Eleocharis obtusa [ELEOOBT] Cyperaceae (V:M)
Eleocharis obtusa var. *gigantea* (C.B. Clarke) Fern. = Eleocharis obtusa [ELEOOBT] Cyperaceae (V:M)
Eleocharis obtusa var. *jejuna* Fern. = Eleocharis obtusa [ELEOOBT] Cyperaceae (V:M)
Eleocharis obtusa var. *peasei* Svens. = Eleocharis obtusa [ELEOOBT] Cyperaceae (V:M)
Eleocharis ovata var. *obtusa* (Willd.) Kükenth. ex Skottsberg = Eleocharis obtusa [ELEOOBT] Cyperaceae (V:M)
Eleocharis palustris (L.) Roemer & J.A. Schultes [ELEOPAL] Cyperaceae (V:M)
Eleocharis palustris var. *australis* Nees = Eleocharis palustris [ELEOPAL] Cyperaceae (V:M)
Eleocharis palustris var. *major* Sonder = Eleocharis palustris [ELEOPAL] Cyperaceae (V:M)
Eleocharis parvula (Roemer & J.A. Schultes) Link ex Bluff, Nees & Schauer [ELEOPAR] Cyperaceae (V:M)
Eleocharis parvula var. *anachaeta* (Torr.) Svens. = Eleocharis parvula [ELEOPAR] Cyperaceae (V:M)
Eleocharis parvula var. *coloradoensis* (Britt.) Beetle = Eleocharis parvula [ELEOPAR] Cyperaceae (V:M)
Eleocharis pauciflora (Lightf.) Link = Eleocharis quinqueflora [ELEOQUI] Cyperaceae (V:M)
Eleocharis pauciflora var. *bernardina* (Munz & Johnston) Svens. = Eleocharis quinqueflora [ELEOQUI] Cyperaceae (V:M)
Eleocharis pauciflora var. *fernaldii* Svens. = Eleocharis quinqueflora [ELEOQUI] Cyperaceae (V:M)
Eleocharis pauciflora var. *suksdorfiana* (Beauv.) Svens. = Eleocharis quinqueflora [ELEOQUI] Cyperaceae (V:M)
Eleocharis perlonga Fern. & Brack. = Eleocharis palustris [ELEOPAL] Cyperaceae (V:M)
Eleocharis quinqueflora (F.X. Hartmann) Schwarz [ELEOQUI] Cyperaceae (V:M)
Eleocharis quinqueflora ssp. *fernaldii* (Svens.) Hultén = Eleocharis quinqueflora [ELEOQUI] Cyperaceae (V:M)
Eleocharis quinqueflora ssp. *suksdorfiana* (Beauv.) Hultén = Eleocharis quinqueflora [ELEOQUI] Cyperaceae (V:M)

Eleocharis quinqueflora var. *suksdorfiana* (Beauv.) J.T. Howell = Eleocharis quinqueflora [ELEOQUI] Cyperaceae (V:M)
Eleocharis rostellata (Torr.) Torr. [ELEOROS] Cyperaceae (V:M)
Eleocharis rostellata var. *congdonii* Jepson = Eleocharis rostellata [ELEOROS] Cyperaceae (V:M)
Eleocharis rostellata var. *occidentalis* S. Wats. = Eleocharis rostellata [ELEOROS] Cyperaceae (V:M)
Eleocharis smallii var. *major* (Sonder) Seymour = Eleocharis palustris [ELEOPAL] Cyperaceae (V:M)
Eleocharis suksdorfiana Beauv. = Eleocharis quinqueflora [ELEOQUI] Cyperaceae (V:M)
Eleocharis tenuis (Willd.) J.A. Schultes [ELEOTEN] Cyperaceae (V:M)
Eleocharis tenuis var. *atrata* (Svens.) Boivin = Eleocharis compressa [ELEOCOM] Cyperaceae (V:M)
Eleocharis tenuis var. *borealis* (Svens.) Gleason = Eleocharis elliptica [ELEOELL] Cyperaceae (V:M)
Eleocharis uniglumis (Link) J.A. Schultes [ELEOUNI] Cyperaceae (V:M)
Eleocharis xyridiformis Fern. & Brack. = Eleocharis palustris [ELEOPAL] Cyperaceae (V:M)
Elliottia Muhl. ex Ell. [ELLIOTT$] Ericaceae (V:D)
Elliottia pyroliflorus (Bong.) S.W. Brim & P.F. Stevens [ELLIPYR] Ericaceae (V:D)
Ellisia L. [ELLISIA$] Hydrophyllaceae (V:D)
Ellisia nyctelea (L.) L. [ELLINYC] Hydrophyllaceae (V:D)
Ellisia nyctelea var. *coloradensis* Brand = Ellisia nyctelea [ELLINYC] Hydrophyllaceae (V:D)
Elmera Rydb. [ELMERA$] Saxifragaceae (V:D)
Elmera racemosa (S. Wats.) Rydb. [ELMERAC] Saxifragaceae (V:D)
Elodea Michx. [ELODEA$] Hydrocharitaceae (V:M)
Elodea brandegae St. John = Elodea canadensis [ELODCAN] Hydrocharitaceae (V:M)
Elodea canadensis Michx. [ELODCAN] Hydrocharitaceae (V:M)
Elodea columbiana St. John = Elodea nuttallii [ELODNUT] Hydrocharitaceae (V:M)
Elodea densa (Planch.) Caspary = Egeria densa [EGERDEN] Hydrocharitaceae (V:M)
Elodea ioensis Wylie = Elodea canadensis [ELODCAN] Hydrocharitaceae (V:M)
Elodea linearis (Rydb.) St. John = Elodea canadensis [ELODCAN] Hydrocharitaceae (V:M)
Elodea minor (Engelm. ex Caspary) Farw. = Elodea nuttallii [ELODNUT] Hydrocharitaceae (V:M)
Elodea nuttallii (Planch.) St. John [ELODNUT] Hydrocharitaceae (V:M)
Elodea occidentalis (Pursh) St. John = Elodea nuttallii [ELODNUT] Hydrocharitaceae (V:M)
Elodea planchonii Caspary = Elodea canadensis [ELODCAN] Hydrocharitaceae (V:M)
Elodium blandowii (Web. & Mohr) Broth. = Helodium blandowii [HELOBLA] Helodiaceae (B:M)
× **Elyhordeum** Mansf. ex Zizin & Petrowa [ELYHORD$] Poaceae (V:M)
× **Elyhordeum macounii** (Vasey) Barkworth & Dewey [ELYHMAC] Poaceae (V:M)
× *Elymordeum macounii* (Vasey) Barkworth = × Elyhordeum macounii [ELYHMAC] Poaceae (V:M)
Elymus L. [ELYMUS$] Poaceae (V:M)
Elymus alaskanus (Scribn. & Merr.) A. Löve [ELYMALA] Poaceae (V:M)
Elymus alaskanus ssp. **latiglumis** (Scribn. & J.G. Sm.) A. Löve [ELYMALA1] Poaceae (V:M)
Elymus albicans (Scribn. & J.G. Sm.) A. Löve = Elymus lanceolatus ssp. albicans [ELYMLAN1] Poaceae (V:M)
Elymus albicans var. *griffithii* (Scribn. & J.G. Sm. ex Piper) Dorn = Elymus lanceolatus ssp. albicans [ELYMLAN1] Poaceae (V:M)
Elymus brachystachys Scribn. & Ball = Elymus canadensis [ELYMCAN] Poaceae (V:M)
Elymus brownii Scribn. & J.G. Sm. = Leymus innovatus [LEYMINN] Poaceae (V:M)
Elymus canadensis L. [ELYMCAN] Poaceae (V:M)
Elymus cinereus Scribn. & Merr. = Leymus cinereus [LEYMCIN] Poaceae (V:M)
Elymus cinereus var. *pubens* (Piper) C.L. Hitchc. = Leymus cinereus [LEYMCIN] Poaceae (V:M)
Elymus condensatus var. *pubens* Piper = Leymus cinereus [LEYMCIN] Poaceae (V:M)
Elymus condensatus var. *triticoides* (Buckl.) Thurb. = Leymus triticoides [LEYMTRI] Poaceae (V:M)
Elymus elongatus ssp. *ponticus* (Podp.) Melderis = Elytrigia pontica [ELYTPON] Poaceae (V:M)
Elymus elongatus var. *ponticus* (Podp.) Dorn = Elytrigia pontica [ELYTPON] Poaceae (V:M)
Elymus elymoides (Raf.) Swezey [ELYMELY] Poaceae (V:M)
Elymus elymoides ssp. *californicus* Barkworth = Elymus elymoides [ELYMELY] Poaceae (V:M)
Elymus elymoides var. *brevifolius* (J.G. Sm.) Dorn = Elymus elymoides [ELYMELY] Poaceae (V:M)
Elymus glaucus Buckl. [ELYMGLA] Poaceae (V:M)
Elymus glaucus ssp. **glaucus** [ELYMGLA0] Poaceae (V:M)
Elymus glaucus ssp. **virescens** (Piper) Gould [ELYMGLA5] Poaceae (V:M)
Elymus glaucus var. *breviaristatus* Burtt-Davy = Elymus glaucus ssp. glaucus [ELYMGLA0] Poaceae (V:M)
Elymus glaucus var. *virescens* (Piper) Bowden = Elymus glaucus ssp. virescens [ELYMGLA5] Poaceae (V:M)
Elymus griffithii (Scribn. & J.G. Sm. ex Piper) A. Löve = Elymus lanceolatus ssp. albicans [ELYMLAN1] Poaceae (V:M)
Elymus hirsutus J. Presl [ELYMHIR] Poaceae (V:M)
Elymus hispidus (Opiz) Melderis = Elytrigia intermedia [ELYTINT] Poaceae (V:M)
Elymus hispidus ssp. *barbulatus* (Schur) Melderis = Elytrigia intermedia [ELYTINT] Poaceae (V:M)
Elymus hispidus var. *ruthenicus* (Griseb.) Dorn = Elytrigia intermedia [ELYTINT] Poaceae (V:M)
Elymus innovatus Beal = Leymus innovatus [LEYMINN] Poaceae (V:M)
Elymus innovatus var. *glabratus* Bowden = Leymus innovatus [LEYMINN] Poaceae (V:M)
Elymus innovatus var. *velutinus* Bowden = Leymus innovatus [LEYMINN] Poaceae (V:M)
Elymus lanceolatus (Scribn. & J.G. Sm.) Gould [ELYMLAN] Poaceae (V:M)
Elymus lanceolatus ssp. **albicans** (Scribn. & J.G. Sm.) Barkworth & Dewey [ELYMLAN1] Poaceae (V:M)
Elymus lanceolatus ssp. **lanceolatus** [ELYMLAN0] Poaceae (V:M)
Elymus lanceolatus ssp. *yukonensis* (Scribn. & Merr.) A. Löve = Elymus × yukonensis [ELYMYUK] Poaceae (V:M)
Elymus lanceolatus var. *riparius* (Scribn. & J.G. Sm.) Dorn = Elymus lanceolatus ssp. lanceolatus [ELYMLAN0] Poaceae (V:M)
Elymus longifolius (J.G. Sm.) Gould = Elymus elymoides [ELYMELY] Poaceae (V:M)
Elymus macounii Vasey = × Elyhordeum macounii [ELYHMAC] Poaceae (V:M)
Elymus mollis Trin. = Leymus mollis [LEYMMOL] Poaceae (V:M)
Elymus orcuttianus Vasey = Leymus triticoides [LEYMTRI] Poaceae (V:M)
Elymus pauciflorum var. *subsecundus* Gould = Elymus trachycaulus ssp. subsecundus [ELYMTRA1] Poaceae (V:M)
Elymus pauciflorus (Schwein.) Gould = Elymus trachycaulus ssp. trachycaulus [ELYMTRA0] Poaceae (V:M)
Elymus piperi Bowden = Leymus cinereus [LEYMCIN] Poaceae (V:M)
Elymus repens (L.) Gould = Elytrigia repens [ELYTREP] Poaceae (V:M)
Elymus × **saundersii** Vasey (pro sp.) [ELYMXSA] Poaceae (V:M)
Elymus saundersii var. *californicus* Hoover = Elymus × saundersii [ELYMXSA] Poaceae (V:M)
Elymus scribneri (Vasey) M.E. Jones [ELYMSCR] Poaceae (V:M)
Elymus sitanion J.A. Schultes = Elymus elymoides [ELYMELY] Poaceae (V:M)
Elymus smithii (Rydb.) Gould = Pascopyrum smithii [PASCSMI] Poaceae (V:M)

Elymus spicatus (Pursh) Gould = Pseudoroegneria spicata [PSEUSPI] Poaceae (V:M)
Elymus trachycaulus (Link) Gould ex Shinners [ELYMTRA] Poaceae (V:M)
Elymus trachycaulus ssp. *andinus* (Scribn. & J.G. Sm.) A. & D. Löve = Elymus trachycaulus ssp. trachycaulus [ELYMTRA0] Poaceae (V:M)
Elymus trachycaulus ssp. *latiglumis* (Scribn. & J.G. Sm.) Barkworth & Dewey = Elymus alaskanus ssp. latiglumis [ELYMALA1] Poaceae (V:M)
Elymus trachycaulus ssp. *novae-angliae* (Scribn.) Tzvelev = Elymus trachycaulus ssp. trachycaulus [ELYMTRA0] Poaceae (V:M)
Elymus trachycaulus ssp. **subsecundus** (Link) A. & D. Löve [ELYMTRA1] Poaceae (V:M)
Elymus trachycaulus ssp. *teslinensis* (Porsild & Senn) A. Löve = Elymus trachycaulus ssp. trachycaulus [ELYMTRA0] Poaceae (V:M)
Elymus trachycaulus ssp. **trachycaulus** [ELYMTRA0] Poaceae (V:M)
Elymus trachycaulus ssp. *violaceus* (Hornem.) A. & D. Löve = Elymus alaskanus [ELYMALA] Poaceae (V:M)
Elymus trachycaulus var. *andinus* (Scribn. & J.G. Sm.) Dorn = Elymus trachycaulus ssp. trachycaulus [ELYMTRA0] Poaceae (V:M)
Elymus trachycaulus var. *latiglumis* (Scribn. & J.G. Sm.) Beetle = Elymus alaskanus ssp. latiglumis [ELYMALA1] Poaceae (V:M)
Elymus trachycaulus var. *majus* (Vasey) Beetle = Elymus trachycaulus ssp. trachycaulus [ELYMTRA0] Poaceae (V:M)
Elymus trachycaulus var. *unilateralis* (Cassidy) Beetle = Elymus trachycaulus ssp. subsecundus [ELYMTRA1] Poaceae (V:M)
Elymus triticoides Buckl. = Leymus triticoides [LEYMTRI] Poaceae (V:M)
Elymus triticoides var. *pubescens* A.S. Hitchc. = Leymus triticoides [LEYMTRI] Poaceae (V:M)
Elymus varnensis (Velen.) Runemark = Elytrigia pontica [ELYTPON] Poaceae (V:M)
Elymus virescens Piper = Elymus glaucus ssp. virescens [ELYMGLA5] Poaceae (V:M)
Elymus × yukonensis (Scribn. & Merr.) A. Löve (pro sp.) [ELYMYUK] Poaceae (V:M)
Elymus yukonensis (Scribn. & Merr.) A. Löve = Elymus × yukonensis [ELYMYUK] Poaceae (V:M)
Elyna bellardii (All.) Degl. = Kobresia myosuroides [KOBRMYO] Cyperaceae (V:M)
Elyna sibirica Turcz. ex Ledeb. = Kobresia sibirica [KOBRSIB] Cyperaceae (V:M)
× *Elytesion macounii* (Vasey) Barkworth & Dewey = × Elyhordeum macounii [ELYHMAC] Poaceae (V:M)
Elytrigia Desv. [ELYTRIG$] Poaceae (V:M)
Elytrigia dasystachya (Hook.) A. & D. Löve = Elymus lanceolatus ssp. lanceolatus [ELYMLAN0] Poaceae (V:M)
Elytrigia dasystachya ssp. *albicans* (Scribn. & J.G. Sm.) Dewey = Elymus lanceolatus ssp. albicans [ELYMLAN1] Poaceae (V:M)
Elytrigia dasystachya ssp. *yukonensis* (Scribn. & Merr.) Dewey = Elymus × yukonensis [ELYMYUK] Poaceae (V:M)
***Elytrigia intermedia** (Host) Nevski [ELYTINT] Poaceae (V:M)
Elytrigia intermedia ssp. *barbulata* (Schur) A. Löve = Elytrigia intermedia [ELYTINT] Poaceae (V:M)
***Elytrigia pontica** (Podp.) Holub [ELYTPON] Poaceae (V:M)
***Elytrigia repens** (L.) Desv. ex B.D. Jackson [ELYTREP] Poaceae (V:M)
Elytrigia ripara (Scribn. & J.G. Sm.) Beetle = Elymus lanceolatus ssp. lanceolatus [ELYMLAN0] Poaceae (V:M)
Elytrigia smithii (Rydb.) Nevski = Pascopyrum smithii [PASCSMI] Poaceae (V:M)
Elytrigia smithii var. *mollis* (Scribn. & J.G. Sm.) Beetle = Pascopyrum smithii [PASCSMI] Poaceae (V:M)
Elytrigia spicata (Pursh) Dewey = Pseudoroegneria spicata [PSEUSPI] Poaceae (V:M)
Empetrum L. [EMPETRU$] Empetraceae (V:D)
Empetrum nigrum L. [EMPENIG] Empetraceae (V:D)
Encalypta Hedw. [ENCALYP$] Encalyptaceae (B:M)
Encalypta affinis Hedw. f. in Web. & Mohr [ENCAAFF] Encalyptaceae (B:M)
Encalypta affinis ssp. *macounii* (Aust.) Horton = Encalypta affinis var. macounii [ENCAAFF1] Encalyptaceae (B:M)
Encalypta affinis var. **affinis** [ENCAAFF0] Encalyptaceae (B:M)
Encalypta affinis var. **macounii** (Aust.) Crum & Anderson [ENCAAFF1] Encalyptaceae (B:M)
Encalypta alpina Sm. [ENCAALP] Encalyptaceae (B:M)
Encalypta apophysata Nees & Hornsch. = Encalypta affinis var. affinis [ENCAAFF0] Encalyptaceae (B:M)
Encalypta brevicolla (Bruch & Schimp. in B.S.G.) Bruch ex Ångstr. [ENCABRE] Encalyptaceae (B:M)
Encalypta brevipes Schljak. [ENCABRV] Encalyptaceae (B:M)
Encalypta ciliata Hedw. [ENCACIL] Encalyptaceae (B:M)
Encalypta ciliata var. *microstoma* Schimp. = Encalypta ciliata [ENCACIL] Encalyptaceae (B:M)
Encalypta intermedia Jur. in Jur. & Milde [ENCAINT] Encalyptaceae (B:M)
Encalypta labradorica Kindb. = Encalypta brevicolla [ENCABRE] Encalyptaceae (B:M)
Encalypta laciniata Lindb. = Encalypta ciliata [ENCACIL] Encalyptaceae (B:M)
Encalypta leiocarpa Kindb. in Mac. = Encalypta affinis var. macounii [ENCAAFF1] Encalyptaceae (B:M)
Encalypta longicolla Bruch [ENCALON] Encalyptaceae (B:M)
Encalypta longipes Mitt. = Encalypta procera [ENCAPRO] Encalyptaceae (B:M)
Encalypta macounii Aust. = Encalypta affinis var. macounii [ENCAAFF1] Encalyptaceae (B:M)
Encalypta mutica Hag. [ENCAMUT] Encalyptaceae (B:M)
Encalypta procera Bruch [ENCAPRO] Encalyptaceae (B:M)
Encalypta rhaptocarpa Schwaegr. [ENCARHA] Encalyptaceae (B:M)
Encalypta rhaptocarpa var. *leiomitra* Kindb. = Encalypta rhaptocarpa [ENCARHA] Encalyptaceae (B:M)
Encalypta rhaptocarpa var. *subspathulata* (C. Müll. & Kindb. in Mac. & Kindb.) Flow. = Encalypta rhaptocarpa [ENCARHA] Encalyptaceae (B:M)
Encalypta selwynii Aust. = Encalypta procera [ENCAPRO] Encalyptaceae (B:M)
Encalypta spathulata C. Müll. [ENCASPA] Encalyptaceae (B:M)
Encalypta subbrevicolla Kindb. = Encalypta brevicolla [ENCABRE] Encalyptaceae (B:M)
Encalypta subspathulata C. Müll. & Kindb. in Mac. & Kindb. = Encalypta rhaptocarpa [ENCARHA] Encalyptaceae (B:M)
Encalypta vulgaris Hedw. [ENCAVUL] Encalyptaceae (B:M)
Encalypta vulgaris var. *apiculata* Wahlenb. = Encalypta vulgaris [ENCAVUL] Encalyptaceae (B:M)
Encalypta vulgaris var. *mutica* Brid. = Encalypta vulgaris [ENCAVUL] Encalyptaceae (B:M)
Encalypta vulgaris var. *rhaptocarpa* (Schwaegr.) Lawt. = Encalypta rhaptocarpa [ENCARHA] Encalyptaceae (B:M)
Endocarpon Hedwig [ENDOCAR$] Verrucariaceae (L)
Endocarpon cinereum Pers. = Catapyrenium cinereum [CATACIN] Verrucariaceae (L)
Endocarpon drummondii (Tuck.) Choisy = Staurothele drummondii [STAUDRU] Verrucariaceae (L)
Endocarpon hepaticum Ach. = Catapyrenium lachneum [CATALAC] Verrucariaceae (L)
Endocarpon lachneum (Ach.) Ach. = Catapyrenium lachneum [CATALAC] Verrucariaceae (L)
Endocarpon pallidum Ach. [ENDOPAL] Verrucariaceae (L)
Endocarpon pulvinatum Th. Fr. [ENDOPUL] Verrucariaceae (L)
Endocarpon pusillum Hedwig [ENDOPUS] Verrucariaceae (L)
Endocarpon pusillum var. *pallidum* (Ach.) Körber = Endocarpon pallidum [ENDOPAL] Verrucariaceae (L)
Endocarpon tephroides (Ach.) Ach. = Catapyrenium cinereum [CATACIN] Verrucariaceae (L)
Endocarpon wilmsoides Zahlbr. = Staurothele drummondii [STAUDRU] Verrucariaceae (L)
Endopyrenium americanum de Lesd. = Catapyrenium compactum [CATACOM] Verrucariaceae (L)

Endopyrenium hepaticum (Ach.) Körber = Catapyrenium lachneum [CATALAC] Verrucariaceae (L)
Endopyrenium rupicola de Lesd. = Catapyrenium compactum [CATACOM] Verrucariaceae (L)
Endymion hispanicus (P. Mill.) Chouard = Hyacinthoides hispanica [HYACHIS] Liliaceae (V:M)
Enemion Raf. [ENEMION$] Ranunculaceae (V:D)
Enemion savilei (Calder & Taylor) Keener [ENEMSAV] Ranunculaceae (V:D)
Enterographa Fée [ENTEROG$] Opegraphaceae (L)
Enterographa zonata (Körber) Källsten [ENTEZON] Opegraphaceae (L)
Entodon C. Müll. [ENTODON$] Entodontaceae (B:M)
Entodon concinnus (De Not.) Par. [ENTOCON] Entodontaceae (B:M)
Entodon expallens C. Müll. & Kindb. in Mac. & Kindb. = Hypnum pratense [HYPNPRA] Hypnaceae (B:M)
Entodon macounii C. Müll. & Kindb. in Mac. & Kindb. = Hypnum pratense [HYPNPRA] Hypnaceae (B:M)
Entodon orthocarpus (Brid.) Lindb. = Entodon concinnus [ENTOCON] Entodontaceae (B:M)
Entodon repens var. *orthoclados* (Kindb.) Grout = Platygyrium repens [PLATREP] Hypnaceae (B:M)
Entosthodon Schwaegr. [ENTOSTH$] Funariaceae (B:M)
Entosthodon fascicularis (Hedw.) C. Müll. [ENTOFAS] Funariaceae (B:M)
Entosthodon leibergii Britt. = Entosthodon fascicularis [ENTOFAS] Funariaceae (B:M)
Entosthodon muhlenbergii (Turn.) Fife = Funaria muhlenbergii [FUNAMUH] Funariaceae (B:M)
Entosthodon neoscoticus M.S. Brown = Desmatodon randii [DESMRAN] Pottiaceae (B:M)
Entosthodon rubiginosus (Williams) Grout [ENTORUB] Funariaceae (B:M)
Enydria aquatica Vell. = Myriophyllum aquaticum [MYRIAQU] Haloragaceae (V:D)
Eopyrenula R.C. Harris [EOPYREN$] Pyrenulaceae (L)
Eopyrenula intermedia Coppins [EOPYINT] Pyrenulaceae (L)
Ephebe Fr. [EPHEBE$] Lichinaceae (L)
Ephebe hispidula (Ach.) Horwood [EPHEHIS] Lichinaceae (L)
Ephebe lanata (L.) Vainio [EPHELAN] Lichinaceae (L)
Ephebe pubescens (L.) Fr. = Pseudephebe pubescens [PSEUPUB] Parmeliaceae (L)
Ephebeia hispidula (Ach.) Nyl. = Ephebe hispidula [EPHEHIS] Lichinaceae (L)
Epilobium L. [EPILOBI$] Onagraceae (V:D)
Epilobium adenocaulon Hausskn. = Epilobium ciliatum ssp. ciliatum [EPILCIL0] Onagraceae (V:D)
Epilobium adenocaulon var. *cinerascens* (Piper) M.E. Peck = Epilobium ciliatum ssp. glandulosum [EPILCIL2] Onagraceae (V:D)
Epilobium adenocaulon var. *ecomosum* (Fassett) Munz = Epilobium ciliatum ssp. ciliatum [EPILCIL0] Onagraceae (V:D)
Epilobium adenocaulon var. *holosericeum* (Trel.) Munz = Epilobium ciliatum ssp. ciliatum [EPILCIL0] Onagraceae (V:D)
Epilobium adenocaulon var. *occidentale* Trel. = Epilobium ciliatum ssp. glandulosum [EPILCIL2] Onagraceae (V:D)
Epilobium adenocaulon var. *parishii* (Trel.) Munz = Epilobium ciliatum ssp. ciliatum [EPILCIL0] Onagraceae (V:D)
Epilobium adenocaulon var. *perplexans* Trel. = Epilobium ciliatum ssp. ciliatum [EPILCIL0] Onagraceae (V:D)
Epilobium alpinum L. p.p. = Epilobium anagallidifolium [EPILANA] Onagraceae (V:D)
Epilobium alpinum var. *albiflorum* (Suksdorf) C.L. Hitchc. = Epilobium clavatum [EPILCLA] Onagraceae (V:D)
Epilobium alpinum var. *clavatum* (Trel.) C.L. Hitchc. = Epilobium clavatum [EPILCLA] Onagraceae (V:D)
Epilobium alpinum var. *gracillimum* (Trel.) C.L. Hitchc. = Epilobium oregonense [EPILORE] Onagraceae (V:D)
Epilobium alpinum var. *lactiflorum* (Hausskn.) C.L. Hitchc. = Epilobium lactiflorum [EPILLAC] Onagraceae (V:D)
Epilobium alpinum var. *nutans* Hornem. = Epilobium hornemannii ssp. hornemannii [EPILHOR0] Onagraceae (V:D)
Epilobium alpinum var. *sertulatum* (Hausskn.) Welsh = Epilobium hornemannii ssp. behringianum [EPILHOR1] Onagraceae (V:D)
Epilobium americanum Hausskn. = Epilobium ciliatum ssp. ciliatum [EPILCIL0] Onagraceae (V:D)
Epilobium anagallidifolium Lam. [EPILANA] Onagraceae (V:D)
Epilobium anagallidifolium var. *pseudoscaposum* (Hausskn.) Hultén = Epilobium anagallidifolium [EPILANA] Onagraceae (V:D)
Epilobium angustifolium L. [EPILANG] Onagraceae (V:D)
Epilobium angustifolium ssp. **angustifolium** [EPILANG0] Onagraceae (V:D)
Epilobium angustifolium ssp. **circumvagum** Mosquin [EPILANG2] Onagraceae (V:D)
Epilobium angustifolium ssp. *macrophyllum* (Hausskn.) Hultén = Epilobium angustifolium ssp. circumvagum [EPILANG2] Onagraceae (V:D)
Epilobium angustifolium var. *abbreviatum* (Lunell) Munz = Epilobium angustifolium ssp. circumvagum [EPILANG2] Onagraceae (V:D)
Epilobium angustifolium var. *canescens* Wood = Epilobium angustifolium ssp. circumvagum [EPILANG2] Onagraceae (V:D)
Epilobium angustifolium var. *intermedium* (Lange) Fern. = Epilobium angustifolium ssp. angustifolium [EPILANG0] Onagraceae (V:D)
Epilobium angustifolium var. *macrophyllum* (Hausskn.) Fern. = Epilobium angustifolium ssp. circumvagum [EPILANG2] Onagraceae (V:D)
Epilobium angustifolium var. *platyphyllum* (Daniels) Fern. = Epilobium angustifolium ssp. circumvagum [EPILANG2] Onagraceae (V:D)
Epilobium behringianum Hausskn. = Epilobium hornemannii ssp. behringianum [EPILHOR1] Onagraceae (V:D)
Epilobium boreale Hausskn. = Epilobium ciliatum ssp. glandulosum [EPILCIL2] Onagraceae (V:D)
Epilobium brachycarpum K. Presl [EPILBRA] Onagraceae (V:D)
Epilobium brevistylum Barbey = Epilobium ciliatum ssp. ciliatum [EPILCIL0] Onagraceae (V:D)
Epilobium brevistylum var. *subfalcatum* (Trel.) Munz = Epilobium halleanum [EPILHAL] Onagraceae (V:D)
Epilobium brevistylum var. *tenue* (Trel.) Jepson = Epilobium halleanum [EPILHAL] Onagraceae (V:D)
Epilobium brevistylum var. *ursinum* (Parish ex Trel.) Jepson = Epilobium ciliatum ssp. ciliatum [EPILCIL0] Onagraceae (V:D)
Epilobium californicum Hausskn. = Epilobium ciliatum ssp. ciliatum [EPILCIL0] Onagraceae (V:D)
Epilobium californicum var. *holosericeum* (Trel.) Munz = Epilobium ciliatum ssp. ciliatum [EPILCIL0] Onagraceae (V:D)
Epilobium ciliatum Raf. [EPILCIL] Onagraceae (V:D)
Epilobium ciliatum ssp. **ciliatum** [EPILCIL0] Onagraceae (V:D)
Epilobium ciliatum ssp. **glandulosum** (Lehm.) Hoch & Raven [EPILCIL2] Onagraceae (V:D)
Epilobium ciliatum ssp. **watsonii** (Barbey) Hoch & Raven [EPILCIL3] Onagraceae (V:D)
Epilobium ciliatum var. *ecomosum* (Fassett) Boivin = Epilobium ciliatum ssp. ciliatum [EPILCIL0] Onagraceae (V:D)
Epilobium ciliatum var. *glandulosum* (Lehm.) Dorn = Epilobium ciliatum ssp. glandulosum [EPILCIL2] Onagraceae (V:D)
Epilobium clavatum Trel. [EPILCLA] Onagraceae (V:D)
Epilobium clavatum var. *glareosum* (G.N. Jones) Munz = Epilobium clavatum [EPILCLA] Onagraceae (V:D)
Epilobium davuricum Fisch. ex Hornem. [EPILDAV] Onagraceae (V:D)
Epilobium delicatum Trel. = Epilobium ciliatum ssp. ciliatum [EPILCIL0] Onagraceae (V:D)
Epilobium drummondii Hausskn. = Epilobium saximontanum [EPILSAX] Onagraceae (V:D)
Epilobium ecomosum (Fassett) Fern. = Epilobium ciliatum ssp. ciliatum [EPILCIL0] Onagraceae (V:D)
Epilobium foliosum (Torr. & Gray) Suksdorf [EPILFOL] Onagraceae (V:D)
Epilobium franciscanum Barbey = Epilobium ciliatum ssp. watsonii [EPILCIL3] Onagraceae (V:D)
Epilobium glaberrimum Barbey [EPILGLA] Onagraceae (V:D)
Epilobium glaberrimum ssp. **fastigiatum** (Nutt.) Hoch & Raven [EPILGLA1] Onagraceae (V:D)

Epilobium glaberrimum var. *fastigiatum* (Nutt.) Trel. ex Jepson = Epilobium glaberrimum ssp. fastigiatum [EPILGLA1] Onagraceae (V:D)
Epilobium glandulosum Lehm. = Epilobium ciliatum ssp. glandulosum [EPILCIL2] Onagraceae (V:D)
Epilobium glandulosum var. *adenocaulon* (Hausskn.) Fern. = Epilobium ciliatum ssp. ciliatum [EPILCIL0] Onagraceae (V:D)
Epilobium glandulosum var. *brionense* Fern. = Epilobium saximontanum [EPILSAX] Onagraceae (V:D)
Epilobium glandulosum var. *cardiophyllum* Fern. = Epilobium ciliatum ssp. glandulosum [EPILCIL2] Onagraceae (V:D)
Epilobium glandulosum var. *macounii* (Trel.) C.L. Hitchc. = Epilobium ciliatum ssp. ciliatum [EPILCIL0] Onagraceae (V:D)
Epilobium glandulosum var. *occidentale* (Trel.) Fern. = Epilobium ciliatum ssp. glandulosum [EPILCIL2] Onagraceae (V:D)
Epilobium glandulosum var. *tenue* (Trel.) C.L. Hitchc. = Epilobium halleanum [EPILHAL] Onagraceae (V:D)
Epilobium glareosum G.N. Jones = Epilobium clavatum [EPILCLA] Onagraceae (V:D)
Epilobium halleanum Hausskn. [EPILHAL] Onagraceae (V:D)
*Epilobium hirsutum** L. [EPILHIR] Onagraceae (V:D)
Epilobium hornemannii Reichenb. [EPILHOR] Onagraceae (V:D)
Epilobium hornemannii ssp. **behringianum** (Hausskn.) Hoch & Raven [EPILHOR1] Onagraceae (V:D)
Epilobium hornemannii ssp. **hornemannii** [EPILHOR0] Onagraceae (V:D)
Epilobium hornemannii var. *lactiflorum* (Hausskn.) D. Löve = Epilobium lactiflorum [EPILLAC] Onagraceae (V:D)
Epilobium lactiflorum Hausskn. [EPILLAC] Onagraceae (V:D)
Epilobium latifolium L. [EPILLAT] Onagraceae (V:D)
Epilobium leptocarpum Hausskn. [EPILLEP] Onagraceae (V:D)
Epilobium leptocarpum var. *macounii* Trel. = Epilobium ciliatum ssp. ciliatum [EPILCIL0] Onagraceae (V:D)
Epilobium leptophyllum Raf. [EPILLET] Onagraceae (V:D)
Epilobium lineare Muhl. = Epilobium palustre [EPILPAL] Onagraceae (V:D)
Epilobium luteum Pursh [EPILLUT] Onagraceae (V:D)
Epilobium minutum Lindl. ex Lehm. [EPILMIN] Onagraceae (V:D)
Epilobium minutum var. *foliosum* Torr. & Gray = Epilobium foliosum [EPILFOL] Onagraceae (V:D)
Epilobium mirabile Trel. ex Piper [EPILMIR] Onagraceae (V:D)
Epilobium nesophilum (Fern.) Fern. = Epilobium leptophyllum [EPILLET] Onagraceae (V:D)
Epilobium nesophilum var. *sabulonense* Fern. = Epilobium leptophyllum [EPILLET] Onagraceae (V:D)
Epilobium oliganthum Michx. = Epilobium palustre [EPILPAL] Onagraceae (V:D)
Epilobium oregonense Hausskn. [EPILORE] Onagraceae (V:D)
Epilobium palustre L. [EPILPAL] Onagraceae (V:D)
Epilobium palustre var. *davuricum* (Fisch. ex Hornem.) Welsh = Epilobium davuricum [EPILDAV] Onagraceae (V:D)
Epilobium palustre var. *gracile* (Farw.) Dorn = Epilobium leptophyllum [EPILLET] Onagraceae (V:D)
Epilobium palustre var. *grammadophyllum* Hausskn. = Epilobium palustre [EPILPAL] Onagraceae (V:D)
Epilobium palustre var. *labradoricum* Hausskn. = Epilobium palustre [EPILPAL] Onagraceae (V:D)
Epilobium palustre var. *lapponicum* Wahlenb. = Epilobium palustre [EPILPAL] Onagraceae (V:D)
Epilobium palustre var. *longirameum* Fern. & Wieg. = Epilobium palustre [EPILPAL] Onagraceae (V:D)
Epilobium palustre var. *oliganthum* (Michx.) Fern. = Epilobium palustre [EPILPAL] Onagraceae (V:D)
Epilobium palustre var. *sabulonense* (Fern.) Boivin = Epilobium leptophyllum [EPILLET] Onagraceae (V:D)
Epilobium paniculatum Nutt. ex Torr. & Gray = Epilobium brachycarpum [EPILBRA] Onagraceae (V:D)
Epilobium paniculatum var. *hammondii* (T.J. Howell) M.E. Peck = Epilobium brachycarpum [EPILBRA] Onagraceae (V:D)
Epilobium paniculatum var. *juncundum* (Gray) Trel. = Epilobium brachycarpum [EPILBRA] Onagraceae (V:D)
Epilobium paniculatum var. *laevicaule* (Rydb.) Munz = Epilobium brachycarpum [EPILBRA] Onagraceae (V:D)
Epilobium paniculatum var. *subulatum* (Hausskn.) Fern. = Epilobium brachycarpum [EPILBRA] Onagraceae (V:D)
Epilobium paniculatum var. *tracyi* (Rydb.) Munz = Epilobium brachycarpum [EPILBRA] Onagraceae (V:D)
Epilobium platyphyllum Rydb. = Epilobium glaberrimum ssp. fastigiatum [EPILGLA1] Onagraceae (V:D)
Epilobium pringleanum Hausskn. = Epilobium halleanum [EPILHAL] Onagraceae (V:D)
Epilobium pringleanum var. *tenue* (Trel.) Munz = Epilobium halleanum [EPILHAL] Onagraceae (V:D)
Epilobium pylaieanum Fern. = Epilobium palustre [EPILPAL] Onagraceae (V:D)
Epilobium rosmarinifolium Pursh = Epilobium leptophyllum [EPILLET] Onagraceae (V:D)
Epilobium saximontanum Hausskn. [EPILSAX] Onagraceae (V:D)
Epilobium scalare Fern. = Epilobium saximontanum [EPILSAX] Onagraceae (V:D)
Epilobium sertulatum Hausskn. = Epilobium hornemannii ssp. behringianum [EPILHOR1] Onagraceae (V:D)
Epilobium steckerianum Fern. = Epilobium saximontanum [EPILSAX] Onagraceae (V:D)
Epilobium treleaseanum Levl. p.p. = Epilobium luteum [EPILLUT] Onagraceae (V:D)
Epilobium ursinum Parish ex Trel. = Epilobium ciliatum ssp. ciliatum [EPILCIL0] Onagraceae (V:D)
Epilobium watsonii Barbey = Epilobium ciliatum ssp. watsonii [EPILCIL3] Onagraceae (V:D)
Epilobium watsonii var. *franciscanum* (Barbey) Jepson = Epilobium ciliatum ssp. watsonii [EPILCIL3] Onagraceae (V:D)
Epilobium watsonii var. *occidentale* (Trel.) C.L. Hitchc. = Epilobium ciliatum ssp. glandulosum [EPILCIL2] Onagraceae (V:D)
Epilobium watsonii var. *parishii* (Trel.) C.L. Hitchc. = Epilobium ciliatum ssp. ciliatum [EPILCIL0] Onagraceae (V:D)
Epilobium wyomingense A. Nels. = Epilobium palustre [EPILPAL] Onagraceae (V:D)
Epipactis Zinn [EPIPACT$] Orchidaceae (V:M)
Epipactis gigantea Dougl. ex Hook. [EPIPGIG] Orchidaceae (V:M)
***Epipactis helleborine** (L.) Crantz [EPIPHEL] Orchidaceae (V:M)
Epipactis latifolia (L.) All. = Epipactis helleborine [EPIPHEL] Orchidaceae (V:M)
Epipterygium Lindb. [EPIPTER$] Bryaceae (B:M)
Epipterygium tozeri (Grev.) Lindb. [EPIPTOZ] Bryaceae (B:M)
Equisetum L. [EQUISET$] Equisetaceae (V:F)
Equisetum affine Engelm. = Equisetum hyemale var. affine [EQUIHYE2] Equisetaceae (V:F)
Equisetum arvense L. [EQUIARV] Equisetaceae (V:F)
Equisetum arvense var. *alpestre* Wahlenb. = Equisetum arvense [EQUIARV] Equisetaceae (V:F)
Equisetum arvense var. *boreale* (Bong.) Rupr. = Equisetum arvense [EQUIARV] Equisetaceae (V:F)
Equisetum arvense var. *riparium* Farw. = Equisetum arvense [EQUIARV] Equisetaceae (V:F)
Equisetum braunii Milde = Equisetum telmateia var. braunii [EQUITEL1] Equisetaceae (V:F)
Equisetum calderi Boivin = Equisetum arvense [EQUIARV] Equisetaceae (V:F)
Equisetum fluviatile L. [EQUIFLU] Equisetaceae (V:F)
Equisetum fluviatile var. *limosum* (L.) Gilbert = Equisetum fluviatile [EQUIFLU] Equisetaceae (V:F)
Equisetum funstonii A.A. Eat. = Equisetum laevigatum [EQUILAE] Equisetaceae (V:F)
Equisetum hyemale L. [EQUIHYE] Equisetaceae (V:F)
Equisetum hyemale ssp. *affine* (Engelm.) Calder & Taylor = Equisetum hyemale var. affine [EQUIHYE2] Equisetaceae (V:F)
Equisetum hyemale var. **affine** (Engelm.) A.A. Eat. [EQUIHYE2] Equisetaceae (V:F)
Equisetum hyemale var. *californicum* Milde = Equisetum hyemale var. affine [EQUIHYE2] Equisetaceae (V:F)
Equisetum hyemale var. *elatum* (Engelm.) Morton = Equisetum hyemale var. affine [EQUIHYE2] Equisetaceae (V:F)
Equisetum hyemale var. *pseudohyemale* (Farw.) Morton = Equisetum hyemale var. affine [EQUIHYE2] Equisetaceae (V:F)

Equisetum hyemale var. *robustum* (A. Braun) A.A. Eat. = Equisetum hyemale var. affine [EQUIHYE2] Equisetaceae (V:F)
Equisetum kansanum Schaffn. = Equisetum laevigatum [EQUILAE] Equisetaceae (V:F)
Equisetum laevigatum A. Braun [EQUILAE] Equisetaceae (V:F)
Equisetum laevigatum ssp. *funstonii* (A.A. Eat.) Hartman = Equisetum laevigatum [EQUILAE] Equisetaceae (V:F)
Equisetum limosum L. = Equisetum fluviatile [EQUIFLU] Equisetaceae (V:F)
Equisetum × mackaii (Newm.) Brichan [EQUIMAC] Equisetaceae (V:F)
Equisetum maximum auct. = Equisetum telmateia var. braunii [EQUITEL1] Equisetaceae (V:F)
Equisetum palustre L. [EQUIPAL] Equisetaceae (V:F)
Equisetum palustre var. *americanum* Victorin = Equisetum palustre [EQUIPAL] Equisetaceae (V:F)
Equisetum palustre var. *simplicissimum* A. Braun = Equisetum palustre [EQUIPAL] Equisetaceae (V:F)
Equisetum praealtum Raf. = Equisetum hyemale var. affine [EQUIHYE2] Equisetaceae (V:F)
Equisetum pratense Ehrh. [EQUIPRA] Equisetaceae (V:F)
Equisetum robustum A. Braun = Equisetum hyemale var. affine [EQUIHYE2] Equisetaceae (V:F)
Equisetum scirpoides Michx. [EQUISCI] Equisetaceae (V:F)
Equisetum sylvaticum L. [EQUISYL] Equisetaceae (V:F)
Equisetum sylvaticum var. *multiramosum* (Fern.) Wherry = Equisetum sylvaticum [EQUISYL] Equisetaceae (V:F)
Equisetum sylvaticum var. *pauciramosum* Milde = Equisetum sylvaticum [EQUISYL] Equisetaceae (V:F)
Equisetum telmateia Ehrh. [EQUITEL] Equisetaceae (V:F)
Equisetum telmateia ssp. *braunii* (Milde) Hauke = Equisetum telmateia var. braunii [EQUITEL1] Equisetaceae (V:F)
Equisetum telmateia var. **braunii** (Milde) Milde [EQUITEL1] Equisetaceae (V:F)
Equisetum × trachyodon A. Br. = Equisetum × mackaii [EQUIMAC] Equisetaceae (V:F)
Equisetum trachyodon (A. Braun) W.D.J. Koch = Equisetum × mackaii [EQUIMAC] Equisetaceae (V:F)
Equisetum variegatum Schleich. ex F. Weber & D.M.H. Mohr [EQUIVAR] Equisetaceae (V:F)
Equisetum variegatum ssp. *alaskanum* (A.A. Eat.) Hultén = Equisetum variegatum var. alaskanum [EQUIVAR1] Equisetaceae (V:F)
Equisetum variegatum var. **alaskanum** A.A. Eat. [EQUIVAR1] Equisetaceae (V:F)
Equisetum variegatum var. *anceps* Milde = Equisetum variegatum var. variegatum [EQUIVAR0] Equisetaceae (V:F)
Equisetum variegatum var. *jesupii* A.A. Eat. = Equisetum × mackaii [EQUIMAC] Equisetaceae (V:F)
Equisetum variegatum var. **variegatum** [EQUIVAR0] Equisetaceae (V:F)
Eragrostis von Wolf [ERAGROS$] Poaceae (V:M)
Eragrostis cilianensis* (All.) Lut. ex Janchen [ERAGCIL] Poaceae (V:M)
Eragrostis eragrostis (L.) Beauv. = Eragrostis minor [ERAGMIN] Poaceae (V:M)
Eragrostis major Host = Eragrostis cilianensis [ERAGCIL] Poaceae (V:M)
Eragrostis megastachya (Koel.) Link = Eragrostis cilianensis [ERAGCIL] Poaceae (V:M)
***Eragrostis minor** Host [ERAGMIN] Poaceae (V:M)
Eragrostis pectinacea (Michx.) Nees ex Steud. [ERAGPEC] Poaceae (V:M)
***Eragrostis pilosa** (L.) Beauv. [ERAGPIL] Poaceae (V:M)
Eragrostis poaeoides Beauv. ex Roemer & J.A. Schultes = Eragrostis minor [ERAGMIN] Poaceae (V:M)
Eremogone americana (Maguire) S. Ikonnikov = Arenaria capillaris ssp. americana [ARENCAP1] Caryophyllaceae (V:D)
Eremonotus Lindb. & Kaal. ex Pears. [EREMONO$] Gymnomitriaceae (B:H)
Eremonotus hellerianus (Meyl.) Schust. = Anastrophyllum helleranum [ANASHEL] Jungermanniaceae (B:H)
Eremonotus minutus (schreb.) Schust. = Anastrophyllum minutum var. minutum [ANASMIN0] Jungermanniaceae (B:H)
Eremonotus myriocarpus (Carring.) Pears. [EREMMYR] Gymnomitriaceae (B:H)
Eremopyrum (Ledeb.) Jaubert & Spach [EREMOPY$] Poaceae (V:M)
***Eremopyrum triticeum** (Gaertn.) Nevski [EREMTRI] Poaceae (V:M)
Ericameria Nutt. [ERICAME$] Asteraceae (V:D)
Ericameria bloomeri (Gray) J.F. Macbr. [ERICBLO] Asteraceae (V:D)
Erigeron L. [ERIGERO$] Asteraceae (V:D)
Erigeron acris L. = Trimorpha acris var. asteroides [TRIMACR1] Asteraceae (V:D)
Erigeron acris ssp. *debilis* (Gray) Piper = Trimorpha acris var. debilis [TRIMACR2] Asteraceae (V:D)
Erigeron acris ssp. *politus* (Fries) Schinz & R. Keller = Trimorpha acris var. asteroides [TRIMACR1] Asteraceae (V:D)
Erigeron acris var. *asteroides* (Andrz. ex Bess.) DC. = Trimorpha acris var. asteroides [TRIMACR1] Asteraceae (V:D)
Erigeron acris var. *debilis* Gray = Trimorpha acris var. debilis [TRIMACR2] Asteraceae (V:D)
Erigeron acris var. *elatus* (Hook.) Cronq. = Trimorpha elata [TRIMELA] Asteraceae (V:D)
Erigeron alpinus var. *elatus* Hook. = Trimorpha elata [TRIMELA] Asteraceae (V:D)
Erigeron angulosus ssp. *debilis* (Gray) Piper = Trimorpha acris var. debilis [TRIMACR2] Asteraceae (V:D)
***Erigeron annuus** (L.) Pers. [ERIGANN] Asteraceae (V:D)
Erigeron annuus ssp. *strigosus* (Muhl. ex Willd.) Wagenitz = Erigeron strigosus [ERIGSTR] Asteraceae (V:D)
Erigeron annuus var. *discoideus* (Victorin & Rouss.) Cronq. = Erigeron annuus [ERIGANN] Asteraceae (V:D)
Erigeron anodonta Lunell = Erigeron glabellus var. pubescens [ERIGGLA1] Asteraceae (V:D)
Erigeron asperum var. *pubescens* (Hook.) Breitung = Erigeron glabellus var. pubescens [ERIGGLA1] Asteraceae (V:D)
Erigeron aureus Greene [ERIGAUR] Asteraceae (V:D)
Erigeron caespitosus Nutt. [ERIGCAE] Asteraceae (V:D)
Erigeron caespitosus var. *laccoliticus* M.E. Jones = Erigeron caespitosus [ERIGCAE] Asteraceae (V:D)
Erigeron canadensis L. = Conyza canadensis [CONYCAN] Asteraceae (V:D)
Erigeron canadensis var. *glabratus* Gray = Conyza canadensis var. glabrata [CONYCAN2] Asteraceae (V:D)
Erigeron compositus Pursh [ERIGCOM] Asteraceae (V:D)
Erigeron compositus var. *discoideus* Gray = Erigeron compositus [ERIGCOM] Asteraceae (V:D)
Erigeron compositus var. *glabratus* Macoun = Erigeron compositus [ERIGCOM] Asteraceae (V:D)
Erigeron compositus var. *multifidus* (Rydb.) J.F. Macbr. & Payson = Erigeron compositus [ERIGCOM] Asteraceae (V:D)
Erigeron compositus var. *typicus* Hook. = Erigeron compositus [ERIGCOM] Asteraceae (V:D)
Erigeron conspicuus Rydb. = Erigeron subtrinervis var. conspicuus [ERIGSUB1] Asteraceae (V:D)
Erigeron corymbosus Nutt. [ERIGCOR] Asteraceae (V:D)
Erigeron debilis (Gray) Rydb. = Trimorpha acris var. debilis [TRIMACR2] Asteraceae (V:D)
Erigeron divergens Torr. & Gray [ERIGDIV] Asteraceae (V:D)
Erigeron divergens var. *typicus* Cronq. = Erigeron divergens [ERIGDIV] Asteraceae (V:D)
Erigeron droebachianus O.F. Muell. ex Retz. = Trimorpha acris var. asteroides [TRIMACR1] Asteraceae (V:D)
Erigeron elatus (Hook.) Greene = Trimorpha elata [TRIMELA] Asteraceae (V:D)
Erigeron elatus var. *oligocephalus* (Fern. & Wieg.) Fern. = Trimorpha elata [TRIMELA] Asteraceae (V:D)
Erigeron elongatus Ledeb. = Trimorpha acris var. asteroides [TRIMACR1] Asteraceae (V:D)
Erigeron eriocephalus J. Vahl = Erigeron uniflorus ssp. eriocephalus [ERIGUNI1] Asteraceae (V:D)
Erigeron filifolius (Hook.) Nutt. [ERIGFIL] Asteraceae (V:D)

Erigeron filifolius var. *typicus* Cronq. = Erigeron filifolius [ERIGFIL] Asteraceae (V:D)
Erigeron flagellaris Gray [ERIGFLA] Asteraceae (V:D)
Erigeron flagellaris var. *typicus* Cronq. = Erigeron flagellaris [ERIGFLA] Asteraceae (V:D)
Erigeron glabellus Nutt. [ERIGGLA] Asteraceae (V:D)
Erigeron glabellus ssp. *pubescens* (Hook.) Cronq. = Erigeron glabellus var. pubescens [ERIGGLA1] Asteraceae (V:D)
Erigeron glabellus var. **pubescens** Hook. [ERIGGLA1] Asteraceae (V:D)
Erigeron grandiflorus Hook. [ERIGGRA] Asteraceae (V:D)
Erigeron humilis Graham [ERIGHUM] Asteraceae (V:D)
Erigeron jucundus Greene = Trimorpha acris var. debilis [TRIMACR2] Asteraceae (V:D)
Erigeron lanatus Hook. [ERIGLAN] Asteraceae (V:D)
Erigeron leibergii Piper [ERIGLEI] Asteraceae (V:D)
Erigeron linearis (Hook.) Piper [ERIGLIN] Asteraceae (V:D)
Erigeron lonchophyllus Hook. = Trimorpha lonchophylla [TRIMLON] Asteraceae (V:D)
Erigeron lonchophyllus var. *laurentianus* Victorin = Trimorpha lonchophylla [TRIMLON] Asteraceae (V:D)
Erigeron minor (Hook.) Rydb. = Trimorpha lonchophylla [TRIMLON] Asteraceae (V:D)
Erigeron nivalis Nutt. = Trimorpha acris var. debilis [TRIMACR2] Asteraceae (V:D)
Erigeron nudiflorus Buckl. = Erigeron flagellaris [ERIGFLA] Asteraceae (V:D)
Erigeron oligodontus Lunell = Erigeron glabellus var. pubescens [ERIGGLA1] Asteraceae (V:D)
Erigeron pallens Cronq. [ERIGPAL] Asteraceae (V:D)
Erigeron peregrinus (Banks ex Pursh) Greene [ERIGPER] Asteraceae (V:D)
Erigeron peregrinus ssp. **callianthemus** (Greene) Cronq. [ERIGPER2] Asteraceae (V:D)
Erigeron peregrinus ssp. *callianthemus* var. *angustifolius* (Gray) Cronq. = Erigeron peregrinus ssp. callianthemus [ERIGPER2] Asteraceae (V:D)
Erigeron peregrinus ssp. *callianthemus* var. *scaposus* (Torr. & Gray) Cronq. = Erigeron peregrinus ssp. callianthemus [ERIGPER2] Asteraceae (V:D)
Erigeron peregrinus ssp. **peregrinus** [ERIGPER0] Asteraceae (V:D)
Erigeron peregrinus ssp. **peregrinus** var. **dawsonii** Greene [ERIGPER1] Asteraceae (V:D)
Erigeron peregrinus var. *thompsonii* (Blake ex J.W. Thompson) Cronq. = Erigeron peregrinus ssp. peregrinus [ERIGPER0] Asteraceae (V:D)
Erigeron peucephyllus Gray = Erigeron linearis [ERIGLIN] Asteraceae (V:D)
Erigeron philadelphicus L. [ERIGPHI] Asteraceae (V:D)
Erigeron poliospermus Gray [ERIGPOL] Asteraceae (V:D)
Erigeron poliospermus var. *typicus* Cronq. = Erigeron poliospermus [ERIGPOL] Asteraceae (V:D)
Erigeron politus Fries = Trimorpha acris var. asteroides [TRIMACR1] Asteraceae (V:D)
Erigeron pumilus Nutt. [ERIGPUM] Asteraceae (V:D)
Erigeron pumilus ssp. **intermedius** Cronq. [ERIGPUM1] Asteraceae (V:D)
Erigeron purpuratus Greene [ERIGPUR] Asteraceae (V:D)
Erigeron purpuratus ssp. *pallens* (Cronq.) G.W. Douglas = Erigeron pallens [ERIGPAL] Asteraceae (V:D)
Erigeron ramosus (Walt.) B.S.P. = Erigeron strigosus [ERIGSTR] Asteraceae (V:D)
Erigeron salishii G.W. Douglas & Packer [ERIGSAL] Asteraceae (V:D)
Erigeron speciosus (Lindl.) DC. [ERIGSPE] Asteraceae (V:D)
Erigeron speciosus var. *conspicuus* (Rydb.) Breitung = Erigeron subtrinervis var. conspicuus [ERIGSUB1] Asteraceae (V:D)
Erigeron speciosus var. *typicus* Cronq. = Erigeron speciosus [ERIGSPE] Asteraceae (V:D)
Erigeron strigosus Muhl. ex Willd. [ERIGSTR] Asteraceae (V:D)
Erigeron strigosus var. *discoideus* Robbins ex Gray = Erigeron strigosus [ERIGSTR] Asteraceae (V:D)
Erigeron strigosus var. *eligulatus* Cronq. = Erigeron strigosus [ERIGSTR] Asteraceae (V:D)
Erigeron strigosus var. *typicus* Cronq. = Erigeron strigosus [ERIGSTR] Asteraceae (V:D)
Erigeron subtrinervis Rydb. ex Porter & Britt. [ERIGSUB] Asteraceae (V:D)
Erigeron subtrinervis ssp. *conspicuus* (Rydb.) Cronq. = Erigeron subtrinervis var. conspicuus [ERIGSUB1] Asteraceae (V:D)
Erigeron subtrinervis var. **conspicuus** (Rydb.) Cronq. [ERIGSUB1] Asteraceae (V:D)
Erigeron traversii Shinners = Erigeron strigosus [ERIGSTR] Asteraceae (V:D)
Erigeron trifidus Hook. [ERIGTRI] Asteraceae (V:D)
Erigeron unalaschkensis (DC.) Vierh. = Erigeron humilis [ERIGHUM] Asteraceae (V:D)
Erigeron uniflorus L. [ERIGUNI] Asteraceae (V:D)
Erigeron uniflorus ssp. **eriocephalus** (J. Vahl) Cronq. [ERIGUNI1] Asteraceae (V:D)
Erigeron uniflorus var. *eriocephalus* (J. Vahl) Boivin = Erigeron uniflorus ssp. eriocephalus [ERIGUNI1] Asteraceae (V:D)
Erigeron uniflorus var. *unalaschkensis* (DC.) Ostenf. = Erigeron humilis [ERIGHUM] Asteraceae (V:D)
Eriocoma cuspidata Nutt. = Oryzopsis hymenoides [ORYZHYM] Poaceae (V:M)
Erioderma Fée [ERIODER$] Pannariaceae (L)
Erioderma mollissimum (Samp.) Du Rietz [ERIOMOL] Pannariaceae (L)
Erioderma sorediatum D.J. Galloway & P.M. Jørg. [ERIOSOR] Pannariaceae (L)
Eriogonum Michx. [ERIOGON$] Polygonaceae (V:D)
Eriogonum androsaceum Benth. [ERIOAND] Polygonaceae (V:D)
Eriogonum angustifolium Nutt. = Eriogonum heracleoides var. angustifolium [ERIOHER1] Polygonaceae (V:D)
Eriogonum depauperatum Small = Eriogonum pauciflorum [ERIOPAU] Polygonaceae (V:D)
Eriogonum flavum Nutt. [ERIOFLA] Polygonaceae (V:D)
Eriogonum flavum ssp. *piperi* (Greene) S. Stokes = Eriogonum flavum var. piperi [ERIOFLA1] Polygonaceae (V:D)
Eriogonum flavum var. *androsaceum* (Benth.) M.E. Jones = Eriogonum androsaceum [ERIOAND] Polygonaceae (V:D)
Eriogonum flavum var. *linguifolium* Gandog. = Eriogonum flavum var. piperi [ERIOFLA1] Polygonaceae (V:D)
Eriogonum flavum var. **piperi** (Greene) M.E. Jones [ERIOFLA1] Polygonaceae (V:D)
Eriogonum heracleoides Nutt. [ERIOHER] Polygonaceae (V:D)
Eriogonum heracleoides var. **angustifolium** (Nutt.) Torr. & Gray [ERIOHER1] Polygonaceae (V:D)
Eriogonum heracleoides var. *subalpinum* (Greene) S. Stokes = Eriogonum umbellatum var. majus [ERIOUMB1] Polygonaceae (V:D)
Eriogonum multiceps Nees = Eriogonum pauciflorum [ERIOPAU] Polygonaceae (V:D)
Eriogonum nivale Canby = Eriogonum ovalifolium var. nivale [ERIOOVA3] Polygonaceae (V:D)
Eriogonum niveum Dougl. ex Benth. [ERIONIV] Polygonaceae (V:D)
Eriogonum niveum ssp. *decumbens* (Benth.) S. Stokes = Eriogonum niveum [ERIONIV] Polygonaceae (V:D)
Eriogonum niveum var. *decumbens* (Benth.) Torr. & Gray = Eriogonum niveum [ERIONIV] Polygonaceae (V:D)
Eriogonum niveum var. *dichotomum* (Dougl. ex Benth.) M.E. Jones = Eriogonum niveum [ERIONIV] Polygonaceae (V:D)
Eriogonum ovalifolium Nutt. [ERIOOVA] Polygonaceae (V:D)
Eriogonum ovalifolium var. **nivale** (Canby) M.E. Jones [ERIOOVA3] Polygonaceae (V:D)
Eriogonum pauciflorum Pursh [ERIOPAU] Polygonaceae (V:D)
Eriogonum piperi Greene = Eriogonum flavum var. piperi [ERIOFLA1] Polygonaceae (V:D)
Eriogonum pyrolifolium Hook. [ERIOPYR] Polygonaceae (V:D)
Eriogonum pyrolifolium var. *bellingerianum* M.E. Peck = Eriogonum pyrolifolium var. coryphaeum [ERIOPYR1] Polygonaceae (V:D)
Eriogonum pyrolifolium var. **coryphaeum** Torr. & Gray [ERIOPYR1] Polygonaceae (V:D)

Eriogonum rhodanthum A. Nels. & Kennedy = Eriogonum ovalifolium var. nivale [ERIOOVA3] Polygonaceae (V:D)
Eriogonum strictum Benth. [ERIOSTR] Polygonaceae (V:D)
Eriogonum strictum ssp. **proliferum** (Torr. & Gray) S. Stokes [ERIOSTR1] Polygonaceae (V:D)
Eriogonum strictum var. *lachnostegium* Benth. = Eriogonum niveum [ERIONIV] Polygonaceae (V:D)
Eriogonum subalpinum Greene = Eriogonum umbellatum var. majus [ERIOUMB1] Polygonaceae (V:D)
Eriogonum umbellatum Torr. [ERIOUMB] Polygonaceae (V:D)
Eriogonum umbellatum ssp. *majus* (Hook.) Piper = Eriogonum umbellatum var. majus [ERIOUMB1] Polygonaceae (V:D)
Eriogonum umbellatum var. **majus** Hook. [ERIOUMB1] Polygonaceae (V:D)
Eriogonum umbellatum var. *subalpinum* (Greene) M.E. Jones = Eriogonum umbellatum var. majus [ERIOUMB1] Polygonaceae (V:D)
Eriogonum umbellatum var. **umbellatum** [ERIOUMB0] Polygonaceae (V:D)
Eriophorum L. [ERIOPHO$] Cyperaceae (V:M)
Eriophorum alpinum L. [ERIOALP] Cyperaceae (V:M)
Eriophorum altaicum Meinsh. [ERIOALT] Cyperaceae (V:M)
Eriophorum altaicum var. **neogaeum** Raymond [ERIOALT1] Cyperaceae (V:M)
Eriophorum angustifolium Honckeny [ERIOANG] Cyperaceae (V:M)
Eriophorum angustifolium ssp. **scabriusculum** Hultén [ERIOANG1] Cyperaceae (V:M)
Eriophorum angustifolium ssp. **subarcticum** (Vassiljev) Hultén ex Kartesz & Gandhi [ERIOANG2] Cyperaceae (V:M)
Eriophorum angustifolium ssp. **triste** (T. Fries) Hultén [ERIOANG3] Cyperaceae (V:M)
Eriophorum brachyantherum Trautv. & C.A. Mey. [ERIOBRA] Cyperaceae (V:M)
Eriophorum callitrix Cham. ex C.A. Mey. [ERIOCAL] Cyperaceae (V:M)
Eriophorum chamissonis C.A. Mey. [ERIOCHA] Cyperaceae (V:M)
Eriophorum gracile W.D.J. Koch [ERIOGRA] Cyperaceae (V:M)
Eriophorum polystachion var. *viridicarinatum* Engelm. = Eriophorum viridicarinatum [ERIOVIR] Cyperaceae (V:M)
Eriophorum russeolum Fries ex Hartman [ERIORUS] Cyperaceae (V:M)
Eriophorum russeolum ssp. *rufescens* (E. Anders.) Hyl. = Eriophorum chamissonis [ERIOCHA] Cyperaceae (V:M)
Eriophorum scheuchzeri Hoppe [ERIOSCH] Cyperaceae (V:M)
Eriophorum scheuchzeri var. *tenuifolium* Ohwi = Eriophorum scheuchzeri [ERIOSCH] Cyperaceae (V:M)
Eriophorum spissum Fern. = Eriophorum vaginatum var. spissum [ERIOVAG2] Cyperaceae (V:M)
Eriophorum spissum var. *erubescens* (Fern.) Fern. = Eriophorum vaginatum var. spissum [ERIOVAG2] Cyperaceae (V:M)
Eriophorum triste (T. Fries) Hadac & A. Löve = Eriophorum angustifolium ssp. triste [ERIOANG3] Cyperaceae (V:M)
Eriophorum vaginatum L. [ERIOVAG] Cyperaceae (V:M)
Eriophorum vaginatum ssp. *spissum* (Fern.) Hultén = Eriophorum vaginatum var. spissum [ERIOVAG2] Cyperaceae (V:M)
Eriophorum vaginatum var. **spissum** (Fern.) Boivin [ERIOVAG2] Cyperaceae (V:M)
****Eriophorum virginicum** L. [ERIOVIG] Cyperaceae (V:M)
Eriophorum viridicarinatum (Engelm.) Fern. [ERIOVIR] Cyperaceae (V:M)
Eriophyllum Lag. [ERIOPHY$] Asteraceae (V:D)
Eriophyllum lanatum (Pursh) Forbes [ERIOLAN] Asteraceae (V:D)
Eriophyllum lanatum var. *typicum* Constance = Eriophyllum lanatum [ERIOLAN] Asteraceae (V:D)
Erocallis triphylla (S. Wats.) Rydb. = Lewisia triphylla [LEWITRI] Portulacaceae (V:D)
Erodium L'Hér. ex Ait. [ERODIUM$] Geraniaceae (V:D)
****Erodium cicutarium** (L.) L'Hér. ex Ait. [ERODCIC] Geraniaceae (V:D)
Erophila spathulata A.F. Lang = Draba verna [DRABVER] Brassicaceae (V:D)
Erophila verna (L.) Bess. = Draba verna [DRABVER] Brassicaceae (V:D)
Erophila verna ssp. *praecox* (Stev.) S.M. Walters = Draba verna [DRABVER] Brassicaceae (V:D)
Erophila verna ssp. *spathulata* (A.F. Lang) S.M. Walters = Draba verna [DRABVER] Brassicaceae (V:D)
Erucastrum K. Presl [ERUCAST$] Brassicaceae (V:D)
****Erucastrum gallicum** (Willd.) O.E. Schulz [ERUCGAL] Brassicaceae (V:D)
Erxlebenia minor (L.) Rydb. = Pyrola minor [PYROMIN] Pyrolaceae (V:D)
Eryngium L. [ERYNGIU$] Apiaceae (V:D)
****Eryngium planum** L. [ERYNPLA] Apiaceae (V:D)
Erysimum L. [ERYSIMU$] Brassicaceae (V:D)
Erysimum alliaria L. = Alliaria petiolata [ALLIPET] Brassicaceae (V:D)
Erysimum arenicola S. Wats. [ERYSARE] Brassicaceae (V:D)
Erysimum arenicola var. **torulosum** (Piper) C.L. Hitchc. [ERYSARE1] Brassicaceae (V:D)
Erysimum capitatum (Dougl. ex Hook.) Greene [ERYSCAP] Brassicaceae (V:D)
****Erysimum cheiranthoides** L. [ERYSCHE] Brassicaceae (V:D)
Erysimum cheiranthoides ssp. *altum* Ahti = Erysimum cheiranthoides [ERYSCHE] Brassicaceae (V:D)
****Erysimum cheiri** (L.) Crantz [ERYSCHI] Brassicaceae (V:D)
Erysimum drummondii (Gray) Kuntze = Arabis drummondii [ARABDRU] Brassicaceae (V:D)
Erysimum inconspicuum (S. Wats.) MacM. [ERYSINC] Brassicaceae (V:D)
Erysimum officinale L. = Sisymbrium officinale [SISYOFF] Brassicaceae (V:D)
Erysimum pallasii (Pursh) Fern. [ERYSPAL] Brassicaceae (V:D)
Erysimum torulosum Piper = Erysimum arenicola var. torulosum [ERYSARE1] Brassicaceae (V:D)
Erysimum vernum P. Mill. = Barbarea verna [BARBVER] Brassicaceae (V:D)
Erythrobarbula recurvirostris (Hedw.) Steere = Bryoerythrophyllum recurvirostre [BRYOREC] Pottiaceae (B:M)
Erythrocoma ciliata Pursh = Geum triflorum var. ciliatum [GEUMTRI1] Rosaceae (V:D)
Erythrocoma triflora (Pursh) Greene = Geum triflorum var. triflorum [GEUMTRI0] Rosaceae (V:D)
Erythronium L. [ERYTHRO$] Liliaceae (V:M)
Erythronium grandiflorum Pursh [ERYTGRA] Liliaceae (V:M)
Erythronium montanum S. Wats. [ERYTMON] Liliaceae (V:M)
Erythronium oregonum Applegate [ERYTORE] Liliaceae (V:M)
Erythronium revolutum Sm. [ERYTREV] Liliaceae (V:M)
Eschscholzia Cham. [ESCHSCH$] Papaveraceae (V:D)
****Eschscholzia californica** Cham. [ESCHCAL] Papaveraceae (V:D)
Esslingeriana Hale & M.J. Lai [ESSLING$] Parmeliaceae (L)
Esslingeriana idahoensis (Essl.) Hale & M.J. Lai [ESSLIDA] Parmeliaceae (L)
Eucephalus engelmannii (D.C. Eat.) Greene = Aster engelmannii [ASTEENG] Asteraceae (V:D)
Eucladium Bruch & Schimp. in B.S.G. [EUCLADI$] Pottiaceae (B:M)
Eucladium verticillatum (Brid.) Bruch & Schimp. in B.S.G. [EUCLVER] Pottiaceae (B:M)
Eunanus breweri Greene = Mimulus breweri [MIMUBRW] Scrophulariaceae (V:D)
Euonymus L. [EUONYMU$] Celastraceae (V:D)
Euonymus occidentalis Nutt. ex Torr. [EUONOCC] Celastraceae (V:D)
Euopsis Nyl. [EUOPSIS$] Lichinaceae (L)
Euopsis pulvinata (Schaerer) Nyl. [EUOPPUL] Lichinaceae (L)
Eupatoriadelphus maculatus var. *bruneri* (Gray) King & H.E. Robins. = Eupatorium maculatum var. bruneri [EUPAMAC1] Asteraceae (V:D)
Eupatorium L. [EUPATOR$] Asteraceae (V:D)
Eupatorium bruneri Gray = Eupatorium maculatum var. bruneri [EUPAMAC1] Asteraceae (V:D)
Eupatorium maculatum L. [EUPAMAC] Asteraceae (V:D)

Eupatorium maculatum ssp. *bruneri* (Gray) G.W. Douglas = Eupatorium maculatum var. bruneri [EUPAMAC1] Asteraceae (V:D)
Eupatorium maculatum var. **bruneri** (Gray) Breitung [EUPAMAC1] Asteraceae (V:D)
Euphorbia L. [EUPHORB$] Euphorbiaceae (V:D)
*****Euphorbia cyparissias** L. [EUPHCYP] Euphorbiaceae (V:D)
*****Euphorbia esula** L. [EUPHESU] Euphorbiaceae (V:D)
*****Euphorbia exigua** L. [EUPHEXI] Euphorbiaceae (V:D)
Euphorbia glyptosperma Engelm. = Chamaesyce glyptosperma [CHAMGLY] Euphorbiaceae (V:D)
*****Euphorbia helioscopia** L. [EUPHHEL] Euphorbiaceae (V:D)
*****Euphorbia lathyris** L. [EUPHLAT] Euphorbiaceae (V:D)
Euphorbia maculata L. = Chamaesyce maculata [CHAMMAC] Euphorbiaceae (V:D)
Euphorbia neomexicana Greene = Chamaesyce serpyllifolia [CHAMSER] Euphorbiaceae (V:D)
*****Euphorbia peplus** L. [EUPHPEP] Euphorbiaceae (V:D)
Euphorbia serpyllifolia Pers. = Chamaesyce serpyllifolia [CHAMSER] Euphorbiaceae (V:D)
Euphorbia supina Raf. = Chamaesyce maculata [CHAMMAC] Euphorbiaceae (V:D)
Euphrasia L. [EUPHRAS$] Scrophulariaceae (V:D)
Euphrasia americana Wettst. = Euphrasia nemorosa [EUPHNEM] Scrophulariaceae (V:D)
Euphrasia arctica ssp. *borealis* (Townsend) Yeo = Euphrasia nemorosa [EUPHNEM] Scrophulariaceae (V:D)
Euphrasia arctica var. *disjuncta* (Fern. & Wieg.) Cronq. = Euphrasia disjuncta [EUPHDIS] Scrophulariaceae (V:D)
Euphrasia borealis (Townsend) Wettst. = Euphrasia nemorosa [EUPHNEM] Scrophulariaceae (V:D)
Euphrasia disjuncta Fern. & Wieg. [EUPHDIS] Scrophulariaceae (V:D)
*****Euphrasia nemorosa** (Pers.) Wallr. [EUPHNEM] Scrophulariaceae (V:D)
Eurhynchium Schimp. in B.S.G. [EURHYNC$] Brachytheciaceae (B:M)
Eurhynchium acutifolium Kindb. = Eurhynchium praelongum [EURHPRA] Brachytheciaceae (B:M)
Eurhynchium brittoniae Grout = Eurhynchium praelongum [EURHPRA] Brachytheciaceae (B:M)
Eurhynchium diversifolium Schimp. in B.S.G. = Eurhynchium pulchellum var. pulchellum [EURHPUL0] Brachytheciaceae (B:M)
Eurhynchium fallax (Ren. & Card.) Grout = Eurhynchium pulchellum var. pulchellum [EURHPUL0] Brachytheciaceae (B:M)
Eurhynchium fallax var. *taylorae* (Williams) Grout = Eurhynchium pulchellum var. barnesii [EURHPUL1] Brachytheciaceae (B:M)
Eurhynchium fasciculosum (Hedw.) Dix. = Eurhynchium pulchellum var. pulchellum [EURHPUL0] Brachytheciaceae (B:M)
Eurhynchium labradoricum Kindb. = Brachythecium calcareum [BRACCAL] Brachytheciaceae (B:M)
Eurhynchium macounii Kindb. = Scleropodium touretii var. colpophyllum [SCLETOU1] Brachytheciaceae (B:M)
Eurhynchium myosuroides (Brid.) Schimp. = Isothecium myosuroides [ISOTMYO] Brachytheciaceae (B:M)
Eurhynchium oreganum (Sull.) Jaeg. [EURHORE] Brachytheciaceae (B:M)
Eurhynchium piliferum (Hedw.) B.S.G. = Cirriphyllum piliferum [CIRRPIL] Brachytheciaceae (B:M)
Eurhynchium praecox (Hedw.) De Not. in Picc. = Eurhynchium pulchellum var. pulchellum [EURHPUL0] Brachytheciaceae (B:M)
Eurhynchium praelongum (Hedw.) Schimp. in B.S.G. [EURHPRA] Brachytheciaceae (B:M)
Eurhynchium praelongum var. *californicum* Grout = Eurhynchium praelongum [EURHPRA] Brachytheciaceae (B:M)
Eurhynchium praelongum var. *stokesii* (Turn.) Dix. = Eurhynchium praelongum [EURHPRA] Brachytheciaceae (B:M)
Eurhynchium pseudoserrulatum Kindb. = Brachythecium starkei [BRACSTA] Brachytheciaceae (B:M)
Eurhynchium pulchellum (Hedw.) Jenn. [EURHPUL] Brachytheciaceae (B:M)
Eurhynchium pulchellum var. **barnesii** (Ren. & Card.) Grout [EURHPUL1] Brachytheciaceae (B:M)
Eurhynchium pulchellum var. *diversifolium* (B.S.G.) C. Jens. = Eurhynchium pulchellum [EURHPUL] Brachytheciaceae (B:M)
Eurhynchium pulchellum var. *praecox* (Hedw.) Dix. = Eurhynchium pulchellum var. pulchellum [EURHPUL0] Brachytheciaceae (B:M)
Eurhynchium pulchellum var. **pulchellum** [EURHPUL0] Brachytheciaceae (B:M)
Eurhynchium pulchellum var. *robustum* (Röll) Amann = Eurhynchium pulchellum var. pulchellum [EURHPUL0] Brachytheciaceae (B:M)
Eurhynchium riparioides (Hedw.) P. Rich. = Platyhypnidium riparioides [PLATRIP] Brachytheciaceae (B:M)
Eurhynchium rusciforme (Brid.) Milde = Platyhypnidium riparioides [PLATRIP] Brachytheciaceae (B:M)
Eurhynchium rusciforme var. *complanatum* H. Schultz ex Limpr. = Platyhypnidium riparioides [PLATRIP] Brachytheciaceae (B:M)
Eurhynchium semiasperum C. Müll. & Kindb. in Mac. & Kindb. = Brachythecium plumosum [BRACPLU] Brachytheciaceae (B:M)
Eurhynchium serrulatum (Hedw.) Kindb. = Steerecleus serrulatus [STEESER] Brachytheciaceae (B:M)
Eurhynchium serrulatum ssp. *ericense* Kindb. = Steerecleus serrulatus [STEESER] Brachytheciaceae (B:M)
Eurhynchium stokesii (Turn.) Schimp. in B.S.G. = Eurhynchium praelongum [EURHPRA] Brachytheciaceae (B:M)
Eurhynchium stokesii ssp. *pseudospeciosum* Kindb. = Eurhynchium praelongum [EURHPRA] Brachytheciaceae (B:M)
Eurhynchium stokesii var. *californicum* (Grout) Grout = Eurhynchium praelongum [EURHPRA] Brachytheciaceae (B:M)
Eurhynchium strigosum (Web. & Mohr) Schimp. in B.S.G. = Eurhynchium pulchellum var. pulchellum [EURHPUL0] Brachytheciaceae (B:M)
Eurhynchium strigosum var. *praecox* (Hedw.) Husn. = Eurhynchium pulchellum var. pulchellum [EURHPUL0] Brachytheciaceae (B:M)
Eurhynchium strigosum var. *robustum* Röll = Eurhynchium pulchellum var. pulchellum [EURHPUL0] Brachytheciaceae (B:M)
Eurhynchium strigosum var. *scabrisetum* Grout = Eurhynchium pulchellum var. pulchellum [EURHPUL0] Brachytheciaceae (B:M)
Eurhynchium subintegrifolium Kindb. = Brachythecium oedipodium [BRACOED] Brachytheciaceae (B:M)
Eurhynchium substrigosum Kindb. in Mac. & Kindb. = Eurhynchium pulchellum var. pulchellum [EURHPUL0] Brachytheciaceae (B:M)
Eurhynchium substrigosum var. *taylorae* (Ren. & Card.) Grout = Eurhynchium pulchellum var. barnesii [EURHPUL1] Brachytheciaceae (B:M)
Eurhynchium vaucheri (Lesq.) B.S.G. = Hypnum vaucheri [HYPNVAU] Hypnaceae (B:M)
Euthamia Nutt. ex Cass. [EUTHAMI$] Asteraceae (V:D)
Euthamia californica Gandog. = Euthamia occidentalis [EUTHOCC] Asteraceae (V:D)
Euthamia graminifolia (L.) Nutt. [EUTHGRA] Asteraceae (V:D)
Euthamia graminifolia var. *major* (Michx.) Moldenke = Euthamia graminifolia [EUTHGRA] Asteraceae (V:D)
Euthamia linearifolia Gandog. = Euthamia occidentalis [EUTHOCC] Asteraceae (V:D)
Euthamia occidentalis Nutt. [EUTHOCC] Asteraceae (V:D)
Eutrema R. Br. [EUTREMA$] Brassicaceae (V:D)
Eutrema edwardsii R. Br. [EUTREDW] Brassicaceae (V:D)
Evernia Ach. [EVERNIA$] Parmeliaceae (L)
Evernia divaricata (L.) Ach. [EVERDIV] Parmeliaceae (L)
Evernia mesomorpha Nyl. [EVERMES] Parmeliaceae (L)
Evernia perfragilis Llano [EVERPER] Parmeliaceae (L)
Evernia prunastri (L.) Ach. [EVERPRU] Parmeliaceae (L)
Evernia thamnodes (Flotow) Arnold = Evernia mesomorpha [EVERMES] Parmeliaceae (L)
Evernia vulpina (L.) Ach. = Letharia vulpina [LETHVUL] Parmeliaceae (L)

Evonymus occidentalis Nutt. ex Torr. = Euonymus occidentalis [EUONOCC] Celastraceae (V:D)

F

Fabronia Raddi [FABRONI$] Fabroniaceae (B:M)
Fabronia pusilla Raddi [FABRPUS] Fabroniaceae (B:M)
Fagopyrum P. Mill. [FAGOPYR$] Polygonaceae (V:D)
***Fagopyrum esculentum** Moench [FAGOESC] Polygonaceae (V:D)
Fagopyrum fagopyrum (L.) Karst. = Fagopyrum esculentum [FAGOESC] Polygonaceae (V:D)
Fagopyrum sagittatum Gilib. = Fagopyrum esculentum [FAGOESC] Polygonaceae (V:D)
Fagopyrum vulgare Hill = Fagopyrum esculentum [FAGOESC] Polygonaceae (V:D)
Fallopia convolvulus (L.) A. Löve = Polygonum convolvulus [POLYCON] Polygonaceae (V:D)
Farnoldia Hertel [FARNOLD$] Porpidiaceae (L)
Farnoldia hypocrita (A. Massal.) Frøberg [FARNHYP] Porpidiaceae (L)
Farnoldia jurana (Schaerer) Hertel [FARNJUR] Porpidiaceae (L)
Fauria Franch. [FAURIA$] Menyanthaceae (V:D)
Fauria crista-galli (Menzies ex Hook.) Makino [FAURCRI] Menyanthaceae (V:D)
Fellhanera Vezda [FELLHAN$] Pilocarpaceae (L)
Fellhanera bouteillei (Desmaz.) Vezda [FELLBOU] Pilocarpaceae (L)
Fellhanera subtilis (Vezda) Diederich & Sérusiaux [FELLSUB] Pilocarpaceae (L)
Ferula macrocarpa Hook. & Arn. = Lomatium macrocarpum [LOMAMAC] Apiaceae (V:D)
Festuca L. [FESTUCA$] Poaceae (V:M)
Festuca altaica Trin. [FESTALT] Poaceae (V:M)
Festuca altaica ssp. *scabrella* (Torr. ex Hook.) Hultén = Festuca campestris [FESTCAM] Poaceae (V:M)
Festuca altaica var. *major* (Vasey) Gleason = Festuca campestris [FESTCAM] Poaceae (V:M)
Festuca altaica var. *scabrella* (Torr. ex Hook.) Breitung = Festuca campestris [FESTCAM] Poaceae (V:M)
***Festuca arundinacea** Schreb. [FESTARU] Poaceae (V:M)
Festuca aucta Krecz. & Bobr. = Festuca rubra ssp. aucta [FESTRUB5] Poaceae (V:M)
Festuca baffinensis Polunin [FESTBAF] Poaceae (V:M)
Festuca brachyphylla J.A. Schultes ex J.A. & J.H. Schultes [FESTBRA] Poaceae (V:M)
Festuca brachyphylla var. *rydbergii* (St.-Yves) Cronq. = Festuca saximontana [FESTSAX] Poaceae (V:M)
Festuca brevifolia var. *utahensis* St.-Yves = Festuca minutiflora [FESTMIN] Poaceae (V:M)
Festuca brevipila Tracey = Festuca trachyphylla [FESTTRA] Poaceae (V:M)
Festuca bromoides L. = Vulpia bromoides [VULPBRO] Poaceae (V:M)
Festuca campestris Rydb. [FESTCAM] Poaceae (V:M)
Festuca capillata Lam. = Festuca filiformis [FESTFIL] Poaceae (V:M)
Festuca decumbens L. = Danthonia decumbens [DANTDEC] Poaceae (V:M)
Festuca densiuscula (Hack. ex Piper) Alexeev = Festuca rubra ssp. pruinosa [FESTRUB1] Poaceae (V:M)
Festuca detonensis (All.) Aschers. & Graebn. = Vulpia bromoides [VULPBRO] Poaceae (V:M)
Festuca duriuscula auct. = Festuca trachyphylla [FESTTRA] Poaceae (V:M)
Festuca elatior L. = Festuca pratensis [FESTPRA] Poaceae (V:M)
Festuca elatior ssp. *arundinacea* (Schreb.) Hack. = Festuca arundinacea [FESTARU] Poaceae (V:M)
Festuca elatior var. *arundinacea* (Schreb.) C.F.H. Wimmer = Festuca arundinacea [FESTARU] Poaceae (V:M)
***Festuca filiformis** Pourret [FESTFIL] Poaceae (V:M)
Festuca gracilenta Buckl. = Vulpia octoflora var. glauca [VULPOCT1] Poaceae (V:M)
Festuca howellii Hack. ex Beal = Festuca viridula [FESTVIR] Poaceae (V:M)
Festuca idahoensis Elmer [FESTIDA] Poaceae (V:M)
Festuca idahoensis var. **roemeri** Pavlick [FESTIDA2] Poaceae (V:M)
Festuca megalura Nutt. = Vulpia myuros [VULPMYU] Poaceae (V:M)
Festuca megalura var. *hirsuta* (Hack.) Aschers. & Graebn. = Vulpia myuros [VULPMYU] Poaceae (V:M)
Festuca microstachys var. *pauciflora* Scribn. ex Beal = Vulpia microstachys var. pauciflora [VULPMIC1] Poaceae (V:M)
Festuca microstachys var. *simulans* (Hoover) Hoover = Vulpia microstachys var. pauciflora [VULPMIC1] Poaceae (V:M)
Festuca minutiflora Rydb. [FESTMIN] Poaceae (V:M)
Festuca myuros L. = Vulpia myuros [VULPMYU] Poaceae (V:M)
Festuca occidentalis Hook. [FESTOCC] Poaceae (V:M)
Festuca octoflora Walt. = Vulpia octoflora var. octoflora [VULPOCT0] Poaceae (V:M)
Festuca octoflora ssp. *hirtella* Piper = Vulpia octoflora var. hirtella [VULPOCT2] Poaceae (V:M)
Festuca octoflora var. *aristulata* Torr. ex L.H. Dewey = Vulpia octoflora var. octoflora [VULPOCT0] Poaceae (V:M)
Festuca octoflora var. *glauca* (Nutt.) Fern. = Vulpia octoflora var. glauca [VULPOCT1] Poaceae (V:M)
Festuca octoflora var. *hirtella* (Piper) Piper ex A.S. Hitchc. = Vulpia octoflora var. hirtella [VULPOCT2] Poaceae (V:M)
Festuca octoflora var. *tenella* (Willd.) Fern. = Vulpia octoflora var. glauca [VULPOCT1] Poaceae (V:M)
Festuca ovina var. *brachyphylla* (J.A. Schultes ex J.A. & J.H. Schultes) Piper ex A.S. Hitchc. = Festuca brachyphylla [FESTBRA] Poaceae (V:M)
Festuca ovina var. *brevifolia* S. Wats. = Festuca brachyphylla [FESTBRA] Poaceae (V:M)
Festuca ovina var. *capillata* (Lam.) Alef. = Festuca filiformis [FESTFIL] Poaceae (V:M)
Festuca ovina var. *duriuscula* auct. = Festuca trachyphylla [FESTTRA] Poaceae (V:M)
Festuca ovina var. *minutiflora* (Rydb.) J.T. Howell = Festuca minutiflora [FESTMIN] Poaceae (V:M)
Festuca ovina var. *polyphylla* Vasey ex Beal = Festuca occidentalis [FESTOCC] Poaceae (V:M)
Festuca ovina var. *purpusiana* St.-Yves = Festuca saximontana var. purpusiana [FESTSAX1] Poaceae (V:M)
Festuca ovina var. *saximontana* (Rydb.) Gleason = Festuca saximontana [FESTSAX] Poaceae (V:M)
Festuca ovina var. *tenuifolia* (Sibthorp) Sm. = Festuca filiformis [FESTFIL] Poaceae (V:M)
Festuca pacifica Piper = Vulpia microstachys var. pauciflora [VULPMIC1] Poaceae (V:M)
Festuca pacifica var. *simulans* Hoover = Vulpia microstachys var. pauciflora [VULPMIC1] Poaceae (V:M)
***Festuca pratensis** Huds. [FESTPRA] Poaceae (V:M)
Festuca prolifera (Piper) Fern. = Festuca rubra ssp. arctica [FESTRUB3] Poaceae (V:M)
Festuca prolifera var. *lasiolepis* Fern. = Festuca rubra ssp. arctica [FESTRUB3] Poaceae (V:M)
Festuca reflexa Buckl. = Vulpia microstachys var. pauciflora [VULPMIC1] Poaceae (V:M)
Festuca richardsonii Hook. = Festuca rubra ssp. arctica [FESTRUB3] Poaceae (V:M)
Festuca richardsonii ssp. *cryophila* (Krecz. & Bobr.) A. & D. Löve = Festuca rubra ssp. arctica [FESTRUB3] Poaceae (V:M)

Festuca roemeri (Pavlick) Alexeev = Festuca idahoensis var. roemeri [FESTIDA2] Poaceae (V:M)
Festuca rubra L. [FESTRUB] Poaceae (V:M)
Festuca rubra ssp. **arctica** (Hack.) Govor. [FESTRUB3] Poaceae (V:M)
Festuca rubra ssp. **arnicola** Alexeev [FESTRUB4] Poaceae (V:M)
Festuca rubra ssp. **aucta** (Krecz. & Bobr.) Hultén [FESTRUB5] Poaceae (V:M)
Festuca rubra ssp. *aucta* f. *pseudovivipara* Pavl. = Festuca rubra ssp. aucta [FESTRUB5] Poaceae (V:M)
Festuca rubra ssp. *cryophila* (Krecz. & Bobr.) Hultén = Festuca rubra ssp. arctica [FESTRUB3] Poaceae (V:M)
Festuca rubra ssp. *densiuscula* Hack. ex Piper = Festuca rubra ssp. pruinosa [FESTRUB1] Poaceae (V:M)
Festuca rubra ssp. **pruinosa** (Hack.) Piper [FESTRUB1] Poaceae (V:M)
Festuca rubra ssp. *richardsonii* (Hook.) Hultén = Festuca rubra ssp. arctica [FESTRUB3] Poaceae (V:M)
Festuca rubra ssp. **secunda** (J. Presl) Pavlick [FESTRUB7] Poaceae (V:M)
Festuca rubra ssp. **secunda** var. **mediana** Pavlick [FESTRUB8] Poaceae (V:M)
Festuca rubra ssp. **vallicola** (Rydb.) Pavlick [FESTRUB9] Poaceae (V:M)
Festuca rubra var. *littoralis* Vasey ex Beal = Festuca rubra ssp. secunda var. mediana [FESTRUB8] Poaceae (V:M)
Festuca rubra var. *mutica* Hartman = Festuca rubra ssp. arctica [FESTRUB3] Poaceae (V:M)
Festuca rubra var. *prolifera* (Piper) Piper = Festuca rubra ssp. arctica [FESTRUB3] Poaceae (V:M)
Festuca rubra var. *richardsonii* (Hook.) Hultén = Festuca rubra ssp. arctica [FESTRUB3] Poaceae (V:M)
Festuca saximontana Rydb. [FESTSAX] Poaceae (V:M)
Festuca saximontana var. **purpusiana** (St.-Yves) Frederiksen & Pavlick [FESTSAX1] Poaceae (V:M)
Festuca saximontana var. **robertsiana** Pavlick [FESTSAX2] Poaceae (V:M)
Festuca scabrella Torr. ex Hook. = Festuca campestris [FESTCAM] Poaceae (V:M)
Festuca scabrella var. *major* Vasey = Festuca campestris [FESTCAM] Poaceae (V:M)
Festuca shortii Kunth ex Wood = Festuca pratensis [FESTPRA] Poaceae (V:M)
Festuca subulata Trin. [FESTSUB] Poaceae (V:M)
Festuca subuliflora Scribn. [FESTSUU] Poaceae (V:M)
Festuca tenella Willd. = Vulpia octoflora var. glauca [VULPOCT1] Poaceae (V:M)
Festuca tenella var. *glauca* Nutt. = Vulpia octoflora var. glauca [VULPOCT1] Poaceae (V:M)
Festuca tenuifolia Sibthorp = Festuca filiformis [FESTFIL] Poaceae (V:M)
*****Festuca trachyphylla** (Hack.) Krajina [FESTTRA] Poaceae (V:M)
Festuca vallicola Rydb. = Festuca rubra ssp. vallicola [FESTRUB9] Poaceae (V:M)
Festuca viridula Vasey [FESTVIR] Poaceae (V:M)
Festuca vivipara (L.) Sm. [FESTVIV] Poaceae (V:M)
Festuca vivipara ssp. **glabra** Frederiksen [FESTVIV2] Poaceae (V:M)
Festuca viviparoidea Krajina ex Pavlick [FESTVII] Poaceae (V:M)
Festuca viviparoidea var. **krajinae** Pavlick [FESTVII1] Poaceae (V:M)
Filaginella palustris (Nutt.) Holub = Gnaphalium palustre [GNAPPAL] Asteraceae (V:D)
Filaginella uliginosa (L.) Opiz = Gnaphalium uliginosum [GNAPULI] Asteraceae (V:D)
Filago L. [FILAGO$] Asteraceae (V:D)
*****Filago arvensis** L. [FILAARV] Asteraceae (V:D)
Filago germanica L. = Filago vulgaris [FILAVUL] Asteraceae (V:D)
*****Filago vulgaris** Lam. [FILAVUL] Asteraceae (V:D)
Filix fragilis (L.) Underwood = Cystopteris fragilis [CYSTFRA] Dryopteridaceae (V:F)
Filix montana (Lam.) Underwood = Cystopteris montana [CYSTMON] Dryopteridaceae (V:F)
Fissidens Hedw. [FISSIDE$] Fissidentaceae (B:M)
Fissidens adianthoides Hedw. [FISSADI] Fissidentaceae (B:M)
Fissidens adianthoides var. *immarginatus* Lindb. ex Lesq. & James = Fissidens adianthoides [FISSADI] Fissidentaceae (B:M)
Fissidens andersonii Grout = Fissidens bryoides [FISSBRY] Fissidentaceae (B:M)
Fissidens aphelotaxifolius Purs. [FISSAPH] Fissidentaceae (B:M)
Fissidens bryoides Hedw. [FISSBRY] Fissidentaceae (B:M)
Fissidens bryoides var. *brevifolius* (Ren. & Card.) Wijk & Marg. = Fissidens bryoides [FISSBRY] Fissidentaceae (B:M)
Fissidens bryoides var. *incurvus* (Röhl.) Hüb. = Fissidens bryoides [FISSBRY] Fissidentaceae (B:M)
Fissidens bryoides var. *viridulus* (Sw.) Broth. = Fissidens bryoides [FISSBRY] Fissidentaceae (B:M)
Fissidens debilis Schwaegr. = Fissidens fontanus [FISSFON] Fissidentaceae (B:M)
Fissidens exiguus Sull. = Fissidens bryoides [FISSBRY] Fissidentaceae (B:M)
Fissidens exiguus var. *falcatulus* (Ren. & Card.) Grout = Fissidens bryoides [FISSBRY] Fissidentaceae (B:M)
Fissidens falcatulus Ren. & Card. = Fissidens bryoides [FISSBRY] Fissidentaceae (B:M)
Fissidens fontanus (B. Pyl.) Steud. [FISSFON] Fissidentaceae (B:M)
Fissidens grandifrons Brid. [FISSGRA] Fissidentaceae (B:M)
Fissidens incurvus Starke ex Röhl. = Fissidens bryoides [FISSBRY] Fissidentaceae (B:M)
Fissidens incurvus var. *brevifolius* Ren. & Card. = Fissidens bryoides [FISSBRY] Fissidentaceae (B:M)
Fissidens julianus (Savi ex DC.) Schimp. = Fissidens fontanus [FISSFON] Fissidentaceae (B:M)
Fissidens julianus var. *americanus* Kindb. = Fissidens fontanus [FISSFON] Fissidentaceae (B:M)
Fissidens limbatus Sull. [FISSLIM] Fissidentaceae (B:M)
Fissidens longifolius Brid. = Fissidens bryoides [FISSBRY] Fissidentaceae (B:M)
Fissidens minutulus Sull. = Fissidens bryoides [FISSBRY] Fissidentaceae (B:M)
Fissidens osmundioides Hedw. [FISSOSM] Fissidentaceae (B:M)
Fissidens pauperculus Howe [FISSPAU] Fissidentaceae (B:M)
Fissidens pusillus (Wils.) Milde = Fissidens bryoides [FISSBRY] Fissidentaceae (B:M)
Fissidens repandus Wils. = Fissidens bryoides [FISSBRY] Fissidentaceae (B:M)
Fissidens rufulus auct. = Fissidens ventricosus [FISSVEN] Fissidentaceae (B:M)
Fissidens tamarindifolius (Turn.) Brid. = Fissidens bryoides [FISSBRY] Fissidentaceae (B:M)
Fissidens texanus Lesq. & James = Fissidens bryoides [FISSBRY] Fissidentaceae (B:M)
Fissidens tortilis Hampe & C. Müll. = Fissidens bryoides [FISSBRY] Fissidentaceae (B:M)
Fissidens ventricosus Lesq. [FISSVEN] Fissidentaceae (B:M)
Fissidens viridulus (Sw.) Wahlenb. = Fissidens bryoides [FISSBRY] Fissidentaceae (B:M)
Fissidens viridulus var. *brevifolius* (Ren. & Card.) Grout = Fissidens bryoides [FISSBRY] Fissidentaceae (B:M)
Fissidens viridulus var. *tamarindifolius* (Turn.) Grout = Fissidens bryoides [FISSBRY] Fissidentaceae (B:M)
Fissidens viridulus var. *texanus* (Lesq. & James) Grout = Fissidens bryoides [FISSBRY] Fissidentaceae (B:M)
Fistulariella dilacerata (Hoffm.) Bowler & Rundel = Ramalina dilacerata [RAMADIL] Ramalinaceae (L)
Fistulariella geniculata (Hook. f. & Taylor) Bowler & Rundel = Ramalina geniculata [RAMAGEN] Ramalinaceae (L)
Fistulariella inflata (Hook. f. & Taylor) Bowler & Rundel = Ramalina inflata [RAMAINF] Ramalinaceae (L)
Fistulariella minuscula (Nyl.) Bowler & Rundel = Ramalina dilacerata [RAMADIL] Ramalinaceae (L)
Fistulariella roesleri (Hochst. ex Schaerer) Bowler & Rundel = Ramalina roesleri [RAMAROE] Ramalinaceae (L)
Flavocetraria Kärnefelt & Thell [FLAVOCE$] Parmeliaceae (L)
Flavocetraria cucullata (Bellardi) Kärnefelt & Thell [FLAVCUC] Parmeliaceae (L)

Flavocetraria nivalis (L.) Kärnefelt & Thell [FLAVNIV] Parmeliaceae (L)
Flavopunctelia (Krog) Hale [FLAVOPU$] Parmeliaceae (L)
Flavopunctelia flaventior (Stirton) Hale [FLAVFLA] Parmeliaceae (L)
Fluminia festucacea (Willd.) A.S. Hitchc. = Scolochloa festucacea [SCOLFES] Poaceae (V:M)
Foeniculum P. Mill. [FOENICU$] Apiaceae (V:D)
Foeniculum foeniculum (L.) Karst. = Foeniculum vulgare [FOENVUL] Apiaceae (V:D)
*****Foeniculum vulgare** P. Mill. [FOENVUL] Apiaceae (V:D)
Fontinalis Hedw. [FONTINA$] Fontinalaceae (B:M)
Fontinalis allenii Card. in Nichols = Fontinalis antipyretica var. oregonensis [FONTANT3] Fontinalaceae (B:M)
Fontinalis antipyretica Hedw. [FONTANT] Fontinalaceae (B:M)
Fontinalis antipyretica var. **antipyretica** [FONTANT0] Fontinalaceae (B:M)
Fontinalis antipyretica var. **gigantea** (Sull.) Sull. [FONTANT1] Fontinalaceae (B:M)
Fontinalis antipyretica var. *mollis* (C. Müll.) Welch in Grout = Fontinalis antipyretica var. antipyretica [FONTANT0] Fontinalaceae (B:M)
Fontinalis antipyretica var. **oregonensis** Ren. & Card. [FONTANT3] Fontinalaceae (B:M)
Fontinalis antipyretica var. *patula* (Card.) Welch = Fontinalis antipyretica var. antipyretica [FONTANT0] Fontinalaceae (B:M)
Fontinalis chrysophylla Card. = Fontinalis antipyretica var. oregonensis [FONTANT3] Fontinalaceae (B:M)
Fontinalis gigantea Sull. in Sull. & Lesq. = Fontinalis antipyretica var. gigantea [FONTANT1] Fontinalaceae (B:M)
Fontinalis howellii Ren. & Card. = Fontinalis antipyretica var. oregonensis [FONTANT3] Fontinalaceae (B:M)
Fontinalis hypnoides Hartm. [FONTHYP] Fontinalaceae (B:M)
Fontinalis kindbergii Ren. & Card. = Fontinalis antipyretica var. oregonensis [FONTANT3] Fontinalaceae (B:M)
Fontinalis neomexicana Sull. & Lesq. [FONTNEO] Fontinalaceae (B:M)
Fontinalis patula Card. = Fontinalis antipyretica var. antipyretica [FONTANT0] Fontinalaceae (B:M)
Foraminella ambigua (Wulfen) S.F. Meyer = Parmeliopsis ambigua [PARMAMB] Parmeliaceae (L)
Foraminella hyperopta (Ach.) S.F. Meyer = Parmeliopsis hyperopta [PARMHYE] Parmeliaceae (L)
Fossombronia Raddi [FOSSOMB$] Codoniaceae (B:H)
Fossombronia dumortieri (Hüb. & Genth.) Lindb. = Fossombronia foveolata [FOSSFOV] Codoniaceae (B:H)
Fossombronia foveolata Lindb. [FOSSFOV] Codoniaceae (B:H)
Fossombronia longiseta Aust. [FOSSLON] Codoniaceae (B:H)
Fragaria L. [FRAGARI$] Rosaceae (V:D)
Fragaria americana (Porter) Britt. = Fragaria vesca ssp. americana [FRAGVES1] Rosaceae (V:D)
Fragaria bracteata Heller = Fragaria vesca ssp. bracteata [FRAGVES2] Rosaceae (V:D)
Fragaria chiloensis (L.) P. Mill. [FRAGCHI] Rosaceae (V:D)
Fragaria chiloensis ssp. **lucida** (Vilm.) Staudt [FRAGCHI1] Rosaceae (V:D)
Fragaria chiloensis ssp. **pacifica** Staudt [FRAGCHI2] Rosaceae (V:D)
Fragaria glauca (S. Wats.) Rydb. = Fragaria virginiana ssp. glauca [FRAGVIR1] Rosaceae (V:D)
Fragaria helleri Holz. = Fragaria vesca ssp. bracteata [FRAGVES2] Rosaceae (V:D)
Fragaria indica Andr. = Duchesnea indica [DUCHIND] Rosaceae (V:D)
Fragaria pauciflora Rydb. = Fragaria virginiana ssp. glauca [FRAGVIR1] Rosaceae (V:D)
Fragaria platypetala Rydb. = Fragaria virginiana ssp. platypetala [FRAGVIR2] Rosaceae (V:D)
Fragaria platypetala var. *sibbaldifolia* (Rydb.) Jepson = Fragaria virginiana ssp. platypetala [FRAGVIR2] Rosaceae (V:D)
Fragaria sibbaldifolia Rydb. = Fragaria virginiana ssp. platypetala [FRAGVIR2] Rosaceae (V:D)
Fragaria suksdorfii Rydb. = Fragaria virginiana ssp. platypetala [FRAGVIR2] Rosaceae (V:D)
Fragaria truncata Rydb. = Fragaria virginiana ssp. platypetala [FRAGVIR2] Rosaceae (V:D)
Fragaria vesca L. [FRAGVES] Rosaceae (V:D)
Fragaria vesca ssp. **americana** (Porter) Staudt [FRAGVES1] Rosaceae (V:D)
Fragaria vesca ssp. **bracteata** (Heller) Staudt [FRAGVES2] Rosaceae (V:D)
Fragaria vesca var. *americana* Porter = Fragaria vesca ssp. americana [FRAGVES1] Rosaceae (V:D)
Fragaria vesca var. *bracteata* (Heller) R.J. Davis = Fragaria vesca ssp. bracteata [FRAGVES2] Rosaceae (V:D)
Fragaria virginiana Duchesne [FRAGVIR] Rosaceae (V:D)
Fragaria virginiana ssp. **glauca** (S. Wats.) Staudt [FRAGVIR1] Rosaceae (V:D)
Fragaria virginiana ssp. **platypetala** (Rydb.) Staudt [FRAGVIR2] Rosaceae (V:D)
Fragaria virginiana var. *glauca* S. Wats. = Fragaria virginiana ssp. glauca [FRAGVIR1] Rosaceae (V:D)
Fragaria virginiana var. *platypetala* (Rydb.) Hall = Fragaria virginiana ssp. platypetala [FRAGVIR2] Rosaceae (V:D)
Fragaria virginiana var. *terrae-novae* (Rydb.) Fern. & Wieg. = Fragaria virginiana ssp. glauca [FRAGVIR1] Rosaceae (V:D)
Frangula P. Mill. [FRANGUL$] Rhamnaceae (V:D)
Frangula purshiana (DC.) Cooper [FRANPUR] Rhamnaceae (V:D)
Franseria chamissonis Less. = Ambrosia chamissonis [AMBRCHA] Asteraceae (V:D)
Franseria chamissonis ssp. *bipinnatisecta* (Less.) Wiggins & Stockwell = Ambrosia chamissonis [AMBRCHA] Asteraceae (V:D)
Franseria chamissonis var. *bipinnatisecta* Less. = Ambrosia chamissonis [AMBRCHA] Asteraceae (V:D)
Fritillaria L. [FRITILL$] Liliaceae (V:M)
Fritillaria affinis (Schultes) Sealy = Fritillaria lanceolata [FRITLAN] Liliaceae (V:M)
Fritillaria camschatcensis (L.) Ker-Gawl. [FRITCAM] Liliaceae (V:M)
Fritillaria camschatcensis var. *floribunda* (Benth.) Boivin = Fritillaria lanceolata [FRITLAN] Liliaceae (V:M)
Fritillaria eximia Eastw. = Fritillaria lanceolata [FRITLAN] Liliaceae (V:M)
Fritillaria lanceolata Pursh [FRITLAN] Liliaceae (V:M)
Fritillaria lanceolata var. *tristulis* A.L. Grant = Fritillaria lanceolata [FRITLAN] Liliaceae (V:M)
Fritillaria mutica Lindl. = Fritillaria lanceolata [FRITLAN] Liliaceae (V:M)
Fritillaria mutica var. *gracilis* (S. Wats.) Jepson = Fritillaria lanceolata [FRITLAN] Liliaceae (V:M)
Fritillaria pudica (Pursh) Spreng. [FRITPUD] Liliaceae (V:M)
Frullania Raddi [FRULLAN$] Jubulaceae (B:H)
Frullania bolanderi Aust. [FRULBOL] Jubulaceae (B:H)
Frullania californica (Aust.) Evans [FRULCAL] Jubulaceae (B:H)
Frullania franciscana M.A. Howe [FRULFRA] Jubulaceae (B:H)
Frullania hattoriana J.D. Godfr. & G.A. Godfr. [FRULHAT] Jubulaceae (B:H)
Frullania nisquallensis Sull. = Frullania tamarisci ssp. nisquallensis [FRULTAM2] Jubulaceae (B:H)
Frullania tamarisci (L.) Dum. [FRULTAM] Jubulaceae (B:H)
Frullania tamarisci ssp. **nisquallensis** (Sull.) Hatt. [FRULTAM2] Jubulaceae (B:H)
Frullania tamarisci ssp. **tamarisci** [FRULTAM0] Jubulaceae (B:H)
Fulgensia A. Massal. & De Not. [FULGENS$] Teloschistaceae (L)
Fulgensia bracteata (Hoffm.) Räsänen [FULGBRA] Teloschistaceae (L)
Fulgensia desertorum (Tomin) Poelt [FULGDES] Teloschistaceae (L)
Fumaria L. [FUMARIA$] Fumariaceae (V:D)
*****Fumaria bastardii** Boreau [FUMABAS] Fumariaceae (V:D)
*****Fumaria officinalis** L. [FUMAOFF] Fumariaceae (V:D)
Funaria Hedw. [FUNARIA$] Funariaceae (B:M)
Funaria calcarea Wahlenb. = Funaria muhlenbergii [FUNAMUH] Funariaceae (B:M)

Funaria hygrometrica Hedw. [FUNAHYG] Funariaceae (B:M)
Funaria hygrometrica var. *convoluta* (Hampe) Grout = Funaria hygrometrica [FUNAHYG] Funariaceae (B:M)
Funaria hygrometrica var. *patula* Bruch & Schimp. in B.S.G. = Funaria hygrometrica [FUNAHYG] Funariaceae (B:M)
Funaria hygrometrica var. *utahensis* Grout = Funaria hygrometrica [FUNAHYG] Funariaceae (B:M)
Funaria mediterranea Lindb. = Funaria muhlenbergii [FUNAMUH] Funariaceae (B:M)
Funaria muhlenbergii Turn. [FUNAMUH] Funariaceae (B:M)
Funaria muhlenbergii var. *lineata* Grout = Funaria muhlenbergii [FUNAMUH] Funariaceae (B:M)
Funaria muhlenbergii var. *patula* (Bruch & Schimp. in B.S.G.) Schimp. = Funaria muhlenbergii [FUNAMUH] Funariaceae (B:M)
Fuscidea V. Wirth & Vezda [FUSCIDE$] Fuscideaceae (L)
Fuscidea lightfootii (Sm.) Coppins & P. James [FUSCLIG] Fuscideaceae (L)
Fuscidea mollis (Wahlenb.) V. Wirth & Vezda [FUSCMOL] Fuscideaceae (L)
Fuscidea viridis Tønsberg = Ropalospora viridis [ROPAVIR] Fuscideaceae (L)
Fuscopannaria P.M. Jørg. [FUSCOPA$] Uncertain-family (L)
Fuscopannaria ahlneri (P.M. Jørg.) P.M. Jørg. [FUSCAHL] Uncertain-family (L)
Fuscopannaria laceratula (Hue) P.M. Jørg. [FUSCLAC] Uncertain-family (L)
Fuscopannaria leucostictoides (Ohlsson) P.M. Jørg. [FUSCLEU] Uncertain-family (L)
Fuscopannaria maritima (P.M. Jørg.) P.M. Jørg. [FUSCMAR] Uncertain-family (L)
Fuscopannaria mediterranea (Tav.) P.M. Jørg. [FUSCMED] Uncertain-family (L)
Fuscopannaria praetermissa (Nyl.) P.M. Jørg. [FUSCPRE] Uncertain-family (L)
Fuscopannaria saubinetii (Mont.) P.M. Jørg. [FUSCSAU] Uncertain-family (L)

G

Gaillardia Foug. [GAILLAR$] Asteraceae (V:D)
Gaillardia aristata Pursh [GAILARI] Asteraceae (V:D)
Galarhoeus cyparissias (L.) Small ex Rydb. = Euphorbia cyparissias [EUPHCYP] Euphorbiaceae (V:D)
Galarhoeus helioscopius (L.) Haw. = Euphorbia helioscopia [EUPHHEL] Euphorbiaceae (V:D)
Galarhoeus lathyris (L.) Haw. = Euphorbia lathyris [EUPHLAT] Euphorbiaceae (V:D)
Galarhoeus peplus (L.) Haw. = Euphorbia peplus [EUPHPEP] Euphorbiaceae (V:D)
Galeopsis L. [GALEOPS$] Lamiaceae (V:D)
***Galeopsis bifida** Boenn. [GALEBIF] Lamiaceae (V:D)
***Galeopsis tetrahit** L. [GALETET] Lamiaceae (V:D)
Galeopsis tetrahit var. *bifida* (Boenn.) Lej. & Court. = Galeopsis bifida [GALEBIF] Lamiaceae (V:D)
Galinsoga Ruiz & Pavón [GALINSO$] Asteraceae (V:D)
Galinsoga aristulata Bickn. = Galinsoga quadriradiata [GALIQUA] Asteraceae (V:D)
Galinsoga bicolorata St. John & White = Galinsoga quadriradiata [GALIQUA] Asteraceae (V:D)
Galinsoga caracasana (DC.) Schultz-Bip. = Galinsoga quadriradiata [GALIQUA] Asteraceae (V:D)
Galinsoga ciliata (Raf.) Blake = Galinsoga quadriradiata [GALIQUA] Asteraceae (V:D)
***Galinsoga parviflora** Cav. [GALIPAR] Asteraceae (V:D)
Galinsoga parviflora var. *semicalva* Gray = Galinsoga parviflora [GALIPAR] Asteraceae (V:D)
***Galinsoga quadriradiata** Ruiz & Pavón [GALIQUA] Asteraceae (V:D)
Galinsoga semicalva (Gray) St. John & White = Galinsoga parviflora [GALIPAR] Asteraceae (V:D)
Galinsoga semicalva var. *percalva* Blake = Galinsoga parviflora [GALIPAR] Asteraceae (V:D)
Galium L. [GALIUM$] Rubiaceae (V:D)
Galium agreste var. *echinospermum* Wallr. = Galium aparine [GALIAPA] Rubiaceae (V:D)
Galium aparine L. [GALIAPA] Rubiaceae (V:D)
Galium aparine var. *echinospermum* (Wallr.) Farw. = Galium aparine [GALIAPA] Rubiaceae (V:D)
Galium aparine var. *intermedium* (Merr.) Briq. = Galium aparine [GALIAPA] Rubiaceae (V:D)
Galium aparine var. *minor* Hook. = Galium aparine [GALIAPA] Rubiaceae (V:D)
Galium aparine var. *vaillantii* (DC.) Koch = Galium aparine [GALIAPA] Rubiaceae (V:D)
Galium asperrimum var. *asperulum* Gray = Galium mexicanum ssp. asperulum [GALIMEX1] Rubiaceae (V:D)
Galium asperulum (Gray) Rydb. = Galium mexicanum ssp. asperulum [GALIMEX1] Rubiaceae (V:D)
Galium bifolium S. Wats. [GALIBIF] Rubiaceae (V:D)
Galium boreale L. [GALIBOR] Rubiaceae (V:D)
Galium boreale ssp. *septentrionale* (Roemer & J.A. Schultes) Iltis = Galium boreale [GALIBOR] Rubiaceae (V:D)
Galium boreale ssp. *septentrionale* (Roemer & J.A. Schultes) Hara = Galium boreale [GALIBOR] Rubiaceae (V:D)
Galium boreale var. *hyssopifolium* (Hoffmann) DC. = Galium boreale [GALIBOR] Rubiaceae (V:D)
Galium boreale var. *intermedium* DC. = Galium boreale [GALIBOR] Rubiaceae (V:D)
Galium boreale var. *linearifolium* Rydb. = Galium boreale [GALIBOR] Rubiaceae (V:D)
Galium boreale var. *scabrum* DC. = Galium boreale [GALIBOR] Rubiaceae (V:D)
Galium boreale var. *typicum* G. Beck = Galium boreale [GALIBOR] Rubiaceae (V:D)
Galium brachiatum Pursh = Galium triflorum [GALITRF] Rubiaceae (V:D)
Galium brandegei Gray p.p. = Galium trifidum ssp. trifidum [GALITRI0] Rubiaceae (V:D)
Galium claytonii var. *subbiflorum* (Wieg.) Wieg. = Galium trifidum ssp. subbiflorum [GALITRI4] Rubiaceae (V:D)
Galium columbianum Rydb. = Galium trifidum ssp. columbianum [GALITRI3] Rubiaceae (V:D)
Galium cymosum Wieg. = Galium trifidum ssp. columbianum [GALITRI3] Rubiaceae (V:D)
Galium erectum Huds. = Galium mollugo [GALIMOL] Rubiaceae (V:D)
Galium hyssopifolium Hoffmann = Galium boreale [GALIBOR] Rubiaceae (V:D)
Galium kamtschaticum Steller ex J.A. & J.H. Schultes [GALIKAM] Rubiaceae (V:D)
Galium mexicanum Kunth [GALIMEX] Rubiaceae (V:D)
Galium mexicanum ssp. **asperulum** (Gray) Dempster [GALIMEX1] Rubiaceae (V:D)
Galium mexicanum var. *asperulum* (Gray) Dempster = Galium mexicanum ssp. asperulum [GALIMEX1] Rubiaceae (V:D)
***Galium mollugo** L. [GALIMOL] Rubiaceae (V:D)
Galium mollugo ssp. *erectum* (Huds.) Briq. = Galium mollugo [GALIMOL] Rubiaceae (V:D)
Galium mollugo var. *erectum* (Huds.) Domin = Galium mollugo [GALIMOL] Rubiaceae (V:D)
***Galium odoratum** (L.) Scop. [GALIODO] Rubiaceae (V:D)
Galium pennsylvanicum W. Bart. = Galium triflorum [GALITRF] Rubiaceae (V:D)

Galium septentrionale Roemer & J.A. Schultes = Galium boreale [GALIBOR] Rubiaceae (V:D)
Galium spurium var. *echinospermum* (Wallr.) Hayek = Galium aparine [GALIAPA] Rubiaceae (V:D)
Galium spurium var. *vaillantii* (DC.) G. Beck = Galium aparine [GALIAPA] Rubiaceae (V:D)
Galium spurium var. *vaillantii* (DC.) Gren. & Godr. = Galium aparine [GALIAPA] Rubiaceae (V:D)
Galium strictum Torr. = Galium boreale [GALIBOR] Rubiaceae (V:D)
Galium subbiflorum (Wieg.) Rydb. = Galium trifidum ssp. subbiflorum [GALITRI4] Rubiaceae (V:D)
Galium tinctorium var. *subbiflorum* (Wieg.) Fern. = Galium trifidum ssp. subbiflorum [GALITRI4] Rubiaceae (V:D)
Galium trifidum L. [GALITRI] Rubiaceae (V:D)
Galium trifidum ssp. **columbianum** (Rydb.) Hultén [GALITRI3] Rubiaceae (V:D)
Galium trifidum ssp. *pacificum* (Wieg.) Piper = Galium trifidum ssp. columbianum [GALITRI3] Rubiaceae (V:D)
Galium trifidum ssp. **subbiflorum** (Wieg.) Piper [GALITRI4] Rubiaceae (V:D)
Galium trifidum ssp. **trifidum** [GALITRI0] Rubiaceae (V:D)
Galium trifidum var. *pacificum* Wieg. = Galium trifidum ssp. columbianum [GALITRI3] Rubiaceae (V:D)
Galium trifidum var. *pusillum* Gray = Galium trifidum ssp. subbiflorum [GALITRI4] Rubiaceae (V:D)
Galium trifidum var. *subbiflorum* Wieg. = Galium trifidum ssp. subbiflorum [GALITRI4] Rubiaceae (V:D)
Galium triflorum Michx. [GALITRF] Rubiaceae (V:D)
Galium triflorum var. *asprelliforme* Fern. = Galium triflorum [GALITRF] Rubiaceae (V:D)
Galium triflorum var. *viridiflorum* DC. = Galium triflorum [GALITRF] Rubiaceae (V:D)
Galium vaillantii DC. = Galium aparine [GALIAPA] Rubiaceae (V:D)
***Galium verum** L. [GALIVER] Rubiaceae (V:D)
Gamochaeta Weddell [GAMOCHA$] Asteraceae (V:D)
***Gamochaeta purpurea** (L.) Cabrera [GAMOPUR] Asteraceae (V:D)
Gastrolychnis affinis (J. Vahl ex Fries) Tolm. & Kozh. = Silene involucrata [SILEINV] Caryophyllaceae (V:D)
Gastrolychnis drummondii (Hook.) A. & D. Löve = Silene drummondii [SILEDRU] Caryophyllaceae (V:D)
Gastrolychnis involucrata (Cham. & Schlecht.) A. & D. Löve = Silene involucrata [SILEINV] Caryophyllaceae (V:D)
Gastrolychnis ostenfeldii (Porsild) Petrovsky = Silene taimyrensis [SILETAI] Caryophyllaceae (V:D)
Gastrolychnis taimyrensis (Tolm.) S.K. Czer. = Silene taimyrensis [SILETAI] Caryophyllaceae (V:D)
Gastrolychnis triflora ssp. *dawsonii* (B.L. Robins.) A. & D. Löve = Silene taimyrensis [SILETAI] Caryophyllaceae (V:D)
Gaultheria L. [GAULTHE$] Ericaceae (V:D)
Gaultheria hispidula (L.) Muhl. ex Bigelow [GAULHIS] Ericaceae (V:D)
Gaultheria humifusa (Graham) Rydb. [GAULHUM] Ericaceae (V:D)
Gaultheria ovatifolia Gray [GAULOVA] Ericaceae (V:D)
Gaultheria shallon Pursh [GAULSHA] Ericaceae (V:D)
Gaura L. [GAURA$] Onagraceae (V:D)
Gaura coccinea Nutt. ex Pursh [GAURCOC] Onagraceae (V:D)
Gaura coccinea var. *arizonica* Munz = Gaura coccinea [GAURCOC] Onagraceae (V:D)
Gaura coccinea var. *epilobioides* (Kunth) Munz = Gaura coccinea [GAURCOC] Onagraceae (V:D)
Gaura coccinea var. *glabra* (Lehm.) Munz = Gaura coccinea [GAURCOC] Onagraceae (V:D)
Gaura coccinea var. *parvifolia* (Torr.) Rickett = Gaura coccinea [GAURCOC] Onagraceae (V:D)
Gaura coccinea var. *typica* Munz = Gaura coccinea [GAURCOC] Onagraceae (V:D)
Gaura glabra Lehm. = Gaura coccinea [GAURCOC] Onagraceae (V:D)
Gaura odorata Sessé ex Lag. = Gaura coccinea [GAURCOC] Onagraceae (V:D)
Gayophytum A. Juss. [GAYOPHY$] Onagraceae (V:D)
Gayophytum diffusum Torr. & Gray [GAYODIF] Onagraceae (V:D)
Gayophytum diffusum ssp. **parviflorum** Lewis & Szweykowski [GAYODIF1] Onagraceae (V:D)
Gayophytum diffusum var. *strictipes* (Hook.) Dorn = Gayophytum diffusum ssp. parviflorum [GAYODIF1] Onagraceae (V:D)
Gayophytum helleri var. *erosulatum* Jepson = Gayophytum diffusum ssp. parviflorum [GAYODIF1] Onagraceae (V:D)
Gayophytum humile Juss. [GAYOHUM] Onagraceae (V:D)
Gayophytum intermedium Rydb. = Gayophytum diffusum ssp. parviflorum [GAYODIF1] Onagraceae (V:D)
Gayophytum lasiospermum Greene = Gayophytum diffusum ssp. parviflorum [GAYODIF1] Onagraceae (V:D)
Gayophytum lasiospermum var. *hoffmannii* Munz = Gayophytum diffusum ssp. parviflorum [GAYODIF1] Onagraceae (V:D)
Gayophytum nuttallii Torr. & Gray = Gayophytum humile [GAYOHUM] Onagraceae (V:D)
Gayophytum nuttallii var. *abramsii* Munz = Gayophytum diffusum ssp. parviflorum [GAYODIF1] Onagraceae (V:D)
Gayophytum nuttallii var. *intermedium* (Rydb.) Munz = Gayophytum diffusum ssp. parviflorum [GAYODIF1] Onagraceae (V:D)
Gayophytum ramosissimum Torr. & Gray [GAYORAM] Onagraceae (V:D)
Gayophytum strictum Gray = Boisduvalia stricta [BOISSTR] Onagraceae (V:D)
Geheebia Schimp. [GEHEEBI$] Pottiaceae (B:M)
Geheebia gigantea (Funck) Boul. [GEHEGIG] Pottiaceae (B:M)
Gentiana L. [GENTIAN$] Gentianaceae (V:D)
Gentiana acuta Michx. = Gentianella amarella ssp. acuta [GENTAMA1] Gentianaceae (V:D)
Gentiana affinis Griseb. [GENTAFF] Gentianaceae (V:D)
Gentiana affinis var. *bigelovii* (Gray) Kusnez. = Gentiana affinis [GENTAFF] Gentianaceae (V:D)
Gentiana affinis var. *forwoodii* (Gray) Kusnez. = Gentiana affinis [GENTAFF] Gentianaceae (V:D)
Gentiana affinis var. *major* A. Nels. & Kennedy = Gentiana affinis [GENTAFF] Gentianaceae (V:D)
Gentiana affinis var. *ovata* Gray = Gentiana affinis [GENTAFF] Gentianaceae (V:D)
Gentiana affinis var. *parvidentata* Kusnez. = Gentiana affinis [GENTAFF] Gentianaceae (V:D)
Gentiana amarella L. auct. p.p. = Gentianella amarella ssp. acuta [GENTAMA1] Gentianaceae (V:D)
Gentiana amarella ssp. *acuta* (Michx.) Hultén = Gentianella amarella ssp. acuta [GENTAMA1] Gentianaceae (V:D)
Gentiana amarella var. *acuta* (Michx.) Herder = Gentianella amarella ssp. acuta [GENTAMA1] Gentianaceae (V:D)
Gentiana amarella var. *plebeja* (Ledeb. ex Spreng.) Hultén = Gentianella amarella ssp. acuta [GENTAMA1] Gentianaceae (V:D)
Gentiana amarella var. *stricta* (Griseb.) S. Wats. = Gentianella amarella ssp. acuta [GENTAMA1] Gentianaceae (V:D)
Gentiana arctophila Griseb. = Gentianella propinqua [GENTPRP] Gentianaceae (V:D)
Gentiana bigelovii Gray = Gentiana affinis [GENTAFF] Gentianaceae (V:D)
Gentiana calycosa Griseb. [GENTCAL] Gentianaceae (V:D)
Gentiana calycosa var. *obtusiloba* (Rydb.) C.L. Hitchc. = Gentiana calycosa [GENTCAL] Gentianaceae (V:D)
Gentiana calycosa var. *xantha* A. Nels. = Gentiana calycosa [GENTCAL] Gentianaceae (V:D)
Gentiana crinita var. *tonsa* (Lunell) Boivin = Gentianopsis macounii [GENTMAC] Gentianaceae (V:D)
Gentiana douglasiana Bong. [GENTDOU] Gentianaceae (V:D)
Gentiana forwoodii Gray = Gentiana affinis [GENTAFF] Gentianaceae (V:D)
Gentiana gaspensis Victorin = Gentianopsis macounii [GENTMAC] Gentianaceae (V:D)
Gentiana glauca Pallas [GENTGLA] Gentianaceae (V:D)
Gentiana interrupta Greene = Gentiana affinis [GENTAFF] Gentianaceae (V:D)
Gentiana macounii Holm = Gentianopsis macounii [GENTMAC] Gentianaceae (V:D)
Gentiana menziesii Griseb. = Gentiana sceptrum [GENTSCE] Gentianaceae (V:D)

Gentiana oregana Engelm. ex Gray = Gentiana affinis [GENTAFF] Gentianaceae (V:D)
Gentiana platypetala Griseb. [GENTPLA] Gentianaceae (V:D)
Gentiana plebeja Ledeb. ex Spreng. = Gentianella amarella ssp. acuta [GENTAMA1] Gentianaceae (V:D)
Gentiana plebeja var. *holmii* Wettst. = Gentianella amarella ssp. acuta [GENTAMA1] Gentianaceae (V:D)
Gentiana propinqua Richards. = Gentianella propinqua [GENTPRP] Gentianaceae (V:D)
Gentiana propinqua ssp. *arctophila* (Griseb.) Hultén = Gentianella propinqua [GENTPRP] Gentianaceae (V:D)
Gentiana prostrata Haenke [GENTPRO] Gentianaceae (V:D)
Gentiana prostrata var. *americana* Engelm. = Gentiana prostrata [GENTPRO] Gentianaceae (V:D)
Gentiana rusbyi Greene = Gentiana affinis [GENTAFF] Gentianaceae (V:D)
Gentiana sceptrum Griseb. [GENTSCE] Gentianaceae (V:D)
Gentiana sceptrum var. *cascadensis* M.E. Peck = Gentiana sceptrum [GENTSCE] Gentianaceae (V:D)
Gentiana sceptrum var. *humilis* Engelm. ex Gray = Gentiana sceptrum [GENTSCE] Gentianaceae (V:D)
Gentiana strictiflora (Rydb.) A. Nels. = Gentianella amarella ssp. acuta [GENTAMA1] Gentianaceae (V:D)
Gentiana tenella Rottb. = Gentianella tenella [GENTTEN] Gentianaceae (V:D)
Gentiana tonsa (Lunell) Victorin = Gentianopsis macounii [GENTMAC] Gentianaceae (V:D)
Gentianella Moench [GENTIAE$] Gentianaceae (V:D)
Gentianella acuta (Michx.) Hiitonen = Gentianella amarella ssp. acuta [GENTAMA1] Gentianaceae (V:D)
Gentianella amarella (L.) Boerner [GENTAMA] Gentianaceae (V:D)
Gentianella amarella ssp. **acuta** (Michx.) J. Gillett [GENTAMA1] Gentianaceae (V:D)
Gentianella amarella var. *acuta* (Michx.) Herder = Gentianella amarella ssp. acuta [GENTAMA1] Gentianaceae (V:D)
Gentianella crinita ssp. *macounii* (Holm) J. Gillett = Gentianopsis macounii [GENTMAC] Gentianaceae (V:D)
Gentianella propinqua (Richards.) J. Gillett [GENTPRP] Gentianaceae (V:D)
Gentianella strictiflora (Rydb.) W.A. Weber = Gentianella amarella ssp. acuta [GENTAMA1] Gentianaceae (V:D)
Gentianella tenella (Rottb.) Boerner [GENTTEN] Gentianaceae (V:D)
Gentianodes glauca (Pallas) A. & D. Löve = Gentiana glauca [GENTGLA] Gentianaceae (V:D)
Gentianopsis Ma [GENTIAO$] Gentianaceae (V:D)
Gentianopsis macounii (Holm) Iltis [GENTMAC] Gentianaceae (V:D)
Gentianopsis procera ssp. *macounii* (Holm) Iltis = Gentianopsis macounii [GENTMAC] Gentianaceae (V:D)
Geocalyx Nees [GEOCALY$] Geocalycaceae (B:H)
Geocalyx graveolens (Schrad.) Nees [GEOCGRA] Geocalycaceae (B:H)
Geocaulon Fern. [GEOCAUL$] Santalaceae (V:D)
Geocaulon lividum (Richards.) Fern. [GEOCLIV] Santalaceae (V:D)
Georgia cuspidata Kindb. = Tetraphis pellucida [TETRPEL] Tetraphidaceae (B:M)
Georgia geniculata (Milde) Brockm. = Tetraphis geniculata [TETRGEN] Tetraphidaceae (B:M)
Georgia pellucida (Hedw.) Rabenh. = Tetraphis pellucida [TETRPEL] Tetraphidaceae (B:M)
Georgia trachypoda Kindb. = Tetraphis geniculata [TETRGEN] Tetraphidaceae (B:M)
Geranium L. [GERANIU$] Geraniaceae (V:D)
Geranium bicknellii Britt. [GERABIC] Geraniaceae (V:D)
Geranium carolinianum L. [GERACAR] Geraniaceae (V:D)
Geranium carolinianum var. **sphaerospermum** (Fern.) Breitung [GERACAR2] Geraniaceae (V:D)
*__Geranium dissectum__ L. [GERADIS] Geraniaceae (V:D)
Geranium erianthum DC. [GERAERI] Geraniaceae (V:D)
Geranium laxum Hanks = Geranium dissectum [GERADIS] Geraniaceae (V:D)
*__Geranium molle__ L. [GERAMOL] Geraniaceae (V:D)
Geranium pratense var. *erianthum* (DC.) Boivin = Geranium erianthum [GERAERI] Geraniaceae (V:D)
*__Geranium pusillum__ L. [GERAPUS] Geraniaceae (V:D)
Geranium richardsonii Fisch. & Trautv. [GERARIC] Geraniaceae (V:D)
Geranium robertianum L. [GERAROB] Geraniaceae (V:D)
Geranium sphaerospermum Fern. = Geranium carolinianum var. sphaerospermum [GERACAR2] Geraniaceae (V:D)
Geranium viscosissimum Fisch. & C.A. Mey. ex C.A. Mey. [GERAVIS] Geraniaceae (V:D)
Geum L. [GEUM$] Rosaceae (V:D)
Geum aleppicum Jacq. [GEUMALE] Rosaceae (V:D)
Geum aleppicum ssp. *strictum* (Ait.) Clausen = Geum aleppicum [GEUMALE] Rosaceae (V:D)
Geum aleppicum var. *strictum* (Ait.) Fern. = Geum aleppicum [GEUMALE] Rosaceae (V:D)
Geum calthifolium Menzies ex Sm. [GEUMCAL] Rosaceae (V:D)
Geum ciliatum Pursh = Geum triflorum var. ciliatum [GEUMTRI1] Rosaceae (V:D)
Geum ciliatum var. *griseum* (Greene) Kearney & Peebles = Geum triflorum var. triflorum [GEUMTRI0] Rosaceae (V:D)
Geum × macranthum (Kearney) Boivin [GEUMMAR] Rosaceae (V:D)
Geum macrophyllum Willd. [GEUMMAC] Rosaceae (V:D)
Geum macrophyllum ssp. *perincisum* (Rydb.) Hultén = Geum macrophyllum var. perincisum [GEUMMAC1] Rosaceae (V:D)
Geum macrophyllum var. **macrophyllum** [GEUMMAC0] Rosaceae (V:D)
Geum macrophyllum var. **perincisum** (Rydb.) Raup [GEUMMAC1] Rosaceae (V:D)
Geum macrophyllum var. *rydbergii* Farw. = Geum macrophyllum var. perincisum [GEUMMAC1] Rosaceae (V:D)
Geum oregonense (Scheutz) Rydb. = Geum macrophyllum var. perincisum [GEUMMAC1] Rosaceae (V:D)
Geum perincisum Rydb. = Geum macrophyllum var. perincisum [GEUMMAC1] Rosaceae (V:D)
Geum perincisum var. *intermedium* Boivin = Geum macrophyllum var. perincisum [GEUMMAC1] Rosaceae (V:D)
Geum rivale L. [GEUMRIV] Rosaceae (V:D)
Geum rossii (R. Br.) Ser. [GEUMROS] Rosaceae (V:D)
Geum schofieldii Calder & Taylor = Geum × macranthum [GEUMMAR] Rosaceae (V:D)
Geum strictum Ait. = Geum aleppicum [GEUMALE] Rosaceae (V:D)
Geum strictum var. *decurrens* (Rydb.) Kearney & Peebles = Geum aleppicum [GEUMALE] Rosaceae (V:D)
Geum triflorum Pursh [GEUMTRI] Rosaceae (V:D)
Geum triflorum var. **ciliatum** (Pursh) Fassett [GEUMTRI1] Rosaceae (V:D)
Geum triflorum var. **triflorum** [GEUMTRI0] Rosaceae (V:D)
Gifola germanica Dumort. = Filago vulgaris [FILAVUL] Asteraceae (V:D)
Gilia Ruiz & Pavón [GILIA$] Polemoniaceae (V:D)
Gilia aggregata (Pursh) Spreng. = Ipomopsis aggregata [IPOMAGG] Polemoniaceae (V:D)
Gilia capitata Sims [GILICAP] Polemoniaceae (V:D)
Gilia gracilis Hook. = Phlox gracilis [PHLOGRA] Polemoniaceae (V:D)
Gilia hallii Parish = Leptodactylon pungens [LEPTPUN] Polemoniaceae (V:D)
Gilia inconspicua var. *sinuata* (Dougl. ex Benth.) Gray = Gilia sinuata [GILISIN] Polemoniaceae (V:D)
Gilia minutiflora Benth. = Ipomopsis minutiflora [IPOMMIN] Polemoniaceae (V:D)
Gilia pungens (Torr.) Benth. = Leptodactylon pungens [LEPTPUN] Polemoniaceae (V:D)
Gilia pungens var. *hookeri* (Dougl. ex Hook.) Gray = Leptodactylon pungens [LEPTPUN] Polemoniaceae (V:D)
Gilia septentrionalis (Mason) St. John = Linanthus septentrionalis [LINASEP] Polemoniaceae (V:D)
Gilia sinuata Dougl. ex Benth. [GILISIN] Polemoniaceae (V:D)
Gilia squarrosa (Eschsch.) Hook. & Arn. = Navarretia squarrosa [NAVASQU] Polemoniaceae (V:D)

Girgensohnia ruthenica (Weinm.) Kindb. = Pleuroziopsis ruthenica [PLEURUT] Pleuroziopsidaceae (B:M)
Githopsis Nutt. [GITHOPS$] Campanulaceae (V:D)
Githopsis calycina Benth. = Githopsis specularioides [GITHSPE] Campanulaceae (V:D)
Githopsis calycina var. *hirsuta* Benth. = Githopsis specularioides [GITHSPE] Campanulaceae (V:D)
Githopsis specularioides Nutt. [GITHSPE] Campanulaceae (V:D)
Githopsis specularioides var. *hirsuta* Nutt. = Githopsis specularioides [GITHSPE] Campanulaceae (V:D)
Glaucomaria rupicola (L.) Choisy = Lecanora rupicola [LECARUP] Lecanoraceae (L)
Glaux L. [GLAUX$] Primulaceae (V:D)
Glaux maritima L. [GLAUMAR] Primulaceae (V:D)
Glaux maritima ssp. **maritima** [GLAUMAR0] Primulaceae (V:D)
Glaux maritima ssp. **obtusifolia** (Fern.) Boivin [GLAUMAR2] Primulaceae (V:D)
Glaux maritima var. *angustifolia* Boivin = Glaux maritima ssp. obtusifolia [GLAUMAR2] Primulaceae (V:D)
Glaux maritima var. *macrophylla* Boivin = Glaux maritima ssp. obtusifolia [GLAUMAR2] Primulaceae (V:D)
Glaux maritima var. *obtusifolia* Fern. = Glaux maritima ssp. obtusifolia [GLAUMAR2] Primulaceae (V:D)
Glechoma L. [GLECHOM$] Lamiaceae (V:D)
*Glechoma hederacea** L. [GLECHED] Lamiaceae (V:D)
Glehnia F. Schmidt ex Miq. [GLEHNIA$] Apiaceae (V:D)
Glehnia leiocarpa Mathias = Glehnia littoralis ssp. leiocarpa [GLEHLIT1] Apiaceae (V:D)
Glehnia littoralis F. Schmidt ex Miq. [GLEHLIT] Apiaceae (V:D)
Glehnia littoralis ssp. **leiocarpa** (Mathias) Hultén [GLEHLIT1] Apiaceae (V:D)
Glehnia littoralis var. *leiocarpa* (Mathias) Boivin = Glehnia littoralis ssp. leiocarpa [GLEHLIT1] Apiaceae (V:D)
Glyceria R. Br. [GLYCERI$] Poaceae (V:M)
Glyceria borealis (Nash) Batchelder [GLYCBOR] Poaceae (V:M)
Glyceria canadensis (Michx.) Trin. [GLYCCAN] Poaceae (V:M)
Glyceria elata (Nash ex Rydb.) M.E. Jones [GLYCELA] Poaceae (V:M)
Glyceria fernaldii (A.S. Hitchc.) St. John = Torreyochloa pallida var. fernaldii [TORRPAL1] Poaceae (V:M)
Glyceria grandis S. Wats. [GLYCGRA] Poaceae (V:M)
Glyceria leptostachya Buckl. [GLYCLEP] Poaceae (V:M)
*Glyceria maxima** (Hartman) Holmb. [GLYCMAX] Poaceae (V:M)
Glyceria occidentalis (Piper) J.C. Nels. [GLYCOCC] Poaceae (V:M)
Glyceria pallida (Torr.) Trin. = Torreyochloa pallida [TORRPAL] Poaceae (V:M)
Glyceria pallida var. *fernaldii* A.S. Hitchc. = Torreyochloa pallida var. fernaldii [TORRPAL1] Poaceae (V:M)
Glyceria pauciflora J. Presl = Torreyochloa pallida var. pauciflora [TORRPAL2] Poaceae (V:M)
Glyceria spectabilis Mert. & Koch = Glyceria maxima [GLYCMAX] Poaceae (V:M)
Glyceria striata (Lam.) A.S. Hitchc. [GLYCSTR] Poaceae (V:M)
Glyceria striata ssp. *stricta* (Scribn.) Hultén = Glyceria striata var. stricta [GLYCSTR1] Poaceae (V:M)
Glyceria striata var. **stricta** (Scribn.) Fern. [GLYCSTR1] Poaceae (V:M)
Glycosma occidentalis Nutt. ex Torr. & Gray = Osmorhiza occidentalis [OSMOOCC] Apiaceae (V:D)
Glycyrrhiza L. [GLYCYRR$] Fabaceae (V:D)
Glycyrrhiza glutinosa Nutt. = Glycyrrhiza lepidota [GLYCLEI] Fabaceae (V:D)
Glycyrrhiza lepidota Pursh [GLYCLEI] Fabaceae (V:D)
Glycyrrhiza lepidota var. *glutinosa* (Nutt.) S. Wats. = Glycyrrhiza lepidota [GLYCLEI] Fabaceae (V:D)
Glypholecia Nyl. [GLYPHOL$] Acarosporaceae (L)
Glypholecia scabra (Pers.) Müll. Arg. [GLYPSCA] Acarosporaceae (L)
Gnaphalium L. [GNAPHAL$] Asteraceae (V:D)
Gnaphalium arvense L. = Filago arvensis [FILAARV] Asteraceae (V:D)
Gnaphalium beneolens A. Davids. = Gnaphalium microcephalum [GNAPMIC] Asteraceae (V:D)
Gnaphalium chilense Spreng. = Gnaphalium stramineum [GNAPSTR] Asteraceae (V:D)
Gnaphalium chilense var. *confertifolium* Greene = Gnaphalium stramineum [GNAPSTR] Asteraceae (V:D)
Gnaphalium decurrens Ives = Gnaphalium viscosum [GNAPVIS] Asteraceae (V:D)
Gnaphalium dimorphum Nutt. = Antennaria dimorpha [ANTEDIM] Asteraceae (V:D)
Gnaphalium germanicum Scopoli = Filago vulgaris [FILAVUL] Asteraceae (V:D)
Gnaphalium macounii Greene = Gnaphalium viscosum [GNAPVIS] Asteraceae (V:D)
Gnaphalium margaritaceum L. = Anaphalis margaritacea [ANAPMAR] Asteraceae (V:D)
Gnaphalium microcephalum Nutt. [GNAPMIC] Asteraceae (V:D)
Gnaphalium microcephalum ssp. *thermale* (E. Nels.) G.W. Douglas = Gnaphalium microcephalum [GNAPMIC] Asteraceae (V:D)
Gnaphalium microcephalum var. *thermale* (E. Nels.) Cronq. = Gnaphalium microcephalum [GNAPMIC] Asteraceae (V:D)
Gnaphalium palustre Nutt. [GNAPPAL] Asteraceae (V:D)
Gnaphalium purpureum L. = Gamochaeta purpurea [GAMOPUR] Asteraceae (V:D)
Gnaphalium stramineum Kunth [GNAPSTR] Asteraceae (V:D)
Gnaphalium sylvaticum L. = Omalotheca sylvatica [OMALSYL] Asteraceae (V:D)
Gnaphalium thermale E. Nels. = Gnaphalium microcephalum [GNAPMIC] Asteraceae (V:D)
*Gnaphalium uliginosum** L. [GNAPULI] Asteraceae (V:D)
Gnaphalium viscosum Kunth [GNAPVIS] Asteraceae (V:D)
Godetia pacifica M.E. Peck = Clarkia amoena ssp. caurina [CLARAMO1] Onagraceae (V:D)
Gollania Broth. [GOLLANI$] Hypnaceae (B:M)
Gollania densepinnata Dix. = Gollania turgens [GOLLTUR] Hypnaceae (B:M)
Gollania turgens (C. Müll.) Ando [GOLLTUR] Hypnaceae (B:M)
Gonohymenia nigritella (Lettau) Henssen = Lichinella nigritella [LICHNIG] Lichinaceae (L)
Goodyera R. Br. ex Ait. f. [GOODYER$] Orchidaceae (V:M)
Goodyera decipiens (Hook.) F.T. Hubbard = Goodyera oblongifolia [GOODOBL] Orchidaceae (V:M)
Goodyera oblongifolia Raf. [GOODOBL] Orchidaceae (V:M)
Goodyera oblongifolia var. *reticulata* Boivin = Goodyera oblongifolia [GOODOBL] Orchidaceae (V:M)
Goodyera ophioides (Fern.) Rydb. = Goodyera repens [GOODREP] Orchidaceae (V:M)
Goodyera repens (L.) R. Br. ex Ait. f. [GOODREP] Orchidaceae (V:M)
Goodyera repens ssp. *ophioides* (Fern.) A. Löve & Simon = Goodyera repens [GOODREP] Orchidaceae (V:M)
Goodyera repens var. *ophioides* Fern. = Goodyera repens [GOODREP] Orchidaceae (V:M)
Gordonia spathulifolia (Hook.) A. & D. Löve = Sedum spathulifolium ssp. spathulifolium [SEDUSPA0] Crassulaceae (V:D)
Grammica cephalanthi (Engelm.) Hadac & Chrtek = Cuscuta cephalanthi [CUSCCEP] Cuscutaceae (V:D)
Grammica pentagona (Engelm.) W.A. Weber = Cuscuta pentagona [CUSCPEN] Cuscutaceae (V:D)
Grammica salina (Engelm.) Taylor & MacBryde = Cuscuta salina [CUSCSAL] Cuscutaceae (V:D)
Graphina Müll. Arg. [GRAPHIN$] Graphidaceae (L)
Graphina anguina (Mont.) Müll. Arg. [GRAPANG] Graphidaceae (L)
Graphis Adans. [GRAPHIS$] Graphidaceae (L)
Graphis elegans (Borrer ex Sm.) Ach. [GRAPELE] Graphidaceae (L)
Graphis scripta (L.) Ach. [GRAPSCR] Graphidaceae (L)
Gratiola L. [GRATIOL$] Scrophulariaceae (V:D)
Gratiola ebracteata Benth. ex A. DC. [GRATEBR] Scrophulariaceae (V:D)
Gratiola neglecta Torr. [GRATNEG] Scrophulariaceae (V:D)
Gratiola neglecta var. *glaberrima* Fern. = Gratiola neglecta [GRATNEG] Scrophulariaceae (V:D)
Gregoria montana (Gray) House = Douglasia montana [DOUGMON] Primulaceae (V:D)

Grevilleanum serratum Beck & Emmons = Timmia megapolitana var. megapolitana [TIMMMEG0] Timmiaceae (B:M)
Grimaldia barbifrons Bisch. = Mannia fragrans [MANNFRA] Aytoniaceae (B:H)
Grimmia Hedw. [GRIMMIA$] Grimmiaceae (B:M)
Grimmia affinis Hoppe & Hornsch. ex Hornsch. [GRIMAFF] Grimmiaceae (B:M)
Grimmia agassizii (Sull. & Lesq. in Sull.) Jaeg. = Schistidium agassizii [SCHIAGA] Grimmiaceae (B:M)
Grimmia alpestris (Web. & Mohr) Schleich. ex Nees = Grimmia tenerrima [GRIMTEE] Grimmiaceae (B:M)
Grimmia alpestris var. *caespiticia* (Brid.) G. Jones in Grout = Grimmia tenerrima [GRIMTEE] Grimmiaceae (B:M)
Grimmia alpestris var. *holzingeri* (Card. & Thér.) G. Jones in Grout = Grimmia tenerrima [GRIMTEE] Grimmiaceae (B:M)
Grimmia alpestris var. *manniae* (C. Müll.) G. Jones in Grout = Grimmia tenerrima [GRIMTEE] Grimmiaceae (B:M)
Grimmia alpestris var. *microstoma* Bruch & Schimp. in B.S.G. = Grimmia tenerrima [GRIMTEE] Grimmiaceae (B:M)
Grimmia alpicola Sw. ex Hedw. = Schistidium agassizii [SCHIAGA] Grimmiaceae (B:M)
Grimmia alpicola var. *dupretii* (Thér.) Crum = Schistidium apocarpum [SCHIAPO] Grimmiaceae (B:M)
Grimmia alpicola var. *papillosa* (G. Jones) Hab. = Schistidium rivulare [SCHIRIV] Grimmiaceae (B:M)
Grimmia alpicola var. *rivularis* f. *papillosa* G. Jones in Grout = Schistidium rivulare [SCHIRIV] Grimmiaceae (B:M)
Grimmia angusta (Hag.) Broth. = Schistidium agassizii [SCHIAGA] Grimmiaceae (B:M)
Grimmia anodon Bruch & Schimp. in B.S.G. [GRIMANO] Grimmiaceae (B:M)
Grimmia anomala Hampe ex Schimp. [GRIMANM] Grimmiaceae (B:M)
Grimmia apocarpa Hedw. = Schistidium apocarpum [SCHIAPO] Grimmiaceae (B:M)
Grimmia apocarpa f. *canadensis* Loeske = Schistidium apocarpum [SCHIAPO] Grimmiaceae (B:M)
Grimmia apocarpa f. *epilosa* Warnst. = Schistidium apocarpum [SCHIAPO] Grimmiaceae (B:M)
Grimmia apocarpa var. *alpicola* (Hedw.) Röhl. = Schistidium agassizii [SCHIAGA] Grimmiaceae (B:M)
Grimmia apocarpa var. *ambigua* (Sull.) G. Jones in Grout = Schistidium apocarpum [SCHIAPO] Grimmiaceae (B:M)
Grimmia apocarpa var. *atrofusca* (Schimp.) Husn. = Schistidium apocarpum [SCHIAPO] Grimmiaceae (B:M)
Grimmia apocarpa var. *brunnescens* (Limpr.) Mönk. = Schistidium apocarpum [SCHIAPO] Grimmiaceae (B:M)
Grimmia apocarpa var. *canadensis* Dupret = Schistidium apocarpum [SCHIAPO] Grimmiaceae (B:M)
Grimmia apocarpa var. *conferta* (Funck) Spreng. = Schistidium apocarpum [SCHIAPO] Grimmiaceae (B:M)
Grimmia apocarpa var. *conferta* f. *obtusifolia* (Schimp.) Mönk. = Schistidium apocarpum [SCHIAPO] Grimmiaceae (B:M)
Grimmia apocarpa var. *dupretii* (Thér.) Sayre = Schistidium apocarpum [SCHIAPO] Grimmiaceae (B:M)
Grimmia apocarpa var. *gracilis* Web. & Mohr = Schistidium apocarpum [SCHIAPO] Grimmiaceae (B:M)
Grimmia apocarpa var. *nigrescens* Mol. = Schistidium apocarpum [SCHIAPO] Grimmiaceae (B:M)
Grimmia apocarpa var. *obscuroviridis* Crum = Schistidium apocarpum [SCHIAPO] Grimmiaceae (B:M)
Grimmia apocarpa var. *rivularis* (Brid.) Nees & Hornsch. = Schistidium rivulare [SCHIRIV] Grimmiaceae (B:M)
Grimmia apocarpa var. *stricta* (Turn.) Hook. & Tayl. = Schistidium apocarpum [SCHIAPO] Grimmiaceae (B:M)
Grimmia apocarpa var. *tenerrima* Nees & Hornsch. = Schistidium apocarpum [SCHIAPO] Grimmiaceae (B:M)
Grimmia arctophila Kindb. = Grimmia laevigata [GRIMLAE] Grimmiaceae (B:M)
Grimmia arctophila ssp. *labradorica* Kindb. = Grimmia affinis [GRIMAFF] Grimmiaceae (B:M)
Grimmia arcuatifolia Kindb. = Racomitrium heterostichum [RACOHET] Grimmiaceae (B:M)
Grimmia atricha C. Müll. & Kindb. in Mac. & Kindb. = Schistidium apocarpum [SCHIAPO] Grimmiaceae (B:M)
Grimmia brittoniae Williams [GRIMBRI] Grimmiaceae (B:M)
Grimmia calyptrata Hook. in Drumm. = Coscinodon calyptratus [COSCCAL] Grimmiaceae (B:M)
Grimmia canadensis Kindb. = Grimmia trichophylla [GRIMTRI] Grimmiaceae (B:M)
Grimmia chloroblasta Kindb. in Mac. & Kindb. = Schistidium heterophyllum [SCHIHET] Grimmiaceae (B:M)
Grimmia coloradensis Aust. = Schistidium apocarpum [SCHIAPO] Grimmiaceae (B:M)
Grimmia commutata Hüb. = Grimmia ovalis [GRIMOVA] Grimmiaceae (B:M)
Grimmia conferta Funck = Schistidium apocarpum [SCHIAPO] Grimmiaceae (B:M)
Grimmia conferta var. *compacta* Lesq. & James = Schistidium apocarpum [SCHIAPO] Grimmiaceae (B:M)
Grimmia crassinervia (C. Müll.) Mac. & Kindb. = Schistidium maritimum [SCHIMAR] Grimmiaceae (B:M)
Grimmia cribrosa Hedw. = Coscinodon cribrosus [COSCCRI] Grimmiaceae (B:M)
Grimmia curvata (Brid.) De Sloover = Dryptodon patens [DRYPPAT] Grimmiaceae (B:M)
Grimmia decipiens auct. = Grimmia trichophylla [GRIMTRI] Grimmiaceae (B:M)
Grimmia densa Kindb. = Grimmia trichophylla [GRIMTRI] Grimmiaceae (B:M)
Grimmia depilata Kindb. in Mac. & Kindb. = Grimmia trichophylla [GRIMTRI] Grimmiaceae (B:M)
Grimmia donniana Sm. [GRIMDON] Grimmiaceae (B:M)
Grimmia donniana var. *holzingeri* (Card. & Thér.) Wijk & Marg. = Grimmia tenerrima [GRIMTEE] Grimmiaceae (B:M)
Grimmia donniana var. *manniae* (C. Müll.) Wijk & Marg. = Grimmia tenerrima [GRIMTEE] Grimmiaceae (B:M)
Grimmia donniana var. *triformis* (Carest. & De Not. in De Not.) Loeske = Grimmia donniana [GRIMDON] Grimmiaceae (B:M)
Grimmia dupretii Thér. = Schistidium apocarpum [SCHIAPO] Grimmiaceae (B:M)
Grimmia elata Kindb. = Grimmia elatior [GRIMELA] Grimmiaceae (B:M)
Grimmia elatior Bruch ex Bals. & De Not. [GRIMELA] Grimmiaceae (B:M)
Grimmia elongata Kaulf. in Sturm [GRIMELO] Grimmiaceae (B:M)
Grimmia glauca auct. = Grimmia laevigata [GRIMLAE] Grimmiaceae (B:M)
Grimmia gracilis (Web. & Mohr) Schleich. ex Limpr. = Schistidium apocarpum [SCHIAPO] Grimmiaceae (B:M)
Grimmia hartmanii var. *anomala* (Hampe ex Schimp.) Mönk. = Grimmia anomala [GRIMANM] Grimmiaceae (B:M)
Grimmia heterophylla Kindb. in Mac. & Kindb. = Schistidium heterophyllum [SCHIHET] Grimmiaceae (B:M)
Grimmia holzingeri Card. & Thér. = Grimmia tenerrima [GRIMTEE] Grimmiaceae (B:M)
Grimmia laevigata (Brid.) Brid. [GRIMLAE] Grimmiaceae (B:M)
Grimmia leibergii Par. = Grimmia trichophylla [GRIMTRI] Grimmiaceae (B:M)
Grimmia manniae C. Müll. = Grimmia donniana [GRIMDON] Grimmiaceae (B:M)
Grimmia maritima Turn. = Schistidium maritimum [SCHIMAR] Grimmiaceae (B:M)
Grimmia microtricha C. Müll. & Kindb. in Mac. & Kindb. = Grimmia donniana [GRIMDON] Grimmiaceae (B:M)
Grimmia mollis Bruch & Schimp. in B.S.G. [GRIMMOL] Grimmiaceae (B:M)
Grimmia montana Bruch & Schimp. in B.S.G. [GRIMMON] Grimmiaceae (B:M)
Grimmia nivalis Kindb. = Grimmia donniana [GRIMDON] Grimmiaceae (B:M)
Grimmia olympica Britt. in Frye = Brachydontium olympicum [BRACOLY] Seligeriaceae (B:M)
Grimmia ovalis (Hedw.) Lindb. [GRIMOVA] Grimmiaceae (B:M)
Grimmia ovalis f. *affinis* (Hoppe & Hornsch. ex Hornsch.) G. Jones in Grout = Grimmia affinis [GRIMAFF] Grimmiaceae (B:M)

Grimmia ovata Web. & Mohr = Grimmia ovalis [GRIMOVA] Grimmiaceae (B:M)
Grimmia patens (Hedw.) Bruch & Schimp. in B.S.G. = Dryptodon patens [DRYPPAT] Grimmiaceae (B:M)
Grimmia plagiopodia Hedw. [GRIMPLA] Grimmiaceae (B:M)
Grimmia plagiopodia ssp. *brandegei* (Aust.) Kindb. = Schistidium apocarpum [SCHIAPO] Grimmiaceae (B:M)
Grimmia plagiopodia ssp. *pilifera* Lesq. & James = Schistidium apocarpum [SCHIAPO] Grimmiaceae (B:M)
Grimmia procera Kindb. = Grimmia trichophylla [GRIMTRI] Grimmiaceae (B:M)
Grimmia prolifera C. Müll. & Kindb. in Mac. & Kindb. = Grimmia torquata [GRIMTOR] Grimmiaceae (B:M)
Grimmia pseudorivularis ssp. *lancifolia* Kindb. = Schistidium rivulare [SCHIRIV] Grimmiaceae (B:M)
Grimmia pulvinata (Hedw.) Sm. [GRIMPUL] Grimmiaceae (B:M)
Grimmia sarcocalyx Kindb. in Mac. & Kindb. = Grimmia laevigata [GRIMLAE] Grimmiaceae (B:M)
Grimmia sessitana var. *subsulcata* (Limpr.) Breidl. = Grimmia tenerrima [GRIMTEE] Grimmiaceae (B:M)
Grimmia stricta Turn. = Schistidium apocarpum [SCHIAPO] Grimmiaceae (B:M)
Grimmia subpapillinervis Kindb. = Grimmia tenerrima [GRIMTEE] Grimmiaceae (B:M)
Grimmia tenella (C. Müll.) Mac. & Kindb. = Grimmia donniana [GRIMDON] Grimmiaceae (B:M)
Grimmia tenera Zett. = Schistidium tenerum [SCHITEN] Grimmiaceae (B:M)
Grimmia tenerrima Ren. & Card. [GRIMTEE] Grimmiaceae (B:M)
Grimmia tenuicaulis Williams = Schistidium tenerum [SCHITEN] Grimmiaceae (B:M)
Grimmia teretinervis Limpr. [GRIMTER] Grimmiaceae (B:M)
Grimmia torquata Hornsch. in Grev. [GRIMTOR] Grimmiaceae (B:M)
Grimmia tortifolia Kindb. = Grimmia torquata [GRIMTOR] Grimmiaceae (B:M)
Grimmia tortifolia ssp. *pellucida* Kindb. in Röll = Grimmia torquata [GRIMTOR] Grimmiaceae (B:M)
Grimmia trichophylla Grev. [GRIMTRI] Grimmiaceae (B:M)
Grimmia trichophylla var. *brachycarpa* De Not. = Grimmia trichophylla [GRIMTRI] Grimmiaceae (B:M)
Grimmia trichophylla var. *meridionalis* C. Müll. = Grimmia trichophylla [GRIMTRI] Grimmiaceae (B:M)
Grimmia unicolor Hook. in Grev. [GRIMUNI] Grimmiaceae (B:M)
Grimmia velutina Kindb. = Grimmia donniana [GRIMDON] Grimmiaceae (B:M)
Grindelia Willd. [GRINDEL$] Asteraceae (V:D)
Grindelia arenicola Steyermark = Grindelia integrifolia var. macrophylla [GRININT1] Asteraceae (V:D)
Grindelia integrifolia DC. [GRININT] Asteraceae (V:D)
Grindelia integrifolia var. **macrophylla** (Greene) Cronq. [GRININT1] Asteraceae (V:D)
Grindelia macrophylla Greene = Grindelia integrifolia var. macrophylla [GRININT1] Asteraceae (V:D)
Grindelia perennis A. Nels. = Grindelia squarrosa var. quasiperennis [GRINSQU1] Asteraceae (V:D)
Grindelia serrulata Rydb. = Grindelia squarrosa var. serrulata [GRINSQU2] Asteraceae (V:D)
Grindelia squarrosa (Pursh) Dunal [GRINSQU] Asteraceae (V:D)
Grindelia squarrosa var. **quasiperennis** Lunell [GRINSQU1] Asteraceae (V:D)
*__Grindelia squarrosa__ var. **serrulata** (Rydb.) Steyermark [GRINSQU2] Asteraceae (V:D)
*__Grindelia squarrosa__ var. **squarrosa** [GRINSQU0] Asteraceae (V:D)
Grindelia stricta DC. = Grindelia integrifolia var. macrophylla [GRININT1] Asteraceae (V:D)
Grindelia stricta ssp. *blakei* (Steyermark) Keck = Grindelia integrifolia var. macrophylla [GRININT1] Asteraceae (V:D)
Grindelia stricta ssp. *venulosa* (Jepson) Keck = Grindelia integrifolia var. macrophylla [GRININT1] Asteraceae (V:D)

Grossularia cognata (Greene) Coville & Britt. = Ribes oxyacanthoides ssp. cognatum [RIBEOXY1] Grossulariaceae (V:D)
Grossularia irrigua (Dougl.) Coville & Britt. = Ribes oxyacanthoides ssp. irriguum [RIBEOXY2] Grossulariaceae (V:D)
Grossularia lobbii (Gray) Coville & Britt. = Ribes lobbii [RIBELOB] Grossulariaceae (V:D)
Grossularia nonscripta Berger = Ribes oxyacanthoides ssp. irriguum [RIBEOXY2] Grossulariaceae (V:D)
Grossularia oxyacanthoides (L.) P. Mill. = Ribes oxyacanthoides ssp. oxyacanthoides [RIBEOXY0] Grossulariaceae (V:D)
Gyalecta Ach. [GYALECT$] Gyalectaceae (L)
Gyalecta cupularis (Hedw.) Schaerer = Gyalecta jenensis [GYALJEN] Gyalectaceae (L)
Gyalecta friesii Flotow ex Körber [GYALFRI] Gyalectaceae (L)
Gyalecta jenensis (Batsch) Zahlbr. [GYALJEN] Gyalectaceae (L)
Gyalidea Lettau ex Vezda [GYALIDE$] Solorinellaceae (L)
Gyalidea dodgei Vezda = Gyalidea hyalinescens [GYALHYA] Solorinellaceae (L)
Gyalidea hyalinescens (Nyl.) Vezda [GYALHYA] Solorinellaceae (L)
Gyalideopsis Vezda [GYALIDO$] Gomphillaceae (L)
Gyalideopsis alnicola W. Noble & Vezda = Gyalideopsis piceicola [GYALPIC] Gomphillaceae (L)
Gyalideopsis anastomosans P. James & Vezda [GYALANA] Gomphillaceae (L)
Gyalideopsis piceicola (Nyl.) Vezda [GYALPIC] Gomphillaceae (L)
Gymnocarpium Newman [GYMNOCA$] Dryopteridaceae (V:F)
Gymnocarpium continentale (V. Petrov) Pojark. = Gymnocarpium jessoense ssp. parvulum [GYMNJES1] Dryopteridaceae (V:F)
Gymnocarpium disjunctum (Rupr.) Ching [GYMNDIS] Dryopteridaceae (V:F)
Gymnocarpium dryopteris (L.) Newman [GYMNDRY] Dryopteridaceae (V:F)
Gymnocarpium dryopteris ssp. *disjunctum* (Rupr.) Sarvela = Gymnocarpium disjunctum [GYMNDIS] Dryopteridaceae (V:F)
Gymnocarpium dryopteris var. *disjunctum* (Rupr.) Ching = Gymnocarpium disjunctum [GYMNDIS] Dryopteridaceae (V:F)
Gymnocarpium jessoense (Koidzumi) Koidzumi [GYMNJES] Dryopteridaceae (V:F)
Gymnocarpium jessoense ssp. **parvulum** Sarvela [GYMNJES1] Dryopteridaceae (V:F)
Gymnocolea (Dum.) Dum. [GYMNOCO$] Jungermanniaceae (B:H)
Gymnocolea inflata (Huds.) Dum. [GYMNINF] Jungermanniaceae (B:H)
Gymnomitrion Corda [GYMNOMI$] Gymnomitriaceae (B:H)
Gymnomitrion apiculatum (Schiffn.) K. Müll. [GYMNAPI] Gymnomitriaceae (B:H)
Gymnomitrion concinnatum (Lightf.) Corda [GYMNCON] Gymnomitriaceae (B:H)
Gymnomitrion corallioides Nees [GYMNCOR] Gymnomitriaceae (B:H)
Gymnomitrion obtusum (Lindb.) Pears. [GYMNOBT] Gymnomitriaceae (B:H)
Gymnomitrion pacificum Grolle [GYMNPAC] Gymnomitriaceae (B:H)
Gymnomitrium crenulatum auct. = Gymnomitrion pacificum [GYMNPAC] Gymnomitriaceae (B:H)
Gymnomitrium revolutum (Nees) Philib. = Marsupella revoluta [MARSREV] Gymnomitriaceae (B:H)
Gymnomitrium varians (Lindb.) K. Müll. = Marsupella brevissima [MARSBRE] Gymnomitriaceae (B:H)
Gymnostomum Nees & Hornsch. [GYMNOSO$] Pottiaceae (B:M)
Gymnostomum aeruginosum Sm. [GYMNAER] Pottiaceae (B:M)
Gymnostomum curvirostre Hedw. ex Brid. = Hymenostylium recurvirostre [HYMEREC] Pottiaceae (B:M)
Gymnostomum curvirostre var. *commutatum* (Mitt.) Card. & Thér. = Hymenostylium recurvirostre [HYMEREC] Pottiaceae (B:M)
Gymnostomum insigne (Dix.) A.J.E. Sm. = Hymenostylium insigne [HYMEINS] Pottiaceae (B:M)
Gymnostomum latifolium (Zett.) Flow. in Crum = Hymenostylium recurvirostre [HYMEREC] Pottiaceae (B:M)

Gymnostomum recurvirostre Hedw. = Hymenostylium recurvirostre [HYMEREC] Pottiaceae (B:M)
Gymnostomum recurvirostre var. *commutatum* (Mitt.) Grout = Hymenostylium recurvirostre [HYMEREC] Pottiaceae (B:M)
Gymnostomum recurvirostre var. *latifolium* Zett. = Hymenostylium recurvirostre [HYMEREC] Pottiaceae (B:M)
Gymnostomum recurvirostre var. *scabrum* (Lindb.) Grout = Hymenostylium recurvirostre [HYMEREC] Pottiaceae (B:M)
Gymnostomum rupestre Schleich. ex Schwaegr. = Gymnostomum aeruginosum [GYMNAER] Pottiaceae (B:M)
Gypsophila L. [GYPSOPH$] Caryophyllaceae (V:D)
***Gypsophila paniculata** L. [GYPSPAN] Caryophyllaceae (V:D)
Gypsoplaca Timdal [GYPSOPL$] Gypsoplacaceae (L)
Gypsoplaca macrophylla (Zahlbr.) Timdal [GYPSMAC] Gypsoplacaceae (L)
Gyrophora angulata (Tuck.) Herre = Umbilicaria angulata [UMBIANG] Umbilicariaceae (L)
Gyrophora anthracina (Wulfen) Körber = Umbilicaria rigida [UMBIRIG] Umbilicariaceae (L)
Gyrophora arctica Ach. = Umbilicaria arctica [UMBIARC] Umbilicariaceae (L)
Gyrophora cylindrica (L.) Ach. = Umbilicaria cylindrica [UMBICYL] Umbilicariaceae (L)
Gyrophora decussata (Vill.) Zahlbr. = Umbilicaria decussata [UMBIDEC] Umbilicariaceae (L)
Gyrophora deusta (L.) Ach. = Umbilicaria deusta [UMBIDEU] Umbilicariaceae (L)
Gyrophora erosa (G. Weber) Ach. = Umbilicaria torrefacta [UMBITOR] Umbilicariaceae (L)
Gyrophora hyperborea Ach. = Umbilicaria hyperborea var. hyperborea [UMBIHYP0] Umbilicariaceae (L)
Gyrophora muehlenbergii Ach. = Umbilicaria muehlenbergii [UMBIMUE] Umbilicariaceae (L)
Gyrophora phaea (Tuck.) Nyl. = Umbilicaria phaea [UMBIPHA] Umbilicariaceae (L)
Gyrophora polyphylla (L.) Fink = Umbilicaria polyphylla [UMBIPOL] Umbilicariaceae (L)
Gyrophora polyrrhiza (L.) Körber = Umbilicaria polyrrhiza [UMBIPOY] Umbilicariaceae (L)
Gyrophora vellea (L.) Ach. = Umbilicaria vellea [UMBIVEL] Umbilicariaceae (L)
Gyrothyra M.A. Howe [GYROTHY$] Gyrothyraceae (B:H)
Gyrothyra underwoodiana M.A. Howe [GYROUND] Gyrothyraceae (B:H)

H

Habenaria bracteata (Muhl. ex Willd.) R. Br. ex Ait. f. = Coeloglossum viride var. virescens [COELVIR1] Orchidaceae (V:M)
Habenaria chorisiana Cham. = Platanthera chorisiana [PLATCHO] Orchidaceae (V:M)
Habenaria dilatata (Pursh) Hook. = Platanthera dilatata var. dilatata [PLATDIL0] Orchidaceae (V:M)
Habenaria dilatata var. *albiflora* (Cham.) Correll = Platanthera dilatata var. albiflora [PLATDIL1] Orchidaceae (V:M)
Habenaria dilatata var. *leucostachys* (Lindl.) Ames = Platanthera leucostachys [PLATLEU] Orchidaceae (V:M)
Habenaria elegans (Lindl.) Boland. = Piperia elegans [PIPEELE] Orchidaceae (V:M)
Habenaria elegans var. *maritima* (Greene) Ames = Piperia elegans [PIPEELE] Orchidaceae (V:M)
Habenaria greenei Jepson = Piperia elegans [PIPEELE] Orchidaceae (V:M)
Habenaria hyperborea (L.) R. Br. ex Ait. f. = Platanthera hyperborea [PLATHYP] Orchidaceae (V:M)
Habenaria leucostachys (Lindl.) S. Wats. = Platanthera leucostachys [PLATLEU] Orchidaceae (V:M)
Habenaria maritima Greene = Piperia elegans [PIPEELE] Orchidaceae (V:M)
Habenaria obtusata (Banks ex Pursh) Richards. = Platanthera obtusata [PLATOBT] Orchidaceae (V:M)
Habenaria obtusata var. *collectanea* Fern. = Platanthera obtusata [PLATOBT] Orchidaceae (V:M)
Habenaria orbiculata (Pursh) Torr. = Platanthera orbiculata [PLATORB] Orchidaceae (V:M)
Habenaria saccata Greene = Platanthera stricta [PLATSTR] Orchidaceae (V:M)
Habenaria stricta (Lindl.) Rydb. = Platanthera stricta [PLATSTR] Orchidaceae (V:M)
Habenaria unalascensis (Spreng.) S. Wats. = Piperia unalascensis [PIPEUNA] Orchidaceae (V:M)
Habenaria unalascensis var. *maritima* (Greene) Correll = Piperia elegans [PIPEELE] Orchidaceae (V:M)
Habenaria viridis var. *bracteata* (Muhl. ex Willd.) Reichenb. ex Gray = Coeloglossum viride var. virescens [COELVIR1] Orchidaceae (V:M)
Habrodon leucotrichus (Mitt.) Perss. = Iwatsukiella leucotricha [IWATLEU] Pterigynandraceae (B:M)
Hackelia Opiz [HACKELI$] Boraginaceae (V:D)
Hackelia americana (Gray) Fern. = Hackelia deflexa var. americana [HACKDEF2] Boraginaceae (V:D)
Hackelia ciliata (Dougl. ex Lehm.) I.M. Johnston [HACKCIL] Boraginaceae (V:D)
Hackelia deflexa (Wahlenb.) Opiz [HACKDEF] Boraginaceae (V:D)
Hackelia deflexa ssp. *americana* (Gray) A. & D. Löve = Hackelia deflexa var. americana [HACKDEF2] Boraginaceae (V:D)
Hackelia deflexa var. **americana** (Gray) Fern. & I.M. Johnston [HACKDEF2] Boraginaceae (V:D)
Hackelia diffusa (Lehm.) I.M. Johnston [HACKDIF] Boraginaceae (V:D)
Hackelia floribunda (Lehm.) I.M. Johnston [HACKFLO] Boraginaceae (V:D)
Hackelia jessicae (McGregor) Brand = Hackelia micrantha [HACKMIC] Boraginaceae (V:D)
Hackelia leptophylla (Rydb.) I.M. Johnston = Hackelia floribunda [HACKFLO] Boraginaceae (V:D)
Hackelia micrantha (Eastw.) J.L. Gentry [HACKMIC] Boraginaceae (V:D)
Haematomma caesium Coppins & P. James = Mycoblastus caesius [MYCOCAE] Mycoblastaceae (L)
Haematomma californicum Sigal & D. Toren = Ophioparma rubricosa [OPHIRUB] Ophioparmaceae (L)
Haematomma elatinum (Ach.) A. Massal. = Loxospora elatina [LOXOELA] Haematommataceae (L)
Haematomma lapponicum Räsänen = Ophioparma lapponica [OPHILAP] Ophioparmaceae (L)
Haematomma ochrophaeum (Tuck.) A. Massal. = Loxospora ochrophaea [LOXOOCH] Haematommataceae (L)
Haematomma ventosum (L.) A. Massal. = Ophioparma lapponica [OPHILAP] Ophioparmaceae (L)
Hafellia Kalb, H. Mayrh. & Scheid. [HAFELLI$] Physciaceae (L)
Hafellia callispora (C. Knight) H. Mayrh. & Sheard [HAFECAL] Physciaceae (L)
Hafellia fosteri Imshaug & Sheard [HAFEFOS] Physciaceae (L)
Halecania M. Mayrh. [HALECAN$] Catillariaceae (L)
Halecania alpivaga (Th. Fr.) M. Mayrh. [HALEALP] Catillariaceae (L)
Halenia Borkh. [HALENIA$] Gentianaceae (V:D)
Halenia deflexa (Sm.) Griseb. [HALEDEF] Gentianaceae (V:D)
Halimolobos Tausch [HALIMOL$] Brassicaceae (V:D)
Halimolobos mollis (Hook.) Rollins [HALIMOI] Brassicaceae (V:D)

Halimolobos whitedii (Piper) Rollins [HALIWHI] Brassicaceae (V:D)
Haloscias hultenii (Fern.) Holub = Ligusticum scoticum ssp. hultenii [LIGUSCO1] Apiaceae (V:D)
Hamatocaulis Hedenäs [HAMATOC$] Amblystegiaceae (B:M)
Hamatocaulis lapponicus (Norrl.) Hedenäs [HAMALAP] Amblystegiaceae (B:M)
Hamatocaulis vernicosus (Mitt.) Hedenäs [HAMAVER] Amblystegiaceae (B:M)
Hammarbya paludosa (L.) Kuntze = Malaxis paludosa [MALAPAL] Orchidaceae (V:M)
Haplocladium microphyllum (Hedw.) Broth. = Bryohaplocladium microphyllum [BRYOMIC] Leskeaceae (B:M)
Haplocladium microphyllum var. *lignicola* (Kindb.) Reim. = Bryohaplocladium microphyllum [BRYOMIC] Leskeaceae (B:M)
Haplocladium microphyllum var. *obtusum* (Grout) Crum = Bryohaplocladium microphyllum [BRYOMIC] Leskeaceae (B:M)
Haplomitrium Nees [HAPLOMI$] Haplomitriaceae (B:H)
Haplomitrium hookeri (Sm.) Nees [HAPLHOO] Haplomitriaceae (B:H)
Haplopappus bloomeri Gray = Ericameria bloomeri [ERICBLO] Asteraceae (V:D)
Haplopappus bloomeri ssp. *sonnei* (Gray) Hall = Ericameria bloomeri [ERICBLO] Asteraceae (V:D)
Haplopappus bloomeri var. *angustatus* Gray = Ericameria bloomeri [ERICBLO] Asteraceae (V:D)
Haplopappus bloomeri var. *sonnei* (Gray) Greene = Ericameria bloomeri [ERICBLO] Asteraceae (V:D)
Haplopappus carthamoides (Hook.) Gray = Pyrrocoma carthamoides [PYRRCAR] Asteraceae (V:D)
Haplopappus carthamoides ssp. *rigidus* (Rydb.) Hall = Pyrrocoma carthamoides [PYRRCAR] Asteraceae (V:D)
Haplopappus carthamoides var. *erythropappus* (Rydb.) St. John = Pyrrocoma carthamoides [PYRRCAR] Asteraceae (V:D)
Haplopappus carthamoides var. *rigidus* (Rydb.) M.E. Peck = Pyrrocoma carthamoides [PYRRCAR] Asteraceae (V:D)
Haplopappus carthamoides var. *typicus* (Hall) Cronq. = Pyrrocoma carthamoides [PYRRCAR] Asteraceae (V:D)
Haplopappus lyallii Gray = Tonestus lyallii [TONELYA] Asteraceae (V:D)
Harpanthus Nees [HARPANT$] Geocalycaceae (B:H)
Harpanthus flotovianus (Nees) Nees [HARPFLO] Geocalycaceae (B:H)
Harpidium polycarpum (Voit) Williams = Drepanocladus aduncus var. polycarpus [DREPADU4] Amblystegiaceae (B:M)
Harrimanella Coville [HARRIMA$] Ericaceae (V:D)
Harrimanella stelleriana (Pallas) Coville [HARRSTE] Ericaceae (V:D)
Hawksworthiana U. Braun [HAWKSWO$] Uncertain-family (L)
Hawksworthiana peltigericola (D. Hawksw.) U. Braun [HAWKPEL] Uncertain-family (L)
Hedera L. [HEDERA$] Araliaceae (V:D)
***Hedera helix** L. [HEDEHEL] Araliaceae (V:D)
Hedwigia P. Beauv. [HEDWIGI$] Hedwigiaceae (B:M)
Hedwigia albicans Lindb. = Hedwigia ciliata [HEDWCIL] Hedwigiaceae (B:M)
Hedwigia albicans var. *subnuda* (Kindb.) C. Mohr = Hedwigia ciliata [HEDWCIL] Hedwigiaceae (B:M)
Hedwigia ciliata (Hedw.) P. Beauv. [HEDWCIL] Hedwigiaceae (B:M)
Hedwigia ciliata f. *secunda* (B.S.G.) G. Jones in Grout = Hedwigia ciliata [HEDWCIL] Hedwigiaceae (B:M)
Hedwigia ciliata f. *viridis* (B.S.G.) G. Jones in Grout = Hedwigia ciliata [HEDWCIL] Hedwigiaceae (B:M)
Hedwigia ciliata ssp. *subnuda* Kindb. in Mac. & Kindb. = Hedwigia ciliata [HEDWCIL] Hedwigiaceae (B:M)
Hedwigia ciliata var. *leucophaea* B.S.G. = Hedwigia ciliata [HEDWCIL] Hedwigiaceae (B:M)
Hedwigia ciliata var. *secunda* B.S.G. = Hedwigia ciliata [HEDWCIL] Hedwigiaceae (B:M)
Hedwigia ciliata var. *subnuda* Kindb. ex G. Jones in Grout = Hedwigia ciliata [HEDWCIL] Hedwigiaceae (B:M)
Hedwigia ciliata var. *viridis* B.S.G. = Hedwigia ciliata [HEDWCIL] Hedwigiaceae (B:M)
Hedwigia stellata Hedenäs [HEDWSTE] Hedwigiaceae (B:M)
Hedysarum L. [HEDYSAR$] Fabaceae (V:D)
Hedysarum alpinum L. [HEDYALP] Fabaceae (V:D)
Hedysarum alpinum ssp. *americanum* (Michx.) Fedtsch. = Hedysarum alpinum var. americanum [HEDYALP1] Fabaceae (V:D)
Hedysarum alpinum var. **americanum** Michx. [HEDYALP1] Fabaceae (V:D)
Hedysarum americanum (Michx.) Britt. = Hedysarum alpinum var. americanum [HEDYALP1] Fabaceae (V:D)
Hedysarum boreale Nutt. [HEDYBOR] Fabaceae (V:D)
Hedysarum boreale ssp. **boreale** [HEDYBOR0] Fabaceae (V:D)
Hedysarum boreale ssp. **mackenzii** (Richards.) Welsh [HEDYBOR2] Fabaceae (V:D)
Hedysarum boreale var. *mackenzii* (Richards.) C.L. Hitchc. = Hedysarum boreale ssp. mackenzii [HEDYBOR2] Fabaceae (V:D)
Hedysarum mackenzii Richards. = Hedysarum boreale ssp. mackenzii [HEDYBOR2] Fabaceae (V:D)
Hedysarum occidentale Greene [HEDYOCC] Fabaceae (V:D)
Hedysarum occidentale var. *canone* Welsh = Hedysarum occidentale [HEDYOCC] Fabaceae (V:D)
Hedysarum sulphurescens Rydb. [HEDYSUL] Fabaceae (V:D)
Hedysarum uintahense A. Nels. = Hedysarum occidentale [HEDYOCC] Fabaceae (V:D)
Helenium L. [HELENIU$] Asteraceae (V:D)
Helenium autumnale L. [HELEAUT] Asteraceae (V:D)
Helenium autumnale var. **grandiflorum** (Nutt.) Torr. & Gray [HELEAUT1] Asteraceae (V:D)
Helenium autumnale var. **montanum** (Nutt.) Fern. [HELEAUT2] Asteraceae (V:D)
Helenium macranthum Rydb. = Helenium autumnale var. grandiflorum [HELEAUT1] Asteraceae (V:D)
Helenium montanum Nutt. = Helenium autumnale var. montanum [HELEAUT2] Asteraceae (V:D)
Helianthella Torr. & Gray [HELIANT$] Asteraceae (V:D)
Helianthella uniflora (Nutt.) Torr. & Gray [HELIUNI] Asteraceae (V:D)
Helianthella uniflora var. **douglasii** (Torr. & Gray) W.A. Weber [HELIUNI1] Asteraceae (V:D)
Helianthus L. [HELIANH$] Asteraceae (V:D)
***Helianthus annuus** L. [HELIANN] Asteraceae (V:D)
Helianthus annuus ssp. *jaegeri* (Heiser) Heiser = Helianthus annuus [HELIANN] Asteraceae (V:D)
Helianthus annuus ssp. *lenticularis* (Dougl. ex Lindl.) Cockerell = Helianthus annuus [HELIANN] Asteraceae (V:D)
Helianthus annuus ssp. *texanus* Heiser = Helianthus annuus [HELIANN] Asteraceae (V:D)
Helianthus annuus var. *lenticularis* (Dougl. ex Lindl.) Steyermark = Helianthus annuus [HELIANN] Asteraceae (V:D)
Helianthus annuus var. *macrocarpus* (DC.) Cockerell = Helianthus annuus [HELIANN] Asteraceae (V:D)
Helianthus annuus var. *texanus* (Heiser) Shinners = Helianthus annuus [HELIANN] Asteraceae (V:D)
Helianthus aridus Rydb. = Helianthus annuus [HELIANN] Asteraceae (V:D)
Helianthus fascicularis Greene = Helianthus nuttallii [HELINUT] Asteraceae (V:D)
Helianthus laetiflorus var. *subrhomboideus* (Rydb.) Fern. = Helianthus pauciflorus ssp. subrhomboideus [HELIPAU1] Asteraceae (V:D)
Helianthus lenticularis Dougl. ex Lindl. = Helianthus annuus [HELIANN] Asteraceae (V:D)
Helianthus nuttallii Torr. & Gray [HELINUT] Asteraceae (V:D)
Helianthus nuttallii ssp. *canadensis* R.W. Long = Helianthus nuttallii [HELINUT] Asteraceae (V:D)
Helianthus nuttallii ssp. *coloradensis* (Cockerell) R.W. Long = Helianthus nuttallii [HELINUT] Asteraceae (V:D)
Helianthus pauciflorus Nutt. [HELIPAU] Asteraceae (V:D)
Helianthus pauciflorus ssp. **subrhomboideus** (Rydb.) O. Spring & E. Schilling [HELIPAU1] Asteraceae (V:D)

Helianthus pauciflorus var. *subrhomboideus* (Rydb.) Cronq. = Helianthus pauciflorus ssp. subrhomboideus [HELIPAU1] Asteraceae (V:D)
Helianthus rigidus ssp. *laetiflorus* (Rydb.) Heiser = Helianthus pauciflorus ssp. subrhomboideus [HELIPAU1] Asteraceae (V:D)
Helianthus rigidus ssp. *subrhomboideus* (Rydb.) Heiser = Helianthus pauciflorus ssp. subrhomboideus [HELIPAU1] Asteraceae (V:D)
Helianthus rigidus var. *subrhomboideus* (Rydb.) Cronq. = Helianthus pauciflorus ssp. subrhomboideus [HELIPAU1] Asteraceae (V:D)
Helianthus subrhomboideus Rydb. = Helianthus pauciflorus ssp. subrhomboideus [HELIPAU1] Asteraceae (V:D)
Helictotrichon Bess. ex J.A. & J.H. Schultes [HELICTO$] Poaceae (V:M)
Helictotrichon hookeri (Scribn.) Henr. [HELIHOO] Poaceae (V:M)
Helocarpon Th. Fr. [HELOCAR$] Micareaceae (L)
Helocarpon crassipes Th. Fr. [HELOCRA] Micareaceae (L)
Helodium Warnst. [HELODIU$] Helodiaceae (B:M)
Helodium blandowii (Web. & Mohr) Warnst. [HELOBLA] Helodiaceae (B:M)
Helodium lanatum (Brid.) Broth. = Helodium blandowii [HELOBLA] Helodiaceae (B:M)
Helodium pratense W. Koch ex Moxley = Hypnum pratense [HYPNPRA] Hypnaceae (B:M)
Helodium reptile Michx. ex Moxley = Hypnum pallescens [HYPNPAL] Hypnaceae (B:M)
Helonias tenax Pursh = Xerophyllum tenax [XEROTEN] Liliaceae (V:M)
Hemicarpha micrantha (Vahl) Pax = Lipocarpha micrantha [LIPOMIC] Cyperaceae (V:M)
Hemicarpha micrantha var. *minor* (Schrad.) Friedland = Lipocarpha micrantha [LIPOMIC] Cyperaceae (V:M)
Hemieva ranunculifolia (Hook.) Raf. = Suksdorfia ranunculifolia [SUKSRAN] Saxifragaceae (V:D)
Hemitomes Gray [HEMITOM$] Monotropaceae (V:D)
Hemitomes congestum Gray [HEMICON] Monotropaceae (V:D)
Heppia Nägeli [HEPPIA$] Heppiaceae (L)
Heppia despreauxii (Mont.) Tuck. = Heppia lutosa [HEPPLUT] Heppiaceae (L)
Heppia euploca (Ach.) Vainio = Peltula euploca [PELTEUP] Peltulaceae (L)
Heppia guepinii (Delise) Nyl. = Peltula euploca [PELTEUP] Peltulaceae (L)
Heppia lutosa (Ach.) Nyl. [HEPPLUT] Heppiaceae (L)
Heppia polyphylla de Lesd. = Peltula euploca [PELTEUP] Peltulaceae (L)
Heppia virescens (Despr.) Nyl. = Heppia lutosa [HEPPLUT] Heppiaceae (L)
Heracleum L. [HERACLE$] Apiaceae (V:D)
Heracleum lanatum Michx. = Heracleum maximum [HERAMAX] Apiaceae (V:D)
***Heracleum mantegazzianum** Sommier & Levier [HERAMAN] Apiaceae (V:D)
Heracleum maximum Bartr. [HERAMAX] Apiaceae (V:D)
Heracleum sphondylium ssp. *lanatum* (Michx.) A. & D. Löve = Heracleum maximum [HERAMAX] Apiaceae (V:D)
Heracleum sphondylium ssp. *montanum* (Schleich. ex Gaudin) Briq. = Heracleum maximum [HERAMAX] Apiaceae (V:D)
Heracleum sphondylium var. *lanatum* (Michx.) Dorn = Heracleum maximum [HERAMAX] Apiaceae (V:D)
Herberta Gray = Herbertus [HERBERT$] Herbertaceae (B:H)
Herbertus Gray [HERBERT$] Herbertaceae (B:H)
Herbertus aduncus (Dicks.) Gray [HERBADU] Herbertaceae (B:H)
Herbertus hawaiiensis Mill. [HERBHAW] Herbertaceae (B:H)
Herbertus himalayanus auct. = Herbertus sakuraii [HERBSAK] Herbertaceae (B:H)
Herbertus sakuraii Steph. [HERBSAK] Herbertaceae (B:H)
Herbertus sendtneri (Nees) Lindb. [HERBSEN] Herbertaceae (B:H)
Herzogiella Broth. [HERZOGI$] Hypnaceae (B:M)
Herzogiella adscendens (Lindb.) Iwats. & Schof. [HERZADS] Hypnaceae (B:M)
Herzogiella seligeri (Brid.) Iwats. [HERZSEL] Hypnaceae (B:M)
Herzogiella striatella (Brid.) Iwats. [HERZSTR] Hypnaceae (B:M)
Hesperis L. [HESPERI$] Brassicaceae (V:D)
***Hesperis matronalis** L. [HESPMAT] Brassicaceae (V:D)
Hesperochiron S. Wats. [HESPERO$] Hydrophyllaceae (V:D)
Hesperochiron pumilus (Dougl. ex Griseb.) Porter [HESPPUM] Hydrophyllaceae (V:D)
Hesperochiron villosulus (Greene) Suksdorf = Hesperochiron pumilus [HESPPUM] Hydrophyllaceae (V:D)
Heterocladium Schimp. in B.S.G. [HETEROL$] Pterigynandraceae (B:M)
Heterocladium aberrans Ren. & Card. = Heterocladium procurrens [HETEPRO] Pterigynandraceae (B:M)
Heterocladium dimorphum (Brid.) Schimp. in B.S.G. [HETEDIM] Pterigynandraceae (B:M)
Heterocladium frullaniopsis C. Müll. & Kindb. in Mac. & Kindb. = Pterigynandrum filiforme [PTERFIL] Pterigynandraceae (B:M)
Heterocladium heteropteroides Best = Heterocladium macounii [HETEMAC] Pterigynandraceae (B:M)
Heterocladium heteropteroides var. *filescens* Best = Heterocladium macounii [HETEMAC] Pterigynandraceae (B:M)
Heterocladium heteropterum (Brid.) B.S.G. = Heterocladium macounii [HETEMAC] Pterigynandraceae (B:M)
Heterocladium macounii Best [HETEMAC] Pterigynandraceae (B:M)
Heterocladium papillosum (Lindb.) Lindb. = Pseudoleskeella papillosa [PSEUPAP] Leskeaceae (B:M)
Heterocladium procurrens (Mitt.) Jaeg. [HETEPRO] Pterigynandraceae (B:M)
Heterocladium squarrosulum Lindb. = Heterocladium dimorphum [HETEDIM] Pterigynandraceae (B:M)
Heterocladium vancouveriense (Kindb.) Kindb. in Mac. & Kindb. = Porotrichum vancouveriense [POROVAN] Thamnobryaceae (B:M)
Heterocodon Nutt. [HETEROC$] Campanulaceae (V:D)
Heterocodon rariflorum Nutt. [HETERAR] Campanulaceae (V:D)
Heterodermia Trevisan [HETEROD$] Physciaceae (L)
Heterodermia leucomelaena (L.) Poelt = Heterodermia leucomelos [HETELEU] Physciaceae (L)
Heterodermia leucomelos (L.) Poelt [HETELEU] Physciaceae (L)
Heterodermia sitchensis Goward & W. Noble [HETESIT] Physciaceae (L)
Heterodermia speciosa (Wulfen) Trevisan [HETESPE] Physciaceae (L)
Heterodermia tremulans (Müll. Arg.) Culb. = Heterodermia speciosa [HETESPE] Physciaceae (L)
Heterophyllium haldanianum (Grev.) Fleisch. = Callicladium haldanianum [CALLHAL] Hypnaceae (B:M)
Heterotheca Cass. [HETEROT$] Asteraceae (V:D)
Heterotheca horrida (Rydb.) Harms = Heterotheca villosa var. hispida [HETEVIL1] Asteraceae (V:D)
Heterotheca horrida ssp. *cinerascens* (Blake) Semple = Heterotheca villosa var. hispida [HETEVIL1] Asteraceae (V:D)
Heterotheca villosa (Pursh) Shinners [HETEVIL] Asteraceae (V:D)
Heterotheca villosa var. *angustifolia* (Rydb.) Harms = Heterotheca villosa var. villosa [HETEVIL0] Asteraceae (V:D)
Heterotheca villosa var. *foliosa* (Nutt.) Harms = Heterotheca villosa var. villosa [HETEVIL0] Asteraceae (V:D)
Heterotheca villosa var. **hispida** (Hook.) Harms [HETEVIL1] Asteraceae (V:D)
Heterotheca villosa var. *pedunculata* (Greene) Harms ex Semple = Heterotheca villosa var. villosa [HETEVIL0] Asteraceae (V:D)
Heterotheca villosa var. **villosa** [HETEVIL0] Asteraceae (V:D)
Heterotheca wisconsinensis (Shinners) Shinners = Heterotheca villosa var. hispida [HETEVIL1] Asteraceae (V:D)
Heuchera L. [HEUCHER$] Saxifragaceae (V:D)
Heuchera chlorantha Piper [HEUCCHL] Saxifragaceae (V:D)
Heuchera cylindrica Dougl. ex Hook. [HEUCCYL] Saxifragaceae (V:D)
Heuchera cylindrica var. **cylindrica** [HEUCCYL0] Saxifragaceae (V:D)
Heuchera cylindrica var. **glabella** (Torr. & Gray) Wheelock [HEUCCYL2] Saxifragaceae (V:D)
Heuchera cylindrica var. **orbicularis** (Rosendahl, Butters & Lakela) Calder & Savile [HEUCCYL3] Saxifragaceae (V:D)
Heuchera cylindrica var. **septentrionalis** Rosendahl, Butters & Lakela [HEUCCYL4] Saxifragaceae (V:D)

Heuchera cylindrica var. *suksdorfii* (Rydb.) Dorn = Heuchera cylindrica var. cylindrica [HEUCCYL0] Saxifragaceae (V:D)
Heuchera diversifolia Rydb. = Heuchera micrantha var. diversifolia [HEUCMIC1] Saxifragaceae (V:D)
Heuchera glabella Torr. & Gray = Heuchera cylindrica var. glabella [HEUCCYL2] Saxifragaceae (V:D)
Heuchera glabra Willd. ex Roemer & J.A. Schultes [HEUCGLA] Saxifragaceae (V:D)
Heuchera micrantha Dougl. ex Lindl. [HEUCMIC] Saxifragaceae (V:D)
Heuchera micrantha var. **diversifolia** (Rydb.) Rosendahl, Butters & Lakela [HEUCMIC1] Saxifragaceae (V:D)
Heuchera ovalifolia var. *orbicularis* Rosendahl, Butters & Lakela = Heuchera cylindrica var. orbicularis [HEUCCYL3] Saxifragaceae (V:D)
Heuchera racemosa S. Wats. = Elmera racemosa [ELMERAC] Saxifragaceae (V:D)
Heuchera richardsonii R. Br. [HEUCRIC] Saxifragaceae (V:D)
Heuchera richardsonii var. *affinis* Rosendahl, Butters & Lakela = Heuchera richardsonii [HEUCRIC] Saxifragaceae (V:D)
Heuchera richardsonii var. *grayana* Rosendahl, Butters & Lakela = Heuchera richardsonii [HEUCRIC] Saxifragaceae (V:D)
Heuchera richardsonii var. *hispidior* Rosendahl, Butters & Lakela = Heuchera richardsonii [HEUCRIC] Saxifragaceae (V:D)
Heuchera saxicola E. Nels. = Heuchera cylindrica var. cylindrica [HEUCCYL0] Saxifragaceae (V:D)
Heuchera suksdorfii Rydb. = Heuchera cylindrica var. cylindrica [HEUCCYL0] Saxifragaceae (V:D)
Hieracium L. [HIERACI$] Asteraceae (V:D)
Hieracium albertinum Farr = Hieracium cynoglossoides [HIERCYN] Asteraceae (V:D)
Hieracium albiflorum Hook. [HIERALI] Asteraceae (V:D)
***Hieracium aurantiacum** L. [HIERAUR] Asteraceae (V:D)
Hieracium chapacanum Zahn = Hieracium scouleri [HIERSCO] Asteraceae (V:D)
Hieracium cusickii Gandog. = Hieracium cynoglossoides [HIERCYN] Asteraceae (V:D)
Hieracium cynoglossoides Arv.-Touv. [HIERCYN] Asteraceae (V:D)
Hieracium florentinum All. = Hieracium piloselloides [HIERPIO] Asteraceae (V:D)
Hieracium gracile Hook. [HIERGRA] Asteraceae (V:D)
Hieracium helleri Gandog. = Hieracium albiflorum [HIERALI] Asteraceae (V:D)
Hieracium lachenalii K.C. Gmel. [HIERLAC] Asteraceae (V:D)
***Hieracium murorum** L. [HIERMUR] Asteraceae (V:D)
Hieracium parryi Zahn = Hieracium scouleri [HIERSCO] Asteraceae (V:D)
***Hieracium pilosella** L. [HIERPIL] Asteraceae (V:D)
***Hieracium piloselloides** Vill. [HIERPIO] Asteraceae (V:D)
Hieracium scabriusculum Schwein. = Hieracium umbellatum [HIERUMB] Asteraceae (V:D)
Hieracium scabriusculum var. *perhirsutum* Lepage = Hieracium umbellatum [HIERUMB] Asteraceae (V:D)
Hieracium scabriusculum var. *saximontanum* Lepage = Hieracium umbellatum [HIERUMB] Asteraceae (V:D)
Hieracium scabriusculum var. *scabrum* (Schwein.) Lepage = Hieracium umbellatum [HIERUMB] Asteraceae (V:D)
Hieracium scouleri Hook. [HIERSCO] Asteraceae (V:D)
Hieracium scouleri var. *albertinum* (Farr) G.W. Douglas & G.A. Allen = Hieracium cynoglossoides [HIERCYN] Asteraceae (V:D)
Hieracium scouleri var. *griseum* (Rydb.) A. Nels. = Hieracium cynoglossoides [HIERCYN] Asteraceae (V:D)
Hieracium triste Willd. ex Spreng. [HIERTRI] Asteraceae (V:D)
Hieracium triste var. *gracile* (Hook.) Gray = Hieracium gracile [HIERGRA] Asteraceae (V:D)
***Hieracium umbellatum** L. [HIERUMB] Asteraceae (V:D)
Hieracium vulgatum Fries = Hieracium lachenalii [HIERLAC] Asteraceae (V:D)
Hierochloe R. Br. [HIEROCH$] Poaceae (V:M)
Hierochloe alpina (Sw. ex Willd.) Roemer & J.A. Schultes [HIERALP] Poaceae (V:M)
Hierochloe odorata (L.) Beauv. [HIERODO] Poaceae (V:M)
Hilpertia scotteri (Zand. & Steere) Zand. = Tortula scotteri [TORTSCO] Pottiaceae (B:M)
Hippochaete hyemalis (L.) Brubin = Equisetum hyemale var. affine [EQUIHYE2] Equisetaceae (V:F)
Hippochaete hyemalis ssp. *affinis* (Engelm.) W.A. Weber = Equisetum hyemale var. affine [EQUIHYE2] Equisetaceae (V:F)
Hippochaete laevigata (A. Braun) Farw. = Equisetum laevigatum [EQUILAE] Equisetaceae (V:F)
Hippochaete variegata (Schleich. ex F. Weber & D.M.H. Mohr) Bruhin = Equisetum variegatum var. variegatum [EQUIVAR0] Equisetaceae (V:F)
Hippuris L. [HIPPURI$] Hippuridaceae (V:D)
Hippuris montana Ledeb. [HIPPMON] Hippuridaceae (V:D)
Hippuris tetraphylla L. f. [HIPPTET] Hippuridaceae (V:D)
Hippuris vulgaris L. [HIPPVUL] Hippuridaceae (V:D)
Hirculus serpyllifolius (Pursh) W.A. Weber = Saxifraga serpyllifolia [SAXISER] Saxifragaceae (V:D)
Holcus L. [HOLCUS$] Poaceae (V:M)
***Holcus lanatus** L. [HOLCLAN] Poaceae (V:M)
***Holcus mollis** L. [HOLCMOL] Poaceae (V:M)
Holmgrenia chrysea (Schwaegr. in Schultes) Lindb. = Orthothecium chryseum [ORTHCHR] Hypnaceae (B:M)
Holmgrenia intricata (Hartm.) Lindb. = Orthothecium intricatum [ORTHINT] Hypnaceae (B:M)
Holmgrenia stricta (Lor.) Grout = Orthothecium strictum [ORTHSTR] Hypnaceae (B:M)
Holodiscus (K. Koch) Maxim. [HOLODIS$] Rosaceae (V:D)
Holodiscus boursieri (Carr.) Rehd. = Holodiscus discolor [HOLODIC] Rosaceae (V:D)
Holodiscus discolor (Pursh) Maxim. [HOLODIC] Rosaceae (V:D)
Holodiscus discolor ssp. *franciscanus* (Rydb.) Taylor & MacBryde = Holodiscus discolor [HOLODIC] Rosaceae (V:D)
Holodiscus discolor var. *delnortensis* Ley = Holodiscus discolor [HOLODIC] Rosaceae (V:D)
Holodiscus discolor var. *franciscanus* (Rydb.) Jepson = Holodiscus discolor [HOLODIC] Rosaceae (V:D)
Holodiscus discolor var. *glabrescens* (Greenm.) Jepson = Holodiscus discolor [HOLODIC] Rosaceae (V:D)
Holodiscus dumosus ssp. *saxicola* (Heller) Abrams = Holodiscus discolor [HOLODIC] Rosaceae (V:D)
Holodiscus dumosus var. *australis* (Heller) Ley = Holodiscus discolor [HOLODIC] Rosaceae (V:D)
Holodiscus dumosus var. *glabrescens* (Greenm.) C.L. Hichc. = Holodiscus discolor [HOLODIC] Rosaceae (V:D)
Holodiscus glabrescens (Greenm.) Heller ex Jepson = Holodiscus discolor [HOLODIC] Rosaceae (V:D)
Holodiscus microphyllus Rydb. = Holodiscus discolor [HOLODIC] Rosaceae (V:D)
Holodiscus microphyllus var. *glabrescens* (Greenm.) Ley = Holodiscus discolor [HOLODIC] Rosaceae (V:D)
Holodiscus microphyllus var. *sericeus* Ley = Holodiscus discolor [HOLODIC] Rosaceae (V:D)
Holodiscus microphyllus var. *typicus* Ley = Holodiscus discolor [HOLODIC] Rosaceae (V:D)
Holosteum L. [HOLOSTE$] Caryophyllaceae (V:D)
***Holosteum umbellatum** L. [HOLOUMB] Caryophyllaceae (V:D)
Homalia (Brid.) Schimp. in B.S.G. [HOMALIA$] Neckeraceae (B:M)
Homalia gracilis James in Peck = Homalia trichomanoides [HOMATRI] Neckeraceae (B:M)
Homalia jamesii Schimp. = Homalia trichomanoides [HOMATRI] Neckeraceae (B:M)
Homalia jamesii var. *gracilis* (James in Peck) Wagn. = Homalia trichomanoides [HOMATRI] Neckeraceae (B:M)
Homalia macounii C. Müll. & Kindb. in Mac. & Kindb. = Homalia trichomanoides [HOMATRI] Neckeraceae (B:M)
Homalia trichomanoides (Hedw.) Schimp. in B.S.G. [HOMATRI] Neckeraceae (B:M)
Homalobus tenellus (Pursh) Britt. = Astragalus tenellus [ASTRTEN] Fabaceae (V:D)
Homalocenchrus oryzoides (L.) Pollich = Leersia oryzoides [LEERORY] Poaceae (V:M)
Homalothecium Schimp. in B.S.G. [HOMALOH$] Brachytheciaceae (B:M)

Homalothecium aeneum (Mitt.) Lawt. [HOMAAEN] Brachytheciaceae (B:M)
Homalothecium arenarium (Lesq.) Lawt. [HOMAARE] Brachytheciaceae (B:M)
Homalothecium corticola Kindb. in Mac. & Kindb. = Homalothecium nevadense [HOMANEV] Brachytheciaceae (B:M)
Homalothecium fulgescens (Mitt. ex C. Müll.) Lawt. [HOMAFUL] Brachytheciaceae (B:M)
Homalothecium megaptilum (Sull.) Robins. = Trachybryum megaptilum [TRACMEG] Brachytheciaceae (B:M)
Homalothecium nevadense (Lesq.) Ren. & Card. [HOMANEV] Brachytheciaceae (B:M)
Homalothecium nevadense var. *subulatum* Ren. & Card. in Röll = Homalothecium nevadense [HOMANEV] Brachytheciaceae (B:M)
Homalothecium nitens (Hedw.) Robins. = Tomentypnum nitens [TOMENIT] Brachytheciaceae (B:M)
Homalothecium nitens var. *falcifolium* (Ren. ex Nichols) Wijk & Marg. = Tomentypnum falcifolium [TOMEFAL] Brachytheciaceae (B:M)
Homalothecium nuttallii (Wils.) Jaeg. [HOMANUT] Brachytheciaceae (B:M)
Homalothecium nuttallii var. *hamatidens* (Kindb.) Grout = Homalothecium nuttallii [HOMANUT] Brachytheciaceae (B:M)
Homalothecium nuttallii var. *stoloniferum* (Lesq.) L. Koch = Homalothecium nuttallii [HOMANUT] Brachytheciaceae (B:M)
Homalothecium nuttallii var. *tenue* (Kindb.) Grout = Homalothecium nuttallii [HOMANUT] Brachytheciaceae (B:M)
Homalothecium pinnatifidum (Sull. & Lesq.) Lawt. [HOMAPIN] Brachytheciaceae (B:M)
Homalothecium pseudosericeum (C. Müll.) Jaeg. & Sauerb. = Homalothecium fulgescens [HOMAFUL] Brachytheciaceae (B:M)
Homalothecium sericeoides C. Müll. & Kindb. in Mac. & Kindb. = Homalothecium nevadense [HOMANEV] Brachytheciaceae (B:M)
Honkenya Ehrh. [HONKENY$] Caryophyllaceae (V:D)
Honkenya oblongifolia Torr. & Gray = Honkenya peploides ssp. major [HONKPEP1] Caryophyllaceae (V:D)
Honkenya peploides (L.) Ehrh. [HONKPEP] Caryophyllaceae (V:D)
Honkenya peploides ssp. **major** (Hook.) Hultén [HONKPEP1] Caryophyllaceae (V:D)
Honkenya peploides var. *major* (Hook.) Abrams = Honkenya peploides ssp. major [HONKPEP1] Caryophyllaceae (V:D)
Hookeria Sm. [HOOKERI$] Hookeriaceae (B:M)
Hookeria acutifolia Hook. & Grev. [HOOKACU] Hookeriaceae (B:M)
Hookeria lucens (Hedw.) Sm. [HOOKLUC] Hookeriaceae (B:M)
Hordeum L. [HORDEUM$] Poaceae (V:M)
Hordeum aegiceras Nees ex Royle = Hordeum vulgare [HORDVUL] Poaceae (V:M)
Hordeum boreale Scribn. & J.G. Sm. = Hordeum brachyantherum [HORDBRA] Poaceae (V:M)
Hordeum brachyantherum Nevski [HORDBRA] Poaceae (V:M)
Hordeum caespitosum Scribn. ex Pammel = Hordeum jubatum ssp. jubatum [HORDJUB0] Poaceae (V:M)
Hordeum × *caespitosum* Scribn. ex Pammel = Hordeum jubatum ssp. jubatum [HORDJUB0] Poaceae (V:M)
Hordeum distichon L. = Hordeum vulgare [HORDVUL] Poaceae (V:M)
Hordeum geniculatum All. = Hordeum marinum ssp. gussonianum [HORDMAR1] Poaceae (V:M)
Hordeum gussonianum Parl. = Hordeum marinum ssp. gussonianum [HORDMAR1] Poaceae (V:M)
Hordeum hexastichum L. = Hordeum vulgare [HORDVUL] Poaceae (V:M)
Hordeum hystrix Roth = Hordeum marinum ssp. gussonianum [HORDMAR1] Poaceae (V:M)
Hordeum jubatum L. [HORDJUB] Poaceae (V:M)
Hordeum jubatum ssp. **breviaristatum** Bowden [HORDJUB1] Poaceae (V:M)
Hordeum jubatum ssp. × **intermedium** Bowden [HORDJUB2] Poaceae (V:M)
Hordeum jubatum ssp. **jubatum** [HORDJUB0] Poaceae (V:M)
Hordeum jubatum var. *boreale* (Scribn. & J.G. Sm.) Boivin = Hordeum brachyantherum [HORDBRA] Poaceae (V:M)
Hordeum jubatum var. *breviaristatum* (Nevski) Bowden = Hordeum jubatum ssp. breviaristatum [HORDJUB1] Poaceae (V:M)
Hordeum jubatum var. *caespitosum* (Scribn. ex Pammel) A.S. Hitchc. = Hordeum jubatum ssp. jubatum [HORDJUB0] Poaceae (V:M)
Hordeum leporinum Link = Hordeum murinum ssp. leporinum [HORDMUR1] Poaceae (V:M)
***Hordeum marinum** Huds. [HORDMAR] Poaceae (V:M)
***Hordeum marinum** ssp. **gussonianum** (Parl.) Thellung [HORDMAR1] Poaceae (V:M)
***Hordeum murinum** L. [HORDMUR] Poaceae (V:M)
***Hordeum murinum** ssp. **leporinum** (Link) Arcang. [HORDMUR1] Poaceae (V:M)
***Hordeum murinum** ssp. **murinum** [HORDMUR0] Poaceae (V:M)
Hordeum nodosum L. = Hordeum brachyantherum [HORDBRA] Poaceae (V:M)
Hordeum nodosum var. *boreale* (Scribn. & J.G. Sm.) A.S. Hitchc. = Hordeum brachyantherum [HORDBRA] Poaceae (V:M)
***Hordeum vulgare** L. [HORDVUL] Poaceae (V:M)
Hordeum vulgare var. *trifurcatum* (Schlecht.) Alef. = Hordeum vulgare [HORDVUL] Poaceae (V:M)
Hosackia americana (Nutt.) Piper = Lotus unifoliolatus [LOTUUNI] Fabaceae (V:D)
Hosackia decumbens Benth. = Lotus nevadensis var. douglasii [LOTUNEV1] Fabaceae (V:D)
Hosackia denticulata E. Drew = Lotus denticulatus [LOTUDEN] Fabaceae (V:D)
Hosackia gracilis Benth. = Lotus formosissimus [LOTUFOR] Fabaceae (V:D)
Hosackia parviflora Benth. = Lotus micranthus [LOTUMIC] Fabaceae (V:D)
Hosackia pinnata (Hook.) Abrams = Lotus pinnatus [LOTUPIN] Fabaceae (V:D)
Huilia albocaerulescens (Wulfen) Hertel = Porpidia albocaerulescens [PORPALB] Porpidiaceae (L)
Huilia consentiens (Nyl.) Hertel = Amygdalaria consentiens [AMYGCON] Porpidiaceae (L)
Huilia crustulata (Ach.) Hertel = Porpidia crustulata [PORPCRU] Porpidiaceae (L)
Huilia macrocarpa (DC.) Hertel = Porpidia macrocarpa [PORPMAC] Porpidiaceae (L)
Huilia nigrocruenta (Anzi) Hertel = Porpidia macrocarpa [PORPMAC] Porpidiaceae (L)
Huilia panaeola (Ach.) Hertel = Amygdalaria panaeola [AMYGPAN] Porpidiaceae (L)
Huilia soredizodes (Lamy ex Nyl.) Hertel = Porpidia soredizodes [PORPSOR] Porpidiaceae (L)
Huilia tuberculosa (Sm.) P. James = Porpidia tuberculosa [PORPTUB] Porpidiaceae (L)
Humulus L. [HUMULUS$] Cannabaceae (V:D)
***Humulus lupulus** L. [HUMULUP] Cannabaceae (V:D)
Huperzia Bernh. [HUPERZI$] Lycopodiaceae (V:F)
Huperzia chinensis (Christ) Czern. [HUPECHI] Lycopodiaceae (V:F)
Huperzia haleakalae (Brack.) Holub [HUPEHAL] Lycopodiaceae (V:F)
Huperzia miyoshiana (Makino) Ching = Huperzia chinensis [HUPECHI] Lycopodiaceae (V:F)
Huperzia occidentalis (Clute) Kartesz & Gandhi [HUPEOCC] Lycopodiaceae (V:F)
Huperzia selago ssp. *chinensis* (Christ) A. & D. Löve = Huperzia chinensis [HUPECHI] Lycopodiaceae (V:F)
Huperzia selago var. *miyoshiana* (Makino) Taylor & MacBryde = Huperzia chinensis [HUPECHI] Lycopodiaceae (V:F)
Husnotiella pringlei (Card.) Grout = Didymodon tophaceus [DIDYTOP] Pottiaceae (B:M)
Husnotiella revoluta Card. = Didymodon revolutus [DIDYREV] Pottiaceae (B:M)
Husnotiella revoluta var. *palmeri* (Card.) Bartr. = Didymodon revolutus [DIDYREV] Pottiaceae (B:M)
Husnotiella torquescens (Card.) Bartr. = Trichostomopsis australasiae [TRICAUS] Pottiaceae (B:M)
Hutchinsia Ait. f. [HUTCHIN$] Brassicaceae (V:D)
Hutchinsia procumbens (L.) Desv. [HUTCPRO] Brassicaceae (V:D)
Hyacinthoides Medik. [HYACINT$] Liliaceae (V:M)

•**Hyacinthoides hispanica** (P. Mill.) Rothm. [HYACHIS] Liliaceae (V:M)
Hydastylus borealis Bickn. = Sisyrinchium californicum [SISYCAL] Iridaceae (V:M)
Hydastylus brachypus Bickn. = Sisyrinchium californicum [SISYCAL] Iridaceae (V:M)
Hydastylus californicus (Ker-Gawl. ex Sims) Salisb. = Sisyrinchium californicum [SISYCAL] Iridaceae (V:M)
Hydrocotyle L. [HYDROCO$] Apiaceae (V:D)
Hydrocotyle ranunculoides L. f. [HYDRRAN] Apiaceae (V:D)
Hydrocotyle verticillata Thunb. [HYDRVER] Apiaceae (V:D)
Hydrogrimmia mollis (Bruch & Schimp. in B.S.G.) Loeske = Grimmia mollis [GRIMMOL] Grimmiaceae (B:M)
Hydrophila aquatica (L.) House = Crassula aquatica [CRASAQU] Crassulaceae (V:D)
Hydrophyllum L. [HYDROPH$] Hydrophyllaceae (V:D)
Hydrophyllum capitatum Dougl. ex Benth. [HYDRCAP] Hydrophyllaceae (V:D)
Hydrophyllum fendleri (Gray) Heller [HYDRFEN] Hydrophyllaceae (V:D)
Hydrophyllum fendleri var. **albifrons** (Heller) J.F. Macbr. [HYDRFEN1] Hydrophyllaceae (V:D)
Hydrophyllum tenuipes Heller [HYDRTEN] Hydrophyllaceae (V:D)
Hydrothyria J.L. Russell [HYDROTH$] Peltigeraceae (L)
Hydrothyria venosa J.L. Russell [HYDRVEN] Peltigeraceae (L)
Hygroamblystegium Loeske [HYGROAM$] Amblystegiaceae (B:M)
Hygroamblystegium filicinum (Hedw.) Loeske = Cratoneuron filicinum [CRATFIL] Amblystegiaceae (B:M)
Hygroamblystegium fluviatile (Hedw.) Loeske [HYGRFLU] Amblystegiaceae (B:M)
Hygroamblystegium fluviatile var. *ovatum* Grout = Hygroamblystegium fluviatile [HYGRFLU] Amblystegiaceae (B:M)
Hygroamblystegium irriguum f. *marianopolitana* Dupret = Hygroamblystegium tenax [HYGRTEN] Amblystegiaceae (B:M)
Hygroamblystegium noterophilum (Sull. & Lesq. in Sull.) Warnst. [HYGRNOT] Amblystegiaceae (B:M)
Hygroamblystegium tenax (Hedw.) Jenn. [HYGRTEN] Amblystegiaceae (B:M)
Hygroamblystegium varium (Hedw.) Mönk. = Amblystegium varium [AMBLVAR] Amblystegiaceae (B:M)
Hygrobiella Spruce [HYGROBI$] Cephaloziaceae (B:H)
Hygrobiella laxifolia (Hook.) Spruce [HYGRLAX] Cephaloziaceae (B:H)
Hygrohypnum Lindb. [HYGROHY$] Amblystegiaceae (B:M)
Hygrohypnum alpestre (Hedw.) Loeske [HYGRALP] Amblystegiaceae (B:M)
Hygrohypnum alpinum (Lindb.) Loeske [HYGRALI] Amblystegiaceae (B:M)
Hygrohypnum arcticum (Somm.) Loeske = Hygrohypnum smithii [HYGRSMI] Amblystegiaceae (B:M)
Hygrohypnum bestii (Ren. & Bryhn in Ren.) Broth. [HYGRBES] Amblystegiaceae (B:M)
Hygrohypnum cochlearifolium (Vent. ex De Not.) Broth. [HYGRCOC] Amblystegiaceae (B:M)
Hygrohypnum dilitatum (Wils.) Loeske = Hygrohypnum molle [HYGRMOL] Amblystegiaceae (B:M)
Hygrohypnum duriusculum (De Not.) Jamieson [HYGRDUR] Amblystegiaceae (B:M)
Hygrohypnum luridum (Hedw.) Jenn. [HYGRLUR] Amblystegiaceae (B:M)
Hygrohypnum luridum ssp. *pseudomontanum* (Kindb.) Wijk & Marg. = Hygrohypnum luridum [HYGRLUR] Amblystegiaceae (B:M)
Hygrohypnum luridum var. *ehlei* (Arnell) Wijk & Marg. = Hygrohypnum luridum [HYGRLUR] Amblystegiaceae (B:M)
Hygrohypnum luridum var. *julaceum* (Schimp. in B.S.G.) Podp. = Hygrohypnum luridum [HYGRLUR] Amblystegiaceae (B:M)
Hygrohypnum luridum var. *subsphaericarpon* (Brid.) C. Jens. in Podp. = Hygrohypnum luridum [HYGRLUR] Amblystegiaceae (B:M)
Hygrohypnum micans (Mitt.) Broth. [HYGRMIC] Amblystegiaceae (B:M)
Hygrohypnum molle (Hedw.) Loeske [HYGRMOL] Amblystegiaceae (B:M)
Hygrohypnum molle var. *bestii* (Ren. & Bryhn) Hab. = Hygrohypnum bestii [HYGRBES] Amblystegiaceae (B:M)
Hygrohypnum norvegicum (Schimp. in B.S.G.) Amann [HYGRNOR] Amblystegiaceae (B:M)
Hygrohypnum novae-caesareae (Aust.) Grout = Hygrohypnum micans [HYGRMIC] Amblystegiaceae (B:M)
Hygrohypnum ochraceum (Turn. ex Wils.) Loeske [HYGROCH] Amblystegiaceae (B:M)
Hygrohypnum ochraceum var. *filiforme* (Limpr.) Amann = Hygrohypnum ochraceum [HYGROCH] Amblystegiaceae (B:M)
Hygrohypnum ochraceum var. *flaccidum* (Milde) Amann = Hygrohypnum ochraceum [HYGROCH] Amblystegiaceae (B:M)
Hygrohypnum ochraceum var. *uncinatum* (Milde) Loeske = Hygrohypnum ochraceum [HYGROCH] Amblystegiaceae (B:M)
Hygrohypnum palustre Loeske = Hygrohypnum luridum [HYGRLUR] Amblystegiaceae (B:M)
Hygrohypnum palustre var. *ehlei* (Arnell) Grout = Hygrohypnum luridum [HYGRLUR] Amblystegiaceae (B:M)
Hygrohypnum palustre var. *julaceum* (Schimp. in B.S.G.) Loeske = Hygrohypnum luridum [HYGRLUR] Amblystegiaceae (B:M)
Hygrohypnum palustre var. *subsphaericarpon* (Brid.) Loeske = Hygrohypnum luridum [HYGRLUR] Amblystegiaceae (B:M)
Hygrohypnum polare (Lindb.) Loeske [HYGRPOL] Amblystegiaceae (B:M)
Hygrohypnum polare var. *falcatum* (Bryhn) Broth. = Hygrohypnum polare [HYGRPOL] Amblystegiaceae (B:M)
Hygrohypnum pseudomontanum (Kindb.) Grout = Hygrohypnum luridum [HYGRLUR] Amblystegiaceae (B:M)
Hygrohypnum smithii (Sw. in Lilj.) Broth. [HYGRSMI] Amblystegiaceae (B:M)
Hygrohypnum smithii var. *goulardii* (Schimp.) Wijk & Marg. = Hygrohypnum molle [HYGRMOL] Amblystegiaceae (B:M)
Hygrohypnum styriacum (Limpr.) Broth. [HYGRSTY] Amblystegiaceae (B:M)
Hygrohypnum subeugyrium var. *occidentale* (Card. & Thér.) Grout = Hygrohypnum luridum [HYGRLUR] Amblystegiaceae (B:M)
Hylocomiastrum Fleisch. [HYLOCOI$] Hylocomiaceae (B:M)
Hylocomiastrum pyrenaicum (Spruce) Fleisch. in Broth. [HYLOPYE] Hylocomiaceae (B:M)
Hylocomiastrum umbratum (Hedw.) Fleisch. in Broth. [HYLOUMR] Hylocomiaceae (B:M)
Hylocomium Schimp. in B.S.G. [HYLOCOM$] Hylocomiaceae (B:M)
Hylocomium alaskanum (Lesq. & James) Aust. = Hylocomium splendens [HYLOSPL] Hylocomiaceae (B:M)
Hylocomium calvescens Lindb. = Rhytidiadelphus squarrosus [RHYTSQU] Hylocomiaceae (B:M)
Hylocomium giganteum Perss., nom. nud., non Bartr. = Hylocomium splendens [HYLOSPL] Hylocomiaceae (B:M)
Hylocomium loreum (Hedw.) B.S.G. = Rhytidiadelphus loreus [RHYTLOR] Hylocomiaceae (B:M)
Hylocomium proliferum (Brid.) Lindb. = Hylocomium splendens [HYLOSPL] Hylocomiaceae (B:M)
Hylocomium pyrenaicum (Spruce) Lindb. = Hylocomiastrum pyrenaicum [HYLOPYE] Hylocomiaceae (B:M)
Hylocomium robustum (Hook.) Kindb. = Rhytidiopsis robusta [RHYTROB] Hylocomiaceae (B:M)
Hylocomium splendens (Hedw.) Schimp. in B.S.G. [HYLOSPL] Hylocomiaceae (B:M)
Hylocomium splendens var. *alaskanum* (Lesq. & James) Limpr. = Hylocomium splendens [HYLOSPL] Hylocomiaceae (B:M)
Hylocomium splendens var. *compactum* (Lesq. & James) Mac. & Kindb. = Hylocomium splendens [HYLOSPL] Hylocomiaceae (B:M)
Hylocomium splendens var. *gracilius* (Boul.) Husn. = Hylocomium splendens [HYLOSPL] Hylocomiaceae (B:M)
Hylocomium splendens var. *obtusifolium* (Geh.) Par. = Hylocomium splendens [HYLOSPL] Hylocomiaceae (B:M)
Hylocomium squarrosum ssp. *calvescens* Kindb. = Rhytidiadelphus squarrosus [RHYTSQU] Hylocomiaceae (B:M)

Hylocomium squarrosum var. *calvescens* (Kindb.) Hobk. = Rhytidiadelphus squarrosus [RHYTSQU] Hylocomiaceae (B:M)
Hylocomium triquetrum (Hedw.) B.S.G. = Rhytidiadelphus triquetrus [RHYTTRI] Hylocomiaceae (B:M)
Hylocomium umbratum (Hedw.) Schimp. in B.S.G. = Hylocomiastrum umbratum [HYLOUMR] Hylocomiaceae (B:M)
Hymenelia Kremp. [HYMENEL$] Hymeneliaceae (L)
Hymenelia epulotica (Ach.) Lutzoni [HYMEEPU] Hymeneliaceae (L)
Hymenelia heteromorpha (Kremp.) Lutzoni [HYMEHET] Hymeneliaceae (L)
Hymenelia lacustris (With.) Choisy = Ionaspis lacustris [IONALAC] Hymeneliaceae (L)
Hymenelia melanocarpa (Kremp.) Arnold [HYMEMEL] Hymeneliaceae (L)
Hymenelia ochrolemma (Vainio) Gowan & Ahti = Porpidia ochrolemma [PORPOCH] Porpidiaceae (L)
Hymenelia prevostii (Duby) Kremp. = Hymenelia epulotica [HYMEEPU] Hymeneliaceae (L)
Hymenolobus procumbens (L.) Nutt. ex Schinz & Thellung = Hutchinsia procumbens [HUTCPRO] Brassicaceae (V:D)
Hymenophyllum Sm. [HYMENOP$] Hymenophyllaceae (V:F)
Hymenophyllum wrightii Bosch [HYMEWRI] Hymenophyllaceae (V:F)
Hymenophysa pubescens C.A. Mey. = Cardaria pubescens [CARDPUB] Brassicaceae (V:D)
Hymenostomum microstomum (Hedw.) R. Br. = Weissia hedwigii [WEISHED] Pottiaceae (B:M)
Hymenostylium Brid. [HYMENOS$] Pottiaceae (B:M)
Hymenostylium curvirostre Mitt. = Hymenostylium recurvirostre [HYMEREC] Pottiaceae (B:M)
Hymenostylium insigne (Dix.) Podp. [HYMEINS] Pottiaceae (B:M)
Hymenostylium recurvirostre (Hedw.) Dix. [HYMEREC] Pottiaceae (B:M)
Hymenostylium recurvirostre var. *insigne* (Dix.) Bartr. = Hymenostylium insigne [HYMEINS] Pottiaceae (B:M)
Hymenostylium recurvirostre var. *latifolium* (Zett.) Wijk & Marg. = Hymenostylium recurvirostre [HYMEREC] Pottiaceae (B:M)
Hymenostylium scabrum (Lindb.) Loeske = Hymenostylium recurvirostre [HYMEREC] Pottiaceae (B:M)
Hyoseris virginica L. = Krigia virginica [KRIGVIR] Asteraceae (V:D)
Hypericum L. [HYPERIC$] Clusiaceae (V:D)
Hypericum anagalloides Cham. & Schlecht. [HYPEANA] Clusiaceae (V:D)
Hypericum canadense var. *majus* Gray = Hypericum majus [HYPEMAJ] Clusiaceae (V:D)
Hypericum formosum auct. = Hypericum scouleri [HYPESCO] Clusiaceae (V:D)
Hypericum formosum ssp. *scouleri* (Hook.) C.L. Hitchc. = Hypericum scouleri ssp. scouleri [HYPESCO0] Clusiaceae (V:D)
Hypericum formosum var. *nortoniae* (M.E. Jones) C.L. Hitchc. = Hypericum scouleri ssp. nortoniae [HYPESCO4] Clusiaceae (V:D)
Hypericum formosum var. *scouleri* (Hook.) Coult. = Hypericum scouleri ssp. scouleri [HYPESCO0] Clusiaceae (V:D)
Hypericum majus (Gray) Britt. [HYPEMAJ] Clusiaceae (V:D)
Hypericum nortoniae M.E. Jones = Hypericum scouleri ssp. nortoniae [HYPESCO4] Clusiaceae (V:D)
***Hypericum perforatum** L. [HYPEPER] Clusiaceae (V:D)
Hypericum scouleri Hook. [HYPESCO] Clusiaceae (V:D)
Hypericum scouleri ssp. **nortoniae** (M.E. Jones) C.L. Hitchc. [HYPESCO4] Clusiaceae (V:D)
Hypericum scouleri ssp. **scouleri** [HYPESCO0] Clusiaceae (V:D)
Hypnum Hedw. [HYPNUM$] Hypnaceae (B:M)
Hypnum abietinum Hedw. = Abietinella abietina [ABIEABI] Thuidiaceae (B:M)
Hypnum aduncum Hedw. = Drepanocladus aduncus [DREPADU] Amblystegiaceae (B:M)
Hypnum aduncum var. *aquaticum* (Sanio) Sanio = Drepanocladus aduncus [DREPADU] Amblystegiaceae (B:M)
Hypnum aduncum var. *platyphyllum* (Kindb.) Par. = Drepanocladus aduncus [DREPADU] Amblystegiaceae (B:M)
Hypnum alaskae Kindb. = Hypnum callichroum [HYPNCAL] Hypnaceae (B:M)
Hypnum albicans Neck. ex Hedw. = Brachythecium albicans [BRACALB] Brachytheciaceae (B:M)
Hypnum alpestre Sw. ex Hedw. = Hygrohypnum alpestre [HYGRALP] Amblystegiaceae (B:M)
Hypnum amblyphyllum Williams = Warnstorfia tundrae [WARNTUN] Amblystegiaceae (B:M)
Hypnum arcticum Somm. in Wahlenb. = Hygrohypnum smithii [HYGRSMI] Amblystegiaceae (B:M)
Hypnum arcuatiforme Kindb. in Mac. & Kindb. = Hypnum lindbergii [HYPNLIN] Hypnaceae (B:M)
Hypnum arcuatum Lindb. = Hypnum lindbergii [HYPNLIN] Hypnaceae (B:M)
Hypnum arcuatum var. *americanum* Ren. & Card. = Hypnum lindbergii [HYPNLIN] Hypnaceae (B:M)
Hypnum arcuatum var. *demissum* Schimp. = Hypnum lindbergii [HYPNLIN] Hypnaceae (B:M)
Hypnum arcuatum var. *elatum* Schimp. = Hypnum lindbergii [HYPNLIN] Hypnaceae (B:M)
Hypnum atrovirens Dicks. = Pseudoleskea patens [PSEUPAT] Leskeaceae (B:M)
Hypnum badium C.J. Hartm. = Loeskypnum badium [LOESBAD] Amblystegiaceae (B:M)
Hypnum bambergeri Schimp. [HYPNBAM] Hypnaceae (B:M)
Hypnum bergenense Aust. = Campylium radicale [CAMPRAD] Amblystegiaceae (B:M)
Hypnum blandowii Web. & Mohr = Helodium blandowii [HELOBLA] Helodiaceae (B:M)
Hypnum bridelianum Crum, Steere & Anderson = Hypnum recurvatum [HYPNREC] Hypnaceae (B:M)
Hypnum byssirameum C. Müll. & Kindb. in Mac. & Kindb. = Campylium hispidulum [CAMPHIS] Amblystegiaceae (B:M)
Hypnum callichroum Funck ex Brid. [HYPNCAL] Hypnaceae (B:M)
Hypnum campestre (C. Müll.) Mitt. = Brachythecium campestre [BRACCAM] Brachytheciaceae (B:M)
Hypnum canadense Kindb. = Hypnum dieckii [HYPNDIE] Hypnaceae (B:M)
Hypnum capillifolium Warnst. = Drepanocladus capillifolius [DREPCAP] Amblystegiaceae (B:M)
Hypnum chrysophyllum Brid. = Campylium chrysophyllum [CAMPCHR] Amblystegiaceae (B:M)
Hypnum chrysophyllum var. *brevifolium* Ren. & Card. = Campylium chrysophyllum [CAMPCHR] Amblystegiaceae (B:M)
Hypnum circinale Hook. [HYPNCIR] Hypnaceae (B:M)
Hypnum circulifolium C. Müll. & Kindb. in Mac. & Kindb. = Hygrohypnum molle [HYGRMOL] Amblystegiaceae (B:M)
Hypnum columbicopalustre C. Müll. & Kindb. in Mac. & Kindb. = Hygrohypnum luridum [HYGRLUR] Amblystegiaceae (B:M)
Hypnum commutatum Hedw. = Palustriella commutata [PALUCOM] Amblystegiaceae (B:M)
Hypnum commutatum ssp. *sulcatum* (Lindb.) Boul. = Palustriella commutata [PALUCOM] Amblystegiaceae (B:M)
Hypnum compactum (Hook.) C. Müll. = Conardia compacta [CONACOM] Amblystegiaceae (B:M)
Hypnum complexum (Mitt.) Jaeg. & Sauerb. = Hypnum cupressiforme [HYPNCUP] Hypnaceae (B:M)
Hypnum conflatum C. Müll. & Kindb. in Mac. & Kindb. = Drepanocladus aduncus [DREPADU] Amblystegiaceae (B:M)
Hypnum cordifolium Hedw. = Calliergon cordifolium [CALLCOR] Amblystegiaceae (B:M)
Hypnum cossonii Schimp. = Limprichtia revolvens [LIMPREV] Amblystegiaceae (B:M)
Hypnum crista-castrensis Hedw. = Ptilium crista-castrensis [PTILCRI] Hypnaceae (B:M)
Hypnum cupressiforme Hedw. [HYPNCUP] Hypnaceae (B:M)
Hypnum cupressiforme var. *brevisetum* Schimp. = Hypnum cupressiforme var. cupressiforme [HYPNCUP0] Hypnaceae (B:M)
Hypnum cupressiforme var. **cupressiforme** [HYPNCUP0] Hypnaceae (B:M)
Hypnum cupressiforme var. **filiforme** Brid. [HYPNCUP2] Hypnaceae (B:M)
Hypnum cupressiforme var. **lacunosum** Brid. [HYPNCUP1] Hypnaceae (B:M)

Hypnum cupressiforme var. **resupinatum** (Tayl.) Schimp. in Spruce [HYPNCUP3] Hypnaceae (B:M)
Hypnum cuspidatum Hedw. = Calliergonella cuspidata [CALLCUS] Amblystegiaceae (B:M)
Hypnum decursivulum C. Müll. & Kindb. in Mac. & Kindb. = Campylium radicale [CAMPRAD] Amblystegiaceae (B:M)
Hypnum denticulatum Hedw. = Plagiothecium denticulatum [PLAGDEN] Plagiotheciaceae (B:M)
Hypnum dieckii Ren. & Card. in Röll [HYPNDIE] Hypnaceae (B:M)
Hypnum dilatatum Wils. ex Schimp. = Hygrohypnum molle [HYGRMOL] Amblystegiaceae (B:M)
Hypnum dimorphum Brid. = Heterocladium dimorphum [HETEDIM] Pterigynandraceae (B:M)
Hypnum exannulatum B.S.G. = Warnstorfia exannulata var. exannulata [WARNEXA0] Amblystegiaceae (B:M)
Hypnum exannulatum ssp. *pseudolycopodioides* Kindb. = Hygrohypnum ochraceum [HYGROCH] Amblystegiaceae (B:M)
Hypnum exannulatum ssp. *pseudostramineum* (C. Müll.) Kindb. = Warnstorfia pseudostraminea [WARNPSE] Amblystegiaceae (B:M)
Hypnum exannulatum var. *purpurascens* (Schimp.) Milde = Warnstorfia exannulata var. exannulata [WARNEXA0] Amblystegiaceae (B:M)
Hypnum falcatum Brid. = Palustriella commutata [PALUCOM] Amblystegiaceae (B:M)
Hypnum falcatum var. *microphyllum* Kindb. in Mac. & Kindb. = Drepanocladus sendtneri [DREPSEN] Amblystegiaceae (B:M)
Hypnum fastigiatum Brid. = Hypnum recurvatum [HYPNREC] Hypnaceae (B:M)
Hypnum filicinum Hedw. = Cratoneuron filicinum [CRATFIL] Amblystegiaceae (B:M)
Hypnum filicinum var. *aciculinum* C. Müll. & Kindb. in Mac. & Kindb. = Cratoneuron filicinum [CRATFIL] Amblystegiaceae (B:M)
Hypnum filicinum var. *filiforme* Berggr. = Cratoneuron filicinum [CRATFIL] Amblystegiaceae (B:M)
Hypnum flaccum C. Müll. & Kindb. in Mac. & Kindb. = Callicladium haldanianum [CALLHAL] Hypnaceae (B:M)
Hypnum fluitans Hedw. = Warnstorfia fluitans [WARNFLU] Amblystegiaceae (B:M)
Hypnum fluitans var. *falcifolius* f. *viridis* Boul. = Warnstorfia exannulata var. exannulata [WARNEXA0] Amblystegiaceae (B:M)
Hypnum fluitans var. *gracilis* Boul. in Ren. = Warnstorfia fluitans [WARNFLU] Amblystegiaceae (B:M)
Hypnum fluitans var. *jeanbernatii* Ren. = Warnstorfia fluitans [WARNFLU] Amblystegiaceae (B:M)
Hypnum fluviatile Hedw. = Hygroamblystegium fluviatile [HYGRFLU] Amblystegiaceae (B:M)
Hypnum geminum (Mitt.) Lesq. & James [HYPNGEM] Hypnaceae (B:M)
Hypnum giganteum Schimp. = Calliergon giganteum [CALLGIG] Amblystegiaceae (B:M)
Hypnum giganteum var. *fluitans* (Rabenh.) Heug. = Calliergon giganteum [CALLGIG] Amblystegiaceae (B:M)
Hypnum goulardii Schimp. = Hygrohypnum smithii [HYGRSMI] Amblystegiaceae (B:M)
Hypnum gracile Bruch & Schimp. in Hook. = Bryohaplocladium microphyllum [BRYOMIC] Leskeaceae (B:M)
Hypnum haldanianum Grev. = Callicladium haldanianum [CALLHAL] Hypnaceae (B:M)
Hypnum hamulosum Schimp. in B.S.G. [HYPNHAM] Hypnaceae (B:M)
Hypnum hispidulum Brid. = Campylium hispidulum [CAMPHIS] Amblystegiaceae (B:M)
Hypnum implexum Ren. & Card. = Hypnum plicatulum [HYPNPLI] Hypnaceae (B:M)
Hypnum julaceum (Schwaegr.) Vill. ex Schwaegr. = Myurella julacea [MYURJUL] Pterigynandraceae (B:M)
Hypnum kneiffii (B.S.G.) Schimp. in Wils. = Drepanocladus aduncus var. kneiffii [DREPADU3] Amblystegiaceae (B:M)
Hypnum kneiffii var. *platyphyllum* Kindb. in Mac. & Kindb. = Drepanocladus aduncus [DREPADU] Amblystegiaceae (B:M)
Hypnum lindbergii Mitt. [HYPNLIN] Hypnaceae (B:M)
Hypnum lindbergii var. *americanum* (Ren. & Card.) E. Whiteh. = Hypnum lindbergii [HYPNLIN] Hypnaceae (B:M)
Hypnum lindbergii var. *demissum* (Schimp.) Loeske = Hypnum lindbergii [HYPNLIN] Hypnaceae (B:M)
Hypnum lindbergii var. *elatum* (Schimp.) Williams = Hypnum lindbergii [HYPNLIN] Hypnaceae (B:M)
Hypnum longinerve (Kindb.) Kindb. = Warnstorfia exannulata var. exannulata [WARNEXA0] Amblystegiaceae (B:M)
Hypnum molluscoides Kindb. = Hypnum plicatulum [HYPNPLI] Hypnaceae (B:M)
Hypnum molluscum var. *molluscoides* (Par.) Grout = Hypnum plicatulum [HYPNPLI] Hypnaceae (B:M)
Hypnum moseri Kindb. = Sanionia uncinata [SANIUNC] Amblystegiaceae (B:M)
Hypnum muehlenbeckii Schimp. ex Hartm. = Herzogiella striatella [HERZSTR] Hypnaceae (B:M)
Hypnum myosuroides Brid. = Isothecium myosuroides [ISOTMYO] Brachytheciaceae (B:M)
Hypnum nitens Hedw. = Tomentypnum nitens [TOMENIT] Brachytheciaceae (B:M)
Hypnum nitidulum Wahlenb. = Isopterygiopsis pulchella [ISOPPUC] Hypnaceae (B:M)
Hypnum oakesii Sull. in Gray = Hylocomiastrum pyrenaicum [HYLOPYE] Hylocomiaceae (B:M)
Hypnum ochraceum Turn. ex Wils. = Hygrohypnum ochraceum [HYGROCH] Amblystegiaceae (B:M)
Hypnum oreganum Sull. = Eurhynchium oreganum [EURHORE] Brachytheciaceae (B:M)
Hypnum orthothecioides (Lindb.) Par. = Sanionia uncinata [SANIUNC] Amblystegiaceae (B:M)
Hypnum pallescens (Hedw.) P. Beauv. [HYPNPAL] Hypnaceae (B:M)
Hypnum palustre Brid. = Hygrohypnum luridum [HYGRLUR] Amblystegiaceae (B:M)
Hypnum passaicense (Aust.) Lesq. & James = Isopterygiopsis pulchella [ISOPPUC] Hypnaceae (B:M)
Hypnum patientiae Lindb. ex Milde = Hypnum lindbergii [HYPNLIN] Hypnaceae (B:M)
Hypnum patientiae var. *americanum* (Ren. & Card.) Ren. & Card. = Hypnum lindbergii [HYPNLIN] Hypnaceae (B:M)
Hypnum patientiae var. *demissum* (Schimp.) Jaeg. = Hypnum lindbergii [HYPNLIN] Hypnaceae (B:M)
Hypnum patientiae var. *elatum* (Schimp.) Jaeg. = Hypnum lindbergii [HYPNLIN] Hypnaceae (B:M)
Hypnum plicatile (Mitt.) Lesq. & James = Hypnum revolutum [HYPNREV] Hypnaceae (B:M)
Hypnum plicatulum (Lindb.) Jaeg. [HYPNPLI] Hypnaceae (B:M)
Hypnum plumosum Hedw. = Brachythecium plumosum [BRACPLU] Brachytheciaceae (B:M)
Hypnum polare Lindb. = Hygrohypnum polare [HYGRPOL] Amblystegiaceae (B:M)
Hypnum polygamum (B.S.G.) Wils. = Campylium polygamum [CAMPPOL] Amblystegiaceae (B:M)
Hypnum populeum Hedw. = Brachythecium populeum [BRACPOP] Brachytheciaceae (B:M)
Hypnum praelongum Hedw. = Eurhynchium praelongum [EURHPRA] Brachytheciaceae (B:M)
Hypnum pratense (Rabenh.) W. Koch ex Spruce [HYPNPRA] Hypnaceae (B:M)
Hypnum procerrimum Mol. [HYPNPRO] Hypnaceae (B:M)
Hypnum pseudo-arcticum Kindb. = Hygrohypnum luridum [HYGRLUR] Amblystegiaceae (B:M)
Hypnum pseudo-nemorosum (Kindb.) Par. = Hypnum cupressiforme var. resupinatum [HYPNCUP3] Hypnaceae (B:M)
Hypnum pseudodrepanium C. Müll. & Kindb. in Mac. & Kindb. = Hypnum lindbergii [HYPNLIN] Hypnaceae (B:M)
Hypnum pseudofastigiatum C. Müll. & Kindb. in Mac. & Kindb. = Hypnum cupressiforme [HYPNCUP] Hypnaceae (B:M)
Hypnum pseudolycopodioides Kindb. ex Nichols = Hygrohypnum ochraceum [HYGROCH] Amblystegiaceae (B:M)
Hypnum pseudopratense Kindb. in Mac. & Kindb. = Hypnum pratense [HYPNPRA] Hypnaceae (B:M)

Hypnum pseudorecurvans Kindb. = Hypnum circinale [HYPNCIR] Hypnaceae (B:M)
Hypnum pulchellum (Hedw.) Dicks. ex Brid. = Isopterygiopsis pulchella [ISOPPUC] Hypnaceae (B:M)
Hypnum pyrenaicum Spruce = Hylocomiastrum pyrenaicum [HYLOPYE] Hylocomiaceae (B:M)
Hypnum radicale (P. Beauv.) Brid. = Campylium radicale [CAMPRAD] Amblystegiaceae (B:M)
Hypnum ravaudii ssp. *fastigiatum* (Brid.) Wijk & Marg. = Hypnum recurvatum [HYPNREC] Hypnaceae (B:M)
Hypnum recognitum Hedw. = Thuidium recognitum [THUIREC] Thuidiaceae (B:M)
Hypnum recurvatum (Lindb. & Arnell) Kindb. [HYPNREC] Hypnaceae (B:M)
Hypnum reflexum Stark in Web. & Mohr = Brachythecium reflexum [BRACREF] Brachytheciaceae (B:M)
Hypnum renauldii Kindb. in Mac. & Kindb. = Hypnum lindbergii [HYPNLIN] Hypnaceae (B:M)
Hypnum revolutum (Mitt.) Lindb. [HYPNREV] Hypnaceae (B:M)
Hypnum revolutum var. *dolomiticum* (Milde) Mönk. = Hypnum revolutum [HYPNREV] Hypnaceae (B:M)
Hypnum revolutum var. *subjulaceum* Bryhn = Hypnum revolutum [HYPNREV] Hypnaceae (B:M)
Hypnum revolvens Sw. = Limprichtia revolvens [LIMPREV] Amblystegiaceae (B:M)
Hypnum richardsonii (Mitt.) Lesq. & James = Calliergon richardsonii [CALLRIC] Amblystegiaceae (B:M)
Hypnum riparium Hedw. = Leptodictyum riparium [LEPTRIP] Amblystegiaceae (B:M)
Hypnum rivulare (B.S.G.) Bruch in Wils. = Brachythecium rivulare [BRACRIV] Brachytheciaceae (B:M)
Hypnum rugosum Hedw. = Rhytidium rugosum [RHYTRUG] Rhytidiaceae (B:M)
Hypnum rusciforme Weiss ex Brid. = Platyhypnidium riparioides [PLATRIP] Brachytheciaceae (B:M)
Hypnum rutabulum Hedw. = Brachythecium rutabulum [BRACRUT] Brachytheciaceae (B:M)
Hypnum salebrosum Hoffm. ex Web. & Mohr = Brachythecium salebrosum [BRACSAL] Brachytheciaceae (B:M)
Hypnum sarmentosum Wahlenb. = Sarmenthypnum sarmentosum [SARMSAR] Amblystegiaceae (B:M)
Hypnum sarmentosum var. *acuminatum* Bryhn = Sarmenthypnum sarmentosum [SARMSAR] Amblystegiaceae (B:M)
Hypnum sarmentosum var. *fontinaloides* Berggr. = Sarmenthypnum sarmentosum [SARMSAR] Amblystegiaceae (B:M)
Hypnum schreberi Brid. = Pleurozium schreberi [PLEUSCH] Hylocomiaceae (B:M)
Hypnum scorpioides Hedw. = Scorpidium scorpioides [SCORSCO] Amblystegiaceae (B:M)
Hypnum scorpioides var. *gracilescens* Sanio in Klinggr. = Scorpidium scorpioides [SCORSCO] Amblystegiaceae (B:M)
Hypnum sendtneri Schimp. in H. Müll. = Drepanocladus sendtneri [DREPSEN] Amblystegiaceae (B:M)
Hypnum sequoieti C. Müll. = Hypnum circinale [HYPNCIR] Hypnaceae (B:M)
Hypnum serpens Hedw. = Amblystegium serpens [AMBLSER] Amblystegiaceae (B:M)
Hypnum serrulatum Hedw. = Steerecleus serrulatus [STEESER] Brachytheciaceae (B:M)
Hypnum sinuolatum (Kindb.) Kindb. in Röll = Campylium chrysophyllum [CAMPCHR] Amblystegiaceae (B:M)
Hypnum sommerfeltii Myr. = Campylium hispidulum [CAMPHIS] Amblystegiaceae (B:M)
Hypnum splendens Hedw. = Hylocomium splendens [HYLOSPL] Hylocomiaceae (B:M)
Hypnum squarrosum Hedw. = Rhytidiadelphus squarrosus [RHYTSQU] Hylocomiaceae (B:M)
Hypnum starkei Brid. = Brachythecium starkei [BRACSTA] Brachytheciaceae (B:M)
Hypnum stellatum Hedw. = Campylium stellatum [CAMPSTE] Amblystegiaceae (B:M)
Hypnum stellatum var. *protensum* (Brid.) Röhl. = Campylium stellatum var. protensum [CAMPSTE2] Amblystegiaceae (B:M)
Hypnum stokessii Turn. = Eurhynchium praelongum [EURHPRA] Brachytheciaceae (B:M)
Hypnum stramineum Dicks. ex Brid. = Calliergon stramineum [CALLSTR] Amblystegiaceae (B:M)
Hypnum stramineum var. *patens* (Lindb.) Par. = Calliergon stramineum [CALLSTR] Amblystegiaceae (B:M)
Hypnum strigosum Hoffm. ex Web. & Mohr = Eurhynchium pulchellum [EURHPUL] Brachytheciaceae (B:M)
Hypnum subcomplexum Kindb. = Hypnum vaucheri [HYPNVAU] Hypnaceae (B:M)
Hypnum subflaccum C. Müll. & Kindb. in Mac. & Kindb. = Hypnum pratense [HYPNPRA] Hypnaceae (B:M)
Hypnum subgiganteum Kindb. ex Grout = Calliergon richardsonii [CALLRIC] Amblystegiaceae (B:M)
Hypnum subimponens Lesq. [HYPNSUB] Hypnaceae (B:M)
Hypnum subimponens var. *cristulum* Kindb. in Mac. = Hypnum subimponens [HYPNSUB] Hypnaceae (B:M)
Hypnum subplicatile Limpr. = Hypnum plicatulum [HYPNPLI] Hypnaceae (B:M)
Hypnum subsecundum Kindb. = Campylium chrysophyllum [CAMPCHR] Amblystegiaceae (B:M)
Hypnum sullivantiae Schimp. ex Sull. = Plagiothecium cavifolium [PLAGCAV] Plagiotheciaceae (B:M)
Hypnum torrentis C. Müll. & Kindb. in Mac. & Kindb. = Hygrohypnum smithii [HYGRSMI] Amblystegiaceae (B:M)
Hypnum trifarium Web. & Mohr = Calliergon trifarium [CALLTRI] Amblystegiaceae (B:M)
Hypnum triquetrum Hedw. = Rhytidiadelphus triquetrus [RHYTTRI] Hylocomiaceae (B:M)
Hypnum turgescens T. Jens. = Pseudocalliergon turgescens [PSEUTUR] Amblystegiaceae (B:M)
Hypnum umbratum Ehrh. ex Hedw. = Hylocomiastrum umbratum [HYLOUMR] Hylocomiaceae (B:M)
Hypnum uncinatum Hedw. = Sanionia uncinata [SANIUNC] Amblystegiaceae (B:M)
Hypnum uncinatum var. *alpinum* (Ren.) Hamm. = Sanionia uncinata [SANIUNC] Amblystegiaceae (B:M)
Hypnum uncinatum var. *gracilescens* B.S.G. = Sanionia uncinata [SANIUNC] Amblystegiaceae (B:M)
Hypnum uncinatum var. *gracillimum* Berggr. = Sanionia uncinata [SANIUNC] Amblystegiaceae (B:M)
Hypnum uncinatum var. *micropterum* Kindb. in Mac. & Kindb. = Sanionia uncinata [SANIUNC] Amblystegiaceae (B:M)
Hypnum uncinatum var. *plumosum* B.S.G. = Sanionia uncinata [SANIUNC] Amblystegiaceae (B:M)
Hypnum uncinatum var. *symmetricum* (Ren. & Card.) Ren. & Card. = Sanionia uncinata var. symmetrica [SANIUNC1] Amblystegiaceae (B:M)
Hypnum unicostatum C. Müll. & Kindb. in Mac. & Kindb. = Campylium chrysophyllum [CAMPCHR] Amblystegiaceae (B:M)
Hypnum vacillans Lesq. & James = Leptodictyum riparium [LEPTRIP] Amblystegiaceae (B:M)
Hypnum vacium (Hedw.) P. Beauv. = Amblystegium varium [AMBLVAR] Amblystegiaceae (B:M)
Hypnum vaucheri Lesq. [HYPNVAU] Hypnaceae (B:M)
Hypnum velutinum Hedw. = Brachythecium velutinum [BRACVEL] Brachytheciaceae (B:M)
Hypnum vernicosus Lindb. in Hartm. = Hamatocaulis vernicosus [HAMAVER] Amblystegiaceae (B:M)
Hypnum wilsonii (Lindb.) Schimp. ex Ren. = Drepanocladus sendtneri [DREPSEN] Amblystegiaceae (B:M)
Hypocenomyce Choisy [HYPOCEN$] Lecideaceae (L)
Hypocenomyce anthracophila (Nyl.) P. James & Gotth. Schneider = Biatora anthracophila [BIATANT] Bacidiaceae (L)
Hypocenomyce castaneocinerea (Räsänen) Timdal [HYPOCAS] Lecideaceae (L)
Hypocenomyce friesii (Ach.) P. James & Gotth. Schneider [HYPOFRI] Lecideaceae (L)
Hypocenomyce leucococca R. Sant. [HYPOLEU] Lecideaceae (L)
Hypocenomyce scalaris (Ach.) Choisy [HYPOSCA] Lecideaceae (L)
Hypochaeris L. [HYPOCHA$] Asteraceae (V:D)
*__Hypochaeris glabra__ L. [HYPOGLA] Asteraceae (V:D)
*__Hypochaeris radicata__ L. [HYPORAD] Asteraceae (V:D)

Hypochoeris L. = Hypochaeris [HYPOCHA$] Asteraceae (V:D)
Hypogymnia (Nyl.) Nyl. [HYPOGYM$] Parmeliaceae (L)
Hypogymnia apinnata Goward & McCune [HYPOAPI] Parmeliaceae (L)
Hypogymnia austerodes (Nyl.) Räsänen [HYPOAUS] Parmeliaceae (L)
Hypogymnia bitteri (Lynge) Ahti [HYPOBIT] Parmeliaceae (L)
Hypogymnia duplicata (Ach.) Rass. [HYPODUP] Parmeliaceae (L)
Hypogymnia elongata (Hillm.) Rass. = Hypogymnia duplicata [HYPODUP] Parmeliaceae (L)
Hypogymnia enteromorpha (Ach.) Nyl. [HYPOENT] Parmeliaceae (L)
Hypogymnia heterophylla L. Pike [HYPOHET] Parmeliaceae (L)
Hypogymnia imshaugii Krog [HYPOIMS] Parmeliaceae (L)
Hypogymnia imshaugii var. *inactiva* Krog = Hypogymnia inactiva [HYPOINA] Parmeliaceae (L)
Hypogymnia inactiva (Krog) Ohlsson [HYPOINA] Parmeliaceae (L)
Hypogymnia metaphysodes (Asah.) Rass. [HYPOMET] Parmeliaceae (L)
Hypogymnia occidentalis L. Pike [HYPOOCC] Parmeliaceae (L)
Hypogymnia oceanica Goward [HYPOOCE] Parmeliaceae (L)
Hypogymnia oroarctica Krog = Brodoa oroarctica [BRODORO] Parmeliaceae (L)
Hypogymnia physodes (L.) Nyl. [HYPOPHY] Parmeliaceae (L)
Hypogymnia pseudophysodes auct. = Hypogymnia oceanica [HYPOOCE] Parmeliaceae (L)
Hypogymnia rugosa (G. Merr.) L. Pike [HYPORUG] Parmeliaceae (L)
Hypogymnia subobscura (Vainio) Poelt [HYPOSUO] Parmeliaceae (L)
Hypogymnia tubulosa (Schaerer) Hav. [HYPOTUB] Parmeliaceae (L)
Hypogymnia vittata (Ach.) Parrique [HYPOVIT] Parmeliaceae (L)
Hypopitys americana (DC.) Small = Monotropa hypopithys [MONOHYP] Monotropaceae (V:D)
Hypopitys fimbriata (Gray) T.J. Howell = Monotropa hypopithys [MONOHYP] Monotropaceae (V:D)
Hypopitys insignata Bickn. = Monotropa hypopithys [MONOHYP] Monotropaceae (V:D)
Hypopitys lanuginosa (Michx.) Nutt. = Monotropa hypopithys [MONOHYP] Monotropaceae (V:D)
Hypopitys latisquama Rydb. = Monotropa hypopithys [MONOHYP] Monotropaceae (V:D)
Hypopitys monotropa Crantz = Monotropa hypopithys [MONOHYP] Monotropaceae (V:D)
Hypopterygium Brid. [HYPOPTE$] Hypopterygiaceae (B:M)
Hypopterygium canadense Kindb. = Hypopterygium fauriei [HYPOFAU] Hypopterygiaceae (B:M)
Hypopterygium fauriei Besch. [HYPOFAU] Hypopterygiaceae (B:M)
Hypotrachyna (Vainio) Hale [HYPOTRA$] Parmeliaceae (L)
Hypotrachyna revoluta (Flörke) Hale [HYPOREV] Parmeliaceae (L)
Hypotrachyna sinuosa (Sm.) Hale [HYPOSIN] Parmeliaceae (L)

I

Iberis nudicaulis L. = Teesdalia nudicaulis [TEESNUD] Brassicaceae (V:D)
Ibidium strictum (Rydb.) House = Spiranthes romanzoffiana [SPIRROM] Orchidaceae (V:M)
Icmadophila Trevisan [ICMADOP$] Baeomycetaceae (L)
Icmadophila ericetorum (L.) Zahlbr. [ICMAERI] Baeomycetaceae (L)
Idahoa A. Nels. & J.F. Macbr. [IDAHOA$] Brassicaceae (V:D)
Idahoa scapigera (Hook.) A. Nels. & J.F. Macbr. [IDAHSCA] Brassicaceae (V:D)
Ilex L. [ILEX$] Aquifoliaceae (V:D)
•**Ilex aquifolium** L. [ILEXAQU] Aquifoliaceae (V:D)
Iliamna Greene [ILIAMNA$] Malvaceae (V:D)
Iliamna rivularis (Dougl. ex Hook.) Greene [ILIARIV] Malvaceae (V:D)
Illosporium Martius [ILLOSPO$] Uncertain-family (L)
Illosporium carneum Fr. [ILLOCAR] Uncertain-family (L)
Ilysanthes inequalis (Walt.) Pennell = Lindernia dubia var. anagallidea [LINDDUB1] Scrophulariaceae (V:D)
Impatiens L. [IMPATIE$] Balsaminaceae (V:D)
Impatiens aurella Rydb. [IMPAAUR] Balsaminaceae (V:D)
Impatiens biflora Walt. = Impatiens capensis [IMPACAP] Balsaminaceae (V:D)
Impatiens biflora var. *ecalcarata* (Blank) M.E. Jones = Impatiens ecalcarata [IMPAECA] Balsaminaceae (V:D)
Impatiens capensis Meerb. [IMPACAP] Balsaminaceae (V:D)
Impatiens ecalcarata Blank. [IMPAECA] Balsaminaceae (V:D)
Impatiens fulva Nutt. = Impatiens capensis [IMPACAP] Balsaminaceae (V:D)
•**Impatiens glandulifera** Royle [IMPAGLA] Balsaminaceae (V:D)
•**Impatiens noli-tangere** L. [IMPANOL] Balsaminaceae (V:D)
Impatiens noli-tangere ssp. *biflora* (Walt.) Hultén = Impatiens capensis [IMPACAP] Balsaminaceae (V:D)
Impatiens nortonii Rydb. = Impatiens capensis [IMPACAP] Balsaminaceae (V:D)
Impatiens occidentalis Rydb. = Impatiens noli-tangere [IMPANOL] Balsaminaceae (V:D)
•**Impatiens parviflora** DC. [IMPAPAR] Balsaminaceae (V:D)
Impatiens roylei Walp. = Impatiens glandulifera [IMPAGLA] Balsaminaceae (V:D)
Imshaugia S.F. Meyer [IMSHAUG$] Parmeliaceae (L)
Imshaugia aleurites (Ach.) S.F. Meyer [IMSHALE] Parmeliaceae (L)
Inula L. [INULA$] Asteraceae (V:D)
•**Inula helenium** L. [INULHEL] Asteraceae (V:D)
Ionactis Greene [IONACTI$] Asteraceae (V:D)
Ionactis stenomeres (Gray) Greene [IONASTE] Asteraceae (V:D)
Ionaspis Th. Fr. [IONASPI$] Hymeneliaceae (L)
Ionaspis epulotica (Ach.) Blomb. & Forssell = Hymenelia epulotica [HYMEEPU] Hymeneliaceae (L)
Ionaspis heteromorpha (Kremp.) Arnold = Hymenelia heteromorpha [HYMEHET] Hymeneliaceae (L)
Ionaspis lacustris (With.) Lutzoni [IONALAC] Hymeneliaceae (L)
Ionaspis melanocarpa (Kremp.) Arnold = Hymenelia melanocarpa [HYMEMEL] Hymeneliaceae (L)
Ionaspis ochracella (Nyl.) H. Magn. = Hymenelia heteromorpha [HYMEHET] Hymeneliaceae (L)
Ionaspis odora (Ach.) Th. Fr. ex Stein [IONAODO] Hymeneliaceae (L)
Ionaspis reducta H. Magn. = Hymenelia heteromorpha [HYMEHET] Hymeneliaceae (L)
Ionaspis schismatopsis (Nyl.) Hue = Hymenelia heteromorpha [HYMEHET] Hymeneliaceae (L)
Ipomopsis Michx. [IPOMOPS$] Polemoniaceae (V:D)
Ipomopsis aggregata (Pursh) V. Grant [IPOMAGG] Polemoniaceae (V:D)
Ipomopsis minutiflora (Benth.) V. Grant [IPOMMIN] Polemoniaceae (V:D)
Iris L. [IRIS$] Iridaceae (V:M)
Iris longipetala Herbert = Iris missouriensis [IRISMIS] Iridaceae (V:M)
Iris missouriensis Nutt. [IRISMIS] Iridaceae (V:M)
Iris missouriensis var. *arizonica* (Dykes) R.C. Foster = Iris missouriensis [IRISMIS] Iridaceae (V:M)
Iris missouriensis var. *pelogonus* (Goodding) R.C. Foster = Iris missouriensis [IRISMIS] Iridaceae (V:M)
Iris pariensis Welsh = Iris missouriensis [IRISMIS] Iridaceae (V:M)

*__Iris pseudacorus__ L. [IRISPSE] Iridaceae (V:M)
*__Iris sibirica__ L. [IRISSIB] Iridaceae (V:M)
Iris tolmieana Herbert = Iris missouriensis [IRISMIS] Iridaceae (V:M)
Isatis L. [ISATIS$] Brassicaceae (V:D)
*__Isatis tinctoria__ L. [ISATTIN] Brassicaceae (V:D)
Isnardia palustris L. = Ludwigia palustris [LUDWPAL] Onagraceae (V:D)
Isoetes L. [ISOETES$] Isoetaceae (V:F)
Isoetes beringensis Komarov = Isoetes maritima [ISOEMAR] Isoetaceae (V:F)
Isoetes bolanderi Engelm. [ISOEBOL] Isoetaceae (V:F)
Isoetes braunii Durieu = Isoetes echinospora [ISOEECH] Isoetaceae (V:F)
Isoetes echinospora Durieu [ISOEECH] Isoetaceae (V:F)
Isoetes echinospora ssp. *asiatica* (Makino) A. Löve = Isoetes echinospora [ISOEECH] Isoetaceae (V:F)
Isoetes echinospora ssp. *maritima* (Underwood) A. Löve = Isoetes maritima [ISOEMAR] Isoetaceae (V:F)
Isoetes echinospora ssp. *muricata* (Durieu) A. & D. Löve = Isoetes echinospora [ISOEECH] Isoetaceae (V:F)
Isoetes echinospora var. *asiatica* Makino = Isoetes echinospora [ISOEECH] Isoetaceae (V:F)
Isoetes echinospora var. *braunii* (Durieu) Engelm. = Isoetes echinospora [ISOEECH] Isoetaceae (V:F)
Isoetes echinospora var. *hesperia* (C.F. Reed) A. Löve = Isoetes echinospora [ISOEECH] Isoetaceae (V:F)
Isoetes echinospora var. *maritima* (Underwood) A.A. Eat. = Isoetes maritima [ISOEMAR] Isoetaceae (V:F)
Isoetes echinospora var. *muricata* (Durieu) Engelm. = Isoetes echinospora [ISOEECH] Isoetaceae (V:F)
Isoetes echinospora var. *robusta* Engelm. = Isoetes echinospora [ISOEECH] Isoetaceae (V:F)
Isoetes echinospora var. *savilei* Boivin = Isoetes echinospora [ISOEECH] Isoetaceae (V:F)
Isoetes flettii (A.A. Eat.) N.E. Pfeiffer [ISOEFLE] Isoetaceae (V:F)
Isoetes howellii Engelm. [ISOEHOW] Isoetaceae (V:F)
Isoetes howellii var. *minima* (A.A. Eat.) Pfeiffer = Isoetes howellii [ISOEHOW] Isoetaceae (V:F)
Isoetes lacustris ssp. *paupercula* (Engelm.) J. Feilberg = Isoetes occidentalis [ISOEOCC] Isoetaceae (V:F)
Isoetes lacustris var. *paupercula* Engelm. = Isoetes occidentalis [ISOEOCC] Isoetaceae (V:F)
Isoetes macounii A.A. Eaton = Isoetes maritima [ISOEMAR] Isoetaceae (V:F)
Isoetes maritima Underwood [ISOEMAR] Isoetaceae (V:F)
Isoetes muricata Durieu = Isoetes echinospora [ISOEECH] Isoetaceae (V:F)
Isoetes muricata ssp. *maritima* (Underwood) Hultén = Isoetes maritima [ISOEMAR] Isoetaceae (V:F)
Isoetes muricata var. *braunii* (Durieu) C.F. Reed = Isoetes echinospora [ISOEECH] Isoetaceae (V:F)
Isoetes muricata var. *hesperia* C.F. Reed = Isoetes echinospora [ISOEECH] Isoetaceae (V:F)
Isoetes nuttallii A. Braun ex Engelm. [ISOENUT] Isoetaceae (V:F)
Isoetes occidentalis Henderson [ISOEOCC] Isoetaceae (V:F)
Isoetes paupercula (Engelm.) A.A. Eat. = Isoetes occidentalis [ISOEOCC] Isoetaceae (V:F)
Isoetes piperi A.A. Eat. = Isoetes occidentalis [ISOEOCC] Isoetaceae (V:F)
Isoetes × pseudotruncata D.M. Britton & D.F. Brunt. [ISOEPSE] Isoetaceae (V:F)
Isoetes setacea Lam. p.p. = Isoetes echinospora [ISOEECH] Isoetaceae (V:F)
Isoetes setacea ssp. *muricata* (Durieu) Holub = Isoetes echinospora [ISOEECH] Isoetaceae (V:F)
Isoetes × truncata (A.A. Eat.) Clute (pro sp.) [ISOETRU] Isoetaceae (V:F)
Isoetes truncata (A.A. Eat.) Clute = Isoetes × truncata [ISOETRU] Isoetaceae (V:F)
Isolepis cernua (Vahl) Roemer & J.A. Schultes p.p. = Scirpus cernuus var. californicus [SCIRCER1] Cyperaceae (V:M)
Isolepis setaceus (L.) R. Br. = Scirpus setaceus [SCIRSET] Cyperaceae (V:M)
Isopaches bicrenatus (Schmid.) Buch = Lophozia bicrenata [LOPHBIC] Jungermanniaceae (B:H)
Isopaches decolorans Buch = Lophozia decolorans [LOPHDEC] Jungermanniaceae (B:H)
Isopaches hellerianus (Nees) Buch = Anastrophyllum helleranum [ANASHEL] Jungermanniaceae (B:H)
Isopterygiopsis Iwats. [ISOPTER$] Hypnaceae (B:M)
Isopterygiopsis muelleriana (Schimp.) Iwats. [ISOPMUE] Hypnaceae (B:M)
Isopterygiopsis pulchella (Hedw.) Iwats. [ISOPPUC] Hypnaceae (B:M)
Isopterygium borrerianum (C. Müll.) Lindb. = Pseudotaxiphyllum elegans [PSEUELE] Hypnaceae (B:M)
Isopterygium borrerianum var. *gracilens* (Aust. ex Grout) Crum, Steere & Anderson = Pseudotaxiphyllum elegans [PSEUELE] Hypnaceae (B:M)
Isopterygium borrerianum var. *terrestre* (Lindb.) Crum, Steere & Anderson = Pseudotaxiphyllum elegans [PSEUELE] Hypnaceae (B:M)
Isopterygium elegans (Brid.) Lindb. = Pseudotaxiphyllum elegans [PSEUELE] Hypnaceae (B:M)
Isopterygium elegans var. *gracilens* Aust. ex Grout = Pseudotaxiphyllum elegans [PSEUELE] Hypnaceae (B:M)
Isopterygium elegans var. *schimperi* (Jur. & Milde) Limpr. = Pseudotaxiphyllum elegans [PSEUELE] Hypnaceae (B:M)
Isopterygium elegans var. *terrestre* (Lindb.) Wijk & Marg. = Pseudotaxiphyllum elegans [PSEUELE] Hypnaceae (B:M)
Isopterygium muellerianum (Schimp.) Jaeg. = Isopterygiopsis muelleriana [ISOPMUE] Hypnaceae (B:M)
Isopterygium nitidulum (Wahlenb.) Kindb. = Isopterygiopsis pulchella [ISOPPUC] Hypnaceae (B:M)
Isopterygium nitidum Lindb. = Isopterygiopsis pulchella [ISOPPUC] Hypnaceae (B:M)
Isopterygium piliferum (Sw. ex Hartm.) Loeske = Plagiothecium piliferum [PLAGPIL] Plagiotheciaceae (B:M)
Isopterygium pulchellum (Hedw.) Jaeg. = Isopterygiopsis pulchella [ISOPPUC] Hypnaceae (B:M)
Isopterygium pulchellum var. *nitidulum* (Wahlenb.) G. Roth = Isopterygiopsis pulchella [ISOPPUC] Hypnaceae (B:M)
Isopterygium seligeri (Brid.) Dix. in C. Jens. = Herzogiella seligeri [HERZSEL] Hypnaceae (B:M)
Isopterygium silesiacum (Selig. ex P. Beauv.) Kindb. = Herzogiella seligeri [HERZSEL] Hypnaceae (B:M)
Isopterygium striatellum (Brid.) Loeske = Herzogiella striatella [HERZSTR] Hypnaceae (B:M)
Isopyrum savilei Calder & Taylor = Enemion savilei [ENEMSAV] Ranunculaceae (V:D)
Isothecium Brid. [ISOTHEC$] Brachytheciaceae (B:M)
Isothecium acuticuspis (Mitt.) Mac. & Kindb. = Isothecium myosuroides [ISOTMYO] Brachytheciaceae (B:M)
Isothecium aggredatum (Mitt.) Jaeg. & Sauerb. = Isothecium cristatum [ISOTCRI] Brachytheciaceae (B:M)
Isothecium aplocladum (Mitt.) Kindb. = Scleropodium cespitans var. sublaeve [SCLECES2] Brachytheciaceae (B:M)
Isothecium brachycladon Kindb. = Isothecium myosuroides [ISOTMYO] Brachytheciaceae (B:M)
Isothecium brewerianum (Lesq.) Kindb. in Mac. & Kindb. = Isothecium cristatum [ISOTCRI] Brachytheciaceae (B:M)
Isothecium cardotii Kindb. = Isothecium myosuroides [ISOTMYO] Brachytheciaceae (B:M)
Isothecium cristatum (Hampe) Robins. [ISOTCRI] Brachytheciaceae (B:M)
Isothecium cristatum var. *howei* (Kindb.) Crum, Steere & Anderson = Isothecium cristatum [ISOTCRI] Brachytheciaceae (B:M)
Isothecium cristatum var. *lutescens* (Lesq. & James) Crum, Steere & Anderson = Isothecium cristatum [ISOTCRI] Brachytheciaceae (B:M)
Isothecium eumyosuroides Dix. = Isothecium myosuroides [ISOTMYO] Brachytheciaceae (B:M)
Isothecium eumyosuroides var. *cavernarum* (Mol.) Crum, Steere & Anderson = Isothecium myosuroides [ISOTMYO] Brachytheciaceae (B:M)

Isothecium hylocomioides (Kindb.) Kindb. = Isothecium cristatum [ISOTCRI] Brachytheciaceae (B:M)
Isothecium lentum (Mitt.) Kindb. = Scleropodium cespitans [SCLECES] Brachytheciaceae (B:M)
Isothecium myosuroides Brid. [ISOTMYO] Brachytheciaceae (B:M)
Isothecium myosuroides var. *cavernarum* Mol. = Isothecium myosuroides [ISOTMYO] Brachytheciaceae (B:M)
Isothecium spiculiferum (Mitt.) Ren. & Card. = Isothecium myosuroides [ISOTMYO] Brachytheciaceae (B:M)
Isothecium spiculiferum var. *cardotii* (Kindb. in Mac. & Kindb.) Crum, Steere & Anderson = Isothecium myosuroides [ISOTMYO] Brachytheciaceae (B:M)
Isothecium spiculiferum var. *myurellum* (Kindb.) Wijk & Marg. = Isothecium myosuroides [ISOTMYO] Brachytheciaceae (B:M)
Isothecium stoloniferum Brid. = Isothecium myosuroides [ISOTMYO] Brachytheciaceae (B:M)
Isothecium stoloniferum var. *cardotii* (Kindb. in Mac. & Kindb.) Wijk & Marg. = Isothecium myosuroides [ISOTMYO] Brachytheciaceae (B:M)
Isothecium stoloniferum var. *myurellum* (Kindb.) Wijk & Marg. = Isothecium myosuroides [ISOTMYO] Brachytheciaceae (B:M)
Isothecium stoloniferum var. *spiculiferum* (Mitt.) R.D. Williams & Schof. = Isothecium myosuroides [ISOTMYO] Brachytheciaceae (B:M)
Isothecium stoloniferum var. *thamnoides* (Kindb.) Wijk & Marg. = Isothecium myosuroides [ISOTMYO] Brachytheciaceae (B:M)
Isothecium thamnoides Kindb. = Isothecium myosuroides [ISOTMYO] Brachytheciaceae (B:M)
Iva L. [IVA$] Asteraceae (V:D)
Iva axillaris Pursh [IVA AXI] Asteraceae (V:D)
Iva axillaris ssp. *robustior* (Hook.) Bassett = Iva axillaris [IVA AXI] Asteraceae (V:D)
Iva axillaris var. *robustior* Hook. = Iva axillaris [IVA AXI] Asteraceae (V:D)
*****Iva xanthifolia** Nutt. [IVA XAN] Asteraceae (V:D)
Iwatsukiella Buck & Crum [IWATSUK$] Pterigynandraceae (B:M)
Iwatsukiella leucotricha (Mitt.) Buck & Crum [IWATLEU] Pterigynandraceae (B:M)

J

Jamesoniella (Spruce) Carring. [JAMESON$] Jungermanniaceae (B:H)
Jamesoniella autumnalis (DC.) Steph. [JAMEAUT] Jungermanniaceae (B:H)
Japewia Tønsberg [JAPEWIA$] Bacidiaceae (L)
Japewia carrollii (Coppins & P. James) Tønsberg [JAPECAR] Bacidiaceae (L)
Japewia subaurifera Muhr & Tønsberg [JAPESUB] Bacidiaceae (L)
Japewia tornoënsis (Nyl.) Tønsberg [JUPETOR] Bacidiaceae (L)
Jaumea Pers. [JAUMEA$] Asteraceae (V:D)
Jaumea carnosa (Less.) Gray [JAUMCAR] Asteraceae (V:D)
Juncoides hyperboreum (R. Br.) Sheldon p.p. = Luzula confusa [LUZUCON] Juncaceae (V:M)
Juncoides piperi Coville = Luzula piperi [LUZUPIP] Juncaceae (V:M)
Juncoides spicatum (L.) Kuntze = Luzula spicata [LUZUSPI] Juncaceae (V:M)
Juncus L. [JUNCUS$] Juncaceae (V:M)
Juncus acuminatus Michx. [JUNCACU] Juncaceae (V:M)
Juncus albescens (Lange) Fern. [JUNCALB] Juncaceae (V:M)
Juncus alpinoarticulatus Chaix [JUNCALP] Juncaceae (V:M)
Juncus arcticus Willd. [JUNCARC] Juncaceae (V:M)
Juncus arcticus ssp. **alaskanus** Hultén [JUNCARC1] Juncaceae (V:M)
Juncus arcticus ssp. *ater* (Rydb.) Hultén = Juncus balticus var. montanus [JUNCBAL1] Juncaceae (V:M)
Juncus arcticus ssp. *balticus* (Willd.) Hyl. = Juncus balticus [JUNCBAL] Juncaceae (V:M)
Juncus arcticus ssp. *littoralis* (Engelm.) Hultén = Juncus balticus var. littoralis [JUNCBAL2] Juncaceae (V:M)
Juncus arcticus ssp. *sitchensis* Engelm. = Juncus haenkei [JUNCHAE] Juncaceae (V:M)
Juncus arcticus var. *alaskanus* (Hultén) Welsh = Juncus arcticus ssp. alaskanus [JUNCARC1] Juncaceae (V:M)
Juncus arcticus var. *montanus* (Engelm.) Welsh = Juncus balticus var. montanus [JUNCBAL1] Juncaceae (V:M)
Juncus arcticus var. *sitchensis* Engelm. = Juncus haenkei [JUNCHAE] Juncaceae (V:M)
Juncus articulatus L. [JUNCART] Juncaceae (V:M)
Juncus articulatus var. *obtusatus* Engelm. = Juncus articulatus [JUNCART] Juncaceae (V:M)
Juncus ater Rydb. = Juncus balticus var. montanus [JUNCBAL1] Juncaceae (V:M)
Juncus balticus Willd. [JUNCBAL] Juncaceae (V:M)
Juncus balticus var. *alaskanus* (Hultén) Porsild = Juncus arcticus ssp. alaskanus [JUNCARC1] Juncaceae (V:M)
Juncus balticus var. *haenkei* (E. Mey.) Buch. = Juncus haenkei [JUNCHAE] Juncaceae (V:M)
Juncus balticus var. **littoralis** Engelm. [JUNCBAL2] Juncaceae (V:M)
Juncus balticus var. **montanus** Engelm. [JUNCBAL1] Juncaceae (V:M)
Juncus biglumis L. [JUNCBIG] Juncaceae (V:M)
Juncus bolanderi Engelm. [JUNCBOL] Juncaceae (V:M)
Juncus brachystylus (Engelm.) Piper = Juncus kelloggii [JUNCKEL] Juncaceae (V:M)
Juncus brevicaudatus (Engelm.) Fern. [JUNCBRE] Juncaceae (V:M)
Juncus breweri Engelm. [JUNCBRW] Juncaceae (V:M)
Juncus bufonius L. [JUNCBUF] Juncaceae (V:M)
*****Juncus bulbosus** L. [JUNCBUL] Juncaceae (V:M)
Juncus castaneus Sm. [JUNCCAS] Juncaceae (V:M)
Juncus castaneus ssp. **castaneus** [JUNCCAS0] Juncaceae (V:M)
Juncus castaneus ssp. **leucochlamys** (Zing. ex Krecz.) Hultén [JUNCCAS2] Juncaceae (V:M)
Juncus confusus Coville [JUNCCON] Juncaceae (V:M)
Juncus conglomeratus L. = Juncus effusus var. conglomeratus [JUNCEFF6] Juncaceae (V:M)
Juncus covillei Piper [JUNCCOV] Juncaceae (V:M)
Juncus drummondii E. Mey. [JUNCDRU] Juncaceae (V:M)
Juncus drummondii var. **drummondii** [JUNCDRU0] Juncaceae (V:M)
Juncus drummondii var. *logifructus* St. John = Juncus drummondii var. subtriflorus [JUNCDRU2] Juncaceae (V:M)
Juncus drummondii var. **subtriflorus** (E. Mey.) C.L. Hitchc. [JUNCDRU2] Juncaceae (V:M)
Juncus effusus L. [JUNCEFF] Juncaceae (V:M)
Juncus effusus var. **brunneus** Engelm. [JUNCEFF1] Juncaceae (V:M)
Juncus effusus var. *caeruleomontanus* St. John = Juncus effusus var. conglomeratus [JUNCEFF6] Juncaceae (V:M)
Juncus effusus var. *compactus* auct. = Juncus effusus var. conglomeratus [JUNCEFF6] Juncaceae (V:M)
*****Juncus effusus** var. **conglomeratus** (L.) Engelm. [JUNCEFF6] Juncaceae (V:M)
Juncus effusus var. **gracilis** Hook. [JUNCEFF2] Juncaceae (V:M)
Juncus effusus var. **pacificus** Fern. & Wieg. [JUNCEFF3] Juncaceae (V:M)
Juncus ensifolius Wikstr. [JUNCENS] Juncaceae (V:M)
Juncus ensifolius var. *brunnescens* (Rydb.) Cronq. = Juncus saximontanus [JUNCSAX] Juncaceae (V:M)
Juncus ensifolius var. *major* Hook. = Juncus ensifolius [JUNCENS] Juncaceae (V:M)

Juncus ensifolius var. *montanus* (Engelm.) C.L. Hitchc. = Juncus saximontanus [JUNCSAX] Juncaceae (V:M)
Juncus exilis Osterhout = Juncus confusus [JUNCCON] Juncaceae (V:M)
Juncus falcatus E. Mey. [JUNCFAL] Juncaceae (V:M)
Juncus falcatus ssp. *sitchensis* (Buch.) Hultén = Juncus prominens [JUNCPRO] Juncaceae (V:M)
Juncus falcatus var. *sitchensis* Buch. = Juncus prominens [JUNCPRO] Juncaceae (V:M)
Juncus filiformis L. [JUNCFIL] Juncaceae (V:M)
Juncus geniculatus Schrank = Juncus alpinoarticulatus [JUNCALP] Juncaceae (V:M)
Juncus gerardii Loisel. [JUNCGER] Juncaceae (V:M)
Juncus greenei var. *vaseyi* (Engelm.) Boivin = Juncus vaseyi [JUNCVAS] Juncaceae (V:M)
Juncus haenkei E. Mey. [JUNCHAE] Juncaceae (V:M)
Juncus jonesii Rydb. = Juncus regelii [JUNCREG] Juncaceae (V:M)
Juncus kelloggii Engelm. [JUNCKEL] Juncaceae (V:M)
Juncus lesuerii Boland. [JUNCLES] Juncaceae (V:M)
Juncus lesuerii var. *tracyi* Jepson = Juncus tracyi [JUNCTRA] Juncaceae (V:M)
Juncus leucochlamys Zing. ex Krecz. = Juncus castaneus ssp. leucochlamys [JUNCCAS2] Juncaceae (V:M)
Juncus longistylis Torr. [JUNCLON] Juncaceae (V:M)
Juncus macer S.F. Gray = Juncus tenuis [JUNCTEN] Juncaceae (V:M)
Juncus melanocarpus Michx. = Luzula parviflora ssp. melanocarpa [LUZUPAR4] Juncaceae (V:M)
Juncus mertensianus Bong. [JUNCMER] Juncaceae (V:M)
Juncus nevadensis S. Wats. [JUNCNEV] Juncaceae (V:M)
Juncus nodosus L. [JUNCNOD] Juncaceae (V:M)
Juncus occidentalis Wieg. [JUNCOCC] Juncaceae (V:M)
Juncus oreganus S. Wats. = Juncus supiniformis [JUNCSUP] Juncaceae (V:M)
Juncus oxymeris Engelm. [JUNCOXY] Juncaceae (V:M)
Juncus parryi Engelm. [JUNCPAR] Juncaceae (V:M)
Juncus prominens (Buch.) Miyabe & Kudo [JUNCPRO] Juncaceae (V:M)
Juncus regelii Buch. [JUNCREG] Juncaceae (V:M)
Juncus saximontanus A. Nels. [JUNCSAX] Juncaceae (V:M)
Juncus saximontanus var. *robustior* M.E. Peck = Juncus saximontanus [JUNCSAX] Juncaceae (V:M)
Juncus slwookoorum S.B. Young = Juncus mertensianus [JUNCMER] Juncaceae (V:M)
Juncus stygius L. [JUNCSTY] Juncaceae (V:M)
Juncus stygius ssp. **americanus** (Buch.) Hultén [JUNCSTY1] Juncaceae (V:M)
Juncus stygius var. *americanus* Buch. = Juncus stygius ssp. americanus [JUNCSTY1] Juncaceae (V:M)
Juncus supiniformis Engelm. [JUNCSUP] Juncaceae (V:M)
Juncus supinus Moench = Juncus bulbosus [JUNCBUL] Juncaceae (V:M)
Juncus tenuis Willd. [JUNCTEN] Juncaceae (V:M)
Juncus tenuis var. *anthelatus* Wieg. = Juncus tenuis [JUNCTEN] Juncaceae (V:M)
Juncus tenuis var. *congestus* Engelm. = Juncus occidentalis [JUNCOCC] Juncaceae (V:M)
Juncus tenuis var. *multicornis* E. Mey. = Juncus tenuis [JUNCTEN] Juncaceae (V:M)
Juncus tenuis var. *williamsii* Fern. = Juncus tenuis [JUNCTEN] Juncaceae (V:M)
Juncus torreyi Coville [JUNCTOR] Juncaceae (V:M)
Juncus tracyi Rydb. [JUNCTRA] Juncaceae (V:M)
Juncus triglumis L. [JUNCTRI] Juncaceae (V:M)
Juncus triglumis ssp. *albescens* (Lange) Hultén = Juncus albescens [JUNCALB] Juncaceae (V:M)
Juncus triglumis var. *albescens* Lange = Juncus albescens [JUNCALB] Juncaceae (V:M)
Juncus vallicola Rydb. = Juncus balticus [JUNCBAL] Juncaceae (V:M)
Juncus vaseyi Engelm. [JUNCVAS] Juncaceae (V:M)
Juncus xiphioides var. *macranthus* Engelm. = Juncus saximontanus [JUNCSAX] Juncaceae (V:M)
Juncus xiphioides var. *montanus* Engelm. = Juncus saximontanus [JUNCSAX] Juncaceae (V:M)
Juncus xiphioides var. *triandrus* Engelm. = Juncus ensifolius [JUNCENS] Juncaceae (V:M)
Jungermannia L. emend. Dum. [JUNGERM$] Jungermanniaceae (B:H)
Jungermannia allenii Clark = Jamesoniella autumnalis [JAMEAUT] Jungermanniaceae (B:H)
Jungermannia atrovirens Dum. [JUNGATR] Jungermanniaceae (B:H)
Jungermannia bolanderi Gott. ex Underw. = Jungermannia confertissima [JUNGCON] Jungermanniaceae (B:H)
Jungermannia borealis Damscholdt & Vana [JUNGBOR] Jungermanniaceae (B:H)
Jungermannia confertissima Nees [JUNGCON] Jungermanniaceae (B:H)
Jungermannia confervoides Raddi = Cephaloziella divaricata [CEPHDIV] Cephaloziellaceae (B:H)
Jungermannia danicola Gott. ex Underw. = Jungermannia confertissima [JUNGCON] Jungermanniaceae (B:H)
Jungermannia decolorans Limpr. = Lophozia decolorans [LOPHDEC] Jungermanniaceae (B:H)
Jungermannia exsertifolia Steph. [JUNGEXS] Jungermanniaceae (B:H)
Jungermannia exsertifolia ssp. **cordifolia** (Dum.) Vana [JUNGEXS1] Jungermanniaceae (B:H)
Jungermannia exsertifolia ssp. **exsertifolia** [JUNGEXS0] Jungermanniaceae (B:H)
Jungermannia hyalina Lyell [JUNGHYA] Jungermanniaceae (B:H)
Jungermannia lanceolata L. = Jungermannia atrovirens [JUNGATR] Jungermanniaceae (B:H)
Jungermannia lanceolata auct. = Jungermannia leiantha [JUNGLEI] Jungermanniaceae (B:H)
Jungermannia leiantha Grolle [JUNGLEI] Jungermanniaceae (B:H)
Jungermannia obovata Nees [JUNGOBO] Jungermanniaceae (B:H)
Jungermannia pumila With. [JUNGPUM] Jungermanniaceae (B:H)
Jungermannia pumila var. *rivularis* (Schiffn.) Frye & Clark = Jungermannia pumila [JUNGPUM] Jungermanniaceae (B:H)
Jungermannia rubra Gott. ex Underw. [JUNGRUB] Jungermanniaceae (B:H)
Jungermannia schusterana J.D. Godfr. & G.A. Godfr. [JUNGSCH] Jungermanniaceae (B:H)
Jungermannia sphaerocarpa Hook. [JUNGSPH] Jungermanniaceae (B:H)
Jungermannia sphaerocarpa auct. = Jungermannia confertissima [JUNGCON] Jungermanniaceae (B:H)
Jungermannia sphaerocarpa var. *amplexicaulis* (Dum.) Frye & Clark = Jungermannia sphaerocarpa [JUNGSPH] Jungermanniaceae (B:H)
Jungermannia sphaerocarpa var. *nana* (Ness) Frye & Clark = Jungermannia sphaerocarpa [JUNGSPH] Jungermanniaceae (B:H)
Jungermannia subelliptica (Lindb. ex Kaal.) Lev. [JUNGSUB] Jungermanniaceae (B:H)
Jungermannia tristis Nees = Jungermannia atrovirens [JUNGATR] Jungermanniaceae (B:H)
Juniperus L. [JUNIPER$] Cupressaceae (V:G)
Juniperus alpina (Sm.) S.F. Gray = Juniperus communis var. montana [JUNICOM1] Cupressaceae (V:G)
Juniperus communis L. [JUNICOM] Cupressaceae (V:G)
Juniperus communis ssp. *alpina* (Sm.) Celak. = Juniperus communis var. montana [JUNICOM1] Cupressaceae (V:G)
Juniperus communis ssp. *nana* (Willd.) Syme = Juniperus communis var. montana [JUNICOM1] Cupressaceae (V:G)
Juniperus communis ssp. *saxitilis* (Pallas) E. Murr. = Juniperus communis var. montana [JUNICOM1] Cupressaceae (V:G)
Juniperus communis var. *alpina* Sm. = Juniperus communis var. montana [JUNICOM1] Cupressaceae (V:G)
Juniperus communis var. *jackii* Rehd. = Juniperus communis var. montana [JUNICOM1] Cupressaceae (V:G)
Juniperus communis var. **montana** Ait. [JUNICOM1] Cupressaceae (V:G)
Juniperus communis var. *saxatilis* Pallas = Juniperus communis var. montana [JUNICOM1] Cupressaceae (V:G)

Juniperus horizontalis Moench [JUNIHOR] Cupressaceae (V:G)
Juniperus horizontalis var. *douglasii* hort. = Juniperus horizontalis [JUNIHOR] Cupressaceae (V:G)
Juniperus horizontalis var. *variegata* Beissn. = Juniperus horizontalis [JUNIHOR] Cupressaceae (V:G)
Juniperus hudsonica Forbes = Juniperus horizontalis [JUNIHOR] Cupressaceae (V:G)
Juniperus nana Willd. = Juniperus communis var. montana [JUNICOM1] Cupressaceae (V:G)
Juniperus prostrata Pers. = Juniperus horizontalis [JUNIHOR] Cupressaceae (V:G)
Juniperus repens Nutt. = Juniperus horizontalis [JUNIHOR] Cupressaceae (V:G)
Juniperus scopulorum Sarg. [JUNISCO] Cupressaceae (V:G)
Juniperus scopulorum var. *columnaris* Fassett = Juniperus scopulorum [JUNISCO] Cupressaceae (V:G)
Juniperus sibirica Burgsd. = Juniperus communis var. montana [JUNICOM1] Cupressaceae (V:G)
Juniperus virginiana ssp. *scopulorum* (Sarg.) E. Murr. = Juniperus scopulorum [JUNISCO] Cupressaceae (V:G)
Juniperus virginiana var. *montana* Vasey = Juniperus scopulorum [JUNISCO] Cupressaceae (V:G)
Juniperus virginiana var. *prostrata* (Pers.) Torr. = Juniperus horizontalis [JUNIHOR] Cupressaceae (V:G)
Juniperus virginiana var. *scopulorum* (Sarg.) Lemmon = Juniperus scopulorum [JUNISCO] Cupressaceae (V:G)
Jurtsevia richardsonii (Hook.) A. & D. Löve = Anemone richardsonii [ANEMRIC] Ranunculaceae (V:D)

K

Kaernefeltia Thell & Goward [KAERNEF$] Parmeliaceae (L)
Kaernefeltia californica (Tuck.) Thell & Goward [KAERCAL] Parmeliaceae (L)
Kaernefeltia merrillii (Du Rietz) Thell & Goward [KAERMER] Parmeliaceae (L)
Kalmia L. [KALMIA$] Ericaceae (V:D)
Kalmia microphylla (Hook.) Heller [KALMMIC] Ericaceae (V:D)
Kalmia microphylla ssp. *occidentalis* (Small) Taylor & MacBryde = Kalmia microphylla [KALMMIC] Ericaceae (V:D)
Kalmia microphylla var. *occidentalis* (Small) Ebinger = Kalmia microphylla [KALMMIC] Ericaceae (V:D)
Kalmia occidentalis Small = Kalmia microphylla [KALMMIC] Ericaceae (V:D)
Kalmia polifolia ssp. *microphylla* (Hook.) Calder & Taylor = Kalmia microphylla [KALMMIC] Ericaceae (V:D)
Kalmia polifolia ssp. *occidentalis* (Small) Abrams = Kalmia microphylla [KALMMIC] Ericaceae (V:D)
Kalmia polifolia var. *microphylla* (Hook.) Hall = Kalmia microphylla [KALMMIC] Ericaceae (V:D)
Kiaeria Hag. [KIAERIA$] Dicranaceae (B:M)
Kiaeria blyttii (Schimp.) Broth. [KIAEBLY] Dicranaceae (B:M)
Kiaeria blyttii var. *hispidula* (Williams) Wijk & Marg. = Kiaeria blyttii [KIAEBLY] Dicranaceae (B:M)
Kiaeria falcata (Hedw.) Hag. [KIAEFAL] Dicranaceae (B:M)
Kiaeria falcata var. *pumila* (Saut.) Podp. = Kiaeria falcata [KIAEFAL] Dicranaceae (B:M)
Kiaeria glacialis (Berggr.) Hag. [KIAEGLA] Dicranaceae (B:M)
Kiaeria starkei (Web. & Mohr) Hag. [KIAESTA] Dicranaceae (B:M)
Kickxia Dumort. [KICKXIA$] Scrophulariaceae (V:D)
***Kickxia elatine** (L.) Dumort. [KICKELA] Scrophulariaceae (V:D)
Kiliasia athallina (Hepp) Hafellner = Toninia athallina [TONIATH] Catillariaceae (L)
Kiliasia philippea (Mont.) Hafellner = Toninia philippea [TONIPHI] Catillariaceae (L)
Kindbergia brittoniae (Grout) Ochyra = Eurhynchium praelongum [EURHPRA] Brachytheciaceae (B:M)
Kindbergia oregana (Sull.) Ochyra = Eurhynchium oreganum [EURHORE] Brachytheciaceae (B:M)
Kindbergia praelonga (Hedw.) Ochyra = Eurhynchium praelongum [EURHPRA] Brachytheciaceae (B:M)
Kindbergia praelonga var. *californica* (Grout) Ochyra = Eurhynchium praelongum [EURHPRA] Brachytheciaceae (B:M)
Kindbergia praelonga var. *stokesii* (Turn.) Ochyra = Eurhynchium praelongum [EURHPRA] Brachytheciaceae (B:M)
Knautia L. [KNAUTIA$] Dipsacaceae (V:D)
***Knautia arvensis** (L.) Coult. [KNAUARV] Dipsacaceae (V:D)
Kneiffia perennis (L.) Pennell = Oenothera perennis [OENOPER] Onagraceae (V:D)
Kneiffia pumila (L.) Spach = Oenothera perennis [OENOPER] Onagraceae (V:D)
Kobresia Willd. [KOBRESI$] Cyperaceae (V:M)
Kobresia bellardii (All.) K. Koch = Kobresia myosuroides [KOBRMYO] Cyperaceae (V:M)
Kobresia bellardii var. *macrocarpa* (Clokey ex Mackenzie) Harrington = Kobresia sibirica [KOBRSIB] Cyperaceae (V:M)
Kobresia bipartita (All.) Dalla Torre = Kobresia simpliciuscula [KOBRSIM] Cyperaceae (V:M)
Kobresia hyperborea Porsild = Kobresia sibirica [KOBRSIB] Cyperaceae (V:M)
Kobresia hyperborea var. *alaskana* Duman = Kobresia sibirica [KOBRSIB] Cyperaceae (V:M)
Kobresia macrocarpa Clokey ex Mackenzie = Kobresia sibirica [KOBRSIB] Cyperaceae (V:M)
Kobresia myosuroides (Vill.) Fiori [KOBRMYO] Cyperaceae (V:M)
Kobresia schoenoides (C.A. Mey.) Steud. p.p. = Kobresia sibirica [KOBRSIB] Cyperaceae (V:M)
Kobresia schoenoides var. *lepagei* (Duman) Boivin = Kobresia sibirica [KOBRSIB] Cyperaceae (V:M)
Kobresia sibirica (Turcz. ex Ledeb.) Boeckl. [KOBRSIB] Cyperaceae (V:M)
Kobresia simpliciuscula (Wahlenb.) Mackenzie [KOBRSIM] Cyperaceae (V:M)
Kobresia simpliciuscula var. *americana* Duman = Kobresia simpliciuscula [KOBRSIM] Cyperaceae (V:M)
Kochia Roth [KOCHIA$] Chenopodiaceae (V:D)
Kochia alata Bates = Kochia scoparia [KOCHSCO] Chenopodiaceae (V:D)
Kochia hyssopifolia (Pallas) Schrad. = Bassia hyssopifolia [BASSHYS] Chenopodiaceae (V:D)
***Kochia scoparia** (L.) Schrad. [KOCHSCO] Chenopodiaceae (V:D)
Kochia scoparia var. *culta* Farw. = Kochia scoparia [KOCHSCO] Chenopodiaceae (V:D)
Kochia scoparia var. *pubescens* Fenzl = Kochia scoparia [KOCHSCO] Chenopodiaceae (V:D)
Kochia scoparia var. *subvillosa* Moq. = Kochia scoparia [KOCHSCO] Chenopodiaceae (V:D)
Kochia scoparia var. *trichophila* (Stapf) Bailey = Kochia scoparia [KOCHSCO] Chenopodiaceae (V:D)
Kochia sieversiana (Pallas) C.A. Mey. = Kochia scoparia [KOCHSCO] Chenopodiaceae (V:D)
Kochia trichophila Stapf = Kochia scoparia [KOCHSCO] Chenopodiaceae (V:D)
Koeleria Pers. [KOELERI$] Poaceae (V:M)
Koeleria cristata auct. = Koeleria macrantha [KOELMAC] Poaceae (V:M)
Koeleria cristata var. *longifolia* Vasey ex Burtt-Davy = Koeleria macrantha [KOELMAC] Poaceae (V:M)
Koeleria cristata var. *pinetorum* Abrams = Koeleria macrantha [KOELMAC] Poaceae (V:M)
Koeleria gracilis Pers. = Koeleria macrantha [KOELMAC] Poaceae (V:M)
Koeleria macrantha (Ledeb.) J.A. Schultes [KOELMAC] Poaceae (V:M)

Koeleria nitida Nutt. = Koeleria macrantha [KOELMAC] Poaceae (V:M)
Koeleria pyramidata auct. = Koeleria macrantha [KOELMAC] Poaceae (V:M)
Koeleria yukonensis Hultén = Koeleria macrantha [KOELMAC] Poaceae (V:M)
Koenigia L. [KOENIGI$] Polygonaceae (V:D)
Koenigia islandica L. [KOENISL] Polygonaceae (V:D)
Koerberia A. Massal. [KOERBER$] Placynthiaceae (L)
Koerberia sonomensis (Tuck.) Henssen [KOERSON] Placynthiaceae (L)
Koerberiella Stein [KOERBEI$] Porpidiaceae (L)
Koerberiella wimmeriana (Körber) Stein [KOERWIM] Porpidiaceae (L)
Kohlmeyera Schatz [KOHLMEY$] Mastodiaceae (L)
Kohlmeyera complicatula (Nyl.) Schatz [KOHLCOM] Mastodiaceae (L)
Koniga maritima (L.) R. Br. = Lobularia maritima [LOBUMAR] Brassicaceae (V:D)
Krigia Schreb. [KRIGIA$] Asteraceae (V:D)
*****Krigia virginica** (L.) Willd. [KRIGVIR] Asteraceae (V:D)
Kruhsea streptopoides (Ledeb.) Kearney p.p. = Streptopus streptopoides var. brevipes [STRESTR1] Liliaceae (V:M)
Kumlienia Greene [KUMLIEN$] Ranunculaceae (V:D)
Kumlienia cooleyae (Vasey & Rose) Greene [KUMLCOO] Ranunculaceae (V:D)
Kurzia G. Martens [KURZIA$] Lepidoziaceae (B:H)
Kurzia makinoana (Steph.) Grolle = Kurzia sylvatica [KURZSYL] Lepidoziaceae (B:H)
Kurzia pauciflora (Dicks.) Grolle [KURZPAU] Lepidoziaceae (B:H)
Kurzia setacea (Web.) Grolle [KURZSET] Lepidoziaceae (B:H)
Kurzia sylvatica (Evans) Grolle [KURZSYL] Lepidoziaceae (B:H)
Kurzia trichoclados (K. Müll.) Grolle [KURZTRI] Lepidoziaceae (B:H)

L

Laburnum Medik. [LABURNU$] Fabaceae (V:D)
*****Laburnum anagyroides** Medik. [LABUANA] Fabaceae (V:D)
Lactuca L. [LACTUCA$] Asteraceae (V:D)
Lactuca biennis (Moench) Fern. [LACTBIE] Asteraceae (V:D)
*****Lactuca canadensis** L. [LACTCAN] Asteraceae (V:D)
Lactuca canadensis var. *integrifolia* (Bigelow) Torr. & Gray = Lactuca canadensis [LACTCAN] Asteraceae (V:D)
Lactuca canadensis var. *typica* Wieg. = Lactuca canadensis [LACTCAN] Asteraceae (V:D)
Lactuca muralis (L.) Fresen. = Mycelis muralis [MYCEMUR] Asteraceae (V:D)
Lactuca oblongifolia Nutt. = Lactuca tatarica var. pulchella [LACTTAT1] Asteraceae (V:D)
Lactuca pulchella (Pursh) DC. = Lactuca tatarica var. pulchella [LACTTAT1] Asteraceae (V:D)
Lactuca sagittifolia Ell. = Lactuca canadensis [LACTCAN] Asteraceae (V:D)
Lactuca scariola L. = Lactuca serriola [LACTSER] Asteraceae (V:D)
*****Lactuca serriola** L. [LACTSER] Asteraceae (V:D)
Lactuca spicata var. *integrifolia* (Torr. & Gray) Britt. = Lactuca biennis [LACTBIE] Asteraceae (V:D)
Lactuca tatarica (L.) C.A. Mey. [LACTTAT] Asteraceae (V:D)
Lactuca tatarica ssp. *pulchella* (Pursh) Stebbins = Lactuca tatarica var. pulchella [LACTTAT1] Asteraceae (V:D)
Lactuca tatarica var. *heterophylla* (Nutt.) Boivin = Lactuca tatarica var. pulchella [LACTTAT1] Asteraceae (V:D)
Lactuca tatarica var. **pulchella** (Pursh) Breitung [LACTTAT1] Asteraceae (V:D)
Lamium L. [LAMIUM$] Lamiaceae (V:D)
*****Lamium amplexicaule** L. [LAMIAMP] Lamiaceae (V:D)
Lamium amplexicaule var. *album* A.L. & M.C. Pickens = Lamium amplexicaule [LAMIAMP] Lamiaceae (V:D)
Lamium hybridum auct. = Lamium purpureum var. incisum [LAMIPUR1] Lamiaceae (V:D)
*****Lamium maculatum** L. [LAMIMAC] Lamiaceae (V:D)
*****Lamium purpureum** L. [LAMIPUR] Lamiaceae (V:D)
*****Lamium purpureum** var. **incisum** (Willd.) Pers. [LAMIPUR1] Lamiaceae (V:D)
Lappa minor Hill = Arctium minus [ARCTMIN] Asteraceae (V:D)
Lappula Moench [LAPPULA$] Boraginaceae (V:D)
Lappula americana (Gray) Rydb. = Hackelia deflexa var. americana [HACKDEF2] Boraginaceae (V:D)
Lappula deflexa (Wahlenb.) Garcke p.p. = Hackelia deflexa var. americana [HACKDEF2] Boraginaceae (V:D)
Lappula deflexa ssp. *americana* (Gray) Hultén = Hackelia deflexa var. americana [HACKDEF2] Boraginaceae (V:D)
Lappula deflexa var. *americana* (Gray) Greene = Hackelia deflexa var. americana [HACKDEF2] Boraginaceae (V:D)
Lappula echinata Gilib. = Lappula squarrosa [LAPPSQU] Boraginaceae (V:D)
Lappula echinata var. *occidentalis* (S. Wats.) Boivin = Lappula occidentalis var. occidentalis [LAPPOCC0] Boraginaceae (V:D)
Lappula erecta A. Nels. = Lappula squarrosa [LAPPSQU] Boraginaceae (V:D)
Lappula floribunda (Lehm.) Greene = Hackelia floribunda [HACKFLO] Boraginaceae (V:D)
Lappula fremontii (Torr.) Greene = Lappula squarrosa [LAPPSQU] Boraginaceae (V:D)
Lappula lappula (L.) Karst. = Lappula squarrosa [LAPPSQU] Boraginaceae (V:D)
Lappula myosotis Moench = Lappula squarrosa [LAPPSQU] Boraginaceae (V:D)
Lappula occidentalis (S. Wats.) Greene [LAPPOCC] Boraginaceae (V:D)
Lappula occidentalis var. **cupulata** (Gray) Higgins [LAPPOCC1] Boraginaceae (V:D)
Lappula occidentalis var. **occidentalis** [LAPPOCC0] Boraginaceae (V:D)
Lappula redowskii auct. = Lappula occidentalis var. occidentalis [LAPPOCC0] Boraginaceae (V:D)
Lappula redowskii var. *cupulata* (Gray) M.E. Jones = Lappula occidentalis var. cupulata [LAPPOCC1] Boraginaceae (V:D)
Lappula redowskii var. *desertorum* (Greene) I.M. Johnston = Lappula occidentalis var. occidentalis [LAPPOCC0] Boraginaceae (V:D)
Lappula redowskii var. *occidentalis* (S. Wats.) Rydb. = Lappula occidentalis var. occidentalis [LAPPOCC0] Boraginaceae (V:D)
Lappula redowskii var. *texana* (Scheele) Brand = Lappula occidentalis var. cupulata [LAPPOCC1] Boraginaceae (V:D)
*****Lappula squarrosa** (Retz.) Dumort. [LAPPSQU] Boraginaceae (V:D)
Lappula squarrosa var. *erecta* (A. Nels.) Dorn = Lappula squarrosa [LAPPSQU] Boraginaceae (V:D)
Lappula texana (Scheele) Britt. = Lappula occidentalis var. cupulata [LAPPOCC1] Boraginaceae (V:D)
Lappula texana var. *coronata* (Greene) A. Nels. & J.F. Macbr. = Lappula occidentalis var. cupulata [LAPPOCC1] Boraginaceae (V:D)
Lappula texana var. *heterosperma* (Greene) A. Nels. & J.F. Macbr. = Lappula occidentalis var. cupulata [LAPPOCC1] Boraginaceae (V:D)
Lappula texana var. *homosperma* (A. Nels.) A. Nels. & J.F. Macbr. = Lappula occidentalis var. cupulata [LAPPOCC1] Boraginaceae (V:D)
Lapsana L. [LAPSANA$] Asteraceae (V:D)
*****Lapsana communis** L. [LAPSCOM] Asteraceae (V:D)
Larix P. Mill. [LARIX$] Pinaceae (V:G)

Larix alaskensis W. Wight = Larix laricina [LARILAR] Pinaceae (V:G)
Larix laricina (Du Roi) K. Koch [LARILAR] Pinaceae (V:G)
Larix laricina var. *alaskensis* (W. Wight) Raup = Larix laricina [LARILAR] Pinaceae (V:G)
Larix lyallii Parl. [LARILYA] Pinaceae (V:G)
Larix occidentalis Nutt. [LARIOCC] Pinaceae (V:G)
Lasallea falcata (Lindl.) Semple & Brouillet = Aster falcatus [ASTEFAL] Asteraceae (V:D)
Lasallia Mérat [LASALLI$] Umbilicariaceae (L)
Lasallia papulosa (Ach.) Llano [LASAPAP] Umbilicariaceae (L)
Lasallia pensylvanica (Hoffm.) Llano [LASAPEN] Umbilicariaceae (L)
Lasallia pustulata ssp. *papulosa* (Ach.) W.A. Weber = Lasallia papulosa [LASAPAP] Umbilicariaceae (L)
Lasthenia Cass. [LASTHEN$] Asteraceae (V:D)
Lasthenia maritima (Gray) Ornduff [LASTMAR] Asteraceae (V:D)
Lasthenia minor ssp. *maritima* (Gray) Ornduff = Lasthenia maritima [LASTMAR] Asteraceae (V:D)
Lasthenia minor var. *maritima* (Gray) Cronq. = Lasthenia maritima [LASTMAR] Asteraceae (V:D)
Lastrea oregana (C. Christens.) Copeland = Thelypteris nevadensis [THELNEV] Thelypteridaceae (V:F)
Lathyrus L. [LATHYRU$] Fabaceae (V:D)
Lathyrus bijugatus White [LATHBIJ] Fabaceae (V:D)
Lathyrus bijugatus var. *sandbergii* White = Lathyrus bijugatus [LATHBIJ] Fabaceae (V:D)
Lathyrus japonicus Willd. [LATHJAP] Fabaceae (V:D)
Lathyrus japonicus ssp. *maritimus* (L.) P.W. Ball = Lathyrus japonicus var. maritimus [LATHJAP1] Fabaceae (V:D)
Lathyrus japonicus var. *glaber* (Ser.) Fern. = Lathyrus japonicus var. maritimus [LATHJAP1] Fabaceae (V:D)
Lathyrus japonicus var. **maritimus** (L.) Kartesz & Gandhi [LATHJAP1] Fabaceae (V:D)
***Lathyrus latifolius** L. [LATHLAT] Fabaceae (V:D)
Lathyrus littoralis (Nutt.) Endl. [LATHLIT] Fabaceae (V:D)
Lathyrus maritimus Bigelow = Lathyrus japonicus var. maritimus [LATHJAP1] Fabaceae (V:D)
Lathyrus maritimus var. *glaber* (Ser.) Eames = Lathyrus japonicus var. maritimus [LATHJAP1] Fabaceae (V:D)
Lathyrus myrtifolius Muhl. ex Willd. = Lathyrus palustris [LATHPAL] Fabaceae (V:D)
Lathyrus nevadensis S. Wats. [LATHNEV] Fabaceae (V:D)
Lathyrus nevadensis ssp. **lanceolatus** var. **pilosellus** (M.E. Peck) C.L. Hitchc. [LATHNEV2] Fabaceae (V:D)
Lathyrus nevadensis var. *pilosellus* (Peck) C.L. Hitchc. = Lathyrus nevadensis ssp. lanceolatus var. pilosellus [LATHNEV2] Fabaceae (V:D)
Lathyrus ochroleucus Hook. [LATHOCH] Fabaceae (V:D)
Lathyrus palustris L. [LATHPAL] Fabaceae (V:D)
Lathyrus palustris ssp. *pilosus* (Cham.) Hultén = Lathyrus palustris [LATHPAL] Fabaceae (V:D)
Lathyrus palustris var. *linearifolius* Ser. = Lathyrus palustris [LATHPAL] Fabaceae (V:D)
Lathyrus palustris var. *macranthus* (White) Fern. = Lathyrus palustris [LATHPAL] Fabaceae (V:D)
Lathyrus palustris var. *meridionalis* Butters & St. John = Lathyrus palustris [LATHPAL] Fabaceae (V:D)
Lathyrus palustris var. *myrtifolius* (Muhl. ex Willd.) Gray = Lathyrus palustris [LATHPAL] Fabaceae (V:D)
Lathyrus palustris var. *pilosus* (Cham.) Ledeb. = Lathyrus palustris [LATHPAL] Fabaceae (V:D)
Lathyrus palustris var. *retusus* Fern. & St. John = Lathyrus palustris [LATHPAL] Fabaceae (V:D)
***Lathyrus pratensis** L. [LATHPRA] Fabaceae (V:D)
***Lathyrus sphaericus** Retz. [LATHSPH] Fabaceae (V:D)
***Lathyrus sylvestris** L. [LATHSYL] Fabaceae (V:D)
Lecanactis Körber [LECANAC$] Opegraphaceae (L)
Lecanactis abietina (Ach.) Körber [LECAABI] Opegraphaceae (L)
Lecanactis akompsa (Tuck.) Essl. & Egan [LECAAKO] Opegraphaceae (L)
Lecanactis megaspora (G. Merr.) Brodo = Lecanactis abietina [LECAABI] Opegraphaceae (L)
Lecanactis patellarioides (Nyl.) Vainio = Bactrospora patellarioides [BACTPAT] Opegraphaceae (L)
Lecania A. Massal. [LECANIA$] Bacidiaceae (L)
Lecania alpivaga Th. Fr. = Halecania alpivaga [HALEALP] Catillariaceae (L)
Lecania arctica Lynge = Caloplaca diphyodes [CALODIP] Teloschistaceae (L)
Lecania curvescens (Mudd) A.L. Sm. = Bryonora curvescens [BRYOCUR] Lecanoraceae (L)
Lecania cyrtella (Ach.) Th. Fr. [LECACYR] Bacidiaceae (L)
Lecania cyrtellina (Nyl.) Sandst. = Lecania cyrtella [LECACYR] Bacidiaceae (L)
Lecania dimera (Nyl.) Th. Fr. = Lecania dubitans [LECADUB] Bacidiaceae (L)
Lecania dubitans (Nyl.) A.L. Sm. [LECADUB] Bacidiaceae (L)
Lecania dudleyi Herre [LECADUD] Bacidiaceae (L)
Lecania erysibe (Ach.) Mudd [LECAERY] Bacidiaceae (L)
Lecania fructigena Zahlbr. [LECAFRU] Bacidiaceae (L)
Lecania hassei (Zahlbr.) W. Noble [LECAHAS] Bacidiaceae (L)
Lecania nylanderiana A. Massal. [LECANYL] Bacidiaceae (L)
Lecania thallophila H. Magn. = Halecania alpivaga [HALEALP] Catillariaceae (L)
Lecanora Ach. [LECANOR$] Lecanoraceae (L)
Lecanora albescens (Hoffm.) Branth & Rostrup [LECAALE] Lecanoraceae (L)
Lecanora allophana Nyl. [LECAALL] Lecanoraceae (L)
Lecanora alphoplaca (Wahlenb.) Ach. = Lobothallia alphoplaca [LOBOALP] Hymeneliaceae (L)
Lecanora alpina Sommerf. = Bellemerea alpina [BELLALP] Porpidiaceae (L)
Lecanora anseris Lynge = Aspicilia anseris [ASPIANS] Hymeneliaceae (L)
Lecanora aquatica (Körber) Hepp = Aspicilia aquatica [ASPIAQU] Hymeneliaceae (L)
Lecanora argopholis (Ach.) Ach. [LECAARO] Lecanoraceae (L)
Lecanora atra (Hudson) Ach. = Tephromela atra [TEPHATR] Bacidiaceae (L)
Lecanora atrynea (Ach.) Röhl. = Lecanora cenisia [LECACEN] Lecanoraceae (L)
Lecanora badia (Hoffm.) Ach. = Protoparmelia badia [PROTBAD] Lecanoraceae (L)
Lecanora beringii Nyl. [LECABER] Lecanoraceae (L)
Lecanora boligera (Norman ex Th. Fr.) Hedl. [LECABOL] Lecanoraceae (L)
Lecanora cadubriae (A. Massal.) Hedl. [LECACAD] Lecanoraceae (L)
Lecanora caesiocinerea Nyl. ex Malbr. = Aspicilia caesiocinerea [ASPICAE] Hymeneliaceae (L)
Lecanora caesiopruinosa H. Magn. = Aspicilia caesiopruinosa [ASPICAS] Hymeneliaceae (L)
Lecanora calcarea (L.) Sommerf. = Aspicilia calcarea [ASPICAL] Hymeneliaceae (L)
Lecanora candida (Anzi) Nyl. = Aspicilia candida [ASPICAN] Hymeneliaceae (L)
Lecanora cascadensis H. Magn. = Lecanora garovaglii [LECAGAR] Lecanoraceae (L)
Lecanora castanea (Hepp) Th. Fr. = Bryonora castanea [BRYOCAS] Lecanoraceae (L)
Lecanora cateilea (Ach.) A. Massal. [LECACTE] Lecanoraceae (L)
Lecanora cenisia Ach. [LECACEN] Lecanoraceae (L)
Lecanora chlarona (Ach.) Nyl. = Lecanora pulicaris [LECAPUL] Lecanoraceae (L)
Lecanora chlarotera Nyl. [LECACHA] Lecanoraceae (L)
Lecanora chlorina (Flotow) Lamy = Caloplaca chlorina [CALOCHL] Teloschistaceae (L)
Lecanora chloropolia (Erichsen) Almb. = Lecanora impudens [LECAIMP] Lecanoraceae (L)
Lecanora christoi W.A. Weber = Lecanora phaedrophthalma [LECAPHA] Lecanoraceae (L)
Lecanora chrysoleuca (Sm.) Ach. = Rhizoplaca chrysoleuca [RHIZCHR] Lecanoraceae (L)
Lecanora cinerea (L.) Sommerf. = Aspicilia cinerea [ASPICIN] Hymeneliaceae (L)

Lecanora cinereofusca H. Magn. [LECACIN] Lecanoraceae (L)
Lecanora cinereorufescens (Ach.) Hepp = Bellemerea cinereorufescens [BELLCIN] Porpidiaceae (L)
Lecanora circumborealis Brodo & Vitik. [LECACIC] Lecanoraceae (L)
Lecanora coilocarpa (Ach.) Nyl. = Lecanora pulicaris [LECAPUL] Lecanoraceae (L)
Lecanora coilocarpa auct. = Lecanora circumborealis [LECACIC] Lecanoraceae (L)
Lecanora confusa Almb. [LECACOF] Lecanoraceae (L)
Lecanora congesta Lynge [LECACON] Lecanoraceae (L)
Lecanora conizaea (Ach.) Nyl. ex Crombie = Lecanora expallens [LECAEXP] Lecanoraceae (L)
Lecanora conizaea auct. = Lecanora strobilina [LECASTO] Lecanoraceae (L)
Lecanora conizaeoides Nyl. ex Crombie [LECACOA] Lecanoraceae (L)
Lecanora contorta (Hoffm.) J. Steiner = Aspicilia contorta [ASPICON] Hymeneliaceae (L)
Lecanora contractula Nyl. [LECACOR] Lecanoraceae (L)
Lecanora curvescens (Mudd) Nyl. = Bryonora curvescens [BRYOCUR] Lecanoraceae (L)
Lecanora demissa (Flotow) Zahlbr. [LECADEM] Lecanoraceae (L)
Lecanora desertorum Kremp. = Aspicilia desertorum [ASPIDES] Hymeneliaceae (L)
Lecanora diamarta (Ach.) Vainio = Bellemerea diamarta [BELLDIA] Porpidiaceae (L)
Lecanora diffracta Ach. = Lecanora muralis [LECAMUR] Lecanoraceae (L)
Lecanora dispersa (Pers.) Sommerf. [LECADIP] Lecanoraceae (L)
Lecanora disserpens (Zahlbr.) H. Magn. = Aspicilia disserpens [ASPIDIS] Hymeneliaceae (L)
Lecanora effusa (Hoffm.) Ach. = Lecanora saligna [LECASAI] Lecanoraceae (L)
Lecanora epanora (Ach.) Ach. [LECAEPA] Lecanoraceae (L)
Lecanora epibryon (Ach.) Ach. [LECAEPI] Lecanoraceae (L)
Lecanora epulotica (Ach.) Nyl. = Hymenelia epulotica [HYMEEPU] Hymeneliaceae (L)
Lecanora expallens Ach. [LECAEXP] Lecanoraceae (L)
Lecanora eyerdamii Herre = Lecanora xylophila [LECAXYL] Lecanoraceae (L)
Lecanora fimbriata H. Magn. = Aspicilia fimbriata [ASPIFIM] Hymeneliaceae (L)
Lecanora flavopunctata Tønsberg = Biatora flavopunctata [BIATFLA] Bacidiaceae (L)
Lecanora frustulosa (Dickson) Ach. = Lecanora argopholis [LECAARO] Lecanoraceae (L)
Lecanora fuscescens (Sommerf.) Nyl. [LECAFUE] Lecanoraceae (L)
Lecanora garovaglii (Körber) Zahlbr. [LECAGAR] Lecanoraceae (L)
Lecanora gelida (L.) Ach. = Placopsis gelida [PLACGEL] Trapeliaceae (L)
Lecanora gibbosa (Ach.) Nyl. = Aspicilia gibbosa [ASPIGIB] Hymeneliaceae (L)
Lecanora gibbosula H. Magn. = Aspicilia gibbosa [ASPIGIB] Hymeneliaceae (L)
Lecanora glaucomela Tuck. = Pertusaria glaucomela [PERTGLA] Pertusariaceae (L)
Lecanora grandis H. Magn. = Protoparmelia badia [PROTBAD] Lecanoraceae (L)
Lecanora grantii H. Magn. = Lecanora xylophila [LECAXYL] Lecanoraceae (L)
Lecanora hagenii (Ach.) Ach. [LECAHAG] Lecanoraceae (L)
Lecanora helicopis (Wahlenb.) Ach. [LECAHEL] Lecanoraceae (L)
Lecanora hypopta (Ach.) Vainio [LECAHYA] Lecanoraceae (L)
Lecanora impudens Degel. [LECAIMP] Lecanoraceae (L)
Lecanora incusa (Fr.) Vainio = Lecanora demissa [LECADEM] Lecanoraceae (L)
Lecanora intricata (Ach.) Ach. [LECAINT] Lecanoraceae (L)
Lecanora intumescens (Rebent.) Rabenh. [LECAINU] Lecanoraceae (L)
Lecanora karelica H. Magn. = Aspicilia karelica [ASPIKAR] Hymeneliaceae (L)
Lecanora lacustris (With.) Nyl. = Ionaspis lacustris [IONALAC] Hymeneliaceae (L)
Lecanora laxula H. Magn. = Aspicilia laxula [ASPILAX] Hymeneliaceae (L)
Lecanora lentigera (Weber) Ach. = Squamarina lentigera [SQUALEN] Bacidiaceae (L)
Lecanora leprosescens Sandst. = Aspicilia leprosescens [ASPILEP] Hymeneliaceae (L)
Lecanora luteovernalis Brodo [LECALUT] Lecanoraceae (L)
Lecanora marginata (Schaerer) Hertel & Rambold [LECAMAG] Lecanoraceae (L)
Lecanora marina Wedd. = Caloplaca marina [CALOMAR] Teloschistaceae (L)
Lecanora melanaspis (Ach.) Ach. = Lobothallia melanaspis [LOBOMEL] Hymeneliaceae (L)
Lecanora melanophthalma (DC.) Ramond = Rhizoplaca melanophthalma [RHIZMEL] Lecanoraceae (L)
Lecanora mughicola Nyl. [LECAMUG] Lecanoraceae (L)
Lecanora muralis (Schreber) Rabenh. [LECAMUR] Lecanoraceae (L)
Lecanora mutabilis (Ach.) Nyl. = Megaspora verrucosa [MEGAVEU] Hymeneliaceae (L)
Lecanora mutabilis Sommerf. = Lecanora intricata [LECAINT] Lecanoraceae (L)
Lecanora myrinii (Fr.) Tuck. = Aspicilia myrinii [ASPIMYR] Hymeneliaceae (L)
Lecanora nevadensis H. Magn. = Lecanora garovaglii [LECAGAR] Lecanoraceae (L)
Lecanora occidentalis (Lynge) Lynge = Lecanora argopholis [LECAARO] Lecanoraceae (L)
Lecanora ochrococca (Nyl.) Clauzade & Roux = Protoparmelia ochrococca [PROTOCH] Lecanoraceae (L)
Lecanora orae-frigidae R. Sant. [LECAORA] Lecanoraceae (L)
Lecanora oregana Tuck. = Lecanora argopholis [LECAARO] Lecanoraceae (L)
Lecanora pacifica Tuck. [LECAPAI] Lecanoraceae (L)
Lecanora palanderi Vainio = Lecanora zosterae [LECAZOS] Lecanoraceae (L)
Lecanora pelobotrya (Wahlenb.) Sommerf. = Amygdalaria pelobotryon [AMYGPEL] Porpidiaceae (L)
Lecanora peltata (Ramond) Steudel = Rhizoplaca peltata [RHIZPEL] Lecanoraceae (L)
Lecanora persimilis (Th. Fr.) Nyl. [LECAPER] Lecanoraceae (L)
Lecanora phaedrophthalma Poelt [LECAPHA] Lecanoraceae (L)
Lecanora phaeobola Tuck. = Protoparmelia ochrococca [PROTOCH] Lecanoraceae (L)
Lecanora pinastri (Schaerer) H. Magn. = Lecanora pulicaris [LECAPUL] Lecanoraceae (L)
Lecanora polytropa (Hoffm.) Rabenh. [LECAPOY] Lecanoraceae (L)
Lecanora praeradiosa Nyl. = Lobothallia praeradiosa [LOBOPRA] Hymeneliaceae (L)
Lecanora pringlei (Tuck.) Lamb [LECAPRI] Lecanoraceae (L)
Lecanora pulicaris (Pers.) Ach. [LECAPUL] Lecanoraceae (L)
Lecanora reagens Norman [LECAREA] Lecanoraceae (L)
Lecanora reptans Looman = Aspicilia reptans [ASPIREP] Hymeneliaceae (L)
Lecanora riparia G. Merr. = Lecanora xylophila [LECAXYL] Lecanoraceae (L)
Lecanora rubina (Vill.) Ach. = Rhizoplaca chrysoleuca [RHIZCHR] Lecanoraceae (L)
Lecanora rugosella Zahlbr. [LECARUG] Lecanoraceae (L)
Lecanora rupicola (L.) Zahlbr. [LECARUP] Lecanoraceae (L)
Lecanora ryrkaipiae H. Magn. = Aspicilia ryrkaipiae [ASPIRYR] Hymeneliaceae (L)
Lecanora salicicola H. Magn. [LECASAL] Lecanoraceae (L)
Lecanora saligna (Schrader) Zahlbr. [LECASAI] Lecanoraceae (L)
Lecanora sambuci (Pers.) Nyl. [LECASAM] Lecanoraceae (L)
Lecanora scotopholis (Tuck.) Timdal [LECASCO] Lecanoraceae (L)
Lecanora semitensis Tuck. [LECASEM] Lecanoraceae (L)
Lecanora sordida (Pers.) Th. Fr. = Lecanora rupicola [LECARUP] Lecanoraceae (L)
Lecanora straminea Ach. [LECASTR] Lecanoraceae (L)

Lecanora strobilina (Sprengel) Kieffer [LECASTO] Lecanoraceae (L)
Lecanora subfusca (L.) Ach. = Lecanora allophana [LECAALL] Lecanoraceae (L)
Lecanora subintricata (Nyl.) Th. Fr. [LECASUI] Lecanoraceae (L)
Lecanora subolivascens Nyl. = Lecanora demissa [LECADEM] Lecanoraceae (L)
Lecanora supertegens (Arnold) Zahlbr. = Aspicilia supertegens [ASPISUP] Hymeneliaceae (L)
Lecanora symmicta (Ach.) Ach. [LECASYM] Lecanoraceae (L)
Lecanora symmictera Nyl. = Lecanora symmicta [LECASYM] Lecanoraceae (L)
Lecanora tetraspora H. Magn. = Lecanora pacifica [LECAPAI] Lecanoraceae (L)
Lecanora thamnoplaca Tuck. = Lobothallia alphoplaca [LOBOALP] Hymeneliaceae (L)
Lecanora torrida Vainio [LECATOR] Lecanoraceae (L)
Lecanora urceolata (Fr.) Wetmore = Megaspora verrucosa [MEGAVEU] Hymeneliaceae (L)
Lecanora varia (Hoffm.) Ach. [LECAVAR] Lecanoraceae (L)
Lecanora variolascens auct. = Lecanora impudens [LECAIMP] Lecanoraceae (L)
Lecanora verrucigera (Hue) Zahlbr. = Aspicilia verrucigera [ASPIVER] Hymeneliaceae (L)
Lecanora verrucosa (Ach.) Laurer = Megaspora verrucosa [MEGAVEU] Hymeneliaceae (L)
Lecanora weberi B.D. Ryan [LECAWEB] Lecanoraceae (L)
Lecanora wisconsinensis H. Magn. [LECAWIS] Lecanoraceae (L)
Lecanora xylophila Hue [LECAXYL] Lecanoraceae (L)
Lecanora zosterae (Ach.) Nyl. [LECAZOS] Lecanoraceae (L)
Lecanoropsis saligna (Schrader) Choisy = Lecanora saligna [LECASAI] Lecanoraceae (L)
Lecidea Ach. [LECIDEA$] Lecideaceae (L)
Lecidea absistens Nyl. = Bacidia absistens [BACIABS] Bacidiaceae (L)
Lecidea acrocyanea (Th. Fr.) H. Magn. = Lecidella patavina [LECIPAA] Lecanoraceae (L)
Lecidea aeruginosa Borrer = Trapeliopsis flexuosa [TRAPFLE] Trapeliaceae (L)
Lecidea alaiensis Vainio = Lecidella patavina [LECIPAA] Lecanoraceae (L)
Lecidea alaskensis Nyl. = Bacidia alaskensis [BACIALA] Bacidiaceae (L)
Lecidea albocaerulescens (Wulfen) Ach. = Porpidia albocaerulescens [PORPALB] Porpidiaceae (L)
Lecidea albofuscescens Nyl. [LECIALO] Lecideaceae (L)
Lecidea albosuffusa Th. Fr. = Farnoldia jurana [FARNJUR] Porpidiaceae (L)
Lecidea alpestris Sommerf. [LECIALP] Lecideaceae (L)
Lecidea aniptiza Stirton = Micarea denigrata [MICADEN] Micareaceae (L)
Lecidea anthracophila Nyl. = Biatora anthracophila [BIATANT] Bacidiaceae (L)
Lecidea arctica Sommerf. = Lecidea caesioatra [LECICAS] Lecideaceae (L)
Lecidea armeniaca (DC.) Fr. = Tephromela armeniaca [TEPHARM] Bacidiaceae (L)
Lecidea assimilata Nyl. = Micarea assimilata [MICAASS] Micareaceae (L)
Lecidea atrata (Ach.) Wahlenb. = Tremolecia atrata [TREMATR] Hymeneliaceae (L)
Lecidea atrobrunnea (Ramond ex Lam. & DC.) Schaerer [LECIATR] Lecideaceae (L)
Lecidea atroflava Turn. = Caloplaca atroflava [CALOATR] Teloschistaceae (L)
Lecidea atrofusca (Hepp) Mudd = Mycobilimbia hypnorum [MYCOHYP] Porpidiaceae (L)
Lecidea atrolutescens Nyl. = Lecidea mannii [LECIMAN] Lecideaceae (L)
Lecidea atroviridis (Arnold) Th. Fr. [LECIATO] Lecideaceae (L)
Lecidea auerswaldii Hepp ex Stizenb. = Bacidia auerswaldii [BACIAUE] Bacidiaceae (L)
Lecidea auriculata Th. Fr. [LECIAUR] Lecideaceae (L)
Lecidea bacillifera f. *circumspecta* Nyl., nom. nud. = Bacidia circumspecta [BACICIR] Bacidiaceae (L)
Lecidea bacillifera var. *circumspecta* Nyl. ex Vainio = Bacidia circumspecta [BACICIR] Bacidiaceae (L)
Lecidea berengeriana (A. Massal.) Nyl. = Mycobilimbia berengeriana [MYCOBER] Porpidiaceae (L)
Lecidea botryosa (Fr.) Th. Fr. = Biatora botryosa [BIATBOT] Bacidiaceae (L)
Lecidea brandegei Tuck. = Rhizoplaca melanophthalma [RHIZMEL] Lecanoraceae (L)
Lecidea brouardii (de Lesd.) Zahlbr. = Psorula rufonigra [PSORRUF] Psoraceae (L)
Lecidea cadubriae (A. Massal.) Nyl. = Lecanora cadubriae [LECACAD] Lecanoraceae (L)
Lecidea caesioatra Schaerer [LECICAS] Lecideaceae (L)
Lecidea callispora Knight = Hafellia callispora [HAFECAL] Physciaceae (L)
Lecidea carpathica (Körber) Szat. = Lecidella carpathica [LECICAA] Lecanoraceae (L)
Lecidea catalinaria Stizenb. = Lecidella asema [LECIASE] Lecanoraceae (L)
Lecidea cinnabarina Sommerf. = Pyrrhospora cinnabarina [PYRRCIN] Lecanoraceae (L)
Lecidea circumspecta (Nyl. ex Vainio) Hedl. = Bacidia circumspecta [BACICIR] Bacidiaceae (L)
Lecidea coarctata (Sm.) Nyl. = Trapelia coarctata [TRAPCOA] Trapeliaceae (L)
Lecidea columbiana H. Magn. = Lecidea tessellata [LECITEL] Lecideaceae (L)
Lecidea confluens (Weber) Ach. [LECICOF] Lecideaceae (L)
Lecidea contigua Fr. = Porpidia macrocarpa [PORPMAC] Porpidiaceae (L)
Lecidea contraponenda Arnold = Porpidia contraponenda [PORPCON] Porpidiaceae (L)
Lecidea crassipes (Th. Fr.) Nyl. = Helocarpon crassipes [HELOCRA] Micareaceae (L)
Lecidea cruciaria Tuck. [LECICRU] Lecideaceae (L)
Lecidea crustulata (Ach.) Sprengel = Porpidia crustulata [PORPCRU] Porpidiaceae (L)
Lecidea cyanea (Ach.) Rohl. = Lecidea tessellata [LECITEL] Lecideaceae (L)
Lecidea cyanescens Lynge = Lecidea lapicida [LECILAP] Lecideaceae (L)
Lecidea decipiens (Hedwig) Ach. = Psora decipiens [PSORDEC] Psoraceae (L)
Lecidea demissa (Rutstr.) Ach. = Lecidoma demissum [LECIDEU] Psoraceae (L)
Lecidea dicksonii auct. = Tremolecia atrata [TREMATR] Hymeneliaceae (L)
Lecidea effusa (Auersw. ex Rabenh.) Stizenb. = Bacidia auerswaldii [BACIAUE] Bacidiaceae (L)
Lecidea elabens Fr. = Pyrrhospora elabens [PYRRELA] Lecanoraceae (L)
Lecidea elaeochroma (Ach.) Ach. = Lecidella elaeochroma [LECIELO] Lecanoraceae (L)
Lecidea elata Schaerer = Lecanora marginata [LECAMAG] Lecanoraceae (L)
Lecidea endoleucula Nyl. = Bacidia laurocerasi [BACILAU] Bacidiaceae (L)
Lecidea endolithea Lynge = Lecidella patavina [LECIPAA] Lecanoraceae (L)
Lecidea euphorea (Flörke) Nyl. = Lecidella euphorea [LECIEUP] Lecanoraceae (L)
Lecidea evansii H. Magn. = Lecidella carpathica [LECICAA] Lecanoraceae (L)
Lecidea flexuosa (Fr.) Nyl. = Trapeliopsis flexuosa [TRAPFLE] Trapeliaceae (L)
Lecidea friesii Ach. = Hypocenomyce friesii [HYPOFRI] Lecideaceae (L)
Lecidea fusca (Schaerer) Th. Fr. = Mycobilimbia hypnorum [MYCOHYP] Porpidiaceae (L)
Lecidea fuscescens Sommerf. = Lecanora fuscescens [LECAFUE] Lecanoraceae (L)

Lecidea fuscoatra (L.) Ach. [LECIFUA] Lecideaceae (L)
Lecidea gelatinosa Flörke = Trapeliopsis gelatinosa [TRAPGEL] Trapeliaceae (L)
Lecidea geophana Nyl. = Steinia geophana [STEIGEO] Lecideaceae (L)
Lecidea glabrescens Nyl. = Mycoblastus affinis [MYCOAFF] Mycoblastaceae (L)
Lecidea glebulosa (Fr.) Clem. = Trapeliopsis wallrothii [TRAPWAL] Trapeliaceae (L)
Lecidea globifera Ach. = Psora globifera [PSORGLO] Psoraceae (L)
Lecidea glomerulosa (DC.) Steudel = Lecidella euphorea [LECIEUP] Lecanoraceae (L)
Lecidea granulosa (Hoffm.) Ach. = Trapeliopsis granulosa [TRAPGRA] Trapeliaceae (L)
Lecidea gregaria G. Merr. = Trapelia involuta [TRAPINV] Trapeliaceae (L)
Lecidea grisella Flörke ex Schaerer = Lecidea fuscoatra [LECIFUA] Lecideaceae (L)
Lecidea hegetschweileri Hepp = Bacidia subincompta [BACISUI] Bacidiaceae (L)
Lecidea helvola (Körber) Th. Fr. = Biatora helvola [BIATHEL] Bacidiaceae (L)
Lecidea heppii R. Anderson & W.A. Weber = Lecidella wulfenii [LECIWUL] Lecanoraceae (L)
Lecidea humosa (Hoffm.) Nyl. = Placynthiella uliginosa [PLACULI] Trapeliaceae (L)
Lecidea hypnorum Lib. = Mycobilimbia hypnorum [MYCOHYP] Porpidiaceae (L)
Lecidea hypocrita A. Massal. = Farnoldia hypocrita [FARNHYP] Porpidiaceae (L)
Lecidea hypopta Ach. = Lecanora hypopta [LECAHYA] Lecanoraceae (L)
Lecidea insularis Nyl. = Rimularia insularis [RIMUINS] Rimulariaceae (L)
Lecidea intumescens (Flotow) Nyl. = Rimularia insularis [RIMUINS] Rimulariaceae (L)
Lecidea invertens Nyl. = Bacidia laurocerasi [BACILAU] Bacidiaceae (L)
Lecidea jurana Schaerer = Farnoldia jurana [FARNJUR] Porpidiaceae (L)
Lecidea lactea Flörke ex Schaerer = Lecidea lapicida [LECILAP] Lecideaceae (L)
Lecidea lacus-crateris H. Magn. = Lecidella stigmatea [LECISTI] Lecanoraceae (L)
Lecidea lapicida (Ach.) Ach. [LECILAP] Lecideaceae (L)
Lecidea latypea Ach. = Lecidea plana [LECIPLN] Lecideaceae (L)
Lecidea latypea auct. = Lecidella carpathica [LECICAA] Lecanoraceae (L)
Lecidea latypiza Nyl. = Lecidella carpathica [LECICAA] Lecanoraceae (L)
Lecidea leprarioides Tønsberg [LECILEP] Lecideaceae (L)
Lecidea leucothallina Arnold [LECILEC] Lecideaceae (L)
Lecidea limitata auct. = Lecidella elaeochroma [LECIELO] Lecanoraceae (L)
Lecidea lithophila (Ach.) Ach. [LECILIT] Lecideaceae (L)
Lecidea lithospersa Zahlbr. = Farnoldia hypocrita [FARNHYP] Porpidiaceae (L)
Lecidea lucida (Ach.) Ach. = Psilolechia lucida [PSILLUC] Micareaceae (L)
Lecidea lurida (Ach.) DC. [LECILUR] Lecideaceae (L)
Lecidea luridella Tuck. = Psora luridella [PSORLUR] Psoraceae (L)
Lecidea luteola (Schrad.) Ach. = Bacidia rubella [BACIRUB] Bacidiaceae (L)
Lecidea lyngeana Zahlbr. = Adelolecia pilati [ADELPIL] Bacidiaceae (L)
Lecidea macrocarpa (DC.) Steudel = Porpidia macrocarpa [PORPMAC] Porpidiaceae (L)
Lecidea mannii Tuck. [LECIMAN] Lecideaceae (L)
Lecidea marginata Schaerer = Lecanora marginata [LECAMAG] Lecanoraceae (L)
Lecidea melancheima Tuck. = Pyrrhospora elabens [PYRRELA] Lecanoraceae (L)
Lecidea micacea Körber = Lecidella stigmatea [LECISTI] Lecanoraceae (L)
Lecidea misella (Nyl.) Nyl. = Micarea misella [MICAMIS] Micareaceae (L)
Lecidea mollis (Wahlenb.) Nyl. = Fuscidea mollis [FUSCMOL] Fuscideaceae (L)
Lecidea monticola Ach. = Clauzadea monticola [CLAUMON] Porpidiaceae (L)
Lecidea nigrocruenta Anzi = Porpidia macrocarpa [PORPMAC] Porpidiaceae (L)
Lecidea novomexicana (de Lesd.) W.A. Weber ex R. Anderson = Psora nipponica [PSORNIP] Psoraceae (L)
Lecidea nylanderi (Anzi) Th. Fr. [LECINYL] Lecideaceae (L)
Lecidea obtegens Th. Fr. = Trapelia obtegens [TRAPOBT] Trapeliaceae (L)
Lecidea ochrococca Nyl. = Protoparmelia ochrococca [PROTOCH] Lecanoraceae (L)
Lecidea oligotropha J.R. Laundon = Placynthiella oligotropha [PLACOLI] Trapeliaceae (L)
Lecidea olivacea (Hoffm.) A. Massal. = Lecidella elaeochroma [LECIELO] Lecanoraceae (L)
Lecidea ornata (Sommerf.) Hue = Trapelia involuta [TRAPINV] Trapeliaceae (L)
Lecidea ostreata (Hoffm.) Schaerer = Hypocenomyce scalaris [HYPOSCA] Lecideaceae (L)
Lecidea oxyspora (Tul.) Nyl. = Phacopsis oxyspora [PHACOXY] Lecanoraceae (L)
Lecidea pallida Th. Fr. = Pilophorus dovrensis [PILODOV] Stereocaulaceae (L)
Lecidea panaeola (Ach.) Ach. = Amygdalaria panaeola [AMYGPAN] Porpidiaceae (L)
Lecidea pantherina (Ach.) Th. Fr. = Lecidea lapicida [LECILAP] Lecideaceae (L)
Lecidea parasema var. *crustulata* Ach. = Porpidia crustulata [PORPCRU] Porpidiaceae (L)
Lecidea paupercula Th. Fr. = Lecidea praenubila [LECIPRE] Lecideaceae (L)
Lecidea pelobotrya (Wahlenb.) Leighton = Amygdalaria pelobotryon [AMYGPEL] Porpidiaceae (L)
Lecidea petri (Tuck.) Zahlbr. = Lecidea lurida [LECILUR] Lecideaceae (L)
Lecidea phaeops Nyl. [LECIPHA] Lecideaceae (L)
Lecidea phylliscina Nyl. = Porpidia macrocarpa [PORPMAC] Porpidiaceae (L)
Lecidea pilati (Hepp) Körber = Adelolecia pilati [ADELPIL] Bacidiaceae (L)
Lecidea plana (J. Lahm) Nyl. [LECIPLN] Lecideaceae (L)
Lecidea platycarpa Ach. = Porpidia macrocarpa [PORPMAC] Porpidiaceae (L)
Lecidea praenubila Nyl. [LECIPRE] Lecideaceae (L)
Lecidea prasinula (Wedd.) de Lesd. = Lecidella scabra [LECISCB] Lecanoraceae (L)
Lecidea pringlei Tuck. = Lecanora pringlei [LECAPRI] Lecanoraceae (L)
Lecidea pruinosa Ach. = Lecidea lithophila [LECILIT] Lecideaceae (L)
Lecidea pullata (Norman) Th. Fr. = Biatora pullata [BIATPUL] Bacidiaceae (L)
Lecidea purissima Darbish. = Lecanora marginata [LECAMAG] Lecanoraceae (L)
Lecidea quadricolor (Dickson) Borrer = Trapeliopsis granulosa [TRAPGRA] Trapeliaceae (L)
Lecidea quernea (Dickson) Ach. = Pyrrhospora quernea [PYRRQUE] Lecanoraceae (L)
Lecidea ramulicola H. Magn. = Lecanora cadubriae [LECACAD] Lecanoraceae (L)
Lecidea restituta Stirton = Hafellia callispora [HAFECAL] Physciaceae (L)
Lecidea rubiformis (Ach.) Wahlenb. = Psora rubiformis [PSORRUB] Psoraceae (L)
Lecidea rufofusca (Anzi) Nyl. [LECIRUF] Lecideaceae (L)
Lecidea rufonigra (Tuck.) Nyl. = Psorula rufonigra [PSORRUF] Psoraceae (L)

Lecidea russellii Tuck. = Psora russellii [PSORRUS] Psoraceae (L)
Lecidea russula Ach. = Pyrrhospora russula [PYRRRUS] Lecanoraceae (L)
Lecidea sanguineoatra auct. = Mycobilimbia hypnorum [MYCOHYP] Porpidiaceae (L)
Lecidea scabra Taylor = Lecidella scabra [LECISCB] Lecanoraceae (L)
Lecidea scalaris (Ach. ex Lilj.) Ach. = Hypocenomyce scalaris [HYPOSCA] Lecideaceae (L)
Lecidea scotopholis (Tuck.) Herre = Lecanora scotopholis [LECASCO] Lecanoraceae (L)
Lecidea separabilis (Nyl.) Arnold = Bacidia subincompta [BACISUI] Bacidiaceae (L)
Lecidea silacea Ach. [LECISIL] Lecideaceae (L)
Lecidea sorediata Lynge = Lecanora orae-frigidae [LECAORA] Lecanoraceae (L)
Lecidea soredifera Lowe = Porpidia crustulata [PORPCRU] Porpidiaceae (L)
Lecidea sorediza Nyl. = Porpidia tuberculosa [PORPTUB] Porpidiaceae (L)
Lecidea soredizodes (Lamy ex Nyl.) Vainio = Porpidia soredizodes [PORPSOR] Porpidiaceae (L)
Lecidea steriza (Ach.) Vainio = Porpidia macrocarpa [PORPMAC] Porpidiaceae (L)
Lecidea stigmatea Ach. = Lecidella stigmatea [LECISTI] Lecanoraceae (L)
Lecidea subauriculata Lynge = Adelolecia pilati [ADELPIL] Bacidiaceae (L)
Lecidea subcinnabarina Tønsberg = Pyrrhospora subcinnabarina [PYRRSUB] Lecanoraceae (L)
Lecidea subconnexa Stirton = Hafellia callispora [HAFECAL] Physciaceae (L)
Lecidea subcontinuior de Lesd. = Lecidella carpathica [LECICAA] Lecanoraceae (L)
Lecidea subincompta Nyl. = Bacidia subincompta [BACISUI] Bacidiaceae (L)
Lecidea subsorediza Lynge = Bellemerea subsorediza [BELLSUB] Porpidiaceae (L)
Lecidea sylvicola Flotow = Micarea sylvicola [MICASYL] Micareaceae (L)
Lecidea symmicta (Ach.) Ach. = Lecanora symmicta [LECASYM] Lecanoraceae (L)
Lecidea templetonii Taylor = Mycobilimbia hypnorum [MYCOHYP] Porpidiaceae (L)
Lecidea tessellata Flörke [LECITEL] Lecideaceae (L)
Lecidea tornoënsis Nyl. = Japewia tornoënsis [JUPETOR] Bacidiaceae (L)
Lecidea tumida A. Massal. = Porpidia tuberculosa [PORPTUB] Porpidiaceae (L)
Lecidea uliginosa (Schrader) Ach. = Placynthiella uliginosa [PLACULI] Trapeliaceae (L)
Lecidea umbonata (Hepp) Mudd [LECIUMB] Lecideaceae (L)
Lecidea vacciniicola Tønsberg [LECIVAC] Lecideaceae (L)
Lecidea varians Ach. = Pyrrhospora varians [PYRRVAR] Lecanoraceae (L)
Lecidea vernalis (L.) Ach. = Biatora vernalis [BIATVER] Bacidiaceae (L)
Lecidea viridescens (Schrader) Ach. = Trapeliopsis viridescens [TRAPVIR] Trapeliaceae (L)
Lecidea vorticosa (Flörke) Körber = Carbonea vorticosa [CARBVOR] Lecanoraceae (L)
Lecidea vulgata Zahlbr. = Lecidella stigmatea [LECISTI] Lecanoraceae (L)
Lecidea wallrothii Flörke ex Sprengel = Trapeliopsis wallrothii [TRAPWAL] Trapeliaceae (L)
Lecidea wulfenii (Hepp) Arnold = Lecidella wulfenii [LECIWUL] Lecanoraceae (L)
Lecidea ypocrita A. Massal. = Farnoldia hypocrita [FARNHYP] Porpidiaceae (L)
Lecidella Körber [LECIDLL$] Lecanoraceae (L)
Lecidella alaiensis (Vainio) Hertel = Lecidella patavina [LECIPAA] Lecanoraceae (L)
Lecidella asema (Nyl.) Knoph & Hertel [LECIASE] Lecanoraceae (L)
Lecidella carpathica Körber [LECICAA] Lecanoraceae (L)
Lecidella elaeochroma (Ach.) Hazsl. [LECIELO] Lecanoraceae (L)
Lecidella elaeochromoides (Nyl.) Knoph & Hertel = Lecidella asema [LECIASE] Lecanoraceae (L)
Lecidella euphorea (Flörke) Hertel [LECIEUP] Lecanoraceae (L)
Lecidella glomerulosa (DC.) Choisy = Lecidella euphorea [LECIEUP] Lecanoraceae (L)
Lecidella inamoena (Müll. Arg.) Hertel = Lecidella patavina [LECIPAA] Lecanoraceae (L)
Lecidella laureri (Hepp) Körber [LECILAU] Lecanoraceae (L)
Lecidella patavina (A. Massal.) Knoph & Leuckert [LECIPAA] Lecanoraceae (L)
Lecidella prasinula (Wedd.) Hertel = Lecidella scabra [LECISCB] Lecanoraceae (L)
Lecidella scabra (Taylor) Hertel & Leuckert [LECISCB] Lecanoraceae (L)
Lecidella spitsbergensis (Lynge) Hertel & Leuckert = Lecidella patavina [LECIPAA] Lecanoraceae (L)
Lecidella stigmatea (Ach.) Hertel & Leuckert [LECISTI] Lecanoraceae (L)
Lecidella subincongrua (Nyl.) Hertel & Leuckert = Lecidella asema [LECIASE] Lecanoraceae (L)
Lecidella subincongrua var. *elaeochromoides* (Nyl.) Hertel & Leuckert = Lecidella asema [LECIASE] Lecanoraceae (L)
Lecidella wulfenii (Hepp) Körber [LECIWUL] Lecanoraceae (L)
Lecidoma Gotth. Schneider & Hertel [LECIDOM$] Psoraceae (L)
Lecidoma demissum (Rutstr.) Gotth. Schneider & Hertel [LECIDEU] Psoraceae (L)
Leciophysma Th. Fr. [LECIOPH$] Collemataceae (L)
Leciophysma finmarkicum Th. Fr. [LECIFIN] Collemataceae (L)
Leciophysma furfurascens (Nyl.) Gyelnik [LECIFUU] Collemataceae (L)
Ledum L. [LEDUM$] Ericaceae (V:D)
Ledum decumbens (Ait.) Lodd. ex Steud. = Ledum palustre ssp. decumbens [LEDUPAL1] Ericaceae (V:D)
Ledum glandulosum Nutt. [LEDUGLA] Ericaceae (V:D)
Ledum glandulosum var. *californicum* (Kellogg) C.L. Hitchc. = Ledum glandulosum [LEDUGLA] Ericaceae (V:D)
Ledum groenlandicum Oeder [LEDUGRO] Ericaceae (V:D)
Ledum palustre L. [LEDUPAL] Ericaceae (V:D)
Ledum palustre ssp. **decumbens** (Ait.) Hultén [LEDUPAL1] Ericaceae (V:D)
Ledum palustre ssp. *groenlandicum* (Oeder) Hultén = Ledum groenlandicum [LEDUGRO] Ericaceae (V:D)
Ledum palustre var. *decumbens* Ait. = Ledum palustre ssp. decumbens [LEDUPAL1] Ericaceae (V:D)
Ledum palustre var. *latifolium* (Jacq.) Michx. = Ledum groenlandicum [LEDUGRO] Ericaceae (V:D)
Leersia Sw. [LEERSIA$] Poaceae (V:M)
Leersia oryzoides (L.) Sw. [LEERORY] Poaceae (V:M)
Legousia perfoliata (L.) Britt. = Triodanis perfoliata [TRIOPER] Campanulaceae (V:D)
Leiocolea badensis (Gott. ex Rabenh.) Joerg. = Lophozia badensis [LOPHBAD] Jungermanniaceae (B:H)
Leiocolea bantriensis (Hook.) Joerg. = Lophozia bantriensis [LOPHBAN] Jungermanniaceae (B:H)
Leiocolea gillmanii (Aust.) Evans = Lophozia gillmanii [LOPHGIL] Jungermanniaceae (B:H)
Leiocolea heterocolpa (Thed.) Buch = Lophozia heterocolpos [LOPHHEE] Jungermanniaceae (B:H)
Leiocolea heterocolpos (Thed.) Buch = Lophozia heterocolpos [LOPHHEE] Jungermanniaceae (B:H)
Leiocolea muelleri (Nees) Joerg. = Lophozia collaris [LOPHCOL] Jungermanniaceae (B:H)
Leiocolea obtusa (Lindb.) Buch = Lophozia obtusa [LOPHOBT] Jungermanniaceae (B:H)
Leiocolea rutheana (Lindb.) K. Müll. = Lophozia rutheana [LOPHRUT] Jungermanniaceae (B:H)
Leiocolea setacea (Web.) Mitt. = Kurzia setacea [KURZSET] Lepidoziaceae (B:H)
Leiocolea sylvatica Evans = Kurzia sylvatica [KURZSYL] Lepidoziaceae (B:H)
Leioderma Nyl. [LEIODER$] Pannariaceae (L)

Leioderma sorediatum D.J. Galloway & P.M. Jørg. [LEIOSOR] Pannariaceae (L)
Lemna L. [LEMNA$] Lemnaceae (V:M)
Lemna cyclostasa Ell. ex Schleid. = Lemna minor [LEMNMIN] Lemnaceae (V:M)
Lemna minima Chev. ex Schleid. = Lemna minor [LEMNMIN] Lemnaceae (V:M)
Lemna minor L. [LEMNMIN] Lemnaceae (V:M)
Lemna trisulca L. [LEMNTRI] Lemnaceae (V:M)
Lemna turionifera Landolt [LEMNTUR] Lemnaceae (V:M)
Lempholemma Körber [LEMPHOL$] Lichinaceae (L)
Lempholemma isidiodes (Nyl. ex Arnold) H. Magn. [LEMPISI] Lichinaceae (L)
Lempholemma myriococcum (Ach.) Th. Fr. = Lempholemma polyanthes [LEMPPOL] Lichinaceae (L)
Lempholemma polyanthes (Bernh.) Malme [LEMPPOL] Lichinaceae (L)
Lempholemma radiatum (Sommerf.) Henssen [LEMPRAD] Lichinaceae (L)
Leontodon L. [LEONTOD$] Asteraceae (V:D)
*****Leontodon autumnalis** L. [LEONAUT] Asteraceae (V:D)
Leontodon erythrospermum (Andrz. ex Bess.) Britt. = Taraxacum laevigatum [TARALAE] Asteraceae (V:D)
Leontodon leysseri (Wallr.) G. Beck = Leontodon taraxacoides [LEONTAR] Asteraceae (V:D)
Leontodon nudicaulis ssp. *taraxacoides* (Vill.) Schinz & Thellung = Leontodon taraxacoides [LEONTAR] Asteraceae (V:D)
Leontodon saxatilis ssp. *taraxacoides* (Vill.) Holub & Moravec = Leontodon taraxacoides [LEONTAR] Asteraceae (V:D)
*****Leontodon taraxacoides** (Vill.) Mérat [LEONTAR] Asteraceae (V:D)
Leonurus L. [LEONURU$] Lamiaceae (V:D)
*****Leonurus cardiaca** L. [LEONCAR] Lamiaceae (V:D)
Lepachys columnifera (Nutt.) Rydb. = Ratibida columnifera [RATICOL] Asteraceae (V:D)
Lepargyrea argentea (Pursh) Greene = Shepherdia argentea [SHEPARG] Elaeagnaceae (V:D)
Lepargyrea canadensis (L.) Greene = Shepherdia canadensis [SHEPCAN] Elaeagnaceae (V:D)
Lepidanthus suaveolens (Pursh) Nutt. = Matricaria discoidea [MATRDIS] Asteraceae (V:D)
Lepidium L. [LEPIDIU$] Brassicaceae (V:D)
Lepidium bourgeauanum Thellung [LEPIBOU] Brassicaceae (V:D)
*****Lepidium campestre** (L.) Ait. f. [LEPICAM] Brassicaceae (V:D)
Lepidium densiflorum Schrad. [LEPIDEN] Brassicaceae (V:D)
Lepidium densiflorum var. *bourgeauanum* (Thellung) C.L. Hitchc. = Lepidium bourgeauanum [LEPIBOU] Brassicaceae (V:D)
Lepidium densiflorum var. **densiflorum** [LEPIDEN0] Brassicaceae (V:D)
Lepidium densiflorum var. **elongatum** (Rydb.) Thellung [LEPIDEN2] Brassicaceae (V:D)
Lepidium densiflorum var. **macrocarpum** Mulligan [LEPIDEN3] Brassicaceae (V:D)
Lepidium densiflorum var. **pubicarpum** (A. Nels.) Thellung [LEPIDEN4] Brassicaceae (V:D)
Lepidium densiflorum var. *typicum* Thellung = Lepidium densiflorum var. densiflorum [LEPIDEN0] Brassicaceae (V:D)
Lepidium didymum L. = Coronopus didymus [CORODID] Brassicaceae (V:D)
Lepidium divergens Osterhout = Lepidium ramosissimum [LEPIRAM] Brassicaceae (V:D)
Lepidium elongatum Rydb. = Lepidium densiflorum var. elongatum [LEPIDEN2] Brassicaceae (V:D)
*****Lepidium heterophyllum** (DC.) Benth. [LEPIHET] Brassicaceae (V:D)
Lepidium neglectum Thellung = Lepidium densiflorum var. densiflorum [LEPIDEN0] Brassicaceae (V:D)
*****Lepidium perfoliatum** L. [LEPIPER] Brassicaceae (V:D)
Lepidium pubecarpum A. Nels. = Lepidium densiflorum var. pubicarpum [LEPIDEN4] Brassicaceae (V:D)
Lepidium ramosissimum A. Nels. [LEPIRAM] Brassicaceae (V:D)
Lepidium repens (Schrenk) Boiss. = Cardaria draba ssp. chalapensis [CARDDRA1] Brassicaceae (V:D)
*****Lepidium sativum** L. [LEPISAT] Brassicaceae (V:D)
Lepidium smithii Hook. = Lepidium heterophyllum [LEPIHET] Brassicaceae (V:D)
Lepidium texanum Buckl. = Lepidium densiflorum var. densiflorum [LEPIDEN0] Brassicaceae (V:D)
Lepidium virginicum L. [LEPIVIR] Brassicaceae (V:D)
Lepidoma demissum (Rutstr.) Choisy = Lecidoma demissum [LECIDEU] Psoraceae (L)
Lepidotheca suaveolens (Pursh) Nutt. = Matricaria discoidea [MATRDIS] Asteraceae (V:D)
Lepidotis inundata (L.) C. Borner = Lycopodiella inundata [LYCOINU] Lycopodiaceae (V:F)
Lepidozia (Dum.) Dum. [LEPIDOZ$] Lepidoziaceae (B:H)
Lepidozia filamentosa (Lehm. & Lindenb.) Lindenb. [LEPIFIL] Lepidoziaceae (B:H)
Lepidozia reptans (L.) Dum. [LEPIREP] Lepidoziaceae (B:H)
Lepidozia sandvicensis Lindenb. [LEPISAN] Lepidoziaceae (B:H)
Lepraria Ach. [LEPRARI$] Uncertain-family (L)
Lepraria aeruginosa auct. = Lepraria incana [LEPRINC] Uncertain-family (L)
Lepraria arctica (Lynge) Wetmore = Leproloma vouauxii [LEPRVOU] Pannariaceae (L)
Lepraria cacuminum (A. Massal.) Lohtander [LEPRCAC] Uncertain-family (L)
Lepraria candelaris (L.) Fr. = Chrysothrix candelaris [CHRYCAN] Chrysotrichaceae (L)
Lepraria chlorina (Ach.) Ach. = Chrysothrix chlorina [CHRYCHL] Chrysotrichaceae (L)
Lepraria citrina (Schaerer) Rabenh. = Chrysothrix candelaris [CHRYCAN] Chrysotrichaceae (L)
Lepraria crassissima (Hue) Lettau [LEPRCRA] Uncertain-family (L)
Lepraria eburnea J.R. Laundon [LEPREBU] Uncertain-family (L)
Lepraria flava (Schreber) Sm. = Chrysothrix candelaris [CHRYCAN] Chrysotrichaceae (L)
Lepraria frigida J.R. Laundon [LEPRFRI] Uncertain-family (L)
Lepraria incana (L.) Ach. [LEPRINC] Uncertain-family (L)
Lepraria lobificans Nyl. [LEPRLOB] Uncertain-family (L)
Lepraria membranacea (Dickson) Vainio = Leproloma membranaceum [LEPRMEB] Pannariaceae (L)
Lepraria neglecta (Nyl.) Erichsen [LEPRNEG] Uncertain-family (L)
Lepraria rigidula (de Lesd.) Tønsberg [LEPRRIG] Uncertain-family (L)
Lepraria vouauxii (Hue) R.C. Harris = Leproloma vouauxii [LEPRVOU] Pannariaceae (L)
Leprocaulon Nyl. ex Lamy [LEPROCA$] Uncertain-family (L)
Leprocaulon albicans (Th. Fr.) Nyl. ex Hue [LEPRALB] Uncertain-family (L)
Leprocaulon microscopicum (Vill.) Gams ex D. Hawksw. [LEPRMIC] Uncertain-family (L)
Leprocaulon subalbicans (Lamb) Lamb & Ward [LEPRSUB] Uncertain-family (L)
Leproloma Nyl. ex Crombie [LEPROLO$] Pannariaceae (L)
Leproloma angardianum (Ovstedal) J.R. Laundon = Lepraria cacuminum [LEPRCAC] Uncertain-family (L)
Leproloma cacuminum (A. Massal.) J.R. Laundon = Lepraria cacuminum [LEPRCAC] Uncertain-family (L)
Leproloma diffusum J.R. Laundon [LEPRDIF] Pannariaceae (L)
Leproloma diffusum var. **chrysodetoides** J.R. Laundon [LEPRDIF1] Pannariaceae (L)
Leproloma diffusum var. **diffusum** [LEPRDIF0] Pannariaceae (L)
Leproloma membranaceum (Dickson) Vainio [LEPRMEB] Pannariaceae (L)
Leproloma vouauxii (Hue) J.R. Laundon [LEPRVOU] Pannariaceae (L)
Leptarrhena R. Br. [LEPTARR$] Saxifragaceae (V:D)
Leptarrhena pyrolifolia (D. Don) R. Br. ex Ser. [LEPTPYR] Saxifragaceae (V:D)
Leptasea aizoides (L.) Haw. = Saxifraga aizoides [SAXIAIZ] Saxifragaceae (V:D)
Leptasea tricuspidata (Rottb.) Haw. = Saxifraga tricuspidata [SAXITRI] Saxifragaceae (V:D)
Leptobryum (Bruch & Schimp. in B.S.G.) Wils. [LEPTOBR$] Bryaceae (B:M)

Leptobryum pyriforme (Hedw.) Wils. [LEPTPYI] Bryaceae (B:M)
Leptobryum pyriforme var. *flagelliferum* Holz. = Leptobryum pyriforme [LEPTPYI] Bryaceae (B:M)
Leptochidium Choisy [LEPTOCI$] Placynthiaceae (L)
Leptochidium albociliatum (Desmaz.) Choisy [LEPTALB] Placynthiaceae (L)
Leptochloa Beauv. [LEPTOCH$] Poaceae (V:M)
*__Leptochloa fascicularis__ (Lam.) Gray [LEPTFAS] Poaceae (V:M)
Leptodactylon Hook. & Arn. [LEPTODA$] Polemoniaceae (V:D)
Leptodactylon hazeliae M.E. Peck = Leptodactylon pungens [LEPTPUN] Polemoniaceae (V:D)
Leptodactylon lilacinum Greene ex Baker = Leptodactylon pungens [LEPTPUN] Polemoniaceae (V:D)
Leptodactylon pungens (Torr.) Torr. ex Nutt. [LEPTPUN] Polemoniaceae (V:D)
Leptodactylon pungens ssp. *brevifolium* (Rydb.) Wherry = Leptodactylon pungens [LEPTPUN] Polemoniaceae (V:D)
Leptodactylon pungens ssp. *eupungens* (Brand) Wherry = Leptodactylon pungens [LEPTPUN] Polemoniaceae (V:D)
Leptodactylon pungens ssp. *hallii* (Parish) Mason = Leptodactylon pungens [LEPTPUN] Polemoniaceae (V:D)
Leptodactylon pungens ssp. *hazeliae* (M.E. Peck) Meinke = Leptodactylon pungens [LEPTPUN] Polemoniaceae (V:D)
Leptodactylon pungens ssp. *hookeri* (Dougl. ex Hook.) Wherry = Leptodactylon pungens [LEPTPUN] Polemoniaceae (V:D)
Leptodactylon pungens ssp. *pulchriflorum* (Brand) Mason = Leptodactylon pungens [LEPTPUN] Polemoniaceae (V:D)
Leptodactylon pungens ssp. *squarrosum* (Gray) Tidestrom = Leptodactylon pungens [LEPTPUN] Polemoniaceae (V:D)
Leptodactylon pungens var. *hallii* (Parish) Jepson = Leptodactylon pungens [LEPTPUN] Polemoniaceae (V:D)
Leptodactylon pungens var. *hookeri* (Dougl. ex Hook.) Jepson = Leptodactylon pungens [LEPTPUN] Polemoniaceae (V:D)
Leptodictyum (Schimp.) Warnst. [LEPTODI$] Amblystegiaceae (B:M)
Leptodictyum brevipes (Card. & Thér. ex Holz.) Broth. = Leptodictyum riparium [LEPTRIP] Amblystegiaceae (B:M)
Leptodictyum laxirete (Card. & Thér.) Broth. = Leptodictyum riparium [LEPTRIP] Amblystegiaceae (B:M)
Leptodictyum pennellii (Bartr.) Robins. = Leptodictyum riparium [LEPTRIP] Amblystegiaceae (B:M)
Leptodictyum radicale (P. Beauv.) Kanda = Campylium radicale [CAMPRAD] Amblystegiaceae (B:M)
Leptodictyum riparium (Hedw.) Warnst. [LEPTRIP] Amblystegiaceae (B:M)
Leptodictyum riparium f. *abbreviatum* (Bruch & Schimp.) Grout = Leptodictyum riparium [LEPTRIP] Amblystegiaceae (B:M)
Leptodictyum riparium f. *fluitans* (Lesq. & James) Grout = Leptodictyum riparium [LEPTRIP] Amblystegiaceae (B:M)
Leptodictyum riparium f. *longifolium* (Schultz) Grout = Leptodictyum riparium [LEPTRIP] Amblystegiaceae (B:M)
Leptodictyum riparium f. *obtusum* (Grout) Grout = Leptodictyum riparium [LEPTRIP] Amblystegiaceae (B:M)
Leptodictyum riparium var. *abbreviatum* (Schimp. in B.S.G.) Grout = Leptodictyum riparium [LEPTRIP] Amblystegiaceae (B:M)
Leptodictyum riparium var. *brachyphyllum* (Card. & Thér.) Grout = Leptodictyum riparium [LEPTRIP] Amblystegiaceae (B:M)
Leptodictyum riparium var. *elongatum* (Schimp. in B.S.G.) Warnst. = Leptodictyum riparium [LEPTRIP] Amblystegiaceae (B:M)
Leptodictyum riparium var. *flaccidum* (Lesq. & James) Grout = Leptodictyum riparium [LEPTRIP] Amblystegiaceae (B:M)
Leptodictyum riparium var. *longifolium* (Schultz) Warnst. = Leptodictyum riparium [LEPTRIP] Amblystegiaceae (B:M)
Leptodictyum riparium var. *nigrescens* Wynne in E. Whiteh. = Leptodictyum riparium [LEPTRIP] Amblystegiaceae (B:M)
Leptodictyum riparium var. *obtusum* (Grout) Grout = Leptodictyum riparium [LEPTRIP] Amblystegiaceae (B:M)
Leptodictyum sipho (P. Beauv.) Broth. = Leptodictyum riparium [LEPTRIP] Amblystegiaceae (B:M)
Leptodictyum vacillans (Sull.) Broth. = Leptodictyum riparium [LEPTRIP] Amblystegiaceae (B:M)
Leptodontium recurvifolium (Tayl.) Lindb. = Paraleptodontium recurvifolium [PARAREC] Pottiaceae (B:M)
Leptogidium dendriscum (Nyl.) Nyl. = Polychidium dendriscum [POLYDEN] Placynthiaceae (L)
Leptogidium intricatulum Nyl. = Dendriscocaulon intricatulum [DENDINT] Lobariaceae (L)
Leptogiopsis complicatula Nyl. = Kohlmeyera complicatula [KOHLCOM] Mastodiaceae (L)
Leptogium (Ach.) Gray [LEPTOGI$] Collemataceae (L)
Leptogium albociliatum Desmaz. = Leptochidium albociliatum [LEPTALB] Placynthiaceae (L)
Leptogium brebissonii Mont. [LEPTBRE] Collemataceae (L)
Leptogium burgessii (L.) Mont. [LEPTBUR] Collemataceae (L)
Leptogium burnetiae C. W. Dodge [LEPTBUN] Collemataceae (L)
Leptogium burnetiae var. *hirsutum* (Sierk) P.M. Jørg. = Leptogium hirsutum [LEPTHIR] Collemataceae (L)
Leptogium caesium (Ach.) Vainio = Leptogium cyanescens [LEPTCYA] Collemataceae (L)
Leptogium californicum Tuck. [LEPTCAL] Collemataceae (L)
Leptogium corniculatum (Hoffm.) Minks [LEPTCOI] Collemataceae (L)
Leptogium cyanescens (Rabenh.) Körber [LEPTCYA] Collemataceae (L)
Leptogium furfuraceum (Harm.) Sierk [LEPTFUR] Collemataceae (L)
Leptogium gelatinosum (With.) J.R. Laundon [LEPTGEL] Collemataceae (L)
Leptogium hirsutum Sierk [LEPTHIR] Collemataceae (L)
Leptogium inflexum Nyl. = Leptogium burgessii [LEPTBUR] Collemataceae (L)
Leptogium lichenoides (L.) Zahlbr. [LEPTLIC] Collemataceae (L)
Leptogium lividofuscum (Florke ex Schlecht.) Flotow = Leptogium tenuissimum [LEPTTEN] Collemataceae (L)
Leptogium minutissimum (Flörke) Fr. = Leptogium subtile [LEPTSUB] Collemataceae (L)
Leptogium muscicola (Sw.) Fr. = Polychidium muscicola [POLYMUS] Placynthiaceae (L)
Leptogium nanum Herre = Leptogium tenuissimum [LEPTTEN] Collemataceae (L)
Leptogium palmatum (Hudson) Mont. = Leptogium corniculatum [LEPTCOI] Collemataceae (L)
Leptogium papillosum auct. = Leptogium furfuraceum [LEPTFUR] Collemataceae (L)
Leptogium perminutum Fink = Leptogium tenuissimum [LEPTTEN] Collemataceae (L)
Leptogium pilosellum G. Merr. = Leptochidium albociliatum [LEPTALB] Placynthiaceae (L)
Leptogium platynum (Tuck.) Herre [LEPTPLA] Collemataceae (L)
Leptogium polycarpum P.M. Jørg. & Goward [LEPTPOL] Collemataceae (L)
Leptogium rivale Tuck. [LEPTRIV] Collemataceae (L)
Leptogium saturninum (Dickson) Nyl. [LEPTSAT] Collemataceae (L)
Leptogium schraderi (Ach.) Nyl. [LEPTSCH] Collemataceae (L)
Leptogium sinuatum (Hudson) A. Massal. = Leptogium gelatinosum [LEPTGEL] Collemataceae (L)
Leptogium subaridum Jørgensen & Goward [LEPTSUA] Collemataceae (L)
Leptogium subtile (Schrader) Torss. [LEPTSUB] Collemataceae (L)
Leptogium tenuissimum (Dickson) Körber [LEPTTEN] Collemataceae (L)
Leptogium teretiusculum (Wallr.) Arnold [LEPTTER] Collemataceae (L)
Leptogium tremelloides auct. = Leptogium cyanescens [LEPTCYA] Collemataceae (L)
Leptotaenia dissecta Nutt. = Lomatium dissectum var. dissectum [LOMADIS0] Apiaceae (V:D)
Leptotaenia multifida Nutt. = Lomatium dissectum var. multifidum [LOMADIS2] Apiaceae (V:D)
Leptotrichum brevifolium Kindb. ex Par. = Ditrichum flexicaule [DITRFLE] Ditrichaceae (B:M)
Leptotrichum brevifolium var. *densum* Schimp. = Ditrichum flexicaule [DITRFLE] Ditrichaceae (B:M)
Leptotrichum glaucescens (Hedw.) Hampe in Schimp. = Saelania glaucescens [SAELGLA] Ditrichaceae (B:M)

Leptotrichum tortile (Schrad.) Hampe in C. Müll. = Ditrichum pusillum [DITRPUS] Ditrichaceae (B:M)
Lescuraea Schimp. in B.S.G. [LESCURA$] Leskeaceae (B:M)
Lescuraea atricha (Kindb. in Mac. & Kindb.) Lawt. = Pseudoleskea atricha [PSEUATR] Leskeaceae (B:M)
Lescuraea baileyi (Best & Grout in Grout) Lawt. = Pseudoleskea baileyi [PSEUBAI] Leskeaceae (B:M)
Lescuraea frigida Kindb. = Lescuraea saxicola [LESCSAX] Leskeaceae (B:M)
Lescuraea iliamniana Lawt. = Pseudoleskea julacea [PSEUJUL] Leskeaceae (B:M)
Lescuraea imperfecta C. Müll. & Kindb. in Mac. & Kindb. = Pseudoleskea stenophylla [PSEUSTE] Leskeaceae (B:M)
Lescuraea incurvata (Hedw.) Lawt. = Pseudoleskea incurvata [PSEUINC] Leskeaceae (B:M)
Lescuraea incurvata var. *gigantea* Lawt. = Pseudoleskea incurvata var. gigantea [PSEUINC1] Leskeaceae (B:M)
Lescuraea julacea Besch. & Card. in Card. = Pseudoleskea julacea [PSEUJUL] Leskeaceae (B:M)
Lescuraea mutabilis var. *saxicola* (Schimp. in B.S.G.) Hag. = Lescuraea saxicola [LESCSAX] Leskeaceae (B:M)
Lescuraea patens (Lindb.) Arnell & C. Jens. = Pseudoleskea patens [PSEUPAT] Leskeaceae (B:M)
Lescuraea radicosa (Mitt.) Mönk. = Pseudoleskea radicosa [PSEURAD] Leskeaceae (B:M)
Lescuraea radicosa var. *compacta* (Best) Lawt. = Pseudoleskea radicosa var. compacta [PSEURAD1] Leskeaceae (B:M)
Lescuraea radicosa var. *denudata* (Kindb. in Mac. & Kindb.) Lawt. = Pseudoleskea radicosa var. denudata [PSEURAD2] Leskeaceae (B:M)
Lescuraea radicosa var. *pallida* (Best) Lawt. = Pseudoleskea radicosa var. pallida [PSEURAD3] Leskeaceae (B:M)
Lescuraea saxicola (Schimp. in B.S.G.) Milde [LESCSAX] Leskeaceae (B:M)
Lescuraea stenophylla (Ren. & Card. in Röll) Kindb. = Pseudoleskea stenophylla [PSEUSTE] Leskeaceae (B:M)
Leskea Hedw. [LESKEA$] Leskeaceae (B:M)
Leskea arenicola Best = Leskea polycarpa [LESKPOL] Leskeaceae (B:M)
Leskea cyrtophylla Kindb. in Mac. & Kindb. = Pseudoleskeella tectorum [PSEUTEC] Leskeaceae (B:M)
Leskea montanae (Bryhn) Grout = Leskeella nervosa [LESKNER] Leskeaceae (B:M)
Leskea nervosa (Brid.) Myr. = Leskeella nervosa [LESKNER] Leskeaceae (B:M)
Leskea nervosa var. *bulbifera* (Brid.) Best in Mac. = Leskeella nervosa [LESKNER] Leskeaceae (B:M)
Leskea nervosa var. *flagellifera* Kindb. = Leskeella nervosa [LESKNER] Leskeaceae (B:M)
Leskea nervosa var. *nigrescens* (Kindb.) Best = Leskeella nervosa [LESKNER] Leskeaceae (B:M)
Leskea nigrescens Kindb. = Leskeella nervosa [LESKNER] Leskeaceae (B:M)
Leskea polycarpa Hedw. [LESKPOL] Leskeaceae (B:M)
Leskea polycarpa var. *paludosa* (Hedw.) Schimp. = Leskea polycarpa [LESKPOL] Leskeaceae (B:M)
Leskea pulvinata Wahlenb. = Myrinia pulvinata [MYRIPUL] Myriniaceae (B:M)
Leskea seligeri Brid. = Herzogiella seligeri [HERZSEL] Hypnaceae (B:M)
Leskea striatella Brid. = Herzogiella striatella [HERZSTR] Hypnaceae (B:M)
Leskea subobtusifolia C. Müll. & Kindb. in Mac. & Kindb. = Leskea polycarpa [LESKPOL] Leskeaceae (B:M)
Leskea tectorum (Funck ex Brid.) Lindb. = Pseudoleskeella tectorum [PSEUTEC] Leskeaceae (B:M)
Leskea tectorum var. *flagellifera* Best = Pseudoleskeella tectorum [PSEUTEC] Leskeaceae (B:M)
Leskea williamsii Best = Pseudoleskeella tectorum [PSEUTEC] Leskeaceae (B:M)
Leskea williamsii var. *filamentosa* Best = Pseudoleskeella tectorum [PSEUTEC] Leskeaceae (B:M)
Leskea wollei Aust. = Pseudoleskeella tectorum [PSEUTEC] Leskeaceae (B:M)
Leskeella (Limpr.) Loeske [LESKEEL$] Leskeaceae (B:M)
Leskeella nervosa (Brid.) Loeske [LESKNER] Leskeaceae (B:M)
Leskeella nervosa f. *bulbifera* Möll. = Leskeella nervosa [LESKNER] Leskeaceae (B:M)
Leskeella nervosa var. *flagellifera* (Kindb.) Wijk & Marg. = Leskeella nervosa [LESKNER] Leskeaceae (B:M)
Leskeella nervosa var. *sibirica* (Arnell) Broth. = Leskeella nervosa [LESKNER] Leskeaceae (B:M)
Leskeella nervosa var. *subrigidula* (Kindb.) Podp. = Leskeella nervosa [LESKNER] Leskeaceae (B:M)
Leskeella tectorum (Funck ex Brid.) Hag. = Pseudoleskeella tectorum [PSEUTEC] Leskeaceae (B:M)
Leskeella tectorum ssp. *cyrtophylla* (Kindb. in Mac. & Kindb.) Wijk & Marg. = Pseudoleskeella tectorum [PSEUTEC] Leskeaceae (B:M)
Lesquerella S. Wats. [LESQUER$] Brassicaceae (V:D)
Lesquerella arctica (Wormsk. ex Hornem.) S. Wats. [LESQARC] Brassicaceae (V:D)
Lesquerella arctica ssp. *purshii* (S. Wats.) Porsild = Lesquerella arctica [LESQARC] Brassicaceae (V:D)
Lesquerella arctica var. *purshii* S. Wats. = Lesquerella arctica [LESQARC] Brassicaceae (V:D)
Lesquerella arctica var. *scammaniae* Rollins = Lesquerella arctica [LESQARC] Brassicaceae (V:D)
Lesquerella douglasii S. Wats. [LESQDOU] Brassicaceae (V:D)
Lesquerella purshii (S. Wats.) Fern. = Lesquerella arctica [LESQARC] Brassicaceae (V:D)
Letharia (Th. Fr.) Zahlbr. [LETHARI$] Parmeliaceae (L)
Letharia californica (Lév.) Hue = Letharia columbiana [LETHCOL] Parmeliaceae (L)
Letharia columbiana (Nutt.) J.W. Thomson [LETHCOL] Parmeliaceae (L)
Letharia vulpina (L.) Hue [LETHVUL] Parmeliaceae (L)
Letharia vulpina f. *californica* (Lév.) W.A. Weber = Letharia columbiana [LETHCOL] Parmeliaceae (L)
Lethariicola Grummann [LETHARC$] Odontotremataceae (L)
Lethariicola cucularis (Norman) Lumbsch & D. Hawksw. [LETHCUC] Odontotremataceae (L)
Lethariicola sipei Grummann = Lethariicola cucularis [LETHCUC] Odontotremataceae (L)
Leucacantha cyanus (L.) Nieuwl. & Lunell = Centaurea cyanus [CENTCYA] Asteraceae (V:D)
Leucanthemum P. Mill. [LEUCANT$] Asteraceae (V:D)
Leucanthemum arcticum auct. = Dendranthema arcticum ssp. polare [DENDARC1] Asteraceae (V:D)
Leucanthemum integrifolium (Richards.) DC. [LEUCINT] Asteraceae (V:D)
Leucanthemum leucanthemum (L.) Rydb. = Leucanthemum vulgare [LEUCVUL] Asteraceae (V:D)
*__Leucanthemum vulgare__ Lam. [LEUCVUL] Asteraceae (V:D)
Leucanthemum vulgare var. *pinnatifidum* (Lecoq & Lamotte) Moldenke = Leucanthemum vulgare [LEUCVUL] Asteraceae (V:D)
Leucocoma alpina (L.) Rydb. = Eriophorum alpinum [ERIOALP] Cyperaceae (V:M)
Leucolepis Lindb. [LEUCOLE$] Mniaceae (B:M)
Leucolepis acanthoneuron (Schwaegr.) Lindb. [LEUCACA] Mniaceae (B:M)
Leucolepis menziesii (Hook.) Steere in L. Koch = Leucolepis acanthoneuron [LEUCACA] Mniaceae (B:M)
Lewisia Pursh [LEWISIA$] Portulacaceae (V:D)
Lewisia columbiana (T.J. Howell ex Gray) B.L. Robins. [LEWICOL] Portulacaceae (V:D)
Lewisia glandulosa (Rydb.) Dempster = Lewisia pygmaea [LEWIPYG] Portulacaceae (V:D)
Lewisia minima (A. Nels.) A. Nels. = Lewisia pygmaea [LEWIPYG] Portulacaceae (V:D)
Lewisia pygmaea (Gray) B.L. Robins. [LEWIPYG] Portulacaceae (V:D)
Lewisia pygmaea ssp. *glandulosa* (Rydb.) Ferris = Lewisia pygmaea [LEWIPYG] Portulacaceae (V:D)

Lewisia pygmaea var. *aridorum* Bartlett = Lewisia pygmaea [LEWIPYG] Portulacaceae (V:D)
Lewisia rediviva Pursh [LEWIRED] Portulacaceae (V:D)
Lewisia triphylla (S. Wats.) B.L. Robins. [LEWITRI] Portulacaceae (V:D)
Lewisia tweedyi (Gray) B.L. Robins. = Cistanthe tweedyi [CISTTWE] Portulacaceae (V:D)
Leymus Hochst. [LEYMUS$] Poaceae (V:M)
Leymus cinereus (Scribn. & Merr.) A. Löve [LEYMCIN] Poaceae (V:M)
Leymus innovatus (Beal) Pilger [LEYMINN] Poaceae (V:M)
Leymus mollis (Trin.) Hara [LEYMMOL] Poaceae (V:M)
Leymus triticoides (Buckl.) Pilger [LEYMTRI] Poaceae (V:M)
Leymus velutinus (Bowden) A. & D. Löve = Leymus innovatus [LEYMINN] Poaceae (V:M)
Lichen albocaerulescens Wulfen = Porpidia albocaerulescens [PORPALB] Porpidiaceae (L)
Lichen intestiniformis Vill. = Brodoa oroarctica [BRODORO] Parmeliaceae (L)
Lichen lachneus Ach. = Catapyrenium lachneum [CATALAC] Verrucariaceae (L)
Lichen luteolus Schrad. = Bacidia rubella [BACIRUB] Bacidiaceae (L)
Lichen lutereus J.F. Gmel. = Bacidia rubella [BACIRUB] Bacidiaceae (L)
Lichen rubellus Ehrh., nom. nud. = Bacidia rubella [BACIRUB] Bacidiaceae (L)
Lichen tephroides Ach. = Catapyrenium cinereum [CATACIN] Verrucariaceae (L)
Lichenochora Hafellner [LICHENH$] Phyllachoraceae (L)
Lichenochora thallina (Cooke) Hafellner [LICHTHA] Phyllachoraceae (L)
Lichenoconium Petrak & H. Sydow [LICHENC$] Uncertain-family (L)
Lichenoconium usneae (Anzi) D. Hawksw. [LICHUSN] Uncertain-family (L)
Lichenomyces lichenum (Sommerf.) R. Sant. = Plectocarpon lichenum [PLECLIC] Uncertain-family (L)
Lichenosticta Zopf [LICHENS$] Uncertain-family (L)
Lichenosticta alcicorniaria (Lindsay) D. Hawksw. [LICHALC] Uncertain-family (L)
Lichenothelia D. Hawksw. [LICHENT$] Lichenotheliaceae (L)
Lichenothelia metzleri (J. Lahm) D. Hawksw. [LICHMET] Lichenotheliaceae (L)
Lichinella Nyl. [LICHINE$] Lichinaceae (L)
Lichinella nigritella (Lettau) Moreno & Egea [LICHNIG] Lichinaceae (L)
Lichinodium Nyl. [LICHINO$] Lichinaceae (L)
Lichinodium canadense Henssen [LICHCAN] Lichinaceae (L)
Lidia biflora (L.) A. & D. Löve = Minuartia biflora [MINUBIF] Caryophyllaceae (V:D)
Lidia obtusiloba (Rydb.) A. & D. Löve = Minuartia obtusiloba [MINUOBT] Caryophyllaceae (V:D)
Ligusticum L. [LIGUSTI$] Apiaceae (V:D)
Ligusticum caeruleimontanum St. John = Ligusticum canbyi [LIGUCAN] Apiaceae (V:D)
Ligusticum calderi Mathias & Constance [LIGUCAL] Apiaceae (V:D)
Ligusticum canbyi Coult. & Rose [LIGUCAN] Apiaceae (V:D)
Ligusticum hultenii Fern. = Ligusticum scoticum ssp. hultenii [LIGUSCO1] Apiaceae (V:D)
Ligusticum leibergii Coult. & Rose = Ligusticum canbyi [LIGUCAN] Apiaceae (V:D)
Ligusticum scoticum L. [LIGUSCO] Apiaceae (V:D)
Ligusticum scoticum ssp. **hultenii** (Fern.) Calder & Taylor [LIGUSCO1] Apiaceae (V:D)
Ligusticum scoticum var. *hultenii* (Fern.) Boivin = Ligusticum scoticum ssp. hultenii [LIGUSCO1] Apiaceae (V:D)
Ligusticum verticillatum (Hook.) Coult. & Rose ex Rose [LIGUVER] Apiaceae (V:D)
Ligustrum L. [LIGUSTR$] Oleaceae (V:D)
*__Ligustrum vulgare__ L. [LIGUVUL] Oleaceae (V:D)
Lilaea Bonpl. [LILAEA$] Juncaginaceae (V:M)
Lilaea scilloides (Poir.) Hauman [LILASCI] Juncaginaceae (V:M)
Lilaea subulata Bonpl. = Lilaea scilloides [LILASCI] Juncaginaceae (V:M)
Lilaeopsis Greene [LILAEOP$] Apiaceae (V:D)
Lilaeopsis occidentalis Coult. & Rose [LILAOCC] Apiaceae (V:D)
Lilium L. [LILIUM$] Liliaceae (V:M)
Lilium andinum Nutt. = Lilium philadelphicum var. andinum [LILIPHI1] Liliaceae (V:M)
Lilium camschatcense L. = Fritillaria camschatcensis [FRITCAM] Liliaceae (V:M)
Lilium canadense var. *parviflorum* Hook. = Lilium columbianum [LILICOL] Liliaceae (V:M)
Lilium columbianum hort. ex Baker [LILICOL] Liliaceae (V:M)
Lilium montanum A. Nels. = Lilium philadelphicum var. andinum [LILIPHI1] Liliaceae (V:M)
Lilium philadelphicum L. [LILIPHI] Liliaceae (V:M)
Lilium philadelphicum var. **andinum** (Nutt.) Ker-Gawl. [LILIPHI1] Liliaceae (V:M)
Lilium philadelphicum var. *montanum* (A. Nels.) Wherry = Lilium philadelphicum var. andinum [LILIPHI1] Liliaceae (V:M)
Lilium pudicum Pursh = Fritillaria pudica [FRITPUD] Liliaceae (V:M)
Lilium umbellatum Pursh = Lilium philadelphicum var. andinum [LILIPHI1] Liliaceae (V:M)
Limnanthes R. Br. [LIMNANT$] Limnanthaceae (V:D)
Limnanthes macounii Trel. [LIMNMAC] Limnanthaceae (V:D)
Limnobotrya lacustris (Pers.) Rydb. = Ribes lacustre [RIBELAC] Grossulariaceae (V:D)
Limnobotrya montigena (McClatchie) Rydb. = Ribes montigenum [RIBEMON] Grossulariaceae (V:D)
Limnorchis dilatata (Pursh) Rydb. = Platanthera dilatata var. dilatata [PLATDIL0] Orchidaceae (V:M)
Limnorchis dilatata ssp. *albiflora* (Cham.) A. Löve & Simon = Platanthera dilatata var. albiflora [PLATDIL1] Orchidaceae (V:M)
Limnorchis graminifolia Rydb. = Platanthera leucostachys [PLATLEU] Orchidaceae (V:M)
Limnorchis saccata (Greene) A. Löve & Simon = Platanthera stricta [PLATSTR] Orchidaceae (V:M)
Limnorchis stricta (Lindl.) Rydb. = Platanthera stricta [PLATSTR] Orchidaceae (V:M)
Limosella L. [LIMOSEL$] Scrophulariaceae (V:D)
Limosella aquatica L. [LIMOAQU] Scrophulariaceae (V:D)
Limprichtia Loeske [LIMPRIC$] Amblystegiaceae (B:M)
Limprichtia cossonii (Schimp.) Anderson, Crum & Buck [LIMPCOS] Amblystegiaceae (B:M)
Limprichtia intermedia (Lindb. in Hartm.) Loeske = Limprichtia cossonii [LIMPCOS] Amblystegiaceae (B:M)
Limprichtia revolvens (Sw.) Loeske [LIMPREV] Amblystegiaceae (B:M)
Limprichtia vernicosa (Mitt.) Loeske = Hamatocaulis vernicosus [HAMAVER] Amblystegiaceae (B:M)
Linanthus Benth. [LINANTH$] Polemoniaceae (V:D)
Linanthus bicolor (Nutt.) Greene [LINABIC] Polemoniaceae (V:D)
Linanthus harknessii (Curran) Greene [LINAHAR] Polemoniaceae (V:D)
Linanthus harknessii var. *septentrionalis* (Mason) Jepson & V. Bailey = Linanthus septentrionalis [LINASEP] Polemoniaceae (V:D)
Linanthus septentrionalis Mason [LINASEP] Polemoniaceae (V:D)
Linaria P. Mill. [LINARIA$] Scrophulariaceae (V:D)
Linaria canadensis var. *texana* (Scheele) Pennell = Nuttallanthus texanus [NUTTTEX] Scrophulariaceae (V:D)
Linaria cymbalaria (L.) P. Mill. = Cymbalaria muralis [CYMBMUR] Scrophulariaceae (V:D)
*__Linaria dalmatica__ (L.) P. Mill. [LINADAL] Scrophulariaceae (V:D)
Linaria elatine (L.) P. Mill. = Kickxia elatine [KICKELA] Scrophulariaceae (V:D)
*__Linaria genistifolia__ (L.) P. Mill. [LINAGEN] Scrophulariaceae (V:D)
Linaria genistifolia ssp. *dalmatica* (L.) Maire & Petitm. = Linaria dalmatica [LINADAL] Scrophulariaceae (V:D)
Linaria linaria (L.) Karst. = Linaria vulgaris [LINAVUL] Scrophulariaceae (V:D)
*__Linaria purpurea__ (L.) P. Mill. [LINAPUR] Scrophulariaceae (V:D)

Linaria texana Scheele = Nuttallanthus texanus [NUTTTEX] Scrophulariaceae (V:D)
•**Linaria vulgaris** P. Mill. [LINAVUL] Scrophulariaceae (V:D)
Lindernia All. [LINDERN$] Scrophulariaceae (V:D)
Lindernia anagallidea (Michx.) Pennell = Lindernia dubia var. anagallidea [LINDDUB1] Scrophulariaceae (V:D)
Lindernia dubia (L.) Pennell [LINDDUB] Scrophulariaceae (V:D)
Lindernia dubia var. **anagallidea** (Michx.) Cooperrider [LINDDUB1] Scrophulariaceae (V:D)
Linnaea L. [LINNAEA$] Caprifoliaceae (V:D)
Linnaea americana Forbes = Linnaea borealis ssp. longiflora [LINNBOR2] Caprifoliaceae (V:D)
Linnaea borealis L. [LINNBOR] Caprifoliaceae (V:D)
Linnaea borealis ssp. *americana* (Forbes) Hultén ex Clausen = Linnaea borealis ssp. longiflora [LINNBOR2] Caprifoliaceae (V:D)
Linnaea borealis ssp. **borealis** [LINNBOR0] Caprifoliaceae (V:D)
Linnaea borealis ssp. **longiflora** (Torr.) Hultén [LINNBOR2] Caprifoliaceae (V:D)
Linnaea borealis var. *americana* (Forbes) Rehd. = Linnaea borealis ssp. longiflora [LINNBOR2] Caprifoliaceae (V:D)
Linnaea borealis var. *longiflora* Torr. = Linnaea borealis ssp. longiflora [LINNBOR2] Caprifoliaceae (V:D)
Linum L. [LINUM$] Linaceae (V:D)
Linum angustifolium Huds. = Linum bienne [LINUBIE] Linaceae (V:D)
•**Linum bienne** P. Mill. [LINUBIE] Linaceae (V:D)
Linum humile P. Mill. = Linum usitatissimum [LINUUSI] Linaceae (V:D)
Linum lewisii Pursh [LINULEW] Linaceae (V:D)
Linum perenne L. [LINUPER] Linaceae (V:D)
Linum perenne ssp. *lewisii* (Pursh) Hultén = Linum lewisii [LINULEW] Linaceae (V:D)
Linum perenne var. *lewisii* (Pursh) Eat. & J. Wright = Linum lewisii [LINULEW] Linaceae (V:D)
•**Linum usitatissimum** L. [LINUUSI] Linaceae (V:D)
Linum usitatissimum var. *humile* (P. Mill.) Pers. = Linum usitatissimum [LINUUSI] Linaceae (V:D)
Liparis L.C. Rich. [LIPARIS$] Orchidaceae (V:M)
Liparis loeselii (L.) L.C. Rich. [LIPALOE] Orchidaceae (V:M)
Lipocarpha R. Br. [LIPOCAR$] Cyperaceae (V:M)
Lipocarpha micrantha (Vahl) G. Tucker [LIPOMIC] Cyperaceae (V:M)
Listera R. Br. ex Ait. f. [LISTERA$] Orchidaceae (V:M)
Listera banksiana auct. = Listera caurina [LISTCAU] Orchidaceae (V:M)
Listera borealis Morong [LISTBOR] Orchidaceae (V:M)
Listera caurina Piper [LISTCAU] Orchidaceae (V:M)
Listera convallarioides (Sw.) Nutt. ex Ell. [LISTCON] Orchidaceae (V:M)
Listera cordata (L.) R. Br. ex Ait. f. [LISTCOR] Orchidaceae (V:M)
Lithophragma (Nutt.) Torr. & Gray [LITHOPH$] Saxifragaceae (V:D)
Lithophragma australe Rydb. = Lithophragma tenellum [LITHTEN] Saxifragaceae (V:D)
Lithophragma brevilobum Rydb. = Lithophragma tenellum [LITHTEN] Saxifragaceae (V:D)
Lithophragma bulbiferum Rydb. = Lithophragma glabrum [LITHGLA] Saxifragaceae (V:D)
Lithophragma glabrum Nutt. [LITHGLA] Saxifragaceae (V:D)
Lithophragma glabrum var. *bulbiferum* (Rydb.) Jepson = Lithophragma glabrum [LITHGLA] Saxifragaceae (V:D)
Lithophragma glabrum var. *ramulosum* (Suksdorf) Boivin = Lithophragma glabrum [LITHGLA] Saxifragaceae (V:D)
Lithophragma parviflorum (Hook.) Nutt. ex Torr. & Gray [LITHPAR] Saxifragaceae (V:D)
Lithophragma rupicola Greene = Lithophragma tenellum [LITHTEN] Saxifragaceae (V:D)
Lithophragma tenellum Nutt. [LITHTEN] Saxifragaceae (V:D)
Lithophragma tenellum var. *thompsonii* (Hoover) C.L. Hitchc. = Lithophragma tenellum [LITHTEN] Saxifragaceae (V:D)
Lithophragma thompsonii Hoover = Lithophragma tenellum [LITHTEN] Saxifragaceae (V:D)
Lithospermum L. [LITHOSP$] Boraginaceae (V:D)
Lithospermum angustifolium Michx. = Lithospermum incisum [LITHINC] Boraginaceae (V:D)
Lithospermum arvense L. = Buglossoides arvensis [BUGLARV] Boraginaceae (V:D)
Lithospermum incisum Lehm. [LITHINC] Boraginaceae (V:D)
Lithospermum linearifolium Goldie = Lithospermum incisum [LITHINC] Boraginaceae (V:D)
Lithospermum mandanense Spreng. = Lithospermum incisum [LITHINC] Boraginaceae (V:D)
Lithospermum pilosum Nutt. = Lithospermum ruderale [LITHRUD] Boraginaceae (V:D)
Lithospermum ruderale Dougl. ex Lehm. [LITHRUD] Boraginaceae (V:D)
Lloydia Salisb. ex Reichenb. [LLOYDIA$] Liliaceae (V:M)
Lloydia serotina (L.) Reichenb. [LLOYSER] Liliaceae (V:M)
Lloydia serotina ssp. **flava** Calder & Taylor [LLOYSER1] Liliaceae (V:M)
Lloydia serotina ssp. **serotina** [LLOYSER0] Liliaceae (V:M)
Lloydia serotina var. *flava* (Calder & Taylor) Boivin = Lloydia serotina ssp. flava [LLOYSER1] Liliaceae (V:M)
Lobaria (Schreber) Hoffm. [LOBARIA$] Lobariaceae (L)
Lobaria hallii (Tuck.) Zahlbr. [LOBAHAL] Lobariaceae (L)
Lobaria linita (Ach.) Rabenh. [LOBALIN] Lobariaceae (L)
Lobaria linita var. **linita** [LOBALIN0] Lobariaceae (L)
Lobaria linita var. **tenuior** (Del.) Asah. [LOBALIN1] Lobariaceae (L)
Lobaria oregana (Tuck.) Müll. Arg. [LOBAORE] Lobariaceae (L)
Lobaria pulmonaria (L.) Hoffm. [LOBAPUL] Lobariaceae (L)
Lobaria retigera (Bory) Trevisan [LOBARET] Lobariaceae (L)
Lobaria scrobiculata (Scop.) DC. [LOBASCR] Lobariaceae (L)
Lobaria verrucosa (Hudson) Hoffm. = Lobaria scrobiculata [LOBASCR] Lobariaceae (L)
Lobelia L. [LOBELIA$] Campanulaceae (V:D)
Lobelia dortmanna L. [LOBEDOR] Campanulaceae (V:D)
Lobelia inflata L. [LOBEINF] Campanulaceae (V:D)
Lobelia kalmii L. [LOBEKAL] Campanulaceae (V:D)
Lobelia kalmii var. *strictiflora* Rydb. = Lobelia kalmii [LOBEKAL] Campanulaceae (V:D)
Lobelia strictiflora (Rydb.) Lunell = Lobelia kalmii [LOBEKAL] Campanulaceae (V:D)
Lobothallia (Clauzade & Roux) Hafellner [LOBOTHA$] Hymeneliaceae (L)
Lobothallia alphoplaca (Wahlenb.) Hafellner [LOBOALP] Hymeneliaceae (L)
Lobothallia melanaspis (Ach.) Hafellner [LOBOMEL] Hymeneliaceae (L)
Lobothallia praeradiosa (Nyl.) Hafellner [LOBOPRA] Hymeneliaceae (L)
Lobularia Desv. [LOBULAR$] Brassicaceae (V:D)
•**Lobularia maritima** (L.) Desv. [LOBUMAR] Brassicaceae (V:D)
Loeskypnum Paul [LOESKYP$] Amblystegiaceae (B:M)
Loeskypnum badium (Hartm.) Paul [LOESBAD] Amblystegiaceae (B:M)
Loeskypnum wickesiae (Grout) Tuom. [LOESWIC] Amblystegiaceae (B:M)
Logfia arvensis (L.) Holub = Filago arvensis [FILAARV] Asteraceae (V:D)
Loiseleuria Desv. [LOISELE$] Ericaceae (V:D)
Loiseleuria procumbens (L.) Desv. [LOISPRO] Ericaceae (V:D)
Lolium L. [LOLIUM$] Poaceae (V:M)
Lolium arvense With. = Lolium temulentum [LOLITEM] Poaceae (V:M)
Lolium multiflorum Lam. = Lolium perenne ssp. multiflorum [LOLIPER1] Poaceae (V:M)
Lolium multiflorum var. *diminutum* Mutel = Lolium perenne ssp. multiflorum [LOLIPER1] Poaceae (V:M)
Lolium multiflorum var. *muticum* DC. = Lolium perenne ssp. multiflorum [LOLIPER1] Poaceae (V:M)
Lolium multiflorum var. *ramosum* Guss. ex Arcang. = Lolium perenne ssp. multiflorum [LOLIPER1] Poaceae (V:M)
•**Lolium perenne** L. [LOLIPER] Poaceae (V:M)

*•**Lolium perenne** ssp. **multiflorum** (Lam.) Husnot [LOLIPER1] Poaceae (V:M)
Lolium perenne var. *multiflorum* (Lam.) Parnell = Lolium perenne ssp. multiflorum [LOLIPER1] Poaceae (V:M)
•**Lolium temulentum** L. [LOLITEM] Poaceae (V:M)
Lolium temulentum var. *arvense* (With.) Lilja = Lolium temulentum [LOLITEM] Poaceae (V:M)
Lolium temulentum var. *leptochaeton* A. Braun = Lolium temulentum [LOLITEM] Poaceae (V:M)
Lolium temulentum var. *macrochaeton* A. Braun = Lolium temulentum [LOLITEM] Poaceae (V:M)
Lomaria spicant (L.) Desv. = Blechnum spicant [BLECSPI] Blechnaceae (V:F)
Lomatium Raf. [LOMATIU$] Apiaceae (V:D)
Lomatium ambiguum (Nutt.) Coult. & Rose [LOMAAMB] Apiaceae (V:D)
Lomatium angustatum (Coult. & Rose) St. John = Lomatium martindalei [LOMAMAR] Apiaceae (V:D)
Lomatium angustatum var. *flavum* G.N. Jones = Lomatium martindalei [LOMAMAR] Apiaceae (V:D)
Lomatium brandegei (Coult. & Rose) J.F. Macbr. [LOMABRA] Apiaceae (V:D)
Lomatium dissectum (Nutt.) Mathias & Constance [LOMADIS] Apiaceae (V:D)
Lomatium dissectum var. **dissectum** [LOMADIS0] Apiaceae (V:D)
Lomatium dissectum var. *eatonii* (Coult. & Rose) Cronq. = Lomatium dissectum var. multifidum [LOMADIS2] Apiaceae (V:D)
Lomatium dissectum var. **multifidum** (Nutt.) Mathias & Constance [LOMADIS2] Apiaceae (V:D)
Lomatium foeniculaceum (Nutt.) Coult. & Rose [LOMAFOE] Apiaceae (V:D)
Lomatium geyeri (S. Wats.) Coult. & Rose [LOMAGEY] Apiaceae (V:D)
Lomatium grayi (Coult. & Rose) Coult. & Rose [LOMAGRA] Apiaceae (V:D)
Lomatium macrocarpum (Nutt. ex Torr. & Gray) Coult. & Rose [LOMAMAC] Apiaceae (V:D)
Lomatium macrocarpum var. *artemisiarum* Piper = Lomatium macrocarpum [LOMAMAC] Apiaceae (V:D)
Lomatium macrocarpum var. *ellipticum* (Torr. & Gray) Jepson = Lomatium macrocarpum [LOMAMAC] Apiaceae (V:D)
Lomatium martindalei (Coult. & Rose) Coult. & Rose [LOMAMAR] Apiaceae (V:D)
Lomatium martindalei var. *angustatum* (Coult. & Rose) Coult. & Rose = Lomatium martindalei [LOMAMAR] Apiaceae (V:D)
Lomatium martindalei var. *flavum* (G.N. Jones) Cronq. = Lomatium martindalei [LOMAMAR] Apiaceae (V:D)
Lomatium nudicaule (Pursh) Coult. & Rose [LOMANUD] Apiaceae (V:D)
Lomatium platycarpum (Torr.) Coult. & Rose = Lomatium simplex [LOMASIM] Apiaceae (V:D)
Lomatium sandbergii (Coult. & Rose) Coult. & Rose [LOMASAN] Apiaceae (V:D)
Lomatium simplex (Nutt.) J.F. Macbr. [LOMASIM] Apiaceae (V:D)
Lomatium triternatum (Pursh) Coult. & Rose [LOMATRI] Apiaceae (V:D)
Lomatium triternatum ssp. *platycarpum* (Torr.) Cronq. = Lomatium simplex [LOMASIM] Apiaceae (V:D)
Lomatium utriculatum (Nutt. ex Torr. & Gray) Coult. & Rose [LOMAUTR] Apiaceae (V:D)
Lomatium vaseyi (Coult. & Rose) Coult. & Rose = Lomatium utriculatum [LOMAUTR] Apiaceae (V:D)
Lomatogonium A. Braun [LOMATOG$] Gentianaceae (V:D)
Lomatogonium rotatum (L.) Fries ex Fern. [LOMAROT] Gentianaceae (V:D)
Lomatogonium rotatum ssp. *tenuifolium* (Griseb.) Porsild = Lomatogonium rotatum [LOMAROT] Gentianaceae (V:D)
Lomatogonium tenellum (Rottb.) A. Löve & D. Löve = Gentianella tenella [GENTTEN] Gentianaceae (V:D)
Lonicera L. [LONICER$] Caprifoliaceae (V:D)
•**Lonicera caerulea** L. [LONICAE] Caprifoliaceae (V:D)
Lonicera ciliosa (Pursh) Poir. ex DC. [LONICIL] Caprifoliaceae (V:D)
Lonicera dioica L. [LONIDIO] Caprifoliaceae (V:D)
Lonicera dioica var. **glaucescens** (Rydb.) Butters [LONIDIO1] Caprifoliaceae (V:D)
•**Lonicera etrusca** Santi [LONIETR] Caprifoliaceae (V:D)
Lonicera glaucescens (Rydb.) Rydb. = Lonicera dioica var. glaucescens [LONIDIO1] Caprifoliaceae (V:D)
Lonicera hispidula (Lindl.) Dougl. ex Torr. & Gray [LONIHIS] Caprifoliaceae (V:D)
Lonicera involucrata Banks ex Spreng. [LONIINV] Caprifoliaceae (V:D)
Lonicera involucrata var. *flavescens* (Dippel) Rehd. = Lonicera involucrata [LONIINV] Caprifoliaceae (V:D)
Lonicera utahensis S. Wats. [LONIUTA] Caprifoliaceae (V:D)
Lopadium Körber [LOPADIU$] Ectolechiaceae (L)
Lopadium alpinum (Körber) R. Sant. = Schadonia alpina [SCHAALP] Bacidiaceae (L)
Lopadium disciforme (Flotow) Kullhem [LOPADIS] Ectolechiaceae (L)
Lopadium gemellum (Anzi) Stizenb. = Schadonia alpina [SCHAALP] Bacidiaceae (L)
Lopadium pezizoideum (Ach.) Körber [LOPAPEZ] Ectolechiaceae (L)
Lopadium pezizoideum var. *disciforme* (Flotow) Körber = Lopadium disciforme [LOPADIS] Ectolechiaceae (L)
Lophochlaena Nees [LOPHOCH$] Poaceae (V:M)
Lophochlaena refracta Gray [LOPHREF] Poaceae (V:M)
Lophocolea (Dum.) Dum. [LOPHOCO$] Lophocoleaceae (B:H)
Lophocolea austinii Lindb. = Lophocolea heterophylla [LOPHHET] Lophocoleaceae (B:H)
Lophocolea bidentata (L.) Dum. [LOPHBID] Lophocoleaceae (B:H)
Lophocolea heterophylla (Schrad.) Dum. [LOPHHET] Lophocoleaceae (B:H)
Lophocolea macounii Aust. = Lophocolea heterophylla [LOPHHET] Lophocoleaceae (B:H)
Lophocolea minor Nees [LOPHMIN] Lophocoleaceae (B:H)
Lophozia (Dum.) Dum. [LOPHOZI$] Jungermanniaceae (B:H)
Lophozia alpestris (Schleich. ex Web.) Evans [LOPHALP] Jungermanniaceae (B:H)
Lophozia ascendens (Warnst.) Schust. [LOPHASC] Jungermanniaceae (B:H)
Lophozia attenuata (Mart.) Dum. = Barbilophozia attenuata [BARBATT] Jungermanniaceae (B:H)
Lophozia badensis (Gott. ex Rabenh.) Schiffn. [LOPHBAD] Jungermanniaceae (B:H)
Lophozia bantriensis (Hook.) Steph. [LOPHBAN] Jungermanniaceae (B:H)
Lophozia barbata (Schmid.) Dum. = Barbilophozia barbata [BARBBAR] Jungermanniaceae (B:H)
Lophozia baueriana Schiffn. = Barbilophozia hatcheri [BARBHAT] Jungermanniaceae (B:H)
Lophozia bicrenata (Schmid. ex Hoffm.) Dum. [LOPHBIC] Jungermanniaceae (B:H)
Lophozia binsteadii (Kaal.) Evans = Barbilophozia binsteadii [BARBBIN] Jungermanniaceae (B:H)
Lophozia collaris (Nees) Dum. [LOPHCOL] Jungermanniaceae (B:H)
Lophozia confertifolia Schiffn. = Lophozia wenzelii [LOPHWEN] Jungermanniaceae (B:H)
Lophozia decolorans (Limpr.) Steph. [LOPHDEC] Jungermanniaceae (B:H)
Lophozia excisa (Dicks.) Dum. [LOPHEXC] Jungermanniaceae (B:H)
Lophozia fauriana Steph. = Lophozia guttulata [LOPHGUT] Jungermanniaceae (B:H)
Lophozia floerkei (Web. & Mohr) Schiffn. = Barbilophozia floerkei [BARBFLO] Jungermanniaceae (B:H)
Lophozia gillmanii (Aust.) Schust. [LOPHGIL] Jungermanniaceae (B:H)
Lophozia gracilis (Schleich.) Steph. = Barbilophozia attenuata [BARBATT] Jungermanniaceae (B:H)
Lophozia gracillima Buch = Lophozia ascendens [LOPHASC] Jungermanniaceae (B:H)

Lophozia grandiretis (Lindb.) Schiffn. [LOPHGRA] Jungermanniaceae (B:H)
Lophozia guttulata (Lindb. & Arnell) Evans [LOPHGUT] Jungermanniaceae (B:H)
Lophozia hatcheri (Evans) Steph. = Barbilophozia hatcheri [BARBHAT] Jungermanniaceae (B:H)
Lophozia heterocolpos (Thed.) M.A. Howe [LOPHHEE] Jungermanniaceae (B:H)
Lophozia hornschuchiana Macoun = Lophozia bantriensis [LOPHBAN] Jungermanniaceae (B:H)
Lophozia incisa (Schrad.) Dum. [LOPHINC] Jungermanniaceae (B:H)
Lophozia kaurini (Limpr.) Steph. = Lophozia gillmanii [LOPHGIL] Jungermanniaceae (B:H)
Lophozia kunzeana (Hüb.) Buch = Barbilophozia kunzeana [BARBKUN] Jungermanniaceae (B:H)
Lophozia longidens (Lindb.) Macoun [LOPHLON] Jungermanniaceae (B:H)
Lophozia longidens auct. = Lophozia ascendens [LOPHASC] Jungermanniaceae (B:H)
Lophozia longiflora (Nees) Schiffn. = Lophozia ventricosa var. longiflora [LOPHVEN3] Jungermanniaceae (B:H)
Lophozia lycopodioides (Wallr.) Cogn. = Barbilophozia lycopodioides [BARBLYC] Jungermanniaceae (B:H)
Lophozia muelleri (Nees) Dum. = Lophozia collaris [LOPHCOL] Jungermanniaceae (B:H)
Lophozia murmanica Kaal. = Lophozia grandiretis [LOPHGRA] Jungermanniaceae (B:H)
Lophozia obtusa (Lindb.) Evans [LOPHOBT] Jungermanniaceae (B:H)
Lophozia opacifolia Culm. [LOPHOPA] Jungermanniaceae (B:H)
Lophozia porphyroleuca (Nees) Schiffn. = Lophozia guttulata [LOPHGUT] Jungermanniaceae (B:H)
Lophozia quadriloba (Lindb.) Evans = Barbilophozia quadriloba [BARBQUA] Jungermanniaceae (B:H)
Lophozia quinquedentata (Huds.) Cogn. = Tritomaria quinquedentata [TRITQUI] Jungermanniaceae (B:H)
Lophozia rotundiloba nom. herb. = Barbilophozia kunzeana [BARBKUN] Jungermanniaceae (B:H)
Lophozia rutheana (Limpr.) M.A. Howe [LOPHRUT] Jungermanniaceae (B:H)
Lophozia schultzii (Nees) Schiffn. = Lophozia rutheana [LOPHRUT] Jungermanniaceae (B:H)
Lophozia silvicola Buch = Lophozia ventricosa var. silvicola [LOPHVEN5] Jungermanniaceae (B:H)
Lophozia silvicoloides Kitag. = Lophozia ventricosa var. silvicola [LOPHVEN5] Jungermanniaceae (B:H)
Lophozia ventricosa (Dicks.) Dum. [LOPHVEN] Jungermanniaceae (B:H)
Lophozia ventricosa var. **longiflora** (Nees) Macoun [LOPHVEN3] Jungermanniaceae (B:H)
Lophozia ventricosa var. **silvicola** (Buch) Jones [LOPHVEN5] Jungermanniaceae (B:H)
Lophozia ventricosa var. **ventricosa** [LOPHVEN0] Jungermanniaceae (B:H)
Lophozia wenzelii (Nees) Steph. [LOPHWEN] Jungermanniaceae (B:H)
Lotus L. [LOTUS$] Fabaceae (V:D)
Lotus americanus (Nutt.) Bisch. = Lotus unifoliolatus [LOTUUNI] Fabaceae (V:D)
•**Lotus corniculatus** L. [LOTUCOR] Fabaceae (V:D)
Lotus corniculatus var. *arvensis* (Schkuhr) Ser. ex DC. = Lotus corniculatus [LOTUCOR] Fabaceae (V:D)
Lotus corniculatus var. *tenuifolius* L. = Lotus tenuis [LOTUTEN] Fabaceae (V:D)
Lotus denticulatus (E. Drew) Greene [LOTUDEN] Fabaceae (V:D)
Lotus douglasii Greene = Lotus nevadensis var. douglasii [LOTUNEV1] Fabaceae (V:D)
Lotus formosissimus Greene [LOTUFOR] Fabaceae (V:D)
Lotus micranthus Benth. [LOTUMIC] Fabaceae (V:D)
Lotus nevadensis (S. Wats.) Greene [LOTUNEV] Fabaceae (V:D)
Lotus nevadensis var. **douglasii** (Greene) Ottley [LOTUNEV1] Fabaceae (V:D)
•**Lotus pedunculatus** Cav. [LOTUPED] Fabaceae (V:D)
Lotus pinnatus Hook. [LOTUPIN] Fabaceae (V:D)
Lotus purshianus F.E. & E.G. Clem. = Lotus unifoliolatus [LOTUUNI] Fabaceae (V:D)
Lotus purshianus var. *glaber* (Nutt.) Munz = Lotus unifoliolatus [LOTUUNI] Fabaceae (V:D)
Lotus sericeus Pursh = Lotus unifoliolatus [LOTUUNI] Fabaceae (V:D)
•**Lotus tenuis** Waldst. & Kit. ex Willd. [LOTUTEN] Fabaceae (V:D)
Lotus uliginosus Schkuhr = Lotus pedunculatus [LOTUPED] Fabaceae (V:D)
Lotus unifoliolatus (Hook.) Benth. [LOTUUNI] Fabaceae (V:D)
Loxospora A. Massal. [LOXOSPO$] Haematommataceae (L)
Loxospora elatina (Ach.) A. Massal. [LOXOELA] Haematommataceae (L)
Loxospora ochrophaea (Tuck.) R.C. Harris [LOXOOCH] Haematommataceae (L)
Ludwigia L. [LUDWIGI$] Onagraceae (V:D)
Ludwigia palustris (L.) Ell. [LUDWPAL] Onagraceae (V:D)
Ludwigia palustris var. *americana* (DC.) Fern. & Grisc. = Ludwigia palustris [LUDWPAL] Onagraceae (V:D)
Ludwigia palustris var. *nana* Fern. & Grisc. = Ludwigia palustris [LUDWPAL] Onagraceae (V:D)
Ludwigia palustris var. *pacifica* Fern. & Grisc. = Ludwigia palustris [LUDWPAL] Onagraceae (V:D)
Luetkea Bong. [LUETKEA$] Rosaceae (V:D)
Luetkea pectinata (Pursh) Kuntze [LUETPEC] Rosaceae (V:D)
Luina Benth. [LUINA$] Asteraceae (V:D)
Luina hypoleuca Benth. [LUINHYP] Asteraceae (V:D)
Luina nardosmia (Gray) Cronq. = Cacaliopsis nardosmia [CACANAR] Asteraceae (V:D)
Luina nardosmia var. *glabrata* (Piper) Cronq. = Cacaliopsis nardosmia [CACANAR] Asteraceae (V:D)
Lunaria L. [LUNARIA$] Brassicaceae (V:D)
•**Lunaria annua** L. [LUNAANN] Brassicaceae (V:D)
Lunularia Adans. [LUNULAR$] Lunulariaceae (B:H)
Lunularia cruciata (L.) Dum. [LUNUCRU] Lunulariaceae (B:H)
Lupinaster wormskioldii (Lehm.) K. Presl = Trifolium wormskjoldii [TRIFWOR] Fabaceae (V:D)
Lupinus L. [LUPINUS$] Fabaceae (V:D)
Lupinus albertensis C.P. Sm. = Lupinus nootkatensis var. nootkatensis [LUPINOO0] Fabaceae (V:D)
Lupinus amniculi-putori C.P. Sm. = Lupinus arbustus ssp. neolaxiflorus [LUPIARU1] Fabaceae (V:D)
Lupinus apricus Greene = Lupinus vallicola ssp. apricus [LUPIVAL1] Fabaceae (V:D)
•**Lupinus arboreus** Sims [LUPIARB] Fabaceae (V:D)
Lupinus arboreus var. *fruticosus* (Sims) S. Wats. = Lupinus nootkatensis var. fruticosus [LUPINOO1] Fabaceae (V:D)
Lupinus arbustus Dougl. ex Lindl. [LUPIARU] Fabaceae (V:D)
Lupinus arbustus ssp. **neolaxiflorus** D. Dunn [LUPIARU1] Fabaceae (V:D)
Lupinus arbustus ssp. **pseudoparviflorus** (Rydb.) D. Dunn [LUPIARU2] Fabaceae (V:D)
Lupinus arcticus S. Wats. [LUPIARC] Fabaceae (V:D)
Lupinus arcticus ssp. **arcticus** [LUPIARC0] Fabaceae (V:D)
Lupinus arcticus ssp. **canadensis** (C.P. Sm.) D. Dunn [LUPIARC2] Fabaceae (V:D)
Lupinus arcticus ssp. **subalpinus** (Piper & B.L. Robins.) D. Dunn [LUPIARC3] Fabaceae (V:D)
Lupinus arcticus var. *subalpinus* (Piper & B.L. Robins.) C.P. Sm. = Lupinus arcticus ssp. subalpinus [LUPIARC3] Fabaceae (V:D)
Lupinus argenteus Pursh [LUPIARG] Fabaceae (V:D)
Lupinus argenteus ssp. **argenteus** [LUPIARG0] Fabaceae (V:D)
Lupinus augustii C.P. Sm. = Lupinus arbustus ssp. neolaxiflorus [LUPIARU1] Fabaceae (V:D)
Lupinus bicolor Lindl. [LUPIBIC] Fabaceae (V:D)
Lupinus bingenensis Suksdorf [LUPIBIN] Fabaceae (V:D)
Lupinus bingenensis var. **subsaccatus** Suksdorf [LUPIBIN1] Fabaceae (V:D)
Lupinus borealis Heller = Lupinus arcticus ssp. arcticus [LUPIARC0] Fabaceae (V:D)
Lupinus burkei S. Wats. [LUPIBUR] Fabaceae (V:D)

Lupinus caudatus var. *submanens* C.P. Sm. = Lupinus arbustus ssp. neolaxiflorus [LUPIARU1] Fabaceae (V:D)
Lupinus densiflorus Benth. [LUPIDEN] Fabaceae (V:D)
Lupinus densiflorus var. *latilabris* C.P. Sm. = Lupinus densiflorus [LUPIDEN] Fabaceae (V:D)
Lupinus densiflorus var. *scopulorum* C.P. Sm. = Lupinus densiflorus [LUPIDEN] Fabaceae (V:D)
Lupinus densiflorus var. *stenopetalus* C.P. Sm. = Lupinus densiflorus [LUPIDEN] Fabaceae (V:D)
Lupinus densiflorus var. *tracyi* C.P. Sm. = Lupinus densiflorus [LUPIDEN] Fabaceae (V:D)
Lupinus donnellyensis C.P. Sm. = Lupinus arcticus ssp. arcticus [LUPIARC0] Fabaceae (V:D)
Lupinus gakonensis C.P. Sm. = Lupinus arcticus ssp. arcticus [LUPIARC0] Fabaceae (V:D)
Lupinus glacialis C.P. Sm. = Lupinus arcticus ssp. subalpinus [LUPIARC3] Fabaceae (V:D)
Lupinus kiskensis C.P. Sm. = Lupinus nootkatensis var. nootkatensis [LUPINOO0] Fabaceae (V:D)
Lupinus kuschei Eastw. [LUPIKUS] Fabaceae (V:D)
Lupinus latifolius var. *canadensis* C.P. Sm. = Lupinus arcticus ssp. canadensis [LUPIARC2] Fabaceae (V:D)
Lupinus latifolius var. *subalpinus* (Piper & B.L. Robins.) C.P. Sm. = Lupinus arcticus ssp. subalpinus [LUPIARC3] Fabaceae (V:D)
Lupinus laxiflorus var. *elmerianus* C.P. Sm. = Lupinus arbustus ssp. pseudoparviflorus [LUPIARU2] Fabaceae (V:D)
Lupinus laxiflorus var. *lyleanus* C.P. Sm. = Lupinus arbustus ssp. neolaxiflorus [LUPIARU1] Fabaceae (V:D)
Lupinus laxiflorus var. *pseudoparviflorus* (Rydb.) C.P. Sm. & St. John = Lupinus arbustus ssp. pseudoparviflorus [LUPIARU2] Fabaceae (V:D)
Lupinus laxispicatus Rydb. = Lupinus arbustus ssp. pseudoparviflorus [LUPIARU2] Fabaceae (V:D)
Lupinus laxispicatus var. *whithamii* C.P. Sm. = Lupinus arbustus ssp. pseudoparviflorus [LUPIARU2] Fabaceae (V:D)
Lupinus lepidus Dougl. ex Lindl. [LUPILEP] Fabaceae (V:D)
Lupinus lepidus ssp. *lyallii* (Gray) Detling = Lupinus lyallii [LUPILYA] Fabaceae (V:D)
Lupinus leucophyllus Dougl. ex Lindl. [LUPILEU] Fabaceae (V:D)
Lupinus lignipes Heller = Lupinus rivularis [LUPIRIV] Fabaceae (V:D)
Lupinus littoralis Dougl. [LUPILIT] Fabaceae (V:D)
Lupinus lyallii Gray [LUPILYA] Fabaceae (V:D)
Lupinus lyleanus C.P. Sm. = Lupinus arbustus ssp. neolaxiflorus [LUPIARU1] Fabaceae (V:D)
Lupinus mackeyi C.P. Sm. = Lupinus arbustus ssp. neolaxiflorus [LUPIARU1] Fabaceae (V:D)
Lupinus micranthus Dougl. = Lupinus polycarpus [LUPIPOL] Fabaceae (V:D)
Lupinus micranthus var. *bicolor* (Lindl.) S. Wats. = Lupinus bicolor [LUPIBIC] Fabaceae (V:D)
Lupinus microcarpus ssp. *scopulorum* (C.P. Sm.) C.P. Sm. = Lupinus densiflorus [LUPIDEN] Fabaceae (V:D)
Lupinus microcarpus var. *scopulorum* C.P. Sm. = Lupinus densiflorus [LUPIDEN] Fabaceae (V:D)
Lupinus minimus Dougl. ex Hook. [LUPIMIN] Fabaceae (V:D)
Lupinus mucronulatus var. *umatillensis* C.P. Sm. = Lupinus arbustus ssp. pseudoparviflorus [LUPIARU2] Fabaceae (V:D)
Lupinus multicaulis C.P. Sm. = Lupinus arcticus ssp. arcticus [LUPIARC0] Fabaceae (V:D)
Lupinus multifolius C.P. Sm. = Lupinus arcticus ssp. arcticus [LUPIARC0] Fabaceae (V:D)
Lupinus nanus var. *apricus* (Greene) C.P. Sm. = Lupinus vallicola ssp. apricus [LUPIVAL1] Fabaceae (V:D)
Lupinus nootkatensis Donn ex Sims [LUPINOO] Fabaceae (V:D)
Lupinus nootkatensis var. *ethel-looffii* C.P. Sm. = Lupinus nootkatensis var. nootkatensis [LUPINOO0] Fabaceae (V:D)
Lupinus nootkatensis var. **fruticosus** Sims [LUPINOO1] Fabaceae (V:D)
Lupinus nootkatensis var. *glaber* Hook. = Lupinus nootkatensis var. fruticosus [LUPINOO1] Fabaceae (V:D)
Lupinus nootkatensis var. *henry-looffii* C.P. Sm. = Lupinus nootkatensis var. nootkatensis [LUPINOO0] Fabaceae (V:D)
Lupinus nootkatensis var. *kiellmannii* Ostenf. = Lupinus arcticus ssp. arcticus [LUPIARC0] Fabaceae (V:D)
Lupinus nootkatensis var. **nootkatensis** [LUPINOO0] Fabaceae (V:D)
Lupinus nootkatensis var. *perlanatus* C.P. Sm. = Lupinus nootkatensis var. nootkatensis [LUPINOO0] Fabaceae (V:D)
Lupinus nootkatensis var. *unalaskensis* S. Wats. = Lupinus nootkatensis var. fruticosus [LUPINOO1] Fabaceae (V:D)
Lupinus oreganus Heller [LUPIORE] Fabaceae (V:D)
Lupinus oreganus var. **kincaidii** C.P. Sm. [LUPIORE1] Fabaceae (V:D)
Lupinus ovinus Greene = Lupinus minimus [LUPIMIN] Fabaceae (V:D)
Lupinus perennis ssp. *nootkatensis* (Donn ex Sims) L. Phillips = Lupinus nootkatensis var. nootkatensis [LUPINOO0] Fabaceae (V:D)
Lupinus piperi B.L. Robins. = Lupinus minimus [LUPIMIN] Fabaceae (V:D)
Lupinus polycarpus Greene [LUPIPOL] Fabaceae (V:D)
Lupinus polyphyllus Lindl. [LUPIPOY] Fabaceae (V:D)
Lupinus polyphyllus ssp. *arcticus* (S. Wats.) L. Phillips = Lupinus arcticus ssp. arcticus [LUPIARC0] Fabaceae (V:D)
Lupinus polyphyllus var. *burkei* (S. Wats.) C.L. Hitchc. = Lupinus burkei [LUPIBUR] Fabaceae (V:D)
Lupinus relictus Hultén = Lupinus arcticus ssp. arcticus [LUPIARC0] Fabaceae (V:D)
Lupinus rivularis Dougl. ex Lindl. [LUPIRIV] Fabaceae (V:D)
Lupinus scheuberae Rydb. = Lupinus arbustus ssp. pseudoparviflorus [LUPIARU2] Fabaceae (V:D)
Lupinus sericeus Pursh [LUPISER] Fabaceae (V:D)
Lupinus sericeus var. *kuschei* (Eastw.) Boivin = Lupinus kuschei [LUPIKUS] Fabaceae (V:D)
Lupinus standingii C.P. Sm. = Lupinus arbustus ssp. neolaxiflorus [LUPIARU1] Fabaceae (V:D)
Lupinus stipaphilus C.P. Sm. = Lupinus arbustus ssp. neolaxiflorus [LUPIARU1] Fabaceae (V:D)
Lupinus stockii C.P. Sm. = Lupinus arbustus ssp. neolaxiflorus [LUPIARU1] Fabaceae (V:D)
Lupinus subalpinus Piper & B.L. Robins. = Lupinus arcticus ssp. subalpinus [LUPIARC3] Fabaceae (V:D)
Lupinus sulphureus Dougl. ex Hook. [LUPISUL] Fabaceae (V:D)
Lupinus sulphureus ssp. *kincaidii* (C.P. Sm.) L. Phillips = Lupinus oreganus var. kincaidii [LUPIORE1] Fabaceae (V:D)
Lupinus sulphureus ssp. *subsaccatus* (Suksdorf) L. Phillips = Lupinus bingenensis var. subsaccatus [LUPIBIN1] Fabaceae (V:D)
Lupinus sulphureus var. *applegateanus* C.P. Sm. = Lupinus sulphureus [LUPISUL] Fabaceae (V:D)
Lupinus sulphureus var. *echleranus* C.P. Sm. = Lupinus sulphureus [LUPISUL] Fabaceae (V:D)
Lupinus sulphureus var. *kincaidii* (C.P. Sm.) C.L. Hitchc. = Lupinus oreganus var. kincaidii [LUPIORE1] Fabaceae (V:D)
Lupinus sulphureus var. *subsaccatus* (Suksdorf) C.L. Hitchc. = Lupinus bingenensis var. subsaccatus [LUPIBIN1] Fabaceae (V:D)
Lupinus trifurcatus C.P. Sm. = Lupinus nootkatensis var. nootkatensis [LUPINOO0] Fabaceae (V:D)
Lupinus vallicola Heller [LUPIVAL] Fabaceae (V:D)
Lupinus vallicola ssp. **apricus** (Greene) D. Dunn [LUPIVAL1] Fabaceae (V:D)
Lupinus vallicola var. *apricus* (Greene) C.P. Sm. = Lupinus vallicola ssp. apricus [LUPIVAL1] Fabaceae (V:D)
Lupinus volcanicus Greene = Lupinus arcticus ssp. subalpinus [LUPIARC3] Fabaceae (V:D)
Lupinus volcanicus var. *rupestricola* C.P. Sm. = Lupinus arcticus ssp. subalpinus [LUPIARC3] Fabaceae (V:D)
Lupinus wenachensis Eastw. = Lupinus arbustus ssp. neolaxiflorus [LUPIARU1] Fabaceae (V:D)
Lupinus wyethii S. Wats. [LUPIWYE] Fabaceae (V:D)
Lupinus yakimensis C.P. Sm. = Lupinus arbustus ssp. neolaxiflorus [LUPIARU1] Fabaceae (V:D)
Lupinus yukonensis Greene = Lupinus arcticus ssp. arcticus [LUPIARC0] Fabaceae (V:D)
Luzula DC. [LUZULA$] Juncaceae (V:M)
Luzula arctica Blytt [LUZUARC] Juncaceae (V:M)

Luzula arcuata (Wahlenb.) Sw. [LUZUARU] Juncaceae (V:M)
Luzula arcuata ssp. **unalaschcensis** (Buch.) Hultén [LUZUARU1] Juncaceae (V:M)
Luzula arcuata var. *unalaschcensis* Buch. = Luzula arcuata ssp. unalaschcensis [LUZUARU1] Juncaceae (V:M)
Luzula campestris var. *congesta* (Thuill.) E. Mey. = Luzula congesta [LUZUCOG] Juncaceae (V:M)
Luzula campestris var. *frigida* Buch. = Luzula multiflora ssp. frigida [LUZUMUL7] Juncaceae (V:M)
Luzula campestris var. *multiflora* (Ehrh.) Celak. = Luzula multiflora [LUZUMUL] Juncaceae (V:M)
Luzula comosa E. Mey. = Luzula congesta [LUZUCOG] Juncaceae (V:M)
Luzula comosa var. *congesta* (Thuill.) S. Wats. = Luzula congesta [LUZUCOG] Juncaceae (V:M)
Luzula confusa Lindeberg [LUZUCON] Juncaceae (V:M)
Luzula congesta (Thuill.) Lej. [LUZUCOG] Juncaceae (V:M)
Luzula fastigiata E. Mey. = Luzula parviflora ssp. fastigiata [LUZUPAR1] Juncaceae (V:M)
Luzula frigida (Buch.) Samuelsson = Luzula multiflora ssp. frigida [LUZUMUL7] Juncaceae (V:M)
Luzula glabrata (Hoppe ex Rostk.) Desv. [LUZUGLA] Juncaceae (V:M)
Luzula glabrata var. **hitchcockii** (Hämet-Ahti) Dorn [LUZUGLA1] Juncaceae (V:M)
Luzula groenlandica Böcher [LUZUGRO] Juncaceae (V:M)
Luzula hitchcockii Hämet-Ahti = Luzula glabrata var. hitchcockii [LUZUGLA1] Juncaceae (V:M)
Luzula hyperborea R. Br. p.p. = Juncus articulatus [JUNCART] Juncaceae (V:M)
Luzula intermedia (Thuill.) A. Nels. = Luzula congesta [LUZUCOG] Juncaceae (V:M)
Luzula melanocarpa (Michx.) Tolm. = Luzula parviflora ssp. melanocarpa [LUZUPAR4] Juncaceae (V:M)
Luzula multiflora (Ehrh.) Lej. [LUZUMUL] Juncaceae (V:M)
Luzula multiflora ssp. *comosa* (E. Mey.) Hultén = Luzula congesta [LUZUCOG] Juncaceae (V:M)
Luzula multiflora ssp. *congesta* (Thuill.) Hyl. = Luzula congesta [LUZUCOG] Juncaceae (V:M)
Luzula multiflora ssp. **frigida** (Buch.) Krecz. [LUZUMUL7] Juncaceae (V:M)
Luzula multiflora ssp. *kjellmaniana* (Miyabe & Kudo) Tolm. = Luzula multiflora ssp. multiflora var. kjellmanioides [LUZUMUL4] Juncaceae (V:M)
Luzula multiflora ssp. **multiflora** [LUZUMUL0] Juncaceae (V:M)
Luzula multiflora ssp. **multiflora** var. **kjellmanioides** Taylor & MacBryde [LUZUMUL4] Juncaceae (V:M)
Luzula multiflora var. *acadiensis* Fern. = Luzula congesta [LUZUCOG] Juncaceae (V:M)
Luzula multiflora var. *comosa* (E. Mey.) Fern. & Wieg. = Luzula congesta [LUZUCOG] Juncaceae (V:M)
Luzula multiflora var. *congesta* (Thuill.) Koch = Luzula congesta [LUZUCOG] Juncaceae (V:M)
Luzula multiflora var. *frigida* (Buch.) Samuelsson = Luzula multiflora ssp. frigida [LUZUMUL7] Juncaceae (V:M)
Luzula multiflora var. *fusconigra* Celak. = Luzula multiflora ssp. frigida [LUZUMUL7] Juncaceae (V:M)
Luzula nivalis (Laestad.) Beurling = Luzula arctica [LUZUARC] Juncaceae (V:M)
Luzula orestera C.W. Sharsmith = Luzula congesta [LUZUCOG] Juncaceae (V:M)
Luzula parviflora (Ehrh.) Desv. [LUZUPAR] Juncaceae (V:M)
Luzula parviflora ssp. **fastigiata** (E. Mey.) Hämet-Ahti [LUZUPAR1] Juncaceae (V:M)
Luzula parviflora ssp. **melanocarpa** (Michx.) Hämet-Ahti [LUZUPAR4] Juncaceae (V:M)
Luzula parviflora ssp. *melanocarpa* (Michx.) Tolm. = Luzula parviflora ssp. melanocarpa [LUZUPAR4] Juncaceae (V:M)
Luzula parviflora var. *melanocarpa* (Michx.) Buch. = Luzula parviflora ssp. melanocarpa [LUZUPAR4] Juncaceae (V:M)
Luzula pilosa var. *rufescens* (Fisch. ex E. Mey.) Boivin = Luzula rufescens [LUZURUF] Juncaceae (V:M)
Luzula piperi (Coville) M.E. Jones [LUZUPIP] Juncaceae (V:M)
Luzula rufescens Fisch. ex E. Mey. [LUZURUF] Juncaceae (V:M)
Luzula spicata (L.) DC. [LUZUSPI] Juncaceae (V:M)
Luzula spicata ssp. *saximontana* A. & D. Löve = Luzula congesta [LUZUCOG] Juncaceae (V:M)
Luzula sudetica var. *frigida* (Buch.) Fern. = Luzula multiflora ssp. frigida [LUZUMUL7] Juncaceae (V:M)
Luzula unalaschcensis (Buch.) Satake = Luzula arcuata ssp. unalaschcensis [LUZUARU1] Juncaceae (V:M)
Luzula wahlenbergii ssp. *piperi* (Coville) Hultén = Luzula piperi [LUZUPIP] Juncaceae (V:M)
Lychnis L. [LYCHNIS$] Caryophyllaceae (V:D)
Lychnis affinis J. Vahl ex Fries = Silene involucrata [SILEINV] Caryophyllaceae (V:D)
Lychnis alba P. Mill. = Silene latifolia ssp. alba [SILELAT1] Caryophyllaceae (V:D)
Lychnis apetala ssp. *attenuata* (Farr) Maguire = Silene uralensis ssp. attenuata [SILEURA1] Caryophyllaceae (V:D)
Lychnis apetala var. *attenuata* (Farr) C.L. Hitchc. = Silene uralensis ssp. attenuata [SILEURA1] Caryophyllaceae (V:D)
Lychnis attenuata Farr = Silene uralensis ssp. attenuata [SILEURA1] Caryophyllaceae (V:D)
•**Lychnis coronaria** (L.) Desr. [LYCHCOR] Caryophyllaceae (V:D)
Lychnis dawsonii (B.L. Robins.) J.P. Andrs. = Silene taimyrensis [SILETAI] Caryophyllaceae (V:D)
Lychnis dioica L. = Silene dioica [SILEDIO] Caryophyllaceae (V:D)
Lychnis drummondii (Hook.) S. Wats. = Silene drummondii [SILEDRU] Caryophyllaceae (V:D)
Lychnis furcata (Raf.) Fern. = Silene involucrata [SILEINV] Caryophyllaceae (V:D)
Lychnis gillettii Boivin = Silene involucrata [SILEINV] Caryophyllaceae (V:D)
Lychnis loveae Boivin = Silene latifolia ssp. alba [SILELAT1] Caryophyllaceae (V:D)
Lychnis × *loveae* Boivin = Silene latifolia ssp. alba [SILELAT1] Caryophyllaceae (V:D)
Lychnis pudica Boivin = Silene drummondii [SILEDRU] Caryophyllaceae (V:D)
Lychnis saponaria Jessen = Saponaria officinalis [SAPOOFF] Caryophyllaceae (V:D)
Lychnis taimyrensis (Tolm.) Polunin = Silene taimyrensis [SILETAI] Caryophyllaceae (V:D)
Lychnis triflora ssp. *dawsonii* (B.L. Robins.) Maguire = Silene taimyrensis [SILETAI] Caryophyllaceae (V:D)
Lychnis triflora var. *dawsonii* B.L. Robins. = Silene taimyrensis [SILETAI] Caryophyllaceae (V:D)
Lychnis vespertina Sibthorp = Silene latifolia ssp. alba [SILELAT1] Caryophyllaceae (V:D)
Lycium L. [LYCIUM$] Solanaceae (V:D)
•**Lycium barbarum** L. [LYCIBAR] Solanaceae (V:D)
Lycium halimifolium P. Mill. = Lycium barbarum [LYCIBAR] Solanaceae (V:D)
Lycopodiella Holub [LYCOPOD$] Lycopodiaceae (V:F)
Lycopodiella inundata (L.) Holub [LYCOINU] Lycopodiaceae (V:F)
Lycopodium L. [LYCOPOI$] Lycopodiaceae (V:F)
Lycopodium alpinum L. [LYCOALP] Lycopodiaceae (V:F)
Lycopodium anceps Wallr. = Lycopodium complanatum [LYCOCOM] Lycopodiaceae (V:F)
Lycopodium annotinum L. [LYCOANN] Lycopodiaceae (V:F)
Lycopodium chinense Christ = Huperzia chinensis [HUPECHI] Lycopodiaceae (V:F)
Lycopodium clavatum L. [LYCOCLA] Lycopodiaceae (V:F)
Lycopodium clavatum ssp. *megastachyon* (Fern. & Bissell) Selin = Lycopodium clavatum var. monostachyon [LYCOCLA3] Lycopodiaceae (V:F)
Lycopodium clavatum ssp. *monostachyon* (Grev. & Hook.) Seland. = Lycopodium clavatum var. monostachyon [LYCOCLA3] Lycopodiaceae (V:F)
Lycopodium clavatum var. **clavatum** [LYCOCLA0] Lycopodiaceae (V:F)
Lycopodium clavatum var. **integerrimum** Spring [LYCOCLA2] Lycopodiaceae (V:F)
Lycopodium clavatum var. *laurentianum* Victorin = Lycopodium clavatum var. clavatum [LYCOCLA0] Lycopodiaceae (V:F)

Lycopodium clavatum var. *megastachyon* Fern. & Bissell = Lycopodium clavatum var. monostachyon [LYCOCLA3] Lycopodiaceae (V:F)
Lycopodium clavatum var. **monostachyon** Grev. & Hook. [LYCOCLA3] Lycopodiaceae (V:F)
Lycopodium clavatum var. *subremotum* Victorin = Lycopodium clavatum var. clavatum [LYCOCLA0] Lycopodiaceae (V:F)
Lycopodium clavatum var. *tristachyum* Hook. = Lycopodium clavatum var. clavatum [LYCOCLA0] Lycopodiaceae (V:F)
Lycopodium complanatum L. [LYCOCOM] Lycopodiaceae (V:F)
Lycopodium complanatum ssp. *anceps* (Wallr.) Aschers. = Lycopodium complanatum [LYCOCOM] Lycopodiaceae (V:F)
Lycopodium complanatum var. *canadense* Victorin = Lycopodium complanatum [LYCOCOM] Lycopodiaceae (V:F)
Lycopodium dendroideum Michx. [LYCODEN] Lycopodiaceae (V:F)
Lycopodium haleakalae Brack. = Huperzia haleakalae [HUPEHAL] Lycopodiaceae (V:F)
Lycopodium inundatum L. = Lycopodiella inundata [LYCOINU] Lycopodiaceae (V:F)
Lycopodium lagopus (Laestad.) Zinserl. ex Kuzen = Lycopodium clavatum var. monostachyon [LYCOCLA3] Lycopodiaceae (V:F)
Lycopodium lucidulum var. *occidentale* (Clute) L.R. Wilson = Huperzia occidentalis [HUPEOCC] Lycopodiaceae (V:F)
Lycopodium lucidulum var. *tryonii* Mohlenbrock = Huperzia occidentalis [HUPEOCC] Lycopodiaceae (V:F)
Lycopodium miyoshianum Makino = Huperzia chinensis [HUPECHI] Lycopodiaceae (V:F)
Lycopodium obscurum var. *dendroideum* (Michx.) D.C. Eat. = Lycopodium dendroideum [LYCODEN] Lycopodiaceae (V:F)
Lycopodium obscurum var. *hybridum* Farw. = Lycopodium dendroideum [LYCODEN] Lycopodiaceae (V:F)
Lycopodium pungens La Pylaie ex Iljin = Lycopodium annotinum [LYCOANN] Lycopodiaceae (V:F)
Lycopodium sabinifolium ssp. *sitchense* (Rupr.) Calder & Taylor = Lycopodium sitchense [LYCOSIT] Lycopodiaceae (V:F)
Lycopodium sabinifolium var. *sitchense* (Rupr.) Fern. = Lycopodium sitchense [LYCOSIT] Lycopodiaceae (V:F)
Lycopodium selago ssp. *chinense* (Christ) Hultén = Huperzia chinensis [HUPECHI] Lycopodiaceae (V:F)
Lycopodium selago ssp. *miyoshianum* (Makino) Calder & Taylor = Huperzia chinensis [HUPECHI] Lycopodiaceae (V:F)
Lycopodium selago var. *miyoshianum* (Makino) Makino = Huperzia chinensis [HUPECHI] Lycopodiaceae (V:F)
Lycopodium sitchense Rupr. [LYCOSIT] Lycopodiaceae (V:F)
Lycopus L. [LYCOPUS$] Lamiaceae (V:D)
Lycopus americanus Muhl. ex W. Bart. [LYCOAME] Lamiaceae (V:D)
Lycopus americanus var. *longii* Benner = Lycopus americanus [LYCOAME] Lamiaceae (V:D)
Lycopus americanus var. *scabrifolius* Fern. = Lycopus americanus [LYCOAME] Lamiaceae (V:D)
Lycopus asper Greene [LYCOASP] Lamiaceae (V:D)
Lycopus lucidus ssp. *americanus* (Gray) Hultén = Lycopus asper [LYCOASP] Lamiaceae (V:D)
Lycopus lucidus var. *americanus* Gray = Lycopus asper [LYCOASP] Lamiaceae (V:D)
Lycopus sinuatus Ell. = Lycopus americanus [LYCOAME] Lamiaceae (V:D)
Lycopus uniflorus Michx. [LYCOUNI] Lamiaceae (V:D)
Lyellia lescurii (James) Salm. = Bartramiopsis lescurii [BARTLES] Polytrichaceae (B:M)
Lygodesmia D. Don [LYGODES$] Asteraceae (V:D)
Lygodesmia juncea (Pursh) D. Don ex Hook. [LYGOJUN] Asteraceae (V:D)
Lysichiton Schott [LYSICHI$] Araceae (V:M)
Lysichiton americanum Hultén & St. John [LYSIAME] Araceae (V:M)
Lysiella obtusata (Banks ex Pursh) Rydb. = Platanthera obtusata [PLATOBT] Orchidaceae (V:M)
Lysimachia L. [LYSIMAC$] Primulaceae (V:D)
Lysimachia ciliata L. [LYSICIL] Primulaceae (V:D)
***Lysimachia nummularia** L. [LYSINUM] Primulaceae (V:D)
***Lysimachia punctata** L. [LYSIPUN] Primulaceae (V:D)
Lysimachia punctata var. *verticillata* (Bieb.) Klatt = Lysimachia punctata [LYSIPUN] Primulaceae (V:D)
***Lysimachia terrestris** (L.) B.S.P. [LYSITER] Primulaceae (V:D)
Lysimachia terrestris var. *ovata* (Rand & Redf.) Fern. = Lysimachia terrestris [LYSITER] Primulaceae (V:D)
Lysimachia thyrsiflora L. [LYSITHY] Primulaceae (V:D)
***Lysimachia vulgaris** L. [LYSIVUL] Primulaceae (V:D)
Lythrum L. [LYTHRUM$] Lythraceae (V:D)
Lythrum alatum Pursh [LYTHALA] Lythraceae (V:D)
***Lythrum portula** (L.) D.A. Webber [LYTHPOR] Lythraceae (V:D)
***Lythrum salicaria** L. [LYTHSAL] Lythraceae (V:D)
Lythrum salicaria var. *gracilior* Turcz. = Lythrum salicaria [LYTHSAL] Lythraceae (V:D)
Lythrum salicaria var. *tomentosum* (P. Mill.) DC. = Lythrum salicaria [LYTHSAL] Lythraceae (V:D)
Lythrum salicaria var. *vulgare* DC. = Lythrum salicaria [LYTHSAL] Lythraceae (V:D)

M

Machaeranthera Nees [MACHAER$] Asteraceae (V:D)
Machaeranthera canescens (Pursh) Gray [MACHCAN] Asteraceae (V:D)
Macounastrum islandicum (L.) Small = Koenigia islandica [KOENISL] Polygonaceae (V:D)
Macrodiplophyllum imbricatum (M.A. Howe) Perss. = Diplophyllum imbricatum [DIPLIMB] Scapaniaceae (B:H)
Macrodiplophyllum plicatum (Lindb.) Perss. = Diplophyllum plicatum [DIPLPLI] Scapaniaceae (B:H)
Macrorhynchus heterophyllus Nutt. = Agoseris heterophylla [AGOSHET] Asteraceae (V:D)
Madia Molina [MADIA$] Asteraceae (V:D)
Madia capitata Nutt. = Madia sativa [MADISAT] Asteraceae (V:D)
Madia dissitiflora (Nutt.) Torr. & Gray = Madia gracilis [MADIGRA] Asteraceae (V:D)
Madia exigua (Sm.) Gray [MADIEXI] Asteraceae (V:D)
Madia glomerata Hook. [MADIGLO] Asteraceae (V:D)
Madia gracilis (Sm.) Keck & J. Clausen ex Applegate [MADIGRA] Asteraceae (V:D)
Madia gracilis ssp. *collina* Keck = Madia gracilis [MADIGRA] Asteraceae (V:D)
Madia gracilis ssp. *pilosa* Keck = Madia gracilis [MADIGRA] Asteraceae (V:D)
Madia madioides (Nutt.) Greene [MADIMAD] Asteraceae (V:D)
Madia minima (Gray) Keck [MADIMIN] Asteraceae (V:D)
Madia sativa Molina [MADISAT] Asteraceae (V:D)
Madia sativa ssp. *capitata* (Nutt.) Piper = Madia sativa [MADISAT] Asteraceae (V:D)
Madia sativa var. *congesta* Torr. & Gray = Madia sativa [MADISAT] Asteraceae (V:D)
Madotheca cordaeana (Hüb.) Dum. = Porella cordaeana [PORECOR] Porellaceae (B:H)
Madotheca navicularis (Lehm. & Lindenb.) Dum. = Porella navicularis [PORENAV] Porellaceae (B:H)
Madotheca roellii Steph. = Porella roellii [POREROE] Porellaceae (B:H)
Mahonia Nutt. [MAHONIA$] Berberidaceae (V:D)
Mahonia aquifolium (Pursh) Nutt. [MAHOAQU] Berberidaceae (V:D)

Mahonia nervosa (Pursh) Nutt. [MAHONER] Berberidaceae (V:D)
Mahonia repens (Lindl.) G. Don [MAHOREP] Berberidaceae (V:D)
Mahonia sonnei Abrams = Mahonia repens [MAHOREP] Berberidaceae (V:D)
Maianthemum G.H. Weber ex Wiggers [MAIANTH$] Liliaceae (V:M)
Maianthemum amplexicaule (Nutt.) W.A. Weber = Maianthemum racemosum ssp. amplexicaule [MAIARAC1] Liliaceae (V:M)
Maianthemum bifolium ssp. *kamtschaticum* (J.F. Gmel. ex Cham.) E. Murr. = Maianthemum dilatatum [MAIADIL] Liliaceae (V:M)
Maianthemum bifolium var. *kamtschaticum* (J.F. Gmel. ex Cham.) Trautv. & C.A. Mey. = Maianthemum dilatatum [MAIADIL] Liliaceae (V:M)
Maianthemum canadense Desf. [MAIACAN] Liliaceae (V:M)
Maianthemum canadense var. *interius* Fern. = Maianthemum canadense [MAIACAN] Liliaceae (V:M)
Maianthemum canadense var. *pubescens* Gates & Ehlers = Maianthemum canadense [MAIACAN] Liliaceae (V:M)
Maianthemum dilatatum (Wood) A. Nels. & J.F. Macbr. [MAIADIL] Liliaceae (V:M)
Maianthemum kamtschaticum (J.F. Gmel. ex Cham.) Nakai = Maianthemum dilatatum [MAIADIL] Liliaceae (V:M)
Maianthemum racemosum (L.) Link [MAIARAC] Liliaceae (V:M)
Maianthemum racemosum ssp. **amplexicaule** (Nutt.) LaFrankie [MAIARAC1] Liliaceae (V:M)
Maianthemum racemosum var. *amplexicaule* (Nutt.) Dorn = Maianthemum racemosum ssp. amplexicaule [MAIARAC1] Liliaceae (V:M)
Maianthemum stellatum (L.) Link [MAIASTE] Liliaceae (V:M)
Maianthemum trifolium (L.) Sloboda [MAIATRI] Liliaceae (V:M)
Mairania alpina (L.) Desv. = Arctostaphylos alpina [ARCTALP] Ericaceae (V:D)
Malachium aquaticum (L.) Fries = Myosoton aquaticum [MYOSAQU] Caryophyllaceae (V:D)
Malaxis Soland. ex Sw. [MALAXIS$] Orchidaceae (V:M)
Malaxis brachypoda (Gray) Fern. [MALABRA] Orchidaceae (V:M)
Malaxis diphyllos Cham. [MALADIP] Orchidaceae (V:M)
Malaxis monophyllos auct. = Malaxis brachypoda [MALABRA] Orchidaceae (V:M)
Malaxis monophyllos ssp. *brachypoda* (Gray) A. & D. Löve = Malaxis brachypoda [MALABRA] Orchidaceae (V:M)
Malaxis monophyllos var. *brachypoda* (Gray) F. Morris & Eames = Malaxis brachypoda [MALABRA] Orchidaceae (V:M)
Malaxis monophyllos var. *diphyllos* (Cham.) Luer = Malaxis diphyllos [MALADIP] Orchidaceae (V:M)
Malaxis paludosa (L.) Sw. [MALAPAL] Orchidaceae (V:M)
Malus P. Mill. [MALUS$] Rosaceae (V:D)
Malus communis Poir. = Malus pumila [MALUPUM] Rosaceae (V:D)
Malus diversifolia (Bong.) M. Roemer = Malus fusca [MALUFUS] Rosaceae (V:D)
Malus domestica (Borkh.) Borkh. = Malus pumila [MALUPUM] Rosaceae (V:D)
Malus fusca (Raf.) Schneid. [MALUFUS] Rosaceae (V:D)
Malus fusca var. *levipes* (Nutt.) Schneid. = Malus fusca [MALUFUS] Rosaceae (V:D)
Malus malus (L.) Britt. = Malus sylvestris [MALUSYL] Rosaceae (V:D)
*__Malus pumila__ P. Mill. [MALUPUM] Rosaceae (V:D)
Malus sylvestris P. Mill. [MALUSYL] Rosaceae (V:D)
Malva L. [MALVA$] Malvaceae (V:D)
Malva mauritiana L. = Malva sylvestris [MALVSYL] Malvaceae (V:D)
*__Malva moschata__ L. [MALVMOS] Malvaceae (V:D)
*__Malva neglecta__ Wallr. [MALVNEG] Malvaceae (V:D)
*__Malva parviflora__ L. [MALVPAR] Malvaceae (V:D)
Malva pusilla auct. = Malva rotundifolia [MALVROT] Malvaceae (V:D)
*__Malva rotundifolia__ L. [MALVROT] Malvaceae (V:D)
*__Malva sylvestris__ L. [MALVSYL] Malvaceae (V:D)
Malva sylvestris ssp. *mauritiana* (L.) Thellung = Malva sylvestris [MALVSYL] Malvaceae (V:D)
Malva sylvestris var. *mauritiana* (L.) Boiss. = Malva sylvestris [MALVSYL] Malvaceae (V:D)
Mannia Opiz [MANNIA$] Aytoniaceae (B:H)
Mannia fragrans (Balbis) Frye & Clark [MANNFRA] Aytoniaceae (B:H)
Marah Kellogg [MARAH$] Cucurbitaceae (V:D)
Marah oreganus (Torr. ex S. Wats.) T.J. Howell [MARAORE] Cucurbitaceae (V:D)
Marchantia L. [MARCHAN$] Marchantiaceae (B:H)
Marchantia alpestris (Nees) Burgeff [MARCALP] Marchantiaceae (B:H)
Marchantia aquatica (Nees) Burgeff = Marchantia polymorpha [MARCPOL] Marchantiaceae (B:H)
Marchantia polymorpha L. [MARCPOL] Marchantiaceae (B:H)
Mariana mariana (L.) Hill = Silybum marianum [SILYMAR] Asteraceae (V:D)
Marrubium L. [MARRUBI$] Lamiaceae (V:D)
*__Marrubium vulgare__ L. [MARRVUL] Lamiaceae (V:D)
Marsilea L. [MARSILE$] Marsileaceae (V:F)
Marsilea vestita Hook. & Grev. [MARSVES] Marsileaceae (V:F)
Marsupella Dum. [MARSUPE$] Gymnomitriaceae (B:H)
Marsupella alpina (Mass. & Carest.) H. Bern. [MARSALP] Gymnomitriaceae (B:H)
Marsupella apiculata Schiffn. = Gymnomitrion apiculatum [GYMNAPI] Gymnomitriaceae (B:H)
Marsupella aquatica (Lindb.) Schiffn. = Marsupella emarginata ssp. emarginata var. aquatica [MARSEMA2] Gymnomitriaceae (B:H)
Marsupella boeckii (Aust.) Kaal. [MARSBOE] Gymnomitriaceae (B:H)
Marsupella brevissima (Dum.) Grolle [MARSBRE] Gymnomitriaceae (B:H)
Marsupella commutata (Limpr.) H. Bern. [MARSCOM] Gymnomitriaceae (B:H)
Marsupella condensata (Angstr.) Schiffn. [MARSCON] Gymnomitriaceae (B:H)
Marsupella emarginata (Ehrh.) Dum. [MARSEMA] Gymnomitriaceae (B:H)
Marsupella emarginata ssp. **emarginata** [MARSEMA0] Gymnomitriaceae (B:H)
Marsupella emarginata ssp. **emarginata** var. **aquatica** (Lindenb.) Dum. [MARSEMA2] Gymnomitriaceae (B:H)
Marsupella emarginata ssp. **emarginata** var. **emarginata** [MARSEMA3] Gymnomitriaceae (B:H)
Marsupella emarginata ssp. **tubulosa** (Steph.) Kitag. [MARSEMA4] Gymnomitriaceae (B:H)
Marsupella emarginata var. *pearsonii* (Schiffn.) Joerg. = Marsupella emarginata ssp. emarginata var. aquatica [MARSEMA2] Gymnomitriaceae (B:H)
Marsupella revoluta (Nees) Dum. [MARSREV] Gymnomitriaceae (B:H)
Marsupella sparsifolia (Lindb.) Dum. [MARSSPA] Gymnomitriaceae (B:H)
Marsupella sphacelata (Gieseke) Dum. [MARSSPH] Gymnomitriaceae (B:H)
Marsupella sphacelata f. **media** (Gott.) Schust. [MARSSPH1] Gymnomitriaceae (B:H)
Marsupella sphacelata f. **sphacelata** [MARSSPH0] Gymnomitriaceae (B:H)
Marsupella sphacelata var. *erythrorhiza* (Limpr.) Schiffn. = Marsupella sphacelata [MARSSPH] Gymnomitriaceae (B:H)
Marsupella sprucei (Limpr.) H. Bern. = Marsupella ustulata var. sprucei [MARSUST1] Gymnomitriaceae (B:H)
Marsupella stableri Spruce = Marsupella boeckii [MARSBOE] Gymnomitriaceae (B:H)
Marsupella sullivantii (De Not.) Evans = Marsupella sphacelata [MARSSPH] Gymnomitriaceae (B:H)
Marsupella ustulata (Hub.) Spruce [MARSUST] Gymnomitriaceae (B:H)
Marsupella ustulata var. **sprucei** (Limpr.) Schust. [MARSUST1] Gymnomitriaceae (B:H)
Maruta cotula (L.) DC. = Anthemis cotula [ANTHCOT] Asteraceae (V:D)
Masonhalea Kärnefelt [MASONHA$] Parmeliaceae (L)
Masonhalea richardsonii (Hook.) Kärnefelt [MASORIC] Parmeliaceae (L)

Massalongia Körber [MASSALO$] Peltigeraceae (L)
Massalongia carnosa (Dickson) Körber [MASSCAR] Peltigeraceae (L)
Massalongia microphylliza (Nyl. ex Hasse) Henssen [MASSMIC] Peltigeraceae (L)
Mastigophora Nees [MASTIGP$] Mastigophoraceae (B:H)
Mastigophora woodsii (Hook.) Nees [MASTWOO] Mastigophoraceae (B:H)
Mastodia tessellata Hook. f. & Harvey = Kohlmeyera complicatula [KOHLCOM] Mastodiaceae (L)
Matricaria L. [MATRICA$] Asteraceae (V:D)
Matricaria chamomilla L. 1755 & 1763, non 1753 = Matricaria recutita [MATRREC] Asteraceae (V:D)
Matricaria chamomilla var. *coronata* (J. Gay) Coss. & Germ. = Matricaria recutita [MATRREC] Asteraceae (V:D)
Matricaria discoidea DC. [MATRDIS] Asteraceae (V:D)
Matricaria inodora L. = Matricaria perforata [MATRPER] Asteraceae (V:D)
Matricaria inodora var. *agrestis* (Knaf.) Willmot = Matricaria perforata [MATRPER] Asteraceae (V:D)
Matricaria maritima ssp. *inodora* (L.) Clapham = Matricaria perforata [MATRPER] Asteraceae (V:D)
Matricaria maritima var. *agrestis* (Knaf) Wilmott = Matricaria perforata [MATRPER] Asteraceae (V:D)
Matricaria matricarioides auct. = Matricaria discoidea [MATRDIS] Asteraceae (V:D)
Matricaria parthenium L. = Tanacetum parthenium [TANAPAR] Asteraceae (V:D)
*__Matricaria perforata__ Mérat [MATRPER] Asteraceae (V:D)
*__Matricaria recutita__ L. [MATRREC] Asteraceae (V:D)
Matricaria suaveolens (Pursh) Buch. = Matricaria discoidea [MATRDIS] Asteraceae (V:D)
Matricaria suaveolens L. = Matricaria recutita [MATRREC] Asteraceae (V:D)
Matteuccia Todaro [MATTEUC$] Dryopteridaceae (V:F)
Matteuccia pensylvanica (Willd.) Raymond = Matteuccia struthiopteris [MATTSTR] Dryopteridaceae (V:F)
Matteuccia struthiopteris (L.) Todaro [MATTSTR] Dryopteridaceae (V:F)
Matteuccia struthiopteris var. *pensylvanica* (Willd.) Morton = Matteuccia struthiopteris [MATTSTR] Dryopteridaceae (V:F)
Matteuccia struthiopteris var. *pubescens* (Terry) Clute = Matteuccia struthiopteris [MATTSTR] Dryopteridaceae (V:F)
Mecodium wrightii (Bosch) Copeland = Hymenophyllum wrightii [HYMEWRI] Hymenophyllaceae (V:F)
Meconella Nutt. [MECONEL$] Papaveraceae (V:D)
Meconella oregana Nutt. [MECOORE] Papaveraceae (V:D)
Medicago L. [MEDICAG$] Fabaceae (V:D)
*__Medicago arabica__ (L.) Huds. [MEDIARA] Fabaceae (V:D)
Medicago arabica ssp. *inermis* Ricker = Medicago arabica [MEDIARA] Fabaceae (V:D)
Medicago falcata L. = Medicago sativa ssp. falcata [MEDISAT1] Fabaceae (V:D)
Medicago hispida Gaertn. = Medicago polymorpha [MEDIPOL] Fabaceae (V:D)
Medicago hispida var. *apiculata* (Willd.) Urban = Medicago polymorpha [MEDIPOL] Fabaceae (V:D)
Medicago hispida var. *confinis* (W.D.J. Koch) Burnat = Medicago polymorpha [MEDIPOL] Fabaceae (V:D)
*__Medicago lupulina__ L. [MEDILUP] Fabaceae (V:D)
Medicago lupulina var. *cupaniana* (Guss.) Boiss. = Medicago lupulina [MEDILUP] Fabaceae (V:D)
Medicago lupulina var. *glandulosa* Neilr. = Medicago lupulina [MEDILUP] Fabaceae (V:D)
*__Medicago polymorpha__ L. [MEDIPOL] Fabaceae (V:D)
Medicago polymorpha var. *brevispina* (Benth.) Heyn = Medicago polymorpha [MEDIPOL] Fabaceae (V:D)
Medicago polymorpha var. *ciliaris* (Ser.) Shinners = Medicago polymorpha [MEDIPOL] Fabaceae (V:D)
Medicago polymorpha var. *nigra* L. = Medicago polymorpha [MEDIPOL] Fabaceae (V:D)
Medicago polymorpha var. *polygyra* (Urban) Shinners = Medicago polymorpha [MEDIPOL] Fabaceae (V:D)
Medicago polymorpha var. *tricycla* (Gren. & Godr.) Shinners = Medicago polymorpha [MEDIPOL] Fabaceae (V:D)
Medicago polymorpha var. *vulgaris* (Benth.) Shinners = Medicago polymorpha [MEDIPOL] Fabaceae (V:D)
*__Medicago sativa__ L. [MEDISAT] Fabaceae (V:D)
*__Medicago sativa__ ssp. __falcata__ (L.) Arcang. [MEDISAT1] Fabaceae (V:D)
Meesia Hedw. [MEESIA$] Meesiaceae (B:M)
Meesia albertini B.S.G. = Meesia longiseta [MEESLON] Meesiaceae (B:M)
Meesia hexasticha auct. = Meesia longiseta [MEESLON] Meesiaceae (B:M)
Meesia longiseta Hedw. [MEESLON] Meesiaceae (B:M)
Meesia longiseta var. *macounii* Grout = Meesia longiseta [MEESLON] Meesiaceae (B:M)
Meesia trichodes Spruce = Meesia uliginosa [MEESULI] Meesiaceae (B:M)
Meesia trifaria Crum, Steere & Anderson = Meesia triquetra [MEESTRI] Meesiaceae (B:M)
Meesia triquetra (Richt.) Ångstr. [MEESTRI] Meesiaceae (B:M)
Meesia tristicha Bruch = Meesia triquetra [MEESTRI] Meesiaceae (B:M)
Meesia uliginosa Hedw. [MEESULI] Meesiaceae (B:M)
Meesia uliginosa var. *alpina* (Bruch) Hampe = Meesia uliginosa [MEESULI] Meesiaceae (B:M)
Meesia uliginosa var. *angustifolia* (Brid.) Hampe = Meesia uliginosa [MEESULI] Meesiaceae (B:M)
Meesia uliginosa var. *minor* (Brid.) Web. & Mohr = Meesia uliginosa [MEESULI] Meesiaceae (B:M)
Megalaria Hafellner [MEGALAR$] Lecanoraceae (L)
Megalaria brodoana S. Ekman & Tønsberg [MEGABRO] Lecanoraceae (L)
Megalodonta Greene [MEGALOD$] Asteraceae (V:D)
Megalodonta beckii (Torr. ex Spreng.) Greene [MEGABEC] Asteraceae (V:D)
Megaspora (Clauz. & Roux) Hafellner & V. Wirth [MEGASPO$] Hymeneliaceae (L)
Megaspora verrucosa (Ach.) Hafellner & V. Wirth [MEGAVEU] Hymeneliaceae (L)
Melampyrum L. [MELAMPY$] Scrophulariaceae (V:D)
*__Melampyrum lineare__ Desr. [MELALIN] Scrophulariaceae (V:D)
Melampyrum lineare var. *americanum* (Michx.) Beauverd = Melampyrum lineare [MELALIN] Scrophulariaceae (V:D)
Melandrium affine (J. Vahl ex Fries) J. Vahl = Silene involucrata [SILEINV] Caryophyllaceae (V:D)
Melandrium album (P. Mill.) Garcke = Silene latifolia ssp. alba [SILELAT1] Caryophyllaceae (V:D)
Melandrium apetalum ssp. *attenuatum* (Farr) Hara = Silene uralensis ssp. attenuata [SILEURA1] Caryophyllaceae (V:D)
Melandrium dioicum (L.) Coss. & Germ. = Silene dioica [SILEDIO] Caryophyllaceae (V:D)
Melandrium dioicum ssp. *rubrum* (Wieg.) D. Löve = Silene dioica [SILEDIO] Caryophyllaceae (V:D)
Melandrium drummondii (Hook.) Hultén = Silene drummondii [SILEDRU] Caryophyllaceae (V:D)
Melandrium furcatum Raf.) Hadac = Silene involucrata [SILEINV] Caryophyllaceae (V:D)
Melandrium noctiflorum (L.) Fries = Silene noctiflora [SILENOC] Caryophyllaceae (V:D)
Melandrium ostenfeldii Porsild = Silene taimyrensis [SILETAI] Caryophyllaceae (V:D)
Melandrium taimyrense Tolm. = Silene taimyrensis [SILETAI] Caryophyllaceae (V:D)
Melanelia Essl. [MELANEL$] Parmeliaceae (L)
Melanelia agnata (Nyl.) Thell [MELAAGN] Parmeliaceae (L)
Melanelia albertana (Ahti) Essl. [MELAALE] Parmeliaceae (L)
Melanelia commixta (Nyl.) Thell [MELACOM] Parmeliaceae (L)
Melanelia disjuncta (Erichsen) Essl. [MELADIS] Parmeliaceae (L)
Melanelia elegantula (Zahlbr.) Essl. [MELAELE] Parmeliaceae (L)
Melanelia exasperatula (Nyl.) Essl. [MELAEXS] Parmeliaceae (L)
Melanelia fuliginosa (Fr. ex Duby) Essl. [MELAFUL] Parmeliaceae (L)

Melanelia glabratula (Lamy) Essl. = Melanelia fuliginosa [MELAFUL] Parmeliaceae (L)
Melanelia granulosa (Lynge) Essl. = Melanelia disjuncta [MELADIS] Parmeliaceae (L)
Melanelia hepatizon (Ach.) Thell [MELAHEP] Parmeliaceae (L)
Melanelia incolorata (Parr.) Essl. = Melanelia elegantula [MELAELE] Parmeliaceae (L)
Melanelia infumata (Nyl.) Essl. [MELAINF] Parmeliaceae (L)
Melanelia loxodes (Nyl.) Essl. = Neofuscelia loxodes [NEOFLOX] Parmeliaceae (L)
Melanelia multispora (A. Schneider) Essl. [MELAMUL] Parmeliaceae (L)
Melanelia olivaceoides (Krog) Essl. [MELAOLI] Parmeliaceae (L)
Melanelia panniformis (Nyl.) Essl. [MELAPAN] Parmeliaceae (L)
Melanelia septentrionalis (Lynge) Essl. [MELASET] Parmeliaceae (L)
Melanelia sorediata (Ach.) Goward & Ahti [MELASOR] Parmeliaceae (L)
Melanelia sorediosa (Almb.) Essl. = Melanelia sorediata [MELASOR] Parmeliaceae (L)
Melanelia stygia (L.) Essl. [MELASTY] Parmeliaceae (L)
Melanelia subargentifera (Nyl.) Essl. [MELASUA] Parmeliaceae (L)
Melanelia subaurifera (Nyl.) Essl. [MELASUU] Parmeliaceae (L)
Melanelia subelegantula (Essl.) Essl. [MELASUB] Parmeliaceae (L)
Melanelia subolivacea (Nyl.) Essl. [MELASUO] Parmeliaceae (L)
Melanelia substygia (Räsänen) Essl. = Melanelia tominii [MELATOM] Parmeliaceae (L)
Melanelia tominii (Oksner) Essl. [MELATOM] Parmeliaceae (L)
Melanelia trabeculata (Ahti) Essl. [MELATRA] Parmeliaceae (L)
Melanolecia jurana (Schaerer) Hertel = Farnoldia jurana [FARNJUR] Porpidiaceae (L)
Melica L. [MELICA$] Poaceae (V:M)
Melica bella Piper = Melica bulbosa [MELIBUL] Poaceae (V:M)
Melica bella ssp. *intonsa* Piper = Melica bulbosa [MELIBUL] Poaceae (V:M)
Melica bulbosa Geyer ex Porter & Coult. [MELIBUL] Poaceae (V:M)
Melica bulbosa var. *intonsa* (Piper) M.E. Peck = Melica bulbosa [MELIBUL] Poaceae (V:M)
Melica harfordii Boland. [MELIHAR] Poaceae (V:M)
Melica harfordii var. *minor* Vasey = Melica harfordii [MELIHAR] Poaceae (V:M)
Melica purpurascens (Torr.) A.S. Hitchc. = Schizachne purpurascens [SCHIPUR] Poaceae (V:M)
Melica smithii (Porter ex Gray) Vasey [MELISMI] Poaceae (V:M)
Melica spectabilis Scribn. [MELISPE] Poaceae (V:M)
Melica subulata (Griseb.) Scribn. [MELISUB] Poaceae (V:M)
Melilotus P. Mill. [MELILOT$] Fabaceae (V:D)
Melilotus alba Desr. = Melilotus officinalis [MELIOFF] Fabaceae (V:D)
Melilotus albus Medik. = Melilotus officinalis [MELIOFF] Fabaceae (V:D)
Melilotus albus var. *annuus* Coe = Melilotus officinalis [MELIOFF] Fabaceae (V:D)
***Melilotus officinalis** (L.) Lam. [MELIOFF] Fabaceae (V:D)
Melissa L. [MELISSA$] Lamiaceae (V:D)
***Melissa officinalis** L. [MELIOFI] Lamiaceae (V:D)
Menegazzia A. Massal. [MENEGAZ$] Parmeliaceae (L)
Menegazzia pertusa (Schrank) Stein = Menegazzia terebrata [MENETER] Parmeliaceae (L)
Menegazzia terebrata (Hoffm.) A. Massal. [MENETER] Parmeliaceae (L)
Mentha L. [MENTHA$] Lamiaceae (V:D)
***Mentha aquatica** L. [MENTAQU] Lamiaceae (V:D)
Mentha aquatica var. *crispa* (L.) Benth. = Mentha × piperita [MENTXPI] Lamiaceae (V:D)
Mentha arvensis L. [MENTARV] Lamiaceae (V:D)
Mentha arvensis ssp. *borealis* (Michx.) Taylor & MacBryde = Mentha canadensis [MENTCAN] Lamiaceae (V:D)
Mentha arvensis ssp. *haplocalyx* Briq. = Mentha canadensis [MENTCAN] Lamiaceae (V:D)
Mentha arvensis var. *canadensis* (L.) Kuntze = Mentha canadensis [MENTCAN] Lamiaceae (V:D)
Mentha arvensis var. *glabrata* (Benth.) Fern. = Mentha canadensis [MENTCAN] Lamiaceae (V:D)
Mentha arvensis var. *lanata* Piper = Mentha canadensis [MENTCAN] Lamiaceae (V:D)
Mentha arvensis var. *villosa* (Benth.) S.R. Stewart = Mentha canadensis [MENTCAN] Lamiaceae (V:D)
Mentha canadensis L. [MENTCAN] Lamiaceae (V:D)
Mentha citrata Ehrh. = Mentha aquatica [MENTAQU] Lamiaceae (V:D)
Mentha crispa L. = Mentha × piperita [MENTXPI] Lamiaceae (V:D)
Mentha dumetorum Schultes = Mentha × piperita [MENTXPI] Lamiaceae (V:D)
Mentha × gentilis L. = Mentha arvensis [MENTARV] Lamiaceae (V:D)
Mentha gentilis L. = Mentha arvensis [MENTARV] Lamiaceae (V:D)
Mentha glabrior (Hook.) Rydb. = Mentha canadensis [MENTCAN] Lamiaceae (V:D)
Mentha penardii (Briq.) Rydb. = Mentha canadensis [MENTCAN] Lamiaceae (V:D)
***Mentha × piperita** L. (pro sp.) [MENTXPI] Lamiaceae (V:D)
Mentha piperita ssp. *citrata* (Ehrh.) Briq. = Mentha aquatica [MENTAQU] Lamiaceae (V:D)
Mentha × piperita var. *citrata* (Ehrh.) Boivin (pro nm.) = Mentha aquatica [MENTAQU] Lamiaceae (V:D)
***Mentha pulegium** L. [MENTPUL] Lamiaceae (V:D)
Mentha rotundifolia auct. = Mentha suaveolens [MENTSUA] Lamiaceae (V:D)
***Mentha spicata** L. [MENTSPI] Lamiaceae (V:D)
***Mentha suaveolens** Ehrh. [MENTSUA] Lamiaceae (V:D)
Mentha viridis L. = Mentha spicata [MENTSPI] Lamiaceae (V:D)
Mentzelia L. [MENTZEL$] Loasaceae (V:D)
Mentzelia albicaulis (Dougl. ex Hook.) Dougl. ex Torr. & Gray [MENTALB] Loasaceae (V:D)
Mentzelia albicaulis var. *ctenophora* (Rydb.) St. John = Mentzelia albicaulis [MENTALB] Loasaceae (V:D)
Mentzelia albicaulis var. *gracilis* J. Darl. = Mentzelia albicaulis [MENTALB] Loasaceae (V:D)
Mentzelia albicaulis var. *tenerrima* (Rydb.) St. John = Mentzelia albicaulis [MENTALB] Loasaceae (V:D)
Mentzelia brandegei S. Wats. = Mentzelia laevicaulis var. parviflora [MENTLAE2] Loasaceae (V:D)
Mentzelia dispersa S. Wats. [MENTDIS] Loasaceae (V:D)
Mentzelia douglasii St. John = Mentzelia laevicaulis var. parviflora [MENTLAE2] Loasaceae (V:D)
Mentzelia gracilis H.J. Thompson & Lewis = Mentzelia albicaulis [MENTALB] Loasaceae (V:D)
Mentzelia laevicaulis (Dougl. ex Hook.) Torr. & Gray [MENTLAE] Loasaceae (V:D)
Mentzelia laevicaulis var. *acuminata* (Rydb.) A. Nels. & J.F. Macbr. = Mentzelia laevicaulis var. laevicaulis [MENTLAE0] Loasaceae (V:D)
Mentzelia laevicaulis var. **laevicaulis** [MENTLAE0] Loasaceae (V:D)
Mentzelia laevicaulis var. **parviflora** (Dougl. ex Hook.) C.L. Hitchc. [MENTLAE2] Loasaceae (V:D)
Menyanthes L. [MENYANT$] Menyanthaceae (V:D)
Menyanthes trifoliata L. [MENYTRI] Menyanthaceae (V:D)
Menyanthes trifoliata var. *minor* Raf. = Menyanthes trifoliata [MENYTRI] Menyanthaceae (V:D)
Menziesia Sm. [MENZIES$] Ericaceae (V:D)
Menziesia ferruginea Sm. [MENZFER] Ericaceae (V:D)
Menziesia ferruginea ssp. *glabella* (Gray) Calder & Taylor = Menziesia ferruginea [MENZFER] Ericaceae (V:D)
Menziesia ferruginea var. *glabella* (Gray) M.E. Peck = Menziesia ferruginea [MENZFER] Ericaceae (V:D)
Menziesia glabella Gray = Menziesia ferruginea [MENZFER] Ericaceae (V:D)
Merceya latifolia Kindb. in Mac. = Crumia latifolia [CRUMLAT] Pottiaceae (B:M)
Mertensia Roth [MERTENS$] Boraginaceae (V:D)
Mertensia longiflora Greene [MERTLON] Boraginaceae (V:D)
Mertensia maritima (L.) S.F. Gray [MERTMAR] Boraginaceae (V:D)

Mertensia palmeri A. Nels. & J.F. Macbr. = Mertensia paniculata var. paniculata [MERTPAN0] Boraginaceae (V:D)
Mertensia paniculata (Ait.) G. Don [MERTPAN] Boraginaceae (V:D)
Mertensia paniculata var. **borealis** (J.F. Macbr.) L.O. Williams [MERTPAN1] Boraginaceae (V:D)
Mertensia paniculata var. **paniculata** [MERTPAN0] Boraginaceae (V:D)
Metaneckera Steere [METANEC$] Neckeraceae (B:M)
Metaneckera menziesii (Hook. in Drumm.) Steere [METAMEN] Neckeraceae (B:M)
Metzgeria Raddi [METZGER$] Metzgeriaceae (B:H)
Metzgeria conjugata Lindb. [METZCON] Metzgeriaceae (B:H)
Metzgeria fruticulosa auct. = Metzgeria temperata [METZTEM] Metzgeriaceae (B:H)
Metzgeria hamata Lindb. = Metzgeria leptoneura [METZLEP] Metzgeriaceae (B:H)
Metzgeria leptoneura Spruce [METZLEP] Metzgeriaceae (B:H)
Metzgeria pubescens (Schrank) Raddi = Apometzgeria pubescens [APOMPUB] Metzgeriaceae (B:H)
Metzgeria temperata Kuwah. [METZTEM] Metzgeriaceae (B:H)
Micarea Fr. [MICAREA$] Micareaceae (L)
Micarea assimilata (Nyl.) Coppins [MICAASS] Micareaceae (L)
Micarea cinerea (Schaerer) Hedl. [MICACIN] Micareaceae (L)
Micarea crassipes (Th. Fr.) Coppins = Helocarpon crassipes [HELOCRA] Micareaceae (L)
Micarea denigrata (Fr.) Hedl. [MICADEN] Micareaceae (L)
Micarea gelatinosa (Flörke) Brodo = Trapeliopsis gelatinosa [TRAPGEL] Trapeliaceae (L)
Micarea globularis (Ach. ex Nyl.) Hedl. = Micarea misella [MICAMIS] Micareaceae (L)
Micarea hemipoliella (Nyl.) Vezda = Micarea denigrata [MICADEN] Micareaceae (L)
Micarea lithinella (Nyl.) Hedl. [MICALIT] Micareaceae (L)
Micarea lutulata (Nyl.) Coppins [MICALUT] Micareaceae (L)
Micarea melaena (Nyl.) Hedl. [MICAMEL] Micareaceae (L)
Micarea misella (Nyl.) Hedl. [MICAMIS] Micareaceae (L)
Micarea peliocarpa (Anzi) Coppins & R. Sant. [MICAPEL] Micareaceae (L)
Micarea prasina Fr. [MICAPRA] Micareaceae (L)
Micarea sylvicola (Flotow) Vezda & V. Wirth [MICASYL] Micareaceae (L)
Micarea trisepta (Hellbom) Wetmore = Micarea peliocarpa [MICAPEL] Micareaceae (L)
Micarea violacea (Crouan ex Nyl.) Hedl. = Micarea peliocarpa [MICAPEL] Micareaceae (L)
Micarea viridescens (Schrader) Brodo = Trapeliopsis viridescens [TRAPVIR] Trapeliaceae (L)
Micrampelis lobata (Michx.) Greene = Echinocystis lobata [ECHILOB] Cucurbitaceae (V:D)
Micranthes bidens Small = Saxifraga integrifolia [SAXIINT] Saxifragaceae (V:D)
Micranthes lata Small = Saxifraga occidentalis [SAXIOCC] Saxifragaceae (V:D)
Micranthes lyallii (Engl.) Small = Saxifraga lyallii ssp. lyallii [SAXILYA0] Saxifragaceae (V:D)
Micranthes montana Small = Saxifraga nidifica [SAXINID] Saxifragaceae (V:D)
Micranthes nivalis (L.) Small = Saxifraga nivalis [SAXINIV] Saxifragaceae (V:D)
Micranthes occidentalis (S. Wats.) Small = Saxifraga occidentalis [SAXIOCC] Saxifragaceae (V:D)
Micranthes odontoloma (Piper) W.A. Weber = Saxifraga odontoloma [SAXIODO] Saxifragaceae (V:D)
Micranthes saximontana Small = Saxifraga occidentalis [SAXIOCC] Saxifragaceae (V:D)
Microcalicium Vainio [MICROCA$] Microcaliciaceae (L)
Microcalicium arenarium (Hampe ex Massal.) Tibell [MICRARE] Microcaliciaceae (L)
Microcalicium disseminatum (Ach.) Vainio [MICRDIS] Microcaliciaceae (L)
Microcalicium subpedicellatum (Schaerer) Tibell = Microcalicium disseminatum [MICRDIS] Microcaliciaceae (L)
Microglaena sphinctrinoides (Nyl.) Lönnr. = Protothelenella sphinctrinoides [PROTSPH] Protothelenellaceae (L)
Microlepidozia makinoana Hatt. & Mizut. = Kurzia sylvatica [KURZSYL] Lepidoziaceae (B:H)
Microlepidozia setacea (Web.) Joerg. = Kurzia setacea [KURZSET] Lepidoziaceae (B:H)
Microlepidozia sylvatica (Evans) Joerg. = Kurzia sylvatica [KURZSYL] Lepidoziaceae (B:H)
Microlychnus A. Funk [MICROLY$] Uncertain-family (L)
Microlychnus epicorticis A. Funk [MICREPI] Uncertain-family (L)
Micromeria chamissonis (Benth.) Greene = Satureja douglasii [SATUDOU] Lamiaceae (V:D)
Micromitrium Aust. [MICROMI$] Ephemeraceae (B:M)
*__Micromitrium tenerum__ (Bruch & Schimp. in B.S.G.) Crosby [MICRTEN] Ephemeraceae (B:M)
Microseris D. Don [MICROSE$] Asteraceae (V:D)
Microseris bigelovii (Gray) Schultz-Bip. [MICRBIG] Asteraceae (V:D)
Microseris borealis (Bong.) Schultz-Bip. [MICRBOR] Asteraceae (V:D)
Microseris nutans (Hook.) Schultz-Bip. [MICRNUT] Asteraceae (V:D)
Microseris troximoides Gray = Nothocalais troximoides [NOTHTRO] Asteraceae (V:D)
Microsteris gracilis (Hook.) Greene = Phlox gracilis [PHLOGRA] Polemoniaceae (V:D)
Microthelia metzleri J. Lahm = Lichenothelia metzleri [LICHMET] Lichenotheliaceae (L)
Mielichhoferia Nees & Hornsch. [MIELICH$] Bryaceae (B:M)
Mielichhoferia macrocarpa (Hook. in Drumm.) Bruch & Schimp. ex Jaeg. [MIELMAC] Bryaceae (B:M)
Mielichhoferia macrocarpa var. *pungens* Bartr. = Mielichhoferia macrocarpa [MIELMAC] Bryaceae (B:M)
Mielichhoferia mielichhoferi (Funck ex Hook.) Wijk & Marg. = Mielichhoferia mielichhoferiana [MIELMIE] Bryaceae (B:M)
Mielichhoferia mielichhoferiana (Funck in Hook.) Loeske [MIELMIE] Bryaceae (B:M)
Mielichhoferia recurvifolium Kindb. = Ceratodon purpureus [CERAPUR] Ditrichaceae (B:M)
Mimulus L. [MIMULUS$] Scrophulariaceae (V:D)
Mimulus alpinus (Gray) Piper = Mimulus tilingii var. caespitosus [MIMUTIL1] Scrophulariaceae (V:D)
Mimulus alsinoides Dougl. ex Benth. [MIMUALS] Scrophulariaceae (V:D)
Mimulus arvensis Greene = Mimulus guttatus [MIMUGUT] Scrophulariaceae (V:D)
Mimulus bakeri Gandog. = Mimulus guttatus [MIMUGUT] Scrophulariaceae (V:D)
Mimulus brachystylis Edwin = Mimulus guttatus [MIMUGUT] Scrophulariaceae (V:D)
Mimulus breviflorus Piper [MIMUBRE] Scrophulariaceae (V:D)
Mimulus breweri (Greene) Coville [MIMUBRW] Scrophulariaceae (V:D)
Mimulus caespitosus (Greene) Greene = Mimulus tilingii var. caespitosus [MIMUTIL1] Scrophulariaceae (V:D)
Mimulus caespitosus var. *implexus* (Greene) M.E. Peck = Mimulus tilingii var. caespitosus [MIMUTIL1] Scrophulariaceae (V:D)
Mimulus clementinus Greene = Mimulus guttatus [MIMUGUT] Scrophulariaceae (V:D)
Mimulus corallinus (Greene) A.L. Grant = Mimulus tilingii var. tilingii [MIMUTIL0] Scrophulariaceae (V:D)
Mimulus cordatus Greene = Mimulus guttatus [MIMUGUT] Scrophulariaceae (V:D)
Mimulus cuspidata Greene = Mimulus guttatus [MIMUGUT] Scrophulariaceae (V:D)
Mimulus decorus (A.L. Grant) Suksdorf = Mimulus guttatus [MIMUGUT] Scrophulariaceae (V:D)
Mimulus deltoides Gandog. = Mimulus floribundus [MIMUFLO] Scrophulariaceae (V:D)
Mimulus dentatus Nutt. ex Benth. [MIMUDEN] Scrophulariaceae (V:D)
Mimulus equinnus Greene = Mimulus guttatus [MIMUGUT] Scrophulariaceae (V:D)

Mimulus floribundus Lindl. [MIMUFLO] Scrophulariaceae (V:D)
Mimulus floribundus ssp. *moorei* Iltis = Mimulus floribundus [MIMUFLO] Scrophulariaceae (V:D)
Mimulus floribundus var. *geniculatus* (Greene) A.L. Grant = Mimulus floribundus [MIMUFLO] Scrophulariaceae (V:D)
Mimulus floribundus var. *membranaceus* (A. Nels.) A.L. Grant = Mimulus floribundus [MIMUFLO] Scrophulariaceae (V:D)
Mimulus floribundus var. *subulatus* A.L. Grant = Mimulus floribundus [MIMUFLO] Scrophulariaceae (V:D)
Mimulus geniculatus Greene = Mimulus floribundus [MIMUFLO] Scrophulariaceae (V:D)
Mimulus glabratus var. *ascendens* Gray = Mimulus guttatus [MIMUGUT] Scrophulariaceae (V:D)
Mimulus glareosus Greene = Mimulus guttatus [MIMUGUT] Scrophulariaceae (V:D)
Mimulus grandiflorus J.T. Howell = Mimulus guttatus [MIMUGUT] Scrophulariaceae (V:D)
Mimulus grandis (Greene) Heller = Mimulus guttatus [MIMUGUT] Scrophulariaceae (V:D)
Mimulus guttatus DC. [MIMUGUT] Scrophulariaceae (V:D)
Mimulus guttatus ssp. *arenicola* Pennell = Mimulus guttatus [MIMUGUT] Scrophulariaceae (V:D)
Mimulus guttatus ssp. *arvensis* (Greene) Munz = Mimulus guttatus [MIMUGUT] Scrophulariaceae (V:D)
Mimulus guttatus ssp. *haidensis* Calder & Taylor = Mimulus guttatus [MIMUGUT] Scrophulariaceae (V:D)
Mimulus guttatus ssp. *littoralis* Pennell = Mimulus guttatus [MIMUGUT] Scrophulariaceae (V:D)
Mimulus guttatus ssp. *micranthus* (Heller) Munz = Mimulus guttatus [MIMUGUT] Scrophulariaceae (V:D)
Mimulus guttatus ssp. *scouleri* (Hook.) Pennell = Mimulus guttatus [MIMUGUT] Scrophulariaceae (V:D)
Mimulus guttatus var. *arvensis* (Greene) A.L. Grant = Mimulus guttatus [MIMUGUT] Scrophulariaceae (V:D)
Mimulus guttatus var. *decorus* A.L. Grant = Mimulus guttatus [MIMUGUT] Scrophulariaceae (V:D)
Mimulus guttatus var. *depauperatus* (Gray) A.L. Grant = Mimulus guttatus [MIMUGUT] Scrophulariaceae (V:D)
Mimulus guttatus var. *gracilis* (Gray) Campbell = Mimulus guttatus [MIMUGUT] Scrophulariaceae (V:D)
Mimulus guttatus var. *grandis* Greene = Mimulus guttatus [MIMUGUT] Scrophulariaceae (V:D)
Mimulus guttatus var. *hallii* (Greene) A.L. Grant = Mimulus guttatus [MIMUGUT] Scrophulariaceae (V:D)
Mimulus guttatus var. *insignis* Greene = Mimulus guttatus [MIMUGUT] Scrophulariaceae (V:D)
Mimulus guttatus var. *laxus* (Pennell ex M.E. Peck) M.E. Peck = Mimulus guttatus [MIMUGUT] Scrophulariaceae (V:D)
Mimulus guttatus var. *lyratus* (Benth.) Pennell ex M.E. Peck = Mimulus guttatus [MIMUGUT] Scrophulariaceae (V:D)
Mimulus guttatus var. *microphyllus* (Benth.) Pennell ex M.E. Peck = Mimulus guttatus [MIMUGUT] Scrophulariaceae (V:D)
Mimulus guttatus var. *nasutus* (Greene) Jepson = Mimulus guttatus [MIMUGUT] Scrophulariaceae (V:D)
Mimulus guttatus var. *puberulus* (Greene ex Rydb.) A.L. Grant = Mimulus guttatus [MIMUGUT] Scrophulariaceae (V:D)
Mimulus hallii Greene = Mimulus guttatus [MIMUGUT] Scrophulariaceae (V:D)
Mimulus hirsutus J.T. Howell = Mimulus guttatus [MIMUGUT] Scrophulariaceae (V:D)
Mimulus implicatus Greene = Mimulus tilingii var. tilingii [MIMUTIL0] Scrophulariaceae (V:D)
Mimulus inflatulus Suksdorf = Mimulus breviflorus [MIMUBRE] Scrophulariaceae (V:D)
Mimulus langsdorfii Donn ex Greene = Mimulus guttatus [MIMUGUT] Scrophulariaceae (V:D)
Mimulus langsdorfii var. *argutus* Greene = Mimulus guttatus [MIMUGUT] Scrophulariaceae (V:D)
Mimulus langsdorfii var. *arvensis* (Greene) Jepson = Mimulus guttatus [MIMUGUT] Scrophulariaceae (V:D)
Mimulus langsdorfii var. *californicus* Jepson = Mimulus guttatus [MIMUGUT] Scrophulariaceae (V:D)
Mimulus langsdorfii var. *grandis* (Greene) Greene = Mimulus guttatus [MIMUGUT] Scrophulariaceae (V:D)
Mimulus langsdorfii var. *guttatus* (Fisch. ex DC.) Jepson = Mimulus guttatus [MIMUGUT] Scrophulariaceae (V:D)
Mimulus langsdorfii var. *insignis* (Greene) A.L. Grant = Mimulus guttatus [MIMUGUT] Scrophulariaceae (V:D)
Mimulus langsdorfii var. *microphyllus* (Benth.) A. Nels. & J.F. Macbr. = Mimulus guttatus [MIMUGUT] Scrophulariaceae (V:D)
Mimulus langsdorfii var. *minimus* Henry = Mimulus guttatus [MIMUGUT] Scrophulariaceae (V:D)
Mimulus langsdorfii var. *minor* (A. Nels.) Cockerell = Mimulus tilingii var. tilingii [MIMUTIL0] Scrophulariaceae (V:D)
Mimulus langsdorfii var. *nasutus* (Greene) Jepson = Mimulus guttatus [MIMUGUT] Scrophulariaceae (V:D)
Mimulus langsdorfii var. *platyphyllus* Greene = Mimulus guttatus [MIMUGUT] Scrophulariaceae (V:D)
Mimulus langsdorfii var. *tilingii* (Regel) Greene = Mimulus tilingii var. tilingii [MIMUTIL0] Scrophulariaceae (V:D)
Mimulus laxus Pennell ex M.E. Peck = Mimulus guttatus [MIMUGUT] Scrophulariaceae (V:D)
Mimulus lewisii Pursh [MIMULEW] Scrophulariaceae (V:D)
Mimulus longulus Greene = Mimulus guttatus [MIMUGUT] Scrophulariaceae (V:D)
Mimulus lucens Greene = Mimulus tilingii var. tilingii [MIMUTIL0] Scrophulariaceae (V:D)
Mimulus luteus var. *alpinus* Gray = Mimulus tilingii var. caespitosus [MIMUTIL1] Scrophulariaceae (V:D)
Mimulus luteus var. *depauperatus* Gray = Mimulus guttatus [MIMUGUT] Scrophulariaceae (V:D)
Mimulus luteus var. *gracilis* Gray = Mimulus guttatus [MIMUGUT] Scrophulariaceae (V:D)
Mimulus lyratus Benth. = Mimulus guttatus [MIMUGUT] Scrophulariaceae (V:D)
Mimulus maguirei Pennell = Mimulus guttatus [MIMUGUT] Scrophulariaceae (V:D)
Mimulus marmoratus Greene = Mimulus guttatus [MIMUGUT] Scrophulariaceae (V:D)
Mimulus membranaceus A. Nels. = Mimulus floribundus [MIMUFLO] Scrophulariaceae (V:D)
Mimulus micranthus Heller = Mimulus guttatus [MIMUGUT] Scrophulariaceae (V:D)
Mimulus microphyllus Benth. = Mimulus guttatus [MIMUGUT] Scrophulariaceae (V:D)
Mimulus minor A. Nels. = Mimulus tilingii var. tilingii [MIMUTIL0] Scrophulariaceae (V:D)
Mimulus minusculus Greene = Mimulus tilingii var. tilingii [MIMUTIL0] Scrophulariaceae (V:D)
Mimulus moschatus Dougl. ex Lindl. [MIMUMOS] Scrophulariaceae (V:D)
Mimulus multiflorus Pennell = Mimulus floribundus [MIMUFLO] Scrophulariaceae (V:D)
Mimulus nasutus Greene = Mimulus guttatus [MIMUGUT] Scrophulariaceae (V:D)
Mimulus nasutus var. *micranthus* (Heller) A.L. Grant = Mimulus guttatus [MIMUGUT] Scrophulariaceae (V:D)
Mimulus paniculatus Greene = Mimulus guttatus [MIMUGUT] Scrophulariaceae (V:D)
Mimulus pardalis Pennell = Mimulus guttatus [MIMUGUT] Scrophulariaceae (V:D)
Mimulus parishii Gandog. = Mimulus guttatus [MIMUGUT] Scrophulariaceae (V:D)
Mimulus peduncularis Dougl. ex Benth. = Mimulus floribundus [MIMUFLO] Scrophulariaceae (V:D)
Mimulus petiolaris Greene = Mimulus guttatus [MIMUGUT] Scrophulariaceae (V:D)
Mimulus prionophyllus Greene = Mimulus guttatus [MIMUGUT] Scrophulariaceae (V:D)
Mimulus procerus Greene = Mimulus guttatus [MIMUGUT] Scrophulariaceae (V:D)
Mimulus puberulus Greene ex Rydb. = Mimulus guttatus [MIMUGUT] Scrophulariaceae (V:D)
Mimulus pubescens Benth. = Mimulus floribundus [MIMUFLO] Scrophulariaceae (V:D)

Mimulus puncticalyx Gandog. = Mimulus guttatus [MIMUGUT] Scrophulariaceae (V:D)
Mimulus rivularis Nutt. = Mimulus guttatus [MIMUGUT] Scrophulariaceae (V:D)
Mimulus rubellus var. *breweri* (Greene) Jepson = Mimulus breweri [MIMUBRW] Scrophulariaceae (V:D)
Mimulus scouleri Hook. = Mimulus guttatus [MIMUGUT] Scrophulariaceae (V:D)
Mimulus scouleri var. *caespitosus* Greene = Mimulus tilingii var. caespitosus [MIMUTIL1] Scrophulariaceae (V:D)
Mimulus serotinus Suksdorf = Mimulus floribundus [MIMUFLO] Scrophulariaceae (V:D)
Mimulus subreniformis Greene = Mimulus guttatus [MIMUGUT] Scrophulariaceae (V:D)
Mimulus subulatus (A.L. Grant) Pennell = Mimulus floribundus [MIMUFLO] Scrophulariaceae (V:D)
Mimulus tenellus Nutt. ex Gray = Mimulus guttatus [MIMUGUT] Scrophulariaceae (V:D)
Mimulus thermalis A. Nels. = Mimulus guttatus [MIMUGUT] Scrophulariaceae (V:D)
Mimulus tilingii Regel [MIMUTIL] Scrophulariaceae (V:D)
Mimulus tilingii var. **caespitosus** (Greene) A.L. Grant [MIMUTIL1] Scrophulariaceae (V:D)
Mimulus tilingii var. *corallinus* (Greene) A.L. Grant = Mimulus tilingii var. tilingii [MIMUTIL0] Scrophulariaceae (V:D)
Mimulus tilingii var. **tilingii** [MIMUTIL0] Scrophulariaceae (V:D)
Mimulus trisulcatus Pennell = Mimulus floribundus [MIMUFLO] Scrophulariaceae (V:D)
Mimulus unimaculatus Pennell = Mimulus guttatus [MIMUGUT] Scrophulariaceae (V:D)
Mimulus veronicifolius Greene = Mimulus tilingii var. tilingii [MIMUTIL0] Scrophulariaceae (V:D)
Minuartia L. [MINUART$] Caryophyllaceae (V:D)
Minuartia austromontana S.J. Wolf & Packer [MINUAUS] Caryophyllaceae (V:D)
Minuartia biflora (L.) Schinz & Thellung [MINUBIF] Caryophyllaceae (V:D)
Minuartia dawsonensis (Britt.) House [MINUDAW] Caryophyllaceae (V:D)
Minuartia elegans (Cham. & Schlecht.) Schischkin [MINUELE] Caryophyllaceae (V:D)
Minuartia nuttallii (Pax) Briq. [MINUNUT] Caryophyllaceae (V:D)
Minuartia obtusiloba (Rydb.) House [MINUOBT] Caryophyllaceae (V:D)
Minuartia orthotrichoides Schischkin = Minuartia elegans [MINUELE] Caryophyllaceae (V:D)
Minuartia pungens (Nutt.) Mattf. = Minuartia nuttallii [MINUNUT] Caryophyllaceae (V:D)
Minuartia pusilla (S. Wats.) Mattf. [MINUPUS] Caryophyllaceae (V:D)
Minuartia rossii var. *elegans* (Cham. & Schlecht.) Hultén = Minuartia elegans [MINUELE] Caryophyllaceae (V:D)
Minuartia rossii var. *orthotrichoides* (Schischkin) Hultén = Minuartia elegans [MINUELE] Caryophyllaceae (V:D)
Minuartia rubella (Wahlenb.) Hiern [MINURUB] Caryophyllaceae (V:D)
Minuartia tenella (Nutt.) Mattf. [MINUTEN] Caryophyllaceae (V:D)
Minuopsis nuttallii (Pax) W.A. Weber = Minuartia nuttallii [MINUNUT] Caryophyllaceae (V:D)
Mirabilis L. [MIRABIL$] Nyctaginaceae (V:D)
Mirabilis hirsuta (Pursh) MacM. [MIRAHIR] Nyctaginaceae (V:D)
•**Mirabilis nyctaginea** (Michx.) MacM. [MIRANYC] Nyctaginaceae (V:D)
Mitella L. [MITELLA$] Saxifragaceae (V:D)
Mitella breweri Gray [MITEBRE] Saxifragaceae (V:D)
Mitella caulescens Nutt. [MITECAU] Saxifragaceae (V:D)
Mitella nuda L. [MITENUD] Saxifragaceae (V:D)
Mitella ovalis Greene [MITEOVA] Saxifragaceae (V:D)
Mitella pentandra Hook. [MITEPEN] Saxifragaceae (V:D)
Mitella trifida Graham [MITETRI] Saxifragaceae (V:D)
Mitellastra caulescens (Nutt.) T.J. Howell = Mitella caulescens [MITECAU] Saxifragaceae (V:D)
Mniobryum albicans (Wahlenb.) Limpr. = Pohlia wahlenbergii [POHLWAH] Bryaceae (B:M)
Mniobryum longibracteatum (Broth. in Röll) Broth. = Pohlia longibracteata [POHLLON] Bryaceae (B:M)
Mniobryum wahlenbergii (Web. & Mohr) Jenn. = Pohlia wahlenbergii [POHLWAH] Bryaceae (B:M)
Mniobryum wahlenbergii var. *glacialis* (Brid.) Wijk & Marg. = Pohlia wahlenbergii [POHLWAH] Bryaceae (B:M)
Mnium Hedw. [MNIUM$] Mniaceae (B:M)
Mnium affine var. *ciliare* C. Müll. = Plagiomnium ciliare [PLAGCIL] Mniaceae (B:M)
Mnium affine var. *integrifolium* (Lindb.) Milde = Plagiomnium ellipticum [PLAGELL] Mniaceae (B:M)
Mnium affine var. *rugicum* (Laur.) Bruch & Schimp. in B.S.G. = Plagiomnium ellipticum [PLAGELL] Mniaceae (B:M)
Mnium ambiguum H. Müll. [MNIUAMB] Mniaceae (B:M)
Mnium arizonicum Amann [MNIUARI] Mniaceae (B:M)
Mnium blyttii Bruch & Schimp in B.S.G. [MNIUBLY] Mniaceae (B:M)
Mnium ciliare (C. Müll.) Schimp. = Plagiomnium ciliare [PLAGCIL] Mniaceae (B:M)
Mnium cinclidioides Hüb. = Pseudobryum cinclidioides [PSEUCIN] Mniaceae (B:M)
Mnium cuspidatum Hedw. = Plagiomnium cuspidatum [PLAGCUS] Mniaceae (B:M)
Mnium cuspidatum var. *tenellum* Kindb. in Mac. & Kindb. = Plagiomnium cuspidatum [PLAGCUS] Mniaceae (B:M)
Mnium decurrens C. Müll. & Kindb. in Mac. & Kindb. = Mnium thomsonii [MNIUTHO] Mniaceae (B:M)
Mnium drummondii Bruch & Schimp. = Plagiomnium drummondii [PLAGDRU] Mniaceae (B:M)
Mnium ellipticum Brid. = Plagiomnium ellipticum [PLAGELL] Mniaceae (B:M)
Mnium glabrescens Kindb. = Rhizomnium glabrescens [RHIZGLB] Mniaceae (B:M)
Mnium glabrescens ssp. *chlorophyllosum* Kindb. = Rhizomnium punctatum [RHIZPUN] Mniaceae (B:M)
Mnium gracile (T. Kop.) Crum & Anderson = Rhizomnium gracile [RHIZGRC] Mniaceae (B:M)
Mnium hymenophylloides Hüb. = Cyrtomnium hymenophylloides [CYRTHYM] Mniaceae (B:M)
Mnium hymenophyllum Bruch & Schimp. in B.S.G. = Cyrtomnium hymenophyllum [CYRTHYE] Mniaceae (B:M)
Mnium inclinatum Lindb. = Mnium ambiguum [MNIUAMB] Mniaceae (B:M)
Mnium insigne Mitt. = Plagiomnium insigne [PLAGINS] Mniaceae (B:M)
Mnium insigne var. *intermedium* Kindb. in Mac. = Plagiomnium insigne [PLAGINS] Mniaceae (B:M)
Mnium koponenii Crum = Rhizomnium gracile [RHIZGRC] Mniaceae (B:M)
Mnium longirostre Brid. = Plagiomnium rostratum [PLAGROS] Mniaceae (B:M)
Mnium lycopodioides var. *inclinatum* (Lindb.) Wijk & Marg. = Mnium ambiguum [MNIUAMB] Mniaceae (B:M)
Mnium macrociliare C. Müll. & Kindb. in Mac. & Kindb. = Plagiomnium ciliare [PLAGCIL] Mniaceae (B:M)
Mnium magnifolium Horik. = Rhizomnium magnifolium [RHIZMAG] Mniaceae (B:M)
Mnium marginatum (With.) Brid. ex P. Beauv. [MNIUMAR] Mniaceae (B:M)
Mnium medium Bruch & Schimp. in B.S.G. = Plagiomnium medium [PLAGMED] Mniaceae (B:M)
Mnium menziesii Hook. = Leucolepis acanthoneuron [LEUCACA] Mniaceae (B:M)
Mnium nudum Britt. & Williams = Rhizomnium nudum [RHIZNUD] Mniaceae (B:M)
Mnium orthorrhynchum Brid. = Atrichum tenellum [ATRITEN] Polytrichaceae (B:M)
Mnium orthorrhynchum auct. = Mnium thomsonii [MNIUTHO] Mniaceae (B:M)
Mnium pseudolycopodioides C. Müll. & Kindb. in Mac. & Kindb. = Mnium ambiguum [MNIUAMB] Mniaceae (B:M)

Mnium pseudopunctatum Bruch & Schimp. = Rhizomnium pseudopunctatum [RHIZPSE] Mniaceae (B:M)
Mnium punctatum Hedw. = Rhizomnium punctatum [RHIZPUN] Mniaceae (B:M)
Mnium punctatum var. *elatum* Schimp. = Rhizomnium magnifolium [RHIZMAG] Mniaceae (B:M)
Mnium riparium Mitt. = Mnium ambiguum [MNIUAMB] Mniaceae (B:M)
Mnium robustum (Kindb.) Kindb. = Plagiomnium insigne [PLAGINS] Mniaceae (B:M)
Mnium rostratum Schrad. = Plagiomnium rostratum [PLAGROS] Mniaceae (B:M)
Mnium rugicum Laur. = Plagiomnium ellipticum [PLAGELL] Mniaceae (B:M)
Mnium saximontanum M. Bowers = Mnium arizonicum [MNIUARI] Mniaceae (B:M)
Mnium serratum Schrad. ex Brid. = Mnium marginatum [MNIUMAR] Mniaceae (B:M)
Mnium serratum var. *macounii* Kindb. in Mac. & Kindb. = Mnium spinulosum [MNIUSPN] Mniaceae (B:M)
Mnium spinulosum Bruch & Schimp. in B.S.G. [MNIUSPN] Mniaceae (B:M)
Mnium subglobosum Bruch & Schimp. in B.S.G. = Rhizomnium pseudopunctatum [RHIZPSE] Mniaceae (B:M)
Mnium sylvaticum Lindb. = Plagiomnium cuspidatum [PLAGCUS] Mniaceae (B:M)
Mnium thomsonii Schimp. [MNIUTHO] Mniaceae (B:M)
Mnium umbratile Mitt. = Mnium ambiguum [MNIUAMB] Mniaceae (B:M)
Mnium venustum Mitt. = Plagiomnium venustum [PLAGVEN] Mniaceae (B:M)
Moehringia L. [MOEHRIN$] Caryophyllaceae (V:D)
Moehringia lateriflora (L.) Fenzl [MOEHLAT] Caryophyllaceae (V:D)
Moehringia macrophylla (Hook.) Fenzl [MOEHMAC] Caryophyllaceae (V:D)
Moenchia Ehrh. [MOENCHI$] Caryophyllaceae (V:D)
***Moenchia erecta** (L.) P.G. Gaertn., B. Mey. & Scherb. [MOENERE] Caryophyllaceae (V:D)
Moerckia Gott. [MOERCKI$] Pallaviciniaceae (B:H)
Moerckia blyttii (Moerck) Brockm. [MOERBLY] Pallaviciniaceae (B:H)
Moerckia flotoviana (Nees) Schiffn. = Moerckia hibernica [MOERHIB] Pallaviciniaceae (B:H)
Moerckia hibernica (Hook.) Gott. [MOERHIB] Pallaviciniaceae (B:H)
Moldavica parviflora (Nutt.) Britt. = Dracocephalum parviflorum [DRACPAR] Lamiaceae (V:D)
Moldavica thymiflora (L.) Rydb. = Dracocephalum thymiflorum [DRACTHY] Lamiaceae (V:D)
Molendoa Lindb. [MOLENDO$] Pottiaceae (B:M)
Molendoa hornschuchiana (Hook.) Lindb. ex Limpr. [MOLEHOR] Pottiaceae (B:M)
Molendoa obtusifolia Broth. & Par. in Card. = Molendoa sendtneriana [MOLESEN] Pottiaceae (B:M)
Molendoa sendtneriana (Bruch & Schimp. in B.S.G.) Limpr. [MOLESEN] Pottiaceae (B:M)
Molinia maxima Hartman = Glyceria maxima [GLYCMAX] Poaceae (V:M)
Mollugo L. [MOLLUGO$] Molluginaceae (V:D)
Mollugo berteriana Ser. = Mollugo verticillata [MOLLVER] Molluginaceae (V:D)
Mollugo tetraphylla L. = Polycarpon tetraphyllum [POLYTET] Caryophyllaceae (V:D)
***Mollugo verticillata** L. [MOLLVER] Molluginaceae (V:D)
Monarda L. [MONARDA$] Lamiaceae (V:D)
Monarda fistulosa L. [MONAFIS] Lamiaceae (V:D)
Monarda fistulosa var. **menthifolia** (Graham) Fern. [MONAFIS2] Lamiaceae (V:D)
Monarda fistulosa var. **mollis** (L.) Benth. [MONAFIS1] Lamiaceae (V:D)
Monarda menthifolia Graham = Monarda fistulosa var. menthifolia [MONAFIS2] Lamiaceae (V:D)
Monarda mollis L. = Monarda fistulosa var. mollis [MONAFIS1] Lamiaceae (V:D)
Monarda scabra Beck = Monarda fistulosa var. mollis [MONAFIS1] Lamiaceae (V:D)
Monardella Benth. [MONARDE$] Lamiaceae (V:D)
Monardella odoratissima Benth. [MONAODO] Lamiaceae (V:D)
Moneses Salisb. ex S.F. Gray [MONESES$] Pyrolaceae (V:D)
Moneses uniflora (L.) Gray [MONEUNI] Pyrolaceae (V:D)
Monilistus monilifera (Ait.) Raf. ex B.D. Jackson = Populus deltoides ssp. monilifera [POPUDEL2] Salicaceae (V:D)
Monolepis Schrad. [MONOLEP$] Chenopodiaceae (V:D)
***Monolepis nuttalliana** (J.A. Schultes) Greene [MONONUT] Chenopodiaceae (V:D)
Monotropa L. [MONOTRO$] Monotropaceae (V:D)
Monotropa brittonii Small = Monotropa uniflora [MONOUNI] Monotropaceae (V:D)
Monotropa hypopithys L. [MONOHYP] Monotropaceae (V:D)
Monotropa hypopithys ssp. *lanuginosa* (Michx.) Hara = Monotropa hypopithys [MONOHYP] Monotropaceae (V:D)
Monotropa hypopithys var. *americana* (DC.) Domin = Monotropa hypopithys [MONOHYP] Monotropaceae (V:D)
Monotropa hypopithys var. *latisquama* (Rydb.) Kearney & Peebles = Monotropa hypopithys [MONOHYP] Monotropaceae (V:D)
Monotropa hypopithys var. *rubra* (Torr.) Farw. = Monotropa hypopithys [MONOHYP] Monotropaceae (V:D)
Monotropa lanuginosa Michx. = Monotropa hypopithys [MONOHYP] Monotropaceae (V:D)
Monotropa latisquama (Rydb.) Hultén = Monotropa hypopithys [MONOHYP] Monotropaceae (V:D)
Monotropa uniflora L. [MONOUNI] Monotropaceae (V:D)
Montia L. [MONTIA$] Portulacaceae (V:D)
Montia bostockii (Porsild) Welsh [MONTBOS] Portulacaceae (V:D)
Montia chamissoi (Ledeb. ex Spreng.) Greene [MONTCHA] Portulacaceae (V:D)
Montia cordifolia (S. Wats.) Pax & K. Hoffmann = Claytonia cordifolia [CLAYCOR] Portulacaceae (V:D)
Montia dichotoma (Nutt.) T.J. Howell [MONTDIC] Portulacaceae (V:D)
Montia diffusa (Nutt.) Greene [MONTDIF] Portulacaceae (V:D)
Montia flagellaris (Bong.) B.L. Robins. = Montia parvifolia ssp. flagellaris [MONTPAR1] Portulacaceae (V:D)
Montia fontana L. [MONTFON] Portulacaceae (V:D)
Montia fontana ssp. *amporitana* auct. = Montia fontana [MONTFON] Portulacaceae (V:D)
Montia fontana ssp. **variabilis** S.M. Walters [MONTFON3] Portulacaceae (V:D)
Montia fontana var. *tenerrima* (Gray) Fern. & Wieg. = Montia fontana [MONTFON] Portulacaceae (V:D)
Montia funstonii Rydb. = Montia fontana ssp. variabilis [MONTFON3] Portulacaceae (V:D)
Montia heterophylla (Torr. & Gray) Jepson = Claytonia heterophylla [CLAYHET] Portulacaceae (V:D)
Montia howellii S. Wats. [MONTHOW] Portulacaceae (V:D)
Montia lamprosperma Cham. = Montia fontana [MONTFON] Portulacaceae (V:D)
Montia linearis (Dougl. ex Hook.) Greene [MONTLIN] Portulacaceae (V:D)
Montia parviflora (Dougl. ex Hook.) T.J. Howell = Claytonia parviflora [CLAYPAR] Portulacaceae (V:D)
Montia parvifolia (Moc. ex DC.) Greene [MONTPAR] Portulacaceae (V:D)
Montia parvifolia ssp. **flagellaris** (Bong.) Ferris [MONTPAR1] Portulacaceae (V:D)
Montia parvifolia ssp. **parvifolia** [MONTPAR0] Portulacaceae (V:D)
Montia parvifolia var. *flagellaris* (Bong.) C.L. Hitchc. = Montia parvifolia ssp. flagellaris [MONTPAR1] Portulacaceae (V:D)
Montia perfoliata (Donn ex Willd.) T.J. Howell = Claytonia perfoliata [CLAYPER] Portulacaceae (V:D)
Montia perfoliata ssp. *glauca* (Nutt. ex Torr. & Gray) Ferris = Claytonia rubra [CLAYRUB] Portulacaceae (V:D)
Montia perfoliata var. *depressa* (Gray) Jepson = Claytonia rubra [CLAYRUB] Portulacaceae (V:D)

Montia perfoliata var. *parviflora* (Dougl. ex Hook.) Jepson = Claytonia parviflora [CLAYPAR] Portulacaceae (V:D)
Montia sarmentosa (C.A. Mey.) B.L. Robins. = Claytonia sarmentosa [CLAYSAR] Portulacaceae (V:D)
Montia sibirica (L.) T.J. Howell = Claytonia sibirica [CLAYSIB] Portulacaceae (V:D)
Montia sibirica var. *heterophylla* (Torr. & Gray) B.L. Robins. = Claytonia heterophylla [CLAYHET] Portulacaceae (V:D)
Montia spathulata (Dougl. ex Hook.) T.J. Howell = Claytonia spathulata [CLAYSPA] Portulacaceae (V:D)
Montiastrum dichotomum (Nutt.) Rydb. = Montia dichotoma [MONTDIC] Portulacaceae (V:D)
Montiastrum lineare (Dougl. ex Hook.) Rydb. = Montia linearis [MONTLIN] Portulacaceae (V:D)
Morus L. [MORUS$] Moraceae (V:D)
•**Morus alba** L. [MORUALB] Moraceae (V:D)
Morus alba var. *tatarica* (L.) Ser. = Morus alba [MORUALB] Moraceae (V:D)
Morus tatarica L. = Morus alba [MORUALB] Moraceae (V:D)
Muellerella Hepp ex Müll. Arg. [MUELLER$] Verrucariaceae (L)
Muellerella pygmaea (Körber) D. Hawksw. [MUELPYG] Verrucariaceae (L)
Muhlenbergia Schreb. [MUHLENB$] Poaceae (V:M)
Muhlenbergia ambigua Torr. = Muhlenbergia mexicana [MUHLMEX] Poaceae (V:M)
Muhlenbergia andina (Nutt.) A.S. Hitchc. [MUHLAND] Poaceae (V:M)
Muhlenbergia asperifolia (Nees & Meyen ex Trin.) Parodi [MUHLASP] Poaceae (V:M)
Muhlenbergia comata (Thurb.) Thurb. ex Benth. = Muhlenbergia andina [MUHLAND] Poaceae (V:M)
Muhlenbergia filiformis (Thurb. ex S. Wats.) Rydb. [MUHLFIL] Poaceae (V:M)
Muhlenbergia filiformis var. *fortis* E.H. Kelso = Muhlenbergia filiformis [MUHLFIL] Poaceae (V:M)
Muhlenbergia foliosa (Roemer & J.A. Schultes) Trin. = Muhlenbergia mexicana [MUHLMEX] Poaceae (V:M)
Muhlenbergia foliosa ssp. *ambigua* (Torr.) Scribn. = Muhlenbergia mexicana [MUHLMEX] Poaceae (V:M)
Muhlenbergia foliosa ssp. *setiglumis* (S. Wats.) Scribn. = Muhlenbergia mexicana [MUHLMEX] Poaceae (V:M)
Muhlenbergia glomerata (Willd.) Trin. [MUHLGLO] Poaceae (V:M)
Muhlenbergia glomerata var. *cinnoides* (Link) F.J. Herm. = Muhlenbergia glomerata [MUHLGLO] Poaceae (V:M)
Muhlenbergia idahoensis St. John = Muhlenbergia filiformis [MUHLFIL] Poaceae (V:M)
Muhlenbergia mexicana (L.) Trin. [MUHLMEX] Poaceae (V:M)
Muhlenbergia mexicana var. *filiformis* (Willd.) Scribn. = Muhlenbergia mexicana [MUHLMEX] Poaceae (V:M)
Muhlenbergia racemosa (Michx.) B.S.P. [MUHLRAC] Poaceae (V:M)
Muhlenbergia racemosa var. *cinnoides* (Link) Boivin = Muhlenbergia glomerata [MUHLGLO] Poaceae (V:M)
Muhlenbergia richardsonis (Trin.) Rydb. [MUHLRIC] Poaceae (V:M)
Muhlenbergia simplex (Scribn.) Rydb. = Muhlenbergia filiformis [MUHLFIL] Poaceae (V:M)
Muhlenbergia squarrosa (Trin.) Rydb. = Muhlenbergia richardsonis [MUHLRIC] Poaceae (V:M)
Mulgedium spicatum var. *integrifolium* (Torr. & Gray) Small = Lactuca biennis [LACTBIE] Asteraceae (V:D)
Multiclavula R. Petersen [MULTICL$] Clavariaceae (L)
Multiclavula corynoides (Peck) R. Petersen [MULTCOY] Clavariaceae (L)
Multiclavula sharpii R. Petersen [MULTSHA] Clavariaceae (L)
Muscaria adscendens (L.) Small p.p. = Saxifraga adscendens ssp. oregonensis [SAXIADS1] Saxifragaceae (V:D)
Muscaria micropetala Small = Saxifraga cespitosa ssp. monticola [SAXICES1] Saxifragaceae (V:D)
Muscaria monticola Small = Saxifraga cespitosa ssp. monticola [SAXICES1] Saxifragaceae (V:D)
Myagrum paniculatum L. = Neslia paniculata [NESLPAN] Brassicaceae (V:D)
Mycelis Cass. [MYCELIS$] Asteraceae (V:D)
•**Mycelis muralis** (L.) Dumort. [MYCEMUR] Asteraceae (V:D)
Mycobilimbia Rehm [MYCOBIL$] Porpidiaceae (L)
Mycobilimbia accedens (Arnold) V. Wirth ex Hafellner = Mycobilimbia sabuletorum [MYCOSAB] Porpidiaceae (L)
Mycobilimbia berengeriana (A. Massal.) Hafellner & V. Wirth [MYCOBER] Porpidiaceae (L)
Mycobilimbia fusca (A. Massal.) Hafellner & V. Wirth = Mycobilimbia tetramera [MYCOTET] Porpidiaceae (L)
Mycobilimbia hypnorum (Lib.) Kalb & Hafellner [MYCOHYP] Porpidiaceae (L)
Mycobilimbia lobulata (Sommerf.) Hafellner [MYCOLOB] Porpidiaceae (L)
Mycobilimbia obscurata (Sommerf.) Rehm = Mycobilimbia tetramera [MYCOTET] Porpidiaceae (L)
Mycobilimbia sabuletorum (Schreber) Hafellner [MYCOSAB] Porpidiaceae (L)
Mycobilimbia tetramera (De Not.) W. Brunnbauer [MYCOTET] Porpidiaceae (L)
Mycoblastus Norman [MYCOBLA$] Mycoblastaceae (L)
Mycoblastus affinis (Schaerer) Schauer [MYCOAFF] Mycoblastaceae (L)
Mycoblastus alpinus (Fr.) Kernst. [MYCOALP] Mycoblastaceae (L)
Mycoblastus caesius (Coppins & P. James) Tønsberg [MYCOCAE] Mycoblastaceae (L)
Mycoblastus fucatus (Stirton) Zahlbr. [MYCOFUC] Mycoblastaceae (L)
Mycoblastus glabrescens (Nyl.) Zahlbr. = Mycoblastus fucatus [MYCOFUC] Mycoblastaceae (L)
Mycoblastus melinus (Kremp. ex Nyl.) Hellbom = Mycoblastus affinis [MYCOAFF] Mycoblastaceae (L)
Mycoblastus sanguinarius (L.) Norman [MYCOSAN] Mycoblastaceae (L)
Mycoblastus sanguinarius var. *dodgeanus* Räsänen = Mycoblastus sanguinarius [MYCOSAN] Mycoblastaceae (L)
Mycoblastus tornoënsis (Nyl.) R. Anderson = Japewia tornoënsis [JUPETOR] Bacidiaceae (L)
Mycocalicium Vainio [MYCOCAC$] Mycocaliciaceae (L)
Mycocalicium compressulum Nyl. ex Szat. = Phaeocalicium compressulum [PHAECOM] Mycocaliciaceae (L)
Mycocalicium disseminatum (Ach.) Fink = Microcalicium disseminatum [MICRDIS] Microcaliciaceae (L)
Mycocalicium parietinum (Ach. ex Schaerer) D. Hawksw. = Mycocalicium subtile [MYCOSUB] Mycocaliciaceae (L)
Mycocalicium pusiolum (Ach.) Räsänen = Chaenothecopsis pusiola [CHAEPUI] Mycocaliciaceae (L)
Mycocalicium subtile (Pers.) Szat. [MYCOSUB] Mycocaliciaceae (L)
Mylia Gray [MYLIA$] Jungermanniaceae (B:H)
Mylia anomala (Hook.) Gray [MYLIANO] Jungermanniaceae (B:H)
Mylia taylorii (Hook.) Gray [MYLITAY] Jungermanniaceae (B:H)
Myosotis L. [MYOSOTI$] Boraginaceae (V:D)
Myosotis alpestris auct. = Myosotis asiatica [MYOSASI] Boraginaceae (V:D)
Myosotis alpestris ssp. *asiatica* Vesterg. = Myosotis asiatica [MYOSASI] Boraginaceae (V:D)
•**Myosotis arvensis** (L.) Hill [MYOSARV] Boraginaceae (V:D)
Myosotis asiatica (Vesterg.) Schischkin & Sergievskaja [MYOSASI] Boraginaceae (V:D)
•**Myosotis discolor** Pers. [MYOSDIS] Boraginaceae (V:D)
Myosotis laxa Lehm. [MYOSLAX] Boraginaceae (V:D)
Myosotis micrantha auct. = Myosotis stricta [MYOSSTR] Boraginaceae (V:D)
Myosotis palustris (L.) Hill = Myosotis scorpioides [MYOSSCO] Boraginaceae (V:D)
•**Myosotis scorpioides** L. [MYOSSCO] Boraginaceae (V:D)
Myosotis scorpioides var. *arvensis* L. = Myosotis arvensis [MYOSARV] Boraginaceae (V:D)
•**Myosotis stricta** Link ex Roemer & J.A. Schultes [MYOSSTR] Boraginaceae (V:D)

*Myosotis sylvatica** Ehrh. ex Hoffmann [MYOSSYL] Boraginaceae (V:D)
Myosotis sylvatica var. *alpestris* auct. = Myosotis asiatica [MYOSASI] Boraginaceae (V:D)
*Myosotis verna** Nutt. [MYOSVER] Boraginaceae (V:D)
Myosotis versicolor (Pers.) Sm. = Myosotis discolor [MYOSDIS] Boraginaceae (V:D)
Myosoton Moench [MYOSOTO$] Caryophyllaceae (V:D)
*Myosoton aquaticum** (L.) Moench [MYOSAQU] Caryophyllaceae (V:D)
Myosurus L. [MYOSURU$] Ranunculaceae (V:D)
Myosurus aristatus Benth. [MYOSARI] Ranunculaceae (V:D)
Myosurus minimus L. [MYOSMIN] Ranunculaceae (V:D)
Myosurus minimus var. *aristatus* (Benth.) Boivin = Myosurus aristatus [MYOSARI] Ranunculaceae (V:D)
Myrica L. [MYRICA$] Myricaceae (V:D)
Myrica californica Cham. [MYRICAL] Myricaceae (V:D)
Myrica gale L. [MYRIGAL] Myricaceae (V:D)
Myrinia Schimp. [MYRINIA$] Myriniaceae (B:M)
Myrinia dieckii Ren. & Card. in Röll = Scleropodium cespitans var. sublaeve [SCLECES2] Brachytheciaceae (B:M)
Myrinia pulvinata (Wahlenb.) Schimp. [MYRIPUL] Myriniaceae (B:M)
Myriophyllum L. [MYRIOPH$] Haloragaceae (V:D)
*Myriophyllum aquaticum** (Vell.) Verdc. [MYRIAQU] Haloragaceae (V:D)
Myriophyllum brasiliense Camb. = Myriophyllum aquaticum [MYRIAQU] Haloragaceae (V:D)
Myriophyllum elatinoides Gaud. = Myriophyllum quitense [MYRIQUI] Haloragaceae (V:D)
Myriophyllum exalbescens Fern. = Myriophyllum sibiricum [MYRISIB] Haloragaceae (V:D)
Myriophyllum exalbescens var. *magdalenense* (Fern.) A. Löve = Myriophyllum sibiricum [MYRISIB] Haloragaceae (V:D)
Myriophyllum farwellii Morong [MYRIFAR] Haloragaceae (V:D)
*Myriophyllum heterophyllum** Michx. [MYRIHET] Haloragaceae (V:D)
Myriophyllum hippuroides Nutt. ex Torr. & Gray [MYRIHIP] Haloragaceae (V:D)
Myriophyllum magdalenense Fern. = Myriophyllum sibiricum [MYRISIB] Haloragaceae (V:D)
Myriophyllum proserpinacoides Gillies ex Hook. & Arn. = Myriophyllum aquaticum [MYRIAQU] Haloragaceae (V:D)
Myriophyllum quitense Kunth [MYRIQUI] Haloragaceae (V:D)
Myriophyllum sibiricum Komarov [MYRISIB] Haloragaceae (V:D)
*Myriophyllum spicatum** L. [MYRISPI] Haloragaceae (V:D)
Myriophyllum spicatum ssp. *exalbescens* (Fern.) Hultén = Myriophyllum sibiricum [MYRISIB] Haloragaceae (V:D)
Myriophyllum spicatum ssp. *squamosum* Laestad. ex Hartman = Myriophyllum sibiricum [MYRISIB] Haloragaceae (V:D)
Myriophyllum spicatum var. *capillaceum* Lange = Myriophyllum sibiricum [MYRISIB] Haloragaceae (V:D)
Myriophyllum spicatum var. *exalbescens* (Fern.) Jepson = Myriophyllum sibiricum [MYRISIB] Haloragaceae (V:D)
Myriophyllum spicatum var. *squamosum* (Laestad. ex Hartman) Hartman = Myriophyllum sibiricum [MYRISIB] Haloragaceae (V:D)
Myriophyllum ussuriense (Regel) Maxim. [MYRIUSS] Haloragaceae (V:D)
*Myriophyllum verticillatum** L. [MYRIVER] Haloragaceae (V:D)
Myriophyllum verticillatum var. *cheneyi* Fassett = Myriophyllum verticillatum [MYRIVER] Haloragaceae (V:D)
Myriophyllum verticillatum var. *intermedium* W.D.J. Koch = Myriophyllum verticillatum [MYRIVER] Haloragaceae (V:D)
Myriophyllum verticillatum var. *pectinatum* Wallr. = Myriophyllum verticillatum [MYRIVER] Haloragaceae (V:D)
Myriophyllum verticillatum var. *pinnatifidum* Wallr. = Myriophyllum verticillatum [MYRIVER] Haloragaceae (V:D)
Myurella Schimp. in B.S.G. [MYURELL$] Pterigynandraceae (B:M)
Myurella apiculata (Somm.) Schimp. in B.S.G. = Myurella tenerrima [MYURTEN] Pterigynandraceae (B:M)
Myurella careyana Sull. in Sull. & Lesq. = Myurella sibirica [MYURSIB] Pterigynandraceae (B:M)
Myurella careyana var. *tenella* Hab. = Myurella sibirica [MYURSIB] Pterigynandraceae (B:M)
Myurella gracilis Lindb. = Myurella sibirica [MYURSIB] Pterigynandraceae (B:M)
Myurella julacea (Schwaegr.) Schimp. in B.S.G. [MYURJUL] Pterigynandraceae (B:M)
Myurella julacea var. *scabrifolia* Lindb. ex Limpr. = Myurella julacea [MYURJUL] Pterigynandraceae (B:M)
Myurella sibirica (C. Müll.) Reim. [MYURSIB] Pterigynandraceae (B:M)
Myurella sibirica var. *tenella* (Hab.) Crum, Steere & Anderson = Myurella sibirica [MYURSIB] Pterigynandraceae (B:M)
Myurella squarrosa Grout = Campylium hispidulum [CAMPHIS] Amblystegiaceae (B:M)
Myurella tenerrima (Brid.) Lindb. [MYURTEN] Pterigynandraceae (B:M)
Myzorrhiza californica (Cham. & Schlecht.) Rydb. = Orobanche californica [OROBCAL] Orobanchaceae (V:D)
Myzorrhiza pinorum (Geyer ex Hook.) Rydb. = Orobanche pinorum [OROBPIN] Orobanchaceae (V:D)

N

Najas L. [NAJAS$] Najadaceae (V:M)
Najas caespitosus (Maguire) Reveal = Najas flexilis [NAJAFLE] Najadaceae (V:M)
Najas flexilis (Willd.) Rostk. & Schmidt [NAJAFLE] Najadaceae (V:M)
Najas flexilis ssp. *caespitosus* Maguire = Najas flexilis [NAJAFLE] Najadaceae (V:M)
Najas flexilis var. *congesta* Farw. = Najas flexilis [NAJAFLE] Najadaceae (V:M)
Najas flexilis var. *robusta* Morong = Najas flexilis [NAJAFLE] Najadaceae (V:M)
Nanomitrium tenerum (Bruch & Schimp. in B.S.G.) Lindb. = Micromitrium tenerum [MICRTEN] Ephemeraceae (B:M)
Nanostictis M.S. Christ. [NANOSTI$] Stictidaceae (L)
Nanostictis peltigerae Sherwood [NANOPEL] Stictidaceae (L)
Narcissus L. [NARCISS$] Liliaceae (V:M)
*Narcissus poeticus** L. [NARCPOE] Liliaceae (V:M)
*Narcissus pseudonarcissus** L. [NARCPSE] Liliaceae (V:M)
Nardia Gray [NARDIA$] Jungermanniaceae (B:H)
Nardia breidleri (Limpr.) Lindb. [NARDBRE] Jungermanniaceae (B:H)
Nardia cavana Clark = Nardia scalaris [NARDSCA] Jungermanniaceae (B:H)
Nardia compressa (Hook.) Gray [NARDCOM] Jungermanniaceae (B:H)
Nardia geoscyphus (DeNot.) Lindb. [NARDGEO] Jungermanniaceae (B:H)
Nardia geoscyphus var. *insecta* (K. Müll.) Schiffn. = Nardia insecta [NARDINS] Jungermanniaceae (B:H)
Nardia insecta Lindb. [NARDINS] Jungermanniaceae (B:H)
Nardia japonica Steph. [NARDJAP] Jungermanniaceae (B:H)
Nardia rubra (Gott.) Evans = Jungermannia rubra [JUNGRUB] Jungermanniaceae (B:H)
Nardia scalaris Gray [NARDSCA] Jungermanniaceae (B:H)
Nardosmia arctica (Porsild) A. & D. Löve = Petasites frigidus var. palmatus [PETAFRI3] Asteraceae (V:D)
Nardosmia japonica Sieb. & Zucc. = Petasites japonicus [PETAJAP] Asteraceae (V:D)

Narthecium pusillum Michx. = Tofieldia pusilla [TOFIPUS] Liliaceae (V:M)
Nassella (Trin.) Desv. [NASSELL$] Poaceae (V:M)
Nassella viridula (Trin.) Barkworth [NASSVIR] Poaceae (V:M)
Nasturtium microphyllum Boenn. ex Reichenb. = Rorippa microphylla [RORIMIR] Brassicaceae (V:D)
Nasturtium officinale Ait. f. = Rorippa nasturtium-aquaticum [RORINAS] Brassicaceae (V:D)
Nasturtium officinale var. *microphyllum* (Boenn. ex Reichenb.) Thellung = Rorippa microphylla [RORIMIR] Brassicaceae (V:D)
Nasturtium officinale var. *siifolium* (Reichenb.) W.D.J. Koch = Rorippa nasturtium-aquaticum [RORINAS] Brassicaceae (V:D)
Naumburgia thyrsiflora (L.) Duby = Lysimachia thyrsiflora [LYSITHY] Primulaceae (V:D)
Navarretia Ruiz & Pavón [NAVARRE$] Polemoniaceae (V:D)
Navarretia intertexta (Benth.) Hook. [NAVAINT] Polemoniaceae (V:D)
Navarretia minima var. *intertexta* (Benth.) Boivin = Navarretia intertexta [NAVAINT] Polemoniaceae (V:D)
Navarretia squarrosa (Eschsch.) Hook. & Arn. [NAVASQU] Polemoniaceae (V:D)
Neckera Hedw. [NECKERA$] Neckeraceae (B:M)
Neckera douglasii Hook. [NECKDOU] Neckeraceae (B:M)
Neckera douglasii var. *macounii* Kindb. = Neckera douglasii [NECKDOU] Neckeraceae (B:M)
Neckera gracilis (James in Peck) Kindb. = Homalia trichomanoides [HOMATRI] Neckeraceae (B:M)
Neckera menziesii Hook. in Drumm. = Metaneckera menziesii [METAMEN] Neckeraceae (B:M)
Neckera menziesii var. *amblyclada* Kindb. in Mac. & Kindb. = Metaneckera menziesii [METAMEN] Neckeraceae (B:M)
Neckera neomexicana (Card.) Grout = Metaneckera menziesii [METAMEN] Neckeraceae (B:M)
Neckera oligocarpa Bruch in Ångstr. = Neckera pennata [NECKPEN] Neckeraceae (B:M)
Neckera pennata Hedw. [NECKPEN] Neckeraceae (B:M)
Neckera pennata var. *oligocarpa* (Bruch in Ångstr.) C. Müll. = Neckera pennata [NECKPEN] Neckeraceae (B:M)
Neckera pennata var. *tenera* C. Müll. = Neckera pennata [NECKPEN] Neckeraceae (B:M)
Neckera pterantha (C. Müll. & Kindb.) Kindb. = Neckera pennata [NECKPEN] Neckeraceae (B:M)
Neckeradelphus menziesii (Hook. in Drumm.) Steere = Metaneckera menziesii [METAMEN] Neckeraceae (B:M)
Nemophila Nutt. [NEMOPHI$] Hydrophyllaceae (V:D)
Nemophila breviflora Gray [NEMOBRE] Hydrophyllaceae (V:D)
Nemophila parviflora Dougl. ex Benth. [NEMOPAR] Hydrophyllaceae (V:D)
Nemophila pedunculata Dougl. ex Benth. [NEMOPED] Hydrophyllaceae (V:D)
Neofuscelia Essl. [NEOFUSC$] Parmeliaceae (L)
Neofuscelia loxodes (Nyl.) Essl. [NEOFLOX] Parmeliaceae (L)
Neofuscelia subhosseana (Essl.) Essl. [NEOFSUB] Parmeliaceae (L)
Neofuscelia verruculifera (Nyl.) Essl. [NEOFVER] Parmeliaceae (L)
Neolepia campestris (L.) W.A. Weber = Lepidium campestre [LEPICAM] Brassicaceae (V:D)
Nepeta L. [NEPETA$] Lamiaceae (V:D)
***Nepeta cataria** L. [NEPECAT] Lamiaceae (V:D)
Nephroma Ach. [NEPHROM$] Nephromataceae (L)
Nephroma arcticum (L.) Torss. [NEPHARC] Nephromataceae (L)
Nephroma bellum (Sprengel) Tuck. [NEPHBEL] Nephromataceae (L)
Nephroma expallidum (Nyl.) Nyl. [NEPHEXP] Nephromataceae (L)
Nephroma helveticum Ach. [NEPHHEL] Nephromataceae (L)
Nephroma helveticum ssp. **helveticum** [NEPHHEL0] Nephromataceae (L)
Nephroma helveticum ssp. **sipeanum** (Gyelnik) Goward & Ahti [NEPHHEL1] Nephromataceae (L)
Nephroma helveticum var. *sipeanum* (Gyelnik) Wetmore = Nephroma helveticum ssp. sipeanum [NEPHHEL1] Nephromataceae (L)
Nephroma isidiosum (Nyl.) Gyelnik [NEPHISI] Nephromataceae (L)
Nephroma laevigatum Ach. [NEPHLAE] Nephromataceae (L)
Nephroma laevigatum auct. = Nephroma bellum [NEPHBEL] Nephromataceae (L)
Nephroma lusitanicum auct. = Nephroma laevigatum [NEPHLAE] Nephromataceae (L)
Nephroma occultum Wetmore [NEPHOCC] Nephromataceae (L)
Nephroma parile (Ach.) Ach. [NEPHPAR] Nephromataceae (L)
Nephroma resupinatum (L.) Ach. [NEPHRES] Nephromataceae (L)
Nephroma silvae-veteris Goward & Goffinet [NEPHSIL] Nephromataceae (L)
Nephroma subtomentellum (Nyl.) Gyelnik = Nephroma bellum [NEPHBEL] Nephromataceae (L)
Nephromium canadense Räsänen = Nephroma helveticum ssp. sipeanum [NEPHHEL1] Nephromataceae (L)
Nephromopsis platyphylla (Tuck.) Herre = Tuckermannopsis platyphylla [TUCKPLA] Parmeliaceae (L)
Nephrophyllidium crista-galli (Menzies ex Hook.) Gilg = Fauria crista-galli [FAURCRI] Menyanthaceae (V:D)
Neslia Desv. [NESLIA$] Brassicaceae (V:D)
***Neslia paniculata** (L.) Desv. [NESLPAN] Brassicaceae (V:D)
Nesolechia oxyspora (Tul.) A. Massal. = Phacopsis oxyspora [PHACOXY] Lecanoraceae (L)
Neuroloma nudicaule (L.) DC. = Parrya nudicaulis [PARRNUD] Brassicaceae (V:D)
Niebla Rundel & Bowler [NIEBLA$] Ramalinaceae (L)
Niebla cephalota (Tuck.) Rundel & Bowler [NIEBCEP] Ramalinaceae (L)
Nodobryoria Common & Brodo [NODOBRY$] Parmeliaceae (L)
Nodobryoria abbreviata (Müll. Arg.) Common & Brodo [NODOABB] Parmeliaceae (L)
Nodobryoria oregana (Tuck.) Common & Brodo [NODOORE] Parmeliaceae (L)
Nodobryoria subdivergens (E. Dahl) Common & Brodo [NODOSUB] Parmeliaceae (L)
Normandina Nyl. [NORMAND$] Verrucariaceae (L)
Normandina pulchella (Borrer) Nyl. [NORMPUL] Verrucariaceae (L)
Norta altissima (L.) Britt. = Sisymbrium altissimum [SISYALT] Brassicaceae (V:D)
Nothocalais (Gray) Greene [NOTHOCA$] Asteraceae (V:D)
Nothocalais troximoides (Gray) Greene [NOTHTRO] Asteraceae (V:D)
Nothochelone (Gray) Straw [NOTHOCH$] Scrophulariaceae (V:D)
Nothochelone nemorosa (Dougl. ex Lindl.) Straw [NOTHNEM] Scrophulariaceae (V:D)
Nothoholcus lanatus (L.) Nash = Holcus lanatus [HOLCLAN] Poaceae (V:M)
Nuphar Sm. [NUPHAR$] Nymphaeaceae (V:D)
Nuphar advena var. *fraterna* (Mill. & Standl.) Standl. = Nuphar lutea ssp. variegata [NUPHLUT2] Nymphaeaceae (V:D)
Nuphar lutea (L.) Sm. [NUPHLUT] Nymphaeaceae (V:D)
Nuphar lutea ssp. **polysepala** (Engelm.) E.O. Beal [NUPHLUT1] Nymphaeaceae (V:D)
Nuphar lutea ssp. **variegata** (Dur.) E.O. Beal [NUPHLUT2] Nymphaeaceae (V:D)
Nuphar polysepala Engelm. = Nuphar lutea ssp. polysepala [NUPHLUT1] Nymphaeaceae (V:D)
Nuphar variegata Dur. = Nuphar lutea ssp. variegata [NUPHLUT2] Nymphaeaceae (V:D)
Nuttallanthus D.A. Sutton [NUTTALL$] Scrophulariaceae (V:D)
Nuttallanthus texanus (Scheele) D.A. Sutton [NUTTTEX] Scrophulariaceae (V:D)
Nuttallia cerasiformis Torr. & Gray ex Hook. & Arn. = Oemleria cerasiformis [OEMLCER] Rosaceae (V:D)
Nuttallia laevicaulis (Dougl. ex Hook.) Greene = Mentzelia laevicaulis var. laevicaulis [MENTLAE0] Loasaceae (V:D)
Nyctelea nyctelea (L.) Britt. = Ellisia nyctelea [ELLINYC] Hydrophyllaceae (V:D)
Nyholmiella obtusifolia (Brid.) Holm. & Warncke = Orthotrichum obtusifolium [ORTHOBT] Orthotrichaceae (B:M)
Nymphaea L. [NYMPHAE$] Nymphaeaceae (V:D)
***Nymphaea alba** L. [NYMPALB] Nymphaeaceae (V:D)
Nymphaea fraterna Mill. & Standl. = Nuphar lutea ssp. variegata [NUPHLUT2] Nymphaeaceae (V:D)

*·**Nymphaea mexicana** Zucc. [NYMPMEX] Nymphaeaceae (V:D)
Nymphaea minor (Sims) DC. = Nymphaea odorata [NYMPODO] Nymphaeaceae (V:D)
·**Nymphaea odorata** Ait. [NYMPODO] Nymphaeaceae (V:D)
Nymphaea odorata var. *gigantea* Tricker = Nymphaea odorata [NYMPODO] Nymphaeaceae (V:D)
Nymphaea odorata var. *godfreyi* Ward = Nymphaea odorata [NYMPODO] Nymphaeaceae (V:D)
Nymphaea odorata var. *maxima* (Conrad) Boivin = Nymphaea odorata [NYMPODO] Nymphaeaceae (V:D)
Nymphaea odorata var. *minor* Sims = Nymphaea odorata [NYMPODO] Nymphaeaceae (V:D)
Nymphaea odorata var. *rosea* Pursh = Nymphaea odorata [NYMPODO] Nymphaeaceae (V:D)
Nymphaea odorata var. *stenopetala* Fern. = Nymphaea odorata [NYMPODO] Nymphaeaceae (V:D)
Nymphaea odorata var. *villosa* Caspary = Nymphaea odorata [NYMPODO] Nymphaeaceae (V:D)
Nymphaea polysepala (Engelm.) Greene = Nuphar lutea ssp. polysepala [NUPHLUT1] Nymphaeaceae (V:D)
Nymphaea tetragona Georgi [NYMPTET] Nymphaeaceae (V:D)
Nymphaea tetragona ssp. *leibergii* (Morong) Porsild = Nymphaea tetragona [NYMPTET] Nymphaeaceae (V:D)
Nymphaea tetragona var. *leibergii* (Morong) Boivin = Nymphaea tetragona [NYMPTET] Nymphaeaceae (V:D)
Nymphaea tuberosa Paine = Nymphaea odorata [NYMPODO] Nymphaeaceae (V:D)
Nymphozanthus polysepalus (Engelm.) Fern. = Nuphar lutea ssp. polysepala [NUPHLUT1] Nymphaeaceae (V:D)

O

Oberna commutata (Guss.) S. Ikonnikov = Silene vulgaris [SILEVUL] Caryophyllaceae (V:D)
Obtusifolium obtusum (Lindb.) S. Arnell = Lophozia obtusa [LOPHOBT] Jungermanniaceae (B:H)
Ochrocodon pudicus (Pursh) Rydb. = Fritillaria pudica [FRITPUD] Liliaceae (V:M)
Ochrolechia A. Massal. [OCHROLE$] Pertusariaceae (L)
Ochrolechia androgyna (Hoffm.) Arnold [OCHRAND] Pertusariaceae (L)
Ochrolechia androgyna var. *pergranulosa* Räsänen = Ochrolechia androgyna [OCHRAND] Pertusariaceae (L)
Ochrolechia arborea (Kreyer) Almb. [OCHRARB] Pertusariaceae (L)
Ochrolechia californica Vers. = Ochrolechia oregonensis [OCHRORE] Pertusariaceae (L)
Ochrolechia elisabethae-kolae Vers. = Ochrolechia frigida [OCHRFRI] Pertusariaceae (L)
Ochrolechia farinacea Howard [OCHRFAR] Pertusariaceae (L)
Ochrolechia frigida (Sw.) Lynge [OCHRFRI] Pertusariaceae (L)
Ochrolechia geminipara (Th. Fr.) Vainio = Pertusaria geminipara [PERTGEM] Pertusariaceae (L)
Ochrolechia gonatodes (Ach.) Räsänen = Ochrolechia frigida [OCHRFRI] Pertusariaceae (L)
Ochrolechia gowardii Brodo [OCHRGOW] Pertusariaceae (L)
Ochrolechia gyalectina (Nyl.) Zahlbr. [OCHRGYA] Pertusariaceae (L)
Ochrolechia juvenalis Brodo [OCHRJUV] Pertusariaceae (L)
Ochrolechia laevigata (Räsänen) Vers. ex Brodo [OCHRLAE] Pertusariaceae (L)
Ochrolechia mahluensis Räsänen = Ochrolechia androgyna [OCHRAND] Pertusariaceae (L)
Ochrolechia montana Brodo [OCHRMON] Pertusariaceae (L)
Ochrolechia oregonensis H. Magn. [OCHRORE] Pertusariaceae (L)
Ochrolechia pacifica H. Magn. = Coccotrema pocillarium [COCCPOC] Coccotremataceae (L)
Ochrolechia pallescens var. *laevigata* Räsänen = Ochrolechia laevigata [OCHRLAE] Pertusariaceae (L)
Ochrolechia parella (L.) A. Massal. [OCHRPAR] Pertusariaceae (L)
Ochrolechia pterulina (Nyl.) Howard = Ochrolechia frigida [OCHRFRI] Pertusariaceae (L)
Ochrolechia rhodoleuca (Th. Fr.) Brodo [OCHRRHO] Pertusariaceae (L)
Ochrolechia sorediosa Howard = Ochrolechia szatalaënsis [OCHRSZA] Pertusariaceae (L)
Ochrolechia subathallina H. Magn. [OCHRSUB] Pertusariaceae (L)
Ochrolechia subpallescens Vers. [OCHRSUP] Pertusariaceae (L)
Ochrolechia subplicans (Nyl.) Brodo [OCHRSUL] Pertusariaceae (L)
Ochrolechia subplicans ssp. **hultenii** (Erichsen) Brodo [OCHRSUL1] Pertusariaceae (L)
Ochrolechia szatalaënsis Vers. [OCHRSZA] Pertusariaceae (L)
Ochrolechia tartarea (L.) A. Massal. [OCHRTAR] Pertusariaceae (L)
Ochrolechia turneri (Sm.) Hasselrot [OCHRTUR] Pertusariaceae (L)
Ochrolechia upsaliensis (L.) A. Massal. [OCHRUPS] Pertusariaceae (L)
Ochrolechia xanthostoma (Sommerf.) K. Schmitz & Lumbsch [OCHRXAN] Pertusariaceae (L)
Odontoschisma (Dum.) Dum. [ODONTOS$] Adelanthaceae (B:H)
Odontoschisma denudatum (Nees ex Mart.) Dum. [ODONDEN] Adelanthaceae (B:H)
Odontoschisma elongatum (Lindb.) Evans [ODONELO] Adelanthaceae (B:H)
Odontoschisma gibbsiae Evans = Odontoschisma denudatum [ODONDEN] Adelanthaceae (B:H)
Odontoschisma macounii (Aust.) Underw. [ODONMAC] Adelanthaceae (B:H)
Odontoschisma sphagni auct. = Odontoschisma denudatum [ODONDEN] Adelanthaceae (B:H)
Odostemon aquifolium (Pursh) Rydb. = Mahonia aquifolium [MAHOAQU] Berberidaceae (V:D)
Odostemon repens (Lindl.) Cockerell = Mahonia repens [MAHOREP] Berberidaceae (V:D)
Oedipodium Schwaegr. [OEDIPOD$] Oedipodiaceae (B:M)
Oedipodium griffithianum (Dicks.) Schwaegr. [OEDIGRI] Oedipodiaceae (B:M)
Oemleria Reichenb. [OEMLERI$] Rosaceae (V:D)
Oemleria cerasiformis (Torr. & Gray ex Hook. & Arn.) Landon [OEMLCER] Rosaceae (V:D)
Oenanthe L. [OENANTH$] Apiaceae (V:D)
Oenanthe sarmentosa K. Presl ex DC. [OENASAR] Apiaceae (V:D)
Oenothera L. [OENOTHE$] Onagraceae (V:D)
Oenothera andina Nutt. = Camissonia andina [CAMIAND] Onagraceae (V:D)
·**Oenothera biennis** L. [OENOBIE] Onagraceae (V:D)
Oenothera biennis ssp. *caeciarum* Munz = Oenothera biennis [OENOBIE] Onagraceae (V:D)
Oenothera biennis ssp. *centralis* Munz = Oenothera biennis [OENOBIE] Onagraceae (V:D)
Oenothera biennis var. *pycnocarpa* (Atkinson & Bartlett) Wieg. = Oenothera biennis [OENOBIE] Onagraceae (V:D)
Oenothera biennis var. *strigosa* (Rydb.) Piper = Oenothera villosa ssp. strigosa [OENOVIL1] Onagraceae (V:D)
Oenothera breviflora Torr. & Gray = Camissonia breviflora [CAMIBRE] Onagraceae (V:D)
Oenothera cheradophila Bartlett = Oenothera villosa ssp. strigosa [OENOVIL1] Onagraceae (V:D)
Oenothera contorta Dougl. ex Lehm. = Camissonia contorta [CAMICON] Onagraceae (V:D)
Oenothera cruciata (S. Wats.) Munz = Camissonia contorta [CAMICON] Onagraceae (V:D)

Oenothera densiflora Lindl. = Boisduvalia densiflora [BOISDEN] Onagraceae (V:D)
Oenothera depressa ssp. *strigosa* (Rydb.) Taylor & MacBryde = Oenothera villosa ssp. strigosa [OENOVIL1] Onagraceae (V:D)
Oenothera erythrosepala Borbás = Oenothera glazioviana [OENOGLA] Onagraceae (V:D)
Oenothera glabella Nutt. = Boisduvalia glabella [BOISGLA] Onagraceae (V:D)
*****Oenothera glazioviana** Micheli [OENOGLA] Onagraceae (V:D)
Oenothera muricata L. = Oenothera biennis [OENOBIE] Onagraceae (V:D)
Oenothera pallida Lindl. [OENOPAL] Onagraceae (V:D)
Oenothera pallida var. *idahoensis* Munz = Oenothera pallida [OENOPAL] Onagraceae (V:D)
Oenothera pallida var. *typica* Munz = Oenothera pallida [OENOPAL] Onagraceae (V:D)
*****Oenothera perennis** L. [OENOPER] Onagraceae (V:D)
Oenothera perennis var. *rectipilis* (Blake) Blake = Oenothera perennis [OENOPER] Onagraceae (V:D)
Oenothera perennis var. *typica* Munz = Oenothera perennis [OENOPER] Onagraceae (V:D)
Oenothera procera Woot. & Standl. = Oenothera villosa ssp. strigosa [OENOVIL1] Onagraceae (V:D)
Oenothera pycnocarpa Atkinson & Bartlett = Oenothera biennis [OENOBIE] Onagraceae (V:D)
Oenothera rydbergii House = Oenothera villosa ssp. strigosa [OENOVIL1] Onagraceae (V:D)
Oenothera strigosa (Rydb.) Mackenzie & Bush = Oenothera villosa ssp. strigosa [OENOVIL1] Onagraceae (V:D)
Oenothera strigosa ssp. *cheradophila* (Bartlett) Munz = Oenothera villosa ssp. strigosa [OENOVIL1] Onagraceae (V:D)
Oenothera villosa Thunb. [OENOVIL] Onagraceae (V:D)
Oenothera villosa ssp. *cheradophila* (Bartlett) W. Dietr. & Raven = Oenothera villosa ssp. strigosa [OENOVIL1] Onagraceae (V:D)
Oenothera villosa ssp. **strigosa** (Rydb.) W. Dietr. & Raven [OENOVIL1] Onagraceae (V:D)
Oenothera villosa var. *strigosa* (Rydb.) Dorn = Oenothera villosa ssp. strigosa [OENOVIL1] Onagraceae (V:D)
Oglifa arvensis (L.) Cass. = Filago arvensis [FILAARV] Asteraceae (V:D)
Oligosporus campestris ssp. *pacificus* (Nutt.) W.A. Weber = Artemisia campestris ssp. pacifica [ARTECAM2] Asteraceae (V:D)
Oligotrichum Lam. & DC. [OLIGOTR$] Polytrichaceae (B:M)
Oligotrichum aligerum Mitt. [OLIGALI] Polytrichaceae (B:M)
Oligotrichum hercynicum (Hedw.) Lam. & DC. [OLIGHER] Polytrichaceae (B:M)
Oligotrichum hercynicum var. *latifolium* C. Müll. & Kindb. in Mac. & Kindb. = Oligotrichum hercynicum [OLIGHER] Polytrichaceae (B:M)
Oligotrichum laevigatum var. *cavifolium* (Wils.) Frye in Grout = Psilopilum cavifolium [PSILCAV] Polytrichaceae (B:M)
Oligotrichum lyallii (Mitt.) Lindb. = Polytrichum lyallii [POLYLYA] Polytrichaceae (B:M)
Oligotrichum parallelum (Mitt.) Kindb. [OLIGPAR] Polytrichaceae (B:M)
Olsynium Raf. [OLSYNIU$] Iridaceae (V:M)
Olsynium douglasii (A. Dietr.) Bickn. [OLSYDOU] Iridaceae (V:M)
Olsynium douglasii var. **douglasii** [OLSYDOU0] Iridaceae (V:M)
Olsynium douglasii var. **inflatum** (Suksdorf) Cholewa & D. Henderson [OLSYDOU2] Iridaceae (V:M)
Omalia trichomanoides (Brid.) B.S.G. = Homalia trichomanoides [HOMATRI] Neckeraceae (B:M)
Omalotheca Cass. [OMALOTH$] Asteraceae (V:D)
*****Omalotheca sylvatica** (L.) Schultz-Bip. & F.W. Schultz [OMALSYL] Asteraceae (V:D)
Omphalina Quélet [OMPHALI$] Tricholomataceae (L)
Omphalina ericetorum (Pers.) M.T. Lange = Omphalina umbellifera [OMPHUMB] Tricholomataceae (L)
Omphalina hudsoniana (H.S. Jenn.) H.E. Bigelow [OMPHHUD] Tricholomataceae (L)
Omphalina umbellifera (L.) Quélet [OMPHUMB] Tricholomataceae (L)
Omphalodiscus decussatus (Vill.) Schol. = Umbilicaria decussata [UMBIDEC] Umbilicariaceae (L)
Omphalodiscus krascheninnikovii (Savicz) Schol. = Umbilicaria krascheninnikovii [UMBIKRA] Umbilicariaceae (L)
Omphalodiscus virginis (Schaerer) Schol. = Umbilicaria virginis [UMBIVIR] Umbilicariaceae (L)
Oncophorus (Brid.) Brid. [ONCOPHO$] Dicranaceae (B:M)
Oncophorus alpestris (Wahlenb.) Lindb. = Cynodontium alpestre [CYNOALP] Dicranaceae (B:M)
Oncophorus glaucescens Lindb. & Arnell = Cynodontium glaucescens [CYNOGLA] Dicranaceae (B:M)
Oncophorus jenneri (Schimp. in Howie) Williams = Cynodontium jenneri [CYNOJEN] Dicranaceae (B:M)
Oncophorus polycarpus (Hedw.) Brid. = Cynodontium polycarpon [CYNOPOL] Dicranaceae (B:M)
Oncophorus polycarpus var. *strumiferus* (Hedw.) Brid. = Cynodontium strumiferum [CYNOSTR] Dicranaceae (B:M)
Oncophorus schisti (Web. & Mohr) Lindb. = Cynodontium schisti [CYNOSCH] Dicranaceae (B:M)
Oncophorus strumulosus (C. Müll. & Kindb.) Britt. in Williams = Cynodontium alpestre [CYNOALP] Dicranaceae (B:M)
Oncophorus tenellus (Bruch & Schimp. in B.S.G.) Williams = Cynodontium tenellum [CYNOTEN] Dicranaceae (B:M)
Oncophorus virens (Hedw.) Brid. [ONCOVIR] Dicranaceae (B:M)
Oncophorus virens var. *nigrescens* Williams in Mac. = Oncophorus virens [ONCOVIR] Dicranaceae (B:M)
Oncophorus virens var. *serratus* (Bruch & Schimp. in B.S.G.) Braithw. = Oncophorus virens [ONCOVIR] Dicranaceae (B:M)
Oncophorus wahlenbergii Brid. [ONCOWAH] Dicranaceae (B:M)
Oncophorus wahlenbergii var. *compactus* (Bruch & Schimp. in B.S.G.) Braithw. = Oncophorus wahlenbergii [ONCOWAH] Dicranaceae (B:M)
Oncophorus wahlenbergii var. *gracilis* (Broth.) Arnell & C. Jens. = Oncophorus wahlenbergii [ONCOWAH] Dicranaceae (B:M)
Onobrychis P. Mill. [ONOBRYC$] Fabaceae (V:D)
*****Onobrychis viciifolia** Scop. [ONOBVIC] Fabaceae (V:D)
Onoclea struthiopteris (L.) Hoffmann p.p. = Matteuccia struthiopteris [MATTSTR] Dryopteridaceae (V:F)
Onoclea struthiopteris var. *pensylvanica* (Willd.) Boivin = Matteuccia struthiopteris [MATTSTR] Dryopteridaceae (V:F)
Onopordum L. [ONOPORD$] Asteraceae (V:D)
*****Onopordum acanthium** L. [ONOPACA] Asteraceae (V:D)
Onychium densum Brack. = Aspidotis densa [ASPIDEN] Pteridaceae (V:F)
Opegrapha Ach. [OPEGRAP$] Opegraphaceae (L)
Opegrapha betulina Sm. = Opegrapha herbarum [OPEGHER] Opegraphaceae (L)
Opegrapha diaphora (Ach.) Ach. = Opegrapha varia [OPEGVAR] Opegraphaceae (L)
Opegrapha gyrocarpa Flotow [OPEGGYR] Opegraphaceae (L)
Opegrapha herbarum Mont. [OPEGHER] Opegraphaceae (L)
Opegrapha herpetica (Ach.) Ach. = Opegrapha rufescens [OPEGRUF] Opegraphaceae (L)
Opegrapha lichenoides Pers. = Opegrapha varia [OPEGVAR] Opegraphaceae (L)
Opegrapha ochrocheila Nyl. [OPEGOCH] Opegraphaceae (L)
Opegrapha pulicaris auct. = Opegrapha varia [OPEGVAR] Opegraphaceae (L)
Opegrapha rimalis Pers. ex Ach. = Opegrapha varia [OPEGVAR] Opegraphaceae (L)
Opegrapha rufescens Pers. [OPEGRUF] Opegraphaceae (L)
Opegrapha sphaerophoricola Isbrand & Alstrup [OPEGSPH] Opegraphaceae (L)
Opegrapha varia Pers. [OPEGVAR] Opegraphaceae (L)
Opegrapha zonata Körber = Enterographa zonata [ENTEZON] Opegraphaceae (L)
Ophioglossum L. [OPHIOGL$] Ophioglossaceae (V:F)
Ophioglossum alaskanum Britton = Ophioglossum pusillum [OPHIPUS] Ophioglossaceae (V:F)
Ophioglossum pusillum Raf. [OPHIPUS] Ophioglossaceae (V:F)
Ophioglossum vulgatum auct. = Ophioglossum pusillum [OPHIPUS] Ophioglossaceae (V:F)

Ophioglossum vulgatum var. *alaskanum* (E.G. Britt.) C. Christens. = Ophioglossum pusillum [OPHIPUS] Ophioglossaceae (V:F)
Ophioglossum vulgatum var. *pseudopodum* (Blake) Farw. = Ophioglossum pusillum [OPHIPUS] Ophioglossaceae (V:F)
Ophioparma Norman [OPHIOPA$] Ophioparmaceae (L)
Ophioparma lapponica (Räsänen) Hafellner & R.W. Rogers [OPHILAP] Ophioparmaceae (L)
Ophioparma rubricosa (Müll. Arg.) S. Ekman [OPHIRUB] Ophioparmaceae (L)
Ophrys caurina (Piper) Rydb. = Listera caurina [LISTCAU] Orchidaceae (V:M)
Ophrys convallarioides (Sw.) W. Wight = Listera convallarioides [LISTCON] Orchidaceae (V:M)
Oplopanax (Torr. & Gray) Miq. [OPLOPAN$] Araliaceae (V:D)
Oplopanax horridus Miq. [OPLOHOR] Araliaceae (V:D)
Opuntia P. Mill. [OPUNTIA$] Cactaceae (V:D)
Opuntia fragilis (Nutt.) Haw. [OPUNFRA] Cactaceae (V:D)
Opuntia polyacantha Haw. [OPUNPOL] Cactaceae (V:D)
Orchis rotundifolia Banks ex Pursh = Amerorchis rotundifolia [AMERROT] Orchidaceae (V:M)
Orchis rotundifolia var. *lineata* Mousley = Amerorchis rotundifolia [AMERROT] Orchidaceae (V:M)
Oreas Brid. [OREAS$] Dicranaceae (B:M)
Oreas martiana (Hoppe & Hornsch. in Hornsch.) Brid. [OREAMAR] Dicranaceae (B:M)
Oreobroma pygmaeum (Gray) T.J. Howell = Lewisia pygmaea [LEWIPYG] Portulacaceae (V:D)
Oreocarya celosioides Eastw. = Cryptantha celosioides [CRYPCEL] Boraginaceae (V:D)
Oreocarya glomerata (Pursh) Greene = Cryptantha celosioides [CRYPCEL] Boraginaceae (V:D)
Oreocarya macounii Eastw. = Cryptantha celosioides [CRYPCEL] Boraginaceae (V:D)
Oreocarya nubigena Greene = Cryptantha nubigena [CRYPNUB] Boraginaceae (V:D)
Oreocarya sheldonii Brand = Cryptantha celosioides [CRYPCEL] Boraginaceae (V:D)
Oreopteris quelpaertensis (Christ) Holub = Thelypteris quelpaertensis [THELQUE] Thelypteridaceae (V:F)
Oreosedum album (L.) Grulich = Sedum album [SEDUALB] Crassulaceae (V:D)
Origanum L. [ORIGANU$] Lamiaceae (V:D)
***Origanum vulgare** L. [ORIGVUL] Lamiaceae (V:D)
Ornithogalum L. [ORNITHO$] Liliaceae (V:M)
***Ornithogalum umbellatum** L. [ORNIUMB] Liliaceae (V:M)
Orobanche L. [OROBANC$] Orobanchaceae (V:D)
Orobanche californica Cham. & Schlecht. [OROBCAL] Orobanchaceae (V:D)
Orobanche corymbosa (Rydb.) Ferris [OROBCOR] Orobanchaceae (V:D)
Orobanche corymbosa ssp. **mutabilis** Heckard [OROBCOR1] Orobanchaceae (V:D)
Orobanche fasciculata Nutt. [OROBFAS] Orobanchaceae (V:D)
Orobanche fasciculata var. *franciscana* Achey = Orobanche fasciculata [OROBFAS] Orobanchaceae (V:D)
Orobanche fasciculata var. *lutea* (Parry) Achey = Orobanche fasciculata [OROBFAS] Orobanchaceae (V:D)
Orobanche fasciculata var. *subulata* Goodman = Orobanche fasciculata [OROBFAS] Orobanchaceae (V:D)
Orobanche fasciculata var. *typica* Achey = Orobanche fasciculata [OROBFAS] Orobanchaceae (V:D)
Orobanche grayana var. *nelsonii* Munz = Orobanche californica [OROBCAL] Orobanchaceae (V:D)
Orobanche grayana var. *violacea* (Eastw.) Munz = Orobanche californica [OROBCAL] Orobanchaceae (V:D)
Orobanche pinorum Geyer ex Hook. [OROBPIN] Orobanchaceae (V:D)
Orobanche porphyrantha G. Beck = Orobanche uniflora [OROBUNI] Orobanchaceae (V:D)
Orobanche purpurea Jacq. = Orobanche uniflora [OROBUNI] Orobanchaceae (V:D)
Orobanche terrae-novae Fern. = Orobanche uniflora [OROBUNI] Orobanchaceae (V:D)
Orobanche uniflora L. [OROBUNI] Orobanchaceae (V:D)
Orobanche uniflora ssp. *occidentalis* (Greene) Abrams ex Ferris = Orobanche uniflora [OROBUNI] Orobanchaceae (V:D)
Orobanche uniflora var. *minuta* (Suksdorf) G. Beck = Orobanche uniflora [OROBUNI] Orobanchaceae (V:D)
Orobanche uniflora var. *occidentalis* (Greene) Taylor & MacBryde = Orobanche uniflora [OROBUNI] Orobanchaceae (V:D)
Orobanche uniflora var. *purpurea* (Heller) Achey = Orobanche uniflora [OROBUNI] Orobanchaceae (V:D)
Orobanche uniflora var. *sedii* (Suksdorf) Achey = Orobanche uniflora [OROBUNI] Orobanchaceae (V:D)
Orobanche uniflora var. *terrae-novae* (Fern.) Munz = Orobanche uniflora [OROBUNI] Orobanchaceae (V:D)
Orobanche uniflora var. *typica* Achey = Orobanche uniflora [OROBUNI] Orobanchaceae (V:D)
Orobus myrtifolius (Muhl. ex Willd.) A. Hall = Lathyrus palustris [LATHPAL] Fabaceae (V:D)
Orphniospora Körber [ORPHNIO$] Fuscideaceae (L)
Orphniospora atrata (Sm.) Poelt = Orphniospora moriopsis [ORPHMOR] Fuscideaceae (L)
Orphniospora moriopsis (A. Massal.) D. Hawksw. [ORPHMOR] Fuscideaceae (L)
Orthilia Raf. [ORTHILI$] Pyrolaceae (V:D)
Orthilia secunda (L.) House [ORTHSEC] Pyrolaceae (V:D)
Orthilia secunda ssp. *obtusata* (Turcz.) Böcher = Orthilia secunda [ORTHSEC] Pyrolaceae (V:D)
Orthilia secunda var. *obtusata* (Turcz.) House = Orthilia secunda [ORTHSEC] Pyrolaceae (V:D)
Orthocarpus Nutt. [ORTHOCA$] Scrophulariaceae (V:D)
Orthocarpus attenuatus Gray = Castilleja attenuata [CASTATT] Scrophulariaceae (V:D)
Orthocarpus bracteosus Benth. [ORTHBRA] Scrophulariaceae (V:D)
Orthocarpus bracteosus var. *albus* Keck = Orthocarpus bracteosus [ORTHBRA] Scrophulariaceae (V:D)
Orthocarpus castillejoides Benth. = Castilleja ambigua [CASTAMB] Scrophulariaceae (V:D)
Orthocarpus faucibarbatus ssp. *albidus* Keck = Triphysaria versicolor [TRIPVER] Scrophulariaceae (V:D)
Orthocarpus faucibarbatus var. *albidus* (Keck) J.T. Howell = Triphysaria versicolor [TRIPVER] Scrophulariaceae (V:D)
Orthocarpus hispidus Benth. = Castilleja tenuis [CASTTEN] Scrophulariaceae (V:D)
Orthocarpus imbricatus Torr. ex S. Wats. [ORTHIMB] Scrophulariaceae (V:D)
Orthocarpus luteus Nutt. [ORTHLUT] Scrophulariaceae (V:D)
Orthocarpus pusillus Benth. = Triphysaria pusilla [TRIPPUS] Scrophulariaceae (V:D)
Orthocarpus pusillus var. *densiuscuus* (Gandog.) Keck = Triphysaria pusilla [TRIPPUS] Scrophulariaceae (V:D)
Orthocarpus tenuifolius (Pursh) Benth. [ORTHTEN] Scrophulariaceae (V:D)
Orthocaulis attenuatus (Mart.) Evans = Barbilophozia attenuata [BARBATT] Jungermanniaceae (B:H)
Orthocaulis binsteadii (Kaal.) Buch = Barbilophozia binsteadii [BARBBIN] Jungermanniaceae (B:H)
Orthocaulis floerkei (Web. & Mohr) Buch = Barbilophozia floerkei [BARBFLO] Jungermanniaceae (B:H)
Orthocaulis gracilis (Schleich.) Buch = Barbilophozia attenuata [BARBATT] Jungermanniaceae (B:H)
Orthocaulis kunzeana (Hüb.) Buch = Barbilophozia kunzeana [BARBKUN] Jungermanniaceae (B:H)
Orthocaulis quadrilobus (Lindb.) Buch = Barbilophozia quadriloba [BARBQUA] Jungermanniaceae (B:H)
Orthodicranum flagellare (Hedw.) Loeske = Dicranum flagellare [DICRFLA] Dicranaceae (B:M)
Orthodicranum montanum (Hedw.) Loeske = Dicranum montanum [DICRMON] Dicranaceae (B:M)
Orthothecium Schimp. in B.S.G. [ORTHOTH$] Hypnaceae (B:M)
Orthothecium chryseum (Schwaegr. in Schultes) Schimp. in B.S.G. [ORTHCHR] Hypnaceae (B:M)
Orthothecium intricatum (Hartm.) Schimp. in B.S.G. [ORTHINT] Hypnaceae (B:M)

Orthothecium rubellum (Mitt.) Kindb. = Orthothecium strictum [ORTHSTR] Hypnaceae (B:M)
Orthothecium strictum Lor. [ORTHSTR] Hypnaceae (B:M)
Orthotrichum Hedw. [ORTHOTR$] Orthotrichaceae (B:M)
Orthotrichum affine Brid. [ORTHAFF] Orthotrichaceae (B:M)
Orthotrichum alpestre Hornsch. in B.S.G. [ORTHALP] Orthotrichaceae (B:M)
Orthotrichum alpestre var. *majus* Lesq. & James = Orthotrichum alpestre [ORTHALP] Orthotrichaceae (B:M)
Orthotrichum alpestre var. *occidentale* (James) Grout = Orthotrichum alpestre [ORTHALP] Orthotrichaceae (B:M)
Orthotrichum alpestre var. *watsonii* (James) Grout = Orthotrichum alpestre [ORTHALP] Orthotrichaceae (B:M)
Orthotrichum anomalum Hedw. [ORTHANO] Orthotrichaceae (B:M)
Orthotrichum anomalum var. *americanum* Vent. in Mac. & Kindb. = Orthotrichum anomalum [ORTHANO] Orthotrichaceae (B:M)
Orthotrichum anomalum var. *saxatile* Milde = Orthotrichum anomalum [ORTHANO] Orthotrichaceae (B:M)
Orthotrichum arcticum Schimp. = Orthotrichum pylaisii [ORTHPYL] Orthotrichaceae (B:M)
Orthotrichum blyttii Schimp. = Orthotrichum pylaisii [ORTHPYL] Orthotrichaceae (B:M)
Orthotrichum canadense Bruch & Schimp. = Orthotrichum anomalum [ORTHANO] Orthotrichaceae (B:M)
Orthotrichum columbicum Mitt. = Orthotrichum consimile [ORTHCON] Orthotrichaceae (B:M)
Orthotrichum consimile Mitt. [ORTHCON] Orthotrichaceae (B:M)
Orthotrichum consimile ssp. *anomaloides* Kindb. = Orthotrichum anomalum [ORTHANO] Orthotrichaceae (B:M)
Orthotrichum cupulatum Brid. [ORTHCUP] Orthotrichaceae (B:M)
Orthotrichum cupulatum var. *jamesianum* (Sull. in James in Watson) Lawt. = Orthotrichum pellucidum [ORTHPEU] Orthotrichaceae (B:M)
Orthotrichum cylindricarpum Lesq. in Jaeg. = Orthotrichum tenellum [ORTHTNE] Orthotrichaceae (B:M)
Orthotrichum diaphanum Brid. [ORTHDIA] Orthotrichaceae (B:M)
Orthotrichum elegans Schwaegr. ex Grev. = Orthotrichum speciosum var. elegans [ORTHSPE1] Orthotrichaceae (B:M)
Orthotrichum euryphyllum Vent. in Röll = Orthotrichum rivulare [ORTHRIV] Orthotrichaceae (B:M)
Orthotrichum garrettii Grout & Flow. in Grout = Orthotrichum diaphanum [ORTHDIA] Orthotrichaceae (B:M)
Orthotrichum hallii Sull. & Lesq. in Sull. [ORTHHAL] Orthotrichaceae (B:M)
Orthotrichum jamesianum Sull. in James in Watson = Orthotrichum pellucidum [ORTHPEU] Orthotrichaceae (B:M)
Orthotrichum killiasii C. Müll. = Orthotrichum speciosum var. speciosum [ORTHSPE0] Orthotrichaceae (B:M)
Orthotrichum kingianum Lesq. = Orthotrichum laevigatum [ORTHLAE] Orthotrichaceae (B:M)
Orthotrichum laevigatum Zett. [ORTHLAE] Orthotrichaceae (B:M)
Orthotrichum laevigatum f. *macounii* (Aust.) Lawt. & Vitt in Lawt. = Orthotrichum laevigatum [ORTHLAE] Orthotrichaceae (B:M)
Orthotrichum laevigatum var. *kingianum* (Lesq.) Grout = Orthotrichum laevigatum [ORTHLAE] Orthotrichaceae (B:M)
Orthotrichum lonchothecium C. Müll. & Kindb. in Mac. & Kindb. = Orthotrichum laevigatum [ORTHLAE] Orthotrichaceae (B:M)
Orthotrichum lyellii Hook. & Tayl. [ORTHLYE] Orthotrichaceae (B:M)
Orthotrichum lyellii var. *papillosum* (Hampe) Sull. = Orthotrichum lyellii [ORTHLYE] Orthotrichaceae (B:M)
Orthotrichum lyelloides Kindb. = Orthotrichum lyellii [ORTHLYE] Orthotrichaceae (B:M)
Orthotrichum macounii Aust. = Orthotrichum laevigatum [ORTHLAE] Orthotrichaceae (B:M)
Orthotrichum macounii var. *lonchothecium* (C. Müll. & Kindb. in Mac. & Kindb.) Grout = Orthotrichum laevigatum [ORTHLAE] Orthotrichaceae (B:M)
Orthotrichum microblepharum Schimp. = Orthotrichum pylaisii [ORTHPYL] Orthotrichaceae (B:M)
Orthotrichum nudum var. *rudolphianum* Vent. in Husn. = Orthotrichum cupulatum [ORTHCUP] Orthotrichaceae (B:M)
Orthotrichum obtusifolium Brid. [ORTHOBT] Orthotrichaceae (B:M)
Orthotrichum pallens Bruch ex Brid. [ORTHPAL] Orthotrichaceae (B:M)
Orthotrichum papillosum Hampe = Orthotrichum lyellii [ORTHLYE] Orthotrichaceae (B:M)
Orthotrichum papillosum ssp. *strictum* (Vent.) Kindb. = Orthotrichum lyellii [ORTHLYE] Orthotrichaceae (B:M)
Orthotrichum pellucidum Lindb. [ORTHPEU] Orthotrichaceae (B:M)
Orthotrichum psilothecium C. Müll. & Kindb. in Mac. & Kindb. = Orthotrichum speciosum var. elegans [ORTHSPE1] Orthotrichaceae (B:M)
Orthotrichum pulchellum Brunt. in Winch. & Gateh. [ORTHPUL] Orthotrichaceae (B:M)
Orthotrichum pulchellum ssp. *ulotaeforme* (Ren. & Card.) Kindb. = Orthotrichum consimile [ORTHCON] Orthotrichaceae (B:M)
Orthotrichum pulchellum var. *columbicum* Grout = Orthotrichum consimile [ORTHCON] Orthotrichaceae (B:M)
Orthotrichum pulchellum var. *groutii* Lawt. = Orthotrichum pulchellum [ORTHPUL] Orthotrichaceae (B:M)
Orthotrichum pylaisii Brid. [ORTHPYL] Orthotrichaceae (B:M)
Orthotrichum rivulare Turn. [ORTHRIV] Orthotrichaceae (B:M)
Orthotrichum roellii Vent. in Röll = Orthotrichum laevigatum [ORTHLAE] Orthotrichaceae (B:M)
Orthotrichum rupestre Schleich. ex Schwaegr. [ORTHRUP] Orthotrichaceae (B:M)
Orthotrichum rupestre ssp. *sturmii* (Hoppe & Hornsch.) Boul. = Orthotrichum rupestre [ORTHRUP] Orthotrichaceae (B:M)
Orthotrichum rupestre var. *globosum* (Lesq.) Grout = Orthotrichum rupestre [ORTHRUP] Orthotrichaceae (B:M)
Orthotrichum rupestre var. *macfaddenae* (Williams) Grout = Orthotrichum rupestre [ORTHRUP] Orthotrichaceae (B:M)
Orthotrichum speciosum Nees in Sturm [ORTHSPE] Orthotrichaceae (B:M)
Orthotrichum speciosum var. **elegans** (Schwaegr. ex Hook. & Grev.) Warnst. [ORTHSPE1] Orthotrichaceae (B:M)
Orthotrichum speciosum var. *hainesiae* (Aust.) Par. = Orthotrichum laevigatum [ORTHLAE] Orthotrichaceae (B:M)
Orthotrichum speciosum var. *killiasii* (C. Müll.) Vent. = Orthotrichum speciosum var. speciosum [ORTHSPE0] Orthotrichaceae (B:M)
Orthotrichum speciosum var. **speciosum** [ORTHSPE0] Orthotrichaceae (B:M)
Orthotrichum sprucei auct. = Orthotrichum rivulare [ORTHRIV] Orthotrichaceae (B:M)
Orthotrichum striatum Hedw. [ORTHSTI] Orthotrichaceae (B:M)
Orthotrichum tenellum Bruch ex Brid. [ORTHTNE] Orthotrichaceae (B:M)
Orthotrichum tenellum var. *coulteri* (Mitt.) Grout = Orthotrichum tenellum [ORTHTNE] Orthotrichaceae (B:M)
Orthotrichum texanum Sull. & Lesq. = Orthotrichum rupestre [ORTHRUP] Orthotrichaceae (B:M)
Orthotrichum winteri Schimp. = Orthotrichum consimile [ORTHCON] Orthotrichaceae (B:M)
Oryzopsis Michx. [ORYZOPS$] Poaceae (V:M)
Oryzopsis asperifolia Michx. [ORYZASP] Poaceae (V:M)
Oryzopsis canadensis (Poir.) Torr. [ORYZCAN] Poaceae (V:M)
Oryzopsis exigua Thurb. [ORYZEXI] Poaceae (V:M)
Oryzopsis hymenoides (Roemer & J.A. Schultes) Ricker ex Piper [ORYZHYM] Poaceae (V:M)
Oryzopsis micrantha (Trin. & Rupr.) Thurb. [ORYZMIC] Poaceae (V:M)
Oryzopsis pungens (Torr. ex Spreng.) A.S. Hitchc. [ORYZPUN] Poaceae (V:M)
Osmaronia cerasiformis (Torr. & Gray ex Hook. & Arn.) Greene = Oemleria cerasiformis [OEMLCER] Rosaceae (V:D)
Osmorhiza Raf. [OSMORHI$] Apiaceae (V:D)
Osmorhiza berteroi DC. [OSMOBER] Apiaceae (V:D)
Osmorhiza chilensis Hook. & Arn. = Osmorhiza berteroi [OSMOBER] Apiaceae (V:D)
Osmorhiza chilensis var. *cupressimontana* (Boivin) Boivin = Osmorhiza depauperata [OSMODEP] Apiaceae (V:D)

Osmorhiza chilensis var. *purpurea* (Coult. & Rose) Boivin = Osmorhiza purpurea [OSMOPUR] Apiaceae (V:D)
Osmorhiza depauperata Phil. [OSMODEP] Apiaceae (V:D)
Osmorhiza divaricata (Britt.) Suksdorf = Osmorhiza berteroi [OSMOBER] Apiaceae (V:D)
Osmorhiza nuda Torr. = Osmorhiza berteroi [OSMOBER] Apiaceae (V:D)
Osmorhiza obtusa (Coult. & Rose) Fern. = Osmorhiza depauperata [OSMODEP] Apiaceae (V:D)
Osmorhiza occidentalis (Nutt. ex Torr. & Gray) Torr. [OSMOOCC] Apiaceae (V:D)
Osmorhiza purpurea (Coult. & Rose) Suksdorf [OSMOPUR] Apiaceae (V:D)
Osmunda spicant L. = Blechnum spicant [BLECSPI] Blechnaceae (V:F)
Oxalis L. [OXALIS$] Oxalidaceae (V:D)
Oxalis acetosella ssp. *oregana* (Nutt.) D. Löve = Oxalis oregana [OXALORE] Oxalidaceae (V:D)
Oxalis bushii (Small) Small = Oxalis stricta [OXALSTR] Oxalidaceae (V:D)
Oxalis coloradensis Rydb. = Oxalis stricta [OXALSTR] Oxalidaceae (V:D)
*****Oxalis corniculata** L. [OXALCOR] Oxalidaceae (V:D)
Oxalis corniculata var. *atropurpurea* Planch. = Oxalis corniculata [OXALCOR] Oxalidaceae (V:D)
Oxalis corniculata var. *langloisii* (Small) Wieg. = Oxalis corniculata [OXALCOR] Oxalidaceae (V:D)
Oxalis corniculata var. *lupulina* (R. Knuth) Zucc. = Oxalis corniculata [OXALCOR] Oxalidaceae (V:D)
Oxalis corniculata var. *macrophylla* Arsene ex R. Knuth = Oxalis corniculata [OXALCOR] Oxalidaceae (V:D)
Oxalis corniculata var. *minor* Laing = Oxalis corniculata [OXALCOR] Oxalidaceae (V:D)
Oxalis corniculata var. *reptans* Laing = Oxalis corniculata [OXALCOR] Oxalidaceae (V:D)
Oxalis corniculata var. *villosa* (Bieb.) Hohen. = Oxalis corniculata [OXALCOR] Oxalidaceae (V:D)
Oxalis corniculata var. *viscidula* Wieg. = Oxalis corniculata [OXALCOR] Oxalidaceae (V:D)
Oxalis cymosa Small = Oxalis stricta [OXALSTR] Oxalidaceae (V:D)
Oxalis europaea Jord. = Oxalis stricta [OXALSTR] Oxalidaceae (V:D)
Oxalis europaea var. *bushii* (Small) Wieg. = Oxalis stricta [OXALSTR] Oxalidaceae (V:D)
Oxalis europaea var. *rufa* (Small) Young = Oxalis stricta [OXALSTR] Oxalidaceae (V:D)
Oxalis fontana Bunge = Oxalis stricta [OXALSTR] Oxalidaceae (V:D)
Oxalis fontana var. *bushii* (Small) Hara = Oxalis stricta [OXALSTR] Oxalidaceae (V:D)
Oxalis interior (Small) Fedde = Oxalis stricta [OXALSTR] Oxalidaceae (V:D)
Oxalis langloisii (Small) Fedde = Oxalis corniculata [OXALCOR] Oxalidaceae (V:D)
Oxalis oregana Nutt. [OXALORE] Oxalidaceae (V:D)
Oxalis oregana var. *smallii* (R. Knuth) M.E. Peck = Oxalis oregana [OXALORE] Oxalidaceae (V:D)
Oxalis pusilla Salisb. = Oxalis corniculata [OXALCOR] Oxalidaceae (V:D)
Oxalis repens Thunb. = Oxalis corniculata [OXALCOR] Oxalidaceae (V:D)
Oxalis rufa Small = Oxalis stricta [OXALSTR] Oxalidaceae (V:D)
*****Oxalis stricta** L. [OXALSTR] Oxalidaceae (V:D)
Oxalis stricta var. *decumbens* Bitter = Oxalis stricta [OXALSTR] Oxalidaceae (V:D)
Oxalis stricta var. *piletocarpa* Wieg. = Oxalis stricta [OXALSTR] Oxalidaceae (V:D)
Oxalis stricta var. *rufa* (Small) Farw. = Oxalis stricta [OXALSTR] Oxalidaceae (V:D)
Oxalis stricta var. *villicaulis* (Wieg.) Farw. = Oxalis stricta [OXALSTR] Oxalidaceae (V:D)
Oxalis villosa Bieb. = Oxalis corniculata [OXALCOR] Oxalidaceae (V:D)
Oxybaphus hirsutus (Pursh) Sweet = Mirabilis hirsuta [MIRAHIR] Nyctaginaceae (V:D)
Oxybaphus nyctagineus (Michx.) Sweet = Mirabilis nyctaginea [MIRANYC] Nyctaginaceae (V:D)
Oxycoccus hagerupii A. & D. Löve = Vaccinium oxycoccos [VACCOXY] Ericaceae (V:D)
Oxycoccus intermedius (Gray) Rydb. = Vaccinium oxycoccos [VACCOXY] Ericaceae (V:D)
Oxycoccus microcarpos Turcz. ex Rupr. = Vaccinium oxycoccos [VACCOXY] Ericaceae (V:D)
Oxycoccus ovalifolius (Michx.) Porsild = Vaccinium oxycoccos [VACCOXY] Ericaceae (V:D)
Oxycoccus oxycoccos (L.) Adolphi = Vaccinium oxycoccos [VACCOXY] Ericaceae (V:D)
Oxycoccus oxycoccos (L.) MacM. = Vaccinium oxycoccos [VACCOXY] Ericaceae (V:D)
Oxycoccus palustris Pers. = Vaccinium oxycoccos [VACCOXY] Ericaceae (V:D)
Oxycoccus palustris ssp. *microphyllus* (Lange) A. & D. Löve = Vaccinium oxycoccos [VACCOXY] Ericaceae (V:D)
Oxycoccus palustris var. *intermedius* (Gray) T.J. Howell = Vaccinium oxycoccos [VACCOXY] Ericaceae (V:D)
Oxycoccus palustris var. *ovalifolius* (Michx.) Seymour = Vaccinium oxycoccos [VACCOXY] Ericaceae (V:D)
Oxycoccus quadripetalus Gilib. = Vaccinium oxycoccos [VACCOXY] Ericaceae (V:D)
Oxycoccus quadripetalus var. *microphyllus* (Lange) Porsild = Vaccinium oxycoccos [VACCOXY] Ericaceae (V:D)
Oxyria Hill [OXYRIA$] Polygonaceae (V:D)
Oxyria digyna (L.) Hill [OXYRDIG] Polygonaceae (V:D)
Oxyrrhynchium praelongum (Hedw.) Warnst. = Eurhynchium praelongum [EURHPRA] Brachytheciaceae (B:M)
Oxyrrhynchium praelongum var. *californicum* (Grout) Wijk & Marg. = Eurhynchium praelongum [EURHPRA] Brachytheciaceae (B:M)
Oxyrrhynchium praelongum var. *stokesii* (Turn.) Podp. = Eurhynchium praelongum [EURHPRA] Brachytheciaceae (B:M)
Oxyrrhynchium riparioides (Hedw.) Jenn. = Platyhypnidium riparioides [PLATRIP] Brachytheciaceae (B:M)
Oxyrrhynchium rusciforme Warnst. = Platyhypnidium riparioides [PLATRIP] Brachytheciaceae (B:M)
Oxyrrhynchium rusciforme var. *complanatum* (Limpr.) Warnst. = Platyhypnidium riparioides [PLATRIP] Brachytheciaceae (B:M)
Oxystegus (Limpr.) Hilp. [OXYSTEG$] Pottiaceae (B:M)
Oxystegus recurvifolius (Tayl.) Zand. = Paraleptodontium recurvifolium [PARAREC] Pottiaceae (B:M)
Oxystegus tenuirostris (Hook. & Tayl.) A. J. E. Sm. [OXYSTEN] Pottiaceae (B:M)
Oxytropis DC. [OXYTROP$] Fabaceae (V:D)
Oxytropis alaskana A. Nels. = Oxytropis campestris var. varians [OXYTCAM7] Fabaceae (V:D)
Oxytropis alpicola (Rydb.) M.E. Jones = Oxytropis campestris var. cusickii [OXYTCAM2] Fabaceae (V:D)
Oxytropis arctica R. Br. [OXYTARC] Fabaceae (V:D)
Oxytropis campestris (L.) DC. [OXYTCAM] Fabaceae (V:D)
Oxytropis campestris ssp. *gracilis* (A. Nels.) Hultén = Oxytropis monticola [OXYTMON] Fabaceae (V:D)
Oxytropis campestris ssp. *jordalii* (Porsild) Hultén = Oxytropis jordalii ssp. jordalii [OXYTJOR0] Fabaceae (V:D)
Oxytropis campestris ssp. *melanocephala* Hook. = Oxytropis maydelliana [OXYTMAY] Fabaceae (V:D)
Oxytropis campestris var. *cervinus* (Greene) Boivin = Oxytropis monticola [OXYTMON] Fabaceae (V:D)
Oxytropis campestris var. **columbiana** (St. John) Barneby [OXYTCAM8] Fabaceae (V:D)
Oxytropis campestris var. **cusickii** (Greenm.) Barneby [OXYTCAM2] Fabaceae (V:D)
Oxytropis campestris var. *davisii* Welsh = Oxytropis jordalii ssp. davisii [OXYTJOR1] Fabaceae (V:D)
Oxytropis campestris var. *glabrata* Hook. = Oxytropis maydelliana [OXYTMAY] Fabaceae (V:D)
Oxytropis campestris var. *gracilis* (A. Nels.) Barneby = Oxytropis monticola [OXYTMON] Fabaceae (V:D)

Oxytropis campestris var. *jordalii* (Porsild) Welsh = Oxytropis jordalii ssp. jordalii [OXYTJOR0] Fabaceae (V:D)
Oxytropis campestris var. *rydbergii* (A. Nels.) R.J. Davis = Oxytropis campestris var. cusickii [OXYTCAM2] Fabaceae (V:D)
Oxytropis campestris var. **varians** (Rydb.) Barneby [OXYTCAM7] Fabaceae (V:D)
Oxytropis columbiana St. John = Oxytropis campestris var. columbiana [OXYTCAM8] Fabaceae (V:D)
Oxytropis cusickii Greenm. = Oxytropis campestris var. cusickii [OXYTCAM2] Fabaceae (V:D)
Oxytropis deflexa (Pallas) DC. [OXYTDEF] Fabaceae (V:D)
Oxytropis glabrata (Hook.) A. Nels. = Oxytropis maydelliana [OXYTMAY] Fabaceae (V:D)
Oxytropis gracilis (A. Nels.) K. Schum. = Oxytropis monticola [OXYTMON] Fabaceae (V:D)
Oxytropis huddelsonii Porsild [OXYTHUD] Fabaceae (V:D)
Oxytropis hyperborea Porsild = Oxytropis campestris var. varians [OXYTCAM7] Fabaceae (V:D)
Oxytropis jordalii Porsild [OXYTJOR] Fabaceae (V:D)
Oxytropis jordalii ssp. **davisii** (Welsh) Elisens & Packer [OXYTJOR1] Fabaceae (V:D)
Oxytropis jordalii ssp. **jordalii** [OXYTJOR0] Fabaceae (V:D)
Oxytropis leucantha var. *depressus* (Rydb.) Boivin = Oxytropis viscida [OXYTVIS] Fabaceae (V:D)
Oxytropis luteola (Greene) Piper & Beattie = Oxytropis monticola [OXYTMON] Fabaceae (V:D)
Oxytropis macounii (Greene) Rydb. = Oxytropis sericea var. spicata [OXYTSER1] Fabaceae (V:D)
Oxytropis maydelliana Trautv. [OXYTMAY] Fabaceae (V:D)
Oxytropis maydelliana ssp. *melanocephala* (Hook.) Porsild = Oxytropis maydelliana [OXYTMAY] Fabaceae (V:D)
Oxytropis monticola Gray [OXYTMON] Fabaceae (V:D)
Oxytropis nigrescens (Pallas) Fisch. ex DC. [OXYTNIG] Fabaceae (V:D)
Oxytropis podocarpa Gray [OXYTPOD] Fabaceae (V:D)
Oxytropis podocarpa var. *inflata* (Hook.) Boivin = Oxytropis podocarpa [OXYTPOD] Fabaceae (V:D)
Oxytropis richardsonii (Hook.) K. Schum. = Oxytropis splendens [OXYTSPL] Fabaceae (V:D)
Oxytropis rydbergii A. Nels. = Oxytropis campestris var. cusickii [OXYTCAM2] Fabaceae (V:D)
Oxytropis scammaniana Hultén [OXYTSCA] Fabaceae (V:D)
Oxytropis sericea Nutt. [OXYTSER] Fabaceae (V:D)
Oxytropis sericea var. **spicata** (Hook.) Barneby [OXYTSER1] Fabaceae (V:D)
Oxytropis spicata (Hook.) Standl. = Oxytropis sericea var. spicata [OXYTSER1] Fabaceae (V:D)
Oxytropis splendens Dougl. ex Hook. [OXYTSPL] Fabaceae (V:D)
Oxytropis splendens var. *richardsonii* Hook. = Oxytropis splendens [OXYTSPL] Fabaceae (V:D)
Oxytropis splendens var. *vestita* Hook. = Oxytropis splendens [OXYTSPL] Fabaceae (V:D)
Oxytropis varians (Rydb.) K. Schum. = Oxytropis campestris var. varians [OXYTCAM7] Fabaceae (V:D)
Oxytropis villosa (Rydb.) K. Schum. = Oxytropis monticola [OXYTMON] Fabaceae (V:D)
Oxytropis viscida Nutt. [OXYTVIS] Fabaceae (V:D)

P

Pachistima myrsinites (Pursh) Raf. = Paxistima myrsinites [PAXIMYR] Celastraceae (V:D)
Pachyospora mutabilis (Ach.) A. Massal. = Megaspora verrucosa [MEGAVEU] Hymeneliaceae (L)
Pachyospora verrucosa (Ach.) A. Massal. = Megaspora verrucosa [MEGAVEU] Hymeneliaceae (L)
Packera cana (Hook.) W.A. Weber & A. Löve = Senecio canus [SENECAN] Asteraceae (V:D)
Packera cymbalarioides (Buek) W.A. Weber & A. Löve = Senecio cymbalarioides [SENECYM] Asteraceae (V:D)
Packera hyperborealis (Greenm.) A. & D. Löve = Senecio hyperborealis [SENEHYP] Asteraceae (V:D)
Packera indecora (Greene) A. & D. Löve = Senecio indecorus [SENEIND] Asteraceae (V:D)
Packera ogotorukensis (Packer) A. & D. Löve = Senecio ogotorukensis [SENEOGO] Asteraceae (V:D)
Packera pauciflora (Pursh) A. & D. Löve = Senecio pauciflorus [SENEPAU] Asteraceae (V:D)
Packera paupercula (Michx.) A. & D. Löve = Senecio pauperculus [SENEPAP] Asteraceae (V:D)
Packera pseudaurea (Rydb.) W.A. Weber & A. Löve = Senecio pseudaureus [SENEPSE] Asteraceae (V:D)
Padus melanocarpa (A. Nels.) Shafer = Prunus virginiana var. melanocarpa [PRUNVIR2] Rosaceae (V:D)
Padus virginiana ssp. *melanocarpa* (A. Nels.) W.A. Weber = Prunus virginiana var. melanocarpa [PRUNVIR2] Rosaceae (V:D)
Paludella Brid. [PALUDEL$] Meesiaceae (B:M)
Paludella squarrosa (Hedw.) Brid. [PALUSQU] Meesiaceae (B:M)
Palustriella Ochyra [PALUSTR$] Amblystegiaceae (B:M)
Palustriella commutata (Brid.) Ochyra [PALUCOM] Amblystegiaceae (B:M)
Panicularia borealis Nash = Glyceria borealis [GLYCBOR] Poaceae (V:M)
Panicularia canadensis (Michx.) Kuntze = Glyceria canadensis [GLYCCAN] Poaceae (V:M)
Panicum L. [PANICUM$] Poaceae (V:M)
Panicum acuminatum var. *fasciculatum* (Torr.) Lelong = Dichanthelium acuminatum var. fasciculatum [DICHACU1] Poaceae (V:M)
Panicum acuminatum var. *implicatum* (Scribn.) C.F. Reed = Dichanthelium acuminatum var. fasciculatum [DICHACU1] Poaceae (V:M)
Panicum barbipulvinatum Nash = Panicum capillare [PANICAP] Poaceae (V:M)
Panicum brodiei St. John = Dichanthelium acuminatum var. fasciculatum [DICHACU1] Poaceae (V:M)
Panicum capillare L. [PANICAP] Poaceae (V:M)
Panicum capillare ssp. *barbipulvinatum* (Nash) Tzvelev = Panicum capillare [PANICAP] Poaceae (V:M)
Panicum capillare var. *agreste* Gattinger = Panicum capillare [PANICAP] Poaceae (V:M)
Panicum capillare var. *barbipulvinatum* (Nash) R.L. McGregor = Panicum capillare [PANICAP] Poaceae (V:M)
Panicum capillare var. *brevifolium* Vasey ex Rydb. & Shear = Panicum capillare [PANICAP] Poaceae (V:M)
Panicum capillare var. *occidentale* Rydb. = Panicum capillare [PANICAP] Poaceae (V:M)
Panicum curtifolium Nash = Dichanthelium acuminatum var. fasciculatum [DICHACU1] Poaceae (V:M)
***Panicum dichotomiflorum** Michx. [PANIDIC] Poaceae (V:M)
Panicum glaucum L. = Setaria glauca [SETAGLA] Poaceae (V:M)
Panicum glutinoscabrum Fern. = Dichanthelium acuminatum var. fasciculatum [DICHACU1] Poaceae (V:M)
Panicum helleri Nash = Dichanthelium oligosanthes var. scribnerianum [DICHOLI1] Poaceae (V:M)
Panicum huachucae Ashe = Dichanthelium acuminatum var. fasciculatum [DICHACU1] Poaceae (V:M)
Panicum huachucae var. *fasciculatum* (Torr.) F.T. Hubbard = Dichanthelium acuminatum var. fasciculatum [DICHACU1] Poaceae (V:M)
Panicum implicatum Scribn. = Dichanthelium acuminatum var. fasciculatum [DICHACU1] Poaceae (V:M)
Panicum italicum L. = Setaria italica [SETAITA] Poaceae (V:M)

Panicum languidum A.S. Hitchc. = Dichanthelium acuminatum var. fasciculatum [DICHACU1] Poaceae (V:M)
Panicum lanuginosum Ell. = Dichanthelium acuminatum var. fasciculatum [DICHACU1] Poaceae (V:M)
Panicum lanuginosum var. *fasciculatum* (Torr.) Fern. = Dichanthelium acuminatum var. fasciculatum [DICHACU1] Poaceae (V:M)
Panicum lanuginosum var. *huachucae* (Ashe) A.S. Hitchc. = Dichanthelium acuminatum var. fasciculatum [DICHACU1] Poaceae (V:M)
Panicum lanuginosum var. *implicatum* (Scribn.) Fern. = Dichanthelium acuminatum var. fasciculatum [DICHACU1] Poaceae (V:M)
Panicum lanuginosum var. *tennesseense* (Ashe) Gleason = Dichanthelium acuminatum var. fasciculatum [DICHACU1] Poaceae (V:M)
Panicum lassenianum Schmoll = Dichanthelium acuminatum var. fasciculatum [DICHACU1] Poaceae (V:M)
Panicum lindheimeri var. *fasciculatum* (Torr.) Fern. = Dichanthelium acuminatum var. fasciculatum [DICHACU1] Poaceae (V:M)
Panicum macrocarpon Le Conte ex Torr. = Dichanthelium oligosanthes var. scribnerianum [DICHOLI1] Poaceae (V:M)
*****Panicum miliaceum** L. [PANIMIL] Poaceae (V:M)
Panicum occidentale Scribn. = Dichanthelium acuminatum var. fasciculatum [DICHACU1] Poaceae (V:M)
Panicum oligosanthes var. *helleri* (Nash) Fern. = Dichanthelium oligosanthes var. scribnerianum [DICHOLI1] Poaceae (V:M)
Panicum oligosanthes var. *scribnerianum* (Nash) Fern. = Dichanthelium oligosanthes var. scribnerianum [DICHOLI1] Poaceae (V:M)
Panicum pacificum A.S. Hitchc. & Chase = Dichanthelium acuminatum var. fasciculatum [DICHACU1] Poaceae (V:M)
Panicum sanguinale L. = Digitaria sanguinalis [DIGISAN] Poaceae (V:M)
Panicum scoparium S. Wats. ex Nash = Dichanthelium oligosanthes var. scribnerianum [DICHOLI1] Poaceae (V:M)
Panicum scribnerianum Nash = Dichanthelium oligosanthes var. scribnerianum [DICHOLI1] Poaceae (V:M)
Panicum subvillosum Ashe = Dichanthelium acuminatum var. fasciculatum [DICHACU1] Poaceae (V:M)
Panicum tennesseense Ashe = Dichanthelium acuminatum var. fasciculatum [DICHACU1] Poaceae (V:M)
Panicum thermale Boland. = Dichanthelium acuminatum var. thermale [DICHACU2] Poaceae (V:M)
Panicum verticillatum var. *ambiguum* Guss. = Setaria verticillata var. ambigua [SETAVER1] Poaceae (V:M)
Pannaria Delise [PANNARI$] Pannariaceae (L)
Pannaria ahlneri P.M. Jørg. = Fuscopannaria ahlneri [FUSCAHL] Pannariaceae (L)
Pannaria cheiroloba (Müll. Arg.) Essl. & Egan [PANNCHE] Pannariaceae (L)
Pannaria cyanolepra Tuck. [PANNCYA] Pannariaceae (L)
Pannaria hypnorum (Vahl) Körber = Psoroma hypnorum [PSORHYP] Pannariaceae (L)
Pannaria isidiata Degel. = Vestergrenopsis isidiata [VESTISI] Placynthiaceae (L)
Pannaria laceratula Hue = Fuscopannaria laceratula [FUSCLAC] Pannariaceae (L)
Pannaria lepidiota (Sommerf.) Th. Fr. = Fuscopannaria praetermissa [FUSCPRE] Uncertain-family (L)
Pannaria leucostictoides Ohlsson = Fuscopannaria leucostictoides [FUSCLEU] Pannariaceae (L)
Pannaria maritima P.M. Jørg. = Fuscopannaria maritima [FUSCMAR] Pannariaceae (L)
Pannaria mediterranea Tav. = Fuscopannaria mediterranea [FUSCMED] Pannariaceae (L)
Pannaria pezizoides (Weber) Trevisan [PANNPEZ] Pannariaceae (L)
Pannaria praetermissa Nyl. = Fuscopannaria praetermissa [FUSCPRE] Pannariaceae (L)
Pannaria rubiginosa (Ach.) Bory [PANNRUB] Pannariaceae (L)
Pannaria saubinetii (Mont.) Nyl. = Fuscopannaria saubinetii [FUSCSAU] Pannariaceae (L)
Pannaria sonomensis Tuck. = Koerberia sonomensis [KOERSON] Placynthiaceae (L)
Papaver L. [PAPAVER$] Papaveraceae (V:D)
Papaver alaskanum Hultén = Papaver lapponicum ssp. occidentale [PAPALAP1] Papaveraceae (V:D)
Papaver alaskanum var. *macranthum* Hultén = Papaver macounii [PAPAMAC] Papaveraceae (V:D)
Papaver alboroseum Hultén [PAPAALB] Papaveraceae (V:D)
Papaver alpinum L. [PAPAALP] Papaveraceae (V:D)
Papaver cornwallisense D. Löve = Papaver lapponicum ssp. occidentale [PAPALAP1] Papaveraceae (V:D)
Papaver denalii Gjaerevoll = Papaver lapponicum ssp. occidentale [PAPALAP1] Papaveraceae (V:D)
Papaver freedmanianum D. Löve = Papaver lapponicum ssp. occidentale [PAPALAP1] Papaveraceae (V:D)
Papaver hultenii Knaben = Papaver macounii [PAPAMAC] Papaveraceae (V:D)
Papaver hultenii var. *salmonicolor* Hultén = Papaver macounii [PAPAMAC] Papaveraceae (V:D)
Papaver keelei Porsild = Papaver macounii [PAPAMAC] Papaveraceae (V:D)
Papaver kluanense D. Löve = Papaver lapponicum ssp. occidentale [PAPALAP1] Papaveraceae (V:D)
Papaver lapponicum (Tolm.) Nordh. [PAPALAP] Papaveraceae (V:D)
Papaver lapponicum ssp. **occidentale** (Lundstr.) Knaben [PAPALAP1] Papaveraceae (V:D)
Papaver lapponicum ssp. *porsildii* Knaben = Papaver lapponicum ssp. occidentale [PAPALAP1] Papaveraceae (V:D)
Papaver macounii Greene [PAPAMAC] Papaveraceae (V:D)
Papaver macounii var. *discolor* Hultén = Papaver macounii [PAPAMAC] Papaveraceae (V:D)
Papaver nigroflavum D. Löve = Papaver lapponicum ssp. occidentale [PAPALAP1] Papaveraceae (V:D)
*****Papaver nudicaule** L. [PAPANUD] Papaveraceae (V:D)
Papaver nudicaule var. *coloradense* Fedde = Papaver lapponicum ssp. occidentale [PAPALAP1] Papaveraceae (V:D)
Papaver nudicaule var. *columbianum* Fedde = Papaver lapponicum ssp. occidentale [PAPALAP1] Papaveraceae (V:D)
Papaver nudicaule var. *pseudocorylatifolium* Fedde = Papaver alpinum [PAPAALP] Papaveraceae (V:D)
Papaver pygmaeum Rydb. [PAPAPYG] Papaveraceae (V:D)
Papaver radicatum ssp. *lapponicum* Tolm. = Papaver lapponicum ssp. occidentale [PAPALAP1] Papaveraceae (V:D)
Papaver radicatum ssp. *occidentale* Lundstr. = Papaver lapponicum ssp. occidentale [PAPALAP1] Papaveraceae (V:D)
Papaver radicatum ssp. *porsildii* (Knaben) D. Löve = Papaver lapponicum ssp. occidentale [PAPALAP1] Papaveraceae (V:D)
Papaver radicatum var. *pygmaeum* (Rydb.) Welsh = Papaver pygmaeum [PAPAPYG] Papaveraceae (V:D)
*****Papaver rhoeas** L. [PAPARHO] Papaveraceae (V:D)
Papaver scammianum D. Löve = Papaver macounii [PAPAMAC] Papaveraceae (V:D)
*****Papaver somniferum** L. [PAPASOM] Papaveraceae (V:D)
Paraleptodontium Long [PARALEP$] Pottiaceae (B:M)
Paraleptodontium recurvifolium (Tayl.) Long [PARAREC] Pottiaceae (B:M)
Paraleucobryum (Lindb.) Loeske [PARALEU$] Dicranaceae (B:M)
Paraleucobryum enerve (Thed. in Hartm.) Loeske [PARAENE] Dicranaceae (B:M)
Paraleucobryum longifolium (Hedw.) Loeske [PARALON] Dicranaceae (B:M)
Paraleucobryum longifolium ssp. *serratum* (Kindb.) Podp. = Paraleucobryum longifolium [PARALON] Dicranaceae (B:M)
Paraleucobryum longifolium var. *subalpinum* (Milde) Cas.-Gil = Paraleucobryum longifolium [PARALON] Dicranaceae (B:M)
Paraleucobryum sauteri (Bruch & Schimp. in B.S.G.) Loeske = Paraleucobryum enerve [PARAENE] Dicranaceae (B:M)
Parathelium microcarpum Riddle = Pyrenula microtheca [PYREMIC] Pyrenulaceae (L)
Parentucellia Viviani [PARENTU$] Scrophulariaceae (V:D)
*****Parentucellia viscosa** (L.) Caruel [PAREVIS] Scrophulariaceae (V:D)

Parietaria L. [PARIETA$] Urticaceae (V:D)
Parietaria pensylvanica Muhl. ex Willd. [PARIPEN] Urticaceae (V:D)
Parmelia Ach. [PARMELI$] Parmeliaceae (L)
Parmelia albertana Ahti = Melanelia albertana [MELAALE] Parmeliaceae (L)
Parmelia aleuritica Nyl. = Arctoparmelia centrifuga [ARCTCEN] Parmeliaceae (L)
Parmelia almquistii Vainio = Allantoparmelia almquistii [ALLAALM] Parmeliaceae (L)
Parmelia alpicola Th. Fr. = Allantoparmelia alpicola [ALLAALP] Parmeliaceae (L)
Parmelia andreana Müll. Arg. = Flavopunctelia flaventior [FLAVFLA] Parmeliaceae (L)
Parmelia arnoldii Du Rietz = Parmotrema arnoldii [PARMARO] Parmeliaceae (L)
Parmelia austerodes Nyl. = Hypogymnia austerodes [HYPOAUS] Parmeliaceae (L)
Parmelia birulae var. *grumosa* Llano = Arctoparmelia separata [ARCTSEP] Parmeliaceae (L)
Parmelia bitteri Lynge = Hypogymnia bitteri [HYPOBIT] Parmeliaceae (L)
Parmelia centrifuga (L.) Ach. = Arctoparmelia centrifuga [ARCTCEN] Parmeliaceae (L)
Parmelia cetrarioides (Delise ex Duby) Nyl. = Cetrelia cetrarioides [CETRCET] Parmeliaceae (L)
Parmelia conspurcata (Schaerer) Vainio = Melanelia subargentifera [MELASUA] Parmeliaceae (L)
Parmelia crinita Ach. = Parmotrema crinitum [PARMCRN] Parmeliaceae (L)
Parmelia cumberlandia (Gyelnik) Hale = Xanthoparmelia cumberlandia [XANTCUM] Parmeliaceae (L)
Parmelia denalii Krog = Melanelia disjuncta [MELADIS] Parmeliaceae (L)
Parmelia disjuncta Erichsen = Melanelia disjuncta [MELADIS] Parmeliaceae (L)
Parmelia dubia (Wulfen) Schaerer = Punctelia subrudecta [PUNCSUR] Parmeliaceae (L)
Parmelia duplicata var. *douglasicola* Gyelnik = Hypogymnia physodes [HYPOPHY] Parmeliaceae (L)
Parmelia elegantula (Zahlbr.) Szat. = Melanelia elegantula [MELAELE] Parmeliaceae (L)
Parmelia elegantula var. *americana* Räsänen = Melanelia elegantula [MELAELE] Parmeliaceae (L)
Parmelia exasperatula Nyl. = Melanelia exasperatula [MELAEXS] Parmeliaceae (L)
Parmelia flaventior Stirton = Flavopunctelia flaventior [FLAVFLA] Parmeliaceae (L)
Parmelia fraudans (Nyl.) Nyl. [PARMFRA] Parmeliaceae (L)
Parmelia glabratula (Lamy) Nyl. = Melanelia fuliginosa [MELAFUL] Parmeliaceae (L)
Parmelia graminicola de Lesd. = Punctelia subrudecta [PUNCSUR] Parmeliaceae (L)
Parmelia granulosa Lynge = Melanelia disjuncta [MELADIS] Parmeliaceae (L)
Parmelia hygrophila Goward & Ahti [PARMHYG] Parmeliaceae (L)
Parmelia incurva (Pers.) Fr. = Arctoparmelia incurva [ARCTINC] Parmeliaceae (L)
Parmelia infumata Nyl. = Melanelia infumata [MELAINF] Parmeliaceae (L)
Parmelia isidiotyla Nyl. = Neofuscelia loxodes [NEOFLOX] Parmeliaceae (L)
Parmelia kerguelensis A. Wilson = Parmelia saxatilis [PARMSAX] Parmeliaceae (L)
Parmelia kerguelensis auct. = Parmelia pseudosulcata [PARMPSE] Parmeliaceae (L)
Parmelia kernstockii (Lynge) Zahlbr. = Flavopunctelia flaventior [FLAVFLA] Parmeliaceae (L)
Parmelia lanata (L.) Wallr. = Pseudephebe pubescens [PSEUPUB] Parmeliaceae (L)
Parmelia lineola E.C. Berry = Xanthoparmelia lineola [XANTLIN] Parmeliaceae (L)
Parmelia loxodes Nyl. = Neofuscelia loxodes [NEOFLOX] Parmeliaceae (L)
Parmelia mexicana Gyelnik = Xanthoparmelia mexicana [XANTMEX] Parmeliaceae (L)
Parmelia mougeotii Schaerer = Xanthoparmelia mougeotii [XANTMOU] Parmeliaceae (L)
Parmelia multispora A. Schneider = Melanelia multispora [MELAMUL] Parmeliaceae (L)
Parmelia neoconspersa Gyelnik = Xanthoparmelia neoconspersa [XANTNEO] Parmeliaceae (L)
Parmelia olivaceoides Krog = Melanelia olivaceoides [MELAOLI] Parmeliaceae (L)
Parmelia omphalodes (L.) Ach. [PARMOMP] Parmeliaceae (L)
Parmelia oregana Gyelnik = Hypogymnia physodes [HYPOPHY] Parmeliaceae (L)
Parmelia panniformis (Nyl.) Vainio = Melanelia panniformis [MELAPAN] Parmeliaceae (L)
Parmelia perlata (Hudson) Ach. = Parmotrema chinense [PARMCHN] Parmeliaceae (L)
Parmelia pertusa (Schrank) Schaerer = Menegazzia terebrata [MENETER] Parmeliaceae (L)
Parmelia physodes (L.) Ach. = Hypogymnia physodes [HYPOPHY] Parmeliaceae (L)
Parmelia plittii Gyelnik = Xanthoparmelia plittii [XANTPLI] Parmeliaceae (L)
Parmelia pseudosulcata Gyelnik [PARMPSE] Parmeliaceae (L)
Parmelia pubescens (L.) Vainio = Pseudephebe pubescens [PSEUPUB] Parmeliaceae (L)
Parmelia pubescens var. *reticulata* Cromb. = Pseudephebe pubescens [PSEUPUB] Parmeliaceae (L)
Parmelia revoluta Flörke = Hypotrachyna revoluta [HYPOREV] Parmeliaceae (L)
Parmelia saxatilis (L.) Ach. [PARMSAX] Parmeliaceae (L)
Parmelia saxatilis var. *divaricata* (Nyl.) Hue = Parmelia squarrosa [PARMSQU] Parmeliaceae (L)
Parmelia saximontana R. Anderson & W.A. Weber = Melanelia tominii [MELATOM] Parmeliaceae (L)
Parmelia separata Th. Fr. = Arctoparmelia separata [ARCTSEP] Parmeliaceae (L)
Parmelia septentrionalis (Lynge) Ahti = Melanelia septentrionalis [MELASET] Parmeliaceae (L)
Parmelia sinuosa (Sm.) Ach. = Hypotrachyna sinuosa [HYPOSIN] Parmeliaceae (L)
Parmelia sipeana Gyelnik = Menegazzia terebrata [MENETER] Parmeliaceae (L)
Parmelia somloënsis Gyelnik = Xanthoparmelia somloënsis [XANTSOM] Parmeliaceae (L)
Parmelia sorediata (Ach.) Th. Fr. = Melanelia sorediata [MELASOR] Parmeliaceae (L)
Parmelia sorediosa Almb. = Melanelia sorediata [MELASOR] Parmeliaceae (L)
Parmelia sphaerosporella Müll. Arg. = Ahtiana sphaerosporella [AHTISPH] Parmeliaceae (L)
Parmelia squarrosa Hale [PARMSQU] Parmeliaceae (L)
Parmelia stenophylla (Ach.) Du Rietz = Xanthoparmelia somloënsis [XANTSOM] Parmeliaceae (L)
Parmelia stictica (Duby) Nyl. = Punctelia stictica [PUNCSTI] Parmeliaceae (L)
Parmelia stygia (L.) Ach. = Melanelia stygia [MELASTY] Parmeliaceae (L)
Parmelia subargentifera Nyl. = Melanelia subargentifera [MELASUA] Parmeliaceae (L)
Parmelia subaurifera Nyl. = Melanelia subaurifera [MELASUU] Parmeliaceae (L)
Parmelia subcentrifuga Oksner = Arctoparmelia subcentrifuga [ARCTSUB] Parmeliaceae (L)
Parmelia subelegantula Essl. = Melanelia subelegantula [MELASUB] Parmeliaceae (L)
Parmelia subhosseana Essl. = Neofuscelia subhosseana [NEOFSUB] Parmeliaceae (L)
Parmelia subobscura Vainio = Hypogymnia subobscura [HYPOSUO] Parmeliaceae (L)

Parmelia subolivacea Nyl. = Melanelia subolivacea [MELASUO] Parmeliaceae (L)
Parmelia subrudecta Nyl. = Punctelia subrudecta [PUNCSUR] Parmeliaceae (L)
Parmelia substygia Räsänen = Melanelia tominii [MELATOM] Parmeliaceae (L)
Parmelia sulcata Taylor [PARMSUL] Parmeliaceae (L)
Parmelia tasmanica Hook. f. & Taylor = Xanthoparmelia tasmanica [XANTTAS] Parmeliaceae (L)
Parmelia trabeculata Ahti = Melanelia trabeculata [MELATRA] Parmeliaceae (L)
Parmelia trichotera Hue = Parmotrema chinense [PARMCHN] Parmeliaceae (L)
Parmelia tubulosa (Schaerer) Bitter = Hypogymnia tubulosa [HYPOTUB] Parmeliaceae (L)
Parmelia ulophylla (Ach.) G. Merr. = Punctelia subrudecta [PUNCSUR] Parmeliaceae (L)
Parmelia verruculifera Nyl. = Neofuscelia verruculifera [NEOFVER] Parmeliaceae (L)
Parmelia vittata (Ach.) Nyl. = Hypogymnia vittata [HYPOVIT] Parmeliaceae (L)
Parmelia wyomingica (Gyelnik) Hale = Xanthoparmelia wyomingica [XANTWYO] Parmeliaceae (L)
Parmeliella Müll. Arg. [PARMELL$] Pannariaceae (L)
Parmeliella cheiroloba Müll. Arg. = Pannaria cheiroloba [PANNCHE] Pannariaceae (L)
Parmeliella saubinetii (Mont.) Zahlbr. = Fuscopannaria saubinetii [FUSCSAU] Uncertain-family (L)
Parmeliella triptophylla (Ach.) Müll. Arg. [PARMTRI] Pannariaceae (L)
Parmeliopsis Nyl. [PARMELO$] Parmeliaceae (L)
Parmeliopsis aleurites (Ach.) Nyl. = Imshaugia aleurites [IMSHALE] Parmeliaceae (L)
Parmeliopsis ambigua (Wulfen) Nyl. [PARMAMB] Parmeliaceae (L)
Parmeliopsis diffusa (Weber) Riddle = Parmeliopsis hyperopta [PARMHYE] Parmeliaceae (L)
Parmeliopsis hyperopta (Ach.) Arnold [PARMHYE] Parmeliaceae (L)
Parmotrema A. Massal. [PARMOTR$] Parmeliaceae (L)
Parmotrema arnoldii (Du Rietz) Hale [PARMARO] Parmeliaceae (L)
Parmotrema chinense (Osbeck) Hale & Ahti [PARMCHN] Parmeliaceae (L)
Parmotrema crinitum (Ach.) Choisy [PARMCRN] Parmeliaceae (L)
Parmotrema perlatum (Hudson) Choisy = Parmotrema chinense [PARMCHN] Parmeliaceae (L)
Parmotrema sinense (Osbeck) Hale & Ahti [PARMSIN] Parmeliaceae (L)
Parnassia L. [PARNASS$] Saxifragaceae (V:D)
Parnassia fimbriata Koenig [PARNFIM] Saxifragaceae (V:D)
Parnassia kotzebuei Cham. ex Spreng. [PARNKOT] Saxifragaceae (V:D)
Parnassia kotzebuei var. *pumila* C.L. Hitchc. & Ownbey = Parnassia kotzebuei [PARNKOT] Saxifragaceae (V:D)
Parnassia montanensis Fern. & Rydb. ex Rydb. = Parnassia palustris var. montanensis [PARNPAL2] Saxifragaceae (V:D)
Parnassia palustris L. [PARNPAL] Saxifragaceae (V:D)
Parnassia palustris var. **montanensis** (Fern. & Rydb. ex Rydb.) C.L. Hitchc. [PARNPAL2] Saxifragaceae (V:D)
Parnassia palustris var. *parviflora* (DC.) Boivin = Parnassia parviflora [PARNPAR] Saxifragaceae (V:D)
Parnassia parviflora DC. [PARNPAR] Saxifragaceae (V:D)
Parrya R. Br. [PARRYA$] Brassicaceae (V:D)
Parrya nudicaulis (L.) Boiss. [PARRNUD] Brassicaceae (V:D)
Parrya nudicaulis ssp. *interior* Hultén = Parrya nudicaulis [PARRNUD] Brassicaceae (V:D)
Parrya nudicaulis ssp. *septentrionalis* Hultén = Parrya nudicaulis [PARRNUD] Brassicaceae (V:D)
Parrya nudicaulis var. *grandiflora* Hultén = Parrya nudicaulis [PARRNUD] Brassicaceae (V:D)
Parrya nudicaulis var. *interior* (Hultén) Boivin = Parrya nudicaulis [PARRNUD] Brassicaceae (V:D)
Parrya platycarpa Rydb. = Parrya nudicaulis [PARRNUD] Brassicaceae (V:D)
Parrya rydbergii Botsch. = Parrya nudicaulis [PARRNUD] Brassicaceae (V:D)
Pascopyrum A. Löve [PASCOPY$] Poaceae (V:M)
Pascopyrum smithii (Rydb.) A. Löve [PASCSMI] Poaceae (V:M)
Pastinaca L. [PASTINA$] Apiaceae (V:D)
Pastinaca sativa L. [PASTSAT] Apiaceae (V:D)
Patellaria laurocerasi Delise ex Duby = Bacidia laurocerasi [BACILAU] Bacidiaceae (L)
Patellaria macrocarpa DC. = Porpidia macrocarpa [PORPMAC] Porpidiaceae (L)
Paxistima Raf. [PAXISTI$] Celastraceae (V:D)
Paxistima myrsinites (Pursh) Raf. [PAXIMYR] Celastraceae (V:D)
Pectiantia breweri (Gray) Rydb. = Mitella breweri [MITEBRE] Saxifragaceae (V:D)
Pectiantia ovalis (Greene) Rydb. = Mitella ovalis [MITEOVA] Saxifragaceae (V:D)
Pectiantia pentandra (Hook.) Rydb. = Mitella pentandra [MITEPEN] Saxifragaceae (V:D)
Pectocarya DC. ex Meisn. [PECTOCA$] Boraginaceae (V:D)
Pectocarya linearis var. *penicillata* (Hook. & Arn.) M.E. Jones = Pectocarya penicillata [PECTPEN] Boraginaceae (V:D)
Pectocarya penicillata (Hook. & Arn.) A. DC. [PECTPEN] Boraginaceae (V:D)
Pedicularioposis verticillata (L.) A. & D. Löve = Pedicularis verticillata [PEDIVER] Scrophulariaceae (V:D)
Pedicularis L. [PEDICUL$] Scrophulariaceae (V:D)
Pedicularis arctica R. Br. = Pedicularis langsdorfii ssp. arctica [PEDILAG1] Scrophulariaceae (V:D)
Pedicularis bracteosa Benth. [PEDIBRA] Scrophulariaceae (V:D)
Pedicularis bracteosa var. **bracteosa** [PEDIBRA0] Scrophulariaceae (V:D)
Pedicularis bracteosa var. **latifolia** (Pennell) Cronq. [PEDIBRA2] Scrophulariaceae (V:D)
Pedicularis capitata M.F. Adams [PEDICAP] Scrophulariaceae (V:D)
Pedicularis contorta Benth. [PEDICON] Scrophulariaceae (V:D)
Pedicularis groenlandica Retz. [PEDIGRO] Scrophulariaceae (V:D)
Pedicularis hians Eastw. = Pedicularis langsdorfii ssp. arctica [PEDILAG1] Scrophulariaceae (V:D)
Pedicularis kanei Dur. = Pedicularis lanata [PEDILAN] Scrophulariaceae (V:D)
Pedicularis labradorica Wirsing [PEDILAB] Scrophulariaceae (V:D)
Pedicularis lanata Cham. & Schlecht. [PEDILAN] Scrophulariaceae (V:D)
Pedicularis langsdorfii Fisch. ex Stev. [PEDILAG] Scrophulariaceae (V:D)
Pedicularis langsdorfii ssp. **arctica** (R. Br.) Pennell [PEDILAG1] Scrophulariaceae (V:D)
Pedicularis langsdorfii var. *arctica* (R. Br.) Polunin = Pedicularis langsdorfii ssp. arctica [PEDILAG1] Scrophulariaceae (V:D)
Pedicularis latifolia Pennell = Pedicularis bracteosa var. latifolia [PEDIBRA2] Scrophulariaceae (V:D)
Pedicularis macrodonta Richards. [PEDIMAC] Scrophulariaceae (V:D)
Pedicularis montanensis Rydb. = Pedicularis bracteosa var. bracteosa [PEDIBRA0] Scrophulariaceae (V:D)
Pedicularis nelsonii R. Br. = Pedicularis capitata [PEDICAP] Scrophulariaceae (V:D)
Pedicularis oederi Vahl ex Hornem. [PEDIOED] Scrophulariaceae (V:D)
Pedicularis ornithorhyncha Benth. [PEDIORN] Scrophulariaceae (V:D)
Pedicularis paddoensis Pennell = Pedicularis bracteosa var. bracteosa [PEDIBRA0] Scrophulariaceae (V:D)
Pedicularis parviflora Sm. ex Rees [PEDIPAR] Scrophulariaceae (V:D)
Pedicularis parviflora var. *macrodonta* (Richards.) Welsh = Pedicularis macrodonta [PEDIMAC] Scrophulariaceae (V:D)
Pedicularis pedicellata Bunge = Pedicularis ornithorhyncha [PEDIORN] Scrophulariaceae (V:D)

Pedicularis pennellii ssp. *insularis* Calder & Taylor = Pedicularis parviflora [PEDIPAR] Scrophulariaceae (V:D)
Pedicularis pennellii var. *insularis* (Calder & Taylor) Boivin = Pedicularis parviflora [PEDIPAR] Scrophulariaceae (V:D)
Pedicularis racemosa Dougl. ex Benth. [PEDIRAC] Scrophulariaceae (V:D)
Pedicularis subnuda Benth. = Pedicularis ornithorhyncha [PEDIORN] Scrophulariaceae (V:D)
Pedicularis sudetica Willd. [PEDISUD] Scrophulariaceae (V:D)
Pedicularis sudetica ssp. **interior** (Hultén) Hultén [PEDISUD1] Scrophulariaceae (V:D)
Pedicularis sudetica var. *gymnocephala* Trautv. = Pedicularis sudetica ssp. interior [PEDISUD1] Scrophulariaceae (V:D)
Pedicularis thompsonii Pennell = Pedicularis bracteosa var. bracteosa [PEDIBRA0] Scrophulariaceae (V:D)
Pedicularis versicolor Wahlenb. = Pedicularis oederi [PEDIOED] Scrophulariaceae (V:D)
Pedicularis verticillata L. [PEDIVER] Scrophulariaceae (V:D)
Pedicularis willdenowii Vved. = Pedicularis lanata [PEDILAN] Scrophulariaceae (V:D)
Pellaea Link [PELLAEA$] Pteridaceae (V:F)
Pellaea atropurpurea var. *simplex* (Butters) Morton = Pellaea glabella ssp. simplex [PELLGLA2] Pteridaceae (V:F)
Pellaea densa (Brack.) Hook. = Aspidotis densa [ASPIDEN] Pteridaceae (V:F)
Pellaea glabella Mett. ex Kuhn [PELLGLA] Pteridaceae (V:F)
Pellaea glabella ssp. **occidentalis** (E. Nels.) Windham [PELLGLA1] Pteridaceae (V:F)
Pellaea glabella ssp. **simplex** (Butters) A. & D. Löve [PELLGLA2] Pteridaceae (V:F)
Pellaea glabella var. *nana* (L.C. Rich.) Cody = Pellaea glabella ssp. occidentalis [PELLGLA1] Pteridaceae (V:F)
Pellaea glabella var. *occidentalis* (E. Nels.) Butters = Pellaea glabella ssp. occidentalis [PELLGLA1] Pteridaceae (V:F)
Pellaea glabella var. *simplex* Butters = Pellaea glabella ssp. simplex [PELLGLA2] Pteridaceae (V:F)
Pellaea occidentalis (E. Nels.) Rydb. = Pellaea glabella ssp. occidentalis [PELLGLA1] Pteridaceae (V:F)
Pellaea occidentalis ssp. *simplex* (Butters) Gastony = Pellaea glabella ssp. simplex [PELLGLA2] Pteridaceae (V:F)
Pellaea pumila Rydb. = Pellaea glabella ssp. occidentalis [PELLGLA1] Pteridaceae (V:F)
Pellaea suksdorfiana Butters = Pellaea glabella ssp. simplex [PELLGLA2] Pteridaceae (V:F)
Pellia Raddi [PELLIA$] Pelliaceae (B:H)
Pellia borealis Lorbeer = Pellia epiphylla [PELLEPI] Pelliaceae (B:H)
Pellia columbiana Krajina & Brayshaw = Pellia neesiana [PELLNEE] Pelliaceae (B:H)
Pellia endiviifolia (Dicks.) Dum. [PELLEND] Pelliaceae (B:H)
Pellia epiphylla (L.) Corda [PELLEPI] Pelliaceae (B:H)
Pellia fabbroniana Raddi = Pellia endiviifolia [PELLEND] Pelliaceae (B:H)
Pellia neesiana (Gott.) Limpr. [PELLNEE] Pelliaceae (B:H)
Peltigera Willd. [PELTIGE$] Peltigeraceae (L)
Peltigera aphthosa (L.) Willd. [PELTAPH] Peltigeraceae (L)
Peltigera aphthosa f. *complicata* (Th. Fr.) Zahlbr. = Peltigera leucophlebia [PELTLEU] Peltigeraceae (L)
Peltigera aphthosa var. *variolosa* A. Massal. = Peltigera leucophlebia [PELTLEU] Peltigeraceae (L)
Peltigera britannica (Gyelnik) Holt.-Hartw. & Tønsberg [PELTBRI] Peltigeraceae (L)
Peltigera canina (L.) Willd. [PELTCAN] Peltigeraceae (L)
Peltigera canina var. *extenuata* Nyl. ex Vainio = Peltigera didactyla var. extenuata [PELTDID1] Peltigeraceae (L)
Peltigera canina var. *rufescens* (Weiss) Mudd = Peltigera rufescens [PELTRUF] Peltigeraceae (L)
Peltigera canina var. *rufescens* f. *innovans* (Körber) J.W. Thomson = Peltigera praetextata [PELTPRA] Peltigeraceae (L)
Peltigera canina var. *spuria* (Ach.) Schaerer = Peltigera didactyla [PELTDID] Peltigeraceae (L)
Peltigera chionophila Goward [PELTCHI] Peltigeraceae (L)
Peltigera cinnamomea Goward [PELTCIN] Peltigeraceae (L)
Peltigera collina (Ach.) Schrader [PELTCOL] Peltigeraceae (L)
Peltigera degenii Gyelnik [PELTDEG] Peltigeraceae (L)
Peltigera didactyla (With.) J.R. Laundon [PELTDID] Peltigeraceae (L)
Peltigera didactyla var. **didactyla** [PELTDID0] Peltigeraceae (L)
Peltigera didactyla var. **extenuata** (Nyl. ex Vainio) Goffinet & Hastings [PELTDID1] Peltigeraceae (L)
Peltigera elisabethae Gyelnik [PELTELI] Peltigeraceae (L)
Peltigera erumpens (Taylor) Elenkin = Peltigera didactyla [PELTDID] Peltigeraceae (L)
Peltigera evansiana Gyelnik [PELTEVA] Peltigeraceae (L)
Peltigera horizontalis (Hudson) Baumg. [PELTHOR] Peltigeraceae (L)
Peltigera horizontalis f. *zopfii* auct. = Peltigera elisabethae [PELTELI] Peltigeraceae (L)
Peltigera hymenina (Ach.) Delise [PELTHYM] Peltigeraceae (L)
Peltigera kristinssonii Vitik. [PELTKRI] Peltigeraceae (L)
Peltigera lactucifolia (With.) J.R. Laundon = Peltigera hymenina [PELTHYM] Peltigeraceae (L)
Peltigera lepidophora (Vainio) Bitter [PELTLEP] Peltigeraceae (L)
Peltigera leucophlebia (Nyl.) Gyelnik [PELTLEU] Peltigeraceae (L)
Peltigera malacea (Ach.) Funck [PELTMAL] Peltigeraceae (L)
Peltigera membranacea (Ach.) Nyl. [PELTMEM] Peltigeraceae (L)
Peltigera neckeri Hepp ex Müll. Arg. [PELTNEC] Peltigeraceae (L)
Peltigera neopolydactyla (Gyelnik) Gyelnik [PELTNEO] Peltigeraceae (L)
Peltigera occidentalis (E. Dahl) Kristinsson = Peltigera neopolydactyla [PELTNEO] Peltigeraceae (L)
Peltigera occidentalis auct. = Peltigera kristinssonii [PELTKRI] Peltigeraceae (L)
Peltigera pacifica Vitik. [PELTPAC] Peltigeraceae (L)
Peltigera polydactyla (Necker) Hoffm. = Peltigera polydactylon [PELTPOL] Peltigeraceae (L)
Peltigera polydactylon (Necker) Hoffm. [PELTPOL] Peltigeraceae (L)
Peltigera polydactyla var. *hymenina* (Ach.) Flotow = Peltigera hymenina [PELTHYM] Peltigeraceae (L)
Peltigera polydactyla var. *neopolydactyla* Gyelnik = Peltigera neopolydactyla [PELTNEO] Peltigeraceae (L)
Peltigera ponojensis Gyelnik [PELTPON] Peltigeraceae (L)
Peltigera praetextata (Flörke ex Sommerf.) Zopf [PELTPRA] Peltigeraceae (L)
Peltigera pulverulenta (Taylor) Kremp. = Peltigera scabrosa [PELTSCA] Peltigeraceae (L)
Peltigera retifoveata Vitik. [PELTRET] Peltigeraceae (L)
Peltigera rufescens (Weiss) Humb. [PELTRUF] Peltigeraceae (L)
Peltigera scabrosa Th. Fr. [PELTSCA] Peltigeraceae (L)
Peltigera scutata (Dickson) Duby = Peltigera collina [PELTCOL] Peltigeraceae (L)
Peltigera spuria (Ach.) DC. = Peltigera didactyla [PELTDID] Peltigeraceae (L)
Peltigera variolosa (A. Massal.) Gyelnik = Peltigera leucophlebia [PELTLEU] Peltigeraceae (L)
Peltigera venosa (L.) Hoffm. [PELTVEN] Peltigeraceae (L)
Peltolepis Lindb. [PELTOLE$] Cleveaceae (B:H)
Peltolepis grandis Lindb. = Peltolepis quadrata [PELTQUA] Cleveaceae (B:H)
Peltolepis quadrata (Saut.) K. Müll. [PELTQUA] Cleveaceae (B:H)
Peltula Nyl. [PELTULA$] Peltulaceae (L)
Peltula euploca (Ach.) Poelt [PELTEUP] Peltulaceae (L)
Pennisetum glaucum (L.) R. Br. = Setaria glauca [SETAGLA] Poaceae (V:M)
Penstemon Schmidel [PENSTEM$] Scrophulariaceae (V:D)
Penstemon albertinus Greene [PENSALB] Scrophulariaceae (V:D)
Penstemon confertus Dougl. ex Lindl. [PENSCON] Scrophulariaceae (V:D)
Penstemon confertus ssp. *procerus* (Dougl. ex Graham) D.V. Clark = Penstemon procerus var. procerus [PENSPRO0] Scrophulariaceae (V:D)
Penstemon confertus var. *procerus* (Dougl. ex Graham) Coville = Penstemon procerus var. procerus [PENSPRO0] Scrophulariaceae (V:D)
Penstemon cusickii Gray [PENSCUS] Scrophulariaceae (V:D)
Penstemon davidsonii Greene [PENSDAV] Scrophulariaceae (V:D)

Penstemon davidsonii ssp. *menziesii* Keck = Penstemon davidsonii var. menziesii [PENSDAV2] Scrophulariaceae (V:D)
Penstemon davidsonii var. **davidsonii** [PENSDAV0] Scrophulariaceae (V:D)
Penstemon davidsonii var. *ellipticus* (Coult. & Fisher) Boivin = Penstemon ellipticus [PENSELL] Scrophulariaceae (V:D)
Penstemon davidsonii var. **menziesii** (Keck) Cronq. [PENSDAV2] Scrophulariaceae (V:D)
Penstemon ellipticus Coult. & Fisher [PENSELL] Scrophulariaceae (V:D)
Penstemon eriantherus Pursh [PENSERI] Scrophulariaceae (V:D)
Penstemon fruticosus (Pursh) Greene [PENSFRU] Scrophulariaceae (V:D)
Penstemon fruticosus ssp. *scouleri* (Lindl.) Pennell & Keck = Penstemon fruticosus var. scouleri [PENSFRU2] Scrophulariaceae (V:D)
Penstemon fruticosus var. **scouleri** (Lindl.) Cronq. [PENSFRU2] Scrophulariaceae (V:D)
Penstemon gormanii Greene [PENSGOR] Scrophulariaceae (V:D)
Penstemon gracilis Nutt. [PENSGRA] Scrophulariaceae (V:D)
Penstemon lyallii (Gray) Gray [PENSLYA] Scrophulariaceae (V:D)
Penstemon menziesii Hook. p.p. = Penstemon davidsonii var. menziesii [PENSDAV2] Scrophulariaceae (V:D)
Penstemon menziesii ssp. *davidsonii* (Greene) Piper = Penstemon davidsonii var. davidsonii [PENSDAV0] Scrophulariaceae (V:D)
Penstemon menziesii ssp. *thompsonii* Pennell = Penstemon davidsonii var. davidsonii [PENSDAV0] Scrophulariaceae (V:D)
Penstemon nemorosus (Dougl. ex Lindl.) Trautv. = Nothochelone nemorosa [NOTHNEM] Scrophulariaceae (V:D)
Penstemon nitidus Dougl. ex Benth. [PENSNIT] Scrophulariaceae (V:D)
Penstemon ovatus Dougl. ex Hook. [PENSOVA] Scrophulariaceae (V:D)
Penstemon procerus Dougl. ex Graham [PENSPRO] Scrophulariaceae (V:D)
Penstemon procerus var. **procerus** [PENSPRO0] Scrophulariaceae (V:D)
Penstemon procerus var. **tolmiei** (Hook.) Cronq. [PENSPRO2] Scrophulariaceae (V:D)
Penstemon pruinosus Dougl. ex Lindl. [PENSPRU] Scrophulariaceae (V:D)
Penstemon richardsonii Dougl. ex Lindl. [PENSRIC] Scrophulariaceae (V:D)
Penstemon serrulatus Menzies ex Sm. [PENSSER] Scrophulariaceae (V:D)
Penstemon tolmiei Hook. = Penstemon procerus var. tolmiei [PENSPRO2] Scrophulariaceae (V:D)
Pentagramma Yatsk. [PENTAGR$] Pteridaceae (V:F)
Pentagramma triangularis (Kaulfuss) Yatskievych, Windham & Wollenweber [PENTTRI] Pteridaceae (V:F)
Pentaphylloides Duham. [PENTAPH$] Rosaceae (V:D)
Pentaphylloides floribunda (Pursh) A. Löve [PENTFLO] Rosaceae (V:D)
Peplis portula L. = Lythrum portula [LYTHPOR] Lythraceae (V:D)
Peramium decipiens (Hook.) Piper = Goodyera oblongifolia [GOODOBL] Orchidaceae (V:M)
Peramium ophioides (Fern.) Rydb. = Goodyera repens [GOODREP] Orchidaceae (V:M)
Perforaria minuta Degel. = Coccotrema pocillarium [COCCPOC] Coccotremataceae (L)
Perideridia Reichenb. [PERIDER$] Apiaceae (V:D)
Perideridia gairdneri (Hook. & Arn.) Mathias [PERIGAI] Apiaceae (V:D)
Persicaria amphibia (L.) S.F. Gray p.p. = Polygonum amphibium var. emersum [POLYAMP1] Polygonaceae (V:D)
Persicaria amphibia var. *emersa* (Michx.) Hickman = Polygonum amphibium var. emersum [POLYAMP1] Polygonaceae (V:D)
Persicaria amphibia var. *stipulacea* (Coleman) Hara = Polygonum amphibium var. stipulaceum [POLYAMP2] Polygonaceae (V:D)
Persicaria coccinea (Muhl. ex Willd.) Greene = Polygonum amphibium var. emersum [POLYAMP1] Polygonaceae (V:D)
Persicaria hydropiper (L.) Opiz = Polygonum hydropiper [POLYHYD] Polygonaceae (V:D)
Persicaria hydropiperoides (Michx.) Small = Polygonum hydropiperoides [POLYHYR] Polygonaceae (V:D)
Persicaria hydropiperoides var. *breviciliata* (Fern.) C.F. Reed = Polygonum hydropiperoides [POLYHYR] Polygonaceae (V:D)
Persicaria hydropiperoides var. *euronotora* (Fern.) C.F. Reed = Polygonum hydropiperoides [POLYHYR] Polygonaceae (V:D)
Persicaria hydropiperoides var. *opelousana* (Riddell ex Small) J.S. Wilson = Polygonum hydropiperoides [POLYHYR] Polygonaceae (V:D)
Persicaria lapathifolia (L.) S.F. Gray = Polygonum lapathifolium [POLYLAP] Polygonaceae (V:D)
Persicaria maculata (Raf.) S.F. Gray = Polygonum persicaria [POLYPES] Polygonaceae (V:D)
Persicaria mesochora Greene = Polygonum amphibium var. stipulaceum [POLYAMP2] Polygonaceae (V:D)
Persicaria muehlenbergii (S. Wats.) Small = Polygonum amphibium var. emersum [POLYAMP1] Polygonaceae (V:D)
Persicaria nebraskensis Greene = Polygonum amphibium var. stipulaceum [POLYAMP2] Polygonaceae (V:D)
Persicaria nepalense (Meisn.) Miyabe = Polygonum nepalense [POLYNEP] Polygonaceae (V:D)
Persicaria opelousana (Riddell ex Small) Small = Polygonum hydropiperoides [POLYHYR] Polygonaceae (V:D)
Persicaria paludicola Small = Polygonum hydropiperoides [POLYHYR] Polygonaceae (V:D)
Persicaria persicaria (L.) Small = Polygonum persicaria [POLYPES] Polygonaceae (V:D)
Persicaria persicarioides (Kunth) Small = Polygonum hydropiperoides [POLYHYR] Polygonaceae (V:D)
Persicaria punctata (Ell.) Small = Polygonum punctatum [POLYPUN] Polygonaceae (V:D)
Persicaria ruderalis (Salisb.) C.F. Reed = Polygonum persicaria [POLYPES] Polygonaceae (V:D)
Persicaria ruderalis var. *vulgaris* (Webb & Moq.) C.F. Reed = Polygonum persicaria [POLYPES] Polygonaceae (V:D)
Persicaria vulgaris Webb & Moq. = Polygonum persicaria [POLYPES] Polygonaceae (V:D)
Pertusaria DC. [PERTUSA$] Pertusariaceae (L)
Pertusaria alaskensis Erichsen [PERTALA] Pertusariaceae (L)
Pertusaria albescens (Hudson) Choisy & Werner [PERTALB] Pertusariaceae (L)
Pertusaria aleutensis Erichsen = Pertusaria alaskensis [PERTALA] Pertusariaceae (L)
Pertusaria amara (Ach.) Nyl. [PERTAMA] Pertusariaceae (L)
Pertusaria ambigens (Nyl.) Tuck. = Pertusaria subambigens [PERTSUA] Pertusariaceae (L)
Pertusaria borealis Erichsen [PERTBOR] Pertusariaceae (L)
Pertusaria carneopallida (Nyl.) Anzi [PERTCAR] Pertusariaceae (L)
Pertusaria chiodectonoides Bagl. ex A. Massal. [PERTCHI] Pertusariaceae (L)
Pertusaria concentrica Erichsen = Pertusaria multipunctoides [PERTMUT] Pertusariaceae (L)
Pertusaria coriacea (Th. Fr.) Th. Fr. [PERTCOR] Pertusariaceae (L)
Pertusaria coriacea var. *obducens* (Nyl.) Vainio = Pertusaria coriacea [PERTCOR] Pertusariaceae (L)
Pertusaria dactylina (Ach.) Nyl. [PERTDAC] Pertusariaceae (L)
Pertusaria discoidea (Pers.) Malme = Pertusaria albescens [PERTALB] Pertusariaceae (L)
Pertusaria freyi Erichsen = Megaspora verrucosa [MEGAVEU] Hymeneliaceae (L)
Pertusaria geminipara (Th. Fr.) C. Knight ex Brodo [PERTGEM] Pertusariaceae (L)
Pertusaria glaucomela (Tuck.) Nyl. [PERTGLA] Pertusariaceae (L)
Pertusaria gyalectina (Nyl.) Zahlbr. [PERTGYA] Pertusariaceae (L)
Pertusaria hakkodensis Yasuda ex Räsänen [PERTHAK] Pertusariaceae (L)
Pertusaria hultenii Erichsen = Ochrolechia subplicans ssp. hultenii [OCHRSUL1] Pertusariaceae (L)
Pertusaria leioplaca DC. [PERTLEI] Pertusariaceae (L)
Pertusaria leucostoma A. Massal. = Pertusaria leioplaca [PERTLEI] Pertusariaceae (L)
Pertusaria multipunctoides Dibben [PERTMUT] Pertusariaceae (L)

Pertusaria nolens Nyl. = Pertusaria chiodectonoides [PERTCHI] Pertusariaceae (L)
Pertusaria octomela (Norman) Erichsen [PERTOCT] Pertusariaceae (L)
Pertusaria oculata (Dickson) Th. Fr. [PERTOCU] Pertusariaceae (L)
Pertusaria ophthalmiza (Nyl.) Nyl. [PERTOPH] Pertusariaceae (L)
Pertusaria panyrga (Ach.) A. Massal. [PERTPAN] Pertusariaceae (L)
Pertusaria protuberans (Sommerf. ex Th. Fr.) Th. Fr. = Pertusaria carneopallida [PERTCAR] Pertusariaceae (L)
Pertusaria pupillaris (Nyl.) Th. Fr. [PERTPUP] Pertusariaceae (L)
Pertusaria rhodoleuca Th. Fr. = Ochrolechia rhodoleuca [OCHRRHO] Pertusariaceae (L)
Pertusaria sommerfeltii (Flörke ex Sommerf.) Fr. [PERTSOM] Pertusariaceae (L)
Pertusaria stenhammarii Hellb. [PERTSTE] Pertusariaceae (L)
Pertusaria subambigens Dibben [PERTSUA] Pertusariaceae (L)
Pertusaria subdactylina Nyl. [PERTSUB] Pertusariaceae (L)
Pertusaria suboculata Brodo & Dibben [PERTSUO] Pertusariaceae (L)
Pertusaria subplicans Nyl. = Ochrolechia subplicans [OCHRSUL] Pertusariaceae (L)
Pertusaria subpupillaris Vezda = Pertusaria glaucomela [PERTGLA] Pertusariaceae (L)
Pertusaria tabuliformis Erichsen = Pertusaria leioplaca [PERTLEI] Pertusariaceae (L)
Pertusaria xanthodes Müll. Arg. [PERTXAN] Pertusariaceae (L)
Pertusaria xanthostoma (Sommerf.) Fr. = Ochrolechia xanthostoma [OCHRXAN] Pertusariaceae (L)
Petasites P. Mill. [PETASIT$] Asteraceae (V:D)
Petasites alaskanus Rydb. = Petasites frigidus var. frigidus [PETAFRI0] Asteraceae (V:D)
Petasites arcticus Porsild = Petasites frigidus var. palmatus [PETAFRI3] Asteraceae (V:D)
Petasites corymbosus (R. Br.) Rydb. = Petasites frigidus var. frigidus [PETAFRI0] Asteraceae (V:D)
Petasites dentatus Blank. = Petasites sagittatus [PETASAG] Asteraceae (V:D)
Petasites frigidus (L.) Fries [PETAFRI] Asteraceae (V:D)
Petasites frigidus var. *corymbosus* (R. Br.) Cronq. = Petasites frigidus var. frigidus [PETAFRI0] Asteraceae (V:D)
Petasites frigidus var. **frigidus** [PETAFRI0] Asteraceae (V:D)
Petasites frigidus var. *hyperboreoides* Hultén = Petasites frigidus var. frigidus [PETAFRI0] Asteraceae (V:D)
Petasites frigidus var. **nivalis** (Greene) Cronq. [PETAFRI2] Asteraceae (V:D)
Petasites frigidus var. **palmatus** (Ait.) Cronq. [PETAFRI3] Asteraceae (V:D)
Petasites gracilis Britt. = Petasites frigidus var. frigidus [PETAFRI0] Asteraceae (V:D)
Petasites hookerianus (Nutt.) Rydb. = Petasites frigidus var. palmatus [PETAFRI3] Asteraceae (V:D)
Petasites hyperboreus Rydb. = Petasites frigidus var. nivalis [PETAFRI2] Asteraceae (V:D)
*__Petasites japonicus__ (Sieb. & Zucc.) Maxim. [PETAJAP] Asteraceae (V:D)
Petasites nivalis Greene = Petasites frigidus var. nivalis [PETAFRI2] Asteraceae (V:D)
Petasites palmatus (Ait.) Gray = Petasites frigidus var. palmatus [PETAFRI3] Asteraceae (V:D)
Petasites palmatus var. *frigidus* Macoun = Petasites frigidus var. nivalis [PETAFRI2] Asteraceae (V:D)
Petasites sagittatus (Banks ex Pursh) Gray [PETASAG] Asteraceae (V:D)
Petasites speciosus (Nutt.) Piper = Petasites frigidus var. palmatus [PETAFRI3] Asteraceae (V:D)
Petasites vitifolius Greene = Petasites frigidus var. nivalis [PETAFRI2] Asteraceae (V:D)
Petrorhagia (Ser.) Link [PETRORH$] Caryophyllaceae (V:D)
*__Petrorhagia prolifera__ (L.) P.W. Ball & Heywood [PETRPRO] Caryophyllaceae (V:D)
*__Petrorhagia saxifraga__ (L.) Link [PETRSAX] Caryophyllaceae (V:D)
Peucedanum ambiguum (Nutt.) Nutt. ex Torr. & Gray = Lomatium ambiguum [LOMAAMB] Apiaceae (V:D)
Peucedanum macrocarpum Nutt. ex Torr. & Gray = Lomatium macrocarpum [LOMAMAC] Apiaceae (V:D)
Peucedanum sandbergii Coult. & Rose = Lomatium sandbergii [LOMASAN] Apiaceae (V:D)
Phaca americana (Hook.) Rydb. ex Small = Astragalus americanus [ASTRAME] Fabaceae (V:D)
Phacelia Juss. [PHACELI$] Hydrophyllaceae (V:D)
Phacelia franklinii (R. Br.) Gray [PHACFRA] Hydrophyllaceae (V:D)
Phacelia hastata Dougl. ex Lehm. [PHACHAS] Hydrophyllaceae (V:D)
Phacelia hastata var. *leptosepala* (Rydb.) Cronq. = Phacelia leptosepala [PHACLEP] Hydrophyllaceae (V:D)
Phacelia hastata var. *leucophylla* (Torr.) Cronq. = Phacelia hastata [PHACHAS] Hydrophyllaceae (V:D)
Phacelia leptosepala Rydb. [PHACLEP] Hydrophyllaceae (V:D)
Phacelia linearis (Pursh) Holz. [PHACLIN] Hydrophyllaceae (V:D)
Phacelia lyallii (Gray) Rydb. [PHACLYA] Hydrophyllaceae (V:D)
Phacelia mollis J.F. Macbr. [PHACMOL] Hydrophyllaceae (V:D)
Phacelia ramosissima Dougl. ex Lehm. [PHACRAM] Hydrophyllaceae (V:D)
Phacelia sericea (Graham) Gray [PHACSER] Hydrophyllaceae (V:D)
Phacopsis Tul. [PHACOPS$] Lecanoraceae (L)
Phacopsis huuskonenii Räsänen [PHACHUU] Lecanoraceae (L)
Phacopsis oxyspora (Tul.) Triebel & Rambold [PHACOXY] Lecanoraceae (L)
Phacopsis vulpina Tul. [PHACVUL] Lecanoraceae (L)
Phaeocalicium A.F.W. Schmidt [PHAEOCA$] Mycocaliciaceae (L)
Phaeocalicium compressulum (Nyl. ex Szat.) A.F.W. Schmidt [PHAECOM] Mycocaliciaceae (L)
Phaeocalicium populneum (Brond. ex Duby) A.F.W. Schmidt [PHAEPOP] Mycocaliciaceae (L)
Phaeoceros Prosk. [PHAEOCE$] Anthocerotaceae (B:H)
Phaeoceros hallii (Aust.) Prosk. [PHAEHAL] Anthocerotaceae (B:H)
Phaeoceros laevis (L.) Prosk. [PHAELAE] Anthocerotaceae (B:H)
Phaeophyscia Moberg [PHAEOPH$] Physciaceae (L)
Phaeophyscia adiastola (Essl.) Essl. [PHAEADI] Physciaceae (L)
Phaeophyscia ciliata (Hoffm.) Moberg [PHAECIL] Physciaceae (L)
Phaeophyscia constipata (Norrlin & Nyl.) Moberg [PHAECON] Physciaceae (L)
Phaeophyscia endococcina (Körber) Moberg [PHAEEND] Physciaceae (L)
Phaeophyscia hirsuta (Mereschk.) Essl. [PHAEHIR] Physciaceae (L)
Phaeophyscia hirtella Essl. [PHAEHIT] Physciaceae (L)
Phaeophyscia hispidula (Ach.) Essl. [PHAEHIS] Physciaceae (L)
Phaeophyscia hispidula ssp. **hispidula** [PHAEHIS0] Physciaceae (L)
Phaeophyscia hispidula ssp. **limbata** Poelt [PHAEHIS1] Physciaceae (L)
Phaeophyscia kairamoi (Vainio) Moberg [PHAEKAI] Physciaceae (L)
Phaeophyscia nigricans (Flörke) Moberg [PHAENIG] Physciaceae (L)
Phaeophyscia orbicularis (Necker) Moberg [PHAEORB] Physciaceae (L)
Phaeophyscia sciastra (Ach.) Moberg [PHAESCI] Physciaceae (L)
Phaeorrhiza H. Mayrh. & Poelt [PHAEORR$] Physciaceae (L)
Phaeorrhiza nimbosa (Fr.) H. Mayrh. & Poelt [PHAENIM] Physciaceae (L)
Phaeorrhiza sareptana (Tomin) H. Mayrh. & Poelt [PHAESAR] Physciaceae (L)
Phalangium quamash Pursh = Camassia quamash ssp. quamash [CAMAQUA0] Liliaceae (V:M)
Phalaris L. [PHALARI$] Poaceae (V:M)
Phalaris arundinacea L. [PHALARU] Poaceae (V:M)
Phalaris arundinacea var. *picta* L. = Phalaris arundinacea [PHALARU] Poaceae (V:M)
*__Phalaris canariensis__ L. [PHALCAN] Poaceae (V:M)
Phalaris oryzoides L. = Leersia oryzoides [LEERORY] Poaceae (V:M)
Phalaroides arundinacea (L.) Raeusch. = Phalaris arundinacea [PHALARU] Poaceae (V:M)

Phalaroides arundinacea var. *picta* (L.) Tzvelev = Phalaris arundinacea [PHALARU] Poaceae (V:M)
Pharomitrium subsessile (Brid.) Schimp. = Pterygoneurum subsessile [PTERSUB] Pottiaceae (B:M)
Phascum Hedw. [PHASCUM$] Pottiaceae (B:M)
Phascum acaulon With. = Phascum cuspidatum [PHASCUS] Pottiaceae (B:M)
Phascum cuspidatum Hedw. [PHASCUS] Pottiaceae (B:M)
Phascum cuspidatum var. *americanum* Ren. & Card. = Phascum cuspidatum [PHASCUS] Pottiaceae (B:M)
Phascum cuspidatum var. *henricii* (Ren. & Card.) Wijk & Marg. = Phascum cuspidatum [PHASCUS] Pottiaceae (B:M)
Phascum cuspidatum var. *piliferum* (Hedw.) Amann = Phascum cuspidatum [PHASCUS] Pottiaceae (B:M)
Phascum cuspidatum var. *schreberianum* (Dicks.) Brid. = Phascum cuspidatum [PHASCUS] Pottiaceae (B:M)
Phascum piliferum Schreb. ex Hedw. = Phascum cuspidatum [PHASCUS] Pottiaceae (B:M)
Phascum subexsertum Hook. in Drumm. = Pottia bryoides [POTTBRY] Pottiaceae (B:M)
Phascum vlassovii Laz. [PHASVLA] Pottiaceae (B:M)
Phegopteris Fée [PHEGOPT$] Thelypteridaceae (V:F)
Phegopteris connectilis (Michx.) Watt [PHEGCON] Thelypteridaceae (V:F)
Phegopteris dryopteris (L.) Fée = Gymnocarpium dryopteris [GYMNDRY] Dryopteridaceae (V:F)
Phegopteris polypodioides Fée = Phegopteris connectilis [PHEGCON] Thelypteridaceae (V:F)
Phelipaea pinorum (Geyer ex Hook.) Gray = Orobanche pinorum [OROBPIN] Orobanchaceae (V:D)
Philadelphus L. [PHILADE$] Hydrangeaceae (V:D)
Philadelphus lewisii Pursh [PHILLEW] Hydrangeaceae (V:D)
Philonotis Brid. [PHILONO$] Bartramiaceae (B:M)
Philonotis acutiflora Kindb. in Röll = Philonotis fontana var. pumila [PHILFON4] Bartramiaceae (B:M)
Philonotis americana Dism. = Philonotis fontana var. americana [PHILFON1] Bartramiaceae (B:M)
Philonotis americana var. *gracilescens* (Dism.) Flow. in Grout = Philonotis fontana var. americana [PHILFON1] Bartramiaceae (B:M)
Philonotis americana var. *torquata* (Ren. & Geh. in Geh.) Flow. in Grout = Philonotis fontana var. americana [PHILFON1] Bartramiaceae (B:M)
Philonotis arnellii Husn. = Philonotis capillaris [PHILCAP] Bartramiaceae (B:M)
Philonotis calcarea f. *occidentalis* Flow. in Grout = Philonotis fontana var. fontana [PHILFON0] Bartramiaceae (B:M)
Philonotis capillaris Lindb. in Hartm. [PHILCAP] Bartramiaceae (B:M)
Philonotis fontana (Hedw.) Brid. [PHILFON] Bartramiaceae (B:M)
Philonotis fontana var. *adpressa* (Ferg.) Limpr. = Philonotis fontana var. fontana [PHILFON0] Bartramiaceae (B:M)
Philonotis fontana var. **americana** (Dism.) Flow. [PHILFON1] Bartramiaceae (B:M)
Philonotis fontana var. *borealis* Hag. = Philonotis fontana var. fontana [PHILFON0] Bartramiaceae (B:M)
Philonotis fontana var. *brachyphylla* Kindb. in Mac. = Philonotis fontana var. fontana [PHILFON0] Bartramiaceae (B:M)
Philonotis fontana var. *columbiae* Kindb. in Mac. & Kindb. = Philonotis fontana var. fontana [PHILFON0] Bartramiaceae (B:M)
Philonotis fontana var. *falcata* (Hook.) Brid. = Philonotis fontana var. fontana [PHILFON0] Bartramiaceae (B:M)
Philonotis fontana var. **fontana** [PHILFON0] Bartramiaceae (B:M)
Philonotis fontana var. *heterophylla* Dism. = Philonotis fontana var. fontana [PHILFON0] Bartramiaceae (B:M)
Philonotis fontana var. *laxa* Vent. = Philonotis fontana var. fontana [PHILFON0] Bartramiaceae (B:M)
Philonotis fontana var. *microblasta* C. Müll. & Kindb. in Mac. & Kindb. = Philonotis fontana [PHILFON] Bartramiaceae (B:M)
Philonotis fontana var. *microthamnia* Kindb. = Philonotis fontana [PHILFON] Bartramiaceae (B:M)
Philonotis fontana var. **pumila** (Turn.) Brid. [PHILFON4] Bartramiaceae (B:M)
Philonotis fontana var. *seriata* (Mitt.) Kindb. = Philonotis yezoana [PHILYEZ] Bartramiaceae (B:M)
Philonotis fontana var. *seriata* Breidl. = Philonotis fontana var. americana [PHILFON1] Bartramiaceae (B:M)
Philonotis fontana var. *seriata* f. *dimorphophylla* Flow. in Grout = Philonotis yezoana [PHILYEZ] Bartramiaceae (B:M)
Philonotis fontana var. *seriata* f. *occidentalis* Flow. in Grout = Philonotis yezoana [PHILYEZ] Bartramiaceae (B:M)
Philonotis fontana var. *serrata* Kindb. in Mac. & Kindb. = Philonotis fontana var. americana [PHILFON1] Bartramiaceae (B:M)
Philonotis macounii Lesq. & James = Philonotis capillaris [PHILCAP] Bartramiaceae (B:M)
Philonotis marchica (Hedw.) Brid. [PHILMAR] Bartramiaceae (B:M)
Philonotis marchica var. *laxa* (Limpr.) Loeske & Warnst. = Philonotis marchica [PHILMAR] Bartramiaceae (B:M)
Philonotis tomentella Mol. in Lor. = Philonotis fontana var. pumila [PHILFON4] Bartramiaceae (B:M)
Philonotis vancouverensis Kindb. = Philonotis capillaris [PHILCAP] Bartramiaceae (B:M)
Philonotis yezoana Besch. & Card. in Card. [PHILYEZ] Bartramiaceae (B:M)
Philotria angustifolia (Muhl.) Britt. ex Rydb. = Elodea nuttallii [ELODNUT] Hydrocharitaceae (V:M)
Philotria canadensis (Michx.) Britt. = Elodea canadensis [ELODCAN] Hydrocharitaceae (V:M)
Philotria densa (Planch.) Small & St. John = Egeria densa [EGERDEN] Hydrocharitaceae (V:M)
Philotria linearis Rydb. = Elodea canadensis [ELODCAN] Hydrocharitaceae (V:M)
Philotria minor (Engelm. ex Caspary) Small = Elodea nuttallii [ELODNUT] Hydrocharitaceae (V:M)
Philotria nuttallii (Planch.) Rydb. = Elodea nuttallii [ELODNUT] Hydrocharitaceae (V:M)
Philotria occidentalis (Pursh) House = Elodea nuttallii [ELODNUT] Hydrocharitaceae (V:M)
Phippsia arctica (Hook.) A. & D. Löve = Puccinellia arctica [PUCCARC] Poaceae (V:M)
Phippsia borealis (Swallen) A. & D. Löve = Puccinellia arctica [PUCCARC] Poaceae (V:M)
Phippsia interior (Sorensen) A. & D. Löve = Puccinellia interior [PUCCINT] Poaceae (V:M)
Phippsia nutkaensis (J. Presl) A. & D. Löve = Puccinellia nutkaensis [PUCCNUT] Poaceae (V:M)
Phleum L. [PHLEUM$] Poaceae (V:M)
Phleum alpinum L. [PHLEALP] Poaceae (V:M)
Phleum alpinum var. *commutatum* (Gaudin) Griseb. = Phleum alpinum [PHLEALP] Poaceae (V:M)
Phleum commutatum Gaudin = Phleum alpinum [PHLEALP] Poaceae (V:M)
Phleum commutatum var. *americanum* (Fourn.) Hultén = Phleum alpinum [PHLEALP] Poaceae (V:M)
***Phleum pratense** L. [PHLEPRA] Poaceae (V:M)
Phlox L. [PHLOX$] Polemoniaceae (V:D)
Phlox alyssifolia Greene [PHLOALY] Polemoniaceae (V:D)
Phlox caespitosa Nutt. [PHLOCAE] Polemoniaceae (V:D)
Phlox caespitosa ssp. *eucaespitosa* Brand = Phlox caespitosa [PHLOCAE] Polemoniaceae (V:D)
Phlox diffusa Benth. [PHLODIF] Polemoniaceae (V:D)
Phlox diffusa ssp. **longistylis** Wherry [PHLODIF1] Polemoniaceae (V:D)
Phlox diffusa var. *longistylis* (Wherry) M.E. Peck = Phlox diffusa ssp. longistylis [PHLODIF1] Polemoniaceae (V:D)
Phlox douglasii Hook. = Phlox caespitosa [PHLOCAE] Polemoniaceae (V:D)
Phlox douglasii ssp. *eudouglasii* Brand = Phlox caespitosa [PHLOCAE] Polemoniaceae (V:D)
Phlox gracilis (Hook.) Greene [PHLOGRA] Polemoniaceae (V:D)
Phlox hoodii Richards. [PHLOHOO] Polemoniaceae (V:D)
Phlox longifolia Nutt. [PHLOLON] Polemoniaceae (V:D)
Phlox speciosa Pursh [PHLOSPE] Polemoniaceae (V:D)
Phlyctis Wallr. [PHLYCTS$] Phlyctidaceae (L)
Phlyctis argena (Sprengel) Flotow [PHLYARG] Phlyctidaceae (L)

Phlyctis speirea G. Merr. [PHLYSPE] Phlyctidaceae (L)
Phragmites Adans. [PHRAGMI$] Poaceae (V:M)
Phragmites australis (Cav.) Trin. ex Steud. [PHRAAUS] Poaceae (V:M)
Phragmites australis var. *berlandieri* (Fourn.) C.F. Reed = Phragmites australis [PHRAAUS] Poaceae (V:M)
Phragmites communis Trin. = Phragmites australis [PHRAAUS] Poaceae (V:M)
Phragmites communis ssp. *berlandieri* (Fourn.) A. & D. Löve = Phragmites australis [PHRAAUS] Poaceae (V:M)
Phragmites communis var. *berlandieri* (Fourn.) Fern. = Phragmites australis [PHRAAUS] Poaceae (V:M)
Phragmites phragmites (L.) Karst. = Phragmites australis [PHRAAUS] Poaceae (V:M)
Phragmonaevia fuckelii Rehm = Corticifraga fuckelii [CORTFUC] Uncertain-family (L)
Phylliscum Nyl. [PHYLLIS$] Lichinaceae (L)
Phylliscum demangeonii (Moug. & Mont.) Nyl. [PHYLDEM] Lichinaceae (L)
Phyllodoce Salisb. [PHYLLOD$] Ericaceae (V:D)
Phyllodoce aleutica ssp. *glanduliflora* (Hook.) Hultén = Phyllodoce glanduliflora [PHYLGLA] Ericaceae (V:D)
Phyllodoce empetriformis (Sm.) D. Don [PHYLEMP] Ericaceae (V:D)
Phyllodoce glanduliflora (Hook.) Coville [PHYLGLA] Ericaceae (V:D)
Phyllospadix Hook. [PHYLLOS$] Zosteraceae (V:M)
Phyllospadix scouleri Hook. [PHYLSCO] Zosteraceae (V:M)
Phyllospadix serrulatus Rupr. ex Aschers. [PHYLSER] Zosteraceae (V:M)
Phyllospadix torreyi S. Wats. [PHYLTOR] Zosteraceae (V:M)
Physaria (Nutt. ex Torr. & Gray) Gray [PHYSARI$] Brassicaceae (V:D)
Physaria didymocarpa (Hook.) Gray [PHYSDID] Brassicaceae (V:D)
Physaria didymocarpa var. *normalis* Kuntze = Physaria didymocarpa [PHYSDID] Brassicaceae (V:D)
Physcia (Schreber) Michaux [PHYSCIA$] Physciaceae (L)
Physcia adiastola Essl. = Phaeophyscia adiastola [PHAEADI] Physciaceae (L)
Physcia adscendens (Fr.) H. Olivier [PHYSADS] Physciaceae (L)
Physcia aipolia (Ehrh. ex Humb.) Fürnr. [PHYSAIP] Physciaceae (L)
Physcia aipolia var. **aipolia** [PHYSAIP0] Physciaceae (L)
Physcia aipolia var. **alnophila** (Vainio) Lynge [PHYSAIP1] Physciaceae (L)
Physcia biziana (A. Massal.) Zahlbr. [PHYSBIZ] Physciaceae (L)
Physcia caesia (Hoffm.) Fürnr. [PHYSCAE] Physciaceae (L)
Physcia cainii Räsänen = Physcia aipolia [PHYSAIP] Physciaceae (L)
Physcia callosa Nyl. [PHYSCAO] Physciaceae (L)
Physcia cascadensis H. Magn. [PHYSCAS] Physciaceae (L)
Physcia ciliata (Hoffm.) Du Rietz = Phaeophyscia ciliata [PHAECIL] Physciaceae (L)
Physcia columbiana de Lesd. = Phaeophyscia endococcina [PHAEEND] Physciaceae (L)
Physcia constipata Norrlin & Nyl. = Phaeophyscia constipata [PHAECON] Physciaceae (L)
Physcia detersa (Nyl.) Nyl. = Physconia detersa [PHYSDET] Physciaceae (L)
Physcia dimidiata (Arnold) Nyl. [PHYSDIM] Physciaceae (L)
Physcia dubia (Hoffm.) Lettau [PHYSDUB] Physciaceae (L)
Physcia endococcina (Körber) Th. Fr. = Phaeophyscia endococcina [PHAEEND] Physciaceae (L)
Physcia hirsuta Mereschk. = Phaeophyscia hirsuta [PHAEHIR] Physciaceae (L)
Physcia hispida auct. = Physcia tenella [PHYSTEN] Physciaceae (L)
Physcia hispidula (Ach.) Frey = Phaeophyscia hispidula [PHAEHIS] Physciaceae (L)
Physcia intermedia Vainio = Physcia dubia [PHYSDUB] Physciaceae (L)
Physcia kairamoi Vainio = Phaeophyscia kairamoi [PHAEKAI] Physciaceae (L)
Physcia leptalea (Ach.) DC. [PHYSLEP] Physciaceae (L)
Physcia lithotea (Ach.) Nyl. = Phaeophyscia sciastra [PHAESCI] Physciaceae (L)
Physcia lithotodes Nyl. = Phaeophyscia endococcina [PHAEEND] Physciaceae (L)
Physcia melops Dufour = Physcia phaea [PHYSPHA] Physciaceae (L)
Physcia muscigena (Ach.) Nyl. = Physconia muscigena [PHYSMUS] Physciaceae (L)
Physcia nigricans (Flörke) Stizenb. = Phaeophyscia nigricans [PHAENIG] Physciaceae (L)
Physcia obscura auct. = Phaeophyscia ciliata [PHAECIL] Physciaceae (L)
Physcia orbicularis (Necker) Poetsch = Phaeophyscia orbicularis [PHAEORB] Physciaceae (L)
Physcia phaea (Tuck.) J.W. Thomson [PHYSPHA] Physciaceae (L)
Physcia sciastra (Ach.) Du Rietz = Phaeophyscia sciastra [PHAESCI] Physciaceae (L)
Physcia semipinnata (J.F. Gmelin) Moberg = Physcia leptalea [PHYSLEP] Physciaceae (L)
Physcia setosa (Ach.) Nyl. = Phaeophyscia hispidula [PHAEHIS] Physciaceae (L)
Physcia stellaris (L.) Nyl. [PHYSSTE] Physciaceae (L)
Physcia tenella (Scop.) DC. [PHYSTEN] Physciaceae (L)
Physcia teretiuscula (Ach.) Lynge = Physcia dubia [PHYSDUB] Physciaceae (L)
Physcia wainioi Räsänen [PHYSWAI] Physciaceae (L)
Physcomitrella Bruch & Schimp. in B.S.G. [PHYSCOM$] Funariaceae (B:M)
Physcomitrella patens (Hedw.) Bruch & Schimp. in B.S.G. [PHYSPAT] Funariaceae (B:M)
Physcomitrium (Brid.) Brid. [PHYSCOI$] Funariaceae (B:M)
Physcomitrium australe Britt. = Physcomitrium pyriforme [PHYSPYR] Funariaceae (B:M)
Physcomitrium californicum Britt. = Physcomitrium pyriforme [PHYSPYR] Funariaceae (B:M)
Physcomitrium drummondii Britt. = Physcomitrium pyriforme [PHYSPYR] Funariaceae (B:M)
Physcomitrium immersum Sull. [PHYSIMM] Funariaceae (B:M)
Physcomitrium kellermanii Britt. = Physcomitrium pyriforme [PHYSPYR] Funariaceae (B:M)
Physcomitrium kellermanii var. *drummondii* (Britt.) Grout = Physcomitrium pyriforme [PHYSPYR] Funariaceae (B:M)
Physcomitrium megalocarpum Kindb. = Physcomitrium pyriforme [PHYSPYR] Funariaceae (B:M)
Physcomitrium megalocarpum var. *californicum* (Britt.) Grout = Physcomitrium pyriforme [PHYSPYR] Funariaceae (B:M)
Physcomitrium pyriforme (Hedw.) Hampe [PHYSPYR] Funariaceae (B:M)
Physcomitrium pyriforme var. *serratum* (Ren. & Card.) Crum & Anderson = Physcomitrium pyriforme [PHYSPYR] Funariaceae (B:M)
Physcomitrium strangulatum Kindb. = Physcomitrium pyriforme [PHYSPYR] Funariaceae (B:M)
Physcomitrium turbinatum C. Müll. ex Lesq. & James = Physcomitrium pyriforme [PHYSPYR] Funariaceae (B:M)
Physcomitrium turbinatum (Michx.) Britt. = Physcomitrium pyriforme [PHYSPYR] Funariaceae (B:M)
Physcomitrium turbinatum f. *australe* (Britt.) Grout = Physcomitrium pyriforme [PHYSPYR] Funariaceae (B:M)
Physcomitrium turbinatum var. *langloisii* (Ren. & Card.) Britt. = Physcomitrium pyriforme [PHYSPYR] Funariaceae (B:M)
Physconia Poelt [PHYSCOA$] Physciaceae (L)
Physconia americana Essl. [PHYSAMR] Physciaceae (L)
Physconia detersa (Nyl.) Poelt [PHYSDET] Physciaceae (L)
Physconia enteroxantha (Nyl.) Poelt [PHYSENT] Physciaceae (L)
Physconia farrea auct. = Physconia perisidiosa [PHYSPER] Physciaceae (L)
Physconia muscigena (Ach.) Poelt [PHYSMUS] Physciaceae (L)
Physconia perisidiosa (Erichsen) Moberg [PHYSPER] Physciaceae (L)
Physconia thomsonii Essl. [PHYSTHO] Physciaceae (L)
Physocarpus Maxim. [PHYSOCA$] Rosaceae (V:D)
Physocarpus capitatus (Pursh) Kuntze [PHYSCAP] Rosaceae (V:D)

Physocarpus malvaceus (Greene) Kuntze [PHYSMAL] Rosaceae (V:D)
Physocarpus opulifolius var. *tomentellus* (Ser.) Boivin = Physocarpus capitatus [PHYSCAP] Rosaceae (V:D)
Physostegia Benth. [PHYSOST$] Lamiaceae (V:D)
Physostegia nuttallii (Britt.) Fassett = Physostegia parviflora [PHYSPAR] Lamiaceae (V:D)
Physostegia parviflora Nutt. ex Gray [PHYSPAR] Lamiaceae (V:D)
Physostegia virginiana var. *parviflora* (Nutt. ex Gray) Boivin = Physostegia parviflora [PHYSPAR] Lamiaceae (V:D)
Phytoconis ericetorum (Pers.) Redhead & Kuyper = Omphalina umbellifera [OMPHUMB] Tricholomataceae (L)
Phytoconis viridis (Ach.) Redhead & Kuyper = Omphalina hudsoniana [OMPHHUD] Tricholomataceae (L)
Picea A. Dietr. [PICEA$] Pinaceae (V:G)
Picea canadensis (P. Mill.) B.S.P. = Picea glauca [PICEGLA] Pinaceae (V:G)
Picea engelmannii Parry ex Engelm. [PICEENG] Pinaceae (V:G)
Picea engelmannii × glauca [PICEENE] Pinaceae (V:G)
Picea glauca (Moench) Voss [PICEGLA] Pinaceae (V:G)
Picea glauca ssp. *engelmannii* (Parry ex Engelm.) T.M.C. Taylor = Picea engelmannii [PICEENG] Pinaceae (V:G)
Picea glauca var. *albertiana* (S. Br.) Sarg. = Picea glauca [PICEGLA] Pinaceae (V:G)
Picea glauca var. *densata* Bailey = Picea glauca [PICEGLA] Pinaceae (V:G)
Picea glauca var. *engelmannii* (Parry ex Engelm.) Boivin = Picea engelmannii [PICEENG] Pinaceae (V:G)
Picea glauca var. *porsildii* Raup = Picea glauca [PICEGLA] Pinaceae (V:G)
Picea × lutzii Little [PICEXLU] Pinaceae (V:G)
Picea mariana (P. Mill.) B.S.P. [PICEMAR] Pinaceae (V:G)
Picea sitchensis (Bong.) Carr. [PICESIT] Pinaceae (V:G)
Pilophorus Th. Fr. [PILOPHO$] Stereocaulaceae (L)
Pilophorus acicularis (Ach.) Th. Fr. [PILOACI] Stereocaulaceae (L)
Pilophorus cereolus (Ach.) Th. Fr. [PILOCER] Stereocaulaceae (L)
Pilophorus cereolus var. *acicularis* (Ach.) Tuck. = Pilophorus acicularis [PILOACI] Stereocaulaceae (L)
Pilophorus clavatus Th. Fr. [PILOCLA] Stereocaulaceae (L)
Pilophorus dovrensis (Nyl.) Timdal, Hertel & Rambold [PILODOV] Stereocaulaceae (L)
Pilophorus hallii (Tuck.) Vainio = Pilophorus clavatus [PILOCLA] Stereocaulaceae (L)
Pilophorus nigricaulis Satô [PILONIG] Stereocaulaceae (L)
Pilophorus pallidus (Th. Fr.) Timdal = Pilophorus dovrensis [PILODOV] Stereocaulaceae (L)
Pilophorus vegae Krog [PILOVEG] Stereocaulaceae (L)
Pilosella novae-angliae Rydb. = Braya humilis [BRAYHUM] Brassicaceae (V:D)
Pilosella richardsonii Rydb. = Braya humilis [BRAYHUM] Brassicaceae (V:D)
Pinguicula L. [PINGUIC$] Lentibulariaceae (V:D)
Pinguicula arctica Eastw. = Pinguicula macroceras [PINGMAC] Lentibulariaceae (V:D)
Pinguicula macroceras Link [PINGMAC] Lentibulariaceae (V:D)
Pinguicula villosa L. [PINGVIL] Lentibulariaceae (V:D)
Pinguicula vulgaris L. [PINGVUL] Lentibulariaceae (V:D)
Pinguicula vulgaris ssp. *macroceras* (Link) Calder & Taylor = Pinguicula macroceras [PINGMAC] Lentibulariaceae (V:D)
Pinguicula vulgaris var. *americana* Gray = Pinguicula vulgaris [PINGVUL] Lentibulariaceae (V:D)
Pinguicula vulgaris var. *macroceras* (Link) Herder = Pinguicula macroceras [PINGMAC] Lentibulariaceae (V:D)
Pinus L. [PINUS$] Pinaceae (V:G)
Pinus albicaulis Engelm. [PINUALB] Pinaceae (V:G)
Pinus balsamea L. = Abies balsemea [ABIEBAL] Pinaceae (V:G)
Pinus banksiana Lamb. [PINUBAN] Pinaceae (V:G)
Pinus contorta Dougl. ex Loud. [PINUCON] Pinaceae (V:G)
Pinus contorta ssp. *latifolia* (Engelm. ex S. Wats.) Critchfield = Pinus contorta var. latifolia [PINUCON2] Pinaceae (V:G)
Pinus contorta var. **contorta** [PINUCON0] Pinaceae (V:G)
Pinus contorta var. **latifolia** Engelm. ex S. Wats. [PINUCON2] Pinaceae (V:G)
Pinus divaricata (Ait.) Dum.-Cours. = Pinus banksiana [PINUBAN] Pinaceae (V:G)
Pinus divaricata var. *hendersonii* (Lemmon) Boivin = Pinus contorta var. latifolia [PINUCON2] Pinaceae (V:G)
Pinus divaricata var. *latifolia* (Engelm. ex S. Wats.) Boivin = Pinus contorta var. latifolia [PINUCON2] Pinaceae (V:G)
Pinus divaricata var. × *musci* Boivin = Pinus × murraybanksiana [PINUXMU] Pinaceae (V:G)
Pinus flexilis James [PINUFLE] Pinaceae (V:G)
Pinus monticola Dougl. ex D. Don [PINUMON] Pinaceae (V:G)
Pinus × murraybanksiana Righter & Stockwell [PINUXMU] Pinaceae (V:G)
Pinus ponderosa P.& C. Lawson [PINUPON] Pinaceae (V:G)
Pinus strobus var. *monticola* (Dougl. ex D. Don) Nutt. = Pinus monticola [PINUMON] Pinaceae (V:G)
Piperia Rydb. [PIPERIA$] Orchidaceae (V:M)
Piperia elegans (Lindl.) Rydb. [PIPEELE] Orchidaceae (V:M)
Piperia maritima (Greene) Rydb. = Piperia elegans [PIPEELE] Orchidaceae (V:M)
Piperia multiflora Rydb. = Piperia elegans [PIPEELE] Orchidaceae (V:M)
Piperia unalascensis (Spreng.) Rydb. [PIPEUNA] Orchidaceae (V:M)
Pisum maritimum L. = Lathyrus japonicus var. maritimus [LATHJAP1] Fabaceae (V:D)
Pisum maritimum var. *glabrum* Ser. = Lathyrus japonicus var. maritimus [LATHJAP1] Fabaceae (V:D)
Pityrogramma triangularis (Kaulfuss) Maxon = Pentagramma triangularis [PENTTRI] Pteridaceae (V:F)
Placidiopsis Beltr. [PLACIDI$] Verrucariaceae (L)
Placidiopsis pseudocinerea Breuss [PLACPSE] Verrucariaceae (L)
Placidium hepaticum var. *pruinosum* de Lesd. = Catapyrenium lachneum [CATALAC] Verrucariaceae (L)
Placidium hepaticum var. *reticulatum* de Lesd. = Catapyrenium lachneum [CATALAC] Verrucariaceae (L)
Placodium bolacinum Tuck. = Caloplaca bolacina [CALOBOL] Teloschistaceae (L)
Placodium luteominium Tuck. = Caloplaca luteominia [CALOLUT] Teloschistaceae (L)
Placopsis (Nyl.) Lindsay [PLACOPS$] Trapeliaceae (L)
Placopsis cribellans (Nyl.) Räsänen [PLACCRI] Trapeliaceae (L)
Placopsis effusa Lamb [PLACEFF] Trapeliaceae (L)
Placopsis gelida (L.) Lindsay [PLACGEL] Trapeliaceae (L)
Placynthiella Elenkin [PLACYNH$] Trapeliaceae (L)
Placynthiella icmalea (Ach.) Coppins & P. James [PLACICM] Trapeliaceae (L)
Placynthiella oligotropha (J.R. Laundon) Coppins & P. James [PLACOLI] Trapeliaceae (L)
Placynthiella uliginosa (Schrader) Coppins & P. James [PLACULI] Trapeliaceae (L)
Placynthium (Ach.) Gray [PLACYNT$] Placynthiaceae (L)
Placynthium asperellum (Ach.) Trevisan [PLACASE] Placynthiaceae (L)
Placynthium aspratile (Ach.) Henssen = Placynthium asperellum [PLACASE] Placynthiaceae (L)
Placynthium dubium Herre = Massalongia microphylliza [MASSMIC] Peltigeraceae (L)
Placynthium flabellosum (Tuck.) Zahlbr. [PLACFLA] Placynthiaceae (L)
Placynthium microphyllizum (Nyl. ex Hasse) Hasse = Massalongia microphylliza [MASSMIC] Peltigeraceae (L)
Placynthium nigrum (Hudson) Gray [PLACNIG] Placynthiaceae (L)
Placynthium nigrum var. **nigrum** [PLACNIG0] Placynthiaceae (L)
Placynthium nigrum var. **tantaleum** (Hepp) Arnold [PLACNIG1] Placynthiaceae (L)
Placynthium stenophyllum (Tuck.) Fink [PLACSTE] Placynthiaceae (L)
Placynthium stenophyllum var. **isidiatum** Henssen [PLACSTE1] Placynthiaceae (L)
Placynthium subradiatum (Nyl.) Arnold [PLACSUR] Placynthiaceae (L)
Plagiobothrys Fisch. & C.A. Mey. [PLAGIOB$] Boraginaceae (V:D)

Plagiobothrys asper Greene = Plagiobothrys tenellus [PLAGTEN] Boraginaceae (V:D)
Plagiobothrys figuratus (Piper) I.M. Johnston ex M.E. Peck [PLAGFIG] Boraginaceae (V:D)
Plagiobothrys scouleri (Hook. & Arn.) I.M. Johnston [PLAGSCO] Boraginaceae (V:D)
Plagiobothrys tenellus (Nutt. ex Hook.) Gray [PLAGTEN] Boraginaceae (V:D)
Plagiobryum Lindb. [PLAGIOR$] Bryaceae (B:M)
Plagiobryum demissum (Hook.) Lindb. [PLAGDEM] Bryaceae (B:M)
Plagiobryum zieri (Hedw.) Lindb. [PLAGZIE] Bryaceae (B:M)
Plagiochila (Dum.) Dum. [PLAGIOH$] Plagiochilaceae (B:H)
Plagiochila alaskana Evans = Plagiochila semidecurrens var. alaskana [PLAGSEM1] Plagiochilaceae (B:H)
Plagiochila arctica auct. = Plagiochila asplenioides [PLAGASL] Plagiochilaceae (B:H)
Plagiochila asplenioides (L.) Dum. [PLAGASL] Plagiochilaceae (B:H)
Plagiochila asplenioides ssp. *porelloides* (Torrey ex Nees) Schust. = Plagiochila porelloides [PLAGPOR] Plagiochilaceae (B:H)
Plagiochila firma ssp. *confusa* (Schust.) H. Inoue = Plagiochila schofieldiana [PLAGSCH] Plagiochilaceae (B:H)
Plagiochila fryei Evans = Plagiochila semidecurrens var. alaskana [PLAGSEM1] Plagiochilaceae (B:H)
Plagiochila major (Nees) S. Arnell = Plagiochila asplenioides [PLAGASL] Plagiochilaceae (B:H)
Plagiochila porelloides (Torrey ex Nees) Lindenb. [PLAGPOR] Plagiochilaceae (B:H)
Plagiochila rhizophora ssp. *confusa* Schust. = Plagiochila schofieldiana [PLAGSCH] Plagiochilaceae (B:H)
Plagiochila satoi Hatt. [PLAGSAT] Plagiochilaceae (B:H)
Plagiochila satoi var. **magnum** Schof. & Hong [PLAGSAT1] Plagiochilaceae (B:H)
Plagiochila satoi var. **satoi** [PLAGSAT0] Plagiochilaceae (B:H)
Plagiochila schofieldiana H. Inoue [PLAGSCH] Plagiochilaceae (B:H)
Plagiochila semidecurrens Lehm. & Lindenb. [PLAGSEM] Plagiochilaceae (B:H)
Plagiochila semidecurrens ssp. *grossidens* (Herz.) Schust. p.p. = Plagiochila semidecurrens var. alaskana [PLAGSEM1] Plagiochilaceae (B:H)
Plagiochila semidecurrens ssp. *grossidens* (Herz.) Schust. p.p. = Plagiochila semidecurrens var. semidecurrens [PLAGSEM0] Plagiochilaceae (B:H)
Plagiochila semidecurrens var. **alaskana** (Evans) H. Inoue [PLAGSEM1] Plagiochilaceae (B:H)
Plagiochila semidecurrens var. **semidecurrens** [PLAGSEM0] Plagiochilaceae (B:H)
Plagiomnium T. Kop. [PLAGIOM$] Mniaceae (B:M)
Plagiomnium ciliare (C. Müll.) T. Kop. [PLAGCIL] Mniaceae (B:M)
Plagiomnium cinclidioides (Hüb.) M. Bowers = Pseudobryum cinclidioides [PSEUCIN] Mniaceae (B:M)
Plagiomnium cuspidatum (Hedw.) T. Kop. [PLAGCUS] Mniaceae (B:M)
Plagiomnium drummondii (Bruch & Schimp.) T. Kop. [PLAGDRU] Mniaceae (B:M)
Plagiomnium ellipticum (Brid.) T. Kop. [PLAGELL] Mniaceae (B:M)
Plagiomnium insigne (Mitt.) T. Kop. [PLAGINS] Mniaceae (B:M)
Plagiomnium medium (Bruch & Schimp. in B.S.G.) T. Kop. [PLAGMED] Mniaceae (B:M)
Plagiomnium rostratum (Schrad.) T. Kop. [PLAGROS] Mniaceae (B:M)
Plagiomnium rugicum (Laur.) T. Kop. = Plagiomnium ellipticum [PLAGELL] Mniaceae (B:M)
Plagiomnium venustum (Mitt.) T. Kop. [PLAGVEN] Mniaceae (B:M)
Plagiopus Brid. [PLAGIOP$] Bartramiaceae (B:M)
Plagiopus oederi (Brid.) Limpr. = Plagiopus oederiana [PLAGOED] Bartramiaceae (B:M)
Plagiopus oederi f. *alpinus* (Schwaegr.) Hab. = Plagiopus oederiana [PLAGOED] Bartramiaceae (B:M)
Plagiopus oederi var. *alpina* (Schwaegr.) Torre & Sarnth. = Plagiopus oederiana [PLAGOED] Bartramiaceae (B:M)
Plagiopus oederiana (Sw.) Crum & Anderson [PLAGOED] Bartramiaceae (B:M)
Plagiotheciella pilifera (Sw. ex Hartm.) Fleisch. = Plagiothecium piliferum [PLAGPIL] Plagiotheciaceae (B:M)
Plagiothecium Schimp. in B.S.G. [PLAGIOT$] Plagiotheciaceae (B:M)
Plagiothecium aciculari-pungens C. Müll. & Kindb. in Mac. & Kindb. = Plagiothecium cavifolium [PLAGCAV] Plagiotheciaceae (B:M)
Plagiothecium bifariellum Kindb. = Eurhynchium praelongum [EURHPRA] Brachytheciaceae (B:M)
Plagiothecium brevipungens Kindb. in Mac. & Kindb. = Callicladium haldanianum [CALLHAL] Hypnaceae (B:M)
Plagiothecium cavifolium (Brid.) Iwats. [PLAGCAV] Plagiotheciaceae (B:M)
Plagiothecium cavifolium var. *fallax* (Card. & Thér.) Iwats. = Plagiothecium cavifolium [PLAGCAV] Plagiotheciaceae (B:M)
Plagiothecium curvifolium auct. = Plagiothecium laetum [PLAGLAE] Plagiotheciaceae (B:M)
Plagiothecium decursivifolium Kindb. in Mac. & Kindb. = Plagiothecium laetum [PLAGLAE] Plagiotheciaceae (B:M)
Plagiothecium denticulatum (Hedw.) Schimp. in B.S.G. [PLAGDEN] Plagiotheciaceae (B:M)
Plagiothecium denticulatum var. *aptychus* (Spruce) Lees in Dix. = Plagiothecium laetum [PLAGLAE] Plagiotheciaceae (B:M)
Plagiothecium denticulatum var. *bullulae* Grout = Plagiothecium denticulatum [PLAGDEN] Plagiotheciaceae (B:M)
Plagiothecium denticulatum var. *donii* Lindb. = Plagiothecium denticulatum [PLAGDEN] Plagiotheciaceae (B:M)
Plagiothecium denticulatum var. *laetum* (Schimp. in B.S.G.) Lindb. = Plagiothecium laetum [PLAGLAE] Plagiotheciaceae (B:M)
Plagiothecium denticulatum var. *majus* (Boul.) Delogne = Plagiothecium denticulatum [PLAGDEN] Plagiotheciaceae (B:M)
Plagiothecium denticulatum var. *obtusifolium* (Turn.) Moore = Plagiothecium denticulatum [PLAGDEN] Plagiotheciaceae (B:M)
Plagiothecium denticulatum var. *squarrosum* Kindb. = Plagiothecium denticulatum [PLAGDEN] Plagiotheciaceae (B:M)
Plagiothecium denticulatum var. *tenellum* Schimp. in B.S.G. = Plagiothecium laetum [PLAGLAE] Plagiotheciaceae (B:M)
Plagiothecium denticulatum var. *undulatum* Ruthe ex Geh. = Plagiothecium denticulatum [PLAGDEN] Plagiotheciaceae (B:M)
Plagiothecium elegans (Brid.) Sull. in Gray = Pseudotaxiphyllum elegans [PSEUELE] Hypnaceae (B:M)
Plagiothecium elegans var. *gracilens* Aust. ex Grout = Pseudotaxiphyllum elegans [PSEUELE] Hypnaceae (B:M)
Plagiothecium fallax Card. & Thér. = Plagiothecium cavifolium [PLAGCAV] Plagiotheciaceae (B:M)
Plagiothecium geminum (Mitt.) Jaeg. & Sauerb. = Isopterygiopsis pulchella [ISOPPUC] Hypnaceae (B:M)
Plagiothecium laetum Schimp. in B.S.G. [PLAGLAE] Plagiotheciaceae (B:M)
Plagiothecium muehlenbeckii B.S.G. = Herzogiella striatella [HERZSTR] Hypnaceae (B:M)
Plagiothecium muellerianum Schimp. = Isopterygiopsis muelleriana [ISOPMUE] Hypnaceae (B:M)
Plagiothecium neglectum Mönk. = Plagiothecium cavifolium [PLAGCAV] Plagiotheciaceae (B:M)
Plagiothecium passaicence Aust. = Isopterygiopsis pulchella [ISOPPUC] Hypnaceae (B:M)
Plagiothecium piliferum (Sw. ex Hartm.) Schimp. in B.S.G. [PLAGPIL] Plagiotheciaceae (B:M)
Plagiothecium platyphyllum auct. = Plagiothecium cavifolium [PLAGCAV] Plagiotheciaceae (B:M)
Plagiothecium pseudolatebricola Kindb. in Mac. & Kindb. = Isopterygiopsis pulchella [ISOPPUC] Hypnaceae (B:M)
Plagiothecium pulchellum (Hedw.) Schimp. in B.S.G. = Isopterygiopsis pulchella [ISOPPUC] Hypnaceae (B:M)
Plagiothecium pulchellum var. *nitidulum* (Wahlenb.) Ren. & Card. = Isopterygiopsis pulchella [ISOPPUC] Hypnaceae (B:M)
Plagiothecium roeseanum Schimp. in B.S.G. = Plagiothecium cavifolium [PLAGCAV] Plagiotheciaceae (B:M)

Plagiothecium ruthei auct. = Plagiothecium denticulatum [PLAGDEN] Plagiotheciaceae (B:M)
Plagiothecium seligeri (Brid.) Lindb. = Herzogiella seligeri [HERZSEL] Hypnaceae (B:M)
Plagiothecium silesiacum (Selig. ex P. Beauv.) Schimp. in B.S.G. = Herzogiella seligeri [HERZSEL] Hypnaceae (B:M)
Plagiothecium striatellum (Brid.) Lindb. = Herzogiella striatella [HERZSTR] Hypnaceae (B:M)
Plagiothecium succulentum auct. = Plagiothecium cavifolium [PLAGCAV] Plagiotheciaceae (B:M)
Plagiothecium sylvaticum var. *orthocladium* (Schimp. in B.S.G.) Schimp. = Plagiothecium cavifolium [PLAGCAV] Plagiotheciaceae (B:M)
Plagiothecium sylvaticum var. *roeseanum* (Schimp. in B.S.G.) Lindb. = Plagiothecium cavifolium [PLAGCAV] Plagiotheciaceae (B:M)
Plagiothecium sylvaticum var. *succulentum* auct. = Plagiothecium cavifolium [PLAGCAV] Plagiotheciaceae (B:M)
Plagiothecium trichophorum (Spruce) Vent. & Bott. = Plagiothecium piliferum [PLAGPIL] Plagiotheciaceae (B:M)
Plagiothecium undulatum (Hedw.) Schimp. in B.S.G. [PLAGUND] Plagiotheciaceae (B:M)
Plagiothecium undulatum var. *myurum* Card. & Thér. = Plagiothecium undulatum [PLAGUND] Plagiotheciaceae (B:M)
Plantago L. [PLANTAG$] Plantaginaceae (V:D)
Plantago arenaria Waldst. & Kit. = Plantago psyllium [PLANPSY] Plantaginaceae (V:D)
Plantago bigelovii Gray [PLANBIG] Plantaginaceae (V:D)
Plantago canescens M.F. Adams [PLANCAN] Plantaginaceae (V:D)
Plantago canescens var. *cylindrica* (Macoun) Boivin = Plantago canescens [PLANCAN] Plantaginaceae (V:D)
*****Plantago coronopus** L. [PLANCOR] Plantaginaceae (V:D)
Plantago coronopus ssp. *commutata* (Guss.) Pilger = Plantago coronopus [PLANCOR] Plantaginaceae (V:D)
Plantago elongata Pursh [PLANELO] Plantaginaceae (V:D)
Plantago elongata ssp. **pentasperma** Bassett [PLANELO2] Plantaginaceae (V:D)
Plantago eriopoda Torr. [PLANERI] Plantaginaceae (V:D)
Plantago gnaphalioides Nutt. = Plantago patagonica [PLANPAT] Plantaginaceae (V:D)
Plantago indica L. = Plantago psyllium [PLANPSY] Plantaginaceae (V:D)
Plantago juncoides Lam. = Plantago maritima var. juncoides [PLANMAR1] Plantaginaceae (V:D)
Plantago juncoides var. *decipiens* (Barneoud) Fern. = Plantago maritima var. juncoides [PLANMAR1] Plantaginaceae (V:D)
Plantago juncoides var. *glauca* (Hornem.) Fern. = Plantago maritima var. juncoides [PLANMAR1] Plantaginaceae (V:D)
Plantago juncoides var. *laurentiana* Fern. = Plantago maritima var. juncoides [PLANMAR1] Plantaginaceae (V:D)
*****Plantago lanceolata** L. [PLANLAN] Plantaginaceae (V:D)
Plantago lanceolata var. *sphaerostachya* Mert. & Koch = Plantago lanceolata [PLANLAN] Plantaginaceae (V:D)
Plantago macrocarpa Cham. & Schlecht. [PLANMAC] Plantaginaceae (V:D)
*****Plantago major** L. [PLANMAJ] Plantaginaceae (V:D)
Plantago maritima L. [PLANMAR] Plantaginaceae (V:D)
Plantago maritima ssp. *borealis* (Lange) Blytt & O. Dahl = Plantago maritima var. juncoides [PLANMAR1] Plantaginaceae (V:D)
Plantago maritima ssp. *juncoides* (Lam.) Hultén = Plantago maritima var. juncoides [PLANMAR1] Plantaginaceae (V:D)
Plantago maritima var. **juncoides** (Lam.) Gray [PLANMAR1] Plantaginaceae (V:D)
*****Plantago media** L. [PLANMED] Plantaginaceae (V:D)
Plantago media var. *monnieri* (Giraud.) Roug. = Plantago media [PLANMED] Plantaginaceae (V:D)
Plantago oliganthos Roemer & J.A. Schultes = Plantago maritima var. juncoides [PLANMAR1] Plantaginaceae (V:D)
Plantago oliganthos var. *fallax* Fern. = Plantago maritima var. juncoides [PLANMAR1] Plantaginaceae (V:D)
Plantago patagonica Jacq. [PLANPAT] Plantaginaceae (V:D)
Plantago patagonica var. *breviscapa* (Shinners) Shinners = Plantago patagonica [PLANPAT] Plantaginaceae (V:D)
Plantago patagonica var. *gnaphalioides* (Nutt.) Gray = Plantago patagonica [PLANPAT] Plantaginaceae (V:D)
Plantago patagonica var. *oblonga* (Morris) Shinners = Plantago patagonica [PLANPAT] Plantaginaceae (V:D)
Plantago patagonica var. *spinulosa* (Dcne.) Gray = Plantago patagonica [PLANPAT] Plantaginaceae (V:D)
Plantago picta Morris = Plantago patagonica [PLANPAT] Plantaginaceae (V:D)
*****Plantago psyllium** L. [PLANPSY] Plantaginaceae (V:D)
Plantago purshii Roemer & J.A. Schultes = Plantago patagonica [PLANPAT] Plantaginaceae (V:D)
Plantago purshii var. *breviscapa* Shinners = Plantago patagonica [PLANPAT] Plantaginaceae (V:D)
Plantago purshii var. *oblonga* (Morris) Shinners = Plantago patagonica [PLANPAT] Plantaginaceae (V:D)
Plantago purshii var. *picta* Pilger = Plantago patagonica [PLANPAT] Plantaginaceae (V:D)
Plantago purshii var. *spinulosa* (Dcne.) Shinners = Plantago patagonica [PLANPAT] Plantaginaceae (V:D)
Plantago scabra Moench = Plantago psyllium [PLANPSY] Plantaginaceae (V:D)
Plantago septata Morris ex Rydb. = Plantago canescens [PLANCAN] Plantaginaceae (V:D)
Plantago shastensis Greene = Plantago eriopoda [PLANERI] Plantaginaceae (V:D)
Plantago spinulosa Dcne. = Plantago patagonica [PLANPAT] Plantaginaceae (V:D)
Plantago wyomingensis Gandog. = Plantago patagonica [PLANPAT] Plantaginaceae (V:D)
Platanthera L.C. Rich. [PLATANT$] Orchidaceae (V:M)
Platanthera chorisiana (Cham.) Reichenb. [PLATCHO] Orchidaceae (V:M)
Platanthera cooperi (S. Wats.) Rydb. = Piperia unalascensis [PIPEUNA] Orchidaceae (V:M)
Platanthera dilatata (Pursh) Lindl. ex Beck [PLATDIL] Orchidaceae (V:M)
Platanthera dilatata var. **albiflora** (Cham.) Ledeb. [PLATDIL1] Orchidaceae (V:M)
Platanthera dilatata var. *angustifolia* Hook. = Platanthera dilatata var. dilatata [PLATDIL0] Orchidaceae (V:M)
Platanthera dilatata var. **dilatata** [PLATDIL0] Orchidaceae (V:M)
Platanthera dilatata var. *gracilis* Ledeb. = Platanthera stricta [PLATSTR] Orchidaceae (V:M)
Platanthera dilatata var. *leucostachys* (Lindl.) Luer = Platanthera leucostachys [PLATLEU] Orchidaceae (V:M)
Platanthera hyperborea (L.) Lindl. [PLATHYP] Orchidaceae (V:M)
Platanthera hyperborea var. *purpurascens* (Rydb.) Luer = Platanthera stricta [PLATSTR] Orchidaceae (V:M)
Platanthera leucostachys Lindl. [PLATLEU] Orchidaceae (V:M)
Platanthera obtusata (Banks ex Pursh) Lindl. [PLATOBT] Orchidaceae (V:M)
Platanthera obtusata ssp. *oligantha* (Turcz.) Hultén = Platanthera obtusata [PLATOBT] Orchidaceae (V:M)
Platanthera orbiculata (Pursh) Lindl. [PLATORB] Orchidaceae (V:M)
Platanthera saccata (Greene) Hultén = Platanthera stricta [PLATSTR] Orchidaceae (V:M)
Platanthera stricta Lindl. [PLATSTR] Orchidaceae (V:M)
Platanthera unalascensis (Spreng.) Kurtz = Piperia unalascensis [PIPEUNA] Orchidaceae (V:M)
Platanthera unalascensis ssp. *maritima* (Greene) de Filipps = Piperia elegans [PIPEELE] Orchidaceae (V:M)
Platanthera unalascensis var. *maritima* (Greene) Correll = Piperia elegans [PIPEELE] Orchidaceae (V:M)
Platismatia Culb. & C. Culb. [PLATISM$] Parmeliaceae (L)
Platismatia glauca (L.) Culb. & C. Culb. [PLATGLA] Parmeliaceae (L)
Platismatia herrei (Imshaug) Culb. & C. Culb. [PLATHER] Parmeliaceae (L)
Platismatia lacunosa (Ach.) Culb. & C. Culb. [PLATLAC] Parmeliaceae (L)
Platismatia lacunosa var. *macounii* (Du Rietz) Culb. & C. Culb. = Platismatia lacunosa [PLATLAC] Parmeliaceae (L)

Platismatia norvegica (Lynge) Culb. & C. Culb. [PLATNOR] Parmeliaceae (L)
Platismatia stenophylla (Tuck.) Culb. & C. Culb. [PLATSTE] Parmeliaceae (L)
Platydictya Berk. [PLATYDI$] Hypnaceae (B:M)
Platydictya jungermannioides (Brid.) Crum [PLATJUN] Hypnaceae (B:M)
Platygyrium Schimp. in B.S.G. [PLATYGY$] Hypnaceae (B:M)
Platygyrium orthoclados Kindb. in Mac. & Kindb. = Platygyrium repens [PLATREP] Hypnaceae (B:M)
Platygyrium repens (Brid.) Schimp. in B.S.G. [PLATREP] Hypnaceae (B:M)
Platygyrium repens var. *ascendens* (Schwaegr.) Grout = Platygyrium repens [PLATREP] Hypnaceae (B:M)
Platygyrium repens var. *orthoclados* Kindb. = Platygyrium repens [PLATREP] Hypnaceae (B:M)
Platyhypnidium Fleisch. [PLATYHY$] Brachytheciaceae (B:M)
Platyhypnidium riparioides (Hedw.) Dix. [PLATRIP] Brachytheciaceae (B:M)
Plectocarpon Fée [PLECTOC$] Uncertain-family (L)
Plectocarpon lichenum (Sommerf.) D. Hawksw. [PLECLIC] Uncertain-family (L)
Plectocarpon nephromeum (Norman) Sant. [PLECNEP] Uncertain-family (L)
Plectocolea hyalina (Lyell) Mitt. = Jungermannia hyalina [JUNGHYA] Jungermanniaceae (B:H)
Plectocolea obovata (Nees) Lindb. = Jungermannia obovata [JUNGOBO] Jungermanniaceae (B:H)
Plectocolea rubra (Gott. ex Underw.) Buch et al. = Jungermannia rubra [JUNGRUB] Jungermanniaceae (B:H)
Plectocolea subelliptica (Lindb. ex Kaal.) Evans = Jungermannia subelliptica [JUNGSUB] Jungermanniaceae (B:H)
Plectritis (Lindl.) DC. [PLECTRI$] Valerianaceae (V:D)
Plectritis anomala (Gray) Suksdorf = Plectritis congesta ssp. brachystemon [PLECCON1] Valerianaceae (V:D)
Plectritis anomala var. *gibbosa* (Suksdorf) Dyal = Plectritis congesta ssp. brachystemon [PLECCON1] Valerianaceae (V:D)
Plectritis aphanoptera (Gray) Suksdorf = Plectritis congesta ssp. brachystemon [PLECCON1] Valerianaceae (V:D)
Plectritis congesta (Lindl.) DC. [PLECCON] Valerianaceae (V:D)
Plectritis congesta ssp. **brachystemon** (Fisch. & C.A. Mey.) Morey [PLECCON1] Valerianaceae (V:D)
Plectritis congesta var. *major* (Fisch. & C.A. Mey.) Dyal = Plectritis congesta ssp. brachystemon [PLECCON1] Valerianaceae (V:D)
Plectritis macrocera Torr. & Gray [PLECMAC] Valerianaceae (V:D)
Plectritis magna (Greene) Suksdorf = Plectritis congesta ssp. brachystemon [PLECCON1] Valerianaceae (V:D)
Plectritis samolifolia (DC.) Hoeck = Plectritis congesta ssp. brachystemon [PLECCON1] Valerianaceae (V:D)
Plectritis samolifolia var. *involuta* (Suksdorf) Dyal = Plectritis congesta ssp. brachystemon [PLECCON1] Valerianaceae (V:D)
Pleopsidium Körber [PLEOPSI$] Acarosporaceae (L)
Pleopsidium chlorophanum (Wahlenb.) Zopf [PLEOCHL] Acarosporaceae (L)
Pleuricospora Gray [PLEURIC$] Monotropaceae (V:D)
Pleuricospora fimbriolata Gray [PLEUFIM] Monotropaceae (V:D)
Pleuricospora longipetala T.J. Howell = Pleuricospora fimbriolata [PLEUFIM] Monotropaceae (V:D)
Pleuridium Rabenh. [PLEURID$] Ditrichaceae (B:M)
Pleuridium alternifolium auct. = Pleuridium subulatum [PLEUSUB] Ditrichaceae (B:M)
Pleuridium axillare (Sm.) Lindb. = Pseudephemerum nitidum [PSEUNIT] Dicranaceae (B:M)
Pleuridium subulatum (Hedw.) Rabenh. [PLEUSUB] Ditrichaceae (B:M)
Pleuroclada albescens (Hook.) Spruce = Pleurocladula albescens [PLEUALB] Cephaloziaceae (B:H)
Pleuroclada albescens var. *islandica* (Nees) Spruce = Pleurocladula albescens [PLEUALB] Cephaloziaceae (B:H)
Pleurocladula Grolle [PLEUROL$] Cephaloziaceae (B:H)
Pleurocladula albescens (Hook.) Grolle [PLEUALB] Cephaloziaceae (B:H)
Pleurocladula islandica (Nees) Grolle = Pleurocladula albescens [PLEUALB] Cephaloziaceae (B:H)
Pleurogyne rotata (L.) Griseb. = Lomatogonium rotatum [LOMAROT] Gentianaceae (V:D)
Pleuropogon refractus (Gray) Benth. ex Vasey = Lophochlaena refracta [LOPHREF] Poaceae (V:M)
Pleuropterus cuspidatus (Sieb. & Zucc.) Moldenke = Polygonum cuspidatum [POLYCUS] Polygonaceae (V:D)
Pleuropterus zuccarinii (Small) Small = Polygonum cuspidatum [POLYCUS] Polygonaceae (V:D)
Pleurozia Dum. [PLEUROA$] Pleuroziaceae (B:H)
Pleurozia purpurea Lindb. [PLEUPUR] Pleuroziaceae (B:H)
Pleuroziopsis Kindb. ex Britt. [PLEUROZ$] Pleuroziopsidaceae (B:M)
Pleuroziopsis ruthenica (Weinm.) Kindb. ex Britt. [PLEURUT] Pleuroziopsidaceae (B:M)
Pleurozium Mitt. [PLEUROI$] Hylocomiaceae (B:M)
Pleurozium schreberi (Brid.) Mitt. [PLEUSCH] Hylocomiaceae (B:M)
Pleurozium schreberi var. *tananae* (Grout) Wijk & Marg. = Pleurozium schreberi [PLEUSCH] Hylocomiaceae (B:M)
Pneumonanthe affinis (Griseb.) W.A. Weber = Gentiana affinis [GENTAFF] Gentianaceae (V:D)
Pneumonanthe calycosa (Griseb.) Greene = Gentiana calycosa [GENTCAL] Gentianaceae (V:D)
Poa L. [POA$] Poaceae (V:M)
Poa abbreviata R. Br. [POA ABB] Poaceae (V:M)
Poa abbreviata ssp. *jordalii* (Porsild) Hultén = Poa abbreviata ssp. pattersonii [POA ABB3] Poaceae (V:M)
Poa abbreviata ssp. **pattersonii** (Vasey) A. & D. Löve & Kapoor [POA ABB3] Poaceae (V:M)
Poa agassizensis Boivin & D. Löve = Poa pratensis [POA PRA] Poaceae (V:M)
Poa alpigena (Fries ex Blytt) Lindm. f. = Poa pratensis [POA PRA] Poaceae (V:M)
Poa alpigena ssp. *colpodea* (Fries ex Blytt) Tzvelev = Poa pratensis [POA PRA] Poaceae (V:M)
Poa alpigena var. *colpodea* (Fries ex Blytt) Schol. = Poa pratensis [POA PRA] Poaceae (V:M)
Poa alpina L. [POA ALP] Poaceae (V:M)
Poa alpina var. *purpurascens* Vassey = Poa fendleriana [POA FEN] Poaceae (V:M)
Poa ampla Merr. = Poa secunda [POA SEC] Poaceae (V:M)
Poa angustifolia L. = Poa pratensis [POA PRA] Poaceae (V:M)
***Poa annua** L. [POA ANN] Poaceae (V:M)
Poa annua var. *aquatica* Aschers. = Poa annua [POA ANN] Poaceae (V:M)
Poa annua var. *reptans* Hausskn. = Poa annua [POA ANN] Poaceae (V:M)
Poa arctica R. Br. [POA ARC] Poaceae (V:M)
Poa arctica ssp. **arctica** [POA ARC0] Poaceae (V:M)
Poa arctica ssp. **lanata** (Scribn. & Merr.) Soreng [POA ARC5] Poaceae (V:M)
Poa arctica ssp. **longiculmis** Hultén [POA ARC3] Poaceae (V:M)
Poa arctica ssp. **williamsii** (Nash) Hultén [POA ARC4] Poaceae (V:M)
Poa arctica var. *glabriflora* Rosh. = Poa arctica ssp. arctica [POA ARC0] Poaceae (V:M)
Poa arctica var. *lanata* (Scribn. & Merr.) Bovin = Poa arctica ssp. lanata [POA ARC5] Poaceae (V:M)
Poa arctica var. *vivipara* Hultén = Poa arctica ssp. lanata [POA ARC5] Poaceae (V:M)
Poa arctica var. *vivipara* Hook. = Poa arctica ssp. arctica [POA ARC0] Poaceae (V:M)
Poa bolanderi Vasey [POA BOL] Poaceae (V:M)
Poa bolanderi ssp. **howellii** (Vasey & Scribn.) Keck [POA BOL1] Poaceae (V:M)
Poa bolanderi var. *howellii* (Vasey & Scribn.) M.E. Jones = Poa bolanderi ssp. howellii [POA BOL1] Poaceae (V:M)
Poa brachyanthera Hultén = Poa pseudoabbreviata [POA PSE] Poaceae (V:M)
Poa brachyglossa Piper = Poa secunda [POA SEC] Poaceae (V:M)
Poa buckleyana Nash = Poa secunda [POA SEC] Poaceae (V:M)

*•**Poa bulbosa** L. [POA BUL] Poaceae (V:M)
Poa canbyi (Scribn.) T.J. Howell = Poa secunda [POA SEC] Poaceae (V:M)
Poa cenisia var. *arctica* (R. Br.) Richter = Poa arctica ssp. arctica [POA ARC0] Poaceae (V:M)
Poa cilianensis All. = Eragrostis cilianensis [ERAGCIL] Poaceae (V:M)
•**Poa compressa** L. [POA COM] Poaceae (V:M)
Poa confinis Vasey [POA CON] Poaceae (V:M)
Poa confusa Rydb. = Poa secunda [POA SEC] Poaceae (V:M)
Poa crocata Michx. = Poa palustris [POA PAL] Poaceae (V:M)
Poa cusickii Vasey = Poa fendleriana [POA FEN] Poaceae (V:M)
Poa cusickii ssp. *epilis* (Scribn.) W.A. Weber = Poa fendleriana [POA FEN] Poaceae (V:M)
Poa cusickii ssp. *pallida* Soreng = Poa fendleriana [POA FEN] Poaceae (V:M)
Poa cusickii ssp. *pubens* Keck = Poa fendleriana [POA FEN] Poaceae (V:M)
Poa cusickii ssp. *purpurascens* (Vasey) Soreng = Poa fendleriana [POA FEN] Poaceae (V:M)
Poa cusickii var. *epilis* (Scribn.) C.L. Hitchc. = Poa fendleriana [POA FEN] Poaceae (V:M)
Poa cusickii var. *purpurascens* (Vasey) C.L. Hitchc. = Poa fendleriana [POA FEN] Poaceae (V:M)
Poa douglasii Nees [POA DOU] Poaceae (V:M)
Poa douglasii ssp. **macrantha** (Vasey) Keck [POA DOU1] Poaceae (V:M)
Poa douglasii var. *macrantha* (Vasey) Boivin = Poa douglasii ssp. macrantha [POA DOU1] Poaceae (V:M)
Poa eminens J. Presl [POA EMI] Poaceae (V:M)
Poa englishii St. John & Hardin = Poa secunda [POA SEC] Poaceae (V:M)
Poa epilis Scribn. = Poa fendleriana [POA FEN] Poaceae (V:M)
Poa fendleriana (Steud.) Vasey [POA FEN] Poaceae (V:M)
Poa fendleriana ssp. *longiligula* (Scribn. & Williams) Soreng = Poa fendleriana [POA FEN] Poaceae (V:M)
Poa fendleriana var. *longiligula* (Scribn. & Williams) Gould = Poa fendleriana [POA FEN] Poaceae (V:M)
Poa fendleriana var. *wyomingensis* T.A. Williams = Poa fendleriana [POA FEN] Poaceae (V:M)
Poa filifolia Vasey = Poa secunda [POA SEC] Poaceae (V:M)
Poa glacialis A. Hitchc. = Poa leptocoma ssp. paucispicula [POA LEP3] Poaceae (V:M)
Poa glauca Vahl [POA GLA] Poaceae (V:M)
Poa glauca ssp. *conferta* (Blytt) Lindm. = Poa glauca ssp. glauca [POA GLA0] Poaceae (V:M)
Poa glauca ssp. **glauca** [POA GLA0] Poaceae (V:M)
Poa glauca ssp. **rupicola** (Nash ex Rydb.) W.A. Weber [POA GLA2] Poaceae (V:M)
Poa glauca var. *conferta* (Blytt) Nannf. = Poa glauca ssp. glauca [POA GLA0] Poaceae (V:M)
Poa glauca var. *laxiuscula* (Blytt) Lindm. = Poa glauca ssp. glauca [POA GLA0] Poaceae (V:M)
Poa glauca var. *rupicola* (Nash ex Rydb.) Boivin = Poa glauca ssp. rupicola [POA GLA2] Poaceae (V:M)
Poa gracillima Vasey = Poa secunda [POA SEC] Poaceae (V:M)
Poa gracillima var. *multnomae* (Piper) C.L. Hitchc. = Poa secunda [POA SEC] Poaceae (V:M)
Poa howellii Vasey & Scribn. = Poa bolanderi ssp. howellii [POA BOL1] Poaceae (V:M)
Poa incurva Scribn. & Williams = Poa secunda [POA SEC] Poaceae (V:M)
•**Poa infirma** Kunth [POA INF] Poaceae (V:M)
Poa interior Rydb., p.p. [POA INT] Poaceae (V:M)
Poa irrigata Lindm. = Poa subcaerulea [POA SUB] Poaceae (V:M)
Poa jordalii Porsild = Poa abbreviata ssp. pattersonii [POA ABB3] Poaceae (V:M)
Poa juncifolia Scribn. = Poa secunda [POA SEC] Poaceae (V:M)
Poa juncifolia ssp. *porteri* Keck = Poa secunda [POA SEC] Poaceae (V:M)
Poa juncifolia var. *ampla* (Merr.) Dorn = Poa secunda [POA SEC] Poaceae (V:M)
Poa laevigata Scribn. = Poa secunda [POA SEC] Poaceae (V:M)
Poa lanata Scribn. & Merr. = Poa arctica ssp. lanata [POA ARC5] Poaceae (V:M)
Poa laxa Haenke [POA LAA] Poaceae (V:M)
Poa laxa ssp. **banffiana** Soreng [POA LAA1] Poaceae (V:M)
Poa laxiflora Buckl. [POA LAX] Poaceae (V:M)
Poa leptocoma Trin. [POA LEP] Poaceae (V:M)
Poa leptocoma ssp. **paucispicula** (Scribn. & Merr.) Tzvelev [POA LEP3] Poaceae (V:M)
Poa leptocoma var. *paucispicula* (Scribn. & Merr.) C.L. Hitchc. = Poa leptocoma ssp. paucispicula [POA LEP3] Poaceae (V:M)
Poa lettermanii Vasey [POA LET] Poaceae (V:M)
Poa longiligula Scribn. & Williams = Poa fendleriana [POA FEN] Poaceae (V:M)
Poa longipila Nash = Poa arctica ssp. arctica [POA ARC0] Poaceae (V:M)
Poa macrantha Vasey = Poa douglasii ssp. macrantha [POA DOU1] Poaceae (V:M)
Poa marcida A.S. Hitchc. [POA MAR] Poaceae (V:M)
Poa merrilliana Scribn. & Merr. = Poa leptocoma ssp. paucispicula [POA LEP3] Poaceae (V:M)
Poa montevansii L. Kelso = Poa lettermanii [POA LET] Poaceae (V:M)
Poa nematophylla Rydb. = Poa fendleriana [POA FEN] Poaceae (V:M)
•**Poa nemoralis** L. [POA NEM] Poaceae (V:M)
Poa nemoralis ssp. *interior* (Rydb.) W.A. Weber = Poa interior [POA INT] Poaceae (V:M)
Poa nemoralis var. *interior* (Rydb.) Butters & Abbe = Poa interior [POA INT] Poaceae (V:M)
Poa nervosa (Hook.) Vasey [POA NER] Poaceae (V:M)
Poa nervosa var. **wheeleri** (Vasey) C.L. Hitchc. [POA NER1] Poaceae (V:M)
Poa nevadensis Vasey ex Scribn. = Poa secunda [POA SEC] Poaceae (V:M)
Poa nevadensis var. *juncifolia* (Scribn.) Beetle = Poa secunda [POA SEC] Poaceae (V:M)
Poa orcuttiana Vasey = Poa secunda [POA SEC] Poaceae (V:M)
Poa palustris L. [POA PAL] Poaceae (V:M)
Poa pattersonii Vasey = Poa abbreviata ssp. pattersonii [POA ABB3] Poaceae (V:M)
Poa paucispicula Scribn. & Merr. = Poa leptocoma ssp. paucispicula [POA LEP3] Poaceae (V:M)
•**Poa pratensis** L. [POA PRA] Poaceae (V:M)
Poa pratensis ssp. *agassizensis* (Boivin & D. Löve) Taylor & MacBryde = Poa pratensis [POA PRA] Poaceae (V:M)
Poa pratensis ssp. *alpigena* (Fries ex Blytt) Hiitonen = Poa pratensis [POA PRA] Poaceae (V:M)
Poa pratensis ssp. *angustifolia* (L.) Lej. = Poa pratensis [POA PRA] Poaceae (V:M)
Poa pratensis ssp. *colpodea* (Fries ex Blytt) Tzvelev = Poa pratensis [POA PRA] Poaceae (V:M)
Poa pratensis ssp. *irrigata* (Lindm.) Lindb. f. = Poa subcaerulea [POA SUB] Poaceae (V:M)
Poa pratensis ssp. *rigens* (Hartman) Tzvelev = Poa subcaerulea [POA SUB] Poaceae (V:M)
Poa pratensis var. *angustifolia* (L.) Gaudin = Poa pratensis [POA PRA] Poaceae (V:M)
Poa pratensis var. *apligena* Fries ex Blytt = Poa pratensis [POA PRA] Poaceae (V:M)
Poa pratensis var. *colpodea* (Fries ex Blytt) Soreng = Poa pratensis [POA PRA] Poaceae (V:M)
Poa pratensis var. *domestica* Laestad. = Poa pratensis [POA PRA] Poaceae (V:M)
Poa pratensis var. *gelida* (Roemer & J.A. Schultes) Böcher = Poa pratensis [POA PRA] Poaceae (V:M)
Poa pratensis var. *iantha* Wahlenb. = Poa pratensis [POA PRA] Poaceae (V:M)
Poa pratensis var. *rigens* (Hartman) Wahlenb. = Poa subcaerulea [POA SUB] Poaceae (V:M)
Poa pratensis var. *vivipara* (Malmgr.) Boivin = Poa pratensis [POA PRA] Poaceae (V:M)
Poa pseudoabbreviata Rosh. [POA PSE] Poaceae (V:M)

Poa purpurascens Vasey = Poa fendleriana [POA FEN] Poaceae (V:M)
Poa purpurascens var. *epilis* (Scribn.) M.E. Jones = Poa fendleriana [POA FEN] Poaceae (V:M)
Poa rigens Hartman = Poa subcaerulea [POA SUB] Poaceae (V:M)
Poa rupicola Nash ex Rydb. = Poa glauca ssp. rupicola [POA GLA2] Poaceae (V:M)
Poa saltuensis var. *marcida* (A.S. Hitchc.) Boivin = Poa marcida [POA MAR] Poaceae (V:M)
Poa sandbergii Vasey = Poa secunda [POA SEC] Poaceae (V:M)
Poa scabrella (Thurb.) Benth. ex Vasey = Poa secunda [POA SEC] Poaceae (V:M)
Poa secunda J. Presl [POA SEC] Poaceae (V:M)
Poa secunda ssp. *juncifolia* (Scribn.) Soreng = Poa secunda [POA SEC] Poaceae (V:M)
Poa secunda var. *elongata* (Vasey) Dorn = Poa secunda [POA SEC] Poaceae (V:M)
Poa secunda var. *incurva* (Scribn. & Williams) Beetle = Poa secunda [POA SEC] Poaceae (V:M)
Poa secunda var. *stenophylla* (Vasey) Beetle = Poa secunda [POA SEC] Poaceae (V:M)
Poa stenantha Trin. [POA STE] Poaceae (V:M)
Poa subaristata Scribn. = Poa fendleriana [POA FEN] Poaceae (V:M)
***Poa subcaerulea** Sm. [POA SUB] Poaceae (V:M)
Poa tenerrima Scribn. = Poa secunda [POA SEC] Poaceae (V:M)
Poa triflora Gilib. = Poa palustris [POA PAL] Poaceae (V:M)
***Poa trivialis** L. [POA TRI] Poaceae (V:M)
Poa wheeleri Vasey = Poa nervosa var. wheeleri [POA NER1] Poaceae (V:M)
Poa williamsii Nash = Poa arctica ssp. williamsii [POA ARC4] Poaceae (V:M)
Podagrostis aequivalvis (Trin.) Scribn. & Merr. = Agrostis aequivalvis [AGROAEQ] Poaceae (V:M)
Poeltinula Hafellner [POELTIN$] Rhizocarpaceae (L)
Poeltinula cerebrina (DC.) Hafellner [POELCER] Rhizocarpaceae (L)
Pogonatum P. Beauv. [POGONAT$] Polytrichaceae (B:M)
Pogonatum atrovirens Mitt. = Pogonatum contortum [POGOCON] Polytrichaceae (B:M)
Pogonatum capillare (Michx.) Brid. = Pogonatum dentatum [POGODEN] Polytrichaceae (B:M)
Pogonatum capillare var. *minus* (Wahlenb.) Par. = Pogonatum dentatum [POGODEN] Polytrichaceae (B:M)
Pogonatum contortum (Brid.) Lesq. [POGOCON] Polytrichaceae (B:M)
Pogonatum dentatum (Brid.) Brid. [POGODEN] Polytrichaceae (B:M)
Pogonatum erythrodontium Kindb. in Mac. & Kindb. = Pogonatum contortum [POGOCON] Polytrichaceae (B:M)
Pogonatum laterale Schimp. in Jard. = Pogonatum contortum [POGOCON] Polytrichaceae (B:M)
Pogonatum sphaerothecium Besch. = Polytrichum sphaerothecium [POLYSPH] Polytrichaceae (B:M)
Pogonatum urnigerum (Hedw.) P. Beauv. [POGOURN] Polytrichaceae (B:M)
Pogonatum urnigerum var. *fasciculatum* (Michx.) Brid. = Pogonatum urnigerum [POGOURN] Polytrichaceae (B:M)
Pogonatum urnigerum var. *humile* (Wahlenb.) Brid. = Pogonatum urnigerum [POGOURN] Polytrichaceae (B:M)
Pogonatum urnigerum var. *subintegrifolium* (Arnell & C. Jens.) Möll. = Pogonatum urnigerum [POGOURN] Polytrichaceae (B:M)
Pohlia Hedw. [POHLIA$] Bryaceae (B:M)
Pohlia acuminata Hoppe & Hornsch. = Pohlia elongata var. elongata [POHLELO0] Bryaceae (B:M)
Pohlia albicans Lindb. = Pohlia wahlenbergii [POHLWAH] Bryaceae (B:M)
Pohlia andalusica (Hohn.) Broth. [POHLAND] Bryaceae (B:M)
Pohlia annotina (Hedw.) Lindb. [POHLANN] Bryaceae (B:M)
Pohlia annotina var. *decipiens* Loeske = Pohlia annotina [POHLANN] Bryaceae (B:M)
Pohlia annotina var. *loeskei* Crum, Steere & Anderson = Pohlia annotina [POHLANN] Bryaceae (B:M)
Pohlia atropurpurea (Wahlenb. in Fürnr.) H. Lindb. [POHLATR] Bryaceae (B:M)
Pohlia bolanderi (Sull.) Broth. [POHLBOL] Bryaceae (B:M)
Pohlia brevinervis Lindb. & Arnell [POHLBRE] Bryaceae (B:M)
Pohlia bulbifera (Warnst.) Warnst. [POHLBUL] Bryaceae (B:M)
Pohlia camptotrachela (Ren. & Card.) Broth. [POHLCAM] Bryaceae (B:M)
Pohlia camptotrachela var. *bulbifera* (Lam. & DC.) Wijk & Marg. = Pohlia bulbifera [POHLBUL] Bryaceae (B:M)
Pohlia camptotrachela var. *decipiens* (Loeske) Nyh. = Pohlia annotina [POHLANN] Bryaceae (B:M)
Pohlia cardotii (Ren. in Ren. & Card.) Broth. [POHLCAR] Bryaceae (B:M)
Pohlia carnea (Schimp.) Lindb. = Pohlia melanodon [POHLMEL] Bryaceae (B:M)
Pohlia columbica (Kindb. in Mac. & Kindb.) Andrews [POHLCOL] Bryaceae (B:M)
Pohlia commutata (Schimp.) Lindb. = Pohlia drummondii [POHLDRU] Bryaceae (B:M)
Pohlia commutata var. *filum* (Schimp.) Dus. = Pohlia filum [POHLFIU] Bryaceae (B:M)
Pohlia cruda (Hedw.) Lindb. [POHLCRU] Bryaceae (B:M)
Pohlia cruda var. *oregonensis* (Par.) Wijk & Marg. = Pohlia cruda [POHLCRU] Bryaceae (B:M)
Pohlia crudoides (Sull. & Lesq.) Broth. [POHLCRD] Bryaceae (B:M)
Pohlia cucullata (Schwaegr.) Lindb. = Pohlia obtusifolia [POHLOBT] Bryaceae (B:M)
Pohlia defecta (Sanio) Andrews in Grout = Pohlia erecta [POHLERE] Bryaceae (B:M)
Pohlia delicatula (Hedw.) Grout = Pohlia melanodon [POHLMEL] Bryaceae (B:M)
Pohlia drummondii (C. Müll.) Andrews [POHLDRU] Bryaceae (B:M)
Pohlia drummondii var. *gracilis* (Bruch & Schimp. in B.S.G.) Podp. = Pohlia filum [POHLFIU] Bryaceae (B:M)
Pohlia elongata Hedw. [POHLELO] Bryaceae (B:M)
Pohlia elongata var. **elongata** [POHLELO0] Bryaceae (B:M)
Pohlia elongata var. **greenii** (Brid.) Shaw [POHLELO1] Bryaceae (B:M)
Pohlia elongata var. *minor* Hartm. = Pohlia elongata var. greenii [POHLELO1] Bryaceae (B:M)
Pohlia erecta Lindb. [POHLERE] Bryaceae (B:M)
Pohlia filiformis (Dicks.) Andrews in Grout = Anomobryum filiforme [ANOMFIL] Bryaceae (B:M)
Pohlia filiformis var. *concinnata* (Spruce) Grout = Anomobryum filiforme [ANOMFIL] Bryaceae (B:M)
Pohlia filum (Schimp.) Mårt. [POHLFIU] Bryaceae (B:M)
Pohlia gracilis (Bruch & Schimp. in B.S.G.) Lindb. = Pohlia filum [POHLFIU] Bryaceae (B:M)
Pohlia grandiflora ssp. *proligera* (Kindb. ex Breidl.) Kuc = Pohlia proligera [POHLPRO] Bryaceae (B:M)
Pohlia grandiflora ssp. *proligera* var. *decipiens* (Loeske) Kuc = Pohlia annotina [POHLANN] Bryaceae (B:M)
Pohlia lescuriana (Sull.) Grout [POHLLES] Bryaceae (B:M)
Pohlia longibracteata Broth. in Röll [POHLLON] Bryaceae (B:M)
Pohlia longicolla (Hedw.) Lindb. [POHLLOG] Bryaceae (B:M)
Pohlia ludwigii (Spreng. ex Schwaegr.) Broth. [POHLLUD] Bryaceae (B:M)
Pohlia melanodon (Brid.) Shaw [POHLMEL] Bryaceae (B:M)
Pohlia minor Schleich ex Schwaegr. = Pohlia elongata var. greenii [POHLELO1] Bryaceae (B:M)
Pohlia minor ssp. *acuminata* (Hoppe & Hornsch.) Wijk & Marg. = Pohlia elongata var. elongata [POHLELO0] Bryaceae (B:M)
Pohlia minor var. *brachycarpa* (Hoppe & Hornsch.) Wijk & Marg. = Pohlia elongata var. elongata [POHLELO0] Bryaceae (B:M)
Pohlia nutans (Hedw.) Lindb. [POHLNUT] Bryaceae (B:M)
Pohlia nutans ssp. *schimperi* (C. Müll.) Nyh. = Pohlia nutans [POHLNUT] Bryaceae (B:M)
Pohlia obtusifolia (Brid.) L. Koch [POHLOBT] Bryaceae (B:M)
Pohlia pacifica Shaw [POHLPAC] Bryaceae (B:M)
Pohlia proligera (Kindb. ex Breidl.) Lindb. ex Arnell [POHLPRO] Bryaceae (B:M)

Pohlia pulchella (Hedw.) Lindb. = Pohlia lescuriana [POHLLES] Bryaceae (B:M)
Pohlia rothii (Corr. ex Limpr.) Broth. = Pohlia filum [POHLFIU] Bryaceae (B:M)
Pohlia rothii auct. = Pohlia andalusica [POHLAND] Bryaceae (B:M)
Pohlia rutilans (Schimp.) Lindb. = Pohlia nutans [POHLNUT] Bryaceae (B:M)
Pohlia schimperi (C. Müll.) Andrews in Grout = Pohlia nutans [POHLNUT] Bryaceae (B:M)
Pohlia schleicheri Crum = Pohlia filum [POHLFIU] Bryaceae (B:M)
Pohlia sphagnicola (Bruch & Schimp.) Lindb. & Arnell [POHLSPH] Bryaceae (B:M)
Pohlia tozeri (Grev.) Delogne = Epipterygium tozeri [EPIPTOZ] Bryaceae (B:M)
Pohlia tundrae Shaw [POHLTUN] Bryaceae (B:M)
Pohlia vexans (Limpr.) H. Lindb. [POHLVEX] Bryaceae (B:M)
Pohlia wahlenbergii (Web. & Mohr) Andrews [POHLWAH] Bryaceae (B:M)
Polemoniella micrantha (Benth.) Heller = Polemonium micranthum [POLEMIC] Polemoniaceae (V:D)
Polemonium L. [POLEMON$] Polemoniaceae (V:D)
Polemonium acutiflorum Willd. ex Roemer & J.A. Schultes [POLEACU] Polemoniaceae (V:D)
Polemonium boreale M.F. Adams [POLEBOR] Polemoniaceae (V:D)
•**Polemonium caeruleum** L. [POLECAE] Polemoniaceae (V:D)
Polemonium caeruleum ssp. *amygdalinum* (Wherry) Munz = Polemonium occidentale [POLEOCC] Polemoniaceae (V:D)
Polemonium caeruleum ssp. *occidentale* (Greene) J.F. Davids. = Polemonium occidentale [POLEOCC] Polemoniaceae (V:D)
Polemonium caeruleum ssp. *villosum* (J.H. Rudolph ex Georgi) Brand = Polemonium acutiflorum [POLEACU] Polemoniaceae (V:D)
Polemonium caeruleum var. *pterospermum* Benth. = Polemonium occidentale [POLEOCC] Polemoniaceae (V:D)
Polemonium elegans Greene [POLEELE] Polemoniaceae (V:D)
Polemonium helleri Brand = Polemonium occidentale [POLEOCC] Polemoniaceae (V:D)
Polemonium intermedium (Brand) Rydb. = Polemonium occidentale [POLEOCC] Polemoniaceae (V:D)
Polemonium micranthum Benth. [POLEMIC] Polemoniaceae (V:D)
Polemonium occidentale Greene [POLEOCC] Polemoniaceae (V:D)
Polemonium occidentale ssp. *amygdalium* Wherry = Polemonium occidentale [POLEOCC] Polemoniaceae (V:D)
Polemonium occidentale ssp. *typicum* Wherry = Polemonium occidentale [POLEOCC] Polemoniaceae (V:D)
Polemonium pulcherrimum Hook. [POLEPUL] Polemoniaceae (V:D)
Polemonium viscosum Nutt. [POLEVIS] Polemoniaceae (V:D)
Polemonium viscosum ssp. *genuinum* Wherry = Polemonium viscosum [POLEVIS] Polemoniaceae (V:D)
Polemonium viscosum ssp. *lemmonii* (Brand) Wherry = Polemonium viscosum [POLEVIS] Polemoniaceae (V:D)
Polyblastia A. Massal. [POLYBLA$] Verrucariaceae (L)
Polyblastia gelatinosa (Ach.) Th. Fr. [POLYGEL] Verrucariaceae (L)
Polyblastia hyperborea Th. Fr. [POLYHYP] Verrucariaceae (L)
Polyblastia integrascens (Nyl.) Vainio = Polyblastia hyperborea [POLYHYP] Verrucariaceae (L)
Polyblastia obsoleta Arnold [POLYOBS] Verrucariaceae (L)
Polyblastia rupifraga A. Massal. = Staurothele rupifraga [STAURUP] Verrucariaceae (L)
Polyblastia sommerfeltii Lynge = Polyblastia terrestris [POLYTER] Verrucariaceae (L)
Polyblastia terrestris Th. Fr. [POLYTER] Verrucariaceae (L)
Polyblastia theleodes (Sommerf.) Th. Fr. [POLYTHE] Verrucariaceae (L)
Polyblastia tristicula (Nyl.) Arnold = Agonimia tristicula [AGONTRI] Verrucariaceae (L)
Polyblastia umbrina var. *areolata* (Ach.) Boist. = Staurothele areolata [STAUARE] Verrucariaceae (L)
Polyblastiopsis fallax (Nyl.) Fink = Arthopyrenia lapponina [ARTHLAO] Arthopyreniaceae (L)
Polycarpon L. [POLYCAR$] Caryophyllaceae (V:D)
•**Polycarpon tetraphyllum** (L.) L. [POLYTET] Caryophyllaceae (V:D)
Polychidium (Ach.) Gray [POLYCHI$] Placynthiaceae (L)
Polychidium albociliatum (Desmaz.) Zahlbr. = Leptochidium albociliatum [LEPTALB] Placynthiaceae (L)
Polychidium contortum Henssen [POLYCOT] Placynthiaceae (L)
Polychidium dendriscum (Nyl.) Henssen [POLYDEN] Placynthiaceae (L)
Polychidium intricatulum (Nyl.) Henssen = Dendriscocaulon intricatulum [DENDINT] Lobariaceae (L)
Polychidium muscicola (Sw.) Gray [POLYMUS] Placynthiaceae (L)
Polychidium rivale (Tuck.) Fink = Leptogium rivale [LEPTRIV] Collemataceae (L)
Polycnemum L. [POLYCNE$] Chenopodiaceae (V:D)
•**Polycnemum arvense** L. [POLYARV] Chenopodiaceae (V:D)
Polygala L. [POLYGAL$] Polygalaceae (V:D)
Polygala senega L. [POLYSEN] Polygalaceae (V:D)
Polygonum L. [POLYGON$] Polygonaceae (V:D)
Polygonum achoreum Blake [POLYACH] Polygonaceae (V:D)
Polygonum aequale Lindm. = Polygonum arenastrum [POLYARE] Polygonaceae (V:D)
Polygonum alatum Hamilton ex D. Don = Polygonum nepalense [POLYNEP] Polygonaceae (V:D)
Polygonum amphibium L. [POLYAMP] Polygonaceae (V:D)
Polygonum amphibium ssp. *laevimarginatum* Hultén = Polygonum amphibium var. stipulaceum [POLYAMP2] Polygonaceae (V:D)
Polygonum amphibium var. **emersum** Michx. [POLYAMP1] Polygonaceae (V:D)
Polygonum amphibium var. *hartwrightii* (Gray) Bissell = Polygonum amphibium var. stipulaceum [POLYAMP2] Polygonaceae (V:D)
Polygonum amphibium var. **stipulaceum** Coleman [POLYAMP2] Polygonaceae (V:D)
•**Polygonum arenastrum** Jord. ex Boreau [POLYARE] Polygonaceae (V:D)
Polygonum austiniae Greene = Polygonum douglasii ssp. austiniae [POLYDOU3] Polygonaceae (V:D)
•**Polygonum aviculare** L. [POLYAVI] Polygonaceae (V:D)
Polygonum aviculare var. *arenastrum* (Jord. ex Boreau) Rouy = Polygonum arenastrum [POLYARE] Polygonaceae (V:D)
Polygonum aviculare var. *littorale* (Link) Mert. = Polygonum buxiforme [POLYBUX] Polygonaceae (V:D)
Polygonum aviculare var. *vegetum* Ledeb. = Polygonum aviculare [POLYAVI] Polygonaceae (V:D)
Polygonum bistortoides Pursh [POLYBIS] Polygonaceae (V:D)
Polygonum bistortoides var. *linearifolium* (S. Wats.) Small = Polygonum bistortoides [POLYBIS] Polygonaceae (V:D)
Polygonum bistortoides var. *oblongifolium* (Meisn.) St. John = Polygonum bistortoides [POLYBIS] Polygonaceae (V:D)
Polygonum buxifolium Nutt. ex Bong. = Polygonum fowleri [POLYFOW] Polygonaceae (V:D)
•**Polygonum buxiforme** Small [POLYBUX] Polygonaceae (V:D)
Polygonum buxiforme var. *montanum* (Small) R.J. Davis = Polygonum douglasii ssp. douglasii [POLYDOU0] Polygonaceae (V:D)
Polygonum cephalophorum Greene = Polygonum bistortoides [POLYBIS] Polygonaceae (V:D)
Polygonum coarctatum var. *majus* Meisn. = Polygonum douglasii ssp. majus [POLYDOU5] Polygonaceae (V:D)
Polygonum coccineum Muhl. ex Willd. = Polygonum amphibium var. emersum [POLYAMP1] Polygonaceae (V:D)
Polygonum coccineum var. *pratincola* (Greene) Stanford = Polygonum amphibium var. emersum [POLYAMP1] Polygonaceae (V:D)
Polygonum coccineum var. *rigidulum* (Sheldon) Stanford = Polygonum amphibium var. stipulaceum [POLYAMP2] Polygonaceae (V:D)
Polygonum coccineum var. *terrestre* Willd. = Polygonum amphibium var. emersum [POLYAMP1] Polygonaceae (V:D)
•**Polygonum convolvulus** L. [POLYCON] Polygonaceae (V:D)
•**Polygonum cuspidatum** Sieb. & Zucc. [POLYCUS] Polygonaceae (V:D)
Polygonum cuspidatum var. *compactum* (Hook f.) Bailey = Polygonum cuspidatum [POLYCUS] Polygonaceae (V:D)

Polygonum douglasii Greene [POLYDOU] Polygonaceae (V:D)
Polygonum douglasii ssp. **austiniae** (Greene) E. Murr. [POLYDOU3] Polygonaceae (V:D)
Polygonum douglasii ssp. **douglasii** [POLYDOU0] Polygonaceae (V:D)
Polygonum douglasii ssp. **engelmannii** (Greene) Kartesz & Gandhi [POLYDOU1] Polygonaceae (V:D)
Polygonum douglasii ssp. **majus** (Meisn.) Hickman [POLYDOU5] Polygonaceae (V:D)
Polygonum douglasii ssp. **nuttallii** (Small) Hickman [POLYDOU6] Polygonaceae (V:D)
Polygonum douglasii ssp. **spergulariiforme** (Meisn. ex Small) Hickman [POLYDOU7] Polygonaceae (V:D)
Polygonum douglasii var. *austiniae* (Greene) M.E. Jones = Polygonum douglasii ssp. austiniae [POLYDOU3] Polygonaceae (V:D)
Polygonum douglasii var. *latifolium* (Engelm.) Greene = Polygonum douglasii ssp. douglasii [POLYDOU0] Polygonaceae (V:D)
Polygonum douglasii var. *microspermum* (Engelm.) Dorn = Polygonum douglasii ssp. engelmannii [POLYDOU1] Polygonaceae (V:D)
Polygonum dubium Stein = Polygonum persicaria [POLYPES] Polygonaceae (V:D)
Polygonum emaciatum A. Nels. = Polygonum douglasii ssp. douglasii [POLYDOU0] Polygonaceae (V:D)
Polygonum engelmannii Greene = Polygonum douglasii ssp. engelmannii [POLYDOU1] Polygonaceae (V:D)
Polygonum erectum ssp. *achoreum* Blaka = Polygonum achoreum [POLYACH] Polygonaceae (V:D)
Polygonum fagopyrum L. = Fagopyrum esculentum [FAGOESC] Polygonaceae (V:D)
Polygonum fluitans Eat. = Polygonum amphibium var. stipulaceum [POLYAMP2] Polygonaceae (V:D)
Polygonum fowleri B.L. Robins. [POLYFOW] Polygonaceae (V:D)
Polygonum fugax Small = Polygonum viviparum [POLYVIV] Polygonaceae (V:D)
Polygonum glastifolium Greene = Polygonum bistortoides [POLYBIS] Polygonaceae (V:D)
Polygonum hartwrightii Gray = Polygonum amphibium var. stipulaceum [POLYAMP2] Polygonaceae (V:D)
Polygonum heterophyllum Lindl. = Polygonum aviculare [POLYAVI] Polygonaceae (V:D)
***Polygonum hydropiper** L. [POLYHYD] Polygonaceae (V:D)
Polygonum hydropiper var. *projectum* Stanford = Polygonum hydropiper [POLYHYD] Polygonaceae (V:D)
Polygonum hydropiperoides Michx. [POLYHYR] Polygonaceae (V:D)
Polygonum hydropiperoides var. *adenocalyx* (Stanford) Gleason = Polygonum hydropiperoides [POLYHYR] Polygonaceae (V:D)
Polygonum hydropiperoides var. *asperifolium* Stanford = Polygonum hydropiperoides [POLYHYR] Polygonaceae (V:D)
Polygonum hydropiperoides var. *breviciliatum* Fern. = Polygonum hydropiperoides [POLYHYR] Polygonaceae (V:D)
Polygonum hydropiperoides var. *bushianum* Stanford = Polygonum hydropiperoides [POLYHYR] Polygonaceae (V:D)
Polygonum hydropiperoides var. *digitatum* Fern. = Polygonum hydropiperoides [POLYHYR] Polygonaceae (V:D)
Polygonum hydropiperoides var. *euronotorum* Fern. = Polygonum hydropiperoides [POLYHYR] Polygonaceae (V:D)
Polygonum hydropiperoides var. *opelousanum* (Riddell ex Small) Riddell ex W. Stone = Polygonum hydropiperoides [POLYHYR] Polygonaceae (V:D)
Polygonum hydropiperoides var. *psilostachyum* St. John = Polygonum hydropiperoides [POLYHYR] Polygonaceae (V:D)
Polygonum hydropiperoides var. *strigosum* (Small) Stanford = Polygonum hydropiperoides [POLYHYR] Polygonaceae (V:D)
Polygonum inundatum Raf. = Polygonum amphibium var. stipulaceum [POLYAMP2] Polygonaceae (V:D)
Polygonum kelloggii Greene = Polygonum polygaloides ssp. kelloggii [POLYPOL1] Polygonaceae (V:D)
Polygonum lapathifolium L. [POLYLAP] Polygonaceae (V:D)
Polygonum littorale Link = Polygonum buxiforme [POLYBUX] Polygonaceae (V:D)
Polygonum macounii Small ex Macoun = Polygonum viviparum [POLYVIV] Polygonaceae (V:D)
Polygonum majus (Meisn.) Piper = Polygonum douglasii ssp. majus [POLYDOU5] Polygonaceae (V:D)
Polygonum microspermum (Engelm.) Small = Polygonum douglasii ssp. engelmannii [POLYDOU1] Polygonaceae (V:D)
Polygonum minimum S. Wats. [POLYMIN] Polygonaceae (V:D)
Polygonum minus auct. = Polygonum persicaria [POLYPES] Polygonaceae (V:D)
Polygonum minus var. *subcontinuum* (Meisn.) Fern. = Polygonum persicaria [POLYPES] Polygonaceae (V:D)
Polygonum monspeliense Pers. = Polygonum aviculare [POLYAVI] Polygonaceae (V:D)
Polygonum montanum Small = Polygonum douglasii ssp. douglasii [POLYDOU0] Polygonaceae (V:D)
Polygonum montereyense Brenckle = Polygonum arenastrum [POLYARE] Polygonaceae (V:D)
Polygonum muehlenbergii S. Wats. = Polygonum amphibium var. emersum [POLYAMP1] Polygonaceae (V:D)
Polygonum muehlenbergii var. *terrestre* (Willd.) Trel. = Polygonum amphibium var. emersum [POLYAMP1] Polygonaceae (V:D)
Polygonum natans Eat. = Polygonum amphibium var. stipulaceum [POLYAMP2] Polygonaceae (V:D)
***Polygonum nepalense** Meisn. [POLYNEP] Polygonaceae (V:D)
Polygonum nuttallii Small = Polygonum douglasii ssp. nuttallii [POLYDOU6] Polygonaceae (V:D)
Polygonum opelousanum Riddell ex Small = Polygonum hydropiperoides [POLYHYR] Polygonaceae (V:D)
Polygonum opelousanum var. *adenocalyx* Stanford = Polygonum hydropiperoides [POLYHYR] Polygonaceae (V:D)
Polygonum paronychia Cham. & Schlecht. [POLYPAR] Polygonaceae (V:D)
***Polygonum persicaria** L. [POLYPES] Polygonaceae (V:D)
Polygonum persicaria var. *angustifolium* Beckh. = Polygonum persicaria [POLYPES] Polygonaceae (V:D)
Polygonum persicaria var. *ruderale* (Salisb.) Meisn. = Polygonum persicaria [POLYPES] Polygonaceae (V:D)
Polygonum persicarioides Kunth = Polygonum hydropiperoides [POLYHYR] Polygonaceae (V:D)
Polygonum polygaloides Meisn. [POLYPOL] Polygonaceae (V:D)
Polygonum polygaloides ssp. **kelloggii** (Greene) Hickman [POLYPOL1] Polygonaceae (V:D)
***Polygonum polystachyum** Wallich ex Meisn. [POLYPOY] Polygonaceae (V:D)
Polygonum prolificum (Small) B.L. Robins. = Polygonum ramosissimum var. prolificum [POLYRAM1] Polygonaceae (V:D)
Polygonum punctatum Ell. [POLYPUN] Polygonaceae (V:D)
Polygonum puritanorum Fern. = Polygonum persicaria [POLYPES] Polygonaceae (V:D)
Polygonum ramosissimum Michx. [POLYRAM] Polygonaceae (V:D)
Polygonum ramosissimum var. **prolificum** Small [POLYRAM1] Polygonaceae (V:D)
***Polygonum sachalinense** F. Schmidt ex Maxim. [POLYSAC] Polygonaceae (V:D)
Polygonum scabrum Moench = Polygonum lapathifolium [POLYLAP] Polygonaceae (V:D)
Polygonum spergulariiforme Meisn. ex Small = Polygonum douglasii ssp. spergulariiforme [POLYDOU7] Polygonaceae (V:D)
Polygonum tenue var. *microspermum* Engelm. = Polygonum douglasii ssp. engelmannii [POLYDOU1] Polygonaceae (V:D)
Polygonum unifolium Greene = Polygonum polygaloides ssp. kelloggii [POLYPOL1] Polygonaceae (V:D)
Polygonum viviparum L. [POLYVIV] Polygonaceae (V:D)
Polygonum viviparum var. *alpinum* Wahlenb. = Polygonum viviparum [POLYVIV] Polygonaceae (V:D)
Polygonum viviparum var. *macounii* (Small ex Macoun) Hultén = Polygonum viviparum [POLYVIV] Polygonaceae (V:D)
Polygonum vulcanicum Greene = Polygonum bistortoides [POLYBIS] Polygonaceae (V:D)
Polygonum zuccarinii Small = Polygonum cuspidatum [POLYCUS] Polygonaceae (V:D)
Polypodium L. [POLYPOD$] Polypodiaceae (V:F)

Polypodium amorphum Suksdorf [POLYAMO] Polypodiaceae (V:F)
Polypodium glycyrrhiza D.C. Eat. [POLYGLY] Polypodiaceae (V:F)
Polypodium hesperium Maxon [POLYHES] Polypodiaceae (V:F)
Polypodium montense F.A. Lang = Polypodium amorphum [POLYAMO] Polypodiaceae (V:F)
Polypodium occidentale (Hook.) Maxon = Polypodium glycyrrhiza [POLYGLY] Polypodiaceae (V:F)
Polypodium scouleri Hook. & Grev. [POLYSCO] Polypodiaceae (V:F)
Polypodium sibiricum Sipl. [POLYSIB] Polypodiaceae (V:F)
Polypodium vulgare ssp. *columbianum* (Gilbert) Hultén = Polypodium hesperium [POLYHES] Polypodiaceae (V:F)
Polypodium vulgare ssp. *occidentale* (Hook.) Hultén = Polypodium glycyrrhiza [POLYGLY] Polypodiaceae (V:F)
Polypodium vulgare var. *columbianum* Gilbert = Polypodium hesperium [POLYHES] Polypodiaceae (V:F)
Polypodium vulgare var. *commune* Milde = Polypodium glycyrrhiza [POLYGLY] Polypodiaceae (V:F)
Polypodium vulgare var. *hesperium* (Maxon) A. Nels. & J.F. Macbr. = Polypodium hesperium [POLYHES] Polypodiaceae (V:F)
Polypodium vulgare var. *occidentale* Hook. = Polypodium glycyrrhiza [POLYGLY] Polypodiaceae (V:F)
Polypogon Desf. [POLYPOG$] Poaceae (V:M)
•**Polypogon monspeliensis** (L.) Desf. [POLYMOS] Poaceae (V:M)
Polysporina Vezda [POLYSPO$] Acarosporaceae (L)
Polysporina simplex (Davies) Vezda [POLYSIM] Acarosporaceae (L)
Polysporina urceolata (Anzi) Brodo [POLYURC] Acarosporaceae (L)
Polystichum Roth [POLYSTI$] Dryopteridaceae (V:F)
Polystichum alaskense Maxon = Polystichum setigerum [POLYSET] Dryopteridaceae (V:F)
Polystichum andersonii Hopkins [POLYAND] Dryopteridaceae (V:F)
Polystichum braunii (Spenner) Fée [POLYBRA] Dryopteridaceae (V:F)
Polystichum braunii ssp. *alaskense* (Maxon) Calder & Taylor = Polystichum setigerum [POLYSET] Dryopteridaceae (V:F)
Polystichum braunii ssp. *andersonii* (Hopkins) Calder & Taylor = Polystichum andersonii [POLYAND] Dryopteridaceae (V:F)
Polystichum braunii ssp. *purshii* (Fern.) Calder & Taylor = Polystichum braunii [POLYBRA] Dryopteridaceae (V:F)
Polystichum braunii var. *alaskense* (Maxon) Hultén = Polystichum setigerum [POLYSET] Dryopteridaceae (V:F)
Polystichum braunii var. *purshii* Fern. = Polystichum braunii [POLYBRA] Dryopteridaceae (V:F)
Polystichum imbricans (D.C. Eat.) D.H. Wagner [POLYIMB] Dryopteridaceae (V:F)
Polystichum kruckebergii W.H. Wagner [POLYKRU] Dryopteridaceae (V:F)
Polystichum lemmonii Underwood [POLYLEM] Dryopteridaceae (V:F)
Polystichum lonchitis (L.) Roth [POLYLOC] Dryopteridaceae (V:F)
Polystichum mohrioides var. *lemmonii* (Underwood) Fern. = Polystichum lemmonii [POLYLEM] Dryopteridaceae (V:F)
Polystichum mohrioides var. *scopulinum* (D.C. Eat.) Fern. = Polystichum scopulinum [POLYSCP] Dryopteridaceae (V:F)
Polystichum munitum (Kaulfuss) K. Presl [POLYMUN] Dryopteridaceae (V:F)
Polystichum munitum var. *incisoserratum* (D.C. Eat.) Underwood = Polystichum munitum [POLYMUN] Dryopteridaceae (V:F)
Polystichum scopulinum (D.C. Eat.) Maxon [POLYSCP] Dryopteridaceae (V:F)
Polystichum setigerum (K. Presl) K. Presl [POLYSET] Dryopteridaceae (V:F)
Polytrichadelphus lyallii Mitt. = Polytrichum lyallii [POLYLYA] Polytrichaceae (B:M)
Polytrichastrum G.L. Sm. [POLYTRC$] Polytrichaceae (B:M)
Polytrichastrum alpinum (Hedw.) G.L. Sm. [POLYALP] Polytrichaceae (B:M)
Polytrichastrum formosum (Hedw.) G.L. Sm. = Polytrichum formosum [POLYFOR] Polytrichaceae (B:M)
Polytrichastrum longisetum (Brid.) G. L. Sm. = Polytrichum longisetum [POLYLOG] Polytrichaceae (B:M)
Polytrichastrum lyallii (Mitt.) G.L. Sm. = Polytrichum lyallii [POLYLYA] Polytrichaceae (B:M)
Polytrichastrum sexangulare (Brid.) G. L. Sm. = Polytrichum sexangulare [POLYSEX] Polytrichaceae (B:M)
Polytrichum Hedw. [POLYTRI$] Polytrichaceae (B:M)
Polytrichum affine Funck = Polytrichum strictum [POLYSTR] Polytrichaceae (B:M)
Polytrichum alpestre Hoppe = Polytrichum strictum [POLYSTR] Polytrichaceae (B:M)
Polytrichum anomalum (Milde) Milde = Polytrichum longisetum [POLYLOG] Polytrichaceae (B:M)
Polytrichum apiculatum Kindb. = Polytrichum juniperinum [POLYJUN] Polytrichaceae (B:M)
Polytrichum attenuatum Menz. ex Brid. = Polytrichum formosum [POLYFOR] Polytrichaceae (B:M)
Polytrichum aurantiacum Hoppe ex Brid. = Polytrichum longisetum [POLYLOG] Polytrichaceae (B:M)
Polytrichum capillare Michx. = Pogonatum dentatum [POGODEN] Polytrichaceae (B:M)
Polytrichum commune Hedw. [POLYCOM] Polytrichaceae (B:M)
Polytrichum commune var. *canadense* Kindb. in Mac. & Kindb. = Polytrichum commune [POLYCOM] Polytrichaceae (B:M)
Polytrichum commune var. **commune** [POLYCOM0] Polytrichaceae (B:M)
Polytrichum commune var. *maximoviczii* Lindb. = Polytrichum commune var. commune [POLYCOM0] Polytrichaceae (B:M)
Polytrichum commune var. *nigrescens* Warnst. = Polytrichum commune var. commune [POLYCOM0] Polytrichaceae (B:M)
Polytrichum commune var. **perigoniale** (Michx.) Hampe [POLYCOM4] Polytrichaceae (B:M)
Polytrichum commune var. *uliginosum* Wallr. = Polytrichum commune var. commune [POLYCOM0] Polytrichaceae (B:M)
Polytrichum conorhynchum Kindb. in Mac. & Kindb. = Polytrichum formosum [POLYFOR] Polytrichaceae (B:M)
Polytrichum formosum Hedw. [POLYFOR] Polytrichaceae (B:M)
Polytrichum formosum var. *aurantiacum* (Hoppe ex Brid.) Hartm. = Polytrichum longisetum [POLYLOG] Polytrichaceae (B:M)
Polytrichum gracile Bryhn = Polytrichum longisetum [POLYLOG] Polytrichaceae (B:M)
Polytrichum gracile var. *anomalum* (Milde) Hag. = Polytrichum longisetum [POLYLOG] Polytrichaceae (B:M)
Polytrichum juniperinum Hedw. [POLYJUN] Polytrichaceae (B:M)
Polytrichum juniperinum var. *affine* (Funck) Brid. = Polytrichum strictum [POLYSTR] Polytrichaceae (B:M)
Polytrichum juniperinum var. *alpestre* (Hoppe) Röhl. = Polytrichum strictum [POLYSTR] Polytrichaceae (B:M)
Polytrichum juniperinum var. *gracilius* Wahlenb. = Polytrichum strictum [POLYSTR] Polytrichaceae (B:M)
Polytrichum juniperinum var. *waghornei* Kindb. in Mac. = Polytrichum juniperinum [POLYJUN] Polytrichaceae (B:M)
Polytrichum longisetum Brid. [POLYLOG] Polytrichaceae (B:M)
Polytrichum longisetum var. *anomalum* (Milde) Hag. = Polytrichum longisetum [POLYLOG] Polytrichaceae (B:M)
Polytrichum longisetum var. *aurantiacum* (Hoppe ex Brid.) = Polytrichum longisetum [POLYLOG] Polytrichaceae (B:M)
Polytrichum lyallii (Mitt.) Kindb. [POLYLYA] Polytrichaceae (B:M)
Polytrichum norvegicum auct. = Polytrichum sexangulare [POLYSEX] Polytrichaceae (B:M)
Polytrichum norvegicum var. *vulcanicum* (C. Jens.) Podp. = Polytrichum sphaerothecium [POLYSPH] Polytrichaceae (B:M)
Polytrichum piliferum Hedw. [POLYPIL] Polytrichaceae (B:M)
Polytrichum sexangulare Brid. [POLYSEX] Polytrichaceae (B:M)
Polytrichum sexangulare var. *vulcanicum* C. Jens. = Polytrichum sphaerothecium [POLYSPH] Polytrichaceae (B:M)
Polytrichum sphaerothecium (Besch.) C. Müll. [POLYSPH] Polytrichaceae (B:M)
Polytrichum strictum Brid. [POLYSTR] Polytrichaceae (B:M)
Populus L. [POPULUS$] Salicaceae (V:D)
Populus angulata Ait. = Populus deltoides ssp. deltoides [POPUDEL0] Salicaceae (V:D)

Populus angulata var. *missouriensis* A. Henry = Populus deltoides ssp. deltoides [POPUDEL0] Salicaceae (V:D)
Populus aurea Tidestrom = Populus tremuloides [POPUTRE] Salicaceae (V:D)
Populus balsamifera L. [POPUBAL] Salicaceae (V:D)
Populus balsamifera ssp. **balsamifera** [POPUBAL0] Salicaceae (V:D)
Populus balsamifera ssp. **trichocarpa** (Torr. & Gray ex Hook.) Brayshaw [POPUBAL2] Salicaceae (V:D)
Populus balsamifera var. *californica* S. Wats. = Populus balsamifera ssp. trichocarpa [POPUBAL2] Salicaceae (V:D)
Populus balsamifera var. *candicans* (Ait.) Gray = Populus balsamifera ssp. balsamifera [POPUBAL0] Salicaceae (V:D)
Populus balsamifera var. *fernaldiana* Rouleau = Populus balsamifera ssp. balsamifera [POPUBAL0] Salicaceae (V:D)
Populus balsamifera var. *lanceolata* Marsh. = Populus balsamifera ssp. balsamifera [POPUBAL0] Salicaceae (V:D)
Populus balsamifera var. *michauxii* (Dode) A. Henry = Populus balsamifera ssp. balsamifera [POPUBAL0] Salicaceae (V:D)
Populus balsamifera var. *missouriensis* (A. Henry) Rehd. = Populus deltoides ssp. deltoides [POPUDEL0] Salicaceae (V:D)
Populus balsamifera var. *pilosa* Sarg. = Populus deltoides ssp. deltoides [POPUDEL0] Salicaceae (V:D)
Populus balsamifera var. *subcordata* Hyl. = Populus balsamifera ssp. balsamifera [POPUBAL0] Salicaceae (V:D)
Populus balsamifera var. *virginiana* (Foug.) Sarg. = Populus deltoides ssp. deltoides [POPUDEL0] Salicaceae (V:D)
Populus besseyana Dode = Populus deltoides ssp. monilifera [POPUDEL2] Salicaceae (V:D)
Populus canadensis var. *virginiana* (Foug.) Fiori = Populus deltoides ssp. deltoides [POPUDEL0] Salicaceae (V:D)
Populus candicans Ait. = Populus balsamifera ssp. balsamifera [POPUBAL0] Salicaceae (V:D)
Populus cercidiphylla Britt. = Populus tremuloides [POPUTRE] Salicaceae (V:D)
*****Populus deltoides** Bartr. ex Marsh. [POPUDEL] Salicaceae (V:D)
*****Populus deltoides** ssp. **deltoides** [POPUDEL0] Salicaceae (V:D)
*****Populus deltoides** ssp. **monilifera** (Ait.) Eckenwalder [POPUDEL2] Salicaceae (V:D)
Populus deltoides var. *angulata* (Ait.) Sarg. = Populus deltoides ssp. deltoides [POPUDEL0] Salicaceae (V:D)
Populus deltoides var. *missouriensis* (A. Henry) A. Henry = Populus deltoides ssp. deltoides [POPUDEL0] Salicaceae (V:D)
Populus deltoides var. *occidentalis* Rydb. = Populus deltoides ssp. monilifera [POPUDEL2] Salicaceae (V:D)
Populus deltoides var. *pilosa* (Sarg.) Sudworth = Populus deltoides ssp. deltoides [POPUDEL0] Salicaceae (V:D)
Populus deltoides var. *virginiana* (Foug.) Sudworth = Populus deltoides ssp. deltoides [POPUDEL0] Salicaceae (V:D)
Populus hastata Dode p.p. = Populus balsamifera ssp. trichocarpa [POPUBAL2] Salicaceae (V:D)
Populus michauxii Dode = Populus balsamifera ssp. balsamifera [POPUBAL0] Salicaceae (V:D)
Populus monilifera Ait. = Populus deltoides ssp. monilifera [POPUDEL2] Salicaceae (V:D)
Populus nigra var. *virginiana* (Foug.) Castigl. = Populus deltoides ssp. deltoides [POPUDEL0] Salicaceae (V:D)
Populus occidentalis (Rydb.) Britt. ex Rydb. = Populus deltoides ssp. monilifera [POPUDEL2] Salicaceae (V:D)
Populus palmeri Sarg. = Populus deltoides ssp. deltoides [POPUDEL0] Salicaceae (V:D)
Populus × polygonifolia Bernard = Populus tremuloides [POPUTRE] Salicaceae (V:D)
Populus sargentii Dode = Populus deltoides ssp. monilifera [POPUDEL2] Salicaceae (V:D)
Populus sargentii var. *texana* (Sarg.) Correll = Populus deltoides ssp. monilifera [POPUDEL2] Salicaceae (V:D)
Populus tacamahaca P. Mill. = Populus balsamifera ssp. balsamifera [POPUBAL0] Salicaceae (V:D)
Populus tacamahaca var. *candicans* (Ait.) Stout = Populus balsamifera ssp. balsamifera [POPUBAL0] Salicaceae (V:D)
Populus tacamahaca var. *lanceolata* (Marsh.) Farw. = Populus balsamifera ssp. balsamifera [POPUBAL0] Salicaceae (V:D)
Populus tacamahaca var. *michauxii* (Dode) Farw. = Populus balsamifera ssp. balsamifera [POPUBAL0] Salicaceae (V:D)
Populus texana Sarg. = Populus deltoides ssp. monilifera [POPUDEL2] Salicaceae (V:D)
Populus tremula ssp. *tremuloides* (Michx.) A. & D. Löve = Populus tremuloides [POPUTRE] Salicaceae (V:D)
Populus tremuloides Michx. [POPUTRE] Salicaceae (V:D)
Populus tremuloides var. *aurea* (Tidestrom) Daniels = Populus tremuloides [POPUTRE] Salicaceae (V:D)
Populus tremuloides var. *cercidiphylla* (Britt.) Sudworth = Populus tremuloides [POPUTRE] Salicaceae (V:D)
Populus tremuloides var. *intermedia* Victorin = Populus tremuloides [POPUTRE] Salicaceae (V:D)
Populus tremuloides var. *magnifica* Victorin = Populus tremuloides [POPUTRE] Salicaceae (V:D)
Populus tremuloides var. *rhomboidea* Victorin = Populus tremuloides [POPUTRE] Salicaceae (V:D)
Populus tremuloides var. *vancouveriana* (Trel.) Sarg. = Populus tremuloides [POPUTRE] Salicaceae (V:D)
Populus trichocarpa Torr. & Gray ex Hook. = Populus balsamifera ssp. trichocarpa [POPUBAL2] Salicaceae (V:D)
Populus trichocarpa ssp. *hastata* (Dode) Dode p.p. = Populus balsamifera ssp. trichocarpa [POPUBAL2] Salicaceae (V:D)
Populus trichocarpa var. *cupulata* S. Wats. = Populus balsamifera ssp. trichocarpa [POPUBAL2] Salicaceae (V:D)
Populus trichocarpa var. *hastata* (Dode) A. Henry p.p. = Populus balsamifera ssp. trichocarpa [POPUBAL2] Salicaceae (V:D)
Populus trichocarpa var. *ingrata* (Jepson) Jepson = Populus balsamifera ssp. trichocarpa [POPUBAL2] Salicaceae (V:D)
Populus vancouveriana Trel. = Populus tremuloides [POPUTRE] Salicaceae (V:D)
Populus virginiana Foug. = Populus deltoides ssp. deltoides [POPUDEL0] Salicaceae (V:D)
Populus virginiana var. *pilosa* (Sarg.) F.C. Gates = Populus deltoides ssp. deltoides [POPUDEL0] Salicaceae (V:D)
Porella L. [PORELLA$] Porellaceae (B:H)
Porella cordaeana (Hub.) Moore [PORECOR] Porellaceae (B:H)
Porella navicularis (Lehm. & Lindenb.) Lindb. [PORENAV] Porellaceae (B:H)
Porella platyphylla (L.) Pfeiff. [POREPLA] Porellaceae (B:H)
Porella platyphylla var. *platyphylloidea* (Schwein.) Frye & Clark = Porella platyphylloidea [POREPLT] Porellaceae (B:H)
Porella platyphylloidea (Schwein.) Lindb. [POREPLT] Porellaceae (B:H)
Porella roellii Steph. [PORERОЕ] Porellaceae (B:H)
Porella roellii f. **crispata** Hong [PORERОЕ1] Porellaceae (B:H)
Porella roellii f. **schofieldiana** Hong [PORERОЕ2] Porellaceae (B:H)
Porina Müll. Arg. [PORINA$] Trichotheliaceae (L)
Porina aenea (Wallr.) Zahlbr. = Pseudosagedia aenea [PSEUAEN] Trichotheliaceae (L)
Porina carpinea (Pers. ex Ach.) Zahlbr. = Pseudosagedia aenea [PSEUAEN] Trichotheliaceae (L)
Porina chlorotica (Ach.) Müll. Arg. = Pseudosagedia chlorotica [PSEUCHL] Trichotheliaceae (L)
Porina guentheri (Flotow) Zahlbr. = Pseudosagedia guentheri [PSEUGUE] Trichotheliaceae (L)
Porina lectissima (Fr.) Zahlbr. [PORILEC] Trichotheliaceae (L)
Porina leptalea (Durieu & Mont.) A.L. Sm. [PORILEP] Trichotheliaceae (L)
Porina linearis (Leighton) Zahlbr. = Pseudosagedia linearis [PSEULIN] Trichotheliaceae (L)
Porina plumbaria (Stizenb.) Hasse = Arthopyrenia plumbaria [ARTHPLU] Arthopyreniaceae (L)
Porocyphus Körber [POROCYP$] Lichinaceae (L)
Porocyphus kenmorensis (Holl ex Nyl.) Henssen [POROKEN] Lichinaceae (L)
Porothamnium bigelovii (Sull.) Fleisch. in Broth. = Porotrichum bigelovii [POROBIG] Thamnobryaceae (B:M)
Porotrichum (Brid.) Hampe [POROTRI$] Thamnobryaceae (B:M)
Porotrichum bigelovii (Sull.) Kindb. [POROBIG] Thamnobryaceae (B:M)

Porotrichum neckeroides (Hook.) Williams = Thamnobryum neckeroides [THAMNEC] Thamnobryaceae (B:M)
Porotrichum neomexicanum (Card.) Wagn. = Metaneckera menziesii [METAMEN] Neckeraceae (B:M)
Porotrichum vancouveriense (Kindb. in Mac.) Crum [POROVAN] Thamnobryaceae (B:M)
Porpidia Körber [PORPIDI$] Porpidiaceae (L)
Porpidia albocaerulescens (Wulfen) Hertel & Knoph [PORPALB] Porpidiaceae (L)
Porpidia carlottiana Gowan [PORPCAR] Porpidiaceae (L)
Porpidia contraponenda (Arnold) Knoph & Hertel [PORPCON] Porpidiaceae (L)
Porpidia crustulata (Ach.) Hertel & Knoph [PORPCRU] Porpidiaceae (L)
Porpidia macrocarpa (DC.) Hertel & A.J. Schwab [PORPMAC] Porpidiaceae (L)
Porpidia nigrocruenta (Anzi) Diederich & Sérus. = Porpidia macrocarpa [PORPMAC] Porpidiaceae (L)
Porpidia ochrolemma (Vainio) Brodo & R. Sant. [PORPOCH] Porpidiaceae (L)
Porpidia pseudomelinodes A.J. Schwab = Porpidia ochrolemma [PORPOCH] Porpidiaceae (L)
Porpidia soredizodes (Lamy ex Nyl.) J.R. Laundon [PORPSOR] Porpidiaceae (L)
Porpidia thomsonii Gowan [PORPTHO] Porpidiaceae (L)
Porpidia tuberculosa (Sm.) Hertel & Knoph [PORPTUB] Porpidiaceae (L)
Portulaca L. [PORTULA$] Portulacaceae (V:D)
*__Portulaca oleracea__ L. [PORTOLE] Portulacaceae (V:D)
Potamogeton L. [POTAMOG$] Potamogetonaceae (V:M)
Potamogeton alpinus Balbis [POTAALP] Potamogetonaceae (V:M)
Potamogeton alpinus ssp. *tenuifolius* (Raf.) Hultén = Potamogeton alpinus [POTAALP] Potamogetonaceae (V:M)
Potamogeton alpinus var. *subellipticus* (Fern.) Ogden = Potamogeton alpinus [POTAALP] Potamogetonaceae (V:M)
Potamogeton alpinus var. *tenuifolius* (Raf.) Ogden = Potamogeton alpinus [POTAALP] Potamogetonaceae (V:M)
Potamogeton americanus Cham. & Schlecht. = Potamogeton nodosus [POTANOD] Potamogetonaceae (V:M)
Potamogeton amplexicaulis Kar. = Potamogeton perfoliatus [POTAPER] Potamogetonaceae (V:M)
Potamogeton amplifolius Tuckerman [POTAAMP] Potamogetonaceae (V:M)
Potamogeton angustifolius Bercht. & K. Presl = Potamogeton illinoensis [POTAILL] Potamogetonaceae (V:M)
Potamogeton berchtoldii Fieber = Potamogeton pusillus var. tenuissimus [POTAPUS2] Potamogetonaceae (V:M)
Potamogeton berchtoldii var. *acuminatus* Fieber = Potamogeton pusillus var. tenuissimus [POTAPUS2] Potamogetonaceae (V:M)
Potamogeton berchtoldii var. *colpophilus* (Fern.) Fern. = Potamogeton pusillus var. tenuissimus [POTAPUS2] Potamogetonaceae (V:M)
Potamogeton berchtoldii var. *lacunatus* (Hagstr.) Fern. = Potamogeton pusillus var. tenuissimus [POTAPUS2] Potamogetonaceae (V:M)
Potamogeton berchtoldii var. *mucronatus* Fieber = Potamogeton pusillus var. tenuissimus [POTAPUS2] Potamogetonaceae (V:M)
Potamogeton berchtoldii var. *polyphyllus* (Morong) Fern. = Potamogeton pusillus var. tenuissimus [POTAPUS2] Potamogetonaceae (V:M)
Potamogeton berchtoldii var. *tenuissimus* (Mert. & Koch) Fern. = Potamogeton pusillus var. tenuissimus [POTAPUS2] Potamogetonaceae (V:M)
Potamogeton bupleuroides Fern. = Potamogeton perfoliatus [POTAPER] Potamogetonaceae (V:M)
Potamogeton compressus auct. = Potamogeton zosteriformis [POTAZOS] Potamogetonaceae (V:M)
*__Potamogeton crispus__ L. [POTACRI] Potamogetonaceae (V:M)
Potamogeton epihydrus Raf. [POTAEPI] Potamogetonaceae (V:M)
Potamogeton epihydrus ssp. *nuttallii* (Cham. & Schlecht.) Calder & Taylor = Potamogeton epihydrus [POTAEPI] Potamogetonaceae (V:M)
Potamogeton epihydrus var. *nuttallii* (Cham. & Schlecht.) Fern. = Potamogeton epihydrus [POTAEPI] Potamogetonaceae (V:M)
Potamogeton epihydrus var. *ramosus* (Peck) House = Potamogeton epihydrus [POTAEPI] Potamogetonaceae (V:M)
Potamogeton filiformis Pers. [POTAFIL] Potamogetonaceae (V:M)
Potamogeton fluitans Roth = Potamogeton nodosus [POTANOD] Potamogetonaceae (V:M)
Potamogeton foliosus Raf. [POTAFOL] Potamogetonaceae (V:M)
Potamogeton foliosus var. *macellus* Fern. = Potamogeton foliosus [POTAFOL] Potamogetonaceae (V:M)
Potamogeton friesii Rupr. [POTAFRI] Potamogetonaceae (V:M)
Potamogeton gramineus L. [POTAGRA] Potamogetonaceae (V:M)
Potamogeton gramineus var. *graminifolius* Fries = Potamogeton gramineus [POTAGRA] Potamogetonaceae (V:M)
Potamogeton gramineus var. *maximus* Morong = Potamogeton gramineus [POTAGRA] Potamogetonaceae (V:M)
Potamogeton gramineus var. *myriophyllus* J.W. Robbins = Potamogeton gramineus [POTAGRA] Potamogetonaceae (V:M)
Potamogeton gramineus var. *typicus* Ogden = Potamogeton gramineus [POTAGRA] Potamogetonaceae (V:M)
Potamogeton heterophyllus Schreb. = Potamogeton illinoensis [POTAILL] Potamogetonaceae (V:M)
Potamogeton illinoensis Morong [POTAILL] Potamogetonaceae (V:M)
Potamogeton interruptus Kit. = Potamogeton vaginatus [POTAVAG] Potamogetonaceae (V:M)
Potamogeton longiligulatus Fern. = Potamogeton strictifolius [POTASTR] Potamogetonaceae (V:M)
Potamogeton natans L. [POTANAT] Potamogetonaceae (V:M)
Potamogeton nodosus Poir. [POTANOD] Potamogetonaceae (V:M)
Potamogeton nuttallii Cham. & Schlecht. = Potamogeton epihydrus [POTAEPI] Potamogetonaceae (V:M)
Potamogeton oakesianus J.W. Robbins [POTAOAK] Potamogetonaceae (V:M)
Potamogeton obtusifolius Mert. & Koch [POTAOBT] Potamogetonaceae (V:M)
Potamogeton panormitanus Biv. = Potamogeton pusillus [POTAPUS] Potamogetonaceae (V:M)
Potamogeton pectinatus L. [POTAPEC] Potamogetonaceae (V:M)
Potamogeton perfoliatus L. [POTAPER] Potamogetonaceae (V:M)
Potamogeton perfoliatus ssp. *bupleuroides* (Fern.) Hultén = Potamogeton perfoliatus [POTAPER] Potamogetonaceae (V:M)
Potamogeton perfoliatus ssp. *richardsonii* (Benn.) Hultén = Potamogeton richardsonii [POTARIC] Potamogetonaceae (V:M)
Potamogeton perfoliatus var. *bupleuroides* (Fern.) Farw. = Potamogeton perfoliatus [POTAPER] Potamogetonaceae (V:M)
Potamogeton perfoliatus var. *richardsonii* Benn. = Potamogeton richardsonii [POTARIC] Potamogetonaceae (V:M)
Potamogeton praelongus Wulfen [POTAPRA] Potamogetonaceae (V:M)
Potamogeton praelongus var. *angustifolius* Graebn. = Potamogeton praelongus [POTAPRA] Potamogetonaceae (V:M)
Potamogeton pusillus L. [POTAPUS] Potamogetonaceae (V:M)
Potamogeton pusillus var. *minor* (Biv.) Fern. & Schub. = Potamogeton pusillus [POTAPUS] Potamogetonaceae (V:M)
Potamogeton pusillus var. *mucronatus* (Fieber) Graebn. = Potamogeton pusillus var. tenuissimus [POTAPUS2] Potamogetonaceae (V:M)
Potamogeton pusillus var. *rutiloides* (Fern.) Boivin = Potamogeton strictifolius [POTASTR] Potamogetonaceae (V:M)
Potamogeton pusillus var. **tenuissimus** Mert. & Koch [POTAPUS2] Potamogetonaceae (V:M)
Potamogeton richardsonii (Benn.) Rydb. [POTARIC] Potamogetonaceae (V:M)
Potamogeton robbinsii Oakes [POTAROB] Potamogetonaceae (V:M)
Potamogeton strictifolius Benn. [POTASTR] Potamogetonaceae (V:M)
Potamogeton strictifolius var. *rutiloides* Fern. = Potamogeton strictifolius [POTASTR] Potamogetonaceae (V:M)
Potamogeton strictifolius var. *typicus* Fern. = Potamogeton strictifolius [POTASTR] Potamogetonaceae (V:M)

Potamogeton tenuifolius Raf. = Potamogeton alpinus [POTAALP] Potamogetonaceae (V:M)
Potamogeton vaginatus Turcz. [POTAVAG] Potamogetonaceae (V:M)
Potamogeton zosterifolius ssp. *zosteriformis* (Fern.) Hultén = Potamogeton zosteriformis [POTAZOS] Potamogetonaceae (V:M)
Potamogeton zosteriformis Fern. [POTAZOS] Potamogetonaceae (V:M)
Potentilla L. [POTENTI$] Rosaceae (V:D)
Potentilla altaica Bunge = Potentilla nivea var. pentaphylla [POTENIV1] Rosaceae (V:D)
Potentilla angustata Rydb. = Potentilla gracilis var. nuttallii [POTEGRA1] Rosaceae (V:D)
Potentilla anomalofolia M.E. Peck = Potentilla drummondii [POTEDRU] Rosaceae (V:D)
Potentilla anserina L. = Argentina anserina [ARGEANS] Rosaceae (V:D)
Potentilla anserina ssp. *egedii* (Wormsk.) Hiitonen = Argentina egedii [ARGEEGE] Rosaceae (V:D)
Potentilla anserina ssp. *pacifica* (T.J. Howell) Rousi = Argentina egedii [ARGEEGE] Rosaceae (V:D)
Potentilla anserina var. *concolor* Ser. = Argentina anserina [ARGEANS] Rosaceae (V:D)
Potentilla anserina var. *grandis* Torr. & Gray = Argentina egedii [ARGEEGE] Rosaceae (V:D)
Potentilla anserina var. *lanata* Boivin = Argentina egedii [ARGEEGE] Rosaceae (V:D)
Potentilla anserina var. *rolandii* (Boivin) Boivin = Argentina egedii [ARGEEGE] Rosaceae (V:D)
Potentilla anserina var. *sericea* (L.) Hayne = Argentina anserina [ARGEANS] Rosaceae (V:D)
Potentilla anserina var. *yukonensis* (Hultén) Boivin = Argentina anserina [ARGEANS] Rosaceae (V:D)
*__Potentilla argentea__ L. [POTEARG] Rosaceae (V:D)
Potentilla arguta Pursh [POTEARU] Rosaceae (V:D)
Potentilla arguta ssp. **convallaria** (Rydb.) Keck [POTEARU2] Rosaceae (V:D)
Potentilla arguta var. *convallaria* (Rydb.) T. Wolf = Potentilla arguta ssp. convallaria [POTEARU2] Rosaceae (V:D)
Potentilla biennis Greene [POTEBIE] Rosaceae (V:D)
Potentilla biflora Willd. ex Schlecht. [POTEBIF] Rosaceae (V:D)
Potentilla bipinnatifida Dougl. ex Hook. [POTEBIP] Rosaceae (V:D)
Potentilla bipinnatifida var. **glabrata** (Lehm. ex Hook.) Kohli & Packer [POTEBIP2] Rosaceae (V:D)
Potentilla blasckeana Turcz. ex Lehm. = Potentilla gracilis var. nuttallii [POTEGRA1] Rosaceae (V:D)
Potentilla blasckeana var. *permollis* (Rydb.) Wolf = Potentilla gracilis var. nuttallii [POTEGRA1] Rosaceae (V:D)
Potentilla camporum Rydb. = Potentilla pulcherrima [POTEPUC] Rosaceae (V:D)
Potentilla convallaria Rydb. = Potentilla arguta ssp. convallaria [POTEARU2] Rosaceae (V:D)
Potentilla diversifolia Lehm. [POTEDIV] Rosaceae (V:D)
Potentilla diversifolia ssp. *glaucophylla* Lehm. = Potentilla diversifolia var. diversifolia [POTEDIV0] Rosaceae (V:D)
Potentilla diversifolia var. **diversifolia** [POTEDIV0] Rosaceae (V:D)
Potentilla diversifolia var. *glaucophylla* (Lehm.) S. Wats. = Potentilla diversifolia var. diversifolia [POTEDIV0] Rosaceae (V:D)
Potentilla diversifolia var. **perdissecta** (Rydb.) C.L. Hitchc. [POTEDIV2] Rosaceae (V:D)
Potentilla drummondii Lehm. [POTEDRU] Rosaceae (V:D)
Potentilla egedii Wormsk. = Argentina egedii [ARGEEGE] Rosaceae (V:D)
Potentilla egedii ssp. *grandis* (Torr. & Gray) Hultén = Argentina egedii [ARGEEGE] Rosaceae (V:D)
Potentilla egedii ssp. *yukonensis* (Hultén) Hultén = Argentina anserina [ARGEANS] Rosaceae (V:D)
Potentilla egedii var. *grandis* (Torr. & Gray) J.T. Howell = Argentina egedii [ARGEEGE] Rosaceae (V:D)
Potentilla elegans Cham. & Schlecht. [POTEELE] Rosaceae (V:D)
Potentilla emarginata Pursh = Potentilla nana [POTENAN] Rosaceae (V:D)
Potentilla etomentosa Rydb. = Potentilla gracilis var. nuttallii [POTEGRA1] Rosaceae (V:D)
Potentilla etomentosa var. *hallii* (Rydb.) Abrams = Potentilla gracilis var. nuttallii [POTEGRA1] Rosaceae (V:D)
Potentilla fastigiata Nutt. = Potentilla gracilis var. nuttallii [POTEGRA1] Rosaceae (V:D)
Potentilla flabellifolia Hook. ex Torr. & Gray [POTEFLA] Rosaceae (V:D)
Potentilla flabellifolia var. *emarginata* (Pursh) Boivin = Potentilla nana [POTENAN] Rosaceae (V:D)
Potentilla flabelliformis Lehm. [POTEFLB] Rosaceae (V:D)
Potentilla floribunda Pursh = Pentaphylloides floribunda [PENTFLO] Rosaceae (V:D)
Potentilla fruticosa auct. = Pentaphylloides floribunda [PENTFLO] Rosaceae (V:D)
Potentilla fruticosa ssp. *floribunda* (Pursh) Elkington = Pentaphylloides floribunda [PENTFLO] Rosaceae (V:D)
Potentilla fruticosa var. *tenuifolia* Lehm. = Pentaphylloides floribunda [PENTFLO] Rosaceae (V:D)
Potentilla glabella Rydb. = Potentilla bipinnatifida var. glabrata [POTEBIP2] Rosaceae (V:D)
Potentilla glandulosa Lindl. [POTEGLA] Rosaceae (V:D)
Potentilla glandulosa ssp. **glandulosa** [POTEGLA0] Rosaceae (V:D)
Potentilla glandulosa ssp. **pseudorupestris** (Rydb.) Keck [POTEGLA2] Rosaceae (V:D)
Potentilla glandulosa ssp. *typica* Keck = Potentilla glandulosa ssp. glandulosa [POTEGLA0] Rosaceae (V:D)
Potentilla glandulosa var. *campanulata* C.L. Hitchc. = Potentilla glandulosa ssp. glandulosa [POTEGLA0] Rosaceae (V:D)
Potentilla glandulosa var. *incisa* Lindl. = Potentilla glandulosa ssp. glandulosa [POTEGLA0] Rosaceae (V:D)
Potentilla glandulosa var. *pseudorupestris* (Rydb.) Breitung = Potentilla glandulosa ssp. pseudorupestris [POTEGLA2] Rosaceae (V:D)
Potentilla glomerata A. Nels. = Potentilla gracilis var. nuttallii [POTEGRA1] Rosaceae (V:D)
Potentilla gracilis Dougl. ex Hook. [POTEGRA] Rosaceae (V:D)
Potentilla gracilis ssp. *nuttallii* (Lehm.) Keck = Potentilla gracilis var. nuttallii [POTEGRA1] Rosaceae (V:D)
Potentilla gracilis var. *blasckeana* (Turcz. ex Lehm.) Jepson = Potentilla gracilis var. nuttallii [POTEGRA1] Rosaceae (V:D)
Potentilla gracilis var. *flabelliformis* (Lehm.) Nutt. ex Torr. & Gray = Potentilla flabelliformis [POTEFLB] Rosaceae (V:D)
Potentilla gracilis var. *glabrata* (Lehm.) C.L. Hitchc. = Potentilla gracilis var. nuttallii [POTEGRA1] Rosaceae (V:D)
Potentilla gracilis var. **gracilis** [POTEGRA0] Rosaceae (V:D)
Potentilla gracilis var. **nuttallii** (Lehm.) Sheldon [POTEGRA1] Rosaceae (V:D)
Potentilla gracilis var. *permollis* (Rydb.) C.L. Hitchc. = Potentilla gracilis var. nuttallii [POTEGRA1] Rosaceae (V:D)
Potentilla gracilis var. *pulcherrima* (Lehm.) Fern. = Potentilla pulcherrima [POTEPUC] Rosaceae (V:D)
Potentilla gracilis var. *rigida* S. Wats. = Potentilla gracilis var. nuttallii [POTEGRA1] Rosaceae (V:D)
Potentilla hippiana Lehm. [POTEHIP] Rosaceae (V:D)
Potentilla hookeriana Lehm. [POTEHOO] Rosaceae (V:D)
Potentilla hyparctica Malte = Potentilla nana [POTENAN] Rosaceae (V:D)
Potentilla hyparctica ssp. *nana* (Willd. ex Schlecht.) Hultén = Potentilla nana [POTENAN] Rosaceae (V:D)
Potentilla hyparctica var. *elatior* (Abrom.) Fern. = Potentilla nana [POTENAN] Rosaceae (V:D)
Potentilla indiges M.E. Peck = Potentilla gracilis var. nuttallii [POTEGRA1] Rosaceae (V:D)
Potentilla jucunda A. Nels. = Potentilla gracilis var. nuttallii [POTEGRA1] Rosaceae (V:D)
Potentilla ledebouriana Porsild = Potentilla uniflora [POTEUNI] Rosaceae (V:D)
Potentilla longipedunculata Rydb. = Potentilla gracilis var. gracilis [POTEGRA0] Rosaceae (V:D)
Potentilla macropetala Rydb. = Potentilla gracilis var. gracilis [POTEGRA0] Rosaceae (V:D)
Potentilla nana Willd. ex Schlecht. [POTENAN] Rosaceae (V:D)

Potentilla nicolletii (S. Wats.) Sheldon = Potentilla paradoxa [POTEPAR] Rosaceae (V:D)
Potentilla nivea L. [POTENIV] Rosaceae (V:D)
Potentilla nivea ssp. *chionodes* Hiitonen = Potentilla nivea var. pentaphylla [POTENIV1] Rosaceae (V:D)
Potentilla nivea ssp. *subquinata* (Lange) Hultén = Potentilla nivea var. pentaphylla [POTENIV1] Rosaceae (V:D)
Potentilla nivea var. *macrophylla* Ser. = Potentilla nivea var. pentaphylla [POTENIV1] Rosaceae (V:D)
Potentilla nivea var. **pentaphylla** Lehm. [POTENIV1] Rosaceae (V:D)
Potentilla nivea var. *subquinata* Lange = Potentilla nivea var. pentaphylla [POTENIV1] Rosaceae (V:D)
Potentilla nivea var. *villosa* (Pallas ex Pursh) Regel & Tiling = Potentilla villosa [POTEVIL] Rosaceae (V:D)
Potentilla norvegica L. [POTENOR] Rosaceae (V:D)
Potentilla nuttallii Lehm. = Potentilla gracilis var. nuttallii [POTEGRA1] Rosaceae (V:D)
Potentilla ovina Macoun [POTEOVI] Rosaceae (V:D)
Potentilla pacifica T.J. Howell = Argentina egedii [ARGEEGE] Rosaceae (V:D)
Potentilla palustris (L.) Scop. = Comarum palustre [COMAPAL] Rosaceae (V:D)
Potentilla palustris var. *parvifolia* (Raf.) Fern. & Long = Comarum palustre [COMAPAL] Rosaceae (V:D)
Potentilla palustris var. *villosa* (Pers.) Lehm. = Comarum palustre [COMAPAL] Rosaceae (V:D)
Potentilla paradoxa Nutt. [POTEPAR] Rosaceae (V:D)
Potentilla pensylvanica L. [POTEPEN] Rosaceae (V:D)
Potentilla pensylvanica var. *bipinnatifida* (Dougl. ex Hook.) Torr. & Gray = Potentilla bipinnatifida [POTEBIP] Rosaceae (V:D)
Potentilla pensylvanica var. *glabrata* (Lehm. ex Hook.) S. Wats. = Potentilla bipinnatifida var. glabrata [POTEBIP2] Rosaceae (V:D)
Potentilla permollis Rydb. = Potentilla gracilis var. nuttallii [POTEGRA1] Rosaceae (V:D)
Potentilla pulcherrima Lehm. [POTEPUC] Rosaceae (V:D)
Potentilla quinquefolia (Rydb.) Rydb. = Potentilla nivea var. pentaphylla [POTENIV1] Rosaceae (V:D)
***Potentilla recta** L. [POTEREC] Rosaceae (V:D)
Potentilla recta var. *obscura* (Nestler) W.D.J. Koch = Potentilla recta [POTEREC] Rosaceae (V:D)
Potentilla recta var. *pilosa* (Willd.) Ledeb. = Potentilla recta [POTEREC] Rosaceae (V:D)
Potentilla recta var. *sulphurea* (Lam. & DC.) Peyr. = Potentilla recta [POTEREC] Rosaceae (V:D)
Potentilla rectiformis Rydb. = Potentilla gracilis var. nuttallii [POTEGRA1] Rosaceae (V:D)
Potentilla rhomboidea Rydb. = Potentilla glandulosa ssp. glandulosa [POTEGLA0] Rosaceae (V:D)
Potentilla rivalis Nutt. [POTERIV] Rosaceae (V:D)
Potentilla rolandii Boivin = Argentina egedii [ARGEEGE] Rosaceae (V:D)
Potentilla sibbaldii Haller f. = Sibbaldia procumbens [SIBBPRO] Rosaceae (V:D)
Potentilla supina ssp. *paradoxa* (Nutt.) Soják = Potentilla paradoxa [POTEPAR] Rosaceae (V:D)
Potentilla uniflora Ledeb. [POTEUNI] Rosaceae (V:D)
Potentilla villosa Pallas ex Pursh [POTEVIL] Rosaceae (V:D)
Potentilla villosa var. *parviflora* C.L. Hitchc. = Potentilla villosa [POTEVIL] Rosaceae (V:D)
Potentilla viridescens Rydb. = Potentilla gracilis var. nuttallii [POTEGRA1] Rosaceae (V:D)
Potentilla yukonensis Hultén = Argentina anserina [ARGEANS] Rosaceae (V:D)
Pottia (Reichenb.) Fürnr. [POTTIA$] Pottiaceae (B:M)
Pottia bryoides (Dicks.) Mitt. [POTTBRY] Pottiaceae (B:M)
Pottia cavifolia Ehrh. ex Fürnr. = Pterygoneurum ovatum [PTEROVA] Pottiaceae (B:M)
Pottia latifolia (Schwaegr. in Schultes) C. Müll. = Stegonia latifolia [STEGLAT] Pottiaceae (B:M)
Pottia latifolia var. *pilifera* (Brid.) C. Müll. = Stegonia pilifera [STEGPIL] Pottiaceae (B:M)
Pottia littoralis Mitt. = Pottia truncata [POTTTRU] Pottiaceae (B:M)
Pottia nevadensis Card. & Thér. [POTTNEV] Pottiaceae (B:M)
Pottia randii Kenn. = Desmatodon randii [DESMRAN] Pottiaceae (B:M)
Pottia truncata (Hedw.) Fürnr. ex B.S.G. [POTTTRU] Pottiaceae (B:M)
Pottia truncata var. *littoralis* (Mitt.) Warnst. = Pottia truncata [POTTTRU] Pottiaceae (B:M)
Pottia truncatula (With.) Buse = Pottia truncata [POTTTRU] Pottiaceae (B:M)
Pottia wilsonii (Hook.) Bruch & Schimp. in B.S.G. [POTTWIL] Pottiaceae (B:M)
Preissia Corda [PREISSI$] Marchantiaceae (B:H)
Preissia quadrata (Scop.) Nees [PREIQUA] Marchantiaceae (B:H)
Prenanthes L. [PRENANT$] Asteraceae (V:D)
Prenanthes alata (Hook.) D. Dietr. [PRENALA] Asteraceae (V:D)
Prenanthes lessingii Hultén = Prenanthes alata [PRENALA] Asteraceae (V:D)
Prenanthes racemosa Michx. [PRENRAC] Asteraceae (V:D)
Prenanthes racemosa ssp. **multiflora** Cronq. [PRENRAC1] Asteraceae (V:D)
Prenanthes racemosa var. *multiflora* (Cronq.) Dorn = Prenanthes racemosa ssp. multiflora [PRENRAC1] Asteraceae (V:D)
Pressia commutata Nees = Preissia quadrata [PREIQUA] Marchantiaceae (B:H)
Primula L. [PRIMULA$] Primulaceae (V:D)
Primula clusiana auct. = Primula nutans [PRIMNUT] Primulaceae (V:D)
Primula cuneifolia Ledeb. [PRIMCUN] Primulaceae (V:D)
Primula cuneifolia ssp. *saxifragifolia* (Lehm.) W.W. Sm. & G. Forrest = Primula cuneifolia var. saxifragifolia [PRIMCUN1] Primulaceae (V:D)
Primula cuneifolia var. **saxifragifolia** (Lehm.) Pax & R. Knuth [PRIMCUN1] Primulaceae (V:D)
Primula egaliksensis Wormsk. ex Hornem. [PRIMEGA] Primulaceae (V:D)
Primula farinosa var. *incana* Fern. = Primula incana [PRIMINC] Primulaceae (V:D)
Primula groenlandica (Warming) W.W. Sm. & G. Forrest = Primula egaliksensis [PRIMEGA] Primulaceae (V:D)
Primula incana M.E. Jones [PRIMINC] Primulaceae (V:D)
Primula intercedens Fern. = Primula mistassinica [PRIMMIS] Primulaceae (V:D)
Primula maccalliana Wieg. = Primula mistassinica [PRIMMIS] Primulaceae (V:D)
Primula mistassinica Michx. [PRIMMIS] Primulaceae (V:D)
Primula mistassinica var. *intercedens* (Fern.) Boivin = Primula mistassinica [PRIMMIS] Primulaceae (V:D)
Primula mistassinica var. *noveboracensis* Fern. = Primula mistassinica [PRIMMIS] Primulaceae (V:D)
Primula nutans Georgi [PRIMNUT] Primulaceae (V:D)
Primula sibirica Jacq. = Primula nutans [PRIMNUT] Primulaceae (V:D)
Primula stricta Hornem. [PRIMSTR] Primulaceae (V:D)
Prionolobus turneri (Hook.) Spruce = Cephaloziella turneri [CEPHTUR] Cephaloziellaceae (B:H)
Prosartes trachycarpa S. Wats. = Disporum trachycarpum [DISPTRA] Liliaceae (V:M)
Protoblastenia (Zahlbr.) J. Steiner [PROTOBL$] Psoraceae (L)
Protoblastenia cinnabarina (Sommerf.) Räsänen = Pyrrhospora cinnabarina [PYRRCIN] Lecanoraceae (L)
Protoblastenia incrustans (DC.) J. Steiner [PROTINC] Psoraceae (L)
Protoblastenia monticola (Ach.) J. Steiner = Clauzadea monticola [CLAUMON] Porpidiaceae (L)
Protoblastenia quernea (Dickson) Clauzade = Pyrrhospora quernea [PYRRQUE] Lecanoraceae (L)
Protoblastenia rupestris (Scop.) J. Steiner [PROTRUP] Psoraceae (L)
Protoblastenia russula (Ach.) Räsänen = Pyrrhospora russula [PYRRRUS] Lecanoraceae (L)
Protoblastenia terricola (Anzi) Lynge [PROTTER] Psoraceae (L)
Protoparmelia Choisy [PROTOPA$] Lecanoraceae (L)
Protoparmelia badia (Hoffm.) Hafellner [PROTBAD] Lecanoraceae (L)

Protoparmelia ochrococca (Nyl.) P.M. Jørg., Rambold & Hertel [PROTOCH] Lecanoraceae (L)
Protoparmeliopsis muralis (Schreber) Choisy = Lecanora muralis [LECAMUR] Lecanoraceae (L)
Protothelenella Räsänen [PROTOTH$] Protothelenellaceae (L)
Protothelenella leucothelia (Nyl.) H. Mayrh. & Poelt [PROTLEU] Protothelenellaceae (L)
Protothelenella sphinctrinoides (Nyl.) H. Mayrh. & Poelt [PROTSPH] Protothelenellaceae (L)
Prunella L. [PRUNELL$] Lamiaceae (V:D)
Prunella vulgaris L. [PRUNVUL] Lamiaceae (V:D)
Prunella vulgaris ssp. **lanceolata** (W. Bart.) Hultén [PRUNVUL1] Lamiaceae (V:D)
****Prunella vulgaris** ssp. **vulgaris** [PRUNVUL0] Lamiaceae (V:D)
Prunella vulgaris var. *atropurpurea* Fern. = Prunella vulgaris ssp. vulgaris [PRUNVUL0] Lamiaceae (V:D)
Prunella vulgaris var. *calvescens* Fern. = Prunella vulgaris ssp. vulgaris [PRUNVUL0] Lamiaceae (V:D)
Prunella vulgaris var. *elongata* Benth. = Prunella vulgaris ssp. lanceolata [PRUNVUL1] Lamiaceae (V:D)
Prunella vulgaris var. *hispida* Benth. = Prunella vulgaris ssp. vulgaris [PRUNVUL0] Lamiaceae (V:D)
Prunella vulgaris var. *lanceolata* (W. Bart.) Fern. = Prunella vulgaris ssp. lanceolata [PRUNVUL1] Lamiaceae (V:D)
Prunella vulgaris var. *minor* Sm. = Prunella vulgaris ssp. vulgaris [PRUNVUL0] Lamiaceae (V:D)
Prunella vulgaris var. *nana* Clute = Prunella vulgaris ssp. vulgaris [PRUNVUL0] Lamiaceae (V:D)
Prunella vulgaris var. *parviflora* (Poir.) DC. = Prunella vulgaris ssp. vulgaris [PRUNVUL0] Lamiaceae (V:D)
Prunella vulgaris var. *rouleauiana* Victorin = Prunella vulgaris ssp. vulgaris [PRUNVUL0] Lamiaceae (V:D)
Prunus L. [PRUNUS$] Rosaceae (V:D)
***Prunus avium** (L.) L. [PRUNAVI] Rosaceae (V:D)
Prunus demissa (Nutt.) Walp. = Prunus virginiana var. demissa [PRUNVIR1] Rosaceae (V:D)
Prunus emarginata (Dougl. ex Hook.) Walp. [PRUNEMA] Rosaceae (V:D)
***Prunus laurocerasus** L. [PRUNLAU] Rosaceae (V:D)
***Prunus mahaleb** L. [PRUNMAH] Rosaceae (V:D)
Prunus melanocarpa (A. Nels.) Rydb. = Prunus virginiana var. melanocarpa [PRUNVIR2] Rosaceae (V:D)
Prunus pensylvanica L. f. [PRUNPEN] Rosaceae (V:D)
***Prunus spinosa** L. [PRUNSPI] Rosaceae (V:D)
Prunus virginiana L. [PRUNVIR] Rosaceae (V:D)
Prunus virginiana ssp. *demissa* (Nutt.) Taylor & MacBryde = Prunus virginiana var. demissa [PRUNVIR1] Rosaceae (V:D)
Prunus virginiana ssp. *melanocarpa* (A. Nels.) Taylor & MacBryde = Prunus virginiana var. melanocarpa [PRUNVIR2] Rosaceae (V:D)
Prunus virginiana var. **demissa** (Nutt.) Torr. [PRUNVIR1] Rosaceae (V:D)
Prunus virginiana var. **melanocarpa** (A. Nels.) Sarg. [PRUNVIR2] Rosaceae (V:D)
Pseudephebe Choisy [PSEUDEH$] Parmeliaceae (L)
Pseudephebe minuscula (Nyl. ex Arnold) Brodo & D. Hawksw. [PSEUMIN] Parmeliaceae (L)
Pseudephebe pubescens (L.) Choisy [PSEUPUB] Parmeliaceae (L)
Pseudephemerum (Lindb.) Loeske [PSEUDEP$] Dicranaceae (B:M)
***Pseudephemerum nitidum** (Hedw.) Loeske [PSEUNIT] Dicranaceae (B:M)
Pseudisothecium myosuroides (Brid.) Grout = Isothecium myosuroides [ISOTMYO] Brachytheciaceae (B:M)
Pseudisothecium myosuroides var. *filescens* (Ren.) Grout = Isothecium myosuroides [ISOTMYO] Brachytheciaceae (B:M)
Pseudisothecium stoloniferum (Brid.) Grout = Isothecium myosuroides [ISOTMYO] Brachytheciaceae (B:M)
Pseudisothecium stoloniferum var. *cardotii* (Kindb. in Mac. & Kindb.) Grout = Isothecium myosuroides [ISOTMYO] Brachytheciaceae (B:M)
Pseudisothecium stoloniferum var. *myurellum* (Kindb.) Grout = Isothecium myosuroides [ISOTMYO] Brachytheciaceae (B:M)
Pseudobraunia (Lesq. & James) Broth. [PSEUDOB$] Hedwigiaceae (B:M)
Pseudobraunia californica (Lesq.) Broth. [PSEUCAL] Hedwigiaceae (B:M)
Pseudobryum (Kindb.) T. Kop. [PSEUDBY$] Mniaceae (B:M)
Pseudobryum cinclidioides (Hüb.) T. Kop. [PSEUCIN] Mniaceae (B:M)
Pseudocalliergon (Limpr.) Loeske [PSEUDOG$] Amblystegiaceae (B:M)
Pseudocalliergon turgescens (T. Jens.) Loeske [PSEUTUR] Amblystegiaceae (B:M)
Pseudocrossidium Williams [PSEUDOO$] Pottiaceae (B:M)
Pseudocrossidium hornschuchianum (Schultz) Zand. [PSEUHOR] Pottiaceae (B:M)
Pseudocrossidium revolutum (Brid. in Schrad.) Zand. [PSEUREV] Pottiaceae (B:M)
Pseudocyphellaria Vainio [PSEUDOY$] Lobariaceae (L)
Pseudocyphellaria anomala Brodo & Ahti [PSEUANO] Lobariaceae (L)
Pseudocyphellaria anthraspis (Ach.) H. Magn. [PSEUANT] Lobariaceae (L)
Pseudocyphellaria aurata (Ach.) Vainio [PSEUAUR] Lobariaceae (L)
Pseudocyphellaria crocata (L.) Vainio [PSEUCRO] Lobariaceae (L)
Pseudocyphellaria mougeotiana (Delise) Vainio = Pseudocyphellaria crocata [PSEUCRO] Lobariaceae (L)
Pseudocyphellaria rainierensis Imshaug [PSEURAI] Lobariaceae (L)
Pseudognaphalium stramineum (Kunth) W.A. Weber = Gnaphalium stramineum [GNAPSTR] Asteraceae (V:D)
Pseudognaphalium viscosum (Kunth) W.A. Weber = Gnaphalium viscosum [GNAPVIS] Asteraceae (V:D)
Pseudoleskea Schimp. in B.S.G. [PSEUDOL$] Leskeaceae (B:M)
Pseudoleskea atricha (Kindb. in Mac. & Kindb.) Kindb. [PSEUATR] Leskeaceae (B:M)
Pseudoleskea atrovirens (Brid.) Schimp. in B.S.G. = Pseudoleskea incurvata var. incurvata [PSEUINC0] Leskeaceae (B:M)
Pseudoleskea atrovirens auct. = Pseudoleskea patens [PSEUPAT] Leskeaceae (B:M)
Pseudoleskea baileyi Best & Grout in Grout [PSEUBAI] Leskeaceae (B:M)
Pseudoleskea brachyclados (Schwaegr.) Kindb. = Pseudoleskea radicosa var. radicosa [PSEURAD0] Leskeaceae (B:M)
Pseudoleskea breidleri Kindb. = Pseudoleskea radicosa var. denudata [PSEURAD2] Leskeaceae (B:M)
Pseudoleskea denudata (Kindb. in Mac. & Kindb.) Kindb. in Best = Pseudoleskea radicosa var. denudata [PSEURAD2] Leskeaceae (B:M)
Pseudoleskea denudata var. *holzingeri* Best = Pseudoleskea radicosa var. radicosa [PSEURAD0] Leskeaceae (B:M)
Pseudoleskea falcicuspis C. Müll. & Kindb. in Mac. & Kindb. = Pseudoleskea incurvata var. incurvata [PSEUINC0] Leskeaceae (B:M)
Pseudoleskea frigida (Kindb.) Sharp in Grout = Lescuraea saxicola [LESCSAX] Leskeaceae (B:M)
Pseudoleskea howei (Best) L. Koch = Pseudoleskea radicosa var. radicosa [PSEURAD0] Leskeaceae (B:M)
Pseudoleskea iliamniana (Lawt.) Crum, Steere & Anderson = Pseudoleskea julacea [PSEUJUL] Leskeaceae (B:M)
Pseudoleskea incurvata (Hedw.) Loeske [PSEUINC] Leskeaceae (B:M)
Pseudoleskea incurvata var. **gigantea** (Lawt.) Crum, Steere & Anderson [PSEUINC1] Leskeaceae (B:M)
Pseudoleskea incurvata var. **incurvata** [PSEUINC0] Leskeaceae (B:M)
Pseudoleskea julacea (Besch. & Card. in Card.) Crum, Steere & Anderson [PSEUJUL] Leskeaceae (B:M)
Pseudoleskea oligoclada Kindb. = Pseudoleskea incurvata var. incurvata [PSEUINC0] Leskeaceae (B:M)
Pseudoleskea pallida Best = Pseudoleskea radicosa var. pallida [PSEURAD3] Leskeaceae (B:M)
Pseudoleskea patens (Lindb.) Kindb. [PSEUPAT] Leskeaceae (B:M)
Pseudoleskea radicosa (Mitt.) Mac. & Kindb. [PSEURAD] Leskeaceae (B:M)

Pseudoleskea radicosa var. **compacta** Best [PSEURAD1] Leskeaceae (B:M)
Pseudoleskea radicosa var. **denudata** (Kindb. in Mac. & Kindb.) Wijk & Marg. [PSEURAD2] Leskeaceae (B:M)
Pseudoleskea radicosa var. *holzingeri* (Best) Hag. = Pseudoleskea radicosa var. radicosa [PSEURAD0] Leskeaceae (B:M)
Pseudoleskea radicosa var. **pallida** (Best) Crum, Steere & Anderson [PSEURAD3] Leskeaceae (B:M)
Pseudoleskea radicosa var. **radicosa** [PSEURAD0] Leskeaceae (B:M)
Pseudoleskea rigescens Lindb. = Pseudoleskea stenophylla [PSEUSTE] Leskeaceae (B:M)
Pseudoleskea saxicola (Schimp. in B.S.G.) Crum, Steere & Anderson = Lescuraea saxicola [LESCSAX] Leskeaceae (B:M)
Pseudoleskea sciuroides Kindb. = Pseudoleskea radicosa var. radicosa [PSEURAD0] Leskeaceae (B:M)
Pseudoleskea sciuroides var. *denudata* Kindb. in Mac. & Kindb. = Pseudoleskea radicosa var. denudata [PSEURAD2] Leskeaceae (B:M)
Pseudoleskea stenophylla Ren. & Card. in Röll [PSEUSTE] Leskeaceae (B:M)
Pseudoleskea substriata Best = Lescuraea saxicola [LESCSAX] Leskeaceae (B:M)
Pseudoleskeella Kindb. [PSEUDOE$] Leskeaceae (B:M)
Pseudoleskeella nervosa (Brid.) Nyh. = Leskeella nervosa [LESKNER] Leskeaceae (B:M)
Pseudoleskeella occidentalis (Kindb.) Kindb. = Porotrichum vancouveriense [POROVAN] Thamnobryaceae (B:M)
Pseudoleskeella papillosa (Lindb.) Lindb. [PSEUPAP] Leskeaceae (B:M)
Pseudoleskeella tectorum (Funck ex Brid.) Kindb. in Broth. [PSEUTEC] Leskeaceae (B:M)
Pseudoleskeella tectorum var. *cyrtophylla* (Kindb. in Mac. & Kindb.) Crum, Steere & Anderson = Pseudoleskeella tectorum [PSEUTEC] Leskeaceae (B:M)
Pseudoleskeella tectorum var. *flagellifera* (Best) Amann = Pseudoleskeella tectorum [PSEUTEC] Leskeaceae (B:M)
Pseudoleskeella williamsii (Best) Crum, Steere & Anderson = Pseudoleskeella tectorum [PSEUTEC] Leskeaceae (B:M)
Pseudoleskeella williamsii var. *filamentosa* (Best) Crum, Steere & Anderson = Pseudoleskeella tectorum [PSEUTEC] Leskeaceae (B:M)
Pseudoroegneria (Nevski) A. Löve [PSEUDOR$] Poaceae (V:M)
Pseudoroegneria spicata (Pursh) A. Löve [PSEUSPI] Poaceae (V:M)
Pseudoroegneria spicata ssp. **inermis** (Scribn. & J.G. Sm.) A. Löve [PSEUSPI1] Poaceae (V:M)
Pseudosagedia (Müll. Arg.) Choisy [PSEUDOD$] Trichotheliaceae (L)
Pseudosagedia aenea (Wallr.) Hafellner & Kalb [PSEUAEN] Trichotheliaceae (L)
Pseudosagedia chlorotica (Ach.) Hafellner & Kalb [PSEUCHL] Trichotheliaceae (L)
Pseudosagedia guentheri (Flotow) Hafellner & Kalb [PSEUGUE] Trichotheliaceae (L)
Pseudosagedia linearis (Leighton) Hafellner & Kalb [PSEULIN] Trichotheliaceae (L)
Pseudoscleropodium (Limpr.) Fleisch. in Broth. [PSEUDOS$] Brachytheciaceae (B:M)
***Pseudoscleropodium purum** (Hedw.) Fleisch. in Broth. [PSEUPUR] Brachytheciaceae (B:M)
Pseudostereodon procerrimum (Mol.) Fleisch. in Broth. = Hypnum procerrimum [HYPNPRO] Hypnaceae (B:M)
Pseudotaxiphyllum Iwats. [PSEUDOH$] Hypnaceae (B:M)
Pseudotaxiphyllum elegans (Brid.) Iwats. [PSEUELE] Hypnaceae (B:M)
Pseudotsuga Carr. [PSEUDOT$] Pinaceae (V:G)
Pseudotsuga menziesii (Mirbel) Franco [PSEUMEN] Pinaceae (V:G)
Pseudotsuga menziesii var. **glauca** (Beissn.) Franco [PSEUMEN1] Pinaceae (V:G)
Pseudotsuga menziesii var. **menziesii** [PSEUMEN0] Pinaceae (V:G)
Pseudotsuga taxifolia (Lamb.) Britt. = Pseudotsuga menziesii var. menziesii [PSEUMEN0] Pinaceae (V:G)
Pseudotsuga taxifolia var. *glauca* (Beissn.) Sudworth = Pseudotsuga menziesii var. glauca [PSEUMEN1] Pinaceae (V:G)
Psilocarphus Nutt. [PSILOCA$] Asteraceae (V:D)
Psilocarphus elatior (Gray) Gray [PSILELA] Asteraceae (V:D)
Psilocarphus tenellus Nutt. [PSILTEN] Asteraceae (V:D)
Psilochenia atribarba (Heller) W.A. Weber = Crepis atrabarba ssp. atrabarba [CREPATR0] Asteraceae (V:D)
Psilochenia intermedia (Gray) W.A. Weber = Crepis intermedia [CREPINT] Asteraceae (V:D)
Psilochenia modocensis (Greene) W.A. Weber = Crepis modocensis ssp. modocensis [CREPMOD0] Asteraceae (V:D)
Psilochenia modocensis ssp. *rostrata* (Coville) W.A. Weber = Crepis modocensis ssp. rostrata [CREPMOD2] Asteraceae (V:D)
Psilochenia occidentalis (Nutt.) Nutt. = Crepis occidentalis ssp. occidentalis [CREPOCC0] Asteraceae (V:D)
Psilochenia occidentalis ssp. *costata* (Gray) W.A. Weber = Crepis occidentalis ssp. costata [CREPOCC1] Asteraceae (V:D)
Psilochenia runcinata (James) A. & D. Löve = Crepis runcinata [CREPRUN] Asteraceae (V:D)
Psilolechia A. Massal. [PSILOLE$] Micareaceae (L)
Psilolechia lucida (Ach.) Choisy [PSILLUC] Micareaceae (L)
Psilopilum Brid. [PSILOPI$] Polytrichaceae (B:M)
Psilopilum cavifolium (Wils.) Hag. [PSILCAV] Polytrichaceae (B:M)
Psora Hoffm. [PSORA$] Psoraceae (L)
Psora anthracophila (Nyl.) Arnold = Biatora anthracophila [BIATANT] Bacidiaceae (L)
Psora cerebriformis W.A. Weber [PSORCER] Psoraceae (L)
Psora decipiens (Hedwig) Hoffm. [PSORDEC] Psoraceae (L)
Psora demissa (Rutstr.) Hepp = Lecidoma demissum [LECIDEU] Psoraceae (L)
Psora elenkinii Rass. = Psora himalayana [PSORHIM] Psoraceae (L)
Psora friesii (Ach.) Hellbom = Hypocenomyce friesii [HYPOFRI] Lecideaceae (L)
Psora globifera (Ach.) A. Massal. [PSORGLO] Psoraceae (L)
Psora himalayana (Church. Bab.) Timdal [PSORHIM] Psoraceae (L)
Psora lurida (Ach.) DC. = Lecidea lurida [LECILUR] Lecideaceae (L)
Psora luridella (Tuck.) Fink [PSORLUR] Psoraceae (L)
Psora montana Timdal [PSORMON] Psoraceae (L)
Psora nicolai de Lesd. = Psora russellii [PSORRUS] Psoraceae (L)
Psora nipponica (Zahlbr.) Gotth. Schneider [PSORNIP] Psoraceae (L)
Psora novomexicana de Lesd. = Psora nipponica [PSORNIP] Psoraceae (L)
Psora ostreata Hoffm. = Hypocenomyce scalaris [HYPOSCA] Lecideaceae (L)
Psora petri (Tuck.) Fink = Lecidea lurida [LECILUR] Lecideaceae (L)
Psora rubiformis (Ach.) Hook. [PSORRUB] Psoraceae (L)
Psora rufonigra (Tuck.) A. Schneider = Psorula rufonigra [PSORRUF] Psoraceae (L)
Psora russellii (Tuck.) A. Schneider [PSORRUS] Psoraceae (L)
Psora scalaris (Ach. ex Lilj.) Hook. = Hypocenomyce scalaris [HYPOSCA] Lecideaceae (L)
Psora scholanderi (Lynge) R. Anderson = Toninia tristis [TONITRI] Catillariaceae (L)
Psora scotopholia (Tuck.) Fink = Lecanora scotopholis [LECASCO] Lecanoraceae (L)
Psora tuckermanii R. Anderson ex Timdal [PSORTUC] Psoraceae (L)
Psoralea physodes Dougl. ex Hook. = Rupertia physodes [RUPEPHY] Fabaceae (V:D)
Psoroma Michaux [PSOROMA$] Pannariaceae (L)
Psoroma hypnorum (Vahl) Gray [PSORHYP] Pannariaceae (L)
Psorula Gotth. Schneider [PSORULA$] Psoraceae (L)
Psorula rufonigra (Tuck.) Gotth. Schneider [PSORRUF] Psoraceae (L)
Psorula scotopholis (Tuck.) Gotth. Schneider = Lecanora scotopholis [LECASCO] Lecanoraceae (L)
Pteretis nodulosa (Michx.) Nieuwl. = Matteuccia struthiopteris [MATTSTR] Dryopteridaceae (V:F)
Pteretis pensylvanica (Willd.) Fern. = Matteuccia struthiopteris [MATTSTR] Dryopteridaceae (V:F)
Pteridium Gleditsch ex Scop. [PTERIDI$] Dennstaedtiaceae (V:F)

Pteridium aquilinum (L.) Kuhn [PTERAQU] Dennstaedtiaceae (V:F)
Pteridium aquilinum ssp. *lanuginosum* (Bong.) Hultén = Pteridium aquilinum var. pubescens [PTERAQU7] Dennstaedtiaceae (V:F)
Pteridium aquilinum ssp. *latiusculum* (Desv.) C.N. Page = Pteridium aquilinum var. latiusculum [PTERAQU6] Dennstaedtiaceae (V:F)
Pteridium aquilinum var. *lanuginosum* (Bong.) Fern. = Pteridium aquilinum var. pubescens [PTERAQU7] Dennstaedtiaceae (V:F)
Pteridium aquilinum var. **latiusculum** (Desv.) Underwood ex Heller [PTERAQU6] Dennstaedtiaceae (V:F)
Pteridium aquilinum var. **pubescens** Underwood [PTERAQU7] Dennstaedtiaceae (V:F)
Pteridium latiusculum (Desv.) Hieron. = Pteridium aquilinum var. latiusculum [PTERAQU6] Dennstaedtiaceae (V:F)
Pterigynandrum Hedw. [PTERIGY$] Pterigynandraceae (B:M)
Pterigynandrum filiforme Hedw. [PTERFIL] Pterigynandraceae (B:M)
Pterigynandrum filiforme f. *majus* (De Not.) De Not. = Pterigynandrum filiforme [PTERFIL] Pterigynandraceae (B:M)
Pterigynandrum filiforme f. *minus* Lesq. & James = Pterigynandrum filiforme [PTERFIL] Pterigynandraceae (B:M)
Pterigynandrum filiforme f. *papillosulum* (C. Müll. & Kindb. in Mac. & Kindb.) Grout = Pterigynandrum filiforme [PTERFIL] Pterigynandraceae (B:M)
Pterigynandrum filiforme ssp. *decipiens* (Web. & Mohr) Kindb. = Pterigynandrum filiforme [PTERFIL] Pterigynandraceae (B:M)
Pterigynandrum filiforme var. *decipiens* (Web. & Mohr) Limpr. = Pterigynandrum filiforme [PTERFIL] Pterigynandraceae (B:M)
Pterigynandrum filiforme var. *majus* (De Not.) De Not. = Pterigynandrum filiforme [PTERFIL] Pterigynandraceae (B:M)
Pterigynandrum filiforme var. *minus* Lesq. & James = Pterigynandrum filiforme [PTERFIL] Pterigynandraceae (B:M)
Pterigynandrum filiforme var. *papillosulum* (C. Müll. & Kindb. in Mac. & Kindb.) Thér. = Pterigynandrum filiforme [PTERFIL] Pterigynandraceae (B:M)
Pterigynandrum papillosulum C. Müll. & Kindb. in Mac. & Kindb. = Pterigynandrum filiforme [PTERFIL] Pterigynandraceae (B:M)
Pterogonium Sm. [PTEROGO$] Anomodontaceae (B:M)
Pterogonium gracile (Hedw.) Sm. [PTERGRA] Anomodontaceae (B:M)
Pterogonium peregrinum Schimp. = Isothecium cristatum [ISOTCRI] Brachytheciaceae (B:M)
Pterospora Nutt. [PTEROSP$] Monotropaceae (V:D)
Pterospora andromedea Nutt. [PTERAND] Monotropaceae (V:D)
Pterygoneurum Jur. [PTERYGO$] Pottiaceae (B:M)
Pterygoneurum arcticum Steere = Pterygoneurum lamellatum [PTERLAM] Pottiaceae (B:M)
Pterygoneurum kozlovii Laz. [PTERKOZ] Pottiaceae (B:M)
Pterygoneurum lamellatum (Lindb.) Jur. [PTERLAM] Pottiaceae (B:M)
Pterygoneurum ovatum (Hedw.) Dix. [PTEROVA] Pottiaceae (B:M)
Pterygoneurum smardaeanum Vanek = Pterygoneurum kozlovii [PTERKOZ] Pottiaceae (B:M)
Pterygoneurum subsessile (Brid.) Jur. [PTERSUB] Pottiaceae (B:M)
Pterygoneurum subsessile var. *henricii* (Rau) Wareh. in Grout = Pterygoneurum subsessile [PTERSUB] Pottiaceae (B:M)
Pterygoneurum subsessile var. *kieneri* Hab. = Pterygoneurum subsessile [PTERSUB] Pottiaceae (B:M)
Pterygophyllum lucens (Hedw.) Brid. = Hookeria lucens [HOOKLUC] Hookeriaceae (B:M)
Ptilidium Nees [PTILIDI$] Ptilidiaceae (B:H)
Ptilidium californicum (Aust.) Underw. [PTILCAL] Ptilidiaceae (B:H)
Ptilidium ciliare (L.) Hampe [PTILCIL] Ptilidiaceae (B:H)
Ptilidium pulcherrimum (G. Web.) Hampe [PTILPUL] Ptilidiaceae (B:H)
Ptilium De Not. [PTILIUM$] Hypnaceae (B:M)
Ptilium crista-castrensis (Hedw.) De Not. [PTILCRI] Hypnaceae (B:M)
Ptilocalais nutans (Hook.) Greene = Microseris nutans [MICRNUT] Asteraceae (V:D)
Ptychomitrium Fürnr. [PTYCHOM$] Ptychomitriaceae (B:M)
Ptychomitrium gardneri Lesq. [PTYCGAR] Ptychomitriaceae (B:M)
Ptychomitrium kiusiuense Sak. = Ptychomitrium gardneri [PTYCGAR] Ptychomitriaceae (B:M)
Ptychomitrium longisetum Reim. & Sak. = Ptychomitrium gardneri [PTYCGAR] Ptychomitriaceae (B:M)
Ptychomitrium polyphylloides (C. Müll.) Par. = Ptychomitrium gardneri [PTYCGAR] Ptychomitriaceae (B:M)
Ptychomitrium robustum Broth. = Ptychomitrium gardneri [PTYCGAR] Ptychomitriaceae (B:M)
Ptychomitrium viride Sak. = Ptychomitrium gardneri [PTYCGAR] Ptychomitriaceae (B:M)
Puccinellia Parl. [PUCCINE$] Poaceae (V:M)
Puccinellia airoides (Nutt.) S. Wats. & Coult. = Puccinellia nuttalliana [PUCCNUA] Poaceae (V:M)
Puccinellia arctica (Hook.) Fern. & Weatherby [PUCCARC] Poaceae (V:M)
Puccinellia borealis Swallen = Puccinellia arctica [PUCCARC] Poaceae (V:M)
Puccinellia cusickii Weatherby = Puccinellia nuttalliana [PUCCNUA] Poaceae (V:M)
*__Puccinellia distans__ (Jacq.) Parl. [PUCCDIS] Poaceae (V:M)
Puccinellia fernaldii (A.S. Hitchc.) E.G. Voss = Torreyochloa pallida var. fernaldii [TORRPAL1] Poaceae (V:M)
Puccinellia grandis Swallen [PUCCGRA] Poaceae (V:M)
Puccinellia interior Sorensen [PUCCINT] Poaceae (V:M)
Puccinellia kurilensis (Takeda) Honda [PUCCKUR] Poaceae (V:M)
Puccinellia lettermanii (Vasey) Ponert = Poa lettermanii [POA LET] Poaceae (V:M)
Puccinellia nutkaensis (J. Presl) Fern. & Weatherby [PUCCNUT] Poaceae (V:M)
Puccinellia nuttalliana (J.A. Schultes) A.S. Hitchc. [PUCCNUA] Poaceae (V:M)
Puccinellia pauciflora (J. Presl) Munz = Torreyochloa pallida var. pauciflora [TORRPAL2] Poaceae (V:M)
Puccinellia pauciflora var. *holmii* (Beal) C.L. Hitchc. = Torreyochloa pallida var. pauciflora [TORRPAL2] Poaceae (V:M)
Puccinellia pauciflora var. *microtheca* (Buckl.) C.L. Hitchc. = Torreyochloa pallida var. pauciflora [TORRPAL2] Poaceae (V:M)
Puccinellia pumila (Vasey) A.S. Hitchc. = Puccinellia kurilensis [PUCCKUR] Poaceae (V:M)
Pulsatilla P. Mill. [PULSATI$] Ranunculaceae (V:D)
Pulsatilla hirsutissima (Pursh) Britt. = Pulsatilla patens ssp. multifida [PULSPAT1] Ranunculaceae (V:D)
Pulsatilla ludoviciana Heller = Pulsatilla patens ssp. multifida [PULSPAT1] Ranunculaceae (V:D)
Pulsatilla nuttalliana (DC.) Spreng. = Pulsatilla patens ssp. multifida [PULSPAT1] Ranunculaceae (V:D)
Pulsatilla occidentalis (S. Wats.) Freyn [PULSOCC] Ranunculaceae (V:D)
Pulsatilla patens (L.) P. Mill. [PULSPAT] Ranunculaceae (V:D)
Pulsatilla patens ssp. *hirsutissima* (Pursh) Zamels = Pulsatilla patens ssp. multifida [PULSPAT1] Ranunculaceae (V:D)
Pulsatilla patens ssp. **multifida** (Pritz.) Zamels [PULSPAT1] Ranunculaceae (V:D)
Punctelia Krog [PUNCTEL$] Parmeliaceae (L)
Punctelia flaventior (Stirton) Krog = Flavopunctelia flaventior [FLAVFLA] Parmeliaceae (L)
Punctelia stictica (Duby) Krog [PUNCSTI] Parmeliaceae (L)
Punctelia subrudecta (Nyl.) Krog [PUNCSUR] Parmeliaceae (L)
Purshia DC. ex Poir. [PURSHIA$] Rosaceae (V:D)
Purshia tridentata (Pursh) DC. [PURSTRI] Rosaceae (V:D)
Pylaisia filari-acuminata C. Müll. & Kindb. in Mac. & Kindb. = Pylaisiella polyantha [PYLAPOL] Hypnaceae (B:M)
Pylaisia heteromalla Bruch & Schimp. = Pylaisiella polyantha [PYLAPOL] Hypnaceae (B:M)
Pylaisia intricata (Hedw.) Schimp. in B.S.G. = Pylaisiella intricata [PYLAINT] Hypnaceae (B:M)
Pylaisia jamesii Sull. & Lesq. = Pylaisiella polyantha [PYLAPOL] Hypnaceae (B:M)
Pylaisia ontariensis C. Müll. & Kindb. in Mac. & Kindb. = Pylaisiella polyantha [PYLAPOL] Hypnaceae (B:M)
Pylaisia polyantha (Hedw.) Schimp. in B.S.G. = Pylaisiella polyantha [PYLAPOL] Hypnaceae (B:M)

Pylaisia polyantha var. *brevifolia* (Lindb. & Arnell) Limpr. = Pylaisiella polyantha [PYLAPOL] Hypnaceae (B:M)
Pylaisia polyantha var. *jamesii* (Sull. & Lesq.) Rau & Herv. = Pylaisiella polyantha [PYLAPOL] Hypnaceae (B:M)
Pylaisia polyantha var. *pseudoplatygyria* (Kindb.) Grout = Pylaisiella polyantha [PYLAPOL] Hypnaceae (B:M)
Pylaisia subdenticulata Schimp. in B.S.G. = Pylaisiella polyantha [PYLAPOL] Hypnaceae (B:M)
Pylaisia velutina Schimp. in B.S.G. = Pylaisiella intricata [PYLAINT] Hypnaceae (B:M)
Pylaisiadelpha roellii (Ren. & Card. in Röll) Buck = Brotherella roellii [BROTROE] Sematophyllaceae (B:M)
Pylaisiella Kindb. [PYLAISI$] Hypnaceae (B:M)
Pylaisiella intricata (Hedw.) Grout [PYLAINT] Hypnaceae (B:M)
Pylaisiella polyantha (Hedw.) Grout [PYLAPOL] Hypnaceae (B:M)
Pylaisiella polyantha var. *pseudoplatygyria* (Kindb.) Grout = Pylaisiella polyantha [PYLAPOL] Hypnaceae (B:M)
Pyrenocollema Reinke [PYRENOC$] Pyrenulaceae (L)
Pyrenocollema halodytes (Nyl.) R.C. Harris [PYREHAL] Pyrenulaceae (L)
Pyrenopsis pulvinata (Schaerer) Th. Fr. = Euopsis pulvinata [EUOPPUL] Lichinaceae (L)
Pyrenothamnia brandegei (Tuck.) Zahlbr. = Endocarpon pulvinatum [ENDOPUL] Verrucariaceae (L)
Pyrenothamnia spraguei Tuck. = Endocarpon pulvinatum [ENDOPUL] Verrucariaceae (L)
Pyrenula A. Massal. [PYRENUL$] Pyrenulaceae (L)
Pyrenula acutispora Kalb & Hafellner [PYREACU] Pyrenulaceae (L)
Pyrenula areolata Ach. = Staurothele areolata [STAUARE] Verrucariaceae (L)
Pyrenula herrei Fink = Arthopyrenia plumbaria [ARTHPLU] Arthopyreniaceae (L)
Pyrenula leucoplaca var. *pluriloculata* Fink = Eopyrenula intermedia [EOPYINT] Pyrenulaceae (L)
Pyrenula microtheca R.C. Harris [PYREMIC] Pyrenulaceae (L)
Pyrenula neglecta R.C. Harris = Pyrenula pseudobufonia [PYREPSE] Pyrenulaceae (L)
Pyrenula occidentalis (R.C. Harris) R.C. Harris [PYREOCC] Pyrenulaceae (L)
Pyrenula pseudobufonia (Rehm) R.C. Harris [PYREPSE] Pyrenulaceae (L)
Pyrola L. [PYROLA$] Pyrolaceae (V:D)
Pyrola aphylla Sm. = Pyrola picta [PYROPIC] Pyrolaceae (V:D)
Pyrola aphylla var. *leptosepala* Nutt. = Pyrola picta [PYROPIC] Pyrolaceae (V:D)
Pyrola aphylla var. *paucifolia* T.J. Howell = Pyrola picta [PYROPIC] Pyrolaceae (V:D)
Pyrola asarifolia Michx. [PYROASA] Pyrolaceae (V:D)
Pyrola asarifolia ssp. **bracteata** (Hook.) Haber [PYROASA1] Pyrolaceae (V:D)
Pyrola asarifolia var. *bracteata* (Hook.) Jepson = Pyrola asarifolia ssp. bracteata [PYROASA1] Pyrolaceae (V:D)
Pyrola asarifolia var. *incarnata* (DC.) Fern. = Pyrola asarifolia [PYROASA] Pyrolaceae (V:D)
Pyrola asarifolia var. *ovata* Farw. = Pyrola asarifolia [PYROASA] Pyrolaceae (V:D)
Pyrola asarifolia var. *purpurea* (Bunge) Fern. = Pyrola asarifolia [PYROASA] Pyrolaceae (V:D)
Pyrola blanda Andres = Pyrola picta [PYROPIC] Pyrolaceae (V:D)
Pyrola borealis Rydb. = Pyrola grandiflora [PYROGRA] Pyrolaceae (V:D)
Pyrola bracteata Hook. = Pyrola asarifolia ssp. bracteata [PYROASA1] Pyrolaceae (V:D)
Pyrola bracteata var. *hillii* J.K. Henry = Pyrola asarifolia ssp. bracteata [PYROASA1] Pyrolaceae (V:D)
Pyrola californica Krísa = Pyrola asarifolia [PYROASA] Pyrolaceae (V:D)
Pyrola canadensis Andres = Pyrola grandiflora [PYROGRA] Pyrolaceae (V:D)
Pyrola chlorantha Sw. [PYROCHL] Pyrolaceae (V:D)
Pyrola chlorantha var. *convoluta* (W. Bart.) Fern. = Pyrola chlorantha [PYROCHL] Pyrolaceae (V:D)
Pyrola chlorantha var. *paucifolia* Fern. = Pyrola chlorantha [PYROCHL] Pyrolaceae (V:D)
Pyrola chlorantha var. *revoluta* Jennings = Pyrola chlorantha [PYROCHL] Pyrolaceae (V:D)
Pyrola compacta Jennings = Pyrola elliptica [PYROELL] Pyrolaceae (V:D)
Pyrola conardiana Andres = Pyrola picta [PYROPIC] Pyrolaceae (V:D)
Pyrola convoluta W. Bart. = Pyrola chlorantha [PYROCHL] Pyrolaceae (V:D)
Pyrola dentata Sm. = Pyrola picta [PYROPIC] Pyrolaceae (V:D)
Pyrola dentata var. *apophylla* Copeland = Pyrola picta [PYROPIC] Pyrolaceae (V:D)
Pyrola dentata var. *integra* Gray = Pyrola picta [PYROPIC] Pyrolaceae (V:D)
Pyrola elata Nutt. = Pyrola asarifolia [PYROASA] Pyrolaceae (V:D)
Pyrola elliptica Nutt. [PYROELL] Pyrolaceae (V:D)
Pyrola gormanii Rydb. = Pyrola grandiflora [PYROGRA] Pyrolaceae (V:D)
Pyrola grandiflora Radius [PYROGRA] Pyrolaceae (V:D)
Pyrola grandiflora var. *canadensis* (Andres) Porsild = Pyrola grandiflora [PYROGRA] Pyrolaceae (V:D)
Pyrola minor L. [PYROMIN] Pyrolaceae (V:D)
Pyrola minor var. *parviflora* Boivin = Pyrola minor [PYROMIN] Pyrolaceae (V:D)
Pyrola occidentalis R. Br. ex D. Don = Pyrola grandiflora [PYROGRA] Pyrolaceae (V:D)
Pyrola oxypetala Austin ex Gray = Pyrola chlorantha [PYROCHL] Pyrolaceae (V:D)
Pyrola pallida Greene = Pyrola picta [PYROPIC] Pyrolaceae (V:D)
Pyrola paradoxa Andres = Pyrola picta [PYROPIC] Pyrolaceae (V:D)
Pyrola picta Sm. [PYROPIC] Pyrolaceae (V:D)
Pyrola picta ssp. *dentata* (Sm.) Piper = Pyrola picta [PYROPIC] Pyrolaceae (V:D)
Pyrola picta ssp. *integra* (Gray) Piper = Pyrola picta [PYROPIC] Pyrolaceae (V:D)
Pyrola picta ssp. *pallida* Andres = Pyrola picta [PYROPIC] Pyrolaceae (V:D)
Pyrola picta var. *dentata* (Sm.) Dorn = Pyrola picta [PYROPIC] Pyrolaceae (V:D)
Pyrola rotundifolia ssp. *asarifolia* (Michx.) A. & D. Löve = Pyrola asarifolia [PYROASA] Pyrolaceae (V:D)
Pyrola secunda L. = Orthilia secunda [ORTHSEC] Pyrolaceae (V:D)
Pyrola secunda ssp. *obtusata* (Turcz.) Hultén = Orthilia secunda [ORTHSEC] Pyrolaceae (V:D)
Pyrola secunda var. *obtusata* Turcz. = Orthilia secunda [ORTHSEC] Pyrolaceae (V:D)
Pyrola septentrionalis Andres = Pyrola picta [PYROPIC] Pyrolaceae (V:D)
Pyrola sparsifolia Suksdorf = Pyrola picta [PYROPIC] Pyrolaceae (V:D)
Pyrola uliginosa Torr. & Gray ex Torr. = Pyrola asarifolia [PYROASA] Pyrolaceae (V:D)
Pyrola uliginosa var. *gracilis* Jennings = Pyrola asarifolia [PYROASA] Pyrolaceae (V:D)
Pyrola uniflora L. = Moneses uniflora [MONEUNI] Pyrolaceae (V:D)
Pyrola virens Schreb. = Pyrola chlorantha [PYROCHL] Pyrolaceae (V:D)
Pyrola virens var. *convoluta* (W. Bart.) Fern. = Pyrola chlorantha [PYROCHL] Pyrolaceae (V:D)
Pyrola virens var. *saximontana* Fern. = Pyrola chlorantha [PYROCHL] Pyrolaceae (V:D)
Pyrrhospora Körber [PYRRHOS$] Lecanoraceae (L)
Pyrrhospora cinnabarina (Sommerf.) Choisy [PYRRCIN] Lecanoraceae (L)
Pyrrhospora elabens (Fr.) Hafellner [PYRRELA] Lecanoraceae (L)
Pyrrhospora quernea (Dickson) Körber [PYRRQUE] Lecanoraceae (L)
Pyrrhospora russula (Ach.) Hafellner [PYRRRUS] Lecanoraceae (L)
Pyrrhospora subcinnabarina (Tønsberg) Hafellner [PYRRSUB] Lecanoraceae (L)
Pyrrhospora varians (Ach.) R.C. Harris [PYRRVAR] Lecanoraceae (L)

Pyrrocoma Hook. [PYRROCO$] Asteraceae (V:D)
Pyrrocoma carthamoides Hook. [PYRRCAR] Asteraceae (V:D)
Pyrus L. [PYRUS$] Rosaceae (V:D)
Pyrus aucuparia (L.) Gaertn. = Sorbus aucuparia [SORBAUC] Rosaceae (V:D)
*__Pyrus communis__ L. [PYRUCOM] Rosaceae (V:D)
Pyrus diversifolia Bong. = Malus fusca [MALUFUS] Rosaceae (V:D)
Pyrus fusca Raf. = Malus fusca [MALUFUS] Rosaceae (V:D)
Pyrus malus L. = Malus sylvestris [MALUSYL] Rosaceae (V:D)
Pyrus pumila (P. Mill.) K. Koch = Malus pumila [MALUPUM] Rosaceae (V:D)
Pyrus rivularis Dougl. ex Hook. = Malus fusca [MALUFUS] Rosaceae (V:D)
Pyrus sitchensis (M. Roemer) Piper = Sorbus sitchensis var. sitchensis [SORBSIT0] Rosaceae (V:D)

Q

Quercus L. [QUERCUS$] Fagaceae (V:D)
Quercus garryana Dougl. ex Hook. [QUERGAR] Fagaceae (V:D)
*__Quercus robur__ L. [QUERROB] Fagaceae (V:D)

R

Racodium Fr. [RACODIU$] Uncertain-family (L)
Racodium rupestre Pers. [RACORUP] Uncertain-family (L)
Racomitrium Brid. [RACOMIT$] Grimmiaceae (B:M)
Racomitrium aciculare (Hedw.) Brid. [RACOACI] Grimmiaceae (B:M)
Racomitrium affine (Schleich. ex Web. & Mohr) Lindb. [RACOAFF] Grimmiaceae (B:M)
Racomitrium alternuatum C. Müll. & Kindb. in Mac. & Kindb. = Racomitrium macounii [RACOMAC] Grimmiaceae (B:M)
Racomitrium aquaticum (Brid. ex Schrad.) Brid. [RACOAQU] Grimmiaceae (B:M)
Racomitrium brevipes Kindb. in Mac. [RACOBRE] Grimmiaceae (B:M)
Racomitrium canescens (Hedw.) Brid. [RACOCAN] Grimmiaceae (B:M)
Racomitrium canescens f. *epilosum* (H. Müll. ex Milde) G. Jones in Grout = Racomitrium ericoides [RACOERI] Grimmiaceae (B:M)
Racomitrium canescens f. *ericoides* (Web. ex Brid.) Mönk. = Racomitrium ericoides [RACOERI] Grimmiaceae (B:M)
Racomitrium canescens ssp. *latifolium* (C. Jens. in Lange & C. Jens.) Frisv. = Racomitrium canescens var. latifolium [RACOCAN1] Grimmiaceae (B:M)
Racomitrium canescens var. **canescens** [RACOCAN0] Grimmiaceae (B:M)
Racomitrium canescens var. *epilosum* H. Müll. ex Milde = Racomitrium ericoides [RACOERI] Grimmiaceae (B:M)
Racomitrium canescens var. *ericoides* (Hedw.) Hampe = Racomitrium ericoides [RACOERI] Grimmiaceae (B:M)
Racomitrium canescens var. **latifolium** C. Jens. [RACOCAN1] Grimmiaceae (B:M)
Racomitrium canescens var. *muticum* (Kindb. in Mac.) Mac. & Kindb. = Racomitrium muticum [RACOMUT] Grimmiaceae (B:M)
Racomitrium canescens var. *strictum* Schlieph. in Limpr. = Racomitrium ericoides [RACOERI] Grimmiaceae (B:M)
Racomitrium elongatum Ehrh. ex Frisv. [RACOELO] Grimmiaceae (B:M)
Racomitrium ericoides (Web. ex Brid.) Brid. [RACOERI] Grimmiaceae (B:M)
Racomitrium fasciculare (Hedw.) Brid. [RACOFAS] Grimmiaceae (B:M)
Racomitrium heterostichum (Hedw.) Brid. [RACOHET] Grimmiaceae (B:M)
Racomitrium heterostichum ssp. *micropoides* (Kindb.) Kindb. = Racomitrium affine [RACOAFF] Grimmiaceae (B:M)
Racomitrium heterostichum var. *affine* (Schleich. ex Web. & Mohr) Lesq. = Racomitrium affine [RACOAFF] Grimmiaceae (B:M)
Racomitrium heterostichum var. *alopecurum* Hüb. = Racomitrium affine [RACOAFF] Grimmiaceae (B:M)
Racomitrium heterostichum var. *gracilescens* Bruch & Schimp. in B.S.G. = Racomitrium heterostichum [RACOHET] Grimmiaceae (B:M)
Racomitrium heterostichum var. *macounii* (Kindb. in Mac.) G. Jones in Grout = Racomitrium macounii [RACOMAC] Grimmiaceae (B:M)
Racomitrium heterostichum var. *microcarpon* (Hedw.) Boul. = Racomitrium microcarpon [RACOMIC] Grimmiaceae (B:M)
Racomitrium heterostichum var. *occidentale* Ren. & Card. = Racomitrium occidentale [RACOOCC] Grimmiaceae (B:M)
Racomitrium heterostichum var. *ramulosum* (Lindb.) Corb. = Racomitrium microcarpon [RACOMIC] Grimmiaceae (B:M)
Racomitrium heterostichum var. *sudeticum* (Funck) Bauer = Racomitrium sudeticum [RACOSUD] Grimmiaceae (B:M)
Racomitrium hypnoides Lindb. = Racomitrium lanuginosum [RACOLAN] Grimmiaceae (B:M)
Racomitrium lanuginosum (Hedw.) Brid. [RACOLAN] Grimmiaceae (B:M)
Racomitrium lanuginosum var. *subimberbe* (Hartm.) Lindb. = Racomitrium lanuginosum [RACOLAN] Grimmiaceae (B:M)
Racomitrium lawtonae Irel. [RACOLAW] Grimmiaceae (B:M)
Racomitrium macounii Kindb. in Mac. [RACOMAC] Grimmiaceae (B:M)
Racomitrium macounii ssp. *alpinum* (Lawt.) Frisv. = Racomitrium macounii [RACOMAC] Grimmiaceae (B:M)
Racomitrium microcarpon (Hedw.) Brid. [RACOMIC] Grimmiaceae (B:M)
Racomitrium muticum (Kindb. in Mac.) Frisv. [RACOMUT] Grimmiaceae (B:M)
Racomitrium neevii (C. Müll.) Watts = Racomitrium aciculare [RACOACI] Grimmiaceae (B:M)
Racomitrium obesum Frisv. [RACOOBE] Grimmiaceae (B:M)
Racomitrium occidentale (Ren. & Card.) Ren. & Card. [RACOOCC] Grimmiaceae (B:M)
Racomitrium pacificum Irel. & Spence [RACOPAC] Grimmiaceae (B:M)
Racomitrium palmeri (Kindb.) Kindb. = Racomitrium fasciculare [RACOFAS] Grimmiaceae (B:M)
Racomitrium panschii (C. Müll.) Kindb. [RACOPAN] Grimmiaceae (B:M)
Racomitrium patens (Hedw.) Hüb. = Dryptodon patens [DRYPPAT] Grimmiaceae (B:M)
Racomitrium protensum (Braun) Hüb. = Racomitrium aquaticum [RACOAQU] Grimmiaceae (B:M)
Racomitrium pygmaeum Frisv. [RACOPYG] Grimmiaceae (B:M)

Racomitrium ramulosum Lindb. = Racomitrium microcarpon [RACOMIC] Grimmiaceae (B:M)
Racomitrium robustifolium Kindb. = Racomitrium macounii [RACOMAC] Grimmiaceae (B:M)
Racomitrium sudeticum (Funck) Bruch & Schimp. in B.S.G. [RACOSUD] Grimmiaceae (B:M)
Racomitrium sudeticum f. *alpinum* Lawt. = Racomitrium macounii [RACOMAC] Grimmiaceae (B:M)
Racomitrium sudeticum f. *americanum* Lawt. = Racomitrium macounii [RACOMAC] Grimmiaceae (B:M)
Racomitrium sudeticum f. *brevipes* (Kindb. in Mac.) Lawt. = Racomitrium brevipes [RACOBRE] Grimmiaceae (B:M)
Racomitrium sudeticum f. *macounii* (Kindb. in Mac.) Lawt. = Racomitrium macounii [RACOMAC] Grimmiaceae (B:M)
Racomitrium sudeticum var. *validius* Jur. = Racomitrium sudeticum [RACOSUD] Grimmiaceae (B:M)
Racomitrium tenuinerve Kindb. = Racomitrium fasciculare [RACOFAS] Grimmiaceae (B:M)
Racomitrium varium (Mitt.) Jaeg. [RACOVAR] Grimmiaceae (B:M)
Radicula armoracia (L.) B.L. Robins. = Armoracia rusticana [ARMORUS] Brassicaceae (V:D)
Radicula sylvestris (L.) Druce = Rorippa sylvestris [RORISYL] Brassicaceae (V:D)
Radula Dum. [RADULA$] Radulaceae (B:H)
Radula auriculata Steph. [RADUAUR] Radulaceae (B:H)
Radula bolanderi Gott. [RADUBOL] Radulaceae (B:H)
Radula complanata (L.) Dum. [RADUCOM] Radulaceae (B:H)
Radula hallii Aust. = Radula complanata [RADUCOM] Radulaceae (B:H)
Radula obtusiloba Steph. [RADUOBT] Radulaceae (B:H)
Radula obtusiloba ssp. **obtusiloba** [RADUOBT0] Radulaceae (B:H)
Radula obtusiloba ssp. **polyclada** (Evans) Hatt. [RADUOBT1] Radulaceae (B:H)
Radula polyclada Evans = Radula obtusiloba ssp. polyclada [RADUOBT1] Radulaceae (B:H)
Radula prolifera S. Arnell [RADUPRO] Radulaceae (B:H)
Ramalina Ach. [RAMALIN$] Ramalinaceae (L)
Ramalina calicaris var. *farinacea* Schaerer = Ramalina farinacea [RAMAFAR] Ramalinaceae (L)
Ramalina cephalota Tuck. = Niebla cephalota [NIEBCEP] Ramalinaceae (L)
Ramalina crinalis (Ach.) Gyelnik = Ramalina thrausta [RAMATHR] Ramalinaceae (L)
Ramalina dilacerata (Hoffm.) Hoffm. [RAMADIL] Ramalinaceae (L)
Ramalina dilacerata f. **pollinariella** Arnold [RAMADIL1] Ramalinaceae (L)
Ramalina farinacea (L.) Ach. [RAMAFAR] Ramalinaceae (L)
Ramalina farinacea Schaerer = Ramalina farinacea [RAMAFAR] Ramalinaceae (L)
Ramalina geniculata Hook. f. & Taylor [RAMAGEN] Ramalinaceae (L)
Ramalina hypoprotocetrarica Culb. = Ramalina farinacea [RAMAFAR] Ramalinaceae (L)
Ramalina inflata (Hook. f. & Taylor) Hook. f. & Taylor [RAMAINF] Ramalinaceae (L)
Ramalina menziesii Taylor [RAMAMEN] Ramalinaceae (L)
Ramalina minuscula (Nyl.) Nyl. = Ramalina dilacerata [RAMADIL] Ramalinaceae (L)
Ramalina obtusata (Arnold) Bitter [RAMAOBT] Ramalinaceae (L)
Ramalina pollinaria (Westr.) Ach. [RAMAPOL] Ramalinaceae (L)
Ramalina pollinariella Nyl. = Ramalina roesleri [RAMAROE] Ramalinaceae (L)
Ramalina reagens (de Lesd.) Culb. = Ramalina farinacea [RAMAFAR] Ramalinaceae (L)
Ramalina reticulata (Nöhden) Kremp. = Ramalina menziesii [RAMAMEN] Ramalinaceae (L)
Ramalina roesleri (Hochst. ex Schaerer) Hue [RAMAROE] Ramalinaceae (L)
Ramalina sinensis Jatta [RAMASIN] Ramalinaceae (L)
Ramalina subleptocarpha Rundel & Bowler [RAMASUB] Ramalinaceae (L)
Ramalina thrausta (Ach.) Nyl. [RAMATHR] Ramalinaceae (L)
Ramischia elatior Rydb. = Orthilia secunda [ORTHSEC] Pyrolaceae (V:D)
Ramischia secunda (L.) Garcke = Orthilia secunda [ORTHSEC] Pyrolaceae (V:D)
Ramularia peltigericola D. Hawksw. = Hawksworthiana peltigericola [HAWKPEL] Uncertain-family (L)
Ranunculus L. [RANUNCU$] Ranunculaceae (V:D)
Ranunculus abortivus L. [RANUABO] Ranunculaceae (V:D)
Ranunculus abortivus ssp. *acrolasius* (Fern.) Kapoor & A. & D. Löve = Ranunculus abortivus [RANUABO] Ranunculaceae (V:D)
Ranunculus abortivus var. *acrolasius* Fern. = Ranunculus abortivus [RANUABO] Ranunculaceae (V:D)
Ranunculus abortivus var. *eucyclus* Fern. = Ranunculus abortivus [RANUABO] Ranunculaceae (V:D)
Ranunculus abortivus var. *indivisus* Fern. = Ranunculus abortivus [RANUABO] Ranunculaceae (V:D)
Ranunculus abortivus var. *typicus* Fern. = Ranunculus abortivus [RANUABO] Ranunculaceae (V:D)
***Ranunculus acris** L. [RANUACR] Ranunculaceae (V:D)
Ranunculus affinis R. Br. = Ranunculus pedatifidus var. affinis [RANUPED1] Ranunculaceae (V:D)
Ranunculus alismifolius Geyer ex Benth. [RANUALI] Ranunculaceae (V:D)
Ranunculus alismifolius var. *hartwegii* (Greene) Jepson = Ranunculus alismifolius [RANUALI] Ranunculaceae (V:D)
Ranunculus alismifolius var. *lemmonii* (Gray) L. Benson = Ranunculus alismifolius [RANUALI] Ranunculaceae (V:D)
Ranunculus alismifolius var. *typicus* L. Benson = Ranunculus alismifolius [RANUALI] Ranunculaceae (V:D)
Ranunculus alpeophilus A. Nels. = Ranunculus inamoenus var. alpeophilus [RANUINA1] Ranunculaceae (V:D)
Ranunculus amphibius James = Ranunculus longirostris [RANULON] Ranunculaceae (V:D)
Ranunculus aquatilis L. [RANUAQU] Ranunculaceae (V:D)
Ranunculus aquatilis var. *calvescens* (W. Drew) L. Benson = Ranunculus trichophyllus [RANUTRI] Ranunculaceae (V:D)
Ranunculus aquatilis var. *capillaceus* (Thuill.) DC. = Ranunculus trichophyllus [RANUTRI] Ranunculaceae (V:D)
Ranunculus aquatilis var. *harrisii* L. Benson = Ranunculus trichophyllus [RANUTRI] Ranunculaceae (V:D)
Ranunculus aquatilis var. *hispidulus* E. Drew = Ranunculus aquatilis [RANUAQU] Ranunculaceae (V:D)
Ranunculus aquatilis var. *lalondei* L. Benson = Ranunculus trichophyllus [RANUTRI] Ranunculaceae (V:D)
Ranunculus aquatilis var. *lobbii* (Hiern) S. Wats. = Ranunculus lobbii [RANULOB] Ranunculaceae (V:D)
Ranunculus aquatilis var. *longirostris* (Godr.) Lawson = Ranunculus longirostris [RANULON] Ranunculaceae (V:D)
Ranunculus aquatilis var. *porteri* (Britt.) L. Benson = Ranunculus trichophyllus [RANUTRI] Ranunculaceae (V:D)
Ranunculus aquatilis var. *subrigidus* (W. Drew) Breitung = Ranunculus longirostris [RANULON] Ranunculaceae (V:D)
Ranunculus aquatilis var. *trichophyllus* (Chaix) Gray = Ranunculus trichophyllus [RANUTRI] Ranunculaceae (V:D)
Ranunculus aquatilis var. *typicus* L. Benson = Ranunculus aquatilis [RANUAQU] Ranunculaceae (V:D)
Ranunculus bongardii Greene = Ranunculus uncinatus var. parviflorus [RANUUNC1] Ranunculaceae (V:D)
Ranunculus californicus Benth. [RANUCAL] Ranunculaceae (V:D)
Ranunculus californicus var. *austromontanus* L. Benson = Ranunculus californicus [RANUCAL] Ranunculaceae (V:D)
Ranunculus californicus var. *cuneatus* Greene = Ranunculus californicus [RANUCAL] Ranunculaceae (V:D)
Ranunculus californicus var. *gratus* Jepson = Ranunculus californicus [RANUCAL] Ranunculaceae (V:D)
Ranunculus californicus var. *rugulosus* (Greene) L. Benson = Ranunculus californicus [RANUCAL] Ranunculaceae (V:D)
Ranunculus californicus var. *typicus* L. Benson = Ranunculus californicus [RANUCAL] Ranunculaceae (V:D)
Ranunculus cardiophyllus Hook. [RANUCAR] Ranunculaceae (V:D)
Ranunculus cardiophyllus var. *subsagittatus* (Gray) L. Benson = Ranunculus cardiophyllus [RANUCAR] Ranunculaceae (V:D)

Ranunculus cardiophyllus var. *typicus* L. Benson = Ranunculus cardiophyllus [RANUCAR] Ranunculaceae (V:D)
Ranunculus circinatus auct. = Ranunculus longirostris [RANULON] Ranunculaceae (V:D)
Ranunculus circinatus var. *subrigidus* (W. Drew) L. Benson = Ranunculus longirostris [RANULON] Ranunculaceae (V:D)
Ranunculus cooleyae Vasey & Rose = Kumlienia cooleyae [KUMLCOO] Ranunculaceae (V:D)
Ranunculus cymbalaria Pursh [RANUCYM] Ranunculaceae (V:D)
Ranunculus cymbalaria ssp. *saximontanus* (Fern.) Thorne = Ranunculus cymbalaria var. saximontanus [RANUCYM2] Ranunculaceae (V:D)
Ranunculus cymbalaria var. **saximontanus** Fern. [RANUCYM2] Ranunculaceae (V:D)
Ranunculus delphiniifolius Torr. ex Eat. = Ranunculus flabellaris [RANUFLA] Ranunculaceae (V:D)
Ranunculus ellipticus Greene = Ranunculus glaberrimus var. ellipticus [RANUGLA1] Ranunculaceae (V:D)
Ranunculus eschscholtzii Schlecht. [RANUESC] Ranunculaceae (V:D)
Ranunculus eschscholtzii var. *suksdorfii* (Gray) L. Benson = Ranunculus suksdorfii [RANUSUK] Ranunculaceae (V:D)
Ranunculus eschscholtzii var. *typicus* L. Benson = Ranunculus eschscholtzii [RANUESC] Ranunculaceae (V:D)
****Ranunculus ficaria** L. [RANUFIC] Ranunculaceae (V:D)
Ranunculus filiformis Michx. = Ranunculus flammula var. filiformis [RANUFLM1] Ranunculaceae (V:D)
Ranunculus flabellaris Raf. [RANUFLA] Ranunculaceae (V:D)
Ranunculus flaccidus Pers. = Ranunculus trichophyllus [RANUTRI] Ranunculaceae (V:D)
Ranunculus flammula L. [RANUFLM] Ranunculaceae (V:D)
Ranunculus flammula var. **filiformis** (Michx.) Hook. [RANUFLM1] Ranunculaceae (V:D)
Ranunculus flammula var. *ovalis* (Bigelow) L. Benson = Ranunculus flammula [RANUFLM] Ranunculaceae (V:D)
Ranunculus flammula var. *reptans* (L.) E. Mey. = Ranunculus flammula var. filiformis [RANUFLM1] Ranunculaceae (V:D)
Ranunculus gelidus Kar. & Kir. = Ranunculus karelinii [RANUKAR] Ranunculaceae (V:D)
Ranunculus gelidus ssp. *grayi* (Britt.) Hultén = Ranunculus karelinii [RANUKAR] Ranunculaceae (V:D)
Ranunculus gelidus var. *shumaginensis* Hultén = Ranunculus karelinii [RANUKAR] Ranunculaceae (V:D)
Ranunculus glaberrimus Hook. [RANUGLA] Ranunculaceae (V:D)
Ranunculus glaberrimus var. *buddii* Boivin = Ranunculus glaberrimus var. ellipticus [RANUGLA1] Ranunculaceae (V:D)
Ranunculus glaberrimus var. **ellipticus** (Greene) Greene [RANUGLA1] Ranunculaceae (V:D)
Ranunculus glaberrimus var. **glaberrimus** [RANUGLA0] Ranunculaceae (V:D)
Ranunculus glaberrimus var. *typicus* L. Benson = Ranunculus glaberrimus var. glaberrimus [RANUGLA0] Ranunculaceae (V:D)
Ranunculus gmelinii DC. [RANUGME] Ranunculaceae (V:D)
Ranunculus gmelinii ssp. *purshii* (Richards.) Hultén = Ranunculus gmelinii var. purshii [RANUGME1] Ranunculaceae (V:D)
Ranunculus gmelinii var. *hookeri* (D. Don) L. Benson = Ranunculus gmelinii var. purshii [RANUGME1] Ranunculaceae (V:D)
Ranunculus gmelinii var. **limosus** (Nutt.) Hara [RANUGME4] Ranunculaceae (V:D)
Ranunculus gmelinii var. *prolificus* (Fern.) Hara = Ranunculus gmelinii var. purshii [RANUGME1] Ranunculaceae (V:D)
Ranunculus gmelinii var. **purshii** (Richards.) Hara [RANUGME1] Ranunculaceae (V:D)
Ranunculus gmelinii var. *terrestris* (Ledeb.) L. Benson = Ranunculus gmelinii var. purshii [RANUGME1] Ranunculaceae (V:D)
Ranunculus grayi Britt. = Ranunculus karelinii [RANUKAR] Ranunculaceae (V:D)
Ranunculus hartwegii Greene = Ranunculus alismifolius [RANUALI] Ranunculaceae (V:D)
Ranunculus hexasepalus (L. Benson) L. Benson [RANUHEX] Ranunculaceae (V:D)
Ranunculus hyperboreus Rottb. [RANUHYP] Ranunculaceae (V:D)
Ranunculus inamoenus Greene [RANUINA] Ranunculaceae (V:D)
Ranunculus inamoenus var. **alpeophilus** (A. Nels.) L. Benson [RANUINA1] Ranunculaceae (V:D)
Ranunculus karelinii Czern. [RANUKAR] Ranunculaceae (V:D)
Ranunculus lapponicus L. [RANULAP] Ranunculaceae (V:D)
Ranunculus lemmonii Gray = Ranunculus alismifolius [RANUALI] Ranunculaceae (V:D)
Ranunculus limosus Nutt. = Ranunculus gmelinii var. limosus [RANUGME4] Ranunculaceae (V:D)
Ranunculus lobbii (Hiern) Gray [RANULOB] Ranunculaceae (V:D)
Ranunculus longirostris Godr. [RANULON] Ranunculaceae (V:D)
Ranunculus macounii Britt. [RANUMAC] Ranunculaceae (V:D)
Ranunculus macounii var. *oreganus* (Gray) Davis = Ranunculus macounii [RANUMAC] Ranunculaceae (V:D)
Ranunculus natans var. *intertextus* (Greene) L. Benson = Ranunculus hyperboreus [RANUHYP] Ranunculaceae (V:D)
Ranunculus nivalis L. [RANUNIV] Ranunculaceae (V:D)
Ranunculus nivalis var. *eschscholtzii* (Schlecht.) S. Wats. = Ranunculus eschscholtzii [RANUESC] Ranunculaceae (V:D)
Ranunculus nivalis var. *sulphureus* (Soland. ex Phipps) Wahlenb. = Ranunculus sulphureus [RANUSUL] Ranunculaceae (V:D)
Ranunculus occidentalis Nutt. [RANUOCC] Ranunculaceae (V:D)
Ranunculus occidentalis var. **brevistylis** Greene [RANUOCC2] Ranunculaceae (V:D)
Ranunculus occidentalis var. *hexasepalus* L. Benson = Ranunculus hexasepalus [RANUHEX] Ranunculaceae (V:D)
Ranunculus occidentalis var. *parviflorus* Torr. = Ranunculus uncinatus var. parviflorus [RANUUNC1] Ranunculaceae (V:D)
Ranunculus orthorhynchus Hook. [RANUORT] Ranunculaceae (V:D)
Ranunculus orthorhynchus ssp. *alaschensis* (L. Benson) Hultén = Ranunculus orthorhynchus [RANUORT] Ranunculaceae (V:D)
Ranunculus orthorhynchus ssp. *platyphyllus* (Gray) Taylor & MacBryde = Ranunculus orthorhynchus [RANUORT] Ranunculaceae (V:D)
Ranunculus orthorhynchus var. *alaschensis* L. Benson = Ranunculus orthorhynchus [RANUORT] Ranunculaceae (V:D)
Ranunculus orthorhynchus var. *hallii* Jepson = Ranunculus orthorhynchus [RANUORT] Ranunculaceae (V:D)
Ranunculus orthorhynchus var. *platyphyllus* Gray = Ranunculus orthorhynchus [RANUORT] Ranunculaceae (V:D)
Ranunculus orthorhynchus var. *typicus* L. Benson = Ranunculus orthorhynchus [RANUORT] Ranunculaceae (V:D)
Ranunculus ovalis Raf. = Ranunculus rhomboideus [RANURHO] Ranunculaceae (V:D)
Ranunculus parvulus L. = Ranunculus sardous [RANUSAR] Ranunculaceae (V:D)
Ranunculus pedatifidus Sm. [RANUPED] Ranunculaceae (V:D)
Ranunculus pedatifidus ssp. *affinis* (R. Br.) Hultén = Ranunculus pedatifidus var. affinis [RANUPED1] Ranunculaceae (V:D)
Ranunculus pedatifidus var. **affinis** (R. Br.) L. Benson [RANUPED1] Ranunculaceae (V:D)
Ranunculus pedatifidus var. *cardiophyllus* (Hook.) Britt. = Ranunculus cardiophyllus [RANUCAR] Ranunculaceae (V:D)
Ranunculus pedatifidus var. *leiocarpus* (Trautv.) Fern. = Ranunculus pedatifidus var. affinis [RANUPED1] Ranunculaceae (V:D)
Ranunculus pensylvanicus L. f. [RANUPEN] Ranunculaceae (V:D)
Ranunculus purshii Richards. = Ranunculus gmelinii var. purshii [RANUGME1] Ranunculaceae (V:D)
Ranunculus pygmaeus Wahlenb. [RANUPYG] Ranunculaceae (V:D)
****Ranunculus repens** L. [RANUREP] Ranunculaceae (V:D)
Ranunculus repens var. **degeneratus** Schur [RANUREP2] Ranunculaceae (V:D)
Ranunculus repens var. *erectus* DC. = Ranunculus repens var. glabratus [RANUREP1] Ranunculaceae (V:D)
Ranunculus repens var. **glabratus** DC. [RANUREP1] Ranunculaceae (V:D)
Ranunculus repens var. *pleniflorus* Fern. = Ranunculus repens var. degeneratus [RANUREP2] Ranunculaceae (V:D)
Ranunculus repens var. *villosus* Lamotte = Ranunculus repens [RANUREP] Ranunculaceae (V:D)
Ranunculus reptans L. = Ranunculus flammula var. filiformis [RANUFLM1] Ranunculaceae (V:D)

Ranunculus reptans var. *filiformis* (Michx.) DC. = Ranunculus flammula var. filiformis [RANUFLM1] Ranunculaceae (V:D)
Ranunculus reptans var. *intermedius* (Hook.) Torr. & Gray = Ranunculus flammula var. filiformis [RANUFLM1] Ranunculaceae (V:D)
Ranunculus rhomboideus Goldie [RANURHO] Ranunculaceae (V:D)
Ranunculus rugulosus Greene = Ranunculus californicus [RANUCAL] Ranunculaceae (V:D)
•**Ranunculus sardous** Crantz [RANUSAR] Ranunculaceae (V:D)
Ranunculus sceleratus L. [RANUSCE] Ranunculaceae (V:D)
Ranunculus sceleratus ssp. *multifidus* (Nutt.) Hultén = Ranunculus sceleratus var. multifidus [RANUSCE1] Ranunculaceae (V:D)
Ranunculus sceleratus var. **multifidus** Nutt. [RANUSCE1] Ranunculaceae (V:D)
Ranunculus subrigidus W. Drew = Ranunculus longirostris [RANULON] Ranunculaceae (V:D)
Ranunculus suksdorfii Gray [RANUSUK] Ranunculaceae (V:D)
Ranunculus sulphureus Soland. ex Phipps [RANUSUL] Ranunculaceae (V:D)
Ranunculus sulphureus var. *intercedens* Hultén = Ranunculus sulphureus [RANUSUL] Ranunculaceae (V:D)
Ranunculus testiculatus Crantz = Ceratocephala testiculatus [CERATES] Ranunculaceae (V:D)
Ranunculus trichophyllus Chaix [RANUTRI] Ranunculaceae (V:D)
Ranunculus trichophyllus var. *calvescens* W. Drew = Ranunculus trichophyllus [RANUTRI] Ranunculaceae (V:D)
Ranunculus trichophyllus var. *hispidulus* (E. Drew) W. Drew = Ranunculus aquatilis [RANUAQU] Ranunculaceae (V:D)
Ranunculus trichophyllus var. *typicus* W. Drew = Ranunculus trichophyllus [RANUTRI] Ranunculaceae (V:D)
Ranunculus uncinatus D. Don ex G. Don [RANUUNC] Ranunculaceae (V:D)
Ranunculus uncinatus var. **parviflorus** (Torr.) L. Benson [RANUUNC1] Ranunculaceae (V:D)
Ranunculus usneoides Greene = Ranunculus longirostris [RANULON] Ranunculaceae (V:D)
Ranunculus verecundus B.L. Robins. ex Piper [RANUVER] Ranunculaceae (V:D)
Raphanus L. [RAPHANU$] Brassicaceae (V:D)
•**Raphanus raphanistrum** L. [RAPHRAP] Brassicaceae (V:D)
Raphanus raphanistrum var. *sativus* (L.) G. Beck = Raphanus sativus [RAPHSAT] Brassicaceae (V:D)
•**Raphanus sativus** L. [RAPHSAT] Brassicaceae (V:D)
Ratibida Raf. [RATIBID$] Asteraceae (V:D)
Ratibida columnaris (Sims) D. Don = Ratibida columnifera [RATICOL] Asteraceae (V:D)
Ratibida columnaris var. *pulcherrima* (DC.) D. Don = Ratibida columnifera [RATICOL] Asteraceae (V:D)
Ratibida columnifera (Nutt.) Woot. & Standl. [RATICOL] Asteraceae (V:D)
Reboulia Raddi [REBOULI$] Aytoniaceae (B:H)
Reboulia hemisphaerica (L.) Raddi [REBOHEM] Aytoniaceae (B:H)
Refractohilum D. Hawksw. [REFRACT$] Uncertain-family (L)
Refractohilum peltigerae (Keissler) D. Hawksw. [REFRPEL] Uncertain-family (L)
Reseda L. [RESEDA$] Resedaceae (V:D)
•**Reseda alba** L. [RESEALB] Resedaceae (V:D)
•**Reseda lutea** L. [RESELUT] Resedaceae (V:D)
Reynoutria japonica Houtt. = Polygonum cuspidatum [POLYCUS] Polygonaceae (V:D)
Reynoutria sachalinensis (F. Schmidt ex Maxim.) Nakai = Polygonum sachalinense [POLYSAC] Polygonaceae (V:D)
Rhabdoweisia Bruch & Schimp. in B.S.G. [RHABDOW$] Dicranaceae (B:M)
Rhabdoweisia crispata (With.) Lindb. [RHABCRI] Dicranaceae (B:M)
Rhabdoweisia denticulata Bruch & Schimp. in B.S.G. = Rhabdoweisia crispata [RHABCRI] Dicranaceae (B:M)
Rhacomitrium Brid. = Racomitrium [RACOMIT$] Grimmiaceae (B:M)
Rhamnus L. [RHAMNUS$] Rhamnaceae (V:D)
Rhamnus alnifolia L'Hér. [RHAMALN] Rhamnaceae (V:D)
Rhamnus purshiana DC. = Frangula purshiana [FRANPUR] Rhamnaceae (V:D)
Rhaphidostegium demissum (Wils.) De Not. = Sematophyllum demissum [SEMADEM] Sematophyllaceae (B:M)
Rhaphidostegium novae-caesareae (Aust.) Kindb. = Hygrohypnum micans [HYGRMIC] Amblystegiaceae (B:M)
Rhaphidostegium roellii Ren. & Card. in Röll = Brotherella roellii [BROTROE] Sematophyllaceae (B:M)
Rhinanthus L. [RHINANT$] Scrophulariaceae (V:D)
Rhinanthus borealis ssp. *kyrolliae* (Chabert) Pennell = Rhinanthus minor [RHINMIN] Scrophulariaceae (V:D)
Rhinanthus crista-galli L. = Rhinanthus minor [RHINMIN] Scrophulariaceae (V:D)
Rhinanthus minor L. [RHINMIN] Scrophulariaceae (V:D)
Rhinanthus stenophyllus (Schur) Schinz & Thellung = Rhinanthus minor [RHINMIN] Scrophulariaceae (V:D)
Rhizocarpon Ramond ex DC. [RHIZOCA$] Rhizocarpaceae (L)
Rhizocarpon albineum (Tuck.) Fink = Rhizocarpon obscuratum [RHIZOBS] Rhizocarpaceae (L)
Rhizocarpon alboatrum (Hoffm.) Anzi = Diplotomma alboatrum [DIPLALO] Physciaceae (L)
Rhizocarpon alpicola (Anzi) Rabenh. [RHIZALP] Rhizocarpaceae (L)
Rhizocarpon ambiguum (Schaerer) Zahlbr. = Rhizocarpon distinctum [RHIZDIT] Rhizocarpaceae (L)
Rhizocarpon atroalbescens (Nyl.) Zahlbr. = Rhizocarpon eupetraeoides [RHIZEUE] Rhizocarpaceae (L)
Rhizocarpon atroflavescens Lynge [RHIZATR] Rhizocarpaceae (L)
Rhizocarpon badioatrum (Flörke ex Sprengel) Th. Fr. [RHIZBAD] Rhizocarpaceae (L)
Rhizocarpon bolanderi (Tuck.) Herre [RHIZBOL] Rhizocarpaceae (L)
Rhizocarpon chioneum (Norman) Th. Fr. [RHIZCHI] Rhizocarpaceae (L)
Rhizocarpon chionophilum Th. Fr. = Rhizocarpon alpicola [RHIZALP] Rhizocarpaceae (L)
Rhizocarpon chlorophaeum Hepp ex Leighton = Diplotomma chlorophaeum [DIPLCHL] Physciaceae (L)
Rhizocarpon cinereonigrum Vainio [RHIZCIE] Rhizocarpaceae (L)
Rhizocarpon cinereovirens (Müll. Arg.) Vainio [RHIZCIN] Rhizocarpaceae (L)
Rhizocarpon concentricum (Davies) Beltr. [RHIZCON] Rhizocarpaceae (L)
Rhizocarpon concretum (Ach.) Elenkin = Rhizocarpon geminatum [RHIZGEM] Rhizocarpaceae (L)
Rhizocarpon crystalligenum Lynge = Rhizocarpon superficiale ssp. boreale [RHIZSUP1] Rhizocarpaceae (L)
Rhizocarpon disporum auct. = Rhizocarpon geminatum [RHIZGEM] Rhizocarpaceae (L)
Rhizocarpon distinctum Th. Fr. [RHIZDIT] Rhizocarpaceae (L)
Rhizocarpon eupetraeoides (Nyl.) Blomb. & Forss. [RHIZEUE] Rhizocarpaceae (L)
Rhizocarpon eupetraeum (Nyl.) Arnold [RHIZEUP] Rhizocarpaceae (L)
Rhizocarpon geminatum Körber [RHIZGEM] Rhizocarpaceae (L)
Rhizocarpon geographicum (L.) DC. [RHIZGEO] Rhizocarpaceae (L)
Rhizocarpon grande (Flörke ex Flotow) Arnold [RHIZGRA] Rhizocarpaceae (L)
Rhizocarpon hensseniae Brodo [RHIZHEN] Rhizocarpaceae (L)
Rhizocarpon hochstetteri (Körber) Vainio [RHIZHOC] Rhizocarpaceae (L)
Rhizocarpon inarense (Vainio) Vainio [RHIZINA] Rhizocarpaceae (L)
Rhizocarpon intermediellum Räsänen [RHIZINT] Rhizocarpaceae (L)
Rhizocarpon intermedium Degel. = Rhizocarpon eupetraeum [RHIZEUP] Rhizocarpaceae (L)
Rhizocarpon interponens (Nyl.) Zahlbr. = Rhizocarpon obscuratum [RHIZOBS] Rhizocarpaceae (L)
Rhizocarpon lavatum (Fr.) Hazsl. [RHIZLAV] Rhizocarpaceae (L)
Rhizocarpon lecanorinum Anders [RHIZLEC] Rhizocarpaceae (L)

Rhizocarpon macrosporum Räsänen [RHIZMAC] Rhizocarpaceae (L)
Rhizocarpon obscuratum (Ach.) A. Massal. [RHIZOBS] Rhizocarpaceae (L)
Rhizocarpon occidentale Lynge = Rhizocarpon superficiale [RHIZSUP] Rhizocarpaceae (L)
Rhizocarpon oidaleum (Nyl.) Fink = Buellia oidalea [BUELOID] Physciaceae (L)
Rhizocarpon oreites (Vainio) Zahlbr. = Rhizocarpon alpicola [RHIZALP] Rhizocarpaceae (L)
Rhizocarpon penichrum (Tuck.) G. Merr. = Diplotomma penichrum [DIPLPEN] Physciaceae (L)
Rhizocarpon perlutum (Nyl.) Zahlbr. = Rhizocarpon concentricum [RHIZCON] Rhizocarpaceae (L)
Rhizocarpon permodestum Arnold = Rhizocarpon obscuratum [RHIZOBS] Rhizocarpaceae (L)
Rhizocarpon polare Räsänen = Rhizocarpon superficiale [RHIZSUP] Rhizocarpaceae (L)
Rhizocarpon pusillum Runem. [RHIZPUS] Rhizocarpaceae (L)
Rhizocarpon reductum Th. Fr. = Rhizocarpon obscuratum [RHIZOBS] Rhizocarpaceae (L)
Rhizocarpon riparium Räsänen [RHIZRIP] Rhizocarpaceae (L)
Rhizocarpon sphaerosporum Räsänen = Rhizocarpon macrosporum [RHIZMAC] Rhizocarpaceae (L)
Rhizocarpon subtile Runem. = Rhizocarpon viridiatrum [RHIZVIR] Rhizocarpaceae (L)
Rhizocarpon superficiale (Schaerer) Vainio [RHIZSUP] Rhizocarpaceae (L)
Rhizocarpon superficiale ssp. **boreale** Runem. [RHIZSUP1] Rhizocarpaceae (L)
Rhizocarpon umbilicatum (Ramond) Flagey [RHIZUMB] Rhizocarpaceae (L)
Rhizocarpon umbilicatum f. pseudospeireum (Th. Fr.) Magnusson [RHIZUMB] Rhizocarpaceae (L)
Rhizocarpon viridiatrum (Wulfen) Körber [RHIZVIR] Rhizocarpaceae (L)
Rhizomatopteris montana (Lam.) Khokhr. = Cystopteris montana [CYSTMON] Dryopteridaceae (V:F)
Rhizomnium (Broth.) T. Kop. [RHIZOMN$] Mniaceae (B:M)
Rhizomnium glabrescens (Kindb.) T. Kop. [RHIZGLB] Mniaceae (B:M)
Rhizomnium gracile T. Kop. [RHIZGRC] Mniaceae (B:M)
Rhizomnium magnifolium (Horik.) T. Kop. [RHIZMAG] Mniaceae (B:M)
Rhizomnium nudum (Britt. & Williams) T. Kop. [RHIZNUD] Mniaceae (B:M)
Rhizomnium perssonii T. Kop = Rhizomnium magnifolium [RHIZMAG] Mniaceae (B:M)
Rhizomnium pseudopunctatum (Bruch & Schimp.) T. Kop. [RHIZPSE] Mniaceae (B:M)
Rhizomnium punctatum (Hedw.) T. Kop. [RHIZPUN] Mniaceae (B:M)
Rhizomnium punctatum ssp. *chlorophyllosum* (Kindb.) T. Kop. = Rhizomnium punctatum [RHIZPUN] Mniaceae (B:M)
Rhizomnium punctatum var. *elatum* (Schimp.) T. Kop. = Rhizomnium magnifolium [RHIZMAG] Mniaceae (B:M)
Rhizoplaca Zopf [RHIZOPL$] Lecanoraceae (L)
Rhizoplaca chrysoleuca (Sm.) Zopf [RHIZCHR] Lecanoraceae (L)
Rhizoplaca melanophthalma (DC.) Leuckert & Poelt [RHIZMEL] Lecanoraceae (L)
Rhizoplaca peltata (Ramond) Leuckert & Poelt [RHIZPEL] Lecanoraceae (L)
Rhizoplaca subdiscrepans (Nyl.) R. Sant. [RHIZSUB] Lecanoraceae (L)
Rhodiola integrifolia Raf. = Sedum integrifolium [SEDUINT] Crassulaceae (V:D)
Rhodobryum (Schimp.) Hampe [RHODOBR$] Bryaceae (B:M)
Rhodobryum roseum (Hedw.) Limpr. [RHODROS] Bryaceae (B:M)
Rhododendron L. [RHODODE$] Ericaceae (V:D)
Rhododendron albiflorum Hook. [RHODALB] Ericaceae (V:D)
Rhododendron californicum Hook. = Rhododendron macrophyllum [RHODMAC] Ericaceae (V:D)
Rhododendron groenlandicum (Oeder) Kron & Judd = Ledum groenlandicum [LEDUGRO] Ericaceae (V:D)
Rhododendron lapponicum (L.) Wahlenb. [RHODLAP] Ericaceae (V:D)
Rhododendron macrophyllum D. Don ex G. Don [RHODMAC] Ericaceae (V:D)
Rhododendron tomentosum ssp. *subarcticum* (Harmaja) G. Wallace = Ledum palustre ssp. decumbens [LEDUPAL1] Ericaceae (V:D)
Rhus L. [RHUS$] Anacardiaceae (V:D)
Rhus borealis Greene = Rhus glabra [RHUSGLA] Anacardiaceae (V:D)
Rhus calophylla Greene = Rhus glabra [RHUSGLA] Anacardiaceae (V:D)
Rhus diversiloba Torr. & Gray = Toxicodendron diversilobum [TOXIDIV] Anacardiaceae (V:D)
Rhus glabra L. [RHUSGLA] Anacardiaceae (V:D)
Rhus glabra var. *laciniata* Carr. = Rhus glabra [RHUSGLA] Anacardiaceae (V:D)
Rhus glabra var. *occidentalis* Torr. = Rhus glabra [RHUSGLA] Anacardiaceae (V:D)
Rhus radicans var. *rydbergii* (Small ex Rydb.) Rehd. = Toxicodendron rydbergii [TOXIRYD] Anacardiaceae (V:D)
Rhus radicans var. *vulgaris* (Michx.) DC. = Toxicodendron rydbergii [TOXIRYD] Anacardiaceae (V:D)
Rhus toxicodendron var. *vulgaris* Michx. = Toxicodendron rydbergii [TOXIRYD] Anacardiaceae (V:D)
Rhynchospora Vahl [RHYNCHO$] Cyperaceae (V:M)
Rhynchospora alba (L.) Vahl [RHYNALB] Cyperaceae (V:M)
Rhynchospora capillacea Torr. [RHYNCAP] Cyperaceae (V:M)
Rhynchospora capillacea var. *leviseta* E.J. Hill ex Gray = Rhynchospora capillacea [RHYNCAP] Cyperaceae (V:M)
Rhynchospora luquillensis Britt. = Rhynchospora alba [RHYNALB] Cyperaceae (V:M)
Rhynchospora smallii Britt. = Rhynchospora capillacea [RHYNCAP] Cyperaceae (V:M)
Rhynchostegiella compacta (C. Müll.) Loeske = Conardia compacta [CONACOM] Amblystegiaceae (B:M)
Rhynchostegiella compacta var. *americana* (Grout) Crum, Steere & Anderson = Conardia compacta [CONACOM] Amblystegiaceae (B:M)
Rhynchostegiella georgiana Dix. & Grout in Grout = Leptodictyum riparium [LEPTRIP] Amblystegiaceae (B:M)
Rhynchostegium pseudo-serrulatum (Kindb.) Kindb. = Brachythecium starkei [BRACSTA] Brachytheciaceae (B:M)
Rhynchostegium pulchellum (Hedw.) Robins. = Eurhynchium pulchellum [EURHPUL] Brachytheciaceae (B:M)
Rhynchostegium revelstokense (Kindb.) Kindb. = Brachythecium acutum [BRACACT] Brachytheciaceae (B:M)
Rhynchostegium riparioides (Hedw.) Card. in Tourr. = Platyhypnidium riparioides [PLATRIP] Brachytheciaceae (B:M)
Rhynchostegium rusciforme (Brid.) Schimp. in B.S.G. = Platyhypnidium riparioides [PLATRIP] Brachytheciaceae (B:M)
Rhynchostegium serrulatum (Hedw.) Jaeg. = Steerecleus serrulatus [STEESER] Brachytheciaceae (B:M)
Rhynchostegium serrulatum ssp. *eriense* (Kindb.) Kindb. = Steerecleus serrulatus [STEESER] Brachytheciaceae (B:M)
Rhynchostegium serrulatum ssp. *hispidifolium* (Kindb.) Kindb. = Steerecleus serrulatus [STEESER] Brachytheciaceae (B:M)
Rhynchostegium strigosum (Web. & Mohr) De Not. = Eurhynchium pulchellum var. pulchellum [EURHPUL0] Brachytheciaceae (B:M)
Rhynchostegium subintegrifolium (Kindb.) Kindb. = Brachythecium oedipodium [BRACOED] Brachytheciaceae (B:M)
Rhytidiadelphus (Lindb. ex Limpr.) Warnst. [RHYTIDI$] Hylocomiaceae (B:M)
Rhytidiadelphus calvescens (Kindb.) Broth. = Rhytidiadelphus squarrosus [RHYTSQU] Hylocomiaceae (B:M)
Rhytidiadelphus loreus (Hedw.) Warnst. [RHYTLOR] Hylocomiaceae (B:M)
Rhytidiadelphus squarrosus (Hedw.) Warnst. [RHYTSQU] Hylocomiaceae (B:M)
Rhytidiadelphus squarrosus var. *calvescens* (Kindb.) Warnst. = Rhytidiadelphus squarrosus [RHYTSQU] Hylocomiaceae (B:M)

Rhytidiadelphus subpinnatus (Lindb.) T. Kop. = Rhytidiadelphus squarrosus [RHYTSQU] Hylocomiaceae (B:M)
Rhytidiadelphus triquetrus (Hedw.) Warnst. [RHYTTRI] Hylocomiaceae (B:M)
Rhytidiadelphus triquetrus var. *beringianus* (Card. & Thér.) Grout = Rhytidiadelphus triquetrus [RHYTTRI] Hylocomiaceae (B:M)
Rhytidiopsis Broth. [RHYTIDO$] Hylocomiaceae (B:M)
Rhytidiopsis robusta (Hook.) Broth. [RHYTROB] Hylocomiaceae (B:M)
Rhytidium (Sull.) Kindb. [RHYTIDU$] Rhytidiaceae (B:M)
Rhytidium rugosum (Hedw.) Kindb. [RHYTRUG] Rhytidiaceae (B:M)
Ribes L. [RIBES$] Grossulariaceae (V:D)
Ribes acerifolium T.J. Howell [RIBEACE] Grossulariaceae (V:D)
Ribes bracteosum Dougl. ex Hook. [RIBEBRA] Grossulariaceae (V:D)
Ribes cereum Dougl. [RIBECER] Grossulariaceae (V:D)
Ribes cognatum Greene = Ribes oxyacanthoides ssp. cognatum [RIBEOXY1] Grossulariaceae (V:D)
Ribes divaricatum Dougl. [RIBEDIV] Grossulariaceae (V:D)
Ribes divaricatum var. *irriguum* (Dougl.) Gray = Ribes oxyacanthoides ssp. irriguum [RIBEOXY2] Grossulariaceae (V:D)
Ribes glandulosum Grauer [RIBEGLA] Grossulariaceae (V:D)
Ribes howellii Greene = Ribes acerifolium [RIBEACE] Grossulariaceae (V:D)
Ribes hudsonianum Richards. [RIBEHUD] Grossulariaceae (V:D)
Ribes hudsonianum var. **hudsonianum** [RIBEHUD0] Grossulariaceae (V:D)
Ribes hudsonianum var. **petiolare** (Dougl.) Jancz. [RIBEHUD2] Grossulariaceae (V:D)
Ribes inerme Rydb. [RIBEINE] Grossulariaceae (V:D)
Ribes irriguum Dougl. = Ribes oxyacanthoides ssp. irriguum [RIBEOXY2] Grossulariaceae (V:D)
Ribes lacustre (Pers.) Poir. [RIBELAC] Grossulariaceae (V:D)
Ribes lacustre var. *molle* Gray = Ribes montigenum [RIBEMON] Grossulariaceae (V:D)
Ribes lacustre var. *parvulum* Gray = Ribes lacustre [RIBELAC] Grossulariaceae (V:D)
Ribes laxiflorum Pursh [RIBELAX] Grossulariaceae (V:D)
Ribes lentum Coville & Rose = Ribes montigenum [RIBEMON] Grossulariaceae (V:D)
Ribes leucoderme Heller = Ribes oxyacanthoides ssp. irriguum [RIBEOXY2] Grossulariaceae (V:D)
Ribes lobbii Gray [RIBELOB] Grossulariaceae (V:D)
Ribes montigenum McClatchie [RIBEMON] Grossulariaceae (V:D)
Ribes nonscripta (Berger) Standl. = Ribes oxyacanthoides ssp. irriguum [RIBEOXY2] Grossulariaceae (V:D)
Ribes nubigenum McClatchie = Ribes montigenum [RIBEMON] Grossulariaceae (V:D)
Ribes oxyacanthoides L. [RIBEOXY] Grossulariaceae (V:D)
Ribes oxyacanthoides ssp. **cognatum** (Greene) Sinnott [RIBEOXY1] Grossulariaceae (V:D)
Ribes oxyacanthoides ssp. **irriguum** (Dougl.) Sinnott [RIBEOXY2] Grossulariaceae (V:D)
Ribes oxyacanthoides ssp. **oxyacanthoides** [RIBEOXY0] Grossulariaceae (V:D)
Ribes oxyacanthoides var. *irriguum* (Dougl.) Jancz. = Ribes oxyacanthoides ssp. irriguum [RIBEOXY2] Grossulariaceae (V:D)
Ribes oxyacanthoides var. *leucoderme* (Heller) Jancz. = Ribes oxyacanthoides ssp. irriguum [RIBEOXY2] Grossulariaceae (V:D)
Ribes oxycanthoides var. *lacustre* Pers. = Ribes lacustre [RIBELAC] Grossulariaceae (V:D)
Ribes petiolare Dougl. = Ribes hudsonianum var. petiolare [RIBEHUD2] Grossulariaceae (V:D)
Ribes prostratum L'Hér. = Ribes glandulosum [RIBEGLA] Grossulariaceae (V:D)
Ribes reniforme Nutt. = Ribes cereum [RIBECER] Grossulariaceae (V:D)
Ribes resinosum Pursh = Ribes glandulosum [RIBEGLA] Grossulariaceae (V:D)
Ribes rubrum var. *alaskanum* (Berger) Boivin = Ribes triste [RIBETRI] Grossulariaceae (V:D)
Ribes rubrum var. *propinquum* (Turcz.) Trautv. & C.A. Mey. = Ribes triste [RIBETRI] Grossulariaceae (V:D)
Ribes sanguineum Pursh [RIBESAN] Grossulariaceae (V:D)
Ribes triste Pallas [RIBETRI] Grossulariaceae (V:D)
Ribes triste var. *albinervium* (Michx.) Fern. = Ribes triste [RIBETRI] Grossulariaceae (V:D)
Ribes viscidulum Berger = Ribes cereum [RIBECER] Grossulariaceae (V:D)
Ribes viscosissimum Pursh [RIBEVIS] Grossulariaceae (V:D)
Ribes viscosissimum var. *hallii* Jancz. = Ribes viscosissimum [RIBEVIS] Grossulariaceae (V:D)
Riccardia Gray [RICCARD$] Aneuraceae (B:H)
Riccardia chamedryfolia (With.) Grolle [RICCCHA] Aneuraceae (B:H)
Riccardia latifrons Lindb. [RICCLAT] Aneuraceae (B:H)
Riccardia multifida (L.) Gray [RICCMUL] Aneuraceae (B:H)
Riccardia palmata (Hedw.) Carruth. [RICCPAL] Aneuraceae (B:H)
Riccardia pinguis (L.) Gray = Aneura pinguis [ANEUPIN] Aneuraceae (B:H)
Riccardia sinuata (Dicks.) Trev. = Riccardia chamedryfolia [RICCCHA] Aneuraceae (B:H)
Riccia L. [RICCIA$] Ricciaceae (B:H)
Riccia beyrichiana Hampe ex Lehm. [RICCBEY] Ricciaceae (B:H)
Riccia cavernosa Hoffm. [RICCCAV] Ricciaceae (B:H)
Riccia crystallina auct. = Riccia cavernosa [RICCCAV] Ricciaceae (B:H)
Riccia fluitans L. [RICCFLU] Ricciaceae (B:H)
Riccia rhenana Lorbeer = Riccia fluitans [RICCFLU] Ricciaceae (B:H)
Riccia sorocarpa Bisch. [RICCSOR] Ricciaceae (B:H)
Ricciocarpos Corda [RICCIOC$] Ricciaceae (B:H)
Ricciocarpos natans (L.) Corda [RICCNAT] Ricciaceae (B:H)
Rimularia Nyl. [RIMULAR$] Rimulariaceae (L)
Rimularia insularis (Nyl.) Rambold & Hertel [RIMUINS] Rimulariaceae (L)
Rinodina (Ach.) Gray [RINODIN$] Physciaceae (L)
Rinodina archaea (Ach.) Arnold [RINOARC] Physciaceae (L)
Rinodina bolanderi H. Magn. [RINOBOL] Physciaceae (L)
Rinodina calcigena (Th. Fr.) Lynge [RINOCAL] Physciaceae (L)
Rinodina colobina (Ach.) Th. Fr. [RINOCOL] Physciaceae (L)
Rinodina confragosa (Ach.) Körber [RINOCON] Physciaceae (L)
Rinodina conradii Körber [RINOCOR] Physciaceae (L)
Rinodina disjuncta Sheard & Tønsberg [RINODIS] Physciaceae (L)
Rinodina efflorescens Malme [RINOEFF] Physciaceae (L)
Rinodina excrescens Vainio [RINOEXC] Physciaceae (L)
Rinodina exigua (Ach.) Gray [RINOEXI] Physciaceae (L)
Rinodina gennarii Bagl. [RINOGEN] Physciaceae (L)
Rinodina hallii Tuck. [RINOHAI] Physciaceae (L)
Rinodina hueana Vainio = Dimelaena oreina [DIMEORE] Physciaceae (L)
Rinodina intermedia Bagl. = Rinodina conradii [RINOCOR] Physciaceae (L)
Rinodina marysvillensis H. Magn. [RINOMAR] Physciaceae (L)
Rinodina mniaraea (Ach.) Körber [RINOMNI] Physciaceae (L)
Rinodina mucronatula H. Magn. [RINOMUC] Physciaceae (L)
Rinodina nimbosa (Fr.) Th. Fr. = Phaeorrhiza nimbosa [PHAENIM] Physciaceae (L)
Rinodina novomexicana de Lesd. = Dimelaena oreina [DIMEORE] Physciaceae (L)
Rinodina occidentalis Lynge = Rinodina calcigena [RINOCAL] Physciaceae (L)
Rinodina orbata (Ach.) Vainio = Rinodina turfacea [RINOTUR] Physciaceae (L)
Rinodina oreina (Ach.) A. Massal. = Dimelaena oreina [DIMEORE] Physciaceae (L)
Rinodina phaeocarpa (Sommerf.) Vainio = Phaeorrhiza nimbosa [PHAENIM] Physciaceae (L)
Rinodina pyrina (Ach.) Arnold [RINOPYR] Physciaceae (L)
Rinodina roscida (Sommerf.) Arnold [RINOROS] Physciaceae (L)
Rinodina sabulosa Tuck. = Rinodina conradii [RINOCOR] Physciaceae (L)
Rinodina salina Degel. = Rinodina gennarii [RINOGEN] Physciaceae (L)

Rinodina sareptana (Tomin) H. Magn. = Phaeorrhiza sareptana [PHAESAR] Physciaceae (L)
Rinodina septentrionalis Malme [RINOSEP] Physciaceae (L)
Rinodina suboreina de Lesd. = Dimelaena oreina [DIMEORE] Physciaceae (L)
Rinodina thysanota Tuck. = Dimelaena thysanota [DIMETHY] Physciaceae (L)
Rinodina turfacea (Wahlenb.) Körber [RINOTUR] Physciaceae (L)
Robertiella robertiana (L.) Hanks = Geranium robertianum [GERAROB] Geraniaceae (V:D)
Robinia L. [ROBINIA$] Fabaceae (V:D)
****Robinia pseudoacacia** L. [ROBIPSE] Fabaceae (V:D)
Robinia pseudoacacia var. *rectissima* (L.) Raber = Robinia pseudoacacia [ROBIPSE] Fabaceae (V:D)
Robinsonia haldaniana (Grev.) Crum = Callicladium haldanianum [CALLHAL] Hypnaceae (B:M)
Roegneria albicans (Scribn. & J.G. Sm.) Beetle = Elymus lanceolatus ssp. albicans [ELYMLAN1] Poaceae (V:M)
Roegneria albicans var. *griffithii* (Scribn. & J.G. Sm. ex Piper) Beetle = Elymus lanceolatus ssp. albicans [ELYMLAN1] Poaceae (V:M)
Roegneria pauciflora (Schwein.) Hyl. = Elymus trachycaulus ssp. trachycaulus [ELYMTRA0] Poaceae (V:M)
Roegneria spicata (Pursh) Beetle = Pseudoroegneria spicata [PSEUSPI] Poaceae (V:M)
Roegneria trachycaula (Link) Nevski = Elymus trachycaulus ssp. trachycaulus [ELYMTRA0] Poaceae (V:M)
Roellia Kindb. [ROELLIA$] Bryaceae (B:M)
Roellia roellii (Broth. in Röll) Andrews ex Crum [ROELROE] Bryaceae (B:M)
Roellia simplex Kindb. = Roellia roellii [ROELROE] Bryaceae (B:M)
Romanzoffia Cham. [ROMANZO$] Hydrophyllaceae (V:D)
Romanzoffia sitchensis Bong. [ROMASIT] Hydrophyllaceae (V:D)
Romanzoffia tracyi Jepson [ROMATRA] Hydrophyllaceae (V:D)
Ropalospora A. Massal. [ROPALOS$] Fuscideaceae (L)
Ropalospora viridis (Tønsberg) Tønsberg [ROPAVIR] Fuscideaceae (L)
Rorippa Scop. [RORIPPA$] Brassicaceae (V:D)
Rorippa armoracia (L.) A.S. Hitchc. = Armoracia rusticana [ARMORUS] Brassicaceae (V:D)
Rorippa barbareifolia (DC.) Kitagawa [RORIBAR] Brassicaceae (V:D)
Rorippa curvipes Greene [RORICUR] Brassicaceae (V:D)
Rorippa curvipes var. **integra** (Rydb.) R. Stuckey [RORICUR2] Brassicaceae (V:D)
Rorippa curvisiliqua (Hook.) Bess. ex Britt. [RORICUV] Brassicaceae (V:D)
Rorippa hispida var. *barbareifolia* (DC.) Hultén = Rorippa barbareifolia [RORIBAR] Brassicaceae (V:D)
Rorippa islandica var. *barbareifolia* (DC.) Welsh = Rorippa barbareifolia [RORIBAR] Brassicaceae (V:D)
****Rorippa microphylla** (Boenn. ex Reichenb.) Hyl. ex A. & D. Löve [RORIMIR] Brassicaceae (V:D)
****Rorippa nasturtium-aquaticum** (L.) Hayek [RORINAS] Brassicaceae (V:D)
Rorippa nasturtium-aquaticum var. *longisiliqua* (Irmisch) Boivin = Rorippa microphylla [RORIMIR] Brassicaceae (V:D)
Rorippa obtusa var. *integra* (Rydb.) Victorin = Rorippa curvipes var. integra [RORICUR2] Brassicaceae (V:D)
Rorippa palustris (L.) Bess. [RORIPAL] Brassicaceae (V:D)
****Rorippa sylvestris** (L.) Bess. [RORISYL] Brassicaceae (V:D)
Rosa L. [ROSA$] Rosaceae (V:D)
Rosa acicularis Lindl. [ROSAACI] Rosaceae (V:D)
Rosa acicularis ssp. **sayi** (Schwein.) W.H. Lewis [ROSAACI1] Rosaceae (V:D)
Rosa acicularis var. *bourgeauiana* (Crépin) Crépin = Rosa acicularis ssp. sayi [ROSAACI1] Rosaceae (V:D)
Rosa acicularis var. *sayana* Erlanson = Rosa acicularis ssp. sayi [ROSAACI1] Rosaceae (V:D)
Rosa anatonensis St. John = Rosa nutkana var. hispida [ROSANUT1] Rosaceae (V:D)
Rosa arizonica Rydb. = Rosa woodsii var. ultramontana [ROSAWOO1] Rosaceae (V:D)
Rosa arizonica var. *granulifera* (Rydb.) Kearney & Peebles = Rosa woodsii var. ultramontana [ROSAWOO1] Rosaceae (V:D)
Rosa arkansana Porter [ROSAARK] Rosaceae (V:D)
Rosa bourgeauiana Crépin = Rosa acicularis ssp. sayi [ROSAACI1] Rosaceae (V:D)
Rosa caeruleimontana St. John = Rosa nutkana var. hispida [ROSANUT1] Rosaceae (V:D)
****Rosa canina** L. [ROSACAN] Rosaceae (V:D)
Rosa canina var. *dumetorum* Baker = Rosa canina [ROSACAN] Rosaceae (V:D)
Rosa collaris Rydb. = Rosa acicularis ssp. sayi [ROSAACI1] Rosaceae (V:D)
Rosa covillei Greene = Rosa woodsii var. ultramontana [ROSAWOO1] Rosaceae (V:D)
Rosa durandii Crépin = Rosa nutkana var. nutkana [ROSANUT0] Rosaceae (V:D)
****Rosa eglanteria** L. [ROSAEGL] Rosaceae (V:D)
Rosa engelmannii S. Wats. = Rosa acicularis ssp. sayi [ROSAACI1] Rosaceae (V:D)
Rosa gymnocarpa Nutt. [ROSAGYM] Rosaceae (V:D)
Rosa jonesii St. John = Rosa nutkana var. hispida [ROSANUT1] Rosaceae (V:D)
Rosa lapwaiensis St. John = Rosa woodsii var. ultramontana [ROSAWOO1] Rosaceae (V:D)
Rosa macdougalii Holz. = Rosa nutkana var. hispida [ROSANUT1] Rosaceae (V:D)
Rosa macounii Greene = Rosa woodsii var. ultramontana [ROSAWOO1] Rosaceae (V:D)
Rosa megalantha G.N. Jones = Rosa nutkana var. hispida [ROSANUT1] Rosaceae (V:D)
Rosa nutkana K. Presl [ROSANUT] Rosaceae (V:D)
Rosa nutkana var. **hispida** Fern. [ROSANUT1] Rosaceae (V:D)
Rosa nutkana var. **nutkana** [ROSANUT0] Rosaceae (V:D)
Rosa pecosensis Cockerell = Rosa woodsii var. ultramontana [ROSAWOO1] Rosaceae (V:D)
Rosa pisocarpa Gray [ROSAPIS] Rosaceae (V:D)
Rosa pisocarpa var. *rivalis* (Eastw.) Jepson = Rosa pisocarpa [ROSAPIS] Rosaceae (V:D)
Rosa rivalis Eastw. = Rosa pisocarpa [ROSAPIS] Rosaceae (V:D)
Rosa rubiginosa L. = Rosa eglanteria [ROSAEGL] Rosaceae (V:D)
Rosa sayi Schwein. = Rosa acicularis ssp. sayi [ROSAACI1] Rosaceae (V:D)
Rosa spaldingii Crépin = Rosa nutkana var. hispida [ROSANUT1] Rosaceae (V:D)
Rosa spaldingii var. *alta* (Suksdorf) G.N. Jones = Rosa nutkana var. hispida [ROSANUT1] Rosaceae (V:D)
Rosa spaldingii var. *hispida* (Fern.) G.N. Jones = Rosa nutkana var. hispida [ROSANUT1] Rosaceae (V:D)
Rosa spaldingii var. *parkeri* (S. Wats.) St. John = Rosa nutkana var. hispida [ROSANUT1] Rosaceae (V:D)
Rosa ultramontana (S. Wats.) Heller = Rosa woodsii var. ultramontana [ROSAWOO1] Rosaceae (V:D)
Rosa woodsii Lindl. [ROSAWOO] Rosaceae (V:D)
Rosa woodsii ssp. *ultramontana* (S. Wats.) Taylor & MacBryde = Rosa woodsii var. ultramontana [ROSAWOO1] Rosaceae (V:D)
Rosa woodsii var. *arizonica* (Rydb.) W.C. Martin & C.R. Hutchins = Rosa woodsii var. ultramontana [ROSAWOO1] Rosaceae (V:D)
Rosa woodsii var. *granulifera* (Rydb.) W.C. Martin & C.R. Hutchins = Rosa woodsii var. ultramontana [ROSAWOO1] Rosaceae (V:D)
Rosa woodsii var. *macounii* (Greene) W.C. Martin & C.R. Hutchins = Rosa woodsii var. ultramontana [ROSAWOO1] Rosaceae (V:D)
Rosa woodsii var. **ultramontana** (S. Wats.) Jepson [ROSAWOO1] Rosaceae (V:D)
Roselliniella Vainio [ROSELLI$] Uncertain-family (L)
Roselliniella nephromatis (Crouan) Matzer & Hafellner [ROSENEP] Uncertain-family (L)
Rotala L. [ROTALA$] Lythraceae (V:D)
Rotala catholica (Cham. & Schlecht.) van Leeuwen = Rotala ramosior [ROTARAM] Lythraceae (V:D)
Rotala dentifera (Gray) Koehne = Rotala ramosior [ROTARAM] Lythraceae (V:D)
Rotala ramosior (L.) Koehne [ROTARAM] Lythraceae (V:D)

Rotala ramosior var. *interior* Fern. & Grisc. = Rotala ramosior [ROTARAM] Lythraceae (V:D)
Rotala ramosior var. *typica* Fern. & Grisc. = Rotala ramosior [ROTARAM] Lythraceae (V:D)
Rubus L. [RUBUS$] Rosaceae (V:D)
Rubus acaulis Michx. = Rubus arcticus ssp. acaulis [RUBUARC1] Rosaceae (V:D)
*****Rubus allegheniensis** Porter [RUBUALL] Rosaceae (V:D)
Rubus arcticus L. [RUBUARC] Rosaceae (V:D)
Rubus arcticus ssp. **acaulis** (Michx.) Focke [RUBUARC1] Rosaceae (V:D)
Rubus arcticus ssp. **stellatus** (Sm.) Boivin [RUBUARC2] Rosaceae (V:D)
Rubus arcticus var. *acaulis* (Michx.) Boivin = Rubus arcticus ssp. acaulis [RUBUARC1] Rosaceae (V:D)
Rubus arcticus var. *stellatus* (Sm.) Boivin = Rubus arcticus ssp. stellatus [RUBUARC2] Rosaceae (V:D)
Rubus carolinianus Rydb. = Rubus idaeus ssp. strigosus [RUBUIDA1] Rosaceae (V:D)
Rubus chamaemorus L. [RUBUCHA] Rosaceae (V:D)
*****Rubus discolor** Weihe & Nees [RUBUDIS] Rosaceae (V:D)
Rubus idaeus L. [RUBUIDA] Rosaceae (V:D)
Rubus idaeus ssp. *melanolasius* (Dieck) Focke = Rubus idaeus ssp. strigosus [RUBUIDA1] Rosaceae (V:D)
Rubus idaeus ssp. *sachalinensis* (Levl.) Focke = Rubus idaeus ssp. strigosus [RUBUIDA1] Rosaceae (V:D)
Rubus idaeus ssp. **strigosus** (Michx.) Focke [RUBUIDA1] Rosaceae (V:D)
Rubus idaeus var. *aculeatissimus* Regel & Tiling = Rubus idaeus ssp. strigosus [RUBUIDA1] Rosaceae (V:D)
Rubus idaeus var. *canadensis* Richards. = Rubus idaeus ssp. strigosus [RUBUIDA1] Rosaceae (V:D)
Rubus idaeus var. *gracilipes* M.E. Jones = Rubus idaeus ssp. strigosus [RUBUIDA1] Rosaceae (V:D)
Rubus idaeus var. *melanolasius* (Dieck) R.J. Davis = Rubus idaeus ssp. strigosus [RUBUIDA1] Rosaceae (V:D)
Rubus idaeus var. *melanotrachys* (Focke) Fern. = Rubus idaeus ssp. strigosus [RUBUIDA1] Rosaceae (V:D)
Rubus idaeus var. *strigosus* (Michx.) Maxim. = Rubus idaeus ssp. strigosus [RUBUIDA1] Rosaceae (V:D)
*****Rubus laciniatus** Willd. [RUBULAC] Rosaceae (V:D)
Rubus lasiococcus Gray [RUBULAS] Rosaceae (V:D)
Rubus leucodermis Dougl. ex Torr. & Gray [RUBULEU] Rosaceae (V:D)
Rubus macropetalus Dougl. ex Hook. = Rubus ursinus ssp. macropetalus [RUBUURS1] Rosaceae (V:D)
Rubus melanolasius Dieck = Rubus idaeus ssp. strigosus [RUBUIDA1] Rosaceae (V:D)
Rubus neglectus Peck = Rubus idaeus ssp. strigosus [RUBUIDA1] Rosaceae (V:D)
Rubus nivalis Dougl. ex Hook. [RUBUNIV] Rosaceae (V:D)
Rubus parviflorus Nutt. [RUBUPAR] Rosaceae (V:D)
Rubus pedatus Sm. [RUBUPED] Rosaceae (V:D)
Rubus procerus P.J. Muell. = Rubus discolor [RUBUDIS] Rosaceae (V:D)
Rubus pubescens Raf. [RUBUPUB] Rosaceae (V:D)
Rubus spectabilis Pursh [RUBUSPE] Rosaceae (V:D)
Rubus stellatus Sm. = Rubus arcticus ssp. stellatus [RUBUARC2] Rosaceae (V:D)
Rubus strigosus Michx. = Rubus idaeus ssp. strigosus [RUBUIDA1] Rosaceae (V:D)
Rubus strigosus var. *acalyphacea* (Greene) Bailey = Rubus idaeus ssp. strigosus [RUBUIDA1] Rosaceae (V:D)
Rubus strigosus var. *arizonicus* (Greene) Kearney & Peebles = Rubus idaeus ssp. strigosus [RUBUIDA1] Rosaceae (V:D)
Rubus strigosus var. *canadensis* (Richards.) House = Rubus idaeus ssp. strigosus [RUBUIDA1] Rosaceae (V:D)
Rubus ursinus Cham. & Schlecht. [RUBUURS] Rosaceae (V:D)
Rubus ursinus ssp. **macropetalus** (Dougl. ex Hook.) Taylor & MacBryde [RUBUURS1] Rosaceae (V:D)
Rubus ursinus var. *macropetalus* (Dougl. ex Hook.) S.W. Br. = Rubus ursinus ssp. macropetalus [RUBUURS1] Rosaceae (V:D)
Rudbeckia L. [RUDBECK$] Asteraceae (V:D)
Rudbeckia columnaris Sims = Ratibida columnifera [RATICOL] Asteraceae (V:D)
Rudbeckia columnaris Pursh = Ratibida columnifera [RATICOL] Asteraceae (V:D)
Rudbeckia columnifera Nutt. = Ratibida columnifera [RATICOL] Asteraceae (V:D)
*****Rudbeckia hirta** L. [RUDBHIR] Asteraceae (V:D)
Rumex L. [RUMEX$] Polygonaceae (V:D)
Rumex acetosa L. [RUMEACE] Polygonaceae (V:D)
*****Rumex acetosa** ssp. **acetosa** [RUMEACE0] Polygonaceae (V:D)
Rumex acetosa ssp. **alpestris** (Jacq.) A. Löve [RUMEACE3] Polygonaceae (V:D)
Rumex acetosa ssp. **arifolius** (All.) Blytt & O. Dahl [RUMEACE2] Polygonaceae (V:D)
*****Rumex acetosella** L. [RUMEACT] Polygonaceae (V:D)
Rumex acetosella ssp. *angiocarpus* (Murb.) Murb. = Rumex acetosella [RUMEACT] Polygonaceae (V:D)
Rumex acetosella var. *pyrenaeus* (Pourret) Timbal-Lagrave = Rumex acetosella [RUMEACT] Polygonaceae (V:D)
Rumex acetosella var. *tenuifolius* Wallr. = Rumex acetosella [RUMEACT] Polygonaceae (V:D)
Rumex alpestris Jacq. = Rumex acetosa ssp. alpestris [RUMEACE3] Polygonaceae (V:D)
Rumex angiocarpus Murb. = Rumex acetosella [RUMEACT] Polygonaceae (V:D)
Rumex aquaticus L. [RUMEAQU] Polygonaceae (V:D)
Rumex aquaticus ssp. *fenestratus* (Greene) Hultén = Rumex aquaticus var. fenestratus [RUMEAQU1] Polygonaceae (V:D)
Rumex aquaticus ssp. *occidentalis* (S. Wats.) Hultén = Rumex aquaticus var. fenestratus [RUMEAQU1] Polygonaceae (V:D)
Rumex aquaticus var. **fenestratus** (Greene) Dorn [RUMEAQU1] Polygonaceae (V:D)
Rumex arcticus Trautv. [RUMEARC] Polygonaceae (V:D)
*****Rumex conglomeratus** Murr. [RUMECON] Polygonaceae (V:D)
*****Rumex crispus** L. [RUMECRI] Polygonaceae (V:D)
Rumex digyna L. = Oxyria digyna [OXYRDIG] Polygonaceae (V:D)
Rumex fenestratus Greene = Rumex aquaticus var. fenestratus [RUMEAQU1] Polygonaceae (V:D)
Rumex fueginus Phil. = Rumex maritimus [RUMEMAR] Polygonaceae (V:D)
Rumex maritimus L. [RUMEMAR] Polygonaceae (V:D)
Rumex maritimus ssp. *fueginus* (Phil.) Hultén = Rumex maritimus [RUMEMAR] Polygonaceae (V:D)
Rumex maritimus var. *athrix* St. John = Rumex maritimus [RUMEMAR] Polygonaceae (V:D)
Rumex maritimus var. *fueginus* (Phil.) Dusen = Rumex maritimus [RUMEMAR] Polygonaceae (V:D)
Rumex maritimus var. *persicarioides* (L.) R.S. Mitchell = Rumex maritimus [RUMEMAR] Polygonaceae (V:D)
Rumex mexicanus Meisn. = Rumex salicifolius var. mexicanus [RUMESAL1] Polygonaceae (V:D)
Rumex mexicanus var. *angustifolius* (Meisn.) Boivin = Rumex salicifolius var. mexicanus [RUMESAL1] Polygonaceae (V:D)
Rumex mexicanus var. *subarcticus* (Lepage) Boivin = Rumex salicifolius var. mexicanus [RUMESAL1] Polygonaceae (V:D)
Rumex mexicanus var. *transitorius* (Rech. f.) Boivin = Rumex salicifolius var. transitorius [RUMESAL2] Polygonaceae (V:D)
*****Rumex obtusifolius** L. [RUMEOBT] Polygonaceae (V:D)
Rumex obtusifolius ssp. *agrestis* (Fries) Danser = Rumex obtusifolius [RUMEOBT] Polygonaceae (V:D)
Rumex obtusifolius ssp. *sylvestris* (Wallr.) Rech. f. = Rumex obtusifolius [RUMEOBT] Polygonaceae (V:D)
Rumex obtusifolius var. *sylvestris* (Wallr.) Koch = Rumex obtusifolius [RUMEOBT] Polygonaceae (V:D)
Rumex occidentalis S. Wats. = Rumex aquaticus var. fenestratus [RUMEAQU1] Polygonaceae (V:D)
Rumex occidentalis var. *fenestratus* (Greene) Lepage = Rumex aquaticus var. fenestratus [RUMEAQU1] Polygonaceae (V:D)
Rumex occidentalis var. *labradoricus* (Rech. f.) Lepage = Rumex aquaticus var. fenestratus [RUMEAQU1] Polygonaceae (V:D)
Rumex occidentalis var. *procerus* (Greene) J.T. Howell = Rumex aquaticus var. fenestratus [RUMEAQU1] Polygonaceae (V:D)
*****Rumex patientia** L. [RUMEPAT] Polygonaceae (V:D)

Rumex paucifolius Nutt. [RUMEPAU] Polygonaceae (V:D)
Rumex persicarioides L. = Rumex maritimus [RUMEMAR] Polygonaceae (V:D)
Rumex quadrangulivalvis (Danser) Rech. f. = Rumex salicifolius var. mexicanus [RUMESAL1] Polygonaceae (V:D)
Rumex salicifolius Weinm. [RUMESAL] Polygonaceae (V:D)
Rumex salicifolius ssp. *triangulivalvis* Danser = Rumex salicifolius var. mexicanus [RUMESAL1] Polygonaceae (V:D)
Rumex salicifolius ssp. *triangulivalvis* var. *mexicanus* (Meisn.) C.L. Hitchc. = Rumex salicifolius var. mexicanus [RUMESAL1] Polygonaceae (V:D)
Rumex salicifolius var. **mexicanus** (Meisn.) C.L. Hitchc. [RUMESAL1] Polygonaceae (V:D)
Rumex salicifolius var. **salicifolius** [RUMESAL0] Polygonaceae (V:D)
Rumex salicifolius var. **transitorius** (Rech. f.) Hickman [RUMESAL2] Polygonaceae (V:D)
Rumex salicifolius var. *triangulivalvis* (Danser) C.L. Hitchc. = Rumex salicifolius var. mexicanus [RUMESAL1] Polygonaceae (V:D)
Rumex tenuifolius (Wallr.) A. Löve = Rumex acetosella [RUMEACT] Polygonaceae (V:D)
Rumex transitorius Rech. f. = Rumex salicifolius var. transitorius [RUMESAL2] Polygonaceae (V:D)
Rumex triangulivalvis (Danser) Rech. f. = Rumex salicifolius var. mexicanus [RUMESAL1] Polygonaceae (V:D)
Rumex triangulivalvis var. *oreolapathum* Rech. f. = Rumex salicifolius var. mexicanus [RUMESAL1] Polygonaceae (V:D)
Rupertia J. Grimes [RUPERTI$] Fabaceae (V:D)
Rupertia physodes (Dougl. ex Hook.) J. Grimes [RUPEPHY] Fabaceae (V:D)
Ruppia L. [RUPPIA$] Ruppiaceae (V:M)
Ruppia cirrhosa ssp. *occidentalis* (S. Wats.) A. & D. Löve = Ruppia maritima [RUPPMAR] Ruppiaceae (V:M)
Ruppia maritima L. [RUPPMAR] Ruppiaceae (V:M)
Ruppia maritima var. *brevirostris* Agardh = Ruppia maritima [RUPPMAR] Ruppiaceae (V:M)
Ruppia maritima var. *exigua* Fern. & Wieg. = Ruppia maritima [RUPPMAR] Ruppiaceae (V:M)
Ruppia maritima var. *intermedia* (Thed.) Aschers. & Graebn. = Ruppia maritima [RUPPMAR] Ruppiaceae (V:M)
Ruppia maritima var. *longipes* Hagstr. = Ruppia maritima [RUPPMAR] Ruppiaceae (V:M)
Ruppia maritima var. *obliqua* (Schur) Aschers. & Graebn. = Ruppia maritima [RUPPMAR] Ruppiaceae (V:M)
Ruppia maritima var. *occidentalis* (S. Wats.) Graebn. = Ruppia maritima [RUPPMAR] Ruppiaceae (V:M)
Ruppia maritima var. *pacifica* St. John & Fosberg = Ruppia maritima [RUPPMAR] Ruppiaceae (V:M)
Ruppia maritima var. *rostrata* Agardh = Ruppia maritima [RUPPMAR] Ruppiaceae (V:M)
Ruppia maritima var. *spiralis* Morris = Ruppia maritima [RUPPMAR] Ruppiaceae (V:M)
Ruppia maritima var. *subcapitata* Fern. & Wieg. = Ruppia maritima [RUPPMAR] Ruppiaceae (V:M)
Ruppia occidentalis S. Wats. = Ruppia maritima [RUPPMAR] Ruppiaceae (V:M)
Ruppia pectinata Rydb. = Ruppia maritima [RUPPMAR] Ruppiaceae (V:M)

S

Sabina horizontalis (Moench) Rydb. = Juniperus horizontalis [JUNIHOR] Cupressaceae (V:G)
Sabina prostrata (Pers.) Antoine = Juniperus horizontalis [JUNIHOR] Cupressaceae (V:G)
Sabina scopulorum (Sarg.) Rydb. = Juniperus scopulorum [JUNISCO] Cupressaceae (V:G)
Saccobasis polita (Nees) Buch = Tritomaria polita [TRITPOL] Jungermanniaceae (B:H)
Saccomorpha icmalea (Ach.) Clauzade & Roux = Placynthiella icmalea [PLACICM] Trapeliaceae (L)
Saccomorpha oligotropha (J.R. Laundon) Clauzade & Roux = Placynthiella oligotropha [PLACOLI] Trapeliaceae (L)
Saccomorpha uliginosa (Schrader) Hafellner = Placynthiella uliginosa [PLACULI] Trapeliaceae (L)
Saelania Lindb. [SAELANI$] Ditrichaceae (B:M)
Saelania glaucescens (Hedw.) Broth. in Bomanss. & Broth. [SAELGLA] Ditrichaceae (B:M)
Sagedia cinerea (Pers.) Fr. = Catapyrenium cinereum [CATACIN] Verrucariaceae (L)
Sagina L. [SAGINA$] Caryophyllaceae (V:D)
Sagina crassicaulis S. Wats. = Sagina maxima ssp. crassicaulis [SAGIMAX1] Caryophyllaceae (V:D)
Sagina crassicaulis var. *littoralis* (Hultén) Hultén = Sagina maxima ssp. maxima [SAGIMAX0] Caryophyllaceae (V:D)
Sagina decumbens (Ell.) Torr. & Gray [SAGIDEC] Caryophyllaceae (V:D)
Sagina decumbens ssp. **occidentalis** (S. Wats.) Crow [SAGIDEC1] Caryophyllaceae (V:D)
Sagina intermedia Fenzl = Sagina nivalis [SAGINIV] Caryophyllaceae (V:D)
***Sagina japonica** (Sw.) Ohwi [SAGIJAP] Caryophyllaceae (V:D)
Sagina linnaei K. Presl = Sagina saginoides [SAGISAG] Caryophyllaceae (V:D)
Sagina littoralis Hultén = Sagina maxima ssp. maxima [SAGIMAX0] Caryophyllaceae (V:D)
Sagina maxima Gray [SAGIMAX] Caryophyllaceae (V:D)
Sagina maxima ssp. **crassicaulis** (S. Wats.) Crow [SAGIMAX1] Caryophyllaceae (V:D)
Sagina maxima ssp. **maxima** [SAGIMAX0] Caryophyllaceae (V:D)
Sagina maxima var. *littorea* (Mackenzie) Hara = Sagina maxima ssp. maxima [SAGIMAX0] Caryophyllaceae (V:D)
Sagina micrantha (Bunge) Fern. = Sagina saginoides [SAGISAG] Caryophyllaceae (V:D)
Sagina nivalis (Lindbl.) Fries [SAGINIV] Caryophyllaceae (V:D)
Sagina occidentalis S. Wats. = Sagina decumbens ssp. occidentalis [SAGIDEC1] Caryophyllaceae (V:D)
Sagina procumbens L. [SAGIPRO] Caryophyllaceae (V:D)
Sagina procumbens var. *compacta* Lange = Sagina procumbens [SAGIPRO] Caryophyllaceae (V:D)
Sagina saginoides (L.) Karst. [SAGISAG] Caryophyllaceae (V:D)
Sagina saginoides var. *hesperia* Fern. = Sagina saginoides [SAGISAG] Caryophyllaceae (V:D)
Sagittaria L. [SAGITTA$] Alismataceae (V:M)
Sagittaria arifolia Nutt. ex J.G. Sm. = Sagittaria cuneata [SAGICUN] Alismataceae (V:M)
Sagittaria cuneata Sheldon [SAGICUN] Alismataceae (V:M)
Sagittaria engelmanniana ssp. *longirostra* (Micheli) Bogin = Sagittaria latifolia [SAGILAT] Alismataceae (V:M)
Sagittaria esculenta T.J. Howell = Sagittaria latifolia [SAGILAT] Alismataceae (V:M)
Sagittaria latifolia Willd. [SAGILAT] Alismataceae (V:M)
Sagittaria latifolia var. *obtusa* (Engelm.) Wieg. = Sagittaria latifolia [SAGILAT] Alismataceae (V:M)
Sagittaria longirostra (Micheli) J.G. Sm. = Sagittaria latifolia [SAGILAT] Alismataceae (V:M)
Sagittaria obtusa Muhl. ex Willd. = Sagittaria latifolia [SAGILAT] Alismataceae (V:M)
Sagittaria ornithorhyncha Small = Sagittaria latifolia [SAGILAT] Alismataceae (V:M)
Sagittaria planipes Fern. = Sagittaria latifolia [SAGILAT] Alismataceae (V:M)

Sagittaria variabilis var. *obtusa* Engelm. = Sagittaria latifolia [SAGILAT] Alismataceae (V:M)
Sagittaria viscosa C. Mohr p.p. = Sagittaria latifolia [SAGILAT] Alismataceae (V:M)
Salicornia L. [SALICOR$] Chenopodiaceae (V:D)
Salicornia depressa Standl. = Salicornia virginica [SALIVIR] Chenopodiaceae (V:D)
Salicornia europaea auct. = Salicornia maritima [SALIMAR] Chenopodiaceae (V:D)
Salicornia europaea ssp. *rubra* (A. Nels.) Breitung = Salicornia rubra [SALIRUB] Chenopodiaceae (V:D)
Salicornia europaea var. *prona* (Lunell) Boivin = Salicornia rubra [SALIRUB] Chenopodiaceae (V:D)
Salicornia maritima Wolff & Jefferies [SALIMAR] Chenopodiaceae (V:D)
Salicornia pacifica Standl. = Sarcocornia pacifica [SARCPAC] Chenopodiaceae (V:D)
Salicornia rubra A. Nels. [SALIRUB] Chenopodiaceae (V:D)
Salicornia virginica L. [SALIVIR] Chenopodiaceae (V:D)
Salix L. [SALIX$] Salicaceae (V:D)
Salix alaxensis (Anderss.) Coville [SALIALA] Salicaceae (V:D)
Salix alaxensis ssp. *longistylis* (Rydb.) Hultén = Salix alaxensis var. longistylis [SALIALA2] Salicaceae (V:D)
Salix alaxensis var. **alaxensis** [SALIALA0] Salicaceae (V:D)
Salix alaxensis var. **longistylis** (Rydb.) Schneid. [SALIALA2] Salicaceae (V:D)
Salix alaxensis var. *obovalifolia* Ball = Salix alaxensis var. alaxensis [SALIALA0] Salicaceae (V:D)
*__Salix alba__ L. [SALIALB] Salicaceae (V:D)
Salix alba ssp. *vitellina* (L.) Arcang. = Salix alba var. vitellina [SALIALB1] Salicaceae (V:D)
*__Salix alba__ var. **vitellina** (L.) Stokes [SALIALB1] Salicaceae (V:D)
Salix albertana Rowlee = Salix barrattiana [SALIBAA] Salicaceae (V:D)
Salix aliena Flod. = Salix setchelliana [SALISET] Salicaceae (V:D)
Salix amplifolia Coville = Salix hookeriana [SALIHOO] Salicaceae (V:D)
Salix amygdaloides Anderss. [SALIAMY] Salicaceae (V:D)
Salix amygdaloides var. *wrightii* (Anderss.) Schneid. = Salix amygdaloides [SALIAMY] Salicaceae (V:D)
Salix ancorifera Fern. = Salix discolor [SALIDIS] Salicaceae (V:D)
Salix anglorum var. *antiplasta* Schneid. = Salix arctica [SALIARC] Salicaceae (V:D)
Salix anglorum var. *araioclada* Schneid. = Salix arctica [SALIARC] Salicaceae (V:D)
Salix anglorum var. *kophophylla* Schneid. = Salix arctica [SALIARC] Salicaceae (V:D)
Salix arbusculoides Anderss. [SALIARB] Salicaceae (V:D)
Salix arbusculoides var. *glabra* (Anderss.) Anderss. ex Schneid. = Salix planifolia ssp. pulchra [SALIPLA2] Salicaceae (V:D)
Salix arctica Pallas [SALIARC] Salicaceae (V:D)
Salix arctica R. Br. ex Richards. = Salix arctica [SALIARC] Salicaceae (V:D)
Salix arctica ssp. *crassijulis* (Trautv.) Skvort. = Salix arctica [SALIARC] Salicaceae (V:D)
Salix arctica ssp. *petraea* (Anderss.) A. & D. Löve & Kapoor = Salix arctica [SALIARC] Salicaceae (V:D)
Salix arctica ssp. *tortulosa* (Trautv.) Hultén = Salix arctica [SALIARC] Salicaceae (V:D)
Salix arctica var. *antiplasta* (Schneid.) Fern. = Salix arctica [SALIARC] Salicaceae (V:D)
Salix arctica var. *araioclada* (Schneid.) Raup = Salix arctica [SALIARC] Salicaceae (V:D)
Salix arctica var. *brownei* Anderss. = Salix arctica [SALIARC] Salicaceae (V:D)
Salix arctica var. *caespitosa* (Kennedy) L. Kelso = Salix arctica [SALIARC] Salicaceae (V:D)
Salix arctica var. *graminifolia* (E.H. Kelso) L. Kelso = Salix arctica [SALIARC] Salicaceae (V:D)
Salix arctica var. *kophophylla* (Schneid.) Polunin = Salix arctica [SALIARC] Salicaceae (V:D)
Salix arctica var. *pallasii* (Anderss.) Kurtz = Salix arctica [SALIARC] Salicaceae (V:D)
Salix arctica var. *petraea* Anderss. = Salix arctica [SALIARC] Salicaceae (V:D)
Salix arctica var. *petrophila* (Rydb.) L. Kelso = Salix arctica [SALIARC] Salicaceae (V:D)
Salix arctica var. *subcordata* (Anderss.) Schneid. = Salix glauca var. villosa [SALIGLA2] Salicaceae (V:D)
Salix arctica var. *tortulosa* (Trautv.) Raup = Salix arctica [SALIARC] Salicaceae (V:D)
Salix argophylla Nutt. = Salix exigua [SALIEXI] Salicaceae (V:D)
Salix arguta Anderss. = Salix lucida ssp. lasiandra [SALILUC2] Salicaceae (V:D)
Salix arguta var. *alpigena* Anderss. = Salix serissima [SALISER] Salicaceae (V:D)
Salix arguta var. *erythrocoma* (Anderss.) Anderss. = Salix lucida ssp. lasiandra [SALILUC2] Salicaceae (V:D)
Salix arguta var. *lasiandra* (Benth.) Anderss. = Salix lucida ssp. lasiandra [SALILUC2] Salicaceae (V:D)
Salix arguta var. *pallescens* Anderss. = Salix serissima [SALISER] Salicaceae (V:D)
Salix athabascensis Raup [SALIATH] Salicaceae (V:D)
Salix austinae Bebb = Salix lemmonii [SALILEM] Salicaceae (V:D)
Salix balsamifera (Hook.) Barratt ex Anderss. = Salix pyrifolia [SALIPYR] Salicaceae (V:D)
Salix balsamifera var. *alpestris* Bebb = Salix pyrifolia [SALIPYR] Salicaceae (V:D)
Salix balsamifera var. *lanceolata* Bebb = Salix pyrifolia [SALIPYR] Salicaceae (V:D)
Salix balsamifera var. *vegeta* Bebb = Salix pyrifolia [SALIPYR] Salicaceae (V:D)
Salix barclayi Anderss. [SALIBAR] Salicaceae (V:D)
Salix barclayi var. *angustifolia* (Anderss.) Anderss. ex Schneid. = Salix barclayi [SALIBAR] Salicaceae (V:D)
Salix barclayi var. *commutata* (Bebb) L. Kelso = Salix commutata [SALICOM] Salicaceae (V:D)
Salix barclayi var. *conjuncta* (Bebb) Ball ex Schneid. = Salix barclayi [SALIBAR] Salicaceae (V:D)
Salix barclayi var. *hebecarpa* Anderss. = Salix planifolia ssp. pulchra [SALIPLA2] Salicaceae (V:D)
Salix barclayi var. *pseudomonticola* (Ball) L. Kelso = Salix pseudomonticola [SALIPSE] Salicaceae (V:D)
Salix barrattiana Hook. [SALIBAA] Salicaceae (V:D)
Salix barrattiana var. *angustifolia* Anderss. = Salix barrattiana [SALIBAA] Salicaceae (V:D)
Salix barrattiana var. *latifolia* Anderss. = Salix barrattiana [SALIBAA] Salicaceae (V:D)
Salix barrattiana var. *marcescens* Raup = Salix barrattiana [SALIBAA] Salicaceae (V:D)
Salix barrattiana var. *tweedyi* Bebb ex Rose = Salix tweedyi [SALITWE] Salicaceae (V:D)
Salix bebbiana Sarg. [SALIBEB] Salicaceae (V:D)
Salix bebbiana var. *capreifolia* (Fern.) Fern. = Salix bebbiana [SALIBEB] Salicaceae (V:D)
Salix bebbiana var. *depilis* Raup = Salix bebbiana [SALIBEB] Salicaceae (V:D)
Salix bebbiana var. *luxurians* (Fern.) Fern. = Salix bebbiana [SALIBEB] Salicaceae (V:D)
Salix bebbiana var. *perrostrata* (Rydb.) Schneid. = Salix bebbiana [SALIBEB] Salicaceae (V:D)
Salix bebbiana var. *projecta* (Fern.) Schneid. = Salix bebbiana [SALIBEB] Salicaceae (V:D)
Salix bella Piper = Salix drummondiana [SALIDRU] Salicaceae (V:D)
Salix bolanderiana Rowlee = Salix melanopsis [SALIMEL] Salicaceae (V:D)
Salix boothii Dorn [SALIBOO] Salicaceae (V:D)
Salix brachycarpa Nutt. [SALIBRA] Salicaceae (V:D)
Salix brachycarpa ssp. **brachycarpa** [SALIBRA0] Salicaceae (V:D)
Salix brachycarpa ssp. **niphoclada** (Rydb.) Argus [SALIBRA2] Salicaceae (V:D)
Salix brachycarpa var. *mexiae* Ball = Salix brachycarpa ssp. niphoclada [SALIBRA2] Salicaceae (V:D)
Salix brachystachys Benth. = Salix scouleriana [SALISCO] Salicaceae (V:D)

Salix brachystachys var. *scouleriana* (Barratt ex Hook.) Anderss. = Salix scouleriana [SALISCO] Salicaceae (V:D)
Salix brownei (Anderss.) Bebb = Salix arctica [SALIARC] Salicaceae (V:D)
Salix brownei var. *petraea* (Anderss.) Bebb = Salix arctica [SALIARC] Salicaceae (V:D)
Salix brownii var. *tenera* (Anderss.) M.E. Jones = Salix cascadensis [SALICAS] Salicaceae (V:D)
Salix caespitosa Kennedy = Salix arctica [SALIARC] Salicaceae (V:D)
Salix candida Fluegge ex Willd. [SALICAN] Salicaceae (V:D)
Salix candida var. *denudata* Anderss. = Salix candida [SALICAN] Salicaceae (V:D)
Salix candida var. *tomentosa* Anderss. = Salix candida [SALICAN] Salicaceae (V:D)
Salix candidula Nieuwl. = Salix candida [SALICAN] Salicaceae (V:D)
Salix capreoides Anderss. = Salix scouleriana [SALISCO] Salicaceae (V:D)
Salix cascadensis Cockerell [SALICAS] Salicaceae (V:D)
Salix cascadensis var. *thompsonii* Brayshaw = Salix cascadensis [SALICAS] Salicaceae (V:D)
Salix caudata (Nutt.) Heller = Salix lucida ssp. caudata [SALILUC1] Salicaceae (V:D)
Salix caudata var. *bryantiana* Ball & Braceline = Salix lucida ssp. caudata [SALILUC1] Salicaceae (V:D)
Salix caudata var. *parvifolia* Ball = Salix lucida ssp. caudata [SALILUC1] Salicaceae (V:D)
Salix chlorophylla Anderss. = Salix planifolia ssp. planifolia [SALIPLA0] Salicaceae (V:D)
Salix chlorophylla var. *monica* (Bebb) Flod. = Salix planifolia ssp. planifolia [SALIPLA0] Salicaceae (V:D)
Salix chlorophylla var. *nelsonii* (Ball) Flod. = Salix planifolia ssp. planifolia [SALIPLA0] Salicaceae (V:D)
Salix chlorophylla var. *pychnocarpa* (Anderss.) Anderss. = Salix planifolia ssp. planifolia [SALIPLA0] Salicaceae (V:D)
Salix commutata Bebb [SALICOM] Salicaceae (V:D)
Salix commutata var. *denudata* Bebb = Salix commutata [SALICOM] Salicaceae (V:D)
Salix commutata var. *mixta* Piper = Salix commutata [SALICOM] Salicaceae (V:D)
Salix commutata var. *puberula* Bebb = Salix commutata [SALICOM] Salicaceae (V:D)
Salix commutata var. *sericea* Bebb = Salix commutata [SALICOM] Salicaceae (V:D)
Salix conformis Forbes = Salix discolor [SALIDIS] Salicaceae (V:D)
Salix conjuncta Bebb = Salix barclayi [SALIBAR] Salicaceae (V:D)
Salix cordata var. *balsamifera* Hook. = Salix pyrifolia [SALIPYR] Salicaceae (V:D)
Salix cordata var. *mackenzieana* Hook. = Salix prolixa [SALIPRO] Salicaceae (V:D)
Salix coulteri Anderss. = Salix sitchensis [SALISIT] Salicaceae (V:D)
Salix covillei Eastw. = Salix drummondiana [SALIDRU] Salicaceae (V:D)
Salix crassa Barratt = Salix discolor [SALIDIS] Salicaceae (V:D)
Salix crassijulis Trautv. = Salix arctica [SALIARC] Salicaceae (V:D)
Salix cuneata Nutt. = Salix sitchensis [SALISIT] Salicaceae (V:D)
Salix depressa ssp. *rostrata* (Richards.) Hiitonen = Salix bebbiana [SALIBEB] Salicaceae (V:D)
Salix desertorum Richards. = Salix glauca var. villosa [SALIGLA2] Salicaceae (V:D)
Salix desertorum var. *elata* Anderss. = Salix glauca var. villosa [SALIGLA2] Salicaceae (V:D)
Salix discolor Muhl. [SALIDIS] Salicaceae (V:D)
Salix discolor var. *overi* Ball = Salix discolor [SALIDIS] Salicaceae (V:D)
Salix discolor var. *prinoides* (Pursh) Anderss. = Salix discolor [SALIDIS] Salicaceae (V:D)
Salix discolor var. *rigidior* (Anderss.) Schneid. = Salix discolor [SALIDIS] Salicaceae (V:D)
Salix drummondiana Barratt ex Hook. [SALIDRU] Salicaceae (V:D)
Salix drummondiana ssp. *subcoerulea* (Piper) E. Murr. = Salix drummondiana [SALIDRU] Salicaceae (V:D)
Salix drummondiana var. *bella* (Piper) Ball = Salix drummondiana [SALIDRU] Salicaceae (V:D)
Salix drummondiana var. *subcoerulea* (Piper) Ball = Salix drummondiana [SALIDRU] Salicaceae (V:D)
Salix eriocephala ssy. prolixa (Anderss.) Argus = Salix prolixa [SALIPRO] Salicaceae (V:D)
Salix exigua Nutt. [SALIEXI] Salicaceae (V:D)
Salix exigua ssp. *interior* (Rowlee) Cronq. = Salix exigua [SALIEXI] Salicaceae (V:D)
Salix exigua ssp. *melanopsis* (Nutt.) Cronq. = Salix melanopsis [SALIMEL] Salicaceae (V:D)
Salix exigua var. *angustissima* (Anderss.) Reveal & Broome = Salix exigua [SALIEXI] Salicaceae (V:D)
Salix exigua var. *exterior* (Fern.) C.F. Reed = Salix exigua [SALIEXI] Salicaceae (V:D)
Salix exigua var. *gracilipes* (Ball) Cronq. = Salix melanopsis [SALIMEL] Salicaceae (V:D)
Salix exigua var. *luteosericea* (Rydb.) Schneid. = Salix exigua [SALIEXI] Salicaceae (V:D)
Salix exigua var. *nevadensis* (S. Wats.) Schneid. = Salix exigua [SALIEXI] Salicaceae (V:D)
Salix exigua var. *parishiana* (Rowlee) Jepson = Salix sessilifolia [SALISES] Salicaceae (V:D)
Salix exigua var. *pedicellata* (Anderss.) Cronq. = Salix exigua [SALIEXI] Salicaceae (V:D)
Salix exigua var. *stenophylla* (Rydb.) Schneid. = Salix exigua [SALIEXI] Salicaceae (V:D)
Salix exigua var. *tenerrima* (Henderson) Schneid. = Salix melanopsis [SALIMEL] Salicaceae (V:D)
Salix exigua var. *virens* Rowlee = Salix exigua [SALIEXI] Salicaceae (V:D)
Salix fallax Raup = Salix athabascensis [SALIATH] Salicaceae (V:D)
Salix farriae Ball [SALIFAR] Salicaceae (V:D)
Salix farriae var. *microserrulata* Ball = Salix farriae [SALIFAR] Salicaceae (V:D)
Salix fendleriana Anderss. = Salix lucida ssp. caudata [SALILUC1] Salicaceae (V:D)
Salix fernaldii Blank. = Salix vestita [SALIVES] Salicaceae (V:D)
Salix flavescens Nutt. = Salix scouleriana [SALISCO] Salicaceae (V:D)
Salix flavescens var. *capreoides* (Anderss.) Bebb = Salix scouleriana [SALISCO] Salicaceae (V:D)
Salix flavescens var. *scouleriana* (Barratt ex Hook.) Bebb = Salix scouleriana [SALISCO] Salicaceae (V:D)
Salix fluviatilis var. *argophylla* (Nutt.) Sarg. = Salix exigua [SALIEXI] Salicaceae (V:D)
Salix fluviatilis var. *sericans* (Nees) Boivin = Salix exigua [SALIEXI] Salicaceae (V:D)
Salix fluviatilis var. *sessilifolia* (Nutt.) Scoggan = Salix sessilifolia [SALISES] Salicaceae (V:D)
Salix fluviatilis var. *tenerrima* (Henderson) T.J. Howell = Salix melanopsis [SALIMEL] Salicaceae (V:D)
Salix fulcrata var. *subglauca* Anderss. = Salix planifolia ssp. pulchra [SALIPLA2] Salicaceae (V:D)
Salix fuscata Pursh = Salix discolor [SALIDIS] Salicaceae (V:D)
Salix fuscescens var. *hebecarpa* Fern. = Salix pedicellaris [SALIPED] Salicaceae (V:D)
Salix geyeriana Anderss. [SALIGEY] Salicaceae (V:D)
Salix geyeriana ssp. *argentea* (Bebb) E. Murr. = Salix geyeriana [SALIGEY] Salicaceae (V:D)
Salix geyeriana var. *argentea* (Bebb) Schneid. = Salix geyeriana [SALIGEY] Salicaceae (V:D)
Salix geyeriana var. *meleina* J.K. Henry = Salix geyeriana [SALIGEY] Salicaceae (V:D)
Salix glauca L. [SALIGLA] Salicaceae (V:D)
Salix glauca ssp. *acutifolia* (Hook.) Hultén = Salix glauca var. acutifolia [SALIGLA1] Salicaceae (V:D)
Salix glauca ssp. *desertorum* (Richards.) Hultén = Salix glauca var. villosa [SALIGLA2] Salicaceae (V:D)
Salix glauca ssp. *glabrescens* (Anderss.) Hultén = Salix glauca var. villosa [SALIGLA2] Salicaceae (V:D)
Salix glauca var. **acutifolia** (Hook.) Schneid. [SALIGLA1] Salicaceae (V:D)

Salix glauca var. *alicea* Ball = Salix glauca var. acutifolia [SALIGLA1] Salicaceae (V:D)
Salix glauca var. *glabrescens* (Anderss.) Schneid. = Salix glauca var. villosa [SALIGLA2] Salicaceae (V:D)
Salix glauca var. *kenosha* (L. Kelso) L. Kelso = Salix glauca var. villosa [SALIGLA2] Salicaceae (V:D)
Salix glauca var. *niphoclada* (Rydb.) Wiggins = Salix brachycarpa ssp. niphoclada [SALIBRA2] Salicaceae (V:D)
Salix glauca var. *perstipulata* Raup = Salix glauca var. acutifolia [SALIGLA1] Salicaceae (V:D)
Salix glauca var. *poliophylla* (Schneid.) Raup = Salix glauca var. acutifolia [SALIGLA1] Salicaceae (V:D)
Salix glauca var. *pseudolapponum* (von Seem.) L. Kelso = Salix glauca var. villosa [SALIGLA2] Salicaceae (V:D)
Salix glauca var. *seemanii* (Rydb.) Ostenf. = Salix glauca var. acutifolia [SALIGLA1] Salicaceae (V:D)
Salix glauca var. *sericea* Hultén = Salix glauca var. villosa [SALIGLA2] Salicaceae (V:D)
Salix glauca var. *subincurva* (E.H. Kelso) L. Kelso = Salix glauca var. villosa [SALIGLA2] Salicaceae (V:D)
Salix glauca var. **villosa** (D. Don ex Hook.) Anderss. [SALIGLA2] Salicaceae (V:D)
Salix glaucops Anderss. = Salix glauca var. villosa [SALIGLA2] Salicaceae (V:D)
Salix × *glaucops* Anderss. = Salix glauca var. villosa [SALIGLA2] Salicaceae (V:D)
Salix glaucops var. *glabrescens* Anderss. = Salix glauca var. villosa [SALIGLA2] Salicaceae (V:D)
Salix glaucops var. *villosa* (D. Don ex Hook.) Anderss. = Salix glauca var. villosa [SALIGLA2] Salicaceae (V:D)
Salix gracilis Anderss. = Salix petiolaris [SALIPET] Salicaceae (V:D)
Salix gracilis var. *rosmarinoides* Anderss. = Salix petiolaris [SALIPET] Salicaceae (V:D)
Salix gracilis var. *textoris* Fern. = Salix petiolaris [SALIPET] Salicaceae (V:D)
Salix hastata var. *farriae* (Ball) Hultén = Salix farriae [SALIFAR] Salicaceae (V:D)
Salix hebecarpa (Fern.) Fern. = Salix pedicellaris [SALIPED] Salicaceae (V:D)
Salix hindsiana var. *parishiana* (Rowlee) Ball = Salix sessilifolia [SALISES] Salicaceae (V:D)
Salix hindsiana var. *tenuifolia* (Anderss.) Anderss. = Salix exigua [SALIEXI] Salicaceae (V:D)
Salix hookeriana Barratt ex Hook. [SALIHOO] Salicaceae (V:D)
Salix hookeriana var. *laurifolia* J.K. Henry = Salix hookeriana [SALIHOO] Salicaceae (V:D)
Salix hookeriana var. *tomentosa* J.K. Henry ex Schneid. = Salix hookeriana [SALIHOO] Salicaceae (V:D)
Salix hoyeriana Dieck = Salix barclayi [SALIBAR] Salicaceae (V:D)
Salix hudsonensis Schneid. = Salix arctica [SALIARC] Salicaceae (V:D)
Salix humillima Anderss. = Salix arbusculoides [SALIARB] Salicaceae (V:D)
Salix humillima var. *puberula* (Anderss.) Anderss. = Salix arbusculoides [SALIARB] Salicaceae (V:D)
Salix interior Rowlee = Salix exigua [SALIEXI] Salicaceae (V:D)
Salix interior var. *angustissima* (Anderss.) Dayton = Salix exigua [SALIEXI] Salicaceae (V:D)
Salix interior var. *exterior* Fern. = Salix exigua [SALIEXI] Salicaceae (V:D)
Salix interior var. *luteosericea* (Rydb.) Schneid. = Salix exigua [SALIEXI] Salicaceae (V:D)
Salix interior var. *pedicellata* (Anderss.) Ball = Salix exigua [SALIEXI] Salicaceae (V:D)
Salix interior var. *wheeleri* Rowlee = Salix exigua [SALIEXI] Salicaceae (V:D)
Salix lanata L. [SALILAN] Salicaceae (V:D)
Salix lanata ssp. **richardsonii** (Hook.) Skvort. [SALILAN1] Salicaceae (V:D)
Salix lancifolia Anderss. = Salix lucida ssp. lasiandra [SALILUC2] Salicaceae (V:D)
Salix lasiandra Benth. = Salix lucida ssp. lasiandra [SALILUC2] Salicaceae (V:D)
Salix lasiandra ssp. *caudata* (Nutt.) E. Murr. = Salix lucida ssp. caudata [SALILUC1] Salicaceae (V:D)
Salix lasiandra var. *abramsii* Ball = Salix lucida ssp. lasiandra [SALILUC2] Salicaceae (V:D)
Salix lasiandra var. *caudata* (Nutt.) Sudworth = Salix lucida ssp. caudata [SALILUC1] Salicaceae (V:D)
Salix lasiandra var. *fendleriana* (Anderss.) Bebb = Salix lucida ssp. caudata [SALILUC1] Salicaceae (V:D)
Salix lasiandra var. *lancifolia* (Anderss.) Bebb = Salix lucida ssp. lasiandra [SALILUC2] Salicaceae (V:D)
Salix lasiandra var. *lyallii* Sarg. = Salix lucida ssp. lasiandra [SALILUC2] Salicaceae (V:D)
Salix lasiandra var. *macrophylla* (Anderss.) Little = Salix lucida ssp. lasiandra [SALILUC2] Salicaceae (V:D)
Salix lasiandra var. *recomponens* Raup = Salix lucida ssp. lasiandra [SALILUC2] Salicaceae (V:D)
Salix leiolepis Fern. = Salix vestita [SALIVES] Salicaceae (V:D)
Salix lemmonii Bebb [SALILEM] Salicaceae (V:D)
Salix lemmonii var. *austinae* (Bebb) Schneid. = Salix lemmonii [SALILEM] Salicaceae (V:D)
Salix lemmonii var. *macrostachya* Bebb = Salix lemmonii [SALILEM] Salicaceae (V:D)
Salix lemmonii var. *melanopsis* Bebb = Salix lemmonii [SALILEM] Salicaceae (V:D)
Salix lemmonii var. *sphaerostachya* Bebb = Salix lemmonii [SALILEM] Salicaceae (V:D)
Salix linearifolia Rydb. = Salix exigua [SALIEXI] Salicaceae (V:D)
Salix lingulata Anderss. = Salix myrtillifolia var. myrtillifolia [SALIMYR0] Salicaceae (V:D)
Salix livida var. *occidentalis* (Anderss.) Gray = Salix bebbiana [SALIBEB] Salicaceae (V:D)
Salix livida var. *rostrata* (Richards.) Dippel = Salix bebbiana [SALIBEB] Salicaceae (V:D)
Salix longifolia Muhl. non Lam. = Salix exigua [SALIEXI] Salicaceae (V:D)
Salix longifolia var. *argophylla* (Nutt.) Anderss. = Salix exigua [SALIEXI] Salicaceae (V:D)
Salix longifolia var. *exigua* (Nutt.) Bebb = Salix exigua [SALIEXI] Salicaceae (V:D)
Salix longifolia var. *interior* (Rowlee) M.E. Jones = Salix exigua [SALIEXI] Salicaceae (V:D)
Salix longifolia var. *opaca* Anderss. = Salix exigua [SALIEXI] Salicaceae (V:D)
Salix longifolia var. *pedicellata* Anderss. = Salix exigua [SALIEXI] Salicaceae (V:D)
Salix longifolia var. *sericans* Nees ex Wied-Neuw. = Salix exigua [SALIEXI] Salicaceae (V:D)
Salix longifolia var. *sessilifolia* (Nutt.) M.E. Jones = Salix sessilifolia [SALISES] Salicaceae (V:D)
Salix longifolia var. *tenerrima* Henderson = Salix melanopsis [SALIMEL] Salicaceae (V:D)
Salix longifolia var. *wheeleri* (Rowlee) Schneid. = Salix exigua [SALIEXI] Salicaceae (V:D)
Salix longistylis Rydb. = Salix alaxensis var. longistylis [SALIALA2] Salicaceae (V:D)
Salix lucida Muhl. [SALILUC] Salicaceae (V:D)
Salix lucida ssp. **caudata** (Nutt.) E. Murr. [SALILUC1] Salicaceae (V:D)
Salix lucida ssp. **lasiandra** (Benth.) E. Murr. [SALILUC2] Salicaceae (V:D)
Salix lucida var. *macrophylla* Anderss. = Salix lucida ssp. lasiandra [SALILUC2] Salicaceae (V:D)
Salix lucida var. *serissima* Bailey = Salix serissima [SALISER] Salicaceae (V:D)
Salix luteosericea Rydb. = Salix exigua [SALIEXI] Salicaceae (V:D)
Salix lyallii (Sarg.) Heller = Salix lucida ssp. lasiandra [SALILUC2] Salicaceae (V:D)
Salix maccalliana Rowlee [SALIMAC] Salicaceae (V:D)
Salix mackenzieana (Hook.) Barratt ex Anderss. = Salix prolixa [SALIPRO] Salicaceae (V:D)
Salix mackenzieana var. *macrogemma* Ball = Salix prolixa [SALIPRO] Salicaceae (V:D)

Salix macrocarpa Nutt. = Salix geyeriana [SALIGEY] Salicaceae (V:D)
Salix macrocarpa var. *argentea* Bebb = Salix geyeriana [SALIGEY] Salicaceae (V:D)
Salix macrostachya Nutt. = Salix sessilifolia [SALISES] Salicaceae (V:D)
Salix macrostachya var. *cusickii* Rowlee = Salix sessilifolia [SALISES] Salicaceae (V:D)
Salix malacophylla Nutt. ex Ball = Salix exigua [SALIEXI] Salicaceae (V:D)
Salix melanopsis Nutt. [SALIMEL] Salicaceae (V:D)
Salix melanopsis var. *bolanderiana* (Rowlee) Schneid. = Salix melanopsis [SALIMEL] Salicaceae (V:D)
Salix melanopsis var. *gracilipes* Ball = Salix melanopsis [SALIMEL] Salicaceae (V:D)
Salix melanopsis var. *kronkheittii* L. Kelso = Salix melanopsis [SALIMEL] Salicaceae (V:D)
Salix melanopsis var. *tenerrima* (Henderson) Ball = Salix melanopsis [SALIMEL] Salicaceae (V:D)
Salix meleina (J.K. Henry) G.N. Jones = Salix geyeriana [SALIGEY] Salicaceae (V:D)
Salix monica Bebb = Salix planifolia ssp. planifolia [SALIPLA0] Salicaceae (V:D)
Salix muriei Hultén = Salix brachycarpa ssp. niphoclada [SALIBRA2] Salicaceae (V:D)
Salix myrsinites var. *curtifolia* Anderss. = Salix myrtillifolia var. myrtillifolia [SALIMYR0] Salicaceae (V:D)
Salix myrtillifolia Anderss. [SALIMYR] Salicaceae (V:D)
Salix myrtillifolia var. **cordata** (Anderss.) Dorn [SALIMYR1] Salicaceae (V:D)
Salix myrtillifolia var. *curtifolia* (Anderss.) Bebb ex Rose = Salix myrtillifolia var. myrtillifolia [SALIMYR0] Salicaceae (V:D)
Salix myrtillifolia var. *lingulata* (Anderss.) Ball = Salix myrtillifolia var. myrtillifolia [SALIMYR0] Salicaceae (V:D)
Salix myrtillifolia var. **myrtillifolia** [SALIMYR0] Salicaceae (V:D)
Salix myrtillifolia var. *pseudomyrsinites* (Anderss.) Ball ex Hultén = Salix myrtillifolia var. cordata [SALIMYR1] Salicaceae (V:D)
Salix myrtilloides var. *hypoglauca* (Fern.) Ball = Salix pedicellaris [SALIPED] Salicaceae (V:D)
Salix myrtilloides var. *pedicellaris* (Pursh) Anderss. = Salix pedicellaris [SALIPED] Salicaceae (V:D)
Salix nelsonii Ball = Salix planifolia ssp. planifolia [SALIPLA0] Salicaceae (V:D)
Salix neoforbesii Toepffer = Salix petiolaris [SALIPET] Salicaceae (V:D)
Salix nevadensis S. Wats. = Salix exigua [SALIEXI] Salicaceae (V:D)
Salix nigra var. *amygdaloides* (Anderss.) Anderss. = Salix amygdaloides [SALIAMY] Salicaceae (V:D)
Salix nigra var. *wrightii* (Anderss.) Anderss. = Salix amygdaloides [SALIAMY] Salicaceae (V:D)
Salix niphoclada Rydb. = Salix brachycarpa ssp. niphoclada [SALIBRA2] Salicaceae (V:D)
Salix niphoclada var. *mexiae* (Ball) Hultén = Salix brachycarpa ssp. niphoclada [SALIBRA2] Salicaceae (V:D)
Salix niphoclada var. *muriei* (Hultén) Raup = Salix brachycarpa ssp. niphoclada [SALIBRA2] Salicaceae (V:D)
Salix nivalis Hook. = Salix reticulata ssp. nivalis [SALIRET2] Salicaceae (V:D)
Salix nivalis var. *saximontana* (Rydb.) Schneid. = Salix reticulata ssp. nivalis [SALIRET2] Salicaceae (V:D)
Salix novae-angliae auct. = Salix myrtillifolia var. cordata [SALIMYR1] Salicaceae (V:D)
Salix novae-angliae var. *myrtillifolia* (Anderss.) Anderss. = Salix myrtillifolia var. myrtillifolia [SALIMYR0] Salicaceae (V:D)
Salix novae-angliae var. *pseudocordata* Anderss. = Salix myrtillifolia var. myrtillifolia [SALIMYR0] Salicaceae (V:D)
Salix novae-angliae var. *pseudomyrsinites* (Anderss.) Anderss. = Salix myrtillifolia var. cordata [SALIMYR1] Salicaceae (V:D)
Salix nudescens Rydb. = Salix glauca var. villosa [SALIGLA2] Salicaceae (V:D)
Salix nuttallii Sarg. = Salix scouleriana [SALISCO] Salicaceae (V:D)
Salix nuttallii var. *capreoides* (Anderss.) Sarg. = Salix scouleriana [SALISCO] Salicaceae (V:D)
Salix orbicularis Anderss. = Salix reticulata ssp. reticulata [SALIRET0] Salicaceae (V:D)
Salix pachnophora Rydb. = Salix drummondiana [SALIDRU] Salicaceae (V:D)
Salix pallasii Anderss. = Salix arctica [SALIARC] Salicaceae (V:D)
Salix pallasii var. *crassijulis* (Trautv.) Anderss. = Salix arctica [SALIARC] Salicaceae (V:D)
Salix parishiana Rowlee = Salix sessilifolia [SALISES] Salicaceae (V:D)
Salix parksiana Ball = Salix melanopsis [SALIMEL] Salicaceae (V:D)
Salix pedicellaris Pursh [SALIPED] Salicaceae (V:D)
Salix pedicellaris var. *athabascensis* (Raup) Boivin = Salix athabascensis [SALIATH] Salicaceae (V:D)
Salix pedicellaris var. *hypoglauca* Fern. = Salix pedicellaris [SALIPED] Salicaceae (V:D)
Salix pedicellaris var. *tenuescens* Fern. = Salix pedicellaris [SALIPED] Salicaceae (V:D)
Salix pennata Ball = Salix planifolia ssp. planifolia [SALIPLA0] Salicaceae (V:D)
Salix pentandra var. *caudata* Nutt. = Salix lucida ssp. caudata [SALILUC1] Salicaceae (V:D)
Salix perrostrata Rydb. = Salix bebbiana [SALIBEB] Salicaceae (V:D)
Salix petiolaris Sm. [SALIPET] Salicaceae (V:D)
Salix petiolaris var. *angustifolia* Anderss. = Salix petiolaris [SALIPET] Salicaceae (V:D)
Salix petiolaris var. *gracilis* (Anderss.) Anderss. = Salix petiolaris [SALIPET] Salicaceae (V:D)
Salix petiolaris var. *rosmarinoides* (Anderss.) Schneid. = Salix petiolaris [SALIPET] Salicaceae (V:D)
Salix petiolaris var. *subsericea* Anderss. = Salix petiolaris [SALIPET] Salicaceae (V:D)
Salix petrophila Rydb. = Salix arctica [SALIARC] Salicaceae (V:D)
Salix petrophila var. *caespitosa* (Kennedy) Schneid. = Salix arctica [SALIARC] Salicaceae (V:D)
Salix phylicifolia ssp. *planifolia* (Pursh) Hiitonen = Salix planifolia ssp. planifolia [SALIPLA0] Salicaceae (V:D)
Salix phylicifolia ssp. *pulchra* (Cham.) Hultén = Salix planifolia ssp. pulchra [SALIPLA2] Salicaceae (V:D)
Salix phylicifolia var. *monica* (Bebb) Jepson = Salix planifolia ssp. planifolia [SALIPLA0] Salicaceae (V:D)
Salix phylicifolia var. *pennata* (Ball) Cronq. = Salix planifolia ssp. planifolia [SALIPLA0] Salicaceae (V:D)
Salix phylicifolia var. *subglauca* (Anderss.) Boivin = Salix planifolia ssp. pulchra [SALIPLA2] Salicaceae (V:D)
Salix phylicoides Anderss. = Salix planifolia ssp. pulchra [SALIPLA2] Salicaceae (V:D)
Salix piperi Bebb = Salix hookeriana [SALIHOO] Salicaceae (V:D)
Salix planifolia Pursh [SALIPLA] Salicaceae (V:D)
Salix planifolia ssp. **planifolia** [SALIPLA0] Salicaceae (V:D)
Salix planifolia ssp. **pulchra** (Cham.) Argus [SALIPLA2] Salicaceae (V:D)
Salix planifolia var. *monica* (Bebb) Schneid. = Salix planifolia ssp. planifolia [SALIPLA0] Salicaceae (V:D)
Salix planifolia var. *nelsonii* (Ball) Ball ex E.C. Sm. = Salix planifolia ssp. planifolia [SALIPLA0] Salicaceae (V:D)
Salix planifolia var. *pennata* (Ball) Ball ex Dutilly, Lepage & Daman = Salix planifolia ssp. planifolia [SALIPLA0] Salicaceae (V:D)
Salix planifolia var. *yukonensis* (Schneid.) Argus = Salix planifolia ssp. pulchra [SALIPLA2] Salicaceae (V:D)
Salix polaris Wahlenb. [SALIPOL] Salicaceae (V:D)
Salix polaris ssp. *pseudopolaris* (Flod.) Hultén = Salix polaris [SALIPOL] Salicaceae (V:D)
Salix polaris var. *glabrata* Hultén = Salix polaris [SALIPOL] Salicaceae (V:D)
Salix polaris var. *selwynensis* Raup = Salix polaris [SALIPOL] Salicaceae (V:D)
Salix prinoides Pursh = Salix discolor [SALIDIS] Salicaceae (V:D)
Salix prolixa Anderss. [SALIPRO] Salicaceae (V:D)
Salix pseudocordata (Anderss.) Rydb. = Salix myrtillifolia var. myrtillifolia [SALIMYR0] Salicaceae (V:D)
Salix pseudocordata auct. = Salix boothii [SALIBOO] Salicaceae (V:D)

Salix pseudocordata var. *aequalis* (Anderss.) Ball ex Schneid. = Salix boothii [SALIBOO] Salicaceae (V:D)
Salix pseudocordata var. *cordata* (Anderss.) Ball = Salix myrtillifolia var. cordata [SALIMYR1] Salicaceae (V:D)
Salix pseudolapponum von Seem. = Salix glauca var. villosa [SALIGLA2] Salicaceae (V:D)
Salix pseudolapponum var. *kenosha* L. Kelso = Salix glauca var. villosa [SALIGLA2] Salicaceae (V:D)
Salix pseudolapponum var. *subincurva* E.H. Kelso = Salix glauca var. villosa [SALIGLA2] Salicaceae (V:D)
Salix pseudomonticola Ball [SALIPSE] Salicaceae (V:D)
Salix pseudomyrsinites Anderss. = Salix myrtillifolia var. cordata [SALIMYR1] Salicaceae (V:D)
Salix pseudomyrsinites auct. = Salix boothii [SALIBOO] Salicaceae (V:D)
Salix pseudomyrsinites var. *aequalis* (Anderss.) Anderss. ex Ball = Salix boothii [SALIBOO] Salicaceae (V:D)
Salix pseudopolaris Flod. = Salix polaris [SALIPOL] Salicaceae (V:D)
Salix pulchra Cham. = Salix planifolia ssp. pulchra [SALIPLA2] Salicaceae (V:D)
Salix pulchra var. *looffiae* Ball = Salix planifolia ssp. pulchra [SALIPLA2] Salicaceae (V:D)
Salix pulchra var. *palmeri* Ball = Salix planifolia ssp. pulchra [SALIPLA2] Salicaceae (V:D)
Salix pulchra var. *yukonensis* Schneid. = Salix planifolia ssp. pulchra [SALIPLA2] Salicaceae (V:D)
Salix pychnocarpa Anderss. = Salix planifolia ssp. planifolia [SALIPLA0] Salicaceae (V:D)
Salix pyrifolia Anderss. [SALIPYR] Salicaceae (V:D)
Salix pyrifolia var. *lanceolata* (Bebb) Fern. = Salix pyrifolia [SALIPYR] Salicaceae (V:D)
Salix pyrolifolia var. *hoyeriana* (Dieck) Dippel = Salix barclayi [SALIBAR] Salicaceae (V:D)
Salix raupii Argus [SALIRAU] Salicaceae (V:D)
Salix regelii Anderss. = Salix barclayi [SALIBAR] Salicaceae (V:D)
Salix reticulata L. [SALIRET] Salicaceae (V:D)
Salix reticulata ssp. **glabellicarpa** Argus [SALIRET1] Salicaceae (V:D)
Salix reticulata ssp. **nivalis** (Hook.) A. & D. Löve & Kapoor [SALIRET2] Salicaceae (V:D)
Salix reticulata ssp. *orbicularis* (Anderss.) Flod. = Salix reticulata ssp. reticulata [SALIRET0] Salicaceae (V:D)
Salix reticulata ssp. **reticulata** [SALIRET0] Salicaceae (V:D)
Salix reticulata var. *gigantifolia* Ball = Salix reticulata ssp. reticulata [SALIRET0] Salicaceae (V:D)
Salix reticulata var. *glabra* Trautv. = Salix reticulata ssp. reticulata [SALIRET0] Salicaceae (V:D)
Salix reticulata var. *nana* (Hook.) Anderss. = Salix reticulata ssp. nivalis [SALIRET2] Salicaceae (V:D)
Salix reticulata var. *nivalis* (Hook.) Anderss. = Salix reticulata ssp. nivalis [SALIRET2] Salicaceae (V:D)
Salix reticulata var. *orbicularis* (Anderss.) Komarov = Salix reticulata ssp. reticulata [SALIRET0] Salicaceae (V:D)
Salix reticulata var. *saximontana* (Rydb.) L. Kelso = Salix reticulata ssp. nivalis [SALIRET2] Salicaceae (V:D)
Salix reticulata var. *semicalva* Fern. = Salix reticulata ssp. reticulata [SALIRET0] Salicaceae (V:D)
Salix reticulata var. *vestita* (Pursh) Anderss. = Salix vestita [SALIVES] Salicaceae (V:D)
Salix richardsonii Hook. = Salix lanata ssp. richardsonii [SALILAN1] Salicaceae (V:D)
Salix richardsonii var. *mckeandii* Polunin = Salix lanata ssp. richardsonii [SALILAN1] Salicaceae (V:D)
Salix rigida ssp. *mackenzieana* (Hook.) E. Murr. = Salix prolixa [SALIPRO] Salicaceae (V:D)
Salix rigida var. *mackenzieana* (Hook.) Cronq. = Salix prolixa [SALIPRO] Salicaceae (V:D)
Salix rigida var. *macrogemma* (Ball) Cronq. = Salix prolixa [SALIPRO] Salicaceae (V:D)
Salix rostrata Richards. = Salix bebbiana [SALIBEB] Salicaceae (V:D)
Salix rostrata var. *capreifolia* Fern. = Salix bebbiana [SALIBEB] Salicaceae (V:D)
Salix rostrata var. *luxurians* Fern. = Salix bebbiana [SALIBEB] Salicaceae (V:D)
Salix rostrata var. *perrostrata* (Rydb.) Fern. = Salix bebbiana [SALIBEB] Salicaceae (V:D)
Salix rostrata var. *projecta* Fern. = Salix bebbiana [SALIBEB] Salicaceae (V:D)
Salix rotundifolia Nutt. = Salix tweedyi [SALITWE] Salicaceae (V:D)
Salix rubra Richards. = Salix exigua [SALIEXI] Salicaceae (V:D)
Salix saskatchevana von Seem. = Salix arbusculoides [SALIARB] Salicaceae (V:D)
Salix saximontana Rydb. = Salix reticulata ssp. nivalis [SALIRET2] Salicaceae (V:D)
Salix scouleriana Barratt ex Hook. [SALISCO] Salicaceae (V:D)
Salix scouleriana f. *poikila* Schneid. = Salix scouleriana [SALISCO] Salicaceae (V:D)
Salix scouleriana var. *brachystachys* (Benth.) M.E. Jones = Salix scouleriana [SALISCO] Salicaceae (V:D)
Salix scouleriana var. *coetanea* Ball = Salix scouleriana [SALISCO] Salicaceae (V:D)
Salix scouleriana var. *crassijulis* (Anderss.) Schneid. = Salix scouleriana [SALISCO] Salicaceae (V:D)
Salix scouleriana var. *flavescens* (Nutt.) J.K. Henry = Salix scouleriana [SALISCO] Salicaceae (V:D)
Salix scouleriana var. *poikila* Schneid. = Salix scouleriana [SALISCO] Salicaceae (V:D)
Salix scouleriana var. *thompsonii* Ball = Salix scouleriana [SALISCO] Salicaceae (V:D)
Salix seemanii Rydb. = Salix glauca var. acutifolia [SALIGLA1] Salicaceae (V:D)
Salix sensitiva Barratt = Salix discolor [SALIDIS] Salicaceae (V:D)
Salix sericea var. *subsericea* (Anderss.) Rydb. = Salix petiolaris [SALIPET] Salicaceae (V:D)
Salix serissima (Bailey) Fern. [SALISER] Salicaceae (V:D)
Salix sessilifolia Nutt. [SALISES] Salicaceae (V:D)
Salix sessilifolia var. *vancouverensis* Brayshaw = Salix melanopsis [SALIMEL] Salicaceae (V:D)
Salix sessilifolia var. *villosa* Anderss. = Salix sessilifolia [SALISES] Salicaceae (V:D)
Salix setchelliana Ball [SALISET] Salicaceae (V:D)
Salix sitchensis Sanson ex Bong. [SALISIT] Salicaceae (V:D)
Salix sitchensis var. *congesta* (Anderss.) Anderss. = Salix sitchensis [SALISIT] Salicaceae (V:D)
Salix sitchensis var. *denudata* (Anderss.) Anderss. = Salix sitchensis [SALISIT] Salicaceae (V:D)
Salix sitchensis var. *parviflora* (Jepson) Jepson = Salix sitchensis [SALISIT] Salicaceae (V:D)
Salix sitchensis var. *ralphiana* (Jepson) Jepson = Salix sitchensis [SALISIT] Salicaceae (V:D)
Salix speciosa Hook. & Arn. = Salix alaxensis var. alaxensis [SALIALA0] Salicaceae (V:D)
Salix speciosa Nutt. = Salix lucida ssp. lasiandra [SALILUC2] Salicaceae (V:D)
Salix speciosa var. *alaxensis* Anderss. = Salix alaxensis var. alaxensis [SALIALA0] Salicaceae (V:D)
Salix squamata Rydb. = Salix discolor [SALIDIS] Salicaceae (V:D)
Salix stagnalis Nutt. = Salix scouleriana [SALISCO] Salicaceae (V:D)
Salix starkeana ssp. *bebbiana* (Sarg.) Youngberg = Salix bebbiana [SALIBEB] Salicaceae (V:D)
Salix stenophylla Rydb. = Salix exigua [SALIEXI] Salicaceae (V:D)
Salix stolonifera Coville [SALISTO] Salicaceae (V:D)
Salix subcoerulea Piper = Salix drummondiana [SALIDRU] Salicaceae (V:D)
Salix subcordata Anderss. = Salix glauca var. villosa [SALIGLA2] Salicaceae (V:D)
Salix × subsericea (Anderss.) Schneid. = Salix petiolaris [SALIPET] Salicaceae (V:D)
Salix tenera Anderss. = Salix cascadensis [SALICAS] Salicaceae (V:D)
Salix tenerrima (Henderson) Heller = Salix melanopsis [SALIMEL] Salicaceae (V:D)
Salix thurberi Rowlee = Salix exigua [SALIEXI] Salicaceae (V:D)

Salix tortulosa Trautv. = Salix arctica [SALIARC] Salicaceae (V:D)
Salix tweedyi (Bebb ex Rose) Ball [SALITWE] Salicaceae (V:D)
Salix vagans var. *occidentalis* Anderss. = Salix bebbiana [SALIBEB] Salicaceae (V:D)
Salix vagans var. *rostrata* (Richards.) Anderss. = Salix bebbiana [SALIBEB] Salicaceae (V:D)
Salix venusta Anderss. = Salix reticulata ssp. nivalis [SALIRET2] Salicaceae (V:D)
Salix vestita Pursh [SALIVES] Salicaceae (V:D)
Salix vestita ssp. *leiolepis* (Fern.) Argus = Salix vestita [SALIVES] Salicaceae (V:D)
Salix vestita var. *erecta* Anderss. = Salix vestita [SALIVES] Salicaceae (V:D)
Salix vestita var. *humilior* Anderss. = Salix vestita [SALIVES] Salicaceae (V:D)
Salix vestita var. *nana* Hook. = Salix reticulata ssp. nivalis [SALIRET2] Salicaceae (V:D)
Salix vestita var. *psilophylla* Fern. & St. John = Salix vestita [SALIVES] Salicaceae (V:D)
Salix villosa D. Don ex Hook. = Salix glauca var. villosa [SALIGLA2] Salicaceae (V:D)
Salix villosa var. *acutifolia* Hook. = Salix glauca var. acutifolia [SALIGLA1] Salicaceae (V:D)
Salix vitellina L. = Salix alba var. vitellina [SALIALB1] Salicaceae (V:D)
Salix wheeleri (Rowlee) Rydb. = Salix exigua [SALIEXI] Salicaceae (V:D)
Salix wolfii var. *pseudolapponum* (von Seem.) M.E. Jones = Salix glauca var. villosa [SALIGLA2] Salicaceae (V:D)
Salix wrightii Anderss. = Salix amygdaloides [SALIAMY] Salicaceae (V:D)
Salix wyomingensis Rydb. = Salix glauca var. villosa [SALIGLA2] Salicaceae (V:D)
Salsola L. [SALSOLA$] Chenopodiaceae (V:D)
•**Salsola kali** L. [SALSKAL] Chenopodiaceae (V:D)
Salvia L. [SALVIA$] Lamiaceae (V:D)
•**Salvia nemorosa** L. [SALVNEM] Lamiaceae (V:D)
Sambucus L. [SAMBUCU$] Caprifoliaceae (V:D)
Sambucus callicarpa Greene = Sambucus racemosa ssp. pubens var. arborescens [SAMBRAC2] Caprifoliaceae (V:D)
Sambucus cerulea Raf. [SAMBCER] Caprifoliaceae (V:D)
Sambucus melanocarpa Gray = Sambucus racemosa ssp. pubens var. melanocarpa [SAMBRAC4] Caprifoliaceae (V:D)
Sambucus pubens Michx. = Sambucus racemosa ssp. pubens var. pubens [SAMBRAC5] Caprifoliaceae (V:D)
Sambucus pubens var. *arborescens* Torr. & Gray = Sambucus racemosa ssp. pubens var. arborescens [SAMBRAC2] Caprifoliaceae (V:D)
Sambucus racemosa L. [SAMBRAC] Caprifoliaceae (V:D)
Sambucus racemosa ssp. **pubens** (Michx.) House [SAMBRAC1] Caprifoliaceae (V:D)
Sambucus racemosa ssp. **pubens** var. **arborescens** (Torr. & Gray) Gray [SAMBRAC2] Caprifoliaceae (V:D)
Sambucus racemosa ssp. *pubens* var. *leucocarpa* (Torr. & Gray) Cronq. = Sambucus racemosa ssp. pubens var. pubens [SAMBRAC5] Caprifoliaceae (V:D)
Sambucus racemosa ssp. **pubens** var. **melanocarpa** (Gray) McMinn [SAMBRAC4] Caprifoliaceae (V:D)
Sambucus racemosa ssp. **pubens** var. **pubens** (Michx.) Koehne [SAMBRAC5] Caprifoliaceae (V:D)
Sambucus racemosa var. *leucocarpa* (Torr. & Gray) Cronq. = Sambucus racemosa ssp. pubens var. pubens [SAMBRAC5] Caprifoliaceae (V:D)
Samolus L. [SAMOLUS$] Primulaceae (V:D)
Samolus floribundus Kunth = Samolus valerandi ssp. parviflorus [SAMOVAL1] Primulaceae (V:D)
Samolus parviflorus Raf. = Samolus valerandi ssp. parviflorus [SAMOVAL1] Primulaceae (V:D)
Samolus valerandi L. [SAMOVAL] Primulaceae (V:D)
Samolus valerandi ssp. **parviflorus** (Raf.) Hultén [SAMOVAL1] Primulaceae (V:D)
Sanguisorba L. [SANGUIS$] Rosaceae (V:D)
Sanguisorba canadensis L. [SANGCAN] Rosaceae (V:D)
Sanguisorba canadensis ssp. *latifolia* (Hook.) Calder & Taylor = Sanguisorba canadensis [SANGCAN] Rosaceae (V:D)
Sanguisorba canadensis var. *latifolia* Hook. = Sanguisorba canadensis [SANGCAN] Rosaceae (V:D)
Sanguisorba menziesii Rydb. [SANGMEN] Rosaceae (V:D)
Sanguisorba microcephala K. Presl = Sanguisorba officinalis [SANGOFF] Rosaceae (V:D)
•**Sanguisorba minor** Scop. [SANGMIN] Rosaceae (V:D)
Sanguisorba occidentalis Nutt. [SANGOCC] Rosaceae (V:D)
Sanguisorba officinalis L. [SANGOFF] Rosaceae (V:D)
Sanguisorba officinalis ssp. *microcephala* (K. Presl) Calder & Taylor = Sanguisorba officinalis [SANGOFF] Rosaceae (V:D)
Sanguisorba officinalis var. *polygama* (W. Nyl.) Mela & Caj. = Sanguisorba officinalis [SANGOFF] Rosaceae (V:D)
Sanguisorba sitchensis C.A. Mey. = Sanguisorba canadensis [SANGCAN] Rosaceae (V:D)
Sanguisorba stipulata Raf. = Sanguisorba canadensis [SANGCAN] Rosaceae (V:D)
Sanicula L. [SANICUL$] Apiaceae (V:D)
Sanicula arctopoides Hook. & Arn. [SANIARC] Apiaceae (V:D)
Sanicula bipinnatifida Dougl. ex Hook. [SANIBIP] Apiaceae (V:D)
Sanicula bipinnatifida var. *flava* Jepson = Sanicula bipinnatifida [SANIBIP] Apiaceae (V:D)
Sanicula crassicaulis Poepp. ex DC. [SANICRA] Apiaceae (V:D)
Sanicula crassicaulis var. *howellii* (Coult. & Rose) Mathias = Sanicula arctopoides [SANIARC] Apiaceae (V:D)
Sanicula graveolens Poepp. ex DC. [SANIGRA] Apiaceae (V:D)
Sanicula graveolens var. *septentrionalis* (Greene) St. John = Sanicula graveolens [SANIGRA] Apiaceae (V:D)
Sanicula x *howellii* (Coult. & Rose) Shan & Constance = Sanicula arctopoides [SANIARC] Apiaceae (V:D)
Sanicula marilandica L. [SANIMAR] Apiaceae (V:D)
Sanicula marilandica var. *petiolulata* Fern. = Sanicula marilandica [SANIMAR] Apiaceae (V:D)
Sanicula nevadensis S. Wats. = Sanicula graveolens [SANIGRA] Apiaceae (V:D)
Sanicula nevadensis var. *septentrionalis* (Greene) Mathias = Sanicula graveolens [SANIGRA] Apiaceae (V:D)
Sanionia Loeske [SANIONI$] Amblystegiaceae (B:M)
Sanionia uncinata (Hedw.) Loeske [SANIUNC] Amblystegiaceae (B:M)
Sanionia uncinata var. **symmetrica** (Ren. & Card.) Crum & Anderson [SANIUNC1] Amblystegiaceae (B:M)
Sanionia uncinata var. **uncinata** [SANIUNC0] Amblystegiaceae (B:M)
Santolina suaveolens Pursh = Matricaria discoidea [MATRDIS] Asteraceae (V:D)
Saponaria L. [SAPONAR$] Caryophyllaceae (V:D)
•**Saponaria officinalis** L. [SAPOOFF] Caryophyllaceae (V:D)
Saponaria vaccaria L. = Vaccaria hispanica [VACCHIS] Caryophyllaceae (V:D)
Sarcocornia A.J. Scott [SARCOCO$] Chenopodiaceae (V:D)
Sarcocornia pacifica (Standl.) A.J. Scott [SARCPAC] Chenopodiaceae (V:D)
Sarcogyne Flotow [SARCOGY$] Acarosporaceae (L)
Sarcogyne pruinosa auct. = Sarcogyne regularis [SARCREG] Acarosporaceae (L)
Sarcogyne regularis Körber [SARCREG] Acarosporaceae (L)
Sarcogyne simplex (Davies) Nyl. = Polysporina simplex [POLYSIM] Acarosporaceae (L)
Sarcogyne urceolata Anzi = Polysporina urceolata [POLYURC] Acarosporaceae (L)
Sarmenthypnum Tuom. & T. Kop. [SARMENT$] Amblystegiaceae (B:M)
Sarmenthypnum sarmentosum (Wahlenb.) Tuom. & T. Kop. [SARMSAR] Amblystegiaceae (B:M)
Sarracenia L. [SARRACE$] Sarraceniaceae (V:D)
Sarracenia heterophylla Eat. = Sarracenia purpurea [SARRPUR] Sarraceniaceae (V:D)
Sarracenia purpurea L. [SARRPUR] Sarraceniaceae (V:D)
Sarracenia purpurea ssp. *gibbosa* (Raf.) Wherry = Sarracenia purpurea [SARRPUR] Sarraceniaceae (V:D)

Sarracenia purpurea ssp. *heterophylla* (Eat.) Torr. = Sarracenia purpurea [SARRPUR] Sarraceniaceae (V:D)
Sarracenia purpurea var. *ripicola* Boivin = Sarracenia purpurea [SARRPUR] Sarraceniaceae (V:D)
Sarracenia purpurea var. *stolonifera* Macfarlane & D.W. Steckbeck = Sarracenia purpurea [SARRPUR] Sarraceniaceae (V:D)
Sarracenia purpurea var. *terrae-novae* La Pylaie = Sarracenia purpurea [SARRPUR] Sarraceniaceae (V:D)
Satureja L. [SATUREJ$] Lamiaceae (V:D)
Satureja acinos (L.) Scheele = Acinos arvensis [ACINARV] Lamiaceae (V:D)
Satureja chamissonis (Benth.) Briq. = Satureja douglasii [SATUDOU] Lamiaceae (V:D)
Satureja douglasii (Benth.) Briq. [SATUDOU] Lamiaceae (V:D)
Satureja vulgaris (L.) Fritsch = Clinopodium vulgare [CLINVUL] Lamiaceae (V:D)
Satureja vulgaris var. *diminuta* (Simon) Fern. & Wieg. = Clinopodium vulgare [CLINVUL] Lamiaceae (V:D)
Satureja vulgaris var. *neogaea* Fern. = Clinopodium vulgare [CLINVUL] Lamiaceae (V:D)
Saussurea DC. [SAUSSUR$] Asteraceae (V:D)
Saussurea americana D.C. Eat. [SAUSAME] Asteraceae (V:D)
Saussurea angustifolia (Willd.) DC. [SAUSANG] Asteraceae (V:D)
Saussurea densa (Hook.) Rydb. [SAUSDEN] Asteraceae (V:D)
Saussurea nuda ssp. *densa* (Hook.) G.W. Douglas = Saussurea densa [SAUSDEN] Asteraceae (V:D)
Saussurea nuda var. *densa* (Hook.) Hultén = Saussurea densa [SAUSDEN] Asteraceae (V:D)
Saussurea weberi Hultén = Saussurea densa [SAUSDEN] Asteraceae (V:D)
Sauteria Nees [SAUTERI$] Cleveaceae (B:H)
Sauteria alpina (Nees) Nees [SAUTALP] Cleveaceae (B:H)
Saxifraga L. [SAXIFRA$] Saxifragaceae (V:D)
Saxifraga adscendens L. [SAXIADS] Saxifragaceae (V:D)
Saxifraga adscendens ssp. **oregonensis** (Raf.) Bacig. [SAXIADS1] Saxifragaceae (V:D)
Saxifraga adscendens var. *oregonensis* (Raf.) Breitung = Saxifraga adscendens ssp. oregonensis [SAXIADS1] Saxifragaceae (V:D)
Saxifraga aequidentata (Small) Rosendahl = Saxifraga rufidula [SAXIRUF] Saxifragaceae (V:D)
Saxifraga aestivalis auct. = Saxifraga odontoloma [SAXIODO] Saxifragaceae (V:D)
Saxifraga aizoides L. [SAXIAIZ] Saxifragaceae (V:D)
Saxifraga arguta auct. = Saxifraga odontoloma [SAXIODO] Saxifragaceae (V:D)
Saxifraga austromontana Wieg. = Saxifraga bronchialis ssp. austromontana [SAXIBRO1] Saxifragaceae (V:D)
Saxifraga bracteosa Suksdorf = Saxifraga integrifolia [SAXIINT] Saxifragaceae (V:D)
Saxifraga bracteosa var. *angustifolia* Suksdorf = Saxifraga integrifolia [SAXIINT] Saxifragaceae (V:D)
Saxifraga bracteosa var. *micropetala* Suksdorf = Saxifraga nidifica [SAXINID] Saxifragaceae (V:D)
Saxifraga bronchialis L. [SAXIBRO] Saxifragaceae (V:D)
Saxifraga bronchialis ssp. **austromontana** (Wieg.) Piper [SAXIBRO1] Saxifragaceae (V:D)
Saxifraga bronchialis ssp. **funstonii** (Small) Hultén [SAXIBRO2] Saxifragaceae (V:D)
Saxifraga bronchialis var. *austromontana* (Wieg.) Piper ex G.N. Jones = Saxifraga bronchialis ssp. austromontana [SAXIBRO1] Saxifragaceae (V:D)
Saxifraga bronchialis var. *purpureomaculata* Hultén = Saxifraga bronchialis ssp. funstonii [SAXIBRO2] Saxifragaceae (V:D)
Saxifraga calycina Sternb. [SAXICAL] Saxifragaceae (V:D)
Saxifraga cernua L. [SAXICER] Saxifragaceae (V:D)
Saxifraga cernua var. *exilioides* Polunin = Saxifraga cernua [SAXICER] Saxifragaceae (V:D)
Saxifraga cespitosa L. [SAXICES] Saxifragaceae (V:D)
Saxifraga cespitosa ssp. **monticola** (Small) Porsild [SAXICES1] Saxifragaceae (V:D)
Saxifraga cespitosa ssp. **sileneflora** (Sternb. ex Cham.) Hultén [SAXICES2] Saxifragaceae (V:D)
Saxifraga cespitosa ssp. **subgemmifera** Engl. & Irmsch. [SAXICES3] Saxifragaceae (V:D)
Saxifraga cespitosa var. **emarginata** (Small) Rosendahl [SAXICES4] Saxifragaceae (V:D)
Saxifraga cespitosa var. *minima* Blank. = Saxifraga cespitosa ssp. monticola [SAXICES1] Saxifragaceae (V:D)
Saxifraga cespitosa var. *subgemmifera* (Engl. & Irmsch.) C.L. Hitchc. = Saxifraga cespitosa ssp. subgemmifera [SAXICES3] Saxifragaceae (V:D)
Saxifraga columbiana Piper = Saxifraga nidifica [SAXINID] Saxifragaceae (V:D)
Saxifraga crenatifolia (Small) Fedde = Saxifraga nidifica [SAXINID] Saxifragaceae (V:D)
Saxifraga davurica ssp. *grandipetala* (Engl. & Irmsch.) Hultén = Saxifraga calycina [SAXICAL] Saxifragaceae (V:D)
Saxifraga davurica var. *grandipetala* (Engl. & Irmsch.) Boivin = Saxifraga calycina [SAXICAL] Saxifragaceae (V:D)
Saxifraga debilis Engelm. ex Gray = Saxifraga rivularis [SAXIRIV] Saxifragaceae (V:D)
Saxifraga ferruginea Graham [SAXIFER] Saxifragaceae (V:D)
Saxifraga firma Litv. ex Losinsk. = Saxifraga bronchialis ssp. funstonii [SAXIBRO2] Saxifragaceae (V:D)
Saxifraga flagellaris Willd. ex Sternb. [SAXIFLA] Saxifragaceae (V:D)
Saxifraga flagellaris ssp. **setigera** (Pursh) Tolm. [SAXIFLA1] Saxifragaceae (V:D)
Saxifraga fragosa var. *leucandra* Suksdorf = Saxifraga integrifolia [SAXIINT] Saxifragaceae (V:D)
Saxifraga funstonii (Small) Fedde = Saxifraga bronchialis ssp. funstonii [SAXIBRO2] Saxifragaceae (V:D)
Saxifraga hieraciifolia Waldst. & Kit. [SAXIHIE] Saxifragaceae (V:D)
Saxifraga hirculus L. [SAXIHIR] Saxifragaceae (V:D)
Saxifraga hyperborea R. Br. = Saxifraga rivularis [SAXIRIV] Saxifragaceae (V:D)
Saxifraga hyperborea ssp. *debilis* (Engelm. ex Gray) A. & D. Löve & Kapoor = Saxifraga rivularis [SAXIRIV] Saxifragaceae (V:D)
Saxifraga integrifolia Hook. [SAXIINT] Saxifragaceae (V:D)
Saxifraga integrifolia var. *columbiana* (Piper) C.L. Hitchc. = Saxifraga nidifica [SAXINID] Saxifragaceae (V:D)
Saxifraga integrifolia var. *leptopetala* (Suksdorf) Engl. & Irmsch. = Saxifraga nidifica [SAXINID] Saxifragaceae (V:D)
Saxifraga integrifolia var. *micropetala* (Suksdorf) Engl. & Irmsch. = Saxifraga nidifica [SAXINID] Saxifragaceae (V:D)
Saxifraga klickitatensis A.M. Johnson = Saxifraga rufidula [SAXIRUF] Saxifragaceae (V:D)
Saxifraga laevicarpa A.M. Johnson = Saxifraga integrifolia [SAXIINT] Saxifragaceae (V:D)
Saxifraga lyallii Engl. [SAXILYA] Saxifragaceae (V:D)
Saxifraga lyallii ssp. **hultenii** (Calder & Savile) Calder & Taylor [SAXILYA1] Saxifragaceae (V:D)
Saxifraga lyallii ssp. **lyallii** [SAXILYA0] Saxifragaceae (V:D)
Saxifraga lyallii var. *hultenii* Calder & Savile = Saxifraga lyallii ssp. hultenii [SAXILYA1] Saxifragaceae (V:D)
Saxifraga lyallii var. *laxa* Engl. = Saxifraga lyallii ssp. lyallii [SAXILYA0] Saxifragaceae (V:D)
Saxifraga mertensiana Bong. [SAXIMER] Saxifragaceae (V:D)
Saxifraga mertensiana var. *eastwoodiae* (Small) Engl. & Irmsch. = Saxifraga mertensiana [SAXIMER] Saxifragaceae (V:D)
Saxifraga micropetala (Small) Fedde = Saxifraga cespitosa ssp. monticola [SAXICES1] Saxifragaceae (V:D)
Saxifraga montana (Small) Fedde = Saxifraga nidifica [SAXINID] Saxifragaceae (V:D)
Saxifraga monticola (Small) A. & D. Löve = Saxifraga cespitosa ssp. monticola [SAXICES1] Saxifragaceae (V:D)
Saxifraga nelsoniana D. Don [SAXINEL] Saxifragaceae (V:D)
Saxifraga nelsoniana ssp. **carlottae** (Calder & Savile) Hultén [SAXINEL1] Saxifragaceae (V:D)
Saxifraga nelsoniana ssp. **cascadensis** (Calder & Savile) Hultén [SAXINEL2] Saxifragaceae (V:D)
Saxifraga nelsoniana ssp. **pacifica** (Hultén) Hultén [SAXINEL3] Saxifragaceae (V:D)

Saxifraga nelsoniana ssp. **porsildiana** (Calder & Savile) Hultén [SAXINEL4] Saxifragaceae (V:D)
Saxifraga newcombei (Small) Engl. & Irmsch. = Saxifraga ferruginea [SAXIFER] Saxifragaceae (V:D)
Saxifraga nidifica Greene [SAXINID] Saxifragaceae (V:D)
Saxifraga nivalis L. [SAXINIV] Saxifragaceae (V:D)
Saxifraga occidentalis S. Wats. [SAXIOCC] Saxifragaceae (V:D)
Saxifraga occidentalis ssp. *rufidula* (Small) Bacig. = Saxifraga rufidula [SAXIRUF] Saxifragaceae (V:D)
Saxifraga occidentalis var. *aequidentata* (Small) M.E. Peck = Saxifraga rufidula [SAXIRUF] Saxifragaceae (V:D)
Saxifraga occidentalis var. *allenii* (Small) C.L. Hitchc. = Saxifraga occidentalis [SAXIOCC] Saxifragaceae (V:D)
Saxifraga occidentalis var. *rufidula* (Small) C.L. Hitchc. = Saxifraga rufidula [SAXIRUF] Saxifragaceae (V:D)
Saxifraga occidentalis var. *wallowensis* M.E. Peck = Saxifraga occidentalis [SAXIOCC] Saxifragaceae (V:D)
Saxifraga odontoloma Piper [SAXIODO] Saxifragaceae (V:D)
Saxifraga oppositifolia L. [SAXIOPP] Saxifragaceae (V:D)
Saxifraga pectinata Pursh = Luetkea pectinata [LUETPEC] Rosaceae (V:D)
Saxifraga plantaginea Small = Saxifraga nidifica [SAXINID] Saxifragaceae (V:D)
Saxifraga porsildiana (Calder & Savile) Jurtzev & Petrovsky = Saxifraga nelsoniana ssp. porsildiana [SAXINEL4] Saxifragaceae (V:D)
Saxifraga punctata L. = Saxifraga nelsoniana [SAXINEL] Saxifragaceae (V:D)
Saxifraga punctata ssp. *carlottae* Calder & Savile = Saxifraga nelsoniana ssp. carlottae [SAXINEL1] Saxifragaceae (V:D)
Saxifraga punctata ssp. *cascadensis* Calder & Savile = Saxifraga nelsoniana ssp. cascadensis [SAXINEL2] Saxifragaceae (V:D)
Saxifraga punctata ssp. *pacifica* Hultén = Saxifraga nelsoniana ssp. pacifica [SAXINEL3] Saxifragaceae (V:D)
Saxifraga punctata ssp. *porsildiana* Calder & Savile = Saxifraga nelsoniana ssp. porsildiana [SAXINEL4] Saxifragaceae (V:D)
Saxifraga punctata var. *carlottae* (Calder & Savile) Boivin = Saxifraga nelsoniana ssp. carlottae [SAXINEL1] Saxifragaceae (V:D)
Saxifraga punctata var. *cascadensis* (Calder & Savile) Boivin = Saxifraga nelsoniana ssp. cascadensis [SAXINEL2] Saxifragaceae (V:D)
Saxifraga punctata var. *porsildiana* (Calder & Savile) Boivin = Saxifraga nelsoniana ssp. porsildiana [SAXINEL4] Saxifragaceae (V:D)
Saxifraga pyrolifolia D. Don = Leptarrhena pyrolifolia [LEPTPYR] Saxifragaceae (V:D)
Saxifraga reflexa Hook. [SAXIREF] Saxifragaceae (V:D)
Saxifraga reflexa ssp. *occidentalis* (S. Wats.) Hultén = Saxifraga occidentalis [SAXIOCC] Saxifragaceae (V:D)
Saxifraga rivularis L. [SAXIRIV] Saxifragaceae (V:D)
Saxifraga rivularis var. *debilis* (Engelm. ex Gray) Dorn = Saxifraga rivularis [SAXIRIV] Saxifragaceae (V:D)
Saxifraga rivularis var. *flexuosa* (Sternb.) Engl. & Irmsch. = Saxifraga rivularis [SAXIRIV] Saxifragaceae (V:D)
Saxifraga rivularis var. *hyperborea* (R. Br.) Dorn = Saxifraga rivularis [SAXIRIV] Saxifragaceae (V:D)
Saxifraga rufidula (Small) Macoun [SAXIRUF] Saxifragaceae (V:D)
Saxifraga serpyllifolia Pursh [SAXISER] Saxifragaceae (V:D)
Saxifraga serpyllifolia var. *purpurea* Hultén = Saxifraga serpyllifolia [SAXISER] Saxifragaceae (V:D)
Saxifraga setigera Pursh = Saxifraga flagellaris ssp. setigera [SAXIFLA1] Saxifragaceae (V:D)
Saxifraga sileneflora Sternb. ex Cham. = Saxifraga cespitosa ssp. sileneflora [SAXICES2] Saxifragaceae (V:D)
Saxifraga taylori Calder & Savile [SAXITAY] Saxifragaceae (V:D)
Saxifraga tischii Skelly [SAXITIS] Saxifragaceae (V:D)
Saxifraga tolmiei Torr. & Gray [SAXITOL] Saxifragaceae (V:D)
Saxifraga tolmiei var. *ledifolia* (Greene) Engl. & Irmsch. = Saxifraga tolmiei [SAXITOL] Saxifragaceae (V:D)
Saxifraga tricuspidata Rottb. [SAXITRI] Saxifragaceae (V:D)
***Saxifraga tridactylites** L. [SAXITRD] Saxifragaceae (V:D)
Scabiosa L. [SCABIOS$] Dipsacaceae (V:D)
Scabiosa arvensis L. = Knautia arvensis [KNAUARV] Dipsacaceae (V:D)
***Scabiosa ochroleuca** L. [SCABOCH] Dipsacaceae (V:D)
Scandix L. [SCANDIX$] Apiaceae (V:D)
***Scandix pecten-veneris** L. [SCANPEC] Apiaceae (V:D)
Scapania (Dum.) Dum. [SCAPANI$] Scapaniaceae (B:H)
Scapania americana K. Müll. [SCAPAME] Scapaniaceae (B:H)
Scapania apiculata Spruce [SCAPAPI] Scapaniaceae (B:H)
Scapania bolanderi Aust. [SCAPBOL] Scapaniaceae (B:H)
Scapania bolanderi var. *americana* (K. Müll.) Frye & Clark = Scapania americana [SCAPAME] Scapaniaceae (B:H)
Scapania convexula K. Müll. = Scapania spitzbergensis [SCAPSPI] Scapaniaceae (B:H)
Scapania curta (Mart.) Dum. [SCAPCUR] Scapaniaceae (B:H)
Scapania cuspiduligera (Nees) K. Müll. [SCAPCUS] Scapaniaceae (B:H)
Scapania glaucocephala (Tayl.) Aust. [SCAPGLA] Scapaniaceae (B:H)
Scapania granulifera Evans = Scapania americana [SCAPAME] Scapaniaceae (B:H)
Scapania gymnostomophila Kaal. [SCAPGYM] Scapaniaceae (B:H)
Scapania heterophylla M.A. Howe = Scapania undulata var. undulata [SCAPUND0] Scapaniaceae (B:H)
Scapania hians Steph. [SCAPHIA] Scapaniaceae (B:H)
Scapania hians ssp. **hians** [SCAPHIA0] Scapaniaceae (B:H)
Scapania hians ssp. **salishensis** J. Godfrey & G. Godfrey [SCAPHIA1] Scapaniaceae (B:H)
Scapania hollandiae Hong [SCAPHOL] Scapaniaceae (B:H)
Scapania irrigua (Nees) Gott. & al. [SCAPIRR] Scapaniaceae (B:H)
Scapania mucronata Buch [SCAPMUC] Scapaniaceae (B:H)
Scapania oakesii Aust. = Scapania undulata var. oakesii [SCAPUND2] Scapaniaceae (B:H)
Scapania obscura (Arnell & C. Jens.) Schiffn. [SCAPOBS] Scapaniaceae (B:H)
Scapania paludicola Loeske & K. Müll. [SCAPPAL] Scapaniaceae (B:H)
Scapania paludosa (K. Müll.) K. Müll. [SCAPPAU] Scapaniaceae (B:H)
Scapania perlaxa Warnst. = Scapania subalpina [SCAPSUB] Scapaniaceae (B:H)
Scapania rosacea (Corda) Dum. = Scapania curta [SCAPCUR] Scapaniaceae (B:H)
Scapania scandica (Arnell & Buch) Macv. [SCAPSCA] Scapaniaceae (B:H)
Scapania simmonsii Bryhn & Kaal. [SCAPSIM] Scapaniaceae (B:H)
Scapania spitzbergensis (Lindb.) K. Müll. [SCAPSPI] Scapaniaceae (B:H)
Scapania subalpina (Nees) Dum. [SCAPSUB] Scapaniaceae (B:H)
Scapania subalpina var. *haynesiae* Frye & Clark = Scapania subalpina [SCAPSUB] Scapaniaceae (B:H)
Scapania uliginosa (Sw. ex Lindenb.) Dum. [SCAPULI] Scapaniaceae (B:H)
Scapania umbrosa (Schrad.) Dum. [SCAPUMB] Scapaniaceae (B:H)
Scapania undulata (L.) Dum. [SCAPUND] Scapaniaceae (B:H)
Scapania undulata var. *heterophylla* Warnst. = Scapania undulata var. undulata [SCAPUND0] Scapaniaceae (B:H)
Scapania undulata var. **oakesii** (Aust.) Buch [SCAPUND2] Scapaniaceae (B:H)
Scapania undulata var. **undulata** [SCAPUND0] Scapaniaceae (B:H)
Scapaniella glaucocephala (Tayl.) Evans = Scapania glaucocephala [SCAPGLA] Scapaniaceae (B:H)
Sceptridium multifidum (Gmel.) Tagawa = Botrychium multifidum [BOTRMUL] Ophioglossaceae (V:F)
Schadonia Körber [SCHADON$] Bacidiaceae (L)
Schadonia alpina Körber [SCHAALP] Bacidiaceae (L)
Schaereria Körber [SCHAERE$] Schaereriaceae (L)
Schaereria corticola Muhr & Tønsberg [SCHACOR] Schaereriaceae (L)
Scheuchzeria L. [SCHEUCH$] Scheuchzeriaceae (V:M)
Scheuchzeria americana (Fern.) G.N. Jones = Scheuchzeria palustris ssp. americana [SCHEPAL1] Scheuchzeriaceae (V:M)
Scheuchzeria palustris L. [SCHEPAL] Scheuchzeriaceae (V:M)

Scheuchzeria palustris ssp. **americana** (Fern.) Hultén [SCHEPAL1] Scheuchzeriaceae (V:M)
Scheuchzeria palustris var. *americana* Fern. = Scheuchzeria palustris ssp. americana [SCHEPAL1] Scheuchzeriaceae (V:M)
Schistidium Brid. [SCHISTI$] Grimmiaceae (B:M)
Schistidium agassizii Sull. & Lesq. in Sull. [SCHIAGA] Grimmiaceae (B:M)
Schistidium alpicola (Hedw.) Limpr. = Schistidium agassizii [SCHIAGA] Grimmiaceae (B:M)
Schistidium alpicola var. *dupretii* (Thér.) Crum = Schistidium apocarpum [SCHIAPO] Grimmiaceae (B:M)
Schistidium ambiguum Sull. = Schistidium apocarpum [SCHIAPO] Grimmiaceae (B:M)
Schistidium apocarpum (Hedw.) Bruch & Schimp. in B.S.G. [SCHIAPO] Grimmiaceae (B:M)
Schistidium apocarpum var. *ambiguum* (Sull.) G. Jones in Grout = Schistidium apocarpum [SCHIAPO] Grimmiaceae (B:M)
Schistidium apocarpum var. *atrofuscum* (Schimp.) C. Jens. ex Weim. = Schistidium apocarpum [SCHIAPO] Grimmiaceae (B:M)
Schistidium apocarpum var. *boreale* (Poelt) Duell = Schistidium apocarpum [SCHIAPO] Grimmiaceae (B:M)
Schistidium apocarpum var. *brunnescens* (Limpr.) Herz. = Schistidium apocarpum [SCHIAPO] Grimmiaceae (B:M)
Schistidium apocarpum var. *confertum* (Funck) Möll. = Schistidium apocarpum [SCHIAPO] Grimmiaceae (B:M)
Schistidium apocarpum var. *dupretii* (Thér.) Wijk & Marg. = Schistidium apocarpum [SCHIAPO] Grimmiaceae (B:M)
Schistidium apocarpum var. *gracile* (Web. & Mohr) Bruch & Schimp. in B.S.G. = Schistidium apocarpum [SCHIAPO] Grimmiaceae (B:M)
Schistidium apocarpum var. *nigrescens* (Mol.) Loeske = Schistidium apocarpum [SCHIAPO] Grimmiaceae (B:M)
Schistidium apocarpum var. *strictum* (Turn.) Moore = Schistidium apocarpum [SCHIAPO] Grimmiaceae (B:M)
Schistidium atrofuscum (Schimp.) Limpr. = Schistidium apocarpum [SCHIAPO] Grimmiaceae (B:M)
Schistidium boreale Poelt = Schistidium apocarpum [SCHIAPO] Grimmiaceae (B:M)
Schistidium brunnescens Limpr. = Schistidium apocarpum [SCHIAPO] Grimmiaceae (B:M)
Schistidium confertum (Funck) Bruch & Schimp. in B.S.G. = Schistidium apocarpum [SCHIAPO] Grimmiaceae (B:M)
Schistidium dupretii (Thér.) Web. = Schistidium apocarpum [SCHIAPO] Grimmiaceae (B:M)
Schistidium gracile (Web. & Mohr) Schleich. ex Limpr. = Schistidium apocarpum [SCHIAPO] Grimmiaceae (B:M)
Schistidium heterophyllum (Kindb. in Mac. & Kindb.) McIntosh [SCHIHET] Grimmiaceae (B:M)
Schistidium maritimum (Turn.) Bruch & Schimp. in B.S.G. [SCHIMAR] Grimmiaceae (B:M)
Schistidium papillosum Culm. = Schistidium apocarpum [SCHIAPO] Grimmiaceae (B:M)
Schistidium rivulare (Brid.) Podp. [SCHIRIV] Grimmiaceae (B:M)
Schistidium strictum (Turn.) T. Kop. & Isov. = Schistidium apocarpum [SCHIAPO] Grimmiaceae (B:M)
Schistidium tenerrimum (Ren. & Card.) G. Roth = Grimmia tenerrima [GRIMTEE] Grimmiaceae (B:M)
Schistidium tenerum (Zett.) Nyh. [SCHITEN] Grimmiaceae (B:M)
Schistidium teretinerve (Limpr.) Limpr. = Grimmia teretinervis [GRIMTER] Grimmiaceae (B:M)
Schistidium trichodon (Brid.) Poelt [SCHITRI] Grimmiaceae (B:M)
Schistostega Mohr [SCHISTO$] Schistostegaceae (B:M)
Schistostega osmundacea Mohr = Schistostega pennata [SCHIPEN] Schistostegaceae (B:M)
Schistostega pennata (Hedw.) Web. & Mohr [SCHIPEN] Schistostegaceae (B:M)
Schizachne Hack. [SCHIZAC$] Poaceae (V:M)
Schizachne purpurascens (Torr.) Swallen [SCHIPUR] Poaceae (V:M)
Schizachne purpurascens var. *pubescens* Dore = Schizachne purpurascens [SCHIPUR] Poaceae (V:M)
Schizachne stricta (Michx.) Hultén = Schizachne purpurascens [SCHIPUR] Poaceae (V:M)
Schizachyrium Nees [SCHIZAH$] Poaceae (V:M)
Schizachyrium scoparium (Michx.) Nash [SCHISCO] Poaceae (V:M)
Schobera occidentalis S. Wats. = Suaeda calceoliformis [SUAECAL] Chenopodiaceae (V:D)
Schoenocrambe Greene [SCHOENO$] Brassicaceae (V:D)
Schoenocrambe linifolia (Nutt.) Greene [SCHOLIN] Brassicaceae (V:D)
Schoenoplectus americanus (Pers.) Volk. ex Schinz & R. Keller = Scirpus americanus [SCIRAME] Cyperaceae (V:M)
Schoenoplectus lacustris ssp. *acutus* (Muhl. ex Bigelow) A. & D. Löve = Scirpus acutus [SCIRACU] Cyperaceae (V:M)
Schoenoplectus lacustris ssp. *creber* (Fern.) A. & D. Löve = Scirpus tabernaemontani [SCIRTAB] Cyperaceae (V:M)
Schoenoplectus lacustris ssp. *validus* (Vahl) T. Koyama = Scirpus tabernaemontani [SCIRTAB] Cyperaceae (V:M)
Schoenoplectus maritimus (L.) Lye = Scirpus maritimus [SCIRMAR] Cyperaceae (V:M)
Schoenoplectus pungens (Vahl) Palla = Scirpus pungens [SCIRPUN] Cyperaceae (V:M)
Schoenoplectus validus (Vahl) A. & D. Löve = Scirpus tabernaemontani [SCIRTAB] Cyperaceae (V:M)
Schofieldia Godfr. [SCHOFIE$] Cephaloziaceae (B:H)
Schofieldia monticola Godfr. [SCHOMON] Cephaloziaceae (B:H)
Scilla hispanica P. Mill. = Hyacinthoides hispanica [HYACHIS] Liliaceae (V:M)
Scirpus L. [SCIRPUS$] Cyperaceae (V:M)
Scirpus acutus Muhl. ex Bigelow [SCIRACU] Cyperaceae (V:M)
Scirpus americanus Pers. [SCIRAME] Cyperaceae (V:M)
Scirpus americanus var. *longispicatus* Britt. = Scirpus pungens [SCIRPUN] Cyperaceae (V:M)
Scirpus americanus var. *monophyllus* (J.& K. Presl) T. Koyama = Scirpus americanus [SCIRAME] Cyperaceae (V:M)
Scirpus americanus var. *polyphyllus* (Boeckl.) Beetle = Scirpus pungens [SCIRPUN] Cyperaceae (V:M)
Scirpus atrocinctus Fern. [SCIRATR] Cyperaceae (V:M)
Scirpus atropurpureus Retz. = Eleocharis atropurpurea [ELEOATR] Cyperaceae (V:M)
Scirpus atrovirens var. *pallidus* Britt. = Scirpus pallidus [SCIRPAI] Cyperaceae (V:M)
Scirpus bergsonii Schuyler = Scirpus saximontanus [SCIRSAX] Cyperaceae (V:M)
Scirpus cernuus Vahl [SCIRCER] Cyperaceae (V:M)
Scirpus cernuus ssp. *californicus* (Torr.) Thorne = Scirpus cernuus var. californicus [SCIRCER1] Cyperaceae (V:M)
Scirpus cernuus var. **californicus** (Torr.) Beetle [SCIRCER1] Cyperaceae (V:M)
Scirpus cespitosus L. [SCIRCES] Cyperaceae (V:M)
Scirpus cespitosus var. *austriacus* (Pallas) Aschers. & Graebn. = Scirpus cespitosus [SCIRCES] Cyperaceae (V:M)
Scirpus cespitosus var. *callosus* Bigelow = Scirpus cespitosus [SCIRCES] Cyperaceae (V:M)
Scirpus cespitosus var. *delicatulus* Fern. = Scirpus cespitosus [SCIRCES] Cyperaceae (V:M)
Scirpus chilensis Nees & Meyen ex Kunth = Scirpus americanus [SCIRAME] Cyperaceae (V:M)
Scirpus conglomeratus Kunth = Scirpus americanus [SCIRAME] Cyperaceae (V:M)
Scirpus cyperinus var. *brachypodus* (Fern.) Gilly = Scirpus atrocinctus [SCIRATR] Cyperaceae (V:M)
Scirpus fernaldii Bickn. = Scirpus maritimus [SCIRMAR] Cyperaceae (V:M)
Scirpus fluviatilis (Torr.) Gray [SCIRFLU] Cyperaceae (V:M)
Scirpus hudsonianus (Michx.) Fern. = Eriophorum alpinum [ERIOALP] Cyperaceae (V:M)
Scirpus kamtschaticus C.A. Mey. = Eleocharis kamtschatica [ELEOKAM] Cyperaceae (V:M)
Scirpus lacustris L. p.p. = Scirpus acutus [SCIRACU] Cyperaceae (V:M)
Scirpus lacustris ssp. *creber* (Fern.) T. Koyama = Scirpus tabernaemontani [SCIRTAB] Cyperaceae (V:M)
Scirpus lacustris ssp. *glaucus* (Reichenb.) Hartman = Scirpus tabernaemontani [SCIRTAB] Cyperaceae (V:M)

Scirpus lacustris ssp. *tabernaemontani* (K.C. Gmel.) Syme = Scirpus tabernaemontani [SCIRTAB] Cyperaceae (V:M)
Scirpus lacustris ssp. *validus* (Vahl) T. Koyama = Scirpus tabernaemontani [SCIRTAB] Cyperaceae (V:M)
Scirpus longispicatus (Britt.) Smyth = Scirpus pungens [SCIRPUN] Cyperaceae (V:M)
Scirpus maritimus L. [SCIRMAR] Cyperaceae (V:M)
Scirpus maritimus var. *fernaldii* (Bickn.) Beetle = Scirpus maritimus [SCIRMAR] Cyperaceae (V:M)
Scirpus maritimus var. *paludosus* (A. Nels.) Kükenth. = Scirpus maritimus [SCIRMAR] Cyperaceae (V:M)
Scirpus micranthus Vahl = Lipocarpha micrantha [LIPOMIC] Cyperaceae (V:M)
Scirpus microcarpus J.& K. Presl [SCIRMIC] Cyperaceae (V:M)
Scirpus microcarpus var. *longispicatus* M.E. Peck = Scirpus microcarpus [SCIRMIC] Cyperaceae (V:M)
Scirpus microcarpus var. *rubrotinctus* (Fern.) M.E. Jones = Scirpus microcarpus [SCIRMIC] Cyperaceae (V:M)
Scirpus monophyllus J.& K. Presl = Scirpus americanus [SCIRAME] Cyperaceae (V:M)
Scirpus nanus Spreng. = Eleocharis quinqueflora [ELEOQUI] Cyperaceae (V:M)
Scirpus nevadensis S. Wats. [SCIRNEV] Cyperaceae (V:M)
Scirpus obtusus Willd. = Eleocharis obtusa [ELEOOBT] Cyperaceae (V:M)
Scirpus occidentalis (S. Wats.) Chase = Scirpus acutus [SCIRACU] Cyperaceae (V:M)
Scirpus olneyi Gray = Scirpus americanus [SCIRAME] Cyperaceae (V:M)
Scirpus pacificus Britt. = Scirpus maritimus [SCIRMAR] Cyperaceae (V:M)
Scirpus pallidus (Britt.) Fern. [SCIRPAI] Cyperaceae (V:M)
Scirpus paludosus A. Nels. = Scirpus maritimus [SCIRMAR] Cyperaceae (V:M)
Scirpus paludosus var. *atlanticus* Fern. = Scirpus maritimus [SCIRMAR] Cyperaceae (V:M)
Scirpus pauciflorus Lightf. = Eleocharis quinqueflora [ELEOQUI] Cyperaceae (V:M)
Scirpus pumilus auct. = Scirpus rollandii [SCIRROL] Cyperaceae (V:M)
Scirpus pumilus ssp. *rollandii* (Fern.) Raymond = Scirpus rollandii [SCIRROL] Cyperaceae (V:M)
Scirpus pumilus var. *rollandii* (Fern.) Beetle = Scirpus rollandii [SCIRROL] Cyperaceae (V:M)
Scirpus pungens Vahl [SCIRPUN] Cyperaceae (V:M)
Scirpus pungens ssp. *monophyllus* (J.& K. Presl) Taylor & MacBryde = Scirpus americanus [SCIRAME] Cyperaceae (V:M)
Scirpus pungens var. *longisetus* Benth. & F. Muell. = Scirpus pungens [SCIRPUN] Cyperaceae (V:M)
Scirpus pungens var. *longispicatus* (Britt.) Taylor & MacBryde = Scirpus pungens [SCIRPUN] Cyperaceae (V:M)
Scirpus pungens var. *polyphyllus* Boeckl. = Scirpus pungens [SCIRPUN] Cyperaceae (V:M)
Scirpus quinqueflorus F.X. Hartmann = Eleocharis quinqueflora [ELEOQUI] Cyperaceae (V:M)
Scirpus rollandii Fern. [SCIRROL] Cyperaceae (V:M)
Scirpus rostellatus Torr. = Eleocharis rostellata [ELEOROS] Cyperaceae (V:M)
Scirpus rubrotinctus Fern. = Scirpus microcarpus [SCIRMIC] Cyperaceae (V:M)
Scirpus saximontanus Fern. [SCIRSAX] Cyperaceae (V:M)
*Scirpus setaceus** L. [SCIRSET] Cyperaceae (V:M)
Scirpus steinmetzii Fern. = Scirpus tabernaemontani [SCIRTAB] Cyperaceae (V:M)
Scirpus subterminalis Torr. [SCIRSUB] Cyperaceae (V:M)
Scirpus supinus var. *saximontanus* (Fern.) T. Koyama = Scirpus saximontanus [SCIRSAX] Cyperaceae (V:M)
Scirpus sylvaticus ssp. *digynus* (Boeckl.) Koyama = Scirpus microcarpus [SCIRMIC] Cyperaceae (V:M)
Scirpus tabernaemontani K.C. Gmel. [SCIRTAB] Cyperaceae (V:M)
Scirpus uniglumis Link = Eleocharis uniglumis [ELEOUNI] Cyperaceae (V:M)
Scirpus validus Vahl = Scirpus tabernaemontani [SCIRTAB] Cyperaceae (V:M)
Scirpus validus var. *creber* Fern. = Scirpus tabernaemontani [SCIRTAB] Cyperaceae (V:M)
Scleranthus L. [SCLERAN$] Caryophyllaceae (V:D)
*Scleranthus annuus** L. [SCLEANN] Caryophyllaceae (V:D)
Sclerophora Chevall. [SCLEROI$] Coniocybaceae (L)
Sclerophora amabilis (Tibell) Tibell [SCLEAMA] Coniocybaceae (L)
Sclerophora coniophaea (Norman) J. Mattsson & Middelb. [SCLECON] Coniocybaceae (L)
Sclerophora farinacea (Chevall.) Chevall. [SCLEFAR] Coniocybaceae (L)
Sclerophora peronella (Ach.) Tibell [SCLEPER] Coniocybaceae (L)
Scleropodium Schimp. in B.S.G. [SCLEROP$] Brachytheciaceae (B:M)
Scleropodium apocladum (Mitt.) Grout = Scleropodium cespitans var. sublaeve [SCLECES2] Brachytheciaceae (B:M)
Scleropodium apocladum var. *obtusum* Grout = Scleropodium cespitans var. sublaeve [SCLECES2] Brachytheciaceae (B:M)
Scleropodium caespitosum Schimp. in B.S.G. = Scleropodium cespitans var. cespitans [SCLECES0] Brachytheciaceae (B:M)
Scleropodium caespitosum var. *sublaeve* Ren. & Card. in Röll = Scleropodium cespitans var. sublaeve [SCLECES2] Brachytheciaceae (B:M)
Scleropodium cespitans (C. Müll.) L. Koch [SCLECES] Brachytheciaceae (B:M)
Scleropodium cespitans var. **cespitans** [SCLECES0] Brachytheciaceae (B:M)
Scleropodium cespitans var. *laeve* Ren. & Card. in Röll = Scleropodium cespitans var. sublaeve [SCLECES2] Brachytheciaceae (B:M)
Scleropodium cespitans var. **sublaeve** (Ren. & Card. in Röll) Wijk & Marg. [SCLECES2] Brachytheciaceae (B:M)
Scleropodium colpophyllum (Sull.) Grout = Scleropodium touretii var. colpophyllum [SCLETOU1] Brachytheciaceae (B:M)
Scleropodium colpophyllum var. *attenuatum* Grout = Scleropodium touretii var. colpophyllum [SCLETOU1] Brachytheciaceae (B:M)
Scleropodium illecebrum Schimp. in B.S.G. = Scleropodium touretii [SCLETOU] Brachytheciaceae (B:M)
Scleropodium obtusifolium (Jaeg.) Kindb. in Mac. & Kindb. [SCLEOBT] Brachytheciaceae (B:M)
Scleropodium touretii (Brid.) L. Koch [SCLETOU] Brachytheciaceae (B:M)
Scleropodium touretii var. **colpophyllum** (Sull.) Lawt. ex Crum [SCLETOU1] Brachytheciaceae (B:M)
Scleropodium touretii var. **touretii** [SCLETOU0] Brachytheciaceae (B:M)
Scoliciosporum A. Massal. [SCOLICI$] Lecanoraceae (L)
Scoliciosporum chlorococcum (Stenh.) Vezda [SCOLCHL] Lecanoraceae (L)
Scoliciosporum umbrinum (Ach.) Arnold [SCOLUMB] Lecanoraceae (L)
Scolochloa Link [SCOLOCH$] Poaceae (V:M)
Scolochloa festucacea (Willd.) Link [SCOLFES] Poaceae (V:M)
Scopelophila latifolia (Kindb. in Mac.) Ren. & Card. = Crumia latifolia [CRUMLAT] Pottiaceae (B:M)
Scorpidium (Schimp.) Limpr. [SCORPID$] Amblystegiaceae (B:M)
Scorpidium cossonii (Schimp.) Hedenäs = Limprichtia cossonii [LIMPCOS] Amblystegiaceae (B:M)
Scorpidium revolvens (Sw.) Hedenäs = Limprichtia revolvens [LIMPREV] Amblystegiaceae (B:M)
Scorpidium scorpioides (Hedw.) Limpr. [SCORSCO] Amblystegiaceae (B:M)
Scorpidium scorpioides f. *gracilescens* (Sanio) C. Jens. = Scorpidium scorpioides [SCORSCO] Amblystegiaceae (B:M)
Scorpidium trifarium (Web. & Mohr) Paul = Calliergon trifarium [CALLTRI] Amblystegiaceae (B:M)
Scorpidium turgescens (T. Jens.) Loeske = Pseudocalliergon turgescens [PSEUTUR] Amblystegiaceae (B:M)
Scorpidium vernicosum (Mitt.) Tuom. = Hamatocaulis vernicosus [HAMAVER] Amblystegiaceae (B:M)

Scorzonella nutans Hook. = Microseris nutans [MICRNUT] Asteraceae (V:D)
Scorzonella nutans var. *major* (Gray) M.E. Peck = Microseris nutans [MICRNUT] Asteraceae (V:D)
Scorzonella troximoides (Gray) Jepson = Nothocalais troximoides [NOTHTRO] Asteraceae (V:D)
Scouleria Hook. in Drumm. [SCOULER$] Scouleriaceae (B:M)
Scouleria aquatica Hook. in Drumm. [SCOUAQU] Scouleriaceae (B:M)
Scouleria marginata Britt. [SCOUMAR] Scouleriaceae (B:M)
Scrophularia L. [SCROPHU$] Scrophulariaceae (V:D)
Scrophularia californica var. *oregana* (Pennell) Boivin = Scrophularia oregana [SCROORE] Scrophulariaceae (V:D)
Scrophularia californica var. *oregana* (Pennell) Cronq. = Scrophularia oregana [SCROORE] Scrophulariaceae (V:D)
Scrophularia lanceolata Pursh [SCROLAN] Scrophulariaceae (V:D)
Scrophularia oregana Pennell [SCROORE] Scrophulariaceae (V:D)
Scrophularia pectinata Raf. = Scrophularia lanceolata [SCROLAN] Scrophulariaceae (V:D)
Scutellaria L. [SCUTELL$] Lamiaceae (V:D)
Scutellaria angustifolia Pursh [SCUTANG] Lamiaceae (V:D)
Scutellaria epilobiifolia A. Hamilton = Scutellaria galericulata [SCUTGAL] Lamiaceae (V:D)
Scutellaria galericulata L. [SCUTGAL] Lamiaceae (V:D)
Scutellaria galericulata ssp. *pubescens* (Benth.) A. & D. Löve = Scutellaria galericulata [SCUTGAL] Lamiaceae (V:D)
Scutellaria galericulata var. *epilobiifolia* (A. Hamilton) Jordal = Scutellaria galericulata [SCUTGAL] Lamiaceae (V:D)
Scutellaria galericulata var. *pubescens* Benth. = Scutellaria galericulata [SCUTGAL] Lamiaceae (V:D)
Scutellaria lateriflora L. [SCUTLAT] Lamiaceae (V:D)
Scutula Tul. [SCUTULA$] Micareaceae (L)
Scutula miliaris (Wallr.) Trevisan [SCUTMIL] Micareaceae (L)
Scutula stereocaulorum (Anzi) Körber [SCUTSTE] Micareaceae (L)
Scutula tuberculosa (Th. Fr.) Rehm = Scutula miliaris [SCUTMIL] Micareaceae (L)
Secale L. [SECALE$] Poaceae (V:M)
***Secale cereale** L. [SECACER] Poaceae (V:M)
Securigera varia (L.) Lassen = Coronilla varia [COROVAR] Fabaceae (V:D)
Sedum L. [SEDUM$] Crassulaceae (V:D)
***Sedum acre** L. [SEDUACR] Crassulaceae (V:D)
Sedum alaskanum (Rose) Rose ex Hutchinson = Sedum integrifolium [SEDUINT] Crassulaceae (V:D)
***Sedum album** L. [SEDUALB] Crassulaceae (V:D)
Sedum divergens S. Wats. [SEDUDIV] Crassulaceae (V:D)
Sedum douglasii Hook. = Sedum stenopetalum [SEDUSTE] Crassulaceae (V:D)
Sedum frigidum Rydb. = Sedum integrifolium [SEDUINT] Crassulaceae (V:D)
Sedum integrifolium (Raf.) A. Nels. [SEDUINT] Crassulaceae (V:D)
Sedum lanceolatum Torr. [SEDULAN] Crassulaceae (V:D)
Sedum lanceolatum ssp. **lanceolatum** [SEDULAN0] Crassulaceae (V:D)
Sedum lanceolatum ssp. **nesioticum** (G.N. Jones) Clausen [SEDULAN1] Crassulaceae (V:D)
Sedum lanceolatum var. *nesioticum* (G.N. Jones) C.L. Hitchc. = Sedum lanceolatum ssp. nesioticum [SEDULAN1] Crassulaceae (V:D)
Sedum oreganum Nutt. [SEDUORE] Crassulaceae (V:D)
Sedum pruinosum Britt. = Sedum spathulifolium ssp. pruinosum [SEDUSPA1] Crassulaceae (V:D)
Sedum rosea ssp. *integrifolium* (Raf.) Hultén = Sedum integrifolium [SEDUINT] Crassulaceae (V:D)
Sedum rosea var. *alaskanum* (Rose) Berger = Sedum integrifolium [SEDUINT] Crassulaceae (V:D)
Sedum rosea var. *frigidum* (Rydb.) Hultén = Sedum integrifolium [SEDUINT] Crassulaceae (V:D)
Sedum rosea var. *integrifolium* (Raf.) Berger = Sedum integrifolium [SEDUINT] Crassulaceae (V:D)
Sedum spathulifolium Hook. [SEDUSPA] Crassulaceae (V:D)
Sedum spathulifolium ssp. *anomalum* (Britt.) Clausen & Uhl = Sedum spathulifolium ssp. spathulifolium [SEDUSPA0] Crassulaceae (V:D)
Sedum spathulifolium ssp. **pruinosum** (Britt.) Clausen & Uhl [SEDUSPA1] Crassulaceae (V:D)
Sedum spathulifolium ssp. **spathulifolium** [SEDUSPA0] Crassulaceae (V:D)
Sedum spathulifolium var. *pruinosum* (Britt.) Boivin = Sedum spathulifolium ssp. pruinosum [SEDUSPA1] Crassulaceae (V:D)
Sedum stenopetalum Pursh [SEDUSTE] Crassulaceae (V:D)
Selaginella Beauv. [SELAGIN$] Selaginellaceae (V:F)
Selaginella densa Rydb. [SELADEN] Selaginellaceae (V:F)
Selaginella densa var. **scopulorum** (Maxon) R. Tryon [SELADEN2] Selaginellaceae (V:F)
Selaginella engelmannii var. *scopulorum* (Maxon) C.F. Reed = Selaginella densa var. scopulorum [SELADEN2] Selaginellaceae (V:F)
Selaginella oregana D.C. Eat. [SELAORE] Selaginellaceae (V:F)
Selaginella scopulorum Maxon = Selaginella densa var. scopulorum [SELADEN2] Selaginellaceae (V:F)
Selaginella selaginoides (L.) Link [SELASEL] Selaginellaceae (V:F)
Selaginella sibirica (Milde) Hieron. [SELASIB] Selaginellaceae (V:F)
Selaginella wallacei Hieron. [SELAWAL] Selaginellaceae (V:F)
Seligeria Bruch & Schimp. in B.S.G. [SELIGER$] Seligeriaceae (B:M)
Seligeria acutifolia Lindb. in Hartm. [SELIACU] Seligeriaceae (B:M)
Seligeria campylopoda Kindb. in Mac. & Kindb. [SELICAM] Seligeriaceae (B:M)
Seligeria careyana Vitt. & Schof. [SELICAR] Seligeriaceae (B:M)
Seligeria donniana (Sm.) C. Müll. [SELIDON] Seligeriaceae (B:M)
Seligeria donnii Lindb. = Seligeria donniana [SELIDON] Seligeriaceae (B:M)
Seligeria recurvata (Hedw.) Bruch & Schimp. in B.S.G. [SELIREC] Seligeriaceae (B:M)
Seligeria tristichoides Kindb. [SELITRI] Seligeriaceae (B:M)
Sematophyllum Mitt. [SEMATOP$] Sematophyllaceae (B:M)
Sematophyllum carolinianum (C. Müll.) Britt. = Sematophyllum demissum [SEMADEM] Sematophyllaceae (B:M)
Sematophyllum demissum (Wils.) Mitt. [SEMADEM] Sematophyllaceae (B:M)
Sematophyllum micans (Mitt.) Braithw. = Hygrohypnum micans [HYGRMIC] Amblystegiaceae (B:M)
Sematophyllum novae-caesareae (Aust.) Britt. = Hygrohypnum micans [HYGRMIC] Amblystegiaceae (B:M)
Senecio L. [SENECIO$] Asteraceae (V:D)
Senecio alaskanus Hultén = Senecio yukonensis [SENEYUK] Asteraceae (V:D)
Senecio atropurpureus (Ledeb.) Fedtsch. [SENEATR] Asteraceae (V:D)
Senecio atropurpureus ssp. *frigidus* (Richards.) Hultén = Senecio atropurpureus [SENEATR] Asteraceae (V:D)
Senecio atropurpureus ssp. *tomentosus* (Kjellm.) Hultén = Senecio atropurpureus [SENEATR] Asteraceae (V:D)
Senecio atropurpureus var. *dentatus* (Gray) Hultén = Senecio atropurpureus [SENEATR] Asteraceae (V:D)
Senecio atropurpureus var. *ulmeri* (Steffen) Porsild = Senecio atropurpureus [SENEATR] Asteraceae (V:D)
Senecio balsamitae Muhl. ex Willd. = Senecio pauperculus [SENEPAP] Asteraceae (V:D)
Senecio canus Hook. [SENECAN] Asteraceae (V:D)
Senecio columbianus Greene = Senecio integerrimus var. exaltatus [SENEINT1] Asteraceae (V:D)
Senecio congestus (R. Br.) DC. [SENECON] Asteraceae (V:D)
Senecio congestus ssp. *tonsus* (Fern.) A. & D. Löve = Senecio congestus [SENECON] Asteraceae (V:D)
Senecio congestus var. *palustris* (L.) Fern. = Senecio congestus [SENECON] Asteraceae (V:D)
Senecio congestus var. *tonsus* Fern. = Senecio congestus [SENECON] Asteraceae (V:D)
Senecio conterminus Greenm. [SENECOT] Asteraceae (V:D)

Senecio convallium Greenm. = Senecio canus [SENECAN] Asteraceae (V:D)
Senecio crawfordii (Britt.) G.W. & G.R. Dougl. = Senecio pauperculus [SENEPAP] Asteraceae (V:D)
Senecio crepidineus Greene = Senecio elmeri [SENEELM] Asteraceae (V:D)
Senecio cymbalarioides Buek [SENECYM] Asteraceae (V:D)
Senecio cymbalarioides Nutt. = Senecio streptanthifolius [SENESTR] Asteraceae (V:D)
Senecio cymbalarioides ssp. *moresbiensis* Calder & Taylor = Senecio cymbalarioides [SENECYM] Asteraceae (V:D)
Senecio cymbalarioides var. *moresbiensis* Calder & Taylor = Senecio cymbalarioides [SENECYM] Asteraceae (V:D)
Senecio discoideus (Hook.) Britt. = Senecio pauciflorus [SENEPAU] Asteraceae (V:D)
Senecio elmeri Piper [SENEELM] Asteraceae (V:D)
Senecio eremophilus Richards. [SENEERE] Asteraceae (V:D)
Senecio exaltatus Nutt. = Senecio integerrimus var. exaltatus [SENEINT1] Asteraceae (V:D)
Senecio foetidus J.T. Howell = Senecio hydrophiloides [SENEHYR] Asteraceae (V:D)
Senecio foetidus var. *hydrophiloides* (Rydb.) T.M. Barkl. ex Cronq. = Senecio hydrophiloides [SENEHYR] Asteraceae (V:D)
Senecio fremontii Torr. & Gray [SENEFRE] Asteraceae (V:D)
Senecio frigidus (Richards.) Less. = Senecio atropurpureus [SENEATR] Asteraceae (V:D)
Senecio fuscatus Hayek [SENEFUS] Asteraceae (V:D)
Senecio gaspensis Greenm. = Senecio pauperculus [SENEPAP] Asteraceae (V:D)
Senecio gaspensis var. *firmifolius* (Greenm.) Fern. = Senecio pauperculus [SENEPAP] Asteraceae (V:D)
Senecio gibbsonsii Greene = Senecio triangularis [SENETRI] Asteraceae (V:D)
Senecio glaucifolius Rydb. = Senecio eremophilus [SENEERE] Asteraceae (V:D)
Senecio hallii Britt. = Senecio canus [SENECAN] Asteraceae (V:D)
Senecio hallii var. *discoidea* W.A. Weber = Senecio canus [SENECAN] Asteraceae (V:D)
Senecio harbourii Rydb. = Senecio canus [SENECAN] Asteraceae (V:D)
Senecio hookeri Torr. & Gray = Senecio integerrimus var. exaltatus [SENEINT1] Asteraceae (V:D)
Senecio howellii Greene = Senecio canus [SENECAN] Asteraceae (V:D)
Senecio hydrophiloides Rydb. [SENEHYR] Asteraceae (V:D)
Senecio hydrophilus Nutt. [SENEHYD] Asteraceae (V:D)
Senecio hyperborealis Greenm. [SENEHYP] Asteraceae (V:D)
Senecio indecorus Greene [SENEIND] Asteraceae (V:D)
Senecio integerrimus Nutt. [SENEINT] Asteraceae (V:D)
Senecio integerrimus var. **exaltatus** (Nutt.) Cronq. [SENEINT1] Asteraceae (V:D)
Senecio integerrimus var. *lugens* (Richards.) Boivin = Senecio lugens [SENELUG] Asteraceae (V:D)
Senecio integerrimus var. **ochroleucus** (Gray) Cronq. [SENEINT3] Asteraceae (V:D)
Senecio integerrimus var. *vaseyi* (Greenm.) Cronq. = Senecio integerrimus var. exaltatus [SENEINT1] Asteraceae (V:D)
***Senecio jacobaea** L. [SENEJAC] Asteraceae (V:D)
Senecio kjellmanii Porsild = Senecio atropurpureus [SENEATR] Asteraceae (V:D)
Senecio ligulifolius Greene = Senecio macounii [SENEMAC] Asteraceae (V:D)
Senecio lindstroemii (Ostenf.) Porsild = Senecio fuscatus [SENEFUS] Asteraceae (V:D)
Senecio lugens Richards. [SENELUG] Asteraceae (V:D)
Senecio macounii Greene [SENEMAC] Asteraceae (V:D)
Senecio megacephalus Nutt. [SENEMEG] Asteraceae (V:D)
Senecio moresbiensis (Calder & Taylor) Dougl. & Ruyle-Dougl. = Senecio cymbalarioides [SENECYM] Asteraceae (V:D)
Senecio newcombei Greene = Sinosenecio newcombei [SINONEW] Asteraceae (V:D)
Senecio ogotorukensis Packer [SENEOGO] Asteraceae (V:D)
Senecio palustris (L.) Hook. = Senecio congestus [SENECON] Asteraceae (V:D)
Senecio pauciflorus Pursh [SENEPAU] Asteraceae (V:D)
Senecio pauciflorus var. *fallax* (Greenm.) Greenm. = Senecio indecorus [SENEIND] Asteraceae (V:D)
Senecio pauciflorus var. *jucundulus* Jepson = Senecio pseudaureus [SENEPSE] Asteraceae (V:D)
Senecio pauperculus Michx. [SENEPAP] Asteraceae (V:D)
Senecio pauperculus var. *balsamitae* (Muhl. ex Willd.) Fern. = Senecio pauperculus [SENEPAP] Asteraceae (V:D)
Senecio pauperculus var. *crawfordii* (Britt.) T.M. Barkl. = Senecio pauperculus [SENEPAP] Asteraceae (V:D)
Senecio pauperculus var. *firmifolius* (Greenm.) Greenm. = Senecio pauperculus [SENEPAP] Asteraceae (V:D)
Senecio pauperculus var. *neoscoticus* Fern. = Senecio pauperculus [SENEPAP] Asteraceae (V:D)
Senecio pauperculus var. *praelongus* (Greenm.) House = Senecio pauperculus [SENEPAP] Asteraceae (V:D)
Senecio pauperculus var. *thompsoniensis* (Greenm.) Boivin = Senecio pauperculus [SENEPAP] Asteraceae (V:D)
Senecio pereziifolius Rydb. = Senecio hydrophiloides [SENEHYR] Asteraceae (V:D)
Senecio plattensis Nutt. [SENEPLA] Asteraceae (V:D)
Senecio pseudaureus Rydb. [SENEPSE] Asteraceae (V:D)
Senecio pseudoarnica Less. [SENEPSU] Asteraceae (V:D)
Senecio pseudotomentosus Mackenzie & Bush = Senecio plattensis [SENEPLA] Asteraceae (V:D)
Senecio purshianus Nutt. = Senecio canus [SENECAN] Asteraceae (V:D)
Senecio serra Hook. [SENESER] Asteraceae (V:D)
Senecio sheldonensis Porsild [SENESHE] Asteraceae (V:D)
Senecio streptanthifolius Greene [SENESTR] Asteraceae (V:D)
Senecio streptanthifolius var. *moresbiensis* (Calder & Taylor) Boivin = Senecio cymbalarioides [SENECYM] Asteraceae (V:D)
Senecio subnudus DC. = Senecio cymbalarioides [SENECYM] Asteraceae (V:D)
***Senecio sylvaticus** L. [SENESYL] Asteraceae (V:D)
Senecio triangularis Hook. [SENETRI] Asteraceae (V:D)
Senecio triangularis var. *angustifolius* G.N. Jones = Senecio triangularis [SENETRI] Asteraceae (V:D)
Senecio tundricola Tolm. = Senecio fuscatus [SENEFUS] Asteraceae (V:D)
Senecio tweedyi Rydb. = Senecio pauperculus [SENEPAP] Asteraceae (V:D)
Senecio vaseyi Greenm. = Senecio integerrimus var. exaltatus [SENEINT1] Asteraceae (V:D)
***Senecio viscosus** L. [SENEVIS] Asteraceae (V:D)
***Senecio vulgaris** L. [SENEVUL] Asteraceae (V:D)
Senecio yukonensis Porsild [SENEYUK] Asteraceae (V:D)
Serapias helleborine L. = Epipactis helleborine [EPIPHEL] Orchidaceae (V:M)
Sericocarpus rigidus Lindl. = Aster curtus [ASTECUR] Asteraceae (V:D)
Sericotheca discolor (Pursh) Rydb. = Holodiscus discolor [HOLODIC] Rosaceae (V:D)
Seriphidium tridentatum (Nutt.) W.A. Weber = Artemisia tridentata ssp. tridentata [ARTETRI0] Asteraceae (V:D)
Seriphidium tridentatum ssp. *parishii* (Gray) W.A. Weber = Artemisia tridentata ssp. tridentata [ARTETRI0] Asteraceae (V:D)
Seriphidium tridentatum ssp. *vaseyanum* (Rydb.) W.A. Weber = Artemisia tridentata ssp. vaseyana [ARTETRI2] Asteraceae (V:D)
Seriphidium vaseyanum (Rydb.) W.A. Weber = Artemisia tridentata ssp. vaseyana [ARTETRI2] Asteraceae (V:D)
Serratula arvensis L. = Cirsium arvense [CIRSARV] Asteraceae (V:D)
Setaria Beauv. [SETARIA$] Poaceae (V:M)
Setaria decipiens Schimp. ex Nyman = Setaria verticillata var. ambigua [SETAVER1] Poaceae (V:M)
***Setaria glauca** (L.) Beauv. [SETAGLA] Poaceae (V:M)
***Setaria italica** (L.) Beauv. [SETAITA] Poaceae (V:M)
Setaria italica subvar. metzeri (Koern.) F.T. Hubbard = Setaria italica [SETAITA] Poaceae (V:M)

Setaria italica var. *metzeri* (Koern.) Jáv. = Setaria italica [SETAITA] Poaceae (V:M)
Setaria italica var. *stramineofructa* (F.T. Hubbard) Bailey = Setaria italica [SETAITA] Poaceae (V:M)
Setaria lutescens (Weigel) F.T. Hubbard = Setaria glauca [SETAGLA] Poaceae (V:M)
***Setaria verticillata** (L.) Beauv. [SETAVER] Poaceae (V:M)
Setaria verticillata f. *ambigua* (Guss.) Boivin = Setaria verticillata var. ambigua [SETAVER1] Poaceae (V:M)
Setaria verticillata var. **ambigua** (Guss.) Parl. [SETAVER1] Poaceae (V:M)
***Setaria viridis** (L.) Beauv. [SETAVIR] Poaceae (V:M)
Setaria viridis var. *ambigua* (Guss.) Coss. & Durieu = Setaria verticillata var. ambigua [SETAVER1] Poaceae (V:M)
Sharpiella seligeri (Brid.) Iwats. = Herzogiella seligeri [HERZSEL] Hypnaceae (B:M)
Sharpiella striatella (Brid.) Iwats. = Herzogiella striatella [HERZSTR] Hypnaceae (B:M)
Shepherdia Nutt. [SHEPHER$] Elaeagnaceae (V:D)
Shepherdia argentea (Pursh) Nutt. [SHEPARG] Elaeagnaceae (V:D)
Shepherdia canadensis (L.) Nutt. [SHEPCAN] Elaeagnaceae (V:D)
Sherardia L. [SHERARD$] Rubiaceae (V:D)
***Sherardia arvensis** L. [SHERARV] Rubiaceae (V:D)
Sibbaldia L. [SIBBALD$] Rosaceae (V:D)
Sibbaldia procumbens L. [SIBBPRO] Rosaceae (V:D)
Sicyos lobata Michx. = Echinocystis lobata [ECHILOB] Cucurbitaceae (V:D)
Sidalcea Gray [SIDALCE$] Malvaceae (V:D)
Sidalcea hendersonii S. Wats. [SIDAHEN] Malvaceae (V:D)
Sidalcea oregana (Nutt. ex Torr. & Gray) Gray [SIDAORE] Malvaceae (V:D)
Sidalcea oregana var. **procera** C.L. Hitchc. [SIDAORE1] Malvaceae (V:D)
Sieglingia decumbens (L.) Bernh. = Danthonia decumbens [DANTDEC] Poaceae (V:M)
Siella erecta (Huds.) M. Pimen. = Berula erecta [BERUERE] Apiaceae (V:D)
Sieversia ciliata (Pursh) G. Don = Geum triflorum var. ciliatum [GEUMTRI1] Rosaceae (V:D)
Sieversia × macrantha Kearney = Geum × macranthum [GEUMMAR] Rosaceae (V:D)
Sieversia triflora (Pursh) R. Br. = Geum triflorum var. triflorum [GEUMTRI0] Rosaceae (V:D)
Silene L. [SILENE$] Caryophyllaceae (V:D)
Silene acaulis (L.) Jacq. [SILEACA] Caryophyllaceae (V:D)
Silene acaulis ssp. *subacaulescens* (F.N. Williams) Hultén = Silene acaulis var. subacaulescens [SILEACA5] Caryophyllaceae (V:D)
Silene acaulis var. **acaulis** [SILEACA0] Caryophyllaceae (V:D)
Silene acaulis var. **subacaulescens** (F.N. Williams) Fern. & St. John [SILEACA5] Caryophyllaceae (V:D)
Silene alba (P. Mill.) Krause = Silene latifolia ssp. alba [SILELAT1] Caryophyllaceae (V:D)
Silene anglica L. = Silene gallica [SILEGAL] Caryophyllaceae (V:D)
Silene antirrhina L. [SILEANT] Caryophyllaceae (V:D)
Silene antirrhina var. *confinis* Fern. = Silene antirrhina [SILEANT] Caryophyllaceae (V:D)
Silene antirrhina var. *depauperata* Rydb. = Silene antirrhina [SILEANT] Caryophyllaceae (V:D)
Silene antirrhina var. *divaricata* B.L. Robins. = Silene antirrhina [SILEANT] Caryophyllaceae (V:D)
Silene antirrhina var. *laevigata* Engelm. & Gray = Silene antirrhina [SILEANT] Caryophyllaceae (V:D)
Silene antirrhina var. *subglaber* Engelm. & Gray = Silene antirrhina [SILEANT] Caryophyllaceae (V:D)
Silene antirrhina var. *vaccarifolia* Rydb. = Silene antirrhina [SILEANT] Caryophyllaceae (V:D)
***Silene armeria** L. [SILEARM] Caryophyllaceae (V:D)
Silene attenuata (Farr) Bocquet = Silene uralensis ssp. attenuata [SILEURA1] Caryophyllaceae (V:D)
Silene coronaria (L.) Clairville = Lychnis coronaria [LYCHCOR] Caryophyllaceae (V:D)
***Silene csereii** Baumg. [SILECSE] Caryophyllaceae (V:D)
Silene cucubalus Wibel = Silene vulgaris [SILEVUL] Caryophyllaceae (V:D)
Silene cucubalus var. *latifolia* (P. Mill.) G. Beck = Silene vulgaris [SILEVUL] Caryophyllaceae (V:D)
***Silene dichotoma** Ehrh. [SILEDIC] Caryophyllaceae (V:D)
***Silene dioica** (L.) Clairville [SILEDIO] Caryophyllaceae (V:D)
Silene douglasii Hook. [SILEDOU] Caryophyllaceae (V:D)
Silene douglasii var. *macounii* (S. Wats.) B.L. Robins. = Silene parryi [SILEPAR] Caryophyllaceae (V:D)
Silene douglasii var. *villosa* C.L. Hitchc. & Maguire = Silene douglasii [SILEDOU] Caryophyllaceae (V:D)
Silene drummondii Hook. [SILEDRU] Caryophyllaceae (V:D)
Silene furcata Raf. = Silene involucrata [SILEINV] Caryophyllaceae (V:D)
***Silene gallica** L. [SILEGAL] Caryophyllaceae (V:D)
Silene grandis Eastw. = Silene scouleri [SILESCO] Caryophyllaceae (V:D)
Silene inflata Sm. = Silene vulgaris [SILEVUL] Caryophyllaceae (V:D)
Silene involucrata (Cham. & Schlecht.) Bocquet [SILEINV] Caryophyllaceae (V:D)
Silene latifolia Poir. [SILELAT] Caryophyllaceae (V:D)
Silene latifolia (P. Mill.) Britten & Rendle = Silene vulgaris [SILEVUL] Caryophyllaceae (V:D)
***Silene latifolia** ssp. **alba** (P. Mill.) Greuter & Burdet [SILELAT1] Caryophyllaceae (V:D)
Silene lyallii S. Wats. = Silene douglasii [SILEDOU] Caryophyllaceae (V:D)
Silene macounii S. Wats. = Silene parryi [SILEPAR] Caryophyllaceae (V:D)
Silene menziesii Hook. [SILEMEN] Caryophyllaceae (V:D)
Silene menziesii var. **viscosa** (Greene) C.L. Hitchc. & Maguire [SILEMEN2] Caryophyllaceae (V:D)
***Silene noctiflora** L. [SILENOC] Caryophyllaceae (V:D)
Silene pacifica Eastw. = Silene scouleri ssp. scouleri var. pacifica [SILESCO0] Caryophyllaceae (V:D)
Silene parryi (S. Wats.) C.L. Hitchc. & Maguire [SILEPAR] Caryophyllaceae (V:D)
Silene pratensis (Rafn) Godr. & Gren. = Silene latifolia ssp. alba [SILELAT1] Caryophyllaceae (V:D)
Silene repens Patrin ex Pers. [SILEREP] Caryophyllaceae (V:D)
Silene repens var. *costata* (Williams) Boivin = Silene scouleri [SILESCO] Caryophyllaceae (V:D)
Silene scouleri Hook. [SILESCO] Caryophyllaceae (V:D)
Silene scouleri ssp. *grandis* (Eastw.) C.L. Hitchc. & Maguire = Silene scouleri [SILESCO] Caryophyllaceae (V:D)
Silene scouleri ssp. **scouleri** var. **pacifica** (Eastw.) C.L. Hitchc. [SILESCO0] Caryophyllaceae (V:D)
Silene scouleri var. *macounii* (S. Wats.) Boivin = Silene parryi [SILEPAR] Caryophyllaceae (V:D)
Silene scouleri var. *pacifica* (Eastw.) C.L. Hitchc. = Silene scouleri ssp. scouleri var. pacifica [SILESCO0] Caryophyllaceae (V:D)
Silene taimyrensis (Tolm.) Bocquet [SILETAI] Caryophyllaceae (V:D)
Silene tetonensis A. Nels. = Silene parryi [SILEPAR] Caryophyllaceae (V:D)
Silene uralensis (Rupr.) Bocquet [SILEURA] Caryophyllaceae (V:D)
Silene uralensis ssp. **attenuata** (Farr) McNeill [SILEURA1] Caryophyllaceae (V:D)
***Silene vulgaris** (Moench) Garcke [SILEVUL] Caryophyllaceae (V:D)
Silene wahlbergella ssp. *attenuata* (Farr) Hultén = Silene uralensis ssp. attenuata [SILEURA1] Caryophyllaceae (V:D)
Silybum Adans. [SILYBUM$] Asteraceae (V:D)
***Silybum marianum** (L.) Gaertn. [SILYMAR] Asteraceae (V:D)
Sinapis L. [SINAPIS$] Brassicaceae (V:D)
***Sinapis alba** L. [SINAALB] Brassicaceae (V:D)
***Sinapis arvensis** L. [SINAARV] Brassicaceae (V:D)
Sinapis juncea L. = Brassica juncea [BRASJUN] Brassicaceae (V:D)
Sinapis nigra L. = Brassica nigra [BRASNIG] Brassicaceae (V:D)
Sinosenecio B. Nord. [SINOSEN$] Asteraceae (V:D)
Sinosenecio newcombei (Greene) J.P. Janovec & T.M. Barkley [SINONEW] Asteraceae (V:D)

Siphula Fr. [SIPHULA$] Siphulaceae (L)
Siphula ceratites (Wahlenb.) Fr. [SIPHCER] Siphulaceae (L)
Siphula simplex (Taylor) Nyl. = Siphula ceratites [SIPHCER] Siphulaceae (L)
Sisymbrium L. [SISYMBR$] Brassicaceae (V:D)
Sisymbrium alliaria (L.) Scop. = Alliaria petiolata [ALLIPET] Brassicaceae (V:D)
*Sisymbrium altissimum** L. [SISYALT] Brassicaceae (V:D)
Sisymbrium humile C.A. Mey. = Braya humilis [BRAYHUM] Brassicaceae (V:D)
Sisymbrium incanum Bernh. ex Fisch. & C.A. Mey. = Descurainia incana [DESCINC] Brassicaceae (V:D)
Sisymbrium incisum var. *filipes* Gray = Descurainia pinnata ssp. filipes [DESCPIN1] Brassicaceae (V:D)
Sisymbrium linifolium (Nutt.) Nutt. ex Torr. & Gray = Schoenocrambe linifolia [SCHOLIN] Brassicaceae (V:D)
*Sisymbrium loeselii** L. [SISYLOE] Brassicaceae (V:D)
Sisymbrium longipedicellatum Fourn. = Descurainia pinnata ssp. filipes [DESCPIN1] Brassicaceae (V:D)
Sisymbrium nasturtium-aquaticum L. = Rorippa nasturtium-aquaticum [RORINAS] Brassicaceae (V:D)
*Sisymbrium officinale** (L.) Scop. [SISYOFF] Brassicaceae (V:D)
Sisymbrium officinale var. *leiocarpum* DC. = Sisymbrium officinale [SISYOFF] Brassicaceae (V:D)
Sisymbrium salsugineum Pallas = Arabidopsis salsuginea [ARABSAL] Brassicaceae (V:D)
Sisymbrium sophia L. = Descurainia sophia [DESCSOP] Brassicaceae (V:D)
Sisymbrium sophioides Fisch. ex Hook. = Descurainia sophioides [DESCSOH] Brassicaceae (V:D)
Sisymbrium thalianum (L.) J. Gay & Monn. = Arabidopsis thaliana [ARABTHA] Brassicaceae (V:D)
Sisymbrium viscosum (Rydb.) Blank. = Descurainia incana ssp. viscosa [DESCINC1] Brassicaceae (V:D)
Sisyrinchium L. [SISYRIN$] Iridaceae (V:M)
Sisyrinchium angustifolium P. Mill. [SISYANG] Iridaceae (V:M)
Sisyrinchium birameum Piper = Sisyrinchium idahoense var. idahoense [SISYIDA0] Iridaceae (V:M)
Sisyrinchium boreale (Bickn.) Henry = Sisyrinchium californicum [SISYCAL] Iridaceae (V:M)
Sisyrinchium californicum (Ker-Gawl. ex Sims) Ait. [SISYCAL] Iridaceae (V:M)
Sisyrinchium douglasii A. Dietr. = Olsynium douglasii var. douglasii [OLSYDOU0] Iridaceae (V:M)
Sisyrinchium douglasii var. *inflatum* (Suksdorf) P. Holmgren = Olsynium douglasii var. inflatum [OLSYDOU2] Iridaceae (V:M)
Sisyrinchium graminoides Bickn. = Sisyrinchium angustifolium [SISYANG] Iridaceae (V:M)
Sisyrinchium idahoense Bickn. [SISYIDA] Iridaceae (V:M)
Sisyrinchium idahoense var. **idahoense** [SISYIDA0] Iridaceae (V:M)
Sisyrinchium idahoense var. **macounii** (Bickn.) D. Henderson [SISYIDA2] Iridaceae (V:M)
Sisyrinchium inalatum A. Nels. = Olsynium douglasii var. douglasii [OLSYDOU0] Iridaceae (V:M)
Sisyrinchium inflatum (Suksdorf) St. John = Olsynium douglasii var. inflatum [OLSYDOU2] Iridaceae (V:M)
Sisyrinchium littorale Greene [SISYLIT] Iridaceae (V:M)
Sisyrinchium macounii Bickn. = Sisyrinchium idahoense var. macounii [SISYIDA2] Iridaceae (V:M)
Sisyrinchium montanum Greene [SISYMON] Iridaceae (V:M)
Sisyrinchium septentrionale Bickn. [SISYSEP] Iridaceae (V:M)
Sitanion elymoides Raf. = Elymus elymoides [ELYMELY] Poaceae (V:M)
Sitanion hystrix (Nutt.) J.G. Sm. = Elymus elymoides [ELYMELY] Poaceae (V:M)
Sitanion hystrix var. *brevifolium* (J.G. Sm.) C.L. Hitchc. = Elymus elymoides [ELYMELY] Poaceae (V:M)
Sitanion hystrix var. *californicum* (J.G. Sm.) F.D. Wilson = Elymus elymoides [ELYMELY] Poaceae (V:M)
Sitanion longifolium J.G. Sm. = Elymus elymoides [ELYMELY] Poaceae (V:M)
Sium L. [SIUM$] Apiaceae (V:D)
Sium cicutifolium Schrank = Sium suave [SIUMSUA] Apiaceae (V:D)
Sium floridanum Small = Sium suave [SIUMSUA] Apiaceae (V:D)
Sium incisum Torr. = Berula erecta [BERUERE] Apiaceae (V:D)
Sium suave Walt. [SIUMSUA] Apiaceae (V:D)
Sium suave var. *floridanum* (Small) C.F. Reed = Sium suave [SIUMSUA] Apiaceae (V:D)
Smelowskia C.A. Mey. [SMELOWS$] Brassicaceae (V:D)
Smelowskia calycina (Steph. ex Willd.) C.A. Mey. [SMELCAL] Brassicaceae (V:D)
Smelowskia ovalis M.E. Jones [SMELOVA] Brassicaceae (V:D)
Smilacina amplexicaulis Nutt. = Maianthemum racemosum ssp. amplexicaule [MAIARAC1] Liliaceae (V:M)
Smilacina amplexicaulis var. *glabra* J.F. Macbr. = Maianthemum racemosum ssp. amplexicaule [MAIARAC1] Liliaceae (V:M)
Smilacina amplexicaulis var. *jenkinsii* Boivin = Maianthemum racemosum ssp. amplexicaule [MAIARAC1] Liliaceae (V:M)
Smilacina amplexicaulis var. *ovata* Boivin = Maianthemum racemosum ssp. amplexicaule [MAIARAC1] Liliaceae (V:M)
Smilacina borealis var. *uniflora* Menzies ex J.A. & J.H. Schultes = Clintonia uniflora [CLINUNI] Liliaceae (V:M)
Smilacina liliacea (Greene) Wynd = Maianthemum stellatum [MAIASTE] Liliaceae (V:M)
Smilacina racemosa var. *amplexicaulis* (Nutt.) S. Wats. = Maianthemum racemosum ssp. amplexicaule [MAIARAC1] Liliaceae (V:M)
Smilacina racemosa var. *brachystyla* G. Henderson = Maianthemum racemosum ssp. amplexicaule [MAIARAC1] Liliaceae (V:M)
Smilacina racemosa var. *glabra* (J.F. Macbr.) St. John = Maianthemum racemosum ssp. amplexicaule [MAIARAC1] Liliaceae (V:M)
Smilacina racemosa var. *jenkinsii* (Boivin) Boivin = Maianthemum racemosum ssp. amplexicaule [MAIARAC1] Liliaceae (V:M)
Smilacina sessilifolia Nutt. ex Baker = Maianthemum stellatum [MAIASTE] Liliaceae (V:M)
Smilacina stellata (L.) Desf. = Maianthemum stellatum [MAIASTE] Liliaceae (V:M)
Smilacina stellata var. *crassa* Victorin = Maianthemum stellatum [MAIASTE] Liliaceae (V:M)
Smilacina stellata var. *mollis* Farw. = Maianthemum stellatum [MAIASTE] Liliaceae (V:M)
Smilacina stellata var. *sessilifolia* (Nutt. ex Baker) G. Henderson = Maianthemum stellatum [MAIASTE] Liliaceae (V:M)
Smilacina stellata var. *sylvatica* Victorin & Rouss. = Maianthemum stellatum [MAIASTE] Liliaceae (V:M)
Smilacina trifolia (L.) Desf. = Maianthemum trifolium [MAIATRI] Liliaceae (V:M)
Solanum L. [SOLANUM$] Solanaceae (V:D)
*Solanum americanum** P. Mill. [SOLAAME] Solanaceae (V:D)
Solanum americanum var. *nodiflorum* (Jacq.) Edmonds = Solanum americanum [SOLAAME] Solanaceae (V:D)
Solanum americanum var. *patulum* (L.) Edmonds = Solanum americanum [SOLAAME] Solanaceae (V:D)
Solanum cornutum auct. = Solanum rostratum [SOLAROS] Solanaceae (V:D)
*Solanum dulcamara** L. [SOLADUL] Solanaceae (V:D)
Solanum fistulosum Dunal ex Poir. = Solanum americanum [SOLAAME] Solanaceae (V:D)
Solanum hermannii Dunal = Solanum americanum [SOLAAME] Solanaceae (V:D)
Solanum linnaeanum Hepper & Jaeger = Solanum americanum [SOLAAME] Solanaceae (V:D)
Solanum nigrum L., p.p. = Solanum americanum [SOLAAME] Solanaceae (V:D)
Solanum nigrum var. *americanum* (P. Mill.) O.E. Schulz = Solanum americanum [SOLAAME] Solanaceae (V:D)
Solanum nigrum var. *virginicum* L. = Solanum americanum [SOLAAME] Solanaceae (V:D)
Solanum nodiflorum Jacq. = Solanum americanum [SOLAAME] Solanaceae (V:D)
*Solanum rostratum** Dunal [SOLAROS] Solanaceae (V:D)
*Solanum sarrachoides** Sendtner [SOLASAR] Solanaceae (V:D)
Solanum triflorum Nutt. [SOLATRI] Solanaceae (V:D)
Solenopsora hassei (Zahlbr.) Zahlbr. = Lecania hassei [LECAHAS] Bacidiaceae (L)

Solenostoma atrovirens (Dum.) K. Müll. = Jungermannia atrovirens [JUNGATR] Jungermanniaceae (B:H)
Solenostoma hyalinum (Lyell) Mitt. = Jungermannia hyalina [JUNGHYA] Jungermanniaceae (B:H)
Solenostoma levieri (Steph.) Steph. = Jungermannia confertissima [JUNGCON] Jungermanniaceae (B:H)
Solenostoma obovatum (Nees) Mass. = Jungermannia obovata [JUNGOBO] Jungermanniaceae (B:H)
Solenostoma ontariensis Schust. = Jungermannia hyalina [JUNGHYA] Jungermanniaceae (B:H)
Solenostoma pumilum (With.) K. Müll = Jungermannia pumila [JUNGPUM] Jungermanniaceae (B:H)
Solenostoma pumilum ssp. *anomalum* Schust. & Damsh. = Jungermannia pumila [JUNGPUM] Jungermanniaceae (B:H)
Solenostoma pusillum var. *vinaceum* Schust. = Jungermannia confertissima [JUNGCON] Jungermanniaceae (B:H)
Solenostoma rubrum (Gott. ex Underw.) Schust. = Jungermannia rubra [JUNGRUB] Jungermanniaceae (B:H)
Solenostoma sphaerocarpa (Hook.) Steph. = Jungermannia sphaerocarpa [JUNGSPH] Jungermanniaceae (B:H)
Solenostoma sphaerocarpa var. *nana* (Nees) Schust. = Jungermannia sphaerocarpa [JUNGSPH] Jungermanniaceae (B:H)
Solenostoma subellipticum (Lindb. ex Kaal.) Schust. = Jungermannia subelliptica [JUNGSUB] Jungermanniaceae (B:H)
Solenostoma triste (Nees) K. Müll = Jungermannia atrovirens [JUNGATR] Jungermanniaceae (B:H)
Solidago L. [SOLIDAG$] Asteraceae (V:D)
Solidago altissima var. *gilvocanescens* (Rydb.) Semple = Solidago canadensis var. gilvocanescens [SOLICAN1] Asteraceae (V:D)
Solidago canadensis L. [SOLICAN] Asteraceae (V:D)
Solidago canadensis ssp. *elongata* (Nutt.) Keck = Solidago canadensis var. salebrosa [SOLICAN2] Asteraceae (V:D)
Solidago canadensis ssp. *salebrosa* (Piper) Keck = Solidago canadensis var. salebrosa [SOLICAN2] Asteraceae (V:D)
Solidago canadensis var. *elongata* (Nutt.) M.E. Peck = Solidago canadensis var. salebrosa [SOLICAN2] Asteraceae (V:D)
Solidago canadensis var. **gilvocanescens** Rydb. [SOLICAN1] Asteraceae (V:D)
Solidago canadensis var. **salebrosa** (Piper) M.E. Jones [SOLICAN2] Asteraceae (V:D)
Solidago canadensis var. **subserrata** (DC.) Cronq. [SOLICAN3] Asteraceae (V:D)
Solidago decemflora DC. = Solidago nemoralis var. longipetiolata [SOLINEM1] Asteraceae (V:D)
Solidago elongata Nutt. = Solidago canadensis var. salebrosa [SOLICAN2] Asteraceae (V:D)
Solidago gigantea Ait. [SOLIGIG] Asteraceae (V:D)
Solidago gigantea ssp. *serotina* (Kuntze) McNeill = Solidago gigantea [SOLIGIG] Asteraceae (V:D)
Solidago gigantea var. *leiophylla* Fern. = Solidago gigantea [SOLIGIG] Asteraceae (V:D)
Solidago gigantea var. *pitcheri* (Nutt.) Shinners = Solidago gigantea [SOLIGIG] Asteraceae (V:D)
Solidago gigantea var. *serotina* (Kuntze) Cronq. = Solidago gigantea [SOLIGIG] Asteraceae (V:D)
Solidago gigantea var. *shinnersii* Beaudry = Solidago gigantea [SOLIGIG] Asteraceae (V:D)
Solidago gilvocanescens (Rydb.) Smyth = Solidago canadensis var. gilvocanescens [SOLICAN1] Asteraceae (V:D)
Solidago graminifolia (L.) Salisb. = Euthamia graminifolia [EUTHGRA] Asteraceae (V:D)
Solidago graminifolia var. *major* (Michx.) Fern. = Euthamia graminifolia [EUTHGRA] Asteraceae (V:D)
Solidago × *leiophallax* Friesner = Solidago gigantea [SOLIGIG] Asteraceae (V:D)
Solidago lepida DC. = Solidago canadensis var. subserrata [SOLICAN3] Asteraceae (V:D)
Solidago lepida var. *elongata* (Nutt.) Fern. = Solidago canadensis var. salebrosa [SOLICAN2] Asteraceae (V:D)
Solidago lepida var. *fallax* Fern. = Solidago canadensis var. salebrosa [SOLICAN2] Asteraceae (V:D)
Solidago lepida var. *molina* Fern. = Solidago canadensis var. subserrata [SOLICAN3] Asteraceae (V:D)
Solidago longipetiolata Mackenzie & Bush = Solidago nemoralis var. longipetiolata [SOLINEM1] Asteraceae (V:D)
Solidago missouriensis Nutt. [SOLIMIS] Asteraceae (V:D)
Solidago missouriensis var. *montana* Gray = Solidago missouriensis [SOLIMIS] Asteraceae (V:D)
Solidago multiradiata Ait. [SOLIMUL] Asteraceae (V:D)
Solidago nemoralis Ait. [SOLINEM] Asteraceae (V:D)
Solidago nemoralis ssp. *decemflora* (DC.) Brammall = Solidago nemoralis var. longipetiolata [SOLINEM1] Asteraceae (V:D)
Solidago nemoralis ssp. *longipetiolata* (Mackenzie & Bush) G.W. Douglas = Solidago nemoralis var. longipetiolata [SOLINEM1] Asteraceae (V:D)
Solidago nemoralis var. *decemflora* (DC.) Fern. = Solidago nemoralis var. longipetiolata [SOLINEM1] Asteraceae (V:D)
Solidago nemoralis var. **longipetiolata** (Mackenzie & Bush) Palmer & Steyermark [SOLINEM1] Asteraceae (V:D)
Solidago neomexicana (Gray) Woot. & Standl. = Solidago spathulata var. neomexicana [SOLISPA5] Asteraceae (V:D)
Solidago occidentalis (Nutt.) Torr. & Gray = Euthamia occidentalis [EUTHOCC] Asteraceae (V:D)
Solidago pitcheri Nutt. = Solidago gigantea [SOLIGIG] Asteraceae (V:D)
Solidago pruinosa Greene = Solidago canadensis var. gilvocanescens [SOLICAN1] Asteraceae (V:D)
Solidago pulcherrima A. Nels. = Solidago nemoralis var. longipetiolata [SOLINEM1] Asteraceae (V:D)
Solidago serotina Ait. = Solidago gigantea [SOLIGIG] Asteraceae (V:D)
Solidago serotinoides A. & D. Löve = Solidago gigantea [SOLIGIG] Asteraceae (V:D)
Solidago simplex Kunth [SOLISIM] Asteraceae (V:D)
Solidago spathulata DC. [SOLISPA] Asteraceae (V:D)
Solidago spathulata var. **neomexicana** (Gray) Cronq. [SOLISPA5] Asteraceae (V:D)
Solorina Ach. [SOLORIN$] Peltigeraceae (L)
Solorina bispora Nyl. [SOLOBIS] Peltigeraceae (L)
Solorina crocea (L.) Ach. [SOLOCRO] Peltigeraceae (L)
Solorina octospora (Arnold) Arnold [SOLOOCT] Peltigeraceae (L)
Solorina saccata (L.) Ach. [SOLOSAC] Peltigeraceae (L)
Solorina spongiosa (Ach.) Anzi [SOLOSPO] Peltigeraceae (L)
Solorinaria despreauxii (Mont.) Fink = Heppia lutosa [HEPPLUT] Heppiaceae (L)
Solorinella Anzi [SOLORIE$] Solorinellaceae (L)
Solorinella asteriscus Anzi [SOLOAST] Solorinellaceae (L)
Sonchus L. [SONCHUS$] Asteraceae (V:D)
***Sonchus arvensis** L. [SONCARV] Asteraceae (V:D)
***Sonchus arvensis** ssp. **arvensis** [SONCARV0] Asteraceae (V:D)
***Sonchus arvensis** ssp. **uliginosus** (Bieb.) Nyman [SONCARV1] Asteraceae (V:D)
Sonchus arvensis var. *glabrescens* Guenth., Grab. & Wimmer = Sonchus arvensis ssp. uliginosus [SONCARV1] Asteraceae (V:D)
Sonchus arvensis var. *shumovichii* Boivin = Sonchus arvensis ssp. arvensis [SONCARV0] Asteraceae (V:D)
***Sonchus asper** (L.) Hill [SONCASP] Asteraceae (V:D)
***Sonchus oleraceus** L. [SONCOLE] Asteraceae (V:D)
Sonchus tataricus L. p.p. = Lactuca tatarica var. pulchella [LACTTAT1] Asteraceae (V:D)
Sonchus uliginosus Bieb. = Sonchus arvensis ssp. uliginosus [SONCARV1] Asteraceae (V:D)
Sophia filipes (Gray) Heller = Descurainia pinnata ssp. filipes [DESCPIN1] Brassicaceae (V:D)
Sophia intermedia Rydb. = Descurainia pinnata ssp. intermedia [DESCPIN2] Brassicaceae (V:D)
Sophia richardsonii (O.E. Schulz) Rydb. = Descurainia incana [DESCINC] Brassicaceae (V:D)
Sophia sophia (L.) Britt. = Descurainia sophia [DESCSOP] Brassicaceae (V:D)
Sophia viscosa Rydb. = Descurainia incana ssp. viscosa [DESCINC1] Brassicaceae (V:D)
Sorbus L. [SORBUS$] Rosaceae (V:D)
***Sorbus aucuparia** L. [SORBAUC] Rosaceae (V:D)
Sorbus cascadensis G.N. Jones = Sorbus scopulina var. cascadensis [SORBSCO1] Rosaceae (V:D)

Sorbus occidentalis (S. Wats.) Greene = Sorbus sitchensis var. grayi [SORBSIT3] Rosaceae (V:D)
Sorbus scopulina Greene [SORBSCO] Rosaceae (V:D)
Sorbus scopulina var. **cascadensis** (G.N. Jones) C.L. Hitchc. [SORBSCO1] Rosaceae (V:D)
Sorbus scopulina var. **scopulina** [SORBSCO0] Rosaceae (V:D)
Sorbus sitchensis M. Roemer [SORBSIT] Rosaceae (V:D)
Sorbus sitchensis ssp. *grayi* (Wenzig) Calder & Taylor = Sorbus sitchensis var. grayi [SORBSIT3] Rosaceae (V:D)
Sorbus sitchensis var. **grayi** (Wenzig) C.L. Hitchc. [SORBSIT3] Rosaceae (V:D)
Sorbus sitchensis var. **sitchensis** [SORBSIT0] Rosaceae (V:D)
Sparganium L. [SPARGAN$] Sparganiaceae (V:M)
Sparganium acaule (Beeby) Rydb. = Sparganium angustifolium [SPARANG] Sparganiaceae (V:M)
Sparganium angustifolium Michx. [SPARANG] Sparganiaceae (V:M)
Sparganium angustifolium ssp. *emersum* (Rehmann) T.C. Brayshaw = Sparganium angustifolium [SPARANG] Sparganiaceae (V:M)
Sparganium angustifolium var. *multipedunculatum* (Morong) T.C. Brayshaw = Sparganium angustifolium [SPARANG] Sparganiaceae (V:M)
Sparganium californicum Greene = Sparganium eurycarpum [SPAREUR] Sparganiaceae (V:M)
Sparganium chlorocarpum var. *acaule* (Beeby) Fern. = Sparganium angustifolium [SPARANG] Sparganiaceae (V:M)
Sparganium emersum Rehmann = Sparganium angustifolium [SPARANG] Sparganiaceae (V:M)
Sparganium emersum var. *angustifolium* (Michx.) Taylor & MacBryde = Sparganium angustifolium [SPARANG] Sparganiaceae (V:M)
Sparganium emersum var. *multipedunculatum* (Morong) Reveal = Sparganium angustifolium [SPARANG] Sparganiaceae (V:M)
Sparganium eurycarpum Engelm. ex Gray [SPAREUR] Sparganiaceae (V:M)
Sparganium eurycarpum var. *greenei* (Morong) Graebn. = Sparganium eurycarpum [SPAREUR] Sparganiaceae (V:M)
Sparganium fluctuans (Morong) B.L. Robins. [SPARFLU] Sparganiaceae (V:M)
Sparganium glomeratum Laestad. ex Beurling [SPARGLO] Sparganiaceae (V:M)
Sparganium greenei Morong = Sparganium eurycarpum [SPAREUR] Sparganiaceae (V:M)
Sparganium hyperboreum Laestad. ex Beurling [SPARHYP] Sparganiaceae (V:M)
Sparganium minimum (Hartman) Wallr. = Sparganium nutans [SPARNUT] Sparganiaceae (V:M)
Sparganium multipedunculatum (Morong) Rydb. = Sparganium angustifolium [SPARANG] Sparganiaceae (V:M)
Sparganium nutans L. [SPARNUT] Sparganiaceae (V:M)
Sparganium simplex var. *multipedunculatum* Morong = Sparganium angustifolium [SPARANG] Sparganiaceae (V:M)
Spartina Schreb. [SPARTIN$] Poaceae (V:M)
Spartina gracilis Trin. [SPARGRA] Poaceae (V:M)
*__Spartina patens__ (Ait.) Muhl. [SPARPAT] Poaceae (V:M)
Spartina patens var. *juncea* (Michx.) A.S. Hitchc. = Spartina patens [SPARPAT] Poaceae (V:M)
Spartina patens var. *monogyna* (M.A. Curtis) Fern. = Spartina patens [SPARPAT] Poaceae (V:M)
Specularia perfoliata (L.) A. DC. = Triodanis perfoliata [TRIOPER] Campanulaceae (V:D)
Specularia rariflora (Nutt.) McVaugh = Heterocodon rariflorum [HETERAR] Campanulaceae (V:D)
Spergella intermedia (Fenzl) A. & D. Löve = Sagina nivalis [SAGINIV] Caryophyllaceae (V:D)
Spergella saginoides (L.) Reichenb. = Sagina saginoides [SAGISAG] Caryophyllaceae (V:D)
Spergula L. [SPERGUL$] Caryophyllaceae (V:D)
*__Spergula arvensis__ L. [SPERARV] Caryophyllaceae (V:D)
Spergularia (Pers.) J.& K. Presl [SPERGUA$] Caryophyllaceae (V:D)
Spergularia alata Wieg. = Spergularia salina [SPERSAL] Caryophyllaceae (V:D)
Spergularia canadensis (Pers.) G. Don [SPERCAN] Caryophyllaceae (V:D)
Spergularia canadensis var. **canadensis** [SPERCAN0] Caryophyllaceae (V:D)
Spergularia canadensis var. **occidentalis** R.P. Rossb. [SPERCAN2] Caryophyllaceae (V:D)
Spergularia leiosperma (Kindb.) F. Schmidt = Spergularia salina [SPERSAL] Caryophyllaceae (V:D)
Spergularia macrotheca (Hornem.) Heynh. [SPERMAC] Caryophyllaceae (V:D)
Spergularia marina (L.) Griseb. = Spergularia salina [SPERSAL] Caryophyllaceae (V:D)
Spergularia marina var. *leiosperma* (Kindb.) Guerke = Spergularia salina [SPERSAL] Caryophyllaceae (V:D)
Spergularia marina var. *simonii* O.& I. Deg. = Spergularia salina [SPERSAL] Caryophyllaceae (V:D)
*__Spergularia rubra__ (L.) J.& K. Presl [SPERRUB] Caryophyllaceae (V:D)
Spergularia rubra var. *perennans* (Kindb.) B.L. Robins. = Spergularia rubra [SPERRUB] Caryophyllaceae (V:D)
Spergularia salina J.& K. Presl [SPERSAL] Caryophyllaceae (V:D)
Spergularia sparsiflora (Greene) A. Nels. = Spergularia salina [SPERSAL] Caryophyllaceae (V:D)
Sphaeralcea St.-Hil. [SPHAERA$] Malvaceae (V:D)
Sphaeralcea coccinea (Nutt.) Rydb. [SPHACOC] Malvaceae (V:D)
Sphaeralcea munroana (Dougl. ex Lindl.) Spach ex Gray [SPHAMUN] Malvaceae (V:D)
Sphaeralcea rivularis (Dougl. ex Hook.) Torr. = Iliamna rivularis [ILIARIV] Malvaceae (V:D)
Sphaerellothecium Zopf [SPHAERE$] Mycosphaerellaceae (L)
Sphaerellothecium araneosum (Rehm ex Arnold) Zopf [SPHAARA] Mycosphaerellaceae (L)
Sphaerocarpos Boehmer [SPHAERC$] Sphaerocarpaceae (B:H)
Sphaerocarpos texanus Aust. [SPHATEX] Sphaerocarpaceae (B:H)
Sphaerocarpus terrestris Bisch. = Sphaerocarpos texanus [SPHATEX] Sphaerocarpaceae (B:H)
Sphaeromphale areolata (Ach.) Arnold = Staurothele areolata [STAUARE] Verrucariaceae (L)
Sphaeromphale clopima var. *areolata* (Ach.) Trev. = Staurothele areolata [STAUARE] Verrucariaceae (L)
Sphaeromphale fissa (Tayl.) Körber = Staurothele fissa [STAUFIS] Verrucariaceae (L)
Sphaeromphale hazslinszkyi Körber = Staurothele fissa [STAUFIS] Verrucariaceae (L)
Sphaerophoron coralloides Pers. = Sphaerophorus globosus [SPHAGLO] Sphaerophoraceae (L)
Sphaerophoron globiferum var. *congestum* Müll. Arg. = Sphaerophorus globosus [SPHAGLO] Sphaerophoraceae (L)
Sphaerophorus Pers. [SPHAERO$] Sphaerophoraceae (L)
Sphaerophorus compressum Ach. = Bunodophoron melanocarpum [BUNOMEL] Sphaerophoraceae (L)
Sphaerophorus fragilis (L.) Pers. [SPHAFRA] Sphaerophoraceae (L)
Sphaerophorus globiferis DC. = Sphaerophorus globosus [SPHAGLO] Sphaerophoraceae (L)
Sphaerophorus globiferous (L.) DC. = Sphaerophorus globosus [SPHAGLO] Sphaerophoraceae (L)
Sphaerophorus globiferus (L.) DC. = Sphaerophorus globosus [SPHAGLO] Sphaerophoraceae (L)
Sphaerophorus globosus (Hudson) Vainio [SPHAGLO] Sphaerophoraceae (L)
Sphaerophorus globosus var. **globosus** [SPHAGLO0] Sphaerophoraceae (L)
Sphaerophorus globosus var. **gracilis** (Müll. Arg.) Zahlbr. [SPHAGLO1] Sphaerophoraceae (L)
Sphaerophorus globosus var. *lacunosus* Tuck. = Sphaerophorus globosus [SPHAGLO] Sphaerophoraceae (L)
Sphaerophorus melanocarpus (Sw.) DC. = Bunodophoron melanocarpum [BUNOMEL] Sphaerophoraceae (L)
Sphaerophorus tuckermanii Räsänen = Sphaerophorus globosus var. gracilis [SPHAGLO1] Sphaerophoraceae (L)
Sphagnum L. [SPHAGNU$] Sphagnaceae (B:M)
Sphagnum acutifolium Ehrh. ex Schrad. = Sphagnum capillifolium [SPHACAI] Sphagnaceae (B:M)

Sphagnum acutifolium var. *intermedium* Aust. = Sphagnum warnstorfii [SPHAWAR] Sphagnaceae (B:M)
Sphagnum acutifolium var. *pallescens* Warnst. = Sphagnum capillifolium [SPHACAI] Sphagnaceae (B:M)
Sphagnum acutifolium var. *rubellum* (Wils.) Russ. = Sphagnum rubellum [SPHARUB] Sphagnaceae (B:M)
Sphagnum acutifolium var. *rubrum* Brid. ex Warnst. = Sphagnum capillifolium [SPHACAI] Sphagnaceae (B:M)
Sphagnum acutifolium var. *subtile* Russ. = Sphagnum subtile [SPHASUT] Sphagnaceae (B:M)
Sphagnum acutifolium var. *tenellum* Schimp. = Sphagnum rubellum [SPHARUB] Sphagnaceae (B:M)
Sphagnum acutifolium var. *versicolor* Warnst. = Sphagnum capillifolium [SPHACAI] Sphagnaceae (B:M)
Sphagnum acutiforme var. *robustum* Warnst. = Sphagnum russowii [SPHARUS] Sphagnaceae (B:M)
Sphagnum angustifolium (C. Jens. ex Russ.) C. Jens. in Tolf [SPHAANU] Sphagnaceae (B:M)
Sphagnum annulatum var. *porosum* (Schlieph. & Warnst.) Maass & Isov. in Maass = Sphagnum jensenii [SPHAJEN] Sphagnaceae (B:M)
Sphagnum aongstroemii Hartm. [SPHAAON] Sphagnaceae (B:M)
Sphagnum apiculatum H. Lindb. in Bauer = Sphagnum fallax [SPHAFAL] Sphagnaceae (B:M)
Sphagnum austinii Sull. in Aust. [SPHAAUS] Sphagnaceae (B:M)
Sphagnum balticum (Russ.) C. Jens. [SPHABAL] Sphagnaceae (B:M)
Sphagnum capillaceum (Weiss.) Schrank = Sphagnum capillifolium [SPHACAI] Sphagnaceae (B:M)
Sphagnum capillaceum var. *tenellum* (Schimp.) Andrews = Sphagnum rubellum [SPHARUB] Sphagnaceae (B:M)
Sphagnum capillifolium (Ehrh.) Hedw. [SPHACAI] Sphagnaceae (B:M)
Sphagnum capillifolium var. *tenellum* (Schimp.) Crum = Sphagnum rubellum [SPHARUB] Sphagnaceae (B:M)
Sphagnum carlottae Andrus, nomen nudum = Sphagnum austinii [SPHAAUS] Sphagnaceae (B:M)
Sphagnum centrale C. Jens. in Arnell & C. Jens. [SPHACEN] Sphagnaceae (B:M)
Sphagnum compactum DC. in Lam. & DC. [SPHACOM] Sphagnaceae (B:M)
Sphagnum compactum var. *brachycladum* (Röll) Röll = Sphagnum compactum [SPHACOM] Sphagnaceae (B:M)
Sphagnum compactum var. *expositum* Maass = Sphagnum compactum [SPHACOM] Sphagnaceae (B:M)
Sphagnum compactum var. *imbricatum* Warnst. = Sphagnum compactum [SPHACOM] Sphagnaceae (B:M)
Sphagnum compactum var. *squarrosum* (Russ.) Warnst. = Sphagnum compactum [SPHACOM] Sphagnaceae (B:M)
Sphagnum compactum var. *subsquarrosum* Warnst. = Sphagnum compactum [SPHACOM] Sphagnaceae (B:M)
Sphagnum contortum Schultz [SPHACON] Sphagnaceae (B:M)
Sphagnum cuspidatum Ehrh. ex Hoffm. [SPHACUS] Sphagnaceae (B:M)
Sphagnum cuspidatum var. *dusenii* C. Jens. ex Warnst. = Sphagnum majus [SPHAMAJ] Sphagnaceae (B:M)
Sphagnum cuspidatum var. *plumosum* Nees & Hornsch. = Sphagnum cuspidatum [SPHACUS] Sphagnaceae (B:M)
Sphagnum cuspidatum var. *plumulosum* Schimp. = Sphagnum cuspidatum [SPHACUS] Sphagnaceae (B:M)
Sphagnum cymbifolium (Ehrh.) Hedw. = Sphagnum palustre [SPHAPAL] Sphagnaceae (B:M)
Sphagnum cymbifolium var. *laeve* Warnst. = Sphagnum palustre [SPHAPAL] Sphagnaceae (B:M)
Sphagnum dusenii C. Jens. ex Russ. & Warnst. in Russ. = Sphagnum majus [SPHAMAJ] Sphagnaceae (B:M)
Sphagnum fallax (Klinggr.) Klinggr. [SPHAFAL] Sphagnaceae (B:M)
Sphagnum fallax var. *angustifolium* (C. Jens. ex Russ.) Nyh. = Sphagnum angustifolium [SPHAANU] Sphagnaceae (B:M)
Sphagnum fimbriatum Wils. in Wils. & Hook. f. in Hook. f. [SPHAFIM] Sphagnaceae (B:M)
Sphagnum fimbriatum var. *laxum* (Braithw.) Wijk & Marg. = Sphagnum fimbriatum [SPHAFIM] Sphagnaceae (B:M)
Sphagnum fimbriatum var. *robustum* Braithw. ex Warnst. = Sphagnum fimbriatum [SPHAFIM] Sphagnaceae (B:M)
Sphagnum flexuosum var. *fallax* (Klinggr.) M.O. Hill ex A.J.E. Sm. = Sphagnum fallax [SPHAFAL] Sphagnaceae (B:M)
Sphagnum flexuosum var. *tenue* (Klinggr.) Pilous = Sphagnum angustifolium [SPHAANU] Sphagnaceae (B:M)
Sphagnum fuscum (Schimp.) Klinggr. [SPHAFUS] Sphagnaceae (B:M)
Sphagnum fuscum f. *fuscescens* (Warnst.) Warnst. = Sphagnum fuscum [SPHAFUS] Sphagnaceae (B:M)
Sphagnum fuscum f. *pallescens* (Warnst.) Warnst. = Sphagnum fuscum [SPHAFUS] Sphagnaceae (B:M)
Sphagnum fuscum f. *viride* (Warnst.) Warnst. = Sphagnum fuscum [SPHAFUS] Sphagnaceae (B:M)
Sphagnum fuscum var. *fuscescens* Warnst. = Sphagnum fuscum [SPHAFUS] Sphagnaceae (B:M)
Sphagnum fuscum var. *pallescens* Warnst. = Sphagnum fuscum [SPHAFUS] Sphagnaceae (B:M)
Sphagnum fuscum var. *viride* Warnst. = Sphagnum fuscum [SPHAFUS] Sphagnaceae (B:M)
Sphagnum girgensohnii Russ. [SPHAGIR] Sphagnaceae (B:M)
Sphagnum girgensohnii var. *hygrophyllum* Russ. = Sphagnum girgensohnii [SPHAGIR] Sphagnaceae (B:M)
Sphagnum girgensohnii var. *stachyodes* Russ. in Warnst. = Sphagnum girgensohnii [SPHAGIR] Sphagnaceae (B:M)
Sphagnum girgensohnii var. *strictum* (Lindb.) Russ. = Sphagnum girgensohnii [SPHAGIR] Sphagnaceae (B:M)
Sphagnum henryense Warnst. [SPHAHEN] Sphagnaceae (B:M)
Sphagnum imbricatum ssp. *austinii* (Sull. in Aust.) Flatb. = Sphagnum austinii [SPHAAUS] Sphagnaceae (B:M)
Sphagnum intermedium Hoffm. = Sphagnum capillifolium [SPHACAI] Sphagnaceae (B:M)
Sphagnum jensenii H. Lindb. [SPHAJEN] Sphagnaceae (B:M)
Sphagnum junghuhnianum Dozy & Molk. [SPHAJUN] Sphagnaceae (B:M)
Sphagnum junghuhnianum ssp. *pseudomolle* (Warnst.) Suz. = Sphagnum junghuhnianum var. pseudomolle [SPHAJUN1] Sphagnaceae (B:M)
Sphagnum junghuhnianum var. **junghuhnianum** [SPHAJUN0] Sphagnaceae (B:M)
Sphagnum junghuhnianum var. **pseudomolle** (Warnst.) Warnst. [SPHAJUN1] Sphagnaceae (B:M)
Sphagnum laricinum (Wils.) Spruce = Sphagnum contortum [SPHACON] Sphagnaceae (B:M)
Sphagnum lindbergii Schimp. in Lindb. [SPHALIN] Sphagnaceae (B:M)
Sphagnum lindbergii var. *macrophyllum* Warnst. = Sphagnum lindbergii [SPHALIN] Sphagnaceae (B:M)
Sphagnum lindbergii var. *mesophyllum* Warnst. = Sphagnum lindbergii [SPHALIN] Sphagnaceae (B:M)
Sphagnum lindbergii var. *tenellum* (Lindb.) Röll = Sphagnum lindbergii [SPHALIN] Sphagnaceae (B:M)
Sphagnum magellanicum Brid. [SPHAMAG] Sphagnaceae (B:M)
Sphagnum magellanicum var. *pallescens* (Warnst.) C. Jens. = Sphagnum magellanicum [SPHAMAG] Sphagnaceae (B:M)
Sphagnum magellanicum var. *purpurascens* (Russ.) Warnst. = Sphagnum magellanicum [SPHAMAG] Sphagnaceae (B:M)
Sphagnum magellanicum var. *roseum* (Röll) C. Jens. = Sphagnum magellanicum [SPHAMAG] Sphagnaceae (B:M)
Sphagnum majus (Russ.) C. Jens. [SPHAMAJ] Sphagnaceae (B:M)
Sphagnum medium Limpr. = Sphagnum magellanicum [SPHAMAG] Sphagnaceae (B:M)
Sphagnum medium var. *laeve* f. *purpurascens* Russ. = Sphagnum magellanicum [SPHAMAG] Sphagnaceae (B:M)
Sphagnum medium var. *pallescens* (Warnst.) Warnst. = Sphagnum magellanicum [SPHAMAG] Sphagnaceae (B:M)
Sphagnum medium var. *purpurascens* (Russ.) Warnst. = Sphagnum magellanicum [SPHAMAG] Sphagnaceae (B:M)
Sphagnum medium var. *roseum* (Röll) Warnst. = Sphagnum magellanicum [SPHAMAG] Sphagnaceae (B:M)

Sphagnum mendocinum Sull. & Lesq. in Sull. [SPHAMEN] Sphagnaceae (B:M)
Sphagnum molluscum Bruch = Sphagnum tenellum [SPHATEN] Sphagnaceae (B:M)
Sphagnum nemoreum Scop. auct. plur. = Sphagnum capillifolium [SPHACAI] Sphagnaceae (B:M)
Sphagnum nemoreum f. *versicolor* (Warnst.) Podp. = Sphagnum capillifolium [SPHACAI] Sphagnaceae (B:M)
Sphagnum nemoreum var. *alpinum* (Milde) Wijk & Marg. = Sphagnum capillifolium [SPHACAI] Sphagnaceae (B:M)
Sphagnum nemoreum var. *elegans* (Braithw.) Wijk & Marg. = Sphagnum capillifolium [SPHACAI] Sphagnaceae (B:M)
Sphagnum nemoreum var. *subtile* Russ. = Sphagnum subtile [SPHASUT] Sphagnaceae (B:M)
Sphagnum obtusum var. *dusenii* C. Jens. ex Warnst. = Sphagnum majus [SPHAMAJ] Sphagnaceae (B:M)
Sphagnum pacificum Flatb. [SPHAPAC] Sphagnaceae (B:M)
Sphagnum palustre L. [SPHAPAL] Sphagnaceae (B:M)
Sphagnum papillosum Lindb. [SPHAPAP] Sphagnaceae (B:M)
Sphagnum papillosum var. *intermedium* Warnst. = Sphagnum centrale [SPHACEN] Sphagnaceae (B:M)
Sphagnum papillosum var. *laeve* Warnst. = Sphagnum papillosum [SPHAPAP] Sphagnaceae (B:M)
Sphagnum papillosum var. *sublaeve* Warnst. = Sphagnum papillosum [SPHAPAP] Sphagnaceae (B:M)
Sphagnum parvifolium (Sendtn.) Warnst. = Sphagnum angustifolium [SPHAANU] Sphagnaceae (B:M)
Sphagnum platyphyllum (Lindb. ex Braithw.) Sull. ex Warnst. [SPHAPLA] Sphagnaceae (B:M)
Sphagnum plumulosum Röll = Sphagnum subnitens [SPHASUN] Sphagnaceae (B:M)
Sphagnum pulchricoma C. Müll. = Sphagnum recurvum [SPHAREC] Sphagnaceae (B:M)
Sphagnum pulchrum (Lindb. ex Braithw.) Warnst. [SPHAPUL] Sphagnaceae (B:M)
Sphagnum pulchrum var. *fusco-flavescens* (Warnst.) Warnst. = Sphagnum pulchrum [SPHAPUL] Sphagnaceae (B:M)
Sphagnum pulchrum var. *nigricans* Warnst. = Sphagnum pulchrum [SPHAPUL] Sphagnaceae (B:M)
Sphagnum purpureum C. Jens. = Sphagnum rubellum [SPHARUB] Sphagnaceae (B:M)
Sphagnum quinquefarium (Lindb. ex Braithw.) Warnst. [SPHAQUI] Sphagnaceae (B:M)
Sphagnum recurvum P. Beauv. [SPHAREC] Sphagnaceae (B:M)
Sphagnum recurvum ssp. *angustifolium* C. Jens. ex Russ. = Sphagnum angustifolium [SPHAANU] Sphagnaceae (B:M)
Sphagnum recurvum var. *brevifolium* (Lindb. ex Braithw.) Warnst. = Sphagnum fallax [SPHAFAL] Sphagnaceae (B:M)
Sphagnum recurvum var. *fallax* (Klinggr.) Paul in Koppe = Sphagnum fallax [SPHAFAL] Sphagnaceae (B:M)
Sphagnum recurvum var. *mucronatum* (Russ.) Warnst. = Sphagnum fallax [SPHAFAL] Sphagnaceae (B:M)
Sphagnum recurvum var. *parvifolium* Sendtn. ex Warnst. = Sphagnum angustifolium [SPHAANU] Sphagnaceae (B:M)
Sphagnum recurvum var. *porosum* Schlieph. & Warnst. in Warnst. = Sphagnum jensenii [SPHAJEN] Sphagnaceae (B:M)
Sphagnum recurvum var. *pulchrum* (Lindb.) Warnst. = Sphagnum pulchrum [SPHAPUL] Sphagnaceae (B:M)
Sphagnum recurvum var. *tenue* Klinggr. = Sphagnum angustifolium [SPHAANU] Sphagnaceae (B:M)
Sphagnum rigidum (Nees & Hornsch.) Schimp. = Sphagnum compactum [SPHACOM] Sphagnaceae (B:M)
Sphagnum riparium Ångstr. [SPHARIP] Sphagnaceae (B:M)
Sphagnum robustum (Warnst.) Röll = Sphagnum russowii [SPHARUS] Sphagnaceae (B:M)
Sphagnum rubellum Wils. [SPHARUB] Sphagnaceae (B:M)
Sphagnum rubellum var. *rubescens* Warnst. = Sphagnum rubellum [SPHARUB] Sphagnaceae (B:M)
Sphagnum rubellum var. *rubrum* (Grav.) Horr. = Sphagnum rubellum [SPHARUB] Sphagnaceae (B:M)
Sphagnum rubellum var. *violaceum* C. Jens. = Sphagnum rubellum [SPHARUB] Sphagnaceae (B:M)
Sphagnum rubiginosum Flatb. [SPHARUI] Sphagnaceae (B:M)
Sphagnum russowii Warnst. [SPHARUS] Sphagnaceae (B:M)
Sphagnum russowii var. *girgensohnioides* Russ. ex Warnst. = Sphagnum russowii [SPHARUS] Sphagnaceae (B:M)
Sphagnum russowii var. *poecilum* Russ. ex Warnst. = Sphagnum russowii [SPHARUS] Sphagnaceae (B:M)
Sphagnum russowii var. *rhodochroum* Russ. in Warnst. = Sphagnum russowii [SPHARUS] Sphagnaceae (B:M)
Sphagnum schofieldii Crum [SPHASCH] Sphagnaceae (B:M)
Sphagnum semisquarrosum Russ. = Sphagnum squarrosum [SPHASQU] Sphagnaceae (B:M)
Sphagnum squarrosum Crome [SPHASQU] Sphagnaceae (B:M)
Sphagnum squarrosum var. *cuspidatum* Warnst. = Sphagnum squarrosum [SPHASQU] Sphagnaceae (B:M)
Sphagnum squarrosum var. *imbricatum* Schimp. = Sphagnum squarrosum [SPHASQU] Sphagnaceae (B:M)
Sphagnum squarrosum var. *immersum* Beek ex Warnst. = Sphagnum squarrosum [SPHASQU] Sphagnaceae (B:M)
Sphagnum squarrosum var. *semisquarrosum* Russ. ex Warnst. = Sphagnum squarrosum [SPHASQU] Sphagnaceae (B:M)
Sphagnum squarrosum var. *spectabile* Russ. in Warnst. = Sphagnum squarrosum [SPHASQU] Sphagnaceae (B:M)
Sphagnum squarrosum var. *subsquarrosum* Russ. in Warnst. = Sphagnum squarrosum [SPHASQU] Sphagnaceae (B:M)
Sphagnum subbicolor Hampe = Sphagnum centrale [SPHACEN] Sphagnaceae (B:M)
Sphagnum subnitens Russ. & Warnst. in Warnst. [SPHASUN] Sphagnaceae (B:M)
Sphagnum subobesum Warnst. [SPHASUB] Sphagnaceae (B:M)
Sphagnum subsecundum Nees in Sturm [SPHASUS] Sphagnaceae (B:M)
Sphagnum subsecundum var. *contortum* (Schultz) Hüb. = Sphagnum contortum [SPHACON] Sphagnaceae (B:M)
Sphagnum subsecundum var. *junsaiense* (Warnst.) Crum = Sphagnum subobesum [SPHASUB] Sphagnaceae (B:M)
Sphagnum subsecundum var. *platyphyllum* (Lindb. ex Braithw.) Card. = Sphagnum platyphyllum [SPHAPLA] Sphagnaceae (B:M)
Sphagnum subtile (Russ.) Warnst. [SPHASUT] Sphagnaceae (B:M)
Sphagnum tenellum (Brid.) Bory [SPHATEN] Sphagnaceae (B:M)
Sphagnum tenellum var. *rubellum* (Wils.) Klinggr. = Sphagnum rubellum [SPHARUB] Sphagnaceae (B:M)
Sphagnum tenellum var. *rubrum* Grav. ex Warnst. = Sphagnum rubellum [SPHARUB] Sphagnaceae (B:M)
Sphagnum tenuifolium Warnst. = Sphagnum fuscum [SPHAFUS] Sphagnaceae (B:M)
Sphagnum teres (Schimp.) Ångstr. in Hartm. [SPHATER] Sphagnaceae (B:M)
Sphagnum teres var. *compactum* Warnst. = Sphagnum teres [SPHATER] Sphagnaceae (B:M)
Sphagnum teres var. *imbricatum* Warnst. = Sphagnum teres [SPHATER] Sphagnaceae (B:M)
Sphagnum teres var. *squarrosulum* (Schimp.) Warnst. = Sphagnum teres [SPHATER] Sphagnaceae (B:M)
Sphagnum warnstorfianum Du Rietz = Sphagnum warnstorfii [SPHAWAR] Sphagnaceae (B:M)
Sphagnum warnstorfii Russ. [SPHAWAR] Sphagnaceae (B:M)
Sphagnum warnstorfii var. *purpurascens* Russ. in Warnst. = Sphagnum warnstorfii [SPHAWAR] Sphagnaceae (B:M)
Sphagnum warnstorfii var. *viride* Russ. in Warnst. = Sphagnum warnstorfii [SPHAWAR] Sphagnaceae (B:M)
Sphagnum wilfii Crum [SPHAWIL] Sphagnaceae (B:M)
Sphagnum wulfianum Girg. [SPHAWUL] Sphagnaceae (B:M)
Sphagnum wulfianum var. *squarrulosum* Russ. = Sphagnum wulfianum [SPHAWUL] Sphagnaceae (B:M)
Sphagnum wulfianum var. *versicolor* Warnst. = Sphagnum wulfianum [SPHAWUL] Sphagnaceae (B:M)
Sphagnum wulfianum var. *viride* Warnst. = Sphagnum wulfianum [SPHAWUL] Sphagnaceae (B:M)
Sphenolobopsis Schust. & Kitag. [SPHENOL$] Jungermanniaceae (B:H)
Sphenolobopsis pearsonii (Spruce) Schust. [SPHEPEA] Jungermanniaceae (B:H)
Sphenolobus ascendens Warnst. = Lophozia ascendens [LOPHASC] Jungermanniaceae (B:H)

Sphenolobus exsectiformis (Breidl.) K. Müll. = Tritomaria exsectiformis [TRITEXE] Jungermanniaceae (B:H)
Sphenolobus exsectus (Schmid. ex Schrad.) Steph. = Tritomaria exsecta [TRITEXS] Jungermanniaceae (B:H)
Sphenolobus helleranus (Nees) Steph. = Anastrophyllum helleranum [ANASHEL] Jungermanniaceae (B:H)
Sphenolobus michauxii (Web.) Steph. = Anastrophyllum michauxii [ANASMIC] Jungermanniaceae (B:H)
Sphenolobus minutus (Schreb.) Berggr. = Anastrophyllum minutum [ANASMIN] Jungermanniaceae (B:H)
Sphenolobus minutus var. *grandis* (Lindb.) Lindb. & S. Arnell = Anastrophyllum minutum var. grandis [ANASMIN1] Jungermanniaceae (B:H)
Sphenolobus politus Steph. = Tritomaria polita [TRITPOL] Jungermanniaceae (B:H)
Sphenolobus saccatulus (Lindb.) K. Müll. = Anastrophyllum minutum var. grandis [ANASMIN1] Jungermanniaceae (B:H)
Sphenolobus saxicolus (Schrad.) Steph. = Anastrophyllum saxicola [ANASSAX] Jungermanniaceae (B:H)
Sphenopholis Scribn. [SPHENOP$] Poaceae (V:M)
Sphenopholis intermedia (Rydb.) Rydb. [SPHEINT] Poaceae (V:M)
Sphenopholis intermedia var. *pilosa* Dore = Sphenopholis intermedia [SPHEINT] Poaceae (V:M)
Sphenopholis obtusata (Michx.) Scribn. [SPHEOBT] Poaceae (V:M)
Sphenopholis obtusata var. *lobata* (Trin.) Scribn. = Sphenopholis obtusata [SPHEOBT] Poaceae (V:M)
Sphenopholis obtusata var. *major* (Torr.) K.S. Erdman = Sphenopholis intermedia [SPHEINT] Poaceae (V:M)
Sphenopholis obtusata var. *pubescens* (Scribn. & Merr.) Scribn. = Sphenopholis obtusata [SPHEOBT] Poaceae (V:M)
Spilonema Bornet [SPILONE$] Coccocarpiaceae (L)
Spilonema revertens Nyl. [SPILREV] Coccocarpiaceae (L)
Spilonema tuberculosa Sm. = Porpidia tuberculosa [PORPTUB] Porpidiaceae (L)
Spiraea L. [SPIRAEA$] Rosaceae (V:D)
Spiraea aruncus L. = Aruncus dioicus var. vulgaris [ARUNDIO1] Rosaceae (V:D)
Spiraea beauverdiana auct. = Spiraea stevenii [SPIRSTE] Rosaceae (V:D)
Spiraea beauverdiana var. *stevenii* Schneid. = Spiraea stevenii [SPIRSTE] Rosaceae (V:D)
Spiraea betulifolia Pallas [SPIRBET] Rosaceae (V:D)
Spiraea betulifolia ssp. *lucida* (Dougl. ex Greene) Taylor & MacBryde = Spiraea betulifolia var. lucida [SPIRBET2] Rosaceae (V:D)
Spiraea betulifolia var. **lucida** (Dougl. ex Greene) C.L. Hitchc. [SPIRBET2] Rosaceae (V:D)
Spiraea densiflora Nutt. ex Greenm. = Spiraea splendens [SPIRSPL] Rosaceae (V:D)
Spiraea densiflora ssp. *splendens* (Baumann ex K. Koch) Abrams = Spiraea splendens [SPIRSPL] Rosaceae (V:D)
Spiraea densiflora var. *splendens* (Baumann ex K. Koch) C.L. Hitchc. = Spiraea splendens [SPIRSPL] Rosaceae (V:D)
Spiraea discolor Pursh = Holodiscus discolor [HOLODIC] Rosaceae (V:D)
Spiraea douglasii Hook. [SPIRDOU] Rosaceae (V:D)
Spiraea douglasii ssp. *menziesii* (Hook.) Calder & Taylor = Spiraea douglasii var. menziesii [SPIRDOU1] Rosaceae (V:D)
Spiraea douglasii var. **douglasii** [SPIRDOU0] Rosaceae (V:D)
Spiraea douglasii var. **menziesii** (Hook.) K. Presl [SPIRDOU1] Rosaceae (V:D)
Spiraea douglasii var. *roseata* (Rydb.) C.L. Hitchc. = Spiraea douglasii var. douglasii [SPIRDOU0] Rosaceae (V:D)
Spiraea lucida Dougl. ex Greene = Spiraea betulifolia var. lucida [SPIRBET2] Rosaceae (V:D)
Spiraea menziesii Hook. = Spiraea douglasii var. menziesii [SPIRDOU1] Rosaceae (V:D)
Spiraea × pyramidata Greene (pro sp.) [SPIRPYR] Rosaceae (V:D)
Spiraea pyramidata Greene = Spiraea × pyramidata [SPIRPYR] Rosaceae (V:D)
Spiraea splendens Baumann ex K. Koch [SPIRSPL] Rosaceae (V:D)
Spiraea stevenii (Schneid.) Rydb. [SPIRSTE] Rosaceae (V:D)
Spiraea subvillosa Rydb. = Spiraea douglasii var. menziesii [SPIRDOU1] Rosaceae (V:D)
Spiraea tomentulosa Rydb. = Spiraea × pyramidata [SPIRPYR] Rosaceae (V:D)
Spiranthes L.C. Rich. [SPIRANT$] Orchidaceae (V:M)
Spiranthes romanzoffiana Cham. [SPIRROM] Orchidaceae (V:M)
Spiranthes stricta Rydb. = Spiranthes romanzoffiana [SPIRROM] Orchidaceae (V:M)
Spiranthes unalascensis Spreng. = Piperia unalascensis [PIPEUNA] Orchidaceae (V:M)
Spirodela Schleid. [SPIRODE$] Lemnaceae (V:M)
Spirodela polyrhiza (L.) Schleid. [SPIRPOL] Lemnaceae (V:M)
Spirodela polyrrhiza var. *masonii* Daubs = Spirodela polyrhiza [SPIRPOL] Lemnaceae (V:M)
Splachnobryum kieneri Williams = Bryum turbinatum [BRYUTUR] Bryaceae (B:M)
Splachnum Hedw. [SPLACHU$] Splachnaceae (B:M)
Splachnum ampullaceum Hedw. [SPLAAMP] Splachnaceae (B:M)
Splachnum luteum Hedw. [SPLALUT] Splachnaceae (B:M)
Splachnum ovatum Dicks. ex Hedw. = Splachnum sphaericum [SPLASPH] Splachnaceae (B:M)
Splachnum rubrum Hedw. [SPLARUB] Splachnaceae (B:M)
Splachnum sphaericum Hedw. [SPLASPH] Splachnaceae (B:M)
Splachnum vasculosum Hedw. [SPLAVAS] Splachnaceae (B:M)
Sporastatia A. Massal. [SPORAST$] Acarosporaceae (L)
Sporastatia cinerea (Schaerer) Körber = Sporastatia polyspora [SPORPOL] Acarosporaceae (L)
Sporastatia polyspora (Nyl.) Grummann [SPORPOL] Acarosporaceae (L)
Sporastatia testudinea (Ach.) A. Massal. [SPORTES] Acarosporaceae (L)
Sporobolus R. Br. [SPOROBO$] Poaceae (V:M)
Sporobolus airoides (Torr.) Torr. [SPORAIR] Poaceae (V:M)
Sporobolus asper (Beauv.) Kunth = Sporobolus compositus [SPORCOM] Poaceae (V:M)
Sporobolus asper var. *hookeri* (Trin.) Vasey = Sporobolus compositus [SPORCOM] Poaceae (V:M)
Sporobolus asperifolius (Nees & Meyen ex Trin.) Nees = Muhlenbergia asperifolia [MUHLASP] Poaceae (V:M)
Sporobolus compositus (Poir.) Merr. [SPORCOM] Poaceae (V:M)
Sporobolus cryptandrus (Torr.) Gray [SPORCRY] Poaceae (V:M)
Sporobolus cryptandrus ssp. *fuscicola* (Hook.) E.K. Jones & Fassett = Sporobolus cryptandrus [SPORCRY] Poaceae (V:M)
Sporobolus cryptandrus ssp. *typicus* var. *occidentalis* E.K. Jones & Fassett = Sporobolus cryptandrus [SPORCRY] Poaceae (V:M)
Sporobolus cryptandrus var. *fuscicola* (Hook.) Pohl = Sporobolus cryptandrus [SPORCRY] Poaceae (V:M)
Sporobolus cryptandrus var. *occidentalis* E.K. Jones & Fassett = Sporobolus cryptandrus [SPORCRY] Poaceae (V:M)
*__Sporobolus vaginiflorus__ (Torr. ex Gray) Wood [SPORVAG] Poaceae (V:M)
Sporobolus vaginiflorus var. *inaequalis* Fern. = Sporobolus vaginiflorus [SPORVAG] Poaceae (V:M)
Spraguea umbellata Torr. = Cistanthe umbellata [CISTUMB] Portulacaceae (V:D)
Spraguea umbellata var. *caudicifera* Gray = Cistanthe umbellata var. caudicifera [CISTUMB1] Portulacaceae (V:D)
Squamarina Poelt [SQUAMAR$] Bacidiaceae (L)
Squamarina cartilaginea (With.) P. James [SQUACAR] Bacidiaceae (L)
Squamarina crassa (Hudson) Poelt = Squamarina cartilaginea [SQUACAR] Bacidiaceae (L)
Squamarina lentigera (Weber) Poelt [SQUALEN] Bacidiaceae (L)
Stachys L. [STACHYS$] Lamiaceae (V:D)
*__Stachys arvensis__ (L.) L. [STACARV] Lamiaceae (V:D)
Stachys asperrima Rydb. = Stachys palustris ssp. pilosa [STACPAL1] Lamiaceae (V:D)
Stachys borealis Rydb. = Stachys palustris ssp. pilosa [STACPAL1] Lamiaceae (V:D)
*__Stachys byzantina__ K. Koch ex Scheele [STACBYZ] Lamiaceae (V:D)
Stachys ciliata Epling [STACCIL] Lamiaceae (V:D)

Stachys cooleyae Heller = Stachys ciliata [STACCIL] Lamiaceae (V:D)
Stachys lanata Jacq. = Stachys byzantina [STACBYZ] Lamiaceae (V:D)
Stachys olympica Poir. = Stachys byzantina [STACBYZ] Lamiaceae (V:D)
Stachys palustris L. [STACPAL] Lamiaceae (V:D)
Stachys palustris ssp. **pilosa** (Nutt.) Epling [STACPAL1] Lamiaceae (V:D)
Stachys palustris var. *elliptica* Clos = Stachys palustris ssp. pilosa [STACPAL1] Lamiaceae (V:D)
Stachys palustris var. *petiolata* Clos = Stachys palustris ssp. pilosa [STACPAL1] Lamiaceae (V:D)
Stachys palustris var. *pilosa* (Nutt.) Fern. = Stachys palustris ssp. pilosa [STACPAL1] Lamiaceae (V:D)
Stachys palustris var. *segetum* (Mutel) Grogn. = Stachys palustris ssp. pilosa [STACPAL1] Lamiaceae (V:D)
Stachys pilosa Nutt. = Stachys palustris ssp. pilosa [STACPAL1] Lamiaceae (V:D)
Stachys rigida Nutt. ex Benth. = Stachys palustris ssp. pilosa [STACPAL1] Lamiaceae (V:D)
Stachys rigida ssp. *lanata* Epling = Stachys palustris ssp. pilosa [STACPAL1] Lamiaceae (V:D)
Stachys rigida ssp. *quercetorum* (Heller) Epling = Stachys palustris ssp. pilosa [STACPAL1] Lamiaceae (V:D)
Stachys rigida ssp. *rivularis* (Heller) Epling = Stachys palustris ssp. pilosa [STACPAL1] Lamiaceae (V:D)
Stachys scopulorum Greene = Stachys palustris ssp. pilosa [STACPAL1] Lamiaceae (V:D)
Stachys teucriifolia Rydb. = Stachys palustris ssp. pilosa [STACPAL1] Lamiaceae (V:D)
Stachys teucriiformis Rydb. = Stachys palustris ssp. pilosa [STACPAL1] Lamiaceae (V:D)
Staurothele Norman [STAUROT$] Verrucariaceae (L)
Staurothele ambrosiana (A. Massal.) Zschacke = Staurothele drummondii [STAUDRU] Verrucariaceae (L)
Staurothele areolata (Ach.) Lettau [STAUARE] Verrucariaceae (L)
Staurothele circinata Tuck. = Staurothele fissa [STAUFIS] Verrucariaceae (L)
Staurothele clopima var. *areolata* (Ach.) Vainio = Staurothele areolata [STAUARE] Verrucariaceae (L)
Staurothele clopima var. *mamillata* Vainio = Staurothele areolata [STAUARE] Verrucariaceae (L)
Staurothele diffusilis (Nyl.) Zahlbr. = Staurothele drummondii [STAUDRU] Verrucariaceae (L)
Staurothele drummondii (Tuck.) Tuck. [STAUDRU] Verrucariaceae (L)
Staurothele fissa (Taylor) Zwackh [STAUFIS] Verrucariaceae (L)
Staurothele fuscocuprea (Nyl.) Zschacke = Staurothele drummondii [STAUDRU] Verrucariaceae (L)
Staurothele fuscocuprea f. *rimosior* (Nyl.) Zsch. = Staurothele drummondii [STAUDRU] Verrucariaceae (L)
Staurothele fuscocuprea f. *rimoso-areolata* Zsch. = Staurothele drummondii [STAUDRU] Verrucariaceae (L)
Staurothele glacialis Herre = Staurothele fissa [STAUFIS] Verrucariaceae (L)
Staurothele hazslinskyi (Körber) Blomb. & Forss. = Staurothele fissa [STAUFIS] Verrucariaceae (L)
Staurothele inconversa (Nyl.) Blomb. & Forss. = Staurothele fissa [STAUFIS] Verrucariaceae (L)
Staurothele lithina auct. = Staurothele fissa [STAUFIS] Verrucariaceae (L)
Staurothele perradiata Lynge = Staurothele drummondii [STAUDRU] Verrucariaceae (L)
Staurothele perradiata f. *dissoluta* Lynge = Staurothele drummondii [STAUDRU] Verrucariaceae (L)
Staurothele perradiata var. *hypothallina* Lamb = Staurothele drummondii [STAUDRU] Verrucariaceae (L)
Staurothele rupifraga (A. Massal.) Arnold [STAURUP] Verrucariaceae (L)
Staurothele septentrionalis Lynge = Staurothele drummondii [STAUDRU] Verrucariaceae (L)
Staurothele succedens (Rehm) Arnold = Staurothele drummondii [STAUDRU] Verrucariaceae (L)
Staurothele umbrina (Wahlenb.) Tuck. = Staurothele fissa [STAUFIS] Verrucariaceae (L)
Steerecleus Robins. [STEEREC$] Brachytheciaceae (B:M)
Steerecleus serrulatus (Hedw.) Robins. [STEESER] Brachytheciaceae (B:M)
Stegonia Vent. [STEGONI$] Pottiaceae (B:M)
Stegonia latifolia (Schwaegr. in Schultes) Vent. ex Broth. [STEGLAT] Pottiaceae (B:M)
Stegonia latifolia var. *pilifera* (Brid.) Broth. = Stegonia pilifera [STEGPIL] Pottiaceae (B:M)
Stegonia pilifera (Brid.) Crum & Anderson [STEGPIL] Pottiaceae (B:M)
Steinia Körber [STEINIA$] Lecideaceae (L)
Steinia geophana (Nyl.) Stein [STEIGEO] Lecideaceae (L)
Steironema ciliatum (L.) Baudo = Lysimachia ciliata [LYSICIL] Primulaceae (V:D)
Steironema pumilum Greene = Lysimachia ciliata [LYSICIL] Primulaceae (V:D)
Stellaria L. [STELLAR$] Caryophyllaceae (V:D)
Stellaria alaskana Hultén [STELALA] Caryophyllaceae (V:D)
Stellaria alsine Grimm [STELALS] Caryophyllaceae (V:D)
Stellaria americana (Porter ex B.L. Robins.) Standl. [STELAME] Caryophyllaceae (V:D)
Stellaria aquatica (L.) Scop. = Myosoton aquaticum [MYOSAQU] Caryophyllaceae (V:D)
Stellaria biflora L. = Minuartia biflora [MINUBIF] Caryophyllaceae (V:D)
Stellaria borealis Bigelow [STELBOR] Caryophyllaceae (V:D)
Stellaria borealis ssp. *bongardiana* (Fern.) Piper & Beattie = Stellaria borealis ssp. sitchana [STELBOR1] Caryophyllaceae (V:D)
Stellaria borealis ssp. **sitchana** (Steud.) Piper [STELBOR1] Caryophyllaceae (V:D)
Stellaria borealis var. *bongardiana* Fern. = Stellaria borealis ssp. sitchana [STELBOR1] Caryophyllaceae (V:D)
Stellaria borealis var. *crispa* (Cham. & Schlecht.) Fenzl ex Torr. & Gray = Stellaria crispa [STELCRI] Caryophyllaceae (V:D)
Stellaria borealis var. *simcoei* (T.J. Howell) Fern. = Stellaria calycantha [STELCAL] Caryophyllaceae (V:D)
Stellaria borealis var. *sitchana* (Steud.) Fern. = Stellaria borealis ssp. sitchana [STELBOR1] Caryophyllaceae (V:D)
Stellaria brachypetala var. *bongardiana* Fern. = Stellaria borealis ssp. sitchana [STELBOR1] Caryophyllaceae (V:D)
Stellaria calycantha (Ledeb.) Bong. [STELCAL] Caryophyllaceae (V:D)
Stellaria calycantha ssp. *interior* Hultén = Stellaria borealis [STELBOR] Caryophyllaceae (V:D)
Stellaria calycantha var. *bongardiana* (Fern.) Fern. = Stellaria borealis ssp. sitchana [STELBOR1] Caryophyllaceae (V:D)
Stellaria calycantha var. *isophylla* (Fern.) Fern. = Stellaria borealis [STELBOR] Caryophyllaceae (V:D)
Stellaria calycantha var. *simcoei* (T.J. Howell) Fern. = Stellaria calycantha [STELCAL] Caryophyllaceae (V:D)
Stellaria calycantha var. *sitchana* (Steud.) Fern. = Stellaria borealis ssp. sitchana [STELBOR1] Caryophyllaceae (V:D)
Stellaria crassifolia Ehrh. [STELCRA] Caryophyllaceae (V:D)
Stellaria crispa Cham. & Schlecht. [STELCRI] Caryophyllaceae (V:D)
Stellaria dichotoma var. *americana* Porter ex B.L. Robins. = Stellaria americana [STELAME] Caryophyllaceae (V:D)
Stellaria edwardsii R. Br. = Stellaria longipes [STELLOG] Caryophyllaceae (V:D)
Stellaria gonomischa Boivin = Stellaria umbellata [STELUMB] Caryophyllaceae (V:D)
***Stellaria graminea** L. [STELGRA] Caryophyllaceae (V:D)
Stellaria humifusa Rottb. [STELHUM] Caryophyllaceae (V:D)
Stellaria humifusa var. *oblongifolia* Fenzl = Stellaria humifusa [STELHUM] Caryophyllaceae (V:D)
Stellaria humifusa var. *suberecta* Boivin = Stellaria humifusa [STELHUM] Caryophyllaceae (V:D)
Stellaria laeta Richards. = Stellaria longipes [STELLOG] Caryophyllaceae (V:D)

Stellaria longipes Goldie [STELLOG] Caryophyllaceae (V:D)
Stellaria longipes var. *altocaulis* (Hultén) C.L. Hitchc. = Stellaria longipes [STELLOG] Caryophyllaceae (V:D)
Stellaria longipes var. *edwardsii* (R. Br.) Gray = Stellaria longipes [STELLOG] Caryophyllaceae (V:D)
Stellaria longipes var. *laeta* (Richards.) S. Wats. = Stellaria longipes [STELLOG] Caryophyllaceae (V:D)
Stellaria longipes var. *subvestita* (Greene) Polunin = Stellaria longipes [STELLOG] Caryophyllaceae (V:D)
*Stellaria media** (L.) Vill. [STELMED] Caryophyllaceae (V:D)
Stellaria monantha Hultén = Stellaria longipes [STELLOG] Caryophyllaceae (V:D)
Stellaria monantha var. *altocaulis* Hultén = Stellaria longipes [STELLOG] Caryophyllaceae (V:D)
Stellaria nitens Nutt. [STELNIT] Caryophyllaceae (V:D)
Stellaria obtusa Engelm. [STELOBT] Caryophyllaceae (V:D)
Stellaria praecox A. Nels. = Stellaria nitens [STELNIT] Caryophyllaceae (V:D)
Stellaria simcoei (T.J. Howell) C.L. Hitchc. = Stellaria calycantha [STELCAL] Caryophyllaceae (V:D)
Stellaria sitchana Steud. = Stellaria borealis ssp. sitchana [STELBOR1] Caryophyllaceae (V:D)
Stellaria sitchana var. *bongardiana* (Fern.) Hultén = Stellaria borealis ssp. sitchana [STELBOR1] Caryophyllaceae (V:D)
Stellaria stricta Richards. = Stellaria longipes [STELLOG] Caryophyllaceae (V:D)
Stellaria subvestita Greene = Stellaria longipes [STELLOG] Caryophyllaceae (V:D)
Stellaria uliginosa Murr. = Stellaria alsine [STELALS] Caryophyllaceae (V:D)
Stellaria umbellata Turcz. ex Kar. & Kir. [STELUMB] Caryophyllaceae (V:D)
Stellaria viridula (Piper) St. John = Stellaria obtusa [STELOBT] Caryophyllaceae (V:D)
Stellaria washingtoniana B.L. Robins. = Stellaria obtusa [STELOBT] Caryophyllaceae (V:D)
Stellaria weberi Boivin = Stellaria umbellata [STELUMB] Caryophyllaceae (V:D)
Stellariomnium blyttii (Bruch & Schimp. in B.S.G.) M. Bowers = Mnium blyttii [MNIUBLY] Mniaceae (B:M)
Stenanthella occidentalis (Gray) Rydb. = Stenanthium occidentale [STENOCC] Liliaceae (V:M)
Stenanthium (Gray) Kunth [STENANT$] Liliaceae (V:M)
Stenanthium occidentale Gray [STENOCC] Liliaceae (V:M)
Stenocybe (Nyl.) Körber [STENOCY$] Mycocaliciaceae (L)
Stenocybe byssacea (Fr.) Körber = Stenocybe pullatula [STENPUL] Mycocaliciaceae (L)
Stenocybe clavata Tibell [STENCLA] Mycocaliciaceae (L)
Stenocybe euspora (Nyl.) Anzi = Stenocybe major [STENMAJ] Mycocaliciaceae (L)
Stenocybe major (Nyl.) Körber [STENMAJ] Mycocaliciaceae (L)
Stenocybe pullatula (Ach.) Stein [STENPUL] Mycocaliciaceae (L)
Stephanomeria Nutt. [STEPHAN$] Asteraceae (V:D)
Stephanomeria tenuifolia (Raf.) Hall [STEPTEN] Asteraceae (V:D)
Stereocaulon Hoffm. [STEREOC$] Stereocaulaceae (L)
Stereocaulon albicans Th. Fr. = Leprocaulon albicans [LEPRALB] Uncertain-family (L)
Stereocaulon alpinum Laurer ex Funck [STERALP] Stereocaulaceae (L)
Stereocaulon arenarium (Savicz) Lamb [STERARE] Stereocaulaceae (L)
Stereocaulon botryosum Ach. [STERBOT] Stereocaulaceae (L)
Stereocaulon condensatum Hoffm. [STERCON] Stereocaulaceae (L)
Stereocaulon coniophyllum Lamb [STERCOI] Stereocaulaceae (L)
Stereocaulon coralloides Fr. = Stereocaulon dactylophyllum [STERDAC] Stereocaulaceae (L)
Stereocaulon dactylophyllum Flörke [STERDAC] Stereocaulaceae (L)
Stereocaulon denudatum Flörke = Stereocaulon vesuvianum [STERVES] Stereocaulaceae (L)
Stereocaulon evolutoides (H. Magn.) Frey = Stereocaulon saxatile [STERSAX] Stereocaulaceae (L)
Stereocaulon glareosum (Savicz) H. Magn. [STERGLA] Stereocaulaceae (L)
Stereocaulon grande (H. Magn.) H. Magn. [STERGRA] Stereocaulaceae (L)
Stereocaulon intermedium (Savicz) H. Magn. [STERINT] Stereocaulaceae (L)
Stereocaulon intermedium f. *compactum* Lamb = Stereocaulon intermedium [STERINT] Stereocaulaceae (L)
Stereocaulon microscopicum (Vill.) Frey = Leprocaulon microscopicum [LEPRMIC] Uncertain-family (L)
Stereocaulon myriocarpum Th. Fr. [STERMYR] Stereocaulaceae (L)
Stereocaulon paschale (L.) Hoffm. [STERPAS] Stereocaulaceae (L)
Stereocaulon pseudoarbuscula Asah. = Leprocaulon subalbicans [LEPRSUB] Uncertain-family (L)
Stereocaulon rivulorum H. Magn. [STERRIV] Stereocaulaceae (L)
Stereocaulon sasakii Zahlbr. [STERSAS] Stereocaulaceae (L)
Stereocaulon sasakii var. **simplex** (Riddle) Lamb [STERSAS1] Stereocaulaceae (L)
Stereocaulon sasakii var. **tomentosoides** Lamb [STERSAS2] Stereocaulaceae (L)
Stereocaulon saxatile H. Magn. [STERSAX] Stereocaulaceae (L)
Stereocaulon spathuliferum Vainio [STERSPA] Stereocaulaceae (L)
Stereocaulon sterile (Savicz) Lamb ex Krog [STERSTE] Stereocaulaceae (L)
Stereocaulon subcoralloides (Nyl.) Nyl. [STERSUC] Stereocaulaceae (L)
Stereocaulon tomentosum Fr. [STERTOM] Stereocaulaceae (L)
Stereocaulon tomentosum var. *simplex* Riddle = Stereocaulon sasakii var. simplex [STERSAS1] Stereocaulaceae (L)
Stereocaulon vesuvianum Pers. [STERVES] Stereocaulaceae (L)
Stereocaulon vesuvianum var. *denudatum* (Flörke) Lamb ex Poelt = Stereocaulon vesuvianum [STERVES] Stereocaulaceae (L)
Stereocaulon vesuvianum var. *nodulosum* (Wallr.) Lamb = Stereocaulon vesuvianum [STERVES] Stereocaulaceae (L)
Stereocaulon vesuvianum var. *nodulosum* f. *verrucosum* Lamb = Stereocaulon vesuvianum [STERVES] Stereocaulaceae (L)
Stereodon adscendens Lindb. = Herzogiella adscendens [HERZADS] Hypnaceae (B:M)
Stereodon arcuatus Lindb. = Hypnum lindbergii [HYPNLIN] Hypnaceae (B:M)
Stereodon callichrous (Brid.) Braithw. = Hypnum callichroum [HYPNCAL] Hypnaceae (B:M)
Stereodon cupressiformis (Hedw.) Brid. ex Mitt. = Hypnum cupressiforme [HYPNCUP] Hypnaceae (B:M)
Stereodon lindbergii (Mitt.) Braithw. = Hypnum lindbergii [HYPNLIN] Hypnaceae (B:M)
Stereodon pallescens (Hedw.) Mitt. = Hypnum pallescens [HYPNPAL] Hypnaceae (B:M)
Stereodon pratensis (Rabenh.) Warnst. = Hypnum pratense [HYPNPRA] Hypnaceae (B:M)
Stereodon reptilis (Michx.) Mitt. = Hypnum pallescens [HYPNPAL] Hypnaceae (B:M)
Sticta (Schreber) Ach. [STICTA$] Lobariaceae (L)
Sticta anthraspis Ach. = Pseudocyphellaria anthraspis [PSEUANT] Lobariaceae (L)
Sticta arctica Degel. [STICARC] Lobariaceae (L)
Sticta aurata Ach. = Pseudocyphellaria aurata [PSEUAUR] Lobariaceae (L)
Sticta crocata (L.) Ach. = Pseudocyphellaria crocata [PSEUCRO] Lobariaceae (L)
Sticta fuliginosa (Hoffm.) Ach. [STICFUL] Lobariaceae (L)
Sticta hallii Tuck. = Lobaria hallii [LOBAHAL] Lobariaceae (L)
Sticta limbata (Sm.) Ach. [STICLIM] Lobariaceae (L)
Sticta linita Ach. = Lobaria linita [LOBALIN] Lobariaceae (L)
Sticta oregana Tuck. = Lobaria oregana [LOBAORE] Lobariaceae (L)
Sticta oroborealis Goward & Tonsberg [STICORO] Lobariaceae (L)
Sticta pulmonaria (L.) Biroli = Lobaria pulmonaria [LOBAPUL] Lobariaceae (L)
Sticta verrucosa (Hudson) Fink = Lobaria scrobiculata [LOBASCR] Lobariaceae (L)
Sticta weigelii (Ach.) Vainio [STICWEI] Lobariaceae (L)
Sticta wrightii Tuck. [STICWRI] Lobariaceae (L)

Stigmatomma fissum (Tayl.) Syd. = Staurothele fissa [STAUFIS] Verrucariaceae (L)
Stipa L. [STIPA$] Poaceae (V:M)
Stipa canadensis Poir. = Oryzopsis canadensis [ORYZCAN] Poaceae (V:M)
Stipa columbiana auct. = Stipa nelsonii ssp. dorei [STIPNEL1] Poaceae (V:M)
Stipa comata Trin. & Rupr. [STIPCOM] Poaceae (V:M)
Stipa comata var. **intermedia** Scribn. & Tweedy [STIPCOM1] Poaceae (V:M)
Stipa curtiseta (A.S. Hitchc.) Barkworth [STIPCUR] Poaceae (V:M)
Stipa elmeri Piper & Brodie ex Scribn. = Stipa occidentalis var. pubescens [STIPOCC3] Poaceae (V:M)
Stipa hymenoides Roemer & J.A. Schultes = Oryzopsis hymenoides [ORYZHYM] Poaceae (V:M)
Stipa lemmonii (Vasey) Scribn. [STIPLEM] Poaceae (V:M)
Stipa lemmonii var. *jonesii* Scribn. = Stipa lemmonii [STIPLEM] Poaceae (V:M)
Stipa lemmonii var. *pubescens* Crampton = Stipa lemmonii [STIPLEM] Poaceae (V:M)
Stipa minor (Vasey) Scribn. = Stipa nelsonii ssp. dorei [STIPNEL1] Poaceae (V:M)
Stipa nelsonii Scribn. [STIPNEL] Poaceae (V:M)
Stipa nelsonii ssp. **dorei** Barkworth & Maze [STIPNEL1] Poaceae (V:M)
Stipa nelsonii var. *dorei* (Barkworth & Maze) Dorn = Stipa nelsonii ssp. dorei [STIPNEL1] Poaceae (V:M)
Stipa occidentalis Thurb. ex S. Wats. [STIPOCC] Poaceae (V:M)
Stipa occidentalis var. *minor* (Vasey) C.L. Hitchc. = Stipa nelsonii ssp. dorei [STIPNEL1] Poaceae (V:M)
Stipa occidentalis var. **pubescens** (Vasey) Maze, Taylor & MacBryde [STIPOCC3] Poaceae (V:M)
Stipa richardsonii Link [STIPRIC] Poaceae (V:M)
Stipa spartea Trin. [STIPSPA] Poaceae (V:M)
Stipa spartea var. *curtiseta* A.S. Hitchc. = Stipa curtiseta [STIPCUR] Poaceae (V:M)
Stipa tweedyi Scribn. = Stipa comata [STIPCOM] Poaceae (V:M)
Stipa viridula Trin. = Nassella viridula [NASSVIR] Poaceae (V:M)
Stokesiella brittoniae (Grout) Robins. = Eurhynchium praelongum [EURHPRA] Brachytheciaceae (B:M)
Stokesiella oregana (Sull.) Robins. = Eurhynchium oreganum [EURHORE] Brachytheciaceae (B:M)
Stokesiella praelonga (Hedw.) Robins. = Eurhynchium praelongum [EURHPRA] Brachytheciaceae (B:M)
Stokesiella praelonga var. *californicum* (Grout) Crum & Anderson = Eurhynchium praelongum [EURHPRA] Brachytheciaceae (B:M)
Stokesiella praelonga var. *stokesii* (Turn.) Crum = Eurhynchium praelongum [EURHPRA] Brachytheciaceae (B:M)
Strangospora Körber [STRANGO$] Acarosporaceae (L)
Strangospora moriformis (Ach.) Stein [STRAMOR] Acarosporaceae (L)
Strangospora pinicola (A. Massal.) Körber [STRAPIN] Acarosporaceae (L)
Streblotrichum convolutum (Hedw.) P. Beauv. = Barbula convoluta [BARBCON] Pottiaceae (B:M)
Streptopus Michx. [STREPTO$] Liliaceae (V:M)
Streptopus amplexifolius (L.) DC. [STREAMP] Liliaceae (V:M)
Streptopus amplexifolius var. *americanus* J.A. & J.H. Schultes = Streptopus amplexifolius var. amplexifolius [STREAMP0] Liliaceae (V:M)
Streptopus amplexifolius var. **amplexifolius** [STREAMP0] Liliaceae (V:M)
Streptopus amplexifolius var. **chalazatus** Fassett [STREAMP2] Liliaceae (V:M)
Streptopus amplexifolius var. *denticulatus* Fassett = Streptopus amplexifolius var. amplexifolius [STREAMP0] Liliaceae (V:M)
Streptopus amplexifolius var. *grandiflorus* Fassett = Streptopus amplexifolius var. amplexifolius [STREAMP0] Liliaceae (V:M)
Streptopus curvipes Vail = Streptopus roseus var. curvipes [STREROS1] Liliaceae (V:M)
Streptopus fassettii A. & D. Löve = Streptopus amplexifolius var. chalazatus [STREAMP2] Liliaceae (V:M)
Streptopus roseus Michx. [STREROS] Liliaceae (V:M)
Streptopus roseus ssp. *curvipes* (Vail) Hultén = Streptopus roseus var. curvipes [STREROS1] Liliaceae (V:M)
Streptopus roseus var. **curvipes** (Vail) Fassett [STREROS1] Liliaceae (V:M)
Streptopus streptopoides (Ledeb.) Frye & Rigg [STRESTR] Liliaceae (V:M)
Streptopus streptopoides ssp. *brevipes* (Baker) Calder & Taylor = Streptopus streptopoides var. brevipes [STRESTR1] Liliaceae (V:M)
Streptopus streptopoides var. **brevipes** (Baker) Fassett [STRESTR1] Liliaceae (V:M)
Strigula Fr. [STRIGUL$] Strigulaceae (L)
Strigula stigmatella (Ach.) R.C. Harris [STRISTI] Strigulaceae (L)
Strophocaulos arvensis (L.) Small = Convolvulus arvensis [CONVARV] Convolvulaceae (V:D)
Struthiopteris spicant (L.) Weiss = Blechnum spicant [BLECSPI] Blechnaceae (V:F)
Suaeda Forsk. ex Scop. [SUAEDA$] Chenopodiaceae (V:D)
Suaeda americana (Pers.) Fern. = Suaeda calceoliformis [SUAECAL] Chenopodiaceae (V:D)
Suaeda calceoliformis (Hook.) Moq. [SUAECAL] Chenopodiaceae (V:D)
Suaeda depressa auct. = Suaeda calceoliformis [SUAECAL] Chenopodiaceae (V:D)
Suaeda depressa var. *erecta* S. Wats. = Suaeda calceoliformis [SUAECAL] Chenopodiaceae (V:D)
Suaeda maritima var. *americana* (Pers.) Boivin = Suaeda calceoliformis [SUAECAL] Chenopodiaceae (V:D)
Suaeda minutiflora S. Wats. = Suaeda calceoliformis [SUAECAL] Chenopodiaceae (V:D)
Suaeda occidentalis (S. Wats.) S. Wats. = Suaeda calceoliformis [SUAECAL] Chenopodiaceae (V:D)
Subularia L. [SUBULAR$] Brassicaceae (V:D)
Subularia aquatica L. [SUBUAQU] Brassicaceae (V:D)
Subularia aquatica ssp. *americana* Mulligan & Calder = Subularia aquatica var. americana [SUBUAQU2] Brassicaceae (V:D)
Subularia aquatica var. **americana** (Mulligan & Calder) Boivin [SUBUAQU2] Brassicaceae (V:D)
Suksdorfia Gray [SUKSDOR$] Saxifragaceae (V:D)
Suksdorfia ranunculifolia (Hook.) Engl. [SUKSRAN] Saxifragaceae (V:D)
Suksdorfia violacea Gray [SUKSVIO] Saxifragaceae (V:D)
Swartzia inclinata (Hedw.) P. Beauv. = Distichium inclinatum [DISTINC] Ditrichaceae (B:M)
Swartzia montana Lindb. = Distichium capillaceum [DISTCAP] Ditrichaceae (B:M)
Swertia L. [SWERTIA$] Gentianaceae (V:D)
Swertia perennis L. [SWERPER] Gentianaceae (V:D)
Swertia perennis var. *obtusa* (Ledeb.) Ledeb. ex Griseb. = Swertia perennis [SWERPER] Gentianaceae (V:D)
Swida instolonea (A. Nels.) Rydb. = Cornus sericea [CORNSER] Cornaceae (V:D)
Swida stolonifera (Michx.) Rydb. = Cornus sericea [CORNSER] Cornaceae (V:D)
Swida suecica (L.) Holub = Cornus suecica [CORNSUE] Cornaceae (V:D)
Symphoricarpos Duham. [SYMPHOR$] Caprifoliaceae (V:D)
Symphoricarpos albus (L.) Blake [SYMPALB] Caprifoliaceae (V:D)
Symphoricarpos albus ssp. *laevigatus* (Fern.) Hultén = Symphoricarpos albus var. laevigatus [SYMPALB2] Caprifoliaceae (V:D)
Symphoricarpos albus var. **albus** [SYMPALB0] Caprifoliaceae (V:D)
Symphoricarpos albus var. **laevigatus** (Fern.) Blake [SYMPALB2] Caprifoliaceae (V:D)
Symphoricarpos albus var. *pauciflorus* (W.J. Robins. ex Gray) Blake = Symphoricarpos albus var. albus [SYMPALB0] Caprifoliaceae (V:D)
Symphoricarpos hesperius G.N. Jones [SYMPHES] Caprifoliaceae (V:D)
Symphoricarpos mollis ssp. *hesperius* (G.N. Jones) Abrams ex Ferris = Symphoricarpos hesperius [SYMPHES] Caprifoliaceae (V:D)

Symphoricarpos mollis var. *hesperius* (G.N. Jones) Cronq. = Symphoricarpos hesperius [SYMPHES] Caprifoliaceae (V:D)
Symphoricarpos occidentalis Hook. [SYMPOCC] Caprifoliaceae (V:D)
Symphoricarpos oreophilus Gray [SYMPORE] Caprifoliaceae (V:D)
Symphoricarpos oreophilus var. **utahensis** (Rydb.) A. Nels. [SYMPORE1] Caprifoliaceae (V:D)
Symphoricarpos pauciflorus W.J. Robins. ex Gray = Symphoricarpos albus var. albus [SYMPALB0] Caprifoliaceae (V:D)
Symphoricarpos racemosus Michx. = Symphoricarpos albus var. albus [SYMPALB0] Caprifoliaceae (V:D)
Symphoricarpos rivularis Suksdorf = Symphoricarpos albus var. laevigatus [SYMPALB2] Caprifoliaceae (V:D)
Symphoricarpos tetonensis A. Nels. = Symphoricarpos oreophilus var. utahensis [SYMPORE1] Caprifoliaceae (V:D)
Symphoricarpos utahensis Rydb. = Symphoricarpos oreophilus var. utahensis [SYMPORE1] Caprifoliaceae (V:D)
Symphoricarpos vaccinioides Rydb. = Symphoricarpos oreophilus var. utahensis [SYMPORE1] Caprifoliaceae (V:D)
Symphyotrichum boreale (Torr. & Gray) A. & D. Löve = Aster borealis [ASTEBOR] Asteraceae (V:D)
Symphyotrichum hesperium (Gray) A. & D. Löve = Aster lanceolatus ssp. hesperius [ASTELAN1] Asteraceae (V:D)
Symphytum L. [SYMPHYT$] Boraginaceae (V:D)
Symphytum asperrimum Donn ex Sims = Symphytum asperum [SYMPASP] Boraginaceae (V:D)
***Symphytum asperum** Lepechin [SYMPASP] Boraginaceae (V:D)
***Symphytum officinale** L. [SYMPOFF] Boraginaceae (V:D)
Symphytum officinale ssp. *uliginosum* (Kern.) Nyman = Symphytum officinale [SYMPOFF] Boraginaceae (V:D)
Symphytum uliginosum Kern. = Symphytum officinale [SYMPOFF] Boraginaceae (V:D)
Synechoblastus polycarpus (Hoffm.) Dalla Torre & Sarnth. = Collema polycarpon [COLLPOL] Collemataceae (L)
Synechoblastus rupestris (Sw.) Trevisan = Collema flaccidum [COLLFLA] Collemataceae (L)
Synechoblastus wyomingensis Fink = Collema polycarpon [COLLPOL] Collemataceae (L)
Syntherisma sanguinalis (L.) Dulac = Digitaria sanguinalis [DIGISAN] Poaceae (V:M)
Syntrichia ruralis (Hedw.) Web. & Mohr = Tortula ruralis [TORTRUA] Pottiaceae (B:M)
Szczawinskia A. Funk [SZCZAWI$] Uncertain-family (L)
Szczawinskia tsugae A. Funk [SZCZTSU] Uncertain-family (L)

T

Takakia Hatt. & H. Inoue [TAKAKIA$] Takakiaceae (B:M)
Takakia lepidozioides Hatt. & H. Inoue [TAKALEP] Takakiaceae (B:M)
Talinum Adans. [TALINUM$] Portulacaceae (V:D)
Talinum okanoganense English = Talinum sediforme [TALISED] Portulacaceae (V:D)
Talinum pygmaeum Gray = Lewisia pygmaea [LEWIPYG] Portulacaceae (V:D)
Talinum sediforme Poelln. [TALISED] Portulacaceae (V:D)
Tanacetum L. [TANACET$] Asteraceae (V:D)
Tanacetum bipinnatum (L.) Schultz-Bip. [TANABIP] Asteraceae (V:D)
Tanacetum bipinnatum ssp. **huronense** (Nutt.) Breitung [TANABIP1] Asteraceae (V:D)
Tanacetum douglasii DC. = Tanacetum bipinnatum ssp. huronense [TANABIP1] Asteraceae (V:D)
Tanacetum huronense Nutt. = Tanacetum bipinnatum ssp. huronense [TANABIP1] Asteraceae (V:D)
Tanacetum huronense var. *bifarium* Fern. = Tanacetum bipinnatum ssp. huronense [TANABIP1] Asteraceae (V:D)
Tanacetum huronense var. *floccosum* Raup = Tanacetum bipinnatum ssp. huronense [TANABIP1] Asteraceae (V:D)
Tanacetum huronense var. *johannense* Fern. = Tanacetum bipinnatum ssp. huronense [TANABIP1] Asteraceae (V:D)
Tanacetum huronense var. *terrae-novae* Fern. = Tanacetum bipinnatum ssp. huronense [TANABIP1] Asteraceae (V:D)
***Tanacetum parthenium** (L.) Schultz-Bip. [TANAPAR] Asteraceae (V:D)
Tanacetum suaveolens (Pursh) Hook. = Matricaria discoidea [MATRDIS] Asteraceae (V:D)
***Tanacetum vulgare** L. [TANAVUL] Asteraceae (V:D)
Tanacetum vulgare var. *crispum* DC. = Tanacetum vulgare [TANAVUL] Asteraceae (V:D)
Taraxacum G.H. Weber ex Wiggers [TARAXAC$] Asteraceae (V:D)
Taraxacum ambigens Fern. = Taraxacum officinale ssp. ceratophorum [TARAOFF1] Asteraceae (V:D)
Taraxacum ambigens var. *flutius* Fern. = Taraxacum officinale ssp. ceratophorum [TARAOFF1] Asteraceae (V:D)
Taraxacum amphiphron Böcher = Taraxacum officinale ssp. ceratophorum [TARAOFF1] Asteraceae (V:D)
Taraxacum arctogenum Dahlst. = Taraxacum officinale ssp. ceratophorum [TARAOFF1] Asteraceae (V:D)
Taraxacum brachyceras Dahlst. = Taraxacum officinale ssp. ceratophorum [TARAOFF1] Asteraceae (V:D)
Taraxacum carneocoloratum A. Nels. = Taraxacum officinale ssp. ceratophorum [TARAOFF1] Asteraceae (V:D)
Taraxacum carthamopsis Porsild = Taraxacum officinale ssp. ceratophorum [TARAOFF1] Asteraceae (V:D)
Taraxacum ceratophorum (Ledeb.) DC. = Taraxacum officinale ssp. ceratophorum [TARAOFF1] Asteraceae (V:D)
Taraxacum disseminatum Hagl. = Taraxacum laevigatum [TARALAE] Asteraceae (V:D)
Taraxacum dumetorum Greene = Taraxacum officinale ssp. ceratophorum [TARAOFF1] Asteraceae (V:D)
Taraxacum eriophorum Rydb. [TARAERI] Asteraceae (V:D)
Taraxacum erythrospermum Andrz. ex Bess. = Taraxacum laevigatum [TARALAE] Asteraceae (V:D)
Taraxacum eurylepium Dahlst. = Taraxacum officinale ssp. ceratophorum [TARAOFF1] Asteraceae (V:D)
Taraxacum hyperboreum Dahlst. = Taraxacum officinale ssp. ceratophorum [TARAOFF1] Asteraceae (V:D)
Taraxacum integratum Hagl. = Taraxacum officinale ssp. ceratophorum [TARAOFF1] Asteraceae (V:D)
Taraxacum kamtschaticum Dahlst. = Taraxacum lyratum [TARALYR] Asteraceae (V:D)
Taraxacum lacerum Greene = Taraxacum officinale ssp. ceratophorum [TARAOFF1] Asteraceae (V:D)
Taraxacum lacistophyllum (Dahlst.) Raunk. = Taraxacum laevigatum [TARALAE] Asteraceae (V:D)
***Taraxacum laevigatum** (Willd.) DC. [TARALAE] Asteraceae (V:D)
Taraxacum lapponicum Kihlm. ex Hand.-Maz. = Taraxacum officinale ssp. ceratophorum [TARAOFF1] Asteraceae (V:D)
Taraxacum latispinulosum M.P. Christens. = Taraxacum officinale ssp. ceratophorum [TARAOFF1] Asteraceae (V:D)
Taraxacum laurentianum Fern. = Taraxacum officinale ssp. ceratophorum [TARAOFF1] Asteraceae (V:D)
Taraxacum longii Fern. = Taraxacum officinale ssp. ceratophorum [TARAOFF1] Asteraceae (V:D)
Taraxacum lyratum (Ledeb.) DC. [TARALYR] Asteraceae (V:D)
Taraxacum malteanum Dahlst. = Taraxacum officinale ssp. ceratophorum [TARAOFF1] Asteraceae (V:D)
Taraxacum maurolepium Hagl. = Taraxacum officinale ssp. ceratophorum [TARAOFF1] Asteraceae (V:D)
Taraxacum mitratum Hagl. = Taraxacum officinale ssp. ceratophorum [TARAOFF1] Asteraceae (V:D)

Taraxacum multesimum Hagl. = Taraxacum officinale ssp. ceratophorum [TARAOFF1] Asteraceae (V:D)
Taraxacum naevosum Dahlst. = Taraxacum officinale ssp. ceratophorum [TARAOFF1] Asteraceae (V:D)
•**Taraxacum officinale** G.H. Weber ex Wiggers [TARAOFF] Asteraceae (V:D)
Taraxacum officinale ssp. **ceratophorum** (Ledeb.) Schinz ex Thellung [TARAOFF1] Asteraceae (V:D)
Taraxacum ovinum Rydb. = Taraxacum officinale ssp. ceratophorum [TARAOFF1] Asteraceae (V:D)
Taraxacum paucisquamosum M.E. Peck = Taraxacum officinale ssp. ceratophorum [TARAOFF1] Asteraceae (V:D)
Taraxacum pellianum Porsild = Taraxacum officinale ssp. ceratophorum [TARAOFF1] Asteraceae (V:D)
Taraxacum pseudonorvegicum Dahlst. = Taraxacum officinale ssp. ceratophorum [TARAOFF1] Asteraceae (V:D)
Taraxacum purpuridens Dahlst. = Taraxacum officinale ssp. ceratophorum [TARAOFF1] Asteraceae (V:D)
Taraxacum scanicum Dahlst. = Taraxacum laevigatum [TARALAE] Asteraceae (V:D)
Taraxacum scopulorum (Gray) Rydb. = Taraxacum lyratum [TARALYR] Asteraceae (V:D)
Taraxacum sibiricum Dahlst. = Taraxacum lyratum [TARALYR] Asteraceae (V:D)
Taraxacum torngatense Fern. = Taraxacum officinale ssp. ceratophorum [TARAOFF1] Asteraceae (V:D)
Taraxacum trigonolobum Dahlst. = Taraxacum officinale ssp. ceratophorum [TARAOFF1] Asteraceae (V:D)
Taraxacum umbrinum Dahlst. = Taraxacum officinale ssp. ceratophorum [TARAOFF1] Asteraceae (V:D)
Taraxia breviflora (Torr. & Gray) Nutt. ex Small = Camissonia breviflora [CAMIBRE] Onagraceae (V:D)
Targionia L. [TARGION$] Targioniaceae (B:H)
Targionia hypophylla L. [TARGHYP] Targioniaceae (B:H)
Taxus L. [TAXUS$] Taxaceae (V:G)
Taxus brevifolia Nutt. [TAXUBRE] Taxaceae (V:G)
Tayloria Hook. [TAYLORI$] Splachnaceae (B:M)
Tayloria froelichiana (Hedw.) Mitt. ex Broth. [TAYLFRO] Splachnaceae (B:M)
Tayloria lingulata (Dicks.) Lindb. [TAYLLIN] Splachnaceae (B:M)
Tayloria serrata (Hedw.) Bruch & Schimp. in B.S.G. [TAYLSER] Splachnaceae (B:M)
Tayloria serrata var. *flagellaris* (Brid.) Bruch & Schimp. in B.S.G. = Tayloria serrata [TAYLSER] Splachnaceae (B:M)
Tayloria serrata var. *tenuis* (With.) Bruch & Schimp. in B.S.G. = Tayloria serrata [TAYLSER] Splachnaceae (B:M)
Tayloria splachnoides (Schleich. ex Schwaegr.) Hook. [TAYLSPL] Splachnaceae (B:M)
Tayloria tenuis (With.) Schimp. = Tayloria serrata [TAYLSER] Splachnaceae (B:M)
Teesdalia Ait. f. [TEESDAL$] Brassicaceae (V:D)
•**Teesdalia nudicaulis** (L.) Ait. f. [TEESNUD] Brassicaceae (V:D)
Tellima R. Br. [TELLIMA$] Saxifragaceae (V:D)
Tellima grandiflora (Pursh) Dougl. ex Lindl. [TELLGRA] Saxifragaceae (V:D)
Tellima odorata T.J. Howell = Tellima grandiflora [TELLGRA] Saxifragaceae (V:D)
Teloschistes Norman [TELOSCH$] Teloschistaceae (L)
Teloschistes candelaris (L.) Fink = Xanthoria candelaria [XANTCAN] Teloschistaceae (L)
Teloschistes contortuplicatus (Ach.) Clauzade & Rondon ex Vezda [TELOCON] Teloschistaceae (L)
Teloschistes controversus (Mass.) Müll Arg. = Xanthoria candelaria [XANTCAN] Teloschistaceae (L)
Teloschistes controversus var. *pygmaeus* (Bory) Müll. Arg. = Xanthoria candelaria [XANTCAN] Teloschistaceae (L)
Teloschistes parietinus (L.) Norman = Xanthoria parietina [XANTPAR] Teloschistaceae (L)
Teloschistes polycarpus (Hoffm.) Tuck. = Xanthoria polycarpa [XANTPOL] Teloschistaceae (L)
Teloschistes ramulosus Tuck. = Xanthoria ramulosa [XANTRAM] Teloschistaceae (L)
Teloxys botrys (L.) W.A. Weber = Chenopodium botrys [CHENBOT] Chenopodiaceae (V:D)
Temnoma setiforme (Ehrh.) M.A. Howe = Tetralophozia setiformis [TETRSET] Jungermanniaceae (B:H)
Tephromela Choisy [TEPHROM$] Bacidiaceae (L)
Tephromela armeniaca (DC.) Hertel & Rambold [TEPHARM] Bacidiaceae (L)
Tephromela atra (Hudson) Hafellner [TEPHATR] Bacidiaceae (L)
Tephroseris atropurpurea ssp. *frigida* (Richards.) A. & D. Löve = Senecio atropurpureus [SENEATR] Asteraceae (V:D)
Tephroseris atropurpurea ssp. *tomentosa* (Kjellm.) A. & D. Löve = Senecio atropurpureus [SENEATR] Asteraceae (V:D)
Tephroseris integrifolia ssp. *atropurpurea* (Ledeb.) B. Nordenstam = Senecio atropurpureus [SENEATR] Asteraceae (V:D)
Tetradymia DC. [TETRADY$] Asteraceae (V:D)
Tetradymia canescens DC. [TETRCAN] Asteraceae (V:D)
Tetradymia canescens var. *inermis* (Rydb.) Payson = Tetradymia canescens [TETRCAN] Asteraceae (V:D)
Tetralophozia (Schust.) Schljak. [TETRALO$] Jungermanniaceae (B:H)
Tetralophozia filiformis (Steph.) Lammes [TETRFIL] Jungermanniaceae (B:H)
Tetralophozia setiformis (Ehrh.) Schljak. [TETRSET] Jungermanniaceae (B:H)
Tetraphis Hedw. [TETRAPH$] Tetraphidaceae (B:M)
Tetraphis browniana var. *repanda* (Funck in Sturm) Hampe = Tetrodontium repandum [TETRREP] Tetraphidaceae (B:M)
Tetraphis brownianum (Dicks.) Grev. = Tetrodontium brownianum [TETRBRO] Tetraphidaceae (B:M)
Tetraphis geniculata Girg. ex Milde [TETRGEN] Tetraphidaceae (B:M)
Tetraphis pellucida Hedw. [TETRPEL] Tetraphidaceae (B:M)
Tetraplodon Bruch & Schimp. in B.S.G. [TETRAPL$] Splachnaceae (B:M)
Tetraplodon angustatus (Hedw.) Bruch & Schimp. in B.S.G. [TETRANG] Splachnaceae (B:M)
Tetraplodon bryoides Lindb. = Tetraplodon mnioides [TETRMNI] Splachnaceae (B:M)
Tetraplodon mnioides (Hedw.) Bruch & Schimp. in B.S.G. [TETRMNI] Splachnaceae (B:M)
Tetraplodon mnioides ssp. *breweri* var. *brevicollis* Kindb. = Tetraplodon mnioides [TETRMNI] Splachnaceae (B:M)
Tetraplodon mnioides ssp. *pallidus* (Hag.) Kindb. = Tetraplodon pallidus [TETRPAL] Splachnaceae (B:M)
Tetraplodon mnioides var. *brewerianus* (Hedw.) Bruch & Schimp. in B.S.G. = Tetraplodon mnioides [TETRMNI] Splachnaceae (B:M)
Tetraplodon mnioides var. *cavifolius* Schimp. = Tetraplodon urceolatus [TETRURC] Splachnaceae (B:M)
Tetraplodon mnioides var. *urceolatus* (Hedw.) Steere in Polunin = Tetraplodon urceolatus [TETRURC] Splachnaceae (B:M)
Tetraplodon pallidus Hag. [TETRPAL] Splachnaceae (B:M)
Tetraplodon urceolatus (Hedw.) Bruch & Schimp. in Schimp. [TETRURC] Splachnaceae (B:M)
Tetrodontium Schwaegr. [TETRODO$] Tetraphidaceae (B:M)
Tetrodontium brownianum (Dicks.) Schwaegr. [TETRBRO] Tetraphidaceae (B:M)
Tetrodontium repandum (Funck in Sturm) Schwaegr. [TETRREP] Tetraphidaceae (B:M)
Teucrium L. [TEUCRIU$] Lamiaceae (V:D)
Teucrium boreale Bickn. = Teucrium canadense var. occidentale [TEUCCAN2] Lamiaceae (V:D)
Teucrium canadense L. [TEUCCAN] Lamiaceae (V:D)
Teucrium canadense ssp. *occidentale* (Gray) W.A. Weber = Teucrium canadense var. occidentale [TEUCCAN2] Lamiaceae (V:D)
Teucrium canadense ssp. *viscidum* (Piper) Taylor & MacBryde = Teucrium canadense var. occidentale [TEUCCAN2] Lamiaceae (V:D)
Teucrium canadense var. *boreale* (Bickn.) Shinners = Teucrium canadense var. occidentale [TEUCCAN2] Lamiaceae (V:D)
Teucrium canadense var. **occidentale** (Gray) McClintock & Epling [TEUCCAN2] Lamiaceae (V:D)
Teucrium occidentale Gray = Teucrium canadense var. occidentale [TEUCCAN2] Lamiaceae (V:D)

Teucrium occidentale var. *boreale* (Bickn.) Fern. = Teucrium canadense var. occidentale [TEUCCAN2] Lamiaceae (V:D)
Thacla natans (Pallas ex Georgi) Deyl & Soják = Caltha natans [CALTNAT] Ranunculaceae (V:D)
Thalesia fasciculata (Nutt.) Britt. = Orobanche fasciculata [OROBFAS] Orobanchaceae (V:D)
Thalesia lutea (Parry) Rydb. = Orobanche fasciculata [OROBFAS] Orobanchaceae (V:D)
Thalesia uniflora (L.) Britt. = Orobanche uniflora [OROBUNI] Orobanchaceae (V:D)
Thalictrum L. [THALICT$] Ranunculaceae (V:D)
Thalictrum alpinum L. [THALALP] Ranunculaceae (V:D)
Thalictrum alpinum var. *hebetum* Boivin = Thalictrum alpinum [THALALP] Ranunculaceae (V:D)
Thalictrum breitungii Boivin = Thalictrum occidentale var. occidentale [THALOCC0] Ranunculaceae (V:D)
Thalictrum columbianum Rydb. = Thalictrum venulosum [THALVEN] Ranunculaceae (V:D)
Thalictrum confine Fern. = Thalictrum venulosum [THALVEN] Ranunculaceae (V:D)
Thalictrum confine var. *columbianum* (Rydb.) Boivin = Thalictrum venulosum [THALVEN] Ranunculaceae (V:D)
Thalictrum confine var. *greeneanum* Boivin = Thalictrum venulosum [THALVEN] Ranunculaceae (V:D)
Thalictrum dasycarpum Fisch. & Avé-Lall. [THALDAS] Ranunculaceae (V:D)
Thalictrum dasycarpum var. *hypoglaucum* (Rydb.) Boivin = Thalictrum dasycarpum [THALDAS] Ranunculaceae (V:D)
Thalictrum hypoglaucum Rydb. = Thalictrum dasycarpum [THALDAS] Ranunculaceae (V:D)
Thalictrum occidentale Gray [THALOCC] Ranunculaceae (V:D)
Thalictrum occidentale var. *breitungii* (Bovin) T.C. Brayshaw = Thalictrum occidentale var. occidentale [THALOCC0] Ranunculaceae (V:D)
Thalictrum occidentale var. *columbianum* (Rydb.) M.E. Peck = Thalictrum venulosum [THALVEN] Ranunculaceae (V:D)
Thalictrum occidentale var. **macounii** Boivin [THALOCC4] Ranunculaceae (V:D)
Thalictrum occidentale var. *megacarpum* (Torr.) St. John = Thalictrum occidentale var. occidentale [THALOCC0] Ranunculaceae (V:D)
Thalictrum occidentale var. **occidentale** [THALOCC0] Ranunculaceae (V:D)
Thalictrum occidentale var. *palousense* St. John = Thalictrum occidentale var. occidentale [THALOCC0] Ranunculaceae (V:D)
Thalictrum sparsiflorum Turcz. ex Fisch. & C.A. Mey. [THALSPA] Ranunculaceae (V:D)
Thalictrum sparsiflorum var. **richardsonii** (Gray) Boivin [THALSPA1] Ranunculaceae (V:D)
Thalictrum turneri Boivin = Thalictrum venulosum [THALVEN] Ranunculaceae (V:D)
Thalictrum venulosum Trel. [THALVEN] Ranunculaceae (V:D)
Thalictrum venulosum var. *confine* (Fern.) Boivin = Thalictrum venulosum [THALVEN] Ranunculaceae (V:D)
Thalictrum venulosum var. *fissum* (Greene) Boivin = Thalictrum venulosum [THALVEN] Ranunculaceae (V:D)
Thalictrum venulosum var. *lunellii* (Greene) Boivin = Thalictrum venulosum [THALVEN] Ranunculaceae (V:D)
Thalictrum venulosum var. *turneri* (Boivin) Boivin = Thalictrum venulosum [THALVEN] Ranunculaceae (V:D)
Thamnium bigelovii (Sull.) Jaeg. = Porotrichum bigelovii [POROBIG] Thamnobryaceae (B:M)
Thamnobryum Nieuwl. [THAMNOB$] Thamnobryaceae (B:M)
Thamnobryum leibergii (Britt.) Ren. & Card. = Thamnobryum neckeroides [THAMNEC] Thamnobryaceae (B:M)
Thamnobryum neckeroides (Hook.) Lawt. [THAMNEC] Thamnobryaceae (B:M)
Thamnolia Ach. ex Schaerer [THAMNOL$] Siphulaceae (L)
Thamnolia vermicularis (Sw.) Ach. ex Schaerer [THAMVER] Siphulaceae (L)
Thamnolia vermicularis var. **subuliformis** (Ehrh.) Schaerer [THAMVER1] Siphulaceae (L)
Thamnolia vermicularis var. **vermicularis** [THAMVER0] Siphulaceae (L)
Thelidium A. Massal. [THELIDI$] Verrucariaceae (L)
Thelidium absconditum (Hepp) Rabenh. [THELABS] Verrucariaceae (L)
Thelidium decipiens (Nyl.) Kremp. [THELDEC] Verrucariaceae (L)
Thelidium incavatum Nyl. ex Mudd [THELINC] Verrucariaceae (L)
Thellungiella salsuginea (Pallas) O.E. Schulz = Arabidopsis salsuginea [ARABSAL] Brassicaceae (V:D)
Thelocarpon Nyl. ex Hue [THELOCA$] Acarosporaceae (L)
Thelocarpon epibolum Nyl. [THELEPI] Acarosporaceae (L)
Thelocarpon epibolum var. **epithallinum** (Leighton ex Nyl.) Salisb. [THELEPI1] Acarosporaceae (L)
Thelomma A. Massal. [THELOMM$] Caliciaceae (L)
Thelomma mammosum (Hepp) A. Massal. [THELMAM] Caliciaceae (L)
Thelomma occidentale (Herre) Tibell [THELOCC] Caliciaceae (L)
Thelomma ocellatum (Körber) Tibell [THELOCE] Caliciaceae (L)
Thelotrema Ach. [THELOTR$] Thelotremataceae (L)
Thelotrema lepadinum (Ach.) Ach. [THELLEP] Thelotremataceae (L)
Thelypodium Endl. [THELYPO$] Brassicaceae (V:D)
Thelypodium laciniatum (Hook.) Endl. ex Walp. [THELLAC] Brassicaceae (V:D)
Thelypodium laciniatum var. *milleflorum* (A. Nels.) Payson = Thelypodium milleflorum [THELMIL] Brassicaceae (V:D)
Thelypodium laciniatum var. *streptanthoides* (Leib. ex Piper) Payson = Thelypodium laciniatum [THELLAC] Brassicaceae (V:D)
Thelypodium milleflorum A. Nels. [THELMIL] Brassicaceae (V:D)
Thelypodium streptanthoides Leib. ex Piper = Thelypodium laciniatum [THELLAC] Brassicaceae (V:D)
Thelypteris Schmidel [THELYPT$] Thelypteridaceae (V:F)
Thelypteris dryopteris (L.) Slosson = Gymnocarpium dryopteris [GYMNDRY] Dryopteridaceae (V:F)
Thelypteris limbosperma auct. = Thelypteris quelpaertensis [THELQUE] Thelypteridaceae (V:F)
Thelypteris nevadensis (Baker) Clute ex Morton [THELNEV] Thelypteridaceae (V:F)
Thelypteris phegopteris (L.) Slosson = Phegopteris connectilis [PHEGCON] Thelypteridaceae (V:F)
Thelypteris quelpaertensis (Christ) Ching [THELQUE] Thelypteridaceae (V:F)
Thermopsis R. Br. ex Ait. f. [THERMOP$] Fabaceae (V:D)
Thermopsis montana Nutt. = Thermopsis rhombifolia var. montana [THERRHO1] Fabaceae (V:D)
Thermopsis pinetorum Greene = Thermopsis rhombifolia var. montana [THERRHO1] Fabaceae (V:D)
Thermopsis rhombifolia (Nutt. ex Pursh) Nutt. ex Richards. [THERRHO] Fabaceae (V:D)
Thermopsis rhombifolia var. **montana** (Nutt.) Isely [THERRHO1] Fabaceae (V:D)
Thermutis Fr. [THERMUT$] Lichinaceae (L)
Thermutis velutina (Ach.) Flotow [THERVEL] Lichinaceae (L)
Thinopyrum intermedium (Host) Barkworth & Dewey = Elytrigia intermedia [ELYTINT] Poaceae (V:M)
Thinopyrum intermedium ssp. *barbulatum* (Schur) Barkworth & Dewey = Elytrigia intermedia [ELYTINT] Poaceae (V:M)
Thinopyrum ponticum (Podp.) Barkworth & Dewey = Elytrigia pontica [ELYTPON] Poaceae (V:M)
Thlaspi L. [THLASPI$] Brassicaceae (V:D)
*__Thlaspi arvense__ L. [THLAARV] Brassicaceae (V:D)
Thlaspi bursa-pastoris L. = Capsella bursa-pastoris [CAPSBUR] Brassicaceae (V:D)
Thlaspi campestre L. = Lepidium campestre [LEPICAM] Brassicaceae (V:D)
Tholurna Norman [THOLURN$] Caliciaceae (L)
Tholurna dissimilis (Norman) Norman [THOLDIS] Caliciaceae (L)
Thrombium Wallr. [THROMBI$] Thrombiaceae (L)
Thrombium epigaeum (Pers.) Wallr. [THROEPI] Thrombiaceae (L)
Thuidium Schimp. in B.S.G. [THUIDIU$] Thuidiaceae (B:M)
Thuidium abietinum (Hedw.) Schimp. in B.S.G. = Abietinella abietina [ABIEABI] Thuidiaceae (B:M)

Thuidium blandowii (Web. & Mohr) B.S.G. = Helodium blandowii [HELOBLA] Helodiaceae (B:M)
Thuidium bolanderi (Best) Par. = Claopodium bolanderi [CLAOBOL] Leskeaceae (B:M)
Thuidium crispifolium (Hook.) Lindb. = Claopodium crispifolium [CLAOCRI] Leskeaceae (B:M)
Thuidium delicatulum var. *radicans* (Kindb.) Crum, Steere & Anderson = Thuidium philibertii [THUIPHI] Thuidiaceae (B:M)
Thuidium gracile B.S.G. = Bryohaplocladium microphyllum [BRYOMIC] Leskeaceae (B:M)
Thuidium leuconeuron (Sull. & Lesq.) Lesq. = Claopodium whippleanum [CLAOWHI] Leskeaceae (B:M)
Thuidium lignicola Kindb. = Bryohaplocladium microphyllum [BRYOMIC] Leskeaceae (B:M)
Thuidium microphyllum (Hedw.) Jaeg. = Bryohaplocladium microphyllum [BRYOMIC] Leskeaceae (B:M)
Thuidium microphyllum var. *lignicola* (Kindb.) Best = Bryohaplocladium microphyllum [BRYOMIC] Leskeaceae (B:M)
Thuidium microphyllum var. *obtusum* Grout = Bryohaplocladium microphyllum [BRYOMIC] Leskeaceae (B:M)
Thuidium philibertii Limpr. [THUIPHI] Thuidiaceae (B:M)
Thuidium pseudoabietinum Kindb. = Helodium blandowii [HELOBLA] Helodiaceae (B:M)
Thuidium recognitum (Hedw.) Lindb. [THUIREC] Thuidiaceae (B:M)
Thuidium whippleanum (Sull.) Jaeg. & Sauerb. = Claopodium whippleanum [CLAOWHI] Leskeaceae (B:M)
Thuja L. [THUJA$] Cupressaceae (V:G)
Thuja plicata Donn ex D. Don [THUJPLI] Cupressaceae (V:G)
Thymus L. [THYMUS$] Lamiaceae (V:D)
Thymus arcticus (Dur.) Ronniger = Thymus praecox ssp. arcticus [THYMPRA1] Lamiaceae (V:D)
Thymus praecox Opiz [THYMPRA] Lamiaceae (V:D)
*__Thymus praecox__ ssp. **arcticus** (Dur.) Jalas [THYMPRA1] Lamiaceae (V:D)
Thymus serpyllum auct. = Thymus praecox ssp. arcticus [THYMPRA1] Lamiaceae (V:D)
Thyrea demangeonii (Moug. & Mont.) Fink = Phylliscum demangeonii [PHYLDEM] Lichinaceae (L)
Thyrea nigritella Lettau = Lichinella nigritella [LICHNIG] Lichinaceae (L)
Thysanocarpus Hook. [THYSANO$] Brassicaceae (V:D)
Thysanocarpus amplectens Greene = Thysanocarpus curvipes [THYSCUR] Brassicaceae (V:D)
Thysanocarpus curvipes Hook. [THYSCUR] Brassicaceae (V:D)
Thysanocarpus curvipes var. *elegans* (Fisch. & C.A. Mey.) B.L. Robins. = Thysanocarpus curvipes [THYSCUR] Brassicaceae (V:D)
Thysanocarpus curvipes var. *eradiatus* Jepson = Thysanocarpus curvipes [THYSCUR] Brassicaceae (V:D)
Thysanocarpus curvipes var. *longistylus* Jepson = Thysanocarpus curvipes [THYSCUR] Brassicaceae (V:D)
Thysanocarpus elegans Fisch. & C.A. Mey. = Thysanocarpus curvipes [THYSCUR] Brassicaceae (V:D)
Thysanocarpus pusillus Hook. = Athysanus pusillus [ATHYPUS] Brassicaceae (V:D)
Tiarella L. [TIARELL$] Saxifragaceae (V:D)
Tiarella californica (Kellogg) Rydb. = Tiarella trifoliata var. laciniata [TIARTRI1] Saxifragaceae (V:D)
Tiarella laciniata Hook. = Tiarella trifoliata var. laciniata [TIARTRI1] Saxifragaceae (V:D)
Tiarella menziesii Pursh = Tolmiea menziesii [TOLMMEN] Saxifragaceae (V:D)
Tiarella trifoliata L. [TIARTRI] Saxifragaceae (V:D)
Tiarella trifoliata ssp. *unifoliata* (Hook.) Kern = Tiarella trifoliata var. unifoliata [TIARTRI2] Saxifragaceae (V:D)
Tiarella trifoliata var. **laciniata** (Hook.) Wheelock [TIARTRI1] Saxifragaceae (V:D)
Tiarella trifoliata var. × *laciniata* (Hook.) Wheelock = Tiarella trifoliata var. laciniata [TIARTRI1] Saxifragaceae (V:D)
Tiarella trifoliata var. **trifoliata** [TIARTRI0] Saxifragaceae (V:D)
Tiarella trifoliata var. **unifoliata** (Hook.) Kurtz [TIARTRI2] Saxifragaceae (V:D)
Tiarella unifoliata Hook. = Tiarella trifoliata var. unifoliata [TIARTRI2] Saxifragaceae (V:D)
Tichothecium pygmaeum Körber = Muellerella pygmaea [MUELPYG] Verrucariaceae (L)
Tillaea angustifolia Nutt. = Crassula aquatica [CRASAQU] Crassulaceae (V:D)
Tillaea aquatica L. = Crassula aquatica [CRASAQU] Crassulaceae (V:D)
Tillaea ascendens Eat. = Crassula aquatica [CRASAQU] Crassulaceae (V:D)
Tillaea erecta Hook. & Arn. = Crassula connata [CRASCON] Crassulaceae (V:D)
Tillaea minima Miers ex Hook. & Arn. = Crassula connata [CRASCON] Crassulaceae (V:D)
Tillaea minima Gay = Crassula connata [CRASCON] Crassulaceae (V:D)
Tillaea rubescens Kunth = Crassula connata [CRASCON] Crassulaceae (V:D)
Tillaeastrum aquaticum (L.) Britt. = Crassula aquatica [CRASAQU] Crassulaceae (V:D)
Timmia Hedw. [TIMMIA$] Timmiaceae (B:M)
Timmia austriaca Hedw. [TIMMAUS] Timmiaceae (B:M)
Timmia austriaca var. *arctica* (Lindb.) Arnell = Timmia austriaca [TIMMAUS] Timmiaceae (B:M)
Timmia bavarica Hessl. = Timmia megapolitana var. bavarica [TIMMMEG1] Timmiaceae (B:M)
Timmia cucullata Michx. = Timmia megapolitana [TIMMMEG] Timmiaceae (B:M)
Timmia megapolitana Hedw. [TIMMMEG] Timmiaceae (B:M)
Timmia megapolitana f. *cucullata* (Michx.) Sayre in Grout = Timmia megapolitana [TIMMMEG] Timmiaceae (B:M)
Timmia megapolitana ssp. *bavarica* (Hessl.) Brass. = Timmia megapolitana var. bavarica [TIMMMEG1] Timmiaceae (B:M)
Timmia megapolitana var. **bavarica** (Hessl.) Brid. [TIMMMEG1] Timmiaceae (B:M)
Timmia megapolitana var. **megapolitana** [TIMMMEG0] Timmiaceae (B:M)
Timmia norvegica Zett. [TIMMNOR] Timmiaceae (B:M)
Timmia norvegica var. *crassiretis* Hess. = Timmia sibirica [TIMMSIB] Timmiaceae (B:M)
Timmia sibirica Lindb. & Arnell [TIMMSIB] Timmiaceae (B:M)
Timmiella (De Not.) Limpr. [TIMMIEL$] Pottiaceae (B:M)
Timmiella crassinervis (Hampe) L. Koch [TIMMCRA] Pottiaceae (B:M)
Timmiella flexiseta auct. = Timmiella crassinervis [TIMMCRA] Pottiaceae (B:M)
Timmiella flexiseta var. *vancouveriensis* (Broth.) Grout = Timmiella crassinervis [TIMMCRA] Pottiaceae (B:M)
Tissa canadensis (Pers.) Britt. = Spergularia canadensis var. canadensis [SPERCAN0] Caryophyllaceae (V:D)
Tissa marina (L.) Britt. = Spergularia salina [SPERSAL] Caryophyllaceae (V:D)
Tissa rubra (L.) Britt. = Spergularia rubra [SPERRUB] Caryophyllaceae (V:D)
Tithymalus cyparissias (L.) Hill = Euphorbia cyparissias [EUPHCYP] Euphorbiaceae (V:D)
Tithymalus helioscopius (L.) Hill = Euphorbia helioscopia [EUPHHEL] Euphorbiaceae (V:D)
Tithymalus lathyris (L.) Hill = Euphorbia lathyris [EUPHLAT] Euphorbiaceae (V:D)
Tithymalus peplus (L.) Hill = Euphorbia peplus [EUPHPEP] Euphorbiaceae (V:D)
Tium drummondii (Dougl. ex Hook.) Rydb. = Astragalus drummondii [ASTRDRU] Fabaceae (V:D)
Tofieldia Huds. [TOFIELD$] Liliaceae (V:M)
Tofieldia borealis Wahlenb. = Tofieldia pusilla [TOFIPUS] Liliaceae (V:M)
Tofieldia coccinea Richards. [TOFICOC] Liliaceae (V:M)
Tofieldia glutinosa (Michx.) Pers. [TOFIGLU] Liliaceae (V:M)
Tofieldia pusilla (Michx.) Pers. [TOFIPUS] Liliaceae (V:M)
Tolmachevia integrifolia (Raf.) A. & D. Löve = Sedum integrifolium [SEDUINT] Crassulaceae (V:D)
Tolmiea Torr. & Gray [TOLMIEA$] Saxifragaceae (V:D)

Tolmiea menziesii (Pursh) Torr. & Gray [TOLMMEN] Saxifragaceae (V:D)
Tomasellia A. Massal. [TOMASEL$] Arthopyreniaceae (L)
Tomasellia americana (Minks ex Willey) R.C. Harris [TOMAAME] Arthopyreniaceae (L)
Tomenthypnum Loeske = Tomentypnum [TOMENTY$] Brachytheciaceae (B:M)
Tomentypnum Loeske [TOMENTY$] Brachytheciaceae (B:M)
Tomentypnum falcifolium (Ren. ex Nichols) Tuom. in Ahti & Fagers. [TOMEFAL] Brachytheciaceae (B:M)
Tomentypnum nitens (Hedw.) Loeske [TOMENIT] Brachytheciaceae (B:M)
Tomentypnum nitens var. *falcifolium* Ren. ex Nichols = Tomentypnum falcifolium [TOMEFAL] Brachytheciaceae (B:M)
Tomentypnum nitens var. *insigne* (Milde) C. Jens. = Tomentypnum nitens [TOMENIT] Brachytheciaceae (B:M)
Tomentypnum nitens var. *involutum* (Limpr.) C. Jens. = Tomentypnum nitens [TOMENIT] Brachytheciaceae (B:M)
Tomentypnum paulianum (Grout) Broth. ex Robins. = Sanionia uncinata var. uncinata [SANIUNC0] Amblystegiaceae (B:M)
Tonella Nutt. ex Gray [TONELLA$] Scrophulariaceae (V:D)
Tonella tenella (Benth.) Heller [TONETEN] Scrophulariaceae (V:D)
Tonestus A. Nels. [TONESTU$] Asteraceae (V:D)
Tonestus lyallii (Gray) A. Nels. [TONELYA] Asteraceae (V:D)
Toninia A. Massal. [TONINIA$] Catillariaceae (L)
Toninia aromatica (Sm.) A. Massal. [TONIARO] Catillariaceae (L)
Toninia athallina (Hepp) Timdal [TONIATH] Catillariaceae (L)
Toninia caeruleonigricans (Lightf.) Th. Fr. = Fuscopannaria praetermissa [FUSCPRE] Catillariaceae (L)
Toninia candida (Weber) Th. Fr. [TONICAN] Catillariaceae (L)
Toninia kolax Poelt = Toninia verrucarioides [TONIVER] Catillariaceae (L)
Toninia lobulata (Sommerf.) Lynge = Mycobilimbia lobulata [MYCOLOB] Porpidiaceae (L)
Toninia philippea (Mont.) Timdal [TONIPHI] Catillariaceae (L)
Toninia ruginosa (Tuck.) Herre [TONIRUG] Catillariaceae (L)
Toninia sedifolia (Scop.) Timdal [TONISED] Catillariaceae (L)
Toninia tabacina auct. = Toninia tristis [TONITRI] Catillariaceae (L)
Toninia tristis (Th. Fr.) Th. Fr. [TONITRI] Catillariaceae (L)
Toninia tristis ssp. **asiae-centralis** (H. Magn.) Timdal [TONITRI1] Catillariaceae (L)
Toninia tristis ssp. **canadensis** Timdal [TONITRI2] Catillariaceae (L)
Toninia verrucarioides (Nyl.) Timdal [TONIVER] Catillariaceae (L)
Torreyochloa Church [TORREYO$] Poaceae (V:M)
Torreyochloa fernaldii (A.S. Hitchc.) Church = Torreyochloa pallida var. fernaldii [TORRPAL1] Poaceae (V:M)
Torreyochloa pallida (Torr.) Church [TORRPAL] Poaceae (V:M)
Torreyochloa pallida var. **fernaldii** (A.S. Hitchc.) Dore ex Koyama & Kawano [TORRPAL1] Poaceae (V:M)
Torreyochloa pallida var. **pauciflora** (J. Presl) J.I. Davis [TORRPAL2] Poaceae (V:M)
Torreyochloa pauciflora (J. Presl) Church = Torreyochloa pallida var. pauciflora [TORRPAL2] Poaceae (V:M)
Torreyochloa pauciflora var. *holmii* (Beal) Taylor & MacBryde = Torreyochloa pallida var. pauciflora [TORRPAL2] Poaceae (V:M)
Torreyochloa pauciflora var. *microtheca* (Buckl.) Taylor & MacBryde = Torreyochloa pallida var. pauciflora [TORRPAL2] Poaceae (V:M)
Tortella (Lindb.) Limpr. [TORTELL$] Pottiaceae (B:M)
Tortella arctica (Arnell) Crundw. & Nyh. [TORTARC] Pottiaceae (B:M)
Tortella caespitosa (Schwaegr.) Limpr. = Tortella humilis [TORTHUM] Pottiaceae (B:M)
Tortella fragilis (Hook. & Wils. in Drumm.) Limpr. [TORTFRA] Pottiaceae (B:M)
Tortella humilis (Hedw.) Jenn. [TORTHUM] Pottiaceae (B:M)
Tortella inclinata (Hedw. f.) Limpr. [TORTINC] Pottiaceae (B:M)
Tortella inclinatula (C. Müll. & Kindb. in Mac. & Kindb.) Broth. = Tortella inclinata [TORTINC] Pottiaceae (B:M)
Tortella tortuosa (Hedw.) Limpr. [TORTTOR] Pottiaceae (B:M)
Tortipes amplexifolius (L.) Small = Streptopus amplexifolius var. amplexifolius [STREAMP0] Liliaceae (V:M)
Tortula Hedw. [TORTULA$] Pottiaceae (B:M)
Tortula amplexa (Lesq.) Steere [TORTAMP] Pottiaceae (B:M)
Tortula atrovirens (Sm.) Lindb. = Desmatodon convolutus (Brid.) Grout [DESMCON] Pottiaceae (B:M)
Tortula bistratosa Flow. = Tortula caninervis [TORTCAN] Pottiaceae (B:M)
Tortula bolanderi (Lesq.) Howe [TORTBOL] Pottiaceae (B:M)
Tortula brevipes (Lesq.) Broth. [TORTBRE] Pottiaceae (B:M)
Tortula calcicola Grebe. = Tortula ruralis [TORTRUA] Pottiaceae (B:M)
Tortula calcicolens W. Kramer = Tortula ruralis [TORTRUA] Pottiaceae (B:M)
Tortula caninervis (Mitt.) Broth. [TORTCAN] Pottiaceae (B:M)
Tortula desertorum Broth. = Tortula caninervis [TORTCAN] Pottiaceae (B:M)
Tortula intermedia (Brid.) De Not. = Tortula ruralis [TORTRUA] Pottiaceae (B:M)
Tortula laevipila (Brid.) Schwaegr. [TORTLAE] Pottiaceae (B:M)
Tortula laevipila var. **laevipila** [TORTLAE0] Pottiaceae (B:M)
Tortula laevipila var. **meridionalis** (Schimp.) Wijk & Marg. [TORTLAE2] Pottiaceae (B:M)
Tortula laevipila var. *pagorum* Husn. = Tortula laevipila var. laevipila [TORTLAE0] Pottiaceae (B:M)
Tortula latifolia Bruch ex Hartm. [TORTLAT] Pottiaceae (B:M)
Tortula mucronifolia Schwaegr. [TORTMUC] Pottiaceae (B:M)
Tortula muralis Hedw. [TORTMUR] Pottiaceae (B:M)
Tortula norvegica (Web.) Wahlenb. ex Lindb. [TORTNOR] Pottiaceae (B:M)
Tortula obtusifolia (Schwaegr.) Mathieu = Desmatodon obtusifolius [DESMOBT] Pottiaceae (B:M)
Tortula princeps De Not. [TORTPRI] Pottiaceae (B:M)
Tortula reflexa Brid. = Didymodon fallax var. reflexus [DIDYFAL1] Pottiaceae (B:M)
Tortula ruraliformis (Besch.) Ingh. = Tortula ruralis [TORTRUA] Pottiaceae (B:M)
Tortula ruralis (Hedw.) Gaertn., Meyer & Scherb. [TORTRUA] Pottiaceae (B:M)
Tortula ruralis var. *crinita* De Not. = Tortula ruralis [TORTRUA] Pottiaceae (B:M)
Tortula scotteri Zand. & Steere [TORTSCO] Pottiaceae (B:M)
Tortula subulata Hedw. [TORTSUB] Pottiaceae (B:M)
Tortula subulata var. *angustata* (Schimp.) Lindb. = Tortula subulata [TORTSUB] Pottiaceae (B:M)
Tortula virescens (De Not.) De Not. [TORTVIR] Pottiaceae (B:M)
Torularia arctica (Böcher) A. & D. Löve = Braya humilis [BRAYHUM] Brassicaceae (V:D)
Torularia humilis (C.A. Mey.) O.E. Schulz = Braya humilis [BRAYHUM] Brassicaceae (V:D)
Torularia humilis ssp. *arctica* Böcher = Braya humilis [BRAYHUM] Brassicaceae (V:D)
Tovaria trifolia (L.) Neck. ex Baker = Maianthemum trifolium [MAIATRI] Liliaceae (V:D)
Townsendia Hook. [TOWNSEN$] Asteraceae (V:D)
Townsendia exscapa (Richards.) Porter [TOWNEXS] Asteraceae (V:D)
Townsendia hookeri Beaman [TOWNHOO] Asteraceae (V:D)
Townsendia intermedia Rydb. = Townsendia exscapa [TOWNEXS] Asteraceae (V:D)
Townsendia parryi D.C. Eat. [TOWNPAR] Asteraceae (V:D)
Townsendia sericea Hook. = Townsendia exscapa [TOWNEXS] Asteraceae (V:D)
Toxicodendron P. Mill. [TOXICOD$] Anacardiaceae (V:D)
Toxicodendron desertorum Lunell = Toxicodendron rydbergii [TOXIRYD] Anacardiaceae (V:D)
Toxicodendron diversilobum (Torr. & Gray) Greene [TOXIDIV] Anacardiaceae (V:D)
Toxicodendron radicans ssp. *diversilobum* (Torr. & Gray) Thorne = Toxicodendron diversilobum [TOXIDIV] Anacardiaceae (V:D)
Toxicodendron radicans var. *rydbergii* (Small ex Rydb.) Erskine = Toxicodendron rydbergii [TOXIRYD] Anacardiaceae (V:D)
Toxicodendron rydbergii (Small ex Rydb.) Greene [TOXIRYD] Anacardiaceae (V:D)

Toxicoscordion gramineum (Rydb.) Rydb. = Zigadenus venenosus var. gramineus [ZIGAVEN1] Liliaceae (V:M)
Toxicoscordion venenosum (S. Wats.) Rydb. = Zigadenus venenosus var. venenosus [ZIGAVEN0] Liliaceae (V:M)
Trachybryum (Broth.) Schof. [TRACHYB$] Brachytheciaceae (B:M)
Trachybryum megaptilum (Sull.) Schof. [TRACMEG] Brachytheciaceae (B:M)
Tragopogon L. [TRAGOPO$] Asteraceae (V:D)
*__Tragopogon dubius__ Scop. [TRAGDUB] Asteraceae (V:D)
Tragopogon dubius ssp. *major* (Jacq.) Voll. = Tragopogon dubius [TRAGDUB] Asteraceae (V:D)
Tragopogon major Jacq. = Tragopogon dubius [TRAGDUB] Asteraceae (V:D)
*__Tragopogon porrifolius__ L. [TRAGPOR] Asteraceae (V:D)
*__Tragopogon pratensis__ L. [TRAGPRA] Asteraceae (V:D)
Trapelia Choisy [TRAPELI$] Trapeliaceae (L)
Trapelia coarctata (Sm.) Choisy [TRAPCOA] Trapeliaceae (L)
Trapelia corticola Coppins & P. James [TRAPCOR] Trapeliaceae (L)
Trapelia involuta (Taylor) Hertel [TRAPINV] Trapeliaceae (L)
Trapelia obtegens (Th. Fr.) Hertel [TRAPOBT] Trapeliaceae (L)
Trapelia placodioides Coppins & P. James [TRAPPLA] Trapeliaceae (L)
Trapeliopsis Hertel & Gotth. Schneider [TRAPELO$] Trapeliaceae (L)
Trapeliopsis aeneofusca (Flotow) Coppins & P. James [TRAPAEN] Trapeliaceae (L)
Trapeliopsis flexuosa (Fr.) Coppins & P. James [TRAPFLE] Trapeliaceae (L)
Trapeliopsis gelatinosa (Flörke) Coppins & P. James [TRAPGEL] Trapeliaceae (L)
Trapeliopsis granulosa (Hoffm.) Lumbsch [TRAPGRA] Trapeliaceae (L)
Trapeliopsis pseudogranulosa Coppins & P. James [TRAPPSE] Trapeliaceae (L)
Trapeliopsis viridescens (Schrader) Coppins & P. James [TRAPVIR] Trapeliaceae (L)
Trapeliopsis wallrothii (Flörke) Hertel & Gotth. Schneider [TRAPWAL] Trapeliaceae (L)
Trautvetteria Fisch. & C.A. Mey. [TRAUTVE$] Ranunculaceae (V:D)
Trautvetteria caroliniensis (Walt.) Vail [TRAUCAR] Ranunculaceae (V:D)
Trematodon Michx. [TREMATO$] Bruchiaceae (B:M)
Trematodon ambiguus (Hedw.) Hornsch. [TREMAMB] Bruchiaceae (B:M)
Trematodon boasii Schof. [TREMBOA] Bruchiaceae (B:M)
Trematodon montanus Belland & Brass. [TREMMON] Bruchiaceae (B:M)
Tremolecia Choisy [TREMOLE$] Hymeneliaceae (L)
Tremolecia atrata (Ach.) Hertel [TREMATR] Hymeneliaceae (L)
Tremolecia jurana (Schaerer) Hertel = Farnoldia jurana [FARNJUR] Porpidiaceae (L)
Treubia nana Hatt. & H. Inoue = Apotreubia nana [APOTNAN] Treubiaceae (B:H)
Tribulus L. [TRIBULU$] Zygophyllaceae (V:D)
*__Tribulus terrestris__ L. [TRIBTER] Zygophyllaceae (V:D)
Trichodon Schimp. [TRICHOD$] Ditrichaceae (B:M)
Trichodon cylindricus (Hedw.) Schimp. [TRICCYL] Ditrichaceae (B:M)
Trichophorum alpinum (L.) Pers. = Eriophorum alpinum [ERIOALP] Cyperaceae (V:M)
Trichophorum cespitosum (L.) Hartman = Scirpus cespitosus [SCIRCES] Cyperaceae (V:M)
Trichophorum pumilum auct. = Scirpus rollandii [SCIRROL] Cyperaceae (V:M)
Trichophorum pumilum ssp. *rollandii* (Fern.) Taylor & MacBryde = Scirpus rollandii [SCIRROL] Cyperaceae (V:M)
Trichophorum pumilum var. *rollandii* (Fern.) Hultén = Scirpus rollandii [SCIRROL] Cyperaceae (V:M)
Trichophorum rollandii (Fern.) Hultén = Scirpus rollandii [SCIRROL] Cyperaceae (V:M)
Trichostomopsis Card. [TRICHOS$] Pottiaceae (B:M)
Trichostomopsis australasiae (Grev. & Hook.) Robins. [TRICAUS] Pottiaceae (B:M)
Trichostomopsis australasiae var. *umbrosus* (C. Müll.) Zand. = Trichostomopsis australasiae [TRICAUS] Pottiaceae (B:M)
Trichostomopsis crispifolia Card. = Trichostomopsis australasiae [TRICAUS] Pottiaceae (B:M)
Trichostomopsis diaphanobasis (Card.) Grout = Trichostomopsis australasiae [TRICAUS] Pottiaceae (B:M)
Trichostomopsis fayae Grout = Trichostomopsis australasiae [TRICAUS] Pottiaceae (B:M)
Trichostomopsis umbrosus (Card.) Robins. = Trichostomopsis australasiae [TRICAUS] Pottiaceae (B:M)
Trichostomum Bruch [TRICHOT$] Pottiaceae (B:M)
Trichostomum arcticum Kaal. [TRICARC] Pottiaceae (B:M)
Trichostomum cuspidatissimum Card. & Thér. = Trichostomum arcticum [TRICARC] Pottiaceae (B:M)
Trichostomum cylindricum (Brid.) C. Müll. = Oxystegus tenuirostris [OXYSTEN] Pottiaceae (B:M)
Trichostomum mollissimum auct. = Oxystegus tenuirostris [OXYSTEN] Pottiaceae (B:M)
Trichostomum tenuirostre (Hook. & Tayl.) Lindb. = Oxystegus tenuirostris [OXYSTEN] Pottiaceae (B:M)
Trichostomum tophaceum Brid. = Didymodon tophaceus [DIDYTOP] Pottiaceae (B:M)
Trichostomum tortile Schrad. = Ditrichum pusillum [DITRPUS] Ditrichaceae (B:M)
Trichostomum vancouveriense (Broth.) Kindb. = Timmiella crassinervis [TIMMCRA] Pottiaceae (B:M)
Trientalis L. [TRIENTA$] Primulaceae (V:D)
Trientalis arctica Fisch. ex Hook. = Trientalis europaea ssp. arctica [TRIEEUR1] Primulaceae (V:D)
Trientalis borealis Raf. [TRIEBOR] Primulaceae (V:D)
Trientalis borealis ssp. **latifolia** (Hook.) Hultén [TRIEBOR1] Primulaceae (V:D)
Trientalis europaea L. [TRIEEUR] Primulaceae (V:D)
Trientalis europaea ssp. **arctica** (Fisch. ex Hook.) Hultén [TRIEEUR1] Primulaceae (V:D)
Trientalis europaea var. *arctica* (Fisch. ex Hook.) Ledeb. = Trientalis europaea ssp. arctica [TRIEEUR1] Primulaceae (V:D)
Trientalis europaea var. *latifolia* (Hook.) Torr. = Trientalis borealis ssp. latifolia [TRIEBOR1] Primulaceae (V:D)
Trientalis latifolia Hook. = Trientalis borealis ssp. latifolia [TRIEBOR1] Primulaceae (V:D)
Trifolium L. [TRIFOLI$] Fabaceae (V:D)
Trifolium agrarium L. = Trifolium aureum [TRIFAUR] Fabaceae (V:D)
Trifolium albopurpureum var. *dichotomum* (Hook. & Arn.) Isely = Trifolium dichotomum [TRIFDIC] Fabaceae (V:D)
Trifolium appendiculatum Loja. = Trifolium variegatum [TRIFVAR] Fabaceae (V:D)
*__Trifolium arvense__ L. [TRIFARV] Fabaceae (V:D)
*__Trifolium aureum__ Pollich [TRIFAUR] Fabaceae (V:D)
Trifolium bicephalum Elmer = Trifolium macraei [TRIFMAC] Fabaceae (V:D)
Trifolium bifidum Gray [TRIFBIF] Fabaceae (V:D)
Trifolium bifidum var. *decipiens* Greene = Trifolium bifidum [TRIFBIF] Fabaceae (V:D)
*__Trifolium campestre__ Schreb. [TRIFCAM] Fabaceae (V:D)
Trifolium catalinae S. Wats. = Trifolium macraei [TRIFMAC] Fabaceae (V:D)
Trifolium cyathiferum Lindl. [TRIFCYA] Fabaceae (V:D)
Trifolium depauperatum Desv. [TRIFDEP] Fabaceae (V:D)
Trifolium dianthum Greene = Trifolium variegatum [TRIFVAR] Fabaceae (V:D)
Trifolium dichotomum Hook. & Arn. [TRIFDIC] Fabaceae (V:D)
Trifolium dichotomum var. *turbinatum* Jepson = Trifolium dichotomum [TRIFDIC] Fabaceae (V:D)
*__Trifolium dubium__ Sibthorp [TRIFDUB] Fabaceae (V:D)
Trifolium elegans Savi = Trifolium hybridum [TRIFHYB] Fabaceae (V:D)
Trifolium fendleri Greene = Trifolium wormskjoldii [TRIFWOR] Fabaceae (V:D)

Trifolium fimbriatum Lindl. = Trifolium wormskjoldii [TRIFWOR] Fabaceae (V:D)
Trifolium fragiferum L. [TRIFFRA] Fabaceae (V:D)
Trifolium fragiferum ssp. *bonannii* (K. Presl) Soják = Trifolium fragiferum [TRIFFRA] Fabaceae (V:D)
Trifolium geminiflorum Greene = Trifolium variegatum [TRIFVAR] Fabaceae (V:D)
Trifolium heterodon Torr. & Gray = Trifolium wormskjoldii [TRIFWOR] Fabaceae (V:D)
***Trifolium hybridum** L. [TRIFHYB] Fabaceae (V:D)
Trifolium hybridum ssp. *elegans* (Savi) Aschers. & Graebn. = Trifolium hybridum [TRIFHYB] Fabaceae (V:D)
Trifolium hybridum var. *elegans* (Savi) Boiss. = Trifolium hybridum [TRIFHYB] Fabaceae (V:D)
Trifolium hybridum var. *pratense* Rabenh. = Trifolium hybridum [TRIFHYB] Fabaceae (V:D)
***Trifolium incarnatum** L. [TRIFINC] Fabaceae (V:D)
Trifolium incarnatum var. *elatius* Gibelli & Belli = Trifolium incarnatum [TRIFINC] Fabaceae (V:D)
Trifolium involucratum var. *fendleri* (Greene) McDermott = Trifolium wormskjoldii [TRIFWOR] Fabaceae (V:D)
Trifolium involucratum var. *fimbriatum* (Lindl.) McDermott = Trifolium wormskjoldii [TRIFWOR] Fabaceae (V:D)
Trifolium involucratum var. *heterodon* (Torr. & Gray) S. Wats. = Trifolium wormskjoldii [TRIFWOR] Fabaceae (V:D)
Trifolium involucratum var. *kennedianum* McDermott = Trifolium wormskjoldii [TRIFWOR] Fabaceae (V:D)
Trifolium kennedianum (McDermott) A. Nels. & J.F. Macbr. = Trifolium wormskjoldii [TRIFWOR] Fabaceae (V:D)
Trifolium macraei Hook. & Arn. [TRIFMAC] Fabaceae (V:D)
Trifolium macraei var. *dichotomum* (Hook. & Arn.) Brewer ex S. Wats. = Trifolium dichotomum [TRIFDIC] Fabaceae (V:D)
Trifolium melananthum Hook. & Arn. = Trifolium variegatum [TRIFVAR] Fabaceae (V:D)
Trifolium mercedense Kennedy = Trifolium macraei [TRIFMAC] Fabaceae (V:D)
Trifolium microcephalum Pursh [TRIFMIC] Fabaceae (V:D)
Trifolium microdon Hook. & Arn. [TRIFMIR] Fabaceae (V:D)
Trifolium microdon var. *pilosum* Eastw. = Trifolium microdon [TRIFMIR] Fabaceae (V:D)
Trifolium oliganthum Steud. [TRIFOLG] Fabaceae (V:D)
Trifolium pauciflorum Nutt. = Trifolium oliganthum [TRIFOLG] Fabaceae (V:D)
***Trifolium pratense** L. [TRIFPRA] Fabaceae (V:D)
Trifolium pratense var. *sativum* (P. Mill.) Schreb. = Trifolium pratense [TRIFPRA] Fabaceae (V:D)
Trifolium procumbens L. 1755, non 1753 = Trifolium campestre [TRIFCAM] Fabaceae (V:D)
***Trifolium repens** L. [TRIFREP] Fabaceae (V:D)
Trifolium spinulosum Dougl. ex Hook. = Trifolium wormskjoldii [TRIFWOR] Fabaceae (V:D)
***Trifolium subterraneum** L. [TRIFSUB] Fabaceae (V:D)
Trifolium traskiae Kennedy = Trifolium macraei [TRIFMAC] Fabaceae (V:D)
Trifolium tridentatum Lindl. = Trifolium willdenowii [TRIFWIL] Fabaceae (V:D)
Trifolium tridentatum var. *aciculare* (Nutt.) McDermott = Trifolium willdenowii [TRIFWIL] Fabaceae (V:D)
Trifolium trilobatum Jepson = Trifolium variegatum [TRIFVAR] Fabaceae (V:D)
Trifolium variegatum Nutt. [TRIFVAR] Fabaceae (V:D)
Trifolium variegatum var. *major* Loja. = Trifolium variegatum [TRIFVAR] Fabaceae (V:D)
Trifolium variegatum var. *melananthum* (Hook. & Arn.) Greene = Trifolium variegatum [TRIFVAR] Fabaceae (V:D)
Trifolium variegatum var. *pauciflorum* (Nutt.) McDermott = Trifolium oliganthum [TRIFOLG] Fabaceae (V:D)
Trifolium variegatum var. *rostratum* (Greene) C.L. Hitchc. = Trifolium variegatum [TRIFVAR] Fabaceae (V:D)
Trifolium variegatum var. *trilobatum* (Jepson) McDermott = Trifolium variegatum [TRIFVAR] Fabaceae (V:D)
Trifolium willdenowii Spreng. [TRIFWIL] Fabaceae (V:D)
Trifolium willdenowii var. *fimbriatum* (Lindl.) Ewan = Trifolium wormskjoldii [TRIFWOR] Fabaceae (V:D)
Trifolium willdenowii var. *kennedianum* (McDermott) Ewan = Trifolium wormskjoldii [TRIFWOR] Fabaceae (V:D)
Trifolium wormskjoldii Lehm. [TRIFWOR] Fabaceae (V:D)
Trifolium wormskjoldii var. *fimbriatum* (Lindl.) Jepson = Trifolium wormskjoldii [TRIFWOR] Fabaceae (V:D)
Trifolium wormskjoldii var. *kennedianum* (McDermott) Jepson = Trifolium wormskjoldii [TRIFWOR] Fabaceae (V:D)
Triglochin L. [TRIGLOC$] Juncaginaceae (V:M)
Triglochin concinnum Burtt-Davy [TRIGCON] Juncaginaceae (V:M)
Triglochin concinnum var. *debile* (M.E. Jones) J.T. Howell = Triglochin maritimum [TRIGMAR] Juncaginaceae (V:M)
Triglochin debile (M.E. Jones) A. & D. Löve = Triglochin maritimum [TRIGMAR] Juncaginaceae (V:M)
Triglochin elatum Nutt. = Triglochin maritimum [TRIGMAR] Juncaginaceae (V:M)
Triglochin maritimum L. [TRIGMAR] Juncaginaceae (V:M)
Triglochin maritimum var. *elatum* (Nutt.) Gray = Triglochin maritimum [TRIGMAR] Juncaginaceae (V:M)
Triglochin palustre L. [TRIGPAL] Juncaginaceae (V:M)
Trillium L. [TRILLIU$] Liliaceae (V:M)
Trillium ovatum Pursh [TRILOVA] Liliaceae (V:M)
Trimorpha Cass. [TRIMORP$] Asteraceae (V:D)
Trimorpha acris (L.) Nesom [TRIMACR] Asteraceae (V:D)
Trimorpha acris var. **asteroides** (Andrz. ex Bess.) Nesom [TRIMACR1] Asteraceae (V:D)
Trimorpha acris var. **debilis** (Gray) Nesom [TRIMACR2] Asteraceae (V:D)
Trimorpha elata (Hook.) Nesom [TRIMELA] Asteraceae (V:D)
Trimorpha lonchophylla (Hook.) Nesom [TRIMLON] Asteraceae (V:D)
Triodanis Raf. ex Greene [TRIODAN$] Campanulaceae (V:D)
Triodanis perfoliata (L.) Nieuwl. [TRIOPER] Campanulaceae (V:D)
Triphysaria Fisch. & C.A. Mey. [TRIPHYS$] Scrophulariaceae (V:D)
Triphysaria pusilla (Benth.) Chuang & Heckard [TRIPPUS] Scrophulariaceae (V:D)
Triphysaria versicolor Fisch. & C.A. Mey. [TRIPVER] Scrophulariaceae (V:D)
Tripleurospermum inodorum (L.) Schultz-Bip. = Matricaria perforata [MATRPER] Asteraceae (V:D)
Tripolium angustum Lindl. = Brachyactis ciliata ssp. angusta [BRACCIL1] Asteraceae (V:D)
Tripolium occidentale Nutt. = Aster occidentalis var. occidentalis [ASTEOCC0] Asteraceae (V:D)
Tripterocladium (C. Müll.) Jaeg. [TRIPTER$] Hypnaceae (B:M)
Tripterocladium leucocladulum (C. Müll.) Jaeg. [TRIPLEU] Hypnaceae (B:M)
Tripterocladium rupestris (Kindb.) Kindb. = Tripterocladium leucocladulum [TRIPLEU] Hypnaceae (B:M)
Trisetum Pers. [TRISETU$] Poaceae (V:M)
Trisetum canescens Buckl. = Trisetum cernuum var. canescens [TRISCER3] Poaceae (V:M)
Trisetum cernuum Trin. [TRISCER] Poaceae (V:M)
Trisetum cernuum ssp. *canescens* (Buckl.) Calder & Taylor = Trisetum cernuum var. canescens [TRISCER3] Poaceae (V:M)
Trisetum cernuum var. **canescens** (Buckl.) Beal [TRISCER3] Poaceae (V:M)
Trisetum cernuum var. **cernuum** [TRISCER0] Poaceae (V:M)
Trisetum purpurascens Torr. = Schizachne purpurascens [SCHIPUR] Poaceae (V:M)
Trisetum spicatum (L.) Richter [TRISSPI] Poaceae (V:M)
Trisetum spicatum ssp. *alaskanum* (Nash) Hultén = Trisetum spicatum [TRISSPI] Poaceae (V:M)
Trisetum spicatum ssp. *congdonii* (Scribn. & Merr.) Hultén = Trisetum spicatum [TRISSPI] Poaceae (V:M)
Trisetum spicatum ssp. *majus* (Rydb.) Hultén = Trisetum spicatum [TRISSPI] Poaceae (V:M)
Trisetum spicatum ssp. *molle* (Kunth) Hultén = Trisetum spicatum [TRISSPI] Poaceae (V:M)

Trisetum spicatum ssp. *pilosiglume* (Fern.) Hultén = Trisetum spicatum [TRISSPI] Poaceae (V:M)
Trisetum spicatum var. *alaskanum* (Nash) Malte ex Louis-Marie = Trisetum spicatum [TRISSPI] Poaceae (V:M)
Trisetum spicatum var. *congdonii* (Scribn. & Merr.) A.S. Hitchc. = Trisetum spicatum [TRISSPI] Poaceae (V:M)
Trisetum spicatum var. *maidenii* (Gandog.) Fern. = Trisetum spicatum [TRISSPI] Poaceae (V:M)
Trisetum spicatum var. *majus* (Rydb.) Farw. = Trisetum spicatum [TRISSPI] Poaceae (V:M)
Trisetum spicatum var. *molle* (Kunth) Beal = Trisetum spicatum [TRISSPI] Poaceae (V:M)
Trisetum spicatum var. *pilosiglume* Fern. = Trisetum spicatum [TRISSPI] Poaceae (V:M)
Trisetum spicatum var. *spicatiforme* Hultén = Trisetum spicatum [TRISSPI] Poaceae (V:M)
Trisetum spicatum var. *villosissimum* (Lange) Louis-Marie = Trisetum spicatum [TRISSPI] Poaceae (V:M)
Trisetum subspicatum (L.) Beauv. = Trisetum spicatum [TRISSPI] Poaceae (V:M)
Trisetum triflorum (Bigelow) A. & D. Löve = Trisetum spicatum [TRISSPI] Poaceae (V:M)
Trisetum triflorum ssp. *molle* (Kunth) A. & D. Löve = Trisetum spicatum [TRISSPI] Poaceae (V:M)
Trisetum villosissimum (Lange) Louis-Marie = Trisetum spicatum [TRISSPI] Poaceae (V:M)
Trisetum wolfii Vasey [TRISWOL] Poaceae (V:M)
Triteleia Dougl. ex Lindl. [TRITELE$] Liliaceae (V:M)
Triteleia bicolor (Suksdorf) Abrams = Triteleia howellii [TRITHOW] Liliaceae (V:M)
Triteleia grandiflora Lindl. [TRITGRA] Liliaceae (V:M)
Triteleia grandiflora var. *howellii* (S. Wats.) Hoover = Triteleia howellii [TRITHOW] Liliaceae (V:M)
Triteleia howellii (S. Wats.) Greene [TRITHOW] Liliaceae (V:M)
Triteleia hyacinthina (Lindl.) Greene [TRITHYA] Liliaceae (V:M)
Triticum L. [TRITICU$] Poaceae (V:M)
*****Triticum aestivum** L. [TRITAES] Poaceae (V:M)
Triticum cereale (L.) Salisb. = Secale cereale [SECACER] Poaceae (V:M)
Triticum hybernum L. = Triticum aestivum [TRITAES] Poaceae (V:M)
Triticum macha Dekap. & Menab. = Triticum aestivum [TRITAES] Poaceae (V:M)
Triticum repens L. = Elytrigia repens [ELYTREP] Poaceae (V:M)
Triticum sativum Lam. = Triticum aestivum [TRITAES] Poaceae (V:M)
Triticum sphaerococcum Percival = Triticum aestivum [TRITAES] Poaceae (V:M)
Triticum trachycaulum Link = Elymus trachycaulus ssp. trachycaulus [ELYMTRA0] Poaceae (V:M)
Triticum vulgare Vill. = Triticum aestivum [TRITAES] Poaceae (V:M)
Tritomaria Schiffn. ex Loeske [TRITOMA$] Jungermanniaceae (B:H)
Tritomaria exsecta (Schmid. ex Schrad.) Loeske [TRITEXS] Jungermanniaceae (B:H)
Tritomaria exsectiformis (Breidl.) Loeske [TRITEXE] Jungermanniaceae (B:H)
Tritomaria polita (Nees) Joerg. [TRITPOL] Jungermanniaceae (B:H)
Tritomaria quinquedentata (Huds.) Buch [TRITQUI] Jungermanniaceae (B:H)
Tritomaria scitula (Tayl.) Joerg. [TRITSCI] Jungermanniaceae (B:H)
Tritonia × *crocosmiiflora* (V. Lemoine ex E. Morr.) Nichols. = Crocosmia × crocosmiiflora [CROCXCR] Iridaceae (V:M)
Trollius L. [TROLLIU$] Ranunculaceae (V:D)
Trollius albiflorus (Gray) Rydb. = Trollius laxus ssp. albiflorus [TROLLAX1] Ranunculaceae (V:D)
Trollius laxus Salisb. [TROLLAX] Ranunculaceae (V:D)
Trollius laxus ssp. **albiflorus** (Gray) A. & D. Löve & Kapoor [TROLLAX1] Ranunculaceae (V:D)
Trollius laxus var. *albiflorus* Gray = Trollius laxus ssp. albiflorus [TROLLAX1] Ranunculaceae (V:D)
Troximon aurantiacum Hook. = Agoseris aurantiaca [AGOSAUR] Asteraceae (V:D)
Troximon glaucum var. *dasycephalum* Torr. & Gray = Agoseris glauca var. dasycephala [AGOSGLA1] Asteraceae (V:D)
Troximon grandiflorum Nutt. = Agoseris grandiflora [AGOSGRA] Asteraceae (V:D)
Tryphane rubella (Wahlenb.) Reichenb. = Minuartia rubella [MINURUB] Caryophyllaceae (V:D)
Tsuga Carr. [TSUGA$] Pinaceae (V:G)
Tsuga heterophylla (Raf.) Sarg. [TSUGHET] Pinaceae (V:G)
Tsuga heterophylla × mertensiana [TSUGHEE] Pinaceae (V:G)
Tsuga mertensiana (Bong.) Carr. [TSUGMER] Pinaceae (V:G)
Tuckermannopsis Gyelnik [TUCKERM$] Parmeliaceae (L)
Tuckermannopsis americana (Sprengel) Hale [TUCKAME] Parmeliaceae (L)
Tuckermannopsis canadensis (Räsänen) Hale = Vulpicida canadensis [VULPCAN] Parmeliaceae (L)
Tuckermannopsis chlorophylla (Willd.) Hale [TUCKCHL] Parmeliaceae (L)
Tuckermannopsis halei (Culb. & C. Culb.) M.J. Lai = Tuckermannopsis americana [TUCKAME] Parmeliaceae (L)
Tuckermannopsis juniperina auct. = Vulpicida canadensis [VULPCAN] Parmeliaceae (L)
Tuckermannopsis orbata (Nyl.) M.J. Lai [TUCKORB] Parmeliaceae (L)
Tuckermannopsis pallidula (Tuck. ex Riddle) Hale = Ahtiana pallidula [AHTIPAL] Parmeliaceae (L)
Tuckermannopsis pinastri (Scop.) Hale = Vulpicida pinastri [VULPPIN] Parmeliaceae (L)
Tuckermannopsis platyphylla (Tuck.) Hale [TUCKPLA] Parmeliaceae (L)
Tuckermannopsis sepincola (Ehrh.) Hale [TUCKSEP] Parmeliaceae (L)
Tuckermannopsis subalpina (Imshaug) Kärnefelt [TUCKSUB] Parmeliaceae (L)
Tunica prolifera (L.) Scop. = Petrorhagia prolifera [PETRPRO] Caryophyllaceae (V:D)
Tunica saxifraga (L.) Scop. = Petrorhagia saxifraga [PETRSAX] Caryophyllaceae (V:D)
Turgidosculum complicatulum (Nyl.) Kohlm. & E. Kohlm. = Kohlmeyera complicatula [KOHLCOM] Mastodiaceae (L)
Turritis drummondii (Gray) Lunell = Arabis drummondii [ARABDRU] Brassicaceae (V:D)
Turritis glabra L. = Arabis glabra [ARABGLA] Brassicaceae (V:D)
Turritis mollis Hook. = Halimolobos mollis [HALIMOI] Brassicaceae (V:D)
Tussilago L. [TUSSILA$] Asteraceae (V:D)
*****Tussilago farfara** L. [TUSSFAR] Asteraceae (V:D)
Tylothallia P. James & R. Kilias [TYLOTHA$] Lecanoraceae (L)
Tylothallia biformigera (Leighton) P. James & R. Kilias [TYLOBIF] Lecanoraceae (L)
Typha L. [TYPHA$] Typhaceae (V:M)
Typha angustifolia L. [TYPHANG] Typhaceae (V:M)
Typha angustifolia var. *calumetensis* Peattie = Typha angustifolia [TYPHANG] Typhaceae (V:M)
Typha angustifolia var. *elongata* (Dudley) Wieg. = Typha angustifolia [TYPHANG] Typhaceae (V:M)
Typha latifolia L. [TYPHLAT] Typhaceae (V:M)

U

Udora verticillata var. *minor* Engelm. ex Caspary = Elodea nuttallii [ELODNUT] Hydrocharitaceae (V:M)
Ulex L. [ULEX$] Fabaceae (V:D)
***Ulex europaeus** L. [ULEXEUR] Fabaceae (V:D)
Ulmus L. [ULMUS$] Ulmaceae (V:D)
***Ulmus pumila** L. [ULMUPUM] Ulmaceae (V:D)
Ulota Mohr [ULOTA$] Orthotrichaceae (B:M)
Ulota americanum Mitt. = Ulota curvifolia [ULOTCUR] Orthotrichaceae (B:M)
Ulota cirrata Grout = Ulota curvifolia [ULOTCUR] Orthotrichaceae (B:M)
Ulota crispa var. *alaskana* (Card. & Thér.) Grout = Ulota obtusiuscula [ULOTOBT] Orthotrichaceae (B:M)
Ulota curvifolia (Wahlenb.) Lilj. [ULOTCUR] Orthotrichaceae (B:M)
Ulota drummondii (Hook. & Grev. in Grev.) Brid. [ULOTDRU] Orthotrichaceae (B:M)
Ulota funstonii Grout = Ulota drummondii [ULOTDRU] Orthotrichaceae (B:M)
Ulota maritima C. Müll. & Kindb. in Mac. & Kindb. = Ulota phyllantha [ULOTPHY] Orthotrichaceae (B:M)
Ulota megalospora Vent. in Röll [ULOTMEG] Orthotrichaceae (B:M)
Ulota obtusiuscula C. Müll. & Kindb. in Mac. & Kindb. [ULOTOBT] Orthotrichaceae (B:M)
Ulota phyllantha Brid. [ULOTPHY] Orthotrichaceae (B:M)
Ulota scabrida Kindb. in Mac. & Kindb. = Ulota curvifolia [ULOTCUR] Orthotrichaceae (B:M)
Ulota subulata Kindb. in Mac. & Kindb. = Ulota megalospora [ULOTMEG] Orthotrichaceae (B:M)
Ulota subulifolia C. Müll. & Kindb. in Mac. & Kindb. = Ulota megalospora [ULOTMEG] Orthotrichaceae (B:M)
Umbilicaria Hoffm. [UMBILIC$] Umbilicariaceae (L)
Umbilicaria americana Poelt & T. Nash [UMBIAME] Umbilicariaceae (L)
Umbilicaria angulata Tuck. [UMBIANG] Umbilicariaceae (L)
Umbilicaria aprina Nyl. [UMBIAPR] Umbilicariaceae (L)
Umbilicaria aprina var. **halei** Llano [UMBIAPR1] Umbilicariaceae (L)
Umbilicaria arctica (Ach.) Nyl. [UMBIARC] Umbilicariaceae (L)
Umbilicaria cinereorufescens (Schaerer) Frey [UMBICIN] Umbilicariaceae (L)
Umbilicaria coriacea Imshaug = Umbilicaria rigida [UMBIRIG] Umbilicariaceae (L)
Umbilicaria cylindrica (L.) Delise ex Duby [UMBICYL] Umbilicariaceae (L)
Umbilicaria decussata (Vill.) Zahlbr. [UMBIDEC] Umbilicariaceae (L)
Umbilicaria deusta (L.) Baumg. [UMBIDEU] Umbilicariaceae (L)
Umbilicaria havaasii Llano [UMBIHAV] Umbilicariaceae (L)
Umbilicaria hirsuta (Sw. ex Westr.) Hoffm. [UMBIHIR] Umbilicariaceae (L)
Umbilicaria hyperborea (Ach.) Hoffm. [UMBIHYP] Umbilicariaceae (L)
Umbilicaria hyperborea var. **coccinea** Llano [UMBIHYP1] Umbilicariaceae (L)
Umbilicaria hyperborea var. **hyperborea** [UMBIHYP0] Umbilicariaceae (L)
Umbilicaria hyperborea var. **radicicula** (J.E. Zetterst.) Hasselrot [UMBIHYP2] Umbilicariaceae (L)
Umbilicaria krascheninnikovii (Savicz) Zahlbr. [UMBIKRA] Umbilicariaceae (L)
Umbilicaria lambii Imshaug [UMBILAM] Umbilicariaceae (L)
Umbilicaria leiocarpa DC. [UMBILEI] Umbilicariaceae (L)
Umbilicaria lyngei Schol. [UMBILYN] Umbilicariaceae (L)
Umbilicaria muehlenbergii (Ach.) Tuck. [UMBIMUE] Umbilicariaceae (L)
Umbilicaria nylanderiana (Zahlbr.) H. Magn. [UMBINYL] Umbilicariaceae (L)
Umbilicaria papulosa (Ach.) Nyl. = Lasallia papulosa [LASAPAP] Umbilicariaceae (L)
Umbilicaria pensylvanica Hoffm. = Lasallia pensylvanica [LASAPEN] Umbilicariaceae (L)
Umbilicaria phaea Tuck. [UMBIPHA] Umbilicariaceae (L)
Umbilicaria polyphylla (L.) Baumg. [UMBIPOL] Umbilicariaceae (L)
Umbilicaria polyrrhiza (L.) Fr. [UMBIPOY] Umbilicariaceae (L)
Umbilicaria proboscidea (L.) Schrader [UMBIPRO] Umbilicariaceae (L)
Umbilicaria pustulata var. *papulosa* (Ach.) Tuck. = Lasallia papulosa [LASAPAP] Umbilicariaceae (L)
Umbilicaria rigida (Du Rietz) Frey [UMBIRIG] Umbilicariaceae (L)
Umbilicaria torrefacta (Lightf.) Schrader [UMBITOR] Umbilicariaceae (L)
Umbilicaria vellea (L.) Hoffm. [UMBIVEL] Umbilicariaceae (L)
Umbilicaria virginis Schaerer [UMBIVIR] Umbilicariaceae (L)
Unifolium canadense (Desf.) Greene = Maianthemum canadense [MAIACAN] Liliaceae (V:M)
Uniola spicata L. = Distichlis spicata [DISTSPI] Poaceae (V:M)
Urceolaria scruposa (Schreber) Ach. = Diploschistes scruposus [DIPLSCR] Thelotremataceae (L)
Urostachys chinensis (Christ) Herter ex Nessel = Huperzia chinensis [HUPECHI] Lycopodiaceae (V:F)
Urostachys haleakalae (Brack.) Herter = Huperzia haleakalae [HUPEHAL] Lycopodiaceae (V:F)
Urostachys miyoshiana (Makino) Herter ex Nessel = Huperzia chinensis [HUPECHI] Lycopodiaceae (V:F)
Urtica L. [URTICA$] Urticaceae (V:D)
Urtica californica Greene = Urtica dioica ssp. gracilis [URTIDIO1] Urticaceae (V:D)
Urtica cardiophylla Rydb. = Urtica dioica ssp. gracilis [URTIDIO1] Urticaceae (V:D)
Urtica dioica L. [URTIDIO] Urticaceae (V:D)
Urtica dioica ssp. **gracilis** (Ait.) Seland. [URTIDIO1] Urticaceae (V:D)
Urtica dioica ssp. *gracilis* var. *lyallii* (S. Wats.) C.L. Hitchc. = Urtica dioica ssp. gracilis [URTIDIO1] Urticaceae (V:D)
Urtica dioica var. *angustifolia* Schlecht. = Urtica dioica ssp. gracilis [URTIDIO1] Urticaceae (V:D)
Urtica dioica var. *californica* (Greene) C.L. Hitchc. = Urtica dioica ssp. gracilis [URTIDIO1] Urticaceae (V:D)
Urtica dioica var. *gracilis* (Ait.) C.L. Hitchc. = Urtica dioica ssp. gracilis [URTIDIO1] Urticaceae (V:D)
Urtica dioica var. *lyallii* (S. Wats.) C.L. Hitchc. = Urtica dioica ssp. gracilis [URTIDIO1] Urticaceae (V:D)
Urtica dioica var. *procera* (Muhl. ex Willd.) Weddell = Urtica dioica ssp. gracilis [URTIDIO1] Urticaceae (V:D)
Urtica gracilis Ait. = Urtica dioica ssp. gracilis [URTIDIO1] Urticaceae (V:D)
Urtica lyallii S. Wats. = Urtica dioica ssp. gracilis [URTIDIO1] Urticaceae (V:D)
Urtica lyallii var. *californica* (Greene) Jepson = Urtica dioica ssp. gracilis [URTIDIO1] Urticaceae (V:D)
Urtica major H.P. Fuchs = Urtica dioica ssp. gracilis [URTIDIO1] Urticaceae (V:D)
Urtica procera Muhl. ex Willd. = Urtica dioica ssp. gracilis [URTIDIO1] Urticaceae (V:D)
Urtica strigosissima Rydb. = Urtica dioica ssp. gracilis [URTIDIO1] Urticaceae (V:D)
***Urtica urens** L. [URTIURE] Urticaceae (V:D)
Urtica viridis Rydb. = Urtica dioica ssp. gracilis [URTIDIO1] Urticaceae (V:D)
Usnea Dill. ex Adans. [USNEA$] Parmeliaceae (L)
Usnea alpina Mot. [USNEALP] Parmeliaceae (L)
Usnea angulata Ach. [USNEANG] Parmeliaceae (L)
Usnea barbata (L.) F.H. Wigg. [USNEBAR] Parmeliaceae (L)
Usnea barbata var. *ceratina* Schaerer = Usnea ceratina [USNECER] Parmeliaceae (L)
Usnea barbata var. *dasypoga* Fr. = Usnea filipendula [USNEFIL] Parmeliaceae (L)
Usnea barbata var. *florida* Fr. = Usnea florida [USNEFLO] Parmeliaceae (L)
Usnea barbata var. *hirta* Fr. = Usnea hirta [USNEHIR] Parmeliaceae (L)
Usnea betulina Mot. = Usnea glabrescens [USNEGLB] Parmeliaceae (L)

Usnea capitata Mot. [USNECAP] Parmeliaceae (L)
Usnea catenulata Mot. [USNECAT] Parmeliaceae (L)
Usnea caucasica Vainio = Usnea filipendula [USNEFIL] Parmeliaceae (L)
Usnea cavernosa Tuck. [USNECAV] Parmeliaceae (L)
Usnea ceratina Ach. [USNECER] Parmeliaceae (L)
Usnea chaetophora Stirton [USNECHA] Parmeliaceae (L)
Usnea comosa auct. = Usnea subfloridana [USNESUF] Parmeliaceae (L)
Usnea comosa var. *stuppea* Räsänen = Usnea stuppea [USNESTU] Parmeliaceae (L)
Usnea compacta (Räsänen) Mot. = Usnea glabrescens [USNEGLB] Parmeliaceae (L)
Usnea cornuta Körber [USNECOR] Parmeliaceae (L)
Usnea dasypoga auct. = Usnea filipendula [USNEFIL] Parmeliaceae (L)
Usnea dasypoga ssp. *melanopoga* Mot. = Usnea filipendula [USNEFIL] Parmeliaceae (L)
Usnea dasypoga var. *subscabrata* Vainio = Usnea filipendula [USNEFIL] Parmeliaceae (L)
Usnea diplotypus Vainio [USNEDIP] Parmeliaceae (L)
Usnea distincta Mot. = Usnea glabrescens [USNEGLB] Parmeliaceae (L)
Usnea esthonica Räsänen = Usnea filipendula [USNEFIL] Parmeliaceae (L)
Usnea filipendula Stirton [USNEFIL] Parmeliaceae (L)
Usnea flagellata Mot. = Usnea filipendula [USNEFIL] Parmeliaceae (L)
Usnea florida (L.) F.H. Wigg. [USNEFLO] Parmeliaceae (L)
Usnea fragilescens Hav. ex Lynge [USNEFRA] Parmeliaceae (L)
Usnea fragilescens var. **mollis** (Vainio) Clerc [USNEFRA1] Parmeliaceae (L)
Usnea fulvoreagens (Räsänen) Räsänen [USNEFUL] Parmeliaceae (L)
Usnea glabrata (Ach.) Vainio [USNEGLA] Parmeliaceae (L)
Usnea glabrescens (Nyl. ex Vainio) Vainio [USNEGLB] Parmeliaceae (L)
Usnea graciosa Mot. [USNEGRA] Parmeliaceae (L)
Usnea hesperina Mot. [USNEHES] Parmeliaceae (L)
Usnea hirta (L.) F.H. Wigg. [USNEHIR] Parmeliaceae (L)
Usnea inflata Delise ex Duby = Usnea cornuta [USNECOR] Parmeliaceae (L)
Usnea intexta Stirton = Usnea cornuta [USNECOR] Parmeliaceae (L)
Usnea kujalae Räsänen = Usnea glabrata [USNEGLA] Parmeliaceae (L)
Usnea lapponica Vainio [USNELAP] Parmeliaceae (L)
Usnea laricina Vainio ex Räsänen = Usnea lapponica [USNELAP] Parmeliaceae (L)
Usnea longissima Ach. [USNELON] Parmeliaceae (L)
Usnea longissima var. **corticata** R.H. Howe [USNELON1] Parmeliaceae (L)
Usnea longissima var. **longissima** [USNELON0] Parmeliaceae (L)
Usnea madeirensis Mot. [USNEMAD] Parmeliaceae (L)
Usnea merrillii Mot. [USNEMER] Parmeliaceae (L)
Usnea mollis Stirton = Usnea fragilescens var. mollis [USNEFRA1] Parmeliaceae (L)
Usnea nidulans Mot. [USNENID] Parmeliaceae (L)
Usnea occidentalis Mot. [USNEOCC] Parmeliaceae (L)
Usnea prostrata Vainio ex Räsänen = Usnea barbata [USNEBAR] Parmeliaceae (L)
Usnea rigida (Ach.) Mot. [USNERIG] Parmeliaceae (L)
Usnea rubicunda Stirton [USNERUB] Parmeliaceae (L)
Usnea rugulosa Vainio = Usnea scabrata [USNESCB] Parmeliaceae (L)
Usnea scabiosa Mot. [USNESCA] Parmeliaceae (L)
Usnea scabrata Nyl. [USNESCB] Parmeliaceae (L)
Usnea scabrata ssp. *nylanderiana* Mot. = Usnea scabrata [USNESCB] Parmeliaceae (L)
Usnea silesiaca Mot. [USNESIL] Parmeliaceae (L)
Usnea similis Mot. ex Räsänen = Usnea subfloridana [USNESUF] Parmeliaceae (L)
Usnea sorediifera (Arnold) Lynge = Usnea glabrata [USNEGLA] Parmeliaceae (L)
Usnea sorediifera auct. = Usnea lapponica [USNELAP] Parmeliaceae (L)
Usnea stuppea (Räsänen) Mot. [USNESTU] Parmeliaceae (L)
Usnea subfloridana Stirton [USNESUF] Parmeliaceae (L)
Usnea sublaxa Vainio = Usnea filipendula [USNEFIL] Parmeliaceae (L)
Usnea subpectinata Stirton = Usnea cornuta [USNECOR] Parmeliaceae (L)
Usnea substerilis Mot. [USNESUB] Parmeliaceae (L)
Usnea sylvatica Mot. [USNESYL] Parmeliaceae (L)
Usnea trichodea Ach. [USNETRI] Parmeliaceae (L)
Usnea wasmuthii Räsänen [USNEWAS] Parmeliaceae (L)
Usnea wirthii Clerc [USNEWIR] Parmeliaceae (L)
Utricularia L. [UTRICUL$] Lentibulariaceae (V:D)
Utricularia biflora Lam. = Utricularia gibba [UTRIGIB] Lentibulariaceae (V:D)
Utricularia gibba L. [UTRIGIB] Lentibulariaceae (V:D)
Utricularia intermedia Hayne [UTRIINT] Lentibulariaceae (V:D)
Utricularia macrorhiza Le Conte [UTRIMAC] Lentibulariaceae (V:D)
Utricularia minor L. [UTRIMIN] Lentibulariaceae (V:D)
Utricularia obtusa Sw. = Utricularia gibba [UTRIGIB] Lentibulariaceae (V:D)
Utricularia pumila Walt. = Utricularia gibba [UTRIGIB] Lentibulariaceae (V:D)
Utricularia vulgaris L. = Utricularia macrorhiza [UTRIMAC] Lentibulariaceae (V:D)
Utricularia vulgaris ssp. *macrorhiza* (Le Conte) Clausen = Utricularia macrorhiza [UTRIMAC] Lentibulariaceae (V:D)
Utricularia vulgaris var. *americana* Gray = Utricularia macrorhiza [UTRIMAC] Lentibulariaceae (V:D)
Uva-Ursi uva-ursi (L.) Britt. = Arctostaphylos uva-ursi [ARCTUVA] Ericaceae (V:D)
Uvularia amplexifolia L. = Streptopus amplexifolius var. amplexifolius [STREAMP0] Liliaceae (V:M)
Uvularia smithii Hook. = Disporum smithii [DISPSMI] Liliaceae (V:M)

Vaccaria von Wolf [VACCARI$] Caryophyllaceae (V:D)
•**Vaccaria hispanica** (P. Mill.) Rauschert [VACCHIS] Caryophyllaceae (V:D)
Vaccaria pyramidata Medik. = Vaccaria hispanica [VACCHIS] Caryophyllaceae (V:D)
Vaccaria segetalis Garcke ex Aschers. = Vaccaria hispanica [VACCHIS] Caryophyllaceae (V:D)
Vaccaria vaccaria (L.) Britt. = Vaccaria hispanica [VACCHIS] Caryophyllaceae (V:D)
Vaccaria vulgaris Host = Vaccaria hispanica [VACCHIS] Caryophyllaceae (V:D)
Vaccinium L. [VACCINI$] Ericaceae (V:D)
Vaccinium alaskaense T.J. Howell = Vaccinium ovalifolium [VACCOVA] Ericaceae (V:D)
Vaccinium angustifolium var. *myrtilloides* (Michx.) House = Vaccinium myrtilloides [VACCMYR] Ericaceae (V:D)
Vaccinium caespitosum Michx. [VACCCAE] Ericaceae (V:D)

Vaccinium canadense Kalm ex A. Rich. = Vaccinium myrtilloides [VACCMYR] Ericaceae (V:D)
Vaccinium coccineum Piper = Vaccinium membranaceum [VACCMEM] Ericaceae (V:D)
Vaccinium deliciosum Piper [VACCDEL] Ericaceae (V:D)
Vaccinium globulare Rydb. = Vaccinium membranaceum [VACCMEM] Ericaceae (V:D)
Vaccinium membranaceum Dougl. ex Torr. [VACCMEM] Ericaceae (V:D)
Vaccinium membranaceum var. *rigidum* (Hook.) Fern. = Vaccinium membranaceum [VACCMEM] Ericaceae (V:D)
Vaccinium microcarpum (Turcz. ex Rupr.) Schmalh. = Vaccinium oxycoccos [VACCOXY] Ericaceae (V:D)
Vaccinium myrtilloides Michx. [VACCMYR] Ericaceae (V:D)
Vaccinium myrtillus L. [VACCMYT] Ericaceae (V:D)
Vaccinium occidentale Gray = Vaccinium uliginosum [VACCULI] Ericaceae (V:D)
Vaccinium ovalifolium Sm. [VACCOVA] Ericaceae (V:D)
Vaccinium ovatum Pursh [VACCOVT] Ericaceae (V:D)
Vaccinium oxycoccos L. [VACCOXY] Ericaceae (V:D)
Vaccinium oxycoccos var. *intermedium* Gray = Vaccinium oxycoccos [VACCOXY] Ericaceae (V:D)
Vaccinium oxycoccos var. *microphyllum* (Lange) Rouss. & Raymond = Vaccinium oxycoccos [VACCOXY] Ericaceae (V:D)
Vaccinium oxycoccos var. *ovalifolium* Michx. = Vaccinium oxycoccos [VACCOXY] Ericaceae (V:D)
Vaccinium parvifolium Sm. [VACCPAR] Ericaceae (V:D)
Vaccinium salicinum Cham. = Vaccinium uliginosum [VACCULI] Ericaceae (V:D)
Vaccinium scoparium Leib. ex Coville [VACCSCO] Ericaceae (V:D)
Vaccinium uliginosum L. [VACCULI] Ericaceae (V:D)
Vaccinium uliginosum ssp. *alpinum* (Bigelow) Hultén = Vaccinium uliginosum [VACCULI] Ericaceae (V:D)
Vaccinium uliginosum ssp. *gaultherioides* (Bigelow) S.B. Young = Vaccinium uliginosum [VACCULI] Ericaceae (V:D)
Vaccinium uliginosum ssp. *microphyllum* Lange = Vaccinium uliginosum [VACCULI] Ericaceae (V:D)
Vaccinium uliginosum ssp. *occidentale* (Gray) Hultén = Vaccinium uliginosum [VACCULI] Ericaceae (V:D)
Vaccinium uliginosum ssp. *pedris* (Harshberger) S.B. Young = Vaccinium uliginosum [VACCULI] Ericaceae (V:D)
Vaccinium uliginosum ssp. *pubescens* (Wormsk. ex Hornem.) S.B. Young = Vaccinium uliginosum [VACCULI] Ericaceae (V:D)
Vaccinium uliginosum var. *alpinum* Bigelow = Vaccinium uliginosum [VACCULI] Ericaceae (V:D)
Vaccinium uliginosum var. *langeanum* Malte = Vaccinium uliginosum [VACCULI] Ericaceae (V:D)
Vaccinium uliginosum var. *occidentale* (Gray) Hara = Vaccinium uliginosum [VACCULI] Ericaceae (V:D)
Vaccinium uliginosum var. *salicinum* (Cham.) Hultén = Vaccinium uliginosum [VACCULI] Ericaceae (V:D)
Vaccinium vitis-idaea L. [VACCVIT] Ericaceae (V:D)
Vaccinium vitis-idaea ssp. **minus** (Lodd.) Hultén [VACCVIT1] Ericaceae (V:D)
Vaccinium vitis-idaea var. *minus* Lodd. = Vaccinium vitis-idaea ssp. minus [VACCVIT1] Ericaceae (V:D)
Vaccinium vitis-idaea var. *punctatum* Moench = Vaccinium vitis-idaea ssp. minus [VACCVIT1] Ericaceae (V:D)
Vagnera amplexicaulis (Nutt.) Greene = Maianthemum racemosum ssp. amplexicaule [MAIARAC1] Liliaceae (V:M)
Vagnera amplexicaulis var. *glabra* (J.F. Macbr.) Abrams = Maianthemum racemosum ssp. amplexicaule [MAIARAC1] Liliaceae (V:M)
Vagnera liliacea (Greene) Rydb. = Maianthemum stellatum [MAIASTE] Liliaceae (V:M)
Vagnera sessilifolia (Nutt. ex Baker) Greene = Maianthemum stellatum [MAIASTE] Liliaceae (V:M)
Vagnera stellata (L.) Morong = Maianthemum stellatum [MAIASTE] Liliaceae (V:M)
Vagnera trifolia (L.) Morong = Maianthemum trifolium [MAIATRI] Liliaceae (V:M)
Vahlodea Fries [VAHLODE$] Poaceae (V:M)
Vahlodea atropurpurea (Wahlenb.) Fries ex Hartman [VAHLATR] Poaceae (V:M)
Vahlodea atropurpurea ssp. *latifolia* (Hook.) Porsild = Vahlodea atropurpurea [VAHLATR] Poaceae (V:M)
Vahlodea atropurpurea ssp. *paramushirensis* (Kudo) Hultén = Vahlodea atropurpurea [VAHLATR] Poaceae (V:M)
Vahlodea flexuosa (Honda) Ohwi = Vahlodea atropurpurea [VAHLATR] Poaceae (V:M)
Vahlodea latifolia (Hook.) Hultén = Vahlodea atropurpurea [VAHLATR] Poaceae (V:M)
Valeriana L. [VALERIA$] Valerianaceae (V:D)
Valeriana capitata Pallas ex Link [VALECAP] Valerianaceae (V:D)
Valeriana dioica L. [VALEDIO] Valerianaceae (V:D)
Valeriana dioica ssp. *sylvatica* (S. Wats.) F.G. Mey. = Valeriana dioica var. sylvatica [VALEDIO1] Valerianaceae (V:D)
Valeriana dioica var. **sylvatica** S. Wats. [VALEDIO1] Valerianaceae (V:D)
Valeriana edulis Nutt. ex Torr. & Gray [VALEEDU] Valerianaceae (V:D)
***Valeriana officinalis** L. [VALEOFF] Valerianaceae (V:D)
Valeriana scouleri Rydb. [VALESCO] Valerianaceae (V:D)
Valeriana septentrionalis Rydb. = Valeriana dioica var. sylvatica [VALEDIO1] Valerianaceae (V:D)
Valeriana sitchensis Bong. [VALESIT] Valerianaceae (V:D)
Valeriana sitchensis ssp. *scouleri* (Rydb.) F.G. Mey. = Valeriana scouleri [VALESCO] Valerianaceae (V:D)
Valeriana sitchensis var. *scouleri* (Rydb.) M.E. Jones = Valeriana scouleri [VALESCO] Valerianaceae (V:D)
Valeriana sylvatica Soland. ex Richards. = Valeriana dioica var. sylvatica [VALEDIO1] Valerianaceae (V:D)
Valerianella P. Mill. [VALERIN$] Valerianaceae (V:D)
***Valerianella locusta** (L.) Lat. [VALELOC] Valerianaceae (V:D)
Valerianella olitoria (L.) Pollich = Valerianella locusta [VALELOC] Valerianaceae (V:D)
Vallisneria L. [VALLISN$] Hydrocharitaceae (V:M)
***Vallisneria americana** Michx. [VALLAME] Hydrocharitaceae (V:M)
Vallisneria asiatica Michx. = Vallisneria americana [VALLAME] Hydrocharitaceae (V:M)
Vallisneria neotropicalis Victorin = Vallisneria americana [VALLAME] Hydrocharitaceae (V:M)
Vallisneria spiralis var. *asiatica* (Michx.) Torr. = Vallisneria americana [VALLAME] Hydrocharitaceae (V:M)
Varicellaria Nyl. [VARICEL$] Pertusariaceae (L)
Varicellaria kemensis Räsänen = Varicellaria rhodocarpa [VARIRHO] Pertusariaceae (L)
Varicellaria rhodocarpa (Körber) Th. Fr. [VARIRHO] Pertusariaceae (L)
Variolaria lactea var. *arborea* Kreyer = Ochrolechia arborea [OCHRARB] Pertusariaceae (L)
Ventenata Koel. [VENTENA$] Poaceae (V:M)
Ventenata avenacea Koel. = Ventenata dubia [VENTDUB] Poaceae (V:M)
***Ventenata dubia** (Leers) Coss. & Durieu [VENTDUB] Poaceae (V:M)
Veratrum L. [VERATRU$] Liliaceae (V:M)
Veratrum eschscholtzianum (J.A. & J.H. Schultes) Rydb. ex Heller = Veratrum viride [VERAVIR] Liliaceae (V:M)
Veratrum eschscholtzii Gray = Veratrum viride [VERAVIR] Liliaceae (V:M)
Veratrum eschscholtzii var. *incriminatum* Boivin = Veratrum viride [VERAVIR] Liliaceae (V:M)
Veratrum viride Ait. [VERAVIR] Liliaceae (V:M)
Veratrum viride ssp. *eschscholtzii* (Gray) A. & D. Löve = Veratrum viride [VERAVIR] Liliaceae (V:M)
Veratrum viride var. *eschscholtzii* (Gray) Breitung = Veratrum viride [VERAVIR] Liliaceae (V:M)
Verbascum L. [VERBASC$] Scrophulariaceae (V:D)
***Verbascum blattaria** L. [VERBBLA] Scrophulariaceae (V:D)
***Verbascum phlomoides** L. [VERBPHL] Scrophulariaceae (V:D)
***Verbascum thapsus** L. [VERBTHA] Scrophulariaceae (V:D)
Verbena L. [VERBENA$] Verbenaceae (V:D)
Verbena bracteata Lag. & Rodr. [VERBBRA] Verbenaceae (V:D)

Verbena bracteosa Michx. = Verbena bracteata [VERBBRA] Verbenaceae (V:D)
Verbena hastata L. [VERBHAS] Verbenaceae (V:D)
Verbena imbricata Woot. & Standl. = Verbena bracteata [VERBBRA] Verbenaceae (V:D)
Veronica L. [VERONIC$] Scrophulariaceae (V:D)
Veronica alpina var. *alterniflora* Fern. = Veronica wormskjoldii [VEROWOR] Scrophulariaceae (V:D)
Veronica alpina var. *cascadensis* Fern. = Veronica wormskjoldii [VEROWOR] Scrophulariaceae (V:D)
Veronica alpina var. *geminiflora* Fern. = Veronica wormskjoldii [VEROWOR] Scrophulariaceae (V:D)
Veronica alpina var. *nutans* (Bong.) Boivin = Veronica wormskjoldii [VEROWOR] Scrophulariaceae (V:D)
Veronica alpina var. *terrae-novae* Fern. = Veronica wormskjoldii [VEROWOR] Scrophulariaceae (V:D)
Veronica alpina var. *unalaschcensis* Cham. & Schlecht. = Veronica wormskjoldii [VEROWOR] Scrophulariaceae (V:D)
Veronica alpina var. *wormskjoldii* Hook. = Veronica wormskjoldii [VEROWOR] Scrophulariaceae (V:D)
Veronica americana Schwein. ex Benth. [VEROAME] Scrophulariaceae (V:D)
Veronica anagallis L. = Veronica anagallis-aquatica [VEROANA] Scrophulariaceae (V:D)
*__Veronica anagallis-aquatica__ L. [VEROANA] Scrophulariaceae (V:D)
*__Veronica arvensis__ L. [VEROARV] Scrophulariaceae (V:D)
Veronica beccabunga L. [VEROBEC] Scrophulariaceae (V:D)
Veronica beccabunga ssp. *americana* (Raf.) Sellers = Veronica americana [VEROAME] Scrophulariaceae (V:D)
Veronica catenata Pennell = Veronica anagallis-aquatica [VEROANA] Scrophulariaceae (V:D)
Veronica catenata var. *glandulosa* (Farw.) Pennell = Veronica anagallis-aquatica [VEROANA] Scrophulariaceae (V:D)
*__Veronica chamaedrys__ L. [VEROCHA] Scrophulariaceae (V:D)
Veronica comosa var. *glaberrima* (Pennell) Boivin = Veronica anagallis-aquatica [VEROANA] Scrophulariaceae (V:D)
Veronica comosa var. *glandulosa* (Farw.) Boivin = Veronica anagallis-aquatica [VEROANA] Scrophulariaceae (V:D)
Veronica connata ssp. *glaberrima* Pennell = Veronica anagallis-aquatica [VEROANA] Scrophulariaceae (V:D)
Veronica connata var. *glaberrima* (Pennell) Fern. = Veronica anagallis-aquatica [VEROANA] Scrophulariaceae (V:D)
Veronica connata var. *typica* Pennell = Veronica anagallis-aquatica [VEROANA] Scrophulariaceae (V:D)
Veronica cusickii Gray [VEROCUS] Scrophulariaceae (V:D)
*__Veronica filiformis__ Sm. [VEROFIL] Scrophulariaceae (V:D)
Veronica glandifera Pennell = Veronica anagallis-aquatica [VEROANA] Scrophulariaceae (V:D)
Veronica hederaefolia L. = Veronica hederifolia [VEROHED] Scrophulariaceae (V:D)
*__Veronica hederifolia__ L. [VEROHED] Scrophulariaceae (V:D)
Veronica humifusa Dickson = Veronica serpyllifolia ssp. humifusa [VEROSER1] Scrophulariaceae (V:D)
Veronica × lackschewitzii Keller = Veronica anagallis-aquatica [VEROANA] Scrophulariaceae (V:D)
Veronica nutans Bong. = Veronica wormskjoldii [VEROWOR] Scrophulariaceae (V:D)
*__Veronica officinalis__ L. [VEROOFF] Scrophulariaceae (V:D)
Veronica peregrina L. [VEROPER] Scrophulariaceae (V:D)
Veronica peregrina ssp. **peregrina** [VEROPER0] Scrophulariaceae (V:D)
Veronica peregrina ssp. **xalapensis** (Kunth) Pennell [VEROPER1] Scrophulariaceae (V:D)
Veronica peregrina var. *typica* Pennell = Veronica peregrina ssp. peregrina [VEROPER0] Scrophulariaceae (V:D)
Veronica peregrina var. *xalapensis* (Kunth) Pennell = Veronica peregrina ssp. xalapensis [VEROPER1] Scrophulariaceae (V:D)
*__Veronica persica__ Poir. [VEROPES] Scrophulariaceae (V:D)
Veronica scutellata L. [VEROSCU] Scrophulariaceae (V:D)
Veronica scutellata var. *villosa* Schumacher = Veronica scutellata [VEROSCU] Scrophulariaceae (V:D)
*__Veronica serpyllifolia__ L. [VEROSER] Scrophulariaceae (V:D)
Veronica serpyllifolia ssp. **humifusa** (Dickson) Syme [VEROSER1] Scrophulariaceae (V:D)
Veronica serpyllifolia ssp. **serpyllifolia** [VEROSER0] Scrophulariaceae (V:D)
Veronica serpyllifolia var. *borealis* Laestad. = Veronica serpyllifolia ssp. humifusa [VEROSER1] Scrophulariaceae (V:D)
Veronica serpyllifolia var. *humifusa* (Dickson) Vahl = Veronica serpyllifolia ssp. humifusa [VEROSER1] Scrophulariaceae (V:D)
Veronica sherwoodii M.E. Peck = Veronica peregrina ssp. xalapensis [VEROPER1] Scrophulariaceae (V:D)
Veronica stelleri Pallas ex Link = Veronica wormskjoldii [VEROWOR] Scrophulariaceae (V:D)
Veronica stelleri var. *glabrescens* Hultén = Veronica wormskjoldii [VEROWOR] Scrophulariaceae (V:D)
Veronica tenella All. = Veronica serpyllifolia ssp. humifusa [VEROSER1] Scrophulariaceae (V:D)
*__Veronica verna__ L. [VEROVER] Scrophulariaceae (V:D)
Veronica wormskjoldii Roemer & J.A. Schultes [VEROWOR] Scrophulariaceae (V:D)
Veronica wormskjoldii ssp. *alterniflora* (Fern.) Pennell = Veronica wormskjoldii [VEROWOR] Scrophulariaceae (V:D)
Veronica xalapensis Kunth = Veronica peregrina ssp. xalapensis [VEROPER1] Scrophulariaceae (V:D)
Veronicastrum serpyllifolium (L.) Fourr. = Veronica serpyllifolia ssp. serpyllifolia [VEROSER0] Scrophulariaceae (V:D)
Veronicastrum serpyllifolium ssp. *humifusum* (Dickson) W.A. Weber = Veronica serpyllifolia ssp. humifusa [VEROSER1] Scrophulariaceae (V:D)
Verrucaria Schrader [VERRUCA$] Verrucariaceae (L)
Verrucaria aethiobola Wahlenb. [VERRAET] Verrucariaceae (L)
Verrucaria citrina Hoffm. = Caloplaca citrina [CALOCIT] Teloschistaceae (L)
Verrucaria cuprea var. *fuscocuprea* Nyl. = Staurothele drummondii [STAUDRU] Verrucariaceae (L)
Verrucaria devergens Nyl. [VERRDEV] Verrucariaceae (L)
Verrucaria diffusilis Nyl. = Staurothele drummondii [STAUDRU] Verrucariaceae (L)
Verrucaria drummondii Tuck. = Staurothele drummondii [STAUDRU] Verrucariaceae (L)
Verrucaria erichsenii Zsch. [VERRERI] Verrucariaceae (L)
Verrucaria fissa Tayl. = Staurothele fissa [STAUFIS] Verrucariaceae (L)
Verrucaria funckii (Sprengel) Zahlbr. [VERRFUN] Verrucariaceae (L)
Verrucaria fuscocuprea (Nyl.) Nyl. = Staurothele drummondii [STAUDRU] Verrucariaceae (L)
Verrucaria fuscocuprea f. *rimosior* Nyl. = Staurothele drummondii [STAUDRU] Verrucariaceae (L)
Verrucaria fusconigrescens Nyl. [VERRFUS] Verrucariaceae (L)
Verrucaria inconversa Nyl. = Staurothele fissa [STAUFIS] Verrucariaceae (L)
Verrucaria laevata Ach. = Verrucaria aethiobola [VERRAET] Verrucariaceae (L)
Verrucaria macrostoma Dufour ex DC. [VERRMAC] Verrucariaceae (L)
Verrucaria margacea (Wahlenb.) Wahlenb. [VERRMAR] Verrucariaceae (L)
Verrucaria maura Wahlenb. [VERRMAU] Verrucariaceae (L)
Verrucaria mucosa Wahlenb. [VERRMUC] Verrucariaceae (L)
Verrucaria muralis Ach. [VERRMUR] Verrucariaceae (L)
Verrucaria nigrescens Pers. [VERRNIG] Verrucariaceae (L)
Verrucaria plumbaria Stizenb. = Arthopyrenia plumbaria [ARTHPLU] Arthopyreniaceae (L)
Verrucaria prominula Nyl. [VERRPRO] Verrucariaceae (L)
Verrucaria rubella Hoffm. = Bacidia rubella [BACIRUB] Bacidiaceae (L)
Verrucaria rupestris Schrader [VERRRUP] Verrucariaceae (L)
Verrucaria sphaerospora Anzi [VERRSPH] Verrucariaceae (L)
Verrucaria tectorum (A. Massal.) Körber [VERRTEC] Verrucariaceae (L)
Verrucaria viridula (Schrader) Ach. [VERRVII] Verrucariaceae (L)
Vestergrenopsis Gyelnik [VESTERG$] Placynthiaceae (L)

Vestergrenopsis elaeina (Wahlenb.) Gyelnik [VESTELA] Placynthiaceae (L)
Vestergrenopsis isidiata (Degel.) E. Dahl [VESTISI] Placynthiaceae (L)
Viburnum L. [VIBURNU$] Caprifoliaceae (V:D)
Viburnum edule (Michx.) Raf. [VIBUEDU] Caprifoliaceae (V:D)
Viburnum opulus L. [VIBUOPU] Caprifoliaceae (V:D)
Viburnum opulus ssp. *trilobum* (Marsh.) Clausen = Viburnum opulus var. americanum [VIBUOPU1] Caprifoliaceae (V:D)
Viburnum opulus var. **americanum** Ait. [VIBUOPU1] Caprifoliaceae (V:D)
Viburnum pauciflorum La Pylaie ex Torr. & Gray = Viburnum edule [VIBUEDU] Caprifoliaceae (V:D)
Viburnum trilobum Marsh. = Viburnum opulus var. americanum [VIBUOPU1] Caprifoliaceae (V:D)
Vicia L. [VICIA$] Fabaceae (V:D)
Vicia americana Muhl. ex Willd. [VICIAME] Fabaceae (V:D)
Vicia angustifolia L. = Vicia sativa ssp. nigra [VICISAT1] Fabaceae (V:D)
Vicia angustifolia var. *segetalis* (Thuill.) W.D.J. Koch = Vicia sativa ssp. nigra [VICISAT1] Fabaceae (V:D)
Vicia angustifolia var. *uncinata* (Desv.) Rouy = Vicia sativa ssp. nigra [VICISAT1] Fabaceae (V:D)
•**Vicia cracca** L. [VICICRA] Fabaceae (V:D)
Vicia gigantea Hook. = Vicia nigricans ssp. gigantea [VICINIG1] Fabaceae (V:D)
•**Vicia hirsuta** (L.) S.F. Gray [VICIHIR] Fabaceae (V:D)
Vicia nigricans Hook. & Arn. [VICINIG] Fabaceae (V:D)
Vicia nigricans ssp. **gigantea** (Hook.) Lassetter & Gunn. [VICINIG1] Fabaceae (V:D)
•**Vicia sativa** L. [VICISAT] Fabaceae (V:D)
•**Vicia sativa** ssp. **nigra** (L.) Ehrh. [VICISAT1] Fabaceae (V:D)
•**Vicia sativa** ssp. **sativa** [VICISAT0] Fabaceae (V:D)
Vicia sativa var. *angustifolia* (L.) Ser. = Vicia sativa ssp. nigra [VICISAT1] Fabaceae (V:D)
Vicia sativa var. *linearis* Lange = Vicia sativa ssp. sativa [VICISAT0] Fabaceae (V:D)
Vicia sativa var. *nigra* L. = Vicia sativa ssp. nigra [VICISAT1] Fabaceae (V:D)
Vicia sativa var. *segetalis* (Thuill.) Ser. = Vicia sativa ssp. nigra [VICISAT1] Fabaceae (V:D)
•**Vicia tetrasperma** (L.) Schreb. [VICITET] Fabaceae (V:D)
Vicia tetrasperma var. *tenuissima* Druce = Vicia tetrasperma [VICITET] Fabaceae (V:D)
•**Vicia villosa** Roth [VICIVIL] Fabaceae (V:D)
Vinca L. [VINCA$] Apocynaceae (V:D)
•**Vinca major** L. [VINCMAJ] Apocynaceae (V:D)
Vinca major var. *variegata* Loud. = Vinca major [VINCMAJ] Apocynaceae (V:D)
•**Vinca minor** L. [VINCMIN] Apocynaceae (V:D)
Viola L. [VIOLA$] Violaceae (V:D)
Viola achyrophora Greene = Viola epipsila ssp. repens [VIOLEPI1] Violaceae (V:D)
Viola adunca Sm. [VIOLADU] Violaceae (V:D)
•**Viola arvensis** Murr. [VIOLARV] Violaceae (V:D)
Viola biflora L. [VIOLBIF] Violaceae (V:D)
Viola biflora ssp. **carlottae** Calder & Taylor [VIOLBIF1] Violaceae (V:D)
Viola biflora var. *carlottae* (Calder & Taylor) Boivin = Viola biflora ssp. carlottae [VIOLBIF1] Violaceae (V:D)
Viola canadensis L. [VIOLCAN] Violaceae (V:D)
Viola canadensis ssp. *rydbergii* (Greene) House = Viola canadensis var. corymbosa [VIOLCAN1] Violaceae (V:D)
Viola canadensis var. **corymbosa** Nutt. ex Torr. & Gray [VIOLCAN1] Violaceae (V:D)
Viola cognata Greene = Viola nephrophylla var. cognata [VIOLNEP1] Violaceae (V:D)
Viola epipsila Ledeb. [VIOLEPI] Violaceae (V:D)
Viola epipsila ssp. **repens** Becker [VIOLEPI1] Violaceae (V:D)
Viola epipsiloides A. & D. Löve = Viola epipsila ssp. repens [VIOLEPI1] Violaceae (V:D)
Viola glabella Nutt. [VIOLGLA] Violaceae (V:D)
Viola howellii Gray [VIOLHOW] Violaceae (V:D)
•**Viola lanceolata** L. [VIOLLAN] Violaceae (V:D)
Viola langsdorfii Fisch. ex Gingins [VIOLLAG] Violaceae (V:D)
Viola maccabeiana M.S. Baker [VIOLMAC] Violaceae (V:D)
Viola macloskeyi Lloyd [VIOLMAL] Violaceae (V:D)
Viola macloskeyi ssp. **pallens** (Banks ex DC.) M.S. Baker [VIOLMAL2] Violaceae (V:D)
Viola macloskeyi var. *pallens* (Banks ex DC.) C.L. Hitchc. = Viola macloskeyi ssp. pallens [VIOLMAL2] Violaceae (V:D)
Viola nephrophylla Greene [VIOLNEP] Violaceae (V:D)
Viola nephrophylla var. **cognata** (Greene) C.L. Hitchc. [VIOLNEP1] Violaceae (V:D)
Viola nephrophylla var. **nephrophylla** [VIOLNEP0] Violaceae (V:D)
Viola nuttallii ssp. *praemorsa* (Dougl. ex Lindl.) Piper = Viola praemorsa [VIOLPRA] Violaceae (V:D)
Viola nuttallii var. *major* Hook. = Viola vallicola var. major [VIOLVAL1] Violaceae (V:D)
Viola nuttallii var. *praemorsa* (Dougl. ex Lindl.) S. Wats. = Viola praemorsa [VIOLPRA] Violaceae (V:D)
Viola nuttallii var. *vallicola* (A. Nels.) St. John = Viola vallicola [VIOLVAL] Violaceae (V:D)
•**Viola odorata** L. [VIOLODO] Violaceae (V:D)
Viola orbiculata Geyer ex Holz. [VIOLORB] Violaceae (V:D)
Viola pallens (Banks ex DC.) Brainerd = Viola macloskeyi ssp. pallens [VIOLMAL2] Violaceae (V:D)
Viola pallens var. *subreptans* Rouss. = Viola macloskeyi ssp. pallens [VIOLMAL2] Violaceae (V:D)
Viola palustris L. [VIOLPAL] Violaceae (V:D)
Viola praemorsa Dougl. ex Lindl. [VIOLPRA] Violaceae (V:D)
Viola praemorsa ssp. *major* (Hook.) M.S. Baker ex M.E. Peck = Viola vallicola var. major [VIOLVAL1] Violaceae (V:D)
Viola praemorsa var. *major* (Hook.) M.E. Peck = Viola vallicola var. major [VIOLVAL1] Violaceae (V:D)
Viola purpurea Kellogg [VIOLPUR] Violaceae (V:D)
Viola purpurea ssp. *atriplicifolia* (Greene) M.S. Baker & J.C. Clausen ex M.E. Peck = Viola purpurea ssp. venosa [VIOLPUR1] Violaceae (V:D)
Viola purpurea ssp. **venosa** (S. Wats.) M.S. Baker & J.C. Clausen [VIOLPUR1] Violaceae (V:D)
Viola purpurea var. *atriplicifolia* (Greene) M.E. Peck = Viola purpurea ssp. venosa [VIOLPUR1] Violaceae (V:D)
Viola purpurea var. *venosa* (S. Wats.) Brainerd = Viola purpurea ssp. venosa [VIOLPUR1] Violaceae (V:D)
Viola renifolia Gray [VIOLREN] Violaceae (V:D)
Viola renifolia var. *brainerdii* (Greene) Fern. = Viola renifolia [VIOLREN] Violaceae (V:D)
Viola repens Turcz. ex Trautv. & C.A. Mey. = Viola epipsila ssp. repens [VIOLEPI1] Violaceae (V:D)
Viola sarmentosa Dougl. ex Hook. = Viola sempervirens [VIOLSEM] Violaceae (V:D)
Viola selkirkii Pursh ex Goldie [VIOLSEL] Violaceae (V:D)
Viola sempervirens Greene [VIOLSEM] Violaceae (V:D)
Viola sempervirens var. *orbiculata* (Geyer ex Holz.) J.K. Henry = Viola orbiculata [VIOLORB] Violaceae (V:D)
Viola sempervirens var. *orbiculoides* M.S. Baker = Viola orbiculata [VIOLORB] Violaceae (V:D)
Viola septentrionalis Greene [VIOLSEP] Violaceae (V:D)
Viola simulata M.S. Baker = Viola langsdorfii [VIOLLAG] Violaceae (V:D)
Viola superba M.S. Baker = Viola langsdorfii [VIOLLAG] Violaceae (V:D)
•**Viola tricolor** L. [VIOLTRI] Violaceae (V:D)
Viola tricolor var. *arvensis* (Murr.) Boiss. = Viola arvensis [VIOLARV] Violaceae (V:D)
Viola vallicola A. Nels. [VIOLVAL] Violaceae (V:D)
Viola vallicola var. **major** (Hook.) Fabijan [VIOLVAL1] Violaceae (V:D)
Virgulaster ascendens (Lindl.) Semple = Aster ascendens [ASTEASC] Asteraceae (V:D)
Virgulus campestris (Nutt.) Reveal & Keener = Aster campestris [ASTECAM] Asteraceae (V:D)
Virgulus falcatus (Lindl.) Reveal & Keener = Aster falcatus [ASTEFAL] Asteraceae (V:D)

Vulpia K.C. Gmel. [VULPIA$] Poaceae (V:M)
*Vulpia bromoides** (L.) S.F. Gray [VULPBRO] Poaceae (V:M)
Vulpia dertonensis (All.) Gola = Vulpia bromoides [VULPBRO] Poaceae (V:M)
Vulpia megalura (Nutt.) Rydb. = Vulpia myuros [VULPMYU] Poaceae (V:M)
Vulpia microstachys (Nutt.) Munro [VULPMIC] Poaceae (V:M)
Vulpia microstachys var. **pauciflora** (Scribn. ex Beal) Lonard & Gould [VULPMIC1] Poaceae (V:M)
Vulpia myuros (L.) K.C. Gmel. [VULPMYU] Poaceae (V:M)
Vulpia myuros var. *hirsuta* Hack. = Vulpia myuros [VULPMYU] Poaceae (V:M)
Vulpia octoflora (Walt.) Rydb. [VULPOCT] Poaceae (V:M)
Vulpia octoflora var. **glauca** (Nutt.) Fern. [VULPOCT1] Poaceae (V:M)
Vulpia octoflora var. **hirtella** (Piper) Henr. [VULPOCT2] Poaceae (V:M)
Vulpia octoflora var. **octoflora** [VULPOCT0] Poaceae (V:M)
Vulpia octoflora var. *tenella* (Willd.) Fern. = Vulpia octoflora var. glauca [VULPOCT1] Poaceae (V:M)
Vulpia pacifica (Piper) Rydb. = Vulpia microstachys var. pauciflora [VULPMIC1] Poaceae (V:M)
Vulpia reflexa (Buckl.) Rydb. = Vulpia microstachys var. pauciflora [VULPMIC1] Poaceae (V:M)
Vulpicida J.-E. Mattsson & M.J. Lai [VULPICI$] Parmeliaceae (L)
Vulpicida canadensis (Räsänen) J.-E. Mattsson & M.J. Lai [VULPCAN] Parmeliaceae (L)
Vulpicida pinastri (Scop.) J.-E. Mattsson & M.J. Lai [VULPPIN] Parmeliaceae (L)
Vulpicida tilesii (Ach.) J.-E. Mattsson & M.J. Lai [VULPTIL] Parmeliaceae (L)

W

Wahlbergella attenuata (Farr) Rydb. = Silene uralensis ssp. attenuata [SILEURA1] Caryophyllaceae (V:D)
Wahlbergella drummondii (Hook.) Rydb. = Silene drummondii [SILEDRU] Caryophyllaceae (V:D)
Wahlbergella parryi (S. Wats.) Rydb. = Silene parryi [SILEPAR] Caryophyllaceae (V:D)
Warnstorfia Loeske [WARNSTO$] Amblystegiaceae (B:M)
Warnstorfia crassicostata (Jansssens) Crum & Anderson = Drepanocladus crassicostatus [DREPCRA] Amblystegiaceae (B:M)
Warnstorfia exannulata (Schimp. in B.S.G.) Loeske [WARNEXA] Amblystegiaceae (B:M)
Warnstorfia exannulata var. **exannulata** [WARNEXA0] Amblystegiaceae (B:M)
Warnstorfia exannulata var. **purpurascens** (Schimp.) Tuom. & T. Kop. [WARNEXA1] Amblystegiaceae (B:M)
Warnstorfia fluitans (Hedw.) Loeske [WARNFLU] Amblystegiaceae (B:M)
Warnstorfia fluitans var. **falcata** (Sanio ex C. Jens.) Crum & Anderson [WARNFLU1] Amblystegiaceae (B:M)
Warnstorfia fluitans var. **fluitans** [WARNFLU0] Amblystegiaceae (B:M)
Warnstorfia procera (Ren. & Arnell in Husn.) Tuom. & T. Kop. [WARNPRO] Amblystegiaceae (B:M)
Warnstorfia pseudostraminea (C. Müll.) Tuom. & T. Kop. [WARNPSE] Amblystegiaceae (B:M)
Warnstorfia trichophylla (Warnst.) Tuom. & T. Kop. [WARNTRI] Amblystegiaceae (B:M)
Warnstorfia tundrae (Arnell) Loeske [WARNTUN] Amblystegiaceae (B:M)
Washingtonia divaricata Britt. = Osmorhiza berteroi [OSMOBER] Apiaceae (V:D)
Washingtonia obtusa Coult. & Rose = Osmorhiza depauperata [OSMODEP] Apiaceae (V:D)
Washingtonia purpurea Coult. & Rose = Osmorhiza purpurea [OSMOPUR] Apiaceae (V:D)
Waynea Moberg [WAYNEA$] Bacidiaceae (L)
Waynea californica Moberg = Waynea stoechadiana [WAYNSTO] Bacidiaceae (L)
Waynea stoechadiana (Abbassi Maaf & Roux) Roux & Clerc [WAYNSTO] Bacidiaceae (L)
Webera acuminata (Hoppe & Hornsch.) Schimp. = Pohlia elongata [POHLELO] Bryaceae (B:M)
Webera albicans Schimp. = Pohlia wahlenbergii [POHLWAH] Bryaceae (B:M)
Webera annotina (Hedw.) Bruch in Schwaegr. = Pohlia annotina [POHLANN] Bryaceae (B:M)
Webera canaliculata C. Müll. & Kindb. in Mac. & Kindb. = Pohlia nutans [POHLNUT] Bryaceae (B:M)
Webera carnea Schimp. = Pohlia melanodon [POHLMEL] Bryaceae (B:M)
Webera columbica Kindb. in Mac. & Kindb. = Pohlia columbica [POHLCOL] Bryaceae (B:M)
Webera commutata Schimp. = Pohlia drummondii [POHLDRU] Bryaceae (B:M)
Webera commutata var. *gracile* Schleich. = Pohlia filum [POHLFIU] Bryaceae (B:M)
Webera cruda (Hedw.) Fürnr. = Pohlia cruda [POHLCRU] Bryaceae (B:M)
Webera crudoides (Sull. & Lesq.) Jaeg. & Sauerb. = Pohlia crudoides [POHLCRD] Bryaceae (B:M)
Webera cucullata (Schwaegr.) Schimp. = Pohlia obtusifolia [POHLOBT] Bryaceae (B:M)
Webera elongata (Hedw.) Schwaegr. = Pohlia elongata [POHLELO] Bryaceae (B:M)
Webera gracilis (B.S.G.) De Not. = Pohlia filum [POHLFIU] Bryaceae (B:M)
Webera lescuriana (Sull.) Jaeg. & Sauerb. = Pohlia lescuriana [POHLLES] Bryaceae (B:M)
Webera longicolla Hedw. = Pohlia longicolla [POHLLOG] Bryaceae (B:M)
Webera ludwigii (Schwaegr.) Fürnr. = Pohlia ludwigii [POHLLUD] Bryaceae (B:M)
Webera ludwigii var. *microphylla* Kindb. in Mac. & Kindb. = Pohlia ludwigii [POHLLUD] Bryaceae (B:M)
Webera micro-apiculata C. Müll. & Kindb. in Mac. & Kindb. = Pohlia cruda [POHLCRU] Bryaceae (B:M)
Webera microcaulon C. Müll. & Kindb. in Mac. & Kindb. = Pohlia elongata [POHLELO] Bryaceae (B:M)
Webera microdenticulata C. Müll. & Kindb. in Mac. & Kindb. = Pohlia drummondii [POHLDRU] Bryaceae (B:M)
Webera nutans Hedw. = Pohlia nutans [POHLNUT] Bryaceae (B:M)
Webera nutans var. *macounii* (Kindb.) Mac. = Pohlia nutans [POHLNUT] Bryaceae (B:M)
Webera polymorpha (Hoppe & Hornsch.) Schimp. = Pohlia elongata var. greenii [POHLELO1] Bryaceae (B:M)
Webera polymorphoides Kindb. in Mac. & Kindb. = Pohlia drummondii [POHLDRU] Bryaceae (B:M)
Webera proligera Kindb. = Pohlia proligera [POHLPRO] Bryaceae (B:M)
Webera pseudo-carnea (Kindb.) Mac. = Pohlia bulbifera [POHLBUL] Bryaceae (B:M)
Webera pulchella Schimp. = Pohlia lescuriana [POHLLES] Bryaceae (B:M)
Webera pycnodecurrens C. Müll. & Kindb. in Mac. & Kindb. = Pohlia drummondii [POHLDRU] Bryaceae (B:M)
Webera schimperi (C. Müll.) Schimp. = Pohlia nutans [POHLNUT] Bryaceae (B:M)

Webera sessilis Lindb. = Diphyscium foliosum [DIPHFOL] Buxbaumiaceae (B:M)
Webera sphagnicola (B.S.G.) Schimp. = Pohlia sphagnicola [POHLSPH] Bryaceae (B:M)
Webera subcucullata C. Müll. & Kindb. in Mac. & Kindb. = Pohlia drummondii [POHLDRU] Bryaceae (B:M)
Webera subpolymorpha (Kindb.) Par. = Pohlia drummondii [POHLDRU] Bryaceae (B:M)
Webera tozeri (Grev.) Schimp. = Epipterygium tozeri [EPIPTOZ] Bryaceae (B:M)
Weberaster modestus (Lindl.) A. & D. Löve = Aster modestus [ASTEMOD] Asteraceae (V:D)
Weberaster radulinus (Gray) A. & D. Löve = Aster radulinus [ASTERAD] Asteraceae (V:D)
Weissia Hedw. [WEISSIA$] Pottiaceae (B:M)
Weissia andrewsii Bartr. = Weissia controversa [WEISCON] Pottiaceae (B:M)
Weissia brachycarpa auct. = Weissia hedwigii Crum [WEISHED] Pottiaceae (B:M)
Weissia controversa Hedw. [WEISCON] Pottiaceae (B:M)
Weissia controversa var. *longiseta* (Lesq. & James) Crum, Steere & Anderson = Weissia controversa [WEISCON] Pottiaceae (B:M)
Weissia controversa var. *wolfii* (Lesq. & James) Crum, Steere & Anderson = Weissia controversa [WEISCON] Pottiaceae (B:M)
Weissia curvifolia (Wahlenb.) Lindb. in Braithw. = Ulota curvifolia [ULOTCUR] Orthotrichaceae (B:M)
Weissia curvirostris C. Müll. = Hymenostylium recurvirostre [HYMEREC] Pottiaceae (B:M)
Weissia hedwigii Crum [WEISHED] Pottiaceae (B:M)
Weissia maritima (C. Müll. & Kindb.) Britt. = Ulota phyllantha [ULOTPHY] Orthotrichaceae (B:M)
Weissia microstoma (Hedw.) C. Müll. = Weissia hedwigii [WEISHED] Pottiaceae (B:M)
Weissia rupestris (Schwaegr.) C. Müll. = Gymnostomum aeruginosum [GYMNAER] Pottiaceae (B:M)
Weissia viridula Hedw. ex Brid. = Weissia controversa [WEISCON] Pottiaceae (B:M)
Weissia viridula var. *australis* Aust. = Weissia controversa [WEISCON] Pottiaceae (B:M)
Weissia viridula var. *wolfii* Lesq. & James = Weissia controversa [WEISCON] Pottiaceae (B:M)
Weissia wolfii Lesq. & James = Weissia controversa [WEISCON] Pottiaceae (B:M)
Wijkia Crum [WIJKIA$] Sematophyllaceae (B:M)
Wijkia carlottae (Schof.) Crum [WIJKCAR] Sematophyllaceae (B:M)
Woessia fusarioides D. Hawksw. = Bacidina arnoldiana [BACIARN] Lecanoraceae (L)
Wolffia Horkel ex Schleid. [WOLFFIA$] Lemnaceae (V:M)
Wolffia borealis (Engelm. ex Hegelm.) Landolt ex Landolt & Wildi [WOLFBOR] Lemnaceae (V:M)
Wolffia columbiana Karst. [WOLFCOL] Lemnaceae (V:M)
Wolffia punctata auct. = Wolffia borealis [WOLFBOR] Lemnaceae (V:M)
Woodsia R. Br. [WOODSIA$] Dryopteridaceae (V:F)
Woodsia alpina (Bolton) S.F. Gray [WOODALP] Dryopteridaceae (V:F)
Woodsia alpina var. *bellii* Lawson = Woodsia alpina [WOODALP] Dryopteridaceae (V:F)
Woodsia glabella R. Br. ex Richards. [WOODGLA] Dryopteridaceae (V:F)
Woodsia glabella var. *bellii* (Lawson) Lawson = Woodsia alpina [WOODALP] Dryopteridaceae (V:F)
Woodsia ilvensis (L.) R. Br. [WOODILV] Dryopteridaceae (V:F)
Woodsia oregana D.C. Eat. [WOODORE] Dryopteridaceae (V:F)
Woodsia scopulina D.C. Eat. [WOODSCO] Dryopteridaceae (V:F)
Woodwardia Sm. [WOODWAR$] Blechnaceae (V:F)
Woodwardia chamissoi Brack. = Woodwardia fimbriata [WOODFIM] Blechnaceae (V:F)
Woodwardia fimbriata Sm. [WOODFIM] Blechnaceae (V:F)

X

Xanthium L. [XANTHIU$] Asteraceae (V:D)
Xanthium acerosum Greene = Xanthium strumarium var. canadense [XANTSTR1] Asteraceae (V:D)
Xanthium californicum Greene = Xanthium strumarium var. canadense [XANTSTR1] Asteraceae (V:D)
Xanthium californicum var. *rotundifolium* Widder = Xanthium strumarium var. canadense [XANTSTR1] Asteraceae (V:D)
Xanthium campestre Greene = Xanthium strumarium var. canadense [XANTSTR1] Asteraceae (V:D)
Xanthium canadense P. Mill. = Xanthium strumarium var. canadense [XANTSTR1] Asteraceae (V:D)
Xanthium cavanillesii Schouw = Xanthium strumarium var. canadense [XANTSTR1] Asteraceae (V:D)
Xanthium cenchroides Millsp. & Sherff = Xanthium strumarium var. canadense [XANTSTR1] Asteraceae (V:D)
Xanthium commune Britt. = Xanthium strumarium var. canadense [XANTSTR1] Asteraceae (V:D)
Xanthium echinatum Murr. = Xanthium strumarium var. canadense [XANTSTR1] Asteraceae (V:D)
Xanthium glanduliferum Greene = Xanthium strumarium var. canadense [XANTSTR1] Asteraceae (V:D)
Xanthium italicum Moretti = Xanthium strumarium var. canadense [XANTSTR1] Asteraceae (V:D)
Xanthium macounii Britt. = Xanthium strumarium var. canadense [XANTSTR1] Asteraceae (V:D)
Xanthium oligacanthum Piper = Xanthium strumarium var. canadense [XANTSTR1] Asteraceae (V:D)
Xanthium oviforme Wallr. = Xanthium strumarium var. canadense [XANTSTR1] Asteraceae (V:D)
Xanthium pensylvanicum Wallr. = Xanthium strumarium var. canadense [XANTSTR1] Asteraceae (V:D)
Xanthium saccharatum Wallr. = Xanthium strumarium var. canadense [XANTSTR1] Asteraceae (V:D)
Xanthium speciosum Kearney = Xanthium strumarium var. canadense [XANTSTR1] Asteraceae (V:D)
*__Xanthium strumarium__ L. [XANTSTR] Asteraceae (V:D)
Xanthium strumarium ssp. *italicum* (Moretti) D. Löve = Xanthium strumarium var. canadense [XANTSTR1] Asteraceae (V:D)
*__Xanthium strumarium__ var. **canadense** (P. Mill.) Torr. & Gray [XANTSTR1] Asteraceae (V:D)
Xanthium strumarium var. *oviforme* (Wallr.) M.E. Peck = Xanthium strumarium var. canadense [XANTSTR1] Asteraceae (V:D)
Xanthium strumarium var. *pensylvanicum* (Wallr.) M.E. Peck = Xanthium strumarium var. canadense [XANTSTR1] Asteraceae (V:D)
Xanthium varians Greene = Xanthium strumarium var. canadense [XANTSTR1] Asteraceae (V:D)
Xanthoparmelia (Vainio) Hale [XANTHOP$] Parmeliaceae (L)
Xanthoparmelia angustiphylla (Gyelnik) Hale [XANTANG] Parmeliaceae (L)
Xanthoparmelia camtschadalis (Ach.) Hale [XANTCAM] Parmeliaceae (L)
Xanthoparmelia centrifuga (L.) Hale = Arctoparmelia centrifuga [ARCTCEN] Parmeliaceae (L)
Xanthoparmelia coloradoensis (Gyelnik) Hale [XANTCOL] Parmeliaceae (L)
Xanthoparmelia cumberlandia (Gyelnik) Hale [XANTCUM] Parmeliaceae (L)
Xanthoparmelia hypopsila auct. = Xanthoparmelia angustiphylla [XANTANG] Parmeliaceae (L)

Xanthoparmelia incurva (Pers.) Hale = Arctoparmelia incurva [ARCTINC] Parmeliaceae (L)
Xanthoparmelia lineola (E.C. Berry) Hale [XANTLIN] Parmeliaceae (L)
Xanthoparmelia mexicana (Gyelnik) Hale [XANTMEX] Parmeliaceae (L)
Xanthoparmelia mougeotii (Schaerer) Hale [XANTMOU] Parmeliaceae (L)
Xanthoparmelia neoconspersa (Gyelnik) Hale [XANTNEO] Parmeliaceae (L)
Xanthoparmelia planilobata (Gyelnik) Hale [XANTPLA] Parmeliaceae (L)
Xanthoparmelia plittii (Gyelnk) Hale [XANTPLI] Parmeliaceae (L)
Xanthoparmelia separata (Th. Fr.) Hale = Arctoparmelia separata [ARCTSEP] Parmeliaceae (L)
Xanthoparmelia somloënsis (Gyelnik) Hale [XANTSOM] Parmeliaceae (L)
Xanthoparmelia subcentrifuga (Oksner) Hale = Arctoparmelia subcentrifuga [ARCTSUB] Parmeliaceae (L)
Xanthoparmelia taractica auct. = Xanthoparmelia coloradoensis [XANTCOL] Parmeliaceae (L)
Xanthoparmelia tasmanica (Hook. f. & Taylor) Hale [XANTTAS] Parmeliaceae (L)
Xanthoparmelia wyomingica (Gyelnik) Hale [XANTWYO] Parmeliaceae (L)
Xanthoria (Fr.) Th. Fr. [XANTHOR$] Teloschistaceae (L)
Xanthoria candelaria (L.) Th. Fr. [XANTCAN] Teloschistaceae (L)
Xanthoria elegans (Link) Th. Fr. [XANTELE] Teloschistaceae (L)
Xanthoria fallax (Hepp) Arnold [XANTFAL] Teloschistaceae (L)
Xanthoria fulva (Hoffm.) Poelt & Petutschnig [XANTFUL] Teloschistaceae (L)
Xanthoria oregana Gyelnik = Xanthoria candelaria [XANTCAN] Teloschistaceae (L)
Xanthoria parietina (L.) Th. Fr. [XANTPAR] Teloschistaceae (L)
Xanthoria polycarpa (Hoffm.) Rieber [XANTPOL] Teloschistaceae (L)
Xanthoria ramulosa (Tuck.) Herre [XANTRAM] Teloschistaceae (L)
Xanthoria sorediata (Vainio) Poelt [XANTSOR] Teloschistaceae (L)
Xanthoxalis bushii Small = Oxalis stricta [OXALSTR] Oxalidaceae (V:D)
Xanthoxalis coloradensis (Rydb.) Rydb. = Oxalis stricta [OXALSTR] Oxalidaceae (V:D)
Xanthoxalis corniculata (L.) Small = Oxalis corniculata [OXALCOR] Oxalidaceae (V:D)
Xanthoxalis corniculata var. *atropurpurea* (Planch.) Moldenke = Oxalis corniculata [OXALCOR] Oxalidaceae (V:D)
Xanthoxalis cymosa (Small) Small = Oxalis stricta [OXALSTR] Oxalidaceae (V:D)
Xanthoxalis dillenii var. *piletocarpa* (Wieg.) Holub = Oxalis stricta [OXALSTR] Oxalidaceae (V:D)
Xanthoxalis interior Small = Oxalis stricta [OXALSTR] Oxalidaceae (V:D)
Xanthoxalis langloisii Small = Oxalis corniculata [OXALCOR] Oxalidaceae (V:D)
Xanthoxalis repens (Thunb.) Moldenke = Oxalis corniculata [OXALCOR] Oxalidaceae (V:D)
Xanthoxalis rufa (Small) Small = Oxalis stricta [OXALSTR] Oxalidaceae (V:D)
Xanthoxalis stricta (L.) Small = Oxalis stricta [OXALSTR] Oxalidaceae (V:D)
Xanthoxalis stricta var. *piletocarpa* (Wieg.) Moldenke = Oxalis stricta [OXALSTR] Oxalidaceae (V:D)
Xerophyllum Michx. [XEROPHY$] Liliaceae (V:M)
Xerophyllum tenax (Pursh) Nutt. [XEROTEN] Liliaceae (V:M)
Xylographa (Fr.) Fr. [XYLOGRA$] Agyriaceae (L)
Xylographa abietina (Pers.) Zahlbr. = Xylographa parallela [XYLOPAR] Agyriaceae (L)
Xylographa hians Tuck. [XYLOHIA] Agyriaceae (L)
Xylographa micrographa G. Merr. = Xylographa hians [XYLOHIA] Agyriaceae (L)
Xylographa opegraphella Nyl. ex Rothr. [XYLOOPE] Agyriaceae (L)
Xylographa parallela (Ach.) Behlen & Desberger [XYLOPAR] Agyriaceae (L)
Xylographa spilomatica (Anzi) Th. Fr. = Xylographa vitiligo [XYLOVIT] Agyriaceae (L)
Xylographa vitiligo (Ach.) J.R. Laundon [XYLOVIT] Agyriaceae (L)
Xylosteum involucratum (Banks ex Spreng.) Richards. = Lonicera involucrata [LONIINV] Caprifoliaceae (V:D)

Y

Yabea K.-Pol. [YABEA$] Apiaceae (V:D)
Yabea microcarpa (Hook. & Arn.) K.-Pol. [YABEMIC] Apiaceae (V:D)

Z

Zahlbrucknerella Herre [ZAHLBRU$] Lichinaceae (L)
Zahlbrucknerella calcarea (Herre) Herre [ZAHLCAL] Lichinaceae (L)
Zannichellia L. [ZANNICH$] Zannichelliaceae (V:M)
Zannichellia major (Hartman) Boenn. ex Reichenb. = Zannichellia palustris [ZANNPAL] Zannichelliaceae (V:M)
Zannichellia palustris L. [ZANNPAL] Zannichelliaceae (V:M)
Zannichellia palustris var. *major* (Hartman) W.D.J. Koch = Zannichellia palustris [ZANNPAL] Zannichelliaceae (V:M)
Zannichellia palustris var. *stenophylla* Aschers. & Graebn. = Zannichellia palustris [ZANNPAL] Zannichelliaceae (V:M)
Zeora cerina var. *chlorina* Flotow = Caloplaca chlorina [CALOCHL] Teloschistaceae (L)
Zieria demissa (Hook.) Schimp. = Plagiobryum demissum [PLAGDEM] Bryaceae (B:M)
Zieria julacea Schimp. = Plagiobryum zieri [PLAGZIE] Bryaceae (B:M)
Zieria lapponicus (Hedw.) B.S.G. = Amphidium lapponicum [AMPHLAP] Orthotrichaceae (B:M)
Zieria mougeotii B.S.G. = Amphidium mougeotii [AMPHMOU] Orthotrichaceae (B:M)
Zigadenus Michx. [ZIGADEN$] Liliaceae (V:M)
Zigadenus elegans Pursh [ZIGAELE] Liliaceae (V:M)
Zigadenus gramineus Rydb. = Zigadenus venenosus var. gramineus [ZIGAVEN1] Liliaceae (V:M)
Zigadenus intermedius Rydb. = Zigadenus venenosus var. gramineus [ZIGAVEN1] Liliaceae (V:M)
Zigadenus venenosus S. Wats. [ZIGAVEN] Liliaceae (V:M)
Zigadenus venenosus var. **gramineus** (Rydb.) Walsh ex M.E. Peck [ZIGAVEN1] Liliaceae (V:M)
Zigadenus venenosus var. **venenosus** [ZIGAVEN0] Liliaceae (V:M)

Zizia W.D.J. Koch [ZIZIA$] Apiaceae (V:D)
Zizia aptera (Gray) Fern. [ZIZIAPT] Apiaceae (V:D)
Zizia aptera var. *occidentalis* Fern. = Zizia aptera [ZIZIAPT] Apiaceae (V:D)
Zizia cordata W.D.J. Koch ex DC. = Zizia aptera [ZIZIAPT] Apiaceae (V:D)
Zostera L. [ZOSTERA$] Zosteraceae (V:M)
Zostera americana den Hartog = Zostera japonica [ZOSTJAP] Zosteraceae (V:M)
***Zostera japonica** Aschers. & Graebn. [ZOSTJAP] Zosteraceae (V:M)
Zostera marina L. [ZOSTMAR] Zosteraceae (V:M)
Zostera marina var. **latifolia** Morong [ZOSTMAR1] Zosteraceae (V:M)
Zostera marina var. **stenophylla** Aschers. & Graebn. [ZOSTMAR2] Zosteraceae (V:M)
Zostera nana Roth = Zostera japonica [ZOSTJAP] Zosteraceae (V:M)
Zostera noltii auct. = Zostera japonica [ZOSTJAP] Zosteraceae (V:M)
Zostera oregana S. Wats. = Zostera marina [ZOSTMAR] Zosteraceae (V:M)
Zostera stenophylla Raf. = Zostera marina [ZOSTMAR] Zosteraceae (V:M)
Zwackhiomyces Grube & Hafellner [ZWACKHI$] Arthopyreniaceae (L)
Zwackhiomyces coepulonus (Norman) Grube & R. Sant. [ZWACCOE] Arthopyreniaceae (L)
Zygadenus Endl. = Zigadenus [ZIGADEN$] Liliaceae (V:M)
Zygodon Hook. & Tayl. [ZYGODON$] Orthotrichaceae (B:M)
Zygodon baumgartneri Malta = Zygodon viridissimus var. rupestris [ZYGOVIR2] Orthotrichaceae (B:M)
Zygodon gracilis Wils. in Berk. [ZYGOGRA] Orthotrichaceae (B:M)
Zygodon reinwardtii (Hornsch. in Reinw. & Hornsch.) A. Br. in Bruch & Schimp. [ZYGOREI] Orthotrichaceae (B:M)
Zygodon reinwardtii var. *subintegrifolius* Malta = Zygodon reinwardtii [ZYGOREI] Orthotrichaceae (B:M)
Zygodon rufotomentosus Britt. ex Malta = Zygodon viridissimus var. rupestris [ZYGOVIR2] Orthotrichaceae (B:M)
Zygodon rupestris (Lindb. ex Hartm.) Lindb. ex Britt. = Zygodon viridissimus var. rupestris [ZYGOVIR2] Orthotrichaceae (B:M)
Zygodon viridissimus (Dicks.) Brid. [ZYGOVIR] Orthotrichaceae (B:M)
Zygodon viridissimus var. *dentatus* (Breidl.) Limpr. = Zygodon viridissimus var. viridissimus [ZYGOVIR0] Orthotrichaceae (B:M)
Zygodon viridissimus var. *rufotomentosus* (Britt. ex Malta) Grout = Zygodon viridissimus var. rupestris [ZYGOVIR2] Orthotrichaceae (B:M)
Zygodon viridissimus var. **rupestris** Lindb. ex Hartm. [ZYGOVIR2] Orthotrichaceae (B:M)
Zygodon viridissimus var. **viridissimus** [ZYGOVIR0] Orthotrichaceae (B:M)
Zygodon viridissimus var. *vulgaris* Malta = Zygodon viridissimus var. rupestris [ZYGOVIR2] Orthotrichaceae (B:M)
Zygodon vulgaris (Malta) Nyh. = Zygodon viridissimus var. rupestris [ZYGOVIR2] Orthotrichaceae (B:M)

Part 3

Alphabetical List of Common Names

A

abbreviated bluegrass = Poa abbreviata ssp. pattersonii (V:M)
abraded brown = Melanelia fuliginosa (L)
abraded brown = Melanelia subaurifera (L)
absinthe = Artemisia absinthium (V:D)
acute-leaved peat moss = Sphagnum capillifolium (B:M)
adder's-tongue, northern → northern adder's-tongue
agoseris, annual → annual agoseris
agoseris, large-flowered → large-flowered agoseris
agoseris, orange → orange agoseris
agoseris, pale → pale agoseris
agoseris, pink → pink agoseris
agoseris, short-beaked → short-beaked agoseris
agrimony = Agrimonia striata (V:D)
agrimony, common → common agrimony
agrimony, grooved → grooved agrimony
Alaska alkaligrass = Puccinellia nutkaensis (V:M)
Alaska bentgrass = Agrostis aequivalvis (V:M)
Alaska brome = Bromus sitchensis (V:M)
Alaska cedar = Chamaecyparis nootkatensis (V:G)
Alaska clubmoss = Lycopodium sitchense (V:F)
Alaska cypress = Chamaecyparis nootkatensis (V:G)
Alaska draba = Draba stenoloba (V:D)
Alaska holly fern = Polystichum setigerum (V:F)
Alaska oniongrass = Melica subulata (V:M)
Alaska orache = Atriplex alaskensis (V:D)
Alaska paper birch = Betula neoalaskana (V:D)
Alaska × paper birch hybrid = Betula × winteri (V:D)
Alaska plantain = Plantago macrocarpa (V:D)
Alaska rein orchid = Piperia unalascensis (V:M)
Alaska sagebrush = Artemisia alaskana (V:D)
Alaska saxifrage = Saxifraga ferruginea (V:D)
Alaska spring-beauty = Claytonia sarmentosa (V:D)
Alaska violet = Viola langsdorfii (V:D)
Alaska whitlow-grass = Draba stenoloba (V:D)
Alaska wildrye = Elymus alaskanus ssp. latiglumis (V:M)
Alaska wildrye = Elymus alaskanus (V:M)
Alaska willow = Salix alaxensis (V:D)
Alaskan harebell = Campanula aurita (V:D)
Alberta penstemon = Penstemon albertinus (V:D)
alder-leaved buckthorn = Rhamnus alnifolia (V:D)
alder turtle = Hypocenomyce leucococca (L)
alder, gray → gray alder
alder, green → green alder
alder, mountain → mountain alder
alder, red → red alder
alder, Sitka → Sitka alder
Aleutian mugwort = Artemisia tilesii (V:D)
Alexanders, heart-leaved → heart-leaved Alexanders
alfalfa = Medicago sativa (V:D)
alfalfa dodder = Cuscuta approximata (V:D)
alkali cordgrass = Spartina gracilis (V:M)
alkali dodder = Cuscuta salina (V:D)
alkali-marsh butterweed = Senecio hydrophilus (V:D)
alkali muhly = Muhlenbergia asperifolia (V:M)
alkali plantain = Plantago eriopoda (V:D)
alkaligrass, Alaska → Alaska alkaligrass
alkaligrass, European → European alkaligrass
alkaligrass, inland → inland alkaligrass
alkaligrass, Nuttall's → Nuttall's alkaligrass
alkaligrass, Pacific → Pacific alkaligrass
alkaligrass, weeping → weeping alkaligrass
alkanet bugloss = Anchusa officinalis (V:D)
all-seed, four-leaved → four-leaved all-seed
Allegheny blackberry = Rubus allegheniensis (V:D)
alp lily = Lloydia serotina (V:M)
alpine anemone = Anemone drummondii (V:D)
alpine arnica = Arnica angustifolia (V:D)
alpine aster = Aster alpinus (V:D)
alpine-azalea = Loiseleuria procumbens (V:D)
alpine bearberry = Arctostaphylos alpina (V:D)
alpine bistort = Polygonum viviparum (V:D)
alpine bitter-cress = Cardamine bellidifolia (V:D)
alpine bluegrass = Poa alpina (V:M)
alpine bog swertia = Swertia perennis (V:D)
alpine buckwheat = Eriogonum pyrolifolium var. coryphaeum (V:D)
alpine buckwheat = Eriogonum pyrolifolium (V:D)
alpine chickweed = Cerastium beeringianum (V:D)
alpine cliff fern = Woodsia alpina (V:F)
alpine clubmoss = Lycopodium alpinum (V:F)
alpine draba = Draba alpina (V:D)
alpine enchanter's-nightshade = Circaea alpina (V:D)
alpine fescue = Festuca brachyphylla (V:M)
alpine fir = Abies lasiocarpa (V:G)
alpine golden wild buckwheat = Eriogonum flavum (V:D)
alpine goldenrod = Solidago multiradiata (V:D)
alpine hedysarum = Hedysarum alpinum (V:D)
alpine larch = Larix lyallii (V:G)
alpine laurel = Kalmia microphylla (V:D)
alpine lewisia = Lewisia pygmaea (V:D)
alpine meadow butterweed = Senecio cymbalarioides (V:D)
alpine meadowrue = Thalictrum alpinum (V:D)
alpine milk-vetch = Astragalus alpinus (V:D)
alpine paintbrush = Castilleja rhexifolia (V:D)
alpine paw = Nephroma expallidum (L)
alpine prickly gooseberry = Ribes montigenum (V:D)
alpine rock-cress = Draba alpina (V:D)
alpine rush = Juncus alpinoarticulatus (V:M)
alpine sandwort = Minuartia obtusiloba (V:D)
alpine saxifrage = Saxifraga nivalis (V:D)
alpine smelowskia = Smelowskia calycina (V:D)
alpine sorrel = Rumex paucifolius (V:D)
alpine speedwell = Veronica wormskjoldii (V:D)
alpine spring-beauty = Claytonia megarhiza (V:D)
alpine sweet-vetch = Hedysarum alpinum (V:D)
alpine sweetgrass = Hierochloe alpina (V:M)
alpine timothy = Phleum alpinum (V:M)
alpine white marsh-marigold = Caltha leptosepala (V:D)
alpine whitlow-grass = Draba alpina (V:D)
alpine willowherb = Epilobium anagallidifolium (V:D)
alpine wintergreen = Gaultheria humifusa (V:D)
alsike clover = Trifolium hybridum (V:D)
Altai fescue = Festuca altaica (V:M)
alumroot, meadow → meadow alumroot
alumroot, Richardson's → Richardson's alumroot
alumroot, round-leaved → round-leaved alumroot
alumroot, small-flowered → small-flowered alumroot
alumroot, smooth → smooth alumroot
alyssum-leaved phlox = Phlox alyssifolia (V:D)
alyssum, desert → desert alyssum
alyssum, hoary → hoary alyssum
alyssum, pale → pale alyssum
alyssum, sweet → sweet alyssum
alyssum, wall → wall alyssum
amabilis fir = Abies amabilis (V:G)
amaranth, pigweed → pigweed amaranth
amaranth, Powell's → Powell's amaranth
American bistort = Polygonum bistortoides (V:D)
American brooklime = Veronica americana (V:D)

American bulrush = Scirpus americanus (V:M)
American bush-cranberry = Viburnum opulus (V:D)
American chamaerhodos = Chamaerhodos erecta ssp. nuttallii (V:D)
American dragonhead = Dracocephalum parviflorum (V:D)
American dwarf mistletoe = Arceuthobium americanum (V:D)
American germander = Teucrium canadense (V:D)
American glasswort = Salicornia virginica (V:D)
American glehnia = Glehnia littoralis ssp. leiocarpa (V:D)
American licorice = Glycyrrhiza lepidota (V:D)
American mannagrass = Glyceria grandis (V:M)
American milk-vetch = Astragalus americanus (V:D)
American rattlepod = Astragalus americanus (V:D)
American red raspberry = Rubus idaeus (V:D)
American sawwort = Saussurea americana (V:D)
American searocket = Cakile edentula (V:D)
American silvertop = Glehnia littoralis (V:D)
American sloughgrass = Beckmannia syzigachne (V:M)
American starwort = Stellaria americana (V:D)
American sweet-flag = Acorus americanus (V:M)
American tapegrass = Vallisneria americana (V:M)
American thorough-wax = Bupleurum americanum (V:D)
American vetch = Vicia americana (V:D)
American wild carrot = Daucus pusillus (V:D)
American winter cress = Barbarea orthoceras (V:D)
amphibious tarpaper = Collema glebulentum (L)
analogue sedge = Carex simulata (V:M)
Andean evening-primrose = Camissonia andina (V:D)
Anderson's holly fern = Polystichum andersonii (V:F)
androsace buckwheat = Eriogonum androsaceum (V:D)
anemone, alpine → alpine anemone
anemone, Canada → Canada anemone
anemone, cut-leaved → cut-leaved anemone
anemone, Drummond's → Drummond's anemone
anemone, long-fruited → long-fruited anemone
anemone, long-headed → long-headed anemone
anemone, Lyall's → Lyall's anemone
anemone, narcissus → narcissus anemone
anemone, northern → northern anemone
anemone, Pacific → Pacific anemone
anemone, Piper's → Piper's anemone
anemone, small-flowered → small-flowered anemone
anemone, small wood → small wood anemone
anemone, yellow → yellow anemone
angelica, Dawson's → Dawson's angelica
angelica, kneeling → kneeling angelica
angelica, seacoast → seacoast angelica
angelica, sharptooth → sharptooth angelica
angled bitter-cress = Cardamine angulata (V:D)
annual agoseris = Agoseris heterophylla (V:D)
annual bluegrass = Poa annua (V:M)
annual daisy = Erigeron annuus (V:D)
annual fleabane = Erigeron annuus (V:D)
annual goat-chicory = Agoseris heterophylla (V:D)
annual hairgrass = Deschampsia danthonioides (V:M)
annual hawksbeard = Crepis tectorum (V:D)
annual Jacob's-ladder = Polemonium micranthum (V:D)
annual knawel = Scleranthus annuus (V:D)
annual paintbrush = Castilleja exilis (V:D)
annual ragweed = Ambrosia artemisiifolia (V:D)
annual sow-thistle = Sonchus oleraceus (V:D)
annual wheatgrass = Eremopyrum triticeum (V:M)
annual whitlow-grass = Draba nemorosa (V:D)
antelope-brush = Purshia tridentata (V:D)
antlered perfume = Evernia prunastri (L)
antlered vinyl = Leptogium corniculatum (L)
apargidium = Microseris borealis (V:D)
apetalous campion = Silene uralensis ssp. attenuata (V:D)
apetalous catchfly = Silene uralensis (V:D)
apple moss = Bartramia pomiformis (B:M)
apple pelt = Peltigera malacea (L)
apple, cultivated → cultivated apple
apple, Pacific crab → Pacific crab apple
applemint = Mentha suaveolens (V:D)
appressed vinyl = Leptogium subtile (L)
aquatic apple moss = Philonotis fontana (B:M)
arbutus = Arbutus menziesii (V:D)
arctic aster = Aster sibiricus (V:D)
arctic bladderpod = Lesquerella arctica (V:D)
arctic bluegrass = Poa arctica (V:M)
arctic buttercup = Ranunculus hyperboreus (V:D)
arctic campion = Silene involucrata (V:D)
arctic cotton-grass = Eriophorum callitrix (V:M)
arctic daisy = Erigeron humilis (V:D)
arctic dock = Rumex arcticus (V:D)
arctic harebell = Campanula uniflora (V:D)
arctic kidney lichen = Nephroma arcticum (L)
arctic Langsdorf's lousewort = Pedicularis langsdorfii ssp. arctica (V:D)
arctic locoweed = Oxytropis arctica (V:D)
arctic lupine = Lupinus arcticus (V:D)
arctic moon = Sticta arctica (L)
arctic oxytrope = Oxytropis arctica (V:D)
arctic pearlwort = Sagina saginoides (V:D)
arctic plantain = Plantago canescens (V:D)
arctic poppy = Papaver lapponicum ssp. occidentale (V:D)
arctic rag = Asahinea scholanderi (L)
arctic rush = Juncus arcticus (V:M)
arctic rush = Juncus balticus (V:M)
arctic tumbleweed = Masonhalea richardsonii (L)
arctic willow = Salix arctica (V:D)
arctic wintergreen = Pyrola grandiflora (V:D)
arctic woodrush = Luzula arctica (V:M)
Arkansas rose = Rosa arkansana (V:D)
arnica, alpine → alpine arnica
arnica, hairy → hairy arnica
arnica, heart-leaved → heart-leaved arnica
arnica, high mountain → high mountain arnica
arnica, leafy → leafy arnica
arnica, meadow → meadow arnica
arnica, mountain → mountain arnica
arnica, northern → northern arnica
arnica, orange → orange arnica
arnica, Parry's → Parry's arnica
arnica, purple → purple arnica
arnica, Rydberg's → Rydberg's arnica
arnica, slender → slender arnica
arnica, spear-leaved → spear-leaved arnica
arnica, streambank → streambank arnica
arrow-grass, graceful → graceful arrow-grass
arrow-grass, marsh → marsh arrow-grass
arrow-grass, seaside → seaside arrow-grass
arrow-leaved balsamroot = Balsamorhiza sagittata (V:D)
arrow-leaved coltsfoot = Petasites sagittatus (V:D)
arrow-leaved groundsel = Senecio triangularis (V:D)
arrow-leaved ragwort = Senecio triangularis (V:D)
arrowhead = Sagittaria latifolia (V:M)
arrowhead, arum-leaved → arum-leaved arrowhead
arum-leaved arrowhead = Sagittaria cuneata (V:M)
arum, water → water arum
ascending purple milk-vetch = Astragalus adsurgens var. robustior (V:D)
ash-coloured ground liverwort = Peltigera canina (L)
ashen rocktripe = Umbilicaria aprina (L)
ashen stipplescale = Catapyrenium daedaleum (L)
ashen stipplescale = Catapyrenium cinereum (L)
Asian oak fern = Gymnocarpium jessoense (V:F)
asp-of-Jerusalem = Isatis tinctoria (V:D)
asparagus = Asparagus officinalis (V:M)
asparagus, garden → garden asparagus
aspen = Populus tremuloides (V:D)
aspen comma = Arthonia patellulata (L)
aspen moss = Pylaisiella polyantha (B:M)
aspen, trembling → trembling aspen
asphodel, common false → common false asphodel

asphodel, northern false → northern false asphodel
asphodel, sticky false → sticky false asphodel
aster, alpine → alpine aster
aster, arctic → arctic aster
aster, Douglas' → Douglas' aster
aster, Eaton's → Eaton's aster
aster, Engelmann's → Engelmann's aster
aster, fringed → fringed aster
aster, great northern → great northern aster
aster, hoary → hoary aster
aster, leafy → leafy aster
aster, Lindley's → Lindley's aster
aster, long-leaved → long-leaved aster
aster, marsh → marsh aster
aster, meadow → meadow aster
aster, Olympic Mountain → Olympic Mountain aster
aster, rough-leaved → rough-leaved aster
aster, rough white prairie → rough white prairie aster
aster, rush → rush aster
aster, showy → showy aster
aster, smooth blue → smooth blue aster
aster, western → western aster
aster, western mountain → western mountain aster
aster, white heath → white heath aster
aster, white-top → white-top aster
asterisk rocktripe = Umbilicaria angulata (L)
Athabasca willow = Salix athabascensis (V:D)
auburn lichen = Pannaria pezizoides (L)
Austin's knotweed = Polygonum douglasii ssp. austiniae (V:D)
Austrian draba = Draba fladnizensis (V:D)
Austrian whitlow-grass = Draba fladnizensis (V:D)
autumn hawkbit = Leontodon autumnalis (V:D)
autumn willow = Salix serissima (V:D)
avalanche lily = Erythronium grandiflorum (V:M)
avens, caltha-leaved → caltha-leaved avens
avens, large-leaved → large-leaved avens
avens, purple → purple avens
avens, Ross' → Ross' avens
avens, three-flowered → three-flowered avens
avens, water → water avens
avens, yellow → yellow avens
awl-fruited sedge = Carex stipata (V:M)
awl-shaped stump lichen = Cladonia coniocraea (L)
awlwort = Subularia aquatica (V:D)
awned haircap = Polytrichum piliferum (B:M)
awned haircap moss = Polytrichum piliferum (B:M)
awned sedge = Carex atherodes (V:M)
azalea, false → false azalea

B

baby brown = Melanelia trabeculata (L)
baby's breath = Gypsophila paniculata (V:D)
bachelor's-button = Centaurea cyanus (V:D)
badge moss = Plagiomnium insigne (B:M)
Baffin fescue = Festuca baffinensis (V:M)
Baffin's Bay draba = Draba corymbosa (V:D)
Baffin's Bay whitlow-grass = Draba corymbosa (V:D)
baked-apple berry = Rubus chamaemorus (V:D)
baldhip rose = Rosa gymnocarpa (V:D)
ball mustard = Neslia paniculata (V:D)
ballpoint rocktripe = Umbilicaria polyrrhiza (L)
balm, lemon → lemon balm
balsam fir = Abies balsemea (V:G)
balsam groundsel = Senecio pauperculus (V:D)
balsam poplar = Populus balsamifera (V:D)
balsam poplar = Populus balsamifera ssp. balsamifera (V:D)
balsam willow = Salix pyrifolia (V:D)
balsam, small → small balsam
balsamroot, arrow-leaved → arrow-leaved balsamroot
balsamroot, deltoid → deltoid balsamroot
Baltic rush = Juncus balticus (V:M)
banana water-lily = Nymphaea mexicana (V:D)
baneberry = Actaea rubra (V:D)
baneberry, red and white → red and white baneberry
Banff bluegrass = Poa laxa ssp. banffiana (V:M)
barberry, common → common barberry
Barclay's willow = Salix barclayi (V:D)
bare-stem desert-parsley = Lomatium nudicaule (V:D)
bare-stem lomatium = Lomatium nudicaule (V:D)
bare-stemmed mitrewort = Mitella nuda (V:D)
bark barnacle = Thelotrema lepadinum (L)
barley, common → common barley
barley, foxtail → foxtail barley
barley, meadow → meadow barley
barley, Mediterranean → Mediterranean barley
barley, seaside → seaside barley
barley, wall → wall barley
barnacle, bark → bark barnacle
barnyard-grass, large → large barnyard-grass
Barratt's willow = Salix barrattiana (V:D)
barren brome = Bromus sterilis (V:M)
barren fescue = Vulpia bromoides (V:M)
bassia, five-hooked → five-hooked bassia
bastard fumatory = Fumaria bastardii (V:D)
bastard toad-flax = Geocaulon lividum (V:D)
bastard toad-flax = Comandra umbellata (V:D)
bayberry, California → California bayberry
beach bindweed = Calystegia soldanella (V:D)
beach bluegrass = Poa confinis (V:M)
beach-carrot = Glehnia littoralis ssp. leiocarpa (V:D)
beach groundsel = Senecio pseudoarnica (V:D)
beach knotweed = Polygonum paronychia (V:D)
beach lovage = Ligusticum scoticum ssp. hultenii (V:D)
beach pea = Lathyrus japonicus (V:D)
beach sand-spurry = Spergularia macrotheca (V:D)
beachgrass, European → European beachgrass
beaded bone = Hypogymnia enteromorpha (L)
beaded bone = Hypogymnia apinnata (L)
beaded rosette = Physcia callosa (L)
beak-rush = Rhynchospora alba (V:M)
beak-rush, brown → brown beak-rush
beak-rush, white → white beak-rush
beak-sedge = Rhynchospora alba (V:M)
beaked hawksbeard = Crepis vesicaria (V:D)
beaked hazelnut = Corylus cornuta (V:D)
beaked sedge = Carex rostrata (V:M)
beaked spike-rush = Eleocharis rostellata (V:M)
beaked willow = Salix bebbiana (V:D)
bean, prairie golden → prairie golden bean
bear-grass = Xerophyllum tenax (V:M)
bear's-foot sanicle = Sanicula arctopoides (V:D)
bearberry, alpine → alpine bearberry
bearberry, common → common bearberry
bearberry, red → red bearberry
beard, blood-spattered → blood-spattered beard
beard, Methuselah's → Methuselah's beard
beard, old man's → old man's beard
beard, pitted → pitted beard
beard, pitted old man's → pitted old man's beard
beard, powdery → powdery beard

beard, powdery old man's → powdery old man's beard
beard, powered → powered beard
beard, scruffy → scruffy beard
beard, scruffy old man's → scruffy old man's beard
beard, shaggy old man's → shaggy old man's beard
beard, sugared → sugared beard
beard, sugary → sugary beard
bearded fescue = Festuca subulata (V:M)
bearded ryegrass = Lolium temulentum (V:M)
bearded sedge = Carex comosa (V:M)
beautiful cotton-grass = Eriophorum callitrix (V:M)
beautiful sedge = Carex concinna (V:M)
beauty, river → river beauty
Bebb's sedge = Carex bebbii (V:M)
Bebb's willow = Salix bebbiana (V:D)
Beckwith's milk-vetch = Astragalus beckwithii (V:D)
bedstraw, boreal → boreal bedstraw
bedstraw, Mexican → Mexican bedstraw
bedstraw, northern → northern bedstraw
bedstraw, rough → rough bedstraw
bedstraw, small → small bedstraw
bedstraw, sweet-scented → sweet-scented bedstraw
bedstraw, thin-leaved → thin-leaved bedstraw
bedstraw, white → white bedstraw
bedstraw, yellow → yellow bedstraw
bee-plant, Rocky Mountain → Rocky Mountain bee-plant
bee-plant, spider-flower → spider-flower bee-plant
bee-plant, stinking-clover → stinking-clover bee-plant
beech fern = Phegopteris connectilis (V:F)
beggarticks, common → common beggarticks
beggarticks, nodding → nodding beggarticks
beggarticks, tall → tall beggarticks
beggarticks, Vancouver Island → Vancouver Island beggarticks
bell, yellow → yellow bell
Bellard's kobresia = Kobresia myosuroides (V:M)
bellflower, creeping → creeping bellflower
bellflower, peach-leaved → peach-leaved bellflower
bellflower, Yukon → Yukon bellflower
bent-leaf moss = Rhytidiadelphus squarrosus (B:M)
bent sedge = Carex deflexa (V:M)
bentgrass, Alaska → Alaska bentgrass
bentgrass, colonial → colonial bentgrass
bentgrass, creeping → creeping bentgrass
bentgrass, dense silky → dense silky bentgrass
bentgrass, dune → dune bentgrass
bentgrass, hair → hair bentgrass
bentgrass, mountain → mountain bentgrass
bentgrass, northern → northern bentgrass
bentgrass, Oregon → Oregon bentgrass
bentgrass, small-leaved → small-leaved bentgrass
bentgrass, spike → spike bentgrass
bentgrass, thin → thin bentgrass
bentgrass, winter → winter bentgrass
bergamot, wild → wild bergamot
Bering chickweed = Cerastium beeringianum (V:D)
Bermuda grass = Cynodon dactylon (V:M)
berry, baked-apple → baked-apple berry
Bicknell's crane's-bill = Geranium bicknellii (V:D)
Bicknell's geranium = Geranium bicknellii (V:D)
bicolored flaxflower = Linanthus bicolor (V:D)
bicolored linanthus = Linanthus bicolor (V:D)
biennial campion = Silene csereii (V:D)
biennial cinquefoil = Potentilla biennis (V:D)
biennial sagewort = Artemisia biennis (V:D)
biennial wormwood = Artemisia biennis (V:D)
big-head rush = Juncus vaseyi (V:M)
big-leaved lupine = Lupinus polyphyllus (V:D)
big-leaved sandwort = Moehringia macrophylla (V:D)
big quaking grass = Briza maxima (V:M)
big red stem = Pleurozium schreberi (B:M)
big sagebrush = Artemisia tridentata (V:D)
Bigelow's sedge = Carex bigelowii (V:M)
bigleaf lupine = Lupinus burkei (V:D)
bigleaf maple = Acer macrophyllum (V:D)
bigleaf sedge = Carex amplifolia (V:M)
bigroot = Marah oreganus (V:D)
bilberry willow = Salix myrtillifolia (V:D)
bilberry, bog → bog bilberry
bilberry, dwarf → dwarf bilberry
bilberry, low → low bilberry
bindweed, beach → beach bindweed
bindweed, black → black bindweed
bindweed, field → field bindweed
bindweed, hedge → hedge bindweed
bipinnate cinquefoil = Potentilla bipinnatifida (V:D)
birch-leaved spirea = Spiraea betulifolia (V:D)
birch-leaved spirea = Spiraea betulifolia var. lucida (V:D)
birch, Alaska paper → Alaska paper birch
birch, Alaska × paper birch hybrid → Alaska × paper birch hybrid
birch, black → black birch
birch, bog → bog birch
birch, canoe → canoe birch
birch, European → European birch
birch, low → low birch
birch, mountain → mountain birch
birch, northwestern white → northwestern white birch
birch, paper → paper birch
birch, red → red birch
birch, river → river birch
birch, scrub glandular → scrub glandular birch
birch, silver → silver birch
birch, swamp → swamp birch
birch, water → water birch
birch, white → white birch
bird cherry = Prunus pensylvanica (V:D)
bird rape mustard = Brassica rapa (V:D)
bird's-beak lousewort = Pedicularis ornithorhyncha (V:D)
bird's-eye = Veronica chamaedrys (V:D)
bird's-eye pearlwort = Sagina procumbens (V:D)
bird's-eye speedwell = Veronica persica (V:D)
bird vetch = Vicia cracca (V:D)
birdfoot buttercup = Ranunculus pedatifidus (V:D)
birds-foot trefoil = Lotus corniculatus (V:D)
bishop's-cap = Mitella nuda (V:D)
bistort, alpine → alpine bistort
bistort, American → American bistort
bitter-brush = Purshia tridentata (V:D)
bitter cherry = Prunus emarginata (V:D)
bitter-cress, alpine → alpine bitter-cress
bitter-cress, angled → angled bitter-cress
bitter-cress, Brewer's → Brewer's bitter-cress
bitter-cress, cuckoo → cuckoo bitter-cress
bitter-cress, few-seeded → few-seeded bitter-cress
bitter-cress, hairy → hairy bitter-cress
bitter-cress, little western → little western bitter-cress
bitter-cress, Nuttall's → Nuttall's bitter-cress
bitter-cress, Pennsylvanian → Pennsylvanian bitter-cress
bitter-cress, small-flowered → small-flowered bitter-cress
bitter-cress, western → western bitter-cress
bitter dock = Rumex obtusifolius (V:D)
bitter winter cress = Barbarea vulgaris (V:D)
bitterroot = Lewisia rediviva (V:D)
bittersweet, European → European bittersweet
black alpine sedge = Carex nigricans (V:M)
black bindweed = Polygonum convolvulus (V:D)
black birch = Betula occidentalis (V:D)
black blueberry = Vaccinium membranaceum (V:D)
black brook moss = Racomitrium aciculare (B:M)
black cottonwood = Populus balsamifera ssp. trichocarpa (V:D)
black crottle = Parmelia omphalodes (L)
black elder = Sambucus racemosa ssp. pubens var. melanocarpa (V:D)

black elderberry = Sambucus racemosa ssp. pubens var. melanocarpa (V:D)
black-eyed rockbright = Rhizoplaca melanophthalma (L)
black-eyed rosette = Physcia stellaris (L)
black-eyed rosette = Physcia phaea (L)
black-eyed Susan = Rudbeckia hirta (V:D)
black fish hook moss = Campylopus atrovirens (B:M)
black-foot cladonia = Cladonia phyllophora (L)
black-foot cladonia = Cladonia gracilis ssp. turbinata (L)
black-fruited weissia = Catoscopium nigritum (B:M)
black-fruiting lichen = Nephroma resupinatum (L)
black gooseberry = Ribes lacustre (V:D)
black hawthorn = Crataegus douglasii (V:D)
black huckleberry = Vaccinium membranaceum (V:D)
black knotweed = Polygonum paronychia (V:D)
black lily = Fritillaria camschatcensis (V:M)
black locust = Robinia pseudoacacia (V:D)
black medic = Medicago lupulina (V:D)
black mustard = Brassica nigra (V:D)
black nightshade = Solanum americanum (V:D)
black poplar = Populus balsamifera (V:D)
black raspberry = Rubus leucodermis (V:D)
black rock moss = Andreaea rupestris (B:M)
black rocktripe = Umbilicaria polyphylla (L)
black sanicle = Sanicula marilandica (V:D)
black-scaled sedge = Carex atrosquama (V:M)
black sedge = Carex atratiformis (V:M)
black snake-root = Sanicula marilandica (V:D)
black spruce = Picea mariana (V:G)
black starburst = Melanelia disjuncta (L)
black swamp gooseberry = Ribes lacustre (V:D)
black-tipped groundsel = Senecio lugens (V:D)
black-tufted rock moss = Racomitrium aciculare (B:M)
black twinberry = Lonicera involucrata (V:D)
blackberry = Psora globifera (L)
blackberry, Allegheny → Allegheny blackberry
blackberry, cutleaf evergreen → cutleaf evergreen blackberry
blackberry, dwarf red → dwarf red blackberry
blackberry, evergreen → evergreen blackberry
blackberry, Himalayan → Himalayan blackberry
blackberry, Pacific trailing → Pacific trailing blackberry
blackberry, trailing → trailing blackberry
blackcap = Rubus leucodermis (V:D)
blackcurrant, northern → northern blackcurrant
blackened icelandmoss = Cetraria nigricans (L)
blackened thornbush = Kaernefeltia merrillii (L)
blackish locoweed = Oxytropis nigrescens (V:D)
blackish oxytrope = Oxytropis nigrescens (V:D)
blackthorn = Prunus spinosa (V:D)
bladder campion = Silene vulgaris (V:D)
bladderpod, arctic → arctic bladderpod
bladderpod, Columbia → Columbia bladderpod
bladderwort, flat-leaved → flat-leaved bladderwort
bladderwort, humped → humped bladderwort
bladderwort, lesser → lesser bladderwort
bladderwort, small → small bladderwort
Blake's knotweed = Polygonum achoreum (V:D)
Blandow's feather moss = Helodium blandowii (B:M)
blazing-star = Mentzelia laevicaulis (V:D)
blazing-star mentzelia = Mentzelia laevicaulis (V:D)
bleb tarpaper = Collema fuscovirens (L)
bleeding heart = Dicentra formosa (V:D)
blinks chickweed = Montia fontana (V:D)
blistered paw = Nephroma resupinatum (L)
blistered tarpaper = Collema furfuraceum (L)
blistered brown = Neofuscelia loxodes (L)
blistered brown = Neofuscelia subhosseana (L)
blistered rocktripe = Lasallia pensylvanica (L)
blistered rocktripe = Umbilicaria nylanderiana (L)
blistered rocktripe = Umbilicaria hyperborea (L)
blood-spattered beard = Usnea wirthii (L)
bloody heart = Mycoblastus sanguinarius (L)

blue-bead clintonia = Clintonia uniflora (V:M)
blue clematis = Clematis occidentalis (V:D)
blue columbine = Aquilegia brevistyla (V:D)
blue corydalis = Corydalis pauciflora (V:D)
blue-edged mouse = Fuscopannaria mediterranea (L)
blue elder = Sambucus cerulea (V:D)
blue elderberry = Sambucus cerulea (V:D)
blue-eyed-grass = Sisyrinchium idahoense var. macounii (V:M)
blue-eyed-grass = Sisyrinchium idahoense (V:M)
blue-eyed-grass, common → common blue-eyed-grass
blue-eyed-grass, Idaho → Idaho blue-eyed-grass
blue-eyed-grass, mountain → mountain blue-eyed-grass
blue-eyed-grass, northern → northern blue-eyed-grass
blue-eyed-grass, purple → purple blue-eyed-grass
blue-eyed-grass, shore → shore blue-eyed-grass
blue flax = Linum perenne (V:D)
blue forget-me-not = Myosotis stricta (V:D)
blue grama = Bouteloua gracilis (V:M)
blue-green willow = Salix glauca (V:D)
blue-grey blister lichen = Physcia caesia (L)
blue hackelia = Hackelia micrantha (V:D)
blue-head gily-flower = Gilia capitata (V:D)
blue-leaved cinquefoil = Potentilla diversifolia (V:D)
blue-leaved huckleberry = Vaccinium deliciosum (V:D)
blue-leaved strawberry = Fragaria virginiana ssp. glauca (V:D)
blue mustard = Chorispora tenella (V:D)
blue skullcap = Scutellaria lateriflora (V:D)
blue stickseed = Hackelia micrantha (V:D)
blue vervain = Verbena hastata (V:D)
blue vinyl = Leptogium cyanescens (L)
blue virgin's bower = Clematis occidentalis (V:D)
blue water speedwell = Veronica anagallis-aquatica (V:D)
blue wildrye = Elymus glaucus (V:M)
bluebell = Campanula rotundifolia (V:D)
bluebell, Scouler's → Scouler's bluebell
bluebells of Scotland = Campanula rotundifolia (V:D)
bluebells, long-flowered → long-flowered bluebells
bluebells, sea → sea bluebells
bluebells, Spanish → Spanish bluebells
bluebells, tall → tall bluebells
blueberry willow = Salix myrtillifolia (V:D)
blueberry, black → black blueberry
blueberry, bog → bog blueberry
blueberry, Cascade → Cascade blueberry
blueberry, common → common blueberry
blueberry, dwarf → dwarf blueberry
blueberry, oval-leaved → oval-leaved blueberry
blueberry, velvet-leaved → velvet-leaved blueberry
bluecup, common → common bluecup
bluefield gilia = Gilia capitata (V:D)
bluefly honeysuckle = Lonicera caerulea (V:D)
bluegrass, abbreviated → abbreviated bluegrass
bluegrass, alpine → alpine bluegrass
bluegrass, annual → annual bluegrass
bluegrass, arctic → arctic bluegrass
bluegrass, Banff → Banff bluegrass
bluegrass, beach → beach bluegrass
bluegrass, bog → bog bluegrass
bluegrass, bulbous → bulbous bluegrass
bluegrass, Canada → Canada bluegrass
bluegrass, coastal → coastal bluegrass
bluegrass, diploid annual → diploid annual bluegrass
bluegrass, eminent → eminent bluegrass
bluegrass, Fendler's → Fendler's bluegrass
bluegrass, fowl → fowl bluegrass
bluegrass, glaucous → glaucous bluegrass
bluegrass, inland → inland bluegrass
bluegrass, Kentucky → Kentucky bluegrass
bluegrass, lax-flowered → lax-flowered bluegrass
bluegrass, Letterman's → Letterman's bluegrass
bluegrass, Mt. Washington → Mt. Washington bluegrass

bluegrass, narrow-flowered → narrow-flowered bluegrass
bluegrass, Nevada → Nevada bluegrass
bluegrass, northern → northern bluegrass
bluegrass, polar → polar bluegrass
bluegrass, rough → rough bluegrass
bluegrass, roughstalk → roughstalk bluegrass
bluegrass, Sandberg's → Sandberg's bluegrass
bluegrass, timberline → timberline bluegrass
bluegrass, Trinius' → Trinius' bluegrass
bluegrass, weeping → weeping bluegrass
bluegrass, Wheeler's → Wheeler's bluegrass
bluegrass, wood → wood bluegrass
bluejoint = Calamagrostis canadensis (V:M)
bluejoint reedgrass = Calamagrostis canadensis (V:M)
bluestem, little → little bluestem
bluet, mountain → mountain bluet
blueweed = Echium vulgare (V:D)
bluff mouse = Massalongia carnosa (L)
blunt-fruited sweet-cicely = Osmorhiza depauperata (V:D)
blunt-leaved bog orchid = Platanthera obtusata (V:M)
blunt-leaved bristle moss = Orthotrichum obtusifolium (B:M)
blunt-leaved moss = Scleropodium obtusifolium (B:M)
blunt-leaved pondweed = Potamogeton obtusifolius (V:M)
blunt-leaved sandwort = Moehringia lateriflora (V:D)
blunt-leaved yellow cress = Rorippa curvipes var. integra (V:D)
blunt-leaved yellow cress = Rorippa curvipes (V:D)
blunt sedge = Carex obtusata (V:M)
blunt-sepaled starwort = Stellaria obtusa (V:D)
blush, sea → sea blush
blushing rocktripe = Umbilicaria virginis (L)
bog bilberry = Vaccinium uliginosum (V:D)
bog birch = Betula pumila (V:D)
bog birds-foot trefoil = Lotus pinnatus (V:D)
bog blueberry = Vaccinium uliginosum (V:D)
bog bluegrass = Poa leptocoma (V:M)
bog clubmoss = Lycopodiella inundata (V:F)
bog cranberry = Vaccinium oxycoccos (V:D)
bog cranberry = Vaccinium vitis-idaea (V:D)
bog haircap = Polytrichum strictum (B:M)
bog-laurel, western → western bog-laurel
bog loosestrife = Lysimachia terrestris (V:D)
bog muhly = Muhlenbergia glomerata (V:M)
bog-rosemary = Andromeda polifolia (V:D)
bog rush = Juncus stygius (V:M)
bog sedge = Carex magellanica ssp. irrigua (V:M)
bog St. John's-wort = Hypericum anagalloides (V:D)
bog starwort = Stellaria alsine (V:D)
bog violet = Viola nephrophylla (V:D)
bog willow = Salix pedicellaris (V:D)
bogbean = Menyanthes trifoliata (V:D)
Bolander's quillwort = Isoetes bolanderi (V:F)
Bolander's rush = Juncus bolanderi (V:M)
Bolander's sedge = Carex bolanderi (V:M)
bone, beaded → beaded bone
bone, deflated → deflated bone
bone, dog → dog bone
bone, forking → forking bone
bone, heath → heath bone
bone, hooded → hooded bone
bone, lattice → lattice bone
bone, powdered → powdered bone
bone, puckered → puckered bone
bone, seaside → seaside bone
bone, tickertape → tickertape bone
Booth's willow = Salix boothii (V:D)
borage, common → common borage
bordered frost = Physconia enteroxantha (L)
bordered frost = Physconia perisidiosa (L)
boreal bedstraw = Galium kamtschaticum (V:D)
boreal lousewort = Pedicularis macrodonta (V:D)
boreal paintbrush = Castilleja fulva (V:D)
boreal sandwort = Minuartia rubella (V:D)
born-again pelt = Peltigera praetextata (L)
Bostock's montia = Montia bostockii (V:D)
bottle moss = Amphidium lapponicum (B:M)
bottle sedge = Carex utriculata (V:M)
bottlebrush frost = Physconia detersa (L)
boulder lichen = Parmelia saxatilis (L)
bouncing-bet = Saponaria officinalis (V:D)
Bourgeau's milk-vetch = Astragalus bourgovii (V:D)
Bourgeau's pepper-grass = Lepidium bourgeauanum (V:D)
bower, blue virgin's → blue virgin's bower
bower, Columbia virgin's → Columbia virgin's bower
bower, white virgin's → white virgin's bower
box elder = Acer negundo (V:D)
boxboard felt lichen = Peltigera malacea (L)
boxwood, mountain → mountain boxwood
boxwood, Oregon → Oregon boxwood
boykinia, coast → coast boykinia
bracked honeysuckle = Lonicera involucrata (V:D)
bracken fern = Pteridium aquilinum (V:F)
bracted bog orchid = Coeloglossum viride (V:M)
bracted lousewort = Pedicularis bracteosa (V:D)
bracted vervain = Verbena bracteata (V:D)
bractless hedge-hyssop = Gratiola ebracteata (V:D)
bramble, dwarf → dwarf bramble
bramble, five-leaved → five-leaved bramble
bramble, snow → snow bramble
branched pepper-grass = Lepidium ramosissimum (V:D)
branched pepper-grass = Lepidium bourgeauanum (V:D)
branched phacelia = Phacelia ramosissima (V:D)
Brandegee's lomatium = Lomatium brandegei (V:D)
brass buttons = Cotula coronopifolia (V:D)
Braun's holly fern = Polystichum braunii (V:F)
braya, dwarf → dwarf braya
braya, low → low braya
braya, purple → purple braya
Brazilian water-milfoil = Myriophyllum aquaticum (V:D)
Brazilian waterweed = Egeria densa (V:M)
breaked sedge = Carex utriculata (V:M)
breath, baby's → baby's breath
Brewer's bitter-cress = Cardamine breweri (V:D)
Brewer's mitrewort = Mitella breweri (V:D)
Brewer's monkey-flower = Mimulus breweri (V:D)
Brewer's rush = Juncus breweri (V:M)
brickellia, large-flowered → large-flowered brickellia
brickellia, narrow-leaved → narrow-leaved brickellia
bristle clubrush = Scirpus setaceus (V:M)
bristle-leaf sedge = Carex eburnea (V:M)
bristle-like quillwort = Isoetes echinospora (V:F)
bristle sedge = Carex microglochin (V:M)
bristle-stalked sedge = Carex leporina (V:M)
bristle-stalked sedge = Carex leptalea (V:M)
bristlegrass, bur → bur bristlegrass
bristlegrass, foxtail → foxtail bristlegrass
bristlegrass, green → green bristlegrass
bristlegrass, yellow → yellow bristlegrass
bristly black currant = Ribes lacustre (V:D)
bristly crowfoot = Ranunculus pensylvanicus (V:D)
bristly haircap moss = Polytrichastrum alpinum (B:M)
bristly mousetail = Myosurus aristatus (V:D)
bristly stickseed = Lappula squarrosa (V:D)
brittle horsehair = Bryoria lanestris (L)
brittle horsehair lichen = Bryoria lanestris (L)
brittle prickly-pear cactus = Opuntia fragilis (V:D)
broad-fruited bur-reed = Sparganium eurycarpum (V:M)
broad-leaved cattail = Typha latifolia (V:M)
broad-leaved dock = Rumex obtusifolius (V:D)
broad-leaved maple = Acer macrophyllum (V:D)
broad-leaved peavine = Lathyrus latifolius (V:D)
broad-leaved penstemon = Penstemon ovatus (V:D)
broad-leaved pussytoes = Antennaria neglecta (V:D)
broad-leaved shootingstar = Dodecatheon hendersonii (V:D)

broad-leaved spring-beauty = Claytonia cordifolia (V:D)
broad-leaved stonecrop = Sedum spathulifolium (V:D)
broad-leaved twayblade = Listera convallarioides (V:M)
broad-leaved willowherb = Epilobium latifolium (V:D)
broad-petalled gentian = Gentiana platypetala (V:D)
broad shield = Parmotrema chinense (L)
broad-winged sedge = Carex petasata (V:M)
broadleaf tarpaper = Collema nigrescens (L)
brodiaea, harvest → harvest brodiaea
broken-leaf moss = Dicranum tauricum (B:M)
brome, Alaska → Alaska brome
brome, barren → barren brome
brome, California → California brome
brome, Columbia → Columbia brome
brome, corn → corn brome
brome, European smooth → European smooth brome
brome, fringed → fringed brome
brome, Japanese → Japanese brome
brome, meadow → meadow brome
brome, nodding → nodding brome
brome, northern → northern brome
brome, Pacific → Pacific brome
brome, pumpelly → pumpelly brome
brome, rattle → rattle brome
brome, rye → rye brome
brome, smooth → smooth brome
brome, soft → soft brome
bronze sedge = Carex aenea (V:M)
brook cinquefoil = Potentilla rivalis (V:D)
brook lichen = Dermatocarpon luridum (L)
brook saxifrage = Saxifraga rivularis (V:D)
brook saxifrage = Saxifraga nelsoniana (V:D)
brook spike-primrose = Boisduvalia stricta (V:D)
brooklime, American → American brooklime
brookweed = Samolus valerandi (V:D)
broom-corn millet = Panicum miliaceum (V:M)
broom-leaf toadflax = Linaria genistifolia (V:D)
broom moss = Dicranum scoparium (B:M)
broom, Scotch → Scotch broom
broom, Scots → Scots broom
broomrape, California → California broomrape
broomrape, clustered → clustered broomrape
broomrape, flat-topped → flat-topped broomrape
broomrape, naked → naked broomrape
broomrape, pine → pine broomrape
brow, rock → rock brow
brown bark lichen = Melanelia septentrionalis (L)
brown beak-rush = Rhynchospora capillacea (V:M)
brown cushion lichen = Leptogium lichenoides (L)
brown-eyed rockbright = Rhizoplaca peltata (L)
brown-eyed scale = Psora tuckermanii (L)
brown-eyed scale = Psora montana (L)
brown-eyed sunshine = Vulpicida canadensis (L)
brown-eyed Susan = Gaillardia aristata (V:D)
brown mustard = Brassica juncea (V:D)
brown-stemmed bog moss = Sphagnum lindbergii (B:M)
brown stipplescale = Catapyrenium squamulosum (L)
brown tapering splachnum = Tetraplodon mnioides (B:M)
brown turf lichen = Leptogium tenuissimum (L)
brown, abraded → abraded brown
brown, baby → baby brown
brown, blistered → blistered brown
brown, elegant → elegant brown
brown, eyed → eyed brown
brown, lattice → lattice brown
brown, leather → leather brown
brown, lustrous → lustrous brown
brown, northern → northern brown
brown, powdered → powdered brown
brown, rock → rock brown
brown, subelegant → subelegant brown
brownette = Koerberia sonomensis (L)
brownette = Placynthium stenophyllum (L)
brownette = Placynthium subradiatum (L)
brownette = Placynthium flabellosum (L)
brownette = Placynthium asperellum (L)
brownette, eyed → eyed brownette
brownette, peppered → peppered brownette
brownette, quilted → quilted brownette
brownish sedge = Carex brunnescens (V:M)
bryum, long-necked → long-necked bryum
bryum, red → red bryum
buck's-horn plantain = Plantago coronopus (V:D)
buckbean = Menyanthes trifoliata (V:D)
buckbrush = Symphoricarpos occidentalis (V:D)
buckhorn = Plantago lanceolata (V:D)
buckthorn, alder-leaved → alder-leaved buckthorn
buckwheat = Fagopyrum esculentum (V:D)
buckwheat, alpine → alpine buckwheat
buckwheat, alpine golden wild → alpine golden wild buckwheat
buckwheat, androsace → androsace buckwheat
buckwheat, cushion → cushion buckwheat
buckwheat, few-flowered → few-flowered buckwheat
buckwheat, parsnip-flowered → parsnip-flowered buckwheat
buckwheat, snow → snow buckwheat
buckwheat, strict → strict buckwheat
buckwheat, sulfur → sulfur buckwheat
buckwheat, sulphur → sulphur buckwheat
buckwheat, yellow → yellow buckwheat
buena, yerba → yerba buena
buffalo-berry, Canada → Canada buffalo-berry
buffalo-berry, Canadian → Canadian buffalo-berry
buffalo-berry, thorny → thorny buffalo-berry
buffalo-bur = Solanum rostratum (V:D)
buffalo plum = Astragalus crassicarpus (V:D)
bugbane, false → false bugbane
bugbane, tall → tall bugbane
bugle-weed = Ajuga reptans (V:D)
bugleweed = Lycopus uniflorus (V:D)
bugloss fiddleneck = Amsinckia lycopsoides (V:D)
bugloss, alkanet → alkanet bugloss
bugloss, common → common bugloss
bugloss, viper's → viper's bugloss
bugseed = Corispermum hyssopifolium (V:D)
bulb-bearing water-hemlock = Cicuta bulbifera (V:D)
bulbous bluegrass = Poa bulbosa (V:M)
bulbous rush = Juncus bulbosus (V:M)
bulbous water-hemlock = Cicuta bulbifera (V:D)
bull's-eye = Placopsis gelida (L)
bull thistle = Cirsium vulgare (V:D)
bullate-leaved cotoneaster = Cotoneaster bullatus (V:D)
bullhead waterleaf = Hydrophyllum capitatum (V:D)
bulrush, American → American bulrush
bulrush, common → common bulrush
bulrush, hard-stemmed → hard-stemmed bulrush
bulrush, Nevada → Nevada bulrush
bulrush, pale → pale bulrush
bulrush, river → river bulrush
bulrush, saltmarsh → saltmarsh bulrush
bulrush, small-flowered → small-flowered bulrush
bulrush, small-fruited → small-fruited bulrush
bunchberry = Cornus canadensis (V:D)
bunchberry, Canadian → Canadian bunchberry
bunchberry, Cordilleran → Cordilleran bunchberry
bunchberry, dwarf dog → dwarf dog bunchberry
bur bristlegrass = Setaria verticillata (V:M)
bur chervil = Anthriscus caucalis (V:D)
bur-clover = Medicago polymorpha (V:D)
bur-reed, broad-fruited → broad-fruited bur-reed
bur-reed, giant → giant bur-reed
bur-reed, glomerate → glomerate bur-reed
bur-reed, narrow-leaved → narrow-leaved bur-reed

bur-reed, northern → northern bur-reed
bur-reed, small → small bur-reed
bur-reed, water → water bur-reed
burdock, common → common burdock
burdock, great → great burdock
burgrass = Cenchrus longispinus (V:M)
Burke's larkspur = Delphinium burkei (V:D)
burnet, Canada → Canada burnet
burnet, great → great burnet
burnet, Menzies' → Menzies' burnet
burnet, salad → salad burnet
burnet, Sitka → Sitka burnet
burnet, western → western burnet
burning nettle = Urtica urens (V:D)
burweed, silver → silver burweed
bush-cranberry, American → American bush-cranberry
bush-cranberry, high → high bush-cranberry
bush-cranberry, low → low bush-cranberry
bushy cinquefoil = Potentilla paradoxa (V:D)
bushy knotweed = Polygonum ramosissimum (V:D)
bushy mentzelia = Mentzelia dispersa (V:D)
butter-and-eggs = Linaria vulgaris (V:D)
butterbur, Japanese → Japanese butterbur
buttercup-leaved saxifrage = Suksdorfia ranunculifolia (V:D)
buttercup-leaved suksdorfia = Suksdorfia ranunculifolia (V:D)
buttercup, arctic → arctic buttercup
buttercup, birdfoot → birdfoot buttercup
buttercup, California → California buttercup
buttercup, celery-leaved → celery-leaved buttercup
buttercup, creeping → creeping buttercup
buttercup, cursed → cursed buttercup
buttercup, dwarf → dwarf buttercup
buttercup, far-northern → far-northern buttercup
buttercup, hairy → hairy buttercup
buttercup, heart-leaved → heart-leaved buttercup
buttercup, kidney-leaved → kidney-leaved buttercup
buttercup, Lapland → Lapland buttercup
buttercup, little → little buttercup
buttercup, Macoun's → Macoun's buttercup
buttercup, meadow → meadow buttercup
buttercup, modest → modest buttercup
buttercup, mountain → mountain buttercup
buttercup, Pennsylvania → Pennsylvania buttercup
buttercup, prairie → prairie buttercup
buttercup, pygmy → pygmy buttercup
buttercup, sagebrush → sagebrush buttercup
buttercup, seaside → seaside buttercup
buttercup, shore → shore buttercup
buttercup, small-flowered → small-flowered buttercup
buttercup, snow → snow buttercup
buttercup, snowpatch → snowpatch buttercup
buttercup, straight-beaked → straight-beaked buttercup
buttercup, subalpine → subalpine buttercup
buttercup, sulphur → sulphur buttercup
buttercup, tall → tall buttercup
buttercup, unlovely → unlovely buttercup
buttercup, water-plantain → water-plantain buttercup
buttercup, western → western buttercup
butterfly pelt = Peltigera lepidophora (L)
butterfly-bush = Buddleja davidii (V:D)
butterfly scale = Psora rubiformis (L)
butterfly scale = Psora nipponica (L)
butterfly tarpaper = Collema flaccidum (L)
butterfly vinyl = Leptogium platynum (L)
butterweed, alkali-marsh → alkali-marsh butterweed
butterweed, alpine meadow → alpine meadow butterweed
butterweed, Canadian → Canadian butterweed
butterweed, dwarf mountain → dwarf mountain butterweed
butterweed, Elmer's → Elmer's butterweed
butterweed, high alpine → high alpine butterweed
butterweed, Mount Sheldon → Mount Sheldon butterweed
butterweed, Newcombe's → Newcombe's butterweed
butterweed, Ogotoruk Creek → Ogotoruk Creek butterweed
butterweed, plains → plains butterweed
butterweed, rayless alpine → rayless alpine butterweed
butterweed, rayless mountain → rayless mountain butterweed
butterweed, Rocky Mountain → Rocky Mountain butterweed
butterweed, streambank → streambank butterweed
butterweed, tall → tall butterweed
butterwort = Pinguicula vulgaris (V:D)
butterwort, common → common butterwort
butterwort, hairy → hairy butterwort
button-bush dodder = Cuscuta cephalanthi (V:D)
button lichen = Amandinea punctata (L)
buttons, brass → brass buttons
Buxbaum's sedge = Carex buxbaumii (V:M)

C

cabbage lung = Lobaria linita (L)
cabbage, skunk → skunk cabbage
cactus, brittle prickly-pear → brittle prickly-pear cactus
cactus, plains prickly-pear → plains prickly-pear cactus
Calder's lovage = Ligusticum calderi (V:D)
California bayberry = Myrica californica (V:D)
California brome = Bromus carinatus (V:M)
California broomrape = Orobanche californica (V:D)
California buttercup = Ranunculus californicus (V:D)
California hazelnut = Corylus cornuta var. californica (V:D)
California hazelnut = Corylus cornuta (V:D)
California oatgrass = Danthonia californica (V:M)
California poppy = Eschscholzia californica (V:D)
California rhododendron = Rhododendron macrophyllum (V:D)
California vinyl = Leptogium californicum (L)
California wax-myrtle = Myrica californica (V:D)
calla, wild → wild calla
caltha-leaved avens = Geum calthifolium (V:D)
camas, common → common camas
camas, great → great camas
campion, apetalous → apetalous campion
campion, arctic → arctic campion
campion, biennial → biennial campion
campion, bladder → bladder campion
campion, Douglas' → Douglas' campion
campion, Drummond's → Drummond's campion
campion, Menzies' → Menzies' campion
campion, moss → moss campion
campion, Parry's → Parry's campion
campion, pink → pink campion
campion, red → red campion
campion, rose → rose campion
campion, Scouler's → Scouler's campion
campion, Taimyr → Taimyr campion
Canada anemone = Anemone canadensis (V:D)
Canada bluegrass = Poa compressa (V:M)
Canada buffalo-berry = Shepherdia canadensis (V:D)
Canada burnet = Sanguisorba canadensis (V:D)
Canada goldenrod = Solidago canadensis (V:D)
Canada reedgrass = Calamagrostis canadensis (V:M)

Canada ryegrass = Oryzopsis canadensis (V:M)
Canada thistle = Cirsium arvense (V:D)
Canada violet = Viola canadensis (V:D)
Canada wildrye = Elymus canadensis (V:M)
Canadian buffalo-berry = Shepherdia canadensis (V:D)
Canadian bunchberry = Cornus canadensis (V:D)
Canadian butterweed = Senecio pauperculus (V:D)
Canadian fleabane = Conyza canadensis var. glabrata (V:D)
Canadian fleabane = Conyza canadensis (V:D)
Canadian mayflower = Maianthemum canadense (V:M)
Canadian milk-vetch = Astragalus canadensis (V:D)
Canadian sand-spurry = Spergularia canadensis (V:D)
Canadian waterweed = Elodea canadensis (V:M)
canarygrass = Phalaris canariensis (V:M)
canarygrass, reed → reed canarygrass
Canby's lovage = Ligusticum canbyi (V:D)
candleflame = Candelaria concolor (L)
candlewax, mountain → mountain candlewax
candlewax, whitebark → whitebark candlewax
candystick = Allotropa virgata (V:D)
canoe birch = Betula papyrifera var. papyrifera (V:D)
canoe birch = Betula papyrifera (V:D)
Canterbury-bells = Campanula medium (V:D)
caper spurge = Euphorbia lathyris (V:D)
capillary thread moss = Meesia uliginosa (B:M)
capitate lousewort = Pedicularis capitata (V:D)
capitate sedge = Carex capitata (V:M)
capitate valerian = Valeriana capitata (V:D)
caraway = Carum carvi (V:D)
Carolina crane's-bill = Geranium carolinianum (V:D)
Carolina draba = Draba reptans (V:D)
Carolina geranium = Geranium carolinianum (V:D)
Carolina meadow-foxtail = Alopecurus carolinianus (V:M)
Carolina whitlow-grass = Draba reptans (V:D)
carpet, golden → golden carpet
carpet, Iowa golden → Iowa golden carpet
carpet, northern golden → northern golden carpet
carpet, Wright's golden → Wright's golden carpet
carpetweed, common → common carpetweed
carrot-leaf desert-parsley = Lomatium foeniculaceum (V:D)
carrot, American wild → American wild carrot
carrot, wild → wild carrot
cartilage lichen = Ramalina dilacerata (L)
Cascade blueberry = Vaccinium deliciosum (V:D)
Cascade huckleberry = Vaccinium deliciosum (V:D)
Cascade parsley fern = Cryptogramma cascadensis (V:F)
cascade wallflower = Erysimum arenicola (V:D)
Cascade willow = Salix cascadensis (V:D)
cassandra = Chamaedaphne calyculata (V:D)
cat paw = Nephroma bellum (L)
cat's-breeches = Hydrophyllum capitatum (V:D)
cat's-ear, hairy → hairy cat's-ear
cat's-ear, smooth → smooth cat's-ear
cat-tail moss = Isothecium myosuroides (B:M)
catchfly, apetalous → apetalous catchfly
catchfly, Douglas' → Douglas' catchfly
catchfly, forked → forked catchfly
catchfly, Menzies' → Menzies' catchfly
catchfly, night-flowering → night-flowering catchfly
catchfly, Parry's → Parry's catchfly
catchfly, Scouler's → Scouler's catchfly
catchfly, seabluff → seabluff catchfly
catchfly, sleepy → sleepy catchfly
catchfly, small-flowered → small-flowered catchfly
catchfly, sweet William → sweet William catchfly
catchweed = Asperugo procumbens (V:D)
catnip = Nepeta cataria (V:D)
cattail = Typha latifolia (V:M)
cattail, broad-leaved → broad-leaved cattail
cattail, common → common cattail
cattail, lesser → lesser cattail
ceanothus, redstem → redstem ceanothus
ceanothus, snowbrush → snowbrush ceanothus
cedar-shake liverwort = Plagiochila porelloides (B:H)
cedar-shake liverwort = Plagiochila asplenioides (B:H)
cedar-shake wort = Plagiochila asplenioides (B:H)
cedar, Alaska → Alaska cedar
cedar, yellow → yellow cedar
celandine = Chelidonium majus (V:D)
celandine, lesser → lesser celandine
celery-leaved buttercup = Ranunculus sceleratus (V:D)
centaury, common → common centaury
centaury, European → European centaury
centaury, Muhlenberg's → Muhlenberg's centaury
centaury, western → western centaury
centipede, elegant → elegant centipede
centipede, eyed → eyed centipede
centipede, powdered → powdered centipede
centipede, seaside → seaside centipede
cetraria, curled → curled cetraria
cetraria, moonshine → moonshine cetraria
chalky shield lichen = Parmeliopsis hyperopta (L)
chamaerhodos = Chamaerhodos erecta (V:D)
chamaerhodos, American → American chamaerhodos
Chamisso's cotton-grass = Eriophorum chamissonis (V:M)
Chamisso's montia = Montia chamissoi (V:D)
chamisso's rein orchid = Platanthera chorisiana (V:M)
chamomile, corn → corn chamomile
chamomile, stinking → stinking chamomile
chamomile, wild → wild chamomile
chamomile, yellow → yellow chamomile
charcoal turtle = Hypocenomyce castaneocinerea (L)
charlie, creeping → creeping charlie
cheatgrass = Bromus tectorum (V:M)
checker-mallow, Henderson's → Henderson's checker-mallow
checker-mallow, Oregon → Oregon checker-mallow
Chelan penstemon = Penstemon pruinosus (V:D)
cherry-laurel = Prunus laurocerasus (V:D)
cherry, bird → bird cherry
cherry, bitter → bitter cherry
cherry, choke → choke cherry
cherry, mahaleb → mahaleb cherry
cherry, pin → pin cherry
cherry, sweet → sweet cherry
chervil, bur → bur chervil
chess, rattle → rattle chess
chess, soft → soft chess
chestnut rush = Juncus castaneus (V:M)
chick-pea milk-vetch = Astragalus cicer (V:D)
chickweed = Stellaria media (V:D)
chickweed monkey-flower = Mimulus alsinoides (V:D)
chickweed, alpine → alpine chickweed
chickweed, Bering → Bering chickweed
chickweed, blinks → blinks chickweed
chickweed, common → common chickweed
chickweed, field → field chickweed
chickweed, Fischer's → Fischer's chickweed
chickweed, little → little chickweed
chickweed, long-stalked → long-stalked chickweed
chickweed, long-stalked mouse-ear → long-stalked mouse-ear chickweed
chickweed, mouse-ear → mouse-ear chickweed
chickweed, nodding → nodding chickweed
chickweed, sticky → sticky chickweed
chickweed, umbellate → umbellate chickweed
chickweed, upright → upright chickweed
chickweed, water → water chickweed
chicory = Cichorium intybus (V:D)
Chilean tarweed = Madia sativa (V:D)
Chinese mustard = Brassica juncea (V:D)
Chinese scatter-rug = Parmotrema chinense (L)
chinook licorice = Lupinus littoralis (V:D)

chip, chocolate → chocolate chip
chives, wild → wild chives
chocolate chip = Solorina crocea (L)
chocolate-coloured nephroma = Nephroma parile (L)
chocolate lily = Fritillaria lanceolata (V:M)
chocolate shield = Tuckermannopsis sepincola (L)
choke cherry = Prunus virginiana (V:D)
Christmas-tree, common → common Christmas-tree
cinquefoil, biennial → biennial cinquefoil
cinquefoil, bipinnate → bipinnate cinquefoil
cinquefoil, blue-leaved → blue-leaved cinquefoil
cinquefoil, brook → brook cinquefoil
cinquefoil, bushy → bushy cinquefoil
cinquefoil, diverse-leaved → diverse-leaved cinquefoil
cinquefoil, Drummond's → Drummond's cinquefoil
cinquefoil, elegant → elegant cinquefoil
cinquefoil, fan-leaved → fan-leaved cinquefoil
cinquefoil, graceful → graceful cinquefoil
cinquefoil, Hooker's → Hooker's cinquefoil
cinquefoil, Norwegian → Norwegian cinquefoil
cinquefoil, one-flowered → one-flowered cinquefoil
cinquefoil, Pennsylvania → Pennsylvania cinquefoil
cinquefoil, prairie → prairie cinquefoil
cinquefoil, rough → rough cinquefoil
cinquefoil, sheep → sheep cinquefoil
cinquefoil, silvery → silvery cinquefoil
cinquefoil, snow → snow cinquefoil
cinquefoil, sticky → sticky cinquefoil
cinquefoil, sulphur → sulphur cinquefoil
cinquefoil, tall → tall cinquefoil
cinquefoil, two-flowered → two-flowered cinquefoil
cinquefoil, villous → villous cinquefoil
cinquefoil, white → white cinquefoil
cinquefoil, woolly → woolly cinquefoil
cladonia scales = Cladonia cariosa (L)
cladonia, black-foot → black-foot cladonia
cladonia, dragon → dragon cladonia
cladonia, horn → horn cladonia
cladonia, lipstick → lipstick cladonia
cladonia, orange-foot → orange-foot cladonia
cladonia, pioneer → pioneer cladonia
cladonia, powdered funnel → powdered funnel cladonia
cladonia, prickle → prickle cladonia
cladonia, ribbed → ribbed cladonia
cladonia, shingled → shingled cladonia
cladonia, sieve → sieve cladonia
cladonia, stump → stump cladonia
cladonia, sulphur → sulphur cladonia
cladonia, tiny toothpick → tiny toothpick cladonia
clarkia, common → common clarkia
clasping-leaved pepper-grass = Lepidium perfoliatum (V:D)
clasping-leaved pondweed = Potamogeton richardsonii (V:M)
clasping mullein = Verbascum phlomoides (V:D)
clasping twistedstalk = Streptopus amplexifolius (V:M)
claw-leaved feather moss = Drepanocladus aduncus (B:M)
clay pigtail moss = Hypnum lindbergii (B:M)
clear moss = Hookeria lucens (B:M)
cleavers = Galium aparine (V:D)
clematis, blue → blue clematis
clematis, Columbia → Columbia clematis
clematis, golden → golden clematis
clematis, white → white clematis
cliff-brake, smooth → smooth cliff-brake
cliff paintbrush = Castilleja rupicola (V:D)
clintonia, blue-bead → blue-bead clintonia
close-sheathed cotton-grass = Eriophorum brachyantherum (V:M)
closed-leaved pondweed = Potamogeton foliosus (V:M)
cloudberry = Rubus chamaemorus (V:D)
clover, alsike → alsike clover
clover, crimson → crimson clover
clover, cup → cup clover
clover, few-flowered → few-flowered clover
clover, hare's-foot → hare's-foot clover
clover, Italian → Italian clover
clover, Macrae's → Macrae's clover
clover, pinole → pinole clover
clover, pverty → pverty clover
clover, red → red clover
clover, small-headed → small-headed clover
clover, springbank → springbank clover
clover, strawberry → strawberry clover
clover, subterranean → subterranean clover
clover, thimble → thimble clover
clover, white → white clover
clover, white-tipped → white-tipped clover
clover, yellow → yellow clover
club-fruited willowherb = Epilobium clavatum (V:D)
club-moss mountain-heather = Cassiope lycopodioides (V:D)
club sedge = Carex buxbaumii (V:M)
club, devil's → devil's club
clubmoss, Alaska → Alaska clubmoss
clubmoss, alpine → alpine clubmoss
clubmoss, bog → bog clubmoss
clubmoss, Haleakala fir → Haleakala fir clubmoss
clubmoss, mountain → mountain clubmoss
clubmoss, running → running clubmoss
clubmoss, stiff → stiff clubmoss
clubmoss, western fir → western fir clubmoss
clubrush, bristle → bristle clubrush
clubrush, low → low clubrush
clubrush, Nevada → Nevada clubrush
clubrush, water → water clubrush
clustered broomrape = Orobanche fasciculata (V:D)
clustered dock = Rumex conglomeratus (V:D)
clustered dodder = Cuscuta approximata (V:D)
clustered tarweed = Madia glomerata (V:D)
clustered wild rose = Rosa pisocarpa (V:D)
coal tarpaper = Lichinella nigritella (L)
coast-blite = Chenopodium rubrum (V:D)
coast boykinia = Boykinia occidentalis (V:D)
coast Douglas-fir = Pseudotsuga menziesii var. menziesii (V:G)
Coast Mountain draba = Draba ruaxes (V:D)
Coast Mountain whitlow-grass = Draba ruaxes (V:D)
coast penstemon = Penstemon serrulatus (V:D)
coastal black gooseberry = Ribes divaricatum (V:D)
coastal bluegrass = Poa nervosa (V:M)
coastal leafy moss = Plagiomnium insigne (B:M)
coastal microseris = Microseris bigelovii (V:D)
coastal pearlwort = Sagina maxima (V:D)
coastal quillwort = Isoetes maritima (V:F)
coastal red elder = Sambucus racemosa ssp. pubens var. arborescens (V:D)
coastal red elderberry = Sambucus racemosa ssp. pubens var. arborescens (V:D)
coastal red paintbrush = Castilleja hyetophila (V:D)
coastal reindeer = Cladina portentosa (L)
coastal stellate sedge = Carex echinata ssp. phyllomanica (V:M)
coastal strawberry = Fragaria chiloensis (V:D)
coastal wood fern = Dryopteris arguta (V:F)
cockle, sticky → sticky cockle
cocklebur, common → common cocklebur
cocklebur, rough → rough cocklebur
cockleshell lichen = Hypocenomyce scalaris (L)
cockscomb cryptantha = Cryptantha celosioides (V:D)
coil-beaked lousewort = Pedicularis contorta (V:D)
coiled-leaf moss = Hypnum circinale (B:M)
coiled sedge = Carex circinata (V:M)
collomia, large-flowered → large-flowered collomia
collomia, narrow-leaved → narrow-leaved collomia
collomia, vari-leaved → vari-leaved collomia
colonial bentgrass = Agrostis capillaris (V:M)

Colorado rockfrog = Xanthoparmelia coloradoensis (L)
Colorado rush = Juncus confusus (V:M)
coltsfoot = Tussilago farfara (V:D)
coltsfoot, arrow-leaved → arrow-leaved coltsfoot
coltsfoot, Japanese → Japanese coltsfoot
coltsfoot, sweet → sweet coltsfoot
Columbia bladderpod = Lesquerella douglasii (V:D)
Columbia brome = Bromus vulgaris (V:M)
Columbia clematis = Clematis occidentalis (V:D)
Columbia gromwell = Lithospermum ruderale (V:D)
Columbia hawthorn = Crataegus columbiana (V:D)
Columbia lewisia = Lewisia columbiana (V:D)
Columbia lily = Lilium columbianum (V:M)
Columbia mugwort = Artemisia lindleyana (V:D)
Columbia River mugwort = Artemisia lindleyana (V:D)
Columbia sedge = Carex aperta (V:M)
Columbia virgin's bower = Clematis occidentalis (V:D)
Columbian monkshood = Aconitum columbianum (V:D)
Columbian needlegrass = Stipa nelsonii ssp. dorei (V:M)
columbine, blue → blue columbine
columbine, red → red columbine
columbine, Sitka → Sitka columbine
columbine, small-flowered → small-flowered columbine
columbine, yellow → yellow columbine
comandra, northern → northern comandra
comandra, pale → pale comandra
comb liverwort = Riccardia multifida (B:H)
combseed, winged → winged combseed
comfrey, common → common comfrey
comfrey, prickley → prickley comfrey
comfrey, rough → rough comfrey
comma, aspen → aspen comma
common agrimony = Agrimonia gryposepala (V:D)
common American hedge-hyssop = Gratiola neglecta (V:D)
common barberry = Berberis vulgaris (V:D)
common barley = Hordeum vulgare (V:M)
common beaked moss = Eurhynchium pulchellum (B:M)
common bearberry = Arctostaphylos uva-ursi (V:D)
common beard moss = Schistidium apocarpum (B:M)
common beggarticks = Bidens frondosa (V:D)
common blue-eyed-grass = Sisyrinchium montanum (V:M)
common blue lettuce = Lactuca tatarica (V:D)
common blueberry = Vaccinium myrtilloides (V:D)
common bluecup = Githopsis specularioides (V:D)
common borage = Borago officinalis (V:D)
common brown sphagnum = Sphagnum fuscum (B:M)
common bugloss = Anchusa officinalis (V:D)
common bulrush = Typha latifolia (V:M)
common burdock = Arctium minus (V:D)
common butterwort = Pinguicula vulgaris (V:D)
common camas = Camassia quamash (V:M)
common carpetweed = Mollugo verticillata (V:D)
common cattail = Typha latifolia (V:M)
common centaury = Centaurium erythraea (V:D)
common chickweed = Stellaria media (V:D)
common Christmas-tree = Sphaerophorus globosus (L)
common clarkia = Clarkia rhomboidea (V:D)
common cocklebur = Xanthium strumarium var. canadense (V:D)
common comfrey = Symphytum officinale (V:D)
common coral lichen = Stereocaulon paschale (L)
common crown-vetch = Coronilla varia (V:D)
common cryptantha = Cryptantha affinis (V:D)
common dandelion = Taraxacum officinale (V:D)
common dead-nettle = Lamium amplexicaule (V:D)
common dill = Anethum graveolens (V:D)
common dodder = Cuscuta epithymum (V:D)
common downingia = Downingia elegans (V:D)
common draba = Draba verna (V:D)
common duckweed = Lemna minor (V:M)
common dung moss = Tetraplodon mnioides (B:M)
common eel-grass = Zostera marina (V:M)
common evening-primrose = Oenothera biennis (V:D)
common false asphodel = Tofieldia pusilla (V:M)
common fiddleneck = Amsinckia intermedia (V:D)
common filago = Filago vulgaris (V:D)
common flax = Linum usitatissimum (V:D)
common fold-leaf liverwort = Diplophyllum albicans (B:H)
common forget-me-not = Myosotis discolor (V:D)
common four-tooth moss = Tetraphis pellucida (B:M)
common foxglove = Digitalis purpurea (V:D)
common fumatory = Fumaria officinalis (V:D)
common green sphagnum = Sphagnum girgensohnii (B:M)
common groundsel = Senecio vulgaris (V:D)
common haircap = Polytrichum commune (B:M)
common haircap moss = Polytrichum commune (B:M)
common harebell = Campanula rotundifolia (V:D)
common hawthorn = Crataegus monogyna (V:D)
common hemp-nettle = Galeopsis tetrahit (V:D)
common hook moss = Drepanocladus aduncus (B:M)
common hop = Humulus lupulus (V:D)
common horehound = Marrubium vulgare (V:D)
common hornwort = Ceratophyllum demersum (V:D)
common horseradish = Armoracia rusticana (V:D)
common horsetail = Equisetum arvense (V:F)
common hound's-tongue = Cynoglossum officinale (V:D)
common juniper = Juniperus communis (V:G)
common knotweed = Polygonum aviculare (V:D)
common Labrador tea = Ledum palustre (V:D)
common lantern moss = Andreaea rupestris (B:M)
common lawn moss = Brachythecium albicans (B:M)
common leafy liverwort = Barbilophozia lycopodioides (B:H)
common leafy moss = Plagiomnium medium (B:M)
common lomatium = Lomatium utriculatum (V:D)
common mallow = Malva sylvestris (V:D)
common mare's-tail = Hippuris vulgaris (V:D)
common mitrewort = Mitella nuda (V:D)
common moonwort = Botrychium lunaria (V:F)
common motherwort = Leonurus cardiaca (V:D)
common mugwort = Artemisia vulgaris (V:D)
common mullein = Verbascum thapsus (V:D)
common nettle = Urtica dioica (V:D)
common nodding pohlia = Pohlia nutans (B:M)
common oat = Avena sativa (V:M)
common orache = Atriplex patula (V:D)
common parsnip = Pastinaca sativa (V:D)
common pear = Pyrus communis (V:D)
common pepper-grass = Lepidium densiflorum (V:D)
common periwinkle = Vinca minor (V:D)
common pink wintergreen = Pyrola asarifolia (V:D)
common pitcher-plant = Sarracenia purpurea (V:D)
common plantain = Plantago major (V:D)
common privet = Ligustrum vulgare (V:D)
common purslane = Portulaca oleracea (V:D)
common rabbit-brush = Chrysothamnus nauseosus (V:D)
common ragweed = Ambrosia artemisiifolia (V:D)
common red paintbrush = Castilleja hyetophila (V:D)
common red paintbrush = Castilleja miniata (V:D)
common red sphagnum = Sphagnum capillifolium (B:M)
common reedgrass = Phragmites australis (V:M)
common rush = Juncus effusus (V:M)
common salsify = Tragopogon porrifolius (V:D)
common sandweed = Athysanus pusillus (V:D)
common scissor-leaf liverwort = Herbertus aduncus (B:H)
common scouring-rush = Equisetum hyemale (V:F)
common shingle = Hypocenomyce scalaris (L)
common snapdragon = Chaenorrhinum minus (V:D)
common snowberry = Symphoricarpos albus (V:D)
common sow-thistle = Sonchus oleraceus (V:D)
common speedwell = Veronica officinalis (V:D)
common spike-rush = Eleocharis palustris (V:M)
common St. John's-wort = Hypericum perforatum (V:D)
common starwort = Stellaria media (V:D)
common stork's-bill = Erodium cicutarium (V:D)
common sunflower = Helianthus annuus (V:D)
common sweetgrass = Hierochloe odorata (V:M)

common tansy = Tanacetum vulgare (V:D)
common timothy = Phleum pratense (V:M)
common toadflax = Linaria vulgaris (V:D)
common touch-me-not = Impatiens noli-tangere (V:D)
common tree moss = Climacium dendroides (B:M)
common twinpod = Physaria didymocarpa (V:D)
common valerian = Valeriana officinalis (V:D)
common velvet-grass = Holcus lanatus (V:M)
common verdant moss = Brachythecium rivulare (B:M)
common vetch = Vicia sativa (V:D)
common wallflower = Erysimum cheiri (V:D)
common water lentil = Lemna minor (V:M)
common water moss = Fontinalis antipyretica (B:M)
common whitlow-grass = Draba verna (V:D)
common witch's hair = Alectoria sarmentosa (L)
common witchgrass = Panicum capillare (V:M)
common wormwood = Artemisia absinthium (V:D)
common yarrow = Achillea millefolium (V:D)
compact selaginella = Selaginella densa (V:F)
concentric pelt = Peltigera elisabethae (L)
concentric pelt = Peltigera horizontalis (L)
cone-plant = Hemitomes congestum (V:D)
coneflower, prairie → prairie coneflower
conehead, spicy → spicy conehead
confetti lichen = Normandina pulchella (L)
confused woodrush = Luzula confusa (V:M)
contorted-podded evening-primrose = Camissonia contorta (V:D)
coontail = Ceratophyllum demersum (V:D)
copper wire moss = Pohlia nutans (B:M)
coral, cottontail → cottontail coral
coral, woolly → woolly coral
coralroot, pale → pale coralroot
coralroot, spotted → spotted coralroot
coralroot, striped → striped coralroot
coralroot, yellow → yellow coralroot
cord moss = Funaria hygrometrica (B:M)
cordate-leaved saxifrage = Saxifraga nelsoniana (V:D)
cordgrass, alkali → alkali cordgrass
Cordilleran bunchberry = Cornus unalaschkensis (V:D)
cordroot sedge = Carex chordorrhiza (V:M)
coreopsis, garden → garden coreopsis
corks = Parmelia omphalodes (L)
corn brome = Bromus squarrosus (V:M)
corn chamomile = Anthemis arvensis (V:D)
corn poppy = Papaver rhoeas (V:D)
corn-spurry = Spergula arvensis (V:D)
cornflower = Centaurea cyanus (V:D)
cornsalad = Valerianella locusta (V:D)
corrupt spleenwort = Asplenium adulterinum (V:F)
corydalis, blue → blue corydalis
corydalis, few-flowered → few-flowered corydalis
corydalis, golden → golden corydalis
corydalis, pink → pink corydalis
corydalis, Scouler's → Scouler's corydalis
cotoneaster, bullate-leaved → bullate-leaved cotoneaster
cotoneaster, rock → rock cotoneaster
cotoneaster, Simons' → Simons' cotoneaster
cotton-grass, arctic → arctic cotton-grass
cotton-grass, beautiful → beautiful cotton-grass
cotton-grass, Chamisso's → Chamisso's cotton-grass
cotton-grass, close-sheathed → close-sheathed cotton-grass
cotton-grass, green-keeled → green-keeled cotton-grass
cotton-grass, narrow-leaved → narrow-leaved cotton-grass
cotton-grass, russet → russet cotton-grass
cotton-grass, Scheuchzer's → Scheuchzer's cotton-grass
cotton-grass, sheathed → sheathed cotton-grass
cotton-grass, short-anthered → short-anthered cotton-grass
cotton-grass, slender → slender cotton-grass
cotton-grass, tall → tall cotton-grass
cotton-grass, tawny → tawny cotton-grass
cottontail coral = Stereocaulon paschale (L)
cottonwood, black → black cottonwood
cottonwood, plains → plains cottonwood
cottonwood, southern → southern cottonwood
Coville's rush = Juncus covillei (V:M)
cow-parsnip, giant → giant cow-parsnip
cow pie = Diploschistes muscorum (L)
cow-wheat = Melampyrum lineare (V:D)
cowbane, spotted → spotted cowbane
cowslip = Caltha palustris (V:D)
coyote willow = Salix exigua (V:D)
crabgrass, hairy → hairy crabgrass
crabgrass, smooth → smooth crabgrass
crag holly fern = Polystichum scopulinum (V:F)
cranberry, bog → bog cranberry
cranberry, mountain → mountain cranberry
cranberry, rock → rock cranberry
crane's-bill moss = Atrichum selwynii (B:M)
crane's-bill, Bicknell's → Bicknell's crane's-bill
crane's-bill, Carolina → Carolina crane's-bill
crane's-bill, cut-leaved → cut-leaved crane's-bill
crane's-bill, dovefoot → dovefoot crane's-bill
crane's-bill, northern → northern crane's-bill
crane's-bill, Richardson's → Richardson's crane's-bill
crane's-bill, Robert → Robert crane's-bill
crane's-bill, small-flowered → small-flowered crane's-bill
crane's-bill, sticky purple → sticky purple crane's-bill
crane's-bill, white → white crane's-bill
crap lichen = Melanelia sorediata (L)
Crawe's sedge = Carex crawei (V:M)
Crawford's sedge = Carex crawfordii (V:M)
cream-bush = Holodiscus discolor (V:D)
cream-coloured vetchling = Lathyrus ochroleucus (V:D)
cream-flowered peavine = Lathyrus ochroleucus (V:D)
creamy peavine = Lathyrus ochroleucus (V:D)
creeping bellflower = Campanula rapunculoides (V:D)
creeping bentgrass = Agrostis stolonifera (V:M)
creeping buttercup = Ranunculus repens (V:D)
creeping charlie = Glechoma hederacea (V:D)
creeping jenny = Lysimachia nummularia (V:D)
creeping juniper = Juniperus horizontalis (V:G)
creeping loosestrife = Lysimachia nummularia (V:D)
creeping Oregon-grape = Mahonia repens (V:D)
creeping raspberry = Rubus pedatus (V:D)
creeping snowberry = Gaultheria hispidula (V:D)
creeping softgrass = Holcus mollis (V:M)
creeping spearwort = Ranunculus flammula (V:D)
creeping spike-rush = Eleocharis palustris (V:M)
creeping velvet-grass = Holcus mollis (V:M)
creeping wildrye = Leymus triticoides (V:M)
creeping willow = Salix stolonifera (V:D)
creeping yellow cress = Rorippa sylvestris (V:D)
cress, American winter → American winter cress
cress, bitter winter → bitter winter cress
cress, blunt-leaved yellow → blunt-leaved yellow cress
cress, creeping yellow → creeping yellow cress
cress, early winter → early winter cress
cress, garden → garden cress
cress, hoary yellow → hoary yellow cress
cress, marsh yellow → marsh yellow cress
cress, mouse-ear → mouse-ear cress
cress, shepherd's → shepherd's cress
cress, thale → thale cress
cress, western yellow → western yellow cress
crested dogtail = Cynosurus cristatus (V:M)
crested wheatgrass = Agropyron cristatum (V:M)
crested wood fern = Dryopteris cristata (V:F)
crimson clover = Trifolium incarnatum (V:D)
crinkle-awned fescue = Festuca subuliflora (V:M)
crinkled pulp = Collema crispum (L)
crinkled rag = Platismatia lacunosa (L)
crisp sandwort = Stellaria crispa (V:D)

crisp starwort = Stellaria crispa (V:D)
crocus, prairie → prairie crocus
crottle = Parmelia saxatilis (L)
crottle, black → black crottle
crottle, dark → dark crottle
crowberry = Empetrum nigrum (V:D)
crowfoot, bristly → bristly crowfoot
crown-vetch, common → common crown-vetch
crumpled-leaf moss = Rhytidium rugosum (B:M)
crunch weed = Polycnemum arvense (V:D)
crusted orange lichen = Caloplaca cerina (L)
cryptantha, cockscomb → cockscomb cryptantha
cryptantha, common → common cryptantha
cryptantha, Fendler's → Fendler's cryptantha
cryptantha, obscure → obscure cryptantha
cryptantha, Sierra → Sierra cryptantha
cryptantha, Torrey's → Torrey's cryptantha
cryptantha, Watson's → Watson's cryptantha
cryptic paw = Nephroma occultum (L)
cuckoo bitter-cress = Cardamine pratensis (V:D)
cuckoo bitter-cress = Cardamine pratensis var. angustifolia (V:D)
cuckoo-flower = Cardamine pratensis (V:D)
cucumber, prickly → prickly cucumber
cucumber, wild → wild cucumber
cudweed, lowland → lowland cudweed
cudweed, marsh → marsh cudweed
cudweed, slender → slender cudweed
cudweed, sticky → sticky cudweed
cultivated apple = Malus pumila (V:D)
cup clover = Trifolium cyathiferum (V:D)
cup, queen's → queen's cup
cup, sulphur → sulphur cup
curled cetraria = Flavocetraria cucullata (L)
curled dock = Rumex crispus (V:D)
curled pondweed = Potamogeton crispus (V:M)
curled thistle = Carduus crispus (V:D)
curly-cup gumweed = Grindelia squarrosa (V:D)
curly dock = Rumex crispus (V:D)
curly heron's-bill moss = Dicranum fuscescens (B:M)
curly hypnum = Hypnum subimponens (B:M)
curly sedge = Carex rupestris (V:M)
curly thatch moss = Dicranoweisia cirrata (B:M)
currant, bristly black → bristly black currant
currant, red-flowering → red-flowering currant
currant, red swamp → red swamp currant
currant, skunk → skunk currant
currant, squaw → squaw currant
currant, sticky → sticky currant
currant, stink → stink currant
currant, swamp red → swamp red currant
currant, trailing black → trailing black currant
currant, wild red → wild red currant
cursed buttercup = Ranunculus sceleratus (V:D)
curved alpine woodrush = Luzula arcuata (V:M)
curved-spiked sedge = Carex maritima (V:M)
cushion buckwheat = Eriogonum ovalifolium var. nivale (V:D)
cushion buckwheat = Eriogonum ovalifolium (V:D)
cushion daisy = Erigeron poliospermus (V:D)
cushion fleabane = Erigeron poliospermus (V:D)
cushion mouse = Fuscopannaria laceratula (L)
cushion orange = Xanthoria polycarpa (L)
cushion tarpaper = Collema ceraniscum (L)
Cusick's paintbrush = Castilleja cusickii (V:D)
Cusick's penstemon = Penstemon cusickii (V:D)
Cusick's sedge = Carex cusickii (V:M)
Cusick's speedwell = Veronica cusickii (V:D)
cut-leaved anemone = Anemone multifida (V:D)
cut-leaved crane's-bill = Geranium dissectum (V:D)
cut-leaved daisy = Erigeron compositus (V:D)
cut-leaved geranium = Geranium dissectum (V:D)
cut-leaved nightshade = Solanum triflorum (V:D)
cut-leaved ragwort = Senecio eremophilus (V:D)
cut-leaved water horehound = Lycopus americanus (V:D)
cut-leaved water-parsnip = Berula erecta (V:D)
cutgrass, rice → rice cutgrass
cutleaf evergreen blackberry = Rubus laciniatus (V:D)
cutleaf sagebrush = Artemisia tripartita (V:D)
cypress spurge = Euphorbia cyparissias (V:D)
cypress, Alaska → Alaska cypress
cypress, yellow → yellow cypress

D

daffodil = Narcissus pseudonarcissus (V:M)
dagger-leaved rush = Juncus ensifolius (V:M)
daisy, annual → annual daisy
daisy, arctic → arctic daisy
daisy, cushion → cushion daisy
daisy, cut-leaved → cut-leaved daisy
daisy, easter → easter daisy
daisy, English → English daisy
daisy, entire-leaved → entire-leaved daisy
daisy, fine-leaved → fine-leaved daisy
daisy, golden → golden daisy
daisy, large-flowered → large-flowered daisy
daisy, Leiberg's → Leiberg's daisy
daisy, line-leaved → line-leaved daisy
daisy, long-leaved → long-leaved daisy
daisy, oxeye → oxeye daisy
daisy, Philadelphia → Philadelphia daisy
daisy, purple → purple daisy
daisy, rough-stemmed → rough-stemmed daisy
daisy, Salish → Salish daisy
daisy, shaggy → shaggy daisy
daisy, smooth → smooth daisy
daisy, subalpine → subalpine daisy
daisy, thread-leaved → thread-leaved daisy
daisy, three-lobed → three-lobed daisy
daisy, three-nerve → three-nerve daisy
daisy, trailing → trailing daisy
daisy, triple-nerved → triple-nerved daisy
daisy, tufted → tufted daisy
daisy, woolly → woolly daisy
dalmatian toadflax = Linaria dalmatica (V:D)
dame's-violet = Hesperis matronalis (V:D)
dandelion hawksbeard = Crepis runcinata (V:D)
dandelion, common → common dandelion
dandelion, horned → horned dandelion
dandelion, red-seeded → red-seeded dandelion
dandelion, virginia dwarf → virginia dwarf dandelion
dane's dwarf-gentian = Gentianella tenella (V:D)
dark crottle = Hypogymnia physodes (L)
dark lamb's-quarters = Chenopodium atrovirens (V:D)
Davidson's penstemon = Penstemon davidsonii (V:D)
Dawson's angelica = Angelica dawsonii (V:D)
dead-nettle, common → common dead-nettle
dead-nettle, henbit → henbit dead-nettle
dead-nettle, purple → purple dead-nettle
dead-nettle, spotted → spotted dead-nettle
death-camas, elegant → elegant death-camas

death-camas, grass-leaved → grass-leaved death-camas
death-camas, meadow → meadow death-camas
death-camas, mountain → mountain death-camas
death-camas, white → white death-camas
deer-cabbage = Fauria crista-galli (V:D)
deer fern = Blechnum spicant (V:F)
deer foot = Achlys triphylla (V:D)
deer paintbrush = Castilleja cervina (V:D)
deer-root = Iva axillaris (V:D)
deervetch = Lotus nevadensis (V:D)
deflated bone = Hypogymnia metaphysodes (L)
delight, single → single delight
deltoid balsamroot = Balsamorhiza deltoidea (V:D)
dense silky bentgrass = Apera interrupta (V:M)
dense spike-primrose = Boisduvalia densiflora (V:D)
dentate shootingstar = Dodecatheon dentatum (V:D)
Deptford pink = Dianthus armeria (V:D)
desert alyssum = Alyssum desertorum (V:D)
desert-parsley, bare-stem → bare-stem desert-parsley
desert-parsley, carrot-leaf → carrot-leaf desert-parsley
desert-parsley, fern-leaved → fern-leaved desert-parsley
desert-parsley, fine-leaved → fine-leaved desert-parsley
desert-parsley, Geyer's → Geyer's desert-parsley
desert-parsley, Gray's → Gray's desert-parsley
desert-parsley, large-fruited → large-fruited desert-parsley
desert-parsley, narrow-leaved → narrow-leaved desert-parsley
desert-parsley, nine-leaved → nine-leaved desert-parsley
desert-parsley, Sandberg's → Sandberg's desert-parsley
desert-parsley, swale → swale desert-parsley
desert ragwort = Senecio eremophilus (V:D)
desert rock purslane = Calandrinia ciliata (V:D)
desert shootingstar = Dodecatheon conjugens (V:D)
desert wheatgrass = Agropyron desertorum (V:M)
devil's club = Oplopanax horridus (V:D)
devil's matchstick = Pilophorus acicularis (L)
dewberry = Rubus ursinus (V:D)
dewberry = Rubus nivalis (V:D)
dewberry = Rubus pubescens (V:D)
Dewey's sedge = Carex deweyana (V:M)
diamond-leaved willow = Salix planifolia ssp. pulchra (V:D)
diapensia = Diapensia lapponica (V:D)
dicranum, wavy → wavy dicranum
diffuse knapweed = Centaurea diffusa (V:D)
dill, common → common dill
dimpled lichen = Solorina saccata (L)
dimpled specklebelly = Pseudocyphellaria anthraspis (L)
diploid annual bluegrass = Poa infirma (V:M)
dirty socks = Eriogonum pyrolifolium (V:D)
dissolving rockfrog = Arctoparmelia subcentrifuga (L)
distaff thistle = Carthamus lanatus ssp. baeticus (V:D)
distaff-thistle, woolly → woolly distaff-thistle
ditch-grass = Ruppia maritima (V:M)
diverse-leaved cinquefoil = Potentilla diversifolia (V:D)
diverse-leaved water-starwort = Callitriche heterophylla ssp. bolanderi (V:D)
dock, arctic → arctic dock
dock, bitter → bitter dock
dock, broad-leaved → broad-leaved dock
dock, clustered → clustered dock
dock, curled → curled dock
dock, curly → curly dock
dock, golden → golden dock
dock, narrow-leaved → narrow-leaved dock
dock, patience → patience dock
dock, willow → willow dock
dockleaf smartweed = Polygonum lapathifolium (V:D)
dodder, alfalfa → alfalfa dodder
dodder, alkali → alkali dodder
dodder, button-bush → button-bush dodder
dodder, clustered → clustered dodder
dodder, common → common dodder
dodder, field → field dodder
dodder, five-angled → five-angled dodder
dodder, salt marsh → salt marsh dodder
dodder, thyme → thyme dodder
dog bone = Hypogymnia tubulosa (L)
dog lichen = Peltigera canina (L)
dog mustard = Erucastrum gallicum (V:D)
dog nettle = Urtica urens (V:D)
dog paw = Nephroma helveticum (L)
dog pelt = Peltigera cinnamomea (L)
dog pelt = Peltigera kristinssonii (L)
dog pelt = Peltigera canina (L)
dog pelt = Peltigera membranacea (L)
dog rose = Rosa canina (V:D)
dog tooth lichen = Peltigera canina (L)
dogbane, hemp → hemp dogbane
dogbane, spreading → spreading dogbane
dogtail, crested → crested dogtail
dogtail, hedgehog → hedgehog dogtail
dogwood, dwarf → dwarf dogwood
dogwood, flowering → flowering dogwood
dogwood, mountain → mountain dogwood
dogwood, Pacific → Pacific dogwood
dogwood, red-osier → red-osier dogwood
dogwood, western → western dogwood
dogwood, western flowering → western flowering dogwood
dot lichen = Biatora vernalis (L)
dotted smartweed = Polygonum punctatum (V:D)
doubtful rocktripe = Umbilicaria cinereorufescens (L)
Douglas' aster = Aster subspicatus (V:D)
Douglas' campion = Silene douglasii (V:D)
Douglas' catchfly = Silene douglasii (V:D)
Douglas' dwarf mistletoe = Arceuthobium douglasii (V:D)
Douglas' helianthella = Helianthella uniflora var. douglasii (V:D)
Douglas' knotweed = Polygonum douglasii ssp. douglasii (V:D)
Douglas' knotweed = Polygonum douglasii (V:D)
Douglas' neckera = Neckera douglasii (B:M)
Douglas' sedge = Carex douglasii (V:M)
Douglas' silene = Silene douglasii (V:D)
Douglas' spirea = Spiraea douglasii (V:D)
Douglas' water-hemlock = Cicuta douglasii (V:D)
Douglas-fir = Pseudotsuga menziesii (V:G)
Douglas-fir, coast → coast Douglas-fir
Douglas-fir, interior → interior Douglas-fir
Douglas-fir, Rocky Mountain → Rocky Mountain Douglas-fir
Douglas maple = Acer glabrum var. douglasii (V:D)
douglasia, Gorman's → Gorman's douglasia
douglasia, Rocky Mountain → Rocky Mountain douglasia
douglasia, smooth → smooth douglasia
douglasia, snow → snow douglasia
dovefoot crane's-bill = Geranium molle (V:D)
dovefoot geranium = Geranium molle (V:D)
downingia, common → common downingia
draba, Alaska → Alaska draba
draba, alpine → alpine draba
draba, Austrian → Austrian draba
draba, Baffin's Bay → Baffin's Bay draba
draba, Carolina → Carolina draba
draba, Coast Mountain → Coast Mountain draba
draba, common → common draba
draba, few-seeded → few-seeded draba
draba, golden → golden draba
draba, gray-leaved → gray-leaved draba
draba, lance-fruited → lance-fruited draba
draba, lance-leaved → lance-leaved draba
draba, long-stalked → long-stalked draba
draba, Macoun's → Macoun's draba
draba, milky → milky draba
draba, north Pacific → north Pacific draba
draba, northern → northern draba
draba, Nuttall's → Nuttall's draba

draba, Payson's → Payson's draba
draba, Porsild's → Porsild's draba
draba, Rocky Mountain → Rocky Mountain draba
draba, slender → slender draba
draba, smooth → smooth draba
draba, snow → snow draba
draba, tall → tall draba
draba, thick-leaved → thick-leaved draba
draba, woods → woods draba
draba, Yellowstone → Yellowstone draba
dragon cladonia = Cladonia squamosa (L)
dragonhead, American → American dragonhead
dragonhead, Eurasian → Eurasian dragonhead
dragonhead, purple → purple dragonhead
droping wood-reed = Cinna latifolia (V:M)
dropseed, hairgrass → hairgrass dropseed
dropseed, sand → sand dropseed
Drummond's anemone = Anemone drummondii (V:D)
Drummond's campion = Silene drummondii (V:D)
Drummond's cinquefoil = Potentilla drummondii (V:D)
Drummond's leafy moss = Plagiomnium drummondii (B:M)
Drummond's milk-vetch = Astragalus drummondii (V:D)
Drummond's rockcress = Arabis drummondii (V:D)
Drummond's rush = Juncus drummondii (V:M)
Drummond's thistle = Cirsium drummondii (V:D)
Drummond's willow = Salix drummondiana (V:D)
dryland ragwort = Senecio eremophilus (V:D)
dryland sedge = Carex xerantica (V:M)
duckmeat, great → great duckmeat
duckweed, common → common duckweed
duckweed, ivy-leaved → ivy-leaved duckweed
duckweed, star → star duckweed
dulichium = Dulichium arundinaceum (V:M)
dull Oregon-grape = Mahonia nervosa (V:D)
dull scale feather wort = Porella cordaeana (B:H)
dune bentgrass = Agrostis pallens (V:M)
dune goldenrod = Solidago spathulata (V:D)
dune tansy = Tanacetum bipinnatum ssp. huronense (V:D)
dune tansy = Tanacetum bipinnatum (V:D)
dune wildrye = Leymus mollis (V:M)
dunegrass = Leymus mollis (V:M)
dunhead sedge = Carex phaeocephala (V:M)
dusky fork moss = Dicranum fuscescens (B:M)
dusky willow = Salix melanopsis (V:D)
dust, lime → lime dust
dusty cartilage lichen = Ramalina pollinaria (L)
dusty gristle = Ramalina pollinaria (L)
dusty-margined dog lichen = Peltigera collina (L)
dusty miller = Cerastium tomentosum (V:D)
dwarf bilberry = Vaccinium myrtillus (V:D)
dwarf bilberry = Vaccinium caespitosum (V:D)
dwarf blueberry = Vaccinium caespitosum (V:D)
dwarf bramble = Rubus lasiococcus (V:D)
dwarf braya = Braya humilis (V:D)
dwarf buttercup = Ranunculus pygmaeus (V:D)
dwarf dog bunchberry = Cornus suecica (V:D)
dwarf dogwood = Cornus canadensis (V:D)
dwarf-gentian, dane's → dane's dwarf-gentian
dwarf groundsmoke = Gayophytum humile (V:D)
dwarf hawksbeard = Crepis nana (V:D)
dwarf hesperochiron = Hesperochiron pumilus (V:D)
dwarf larkspur = Delphinium depauperatum (V:D)
dwarf mallow = Malva neglecta (V:D)
dwarf marsh violet = Viola epipsila ssp. repens (V:D)
dwarf marsh violet = Viola epipsila (V:D)
dwarf montia = Montia dichotoma (V:D)
dwarf mountain butterweed = Senecio fremontii (V:D)
dwarf mountain fleabane = Erigeron compositus (V:D)
dwarf mountain lupine = Lupinus lyallii (V:D)
dwarf nagoonberry = Rubus arcticus (V:D)
dwarf poppy = Papaver alpinum (V:D)
dwarf raspberry = Rubus arcticus ssp. acaulis (V:D)
dwarf raspberry = Rubus arcticus (V:D)
dwarf rattlesnake orchid = Goodyera repens (V:M)
dwarf red blackberry = Rubus pubescens (V:D)
dwarf rose = Rosa gymnocarpa (V:D)
dwarf sandwort = Minuartia pusilla (V:D)
dwarf scouring-rush = Equisetum scirpoides (V:F)
dwarf snow willow = Salix reticulata ssp. nivalis (V:D)
dwarf spurge = Euphorbia exigua (V:D)
dyer's woad = Isatis tinctoria (V:D)

E

ear lichen = Pseudocyphellaria anthraspis (L)
early blue violet = Viola adunca (V:D)
early hairgrass = Aira praecox (V:M)
early winter cress = Barbarea verna (V:D)
easter daisy = Townsendia exscapa (V:D)
eastern eyebright = Euphrasia nemorosa (V:D)
eastern knotweed = Polygonum buxiforme (V:D)
Eaton's aster = Aster eatonii (V:D)
edible horsehair = Bryoria fremontii (L)
edible thistle = Cirsium edule (V:D)
edible valerian = Valeriana edulis (V:D)
Edwards' wallflower = Eutrema edwardsii (V:D)
eel-grass pondweed = Potamogeton zosteriformis (V:M)
eel-grass, common → common eel-grass
eel-grass, Japanese → Japanese eel-grass
eels, electric → electric eels
elder, black → black elder
elder, blue → blue elder
elder, box → box elder
elder, coastal red → coastal red elder
elder, red → red elder
elderberry, black → black elderberry
elderberry, blue → blue elderberry
elderberry, coastal red → coastal red elderberry
elderberry, red → red elderberry
elecampane = Inula helenium (V:D)
electric eels = Dicranum polysetum (B:M)
electrified cat's-tail moss = Rhytidiadelphus triquetrus (B:M)
elegant brown = Melanelia infumata (L)
elegant brown = Melanelia elegantula (L)
elegant centipede = Heterodermia leucomelos (L)
elegant cinquefoil = Potentilla elegans (V:D)
elegant death-camas = Zigadenus elegans (V:M)
elegant hawksbeard = Crepis elegans (V:D)
elegant Jacob's-ladder = Polemonium elegans (V:D)
elegant Jacob's-ladder polemonium = Polemonium elegans (V:D)
elegant milk-vetch = Astragalus eucosmus (V:D)
elegant moss = Pseudotaxiphyllum elegans (B:M)
elegant orange = Xanthoria elegans (L)
elegant rein orchid = Piperia elegans (V:M)
elephant's-head lousewort = Pedicularis groenlandica (V:D)
elf-ear = Normandina pulchella (L)
elk sedge = Carex geyeri (V:M)
elk thistle = Cirsium scariosum (V:D)
elkslip = Caltha leptosepala (V:D)
elliptic-leaved penstemon = Penstemon ellipticus (V:D)
ellisia = Ellisia nyctelea (V:D)
elm, Siberian → Siberian elm

Elmer's butterweed = Senecio elmeri (V:D)
Elmer's paintbrush = Castilleja elmeri (V:D)
emery rocktripe = Umbilicaria phaea (L)
eminent bluegrass = Poa eminens (V:M)
enchanter's-nightshade = Circaea alpina (V:D)
enchanter's-nightshade, alpine → alpine enchanter's-nightshade
enchanter's-nightshade, Pacific → Pacific enchanter's-nightshade
enchanter's-nightshade, small → small enchanter's-nightshade
Engelmann's aster = Aster engelmannii (V:D)
Engelmann's sedge = Carex engelmannii (V:M)
Engelmann spruce = Picea engelmannii (V:G)
English daisy = Bellis perennis (V:D)
English holly = Ilex aquifolium (V:D)
English ivy = Hedera helix (V:D)
English oak = Quercus robur (V:D)
English plantain = Plantago lanceolata (V:D)
English ryegrass = Lolium perenne (V:M)
entire-leaf stonecrop = Sedum integrifolium (V:D)
entire-leaved daisy = Leucanthemum integrifolium (V:D)
entire-leaved gumweed = Grindelia integrifolia (V:D)
entire-leaved mountain-avens = Dryas integrifolia (V:D)
entire-leaved white mountain-avens = Dryas integrifolia (V:D)
erect-fruited iris moss = Distichium capillaceum (B:M)
eriophyllum, woolly → woolly eriophyllum
eryngo, plains → plains eryngo
Eschscholtz's little nightmare = Aphragmus eschscholtzianus (V:D)
Etruscan honeysuckle = Lonicera etrusca (V:D)
Eurasian dragonhead = Dracocephalum thymiflorum (V:D)
Eurasian water-milfoil = Myriophyllum spicatum (V:D)
European alkaligrass = Puccinellia distans (V:M)
European beachgrass = Ammophila arenaria (V:M)
European birch = Betula pendula (V:D)
European bittersweet = Solanum dulcamara (V:D)
European centaury = Centaurium erythraea (V:D)
European field pansy = Viola arvensis (V:D)
European mountain-ash = Sorbus aucuparia (V:D)
European searocket = Cakile maritima (V:D)
European smooth brome = Bromus racemosus (V:M)
European speedwell = Veronica beccabunga (V:D)
European starflower = Trientalis europaea (V:D)
European water-hemlock = Cicuta virosa (V:D)
European white water-lily = Nymphaea alba (V:D)
European wild pansy = Viola tricolor (V:D)
even dog lichen = Peltigera malacea (L)
evening-primrose, Andean → Andean evening-primrose
evening-primrose, common → common evening-primrose
evening-primrose, contorted-podded → contorted-podded evening-primrose
evening-primrose, pale → pale evening-primrose
evening-primrose, red-sepaled → red-sepaled evening-primrose
evening-primrose, short-flowered → short-flowered evening-primrose
evening-primrose, white-pole → white-pole evening-primrose
evening-primrose, yellow → yellow evening-primrose
evergreen blackberry = Rubus laciniatus (V:D)
evergreen huckleberry = Vaccinium ovatum (V:D)
evergreen saxifrage = Saxifraga aizoides (V:D)
evergreen violet = Viola sempervirens (V:D)
evergreen yellow violet = Viola orbiculata (V:D)
everlasting, pearly → pearly everlasting
everlasting, small-leaved → small-leaved everlasting
eyebright, eastern → eastern eyebright
eyed brown = Melanelia subolivacea (L)
eyed brown = Melanelia multispora (L)
eyed brownette = Vestergrenopsis elaeina (L)
eyed centipede = Anaptychia setifera (L)
eyed ruffle = Tuckermannopsis sepincola (L)
eyed saguaro = Cavernularia lophyrea (L)

F

fairies, pink → pink fairies
fairy-candelabra, northern → northern fairy-candelabra
fairy-candelabra, slender-flowered → slender-flowered fairy-candelabra
fairy-candelabra, sweet-flowered → sweet-flowered fairy-candelabra
fairy-candelabra, western → western fairy-candelabra
fairy parasols = Splachnum luteum (B:M)
fairy puke = Icmadophila ericetorum (L)
fairy-slipper = Calypso bulbosa (V:M)
fairybells = Disporum trachycarpum (V:M)
fairybells, Hooker's → Hooker's fairybells
fairybells, Oregon → Oregon fairybells
fairybells, rough-fruited → rough-fruited fairybells
fairybells, Smith's → Smith's fairybells
Falkland Island sedge = Carex macloviana (V:M)
fall scouring-rush = Equisetum hyemale (V:F)
false-agoseris = Nothocalais troximoides (V:D)
false azalea = Menziesia ferruginea (V:D)
false bugbane = Trautvetteria caroliniensis (V:D)
false haircap moss = Timmia austriaca (B:M)
false lily-of-the-valley = Maianthemum dilatatum (V:M)
false-manna, Fernald's → Fernald's false-manna
false melic = Schizachne purpurascens (V:M)
false oatgrass = Arrhenatherum elatius (V:M)
false pixie-cup = Cladonia chlorophaea (L)
false-polytrichum = Timmia austriaca (B:M)
false strawberry = Duchesnea indica (V:D)
falsebox = Paxistima myrsinites (V:D)
falseflax = Camelina sativa (V:D)
falseflax, little-podded → little-podded falseflax
fameflower, Okanogan → Okanogan fameflower
fan-leaved cinquefoil = Potentilla flabellifolia (V:D)
fan lichen = Peltigera venosa (L)
fan moss = Rhizomnium glabrescens (B:M)
fan pelt = Peltigera venosa (L)
far-northern buttercup = Ranunculus hyperboreus (V:D)
farewell-to-spring = Clarkia amoena (V:D)
Farr's willow = Salix farriae (V:D)
Farwell's water-milfoil = Myriophyllum farwellii (V:D)
fat bog moss = Sphagnum papillosum (B:M)
feather geranium = Chenopodium botrys (V:D)
feathermoss, red-stemmed → red-stemmed feathermoss
Fees' lip fern = Cheilanthes feei (V:F)
felt-leaf willow = Salix alaxensis (V:D)
felt pelt = Peltigera rufescens (L)
felt pelt = Peltigera ponojensis (L)
felt round moss = Rhizomnium pseudopunctatum (B:M)
felwort = Gentianella amarella (V:D)
felwort, marsh → marsh felwort
Fendler's bluegrass = Poa fendleriana (V:M)
Fendler's cryptantha = Cryptantha fendleri (V:D)
Fendler's muttongrass = Poa fendleriana (V:M)
Fendler's waterleaf = Hydrophyllum fendleri (V:D)
Fendler's waterleaf = Hydrophyllum fendleri var. albifrons (V:D)
fennel-leaved pondweed = Potamogeton pectinatus (V:M)
fennel, sweet → sweet fennel
fern-leaved desert-parsley = Lomatium dissectum (V:D)
fern-leaved goldthread = Coptis aspleniifolia (V:D)
fern-leaved lomatiun = Lomatium dissectum (V:D)
fern moss = Cratoneuron filicinum (B:M)
fern, Alaska holly → Alaska holly fern

fern, alpine cliff → alpine cliff fern
fern, Anderson's holly → Anderson's holly fern
fern, Asian oak → Asian oak fern
fern, beech → beech fern
fern, bracken → bracken fern
fern, Braun's holly → Braun's holly fern
fern, Cascade parsley → Cascade parsley fern
fern, coastal wood → coastal wood fern
fern, crag holly → crag holly fern
fern, crested wood → crested wood fern
fern, deer → deer fern
fern, Fees' lip → Fees' lip fern
fern, fragile → fragile fern
fern, fragile bladder → fragile bladder fern
fern, fragrant shield → fragrant shield fern
fern, fragrant wood → fragrant wood fern
fern, giant chain → giant chain fern
fern, goldenback → goldenback fern
fern, Indian's-dream → Indian's-dream fern
fern, Kruckeberg's holly → Kruckeberg's holly fern
fern, lace → lace fern
fern, lady → lady fern
fern, large mosquito → large mosquito fern
fern, leathery grape → leathery grape fern
fern, Lemmon's holly → Lemmon's holly fern
fern, licorice → licorice fern
fern, maidenhair → maidenhair fern
fern, male → male fern
fern, marginal wood → marginal wood fern
fern, Mexican mosquito → Mexican mosquito fern
fern, mountain → mountain fern
fern, mountain bladder → mountain bladder fern
fern, mountain cliff → mountain cliff fern
fern, mountain holly → mountain holly fern
fern, Nahanni oak → Nahanni oak fern
fern, narrow beech → narrow beech fern
fern, narrow cowboy → narrow cowboy fern
fern, narrow-leaved sword → narrow-leaved sword fern
fern, Nevada marsh → Nevada marsh fern
fern, northern holly → northern holly fern
fern, northern maidenhair → northern maidenhair fern
fern, northern oak → northern oak fern
fern, oak → oak fern
fern, ostrich → ostrich fern
fern, parsley → parsley fern
fern, rattlesnake → rattlesnake fern
fern, rusty cliff → rusty cliff fern
fern, shield → shield fern
fern, Sitka parsley → Sitka parsley fern
fern, slender lip → slender lip fern
fern, smooth cliff → smooth cliff fern
fern, southern maidenhair → southern maidenhair fern
fern, spiny wood → spiny wood fern
fern, sword → sword fern
fern, toothed wood → toothed wood fern
fern, venus-hair → venus-hair fern
fern, Virginia grape → Virginia grape fern
fern, western cliff → western cliff fern
fern, western oak → western oak fern
Fernald's false-manna = Torreyochloa pallida (V:M)
fescue, alpine → alpine fescue
fescue, Altai → Altai fescue
fescue, Baffin → Baffin fescue
fescue, barren → barren fescue
fescue, bearded → bearded fescue
fescue, crinkle-awned → crinkle-awned fescue
fescue, green → green fescue
fescue, hair → hair fescue
fescue, hard → hard fescue
fescue, Idaho → Idaho fescue
fescue, Krajina's → Krajina's fescue
fescue, little → little fescue
fescue, meadow → meadow fescue
fescue, northern rough → northern rough fescue
fescue, rattail → rattail fescue
fescue, red → red fescue
fescue, Rocky Mountain → Rocky Mountain fescue
fescue, rough → rough fescue
fescue, six-weeks → six-weeks fescue
fescue, small → small fescue
fescue, tall → tall fescue
fescue, viviparous → viviparous fescue
fescue, western → western fescue
feverfew = Tanacetum parthenium (V:D)
few-finger lichen = Dactylina arctica (L)
few-flowered buckwheat = Eriogonum pauciflorum (V:D)
few-flowered clover = Trifolium oliganthum (V:D)
few-flowered corydalis = Corydalis pauciflora (V:D)
few-flowered meadowrue = Thalictrum sparsiflorum (V:D)
few-flowered sedge = Carex pauciflora (V:M)
few-flowered shootingstar = Dodecatheon pulchellum (V:D)
few-flowered snowberry = Symphoricarpos albus (V:D)
few-flowered spike-rush = Eleocharis quinqueflora (V:M)
few-fruited lomatium = Lomatium martindalei (V:D)
few-seeded bitter-cress = Cardamine oligosperma (V:D)
few-seeded bog sedge = Carex microglochin (V:M)
few-seeded draba = Draba oligosperma (V:D)
few-seeded whitlow-grass = Draba oligosperma (V:D)
fiddleneck, bugloss → bugloss fiddleneck
fiddleneck, common → common fiddleneck
fiddleneck, rigid → rigid fiddleneck
fiddleneck, seaside → seaside fiddleneck
fiddleneck, small-flowered → small-flowered fiddleneck
field bindweed = Convolvulus arvensis (V:D)
field chickweed = Cerastium arvense (V:D)
field dodder = Cuscuta pentagona (V:D)
field filago = Filago arvensis (V:D)
field forget-me-not = Myosotis arvensis (V:D)
field garlic = Allium vineale (V:M)
field hedge-nettle = Stachys arvensis (V:D)
field horsetail = Equisetum arvense (V:F)
field locoweed = Oxytropis campestris (V:D)
field madder = Sherardia arvensis (V:D)
field milk-vetch = Astragalus agrestis (V:D)
field mint = Mentha arvensis (V:D)
field parsley-piert = Aphanes arvensis (V:D)
field pennycress = Thlaspi arvense (V:D)
field pepper-grass = Lepidium campestre (V:D)
field pussytoes = Antennaria neglecta (V:D)
field sandbur = Cenchrus longispinus (V:M)
field scabious = Knautia arvensis (V:D)
field sedge = Carex praegracilis (V:M)
field woodrush = Luzula multiflora (V:M)
figwort, lance-leaved → lance-leaved figwort
filago, common → common filago
filago, field → field filago
filaree = Erodium cicutarium (V:D)
fine-leaved daisy = Erigeron linearis (V:D)
fine-leaved desert-parsley = Lomatium utriculatum (V:D)
fine-leaved lomatium = Lomatium utriculatum (V:D)
finger felt lichen = Peltigera neopolydactyla (L)
fingered mouse = Parmeliella triptophylla (L)
fingered tarpaper = Collema cristatum (L)
fir, alpine → alpine fir
fir, amabilis → amabilis fir
fir, balsam → balsam fir
fir, grand → grand fir
fir, noble → noble fir
fir, Pacific silver → Pacific silver fir
fir, subalpine → subalpine fir
fire moss = Ceratodon purpureus (B:M)

fireweed = Epilobium angustifolium (V:D)
Fischer's chickweed = Cerastium fischerianum (V:D)
fishnet = Ramalina menziesii (L)
fissured scale = Psora cerebriformis (L)
fissured stippleback = Dermatocarpon intestiniforme (L)
fist lichen = Arctoparmelia incurva (L)
five-angled dodder = Cuscuta pentagona (V:D)
five-hooked bassia = Bassia hyssopifolia (V:D)
five-leaved bramble = Rubus pedatus (V:D)
five o'clock shadow = Phaeophyscia sciastra (L)
five-stamened mitrewort = Mitella pentandra (V:D)
flapper moss = Mnium spinulosum (B:M)
flat dog lichen = Peltigera horizontalis (L)
flat-fruited meadowrue = Thalictrum sparsiflorum (V:D)
flat-leaved bladderwort = Utricularia intermedia (V:D)
flat-leaved liverwort = Radula complanata (B:H)
flat-leaved willow = Salix planifolia (V:D)
flat moss = Plagiothecium undulatum (B:M)
flat-stalked pondweed = Potamogeton friesii (V:M)
flat-top goldentop = Euthamia graminifolia (V:D)
flat-topped broomrape = Orobanche corymbosa (V:D)
flat-topped broomrape = Orobanche corymbosa ssp. mutabilis (V:D)
flattened snow light = Flavocetraria nivalis (L)
flax-leaved plainsmustard = Schoenocrambe linifolia (V:D)
flax, blue → blue flax
flax, common → common flax
flax, hairy → hairy flax
flax, littlepod → littlepod flax
flax, pale → pale flax
flax, wild blue → wild blue flax
flaxflower, bicolored → bicolored flaxflower
flaxflower, Harkness' → Harkness' flaxflower
flaxflower, northern → northern flaxflower
fleabane, annual → annual fleabane
fleabane, Canadian → Canadian fleabane
fleabane, cushion → cushion fleabane
fleabane, dwarf mountain → dwarf mountain fleabane
fleabane, golden → golden fleabane
fleabane, large-flowered → large-flowered fleabane
fleabane, Leiberg's → Leiberg's fleabane
fleabane, line-leaved → line-leaved fleabane
fleabane, long-leaved → long-leaved fleabane
fleabane, marsh → marsh fleabane
fleabane, one-flowered → one-flowered fleabane
fleabane, Philadelphia → Philadelphia fleabane
fleabane, prairie → prairie fleabane
fleabane, rough-stemmed → rough-stemmed fleabane
fleabane, shaggy → shaggy fleabane
fleabane, showy → showy fleabane
fleabane, smooth → smooth fleabane
fleabane, spreading → spreading fleabane
fleabane, streamside → streamside fleabane
fleabane, thread-leaved → thread-leaved fleabane
fleabane, three-nerve → three-nerve fleabane
fleabane, trailing → trailing fleabane
fleabane, triple-nerved → triple-nerved fleabane
fleabane, tufted → tufted fleabane
fleecy rocktripe = Umbilicaria vellea (L)
fleshy jaumea = Jaumea carnosa (V:D)
fleshy stitchwort = Stellaria crassifolia (V:D)
flixweed = Descurainia sophia (V:D)
floating feather moss = Warnstorfia fluitans (B:M)
floating-leaved pondweed = Potamogeton natans (V:M)
floating liverwort = Ricciocarpos natans (B:H)
floating marsh-marigold = Caltha natans (V:D)
floating marsh pennywort = Hydrocotyle ranunculoides (V:D)
Florida blue lettuce = Lactuca canadensis (V:D)
floury funnel lichen = Cladonia cenotea (L)
flower, tunic → tunic flower
flowering dogwood = Cornus nuttallii (V:D)
flowering quillwort = Lilaea scilloides (V:M)
flowering-rush = Butomus umbellatus (V:M)
fluellen, sharp-leaved → sharp-leaved fluellen
foamflower = Tiarella trifoliata (V:D)
foamflower, one-leaved → one-leaved foamflower
foamflower, three-leaved → three-leaved foamflower
fog, Yorkshire → Yorkshire fog
foliose willowherb = Epilobium foliosum (V:D)
fool's huckleberry = Menziesia ferruginea (V:D)
fool's-onion = Triteleia hyacinthina (V:M)
foot, deer → deer foot
foot, whiteman's → whiteman's foot
forest speckleback = Punctelia subrudecta (L)
forget-me-not = Myosotis scorpioides (V:D)
forget-me-not, blue → blue forget-me-not
forget-me-not, common → common forget-me-not
forget-me-not, field → field forget-me-not
forget-me-not, mountain → mountain forget-me-not
forget-me-not, small-flowered → small-flowered forget-me-not
forget-me-not, spring → spring forget-me-not
forget-me-not, wood → wood forget-me-not
forked catchfly = Silene dichotoma (V:D)
forked shore lichen = Cladonia scabriuscula (L)
forking bone = Hypogymnia imshaugii (L)
forking bone = Hypogymnia inactiva (L)
forking tube lichen = Hypogymnia imshaugii (L)
four-angled mountain-heather = Cassiope tetragona (V:D)
four-leaved all-seed = Polycarpon tetraphyllum (V:D)
four-leaved mare's-tail = Hippuris tetraphylla (V:D)
four-parted gentian = Gentianella propinqua (V:D)
four-toothed log moss = Tetraphis pellucida (B:M)
fowl bluegrass = Poa palustris (V:M)
fowl mannagrass = Glyceria striata (V:M)
Fowler's knotweed = Polygonum fowleri (V:D)
fox sedge = Carex vulpinoidea (V:M)
foxglove = Digitalis purpurea (V:D)
foxglove, common → common foxglove
foxtail barley = Hordeum jubatum (V:M)
foxtail bristlegrass = Setaria italica (V:M)
foxtail muhly = Muhlenbergia andina (V:M)
foxtail, green → green foxtail
foxtail, short-awned → short-awned foxtail
fragile bladder fern = Cystopteris fragilis (V:F)
fragile fern = Cystopteris fragilis (V:F)
fragile screw moss = Tortella fragilis (B:M)
fragile sedge = Carex membranacea (V:M)
fragrant popcornflower = Plagiobothrys figuratus (V:D)
fragrant shield fern = Dryopteris fragrans (V:F)
fragrant water-lily = Nymphaea odorata (V:D)
fragrant wood fern = Dryopteris fragrans (V:F)
Franklin's lady's-slipper = Cypripedium passerinum (V:M)
Franklin's phacelia = Phacelia franklinii (V:D)
Franklin's scorpionweed = Phacelia franklinii (V:D)
Franklin's sedge = Carex franklinii (V:M)
freckle pelt = Peltigera britannica (L)
freckle pelt = Peltigera aphthosa (L)
freckle pelt = Peltigera leucophlebia (L)
freckled lichen = Peltigera aphthosa (L)
freckled milk-vetch = Astragalus lentiginosus (V:D)
freckled milk-vetch = Astragalus lentiginosus var. salinus (V:D)
French hawksbeard = Crepis nicaeensis (V:D)
Fries' pondweed = Potamogeton friesii (V:M)
fringecup = Tellima grandiflora (V:D)
fringecup, tall → tall fringecup
fringed aster = Aster ciliolatus (V:D)
fringed brome = Bromus ciliatus (V:M)
fringed grass-of-Parnassus = Parnassia fimbriata (V:D)
fringed loosestrife = Lysimachia ciliata (V:D)
fringed moon = Sticta weigelii (L)
fringed owl = Solorina spongiosa (L)
fringed rocktripe = Umbilicaria cylindrica (L)
fringed rosette = Physcia leptalea (L)
fringed rosette = Physcia tenella (L)

fringed ruffle = Tuckermannopsis americana (L)
fritillary, riceroot → riceroot fritillary
fritillary, yellow → yellow fritillary
frog orchid = Coeloglossum viride (V:M)
frog pelt = Peltigera hymenina (L)
frog pelt = Peltigera degenii (L)
frog pelt = Peltigera polydactylon (L)
frog pelt = Peltigera pacifica (L)
frog pelt = Peltigera neopolydactyla (L)
frog pelt = Peltigera neckeri (L)
frog pelt = Peltigera kristinssonii (L)
frost, bordered → bordered frost
frost, bottlebrush → bottlebrush frost
frost, ground → ground frost
frosted rocktripe = Umbilicaria vellea (L)
frosted rosette = Physcia dimidiata (L)
frosted rosette = Physcia biziana (L)
Fuller's teasel = Dipsacus fullonum (V:D)
fumatory, bastard → bastard fumatory
fumatory, common → common fumatory
furled paperdoll = Flavocetraria cucullata (L)
furrowed shield lichen = Parmelia sulcata (L)
fuzzy-spiked wildrye = Leymus innovatus (V:M)
fuzzy-tongued penstemon = Penstemon eriantherus (V:D)

G

Gairdner's yampah = Perideridia gairdneri (V:D)
gale, sweet → sweet gale
galinsoga, small-flowered → small-flowered galinsoga
Garber's sedge = Carex garberi (V:M)
garden asparagus = Asparagus officinalis (V:M)
garden coreopsis = Coreopsis lanceolata (V:D)
garden cress = Lepidium sativum (V:D)
garden heliotrope = Valeriana officinalis (V:D)
garden orache = Atriplex hortensis (V:D)
garden radish = Raphanus sativus (V:D)
garden sorrel = Rumex acetosa (V:D)
garlic mustard = Alliaria petiolata (V:D)
garlic, field → field garlic
Garry oak = Quercus garryana (V:D)
gaura, scarlet → scarlet gaura
gentian, broad-petalled → broad-petalled gentian
gentian, four-parted → four-parted gentian
gentian, glaucous → glaucous gentian
gentian, king → king gentian
gentian, moss → moss gentian
gentian, mountain bog → mountain bog gentian
gentian, northern → northern gentian
gentian, prairie → prairie gentian
gentian, slender → slender gentian
gentian, spurred → spurred gentian
gentian, swamp → swamp gentian
geranium, Bicknell's → Bicknell's geranium
geranium, Carolina → Carolina geranium
geranium, cut-leaved → cut-leaved geranium
geranium, dovefoot → dovefoot geranium
geranium, feather → feather geranium
geranium, northern → northern geranium
geranium, Richardson's → Richardson's geranium
geranium, Robert → Robert geranium
geranium, small-flowered → small-flowered geranium
geranium, sticky → sticky geranium
geranium, sticky purple → sticky purple geranium
geranium, white → white geranium
Gerard's rush = Juncus gerardii (V:M)
germander speedwell = Veronica chamaedrys (V:D)
germander, American → American germander
Geyer's desert-parsley = Lomatium geyeri (V:D)
Geyer's lomatium = Lomatium geyeri (V:D)
Geyer's onion = Allium geyeri (V:M)
Geyer's sedge = Carex geyeri (V:M)
Geyer's willow = Salix geyeriana (V:D)
giant bur-reed = Sparganium eurycarpum (V:M)
giant chain fern = Woodwardia fimbriata (V:F)
giant cow-parsnip = Heracleum mantegazzianum (V:D)
giant feather moss = Calliergon giganteum (B:M)
giant goldenrod = Solidago gigantea (V:D)
giant helleborine = Epipactis gigantea (V:M)
giant horsetail = Equisetum telmateia (V:F)
giant-hyssop = Agastache foeniculum (V:D)
giant-hyssop, nettle-leaved → nettle-leaved giant-hyssop
giant knotweed = Polygonum sachalinense (V:D)
giant mannagrass = Glyceria maxima (V:M)
giant reedgrass = Phragmites australis (V:M)
giant water moss = Calliergon giganteum (B:M)
giant wildrye = Leymus cinereus (V:M)
gilia, bluefield → bluefield gilia
gilia, globe → globe gilia
gilia, scarlet → scarlet gilia
gilia, shy → shy gilia
gily-flower, blue-head → blue-head gily-flower
ginger, wild → wild ginger
Girgensohn's peat moss = Sphagnum girgensohnii (B:M)
glabrous dwarf willow = Salix reticulata ssp. glabellicarpa (V:D)
glacier lily = Erythronium grandiflorum (V:M)
glacier sedge = Carex glacialis (V:M)
glandular Labrador tea = Ledum glandulosum (V:D)
glasswort, American → American glasswort
glaucous bluegrass = Poa glauca (V:M)
glaucous gentian = Gentiana glauca (V:D)
glaucous-leaved honeysuckle = Lonicera dioica var. glaucescens (V:D)
glehnia, American → American glehnia
globe gilia = Gilia capitata (V:D)
globe-mallow, Munroe's → Munroe's globe-mallow
globe-mallow, scarlet → scarlet globe-mallow
globe-mallow, streambank → streambank globe-mallow
globe-pod hoary-cress = Cardaria pubescens (V:D)
globeflower = Trollius laxus (V:D)
globeflower = Trollius laxus ssp. albiflorus (V:D)
glomerate bur-reed = Sparganium glomeratum (V:M)
glow moss = Aulacomnium palustre (B:M)
Gmelin's orache = Atriplex gmelinii (V:D)
Gmelin's sedge = Carex gmelinii (V:M)
gnome-plant = Hemitomes congestum (V:D)
goat-chicory, annual → annual goat-chicory
goat lichen = Physcia aipolia (L)
goat's-beard = Aruncus dioicus (V:D)
Goblin's gold = Schistostega pennata (B:M)
godetia = Clarkia amoena (V:D)
gold-edge lichen = Pseudocyphellaria crocata (L)
gold-of-pleasure = Camelina sativa (V:D)
gold spoon moss = Calliergon richardsonii (B:M)
gold star = Crocidium multicaule (V:D)
gold, Goblin's → Goblin's gold
gold, spring → spring gold
golden-aster = Heterotheca villosa (V:D)
golden-aster, hairy → hairy golden-aster
golden carpet = Chrysosplenium iowense (V:D)
golden chain tree = Laburnum anagyroides (V:D)

golden clematis = Clematis tangutica (V:D)
golden corydalis = Corydalis aurea (V:D)
golden curls moss = Homalothecium aeneum (B:M)
golden daisy = Erigeron aureus (V:D)
golden dock = Rumex maritimus (V:D)
golden draba = Draba aurea (V:D)
golden-eyed-grass = Sisyrinchium californicum (V:M)
golden fleabane = Erigeron aureus (V:D)
golden fuzzy fen moss = Tomenthypnum nitens (B:M)
golden Indian paintbrush = Castilleja levisecta (V:D)
golden moss = Tomenthypnum nitens (B:M)
golden paintbrush = Castilleja levisecta (V:D)
golden ragged moss = Brachythecium salebrosum (B:M)
golden-saxifrage, Iowa → Iowa golden-saxifrage
golden-saxifrage, northern → northern golden-saxifrage
golden-saxifrage, Wright's → Wright's golden-saxifrage
golden sedge = Carex aurea (V:M)
golden short-capsuled moss = Brachythecium frigidum (B:M)
golden star moss = Campylium stellatum (B:M)
golden whitlow-grass = Draba aurea (V:D)
goldenback fern = Pentagramma triangularis (V:F)
goldenrod, alpine → alpine goldenrod
goldenrod, Canada → Canada goldenrod
goldenrod, dune → dune goldenrod
goldenrod, giant → giant goldenrod
goldenrod, gray → gray goldenrod
goldenrod, late → late goldenrod
goldenrod, Missouri → Missouri goldenrod
goldenrod, mountain → mountain goldenrod
goldenrod, northern → northern goldenrod
goldenrod, smooth → smooth goldenrod
goldenrod, spikelike → spikelike goldenrod
goldenrod, western → western goldenrod
goldentop, flat-top → flat-top goldentop
goldfields, hairy → hairy goldfields
goldmoss stonecrop = Sedum acre (V:D)
goldthread = Coptis trifolia (V:D)
goldthread, fern-leaved → fern-leaved goldthread
goldthread, spleenwort-leaved → spleenwort-leaved goldthread
goldthread, three-leaved → three-leaved goldthread
golf club moss = Catoscopium nigritum (B:M)
goose-foot yellow violet = Viola purpurea (V:D)
goose-necked moss = Rhytidiadelphus triquetrus (B:M)
gooseberry, alpine prickly → alpine prickly gooseberry
gooseberry, black → black gooseberry
gooseberry, black swamp → black swamp gooseberry
gooseberry, coastal black → coastal black gooseberry
gooseberry, gummy → gummy gooseberry
gooseberry, mountain → mountain gooseberry
gooseberry, northern → northern gooseberry
gooseberry, northern smooth → northern smooth gooseberry
gooseberry, swamp black → swamp black gooseberry
gooseberry, white-stemmed → white-stemmed gooseberry
gooseberry, wild → wild gooseberry
goosefoot, red → red goosefoot
goosefoot, upright → upright goosefoot
goosetongue = Plantago maritima (V:D)
Gorman's douglasia = Douglasia gormanii (V:D)
Gorman's penstemon = Penstemon gormanii (V:D)
gorse = Ulex europaeus (V:D)
goutweed = Aegopodium podagraria (V:D)
graceful arrow-grass = Triglochin concinnum (V:M)
graceful cinquefoil = Potentilla gracilis (V:D)
graceful mountain sedge = Carex podocarpa (V:M)
graceful sedge = Carex praegracilis (V:M)
grama, blue → blue grama
grand fir = Abies grandis (V:G)
granite moss = Andreaea rupestris (B:M)
granulated shadow = Phaeophyscia adiastola (L)
granulated shadow = Phaeophyscia orbicularis (L)
granulated shield = Parmotrema crinitum (L)
grass-leaved death-camas = Zigadenus venenosus (V:M)
grass-leaved pondweed = Potamogeton gramineus (V:M)
grass-leaved starwort = Stellaria graminea (V:D)
grass-of-Parnassus, fringed → fringed grass-of-Parnassus
grass-of-Parnassus, Kotzebue's → Kotzebue's grass-of-Parnassus
grass-of-Parnassus, northern → northern grass-of-Parnassus
grass-of-Parnassus, small-flowered → small-flowered grass-of-Parnassus
grass peavine = Lathyrus sphaericus (V:D)
grass, Bermuda → Bermuda grass
grass, big quaking → big quaking grass
grass, little quaking → little quaking grass
grass, marram → marram grass
grass, moss → moss grass
grass, needle-and-thread → needle-and-thread grass
grass, poverty → poverty grass
grass, purple oat → purple oat grass
grass, rough hair → rough hair grass
grass, satin → satin grass
grass, six-weeks → six-weeks grass
grass, squirreltail → squirreltail grass
grass, western needle → western needle grass
grassland lupine = Lupinus arbustus ssp. neolaxiflorus (V:D)
grassland saxifrage = Saxifraga integrifolia (V:D)
gray alder = Alnus incana (V:D)
gray goldenrod = Solidago nemoralis (V:D)
gray hawksbeard = Crepis intermedia (V:D)
gray-leaved draba = Draba cinerea (V:D)
gray-leaved whitlow-grass = Draba cinerea (V:D)
Gray's desert-parsley = Lomatium grayi (V:D)
Gray's lomatium = Lomatium grayi (V:D)
Great Basin nemophila = Nemophila breviflora (V:D)
great burdock = Arctium lappa (V:D)
great burnet = Sanguisorba officinalis (V:D)
great camas = Camassia leichtlinii (V:M)
great duckmeat = Spirodela polyrhiza (V:M)
great leopard's-bane = Doronicum pardalianches (V:D)
great mullein = Verbascum thapsus (V:D)
great northern aster = Aster modestus (V:D)
great spurred violet = Viola selkirkii (V:D)
great sundew = Drosera anglica (V:D)
greater water-starwort = Callitriche heterophylla (V:D)
green alder = Alnus viridis (V:D)
green bristlegrass = Setaria viridis (V:M)
green false hellebore = Veratrum viride (V:M)
green fescue = Festuca viridula (V:M)
green-flowered bog orchid = Platanthera hyperborea (V:M)
green-flowered rein orchid = Platanthera hyperborea (V:M)
green foxtail = Setaria viridis (V:M)
green-fruited sedge = Carex interrupta (V:M)
green-keeled cotton-grass = Eriophorum viridicarinatum (V:M)
green kidney lichen = Nephroma arcticum (L)
green loop = Hypotrachyna sinuosa (L)
green map = Rhizocarpon geographicum (L)
green map lichen = Rhizocarpon geographicum (L)
green moon = Sticta wrightii (L)
green mouse = Psoroma hypnorum (L)
green needlegrass = Nassella viridula (V:M)
green paw = Nephroma arcticum (L)
green pigweed = Amaranthus powellii (V:D)
green rabbit-brush = Chrysothamnus viscidiflorus (V:D)
green reindeer lichen = Cladina mitis (L)
green saxifraga = Chrysosplenium tetrandrum (V:D)
green sedge = Carex viridula (V:M)
green-sheathed sedge = Carex feta (V:M)
green shield = Parmelia fraudans (L)
green sorrel = Rumex acetosa (V:D)
green speckleback = Flavopunctelia flaventior (L)
green starburst = Parmeliopsis ambigua (L)
green stubble = Calicium viride (L)
green-tongue liverwort = Marchantia polymorpha (B:H)
green tongue wort = Marchantia polymorpha (B:H)

green wintergreen = Pyrola chlorantha (V:D)
greenish wintergreen = Pyrola chlorantha (V:D)
Greenland primrose = Primula egaliksensis (V:D)
Greenland woodrush = Luzula groenlandica (V:M)
grey beach peavine = Lathyrus littoralis (V:D)
grey-eyed rosette = Physcia aipolia (L)
grey haircap moss = Pogonatum urnigerum (B:M)
grey horsebrush = Tetradymia canescens (V:D)
grey-leaved willow = Salix glauca (V:D)
grey loop = Hypotrachyna revoluta (L)
grey reindeer lichen = Cladina rangiferina (L)
grey sedge = Carex canescens (V:M)
grey star lichen = Physcia stellaris (L)
grey starburst = Parmeliopsis hyperopta (L)
gristle, dusty → dusty gristle
gristle, punctured → punctured gristle
grizzly shield = Imshaugia aleurites (L)
gromwell, Columbia → Columbia gromwell
gromwell, lemonweed → lemonweed gromwell
gromwell, yellow → yellow gromwell
grooved agrimony = Agrimonia striata (V:D)
grooved gnome-cap moss = Encalypta rhaptocarpa (B:M)
ground-cedar = Lycopodium complanatum (V:F)
ground frost = Physconia muscigena (L)
ground-ivy = Glechoma hederacea (V:D)
ground juniper = Juniperus communis (V:G)
ground-pine = Lycopodium dendroideum (V:F)
ground plum = Astragalus crassicarpus (V:D)
groundcone = Boschniakia hookeri (V:D)
groundcone, northern → northern groundcone
groundcone, Vancouver → Vancouver groundcone
groundsel, arrow-leaved → arrow-leaved groundsel
groundsel, balsam → balsam groundsel
groundsel, beach → beach groundsel
groundsel, black-tipped → black-tipped groundsel
groundsel, common → common groundsel
groundsel, large-headed → large-headed groundsel
groundsel, Macoun's → Macoun's groundsel
groundsel, purple-haired → purple-haired groundsel
groundsel, western → western groundsel
groundsel, wood → wood groundsel
groundsel, woolly → woolly groundsel
groundsel, Yukon → Yukon groundsel
groundsmoke, dwarf → dwarf groundsmoke
groundsmoke, hairstem → hairstem groundsmoke
groundsmoke, spreading → spreading groundsmoke
grouseberry = Vaccinium scoparium (V:D)
gum-succory = Chondrilla juncea (V:D)
gummy gooseberry = Ribes lobbii (V:D)
gumweed, curly-cup → curly-cup gumweed
gumweed, entire-leaved → entire-leaved gumweed
gumweed, Puget Sound → Puget Sound gumweed

H

hackelia, blue → blue hackelia
hackelia, many-flowered → many-flowered hackelia
hackelia, nodding → nodding hackelia
hackelia, Okanogan → Okanogan hackelia
hackelia, spreading → spreading hackelia
hair bentgrass = Agrostis scabra (V:M)
hair fescue = Festuca filiformis (V:M)
hair-like sedge = Carex capillaris (V:M)
hair sedge = Carex capillaris (V:M)
hair, common witch's → common witch's hair
haircap, awned → awned haircap
haircap, bog → bog haircap
haircap, common → common haircap
haircap, juniper → juniper haircap
haircap, slender → slender haircap
hairgrass dropseed = Sporobolus airoides (V:M)
hairgrass, annual → annual hairgrass
hairgrass, early → early hairgrass
hairgrass, mountain → mountain hairgrass
hairgrass, silver → silver hairgrass
hairgrass, slender → slender hairgrass
hairgrass, tufted → tufted hairgrass
hairgrass, water → water hairgrass
hairgrass, wavy → wavy hairgrass
hairstem groundsmoke = Gayophytum ramosissimum (V:D)
hairy arnica = Arnica mollis (V:D)
hairy bitter-cress = Cardamine hirsuta (V:D)
hairy buttercup = Ranunculus sardous (V:D)
hairy butterwort = Pinguicula villosa (V:D)
hairy-cap moss = Polytrichum commune (B:M)
hairy cat's-ear = Hypochoeris radicata (V:D)
hairy crabgrass = Digitaria sanguinalis (V:M)
hairy flax = Camelina microcarpa (V:D)
hairy-fruited sedge = Carex lasiocarpa (V:M)
hairy golden-aster = Heterotheca villosa (V:D)
hairy goldfields = Lasthenia maritima (V:D)
hairy hawkbit = Leontodon taraxacoides (V:D)
hairy honeysuckle = Lonicera hispidula (V:D)
hairy lantern moss = Rhizomnium magnifolium (B:M)
hairy manzanita = Arctostaphylos columbiana (V:D)
hairy nightshade = Solanum sarrachoides (V:D)
hairy rockcress = Arabis hirsuta (V:D)
hairy screw moss = Tortula ruralis (B:M)
hairy-stemmed willowherb = Epilobium mirabile (V:D)
hairy umbrellawort = Mirabilis hirsuta (V:D)
hairy vetch = Vicia villosa (V:D)
hairy vetch = Vicia hirsuta (V:D)
hairy water-clover = Marsilea vestita (V:F)
hairy wildrye = Elymus hirsutus (V:M)
hairy willowherb = Epilobium hirsutum (V:D)
Haleakala fir clubmoss = Huperzia haleakalae (V:F)
halimolobos, soft → soft halimolobos
halimolobos, Whited's → Whited's halimolobos
Hall's willowherb = Epilobium halleanum (V:D)
hanging basket moss = Rhytidiadelphus loreus (B:M)
hanging millipede liverwort = Frullania tamarisci ssp. nisquallensis (B:H)
hanging moss = Antitrichia curtipendula (B:M)
hard fescue = Festuca trachyphylla (V:M)
hard scale liverwort = Mylia anomala (B:H)
hard-stemmed bulrush = Scirpus acutus (V:M)
hardhack = Spiraea douglasii (V:D)
hare's-ear mustard = Conringia orientalis (V:D)
hare's-foot clover = Trifolium arvense (V:D)
harebell, Alaskan → Alaskan harebell
harebell, arctic → arctic harebell
harebell, common → common harebell
harebell, mountain → mountain harebell
harebell, Scouler's → Scouler's harebell
harefoot sedge = Carex leporina (V:M)
Harford's melic = Melica harfordii (V:M)
Harkness' flaxflower = Linanthus harknessii (V:D)
Harkness' linanthus = Linanthus harknessii (V:D)
harsh paintbrush = Castilleja hispida (V:D)
harvest brodiaea = Brodiaea coronaria (V:M)
Hatcher's fan wort = Barbilophozia hatcheri (B:H)

hawkbit, autumn → autumn hawkbit
hawkbit, hairy → hairy hawkbit
hawksbeard, annual → annual hawksbeard
hawksbeard, beaked → beaked hawksbeard
hawksbeard, dandelion → dandelion hawksbeard
hawksbeard, dwarf → dwarf hawksbeard
hawksbeard, elegant → elegant hawksbeard
hawksbeard, French → French hawksbeard
hawksbeard, gray → gray hawksbeard
hawksbeard, low → low hawksbeard
hawksbeard, slender → slender hawksbeard
hawksbeard, smooth → smooth hawksbeard
hawksbeard, western → western hawksbeard
hawkweed-leaved saxifrage = Saxifraga hieraciifolia (V:D)
hawkweed, hound's-tongue → hound's-tongue hawkweed
hawkweed, mouse-ear → mouse-ear hawkweed
hawkweed, narrow-leaved → narrow-leaved hawkweed
hawkweed, orange → orange hawkweed
hawkweed, Scouler's → Scouler's hawkweed
hawkweed, slender → slender hawkweed
hawkweed, tall → tall hawkweed
hawkweed, wall → wall hawkweed
hawkweed, white → white hawkweed
hawkweed, white-flowered → white-flowered hawkweed
hawkweed, woolly → woolly hawkweed
hawthorn, black → black hawthorn
hawthorn, Columbia → Columbia hawthorn
hawthorn, common → common hawthorn
hawthorn, red → red hawthorn
hawthorn, Suksdorf's → Suksdorf's hawthorn
hay sedge = Carex foenea (V:M)
Hayden's sedge = Carex haydeniana (V:M)
hazelnut = Corylus avellana (V:D)
hazelnut, beaked → beaked hazelnut
hazelnut, California → California hazelnut
head, steer's → steer's head
heal-all = Prunella vulgaris (V:D)
heart-leaved Alexanders = Zizia aptera (V:D)
heart-leaved arnica = Arnica cordifolia (V:D)
heart-leaved buttercup = Ranunculus cardiophyllus (V:D)
heart-leaved spring-beauty = Claytonia cordifolia (V:D)
heart-leaved twayblade = Listera cordata (V:M)
heart-podded hoary-cress = Cardaria draba (V:D)
heart, bleeding → bleeding heart
heart, bloody → bloody heart
heart, Pacific bleeding → Pacific bleeding heart
heath bone = Hypogymnia subobscura (L)
heath, spiny → spiny heath
heather = Calluna vulgaris (V:D)
heather-grass = Danthonia decumbens (V:M)
heather rags = Hypogymnia physodes (L)
heather, Mertens' moss → Mertens' moss heather
heather, Scotch → Scotch heather
heather, white moss → white moss heather
hedge bindweed = Calystegia sepium (V:D)
hedge-hyssop, bractless → bractless hedge-hyssop
hedge-hyssop, common American → common American hedge-hyssop
hedge mustard = Sisymbrium officinale (V:D)
hedge-nettle, field → field hedge-nettle
hedge-nettle, marsh → marsh hedge-nettle
hedge-nettle, swamp → swamp hedge-nettle
hedgehog dogtail = Cynosurus echinatus (V:M)
Hedwig's rock moss = Hedwigia ciliata (B:M)
hedysarum, alpine → alpine hedysarum
hedysarum, northern → northern hedysarum
hedysarum, sulphur → sulphur hedysarum
hedysarum, western → western hedysarum
hedysarum, yellow → yellow hedysarum
helianthella, Douglas' → Douglas' helianthella
helianthella, Rocky Mountain → Rocky Mountain helianthella
heliotrope, garden → garden heliotrope
hellebore, green false → green false hellebore
hellebore, Indian → Indian hellebore
helleborine = Epipactis helleborine (V:M)
helleborine, giant → giant helleborine
helmet, policeman's → policeman's helmet
hemlock-parsley = Cnidium cnidiifolium (V:D)
hemlock water-parsnip = Sium suave (V:D)
hemlock, mountain → mountain hemlock
hemlock, western → western hemlock
hemp dogbane = Apocynum cannabinum (V:D)
hemp-nettle = Galeopsis tetrahit (V:D)
hemp-nettle, common → common hemp-nettle
henbit dead-nettle = Lamium amplexicaule (V:D)
Henderson's checker-mallow = Sidalcea hendersonii (V:D)
Henderson's sedge = Carex hendersonii (V:M)
Henderson's shootingstar = Dodecatheon hendersonii (V:D)
herb-Robert = Geranium robertianum (V:D)
hesperochiron, dwarf → dwarf hesperochiron
heterocodon = Heterocodon rariflorum (V:D)
high alpine butterweed = Senecio conterminus (V:D)
high bush-cranberry = Viburnum opulus (V:D)
high mountain arnica = Arnica gracilis (V:D)
highbush-cranberry = Viburnum edule (V:D)
hill milk-vetch = Astragalus collinus (V:D)
hillside milk-vetch = Astragalus collinus (V:D)
Himalayan blackberry = Rubus discolor (V:D)
Himalayan knotweed = Polygonum polystachyum (V:D)
hoary alyssum = Berteroa incana (V:D)
hoary aster = Machaeranthera canescens (V:D)
hoary-cress, globe-pod → globe-pod hoary-cress
hoary-cress, heart-podded → heart-podded hoary-cress
hoary false yarrow = Chaenactis douglasii (V:D)
hoary plantain = Plantago media (V:D)
hoary rock moss = Racomitrium lanuginosum (B:M)
hoary rosette = Physcia aipolia (L)
hoary willow = Salix candida (V:D)
hoary yellow cress = Rorippa barbareifolia (V:D)
Holboell's rockcress = Arabis holboellii (V:D)
holly, English → English holly
hollyhock, mountain → mountain hollyhock
Holm's Rocky Mountain sedge = Carex scopulorum (V:M)
holy-clover = Onobrychis viciifolia (V:D)
honesty = Lunaria annua (V:D)
honeysuckle, bluefly → bluefly honeysuckle
honeysuckle, bracked → bracked honeysuckle
honeysuckle, Etruscan → Etruscan honeysuckle
honeysuckle, glaucous-leaved → glaucous-leaved honeysuckle
honeysuckle, hairy → hairy honeysuckle
honcysuckle, limber → limber honeysuckle
honeysuckle, orange → orange honeysuckle
honeysuckle, purple → purple honeysuckle
honeysuckle, red → red honeysuckle
honeysuckle, twining → twining honeysuckle
honeysuckle, Utah → Utah honeysuckle
honeysuckle, western trumpet → western trumpet honeysuckle
hood lichen = Physcia adscendens (L)
Hood's phlox = Phlox hoodii (V:D)
Hood's sedge = Carex hoodii (V:M)
hooded bone = Hypogymnia physodes (L)
hooded ladies' tresses = Spiranthes romanzoffiana (V:M)
hooded lichen = Physcia adscendens (L)
hooded moss = Orthotrichum speciosum (B:M)
hooded rosette = Physcia adscendens (L)
hooded tube lichen = Hypogymnia physodes (L)
hook-leaf fern moss = Thuidium recognitum (B:M)
hook moss = Sanionia uncinata (B:M)
hooked spring moss = Warnstorfia exannulata var. exannulata (B:M)
Hooker's cinquefoil = Potentilla hookeriana (V:D)
Hooker's fairybells = Disporum hookeri (V:M)

Hooker's onion = Allium acuminatum (V:M)
Hooker's thistle = Cirsium hookerianum (V:D)
Hooker's townsendia = Townsendia hookeri (V:D)
Hooker's willow = Salix hookeriana (V:D)
hop-clover, low → low hop-clover
hop-clover, small → small hop-clover
hop, common → common hop
horehound, common → common horehound
horehound, cut-leaved water → cut-leaved water horehound
horehound, northern water → northern water horehound
horehound, rough water → rough water horehound
horn cladonia = Cladonia cornuta (L)
horned dandelion = Taraxacum lyratum (V:D)
horned pondweed = Zannichellia palustris (V:M)
Hornemann's willowherb = Epilobium hornemannii (V:D)
hornwort = Ceratophyllum demersum (V:D)
hornwort, common → common hornwort
hornwort, spring → spring hornwort
horsebrush, grey → grey horsebrush
horsehair, brittle → brittle horsehair
horsehair, edible → edible horsehair
horsehair, simple → simple horsehair
horsehair, speckled → speckled horsehair
horseradish, common → common horseradish
horsetail, common → common horsetail
horsetail, field → field horsetail
horsetail, giant → giant horsetail
horsetail, marsh → marsh horsetail
horsetail, meadow → meadow horsetail
horsetail, swamp → swamp horsetail
horsetail, water → water horsetail
horsetail, wood → wood horsetail
horsetail, woodland → woodland horsetail
horseweed = Conyza canadensis var. glabrata (V:D)
hound's-tongue hawkweed = Hieracium cynoglossoides (V:D)
hound's-tongue, common → common hound's-tongue
Howell's montia = Montia howellii (V:D)
Howell's quillwort = Isoetes howellii (V:F)
Howell's triteleia = Triteleia howellii (V:M)
Howell's violet = Viola howellii (V:D)
huckleberry, black → black huckleberry
huckleberry, blue-leaved → blue-leaved huckleberry
huckleberry, Cascade → Cascade huckleberry
huckleberry, evergreen → evergreen huckleberry
huckleberry, fool's → fool's huckleberry
huckleberry, red → red huckleberry
Huddelson's locoweed = Oxytropis huddelsonii (V:D)
Huddelson's oxytrope = Oxytropis huddelsonii (V:D)
Hudson Bay sedge = Carex heleonastes (V:M)
humped bladderwort = Utricularia gibba (V:D)
hutchinsia = Hutchinsia procumbens (V:D)
hybrid white spruce = Picea engelmannii × glauca (V:G)
hypnum, curly → curly hypnum

I

Iceland poppy = Papaver nudicaule (V:D)
icelandmoss = Cetraria islandica (L)
icelandmoss = Cetrariella delisei (L)
icelandmoss = Cetraria laevigata (L)
icelandmoss = Cetraria ericetorum (L)
icelandmoss = Tuckermannopsis subalpina (L)
icelandmoss, blackened → blackened icelandmoss
Idaho blue-eyed-grass = Sisyrinchium idahoense var. idahoense (V:M)
Idaho blue-eyed-grass = Sisyrinchium idahoense (V:M)
Idaho fescue = Festuca idahoensis (V:M)
Illinois pondweed = Potamogeton illinoensis (V:M)
Indian consumption plant = Lomatium nudicaule (V:D)
Indian hellebore = Veratrum viride (V:M)
Indian-hemp = Apocynum cannabinum (V:D)
Indian lovegrass = Eragrostis pilosa (V:M)
Indian mustard = Brassica juncea (V:D)
Indian paint = Chenopodium capitatum (V:D)
Indian-paintbrush, northern → northern Indian-paintbrush
Indian-paintbrush, palish → palish Indian-paintbrush
Indian-paintbrush, Raup's → Raup's Indian-paintbrush
Indian-paintbrush, unalaska → unalaska Indian-paintbrush
Indian-pipe = Monotropa uniflora (V:D)
Indian-plum = Oemleria cerasiformis (V:D)
Indian's-dream fern = Aspidotis densa (V:F)
Indian strawberry = Duchesnea indica (V:D)
Indian-tobacco = Lobelia inflata (V:D)
Indian-wheat = Plantago patagonica (V:D)
inflated sedge = Carex exsiccata (V:M)
inflated sedge = Carex vesicaria (V:M)
inland alkaligrass = Puccinellia interior (V:M)
inland bluegrass = Poa interior (V:M)
inland sedge = Carex interior (V:M)
interior Douglas-fir = Pseudotsuga menziesii var. glauca (V:G)
Iowa golden carpet = Chrysosplenium iowense (V:D)
Iowa golden-saxifrage = Chrysosplenium iowense (V:D)
ipomopsis, small-flowered → small-flowered ipomopsis
iris, Siberian → Siberian iris
iris, western blue → western blue iris
iris, yellow → yellow iris
iron lung = Lobaria hallii (L)
irregular polypody = Polypodium amorphum (V:F)
island koenigia = Koenigia islandica (V:D)
Italian clover = Trifolium incarnatum (V:D)
ivy-leaved duckweed = Lemna trisulca (V:M)
ivy-leaved toadflax = Cymbalaria muralis (V:D)
ivy, English → English ivy

J

jack pine = Pinus banksiana (V:G)
Jacob's-ladder, annual → annual Jacob's-ladder
Jacob's-ladder, elegant → elegant Jacob's-ladder
Jacob's-ladder, northern → northern Jacob's-ladder
Jacob's-ladder, showy → showy Jacob's-ladder
Jacob's-ladder, skunk → skunk Jacob's-ladder
Jacob's-ladder, sticky → sticky Jacob's-ladder
Jacob's-ladder, tall → tall Jacob's-ladder
Jameson's liverwort = Jamesoniella autumnalis (B:H)
Japanese brome = Bromus japonicus (V:M)
Japanese butterbur = Petasites japonicus (V:D)
Japanese coltsfoot = Petasites japonicus (V:D)

Japanese eel-grass = Zostera japonica (V:M)
Japanese knotweed = Polygonum cuspidatum (V:D)
Japanese pearlwort = Sagina japonica (V:D)
jaumea, fleshy → fleshy jaumea
Jeffrey's shootingstar = Dodecatheon jeffreyi (V:D)
jellied vinyl = Leptogium brebissonii (L)
jelly tarpaper = Collema auriforme (L)
jenny, creeping → creeping jenny
jersey knapweed = Centaurea paniculata (V:D)
Jerusalem-oak = Chenopodium botrys (V:D)
jewelweed = Impatiens noli-tangere (V:D)
jewelweed, western → western jewelweed
jimsonweed = Datura stramonium (V:D)
joe-pye-weed, spotted → spotted joe-pye-weed
Johnny-jump-up = Viola tricolor (V:D)
jointed rush = Juncus articulatus (V:M)
Jordal's locoweed = Oxytropis jordalii (V:D)
joy, traveler's → traveler's joy
Juneberry = Amelanchier alnifolia (V:D)
Junegrass = Koeleria macrantha (V:M)
juniper haircap = Polytrichum juniperinum (B:M)
juniper haircap moss = Polytrichum juniperinum (B:M)
juniper moss = Polytrichum juniperinum (B:M)
juniper, common → common juniper
juniper, creeping → creeping juniper
juniper, ground → ground juniper
juniper, Rocky Mountain → Rocky Mountain juniper

K

Kalm's lobelia = Lobelia kalmii (V:D)
Kamchatka lyre-leaved rockcress = Arabis lyrata var. kamchatica (V:D)
Kamchatka spike-rush = Eleocharis kamtschatica (V:M)
Kellogg's knotweed = Polygonum polygaloides ssp. kelloggii (V:D)
Kellogg's rush = Juncus kelloggii (V:M)
Kentucky bluegrass = Poa pratensis (V:M)
kidney-leaved buttercup = Ranunculus abortivus (V:D)
kidney-leaved violet = Viola renifolia (V:D)
kidney, pimpled → pimpled kidney
kidney, seaside → seaside kidney
king gentian = Gentiana sceptrum (V:D)
kinnikinnick = Arctostaphylos uva-ursi (V:D)
kitten-tails, Wyoming → Wyoming kitten-tails
knapweed, diffuse → diffuse knapweed
knapweed, jersey → jersey knapweed
knawel, annual → annual knawel
kneeling angelica = Angelica genuflexa (V:D)
knight's plume = Ptilium crista-castrensis (B:M)
knobbed scale = Waynea stoechadiana (L)
knotted rush = Juncus nodosus (V:M)
knotweed, Austin's → Austin's knotweed
knotweed, beach → beach knotweed
knotweed, black → black knotweed
knotweed, Blake's → Blake's knotweed
knotweed, bushy → bushy knotweed
knotweed, common → common knotweed
knotweed, Douglas' → Douglas' knotweed
knotweed, eastern → eastern knotweed
knotweed, Fowler's → Fowler's knotweed
knotweed, giant → giant knotweed
knotweed, Himalayan → Himalayan knotweed
knotweed, Japanese → Japanese knotweed
knotweed, Kellogg's → Kellogg's knotweed
knotweed, leafy dwarf → leafy dwarf knotweed
knotweed, Nepalese → Nepalese knotweed
knotweed, Nuttall's → Nuttall's knotweed
knotweed, oval-leaved → oval-leaved knotweed
knotweed, prostrate → prostrate knotweed
knotweed, Sachaline → Sachaline knotweed
knotweed, short-fringed → short-fringed knotweed
knotweed, spurry → spurry knotweed
knotweed, white-margin → white-margin knotweed
knotweed, wiry → wiry knotweed
knotweed, yellow-flowered → yellow-flowered knotweed
kobresia, Bellard's → Bellard's kobresia
kobresia, Siberian → Siberian kobresia
kobresia, simple → simple kobresia
koenigia, island → island koenigia
Kotzebue's grass-of-Parnassus = Parnassia kotzebuei (V:D)
Krajina's fescue = Festuca viviparoidea var. krajinae (V:M)
Krause's sedge = Carex krausei (V:M)
Kruckeberg's holly fern = Polystichum kruckebergii (V:F)
Kusche's lupine = Lupinus kuschei (V:D)

L

Labrador lousewort = Pedicularis labradorica (V:D)
Labrador tea = Ledum groenlandicum (V:D)
Labrador-tea, marsh → marsh Labrador-tea
laburnum = Laburnum anagyroides (V:D)
lace fern = Cheilanthes gracillima (V:F)
lace, Queen Anne's → Queen Anne's lace
lacepod, sand → sand lacepod
ladies' tresses = Spiranthes romanzoffiana (V:M)
lady fern = Athyrium filix-femina ssp. cyclosorum (V:F)
lady fern = Athyrium filix-femina (V:F)
lady's-mantle = Alchemilla subcrenata (V:D)
lady's-slipper, Franklin's → Franklin's lady's-slipper
lady's-slipper, mountain → mountain lady's-slipper
lady's-slipper, sparrow's-egg → sparrow's-egg lady's-slipper
lady's-smock = Cardamine pratensis (V:D)
lady's-thumb = Polygonum persicaria (V:D)
lamatium, nine-leaved → nine-leaved lamatium
lamb's-ear = Stachys byzantina (V:D)
lamb's-quarters = Chenopodium album (V:D)
lamb's-quarters, dark → dark lamb's-quarters
lance-fruited draba = Draba lonchocarpa (V:D)
lance-fruited whitlow-grass = Draba lonchocarpa (V:D)
lance-leaved draba = Draba cana (V:D)
lance-leaved figwort = Scrophularia lanceolata (V:D)
lance-leaved moonwort = Botrychium lanceolatum (V:F)
lance-leaved stonecrop = Sedum lanceolatum (V:D)
lance-leaved violet = Viola lanceolata (V:D)
lance-leaved whitlow-grass = Draba cana (V:D)
Langsdorf's lousewort = Pedicularis langsdorfii (V:D)
lanky moss = Rhytidiadelphus loreus (B:M)

lantern, swamp → swamp lantern
Lapland buttercup = Ranunculus lapponicus (V:D)
Lapland poppy = Papaver lapponicum (V:D)
Lapland reedgrass = Calamagrostis lapponica (V:M)
Lapland rosebay = Rhododendron lapponicum (V:D)
Lapland sedge = Carex lapponica (V:M)
larch, alpine → alpine larch
larch, subalpine → subalpine larch
larch, western → western larch
large-awned sedge = Carex macrochaeta (V:M)
large barnyard-grass = Echinochloa crusgalli (V:M)
large Canadian St. John's-wort = Hypericum majus (V:D)
large-flowered agoseris = Agoseris grandiflora (V:D)
large-flowered brickellia = Brickellia grandiflora (V:D)
large-flowered collomia = Collomia grandiflora (V:D)
large-flowered daisy = Erigeron grandiflorus (V:D)
large-flowered fleabane = Erigeron grandiflorus (V:D)
large-flowered triteleia = Triteleia grandiflora (V:M)
large-fruited desert-parsley = Lomatium macrocarpum (V:D)
large-fruited lomatium = Lomatium macrocarpum (V:D)
large hair moss = Oligotrichum parallelum (B:M)
large-headed groundsel = Senecio megacephalus (V:D)
large-headed sedge = Carex macrocephala (V:M)
large leafy moss = Rhizomnium glabrescens (B:M)
large-leaved avens = Geum macrophyllum (V:D)
large-leaved lupine = Lupinus polyphyllus (V:D)
large-leaved pondweed = Potamogeton amplifolius (V:M)
large-leaved rattlesnake orchid = Goodyera oblongifolia (V:M)
large mosquito fern = Azolla filiculoides (V:F)
large mountain monkey-flower = Mimulus tilingii (V:D)
large periwinkle = Vinca major (V:D)
large round-leaved rein orchid = Platanthera orbiculata (V:M)
large-sheathed pondweed = Potamogeton vaginatus (V:M)
large wintergreen = Pyrola asarifolia (V:D)
large yellow sedge = Carex flava (V:M)
larkspur, Burke's → Burke's larkspur
larkspur, dwarf → dwarf larkspur
larkspur, Menzies' → Menzies' larkspur
larkspur, Montana → Montana larkspur
larkspur, Nuttall's → Nuttall's larkspur
larkspur, slim → slim larkspur
larkspur, tall → tall larkspur
larkspur, upland → upland larkspur
late goldenrod = Solidago gigantea (V:D)
lattice bone = Hypogymnia oceanica (L)
lattice bone = Hypogymnia occidentalis (L)
lattice brown = Melanelia panniformis (L)
lattice pipe = Hypogymnia occidentalis (L)
laundered rag = Platismatia norvegica (L)
laurel, alpine → alpine laurel
laurel, swamp → swamp laurel
lawn moss = Brachythecium albicans (B:M)
lax-flowered bluegrass = Poa laxiflora (V:M)
leaf mustard = Brassica juncea (V:D)
leafy arnica = Arnica chamissonis (V:D)
leafy aster = Aster foliaceus (V:D)
leafy dwarf knotweed = Polygonum minimum (V:D)
leafy liverwort = Lophozia ventricosa (B:H)
leafy mitrewort = Mitella caulescens (V:D)
leafy spurge = Euphorbia esula (V:D)
leafy thistle = Cirsium foliosum (V:D)
least bladdery milk-vetch = Astragalus microcystis (V:D)
least moonwort = Botrychium simplex (V:F)
least mousetail = Myosurus minimus (V:D)
leather brown = Melanelia agnata (L)
leather brown = Melanelia stygia (L)
leatherleaf = Chamaedaphne calyculata (V:D)
leatherleaf polypody = Polypodium scouleri (V:F)
leatherleaf saxifrage = Leptarrhena pyrolifolia (V:D)
leathery grape fern = Botrychium multifidum (V:F)
leathery polypody = Polypodium scouleri (V:F)
Leiberg's daisy = Erigeron leibergii (V:D)
Leiberg's fleabane = Erigeron leibergii (V:D)
Lemmon's holly fern = Polystichum lemmonii (V:F)
Lemmon's needlegrass = Stipa lemmonii (V:M)
Lemmon's rockcress = Arabis lemmonii (V:D)
Lemmon's willow = Salix lemmonii (V:D)
lemon balm = Melissa officinalis (V:D)
lemon lichen = Candelaria concolor (L)
lemonweed gromwell = Lithospermum ruderale (V:D)
lens-fruited sedge = Carex lenticularis (V:M)
lentil, common water → common water lentil
leopard's-bane, great → great leopard's-bane
lesser bladderwort = Utricularia minor (V:D)
lesser cattail = Typha angustifolia (V:M)
lesser celandine = Ranunculus ficaria (V:D)
lesser panicled sedge = Carex diandra (V:M)
lesser rattlesnake-plantain = Goodyera repens (V:M)
lesser rushy milk-vetch = Astragalus convallarius (V:D)
lesser saltmarsh sedge = Carex glareosa (V:M)
lesser saltmarsh sedge = Carex glareosa var. amphigens (V:M)
lesser spearwort = Ranunculus flammula (V:D)
lesser swine-cress = Coronopus didymus (V:D)
lesser tussock sedge = Carex diandra (V:M)
lesser wintergreen = Pyrola minor (V:D)
Letterman's bluegrass = Poa lettermanii (V:M)
lettuce lichen = Platismatia lacunosa (L)
lettuce lung = Lobaria oregana (L)
lettuce, common blue → common blue lettuce
lettuce, Florida blue → Florida blue lettuce
lettuce, prickly → prickly lettuce
lettuce, Russian blue → Russian blue lettuce
lettuce, tall blue → tall blue lettuce
lettuce, wire → wire lettuce
Lewis' monkey-flower = Mimulus lewisii (V:D)
lewisia, alpine → alpine lewisia
lewisia, Columbia → Columbia lewisia
lewisia, three-leaved → three-leaved lewisia
lichen, arctic kidney → arctic kidney lichen
lichen, auburn → auburn lichen
lichen, awl-shaped stump → awl-shaped stump lichen
lichen, black-fruiting → black-fruiting lichen
lichen, blue-grey blister → blue-grey blister lichen
lichen, boulder → boulder lichen
lichen, boxboard felt → boxboard felt lichen
lichen, brittle horsehair → brittle horsehair lichen
lichen, brook → brook lichen
lichen, brown bark → brown bark lichen
lichen, brown cushion → brown cushion lichen
lichen, brown turf → brown turf lichen
lichen, button → button lichen
lichen, cartilage → cartilage lichen
lichen, chalky shield → chalky shield lichen
lichen, cockleshell → cockleshell lichen
lichen, common coral → common coral lichen
lichen, confetti → confetti lichen
lichen, crap → crap lichen
lichen, crusted orange → crusted orange lichen
lichen, dimpled → dimpled lichen
lichen, dog → dog lichen
lichen, dog tooth → dog tooth lichen
lichen, dot → dot lichen
lichen, dusty cartilage → dusty cartilage lichen
lichen, dusty-margined dog → dusty-margined dog lichen
lichen, ear → ear lichen
lichen, even dog → even dog lichen
lichen, fan → fan lichen
lichen, few-finger → few-finger lichen
lichen, finger felt → finger felt lichen
lichen, fist → fist lichen
lichen, flat dog → flat dog lichen
lichen, floury funnel → floury funnel lichen
lichen, forked shore → forked shore lichen

lichen, forking tube → forking tube lichen
lichen, freckled → freckled lichen
lichen, furrowed shield → furrowed shield lichen
lichen, goat → goat lichen
lichen, gold-edge → gold-edge lichen
lichen, green kidney → green kidney lichen
lichen, green map → green map lichen
lichen, green reindeer → green reindeer lichen
lichen, grey reindeer → grey reindeer lichen
lichen, grey star → grey star lichen
lichen, hood → hood lichen
lichen, hooded → hooded lichen
lichen, hooded tube → hooded tube lichen
lichen, lemon → lemon lichen
lichen, lettuce → lettuce lichen
lichen, lumpy shore → lumpy shore lichen
lichen, lustrous brown → lustrous brown lichen
lichen, mackerel → mackerel lichen
lichen, many-fruited dog → many-fruited dog lichen
lichen, moss trifle → moss trifle lichen
lichen, mouse → mouse lichen
lichen, North American → North American lichen
lichen, northern brown → northern brown lichen
lichen, northern reindeer → northern reindeer lichen
lichen, orange-foot → orange-foot lichen
lichen, pine → pine lichen
lichen, pixie-cup → pixie-cup lichen
lichen, powdered orange → powdered orange lichen
lichen, powdery Swiss → powdery Swiss lichen
lichen, reindeer → reindeer lichen
lichen, rim → rim lichen
lichen, ring → ring lichen
lichen, rock worm → rock worm lichen
lichen, rose-and-gold → rose-and-gold lichen
lichen, rough dog → rough dog lichen
lichen, scurf → scurf lichen
lichen, shingle → shingle lichen
lichen, shrub funnel → shrub funnel lichen
lichen, sieve cup → sieve cup lichen
lichen, slender cup → slender cup lichen
lichen, small felt → small felt lichen
lichen, smoky shield → smoky shield lichen
lichen, smooth Swiss → smooth Swiss lichen
lichen, sooty leather → sooty leather lichen
lichen, speckled horsehair → speckled horsehair lichen
lichen, spike → spike lichen
lichen, spiny heath → spiny heath lichen
lichen, spraypaint → spraypaint lichen
lichen, spring disc → spring disc lichen
lichen, star reindeer → star reindeer lichen
lichen, sticky → sticky lichen
lichen, studded leather → studded leather lichen
lichen, sulphur-dust → sulphur-dust lichen
lichen, sunburst → sunburst lichen
lichen, Swiss → Swiss lichen
lichen, torn club → torn club lichen
lichen, true reindeer → true reindeer lichen
lichen, tumble → tumble lichen
lichen, underwater → underwater lichen
lichen, veined → veined lichen
lichen, velcro → velcro lichen
lichen, wart → wart lichen
lichen, waxpaper → waxpaper lichen
lichen, Weigel's leather → Weigel's leather lichen
lichen, whorled cup → whorled cup lichen
lichen, wolf → wolf lichen
lichen, wreath → wreath lichen
lichen, Wright's leather → Wright's leather lichen
lichen, yellow → yellow lichen
lichen, yellow reindeer → yellow reindeer lichen
licorice fern = Polypodium glycyrrhiza (V:F)
licorice, American → American licorice
licorice, chinook → chinook licorice
light, flattened snow → flattened snow light
lilaeopsis, western → western lilaeopsis
lilliput vinyl = Leptogium tenuissimum (L)
lilliput vinyl = Leptogium subtile (L)
lily-of-the-valley, false → false lily-of-the-valley
lily-of-the-valley, wild → wild lily-of-the-valley
lily, alp → alp lily
lily, avalanche → avalanche lily
lily, black → black lily
lily, chocolate → chocolate lily
lily, Columbia → Columbia lily
lily, glacier → glacier lily
lily, Lyall's mariposa → Lyall's mariposa lily
lily, pink fawn → pink fawn lily
lily, sagebrush mariposa → sagebrush mariposa lily
lily, three-spot mariposa → three-spot mariposa lily
lily, tiger → tiger lily
lily, white fawn → white fawn lily
lily, white glacier → white glacier lily
lily, wood → wood lily
lily, yellow glacier → yellow glacier lily
limber honeysuckle = Lonicera dioica (V:D)
limber pine = Pinus flexilis (V:G)
lime dust = Chrysothrix chlorina (L)
limestone sunshine = Vulpicida tilesii (L)
limy stippleback = Dermatocarpon miniatum (L)
linanthus, bicolored → bicolored linanthus
linanthus, Harkness' → Harkness' linanthus
linanthus, northern → northern linanthus
Lindley's aster = Aster ciliolatus (V:D)
line-leaved daisy = Erigeron linearis (V:D)
line-leaved fleabane = Erigeron linearis (V:D)
Lingbye's sedge = Carex lyngbyei (V:M)
lingonberry = Vaccinium vitis-idaea (V:D)
liparis, Loesel's → Loesel's liparis
lipstick cladonia = Cladonia macilenta (L)
little bluestem = Schizachyrium scoparium (V:M)
little buttercup = Ranunculus uncinatus (V:D)
little chickweed = Cerastium semidecandrum (V:D)
little fescue = Festuca minutiflora (V:M)
little hands liverwort = Lepidozia reptans (B:H)
little lovegrass = Eragrostis minor (V:M)
little meadow-foxtail = Alopecurus aequalis (V:M)
little-podded falseflax = Camelina microcarpa (V:D)
little quaking grass = Briza minor (V:M)
little ricegrass = Oryzopsis exigua (V:M)
little-rose = Chamaerhodos erecta (V:D)
little-seeded ricegrass = Oryzopsis micrantha (V:M)
little tarweed = Madia exigua (V:D)
little-tree willow = Salix arbusculoides (V:D)
little western bitter-cress = Cardamine oligosperma (V:D)
littlebells polemonium = Polemonium micranthum (V:D)
littlepod flax = Camelina microcarpa (V:D)
liverwort, ash-coloured ground → ash-coloured ground liverwort
liverwort, cedar-shake → cedar-shake liverwort
liverwort, comb → comb liverwort
liverwort, common fold-leaf → common fold-leaf liverwort
liverwort, common leafy → common leafy liverwort
liverwort, common scissor-leaf → common scissor-leaf liverwort
liverwort, flat-leaved → flat-leaved liverwort
liverwort, floating → floating liverwort
liverwort, green-tongue → green-tongue liverwort
liverwort, hanging millipede → hanging millipede liverwort
liverwort, hard scale → hard scale liverwort
liverwort, Jameson's → Jameson's liverwort
liverwort, leafy → leafy liverwort
liverwort, little hands → little hands liverwort
liverwort, lung → lung liverwort

liverwort, maple → maple liverwort
liverwort, mountain leafy → mountain leafy liverwort
liverwort, naugehyde → naugehyde liverwort
liverwort, northern naugehyde → northern naugehyde liverwort
liverwort, purple-worm → purple-worm liverwort
liverwort, shiny → shiny liverwort
liverwort, snake → snake liverwort
liverwort, snow-mat → snow-mat liverwort
liverwort, Taylor's → Taylor's liverwort
liverwort, three-ruffle → three-ruffle liverwort
liverwort, three-toothed whip → three-toothed whip liverwort
liverwort, yellow-ladle → yellow-ladle liverwort
livid sedge = Carex livida (V:M)
lizard tripe = Phylliscum demangeonii (L)
Lobb's water-buttercup = Ranunculus lobbii (V:D)
lobelia, Kalm's → Kalm's lobelia
lobelia, water → water lobelia
locoweed, arctic → arctic locoweed
locoweed, blackish → blackish locoweed
locoweed, field → field locoweed
locoweed, Huddelson's → Huddelson's locoweed
locoweed, Jordal's → Jordal's locoweed
locoweed, Maydell's → Maydell's locoweed
locoweed, pendant-pod → pendant-pod locoweed
locoweed, reflexed → reflexed locoweed
locoweed, Rocky Mountain → Rocky Mountain locoweed
locoweed, Scamman's → Scamman's locoweed
locoweed, showy → showy locoweed
locoweed, silky → silky locoweed
locoweed, stalked-pod → stalked-pod locoweed
locoweed, sticky → sticky locoweed
locoweed, yellow-flower → yellow-flower locoweed
locust, black → black locust
lodgepole × jack pine hybrid = Pinus × murraybanksiana (V:G)
lodgepole pine = Pinus contorta var. latifolia (V:G)
lodgepole pine = Pinus contorta (V:G)
Loesel's liparis = Liparis loeselii (V:M)
Loesel's tumble-mustard = Sisymbrium loeselii (V:D)
lomatium, bare-stem → bare-stem lomatium
lomatium, Brandegee's → Brandegee's lomatium
lomatium, common → common lomatium
lomatium, few-fruited → few-fruited lomatium
lomatium, fine-leaved → fine-leaved lomatium
lomatium, Geyer's → Geyer's lomatium
lomatium, Gray's → Gray's lomatium
lomatium, large-fruited → large-fruited lomatium
lomatium, Martindale's → Martindale's lomatium
lomatium, Sandberg's → Sandberg's lomatium
lomatium, swale → swale lomatium
lomatiun, fern-leaved → fern-leaved lomatiun
long-beaked sedge = Carex sychnocephala (V:M)
long-bracted frog orchid = Coeloglossum viride (V:M)
long-bracted orchid = Coeloglossum viride (V:M)
long-bracted sedge = Carex retrorsa (V:M)
long-flowered bluebells = Mertensia longiflora (V:D)
long-flowered mertensia = Mertensia longiflora (V:D)
long-fruited anemone = Anemone cylindrica (V:D)
long-headed anemone = Anemone cylindrica (V:D)
long-leaved aster = Aster ascendens (V:D)
long-leaved daisy = Erigeron corymbosus (V:D)
long-leaved fleabane = Erigeron corymbosus (V:D)
long-leaved mugwort = Artemisia longifolia (V:D)
long-leaved phlox = Phlox longifolia (V:D)
long-leaved pondweed = Potamogeton nodosus (V:M)
long-leaved sundew = Drosera anglica (V:D)
long-necked bryum = Leptobryum pyriforme (B:M)
long-spurred plectritis = Plectritis macrocera (V:D)
long-stalked chickweed = Stellaria longipes (V:D)
long-stalked draba = Draba longipes (V:D)
long-stalked mouse-ear chickweed = Cerastium nutans (V:D)
long-stalked pondweed = Potamogeton praelongus (V:M)
long-stalked starwort = Stellaria longipes (V:D)
long-stalked whitlow-grass = Draba longipes (V:D)
long-stoloned sedge = Carex inops (V:M)
long-styled rush = Juncus longistylis (V:M)
long-styled sedge = Carex stylosa (V:M)
loop, green → green loop
loop, grey → grey loop
loosestrife, bog → bog loosestrife
loosestrife, creeping → creeping loosestrife
loosestrife, fringed → fringed loosestrife
loosestrife, purple → purple loosestrife
loosestrife, spotted → spotted loosestrife
loosestrife, tufted → tufted loosestrife
loosestrife, winged → winged loosestrife
loosestrife, yellow → yellow loosestrife
lotus milk-vetch = Astragalus lotiflorus (V:D)
lotus, meadow → meadow lotus
lotus, seaside → seaside lotus
lotus, small-flowered → small-flowered lotus
lousewort, arctic Langsdorf's → arctic Langsdorf's lousewort
lousewort, bird's-beak → bird's-beak lousewort
lousewort, boreal → boreal lousewort
lousewort, bracted → bracted lousewort
lousewort, capitate → capitate lousewort
lousewort, coil-beaked → coil-beaked lousewort
lousewort, elephant's-head → elephant's-head lousewort
lousewort, Labrador → Labrador lousewort
lousewort, Langsdorf's → Langsdorf's lousewort
lousewort, Oeder's → Oeder's lousewort
lousewort, sickletop → sickletop lousewort
lousewort, small-flowered → small-flowered lousewort
lousewort, Sudeten → Sudeten lousewort
lousewort, swamp → swamp lousewort
lousewort, whorled → whorled lousewort
lousewort, woolly → woolly lousewort
lovage, beach → beach lovage
lovage, Calder's → Calder's lovage
lovage, Canby's → Canby's lovage
lovage, Scottish wild → Scottish wild lovage
lovage, verticillate-umbel → verticillate-umbel lovage
lovegrass, Indian → Indian lovegrass
lovegrass, little → little lovegrass
lovegrass, tufted → tufted lovegrass
lover's moss = Aulacomnium androgynum (B:M)
low bilberry = Vaccinium myrtillus (V:D)
low birch = Betula pumila var. glandulifera (V:D)
low blueberry willow = Salix myrtillifolia (V:D)
low braya = Braya humilis (V:D)
low bush-cranberry = Viburnum edule (V:D)
low clubrush = Scirpus cernuus (V:M)
low hawksbeard = Crepis modocensis (V:D)
low hop-clover = Trifolium campestre (V:D)
low northern sedge = Carex concinna (V:M)
low pussytoes = Antennaria dimorpha (V:D)
low sandwort = Arenaria longipedunculata (V:D)
low selaginella = Selaginella selaginoides (V:F)
lowland cudweed = Gnaphalium palustre (V:D)
luina, silverback → silverback luina
lumpy shore lichen = Xanthoria polycarpa (L)
lung liverwort = Marchantia polymorpha (B:H)
lung, cabbage → cabbage lung
lung, iron → iron lung
lung, lettuce → lettuce lung
lung, smoker's → smoker's lung
lung, textured → textured lung
lungwort = Lobaria pulmonaria (L)
lungwort, sea → sea lungwort
lungwort, tall → tall lungwort
lupine, arctic → arctic lupine
lupine, big-leaved → big-leaved lupine
lupine, bigleaf → bigleaf lupine

lupine, dwarf mountain → dwarf mountain lupine
lupine, grassland → grassland lupine
lupine, Kusche's → Kusche's lupine
lupine, large-leaved → large-leaved lupine
lupine, Montana → Montana lupine
lupine, Nootka → Nootka lupine
lupine, prairie → prairie lupine
lupine, seashore → seashore lupine
lupine, silky → silky lupine
lupine, silvery → silvery lupine
lupine, small-flowered → small-flowered lupine
lupine, spurred → spurred lupine
lupine, stream-bank → stream-bank lupine
lupine, sulphur → sulphur lupine
lupine, tree → tree lupine
lupine, two-coloured → two-coloured lupine
lupine, velvet → velvet lupine
lupine, white-whorl → white-whorl lupine
lupine, Wyeth's → Wyeth's lupine
lupine, Yukon → Yukon lupine
lustrous brown = Melanelia exasperatula (L)
lustrous brown lichen = Melanelia exasperatula (L)
Lyall's anemone = Anemone lyallii (V:D)
Lyall's mariposa lily = Calochortus lyallii (V:M)
Lyall's penstemon = Penstemon lyallii (V:D)
Lyall's phacelia = Phacelia lyallii (V:D)
Lyall's rockcress = Arabis lyallii (V:D)
Lyall's saxifraga = Saxifraga lyallii (V:D)
lyell's bristle moss = Orthotrichum lyellii (B:M)
lyre-leaved rockcress = Arabis lyrata (V:D)

M

MacBryde's phacelia = Phacelia mollis (V:D)
MacCalla's willow = Salix maccalliana (V:D)
Mackenzie's willow = Salix prolixa (V:D)
mackerel lichen = Leptogium subtile (L)
Macoun's buttercup = Ranunculus macounii (V:D)
Macoun's draba = Draba macounii (V:D)
Macoun's groundsel = Senecio macounii (V:D)
Macoun's meadow-foam = Limnanthes macounii (V:D)
Macoun's poppy = Papaver macounii (V:D)
Macoun's whitlow-grass = Draba macounii (V:D)
Macrae's clover = Trifolium macraei (V:D)
madder, field → field madder
madrone = Arbutus menziesii (V:D)
madrone, Pacific → Pacific madrone
madwort = Asperugo procumbens (V:D)
magic treeflute = Menegazzia terebrata (L)
magnificent hook moss = Limprichtia revolvens (B:M)
magnificent moss = Plagiomnium venustum (B:M)
mahaleb cherry = Prunus mahaleb (V:D)
maiden pink = Dianthus deltoides (V:D)
maidenhair fern = Adiantum aleuticum (V:F)
maidenhair spleenwort = Asplenium trichomanes (V:F)
maids, red → red maids
male fern = Dryopteris filix-mas (V:F)
mallow ninebark = Physocarpus malvaceus (V:D)
mallow, common → common mallow
mallow, dwarf → dwarf mallow
mallow, musk → musk mallow
mallow, small-flowered → small-flowered mallow
maltese star-thistle = Centaurea melitensis (V:D)
Manitoba maple = Acer negundo (V:D)
mannagrass, American → American mannagrass
mannagrass, fowl → fowl mannagrass
mannagrass, giant → giant mannagrass
mannagrass, northern → northern mannagrass
mannagrass, reed → reed mannagrass
mannagrass, slender-spiked → slender-spiked mannagrass
mannagrass, tall → tall mannagrass
mannagrass, western → western mannagrass
manroot = Marah oreganus (V:D)
many-flowered hackelia = Hackelia floribunda (V:D)
many-flowered sedge = Carex pluriflora (V:M)
many-flowered stickseed = Hackelia floribunda (V:D)
many-flowered woodrush = Luzula multiflora (V:M)
many-flowered yarrow = Achillea sibirica (V:D)
many-fruited dog lichen = Peltigera polydactylon (L)
many-headed sedge = Carex sychnocephala (V:M)
manzanita, hairy → hairy manzanita
map, green → green map
maple liverwort = Barbilophozia lycopodioides (B:H)
maple, bigleaf → bigleaf maple
maple, broad-leaved → broad-leaved maple
maple, Douglas → Douglas maple
maple, Manitoba → Manitoba maple
maple, norway → norway maple
maple, Oregon → Oregon maple
maple, Rocky Mountain → Rocky Mountain maple
maple, sycamore → sycamore maple
maple, vine → vine maple
mare's-tail = Hippuris vulgaris (V:D)
mare's-tail, common → common mare's-tail
mare's-tail, four-leaved → four-leaved mare's-tail
mare's-tail, mountain → mountain mare's-tail
marginal wood fern = Dryopteris marginalis (V:F)
marjoram, wild → wild marjoram
marram grass = Ammophila arenaria (V:M)
marsh arrow-grass = Triglochin palustre (V:M)
marsh aster = Aster borealis (V:D)
marsh cudweed = Gnaphalium uliginosum (V:D)
marsh-elder = Iva xanthifolia (V:D)
marsh felwort = Lomatogonium rotatum (V:D)
marsh fleabane = Senecio congestus (V:D)
marsh hedge-nettle = Stachys palustris (V:D)
marsh horsetail = Equisetum palustre (V:F)
marsh Labrador-tea = Ledum palustre (V:D)
marsh magnificent moss = Plagiomnium ellipticum (B:M)
marsh-marigold, alpine white → alpine white marsh-marigold
marsh-marigold, floating → floating marsh-marigold
marsh-marigold, mountain → mountain marsh-marigold
marsh-marigold, white → white marsh-marigold
marsh-marigold, yellow → yellow marsh-marigold
marsh muhly = Muhlenbergia glomerata (V:M)
marsh peavine = Lathyrus palustris (V:D)
marsh reedgrass = Calamagrostis canadensis (V:M)
marsh skullcap = Scutellaria galericulata (V:D)
marsh speedwell = Veronica scutellata (V:D)
marsh thistle = Cirsium palustre (V:D)
marsh valerian = Valeriana dioica (V:D)
marsh violet = Viola palustris (V:D)
marsh willowherb = Epilobium palustre (V:D)
marsh yellow cress = Rorippa palustris (V:D)
marshpepper smartweed = Polygonum hydropiper (V:D)
Martindale's lomatium = Lomatium martindalei (V:D)
Mary, small-flowered blue-eyed → small-flowered blue-eyed Mary
mat muhly = Muhlenbergia richardsonis (V:M)
matchstick, devil's → devil's matchstick
Maydell's locoweed = Oxytropis maydelliana (V:D)
Maydell's oxytrope = Oxytropis maydelliana (V:D)

mayflower, Canadian → Canadian mayflower
mayweed = Anthemis cotula (V:D)
mayweed, scentless → scentless mayweed
meadow alumroot = Heuchera chlorantha (V:D)
meadow arnica = Arnica chamissonis (V:D)
meadow aster = Aster campestris (V:D)
meadow barley = Hordeum brachyantherum (V:M)
meadow birds-foot trefoil = Lotus denticulatus (V:D)
meadow brome = Bromus commutatus (V:M)
meadow buttercup = Ranunculus acris (V:D)
meadow death-camas = Zigadenus venenosus (V:M)
meadow fescue = Festuca pratensis (V:M)
meadow-foam, Macoun's → Macoun's meadow-foam
meadow-foam, toothcup → toothcup meadow-foam
meadow-foxtail = Alopecurus pratensis (V:M)
meadow-foxtail, Carolina → Carolina meadow-foxtail
meadow-foxtail, little → little meadow-foxtail
meadow-foxtail, water → water meadow-foxtail
meadow horsetail = Equisetum pratense (V:F)
meadow lotus = Lotus denticulatus (V:D)
meadow nemophila = Nemophila pedunculata (V:D)
meadow peavine = Lathyrus pratensis (V:D)
meadow salsify = Tragopogon pratensis (V:D)
meadow sedge = Carex praticola (V:M)
meadow willow = Salix petiolaris (V:D)
meadowgrass, salt → salt meadowgrass
meadowrue, alpine → alpine meadowrue
meadowrue, few-flowered → few-flowered meadowrue
meadowrue, flat-fruited → flat-fruited meadowrue
meadowrue, purple → purple meadowrue
meadowrue, Richardson's → Richardson's meadowrue
meadowrue, tall → tall meadowrue
meadowrue, veiny → veiny meadowrue
meadowrue, western → western meadowrue
mealberry = Arctostaphylos uva-ursi (V:D)
mealy primrose = Primula incana (V:D)
meconella, white → white meconella
medic, black → black medic
medic, spotted → spotted medic
Mediterranean barley = Hordeum marinum ssp. gussonianum (V:M)
melic, false → false melic
melic, Harford's → Harford's melic
melic, Smith's → Smith's melic
mentzelia, blazing-star → blazing-star mentzelia
mentzelia, bushy → bushy mentzelia
mentzelia, small-flowered → small-flowered mentzelia
mentzelia, white-stemmed → white-stemmed mentzelia
Menzies' burnet = Sanguisorba menziesii (V:D)
Menzies' campion = Silene menziesii (V:D)
Menzies' catchfly = Silene menziesii (V:D)
Menzies' larkspur = Delphinium menziesii (V:D)
Menzies' larkspur = Delphinium menziesii ssp. menziesii (V:D)
Menzies' neckera = Metaneckera menziesii (B:M)
Menzies' pipsissewa = Chimaphila menziesii (V:D)
Menzies' red-mouthed mnium = Mnium spinulosum (B:M)
Menzies' silene = Silene menziesii (V:D)
Menzies' tree moss = Leucolepis acanthoneuron (B:M)
Mertens' moss heather = Cassiope mertensiana (V:D)
Mertens' mountain-heather = Cassiope mertensiana (V:D)
Mertens' rush = Juncus mertensianus (V:M)
Mertens' saxifraga = Saxifraga mertensiana (V:D)
Mertens' sedge = Carex mertensii (V:M)
mertensia, long-flowered → long-flowered mertensia
mertensia, panicled → panicled mertensia
mertensia, sea → sea mertensia
Methuselah's beard = Usnea longissima (L)
Mexican bedstraw = Galium mexicanum (V:D)
Mexican mosquito fern = Azolla mexicana (V:F)
Michaux's mugwort = Artemisia michauxiana (V:D)
microseris, coastal → coastal microseris
microseris, nodding → nodding microseris
midway peat moss = Sphagnum magellanicum (B:M)
mignonette, white → white mignonette
mignonette, yellow → yellow mignonette
milfoil = Achillea millefolium (V:D)
milk thistle = Silybum marianum (V:D)
milk-vetch, alpine → alpine milk-vetch
milk-vetch, American → American milk-vetch
milk-vetch, ascending purple → ascending purple milk-vetch
milk-vetch, Beckwith's → Beckwith's milk-vetch
milk-vetch, Bourgeau's → Bourgeau's milk-vetch
milk-vetch, Canadian → Canadian milk-vetch
milk-vetch, chick-pea → chick-pea milk-vetch
milk-vetch, Drummond's → Drummond's milk-vetch
milk-vetch, elegant → elegant milk-vetch
milk-vetch, field → field milk-vetch
milk-vetch, freckled → freckled milk-vetch
milk-vetch, hill → hill milk-vetch
milk-vetch, hillside → hillside milk-vetch
milk-vetch, least bladdery → least bladdery milk-vetch
milk-vetch, lesser rushy → lesser rushy milk-vetch
milk-vetch, lotus → lotus milk-vetch
milk-vetch, Nutzotin → Nutzotin milk-vetch
milk-vetch, pretty → pretty milk-vetch
milk-vetch, pulse → pulse milk-vetch
milk-vetch, purple → purple milk-vetch
milk-vetch, Robbins' → Robbins' milk-vetch
milk-vetch, Spalding's → Spalding's milk-vetch
milk-vetch, standing → standing milk-vetch
milk-vetch, The Dalles → The Dalles milk-vetch
milk-vetch, threadstalk → threadstalk milk-vetch
milk-vetch, timber → timber milk-vetch
milk-vetch, tundra → tundra milk-vetch
milk-vetch, Weiser → Weiser milk-vetch
milk-vetch, woollypod → woollypod milk-vetch
milkweed, oval-leaved → oval-leaved milkweed
milkweed, showy → showy milkweed
milky draba = Draba lactea (V:D)
milky whitlow-grass = Draba lactea (V:D)
miller, dusty → dusty miller
millet, broom-corn → broom-corn millet
miner's-lettuce = Claytonia perfoliata (V:D)
miner's-lettuce, Siberian → Siberian miner's-lettuce
Mingan moonwort = Botrychium minganense (V:F)
mini rockfrog = Xanthoparmelia planilobata (L)
mint, field → field mint
mint, wild → wild mint
Missouri goldenrod = Solidago missouriensis (V:D)
Mistassini primrose = Primula mistassinica (V:D)
mistletoe, American dwarf → American dwarf mistletoe
mistletoe, Douglas' dwarf → Douglas' dwarf mistletoe
mistletoe, western dwarf → western dwarf mistletoe
mistmaiden, Sitka → Sitka mistmaiden
mistmaiden, Tracy's → Tracy's mistmaiden
mitrewort, bare-stemmed → bare-stemmed mitrewort
mitrewort, Brewer's → Brewer's mitrewort
mitrewort, common → common mitrewort
mitrewort, five-stamened → five-stamened mitrewort
mitrewort, leafy → leafy mitrewort
mitrewort, naked → naked mitrewort
mitrewort, oval-leaved → oval-leaved mitrewort
mitrewort, three-toothed → three-toothed mitrewort
mnium, Menzies' red-mouthed → Menzies' red-mouthed mnium
mnium, red-mouthed → red-mouthed mnium
mnium, woodsy → woodsy mnium
mock-orange = Philadelphus lewisii (V:D)
modest buttercup = Ranunculus verecundus (V:D)
monardella = Monardella odoratissima (V:D)
monk's-hood = Hypogymnia physodes (L)
monk's-hood = Hypogymnia vittata (L)
monkey-flower, Brewer's → Brewer's monkey-flower

monkey-flower, chickweed → chickweed monkey-flower
monkey-flower, large mountain → large mountain monkey-flower
monkey-flower, Lewis' → Lewis' monkey-flower
monkey-flower, mountain → mountain monkey-flower
monkey-flower, pink → pink monkey-flower
monkey-flower, purple-stemmed → purple-stemmed monkey-flower
monkey-flower, short-flowered → short-flowered monkey-flower
monkey-flower, toothed-leaved → toothed-leaved monkey-flower
monkey-flower, yellow → yellow monkey-flower
monkshood, Columbian → Columbian monkshood
monkshood, mountain → mountain monkshood
Montana larkspur = Delphinium bicolor (V:D)
Montana lupine = Lupinus arbustus ssp. pseudoparviflorus (V:D)
montbretia = Crocosmia × crocosmiiflora (V:M)
montia, Bostock's → Bostock's montia
montia, Chamisso's → Chamisso's montia
montia, dwarf → dwarf montia
montia, Howell's → Howell's montia
montia, narrow-leaved → narrow-leaved montia
montia, small-leaved → small-leaved montia
montia, spreading → spreading montia
moon, arctic → arctic moon
moon, fringed → fringed moon
moon, green → green moon
moon, peppered → peppered moon
moon, powdered → powdered moon
moonshine cetraria = Vulpicida pinastri (L)
moonshine ruffle = Vulpicida pinastri (L)
moonwort = Botrychium lunaria (V:F)
moonwort, common → common moonwort
moonwort, lance-leaved → lance-leaved moonwort
moonwort, least → least moonwort
moonwort, Mingan → Mingan moonwort
moonwort, northwestern → northwestern moonwort
moonwort, two-spiked → two-spiked moonwort
mooseberry = Viburnum edule (V:D)
morning-glory, orchard → orchard morning-glory
moschatel = Adoxa moschatellina (V:D)
moss campion = Silene acaulis (V:D)
moss gentian = Gentiana prostrata (V:D)
moss grass = Coleanthus subtilis (V:M)
moss mouse = Fuscopannaria praetermissa (L)
moss trifle lichen = Agonimia tristicula (L)
moss, acute-leaved peat → acute-leaved peat moss
moss, apple → apple moss
moss, aquatic apple → aquatic apple moss
moss, aspen → aspen moss
moss, awned haircap → awned haircap moss
moss, badge → badge moss
moss, bent-leaf → bent-leaf moss
moss, black brook → black brook moss
moss, black fish hook → black fish hook moss
moss, black rock → black rock moss
moss, black-tufted rock → black-tufted rock moss
moss, Blandow's feather → Blandow's feather moss
moss, blunt-leaved → blunt-leaved moss
moss, blunt-leaved bristle → blunt-leaved bristle moss
moss, bottle → bottle moss
moss, bristly haircap → bristly haircap moss
moss, broken-leaf → broken-leaf moss
moss, broom → broom moss
moss, brown-stemmed bog → brown-stemmed bog moss
moss, capillary thread → capillary thread moss
moss, cat-tail → cat-tail moss
moss, claw-leaved feather → claw-leaved feather moss
moss, clay pigtail → clay pigtail moss
moss, clear → clear moss
moss, coastal leafy → coastal leafy moss
moss, coiled-leaf → coiled-leaf moss
moss, common beaked → common beaked moss
moss, common beard → common beard moss
moss, common dung → common dung moss
moss, common four-tooth → common four-tooth moss
moss, common haircap → common haircap moss
moss, common hook → common hook moss
moss, common lantern → common lantern moss
moss, common lawn → common lawn moss
moss, common leafy → common leafy moss
moss, common tree → common tree moss
moss, common verdant → common verdant moss
moss, common water → common water moss
moss, copper wire → copper wire moss
moss, cord → cord moss
moss, crane's-bill → crane's-bill moss
moss, crumpled-leaf → crumpled-leaf moss
moss, curly heron's-bill → curly heron's-bill moss
moss, curly thatch → curly thatch moss
moss, Drummond's leafy → Drummond's leafy moss
moss, dusky fork → dusky fork moss
moss, electrified cat's-tail → electrified cat's-tail moss
moss, elegant → elegant moss
moss, erect-fruited iris → erect-fruited iris moss
moss, false haircap → false haircap moss
moss, fan → fan moss
moss, fat bog → fat bog moss
moss, felt round → felt round moss
moss, fern → fern moss
moss, fire → fire moss
moss, flapper → flapper moss
moss, flat → flat moss
moss, floating feather → floating feather moss
moss, four-toothed log → four-toothed log moss
moss, fragile screw → fragile screw moss
moss, giant feather → giant feather moss
moss, giant water → giant water moss
moss, Girgensohn's peat → Girgensohn's peat moss
moss, glow → glow moss
moss, gold spoon → gold spoon moss
moss, golden → golden moss
moss, golden curls → golden curls moss
moss, golden fuzzy fen → golden fuzzy fen moss
moss, golden ragged → golden ragged moss
moss, golden short-capsuled → golden short-capsuled moss
moss, golden star → golden star moss
moss, golf club → golf club moss
moss, goose-necked → goose-necked moss
moss, granite → granite moss
moss, grey haircap → grey haircap moss
moss, grooved gnome-cap → grooved gnome-cap moss
moss, hairy-cap → hairy-cap moss
moss, hairy lantern → hairy lantern moss
moss, hairy screw → hairy screw moss
moss, hanging → hanging moss
moss, hanging basket → hanging basket moss
moss, Hedwig's rock → Hedwig's rock moss
moss, hoary rock → hoary rock moss
moss, hooded → hooded moss
moss, hook → hook moss
moss, hook-leaf fern → hook-leaf fern moss
moss, hooked spring → hooked spring moss
moss, juniper → juniper moss
moss, juniper haircap → juniper haircap moss
moss, lanky → lanky moss
moss, large hair → large hair moss
moss, large leafy → large leafy moss
moss, lawn → lawn moss
moss, lover's → lover's moss
moss, lyell's bristle → lyell's bristle moss
moss, magnificent → magnificent moss
moss, magnificent hook → magnificent hook moss

moss, marsh magnificent → marsh magnificent moss
moss, Menzies' tree → Menzies' tree moss
moss, midway peat → midway peat moss
moss, mountain curved-back → mountain curved-back moss
moss, Oregon beaked → Oregon beaked moss
moss, pale-stalked broom → pale-stalked broom moss
moss, palm → palm moss
moss, palm tree → palm tree moss
moss, pear → pear moss
moss, pipecleaner → pipecleaner moss
moss, plume → plume moss
moss, poor fen peat → poor fen peat moss
moss, purple horn-toothed → purple horn-toothed moss
moss, red hook → red hook moss
moss, red-mouthed leafy → red-mouthed leafy moss
moss, red roof → red roof moss
moss, red-toothed rock → red-toothed rock moss
moss, ribbed bog → ribbed bog moss
moss, Richardson's water → Richardson's water moss
moss, rigid brook → rigid brook moss
moss, roadside rock → roadside rock moss
moss, rolled-leaf pigtail → rolled-leaf pigtail moss
moss, rough → rough moss
moss, round → round moss
moss, rusty claw → rusty claw moss
moss, rusty mountain heath → rusty mountain heath moss
moss, rusty peat → rusty peat moss
moss, rusty steppe → rusty steppe moss
moss, sausage → sausage moss
moss, scorpion feather → scorpion feather moss
moss, seepage apple → seepage apple moss
moss, shaggy yellow sand → shaggy yellow sand moss
moss, shore-growing peat → shore-growing peat moss
moss, short-leaved ragged → short-leaved ragged moss
moss, showy bristle → showy bristle moss
moss, sickle → sickle moss
moss, sidewalk → sidewalk moss
moss, silky tufted → silky tufted moss
moss, silver → silver moss
moss, slender beaked → slender beaked moss
moss, slender-stemmed hair → slender-stemmed hair moss
moss, small flat → small flat moss
moss, small hair → small hair moss
moss, small mouse-tail → small mouse-tail moss
moss, small red peat → small red peat moss
moss, smooth-stalked yellow feather → smooth-stalked yellow feather moss
moss, spear → spear moss
moss, spread-leaved peat → spread-leaved peat moss
moss, spreading-leaved peat → spreading-leaved peat moss
moss, spring → spring moss
moss, spring claw → spring claw moss
moss, spruce → spruce moss
moss, squarrose peat → squarrose peat moss
moss, stair step → stair step moss
moss, stair-step → stair-step moss
moss, step → step moss
moss, stocking → stocking moss
moss, straw → straw moss
moss, straw-coloured water → straw-coloured water moss
moss, straw-like feather → straw-like feather moss
moss, streamside → streamside moss
moss, swamp → swamp moss
moss, tall clustered thread → tall clustered thread moss
moss, tangle → tangle moss
moss, thick grass → thick grass moss
moss, thick ragged → thick ragged moss
moss, three-angled thread → three-angled thread moss
moss, trailing leafy → trailing leafy moss
moss, tree → tree moss
moss, tufted → tufted moss
moss, velvet feather → velvet feather moss
moss, Warnstorf's peat → Warnstorf's peat moss
moss, water hook → water hook moss
moss, waterside feather → waterside feather moss
moss, wavy-leaved → wavy-leaved moss
moss, wavy-leaved cotton → wavy-leaved cotton moss
moss, wet rock → wet rock moss
moss, whip fork → whip fork moss
moss, white-toothed peat → white-toothed peat moss
moss, wiry fern → wiry fern moss
moss, woodsy leafy → woodsy leafy moss
moss, woodsy ragged → woodsy ragged moss
moss, woolly caterpillar → woolly caterpillar moss
moss, yellow → yellow moss
moss, yellow collar → yellow collar moss
moss, yellow-green cushion → yellow-green cushion moss
moss, yellow-green peat → yellow-green peat moss
moss, yellow-green rock → yellow-green rock moss
moss, yellow star → yellow star moss
moss, yellow starry feather → yellow starry feather moss
moth mullein = Verbascum blattaria (V:D)
moth tarpaper = Collema subflaccidum (L)
motherwort, common → common motherwort
Mount Sheldon butterweed = Senecio sheldonensis (V:D)
mountain alder = Alnus incana ssp. tenuifolia (V:D)
mountain arnica = Arnica latifolia (V:D)
mountain-ash, European → European mountain-ash
mountain-ash, Sitka → Sitka mountain-ash
mountain-ash, western → western mountain-ash
mountain-avens, entire-leaved → entire-leaved mountain-avens
mountain-avens, entire-leaved white → entire-leaved white mountain-avens
mountain-avens, smooth-leaved → smooth-leaved mountain-avens
mountain-avens, white → white mountain-avens
mountain-avens, yellow → yellow mountain-avens
mountain bentgrass = Agrostis variabilis (V:M)
mountain birch = Betula occidentalis (V:D)
mountain bladder fern = Cystopteris montana (V:F)
mountain blue-eyed-grass = Sisyrinchium montanum (V:M)
mountain bluet = Centaurea montana (V:D)
mountain bog gentian = Gentiana calycosa (V:D)
mountain-box = Paxistima myrsinites (V:D)
mountain boxwood = Paxistima myrsinites (V:D)
mountain buttercup = Ranunculus eschscholtzii (V:D)
mountain candlewax = Ahtiana sphaerosporella (L)
mountain cliff fern = Woodsia scopulina (V:F)
mountain clubmoss = Lycopodium sitchense (V:F)
mountain cranberry = Vaccinium vitis-idaea (V:D)
mountain curved-back moss = Oncophorus wahlenbergii (B:M)
mountain death-camas = Zigadenus elegans (V:M)
mountain dogwood = Cornus nuttallii (V:D)
mountain fern = Thelypteris quelpaertensis (V:F)
mountain forget-me-not = Myosotis asiatica (V:D)
mountain goldenrod = Solidago spathulata (V:D)
mountain gooseberry = Ribes montigenum (V:D)
mountain hairgrass = Vahlodea atropurpurea (V:M)
mountain harebell = Campanula lasiocarpa (V:D)
mountain-heather, club-moss → club-moss mountain-heather
mountain-heather, four-angled → four-angled mountain-heather
mountain-heather, Mertens' → Mertens' mountain-heather
mountain-heather, pink → pink mountain-heather
mountain-heather, white → white mountain-heather
mountain-heather, yellow → yellow mountain-heather
mountain hemlock = Tsuga mertensiana (V:G)
mountain holly fern = Polystichum lonchitis (V:F)
mountain hollyhock = Iliamna rivularis (V:D)
mountain lady's-slipper = Cypripedium montanum (V:M)
mountain leafy liverwort = Barbilophozia floerkei (B:H)
mountain-lover = Paxistima myrsinites (V:D)
mountain mare's-tail = Hippuris montana (V:D)

mountain marsh-marigold = Caltha leptosepala (V:D)
mountain monkey-flower = Mimulus tilingii (V:D)
mountain monkshood = Aconitum delphiniifolium (V:D)
mountain-moss = Selaginella selaginoides (V:F)
mountain owl-clover = Orthocarpus imbricatus (V:D)
mountain-parsley = Cryptogramma acrostichoides (V:F)
mountain sagewort = Artemisia arctica (V:D)
mountain sagewort = Artemisia tilesii (V:D)
mountain sandwort = Minuartia biflora (V:D)
mountain scale = Psora himalayana (L)
mountain sneezeweed = Helenium autumnale (V:D)
mountain snowberry = Symphoricarpos oreophilus (V:D)
mountain sorrel = Oxyria digyna (V:D)
mountain spikemoss = Selaginella selaginoides (V:F)
moutain × western hemlock hybrid = Tsuga heterophylla × mertensiana (V:G)
mountain willow = Salix pseudomonticola (V:D)
mountainbells = Stenanthium occidentale (V:M)
mountainbells, western → western mountainbells
mouse-ear chickweed = Cerastium fontanum (V:D)
mouse-ear cress = Arabidopsis thaliana (V:D)
mouse-ear hawkweed = Hieracium pilosella (V:D)
mouse lichen = Leptogium saturninum (L)
mouse-moss, steppe → steppe mouse-moss
mouse, blue-edged → blue-edged mouse
mouse, bluff → bluff mouse
mouse, cushion → cushion mouse
mouse, fingered → fingered mouse
mouse, green → green mouse
mouse, moss → moss mouse
mouse, peacock → peacock mouse
mouse, petalled → petalled mouse
mouse, pink-eyed → pink-eyed mouse
mouse, rock → rock mouse
mouse, roughened → roughened mouse
mouse, seaside → seaside mouse
mouse, soil → soil mouse
mousetail, bristly → bristly mousetail
mousetail, least → least mousetail
mousetail, sedge → sedge mousetail
mousetail, tiny → tiny mousetail
Mt. Washington bluegrass = Poa laxa (V:M)
mud-disk = Cotula coronopifolia (V:D)
mud rush = Juncus gerardii (V:M)
mud sedge = Carex limosa (V:M)
mudwort, water → water mudwort
mugwort, Aleutian → Aleutian mugwort
mugwort, Columbia → Columbia mugwort
mugwort, Columbia River → Columbia River mugwort
mugwort, common → common mugwort
mugwort, long-leaved → long-leaved mugwort
mugwort, Michaux's → Michaux's mugwort
mugwort, Suksdorf's → Suksdorf's mugwort
mugwort, three-forked → three-forked mugwort
mugwort, western → western mugwort
Muhlenberg's centaury = Centaurium muhlenbergii (V:D)
Muhlenberg's rocktripe = Umbilicaria muehlenbergii (L)
muhly, alkali → alkali muhly
muhly, bog → bog muhly
muhly, foxtail → foxtail muhly
muhly, marsh → marsh muhly
muhly, mat → mat muhly
muhly, slender → slender muhly
muhly, wirestem → wirestem muhly
mulberry, white → white mulberry
mullein, clasping → clasping mullein
mullein, common → common mullein
mullein, great → great mullein
mullein, moth → moth mullein
mullein, woolly → woolly mullein
Munroe's globe-mallow = Sphaeralcea munroana (V:D)
musk-flower = Mimulus moschatus (V:D)
musk mallow = Malva moschata (V:D)
musk thistle = Carduus nutans ssp. leiophyllus (V:D)
mustard, ball → ball mustard
mustard, bird rape → bird rape mustard
mustard, black → black mustard
mustard, blue → blue mustard
mustard, brown → brown mustard
mustard, Chinese → Chinese mustard
mustard, dog → dog mustard
mustard, garlic → garlic mustard
mustard, hare's-ear → hare's-ear mustard
mustard, hedge → hedge mustard
mustard, Indian → Indian mustard
mustard, leaf → leaf mustard
mustard, plains → plains mustard
mustard, tower → tower mustard
mustard, wormseed → wormseed mustard
muttongrass, Fendler's → Fendler's muttongrass
myrtle-leaved willow = Salix myrtillifolia (V:D)

N

nagoonberry = Rubus arcticus (V:D)
nagoonberry, dwarf → dwarf nagoonberry
Nahanni oak fern = Gymnocarpium jessoense ssp. parvulum (V:F)
nail-rod = Typha latifolia (V:M)
naked broomrape = Orobanche uniflora (V:D)
naked mitrewort = Mitella nuda (V:D)
narcissus anemone = Anemone narcissiflora (V:D)
narcissus, poet's → poet's narcissus
narrow beech fern = Phegopteris connectilis (V:F)
narrow cowboy fern = Phegopteris connectilis (V:F)
narrow-flowered bluegrass = Poa stenantha (V:M)
narrow-leaved birds-foot trefoil = Lotus tenuis (V:D)
narrow-leaved brickellia = Brickellia oblongifolia (V:D)
narrow-leaved bur-reed = Sparganium angustifolium (V:M)
narrow-leaved collomia = Collomia linearis (V:D)
narrow-leaved cotton-grass = Eriophorum angustifolium (V:M)
narrow-leaved desert-parsley = Lomatium triternatum (V:D)
narrow-leaved dock = Rumex salicifolius (V:D)
narrow-leaved hawkweed = Hieracium umbellatum (V:D)
narrow-leaved montia = Montia linearis (V:D)
narrow-leaved peavine = Lathyrus sylvestris (V:D)
narrow-leaved skeletonweed = Stephanomeria tenuifolia (V:D)
narrow-leaved skullcap = Scutellaria angustifolia (V:D)
narrow-leaved stephanomeria = Stephanomeria tenuifolia (V:D)
narrow-leaved sword fern = Polystichum imbricans (V:F)
narrow-leaved water-plantain = Alisma gramineum (V:M)
narrow-leaved willow = Salix exigua (V:D)
narrow-leaved willowherb = Epilobium leptophyllum (V:D)
narrow-petaled stonecrop = Sedum stenopetalum (V:D)
narrow sedge = Carex arcta (V:M)
narrow-sepaled phacelia = Phacelia leptosepala (V:D)
naugehyde liverwort = Ptilidium pulcherrimum (B:H)
navarretia, needle-leaved → needle-leaved navarretia
neckera, Douglas' → Douglas' neckera
neckera, Menzies' → Menzies' neckera
needle-and-thread grass = Stipa comata (V:M)

needle-leaved navarretia = Navarretia intertexta (V:D)
needle spike-rush = Eleocharis acicularis (V:M)
needlegrass, Columbian → Columbian needlegrass
needlegrass, green → green needlegrass
needlegrass, Lemmon's → Lemmon's needlegrass
needlegrass, Nelson's → Nelson's needlegrass
needlegrass, spreading → spreading needlegrass
needlegrass, stiff → stiff needlegrass
Nelson's needlegrass = Stipa nelsonii (V:M)
nemophila, Great Basin → Great Basin nemophila
nemophila, meadow → meadow nemophila
nemophila, small-flowered → small-flowered nemophila
Nepalese knotweed = Polygonum Nepalense (V:D)
nephroma, chocolate-coloured → chocolate-coloured nephroma
net-veined willow = Salix reticulata (V:D)
net-veined willow = Salix reticulata ssp. reticulata (V:D)
netted rocktripe = Umbilicaria proboscidea (L)
netted rocktripe = Umbilicaria decussata (L)
netted rocktripe = Umbilicaria lyngei (L)
netted rocktripe = Umbilicaria krascheninnikovii (L)
netted specklebelly = Pseudocyphellaria anomala (L)
netted willow = Salix reticulata (V:D)
netted willow = Salix reticulata ssp. reticulata (V:D)
nettle-leaved giant-hyssop = Agastache urticifolia (V:D)
nettle, burning → burning nettle
nettle, common → common nettle
nettle, dog → dog nettle
nettle, stinging → stinging nettle
Nevada birds-food trefoil = Lotus nevadensis var. douglasii (V:D)
Nevada birds-foot trefoil = Lotus nevadensis (V:D)
Nevada bluegrass = Poa secunda (V:M)
Nevada bulrush = Scirpus nevadensis (V:M)
Nevada clubrush = Scirpus nevadensis (V:M)
Nevada marsh fern = Thelypteris nevadensis (V:F)
Newcombe's butterweed = Sinosenecio newcombei (V:D)
night-flowering catchfly = Silene noctiflora (V:D)
nightmare, Eschscholtz's little → Eschscholtz's little nightmare
nightshade, black → black nightshade
nightshade, cut-leaved → cut-leaved nightshade
nightshade, hairy → hairy nightshade
nine-leaved desert-parsley = Lomatium triternatum (V:D)
nine-leaved lamatium = Lomatium triternatum (V:D)
ninebark, mallow → mallow ninebark
ninebark, Pacific → Pacific ninebark
nipplewort = Lapsana communis (V:D)
noble fir = Abies procera (V:G)
nodding beggarticks = Bidens cernua (V:D)
nodding brome = Bromus anomalus (V:M)
nodding chickweed = Cerastium nutans (V:D)
nodding hackelia = Hackelia deflexa (V:D)
nodding microseris = Microseris nutans (V:D)
nodding onion = Allium cernuum (V:M)
nodding pohlia = Pohlia nutans (B:M)
nodding saxifrage = Saxifraga cernua (V:D)
nodding stickseed = Hackelia deflexa (V:D)
nodding thistle = Carduus nutans ssp. leiophyllus (V:D)
nodding trisetum = Trisetum cernuum var. cernuum (V:M)
nodding trisetum = Trisetum cernuum (V:M)
nodding wood-reed = Cinna latifolia (V:M)
Nootka lupine = Lupinus nootkatensis (V:D)
Nootka rose = Rosa nutkana (V:D)
North American lichen = Hydrothyria venosa (L)
North Pacific draba = Draba hyperborea (V:D)
North Pacific whitlow-grass = Draba hyperborea (V:D)
northern brown lichen = Melanelia septentrionalis (L)
northern adder's-tongue = Ophioglossum pusillum (V:F)
northern anemone = Anemone parviflora (V:D)
northern arnica = Arnica frigida (V:D)
northern bastard toad-flax = Geocaulon lividum (V:D)
northern bedstraw = Galium boreale (V:D)
northern bentgrass = Agrostis mertensii (V:M)
northern bentgrass = Agrostis aequivalvis (V:M)
northern blackcurrant = Ribes hudsonianum (V:D)
northern blue-eyed-grass = Sisyrinchium septentrionale (V:M)
northern blue violet = Viola septentrionalis (V:D)
northern bluegrass = Poa abbreviata (V:M)
northern bog sedge = Carex gynocrates (V:M)
northern bog violet = Viola nephrophylla (V:D)
northern brome = Bromus inermis ssp. pumpellianus (V:M)
northern brown = Melanelia septentrionalis (L)
northern bur-reed = Sparganium hyperboreum (V:M)
northern bush willow = Salix arbusculoides (V:D)
northern clustered sedge = Carex arcta (V:M)
northern comandra = Geocaulon lividum (V:D)
northern crane's-bill = Geranium erianthum (V:D)
northern draba = Draba borealis (V:D)
northern fairy-candelabra = Androsace septentrionalis (V:D)
northern false asphodel = Tofieldia coccinea (V:M)
northern flaxflower = Linanthus septentrionalis (V:D)
northern gentian = Gentianella amarella (V:D)
northern gentian = Gentianella amarella ssp. acuta (V:D)
northern geranium = Geranium erianthum (V:D)
northern golden carpet = Chrysosplenium tetrandrum (V:D)
northern golden-saxifrage = Chrysosplenium tetrandrum (V:D)
northern goldenrod = Solidago multiradiata (V:D)
northern gooseberry = Ribes oxyacanthoides (V:D)
northern grass-of-Parnassus = Parnassia palustris (V:D)
northern green bog orchid = Platanthera hyperborea (V:M)
northern groundcone = Boschniakia rossica (V:D)
northern hedysarum = Hedysarum boreale (V:D)
northern holly fern = Polystichum lonchitis (V:F)
northern Indian-paintbrush = Castilleja hyperborea (V:D)
northern Jacob's-ladder = Polemonium boreale (V:D)
northern Labrador tea = Ledum palustre ssp. decumbens (V:D)
northern Labrador tea = Ledum palustre (V:D)
northern linanthus = Linanthus septentrionalis (V:D)
northern maidenhair fern = Adiantum aleuticum (V:F)
northern mannagrass = Glyceria borealis (V:M)
northern naugehyde liverwort = Ptilidium ciliare (B:H)
northern oak fern = Gymnocarpium jessoense (V:F)
northern paintbrush = Castilleja hyperborea (V:D)
northern parrya = Parrya nudicaulis (V:D)
northern perfume = Evernia mesomorpha (L)
northern pondweed = Potamogeton alpinus (V:M)
northern reindeer lichen = Cladina stellaris (L)
northern rice-root = Fritillaria camschatcensis (V:M)
northern ricegrass = Oryzopsis pungens (V:M)
northern rocktripe = Umbilicaria hyperborea (L)
northern rough fescue = Festuca altaica (V:M)
northern sandwort = Minuartia elegans (V:D)
northern sawwort = Saussurea angustifolia (V:D)
northern scouring-rush = Equisetum variegatum (V:F)
northern selaginella = Selaginella sibirica (V:F)
northern shootingstar = Dodecatheon frigidum (V:D)
northern smooth gooseberry = Ribes oxyacanthoides (V:D)
northern starflower = Trientalis borealis (V:D)
northern starwort = Stellaria calycantha (V:D)
northern stitchwort = Stellaria calycantha (V:D)
northern sweet-vetch = Hedysarum boreale (V:D)
northern tansymustard = Descurainia sophioides (V:D)
northern twayblade = Listera borealis (V:M)
northern valerian = Valeriana dioica (V:D)
northern violet = Viola septentrionalis (V:D)
northern water-carpet = Chrysosplenium tetrandrum (V:D)
northern water horehound = Lycopus uniflorus (V:D)
northern water-meal = Wolffia borealis (V:M)
northern water-starwort = Callitriche hermaphroditica (V:D)
northern whitlow-grass = Draba borealis (V:D)
northern wild-licorice = Galium kamtschaticum (V:D)
northern woodsia = Woodsia alpina (V:F)
northern wormwood = Artemisia campestris (V:D)
northwest stippleback = Dermatocarpon reticulatum (L)
northwestern moonwort = Botrychium pinnatum (V:F)
northwestern sedge = Carex concinnoides (V:M)
northwestern twayblade = Listera caurina (V:M)

northwestern white birch = Betula neoalaskana (V:D)
Norway maple = Acer platanoides (V:D)
Norwegian cinquefoil = Potentilla norvegica (V:D)
Nuttall's alkaligrass = Puccinellia nuttalliana (V:M)
Nuttall's bitter-cress = Cardamine nuttallii (V:D)
Nuttall's draba = Draba densifolia (V:D)
Nuttall's knotweed = Polygonum douglasii ssp. nuttallii (V:D)
Nuttall's larkspur = Delphinium nuttallianum (V:D)
Nuttall's orache = Atriplex nuttallii (V:D)
Nuttall's pussytoes = Antennaria parvifolia (V:D)
Nuttall's quillwort = Isoetes nuttallii (V:F)
Nuttall's rockcress = Arabis nuttallii (V:D)
Nuttall's sandwort = Minuartia nuttallii (V:D)
Nuttall's sunflower = Helianthus nuttallii (V:D)
Nuttall's waterweed = Elodea nuttallii (V:M)
Nuttall's whitlow-grass = Draba densifolia (V:D)
Nutzotin milk-vetch = Astragalus nutzotinensis (V:D)
nymph, wavy water → wavy water nymph

O

oak fern = Gymnocarpium dryopteris (V:F)
oak, English → English oak
oak, Garry → Garry oak
oak, Oregon white → Oregon white oak
Oakes' pondweed = Potamogeton oakesianus (V:M)
oat, common → common oat
oat, wild → wild oat
oatgrass, California → California oatgrass
oatgrass, false → false oatgrass
oatgrass, one-spike → one-spike oatgrass
oatgrass, poverty → poverty oatgrass
oatgrass, tall → tall oatgrass
oatgrass, timber → timber oatgrass
oblong-leaved orache = Atriplex oblongifolia (V:D)
oblong-leaved sundew = Drosera anglica (V:D)
obscure cryptantha = Cryptantha ambigua (V:D)
ocean-spray = Holodiscus discolor (V:D)
Oeder's lousewort = Pedicularis oederi (V:D)
oenanthe, Pacific → Pacific oenanthe
Ogotoruk Creek butterweed = Senecio ogotorukensis (V:D)
Okanogan fameflower = Talinum sediforme (V:D)
Okanogan hackelia = Hackelia ciliata (V:D)
Okanogan stickseed = Hackelia ciliata (V:D)
Okanogan talinum = Talinum sediforme (V:D)
old-growth paw = Nephroma silvae-veteris (L)
old-growth specklebelly = Pseudocyphellaria rainierensis (L)
old-growth turtle = Hypocenomyce friesii (L)
old man's beard = Clematis vitalba (V:D)
old man's whiskers = Geum triflorum (V:D)
olive, Russian → Russian olive
Olympic Mountain aster = Aster paucicapitatus (V:D)
Olympic onion = Allium crenulatum (V:M)
one-and-a-half-flowered reedgrass = Calamagrostis sesquiflora (V:M)
one-flowered fleabane = Erigeron uniflorus (V:D)
one-flowered cinquefoil = Potentilla uniflora (V:D)
one-flowered wintergreen = Moneses uniflora (V:D)
one-headed pussytoes = Antennaria monocephala (V:D)
one-horse shadow = Phaeophyscia nigricans (L)
one-leaved foamflower = Tiarella trifoliata var. unifoliata (V:D)
one-leaved rein orchid = Platanthera obtusata (V:M)
one-sided sedge = Carex unilateralis (V:M)
one-sided wintergreen = Orthilia secunda (V:D)
one-spike oatgrass = Danthonia unispicata (V:M)
onion, Geyer's → Geyer's onion
onion, Hooker's → Hooker's onion
onion, nodding → nodding onion
onion, Olympic → Olympic onion
onion, slimleaf → slimleaf onion
onion, swamp → swamp onion
oniongrass = Melica bulbosa (V:M)
oniongrass, Alaska → Alaska oniongrass
oniongrass, purple → purple oniongrass
opium poppy = Papaver somniferum (V:D)
orache = Atriplex patula (V:D)
orache, Alaska → Alaska orache
orache, common → common orache
orache, garden → garden orache
orache, Gmelin's → Gmelin's orache
orache, Nuttall's → Nuttall's orache
orache, oblong-leaved → oblong-leaved orache
orache, red → red orache
orache, saline → saline orache
orache, tumbling → tumbling orache
orache, wedgescalf → wedgescalf orache
orange agoseris = Agoseris aurantiaca (V:D)
orange arnica = Arnica fulgens (V:D)
orange-foot cladonia = Cladonia ecmocyna (L)
orange-foot lichen = Cladonia ecmocyna (L)
orange hawkweed = Hieracium aurantiacum (V:D)
orange honeysuckle = Lonicera ciliosa (V:D)
orange talus wort = Ptilidium ciliare (B:H)
orange touch-me-not = Impatiens aurella (V:D)
orange, cushion → cushion orange
orange, elegant → elegant orange
orange, pincushion → pincushion orange
orange, powdered → powdered orange
orange, rock → rock orange
orange, shrubby → shrubby orange
orange, sugared → sugared orange
orchard morning-glory = Convolvulus arvensis (V:D)
orchardgrass = Dactylis glomerata (V:M)
orchid, Alaska rein → Alaska rein orchid
orchid, blunt-leaved bog → blunt-leaved bog orchid
orchid, bracted bog → bracted bog orchid
orchid, chamisso's rein → chamisso's rein orchid
orchid, dwarf rattlesnake → dwarf rattlesnake orchid
orchid, elegant rein → elegant rein orchid
orchid, frog → frog orchid
orchid, green-flowered bog → green-flowered bog orchid
orchid, green-flowered rein → green-flowered rein orchid
orchid, large-leaved rattlesnake → large-leaved rattlesnake orchid
orchid, large round-leaved rein → large round-leaved rein orchid
orchid, long-bracted → long-bracted orchid
orchid, long-bracted frog → long-bracted frog orchid
orchid, northern green bog → northern green bog orchid
orchid, one-leaved rein → one-leaved rein orchid
orchid, phantom → phantom orchid
orchid, round-leaved bog → round-leaved bog orchid
orchid, round-leaved rein → round-leaved rein orchid
orchid, slender bog → slender bog orchid
orchid, slender rein → slender rein orchid
orchid, small bog → small bog orchid
orchid, white bog → white bog orchid
orchid, white rein → white rein orchid
orchis, round-leaved → round-leaved orchis
Oregon beaked moss = Eurhynchium oreganum (B:M)
Oregon bentgrass = Agrostis oregonensis (V:M)
Oregon boxwood = Paxistima myrsinites (V:D)

Oregon checker-mallow = Sidalcea oregana var. procera (V:D)
Oregon checker-mallow = Sidalcea oregana (V:D)
Oregon fairybells = Disporum hookeri var. oreganum (V:M)
Oregon-grape, creeping → creeping Oregon-grape
Oregon-grape, dull → dull Oregon-grape
Oregon-grape, tall → tall Oregon-grape
Oregon maple = Acer macrophyllum (V:D)
Oregon saxifrage = Saxifraga adscendens ssp. oregonensis (V:D)
Oregon selaginella = Selaginella oregana (V:F)
Oregon stonecrop = Sedum oreganum (V:D)
Oregon sunshine = Eriophyllum lanatum (V:D)
Oregon white oak = Quercus garryana (V:D)
Oregon willowherb = Epilobium oregonense (V:D)
Oregon wintergreen = Gaultheria ovatifolia (V:D)
Oregon woodsia = Woodsia oregana (V:F)
ostrich fern = Matteuccia struthiopteris (V:F)
oval-leaved blueberry = Vaccinium ovalifolium (V:D)
oval-leaved knotweed = Polygonum arenastrum (V:D)
oval-leaved milkweed = Asclepias ovalifolia (V:D)
oval-leaved mitrewort = Mitella ovalis (V:D)
oval-leaved penstemon = Penstemon ellipticus (V:D)
oval penstemon = Penstemon ellipticus (V:D)
owl-clover, mountain → mountain owl-clover
owl-clover, rosy → rosy owl-clover
owl-clover, thin-leaved → thin-leaved owl-clover
owl-clover, yellow → yellow owl-clover
owl, fringed → fringed owl
owl, tundra → tundra owl
owl, woodland → woodland owl
oxalis, upright yellow → upright yellow oxalis
oxalis, yellow → yellow oxalis
oxeye daisy = Leucanthemum vulgare (V:D)
oxytrope, arctic → arctic oxytrope
oxytrope, blackish → blackish oxytrope
oxytrope, Huddelson's → Huddelson's oxytrope
oxytrope, Maydell's → Maydell's oxytrope
oxytrope, Scamman's → Scamman's oxytrope
oyster plant = Tragopogon porrifolius (V:D)
oysterleaf = Mertensia maritima (V:D)

P

Pacific alkaligrass = Puccinellia nutkaensis (V:M)
Pacific anemone = Anemone multifida (V:D)
Pacific bleeding heart = Dicentra formosa (V:D)
Pacific brome = Bromus pacificus (V:M)
Pacific coast strawberry = Fragaria chiloensis (V:D)
Pacific crab apple = Malus fusca (V:D)
Pacific dogwood = Cornus nuttallii (V:D)
Pacific enchanter's-nightshade = Circaea alpina ssp. pacifica (V:D)
Pacific madrone = Arbutus menziesii (V:D)
Pacific ninebark = Physocarpus capitatus (V:D)
Pacific oenanthe = Oenanthe sarmentosa (V:D)
Pacific polypody = Polypodium amorphum (V:F)
Pacific reedgrass = Calamagrostis nutkaensis (V:M)
Pacific rhododendron = Rhododendron macrophyllum (V:D)
Pacific samphire = Salicornia virginica (V:D)
Pacific sanicle = Sanicula crassicaulis (V:D)
Pacific silver fir = Abies amabilis (V:G)
Pacific trailing blackberry = Rubus ursinus ssp. macropetalus (V:D)
Pacific water-parsley = Oenanthe sarmentosa (V:D)
Pacific waterleaf = Hydrophyllum tenuipes (V:D)
Pacific willow = Salix lucida ssp. lasiandra (V:D)
Pacific yew = Taxus brevifolia (V:G)
pahute weed = Suaeda calceoliformis (V:D)
paint, Indian → Indian paint
paintbrush, alpine → alpine paintbrush
paintbrush, annual → annual paintbrush
paintbrush, boreal → boreal paintbrush
paintbrush, cliff → cliff paintbrush
paintbrush, coastal red → coastal red paintbrush
paintbrush, common red → common red paintbrush
paintbrush, Cusick's → Cusick's paintbrush
paintbrush, deer → deer paintbrush
paintbrush, Elmer's → Elmer's paintbrush
paintbrush, golden → golden paintbrush
paintbrush, golden Indian → golden Indian paintbrush
paintbrush, harsh → harsh paintbrush
paintbrush, northern → northern paintbrush
paintbrush, palish → palish paintbrush
paintbrush, purple → purple paintbrush
paintbrush, Raup's → Raup's paintbrush
paintbrush, rhexia-leaved → rhexia-leaved paintbrush
paintbrush, scarlet → scarlet paintbrush
paintbrush, small-flowered → small-flowered paintbrush
paintbrush, sulphur → sulphur paintbrush
paintbrush, Thompson's → Thompson's paintbrush
paintbrush, unalaska → unalaska paintbrush
paintbrush, western → western paintbrush
paintbrush, yellow → yellow paintbrush
pale agoseris = Agoseris glauca (V:D)
pale alyssum = Alyssum alyssoides (V:D)
pale bulrush = Scirpus pallidus (V:M)
pale comandra = Comandra umbellata (V:D)
pale coralroot = Corallorrhiza trifida (V:M)
pale evening-primrose = Oenothera pallida (V:D)
pale flax = Linum bienne (V:D)
pale persicaria = Polygonum lapathifolium (V:D)
pale poppy = Papaver alboroseum (V:D)
pale sedge = Carex livida (V:M)
pale sedge = Carex pallescens (V:M)
pale shield = Platismatia glauca (L)
pale spring-beauty = Claytonia spathulata (V:D)
pale-stalked broom moss = Dicranum pallidisetum (B:M)
palish Indian-paintbrush = Castilleja pallescens (V:D)
palish paintbrush = Castilleja pallescens (V:D)
Pallas' wallflower = Erysimum pallasii (V:D)
pallid ruffle = Ahtiana pallidula (L)
palm moss = Climacium dendroides (B:M)
palm tree moss = Leucolepis acanthoneuron (B:M)
panicled mertensia = Mertensia paniculata (V:D)
pansy, European field → European field pansy
pansy, European wild → European wild pansy
paper birch = Betula papyrifera var. papyrifera (V:D)
paper birch = Betula papyrifera (V:D)
paperdoll, furled → furled paperdoll
paperdoll, ragged → ragged paperdoll
papoose tarpaper = Collema crispum (L)
pappered vinyl = Leptogium saturninum (L)
parasols, fairy → fairy parasols
parentucellia, yellow → yellow parentucellia
Parry's arnica = Arnica parryi (V:D)
Parry's campion = Silene parryi (V:D)
Parry's catchfly = Silene parryi (V:D)
Parry's rush = Juncus parryi (V:M)
Parry's sedge = Carex parryana (V:M)
Parry's townsendia = Townsendia parryi (V:D)
parrya, northern → northern parrya
parsley fern = Cryptogramma acrostichoides (V:F)
parsley-piert, field → field parsley-piert
parsley-piert, small-fruited → small-fruited parsley-piert

parsnip-flowered buckwheat = Eriogonum heracleoides var. angustifolium (V:D)
parsnip-flowered buckwheat = Eriogonum heracleoides (V:D)
parsnip, common → common parsnip
partridgefoot = Luetkea pectinata (V:D)
pasqueflower, western → western pasqueflower
pasture sagewort = Artemisia frigida (V:D)
pasture sedge = Carex petasata (V:M)
pathfinder = Adenocaulon bicolor (V:D)
patience dock = Rumex patientia (V:D)
paw, alpine → alpine paw
paw, blistered → blistered paw
paw, cat → cat paw
paw, cryptic → cryptic paw
paw, dog → dog paw
paw, green → green paw
paw, old-growth → old-growth paw
paw, pepper → pepper paw
paw, powder → powder paw
paw, powdered → powdered paw
paw, seaside → seaside paw
Payson's draba = Draba paysonii (V:D)
Payson's sedge = Carex paysonis (V:M)
Payson's whitlow-grass = Draba paysonii (V:D)
pea, beach → beach pea
peach-leaf willow = Salix amygdaloides (V:D)
peach-leaved bellflower = Campanula persicifolia (V:D)
peacock mouse = Pannaria pezizoides (L)
peacock vinyl = Leptogium polycarpum (L)
peafruit rose = Rosa pisocarpa (V:D)
peak saxifrage = Saxifraga nidifica (V:D)
pear moss = Leptobryum pyriforme (B:M)
pear, common → common pear
pearlwort, arctic → arctic pearlwort
pearlwort, bird's-eye → bird's-eye pearlwort
pearlwort, coastal → coastal pearlwort
pearlwort, Japanese → Japanese pearlwort
pearlwort, snow → snow pearlwort
pearlwort, trailing → trailing pearlwort
pearlwort, western → western pearlwort
pearly everlasting = Anaphalis margaritacea (V:D)
peavine, broad-leaved → broad-leaved peavine
peavine, cream-flowered → cream-flowered peavine
peavine, creamy → creamy peavine
peavine, grass → grass peavine
peavine, grey beach → grey beach peavine
peavine, marsh → marsh peavine
peavine, meadow → meadow peavine
peavine, narrow-leaved → narrow-leaved peavine
peavine, pinewood → pinewood peavine
peavine, purple → purple peavine
Peck's sedge = Carex peckii (V:M)
peduncled sedge = Carex pedunculata (V:M)
pedunculate birds-foot trefoil = Lotus pedunculatus (V:D)
pellia, ring → ring pellia
pellitory, Pennsylvania → Pennsylvania pellitory
pelt, apple → apple pelt
pelt, born-again → born-again pelt
pelt, butterfly → butterfly pelt
pelt, concentric → concentric pelt
pelt, dog → dog pelt
pelt, fan → fan pelt
pelt, felt → felt pelt
pelt, freckle → freckle pelt
pelt, frog → frog pelt
pelt, peppered → peppered pelt
pelt, sponge → sponge pelt
pelt, temporary → temporary pelt
pelt, toad → toad pelt
pelt, tree → tree pelt
pembina = Viburnum opulus (V:D)
pencil scirpt = Graphis scripta (L)
pendant-pod locoweed = Oxytropis deflexa (V:D)
pendantgrass = Arctophila fulva (V:M)
Pennsylvania buttercup = Ranunculus pensylvanicus (V:D)
Pennsylvania cinquefoil = Potentilla pensylvanica (V:D)
Pennsylvania pellitory = Parietaria pensylvanica (V:D)
Pennsylvania rocktripe = Lasallia pensylvanica (L)
Pennsylvanian bitter-cress = Cardamine pensylvanica (V:D)
pennycress, field → field pennycress
pennyroyal = Mentha pulegium (V:D)
pennywort, floating marsh → floating marsh pennywort
pennywort, whorled marsh → whorled marsh pennywort
penstemon, Alberta → Alberta penstemon
penstemon, broad-leaved → broad-leaved penstemon
penstemon, Chelan → Chelan penstemon
penstemon, coast → coast penstemon
penstemon, Cusick's → Cusick's penstemon
penstemon, Davidson's → Davidson's penstemon
penstemon, elliptic-leaved → elliptic-leaved penstemon
penstemon, fuzzy-tongued → fuzzy-tongued penstemon
penstemon, Gorman's → Gorman's penstemon
penstemon, Lyall's → Lyall's penstemon
penstemon, oval → oval penstemon
penstemon, oval-leaved → oval-leaved penstemon
penstemon, Richardson's → Richardson's penstemon
penstemon, shining → shining penstemon
penstemon, shrubby → shrubby penstemon
penstemon, slender → slender penstemon
penstemon, slender blue → slender blue penstemon
penstemon, small-flowered → small-flowered penstemon
penstemon, woodland → woodland penstemon
penstemon, yellow → yellow penstemon
pepper-grass, Bourgeau's → Bourgeau's pepper-grass
pepper-grass, branched → branched pepper-grass
pepper-grass, clasping-leaved → clasping-leaved pepper-grass
pepper-grass, common → common pepper-grass
pepper-grass, field → field pepper-grass
pepper-grass, prairie → prairie pepper-grass
pepper-grass, Smith's → Smith's pepper-grass
pepper-grass, tall → tall pepper-grass
pepper paw = Nephroma isidiosum (L)
peppered brownette = Vestergrenopsis isidiata (L)
peppered moon = Sticta fuliginosa (L)
peppered pelt = Peltigera evansiana (L)
peppered rocktripe = Umbilicaria deusta (L)
peppered vinyl = Leptogium burgessii (L)
peppered vinyl = Leptogium furfuraceum (L)
peppermint = Mentha × piperita (V:D)
perennial ryegrass = Lolium perenne (V:M)
perennial saltwort = Salicornia virginica (V:D)
perennial sow-thistle = Sonchus arvensis (V:D)
perennial sundrops = Oenothera perennis (V:D)
perfoliate pondweed = Potamogeton perfoliatus (V:M)
perfume, antlered → antlered perfume
perfume, northern → northern perfume
periwinkle, common → common periwinkle
periwinkle, large → large periwinkle
persicaria, pale → pale persicaria
petalled mouse = Fuscopannaria leucostictoides (L)
petalled mouse = Pannaria rubiginosa (L)
petalled rocktripe = Umbilicaria polyphylla (L)
petalled vinyl = Leptogium gelatinosum (L)
petrorhagia = Petrorhagia prolifera (V:D)
petty spurge = Euphorbia peplus (V:D)
phacelia, branched → branched phacelia
phacelia, Franklin's → Franklin's phacelia
phacelia, Lyall's → Lyall's phacelia
phacelia, MacBryde's → MacBryde's phacelia
phacelia, narrow-sepaled → narrow-sepaled phacelia
phacelia, silky → silky phacelia
phacelia, silverleaf → silverleaf phacelia

phacelia, thread-leaved → thread-leaved phacelia
phantom orchid = Cephalanthera austiniae (V:M)
Philadelphia daisy = Erigeron philadelphicus (V:D)
Philadelphia fleabane = Erigeron philadelphicus (V:D)
phlox, alyssum-leaved → alyssum-leaved phlox
phlox, Hood's → Hood's phlox
phlox, long-leaved → long-leaved phlox
phlox, prickly → prickly phlox
phlox, showy → showy phlox
phlox, spreading → spreading phlox
phlox, tufted → tufted phlox
phlox, tufyed → tufyed phlox
pie, cow → cow pie
piggy-back plant = Tolmiea menziesii (V:D)
pigmyweed = Crassula aquatica (V:D)
pigweed amaranth = Amaranthus retroflexus (V:D)
pigweed, green → green pigweed
pigweed, prostrate → prostrate pigweed
pigweed, red → red pigweed
pigweed, rough → rough pigweed
pigweed, Russian → Russian pigweed
pigweed, white → white pigweed
pimpernel, scarlet → scarlet pimpernel
pimpled kidney = Nephroma resupinatum (L)
pin cherry = Prunus pensylvanica (V:D)
pincushion orange = Xanthoria ramulosa (L)
pincushion orange = Xanthoria polycarpa (L)
pincushion shadow = Phaeophyscia constipata (L)
pincushion vinyl = Leptogium subaridum (L)
pine broomrape = Orobanche pinorum (V:D)
pine lichen = Vulpicida pinastri (L)
pine, jack → jack pine
pine, limber → limber pine
pine, lodgepole → lodgepole pine
pine, lodgepole × jack → lodgepole × jack pine hybrid
pine, ponderosa → ponderosa pine
pine, prince's → prince's pine
pine, shore → shore pine
pine, western white → western white pine
pine, whitebark → whitebark pine
pine, yellow → yellow pine
pineapple-weed = Matricaria discoidea (V:D)
pinedrops = Pterospora andromedea (V:D)
pinegrass = Calamagrostis rubescens (V:M)
pinewood peavine = Lathyrus bijugatus (V:D)
pink agoseris = Agoseris lackschewitzii (V:D)
pink campion = Silene repens (V:D)
pink corydalis = Corydalis sempervirens (V:D)
pink-eyed mouse = Fuscopannaria saubinetii (L)
pink-eyed rockbright = Rhizoplaca chrysoleuca (L)
pink fairies = Clarkia pulchella (V:D)
pink fawn lily = Erythronium revolutum (V:M)
pink monkey-flower = Mimulus lewisii (V:D)
pink mountain-heather = Phyllodoce empetriformis (V:D)
pink wintergreen = Pyrola asarifolia (V:D)
pink, Deptford → Deptford pink
pink, maiden → maiden pink
pinole clover = Trifolium bifidum (V:D)
pioneer cladonia = Cladonia cornuta (L)
pipe, lattice → lattice pipe
pipecleaner moss = Rhytidiopsis robusta (B:M)
pipecleaner moss = Rhytidium rugosum (B:M)
Piper's anemone = Anemone piperi (V:D)
Piper's woodrush = Luzula piperi (V:M)
pipestems = Clematis ligusticifolia (V:D)
pipsissewa = Chimaphila umbellata (V:D)
pipsissewa, Menzies' → Menzies' pipsissewa
pitcher-plant = Sarracenia purpurea (V:D)
pitcher-plant, common → common pitcher-plant
pitted beard = Usnea cavernosa (L)
pitted old man's beard = Usnea cavernosa (L)
pixie-cup = Cladonia pyxidata (L)
pixie-cup lichen = Cladonia pyxidata (L)
pixie-cup, false → false pixie-cup
pixie-cup, red → red pixie-cup
pixie-eyes = Primula cuneifolia (V:D)
plains butterweed = Senecio plattensis (V:D)
plains cottonwood = Populus deltoides ssp. monilifera (V:D)
plains eryngo = Eryngium planum (V:D)
plains mustard = Schoenocrambe linifolia (V:D)
plains prickly-pear cactus = Opuntia polyacantha (V:D)
plains reedgrass = Calamagrostis montanensis (V:M)
plains wormwood = Artemisia campestris (V:D)
plainsmustard, flax-leaved → flax-leaved plainsmustard
plane-leaved willow = Salix planifolia ssp. planifolia (V:D)
plane-leaved willow = Salix planifolia (V:D)
plant, Indian consumption → Indian consumption plant
plant, oyster → oyster plant
plant, piggy-back → piggy-back plant
plantain, Alaska → Alaska plantain
plantain, alkali → alkali plantain
plantain, arctic → arctic plantain
plantain, buck's-horn → buck's-horn plantain
plantain, common → common plantain
plantain, English → English plantain
plantain, hoary → hoary plantain
plantain, ribwort → ribwort plantain
plantain, saline → saline plantain
plantain, sand → sand plantain
plantain, Siberian → Siberian plantain
plantain, slender → slender plantain
plantain, whorled → whorled plantain
plantain, woolly → woolly plantain
plated rocktripe = Umbilicaria muehlenbergii (L)
plectritis, long-spurred → long-spurred plectritis
plectritis, rosy → rosy plectritis
plum, buffalo → buffalo plum
plum, ground → ground plum
plume moss = Dendroalsia abietina (B:M)
plume, knight's → knight's plume
plumeless thistle = Carduus acanthoides (V:D)
poet's narcissus = Narcissus poeticus (V:M)
pohlia, common nodding → common nodding pohlia
pohlia, nodding → nodding pohlia
pointed broom sedge = Carex scoparia (V:M)
pointed rush = Juncus oxymeris (V:M)
poison-hemlock = Conium maculatum (V:D)
poison-ivy = Toxicodendron rydbergii (V:D)
poison-oak = Toxicodendron diversilobum (V:D)
polar bluegrass = Poa pseudoabbreviata (V:M)
polar willow = Salix polaris (V:D)
polargrass = Arctagrostis latifolia (V:M)
polemonium, elegant Jacob's-ladder → elegant Jacob's-ladder polemonium
polemonium, littlebells → littlebells polemonium
polemonium, skunk → skunk polemonium
polemonium, sticky → sticky polemonium
policeman's helmet = Impatiens glandulifera (V:D)
polypody, irregular → irregular polypody
polypody, leatherleaf → leatherleaf polypody
polypody, leathery → leathery polypody
polypody, Pacific → Pacific polypody
polypody, western → western polypody
polypogon, rabbitfoot → rabbitfoot polypogon
pond water-starwort = Callitriche stagnalis (V:D)
ponderosa pine = Pinus ponderosa (V:G)
pondweed, blunt-leaved → blunt-leaved pondweed
pondweed, clasping-leaved → clasping-leaved pondweed
pondweed, closed-leaved → closed-leaved pondweed
pondweed, curled → curled pondweed
pondweed, eel-grass → eel-grass pondweed
pondweed, fennel-leaved → fennel-leaved pondweed

pondweed, flat-stalked → flat-stalked pondweed
pondweed, floating-leaved → floating-leaved pondweed
pondweed, Fries' → Fries' pondweed
pondweed, grass-leaved → grass-leaved pondweed
pondweed, horned → horned pondweed
pondweed, Illinois → Illinois pondweed
pondweed, large-leaved → large-leaved pondweed
pondweed, large-sheathed → large-sheathed pondweed
pondweed, long-leaved → long-leaved pondweed
pondweed, long-stalked → long-stalked pondweed
pondweed, northern → northern pondweed
pondweed, Oakes' → Oakes' pondweed
pondweed, perfoliate → perfoliate pondweed
pondweed, ribbon-leaved → ribbon-leaved pondweed
pondweed, Richardson's → Richardson's pondweed
pondweed, Robbins' → Robbins' pondweed
pondweed, sheathed → sheathed pondweed
pondweed, sheathing → sheathing pondweed
pondweed, slender-leaved → slender-leaved pondweed
pondweed, small → small pondweed
pondweed, small-leaved → small-leaved pondweed
pondweed, stiff-leaved → stiff-leaved pondweed
pondweed, three-leaved → three-leaved pondweed
pondweed, various-leaved → various-leaved pondweed
poor fen peat moss = Sphagnum angustifolium (B:M)
poor-fen sphagnum = Sphagnum angustifolium (B:M)
poor man's weatherglass = Anagallis arvensis (V:D)
poor sedge = Carex magellanica ssp. irrigua (V:M)
poor sedge = Carex magellanica (V:M)
popcornflower, fragrant → fragrant popcornflower
popcornflower, Scouler's → Scouler's popcornflower
popcornflower, slender → slender popcornflower
poplar, balsam → balsam poplar
poplar, black → black poplar
poplar, white → white poplar
poppy, arctic → arctic poppy
poppy, California → California poppy
poppy, corn → corn poppy
poppy, dwarf → dwarf poppy
poppy, Iceland → Iceland poppy
poppy, Lapland → Lapland poppy
poppy, Macoun's → Macoun's poppy
poppy, opium → opium poppy
poppy, pale → pale poppy
poque = Boschniakia hookeri (V:D)
porcupine sedge = Carex hystericina (V:M)
porcupinegrass = Stipa spartea (V:M)
porcupinegrass, short-awned → short-awned porcupinegrass
Porsild's draba = Draba porsildii (V:D)
Porsild's whitlow-grass = Draba porsildii (V:D)
poverty grass = Sporobolus vaginiflorus (V:M)
poverty oatgrass = Danthonia spicata (V:M)
poverty weed = Monolepis nuttalliana (V:D)
powder paw = Nephroma parile (L)
powdered bone = Hypogymnia austerodes (L)
powdered bone = Hypogymnia bitteri (L)
powdered brown = Melanelia sorediata (L)
powdered brown = Melanelia disjuncta (L)
powdered brown = Melanelia albertana (L)
powdered brown = Melanelia tominii (L)
powdered brown = Neofuscelia verruculifera (L)
powdered brown = Melanelia subargentifera (L)
powdered brown = Melanelia olivaceoides (L)
powdered centipede = Heterodermia speciosa (L)
powdered funnel cladonia = Cladonia cenotea (L)
powdered moon = Sticta limbata (L)
powdered orange = Xanthoria fallax (L)
powdered orange lichen = Xanthoria fallax (L)
powdered paw = Nephroma parile (L)
powdered rockfrog = Arctoparmelia incurva (L)
powdered rockfrog = Xanthoparmelia mougeotii (L)
powdered rosette = Physcia caesia (L)
powdered rosette = Physcia dubia (L)
powdered saguaro = Cavernularia hultenii (L)
powdered scatter-rug = Parmotrema arnoldii (L)
powdered scatter-rug = Parmotrema chinense (L)
powdered shadow = Phaeophyscia hirsuta (L)
powdered shield = Parmelia sulcata (L)
powdered sunshine = Vulpicida pinastri (L)
powdery beard = Usnea lapponica (L)
powdery old man's beard = Usnea lapponica (L)
powdery Swiss lichen = Nephroma parile (L)
Powell's amaranth = Amaranthus powellii (V:D)
powered beard = Usnea lapponica (L)
prairie buttercup = Ranunculus rhomboideus (V:D)
prairie cinquefoil = Potentilla pensylvanica (V:D)
prairie coneflower = Ratibida columnifera (V:D)
prairie crocus = Pulsatilla patens ssp. multifida (V:D)
prairie fleabane = Erigeron strigosus (V:D)
prairie gentian = Gentiana affinis (V:D)
prairie golden bean = Thermopsis rhombifolia (V:D)
prairie lupine = Lupinus lepidus (V:D)
prairie pepper-grass = Lepidium densiflorum (V:D)
prairie rose = Rosa woodsii (V:D)
prairie sagewort = Artemisia frigida (V:D)
prairie sand-reed = Calamovilfa longifolia (V:M)
prairie sandgrass = Calamovilfa longifolia (V:M)
prairie sedge = Carex prairea (V:M)
prairie three-awn = Aristida oligantha (V:M)
prairie violet = Viola praemorsa (V:D)
prairie wedgegrass = Sphenopholis obtusata (V:M)
Presl's sedge = Carex preslii (V:M)
pretty milk-vetch = Astragalus eucosmus (V:D)
pretty shootingstar = Dodecatheon pulchellum (V:D)
prickle cladonia = Cladonia uncialis (L)
prickley comfrey = Symphytum asperum (V:D)
prickly cucumber = Echinocystis lobata (V:D)
prickly lettuce = Lactuca serriola (V:D)
prickly phlox = Leptodactylon pungens (V:D)
prickly rose = Rosa acicularis (V:D)
prickly rose = Rosa acicularis ssp. sayi (V:D)
prickly saxifraga = Saxifraga bronchialis (V:D)
prickly saxifraga = Saxifraga bronchialis ssp. austromontana (V:D)
prickly sow-thistle = Sonchus asper (V:D)
primrose, Greenland → Greenland primrose
primrose, mealy → mealy primrose
primrose, Mistassini → Mistassini primrose
primrose, upright → upright primrose
prince's pine = Chimaphila umbellata (V:D)
prince's pine = Chimaphila umbellata ssp. occidentalis (V:D)
privet, common → common privet
prostrate knotweed = Polygonum aviculare (V:D)
prostrate pigweed = Amaranthus blitoides (V:D)
prostrate sedge = Carex chordorrhiza (V:M)
protean tarpaper = Collema undulatum (L)
protracted tarpaper = Collema multipartitum (L)
psora, sockeye → sockeye psora
puckered bone = Hypogymnia rugosa (L)
puffed shield = Hypogymnia physodes (L)
Puget Sound gumweed = Grindelia integrifolia (V:D)
puke, fairy → fairy puke
pulp, crinkled → crinkled pulp
pulse milk-vetch = Astragalus tenellus (V:D)
pumpelly brome = Bromus inermis ssp. pumpellianus (V:M)
puncture vine = Tribulus terrestris (V:D)
punctured gristle = Ramalina dilacerata (L)
punctured rocktripe = Umbilicaria torrefacta (L)
purple arnica = Arnica lessingii (V:D)
purple avens = Geum rivale (V:D)
purple blue-eyed-grass = Olsynium douglasii var. inflatum (V:M)
purple braya = Braya purpurascens (V:D)
purple daisy = Erigeron purpuratus (V:D)
purple dead-nettle = Lamium purpureum (V:D)

purple dragonhead = Physostegia parviflora (V:D)
purple-haired groundsel = Senecio atropurpureus (V:D)
purple honeysuckle = Lonicera hispidula (V:D)
purple horn-toothed moss = Ceratodon purpureus (B:M)
purple-leaved willowherb = Epilobium ciliatum (V:D)
purple loosestrife = Lythrum salicaria (V:D)
purple meadowrue = Thalictrum dasycarpum (V:D)
purple milk-vetch = Astragalus agrestis (V:D)
purple mountain saxifrage = Saxifraga oppositifolia (V:D)
purple oat grass = Schizachne purpurascens (V:M)
purple oniongrass = Melica spectabilis (V:M)
purple paintbrush = Castilleja raupii (V:D)
purple peavine = Lathyrus nevadensis (V:D)
purple rattlesnake-root = Prenanthes racemosa ssp. multiflora (V:D)
purple rattlesnake-root = Prenanthes racemosa (V:D)
purple reedgrass = Calamagrostis purpurascens (V:M)
purple sanicle = Sanicula bipinnatifida (V:D)
purple spike-rush = Eleocharis atropurpurea (V:M)
purple-stemmed monkey-flower = Mimulus floribundus (V:D)
purple sweet-cicely = Osmorhiza purpurea (V:D)
purple toadflax = Linaria purpurea (V:D)
purple-worm liverwort = Pleurozia purpurea (B:H)
purse, shepherd's → shepherd's purse
purslane speedwell = Veronica peregrina (V:D)
purslane, common → common purslane
purslane, desert rock → desert rock purslane
pussy willow = Salix discolor (V:D)
pussytoes, broad-leaved → broad-leaved pussytoes
pussytoes, field → field pussytoes
pussytoes, low → low pussytoes
pussytoes, Nuttall's → Nuttall's pussytoes
pussytoes, one-headed → one-headed pussytoes
pussytoes, racemose → racemose pussytoes
pussytoes, rosy → rosy pussytoes
pussytoes, showy → showy pussytoes
pussytoes, small-leaved → small-leaved pussytoes
pussytoes, umber → umber pussytoes
pussytoes, woodrush → woodrush pussytoes
pussytoes, woolly → woolly pussytoes
pverty clover = Trifolium depauperatum (V:D)
pygmy buttercup = Ranunculus pygmaeus (V:D)
pygmy water-lily = Nymphaea tetragona (V:D)
Pyrenean sedge = Carex pyrenaica (V:M)
Pyrenean sedge = Carex pyrenaica ssp. micropoda (V:M)

Q

Queen Anne's lace = Daucus carota (V:D)
Queen Charlotte twinflower violet = Viola biflora ssp. carlottae (V:D)
queen's cup = Clintonia uniflora (V:M)
questionable rockfrog = Xanthoparmelia cumberlandia (L)
questionable rockfrog = Xanthoparmelia coloradoensis (L)
quillwort, Bolander's → Bolander's quillwort
quillwort, bristle-like → bristle-like quillwort
quillwort, coastal → coastal quillwort
quillwort, flowering → flowering quillwort
quillwort, Howell's → Howell's quillwort
quillwort, Nuttall's → Nuttall's quillwort
quillwort, western → western quillwort
quilted brownette = Placynthium nigrum (L)

R

rabbit-brush = Chrysothamnus nauseosus (V:D)
rabbit-brush, common → common rabbit-brush
rabbit-brush, green → green rabbit-brush
rabbitfoot polypogon = Polypogon monspeliensis (V:M)
racemose pussytoes = Antennaria racemosa (V:D)
radish, garden → garden radish
radish, wild → wild radish
rag, arctic → arctic rag
rag, crinkled → crinkled rag
rag, laundered → laundered rag
rag, ribbon → ribbon rag
rag, speckled → speckled rag
rag, tattered → tattered rag
rag, yellow → yellow rag
ragbag = Platismatia glauca (L)
ragged paperdoll = Flavocetraria nivalis (L)
ragged rocktripe = Umbilicaria havaasii (L)
ragged snow = Flavocetraria nivalis (L)
rags = Pseudocyphellaria crocata (L)
rags, heather → heather rags
ragweed, annual → annual ragweed
ragweed, common → common ragweed
ragwort, arrow-leaved → arrow-leaved ragwort
ragwort, cut-leaved → cut-leaved ragwort
ragwort, desert → desert ragwort
ragwort, dryland → dryland ragwort
ragwort, rayless → rayless ragwort
ragwort, sticky → sticky ragwort
ragwort, tansy → tansy ragwort
rape, winter → winter rape
raspberry, American red → American red raspberry
raspberry, black → black raspberry
raspberry, creeping → creeping raspberry
raspberry, dwarf → dwarf raspberry
raspberry, red → red raspberry
raspberry, running → running raspberry
raspberry, stemless → stemless raspberry
raspberry, trailing → trailing raspberry
rattail fescue = Vulpia myuros (V:M)
rattle brome = Bromus briziformis (V:M)
rattle chess = Bromus briziformis (V:M)
rattle, yellow → yellow rattle
rattlebox = Rhinanthus minor (V:D)
rattlepod, American → American rattlepod
rattlesnake fern = Botrychium virginianum (V:F)
rattlesnake-grass = Glyceria canadensis (V:M)
rattlesnake-plantain = Goodyera oblongifolia (V:M)
rattlesnake-plantain, lesser → lesser rattlesnake-plantain
rattlesnake-root, purple → purple rattlesnake-root
rattlesnake-root, western → western rattlesnake-root
Raup's Indian-paintbrush = Castilleja raupii (V:D)
Raup's paintbrush = Castilleja raupii (V:D)

Raup's willow = Salix raupii (V:D)
rayless alpine butterweed = Senecio pauciflorus (V:D)
rayless mountain butterweed = Senecio indecorus (V:D)
rayless ragwort = Senecio indecorus (V:D)
Raynolds' sedge = Carex raynoldsii (V:M)
red alder = Alnus rubra (V:D)
red and white baneberry = Actaea rubra (V:D)
red bearberry = Arctostaphylos rubra (V:D)
red birch = Betula occidentalis (V:D)
red bryum = Bryum miniatum (B:M)
red campion = Silene dioica (V:D)
red clover = Trifolium pratense (V:D)
red columbine = Aquilegia formosa (V:D)
red elder = Sambucus racemosa ssp. pubens (V:D)
red elder = Sambucus racemosa (V:D)
red elderberry = Sambucus racemosa (V:D)
red elderberry = Sambucus racemosa ssp. pubens (V:D)
red fescue = Festuca rubra (V:M)
red-flowering currant = Ribes sanguineum (V:D)
red goosefoot = Chenopodium rubrum (V:D)
red hawthorn = Crataegus columbiana (V:D)
red honeysuckle = Lonicera dioica (V:D)
red honeysuckle = Lonicera dioica var. glaucescens (V:D)
red hook moss = Limprichtia revolvens (B:M)
red huckleberry = Vaccinium parvifolium (V:D)
red maids = Calandrinia ciliata (V:D)
red-mounthed mnium = Mnium spinulosum (B:M)
red-mouthed leafy moss = Mnium spinulosum (B:M)
red orache = Atriplex rosea (V:D)
red-osier dogwood = Cornus sericea (V:D)
red pigweed = Amaranthus retroflexus (V:D)
red pixie-cup = Cladonia borealis (L)
red raspberry = Rubus idaeus (V:D)
red roof moss = Ceratodon purpureus (B:M)
red sand-spurry = Spergularia rubra (V:D)
red-seeded dandelion = Taraxacum laevigatum (V:D)
red-sepaled evening-primrose = Oenothera glazioviana (V:D)
red stem, big → big red stem
red stem, Schreber's → Schreber's red stem
red-stemmed feathermoss = Pleurozium schreberi (B:M)
red-stemmed saxifrage = Saxifraga lyallii (V:D)
red swamp currant = Ribes triste (V:D)
red-toothed rock moss = Schistidium apocarpum (B:M)
red twinberry = Lonicera utahensis (V:D)
redcedar, western → western redcedar
reddish sandwort = Minuartia rubella (V:D)
redstem ceanothus = Ceanothus sanguineus (V:D)
redtop = Agrostis gigantea (V:M)
redtop = Agrostis stolonifera (V:M)
redwood sorrel = Oxalis oregana (V:D)
reed canarygrass = Phalaris arundinacea (V:M)
reed mannagrass = Glyceria grandis (V:M)
reedgrass, bluejoint → bluejoint reedgrass
reedgrass, Canada → Canada reedgrass
reedgrass, common → common reedgrass
reedgrass, giant → giant reedgrass
reedgrass, Lapland → Lapland reedgrass
reedgrass, marsh → marsh reedgrass
reedgrass, one-and-a-half-flowered → one-and-a-half-flowered reedgrass
reedgrass, Pacific → Pacific reedgrass
reedgrass, plains → plains reedgrass
reedgrass, purple → purple reedgrass
reedgrass, Scribner's → Scribner's reedgrass
reedgrass, slimstem → slimstem reedgrass
reedgrass, wood → wood reedgrass
reedmace = Typha latifolia (V:M)
reflexed locoweed = Oxytropis deflexa (V:D)
reflexed rockcress = Arabis holboellii (V:D)
Regel's rush = Juncus regelii (V:M)
reindeer lichen = Cladina rangiferina (L)
reindeer, coastal → coastal reindeer
retrorse sedge = Carex retrorsa (V:M)
rhexia-leaved paintbrush = Castilleja rhexifolia (V:D)
rhododendron, California → California rhododendron
rhododendron, Pacific → Pacific rhododendron
rhododendron, white → white rhododendron
rhododendron, white-flowered → white-flowered rhododendron
ribbed bog moss = Aulacomnium palustre (B:M)
ribbed cladonia = Cladonia cariosa (L)
ribbon-leaved pondweed = Potamogeton epihydrus (V:M)
ribbon rag = Platismatia stenophylla (L)
ribwort plantain = Plantago lanceolata (V:D)
rice cutgrass = Leersia oryzoides (V:M)
rice-root, northern → northern rice-root
ricegrass, little → little ricegrass
ricegrass, little-seeded → little-seeded ricegrass
ricegrass, northern → northern ricegrass
ricegrass, rough-leaved → rough-leaved ricegrass
ricegrass, short-awned → short-awned ricegrass
ricegrass, small-flowered → small-flowered ricegrass
ricegrass, white-grained mountain → white-grained mountain ricegrass
riceroot fritillary = Fritillaria camschatcensis (V:M)
Richardson's alumroot = Heuchera richardsonii (V:D)
Richardson's crane's-bill = Geranium richardsonii (V:D)
Richardson's geranium = Geranium richardsonii (V:D)
Richardson's meadowrue = Thalictrum sparsiflorum var. richardsonii (V:D)
Richardson's penstemon = Penstemon richardsonii (V:D)
Richardson's pondweed = Potamogeton richardsonii (V:M)
Richardson's sedge = Carex richardsonii (V:M)
Richardson's water moss = Calliergon richardsonii (B:M)
Richardson's willow = Salix lanata ssp. richardsonii (V:D)
rigid brook moss = Hygrohypnum duriusculum (B:M)
rigid fiddleneck = Amsinckia retrorsa (V:D)
rim lichen = Lecanora circumborealis (L)
ring lichen = Arctoparmelia centrifuga (L)
ring pellia = Pellia neesiana (B:H)
rippled rockfrog = Arctoparmelia separata (L)
rippled rockfrog = Arctoparmelia centrifuga (L)
river beauty = Epilobium latifolium (V:D)
river birch = Betula occidentalis (V:D)
river bulrush = Scirpus fluviatilis (V:M)
rivergrass = Scolochloa festucacea (V:M)
roadside rock moss = Racomitrium canescens (B:M)
Robbins' milk-vetch = Astragalus robbinsii var. minor (V:D)
Robbins' milk-vetch = Astragalus robbinsii (V:D)
Robbins' pondweed = Potamogeton robbinsii (V:M)
Robert crane's-bill = Geranium robertianum (V:D)
Robert geranium = Geranium robertianum (V:D)
Roche spruce = Picea × lutzii (V:G)
rock-brake, slender → slender rock-brake
rock brow = Melanelia hepatizon (L)
rock brown = Melanelia commixta (L)
rock cotoneaster = Cotoneaster horizontalis (V:D)
rock cranberry = Vaccinium vitis-idaea (V:D)
rock-cress, alpine → alpine rock-cress
rock-dwelling sedge = Carex petricosa (V:M)
rock mouse = Pannaria cheiroloba (L)
rock-olive = Peltula euploca (L)
rock orange = Xanthoria elegans (L)
rock sandwort = Minuartia dawsonensis (V:D)
rock sedge = Carex rupestris (V:M)
rock stippleback = Endocarpon pulvinatum (L)
rock willow = Salix vestita (V:D)
rock worm lichen = Thamnolia vermicularis (L)
rockbright, black-eyed → black-eyed rockbright
rockbright, brown-eyed → brown-eyed rockbright
rockbright, pink-eyed → pink-eyed rockbright
rockcress, Drummond's → Drummond's rockcress
rockcress, hairy → hairy rockcress
rockcress, Holboell's → Holboell's rockcress
rockcress, Kamchatka lyre-leaved → Kamchatka lyre-leaved rockcress

rockcress, Lemmon's → Lemmon's rockcress
rockcress, Lyall's → Lyall's rockcress
rockcress, lyre-leaved → lyre-leaved rockcress
rockcress, Nuttall's → Nuttall's rockcress
rockcress, reflexed → reflexed rockcress
rockcress, sickle-pod → sickle-pod rockcress
rockcress, small-leaved → small-leaved rockcress
rocket, small-flowered → small-flowered rocket
rockfrog, Colorado → Colorado rockfrog
rockfrog, dissolving → dissolving rockfrog
rockfrog, mini → mini rockfrog
rockfrog, powdered → powdered rockfrog
rockfrog, questionable → questionable rockfrog
rockfrog, rippled → rippled rockfrog
rockfrog, salted → salted rockfrog
rockfrog, vagabond → vagabond rockfrog
rockfrog, variable → variable rockfrog
rockgrub = Allantoparmelia almquistii (L)
rockgrub = Allantoparmelia alpicola (L)
rockgrub = Brodoa oroarctica (L)
rocktripe, ashen → ashen rocktripe
rocktripe, asterisk → asterisk rocktripe
rocktripe, ballpoint → ballpoint rocktrip
rocktripe, black → black rocktripe
rocktripe, blistered → blistered rocktripe
rocktripe, blushing → blushing rocktripe
rocktripe, doubtful → doubtful rocktripe
rocktripe, emery → emery rocktripe
rocktripe, fleecy → fleecy rocktripe
rocktripe, fringed → fringed rocktripe
rocktripe, frosted → frosted rocktripe
rocktripe, Muhlenberg's → Muhlenberg's rocktripe
rocktripe, netted → netted rocktripe
rocktripe, northern → northern rocktripe
rocktripe, Pennsylvania → Pennsylvania rocktripe
rocktripe, peppered → peppered rocktripe
rocktripe, petalled → petalled rocktripe
rocktripe, plated → plated rocktripe
rocktripe, punctured → punctured rocktripe
rocktripe, ragged → ragged rocktripe
rocktripe, roughened → roughened rocktripe
rocktripe, windward → windward rocktripe
Rocky Mountain bee-plant = Cleome serrulata (V:D)
Rocky Mountain butterweed = Senecio streptanthifolius (V:D)
Rocky Mountain Douglas-fir = Pseudotsuga menziesii var. glauca (V:G)
Rocky Mountain douglasia = Douglasia montana (V:D)
Rocky Mountain draba = Draba crassifolia (V:D)
Rocky Mountain fescue = Festuca saximontana (V:M)
Rocky Mountain helianthella = Helianthella uniflora (V:D)
Rocky Mountain juniper = Juniperus scopulorum (V:G)
Rocky Mountain locoweed = Oxytropis sericea (V:D)
Rocky Mountain maple = Acer glabrum (V:D)
Rocky Mountain sandwort = Minuartia austromontana (V:D)
Rocky Mountain sedge = Carex saximontana (V:M)
Rocky Mountain whitlow-grass = Draba crassifolia (V:D)
Rocky Mountain willowherb = Epilobium saximontanum (V:D)
Rocky Mountain woodsia = Woodsia scopulina (V:F)
rolled-leaf pigtail moss = Hypnum revolutum (B:M)
romanzoffia, Sitka → Sitka romanzoffia
romanzoffia, Tracy's → Tracy's romanzoffia
rose-and-gold lichen = Pseudocyphellaria aurata (L)
rose campion = Lychnis coronaria (V:D)
rose, Arkansas → Arkansas rose
rose, baldhip → baldhip rose
rose, clustered wild → clustered wild rose
rose, dog → dog rose
rose, dwarf → dwarf rose
rose, Nootka → Nootka rose
rose, peafruit → peafruit rose
rose, prairie → prairie rose
rose, prickly → prickly rose
rose, Woods' → Woods' rose
rosebay willowherb = Epilobium angustifolium (V:D)
rosebay, Lapland → Lapland rosebay
roseroot = Sedum integrifolium (V:D)
rosette, beaded → beaded rosette
rosette, black-eyed → black-eyed rosette
rosette, fringed → fringed rosette
rosette, frosted → frosted rosette
rosette, grey-eyed → grey-eyed rosette
rosette, hoary → hoary rosette
rosette, hooded → hooded rosette
rosette, powdered → powdered rosette
Ross' avens = Geum rossii (V:D)
Ross' sedge = Carex rossii (V:M)
rosy owl-clover = Orthocarpus bracteosus (V:D)
rosy plectritis = Plectritis congesta (V:D)
rosy pussytoes = Antennaria microphylla (V:D)
rosy twistedstalk = Streptopus roseus (V:M)
rough bedstraw = Galium mexicanum ssp. asperulum (V:D)
rough bluegrass = Poa trivialis (V:M)
rough cinquefoil = Potentilla norvegica (V:D)
rough cocklebur = Xanthium strumarium (V:D)
rough comfrey = Symphytum asperum (V:D)
rough dog lichen = Peltigera praetextata (L)
rough fescue = Festuca campestris (V:M)
rough-fruited fairybells = Disporum trachycarpum (V:M)
rough hair grass = Agrostis scabra (V:M)
rough-leaved aster = Aster radulinus (V:D)
rough-leaved ricegrass = Oryzopsis asperifolia (V:M)
rough moss = Claopodium crispifolium (B:M)
rough pigweed = Amaranthus retroflexus (V:D)
rough-stemmed daisy = Erigeron strigosus (V:D)
rough-stemmed fleabane = Erigeron strigosus (V:D)
rough water horehound = Lycopus asper (V:D)
rough white prairie aster = Aster falcatus (V:D)
roughened mouse = Fuscopannaria ahlneri (L)
roughened rocktripe = Umbilicaria rigida (L)
roughstalk bluegrass = Poa trivialis (V:M)
round-leaved alumroot = Heuchera cylindrica (V:D)
round-leaved bog orchid = Platanthera orbiculata (V:M)
round-leaved orchis = Amerorchis rotundifolia (V:M)
round-leaved rein orchid = Platanthera orbiculata (V:M)
round-leaved sundew = Drosera rotundifolia (V:D)
round moss = Myurella julacea (B:M)
rounded-leaved violet = Viola orbiculata (V:D)
ruby, soil → soil ruby
rue-leaved saxifrage = Saxifraga tridactylites (V:D)
ruffle, eyed → eyed ruffle
ruffle, fringed → fringed ruffle
ruffle, moonshine → moonshine ruffle
ruffle, pallid → pallid ruffle
ruffle, shadow → shadow ruffle
ruffle, variable → variable ruffle
ruffle, weathered → weathered ruffle
running clubmoss = Lycopodium clavatum (V:F)
running raspberry = Rubus pubescens (V:D)
rush aster = Aster borealis (V:D)
rush-like skeleton-plant = Lygodesmia juncea (V:D)
rush, alpine → alpine rush
rush, arctic → arctic rush
rush, Baltic → Baltic rush
rush, big-head → big-head rush
rush, bog → bog rush
rush, Bolander's → Bolander's rush
rush, Brewer's → Brewer's rush
rush, bulbous → bulbous rush
rush, chestnut → chestnut rush
rush, Colorado → Colorado rush
rush, common → common rush
rush, Coville's → Coville's rush

rush, dagger-leaved → dagger-leaved rush
rush, Drummond's → Drummond's rush
rush, Gerard's → Gerard's rush
rush, jointed → jointed rush
rush, Kellogg's → Kellogg's rush
rush, knotted → knotted rush
rush, long-styled → long-styled rush
rush, Mertens' → Mertens' rush
rush, mud → mud rush
rush, Parry's → Parry's rush
rush, pointed → pointed rush
rush, Regel's → Regel's rush
rush, short-tailed → short-tailed rush
rush, sickle-leaved → sickle-leaved rush
rush, Sierra → Sierra rush
rush, slender → slender rush
rush, spreading → spreading rush
rush, tapered → tapered rush
rush, thread → thread rush
rush, three-flowered → three-flowered rush
rush, toad → toad rush
rush, Torrey's → Torrey's rush
rush, tuberous → tuberous rush
rush, two-flowered → two-flowered rush
rush, Vasey's → Vasey's rush
rush, whitish → whitish rush
rush, wire → wire rush
russet cotton-grass = Eriophorum chamissonis (V:M)
russet sedge = Carex saxatilis (V:M)
Russian blue lettuce = Lactuca tatarica (V:D)
Russian olive = Elaeagnus angustifolia (V:D)
Russian pigweed = Axyris amaranthoides (V:D)
Russian thistle = Salsola kali (V:D)
rusty claw moss = Hypnum revolutum (B:M)
rusty cliff fern = Woodsia ilvensis (V:F)
rusty-haired saxifrage = Saxifraga rufidula (V:D)
rusty mountain heath moss = Pseudoleskea baileyi (B:M)
rusty peat moss = Sphagnum fuscum (B:M)
rusty rock wort = Marsupella emarginata (B:H)
rusty steppe moss = Tortula ruralis (B:M)
rusty woodrush = Luzula rufescens (V:M)
rusty woodsia = Woodsia ilvensis (V:F)
Rydberg's arnica = Arnica rydbergii (V:D)
rye = Secale cereale (V:M)
rye brome = Bromus secalinus (V:M)
ryegrass sedge = Carex loliacea (V:M)
ryegrass, bearded → bearded ryegrass
ryegrass, Canada → Canada ryegrass
ryegrass, English → English ryegrass
ryegrass, perennial → perennial ryegrass

S

Sachaline knotweed = Polygonum sachalinense (V:D)
saffron-yellow solorina = Solorina crocea (L)
sage willow = Salix candida (V:D)
sage, wood → wood sage
sagebrush buttercup = Ranunculus glaberrimus (V:D)
sagebrush mariposa lily = Calochortus macrocarpus (V:M)
sagebrush, Alaska → Alaska sagebrush
sagebrush, big → big sagebrush
sagebrush, cutleaf → cutleaf sagebrush
sagebrush, silver → silver sagebrush
sagebrush, threetip → threetip sagebrush
sagewort, biennial → biennial sagewort
sagewort, mountain → mountain sagewort
sagewort, pasture → pasture sagewort
sagewort, prairie → prairie sagewort
saguaro, eyed → eyed saguaro
saguaro, powdered → powdered saguaro
Sainfoin = Onobrychis viciifolia (V:D)
Saintfoin = Onobrychis viciifolia (V:D)
salad burnet = Sanguisorba minor (V:D)
salal = Gaultheria shallon (V:D)
saline orache = Atriplex subspicata (V:D)
saline plantain = Plantago eriopoda (V:D)
saline shooting star = Dodecatheon pulchellum (V:D)
salish daisy = Erigeron salishii (V:D)
salmonberry = Rubus spectabilis (V:D)
salsify, common → common salsify
salsify, meadow → meadow salsify
salsify, yellow → yellow salsify
salt marsh dodder = Cuscuta salina (V:D)
salt marsh starwort = Stellaria humifusa (V:D)
salt meadowgrass = Spartina patens (V:M)
salted rockfrog = Xanthoparmelia mexicana (L)
salted rockfrog = Xanthoparmelia plittii (L)
salted scatter-rug = Parmotrema crinitum (L)
salted shield = Parmelia hygrophila (L)
salted shield = Parmelia pseudosulcata (L)
salted shield = Parmelia saxatilis (L)
salted shield = Parmelia squarrosa (L)
salted starburst = Imshaugia aleurites (L)
saltgrass, seashore → seashore saltgrass
saltmarsh bulrush = Scirpus maritimus (V:M)
saltwort, perennial → perennial saltwort
samphire, Pacific → Pacific samphire
sand dropseed = Sporobolus cryptandrus (V:M)
sand-dune sedge = Carex pansa (V:M)
sand-dwelling wallflower = Erysimum arenicola var. torulosum (V:D)
sand lacepod = Thysanocarpus curvipes (V:D)
sand plantain = Plantago psyllium (V:D)
sand-reed, prairie → prairie sand-reed
sand-spurry, beach → beach sand-spurry
sand-spurry, Canadian → Canadian sand-spurry
sand-spurry, red → red sand-spurry
sand-verbena, yellow → yellow sand-verbena
sandbar willow = Salix exigua (V:D)
Sandberg's bluegrass = Poa secunda (V:M)
Sandberg's desert-parsley = Lomatium sandbergii (V:D)
Sandberg's lomatium = Lomatium sandbergii (V:D)
sandberry = Arctostaphylos uva-ursi (V:D)
sandbur, field → field sandbur
sandgrass, prairie → prairie sandgrass
sandweed, common → common sandweed
sandwort, alpine → alpine sandwort
sandwort, big-leaved → big-leaved sandwort
sandwort, blunt-leaved → blunt-leaved sandwort
sandwort, boreal → boreal sandwort
sandwort, crisp → crisp sandwort
sandwort, dwarf → dwarf sandwort
sandwort, low → low sandwort
sandwort, mountain → mountain sandwort
sandwort, northern → northern sandwort
sandwort, Nuttall's → Nuttall's sandwort
sandwort, reddish → reddish sandwort
sandwort, rock → rock sandwort
sandwort, Rocky Mountain → Rocky Mountain sandwort
sandwort, seabeach → seabeach sandwort

sandwort, slender → slender sandwort
sandwort, slender mountain → slender mountain sandwort
sandwort, thread-leaved → thread-leaved sandwort
sandwort, thyme-leaved → thyme-leaved sandwort
sanicle, bear's-foot → bear's-foot sanicle
sanicle, black → black sanicle
sanicle, Pacific → Pacific sanicle
sanicle, purple → purple sanicle
sanicle, Sierra → Sierra sanicle
sarsaparilla, wild → wild sarsaparilla
Sartwell's sedge = Carex sartwellii (V:M)
Saskatoon = Amelanchier alnifolia (V:D)
satin grass = Muhlenbergia racemosa (V:M)
satinflower = Olsynium douglasii var. douglasii (V:M)
Saunder's wildrye = Elymus × saundersii (V:M)
sausage moss = Scorpidium scorpioides (B:M)
sawback sedge = Carex stipata (V:M)
sawwort, American → American sawwort
sawwort, northern → northern sawwort
saxifraga, green → green saxifraga
saxifraga, Lyall's → Lyall's saxifraga
saxifraga, Mertens' → Mertens' saxifraga
saxifraga, prickly → prickly saxifraga
saxifraga, yellow → yellow saxifraga
saxifrage, Alaska → Alaska saxifrage
saxifrage, alpine → alpine saxifrage
saxifrage, brook → brook saxifrage
saxifrage, buttercup-leaved → buttercup-leaved saxifrage
saxifrage, cordate-leaved → cordate-leaved saxifrage
saxifrage, evergreen → evergreen saxifrage
saxifrage, grassland → grassland saxifrage
saxifrage, hawkweed-leaved → hawkweed-leaved saxifrage
saxifrage, leatherleaf → leatherleaf saxifrage
saxifrage, nodding → nodding saxifrage
saxifrage, Oregon → Oregon saxifrage
saxifrage, peak → peak saxifrage
saxifrage, purple mountain → purple mountain saxifrage
saxifrage, red-stemmed → red-stemmed saxifrage
saxifrage, rue-leaved → rue-leaved saxifrage
saxifrage, rusty-haired → rusty-haired saxifrage
saxifrage, spotted → spotted saxifrage
saxifrage, stoloniferous → stoloniferous saxifrage
saxifrage, stream → stream saxifrage
saxifrage, Taylor's → Taylor's saxifrage
saxifrage, three-toothed → three-toothed saxifrage
saxifrage, thyme-leaved → thyme-leaved saxifrage
saxifrage, Tolmie's → Tolmie's saxifrage
saxifrage, tufted → tufted saxifrage
saxifrage, violet → violet saxifrage
saxifrage, wedge-leaved → wedge-leaved saxifrage
saxifrage, western → western saxifrage
saxifrage, whiplash → whiplash saxifrage
saxifrage, wood → wood saxifrage
saxifrage, yellow marsh → yellow marsh saxifrage
saxifrage, Yukon → Yukon saxifrage
scabious, field → field scabious
scabious, yellow → yellow scabious
scale, brown-eyed → brown-eyed scale
scale, butterfly → butterfly scale
scale, fissured → fissured scale
scale, knobbed → knobbed scale
scale, mountain → mountain scale
scale, sockeye → sockeye scale
scale, white-edged → white-edged scale
scalepod = Idahoa scapigera (V:D)
scales, cladonia → cladonia scales
Scamman's locoweed = Oxytropis scammaniana (V:D)
Scamman's oxytrope = Oxytropis scammaniana (V:D)
scarlet gaura = Gaura coccinea (V:D)
scarlet gilia = Ipomopsis aggregata (V:D)
scarlet globe-mallow = Sphaeralcea coccinea (V:D)
scarlet paintbrush = Castilleja miniata (V:D)
scarlet pimpernel = Anagallis arvensis (V:D)
scatter-rug, Chinese → Chinese scatter-rug
scatter-rug, powdered → powdered scatter-rug
scatter-rug, salted → salted scatter-rug
scentless mayweed = Matricaria perforata (V:D)
Scheuchzer's cotton-grass = Eriophorum scheuchzeri (V:M)
scheuchzeria = Scheuchzeria palustris (V:M)
Schreber's red stem = Pleurozium schreberi (B:M)
scirpt, pencil → pencil scirpt
scorpion feather moss = Scorpidium scorpioides (B:M)
scorpionweed, Franklin's → Franklin's scorpionweed
Scotch broom = Cytisus scoparius (V:D)
Scotch heather = Calluna vulgaris (V:D)
Scotch thistle = Onopordum acanthium (V:D)
Scotland, bluebells of → bluebells of Scotland
Scots broom = Cytisus scoparius (V:D)
Scottish wild lovage = Ligusticum scoticum (V:D)
Scouler's bluebell = Campanula scouleri (V:D)
Scouler's campion = Silene scouleri (V:D)
Scouler's catchfly = Silene scouleri (V:D)
Scouler's corydalis = Corydalis scouleri (V:D)
Scouler's harebell = Campanula scouleri (V:D)
Scouler's hawkweed = Hieracium scouleri (V:D)
Scouler's popcornflower = Plagiobothrys scouleri (V:D)
Scouler's St. John's-wort = Hypericum scouleri (V:D)
Scouler's surf-grass = Phyllospadix scouleri (V:M)
Scouler's valerian = Valeriana scouleri (V:D)
Scouler's willow = Salix scouleriana (V:D)
scouring-rush = Equisetum hyemale (V:F)
scouring-rush, common → common scouring-rush
scouring-rush, dwarf → dwarf scouring-rush
scouring-rush, fall → fall scouring-rush
scouring-rush, northern → northern scouring-rush
scouring-rush, smooth → smooth scouring-rush
scratchgrass = Muhlenbergia asperifolia (V:M)
Scribner's reedgrass = Calamagrostis scribneri (V:M)
scrub glandular birch = Betula pumila var. glandulifera (V:D)
scruffy beard = Usnea scabrata (L)
scruffy old man's beard = Usnea scabrata (L)
scurf lichen = Psoroma hypnorum (L)
sea bluebells = Mertensia maritima (V:D)
sea blush = Plectritis congesta (V:D)
sea lungwort = Mertensia maritima (V:D)
sea mertensia = Mertensia maritima (V:D)
sea-milkwort = Glaux maritima (V:D)
sea-pink = Armeria maritima (V:D)
sea tar = Verrucaria maura (L)
sea-watch = Angelica lucida (V:D)
seabeach sandwort = Honkenya peploides (V:D)
seabeach sandwort = Honkenya peploides ssp. major (V:D)
seablite = Suaeda calceoliformis (V:D)
seabluff catchfly = Silene douglasii (V:D)
seacoast angelica = Angelica lucida (V:D)
searocket, American → American searocket
searocket, European → European searocket
seashore lupine = Lupinus littoralis (V:D)
seashore saltgrass = Distichlis spicata (V:M)
seaside arrow-grass = Triglochin maritimum (V:M)
seaside barley = Hordeum marinum (V:M)
seaside birds-foot trefoil = Lotus formosissimus (V:D)
seaside bone = Hypogymnia heterophylla (L)
seaside buttercup = Ranunculus cymbalaria (V:D)
seaside centipede = Heterodermia sitchensis (L)
seaside fiddleneck = Amsinckia spectabilis (V:D)
seaside kidney = Nephroma laevigatum (L)
seaside lotus = Lotus formosissimus (V:D)
seaside mouse = Fuscopannaria maritima (L)
seaside paw = Nephroma laevigatum (L)
seaside speckleback = Punctelia stictica (L)
seaside tarpaper = Collema fecundum (L)

seaside thornbush = Kaernefeltia californica (L)
sedge mousetail = Myosurus aristatus (V:D)
sedge, analogue → analogue sedge
sedge, awl-fruited → awl-fruited sedge
sedge, awned → awned sedge
sedge, beaked → beaked sedge
sedge, bearded → bearded sedge
sedge, beautiful → beautiful sedge
sedge, Bebb's → Bebb's sedge
sedge, bent → bent sedge
sedge, Bigelow's → Bigelow's sedge
sedge, bigleaf → bigleaf sedge
sedge, black → black sedge
sedge, black alpine → black alpine sedge
sedge, black-scaled → black-scaled sedge
sedge, blunt → blunt sedge
sedge, bog → bog sedge
sedge, Bolander's → Bolander's sedge
sedge, bottle → bottle sedge
sedge, breaked → breaked sedge
sedge, bristle → bristle sedge
sedge, bristle-leaf → bristle-leaf sedge
sedge, bristle-stalked → bristle-stalked sedge
sedge, broad-winged → broad-winged sedge
sedge, bronze → bronze sedge
sedge, brownish → brownish sedge
sedge, Buxbaum's → Buxbaum's sedge
sedge, capitate → capitate sedge
sedge, club → club sedge
sedge, coastal stellate → coastal stellate sedge
sedge, coiled → coiled sedge
sedge, Columbia → Columbia sedge
sedge, cordroot → cordroot sedge
sedge, Crawe's → Crawe's sedge
sedge, Crawford's → Crawford's sedge
sedge, curly → curly sedge
sedge, curved-spiked → curved-spiked sedge
sedge, Cusick's → Cusick's sedge
sedge, Dewey's → Dewey's sedge
sedge, Douglas' → Douglas' sedge
sedge, dryland → dryland sedge
sedge, dunhead → dunhead sedge
sedge, elk → elk sedge
sedge, Engelmann's → Engelmann's sedge
sedge, Falkland Island → Falkland Island sedge
sedge, few-flowered → few-flowered sedge
sedge, few-seeded bog → few-seeded bog sedge
sedge, field → field sedge
sedge, fox → fox sedge
sedge, fragile → fragile sedge
sedge, Franklin's → Franklin's sedge
sedge, Garber's → Garber's sedge
sedge, Geyer's → Geyer's sedge
sedge, glacier → glacier sedge
sedge, Gmelin's → Gmelin's sedge
sedge, golden → golden sedge
sedge, graceful → graceful sedge
sedge, graceful mountain → graceful mountain sedge
sedge, green → green sedge
sedge, green-fruited → green-fruited sedge
sedge, green-sheathed → green-sheathed sedge
sedge, grey → grey sedge
sedge, hair → hair sedge
sedge, hair-like → hair-like sedge
sedge, hairy-fruited → hairy-fruited sedge
sedge, harefoot → harefoot sedge
sedge, hay → hay sedge
sedge, Hayden's → Hayden's sedge
sedge, Henderson's → Henderson's sedge
sedge, Holm's Rocky Mountain → Holm's Rocky Mountain sedge
sedge, Hood's → Hood's sedge
sedge, Hudson Bay → Hudson Bay sedge
sedge, inflated → inflated sedge
sedge, inland → inland sedge
sedge, Krause's → Krause's sedge
sedge, Lapland → Lapland sedge
sedge, large-awned → large-awned sedge
sedge, large-headed → large-headed sedge
sedge, large yellow → large yellow sedge
sedge, lens-fruited → lens-fruited sedge
sedge, lesser panicled → lesser panicled sedge
sedge, lesser saltmarsh → lesser saltmarsh sedge
sedge, lesser tussock → lesser tussock sedge
sedge, Lingbye's → Lingbye's sedge
sedge, livid → livid sedge
sedge, long-beaked → long-beaked sedge
sedge, long-bracted → long-bracted sedge
sedge, long-stoloned → long-stoloned sedge
sedge, long-styled → long-styled sedge
sedge, low northern → low northern sedge
sedge, many-flowered → many-flowered sedge
sedge, many-headed → many-headed sedge
sedge, meadow → meadow sedge
sedge, Mertens' → Mertens' sedge
sedge, mud → mud sedge
sedge, narrow → narrow sedge
sedge, northern bog → northern bog sedge
sedge, northern clustered → northern clustered sedge
sedge, northwestern → northwestern sedge
sedge, one-sided → one-sided sedge
sedge, pale → pale sedge
sedge, Parry's → Parry's sedge
sedge, pasture → pasture sedge
sedge, Payson's → Payson's sedge
sedge, Peck's → Peck's sedge
sedge, peduncled → peduncled sedge
sedge, pointed broom → pointed broom sedge
sedge, poor → poor sedge
sedge, porcupine → porcupine sedge
sedge, prairie → prairie sedge
sedge, Presl's → Presl's sedge
sedge, prostrate → prostrate sedge
sedge, Pyrenean → Pyrenean sedge
sedge, Raynolds' → Raynolds' sedge
sedge, retrorse → retrorse sedge
sedge, Richardson's → Richardson's sedge
sedge, rock → rock sedge
sedge, rock-dwelling → rock-dwelling sedge
sedge, Rocky Mountain → Rocky Mountain sedge
sedge, Ross' → Ross' sedge
sedge, russet → russet sedge
sedge, ryegrass → ryegrass sedge
sedge, sand-dune → sand-dune sedge
sedge, Sartwell's → Sartwell's sedge
sedge, sawback → sawback sedge
sedge, sheathed → sheathed sedge
sedge, sheep → sheep sedge
sedge, shore → shore sedge
sedge, short → short sedge
sedge, short-beaked → short-beaked sedge
sedge, short-leaved → short-leaved sedge
sedge, short-stemmed → short-stemmed sedge
sedge, showy → showy sedge
sedge, silvertop → silvertop sedge
sedge, silvery-flowered → silvery-flowered sedge
sedge, single-spiked → single-spiked sedge
sedge, slender → slender sedge
sedge, slender-beaked → slender-beaked sedge
sedge, slough → slough sedge
sedge, small-awned → small-awned sedge

sedge, small-winged → small-winged sedge
sedge, smooth-stemmed → smooth-stemmed sedge
sedge, soft-leaved → soft-leaved sedge
sedge, sparse-leaved → sparse-leaved sedge
sedge, spikenard → spikenard sedge
sedge, Sprengel's → Sprengel's sedge
sedge, star → star sedge
sedge, stiff → stiff sedge
sedge, string → string sedge
sedge, sweet → sweet sedge
sedge, tall bog → tall bog sedge
sedge, tender → tender sedge
sedge, thick-headed → thick-headed sedge
sedge, thin-flowered → thin-flowered sedge
sedge, thread-leaved → thread-leaved sedge
sedge, three-seeded → three-seeded sedge
sedge, three-way → three-way sedge
sedge, turned → turned sedge
sedge, two-coloured → two-coloured sedge
sedge, two-parted → two-parted sedge
sedge, two-seeded → two-seeded sedge
sedge, two-stamened → two-stamened sedge
sedge, two-toned → two-toned sedge
sedge, variegated → variegated sedge
sedge, water → water sedge
sedge, weak arctic → weak arctic sedge
sedge, white → white sedge
sedge, woolly → woolly sedge
sedge, yellow → yellow sedge
sedge, yellow bog → yellow bog sedge
sedge, yellow-flowered → yellow-flowered sedge
sedge, yellow-fruited → yellow-fruited sedge
seepage apple moss = Philonotis fontana (B:M)
selaginella, compact → compact selaginella
selaginella, low → low selaginella
selaginella, northern → northern selaginella
selaginella, Oregon → Oregon selaginella
selaginella, Wallace's → Wallace's selaginella
self-heal = Prunella vulgaris (V:D)
Selkirk's violet = Viola selkirkii (V:D)
seneca-root = Polygala senega (V:D)
serviceberry = Amelanchier alnifolia (V:D)
serviceberry willow = Salix pseudomonticola (V:D)
sessile-leaved sandbar willow = Salix sessilifolia (V:D)
Setchell's willow = Salix setchelliana (V:D)
shadow ruffle = Tuckermannopsis chlorophylla (L)
shadow, five o'clock → five o'clock shadow
shadow, granulated → granulated shadow
shadow, one-horse → one-horse shadow
shadow, pincushion → pincushion shadow
shadow, powdered → powdered shadow
shadow, starburst → starburst shadow
shadow, whiskered → whiskered shadow
shag stippleback = Dermatocarpon moulinsii (L)
shaggy daisy = Erigeron pumilus ssp. intermedius (V:D)
shaggy daisy = Erigeron pumilus (V:D)
shaggy fleabane = Erigeron pumilus ssp. intermedius (V:D)
shaggy fleabane = Erigeron pumilus (V:D)
shaggy old man's beard = Usnea hirta (L)
shaggy sphagnum = Sphagnum squarrosum (B:M)
shaggy yellow sand moss = Racomitrium ericoides (B:M)
shaly tarpaper = Collema polycarpon (L)
sharp-leaved fluellen = Kickxia elatine (V:D)
sharptooth angelica = Angelica arguta (V:D)
sheathed cotton-grass = Eriophorum vaginatum (V:M)
sheathed cotton-grass = Eriophorum vaginatum var. spissum (V:M)
sheathed pondweed = Potamogeton vaginatus (V:M)
sheathed sedge = Carex vaginata (V:M)
sheathing pondweed = Potamogeton vaginatus (V:M)
sheep cinquefoil = Potentilla ovina (V:D)
sheep sedge = Carex illota (V:M)
sheep sorrel = Rumex acetosella (V:D)
shepherd's cress = Teesdalia nudicaulis (V:D)
shepherd's-needle = Scandix pecten-veneris (V:D)
shepherd's purse = Capsella bursa-pastoris (V:D)
shield fern = Dryopteris expansa (V:F)
shield, broad → broad shield
shield, chocolate → chocolate shield
shield, granulated → granulated shield
shield, green → green shield
shield, grizzly → grizzly shield
shield, pale → pale shield
shield, powdered → powdered shield
shield, puffed → puffed shield
shield, salted → salted shield
shield, slender → slender shield
shield, unsalted → unsalted shield
shingle lichen = Hypocenomyce scalaris (L)
shingle, common → common shingle
shingled cladonia = Cladonia scabriuscula (L)
shining penstemon = Penstemon nitidus (V:D)
shining willow = Salix lucida (V:D)
shinleaf = Pyrola elliptica (V:D)
shiny liverwort = Pellia neesiana (B:H)
shiny starwort = Stellaria nitens (V:D)
shootingstar, broad-leaved → broad-leaved shootingstar
shootingstar, dentate → dentate shootingstar
shootingstar, desert → desert shootingstar
shootingstar, few-flowered → few-flowered shootingstar
shootingstar, Henderson's → Henderson's shootingstar
shootingstar, Jeffrey's → Jeffrey's shootingstar
shootingstar, northern → northern shootingstar
shootingstar, pretty → pretty shootingstar
shootingstar, slimpod → slimpod shootingstar
shootingstar, tall mountain → tall mountain shootingstar
shootingstar, white → white shootingstar
shore blue-eyed-grass = Sisyrinchium littorale (V:M)
shore buttercup = Ranunculus cymbalaria (V:D)
shore-growing peat moss = Sphagnum riparium (B:M)
shore pine = Pinus contorta var. contorta (V:G)
shore pine = Pinus contorta (V:G)
shore sedge = Carex limosa (V:M)
short-anthered cotton-grass = Eriophorum brachyantherum (V:M)
short-awned foxtail = Alopecurus aequalis (V:M)
short-awned porcupinegrass = Stipa curtiseta (V:M)
short-awned ricegrass = Oryzopsis pungens (V:M)
short-beaked agoseris = Agoseris glauca var. dasycephala (V:D)
short-beaked agoseris = Agoseris glauca (V:D)
short-beaked sedge = Carex brevior (V:M)
short-flowered evening-primrose = Camissonia breviflora (V:D)
short-flowered monkey-flower = Mimulus breviflorus (V:D)
short-fringed knotweed = Centaurea nigrescens (V:D)
short-fruited smelowskia = Smelowskia ovalis (V:D)
short-fruited willow = Salix brachycarpa ssp. brachycarpa (V:D)
short-fruited willow = Salix brachycarpa (V:D)
short-leaved ragged moss = Brachythecium oedipodium (B:M)
short-leaved sedge = Carex misandra (V:M)
short sedge = Carex canescens (V:M)
short-stemmed sedge = Carex brevicaulis (V:M)
short-styled thistle = Cirsium brevistylum (V:D)
short-tailed rush = Juncus brevicaudatus (V:M)
showy aster = Aster conspicuus (V:D)
showy bristle moss = Orthotrichum speciosum (B:M)
showy fleabane = Erigeron speciosus (V:D)
showy Jacob's-ladder = Polemonium pulcherrimum (V:D)
showy locoweed = Oxytropis splendens (V:D)
showy milkweed = Asclepias speciosa (V:D)
showy phlox = Phlox speciosa (V:D)
showy pussytoes = Antennaria pulcherrima (V:D)
showy sedge = Carex spectabilis (V:M)
shrub funnel lichen = Cladonia crispata (L)
shrubby orange = Xanthoria candelaria (L)
shrubby penstemon = Penstemon fruticosus (V:D)

shrubby vinyl = Leptogium teretiusculum (L)
shrubby willow = Salix arbusculoides (V:D)
shy gilia = Gilia sinuata (V:D)
sibbaldia = Sibbaldia procumbens (V:D)
Siberian elm = Ulmus pumila (V:D)
Siberian iris = Iris sibirica (V:M)
Siberian kobresia = Kobresia sibirica (V:M)
Siberian miner's-lettuce = Claytonia sibirica (V:D)
Siberian plantain = Plantago canescens (V:D)
Siberian water-milfoil = Myriophyllum sibiricum (V:D)
Siberian wheatgrass = Agropyron fragile (V:M)
Siberian yarrow = Achillea sibirica (V:D)
sickle-leaved rush = Juncus falcatus (V:M)
sickle moss = Sanionia uncinata (B:M)
sickle-pod rockcress = Arabis sparsiflora (V:D)
sickletop lousewort = Pedicularis racemosa (V:D)
sidewalk moss = Tortula ruralis (B:M)
Sierra cryptantha = Cryptantha nubigena (V:D)
Sierra rush = Juncus nevadensis (V:M)
Sierra sanicle = Sanicula graveolens (V:D)
sieve cladonia = Cladonia multiformis (L)
sieve cup lichen = Cladonia multiformis (L)
silene, Douglas' → Douglas' silene
silene, Menzies' → Menzies' silene
silky locoweed = Oxytropis sericea var. spicata (V:D)
silky locoweed = Oxytropis sericea (V:D)
silky lupine = Lupinus sericeus (V:D)
silky phacelia = Phacelia sericea (V:D)
silky tufted moss = Blindia acuta (B:M)
silver birch = Betula pubescens (V:D)
silver burweed = Ambrosia chamissonis (V:D)
silver hairgrass = Aira caryophyllea (V:M)
silver moss = Bryum argenteum (B:M)
silver sagebrush = Artemisia cana (V:D)
silverback luina = Luina hypoleuca (V:D)
silverberry = Elaeagnus commutata (V:D)
silvercrown = Cacaliopsis nardosmia (V:D)
silverleaf phacelia = Phacelia hastata (V:D)
silverscale = Atriplex argentea (V:D)
silvertop sedge = Carex foenea (V:M)
silvertop, American → American silvertop
silvery cinquefoil = Potentilla argentea (V:D)
silvery-flowered sedge = Carex aenea (V:M)
silvery lupine = Lupinus argenteus (V:D)
Simons' cotoneaster = Cotoneaster simonsii (V:D)
simple horsehair = Bryoria simplicior (L)
simple kobresia = Kobresia simpliciuscula (V:M)
single delight = Moneses uniflora (V:D)
single-spiked sedge = Carex scirpoidea (V:M)
Sitka alder = Alnus viridis ssp. sinuata (V:D)
Sitka burnet = Sanguisorba canadensis (V:D)
Sitka columbine = Aquilegia formosa (V:D)
Sitka mistmaiden = Romanzoffia sitchensis (V:D)
Sitka mountain-ash = Sorbus sitchensis (V:D)
Sitka parsley fern = Cryptogramma sitchensis (V:F)
Sitka romanzoffia = Romanzoffia sitchensis (V:D)
Sitka spruce = Picea sitchensis (V:G)
Sitka valerian = Valeriana sitchensis (V:D)
Sitka willow = Salix sitchensis (V:D)
six-weeks fescue = Vulpia octoflora (V:M)
six-weeks grass = Vulpia octoflora (V:M)
skeleton-plant, rush-like → rush-like skeleton-plant
skeletonweed, narrow-leaved → narrow-leaved skeletonweed
skullcap, blue → blue skullcap
skullcap, marsh → marsh skullcap
skullcap, narrow-leaved → narrow-leaved skullcap
skunk cabbage = Lysichiton americanum (V:M)
skunk currant = Ribes glandulosum (V:D)
skunk Jacob's-ladder = Polemonium viscosum (V:D)
skunk polemonium = Polemonium viscosum (V:D)
skunkweed = Navarretia squarrosa (V:D)
sky-pilot = Phacelia sericea (V:D)
skyrocket = Ipomopsis aggregata (V:D)
sleepy catchfly = Silene antirrhina (V:D)
slender arnica = Arnica gracilis (V:D)
slender beaked moss = Eurhynchium praelongum (B:M)
slender-beaked sedge = Carex athrostachya (V:M)
slender blue penstemon = Penstemon procerus (V:D)
slender bog orchid = Platanthera stricta (V:M)
slender cotton-grass = Eriophorum gracile (V:M)
slender cudweed = Gnaphalium microcephalum (V:D)
slender cup lichen = Cladonia gracilis ssp. turbinata (L)
slender draba = Draba albertina (V:D)
slender-flowered fairy-candelabra = Androsace filiformis (V:D)
slender gentian = Gentianella tenella (V:D)
slender haircap = Polytrichum strictum (B:M)
slender hairgrass = Deschampsia elongata (V:M)
slender hawksbeard = Crepis atrabarba (V:D)
slender hawkweed = Hieracium gracile (V:D)
slender-leaved pondweed = Potamogeton filiformis (V:M)
slender lip fern = Cheilanthes feei (V:F)
slender mountain sandwort = Arenaria capillaris (V:D)
slender muhly = Muhlenbergia filiformis (V:M)
slender penstemon = Penstemon gracilis (V:D)
slender plantain = Plantago elongata (V:D)
slender popcornflower = Plagiobothrys tenellus (V:D)
slender rein orchid = Platanthera stricta (V:M)
slender rock-brake = Cryptogramma stelleri (V:F)
slender rush = Juncus tenuis (V:M)
slender sandwort = Minuartia tenella (V:D)
slender sedge = Carex lasiocarpa (V:M)
slender shield = Platismatia stenophylla (L)
slender speedwell = Veronica filiformis (V:D)
slender spike-rush = Eleocharis tenuis (V:M)
slender-spiked mannagrass = Glyceria leptostachya (V:M)
slender-stemmed hair moss = Ditrichum flexicaule (B:M)
slender tarweed = Madia gracilis (V:D)
slender two-toothed wort = Cephalozia bicuspidata (B:H)
slender vetch = Vicia tetrasperma (V:D)
slender wheatgrass = Elymus trachycaulus (V:M)
slender whitlow-grass = Draba albertina (V:D)
slender woodland star = Lithophragma tenellum (V:D)
slender woolly-heads = Psilocarphus tenellus (V:D)
slim larkspur = Delphinium depauperatum (V:D)
slimleaf onion = Allium amplectens (V:M)
slimpod shootingstar = Dodecatheon conjugens (V:D)
slimstem reedgrass = Calamagrostis stricta (V:M)
sloe = Prunus spinosa (V:D)
slough sedge = Carex obnupta (V:M)
sloughgrass = Beckmannia syzigachne (V:M)
sloughgrass, American → American sloughgrass
small-awned sedge = Carex microchaeta (V:M)
small balsam = Impatiens parviflora (V:D)
small bedstraw = Galium trifidum (V:D)
small bladderwort = Utricularia minor (V:D)
small bog orchid = Platanthera chorisiana (V:M)
small bur-reed = Sparganium nutans (V:M)
small enchanter's-nightshade = Circaea alpina (V:D)
small felt lichen = Peltigera didactyla (L)
small fescue = Vulpia microstachys (V:M)
small fescue = Vulpia microstachys var. pauciflora (V:M)
small flat moss = Pseudotaxiphyllum elegans (B:M)
small-flowered alumroot = Heuchera micrantha (V:D)
small-flowered anemone = Anemone parviflora (V:D)
small-flowered birds-foot trefoil = Lotus micranthus (V:D)
small-flowered bitter-cress = Cardamine parviflora (V:D)
small-flowered blue-eyed Mary = Collinsia parviflora (V:D)
small-flowered bulrush = Scirpus microcarpus (V:M)
small-flowered buttercup = Ranunculus abortivus (V:D)
small-flowered buttercup = Ranunculus uncinatus (V:D)
small-flowered catchfly = Silene gallica (V:D)
small-flowered columbine = Aquilegia brevistyla (V:D)
small-flowered crane's-bill = Geranium pusillum (V:D)
small-flowered evening star = Mentzelia albicaulis (V:D)

small-flowered fiddleneck = Amsinckia menziesii (V:D)
small-flowered forget-me-not = Myosotis laxa (V:D)
small-flowered galinsoga = Galinsoga parviflora (V:D)
small-flowered geranium = Geranium pusillum (V:D)
small-flowered grass-of-Parnassus = Parnassia parviflora (V:D)
small-flowered ipomopsis = Ipomopsis minutiflora (V:D)
small-flowered lotus = Lotus micranthus (V:D)
small-flowered lousewort = Pedicularis parviflora (V:D)
small-flowered lupine = Lupinus polycarpus (V:D)
small-flowered mallow = Malva parviflora (V:D)
small-flowered mentzelia = Mentzelia albicaulis (V:D)
small-flowered nemophila = Nemophila parviflora (V:D)
small-flowered paintbrush = Castilleja parviflora (V:D)
small-flowered penstemon = Penstemon procerus (V:D)
small-flowered ricegrass = Oryzopsis micrantha (V:M)
small-flowered rocket = Erysimum inconspicuum (V:D)
small-flowered tonella = Tonella tenella (V:D)
small-flowered willowherb = Epilobium leptocarpum (V:D)
small-flowered willowherb = Epilobium minutum (V:D)
small-flowered woodland star = Lithophragma parviflorum (V:D)
small-flowered woodrush = Luzula parviflora (V:M)
small-fruited bulrush = Scirpus microcarpus (V:M)
small-fruited parsley-piert = Aphanes microcarpa (V:D)
small hair moss = Oligotrichum aligerum (B:M)
small-headed clover = Trifolium microcephalum (V:D)
small-headed tarweed = Madia minima (V:D)
small hop-clover = Trifolium dubium (V:D)
small-leaved bentgrass = Agrostis microphylla (V:M)
small-leaved everlasting = Antennaria microphylla (V:D)
small-leaved montia = Montia parvifolia (V:D)
small-leaved pondweed = Potamogeton pusillus (V:M)
small-leaved pussytoes = Antennaria microphylla (V:D)
small-leaved rockcress = Arabis microphylla (V:D)
small mouse-tail moss = Myurella julacea (B:M)
small pondweed = Potamogeton pusillus var. tenuissimus (V:M)
small pondweed = Potamogeton pusillus (V:M)
small red peat moss = Sphagnum capillifolium (B:M)
small spike-rush = Eleocharis parvula (V:M)
small touch-me-not = Impatiens parviflora (V:D)
small twistedstalk = Streptopus streptopoides (V:M)
small twistedstalk = Streptopus streptopoides var. brevipes (V:M)
small wallflower = Erysimum inconspicuum (V:D)
small white violet = Viola macloskeyi (V:D)
small-winged sedge = Carex microptera (V:M)
small wood anemone = Anemone parviflora (V:D)
small yellow water-buttercup = Ranunculus gmelinii (V:D)
smartweed, dockleaf → dockleaf smartweed
smartweed, dotted → dotted smartweed
smartweed, marshpepper → marshpepper smartweed
smartweed, swamp → swamp smartweed
smartweed, water → water smartweed
smelowskia, alpine → alpine smelowskia
smelowskia, short-fruited → short-fruited smelowskia
Smith's fairybells = Disporum smithii (V:M)
Smith's melic = Melica smithii (V:M)
Smith's pepper-grass = Lepidium heterophyllum (V:D)
smoker's lung = Lobaria retigera (L)
smoky shield lichen = Parmelia omphalodes (L)
smooth alumroot = Heuchera glabra (V:D)
smooth blue aster = Aster laevis (V:D)
smooth brome = Bromus inermis ssp. inermis (V:M)
smooth brome = Bromus inermis (V:M)
smooth cat's-ear = Hypochoeris glabra (V:D)
smooth cliff-brake = Pellaea glabella (V:F)
smooth cliff fern = Woodsia glabella (V:F)
smooth crabgrass = Digitaria ischaemum (V:M)
smooth daisy = Erigeron glabellus (V:D)
smooth douglasia = Douglasia laevigata (V:D)
smooth draba = Draba glabella (V:D)
smooth fleabane = Erigeron glabellus (V:D)
smooth goldenrod = Solidago gigantea (V:D)
smooth hawksbeard = Crepis capillaris (V:D)
smooth-leaved mountain-avens = Dryas integrifolia (V:D)
smooth scouring-rush = Equisetum laevigatum (V:F)
smooth spike-primrose = Boisduvalia glabella (V:D)
smooth-stalked yellow feather moss = Brachythecium salebrosum (B:M)
smooth-stemmed sedge = Carex laeviculmis (V:M)
smooth sumac = Rhus glabra (V:D)
smooth Swiss lichen = Nephroma laevigatum (L)
smooth whitlow-grass = Draba glabella (V:D)
smooth wild strawberry = Fragaria virginiana (V:D)
smooth willow = Salix glauca (V:D)
smooth willowherb = Epilobium glaberrimum (V:D)
smooth witchgrass = Panicum dichotomiflorum (V:M)
smooth woodland star = Lithophragma glabrum (V:D)
smooth woodsia = Woodsia glabella (V:F)
snake liverwort = Conocephalum conicum (B:H)
snake-root = Sanicula arctopoides (V:D)
snake-root = Sanicula marilandica (V:D)
snake-root, black → black snake-root
snapdragon, common → common snapdragon
sneezeweed, mountain → mountain sneezeweed
snow bramble = Rubus nivalis (V:D)
snow buckwheat = Eriogonum niveum (V:D)
snow buttercup = Ranunculus nivalis (V:D)
snow cinquefoil = Potentilla nivea (V:D)
snow douglasia = Douglasia nivalis (V:D)
snow draba = Draba nivalis (V:D)
snow-in-summer = Cerastium tomentosum (V:D)
snow-mat liverwort = Barbilophozia floerkei (B:H)
snow pearlwort = Sagina nivalis (V:D)
snow whitlow-grass = Draba nivalis (V:D)
snow willow = Salix brachycarpa ssp. niphoclada (V:D)
snow willow = Salix reticulata ssp. nivalis (V:D)
snow, ragged → ragged snow
snowberry, common → common snowberry
snowberry, creeping → creeping snowberry
snowberry, few-flowered → few-flowered snowberry
snowberry, mountain → mountain snowberry
snowberry, Utah mountain → Utah mountain snowberry
snowberry, western → western snowberry
snowbrush = Ceanothus velutinus (V:D)
snowbrush ceanothus = Ceanothus velutinus (V:D)
snowpatch buttercup = Ranunculus eschscholtzii (V:D)
soapberry = Shepherdia canadensis (V:D)
soapwort = Saponaria officinalis (V:D)
sockeye psora = Psora decipiens (L)
sockeye scale = Psora decipiens (L)
socks, dirty → dirty socks
soft brome = Bromus hordeaceus (V:M)
soft chess = Bromus hordeaceus (V:M)
soft halimolobos = Halimolobos mollis (V:D)
soft-leaved sandbar willow = Salix sessilifolia (V:D)
soft-leaved sedge = Carex disperma (V:M)
softgrass, creeping → creeping softgrass
soil mouse = Massalongia microphylliza (L)
soil ruby = Heppia lutosa (L)
soil stippleback = Endocarpon pusillum (L)
Solomon's-seal, two-leaved → two-leaved Solomon's-seal
Solomon's-seal, two-leaved false → two-leaved false Solomon's-seal
solorina, saffron-yellow → saffron-yellow solorina
soopolallie = Shepherdia canadensis (V:D)
sooty leather lichen = Sticta fuliginosa (L)
sorrel, alpine → alpine sorrel
sorrel, garden → garden sorrel
sorrel, green → green sorrel
sorrel, mountain → mountain sorrel
sorrel, redwood → redwood sorrel
sorrel, sheep → sheep sorrel
sour weed = Rumex acetosella (V:D)
southern cottonwood = Populus deltoides ssp. deltoides (V:D)
southern maidenhair fern = Adiantum capillus-veneris (V:F)
sow-thistle, annual → annual sow-thistle

sow-thistle, common → common sow-thistle
sow-thistle, perennial → perennial sow-thistle
sow-thistle, prickly → prickly sow-thistle
Spalding's milk-vetch = Astragalus spaldingii (V:D)
Spanish bluebells = Hyacinthoides hispanica (V:M)
sparrow's-egg lady's-slipper = Cypripedium passerinum (V:M)
sparse-leaved sedge = Carex tenuiflora (V:M)
spear-leaved arnica = Arnica lonchophylla (V:D)
spear moss = Calliergonella cuspidata (B:M)
spearmint = Mentha spicata (V:D)
spearscale = Atriplex patula (V:D)
spearwort, creeping → creeping spearwort
spearwort, lesser → lesser spearwort
speckleback, forest → forest speckleback
speckleback, green → green speckleback
speckleback, seaside → seaside speckleback
specklebelly, dimpled → dimpled specklebelly
specklebelly, netted → netted specklebelly
specklebelly, old-growth → old-growth specklebelly
specklebelly, yellow → yellow specklebelly
speckled horsehair = Bryoria fuscescens (L)
speckled horsehair lichen = Bryoria fuscescens (L)
speckled rag = Cetrelia cetrarioides (L)
speedwell, alpine → alpine speedwell
speedwell, bird's-eye → bird's-eye speedwell
speedwell, blue water → blue water speedwell
speedwell, common → common speedwell
speedwell, Cusick's → Cusick's speedwell
speedwell, European → European speedwell
speedwell, germander → germander speedwell
speedwell, marsh → marsh speedwell
speedwell, purslane → purslane speedwell
speedwell, slender → slender speedwell
speedwell, spring → spring speedwell
speedwell, thyme-leaved → thyme-leaved speedwell
speedwell, wall → wall speedwell
speedwell, water → water speedwell
sphagnum, common brown → common brown sphagnum
sphagnum, common green → common green sphagnum
sphagnum, common red → common red sphagnum
sphagnum, poor-fen → poor-fen sphagnum
sphagnum, shaggy → shaggy sphagnum
spicy conehead = Conocephalum conicum (B:H)
spider-flower bee-plant = Cleome serrulata (V:D)
spike bentgrass = Agrostis exarata (V:M)
spike lichen = Cladonia uncialis (L)
spike-oat = Helictotrichon hookeri (V:M)
spike-primrose, brook → brook spike-primrose
spike-primrose, dense → dense spike-primrose
spike-primrose, smooth → smooth spike-primrose
spike-rush, beaked → beaked spike-rush
spike-rush, common → common spike-rush
spike-rush, creeping → creeping spike-rush
spike-rush, few-flowered → few-flowered spike-rush
spike-rush, Kamchatka → Kamchatka spike-rush
spike-rush, needle → needle spike-rush
spike-rush, purple → purple spike-rush
spike-rush, slender → slender spike-rush
spike-rush, small → small spike-rush
spike trisetum = Trisetum spicatum (V:M)
spiked woodrush = Luzula spicata (V:M)
spikelike goldenrod = Solidago spathulata (V:D)
spikemoss, mountain → mountain spikemoss
spikenard sedge = Carex nardina (V:M)
spiny heath = Cetraria aculeata (L)
spiny heath lichen = Cetraria aculeata (L)
spiny wood fern = Dryopteris expansa (V:F)
spirea, birch-leaved → birch-leaved spirea
spirea, Douglas' → Douglas' spirea
spirea, Steven's → Steven's spirea
splachnum, brown tapering → brown tapering splachnum
spleenwort-leaved goldthread = Coptis aspleniifolia (V:D)
spleenwort, corrupt → corrupt spleenwort
spleenwort, maidenhair → maidenhair spleenwort
sponge pelt = Peltigera retifoveata (L)
spotted coralroot = Corallorrhiza maculata (V:M)
spotted cowbane = Cicuta maculata var. angustifolia (V:D)
spotted dead-nettle = Lamium maculatum (V:D)
spotted joe-pye-weed = Eupatorium maculatum (V:D)
spotted loosestrife = Lysimachia punctata (V:D)
spotted medic = Medicago arabica (V:D)
spotted saxifrage = Saxifraga bronchialis ssp. austromontana (V:D)
spotted saxifrage = Saxifraga bronchialis (V:D)
spotted touch-me-not = Impatiens capensis (V:D)
spotted water-hemlock = Cicuta maculata (V:D)
sprangle-top = Scolochloa festucacea (V:M)
sprangletop = Leptochloa fascicularis (V:M)
spraypaint lichen = Icmadophila ericetorum (L)
spread-leaved peat moss = Sphagnum squarrosum (B:M)
spreading dogbane = Apocynum androsaemifolium (V:D)
spreading fleabane = Erigeron divergens (V:D)
spreading groundsmoke = Gayophytum diffusum (V:D)
spreading hackelia = Hackelia diffusa (V:D)
spreading-leaved peat moss = Sphagnum squarrosum (B:M)
spreading montia = Montia diffusa (V:D)
spreading needlegrass = Stipa richardsonii (V:M)
spreading phlox = Phlox diffusa ssp. longistylis (V:D)
spreading phlox = Phlox diffusa (V:D)
spreading rush = Juncus supiniformis (V:M)
spreading stickseed = Hackelia diffusa (V:D)
spreading stonecrop = Sedum divergens (V:D)
spreading sweet-cicely = Osmorhiza depauperata (V:D)
Sprengel's sedge = Carex sprengelii (V:M)
spring-beauty, Alaska → Alaska spring-beauty
spring-beauty, alpine → alpine spring-beauty
spring-beauty, broad-leaved → broad-leaved spring-beauty
spring-beauty, heart-leaved → heart-leaved spring-beauty
spring-beauty, pale → pale spring-beauty
spring-beauty, streambank → streambank spring-beauty
spring-beauty, tuberous → tuberous spring-beauty
spring-beauty, western → western spring-beauty
spring claw moss = Cratoneuron filicinum (B:M)
spring disc lichen = Biatora vernalis (L)
spring forget-me-not = Myosotis verna (V:D)
spring gold = Lomatium utriculatum (V:D)
spring hornwort = Ceratophyllum echinatum (V:D)
spring moss = Philonotis fontana (B:M)
spring speedwell = Veronica verna (V:D)
spring water-starwort = Callitriche palustris (V:D)
springbank clover = Trifolium wormskjoldii (V:D)
spruce moss = Evernia mesomorpha (L)
spruce, black → black spruce
spruce, Engelmann → Engelmann spruce
spruce, hybrid white → hybrid white spruce
spruce, Roche → Roche spruce
spruce, Sitka → Sitka spruce
spruce, white → white spruce
spurge-laurel = Daphne laureola (V:D)
spurge, caper → caper spurge
spurge, cypress → cypress spurge
spurge, dwarf → dwarf spurge
spurge, leafy → leafy spurge
spurge, petty → petty spurge
spurge, summer → summer spurge
spurless touch-me-not = Impatiens ecalcarata (V:D)
spurred gentian = Halenia deflexa (V:D)
spurred lupine = Lupinus arbustus ssp. neolaxiflorus (V:D)
spurry knotweed = Polygonum douglasii ssp. spergulariiforme (V:D)
squarrose peat moss = Sphagnum squarrosum (B:M)
squashberry = Viburnum edule (V:D)
squaw currant = Ribes cereum (V:D)
squirreltail = Elymus elymoides (V:M)
squirreltail grass = Elymus elymoides (V:M)

St. John's-wort, bog → bog St. John's-wort
St. John's-wort, common → common St. John's-wort
St. John's-wort, large Canadian → large Canadian St. John's-wort
St. John's-wort, Scouler's → Scouler's St. John's-wort
stair step moss = Hylocomium splendens (B:M)
stair-step moss = Hylocomium splendens (B:M)
stalked-pod locoweed = Oxytropis podocarpa (V:D)
standing milk-vetch = Astragalus adsurgens (V:D)
standing milk-vetch = Astragalus adsurgens var. robustior (V:D)
stane-raw = Parmelia saxatilis (L)
star duckweed = Lemna trisulca (V:M)
star-of-Bethlehem = Ornithogalum umbellatum (V:M)
star reindeer lichen = Cladina stellaris (L)
star sedge = Carex echinata (V:M)
star sedge = Carex echinata ssp. echinata (V:M)
star-thistle, maltese → maltese star-thistle
star, gold → gold star
star, saline shooting → saline shooting star
star, slender woodland → slender woodland star
star, small-flowered evening → small-flowered evening star
star, small-flowered woodland → small-flowered woodland star
star, smooth woodland → smooth woodland star
star, white-stemmed evening → white-stemmed evening star
starburst shadow = Phaeophyscia endococcina (L)
starburst shadow = Phaeophyscia ciliata (L)
starburst, black → black starburst
starburst, green → green starburst
starburst, grey → grey starburst
starburst, salted → salted starburst
starflower, European → European starflower
starflower, northern → northern starflower
starwort, American → American starwort
starwort, blunt-sepaled → blunt-sepaled starwort
starwort, bog → bog starwort
starwort, common → common starwort
starwort, crisp → crisp starwort
starwort, grass-leaved → grass-leaved starwort
starwort, long-stalked → long-stalked starwort
starwort, northern → northern starwort
starwort, salt marsh → salt marsh starwort
starwort, shiny → shiny starwort
starwort, thick-leaved → thick-leaved starwort
starwort, umbellate → umbellate starwort
steeplebush = Spiraea douglasii (V:D)
steer's head = Dicentra uniflora (V:D)
stemless raspberry = Rubus arcticus ssp. acaulis (V:D)
stender whitlow-grass = Draba stenoloba (V:D)
step moss = Hylocomium splendens (B:M)
stephanomeria, narrow-leaved → narrow-leaved stephanomeria
steppe mouse-moss = Coscinodon calyptratus (B:M)
Steven's spirea = Spiraea stevenii (V:D)
stickseed, blue → blue stickseed
stickseed, bristly → bristly stickseed
stickseed, many-flowered → many-flowered stickseed
stickseed, nodding → nodding stickseed
stickseed, Okanogan → Okanogan stickseed
stickseed, spreading → spreading stickseed
stickwort = Spergula arvensis (V:D)
sticky chickweed = Cerastium glomeratum (V:D)
sticky cinquefoil = Potentilla glandulosa (V:D)
sticky cockle = Silene noctiflora (V:D)
sticky cudweed = Gnaphalium viscosum (V:D)
sticky currant = Ribes viscosissimum (V:D)
sticky false asphodel = Tofieldia glutinosa (V:M)
sticky geranium = Geranium viscosissimum (V:D)
sticky Jacob's-ladder = Polemonium viscosum (V:D)
sticky lichen = Collema tenax (L)
sticky locoweed = Oxytropis viscida (V:D)
sticky polemonium = Polemonium viscosum (V:D)
sticky purple crane's-bill = Geranium viscosissimum (V:D)
sticky purple geranium = Geranium viscosissimum (V:D)
sticky ragwort = Senecio viscosus (V:D)
stiff clubmoss = Lycopodium annotinum (V:F)
stiff-leaved pondweed = Potamogeton strictifolius (V:M)
stiff needlegrass = Stipa occidentalis var. pubescens (V:M)
stiff needlegrass = Stipa occidentalis (V:M)
stiff sedge = Carex bigelowii (V:M)
stinging nettle = Urtica dioica ssp. gracilis (V:D)
stinging nettle = Urtica dioica (V:D)
stink currant = Ribes bracteosum (V:D)
stinkgrass = Eragrostis cilianensis (V:M)
stinking chamomile = Anthemis cotula (V:D)
stinking-clover bee-plant = Cleome serrulata (V:D)
stinkweed = Thlaspi arvense (V:D)
stippleback, fissured → fissured stippleback
stippleback, limy → limy stippleback
stippleback, northwest → northwest stippleback
stippleback, rock → rock stippleback
stippleback, shag → shag stippleback
stippleback, soil → soil stippleback
stippleback, streamside → streamside stippleback
stipplescale, ashen → ashen stipplescale
stipplescale, brown → brown stipplescale
stitchwort, fleshy → fleshy stitchwort
stitchwort, northern → northern stitchwort
stocking moss = Pylaisiella polyantha (B:M)
stoloniferous saxifrage = Saxifraga flagellaris ssp. setigera (V:D)
stoloniferous saxifrage = Saxifraga flagellaris (V:D)
stoloniferous willow = Salix stolonifera (V:D)
stonecrop, broad-leaved → broad-leaved stonecrop
stonecrop, entire-leaf → entire-leaf stonecrop
stonecrop, goldmoss → goldmoss stonecrop
stonecrop, lance-leaved → lance-leaved stonecrop
stonecrop, narrow-petaled → narrow-petaled stonecrop
stonecrop, Oregon → Oregon stonecrop
stonecrop, spreading → spreading stonecrop
stonecrop, white → white stonecrop
stonecrop, worm-leaved → worm-leaved stonecrop
stork's-bill = Erodium cicutarium (V:D)
stork's-bill, common → common stork's-bill
straight-beaked buttercup = Ranunculus orthorhynchus (V:D)
strapwort = Corrigiola litoralis (V:D)
straw-coloured water moss = Calliergon stramineum (B:M)
straw-like feather moss = Calliergon stramineum (B:M)
straw moss = Calliergon stramineum (B:M)
strawberry-blite = Chenopodium capitatum (V:D)
strawberry clover = Trifolium fragiferum (V:D)
strawberry, blue-leaved → blue-leaved strawberry
strawberry, coastal → coastal strawberry
strawberry, false → false strawberry
strawberry, Indian → Indian strawberry
strawberry, Pacific coast → Pacific coast strawberry
strawberry, smooth wild → smooth wild strawberry
strawberry, wild → wild strawberry
strawberry, wood → wood strawberry
strawberry, woodland → woodland strawberry
stream-bank lupine = Lupinus rivularis (V:D)
stream saxifrage = Saxifraga odontoloma (V:D)
stream violet = Viola glabella (V:D)
streambank arnica = Arnica amplexicaulis (V:D)
streambank butterweed = Senecio pseudaureus (V:D)
streambank globe-mallow = Iliamna rivularis (V:D)
streambank spring-beauty = Montia parvifolia (V:D)
streamside fleabane = Erigeron glabellus (V:D)
streamside moss = Scouleria aquatica (B:M)
streamside stippleback = Dermatocarpon rivulorum (L)
streamside stippleback = Dermatocarpon luridum (L)
streamside vinyl = Leptogium rivale (L)
strict buckwheat = Eriogonum strictum (V:D)
strict buckwheat = Eriogonum strictum ssp. proliferum (V:D)
string sedge = Carex chordorrhiza (V:M)
striped coralroot = Corallorrhiza striata (V:M)

stubble, green → green stubble
stubble, sulphur → sulphur stubble
studded leather lichen = Peltigera aphthosa (L)
stump cladonia = Cladonia botrytes (L)
subalpine buttercup = Ranunculus eschscholtzii (V:D)
subalpine daisy = Erigeron peregrinus (V:D)
subalpine fir = Abies lasiocarpa (V:G)
subalpine larch = Larix lyallii (V:G)
subelegant brown = Melanelia subelegantula (L)
subterranean clover = Trifolium subterraneum (V:D)
Sudeten lousewort = Pedicularis sudetica (V:D)
Sudeten lousewort = Pedicularis sudetica ssp. interior (V:D)
sugared beard = Usnea hirta (L)
sugared orange = Xanthoria sorediata (L)
sugarstick = Allotropa virgata (V:D)
sugary beard = Usnea hirta (L)
Suksdorf's hawthorn = Crataegus suksdorfii (V:D)
Suksdorf's mugwort = Artemisia suksdorfii (V:D)
suksdorfia, buttercup-leaved → buttercup-leaved suksdorfia
suksdorfia, violet → violet suksdorfia
sulfur buckwheat = Eriogonum umbellatum (V:D)
sulphur buckwheat = Eriogonum umbellatum (V:D)
sulphur buttercup = Ranunculus sulphureus (V:D)
sulphur cinquefoil = Potentilla recta (V:D)
sulphur cladonia = Cladonia sulphurina (L)
sulphur cup = Cladonia sulphurina (L)
sulphur-dust lichen = Parmeliopsis ambigua (L)
sulphur hedysarum = Hedysarum sulphurescens (V:D)
sulphur lupine = Lupinus sulphureus (V:D)
sulphur paintbrush = Castilleja sulphurea (V:D)
sulphur stubble = Chaenotheca furfuracea (L)
sulphur sweet-vetch = Hedysarum sulphurescens (V:D)
sumac = Rhus glabra (V:D)
sumac, smooth → smooth sumac
summer-cypress = Kochia scoparia (V:D)
summer spurge = Euphorbia helioscopia (V:D)
sunburst lichen = Arctoparmelia centrifuga (L)
sundew, great → great sundew
sundew, long-leaved → long-leaved sundew
sundew, oblong-leaved → oblong-leaved sundew
sundew, round-leaved → round-leaved sundew
sundrops, perennial → perennial sundrops
sunflower, common → common sunflower
sunflower, Nuttall's → Nuttall's sunflower
sunflower, woolly → woolly sunflower
sunshine, brown-eyed → brown-eyed sunshine
sunshine, limestone → limestone sunshine
sunshine, Oregon → Oregon sunshine
sunshine, powdered → powdered sunshine
surf-grass, Scouler's → Scouler's surf-grass
surf-grass, toothed → toothed surf-grass
surf-grass, Torrey's → Torrey's surf-grass
Susan, black-eyed → black-eyed Susan
Susan, brown-eyed → brown-eyed Susan
swale desert-parsley = Lomatium ambiguum (V:D)
swale lomatium = Lomatium ambiguum (V:D)
swallowwort = Chelidonium majus (V:D)
swamp birch = Betula nana (V:D)
swamp birch = Betula pumila var. glandulifera (V:D)
swamp black gooseberry = Ribes lacustre (V:D)
swamp gentian = Gentiana douglasiana (V:D)
swamp hedge-nettle = Stachys palustris ssp. pilosa (V:D)
swamp hedge-nettle = Stachys palustris (V:D)
swamp horsetail = Equisetum fluviatile (V:F)
swamp lantern = Lysichiton americanum (V:M)
swamp laurel = Kalmia microphylla (V:D)
swamp lousewort = Pedicularis parviflora (V:D)
swamp moss = Philonotis fontana (B:M)
swamp onion = Allium validum (V:M)
swamp red currant = Ribes triste (V:D)
swamp smartweed = Polygonum hydropiperoides (V:D)
swamp willowherb = Epilobium davuricum (V:D)
swamp willowherb = Epilobium palustre (V:D)
sweet alyssum = Lobularia maritima (V:D)
sweet cherry = Prunus avium (V:D)
sweet-cicely, blunt-fruited → blunt-fruited sweet-cicely
sweet-cicely, purple → purple sweet-cicely
sweet-cicely, spreading → spreading sweet-cicely
sweet-cicely, western → western sweet-cicely
sweet-clover, yellow → yellow sweet-clover
sweet coltsfoot = Petasites frigidus (V:D)
sweet fennel = Foeniculum vulgare (V:D)
sweet-flag, American → American sweet-flag
sweet-flowered fairy-candelabra = Androsace chamaejasme (V:D)
sweet gale = Myrica gale (V:D)
sweet-scented bedstraw = Galium triflorum (V:D)
sweet sedge = Carex anthoxanthea (V:M)
sweet vernalgrass = Anthoxanthum odoratum (V:M)
sweet-vetch, alpine → alpine sweet-vetch
sweet-vetch, northern → northern sweet-vetch
sweet-vetch, sulphur → sulphur sweet-vetch
sweet-vetch, western → western sweet-vetch
sweet-vetch, yellow → yellow sweet-vetch
sweet violet = Viola odorata (V:D)
sweet William = Dianthus barbatus (V:D)
sweet William catchfly = Silene armeria (V:D)
sweet woodruff = Galium odoratum (V:D)
sweetbrier = Rosa eglanteria (V:D)
sweetgrass, alpine → alpine sweetgrass
sweetgrass, common → common sweetgrass
swertia, alpine bog → alpine bog swertia
swine-cress, lesser → lesser swine-cress
Swiss lichen = Nephroma helveticum (L)
sword fern = Polystichum munitum (V:F)
sycamore maple = Acer pseudoplatanus (V:D)

T

tail-leaved willow = Salix lucida ssp. caudata (V:D)
Taimyr campion = Silene taimyrensis (V:D)
talinum, Okanogan → Okanogan talinum
tall annual willowherb = Epilobium brachycarpum (V:D)
tall beggarticks = Bidens vulgata (V:D)
tall blue lettuce = Lactuca biennis (V:D)
tall bluebells = Mertensia paniculata (V:D)
tall bog sedge = Carex magellanica ssp. irrigua (V:M)
tall bugbane = Cimicifuga elata (V:D)
tall buttercup = Ranunculus acris (V:D)
tall butterweed = Senecio serra (V:D)
tall cinquefoil = Potentilla arguta ssp. convallaria (V:D)
tall clustered thread moss = Bryum pseudotriquetrum (B:M)
tall cotton-grass = Eriophorum angustifolium (V:M)
tall draba = Draba praealta (V:D)
tall fescue = Festuca arundinacea (V:M)
tall fringecup = Tellima grandiflora (V:D)
tall hawkweed = Hieracium piloselloides (V:D)
tall Jacob's-ladder = Polemonium caeruleum (V:D)
tall larkspur = Delphinium glaucum (V:D)
tall lungwort = Mertensia paniculata (V:D)
tall mannagrass = Glyceria elata (V:M)

tall mannagrass = Glyceria grandis (V:M)
tall meadowrue = Thalictrum dasycarpum (V:D)
tall mountain shootingstar = Dodecatheon jeffreyi (V:D)
tall oatgrass = Arrhenatherum elatius (V:M)
tall Oregon-grape = Mahonia aquifolium (V:D)
tall pepper-grass = Lepidium virginicum (V:D)
tall trisetum = Trisetum cernuum var. canescens (V:M)
tall tumble-mustard = Sisymbrium altissimum (V:D)
tall whitlow-grass = Draba praealta (V:D)
tall woolly-heads = Psilocarphus elatior (V:D)
tamarack = Larix laricina (V:G)
tangle moss = Heterocladium procurrens (B:M)
tansy ragwort = Senecio jacobaea (V:D)
tansy, common → common tansy
tansy, dune → dune tansy
tansy, western dune → western dune tansy
tansymustard = Descurainia sophia (V:D)
tansymustard, northern → northern tansymustard
tansymustard, western → western tansymustard
tapegrass, American → American tapegrass
tapered rush = Juncus acuminatus (V:M)
tar tarpaper = Collema tenax (L)
tar tarpaper = Collema bachmanianum (L)
tar tarpaper = Collema coccophorum (L)
tar, sea → sea tar
tarpaper, amphibious → amphibious tarpaper
tarpaper, bleb → bleb tarpaper
tarpaper, blistered → blistered tarpaper
tarpaper, broadleaf → broadleaf tarpaper
tarpaper, butterfly → butterfly tarpaper
tarpaper, coal → coal tarpaper
tarpaper, cushion → cushion tarpaper
tarpaper, fingered → fingered tarpaper
tarpaper, jelly → jelly tarpaper
tarpaper, moth → moth tarpaper
tarpaper, papoose → papoose tarpaper
tarpaper, protean → protean tarpaper
tarpaper, protracted → protracted tarpaper
tarpaper, seaside → seaside tarpaper
tarpaper, shaly → shaly tarpaper
tarpaper, tar → tar tarpaper
tarpaper, tripe → tripe tarpaper
tarpaper, western → western tarpaper
tarpaper, whiskered → whiskered tarpaper
tarragon = Artemisia dracunculus (V:D)
tarweed, Chilean → Chilean tarweed
tarweed, clustered → clustered tarweed
tarweed, little → little tarweed
tarweed, slender → slender tarweed
tarweed, small-headed → small-headed tarweed
tarweed, woodland → woodland tarweed
tattered rag = Platismatia herrei (L)
tattered vinyl = Leptogium lichenoides (L)
tawny cotton-grass = Eriophorum virginicum (V:M)
Taylor's liverwort = Mylia taylorii (B:H)
Taylor's saxifrage = Saxifraga taylori (V:D)
tea-leaved willow = Salix planifolia (V:D)
tea-leaved willow = Salix planifolia ssp. planifolia (V:D)
tea, common Labrador → common Labrador tea
tea, glandular Labrador → glandular Labrador tea
tea, Labrador → Labrador tea
tea, northern Labrador → northern Labrador tea
tea, trapper's → trapper's tea
teaberry, western → western teaberry
teasel, Fuller's → Fuller's teasel
temporary pelt = Peltigera didactyla (L)
tender sedge = Carex tenera (V:M)
textured lung = Lobaria scrobiculata (L)
thale cress = Arabidopsis thaliana (V:D)
The Dalles milk-vetch = Astragalus sclerocarpus (V:D)
thelypody, thick-leaved → thick-leaved thelypody
thick grass moss = Brachythecium turgidum (B:M)
thick-headed sedge = Carex pachystachya (V:M)
thick-headed sedge = Carex macloviana (V:M)
thick-leaved draba = Draba crassifolia (V:D)
thick-leaved starwort = Stellaria crassifolia (V:D)
thick-leaved thelypody = Thelypodium laciniatum (V:D)
thick-leaved whitlow-grass = Draba crassifolia (V:D)
thick ragged moss = Brachythecium turgidum (B:M)
thickspike wildrye = Elymus lanceolatus (V:M)
thimble clover = Trifolium microdon (V:D)
thimbleberry = Rubus parviflorus (V:D)
thimbleweed = Anemone cylindrica (V:D)
thin bentgrass = Agrostis diegoensis (V:M)
thin-flowered sedge = Carex tenuiflora (V:M)
thin-leaved bedstraw = Galium bifolium (V:D)
thin-leaved owl-clover = Orthocarpus tenuifolius (V:D)
thistle, bull → bull thistle
thistle, Canada → Canada thistle
thistle, curled → curled thistle
thistle, distaff → distaff thistle
thistle, Drummond's → Drummond's thistle
thistle, edible → edible thistle
thistle, elk → elk thistle
thistle, Hooker's → Hooker's thistle
thistle, leafy → leafy thistle
thistle, marsh → marsh thistle
thistle, milk → milk thistle
thistle, musk → musk thistle
thistle, nodding → nodding thistle
thistle, plumeless → plumeless thistle
thistle, Russian → Russian thistle
thistle, Scotch → Scotch thistle
thistle, short-styled → short-styled thistle
thistle, wavy-leaved → wavy-leaved thistle
Thompson's paintbrush = Castilleja thompsonii (V:D)
thornbush, blackened → blackened thornbush
thornbush, seaside → seaside thornbush
thorny buffalo-berry = Shepherdia argentea (V:D)
thorough-wax, American → American thorough-wax
thread-leaved daisy = Erigeron filifolius (V:D)
thread-leaved fleabane = Erigeron filifolius (V:D)
thread-leaved phacelia = Phacelia linearis (V:D)
thread-leaved sandwort = Arenaria capillaris ssp. americana (V:D)
thread-leaved sandwort = Arenaria capillaris (V:D)
thread-leaved sedge = Carex filifolia (V:M)
thread rush = Juncus filiformis (V:M)
threadstalk milk-vetch = Astragalus filipes (V:D)
three-angled thread moss = Meesia triquetra (B:M)
three-awn, prairie → prairie three-awn
three-flowered avens = Geum triflorum (V:D)
three-flowered rush = Juncus triglumis (V:M)
three-flowered waterwort = Elatine rubella (V:D)
three-fork wormwood = Artemisia furcata (V:D)
three-forked mugwort = Artemisia furcata var. heterophylla (V:D)
three-leaved foamflower = Tiarella trifoliata var. trifoliata (V:D)
three-leaved goldthread = Coptis trifolia (V:D)
three-leaved lewisia = Lewisia triphylla (V:D)
three-leaved pondweed = Potamogeton filiformis (V:M)
three-lobed daisy = Erigeron trifidus (V:D)
three-nerve daisy = Erigeron subtrinervis var. conspicuus (V:D)
three-nerve fleabane = Erigeron subtrinervis (V:D)
three-nerve fleabane = Erigeron subtrinervis var. conspicuus (V:D)
three-ruffle liverwort = Porella navicularis (B:H)
three-seeded sedge = Carex trisperma (V:M)
three-spot mariposa lily = Calochortus apiculatus (V:M)
three-toothed mitrewort = Mitella trifida (V:D)
three-toothed saxifrage = Saxifraga tricuspidata (V:D)
three-toothed whip liverwort = Bazzania tricrenata (B:H)
three-way sedge = Dulichium arundinaceum (V:M)
threetip sagebrush = Artemisia tripartita (V:D)
thrift = Armeria maritima (V:D)
thyme dodder = Cuscuta epithymum (V:D)

thyme-leaved sandwort = Arenaria serpyllifolia (V:D)
thyme-leaved saxifrage = Saxifraga serpyllifolia (V:D)
thyme-leaved speedwell = Veronica serpyllifolia (V:D)
tickertape bone = Hypogymnia duplicata (L)
tiger lily = Lilium columbianum (V:M)
timber milk-vetch = Astragalus miser (V:D)
timber oatgrass = Danthonia intermedia (V:M)
timberline bluegrass = Poa glauca (V:M)
timothy = Phleum pratense (V:M)
timothy, alpine → alpine timothy
timothy, common → common timothy
tiny mousetail = Myosurus minimus (V:D)
tiny toothpick cladonia = Cladonia coniocraea (L)
tiny vetch = Vicia hirsuta (V:D)
toad-flax, bastard → bastard toad-flax
toad-flax, northern bastard → northern bastard toad-flax
toad pelt = Peltigera scabrosa (L)
toad rush = Juncus bufonius (V:M)
toadflax, broom-leaf → broom-leaf toadflax
toadflax, common → common toadflax
toadflax, dalmatian → dalmatian toadflax
toadflax, ivy-leaved → ivy-leaved toadflax
toadflax, purple → purple toadflax
Tolmie's saxifrage = Saxifraga tolmiei (V:D)
tonella, small-flowered → small-flowered tonella
toothcup meadow-foam = Rotala ramosior (V:D)
toothed-leaved monkey-flower = Mimulus dentatus (V:D)
toothed surf-grass = Phyllospadix serrulatus (V:M)
toothed wood fern = Dryopteris carthusiana (V:F)
torn club lichen = Cladonia cariosa (L)
Torrey's cryptantha = Cryptantha torreyana (V:D)
Torrey's rush = Juncus torreyi (V:M)
Torrey's surf-grass = Phyllospadix torreyi (V:M)
touch-me-not, common → common touch-me-not
touch-me-not, orange → orange touch-me-not
touch-me-not, small → small touch-me-not
touch-me-not, spotted → spotted touch-me-not
touch-me-not, spurless → spurless touch-me-not
touch-me-not, western → western touch-me-not
tower mustard = Arabis glabra (V:D)
townsendia, Hooker's → Hooker's townsendia
townsendia, Parry's → Parry's townsendia
Tracy's mistmaiden = Romanzoffia tracyi (V:D)
Tracy's romanzoffia = Romanzoffia tracyi (V:D)
trailing black currant = Ribes laxiflorum (V:D)
trailing blackberry = Rubus ursinus ssp. macropetalus (V:D)
trailing blackberry = Rubus ursinus (V:D)
trailing daisy = Erigeron flagellaris (V:D)
trailing fleabane = Erigeron flagellaris (V:D)
trailing leafy moss = Plagiomnium medium (B:M)
trailing pearlwort = Sagina decumbens (V:D)
trailing raspberry = Rubus pubescens (V:D)
trailing yellow violet = Viola sempervirens (V:D)
trapper's tea = Ledum glandulosum (V:D)
traveler's joy = Clematis vitalba (V:D)
tree lupine = Lupinus arboreus (V:D)
tree moss = Climacium dendroides (B:M)
tree pelt = Peltigera collina (L)
tree, golden chain → golden chain tree
treeflute, magic → magic treeflute
treepelt = Erioderma sorediatum (L)
treepelt = Leioderma sorediatum (L)
trefoil, birds-foot → birds-foot trefoil
trefoil, bog birds-foot → bog birds-foot trefoil
trefoil, meadow birds-foot → meadow birds-foot trefoil
trefoil, narrow-leaved birds-foot → narrow-leaved birds-foot trefoil
trefoil, Nevada birds-food → Nevada birds-food trefoil
trefoil, Nevada birds-foot → Nevada birds-foot trefoil
trefoil, pedunculate birds-foot → pedunculate birds-foot trefoil
trefoil, seaside birds-foot → seaside birds-foot trefoil
trefoil, small-flowered birds-foot → small-flowered birds-foot trefoil
trembling aspen = Populus tremuloides (V:D)
tresses, hooded ladies' → hooded ladies' tresses
tresses, ladies' → ladies' tresses
trillium, western → western trillium
Trinius' bluegrass = Poa stenantha (V:M)
tripe tarpaper = Collema callopismum (L)
tripe, lizard → lizard tripe
triple-nerved daisy = Erigeron subtrinervis var. conspicuus (V:D)
triple-nerved fleabane = Erigeron subtrinervis var. conspicuus (V:D)
trisetum, nodding → nodding trisetum
trisetum, spike → spike trisetum
trisetum, tall → tall trisetum
trisetum, Wolf's → Wolf's trisetum
triteleia, Howell's → Howell's triteleia
triteleia, large-flowered → large-flowered triteleia
triteleia, white → white triteleia
true reindeer lichen = Cladina rangiferina (L)
tuberous rush = Juncus nodosus (V:M)
tuberous spring-beauty = Claytonia tuberosa (V:D)
tufted daisy = Erigeron caespitosus (V:D)
tufted fleabane = Erigeron caespitosus (V:D)
tufted hairgrass = Deschampsia cespitosa (V:M)
tufted loosestrife = Lysimachia thyrsiflora (V:D)
tufted lovegrass = Eragrostis pectinacea (V:M)
tufted moss = Aulacomnium palustre (B:M)
tufted phlox = Phlox caespitosa (V:D)
tufted saxifrage = Saxifraga cespitosa (V:D)
tufted vetch = Vicia cracca (V:D)
tufyed phlox = Phlox caespitosa (V:D)
tumble lichen = Masonhalea richardsonii (L)
tumble-mustard, Loesel's → Loesel's tumble-mustard
tumble-mustard, tall → tall tumble-mustard
tumbleweed = Amaranthus albus (V:D)
tumbleweed = Amaranthus blitoides (V:D)
tumbleweed = Salsola kali (V:D)
tumbleweed, arctic → arctic tumbleweed
tumbling orache = Atriplex rosea (V:D)
tundra milk-vetch = Astragalus umbellatus (V:D)
tundra owl = Solorina octospora (L)
tundra owl = Solorina bispora (L)
tunic flower = Petrorhagia saxifraga (V:D)
turned sedge = Carex retrorsa (V:M)
turnip = Brassica napus (V:D)
turtle, alder → alder turtle
turtle, charcoal → charcoal turtle
turtle, old-growth → old-growth turtle
twayblade, broad-leaved → broad-leaved twayblade
twayblade, heart-leaved → heart-leaved twayblade
twayblade, northern → northern twayblade
twayblade, northwestern → northwestern twayblade
Tweedy's willow = Salix tweedyi (V:D)
twinberry, black → black twinberry
twinberry, red → red twinberry
twinflower = Linnaea borealis (V:D)
twinflower violet = Viola biflora (V:D)
twining honeysuckle = Lonicera dioica var. glaucescens (V:D)
twinpod, common → common twinpod
twisted ulota = Ulota obtusiuscula (B:M)
twistedstalk, clasping → clasping twistedstalk
twistedstalk, rosy → rosy twistedstalk
twistedstalk, small → small twistedstalk
two-coloured lupine = Lupinus bicolor (V:D)
two-coloured sedge = Carex bicolor (V:M)
two-flowered cinquefoil = Potentilla biflora (V:D)
two-flowered rush = Juncus biglumis (V:M)
two-leaved false Solomon's-seal = Maianthemum dilatatum (V:M)
two-leaved Solomon's-seal = Maianthemum canadense (V:M)
two-parted sedge = Carex bipartita (V:M)
two-seeded sedge = Carex disperma (V:M)
two-spiked moonwort = Botrychium paradoxum (V:F)
two-stamened sedge = Carex diandra (V:M)

two-toned sedge = Carex albonigra (V:M)

U

ulota, twisted → twisted ulota
umbellate chickweed = Holosteum umbellatum (V:D)
umbellate starwort = Stellaria umbellata (V:D)
umber pussytoes = Antennaria umbrinella (V:D)
umbrellawort = Mirabilis nyctaginea (V:D)
umbrellawort, hairy → hairy umbrellawort
unalaska Indian-paintbrush = Castilleja unalaschcensis (V:D)
unalaska paintbrush = Castilleja unalaschcensis (V:D)
undergreen willow = Salix commutata (V:D)
underwater lichen = Hydrothyria venosa (L)
unlovely buttercup = Ranunculus inamoenus (V:D)
unsalted shield = Parmelia omphalodes (L)
upland larkspur = Delphinium nuttallianum (V:D)
upright chickweed = Moenchia erecta (V:D)
upright goosefoot = Chenopodium urbicum (V:D)
upright primrose = Primula stricta (V:D)
upright yellow oxalis = Oxalis stricta (V:D)
Ussurian water-milfoil = Myriophyllum ussuriense (V:D)
Utah honeysuckle = Lonicera utahensis (V:D)
Utah mountain snowberry = Symphoricarpos oreophilus var. utahensis (V:D)

V

vagabond rockfrog = Xanthoparmelia camtschadalis (L)
valerian, capitate → capitate valerian
valerian, common → common valerian
valerian, edible → edible valerian
valerian, marsh → marsh valerian
valerian, northern → northern valerian
valerian, Scouler's → Scouler's valerian
valerian, Sitka → Sitka valerian
valley violet = Viola vallicola (V:D)
Vancouver groundcone = Boschniakia hookeri (V:D)
Vancouver Island beggarticks = Bidens amplissima (V:D)
vanilla-leaf = Achlys triphylla (V:D)
vari-leaved collomia = Collomia heterophylla (V:D)
vari-leaved water-milfoil = Myriophyllum heterophyllum (V:D)
variable rockfrog = Xanthoparmelia wyomingica (L)
variable ruffle = Tuckermannopsis orbata (L)
variable willow = Salix commutata (V:D)
variegated sedge = Carex stylosa (V:M)
various-leaved pondweed = Potamogeton gramineus (V:M)
Vasey's rush = Juncus vaseyi (V:M)
veined lichen = Lobaria linita (L)
veiny meadowrue = Thalictrum venulosum (V:D)
velcro lichen = Pseudephebe pubescens (L)
velvet feather moss = Brachythecium velutinum (B:M)
velvet-fruited willow = Salix maccalliana (V:D)
velvet-grass, common → common velvet-grass
velvet-grass, creeping → creeping velvet-grass
velvet-leaf = Abutilon theophrasti (V:D)
velvet-leaved blueberry = Vaccinium myrtilloides (V:D)
velvet lupine = Lupinus leucophyllus (V:D)
ventenata = Ventenata dubia (V:M)
Venus'-comb = Scandix pecten-veneris (V:D)
Venus'-slipper = Calypso bulbosa (V:M)
Venus-hair fern = Adiantum capillus-veneris (V:F)
vernalgrass, sweet → sweet vernalgrass
verticillate-umbel lovage = Ligusticum verticillatum (V:D)
verticillate water-milfoil = Myriophyllum verticillatum (V:D)
vervain, blue → blue vervain
vervain, bracted → bracted vervain
vetch, American → American vetch
vetch, bird → bird vetch
vetch, common → common vetch
vetch, hairy → hairy vetch
vetch, slender → slender vetch
vetch, tiny → tiny vetch
vetch, tufted → tufted vetch
vetch, wild → wild vetch
vetch, woolly → woolly vetch
vetchling, cream-coloured → cream-coloured vetchling
villous cinquefoil = Potentilla villosa (V:D)
vine maple = Acer circinatum (V:D)
vine, puncture → puncture vine
vinyl, antlered → antlered vinyl
vinyl, appressed → appressed vinyl
vinyl, blue → blue vinyl
vinyl, butterfly → butterfly vinyl
vinyl, California → California vinyl
vinyl, jellied → jellied vinyl
vinyl, lilliput → lilliput vinyl
vinyl, pappered → pappered vinyl
vinyl, peacock → peacock vinyl
vinyl, petalled → petalled vinyl
vinyl, pincushion → pincushion vinyl
vinyl, shrubby → shrubby vinyl
vinyl, streamside → streamside vinyl
vinyl, tattered → tattered vinyl
vinyl, wrinkled → wrinkled vinyl
violet saxifrage = Suksdorfia violacea (V:D)
violet suksdorfia = Suksdorfia violacea (V:D)
violet, Alaska → Alaska violet
violet, bog → bog violet
violet, Canada → Canada violet
violet, dwarf marsh → dwarf marsh violet
violet, early blue → early blue violet
violet, evergreen → evergreen violet
violet, evergreen yellow → evergreen yellow violet
violet, goose-foot yellow → goose-foot yellow violet
violet, great spurred → great spurred violet
violet, Howell's → Howell's violet
violet, kidney-leaved → kidney-leaved violet
violet, lance-leaved → lance-leaved violet
violet, marsh → marsh violet
violet, northern → northern violet
violet, northern blue → northern blue violet
violet, northern bog → northern bog violet
violet, prairie → prairie violet
violet, Queen Charlotte twinflower → Queen Charlotte twinflower violet
violet, rounded-leaved → rounded-leaved violet

violet, Selkirk's → Selkirk's violet
violet, small white → small white violet
violet, stream → stream violet
violet, sweet → sweet violet
violet, trailing yellow → trailing yellow violet
violet, twinflower → twinflower violet
violet, valley → valley violet
violet, western Canada → western Canada violet
violet, yellow montane → yellow montane violet
violet, yellow prairie → yellow prairie violet
violet, yellow sagebrush → yellow sagebrush violet
violet, yellow wood → yellow wood violet
viper's bugloss = Echium vulgare (V:D)
virginia dwarf dandelion = Krigia virginica (V:D)
Virginia grape fern = Botrychium virginianum (V:F)
viviparous fescue = Festuca viviparoidea (V:M)

W

wahoo, western → western wahoo
wall alyssum = Alyssum murale (V:D)
wall barley = Hordeum murinum (V:M)
wall hawkweed = Hieracium murorum (V:D)
wall-lettuce = Mycelis muralis (V:D)
wall speedwell = Veronica arvensis (V:D)
Wallace's selaginella = Selaginella wallacei (V:F)
wallflower, cascade → cascade wallflower
wallflower, common → common wallflower
wallflower, Edwards' → Edwards' wallflower
wallflower, Pallas' → Pallas' wallflower
wallflower, sand-dwelling → sand-dwelling wallflower
wallflower, small → small wallflower
wapato = Sagittaria latifolia (V:M)
Warnstorf's peat moss = Sphagnum warnstorfii (B:M)
wart lichen = Melanelia exasperatula (L)
water arum = Calla palustris (V:M)
water avens = Geum rivale (V:D)
water birch = Betula occidentalis (V:D)
water bur-reed = Sparganium fluctuans (V:M)
water-buttercup, Lobb's → Lobb's water-buttercup
water-buttercup, small yellow → small yellow water-buttercup
water-buttercup, white → white water-buttercup
water-buttercup, yellow → yellow water-buttercup
water-carpet, northern → northern water-carpet
water chickweed = Myosoton aquaticum (V:D)
water chickweed = Montia fontana (V:D)
water-clover, hairy → hairy water-clover
water clubrush = Scirpus subterminalis (V:M)
water-crowfoot, white → white water-crowfoot
water-crowfoot, yellow → yellow water-crowfoot
water hairgrass = Catabrosa aquatica (V:M)
water-hemlock = Cicuta douglasii (V:D)
water-hemlock = Cicuta maculata var. angustifolia (V:D)
water-hemlock, bulb-bearing → bulb-bearing water-hemlock
water-hemlock, bulbous → bulbous water-hemlock
water-hemlock, Douglas' → Douglas' water-hemlock
water-hemlock, European → European water-hemlock
water-hemlock, spotted → spotted water-hemlock
water hook moss = Warnstorfia fluitans (B:M)
water horsetail = Equisetum fluviatile (V:F)
water-lily, banana → banana water-lily
water-lily, European white → European white water-lily
water-lily, fragrant → fragrant water-lily
water-lily, pygmy → pygmy water-lily
water-lily, yellow → yellow water-lily
water lobelia = Lobelia dortmanna (V:D)
water meadow-foxtail = Alopecurus geniculatus (V:M)
water-meal = Wolffia columbiana (V:M)
water-meal, northern → northern water-meal
water-milfoil, Brazilian → Brazilian water-milfoil
water-milfoil, Eurasian → Eurasian water-milfoil
water-milfoil, Farwell's → Farwell's water-milfoil
water-milfoil, Siberian → Siberian water-milfoil
water-milfoil, Ussurian → Ussurian water-milfoil
water-milfoil, vari-leaved → vari-leaved water-milfoil
water-milfoil, verticillate → verticillate water-milfoil
water-milfoil, waterwort → waterwort water-milfoil
water-milfoil, western → western water-milfoil
water-milfoil, whorled → whorled water-milfoil
water mitten-leaf wort = Scapania undulata (B:H)
water mudwort = Limosella aquatica (V:D)
water-parsley = Oenanthe sarmentosa (V:D)
water-parsley, Pacific → Pacific water-parsley
water-parsnip = Sium suave (V:D)
water-parsnip, cut-leaved → cut-leaved water-parsnip
water-parsnip, hemlock → hemlock water-parsnip
water-pepper = Polygonum hydropiperoides (V:D)
water-plantain buttercup = Ranunculus alismifolius (V:D)
water-plantain, narrow-leaved → narrow-leaved water-plantain
water-purslane = Ludwigia palustris (V:D)
water sedge = Carex aquatilis (V:M)
water-shield = Brasenia schreberi (V:D)
water smartweed = Polygonum amphibium (V:D)
water speedwell = Veronica anagallis-aquatica (V:D)
water-starwort, diverse-leaved → diverse-leaved water-starwort
water-starwort, greater → greater water-starwort
water-starwort, northern → northern water-starwort
water-starwort, pond → pond water-starwort
water-starwort, spring → spring water-starwort
water-starwort, winged → winged water-starwort
waterfan = Hydrothyria venosa (L)
waterleaf, bullhead → bullhead waterleaf
waterleaf, Fendler's → Fendler's waterleaf
waterleaf, Pacific → Pacific waterleaf
waterside feather moss = Brachythecium rivulare (B:M)
waterweed, Brazilian → Brazilian waterweed
waterweed, Canadian → Canadian waterweed
waterweed, Nuttall's → Nuttall's waterweed
waterworm = Siphula ceratites (L)
waterwort water-milfoil = Myriophyllum quitense (V:D)
waterwort, three-flowered → three-flowered waterwort
Watson's cryptantha = Cryptantha watsonii (V:D)
wavy dicranum = Dicranum undulatum (B:M)
wavy dicranum = Dicranum polysetum (B:M)
wavy hairgrass = Deschampsia flexuosa (V:M)
wavy-leaved cotton moss = Plagiothecium undulatum (B:M)
wavy-leaved moss = Dicranum polysetum (B:M)
wavy-leaved thistle = Cirsium undulatum (V:D)
wavy water nymph = Najas flexilis (V:M)
wax-flower = Moneses uniflora (V:D)
wax-myrtle, California → California wax-myrtle
waxberry = Symphoricarpos albus (V:D)
waxpaper lichen = Parmelia sulcata (L)
weak arctic sedge = Carex supina (V:M)
weathered ruffle = Tuckermannopsis platyphylla (L)
weatherglass, poor man's → poor man's weatherglass
wedge-leaved saxifrage = Saxifraga adscendens (V:D)
wedgegrass, prairie → prairie wedgegrass
wedgescalf orache = Atriplex truncata (V:D)
weed, crunch → crunch weed

weed, pahute → pahute weed
weed, poverty → poverty weed
weed, sour → sour weed
weed, willow → willow weed
weeping alkaligrass = Puccinellia distans (V:M)
weeping bluegrass = Poa marcida (V:M)
Weigel's leather lichen = Sticta weigelii (L)
Weiser milk-vetch = Astragalus beckwithii var. weiserensis (V:D)
weissia, black-fruited → black-fruited weissia
western aster = Aster occidentalis (V:D)
western bitter-cress = Cardamine occidentalis (V:D)
western blue iris = Iris missouriensis (V:M)
western bog-laurel = Kalmia microphylla (V:D)
western burnet = Sanguisorba occidentalis (V:D)
western buttercup = Ranunculus occidentalis (V:D)
western Canada violet = Viola canadensis (V:D)
western centaury = Centaurium exaltatum (V:D)
western cliff fern = Woodsia oregana (V:F)
western dogwood = Cornus nuttallii (V:D)
western dune tansy = Tanacetum bipinnatum (V:D)
western dwarf mistletoe = Arceuthobium campylopodum (V:D)
western fairy-candelabra = Androsace occidentalis (V:D)
western fescue = Festuca occidentalis (V:M)
western fir clubmoss = Huperzia occidentalis (V:F)
western flowering dogwood = Cornus nuttallii (V:D)
western goldenrod = Euthamia occidentalis (V:D)
western groundsel = Senecio integerrimus (V:D)
western hawksbeard = Crepis occidentalis (V:D)
western hedysarum = Hedysarum occidentale (V:D)
western hemlock = Tsuga heterophylla (V:G)
western jewelweed = Impatiens noli-tangere (V:D)
western larch = Larix occidentalis (V:G)
western lilaeopsis = Lilaeopsis occidentalis (V:D)
western mannagrass = Glyceria occidentalis (V:M)
western meadowrue = Thalictrum occidentale (V:D)
western mountain-ash = Sorbus scopulina (V:D)
western mountain aster = Aster occidentalis (V:D)
western mountainbells = Stenanthium occidentale (V:M)
western mugwort = Artemisia ludoviciana (V:D)
western needle grass = Stipa occidentalis (V:M)
western oak fern = Gymnocarpium disjunctum (V:F)
western paintbrush = Castilleja occidentalis (V:D)
western pasqueflower = Pulsatilla occidentalis (V:D)
western pearlwort = Sagina decumbens ssp. occidentalis (V:D)
western polypody = Polypodium hesperium (V:F)
western quillwort = Isoetes occidentalis (V:F)
western rattlesnake-root = Prenanthes alata (V:D)
western redcedar = Thuja plicata (V:G)
western saxifrage = Saxifraga occidentalis (V:D)
western snowberry = Symphoricarpos occidentalis (V:D)
western spring-beauty = Claytonia lanceolata (V:D)
western sweet-cicely = Osmorhiza occidentalis (V:D)
western sweet-vetch = Hedysarum occidentale (V:D)
western tansymustard = Descurainia pinnata (V:D)
western tarpaper = Collema subparvum (L)
western teaberry = Gaultheria ovatifolia (V:D)
western touch-me-not = Impatiens ecalcarata (V:D)
western trillium = Trillium ovatum (V:M)
western trumpet honeysuckle = Lonicera ciliosa (V:D)
western wahoo = Euonymus occidentalis (V:D)
western water-milfoil = Myriophyllum hippuroides (V:D)
western white pine = Pinus monticola (V:G)
western yellow cress = Rorippa curvisiliqua (V:D)
western yew = Taxus brevifolia (V:G)
wet rock moss = Dichodontium pellucidum (B:M)
wheat = Triticum aestivum (V:M)
wheatgrass, annual → annual wheatgrass
wheatgrass, crested → crested wheatgrass
wheatgrass, desert → desert wheatgrass
wheatgrass, Siberian → Siberian wheatgrass
wheatgrass, slender → slender wheatgrass
Wheeler's bluegrass = Poa nervosa (V:M)
whip fork moss = Dicranum flagellare (B:M)
whiplash saxifrage = Saxifraga flagellaris (V:D)
whiskered shadow = Phaeophyscia hispidula (L)
whiskered shadow = Phaeophyscia kairamoi (L)
whiskered tarpaper = Leptochidium albociliatum (L)
whiskers, old man's → old man's whiskers
white beak-rush = Rhynchospora alba (V:M)
white bedstraw = Galium mollugo (V:D)
white birch = Betula papyrifera var. papyrifera (V:D)
white birch = Betula papyrifera (V:D)
white bog orchid = Platanthera dilatata (V:M)
white cinquefoil = Potentilla arguta ssp. convallaria (V:D)
white cinquefoil = Potentilla arguta (V:D)
white clematis = Clematis ligusticifolia (V:D)
white clover = Trifolium repens (V:D)
white crane's-bill = Geranium richardsonii (V:D)
white death-camas = Zigadenus elegans (V:M)
white-edged scale = Psora decipiens (L)
white fawn lily = Erythronium oregonum (V:M)
white-flowered hawkweed = Hieracium albiflorum (V:D)
white-flowered rhododendron = Rhododendron albiflorum (V:D)
white-flowered willowherb = Epilobium lactiflorum (V:D)
white geranium = Geranium richardsonii (V:D)
white glacier lily = Erythronium montanum (V:M)
white-grained mountain ricegrass = Oryzopsis asperifolia (V:M)
white hawkweed = Hieracium albiflorum (V:D)
white heath aster = Aster ericoides (V:D)
white-margin knotweed = Polygonum polygaloides (V:D)
white marsh-marigold = Caltha leptosepala (V:D)
white meconella = Meconella oregana (V:D)
white mignonette = Reseda alba (V:D)
white moss heather = Cassiope mertensiana (V:D)
white mountain-avens = Dryas octopetala (V:D)
white mountain-heather = Cassiope mertensiana (V:D)
white mulberry = Morus alba (V:D)
white pigweed = Amaranthus albus (V:D)
white-pole evening-primrose = Oenothera pallida (V:D)
white poplar = Populus tremuloides (V:D)
white rein orchid = Platanthera dilatata (V:M)
white rhododendron = Rhododendron albiflorum (V:D)
white sedge = Carex canescens (V:M)
white shootingstar = Dodecatheon dentatum (V:D)
white spruce = Picea glauca (V:G)
white-stemmed evening star = Mentzelia albicaulis (V:D)
white-stemmed gooseberry = Ribes inerme (V:D)
white-stemmed mentzelia = Mentzelia albicaulis (V:D)
white stonecrop = Sedum album (V:D)
white-tipped clover = Trifolium variegatum (V:D)
white-toothed peat moss = Sphagnum girgensohnii (B:M)
white-top aster = Aster curtus (V:D)
white triteleia = Triteleia hyacinthina (V:M)
white-veined wintergreen = Pyrola picta (V:D)
white virgin's bower = Clematis ligusticifolia (V:D)
white water-buttercup = Ranunculus aquatilis (V:D)
white water-crowfoot = Ranunculus aquatilis (V:D)
white-whorl lupine = Lupinus densiflorus (V:D)
white willow = Salix alba (V:D)
white willow = Salix alba var. vitellina (V:D)
white wintergreen = Pyrola elliptica (V:D)
whitebark candlewax = Ahtiana sphaerosporella (L)
whitebark pine = Pinus albicaulis (V:G)
Whited's halimolobos = Halimolobos whitedii (V:D)
whiteman's foot = Plantago major (V:D)
whitish rush = Juncus albescens (V:M)
whitlow-grass = Draba verna (V:D)
whitlow-grass, Alaska → Alaska whitlow-grass
whitlow-grass, alpine → alpine whitlow-grass
whitlow-grass, annual → annual whitlow-grass
whitlow-grass, Austrian → Austrian whitlow-grass
whitlow-grass, Baffin's Bay → Baffin's Bay whitlow-grass
whitlow-grass, Carolina → Carolina whitlow-grass
whitlow-grass, Coast Mountain → Coast Mountain whitlow-grass

whitlow-grass, common → common whitlow-grass
whitlow-grass, few-seeded → few-seeded whitlow-grass
whitlow-grass, golden → golden whitlow-grass
whitlow-grass, gray-leaved → gray-leaved whitlow-grass
whitlow-grass, lance-fruited → lance-fruited whitlow-grass
whitlow-grass, lance-leaved → lance-leaved whitlow-grass
whitlow-grass, long-stalked → long-stalked whitlow-grass
whitlow-grass, Macoun's → Macoun's whitlow-grass
whitlow-grass, milky → milky whitlow-grass
whitlow-grass, North Pacific → North Pacific whitlow-grass
whitlow-grass, northern → northern whitlow-grass
whitlow-grass, Nuttall's → Nuttall's whitlow-grass
whitlow-grass, Payson's → Payson's whitlow-grass
whitlow-grass, Porsild's → Porsild's whitlow-grass
whitlow-grass, Rocky Mountain → Rocky Mountain whitlow-grass
whitlow-grass, slender → slender whitlow-grass
whitlow-grass, smooth → smooth whitlow-grass
whitlow-grass, snow → snow whitlow-grass
whitlow-grass, stender → stender whitlow-grass
whitlow-grass, tall → tall whitlow-grass
whitlow-grass, thick-leaved → thick-leaved whitlow-grass
whitlow-grass, woods → woods whitlow-grass
whitlow-grass, Yellowstone → Yellowstone whitlow-grass
whorled cup lichen = Cladonia cervicornis ssp. verticillata (L)
whorled lousewort = Pedicularis verticillata (V:D)
whorled marsh pennywort = Hydrocotyle verticillata (V:D)
whorled plantain = Plantago psyllium (V:D)
whorled water-milfoil = Myriophyllum verticillatum (V:D)
widgeon-grass = Ruppia maritima (V:M)
wild bergamot = Monarda fistulosa (V:D)
wild bergamot = Monarda fistulosa var. mollis (V:D)
wild blue flax = Linum lewisii (V:D)
wild calla = Calla palustris (V:M)
wild carrot = Daucus carota (V:D)
wild chamomile = Matricaria recutita (V:D)
wild chives = Allium schoenoprasum (V:M)
wild cucumber = Echinocystis lobata (V:D)
wild ginger = Asarum caudatum (V:D)
wild gooseberry = Ribes divaricatum (V:D)
wild-licorice, northern → northern wild-licorice
wild lily-of-the-valley = Maianthemum canadense (V:M)
wild marjoram = Origanum vulgare (V:D)
wild mint = Mentha arvensis (V:D)
wild oat = Avena fatua (V:M)
wild radish = Raphanus raphanistrum (V:D)
wild red currant = Ribes triste (V:D)
wild sarsaparilla = Aralia nudicaulis (V:D)
wild strawberry = Fragaria virginiana (V:D)
wild vetch = Vicia americana (V:D)
wildrye, Alaska → Alaska wildrye
wildrye, blue → blue wildrye
wildrye, Canada → Canada wildrye
wildrye, creeping → creeping wildrye
wildrye, dune → dune wildrye
wildrye, fuzzy-spiked → fuzzy-spiked wildrye
wildrye, giant → giant wildrye
wildrye, hairy → hairy wildrye
wildrye, Saunder's → Saunder's wildrye
wildrye, thickspike → thickspike wildrye
William, sweet → sweet William
willow dock = Rumex salicifolius (V:D)
willow weed = Polygonum lapathifolium (V:D)
willow, Alaska → Alaska willow
willow, arctic → arctic willow
willow, Athabasca → Athabasca willow
willow, autumn → autumn willow
willow, balsam → balsam willow
willow, Barclay's → Barclay's willow
willow, Barratt's → Barratt's willow
willow, beaked → beaked willow
willow, Bebb's → Bebb's willow
willow, bilberry → bilberry willow
willow, blue-green → blue-green willow
willow, blueberry → blueberry willow
willow, bog → bog willow
willow, Booth's → Booth's willow
willow, Cascade → Cascade willow
willow, coyote → coyote willow
willow, creeping → creeping willow
willow, diamond-leaved → diamond-leaved willow
willow, Drummond's → Drummond's willow
willow, dusky → dusky willow
willow, dwarf snow → dwarf snow willow
willow, Farr's → Farr's willow
willow, felt-leaf → felt-leaf willow
willow, flat-leaved → flat-leaved willow
willow, Geyer's → Geyer's willow
willow, glabrous dwarf → glabrous dwarf willow
willow, grey-leaved → grey-leaved willow
willow, hoary → hoary willow
willow, Hooker's → Hooker's willow
willow, Lemmon's → Lemmon's willow
willow, little-tree → little-tree willow
willow, low blueberry → low blueberry willow
willow, MacCalla's → MacCalla's willow
willow, Mackenzie's → Mackenzie's willow
willow, meadow → meadow willow
willow, mountain → mountain willow
willow, myrtle-leaved → myrtle-leaved willow
willow, narrow-leaved → narrow-leaved willow
willow, net-veined → net-veined willow
willow, netted → netted willow
willow, northern bush → northern bush willow
willow, Pacific → Pacific willow
willow, peach-leaf → peach-leaf willow
willow, plane-leaved → plane-leaved willow
willow, polar → polar willow
willow, pussy → pussy willow
willow, Raup's → Raup's willow
willow, Richardson's → Richardson's willow
willow, rock → rock willow
willow, sage → sage willow
willow, sandbar → sandbar willow
willow, Scouler's → Scouler's willow
willow, serviceberry → serviceberry willow
willow, sessile-leaved sandbar → sessile-leaved sandbar willow
willow, Setchell's → Setchell's willow
willow, shining → shining willow
willow, short-fruited → short-fruited willow
willow, shrubby → shrubby willow
willow, Sitka → Sitka willow
willow, smooth → smooth willow
willow, snow → snow willow
willow, soft-leaved sandbar → soft-leaved sandbar willow
willow, stoloniferous → stoloniferous willow
willow, tail-leaved → tail-leaved willow
willow, tea-leaved → tea-leaved willow
willow, Tweedy's → Tweedy's willow
willow, undergreen → undergreen willow
willow, variable → variable willow
willow, velvet-fruited → velvet-fruited willow
willow, white → white willow
willow, woolly → woolly willow
willowherb, alpine → alpine willowherb
willowherb, broad-leaved → broad-leaved willowherb
willowherb, club-fruited → club-fruited willowherb
willowherb, foliose → foliose willowherb
willowherb, hairy → hairy willowherb
willowherb, hairy-stemmed → hairy-stemmed willowherb
willowherb, Hall's → Hall's willowherb
willowherb, Hornemann's → Hornemann's willowherb

willowherb, marsh → marsh willowherb
willowherb, narrow-leaved → narrow-leaved willowherb
willowherb, Oregon → Oregon willowherb
willowherb, purple-leaved → purple-leaved willowherb
willowherb, Rocky Mountain → Rocky Mountain willowherb
willowherb, rosebay → rosebay willowherb
willowherb, small-flowered → small-flowered willowherb
willowherb, smooth → smooth willowherb
willowherb, swamp → swamp willowherb
willowherb, tall annual → tall annual willowherb
willowherb, white-flowered → white-flowered willowherb
willowherb, yellow → yellow willowherb
windward rocktripe = Umbilicaria lambii (L)
winged combseed = Pectocarya penicillata (V:D)
winged loosestrife = Lythrum alatum (V:D)
winged water-starwort = Callitriche marginata (V:D)
winter bentgrass = Agrostis scabra (V:M)
winter rape = Brassica napus (V:D)
wintergreen, alpine → alpine wintergreen
wintergreen, arctic → arctic wintergreen
wintergreen, common pink → common pink wintergreen
wintergreen, green → green wintergreen
wintergreen, greenish → greenish wintergreen
wintergreen, large → large wintergreen
wintergreen, lesser → lesser wintergreen
wintergreen, one-flowered → one-flowered wintergreen
wintergreen, one-sided → one-sided wintergreen
wintergreen, Oregon → Oregon wintergreen
wintergreen, pink → pink wintergreen
wintergreen, white → white wintergreen
wintergreen, white-veined → white-veined wintergreen
wire lettuce = Stephanomeria tenuifolia (V:D)
wire rush = Juncus balticus (V:M)
wirestem muhly = Muhlenbergia mexicana (V:M)
wiry fern moss = Abietinella abietina (B:M)
wiry knotweed = Polygonum douglasii ssp. majus (V:D)
witchgrass, common → common witchgrass
witchgrass, smooth → smooth witchgrass
woad, dyer's → dyer's woad
wolf lichen = Letharia vulpina (L)
Wolf's trisetum = Trisetum wolfii (V:M)
wolf-willow = Elaeagnus commutata (V:D)
wood bluegrass = Poa nemoralis (V:M)
wood forget-me-not = Myosotis sylvatica (V:D)
wood groundsel = Senecio sylvaticus (V:D)
wood horsetail = Equisetum sylvaticum (V:F)
wood lily = Lilium philadelphicum (V:M)
wood-reed, droping → droping wood-reed
wood-reed, nodding → nodding wood-reed
wood reedgrass = Cinna latifolia (V:M)
wood sage = Salvia nemorosa (V:D)
wood saxifrage = Saxifraga mertensiana (V:D)
wood strawberry = Fragaria vesca (V:D)
woodland horsetail = Equisetum sylvaticum (V:F)
woodland owl = Solorina saccata (L)
woodland penstemon = Nothochelone nemorosa (V:D)
woodland strawberry = Fragaria vesca (V:D)
woodland tarweed = Madia madioides (V:D)
woodruff, sweet → sweet woodruff
woodrush pussytoes = Antennaria luzuloides (V:D)
woodrush, arctic → arctic woodrush
woodrush, confused → confused woodrush
woodrush, curved alpine → curved alpine woodrush
woodrush, field → field woodrush
woodrush, Greenland → Greenland woodrush
woodrush, many-flowered → many-flowered woodrush
woodrush, Piper's → Piper's woodrush
woodrush, rusty → rusty woodrush
woodrush, small-flowered → small-flowered woodrush
woodrush, spiked → spiked woodrush
Woods' rose = Rosa woodsii (V:D)
woods draba = Draba nemorosa (V:D)
woods whitlow-grass = Draba nemorosa (V:D)
woodsia, northern → northern woodsia
woodsia, Oregon → Oregon woodsia
woodsia, Rocky Mountain → Rocky Mountain woodsia
woodsia, rusty → rusty woodsia
woodsia, smooth → smooth woodsia
woodsy leafy moss = Plagiomnium cuspidatum (B:M)
woodsy mnium = Plagiomnium cuspidatum (B:M)
woodsy ragged moss = Brachythecium hylotapetum (B:M)
woolly caterpillar moss = Pterygoneurum subsessile (B:M)
woolly cinquefoil = Potentilla hippiana (V:D)
woolly coral = Stereocaulon tomentosum (L)
woolly daisy = Erigeron lanatus (V:D)
woolly distaff-thistle = Carthamus lanatus (V:D)
woolly eriophyllum = Eriophyllum lanatum (V:D)
woolly groundsel = Senecio canus (V:D)
woolly hawkweed = Hieracium triste (V:D)
woolly-heads, slender → slender woolly-heads
woolly-heads, tall → tall woolly-heads
woolly lousewort = Pedicularis lanata (V:D)
woolly mullein = Verbascum phlomoides (V:D)
woolly plantain = Plantago patagonica (V:D)
woolly pussytoes = Antennaria lanata (V:D)
woolly sedge = Carex lanuginosa (V:M)
woolly sunflower = Eriophyllum lanatum (V:D)
woolly vetch = Vicia villosa (V:D)
woolly willow = Salix lanata (V:D)
woollypod milk-vetch = Astragalus purshii (V:D)
worm-leaved stonecrop = Sedum stenopetalum (V:D)
wormseed mustard = Erysimum cheiranthoides (V:D)
wormwood = Artemisia absinthium (V:D)
wormwood, biennial → biennial wormwood
wormwood, common → common wormwood
wormwood, northern → northern wormwood
wormwood, plains → plains wormwood
wormwood, three-fork → three-fork wormwood
wort, cedar-shake → cedar-shake wort
wort, dull scale feather → dull scale feather wort
wort, green tongue → green tongue wort
wort, Hatcher's fan → Hatcher's fan wort
wort, orange talus → orange talus wort
wort, rusty rock → rusty rock wort
wort, slender two-toothed → slender two-toothed wort
wort, water mitten-leaf → water mitten-leaf wort
wort, yellow double-leaf → yellow double-leaf wort
woundwort = Stachys palustris (V:D)
wreath lichen = Phaeophyscia orbicularis (L)
Wright's golden carpet = Chrysosplenium wrightii (V:D)
Wright's golden-saxifrage = Chrysosplenium wrightii (V:D)
Wright's leather lichen = Sticta wrightii (L)
wrinkled vinyl = Leptogium schraderi (L)
Wyeth's lupine = Lupinus wyethii (V:D)
Wyoming kitten-tails = Besseya wyomingensis (V:D)

yampah = Perideridia gairdneri (V:D)
yampah, Gairdner's → Gairdner's yampah

yarrow = Achillea millefolium (V:D)
yarrow, common → common yarrow
yarrow, hoary false → hoary false yarrow
yarrow, many-flowered → many-flowered yarrow
yarrow, Siberian → Siberian yarrow
yellow anemone = Anemone richardsonii (V:D)
yellow avens = Geum aleppicum (V:D)
yellow bedstraw = Galium verum (V:D)
yellow bell = Fritillaria pudica (V:M)
yellow bog sedge = Carex gynocrates (V:M)
yellow bristlegrass = Setaria glauca (V:M)
yellow buckwheat = Eriogonum flavum (V:D)
yellow cedar = Chamaecyparis nootkatensis (V:G)
yellow chamomile = Anthemis tinctoria (V:D)
yellow clover = Trifolium aureum (V:D)
yellow collar moss = Splachnum luteum (B:M)
yellow columbine = Aquilegia flavescens (V:D)
yellow coralroot = Corallorrhiza trifida (V:M)
yellow cypress = Chamaecyparis nootkatensis (V:G)
yellow double-leaf wort = Diplophyllum taxifolium (B:H)
yellow evening-primrose = Oenothera villosa (V:D)
yellow evening-primrose = Oenothera biennis (V:D)
yellow-flower locoweed = Oxytropis monticola (V:D)
yellow-flowered knotweed = Polygonum ramosissimum (V:D)
yellow-flowered sedge = Carex anthoxanthea (V:M)
yellow fritillary = Fritillaria pudica (V:M)
yellow-fruited sedge = Carex flava (V:M)
yellow glacier lily = Erythronium grandiflorum (V:M)
yellow-green cushion moss = Dicranoweisia crispula (B:M)
yellow-green peat moss = Sphagnum angustifolium (B:M)
yellow-green rock moss = Racomitrium heterostichum (B:M)
yellow gromwell = Lithospermum incisum (V:D)
yellow hedysarum = Hedysarum sulphurescens (V:D)
yellow iris = Iris pseudacorus (V:M)
yellow-ladle liverwort = Scapania bolanderi (B:H)
yellow lichen = Vulpicida tilesii (L)
yellow loosestrife = Lysimachia vulgaris (V:D)
yellow marsh-marigold = Caltha palustris (V:D)
yellow marsh saxifrage = Saxifraga hirculus (V:D)
yellow mignonette = Reseda lutea (V:D)
yellow monkey-flower = Mimulus guttatus (V:D)
yellow montane violet = Viola praemorsa (V:D)
yellow moss = Homalothecium fulgescens (B:M)
yellow mountain-avens = Dryas drummondii (V:D)
yellow mountain-heather = Phyllodoce glanduliflora (V:D)
yellow owl-clover = Orthocarpus luteus (V:D)
yellow oxalis = Oxalis corniculata (V:D)
yellow paintbrush = Castilleja lutescens (V:D)
yellow parentucellia = Parentucellia viscosa (V:D)
yellow penstemon = Penstemon confertus (V:D)
yellow pine = Pinus ponderosa (V:G)
yellow prairie violet = Viola vallicola var. major (V:D)
yellow rag = Esslingeriana idahoensis (L)
yellow rattle = Rhinanthus minor (V:D)
yellow reindeer lichen = Cladina mitis (L)
yellow sagebrush violet = Viola vallicola var. major (V:D)
yellow salsify = Tragopogon dubius (V:D)
yellow sand-verbena = Abronia latifolia (V:D)
yellow saxifraga = Saxifraga aizoides (V:D)
yellow scabious = Scabiosa ochroleuca (V:D)
yellow sedge = Carex flava (V:M)
yellow specklebelly = Pseudocyphellaria aurata (L)
yellow specklebelly = Pseudocyphellaria crocata (L)
yellow star moss = Campylium stellatum (B:M)
yellow starry feather moss = Campylium stellatum (B:M)
yellow sweet-clover = Melilotus officinalis (V:D)
yellow sweet-vetch = Hedysarum sulphurescens (V:D)
yellow water-buttercup = Ranunculus flabellaris (V:D)
yellow water-crowfoot = Ranunculus gmelinii (V:D)
yellow water-lily = Nuphar lutea (V:D)
yellow willowherb = Epilobium luteum (V:D)
yellow wood violet = Viola glabella (V:D)
Yellowstone draba = Draba incerta (V:D)
Yellowstone whitlow-grass = Draba incerta (V:D)
yerba buena = Satureja douglasii (V:D)
yew, Pacific → Pacific yew
yew, western → western yew
Yorkshire fog = Holcus lanatus (V:M)
youth-on-age = Tolmiea menziesii (V:D)
Yukon bellflower = Campanula aurita (V:D)
Yukon groundsel = Senecio yukonensis (V:D)
Yukon lupine = Lupinus kuschei (V:D)
Yukon saxifrage = Saxifraga reflexa (V:D)

Appendix

Excluded Names

Abronia umbellata Lam.
Abronia umbellata ssp. acutalata (Standl.) Tillett
Achillea filipendulina Lam.
Achillea millefolium var. nigrescens E. Mey.
Achillea ptarmica L.
Aconitum columbianum var. howellii (A. Nels. & J.F. Macbr.) C.L. Hitchc.
Aconitum columbianum var. ochroleucum A. Nels.
Adlumia fungosa (Ait.) Greene ex B.S.P.
Aesculus hippocastanum L.
Aethusa cynapium L.
Agoseris aurantiaca var. purpurea (Gray) Cronq.
Agoseris elata (Nutt.) Greene
Agoseris gaspensis Fern.
Agoseris glauca var. laciniata (D.C. Eat.) Smiley
× Agropogon littoralis (Sm.) C.E. Hubbard
Agropyron cristatum ssp. pectinatum (Bieb.) Tzvelev
Agrostemma githago L.
Agrostis alascana Hultén
Agrostis ampla A.S. Hitchc.
Agrostis canina L.
Agrostis humilis Vasey
Agrostis hyemalis (Walt.) B.S.P.
Agrostis idahoensis Nash
Agrostis lepida A.S. Hitchc.
Agrostis rossiae Vasey
Agrostis thurberiana A.S. Hitchc.
Ailanthus altissima (P. Mill.) Swingle
Alcea rosea L.
Alchemilla xanthochlora Rothm.
Alisma plantago-aquatica L.
Alisma subcordatum Raf.
Allium cepa L.
Allium stellatum Nutt. ex Ker-Gawl.
Alnus incana ssp. rugosa (Du Roi) Clausen
Alnus rhombifolia Nutt.
Alopecurus myosuroides Huds.
Alopecurus saccatus Vasey
Althaea hirsuta L.
Amaranthus cruentus L.
Amaranthus hybridus L.
Amaranthus obcordatus (Gray) Standl.
Amberboa moschata (L.) DC.
Ambrosia psilostachya DC.
Ambrosia trifida L.
Ammannia coccinea Rottb.
Ammannia robusta Heer & Regel
Amsinckia tessellata Gray
Anchusa azurea P. Mill.
Andreaea obovata Thed.
Anemone deltoidea Hook.
Anemone multifida var. hudsoniana DC.
Anemone multifida var. sansonii Boivin
Anemone multifida var. saxicola Boivin
Anemone oregana Gray
Anemone virginiana var. cylindroidea Boivin
Antennaria alpina (L.) Gaertn.
Antennaria aromatica Evert
Antennaria corymbosa E. Nels.
Antennaria × erigeroides Greene (pro sp.)
Antennaria howellii ssp. petaloidea (Fern.) Bayer
Antennaria rosea ssp. arida (E. Nels.) Bayer
Antennaria rosea ssp. confinis (Greene) Bayer
Antennaria rosea ssp. pulvinata (Greene) Bayer
Antennaria stenophylla Gray
Anthriscus cerefolium (L.) Hoffmann
Antirrhinum majus L.
Antitrichia curtipendula var. gigantea Sull. & Lesq.
Apocynum androsaemifolium ssp. androsaemifolium var. griseum (Greene) Bég. & Bel.
Apocynum androsaemifolium ssp. androsaemifolium var. incanum A. DC.
Apocynum androsaemifolium ssp. pumilum var. tomentellum (Greene) Boivin
Apocynum androsaemifolium ssp. pumilum var. woodsonii Boivin
Aquilegia flavescens var. miniata A. Nels. & J.F. Macbr.
Aquilegia vulgaris L.
Arabis microphylla var. macounii (S. Wats.) Rollins
Arabis sparsiflora var. columbiana (Macoun) Rollins
Arabis sparsiflora var. subvillosa (S. Wats.) Rollins
Arceuthobium laricis (Piper) St. John
Arceuthobium tsugense (Rosendahl) G.N. Jones
Arctium vulgare (Hill) Evans
Arctostaphylos × media Greene (pro sp.)
Arctostaphylos tomentosa ssp. subcordata (Eastw.) P.V. Wells
Arenaria congesta Nutt.
Arenaria congesta var. subcongesta (S. Wats.) S. Wats.
Arenaria humifusa Wahlenb.
Argemone mexicana L.
Aristida adscensionis L.
Aristida longespica Poir.
Arnica alpina (L.) Olin
Arnica chamissonis ssp. chamissonis var. interior Maguire
Arnica chamissonis ssp. foliosa var. andina (Nutt.) Ediger & Barkl.
Arnica discoidea Benth.
Arnica louiseana Farr
Artemisia arctica ssp. comata (Rydb.) Hultén
Artemisia douglasiana Bess.
Artemisia dracunculus ssp. glauca (Pallas ex Willd.) Hall & Clements
Artemisia hookeriana Bess.
Artemisia stelleriana Bess.
Aruncus dioicus var. acuminatus (Rydb.) Rydb. ex Hara
Asclepias viridiflora Raf.
Asperula arvensis L.
Asplenium polyodon var. subcaudatum (Skottsberg) Morton
Asplenium scolopendrium L.
Asplenium scolopendrium var. americanum (Fern.) Kartesz & Gandhi
Asplenium trichomanes ssp. quadrivalens D.E. Mey.
Aster chilensis Nees
Aster ciliolatus var. wilsonii (Rydb.) A.G. Jones
Aster cusickii Gray
Aster foliaceus var. apricus Gray
Aster foliaceus var. parryi (D.C. Eat.) Gray
Aster hallii Gray
Aster hendersonii Fern.
Aster novi-belgii L.
Aster novi-belgii var. tardiflorus (L.) A.G. Jones
Aster praealtus Poir.
Aster yukonensis Cronq.
Asterella saccata (Wahlenb.) Evans
Astragalus crassicarpus var. paysonii (E.H. Kelso) Barneby
Astragalus flexuosus (Hook.) Dougl. ex G. Don
Astragalus lyallii Gray
Astragalus miser var. decumbens (Nutt. ex Torr. & Gray) Cronq.
Astragalus obcordatus Ell.
Astragalus robbinsii var. occidentalis S. Wats.
Astragalus sinuatus Piper
Astragalus subcinereus Gray
Astragalus trichopodus (Nutt.) Gray
Astragalus trichopodus var. lonchus (M.E. Jones) Barneby
Astragalus vexilliflexus Sheldon
Asyneuma prenanthoides (Dur.) McVaugh
Athyrium filix-femina ssp. angustum (Willd.) Clausen
Atriplex littoralis L.
Atriplex prostrata Bouchér ex DC.
Atriplex semibaccata R. Br.
Avena strigosa Schreb.
Balsamita major Desf.
Balsamorhiza careyana Gray
Balsamorhiza careyana var. intermedia Cronq.

Balsamorhiza hookeri Nutt.
Balsamorhiza hookeri var. hirsuta (Nutt.) A. Nels.
Besseya rubra (Dougl. ex Hook.) Rydb.
Beta vulgaris L.
Betula papyrifera var. kenaica (W.H. Evans) A. Henry
Betula pumila var. renifolia Fern.
Betula × utahensis Britt. (pro sp.)
Blepharipappus scaber Hook.
Botrychium ascendens W.H. Wagner
Botrychium crenulatum W.H. Wagner
Botrychium hesperium (Maxon & Clausen) W.H. Wagner & Lellinger
Botrychium matricariifolium (A. Braun ex Dowell) A. Braun ex Koch
Botrychium montanum W.H. Wagner
Botrychium pedunculosum W.H. Wagner
Boykinia major Gray
Brachyactis ciliata ssp. laurentiana (Fern.) A.G. Jones
Brachyactis frondosa (Nutt.) A.G. Jones
Brachythecium fendleri (Sull.) Jaeg.
Briza media L.
Bromus arvensis L.
Bromus erectus Huds.
Bromus inermis ssp. pumpellianus var. arcticus (Shear ex Scribn. & Merr.) Wagnon
Bromus marginatus var. breviaristatus (Buckl.) Beetle
Bromus orcuttianus Vasey
Bromus rubens L.
Bromus suksdorfii Vasey
Bromus tectorum var. glabratus Spenner
Bunias orientalis L.
Bupleurum rotundifolium L.
Calamagrostis canadensis var. imberbis (Stebbins) C.L. Hitchc.
Calamagrostis lactea Beal
Calamagrostis purpurascens var. laricina Louis-Marie
Calamintha sylvatica Bromf.
Calamintha sylvatica ssp. ascendens (Jord.) P.W. Ball
Calendula arvensis L.
Calendula officinalis L.
Calochortus elegans Pursh
Calypso bulbosa var. americana (R. Br. ex Ait. f.) Luer
Calystegia sepium ssp. americana (Sims) Brummitt
Calystegia sepium ssp. angulata Brummitt
Calystegia silvatica (Kit.) Griseb.
Calystegia silvatica ssp. fraterniflora (Mackenzie & Bush) Brummitt
Camassia leichtlinii ssp. suksdorfii (Greenm.) Gould
Camissonia bistorta (Nutt. ex Torr. & Gray) Raven
Camissonia micrantha (Hornem. ex Spreng.) Raven
Campanula parryi Gray
Cannabis sativa L.
Caragana arborescens Lam.
Cardamine cordifolia Gray
Cardamine flexuosa With.
Cardamine nuttallii var. pulcherrima (Greene) Taylor & MacBryde
Cardamine parviflora var. arenicola (Britt.) O.E. Schulz
Cardionema ramosissimum (Weinm.) A. Nels. & J.F. Macbr.
Carduus nutans ssp. macrolepis (Peterm.) Kazmi
Carex aboriginum M.E. Jones
Carex adelostoma Krecz.
Carex adusta Boott
Carex alma Bailey
Carex backii Boott
Carex barbarae Dewey
Carex bella Bailey
Carex bonanzensis Britt.
Carex breweri Boott
Carex × calderi Boivin
Carex capitata ssp. arctogena (H. Sm.) Hiitonen
Carex eleusinoides Turcz. ex C.A. Mey.
Carex foenea var. tuberculata F.J. Herm.
Carex glareosa ssp. glareosa var. amphigena Fern.
Carex × grahamii Boott (pro sp.)
Carex hassei Bailey
Carex haydenii Dewey
Carex hookeriana Dewey
Carex lacustris Willd.
Carex mackenziei Krecz.
Carex macloviana ssp. subfusca (W. Boott) T. Koyama
Carex macrocephala var. bracteata Holm
Carex merritt-fernaldii Mackenzie
Carex microchaeta ssp. nesophila (Holm) E. Murr.
Carex muricata L.
Carex nebrascensis Dewey
Carex norvegica ssp. inserrulata Kalela
Carex pensylvanica Lam.
Carex × physocarpioides Lepage
Carex piperi Mackenzie
Carex platylepis Mackenzie
Carex praeceptorum Mackenzie
Carex projecta Mackenzie
Carex proposita Mackenzie
Carex rariflora (Wahlenb.) Sm.
Carex rotundata Wahlenb.
Carex rugosperma Mackenzie
Carex tincta (Fern.) Fern.
Carex torreyi Tuckerman
Carex tribuloides Wahlenb.
Carex umbellata Schkuhr ex Willd.
Carex vesicaria var. monile (Tuckerman) Fern.
Carex viridula ssp. oedocarpa (Anderss.) B. Schmid
Carex woodii Dewey
Castanea dentata (Marsh.) Borkh.
Castilleja angustifolia (Nutt.) G. Don
Castilleja applegatei Fern.
Castilleja exserta (Heller) Chuang & Heckard
Castilleja flava S. Wats.
Castilleja miniata var. dixonii (Fern.) A. Nels. & J.F. Macbr.
Castilleja parviflora var. olympica (G.N. Jones) Ownbey
Castilleja parviflora var. oreopola (Greenm.) Ownbey
Castilleja septentrionalis Lindl.
Castilleja suksdorfii Gray
Caulanthus pilosus S. Wats.
Centaurea calcitrapa L.
Centaurea jacea L.
Centaurea macrocephala Puschk. ex Willd.
Centaurea nigra L.
Centaurea scabiosa L.
Centaurea transalpina Schleich. ex DC.
Cephalozia bicuspidata ssp. lammersiana (Hub.) Schust.
Cephaloziella dentata (Raddi) Migula
Cerastium aleuticum Hultén
Cerastium alpinum L.
Cerastium arvense ssp. strictum (L.) Ugborogho
Ceratophyllum muricatum Cham.
Chamaedaphne calyculata var. angustifolia (Ait.) Rehd.
Chamaemelum mixtum (L.) All.
Chamaesyce vermiculata (Raf.) House
Chenopodium album var. missouriense (Aellen) I.J. Bassett & C.W. Crompton
Chenopodium berlandieri Moq.
Chenopodium berlandieri var. zschackii (J. Murr) J. Murr ex Aschers.
Chenopodium carnosulum Moq.
Chenopodium carnosulum var. patagonicum (Phil.) H.A. Wahl
Chenopodium fremontii S. Wats.
Chenopodium glaucum L.
Chenopodium hians Standl.
Chenopodium humile Hook.
Chenopodium leptophyllum (Moq.) Nutt. ex S. Wats.
Chenopodium macrospermum Hook. f.
Chenopodium macrospermum var. farinosum (S. Wats.) J.T. Howell
Chenopodium macrospermum var. halophilum (Phil.) Standl.
Chenopodium murale L.
Chenopodium opulifolium Schrad. ex Koch & Ziz
Chenopodium polyspermum L.
Chenopodium polyspermum var. acutifolium (Sm.) Gaud.
Chenopodium subglabrum (S. Wats.) A. Nels.
Chionophila tweedyi (Canby & Rose) Henderson
Chrysosplenium glechomifolium Nutt.

Chrysothamnus nauseosus ssp. graveolens (Nutt.) Piper
Chrysothamnus viscidiflorus ssp. puberulus (D.C. Eat.) Hall & Clements
Cicer arietinum L.
Cinclidium stygium Sw. in Schrad.
Cinclidium subrotundum Lindb.
Cirsium flodmanii (Rydb.) Arthur
Cirsium × vancouverense Moore & Frankton
Cladonia coccifera (L.) Willd.
Clarkia gracilis (Piper) A. Nels. & J.F. Macbr.
Clarkia purpurea (W. Curtis) A. Nels. & J.F. Macbr.
Clarkia purpurea ssp. quadrivulnera (Dougl. ex Lindl.) H.F. & M.E. Lewis
Claytonia lanceolata var. chrysantha (Greene) C.L. Hitchc.
Claytonia saxosa Brandeg.
Claytonia spathulata var. exigua (Torr. & Gray) Hook. & Arn.
Clematis alpina (L.) P. Mill.
Clematis hirsutissima Pursh
Cnicus benedictus L.
Cochlearia officinalis L.
Collomia tenella Gray
Conioselinum chinense (L.) B.S.P.
Coptis occidentalis (Nutt.) Torr. & Gray
Coptis trifolia ssp. groenlandica (Oeder) Hultén
Cordia subcordata Lam.
Corispermum orientale Lam.
Cornus rugosa Lam.
Cornus sericea ssp. occidentalis (Torr. & Gray) Fosberg
Corydalis aurea ssp. occidentalis (Engelm. ex Gray) G.B. Ownbey
Corynephorus canescens (L.) Beauv.
Cotoneaster franchetii Boiss.
Cotoneaster microphyllus Wallich ex Lindl.
Cotula australis (Sieber) Hook. f.
Crataegus chrysocarpa Ashe
Crataegus curvisepala Lindm.
Crataegus laevigata (Poir.) DC.
Crepis biennis L.
Crepis nana ssp. clivicola Leggett
Crepis occidentalis ssp. conjuncta (Jepson) Babcock & Stebbins
Cryptantha circumscissa (Hook. & Arn.) I.M. Johnston
Cryptantha flaccida (Dougl. ex Lehm.) Greene
Cryptantha leucophaea (Dougl. ex Lehm.) Payson
Cryptantha subcapitata Dorn & Lichvar
Ctenidium molluscum (Hedw.) Mitt.
Cuscuta gronovii Willd. ex J.A. Schultes
Cuscuta salina var. major Yuncker
Cynanchum rossicum (Kleopov) Borhidi
Cynoglossum grande Dougl. ex Lehm.
Cyperus eragrostis Lam.
Cyperus erythrorhizos Muhl.
Cyperus esculentus L.
Cypripedium fasciculatum Kellogg ex S. Wats.
Cypripedium pubescens Willd.
Cystopteris laurentiana (Weatherby) Blasdell
Dalea purpurea Vent.
Danthonia parryi Scribn.
Danthonia sericea Nutt.
Delissea subcordata Gaud.
Delphinium nuttallii Gray
Delphinium occidentale (S. Wats.) S. Wats.
Descurainia incana ssp. incisa (Engelm.) Kartesz & Gandhi
Descurainia pinnata ssp. brachycarpa (Richards.) Detling
Desmatodon laureri (Schultz) Bruch & Schimp. in B.S.G.
Dichanthelium dichotomum (L.) Gould
Dichanthelium dichotomum var. ensifolium (Baldw. ex Ell.) Gould & C.A. Clark
Dichanthelium scoparium (Lam.) Gould
Dichelostemma congestum (Sm.) Kunth
Dichelyma pallescens Schimp. in B.S.G.
Dicranum bonjeanii De Not.
Dicranum howellii Ren. & Card.
Dicranum leioneuron Kindb.
Didymodon vinealis var. brachyphyllus (Whipple) Zand.
Digitalis grandiflora P. Mill.
Digitalis lanata Ehrh.
Diplotaxis muralis (L.) DC.
Diplotaxis tenuifolia (L.) DC.
Dipsacus fullonum ssp. sylvestris (Huds.) Clapham
Dodecatheon meadia L.
Doronicum orientale Hoffmann
Draba borealis var. maxima (Hultén) Welsh
Draba macrocarpa M.F. Adams
Draba nemorosa var. leiocarpa Lindbl.
Draba nivalis var. canadica O.E. Schulz
Draba nivalis var. denudata (O.E. Schulz) C.L. Hitchc.
Draba palanderiana Kjellm.
Draba pectinipila Rollins
Draba stenopetala Trautv.
Draba subcapitata Simm.
Draba ventosa Gray
Drosera intermedia Hayne
Drosera linearis Goldie
Dryopteris campyloptera Clarkson
Dryopteris fragrans var. remotiuscula Komarov
Echinochloa colona (L.) Link
Echinochloa muricata (Beauv.) Fern.
Echinochloa muricata var. microstachya Wieg.
Echinops exaltatus Schrad.
Echinops sphaerocephalus L.
Elaeagnus multiflora Thunb.
Elatine triandra Schkuhr
Eleocharis nitida Fern.
Eleocharis ovata (Roth) Roemer & J.A. Schultes
Eleocharis quadrangulata (Michx.) Roemer & J.A. Schultes
× Elyhordeum schaackianum (Bowden) Bowden
× Elyleymus hirtiflorus (A.S. Hitchc.) Barkworth
× Elyleymus jamesensis (Lepage) Barkworth
× Elyleymus uclueletensis (Bowden) Baum
Elymus californicus (Boland. ex Thurb.) Gould
Elymus caninus (L.) L.
Elymus glaucus ssp. jepsonii (Burtt-Davy) Gould
Elymus × hansenii Scribn. (pro sp.)
Elymus × maltei Bowden
Elymus × pseudorepens (Scribn. & J.G. Sm.) Barkworth & Dewey
Elymus sajanensis ssp. hyperarcticus (Polunin) Tzvelev
Elymus sibiricus L.
Elymus virginicus L.
Elymus virginicus var. submuticus Hook.
Elymus vulpinus Rydb.
Elytrigia elongata (Host) Nevski
Empetrum nigrum ssp. hermaphroditum (Lange ex Hagerup) Böcher
Epilobium obcordatum Gray
Epilobium oreganum Greene
Equisetum × ferrissii Clute (pro sp.)
Equisetum × litorale Kühlewein ex Rupr. (pro sp.)
Equisetum × nelsonii (A.A. Eat.) Schaffn. (pro sp.)
Eragrostis hypnoides (Lam.) B.S.P.
Erigeron asper Nutt.
Erigeron aureus var. acutifolius Raup
Erigeron concinnus (Hook. & Arn.) Torr. & Gray
Erigeron hyssopifolius Michx.
Erigeron melanocephalus (A. Nels.) A. Nels.
Erigeron ochroleucus Nutt.
Erigeron ochroleucus var. scribneri (Canby ex Rydb.) Cronq.
Erigeron philadelphicus var. glaber Henry
Erigeron pumilus ssp. intermedius var. gracilior Cronq.
Erigeron radicatus Hook.
Erigeron speciosus var. macranthus (Nutt.) Cronq.
Erigeron strigosus var. septentrionalis (Fern. & Wieg.) Fern.
Eriogonum ovalifolium var. depressum Blank.
Eriogonum ovalifolium var. ochroleucum (Small ex Rydb.) M.E. Peck
Eriogonum ovalifolium var. vineum (Small) Jepson
Eriogonum sphaerocephalum Dougl. ex Benth.
Eriophorum angustifolium ssp. subarcticum var. coloratum Hultén
Eriophorum angustifolium ssp. subarcticum var. majus F.W. Schultz
Eriophorum russeolum var. albidum W. Nyl.

Eriophyllum lanatum var. achillaeoides (DC.) Jepson
Eriophyllum lanatum var. leucophyllum (DC.) W.R. Carter
Erodium moschatum (L.) L'Hér. ex Ait.
Eryngium articulatum Hook.
Erysimum asperum (Nutt.) DC.
Erysimum capitatum var. angustatum (Greene) G. Rossb.
Erysimum repandum L.
Erythronium grandiflorum ssp. chrysandrum Applegate
Erythronium howellii S. Wats.
Eupatorium cannabinum L.
Eupatorium purpureum L.
Euphorbia esula var. uralensis (Fisch. ex Link) Dorn
Euphorbia marginata Pursh
Euphrasia frigida Pugsley
Euphrasia mollis (Ledeb.) Wettst.
Euphrasia stricta D. Wolff ex J.F. Lehm.
Euphrasia subarctica Raup
Euthamia graminifolia var. nuttallii (Greene) W. Stone
Fagopyrum tataricum (L.) Gaertn.
Falcaria vulgaris Bernh.
Festuca hallii (Vasey) Piper
Festuca rubra ssp. arenaria (Osbeck) Syme
Festuca rubra ssp. falax Thuill.
Filipendula rubra (Hill) B.L. Robins.
Floerkea proserpinacoides Willd.
Fontinalis flaccida Ren. & Card.
Fossombronia pusilla (L.) Dum.
Fragaria × ananassa Duchesne (pro sp.)
Fragaria × ananassa var. cuneifolia (Nutt. ex T.J. Howell) Staudt (pro nm.)
Fragaria crinita Rydb.
Frasera albicaulis Dougl. ex Griseb.
Fraxinus latifolia Benth.
Frullania eboracensis Gott.
Fumaria reuteri Boiss.
Galium labradoricum (Wieg.) Wieg.
Galium oreganum Britt.
Galium palustre L.
Galium tinctorium (L.) Scop.
Gamochaeta ustulata (Nutt.) Nesom
Gayophytum racemosum Torr. & Gray
Genista monspessulana (L.) L. Johnson
Gentianopsis crinita (Froel.) Ma
Gentianopsis thermalis (Kuntze) Iltis
Geranium bicknellii var. longipes (S. Wats.) Fern.
Geranium columbinum L.
Geranium oreganum T.J. Howell
Geranium viscosissimum var. nervosum (Rydb.) C.L. Hitchc.
Geum × aurantiacum Fries
Gilia achilleifolia Benth.
Gilia achilleifolia ssp. multicaulis (Benth.) V.& A. Grant
Glyceria declinata Brébiss.
Glyceria fluitans (L.) R. Br.
Glyceria pulchella (Nash) K. Schum.
Glyceria septentrionalis A.S. Hitchc.
Gnaphalium obtusifolium L.
Grimmia incurva Schwaegr.
Grindelia columbiana (Piper) Rydb.
Grindelia nana Nutt.
Grindelia nana var. integerrima (Rydb.) Steyermark
Gymnocarpium robertianum (Hoffmann) Newman
Gypsophila repens L.
Hackelia diffusa var. arida (Piper) R.L. Carr
Hackelia hispida (Gray) I.M. Johnston
Hackelia patens (Nutt.) I.M. Johnston
Harpanthus scutatus (Web. & Mohr) Spruce
Hedysarum alpinum var. grandiflorum Rollins
Helianthus cusickii Gray
Helianthus giganteus L.
Helianthus grosseserratus Martens
Helianthus × laetiflorus Pers. (pro sp.)
Helianthus maximiliani Schrad.
Helianthus nuttallii ssp. rydbergii (Britt.) R.W. Long
Helianthus petiolaris Nutt.
Heliopsis helianthoides (L.) Sweet
Heliopsis helianthoides var. occidentalis (T.R. Fisher) Steyermark
Heliopsis helianthoides var. scabra (Dunal) Fern.
Herbertus sakuraii ssp. arcticus Inoue & Steere
Herbertus stramineus (Dum.) Trev.
Heteranthera dubia (Jacq.) MacM.
Heuchera cylindrica var. alpina Sw.
Heuchera × easthamii Calder & Savile
Heuchera grossulariifolia Rydb.
Heuchera grossulariifolia var. tenuifolia (Wheelock) C.L. Hitchc.
Hieracium caespitosum Dumort.
Hieracium canadense Michx.
Hieracium × floribundum Wimmer & Grab. (pro sp.)
Hieracium praealtum Vill. ex Gochnat
Hieracium praealtum var. decipiens W.D.J. Koch
Hierochloe alpina ssp. orthantha (Sorensen) G. Weim.
Hierochloe hirta (Schrank) Borbás
Hierochloe hirta ssp. arctica (J. Presl) G. Weim.
Holocarpha obconica (J.C. Clausen & Keck) Keck
Holocarpha obconica ssp. autumnalis Keck
Hordeum depressum (Scribn. & J.G. Sm.) Rydb.
Hordeum pusillum Nutt.
Huperzia porophila (Lloyd & Underwood) Holub
Huperzia selago (L.) Bernh. ex Mart. & Schrank
Huperzia selago var. densa Trevisan
Hyacinthoides nonscripta (L.) Chouard ex Rothm.
Hydrocotyle umbellata L.
Hydrophyllum capitatum var. alpinum S. Wats.
Hyoscyamus niger L.
Hypericum frondosum Michx.
Hypnum jutlandicum Holm. & Warncke
Ionactis alpina (Nutt.) Greene
Ipomopsis aggregata ssp. formosissima (Greene) Wherry
Ipomopsis congesta (Hook.) V. Grant
Iris setosa Pallas ex Link
Iris thompsonii R.C. Foster
Isoetes lacustris L.
Jacquemontia ovalifolia ssp. obcordata (Millsp.) Robertson
Jasione montana L.
Juglans ailanthifolia Carr.
Juncus × alpiniformis Fern.
Juncus alpinoarticulatus ssp. americanus (Farw.) Hämet-Ahti
Juncus alpinoarticulatus ssp. fuscescens (Fern.) Hämet-Ahti
Juncus alpinoarticulatus ssp. nodulosus (Wahlenb.) Hämet-Ahti
Juncus ambiguus Guss.
Juncus canadensis J. Gay ex Laharpe
Juncus dudleyi Wieg.
Juncus effusus var. solutus Fern. & Wieg.
Juncus interior Wieg.
Juncus nevadensis var. badius (Suksdorf) C.L. Hitchc.
Juncus nevadensis var. columbianus (Coville) St. John
Juncus nodosus var. meridianus F.J. Herm.
Juncus orthophyllus Coville
Juncus subcaudatus (Engelm.) Coville & Blake
Juncus subcaudatus var. planisepalus Fern.
Jungermannia gracillima Sm.
Jungermannia hattoriana
Jungermannia karl-muelleri Grolle
Jungermannia polaris Lindb.
Juniperus × fassettii Boivin
Juniperus occidentalis Hook.
Kalmia polifolia Wangenh.
Kickxia spuria (L.) Dumort.
Lactuca canadensis var. longifolia (Michx.) Farw.
Lasthenia minor (DC.) Ornduff
Lathyrus nevadensis ssp. cusickii (S. Wats.) C.L. Hitchc.
Lathyrus nevadensis ssp. lanceolatus (T.J. Howell) C.L. Hitchc.
Lathyrus nevadensis ssp. lanceolatus var. nuttallii (S. Wats.) C.L. Hitchc.
Lathyrus pauciflorus Fern.
Lathyrus polyphyllus Nutt.
Lathyrus pusillus Ell.

Lathyrus tingitanus L.
Lathyrus venosus Muhl. ex Willd.
Layia glandulosa (Hook.) Hook. & Arn.
Lechea maritima var. subcylindrica Hodgdon
Ledum × columbianum Piper (pro sp.)
Lemna gibba L.
Lepidium lasiocarpum Nutt.
Lepidium oxycarpum Torr. & Gray
Lepidium virginicum var. medium (Greene) C.L. Hitchc.
Lepidium virginicum var. menziesii (DC.) C.L. Hitchc.
Lepidium virginicum var. pubescens (Greene) C.L. Hitchc.
Leptodictyum humile (P. Beauv.) Ochyra
Lesquerella alpina (Nutt.) S. Wats.
Leucanthemella serotina (L.) Tzvelev
Lewisia columbiana var. rupicola (English) C.L. Hitchc.
Leymus arenarius (L.) Hochst.
Leymus condensatus (J. Presl) A. Löve
Leymus × vancouverensis (Vasey) Pilger (pro sp.)
Ligusticum apiifolium (Nutt. ex Torr. & Gray) Gray
Ligusticum grayi Coult. & Rose
Lilium canadense L.
Limnanthes douglasii R. Br.
Limosella australis R. Br.
Linanthus bicolor ssp. minimus Mason
Linanthus liniflorus (Benth.) Greene
Linanthus liniflorus ssp. pharnaceoides (Benth.) Mason
Linaria maroccana Hook. f.
Linaria pinifolia (Poir.) Thellung
Lindera subcoriacea B.E. Wofford
Lipocarpha aristulata (Coville) G. Tucker
Lipochaeta subcordata Gray
Lithospermum officinale L.
Lobaria pseudopulmonaria Gyelnik
Lolium persicum Boiss. & Hohen. ex Boiss.
Lomatium farinosum (Hook.) Coult. & Rose
Lomatium gormanii (T.J. Howell) Coult. & Rose
Lomatium simplex var. leptophyllum (Hook.) Mathias
Lomatium triternatum var. macrocarpum (Coult. & Rose) Mathias
Lonicera caerulea var. cauriana (Fern.) Boivin
Lonicera hispidula var. vacillans (Benth.) Gray
Lonicera × notha Zabel
Lonicera periclymenum L.
Lophocolea cuspidata (Nees) Limpr.
Lophozia elongata Steph.
Lophozia laxa (Lindb.) Grolle
Lotus krylovii Schischkin & Sergievskaja
Lotus parviflorus Desf.
Lunaria rediviva L.
Lupinus albicaulis Dougl.
Lupinus argenteus ssp. argenteus var. laxiflorus (Dougl. ex Lindl.) Dorn
Lupinus bicolor ssp. tridentatus (Eastw. ex C.P. Sm.) D. Dunn
Lupinus bingenensis var. dubius C.P. Sm.
Lupinus caespitosus Nutt.
Lupinus caespitosus var. utahensis (S. Wats.) Cox
Lupinus caudatus Kellogg
Lupinus formosus Greene
Lupinus formosus var. robustus C.P. Sm.
Lupinus latifolius Lindl. ex J.G. Agardh
Lupinus polyphyllus ssp. polyphyllus var. pallidipes (Heller) C.P. Sm.
Lupinus prunophilus M.E. Jones
Lupinus sellulus ssp. sellulus var. lobbii (Gray ex S. Wats.) Cox
Lupinus sericeus ssp. sericeus var. egglestonianus C.P. Sm.
Lupinus sericeus ssp. sericeus var. flexuosus (Lindl. ex J.G. Agardh) C.P. Sm.
Lupinus subcarnosus Hook.
Luzula alpinopilosa (Chaix) Breistr.
Luzula arctica ssp. latifolia (Kjellm.) Porsild
Luzula campestris (L.) DC.
Luzula divaricata S. Wats.
Luzula multiflora ssp. multiflora var. kobayasii (Satake) Samuelsson
Luzula subcapitata (Rydb.) Harrington
Luzula subcongesta (S. Wats.) Jepson
Luzula wahlenbergii Rupr.
Lychnis chalcedonica L.
Lycopersicon esculentum P. Mill.
Lycopodium annotinum var. acrifolium Fern.
Lycopodium annotinum var. alpestre Hartman
Lycopodium annotinum var. pungens (La Pylaie) Desv.
Lycopodium obscurum L.
Lycopodium tristachyum Pursh
Lysimachia hybrida Michx.
Machaeranthera canescens ssp. canescens var. incana (Lindl.) Gray
Machaerocarpus californicus (Torr. ex Benth.) Small
Madia elegans D. Don ex Lindl.
Mannia pilosa (Hornem.) Frye & Clark
Mannia triandra (Scop.) Grolle
Matthiola incana (L.) Ait. f.
Medicago orbicularis (L.) Bartalini
Melica aristata Thurb. ex Boland.
Melica geyeri Munro ex Boland.
Melilotus indicus (L.) All.
Mentha × gracilis Sole (pro sp.)
Mentzelia veatchiana Kellogg
Mertensia oblongifolia (Nutt.) G. Don
Mertensia platyphylla Heller
Mertensia platyphylla var. subcordata (Greene) L.O. Williams
Mibora minima (L.) Desv.
Miconia subcorymbosa Britt.
Microseris lindleyi (DC.) Gray
Mimetanthe pilosa (Benth.) Greene
Mimulus cardinalis Dougl. ex Benth.
Mimulus moschatus var. longiflorus Gray
Mimulus moschatus var. sessilifolius Gray
Mimulus suksdorfii Gray
Minuartia michauxii (Fenzl) Farw.
Minuartia rossii (R. Br. ex Richards.) Graebn.
Minuartia stricta (Sw.) Hiern
Minuartia verna (L.) Hiern
Misopates orontium (L.) Raf.
Mitella diversifolia Greene
Mitella trifida var. violacea (Rydb.) Rosendahl
Monarda citriodora Cerv. ex Lag.
Monarda fistulosa ssp. fistulosa var. menthifolia (Graham) Fern.
Monarda fistulosa ssp. fistulosa var. mollis (L.) Benth.
Moneses uniflora ssp. reticulata (Nutt.) Calder & Taylor
Morus rubra L.
Muhlenbergia depauperata Scribn.
Muhlenbergia sylvatica Torr. ex Gray
Muhlenbergia uniflora (Muhl.) Fern.
Myosurus minimus var. major (Greene) Davis
Myrica gale var. tomentosa C. DC.
Myriophyllum alterniflorum DC.
Narcissus jonquilla L.
Navarretia intertexta ssp. propinqua (Suksdorf) Day
Navarretia leucocephala Benth.
Navarretia leucocephala ssp. minima (Nutt.) Day
Nemophila menziesii Hook. & Arn.
Nicandra physalodes (L.) Gaertn.
Nicotiana attenuata Torr. ex S. Wats.
Nigella damascena L.
Nonea rosea (Bieb.) Link
Nothoscordum bivalve (L.) Britt.
Nuttallanthus canadensis (L.) D.A. Sutton
Odontoschisma prostratum (Sw.) Trev.
Oenothera cespitosa Nutt.
Oenothera elata Kunth
Oenothera elata ssp. hirsutissima (Gray ex S. Wats.) W. Dietr.
Oenothera elata ssp. hookeri (Torr. & Gray) W. Dietr. & W.L. Wagner
Oenothera nuttallii Sweet
Orobanche californica ssp. grayana (G. Beck) Heckard
Orobanche ludoviciana Nutt.
Orthotrichum pusillum Mitt.
Oxalis suksdorfii Trel.
Oxytropis campestris var. dispar (A. Nels.) Barneby
Oxytropis deflexa var. foliolosa (Hook.) Barneby

Oxytropis deflexa var. sericea Torr. & Gray
Oxytropis lambertii Pursh
Oxytropis nana Nutt.
Oxytropis nana var. besseyi (Rydb.) Isely
Oxytropis nigrescens var. uniflora (Hook.) Barneby
Oxytropis viscida var. hudsonica (Greene) Barneby
Oxytropis viscida var. subsucculenta (Hook.) Barneby
Paeonia brownii Dougl. ex Hook.
Panicum rigidulum Bosc ex Nees
Panicum sonorum Beal
Parnassia multiseta (Ledeb.) Fern.
Parnassia palustris var. tenuis Wahlenb.
Paronychia jamesii Torr. & Gray
Parthenocissus quinquefolia (L.) Planch.
Pectocarya linearis (Ruiz & Pavón) DC.
Pedicularis groenlandica ssp. surrecta (Benth.) Piper
Pedicularis racemosa ssp. alba Pennell
Pellaea atropurpurea (L.) Link
Pellaea gastonyi Windham
Penstemon attenuatus Dougl. ex Lindl.
Penstemon fruticosus var. serratus (Keck) Cronq.
Penstemon montanus Greene
Penstemon procerus var. formosus (A. Nels.) Cronq.
Penstemon triphyllus Dougl. ex Lindl.
Penstemon wilcoxii Rydb.
Perideridia gairdneri ssp. borealis Chuang & Constance
Perideridia oregana (S. Wats.) Mathias
Petasites hybridus (L.) P.G. Gaertn., B. Mey. & Scherb.
Petroselinum crispum (P. Mill.) Nyman ex A.W. Hill
Phacelia hastata var. compacta (Brand) Cronq.
Phacelia hastata var. dasyphylla (Greene ex J.F. Macbr.) Kartesz & Gandhi
Phacelia heterophylla Pursh
Phacelia nemoralis Greene
Phacelia nemoralis ssp. oregonensis Heckard
Phacelia tanacetifolia Benth.
Phalaris minor Retz.
Philadelphus lewisii var. ellipticus Hu
Philadelphus lewisii var. gordonianus (Lindl.) Jepson
Philadelphus lewisii var. helleri (Rydb.) Hu
Philadelphus trichothecus Hu
Phippsia algida (Phipps) R. Br.
Phlox diffusa ssp. subcarinata Wherry
Phlox gracilis ssp. humilis (Greene) Mason
Phlox hoodii ssp. canescens (Torr. & Gray) Wherry
Phlox hoodii ssp. viscidula (Wherry) Wherry
Phlox speciosa ssp. occidentalis (Dur. ex Torr.) Wherry
Phyllodoce × intermedia (Hook.) Rydb. (pro sp.)
Physalis pubescens L.
Physalis virginiana P. Mill.
Physaria brassicoides Rydb.
Physaria obcordata Rollins
Physocarpus opulifolius (L.) Maxim.
Picea × lutzii Little
Picris hieracioides L.
Pinguicula macroceras var. microceras (Cham.) Casper
Pinus contorta var. murrayana (Grev. & Balf.) Engelm.
Pinus × murraybanksiana Righter & Stockwell
Piperia elongata Rydb.
Piperia transversa Suksdorf
Pisonia subcordata Sw.
Pisum sativum L.
Plagiobothrys hirtus (Greene) I.M. Johnston
Plagiobothrys scouleri var. hispidulus (Greene) Dorn
Plagiochila arctica Bryhn & Kaal.
Plagiochila aspleniformis Schust.
Plagiothecium latebricola Schimp. in B.S.G.
Plantago aristata Michx.
Plantago major var. pilgeri Domin
Plantago pusilla Nutt.
Plantago tweedyi Gray
Platanthera hyperborea var. gracilis (Lindl.) Luer
Platydictya confervoides (Brid.) Crum
Platydictya minutissima (Sull. & Lesq. in Sull.) Crum
Plectritis macrocera ssp. grayi (Suksdorf) Morey
Pleuridium acuminatum Lindb.
Poa arctica ssp. grayana (Vasey) A.& D. Löve & Kapoor
Poa arida Vasey
Poa kelloggii Vasey
Poa leibergii Scribn.
Poa malacantha Komarov
Poa occidentalis Vasey
Poa reflexa Vasey & Scribn. ex Vasey
Poa suksdorfii (Beal) Vasey ex Piper
Poa tracyi Vasey
Polanisia dodecandra (L.) DC.
Polanisia dodecandra ssp. trachysperma (Torr. & Gray) Iltis
Polemonium californicum Eastw.
Polemonium confertum Gray
Polemonium pulcherrimum ssp. lindleyi (Wherry) V. Grant
Polygala vulgaris L.
Polygonum caurianum B.L. Robins.
Polygonum cespitosum Blume
Polygonum cespitosum var. longisetum (de Bruyn) A.N. Steward
Polygonum erectum L.
Polygonum pensylvanicum L.
Polygonum perfoliatum L.
Polygonum phytolaccifolium Meisn. ex Small
Polygonum polygaloides ssp. confertiflorum (Nutt. ex Piper) Hickman
Polygonum punctatum var. confertiflorum (Meisn.) Fassett
Polygonum scandens L.
Polypodium virginianum L.
Polypogon interruptus Kunth
Polypogon viridis (Gouan) Breistr.
Polystichum californicum (D.C. Eat.) Diels
Polystichum kwakiutlii D.H. Wagner
Pontederia cordata L.
Populus alba L.
Populus angustifolia James
Populus × brayshawii Boivin
Populus × canescens (Ait.) Sm. (pro sp.)
Populus grandidentata Michx.
Populus nigra L.
Portulaca oleracea ssp. stellata Danin & Baker
Potamogeton diversifolius Raf.
Potamogeton filiformis var. alpinus (Blytt) Aschers. & Graebn.
Potamogeton × hagstroemii Benn. (pro sp.)
Potamogeton × schreberi Fisch.
Potentilla concinna Richards.
Potentilla glandulosa ssp. glabrata (Rydb.) Keck
Potentilla hookeriana ssp. hookeriana var. furcata (Porsild) Hultén
Potentilla multifida L.
Potentilla multisecta (S. Wats.) Rydb.
Potentilla pectinisecta Rydb.
Potentilla pensylvanica var. ovium Jepson
Potentilla pensylvanica var. strigosa Pallas ex Pursh
Potentilla pulchella R. Br.
Potentilla rivalis var. millegrana (Engelm. ex Lehm.) S. Wats.
Potentilla rivalis var. pentandra (Engelm.) S. Wats.
Potentilla rubricaulis Lehm.
Potentilla vahliana Lehm.
Primula veris L.
Primula vulgaris Huds.
Prunus cerasus L.
Prunus emarginata var. mollis (Dougl. ex Hook.) Brewer
Prunus mexicana S. Wats.
Prunus pensylvanica var. saximontana Rehd.
Prunus subcordata Benth.
Prunus subcordata var. kelloggii J.G. Lemmon
Prunus subcordata var. oregana (Greene) W. Wight ex M.E. Peck
Prunus subcordata var. rubicunda Jepson
Psilocarphus oregonus Nutt.
Pteryxia terebinthina (Hook.) Coult. & Rose
Puccinellia agrostidea Sorensen
Puccinellia ambigua Sorensen
Puccinellia deschampsioides Sorensen

Puccinellia lemmonii (Vasey) Scribn.
Puccinellia lucida Fern. & Weatherby
Puccinellia maritima (Huds.) Parl.
Puccinellia tenella (Lange) Holmb.
Pyracantha coccinea M. Roemer
Pyrrocoma carthamoides var. cusickii (Gray) Kartesz & Gandhi
Pyrrocoma lanceolata (Hook.) Greene
Quercus rubra L.
Quercus × subconvexa Tucker
Radula obconica Sull.
Ranunculus bulbosus L.
Ranunculus eschscholtzii var. trisectus (Eastw.) L. Benson
Ranunculus eximius Greene
Ranunculus ficaria var. bulbifera Marsden-Jones
Ranunculus hebecarpus Hook. & Arn.
Ranunculus hispidus Michx.
Ranunculus hispidus var. nitidus (Chapman) T. Duncan
Ranunculus micranthus Nutt.
Ranunculus occidentalis var. nelsonii (DC.) L. Benson
Ranunculus pygmaeus var. langianus Nathorst
Ranunculus trichophyllus var. eradicatus (Laestad.) W. Drew
Rheum rhabarbarum L.
Rheum rhaponticum L.
Rhinanthus minor ssp. groenlandicus (Ostenf.) L. Neum.
Ribes aureum Pursh
Ribes aureum var. villosum DC.
Ribes hirtellum Michx.
Ribes nigrum L.
Ribes oxyacanthoides ssp. setosum (Lindl.) Sinnott
Ribes rubrum L.
Ribes watsonianum Koehne
Riccardia multifida var. ambrosioides (Nees) Lindb.
Riccia frostii Aust.
Romneya coulteri Harvey
Rorippa cantoniensis (Lour.) Ohwi
Rorippa curvisiliqua var. orientalis R. Stuckey
Rorippa curvisiliqua var. procumbens R. Stuckey
Rorippa indica (L.) Hiern
Rorippa indica var. apetala Hochr.
Rorippa islandica (Oeder) Borbás
Rorippa palustris ssp. fernaldiana (Butters & Abbe) Jonsell
Rorippa palustris ssp. hispida (Desv.) Jonsell
Rorippa palustris ssp. occidentalis (S. Wats.) Abrams
Rorippa sinuata (Nutt.) A.S. Hitchc.
Rorippa teres (Michx.) R. Stuckey
Rosa blanda Ait.
Rosa micrantha Borrer ex Sm.
Rosa nutkana var. muriculata (Greene) G.N. Jones
Rubus alaskensis Bailey
Rubus bifrons Vest ex Tratt.
Rubus ellipticus var. obcordatus Focke
Rubus macrophyllus Weihe & Nees
Rubus phoenicolasius Maxim.
Rubus vitifolius Cham. & Schlecht.
Rudbeckia hirta var. pulcherrima Farw.
Rumex × acutus L. (pro sp.)
Rumex longifolius DC.
Rumex pseudonatronatus (Borbás) Murb.
Rumex pulcher L.
Rumex salicifolius var. angustivalvis Danser
Rumex salicifolius var. denticulatus Torr.
Rumex venosus Pursh
Sagina maritima G. Don
Sairocarpus subcordatus (Gray) D.A. Sutton
Salix babylonica L.
Salix brachycarpa ssp. brachycarpa var. fullertonensis (Schneid.) Argus
Salix fragilis L.
Salix glauca ssp. callicarpaea (Trautv.) Böcher
Salix glauca ssp. glauca var. acutifolia (Hook.) Schneid.
Salix glauca ssp. glauca var. villosa (D. Don ex Hook.) Anderss.
Salix lasiolepis Benth.
Salix lutea Nutt.
Salix monochroma Ball
Salix monticola Bebb
Salix orestera Schneid.
Salix phlebophylla Anderss.
Salix × rubens Schrank (pro sp.)
Salix × sepulcralis Simonkai
Salix × waghornei Rydb. (pro sp.)
Salpichroa origanifolia (Lam.) Baill.
Salsola kali ssp. tragus (L.) Aellen
Sanguisorba annua (Nutt. ex Hook.) Torr. & Gray
Sanguisorba minor ssp. muricata (Spach) Nordborg
Sarcobatus vermiculatus (Hook.) Torr.
Saussurea amara (L.) DC.
Saussurea nuda Ledeb.
Saussurea viscida Hultén
Saxifraga bronchialis ssp. vespertina (Small) Piper
Saxifraga cespitosa ssp. cespitosa var. emarginata (Small) Rosendahl
Saxifraga ferruginea var. vreelandii (Small) Engl. & Irmsch.
Saxifraga foliolosa R. Br.
Saxifraga nidifica var. claytoniifolia (Canby ex Small) Elvander
Saxifraga paniculata P. Mill.
Saxifraga rhomboidea Greene
Saxifraga sibirica L.
Saxifraga spicata D. Don
Saxifraga tenuis (Wahlenb.) H. Sm. ex Lindm.
Scapania aspera M. Bern. & H. Bern.
Scapania nemorosa (L.) Dum.
Schistidium andreaeopsis (C. Müll.) Laz.
Schistidium pulvinatum (Hedw.) Brid.
Schistidium rivulare var. latifolium (Zett.) Crum & Anderson
Scirpus cyperinus (L.) Kunth
Scirpus expansus Fern.
Scirpus hallii Gray
Scirpus robustus Pursh
Scirpus × rubiginosus Beetle (pro sp.)
Scrophularia californica Cham. & Schlecht.
Sedum lanceolatum ssp. subalpinum (Blank.) Clausen
Sedum oreganum ssp. tenue Clausen
Sedum rosea (L.) Scop.
Sedum telephium L.
Selaginella densa var. standleyi (Maxon) R. Tryon
Selaginella douglasii (Hook. & Grev.) Spring
Selaginella rupestris (L.) Spring
Selaginella subcaulescens Baker
Senecio cymbalaria Pursh
Senecio streptanthifolius var. borealis (Torr. & Gray) J.F. Bain
Senecio streptanthifolius var. kluanei J.F. Bain
Senecio werneriifolius (Gray) Gray
Setaria pumila (Poir.) Roemer & J.A. Schultes
Sibara virginica (L.) Rollins
Sidalcea campestris Greene
Sidalcea oregana ssp. oregana var. procera C.L. Hitchc.
Silene acaulis var. exscapa (All.) DC.
Silene involucrata ssp. elatior (Regel) Bocquet
Silene menziesii ssp. menziesii var. viscosa (Greene) C.L. Hitchc. & Maguire
Silene repens ssp. australis C.L. Hitchc. & Maguire
Silene repens ssp. purpurata (Greene) C.L. Hitchc. & Maguire
Silene sorensenis (Boivin) Bocquet
Silene subciliata B.L. Robins.
Sisyrinchium bellum S. Wats.
Sisyrinchium mucronatum Michx.
Sisyrinchium sarmentosum Suksdorf ex Greene
Smelowskia calycina var. americana (Regel & Herder) Drury & Rollins
Solanum tuberosum L.
Solidago missouriensis var. extraria Gray
Solidago multiradiata var. scopulorum Gray
Solidago rugosa P. Mill.
Solidago rugosa ssp. rugosa var. villosa (Pursh) Fern.
Solidago simplex var. nana (Gray) Ringius
Solidago simplex ssp. simplex var. nana (Gray) Ringius
Solidago spathulata var. subcinerea Gray

Sparganium americanum Nutt.
Sparganium erectum L.
Spartina pectinata Link
Spergula arvensis var. sativa (Boenn.) Mert. & Koch
Spergularia diandra (Guss.) Held. & Sart.
Sphagnum lescurii Sull. in Gray
Sphagnum nitidum Warnst.
Sphagnum portoricense Hampe
Sphagnum subsecundum var. andrusii Crum
Sphenopholis pensylvanica (L.) A.S. Hitchc.
Spiraea × subcanescens Rydb. (pro sp.)
Spiranthes porrifolia Lindl.
Stachys ajugoides Benth.
Stachys mexicana Benth.
Stellaria longifolia Muhl. ex Willd.
Stellaria longifolia var. atrata J.W. Moore
Stephanomeria lactucina Gray
Stephanomeria spinosa (Nutt.) S. Tomb
Streptopus streptopoides ssp. streptopoides var. brevipes (Baker) Fassett
Symphoricarpos mollis Nutt.
Symphytum × uplandicum Nyman (pro sp.)
Syringa vulgaris L.
Talinum spinescens Torr.
Tamarix chinensis Lour.
Tamarix parviflora DC.
Tanacetum camphoratum Less.
Taraxacum officinale ssp. vulgare (Lam.) Schinz & R. Keller
Thelesperma subnudum Gray
Thelesperma subnudum var. marginatum (Rydb.) T.E. Melchert ex Cronq.
Thermopsis macrophylla Hook. & Arn.
Thermopsis macrophylla var. venosa (Eastw.) Isely
Thlaspi arcticum Porsild
Thlaspi montanum L.
Thymus pulegioides L.
Tofieldia glutinosa ssp. brevistyla C.L. Hitchc.
Tofieldia glutinosa ssp. montana C.L. Hitchc.
Torilis arvensis (Huds.) Link
Torilis japonica (Houtt.) DC.
Tortella nitida (Lindb.) Broth.
Tortula atrovirens (Sm.) Lindb.
Trautvetteria caroliniensis var. occidentalis (Gray) C.L. Hitchc.
Triadenum fraseri (Spach) Gleason
Triadenum virginicum (L.) Raf.
Trichostomum crispulum Bruch in F. Muell.
Trifolium fucatum Lindl.
Trifolium gracilentum Torr. & Gray
Trifolium longipes Nutt.
Trifolium macrocephalum (Pursh) Poir.
Trifolium medium L.
Trigonella caerulea (L.) Ser.
Trigonotis peduncularis (Trev.) Benth. ex Baker & S. Moore
Triphysaria eriantha (Benth.) Chuang & Heckard
Triphysaria versicolor ssp. faucibarbatus (Gray) Chuang & Heckard
Trisetum flavescens (L.) Beauv.
Tsuga × jeffreyi (Henry) Henry
Ulmus minor P. Mill.
Ulota crispa (Hedw.) Brid.
Ulota japonica (Sull. & Lesq.) Mitt.
Umbilicaria americana Poelt & T. Nash
Urtica dioica ssp. holosericea (Nutt.) Thorne
Utricularia geminiscapa Benj.
Utricularia ochroleuca R.W. Hartman
Vaccinium caespitosum var. paludicola (Camp) Hultén
Vaccinium corymbosum L.
Vaccinium macrocarpon Ait.
Vaccinium myrtillus var. oreophilum (Rydb.) Dorn
Vaccinium × nubigenum Fern. (pro sp.)
Valeriana occidentalis Heller
Vancouveria hexandra (Hook.) Morr. & Dcne.
Verbena hastata var. scabra Moldenke
Veronica agrestis L.
Veronica persica var. aschersoniana (Lehm.) Boivin
Vicia lathyroides L.
Viola affinis Le Conte
Viola canadensis var. rugulosa (Greene) C.L. Hitchc.
Viola labradorica Schrank
Viola nephrophylla var. arizonica (Greene) Kearney & Peebles
Viola nuttallii Pursh
Viola palustris var. brevipes (M.S. Baker) R.J. Davis
Viola sororia Willd.
Vulpia microstachys var. confusa (Piper) Lonard & Gould
Weissia brachycarpa (Nees & Hornsch.) Jur.
Wilhelmsia physodes (Fisch. ex Ser.) McNeill
Woodsia scopulina ssp. laurentiana Windham
Wyethia amplexicaulis (Nutt.) Nutt.
Wyethia angustifolia (DC.) Nutt.
Xanthium spinosum L.
Xanthium strumarium var. glabratum (DC.) Cronq.
Zigadenus paniculatus (Nutt.) S. Wats.
Zizania aquatica L.
Zostera pacifica L.